List of Elements with Their Symbols and Atomic Weights

Element	Symbol	Atomic Number	Atomic Weight
Actinium	Ac	89	227.03[a]
Aluminum	Al	13	26.981538
Americium	Am	95	243.06[a]
Antimony	Sb	51	121.760
Argon	Ar	18	39.948
Arsenic	As	33	74.92160
Astatine	At	85	209.99[a]
Barium	Ba	56	137.327
Berkelium	Bk	97	247.07[a]
Beryllium	Be	4	9.012183
Bismuth	Bi	83	208.98038
Bohrium	Bh	107	264.12[a]
Boron	B	5	10.81
Bromine	Br	35	79.904
Cadmium	Cd	48	112.414
Calcium	Ca	20	40.078
Californium	Cf	98	251.08[a]
Carbon	C	6	12.0107
Cerium	Ce	58	140.116
Cesium	Cs	55	132.905452
Chlorine	Cl	17	35.453
Chromium	Cr	24	51.9961
Cobalt	Co	27	58.933194
Copernicium	Cn	112	285
Copper	Cu	29	63.546
Curium	Cm	96	247.07[a]
Darmstadtium	Ds	110	281.15[a]
Dubnium	Db	105	262.11[a]
Dysprosium	Dy	66	162.50
Einsteinium	Es	99	252.08[a]
Erbium	Er	68	167.259
Europium	Eu	63	151.964
Fermium	Fm	100	257.10[a]
Flerovium	Fl	114	289.2[a]
Fluorine	F	9	18.99840316
Francium	Fr	87	223.02[a]
Gadolinium	Gd	64	157.25
Gallium	Ga	31	69.723
Germanium	Ge	32	72.64
Gold	Au	79	196.966569
Hafnium	Hf	72	178.49
Hassium	Hs	108	269.13[a]
Helium	He	2	4.002602[a]
Holmium	Ho	67	164.93033
Hydrogen	H	1	1.00794
Indium	In	49	114.818
Iodine	I	53	126.90447
Iridium	Ir	77	192.217
Iron	Fe	26	55.845
Krypton	Kr	36	83.80
Lanthanum	La	57	138.9055
Lawrencium	Lr	103	262.11[a]
Lead	Pb	82	207.2
Lithium	Li	3	6.941
Livermorium	Lv	116	293[a]
Lutetium	Lu	71	174.967
Magnesium	Mg	12	24.3050
Manganese	Mn	25	54.938044
Meitnerium	Mt	109	268.14[a]
Mendelevium	Md	101	258.10[a]
Mercury	Hg	80	200.59
Molybdenum	Mo	42	95.95
Neodymium	Nd	60	144.24
Neon	Ne	10	20.1797
Neptunium	Np	93	237.05[a]
Nickel	Ni	28	58.6934
Niobium	Nb	41	92.90637
Nitrogen	N	7	14.0067
Nobelium	No	102	259.10[a]
Osmium	Os	76	190.23
Oxygen	O	8	15.9994
Palladium	Pd	46	106.42
Phosphorus	P	15	30.973762
Platinum	Pt	78	195.078
Plutonium	Pu	94	244.06[a]
Polonium	Po	84	208.98[a]
Potassium	K	19	39.0983
Praseodymium	Pr	59	140.90766
Promethium	Pm	61	145[a]
Protactinium	Pa	91	231.03588
Radium	Ra	88	226.03[a]
Radon	Rn	86	222.02[a]
Rhenium	Re	75	186.207[a]
Rhodium	Rh	45	102.90550
Roentgenium	Rg	111	272.15[a]
Rubidium	Rb	37	85.4678
Ruthenium	Ru	44	101.07
Rutherfordium	Rf	104	261.11[a]
Samarium	Sm	62	150.36
Scandium	Sc	21	44.955908
Seaborgium	Sg	106	266[a]
Selenium	Se	34	78.97
Silicon	Si	14	28.0855
Silver	Ag	47	107.8682
Sodium	Na	11	22.989770
Strontium	Sr	38	87.62
Sulfur	S	16	32.065
Tantalum	Ta	73	180.9479
Technetium	Tc	43	98[a]
Tellurium	Te	52	127.60
Terbium	Tb	65	158.92534
Thallium	Tl	81	204.3833
Thorium	Th	90	232.0377
Thulium	Tm	69	168.93422
Tin	Sn	50	118.710
Titanium	Ti	22	47.867
Tungsten	W	74	183.84
Uranium	U	92	238.02891
Vanadium	V	23	50.9415
Xenon	Xe	54	131.293
Ytterbium	Yb	70	173.04
Yttrium	Y	39	88.90584
Zinc	Zn	30	65.39
Zirconium	Zr	40	91.224
*b		113	284[a]
*b		115	288[a]
*b		117	294[a]
*b		118	294[a]

[a] Mass of longest-lived or most important isotope.
[b] Except for elements 114 and 116, the names and symbols for elements above 113 have not yet been decided.

CHEMISTRY

THE CENTRAL SCIENCE 13TH EDITION

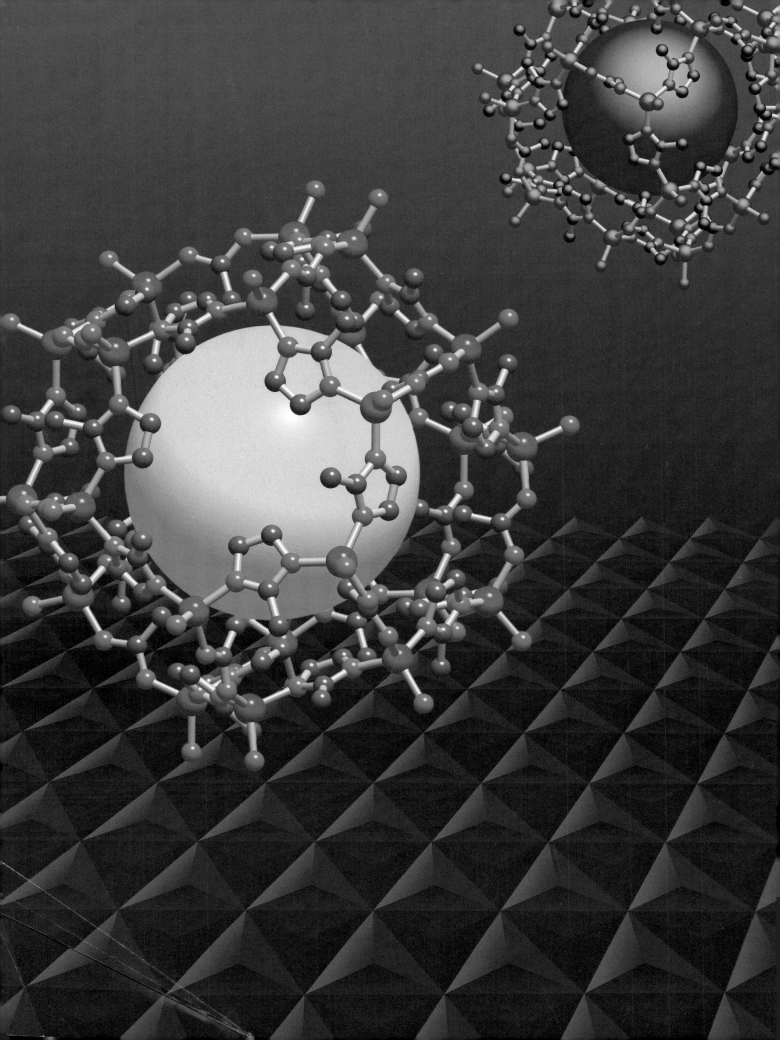

CHEMISTRY

THE CENTRAL SCIENCE 13TH EDITION

Theodore L. Brown

University of Illinois at Urbana-Champaign

H. Eugene LeMay, Jr.

University of Nevada, Reno

Bruce E. Bursten

University of Tennessee, Knoxville

Catherine J. Murphy

University of Illinois at Urbana-Champaign

Patrick M. Woodward

The Ohio State University

Matthew W. Stoltzfus

The Ohio State University

PEARSON

Boston Columbus Indianapolis New York San Francisco Upper Saddle River
Amsterdam Cape Town Dubai London Madrid Milan Munich Paris Montréal Toronto
Delhi Mexico City São Paulo Sydney Hong Kong Seoul Singapore Taipei Tokyo

Editor in Chief, Chemistry: Adam Jaworski
Senior Acquisitions Editor: Terry Haugen
Acquisitions Editor: Chris Hess, Ph.D.
Executive Marketing Manager: Jonathan Cottrell
Associate Team Lead, Program Management, Chemistry and Geoscience: Jessica Moro
Editorial Assistant: Lisa Tarabokjia/Caitlin Falco
Marketing Assistant: Nicola Houston
Director of Development: Jennifer Hart
Development Editor, Text: Carol Pritchard-Martinez
Team Lead, Project Management, Chemistry and Geosciences: Gina M. Cheselka
Project Manager: Beth Sweeten
Full-Service Project Management/Composition: Greg Johnson, PreMediaGlobal
Operations Specialist: Christy Hall
Illustrator: Precision Graphics
Art Director: Mark Ong
Interior / Cover Designer: Tamara Newnam
Image Lead: Maya Melenchuk
Photo Researcher: Kerri Wilson, PreMediaGlobal
Text Permissions Manager: Alison Bruckner
Text Permission Researcher: Jacqueline Bates, GEX Publishing Services
Senior Content Producer: Kristin Mayo
Production Supervisor, Media: Shannon Kong
Electrostatic Potential Maps: Richard Johnson, Chemistry Department, University of New Hampshire
Cover Image Credit: "Metal-Organic Frameworks" by Omar M. Yaghi, University of California, Berkeley

Credits and acknowledgments borrowed from other sources and reproduced, with permission, in this textbook appear on the appropriate page within the text or on pp. P-1–P-2.

Many of the designations used by manufacturers and sellers to distinguish their products are claimed as trademarks. Where those designations appear in this book, and the publisher was aware of a trademark claim, the designations have been printed in initial caps or all caps.

Library of Congress Cataloging-In Publication Data

Brown, Theodore L. (Theodore Lawrence), 1928- author.

    Chemistry the central science.—Thirteenth edition / Theodore L. Brown, University of Illinois at Urbana-Chanmpaign,
    H. Euguene LeMay, Jr., University of Nevada, Reno, Bruce E. Bursten, University of Tennessee, Knoxville,
    Catherine J. Murphy, University of Illinois at Urbana-Chanmpaign, Patrick M. Woodward, The Ohio State University,
    Matthew W. Stoltzfus, The Ohio State University.
       pages cm
    Includes index.
    ISBN-13: 978-0-321-91041-7
    ISBN-10: 0-321-91041-9
    1. Chemistry--Textbooks. I. Title.
    QD31.3.B765 2014
    540—dc23   2013036724

2 3 4 5 6 7 8 9 10—V011— 17 16 15 14

www.pearsonhighered.com

Student Edition: 0-321-91041-9 / 978-0-321-91041-7
Instructor's Resource Copy: 0-321-96239-7 / 978-0-321-96239-3

To our students,
whose enthusiasm and curiosity
have often inspired us,
and whose questions and suggestions
have sometimes taught us.

BRIEF CONTENTS

CONTENTS

6 Electronic Structure of Atoms 212

7 Periodic Properties of the Elements 256

8 Basic Concepts of Chemical Bonding 298

9 Molecular Geometry and Bonding Theories 342

10 Gases 398

11 Liquids and Intermolecular Forces 442

12 Solids and Modern Materials 480

13 Properties of Solutions 530

14 Chemical Kinetics 574

15 Chemical Equilibrium 628

16 Acid–Base Equilibria 670

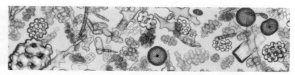

17 Additional Aspects of Aqueous Equilibria 724

18 Chemistry of the Environment 774

19 Chemical Thermodynamics 812

20 Electrochemistry 856

21 Nuclear Chemistry 908

22 Chemistry of the Nonmetals 952

23 Transition Metals and Coordination Chemistry 996

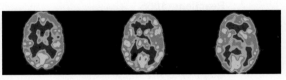

24 The Chemistry of Life: Organic and Biological Chemistry 1040

Appendices

CHEMICAL APPLICATIONS AND ESSAYS

PREFACE

To the Instructor

Philosophy

We authors of *Chemistry: The Central Science* are delighted and honored that you have chosen us as your instructional partners for your general chemistry class. We have all been active researchers who appreciate both the learning and the discovery aspects of the chemical sciences. We have also all taught general chemistry many times. Our varied, wide-ranging experiences have formed the basis of the close collaborations we have enjoyed as coauthors. In writing our book, our focus is on the students: we try to ensure that the text is not only accurate and up-to-date but also clear and readable. We strive to convey the breadth of chemistry and the excitement that scientists experience in making new discoveries that contribute to our understanding of the physical world. We want the student to appreciate that chemistry is not a body of specialized knowledge that is separate from most aspects of modern life, but central to any attempt to address a host of societal concerns, including renewable energy, environmental sustainability, and improved human health.

Publishing the thirteenth edition of this text bespeaks an exceptionally long record of successful textbook writing. We are appreciative of the loyalty and support the book has received over the years, and mindful of our obligation to justify each new edition. We begin our approach to each new edition with an intensive author retreat, in which we ask ourselves the deep questions that we must answer before we can move forward. What justifies yet another edition? What is changing in the world not only of chemistry, but with respect to science education and the qualities of the students we serve? The answer lies only partly in the changing face of chemistry itself. The introduction of many new technologies has changed the landscape in the teaching of sciences at all levels. The use of the Internet in accessing information and presenting learning materials has markedly changed the role of the textbook as one element among many tools for student learning. Our challenge as authors is to maintain the text as the primary source of chemical knowledge and practice, while at the same time integrating it with the new avenues for learning made possible by technology and the Internet. This edition incorporates links to a number of those new methodologies, including use of the Internet, computer-based classroom tools, such as Learning Catalytics™, a cloud-based active learning analytics and assessment system, and web-based tools, particularly MasteringChemistry®, which is continually evolving to provide more effective means of testing and evaluating student performance, while giving the student immediate and helpful feedback. In past versions, MasteringChemistry® provided feedback only on a question level. Now with Knewton-enhanced adaptive follow-up assignments, and Dynamic Study Modules, MasteringChemistry® continually adapts to each student, offering a personalized learning experience.

As authors, we want this text to be a central, indispensable learning tool for students. Whether as a physical book or in electronic form, it can be carried everywhere and used at any time. It is the one place students can go to obtain the information outside of the classroom needed for learning, skill development, reference, and test preparation. The text, more effectively than any other instrument, provides the depth of coverage and coherent background in modern chemistry that students need to serve their professional interests and, as appropriate, to prepare for more advanced chemistry courses.

If the text is to be effective in supporting your role as instructor, it must be addressed to the students. We have done our best to keep our writing clear and interesting and the book attractive and well illustrated. The book has numerous in-text study aids for students, including carefully placed descriptions of problem-solving strategies. We hope that our cumulative experiences as teachers is evident in our pacing, choice of examples, and the kinds of study aids and motivational tools we have employed. We believe students are more enthusiastic about learning chemistry when they see its importance relative to their own goals and interests; therefore, we have highlighted many important applications of chemistry in everyday life. We hope you make use of this material.

It is our philosophy, as authors, that the text and all the supplementary materials provided to support its use must work in concert with you, the instructor. A textbook is only as useful to students as the instructor permits it to be. This book is replete with features that can help students learn and that can guide them as they acquire both conceptual understanding and problem-solving skills. There is a great deal here for the students to use, too much for all of it to be absorbed by any one student. You will be the guide to the best use of the book. Only with your active help will the students be able to utilize most effectively all that the text and its supplements offer. Students care about grades, of course, and with encouragement they will also become interested in the subject matter and care about learning. Please consider emphasizing features of the book that can enhance student appreciation of chemistry, such as the *Chemistry Put to Work* and *Chemistry and Life* boxes that show how chemistry impacts modern life and its relationship to health and life processes. Learn to use, and urge students to use, the rich online resources available. Emphasize conceptual understanding and place less emphasis on simple manipulative, algorithmic problem solving.

What Is New in This Edition?

A great many changes have been made in producing this thirteenth edition. We have continued to improve upon the art program, and new features connected with the art have been introduced. Many figures in the book have undergone modification, and dozens of new figures have been introduced.

A systematic effort has been made to place explanatory labels directly into figures to guide the student. New designs have been employed to more closely integrate photographic materials into figures that convey chemical principles.

We have continued to explore means for more clearly and directly addressing the issue of concept learning. It is well established that conceptual misunderstandings, which impede student learning in many areas, are difficult to correct. We have looked for ways to identify and correct misconceptions via the worked examples in the book, and in the accompanying practice exercises. Among the more important changes made in the new edition, with this in mind, are:

- A major new feature of this edition is the addition of a second Practice Exercise to accompany each Sample Exercise within the chapters. The majority of new *Practice Exercises* are of the multiple-choice variety, which enable feedback via MasteringChemistry®. The correct answers to select Practice Exercises are given in an appendix, and guidance for correcting wrong answers is provided in MasteringChemistry®. The new Practice Exercise feature adds to the aids provided to students for mastering the concepts advanced in the text and rectifying conceptual misunderstandings. The enlarged practice exercise materials also further cement the relationship of the text to the online learning materials. At the same time, they offer a new supportive learning experience for all students, regardless of whether the MasteringChemistry® program is used.

- A second major innovation in this edition is the *Design An Experiment* feature, which appears as a final exercise in all chapters beginning with Chapter 3, as well as in MasteringChemistry®. The *Design an Experiment* exercise is a departure from the usual kinds of end-of-chapter exercises in that it is inquiry based, open ended, and tries to stimulate the student to "think like a scientist." Each exercise presents the student with a scenario in which various unknowns require investigation. The student is called upon to ponder how experiments might be set up to provide answers to particular questions about a system, and/or test plausible hypotheses that might account for a set of observations. The aim of the *Design an Experiment* exercises is to foster critical thinking. We hope that they will be effective in active learning environments, which include classroom-based work and discussions, but they are also suitable for individual student work. There is no one right way to solve these exercises, but we authors offer some ideas in an online Instructor's Resource Manual, which will include results from class testing and analysis of student responses.

- The *Go Figure* exercises introduced in the twelfth edition proved to be a popular innovation, and we have expanded on its use. This feature poses a question that students can answer by examining the figure. These questions encourage students to actually study the figure and understand its primary message. Answers to the *Go Figure* questions are provided in the back of the text.

- The popular *Give It Some Thought (GIST)* questions embedded in the text have been expanded by improvements in some of the existing questions and addition of new ones. The answers to all the GIST items are provided in the back of the text.

- New end-of-chapter exercises have been added, and many of those carried over from the twelfth edition have been significantly revised. Analysis of student responses to the twelfth edition questions in MasteringChemistry® helped us identify and revise or create new questions, prompting improvements and eliminations of some questions. Additionally, analysis of usage of MasteringChemistry® has enhanced our understanding of the ways in which instructors and students have used the end-of-chapter and MasteringChemistry® materials. This, in turn, has led to additional improvements to the content within the text and in the MasteringChemistry® item library. At the end of each chapter, we list the *Learning Outcomes* that students should be able to perform after studying each section. End-of-chapter exercises, both in the text and in MasteringChemistry® offer ample opportunities for students to assess mastery of learning outcomes. We trust the *Learning Outcomes* will help you organize your lectures and tests as the course proceeds.

Organization and Contents

The first five chapters give a largely macroscopic, phenomenological view of chemistry. The basic concepts introduced—such as nomenclature, stoichiometry, and thermochemistry—provide necessary background for many of the laboratory experiments usually performed in general chemistry. We believe that an early introduction to thermochemistry is desirable because so much of our understanding of chemical processes is based on considerations of energy changes. Thermochemistry is also important when we come to a discussion of bond enthalpies. We believe we have produced an effective, balanced approach to teaching thermodynamics in general chemistry, as well as providing students with an introduction to some of the global issues involving energy production and consumption. It is no easy matter to walk the narrow pathway between—on the one hand—trying to teach too much at too high a level and—on the other hand—resorting to oversimplifications. As with the book as a whole, the emphasis has been on imparting *conceptual* understanding, as opposed to presenting equations into which students are supposed to plug numbers.

The next four chapters (Chapters 6–9) deal with electronic structure and bonding. We have largely retained our presentation of atomic orbitals. For more advanced students, *Closer Look* boxes in Chapters 6 and 9 highlight radial probability functions and the phases of orbitals. Our approach of placing this latter discussion in a *Closer Look* box in Chapter 9 enables those who wish to cover this topic to do so, while others may wish to bypass it. In treating this topic and others in Chapters 7 and 9, we have materially enhanced the accompanying figures to more effectively bring home their central messages.

In Chapters 10–13, the focus of the text changes to the next level of the organization of matter: examining the states of matter. Chapters 10 and 11 deal with gases, liquids, and inter-molecular forces, as in earlier editions. Chapter 12 is devoted to solids, presenting an enlarged and more contemporary view of the solid state as well as of modern materials. The chapter provides an opportunity to show how abstract chemical bonding concepts impact real-world applications. The modular organization of the chapter allows you to tailor your coverage to focus on materials (semiconductors, polymers, nanomaterials, and so forth) that are most relevant to your students and your own interests. Chapter 13 treats the formation and properties of solutions in much the same manner as the previous edition.

The next several chapters examine the factors that determine the speed and extent of chemical reactions: kinetics (Chapter 14), equilibria (Chapters 15–17), thermodynamics (Chapter 19), and electrochemistry (Chapter 20). Also in this section is a chapter on environmental chemistry (Chapter 18), in which the concepts developed in preceding chapters are applied to a discussion of the atmosphere and hydrosphere. This chapter has increasingly come to be focused on green chemistry and the impacts of human activities on Earth's water and atmosphere.

After a discussion of nuclear chemistry (Chapter 21), the book ends with three survey chapters. Chapter 22 deals with nonmetals, Chapter 23 with the chemistry of transition metals, including coordination compounds, and Chapter 24 with the chemistry of organic compounds and elementary biochemical themes. These final four chapters are developed in a parallel fashion and can be covered in any order.

Our chapter sequence provides a fairly standard organization, but we recognize that not everyone teaches all the topics in the order we have chosen. We have therefore made sure that instructors can make common changes in teaching sequence with no loss in student comprehension. In particular, many instructors prefer to introduce gases (Chapter 10) after stoichiometry (Chapter 3) rather than with states of matter. The chapter on gases has been written to permit this change with *no* disruption in the flow of material. It is also possible to treat balancing redox equations (Sections 20.1 and 20.2) earlier, after the introduction of redox reactions in Section 4.4. Finally, some instructors like to cover organic chemistry (Chapter 24) right after bonding (Chapters 8 and 9). This, too, is a largely seamless move.

We have brought students into greater contact with descriptive organic and inorganic chemistry by integrating examples throughout the text. You will find pertinent and relevant examples of "real" chemistry woven into all the chapters to illustrate principles and applications. Some chapters, of course, more directly address the "descriptive" properties of elements and their compounds, especially Chapters 4, 7, 11, 18, and 22–24. We also incorporate descriptive organic and inorganic chemistry in the end-of-chapter exercises.

Changes in This Edition

The **What is New in This Edition** section on pp. xx–xxi details changes made throughout the new edition. Beyond a mere listing, however, it is worth dwelling on the general goals we set forth in formulating this new edition. *Chemistry: The Central*

Science has traditionally been valued for its clarity of writing, its scientific accuracy and currency, its strong end-of-chapter exercises, and its consistency in level of coverage. In making changes, we have made sure not to compromise these characteristics, and we have also continued to employ an open, clean design in the layout of the book.

The art program for this thirteenth edition has continued the trajectory set in the twelfth edition: to make greater and more effective use of the figures as learning tools, by drawing the reader more directly into the figure. The art itself has continued to evolve, with modifications of many figures and additions or replacements that teach more effectively. The *Go Figure* feature has been expanded greatly to include a larger number of figures. In the same vein, we have added to the *Give it Some Thought* feature, which stimulates more thoughtful reading of the text and fosters critical thinking.

We provide a valuable overview of each chapter under the *What's Ahead* banner. *Concept links* (∞) continue to provide easy-to-see cross-references to pertinent material covered earlier in the text. The essays titled *Strategies in Chemistry*, which provide advice to students on problem solving and "thinking like a chemist," continue to be an important feature. For example, the new *Strategies in Chemistry* essay at the end of Chapter 3 introduces the new *Design an Experiment* feature and provides a worked out example as guidance.

We have continued to emphasize conceptual exercises in the end-of-chapter exercise materials. The well-received *Visualizing Concepts* exercise category has been continued in this edition. These exercises are designed to facilitate concept understanding through use of models, graphs, and other visual materials. They precede the regular end-of-chapter exercises and are identified in each case with the relevant chapter section number. A generous selection of *Integrative Exercises*, which give students the opportunity to solve problems that integrate concepts from the present chapter with those of previous chapters, is included at the end of each chapter. The importance of integrative problem solving is highlighted by the *Sample Integrative Exercise*, which ends each chapter beginning with Chapter 4. In general, we have included more conceptual end-of-chapter exercises and have made sure that there is a good representation of somewhat more difficult exercises to provide a better mix in terms of topic and level of difficulty. Many of the exercises have been restructured to facilitate their use in MasteringChemistry®. We have made extensive use of the metadata from student use of MasteringChemistry® to analyze end-of-chapter exercises and make appropriate changes, as well as to develop *Learning Outcomes* for each chapter.

New essays in our well-received *Chemistry Put to Work* and *Chemistry and Life* series emphasize world events, scientific discoveries, and medical breakthroughs that bear on topics developed in each chapter. We maintain our focus on the positive aspects of chemistry without neglecting the problems that can arise in an increasingly technological world. Our goal is to help students appreciate the real-world perspective of chemistry and the ways in which chemistry affects their lives.

It is perhaps a natural tendency for chemistry textbooks to grow in length with succeeding editions, but it is one that we have resisted. There are, nonetheless, many new items in this edition, mostly ones that replace other material considered less pertinent. Here is a list of several significant changes in content:

In Chapter 1, the *Closer Look* box on the scientific method has been rewritten. The *Chemistry Put to Work* box, dealing with *Chemistry in the News*, has been completely rewritten, with items that describe diverse ways in which chemistry intersects with the affairs of modern society. The *Chapter Summary* and *Learning Outcomes* sections at the end of the chapter have been rewritten for ease of use by both instructor and student, in this and all chapters in the text. Similarly, the exercises have been thoroughly vetted, modified where this was called for and replaced or added to, here and in all succeeding chapters.

In Chapter 3, graphic elements highlighting the correct approach to problem solving have been added to *Sample Exercises* on calculating an empirical formula from mass percent of the elements present, combustion analysis, and calculating a theoretical yield.

Chapter 5 now presents a more explicit discussion of combined units of measurement, an improved introduction to enthalpy, and more consistent use of color in art.

Changes in Chapter 6 include a significant revision of the discussion of the energy levels of the hydrogen atom, including greater clarity on absorption versus emission processes. There is also a new *Closer Look* box on *Thought Experiments and Schrödinger's Cat*, which gives students a brief glimpse of some of the philosophical issues in quantum mechanics and also connects to the 2012 Nobel Prize in Physics.

In Chapter 7, the emphasis on conceptual thinking was enhanced in several ways: the section on effective nuclear charge was significantly revised to include a classroom-tested analogy, the number of *Go Figure* features was increased substantially, and new end-of-chapter exercises emphasize critical thinking and understanding concepts. In addition, the *Chemistry Put to Work* box on lithium-ion batteries was updated and revised to include discussion of current issues in using these batteries. Finally, the values of ionic radii were revised to be consistent with a recent research study of the best values for these radii.

In Chapter 9, which is one of the most challenging for students, we continue to refine our presentation based on our classroom experience. Twelve new *Go Figure* exercises will stimulate more student thought in a chapter with a large amount of graphic material. The discussion of molecular geometry was made more conceptually oriented. The section on delocalized bonding was completely revised to provide what we believe will be a better introduction that students will find useful in organic chemistry. The *Closer Look* box on phases in orbitals was revamped with improved artwork. We also increased the number of end-of-chapter exercises, especially in the area of molecular orbital theory. The *Design an Experiment* feature in this chapter gives the students the opportunity to explore color and conjugated π systems.

Chapter 10 contains a new *Sample Exercise* that walks the student through the calculations that are needed to understand Torricelli's barometer. Chapter 11 includes an improved definition of hydrogen bonding and updated data for the strengths of intermolecular attractions. Chapter 12 includes the latest updates to materials chemistry, including plastic electronics. New material on the diffusion and mean free path of colloids in solution is added to Chapter 13, making a connection to the diffusion of gas molecules from Chapter 10.

In Chapter 14, ten new *Go Figure* exercises have been added to reinforce many of the concepts presented as figures and graphs in the chapter. The *Design an Experiment* exercise in the chapter connects strongly to the *Closer Look* box on Beer's Law, which is often the basis for spectrometric kinetics experiments performed in the general chemistry laboratory.

The presentation in Chapter 16 was made more closely tied to that in Chapter 15, especially through the use of more initial/change/equilibrium (ICE) charts. The number of conceptual end-of-chapter exercises, including *Visualizing Concepts* features, was increased significantly.

Chapter 17 offers improved clarity on how to make buffers, and when the Henderson–Hasselbalch equation may not be accurate. Chapter 18 has been extensively updated to reflect changes in this rapidly evolving area of chemistry. Two *Closer Look* boxes have been added; one dealing with the shrinking level of water in the Ogallala aquifer and a second with the potential environmental consequences of hydraulic fracking. In Chapter 20, the description of Li-ion batteries has been significantly expanded to reflect the growing importance of these batteries, and a new *Chemistry Put to Work* box on batteries for hybrid and electric vehicles has been added.

Chapter 21 was updated to reflect some of the current issues in nuclear chemistry and more commonly used nomenclature for forms of radiation are now used. Chapter 22 includes an improved discussion of silicates.

In Chapter 23, the section on crystal-field theory (Section 23.6) has undergone considerable revision. The description of how the *d*-orbital energies of a metal ion split in a tetrahedral crystal field has been expanded to put it on par with our treatment of the octahedral geometry, and a new *Sample Exercise* that effectively integrates the links between color, magnetism, and the spectrochemical series has been added. Chapter 24's coverage of organic chemistry and biochemistry now includes oxidation–reduction reactions that organic chemists find most relevant.

To the Student

Chemistry: The Central Science, Thirteenth Edition, has been written to introduce you to modern chemistry. As authors, we have, in effect, been engaged by your instructor to help you learn chemistry. Based on the comments of students and instructors who have used this book in its previous editions, we believe that we have done that job well. Of course, we expect the text to continue to evolve through future editions. We invite you to write to tell us what you like about the book so that we will know where we have helped you most. Also, we would like to learn of any shortcomings so that we might further improve the book in subsequent editions. Our addresses are given at the end of the Preface.

Advice for Learning and Studying Chemistry

Learning chemistry requires both the assimilation of many concepts and the development of analytical skills. In this text, we have provided you with numerous tools to help you succeed in both tasks. If you are going to succeed in your chemistry course, you will have to develop good study habits. Science courses, and chemistry in particular, make different demands on your learning skills than do other types of courses. We offer the following tips for success in your study of chemistry:

Don't fall behind! As the course moves along, new topics will build on material already presented. If you don't keep up in your reading and problem solving, you will find it much harder to follow the lectures and discussions on current topics. Experienced teachers know that students who read the relevant sections of the text *before* coming to a class learn more from the class and retain greater recall. "Cramming" just before an exam has been shown to be an ineffective way to study any subject, chemistry included. So now you know. How important to you, in this competitive world, is a good grade in chemistry?

Focus your study. The amount of information you will be expected to learn can sometimes seem overwhelming. It is essential to recognize those concepts and skills that are particularly important. Pay attention to what your instructor is emphasizing. As you work through the *Sample Exercises* and homework assignments, try to see what general principles and skills they employ. Use the *What's Ahead* feature at the beginning of each chapter to help orient yourself to what is important in each chapter. A single reading of a chapter will simply not be enough for successful learning of chapter concepts and problem-solving skills. You will need to go over assigned materials more than once. Don't skip the *Give It Some Thought* and *Go Figure* features, *Sample Exercises*, and *Practice Exercises*. They are your guides to whether you are learning the material. They are also good preparation for test-taking. The *Learning Outcomes* and *Key Equations* at the end of the chapter should help you focus your study.

Keep good lecture notes. Your lecture notes will provide you with a clear and concise record of what your instructor regards as the most important material to learn. Using your lecture notes in conjunction with this text is the best way to determine which material to study.

Skim topics in the text before they are covered in lecture. Reviewing a topic before lecture will make it easier for you to take good notes. First read the *What's Ahead* points and the end-of-chapter *Summary*; then quickly read through the chapter, skipping Sample Exercises and supplemental sections. Paying attention to the titles of sections and subsections gives you a feeling for the scope of topics. Try to avoid thinking that you must learn and understand everything right away.

You need to do a certain amount of preparation before lecture. More than ever, instructors are using the lecture period not simply as a one-way channel of communication from teacher to student. Rather, they expect students to come to class ready to work on problem solving and critical thinking. Coming to class unprepared is not a good idea for any lecture environment, but it certainly is not an option for an active learning classroom if you aim to do well in the course.

After lecture, carefully read the topics covered in class. As you read, pay attention to the concepts presented and to the application of these concepts in the *Sample Exercises*. Once you think you understand a *Sample Exercise*, test your understanding by working the accompanying *Practice Exercise*.

Learn the language of chemistry. As you study chemistry, you will encounter many new words. It is important to pay attention to these words and to know their meanings or the entities to which they refer. Knowing how to identify chemical substances from their names is an important skill; it can help you avoid painful mistakes on examinations. For example, "chlorine" and "chloride" refer to very different things.

Attempt the assigned end-of-chapter exercises. Working the exercises selected by your instructor provides necessary practice in recalling and using the essential ideas of the chapter. You cannot learn merely by observing; you must be a participant. In particular, try to resist checking the *Student Solutions Manual* (if you have one) until you have made a sincere effort to solve the exercise yourself. If you get stuck on an exercise, however, get help from your instructor, your teaching assistant, or another student. Spending more than 20 minutes on a single exercise is rarely effective unless you know that it is particularly challenging.

Learn to think like a scientist. This book is written by scientists who love chemistry. We encourage you to develop your critical thinking skills by taking advantage of new features in this edition, such as exercises that focus on conceptual learning, and the *Design an Experiment* exercises.

Use online resources. Some things are more easily learned by discovery, and others are best shown in three dimensions. If your instructor has included MasteringChemistry® with your book, take advantage of the unique tools it provides to get the most out of your time in chemistry.

The bottom line is to work hard, study effectively, and use the tools available to you, including this textbook. We want to help you learn more about the world of chemistry and why chemistry is the central science. If you really learn chemistry, you can be the life of the party, impress your friends and parents, and … well, also pass the course with a good grade.

Acknowledgments

The production of a textbook is a team effort requiring the involvement of many people besides the authors who contributed hard work and talent to bring this edition to life. Although their names don't appear on the cover of the book, their creativity, time, and support have been instrumental in all stages of its development and production.

Each of us has benefited greatly from discussions with colleagues and from correspondence with instructors and students both here and abroad. Colleagues have also helped immensely by reviewing our materials, sharing their insights, and providing suggestions for improvements. On this edition, we were particularly blessed with an exceptional group of accuracy checkers who read through our materials looking for both technical inaccuracies and typographical errors.

Thirteenth Edition Reviewers

Yiyan Bai	Houston Community College
Ron Briggs	Arizona State University
Scott Bunge	Kent State University
Jason Coym	University of South Alabama
Ted Clark	The Ohio State University
Michael Denniston	Georgia Perimeter College
Patrick Donoghue	Appalachian State University
Luther Giddings	Salt Lake Community College
Jeffrey Kovac	University of Tennessee
Charity Lovett	Seattle University
Michael Lufaso	University of North Florida
Diane Miller	Marquette University
Gregory Robinson	University of Georgia
Melissa Schultz	The College of Wooster
Mark Schraf	West Virginia University
Richard Spinney	The Ohio State University
Troy Wood	SUNY Buffalo
Kimberly Woznack	California University of Pennsylvania
Edward Zovinka	Saint Francis University

Thirteenth Edition Accuracy Reviewers

Luther Giddings	Salt Lake Community College
Jesudoss Kingston	Iowa State University
Michael Lufaso	University of North Florida
Pamela Marks	Arizona State University
Lee Pedersen	University of North Carolina
Troy Wood	SUNY Buffalo

Thirteenth Edition Focus Group Participants

Tracy Birdwhistle	Xavier University
Cheryl Frech	University of Central Oklahoma
Bridget Gourley	DePauw University
Etta Gravely	North Carolina A&T State University
Thomas J. Greenbowe	Iowa State University
Jason Hofstein	Siena College
Andy Jorgensen	University of Toledo
David Katz	Pima Community College
Sarah Schmidtke	The College of Wooster
Linda Schultz	Tarleton State University
Bob Shelton	Austin Peay State University
Stephen Sieck	Grinnell College
Mark Thomson	Ferris State University

MasteringChemistry® Summit Participants

Phil Bennett	Santa Fe Community College
Jo Blackburn	Richland College
John Bookstaver	St. Charles Community College
David Carter	Angelo State University
Doug Cody	Nassau Community College
Tom Dowd	Harper College
Palmer Graves	Florida International University
Margie Haak	Oregon State University
Brad Herrick	Colorado School of Mines
Jeff Jenson	University of Findlay
Jeff McVey	Texas State University at San Marcos
Gary Michels	Creighton University
Bob Pribush	Butler University
Al Rives	Wake Forest University
Joel Russell	Oakland University
Greg Szulczewski	University of Alabama, Tuscaloosa
Matt Tarr	University of New Orleans
Dennis Taylor	Clemson University
Harold Trimm	Broome Community College
Emanuel Waddell	University of Alabama, Huntsville
Kurt Winklemann	Florida Institute of Technology
Klaus Woelk	University of Missouri, Rolla
Steve Wood	Brigham Young University

Reviewers of Previous Editions of *Chemistry: The Central Science*

S.K. Airee	University of Tennessee
John J. Alexander	University of Cincinnati
Robert Allendoerfer	SUNY Buffalo
Patricia Amateis	Virginia Polytechnic Institute and State University
Sandra Anderson	University of Wisconsin
John Arnold	University of California
Socorro Arteaga	El Paso Community College
Margaret Asirvatham	University of Colorado
Todd L. Austell	University of North Carolina, Chapel Hill
Melita Balch	University of Illinois at Chicago
Rosemary Bartoszek-Loza	The Ohio State University
Rebecca Barlag	Ohio University
Hafed Bascal	University of Findlay

Boyd Beck	Snow College
Kelly Beefus	Anoka-Ramsey Community College
Amy Beilstein	Centre College
Donald Bellew	University of New Mexico
Victor Berner	New Mexico Junior College
Narayan Bhat	University of Texas, Pan American
Merrill Blackman	United States Military Academy
Salah M. Blaih	Kent State University
James A. Boiani	SUNY Geneseo
Leon Borowski	Diablo Valley College
Simon Bott	University of Houston
Kevin L. Bray	Washington State University
Daeg Scott Brenner	Clark University
Gregory Alan Brewer	Catholic University of America
Karen Brewer	Virginia Polytechnic Institute and State University
Edward Brown	Lee University
Gary Buckley	Cameron University
Carmela Byrnes	Texas A&M University
B. Edward Cain	Rochester Institute of Technology
Kim Calvo	University of Akron
Donald L. Campbell	University of Wisconsin
Gene O. Carlisle	Texas A&M University
Elaine Carter	Los Angeles City College
Robert Carter	University of Massachusetts at Boston Harbor
Ann Cartwright	San Jacinto Central College
David L. Cedeño	Illinois State University
Dana Chatellier	University of Delaware
Stanton Ching	Connecticut College
Paul Chirik	Cornell University
Tom Clayton	Knox College
William Cleaver	University of Vermont
Beverly Clement	Blinn College
Robert D. Cloney	Fordham University
John Collins	Broward Community College
Edward Werner Cook	Tunxis Community Technical College
Elzbieta Cook	Louisiana State University
Enriqueta Cortez	South Texas College
Thomas Edgar Crumm	Indiana University of Pennsylvania
Dwaine Davis	Forsyth Tech Community College
Ramón López de la Vega	Florida International University
Nancy De Luca	University of Massachusetts, Lowell North Campus
Angel de Dios	Georgetown University
John M. DeKorte	Glendale Community College
Daniel Domin	Tennessee State University
James Donaldson	University of Toronto
Bill Donovan	University of Akron
Stephen Drucker	University of Wisconsin-Eau Claire
Ronald Duchovic	Indiana University–Purdue University at Fort Wayne
Robert Dunn	University of Kansas
David Easter	Southwest Texas State University
Joseph Ellison	United States Military Academy
George O. Evans II	East Carolina University
James M. Farrar	University of Rochester
Debra Feakes	Texas State University at San Marcos
Gregory M. Ferrence	Illinois State University
Clark L. Fields	University of Northern Colorado
Jennifer Firestine	Lindenwood University
Jan M. Fleischner	College of New Jersey
Paul A. Flowers	University of North Carolina at Pembroke
Michelle Fossum	Laney College
Roger Frampton	Tidewater Community College
Joe Franek	University of Minnesota
David Frank	California State University
Cheryl B. Frech	University of Central Oklahoma
Ewa Fredette	Moraine Valley College
Kenneth A. French	Blinn College
Karen Frindell	Santa Rosa Junior College
John I. Gelder	Oklahoma State University
Robert Gellert	Glendale Community College
Paul Gilletti	Mesa Community College
Peter Gold	Pennsylvania State University
Eric Goll	Brookdale Community College
James Gordon	Central Methodist College
John Gorden	Auburn University
Thomas J. Greenbowe	Iowa State University
Michael Greenlief	University of Missouri
Eric P. Grimsrud	Montana State University
John Hagadorn	University of Colorado
Randy Hall	Louisiana State University
John M. Halpin	New York University
Marie Hankins	University of Southern Indiana
Robert M. Hanson	St. Olaf College
Daniel Haworth	Marquette University
Michael Hay	Pennsylvania State University
Inna Hefley	Blinn College
David Henderson	Trinity College
Paul Higgs	Barry University
Carl A. Hoeger	University of California, San Diego
Gary G. Hoffman	Florida International University
Deborah Hokien	Marywood University
Robin Horner	Fayetteville Tech Community College
Roger K. House	Moraine Valley College
Michael O. Hurst	Georgia Southern University
William Jensen	South Dakota State University
Janet Johannessen	County College of Morris
Milton D. Johnston, Jr.	University of South Florida
Andrew Jones	Southern Alberta Institute of Technology
Booker Juma	Fayetteville State University
Ismail Kady	East Tennessee State University
Siam Kahmis	University of Pittsburgh
Steven Keller	University of Missouri
John W. Kenney	Eastern New Mexico University
Neil Kestner	Louisiana State University
Carl Hoeger	University of California at San Diego
Leslie Kinsland	University of Louisiana
Jesudoss Kingston	Iowa State University
Louis J. Kirschenbaum	University of Rhode Island
Donald Kleinfelter	University of Tennessee, Knoxville
Daniela Kohen	Carleton University
David Kort	George Mason University
George P. Kreishman	University of Cincinnati
Paul Kreiss	Anne Arundel Community College
Manickham Krishnamurthy	Howard University
Sergiy Kryatov	Tufts University
Brian D. Kybett	University of Regina
William R. Lammela	Nazareth College
John T. Landrum	Florida International University
Richard Langley	Stephen F. Austin State University
N. Dale Ledford	University of South Alabama
Ernestine Lee	Utah State University

David Lehmpuhl	University of Southern Colorado
Robley J. Light	Florida State University
Donald E. Linn, Jr.	Indiana University–Purdue University Indianapolis
David Lippmann	Southwest Texas State
Patrick Lloyd	Kingsborough Community College
Encarnacion Lopez	Miami Dade College, Wolfson
Arthur Low	Tarleton State University
Gary L. Lyon	Louisiana State University
Preston J. MacDougall	Middle Tennessee State University
Jeffrey Madura	Duquesne University
Larry Manno	Triton College
Asoka Marasinghe	Moorhead State University
Earl L. Mark	ITT Technical Institute
Pamela Marks	Arizona State University
Albert H. Martin	Moravian College
Przemyslaw Maslak	Pennsylvania State University
Hilary L. Maybaum	ThinkQuest, Inc.
Armin Mayr	El Paso Community College
Marcus T. McEllistrem	University of Wisconsin
Craig McLauchlan	Illinois State University
Jeff McVey	Texas State University at San Marcos
William A. Meena	Valley College
Joseph Merola	Virginia Polytechnic Institute and State University
Stephen Mezyk	California State University
Eric Miller	San Juan College
Gordon Miller	Iowa State University
Shelley Minteer	Saint Louis University
Massoud (Matt) Miri	Rochester Institute of Technology
Mohammad Moharerrzadeh	Bowie State University
Tracy Morkin	Emory University
Barbara Mowery	York College
Kathleen E. Murphy	Daemen College
Kathy Nabona	Austin Community College
Robert Nelson	Georgia Southern University
Al Nichols	Jacksonville State University
Ross Nord	Eastern Michigan University
Jessica Orvis	Georgia Southern University
Mark Ott	Jackson Community College
Jason Overby	College of Charleston
Robert H. Paine	Rochester Institute of Technology
Robert T. Paine	University of New Mexico
Sandra Patrick	Malaspina University College
Mary Jane Patterson	Brazosport College
Tammi Pavelec	Lindenwood University
Albert Payton	Broward Community College
Christopher J. Peeples	University of Tulsa
Kim Percell	Cape Fear Community College
Gita Perkins	Estrella Mountain Community College
Richard Perkins	University of Louisiana
Nancy Peterson	North Central College
Robert C. Pfaff	Saint Joseph's College
John Pfeffer	Highline Community College
Lou Pignolet	University of Minnesota
Bernard Powell	University of Texas
Jeffrey A. Rahn	Eastern Washington University
Steve Rathbone	Blinn College
Scott Reeve	Arkansas State University
John Reissner	University of North Carolina
Helen Richter	University of Akron
Thomas Ridgway	University of Cincinnati
Mark G. Rockley	Oklahoma State University
Lenore Rodicio	Miami Dade College
Amy L. Rogers	College of Charleston
Jimmy R. Rogers	University of Texas at Arlington
Kathryn Rowberg	Purdue University at Calumet
Steven Rowley	Middlesex Community College
James E. Russo	Whitman College
Theodore Sakano	Rockland Community College
Michael J. Sanger	University of Northern Iowa
Jerry L. Sarquis	Miami University
James P. Schneider	Portland Community College
Mark Schraf	West Virginia University
Gray Scrimgeour	University of Toronto
Paula Secondo	Western Connecticut State University
Michael Seymour	Hope College
Kathy Thrush Shaginaw	Villanova University
Susan M. Shih	College of DuPage
David Shinn	University of Hawaii at Hilo
Lewis Silverman	University of Missouri at Columbia
Vince Sollimo	Burlington Community College
David Soriano	University of Pittsburgh-Bradford
Eugene Stevens	Binghamton University
Matthew Stoltzfus	The Ohio State University
James Symes	Cosumnes River College
Iwao Teraoka	Polytechnic University
Domenic J. Tiani	University of North Carolina, Chapel Hill
Edmund Tisko	University of Nebraska at Omaha
Richard S. Treptow	Chicago State University
Michael Tubergen	Kent State University
Claudia Turro	The Ohio State University
James Tyrell	Southern Illinois University
Michael J. Van Stipdonk	Wichita State University
Philip Verhalen	Panola College
Ann Verner	University of Toronto at Scarborough
Edward Vickner	Gloucester County Community College
John Vincent	University of Alabama
Maria Vogt	Bloomfield College
Tony Wallner	Barry University
Lichang Wang	Southern Illinois University
Thomas R. Webb	Auburn University
Clyde Webster	University of California at Riverside
Karen Weichelman	University of Louisiana-Lafayette
Paul G. Wenthold	Purdue University
Laurence Werbelow	New Mexico Institute of Mining and Technology
Wayne Wesolowski	University Of Arizona
Sarah West	University of Notre Dame
Linda M. Wilkes	University at Southern Colorado
Charles A. Wilkie	Marquette University
Darren L. Williams	West Texas A&M University
Troy Wood	SUNY Buffalo
Thao Yang	University of Wisconsin
David Zax	Cornell University
Dr. Susan M. Zirpoli	Slippery Rock University

We would also like to express our gratitude to our many team members at Pearson whose hard work, imagination, and commitment have contributed so greatly to the final form of this edition: Terry Haugen, our senior editor, who has brought energy and imagination to this edition as he has to earlier ones; Chris Hess, our chemistry editor, for many fresh ideas and his unflagging enthusiasm, continuous encouragement, and support; Jennifer Hart, Director of Development, who has brought her experience and insight to oversight of the entire project; Jessica Moro, our project editor, who very effectively coordinated the scheduling and tracked the multidimensional deadlines that come with a project of this magnitude; Jonathan Cottrell our marketing manager, for his energy, enthusiasm, and creative promotion of our text; Carol Pritchard-Martinez, our development editor, whose depth of experience, good judgment, and careful attention to detail were invaluable to this revision, especially in keeping us on task in terms of consistency and student understanding; Donna, our copy editor, for her keen eye; Beth Sweeten, our project manager, and Gina Cheselka, who managed the complex responsibilities of bringing the design, photos, artwork, and writing together with efficiency and good cheer. The Pearson team is a first-class operation.

There are many others who also deserve special recognition, including the following: Greg Johnson, our production editor, who skillfully kept the process moving and us authors on track; Kerri Wilson, our photo researcher, who was so effective in finding photos to bring chemistry to life for students; and Roxy Wilson (University of Illinois), who so ably coordinated the difficult job of working out solutions to the end-of-chapter exercises. Finally, we wish to thank our families and friends for their love, support, encouragement, and patience as we brought this thirteenth edition to completion.

Theodore L. Brown
Department of Chemistry
University of Illinois at
Urbana-Champaign
Urbana, IL 61801
tlbrown@illinois.edu or
tlbrown1@earthlink.net

H. Eugene LeMay, Jr.
Department of Chemistry
University of Nevada
Reno, NV 89557
lemay@unr.edu

Bruce E. Bursten
Department of Chemistry
University of Tennessee
Knoxville, TN 37996
bbursten@utk.edu

Catherine J. Murphy
Department of Chemistry
University of Illinois at
Urbana-Champaign
Urbana, IL 61801
murphycj@illinois.edu.

Patrick M. Woodward
Department of Chemistry
and Biochemistry
The Ohio State University
Columbus, OH 43210
woodward@chemistry.
ohio-state.edu

Matthew W. Stoltzfus
Department of Chemistry
and Biochemistry
The Ohio State University
Columbus, OH 43210
stoltzfus.5@osu.edu

List of Resources

For Students

MasteringChemistry®
(http://www.masteringchemistry.com)
MasteringChemistry® is the most effective, widely used online tutorial, homework and assessment system for chemistry. It helps instructors maximize class time with customizable, easy-to-assign, and automatically graded assessments that motivate students to learn outside of class and arrive prepared for lecture. These assessments can easily be customized and personalized by instructors to suit their individual teaching style. The powerful gradebook provides unique insight into student and class performance even before the first test. As a result, instructors can spend class time where students need it most.

Pearson eText The integration of Pearson eText within MasteringChemistry® gives students with eTexts easy access to the electronic text when they are logged into MasteringChemistry®. Pearson eText pages look exactly like the printed text, offering powerful new functionality for students and instructors. Users can create notes, highlight text in different colors, create bookmarks, zoom, view in single-page or two-page view, and more.

Students Guide (0-321-94928-5) Prepared by James C. Hill of California State University. This book assists students through the text material with chapter overviews, learning objectives, a review of key terms, as well as self-tests with answers and explanations. This edition also features MCAT practice questions.

Solutions to Red Exercises (0-321-94926-9) Prepared by Roxy Wilson of the University of Illinois, Urbana-Champaign. Full solutions to all the red-numbered exercises in the text are provided. (Short answers to red exercises are found in the appendix of the text.)

Solutions to Black Exercises (0-321-94927-7) Prepared by Roxy Wilson of the University of Illinois, Urbana-Champaign. Full solutions to all the black-numbered exercises in the text are provided.

Laboratory Experiments (0-321-94991-9) Prepared by John H. Nelson of the University of Nevada, and Michael Lufaso of the University of North Florida with contributions by Matthew Stoltzfus of The Ohio State University. This manual contains 40 finely tuned experiments chosen to introduce students to basic lab techniques and to illustrate core chemical principles. This new edition has been revised with the addition of four brand new experiments to correlate more tightly with the text. You can also customize these labs through Catalyst, our custom database program. For more information, visit *http://www.pearsoncustom.com/custom-library/*

For Instructors

Solutions to Exercises (0-321-94925-0) Prepared by Roxy Wilson of the University of Illinois, Urbana-Champaign. This manual contains all end-of-chapter exercises in the text. With an instructor's permission, this manual may be made available to students.

Online Instructor Resource Center (0-321-94923-4) This resource provides an integrated collection of resources to help instructors make efficient and effective use of their time. It features all artwork from the text, including figures and tables in PDF format for high-resolution printing, as well as five prebuilt PowerPoint™ presentations. The first presentation contains the images embedded within PowerPoint slides. The second includes a complete lecture outline that is modifiable by the user. The final three presentations contain worked "in-chapter" sample exercises and questions to be used with Classroom Response Systems. The Instructor Resource Center also contains movies, animations, and electronic files of the Instructor Resource Manual, as well as the Test Item File.

TestGen Testbank (0-321-94924-2) Prepared by Andrea Leonard of the University of Louisiana. The Test Item File now provides a selection of more than 4,000 test questions with 200 new questions in the thirteenth edition and 200 additional algorithmic questions.

Online Instructor Resource Manual (0-321-94929-3) Prepared by Linda Brunauer of Santa Clara University and Elzbieta Cook of Louisiana State University. Organized by chapter, this manual offers detailed lecture outlines and complete descriptions of all available lecture demonstrations, interactive media assets, common student misconceptions, and more.

Annotated Instructor's Edition to Laboratory Experiments (0-321-98608-3) Prepared by John H. Nelson of the University of Nevada, and Michael Lufaso of the University of North Florida with contributions by Matthew Stoltzfus of the Ohio State University. This AIE combines the full student lab manual with appendices covering the proper disposal of chemical waste, safety instructions for the lab, descriptions of standard lab equipment, answers to questions, and more.

WebCT Test Item File (IRC download only)
0-321-94931-5

Blackboard Test Item File (IRC download only)
0-321-94930-7

About the Authors

THE BROWN/LEMAY/BURSTEN/ MURPHY/WOODWARD/STOLTZFUS AUTHOR TEAM values collaboration as an integral component to overall success. While each author brings unique talent, research interests, and teaching experiences, the team works together to review and develop the entire text. It is this collaboration that keeps the content ahead of educational trends and contributes to continuous innovations in teaching and learning throughout the text and technology. Some of the new key features in the thirteenth edition and accompanying MasteringChemistry® course are highlighted on the following pages.

THEODORE L. BROWN received his Ph.D. from Michigan State University in 1956. Since then, he has been a member of the faculty of the University of Illinois, Urbana-Champaign, where he is now Professor of Chemistry, Emeritus. He served as Vice Chancellor for Research, and Dean of The Graduate College, from 1980 to 1986, and as Founding Director of the Arnold and Mabel Beckman Institute for Advanced Science and Technology from 1987 to 1993. Professor Brown has been an Alfred P. Sloan Foundation Research Fellow and has been awarded a Guggenheim Fellowship. In 1972 he was awarded the American Chemical Society Award for Research in Inorganic Chemistry and received the American Chemical Society Award for Distinguished Service in the Advancement of Inorganic Chemistry in 1993. He has been elected a Fellow of the American Association for the Advancement of Science, the American Academy of Arts and Sciences, and the American Chemical Society.

H. EUGENE LEMAY, JR., received his B.S. degree in Chemistry from Pacific Lutheran University (Washington) and his Ph.D. in Chemistry in 1966 from the University of Illinois, Urbana-Champaign. He then joined the faculty of the University of Nevada, Reno, where he is currently Professor of Chemistry, Emeritus. He has enjoyed Visiting Professorships at the University of North Carolina at Chapel Hill, at the University College of Wales in Great Britain, and at the University of California, Los Angeles. Professor LeMay is a popular and effective teacher, who has taught thousands of students during more than 40 years of university teaching. Known for the clarity of his lectures and his sense of humor, he has received several teaching awards, including the University Distinguished Teacher of the Year Award (1991) and the first Regents' Teaching Award given by the State of Nevada Board of Regents (1997).

BRUCE E. BURSTEN received his Ph.D. in Chemistry from the University of Wisconsin in 1978. After two years as a National Science Foundation Postdoctoral Fellow at Texas A&M University, he joined the faculty of The Ohio State University, where he rose to the rank of Distinguished University Professor. In 2005, he moved to the University of Tennessee, Knoxville, as Distinguished Professor of Chemistry and Dean of the College of Arts and Sciences. Professor Bursten has been a Camille and Henry Dreyfus Foundation Teacher-Scholar and an Alfred P. Sloan Foundation Research Fellow, and he is a Fellow of both the American Association for the Advancement of Science and the American Chemical Society. At Ohio State he has received the University Distinguished Teaching Award in 1982 and 1996, the Arts and Sciences Student Council Outstanding Teaching Award in 1984, and the University Distinguished Scholar Award in 1990. He received the Spiers Memorial Prize and Medal of the Royal Society of Chemistry in 2003, and the Morley Medal of the Cleveland Section of the American Chemical Society in 2005. He was President of the American Chemical Society for 2008. In addition to his teaching and service activities, Professor Bursten's research program focuses on compounds of the transition-metal and actinide elements.

CATHERINE J. MURPHY received two B.S. degrees, one in Chemistry and one in Biochemistry, from the University of Illinois, Urbana-Champaign, in 1986. She received her Ph.D. in Chemistry from the University of Wisconsin in 1990. She was a National Science Foundation and National Institutes of Health Postdoctoral Fellow at the California Institute of Technology from 1990 to 1993. In 1993, she joined the faculty of the University of South Carolina, Columbia, becoming the Guy F. Lipscomb Professor of Chemistry in 2003. In 2009 she moved to the University of Illinois, Urbana-Champaign, as the Peter C. and Gretchen Miller Markunas Professor of Chemistry. Professor Murphy has been honored for both research and teaching as a Camille Dreyfus Teacher-Scholar, an Alfred P. Sloan Foundation Research Fellow, a Cottrell Scholar of the Research Corporation, a National Science Foundation CAREER Award winner, and a subsequent NSF Award for Special Creativity. She has also received a USC Mortar Board Excellence in Teaching Award, the USC Golden Key Faculty Award for Creative Integration of Research and Undergraduate Teaching, the USC Michael J. Mungo Undergraduate Teaching Award, and the USC Outstanding Undergraduate Research Mentor Award. Since 2006, Professor Murphy has served as a Senior Editor for the Journal of Physical Chemistry. In 2008 she was elected a Fellow of the American Association for the Advancement of Science. Professor Murphy's research program focuses on the synthesis and optical properties of inorganic nanomaterials, and on the local structure and dynamics of the DNA double helix.

PATRICK M. WOODWARD received B.S. degrees in both Chemistry and Engineering from Idaho State University in 1991. He received a M.S. degree in Materials Science and a Ph.D. in Chemistry from Oregon State University in 1996. He spent two years as a postdoctoral researcher in the Department of Physics at Brookhaven National Laboratory. In 1998, he joined the faculty of the Chemistry Department at The Ohio State University where he currently holds the rank of Professor. He has enjoyed visiting professorships at the University of Bordeaux in France and the University of Sydney in Australia. Professor Woodward has been an Alfred P. Sloan Foundation Research Fellow and a National Science Foundation CAREER Award winner. He currently serves as an Associate Editor to the Journal of Solid State Chemistry and as the director of the Ohio REEL program, an NSF-funded center that works to bring authentic research experiments into the laboratories of first- and second-year chemistry classes in 15 colleges and universities across the state of Ohio. Professor Woodward's research program focuses on understanding the links between bonding, structure, and properties of solid-state inorganic functional materials.

MATTHEW W. STOLTZFUS received his B.S. degree in Chemistry from Millersville University in 2002 and his Ph. D. in Chemistry in 2007 from The Ohio State University. He spent two years as a teaching postdoctoral assistant for the Ohio REEL program, an NSF-funded center that works to bring authentic research experiments into the general chemistry lab curriculum in 15 colleges and universities across the state of Ohio. In 2009, he joined the faculty of Ohio State where he currently holds the position of Chemistry Lecturer. In addition to lecturing general chemistry, Stoltzfus accepted the Faculty Fellow position for the Digital First Initiative, inspiring instructors to offer engaging digital learning content to students through emerging technology. Through this initiative, he developed an iTunes U general chemistry course, which has attracted over 120,000 students from all over the world. Stoltzfus has received several teaching awards, including the inaugural Ohio State University 2013 Provost's Award for Distinguished Teaching by a Lecturer and he is recognized as an Apple Distinguished Educator.

Data-Driven Analytics
A New Direction in Chemical Education

Authors traditionally revise roughly 25% of the end of chapter questions when producing a new edition. These changes typically involve modifying numerical variables/identities of chemical formulas to make them "new" to the next batch of students. While these changes are appropriate for the printed version of the text, one of the strengths of MasteringChemistry® is its ability to randomize variables so that every student receives a "different" problem. Hence, the effort which authors have historically put into changing variables can now be used to improve questions.

In order to make informed decisions, the author team consulted the massive reservoir of data available through MasteringChemistry® to revise their question bank. In particular, they analyzed which problems were frequently assigned and why; they paid careful attention to the amount of time it took students to work through a problem (flagging those that took longer than expected) and they observed the wrong answer submissions and hints used (a measure used to calculate the difficulty of problems). This "metadata" served as a starting point for the discussion of which end of chapter questions should be changed.

For example, the breadth of ideas presented in Chapter 9 challenges students to understand three-dimensional visualization while simultaneously introducing several new concepts (particularly VSEPR, hybrids, and Molecular Orbital theory) that challenge their critical thinking skills. In revising the exercises for the chapter, the authors drew on the metadata as well as their own experience in assigning Chapter 9 problems in Mastering Chemistry. From these analyses, we were able to articulate two general revision guidelines.

1. **Improve coverage of topic areas that were underutilized:** In Chapter 9, the authors noticed that there was a particularly low usage rate for questions concerning Molecular Orbital Theory. Based on the metadata and their own teaching experience with Mastering, they recognized an opportunity to expand the coverage of MO theory. Two brand new exercises that emphasize the basics of MO theory were the result of this analysis including the example below. This strategy was replicated throughout the entire book.

9.8 The drawing below shows the overlap of two hybrid orbitals to form a bond in a hydrocarbon. (a) Which of the following types of bonds is being formed: (i) $C\!-\!C\,\sigma$, (ii) $C\!-\!C\,\pi$, or (iii) $C\!-\!H\,\sigma$? (b) Which of the following could be the identity of the hydrocarbon: (i) CH_4, (ii) C_2H_6, (iii) C_2H_4, or (iv) C_2H_2? [Section 9.6]

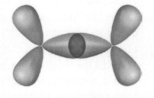

9.10 The following is part of a molecular orbital energy-level diagram for MOs constructed from $1s$ atomic orbitals.

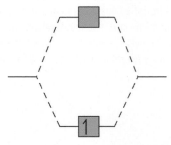

(a) What labels do we use for the two MOs shown? (b) For which of the following molecules or ions could this be the energy-level diagram:

$$H_2, He_2, H_2{}^+, He_2{}^+, \text{ or } H_2{}^-?$$

(c) What is the bond order of the molecule or ion? (d) If an electron is added to the system, into which of the MOs will it be added? [Section 9.7]

2. **Revise the least assigned existing problems.** Much of the appeal of MasteringChemistry®
for students is the immediate feedback they get when they hit submit, which also provides
an opportunity to confront any misconceptions right away. For instructors, the appeal is that
these problems are automatically graded. Essay questions fail to provide these advantages
since they must be graded by an instructor before a student may receive feedback. Wherever
possible, we revised current essay questions to include automatically graded material.

Bottom Line: The revision of the end of chapter questions in this edition is informed by robust
data-driven analytics providing a new level of pedagogically-sound assessments for your students,
all while making the time they spend working these problems even more valuable.

9.93 An AB_5 molecule adopts the geometry shown here. (**a**) What
is the name of this geometry? (**b**) Do you think there are any
nonbonding electron pairs on atom A? (**c**) Suppose the atoms
B are halogen atoms. Of which group in the periodic table is
atom A a member: (i) Group 5A, (ii) Group 6A, (iii) Group
7A, (iv) Group 8A, or (v) More information is needed?

9.4 The molecule shown here is *difluoromethane* (CH_2F_2), which
is used as a refrigerant called R-32. (**a**) Based on the structure, how many electron domains surround the C atom in
this molecule? (**b**) Would the molecule have a nonzero dipole
moment? (**c**) If the molecule is polar, which of the following
describes the direction of the overall dipole moment vector
in the molecule: (i) from the carbon atom toward a fluorine
atom, (ii) from the carbon atom to a point midway between
the fluorine atoms, (iii) from the carbon atom to a point midway between the hydrogen atoms, or (iv) From the carbon
atom toward a hydrogen atom? [Sections 9.2 and 9.3]

Helping Students Think Like Scientists

Design an Experiment

Starting with Chapter 3, every chapter will feature a *Design an Experiment* exercise. The goal of these exercises is to challenge students to think like a scientist, imagining what kind of data needs to be collected and what sort of experimental procedures will provide them the data needed to answer the question. These exercises tend to be integrative, forcing students to draw on many of the skills they have learned in the current and previous chapters.

Strategies in Chemistry

Design an Experiment

One of the most important skills you can learn in school is how to think like a scientist. Questions such as: "What experiment might test this hypothesis?", "How do I interpret these data?", and "Do these data support the hypothesis?" are asked every day by chemists and other scientists as they go about their work.

We want you to become a good critical thinker as well as an active, logical, and curious learner. For this purpose, starting in Chapter 3, we include at the end of each chapter a special exercise called "Design an Experiment." Here is an example:

of the pan? If so, you could weigh it, and calculate the percentage of solids in milk, which would offer good evidence that milk is a mixture. If there is no residue after boiling, then you still cannot distinguish between the two possibilities.

What other experiments might you do to demonstrate that milk is a mixture? You could put a sample of milk in a centrifuge, which you might have used in a biology lab, spin your sample and observe if any solids collect at the bottom of the centrifuge tube; large molecules can be separated in this way from a mixture. Measurement of the mass of the solid at the bottom of the tube is a way to obtain a value for the % solids in milk, and also tells you that milk is indeed a mixture.

Design an Experiment topics include:

Ch 3: Formation of Sulfur Oxides
Ch 4: Identification of Mysterious White Powders
Ch 5: Joule Experiment
Ch 6: Photoelectric Effect and Electron Configurations
Ch 7: Chemistry of Potassium Superoxide
Ch 8: Benzene Resonance
Ch 9: Colors of Organic Dyes
Ch 10: Identification of an Unknown Noble Gas
Ch 11: Hydraulic Fluids
Ch 12: Polymers
Ch 13: Volatile Solvent Molecules

Ch 14: Reaction Kinetics via Spectrophotometry
Ch 15: Beer's Law and Visible-Light Spectroscopy
Ch 16: Acidity/Basicity of an Unknown Liquid
Ch 17: Understanding Differences in pKa
Ch 18: Effects of Fracking on Groundwater
Ch 19: Drug Candidates and the Equilibrium Constant
Ch 20: Voltaic Cells
Ch 21: Discovery and Properties of Radium
Ch 22: Identification of Unknowns
Ch 23: Synthesis and Characterization of a Coordination Compound
Ch 24: Quaternary Structure in Proteins

Go Figure

Go Figure questions encourage students to stop and analyze the artwork in the text, for conceptual understanding. "Voice Balloons" in selected figures help students break down and understand the components of the image. These questions are also available in MasteringChemistry®. The number of Go Figure questions in the thirteenth edition has increased by 25%.

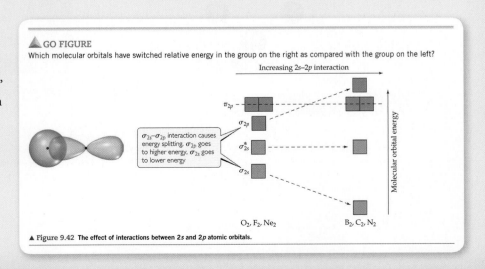

▲ GO FIGURE
Which molecular orbitals have switched relative energy in the group on the right as compared with the group on the left?

Increasing 2s–2p interaction

π_{2p}
σ_{2p}
σ_{2s}^*
σ_{2s}

σ_{2s}–σ_{2p} interaction causes energy splitting. σ_{2p} goes to higher energy, σ_{2s} goes to lower energy

Molecular orbital energy

O_2, F_2, Ne_2 B_2, C_2, N_2

▲ Figure 9.42 The effect of interactions between 2s and 2p atomic orbitals.

Practice Exercises

A major new feature of this edition is the addition of a second Practice Exercise to accompany each Sample Exercise within the chapters. The new Practice Exercises are multiple-choice with correct answers provided for the students in an appendix. Specific wrong answer feedback, written by the authors, will be available in MasteringChemistry® The primary goal of the new Practice Exercise feature is to provide students with an additional problem to test mastery of the concepts in the text and to address the most common conceptual misunderstandings. To ensure the questions touched on the most common student misconceptions, the authors consulted the ACS Chemistry Concept inventory before writing their questions.

SAMPLE EXERCISE 2.6 | Relating Empirical and Molecular Formulas

Write the empirical formulas for **(a)** glucose, a substance also known as either blood sugar or dextrose—molecular formula $C_6H_{12}O_6$; **(b)** nitrous oxide, a substance used as an anesthetic and commonly called laughing gas—molecular formula N_2O.

SOLUTION

(a) The subscripts of an empirical formula are the smallest whole-number ratios. The smallest ratios are obtained by dividing each subscript by the largest common factor, in this case 6. The resultant empirical formula for glucose is CH_2O.

(b) Because the subscripts in N_2O are already the lowest integral numbers, the empirical formula for nitrous oxide is the same as its molecular formula, N_2O.

What are the molecular and empirical formulas of this substance?
(a) C_2O_2, CO_2, **(b)** C_4O, CO, **(c)** CO_2, CO_2, **(d)** C_4O_2, C_2O, **(e)** C_2O, CO_2.

Practice Exercise 1

Tetracarbon dioxide is an unstable oxide of carbon with the following molecular structure:

Practice Exercise 2

Give the empirical formula for *decaborane*, whose molecular formula is $B_{10}H_{14.}$

Give It Some Thought (GIST) questions

These informal, sharply-focused exercises allow students the opportunity to gauge whether they are "getting it" as they read the text. The number of GIST questions has increased throughout the text as well as in MasteringChemistry®.

 Give It Some Thought

A solution of SO_2 in water contains 0.00023 g of SO_2 per liter of solution. What is the concentration of SO_2 in ppm? In ppb?

Active and Visual

The most effective learning happens when students actively participate and interact with material in order to truly internalize key concepts. The Brown/Lemay/Bursten/Murphy/ Woodward/Stoltzfus author team has spent decades refining their text based on educational research to the extent that it has largely defined how the general chemistry course is taught. With the thirteenth edition, these authors have extended this tradition by giving each student a way to personalize their learning experience through MasteringChemistry®. The MasteringChemistry® course for Brown/Lemay/Bursten/Murphy/Woodward/Stoltzfus evolves learning and technology usage far beyond the lecture-homework model. Many of these resources can be used pre-lecture, during class, and for assessment while providing each student with a personalized learning experience which gives them the greatest chance of succeeding.

Learning Catalytics

Learning Catalytics™ is a "bring your own device" student engagement, assessment, and classroom intelligence system. With Learning Catalytics™ you can:

- Assess students in real time, using open-ended tasks to probe student understanding.
- Understand immediately where students are and adjust your lecture accordingly.
- Improve your students' critical-thinking skills.
- Access rich analytics to understand student performance.
- Add your own questions to make Learning Catalytics™ fit your course exactly.
- Manage student interactions with intelligent grouping and timing.

Learning Catalytics™ is a technology that has grown out of twenty years of cutting-edge research, innovation, and implementation of interactive teaching and peer instruction.

Learning Catalytics™ will be included with the purchase of MasteringChemistry® with eText.

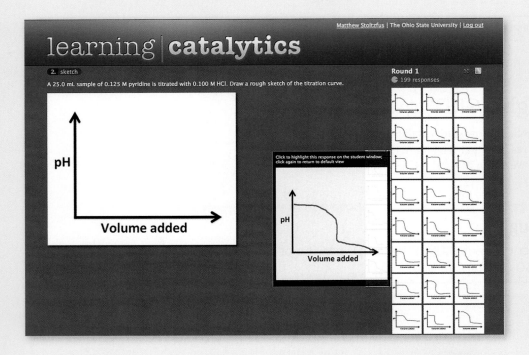

Pause and Predict Videos

Author Dr. Matt Stoltzfus created Pause and Predict Videos. These videos engage students by prompting them to submit a prediction about the outcome of an experiment or demonstration before seeing the final result. A set of assignable tutorials, based on these videos, challenge students to transfer their understanding of the demonstration to related scenarios. These videos are also available in web- and mobile-friendly formats through the study area of MasteringChemistry® and in the Pearson eText.

NEW! **Simulations,** assignable in MasteringChemistry®, include those developed by the PhET Chemistry Group, and the leading authors in simulation development covering some of the most difficult chemistry concepts.

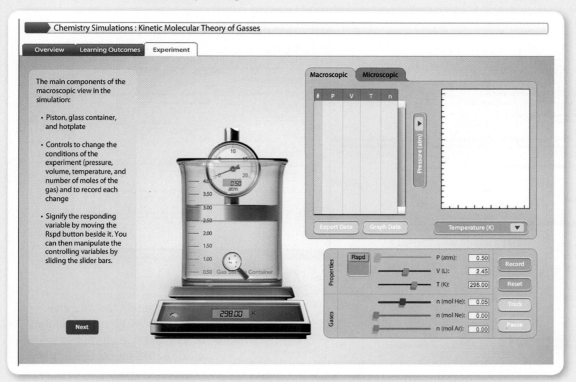

Adaptive

MasteringChemistry® has always been personalized and adaptive on a question level by providing error-specific feedback based on actual student responses; however, Mastering now includes two new adaptive assignment types—Adaptive Follow-Up Assignments and Dynamic Study Modules.

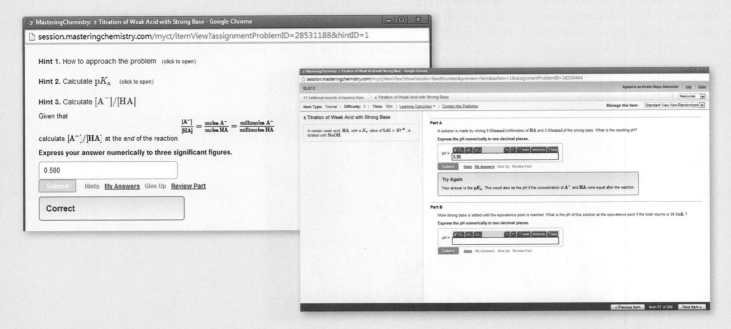

Adaptive Follow-Up Assignments

Instructors have the ability to assign adaptive follow-up assignments. Content delivered to students as part of adaptive learning will be automatically personalized for each individual based on strengths and weaknesses identified by his or her performance on Mastering parent assignments.

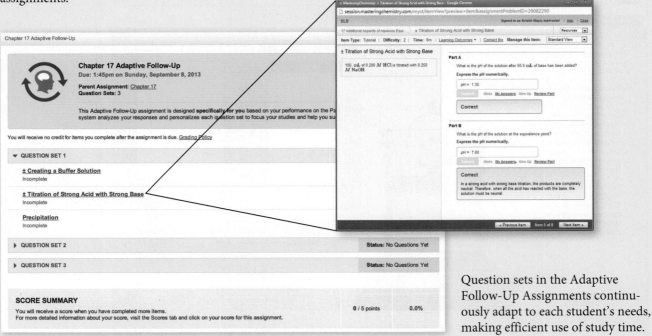

Question sets in the Adaptive Follow-Up Assignments continuously adapt to each student's needs, making efficient use of study time.

Dynamic Study Modules

NEW! **Dynamic Study Modules**, designed to enable students to study effectively on their own as well as help students quickly access and learn the nomenclature they need to be successful in chemistry.

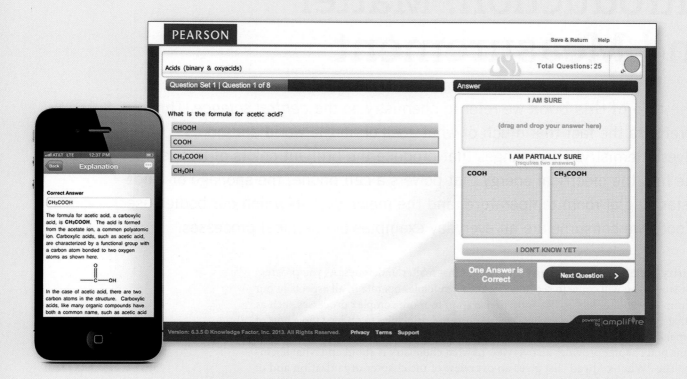

These modules can be accessed on smartphones, tablets, and computers and results can be tracked in the MasteringChemistry® Gradebook. Here's how it works:

1. Students receive an initial set of questions and benefit from the metacognition involved with asking them to indicate how confident they are with their answer.

2. After answering each set of questions, students review their answers.

3. Each question has explanation material that reinforces the correct answer response and addresses the misconceptions found in the wrong answer choices.

4. Once students review the explanations, they are presented with a new set of questions. Students cycle through this dynamic process of test-learn-retest until they achieve mastery of the material.

1

Introduction: Matter and Measurement

In the title of this book we refer to chemistry as the *central science.* This title reflects the fact that much of what goes on in the world around us involves chemistry. The changes that produce the brilliant colors of tree leaves in the fall, the electrical energy that powers a cell phone, the spoilage of foods left standing at room temperature, and the many ways in which our bodies use the foods we consume are all everyday examples of chemical processes.

Chemistry is the study of matter and the changes that matter undergoes. As you progress in your study, you will come to see how chemical principles operate in all aspects of our lives, from everyday activities like food preparation to more complex processes such as those that operate in the environment. We use chemical principles to understand a host of phenomena, from the role of salt in our diet to the workings of a lithium ion battery.

This first chapter provides an overview of what chemistry is about and what chemists do. The "What's Ahead" list gives an overview of the chapter organization and of some of the ideas we will consider.

1.1 | The Study of Chemistry

Chemistry is at the heart of many changes we see in the world around us, and it accounts for the myriad of different properties we see in matter. To understand how these changes and properties arise, we need to look far beneath the surfaces of our everyday observations.

▶ **THE BEAUTIFUL COLORS** that develop in trees in the fall appear when the tree ceases to produce chlorophyll, which imparts the green color to the leaves during the summer. Some of the color we see has been in the leaf all summer, and some develops from the action of sunlight on the leaf as the chlorophyll disappears.

WHAT'S AHEAD ▶

1.1 THE STUDY OF CHEMISTRY We begin with a brief description of what chemistry is, what chemists do, and why it is useful to learn chemistry.

1.2 CLASSIFICATIONS OF MATTER Next, we examine some fundamental ways to classify matter, distinguishing between *pure substances* and *mixtures* and between *elements* and *compounds.*

1.3 PROPERTIES OF MATTER We then consider different characteristics, or *properties*, used to characterize, identify, and separate substances, distinguishing between chemical and physical properties.

1.4 UNITS OF MEASUREMENT We observe that many properties rely on quantitative measurements involving numbers and units. The units of measurement used throughout science are those of the *metric system.*

1.5 UNCERTAINTY IN MEASUREMENT We observe that the uncertainty inherent in all measured quantities is expressed by the number of *significant figures* used to report the quantity. Significant figures are also used to express the uncertainty associated with calculations involving measured quantities.

1.6 DIMENSIONAL ANALYSIS We recognize that units as well as numbers are carried through calculations and that obtaining correct units for the result of a calculation is an important way to check whether the calculation is correct.

The Atomic and Molecular Perspective of Chemistry

Chemistry is the study of the properties and behavior of matter. **Matter** is the physical material of the universe; it is anything that has mass and occupies space. A **property** is any characteristic that allows us to recognize a particular type of matter and to distinguish it from other types. This book, your body, the air you are breathing, and the clothes you are wearing are all samples of matter. We observe a tremendous variety of matter in our world, but countless experiments have shown that all matter is comprised of combinations of only about 100 substances called *elements*. One of our major goals will be to relate the properties of matter to its composition, that is, to the particular elements it contains.

Chemistry also provides a background for understanding the properties of matter in terms of **atoms**, the almost infinitesimally small building blocks of matter. Each element is composed of a unique kind of atom. We will see that the properties of matter relate to both the kinds of atoms the matter contains (*composition*) and the arrangements of these atoms (*structure*).

In **molecules**, two or more atoms are joined in specific shapes. Throughout this text you will see molecules represented using colored spheres to show how the atoms are connected (▼ Figure 1.1). The color provides a convenient way to distinguish between atoms of different elements. For example, notice that the molecules of ethanol and ethylene glycol in Figure 1.1 have different compositions and structures. Ethanol contains one oxygen atom, depicted by one red sphere. In contrast, ethylene glycol contains two oxygen atoms.

Even apparently minor differences in the composition or structure of molecules can cause profound differences in properties. For example, let's compare ethanol and ethylene glycol, which appear in Figure 1.1 to be quite similar. Ethanol is the alcohol in beverages such as beer and wine, whereas ethylene glycol is a viscous liquid used as automobile antifreeze. The properties of these two substances differ in many ways, as do their biological activities. Ethanol is consumed throughout the world, but you should *never* consume ethylene glycol because it is highly toxic. One of the challenges chemists undertake is to alter the composition or structure of molecules in a controlled way, creating new substances with different properties. For example, the common drug aspirin, shown in Figure 1.1, was first synthesized in 1897 in a successful attempt to improve on a natural product extracted from willow bark that had long been used to alleviate pain.

Every change in the observable world—from boiling water to the changes that occur as our bodies combat invading viruses—has its basis in the world of atoms and molecules.

▲ GO FIGURE

Which of the molecules in the figure has the most carbon atoms? How many are there in that molecule?

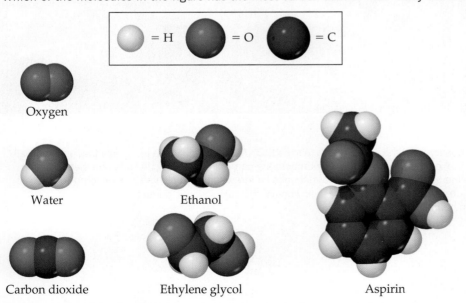

▲ **Figure 1.1 Molecular models.** The white, black, and red spheres represent atoms of hydrogen, carbon, and oxygen, respectively.

Thus, as we proceed with our study of chemistry, we will find ourselves thinking in two realms: the *macroscopic* realm of ordinary-sized objects (*macro* = large) and the *submicroscopic* realm of atoms and molecules. We make our observations in the macroscopic world, but to understand that world, we must visualize how atoms and molecules behave at the submicroscopic level. Chemistry is the science that seeks to understand the properties and behavior of matter by studying the properties and behavior of atoms and molecules.

 Give It Some Thought

(a) Approximately how many elements are there?
(b) What submicroscopic particles are the building blocks of matter?

Why Study Chemistry?

Chemistry lies near the heart of many matters of public concern, such as improvement of health care, conservation of natural resources, protection of the environment, and the supply of energy needed to keep society running. Using chemistry, we have discovered and continually improved upon pharmaceuticals, fertilizers and pesticides, plastics, solar panels, LEDs, and building materials. We have also discovered that some chemicals are potentially harmful to our health or the environment. This means that we must be sure that the materials with which we come into contact are safe. As a citizen and consumer, it is in your best interest to understand the effects, both positive and negative, that chemicals can have, and to arrive at a balanced outlook regarding their uses.

You may be studying chemistry because it is an essential part of your curriculum. Your major might be chemistry, or it could be biology, engineering, pharmacy, agriculture, geology, or some other field. Chemistry is central to a fundamental understanding of governing principles in many science-related fields. For example, our interactions with the material world raise basic questions about the materials around us. ▼ Figure 1.2 illustrates how chemistry is central to several different realms of modern life.

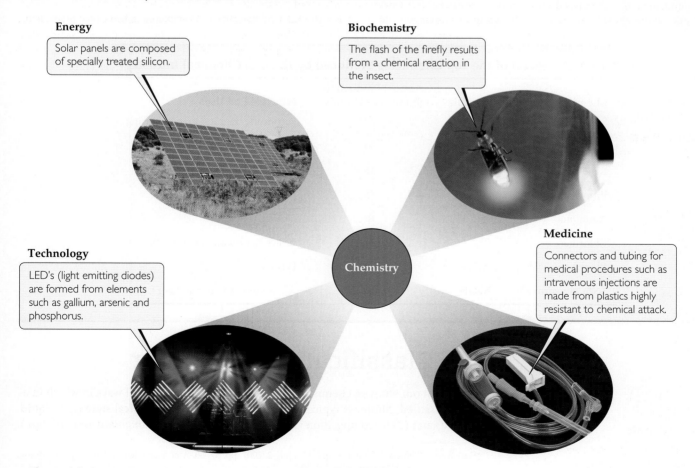

Energy
Solar panels are composed of specially treated silicon.

Biochemistry
The flash of the firefly results from a chemical reaction in the insect.

Technology
LED's (light emitting diodes) are formed from elements such as gallium, arsenic and phosphorus.

Chemistry

Medicine
Connectors and tubing for medical procedures such as intravenous injections are made from plastics highly resistant to chemical attack.

▲ Figure 1.2 **Chemistry is central to our understanding of the world around us.**

Chemistry Put to Work

Chemistry and the Chemical Industry

Chemistry is all around us. Many people are familiar with household chemicals, particularly kitchen chemicals such as those shown in ▶ Figure 1.3. However, few realize the size and importance of the chemical industry. Worldwide sales of chemicals and related products manufactured in the United States total approximately $585 billion annually. Sales of pharmaceuticals total another $180 billion. The chemical industry employs more than 10% of all scientists and engineers and is a major contributor to the U.S. economy.

Vast amounts of industrial chemicals are produced each year. ▼ Table 1.1 lists several of the chemicals produced in highest volumes in the United States. Notice that they all serve as raw materials for a variety of uses, including the manufacture and processing of metals, plastics, fertilizers, and other goods.

Who are chemists, and what do they do? People who have degrees in chemistry hold a variety of positions in industry, government, and academia. Those in industry work as laboratory chemists, developing new products (research and development); analyzing materials (quality control); or assisting customers in using products (sales and service). Those with more experience or training may work as managers or company directors. Chemists are important members of the scientific workforce in government (the National Institutes of Health, Department of Energy, and Environmental Protection Agency all employ chemists) and at universities. A chemistry degree is also good preparation for careers in teaching, medicine, biomedical research, information science, environmental work, technical sales, government regulatory agencies, and patent law.

Fundamentally, chemists do three things: (1) make new types of matter: materials, substances, or combinations of substances with desired properties; (2) measure the properties of matter; and (3) develop models that explain and/or predict the properties of matter. One chemist, for example, may work in the laboratory to discover new drugs. Another may concentrate on the development of new instrumentation to measure properties of matter at the atomic level. Other chemists may use existing materials and methods to understand how pollutants are transported in the environment or how drugs are processed in the body. Yet another chemist will develop theory, write computer code, and run computer simulations to understand how molecules move and react. The collective chemical enterprise is a rich mix of all of these activities.

▲ Figure 1.3 **Common chemicals employed in home food production.**

Table 1.1 **Several of the Top Chemicals Produced by the U.S. Chemical Industry***

Chemical	Formula	Annual Production (Billions of Pounds)	Principal End Uses
Sulfuric acid	H_2SO_4	70	Fertilizers, chemical manufacturing
Ethylene	C_2H_4	50	Plastics, antifreeze
Lime	CaO	45	Paper, cement, steel
Propylene	C_3H_6	35	Plastics
Ammonia	NH_3	18	Fertilizers
Chlorine	Cl_2	21	Bleaches, plastics, water purification
Phosphoric acid	H_3PO_4	20	Fertilizers
Sodium hydroxide	$NaOH$	16	Aluminum production, soap

1.2 | Classifications of Matter

Let's begin our study of chemistry by examining two fundamental ways in which matter is classified. Matter is typically characterized by (1) its physical state (gas, liquid, or solid) and (2) its composition (whether it is an element, a *compound*, or a *mixture*).

*Data from Chemical & Engineering News, July 2, 2007, pp. 57, 60, American Chemical Society; data online from U.S. Geological Survey.

States of Matter

A sample of matter can be a gas, a liquid, or a solid. These three forms, called the **states of matter**, differ in some of their observable properties. A **gas** (also known as *vapor*) has no fixed volume or shape; rather, it uniformly fills its container. A gas can be compressed to occupy a smaller volume, or it can expand to occupy a larger one. A **liquid** has a distinct volume independent of its container, and assumes the shape of the portion of the container it occupies. A **solid** has both a definite shape and a definite volume. Neither liquids nor solids can be compressed to any appreciable extent.

The properties of the states of matter can be understood on the molecular level (▶ Figure 1.4). In a gas the molecules are far apart and moving at high speeds, colliding repeatedly with one another and with the walls of the container. Compressing a gas decreases the amount of space between molecules and increases the frequency of collisions between molecules but does not alter the size or shape of the molecules. In a liquid, the molecules are packed closely together but still move rapidly. The rapid movement allows the molecules to slide over one another; thus, a liquid pours easily. In a solid the molecules are held tightly together, usually in definite arrangements in which the molecules can wiggle only slightly in their otherwise fixed positions. Thus, the distances between molecules are similar in the liquid and solid states, but the two states differ in how free the molecules are to move around. Changes in temperature and/or pressure can lead to conversion from one state of matter to another, illustrated by such familiar processes as ice melting or water vapor condensing.

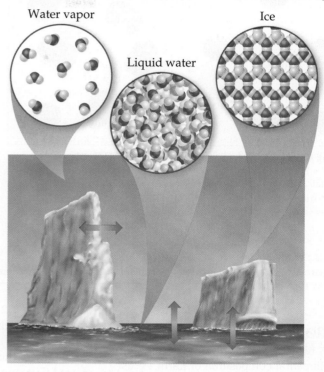

▲ GO FIGURE

In which form of water are the water molecules farthest apart?

Water vapor

Liquid water

Ice

▲ Figure 1.4 **The three physical states of water—water vapor, liquid water, and ice.** We see the liquid and solid states but cannot see the gas (vapor) state. The red arrows show that the three states of matter interconvert.

Pure Substances

Most forms of matter we encounter—the air we breathe (a gas), the gasoline we burn in our cars (a liquid), and the sidewalk we walk on (a solid)—are not chemically pure. We can, however, separate these forms of matter into pure substances. A **pure substance** (usually referred to simply as a *substance*) is matter that has distinct properties and a composition that does not vary from sample to sample. Water and aluminum foil are examples of pure substances.

All substances are either elements or compounds. **Elements** are substances that cannot be decomposed into simpler substances. On the molecular level, each element is composed of only one kind of atom [Figure 1.5(**a** and **b**)]. **Compounds** are substances composed of two or more elements; they contain two or more kinds of atoms [Figure 1.5(**c**)]. Water, for example, is a compound composed of two elements: hydrogen and oxygen. Figure 1.5(**d**) shows a mixture of substances. **Mixtures** are combinations of two or more substances in which each substance retains its chemical identity.

Elements

Currently, 118 elements are known, though they vary widely in abundance. Hydrogen constitutes about 74% of the mass in the Milky Way galaxy, and helium constitutes 24%. Closer to home, only five elements—oxygen, silicon, aluminum, iron, and calcium—account for over 90% of Earth's crust (including oceans and atmosphere), and only three—oxygen, carbon, and hydrogen—account for over 90% of the mass of the human body (**Figure 1.6**).

▲ GO FIGURE

How do the molecules of a compound differ from the molecules of an element?

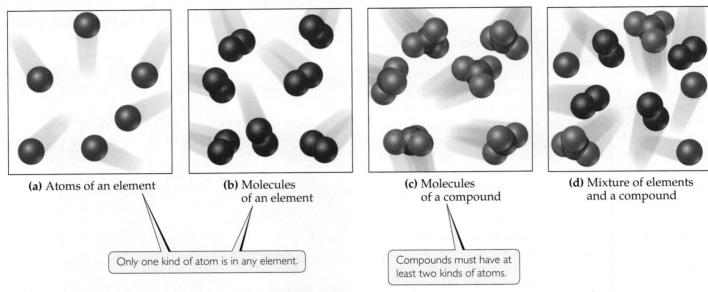

(a) Atoms of an element

(b) Molecules of an element

(c) Molecules of a compound

(d) Mixture of elements and a compound

Only one kind of atom is in any element.

Compounds must have at least two kinds of atoms.

▲ Figure 1.5 Molecular comparison of elements, compounds, and mixtures.

▲ GO FIGURE

Name two significant differences between the elemental composition of Earth's crust and the elemental composition of the human body.

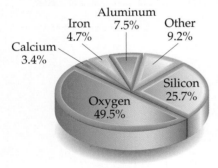

Earth's crust

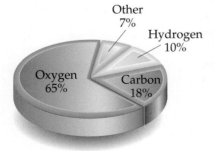

Human body

▲ Figure 1.6 Relative abundances of elements.*
Elements in percent by mass in Earth's crust (including oceans and atmosphere) and the human body.

▼ Table 1.2 lists some common elements, along with the chemical *symbols* used to denote them. The symbol for each element consists of one or two letters, with the first letter capitalized. These symbols are derived mostly from the English names of the elements, but sometimes they are derived from a foreign name instead (last column in Table 1.2). You will need to know these symbols and learn others as we encounter them in the text.

All of the known elements and their symbols are listed on the front inside cover of this text in a table known as the *periodic table*. In the periodic table the elements are arranged in columns so that closely related elements are grouped together. We describe the periodic table in more detail in Section 2.5 and consider the periodically repeating properties of the elements in Chapter 7.

Compounds

Most elements can interact with other elements to form compounds. For example, when hydrogen gas burns in oxygen gas, the elements hydrogen and oxygen combine to form the compound water. Conversely, water can be decomposed into its elements by passing an electrical current through it (▶ Figure 1.7).

Table 1.2   **Some Common Elements and Their Symbols**

Carbon	C	Aluminum	Al	Copper	Cu (from *cuprum*)
Fluorine	F	Bromine	Br	Iron	Fe (from *ferrum*)
Hydrogen	H	Calcium	Ca	Lead	Pb (from *plumbum*)
Iodine	I	Chlorine	Cl	Mercury	Hg (from *hydrargyrum*)
Nitrogen	N	Helium	He	Potassium	K (from *kalium*)
Oxygen	O	Lithium	Li	Silver	Ag (from *argentum*)
Phosphorus	P	Magnesium	Mg	Sodium	Na (from *natrium*)
Sulfur	S	Silicon	Si	Tin	Sn (from *stannum*)

U.S. Geological Survey Circular 285, U.S Department of the Interior.

 **GO FIGURE**

How are the relative gas volumes collected in the two tubes related to the relative number of gas molecules in the tubes?

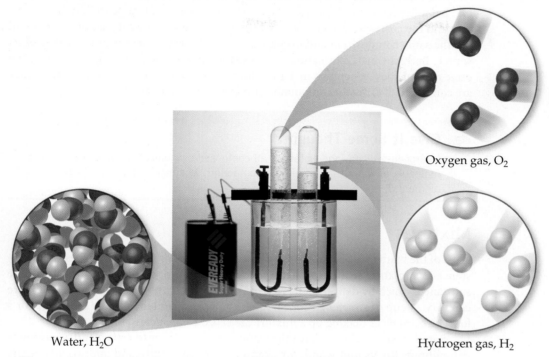

Oxygen gas, O_2

Water, H_2O

Hydrogen gas, H_2

▲ **Figure 1.7 Electrolysis of water.** Water decomposes into its component elements, hydrogen and oxygen, when an electrical current is passed through it. The volume of hydrogen, collected in the right test tube, is twice the volume of oxygen.

Pure water, regardless of its source, consists of 11% hydrogen and 89% oxygen by mass. This macroscopic composition corresponds to the molecular composition, which consists of two hydrogen atoms combined with one oxygen atom:

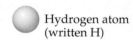 Hydrogen atom (written H)

Oxygen atom (written O)

Water molecule (written H_2O)

The elements hydrogen and oxygen themselves exist naturally as diatomic (two-atom) molecules:

Oxygen molecule (written O_2)

Hydrogen molecule (written H_2)

As seen in ▼ Table 1.3, the properties of water bear no resemblance to the properties of its component elements. Hydrogen, oxygen, and water are each a unique substance, a consequence of the uniqueness of their respective molecules.

Table 1.3 Comparison of Water, Hydrogen, and Oxygen

	Water	Hydrogen	Oxygen
State[a]	Liquid	Gas	Gas
Normal boiling point	100 °C	−253 °C	−183 °C
Density[a]	1000 g/L	0.084 g/L	1.33 g/L
Flammable	No	Yes	No

[a]At room temperature and atmospheric pressure.

The observation that the elemental composition of a compound is always the same is known as the **law of constant composition** (or the **law of definite proportions**). French chemist Joseph Louis Proust (1754–1826) first stated the law in about 1800. Although this law has been known for 200 years, the belief persists among some people that a fundamental difference exists between compounds prepared in the laboratory and the corresponding compounds found in nature. However, a pure compound has the same composition and properties under the same conditions regardless of its source. Both chemists and nature must use the same elements and operate under the same natural laws. When two materials differ in composition or properties, either they are composed of different compounds or they differ in purity.

 Give It Some Thought

Hydrogen, oxygen, and water are all composed of molecules. What is it about a molecule of water that makes it a compound, whereas hydrogen and oxygen are elements?

Mixtures

Most of the matter we encounter consists of mixtures of different substances. Each substance in a mixture retains its chemical identity and properties. In contrast to a pure substance, which by definition has a fixed composition, the composition of a mixture can vary. A cup of sweetened coffee, for example, can contain either a little sugar or a lot. The substances making up a mixture are called *components* of the mixture.

Some mixtures do not have the same composition, properties, and appearance throughout. Rocks and wood, for example, vary in texture and appearance in any typical sample. Such mixtures are *heterogeneous* [▼ Figure 1.8(**a**)]. Mixtures that are uniform throughout are *homogeneous*. Air is a homogeneous mixture of nitrogen, oxygen, and smaller amounts of other gases. The nitrogen in air has all the properties of pure nitrogen because both the pure substance and the mixture contain the same nitrogen molecules. Salt, sugar, and many other substances dissolve in water to form homogeneous mixtures [Figure 1.8(**b**)]. Homogeneous mixtures are also called **solutions**. Although the term *solution* conjures an image of a liquid, solutions can be solids, liquids, or gases.

▶ Figure 1.9 summarizes the classification of matter into elements, compounds, and mixtures.

(**a**)   (**b**)

▲ **Figure 1.8 Mixtures. (a)** Many common materials, including rocks, are heterogeneous mixtures. This photograph of granite shows a heterogeneous mixture of silicon dioxide and other metal oxides. (**b**) Homogeneous mixtures are called solutions. Many substances, including the blue solid shown here [copper(II) sulfate], dissolve in water to form solutions.

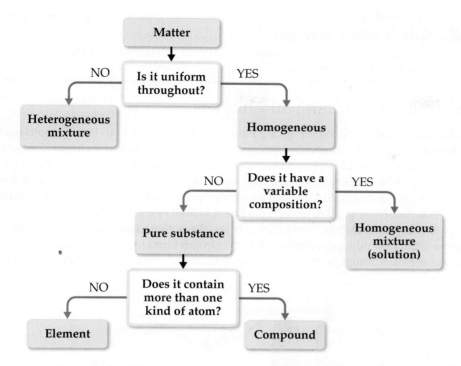

▲ **Figure 1.9 Classification of matter.** All pure matter is classified ultimately as either an element or a compound.

SAMPLE EXERCISE 1.1 | Distinguishing among Elements, Compounds, and Mixtures

"White gold" contains gold and a "white" metal, such as palladium. Two samples of white gold differ in the relative amounts of gold and palladium they contain. Both samples are uniform in composition throughout. Use Figure 1.9 to classify white gold.

SOLUTION

Because the material is uniform throughout, it is homogeneous. Because its composition differs for the two samples, it cannot be a compound. Instead, it must be a homogeneous mixture.

> **Practice Exercise 1**
> Which of the following is the correct description of a cube of material cut from the inside of an apple?
> **(a)** It is a pure compound.
> **(b)** It consists of a homogenous mixture of compounds.
> **(c)** It consists of a heterogeneous mixture of compounds.
> **(d)** It consists of a heterogeneous mixture of elements and compounds.
> **(e)** It consists of a single compound in different states.

> **Practice Exercise 2**
> Aspirin is composed of 60.0% carbon, 4.5% hydrogen, and 35.5% oxygen by mass, regardless of its source. Use Figure 1.9 to classify aspirin.

1.3 | Properties of Matter

Every substance has unique properties. For example, the properties listed in Table 1.3 allow us to distinguish hydrogen, oxygen, and water from one another. The properties of matter can be categorized as physical or chemical. **Physical properties** can be observed without changing the identity and composition of the substance. These properties include color, odor, density, melting point, boiling point, and hardness. **Chemical properties** describe the way a substance may change, or *react*, to form other substances. A common chemical property is flammability, the ability of a substance to burn in the presence of oxygen.

Some properties, such as temperature and melting point, are *intensive properties*. **Intensive properties** do not depend on the amount of sample being examined and are particularly useful in chemistry because many intensive properties can be used to *identify* substances. **Extensive properties** depend on the amount of sample, with two examples being mass and volume. Extensive properties relate to the *amount* of substance present.

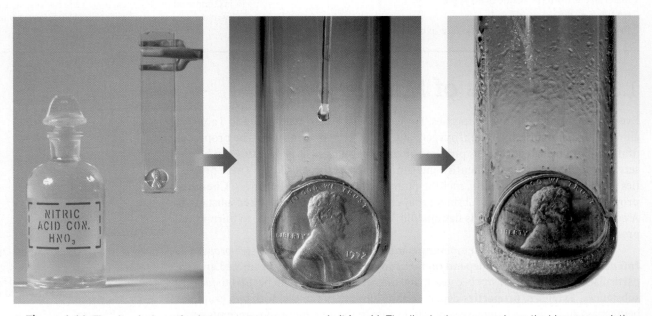

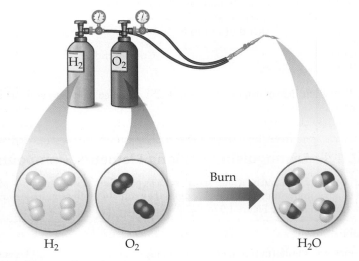

Give It Some Thought

When we say that lead is a denser metal than aluminum, are we talking about an extensive or intensive property?

Physical and Chemical Changes

The changes substances undergo are either physical or chemical. During a **physical change**, a substance changes its physical appearance but not its composition. (That is, it is the same substance before and after the change.) The evaporation of water is a physical change. When water evaporates, it changes from the liquid state to the gas state, but it is still composed of water molecules, as depicted in Figure 1.4. All **changes of state** (for example, from liquid to gas or from liquid to solid) are physical changes.

In a **chemical change** (also called a **chemical reaction**), a substance is transformed into a chemically different substance. When hydrogen burns in air, for example, it undergoes a chemical change because it combines with oxygen to form water (▼ Figure 1.10).

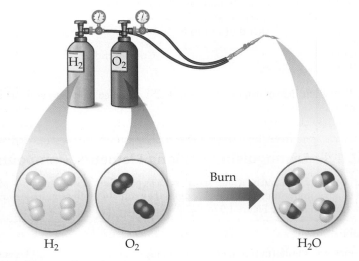

Burn

H₂   O₂   H₂O

▲ Figure 1.10 **A chemical reaction.**

Chemical changes can be dramatic. In the account that follows, Ira Remsen, author of a popular chemistry text published in 1901, describes his first experiences with chemical reactions. The chemical reaction that he observed is shown in ▼ Figure 1.11.

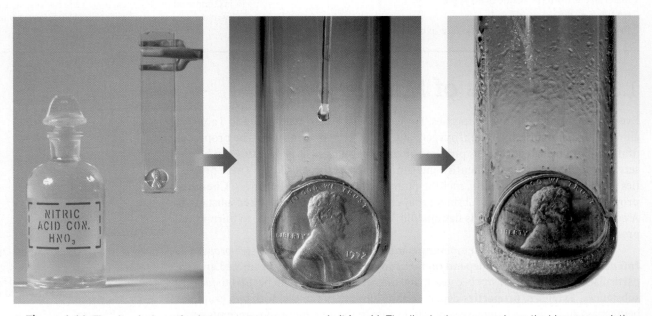

▲ Figure 1.11 **The chemical reaction between a copper penny and nitric acid.** The dissolved copper produces the blue-green solution; the reddish brown gas produced is nitrogen dioxide.

While reading a textbook of chemistry, I came upon the statement "nitric acid acts upon copper," and I determined to see what this meant. Having located some nitric acid, I had only to learn what the words "act upon" meant. In the interest of knowledge I was even willing to sacrifice one of the few copper cents then in my possession. I put one of them on the table, opened a bottle labeled "nitric acid," poured some of the liquid on the copper, and prepared to make an observation. But what was this wonderful thing which I beheld? The cent was already changed, and it was no small change either. A greenish-blue liquid foamed and fumed over the cent and over the table. The air became colored dark red. How could I stop this? I tried by picking the cent up and throwing it out the window. I learned another fact: nitric acid acts upon fingers. The pain led to another unpremeditated experiment. I drew my fingers across my trousers and discovered nitric acid acts upon trousers. That was the most impressive experiment I have ever performed. I tell of it even now with interest. It was a revelation to me. Plainly the only way to learn about such remarkable kinds of action is to see the results, to experiment, to work in the laboratory.*

 Give It Some Thought

Which of these changes are physical and which are chemical? Explain.
(a) Plants make sugar from carbon dioxide and water.
(b) Water vapor in the air forms frost.
(c) A goldsmith melts a nugget of gold and pulls it into a wire.

Separation of Mixtures

We can separate a mixture into its components by taking advantage of differences in their properties. For example, a heterogeneous mixture of iron filings and gold filings could be sorted by color into iron and gold. A less tedious approach would be to use a magnet to attract the iron filings, leaving the gold ones behind. We can also take advantage of an important chemical difference between these two metals: Many acids dissolve iron but not gold. Thus, if we put our mixture into an appropriate acid, the acid would dissolve the iron and the solid gold would be left behind. The two could then be separated by *filtration* (▶ Figure 1.12). We would have to use other chemical reactions, which we will learn about later, to transform the dissolved iron back into metal.

An important method of separating the components of a homogeneous mixture is **distillation**, a process that depends on the different abilities of substances to form gases. For example, if we boil a solution of salt and water, the water evaporates, forming a gas, and the salt is left behind. The gaseous water can be converted back to a liquid on the walls of a condenser, as shown in ▼ Figure 1.13.

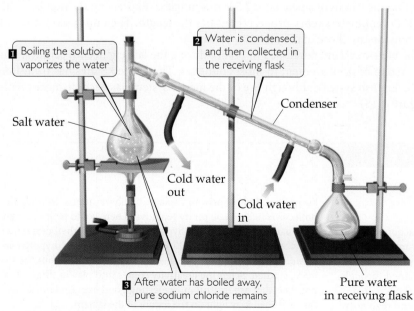

1 Boiling the solution vaporizes the water

2 Water is condensed, and then collected in the receiving flask

Condenser

Salt water

Cold water out

Cold water in

3 After water has boiled away, pure sodium chloride remains

Pure water in receiving flask

▲ **Figure 1.13 Distillation.** Apparatus for separating a sodium chloride solution (salt water) into its components.

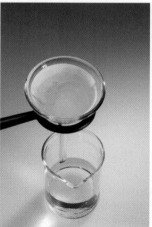

▲ **Figure 1.12 Separation by filtration.** A mixture of a solid and a liquid is poured through filter paper. The liquid passes through the paper while the solid remains on the paper.

*Remsen, Ira, *The Principles of Theoretical Chemistry*, 1887.

▲ GO FIGURE

Is the separation of **a**, **b**, and **c** in Figure 1.14 a physical or chemical process?

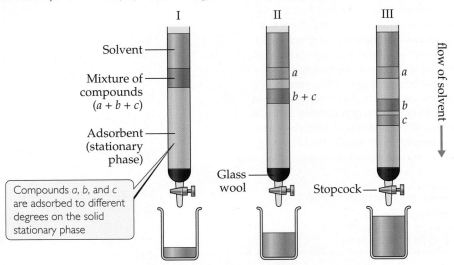

▲ **Figure 1.14 Separation of three substances using column chromatography.**

▲ **Figure 1.15 Metric units.** Metric measurements are increasingly common in the United States, as exemplified by the volume printed on this soda can in both English units (fluid ounces, fl oz) and metric units (milliliters, mL).

The differing abilities of substances to adhere to the surfaces of solids can also be used to separate mixtures. This ability is the basis of *chromatography*, a technique shown in ▲ Figure 1.14.

1.4 | Units of Measurement

Many properties of matter are *quantitative*, that is, associated with numbers. When a number represents a measured quantity, the units of that quantity must be specified. To say that the length of a pencil is 17.5 is meaningless. Expressing the number with its units, 17.5 centimeters (cm), properly specifies the length. The units used for scientific measurements are those of the **metric system**.

The metric system, developed in France during the late eighteenth century, is used as the system of measurement in most countries. The United States has traditionally used the English system, although use of the metric system has become more common (◀ Figure 1.15).

A Closer Look

The Scientific Method

Where does scientific knowledge come from? How is it acquired? How do we know it is reliable? How do scientists add to it, or modify it?

There is nothing mysterious about how scientists work. The first idea to keep in mind is that scientific knowledge is gained through observations of the natural world. A principal aim of the scientist is to organize these observations, by identifying patterns and regularity, making measurements, and associating one set of observations with another. The next step is to ask *why* nature behaves in the manner we observe. To answer this question, the scientist constructs a model,

known as a **hypothesis,** to explain the observations. Initially the hypothesis is likely to be pretty tentative. There could be more than one reasonable hypothesis. If a hypothesis is correct, then certain results and observations should follow from it. In this way hypotheses can stimulate the design of experiments to learn more about the system being studied. Scientific creativity comes into play in thinking of hypotheses that are fruitful in suggesting good experiments to do, ones that will shed new light on the nature of the system.

As more information is gathered, the initial hypotheses get winnowed down. Eventually just one may stand out as most consistent with a body of accumulated evidence. We then begin to call this hypothesis a **theory**, a model that has predictive powers, and that accounts for all the available observations. A theory also generally is consistent with other, perhaps larger and more general theories. For example, a theory of what goes on inside a volcano has to be consistent with more general theories regarding heat transfer, chemistry at high temperature, and so forth.

We will be encountering many theories as we proceed through this book. Some of them have been found over and over again to be consistent with observations. However, no theory can be proven to be absolutely true. We can treat it as though it is, but there always remains a possibility that there is some respect in which a theory is wrong. A famous example is Einstein's theory of relativity. Isaac Newton's theory of mechanics yielded such precise results for the mechanical behavior of matter that no exceptions to it were found before the twentieth century. But Albert Einstein showed that Newton's theory of the nature of space and time is incorrect. Einstein's theory of relativity represented a fundamental shift in how we think of space and time. He predicted where the exceptions to predictions based on Newton's theory might be found. Although only small departures from Newton's theory were predicted, they *were* observed. Einstein's theory of relativity became accepted as the correct model. However, for most uses, Newton's laws of motion are quite accurate enough.

The overall process we have just considered, illustrated in ▶ Figure 1.16, is often referred to as *the scientific method*. But there is no single scientific method. Many factors play a role in advancing scientific knowledge. The one unvarying requirement is that our explanations be consistent with observations, and that they depend solely on natural phenomena.

When nature behaves in a certain way over and over again, under all sorts of different conditions, we can summarize that behavior in a **scientific law**. For example, it has been repeatedly observed that in a chemical reaction there is no change in the total mass of the materials reacting as compared with the materials that are formed; we call this observation the *Law of Conservation of Mass*. It is important to make a distinction between a theory and a scientific law. The latter simply is a statement of what always

happens, to the best of our knowledge. A theory, on the other hand, is an *explanation* for what happens. If we discover some law fails to hold true, then we must assume the theory underlying that law is wrong in some way.

Related Exercises: 1.60, 1.82

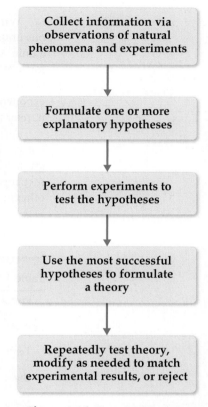

▲ Figure 1.16 **The scientific method.**

SI Units

In 1960 an international agreement was reached specifying a particular choice of metric units for use in scientific measurements. These preferred units are called **SI units**, after the French *Système International d'Unités*. This system has seven *base units* from which all other units are derived (▼ Table 1.4). In this chapter we will consider the base units for length, mass, and temperature.

Table 1.4 **SI Base Units**

Physical Quantity	Name of Unit	Abbreviation
Mass	Kilogram	kg
Length	Meter	m
Time	Second	s or sec
Temperature	Kelvin	K
Amount of substance	Mole	mol
Electric current	Ampere	A or amp
Luminous intensity	Candela	cd

Give It Some Thought

The package of a fluorescent bulb for a table lamp lists the light output in terms of lumens, lm. Which of the seven SI units would you expect to be part of the definition of a lumen?

With SI units, prefixes are used to indicate decimal fractions or multiples of various units. For example, the prefix *milli-* represents a 10^{-3} fraction, one-thousandth, of a unit: A milligram (mg) is 10^{-3} gram (g), a millimeter (mm) is 10^{-3} meter (m), and so forth. ▼ Table 1.5 presents the prefixes commonly encountered in chemistry. In using SI units and in working problems throughout this text, you must be comfortable using exponential notation. If you are unfamiliar with exponential notation or want to review it, refer to Appendix A.1.

Although non–SI units are being phased out, some are still commonly used by scientists. Whenever we first encounter a non–SI unit in the text, the SI unit will also be given. The relations between the non–SI and SI units we will use most frequently in this text appear on the back inside cover. We will discuss how to convert from one to the other in Section 1.6.

Table 1.5   Prefixes Used in the Metric System and with SI Units

Prefix	Abbreviation	Meaning	Example	
Peta	P	10^{15}	1 petawatt (PW)	$= 1 \times 10^{15}$ watts[a]
Tera	T	10^{12}	1 terawatt (TW)	$= 1 \times 10^{12}$ watts
Giga	G	10^{9}	1 gigawatt (GW)	$= 1 \times 10^{9}$ watts
Mega	M	10^{6}	1 megawatt (MW)	$= 1 \times 10^{6}$ watts
Kilo	k	10^{3}	1 kilowatt (kW)	$= 1 \times 10^{3}$ watts
Deci	d	10^{-1}	1 deciwatt (dW)	$= 1 \times 10^{-1}$ watt
Centi	c	10^{-2}	1 centiwatt (cW)	$= 1 \times 10^{-2}$ watt
Milli	m	10^{-3}	1 milliwatt (mW)	$= 1 \times 10^{-3}$ watt
Micro	μ[b]	10^{-6}	1 microwatt (μW)	$= 1 \times 10^{-6}$ watt
Nano	n	10^{-9}	1 nanowatt (nW)	$= 1 \times 10^{-9}$ watt
Pico	p	10^{-12}	1 picowatt (pW)	$= 1 \times 10^{-12}$ watt
Femto	f	10^{-15}	1 femtowatt (fW)	$= 1 \times 10^{-15}$ watt
Atto	a	10^{-18}	1 attowatt (aW)	$= 1 \times 10^{-18}$ watt
Zepto	z	10^{-21}	1 zeptowatt (zW)	$= 1 \times 10^{-21}$ watt

[a]The watt (W) is the SI unit of power, which is the rate at which energy is either generated or consumed. The SI unit of energy is the joule (J); $1 \text{ J} = 1 \text{ kg} \cdot \text{m}^2/\text{s}^2$ and $1 \text{ W} = 1 \text{ J/s}$.
[b]Greek letter mu, pronounced "mew."

Give It Some Thought

How many μg are there in 1 mg?

Length and Mass

The SI base unit of *length* is the meter, a distance slightly longer than a yard. **Mass*** is a measure of the amount of material in an object. The SI base unit of mass is the kilogram (kg), which is equal to about 2.2 pounds (lb). This base unit is unusual because it uses a prefix, *kilo-*, instead of the word *gram* alone. We obtain other units for mass by adding prefixes to the word *gram*.

SAMPLE EXERCISE 1.2 | **Using SI Prefixes**

What is the name of the unit that equals (**a**) 10^{-9} gram, (**b**) 10^{-6} second, (**c**) 10^{-3} meter?

SOLUTION

We can find the prefix related to each power of ten in Table 1.5: (**a**) nanogram, ng; (**b**) microsecond, μs; (**c**) millimeter, mm.

Practice Exercise 1

Which of the following weights would you expect to be suitable for weighing on an ordinary bathroom scale?
(**a**) 2.0×10^7 mg, (**b**) 2500 μg, (**c**) 5×10^{-4} kg, (**d**) 4×10^6 cg, (**e**) 5.5×10^8 dg.

Practice Exercise 2

(**a**) How many picometers are there in 1 m? (**b**) Express 6.0×10^3 m using a prefix to replace the power of ten. (**c**) Use exponential notation to express 4.22 mg in grams. (**d**) Use decimal notation to express 4.22 mg in grams.

Temperature

Temperature, a measure of the hotness or coldness of an object, is a physical property that determines the direction of heat flow. Heat always flows spontaneously from a substance at higher temperature to one at lower temperature. Thus, the influx of heat we feel when we touch a hot object tells us that the object is at a higher temperature than our hand.

The temperature scales commonly employed in science are the Celsius and Kelvin scales. The **Celsius scale** was originally based on the assignment of 0 °C to the freezing point of water and 100 °C to its boiling point at sea level (**Figure 1.17**).

---

*Mass and weight are often incorrectly thought to be the same. The weight of an object is the force that is exerted on its mass by gravity. In space, where gravitational forces are very weak, an astronaut can be weightless, but he or she cannot be massless. The astronaut's mass in space is the same as it is on Earth.

GO FIGURE

True or false: The "size" of a degree on the Celsius scale is the same as the "size" of a degree on the Kelvin scale.

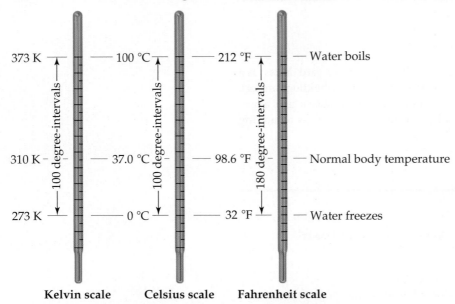

▲ Figure 1.17 **Comparison of the Kelvin, Celsius, and Fahrenheit temperature scales.**

The **Kelvin scale** is the SI temperature scale, and the SI unit of temperature is the *kelvin* (K). Zero on the Kelvin scale is the lowest attainable temperature, referred to as **absolute zero**. On the Celsius scale, absolute zero has the value, $-273.15\,°C$. The Celsius and Kelvin scales have equal-sized units—that is, a kelvin is the same size as a degree Celsius. Thus, the Kelvin and Celsius scales are related according to

$$K = °C + 273.15 \qquad [1.1]$$

The freezing point of water, $0\,°C$, is 273.15 K (Figure 1.17). Notice that we do not use a degree sign (°) with temperatures on the Kelvin scale.

The common temperature scale in the United States is the *Fahrenheit scale*, which is not generally used in science. Water freezes at $32\,°F$ and boils at $212\,°F$. The Fahrenheit and Celsius scales are related according to

$$°C = \frac{5}{9}\,(°F - 32) \quad \text{or} \quad °F = \frac{9}{5}\,(°C) + 32 \qquad [1.2]$$

SAMPLE EXERCISE 1.3 | Converting Units of Temperature

A weather forecaster predicts the temperature will reach 31 °C. What is this temperature **(a)** in K, **(b)** in °F?

SOLUTION

(a) Using Equation 1.1, we have K = 31 + 273 = 304 K.

(b) Using Equation 1.2, we have

$$°F = \frac{9}{5}(31) + 32 = 56 + 32 = 88\,°F.$$

liquid at 525 K (assume samples are protected from air): **(a)** bismuth, Bi; **(b)** platinum, Pt; **(c)** selenium, Se; **(d)** calcium, Ca; **(e)** copper, Cu.

Practice Exercise 1

Using Wolfram Alpha (http://www.wolframalpha.com/) or some other reference, determine which of these elements would be

Practice Exercise 2

Ethylene glycol, the major ingredient in antifreeze, freezes at $-11.5\,°C$. What is the freezing point in **(a)** K, **(b)** °F?

Derived SI Units

The SI base units are used to formulate *derived units*. A **derived unit** is obtained by multiplication or division of one or more of the base units. We begin with the defining equation for a quantity and then substitute the appropriate base units. For example, speed is defined as the ratio of distance traveled to elapsed time. Thus, the SI unit for speed—m/s, read "meters per second"—is a derived unit, the SI unit for distance (length), m, divided by the SI unit for time, s. Two common derived units in chemistry are those for volume and density.

Volume

The *volume* of a cube is its length cubed, length^3. Thus, the derived SI unit of volume is the SI unit of length, m, raised to the third power. The cubic meter, m^3, is the volume of a cube that is 1 m on each edge (▶ **Figure 1.18**). Smaller units, such as cubic centimeters, cm^3 (sometimes written cc), are frequently used in chemistry. Another volume unit used in chemistry is the *liter* (L), which equals a cubic decimeter, dm^3, and is slightly larger than a quart. (The liter is the first metric unit we have encountered that is *not* an SI unit.) There are 1000 milliliters (mL) in a liter, and 1 mL is the same volume as 1 cm^3: 1 mL = 1 cm^3. The devices used most frequently in chemistry to measure volume are illustrated in ▼ **Figure 1.19**.

Syringes, burettes, and pipettes deliver amounts of liquids with more precision than graduated cylinders. Volumetric flasks are used to contain specific volumes of liquid.

Give It Some Thought

Which of the following quantities represents volume measurement: 15 m^2; 2.5 × 10^2 m^3; 5.77 L/s? How do you know?

Density

Density is defined as the amount of mass in a unit volume of a substance:

$$\text{Density} = \frac{\text{mass}}{\text{volume}} \qquad [1.3]$$

 GO FIGURE

How many 1-L bottles are required to contain 1 m^3 of liquid?

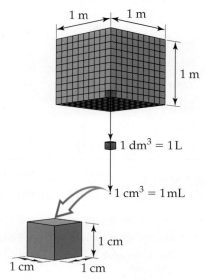

▲ **Figure 1.18 Volume relationships.** The volume occupied by a cube 1 m on each edge is one cubic meter, 1 m^3. Each cubic meter contains 1000 dm^3. One liter is the same volume as one cubic decimeter, 1 L = 1 dm^3. Each cubic decimeter contains 1000 cubic centimeters, 1 dm^3 = 1000 cm^3. One cubic centimeter equals one milliliter, 1 cm^3 = 1 mL.

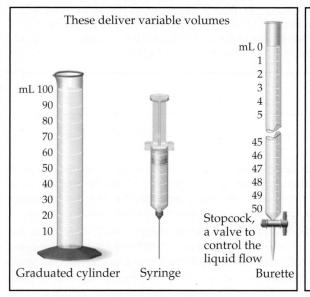

These deliver variable volumes | Pipette delivers a specific volume | Volumetric flask contains a specific volume

Graduated cylinder   Syringe   Burette   Pipette   Volumetric flask

Stopcock, a valve to control the liquid flow

▲ **Figure 1.19 Common volumetric glassware.**

Table 1.6 Densities of Selected Substances at 25 °C

Substance	Density (g/cm³)
Air	0.001
Balsa wood	0.16
Ethanol	0.79
Water	1.00
Ethylene glycol	1.09
Table sugar	1.59
Table salt	2.16
Iron	7.9
Gold	19.32

The densities of solids and liquids are commonly expressed in either grams per cubic centimeter (g/cm^3) or grams per milliliter (g/mL). The densities of some common substances are listed in ◄ Table 1.6. It is no coincidence that the density of water is 1.00 g/mL; the gram was originally defined as the mass of 1 mL of water at a specific temperature. Because most substances change volume when they are heated or cooled, densities are temperature dependent, and so temperature should be specified when reporting densities. If no temperature is reported, we assume 25 °C, close to normal room temperature.

The terms *density* and *weight* are sometimes confused. A person who says that iron weighs more than air generally means that iron has a higher density than air—1 kg of air has the same mass as 1 kg of iron, but the iron occupies a smaller volume, thereby giving it a higher density. If we combine two liquids that do not mix, the less dense liquid will float on the denser liquid.

SAMPLE EXERCISE 1.4 Determining Density and Using Density to Determine Volume or Mass

(a) Calculate the density of mercury if 1.00×10^2 g occupies a volume of 7.36 cm³.

(b) Calculate the volume of 65.0 g of liquid methanol (wood alcohol) if its density is 0.791 g/mL.

(c) What is the mass in grams of a cube of gold (density = 19.32 g/cm³) if the length of the cube is 2.00 cm?

SOLUTION

(a) We are given mass and volume, so Equation 1.3 yields

$$\text{Density} = \frac{\text{mass}}{\text{volume}} = \frac{1.00 \times 10^2 \text{ g}}{7.36 \text{ cm}^3} = 13.6 \text{ g/cm}^3$$

(b) Solving Equation 1.3 for volume and then using the given mass and density gives

$$\text{Volume} = \frac{\text{mass}}{\text{density}} = \frac{65.0 \text{ g}}{0.791 \text{ g/mL}} = 82.2 \text{ mL}$$

(c) We can calculate the mass from the volume of the cube and its density. The volume of a cube is given by its length cubed:

$$\text{Volume} = (2.00 \text{ cm})^3 = (2.00)^3 \text{ cm}^3 = 8.00 \text{ cm}^3$$

Solving Equation 1.3 for mass and substituting the volume and density of the cube, we have

$$\text{Mass} = \text{volume} \times \text{density} = (8.00 \text{ cm}^3)(19.32 \text{ g/cm}^3) = 155 \text{ g}$$

Practice Exercise 1

Platinum, Pt, is one of the rarest of the metals. Worldwide annual production is only about 130 tons. (a) Platinum has a density of 21.4 g/cm³. If thieves were to steal platinum from a bank using a small truck with a maximum payload of 900 lb, how many 1 L bars of the metal could they make off with? (a) 19 bars, (b) 2 bars, (c) 42 bars, (d) 1 bar, (e) 47 bars.

Practice Exercise 2

(a) Calculate the density of a 374.5-g sample of copper if it has a volume of 41.8 cm³. (b) A student needs 15.0 g of ethanol for an experiment. If the density of ethanol is 0.789 g/mL, how many milliliters of ethanol are needed? (c) What is the mass, in grams, of 25.0 mL of mercury (density = 13.6 g/mL)?

Chemistry Put to Work

Chemistry in the News

Because chemistry is so central to our lives, reports on matters of chemical significance appear in the news nearly every day. Some reports tell of breakthroughs in the development of new pharmaceuticals, materials, and processes. Others deal with energy, environmental, and public safety issues. As you study chemistry, you will develop the skills to better understand the importance of chemistry in your life. Here are summaries of a few recent stories in which chemistry plays an important role.

Clean energy from fuel cells. In fuel cells, the energy of a chemical reaction is converted directly into electrical energy. Although fuel cells have long been known as potentially valuable sources of electrical energy, their costs have kept them from widespread use. However, recent advances in technology have brought fuel cells to the fore as sources of reliable and clean electrical power in certain critical situations. They are especially valuable in powering data centers which consume large amounts of electrical power that must be absolutely reliable. For example, failure of electrical power at a major data center for a company such as Amazon, eBay, or Apple could be calamitous for the company and its customers.

eBay recently contracted to build the next phase of its major data center in Utah, utilizing solid–state fuel cells as the source of electrical power. The fuel cells, manufactured by Bloom Energy, a Silicon Valley startup, are large industrial devices about the size of a refrigerator (► Figure 1.20). The eBay installation utilizes biogas, which consists of methane and other fuel gases derived from landfills and farms. The fuel is combined with oxygen, and the mixture run through a special solid–state device to produce electricity. Because the electricity is being produced close to the data center, transmission of the electrical power from source to consumption is more efficient. In contrast to electrical backup systems employed in the past, the new power source will be the *primary* source of power, operating

▲ Figure 1.20 **Solid-state fuel cells manufactured by Bloom Energy.**

24 hours per day, every day of the year. The eBay facility in Utah is the largest nonelectric utility fuel cell installation in the nation. It generates 6 megawatts of power, enough to power about 6000 homes.

Regulation of greenhouse gases. In 2009 the Environmental Protection Agency (EPA) took the position that, under the provisions of the Clean Air Act, it should regulate emissions of "greenhouse" gases. *Greenhouse gases* are substances that have the potential to alter the global climate because of their ability to trap long–wavelength radiation at Earth's surface. (This subject is covered in detail in Section 18.2.) Greenhouse gases include carbon dioxide (CO_2), methane (CH_4), and nitrous oxide (N_2O), as well as other substances. The EPA decision was challenged in the courts by several states, industry organizations, and conservative groups. In a major victory for the EPA, the federal court of appeals of the District of Columbia in July 2012 upheld the agency's position. This case is interesting in part because of the grounds on which the EPA policy was challenged and the way the court responded. The plaintiffs argued that the EPA improperly based its decision on assessments from the Intergovernmental Panel on Climate Change, the U.S. Global Climate Change program, and reports from the National Research Council, rather than on citing the findings of individual research programs in the published literature. The court replied that "it makes no difference that much of the scientific evidence in large part consisted of 'syntheses' of individual studies and research. This is how science works. EPA is not required to re-prove the existence of the atom every time it approaches a scientific question."*

This is an important example of an interaction between science and social policy in our complex, modern society. When other than purely scientific interests are involved, questions about science's reliability and objectivity are bound to arise.

Anesthesia. In the period around the 1840s it became recognized that certain substances, notably ether, chloroform, and nitrous oxide, could induce a state in which the patient had no awareness of bodily pain. You can imagine how joyfully these new discoveries were received by people who had to undergo surgery that would otherwise be unbear-

ably painful. The word *anesthesia* was suggested by Oliver Wendell Holmes, Sr. in 1846 to describe the state in which a person lacks awareness, either total or of a particular part of the body. In time chemists were able to identify certain organic compounds that produced anesthesia without being severely toxic.

More than 40 million patients in North America each year undergo medical procedures that call for anesthesia. The anesthetics used today are most often injected into the blood stream rather than inhaled as a gas. Several organic substances have been identified as effective anesthetics. While modern anesthetics are generally quite safe, they must be administered with care, because they can affect breathing, blood pressure, and heart function. Every drug has a *therapeutic index*, the ratio of the smallest dose that would be fatal to the smallest dose that gives the desired therapeutic effect. Naturally, one wants the therapeutic index for any drug to be as large as possible. Anesthetics have generally low therapeutic indices, which means that they must be administered carefully and with constant monitoring. The death of the entertainer Michael Jackson in June 2009 from an overdose of propofol, a widely used anesthetic (▼ Figure 1.21), illustrates how dangerous such drugs can be when not properly administered. Propofol very quickly renders a patient unconscious and affects breathing. Hence its use must be carefully monitored by a person trained in anesthesiology.

Despite a great deal of research, it is still not clear how anesthetics actually work. It is a near-universal characteristic of life that species ranging from tadpoles to humans can be reversibly immobilized. The search for the mechanisms by which this occurs is important, because it may lead us not only to safer anesthetics, but also to deeper understanding of what we mean by *consciousness* itself.

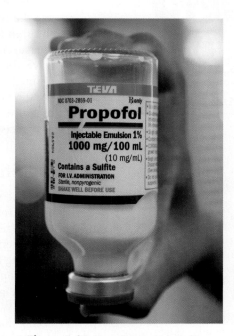

▲ Figure 1.21 **Propofol, an anesthetic.**

---

*U.S. Court of Appeals for the District of Columbia , Case No. 09-1322.

▲ GO FIGURE

How would the darts be positioned on the target for the case of "good accuracy, poor precision"?

Good accuracy
Good precision

Poor accuracy
Good precision

Poor accuracy
Poor precision

▲ Figure 1.22 **Precision and accuracy.**

High precision can be achieved on a scale like this one, which has 0.1 milligram accuracy.

1.5 | Uncertainty in Measurement

Two kinds of numbers are encountered in scientific work: *exact numbers* (those whose values are known exactly) and *inexact numbers* (those whose values have some uncertainty). Most of the exact numbers we will encounter in this book have defined values. For example, there are exactly 12 eggs in a dozen, exactly 1000 g in a kilogram, and exactly 2.54 cm in an inch. The number 1 in any conversion factor, such as 1 m = 100 cm or 1 kg = 2.2046 lb, is an exact number. Exact numbers can also result from counting objects. For example, we can count the exact number of marbles in a jar or the exact number of people in a classroom.

Numbers obtained by measurement are always *inexact*. The equipment used to measure quantities always has inherent limitations (equipment errors), and there are differences in how different people make the same measurement (human errors). Suppose ten students with ten balances are to determine the mass of the same dime. The ten measurements will probably vary slightly for various reasons. The balances might be calibrated slightly differently, and there might be differences in how each student reads the mass from the balance. Remember: *Uncertainties always exist in measured quantities.*

▲ **Give It Some Thought**

Which of the following is an inexact quantity?
(a) the number of people in your chemistry class
(b) the mass of a penny
(c) the number of grams in a kilogram

Precision and Accuracy

The terms *precision* and *accuracy* are often used in discussing the uncertainties of measured values. **Precision** is a measure of how closely individual measurements agree with one another. **Accuracy** refers to how closely individual measurements agree with the correct, or "true," value. The dart analogy in ◀ Figure 1.22 illustrates the difference between these two concepts.

In the laboratory we often perform several "trials" of an experiment and average the results. The precision of the measurements is often expressed in terms of the *standard deviation* (Appendix A.5), which reflects how much the individual measurements differ from the average. We gain confidence in our measurements if we obtain nearly the same value each time—that is, when the standard deviation is small. Figure 1.22 reminds us, however, that precise measurements can be inaccurate. For example, if a very sensitive balance is poorly calibrated, the masses we measure will be consistently either high or low. They will be inaccurate even if they are precise.

Significant Figures

Suppose you determine the mass of a dime on a balance capable of measuring to the nearest 0.0001 g. You could report the mass as 2.2405 ± 0.0001 g. The ± notation (read "plus or minus") expresses the magnitude of the uncertainty of your measurement. In much scientific work we drop the ± notation with the understanding that *there is always some uncertainty in the last digit reported for any measured quantity.*

▶ Figure 1.23 shows a thermometer with its liquid column between two scale marks. We can read the certain digits from the scale and estimate the uncertain one. Seeing that the liquid is between the 25° and 30 °C marks, we estimate the temperature to be 27 °C, being uncertain of the second digit of our measurement. By *uncertain* we mean that the temperature is reliably 27 °C and not 28° or 26 °C, but we can't say that it is *exactly* 27 °C.

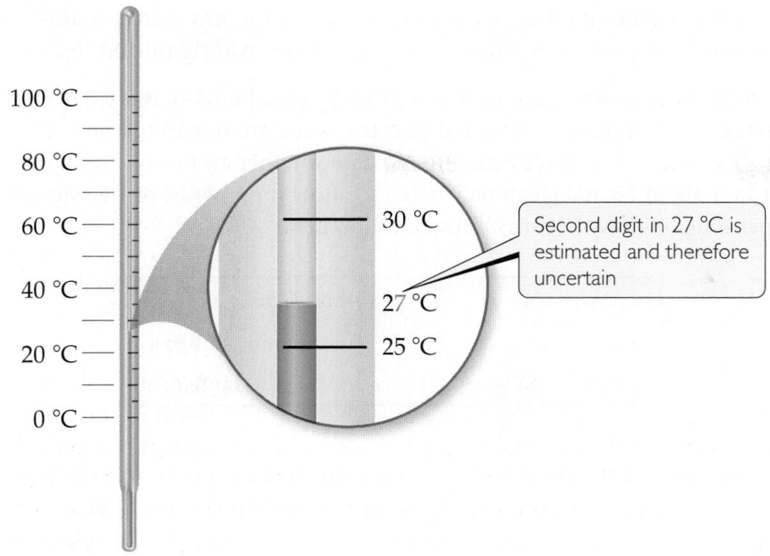

◀ Figure 1.23 **Uncertainty and significant figures in a measurement.**

All digits of a measured quantity, including the uncertain one, are called **significant figures**. A measured mass reported as 2.2 g has two significant figures, whereas one reported as 2.2405 g has five significant figures. The greater the number of significant figures, the greater the precision implied for the measurement.

SAMPLE EXERCISE 1.5 Relating Significant Figures to the Uncertainty of a Measurement

What difference exists between the measured values 4.0 and 4.00 g?

SOLUTION

The value 4.0 has two significant figures, whereas 4.00 has three. This difference implies that 4.0 has more uncertainty. A mass reported as 4.0 g indicates that the uncertainty is in the first decimal place. Thus, the mass is closer to 4.0 than to 3.9 or 4.1 g. We can represent this uncertainty by writing the mass as 4.0 ± 0.1 g. A mass reported as 4.00 g indicates that the uncertainty is in the second decimal place. In this case the mass is closer to 4.00 than 3.99 or 4.01 g, and we can represent it as 4.00 ± 0.01 g. (Without further information, we cannot be sure whether the difference in uncertainties of the two measurements reflects the precision or the accuracy of the measurement.)

Practice Exercise 1

Mo Farah won the 10,000 meter race in the 2012 Olympics with an official time of 27 minutes, 30.42 s. To the correct number of significant figures, what was Farah's average speed in m/sec?
(a) 0. 6059 m/s, **(b)** 1.65042 m/s, **(c)** 6.059064 m/s, **(d)** 0.165042 m/s, **(e)** 6.626192 m/s.

Practice Exercise 2

A sample that has a mass of about 25 g is weighed on a balance that has a precision of ± 0.001 g. How many significant figures should be reported for this measurement?

 Give It Some Thought

A digital bathroom scale gives you the following four readings in a row: 155.2, 154.8, 154.9, 154.8 lbs. How would you record your weight?

To determine the number of significant figures in a reported measurement, read the number from left to right, counting the digits starting with the first digit that is not zero. *In any measurement that is properly reported, all nonzero digits are significant.* Because zeros can be used either as part of the measured value or merely to locate the decimal point, they may or may not be significant:

1. Zeros *between* nonzero digits are always significant—1005 kg (four significant figures); 7.03 cm (three significant figures).

2. Zeros *at the beginning* of a number are never significant; they merely indicate the position of the decimal point—0.02 g (one significant figure); 0.0026 cm (two significant figures).

3. Zeros *at the end* of a number are significant if the number contains a decimal point—0.0200 g (three significant figures); 3.0 cm (two significant figures).

A problem arises when a number ends with zeros but contains no decimal point. In such cases, it is normally assumed that the zeros are not significant. Exponential notation (Appendix A.1) can be used to indicate whether end zeros are significant. For example, a mass of 10,300 g can be written to show three, four, or five significant figures depending on how the measurement is obtained:

1.03×10^4 g	(three significant figures)
1.030×10^4 g	(four significant figures)
1.0300×10^4 g	(five significant figures)

In these numbers all the zeros to the right of the decimal point are significant (rules 1 and 3). (The exponential term 10^4 does not add to the number of significant figures.)

SAMPLE EXERCISE 1.6 | Assigning Appropriate Significant Figures

The state of Colorado is listed in a road atlas as having a population of 4,301,261 and an area of 104,091 square miles. Do the numbers of significant figures in these two quantities seem reasonable? If not, what seems to be wrong with them?

SOLUTION

The population of Colorado must vary from day to day as people move in or out, are born, or die. Thus, the reported number suggests a much higher degree of *accuracy* than is possible. Secondly, it would not be feasible to actually count every individual resident in the state at any given time. Thus, the reported number suggests far greater *precision* than is possible. A reported number of 4,300,000 would better reflect the actual state of knowledge.

The area of Colorado does not normally vary from time to time, so the question here is whether the accuracy of the measurements is good to six significant figures. It would be possible to achieve such accuracy using satellite technology, provided the legal boundaries are known with sufficient accuracy.

Practice Exercise 1

Which of the following numbers in your personal life are exact numbers?
(a) Your cell phone number, (b) your weight, (c) your IQ, (d) your driver's license number, (e) the distance you walked yesterday.

Practice Exercise 2

The back inside cover of the book tells us that there are 5280 ft in 1 mile. Does this make the mile an exact distance?

SAMPLE EXERCISE 1.7 | Determining the Number of Significant Figures in a Measurement

How many significant figures are in each of the following numbers (assume that each number is a measured quantity)? (a) 4.003, (b) 6.023×10^{23}, (c) 5000.

SOLUTION

(a) Four; the zeros are significant figures. (b) Four; the exponential term does not add to the number of significant figures. (c) One; we assume that the zeros are not significant when there is no decimal point shown. If the number has more significant figures, a decimal point should be employed or the number written in exponential notation. Thus, 5000. has four significant figures, whereas 5.00×10^3 has three.

thermometer placed under her tongue and gets a value of 102.8 °F. How many significant figures are in this measurement?
(a) Three, the number of degrees to the left of the decimal point; (b) four, the number of digits in the measured reading; (c) two, the number of digits in the difference between her current reading and her normal body temperature; (d) three, the number of digits in her normal body temperature; (e) one, the number of digits to the right of the decimal point in the measured value.

Practice Exercise 1

Sylvia feels as though she may have a fever. Her normal body temperature is 98.7 °F. She measures her body temperature with a

Practice Exercise 2

How many significant figures are in each of the following measurements? (a) 3.549 g, (b) 2.3×10^4 cm, (c) 0.00134 m³.

Significant Figures in Calculations

When carrying measured quantities through calculations, *the least certain measurement limits the certainty of the calculated quantity and thereby determines the number of significant figures in the final answer*. The final answer should be reported with only one uncertain digit. To keep track of significant figures in calculations, we will make frequent use of two rules: one for addition and subtraction and another for multiplication and division.

1. ***For addition and subtraction*, the result has the same number of decimal places as the measurement with the fewest decimal places**. When the result contains more than the correct number of significant figures, it must be rounded off. Consider the following example in which the uncertain digits appear in color:

This number limits	20.42	⟵ two decimal places
the number of significant	1.322	⟵ three decimal places
figures in the result ⟶	83.1	⟵ one decimal place
	104.842	⟵ round off to one decimal place (104.8)

We report the result as 104.8 because 83.1 has only one decimal place.

2. ***For multiplication and division*, the result contains the same number of significant figures as the measurement with the fewest significant figures.** When the result contains more than the correct number of significant figures, it must be rounded off. For example, the area of a rectangle whose measured edge lengths are 6.221 and 5.2 cm should be reported with two significant figures, 32 cm^2, even though a calculator shows the product to have more digits:

$$\text{Area} = (6.221 \text{ cm})(5.2 \text{ cm}) = 32.3492 \text{ cm}^2 \Rightarrow \text{round off to } 32 \text{ cm}^2$$

because 5.2 has two significant figures.

Notice that for addition and subtraction, decimal places are counted in determining how many digits to report in an answer, whereas for multiplication and division, significant figures are counted in determining how many digits to report.

In determining the final answer for a calculated quantity, *exact numbers* are assumed to have an infinite number of significant figures. Thus, when we say, "There are 12 inches in 1 foot," the number 12 is exact, and we need not worry about the number of significant figures in it.

In *rounding off numbers*, look at the leftmost digit to be removed:

- If the leftmost digit removed is less than 5, the preceding number is left unchanged. Thus, rounding off 7.248 to two significant figures gives 7.2.

- If the leftmost digit removed is 5 or greater, the preceding number is increased by 1. Rounding off 4.735 to three significant figures gives 4.74, and rounding 2.376 to two significant figures gives 2.4.*

 ### Give It Some Thought

A rectangular garden plot is measured to be 25.8 m by 18 m. Which of these dimensions needs to be measured to greater accuracy to provide a more accurate estimate of the area of the plot?

---

*Your instructor may want you to use a slight variation on the rule when the leftmost digit to be removed is exactly 5, with no following digits or only zeros following. One common practice is to round up to the next higher number if that number will be even and down to the next lower number otherwise. Thus, 4.7350 would be rounded to 4.74, and 4.7450 would also be rounded to 4.74.

SAMPLE EXERCISE 1.8 Determining the Number of Significant Figures in a Calculated Quantity

The width, length, and height of a small box are 15.5, 27.3, and 5.4 cm, respectively. Calculate the volume of the box, using the correct number of significant figures in your answer.

SOLUTION

In reporting the volume, we can show only as many significant figures as given in the dimension with the fewest significant figures, which is that for the height (two significant figures):

$$\text{Volume} = \text{width} \times \text{length} \times \text{height}$$
$$= (15.5 \text{ cm})(27.3 \text{ cm})(5.4 \text{ cm})$$
$$= 2285.01 \text{ cm}^3 \Rightarrow 2.3 \times 10^3 \text{ cm}^3$$

A calculator used for this calculation shows 2285.01, which we must round off to two significant figures. Because the resulting number is 2300, it is best reported in exponential notation, 2.3×10^3, to clearly indicate two significant figures.

Practice Exercise 1

Ellen recently purchased a new hybrid car and wants to check her gas mileage. At an odometer setting of 651.1 mi, she fills the tank. At 1314.4 mi she requires 16.1 gal to refill the tank. Assuming that the tank is filled to the same level both times, how is the gas mileage best expressed? (a) 40 mi/gal, (b) 41 mi/gal, (c) 41.2 mi/gal, (d) 41.20 mi/gal.

Practice Exercise 2

It takes 10.5 s for a sprinter to run 100.00 m. Calculate her average speed in meters per second and express the result to the correct number of significant figures.

SAMPLE EXERCISE 1.9 Determining the Number of Significant Figures in a Calculated Quantity

A vessel containing a gas at 25 °C is weighed, emptied, and then reweighed as depicted in ▼ **Figure 1.24**. From the data provided, calculate the density of the gas at 25 °C.

SOLUTION

To calculate the density, we must know both the mass and the volume of the gas. The mass of the gas is just the difference in the masses of the full and empty container:

$$(837.63 - 836.25) \text{ g} = 1.38 \text{ g}$$

In subtracting numbers, we determine the number of significant figures in our result by counting decimal places in each quantity. In this case each quantity has two decimal places. Thus, the mass of the gas, 1.38 g, has two decimal places.

Using the volume given in the question, 1.05×10^3 cm^3, and the definition of density, we have

$$\text{Density} = \frac{\text{mass}}{\text{volume}} = \frac{1.38 \text{ g}}{1.05 \times 10^3 \text{ cm}^3}$$

$$= 1.31 \times 10^{-3} \text{ g/cm}^3 = 0.00131 \text{ g/cm}^3$$

In dividing numbers, we determine the number of significant figures our result should contain by counting the number of significant figures in each quantity. There are three significant figures in our answer, corresponding to the number of significant figures in the two numbers that form the ratio. Notice that in this example, following the rules for determining significant figures gives an answer containing only three significant figures, even though the measured masses contain five significant figures.

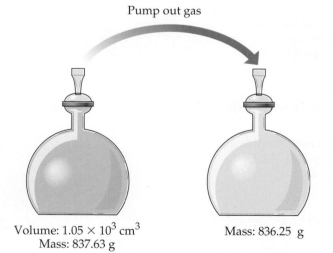

Pump out gas

Volume: 1.05×10^3 cm^3
Mass: 837.63 g

Mass: 836.25 g

▲ **Figure 1.24 Uncertainty and significant figures in a measurement.**

Practice Exercise 1

Which of the following numbers is correctly rounded to three significant figures, as shown in brackets? (a) 12,556 [12,500], (b) 4.5671×10^{-9} [4.567×10^{-9}], (c) 3.00072 [3.001], (d) 0.006739 [0.00674], (e) 5.4589×10^5 [5.459×10^5].

Practice Exercise 2

If the mass of the container in the sample exercise (Figure 1.24) were measured to three decimal places before and after pumping out the gas, could the density of the gas then be calculated to four significant figures?

When a calculation involves two or more steps and you write answers for intermediate steps, retain at least one nonsignificant digit for the intermediate answers. This procedure ensures that small errors from rounding at each step do not combine to affect the final result. When using a calculator, you may enter the numbers one after another, rounding only the final answer. Accumulated rounding-off errors may account for small differences among results you obtain and answers given in the text for numerical problems.

1.6 | Dimensional Analysis

Because measured quantities have units associated with them, it is important to keep track of units as well as numerical values when using the quantities in calculations. Throughout the text we use **dimensional analysis** in solving problems. In **dimensional analysis**, units are multiplied together or divided into each other along with the numerical values. Equivalent units cancel each other. Using dimensional analysis helps ensure that solutions to problems yield the proper units. Moreover, it provides a systematic way of solving many numerical problems and of checking solutions for possible errors.

The key to using dimensional analysis is the correct use of *conversion factors* to change one unit into another. A **conversion factor** is a fraction whose numerator and denominator are the same quantity expressed in different units. For example, 2.54 cm and 1 in. are the same length: 2.54 cm = 1 in. This relationship allows us to write two conversion factors:

$$\frac{2.54 \text{ cm}}{1 \text{ in.}} \quad \text{and} \quad \frac{1 \text{ in.}}{2.54 \text{ cm}}$$

We use the first factor to convert inches to centimeters. For example, the length in centimeters of an object that is 8.50 in. long is

$$\text{Number of centimeters} = (8.50 \text{ in.}) \frac{2.54 \text{ cm}}{1 \text{ in.}} = 21.6 \text{ cm}$$

The unit inches in the denominator of the conversion factor cancels the unit inches in the given data (8.50 *inches*), so that the centimeters unit in the numerator of the conversion factor becomes the unit of the final answer. Because the numerator and denominator of a conversion factor are equal, multiplying any quantity by a conversion factor is equivalent to multiplying by the number 1 and so does not change the intrinsic value of the quantity. The length 8.50 in. is the same as the length 21.6 cm.

In general, we begin any conversion by examining the units of the given data and the units we desire. We then ask ourselves what conversion factors we have available to take us from the units of the given quantity to those of the desired one. When we multiply a quantity by a conversion factor, the units multiply and divide as follows:

$$\text{Given unit} \times \frac{\text{desired unit}}{\text{given unit}} = \text{desired unit}$$

If the desired units are not obtained in a calculation, an error must have been made somewhere. Careful inspection of units often reveals the source of the error.

SAMPLE EXERCISE 1.10 **Converting Units**

If a woman has a mass of 115 lb, what is her mass in grams? (Use the relationships between units given on the back inside cover of the text.)

SOLUTION

Because we want to change from pounds to grams, we look for a relationship between these units of mass. The conversion factor table found on the back inside cover tells us that 1 lb = 453.6 g.

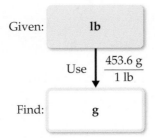

Given: **lb**

Use $\dfrac{453.6 \text{ g}}{1 \text{ lb}}$

Find: **g**

To cancel pounds and leave grams, we write the conversion factor with grams in the numerator and pounds in the denominator:

$$\text{Mass in grams} = (115 \text{ lb})\left(\frac{453.6 \text{ g}}{1 \text{ lb}}\right) = 5.22 \times 10^4 \text{ g}$$

The answer can be given to only three significant figures, the number of significant figures in 115 lb. The process we have used is diagrammed in the margin.

Practice Exercise 1

At a particular instant in time the Earth is judged to be 92,955,000 miles from the Sun. What is the distance in kilometers to four significant figures? (See back inside cover for conversion factor). (a) 5763×10^4 km, (b) 1.496×10^8 km, (c) 1.49596×10^8 km, (d) 1.483×10^4 km, (e) 57,759,000 km.

Practice Exercise 2

By using a conversion factor from the back inside cover, determine the length in kilometers of a 500.0-mi automobile race.

Strategies in Chemistry

Estimating Answers

Calculators are wonderful devices; they enable you to get to the wrong answer very quickly. Of course, that's not the destination you want. You can take certain steps to avoid putting that wrong answer into your homework set or on an exam. One is to keep track of the units in a calculation and use the correct conversion factors. Second, you can do a quick mental check to be sure that your answer is reasonable: you can try to make a "ballpark" estimate.

A ballpark estimate involves making a rough calculation using numbers that are rounded off in such a way that the arithmetic can be done without a calculator. Even though this approach does not give an exact answer, it gives one that is roughly the correct size. By using dimensional analysis and by estimating answers, you can readily check the reasonableness of your calculations.

You can get better at making estimates by practicing in everyday life. How far is it from your dorm room to the chemistry lecture hall? How much do your parents pay for gasoline per year? How many bikes are there on campus? If you respond "I have no idea" to these questions, you're giving up too easily. Try estimating familiar quantities and you'll get better at making estimates in science and in other aspects of your life where a misjudgment can be costly.

Give It Some Thought

How do we determine how many digits to use in conversion factors, such as the one between pounds and grams in Sample Exercise 1.10?

Using Two or More Conversion Factors

It is often necessary to use several conversion factors in solving a problem. As an example, let's convert the length of an 8.00-m rod to inches. The table on the back inside cover does not give the relationship between meters and inches. It *does*, however, give the relationship between centimeters and inches (1 in. = 2.54 cm). From our knowledge of SI prefixes, we know that $1 \text{ cm} = 10^{-2} \text{ m}$. Thus, we can convert step by step, first from meters to centimeters and then from centimeters to inches:

Given: **m** → Use $\dfrac{1 \text{ cm}}{10^{-2} \text{ m}}$ → **cm** → Use $\dfrac{1 \text{ in.}}{2.54 \text{ cm}}$ → Find: **in.**

Combining the given quantity (8.00 m) and the two conversion factors, we have

$$\text{Number of inches} = (8.00 \text{ m})\left(\frac{1 \text{ cm}}{10^{-2} \text{ m}}\right)\left(\frac{1 \text{ in.}}{2.54 \text{ cm}}\right) = 315 \text{ in.}$$

The first conversion factor is used to cancel meters and convert the length to centimeters. Thus, meters are written in the denominator and centimeters in the numerator.

The second conversion factor is used to cancel centimeters and convert the length to inches, so it has centimeters in the denominator and inches, the desired unit, in the numerator.

Note that you could have used 100 cm = 1 m as a conversion factor as well in the second parentheses. As long as you keep track of your given units and cancel them properly to obtain the desired units, you are likely to be successful in your calculations.

SAMPLE EXERCISE 1.11   Converting Units Using Two or More Conversion Factors

The average speed of a nitrogen molecule in air at 25 °C is 515 m/s. Convert this speed to miles per hour.

SOLUTION

To go from the given units, m/s, to the desired units, mi/hr, we must convert meters to miles and seconds to hours. From our knowledge of SI prefixes we know that 1 km = 10^3 m. From the relationships given on the back inside cover of the book, we find that 1 mi = 1.6093 km.

Thus, we can convert m to km and then convert km to mi. From our knowledge of time we know that 60 s = 1 min and 60 min = 1 hr. Thus, we can convert s to min and then convert min to hr. The overall process is

Applying first the conversions for distance and then those for time, we can set up one long equation in which unwanted units are canceled:

$$\text{Speed in mi/hr} = \left(515\frac{m}{s}\right)\left(\frac{1\ km}{10^3\ m}\right)\left(\frac{1\ mi}{1.6093\ km}\right)\left(\frac{60\ s}{1\ min}\right)\left(\frac{60\ min}{1\ hr}\right)$$

$$= 1.15 \times 10^3\ \text{mi/hr}$$

Our answer has the desired units. We can check our calculation, using the estimating procedure described in the "Strategies in Chemistry" box. The given speed is about 500 m/s. Dividing by 1000 converts m to km, giving 0.5 km/s. Because 1 mi is about 1.6 km, this speed corresponds to 0.5/1.6 = 0.3 mi/s. Multiplying by 60 gives about 0.3 × 60 = 20 mi/min. Multiplying again by 60 gives 20 × 60 = 1200 mi/hr. The approximate solution (about 1200 mi/hr) and the detailed solution (1150 mi/hr) are reasonably close. The answer to the detailed solution has three significant figures, corresponding to the number of significant figures in the given speed in m/s.

Practice Exercise 1
Fabiola, who lives in Mexico City, fills her car with gas, paying 357 pesos for 40.0 L. What is her fuel cost in dollars per gallon, if 1 peso = 0.0759 dollars? **(a)** $1.18/gal, **(b)** $3.03/gal, **(c)** $1.47/gal, **(d)** $9.68/gal, **(e)** $2.56/gal.

Practice Exercise 2
A car travels 28 mi per gallon of gasoline. What is the mileage in kilometers per liter?

Conversions Involving Volume

The conversion factors previously noted convert from one unit of a given measure to another unit of the same measure, such as from length to length. We also have conversion factors that convert from one measure to a different one. The density of a substance, for example, can be treated as a conversion factor between mass and volume. Suppose we want to know the mass in grams of 2 cubic inches ($2.00\ \text{in.}^3$) of gold, which has a density of 19.3 g/cm³. The density gives us the conversion factors:

$$\frac{19.3\ g}{1\ cm^3} \quad \text{and} \quad \frac{1\ cm^3}{19.3\ g}$$

Because we want a mass in grams, we use the first factor, which has mass in grams in the numerator. To use this factor, however, we must first convert cubic inches to cubic centimeters. The relationship between in.^{3} and cm^3 is not given on the back inside cover, but the relationship between inches and centimeters is given: 1 in. = 2.54 cm (exactly). Cubing both sides of this equation gives (1 in.)^3 = (2.54 cm)^3, from which we write the desired conversion factor:

$$\frac{(2.54\ \text{cm})^3}{(1\ \text{in.})^3} = \frac{(2.54)^3\ \text{cm}^3}{(1)^3\ \text{in.}^3} = \frac{16.39\ \text{cm}^3}{1\ \text{in.}^3}$$

Notice that both the numbers and the units are cubed. Also, because 2.54 is an exact number, we can retain as many digits of (2.54)^3 as we need. We have used four, one more than the number of digits in the density (19.3 g/cm^3). Applying our conversion factors, we can now solve the problem:

$$\text{Mass in grams} = (2.00\ \text{in.}^3)\left(\frac{16.39\ \text{cm}^3}{1\ \text{in.}^3}\right)\left(\frac{19.3\ \text{g}}{1\ \text{cm}^3}\right) = 633\ \text{g}$$

The procedure is diagrammed here. The final answer is reported to three significant figures, the same number of significant figures as in 2.00 in.^{3} and 19.3 g.

Given: in.^{3} → Use $\left(\dfrac{2.54\ \text{cm}}{1\ \text{in.}}\right)^3$ →   cm^3 → Use $\dfrac{19.3\ \text{g}}{1\ \text{cm}^3}$ →   Find: g

How many liters of water do Earth's oceans contain?

SAMPLE EXERCISE 1.12 Converting Volume Units

Earth's oceans contain approximately 1.36×10^9 km^3 of water. Calculate the volume in liters.

SOLUTION

From the back inside cover, we find 1 L = 10^{-3} m^3, but there is no relationship listed involving km^3. From our knowledge of SI prefixes, however, we know 1 km = 10^3 m and we can use this relationship between lengths to write the desired conversion factor between volumes:

$$\left(\frac{10^3\ \text{m}}{1\ \text{km}}\right)^3 = \frac{10^9\ \text{m}^3}{1\ \text{km}^3}$$

Thus, converting from km^3 to m^3 to L, we have

$$\text{Volume in liters} = (1.36 \times 10^9\ \text{km}^3)\left(\frac{10^9\ \text{m}^3}{1\ \text{km}^3}\right)\left(\frac{1\ \text{L}}{10^{-3}\ \text{m}^3}\right) = 1.36 \times 10^{21}\ \text{L}$$

Practice Exercise 1
A barrel of oil as measured on the oil market is equal to 1.333 U.S. barrels. A U.S. barrel is equal to 31.5 gal. If oil is on the market at $94.0 per barrel, what is the price in dollars per gallon? **(a)** $2.24/gal, **(b)** $3.98/gal, **(c)** $2.98/gal, **(d)** $1.05/gal, **(e)** $8.42/gal.

Practice Exercise 2
The surface area of Earth is 510×10^6 km^2, and 71% of this is ocean. Using the data from the sample exercise, calculate the average depth of the world's oceans in feet.

Strategies in Chemistry

The Importance of Practice

If you have ever played a musical instrument or participated in athletics, you know that the keys to success are practice and discipline. You cannot learn to play a piano merely by listening to music, and you cannot learn how to play basketball merely by watching games on television. Likewise, you cannot learn chemistry by merely watching your instructor give lectures. Simply reading this book, listening to lectures, or reviewing notes will not usually be sufficient when exam time comes around. Your task is to master chemical concepts and practices to a degree that you can put them to use in solving problems and answering questions. Solving problems correctly takes practice—actually, a fair amount of it. You will do well in your chemistry course if you embrace the idea that you need to master the materials presented, and then learn how to apply them in solving problems. Even if you're a brilliant student, this will take time; it's what being a student is all about. Almost no one fully absorbs new material on a first reading, especially when new concepts are being presented. You are

sure to more fully master the content of the chapters by reading them through at least twice, even more for passages that present you with difficulties in understanding.

Throughout the book, we have provided sample exercises in which the solutions are shown in detail. For practice exercises, we supply only the answer, at the back of the book. It is important that you use these exercises to test yourself.

The practice exercises in this text and the homework assignments given by your instructor provide the minimal practice that you will need to succeed in your chemistry course. Only by working all the assigned problems will you face the full range of difficulty and coverage that your instructor expects you to master for exams. There is no substitute for a determined and perhaps lengthy effort to work problems on your own. If you are stuck on a problem, however, ask for help from your instructor, a teaching assistant, a tutor, or a fellow student. Spending an inordinate amount of time on a single exercise is rarely effective unless you know that it is particularly challenging and is expected to require extensive thought and effort.

SAMPLE EXERCISE 1.13   Conversions Involving Density

What is the mass in grams of 1.00 gal of water? The density of water is 1.00 g/mL.

SOLUTION

Before we begin solving this exercise, we note the following:

(1) We are given 1.00 gal of water (the known, or given, quantity) and asked to calculate its mass in grams (the unknown).

(2) We have the following conversion factors either given, commonly known, or available on the back inside cover of the text:

$$\frac{1.00 \text{ g water}}{1 \text{ mL water}} \quad \frac{1 \text{ L}}{1000 \text{ mL}} \quad \frac{1 \text{ L}}{1.057 \text{ qt}} \quad \frac{1 \text{ gal}}{4 \text{ qt}}$$

The first of these conversion factors must be used as written (with grams in the numerator) to give the desired result, whereas the last conversion factor must be inverted in order to cancel gallons:

$$\text{Mass in grams} = (1.00 \text{ gal})\left(\frac{4 \text{ qt}}{1 \text{ gal}}\right)\left(\frac{1 \text{ L}}{1.057 \text{ qt}}\right)\left(\frac{1000 \text{ mL}}{1 \text{ L}}\right)\left(\frac{1.00 \text{ g}}{1 \text{ mL}}\right)$$

$$= 3.78 \times 10^3 \text{ g water}$$

The unit of our final answer is appropriate, and we have taken care of our significant figures. We can further check our calculation by estimating. We can round 1.057 off to 1. Then focusing on the numbers that do not equal 1 gives $4 \times 1000 = 4000$ g, in agreement with the detailed calculation.

You should also use common sense to assess the reasonableness of your answer. In this case we know that most people can lift a gallon of milk with one hand, although it would be tiring to carry it around all day. Milk is mostly water and will have a density not too different from that of water. Therefore, we might estimate that a gallon of water has mass that is more than 5 lb but less than 50 lb. The mass we have calculated, $3.78 \text{ kg} \times 2.2 \text{ lb/kg} = 8.3 \text{ lb}$, is thus reasonable as an order-of-magnitude estimate.

A Trex deck.

Practice Exercise 1

Trex is a manufactured substitute for wood compounded from post-consumer plastic and wood. It is frequently used in outdoor decks. Its density is reported as 60 lb/ft^3. What is the density of Trex in kg/L? **(a)** 138 kg/L, **(b)** 0.960 kg/L, **(c)** 259 kg/L, **(d)** 15.8 kg/L, **(e)** 11.5 kg/L.

Practice Exercise 2

The density of the organic compound benzene is 0.879 g/mL. Calculate the mass in grams of 1.00 qt of benzene.

Strategies in Chemistry

The Features of This Book

If, like most students, you haven't yet read the part of the Preface to this text entitled TO THE STUDENT, *you should do it now.* In less than two pages of reading you will encounter valuable advice on how to navigate your way through this book and through the course. We're serious! This is advice you can use.

The TO THE STUDENT section describes how text features such as "What's Ahead," Key Terms, Learning Outcomes, and Key Equations will help you remember what you have learned. We describe there also how to take advantage of the text's Web site, where many types of online study tools are available. If you have registered for MasteringChemistry®, you will have access to many helpful animations, tutorials, and additional problems correlated to specific topics and sections of each chapter. An interactive eBook is also available online.

As previously mentioned, working exercises is very important—in fact, essential. You will find a large variety of exercises at the end of each chapter that are designed to test your problem-solving skills in chemistry. Your instructor will very likely assign some of these end-of-chapter exercises as homework. The first few exercises called

"Visualizing Concepts" are meant to test how well you understand a concept without plugging a lot of numbers into a formula. The other exercises are grouped in pairs, with the answers given at the back of the book to the odd-numbered exercises (those with red exercise numbers). An exercise with a [bracket] around its number is designed to be more challenging. Additional Exercises appear after the regular exercises; the chapter sections that they cover are not identified, and they are not paired. Integrative Exercises, which start appearing from Chapter 3, are problems that require skills learned in previous chapters. Also first appearing in Chapter 3, are Design an Experiment exercises consisting of problem scenarios that challenge you to design experiments to test hypotheses.

Many chemical databases are available, usually on the Web. The *CRC Handbook of Chemistry and Physics* is the standard reference for many types of data and is available in libraries. The *Merck Index* is a standard reference for the properties of many organic compounds, especially ones of biological interest. WebElements (http://www.webelements .com/) is a good Web site for looking up the properties of the elements. Wolfram Alpha (http://www.wolframalpha.com/) can also be a source of useful information on substances, numerical values, and other data.

Chapter Summary and Key Terms

THE STUDY OF CHEMISTRY (SECTION 1.1) **Chemistry** is the study of the composition, structure, properties, and changes of **matter**. The composition of matter relates to the kinds of **elements** it contains. The structure of matter relates to the ways the **atoms** of these elements are arranged. A **property** is any characteristic that gives a sample of matter its unique identity. A **molecule** is an entity composed of two or more atoms with the atoms attached to one another in a specific way.

CLASSIFICATIONS OF MATTER (SECTION 1.2) Matter exists in three physical states, **gas**, **liquid**, and **solid**, which are known as the **states of matter**. There are two kinds of **pure substances: elements** and **compounds**. Each element has a single kind of atom and is represented by a chemical symbol consisting of one or two letters, with the first letter capitalized. Compounds are composed of two or more elements joined chemically. The **law of constant composition**, also called the **law of definite proportions**, states that the elemental composition of a pure compound is always the same. Most matter consists of a mixture of substances. **Mixtures** have variable compositions and can be either homogeneous or heterogeneous; homogeneous mixtures are called **solutions**.

PROPERTIES OF MATTER (SECTION 1.3) Each substance has a unique set of **physical properties** and **chemical properties** that can be used to identify it. During a **physical change**, matter does not change its composition. **Changes of state** are physical changes. In a **chemical change (chemical reaction)** a substance is transformed into a chemically different substance. **Intensive properties** are independent of the amount of matter examined and are used to identify substances. **Extensive properties** relate to the amount of substance present. Differences in physical and chemical properties are used to separate substances.

The **scientific method** is a dynamic process used to answer questions about the physical world. Observations and experiments lead to tentative explanations or **hypotheses**. As a hypothesis is tested and refined, a **theory** may be developed that can predict the results of future observations and experiments. When observations repeatedly lead to

the same consistent results, we speak of a **scientific law**, a general rule that summarizes how nature behaves.

UNITS OF MEASUREMENT (SECTION 1.4) Measurements in chemistry are made using the **metric system**. Special emphasis is placed on **SI units**, which are based on the meter, the kilogram, and the second as the basic units of length, **mass**, and time, respectively. SI units use prefixes to indicate fractions or multiples of base units. The SI **temperature** scale is the **Kelvin scale**, although the **Celsius scale** is frequently used as well. **Absolute zero** is the lowest temperature attainable. It has the value 0 K. A **derived unit** is obtained by multiplication or division of SI base units. Derived units are needed for defined quantities such as speed or volume. **Density** is an important defined quantity that equals mass divided by volume.

UNCERTAINTY IN MEASUREMENT (SECTION 1.5) All measured quantities are inexact to some extent. The **precision** of a measurement indicates how closely different measurements of a quantity agree with one another. The **accuracy** of a measurement indicates how well a measurement agrees with the accepted or "true" value. The **significant figures** in a measured quantity include one estimated digit, the last digit of the measurement. The significant figures indicate the extent of the uncertainty of the measurement. Certain rules must be followed so that a calculation involving measured quantities is reported with the appropriate number of significant figures.

DIMENSIONAL ANALYSIS (SECTION 1.6) In the **dimensional analysis** approach to problem solving, we keep track of units as we carry measurements through calculations. The units are multiplied together, divided into each other, or canceled like algebraic quantities. Obtaining the proper units for the final result is an important means of checking the method of calculation. When converting units and when carrying out several other types of problems, **conversion factors** can be used. These factors are ratios constructed from valid relations between equivalent quantities.

Learning Outcomes After studying this chapter, you should be able to:

- Distinguish among elements, compounds, and mixtures. (Section 1.2)
- Identify symbols of common elements. (Section 1.2)
- Identify common metric prefixes. (Section 1.4)

- Demonstrate the use of significant figures, scientific notation, and SI units in calculations. (Section 1.5)
- Attach appropriate SI units to defined quantities, and employ dimensional analysis in calculations. (Sections 1.4 and 1.6)

Key Equations

- $K = °C + 273.15$ [1.1]   Converting between Celsius (°C) and Kelvin (K) temperature scales

- $°C = \frac{5}{9}(°F - 32)$ or $°F = \frac{9}{5}(°C) + 32$ [1.2]   Converting between Celsius (°C) and Fahrenheit (°F) temperature scales

- $\text{Density} = \frac{\text{mass}}{\text{volume}}$ [1.3]   Definition of density

Exercises

Visualizing Concepts

1.1 Which of the following figures represents (a) a pure element, (b) a mixture of two elements, (c) a pure compound, (d) a mixture of an element and a compound? (More than one picture might fit each description.) [Section 1.2]

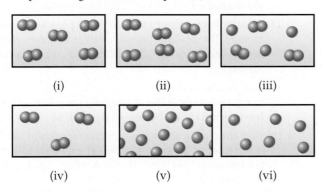

(i)   (ii)   (iii)

(iv)   (v)   (vi)

1.2 Does the following diagram represent a chemical or physical change? How do you know? [Section 1.3]

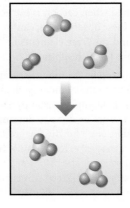

1.3 Describe the separation method(s) involved in brewing a cup of coffee. [Section 1.3]

1.4 Identify each of the following as measurements of length, area, volume, mass, density, time, or temperature: (a) 25 ps, (b) 374.2 mg, (c) 77 K, (d) 100,000 km², (e) 1.06 µm, (f) 16 nm², (g) −78 °C, (h) 2.56 g/cm³, (i) 28 cm³. [Section 1.4]

1.5 (a) Three spheres of equal size are composed of aluminum (density = 2.70 g/cm³), silver (density = 10.49 g/cm³), and nickel (density = 8.90 g/cm³). List the spheres from lightest to heaviest. (b) Three cubes of equal mass are composed of gold (density = 19.32 g/cm³), platinum (density = 21.45 g/cm³), and lead (density = 11.35 g/cm³). List the cubes from smallest to largest. [Section 1.4]

1.6 The three targets from a rifle range shown on the next page were produced by: (A) the instructor firing a newly acquired target rifle; (B) the instructor firing his personal target rifle; and (C) a student who has fired his target rifle only a few times. (a) Comment on the accuracy and precision for each of these three sets of results. (b) For the A and C results in the future to look like those in B, what needs to happen? [Section 1.5]

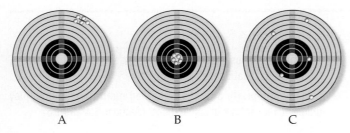

A   B   C

1.7 **(a)** What is the length of the pencil in the following figure if the ruler reads in centimeters? How many significant figures are there in this measurement? **(b)** An automobile speedometer with circular scales reading both miles per hour and kilometers per hour is shown. What speed is indicated, in both units? How many significant figures are in the measurements? [Section 1.5]

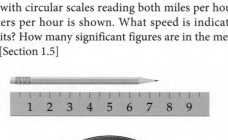

1.8 **(a)** How many significant figures should be reported for the volume of the metal bar shown here? **(b)** If the mass of the bar is 104.72 g, how many significant figures should be reported when its density is determined using the calculated volume? [Section 1.5]

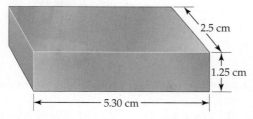

1.9 When you convert units, how do you decide which part of the conversion factor is in the numerator and which is in the denominator? [Section 1.6]

1.10 Show the steps to convert the speed of sound, 344 meters per second, into miles per hour. [Section 1.6]

1.11 Consider the jar of jelly beans in the photo. To get an estimate of the number of beans in the jar you weigh six beans and obtain masses of 3.15, 3.12, 2.98, 3.14, 3.02, and 3.09 g. Then you weigh the jar with all the beans in it, and obtain a mass of 2082 g. The empty jar has a mass of 653 g. Based on these data estimate the number of beans in the jar. Justify the number of significant figures you use in your estimate. [Section 1.5]

1.12 The photo below shows a picture of an agate stone. Jack, who picked up the stone on the Lake Superior shoreline and polished it, insists that agate is a chemical compound. Ellen argues that it cannot be a compound. Discuss the relative merits of their positions. [Section 1.2]

Classification and Properties of Matter (Sections 1.2 and 1.3)

1.13 Classify each of the following as a pure substance or a mixture. If a mixture, indicate whether it is homogeneous or heterogeneous: **(a)** rice pudding, **(b)** seawater, **(c)** magnesium, **(d)** crushed ice.

1.14 Classify each of the following as a pure substance or a mixture. If a mixture, indicate whether it is homogeneous or heterogeneous: **(a)** air, **(b)** tomato juice, **(c)** iodine crystals, **(d)** sand.

1.15 Give the chemical symbol or name for the following elements, as appropriate: **(a)** sulfur, **(b)** gold, **(c)** potassium, **(d)** chlorine, **(e)** copper, **(f)** U, **(g)** Ni, **(h)** Na, **(i)** Al, **(j)** Si.

1.16 Give the chemical symbol or name for each of the following elements, as appropriate: **(a)** carbon, **(b)** nitrogen, **(c)** titanium, **(d)** zinc, **(e)** iron, **(f)** P, **(g)** Ca, **(h)** He, **(i)** Pb, **(j)** Ag.

1.17 A solid white substance A is heated strongly in the absence of air. It decomposes to form a new white substance B and a gas C. The gas has exactly the same properties as the product obtained when carbon is burned in an excess of oxygen. Based on these observations, can we determine whether solids A and B and gas C are elements or compounds? Explain your conclusions for each substance.

1.18 You are hiking in the mountains and find a shiny gold nugget. It might be the element gold, or it might be "fool's gold," which is a nickname for iron pyrite, FeS_2. What kinds of experiments could be done to determine if the shiny nugget is really gold?

1.19 In the process of attempting to characterize a substance, a chemist makes the following observations: The substance is a silvery white, lustrous metal. It melts at 649 °C and boils at 1105 °C. Its density at 20 °C is 1.738 g/cm³. The substance burns in air, producing an intense white light. It reacts with chlorine to give a brittle white solid. The substance can be pounded into thin sheets or drawn into wires. It is a good conductor of electricity. Which of these characteristics are physical properties, and which are chemical properties?

1.20 **(a)** Read the following description of the element zinc and indicate which are physical properties and which are chemical properties.

Zinc melts at 420 °C. When zinc granules are added to dilute sulfuric acid, hydrogen is given off and the metal dissolves. Zinc has a hardness on the Mohs scale of 2.5 and a density of 7.13g/cm³ at 25 °C. It reacts slowly with oxygen gas at elevated temperatures to form zinc oxide, ZnO.

(b) Which properties of zinc can you describe from the photo? Are these physical or chemical properties?

1.21 Label each of the following as either a physical process or a chemical process: **(a)** rusting of a metal can, **(b)** boiling a cup of water, **(c)** pulverizing an aspirin, **(d)** digesting a candy bar, **(e)** exploding of nitroglyerin.

1.22 A match is lit and held under a cold piece of metal. The following observations are made: **(a)** The match burns. **(b)** The metal gets warmer. **(c)** Water condenses on the metal. **(d)** Soot (carbon) is deposited on the metal. Which of these occurrences are due to physical changes, and which are due to chemical changes?

1.23 Suggest a method of separating each of the following mixtures into two components: **(a)** sugar and sand, **(b)** oil and vinegar.

1.24 Three beakers contain clear, colorless liquids. One beaker contains pure water, another contains salt water, and another contains sugar water. How can you tell which beaker is which? (No tasting allowed!)

Units and Measurement (Section 1.4)

1.25 What exponential notation do the following abbreviations represent? **(a)** d, **(b)** c, **(c)** f, **(d)** μ, **(e)** M, **(f)** k, **(g)** n, **(h)** m, **(i)** p.

1.26 Use appropriate metric prefixes to write the following measurements without use of exponents: **(a)** 2.3×10^{-10} L, **(b)** 4.7×10^{-6} g, **(c)** 1.85×10^{-12} m, **(d)** 16.7×10^6 s, **(e)** 15.7×10^3 g, **(f)** 1.34×10^{-3} m, **(g)** 1.84×10^2 cm.

1.27 Make the following conversions: **(a)** 72 °F to °C, **(b)** 216.7 °C to °F, **(c)** 233 °C to K, **(d)** 315 K to °F, **(e)** 2500 °F to K, **(f)** 0 K to °F.

1.28 **(a)** The temperature on a warm summer day is 87 °F. What is the temperature in °C? **(b)** Many scientific data are reported at 25 °C. What is this temperature in kelvins and in degrees Fahrenheit? **(c)** Suppose that a recipe calls for an oven temperature of 400 °F. Convert this temperature to degrees Celsius and to kelvins. **(d)** Liquid nitrogen boils at 77 K. Convert this temperature to degrees Fahrenheit and to degrees Celsius.

1.29 **(a)** A sample of tetrachloroethylene, a liquid used in dry cleaning that is being phased out because of its potential to cause cancer, has a mass of 40.55 g and a volume of 25.0 mL at 25 °C. What is its density at this temperature? Will tetrachloroethylene float on water? (Materials that are less dense than water will float.) **(b)** Carbon dioxide (CO_2) is a gas at room temperature and pressure. However, carbon dioxide can be put under pressure to become a "supercritical fluid" that is a much safer dry-cleaning agent than tetrachloroethylene. At a certain pressure, the density of supercritical CO_2 is 0.469 g/cm³. What is the mass of a 25.0-mL sample of supercritical CO_2 at this pressure?

1.30 **(a)** A cube of osmium metal 1.500 cm on a side has a mass of 76.31 g at 25 °C. What is its density in g/cm³ at this temperature? **(b)** The density of titanium metal is 4.51g/cm³ at 25 °C. What mass of titanium displaces 125.0 mL of water at 25 °C? **(c)** The density of benzene at 15 °C is 0.8787g/mL. Calculate the mass of 0.1500 L of benzene at this temperature.

1.31 **(a)** To identify a liquid substance, a student determined its density. Using a graduated cylinder, she measured out a 45-mL sample of the substance. She then measured the mass of the sample, finding that it weighed 38.5 g. She knew that the substance had to be either isopropyl alcohol (density 0.785 g/mL) or toluene (density 0.866/mL). What are the calculated density and the probable identity of the substance? **(b)** An experiment requires 45.0 g of ethylene glycol, a liquid whose density is 1.114 g/mL. Rather than weigh the sample on a balance, a chemist chooses to dispense the liquid using a graduated cylinder. What volume of the liquid should he use? **(c)** Is a graduated cylinder such as that shown in Figure 1.19 likely to afford the accuracy of measurement needed? **(d)** A cubic piece of metal measures 5.00 cm on each edge. If the metal is nickel, whose density is 8.90 g/cm³, what is the mass of the cube?

1.32 **(a)** After the label fell off a bottle containing a clear liquid believed to be benzene, a chemist measured the density of the liquid to verify its identity. A 25.0-mL portion of the liquid had a mass of 21.95 g. A chemistry handbook lists the density of benzene at 15 °C as 0.8787 g/mL. Is the calculated density in agreement with the tabulated value? **(b)** An experiment requires 15.0 g of cyclohexane, whose density at 25 °C is 0.7781 g/mL. What volume of cyclohexane should be used? **(c)** A spherical ball of lead has a diameter of 5.0 cm. What is the mass of the sphere if lead has a density of 11.34 g/cm³? (The volume of a sphere is $(4/3)\pi r^3$, where r is the radius.)

1.33 In the year 2011, an estimated amount of 35 billion tons of carbon dioxide (CO_2) was emitted worldwide due to fossil fuel combustion and cement production. Express this mass of CO_2 in grams without exponential notation, using an appropriate metric prefix.

1.34 Silicon for computer chips is grown in large cylinders called "boules" that are 300 mm in diameter and 2 m in length, as shown. The density of silicon is 2.33 g/cm^3. Silicon wafers for making integrated circuits are sliced from a 2.0 m boule and are typically 0.75 mm thick and 300 mm in diameter. **(a)** How many wafers can be cut from a single boule? **(b)** What is the mass of a silicon wafer? (The volume of a cylinder is given by $\pi r^2 h$, where r is the radius and h is its height.)

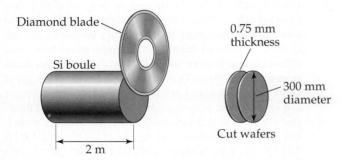

Diamond blade

Si boule

2 m

0.75 mm thickness

300 mm diameter

Cut wafers

Uncertainty in Measurement (Section 1.5)

1.35 Indicate which of the following are exact numbers: **(a)** the mass of a 3 by 5–inch index card, **(b)** the number of ounces in a pound, **(c)** the volume of a cup of Seattle's Best coffee, **(d)** the number of inches in a mile, **(e)** the number of microseconds in a week, **(f)** the number of pages in this book.

1.36 Indicate which of the following are exact numbers: **(a)** the mass of a 32-oz can of coffee, **(b)** the number of students in your chemistry class, **(c)** the temperature of the surface of the Sun, **(d)** the mass of a postage stamp, **(e)** the number of milliliters in a cubic meter of water, **(f)** the average height of NBA basketball players.

1.37 What is the number of significant figures in each of the following measured quantities? **(a)** 601 kg, **(b)** 0.054 s, **(c)** 6.3050 cm, **(d)** 0.0105 L, **(e)** $7.0500 \times 10^{-3} m^3$, **(f)** 400 g.

1.38 Indicate the number of significant figures in each of the following measured quantities: **(a)** 3.774 km, **(b)** 205 m^2, **(c)** 1.700 cm, **(d)** 350.00 K, **(e)** 307.080 g, **(f)** $1.3 \times 10^3 m/s$.

1.39 Round each of the following numbers to four significant figures and express the result in standard exponential notation: **(a)** 102.53070, **(b)** 656.980, **(c)** 0.008543210, **(d)** 0.000257870, **(e)** −0.0357202.

1.40 **(a)** The diameter of Earth at the equator is 7926.381 mi. Round this number to three significant figures and express it in standard exponential notation. **(b)** The circumference of Earth through the poles is 40,008 km. Round this number to four significant figures and express it in standard exponential notation.

1.41 Carry out the following operations and express the answers with the appropriate number of significant figures.

(a) $14.3505 + 2.65$

(b) $952.7 - 140.7389$

(c) $(3.29 \times 10^4)(0.2501)$

(d) $0.0588/0.677$

1.42 Carry out the following operations and express the answer with the appropriate number of significant figures.

(a) $320.5 - (6104.5/2.3)$

(b) $[(285.3 \times 10^5) - (1.200 \times 10^3)] \times 2.8954$

(c) $(0.0045 \times 20,000.0) + (2813 \times 12)$

(d) $863 \times [1255 - (3.45 \times 108)]$

1.43 You weigh an object on a balance and read the mass in grams according to the picture. How many significant figures are in this measurement?

1.44 You have a graduated cylinder that contains a liquid (see photograph). Write the volume of the liquid, in milliliters, using the proper number of significant figures.

Dimensional Analysis (Section 1.6)

1.45 Using your knowledge of metric units, English units, and the information on the back inside cover, write down the conversion factors needed to convert **(a)** mm to nm, **(b)** mg to kg, **(c)** km to ft, **(d)** $in.^3$ to cm^3.

1.46 Using your knowledge of metric units, English units, and the information on the back inside cover, write down the conversion factors needed to convert **(a)** μm to mm, **(b)** ms to ns, **(c)** mi to km, **(d)** ft^3 to L.

1.47 **(a)** A bumblebee flies with a ground speed of 15.2 m/s. Calculate its speed in km/hr. **(b)** The lung capacity of the blue whale is 5.0×10^3 L. Convert this volume into gallons. **(c)** The Statue of Liberty is 151 ft tall. Calculate its height in meters. **(d)** Bamboo can grow up to 60.0 cm/day. Convert this growth rate into inches per hour.

1.48 **(a)** The speed of light in a vacuum is $2.998 \times 10^8 m/s$. Calculate its speed in miles per hour. **(b)** The Sears Tower in Chicago is 1454 ft tall. Calculate its height in meters. **(c)** The Vehicle Assembly Building at the Kennedy Space Center in Florida has a volume of 3,666,500 m^3. Convert this volume to liters and express the result in standard exponential notation. **(d)** An individual suffering from a high cholesterol level in her blood has 242 mg of cholesterol per 100 mL of blood. If the total blood volume of the individual is 5.2 L, how many grams of total blood cholesterol does the individual's body contain?

1.49 The inside dimension of a box that is cubic is 24.8 cm on each edge with an uncertainty of 0.2 cm. What is the volume of the box? What do you estimate to be the uncertainty in the calculated volume?

1.50 The distance from Grand Rapids, Michigan, to Detroit is listed in a road atlas as 153 miles. Describe some of the factors that contribute to the uncertainty in this number. To make the number more precise, what would you need to specify and measure?

1.51 Perform the following conversions: **(a)** 5.00 days to s, **(b)** 0.0550 mi to m, **(c)** $1.89/gal to dollars per liter, **(d)** 0.510 in./ms to km/hr, **(e)** 22.50 gal/min to L/s, **(f)** 0.02500 ft^3 to cm^3.

1.52 Carry out the following conversions: **(a)** 0.105 in. to mm, **(b)** 0.650 qt to mL, **(c)** 8.75 μm/s to km/hr, **(d)** 1.955 m^3 to yd^3, **(e)** $3.99/lb to dollars per kg, **(f)** 8.75 lb/ft^3 to g/mL.

1.53 **(a)** How many liters of wine can be held in a wine barrel whose capacity is 31 gal? **(b)** The recommended adult dose of Elixophyllin®, a drug used to treat asthma, is 6 mg/kg of body mass. Calculate the dose in milligrams for a 185-lb person. **(c)** If an automobile is able to travel 400 km on 47.3 L of gasoline, what is the gas mileage in miles per gallon? **(d)** When the coffee is brewed according to directions, a pound of coffee beans yields 50 cups of coffee (4 cups = 1 qt). How many kg of coffee are required to produce 200 cups of coffee?

1.54 **(a)** If an electric car is capable of going 225 km on a single charge, how many charges will it need to travel from Seattle, Washington, to San Diego, California, a distance of 1257 mi, assuming that the trip begins with a full charge? **(b)** If a migrating loon flies at an average speed of 14 m/s, what is its average speed in mi/hr? **(c)** What is the engine piston displacement in liters of an engine whose displacement is listed as 450 in.^{3}? **(d)** In March 1989 the *Exxon Valdez* ran aground and spilled 240,000 barrels of crude petroleum off the coast of Alaska. One barrel of petroleum is equal to 42 gal. How many liters of petroleum were spilled?

1.55 The density of air at ordinary atmospheric pressure and 25 °C is 1.19 g/L. What is the mass, in kilograms, of the air in a room that measures 14.5 ft × 16.5 ft × 8.0 ft?

1.56 The concentration of carbon monoxide in an urban apartment is 48 μg/m^3. What mass of carbon monoxide in grams is present in a room measuring 10.6 ft × 14.8 ft × 20.5 ft?

1.57 Gold can be hammered into extremely thin sheets called gold leaf. An architect wants to cover a 100 ft × 82 ft ceiling with gold leaf that is five–millionths of an inch thick. The density of gold is 19.32 g/cm^3, and gold costs $1654 per troy ounce (1 troy ounce = 31.1034768 g). How much will it cost the architect to buy the necessary gold?

1.58 A copper refinery produces a copper ingot weighing 150 lb. If the copper is drawn into wire whose diameter is 7.50 mm, how many feet of copper can be obtained from the ingot? The density of copper is 8.94 g/cm^3. (Assume that the wire is a cylinder whose volume $V = \pi r^2 h$, where r is its radius and h is its height or length.)

Additional Exercises

1.59 **(a)** Classify each of the following as a pure substance, a solution, or a heterogeneous mixture: a gold coin, a cup of coffee, a wood plank. **(b)** What ambiguities are there in answering part (a) from the descriptions given?

1.60 **(a)** What is the difference between a hypothesis and a theory? **(b)** Explain the difference between a theory and a scientific law. Which addresses how matter behaves, and which addresses why it behaves that way?

1.61 A sample of ascorbic acid (vitamin C) is synthesized in the laboratory. It contains 1.50 g of carbon and 2.00 g of oxygen. Another sample of ascorbic acid isolated from citrus fruits contains 6.35 g of carbon. How many grams of oxygen does it contain? Which law are you assuming in answering this question?

1.62 Ethyl chloride is sold as a liquid (see photo) under pressure for use as a local skin anesthetic. Ethyl chloride boils at 12 °C at atmospheric pressure. When the liquid is sprayed onto the skin, it boils off, cooling and numbing the skin as it vaporizes. **(a)** What changes of state are involved in this use of ethyl chloride? **(b)** What is the boiling point of ethyl chloride in degrees Fahrenheit? **(c)** The bottle shown contains 103.5 mL of ethyl chloride. The density of ethyl chloride at 25 °C is 0.765 g/cm^3. What is the mass of ethyl chloride in the bottle?

1.63 Two students determine the percentage of lead in a sample as a laboratory exercise. The true percentage is 22.52%. The students' results for three determinations are as follows:

(1) 22.52, 22.48, 22.54

(2) 22.64, 22.58, 22.62

(a) Calculate the average percentage for each set of data and state which set is the more accurate based on the average. (b) Precision can be judged by examining the average of the deviations from the average value for that data set. (Calculate the average value for each data set; then calculate the average value of the absolute deviations of each measurement from the average.) Which set is more precise?

1.64 Is the use of significant figures in each of the following statements appropriate? Why or why not? (a) Apple sold 22,727,000 iPods during the last three months of 2008. (b) New York City receives 49.7 inches of rain, on average, per year. (c) In the United States, 0.621% of the population has the surname Brown. (d) You calculate your grade point average to be 3.87562.

1.65 What type of quantity (for example, length, volume, density) do the following units indicate? (a) mL, (b) cm^2, (c) mm^3, (d) mg/L, (e) ps, (f) nm, (g) K.

1.66 Give the derived SI units for each of the following quantities in base SI units:

(a) acceleration = distance/time^2

(b) force = mass $\times$ acceleration

(c) work = force $\times$ distance

(d) pressure = force/area

(e) power = work/time

(f) velocity = distance/time

(g) energy = mass $\times$ (velocity)^2

1.67 The distance from Earth to the Moon is approximately 240,000 mi. (a) What is this distance in meters? (b) The peregrine falcon has been measured as traveling up to 350 km/hr in a dive. If this falcon could fly to the Moon at this speed, how many seconds would it take? (c) The speed of light is 3.00×10^8 m/s. How long does it take for light to travel from Earth to the Moon and back again? (d) Earth travels around the Sun at an average speed of 29.783 km/s. Convert this speed to miles per hour.

1.68 Which of the following would you characterize as a pure or nearly pure substance? (a) baking powder; (b) lemon juice; (c) propane gas, used in outdoor gas grills; (d) aluminum foil; (e) ibuprofen; (f) bourbon whiskey; (g) helium gas; (h) clear water pumped from a deep aquifer.

1.69 The U.S. quarter has a mass of 5.67 g and is approximately 1.55 mm thick. (a) How many quarters would have to be stacked to reach 575 ft, the height of the Washington Monument? (b) How much would this stack weigh? (c) How much money would this stack contain? (d) The U.S. National Debt Clock showed the outstanding public debt to be $16,213,166,914,811 on October 28, 2012. How many stacks like the one described would be necessary to pay off this debt?

1.70 In the United States, water used for irrigation is measured in acre-feet. An acre-foot of water covers an acre to a depth of exactly 1 ft. An acre is 4840 yd^2. An acre-foot is enough water to supply two typical households for 1.00 yr. (a) If desalinated water costs $1950 per acre-foot, how much does desalinated water cost per liter? (b) How much would it cost one household per day if it were the only source of water?

1.71 By using estimation techniques, determine which of the following is the heaviest and which is the lightest: a 5-lb bag of potatoes, a 5-kg bag of sugar, or 1 gal of water (density = 1.0 g/mL).

1.72 Suppose you decide to define your own temperature scale with units of O, using the freezing point (13 °C) and boiling point (360 °C) of oleic acid, the main component of olive oil. If you set the freezing point of oleic acid as 0 °O and the boiling point as 100 °O, what is the freezing point of water on this new scale?

1.73 The liquid substances mercury (density = 13.6 g/mL), water (1.00 g/mL), and cyclohexane (0.778 g/mL) do not form a solution when mixed but separate in distinct layers. Sketch how the liquids would position themselves in a test tube.

1.74 Two spheres of equal volume are placed on the scales as shown. Which one is more dense?

1.75 Water has a density of 0.997 g/cm^3 at 25 °C; ice has a density of 0.917 g/cm^3 at −10 °C. (a) If a soft-drink bottle whose volume is 1.50 L is completely filled with water and then frozen to −10 °C, what volume does the ice occupy? (b) Can the ice be contained within the bottle?

1.76 A 32.65-g sample of a solid is placed in a flask. Toluene, in which the solid is insoluble, is added to the flask so that the total volume of solid and liquid together is 50.00 mL. The solid and toluene together weigh 58.58 g. The density of toluene at the temperature of the experiment is 0.864 g/mL. What is the density of the solid?

1.77 A thief plans to steal a gold sphere with a radius of 28.9 cm from a museum. If the gold has a density of 19.3 g/cm^3, what is the mass of the sphere in pounds? [The volume of a sphere is $V = (4/3)\pi r^3$.] Is the thief likely to be able to walk off with the gold sphere unassisted?

1.78 Automobile batteries contain sulfuric acid, which is commonly referred to as "battery acid." Calculate the number of grams of sulfuric acid in 1.00 gal of battery acid if the solution has a density of 1.28 g/mL and is 38.1% sulfuric acid by mass.

1.79 A 40-lb container of peat moss measures 14 × 20 × 30 in. A 40-lb container of topsoil has a volume of 1.9 gal. (a) Calculate the average densities of peat moss and topsoil in units of g/cm^3. Would it be correct to say that peat moss is "lighter" than topsoil? Explain. (b) How many bags of peat moss are needed to cover an area measuring 15.0 ft × 20.0 ft to a depth of 3.0 in.?

1.80 A package of aluminum foil contains 50 ft^2 of foil, which weighs approximately 8.0 oz. Aluminum has a density of 2.70 g/cm^3. What is the approximate thickness of the foil in millimeters?

1.81 The total rate at which power used by humans worldwide is approximately 15 TW (terawatts). The solar flux averaged over the sunlit half of Earth is 680 W/m^2. (assuming no clouds). The area of Earth's disc as seen from the sun is 1.28×10^{14} m^2. The surface area of Earth is approximately 197,000,000 square miles. How much of Earth's surface would we need to cover with solar energy collectors to power the planet for use by all humans? Assume that the solar energy collectors can convert only 10% of the available sunlight into useful power.

1.82 In 2005, J. Robin Warren and Barry J. Marshall shared the Nobel Prize in Medicine for discovery of the bacterium *Helicobacter pylori*, and for establishing experimental proof that it plays a major role in gastritis and peptic ulcer disease. The story began when Warren, a pathologist, noticed that bacilli were associated with the tissues taken from patients suffering from ulcers. Look up the history of this case and describe Warren's first hypothesis. What sorts of evidence did it take to create a credible theory based on it?

1.83 A 25.0-cm long cylindrical glass tube, sealed at one end, is filled with ethanol. The mass of ethanol needed to fill the tube is found to be 45.23 g. The density of ethanol is 0.789 g/mL. Calculate the inner diameter of the tube in centimeters.

1.84 Gold is alloyed (mixed) with other metals to increase its hardness in making jewelry. **(a)** Consider a piece of gold jewelry that weighs 9.85 g and has a volume of 0.675 cm³. The jewelry contains only gold and silver, which have densities of 19.3 and 10.5 g/cm³, respectively. If the total volume of the jewelry is the sum of the volumes of the gold and silver that it contains, calculate the percentage of gold (by mass) in the jewelry. **(b)** The relative amount of gold in an alloy is commonly expressed in units of carats. Pure gold is 24 carat, and the percentage of gold in an alloy is given as a percentage of this value. For example, an alloy that is 50% gold is 12 carat. State the purity of the gold jewelry in carats.

1.85 Paper chromatography is a simple but reliable method for separating a mixture into its constituent substances. You have a mixture of two vegetable dyes, one red and one blue, that you are trying to separate. You try two different chromatography procedures and achieve the separations shown in the figure. Which procedure worked better? Can you suggest a method to quantify how good or poor the separation was?

1.86 Judge the following statements as true or false. If you believe a statement to be false, provide a corrected version.

(a) Air and water are both elements.

(b) All mixtures contain at least one element and one compound.

(c) Compounds can be decomposed into two or more other substances; elements cannot.

(d) Elements can exist in any of the three states of matter.

(e) When yellow stains in a kitchen sink are treated with bleach water, the disappearance of the stains is due to a physical change.

(f) A hypothesis is more weakly supported by experimental evidence than a theory.

(g) The number 0.0033 has more significant figures than 0.033.

(h) Conversion factors used in converting units always have a numerical value of one.

(i) Compounds always contain at least two different elements.

1.87 You are assigned the task of separating a desired granular material with a density of 3.62 g/cm³ from an undesired granular material that has a density of 2.04 g/cm³. You want to do this by shaking the mixture in a liquid in which the heavier material will fall to the bottom and the lighter material will float. A solid will float on any liquid that is more dense. Using an Internet-based source or a handbook of chemistry, find the densities of the following substances: carbon tetrachloride, hexane, benzene, and diiodomethane. Which of these liquids will serve your purpose, assuming no chemical interaction between the liquid and the solids?

1.88 In 2009, a team from Northwestern University and Western Washington University reported the preparation of a new "spongy" material composed of nickel, molybdenum, and sulfur that excels at removing mercury from water. The density of this new material is 0.20 g/cm³, and its surface area is 1242 m² per gram of material. **(a)** Calculate the volume of a 10.0-mg sample of this material. **(b)** Calculate the surface area for a 10.0-mg sample of this material. **(c)** A 10.0-mL sample of contaminated water had 7.748 mg of mercury in it. After treatment with 10.0 mg of the new spongy material, 0.001 mg of mercury remained in the contaminated water. What percentage of the mercury was removed from the water? **(d)** What is the final mass of the spongy material after the exposure to mercury?

2

Atoms, Molecules, and Ions

Look around at the great variety of colors, textures, and other properties in the materials that surround you—the colors in a garden, the texture of the fabric in your clothes, the solubility of sugar in a cup of coffee, or the beauty and complexity of a geode like the one shown to the right. How can we explain the striking and seemingly infinite variety of properties of the materials that make up our world? What makes diamonds transparent and hard? A large crystal of sodium chloride, table salt, looks a bit like a diamond, but is brittle and readily dissolves in water. What accounts for the differences? Why does paper burn, and why does water quench fires? The answers to all such questions lie in the structures of atoms, which determine the physical and chemical properties of matter.

Although the materials in our world vary greatly in their properties, everything is formed from only about 100 elements and, therefore, from only about 100 chemically different kinds of atoms. In a sense, these different atoms are like the 26 letters of the English alphabet that join in different combinations to form the immense number of words in our language. But what rules govern the ways in which atoms combine? How do the properties of a substance relate to the kinds of atoms it contains? Indeed, what is an atom like, and what makes the atoms of one element different from those of another?

In this chapter we introduce the basic structure of atoms and discuss the formation of molecules and ions, thereby providing a foundation for exploring chemistry more deeply in later chapters.

▶ **A SECTION THROUGH A GEODE.** A geode is a mass of mineral matter (often containing quartz) that accumulates slowly within the shell of a roughly spherical, hollow rock. Eventually, perfectly formed crystals may develop at a geode's center. The colors of a geode depend upon its composition. Here, agate crystallized out as the geode formed.

WHAT'S ▶ AHEAD

2.1 THE ATOMIC THEORY OF MATTER We begin with a brief history of the notion of *atoms*—the smallest pieces of matter.

2.2 THE DISCOVERY OF ATOMIC STRUCTURE We then look at some key experiments that led to the discovery of *electrons* and to the *nuclear model* of the atom.

2.3 THE MODERN VIEW OF ATOMIC STRUCTURE We explore the modern theory of atomic structure, including the ideas of *atomic numbers*, *mass numbers*, and *isotopes*.

2.4 ATOMIC WEIGHTS We introduce the concept of *atomic weights* and how they relate to the masses of individual atoms.

2.5 THE PERIODIC TABLE We examine the organization of the *periodic table*, in which elements are put in order of increasing atomic number and grouped by chemical similarity.

2.6 MOLECULES AND MOLECULAR COMPOUNDS We discuss the assemblies of atoms called *molecules* and how their compositions are represented by *empirical* and *molecular formulas*.

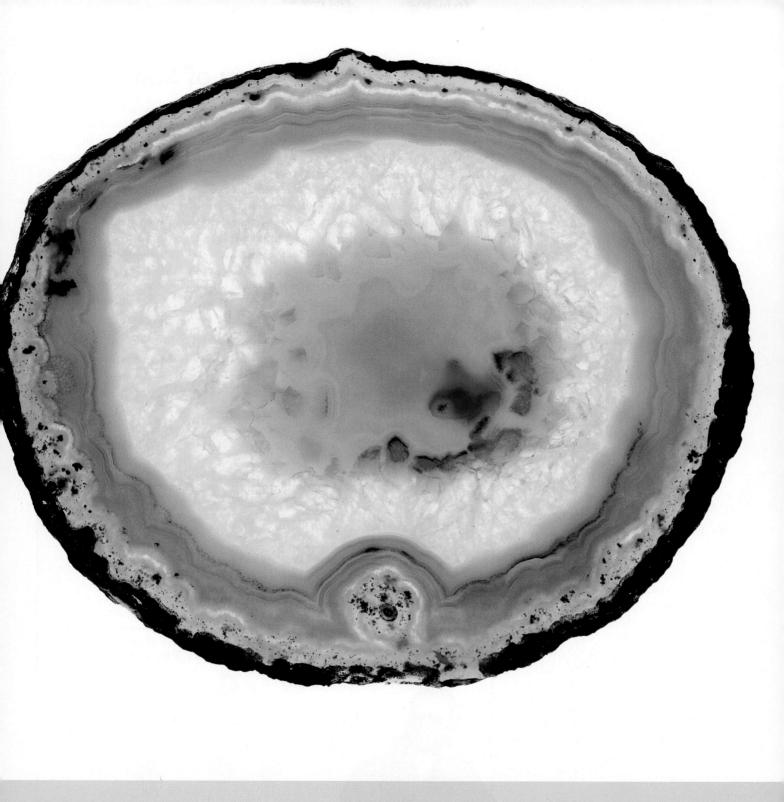

2.7 IONS AND IONIC COMPOUNDS We learn that atoms can gain or lose electrons to form *ions*. We also look at how to use the periodic table to predict the charges on ions and the empirical formulas of *ionic compounds*.

2.8 NAMING INORGANIC COMPOUNDS We consider the systematic way in which substances are named, called *nomenclature*, and how this nomenclature is applied to inorganic compounds.

2.9 SOME SIMPLE ORGANIC COMPOUNDS We introduce *organic chemistry*, the chemistry of the element carbon.

2.1 | The Atomic Theory of Matter

Philosophers from the earliest times speculated about the nature of the fundamental "stuff" from which the world is made. Democritus (460–370 BCE) and other early Greek philosophers described the material world as made up of tiny, indivisible particles that they called *atomos*, meaning "indivisible" or "uncuttable." Later, however, Plato and Aristotle formulated the notion that there can be no ultimately indivisible particles, and the "atomic" view of matter faded for many centuries during which Aristotelean philosophy dominated Western culture.

The notion of **atoms** reemerged in Europe during the seventeenth century. As chemists learned to measure the amounts of elements that reacted with one another to form new substances, the ground was laid for an atomic theory that linked the idea of elements with the idea of atoms. That theory came from the work of John Dalton during the period from 1803 to 1807. Dalton's atomic theory was based on four postulates (see ▼ Figure 2.1).

Dalton's theory explains several laws of chemical combination that were known during his time, including the *law of constant composition* ∞ (Section 1.2),* based on postulate 4:

> In a given compound, the relative numbers and kinds of atoms are constant.

It also explains the **law of conservation of mass**, based on postulate 3:

> The total mass of materials present after a chemical reaction is the same as the total mass present before the reaction.

A good theory explains known facts and predicts new ones. Dalton used his theory to deduce the **law of multiple proportions**:

> If two elements A and B combine to form more than one compound, the masses of B that can combine with a given mass of A are in the ratio of small whole numbers.

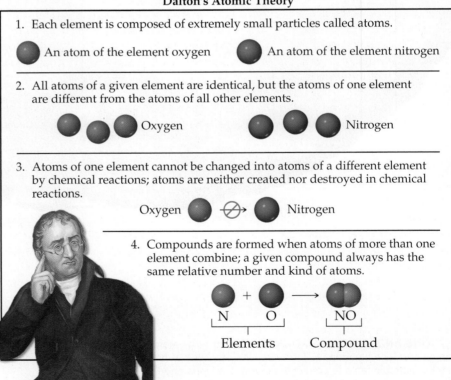

Dalton's Atomic Theory

1. Each element is composed of extremely small particles called atoms.

   An atom of the element oxygen   An atom of the element nitrogen

2. All atoms of a given element are identical, but the atoms of one element are different from the atoms of all other elements.

   Oxygen   Nitrogen

3. Atoms of one element cannot be changed into atoms of a different element by chemical reactions; atoms are neither created nor destroyed in chemical reactions.

   Oxygen ⊘→ Nitrogen

4. Compounds are formed when atoms of more than one element combine; a given compound always has the same relative number and kind of atoms.

   N + O ⟶ NO
   └─ Elements ─┘   └ Compound ┘

▲ Figure 2.1 **Dalton's atomic theory.**[†] John Dalton (1766–1844), the son of a poor English weaver, began teaching at age 12. He spent most of his years in Manchester, where he taught both grammar school and college. His lifelong interest in meteorology led him to study gases, then chemistry, and eventually atomic theory. Despite his humble beginnings, Dalton gained a strong scientific reputation during his lifetime.

---

*The short chainlike symbol (∞) that precedes the section reference indicates a link to ideas presented earlier in the text.
[†]Dalton, John. "Atomic Theory." 1844.

We can illustrate this law by considering water and hydrogen peroxide, both of which consist of the elements hydrogen and oxygen. In forming water, 8.0 g of oxygen combine with 1.0 g of hydrogen. In forming hydrogen peroxide, 16.0 g of oxygen combine with 1.0 g of hydrogen. Thus, the ratio of the masses of oxygen per gram of hydrogen in the two compounds is 2:1. Using Dalton's atomic theory, we conclude that hydrogen peroxide contains twice as many atoms of oxygen per hydrogen atom than does water.

Give It Some Thought

Compound A contains 1.333 g of oxygen per gram of carbon, whereas compound B contains 2.666 g of oxygen per gram of carbon.
(a) What chemical law do these data illustrate?
(b) If compound A has an equal number of oxygen and carbon atoms, what can we conclude about the composition of compound B?

2.2 | The Discovery of Atomic Structure

Dalton based his conclusions about atoms on chemical observations made in the laboratory. By assuming the existence of atoms he was able to account for the laws of constant composition and of multiple proportions. But neither Dalton nor those who followed him during the century after his work was published had any direct evidence for the existence of atoms. Today, however, we can measure the properties of individual atoms and even provide images of them (▶ **Figure 2.2**).

As scientists developed methods for probing the nature of matter, the supposedly indivisible atom began to show signs of a more complex structure, and today we know that the atom is composed of **subatomic particles**. Before we summarize the current model, we briefly consider a few of the landmark discoveries that led to that model. We will see that the atom is composed in part of electrically charged particles, some with a positive charge and some with a negative charge. As we discuss the development of our current model of the atom, keep in mind this fact: *Particles with the same charge repel one another, whereas particles with unlike charges attract one another.*

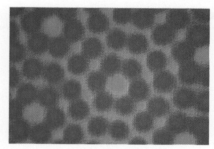

▲ Figure 2.2 **An image of the surface of silicon.** The image was obtained by a technique called scanning tunneling microscopy. The color was added to the image by computer to help distinguish its features. Each red sphere is a silicon atom.

Cathode Rays and Electrons

During the mid-1800s, scientists began to study electrical discharge through a glass tube pumped almost empty of air (**Figure 2.3**). When a high voltage was applied to the electrodes in the tube, radiation was produced between the electrodes. This radiation, called **cathode rays**, originated at the negative electrode and traveled to the positive electrode. Although the rays could not be seen, their presence was detected because they cause certain materials to *fluoresce*, or to give off light.

Experiments showed that cathode rays are deflected by electric or magnetic fields in a way consistent with their being a stream of negative electrical charge. The British scientist J. J. Thomson (1856–1940) observed that cathode rays are the same regardless of the identity of the cathode material. In a paper published in 1897, Thomson described cathode rays as streams of negatively charged particles. His paper is generally accepted as the discovery of what became known as the **electron**.

Thomson constructed a cathode-ray tube having a hole in the anode through which a beam of electrons passed. Electrically charged plates and a magnet were positioned perpendicular to the electron beam, and a fluorescent screen was located at one end (**Figure 2.4**). The electric field deflected the rays in one direction, and the magnetic field deflected them in the opposite direction. Thomson adjusted the strengths of the fields so that the effects balanced each other, allowing the electrons to travel in a straight path to the screen. Knowing the strengths that resulted in the straight path made it possible to calculate a value of 1.76×10^8 coulombs* per gram for the ratio of the electron's electrical charge to its mass.

*The coulomb (C) is the SI unit for electrical charge.

GO FIGURE

How do we deduce from the figure that the cathode rays travel from cathode to anode?

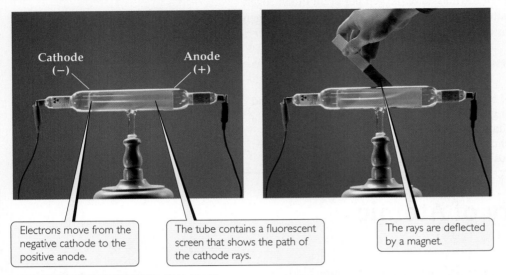

Cathode
(−)

Anode
(+)

Electrons move from the
negative cathode to the
positive anode.

The tube contains a fluorescent
screen that shows the path of
the cathode rays.

The rays are deflected
by a magnet.

▲ **Figure 2.3 Cathode-ray tube.**

 Give It Some Thought

Thomson observed that the cathode rays produced in the cathode–ray tube
behaved identically, regardless of the particular metal used as cathode. What is
the significance of this observation?

 GO FIGURE

If no magnetic field were applied, would you expect the electron beam to be deflected upward or downward by the electric field?

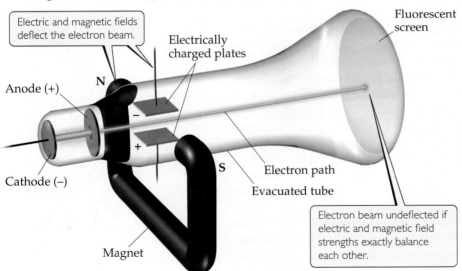

Electric and magnetic fields
deflect the electron beam.

Electrically
charged plates

Fluorescent
screen

Anode (+)

N

−

+

S

Cathode (−)

Electron path

Evacuated tube

Magnet

Electron beam undeflected if
electric and magnetic field
strengths exactly balance
each other.

▲ **Figure 2.4 Cathode-ray tube with perpendicular magnetic and electric fields.** The cathode
rays (electrons) originate at the cathode and are accelerated toward the anode, which has a
hole in its center. A narrow beam of electrons passes through the hole and travels to the
fluorescent screen.

▲ **GO FIGURE**

Would the masses of the oil drops be changed significantly by any electrons that accumulate on them?

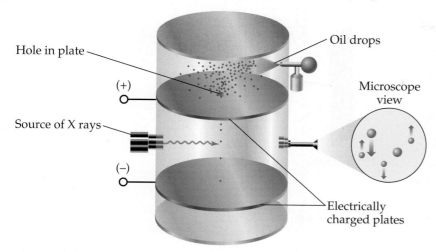

Hole in plate

(+)

Source of X rays

(−)

Oil drops

Microscope view

Electrically charged plates

▲ **Figure 2.5 Millikan's oil-drop experiment to measure the charge of the electron.** Small drops of oil are allowed to fall between electrically charged plates. The drops pick up extra electrons as a result of irradiation by X-rays and so became negatively charged. Millikan measured how varying the voltage between the plates affected the rate of fall. From these data he calculated the negative charge on the drops. Because the charge on any drop was always some integral multiple of 1.602×10^{-19} C, Millikan deduced this value to be the charge of a single electron.

Once the charge-to-mass ratio of the electron was known, measuring either quantity allowed scientists to calculate the other. In 1909 Robert Millikan (1868–1953) of the University of Chicago succeeded in measuring the charge of an electron by performing the experiment described in ▲ **Figure 2.5.** He then calculated the mass of the electron by using his experimental value for the charge, 1.602×10^{-19} C, and Thomson's charge-to-mass ratio, 1.76×10^8 C/g:

$$\text{Electron mass} = \frac{1.602 \times 10^{-19} \text{ C}}{1.76 \times 10^8 \text{ C/g}} = 9.10 \times 10^{-28} \text{ g}$$

This result agrees well with the currently accepted value for the electron mass, 9.10938×10^{-28} g. This mass is about 2000 times smaller than that of hydrogen, the lightest atom.

Radioactivity

In 1896 the French scientist Henri Becquerel (1852–1908) discovered that a compound of uranium spontaneously emits high-energy radiation. This spontaneous emission of radiation is called **radioactivity**. At Becquerel's suggestion, Marie Curie (▶ **Figure 2.6**) and her husband, Pierre, began experiments to identify and isolate the source of radioactivity in the compound. They concluded that it was the uranium atoms.

Further study of radioactivity, principally by the British scientist Ernest Rutherford, revealed three types of radiation: alpha (α), beta (β), and gamma (γ). The paths of α and β radiation are bent by an electric field, although in opposite directions; γ radiation is unaffected by the field (**Figure 2.7**). Rutherford (1871–1937) was a very important figure in this period of atomic science. After working at Cambridge University with J. J. Thomson, he moved to McGill University in Montreal, where he did research on radioactivity that led to his 1908 Nobel Prize in Chemistry. In 1907 he returned to England as a faculty member at Manchester University, where he did his famous α–particle scattering experiments, described below.

Rutherford showed that α and β rays consist of fast-moving particles. In fact, β particles are high-speed electrons and can be considered the radioactive equivalent of cathode rays. They are attracted to a positively charged plate. The α particles have a

▲ **Figure 2.6 Marie Sklodowska Curie (1867–1934).** In 1903 Henri Becquerel, Marie Curie, and her husband, Pierre, were jointly awarded the Nobel Prize in Physics for their pioneering work on radioactivity (a term she introduced). In 1911 Marie Curie won a second Nobel Prize, this time in chemistry for her discovery of the elements polonium and radium.

▲ **GO FIGURE**

Which of the three kinds of radiation shown consists of electrons? Why are these rays deflected to a greater extent than the others?

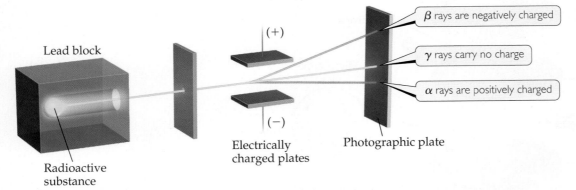

▲ Figure 2.7 **Behavior of alpha (α), beta (β), and gamma (γ) rays in an electric field.**

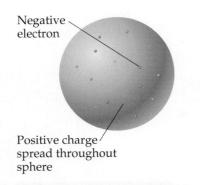

▲ Figure 2.8 **J. J. Thomson's plum-pudding model of the atom.** Ernest Rutherford and Ernest Marsden proved this model wrong.

positive charge and are attracted to a negative plate. In units of the charge of the electron, β particles have a charge of 1− and α particles a charge of 2+. Each α particle has a mass about 7400 times that of an electron. Gamma radiation is high-energy radiation similar to X rays; it does not consist of particles with mass and carries no charge.

The Nuclear Model of the Atom

With growing evidence that the atom is composed of smaller particles, scientist gave attention to how the particles fit together. During the early 1900s, Thomson reasoned that because electrons contribute only a very small fraction of an atom's mass, they probably are responsible for an equally small fraction of the atom's size. He proposed that the atom consists of a uniform positive sphere of matter in which the mass is evenly distributed and in which the electrons are embedded like raisins in a pudding or seeds in a watermelon (◄ Figure 2.8). This *plum-pudding model*, named after a traditional English dessert, was very short-lived.

In 1910 Rutherford was studying the angles at which α particles were deflected, or *scattered*, as they passed through a thin sheet of gold foil (▶ Figure 2.9). He discovered that almost all the particles passed directly through the foil without deflection, with a few particles deflected about 1°, consistent with Thomson's plum-pudding model. For the sake of completeness, Rutherford suggested that Ernest Marsden, an undergraduate student working in the laboratory, look for scattering at large angles. To everyone's surprise, a small amount of scattering was observed at large angles, with some particles scattered back in the direction from which they had come. The explanation for these results was not immediately obvious, but they were clearly inconsistent with Thomson's plum-pudding model.

Rutherford explained the results by postulating the **nuclear model** of the atom, in which most of the mass of each gold atom and all of its positive charge reside in a very small, extremely dense region that he called the **nucleus**. He postulated further that most of the volume of an atom is empty space in which electrons move around the nucleus. In the α-scattering experiment, most of the particles passed through the foil unscattered because they did not encounter the minute nucleus of any gold atom. Occasionally, however, an α particle came close to a gold nucleus. In such encounters, the repulsion between the highly positive charge of the gold nucleus and the positive charge of the α particle was strong enough to deflect the particle, as shown in Figure 2.9.

Subsequent experiments led to the discovery of positive particles (**protons**) and neutral particles (**neutrons**) in the nucleus. Protons were discovered in 1919 by Rutherford and neutrons in 1932 by British scientist James Chadwick (1891–1972). Thus, the atom is composed of electrons, protons, and neutrons.

GO FIGURE

What is the charge on the particles that form the beam?

Experiment   **Interpretation**

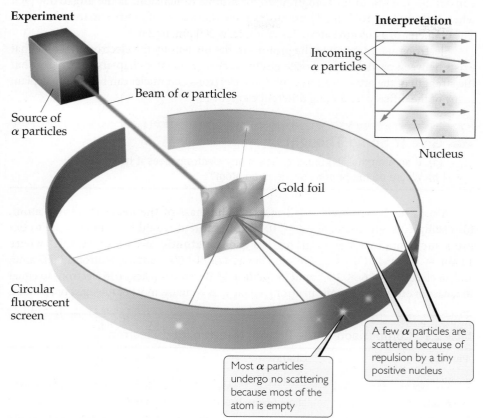

▲ **Figure 2.9 Rutherford's α-Sacttering experiment.** When α particles pass through a gold foil, most pass through undeflected but some are scattered, a few at very large angles. According to the plum-pudding model of the atom, the particles should experience only very minor deflections. The nuclear model of the atom explains why a few α particles are deflected at large angles. Although the nuclear atom has been depicted here as a yellow sphere, it is important to realize that most of the space around the nucleus contains only the low–mass electrons.

 Give It Some Thought

What happens to most of the α particles that strike the gold foil in Rutherford's experiment? Why do they behave that way?

2.3 | The Modern View of Atomic Structure

Since Rutherford's time, as physicists have learned more and more about atomic nuclei, the list of particles that make up nuclei has grown and continues to increase. As chemists, however, we can take a simple view of the atom because only three subatomic particles—the proton, neutron, and electron—have a bearing on chemical behavior.

As noted earlier, the charge of an electron is -1.602×10^{-19} C. The charge of a proton is opposite in sign but equal in magnitude to that of an electron: $+1.602 \times 10^{-19}$ C. The quantity 1.602×10^{-19} C is called the **electronic charge**. For convenience, the charges of atomic and subatomic particles are usually expressed as multiples of this charge rather than in coulombs. Thus, the charge of an electron is $1-$ and that of a proton is $1+$. Neutrons are electrically neutral (which is how they received their name). *Every atom has an equal number of electrons and protons, so atoms have no net electrical charge.*

GO FIGURE

What is the approximate diameter of the nucleus in units of pm?

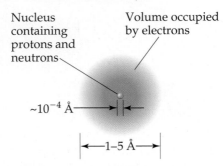

Nucleus containing protons and neutrons

Volume occupied by electrons

~10^{-4} Å

1–5 Å

▲ **Figure 2.10 The structure of the atom.** A cloud of rapidly moving electrons occupies most of the volume of the atom. The nucleus occupies a tiny region at the center of the atom and is composed of the protons and neutrons. The nucleus contains virtually all the mass of the atom.

Protons and neutrons reside in the tiny nucleus of the atom. The vast majority of an atom's volume is the space in which the electrons reside (◄ Figure 2.10). Most atoms have diameters between 1×10^{-10} m (100 pm) and 5×10^{-10} m (500 pm). A convenient non–SI unit of length used for atomic dimensions is the **angstrom** (Å), where 1 Å = 1×10^{-10} m. Thus, atoms have diameters of approximately $1 - 5$ Å. The diameter of a chlorine atom, for example, is 200 pm, or 2.0 Å.

Electrons are attracted to the protons in the nucleus by the electrostatic force that exists between particles of opposite electrical charge. In later chapters we will see that the strength of the attractive forces between electrons and nuclei can be used to explain many of the differences among different elements.

 Give It Some Thought

(a) If an atom has 15 protons, how many electrons does it have?
(b) Where do the protons reside in an atom?

Atoms have extremely small masses. The mass of the heaviest known atom, for example, is approximately 4×10^{-22} g. Because it would be cumbersome to express such small masses in grams, we use the **atomic mass unit** (amu),* where 1 amu = 1.66054×10^{-24} g. A proton has a mass of 1.0073 amu, a neutron 1.0087 amu, and an electron 5.486×10^{-4} amu (▼ Table 2.1). Because it takes 1836 electrons to equal the mass of one proton or one neutron, the nucleus contains most of the mass of an atom.

Table 2.1   Comparison of the Proton, Neutron, and Electron

Particle	Charge	Mass (amu)
Proton	Positive (1+)	1.0073
Neutron	None (neutral)	1.0087
Electron	Negative (1−)	5.486×10^{-4}

SAMPLE EXERCISE 2.1   Atomic Size

The diameter of a U.S. dime is 17.9 mm, and the diameter of a silver atom is 2.88 Å. How many silver atoms could be arranged side by side across the diameter of a dime?

SOLUTION

The unknown is the number of silver (Ag) atoms. Using the relationship 1 Ag atom = 2.88 Å as a conversion factor relating number of atoms and distance, we start with the diameter of the dime, first converting this distance into angstroms and then using the diameter of the Ag atom to convert distance to number of Ag atoms:

$$\text{Ag atoms} = (17.9 \text{ mm})\left(\frac{10^{-3} \text{ m}}{1 \text{ mm}}\right)\left(\frac{1 \text{ Å}}{10^{-10} \text{ m}}\right)\left(\frac{1 \text{ Ag atom}}{2.88 \text{ Å}}\right) = 6.22 \times 10^7 \text{ Ag atoms}$$

That is, 62.2 million silver atoms could sit side by side across a dime!

Practice Exercise 1

Which of the following factors determines the size of an atom?
(a) The volume of the nucleus; (b) the volume of space occupied by the electrons of the atom; (c) the volume of a single electron, multiplied by the number of electrons in the atom; (d) The total nuclear charge; (e) The total mass of the electrons surrounding the nucleus.

Practice Exercise 2

The diameter of a carbon atom is 1.54 Å. (a) Express this diameter in picometers. (b) How many carbon atoms could be aligned side by side across the width of a pencil line that is 0.20 mm wide?

---

*The SI abbreviation for the atomic mass unit is u. We will use the more common abbreviation amu.

The diameter of an atomic nucleus is approximately 10^{-4} Å, only a small fraction of the diameter of the atom as a whole. You can appreciate the relative sizes of the atom and its nucleus by imagining that if the hydrogen atom were as large as a football stadium, the nucleus would be the size of a small marble. Because the tiny nucleus carries most of the mass of the atom in such a small volume, it has an incredibly high density—on the order of 10^{13}–10^{14} g/cm³. A matchbox full of material of such density would weigh over 2.5 billion tons!

A Closer Look

Basic Forces

Four basic forces are known in nature: (1) gravitational, (2) electromagnetic, (3) strong nuclear, and (4) weak nuclear. *Gravitational forces* are attractive forces that act between all objects in proportion to their masses. Gravitational forces between atoms or between subatomic particles are so small that they are of no chemical significance.

Electromagnetic forces are attractive or repulsive forces that act between either electrically charged or magnetic objects. Electric forces are important in understanding the chemical behavior of atoms. The magnitude of the electric force between two charged particles is given by *Coulomb's law:* $F = kQ_1Q_2/d^2$, where Q_1 and Q_2 are the magnitudes of the charges on the two particles, d is the distance between their centers, and k is a constant determined by the units for Q and d.

A negative value for the force indicates attraction, whereas a positive value indicates repulsion. Electric forces are of primary importance in determining the chemical properties of elements.

All nuclei except those of hydrogen atoms contain two or more protons. Because like charges repel, electrical repulsion would cause the protons to fly apart if the *strong nuclear force* did not keep them together. This force acts between subatomic particles, as in the nucleus. At this distance, the attractive strong nuclear force is stronger than the positive–positive repulsive electric force and holds the nucleus together.

The *weak nuclear force* is weaker than the electric force but stronger than the gravitational force. We are aware of its existence only because it shows itself in certain types of radioactivity.

Related Exercise: *2.112*

Figure 2.10 incorporates the features we have just discussed. Electrons play the major role in chemical reactions. The significance of representing the region containing electrons as an indistinct cloud will become clear in later chapters when we consider the energies and spatial arrangements of the electrons. For now we have all the information we need to discuss many topics that form the basis of everyday uses of chemistry.

Atomic Numbers, Mass Numbers, and Isotopes

What makes an atom of one element different from an atom of another element? The atoms of each element have a *characteristic number of protons*. The number of protons in an atom of any particular element is called that element's **atomic number**. Because an atom has no net electrical charge, the number of electrons it contains must equal the number of protons. All atoms of carbon, for example, have six protons and six electrons, whereas all atoms of oxygen have eight protons and eight electrons. Thus, carbon has atomic number 6, and oxygen has atomic number 8. The atomic number of each element is listed with the name and symbol of the element on the front inside cover of the text.

Atoms of a given element can differ in the number of neutrons they contain and, consequently, in mass. For example, while most atoms of carbon have six neutrons, some have more and some have less. The symbol $^{12}_{6}\text{C}$ (read "carbon twelve," carbon-12) represents the carbon atom containing six protons and six neutrons, whereas carbon atoms that contain six protons and eight neutrons have mass number 14, are represented as $^{14}_{6}\text{C}$ or ^{14}C, and are referred to as carbon-14.

The atomic number is indicated by the subscript; the superscript, called the **mass number**, is the number of protons plus neutrons in the atom:

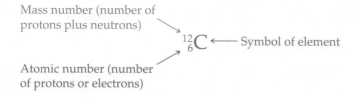

Mass number (number of protons plus neutrons)

$^{12}_{6}\text{C}$ ⟵ Symbol of element

Atomic number (number of protons or electrons)

Table 2.2 Some Isotopes of Carbon[a]

Symbol	Number of Protons	Number of Electrons	Number of Neutrons
^{11}C	6	6	5
^{12}C	6	6	6
^{13}C	6	6	7
^{14}C	6	6	8

[a]Almost 99% of the carbon found in nature is ^{12}C.

Because all atoms of a given element have the same atomic number, the subscript is redundant and is often omitted. Thus, the symbol for carbon-12 can be represented simply as ^{12}C.

Atoms with identical atomic numbers but different mass numbers (that is, same number of protons but different numbers of neutrons) are called **isotopes** of one another. Several isotopes of carbon are listed in ▲ **Table 2.2.** We will generally use the notation with superscripts only when referring to a particular isotope of an element. It is important to keep in mind that the isotopes of any given element are all alike chemically. A carbon dioxide molecule that contains a ^{13}C atom behaves for all practical purposes identically to one that contains a ^{12}C atom.

SAMPLE EXERCISE 2.2 **Determining the Number of Subatomic Particles in Atoms**

How many protons, neutrons, and electrons are in an atom of (**a**) ^{197}Au, (**b**) strontium-90?

SOLUTION

(**a**) The superscript 197 is the mass number (protons + neutrons). According to the list of elements given on the front inside cover, gold has atomic number 79. Consequently, an atom of ^{197}Au has 79 protons, 79 electrons, and $197 - 79 = 118$ neutrons. (**b**) The atomic number of strontium is 38. Thus, all atoms of this element have 38 protons and 38 electrons. The strontium-90 isotope has $90 - 38 = 52$ neutrons.

Practice Exercise 1

Which of these atoms has the largest number of neutrons in the nucleus? (**a**) ^{148}Eu, (**b**) ^{157}Dy, (**c**) ^{149}Nd, (**d**) ^{162}Ho, (**e**) ^{159}Gd.

Practice Exercise 2

How many protons, neutrons, and electrons are in an atom of (**a**) ^{138}Ba, (**b**) phosphorus-31?

SAMPLE EXERCISE 2.3 **Writing Symbols for Atoms**

Magnesium has three isotopes with mass numbers 24, 25, and 26. (**a**) Write the complete chemical symbol (superscript and subscript) for each. (**b**) How many neutrons are in an atom of each isotope?

SOLUTION

(**a**) Magnesium has atomic number 12, so all atoms of magnesium contain 12 protons and 12 electrons. The three isotopes are therefore represented by $^{24}_{12}Mg$, $^{25}_{12}Mg$, and $^{26}_{12}Mg$. (**b**) The number of neutrons in each isotope is the mass number minus the number of protons. The numbers of neutrons in an atom of each isotope are therefore 12, 13, and 14, respectively.

Practice Exercise 1

Which of the following is an incorrect representation for a neutral atom: (**a**) $^{6}_{3}Li$, (**b**) $^{13}_{6}C$, (**c**) $^{63}_{30}Cu$, (**d**) $^{30}_{15}P$, (**e**) $^{108}_{47}Ag$?

Practice Exercise 2

Give the complete chemical symbol for the atom that contains 82 protons, 82 electrons, and 126 neutrons.

2.4 | Atomic Weights

Atoms are small pieces of matter, so they have mass. In this section we discuss the mass scale used for atoms and introduce the concept of *atomic weights*.

The Atomic Mass Scale

Scientists of the nineteenth century were aware that atoms of different elements have different masses. They found, for example, that each 100.0 g of water contains 11.1 g of hydrogen and 88.9 g of oxygen. Thus, water contains $88.9/11.1 = 8$ times as much oxygen, by mass, as hydrogen. Once scientists understood that water contains two hydrogen atoms for each oxygen atom, they concluded that an oxygen atom must have $2 \times 8 = 16$ times as much mass as a hydrogen atom. Hydrogen, the lightest atom, was arbitrarily assigned a relative mass of 1 (no units). Atomic masses of other elements were at first determined relative to this value. Thus, oxygen was assigned an atomic mass of 16.

Today we can determine the masses of individual atoms with a high degree of accuracy. For example, we know that the ^{1}H atom has a mass of 1.6735×10^{-24} g and the ^{16}O atom has a mass of 2.6560×10^{-23} g. As we noted in Section 2.3, it is convenient to use the **atomic mass unit** when dealing with these extremely small masses:

$$1 \text{ amu} = 1.66054 \times 10^{-24} \text{ g and } 1 \text{ g} = 6.02214 \times 10^{23} \text{ amu}$$

The atomic mass unit is presently defined by assigning a mass of exactly 12 amu to a chemically unbound atom of the ^{12}C isotope of carbon. In these units, an ^{1}H atom has a mass of 1.0078 amu and an ^{16}O atom has a mass of 15.9949 amu.

Atomic Weight

Most elements occur in nature as mixtures of isotopes. We can determine the *average atomic mass* of an element, usually called the element's **atomic weight**, by summing (indicated by the Greek sigma, Σ) over the masses of its isotopes multiplied by their relative abundances:

$$\text{Atomic weight} = \sum [(\text{isotope mass}) \times (\text{fractional isotope abundance})]$$
$$\text{over all isotopes of the element} \quad [2.1]$$

Naturally occurring carbon, for example, is composed of 98.93% ^{12}C and 1.07% ^{13}C. The masses of these isotopes are 12 amu (exactly) and 13.00335 amu, respectively, making the atomic weight of carbon

$$(0.9893)(12 \text{ amu}) + (0.0107)(13.00335 \text{ amu}) = 12.01 \text{ amu}$$

The atomic weights of the elements are listed in both the periodic table and the list of elements on the front inside cover of this text.

 Give It Some Thought

A particular atom of chromium has a mass of 52.94 amu, whereas the atomic weight of chromium is given as 51.99 amu. Explain the difference in the two masses.

SAMPLE
EXERCISE 2.4 Calculating the Atomic Weight of an Element from Isotopic Abundances

Naturally occurring chlorine is 75.78% ^{35}Cl (atomic mass 34.969 amu) and 24.22% ^{37}Cl (atomic mass 36.966 amu). Calculate the atomic weight of chlorine.

SOLUTION

We can calculate the atomic weight by multiplying the abundance of each isotope by its atomic mass and summing these products. Because $75.78\% = 0.7578$ and $24.22\% = 0.2422$, we have

$$\text{Atomic weight} = (0.7578)(34.969 \text{ amu}) + (0.2422)(36.966 \text{ amu})$$
$$= 26.50 \text{ amu} + 8.953 \text{ amu}$$
$$= 35.45 \text{ amu}$$

This answer makes sense: The atomic weight, which is actually the average atomic mass, is between the masses of the two isotopes and is closer to the value of ^{35}Cl, the more abundant isotope.

Practice Exercise 1

The atomic weight of copper, Cu, is listed as 63.546. Which of the following statements are untrue?

(a) Not all the atoms of copper have the same number of electrons.
(b) All the copper atoms have 29 protons in the nucleus.
(c) The dominant isotopes of Cu must be ^{63}Cu and ^{64}Cu.
(d) Copper is a mixture of at least two isotopes.
(e) The number of electrons in the copper atoms is independent of atomic mass.

Practice Exercise 2

Three isotopes of silicon occur in nature: ^{28}Si (92.23%), atomic mass 27.97693 amu; ^{29}Si (4.68%), atomic mass 28.97649 amu; and ^{30}Si (3.09%), atomic mass 29.97377 amu. Calculate the atomic weight of silicon.

A Closer Look

The Mass Spectrometer

The most accurate means for determining atomic weights is provided by the **mass spectrometer** (▼ Figure 2.11). A gaseous sample is introduced at *A* and bombarded by a stream of high-energy electrons at *B*. Collisions between the electrons and the atoms or molecules of the gas produce positively charged particles, called *ions*, that are then accelerated toward a negatively charged grid (*C*). After the ions pass through the grid, they encounter two slits that allow only a narrow beam of ions to pass. This beam then passes between the poles of a magnet, which deflects the ions into a curved path. For ions with the same charge, the extent of deflection depends on mass—the more massive the ion, the less the deflection. The ions are thereby separated according to their masses. By changing the strength of the magnetic field or the accelerating voltage on the grid, ions of various masses can be selected to enter the detector.

A graph of the intensity of the detector signal versus ion atomic mass is called a *mass spectrum* (▼ Figure 2.12). Analysis of a mass spectrum gives both the masses of the ions reaching the detector and their relative abundances, which are obtained from the signal intensities. Knowing the atomic mass and the abundance of each isotope allows us to calculate the atomic weight of an element, as shown in Sample Exercise 2.4.

Mass spectrometers are used extensively today to identify chemical compounds and analyze mixtures of substances. Any molecule that loses electrons can fall apart, forming an array of positively charged fragments. The mass spectrometer measures the masses of these fragments, producing a chemical "fingerprint" of the molecule and providing clues about how the atoms were connected in the original molecule. Thus, a chemist might use this technique to determine the molecular structure of a newly synthesized compound or to identify a pollutant in the environment.

Related Exercises: 2.27, 2.38, 2.40, 2.88, 2.98, 2.99

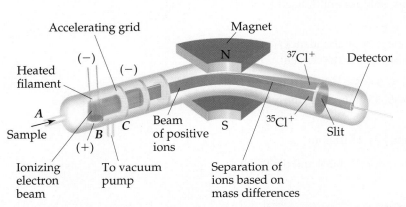

▲ **Figure 2.11 A mass spectrometer.** Cl atoms are introduced at *A* and are ionized to form Cl⁺ ions, which are then directed through a magnetic field. The paths of the ions of the two Cl isotopes diverge as they pass through the field.

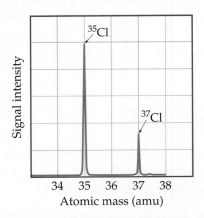

▲ **Figure 2.12 Mass spectrum of atomic chlorine.** The fractional abundances of the isotopes ^{35}Cl and ^{37}Cl are indicated by the relative signal intensities of the beams reaching the detector of the mass spectrometer.

2.5 | The Periodic Table

As the list of known elements expanded during the early 1800s, attempts were made to find patterns in chemical behavior. These efforts culminated in the development of the periodic table in 1869. We will have much to say about the periodic table in later chapters, but it is so important and useful that you should become acquainted with it now. You will quickly learn that *the periodic table is the most significant tool that chemists use for organizing and remembering chemical facts.*

Many elements show strong similarities to one another. The elements lithium (Li), sodium (Na), and potassium (K) are all soft, very reactive metals, for example. The elements helium (He), neon (Ne), and argon (Ar) are all very nonreactive gases. If the elements are arranged in order of increasing atomic number, their chemical and physical properties show a repeating, or *periodic,* pattern. For example, each of the soft, reactive metals—lithium, sodium, and potassium—comes immediately after one of the nonreactive gases—helium, neon, and argon, respectively—as shown in ▶ Figure 2.13.

▲ **GO FIGURE**

If F is a reactive nonmetal, which other element or elements shown here do you expect to also be a reactive nonmetal?

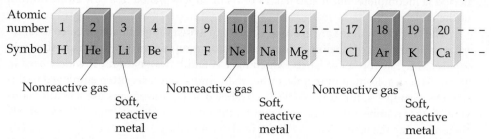

▲ **Figure 2.13 Arranging elements by atomic number reveals a periodic pattern of properties.** This pattern is the basis of the periodic table.

The arrangement of elements in order of increasing atomic number, with elements having similar properties placed in vertical columns, is known as the **periodic table** (▼ Figure 2.14). The table shows the atomic number and atomic symbol for each element, and the atomic weight is often given as well, as in this typical entry for potassium:

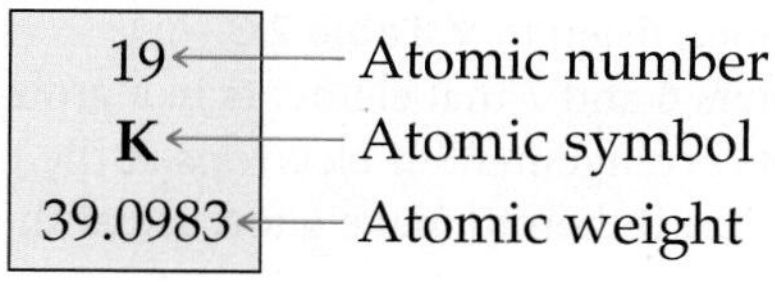

19 ←——— Atomic number
K ←——— Atomic symbol
39.0983 ←——— Atomic weight

You may notice slight variations in periodic tables from one book to another or between those in the lecture hall and in the text. These are simply matters of style, or they might concern the particular information included. There are no fundamental differences.

The horizontal rows of the periodic table are called **periods**. The first period consists of only two elements, hydrogen (H) and helium (He). The second and

▲ **Figure 2.14 Periodic table of elements.**

third periods consist of eight elements each. The fourth and fifth periods contain 18 elements. The sixth period has 32 elements, but for it to fit on a page, 14 of these elements (atomic numbers 57–70) appear at the bottom of the table. The seventh period is incomplete, but it also has 14 of its members placed in a row at the bottom of the table.

The vertical columns are **groups**. The way in which the groups are labeled is somewhat arbitrary. Three labeling schemes are in common use, two of which are shown in Figure 2.14. The top set of labels, which have A and B designations, is widely used in North America. Roman numerals, rather than Arabic ones, are often employed in this scheme. Group 7A, for example, is often labeled VIIA. Europeans use a similar convention that numbers the columns from 1A through 8A and then from 1B through 8B, thereby giving the label 7B (or VIIB) instead of 7A to the group headed by fluorine (F). In an effort to eliminate this confusion, the International Union of Pure and Applied Chemistry (IUPAC) has proposed a convention that numbers the groups from 1 through 18 with no A or B designations, as shown in Figure 2.14. We will use the traditional North American convention with Arabic numerals and the letters A and B.

Elements in a group often exhibit similarities in physical and chemical properties. For example, the "coinage metals"—copper (Cu), silver (Ag), and gold (Au)—belong to group 1B. These elements are less reactive than most metals, which is why they have been traditionally used throughout the world to make coins. Many other groups in the periodic table also have names, listed in ▼ **Table 2.3**.

We will learn in Chapters 6 and 7 that elements in a group have similar properties because they have the same arrangement of electrons at the periphery of their atoms. However, we need not wait until then to make good use of the periodic table; after all,

Table 2.3   **Names of Some Groups in the Periodic Table**

Group	Name	Elements
1A	Alkali metals	Li, Na, K, Rb, Cs, Fr
2A	Alkaline earth metals	Be, Mg, Ca, Sr, Ba, Ra
6A	Chalcogens	O, S, Se, Te, Po
7A	Halogens	F, Cl, Br, I, At
8A	Noble gases (or rare gases)	He, Ne, Ar, Kr, Xe, Rn

A Closer Look

What Are Coins Made Of?

Copper, silver, and gold were traditionally employed to make coins, but modern coins are typically made from other metals. To be useful for coinage, a metal, or combination of metals (called an *alloy*), must be corrosion resistant. It must also be hard enough to withstand rough usage and yet be of a consistency that permits machines to accurately stamp the coins. Some metals that might otherwise make fine coins— for example, manganese (Mn)—are ruled out because they make the coins too hard to stamp. A third criterion is that the value of the metal in the coin should not be as great as the face value of the coin. For example, if pennies were made today

A photo of a coin from the Presidential Dollar coin series.

from pure copper, the metal would be worth more than a penny, thus inviting smelters to melt down the coins for the value of the metal. Pennies today are largely made of zinc with a copper cladding.

One of the traditional alloys for making coins is a mixture of copper and nickel. Today only the U.S. nickel is made from this alloy, called cupronickel, which consists of 75% copper and 25% nickel. The modern U.S. dollar coin, often referred to as the silver dollar, doesn't contain any silver. It consists of copper (88.5%), zinc (6.0%), manganese (3.5%), and nickel (2.0%). In 2007 the U.S. Congress created a new series of $1 coins honoring former U.S. presidents. The coins have not been popular, and supplies stockpiled. The U.S. Treasury secretary suspended production of the coins for general circulation in December 2011.

GO FIGURE

Name two ways in which the metals shown in Figure 2.15 differ in general appearance from the nonmetals.

Metals

Iron (Fe) Copper (Cu) Aluminum (Al)

Silver (Ag) Lead (Pb) Gold (Au)

Nonmetals

Bromine (Br) Carbon (C)

Sulfur (S) Phosphorus (P)

▲ Figure 2.15 **Examples of metals and nonmetals.**

chemists who knew nothing about electrons developed the table! We can use the table, as they intended, to correlate behaviors of elements and to help us remember many facts.

The color code of Figure 2.14 shows that, except for hydrogen, all the elements on the left and in the middle of the table are **metallic elements**, or **metals**. All the metallic elements share characteristic properties, such as luster and high electrical and heat conductivity, and all of them except mercury (Hg) are solid at room temperature. The metals are separated from the **nonmetallic elements**, or **nonmetals**, by a stepped line that runs from boron (B) to astatine (At). (Note that hydrogen, although on the left side of the table, is a nonmetal.) At room temperature some of the nonmetals are gaseous, some are solid, and one is liquid. Nonmetals generally differ from metals in appearance (▲ Figure 2.15) and in other physical properties. Many of the elements that lie along the line that separates metals from nonmetals have properties that fall between those of metals and nonmetals. These elements are often referred to as **metalloids**.

 Give It Some Thought

Chlorine is a halogen (Table 2.3). Locate this element in the periodic table.
(a) What is its symbol?
(b) In which period and in which group is the element located?
(c) What is its atomic number?
(d) Is it a metal or nonmetal?

SAMPLE EXERCISE 2.5 Using the Periodic Table

Which two of these elements would you expect to show the greatest similarity in chemical and physical properties: B, Ca, F, He, Mg, P?

SOLUTION

Elements in the same group of the periodic table are most likely to exhibit similar properties. We therefore expect Ca and Mg to be most alike because they are in the same group (2A, the alkaline earth metals).

Practice Exercise 1

A biochemist who is studying the properties of certain sulfur (S)-containing compounds in the body wonders whether trace

amounts of another nonmetallic element might have similar behavior. To which element should she turn her attention? **(a)** O, **(b)** As, **(c)** Se, **(d)** Cr, **(e)** P.

Practice Exercise 2

Locate Na (sodium) and Br (bromine) in the periodic table. Give the atomic number of each and classify each as metal, metalloid, or nonmetal.

Hydrogen, H_2 Oxygen, O_2

Water, H_2O Hydrogen
                     peroxide, H_2O_2

Carbon   Carbon
monoxide, CO   dioxide, CO_2

Methane, CH_4 Ethylene, C_2H_4

▲ **Figure 2.16 Molecular models.** Notice how the chemical formulas of these simple molecules correspond to their compositions.

2.6 | Molecules and Molecular Compounds

Even though the atom is the smallest representative sample of an element, only the noble-gas elements are normally found in nature as isolated atoms. Most matter is composed of molecules or ions. We examine molecules here and ions in Section 2.7.

Molecules and Chemical Formulas

Several elements are found in nature in molecular form—two or more of the same type of atom bound together. For example, most of the oxygen in air consists of molecules that contain two oxygen atoms. As we saw in Section 1.2, we represent this molecular oxygen by the **chemical formula** O_2 (read "oh two"). The subscript tells us that two oxygen atoms are present in each molecule. A molecule made up of two atoms is called a **diatomic molecule**.

Oxygen also exists in another molecular form known as *ozone*. Molecules of ozone consist of three oxygen atoms, making the chemical formula O_3. Even though "normal" oxygen (O_2) and ozone (O_3) are both composed only of oxygen atoms, they exhibit very different chemical and physical properties. For example, O_2 is essential for life, but O_3 is toxic; O_2 is odorless, whereas O_3 has a sharp, pungent smell.

The elements that normally occur as diatomic molecules are hydrogen, oxygen, nitrogen, and the halogens (H_2, O_2, N_2, F_2, Cl_2, Br_2, and I_2). Except for hydrogen, these diatomic elements are clustered on the right side of the periodic table.

Compounds composed of molecules contain more than one type of atom and are called **molecular compounds**. A molecule of the compound methane, for example, consists of one carbon atom and four hydrogen atoms and is therefore represented by the chemical formula CH_4. Lack of a subscript on the C indicates one atom of C per methane molecule. Several common molecules of both elements and compounds are shown in ◀ **Figure 2.16**. Notice how the composition of each substance is given by its chemical formula. Notice also that these substances are composed only of nonmetallic elements. *Most molecular substances we will encounter contain only nonmetals.*

Molecular and Empirical Formulas

Chemical formulas that indicate the actual numbers of atoms in a molecule are called **molecular formulas**. (The formulas in Figure 2.16 are molecular formulas.) Chemical formulas that give only the relative number of atoms of each type in a molecule are called **empirical formulas**. The subscripts in an empirical formula are always the smallest possible whole-number ratios. The molecular formula for hydrogen peroxide is H_2O_2, for example, whereas its empirical formula is HO. The molecular formula for ethylene is C_2H_4, and its empirical formula is CH_2. For many substances, the molecular formula and the empirical formula are identical, as in the case of water, H_2O.

 Give It Some Thought

Consider the following four formulas: SO_2, B_2H_6, CH, $C_4H_2O_2$. Which of these formulas could be (**a**) only an empirical formula, (**b**) only a molecular formula, (**c**) either a molecular or an empirical formula?

Whenever we know the molecular formula of a compound, we can determine its empirical formula. The converse is not true, however. If we know the empirical formula of a substance, we cannot determine its molecular formula unless we have more information. So why do chemists bother with empirical formulas? As we will see in Chapter 3, certain common methods of analyzing substances lead to the empirical formula only. Once the empirical formula is known, additional experiments can give the information needed to convert the empirical formula to the molecular one. In addition, there are substances that do not exist as isolated molecules. For these substances, we must rely on empirical formulas.

SAMPLE EXERCISE 2.6 | Relating Empirical and Molecular Formulas

Write the empirical formulas for (a) glucose, a substance also known as either blood sugar or dextrose—molecular formula $C_6H_{12}O_6$; (b) nitrous oxide, a substance used as an anesthetic and commonly called laughing gas—molecular formula N_2O.

SOLUTION

(a) The subscripts of an empirical formula are the smallest whole-number ratios. The smallest ratios are obtained by dividing each subscript by the largest common factor, in this case 6. The resultant empirical formula for glucose is CH_2O.

(b) Because the subscripts in N_2O are already the lowest integral numbers, the empirical formula for nitrous oxide is the same as its molecular formula, N_2O.

What are the molecular and empirical formulas of this substance? (a) C_2O_2, CO_2, (b) C_4O, CO, (c) CO_2, CO_2, (d) C_4O_2, C_2O, (e) C_2O, CO_2.

Practice Exercise 1

Tetracarbon dioxide is an unstable oxide of carbon with the following molecular structure:

Practice Exercise 2

Give the empirical formula for *decaborane*, whose molecular formula is $B_{10}H_{14}$.

Picturing Molecules

The molecular formula of a substance summarizes the composition of the substance but does not show how the atoms are joined in the molecule. A **structural formula** shows which atoms are attached to which, as in the following examples:

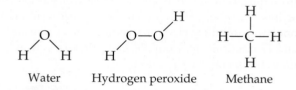

Water   Hydrogen peroxide   Methane

The atoms are represented by their chemical symbols, and lines are used to represent the bonds that hold the atoms together.

A structural formula usually does not depict the actual geometry of the molecule, that is, the actual angles at which atoms are joined. A structural formula can be written as a *perspective drawing* (**Figure 2.17**), however, to portray the three-dimensional shape.

Scientists also rely on various models to help visualize molecules. **Ball-and-stick models** show atoms as spheres and bonds as sticks. This type of model has the advantage of accurately representing the angles at which the atoms are attached to one another in a molecule (**Figure 2.17**). Sometimes the chemical symbols of the elements are superimposed on the balls, but often the atoms are identified simply by color.

A **space-filling model** depicts what a molecule would look like if the atoms were scaled up in size (**Figure 2.17**). These models show the relative sizes of the atoms, but the angles between atoms, which help define their molecular geometry, are often more difficult to see than in ball-and-stick models. As in ball-and-stick models, the identities of the atoms are indicated by color, but they may also be labeled with the element's symbol.

 Give It Some Thought

The structural formula for ethane is

$$H-\overset{\displaystyle H}{\underset{\displaystyle H}{C}}-\overset{\displaystyle H}{\underset{\displaystyle H}{C}}-H$$

(a) What is the molecular formula for ethane?
(b) What is its empirical formula?
(c) Which kind of molecular model would most clearly show the angles between atoms?

GO FIGURE

Which model, the ball-and-stick or the space-filling, more effectively shows the angles between bonds around a central atom?

CH_4

Molecular formula

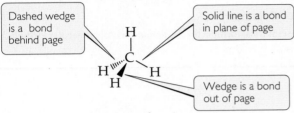

Structural formula

Dashed wedge is a bond behind page

Solid line is a bond in plane of page

Wedge is a bond out of page

Perspective drawing

Ball-and-stick model

Space-filling model

▲ **Figure 2.17 Different representations of the methane (CH_4) molecule.** Structural formulas, perspective drawings, ball-and-stick models, and space-filling models.

2.7 | Ions and Ionic Compounds

The nucleus of an atom is unchanged by chemical processes, but some atoms can readily gain or lose electrons. If electrons are removed from or added to an atom, a charged particle called an **ion** is formed. An ion with a positive charge is a **cation** (pronounced CAT-ion); a negatively charged ion is an **anion** (AN-ion).

To see how ions form, consider the sodium atom, which has 11 protons and 11 electrons. This atom easily loses one electron. The resulting cation has 11 protons and 10 electrons, which means it has a net charge of 1+.

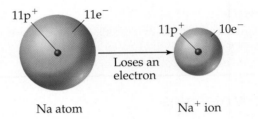

$11p^+$ $11e^-$ $11p^+$ $10e^-$

Loses an electron

Na atom Na^+ ion

The net charge on an ion is represented by a superscript. The superscripts $+, 2+,$ and $3+$, for instance, mean a net charge resulting from the *loss* of one, two, and three electrons, respectively. The superscripts $-, 2-,$ and $3-$ represent net charges resulting from the *gain* of one, two, and three electrons, respectively. Chlorine, with 17 protons and 17 electrons, for example, can gain an electron in chemical reactions, producing the Cl^- anion:

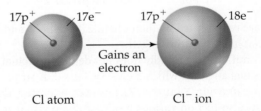

$17p^+$ $17e^-$ $17p^+$ $18e^-$

Gains an electron

Cl atom Cl^- ion

In general, metal atoms tend to lose electrons to form cations and nonmetal atoms tend to gain electrons to form anions. Thus, ionic compounds tend to be composed of metals bonded with nonmetals, as in NaCl.

SAMPLE EXERCISE 2.7 | Writing Chemical Symbols for Ions

Give the chemical symbol, including superscript indicating mass number, for **(a)** the ion with 22 protons, 26 neutrons, and 19 electrons; and **(b)** the ion of sulfur that has 16 neutrons and 18 electrons.

SOLUTION

(a) The number of protons is the atomic number of the element. A periodic table or list of elements tells us that the element with atomic number 22 is titanium (Ti). The mass number (protons plus neutrons) of this isotope of titanium is $22 + 26 = 48$. Because the ion has three more protons than electrons, it has a net charge of $3+$ and is designated $^{48}Ti^{3+}$.

(b) The periodic table tells us that sulfur (S) has an atomic number of 16. Thus, each atom or ion of sulfur contains 16 protons. We are told that the ion also has 16 neutrons, meaning the mass number is $16 + 16 = 32$. Because the ion has 16 protons and 18 electrons, its net charge is $2-$ and the ion symbol is $^{32}S^{2-}$.

In general, we will focus on the net charges of ions and ignore their mass numbers unless the circumstances dictate that we specify a certain isotope.

Practice Exercise 1

In which of the following species is the number of protons less than the number of electrons? (a) Ti^{2+}, (b) P^{3-}, (c) Mn, (d) Se_4^{2-}, (e) Ce^{4+}.

Practice Exercise 2

How many protons, neutrons, and electrons does the $^{79}Se^{2-}$ ion possess?

In addition to simple ions such as Na^+ and Cl^-, there are **polyatomic ions**, such as NH_4^+ (ammonium ion) and SO_4^{2-} (sulfate ion), which consist of atoms joined as in a molecule, but carrying a net positive or negative charge. Polyatomic ions will be discussed in Section 2.8.

It is important to realize that the chemical properties of ions are very different from the chemical properties of the atoms from which the ions are derived. The addition or removal of one or more electrons produces a charged species with behavior very different from that of its associated atom or group of atoms.

Predicting Ionic Charges

As noted in Table 2.3, the elements of group 8A are called the noble–gas elements. The noble gases are chemically nonreactive elements that form very few compounds. Many atoms gain or lose electrons to end up with the same number of electrons as the noble gas closest to them in the periodic table. We might deduce that atoms tend to acquire the electron arrangements of the noble gases because these electron arrangements are very stable. Nearby elements can obtain these same stable arrangements by losing or gaining electrons. For example, the loss of one electron from an atom of sodium leaves it with the same number of electrons as in a neon atom (10). Similarly, when chlorine gains an electron, it ends up with 18, the same number of electrons as in argon. This simple observation will be helpful for now to account for the formation of ions. A deeper explanation awaits us in Chapter 8, where we discuss chemical bonding.

**SAMPLE
EXERCISE 2.8** | **Predicting Ionic Charge**

Predict the charge expected for the most stable ion of barium and the most stable ion of oxygen.

SOLUTION

We will assume that barium and oxygen form ions that have the same number of electrons as the nearest noble-gas atom. From the periodic table, we see that barium has atomic number 56. The nearest noble gas is xenon, atomic number 54. Barium can attain a stable arrangement of 54 electrons by losing two electrons, forming the Ba^{2+} cation.

Oxygen has atomic number 8. The nearest noble gas is neon, atomic number 10. Oxygen can attain this stable electron arrangement by gaining two electrons, forming the O^{2-} anion.

Practice Exercise 1

Although it is helpful to know that many ions have the electron arrangement of a noble gas, many elements, especially among the metals, form ions that do not have a noble–gas electron arrangement. Use the periodic table, Figure 2.14, to determine which of the following ions has a noble–gas electron arrangement, and which do not. For those that do, indicate the noble–gas arrangement they match: (a) Ti^{4+}, (b) Mn^{2+}, (c) Pb^{2+}, (d) Te^{2-}, (e) Zn^{2+}.

Practice Exercise 2

Predict the charge expected for the most stable ion of (a) aluminum and (b) fluorine.

The periodic table is very useful for remembering ionic charges, especially those of elements on the left and right sides of the table. As **Figure 2.18** shows, the charges of these ions relate in a simple way to their positions in the table: The group 1A elements (alkali metals) form 1+ ions, the group 2A elements (alkaline earths) form 2+ ions, the group 7A elements (halogens) form 1− ions, and the group 6A elements form 2− ions. (As noted in Practice Exercise 1 of Sample Exercise 2.8, many of the other groups do not lend themselves to such simple rules.)

GO FIGURE

The most common ions for silver, zinc, and scandium are Ag^+, Zn^{2+}, and Sc^{3+}. Locate the boxes in which you would place these ions in this table. Which of these ions has the same number of electrons as a noble-gas element?

1A														3A	4A	5A	6A	7A	8A
H^+	2A																	H^-	N
Li^+																N^{3-}	O^{2-}	F^-	O
Na^+	Mg^{2+}		Transition metals											Al^{3+}			S^{2-}	Cl^-	B L E
K^+	Ca^{2+}																Se^{2-}	Br^-	G A
Rb^+	Sr^{2+}																Te^{2-}	I^-	S E
Cs^+	Ba^{2+}																		S

▲ **Figure 2.18 Predictable charges of some common ions.** Notice that the red stepped line that divides metals from nonmetals also separates cations from anions. Hydrogen forms both 1+ and 1− ions.

Ionic Compounds

A great deal of chemical activity involves the transfer of electrons from one substance to another. ▼ **Figure 2.19** shows that when elemental sodium is allowed to react with elemental chlorine, an electron transfers from a sodium atom to a chlorine atom, forming a Na^+ ion and a Cl^- ion. Because objects of opposite charges attract, the Na^+ and the Cl^- ions bind together to form the compound sodium chloride (NaCl). Sodium chloride, which we know better as common table salt, is an example of an **ionic compound**, a compound made up of cations and anions.

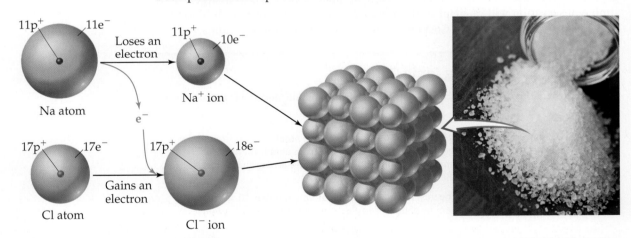

▲ **Figure 2.19 Formation of an ionic compound.** The transfer of an electron from a Na atom to a Cl atom leads to the formation of a Na^+ ion and a Cl^- ion. These ions are arranged in a lattice in solid sodium chloride, NaCl.

We can often tell whether a compound is ionic (consisting of ions) or molecular (consisting of molecules) from its composition. In general, cations are metal ions and anions are nonmetal ions. Consequently, *ionic compounds are generally combinations of metals and nonmetals*, as in NaCl. In contrast, *molecular compounds are generally composed of nonmetals only*, as in H_2O.

SAMPLE EXERCISE 2.9 Identifying Ionic and Molecular Compounds

Which of these compounds would you expect to be ionic: N_2O, Na_2O, $CaCl_2$, SF_4?

SOLUTION

We predict that Na_2O and $CaCl_2$ are ionic compounds because they are composed of a metal combined with a nonmetal. We predict (correctly) that N_2O and SF_4 are molecular compounds because they are composed entirely of nonmetals.

Practice Exercise 1
Which of these compounds are molecular: CBr_4, FeS, P_4O_6, PbF_2?

Practice Exercise 2
Give a reason why each of the following statements is a safe prediction:
(a) Every compound of Rb with a nonmetal is ionic in character.
(b) Every compound of nitrogen with a halogen element is a molecular compound.
(c) The compound $MgKr_2$ does not exist.
(d) Na and K are very similar in the compounds they form with nonmetals.
(e) If contained in an ionic compound, calcium (Ca) will be in the form of the doubly charged ion, Ca^{2+}.

The ions in ionic compounds are arranged in three-dimensional structures, as Figure 2.19 shows for NaCl. Because there is no discrete "molecule" of NaCl, we are able to write only an empirical formula for this substance. This is true for most other ionic compounds.

We can write the empirical formula for an ionic compound if we know the charges of the ions. Because chemical compounds are always electrically neutral, the ions in an ionic compound always occur in such a ratio that the total positive charge equals the total negative charge. Thus, there is one Na^+ to one Cl^- in NaCl, one Ba^{2+} to two Cl^- in $BaCl_2$, and so forth.

As you consider these and other examples, you will see that if the charges on the cation and anion are equal, the subscript on each ion is 1. If the charges are not equal, the charge on one ion (without its sign) will become the subscript on the other ion. For example, the ionic compound formed from Mg (which forms Mg^{2+} ions) and N (which forms N^{3-} ions) is Mg_3N_2:

$$Mg^{2+} \quad N^{3-} \longrightarrow Mg_3N_2$$

 Give It Some Thought

Can you tell from the formula of a substance whether it is ionic or molecular in nature? Why or why not?

Chemistry and Life

Elements Required by Living Organisms

The elements essential to life are highlighted in color in ▼ Figure 2.20. More than 97% of the mass of most organisms is made up of just six of these elements—oxygen, carbon, hydrogen, nitrogen, phosphorus, and sulfur. Water is the most common compound in living organisms, accounting for at least 70% of the mass of most cells. In the solid components of cells, carbon is the most prevalent element by mass. Carbon atoms are found in a vast variety of organic molecules, bonded either to other carbon atoms or to atoms of other elements. All proteins, for example, contain the carbon-based group

which occurs repeatedly in the molecules. (R is either an H atom or a combination of atoms, such as CH_3.)

In addition, 23 other elements have been found in various living organisms. Five are ions required by all organisms: Ca^{2+}, Cl^-, Mg^{2+}, K^+, and Na^+. Calcium ions, for example, are necessary for the formation of bone and transmission of nervous system signals. Many other elements are needed in only very small quantities and consequently are called *trace* elements. For example, trace quantities of copper are required in the diet of humans to aid in the synthesis of hemoglobin.

Related Exercise: 2.102

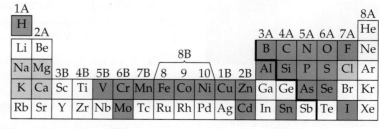

▲ **Figure 2.20 Elements essential to life.**

SAMPLE
EXERCISE 2.10 | Using Ionic Charge to Write Empirical Formulas for Ionic Compounds

Write the empirical formula of the compound formed by **(a)** Al^{3+} and Cl^- ions, **(b)** Al^{3+} and O^{2-} ions, **(c)** Mg^{2+} and NO_3^- ions.

SOLUTION

(a) Three Cl^- ions are required to balance the charge of one Al^{3+} ion, making the empirical formula $AlCl_3$.

(b) Two Al^{3+} ions are required to balance the charge of three O^{2-} ions. A 2:3 ratio is needed to balance the total positive charge of $6+$ and the total negative charge of $6-$. The empirical formula is Al_2O_3.

(c) Two NO_3^- ions are needed to balance the charge of one Mg^{2+}, yielding $Mg(NO_3)_2$. Note that the formula for the polyatomic ion, NO_3^-, must be enclosed in parentheses so that it is clear that the subscript 2 applies to all the atoms of that ion.

> **Practice Exercise 1**
> For the following ionic compounds formed with S^{2-}, what is the empirical formula for the positive ion involved? **(a)** MnS, **(b)** Fe_2S_3, **(c)** MoS_2, **(d)** K_2S, **(e)** Ag_2S.
>
> **Practice Exercise 2**
> Write the empirical formula for the compound formed by **(a)** Na^+ and PO_4^{3-}, **(b)** Zn^{2+} and SO_4^{2-}, **(c)** Fe^{3+} and CO_3^{2-}.

2.8 | Naming Inorganic Compounds

The names and chemical formulas of compounds are essential vocabulary in chemistry. The system used in naming substances is called **chemical nomenclature**, from the Latin words *nomen* (name) and *calare* (to call).

There are more than 50 million known chemical substances. Naming them all would be a hopelessly complicated task if each had a name independent of all others. Many important substances that have been known for a long time, such as water (H_2O) and ammonia (NH_3), do have traditional names (called *common names*). For most substances, however, we rely on a set of rules that leads to an informative and unique name for each substance, one that conveys the composition of the substance.

The rules for chemical nomenclature are based on the division of substances into categories. The major division is between organic and inorganic compounds. *Organic compounds* contain carbon and hydrogen, often in combination with oxygen, nitrogen, or other elements. All others are *inorganic compounds*. Early chemists associated organic compounds with plants and animals and inorganic compounds with the nonliving portion of our world. Although this distinction is no longer pertinent, the classification between organic and inorganic compounds continues to be useful. In this section we consider the basic rules for naming three categories of inorganic compounds: ionic compounds, molecular compounds, and acids.

Names and Formulas of Ionic Compounds

Recall from Section 2.7 that ionic compounds usually consist of metal ions combined with nonmetal ions. The metals form the cations, and the nonmetals form the anions.

1. **Cations**

   a. *Cations formed from metal atoms have the same name as the metal:*

Na^+ sodium ion	Zn^{2+} zinc ion	Al^{3+} aluminum ion

   b. *If a metal can form cations with different charges, the positive charge is indicated by a Roman numeral in parentheses following the name of the metal:*

Fe^{2+} iron(II) ion	Cu^+ copper(I) ion
Fe^{3+} iron(III) ion	Cu^{2+} copper(II) ion

   Ions of the same element that have different charges have different properties, such as different colors (◄ **Figure 2.21**).

 GO FIGURE

Is the difference in properties we see between the two substances in Figure 2.21 a difference in physical or chemical properties?

▲ **Figure 2.21 Different ions of the same element have different properties.** Both substances shown are compounds of iron. The substance on the left is Fe_3O_4, which contains Fe^{2+} and Fe^{3+} ions. The substance on the right is Fe_2O_3, which contains Fe^{3+} ions.

SECTION 2.8 Naming Inorganic Compounds   **63**

Most metals that form cations with different charges are *transition metals*, elements that occur in the middle of the periodic table, from group 3B to group 2B (as indicated on the periodic table on the front inside cover of this book). The metals that form only one cation (only one possible charge) are those of group 1A and group 2A, as well as Al^{3+} (group 3A) and two transition-metal ions: Ag^+ (group 1B) and Zn^{2+} (group 2B). Charges are not expressed when naming these ions. However, if there is any doubt in your mind whether a metal forms more than one cation, use a Roman numeral to indicate the charge. It is never wrong to do so, even though it may be unnecessary.

An older method still widely used for distinguishing between differently charged ions of a metal uses the endings *-ous* and *-ic* added to the root of the element's Latin name:

Fe^{2+}	ferrous ion	Cu^+	cuprous ion
Fe^{3+}	ferric ion	Cu^{2+}	cupric ion

Although we will only rarely use these older names in this text, you might encounter them elsewhere.

c. *Cations formed from nonmetal atoms have names that end in* -ium:

NH_4^+	ammonium ion	H_3O^+	hydronium ion

These two ions are the only ions of this kind that we will encounter frequently in the text.

The names and formulas of some common cations are shown in ▼ Table 2.4 and on the back inside cover of the text. The ions on the left side in Table 2.4 are

Table 2.4 Common Cations[a]

Charge	Formula	Name	Formula	Name
1+	**H⁺**	**hydrogen ion**	**NH₄⁺**	**ammonium ion**
	Li⁺	lithium ion	Cu⁺	copper(I) or cuprous ion
	Na⁺	**sodium ion**		
	K⁺	**potassium ion**		
	Cs⁺	cesium ion		
	Ag⁺	**silver ion**		
2+	**Mg²⁺**	**magnesium ion**	Co²⁺	cobalt(II) or cobaltous ion
	Ca²⁺	**calcium ion**	**Cu²⁺**	**copper(II) or cupric ion**
	Sr²⁺	strontium ion	**Fe²⁺**	**iron(II) or ferrous ion**
	Ba²⁺	barium ion	Mn²⁺	manganese(II) or manganous ion
	Zn²⁺	**zinc ion**	Hg₂²⁺	mercury(I) or mercurous ion
	Cd²⁺	cadmium ion	**Hg²⁺**	**mercury(II) or mercuric ion**
			Ni²⁺	nickel(II) or nickelous ion
			Pb²⁺	**lead(II) or plumbous ion**
			Sn²⁺	tin(II) or stannous ion
3+	**Al³⁺**	**aluminum ion**	Cr³⁺	chromium(III) or chromic ion
			Fe³⁺	**iron(III) or ferric ion**

[a]The ions we use most often in this course are in boldface. Learn them first.

the monatomic ions that do not have more than one possible charge. Those on the right side are either polyatomic cations or cations with more than one possible charge. The Hg_2^{2+} ion is unusual because, even though it is a metal ion, it is not monatomic. It is called the mercury(I) ion because it can be thought of as two Hg^+ ions bound together. The cations that you will encounter most frequently in this text are shown in boldface. You should learn these cations first.

 Give It Some Thought

(a) Why is CrO named using a Roman numeral, chromium(II) oxide, whereas CaO named without a Roman numeral, calcium oxide?

(b) What does the *-ium* ending on the name *ammonium* tell you about the composition of the ion?

2. Anions

a. *The names of monatomic anions are formed by replacing the ending of the name of the element with -ide:*

H^-	hydride ion	O^{2-}	oxide ion	N^{3-}	nitride ion

A few polyatomic anions also have names ending in *-ide*:

OH^-	hydroxide ion	CN^-	cyanide ion	O_2^{2-}	peroxide ion

b. *Polyatomic anions containing oxygen have names ending in either -ate or -ite and are called* **oxyanions**. The *-ate* is used for the most common or representative oxyanion of an element, and *-ite* is used for an oxyanion that has the same charge but one O atom fewer:

NO_3^-	nitrate ion		SO_4^{2-}	sulfate ion
NO_2^-	nitrite ion		SO_3^{2-}	sulfite ion

Prefixes are used when the series of oxyanions of an element extends to four members, as with the halogens. The prefix *per-* indicates one more O atom than the oxyanion ending in *-ate*; *hypo-* indicates one O atom fewer than the oxyanion ending in *-ite*:

ClO_4^-	perchlorate ion (one more O atom than chlorate)
ClO_3^-	chlorate ion
ClO_2^-	chlorite ion (one O atom fewer than chlorate)
ClO^-	hypochlorite ion (one O atom fewer than chlorite)

These rules are summarized in ▼ Figure 2.22.

 GO FIGURE

Name the anion obtained by removing one oxygen atom from the perbromate ion, BrO_4^-.

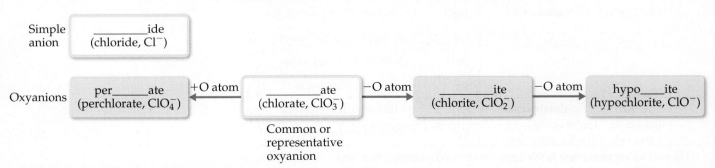

▲ **Figure 2.22 Procedure for naming anions.** The first part of the element's name, such as "chlor" for chlorine or "sulf" for sulfur, goes in the blank.

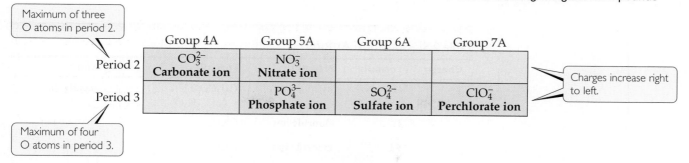

Maximum of three O atoms in period 2.

	Group 4A	Group 5A	Group 6A	Group 7A
Period 2	CO_3^{2-} **Carbonate ion**	NO_3^- **Nitrate ion**		
Period 3		PO_4^{3-} **Phosphate ion**	SO_4^{2-} **Sulfate ion**	ClO_4^- **Perchlorate ion**

Charges increase right to left.

Maximum of four O atoms in period 3.

▲ **Figure 2.23 Common oxyanions.** The composition and charges of common oxyanions are related to their location in the periodic table.

Give It Some Thought

What information is conveyed by the endings *-ide*, *-ate*, and *-ite* in the name of an anion?

▲ Figure 2.23 can help you remember the charge and number of oxygen atoms in the various oxyanions. Notice that C and N, both period 2 elements, have only three O atoms each, whereas the period 3 elements P, S, and Cl have four O atoms each. Beginning at the lower right in Figure 2.23, note that ionic charge increases from right to left, from 1− for ClO_4^- to 3− for PO_4^{3-}. In the second period the charges also increase from right to left, from 1− for NO_3^- to 2− for CO_3^{2-}. Notice also that although each of the anions in Figure 2.23 ends in *-ate*, the ClO_4^- ion also has a *per-* prefix.

Give It Some Thought

Predict the formulas for the borate ion and silicate ion, assuming they contain a single B and Si atom, respectively, and follow the trends shown in Figure 2.23.

SAMPLE EXERCISE 2.11 Determining the Formula of an Oxyanion from Its Name

Based on the formula for the sulfate ion, predict the formula for (**a**) the selenate ion and (**b**) the selenite ion. (Sulfur and selenium are both in group 6A and form analogous oxyanions.)

SOLUTION

(**a**) The sulfate ion is SO_4^{2-}. The analogous selenate ion is therefore SeO_4^{2-}.

(**b**) The ending *-ite* indicates an oxyanion with the same charge but one O atom fewer than the corresponding oxyanion that ends in *-ate*. Thus, the formula for the selenite ion is SeO_3^{2-}.

Practice Exercise 1

Which of the following oxyanions is incorrectly named?
(**a**) ClO_2^-, chlorate; (**b**) IO_4^-, periodate; (**c**) SO_3^{2-}, sulfite; (**d**) IO_3^-, iodate; (**e**) SeO_4^{2-}, selenate.

Practice Exercise 2

The formula for the bromate ion is analogous to that for the chlorate ion. Write the formula for the hypobromite and bromite ions.

c. *Anions derived by adding* H^+ *to an oxyanion are named by adding as a prefix the word* hydrogen *or* dihydrogen, *as appropriate:*

CO_3^{2-}	carbonate ion	PO_4^{3-}	phosphate ion
HCO_3^-	hydrogen carbonate ion	$H_2PO_4^-$	dihydrogen phosphate ion

Notice that each H^+ added reduces the negative charge of the parent anion by one. An older method for naming some of these ions uses the prefix *bi-*. Thus, the HCO_3^- ion is commonly called the bicarbonate ion, and HSO_4^- is sometimes called the bisulfate ion.

The names and formulas of the common anions are listed in **Table 2.5** and on the back inside cover of the text. Those anions whose names end in *-ide* are listed

Table 2.5 Common Anions[a]

Charge	Formula	Name	Formula	Name
1−	H^-	hydride ion	CH_3COO^- (or $C_2H_3O_2^-$)	**acetate ion**
	F^-	**fluoride ion**	ClO_3^-	chlorate ion
	Cl^-	**chloride ion**	ClO_4^-	**perchlorate ion**
	Br^-	**bromide ion**	NO_3^-	**nitrate ion**
	I^-	**iodide ion**	MnO_4^-	permanganate ion
	CN^-	cyanide ion		
	OH^-	**hydroxide ion**		
2−	O^{2-}	**oxide ion**	CO_3^{2-}	**carbonate ion**
	O_2^{2-}	peroxide ion	CrO_4^{2-}	chromate ion
	S^{2-}	**sulfide ion**	$Cr_2O_7^{2-}$	dichromate ion
			SO_4^{2-}	**sulfate ion**
3−	N^{3-}	nitride ion	PO_4^{3-}	**phosphate ion**

[a]The ions we use most often are in boldface. Learn them first.

on the left portion of Table 2.5, and those whose names end in -ate are listed on the right. The most common of these ions are shown in boldface. You should learn names and formulas of these anions first. The formulas of the ions whose names end with -ite can be derived from those ending in -ate by removing an O atom. Notice the location of the monatomic ions in the periodic table. Those of group 7A always have a 1− charge (F^-, Cl^-, Br^-, and I^-), and those of group 6A have a 2− charge (O^{2-} and S^{2-}).

3. Ionic Compounds

Names of ionic compounds consist of the cation name followed by the anion name:

$CaCl_2$	calcium chloride
$Al(NO_3)_3$	aluminum nitrate
$Cu(ClO_4)_2$	copper(II) perchlorate (or cupric perchlorate)

In the chemical formulas for aluminum nitrate and copper(II) perchlorate, parentheses followed by the appropriate subscript are used because the compounds contain two or more polyatomic ions.

SAMPLE EXERCISE 2.12 Determining the Names of Ionic Compounds from Their Formulas

Name the ionic compounds (a) K_2SO_4, (b) $Ba(OH)_2$, (c) $FeCl_3$.

SOLUTION

In naming ionic compounds, it is important to recognize polyatomic ions and to determine the charge of cations with variable charge.

(a) The cation is K^+, the potassium ion, and the anion is SO_4^{2-}, the sulfate ion, making the name potassium sulfate. (If you thought the compound contained S^{2-} and O^{2-} ions, you failed to recognize the polyatomic sulfate ion.)

(b) The cation is Ba^{2+}, the barium ion, and the anion is OH^-, the hydroxide ion: barium hydroxide.

(c) You must determine the charge of Fe in this compound because an iron atom can form more than one cation. Because the compound contains three chloride ions, Cl^-, the cation must be Fe^{3+}, the

iron(III), or ferric, ion. Thus, the compound is iron(III) chloride or ferric chloride.

Practice Exercise 1

Which of the following ionic compounds is incorrectly named? (a) $Zn(NO_3)_2$, zinc nitrate; (b) $TeCl_4$, tellurium(IV) chloride; (c) Fe_2O_3, diiron oxide; (d) BaO, barium oxide; (e) $Mn_3(PO_4)_2$, manganese (II) phosphate.

Practice Exercise 2

Name the ionic compounds (a) NH_4Br, (b) Cr_2O_3, (c) $Co(NO_3)_2$.

Give It Some Thought

Calcium bicarbonate is also called calcium hydrogen carbonate. **(a)** Write the formula for this compound, **(b)** predict the formulas for potassium bisulfate and lithium dihydrogen phosphate.

Names and Formulas of Acids

Acids are an important class of hydrogen-containing compounds, and they are named in a special way. For our present purposes, an *acid* is a substance whose molecules yield hydrogen ions (H^+) when dissolved in water. When we encounter the chemical formula for an acid at this stage of the course, it will be written with H as the first element, as in HCl and H_2SO_4.

An acid is composed of an anion connected to enough H^+ ions to neutralize, or balance, the anion's charge. Thus, the SO_4^{2-} ion requires two H^+ ions, forming H_2SO_4. The name of an acid is related to the name of its anion, as summarized in ▼ Figure 2.24.

1. *Acids containing anions whose names end in -ide are named by changing the -ide ending to -ic, adding the prefix* hydro- *to this anion name, and then following with the word* acid:

Anion	Corresponding Acid
Cl^- (chloride)	HCl (hydrochloric acid)
S^{2-} (sulfide)	H_2S (hydrosulfuric acid)

2. *Acids containing anions whose names end in -ate or -ite are named by changing -ate to -ic and -ite to -ous and then adding the word* acid. *Prefixes in the anion name are retained in the name of the acid:*

Anion	Corresponding Acid
ClO_4^- (perchlorate)	$HClO_4$ (perchloric acid)
ClO_3^- (chlorate)	$HClO_3$ (chloric acid)
ClO_2^- (chlorite)	$HClO_2$ (chlorous acid)
ClO^- (hypochlorite)	HClO (hypochlorous acid)

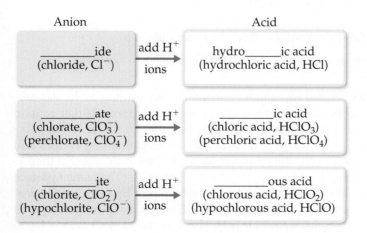

▲ **Figure 2.24 How anion names and acid names relate.** The prefixes *per-* and *hypo-* are retained in going from the anion to the acid.

▲ **Give It Some Thought**

Name the acid obtained by adding H^+ to the iodate ion, IO_3^-.

SAMPLE EXERCISE 2.13 | Relating the Names and Formulas of Acids

Name the acids **(a)** HCN, **(b)** HNO_3, **(c)** H_2SO_4, **(d)** H_2SO_3.

SOLUTION

(a) The anion from which this acid is derived is CN^-, the cyanide ion. Because this ion has an *-ide* ending, the acid is given a *hydro-* prefix and an *-ic* ending: hydrocyanic acid. Only water solutions of HCN are referred to as hydrocyanic acid. The pure compound, which is a gas under normal conditions, is called hydrogen cyanide. Both hydrocyanic acid and hydrogen cyanide are *extremely* toxic.

(b) Because NO_3^- is the nitrate ion, HNO_3 is called nitric acid (the *-ate* ending of the anion is replaced with an *-ic* ending in naming the acid).

(c) Because SO_4^{2-} is the sulfate ion, H_2SO_4 is called sulfuric acid.

(d) Because SO_3^{2-} is the sulfite ion, H_2SO_3 is sulfurous acid (the *-ite* ending of the anion is replaced with an *-ous* ending).

Practice Exercise 1

Which of the following acids are incorrectly named? For those that are, provide a correct name or formula. **(a)** hydrocyanic acid, HCN; **(b)** nitrous acid, HNO_3; **(c)** perbromic acid, $HBrO_4$; **(d)** iodic acid, HI; **(e)** selenic acid, $HSeO_4$.

Practice Exercise 2

Give the chemical formulas for **(a)** hydrobromic acid, **(b)** carbonic acid.

Names and Formulas of Binary Molecular Compounds

The procedures used for naming *binary* (two-element) molecular compounds are similar to those used for naming ionic compounds:

1. *The name of the element farther to the left in the periodic table (closest to the metals) is usually written first.* An exception occurs when the compound contains oxygen and chlorine, bromine, or iodine (any halogen except fluorine), in which case oxygen is written last.

2. *If both elements are in the same group, the one closer to the bottom of the table is named first.*

3. *The name of the second element is given an -ide ending.*

4. *Greek prefixes* (◄ Table 2.6) *indicate the number of atoms of each element.* (Exception: The prefix *mono-* is never used with the first element.) When the prefix ends in *a* or *o* and the name of the second element begins with a vowel, the *a* or *o* of the prefix is often dropped.

The following examples illustrate these rules:

Cl_2O	dichlorine monoxide	NF_3	nitrogen trifluoride
N_2O_4	dinitrogen tetroxide	P_4S_{10}	tetraphosphorus decasulfide

Rule 4 is necessary because we cannot predict formulas for most molecular substances the way we can for ionic compounds. Molecular compounds that contain hydrogen and one other element are an important exception, however. These compounds can be treated as if they were neutral substances containing H^+ ions and anions. Thus, you can predict that the substance named hydrogen chloride has the formula HCl, containing one H^+ to balance the charge of one Cl^-. (The name *hydrogen chloride* is used only for the pure compound; water solutions of HCl are called hydrochloric acid. The distinction, which is important, will be explained in Section 4.1.) Similarly, the formula for hydrogen sulfide is H_2S because two H^+ ions are needed to balance the charge on S^{2-}.

Table 2.6 Prefixes Used in Naming Binary Compounds Formed between Nonmetals

Prefix	Meaning
Mono-	1
Di-	2
Tri-	3
Tetra-	4
Penta-	5
Hexa-	6
Hepta-	7
Octa-	8
Nona-	9
Deca-	10

Give It Some Thought

Is $SOCl_2$ a binary compound?

SAMPLE EXERCISE 2.14 | **Relating the Names and Formulas of Binary Molecular Compounds**

Name the compounds (**a**) SO_2, (**b**) PCl_5, (**c**) Cl_2O_3.

SOLUTION

The compounds consist entirely of nonmetals, so they are molecular rather than ionic. Using the prefixes in Table 2.6, we have (**a**) sulfur dioxide, (**b**) phosphorus pentachloride, (**c**) dichlorine trioxide.

> **Practice Exercise 1**
>
> Give the name for each of the following binary compounds of carbon: (**a**) CS_2, (**b**) CO, (**c**) C_3O_2, (**d**) CBr_4, (**e**) CF.
>
> **Practice Exercise 2**
>
> Give the chemical formulas for (**a**) silicon tetrabromide, (**b**) disulfur dichloride, (**c**) diphosphorus hexaoxide.

2.9 | Some Simple Organic Compounds

The study of compounds of carbon is called **organic chemistry**, and as noted earlier, compounds that contain carbon and hydrogen, often in combination with oxygen, nitrogen, or other elements, are called *organic compounds*. Organic compounds are a very important part of chemistry, far outnumbering all other types of chemical substances. We will examine organic compounds in a systematic way in Chapter 24, but you will encounter many examples of them throughout the text. Here we present a brief introduction to some of the simplest organic compounds and the ways in which they are named.

Alkanes

Compounds that contain only carbon and hydrogen are called **hydrocarbons**. In the simplest class of hydrocarbons, **alkanes**, each carbon is bonded to four other atoms. The three smallest alkanes are methane (CH_4), ethane (C_2H_6), and propane (C_3H_8). The structural formulas of these three alkanes are as follows:

$$
\begin{array}{ccc}
\text{H} & \text{H} \quad \text{H} & \text{H} \quad \text{H} \quad \text{H} \\
| & | \quad\; | & | \quad\; | \quad\; | \\
\text{H}-\text{C}-\text{H} & \text{H}-\text{C}-\text{C}-\text{H} & \text{H}-\text{C}-\text{C}-\text{C}-\text{H} \\
| & | \quad\; | & | \quad\; | \quad\; | \\
\text{H} & \text{H} \quad \text{H} & \text{H} \quad \text{H} \quad \text{H} \\
\text{Methane} & \text{Ethane} & \text{Propane}
\end{array}
$$

Although hydrocarbons are binary molecular compounds, they are not named like the binary inorganic compounds discussed in Section 2.8. Instead, each alkane has a name that ends in *-ane*. The alkane with four carbons is called *butane*. For alkanes with five or more carbons, the names are derived from prefixes like those in Table 2.6. An alkane with eight carbon atoms, for example, is *octane* (C_8H_{18}), where the *octa-* prefix for eight is combined with the *-ane* ending for an alkane.

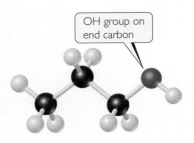

1-Propanol

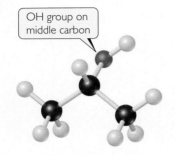

2-Propanol

▲ **Figure 2.25** The two forms (isomers) of propanol.

Some Derivatives of Alkanes

Other classes of organic compounds are obtained when one or more hydrogen atoms in an alkane are replaced with *functional groups*, which are specific groups of atoms. An **alcohol**, for example, is obtained by replacing an H atom of an alkane with an —OH group. The name of the alcohol is derived from that of the alkane by adding an -*ol* ending:

$$
\begin{array}{ccc}
\text{H} & \text{H H} & \text{H H H} \\
| & |\ \ | & |\ \ |\ \ | \\
\text{H---C---OH} & \text{H---C---C---OH} & \text{H---C---C---C---OH} \\
| & |\ \ | & |\ \ |\ \ | \\
\text{H} & \text{H H} & \text{H H H} \\
\text{Methanol} & \text{Ethanol} & \text{1-Propanol}
\end{array}
$$

Alcohols have properties that are very different from those of the alkanes from which the alcohols are obtained. For example, methane, ethane, and propane are all colorless gases under normal conditions, whereas methanol, ethanol, and propanol are colorless liquids. We will discuss the reasons for these differences in Chapter 11.

The prefix "1" in the name 1-propanol indicates that the replacement of H with OH has occurred at one of the "outer" carbon atoms rather than the "middle" carbon atom. A different compound, called either 2-propanol or isopropyl alcohol, is obtained when the OH functional group is attached to the middle carbon atom (◀ **Figure 2.25**).

Compounds with the same molecular formula but different arrangements of atoms are called **isomers**. There are many different kinds of isomers, as we will discover later in this book. What we have here with 1-propanol and 2-propanol are *structural isomers*, compounds having the same molecular formula but different structural formulas.

As already noted, many different functional groups can replace one or more of the hydrogens on an alkane; for example, one or more of the halogens, or a special grouping of carbon and oxygen atoms, such as the carboxylic acid group, —COOH. Here are a few examples of functional groups you will be encountering in the chapters that lie ahead (the functional group is outlined in blue):

$$
\begin{array}{cc}
\text{H H H H} & \text{H H H} \\
|\ \ |\ \ |\ \ | & |\ \ |\ \ | \\
\text{H---C---C---C---C---H} & \text{H---C---C---C---COOH} \\
|\ \ |\ \ |\ \ | & |\ \ |\ \ | \\
\text{H H Br H} & \text{H H H} \\
\text{2-bromobutane} & \text{butyric acid}
\end{array}
$$

$$
\begin{array}{c}
\text{H H H H} \qquad\quad \text{H} \\
|\ \ |\ \ |\ \ | \qquad\ \ | \\
\text{H---C---C---C---C---O---C---H} \\
|\ \ |\ \ |\ \ | \qquad\ \ | \\
\text{H H H H} \qquad\quad \text{H} \\
\text{butyl methyl ether}
\end{array}
$$

 Give It Some Thought

Draw the structural formulas of the two isomers of butane, C_4H_{10}.

Much of the richness of organic chemistry is possible because organic compounds can form long chains of carbon–carbon bonds. The series of alkanes that begins with methane, ethane, and propane and the series of alcohols that begins with methanol, ethanol, and propanol can both be extended for as long as we desire, in principle. The properties of alkanes and alcohols change as the chains get longer. Octanes, which are alkanes with eight carbon atoms, are liquids under normal conditions. If the alkane series is extended to tens of thousands of carbon atoms, we obtain *polyethylene*, a solid substance that is used to make thousands of plastic products, such as plastic bags, food containers, and laboratory equipment.

SAMPLE EXERCISE 2.15 | Writing Structural and Molecular Formulas for Hydrocarbons

Assuming the carbon atoms in *pentane* are in a linear chain, write (**a**) the structural formula and (**b**) the molecular formula for this alkane.

SOLUTION

(**a**) Alkanes contain only carbon and hydrogen, and each carbon is attached to four other atoms. The name pentane contains the prefix *penta-* for five (Table 2.6), and we are told that the carbons are in a linear chain. If we then add enough hydrogen atoms to make four bonds to each carbon, we obtain the structural formula

$$
\begin{array}{ccccc}
H & H & H & H & H \\
| & | & | & | & | \\
H-C-C-C-C-C-H \\
| & | & | & | & | \\
H & H & H & H & H
\end{array}
$$

This form of pentane is often called *n*-pentane, where the *n*- stands for "normal" because all five carbon atoms are in one line in the structural formula.

(**b**) Once the structural formula is written, we determine the molecular formula by counting the atoms present. Thus, *n*-pentane has the molecular formula C_5H_{12}.

Practice Exercise 1

(**a**) What is the molecular formula of hexane, the alkane with six carbons? (**b**) What are the name and molecular formula of an alcohol derived from hexane?

Practice Exercise 2

These two compounds have "butane" in their name. Are they isomers?

$$
\begin{array}{cccc}
H & H & H & H \\
| & | & | & | \\
H-C-C-C-C-H \\
| & | & | & | \\
H & H & H & H
\end{array}
$$
butane

$$
\begin{array}{cc}
H & H \\
| & | \\
H-C-C-H \\
| & | \\
H-C-C-H \\
| & | \\
H & H
\end{array}
$$
cyclobutane

Strategies in Chemistry

How to Take a Test

At about this time in your study of chemistry, you are likely to face your first hour-long examination. The best way to prepare is to study, do homework diligently, and get help from the instructor on any material that is unclear or confusing. (See the advice for learning and studying chemistry presented in the preface of the book.) We present here some general guidelines for taking tests.

Depending on the nature of your course, the exam could consist of a variety of different types of questions.

1. **Multiple-choice questions** In large-enrollment courses, the most common kind of test question is the multiple-choice question. Many of the practice exercise problems in this book are written in this format to give you practice at this style of question. When faced with this type of problem the first thing to realize is that the instructor has written the question so that at first glance all the answers appear to be correct. Thus, you should not jump to the conclusion that because one of the choices looks correct, it must be correct.

   If a multiple-choice question involves a calculation, do the calculation, check your work, and *only then* compare your answer with the choices. Keep in mind, though, that your instructor has anticipated the most common errors you might make in solving a given problem and has probably listed the incorrect answers resulting from those errors. Always double-check your reasoning and use dimensional analysis to arrive at the correct numeric answer and the correct units.

   In multiple-choice questions that do not involve calculations, if you are not sure of the correct choice, eliminate all the choices you know for sure to be incorrect. The reasoning you use in eliminating incorrect choices may offer insight into which of the remaining choices is correct.

2. **Calculations in which you must show your work** In questions of this kind, you may receive partial credit even if you do not arrive at the correct answer, depending on whether the instructor can follow your line of reasoning. It is important, therefore, to be neat and organized in your calculations. Pay particular attention to what information is given and to what your unknown is. Think about how you can get from the given information to your unknown.

   You may want to write a few words or a diagram on the test paper to indicate your approach. Then write out your calculations as neatly as you can. Show the units for every number you write down, and use dimensional analysis as much as you can, showing how units cancel.

3. **Questions requiring drawings** Questions of this kind will come later in the course, but it is useful to talk about them here. (You should review this box before each exam to remind yourself of good exam-taking practices.) Be sure to label your drawing as completely as possible.

Finally, if you find that you simply do not understand how to arrive at a reasoned response to a question, do not linger over the question. Put a check next to it and go on to the next one. If time permits, you can come back to the unanswered questions, but lingering over a question when nothing is coming to mind is wasting time you may need to finish the exam.

Chapter Summary and Key Terms

THE ATOMIC THEORY OF MATTER; THE DISCOVERY OF ATOMIC STRUCTURE (SECTIONS 2.1 AND 2.2) **Atoms** are the basic building blocks of matter. They are the smallest units of an element that can combine with other elements. Atoms are composed of even smaller particles, called **subatomic particles**. Some of these subatomic particles are charged and follow the usual behavior of charged particles: Particles with the same charge repel one another, whereas particles with unlike charges are attracted to one another.

We considered some of the important experiments that led to the discovery and characterization of subatomic particles. Thomson's experiments on the behavior of **cathode rays** in magnetic and electric fields led to the discovery of the electron and allowed its charge-to-mass ratio to be measured. Millikan's oil-drop experiment determined the charge of the electron. Becquerel's discovery of **radioactivity**, the spontaneous emission of radiation by atoms, gave further evidence that the atom has a substructure. Rutherford's studies of how thin metal foils scatter α particles led to the **nuclear model** of the atom, showing that the atom has a dense, positively charged **nucleus**.

THE MODERN VIEW OF ATOMIC STRUCTURE (SECTION 2.3) Atoms have a nucleus that contains **protons** and **neutrons; electrons** move in the space around the nucleus. The magnitude of the charge of the electron, 1.602×10^{-19} C, is called the **electronic charge**. The charges of particles are usually represented as multiples of this charge—an electron has a $1-$ charge, and a proton has a $1+$ charge. The masses of atoms are usually expressed in terms of **atomic mass units** (1 amu $= 1.66054 \times 10^{-24}$ g). The dimensions of atoms are often expressed in units of **angstroms** (1 Å $= 10^{-10}$ m).

Elements can be classified by **atomic number**, the number of protons in the nucleus of an atom. All atoms of a given element have the same atomic number. The **mass number** of an atom is the sum of the numbers of protons and neutrons. Atoms of the same element that differ in mass number are known as **isotopes**.

ATOMIC WEIGHTS (SECTION 2.4) The atomic mass scale is defined by assigning a mass of exactly 12 amu to a ^{12}C atom. The **atomic weight** (average atomic mass) of an element can be calculated from the relative abundances and masses of that element's isotopes. The **mass spectrometer** provides the most direct and accurate means of experimentally measuring atomic (and molecular) weights.

THE PERIODIC TABLE (SECTION 2.5) The **periodic table** is an arrangement of the elements in order of increasing atomic number. Elements with similar properties are placed in vertical columns. The elements in a column are known as a **group**. The elements in a horizontal row are known as a **period**. The **metallic elements** (**metals**), which comprise the majority of the elements, dominate the left side and the middle of the table; the **nonmetallic elements** (**nonmetals**) are located on the upper right side. Many of the elements that lie along the line that separates metals from nonmetals are **metalloids**.

MOLECULES AND MOLECULAR COMPOUNDS (SECTION 2.6) Atoms can combine to form **molecules**. Compounds composed of molecules (**molecular compounds**) usually contain only nonmetallic elements. A molecule that contains two atoms is called a **diatomic molecule**. The composition of a substance is given by its **chemical formula**. A molecular substance can be represented by its **empirical formula**, which gives the relative numbers of atoms of each kind. It is usually represented by its **molecular formula**, however, which gives the actual numbers of each type of atom in a molecule. **Structural formulas** show the order in which the atoms in a molecule are connected. **Ball-and-stick models** and **space-filling models** are often used to represent molecules.

IONS AND IONIC COMPOUNDS (SECTION 2.7) Atoms can either gain or lose electrons, forming charged particles called **ions**. Metals tend to lose electrons, becoming positively charged ions (**cations**). Nonmetals tend to gain electrons, forming negatively charged ions (**anions**). Because **ionic compounds** are electrically neutral, containing both cations and anions, they usually contain both metallic and nonmetallic elements. Atoms that are joined together, as in a molecule, but carry a net charge are called **polyatomic ions**. The chemical formulas used for ionic compounds are empirical formulas, which can be written readily if the charges of the ions are known. The total positive charge of the cations in an ionic compound equals the total negative charge of the anions.

NAMING INORGANIC COMPOUNDS (SECTION 2.8) The set of rules for naming chemical compounds is called **chemical nomenclature**. We studied the systematic rules used for naming three classes of inorganic substances: ionic compounds, acids, and binary molecular compounds. In naming an ionic compound, the cation is named first and then the anion. Cations formed from metal atoms have the same name as the metal. If the metal can form cations of differing charges, the charge is given using Roman numerals. Monatomic anions have names ending in *-ide*. Polyatomic anions containing oxygen and another element (**oxyanions**) have names ending in *-ate* or *-ite*.

SOME SIMPLE ORGANIC COMPOUNDS (SECTION 2.9) **Organic chemistry** is the study of compounds that contain carbon. The simplest class of organic molecules is the **hydrocarbons**, which contain only carbon and hydrogen. Hydrocarbons in which each carbon atom is attached to four other atoms are called **alkanes**. Alkanes have names that end in *-ane,* such as methane and ethane. Other organic compounds are formed when an H atom of a hydrocarbon is replaced with a functional group. An **alcohol**, for example, is a compound in which an H atom of a hydrocarbon is replaced by an OH functional group. Alcohols have names that end in *-ol,* such as methanol and ethanol. Compounds with the same molecular formula but different bonding arrangements of their constituent atoms are called **isomers**.

Learning Outcomes   After studying this chapter, you should be able to:

- List the basic postulates of Dalton's atomic theory. (Section 2.1)

- Describe the key experiments that led to the discovery of electrons and to the nuclear model of the atom. (Section 2.2)

- Describe the structure of the atom in terms of protons, neutrons, and electrons. (Section 2.3)

- Describe the electrical charge and relative masses of protons, neutrons, and electrons. (Section 2.3)

- Use chemical symbols together with atomic number and mass number to express the subatomic composition of isotopes. (Section 2.3)

- Calculate the atomic weight of an element from the masses of individual atoms and a knowledge of natural abundances. (Section 2.4)

- Describe how elements are organized in the periodic table by atomic number and by similarities in chemical behavior, giving rise to periods and groups. (Section 2.5)

- Identify the locations of metals and nonmetals in the periodic table. (Section 2.5)

- Distinguish between molecular substances and ionic substances in terms of their composition. (Sections 2.6 and 2.7)

- Distinguish between empirical formulas and molecular formulas. (Section 2.6)

- Describe how molecular formulas and structural formulas are used to represent the compositions of molecules. (Section 2.6)

- Explain how ions are formed by the gain or loss of electrons and be able to use the periodic table to predict the charges of common ions. (Section 2.7)

- Write the empirical formulas of ionic compounds, given the charges of their component ions. (Section 2.7)

- Write the name of an ionic compound given its chemical formula or write the chemical formula given its name. (Section 2.8)

- Name or write chemical formulas for binary inorganic compounds and for acids. (Section 2.8)

- Identify organic compounds and name simple alkanes and alcohols. (Section 2.9)

Key Equations

$$\text{Atomic weight} = \sum [(\text{isotope mass}) \times (\text{fractional isotope abundance})] \text{ over all isotopes of the element} \qquad [2.1]$$

Calculating atomic weight as a fractionally weighted average of isotopic masses.

Exercises

Visualizing Concepts

These exercises are intended to probe your understanding of key concepts rather than your ability to utilize formulas and perform calculations. Exercises with red numbers have answers in the back of the book.

2.1 A charged particle is caused to move between two electrically charged plates, as shown here.

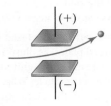

(a) Why does the path of the charged particle bend? (b) What is the sign of the electrical charge on the particle? (c) As the charge on the plates is increased, would you expect the bending to increase, decrease, or stay the same? (d) As the mass of the particle is increased while the speed of the particles remains the same, would you expect the bending to increase, decrease, or stay the same? [Section 2.2]

2.2 The following diagram is a representation of 20 atoms of a fictitious element, which we will call nevadium (Nv). The red spheres are ^{293}Nv, and the blue spheres are ^{295}Nv. (a) Assuming that this sample is a statistically representative sample of the element, calculate the percent abundance of each element. (b) If the mass of ^{293}Nv is 293.15 amu and that of ^{295}Nv is 295.15 amu, what is the atomic weight of Nv? [Section 2.4]

2.3 Four of the boxes in the following periodic table are colored. Which of these are metals and which are nonmetals? Which one is an alkaline earth metal? Which one is a noble gas? [Section 2.5]

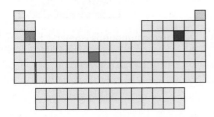

2.4 Does the following drawing represent a neutral atom or an ion? Write its complete chemical symbol including mass number, atomic number, and net charge (if any). [Sections 2.3 and 2.7]

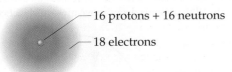

16 protons + 16 neutrons

18 electrons

2.5 Which of the following diagrams most likely represents an ionic compound, and which represents a molecular one? Explain your choice. [Sections 2.6 and 2.7]

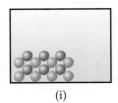
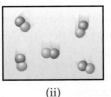

(i)   (ii)

2.6 Write the chemical formula for the following compound. Is the compound ionic or molecular? Name the compound. [Sections 2.6 and 2.8]

2.7 Five of the boxes in the following periodic table are colored. Predict the charge on the ion associated with each of these elements. [Section 2.7]

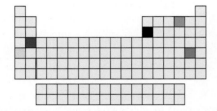

2.8 The following diagram represents an ionic compound in which the red spheres represent cations and blue spheres represent anions. Which of the following formulas is consistent with the drawing? KBr, K_2SO_4, $Ca(NO_3)_2$, $Fe_2(SO_4)_3$. Name the compound. [Sections 2.7 and 2.8]

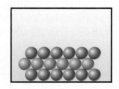

2.9 Are these two compounds isomers? Explain. [Section 2.9]

$$CH_3-CHCl$$
$$|$$
$$CH_2-CH_3$$

$$CH_3-CH_2-CH_2-CH_2Cl$$

2.10 In the Millikan oil–drop experiment (see Figure 2.5) the tiny oil drops are observed through the viewing lens as rising, stationary, or falling, as shown here. (a) What causes their rate of fall to vary from their rate in the absence of an electric field? (b) Why do some drops move upward? [Section 2.2]

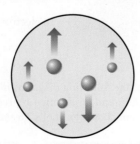

The following exercises are divided into sections that deal with specific topics in the chapter. The exercises are grouped in pairs, with the answers given in the back of the book to the odd-numbered exercises, as indicated by the red exercise numbers. Those exercises whose numbers appear in brackets are more challenging than the nonbracketed exercises.

The Atomic Theory of Matter and the Discovery of Atomic Structure (Sections 2.1 and 2.2)

2.11 How does Dalton's atomic theory account for the fact that when 1.000 g of water is decomposed into its elements, 0.111 g of hydrogen and 0.889 g of oxygen are obtained regardless of the source of the water?

2.12 Hydrogen sulfide is composed of two elements: hydrogen and sulfur. In an experiment, 6.500 g of hydrogen sulfide is fully decomposed into its elements. (a) If 0.384 g of hydrogen is obtained in this experiment, how many grams of sulfur must be obtained? (b) What fundamental law does this experiment demonstrate? (c) How is this law explained by Dalton's atomic theory?

2.13 A chemist finds that 30.82 g of nitrogen will react with 17.60, 35.20, 70.40, or 88.00 g of oxygen to form four different compounds. (a) Calculate the mass of oxygen per gram of nitrogen in each compound. (b) How do the numbers in part (a) support Dalton's atomic theory?

2.14 In a series of experiments, a chemist prepared three different compounds that contain only iodine and fluorine and determined the mass of each element in each compound:

Compound	Mass of Iodine (g)	Mass of Fluorine (g)
1	4.75	3.56
2	7.64	3.43
3	9.41	9.86

(a) Calculate the mass of fluorine per gram of iodine in each compound. (b) How do the numbers in part (a) support the atomic theory?

2.15 Summarize the evidence used by J. J. Thomson to argue that cathode rays consist of negatively charged particles.

2.16 An unknown particle is caused to move between two electrically charged plates, as illustrated in Figure 2.8. Its path is deflected by a smaller magnitude in the opposite direction from that of a beta particle. What can you conclude about the charge and mass of this unknown particle?

2.17 How did Rutherford interpret the following observations made during his α-particle scattering experiments? (a) Most α particles were not appreciably deflected as they passed through the gold foil. (b) A few α particles were deflected at very large angles. (c) What differences would you expect if beryllium foil were used instead of gold foil in the α-particle scattering experiment?

2.18 Millikan determined the charge on the electron by studying the static charges on oil drops falling in an electric field (Figure 2.5). A student carried out this experiment using several oil drops for her measurements and calculated the charges on the drops. She obtained the following data:

Droplet	Calculated Charge (C)
A	1.60×10^{-19}
B	3.15×10^{-19}
C	4.81×10^{-19}
D	6.31×10^{-19}

(a) What is the significance of the fact that the droplets carried different charges? (b) What conclusion can the student draw from these data regarding the charge of the electron? (c) What value (and to how many significant figures) should she report for the electronic charge?

The Modern View of Atomic Structure; Atomic Weights (Sections 2.3 and 2.4)

2.19 The radius of an atom of gold (Au) is about 1.35 Å. (a) Express this distance in nanometers (nm) and in picometers (pm). (b) How many gold atoms would have to be lined up to span 1.0 mm? (c) If the atom is assumed to be a sphere, what is the volume in cm^3 of a single Au atom?

2.20 An atom of rhodium (Rh) has a diameter of about 2.7×10^{-8} cm. (a) What is the radius of a rhodium atom in angstroms (Å) and in meters (m)? (b) How many Rh atoms would have to be placed side by side to span a distance of 6.0 μm? (c) If you assume that the Rh atom is a sphere, what is the volume in m^3 of a single atom?

2.21 Answer the following questions without referring to Table 2.1: (a) What are the main subatomic particles that make up the atom? (b) What is the relative charge (in multiples of the electronic charge) of each of the particles? (c) Which of the particles is the most massive? (d) Which is the least massive?

2.22 Determine whether each of the following statements is true or false. If false, correct the statement to make it true: (a) The nucleus has most of the mass and comprises most of the volume of an atom. (b) Every atom of a given element has the same number of protons. (c) The number of electrons in an atom equals the number of neutrons in the atom. (d) The protons in the nucleus of the helium atom are held together by a force called the strong nuclear force.

2.23 Which of the following pairs of atoms are isotopes of one another? (a) ^{11}B, ^{11}C ; (b) ^{55}Mn, ^{54}Mn; (c) $^{118}_{50}Sn$, $^{120}_{50}Sn$

2.24 What are the differences in the compositions of the following pairs of atomic nuclei? (a) $^{210}_{83}Bi$, $^{210}_{82}Pb$; (b) $^{14}_{7}N$, $^{15}_{7}N$; (c) $^{20}_{10}Ne$, $^{40}_{18}Ar$

2.25 (a) Define atomic number and mass number. (b) Which of these can vary without changing the identity of the element?

2.26 (a) Which two of the following are isotopes of the same element: $^{31}_{16}X$, $^{31}_{15}X$, $^{32}_{16}X$? (b) What is the identity of the element whose isotopes you have selected?

2.27 How many protons, neutrons, and electrons are in the following atoms? (a) ^{40}Ar, (b) ^{65}Zn, (c) ^{70}Ga, (d) ^{80}Br, (e) ^{184}W, (f) ^{243}Am.

2.28 Each of the following isotopes is used in medicine. Indicate the number of protons and neutrons in each isotope: (a) phosphorus-32, (b)– chromium-51, (c) cobalt-60, (d) technetium-99, (e) iodine-131, (f) thallium-201.

2.29 Fill in the gaps in the following table, assuming each column represents a neutral atom.

Symbol	^{79}Br				
Protons		25			82
Neutrons		30	64		
Electrons			48	86	
Mass no.				222	207

2.30 Fill in the gaps in the following table, assuming each column represents a neutral atom.

Symbol	^{112}Cd				
Protons		38			92
Neutrons		58	49		
Electrons			38	36	
Mass no.				81	235

2.31 Write the correct symbol, with both superscript and subscript, for each of the following. Use the list of elements in the front inside cover as needed: (a) the isotope of platinum that contains 118 neutrons, (b) the isotope of krypton with mass number 84, (c) the isotope of arsenic with mass number 75, (d) the isotope of magnesium that has an equal number of protons and neutrons.

2.32 One way in which Earth's evolution as a planet can be understood is by measuring the amounts of certain isotopes in rocks. One quantity recently measured is the ratio of ^{129}Xe to ^{130}Xe in some minerals. In what way do these two isotopes differ from one another? In what respects are they the same?

2.33 (a) What isotope is used as the standard in establishing the atomic mass scale? (b) The atomic weight of boron is reported as 10.81, yet no atom of boron has the mass of 10.81 amu. Explain.

2.34 (a) What is the mass in amu of a carbon-12 atom? (b) Why is the atomic weight of carbon reported as 12.011 in the table of elements and the periodic table in the front inside cover of this text?

2.35 Only two isotopes of copper occur naturally, ^{63}Cu (atomic mass = 62.9296 amu; abundance 69.17%) and ^{65}Cu (atomic mass = 64.9278 amu; abundance 30.83%). Calculate the atomic weight (average atomic mass) of copper.

2.36 Rubidium has two naturally occurring isotopes, rubidium-85 (atomic mass = 84.9118 amu; abundance = 72.15%) and rubidium-87 (atomic mass = 86.9092 amu; abundance = 27.85%). Calculate the atomic weight of rubidium.

2.37 (a) Thomson's cathode–ray tube (Figure 2.4) and the mass spectrometer (Figure 2.11) both involve the use of electric or magnetic fields to deflect charged particles. What are the charged particles involved in each of these experiments? (b) What are the labels on the axes of a mass spectrum? (c) To measure the mass spectrum of an atom, the atom must first lose one or more electrons. Which would you expect to be deflected more by the same setting of the electric and magnetic fields, a Cl^+ or a Cl^{2+} ion?

2.38 (a) The mass spectrometer in Figure 2.11 has a magnet as one of its components. What is the purpose of the magnet? (b) The atomic weight of Cl is 35.5 amu. However, the mass spectrum of Cl (Figure 2.12) does not show a peak at this mass. Explain. (c) A mass spectrum of phosphorus (P) atoms shows only a single peak at a mass of 31. What can you conclude from this observation?

2.39 Naturally occurring magnesium has the following isotopic abundances:

Isotope	Abundance (%)	Atomic mass (amu)
^{24}Mg	78.99	23.98504
^{25}Mg	10.00	24.98584
^{26}Mg	11.01	25.98259

(a) What is the average atomic mass of Mg? (b) Sketch the mass spectrum of Mg.

2.40 Mass spectrometry is more often applied to molecules than to atoms. We will see in Chapter 3 that the *molecular weight* of a molecule is the sum of the atomic weights of the atoms in the molecule. The mass spectrum of H_2 is taken under conditions that prevent decomposition into H atoms. The two naturally occurring isotopes of hydrogen are ^1H (atomic mass = 1.00783 amu; abundance 99.9885%) and ^2H (atomic mass = 2.01410 amu; abundance 0.0115%). **(a)** How many peaks will the mass spectrum have? **(b)** Give the relative atomic masses of each of these peaks. **(c)** Which peak will be the largest, and which the smallest?

The Periodic Table, Molecules and Molecular Compounds, and Ions and Ionic Compounds (Sections 2.5 and 2.7)

2.41 For each of the following elements, write its chemical symbol, locate it in the periodic table, give its atomic number, and indicate whether it is a metal, metalloid, or nonmetal: **(a)** chromium, **(b)** helium, **(c)** phosphorus, **(d)** zinc, **(e)** magnesium, **(f)** bromine, **(g)** arsenic.

2.42 Locate each of the following elements in the periodic table; give its name and atomic number, and indicate whether it is a metal, metalloid, or nonmetal: **(a)** Li, **(b)** Sc, **(c)** Ge, **(d)** Yb, **(e)** Mn, **(f)** Sb, **(g)** Xe.

2.43 For each of the following elements, write its chemical symbol, determine the name of the group to which it belongs (Table 2.3), and indicate whether it is a metal, metalloid, or nonmetal: **(a)** potassium, **(b)** iodine, **(c)** magnesium, **(d)** argon, **(e)** sulfur.

2.44 The elements of group 4A show an interesting change in properties moving down the group. Give the name and chemical symbol of each element in the group and label it as a nonmetal, metalloid, or metal.

2.45 What can we tell about a compound when we know the empirical formula? What additional information is conveyed by the molecular formula? By the structural formula? Explain in each case.

2.46 Two compounds have the same empirical formula. One substance is a gas, whereas the other is a viscous liquid. How is it possible for two substances with the same empirical formula to have markedly different properties?

2.47 What are the molecular and empirical formulas for each of the following compounds?

$$H\!-\!\underset{H}{\overset{H}{N\!-\!N}}\qquad H\!-\!\underset{H}{\overset{H}{N\!=\!N}}\qquad H\!-\!\underset{H}{\overset{H}{N}}\!-\!H$$

2.48 Two substances have the same molecular and empirical formulas. Does this mean that they must be the same compound?

2.49 Write the empirical formula corresponding to each of the following molecular formulas: **(a)** Al_2Br_6, **(b)** C_8H_{10}, **(c)** $C_4H_8O_2$, **(d)** P_4O_{10}, **(e)** $C_6H_4Cl_2$, **(f)** $B_3N_3H_6$.

2.50 Determine the molecular and empirical formulas of the following: **(a)** the organic solvent *benzene*, which has six carbon atoms and six hydrogen atoms; **(b)** the compound *silicon tetrachloride*, which has a silicon atom and four chlorine atoms and is used in the manufacture of computer chips; **(c)** the reactive substance *diborane*, which has two boron atoms and six hydrogen atoms; **(d)** the sugar called *glucose*, which has six carbon atoms, twelve hydrogen atoms, and six oxygen atoms.

2.51 How many hydrogen atoms are in each of the following: **(a)** C_2H_5OH, **(b)** $Ca(C_2H_5COO)_2$, **(c)** $(NH_4)_3PO_4$?

2.52 How many of the indicated atoms are represented by each chemical formula: **(a)** carbon atoms in $C_4H_8COOCH_3$, **(b)** oxygen atoms in $Ca(ClO_3)_2$, **(c)** hydrogen atoms in $(NH_4)_2HPO_4$?

2.53 Write the molecular and structural formulas for the compounds represented by the following molecular models:

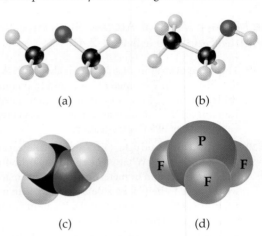

(a)   (b)

(c)   (d)

2.54 Write the molecular and structural formulas for the compounds represented by the following models:

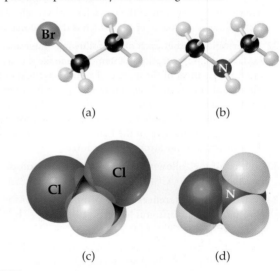

(a)   (b)

(c)   (d)

2.55 Fill in the gaps in the following table:

Symbol	$^{59}Co^{3+}$			
Protons		34	76	80
Neutrons		46	116	120
Electrons		36		78
Net charge			2+	

2.56 Fill in the gaps in the following table:

Symbol	$^{31}P^{3-}$			
Protons		34	50	
Neutrons		45	69	118
Electrons			46	76
Net charge		2−		3+

2.57 Each of the following elements is capable of forming an ion in chemical reactions. By referring to the periodic table, predict the charge of the most stable ion of each: **(a)** Mg, **(b)** Al, **(c)** K, **(d)** S, **(e)** F.

2.58 Using the periodic table, predict the charges of the ions of the following elements: (a) Ga, (b) Sr, (c) As, (d) Br, (e) Se.

2.59 Using the periodic table to guide you, predict the chemical formula and name of the compound formed by the following elements: (a) Ga and F, (b) Li and H, (c) Al and I, (d) K and S.

2.60 The most common charge associated with scandium in its compounds is 3+. Indicate the chemical formulas you would expect for compounds formed between scandium and (a) iodine, (b) sulfur, (c) nitrogen.

2.61 Predict the chemical formula for the ionic compound formed by (a) Ca^{2+} and Br^-, (b) K^+ and CO_3^{2-}, (c) Al^{3+} and CH_3COO^-, (d) NH_4^+ and SO_4^{2-}, (e) Mg^{2+} and PO_4^{3-}.

2.62 Predict the chemical formulas of the compounds formed by the following pairs of ions: (a) Cr^{3+} and Br^-, (b) Fe^{3+} and O^{2-}, (c) Hg_2^{2+} and CO_3^{2-}, (d) Ca^{2+} and ClO_3^-, (e) NH_4^+ and PO_4^{3-}.

2.63 Complete the table by filling in the formula for the ionic compound formed by each pair of cations and anions, as shown for the first pair.

Ion	K^+	NH_4^+	Mg^{2+}	Fe^{3+}
Cl^-	KCl			
OH^-				
CO_3^{2-}				
PO_4^{3-}				

2.64 Complete the table by filling in the formula for the ionic compound formed by each pair of cations and anions, as shown for the first pair.

Ion	Na^+	Ca^{2+}	Fe^{2+}	Al^{3+}
O^{2-}	Na_2O			
NO_3^-				
SO_4^{2-}				
AsO_4^{3-}				

2.65 Predict whether each of the following compounds is molecular or ionic: (a) B_2H_6, (b) CH_3OH, (c) $LiNO_3$, (d) Sc_2O_3, (e) CsBr, (f) NOCl, (g) NF_3, (h) Ag_2SO_4.

2.66 Which of the following are ionic, and which are molecular? (a) PF_5, (b) NaI, (c) SCl_2, (d) $Ca(NO_3)_2$, (e) $FeCl_3$, (f) LaP, (g) $CoCO_3$, (h) N_2O_4.

Naming Inorganic Compounds; Some Simple Organic Compounds (Sections 2.8 and 2.9)

2.67 Give the chemical formula for (a) chlorite ion, (b) chloride ion, (c) chlorate ion, (d) perchlorate ion, (e) hypoite ion.

2.68 Selenium, an element required nutritionally in trace quantities, forms compounds analogous to sulfur. Name the following ions: (a) SeO_4^{2-}, (b) Se^{2-}, (c) HSe^-, (d) $HSeO_3^-$.

2.69 Give the names and charges of the cation and anion in each of the following compounds: (a) CaO, (b) Na_2SO_4, (c) $KClO_4$, (d) $Fe(NO_3)_2$, (e) $Cr(OH)_3$.

2.70 Give the names and charges of the cation and anion in each of the following compounds: (a) CuS, (b) Ag_2SO_4, (c) $Al(ClO_3)_3$, (d) $Co(OH)_2$, (e) $PbCO_3$.

2.71 Name the following ionic compounds: (a) Li_2O, (b) $FeCl_3$, (c) NaClO, (d) $CaSO_3$, (e) $Cu(OH)_2$, (f) $Fe(NO_3)_2$, (g) $Ca(CH_3COO)_2$, (h) $Cr_2(CO_3)_3$, (i) K_2CrO_4, (j) $(NH_4)_2SO_4$.

2.72 Name the following ionic compounds: (a) KCN, (b) $NaBrO_2$, (c) $Sr(OH)_2$, (d) CoTe, (e) $Fe_2(CO_3)_3$, (f) $Cr(NO_3)_3$, (g) $(NH_4)_2SO_3$, (h) NaH_2PO_4, (i) $KMnO_4$, (j) $Ag_2Cr_2O_7$.

2.73 Write the chemical formulas for the following compounds: (a) aluminum hydroxide, (b) potassium sulfate, (c) copper(I) oxide, (d) zinc nitrate, (e) mercury(II) bromide, (f) iron(III) carbonate, (g) sodium hypobromite.

2.74 Give the chemical formula for each of the following ionic compounds: (a) sodium phosphate, (b) zinc nitrate, (c) barium bromate, (d) iron(II) perchlorate, (e) cobalt(II) hydrogen carbonate, (f) chromium(III) acetate, (g) potassium dichromate.

2.75 Give the name or chemical formula, as appropriate, for each of the following acids: (a) $HBrO_3$, (b) HBr, (c) H_3PO_4, (d) hypochlorous acid, (e) iodic acid, (f) sulfurous acid.

2.76 Provide the name or chemical formula, as appropriate, for each of the following acids: (a) hydroiodic acid, (b) chloric acid, (c) nitrous acid, (d) H_2CO_3, (e) $HClO_4$, (f) CH_3COOH.

2.77 Give the name or chemical formula, as appropriate, for each of the following binary molecular substances: (a) SF_6, (b) IF_5, (c) XeO_3, (d) dinitrogen tetroxide, (e) hydrogen cyanide, (f) tetraphosphorus hexasulfide.

2.78 The oxides of nitrogen are very important components in urban air pollution. Name each of the following compounds: (a) N_2O, (b) NO, (c) NO_2, (d) N_2O_5, (e) N_2O_4.

2.79 Write the chemical formula for each substance mentioned in the following word descriptions (use the front inside cover to find the symbols for the elements you do not know). (a) Zinc carbonate can be heated to form zinc oxide and carbon dioxide. (b) On treatment with hydrofluoric acid, silicon dioxide forms silicon tetrafluoride and water. (c) Sulfur dioxide reacts with water to form sulfurous acid. (d) The substance phosphorus trihydride, commonly called phosphine, is a toxic gas. (e) Perchloric acid reacts with cadmium to form cadmium(II) perchlorate. (f) Vanadium(III) bromide is a colored solid.

2.80 Assume that you encounter the following sentences in your reading. What is the chemical formula for each substance mentioned? (a) Sodium hydrogen carbonate is used as a deodorant. (b) Calcium hypochlorite is used in some bleaching solutions. (c) Hydrogen cyanide is a very poisonous gas. (d) Magnesium hydroxide is used as a cathartic. (e) Tin(II) fluoride has been used as a fluoride additive in toothpastes. (f) When cadmium sulfide is treated with sulfuric acid, fumes of hydrogen sulfide are given off.

2.81 (a) What is a hydrocarbon? (b) Pentane is the alkane with a chain of five carbon atoms. Write a structural formula for this compound and determine its molecular and empirical formulas.

2.82 (a) What is meant by the term *isomer*? (b) Among the four alkanes, ethane, propane, butane, and pentane, which is capable of existing in isomeric forms?

2.83 (a) What is a functional group? (b) What functional group characterizes an alcohol? (c) Write a structural formula for 1-pentanol, the alcohol derived from pentane by making a substitution on one of the carbon atoms.

2.84 (a) What do ethane and ethanol have in common? (b) How does 1-propanol differ from propane?

2.85 Chloropropane is derived from propane by substituting Cl for H on one of the carbon atoms. (a) Draw the structural formulas for the two isomers of chloropropane. (b) Suggest names for these two compounds.

2.86 Draw the structural formulas for three isomers of pentane, C_5H_{12}.

Additional Exercises

These exercises are not divided by category, although they are roughly in the order of the topics in the chapter. They are not paired.

2.87 Suppose a scientist repeats the Millikan oil-drop experiment but reports the charges on the drops using an unusual (and imaginary) unit called the *warmomb* (wa). The scientist obtains the following data for four of the drops:

Droplet	Calculated Charge (wa)
A	3.84×10^{-8}
B	4.80×10^{-8}
C	2.88×10^{-8}
D	8.64×10^{-8}

(a) If all the droplets were the same size, which would fall most slowly through the apparatus? (b) From these data, what is the best choice for the charge of the electron in warmombs? (c) Based on your answer to part (b), how many electrons are there on each of the droplets? (d) What is the conversion factor between warmombs and coulombs?

2.88 The natural abundance of ^{3}He is 0.000137%. (a) How many protons, neutrons, and electrons are in an atom of ^{3}He? (b) Based on the sum of the masses of their subatomic particles, which is expected to be more massive, an atom of ^{3}He or an atom of ^{3}H (which is also called *tritium*)? (c) Based on your answer to part (b), what would need to be the precision of a mass spectrometer that is able to differentiate between peaks that are due to ^{3}He$^+$ and ^{3}H$^+$?

2.89 A cube of gold that is 1.00 cm on a side has a mass of 19.3 g. A single gold atom has a mass of 197.0 amu. (a) How many gold atoms are in the cube? (b) From the information given, estimate the diameter in Å of a single gold atom. (c) What assumptions did you make in arriving at your answer for part (b)?

2.90 The diameter of a rubidium atom is 4.95 Å. We will consider two different ways of placing the atoms on a surface. In arrangement A, all the atoms are lined up with one another to form a square grid. Arrangement B is called a *close-packed* arrangement because the atoms sit in the "depressions" formed by the previous row of atoms:

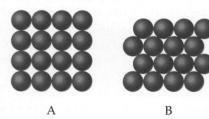

A   B

(a) Using arrangement A, how many Rb atoms could be placed on a square surface that is 1.0 cm on a side? (b) How many Rb atoms could be placed on a square surface that is 1.0 cm on a side, using arrangement B? (c) By what factor has the number of atoms on the surface increased in going to arrangement B from arrangement A? If extended to three dimensions, which arrangement would lead to a greater density for Rb metal?

2.91 (a) Assuming the dimensions of the nucleus and atom shown in Figure 2.11, what fraction of the *volume* of the atom is taken up by the nucleus? (b) Using the mass of the proton from Table 2.1 and assuming its diameter is 1.0×10^{-15} m, calculate the density of a proton in g/cm^3.

2.92 Identify the element represented by each of the following symbols and give the number of protons and neutrons in each: (a) $^{74}_{33}$X, (b) $^{127}_{53}$X, (c) $^{152}_{63}$X, (d) $^{209}_{83}$X.

2.93 The nucleus of ^{6}Li is a powerful absorber of neutrons. It exists in the naturally occurring metal to the extent of 7.5%. In the era of nuclear deterrence, large quantities of lithium were processed to remove ^{6}Li for use in hydrogen bomb production. The lithium metal remaining after removal of ^{6}Li was sold on the market. (a) What are the compositions of the nuclei of ^{6}Li and ^{7}Li? (b) The atomic masses of ^{6}Li and ^{7}Li are 6.015122 and 7.016004 amu, respectively. A sample of lithium depleted in the lighter isotope was found on analysis to contain 1.442% ^{6}Li. What is the average atomic weight of this sample of the metal?

2.94 The element oxygen has three naturally occurring isotopes, with 8, 9, and 10 neutrons in the nucleus, respectively. (a) Write the full chemical symbols for these three isotopes. (b) Describe the similarities and differences between the three kinds of atoms of oxygen.

2.95 The element lead (Pb) consists of four naturally occurring isotopes with atomic masses 203.97302, 205.97444, 206.97587, and 207.97663 amu. The relative abundances of these four isotopes are 1.4, 24.1, 22.1, and 52.4% respectively. From these data, calculate the atomic weight of lead.

2.96 Gallium (Ga) consists of two naturally occurring isotopes with masses of 68.926 and 70.925 amu. (a) How many protons and neutrons are in the nucleus of each isotope? Write the complete atomic symbol for each, showing the atomic number and mass number. (b) The average atomic mass of Ga is 69.72 amu. Calculate the abundance of each isotope.

2.97 Using a suitable reference such as the *CRC Handbook of Chemistry and Physics* or http://www.webelements.com, look up the following information for nickel: (a) the number of known isotopes, (b) the atomic masses (in amu), (c) the natural abundances of the five most abundant isotopes.

2.98 There are two different isotopes of bromine atoms. Under normal conditions, elemental bromine consists of Br$_2$ molecules, and the mass of a Br$_2$ molecule is the sum of the masses of the two atoms in the molecule. The mass spectrum of Br$_2$ consists of three peaks:

Mass (amu)	Relative Size
157.836	0.2569
159.834	0.4999
161.832	0.2431

(a) What is the origin of each peak (of what isotopes does each consist)? (b) What is the mass of each isotope? (c) Determine the average molecular mass of a Br$_2$ molecule. (d) Determine the average atomic mass of a bromine atom. (e) Calculate the abundances of the two isotopes.

2.99 It is common in mass spectrometry to assume that the mass of a cation is the same as that of its parent atom. (a) Using data in Table 2.1, determine the number of significant figures that must be reported before the difference in masses of ^{1}H and ^{1}H$^+$ is significant. (b) What percentage of the mass of an ^{1}H atom does the electron represent?

2.100 From the following list of elements—Ar, H, Ga, Al, Ca, Br, Ge, K, O—pick the one that best fits each description. Use each element only once: (a) an alkali metal, (b) an alkaline earth metal, (**c**) a noble gas, (**d**) a halogen, (**e**) a metalloid, (**f**) a nonmetal listed in group 1A, (**g**) a metal that forms a 3+ ion, (**h**) a nonmetal that forms a 2− ion, (**i**) an element that resembles aluminum.

2.101 The first atoms of seaborgium (Sg) were identified in 1974. The longest-lived isotope of Sg has a mass number of 266. (**a**) How many protons, electrons, and neutrons are in an ^{266}Sg atom? (**b**) Atoms of Sg are very unstable, and it is therefore difficult to study this element's properties. Based on the position of Sg in the periodic table, what element should it most closely resemble in its chemical properties?

2.102 The explosion of an atomic bomb releases many radioactive isotopes, including strontium-90. Considering the location of strontium in the periodic table, suggest a reason for the fact that this isotope is particularly dangerous for human health.

2.103 From the molecular structures shown here, identify the one that corresponds to each of the following species: (**a**) chlorine gas; (**b**) propane; (**c**) nitrate ion; (**d**) sulfur trioxide; (**e**) methyl chloride, CH_3Cl.

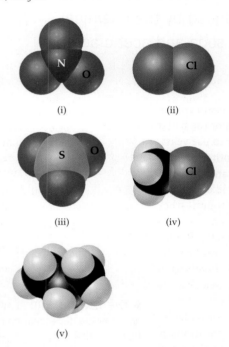

2.104 Name each of the following oxides. Assuming that the compounds are ionic, what charge is associated with the metallic element in each case? (**a**) NiO, (**b**) MnO_2, (**c**) Cr_2O_3, (**d**) MoO_3.

2.105 Fill in the blanks in the following table:

Cation	Anion	Formula	Name
			Lithium oxide
Fe^{2+}	PO_4^{3-}		
		$Al_2(SO_4)_3$	

Cation	Anion	Formula	Name
			Copper(II) nitrate
Cr^{3+}	I^-		
		$MnClO_2$	
			Ammonium carbonate
			Zinc perchlorate

2.106 Cyclopropane is an interesting hydrocarbon. Instead of having three carbons in a row, the three carbons form a ring, as shown in this perspective drawing (see Figure 2.17 for a prior example of this kind of drawing):

Cyclopropane was at one time used as an anesthetic, but its use was discontinued, in part because it is highly inflammable.

(**a**) What is the empirical formula of cyclopropane? How does it differ from that of propane? (**b**) The three carbon atoms are necessarily in a plane. What do the different wedges mean? (**c**) What change would you make to the structure shown to illustrate chlorocyclopropane? Are there isomers of chlorocyclopropane?

2.107 Elements in the same group of the periodic table often form oxyanions with the same general formula. The anions are also named in a similar fashion. Based on these observations, suggest a chemical formula or name, as appropriate, for each of the following ions: (**a**) BrO_4^-, (**b**) SeO_3^{2-}, (**c**) arsenate ion, (**d**) hydrogen tellurate ion.

2.108 Carbonic acid occurs in carbonated beverages. When allowed to react with lithium hydroxide, it produces lithium carbonate. Lithium carbonate is used to treat depression and bipolar disorder. Write chemical formulas for carbonic acid, lithium hydroxide, and lithium carbonate.

2.109 Give the chemical names of each of the following familiar compounds: (**a**) NaCl (table salt), (**b**) $NaHCO_3$ (baking soda), (**c**) NaOCl (in many bleaches), (**d**) NaOH (caustic soda), (**e**) $(NH_4)_2CO_3$ (smelling salts), (**f**) $CaSO_4$ (plaster of Paris).

2.110 Many familiar substances have common, unsystematic names. For each of the following, give the correct systematic name: (**a**) saltpeter, KNO_3; (**b**) soda ash, Na_2CO_3; (**c**) lime, CaO; (**d**) muriatic acid, HCl; (**e**) Epsom salts, $MgSO_4$; (**f**) milk of magnesia, $Mg(OH)_2$.

2.111 Because many ions and compounds have very similar names, there is great potential for confusing them. Write the correct chemical formulas to distinguish between (**a**) calcium sulfide and calcium hydrogen sulfide, (**b**) hydrobromic acid and bromic acid, (**c**) aluminum nitride and aluminum nitrite, (**d**) iron(II) oxide and iron(III) oxide, (**e**) ammonia and ammonium ion, (**f**) potassium sulfite and potassium bisulfite, (**g**) mercurous chloride and mercuric chloride, (**h**) chloric acid and perchloric acid.

2.112 In what part of the atom does the strong nuclear force operate?

3

Chemical Reactions and Reaction Stoichiometry

Have you ever poured vinegar into a vessel containing baking soda? If so, you know the result is an immediate and effervescent cascade of bubbles. The bubbles contain carbon dioxide gas that is produced by the chemical reaction between sodium bicarbonate in the baking soda and acetic acid in the vinegar.

The bubbles released when baking soda reacts with an acid play an important role in baking, where the release of gaseous CO_2 causes the dough in your biscuits or the batter in your pancakes to rise. An alternative way to produce CO_2 in cooking is to use yeasts that rely on chemical reactions to convert sugar into CO_2, ethanol, and other organic compounds. These types of chemical reactions have been used for thousands of years in the baking of breads as well as in the production of alcoholic beverages like beer and wine. Chemical reactions that produce CO_2 are not limited to cooking, though—they occur in places as diverse as the cells in your body and the engine of your car.

In this chapter we explore some important aspects of chemical reactions. Our focus will be both on the use of chemical formulas to represent reactions and on the quantitative information we can obtain about the amounts of substances involved in those reactions. **Stoichiometry** (pronounced stoy-key-OM-uh-tree) is the area of study that examines the quantities of substances consumed and produced in chemical reactions. Stoichiometry (Greek *stoicheion*, "element," and *metron*, "measure") provides an essential set of tools widely used in chemistry, including such diverse applications as measuring ozone concentrations in the atmosphere and assessing different processes for converting coal into gaseous fuels.

▶ **THE TEXTURE AND FLAVORS** of bread and beer are dependent on chemical reactions that occur when yeasts ferment sugars to produce carbon dioxide and ethanol.

WHAT'S AHEAD

3.1 CHEMICAL EQUATIONS We begin by considering how we can use chemical formulas to write equations representing chemical reactions.

3.2 SIMPLE PATTERNS OF CHEMICAL REACTIVITY We then examine some simple chemical reactions: *combination reactions*, *decomposition reactions*, and *combustion reactions*.

3.3 FORMULA WEIGHTS We see how to obtain quantitative information from chemical formulas by using *formula weights*.

3.4 AVOGADRO'S NUMBER AND THE MOLE We use chemical formulas to relate the masses of substances to the numbers of atoms, molecules, or ions contained in the substances, a relationship that leads to the crucially important concept of the *mole*, defined as 6.022×10^{23} objects (atoms, molecules, ions, and so on).

3.5 EMPIRICAL FORMULAS FROM ANALYSES We apply the mole concept to determine chemical formulas from the masses of each element in a given quantity of a compound.

3.6 QUANTITATIVE INFORMATION FROM BALANCED EQUATIONS We use the quantitative information inherent in chemical formulas and equations together with the mole concept to predict the amounts of substances consumed or produced in chemical reactions.

3.7 LIMITING REACTANTS We recognize that one reactant may be used up before others in a chemical reaction. This is the *limiting reactant*. When this happens, the reaction stops, leaving some excess of the other starting materials.

▲ **Figure 3.1 Antoine Lavoisier (1734–1794).** The science career of Lavoisier, who conducted many important studies on combustion reactions, was cut short by the French Revolution. Guillotined in 1794 during the Reign of Terror, he is generally considered the father of modern chemistry because he conducted carefully controlled experiments and used quantitative measurements.

Reactants   Products

$$2\,H_2 + O_2 \longrightarrow 2\,H_2O$$

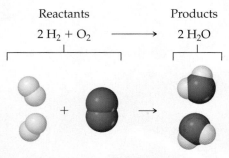

▲ **Figure 3.2 A balanced chemical equation.**

Stoichiometry is built on an understanding of atomic masses ∞∞ (Section 2.4), chemical formulas, and the **law of conservation of mass**. ∞∞ (Section 2.1) The French nobleman and scientist Antoine Lavoisier (◀ Figure 3.1) discovered this important chemical law during the late 1700s. Lavoisier stated the law in this eloquent way: "We may lay it down as an incontestable axiom that, in all the operations of art and nature, nothing is created; an equal quantity of matter exists both before and after the experiment. Upon this principle, the whole art of performing chemical experiments depends."* With the advent of Dalton's atomic theory, chemists came to understand the basis for this law: *Atoms are neither created nor destroyed during a chemical reaction.* The changes that occur during any reaction merely rearrange the atoms. The same collection of atoms is present both before and after the reaction.

3.1 | Chemical Equations

We represent chemical reactions by **chemical equations**. When the gas hydrogen (H_2) burns, for example, it reacts with oxygen (O_2) in the air to form water (H_2O). We write the chemical equation for this reaction as

$$2\,H_2 + O_2 \longrightarrow 2\,H_2O \qquad [3.1]$$

We read the $+$ sign as "reacts with" and the arrow as "produces." The chemical formulas to the left of the arrow represent the starting substances, called **reactants**. The chemical formulas to the right of the arrow represent substances produced in the reaction, called **products**. The numbers in front of the formulas, called coefficients, indicate the relative numbers of molecules of each kind involved in the reaction. (As in algebraic equations, *the coefficient 1 is usually not written.*)

Because atoms are neither created nor destroyed in any reaction, a chemical equation must have an equal number of atoms of each element on each side of the arrow. When this condition is met, the equation is *balanced*. On the right side of Equation 3.1, for example, there are two molecules of H_2O, each composed of two atoms of hydrogen and one atom of oxygen (◀ Figure 3.2). Thus, $2\,H_2O$ (read "two molecules of water") contains $2 \times 2 = 4$ H atoms and $2 \times 1 = 2$ O atoms. Notice that *the number of atoms is obtained by multiplying each subscript in a chemical formula by the coefficient for the formula.* Because there are four H atoms and two O atoms on each side of the equation, the equation is balanced.

▲ Give It Some Thought

How many atoms of Mg, O, and H are represented by the notation $3\,Mg(OH)_2$?

Balancing Equations

To construct a balanced chemical equation we start by writing the formulas for the reactants on the left–hand side of the arrow and the products on the right–hand side. Next we balance the equation by determining the coefficients that provide equal numbers of each type of atom on both sides of the equation. For most purposes, a balanced equation should contain the smallest possible whole-number coefficients.

In balancing an equation, you need to understand the difference between coefficients and subscripts. As ▶ Figure 3.3 illustrates, changing a subscript in a formula—from H_2O to H_2O_2, for example—changes the identity of the substance. The substance H_2O_2, hydrogen peroxide, is quite different from the substance H_2O, water. *Never change subscripts when balancing an equation.* In contrast, placing a coefficient in front of a formula changes only the *amount* of the substance and not its *identity*. Thus, $2\,H_2O$ means two molecules of water, $3\,H_2O$ means three molecules of water, and so forth.

To illustrate the process of balancing an equation, consider the reaction that occurs when methane (CH_4), the principal component of natural gas, burns in air to produce

*Lavoisier, Antoine. "Elements of Chemistry." 1790.

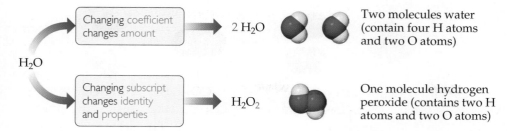

H_2O

Changing coefficient changes amount → $2\,H_2O$ Two molecules water (contain four H atoms and two O atoms)

Changing subscript changes identity and properties → H_2O_2 One molecule hydrogen peroxide (contains two H atoms and two O atoms)

▲ Figure 3.3 **The difference between changing subscripts and changing coefficients in chemical equations.**

carbon dioxide gas (CO_2) and water vapor (H_2O) (▼ Figure 3.4). Both products contain oxygen atoms that come from O_2 in the air. Thus, O_2 is a reactant, and the unbalanced equation is

$$CH_4 + O_2 \longrightarrow CO_2 + H_2O \quad \text{(unbalanced)} \qquad [3.2]$$

It is usually best to balance first those elements that occur in the fewest chemical formulas in the equation. In our example, C appears in only one reactant (CH_4) and one product (CO_2). The same is true for H (CH_4 and H_2O). Notice, however, that O appears in one reactant (O_2) and two products (CO_2 and H_2O). So, let's begin with C. Because one molecule of CH_4 contains the same number of C atoms (one) as one molecule of CO_2, the coefficients for these substances *must* be the same in the balanced equation. Therefore, we start by choosing the coefficient 1 (unwritten) for both CH_4 and CO_2.

Next we focus on H. On the left side of the equation we have CH_4, which has four H atoms, whereas on the right side of the equation we have H_2O, containing two H atoms. To balance the H atoms in the equation we place the coefficient 2 in front of H_2O. Now there are four H atoms on each side of the equation:

$$CH_4 + O_2 \longrightarrow CO_2 + 2\,H_2O \quad \text{(unbalanced)} \qquad [3.3]$$

While the equation is now balanced with respect to hydrogen and carbon, it is not yet balanced for oxygen. Adding the coefficient 2 in front of O_2 balances the equation by giving four O atoms on each side (2×2 left, $2 + 2 \times 1$ right):

$$CH_4 + 2\,O_2 \longrightarrow CO_2 + 2\,H_2O \quad \text{(balanced)} \qquad [3.4]$$

The molecular view of the balanced equation is shown in **Figure 3.5**.

▲ **GO FIGURE**

In the molecular level views shown in the figure how many C, H, and O atoms are present on the reactant side? Are the same number of each type of atom present on the product side?

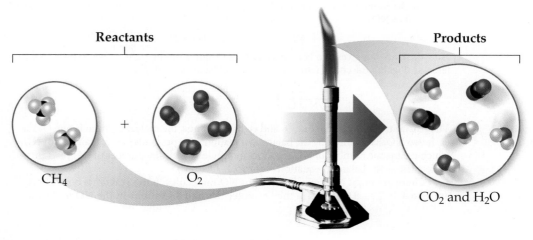

Reactants   Products

CH_4 O_2 CO_2 and H_2O

▲ Figure 3.4 **Methane reacts with oxygen in a Bunsen burner.**

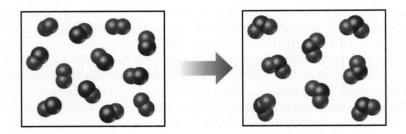

$$CH_4 \quad + \quad 2\,O_2 \quad \longrightarrow \quad CO_2 \quad + \quad 2\,H_2O$$

1 C, 4 H, 4 O   1 C, 4 H, 4 O

▲ **Figure 3.5 Balanced chemical equation for the combustion of CH₄.**

SAMPLE
EXERCISE 3.1 | **Interpreting and Balancing Chemical Equations**

The following diagram represents a chemical reaction in which the red spheres are oxygen atoms and the blue spheres are nitrogen atoms. (**a**) Write the chemical formulas for the reactants and products. (**b**) Write a balanced equation for the reaction. (**c**) Is the diagram consistent with the law of conservation of mass?

SOLUTION

(**a**) The left box, which represents reactants, contains two kinds of molecules, those composed of two oxygen atoms (O_2) and those composed of one nitrogen atom and one oxygen atom (NO). The right box, which represents products, contains only one kind of molecule, which is composed of one nitrogen atom and two oxygen atoms (NO_2).

(**b**) The unbalanced chemical equation is

$$O_2 + NO \longrightarrow NO_2 \quad (\text{unbalanced})$$

An inventory of atoms on each side of the equation shows that there are one N and three O on the left side of the arrow and one N and two O on the right. To balance O we must increase the number of O atoms on the right while keeping the coefficients for NO and NO_2 equal. Sometimes a trial-and-error approach is required; we need to go back and forth several times from one side of an equation to the other, changing coefficients first on one side of the equation and then the other until it is balanced. In our present case, let's start by increasing the number of O atoms on the right side of the equation by placing the coefficient 2 in front of NO_2:

$$O_2 + NO \longrightarrow 2\,NO_2 \quad (\text{unbalanced})$$

Now the equation gives two N atoms and four O atoms on the right, so we go back to the left side. Placing the coefficient 2 in front of NO balances both N and O:

$$O_2 + 2\,NO \longrightarrow 2\,NO_2 \quad (\text{balanced})$$
2 N, 4 O   2 N, 4 O

(**c**) The reactants box contains four O_2 and eight NO. Thus, the molecular ratio is one O_2 for each two NO, as required by the balanced equation. The products box contains eight NO_2, which means the number of NO_2 product molecules equals the number of NO reactant molecules, as the balanced equation requires.

There are eight N atoms in the eight NO molecules in the reactants box. There are also $4 \times 2 = 8$ O atoms in the O_2 molecules and 8 O atoms in the NO molecules, giving a total of 16 O atoms. In the products box, we find eight NO_2 molecules, which contain eight N atoms and $8 \times 2 = 16$ O atoms. Because there are equal numbers of N and O atoms in the two boxes, the drawing is consistent with the law of conservation of mass.

Practice Exercise 1

In the following diagram, the white spheres represent hydrogen atoms and the blue spheres represent nitrogen atoms.

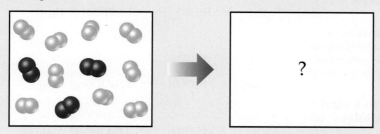

The two reactants combine to form a single product, ammonia, NH_3, which is not shown. Write a balanced chemical equation for the reaction. Based on the equation and the contents of the left (reactants) box, find how many NH_3 molecules should be shown in the right (products) box. **(a)** 2, **(b)** 3, **(c)** 4, **(d)** 6, **(e)** 9.

Practice Exercise 2

In the following diagram, the white spheres represent hydrogen atoms, the black spheres carbon atoms, and the red spheres oxygen atoms.

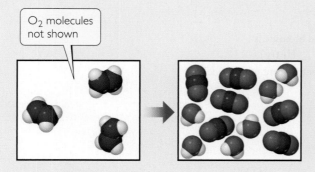

O_2 molecules not shown

In this reaction, there are two reactants, ethylene, C_2H_4, which is shown, and oxygen, O_2, which is not shown, and two products, CO_2 and H_2O, both of which are shown. **(a)** Write a balanced chemical equation for the reaction. **(b)** Determine the number of O_2 molecules that should be shown in the left (reactants) box.

Indicating the States of Reactants and Products

Symbols indicating the physical state of each reactant and product are often shown in chemical equations. We use the symbols (g), (l), (s), and (aq) for substances that are gases, liquids, solids, and dissolved in aqueous (water) solution, respectively. Thus, Equation 3.4 can be written

$$CH_4(g) + 2\,O_2(g) \longrightarrow CO_2(g) + 2\,H_2O(g) \qquad [3.5]$$

Sometimes symbols that represent the conditions under which the reaction proceeds appear above or below the reaction arrow. One example that we will encounter later in this chapter involves the symbol Δ (Greek uppercase delta); a Δ above the reaction arrow indicates the addition of heat.

SAMPLE EXERCISE 3.2 Balancing Chemical Equations

Balance the equation

$$Na(s) + H_2O(l) \longrightarrow NaOH(aq) + H_2(g)$$

SOLUTION

Begin by counting each kind of atom on the two sides of the arrow. There are one Na, one O, and two H on the left side, and one Na, one O, and three H on the right. The Na and O atoms are balanced, but the number of H atoms is not. To increase the number of H atoms on the left, let's try placing the coefficient 2 in front of H_2O:

$$Na(s) + 2\,H_2O(l) \longrightarrow NaOH(aq) + H_2(g)$$

Although beginning this way does not balance H, it does increase the number of reactant H atoms, which we need to do. (Also, adding the coefficient 2 on H_2O unbalances O, but we will take care of that after we balance H.) Now that we have 2 H_2O on the left, we balance H by putting the coefficient 2 in front of NaOH:

$$Na(s) + 2\,H_2O(l) \longrightarrow 2\,NaOH(aq) + H_2(g)$$

Balancing H in this way brings O into balance, but now Na is unbalanced, with one Na on the left and two on the right. To rebalance Na, we put the coefficient 2 in front of the reactant:

$$2\,Na(s) + 2\,H_2O(l) \longrightarrow 2\,NaOH(aq) + H_2(g)$$

We now have two Na atoms, four H atoms, and two O atoms on each side. The equation is balanced.

Comment Notice that we moved back and forth, placing a coefficient in front of H_2O, then NaOH, and finally Na. In balancing equations, we often find ourselves following this pattern of moving back and forth from one side of the arrow to the other, placing coefficients first in front of a formula on one side and then in front of a formula on the other side until the equation is balanced. You can always tell if you have balanced your equation correctly by checking that the number of atoms of each element is the same on the two sides of the arrow, and that you've chosen the smallest set of coefficients that balances the equation.

Practice Exercise 1

The unbalanced equation for the reaction between methane and bromine is

$$\_\_ CH_4(g) + \_\_ Br_2(l) \longrightarrow \_\_ CBr_4(s) + \_\_ HBr(g)$$

Once this equation is balanced what is the value of the coefficient in front of bromine Br_2?
(a) 1, **(b)** 2, **(c)** 3, **(d)** 4, **(e)** 6.

Practice Exercise 2

Balance these equations by providing the missing coefficients:
(a) $\_\_ Fe(s) + \_\_ O_2(g) \longrightarrow \_\_ Fe_2O_3(s)$
(b) $\_\_ Al(s) + \_\_ HCl(aq) \longrightarrow \_\_ AlCl_3(aq) + \_\_ H_2(g)$
(c) $\_\_ CaCO_3(s) + \_\_ HCl(aq) \longrightarrow \_\_ CaCl_2(aq) + \_\_ CO_2(g) + \_\_ H_2O(l)$

3.2 | Simple Patterns of Chemical Reactivity

In this section we examine three types of reactions that we see frequently throughout this chapter: combination reactions, decomposition reactions, and combustion reactions. Our first reason for examining these reactions is to become better acquainted with chemical reactions and their balanced equations. Our second reason is to consider how we might predict the products of some of these reactions knowing only their reactants. The key to predicting the products formed by a given combination of reactants is recognizing general patterns of chemical reactivity. Recognizing a pattern of reactivity for a class of substances gives you a broader understanding than merely memorizing a large number of unrelated reactions.

Combination and Decomposition Reactions

In **combination reactions** two or more substances react to form one product (▶ **Table 3.1**). For example, magnesium metal burns brilliantly in air to produce magnesium oxide (▶ **Figure 3.6**):

$$2\,Mg(s) + O_2(g) \longrightarrow 2\,MgO(s) \qquad\qquad [3.6]$$

This reaction is used to produce the bright flame generated by flares and some fireworks.

A combination reaction between a metal and a nonmetal, as in Equation 3.6, produces an ionic solid. Recall that the formula of an ionic compound can be determined from the charges of its ions. ∞ (Section 2.7) When magnesium reacts with oxygen, the magnesium loses electrons and forms the magnesium ion, Mg^{2+}. The oxygen gains electrons and forms the oxide ion, O^{2-}. Thus, the reaction product is MgO.

You should be able to recognize when a reaction is a combination reaction and to predict the products when the reactants are a metal and a nonmetal.

Table 3.1   Combination and Decomposition Reactions

Combination Reactions

$A + B \longrightarrow C$	Two or more reactants combine to form a single product. Many elements react with one another in this fashion to form compounds.
$C(s) + O_2(g) \longrightarrow CO_2(g)$	
$N_2(g) + 3 H_2(g) \longrightarrow 2 NH_3(g)$	
$CaO(s) + H_2O(l) \longrightarrow Ca(OH)_2(aq)$	

Decomposition Reactions

$C \longrightarrow A + B$	A single reactant breaks apart to form two or more substances. Many compounds react this way when heated.
$2 KClO_3(s) \longrightarrow 2 KCl(s) + 3 O_2(g)$	
$PbCO_3(s) \longrightarrow PbO(s) + CO_2(g)$	
$Cu(OH)_2(s) \longrightarrow CuO(s) + H_2O(g)$	

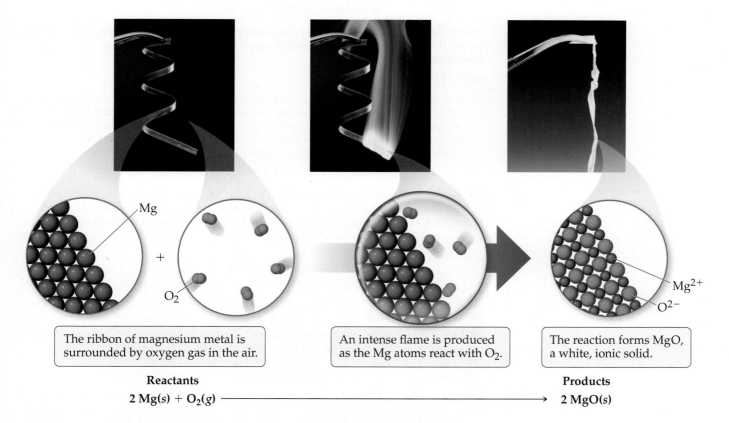

The ribbon of magnesium metal is surrounded by oxygen gas in the air.

An intense flame is produced as the Mg atoms react with O_2.

The reaction forms MgO, a white, ionic solid.

Reactants   Products

$2 Mg(s) + O_2(g) \longrightarrow 2 MgO(s)$

▲ Figure 3.6 Combustion of magnesium metal in air, a combination reaction.

> **Give It Some Thought**
> When Na and S undergo a combination reaction, what is the chemical formula of the product?

In a **decomposition reaction** one substance undergoes a reaction to produce two or more other substances (Table 3.1). For example, many metal carbonates decompose to form metal oxides and carbon dioxide when heated:

$$CaCO_3(s) \xrightarrow{\Delta} CaO(s) + CO_2(g) \qquad [3.7]$$

Decomposition of $CaCO_3$ is an important commercial process. Limestone or seashells, which are both primarily $CaCO_3$, are heated to prepare CaO, known as lime or quick-lime. Tens of millions of tons of CaO is used in the United States each year, in making glass, in metallurgy where it is used to isolate the metals from their ores, and in steel manufacturing where it is used to remove impurities.

The decomposition of sodium azide (NaN_3) rapidly releases $N_2(g)$, so this reaction is used to inflate safety air bags in automobiles (◀ Figure 3.7):

$$2\,NaN_3(s) \longrightarrow 2\,Na(s) + 3\,N_2(g) \qquad [3.8]$$

The system is designed so that an impact ignites a detonator cap, which in turn causes NaN_3 to decompose explosively. A small quantity of NaN_3 (about 100 g) forms a large quantity of gas (about 50 L).

▲ **Figure 3.7 Decomposition of sodium azide, $NaN_3(s)$, is used to inflate air bags in automobiles.**

SAMPLE EXERCISE 3.3   **Writing Balanced Equations for Combination and Decomposition Reactions**

Write a balanced equation for **(a)** the combination reaction between lithium metal and fluorine gas and **(b)** the decomposition reaction that occurs when solid barium carbonate is heated (two products form, a solid and a gas).

SOLUTION

(a) With the exception of mercury, all metals are solids at room temperature. Fluorine occurs as a diatomic molecule. Thus, the reactants are Li(s) and $F_2(g)$. The product will be composed of a metal and a nonmetal, so we expect it to be an ionic solid. Lithium ions have a 1+ charge, Li^+, whereas fluoride ions have a 1− charge, F^-. Thus, the chemical formula for the product is LiF. The balanced chemical equation is

$$2\,Li(s) + F_2(g) \longrightarrow 2\,LiF(s)$$

(b) The chemical formula for barium carbonate is $BaCO_3$. As mentioned, many metal carbonates decompose to metal oxides and carbon dioxide when heated. In Equation 3.7, for example, $CaCO_3$ decomposes to form CaO and CO_2. Thus, we expect $BaCO_3$ to decompose to BaO and CO_2. Barium and calcium are both in group 2A in the periodic table, which further suggests they react in the same way:

$$BaCO_3(s) \longrightarrow BaO(s) + CO_2(g)$$

Practice Exercise 1

Which of the following reactions is the balanced equation that represents the decomposition reaction that occurs when silver (I) oxide is heated? **(a)** $AgO(s) \longrightarrow Ag(s) + O(g)$; **(b)** $2\,AgO(s) \longrightarrow 2\,Ag(s) + O_2(g)$; **(c)** $Ag_2O(s) \longrightarrow 2\,Ag(s) + O(g)$; **(d)** $2\,Ag_2O(s) \longrightarrow 4\,Ag(s) + O_2(g)$; **(e)** $Ag_2O(s) \longrightarrow 2\,Ag(s) + O_2(g)$.

Practice Exercise 2

Write a balanced equation for **(a)** solid mercury (II) sulfide decomposing into its component elements when heated and **(b)** aluminum metal combining with oxygen in the air.

Combustion Reactions

Combustion reactions are rapid reactions that produce a flame. Most combustion reactions we observe involve O_2 from air as a reactant. Equation 3.5 illustrates a general class of reactions involving the burning, or combustion, of hydrocarbons (compounds that contain only carbon and hydrogen, such as CH_4 and C_2H_4). ∞ (Section 2.9)

Hydrocarbons combusted in air react with O_2 to form CO_2 and H_2O.* The number of molecules of O_2 required and the number of molecules of CO_2 and H_2O formed depend on the composition of the hydrocarbon, which acts as the fuel in the reaction. For example, the combustion of propane (C_3H_8, ▶ Figure 3.8), a gas used for cooking and home heating, is described by the equation

$$C_3H_8(g) + 5\,O_2(g) \longrightarrow 3\,CO_2(g) + 4\,H_2O(g) \qquad [3.9]$$

The state of the water in this reaction, $H_2O(g)$ or $H_2O(l)$, depends on the reaction conditions. Water vapor, $H_2O(g)$, is formed at high temperature in an open container.

Combustion of oxygen-containing derivatives of hydrocarbons, such as CH_3OH, also produces CO_2 and H_2O. The rule that hydrocarbons and their oxygen-containing derivatives form CO_2 and H_2O when they burn in air summarizes the reactions of about 3 million compounds with oxygen. Many substances that our bodies use as energy sources, such as the sugar glucose ($C_6H_{12}O_6$), react with O_2 to form CO_2 and H_2O. In our bodies, however, the reactions take place in a series of intermediate steps that occur at body temperature. These reactions that involve intermediate steps are described as *oxidation reactions* instead of combustion reactions.

 GO FIGURE

Does this reaction produce or consume thermal energy (heat)?

▲ Figure 3.8 **Propane burning in air.** Liquid propane in the tank, C_3H_8, vaporizes and mixes with air as it escapes through the nozzle. The combustion reaction of C_3H_8 and O_2 produces a blue flame.

SAMPLE EXERCISE 3.4 | Writing Balanced Equations for Combustion Reactions

Write the balanced equation for the reaction that occurs when methanol, $CH_3OH(l)$, is burned in air.

SOLUTION

When any compound containing C, H, and O is combusted, it reacts with the $O_2(g)$ in air to produce $CO_2(g)$ and $H_2O(g)$. Thus, the unbalanced equation is

$$CH_3OH(l) + O_2(g) \longrightarrow CO_2(g) + H_2O(g)$$

The C atoms are balanced, one on each side of the arrow. Because CH_3OH has four H atoms, we place the coefficient 2 in front of H_2O to balance the H atoms:

$$CH_3OH(l) + O_2(g) \longrightarrow CO_2(g) + 2\,H_2O(g)$$

Adding this coefficient balances H but gives four O atoms in the products. Because there are only three O atoms in the reactants, we are not finished. We can place the coefficient $\frac{3}{2}$ in front of O_2 to give four O atoms in the reactants ($\frac{3}{2} \times 2 = 3$ O atoms in $\frac{3}{2}O_2$):

$$CH_3OH(l) + \tfrac{3}{2}O_2(g) \longrightarrow CO_2(g) + 2\,H_2O(g)$$

Although this equation is balanced, it is not in its most conventional form because it contains a fractional coefficient. However,

multiplying through by 2 removes the fraction and keeps the equation balanced:

$$2\,CH_3OH(l) + 3\,O_2(g) \longrightarrow 2\,CO_2(g) + 4\,H_2O(g)$$

Practice Exercise 1

Write the balanced equation for the reaction that occurs when ethylene glycol, $C_2H_4(OH)_2$, burns in air.
(a) $C_2H_4(OH)_2(l) + 5/2\,O_2(g) \longrightarrow 2\,CO_2(g) + 3\,H_2O(g)$
(b) $2\,C_2H_4(OH)_2(l) + 5\,O_2(g) \longrightarrow 4\,CO_2(g) + 6\,H_2O(g)$
(c) $C_2H_4(OH)_2(l) + 3\,O_2(g) \longrightarrow 2\,CO_2(g) + 3\,H_2O(g)$
(d) $C_2H_4(OH)_2(l) + 5\,O(g) \longrightarrow 2\,CO_2(g) + 3\,H_2O(g)$
(e) $4\,C_2H_4(OH)_2(l) + 10\,O_2(g) \longrightarrow 8\,CO_2(g) + 12\,H_2O(g)$

Practice Exercise 2

Write the balanced equation for the reaction that occurs when ethanol, $C_2H_5OH(l)$, burns in air.

3.3 | Formula Weights

Chemical formulas and chemical equations both have a *quantitative* significance in that the subscripts in formulas and the coefficients in equations represent precise quantities. The formula H_2O indicates that a molecule of this substance (water) contains exactly two atoms of hydrogen and one atom of oxygen. Similarly, the coefficients in a balanced chemical equation indicate the relative quantities of reactants and products. But how do

---

*When there is an insufficient quantity of O_2 present, carbon monoxide (CO) is produced along with CO_2; this is called incomplete combustion. If the quantity of O_2 is severely restricted, the fine particles of carbon we call soot are produced. Complete combustion produces only CO_2 and H_2O. Unless stated to the contrary, we will always take combustion to mean complete combustion.

we relate the numbers of atoms or molecules to the amounts we measure in the laboratory? If you wanted to react hydrogen and oxygen in exactly the right ratio to make H_2O, how would you make sure the reactants contain a 2:1 ratio of hydrogen atoms to oxygen atoms?

It is not possible to count individual atoms or molecules, but we can indirectly determine their numbers if we know their masses. So, if we are to calculate amounts of reactants needed to obtain a given amount of product, or otherwise extrapolate quantitative information from a chemical equation or formula, we need to know more about the masses of atoms and molecules.

Formula and Molecular Weights

The **formula weight** (FW) of a substance is the sum of the atomic weights (AW) of the atoms in the chemical formula of the substance. Using atomic weights, we find, for example, that the formula weight of sulfuric acid (H_2SO_4) is 98.1 amu (atomic mass units):

$$\text{FW of } H_2SO_4 = 2(\text{AW of H}) + (\text{AW of S}) + 4(\text{AW of O})$$
$$= 2(1.0 \text{ amu}) + 32.1 \text{ amu} + 4(16.0 \text{ amu})$$
$$= 98.1 \text{ amu}$$

For convenience, we have rounded off the atomic weights to one decimal place, a practice we will follow in most calculations in this book.

If the chemical formula is the chemical symbol of an element, such as Na, the formula weight equals the atomic weight of the element, in this case 23.0 amu. If the chemical formula is that of a molecule, the formula weight is also called the **molecular weight** (MW). The molecular weight of glucose $(C_6H_{12}O_6)$, for example, is

$$\text{MW of } C_6H_{12}O_6 = 6(12.0 \text{ amu}) + 12(1.0 \text{ amu}) + 6(16.0 \text{ amu}) = 180.0 \text{ amu}$$

Because ionic substances exist as three-dimensional arrays of ions (see Figure 2.21), it is inappropriate to speak of molecules of these substances. Instead we use the empirical formula as the formula unit, and the formula weight of an ionic substance is determined by summing the atomic weights of the atoms in the empirical formula. For example, the formula unit of $CaCl_2$ consists of one Ca^{2+} ion and two Cl^- ions. Thus, the formula weight of $CaCl_2$ is

$$\text{FW of } CaCl_2 = 40.1 \text{ amu} + 2(35.5 \text{ amu}) = 111.1 \text{ amu}$$

SAMPLE EXERCISE 3.5 | Calculating Formula Weights

Calculate the formula weight of (**a**) sucrose, $C_{12}H_{22}O_{11}$ (table sugar); and (**b**) calcium nitrate, $Ca(NO_3)_2$.

SOLUTION

(**a**) By adding the atomic weights of the atoms in sucrose, we find the formula weight to be 342.0 amu:

$$12 \text{ C atoms} = 12(12.0 \text{ amu}) = 144.0 \text{ amu}$$
$$22 \text{ H atoms} = 22(1.0 \text{ amu}) = 22.0 \text{ amu}$$
$$11 \text{ O atoms} = 11(16.0 \text{ amu}) = \frac{176.0 \text{ amu}}{342.0 \text{ amu}}$$

(**b**) If a chemical formula has parentheses, the subscript outside the parentheses is a multiplier for all atoms inside. Thus, for $Ca(NO_3)_2$ we have

$$1 \text{ Ca atom} = 1(40.1 \text{ amu}) = 40.1 \text{ amu}$$
$$2 \text{ N atoms} = 2(14.0 \text{ amu}) = 28.0 \text{ amu}$$
$$6 \text{ O atoms} = 6(16.0 \text{ amu}) = \frac{96.0 \text{ amu}}{164.1 \text{ amu}}$$

Practice Exercise 1

Which of the following is the correct formula weight for calcium phosphate? (**a**) 310.2 amu, (**b**) 135.1 amu, (**c**) 182.2 amu, (**d**) 278.2 amu, (**e**) 175.1 amu.

Practice Exercise 2

Calculate the formula weight of (**a**) $Al(OH)_3$, (**b**) CH_3OH, and (**c**) TaON.

Percentage Composition from Chemical Formulas

Chemists must sometimes calculate the *percentage composition* of a compound—that is, the percentage by mass contributed by each element in the substance. Forensic chemists, for example, can measure the percentage composition of an unknown powder and compare it with the percentage compositions of suspected substances (for example, sugar, salt, or cocaine) to identify the powder.

Calculating the percentage composition of any element in a substance (sometimes called the **elemental composition** of a substance) is straightforward if the chemical formula is known. The calculation depends on the formula weight of the substance, the atomic weight of the element of interest, and the number of atoms of that element in the chemical formula:

$$\% \text{ composition of element} = \frac{\left(\begin{array}{c}\text{number of atoms}\\\text{of element}\end{array}\right)\left(\begin{array}{c}\text{atomic weight}\\\text{of element}\end{array}\right)}{\text{formula weight of substance}} \times 100\% \quad [3.10]$$

SAMPLE EXERCISE 3.6 | Calculating Percentage Composition

Calculate the percentage of carbon, hydrogen, and oxygen (by mass) in $C_{12}H_{22}O_{11}$.

SOLUTION

Let's examine this question using the problem-solving steps in the accompanying "Strategies in Chemistry: Problem Solving" essay.

Analyze We are given a chemical formula and asked to calculate the percentage by mass of each element.

Plan We use Equation 3.10, obtaining our atomic weights from a periodic table. We know the denominator in Equation 3.10, the formula weight of $C_{12}H_{22}O_{11}$, from Sample Exercise 3.5. We must use that value in three calculations, one for each element.

Solve

$$\%C = \frac{(12)(12.0 \text{ amu})}{342.0 \text{ amu}} \times 100\% = 42.1\%$$

$$\%H = \frac{(22)(1.0 \text{ amu})}{342.0 \text{ amu}} \times 100\% = 6.4\%$$

$$\%O = \frac{(11)(16.0 \text{ amu})}{342.0 \text{ amu}} \times 100\% = 51.5\%$$

Check Our calculated percentages must add up to 100%, which they do. We could have used more significant figures for our atomic weights, giving more significant figures for our percentage composition, but we have adhered to our suggested guideline of rounding atomic weights to one digit beyond the decimal point.

Practice Exercise 1

What is the percentage of nitrogen, by mass, in calcium nitrate? (a) 8.54%, (b) 17.1%, (c) 13.7%, (d) 24.4%, (e) 82.9%.

Practice Exercise 2

Calculate the percentage of potassium, by mass, in K_2PtCl_6.

3.4 | Avogadro's Number and the Mole

Even the smallest samples we deal with in the laboratory contain enormous numbers of atoms, ions, or molecules. For example, a teaspoon of water (about 5 mL) contains 2×10^{23} water molecules, a number so large it almost defies comprehension. Chemists therefore have devised a counting unit for describing large numbers of atoms or molecules.

Strategies in Chemistry

Problem Solving

Practice is the key to success in solving problems. As you practice, you can improve your skills by following these steps:

1. **Analyze the problem.** Read the problem carefully. What does it say? Draw a picture or diagram that will help you to visualize the problem. Write down both the data you are given and the quantity you need to obtain (the unknown).

2. **Develop a plan for solving the problem.** Consider a possible path between the given information and the unknown. What principles or equations relate the known data to the unknown? Recognize that some data may not be given explicitly in the problem; you may be expected to know certain quantities (such as Avogadro's number) or look them up in tables (such as atomic weights). Recognize also that your plan may involve either a single step or a series of steps with intermediate answers.

3. **Solve the problem.** Use the known information and suitable equations or relationships to solve for the unknown. Dimensional analysis ∞ (Section 1.6) is a useful tool for solving a great number of problems. Be careful with significant figures, signs, and units.

4. **Check the solution.** Read the problem again to make sure you have found all the solutions asked for in the problem. Does your answer make sense? That is, is the answer outrageously large or small or is it in the ballpark? Finally, are the units and significant figures correct?

In everyday life we use such familiar counting units as dozen (12 objects) and gross (144 objects). In chemistry the counting unit for numbers of atoms, ions, or molecules in a laboratory-size sample is the *mole*, abbreviated mol. One **mole** is the amount of matter that contains as many objects (atoms, molecules, or whatever other objects we are considering) as the number of atoms in exactly 12 g of isotopically pure ^{12}C. From experiments, scientists have determined this number to be $6.02214129 \times 10^{23}$, which we usually round to 6.02×10^{23}. Scientists call this value **Avogadro's number**, N_A, in honor of the Italian scientist Amedeo Avogadro (1776–1856), and it is often cited with units of reciprocal moles, $6.02 \times 10^{23} \text{ mol}^{-1}$.* The unit (read as either "inverse mole" or "per mole") reminds us that there are 6.02×10^{23} objects per one mole. A mole of atoms, a mole of molecules, or a mole of anything else all contain Avogadro's number of objects:

$$1 \text{ mol } ^{12}C \text{ atoms} = 6.02 \times 10^{23} \, ^{12}C \text{ atoms}$$

$$1 \text{ mol } H_2O \text{ molecules} = 6.02 \times 10^{23} \, H_2O \text{ molecules}$$

$$1 \text{ mol } NO_3^- \text{ ions} = 6.02 \times 10^{23} \, NO_3^- \text{ ions}$$

Avogadro's number is so large that it is difficult to imagine. Spreading 6.02×10^{23} marbles over Earth's surface would produce a layer about 3 miles thick. Avogadro's number of pennies placed side by side in a straight line would encircle Earth 300 trillion (3×10^{14}) times.

SAMPLE EXERCISE 3.7 Estimating Numbers of Atoms

Without using a calculator, arrange these samples in order of increasing numbers of carbon atoms: 12 g ^{12}C, 1 mol C_2H_2, 9×10^{23} molecules of CO_2.

SOLUTION

Analyze We are given amounts of three substances expressed in grams, moles, and number of molecules and asked to arrange the samples in order of increasing numbers of C atoms.

Plan To determine the number of C atoms in each sample, we must convert 12 g ^{12}C, 1 mol C_2H_2, and 9×10^{23} molecules CO_2 to numbers of C atoms. To make these conversions, we use the definition of mole and Avogadro's number.

Solve One mole is defined as the amount of matter that contains as many units of the matter as there are C atoms in exactly 12 g of ^{12}C. Thus, 12 g of ^{12}C contains 1 mol of C atoms = 6.02×10^{23} C atoms. One mol of C_2H_2 contains 6.02×10^{23} C_2H_2 molecules. Because there are two C atoms in each molecule, this sample contains 12.04×10^{23} C atoms. Because each CO_2 molecule contains one C atom, the CO_2 sample contains 9×10^{23} C atoms. Hence, the order is 12 g ^{12}C (6×10^{23} C atoms) < 9×10^{23} CO_2 molecules (9×10^{23} C atoms) < 1 mol C_2H_2 (12×10^{23} C atoms).

Check We can check our results by comparing numbers of moles of C atoms in the samples because the number of moles is proportional to the number of atoms. Thus, 12 g of ^{12}C is 1 mol C, 1 mol of C_2H_2 contains 2 mol C, and 9×10^{23} molecules of CO_2 contain 1.5 mol C, giving the same order as stated previously.

Practice Exercise 1

Determine which of the following samples contains the fewest sodium atoms? (**a**) 1 mol sodium oxide, (**b**) 45 g sodium fluoride, (**c**) 50 g sodium chloride, (**d**) 1 mol sodium nitrate?

Practice Exercise 2

Without using a calculator, arrange these samples in order of increasing numbers of O atoms: 1 mol H_2O, 1 mol CO_2, 3×10^{23} molecules of O_3.

---

*Avogadro's number is also referred to as the Avogadro constant. The latter term is the name adopted by agencies such as the National Institute of Standards and Technology (NIST), but Avogadro's number remains in widespread usage and is used in most places in this book.

SAMPLE
EXERCISE 3.8 Converting Moles to Number of Atoms

Calculate the number of H atoms in 0.350 mol of $C_6H_{12}O_6$.

SOLUTION

Analyze We are given the amount of a substance (0.350 mol) and its chemical formula $C_6H_{12}O_6$. The unknown is the number of H atoms in the sample.

Plan Avogadro's number provides the conversion factor between number of moles of $C_6H_{12}O_6$ and number of molecules of $C_6H_{12}O_6$: 1 mol $C_6H_{12}O_6$ = 6.02 × 10^{23} molecules of $C_6H_{12}O_6$. Once we know the number of molecules of $C_6H_{12}O_6$, we can use the chemical formula, which tells us that each molecule of $C_6H_{12}O_6$ contains 12 H atoms. Thus, we convert moles of $C_6H_{12}O_6$ to molecules of $C_6H_{12}O_6$ and then determine the number of atoms of H from the number of molecules of $C_6H_{12}O_6$:

$$\text{Moles } C_6H_{12}O_6 \longrightarrow \text{molecules } C_6H_{12}O_6 \longrightarrow \text{atoms H}$$

Solve

$$\text{H atoms} = (0.350 \text{ mol } C_6H_{12}O_6)\left(\frac{6.02 \times 10^{23} \text{ molecules } C_6H_{12}O_6}{1 \text{ mol } C_6H_{12}O_6}\right)\left(\frac{12 \text{ H atoms}}{1 \text{ molecule } C_6H_{12}O_6}\right)$$

$$= 2.53 \times 10^{24} \text{ H atoms}$$

Check We can do a ballpark calculation, figuring that $0.35(6 \times 10^{23})$ is about 2×10^{23} molecules of $C_6H_{12}O_6$. We know that each one of these molecules contains 12 H atoms. $12(2 \times 10^{23})$ gives $24 \times 10^{23} = 2.4 \times 10^{24}$ H atoms, which is close to our result. Because we were asked for the number of H atoms, the units of our answer are correct. We check, too, for significant figures. The given data had three significant figures, as does our answer.

> **Practice Exercise 1**
> How many sulfur atoms are in (**a**) 0.45 mol $BaSO_4$ and (**b**) 1.10 mol of aluminum sulfide?
>
> **Practice Exercise 2**
> How many oxygen atoms are in (**a**) 0.25 mol $Ca(NO_3)_2$ and (**b**) 1.50 mol of sodium carbonate?

Molar Mass

A dozen is the same number, 12, whether we have a dozen eggs or a dozen elephants. Clearly, however, a dozen eggs does not have the same mass as a dozen elephants. Similarly, a mole is always the *same number* (6.02 × 10^{23}), but 1-mol samples of different substances have *different masses*. Compare, for example, 1 mol of ^{12}C and 1 mol of ^{24}Mg. A single ^{12}C atom has a mass of 12 amu, whereas a single ^{24}Mg atom is twice as massive, 24 amu (to two significant figures). Because a mole of anything always contains the same number of particles, a mole of ^{24}Mg must be twice as massive as a mole of ^{12}C. Because a mole of ^{12}C has a mass of 12 g (by definition), a mole of ^{24}Mg must have a mass of 24 g. This example illustrates a general rule relating the mass of an atom to the mass of Avogadro's number (1 mol) of these atoms: *The atomic weight of an element in atomic mass units is numerically equal to the mass in grams of 1 mol of that element.* For example (the symbol ⇒ means therefore)

Cl has an atomic weight of 35.5 amu ⇒ 1 mol Cl has a mass of 35.5 g.

Au has an atomic weight of 197 amu ⇒ 1 mol Au has a mass of 197 g.

For other kinds of substances, the same numerical relationship exists between formula weight and mass of 1 mol of a substance:

H_2O has a formula weight of 18.0 amu ⇒ 1 mol H_2O has a mass of 18.0 g (► **Figure 3.9**).

NaCl has a formula weight of 58.5 amu ⇒ 1 mol NaCl has a mass of 58.5 g.

▲ GO FIGURE

How many H_2O molecules are in a 9.00-g sample of water?

Single molecule

1 molecule H_2O
(18.0 amu)

Avogadro's number of water molecules in a mole of water.

Laboratory-size sample

1 mol H_2O
(18.0 g)

▲ Figure 3.9 Comparing the mass of 1 molecule and 1 mol of H_2O. Both masses have the same number but different units (atomic mass units and grams). Expressing both masses in grams indicates their huge difference: 1 molecule of H_2O has a mass of 2.99×10^{-23} g, whereas 1 mol H_2O has a mass of 18.0 g.

Give It Some Thought

(a) Which has more mass, a mole of water (H_2O) or a mole of glucose ($C_6H_{12}O_6$)?

(b) Which contains more molecules, a mole of water or a mole of glucose?

The mass in grams of one mole, often abbreviated as 1 mol, of a substance (that is, the mass in grams per mole) is called the **molar mass** of the substance. *The molar mass in grams per mole of any substance is numerically equal to its formula weight in atomic mass units.* For NaCl, for example, the formula weight is 58.5 amu and the molar mass is 58.5 g/mol. Mole relationships for several other substances are shown in ▼ Table 3.2, and ▼ Figure 3.10 shows 1 mol quantities of three common substances.

The entries in Table 3.2 for N and N_2 point out the importance of stating the chemical form of a substance when using the mole concept. Suppose you read that 1 mol of nitrogen is produced in a particular reaction. You might interpret this statement to mean 1 mol of nitrogen atoms (14.0 g). Unless otherwise stated, however, what is probably meant is 1 mol of nitrogen molecules, N_2 (28.0 g), because N_2 is the most

Table 3.2   Mole Relationships

Name of Substance	Formula	Formula Weight (amu)	Molar Mass (g/mol)	Number and Kind of Particles in One Mole
Atomic nitrogen	N	14.0	14.0	6.02×10^{23} N atoms
Molecular nitrogen	N_2	28.0	28.0	$\begin{cases} 6.02 \times 10^{23} \text{ } N_2 \text{ molecules} \\ 2(6.02 \times 10^{23}) \text{ N atoms} \end{cases}$
Silver	Ag	107.9	107.9	6.02×10^{23} Ag atoms
Silver ions	Ag^+	107.9[a]	107.9	6.02×10^{23} Ag^+ ions
Barium chloride	$BaCl_2$	208.2	208.2	$\begin{cases} 6.02 \times 10^{23} \text{ } BaCl_2 \text{ formula units} \\ 6.02 \times 10^{23} \text{ } Ba^{2+} \text{ ions} \\ 2(6.02 \times 10^{23}) \text{ } Cl^- \text{ ions} \end{cases}$

[a]Recall that the mass of an electron is more than 1800 times smaller than the masses of the proton and the neutron; thus, ions and atoms have essentially the same mass.

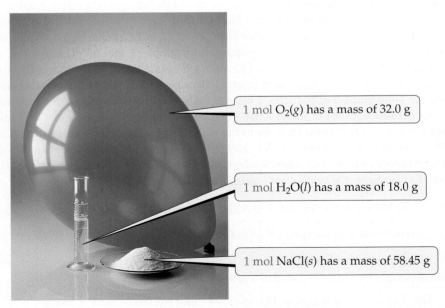

1 mol $O_2(g)$ has a mass of 32.0 g

1 mol $H_2O(l)$ has a mass of 18.0 g

1 mol NaCl(s) has a mass of 58.45 g

▲ **Figure 3.10 One mole each of a solid (NaCl), a liquid (H_2O), and a gas (O_2).** In each case, the mass in grams of 1 mol—that is, the molar mass—is numerically equal to the formula weight in atomic mass units. Each of these samples contains 6.02×10^{23} formula units.

common chemical form of the element. To avoid ambiguity, it is important to state explicitly the chemical form being discussed. Using the chemical formula—N or N_2, for instance—avoids ambiguity.

SAMPLE EXERCISE 3.9 | Calculating Molar Mass

What is the molar mass of glucose, $C_6H_{12}O_6$?

SOLUTION

Analyze We are given a chemical formula and asked to determine its molar mass.

Plan Because the molar mass of any substance is numerically equal to its formula weight, we first determine the formula weight of glucose by adding the atomic weights of its component atoms. The formula weight will have units of amu, whereas the molar mass has units of grams per mole (g/mol).

Solve Our first step is to determine the formula weight of glucose:

$$
\begin{aligned}
6\text{ C atoms} &= 6(12.0\text{ amu}) = 72.0\text{ amu} \\
12\text{ H atoms} &= 12(1.0\text{ amu}) = 12.0\text{ amu} \\
6\text{ O atoms} &= 6(16.0\text{ amu}) = \underline{96.0\text{ amu}} \\
& \phantom{= 6(16.0\text{ amu}) =} 180.0\text{ amu}
\end{aligned}
$$

Because glucose has a formula weight of 180.0 amu, 1 mol of this substance (6.02×10^{23} molecules) has a mass of 180.0 g. In other words, $C_6H_{12}O_6$ has a molar mass of 180.0 g/mol.

Check A molar mass below 250 seems reasonable based on the earlier examples we have encountered, and grams per mole is the appropriate unit for the molar mass.

Practice Exercise 1

A sample of an ionic compound containing iron and chlorine is analyzed and found to have a molar mass of 126.8 g/mol. What is the charge of the iron in this compound? (a) 1+, (b) 2+, (c) 3+, (d) 4+.

Practice Exercise 2

Calculate the molar mass of $Ca(NO_3)_2$.

Chemistry and Life

Glucose Monitoring

Our body converts most of the food we eat into glucose. After digestion, glucose is delivered to cells via the blood. Cells need glucose to live, and the hormone insulin must be present in order for glucose to enter the cells. Normally, the body adjusts the concentration of insulin automatically, in concert with the glucose concentration after eating. However, in a diabetic person, either little or no insulin is produced (Type 1 diabetes) or insulin is produced but the cells cannot take it up properly (Type 2 diabetes). In either case the blood glucose levels are higher than they are in a normal person, typically 70–120 mg/dL. A person who has not eaten for 8 hours or more is diagnosed as diabetic if his or her glucose level is 126 mg/dL or higher.

Glucose meters work by the introduction of blood from a person, usually by a prick of the finger, onto a small strip of paper that contains chemicals that react with glucose. Insertion of the strip into a small battery-operated reader gives the glucose concentration (▼ Figure 3.11). The mechanism of the readout varies from one monitor to another—it may be a measurement of a small electrical current or measurement of light produced in a chemical reaction. Depending on the reading on any given day, a diabetic person may need to receive an injection of insulin or simply limit his or her intake of sugar-rich foods for a while.

▲ Figure 3.11 **Glucose meter.**

Interconverting Masses and Moles

Conversions of mass to moles and of moles to mass are frequently encountered in calculations using the mole concept. These calculations are simplified using dimensional analysis ∞ (Section 1.6), as shown in Sample Exercises 3.10 and 3.11.

SAMPLE EXERCISE 3.10 | Converting Grams to Moles

Calculate the number of moles of glucose $(C_6H_{12}O_6)$ in 5.380 g of $C_6H_{12}O_6$.

SOLUTION

Analyze We are given the number of grams of a substance and its chemical formula and asked to calculate the number of moles.

Plan The molar mass of a substance provides the factor for converting grams to moles. The molar mass of $C_6H_{12}O_6$ is 180.0 g/mol (Sample Exercise 3.9).

Solve Using 1 mol $C_6H_{12}O_6$ = 180.0 g $C_6H_{12}O_6$ to write the appropriate conversion factor, we have

$$\text{Moles } C_6H_{12}O_6 = (5.380 \text{ g } C_6H_{12}O_6)\left(\frac{1 \text{ mol } C_6H_{12}O_6}{180.0 \text{ g } C_6H_{12}O_6}\right) = 0.02989 \text{ mol } C_6H_{12}O_6$$

Check Because 5.380 g is less than the molar mass, an answer less than 1 mol is reasonable. The unit mol is appropriate. The original data had four significant figures, so our answer has four significant figures.

Practice Exercise 1

How many moles of sodium bicarbonate $(NaHCO_3)$ are in 508 g of $NaHCO_3$?

Practice Exercise 2

How many moles of water are in 1.00 L of water, whose density is 1.00 g/mL?

SAMPLE EXERCISE 3.11 | Converting Moles to Grams

Calculate the mass, in grams, of 0.433 mol of calcium nitrate.

SOLUTION

Analyze We are given the number of moles and the name of a substance and asked to calculate the number of grams in the substance.

Plan To convert moles to grams, we need the molar mass, which we can calculate using the chemical formula and atomic weights.

Solve Because the calcium ion is Ca^{2+} and the nitrate ion is NO_3^-, the chemical formula for calcium nitrate is $Ca(NO_3)_2$. Adding the atomic weights of the elements in the compound gives a formula weight of 164.1 amu. Using 1 mol $Ca(NO_3)_2$ = 164.1 g $Ca(NO_3)_2$ to write the appropriate conversion factor, we have

$$\text{Grams } Ca(NO_3)_2 = (0.433 \text{ mol } Ca(NO_3)_2)\left(\frac{164.1 \text{ g } Ca(NO_3)_2}{1 \text{ mol } Ca(NO_3)_2}\right) = 71.1 \text{ g } Ca(NO_3)_2$$

Check The number of moles is less than 1, so the number of grams must be less than the molar mass, 164.1 g. Using rounded numbers to estimate, we have $0.5 \times 150 = 75$ g, which means the magnitude of our answer is reasonable. Both the units (g) and the number of significant figures (3) are correct.

Practice Exercise 1

What is the mass, in grams, of (a) 6.33 mol of $NaHCO_3$ and (b) 3.0×10^{-5} mol of sulfuric acid?

Practice Exercise 2

What is the mass, in grams, of (a) 0.50 mol of diamond (C) and (b) 0.155 mol of ammonium chloride?

Interconverting Masses and Numbers of Particles

The mole concept provides the bridge between mass and number of particles. To illustrate how this bridge works, let's calculate the number of copper atoms in an old copper penny. Such a penny has a mass of about 3 g, and for this illustration we will assume it is 100% copper:

GO FIGURE

What number would you use to convert **(a)** moles of CH_4 to grams of CH_4 and **(b)** number of molecules of CH_4 to moles of CH_4?

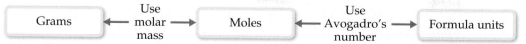

▲ **Figure 3.12 Procedure for interconverting mass and number of formula units.** The number of moles of the substance is central to the calculation. Thus, the mole concept can be thought of as the bridge between the mass of a sample in grams and the number of formula units contained in the sample.

$$\text{Cu atoms} = (3 \text{ g Cu})\left(\frac{1 \text{ mol Cu}}{63.5 \text{ g Cu}}\right)\left(\frac{6.02 \times 10^{23} \text{ Cu atoms}}{1 \text{ mol Cu}}\right)$$

$$= 3 \times 10^{22} \text{ Cu atoms}$$

We have rounded our answer to one significant figure because we used only one significant figure for the mass of the penny. Notice how dimensional analysis provides a straightforward route from grams to numbers of atoms. The molar mass and Avogadro's number are used as conversion factors to convert grams to moles and then moles to atoms. Notice also that our answer is a very large number. Any time you calculate the number of atoms, molecules, or ions in an ordinary sample of matter, you can expect the answer to be very large. In contrast, the number of moles in a sample will usually be small, often less than 1.

The general procedure for interconverting mass and number of formula units (atoms, molecules, ions, or whatever else is represented by the chemical formula) is summarized in ▲ **Figure 3.12.**

SAMPLE EXERCISE 3.12 | Calculating Numbers of Molecules and Atoms from Mass

(a) How many glucose molecules are in 5.23 g of $C_6H_{12}O_6$?

(b) How many oxygen atoms are in this sample?

SOLUTION

Analyze We are given the number of grams and the chemical formula of a substance and asked to calculate **(a)** the number of molecules and **(b)** the number of O atoms in the substance.

Plan (a) The strategy for determining the number of molecules in a given quantity of a substance is summarized in Figure 3.12. We must convert 5.23 g to moles of $C_6H_{12}O_6$ and then convert moles to molecules of $C_6H_{12}O_6$. The first conversion uses the molar mass of $C_6H_{12}O_6$, 180.0 g, and the second conversion uses Avogadro's number.

Solve Molecules $C_6H_{12}O_6$

$$= (5.23 \text{ g } C_6H_{12}O_6)\left(\frac{1 \text{ mol } C_6H_{12}O_6}{180.0 \text{ g } C_6H_{12}O_6}\right)\left(\frac{6.02 \times 10^{23} \text{ molecules } C_6H_{12}O_6}{1 \text{ mol } C_6H_{12}O_6}\right)$$

$$= 1.75 \times 10^{22} \text{ molecules } C_6H_{12}O_6$$

Check Because the mass we began with is less than a mole, there should be fewer than 6.02×10^{23} molecules in the sample, which means the magnitude of our answer is reasonable. A ballpark estimate of the answer comes reasonably close to the answer we derived in this exercise: $5/200 = 2.5 \times 10^{-2}$ mol; $(2.5 \times 10^{-2})(6 \times 10^{23}) = 15 \times 10^{21} = 1.5 \times 10^{22}$ molecules. The units (molecules) and the number of significant figures (three) are appropriate.

Plan (b) To determine the number of O atoms, we use the fact that there are six O atoms in each $C_6H_{12}O_6$ molecule. Thus, multiplying the number of molecules we calculated in **(a)** by the factor (6 atoms $O/1$ molecule $C_6H_{12}O_6$) gives the number of O atoms.

Solve

$$\text{Atoms O} = (1.75 \times 10^{22} \text{ molecules } C_6H_{12}O_6)\left(\frac{6 \text{ atoms O}}{\text{molecule } C_6H_{12}O_6}\right)$$

$$= 1.05 \times 10^{23} \text{ atoms O}$$

Check The answer is six times as large as the answer to part (**a**), exactly what it should be. The number of significant figures (three) and the units (atoms O) are correct.

> **Practice Exercise 1**
>
> How many chlorine atoms are in 12.2 g of CCl_4? (**a**) 4.77×10^{22}, (**b**) 7.34×10^{24}, (**c**) 1.91×10^{23}, (**d**) 2.07×10^{23}.
>
> **Practice Exercise 2**
>
> (**a**) How many nitric acid molecules are in 4.20 g of HNO_3?
> (**b**) How many O atoms are in this sample?

3.5 | Empirical Formulas from Analyses

As we learned in Section 2.6, the empirical formula for a substance tells us the relative number of atoms of each element in the substance. The empirical formula H_2O shows that water contains two H atoms for each O atom. This ratio also applies on the molar level: 1 mol of H_2O contains 2 mol of H atoms and 1 mol of O atoms. Conversely, *the ratio of the numbers of moles of all elements in a compound gives the subscripts in the compound's empirical formula.* Thus, the mole concept provides a way of calculating empirical formulas.

Mercury and chlorine, for example, combine to form a compound that is measured to be 74.0% mercury and 26.0% chlorine by mass. Thus, if we had a 100.0-g sample of the compound, it would contain 74.0 g of mercury and 26.0 g of chlorine. (Samples of any size can be used in problems of this type, but we will generally use 100.0 g to simplify the calculation of mass from percentage.) Using atomic weights to get molar masses, we can calculate the number of moles of each element in the sample:

$$(74.0 \text{ g Hg})\left(\frac{1 \text{ mol Hg}}{200.6 \text{ g Hg}}\right) = 0.369 \text{ mol Hg}$$

$$(26.0 \text{ g Cl})\left(\frac{1 \text{ mol Cl}}{35.5 \text{ g Cl}}\right) = 0.732 \text{ mol Cl}$$

We then divide the larger number of moles by the smaller number to obtain the Cl:Hg mole ratio:

$$\frac{\text{moles of Cl}}{\text{moles of Hg}} = \frac{0.732 \text{ mol Cl}}{0.369 \text{ mol Hg}} = \frac{1.98 \text{ mol Cl}}{1 \text{ mol Hg}}$$

Because of experimental errors, calculated values for a mole ratio may not be whole numbers, as in the calculation here. The number 1.98 is very close to 2, however, and so we can confidently conclude that the empirical formula for the compound is $HgCl_2$. The empirical formula is correct because its subscripts are the smallest integers that express the *ratio* of atoms present in the compound. ∞ (Section 2.6)

The general procedure for determining empirical formulas is outlined in ▼ Figure 3.13.

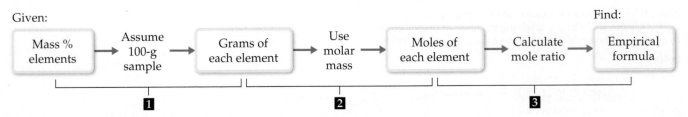

▲ **Figure 3.13 Procedure for calculating an empirical formula from percentage composition.**

⚠ Give It Some Thought
Could the empirical formula determined from chemical analysis be used to tell the difference between acetylene, C_2H_2, and benzene, C_6H_6?

SAMPLE EXERCISE 3.13 | Calculating an Empirical Formula

Ascorbic acid (vitamin C) contains 40.92% C, 4.58% H, and 54.50% O by mass. What is the empirical formula of ascorbic acid?

SOLUTION

Analyze We are to determine the empirical formula of a compound from the mass percentages of its elements.

Plan The strategy for determining the empirical formula involves the three steps given in Figure 3.13.

Solve

(1) For simplicity we assume we have exactly 100 g of material, although any other mass could also be used.

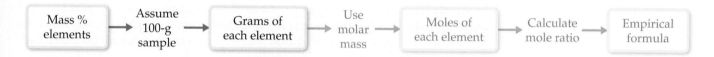

In 100.00 g of ascorbic acid we have 40.92 g C, 4.58 g H, and 54.50 g O.
(2) Next we calculate the number of moles of each element. We use atomic masses with four significant figures to match the precision of our experimental masses.

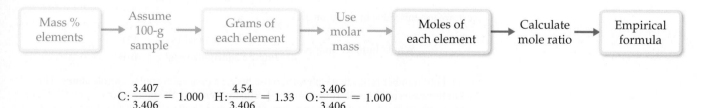

$$\text{Moles C} = (40.92 \text{ g C})\left(\frac{1 \text{ mol C}}{12.01 \text{ g C}}\right) = 3.407 \text{ mol C}$$

$$\text{Moles H} = (4.58 \text{ g H})\left(\frac{1 \text{ mol C}}{1.008 \text{ g H}}\right) = 4.54 \text{ mol H}$$

$$\text{Moles O} = (54.50 \text{ g O})\left(\frac{1 \text{ mol O}}{16.00 \text{ g O}}\right) = 3.406 \text{ mol O}$$

(3) We determine the simplest whole-number ratio of moles by dividing each number of moles by the smallest number of moles.

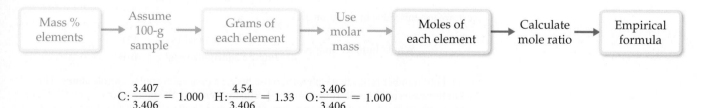

$$\text{C:}\frac{3.407}{3.406} = 1.000 \quad \text{H:}\frac{4.54}{3.406} = 1.33 \quad \text{O:}\frac{3.406}{3.406} = 1.000$$

The ratio for H is too far from 1 to attribute the difference to experimental error; in fact, it is quite close to $1\frac{1}{3}$. This suggests we should multiply the ratios by 3 to obtain whole numbers:

$$C:H:O = (3 \times 1 : 3 \times 1.33 : 3 \times 1) = (3 : 4 : 3)$$

Thus, the empirical formula is $C_3H_4O_3$.

Check It is reassuring that the subscripts are moderate-size whole numbers. Also, calculating the percentage composition of $C_3H_4O_3$ gives values very close to the original percentages.

Practice Exercise 1

A 2.144-g sample of phosgene, a compound used as a chemical warfare agent during World War I, contains 0.260 g of carbon, 0.347 g of oxygen, and 1.537 g of chlorine. What is the empirical formula of this substance? (a) CO_2Cl_6, (b) $COCl_2$, (c) $C_{0.022}O_{0.022}Cl_{0.044}$, (d) C_2OCl_2

Practice Exercise 2

A 5.325-g sample of methyl benzoate, a compound used in the manufacture of perfumes, contains 3.758 g of carbon, 0.316 g of hydrogen, and 1.251 g of oxygen. What is the empirical formula of this substance?

Molecular Formulas from Empirical Formulas

We can obtain the molecular formula for any compound from its empirical formula if we know either the molecular weight or the molar mass of the compound. *The subscripts in the molecular formula of a substance are always whole-number multiples of the subscripts in its empirical formula.* ∞ (Section 2.6) This multiple can be found by dividing the molecular weight by the empirical formula weight:

$$\text{Whole-number multiple} = \frac{\text{molecular weight}}{\text{empirical formula weight}} \qquad [3.11]$$

In Sample Exercise 3.13, for example, the empirical formula of ascorbic acid was determined to be $C_3H_4O_3$. This means the empirical formula weight is 3(12.0 amu) + 4(1.0 amu) + 3(16.0 amu) = 88.0 amu. The experimentally determined molecular weight is 176 amu. Thus, we find the whole-number multiple that converts the empirical formula to the molecular formula by dividing

$$\text{Whole-number multiple} = \frac{\text{molecular weight}}{\text{empirical formula weight}} = \frac{176\ \text{amu}}{88.0\ \text{amu}} = 2$$

Consequently, we multiply the subscripts in the empirical formula by this multiple, giving the molecular formula: $C_6H_8O_6$.

SAMPLE EXERCISE 3.14 | Determining a Molecular Formula

Mesitylene, a hydrocarbon found in crude oil, has an empirical formula of C_3H_4 and an experimentally determined molecular weight of 121 amu. What is its molecular formula?

SOLUTION

Analyze We are given an empirical formula and a molecular weight of a compound and asked to determine its molecular formula.

Plan The subscripts in a compound's molecular formula are whole-number multiples of the subscripts in its empirical formula. We find the appropriate multiple by using Equation 3.11.

Solve The formula weight of the empirical formula C_3H_4 is

$$3(12.0\ \text{amu}) + 4(1.0\ \text{amu}) = 40.0\ \text{amu}$$

Next, we use this value in Equation 3.11:

$$\text{Whole-number multiple} = \frac{\text{molecular weight}}{\text{empirical formula weight}} = \frac{121}{40.0} = 3.03$$

Only whole-number ratios make physical sense because molecules contain whole atoms. The 3.03 in this case could result from a small experimental error in the molecular weight. We therefore multiply each subscript in the empirical formula by 3 to give the molecular formula: C_9H_{12}.

Check We can have confidence in the result because dividing molecular weight by empirical formula weight yields nearly a whole number.

Practice Exercise 1

Cyclohexane, a commonly used organic solvent, is 85.6% C and 14.4% H by mass with a molar mass of 84.2 g/mol. What is its molecular formula? (a) C_6H, (b) CH_2, (c) C_5H_{24}, (d) C_6H_{12}, (e) C_4H_8.

Practice Exercise 2

Ethylene glycol, used in automobile antifreeze, is 38.7% C, 9.7% H, and 51.6% O by mass. Its molar mass is 62.1 g/mol. (**a**) What is the empirical formula of ethylene glycol? (**b**) What is its molecular formula?

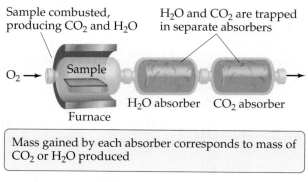

▲ Figure 3.14 **Apparatus for combustion analysis.**

Combustion Analysis

One technique for determining empirical formulas in the laboratory is *combustion analysis*, commonly used for compounds containing principally carbon and hydrogen.

When a compound containing carbon and hydrogen is completely combusted in an apparatus such as that shown in ▲ **Figure 3.14**, the carbon is converted to CO_2 and the hydrogen is converted to H_2O. ∞ (Section 3.2) The amounts of CO_2 and H_2O produced are determined by measuring the mass increase in the CO_2 and H_2O absorbers. From the masses of CO_2 and H_2O we can calculate the number of moles of C and H in the original sample and thereby the empirical formula. If a third element is present in the compound, its mass can be determined by subtracting the measured masses of C and H from the original sample mass.

SAMPLE EXERCISE 3.15 Determining an Empirical Formula by Combustion Analysis

Isopropyl alcohol, sold as rubbing alcohol, is composed of C, H, and O. Combustion of 0.255 g of isopropyl alcohol produces 0.561 g of CO_2 and 0.306 g of H_2O. Determine the empirical formula of isopropyl alcohol.

SOLUTION

Analyze We are told that isopropyl alcohol contains C, H, and O atoms and are given the quantities of CO_2 and H_2O produced when a given quantity of the alcohol is combusted. We must determine the empirical formula for isopropyl alcohol, a task that requires us to calculate the number of moles of C, H, and O in the sample.

Plan We can use the mole concept to calculate grams of C in the CO_2 and grams of H in the H_2O— the masses of C and H in the alcohol before combustion. The mass of O in the compound equals the mass of the original sample minus the sum of the C and H masses. Once we have the C, H, and O masses, we can proceed as in Sample Exercise 3.13.

Solve Because all of the carbon in the sample is converted to CO_2, we can use dimensional analysis and the following steps to calculate the mass C in the sample.

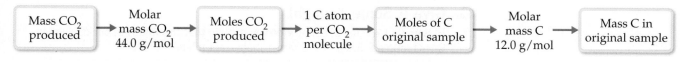

Using the values given in this example, the mass of C is

$$\text{Grams C} = (0.561 \text{ g CO}_2)\left(\frac{1 \text{ mol CO}_2}{44.0 \text{ g CO}_2}\right)\left(\frac{1 \text{ mol C}}{1 \text{ mol CO}_2}\right)\left(\frac{12.0 \text{ g C}}{1 \text{ mol C}}\right)$$

$$= 0.153 \text{ g C}$$

Because all of the hydrogen in the sample is converted to H_2O, we can use dimensional analysis and the following steps to calculate the mass H in the sample. We use three significant figures for the atomic mass of H to match the significant figures in the mass of H_2O produced.

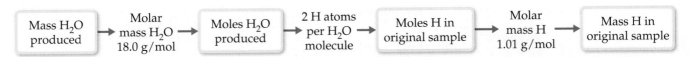

Using the values given in this example, the mass of H is

$$\text{Grams H} = (0.306 \text{ g H}_2\text{O})\left(\frac{1 \text{ mol H}_2\text{O}}{18.0 \text{ g H}_2\text{O}}\right)\left(\frac{2 \text{ mol H}}{1 \text{ mol H}_2\text{O}}\right)\left(\frac{1.01 \text{ g H}}{1 \text{ mol H}}\right)$$

$$= 0.0343 \text{ g H}$$

The mass of the sample, 0.255 g, is the sum of the masses of C, H, and O. Thus, the O mass is

$$\text{Mass of O} = \text{mass of sample} - (\text{mass of C} + \text{mass of H})$$
$$= 0.255 \text{ g} - (0.153 \text{ g} + 0.0343 \text{ g}) = 0.068 \text{ g O}$$

The number of moles of C, H, and O in the sample is therefore

$$\text{Moles C} = (0.153 \text{ g C})\left(\frac{1 \text{ mol C}}{12.0 \text{ g C}}\right) = 0.0128 \text{ mol C}$$

$$\text{Moles H} = (0.0343 \text{ g H})\left(\frac{1 \text{ mol H}}{1.01 \text{ g H}}\right) = 0.0340 \text{ mol H}$$

$$\text{Moles O} = (0.068 \text{ g O})\left(\frac{1 \text{ mol O}}{16.0 \text{ g O}}\right) = 0.0043 \text{ mol O}$$

To find the empirical formula, we must compare the relative number of moles of each element in the sample, as illustrated in Sample Exercise 3.13.

$$\text{C:}\frac{0.0128}{0.0043} = 3.0 \quad \text{H:}\frac{0.0340}{0.0043} = 7.9 \quad \text{O:}\frac{0.0043}{0.0043} = 1.0$$

The first two numbers are very close to the whole numbers 3 and 8, giving the empirical formula C_3H_8O.

Practice Exercise 1

The compound dioxane, which is used as a solvent in various industrial processes, is composed of C, H, and O atoms. Combustion of a 2.203-g sample of this compound produces 4.401 g CO_2 and 1.802 g H_2O. A separate experiment shows that it has a molar mass of 88.1 g/mol. Which of the following is the correct molecular formula for dioxane? (**a**) C_2H_4O, (**b**) $C_4H_4O_2$, (**c**) CH_2, (**d**) $C_4H_8O_2$.

Practice Exercise 2

(**a**) Caproic acid, responsible for the odor of dirty socks, is composed of C, H, and O atoms. Combustion of a 0.225-g sample of this compound produces 0.512 g CO_2 and 0.209 g H_2O. What is the empirical formula of caproic acid? (**b**) Caproic acid has a molar mass of 116 g/mol. What is its molecular formula?

 Give It Some Thought

In Sample Exercise 3.15, how do you explain that the values in our calculated C:H:O ratio are 3.0:7.9:1.0 rather than exact integers 3:8:1?

3.6 | Quantitative Information from Balanced Equations

The coefficients in a chemical equation represent the relative numbers of molecules in a reaction. The mole concept allows us to convert this information to the masses of the substances in the reaction. For instance, the coefficients in the balanced equation

$$2\,H_2(g) + O_2(g) \longrightarrow 2\,H_2O(l) \qquad\qquad [3.12]$$

indicate that two molecules of H_2 react with one molecule of O_2 to form two molecules of H_2O. It follows that the relative numbers of moles are identical to the relative numbers of molecules:

$2\,H_2(g)$	+	$O_2(g)$	$\longrightarrow$	$2\,H_2O(l)$
2 molecules		1 molecule		2 molecules
$2(6.02 \times 10^{23}$ molecules$)$		$1(6.02 \times 10^{23}$ molecules$)$		$2(6.02 \times 10^{23}$ molecules$)$
2 mol		1 mol		2 mol

We can generalize this observation to all balanced chemical equations: *The coefficients in a balanced chemical equation indicate both the relative numbers of molecules (or formula units) in the reaction and the relative numbers of moles.* ▼ **Figure 3.15** shows how this result corresponds to the law of conservation of mass.

The quantities 2 mol H_2, 1 mol O_2, and 2 mol H_2O given by the coefficients in Equation 3.12 are called *stoichiometrically equivalent quantities*. The relationship between these quantities can be represented as

$$2\,\text{mol }H_2 \,\hat{=}\, 1\,\text{mol }O_2 \,\hat{=}\, 2\,\text{mol }H_2O$$

where the $\hat{=}$ symbol means "is stoichiometrically equivalent to." Stoichiometric relations such as these can be used to convert between quantities of reactants and products in a chemical reaction. For example, the number of moles of H_2O produced from 1.57 mol of O_2 is

$$\text{Moles } H_2O = (1.57\,\text{mol } O_2)\left(\frac{2\,\text{mol }H_2O}{1\,\text{mol }O_2}\right) = 3.14\,\text{mol } H_2O$$

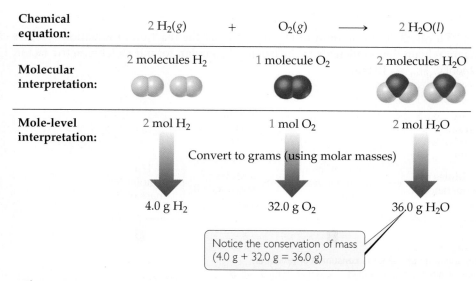

Chemical equation:	$2\,H_2(g)$	+	$O_2(g)$	$\longrightarrow$	$2\,H_2O(l)$
Molecular interpretation:	2 molecules H_2		1 molecule O_2		2 molecules H_2O
Mole-level interpretation:	2 mol H_2		1 mol O_2		2 mol H_2O

Convert to grams (using molar masses)

	4.0 g H_2	32.0 g O_2	36.0 g H_2O

Notice the conservation of mass (4.0 g + 32.0 g = 36.0 g)

▲ **Figure 3.15** Interpreting a balanced chemical equation quantitatively.

▲ **Give It Some Thought**

When 1.57 mol O_2 reacts with H_2 to form H_2O, how many moles of H_2 are consumed in the process?

As an additional example, consider the combustion of butane (C_4H_{10}), the fuel in disposable lighters:

$$2\,C_4H_{10}(l) + 13\,O_2(g) \longrightarrow 8\,CO_2(g) + 10\,H_2O(g) \qquad [3.13]$$

Let's calculate the mass of CO_2 produced when 1.00 g of C_4H_{10} is burned. The coefficients in Equation 3.13 tell us how the amount of C_4H_{10} consumed is related to the amount of CO_2 produced: 2 mol $C_4H_{10} \simeq$ 8 mol CO_2. To use this stoichiometric relationship, we must convert grams of C_4H_{10} to moles using the molar mass of C_4H_{10}, 58.0 g/mol:

$$\text{Moles } C_4H_{10} = (1.00 \text{ g } C_4H_{10})\left(\frac{1 \text{ mol } C_4H_{10}}{58.0 \text{ g } C_4H_{10}}\right)$$

$$= 1.72 \times 10^{-2} \text{ mol } C_4H_{10}$$

We then use the stoichiometric factor from the balanced equation to calculate moles of CO_2:

$$\text{Moles } CO_2 = (1.72 \times 10^{-2} \text{ mol } C_4H_{10})\left(\frac{8 \text{ mol } CO_2}{2 \text{ mol } C_4H_{10}}\right)$$

$$= 6.88 \times 10^{-2} \text{ mol } CO_2$$

Finally, we use the molar mass of CO_2, 44.0 g/mol, to calculate the CO_2 mass in grams:

$$\text{Grams } CO_2 = (6.88 \times 10^{-2} \text{ mol } CO_2)\left(\frac{44.0 \text{ g } CO_2}{1 \text{ mol } CO_2}\right)$$

$$= 3.03 \text{ g } CO_2$$

This conversion sequence involves three steps, as illustrated in ▼ **Figure 3.16.** These three conversions can be combined in a single equation:

$$\text{Grams } CO_2 = (1.00 \text{ g } C_4H_{10})\left(\frac{1 \text{ mol } C_4H_{10}}{58.0 \text{ g } C_4H_{10}}\right)\left(\frac{8 \text{ mol } CO_2}{2 \text{ mol } C_4H_{10}}\right)\left(\frac{44.0 \text{ g } CO_2}{1 \text{ mol } CO_2}\right)$$

$$= 3.03 \text{ g } CO_2$$

To calculate the amount of O_2 consumed in the reaction of Equation 3.13, we again rely on the coefficients in the balanced equation for our stoichiometric factor, 2 mol $C_4H_{10} \simeq$ 13 mol O_2:

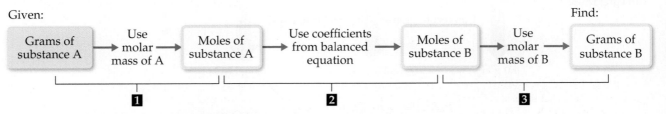

▲ **Figure 3.16 Procedure for calculating amounts of reactants consumed or products formed in a reaction.** The number of grams of a reactant consumed or product formed can be calculated in three steps, starting with the number of grams of any reactant or product.

$$\text{Grams O}_2 = (1.00 \text{ g } \cancel{C_4H_{10}})\left(\frac{1 \text{ mol } \cancel{C_4H_{10}}}{58.0 \text{ g } \cancel{C_4H_{10}}}\right)\left(\frac{13 \text{ mol } \cancel{O_2}}{2 \text{ mol } \cancel{C_4H_{10}}}\right)\left(\frac{32.0 \text{ g } O_2}{1 \text{ mol } \cancel{O_2}}\right)$$

$$= 3.59 \text{ g } O_2$$

 Give It Some Thought

In the previous example, 1.00 g of C_4H_{10} reacts with 3.59 g of O_2 to form 3.03 g of CO_2. Using only addition and subtraction, calculate the amount of H_2O produced.

SAMPLE EXERCISE 3.16 | Calculating Amounts of Reactants and Products

Determine how many grams of water are produced in the oxidation of 1.00 g of glucose, $C_6H_{12}O_6$:

$$C_6H_{12}O_6(s) + 6 O_2(g) \longrightarrow 6 CO_2(g) + 6 H_2O(l)$$

SOLUTION

Analyze We are given the mass of a reactant and must determine the mass of a product in the given reaction.

Plan We follow the general strategy outlined in Figure 3.16:

(1) Convert grams of $C_6H_{12}O_6$ to moles using the molar mass of $C_6H_{12}O_6$.

(2) Convert moles of $C_6H_{12}O_6$ to moles of H_2O using the stoichiometric relationship 1 mol $C_6H_{12}O_6 \triangleq$ 6 mol H_2O.

(3) Convert moles of H_2O to grams using the molar mass of H_2O.

Solve

(1) First we convert grams of $C_6H_{12}O_6$ to moles using the molar mass of $C_6H_{12}O_6$.

Grams of substance A	→ Use molar mass of A →	Moles of substance A	→ Use coefficients from balanced equation →	Moles of substance B	→ Use molar mass of B →	Grams of substance B

$$\text{Moles } C_6H_{12}O_6 = (1.00 \text{ g } \cancel{C_6H_{12}O_6})\left(\frac{1 \text{ mol } C_6H_{12}O_6}{180.0 \text{ g } \cancel{C_6H_{12}O_6}}\right)$$

(2) Next we convert moles of $C_6H_{12}O_6$ to moles of H_2O using the stoichiometric relationship 1 mol $C_6H_{12}O_6 \triangleq$ 6 mol H_2O.

Grams of substance A	→ Use molar mass of A →	Moles of substance A	→ Use coefficients from balanced equation →	Moles of substance B	→ Use molar mass of B →	Grams of substance B

$$\text{Moles } H_2O = (1.00 \text{ g } \cancel{C_6H_{12}O_6})\left(\frac{1 \text{ mol } \cancel{C_6H_{12}O_6}}{180.0 \text{ g } \cancel{C_6H_{12}O_6}}\right)\left(\frac{6 \text{ mol } H_2O}{1 \text{ mol } \cancel{C_6H_{12}O_6}}\right)$$

(3) Finally, we convert moles of H_2O to grams using the molar mass of H_2O.

Grams of substance A	→ Use molar mass of A →	Moles of substance A	→ Use coefficients from balanced equation →	Moles of substance B	→ Use molar mass of B →	Grams of substance B

$$\text{Grams } H_2O = (1.00 \text{ g } \cancel{C_6H_{12}O_6})\left(\frac{1 \text{ mol } \cancel{C_6H_{12}O_6}}{180.0 \text{ g } \cancel{C_6H_{12}O_6}}\right)\left(\frac{6 \text{ mol } \cancel{H_2O}}{1 \text{ mol } \cancel{C_6H_{12}O_6}}\right)\left(\frac{18.0 \text{ g } H_2O}{1 \text{ mol } \cancel{H_2O}}\right)$$

$$= 0.600 \text{ g } H_2O$$

Check We can check how reasonable our result is by doing a ballpark estimate of the mass of H_2O. Because the molar mass of glucose is 180 g/mol, 1 g of glucose equals 1/180 mol. Because 1 mol of glucose yields 6 mol H_2O, we would have 6/180 = 1/30 mol H_2O. The molar mass of water is 18 g/mol, so we have $1/30 \times 18 = 6/10 = 0.6$ g of H_2O, which agrees with the full calculation. The units, grams H_2O, are correct. The initial data had three significant figures, so three significant figures for the answer is correct.

Practice Exercise 1

Sodium hydroxide reacts with carbon dioxide to form sodium carbonate and water:

$$2 \, NaOH(s) + CO_2(g) \longrightarrow Na_2CO_3(s) + H_2O(l)$$

How many grams of Na_2CO_3 can be prepared from 2.40 g of NaOH? **(a)** 3.18 g, **(b)** 6.36 g, **(c)** 1.20 g, **(d)** 0.0300 g.

Practice Exercise 2

Decomposition of $KClO_3$ is sometimes used to prepare small amounts of O_2 in the laboratory: $2 \, KClO_3(s) \longrightarrow 2 \, KCl(s) + 3 \, O_2(g)$. How many grams of O_2 can be prepared from 4.50 g of $KClO_3$?

SAMPLE EXERCISE 3.17 | Calculating Amounts of Reactants and Products

Solid lithium hydroxide is used in space vehicles to remove the carbon dioxide gas exhaled by astronauts. The hydroxide reacts with the carbon dioxide to form solid lithium carbonate and liquid water. How many grams of carbon dioxide can be absorbed by 1.00 g of lithium hydroxide?

SOLUTION

Analyze We are given a verbal description of a reaction and asked to calculate the number of grams of one reactant that reacts with 1.00 g of another.

Plan The verbal description of the reaction can be used to write a balanced equation:

$$2 \, LiOH(s) + CO_2(g) \longrightarrow Li_2CO_3(s) + H_2O(l)$$

We are given the mass in grams of LiOH and asked to calculate the mass in grams of CO_2. We can accomplish this with the three conversion steps in Figure 3.16. The conversion of Step 1 requires the molar mass of LiOH $(6.9 + 16.0 + 1.0 = 23.9 \, g/mol)$. The conversion of Step 2 is based on a stoichiometric relationship from the balanced chemical equation: 2 mol LiOH $\hat{=}$ mol CO_2. For the Step 3 conversion, we use the molar mass of CO_2 $12.0 + 2(16.0) = 44.0 \, g/mol$.

Solve

$$(1.00 \, g \, LiOH)\left(\frac{1 \, mol \, LiOH}{23.9 \, g \, LiOH}\right)\left(\frac{1 \, mol \, CO_2}{2 \, mol \, LiOH}\right)\left(\frac{44.0 \, g \, CO_2}{1 \, mol \, CO_2}\right) = 0.920 \, g \, CO_2$$

Check Notice that 23.9 g LiOH/mol $\approx$ 24 g LiOH/mol, 24 g LiOH/mol $\times$ 2 mol LiOH = 48 g LiOH, and $(44 \, g \, CO_2/mol)/(48 \, g \, LiOH)$ is slightly less than 1. Thus, the magnitude of our answer, 0.919 g CO_2, is reasonable based on the amount of starting LiOH. The number of significant figures and units are also appropriate.

Practice Exercise 1

Propane, C_3H_8 (Figure 3.8), is a common fuel used for cooking and home heating. What mass of O_2 is consumed in the combustion of 1.00 g of propane? **(a)** 5.00 g, **(b)** 0.726 g, **(c)** 2.18 g, **(d)** 3.63 g.

Practice Exercise 2

Methanol, CH_3OH, reacts with oxygen from air in a combustion reaction to form water and carbon dioxide. What mass of water is produced in the combustion of 23.6 g of methanol?

3.7 | Limiting Reactants

Suppose you wish to make several sandwiches using one slice of cheese and two slices of bread for each. Using Bd = bread, Ch = cheese, and Bd_2Ch = sandwich, the recipe for making a sandwich can be represented like a chemical equation:

$$2 \, Bd + Ch \longrightarrow Bd_2Ch$$

If you have ten slices of bread and seven slices of cheese, you can make only five sandwiches and will have two slices of cheese left over. The amount of bread available limits the number of sandwiches.

GO FIGURE

If the amount of H_2 is doubled, how many moles of H_2O would have formed?

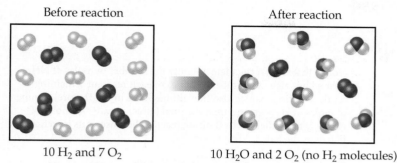

Before reaction

10 H_2 and 7 O_2

After reaction

10 H_2O and 2 O_2 (no H_2 molecules)

▲ **Figure 3.17 Limiting reactant.** Because H_2 is completely consumed, it is the limiting reactant. Because some O_2 is left over after the reaction is complete, it is the excess reactant. The amount of H_2O formed depends on the amount of limiting reactant, H_2.

An analogous situation occurs in chemical reactions when one reactant is used up before the others. The reaction stops as soon as any reactant is totally consumed, leaving the excess reactants as leftovers. Suppose, for example, we have a mixture of 10 mol H_2 and 7 mol O_2, which react to form water:

$$2\,H_2(g) + O_2(g) \longrightarrow 2\,H_2O(g)$$

Because 2 mol $H_2 \backsimeq$ mol O_2, the number of moles of O_2 needed to react with all the H_2 is

$$\text{Moles } O_2 = (10\ \text{mol } H_2)\left(\frac{1\ \text{mol } O_2}{2\ \text{mol } H_2}\right) = 5\ \text{mol } O_2$$

Because 7 mol O_2 is available at the start of the reaction, 7 mol O_2 − 5 mol O_2 = 2 mol O_2 is still present when all the H_2 is consumed.

The reactant that is completely consumed in a reaction is called the **limiting reactant** because it determines, or limits, the amount of product formed. The other reactants are sometimes called *excess reactants*. In our example, shown in ▲ **Figure 3.17,** H_2 is the limiting reactant, which means that once all the H_2 has been consumed, the reaction stops. At that point some of the excess reactant O_2 is left over.

There are no restrictions on the starting amounts of reactants in any reaction. Indeed, many reactions are carried out using an excess of one reactant. The quantities of reactants consumed and products formed, however, are restricted by the quantity of the limiting reactant. For example, when a combustion reaction takes place in the open air, oxygen is plentiful and is therefore the excess reactant. If you run out of gasoline while driving, the car stops because the gasoline is the limiting reactant in the combustion reaction that moves the car.

Before we leave the example illustrated in Figure 3.17, let's summarize the data:

	$2\,H_2(g)$	+	$O_2(g)$	$\longrightarrow$	$2\,H_2O(g)$
Before reaction:	10 mol		7 mol		0 mol
Change (reaction):	−10 mol		−5 mol		+10 mol
After reaction:	0 mol		2 mol		10 mol

The second line in the table (Change) summarizes the amounts of reactants consumed (where this consumption is indicated by the minus signs) and the amount of the product formed (indicated by the plus sign). These quantities are restricted by the quantity of the limiting reactant and depend on the coefficients in the balanced equation. The mole ratio $H_2:O_2:H_2O = 10:5:10$ is a multiple of the ratio of the coefficients in the balanced equation, 2:1:2. The after quantities, which depend on the before quantities and their changes, are found by adding the before quantity and change quantity for each column. The amount of the limiting reactant (H_2) must be zero at the end of the reaction. What remains is 2 mol O_2 (excess reactant) and 10 mol H_2O (product).

SAMPLE EXERCISE 3.18 | Calculating the Amount of Product Formed from a Limiting Reactant

The most important commercial process for converting N_2 from the air into nitrogen-containing compounds is based on the reaction of N_2 and H_2 to form ammonia (NH_3):

$$N_2(g) + 3 H_2(g) \longrightarrow 2 NH_3(g)$$

How many moles of NH_3 can be formed from 3.0 mol of N_2 and 6.0 mol of H_2?

SOLUTION

Analyze We are asked to calculate the number of moles of product, NH_3, given the quantities of each reactant, N_2 and H_2, available in a reaction. This is a limiting reactant problem.

Plan If we assume one reactant is completely consumed, we can calculate how much of the second reactant is needed. By comparing the calculated quantity of the second reactant with the amount available, we can determine which reactant is limiting. We then proceed with the calculation, using the quantity of the limiting reactant.

Solve
The number of moles of H_2 needed for complete consumption of 3.0 mol of N_2 is

$$\text{Moles } H_2 = (3.0 \text{ mol } N_2)\left(\frac{3 \text{ mol } H_2}{1 \text{ mol } N_2}\right) = 9.0 \text{ mol } H_2$$

Because only 6.0 mol H_2 is available, we will run out of H_2 before the N_2 is gone, which tells us that H_2 is the limiting reactant. Therefore, we use the quantity of H_2 to calculate the quantity of NH_3 produced:

$$\text{Moles } NH_3 = (6.0 \text{ mol } H_2)\left(\frac{2 \text{ mol } NH_3}{3 \text{ mol } H_2}\right) = 4.0 \text{ mol } NH_3$$

Comment It is useful to summarize the reaction data in a table:

	$N_2(g)$ +	$3 H_2(g) \longrightarrow$	$2 NH_3(g)$
Before reaction:	3.0 mol	6.0 mol	0 mol
Change (reaction):	−2.0 mol	−6.0 mol	+4.0 mol
After reaction:	1.0 mol	0 mol	4.0 mol

Notice that we can calculate not only the number of moles of NH_3 formed but also the number of moles of each reactant remaining after the reaction. Notice also that although the initial (before) number of moles of H_2 is greater than the final (after) number of moles of N_2, H_2 is nevertheless the limiting reactant because of its larger coefficient in the balanced equation.

Check Examine the Change row of the summary table to see that the mole ratio of reactants consumed and product formed, 2:6:4, is a multiple of the coefficients in the balanced equation, 1:3:2. We confirm that H_2 is the limiting reactant because it is completely consumed in the reaction, leaving 0 mol at the end. Because 6.0 mol H_2 has two significant figures, our answer has two significant figures.

Practice Exercise 1
When 24 mol of methanol and 15 mol of oxygen combine in the combustion reaction $2 CH_3OH(l) + 3 O_2(g) \longrightarrow 2 CO_2(g)+4 H_2O(g)$, what is the excess reactant and how many moles of it remains at the end of the reaction?
(a) 9 mol $CH_3OH(l)$, (b) 10 mol $CO_2(g)$, (c) 10 mol $CH_3OH(l)$, (d) 14 mol $CH_3OH(l)$, (e) 1 mol $O_2(g)$.

Practice Exercise 2
(a) When 1.50 mol of Al and 3.00 mol of Cl_2 combine in the reaction $2 Al(s) + 3 Cl_2(g) \longrightarrow 2 AlCl_3(s)$, which is the limiting reactant? (b) How many moles of $AlCl_3$ are formed? (c) How many moles of the excess reactant remain at the end of the reaction?

SAMPLE EXERCISE 3.19 | Calculating the Amount of Product Formed from a Limiting Reactant

The reaction

$$2 H_2(g) + O_2(g) \longrightarrow 2 H_2O(g)$$

is used to produce electricity in a hydrogen fuel cell. Suppose a fuel cell contains 150 g of $H_2(g)$ and 1500 g of $O_2(g)$ (each measured to two significant figures). How many grams of water can form?

SOLUTION

Analyze We are asked to calculate the amount of a product, given the amounts of two reactants, so this is a limiting reactant problem.

Plan To identify the limiting reactant, we can calculate the number of moles of each reactant and compare their ratio with the ratio of coefficients in the balanced equation. We then use the quantity of the limiting reactant to calculate the mass of water that forms.

Solve From the balanced equation, we have the stoichiometric relations

$$2 \text{ mol } H_2 \simeq \text{ mol } O_2 \simeq 2 \text{ mol } H_2O$$

Using the molar mass of each substance, we calculate the number of moles of each reactant:

$$\text{Moles } H_2 = (150 \text{ g } H_2)\left(\frac{1 \text{ mol } H_2}{2.02 \text{ g } H_2}\right) = 74 \text{ mol } H_2$$

$$\text{Moles } O_2 = (1500 \text{ g } O_2)\left(\frac{1 \text{ mol } O_2}{32.0 \text{ g } O_2}\right) = 47 \text{ mol } O_2$$

The coefficients in the balanced equation indicate that the reaction requires 2 mol of H_2 for every 1 mol of O_2. Therefore, for all the O_2 to completely react, we would need 2 × 47 = 94 mol of H_2. Since there are only 74 mol of H_2, all of the O_2 cannot react, so it is the excess reactant, and H_2 must be the limiting reactant. (Notice that the limiting reactant is not necessarily the one present in the lowest amount.)

We use the given quantity of H_2 (the limiting reactant) to calculate the quantity of water formed. We could begin this calculation with the given H_2 mass, 150 g, but we can save a step by starting with the moles of H_2, 74 mol, we just calculated:

$$\text{Grams } H_2O = (74 \text{ mol } H_2)\left(\frac{2 \text{ mol } H_2O}{2 \text{ mol } H_2}\right)\left(\frac{18.0 \text{ g } H_2O}{1 \text{ mol } H_2O}\right)$$

$$= 1.3 \times 10^2 \text{ g } H_2O$$

Check The magnitude of the answer seems reasonable based on the amounts of the reactants. The units are correct, and the number of significant figures (two) corresponds to those in the values given in the problem statement.

Comment The quantity of the limiting reactant, H_2, can also be used to determine the quantity of O_2 used:

$$\text{Grams } O_2 = (74 \text{ mol } H_2)\left(\frac{1 \text{ mol } O_2}{2 \text{ mol } H_2}\right)\left(\frac{32.0 \text{ g } O_2}{1 \text{ mol } O_2}\right)$$

$$= 1.2 \times 10^3 \text{ g } O_2$$

The mass of O_2 remaining at the end of the reaction equals the starting amount minus the amount consumed:

$$1500 \text{ g} - 1200 \text{ g} = 300 \text{ g}.$$

Practice Exercise 1

Molten gallium reacts with arsenic to form the semiconductor, gallium arsenide, GaAs, used in light–emitting diodes and solar cells :

$$Ga(l) + As(s) \longrightarrow GaAs(s)$$

If 4.00 g of gallium is reacted with 5.50 g of arsenic, how many grams of the excess reactant are left at the end of the reaction? **(a)** 4.94 g As, **(b)** 0.56 g As, **(c)** 8.94 g Ga, or **(d)** 1.50 g As.

Practice Exercise 2

When a 2.00-g strip of zinc metal is placed in an aqueous solution containing 2.50 g of silver nitrate, the reaction is

$$Zn(s) + 2 \, AgNO_3(aq) \longrightarrow 2 \, Ag(s) + Zn(NO_3)_2(aq)$$

(a) Which reactant is limiting? **(b)** How many grams of Ag form? **(c)** How many grams of $Zn(NO_3)_2$ form? **(d)** How many grams of the excess reactant are left at the end of the reaction?

Theoretical and Percent Yields

The quantity of product calculated to form when all of a limiting reactant is consumed is called the **theoretical yield**. The amount of product actually obtained, called the *actual yield*, is almost always less than (and can never be greater than) the theoretical yield. There are many reasons for this difference. Part of the reactants may not react, for example, or they may react in a way different from that desired (side reactions). In addition, it is not always possible to recover all of the product from the reaction mixture. The **percent yield** of a reaction relates actual and theoretical yields:

$$\text{Percent yield} = \frac{\text{actual yield}}{\text{theoretical yield}} \times 100\% \qquad [3.14]$$

SAMPLE EXERCISE 3.20 Calculating Theoretical Yield and Percent Yield

Adipic acid, $H_2C_6H_8O_4$, used to produce nylon, is made commercially by a reaction between cyclohexane (C_6H_{12}) and O_2:

$$2 \, C_6H_{12}(l) + 5 \, O_2(g) \longrightarrow 2 \, H_2C_6H_8O_4(l) + 2 \, H_2O(g)$$

(a) Assume that you carry out this reaction with 25.0 g of cyclohexane and that cyclohexane is the limiting reactant. What is the theoretical yield of adipic acid? **(b)** If you obtain 33.5 g of adipic acid, what is the percent yield for the reaction?

SOLUTION

Analyze We are given a chemical equation and the quantity of the limiting reactant (25.0 g of C_6H_{12}). We are asked to calculate the theoretical yield of a product $H_2C_6H_8O_4$ and the percent yield if only 33.5 g of product is obtained.

Plan

(a) The theoretical yield, which is the calculated quantity of adipic acid formed, can be calculated using the sequence of conversions shown in Figure 3.16.

(b) The percent yield is calculated by using Equation 3.14 to compare the given actual yield (33.5 g) with the theoretical yield.

Solve

(a) The theoretical yield is

$$\text{Grams } H_2C_6H_8O_4 = (25.0 \text{ g } C_6H_{12})\left(\frac{1 \text{ mol } C_6H_{12}}{84.0 \text{ g } C_6H_{12}}\right)\left(\frac{2 \text{ mol } H_2C_6H_8O_4}{2 \text{ mol } C_6H_{12}}\right)\left(\frac{146.0 \text{ g } H_2C_6H_8O_4}{1 \text{ mol } H_2C_6H_8O_4}\right)$$

$$= 43.5 \text{ g } H_2C_6H_8O_4$$

(b) $\text{Percent yield} = \dfrac{\text{actual yield}}{\text{theoretical yield}} \times 100\% = \dfrac{33.5 \text{ g}}{43.5 \text{ g}} \times 100\% = 77.0\%$

Check We can check our answer in (a) by doing a ballpark calculation. From the balanced equation we know that each mole of cyclohexane gives 1 mol adipic acid. We have $25/84 \approx 25/75 = 0.3$ mol hexane, so we expect 0.3 mol adipic acid, which equals about $0.3 \times 150 = 45$ g, about the same magnitude as the 43.5 g obtained in the more detailed calculation given previously. In addition, our answer has the appropriate units and number of significant figures. In (b) the answer is less than 100%, as it must be from the definition of percent yield.

Practice Exercise 1

If 3.00 g of titanium metal is reacted with 6.00 g of chlorine gas, Cl_2, to form 7.7 g of titanium (IV) chloride in a combination reaction, what is the percent yield of the product? (a) 65%, (b) 96%, (c) 48%, or (d) 86%.

Practice Exercise 2

Imagine you are working on ways to improve the process by which iron ore containing Fe_2O_3 is converted into iron:

$$Fe_2O_3(s) + 3 CO(g) \longrightarrow 2 Fe(s) + 3 CO_2(g)$$

(a) If you start with 150 g of Fe_2O_3 as the limiting reactant, what is the theoretical yield of Fe?
(b) If your actual yield is 87.9 g, what is the percent yield?

Strategies in Chemistry

Design an Experiment

One of the most important skills you can learn in school is how to think like a scientist. Questions such as: "What experiment might test this hypothesis?", "How do I interpret these data?", and "Do these data support the hypothesis?" are asked every day by chemists and other scientists as they go about their work.

We want you to become a good critical thinker as well as an active, logical, and curious learner. For this purpose, starting in Chapter 3, we include at the end of each chapter a special exercise called "Design an Experiment." Here is an example:

Is milk a pure liquid or a mixture of chemical components in water? Design an experiment to distinguish between these two possibilities.

You might already know the answer–milk is indeed a mixture of components in water–but the goal is to think of how to demonstrate this in practice. Upon thinking about it, you will likely realize that the key idea for this experiment is separation: You can prove that milk is a mixture of chemical components if you can figure out how to separate these components.

Testing a hypothesis is a creative endeavor. While some experiments may be more efficient than others, there is often more than one good way to test a hypothesis. Our question about milk, for example, might be explored by an experiment in which you boil a known quantity of milk until it is dry. Does a solid residue form in the bottom of the pan? If so, you could weigh it, and calculate the percentage of solids in milk, which would offer good evidence that milk is a mixture. If there is no residue after boiling, then you still cannot distinguish between the two possibilities.

What other experiments might you do to demonstrate that milk is a mixture? You could put a sample of milk in a centrifuge, which you might have used in a biology lab, spin your sample and observe if any solids collect at the bottom of the centrifuge tube; large molecules can be separated in this way from a mixture. Measurement of the mass of the solid at the bottom of the tube is a way to obtain a value for the % solids in milk, and also tells you that milk is indeed a mixture.

Keep an open mind: Lacking a centrifuge, how else might you separate solids in the milk? You could consider using a filter with really tiny holes in it or perhaps even a fine strainer. You could propose that if milk were poured through this filter, some (large) solid components should stay on the top of the filter, while water (and really small molecules or ions) would pass through the filter. That result would be evidence that milk is a mixture. Does such a filter exist? Yes! But for our purposes, the existence of such a filter is not the point: the point is, can you use your imagination and your knowledge of chemistry to design a reasonable experiment? Don't worry too much about the exact apparatus you need for the "Design an Experiment" exercises. The goal is to imagine what you would need to do, or what kind of data would you need to collect, in order to answer the question. If your instructor allows it, you can collaborate with others in your class to develop ideas. Scientists discuss their ideas with other scientists all the time. We find that discussing ideas, and refining them, makes us better scientists and helps us collectively answer important questions.

The design and interpretation of scientific experiments is at the heart of the scientific method. Think of the Design an Experiment exercises as puzzles that can be solved in various ways, and enjoy your explorations!

Chapter Summary and Key Terms

CHEMICAL EQUATIONS (INTRODUCTION AND SECTION 3.1) The study of the quantitative relationships between chemical formulas and chemical equations is known as **stoichiometry**. One of the important concepts of stoichiometry is the **law of conservation of mass**, which states that the total mass of the products of a chemical reaction is the same as the total mass of the reactants. The same numbers of atoms of each type are present before and after a chemical reaction. A balanced **chemical equation** shows equal numbers of atoms of each element on each side of the equation. Equations are balanced by placing coefficients in front of the chemical formulas for the **reactants** and **products** of a reaction, *not* by changing the subscripts in chemical formulas.

SIMPLE PATTERNS OF CHEMICAL REACTIVITY (SECTION 3.2) Among the reaction types described in this chapter are (1) **combination reactions**, in which two reactants combine to form one product; (2) **decomposition reactions**, in which a single reactant forms two or more products; and (3) **combustion reactions** in oxygen, in which a substance, typically a hydrocarbon, reacts rapidly with O_2 to form CO_2 and H_2O.

FORMULA WEIGHTS (SECTION 3.3) Much quantitative information can be determined from chemical formulas and balanced chemical equations by using atomic weights. The **formula weight** of a compound equals the sum of the atomic weights of the atoms in its formula. If the formula is a molecular formula, the formula weight is also called the **molecular weight**. Atomic weights and formula weights can be used to determine the **elemental composition** of a compound.

AVOGADRO'S NUMBER AND THE MOLE (SECTION 3.4) A mole of any substance contains **Avogadro's number** (6.02×10^{23}) of formula

units of that substance. The mass of a **mole** of atoms, molecules, or ions (the **molar mass**) equals the formula weight of that material expressed in grams. The mass of one molecule of H_2O, for example, is 18.0 amu, so the mass of 1 mol of H_2O is 18.0 g. That is, the molar mass of H_2O is 18.0 g/mol.

EMPIRICAL FORMULAS FROM ANALYSIS (SECTION 3.5) The empirical formula of any substance can be determined from its percent composition by calculating the relative number of moles of each atom in 100 g of the substance. If the substance is molecular in nature, its molecular formula can be determined from the empirical formula if the molecular weight is also known. Combustion analysis is a special technique for determining the empirical formulas of compounds containing only carbon, hydrogen, and/or oxygen.

QUANTITATIVE INFORMATION FROM BALANCED EQUATIONS AND LIMITING REACTANTS (SECTIONS 3.6 AND 3.7) The mole concept can be used to calculate the relative quantities of reactants and products in chemical reactions. The coefficients in a balanced equation give the relative numbers of moles of the reactants and products. To calculate the number of grams of a product from the number of grams of a reactant, first convert grams of reactant to moles of reactant. Then use the coefficients in the balanced equation to convert the number of moles of reactant to moles of product. Finally, convert moles of product to grams of product.

A **limiting reactant** is completely consumed in a reaction. When it is used up, the reaction stops, thus limiting the quantities of products formed. The **theoretical yield** of a reaction is the quantity of product calculated to form when all of the limiting reactant reacts. The actual yield of a reaction is always less than the theoretical yield. The **percent yield** compares the actual and theoretical yields.

Learning Outcomes   After studying this chapter, you should be able to:

- Balance chemical equations. (Section 3.1)
- Predict the products of simple combination, decomposition, and combustion reactions. (Section 3.2)
- Calculate formula weights. (Section 3.3)
- Convert grams to moles and vice versa using molar masses. (Section 3.4)
- Convert number of molecules to moles and vice versa using Avogadro's number. (Section 3.4)

- Calculate the empirical and molecular formulas of a compound from percentage composition and molecular weight. (Section 3.5)
- Identify limiting reactants and calculate amounts, in grams or moles, of reactants consumed and products formed for a reaction. (Section 3.6)
- Calculate the percent yield of a reaction. (Section 3.7)

Key Equations

- Elemental composition (%) = $\dfrac{\left(\begin{array}{c}\text{number of atoms}\\\text{of that element}\end{array}\right)\left(\begin{array}{c}\text{atomic weight}\\\text{of element}\end{array}\right)}{\text{formula weight of compound}} \times 100\%$ [3.10] This is the formula to calculate the mass percentage of each element in a compound. The sum of all the percentages of all the elements in a compound should add up to 100%.

- Percent yield = $\dfrac{\text{(actual yield)}}{\text{(theoretical yield)}} \times 100\%$ [3.14] This is the formula to calculate the percent yield of a reaction. The percent yield can never be more than 100%.

Exercises

Visualizing Concepts

3.1 The reaction between reactant A (blue spheres) and reactant B (red spheres) is shown in the following diagram:

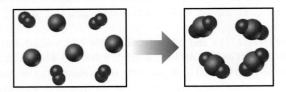

Based on this diagram, which equation best describes the reaction? [Section 3.1]

(a) $A_2 + B \longrightarrow A_2B$

(b) $A_2 + 4 B \longrightarrow 2 AB_2$

(c) $2 A + B_4 \longrightarrow 2 AB_2$

(d) $A + B_2 \longrightarrow AB_2$

3.2 The following diagram shows the combination reaction between hydrogen, H_2, and carbon monoxide, CO, to produce methanol, CH_3OH (white spheres are H, black spheres are C, red spheres are O). The correct number of CO molecules involved in this reaction is not shown. [Section 3.1]

(a) Determine the number of CO molecules that should be shown in the left (reactants) box.

(b) Write a balanced chemical equation for the reaction.

CO molecules not shown

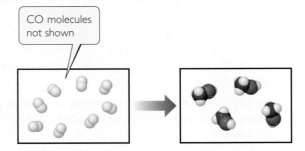

3.3 The following diagram represents the collection of elements formed by a decomposition reaction. (a) If the blue spheres represent N atoms and the red ones represent O atoms, what was the empirical formula of the original compound? (b) Could you draw a diagram representing the molecules of the compound that had been decomposed? Why or why not? [Section 3.2]

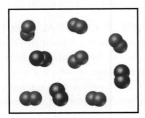

3.4 The following diagram represents the collection of CO_2 and H_2O molecules formed by complete combustion of a hydrocarbon. What is the empirical formula of the hydrocarbon? [Section 3.2]

3.5 Glycine, an amino acid used by organisms to make proteins, is represented by the following molecular model.

(a) Write its molecular formula.

(b) Determine its molar mass.

(c) Calculate the mass of 3 mol of glycine.

(d) Calculate the percent nitrogen by mass in glycine. [Sections 3.3 and 3.5]

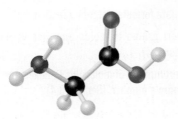

3.6 The following diagram represents a high-temperature reaction between CH_4 and H_2O. Based on this reaction, find how many moles of each product can be obtained starting with 4.0 mol CH_4. [Section 3.6]

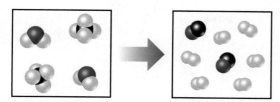

3.7 Nitrogen (N_2) and hydrogen (H_2) react to form ammonia (NH_3). Consider the mixture of N_2 and H_2 shown in the accompanying diagram. The blue spheres represent N, and the white ones represent H. Draw a representation of the product mixture, assuming that the reaction goes to completion. How did you arrive at your representation? What is the limiting reactant in this case? [Section 3.7]

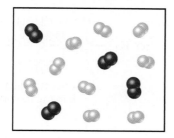

3.8 Nitrogen monoxide and oxygen react to form nitrogen dioxide. Consider the mixture of NO and O_2 shown in the accompanying diagram. The blue spheres represent N, and the red ones represent O. **(a)** Draw a representation of the product mixture, assuming that the reaction goes to completion. What is the limiting reactant in this case? **(b)** How many NO_2 molecules would you draw as products if the reaction had a percent yield of 75%? [Section 3.7]

Chemical Equations and Simple Patterns of Chemical Reactivity (Sections 3.1 and 3.2)

3.9 **(a)** What scientific principle or law is used in the process of balancing chemical equations? **(b)** In balancing equations is it acceptable to change the coefficients, the subscripts in the chemical formula, or both?

3.10 A key step in balancing chemical equations is correctly identifying the formulas of the reactants and products. For example, consider the reaction between calcium oxide, $CaO(s)$, and $H_2O(l)$ to form aqueous calcium hydroxide. **(a)** Write a balanced chemical equation for this combination reaction, having correctly identified the product as $Ca(OH)_2(aq)$. **(b)** Is it possible to balance the equation if you incorrectly identify the product as $CaOH(aq)$, and if so, what is the equation?

3.11 Balance the following equations:
- **(a)** $CO(g) + O_2(g) \longrightarrow CO_2(g)$
- **(b)** $N_2O_5(g) + H_2O(l) \longrightarrow HNO_3(aq)$
- **(c)** $CH_4(g) + Cl_2(g) \longrightarrow CCl_4(l) + HCl(g)$
- **(d)** $Zn(OH)_2(s) + HNO_3(aq) \longrightarrow Zn(NO_3)_2(aq) + H_2O(l)$

3.12 Balance the following equations:
- **(a)** $Li(s) + N_2(g) \longrightarrow Li_3N(s)$
- **(b)** $TiCl_4(l) + H_2O(l) \longrightarrow TiO_2(s) + HCl(aq)$
- **(c)** $NH_4NO_3(s) \longrightarrow N_2(g) + O_2(g) + H_2O(g)$
- **(d)** $AlCl_3(s) + Ca_3N_2(s) \longrightarrow AlN(s) + CaCl_2(s)$

3.13 Balance the following equations:
- **(a)** $Al_4C_3(s) + H_2O(l) \longrightarrow Al(OH)_3(s) + CH_4(g)$
- **(b)** $C_5H_{10}O_2(l) + O_2(g) \longrightarrow CO_2(g) + H_2O(g)$
- **(c)** $Fe(OH)_3(s) + H_2SO_4(aq) \longrightarrow Fe_2(SO_4)_3(aq) + H_2O(l)$
- **(d)** $Mg_3N_2(s) + H_2SO_4(aq) \longrightarrow MgSO_4(aq) + (NH_4)_2SO_4(aq)$

3.14 Balance the following equations:
- **(a)** $Ca_3P_2(s) + H_2O(l) \longrightarrow Ca(OH)_2(aq) + PH_3(g)$
- **(b)** $Al(OH)_3(s) + H_2SO_4(aq) \longrightarrow Al_2(SO_4)_3(aq) + H_2O(l)$
- **(c)** $AgNO_3(aq) + Na_2CO_3(aq) \longrightarrow Ag_2CO_3(s) + NaNO_3(aq)$
- **(d)** $C_2H_5NH_2(g) + O_2(g) \longrightarrow CO_2(g) + H_2O(g) + N_2(g)$

3.15 Write balanced chemical equations corresponding to each of the following descriptions: **(a)** Solid calcium carbide, CaC_2, reacts with water to form an aqueous solution of calcium hydroxide and acetylene gas, C_2H_2. **(b)** When solid potassium chlorate is heated, it decomposes to form solid potassium chloride and oxygen gas. **(c)** Solid zinc metal reacts with sulfuric acid to form hydrogen gas and an aqueous solution of zinc sulfate. **(d)** When liquid phosphorus trichloride is added to water, it reacts to form aqueous phosphorous acid, $H_3PO_3(aq)$, and aqueous hydrochloric acid. **(e)** When hydrogen sulfide gas is passed over solid hot iron(III) hydroxide, the resultant reaction produces solid iron(III) sulfide and gaseous water.

3.16 Write balanced chemical equations to correspond to each of the following descriptions: **(a)** When sulfur trioxide gas reacts with water, a solution of sulfuric acid forms. **b)** Boron sulfide, $B_2S_3(s)$, reacts violently with water to form dissolved boric acid, H_3BO_3, and hydrogen sulfide gas. **(c)** Phosphine, $PH_3(g)$, combusts in oxygen gas to form water vapor and solid tetraphosphorus decaoxide. **(d)** When solid mercury(II) nitrate is heated, it decomposes to form solid mercury(II) oxide, gaseous nitrogen dioxide, and oxygen. **(e)** Copper metal reacts with hot concentrated sulfuric acid solution to form aqueous copper(II) sulfate, sulfur dioxide gas, and water.

Patterns of Chemical Reactivity (Section 3.2)

3.17 **(a)** When the metallic element sodium combines with the nonmetallic element bromine, $Br_2(l)$, what is the chemical formula of the product? **(b)** Is the product a solid, liquid, or gas at room temperature? **(c)** In the balanced chemical equation for this reaction, what is the coefficient in front of the product?

3.18 **(a)** When a compound containing C, H, and O is completely combusted in air, what reactant besides the hydrocarbon is involved in the reaction? **(b)** What products form in this reaction? (c) What is the sum of the coefficients in the balanced chemical equation for the combustion of acetone, $C_3H_6O(l)$, in air?

3.19 Write a balanced chemical equation for the reaction that occurs when (a) $Mg(s)$ reacts with $Cl_2(g)$; (b) barium carbonate decomposes into barium oxide and carbon dioxide gas when heated; (c) the hydrocarbon styrene, $C_8H_8(l)$, is combusted in air; (d) dimethylether, $CH_3OCH_3(g)$, is combusted in air.

3.20 Write a balanced chemical equation for the reaction that occurs when (a) titanium metal undergoes a combination reaction with $O_2(g)$; (b) silver(I) oxide decomposes into silver metal and oxygen gas when heated; (c) propanol, $C_3H_7OH(l)$ burns in air; (d) methyl tert-butyl ether, $C_5H_{12}O(l)$, burns in air.

3.21 Balance the following equations and indicate whether they are combination, decomposition, or combustion reactions:

(a) $C_3H_6(g) + O_2(g) \longrightarrow CO_2(g) + H_2O(g)$

(b) $NH_4NO_3(s) \longrightarrow N_2O(g) + H_2O(g)$

(c) $C_5H_6O(l) + O_2(g) \longrightarrow CO_2(g) + H_2O(g)$

(d) $N_2(g) + H_2(g) \longrightarrow NH_3(g)$

(e) $K_2O(s) + H_2O(l) \longrightarrow KOH(aq)$

3.22 Balance the following equations and indicate whether they are combination, decomposition, or combustion reactions:

(a) $PbCO_3(s) \longrightarrow PbO(s) + CO_2(g)$

(b) $C_2H_4(g) + O_2(g) \longrightarrow CO_2(g) + H_2O(g)$

(c) $Mg(s) + N_2(g) \longrightarrow Mg_3N_2(s)$

(d) $C_7H_8O_2(l) + O_2(g) \longrightarrow CO_2(g) + H_2O(g)$

(e) $Al(s) + Cl_2(g) \longrightarrow AlCl_3(s)$

Formula Weights (Section 3.3)

3.23 Determine the formula weights of each of the following compounds: (a) nitric acid, HNO_3; (b) $KMnO_4$; (c) $Ca_3(PO_4)_2$; (d) quartz, SiO_2; (e) gallium sulfide, (f) chromium(III) sulfate, (g) phosphorus trichloride.

3.24 Determine the formula weights of each of the following compounds: (a) nitrous oxide, N_2O, known as laughing gas and used as an anesthetic in dentistry; (b) benzoic acid; $HC_7H_5O_2$, a substance used as a food preservative; (c) $Mg(OH)_2$, the active ingredient in milk of magnesia; (d) urea, $(NH_2)_2CO$, a compound used as a nitrogen fertilizer; (e) isopentyl acetate, $CH_3CO_2C_5H_{11}$, responsible for the odor of bananas.

3.25 Calculate the percentage by mass of oxygen in the following compounds: (a) morphine, $C_{17}H_{19}NO_3$; (b) codeine, $C_{18}H_{21}NO_3$ (c) cocaine, $C_{17}H_{21}NO_4$; (d) tetracycline, $C_{22}H_{24}N_2O_8$; (e) digitoxin, $C_{41}H_{64}O_{13}$; (f) vancomycin, $C_{66}H_{75}Cl_2N_9O_{24}$.

3.26 Calculate the percentage by mass of the indicated element in the following compounds: (a) carbon in acetylene, C_2H_2, a gas used in welding; (b) hydrogen in ascorbic acid, $HC_6H_7O_6$, also known as vitamin C; (c) hydrogen in ammonium sulfate, $(NH_4)_2SO_4$, a substance used as a nitrogen fertilizer; (d) platinum in $PtCl_2(NH_3)_2$, a chemotherapy agent called cisplatin; (e) oxygen in the female sex hormone estradiol, $C_{18}H_{24}O_2$; (f) carbon in capsaicin, $C_{18}H_{27}NO_3$, the compound that gives the hot taste to chili peppers.

3.27 Based on the following structural formulas, calculate the percentage of carbon by mass present in each compound:

(a) Benzaldehyde (almond fragrance)

(b) Vanillin (vanilla flavor)

(c) Isopentyl acetate (banana flavor)

3.28 Calculate the percentage of carbon by mass in each of the compounds represented by the following models:

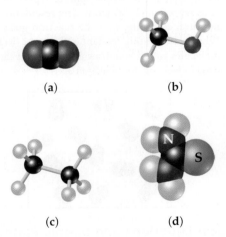

(a)   (b)

(c)   (d)

Avogadro's Number and the Mole (Section 3.4)

3.29 (a) The world population is estimated to be approximately 7 billion people. How many moles of people are there? (b) What units are typically used to express formula weight? (c) What units are typically used to express molar mass?

3.30 (a) What is the mass, in grams, of a mole of ^{12}C? (b) How many carbon atoms are present in a mole of ^{12}C?

3.31 Without doing any detailed calculations (but using a periodic table to give atomic weights), rank the following samples in order of increasing numbers of atoms: 0.50 mol H_2O, 23 g Na, 6.0×10^{23} N_2 molecules.

3.32 Without doing any detailed calculations (but using a periodic table to give atomic weights), rank the following samples in order of increasing numbers of atoms: 42 g of $NaHCO_3$, 1.5 mol CO_2, 6.0×10^{24} Ne atoms.

3.33 What is the mass, in kilograms, of an Avogadro's number of people, if the average mass of a person is 160 lb? How does this compare with the mass of Earth, 5.98×10^{24} kg?

3.34 If Avogadro's number of pennies is divided equally among the 314 million men, women, and children in the United States, how many dollars would each receive? How does this compare with the gross domestic product (GDP) of the United States, which was $15.1 trillion in 2011? (The GDP is the total market value of the nation's goods and services.)

3.35 Calculate the following quantities:

(a) mass, in grams, of 0.105 mol sucrose ($C_{12}H_{22}O_{11}$)

(b) moles of $Zn(NO_3)_2$ in 143.50 g of this substance

(c) number of molecules in 1.0×10^{-6} mol CH_3CH_2OH

(d) number of N atoms in 0.410 mol NH_3

3.36 Calculate the following quantities:

(a) mass, in grams, of 1.50×10^{-2} mol CdS

(b) number of moles of NH_4Cl in 86.6 g of this substance

(c) number of molecules in 8.447×10^{-2} mol C_6H_6

(d) number of O atoms in 6.25×10^{-3} mol $Al(NO_3)_3$

3.37 (a) What is the mass, in grams, of 2.50×10^{-3} mol of ammonium phosphate?

(b) How many moles of chloride ions are in 0.2550 g of aluminum chloride?

(c) What is the mass, in grams, of 7.70×10^{20} molecules of caffeine, $C_8H_{10}N_4O_2$?

(d) What is the molar mass of cholesterol if 0.00105 mol has a mass of 0.406 g?

3.38 (a) What is the mass, in grams, of 1.223 mol of iron (III) sulfate?

(b) How many moles of ammonium ions are in 6.955 g of ammonium carbonate?

(c) What is the mass, in grams, of 1.50×10^{21} molecules of aspirin, $C_9H_8O_4$?

(d) What is the molar mass of diazepam (Valium®) if 0.05570 mol has a mass of 15.86 g?

3.39 The molecular formula of allicin, the compound responsible for the characteristic smell of garlic, is $C_6H_{10}OS_2$. (a) What is the molar mass of allicin? (b) How many moles of allicin are present in 5.00 mg of this substance? (c) How many molecules of allicin are in 5.00 mg of this substance? (d) How many S atoms are present in 5.00 mg of allicin?

3.40 The molecular formula of aspartame, the artificial sweetener marketed as NutraSweet®, is $C_{14}H_{18}N_2O_5$. (a) What is the molar mass of aspartame? (b) How many moles of aspartame are present in 1.00 mg of aspartame? (c) How many molecules of aspartame are present in 1.00 mg of aspartame? (d) How many hydrogen atoms are present in 1.00 mg of aspartame?

3.41 A sample of glucose, $C_6H_{12}O_6$, contains 1.250×10^{21} carbon atoms. (a) How many atoms of hydrogen does it contain? (b) How many molecules of glucose does it contain? (c) How many moles of glucose does it contain? (d) What is the mass of this sample in grams?

3.42 A sample of the male sex hormone testosterone, $C_{19}H_{28}O_2$, contains 3.88×10^{21} hydrogen atoms. (a) How many atoms of carbon does it contain? (b) How many molecules of testosterone does it contain? (c) How many moles of testosterone does it contain? (d) What is the mass of this sample in grams?

3.43 The allowable concentration level of vinyl chloride, C_2H_3Cl, in the atmosphere in a chemical plant is 2.0×10^{-6} g/L. How many moles of vinyl chloride in each liter does this represent? How many molecules per liter?

3.44 At least 25 μg of tetrahydrocannabinol (THC), the active ingredient in marijuana, is required to produce intoxication. The molecular formula of THC is $C_{21}H_{30}O_2$. How many moles of THC does this 25 μg represent? How many molecules?

Empirical Formulas from Analyses (Section 3.5)

3.45 Give the empirical formula of each of the following compounds if a sample contains (a) 0.0130 mol C, 0.0390 mol H, and 0.0065 mol O; (b) 11.66 g iron and 5.01 g oxygen; (c) 40.0% C, 6.7% H, and 53.3% O by mass.

3.46 Determine the empirical formula of each of the following compounds if a sample contains (a) 0.104 mol K, 0.052 mol C, and 0.156 mol O; (b) 5.28 g Sn and 3.37 g F; (c) 87.5% N and 12.5% H by mass.

3.47 Determine the empirical formulas of the compounds with the following compositions by mass:

(a) 10.4% C, 27.8% S, and 61.7% Cl

(b) 21.7% C, 9.6% O, and 68.7% F

(c) 32.79% Na, 13.02% Al, and the remainder F

3.48 Determine the empirical formulas of the compounds with the following compositions by mass:

(a) 55.3% K, 14.6% P, and 30.1% O

(b) 24.5% Na, 14.9% Si, and 60.6% F

(c) 62.1% C, 5.21% H, 12.1% N, and the remainder O

3.49 A compound whose empirical formula is XF_3 consists of 65% F by mass. What is the atomic mass of X?

3.50 The compound XCl_4 contains 75.0% Cl by mass. What is the element X?

3.51 What is the molecular formula of each of the following compounds?

(a) empirical formula CH_2, molar mass = 84 g/mol

(b) empirical formula NH_2Cl, molar mass = 51.5 g/mol

3.52 What is the molecular formula of each of the following compounds?

(a) empirical formula HCO_2, molar mass = 90.0 g/mol

(b) empirical formula C_2H_4O, molar mass = 88 g/mol

3.53 Determine the empirical and molecular formulas of each of the following substances:

(a) Styrene, a compound substance used to make Styrofoam® cups and insulation, contains 92.3% C and 7.7% H by mass and has a molar mass of 104 g/mol.

(b) Caffeine, a stimulant found in coffee, contains 49.5% C, 5.15% H, 28.9% N, and 16.5% O by mass and has a molar mass of 195 g/mol.

(c) Monosodium glutamate (MSG), a flavor enhancer in certain foods, contains 35.51% C, 4.77% H, 37.85% O, 8.29% N, and 13.60% Na, and has a molar mass of 169 g/mol.

3.54 Determine the empirical and molecular formulas of each of the following substances:

(a) Ibuprofen, a headache remedy, contains 75.69% C, 8.80% H, and 15.51% O by mass, and has a molar mass of 206 g/mol.

(b) Cadaverine, a foul-smelling substance produced by the action of bacteria on meat, contains 58.55% C, 13.81% H, and 27.40% N by mass; its molar mass is 102.2 g/mol.

(c) Epinephrine (adrenaline), a hormone secreted into the bloodstream in times of danger or stress, contains 59.0% C, 7.1% H, 26.2% O, and 7.7% N by mass; its MW is about 180 amu.

3.55 (a) Combustion analysis of toluene, a common organic solvent, gives 5.86 mg of CO_2 and 1.37 mg of H_2O. If the compound contains only carbon and hydrogen, what is its empirical formula? **(b)** Menthol, the substance we can smell in mentholated cough drops, is composed of C, H, and O. A 0.1005-g sample of menthol is combusted, producing 0.2829 g of CO_2 and 0.1159 g of H_2O. What is the empirical formula for menthol? If menthol has a molar mass of 156 g/mol, what is its molecular formula?

3.56 (a) The characteristic odor of pineapple is due to ethyl butyrate, a compound containing carbon, hydrogen, and oxygen. Combustion of 2.78 mg of ethyl butyrate produces 6.32 mg of CO_2 and 2.58 mg of H_2O. What is the empirical formula of the compound? **(b)** Nicotine, a component of tobacco, is composed of C, H, and N. A 5.250-mg sample of nicotine was combusted, producing 14.242 mg of CO_2 and 4.083 mg of H_2O. What is the empirical formula for nicotine? If nicotine has a molar mass of 160 ± 5 g/mol, what is its molecular formula?

3.57 Valproic acid, used to treat seizures and bipolar disorder, is composed of C, H, and O. A 0.165-g sample is combusted in an apparatus such as that shown in Figure 3.14. The gain in mass of the H_2O absorber is 0.166 g, whereas that of the CO_2 absorber is 0.403 g. What is the empirical formula for valproic acid? If the molar mass is 144 g/mol what is the molecular formula?

3.58 Propenoic acid is a reactive organic liquid used in the manufacture of plastics, coatings, and adhesives. An unlabeled container is thought to contain this acid. A 0.2033-g sample is combusted in an apparatus such as that shown in Figure 3.14. The gain in mass of the H_2O absorber is 0.102 g, whereas that of the CO_2 absorber is 0.374 g. What is the empirical formula of propenoic acid?

3.59 Washing soda, a compound used to prepare hard water for washing laundry, is a hydrate, which means that a certain number of water molecules are included in the solid structure. Its formula can be written as $Na_2CO_3 \cdot xH_2O$, where x is the number of moles of H_2O per mole of Na_2CO_3. When a 2.558-g sample of washing soda is heated at 125 °C, all the water of hydration is lost, leaving 0.948 g of Na_2CO_3. What is the value of x?

3.60 Epsom salts, a strong laxative used in veterinary medicine, is a hydrate, which means that a certain number of water molecules are included in the solid structure. The formula for Epsom salts can be written as $MgSO_4 \cdot xH_2O$, where x indicates the number of moles of H_2O per mole of $MgSO_4$. When 5.061 g of this hydrate is heated to 250 °C, all the water of hydration is lost, leaving 2.472 g of $MgSO_4$. What is the value of x?

Quantitative Information from Balanced Equations (Section 3.6)

3.61 Hydrofluoric acid, HF(aq), cannot be stored in glass bottles because compounds called silicates in the glass are attacked by the HF(aq). Sodium silicate (Na_2SiO_3), for example, reacts as follows:

$$Na_2SiO_3(s) + 8\,HF(aq) \longrightarrow$$
$$H_2SiF_6(aq) + 2\,NaF(aq) + 3\,H_2O(l)$$

(a) How many moles of HF are needed to react with 0.300 mol of Na_2SiO_3?

(b) How many grams of NaF form when 0.500 mol of HF reacts with excess Na_2SiO_3?

(c) How many grams of Na_2SiO_3 can react with 0.800 g of HF?

3.62 The reaction between potassium superoxide, KO_2, and CO_2,

$$4\,KO_2 + 2\,CO_2 \longrightarrow 2\,K_2CO_3 + 3\,O_2$$

is used as a source of O_2 and absorber of CO_2 in self-contained breathing equipment used by rescue workers.

(a) How many moles of O_2 are produced when 0.400 mol of KO_2 reacts in this fashion?

(b) How many grams of KO_2 are needed to form 7.50 g of O_2?

(c) How many grams of CO_2 are used when 7.50 g of O_2 are produced?

3.63 Several brands of antacids use $Al(OH)_3$ to react with stomach acid, which contains primarily HCl:

$$Al(OH)_3(s) + HCl(aq) \longrightarrow AlCl_3(aq) + H_2O(l)$$

(a) Balance this equation.

(b) Calculate the number of grams of HCl that can react with 0.500 g of $Al(OH)_3$.

(c) Calculate the number of grams of $AlCl_3$ and the number of grams of H_2O formed when 0.500 g of $Al(OH)_3$ reacts.

(d) Show that your calculations in parts (b) and (c) are consistent with the law of conservation of mass.

3.64 An iron ore sample contains Fe_2O_3 together with other substances. Reaction of the ore with CO produces iron metal:

$$Fe_2O_3(s) + CO(g) \longrightarrow Fe(s) + CO_2(g)$$

(a) Balance this equation.

(b) Calculate the number of grams of CO that can react with 0.350 kg of Fe_2O_3.

(c) Calculate the number of grams of Fe and the number of grams of CO_2 formed when 0.350 kg of Fe_2O_3 reacts.

(d) Show that your calculations in parts (b) and (c) are consistent with the law of conservation of mass.

3.65 Aluminum sulfide reacts with water to form aluminum hydroxide and hydrogen sulfide. (**a**) Write the balanced chemical equation for this reaction. (**b**) How many grams of aluminum hydroxide are obtained from 14.2 g of aluminum sulfide?

3.66 Calcium hydride reacts with water to form calcium hydroxide and hydrogen gas. (**a**) Write a balanced chemical equation for the reaction. (**b**) How many grams of calcium hydride are needed to form 4.500 g of hydrogen?

3.67 Automotive air bags inflate when sodium azide, NaN_3, rapidly decomposes to its component elements:

$$2 NaN_3(s) \longrightarrow 2 Na(s) + 3 N_2(g)$$

(**a**) How many moles of N_2 are produced by the decomposition of 1.50 mol of NaN_3?

(**b**) How many grams of NaN_3 are required to form 10.0 g of nitrogen gas?

(**c**) How many grams of NaN_3 are required to produce 10.0 ft^3 of nitrogen gas, about the size of an automotive air bag, if the gas has a density of 1.25 g/L?

3.68 The complete combustion of octane, C_8H_{18}, a component of gasoline, proceeds as follows:

$$2 C_8H_{18}(l) + 25 O_2(g) \longrightarrow 16 CO_2(g) + 18 H_2O(g)$$

(**a**) How many moles of O_2 are needed to burn 1.50 mol of C_8H_{18}?

(**b**) How many grams of O_2 are needed to burn 10.0 g of C_8H_{18}?

(**c**) Octane has a density of 0.692 g/mL at 20 °C. How many grams of O_2 are required to burn 15.0 gal of C_8H_{18} (the capacity of an average fuel tank)?

(**d**) How many grams of CO_2 are produced when 15.0 gal of C_8H_{18} are combusted?

3.69 A piece of aluminum foil 1.00 cm^2 and 0.550-mm thick is allowed to react with bromine to form aluminum bromide.

(**a**) How many moles of aluminum were used? (The density of aluminum is 2.699 g/cm^3.) (**b**) How many grams of aluminum bromide form, assuming the aluminum reacts completely?

3.70 Detonation of nitroglycerin proceeds as follows:

$$4 C_3H_5N_3O_9(l) \longrightarrow$$
$$12 CO_2(g) + 6 N_2(g) + O_2(g) + 10 H_2O(g)$$

(**a**) If a sample containing 2.00 mL of nitroglycerin (density = 1.592 g/mL) is detonated, how many total moles of gas are produced? (**b**) If each mole of gas occupies 55 L under the conditions of the explosion, how many liters of gas are produced? (**c**) How many grams of N_2 are produced in the detonation?

Limiting Reactants (Section 3.7)

3.71 (**a**) Define the terms *limiting reactant* and *excess reactant*. (**b**) Why are the amounts of products formed in a reaction determined only by the amount of the limiting reactant? (**c**) Why should you base your choice of which compound is the limiting reactant on its number of initial moles, not on its initial mass in grams?

3.72 (**a**) Define the terms *theoretical yield*, *actual yield*, and *percent yield*. (**b**) Why is the actual yield in a reaction almost always less than the theoretical yield? (**c**) Can a reaction ever have 110% actual yield?

3.73 A manufacturer of bicycles has 4815 wheels, 2305 frames, and 2255 handlebars. (**a**) How many bicycles can be manufactured using these parts? (**b**) How many parts of each kind are left over? (**c**) Which part limits the production of bicycles?

3.74 A bottling plant has 126,515 bottles with a capacity of 355 mL, 108,500 caps, and 48,775 L of beverage. (**a**) How many bottles can be filled and capped? (**b**) How much of each item is left over? (**c**) Which component limits the production?

3.75 Consider the mixture of ethanol, C_2H_5OH, and O_2 shown in the accompanying diagram. (**a**) Write a balanced equation for the combustion reaction that occurs between ethanol and oxygen. (**b**) Which reactant is the limiting reactant? (**c**) How many molecules of CO_2, H_2O, C_2H_5OH, and O_2 will be present if the reaction goes to completion?

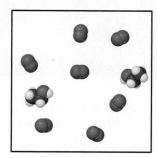

3.76 Consider the mixture of propane, C_3H_8, and O_2 shown below. (**a**) Write a balanced equation for the combustion reaction that occurs between propane and oxygen. (**b**) Which reactant is the limiting reactant? (**c**) How many molecules of CO_2, H_2O, C_3H_8, and O_2 will be present if the reaction goes to completion?

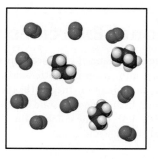

3.77 Sodium hydroxide reacts with carbon dioxide as follows:

$$2\,NaOH(s) + CO_2(g) \longrightarrow Na_2CO_3(s) + H_2O(l)$$

Which is the limiting reactant when 1.85 mol NaOH and 1.00 mol CO_2 are allowed to react? How many moles of Na_2CO_3 can be produced? How many moles of the excess reactant remain after the completion of the reaction?

3.78 Aluminum hydroxide reacts with sulfuric acid as follows:

$$2\,Al(OH)_3(s) + 3\,H_2SO_4(aq) \longrightarrow$$
$$Al_2(SO_4)_3(aq) + 6\,H_2O(l)$$

Which is the limiting reactant when 0.500 mol $Al(OH)_3$ and 0.500 mol H_2SO_4 are allowed to react? How many moles of $Al_2(SO_4)_3$ can form under these conditions? How many moles of the excess reactant remain after the completion of the reaction?

3.79 The fizz produced when an Alka-Seltzer tablet is dissolved in water is due to the reaction between sodium bicarbonate ($NaHCO_3$) and citric acid ($H_3C_6H_5O_7$):

$$3\,NaHCO_3(aq) + H_3C_6H_5O_7(aq) \longrightarrow$$
$$3\,CO_2(g) + 3\,H_2O(l) + Na_3C_6H_5O_7(aq)$$

In a certain experiment 1.00 g of sodium bicarbonate and 1.00 g of citric acid are allowed to react. **(a)** Which is the limiting reactant? **(b)** How many grams of carbon dioxide form? **(c)** How many grams of the excess reactant remain after the limiting reactant is completely consumed?

3.80 One of the steps in the commercial process for converting ammonia to nitric acid is the conversion of NH_3 to NO:

$$4\,NH_3(g) + 5\,O_2(g) \longrightarrow 4\,NO(g) + 6\,H_2O(g)$$

In a certain experiment, 2.00 g of NH_3 reacts with 2.50 g of O_2. **(a)** Which is the limiting reactant? **(b)** How many grams of NO and H_2O form? **(c)** How many grams of the excess reactant remain after the limiting reactant is completely consumed? **(d)** Show that your calculations in parts (b) and (c) are consistent with the law of conservation of mass.

3.81 Solutions of sodium carbonate and silver nitrate react to form solid silver carbonate and a solution of sodium nitrate. A solution containing 3.50 g of sodium carbonate is mixed with one containing 5.00 g of silver nitrate. How many grams of sodium carbonate, silver nitrate, silver carbonate, and sodium nitrate are present after the reaction is complete?

3.82 Solutions of sulfuric acid and lead(II) acetate react to form solid lead(II) sulfate and a solution of acetic acid. If 5.00 g of sulfuric acid and 5.00 g of lead(II) acetate are mixed, calculate the number of grams of sulfuric acid, lead(II) acetate, lead(II) sulfate, and acetic acid present in the mixture after the reaction is complete.

3.83 When benzene (C_6H_6) reacts with bromine (Br_2), bromobenzene (C_6H_5Br) is obtained:

$$C_6H_6 + Br_2 \longrightarrow C_6H_5Br + HBr$$

(a) When 30.0 g of benzene reacts with 65.0 g of bromine, what is the theoretical yield of bromobenzene? **(b)** If the actual yield of bromobenzene is 42.3 g, what is the percentage yield?

3.84 When ethane (C_2H_6) reacts with chlorine (Cl_2), the main product is C_2H_5Cl, but other products containing Cl, such as $C_2H_4Cl_2$, are also obtained in small quantities. The formation of these other products reduces the yield of C_2H_5Cl. **(a)** Calculate the theoretical yield of C_2H_5Cl when 125 g of C_2H_6 reacts with 255 g of Cl_2, assuming that C_2H_6 and Cl_2 react only to form C_2H_2Cl and HCl. **(b)** Calculate the percent yield of C_2H_5Cl if the reaction produces 206 g of C_2H_5Cl.

3.85 Hydrogen sulfide is an impurity in natural gas that must be removed. One common removal method is called the Claus process, which relies on the reaction:

$$8\,H_2S(g) + 4\,O_2(g) \longrightarrow S_8(l) + 8\,H_2O(g)$$

Under optimal conditions the Claus process gives 98% yield of S_8 from H_2S. If you started with 30.0 grams of H_2S and 50.0 grams of O_2, how many grams of S_8 would be produced, assuming 98% yield?

3.86 When hydrogen sulfide gas is bubbled into a solution of sodium hydroxide, the reaction forms sodium sulfide and water. How many grams of sodium sulfide are formed if 1.25 g of hydrogen sulfide is bubbled into a solution containing 2.00 g of sodium hydroxide, assuming that the sodium sulfide is made in 92.0% yield?

Additional Exercises

3.87 Write the balanced chemical equations for **(a)** the complete combustion of acetic acid (CH_3COOH), the main active ingredient in vinegar; **(b)** the decomposition of solid calcium hydroxide into solid calcium (II) oxide (lime) and water vapor; **(c)** the combination reaction between nickel metal and chlorine gas.

3.88 If 1.5 mol C_2H_5OH, 1.5 mol C_3H_8, and 1.5 mol $CH_3CH_2COCH_3$ are completely combusted in oxygen, which produces the largest number of moles of H_2O? Which produces the least? Explain.

3.89 The effectiveness of nitrogen fertilizers depends on both their ability to deliver nitrogen to plants and the amount of nitrogen they can deliver. Four common nitrogen-containing fertilizers are ammonia, ammonium nitrate, ammonium sulfate, and urea $[(NH_2)_2CO]$. Rank these fertilizers in terms of the mass percentage nitrogen they contain.

3.90 (a) The molecular formula of acetylsalicylic acid (aspirin), one of the most common pain relievers, is $C_9H_8O_4$. How many moles of $C_9H_8O_4$ are in a 0.500-g tablet of aspirin? (b) How many molecules of $C_9H_8O_4$ are in this tablet? (c) How many carbon atoms are in the tablet?

3.91 Very small crystals composed of 1000 to 100,000 atoms, called quantum dots, are being investigated for use in electronic devices.

    (a) A quantum dot was made of solid silicon in the shape of a sphere, with a diameter of 4 nm. Calculate the mass of the quantum dot, using the density of silicon (2.3 g/cm^3).

    (b) How many silicon atoms are in the quantum dot?

    (c) The density of germanium is 5.325 g/cm^3. If you made a 4-nm quantum dot of germanium, how many Ge atoms would it contain? Assume the dot is spherical.

3.92 (a) One molecule of the antibiotic penicillin G has a mass of 5.342×10^{-21} g. What is the molar mass of penicillin G?

    (b) Hemoglobin, the oxygen-carrying protein in red blood cells, has four iron atoms per molecule and contains 0.340% iron by mass. Calculate the molar mass of hemoglobin.

3.93 Serotonin is a compound that conducts nerve impulses in the brain. It contains 68.2 mass percent C, 6.86 mass percent H, 15.9 mass percent N, and 9.08 mass percent O. Its molar mass is 176 g/mol. Determine its molecular formula.

3.94 The koala dines exclusively on eucalyptus leaves. Its digestive system detoxifies the eucalyptus oil, a poison to other animals. The chief constituent in eucalyptus oil is a substance called eucalyptol, which contains 77.87% C, 11.76% H, and the remainder O. (a) What is the empirical formula for this substance? (b) A mass spectrum of eucalyptol shows a peak at about 154 amu. What is the molecular formula of the substance?

3.95 Vanillin, the dominant flavoring in vanilla, contains C, H, and O. When 1.05 g of this substance is completely combusted, 2.43 g of CO_2 and 0.50 g of H_2O are produced. What is the empirical formula of vanillin?

3.96 An organic compound was found to contain only C, H, and Cl. When a 1.50-g sample of the compound was completely combusted in air, 3.52 g of CO_2 was formed. In a separate experiment the chlorine in a 1.00-g sample of the compound was converted to 1.27 g of AgCl. Determine the empirical formula of the compound.

3.97 A compound, $KBrO_x$, where x is unknown, is analyzed and found to contain 52.92% Br. What is the value of x?

3.98 An element X forms an iodide (XI_3) and a chloride (XCl_3). The iodide is quantitatively converted to the chloride when it is heated in a stream of chlorine:

$$2 \text{ XI}_3 + 3 \text{ Cl}_2 \longrightarrow 2 \text{ XCl}_3 + 3 \text{ I}_2$$

If 0.5000 g of XI_3 is treated, 0.2360 g of XCl_3 is obtained. (a) Calculate the atomic weight of the element X. (b) Identify the element X.

3.99 A method used by the U.S. Environmental Protection Agency (EPA) for determining the concentration of ozone in air is to pass the air sample through a "bubbler" containing sodium iodide, which removes the ozone according to the following equation:

$$O_3(g) + 2 \text{ NaI}(aq) + H_2O(l) \longrightarrow$$
$$O_2(g) + I_2(s) + 2 \text{ NaOH}(aq)$$

(a) How many moles of sodium iodide are needed to remove 5.95×10^{-6} mol of O_3? (b) How many grams of sodium iodide are needed to remove 1.3 mg of O_3?

3.100 A chemical plant uses electrical energy to decompose aqueous solutions of NaCl to give Cl_2, H_2, and NaOH:

$$2 \text{ NaCl}(aq) + 2 \text{ H}_2O(l) \longrightarrow 2 \text{ NaOH}(aq) + H_2(g) + Cl_2(g)$$

If the plant produces 1.5×10^6 kg (1500 metric tons) of Cl_2 daily, estimate the quantities of H_2 and NaOH produced.

3.101 The fat stored in a camel's hump is a source of both energy and water. Calculate the mass of H_2O produced by the metabolism of 1.0 kg of fat, assuming the fat consists entirely of tristearin $(C_{57}H_{110}O_6)$, a typical animal fat, and assuming that during metabolism, tristearin reacts with O_2 to form only CO_2 and H_2O.

3.102 When hydrocarbons are burned in a limited amount of air, both CO and CO_2 form. When 0.450 g of a particular hydrocarbon was burned in air, 0.467 g of CO, 0.733 g of CO_2, and 0.450 g of H_2O were formed. (a) What is the empirical formula of the compound? (b) How many grams of O_2 were used in the reaction? (c) How many grams would have been required for complete combustion?

3.103 A mixture of $N_2(g)$ and $H_2(g)$ reacts in a closed container to form ammonia, $NH_3(g)$. The reaction ceases before either reactant has been totally consumed. At this stage 3.0 mol N_2, 3.0 mol H_2, and 3.0 mol NH_3 are present. How many moles of N_2 and H_2 were present originally?

3.104 A mixture containing $KClO_3$, K_2CO_3, $KHCO_3$, and KCl was heated, producing CO_2, O_2, and H_2O gases according to the following equations:

$$2 \text{ KClO}_3(s) \longrightarrow 2 \text{ KCl}(s) + 3 \text{ O}_2(g)$$
$$2 \text{ KHCO}_3(s) \longrightarrow K_2O(s) + H_2O(g) + 2 \text{ CO}_2(g)$$
$$K_2CO_3(s) \longrightarrow K_2O(s) + CO_2(g)$$

The KCl does not react under the conditions of the reaction. If 100.0 g of the mixture produces 1.80 g of H_2O, 13.20 g of CO_2, and 4.00 g of O_2, what was the composition of the original mixture? (Assume complete decomposition of the mixture.)

3.105 When a mixture of 10.0 g of acetylene (C_2H_2) and 10.0 g of oxygen (O_2) is ignited, the resultant combustion reaction produces CO_2 and H_2O. (a) Write the balanced chemical equation for this reaction. (b) Which is the limiting reactant? (c) How many grams of C_2H_2, O_2, CO_2, and H_2O are present after the reaction is complete?

Integrative Exercises

These exercises require skills from earlier chapters as well as skills from the present chapter.

3.106   Consider a sample of calcium carbonate in the form of a cube measuring 2.005 in. on each edge. If the sample has a density of 2.71 g/cm^3, how many oxygen atoms does it contain?

3.107   (a) You are given a cube of silver metal that measures 1.000 cm on each edge. The density of silver is 10.5 g/cm^3. How many atoms are in this cube? (b) Because atoms are spherical, they cannot occupy all of the space of the cube. The silver atoms pack in the solid in such a way that 74% of the volume of the solid is actually filled with the silver atoms. Calculate the volume of a single silver atom. (c) Using the volume of a silver atom and the formula for the volume of a sphere, calculate the radius in angstroms of a silver atom.

3.108   (a) If an automobile travels 225 mi with a gas mileage of 20.5 mi/gal, how many kilograms of CO_2 are produced? Assume that the gasoline is composed of octane, $C_8H_{18}(l)$, whose density is 0.69 g/mL. (b) Repeat the calculation for a truck that has a gas mileage of 5 mi/gal.

3.109   ∞ Section 2.9 introduced the idea of structural isomerism, with 1-propanol and 2-propanol as examples. Determine which of these properties would distinguish these two substances: (a) boiling point, (b) combustion analysis results, (c) molecular weight, (d) density at a given temperature and pressure. You can check on the properties of these two compounds in *Wolfram Alpha* (http://www.wolframalpha.com/) or the *CRC Handbook of Chemistry and Physics*.

3.110   A particular coal contains 2.5% sulfur by mass. When this coal is burned at a power plant, the sulfur is converted into sulfur dioxide gas, which is a pollutant. To reduce sulfur dioxide emissions, calcium oxide (lime) is used. The sulfur dioxide reacts with calcium oxide to form solid calcium sulfite. (a) Write the balanced chemical equation for the reaction. (b) If the coal is burned in a power plant that uses 2000 tons of coal per day, what mass of calcium oxide is required daily to eliminate the sulfur dioxide? (c) How many grams of calcium sulfite are produced daily by this power plant?

3.111   Hydrogen cyanide, HCN, is a poisonous gas. The lethal dose is approximately 300 mg HCN per kilogram of air when inhaled. (a) Calculate the amount of HCN that gives the lethal dose in a small laboratory room measuring 12 × 15 × 8.0 ft. The density of air at 26 °C is 0.00118 g/cm^3. (b) If the HCN is formed by reaction of NaCN with an acid such as H_2SO_4, what mass of NaCN gives the lethal dose in the room?

$$2\,NaCN(s) + H_2SO_4(aq) \longrightarrow Na_2SO_4(aq) + 2\,HCN(g)$$

(c) HCN forms when synthetic fibers containing Orlon® or Acrilan® burn. Acrilan® has an empirical formula of CH_2CHCN, so HCN is 50.9% of the formula by mass. A rug measures 12 × 15 ft and contains 30 oz of Acrilan® fibers per square yard of carpet. If the rug burns, will a lethal dose of HCN be generated in the room? Assume that the yield of HCN from the fibers is 20% and that the carpet is 50% consumed.

3.112   The source of oxygen that drives the internal combustion engine in an automobile is air. Air is a mixture of gases, principally N_2(~79%) and O_2(~20%). In the cylinder of an automobile engine, nitrogen can react with oxygen to produce nitric oxide gas, NO. As NO is emitted from the tailpipe of the car, it can react with more oxygen to produce nitrogen dioxide gas. (a) Write balanced chemical equations for both reactions. (b) Both nitric oxide and nitrogen dioxide are pollutants that can lead to acid rain and global warming; collectively, they are called "NO$_x$" gases. In 2007, the United States emitted an estimated 22 million tons of nitrogen dioxide into the atmosphere. How many grams of nitrogen dioxide is this? (c) The production of NO$_x$ gases is an unwanted side reaction of the main engine combustion process that turns octane, C_8H_{18}, into CO_2 and water. If 85% of the oxygen in an engine is used to combust octane, and the remainder used to produce nitrogen dioxide, calculate how many grams of nitrogen dioxide would be produced during the combustion of 500 g of octane.

Design an Experiment

You will learn later in this book that sulfur is capable of forming two common oxides, SO_2 and SO_3. One question that we might ask is whether the direct reaction between sulfur and oxygen leads to the formation of SO_2, SO_3, or a mixture of the two. This question has practical significance because SO_3 can go onto react with water to form sulfuric acid, H_2SO_4, which is produced industrially on a very large scale. Consider also that the answer to this question may depend on the relative amount of each element that is present and the temperature at which the reaction is carried out. For example, carbon and oxygen normally react to form CO_2 but, when there is not enough oxygen present, CO can form. On the other hand, under normal reaction conditions H_2 and O_2 react to form water, H_2O (rather than hydrogen peroxide H_2O_2) regardless of the starting ratio of hydrogen to oxygen.

Suppose you are given a bottle of sulfur, which is a yellow solid, a cylinder of O_2, a transparent reaction vessel that can be evacuated and sealed so that only sulfur, oxygen and the product(s) of the reaction between the two are present, an analytical balance so that you can determine the masses of the reactants and/or products, and a furnace that can be used to heat the reaction vessel to 200 °C where the two elements react. (a) If you start with 0.10 mol of sulfur in the reaction vessel how many moles of oxygen would need to be added to form SO_2, assuming SO_2 forms exclusively? (**b**) How many moles of oxygen would be needed to form SO_3, assuming SO_3 forms exclusively? (**c**) Given the available equipment how would you determine if you added the correct number of moles of each reactant to the reaction vessel? (**d**) What observation or experimental technique would you use to determine the identity of the reaction product(s)? Could differences in the physical properties of SO_2 and SO_3 be used to help identify the product(s)? Have any instruments been described Chapters 1–3 that would allow you to identify the product(s)? (**e**) What experiments would you conduct to determine if the product(s) of this reaction (either SO_2 or SO_3 or a mixture of the two) can be controlled by varying the ratio of sulfur and oxygen that are added to the reaction vessel? What ratio(s) of S to O_2 would you test to answer this question?

4

Reactions in Aqueous Solution

Water covers nearly two-thirds of our planet, and this simple substance has been the key to much of Earth's evolutionary history. Life almost certainly originated in water, and the need for water by all forms of life has helped determine diverse biological structures.

Scientists had studied the chemistry of the ocean for decades before discovering deep sea vents in 1979. The chemical reactions that occur near deep sea vents are, as you might imagine, difficult to study; nonetheless, chemists working in deep-sea submersibles equipped with sampling arms are helping us learn what happens in these hot, toxic waters.

One reaction that occurs in deep sea vents is the conversion of FeS into FeS_2:

$$FeS(s) + H_2S(g) \longrightarrow FeS_2(s) + H_2(g) \qquad [4.1]$$

If we could follow the iron atoms in this reaction as it proceeds, we would learn that they gain and lose electrons and are dissolved in water to different extents at different times (Figure 4.1).

A solution in which water is the dissolving medium is called an **aqueous solution**. In this chapter we examine chemical reactions that take place in aqueous solutions. In addition, we extend the concepts of stoichiometry learned in Chapter 3 by considering how solution concentrations are expressed and used. Although the reactions we will discuss in this chapter are relatively simple, they form the basis for understanding very complex reaction cycles in biology, geology, and oceanography.

▶ DEEP SEA VENTS are amazing places. Superheated water (up to 400 °C) is released from cracks in the bottom of the ocean. Rocks dissolve and reform. The locally high mineral content and sulfur-containing substances in the water provide an environment that favors unusual organisms that are found nowhere else in the world.

WHAT'S AHEAD

4.1 GENERAL PROPERTIES OF AQUEOUS SOLUTIONS We begin by examining whether substances dissolved in water exist as ions, molecules, or a mixture of the two.

4.2 PRECIPITATION REACTIONS We identify reactions in which soluble reactants yield an insoluble product.

4.3 ACIDS, BASES, AND NEUTRALIZATION REACTIONS We explore reactions in which protons, H^+ ions, are transferred from one reactant to another.

4.4 OXIDATION-REDUCTION REACTIONS We examine reactions in which electrons are transferred from one reactant to another.

4.5 CONCENTRATIONS OF SOLUTIONS We learn how the amount of a compound dissolved in a given volume of a solution can be expressed as a *concentration*. Concentration can be defined in many ways; the most common way in chemistry is moles of compound per liter of solution (*molarity*).

4.6 SOLUTION STOICHIOMETRY AND CHEMICAL ANALYSIS We see how the concepts of stoichiometry and concentration can be used to calculate amounts or concentrations of substances in solution through a common chemical practice called *titration*.

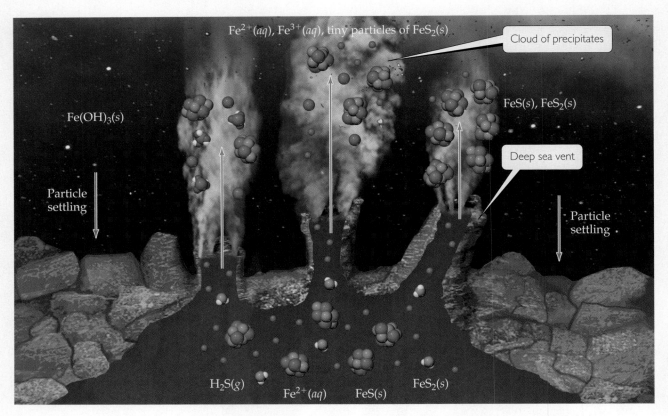

Fe²⁺(aq), Fe³⁺(aq), tiny particles of FeS₂(s)

Cloud of precipitates

Fe(OH)₃(s)

FeS(s), FeS₂(s)

Deep sea vent

Particle settling

Particle settling

H₂S(g)   Fe²⁺(aq)   FeS(s)   FeS₂(s)

▲ Figure 4.1 **Researchers have recently found that compounds containing iron, sulfur, and oxygen are the reactants and products in a deep sea vent.**

4.1 | General Properties of Aqueous Solutions

A *solution* is a homogeneous mixture of two or more substances. ∞ (Section 1.2) The substance present in the greatest quantity is usually called the **solvent**, and the other substances are called **solutes**; they are said to be *dissolved in* the solvent. When a small amount of sodium chloride (NaCl) is dissolved in a large quantity of water, for example, water is the solvent and sodium chloride is the solute.

Electrolytes and Nonelectrolytes

At a young age we learn not to bring electrical devices into the bathtub so as not to electrocute ourselves. That is a useful lesson because most of the water we encounter in daily life is electrically conducting. Pure water, however, is a very poor conductor of electricity. The conductivity of bathwater originates from the substances dissolved in the water, not from the water itself.

Not all substances that dissolve in water make the resulting solution conducting. Experiments show that some solutions conduct electricity better than others. Imagine, for example, preparing two aqueous solutions—one by dissolving a teaspoon of table salt (sodium chloride) in a cup of water and the other by dissolving a teaspoon of table sugar (sucrose) in a cup of water (▶ Figure 4.2). Both solutions are clear and colorless, but they possess very different electrical conductivities: The salt solution is a good conductor of electricity, which we can see from the light bulb turning on. In order for the light bulb in Figure 4.2 to turn on, there must be an electric current (that is, a *flow* of electrically charged particles) between the two electrodes immersed in the solution. The conductivity of pure water is not sufficient to complete the electrical circuit and light

Pure water does not conduct electricity

An **nonelectrolyte** solution does not conduct electricity

An **electrolyte** solution conducts electricity

| Pure water,
$H_2O(l)$ | Sucrose solution,
$C_{12}H_{22}O_{11}(aq)$ | Sodium chloride solution,
$NaCl(aq)$ |

▲ Figure 4.2 **Completion of an electrical circuit with an electrolyte turns on the light.**

the bulb. The situation would change if ions were present in solution, because the ions carry electrical charge from one electrode to the other, completing the circuit. Thus, the conductivity of NaCl solutions indicates the presence of ions. The lack of conductivity of sucrose solutions indicates the absence of ions. When NaCl dissolves in water, the solution contains Na^+ and Cl^- ions, each surrounded by water molecules. When sucrose $(C_{12}H_{22}O_{11})$ dissolves in water, the solution contains only neutral sucrose molecules surrounded by water molecules.

A substance (such as NaCl) whose aqueous solutions contain ions is called an **electrolyte**. A substance (such as $C_{12}H_{22}O_{11}$) that does not form ions in solution is called a **nonelectrolyte**. The different classifications of NaCl and $C_{12}H_{22}O_{11}$ arise largely because NaCl is an ionic compound, whereas $C_{12}H_{22}O_{11}$ is a molecular compound.

How Compounds Dissolve in Water

Recall from Figure 2.19 that solid NaCl consists of an orderly arrangement of Na^+ and Cl^- ions. When NaCl dissolves in water, each ion separates from the solid structure and disperses throughout the solution [**Figure 4.3(a)**]. The ionic solid *dissociates* into its component ions as it dissolves.

Water is a very effective solvent for ionic compounds. Although H_2O is an electrically neutral molecule, the O atom is rich in electrons and has a partial negative charge, while each H atom has a partial positive charge. The lowercase Greek letter delta (δ) is used to denote partial charge: A partial negative charge is denoted δ^- ("delta minus"), and a partial positive charge is denoted by δ^+ ("delta plus"). Cations are attracted by the negative end of H_2O, and anions are attracted by the positive end.

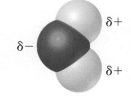

As an ionic compound dissolves, the ions become surrounded by H_2O molecules, as shown in Figure 4.3(**a**). The ions are said to be *solvated*. In chemical equations, we denote solvated ions by writing them as $Na^+(aq)$ and $Cl^-(aq)$, where *aq* is an abbreviation for "aqueous." ∞ (Section 3.1) **Solvation** helps stabilize the ions in solution and prevents cations and anions from recombining. Furthermore, because the ions and their shells of surrounding water molecules are free to move about, the ions become dispersed uniformly throughout the solution.

We can usually predict the nature of the ions in a solution of an ionic compound from the chemical name of the substance. Sodium sulfate (Na_2SO_4), for example, dissociates into sodium ions (Na^+) and sulfate ions (SO_4^{2-}). You must remember the formulas and charges of common ions (Tables 2.4 and 2.5) to understand the forms in which ionic compounds exist in aqueous solution.

▲ GO FIGURE

Which solution, NaCl(*aq*) or CH₃OH(*aq*), conducts electricity?

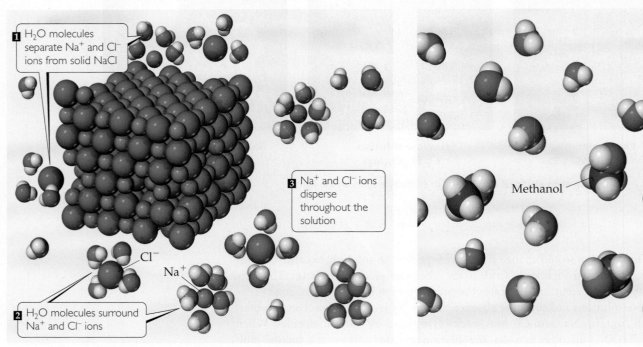

1 H₂O molecules separate Na⁺ and Cl⁻ ions from solid NaCl

3 Na⁺ and Cl⁻ ions disperse throughout the solution

Cl⁻   Na⁺

2 H₂O molecules surround Na⁺ and Cl⁻ ions

Methanol

(a) Ionic compounds like sodium chloride, NaCl, form ions when they dissolve.

(b) Molecular substances like methanol, CH₃OH, dissolve without forming ions.

▲ **Figure 4.3 Dissolution in water.** (**a**) When an ionic compound, such as sodium chloride, NaCl, dissolves in water, H₂O molecules separate, surround, and uniformly disperse the ions into the liquid. (**b**) Molecular substances that dissolve in water, such as methanol, CH₃OH, usually do so without forming ions. We can think of methanol in water as a simple mixing of two molecular species. In both (a) and (b) the water molecules have been moved apart so that the solute particles can be seen clearly.

▲ Give It Some Thought

What dissolved species are present in a solution of
(**a**) KCN?
(**b**) NaClO₄?

When a molecular compound such as sucrose or methanol [Figure 4.3(**b**)] dissolves in water, the solution usually consists of intact molecules dispersed throughout the solution. Consequently, most molecular compounds are nonelectrolytes. A few molecular substances do have aqueous solutions that contain ions. Acids are the most important of these solutions. For example, when HCl(*g*) dissolves in water to form hydrochloric acid, HCl(*aq*), the molecule *ionizes*; that is, it dissociates into H⁺(*aq*) and Cl⁻(*aq*) ions.

Strong and Weak Electrolytes

Electrolytes differ in the extent to which they conduct electricity. **Strong electrolytes** are those solutes that exist in solution completely or nearly completely as ions. Essentially all water-soluble ionic compounds (such as NaCl) and a few molecular compounds (such as HCl) are strong electrolytes. **Weak electrolytes** are those solutes that exist in solution mostly in the form of neutral molecules with only a small fraction in the form of ions. For example, in a solution of acetic acid (CH₃COOH), most of the solute is present as $CH_3COOH(aq)$ molecules. Only a small fraction (about 1%) of the CH_3COOH has dissociated into $H^+(aq)$ and $CH_3COO^-(aq)$ ions.*

We must be careful not to confuse the extent to which an electrolyte dissolves (its solubility) with whether it is strong or weak. For example, CH_3COOH is extremely soluble in water but is a weak electrolyte. $Ca(OH)_2$, on the other hand, is not very soluble in water, but the amount that does dissolve dissociates almost completely. Thus, $Ca(OH)_2$ is a strong electrolyte.

When a weak electrolyte, such as acetic acid, ionizes in solution, we write the reaction in the form

$$CH_3COOH(aq) \rightleftharpoons CH_3COO^-(aq) + H^+(aq) \qquad [4.2]$$

The half-arrows pointing in opposite directions mean that the reaction is significant in both directions. At any given moment some CH_3COOH molecules are ionizing to form H^+ and CH_3COO^- ions but H^+ and CH_3COO^- ions are recombining to form CH_3COOH. The balance between these opposing processes determines the relative numbers of ions and neutral molecules. This balance produces a state of **chemical equilibrium** in which the relative numbers of each type of ion or molecule in the reaction are constant over time. Chemists use half-arrows pointing in opposite directions to represent reactions that go both forward and backward to achieve equilibrium, such as the ionization of weak electrolytes. In contrast, a single reaction arrow is used for reactions that largely go forward, such as the ionization of strong electrolytes. Because HCl is a strong electrolyte, we write the equation for the ionization of HCl as

$$HCl(aq) \longrightarrow H^+(aq) + Cl^-(aq) \qquad [4.3]$$

The absence of a left-pointing half-arrow indicates that the H^+ and Cl^- ions have no tendency to recombine to form HCl molecules.

In the following sections we will look at how a compound's composition lets us predict whether it is a strong electrolyte, weak electrolyte, or nonelectrolyte. For the moment, you need only to remember that *water-soluble ionic compounds are strong electrolytes*. Ionic compounds can usually be identified by the presence of both metals and nonmetals [for example, NaCl, $FeSO_4$, and $Al(NO_3)_3$]. Ionic compounds containing the ammonium ion, NH_4^+ [for example, NH_4Br and $(NH_4)_2CO_3$], are exceptions to this rule of thumb.

 Give It Some Thought

Which solute will cause the light bulb in Figure 4.2 to glow most brightly, CH_3OH, NaOH, or CH_3COOH?

SAMPLE
EXERCISE 4.1 **Relating Relative Numbers of Anions and Cations to Chemical Formulas**

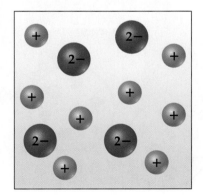

The accompanying diagram represents an aqueous solution of either $MgCl_2$, KCl, or K_2SO_4. Which solution does the drawing best represent?

SOLUTION

Analyze We are asked to associate the charged spheres in the diagram with ions present in a solution of an ionic substance.

Plan We examine each ionic substance given to determine the relative numbers and charges of its ions. We then correlate these ionic species with the ones shown in the diagram.

---

*The chemical formula of acetic acid is sometimes written $HC_2H_3O_2$ so that the formula looks like that of other common acids such as HCl. The formula CH_3COOH conforms to the molecular structure of acetic acid, with the acidic H on the O atom at the end of the formula.

Solve The diagram shows twice as many cations as anions, consistent with the formulation K_2SO_4.

Check Notice that the net charge in the diagram is zero, as it must be if it is to represent an ionic substance.

> **Practice Exercise 1**
>
> If you have an aqueous solution that contains 1.5 moles of HCl, how many moles of ions are in the solution? **(a)** 1.0, **(b)** 1.5, **(c)** 2.0, **(d)** 2.5, **(e)** 3.0
>
> **Practice Exercise 2**
>
> If you were to draw diagrams representing aqueous solutions of **(a)** $NiSO_4$, **(b)** $Ca(NO_3)_2$, **(c)** Na_3PO_4, **(d)** $Al_2(SO_4)_3$, how many anions would you show if each diagram contained six cations?

4.2 | Precipitation Reactions

▼ Figure 4.4 shows two clear solutions being mixed. One solution contains potassium iodide, KI, dissolved in water and the other contains lead nitrate, $Pb(NO_3)_2$, dissolved in water. The reaction between these two solutes produces a water-insoluble yellow solid. Reactions that result in the formation of an insoluble product are called **precipitation reactions**. A **precipitate** is an insoluble solid formed by a reaction in solution.

GO FIGURE

Which ions remain in solution after PbI_2 precipitation is complete?

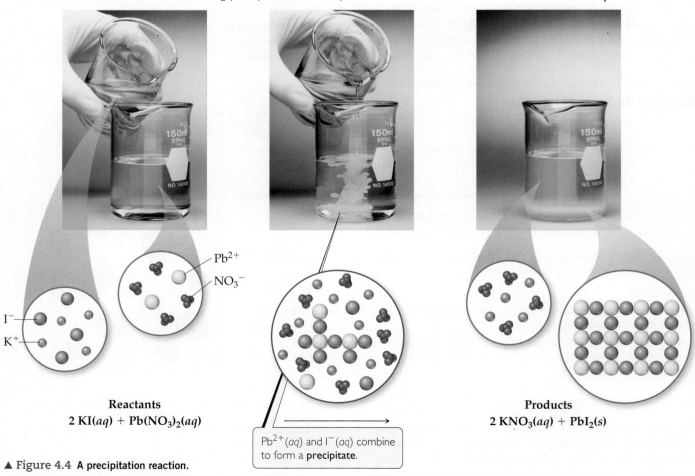

Reactants
$2\ KI(aq) + Pb(NO_3)_2(aq)$

$Pb^{2+}(aq)$ and $I^-(aq)$ combine to form a **precipitate**.

Products
$2\ KNO_3(aq) + PbI_2(s)$

▲ Figure 4.4 **A precipitation reaction.**

In Figure 4.4 the precipitate is lead iodide (PbI_2), a compound that has a very low solubility in water:

$$Pb(NO_3)_2(aq) + 2\,KI(aq) \longrightarrow PbI_2(s) + 2\,KNO_3(aq) \qquad [4.4]$$

The other product of this reaction, potassium nitrate (KNO_3), remains in solution.

Precipitation reactions occur when pairs of oppositely charged ions attract each other so strongly that they form an insoluble ionic solid. These reactions are very common in the ocean, including the deep sea vents we discussed earlier. To predict whether certain combinations of ions form insoluble compounds, we must consider some guidelines concerning the solubilities of common ionic compounds.

Solubility Guidelines for Ionic Compounds

The **solubility** of a substance at a given temperature is the amount of the substance that can be dissolved in a given quantity of solvent at the given temperature. In our discussions, any substance with a solubility less than 0.01 mol/L will be considered *insoluble*. In these cases the attraction between the oppositely charged ions in the solid is too great for the water molecules to separate the ions to any significant extent; the substance remains largely undissolved.

Unfortunately, there are no rules based on simple physical properties such as ionic charge to guide us in predicting whether a particular ionic compound will be soluble. Experimental observations, however, have led to guidelines for predicting solubility for ionic compounds. For example, experiments show that all common ionic compounds that contain the nitrate anion, NO_3^-, are soluble in water. ▼ Table 4.1 summarizes the solubility guidelines for common ionic compounds. The table is organized according to the anion in the compound, but it also reveals many important facts about cations. Note that *all common ionic compounds of the alkali metal ions (group 1A of the periodic table) and of the ammonium ion (NH_4^+) are soluble in water.*

To predict whether a precipitate forms when we mix aqueous solutions of two strong electrolytes, we must (1) note the ions present in the reactants, (2) consider the possible cation–anion combinations, and (3) use Table 4.1 to determine if any of these combinations is insoluble. For example, will a precipitate form when solutions of $Mg(NO_3)_2$ and NaOH are mixed? Both substances are soluble ionic compounds and strong electrolytes. Mixing the solutions first produces a solution containing

Table 4.1   Solubility Guidelines for Common Ionic Compounds in Water

Soluble Ionic Compounds		Important Exceptions
Compounds containing	NO_3^-	None
	CH_3COO^-	None
	Cl^-	Compounds of Ag^+, Hg_2^{2+}, and Pb^{2+}
	Br^-	Compounds of Ag^+, Hg_2^{2+}, and Pb^{2+}
	I^-	Compounds of Ag^+, Hg_2^{2+}, and Pb^{2+}
	SO_4^{2-}	Compounds of Sr^{2+}, Ba^{2+}, Hg_2^{2+}, and Pb^{2+}
Insoluble Ionic Compounds		**Important Exceptions**
Compounds containing	S^{2-}	Compounds of NH_4^+, the alkali metal cations, Ca^{2+}, Sr^{2+}, and Ba^{2+}
	CO_3^{2-}	Compounds of NH_4^+ and the alkali metal cations
	PO_4^{3-}	Compounds of NH_4^+ and the alkali metal cations
	OH^-	Compounds of NH_4^+, the alkali metal cations, Ca^{2+}, Sr^{2+}, and Ba^{2+}

Mg^{2+}, NO_3^-, Na^+, and OH^- ions. Will either cation interact with either anion to form an insoluble compound? Knowing from Table 4.1 that $Mg(NO_3)_2$ and $NaOH$ are both soluble in water, our only possibilities are Mg^{2+} with OH^- and Na^+ with NO_3^-. From Table 4.1 we see that hydroxides are generally insoluble. Because Mg^{2+} is not an exception, $Mg(OH)_2$ is insoluble and thus forms a precipitate. $NaNO_3$, however, is soluble, so Na^+ and NO_3^- remain in solution. The balanced equation for the precipitation reaction is

$$Mg(NO_3)_2(aq) + 2\,NaOH(aq) \longrightarrow Mg(OH)_2(s) + 2\,NaNO_3(aq) \qquad [4.5]$$

SAMPLE EXERCISE 4.2 | Using Solubility Rules

Classify these ionic compounds as soluble or insoluble in water: (**a**) sodium carbonate, Na_2CO_3, (**b**) lead sulfate, $PbSO_4$.

SOLUTION

Analyze We are given the names and formulas of two ionic compounds and asked to predict whether they are soluble or insoluble in water.

Plan We can use Table 4.1 to answer the question. Thus, we need to focus on the anion in each compound because the table is organized by anions.

Solve

(**a**) According to Table 4.1, most carbonates are insoluble. But carbonates of the alkali metal cations (such as sodium ion) are an exception to this rule and are soluble. Thus, Na_2CO_3 is soluble in water.

(**b**) Table 4.1 indicates that although most sulfates are water soluble, the sulfate of Pb^{2+} is an exception. Thus, $PbSO_4$ is insoluble in water.

Practice Exercise 1

Which of the following compounds is insoluble in water? (**a**) $(NH_4)_2S$, (**b**) $CaCO_3$, (**c**) $NaOH$, (**d**) Ag_2SO_4, (**e**) $Pb(CH_3COO)_2$.

Practice Exercise 2

Classify the following compounds as soluble or insoluble in water: (**a**) cobalt(II) hydroxide, (**b**) barium nitrate, (**c**) ammonium phosphate.

Exchange (Metathesis) Reactions

Notice in Equation 4.5 that the reactant cations exchange anions—Mg^{2+} ends up with OH^-, and Na^+ ends up with NO_3^-. The chemical formulas of the products are based on the charges of the ions—two OH^- ions are needed to give a neutral compound with Mg^{2+}, and one NO_3^- ion is needed to give a neutral compound with Na^+. ∞ (Section 2.7) *The equation can be balanced only after the chemical formulas of the products have been determined.*

Reactions in which cations and anions appear to exchange partners conform to the general equation

$$AX + BY \longrightarrow AY + BX \qquad [4.6]$$

Example: $\quad AgNO_3(aq) + KCl(aq) \longrightarrow AgCl(s) + KNO_3(aq)$

Such reactions are called either **exchange reactions** or **metathesis reactions** (meh-TATH-eh-sis, Greek for "to transpose"). Precipitation reactions conform to this pattern, as do many neutralization reactions between acids and bases, as we will see in Section 4.3.

To complete and balance the equation for a metathesis reaction, we follow these steps:

1. Use the chemical formulas of the reactants to determine which ions are present.
2. Write the chemical formulas of the products by combining the cation from one reactant with the anion of the other, using the ionic charges to determine the subscripts in the chemical formulas.
3. Check the water solubilities of the products. For a precipitation reaction to occur, at least one product must be insoluble in water.
4. Balance the equation.

SAMPLE EXERCISE 4.3 | Predicting a Metathesis Reaction

(a) Predict the identity of the precipitate that forms when aqueous solutions of $BaCl_2$ and K_2SO_4 are mixed. **(b)** Write the balanced chemical equation for the reaction.

SOLUTION

Analyze We are given two ionic reactants and asked to predict the insoluble product that they form.

Plan We need to write the ions present in the reactants and exchange the anions between the two cations. Once we have written the chemical formulas for these products, we can use Table 4.1 to determine which is insoluble in water. Knowing the products also allows us to write the equation for the reaction.

Solve

(a) The reactants contain Ba^{2+}, Cl^-, K^+, and SO_4^{2-} ions. Exchanging the anions gives us $BaSO_4$ and KCl. According to Table 4.1, most compounds of SO_4^{2-} are soluble but those of Ba^{2+} are not. Thus, $BaSO_4$ is insoluble and will precipitate from solution. KCl is soluble.

(b) From part (a) we know the chemical formulas of the products, $BaSO_4$ and KCl. The balanced equation is

$$BaCl_2(aq) + K_2SO_4(aq) \longrightarrow BaSO_4(s) + 2\,KCl(aq)$$

Practice Exercise 1

Yes or No: Will a precipitate form when solutions of $Ba(NO_3)_2$ and KOH are mixed?

Practice Exercise 2

(a) What compound precipitates when aqueous solutions of $Fe_2(SO_4)_3$ and LiOH are mixed? **(b)** Write a balanced equation for the reaction.

Ionic Equations and Spectator Ions

In writing equations for reactions in aqueous solution, it is often useful to indicate whether the dissolved substances are present predominantly as ions or as molecules. Let's reconsider the precipitation reaction between $Pb(NO_3)_2$ and 2 KI (Eq. 4.4):

$$Pb(NO_3)_2(aq) + 2\,KI(aq) \longrightarrow PbI_2(s) + 2\,KNO_3(aq)$$

An equation written in this fashion, showing the complete chemical formulas of reactants and products, is called a **molecular equation** because it shows chemical formulas without indicating ionic character. Because $Pb(NO_3)_2$, KI, and KNO_3 are all water-soluble ionic compounds and therefore strong electrolytes, we can write the equation in a form that indicates which species exist as ions in the solution:

$$Pb^{2+}(aq) + 2\,NO_3^-(aq) + 2\,K^+(aq) + 2\,I^-(aq) \longrightarrow$$
$$PbI_2(s) + 2\,K^+(aq) + 2\,NO_3^-(aq) \quad [4.7]$$

An equation written in this form, with all soluble strong electrolytes shown as ions, is called a **complete ionic equation**.

Notice that $K^+(aq)$ and $NO_3^-(aq)$ appear on both sides of Equation 4.7. Ions that appear in identical forms on both sides of a complete ionic equation, called **spectator ions**, play no direct role in the reaction. When spectator ions are omitted from the equation (they cancel out like algebraic quantities), we are left with the **net ionic equation**, which is one that includes only the ions and molecules directly involved in the reaction:

$$Pb^{2+}(aq) + 2\,I^-(aq) \longrightarrow PbI_2(s) \quad [4.8]$$

Because charge is conserved in reactions, the sum of the ionic charges must be the same on both sides of a balanced net ionic equation. In this case the 2+ charge of the cation and the two 1− charges of the anions add to zero, the charge of the electrically neutral product. *If every ion in a complete ionic equation is a spectator, no reaction occurs.*

 Give It Some Thought

Which ions, if any, are spectator ions in this reaction?

$$AgNO_3(aq) + NaCl(aq) \longrightarrow AgCl(s) + NaNO_3(aq)$$

Net ionic equations illustrate the similarities between various reactions involving electrolytes. For example, Equation 4.8 expresses the essential feature of the precipitation reaction between any strong electrolyte containing $Pb^{2+}(aq)$ and any strong electrolyte containing $I^-(aq)$: The ions combine to form a precipitate of PbI_2. Thus, a net ionic equation demonstrates that more than one set of reactants can lead to the same net reaction. For example, aqueous solutions of KI and MgI_2 share many chemical similarities because both contain I^- ions. Either solution when mixed with a $Pb(NO_3)_2$ solution produces $PbI_2(s)$. The complete ionic equation, on the other hand, identifies the actual reactants that participate in a reaction.

The following steps summarize the procedure for writing net ionic equations:

1. Write a balanced molecular equation for the reaction.
2. Rewrite the equation to show the ions that form in solution when each soluble strong electrolyte dissociates into its ions. *Only strong electrolytes dissolved in aqueous solution are written in ionic form.*
3. Identify and cancel spectator ions.

SAMPLE EXERCISE 4.4 | Writing a Net Ionic Equation

Write the net ionic equation for the precipitation reaction that occurs when aqueous solutions of calcium chloride and sodium carbonate are mixed.

SOLUTION

Analyze Our task is to write a net ionic equation for a precipitation reaction, given the names of the reactants present in solution.

Plan We write the chemical formulas of the reactants and products and then determine which product is insoluble. We then write and balance the molecular equation. Next, we write each soluble strong electrolyte as separated ions to obtain the complete ionic equation. Finally, we eliminate the spectator ions to obtain the net ionic equation.

Solve Calcium chloride is composed of calcium ions, Ca^{2+}, and chloride ions, Cl^-; hence, an aqueous solution of the substance is $CaCl_2(aq)$. Sodium carbonate is composed of Na^+ ions and CO_3^{2-} ions; hence, an aqueous solution of the compound is $Na_2CO_3(aq)$. In the molecular equations for precipitation reactions, the anions and cations appear to exchange partners. Thus, we put Ca^{2+} and CO_3^{2-} together to give $CaCO_3$ and Na^+ and Cl^- together to give NaCl. According to the solubility guidelines in Table 4.1, $CaCO_3$ is insoluble and NaCl is soluble. The balanced molecular equation is

$$CaCl_2(aq) + Na_2CO_3(aq) \longrightarrow CaCO_3(s) + 2\,NaCl(aq)$$

In a complete ionic equation, *only* dissolved strong electrolytes (such as soluble ionic compounds) are written as separate ions. As the (aq) designations remind us, $CaCl_2$, Na_2CO_3, and NaCl are all dissolved in the solution. Furthermore, they are all strong electrolytes. $CaCO_3$ is an ionic compound, but it is not soluble. We do not write the formula of any insoluble compound as its component ions. Thus, the complete ionic equation is

$$Ca^{2+}(aq) + 2\,Cl^-(aq) + 2\,Na^+(aq) + CO_3^{2-}(aq) \longrightarrow$$
$$CaCO_3(s) + 2\,Na^+(aq) + 2\,Cl^-(aq)$$

The spectator ions are Na^+ and Cl^-. Canceling them gives the following net ionic equation:

$$Ca^{2+}(aq) + CO_3^{2-}(aq) \longrightarrow CaCO_3(s)$$

Check We can check our result by confirming that both the elements and the electric charge are balanced. Each side has one Ca, one C, and three O, and the net charge on each side equals 0.

Comment If none of the ions in an ionic equation is removed from solution or changed in some way, all ions are spectator ions and a reaction does not occur.

Practice Exercise 1

What happens when you mix an aqueous solution of sodium nitrate with an aqueous solution of barium chloride? **(a)** There is no reaction; all possible products are soluble. **(b)** Only barium nitrate precipitates. **(c)** Only sodium chloride precipitates. **(d)** Both barium nitrate and sodium chloride precipitate. **(e)** Nothing; barium chloride is not soluble and it stays as a precipitate.

Practice Exercise 2

Write the net ionic equation for the precipitation reaction that occurs when aqueous solutions of silver nitrate and potassium phosphate are mixed.

▲ **Figure 4.5** Vinegar and lemon juice are common household acids. Ammonia and baking soda (sodium bicarbonate) are common household bases.

4.3 | Acids, Bases, and Neutralization Reactions

Many acids and bases are industrial and household substances (◄ Figure 4.5), and some are important components of biological fluids. Hydrochloric acid, for example, is an important industrial chemical and the main constituent of gastric juice in your stomach. Acids and bases are also common electrolytes.

Acids

As noted in Section 2.8, **acids** are substances that ionize in aqueous solution to form hydrogen ions $H^+(aq)$. Because a hydrogen atom consists of a proton and an electron,

H^+ is simply a proton. Thus, acids are often called *proton donors*. Molecular models of four common acids are shown in ▶ Figure 4.6.

Protons in aqueous solution are solvated by water molecules, just as other cations are [Figure 4.3(a)]. In writing chemical equations involving protons in water, therefore, we write $H^+(aq)$.

Molecules of different acids ionize to form different numbers of H^+ ions. Both HCl and HNO_3 are *monoprotic* acids, yielding one H^+ per molecule of acid. Sulfuric acid, H_2SO_4, is a *diprotic* acid, one that yields two H^+ per molecule of acid. The ionization of H_2SO_4 and other diprotic acids occurs in two steps:

$$H_2SO_4(aq) \longrightarrow H^+(aq) + HSO_4^-(aq) \qquad [4.9]$$

$$HSO_4^-(aq) \rightleftharpoons H^+(aq) + SO_4^{2-}(aq) \qquad [4.10]$$

Although H_2SO_4 is a strong electrolyte, only the first ionization (Equation 4.9) is complete. Thus, aqueous solutions of sulfuric acid contain a mixture of $H^+(aq)$, $HSO_4^-(aq)$, and $SO_4^{2-}(aq)$.

The molecule CH_3COOH (acetic acid) that we have mentioned frequently is the primary component in vinegar. Acetic acid has four hydrogens, as Figure 4.6 shows, but only one of them, the H that is bonded to an oxygen in the —COOH group, is ionized in water. Thus, the H in the COOH group breaks its O—H bond in water. The three other hydrogens in acetic acid are bound to carbon and do not break their C—H bonds in water. The reasons for this difference are very interesting and will be discussed in Chapter 16.

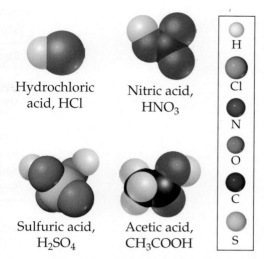

▲ **Figure 4.6 Molecular models of four common acids.**

 Give It Some Thought

The structural formula of citric acid, a main component of citrus fruits, is

How many $H^+(aq)$ can be generated by each citric acid molecule dissolved in water?

Bases

Bases are substances that accept (react with) H^+ ions. Bases produce hydroxide ions (OH^-) when they dissolve in water. Ionic hydroxide compounds, such as NaOH, KOH, and $Ca(OH)_2$, are among the most common bases. When dissolved in water, they dissociate into ions, introducing OH^- ions into the solution.

Compounds that do not contain OH^- ions can also be bases. For example, ammonia (NH_3) is a common base. When added to water, it accepts an H^+ ion from a water molecule and thereby produces an OH^- ion (▶ Figure 4.7):

$$NH_3(aq) + H_2O(l) \rightleftharpoons NH_4^+(aq) + OH^-(aq) \qquad [4.11]$$

Ammonia is a weak electrolyte because only about 1% of the NH_3 forms NH_4^+ and OH^- ions.

Strong and Weak Acids and Bases

Acids and bases that are strong electrolytes (completely ionized in solution) are **strong acids** and **strong bases**. Those that are weak electrolytes (partly ionized) are **weak acids** and **weak bases**. When reactivity depends only on $H^+(aq)$ concentration, strong acids are more reactive

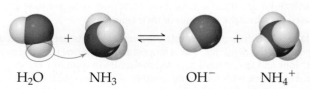

H_2O NH_3 OH^- NH_4^+

▲ **Figure 4.7 Proton transfer.** An H_2O molecule acts as a proton donor (acid), and NH_3 acts as a proton acceptor (base). In aqueous solutions, only a fraction of the NH_3 molecules react with H_2O. Consequently, NH_3 is a weak electrolyte.

Table 4.2 **Common Strong Acids and Bases**

Strong Acids	Strong Bases
Hydrochloric acid, HCl	Group 1A metal hydroxides
Hydrobromic acid, HBr	[LiOH, NaOH, KOH, RbOH, CsOH]
Hydroiodic acid, HI	Heavy group 2A metal hydroxides
Chloric acid, $HClO_3$	$[Ca(OH)_2, Sr(OH)_2, Ba(OH)_2]$
Perchloric acid, $HClO_4$	
Nitric acid, HNO_3	
Sulfuric acid (first proton), H_2SO_4	

than weak acids. The reactivity of an acid, however, can depend on the anion as well as on $H^+(aq)$ concentration. For example, hydrofluoric acid (HF) is a weak acid (only partly ionized in aqueous solution), but it is very reactive and vigorously attacks many substances, including glass. This reactivity is due to the combined action of $H^+(aq)$ and $F^-(aq)$.

▲ Table 4.2 lists the strong acids and bases we are most likely to encounter. You need to commit this information to memory in order to correctly identify strong electrolytes and write net ionic equations. The brevity of this list tells us that most acids are weak. (For H_2SO_4, as we noted earlier, only the first proton completely ionizes.) The only common strong bases are the common soluble metal hydroxides. The most common weak base is NH_3, which reacts with water to form OH^- ions (Equation 4.11).

▲ Give It Some Thought

Why isn't $Al(OH)_3$ classified as a strong base?

SAMPLE EXERCISE 4.5 Comparing Acid Strengths

The following diagrams represent aqueous solutions of acids HX, HY, and HZ, with water molecules omitted for clarity. Rank the acids from strongest to weakest.

HX
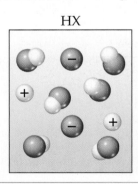

HY
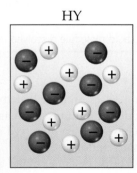

HZ
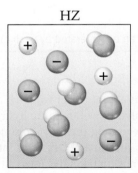

SOLUTION

Analyze We are asked to rank three acids from strongest to weakest, based on schematic drawings of their solutions.

Plan We can determine the relative numbers of uncharged molecular species in the diagrams. The strongest acid is the one with the most H^+ ions and fewest undissociated molecules in solution. The weakest acid is the one with the largest number of undissociated molecules.

Solve The order is HY > HZ > HX. HY is a strong acid because it is totally ionized (no HY molecules in solution), whereas both HX and HZ are weak acids, whose solutions consist of a mixture of molecules and ions. Because HZ contains more H^+ ions and fewer molecules than HX, it is a stronger acid.

Practice Exercise 1

A set of aqueous solutions are prepared containing different acids at the same concentration: acetic acid, chloric acid and hydrobromic acid. Which solution(s) are the most electrically conductive? **(a)** chloric acid, **(b)** hydrobromic acid, **(c)** acetic acid, **(d)** both chloric acid and hydrobromic acid, **(e)** all three solutions have the same electrical conductivity.

Practice Exercise 2

Imagine a diagram showing 10 Na^+ ions and 10 OH^- ions. If this solution were mixed with the one pictured above for HY, what species would be present in a diagram that represents the combined solutions after any possible reaction?

Identifying Strong and Weak Electrolytes

If we remember the common strong acids and bases (Table 4.2) and also remember that NH_3 is a weak base, we can make reasonable predictions about the electrolytic strength of a great number of *water-soluble* substances. ▼ Table 4.3 summarizes our observations about electrolytes. We first ask whether the substance is ionic or molecular. If it is ionic, it is a strong electrolyte. If the substance is molecular, we ask whether it is an acid or a base. (It is an acid if it either has H first in the chemical formula or contains a COOH group.) If it is an acid, we use Table 4.2 to determine whether it is a strong or weak electrolyte: All strong acids are strong electrolytes, and all weak acids are weak electrolytes. If an acid is not listed in Table 4.2, it is probably a weak acid and therefore a weak electrolyte.

If our substance is a base, we use Table 4.2 to determine whether it is a strong base. NH_3 is the only molecular base that we consider in this chapter, and it is a weak base; Table 4.3 tells us it is therefore a weak electrolyte. Finally, any molecular substance that we encounter in this chapter that is not an acid or NH_3 is probably a nonelectrolyte.

Table 4.3 Summary of the Electrolytic Behavior of Common Soluble Ionic and Molecular Compounds

	Strong Electrolyte	Weak Electrolyte	Nonelectrolyte
Ionic	All	None	None
Molecular	Strong acids (see Table 4.2)	Weak acids, weak bases	All other compounds

SAMPLE EXERCISE 4.6 Identifying Strong, Weak, and Nonelectrolytes

Classify these dissolved substances as a strong electrolyte, weak electrolyte, or nonelectrolyte: $CaCl_2$, HNO_3, C_2H_5OH (ethanol), HCOOH (formic acid), KOH.

SOLUTION

Analyze We are given several chemical formulas and asked to classify each substance as a strong electrolyte, weak electrolyte, or nonelectrolyte.

Plan The approach we take is outlined in Table 4.3. We can predict whether a substance is ionic or molecular based on its composition. As we saw in Section 2.7, most ionic compounds we encounter in this text are composed of a metal and a nonmetal, whereas most molecular compounds are composed only of nonmetals.

Solve Two compounds fit the criteria for ionic compounds: $CaCl_2$ and KOH. Because Table 4.3 tells us that all ionic compounds are strong electrolytes, that is how we classify these two substances. The three remaining compounds are molecular. Two of these molecular substances, HNO_3 and HCOOH, are acids. Nitric acid, HNO_3, is a common strong acid, as shown in Table 4.2, and therefore is a strong electrolyte. Because most acids are weak acids, our best guess would be that HCOOH is a weak acid (weak electrolyte), which is in fact the case. The remaining molecular compound, C_2H_5OH, is neither an acid nor a base, so it is a nonelectrolyte.

Comment Although ethanol, C_2H_5OH, has an OH group, it is not a metal hydroxide and therefore not a base. Rather ethanol is a member of a class of organic compounds that have C—OH bonds, which are known as alcohols. ⤻ (Section 2.9) Organic compounds containing the COOH group are called carboxylic acids (Chapter 16). Molecules that have this group are weak acids.

> **Practice Exercise 1**
>
> Which of these substances, when dissolved in water, is a strong electrolyte? (**a**) ammonia, (**b**) hydrofluoric acid, (**c**) folic acid, (**d**) sodium nitrate, (**e**) sucrose.
>
> **Practice Exercise 2**
>
> Consider solutions in which 0.1 mol of each of the following compounds is dissolved in 1 L of water: $Ca(NO_3)_2$ (calcium nitrate), $C_6H_{12}O_6$ (glucose), $NaCH_3COO$ (sodium acetate), and CH_3COOH (acetic acid). Rank the solutions in order of increasing electrical conductivity, knowing that the greater the number of ions in solution, the greater the conductivity.

Neutralization Reactions and Salts

The properties of acidic solutions are quite different from those of basic solutions. Acids have a sour taste, whereas bases have a bitter taste.* Acids change the colors of certain dyes in a way that differs from the way bases affect the same dyes. This is the

---

*Tasting chemical solutions is not a good practice. However, we have all had acids such as ascorbic acid (vitamin C), acetylsalicylic acid (aspirin), and citric acid (in citrus fruits) in our mouths, and we are familiar with their characteristic sour taste. Soaps, which are basic, have the characteristic bitter taste of bases.

Base turns litmus paper blue

Acid turns litmus paper red

▲ Figure 4.8 **Litmus paper.** Litmus paper is coated with dyes that change color in response to exposure to either acids or bases.

principle behind the indicator known as litmus paper (◄ Figure 4.8). In addition, acidic and basic solutions differ in chemical properties in several other important ways that we explore in this chapter and in later chapters.

When a solution of an acid and a solution of a base are mixed, a **neutralization reaction** occurs. The products of the reaction have none of the characteristic properties of either the acidic solution or the basic solution. For example, when hydrochloric acid is mixed with a solution of sodium hydroxide, the reaction is

$$\underset{\text{(acid)}}{HCl(aq)} + \underset{\text{(base)}}{NaOH(aq)} \longrightarrow \underset{\text{(water)}}{H_2O(l)} + \underset{\text{(salt)}}{NaCl(aq)} \qquad [4.12]$$

Water and table salt, NaCl, are the products of the reaction. By analogy to this reaction, the term **salt** has come to mean any ionic compound whose cation comes from a base (for example, Na^+ from NaOH) and whose anion comes from an acid (for example, Cl^- from HCl). In general, *a neutralization reaction between an acid and a metal hydroxide produces water and a salt*.

Because HCl, NaOH, and NaCl are all water-soluble strong electrolytes, the complete ionic equation associated with Equation 4.12 is

$$H^+(aq) + Cl^-(aq) + Na^+(aq) + OH^-(aq) \longrightarrow$$
$$H_2O(l) + Na^+(aq) + Cl^-(aq) \quad [4.13]$$

Therefore, the net ionic equation is

$$H^+(aq) + OH^-(aq) \longrightarrow H_2O(l) \qquad [4.14]$$

Equation 4.14 summarizes the main feature of the neutralization reaction between any strong acid and any strong base: $H^+(aq)$ and $OH^-(aq)$ ions combine to form $H_2O(l)$.

► Figure 4.9 shows the neutralization reaction between hydrochloric acid and the water-insoluble base $Mg(OH)_2$:

Molecular equation:

$$Mg(OH)_2(s) + 2\,HCl(aq) \longrightarrow MgCl_2(aq) + 2\,H_2O(l) \qquad [4.15]$$

Net ionic equation:

$$Mg(OH)_2(s) + 2\,H^+(aq) \longrightarrow Mg^{2+}(aq) + 2\,H_2O(l) \qquad [4.16]$$

Notice that the OH^- ions (this time in a solid reactant) and H^+ ions combine to form H_2O. Because the ions exchange partners, neutralization reactions between acids and metal hydroxides are metathesis reactions.

SAMPLE EXERCISE 4.7 | Writing Chemical Equations for a Neutralization Reaction

For the reaction between aqueous solutions of acetic acid (CH_3COOH) and barium hydroxide, $Ba(OH)_2$, write (**a**) the balanced molecular equation, (**b**) the complete ionic equation, (**c**) the net ionic equation.

SOLUTION

Analyze We are given the chemical formulas for an acid and a base and asked to write a balanced molecular equation, a complete ionic equation, and a net ionic equation for their neutralization reaction.

Plan As Equation 4.12 and the italicized statement that follows it indicate, neutralization reactions form two products, H_2O and a salt. We examine the cation of the base and the anion of the acid to determine the composition of the salt.

Solve

(**a**) The salt contains the cation of the base (Ba^{2+}) and the anion of the acid (CH_3COO^-). Thus, the salt formula is $Ba(CH_3COO)_2$. According to Table 4.1, this compound is soluble in water. The unbalanced molecular equation for the neutralization reaction is

$$CH_3COOH(aq) + Ba(OH)_2(aq) \longrightarrow H_2O(l) + Ba(CH_3COO)_2(aq)$$

To balance this equation, we must provide two molecules of CH_3COOH to furnish the two CH_3COO^- ions and to supply the two H^+ ions needed to combine with the two OH^- ions of the base. The balanced molecular equation is

$$2\,CH_3COOH(aq) + Ba(OH)_2(aq) \longrightarrow$$
$$2\,H_2O(l) + Ba(CH_3COO)_2(aq)$$

(b) To write the complete ionic equation, we identify the strong electrolytes and break them into ions. In this case $Ba(OH)_2$ and $Ba(CH_3COO)_2$ are both water-soluble ionic compounds and hence strong electrolytes. Thus, the complete ionic equation is

$$2\ CH_3COOH(aq) + Ba^{2+}(aq) + 2\ OH^-(aq) \longrightarrow$$
$$2\ H_2O(l) + Ba^{2+}(aq) + 2\ CH_3COO^-(aq)$$

(c) Eliminating the spectator ion, Ba^{2+}, and simplifying coefficients gives the net ionic equation:

$$CH_3COOH(aq) + OH^-(aq) \longrightarrow H_2O(l) + CH_3COO^-(aq)$$

Check We can determine whether the molecular equation is balanced by counting the number of atoms of each kind on both sides of the arrow (10 H, 6 O, 4 C, and 1 Ba on each side). However, it is often easier to check equations by counting groups: There are 2 CH_3COO groups, as well as 1 Ba, and 4 additional H atoms and 2 additional O atoms on each side of the equation. The net ionic equation checks out because the numbers of each kind of element and the net charge are the same on both sides of the equation.

(a) $NH_4^+(aq) + H^+(aq) \longrightarrow NH_5^{2+}(aq)$
(b) $NH_3(aq) + NO_3^-(aq) \longrightarrow NH_2^-(aq) + HNO_3(aq)$
(c) $NH_2^-(aq) + H^+(aq) \longrightarrow NH_3(aq)$
(d) $NH_3(aq) + H^+(aq) \longrightarrow NH_4^+(aq)$
(e) $NH_4^+(aq) + NO_3^-(aq) \longrightarrow NH_4NO_3(aq)$

Practice Exercise 2

For the reaction of phosphorous acid (H_3PO_3) and potassium hydroxide (KOH), write **(a)** the balanced molecular equation and **(b)** the net ionic equation.

Practice Exercise 1

Which is the correct net ionic equation for the reaction of aqueous ammonia with nitric acid?

▲ **GO FIGURE**

Adding just a few drops of hydrochloric acid would not be sufficient to dissolve all the $Mg(OH)_2(s)$. Why not?

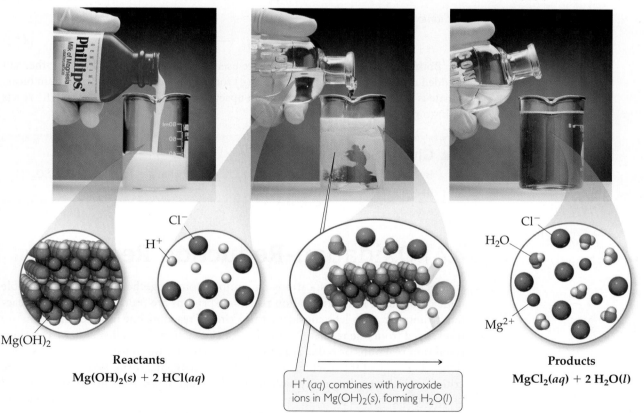

$Mg(OH)_2$

Cl^-

H^+

Reactants
$Mg(OH)_2(s) + 2\ HCl(aq)$

$H^+(aq)$ combines with hydroxide ions in $Mg(OH)_2(s)$, forming $H_2O(l)$

Cl^-

H_2O

Mg^{2+}

Products
$MgCl_2(aq) + 2\ H_2O(l)$

▲ **Figure 4.9 Neutralization reaction between $Mg(OH)_2(s)$ and hydrochloric acid.** Milk of magnesia is a suspension of water-insoluble magnesium hydroxide, $Mg(OH)_2(s)$, in water. When sufficient hydrochloric acid, $HCl(aq)$, is added, a reaction ensues that leads to an aqueous solution containing $Mg^{2+}(aq)$ and $Cl^-(aq)$ ions.

Neutralization Reactions with Gas Formation

Many bases besides OH^- react with H^+ to form molecular compounds. Two of these that you might encounter in the laboratory are the sulfide ion and the carbonate ion. Both of these anions react with acids to form gases that have low solubilities in water. Hydrogen sulfide (H_2S), the substance that gives rotten eggs their foul odor and that is emitted in deep sea vents, forms when an acid such as $HCl(aq)$ reacts with a metal sulfide such as Na_2S:

Molecular equation:

$$2\,HCl(aq) + Na_2S(aq) \longrightarrow H_2S(g) + 2\,NaCl(aq) \qquad [4.17]$$

Net ionic equation:

$$2\,H^+(aq) + S^{2-}(aq) \longrightarrow H_2S(g) \qquad [4.18]$$

Carbonates and bicarbonates react with acids to form $CO_2(g)$. Reaction of CO_3^{2-} or HCO_3^- with an acid first gives carbonic acid (H_2CO_3). For example, when hydrochloric acid is added to sodium bicarbonate, the reaction is

$$HCl(aq) + NaHCO_3(aq) \longrightarrow NaCl(aq) + H_2CO_3(aq) \qquad [4.19]$$

Carbonic acid is unstable. If present in solution in sufficient concentrations, it decomposes to H_2O and CO_2, which escapes from the solution as a gas:

$$H_2CO_3(aq) \longrightarrow H_2O(l) + CO_2(g) \qquad [4.20]$$

The overall reaction is summarized by the following equations:

Molecular equation:

$$HCl(aq) + NaHCO_3(aq) \longrightarrow NaCl(aq) + H_2O(l) + CO_2(g) \qquad [4.21]$$

Net ionic equation:

$$H^+(aq) + HCO_3^-(aq) \longrightarrow H_2O(l) + CO_2(g) \qquad [4.22]$$

Both $NaHCO_3(s)$ and $Na_2CO_3(s)$ are used as neutralizers in acid spills; either salt is added until the fizzing caused by $CO_2(g)$ formation stops. Sometimes sodium bicarbonate is used as an antacid to soothe an upset stomach. In that case the HCO_3^- reacts with stomach acid to form $CO_2(g)$.

 Give It Some Thought

By analogy to examples given in the text, predict what gas forms when $Na_2SO_3(s)$ reacts with $HCl(aq)$.

4.4 | Oxidation-Reduction Reactions

In precipitation reactions, cations and anions come together to form an insoluble ionic compound. In neutralization reactions, protons are transferred from one reactant to another. Now let's consider a third kind of reaction, one in which electrons are transferred from one reactant to another. Such reactions are called either **oxidation-reduction reactions** or **redox reactions**. In this chapter we concentrate on redox reactions where one of the reactants is a metal in its elemental form. Redox reactions are critical in understanding many biological and geological processes in the world around us, including those occurring in deep sea vents; they also form the basis for energy-related technologies such as batteries and fuel cells (Chapter 20).

Oxidation and Reduction

One of the most familiar redox reactions is *corrosion* of a metal (▶ Figure 4.11). In some instances corrosion is limited to the surface of the metal, as is the case with the green coating that forms on copper roofs and statues. In other instances the corrosion

Chemistry Put to Work

Antacids

Your stomach secretes acids to help digest foods. These acids, which include hydrochloric acid, contain about 0.1 mol of H^+ per liter of solution. The stomach and digestive tract are normally protected from the corrosive effects of stomach acid by a mucosal lining. Holes can develop in this lining, however, allowing the acid to attack the underlying tissue, causing painful damage. These holes, known as ulcers, can be caused by the secretion of excess acids and/or by a weakness in the digestive lining. Many peptic ulcers are caused by infection by the bacterium *Helicobacter pylori*. Between 10 and 20% of Americans suffer from ulcers at some point in their lives. Many others experience occasional indigestion, heartburn, or reflux due to digestive acids entering the esophagus.

The problem of excess stomach acid can be addressed by (1) removing the excess acid or (2) decreasing the production of acid. Substances that remove excess acid are called *antacids*, whereas those that decrease acid production are called *acid inhibitors*. ◀ Figure 4.10 shows several common over-the-counter antacids, which usually contain hydroxide, carbonate, or bicarbonate ions (▼ Table 4.4). Antiulcer drugs, such as Tagamet® and Zantac®, are acid inhibitors. They act on acid-producing cells in the lining of the stomach. Formulations that control acid in this way are now available as over-the-counter drugs.

Related Exercise: 4.95

▲ Figure 4.10 Antacids. These products all serve as acid-neutralizing agents in the stomach.

Table 4.4   Some Common Antacids*

Commercial Name	Acid-Neutralizing Agents
Alka-Seltzer®	$NaHCO_3$
Amphojel®	$Al(OH)_3$
Di-Gel®	$Mg(OH)_2$ and $CaCO_3$
Milk of Magnesia	$Mg(OH)_2$
Maalox®	$Mg(OH)_2$ and $Al(OH)_3$
Mylanta®	$Mg(OH)_2$ and $Al(OH)_3$
Rolaids®	$NaAl(OH)_2CO_3$
Tums®	$CaCO_3$

goes deeper, eventually compromising the structural integrity of the metal as happens with the rusting of iron.

Corrosion is the conversion of a metal into a metal compound, by a reaction between the metal and some substance in its environment. When a metal corrodes, each metal atom loses electrons and so forms a cation, which can combine with an anion to form an ionic compound. The green coating on the Statue of Liberty contains Cu^{2+} combined with carbonate and hydroxide anions; rust contains Fe^{3+} combined with oxide and hydroxide anions; and silver tarnish contains Ag^+ combined with sulfide anions.

When an atom, ion, or molecule becomes more positively charged (that is, when it loses electrons), we say that it has been *oxidized*. Loss of electrons by a substance is

(a)   (b)   (c)

▲ Figure 4.11 Familiar corrosion products. (a) A green coating forms when copper is oxidized. (b) Rust forms when iron corrodes. (c) A black tarnish forms as silver corrodes.

*Bunke, B. "Inhibition of pepsin proteolytic activity by some common antacids." NCBI. 1962.

GO FIGURE

How many electrons does each oxygen atom gain during the course of this reaction?

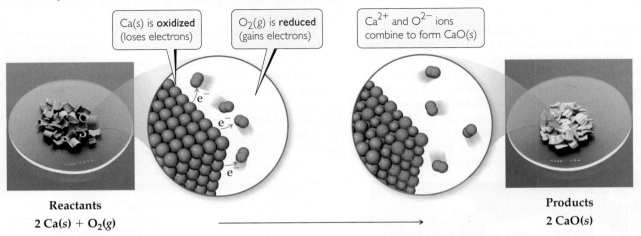

Reactants
$2\,Ca(s) + O_2(g)$

Products
$2\,CaO(s)$

▲ Figure 4.12 **Oxidation of calcium metal by molecular oxygen.**

called **oxidation**. The term *oxidation* is used because the first reactions of this sort to be studied were reactions with oxygen. Many metals react directly with O_2 in air to form metal oxides. In these reactions the metal loses electrons to oxygen, forming an ionic compound of the metal ion and oxide ion. The familiar example of rusting involves the reaction between iron metal and oxygen in the presence of water. In this process Fe is *oxidized* (loses electrons) to form Fe^{3+}.

The reaction between iron and oxygen tends to be relatively slow, but other metals, such as the alkali and alkaline earth metals, react quickly upon exposure to air. ▲ Figure 4.12 shows how the bright metallic surface of calcium tarnishes as CaO forms in the reaction

$$2\,Ca(s) + O_2(g) \longrightarrow 2\,CaO(s) \qquad [4.23]$$

In this reaction Ca is oxidized to Ca^{2+} and neutral O_2 is transformed to O^{2-} ions. When an atom, ion, or molecule becomes more negatively charged (gains electrons), we say that it is *reduced*. The gain of electrons by a substance is called **reduction**. When one reactant loses electrons (that is, when it is oxidized), another reactant must gain them. In other words, oxidation of one substance must be accompanied by reduction of some other substance. The oxidation involves transfer of electrons from the calcium metal to the O_2, leading to formation of CaO.

Oxidation Numbers

Before we can identify an oxidation-reduction reaction, we must have a bookkeeping system—a way of keeping track of electrons gained by the substance being reduced and electrons lost by the substance being oxidized. The concept of oxidation numbers (also called *oxidation states*) was devised as a way of doing this. Each atom in a neutral substance or ion is assigned an **oxidation number**. For monatomic ions the oxidation number is the same as the charge. For neutral molecules and polyatomic ions, the oxidation number of a given atom is a hypothetical charge. This charge is assigned by artificially dividing up the electrons among the atoms in the molecule or ion. We use the following rules for assigning oxidation numbers:

1. *For an atom in its **elemental form**, the oxidation number is always zero.* Thus, each H atom in the H_2 molecule has an oxidation number of 0 and each P atom in the P_4 molecule has an oxidation number of 0.

2. *For any **monatomic ion** the oxidation number equals the ionic charge.* Thus, K^+ has an oxidation number of $+1$, S^{2-} has an oxidation number of -2, and so forth.

In ionic compounds the alkali metal ions (group 1A) always have a 1+ charge and therefore an oxidation number of +1. The alkaline earth metals (group 2A) are always +2, and aluminum (group 3A) is always +3 in ionic compounds. (In writing oxidation numbers we will write the sign before the number to distinguish them from the actual electronic charges, which we write with the number first.)

3. *Nonmetals* usually have negative oxidation numbers, although they can sometimes be positive:

   (a) *The oxidation number of **oxygen** is usually −2 in both ionic and molecular compounds.* The major exception is in compounds called peroxides, which contain the O_2^{2-} ion, giving each oxygen an oxidation number of −1.

   (b) *The oxidation number of **hydrogen** is usually +1 when bonded to nonmetals and −1 when bonded to metals (for example, metal hydrides such as sodium hydride, NaH).*

   (c) *The oxidation number of **fluorine** is −1 in all compounds.* The other **halogens** have an oxidation number of −1 in most binary compounds. When combined with oxygen, as in oxyanions, however, they have positive oxidation states.

4. ***The sum of the oxidation numbers** of all atoms in a neutral compound is zero. The sum of the oxidation numbers in a polyatomic ion equals the charge of the ion.* For example, in the hydronium ion H_3O^+, which is a more accurate description of $H^+(aq)$, the oxidation number of each hydrogen is +1 and that of oxygen is −2. Thus, the sum of the oxidation numbers is $3(+1) + (−2) = +1$, which equals the net charge of the ion. This rule is useful in obtaining the oxidation number of one atom in a compound or ion if you know the oxidation numbers of the other atoms, as illustrated in Sample Exercise 4.8.

 Give It Some Thought

What is the oxidation number of nitrogen (**a**) in aluminum nitride, AlN, and (**b**) in nitric acid, HNO_3?

SAMPLE
EXERCISE 4.8   Determining Oxidation Numbers

Determine the oxidation number of sulfur in (**a**) H_2S, (**b**) S_8, (**c**) SCl_2, (**d**) Na_2SO_3, (**e**) SO_4^{2-}.

SOLUTION

Analyze We are asked to determine the oxidation number of sulfur in two molecular species, in the elemental form, and in two substances containing ions.

Plan In each species the sum of oxidation numbers of all the atoms must equal the charge on the species. We will use the rules outlined previously to assign oxidation numbers.

Solve

(**a**) When bonded to a nonmetal, hydrogen has an oxidation number of +1. Because the H_2S molecule is neutral, the sum of the oxidation numbers must equal zero. Letting x equal the oxidation number of S, we have $2(+1) + x = 0$. Thus, S has an oxidation number of −2.

(**b**) Because S_8 is an elemental form of sulfur, the oxidation number of S is 0.

(**c**) Because SCl_2 is a binary compound, we expect chlorine to have an oxidation number of −1. The sum of the oxidation numbers must equal zero. Letting x equal the oxidation number of S, we have $x + 2(−1) = 0$. Consequently, the oxidation number of S must be +2.

(**d**) Sodium, an alkali metal, always has an oxidation number of +1 in its compounds. Oxygen commonly has an oxidation

state of −2. Letting x equal the oxidation number of S, we have $2(+1) + x + 3(−2) = 0$. Therefore, the oxidation number of S in this compound (Na_2SO_3) is +4.

(**e**) The oxidation state of O is −2. The sum of the oxidation numbers equals −2, the net charge of the SO_4^{2-} ion. Thus, we have $x + 4(−2) = −2$. From this relation we conclude that the oxidation number of S in this ion is +6.

Comment These examples illustrate that the oxidation number of a given element depends on the compound in which it occurs. The oxidation numbers of sulfur, as seen in these examples, range from −2 to +6.

Practice Exercise 1

In which compound is the oxidation state of oxygen −1? (**a**) O_2, (**b**) H_2O, (**c**) H_2SO_4, (**d**) H_2O_2, (**e**) KCH_3COO.

Practice Exercise 2

What is the oxidation state of the boldfaced element in (**a**) $\mathbf{P}_2O_5$, (**b**) Na**H**, (**c**) $\mathbf{Cr}_2O_7^{2-}$, (**d**) $\mathbf{Sn}Br_4$, (**e**) Ba$\mathbf{O}_2$?

GO FIGURE

How many moles of hydrogen gas would be produced for every mole of magnesium added into the HCl solution?

▲ **Figure 4.13 Reaction of magnesium metal with hydrochloric acid.** The metal is readily oxidized by the acid, producing hydrogen gas, $H_2(g)$, and $MgCl_2(aq)$.

Oxidation of Metals by Acids and Salts

The reaction between a metal and either an acid or a metal salt conforms to the general pattern

$$A + BX \longrightarrow AX + B \qquad [4.24]$$

Examples: $Zn(s) + 2\,HBr(aq) \longrightarrow ZnBr_2(aq) + H_2(g)$

$Mn(s) + Pb(NO_3)_2(aq) \longrightarrow Mn(NO_3)_2(aq) + Pb(s)$

These reactions are called **displacement reactions** because the ion in solution is *displaced* (replaced) through oxidation of an element.

Many metals undergo displacement reactions with acids, producing salts and hydrogen gas. For example, magnesium metal reacts with hydrochloric acid to form magnesium chloride and hydrogen gas (▲ **Figure 4.13**):

$$Mg(s) + 2\,HCl(aq) \longrightarrow MgCl_2(aq) + H_2(g) \qquad [4.25]$$

Oxidation number 0   +1 −1   +2 −1   0

The oxidation number of Mg changes from 0 to +2, an increase that indicates the atom has lost electrons and has therefore been oxidized. The oxidation number of H^+ in the acid decreases from +1 to 0, indicating that this ion has gained electrons and has therefore been reduced. Chlorine has an oxidation number of -1 both before and after the reaction, indicating that it is neither oxidized nor reduced. In fact the Cl^- ions are spectator ions, dropping out of the net ionic equation:

$$Mg(s) + 2\,H^+(aq) \longrightarrow Mg^{2+}(aq) + H_2(g) \qquad [4.26]$$

Metals can also be oxidized by aqueous solutions of various salts. Iron metal, for example, is oxidized to Fe^{2+} by aqueous solutions of Ni^{2+} such as $Ni(NO_3)_2(aq)$:

Molecular equation: $\quad Fe(s) + Ni(NO_3)_2(aq) \longrightarrow Fe(NO_3)_2(aq) + Ni(s) \quad [4.27]$

Net ionic equation: $\qquad Fe(s) + Ni^{2+}(aq) \longrightarrow Fe^{2+}(aq) + Ni(s) \qquad [4.28]$

The oxidation of Fe to Fe^{2+} in this reaction is accompanied by the reduction of Ni^{2+} to Ni. Remember: *Whenever one substance is oxidized, another substance must be reduced.*

SAMPLE EXERCISE 4.9 | Writing Equations for Oxidation-Reduction Reactions

Write the balanced molecular and net ionic equations for the reaction of aluminum with hydrobromic acid.

SOLUTION

Analyze We must write two equations—molecular and net ionic—for the redox reaction between a metal and an acid.

Plan Metals react with acids to form salts and H_2 gas. To write the balanced equations, we must write the chemical formulas for the two reactants and then determine the formula of the salt, which is composed of the cation formed by the metal and the anion of the acid.

Solve The reactants are Al and HBr. The cation formed by Al is Al^{3+}, and the anion from hydrobromic acid is Br^-. Thus, the salt formed in the reaction is $AlBr_3$. Writing the reactants and products and then balancing the equation gives the molecular equation:

$$2\,Al(s) + 6\,HBr(aq) \longrightarrow 2\,AlBr_3(aq) + 3\,H_2(g)$$

Both HBr and $AlBr_3$ are soluble strong electrolytes. Thus, the complete ionic equation is

$$2\,Al(s) + 6\,H^+(aq) + 6\,Br^-(aq) \longrightarrow$$
$$2\,Al^{3+}(aq) + 6\,Br^-(aq) + 3\,H_2(g)$$

Because Br^- is a spectator ion, the net ionic equation is

$$2\,Al(s) + 6\,H^+(aq) \longrightarrow 2\,Al^{3+}(aq) + 3\,H_2(g)$$

Comment The substance oxidized is the aluminum metal because its oxidation state changes from 0 in the metal to $+3$ in the cation, thereby increasing in oxidation number. The H^+ is reduced because its oxidation state changes from $+1$ in the acid to 0 in H_2.

Practice Exercise 1

Which of the following statements is true about the reaction between zinc and copper sulfate? **(a)** Zinc is oxidized, and copper ion is reduced. **(b)** Zinc is reduced, and copper ion is oxidized. **(c)** All reactants and products are soluble strong electrolytes. **(d)** The oxidation state of copper in copper sulfate is 0. **(e)** More than one of the previous choices are true.

Practice Exercise 2

(a) Write the balanced molecular and net ionic equations for the reaction between magnesium and cobalt(II) sulfate. **(b)** What is oxidized and what is reduced in the reaction?

The Activity Series

Can we predict whether a certain metal will be oxidized either by an acid or by a particular salt? This question is of practical importance as well as chemical interest. According to Equation 4.27, for example, it would be unwise to store a solution of nickel nitrate in an iron container because the solution would dissolve the container. When a metal is oxidized, it forms various compounds. Extensive oxidation can lead to the failure of metal machinery parts or the deterioration of metal structures.

Different metals vary in the ease with which they are oxidized. Zn is oxidized by aqueous solutions of Cu^{2+}, for example, but Ag is not. Zn, therefore, loses electrons more readily than Ag; that is, Zn is easier to oxidize than Ag.

A list of metals arranged in order of decreasing ease of oxidation, such as in **Table 4.5**, is called an **activity series**. The metals at the top of the table, such as the alkali metals and the alkaline earth metals, are most easily oxidized; that is, they react most readily to form compounds. They are called the *active metals*. The metals at the bottom of the activity series, such as the transition elements from groups 8B and 1B, are very stable and form compounds less readily. These metals, which are used to make coins and jewelry, are called *noble metals* because of their low reactivity.

The activity series can be used to predict the outcome of reactions between metals and either metal salts or acids. *Any metal on the list can be oxidized by the ions of*

Table 4.5 Activity Series of Metals in Aqueous Solution

Metal	Oxidation Reaction
Lithium	$Li(s) \longrightarrow Li^+(aq) + e^-$
Potassium	$K(s) \longrightarrow K^+(aq) + e^-$
Barium	$Ba(s) \longrightarrow Ba^{2+}(aq) + 2e^-$
Calcium	$Ca(s) \longrightarrow Ca^{2+}(aq) + 2e^-$
Sodium	$Na(s) \longrightarrow Na^+(aq) + e^-$
Magnesium	$Mg(s) \longrightarrow Mg^{2+}(aq) + 2e^-$
Aluminum	$Al(s) \longrightarrow Al^{3+}(aq) + 3e^-$
Manganese	$Mn(s) \longrightarrow Mn^{2+}(aq) + 2e^-$
Zinc	$Zn(s) \longrightarrow Zn^{2+}(aq) + 2e^-$
Chromium	$Cr(s) \longrightarrow Cr^{3+}(aq) + 3e^-$
Iron	$Fe(s) \longrightarrow Fe^{2+}(aq) + 2e^-$
Cobalt	$Co(s) \longrightarrow Co^{2+}(aq) + 2e^-$
Nickel	$Ni(s) \longrightarrow Ni^{2+}(aq) + 2e^-$
Tin	$Sn(s) \longrightarrow Sn^{2+}(aq) + 2e^-$
Lead	$Pb(s) \longrightarrow Pb^{2+}(aq) + 2e^-$
Hydrogen	$H_2(g) \longrightarrow 2H^+(aq) + 2e^-$
Copper	$Cu(s) \longrightarrow Cu^{2+}(aq) + 2e^-$
Silver	$Ag(s) \longrightarrow Ag^+(aq) + e^-$
Mercury	$Hg(l) \longrightarrow Hg^{2+}(aq) + 2e^-$
Platinum	$Pt(s) \longrightarrow Pt^{2+}(aq) + 2e^-$
Gold	$Au(s) \longrightarrow Au^{3+}(aq) + 3e^-$

Ease of oxidation increases ↑

elements below it. For example, copper is above silver in the series. Thus, copper metal is oxidized by silver ions:

$$Cu(s) + 2Ag^+(aq) \longrightarrow Cu^{2+}(aq) + 2Ag(s) \qquad [4.29]$$

The oxidation of copper to copper ions is accompanied by the reduction of silver ions to silver metal. The silver metal is evident on the surface of the copper wire in ▶ Figure 4.14. The copper(II) nitrate produces a blue color in the solution, as can be seen most clearly in the photograph on the right of Figure 4.14.

 Give It Some Thought

Does a reaction occur **(a)** when an aqueous solution of $NiCl_2(aq)$ is added to a test tube containing strips of metallic zinc, and **(b)** when $NiCl_2(aq)$ is added to a test tube containing $Zn(NO_3)_2(aq)$?

Only metals above hydrogen in the activity series are able to react with acids to form H_2. For example, Ni reacts with $HCl(aq)$ to form H_2:

$$Ni(s) + 2HCl(aq) \longrightarrow NiCl_2(aq) + H_2(g) \qquad [4.30]$$

Because elements below hydrogen in the activity series are not oxidized by H^+, Cu does not react with $HCl(aq)$. Interestingly, copper does react with nitric acid, as shown in Figure 1.11, but the reaction is not oxidation of Cu by H^+ ions. Instead, the metal is oxidized to Cu^{2+} by the nitrate ion, accompanied by the formation of brown nitrogen dioxide, $NO_2(g)$:

$$Cu(s) + 4HNO_3(aq) \longrightarrow Cu(NO_3)_2(aq) + 2H_2O(l) + 2NO_2(g) \qquad [4.31]$$

As the copper is oxidized in this reaction, NO_3^-, where the oxidation number of nitrogen is +5, is reduced to NO_2, where the oxidation number of nitrogen is +4. We will examine reactions of this type in Chapter 20.

▲ **GO FIGURE**

Why does this solution turn blue?

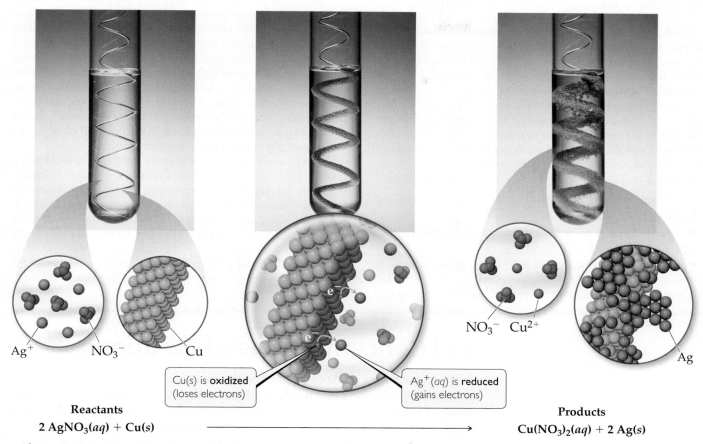

Cu(s) is **oxidized**
(loses electrons)

Ag$^+$(aq) is **reduced**
(gains electrons)

NO$_3^-$ Cu^{2+}

Ag$^+$ NO$_3^-$ Cu   Ag

Reactants

2 AgNO$_3$(aq) + Cu(s)

Products

Cu(NO$_3$)$_2$(aq) + 2 Ag(s)

▲ **Figure 4.14 Reaction of copper metal with silver ion.** When copper metal is placed in a solution of silver nitrate, a redox reaction forms silver metal and a blue solution of copper(II) nitrate.

SAMPLE EXERCISE 4.10 Determining When an Oxidation-Reduction Reaction Can Occur

Will an aqueous solution of iron(II) chloride oxidize magnesium metal? If so, write the balanced molecular and net ionic equations for the reaction.

SOLUTION

Analyze We are given two substances—an aqueous salt, FeCl$_2$, and a metal, Mg—and asked if they react with each other.

Plan A reaction occurs if the reactant that is a metal in its elemental form (Mg) is located above the reactant that is a metal in its oxidized form (Fe^{2+}) in Table 4.5. If the reaction occurs, the Fe^{2+} ion in FeCl$_2$ is reduced to Fe, and the Mg is oxidized to Mg^{2+}.

Solve Because Mg is above Fe in the table, the reaction occurs. To write the formula for the salt produced in the reaction, we must remember the charges on common ions. Magnesium is always present in compounds as Mg^{2+}; the chloride ion is Cl$^-$. The magnesium salt formed in the reaction is MgCl$_2$, meaning the balanced molecular equation is

$$Mg(s) + FeCl_2(aq) \longrightarrow MgCl_2(aq) + Fe(s)$$

Both FeCl$_2$ and MgCl$_2$ are soluble strong electrolytes and can be written in ionic form, which shows us that Cl$^-$ is a spectator ion in the reaction. The net ionic equation is

$$Mg(s) + Fe^{2+}(aq) \longrightarrow Mg^{2+}(aq) + Fe(s)$$

The net ionic equation shows that Mg is oxidized and Fe^{2+} is reduced in this reaction.

Check Note that the net ionic equation is balanced with respect to both charge and mass.

Practice Exercise 1

Which of these metals is the easiest to oxidize? (**a**) gold, (**b**) lithium, (**c**) iron, (**d**) sodium, (**e**) aluminum.

Practice Exercise 2

Which of the following metals will be oxidized by Pb(NO$_3$)$_2$: Zn, Cu, Fe?

Strategies in Chemistry

Analyzing Chemical Reactions

In this chapter you have been introduced to a great number of chemical reactions. It's not easy to get a "feel" for what happens when chemicals react. One goal of this textbook is to help you become more adept at predicting the outcomes of reactions. The key to gaining this "chemical intuition" is to learn how to categorize reactions.

Attempting to memorize individual reactions would be a futile task. It is far more fruitful to recognize patterns to determine the general category of a reaction, such as metathesis or oxidation-reduction. When faced with the challenge of predicting the outcome of a chemical reaction, ask yourself the following questions:

- What are the reactants?
- Are they electrolytes or nonelectrolytes?
- Are they acids or bases?
- If the reactants are electrolytes, will metathesis produce a precipitate? Water? A gas?
- If metathesis cannot occur, can the reactants engage in an oxidation-reduction reaction? This requires that there be both a reactant that can be oxidized and a reactant that can be reduced.

Being able to predict what happens during a reaction follows from asking basic questions like the ones above. Each question narrows the set of possible outcomes, steering you ever closer to a likely outcome. Your prediction might not always be entirely correct, but if you keep your wits about you, you will not be far off. As you gain experience, you will begin to look for reactants that might not be immediately obvious, such as water from the solution or oxygen from the atmosphere. Because proton transfer (acid-base) and electron transfer (oxidation-reduction) are involved in a huge number of chemical reactions, knowing the hallmarks of such reactions will mean you are well on your way to becoming an excellent chemist!

The laboratory is the best place to learn how to think like a chemist. One of the greatest tools available to chemists is experimentation. If you perform an experiment in which two solutions are mixed, you can make observations that help you understand what is happening. Consider, for example, the precipitation experiment in Figure 4.4. Although you might use Table 4.1 to predict whether a precipitate will form, it is much more exciting to actually see the precipitate form! Careful observation in the laboratory portion of the course will make your lecture material both more meaningful and easier to master.

4.5 | Concentrations of Solutions

Scientists use the term **concentration** to designate the amount of solute dissolved in a given quantity of solvent or quantity of solution. The greater the amount of solute dissolved in a certain amount of solvent, the more concentrated the resulting solution. In chemistry we often need to express the concentrations of solutions quantitatively.

Molarity

Molarity (symbol M) expresses the concentration of a solution as the number of moles of solute in a liter of solution (soln):

$$\text{Molarity} = \frac{\text{moles solute}}{\text{volume of solution in liters}} \qquad [4.32]$$

A 1.00 molar solution (written 1.00 M) contains 1.00 mol of solute in every liter of solution. ▶ Figure 4.15 shows the preparation of 0.250 L of a 1.00 M solution of $CuSO_4$. The molarity of the solution is $(0.250 \text{ mol } CuSO_4)/(0.250 \text{ L soln}) = 1.00 \, M$.

 Give It Some Thought

Which is more concentrated, a solution prepared by dissolving 21.0 g of NaF (0.500 mol) in enough water to make 500 mL of solution or a solution prepared by dissolving 10.5 g (0.250 mol) of NaF in enough water to make 100 mL of solution?

SAMPLE EXERCISE 4.11 Calculating Molarity

Calculate the molarity of a solution made by dissolving 23.4 g of sodium sulfate (Na_2SO_4) in enough water to form 125 mL of solution.

SOLUTION

Analyze We are given the number of grams of solute (23.4 g), its chemical formula (Na_2SO_4), and the volume of the solution (125 mL) and asked to calculate the molarity of the solution.

Plan We can calculate molarity using Equation 4.32. To do so, we must convert the number of grams of solute to moles and the volume of the solution from milliliters to liters.

Solve The number of moles of Na_2SO_4 is obtained by using its molar mass:

$$\text{Moles } Na_2SO_4 = (23.4 \text{ g } Na_2SO_4)\left(\frac{1 \text{ mol } Na_2SO_4}{142.1 \text{ g } Na_2SO_4}\right) = 0.165 \text{ mol } Na_2SO_4$$

Converting the volume of the solution to liters:

$$\text{Liters soln} = (125 \text{ mL})\left(\frac{1 \text{ L}}{1000 \text{ mL}}\right) = 0.125 \text{ L}$$

Thus, the molarity is

$$\text{Molarity} = \frac{0.165 \text{ mol } Na_2SO_4}{0.125 \text{ L soln}} = 1.32 \frac{\text{mol } Na_2SO_4}{\text{L soln}} = 1.32 \text{ } M$$

Check Because the numerator is only slightly larger than the denominator, it is reasonable for the answer to be a little over 1 M. The units (mol/L) are appropriate for molarity, and three significant figures are appropriate for the answer because each of the initial pieces of data had three significant figures.

Practice Exercise 1

What is the molarity of a solution that is made by dissolving 3.68 g of sucrose $(C_{12}H_{22}O_{11})$ in sufficient water to form 275.0 mL of

solution? **(a)** 13.4 M, **(b)** $7.43 \times 10^{-2} M$, **(c)** $3.91 \times 10^{-2} M$ **(d)** $7.43 \times 10^{-5} M$ **(e)** $3.91 \times 10^{-5} M$.

Practice Exercise 2

Calculate the molarity of a solution made by dissolving 5.00 g of glucose $(C_6H_{12}O_6)$ in sufficient water to form exactly 100 mL of solution.

▲ **Figure 4.15 Preparing 0.250 L of a 1.00 *M* solution of CuSO₄.**

1 Weigh out 39.9 g (0.250 mol) CuSO₄

2 Put CuSO₄ (solute) into 250-mL volumetric flask; add water and swirl to dissolve solute

3 Add water until solution just reaches calibration mark on neck of flask

Expressing the Concentration of an Electrolyte

In biology, the total concentration of ions in solution is very important in metabolic and cellular processes. When an ionic compound dissolves, the relative concentrations of the ions in the solution depend on the chemical formula of the compound. For example, a 1.0 M solution of NaCl is 1.0 M in Na^+ ions and 1.0 M in Cl^- ions, and a 1.0 M solution of Na_2SO_4 is 2.0 M in Na^+ ions and 1.0 M in SO_4^{2-} ions. Thus, the concentration of an electrolyte solution can be specified either in terms of the compound used to make the solution (1.0 M Na_2SO_4) or in terms of the ions in the solution (2.0 M Na^+ and 1.0 M SO_4^{2-}).

SAMPLE
EXERCISE 4.12 | Calculating Molar Concentrations of Ions

What is the molar concentration of each ion present in a 0.025 M aqueous solution of calcium nitrate?

SOLUTION

Analyze We are given the concentration of the ionic compound used to make the solution and asked to determine the concentrations of the ions in the solution.

Plan We can use the subscripts in the chemical formula of the compound to determine the relative ion concentrations.

Solve Calcium nitrate is composed of calcium ions (Ca^{2+}) and nitrate ions (NO_3^-), so its chemical formula is $Ca(NO_3)_2$. Because there are two NO_3^- ions for each Ca^{2+} ion, each mole of $Ca(NO_3)_2$ that dissolves dissociates into 1 mol of Ca^{2+} and 2 mol of NO_3^-. Thus, a solution that is 0.025 M in $Ca(NO_3)_2$ is 0.025 M in Ca^{2+} and 2 × 0.025 M = 0.050 M in NO_3^-:

$$\frac{mol\ NO_3^-}{L} = \left(\frac{0.025\ mol\ Ca(NO_3)_2}{L}\right)\left(\frac{2\ mol\ NO_3^-}{1\ mol\ Ca(NO_3)_2}\right) = 0.050\ M$$

Check The concentration of NO_3^- ions is twice that of Ca^{2+} ions, as the subscript 2 after the NO_3^- in the chemical formula $Ca(NO_3)_2$ suggests.

Practice Exercise 1
What is the ratio of the concentration of potassium ions to the concentration of carbonate ions in a 0.015 M solution of potassium carbonate? **(a)** 1:0.015, **(b)** 0.015:1, **(c)** 1:1, **(d)** 1:2, **(e)** 2:1.

Practice Exercise 2
What is the molar concentration of K^+ ions in a 0.015 M solution of potassium carbonate?

Interconverting Molarity, Moles, and Volume

If we know any two of the three quantities in Equation 4.32, we can calculate the third. For example, if we know the molarity of an HNO_3 solution to be 0.200 M, which means 0.200 mol of HNO_3 per liter of solution, we can calculate the number of moles of solute in a given volume, say 2.0 L. Molarity therefore is a conversion factor between volume of solution and moles of solute:

$$Moles\ HNO_3 = (2.0\ L\ soln)\left(\frac{0.200\ mol\ HNO_3}{1\ L\ soln}\right) = 0.40\ mol\ HNO_3$$

To illustrate the conversion of moles to volume, let's calculate the volume of 0.30 M HNO_3 solution required to supply 2.0 mol of HNO_3:

$$Liters\ soln = (2.0\ mol\ HNO_3)\left[\frac{1\ L\ soln}{0.30\ mol\ HNO_3}\right] = 6.7\ L\ soln$$

In this case we must use the reciprocal of molarity in the conversion:

$$Liters = moles \times 1/M = moles \times liters/mole.$$

If one of the solutes is a liquid, we can use its density to convert its mass to volume and vice versa. For example, a typical American beer contains 5.0% ethanol (CH_3CH_2OH) by volume in water (along with other components). The density of ethanol is 0.789 g/mL. Therefore, if we wanted to calculate the molarity of ethanol (usually just called "alcohol" in everyday language) in beer, we would first consider 1.00 L of beer.

This 1.00 L of beer contains 0.950 L of water and 0.050 L of ethanol:

$$5.0\% = 5/100 = 0.050$$

Then we can calculate the moles of ethanol by proper cancellation of units, taking into account the density of ethanol and its molar mass (46.0 g/mol):

$$Moles\ ethanol = (0.050\ L)\left(\frac{1000\ mL}{L}\right)\left(\frac{0.789\ g}{mL}\right)\left(\frac{1\ mol}{46.0\ g}\right) = 0.858\ mol$$

Because there are 0.858 moles of ethanol in 1.00 L of beer, the concentration of ethanol in beer is 0.86 M (taking into account significant digits).

SAMPLE EXERCISE 4.13 | **Using Molarity to Calculate Grams of Solute**

How many grams of Na_2SO_4 are required to make 0.350 L of 0.500 M Na_2SO_4?

SOLUTION

Analyze We are given the volume of the solution (0.350 L), its concentration (0.500 M), and the identity of the solute Na_2SO_4 and asked to calculate the number of grams of the solute in the solution.

Plan We can use the definition of molarity (Equation 4.32) to determine the number of moles of solute, and then convert moles to grams using the molar mass of the solute.

$$M_{Na_2SO_4} = \frac{\text{moles } Na_2SO_4}{\text{liters soln}}$$

Solve Calculating the moles of Na_2SO_4 using the molarity and volume of solution gives

$$M_{Na_2SO_4} = \frac{\text{moles } Na_2SO_4}{\text{liters soln}}$$

$$\text{Moles } Na_2SO_4 = \text{liters soln} \times M_{Na_2SO_4}$$

$$= (0.350 \text{ L soln})\left(\frac{0.500 \text{ mol } Na_2SO_4}{1 \text{ L soln}}\right)$$

$$= 0.175 \text{ mol } Na_2SO_4$$

Because each mole of Na_2SO_4 has a mass of 142.1 g, the required number of grams of Na_2SO_4 is

$$\text{Grams } Na_2SO_4 = (0.175 \text{ mol } Na_2SO_4)\left(\frac{142.1 \text{ g } Na_2SO_4}{1 \text{ mol } Na_2SO_4}\right)$$

$$= 24.9 \text{ g } Na_2SO_4$$

Check The magnitude of the answer, the units, and the number of significant figures are all appropriate.

Practice Exercise 1

What is the concentration of ammonia in a solution made by dissolving 3.75 g of ammonia in 120.0 L of water? **(a)** 1.84×10^{-3} M, **(b)** 3.78×10^{-2} M, **(c)** 0.0313 M, **(d)** 1.84 M, **(e)** 7.05 M.

Practice Exercise 2

(a) How many grams of Na_2SO_4 are there in 15 mL of 0.50 M Na_2SO_4? **(b)** How many milliliters of 0.50 M Na_2SO_4 solution are needed to provide 0.038 mol of this salt?

Dilution

Solutions used routinely in the laboratory are often purchased or prepared in concentrated form (called *stock solutions*). Solutions of lower concentrations can then be obtained by adding water, a process called **dilution**.*

Let's see how we can prepare a dilute solution from a concentrated one. Suppose we want to prepare 250.0 mL (that is, 0.2500 L) of 0.100 M $CuSO_4$ solution by diluting a 1.00 M $CuSO_4$ stock solution. The main point to remember is that when solvent is added to a solution, the number of moles of solute remains unchanged:

$$\text{Moles solute before dilution} = \text{moles solute after dilution} \qquad [4.33]$$

Because we know both the volume (250.0 mL) and the concentration (0.100 mol/L) of the dilute solution, we can calculate the number of moles of $CuSO_4$ it contains:

$$\text{Moles } CuSO_4 \text{ in dilute soln} = (0.2500 \text{ L soln})\left(\frac{0.100 \text{ mol } CuSO_4}{\text{L soln}}\right)$$

$$= 0.0250 \text{ mol } CuSO_4$$

The volume of stock solution needed to provide 0.0250 mol $CuSO_4$ is therefore:

$$\text{Liters of conc soln} = (0.0250 \text{ mol } CuSO_4)\left(\frac{1 \text{ L soln}}{1.00 \text{ mol } CuSO_4}\right) = 0.0250 \text{ L}$$

Figure 4.16 shows the dilution carried out in the laboratory. Notice that the diluted solution is less intensely colored than the concentrated one.

---

*In diluting a concentrated acid or base, the acid or base should be added to water and then further diluted by adding more water. Adding water directly to concentrated acid or base can cause spattering because of the intense heat generated.

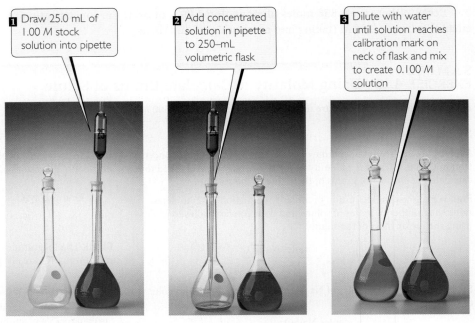

▲ Figure 4.16 **Preparing 250.0 mL of 0.100 M CuSO₄ by dilution of 1.00 M CuSO₄**

 Give It Some Thought

How is the molarity of a 0.50 *M* KBr solution changed when water is added to double its volume?

In laboratory situations, calculations of this sort are often made with an equation derived by remembering that the number of moles of solute is the same in both the concentrated and dilute solutions and that moles = molarity × liters:

$$\text{Moles solute in conc soln} = \text{moles solute in dilute soln}$$

$$M_{\text{conc}} \times V_{\text{conc}} = M_{\text{dil}} \times V_{\text{dil}} \qquad [4.34]$$

Although we derived Equation 4.34 in terms of liters, any volume unit can be used as long as it is used on both sides of the equation. For example, in the calculation we did for the CuSO₄ solution, we have

$$(1.00\ M)(V_{\text{conc}}) = (0.100\ M)(250.0\ \text{mL})$$

Solving for V_{conc} gives $V_{\text{conc}} = 25.0$ mL as before.

SAMPLE
EXERCISE 4.14   **Preparing a Solution by Dilution**

How many milliliters of 3.0 *M* H₂SO₄ are needed to make 450 mL of 0.10 *M* H₂SO₄?

SOLUTION

Analyze We need to dilute a concentrated solution. We are given the molarity of a more concentrated solution (3.0 *M*) and the volume and molarity of a more dilute one containing the same solute (450 mL of 0.10 *M* solution). We must calculate the volume of the concentrated solution needed to prepare the dilute solution.

Plan We can calculate the number of moles of solute, H₂SO₄, in the dilute solution and then calculate the volume of the concentrated solution needed to supply this amount of solute. Alternatively, we can directly apply Equation 4.34. Let's compare the two methods.

Solve Calculating the moles of H₂SO₄ in the dilute solution:

$$\text{Moles H}_2\text{SO}_4 \text{ in dilute solution} = (0.450\ \text{L soln})\left(\frac{0.10\ \text{mol H}_2\text{SO}_4}{1\ \text{L soln}}\right)$$

$$= 0.045\ \text{mol H}_2\text{SO}_4$$

Calculating the volume of the concentrated solution that contains 0.045 mol H₂SO₄:

$$\text{L conc soln} = (0.045\ \text{mol H}_2\text{SO}_4)\left(\frac{1\ \text{L soln}}{3.0\ \text{mol H}_2\text{SO}_4}\right) = 0.015\ \text{L soln}$$

Converting liters to milliliters gives 15 mL.

If we apply Equation 4.34, we get the same result:

$$(3.0\ M)(V_{conc}) = (0.10\ M)(450\ mL)$$

$$(V_{conc}) = \frac{(0.10\ M)(450\ mL)}{3.0\ M} = 15\ mL$$

Either way, we see that if we start with 15 mL of 3.0 M H_2SO_4 and dilute it to a total volume of 450 mL, the desired 0.10 M solution will be obtained.

Check The calculated volume seems reasonable because a small volume of concentrated solution is used to prepare a large volume of dilute solution.

Comment The first approach can also be used to find the final concentration when two solutions of different concentrations are mixed, whereas the second approach, using Equation 4.34,

can be used only for diluting a concentrated solution with pure solvent.

Practice Exercise 1

What volume of a 1.00 M stock solution of glucose must be used to make 500.0 mL of a 1.75×10^{-2} M glucose solution in water? **(a)** 1.75 mL, **(b)** 8.75 mL, **(c)** 48.6 mL, **(d)** 57.1 mL, **(e)** 28,570 mL.

Practice Exercise 2

(a) What volume of 2.50 M lead(II) nitrate solution contains 0.0500 mol of Pb^{2+}? **(b)** How many milliliters of 5.0 M $K_2Cr_2O_7$ solution must be diluted to prepare 250 mL of 0.10 M solution? **(c)** If 10.0 mL of a 10.0 M stock solution of NaOH is diluted to 250 mL, what is the concentration of the resulting stock solution?

4.6 | Solution Stoichiometry and Chemical Analysis

In Chapter 3 we learned that given the chemical equation for a reaction and the amount of one reactant consumed in the reaction, you can calculate the quantities of other reactants and products. In this section we extend this concept to reactions involving solutions.

Recall that the coefficients in a balanced equation give the relative number of moles of reactants and products. ∞ (Section 3.6) To use this information, we must convert the masses of substances involved in a reaction into moles. When dealing with pure substances, as we did in Chapter 3, we use molar mass to convert between grams and moles of the substances. This conversion is not valid when working with a solution because both solute and solvent contribute to its mass. However, if we know the solute concentration, we can use molarity and volume to determine the number of moles (moles solute = $M \times V$). ▼ Figure 4.17 summarizes this approach to using stoichiometry for the reaction between a pure substance and a solution.

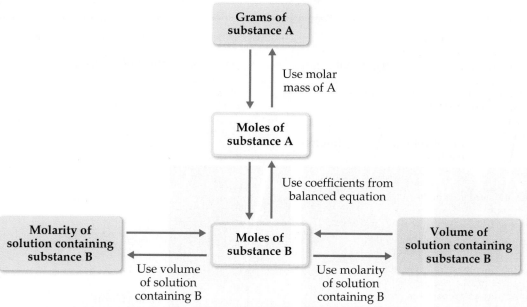

▲ **Figure 4.17 Procedure for solving stoichiometry problems involving reactions between a pure substance A and a solution containing a known concentration of substance B.** Starting from a known mass of substance A, we follow the red arrows to determine either the volume of the solution containing B (if the molarity of B is known) or the molarity of the solution containing B (if the volume of B is known). Starting from either a known volume or known molarity of the solution containing B, we follow the green arrows to determine the mass of substance A.

SAMPLE EXERCISE 4.15 | Using Mass Relations in a Neutralization Reaction

How many grams of $Ca(OH)_2$ are needed to neutralize 25.0 mL of 0.100 M HNO_3?

SOLUTION

Analyze The reactants are an acid, HNO_3, and a base, $Ca(OH)_2$. The volume and molarity of HNO_3 are given, and we are asked how many grams of $Ca(OH)_2$ are needed to neutralize this quantity of HNO_3.

Plan Following the steps outlined by the green arrows in Figure 4.17, we use the molarity and volume of the HNO_3 solution (substance B in Figure 4.17) to calculate the number of moles of HNO_3. We then use the balanced equation to relate moles of HNO_3 to moles of $Ca(OH)_2$ (substance A). Finally, we use the molar mass to convert moles to grams of $Ca(OH)_2$:

$$V_{HNO_3} \times M_{HNO_3} \Rightarrow \text{mol } HNO_3 \Rightarrow \text{mol } Ca(OH)_2 \Rightarrow \text{g } Ca(OH)_2$$

Solve The product of the molar concentration of a solution and its volume in liters gives the number of moles of solute:

$$\text{Moles } HNO_3 = V_{HNO_3} \times M_{HNO_3} = (0.0250 \text{ L})\left(\frac{0.100 \text{ mol } HNO_3}{\text{L}}\right)$$

$$= 2.50 \times 10^{-3} \text{ mol } HNO_3$$

Because this is a neutralization reaction, HNO_3 and $Ca(OH)_2$ react to form H_2O and the salt containing Ca^{2+} and NO_3^-:

$$2 \, HNO_3(aq) + Ca(OH)_2(s) \longrightarrow 2 \, H_2O(l) + Ca(NO_3)_2(aq)$$

Thus, 2 mol HNO_3 ≃ mol $Ca(OH)_2$. Therefore,

$$\text{Grams } Ca(OH)_2 = (2.50 \times 10^{-3} \text{ mol } HNO_3)$$
$$\times \left(\frac{1 \text{ mol } Ca(OH)_2}{2 \text{ mol } HNO_3}\right)\left(\frac{74.1 \text{ g } Ca(OH)_2}{1 \text{ mol } Ca(OH)_2}\right)$$
$$= 0.0926 \text{ g } Ca(OH)_2$$

Check The answer is reasonable because a small volume of dilute acid requires only a small amount of base to neutralize it.

Practice Exercise 1

How many milligrams of sodium sulfide are needed to completely react with 25.00 mL of a 0.0100 M aqueous solution of cadmium nitrate, to form a precipitate of $CdS(s)$? **(a)** 13.8 mg, **(b)** 19.5 mg, **(c)** 23.5 mg, **(d)** 32.1 mg, **(e)** 39.0 mg.

Practice Exercise 2

(a) How many grams of NaOH are needed to neutralize 20.0 mL of 0.150 M H_2SO_4 solution? **(b)** How many liters of 0.500 M $HCl(aq)$ are needed to react completely with 0.100 mol of $Pb(NO_3)_2(aq)$, forming a precipitate of $PbCl_2(s)$?

Titrations

To determine the concentration of a particular solute in a solution, chemists often carry out a **titration**, which involves combining a solution where the solute concentration is not known with a reagent solution of known concentration, called a **standard solution**. Just enough standard solution is added to completely react with the solute in the solution of unknown concentration. The point at which stoichiometrically equivalent quantities are brought together is known as the **equivalence point**.

Titrations can be conducted using neutralization, precipitation, or oxidation-reduction reactions. ▼ Figure 4.18 illustrates a typical neutralization titration, one

▲ GO FIGURE

How would the volume of standard solution added change if that solution were $Ba(OH)_2(aq)$ instead of $NaOH(aq)$?

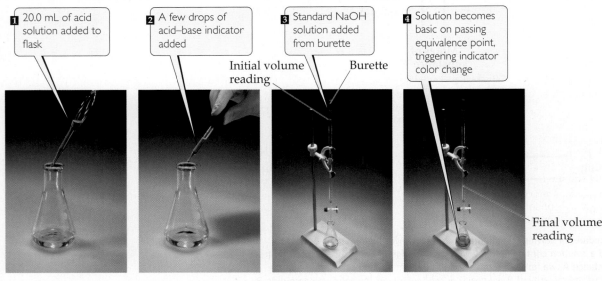

1 20.0 mL of acid solution added to flask

2 A few drops of acid–base indicator added

3 Standard NaOH solution added from burette

4 Solution becomes basic on passing equivalence point, triggering indicator color change

Initial volume reading

Burette

Final volume reading

▲ Figure 4.18 **Procedure for titrating an acid against a standard solution of NaOH.** The acid–base indicator, phenolphthalein, is colorless in acidic solution but takes on a pink color in basic solution.

between an HCl solution of unknown concentration and a standard NaOH solution. To determine the HCl concentration, we first add a specific volume of the HCl solution, 20.0 mL in this example, to a flask. Next we add a few drops of an acid–base indicator. The acid–base indicator is a dye that changes color on passing the equivalence point.* For example, the dye phenolphthalein is colorless in acidic solution but pink in basic solution. The standard solution is then slowly added until the solution turns pink, telling us that the neutralization reaction between HCl and NaOH is complete. The standard solution is added from a *burette* so that we can accurately determine the added volume of NaOH solution. Knowing the volumes of both solutions and the concentration of the standard solution, we can calculate the concentration of the unknown solution as diagrammed in ▼ Figure 4.19.

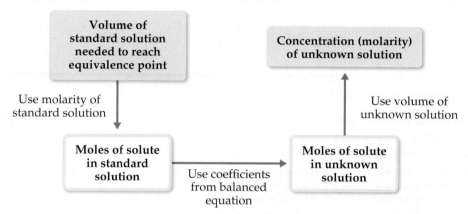

▲ Figure 4.19 **Procedure for determining the concentration of a solution from titration with a standard solution.**

SAMPLE EXERCISE 4.16 | Determining Solution Concentration by an Acid–Base Titration

One commercial method used to peel potatoes is to soak them in a NaOH solution for a short time and then remove the potatoes and spray off the peel. The NaOH concentration is normally 3 to 6 M, and the solution must be analyzed periodically. In one such analysis, 45.7 mL of 0.500 M H_2SO_4 is required to neutralize 20.0 mL of NaOH solution. What is the concentration of the NaOH solution?

SOLUTION

Analyze We are given the volume (45.7 mL) and molarity (0.500 M) of an H_2SO_4 solution (the standard solution) that reacts completely with 20.0 mL of NaOH solution. We are asked to calculate the molarity of the NaOH solution.

Plan Following the steps given in Figure 4.19, we use the H_2SO_4 volume and molarity to calculate the number of moles of H_2SO_4. Then we can use this quantity and the balanced equation for the reaction to calculate moles of NaOH. Finally, we can use moles of NaOH and the NaOH volume to calculate NaOH molarity.

Solve The number of moles of H_2SO_4 is the product of the volume and molarity of this solution:

$$\text{Moles } H_2SO_4 = (45.7 \text{ mL soln})\left(\frac{1 \text{ L soln}}{1000 \text{ mL soln}}\right)\left(\frac{0.500 \text{ mol } H_2SO_4}{\text{L soln}}\right)$$

$$= 2.29 \times 10^{-2} \text{ mol } H_2SO_4$$

Acids react with metal hydroxides to form water and a salt. Thus, the balanced equation for the neutralization reaction is

$$H_2SO_4(aq) + 2 \, NaOH(aq) \longrightarrow 2 \, H_2O(l) + Na_2SO_4(aq)$$

*More precisely, the color change of an indicator signals the end point of the titration, which if the proper indicator is chosen lies very near the equivalence point. Acid–base titrations are discussed in more detail in Section 17.3.

According to the balanced equation, 1 mol $H_2SO_4 \simeq$ 2 mol NaOH. Therefore,

$$\text{Moles NaOH} = (2.28 \times 10^{-2}\ \text{mol}\ H_2SO_4)\left(\frac{2\ \text{mol NaOH}}{1\ \text{mol}\ H_2SO_4}\right)$$

$$= 4.57 \times 10^{-2}\ \text{mol NaOH}$$

Knowing the number of moles of NaOH in 20.0 mL of solution allows us to calculate the molarity of this solution:

$$\text{Molarity NaOH} = \frac{\text{mol NaOH}}{\text{L soln}}$$

$$= \left(\frac{4.56 \times 10^{-2}\ \text{mol NaOH}}{20.0\ \text{mL soln}}\right)\left(\frac{1000\ \text{mL soln}}{1\ \text{L soln}}\right)$$

$$= 2.28\ \frac{\text{mol NaOH}}{\text{L soln}} = 2.29\ M$$

Practice Exercise 1

What is the molarity of an HCl solution if 27.3 mL of it neutralizes 134.5 mL of 0.0165 M $Ba(OH)_2$? (a) 0.0444 M, (b) 0.0813 M, (c) 0.163 M, (d) 0.325 M, (e) 3.35 M.

Practice Exercise 2

What is the molarity of a NaOH solution if 48.0 mL neutralizes 35.0 mL of 0.144 M H_2SO_4?

SAMPLE EXERCISE 4.17 Determining the Quantity of Solute by Titration

The quantity of Cl^- in a municipal water supply is determined by titrating the sample with Ag^+. The precipitation reaction taking place during the titration is

$$Ag^+(aq) + Cl^-(aq) \longrightarrow AgCl(s)$$

The end point in this type of titration is marked by a change in color of a special type of indicator. (a) How many grams of chloride ion are in a sample of the water if 20.2 mL of 0.100 M Ag^+ is needed to react with all the chloride in the sample? (b) If the sample has a mass of 10.0 g, what percentage of Cl^- does it contain?

SOLUTION

Analyze We are given the volume (20.2 mL) and molarity (0.100 M) of a solution of Ag^+ and the chemical equation for reaction of this ion with Cl^-. We are asked to calculate the number of grams of Cl^- in the sample and the mass percentage of Cl^- in the sample.

(a) Plan We can use the procedure outlined by the green arrows in Figure 4.17. We begin by using the volume and molarity of Ag^+ to calculate the number of moles of Ag^+ used in the titration. We then use the balanced equation to determine the moles of Cl^- in the sample and from that the grams of Cl^-.

Solve

$$\text{Moles Ag}^+ = (20.2\ \text{mL soln})\left(\frac{1\ \text{L soln}}{1000\ \text{mL soln}}\right)\left(\frac{0.100\ \text{mol Ag}^+}{\text{L soln}}\right)$$

$$= 2.02 \times 10^{-3}\ \text{mol Ag}^+$$

From the balanced equation we see that 1 mol $Ag^+ \simeq$ 1 mol Cl^-. Using this information and the molar mass of Cl, we have

$$\text{Grams Cl}^- = (2.02 \times 10^{-3}\ \text{mol Ag}^+)\left(\frac{1\ \text{mol Cl}^-}{1\ \text{mol Ag}^+}\right)\left(\frac{35.5\ \text{g Cl}^-}{\text{mol Cl}^-}\right)$$

$$= 7.17 \times 10^{-2}\ \text{g Cl}^-$$

(b) Plan To calculate the percentage of Cl^- in the sample, we compare the number of grams of Cl^- in the sample, 7.17×10^{-2} g, with the original mass of the sample, 10.0 g.

Solve

$$\text{Percent Cl}^- = \frac{7.17 \times 10^{-2}\ \text{g}}{10.0\ \text{g}} \times 100\% = 0.717\%\ \text{Cl}^-$$

Comment Chloride ion is one of the most common ions in water and sewage. Ocean water contains 1.92% Cl^-. Whether water containing Cl^- tastes salty depends on the other ions present. If the only accompanying ions are Na^+, a salty taste may be detected with as little as 0.03% Cl^-.

Practice Exercise 1

A mysterious white powder is found at a crime scene. A simple chemical analysis concludes that the powder is a mixture of sugar and morphine ($C_{17}H_{19}NO_3$), a weak base similar to ammonia. The crime lab takes 10.00 mg of the mysterious white powder, dissolves it in 100.00 mL water, and titrates it to the equivalence point with 2.84 mL of a standard 0.0100 M HCl solution. What is the percentage of morphine in the white powder? (a) 8.10%, (b) 17.3%, (c) 32.6%, (d) 49.7%, (e) 81.0%.

Practice Exercise 2

A sample of an iron ore is dissolved in acid, and the iron is converted to Fe^{2+}. The sample is then titrated with 47.20 mL of 0.02240 M MnO_4^- solution. The oxidation-reduction reaction that occurs during titration is

$$MnO_4^-(aq) + 5\ Fe^{2+}(aq) + 8\ H^+(aq) \longrightarrow$$
$$Mn^{2+}(aq) + 5\ Fe^{3+}(aq) + 4\ H_2O(l)$$

(a) How many moles of MnO_4^- were added to the solution? (b) How many moles of Fe^{2+} were in the sample? (c) How many grams of iron were in the sample? (d) If the sample had a mass of 0.8890 g, what is the percentage of iron in the sample?

SAMPLE INTEGRATIVE EXERCISE | Putting Concepts Together

Note: *Integrative exercises require skills from earlier chapters as well as ones from the present chapter.*

A sample of 70.5 mg of potassium phosphate is added to 15.0 mL of 0.050 M silver nitrate, resulting in the formation of a precipitate. (**a**) Write the molecular equation for the reaction. (**b**) What is the limiting reactant in the reaction? (**c**) Calculate the theoretical yield, in grams, of the precipitate that forms.

SOLUTION

(**a**) Potassium phosphate and silver nitrate are both ionic compounds. Potassium phosphate contains K^+ and PO_4^{3-} ions, so its chemical formula is K_3PO_4. Silver nitrate contains Ag^+ and NO_3^- ions, so its chemical formula is $AgNO_3$. Because both reactants are strong electrolytes, the solution contains, K^+, PO_4^{3-}, Ag^+, and NO_3^- ions before the reaction occurs. According to the solubility guidelines in Table 4.1, Ag^+ and PO_4^{3-} form an insoluble compound, so Ag_3PO_4 will precipitate from the solution. In contrast, K^+ and NO_3^- will remain in solution because KNO_3 is water soluble. Thus, the balanced molecular equation for the reaction is

$$K_3PO_4(aq) + 3\,AgNO_3(aq) \longrightarrow Ag_3PO_4(s) + 3\,KNO_3(aq)$$

(**b**) To determine the limiting reactant, we must examine the number of moles of each reactant. ∞ (Section 3.7) The number of moles of K_3PO_4 is calculated from the mass of the sample using the molar mass as a conversion factor. ∞ (Section 3.4) The molar mass of K_3PO_4 is $3(39.1) + 31.0 + 4(16.0) = 212.3$ g/mol. Converting milligrams to grams and then to moles, we have

$$(70.5\text{ mg K}_3\text{PO}_4)\left(\frac{10^{-3}\text{ g K}_3\text{PO}_4}{1\text{ mg K}_3\text{PO}_4}\right)\left(\frac{1\text{ mol K}_3\text{PO}_4}{212.3\text{ g K}_3\text{PO}_4}\right)$$
$$= 3.32 \times 10^{-4}\text{ mol K}_3\text{PO}_4$$

We determine the number of moles of $AgNO_3$ from the volume and molarity of the solution. (Section 4.5) Converting milliliters to liters and then to moles, we have

$$(15.0\text{ mL})\left(\frac{10^{-3}\text{ L}}{1\text{ mL}}\right)\left(\frac{0.050\text{ mol AgNO}_3}{\text{L}}\right)$$
$$= 7.5 \times 10^{-4}\text{ mol AgNO}_3$$

Comparing the amounts of the two reactants, we find that there are $(7.5 \times 10^{-4})/(3.32 \times 10^{-4}) = 2.3$ times as many moles of $AgNO_3$ as there are moles of K_3PO_4. According to the balanced equation, however, 1 mol K_3PO_4 requires 3 mol $AgNO_3$. Thus, there is insufficient $AgNO_3$ to consume the K_3PO_4, and $AgNO_3$ is the limiting reactant.

(**c**) The precipitate is Ag_3PO_4, whose molar mass is $3(107.9) + 31.0 + 4(16.0) = 418.7$ g/mol. To calculate the number of grams of Ag_3PO_4 that could be produced in this reaction (the theoretical yield), we use the number of moles of the limiting reactant, converting mol $AgNO_3 \Rightarrow$ mol $Ag_3PO_4 \Rightarrow$ g Ag_3PO_4. We use the coefficients in the balanced equation to convert moles of $AgNO_3$ to moles Ag_3PO_4, and we use the molar mass of Ag_3PO_4 to convert the number of moles of this substance to grams.

$$(7.5 \times 10^{-4}\text{ mol AgNO}_3)\left(\frac{1\text{ mol Ag}_3\text{PO}_4}{3\text{ mol AgNO}_3}\right)\left(\frac{418.7\text{ g Ag}_3\text{PO}_4}{1\text{ mol Ag}_3\text{PO}_4}\right)$$
$$= 0.10\text{ g Ag}_3\text{PO}_4$$

The answer has only two significant figures because the quantity of $AgNO_3$ is given to only two significant figures.

Chapter Summary and Key Terms

GENERAL PROPERTIES OF AQUEOUS SOLUTIONS (INTRODUCTION AND SECTION 4.1) Solutions in which water is the dissolving medium are called **aqueous solutions**. The component of the solution that is present in the greatest quantity is the **solvent**. The other components are **solutes**.

Any substance whose aqueous solution contains ions is called an **electrolyte**. Any substance that forms a solution containing no ions is a **nonelectrolyte**. Electrolytes that are present in solution entirely as ions are **strong electrolytes**, whereas those that are present partly as ions and partly as molecules are **weak electrolytes**. Ionic compounds dissociate into ions when they dissolve, and they are strong electrolytes. The solubility of ionic substances is made possible by **solvation**, the interaction of ions with polar solvent molecules. Most molecular compounds are nonelectrolytes, although some are weak electrolytes, and a few are strong electrolytes. When representing the ionization of a weak electrolyte in solution, half-arrows in both directions are used, indicating that the forward and reverse reactions can achieve a chemical balance called a **chemical equilibrium**.

PRECIPITATION REACTIONS (SECTION 4.2) **Precipitation reactions** are those in which an insoluble product, called a **precipitate**, forms. Solubility guidelines help determine whether an ionic compound will be soluble in water. (The **solubility** of a substance is the amount that dissolves in a given quantity of solvent.) Reactions such as precipitation

reactions, in which cations and anions appear to exchange partners, are called **exchange reactions**, or **metathesis reactions**.

Chemical equations can be written to show whether dissolved substances are present in solution predominantly as ions or molecules. When the complete chemical formulas of all reactants and products are used, the equation is called a **molecular equation**. A **complete ionic equation** shows all dissolved strong electrolytes as their component ions. In a **net ionic equation**, those ions that go through the reaction unchanged (**spectator ions**) are omitted.

ACIDS, BASES, AND NEUTRALIZATION REACTIONS (SECTION 4.3) Acids and bases are important electrolytes. **Acids** are proton donors; they increase the concentration of $H^+(aq)$ in aqueous solutions to which they are added. **Bases** are proton acceptors; they increase the concentration of $OH^-(aq)$ in aqueous solutions. Those acids and bases that are strong electrolytes are called **strong acids** and **strong bases**, respectively. Those that are weak electrolytes are **weak acids** and **weak bases**. When solutions of acids and bases are mixed, a neutralization reaction occurs. The **neutralization reaction** between an acid and a metal hydroxide produces water and a **salt**. Gases can also be formed as a result of neutralization reactions. The reaction of a sulfide with an acid forms $H_2S(g)$; the reaction between a carbonate and an acid forms $CO_2(g)$.

OXIDATION-REDUCTION REACTIONS (SECTION 4.4) **Oxidation** is the loss of electrons by a substance, whereas **reduction** is the gain of electrons by a substance. **Oxidation numbers** keep track of electrons during chemical reactions and are assigned to atoms using specific rules. The oxidation of an element results in an increase in its oxidation number, whereas reduction is accompanied by a decrease in oxidation number. Oxidation is always accompanied by reduction, giving **oxidation-reduction**, or **redox, reactions.**

Many metals are oxidized by O_2, acids, and salts. The redox reactions between metals and acids as well as those between metals and salts are called **displacement reactions**. The products of these displacement reactions are always an element (H_2 or a metal) and a salt. Comparing such reactions allows us to rank metals according to their ease of oxidation. A list of metals arranged in order of decreasing ease of oxidation is called an **activity series**. Any metal on the list can be oxidized by ions of metals (or H^+) below it in the series.

CONCENTRATIONS OF SOLUTIONS (SECTION 4.5) The **concentration** of a solution expresses the amount of a solute dissolved in the solution. One of the common ways to express the concentration of a solute is in terms of molarity. The **molarity** of a solution is the number of moles of solute per liter of solution. Molarity makes it possible to interconvert solution volume and number of moles of solute. If the solute is a liquid, its density can be used in molarity calculations to convert between mass, volume, and moles. Solutions of known molarity can be formed either by weighing out the solute and diluting it to a known volume or by the **dilution** of a more concentrated solution of known concentration (a stock solution). Adding solvent to the solution (the process of dilution) decreases the concentration of the solute without changing the number of moles of solute in the solution ($M_{conc} \times V_{conc} = M_{dil} \times V_{dil}$).

SOLUTION STOICHIOMETRY AND CHEMICAL ANALYSIS (SECTION 4.6) In the process called **titration**, we combine a solution of known concentration (a **standard solution**) with a solution of unknown concentration to determine the unknown concentration or the quantity of solute in the unknown. The point in the titration at which stoichiometrically equivalent quantities of reactants are brought together is called the **equivalence point**. An indicator can be used to show the end point of the titration, which coincides closely with the equivalence point.

Learning Outcomes After studying this chapter, you should be able to:

- Identify compounds as acids or bases, and as strong, weak, or nonelectrolytes. (Sections 4.1 and 4.3)

- Recognize reactions by type and be able to predict the products of simple acid–base, precipitation, and redox reactions. (Sections 4.2–4.4)

- Be able to calculate molarity and use it to convert between moles of a substance in solution and volume of the solution. (Section 4.5)

- Describe how to carry out a dilution to achieve a desired solution concentration. (Section 4.5)

- Describe how to perform and interpret the results of a titration. (Section 4.6)

Key Equations

- $$\text{Molarity} = \frac{\text{moles solute}}{\text{volume of solution in liters}} \qquad [4..32]$$

  Molarity is the most commonly used unit of concentration in chemistry.

- $$M_{conc} \times V_{conc} = M_{dil} \times V_{dil} \qquad [4.34]$$

  When adding solvent to a concentrated solution to make a dilute solution, molarities and volumes of both concentrated and dilute solutions can be calculated if three of the quantities are known.

Exercises

Visualizing Concepts

4.1 Which of the following schematic drawings best describes a solution of Li_2SO_4 in water (water molecules not shown for simplicity)? [Section 4.1]

 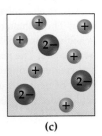 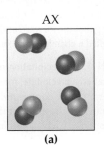

(a)   (b)   (c)

4.2 Aqueous solutions of three different substances, AX, AY, and AZ, are represented by the three accompanying diagrams. Identify each substance as a strong electrolyte, a weak electrolyte, or a nonelectrolyte. [Section 4.1]

AX   AY   AZ

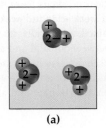 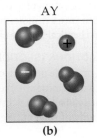 

(a)   (b)   (c)

4.3 Use the molecular representations shown here to classify each compound as a nonelectrolyte, a weak electrolyte, or a strong electrolyte (see Figure 4.6 for element color scheme). [Sections 4.1 and 4.3]

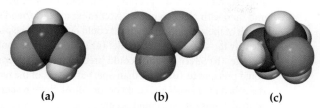

(a)   (b)   (c)

4.4 The concept of chemical equilibrium is very important. Which one of the following statements is the most correct way to think about equilibrium?

(a) If a system is at equilibrium, nothing is happening.

(b) If a system is at equilibrium, the rate of the forward reaction is equal to the rate of the back reaction.

(c) If a system is at equilibrium, the product concentration is changing over time. [Section 4.1]

4.5 You are presented with a white solid and told that due to careless labeling it is not clear if the substance is barium chloride, lead chloride, or zinc chloride. When you transfer the solid to a beaker and add water, the solid dissolves to give a clear solution. Next a $Na_2SO_4(aq)$ solution is added and a white precipitate forms. What is the identity of the unknown white solid? [Section 4.2]

Add
H_2O

Add
$Na_2SO_4(aq)$

4.6 We have seen that ions in aqueous solution are stabilized by the attractions between the ions and the water molecules. Why then do some pairs of ions in solution form precipitates? [Section 4.2]

4.7 Which of the following ions will *always* be a spectator ion in a precipitation reaction? (a) Cl^-, (b) NO_3^-, (c) NH_4^+, (d) S^{2-}, (e) SO_4^{2-}. [Section 4.2]

4.8 The labels have fallen off three bottles containing powdered samples of metals; one contains zinc, one lead, and the other platinum. You have three solutions at your disposal: 1 *M* sodium nitrate, 1 *M* nitric acid, and 1 *M* nickel nitrate. How could you use these solutions to determine the identities of each metal powder? [Section 4.4]

4.9 Explain how a redox reaction involves electrons in the same way that a neutralization reaction involves protons. [Sections 4.3 and 4.4]

4.10 If you want to double the concentration of a solution, how could you do it? [Section 4.5]

4.11 Which data set, of the two graphed here, would you expect to observe from a titration like that shown in Figure 4.18? [Section 4.6]

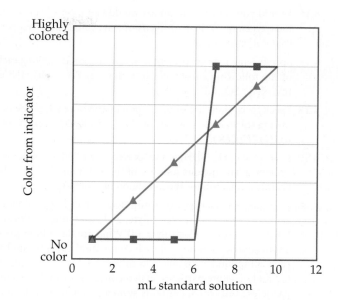

4.12 You are titrating an acidic solution with a basic one, and just realized you forgot to add the indicator that tells you when the equivalence point is reached. In this titration, the indicator turns blue at the equivalence point from an initially colorless solution. You quickly grab a bottle of indicator and throw some into your titration beaker, and the whole solution turns dark blue. What do you do now? [Section 4.6]

General Properties of Aqueous Solutions (Section 4.1)

4.13 State whether each of the statements below is true or false. Justify your answer in each case.

(a) Electrolyte solutions conduct electricity because electrons are moving through the solution.

(b) If you add a nonelectrolyte to an aqueous solution that already contains an electrolyte, the electrical conductivity will not change.

4.14 State whether each of the statements below is true or false. Justify your answer in each case.

(a) When methanol, CH_3OH, is dissolved in water, a conducting solution results.

(b) When acetic acid, CH_3COOH, dissolves in water, the solution is weakly conducting and acidic in nature.

4.15 We have learned in this chapter that many ionic solids dissolve in water as strong electrolytes; that is, as separated ions in solution. Which statement is most correct about this process? (a) Water is a strong acid and therefore is good at dissolving ionic solids. (b) Water is good at solvating ions because the hydrogen and oxygen atoms in water molecules bear partial charges. (c) The hydrogen and oxygen bonds of water are easily broken by ionic solids.

4.16 Would you expect that anions would be physically closer to the oxygen or to the hydrogens of water molecules that surround it in solution?

4.17 Specify what ions are present in solution upon dissolving each of the following substances in water: (a) $FeCl_2$, (b) HNO_3, (c) $(NH_4)_2SO_4$, (d) $Ca(OH)_2$.

4.18 Specify what ions are present upon dissolving each of the following substances in water: (a) MgI_2, (b) K_2CO_3, (c) $HClO_4$, (d) $NaCH_3COO$.

4.19 Formic acid, HCOOH, is a weak electrolyte. What solutes are present in an aqueous solution of this compound? Write the chemical equation for the ionization of HCOOH.

4.20 Acetone, CH_3COCH_3, is a nonelectrolyte; hypochlorous acid, HClO, is a weak electrolyte; and ammonium chloride, NH_4Cl, is a strong electrolyte. (a) What are the solutes present in aqueous solutions of each compound? (b) If 0.1 mol of each compound is dissolved in solution, which one contains 0.2 mol of solute particles, which contains 0.1 mol of solute particles, and which contains somewhere between 0.1 and 0.2 mol of solute particles?

Precipitation Reactions (Section 4.2)

4.21 Using solubility guidelines, predict whether each of the following compounds is soluble or insoluble in water: (a) $MgBr_2$, (b) PbI_2, (c) $(NH_4)_2CO_3$, (d) $Sr(OH)_2$, (e) $ZnSO_4$.

4.22 Predict whether each of the following compounds is soluble in water: (a) AgI, (b) Na_2CO_3, (c) $BaCl_2$, (d) $Al(OH)_3$, (e) $Zn(CH_3COO)_2$.

4.23 Will precipitation occur when the following solutions are mixed? If so, write a balanced chemical equation for the reaction. (a) Na_2CO_3 and $AgNO_3$, (b) $NaNO_3$ and $NiSO_4$, (c) $FeSO_4$ and $Pb(NO_3)_2$.

4.24 Identify the precipitate (if any) that forms when the following solutions are mixed, and write a balanced equation for each reaction. (a) $NaCH_3COO$ and HCl, (b) KOH and $Cu(NO_3)_2$, (c) Na_2S and $CdSO_4$.

4.25 Which ions remain in solution, unreacted, after each of the following pairs of solutions is mixed?

(a) potassium carbonate and magnesium sulfate

(b) lead nitrate and lithium sulfide

(c) ammonium phosphate and calcium chloride

4.26 Write balanced net ionic equations for the reactions that occur in each of the following cases. Identify the spectator ion or ions in each reaction.

(a) $Cr_2(SO_4)_3(aq) + (NH_4)_2CO_3(aq) \longrightarrow$

(b) $Ba(NO_3)_2(aq) + K_2SO_4(aq) \longrightarrow$

(c) $Fe(NO_3)_2(aq) + KOH(aq) \longrightarrow$

4.27 Separate samples of a solution of an unknown salt are treated with dilute solutions of HBr, H_2SO_4, and NaOH. A precipitate forms in all three cases. Which of the following cations could be present in the unknown salt solution: K^+, Pb^{2+}, Ba^{2+}?

4.28 Separate samples of a solution of an unknown ionic compound are treated with dilute $AgNO_3$, $Pb(NO_3)_2$, and $BaCl_2$.

Precipitates form in all three cases. Which of the following could be the anion of the unknown salt: Br^-, CO_3^{2-}, NO_3^-?

4.29 You know that an unlabeled bottle contains an aqueous solution of one of the following: $AgNO_3$, $CaCl_2$, or $Al_2(SO_4)_3$. A friend suggests that you test a portion of the solution with $Ba(NO_3)_2$ and then with NaCl solutions. According to your friend's logic, which of these chemical reactions could occur, thus helping you identify the solution in the bottle? (a) Barium sulfate could precipitate. (b) Silver chloride could precipitate. (c) Silver sulfate could precipitate. (d) More than one, but not all, of the reactions described in answers a–c could occur. (e) All three reactions described in answers a–c could occur.

4.30 Three solutions are mixed together to form a single solution; in the final solution, there are 0.2 mol $Pb(CH_3COO)_2$, 0.1 mol Na_2S, and 0.1 mol $CaCl_2$ present. What solid(s) will precipitate?

Acids, Bases, and Neutralization Reactions (Section 4.3)

4.31 Which of the following solutions is the most acidic? (a) 0.2 M LiOH, (b) 0.2 M HI, (c) 1.0 M methyl alcohol (CH_3OH).

4.32 Which of the following solutions is the most basic? (a) 0.6 M NH_3, (b) 0.150 M KOH, (c) 0.100 M $Ba(OH)_2$.

4.33 State whether each of the following statements is true or false. Justify your answer in each case.

(a) Sulfuric acid is a monoprotic acid.

(b) HCl is a weak acid.

(c) Methanol is a base.

4.34 State whether each of the following statements is true or false. Justify your answer in each case.

(a) NH_3 contains no OH^- ions, and yet its aqueous solutions are basic.

(b) HF is a strong acid.

(c) Although sulfuric acid is a strong electrolyte, an aqueous solution of H_2SO_4 contains more HSO_4^- ions than SO_4^{2-} ions.

4.35 Label each of the following substances as an acid, base, salt, or none of the above. Indicate whether the substance exists in aqueous solution entirely in molecular form, entirely as ions, or as a mixture of molecules and ions. (a) HF, (b) acetonitrile, CH_3CN, (c) $NaClO_4$, (d) $Ba(OH)_2$.

4.36 An aqueous solution of an unknown solute is tested with litmus paper and found to be acidic. The solution is weakly conducting compared with a solution of NaCl of the same concentration. Which of the following substances could the unknown be: KOH, NH_3, HNO_3, $KClO_2$, H_3PO_3, CH_3COCH_3 (acetone)?

4.37 Classify each of the following substances as a nonelectrolyte, weak electrolyte, or strong electrolyte in water: (a) H_2SO_3, (b) C_2H_5OH (ethanol), (c) NH_3, (d) $KClO_3$, (e) $Cu(NO_3)_2$.

4.38 Classify each of the following aqueous solutions as a nonelectrolyte, weak electrolyte, or strong electrolyte: (a) $LiClO_4$, (b) HClO, (c) $CH_3CH_2CH_2OH$ (propanol), (d) $HClO_3$, (e) $CuSO_4$, (f) $C_{12}H_{22}O_{11}$ (sucrose).

4.39 Complete and balance the following molecular equations, and then write the net ionic equation for each:

(a) $HBr(aq) + Ca(OH)_2(aq) \longrightarrow$

(b) $Cu(OH)_2(s) + HClO_4(aq) \longrightarrow$

(c) $Al(OH)_3(s) + HNO_3(aq) \longrightarrow$

4.40 Write the balanced molecular and net ionic equations for each of the following neutralization reactions:

(a) Aqueous acetic acid is neutralized by aqueous barium hydroxide.

(b) Solid chromium(III) hydroxide reacts with nitrous acid.

(c) Aqueous nitric acid and aqueous ammonia react.

4.41 Write balanced molecular and net ionic equations for the following reactions, and identify the gas formed in each: (a) solid cadmium sulfide reacts with an aqueous solution of sulfuric acid; (b) solid magnesium carbonate reacts with an aqueous solution of perchloric acid.

4.42 Because the oxide ion is basic, metal oxides react readily with acids. (a) Write the net ionic equation for the following reaction:

$$FeO(s) + 2\,HClO_4(aq) \longrightarrow Fe(ClO_4)_2(aq) + H_2O(l)$$

(b) Based on the equation in part (a), write the net ionic equation for the reaction that occurs between NiO(s) and an aqueous solution of nitric acid.

4.43 Magnesium carbonate, magnesium oxide, and magnesium hydroxide are all white solids that react with acidic solutions. (a) Write a balanced molecular equation and a net ionic equation for the reaction that occurs when each substance reacts with a hydrochloric acid solution. (b) By observing the reactions in part (a), how could you distinguish any of the three magnesium substances from the other two?

4.44 As K_2O dissolves in water, the oxide ion reacts with water molecules to form hydroxide ions. (a) Write the molecular and net ionic equations for this reaction. (b) Based on the definitions of acid and base, what ion is the base in this reaction? (c) What is the acid in the reaction? (d) What is the spectator ion in the reaction?

Oxidation-Reduction Reactions (Section 4.4)

4.45 True or false:

(a) If a substance is oxidized, it is gaining electrons.

(b) If an ion is oxidized, its oxidation number increases.

4.46 True or false:

(a) Oxidation can occur without oxygen.

(b) Oxidation can occur without reduction.

4.47 (a) Which region of the periodic table shown here contains elements that are easiest to oxidize? (b) Which region contains the least readily oxidized elements?

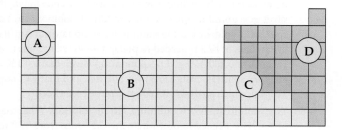

4.48 Determine the oxidation number of sulfur in each of the following substances: (a) barium sulfate, $BaSO_4$, (b) sulfurous acid, H_2SO_3, (c) strontium sulfide, SrS, (d) hydrogen sulfide, H_2S. (e) Locate sulfur in the periodic table in Exercise 4.47; what region is it in? (f) Which region(s) of the period table contains elements that can adopt both positive and negative oxidation numbers?

4.49 Determine the oxidation number for the indicated element in each of the following substances: (a) S in SO_2, (b) C in $COCl_2$, (c) Mn in $KMnO_4$, (d) Br in HBrO, (e) P in PF_3, (f) O in K_2O_2.

4.50 Determine the oxidation number for the indicated element in each of the following compounds: (a) Co in $LiCoO_2$, (b) Al in $NaAlH_4$, (c) C in CH_3OH (methanol), (d) N in GaN, (e) Cl in $HClO_2$, (f) Cr in $BaCrO_4$.

4.51 Which element is oxidized and which is reduced in the following reactions?

(a) $N_2(g) + 3\,H_2(g) \longrightarrow 2\,NH_3(g)$

(b) $3\,Fe(NO_3)_2(aq) + 2\,Al(s) \longrightarrow$
$$3\,Fe(s) + 2\,Al(NO_3)_3(aq)$$

(c) $Cl_2(aq) + 2\,NaI(aq) \longrightarrow I_2(aq) + 2\,NaCl(aq)$

(d) $PbS(s) + 4\,H_2O_2(aq) \longrightarrow PbSO_4(s) + 4\,H_2O(l)$

4.52 Which of the following are redox reactions? For those that are, indicate which element is oxidized and which is reduced. For those that are not, indicate whether they are precipitation or neutralization reactions.

(a) $P_4(s) + 10\,HClO(aq) + 6\,H_2O(l) \longrightarrow$
$$4\,H_3PO_4(aq) + 10\,HCl(aq)$$

(b) $Br_2(l) + 2\,K(s) \longrightarrow 2\,KBr(s)$

(c) $CH_3CH_2OH(l) + 3\,O_2(g) \longrightarrow 3\,H_2O(l) + 2\,CO_2(g)$

(d) $ZnCl_2(aq) + 2\,NaOH(aq) \longrightarrow Zn(OH)_2(s) +$
$$2\,NaCl(aq)$$

4.53 Write balanced molecular and net ionic equations for the reactions of (a) manganese with dilute sulfuric acid, (b) chromium with hydrobromic acid, (c) tin with hydrochloric acid, (d) aluminum with formic acid, HCOOH.

4.54 Write balanced molecular and net ionic equations for the reactions of (a) hydrochloric acid with nickel, (b) dilute sulfuric acid with iron, (c) hydrobromic acid with magnesium, (d) acetic acid, CH_3COOH, with zinc.

4.55 Using the activity series (Table 4.5), write balanced chemical equations for the following reactions. If no reaction occurs, write NR. (a) Iron metal is added to a solution of copper(II) nitrate, (b) zinc metal is added to a solution of magnesium sulfate, (c) hydrobromic acid is added to tin metal, (d) hydrogen gas is bubbled through an aqueous solution of nickel(II) chloride, (e) aluminum metal is added to a solution of cobalt(II) sulfate.

4.56 Using the activity series (Table 4.5), write balanced chemical equations for the following reactions. If no reaction occurs, write NR. (a) Nickel metal is added to a solution of copper(II) nitrate, (b) a solution of zinc nitrate is added to a solution of magnesium sulfate, (c) hydrochloric acid is added to gold metal, (d) chromium metal is immersed in an aqueous solution of cobalt(II) chloride, (e) hydrogen gas is bubbled through a solution of silver nitrate.

4.57 The metal cadmium tends to form Cd^{2+} ions. The following observations are made: (i) When a strip of zinc metal is placed in $CdCl_2(aq)$, cadmium metal is deposited on the strip. (ii) When a strip of cadmium metal is placed in $Ni(NO_3)_2(aq)$, nickel metal is deposited on the strip. (a) Write net ionic equations to explain each of the preceding observations. (b) Which elements more closely define the position of cadmium in the activity series? (c) What experiments would you need to perform to locate more precisely the position of cadmium in the activity series?

4.58 The following reactions (note that the arrows are pointing only one direction) can be used to prepare an activity series for the halogens:

$$Br_2(aq) + 2\,NaI(aq) \longrightarrow 2\,NaBr(aq) + I_2(aq)$$
$$Cl_2(aq) + 2\,NaBr(aq) \longrightarrow 2\,NaCl(aq) + Br_2(aq)$$

(a) Which elemental halogen would you predict is the most stable, upon mixing with other halides? **(b)** Predict whether a reaction will occur when elemental chlorine and potassium iodide are mixed. **(c)** Predict whether a reaction will occur when elemental bromine and lithium chloride are mixed.

Concentrations of Solutions (Section 4.5)

4.59 **(a)** Is the concentration of a solution an intensive or an extensive property? **(b)** What is the difference between 0.50 mol HCl and 0.50 M HCl?

4.60 Your lab partner tells you that he has prepared a solution that contains 1.50 moles of NaOH in 1.50 L of aqueous solution, and therefore that the concentration of NaOH is 1.5 M. **(a)** Is he correct? **(b)** If not, what is the correct concentration?

4.61 **(a)** Calculate the molarity of a solution that contains 0.175 mol $ZnCl_2$ in exactly 150 mL of solution. **(b)** How many moles of protons are present in 35.0 mL of a 4.50 M solution of nitric acid? **(c)** How many milliliters of a 6.00 M NaOH solution are needed to provide 0.350 mol of NaOH?

4.62 **(a)** Calculate the molarity of a solution made by dissolving 12.5 grams of Na_2CrO_4 in enough water to form exactly 750 mL of solution. **(b)** How many moles of KBr are present in 150 mL of a 0.112 M solution? **(c)** How many milliliters of 6.1 M HCl solution are needed to obtain 0.150 mol of HCl?

4.63 The average adult human male has a total blood volume of 5.0 L. If the concentration of sodium ion in this average individual is 0.135 M, what is the mass of sodium ion circulating in the blood?

4.64 A person suffering from hyponatremia has a sodium ion concentration in the blood of 0.118 M and a total blood volume of 4.6 L. What mass of sodium chloride would need to be added to the blood to bring the sodium ion concentration up to 0.138 M, assuming no change in blood volume?

4.65 The concentration of alcohol (CH_3CH_2OH) in blood, called the "blood alcohol concentration" or BAC, is given in units of grams of alcohol per 100 mL of blood. The legal definition of intoxication, in many states of the United States, is that the BAC is 0.08 or higher. What is the concentration of alcohol, in terms of molarity, in blood if the BAC is 0.08?

4.66 The average adult male has a total blood volume of 5.0 L. After drinking a few beers, he has a BAC of 0.10 (see Exercise 4.65). What mass of alcohol is circulating in his blood?

4.67 **(a)** How many grams of ethanol, CH_3CH_2OH, should you dissolve in water to make 1.00 L of vodka (which is an aqueous solution that is 6.86 M ethanol)? **(b)** Using the density of ethanol (0.789 g/mL), calculate the volume of ethanol you need to make 1.00 L of vodka.

4.68 One cup of fresh orange juice contains 124 mg of ascorbic acid (vitamin C, $C_6H_8O_6$). Given that one cup = 236.6 mL, calculate the molarity of vitamin C in organic juice.

4.69 **(a)** Which will have the highest concentration of potassium ion: 0.20 M KCl, 0.15 M K_2CrO_4, or 0.080 M K_3PO_4? **(b)** Which will contain the greater number of moles of potassium ion: 30.0 mL of 0.15 M K_2CrO_4 or 25.0 mL of 0.080 M K_3PO_4?

4.70 In each of the following pairs, indicate which has the higher concentration of I^- ion: **(a)** 0.10 M BaI_2 or 0.25 M KI solution, **(b)** 100 mL of 0.10 M KI solution or 200 mL of 0.040 M ZnI_2 solution, **(c)** 3.2 M HI solution or a solution made by dissolving 145 g of NaI in water to make 150 mL of solution.

4.71 Indicate the concentration of each ion or molecule present in the following solutions: **(a)** 0.25 M $NaNO_3$,

(b) 1.3×10^{-2} M $MgSO_4$, **(c)** 0.0150 M $C_6H_{12}O_6$, **(d)** a mixture of 45.0 mL of 0.272 M NaCl and 65.0 mL of 0.0247 M $(NH_4)_2CO_3$. Assume that the volumes are additive.

4.72 Indicate the concentration of each ion present in the solution formed by mixing **(a)** 42.0 mL of 0.170 M NaOH with 37.6 mL of 0.400 M NaOH, **(b)** 44.0 mL of 0.100 M Na_2SO_4 with 25.0 mL of 0.150 M KCl, **(c)** 3.60 g KCl in 75.0 mL of 0.250 M $CaCl_2$ solution. Assume that the volumes are additive.

4.73 **(a)** You have a stock solution of 14.8 M NH_3. How many milliliters of this solution should you dilute to make 1000.0 mL of 0.250 M NH_3? **(b)** If you take a 10.0-mL portion of the stock solution and dilute it to a total volume of 0.500 L, what will be the concentration of the final solution?

4.74 **(a)** How many milliliters of a stock solution of 6.0 M HNO_3 would you have to use to prepare 110 mL of 0.500 M HNO_3? **(b)** If you dilute 10.0 mL of the stock solution to a final volume of 0.250 L, what will be the concentration of the diluted solution?

4.75 **(a)** Starting with solid sucrose, $C_{12}H_{22}O_{11}$, describe how you would prepare 250 mL of a 0.250 M sucrose solution. **(b)** Describe how you would prepare 350.0 mL of 0.100 M $C_{12}H_{22}O_{11}$ starting with 3.00 L of 1.50 M $C_{12}H_{22}O_{11}$.

4.76 **(a)** How many grams of solid silver nitrate would you need to prepare 200.0 mL of a 0.150 M $AgNO_3$ solution? **(b)** An experiment calls for you to use 100 mL of 0.50 M HNO_3 solution. All you have available is a bottle of 3.6 M HNO_3. How many milliliters of the 3.6 M HNO_3 solution and of water do you need to prepare the desired solution?

4.77 Pure acetic acid, known as glacial acetic acid, is a liquid with a density of 1.049 g/mL at 25 °C. Calculate the molarity of a solution of acetic acid made by dissolving 20.00 mL of glacial acetic acid at 25 °C in enough water to make 250.0 mL of solution.

4.78 Glycerol, $C_3H_8O_3$, is a substance used extensively in the manufacture of cosmetics, foodstuffs, antifreeze, and plastics. Glycerol is a water-soluble liquid with a density of 1.2656 g/mL at 15 °C. Calculate the molarity of a solution of glycerol made by dissolving 50.000 mL glycerol at 15 °C in enough water to make 250.00 mL of solution.

Solution Stoichiometry and Chemical Analysis (Section 4.6)

4.79 You want to analyze a silver nitrate solution. **(a)** You could add HCl(aq) to the solution to precipitate out AgCl(s). What volume of a 0.150 M HCl(aq) solution is needed to precipitate the silver ions from 15.0 mL of a 0.200 M $AgNO_3$ solution? **(b)** You could add solid KCl to the solution to precipitate out AgCl(s). What mass of KCl is needed to precipitate the silver ions from 15.0 mL of 0.200 M $AgNO_3$ solution? **(c)** Given that a 0.150 M HCl(aq) solution costs $39.95 for 500 mL, and that KCl costs $10/ton, which analysis procedure is more cost-effective?

4.80 You want to analyze a cadmium nitrate solution. What mass of NaOH is needed to precipitate the Cd^{2+} ions from 35.0 mL of 0.500 M $Cd(NO_3)_2$ solution?

4.81 **(a)** What volume of 0.115 M $HClO_4$ solution is needed to neutralize 50.00 mL of 0.0875 M NaOH? **(b)** What volume of 0.128 M HCl is needed to neutralize 2.87 g of $Mg(OH)_2$? **(c)** If 25.8 mL of an $AgNO_3$ solution is needed to precipitate all the Cl^- ions in a 785-mg sample of KCl (forming AgCl), what is the molarity of the $AgNO_3$ solution? **(d)** If 45.3 mL of a 0.108 M HCl solution is needed to neutralize a solution of KOH, how many grams of KOH must be present in the solution?

4.82 (a) How many milliliters of 0.120 M HCl are needed to completely neutralize 50.0 mL of 0.101 M Ba(OH)$_2$ solution? (b) How many milliliters of 0.125 M H$_2$SO$_4$ are needed to neutralize 0.200 g of NaOH? (c) If 55.8 mL of a BaCl$_2$ solution is needed to precipitate all the sulfate ion in a 752-mg sample of Na$_2$SO$_4$, what is the molarity of the BaCl$_2$ solution? (d) If 42.7 mL of 0.208 M HCl solution is needed to neutralize a solution of Ca(OH)$_2$, how many grams of Ca(OH)$_2$ must be in the solution?

4.83 Some sulfuric acid is spilled on a lab bench. You can neutralize the acid by sprinkling sodium bicarbonate on it and then mopping up the resultant solution. The sodium bicarbonate reacts with sulfuric acid according to:

$$2\,NaHCO_3(s) + H_2SO_4(aq) \longrightarrow Na_2SO_4(aq) + \\ 2\,H_2O(l) + 2\,CO_2(g)$$

Sodium bicarbonate is added until the fizzing due to the formation of $CO_2(g)$ stops. If 27 mL of 6.0 M H$_2$SO$_4$ was spilled, what is the minimum mass of NaHCO$_3$ that must be added to the spill to neutralize the acid?

4.84 The distinctive odor of vinegar is due to acetic acid, CH$_3$COOH, which reacts with sodium hydroxide according to:

$$CH_3COO(aq) + NaOH(aq) \longrightarrow \\ H_2O(l) + NaCH_3OO(aq)$$

If 3.45 mL of vinegar needs 42.5 mL of 0.115 M NaOH to reach the equivalence point in a titration, how many grams of acetic acid are in a 1.00-qt sample of this vinegar?

4.85 A 4.36-g sample of an unknown alkali metal hydroxide is dissolved in 100.0 mL of water. An acid–base indicator is added and the resulting solution is titrated with 2.50 M HCl(aq) solution. The indicator changes color signaling that the equivalence point has been reached after 17.0 mL of the hydrochloric acid solution has been added. (a) What is the molar mass of the metal hydroxide? (b) What is the identity of the alkali metal cation: Li$^+$, Na$^+$, K$^+$, Rb$^+$, or Cs$^+$?

4.86 An 8.65-g sample of an unknown group 2A metal hydroxide is dissolved in 85.0 mL of water. An acid–base indicator is added and the resulting solution is titrated with 2.50 M HCl(aq) solution. The indicator changes color signaling that the equivalence point has been reached after 56.9 mL of the hydrochloric acid solution has been added. (a) What is the molar mass of the metal hydroxide? (b) What is the identity of the metal cation: Ca^{2+}, Sr^{2+}, Ba^{2+}?

4.87 A solution of 100.0 mL of 0.200 M KOH is mixed with a solution of 200.0 mL of 0.150 M NiSO$_4$. (a) Write the balanced chemical equation for the reaction that occurs. (b) What precipitate forms? (c) What is the limiting reactant? (d) How many grams of this precipitate form? (e) What is the concentration of each ion that remains in solution?

4.88 A solution is made by mixing 15.0 g of Sr(OH)$_2$ and 55.0 mL of 0.200 M HNO$_3$. (a) Write a balanced equation for the reaction that occurs between the solutes. (b) Calculate the concentration of each ion remaining in solution. (c) Is the resultant solution acidic or basic?

4.89 A 0.5895-g sample of impure magnesium hydroxide is dissolved in 100.0 mL of 0.2050 M HCl solution. The excess acid then needs 19.85 mL of 0.1020 M NaOH for neutralization. Calculate the percentage by mass of magnesium hydroxide in the sample, assuming that it is the only substance reacting with the HCl solution.

4.90 A 1.248-g sample of limestone rock is pulverized and then treated with 30.00 mL of 1.035 M HCl solution. The excess acid then requires 11.56 mL of 1.010 M NaOH for neutralization. Calculate the percentage by mass of calcium carbonate in the rock, assuming that it is the only substance reacting with the HCl solution.

Additional Exercises

4.91 Uranium hexafluoride, UF$_6$, is processed to produce fuel for nuclear reactors and nuclear weapons. UF$_6$ is made from the reaction of elemental uranium with ClF$_3$, which also produces Cl$_2$ as a by-product.

(a) Write the balanced molecular equation for the conversion of U and ClF$_3$ into UF$_6$ and Cl$_2$.

(b) Is this a metathesis reaction?

(c) Is this a redox reaction?

4.92 The accompanying photo shows the reaction between a solution of Cd(NO$_3$)$_2$ and one of Na$_2$S. (a) What is the identity of the precipitate? (b) What ions remain in solution? (c) Write the net ionic equation for the reaction. (d) Is this a redox reaction?

4.93 Suppose you have a solution that might contain any or all of the following cations: Ni^{2+}, Ag$^+$, Sr^{2+}, and Mn^{2+}. Addition of HCl solution causes a precipitate to form. After filtering off the precipitate, H$_2$SO$_4$ solution is added to the resulting solution and another precipitate forms. This is filtered off, and a solution of NaOH is added to the resulting solution. No precipitate is observed. Which ions are present in each of the precipitates? Which of the four ions listed above must be absent from the original solution?

4.94 You choose to investigate some of the solubility guidelines for two ions not listed in Table 4.1, the chromate ion (CrO$_4^{2-}$) and the oxalate ion (C$_2$O$_4^{2-}$). You are given 0.01 M solutions (A, B, C, D) of four water-soluble salts:

Solution	Solute	Color of Solution
A	Na$_2$CrO$_4$	Yellow
B	(NH$_4$)$_2$C$_2$O$_4$	Colorless
C	AgNO$_3$	Colorless
D	CaCl$_2$	Colorless

When these solutions are mixed, the following observations are made:

Experiment Number	Solutions Mixed	Result
1	A + B	No precipitate, yellow solution
2	A + C	Red precipitate forms
3	A + D	Yellow precipitate forms
4	B + C	White precipitate forms
5	B + D	White precipitate forms
6	C + D	White precipitate forms

(a) Write a net ionic equation for the reaction that occurs in each of the experiments. (b) Identify the precipitate formed, if any, in each of the experiments.

4.95 Antacids are often used to relieve pain and promote healing in the treatment of mild ulcers. Write balanced net ionic equations for the reactions between the aqueous HCl in the stomach and each of the following substances used in various antacids: (a) $Al(OH)_3(s)$, (b) $Mg(OH)_2(s)$, (c) $MgCO_3(s)$, (d) $NaAl(CO_3)(OH)_2(s)$, (e) $CaCO_3(s)$.

4.96 The commercial production of nitric acid involves the following chemical reactions:

$$4\,NH_3(g) + 5\,O_2(g) \longrightarrow 4\,NO(g) + 6\,H_2O(g)$$
$$2\,NO(g) + O_2(g) \longrightarrow 2\,NO_2(g)$$
$$3\,NO_2(g) + H_2O(l) \longrightarrow 2\,HNO_3(aq) + NO(g)$$

(a) Which of these reactions are redox reactions? (b) In each redox reaction identify the element undergoing oxidation and the element undergoing reduction. (c) How many grams of ammonia must you start with to make 1000.0 L of a 0.150 M aqueous solution of nitric acid? Assume all the reactions give 100% yield.

4.97 Consider the following reagents: zinc, copper, mercury (density 13.6 g/mL), silver nitrate solution, nitric acid solution. (a) Given a 500-mL Erlenmeyer flask and a balloon, can you combine two or more of the foregoing reagents to initiate a chemical reaction that will inflate the balloon? Write a balanced chemical equation to represent this process. What is the identity of the substance that inflates the balloon? (b) What is the theoretical yield of the substance that fills the balloon? (c) Can you combine two or more of the foregoing reagents to initiate a chemical reaction that will produce metallic silver? Write a balanced chemical equation to represent this process. What ions are left behind in solution? (d) What is the theoretical yield of silver?

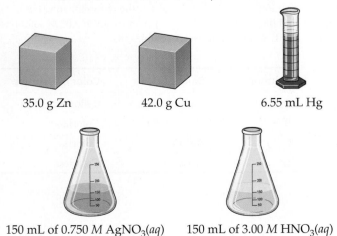

35.0 g Zn   42.0 g Cu   6.55 mL Hg

150 mL of 0.750 M $AgNO_3(aq)$ 150 mL of 3.00 M $HNO_3(aq)$

4.98 Lanthanum metal forms cations with a charge of 3+. Consider the following observations about the chemistry of lanthanum: When lanthanum metal is exposed to air, a white solid (compound A) is formed that contains lanthanum and one other element. When lanthanum metal is added to water, gas bubbles are observed and a different white solid (compound B) is formed. Both A and B dissolve in hydrochloric acid to give a clear solution. When either of these solutions is evaporated, a soluble white solid (compound C) remains. If compound C is dissolved in water and sulfuric acid is added, a white precipitate (compound D) forms. (a) Propose identities for the substances A, B, C, and D. (b) Write net ionic equations for all the reactions described. (c) Based on the preceding observations, what can be said about the position of lanthanum in the activity series (Table 4.5)?

4.99 A 35.0-mL sample of 1.00 M KBr and a 60.0-mL sample of 0.600 M KBr are mixed. The solution is then heated to evaporate water until the total volume is 50.0 mL. How many grams of silver nitrate are required to precipitate out silver bromide in the final solution?

4.100 Using modern analytical techniques, it is possible to detect sodium ions in concentrations as low as 50 pg/mL. What is this detection limit expressed in (a) molarity of Na^+, (b) the number of Na^+ ions per cubic centimeter of solution, (c) the mass of sodium per 1000 L of solution?

4.101 Hard water contains Ca^{2+}, Mg^{2+}, and Fe^{2+}, which interfere with the action of soap and leave an insoluble coating on the insides of containers and pipes when heated. Water softeners replace these ions with Na^+. Keep in mind that charge balance must be maintained. (a) If 1500 L of hard water contains 0.020 M Ca^{2+} and 0.0040 M Mg^{2+}, how many moles of Na^+ is needed to replace these ions? (b) If the sodium is added to the water softener in the form of NaCl, how many grams of sodium chloride are needed?

4.102 Tartaric acid, $H_2C_4H_4O_6$, has two acidic hydrogens. The acid is often present in wines and precipitates from solution as the wine ages. A solution containing an unknown concentration of the acid is titrated with NaOH. It requires 24.65 mL of 0.2500 M NaOH solution to titrate both acidic protons in 50.00 mL of the tartaric acid solution. Write a balanced net ionic equation for the neutralization reaction, and calculate the molarity of the tartaric acid solution.

4.103 (a) A strontium hydroxide solution is prepared by dissolving 12.50 g of $Sr(OH)_2$ in water to make 50.00 mL of solution. What is the molarity of this solution? (b) Next the strontium hydroxide solution prepared in part (a) is used to titrate a nitric acid solution of unknown concentration. Write a balanced chemical equation to represent the reaction between strontium hydroxide and nitric acid solutions. (c) If 23.9 mL of the strontium hydroxide solution was needed to neutralize a 37.5 mL aliquot of the nitric acid solution, what is the concentration (molarity) of the acid?

4.104 A solid sample of $Zn(OH)_2$ is added to 0.350 L of 0.500 M aqueous HBr. The solution that remains is still acidic. It is then titrated with 0.500 M NaOH solution, and it takes 88.5 mL of the NaOH solution to reach the equivalence point. What mass of $Zn(OH)_2$ was added to the HBr solution?

Integrative Exercises

4.105 Suppose you have 5.00 g of powdered magnesium metal, 1.00 L of 2.00 M potassium nitrate solution, and 1.00 L of 2.00 M silver nitrate solution. (a) Which one of the solutions will react with the magnesium powder? (b) What is the net ionic equation that describes this reaction? (c) What volume of solution is needed to completely react with the magnesium? (d) What is the molarity of the Mg^{2+} ions in the resulting solution?

4.106 (a) By titration, 15.0 mL of 0.1008 M sodium hydroxide is needed to neutralize a 0.2053-g sample of a weak acid. What is the molar mass of the acid if it is monoprotic? (b) An elemental analysis of the acid indicates that it is composed of 5.89% H, 70.6% C, and 23.5% O by mass. What is its molecular formula?

4.107 A 3.455-g sample of a mixture was analyzed for barium ion by adding a small excess of sulfuric acid to an aqueous solution of the sample. The resultant reaction produced a precipitate of barium sulfate, which was collected by filtration, washed, dried, and weighed. If 0.2815 g of barium sulfate was obtained, what was the mass percentage of barium in the sample?

4.108 A fertilizer railroad car carrying 34,300 gallons of commercial aqueous ammonia (30% ammonia by mass) tips over and spills. The density of the aqueous ammonia solution is 0.88 g/cm^3. What mass of citric acid, $C(OH)(COOH)(CH_2COOH)_2$, (which contains three acidic protons) is required to neutralize the spill? 1 gallon = 3.785 L.

4.109 A sample of 7.75 g of $Mg(OH)_2$ is added to 25.0 mL of 0.200 M HNO_3. (a) Write the chemical equation for the reaction that occurs. (b) Which is the limiting reactant in the reaction? (c) How many moles of $Mg(OH)_2$, HNO_3, and $Mg(NO_3)_2$ are present after the reaction is complete?

4.110 Lead glass contains 18–40% by mass of PbO (instead of CaO in regular glass). Lead glass is still used industrially, but "lead crystal" drinking goblets are no longer considered safe, as the lead may leach out and cause toxic responses in humans. A particular 286-g lead crystal goblet that holds 450 mL of liquid is 27% PbO by mass, and it leaches 3.4 micrograms of lead every time it is filled. How many grams of sodium sulfide would be required to decontaminate all the lead in the entire goblet?

4.111 The average concentration of gold in seawater is 100 fM (femtomolar). Given that the price of gold is $1764.20 per troy ounce (1 troy ounce = 31.103 g), how many liters of seawater would you need to process to collect $5000 worth of gold,

assuming your processing technique captures only 50% of the gold present in the samples?

4.112 The mass percentage of chloride ion in a 25.00-mL sample of seawater was determined by titrating the sample with silver nitrate, precipitating silver chloride. It took 42.58 mL of 0.2997 M silver nitrate solution to reach the equivalence point in the titration. What is the mass percentage of chloride ion in seawater if its density is 1.025 g/mL?

4.113 The arsenic in a 1.22-g sample of a pesticide was converted to $AsO_4{}^{3-}$ by suitable chemical treatment. It was then titrated using Ag^+ to form Ag_3AsO_4 as a precipitate. (a) What is the oxidation state of As in $AsO_4{}^{3-}$? (b) Name Ag_3AsO_4 by analogy to the corresponding compound containing phosphorus in place of arsenic. (c) If it took 25.0 mL of 0.102 M Ag^+ to reach the equivalence point in this titration, what is the mass percentage of arsenic in the pesticide?

4.114 The U.S. standard for arsenate in drinking water requires that public water supplies must contain no greater than 10 parts per billion (ppb) arsenic. If this arsenic is present as arsenate, $AsO_4{}^{3-}$, what mass of sodium arsenate would be present in a 1.00-L sample of drinking water that just meets the standard? Parts per billion is defined on a mass basis as

$$ppb = \frac{g\ solute}{g\ solution} \times 10^9$$

4.115 Federal regulations set an upper limit of 50 parts per million (ppm) of NH_3 in the air in a work environment [that is, 50 molecules of $NH_3(g)$ for every million molecules in the air]. Air from a manufacturing operation was drawn through a solution containing 1.00×10^2 mL of 0.0105 M HCl. The NH_3 reacts with HCl according to:

$$NH_3(aq) + HCl(aq) \longrightarrow NH_4Cl(aq)$$

After drawing air through the acid solution for 10.0 min at a rate of 10.0 L/min, the acid was titrated. The remaining acid needed 13.1 mL of 0.0588 M NaOH to reach the equivalence point. (a) How many grams of NH_3 were drawn into the acid solution? (b) How many ppm of NH_3 were in the air? (Air has a density of 1.20 g/L and an average molar mass of 29.0 g/mol under the conditions of the experiment.) (c) Is this manufacturer in compliance with regulations?

Design an Experiment

You are cleaning out a chemistry lab and find three unlabeled bottles, each containing white powder. Nearby these bottles are three loose labels: "Sodium sulfide," "Sodium bicarbonate" and "Sodium chloride." Let's design an experiment to figure out which label goes with which bottle.

(a) You could try to use the physical properties of the three solids to distinguish among them. Using an internet resource or the *CRC Handbook of Chemistry and Physics*, look up the melting points, aqueous solubilities, or other properties of these salts. Are the differences among these properties for each salt large enough to distinguish among them? If so, design a set of experiments to

distinguish each salt and therefore figure out which label goes on which bottle.

(b) You could use the chemical reactivity of each salt to distinguish it from the others. Which of these salts, if any, will act as an acid? A base? A strong electrolyte? Can any of these salts be easily oxidized or reduced? Can any of these salts react to produce a gas? Based on your answers to these questions, design a set of experiments to distinguish each salt and thus determine which label goes on which bottle.

5

Thermochemistry

Everything we do is connected in one way or another with energy. Not only our modern society but life itself depends on energy for its existence. The issues surrounding energy—its sources, production, distribution, and consumption—pervade conversations in science, politics, and economics, and relate to environmental concerns and public policy.

With the exception of the energy from the Sun, most of the energy used in our daily lives comes from chemical reactions. The combustion of gasoline, the production of electricity from coal, the heating of homes by natural gas, and the use of batteries to power electronic devices are all examples of how chemistry is used to produce energy. Even solar cells, such as those shown in the chapter-opening photo, rely on chemistry to produce the silicon and other materials that convert solar energy directly to electricity. In addition, chemical reactions provide the energy that sustains living systems. Plants use solar energy to carry out photosynthesis, allowing them to grow. The plants in turn provide food from which we humans derive the energy needed to move, maintain body temperature, and carry out all other bodily functions.

It is evident that the topic of energy is intimately related to chemistry. What exactly is energy, though, and what principles are involved in its production, consumption, and transformation from one form to another?

In this chapter we begin to explore energy and its changes. We are motivated not only by the impact of energy on so many aspects of our daily lives but also because if

▶ **SOLAR PANELS.** Each panel consists of an assembly of solar cells, also known as photovoltaic cells. Various materials have been used in solar cells, but crystalline silicon is most common.

WHAT'S AHEAD ▶

5.1 THE NATURE OF ENERGY We begin by considering the nature of *energy* and the forms it takes, notably *kinetic energy* and *potential energy*. We discuss the units used in measuring energy and the fact that energy can be used to do *work* or to transfer *heat*. To study energy changes, we focus on a particular part of the universe, which we call the *system*. Everything else is called the *surroundings*.

5.2 THE FIRST LAW OF THERMODYNAMICS We then explore the *first law of thermodynamics*: Energy cannot be created or destroyed but can be transformed from one form to another or transferred between systems and surroundings. The energy possessed by a system is called its *internal energy*. Internal energy is a *state function*, a quantity whose value depends only on the current state of a system, not on how the system came to be in that state.

5.3 ENTHALPY Next, we encounter a state function called *enthalpy* that is useful because the change in enthalpy measures the quantity of heat energy gained or lost by a system in a process occurring under constant pressure.

5.4 ENTHALPIES OF REACTION We see that the enthalpy change associated with a chemical reaction is the enthalpies of the products minus the enthalpies of the reactants. This quantity is directly proportional to the amount of reactant consumed in the reaction.

5.5 CALORIMETRY We next examine *calorimetry*, an experimental technique used to measure heat changes in chemical processes.

5.6 HESS'S LAW We observe that the enthalpy change for a given reaction can be calculated using appropriate enthalpy changes for related reactions. To do so, we apply *Hess's law*.

5.7 ENTHALPIES OF FORMATION We discuss how to establish standard values for enthalpy changes in chemical reactions and how to use them to calculate enthalpy changes for reactions.

5.8 FOODS AND FUELS Finally, we examine foods and fuels as sources of energy and discuss some related health and social issues.

we are to properly understand chemistry, we must understand the energy changes that accompany chemical reactions.

The study of energy and its transformations is known as **thermodynamics** (Greek: *thérme-*, "heat"; *dy'namis,* "power"). This area of study began during the Industrial Revolution in order to develop the relationships among heat, work, and fuels in steam engines. In this chapter we will examine the relationships between chemical reactions and energy changes that involve heat. This portion of thermodynamics is called **thermochemistry**. We will discuss additional aspects of thermodynamics in Chapter 19.

5.1 | Energy

Unlike matter, energy does not have mass and cannot be held in our hands, but its effects can be observed and measured. **Energy** is *the capacity to do work or transfer heat.* Before we can make any use of this definition we must understand the concepts of work and heat. **Work** is *the energy used to cause an object to move against a force,* and **heat** is *the energy used to cause the temperature of an object to increase* (◀ **Figure 5.1**). Before we examine these definitions more closely, let's first consider the ways in which matter can possess energy and how that energy can be transferred from one piece of matter to another.

Kinetic Energy and Potential Energy

Objects, whether they are baseballs or molecules, can possess **kinetic energy**, the energy of *motion*. The magnitude of the kinetic energy, E_k, of an object depends on its mass, m, and speed, v:

$$E_k = \tfrac{1}{2} mv^2 \qquad [5.1]$$

Thus, the kinetic energy of an object increases as its speed increases. For example, a car moving at 55 miles per hour (mi/h) has greater kinetic energy than it does at 25 mi/h. For a given speed the kinetic energy increases with increasing mass. Thus, a large truck traveling at 55 mi/h has greater kinetic energy than a small sedan traveling at the same speed because the truck has the greater mass.

In chemistry, we are interested in the kinetic energy of atoms and molecules. Although too small to be seen, these particles have mass and are in motion and, therefore, possess kinetic energy.

All other kinds of energy—the energy stored in a stretched spring, in a weight held above your head, or in a chemical bond, for example—are potential energy. An object has **potential energy** by virtue of its position relative to other objects. Potential energy is, in essence, the "stored" energy that arises from the attractions and repulsions an object experiences in relation to other objects.

We are familiar with many instances in which potential energy is converted into kinetic energy. For example, think of a cyclist poised at the top of a hill (▶ **Figure 5.2**). Because of the attractive force of gravity, the potential energy of the cyclist and her bicycle is greater at the top of the hill than at the bottom. As a result, the bicycle easily moves down the hill with increasing speed. As it does so, potential energy is converted into kinetic energy. The potential energy decreases as the bicycle rolls down the hill, but its kinetic energy increases as the speed increases (Equation 5.1). This example illustrates that forms of energy are interconvertible.

Gravitational forces play a negligible role in the ways that atoms and molecules interact with one another. Forces that arise from electrical charges are more important when dealing with atoms and molecules. One of the most important forms of potential energy in chemistry is *electrostatic potential energy,* E_{el}, which arises from

▲ GO FIGURE

Why is a pitcher able to throw a baseball faster than he could throw a bowling ball?

Work done by pitcher on ball to make ball move

(a)

Heat added by burner to water makes water temperature rise

(b)

▲ **Figure 5.1 Work and heat, two forms of energy. (a)** *Work* is energy used to cause an object to move. **(b)** *Heat* is energy used to cause the temperature of an object to increase.

▲GO FIGURE

Suppose the bicyclist is coasting (not pedaling) at constant speed on a flat road and begins to go up a hill. If she does not start pedaling, what happens to her speed? Why?

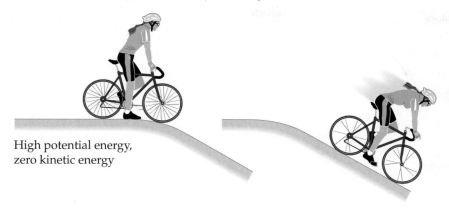

High potential energy, zero kinetic energy

Decreasing potential energy, increasing kinetic energy

▲ **Figure 5.2 Potential energy and kinetic energy.** The potential energy initially stored in the motionless bicycle and rider at the top of the hill is converted to kinetic energy as the bicycle moves down the hill and loses potential energy.

the interactions between charged particles. This energy is proportional to the electrical charges on the two interacting objects, Q_1 and Q_2, and inversely proportional to the distance, d, separating them:

$$E_{el} = \frac{\kappa Q_1 Q_2}{d} \qquad [5.2]$$

In this equation κ is the proportionality constant, 8.99×10^9 J-m/C^2, that relates the units for energy to the units for the charges and their distance of separation. C is the coulomb, a unit of electrical charge ∞ (Section 2.2), and J is the joule, a unit of energy we will discuss soon.* At the molecular level, the electrical charges Q_1 and Q_2 are typically on the order of magnitude of the charge of the electron (1.60×10^{-19} C).

Equation 5.2 shows that the electrostatic potential energy goes to zero as d becomes infinite. Thus, the zero of electrostatic potential energy is defined as infinite separation of the charged particles. **Figure 5.3** illustrates how E_{el} behaves as the distance between two charges changes. When Q_1 and Q_2 have the same sign (for example, both positive), the two charged particles repel each other, and a repulsive force pushes them apart. In this case, E_{el} is positive, and the potential energy decreases as the particles move farther apart. When Q_1 and Q_2 have opposite signs, the particles attract each other, and an attractive force pulls them toward each other. In this case, E_{el} is negative, and the potential energy increases (becomes less negative) as the particles move apart.

One of our goals in chemistry is to relate the energy changes seen in the macroscopic world to the kinetic or potential energy of substances at the molecular level. Many substances—fuels, for example—release energy when they react. The *chemical energy* of a fuel is due to the potential energy stored in the arrangements of its atoms.

---

*We read the combined units J-m/C^2 as joule-meters per coulomb squared. You may see combinations of units such as J-m/C^2 expressed with dots separating units instead of short dashes (J · m/C^2) or with dashes or dots totally absent (J m/C^2).

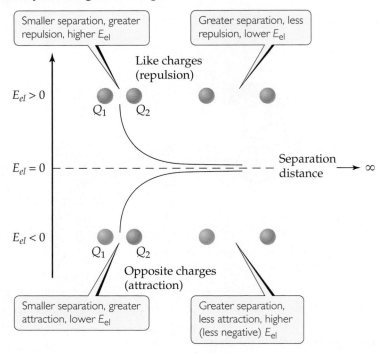

▲ GO FIGURE

A positively charged particle and a negatively charged particle are initially far apart. What happens to their electrostatic potential energy as they are brought closer together?

Like charges (repulsion)

Smaller separation, greater repulsion, higher E_{el}

Greater separation, less repulsion, lower E_{el}

$E_{el} > 0$

Q_1 Q_2

$E_{el} = 0$

Separation distance $\longrightarrow \infty$

$E_{el} < 0$

Q_1 Q_2

Opposite charges (attraction)

Smaller separation, greater attraction, lower E_{el}

Greater separation, less attraction, higher (less negative) E_{el}

▲ **Figure 5.3 Electrostatic potential energy.** At finite separation distances for two charged particles, E_{el} is positive for like charges and negative for opposite charges. As the particles move farther apart, their electrostatic potential energy approaches zero.

When a fuel burns, this chemical energy is converted to *thermal energy*, energy associated with temperature. The increase in thermal energy arises from the increased molecular motion and hence increased kinetic energy at the molecular level.

 Give It Some Thought

The cyclist and bicycle illustrated in Figure 5.2 come to a stop at the bottom of the hill.
(a) Is the potential energy the same as it was at the top of the hill?
(b) Is the kinetic energy the same as it was at the top of the hill?

Units of Energy

The SI unit for energy is the **joule** (pronounced "jool"), J, in honor of James Joule (1818–1889), a British scientist who investigated work and heat: $1\,J = 1\,kg\text{-}m^2/s^2$. Equation 5.1 shows that a mass of 2 kg moving at a speed of 1 m/s possesses a kinetic energy of 1 J:

$$E_k = \tfrac{1}{2} mv^2 = \tfrac{1}{2}(2\,kg)(1\,m/s)^2 = 1\,kg\text{-}m^2/s^2 = 1\,J$$

Because a joule is not a large amount of energy, we often use *kilojoules* (kJ) in discussing the energies associated with chemical reactions.

Traditionally, energy changes accompanying chemical reactions have been expressed in calories, a non–SI unit still widely used in chemistry, biology, and biochemistry. A **calorie** (cal) was originally defined as the amount of energy required to raise the temperature of 1 g of water from 14.5 to 15.5 °C. A calorie is now defined in terms of the joule:

$$1\,cal = 4.184\,J\,(exactly)$$

A related energy unit used in nutrition is the nutritional *Calorie* (note the capital C): 1 Cal = 1000 cal = 1 kcal.

System and Surroundings

When analyzing energy changes, we need to focus on a limited and well-defined part of the universe to keep track of the energy changes that occur. The portion we single out for study is called the **system**; everything else is called the **surroundings**. When we study the energy change that accompanies a chemical reaction in a laboratory, the reactants and products constitute the system. The container and everything beyond it are considered the surroundings.

Systems may be open, closed, or isolated. An *open* system is one in which matter and energy can be exchanged with the surroundings. An uncovered pot of boiling water on a stove, such as that in Figure 5.1(b), is an open system: Heat comes into the system from the stove, and water is released to the surroundings as steam.

The systems we can most readily study in thermochemistry are called *closed systems*—systems that can exchange energy but not matter with their surroundings. For example, consider a mixture of hydrogen gas, H_2, and oxygen gas, O_2, in a cylinder fitted with a piston (▶ Figure 5.4). The system is just the hydrogen and oxygen; the cylinder, piston, and everything beyond them (including us) are the surroundings. If the gases react to form water, energy is liberated:

$$2\,H_2(g) + O_2(g) \longrightarrow 2\,H_2O(g) + \text{energy}$$

Although the chemical form of the hydrogen and oxygen atoms in the system is changed by this reaction, the system has not lost or gained mass, which means it has not exchanged any matter with its surroundings. However, it can exchange energy with its surroundings in the form of *work* and *heat*.

An *isolated* system is one in which neither energy nor matter can be exchanged with the surroundings. An insulated thermos containing hot coffee approximates an isolated system. We know, however, that the coffee eventually cools, so it is not perfectly isolated.

 Give It Some Thought

Is a human being an isolated, closed, or open system?

Transferring Energy: Work and Heat

Figure 5.1 illustrates the two ways we experience energy changes in our everyday lives—in the form of work and in the form of heat. In Figure 5.1(a) work is done as energy is transferred from the pitcher's arm to the ball, directing it toward the plate at high speed. In Figure 5.1(b) energy is transferred in the form of heat. Causing the motion of an object against a force and causing a temperature change are the two general ways that energy can be transferred into or out of a system.

We define *work*, *w*, as the energy transferred when a *force* moves an object. A **force** is any push or pull exerted on an object. The magnitude of the work equals the product of the force, *F*, and the distance, *d*, the object moves:

$$w = F \times d \qquad [5.3]$$

We perform work, for example, when we lift an object against the force of gravity. If we define the object as the system, then we—as part of the surroundings—are performing work on that system, transferring energy to it.

The other way in which energy is transferred is as heat. *Heat* is the energy transferred from a hotter object to a colder one. A combustion reaction, such as the burning of natural gas illustrated in Figure 5.1(b), releases the chemical energy stored in the molecules of the fuel. ⟸ (Section 3.2) If we define the substances involved in the reaction as the system and everything else as the surroundings, we find that the released energy causes the temperature of the system to increase. Energy in the form of heat is then transferred from the hotter system to the cooler surroundings.

▲ **GO FIGURE**

If the piston is pulled upward so that it sits halfway between the position shown and the top of the cylinder, is the system still closed?

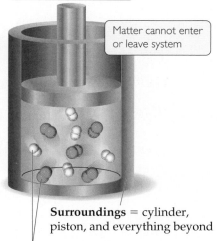

Energy can enter or leave system as heat or as work done on piston

Matter cannot enter or leave system

Surroundings = cylinder, piston, and everything beyond

System = $H_2(g)$ and $O_2(g)$

▲ **Figure 5.4** A closed system.

SAMPLE
EXERCISE 5.1 | Describing and Calculating Energy Changes

A bowler lifts a 5.4-kg (12-lb) bowling ball from ground level to a height of 1.6 m (5.2 ft) and
then drops it. (**a**) What happens to the potential energy of the ball as it is raised? (**b**) What quantity of work, in J, is used to raise the ball? (**c**) After the ball is dropped, it gains kinetic energy. If
all the work done in part (b) has been converted to kinetic energy by the time the ball strikes the
ground, what is the ball's speed just before it hits the ground? (*Note*: The force due to gravity is
$F = m \times g$, where m is the mass of the object and g is the gravitational constant; $g = 9.8 \text{ m/s}^2$.)

SOLUTION

Analyze We need to relate the potential energy of the bowling ball to its position relative to the
ground. We then need to establish the relationship between work and the change in the ball's
potential energy. Finally, we need to connect the change in potential energy when the ball is
dropped with the kinetic energy attained by the ball.

Plan We can calculate the work done in lifting the ball by using Equation 5.3: $w = F \times d$. The
kinetic energy of the ball just before it hits the ground equals its initial potential energy. We can
use the kinetic energy and Equation 5.1 to calculate the speed, v, just before impact.

Solve

(**a**) Because the ball is raised above the ground, its potential energy relative to the ground
increases.

(**b**) The ball has a mass of 5.4 kg and is lifted 1.6 m. To calculate the work performed to raise the
ball, we use Equation 5.3 and $F = m \times g$ for the force that is due to gravity:

$$w = F \times d = m \times g \times d = (5.4 \text{ kg})(9.8 \text{ m/s}^2)(1.6 \text{ m}) = 85 \text{ kg-m}^2/\text{s}^2 = 85 \text{ J}$$

Thus, the bowler has done 85 J of work to lift the ball to a height of 1.6 m.

(**c**) When the ball is dropped, its potential energy is converted to kinetic energy. We assume that the
kinetic energy just before the ball hits the ground is equal to the work done in part (b), 85 J:

$$E_k = \tfrac{1}{2}mv^2 = 85 \text{ J} = 85 \text{ kg-m}^2/\text{s}^2$$

We can now solve this equation for v:

$$v^2 = \left(\frac{2E_k}{m}\right) = \left(\frac{2(85 \text{ kg-m}^2/\text{s}^2)}{5.4 \text{ kg}}\right) = 31.5 \text{ m}^2/\text{s}^2$$

$$v = \sqrt{31.5 \text{ m}^2/\text{s}^2} = 5.6 \text{ m/s}$$

Check Work must be done in (b) to increase the potential energy of the ball, which is in accord
with our experience. The units are appropriate in (b) and (c). The work is in units of J and the
speed in units of m/s. In (c) we carry an additional digit in the intermediate calculation involving the square root, but we report the final value to only two significant figures, as appropriate.

Comment A speed of 1 m/s is roughly 2 mph, so the bowling ball has a speed greater than 10 mph
just before impact.

Practice Exercise 1

Which of the following objects has the greatest kinetic energy? (**a**) a 500-kg motorcycle
moving at 100 km/h, (**b**) a 1,000-kg car moving at 50 km/h, (**c**) a 1,500-kg car moving at
30 km/h, (**d**) a 5,000-kg truck moving at 10 km/h, (**e**) a 10,000-kg truck moving at 5 km/h.

Practice Exercise 2

What is the kinetic energy, in J, of (**a**) an Ar atom moving at a speed of 650 m/s, (**b**) a mole of
Ar atoms moving at 650 m/s? (*Hint*: 1 amu $= 1.66 \times 10^{-27}$ kg.)

5.2 | The First Law of Thermodynamics

We have seen that the potential energy of a system can be converted into kinetic energy,
and vice versa. We have also seen that energy can be transferred back and forth between
a system and its surroundings in the forms of work and heat. All of these conversions
and transfers proceed in accord with one of the most important observations in science:
Energy can be neither created nor destroyed. Any energy that is lost by a system must be
gained by the surroundings, and vice versa. This important observation—that *energy is
conserved*—is known as the **first law of thermodynamics**. To apply this law quantitatively, let's first define the energy of a system more precisely.

Internal Energy

The **internal energy**, E, of a system is the sum of *all* the kinetic and potential energies of the components of the system. For the system in Figure 5.4, for example, the internal energy includes not only the motions and interactions of the H_2 and O_2 molecules but also the motions and interactions of their component nuclei and electrons. We generally do not know the numerical value of a system's internal energy. In thermodynamics, we are mainly concerned with the *change* in E (and, as we shall see, changes in other quantities as well) that accompanies a change in the system.

Imagine that we start with a system with an initial internal energy $E_{initial}$. The system then undergoes a change, which might involve work being done or heat being transferred. After the change, the final internal energy of the system is E_{final}. We define the *change* in internal energy, denoted ΔE (read "delta E"),* as the difference between E_{final} and $E_{initial}$:

$$\Delta E = E_{final} - E_{initial} \qquad [5.4]$$

We generally cannot determine the actual values of E_{final} and $E_{initial}$ for any system of practical interest. Nevertheless, we can determine the value of ΔE experimentally by applying the first law of thermodynamics.

Thermodynamic quantities such as ΔE have three parts: (1) a number and (2) a unit, which together give the magnitude of the change, and (3) a sign that gives the direction. A *positive* value of ΔE results when $E_{final} > E_{initial}$, indicating that the system has gained energy from its surroundings. A *negative* value of ΔE results when $E_{final} < E_{initial}$, indicating that the system has lost energy to its surroundings. Notice that we are taking the point of view of the system rather than that of the surroundings in discussing the energy changes. We need to remember, however, that any increase in the energy of the system is accompanied by a decrease in the energy of the surroundings, and vice versa. These features of energy changes are summarized in ▼ **Figure 5.5**.

 GO FIGURE

What is the value of ΔE if E_{final} equals $E_{initial}$?

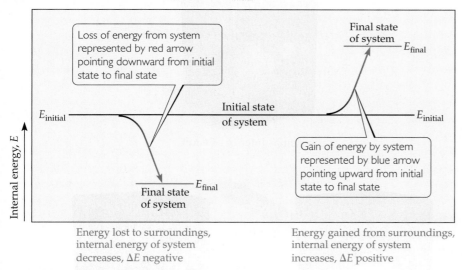

▲ **Figure 5.5 Changes in internal energy.**

In a chemical reaction, the initial state of the system refers to the reactants and the final state refers to the products. In the reaction

$$2\,H_2(g) + O_2(g) \longrightarrow 2\,H_2O(l)$$

for instance, the initial state is the $2\,H_2(g) + O_2(g)$ and the final state is the $2\,H_2O(l)$. When hydrogen and oxygen form water at a given temperature, the system loses energy to the surroundings. Because energy is lost from the system, the internal energy of the products (final state) is less than that of the reactants (initial state), and ΔE for the process is

---

*The symbol Δ is commonly used to denote *change*. For example, a change in height, h, can be represented by Δh.

GO FIGURE

The internal energy for a mixture of Mg(s) and Cl$_2$(g) is greater than that of MgCl$_2$(s). Sketch an energy diagram that represents the reaction MgCl$_2$(s) $\longrightarrow$ Mg(s) + Cl$_2$(g).

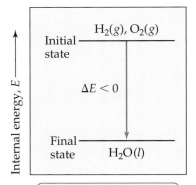

$E_{initial}$ greater than E_{final}; therefore, energy is released from system to surroundings during reaction and $\Delta E < 0$

▲ **Figure 5.6 Energy diagram for the reaction 2 H$_2$(g) + O$_2$(g) $\longrightarrow$ 2 H$_2$O(l).**

negative. Thus, the *energy diagram* in ◄ Figure 5.6 shows that the internal energy of the mixture of H$_2$ and O$_2$ is greater than that of the H$_2$O produced in the reaction.

Relating ΔE to Heat and Work

As we noted in Section 5.1, a system may exchange energy with its surroundings in two general ways: as heat or as work. The internal energy of a system changes in magnitude as heat is added to or removed from the system or as work is done on or by the system. If we think of internal energy as the system's bank account of energy, we see that deposits or withdrawals can be made either in the form of heat or in the form of work. Deposits increase the energy of the system (positive ΔE), whereas withdrawals decrease the energy of the system (negative ΔE).

We can use these ideas to write a useful algebraic expression of the first law of thermodynamics. When a system undergoes any chemical or physical change, the accompanying change in internal energy, ΔE, is the sum of the heat added to or liberated from the system, q, and the work done on or by the system, w:

$$\Delta E = q + w \quad\quad [5.5]$$

When heat is added to a system or work is done on a system, its internal energy increases. Therefore, when heat is transferred to the system from the surroundings, q has a positive value. Adding heat to the system is like making a deposit to the energy account—the energy of the system increases (▼ Figure 5.7). Likewise, when work is done on the

GO FIGURE

Suppose a system receives a "deposit" of 50 J of work from the surroundings and loses a "withdrawal" of 85 J of heat to the surroundings. What is the magnitude and the sign of ΔE for this process?

System is interior of vault

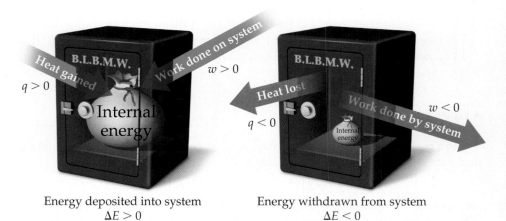

Energy deposited into system
$\Delta E > 0$

Energy withdrawn from system
$\Delta E < 0$

▲ **Figure 5.7 Sign conventions for heat and work.** Heat, q, gained by a system and work, w, done on a system are both positive quantities, corresponding to "deposits" of internal energy into the system. Conversely, heat transferred from the system to the surroundings and work done by the system on the surroundings are both "withdrawals" of internal energy from the system.

Table 5.1 Sign Conventions for q, w, and ΔE

For q	+ means system *gains* heat	− means system *loses* heat
For w	+ means work done *on* system	− means work done *by* system
For ΔE	+ means *net gain* of energy by system	− means *net loss* of energy by system

system by the surroundings, w has a positive value. Conversely, both the heat lost by the system to the surroundings and the work done by the system on the surroundings have negative values; that is, they lower the internal energy of the system. They are energy withdrawals and lower the amount of energy in the system's account.

The sign conventions for q, w, and ΔE are summarized in ▲ **Table 5.1**. Notice that any energy entering the system as either heat or work carries a positive sign.

SAMPLE EXERCISE 5.2 Relating Heat and Work to Changes of Internal Energy

Gases A(g) and B(g) are confined in a cylinder-and-piston arrangement like that in Figure 5.4 and react to form a solid product C(s): A(g) + B(g) $\longrightarrow$ C(s). As the reaction occurs, the system loses 1150 J of heat to the surroundings. The piston moves downward as the gases react to form a solid. As the volume of the gas decreases under the constant pressure of the atmosphere, the surroundings do 480 J of work on the system. What is the change in the internal energy of the system?

SOLUTION

Analyze The question asks us to determine ΔE, given information about q and w.

Plan We first determine the signs of q and w (Table 5.1) and then use Equation 5.5, $\Delta E = q + w$, to calculate ΔE.

Solve Heat is transferred from the system to the surroundings, and work is done on the system by the surroundings, so q is negative and w is positive: $q = -1150$ J and $w = 480$ kJ. Thus,

$$\Delta E = q + w = (-1150 \text{ J}) + (480 \text{ J}) = -670 \text{ J}$$

The negative value of ΔE tells us that a net quantity of 670 J of energy has been transferred from the system to the surroundings.

Comment You can think of this change as a decrease of 670 J in the net value of the system's energy bank account (hence, the negative sign); 1150 J is withdrawn in the form of heat while 480 J is deposited in the form of work. Notice that as the volume of the gases decreases, work is being done *on* the system *by* the surroundings, resulting in a deposit of energy.

Practice Exercise 1

Consider the following four cases: (**i**) A chemical process in which heat is absorbed, (**ii**) A change in which $q = 30$ J, $w = 44$ J, (**iii**) A process in which a system does work on its surroundings with no change in q, (**iv**) A process in which work is done on a system and an equal amount of heat is withdrawn.

In how many of these cases does the internal energy of the system decrease? (**a**) 0, (**b**) 1, (**c**) 2, (**d**) 3, (**e**) 4.

Practice Exercise 2

Calculate the change in the internal energy for a process in which a system absorbs 140 J of heat from the surroundings and does 85 J of work on the surroundings.

Endothermic and Exothermic Processes

Because transfer of heat to and from the system is central to our discussion in this chapter, we have some special terminology to indicate the direction of transfer. When a process occurs in which the system absorbs heat, the process is called **endothermic** (*endo-* means "into"). During an endothermic process, such as the melting of ice, heat flows *into* the system from its surroundings [**Figure 5.8(a)**]. If we, as part of the surroundings, touch a container in which ice is melting, the container feels cold to us because heat has passed from our hand to the container.

A process in which the system loses heat is called **exothermic** (*exo-* means "out of"). During an exothermic process, such as the combustion of gasoline, heat *exits* or flows *out* of the system into the surroundings [**Figure 5.8(b)**].

 Give It Some Thought

When $H_2(g)$ and $O_2(g)$ react to form $H_2O(l)$, heat is released to the surroundings. Consider the reverse reaction, namely, the formation of $H_2(g)$ and $O_2(g)$ from $H_2O(l)$: $2 H_2O(l) \rightarrow 2 H_2(g) + O_2(g)$. Is this reaction exothermic or endothermic? (*Hint:* Refer to Figure 5.6.)

System: reactants + products

Surroundings: solvent, initially at room temperature

(a) An endothermic reaction

Heat flows from surroundings into system, temperature of surroundings drops, thermometer reads temperature well below room temperature

System: reactants + products

Surroundings: air around reactants

(b) An exothermic reaction

Heat flows (violently) from system into surroundings, temperature of surroundings increases

▲ **Figure 5.8 Endothermic and exothermic reactions. (a)** When ammonium thiocyanate and barium hydroxide octahydrate are mixed at room temperature, the temperature drops. **(b)** The reaction of powdered aluminum with Fe_2O_3 (the thermite reaction) proceeds vigorously, releasing heat and forming Al_2O_3 and molten iron.

State Functions

Although we usually have no way of knowing the precise value of the internal energy of a system, E, it does have a fixed value for a given set of conditions. The conditions that influence internal energy include the temperature and pressure. Furthermore, the internal energy of a system is proportional to the total quantity of matter in the system because energy is an extensive property. ∞ (Section 1.3)

Suppose we define our system as 50 g of water at 25 °C (▼ **Figure 5.9**). The system could have reached this state by cooling 50 g of water from 100 to 25 °C or by melting 50 g of ice and subsequently warming the water to 25 °C. The internal energy of the water at 25 °C is the same in either case. Internal energy is an example of a **state function**, a property of a system that is determined by specifying the system's condition, or state (in terms of temperature, pressure, and so forth). *The value of a state function depends only on the present state of the system, not on the path the system took to reach that state.* Because E is a state function, ΔE depends only on the initial and final states of the system, not on how the change occurs.

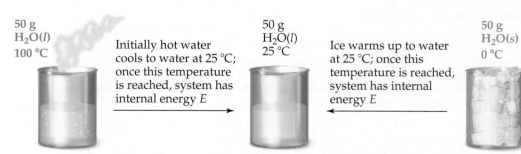

50 g
$H_2O(l)$
100 °C

Initially hot water cools to water at 25 °C; once this temperature is reached, system has internal energy E

50 g
$H_2O(l)$
25 °C

Ice warms up to water at 25 °C; once this temperature is reached, system has internal energy E

50 g
$H_2O(s)$
0 °C

▲ **Figure 5.9 Internal energy, *E*, a state function.** Any state function depends only on the present state of the system and not on the path by which the system arrived at that state.

An analogy may help you understand the difference between quantities that are state functions and those that are not. Suppose you drive from Chicago, which is 596 ft above sea level, to Denver, which is 5280 ft above sea level. No matter which route you take, the altitude change is 4684 ft. The distance you travel, however, depends on your route. Altitude is analogous to a state function because the change in altitude is independent of the path taken. Distance traveled is not a state function.

Some thermodynamic quantities, such as E, are state functions. Other quantities, such as q and w, are not. This means that, although $\Delta E = q + w$ does not depend on how the change occurs, the specific amounts of heat and work depend on the way in which the change occurs. Thus, if changing the path by which a system goes from an initial state to a final state increases the value of q, that path change will also decrease the value of w by exactly the same amount. The result is that ΔE is the same for the two paths.

We can illustrate this principle using a flashlight battery as our system. As the battery is discharged, its internal energy decreases as the energy stored in the battery is released to the surroundings. In ▶ Figure 5.10, we consider two possible ways of discharging the battery at constant temperature. If a wire shorts out the battery, no work is accomplished because nothing is moved against a force. All the energy lost from the battery is in the form of heat. (The wire gets warmer and releases heat to the surroundings.) If the battery is used to make a motor turn, the discharge produces work. Some heat is released, but not as much as when the battery is shorted out. We see that the magnitudes of q and w must be different for these two cases. If the initial and final states of the battery are identical in the two cases, however, then $\Delta E = q + w$ must be the same in both cases because E is a state function. Remember: ΔE depends only on the initial and final states of the system, not on the specific path taken from the initial to the final state.

Give It Some Thought

In what ways is the balance in your checkbook a state function?

5.3 | Enthalpy

The chemical and physical changes that occur around us, such as photosynthesis in the leaves of a plant, evaporation of water from a lake, or a reaction in an open beaker in a laboratory, occur under the essentially constant pressure of Earth's atmosphere.* These changes can result in the release or absorption of heat and can be accompanied by work done by or on the system. In exploring these changes, it is useful to have a thermodynamic function that is a state function and relates mainly to heat flow. Under conditions of constant pressure, a thermodynamic quantity called *enthalpy* (from the Greek *enthalpein*, "to warm") provides such a function.

Enthalpy, which we denote by the symbol H, is defined as the internal energy plus the product of the *pressure*, P, and *volume*, V, of the system:

$$H = E + PV \qquad [5.6]$$

Like internal energy E, both P and V are state functions—they depend only on the current state of the system and not on the path taken to that state. Because energy, pressure, and volume are all state functions, enthalpy is also a state function.

Pressure–Volume Work

To better understand the significance of enthalpy, recall from Equation 5.5 that ΔE involves not only the heat q added to or removed from the system but also the work w done by or on the system. Most commonly, the only kind of work produced by chemical or physical changes open to the atmosphere is the mechanical work associated with a change in volume. For example, when the reaction of zinc metal with hydrochloric acid solution

$$Zn(s) + 2\,H^+(aq) \rightarrow Zn^{2+}(aq) + H_2(g) \qquad [5.7]$$

*You are probably familiar with the notion of atmospheric pressure from a previous course in chemistry. We will discuss it in detail in Chapter 10. Here we need realize only that the atmosphere exerts a pressure on the surface of Earth that is nearly constant.

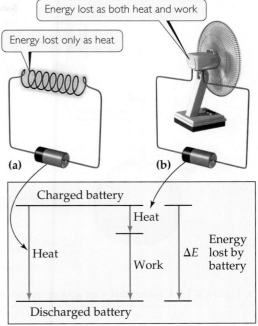

▲ GO FIGURE

If the battery is defined as the system, what is the sign on w in part (b)?

Energy lost as both heat and work

Energy lost only as heat

(a)   (b)

Charged battery

Heat

Heat   Work ΔE Energy lost by battery

Discharged battery

▲ **Figure 5.10 Internal energy is a state function, but heat and work are not. (a)** A battery shorted out by a wire loses energy to the surroundings only as heat; no work is performed. **(b)** A battery discharged through a motor loses energy as work (to make the fan turn) and also loses some energy as heat. The value of ΔE is the same for both processes even though the values of q and w in (a) are different from those in (b).

▲ **GO FIGURE**

If the amount of zinc used in the reaction is increased, will more work be done by the system? Is there additional information you need to answer this question?

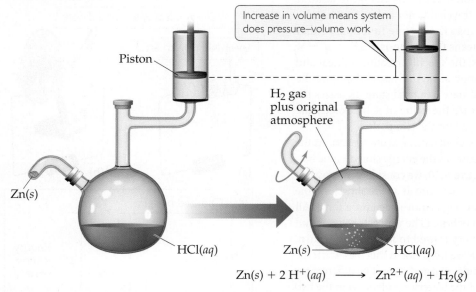

Zn(s) + 2 H$^+$(aq) $\longrightarrow$ Zn^{2+}(aq) + H$_2$(g)

▲ **Figure 5.11 A system that does work on its surroundings.**

is run at constant pressure in the apparatus illustrated in ▲ **Figure 5.11**, the piston moves up or down to maintain a constant pressure in the vessel. If we assume for simplicity that the piston has no mass, the pressure in the apparatus is the same as atmospheric pressure. As the reaction proceeds, H$_2$ gas forms, and the piston rises. The gas within the flask is thus doing work on the surroundings by lifting the piston against the force of atmospheric pressure.

The work involved in the expansion or compression of gases is called **pressure–volume work** (P–V work). When pressure is constant in a process, as in our preceding example, the sign and magnitude of the pressure–volume work are given by

$$w = -P\,\Delta V \qquad [5.8]$$

where P is pressure and $\Delta V = V_{\text{final}} - V_{\text{initial}}$ is the change in volume of the system. The pressure P is always either a positive number or zero. If the volume of the system expands, then ΔV is positive as well. The negative sign in Equation 5.8 is necessary to conform to the sign convention for w (Table 5.1). When a gas expands, the system does work on the surroundings, as indicated by a negative value of w. (The negative sign ✕ positive P ✕ positive ΔV results in a negative number for w.) On the other hand, if the gas is compressed, ΔV is negative (the volume decreases), and Equation 5.8 indicates that w is therefore positive, meaning work is done on the system by the surroundings. "A Closer Look: Energy, Enthalpy, and P–V Work" box discusses pressure–volume work in detail, but all you need to keep in mind for now is Equation 5.8, which applies to processes occurring at constant pressure.

The units of work obtained by using Equation 5.8 will be those of pressure (usually atm) multiplied by those of volume (usually L). To express the work in the more familiar unit of joules, we use the conversion factor 1 L-atm = 101.3 J.

▲ **Give It Some Thought**

If a system does not change its volume during the course of a process, does it do pressure–volume work?

SAMPLE EXERCISE 5.3

A fuel is burned in a cylinder equipped with a piston. The initial volume of the cylinder is 0.250 L, and the final volume is 0.980 L. If the piston expands against a constant pressure of 1.35 atm, how much work (in J) is done? (1 L-atm = 101.3 J)

SOLUTION

Analyze We are given an initial volume and a final volume from which we can calculate ΔV. We are also given the pressure, P. We are asked to calculate work, w.

Plan The equation $w = -P\,\Delta V$ allows us to calculate the work done by the system from the given information.

Solve The volume change is

$$\Delta V = V_{\text{final}} - V_{\text{initial}} = 0.980\ \text{L} - 0.250\ \text{L} = 0.730\ \text{L}$$

Thus, the quantity of work is

$$w = -P\Delta V = -(1.35\ \text{atm})(0.730\ \text{L}) = -0.9855\ \text{L-atm}$$

Converting L-atm to J, we have

$$-(0.9855\ \text{L-atm})\left(\frac{101.3\ \text{J}}{1\ \text{L-atm}}\right) = -99.8\ \text{J}$$

Check The significant figures are correct (3), and the units are the requested ones for energy (J). The negative sign is consistent with an expanding gas doing work on its surroundings.

Practice Exercise 1

If a balloon is expanded from 0.055 to 1.403 L against an external pressure of 1.02 atm, how many L-atm of work is done? (a) −0.056 L-atm, (b) −1.37 L-atm, (c) 1.43 L-atm, (d) 1.49 L-atm, (e) 139 L-atm.

Practice Exercise 2

Calculate the work, in J, if the volume of a system contracts from 1.55 to 0.85 L at a constant pressure of 0.985 atm.

Enthalpy Change

Let's now return to our discussion of enthalpy. When a change occurs at constant pressure, the change in enthalpy, ΔH, is given by the relationship

$$\Delta H = \Delta(E + PV)$$
$$= \Delta E + P\Delta V \quad \text{(constant pressure)} \qquad [5.9]$$

That is, the change in enthalpy equals the change in internal energy plus the product of the constant pressure and the change in volume.

Recall that $\Delta E = q + w$ (Equation 5.5) and that the work involved in the expansion or compression of a gas is $w = -P\,\Delta V$ (at constant pressure). Substituting $-w$ for $P\,\Delta V$ and $q + w$ for ΔE into Equation 5.9, we have

$$\Delta H = \Delta E + P\,\Delta V = (q_P + w) - w = q_P \qquad [5.10]$$

The subscript P on q indicates that the process occurs at constant pressure. Thus, *the change in enthalpy equals the heat q_P gained or lost at constant pressure.* Because q_P is something we can either measure or calculate and because so many physical and chemical changes of interest to us occur at constant pressure, enthalpy is a more useful function for most reactions than is internal energy. In addition, for most reactions the difference in ΔH and ΔE is small because $P\,\Delta V$ is small.

When ΔH is positive (that is, when q_P is positive), the system has gained heat from the surroundings (Table 5.1), which means the process is endothermic. When ΔH is negative, the system has released heat to the surroundings, which means the process is exothermic. To continue the bank analogy of Figure 5.7, under constant pressure, an endothermic process deposits energy in the system in the form of heat and an exothermic process withdraws energy in the form of heat (▶ Figure 5.12).

 Give It Some Thought

When a reaction occurs in a flask, you notice that the flask gets colder. What is the sign of ΔH?

Because H is a state function, ΔH (which equals q_P) depends only on the initial and final states of the system, not on how the change occurs. At first glance this statement might seem to contradict our discussion in Section 5.2, in which we said that q is *not* a state function. There is no contradiction, however, because the relationship between ΔH and q_P has special limitations that only P–V work is involved and that the pressure is constant.

Constant pressure maintained in system

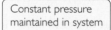 $\Delta H > 0$
Heat gain

(a) An endothermic reaction

$\Delta H < 0$
Heat loss

(b) An exothermic reaction

ΔH is amount of heat that flows into or out of system under constant pressure

▲ **Figure 5.12 Endothermic and exothermic processes. (a)** An endothermic process ($\Delta H > 0$) deposits heat into the system. **(b)** An exothermic process ($\Delta H < 0$) withdraws heat from the system.

SAMPLE EXERCISE 5.4 | Determining the Sign of ΔH

Indicate the sign of the enthalpy change, ΔH, in the following processes carried out under atmospheric pressure and indicate whether each process is endothermic or exothermic: (a) An ice cube melts; (b) 1 g of butane (C_4H_{10}) is combusted in sufficient oxygen to give complete combustion to CO_2 and H_2O.

SOLUTION

Analyze Our goal is to determine whether ΔH is positive or negative for each process. Because each process occurs at constant pressure, the enthalpy change equals the quantity of heat absorbed or released, $\Delta H = q_P$.

Plan We must predict whether heat is absorbed or released by the system in each process. Processes in which heat is absorbed are endothermic and have a positive sign for ΔH; those in which heat is released are exothermic and have a negative sign for ΔH.

Solve In (a) the water that makes up the ice cube is the system. The ice cube absorbs heat from the surroundings as it melts, so ΔH is positive and the process is endothermic. In (b) the system is the 1 g of butane and the oxygen required to combust it. The combustion of butane in oxygen gives off heat, so ΔH is negative and the process is exothermic.

Practice Exercise 1

A chemical reaction that gives off heat to its surroundings is said to be _____ and has a _____ value of ΔH.
(a) endothermic, positive   (c) exothermic, positive
(b) endothermic, negative   (d) exothermic, negative

Practice Exercise 2

Molten gold poured into a mold solidifies at atmospheric pressure. With the gold defined as the system, is the solidification an exothermic or endothermic process?

A Closer Look

Energy, Enthalpy, and *P-V* Work

In chemistry we are interested mainly in two types of work: electrical work and mechanical work done by expanding gases. We focus here on the latter, called pressure–volume, or *P–V*, work. Expanding gases in the cylinder of an automobile engine do *P–V* work on the piston; this work eventually turns the wheels. Expanding gases from an open reaction vessel do *P–V* work on the atmosphere. This work accomplishes nothing in a practical sense, but we must keep track of all work, useful or not, when monitoring energy changes in a system.

Let's consider a gas confined to a cylinder with a movable piston of cross-sectional area A (▼ **Figure 5.13**). A downward force F acts on the piston. The *pressure*, P, on the gas is the force per area: $P = F/A$.

We assume that the piston is massless and that the only pressure acting on it is the *atmospheric pressure* that is due to Earth's atmosphere, which we assume to be constant.

Suppose the gas expands and the piston moves a distance Δh. From Equation 5.3, the magnitude of the work done by the system is

$$\text{Magnitude of work} = \text{force} \times \text{distance} = F \times \Delta h \quad [5.11]$$

We can rearrange the definition of pressure, $P = F/A$, to $F = P \times A$. The volume change, ΔV, resulting from the movement of the piston is the product of the cross-sectional area of the piston and the distance it moves: $\Delta V = A \times \Delta h$. Substituting into Equation 5.11 gives

$$\text{Magnitude of work} = F \times \Delta h = P \times A \times \Delta h$$
$$= P \times \Delta V$$

Because the system (the confined gas) does work on the surroundings, the work is a negative quantity:

$$w = -P\,\Delta V \quad [5.12]$$

Now, if *P–V* work is the only work that can be done, we can substitute Equation 5.12 into Equation 5.5 to give

$$\Delta E = q + w = q - P\,\Delta V \quad [5.13]$$

When a reaction is carried out in a constant-volume container ($\Delta V = 0$), therefore, the heat transferred equals the change in internal energy:

$$\Delta E = q - P\Delta V = q - P(0) = q_V \quad \text{(constant volume)} \quad [5.14]$$

The subscript V indicates that the volume is constant.

Most reactions are run under constant pressure, so that Equation 5.13 becomes

$$\Delta E = q_P - P\,\Delta V$$
$$q_P = \Delta E + P\,\Delta V \quad \text{(constant pressure)} \quad [5.15]$$

We see from Equation 5.9 that the right side of Equation 5.15 is the enthalpy change under constant-pressure conditions. Thus, $\Delta H = q_P$, as we saw in Equation 5.10.

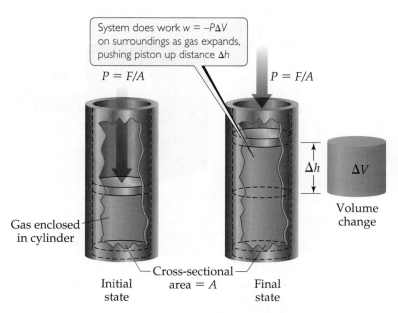

System does work $w = -P\Delta V$ on surroundings as gas expands, pushing piston up distance Δh

$P = F/A$

$P = F/A$

Δh

ΔV

Volume change

Gas enclosed in cylinder

Cross-sectional area = A

Initial state

Final state

▲ **Figure 5.13 Pressure–volume work.** The amount of work done by the system on the surroundings is $w = -P\Delta V$.

In summary, the change in internal energy is equal to the heat gained or lost at constant volume, and the change in enthalpy is equal to the heat gained or lost at constant pressure. The difference between ΔE and ΔH is the amount of P–V work done by the system when the process occurs at constant pressure, $-P\,\Delta V$. The volume change accompanying many reactions is close to zero, which makes $P\,\Delta V$ and, therefore, the difference between ΔE and ΔH small. Under most circumstances, it is generally satisfactory to use ΔH as the measure of energy changes during most chemical processes.

Related Exercises: 5.35, 5.36, 5.37, 5.38

5.4 | Enthalpies of Reaction

Because $\Delta H = H_{final} - H_{initial}$, the enthalpy change for a chemical reaction is given by

$$\Delta H = H_{products} - H_{reactants} \qquad [5.16]$$

The enthalpy change that accompanies a reaction is called either the **enthalpy of reaction** or the *heat of reaction* and is sometimes written ΔH_{rxn}, where "rxn" is a commonly used abbreviation for "reaction."

When we give a numerical value for ΔH_{rxn}, we must specify the reaction involved. For example, when 2 mol $H_2(g)$ burn to form 2 mol $H_2O(g)$ at a constant pressure, the system releases 483.6 kJ of heat. We can summarize this information as

$$2\,H_2(g) + O_2(g) \longrightarrow 2\,H_2O(g) \qquad \Delta H = -483.6\,kJ \qquad [5.17]$$

The negative sign for ΔH tells us that this reaction is exothermic. Notice that ΔH is reported at the end of the balanced equation, without explicitly specifying the amounts of chemicals involved. In such cases the coefficients in the balanced equation represent the number of moles of reactants and products producing the associated enthalpy change. Balanced chemical equations that show the associated enthalpy change in this way are called *thermochemical equations*.

The exothermic nature of this reaction is also shown in the *enthalpy diagram* in ▼ Figure 5.14. Notice that the enthalpy of the reactants is greater (more positive) than the enthalpy of the products. Thus, $\Delta H = H_{products} - H_{reactants}$ is negative.

 Give It Some Thought

If the reaction to form water were written $H_2(g) + \frac{1}{2}O_2(g) \longrightarrow H_2O(g)$, would you expect the same value of ΔH as in Equation 5.17? Why or why not?

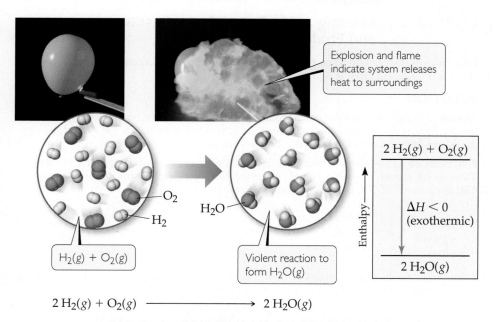

$$2\,H_2(g) + O_2(g) \longrightarrow 2\,H_2O(g)$$

▲ **Figure 5.14 Exothermic reaction of hydrogen with oxygen.** When a mixture of $H_2(g)$ and $O_2(g)$ is ignited to form $H_2O(g)$, the resultant explosion produces a ball of flame. Because the system releases heat to the surroundings, the reaction is exothermic as indicated in the enthalpy diagram.

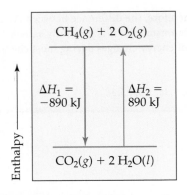

▶ **Figure 5.15 ΔH for a reverse reaction.** Reversing a reaction changes the sign but not the magnitude of the enthalpy change: $\Delta H_2 = -\Delta H_1$.

The following guidelines are helpful when using thermochemical equations and enthalpy diagrams:

1. **Enthalpy is an extensive property.** The magnitude of ΔH is proportional to the amount of reactant consumed in the process. For example, 890 kJ of heat is produced when 1 mol of CH_4 is burned in a constant-pressure system:

$$CH_4(g) + 2\,O_2(g) \longrightarrow CO_2(g) + 2\,H_2O(l) \qquad \Delta H = -890\,\text{kJ} \qquad [5.18]$$

Because the combustion of 1 mol of CH_4 with 2 mol of O_2 releases 890 kJ of heat, the combustion of 2 mol of CH_4 with 4 mol of O_2 releases twice as much heat, 1780 kJ. Although chemical equations are usually written with whole-number coefficients, thermochemical equations sometimes utilize fractions, as in the preceding Give It Some Thought Question.

2. **The enthalpy change for a reaction is equal in magnitude, but opposite in sign, to ΔH for the reverse reaction.** For example, ΔH for the reverse of Equation 5.18 is +890 kJ:

$$CO_2(g) + 2\,H_2O(l) \longrightarrow CH_4(g) + 2\,O_2(g) \qquad \Delta H = +890\,\text{kJ} \qquad [5.19]$$

When we reverse a reaction, we reverse the roles of the products and the reactants. From Equation 5.16, we see that reversing the products and reactants leads to the same magnitude of ΔH, but a change in sign (▶ Figure 5.15).

3. **The enthalpy change for a reaction depends on the states of the reactants and products.** If the product in Equation 5.18 were $H_2O(g)$ instead of $H_2O(l)$, ΔH_{rxn} would be −802 kJ instead of −890 kJ. Less heat would be available for transfer to the surroundings because the enthalpy of $H_2O(g)$ is greater than that of $H_2O(l)$. One way to see this is to imagine that the product is initially liquid water. The liquid water must be converted to water vapor, and the conversion of 2 mol $H_2O(l)$ to 2 mol $H_2O(g)$ is an endothermic process that absorbs 88 kJ:

$$2\,H_2O(l) \longrightarrow 2\,H_2O(g) \qquad \Delta H = +88\,\text{kJ} \qquad [5.20]$$

Thus, it is important to specify the states of the reactants and products in thermochemical equations. In addition, we will generally assume that the reactants and products are both at the same temperature, 25 °C, unless otherwise indicated.

SAMPLE
EXERCISE 5.5   **Relating ΔH to Quantities of Reactants and Products**

How much heat is released when 4.50 g of methane gas is burned in a constant-pressure system? (Use the information given in Equation 5.18.)

SOLUTION

Analyze Our goal is to use a thermochemical equation to calculate the heat produced when a specific amount of methane gas is combusted. According to Equation 5.18, 890 kJ is released by the system when 1 mol CH_4 is burned at constant pressure.

Plan Equation 5.18 provides us with a stoichiometric conversion factor: (1 mol $CH_4 \cong -890$ kJ). Thus, we can convert moles of CH_4 to kJ of energy. First, however, we must convert grams of CH_4 to moles of CH_4. Thus, the conversion sequence is

Grams CH_4 (given) → [Molar mass CH_4 16.0 g/mol] → Moles CH_4 → [$\Delta H = -890$ kJ/mol] → kJ of heat (unkown)

Solve By adding the atomic weights of C and 4 H, we have 1 mol CH_4 = 16.0 g CH_4. We can use the appropriate conversion factors to convert grams of CH_4 to moles of CH_4 to kilojoules:

$$\text{Heat} = (4.50\,\text{g}\,CH_4)\left(\frac{1\,\text{mol}\,CH_4}{16.0\,\text{g}\,CH_4}\right)\left(\frac{-890\,\text{kJ}}{1\,\text{mol}\,CH_4}\right) = -250\,\text{kJ}$$

The negative sign indicates that the system released 250 kJ into the surroundings.

Practice Exercise 1

The complete combustion of ethanol, C_2H_5OH (FW $= 46.0$ g/mol), proceeds as follows:

$$C_2H_5OH(l) + 3O_2(g) \longrightarrow 2CO_2(g) + 3H_2O(l) \qquad \Delta H = -555 \text{ kJ}$$

What is the enthalpy change for combustion of 15.0 g of ethanol?
(a) -12.1 kJ **(b)** -181 kJ **(c)** -422 kJ **(d)** -555 kJ **(e)** -1700 kJ

Practice Exercise 2

Hydrogen peroxide can decompose to water and oxygen by the reaction

$$2\,H_2O_2(l) \longrightarrow 2\,H_2O(l) + O_2(g) \qquad \Delta H = -196 \text{ kJ}$$

Calculate the quantity of heat released when 5.00 g of $H_2O_2(l)$ decomposes at constant pressure.

Strategies in Chemistry

Using Enthalpy as a Guide

If you hold a brick in the air and let it go, you know what happens: It falls as the force of gravity pulls it toward Earth. A process that is thermodynamically favored to happen, such as a brick falling to the ground, is called a *spontaneous* process. A spontaneous process can be either fast or slow; the rate at which processes occur is not governed by thermodynamics.

Chemical processes can be thermodynamically favored, or spontaneous, too. By spontaneous, however, we do not mean that the reaction will form products without any intervention. That can be the case, but often some energy must be imparted to get the process started. The enthalpy change in a reaction gives one indication as to whether the reaction is likely to be spontaneous. The combustion of $H_2(g)$ and $O_2(g)$, for example, is highly exothermic:

$$H_2(g) + \tfrac{1}{2}O_2(g) \longrightarrow H_2O(g) \qquad \Delta H = -242 \text{ kJ}$$

Hydrogen gas and oxygen gas can exist together in a volume indefinitely without noticeable reaction occurring. Once the reaction is initiated, however, energy is rapidly transferred from the system (the reactants) to the surroundings as heat. The system thus loses enthalpy by transferring the heat to the surroundings. (Recall that the first law of thermodynamics tells us that the total energy of the system plus the surroundings does not change; energy is conserved.)

Enthalpy change is not the only consideration in the spontaneity of reactions, however, nor is it a foolproof guide. For example, even though ice melting is an endothermic process,

$$H_2O(s) \longrightarrow H_2O(l) \qquad \Delta H = +6.01 \text{ kJ}$$

this process is spontaneous at temperatures above the freezing point of water $(0 \,^\circ C)$. The reverse process, water freezing, is spontaneous at temperatures below $0 \,^\circ C$. Thus, we know that ice at room temperature melts and water put into a freezer at $-20 \,^\circ C$ turns into ice. Both processes are spontaneous under different conditions even though they are the reverse of one another. In Chapter 19 we will address the spontaneity of processes more fully. We will see why a process can be spontaneous at one temperature but not at another, as is the case for the conversion of water to ice.

Despite these complicating factors, you should pay attention to the enthalpy changes in reactions. As a general observation, when the enthalpy change is large, it is the dominant factor in determining spontaneity. Thus, reactions for which ΔH is *large* (about 100 kJ or more) and *negative* tend to be spontaneous. Reactions for which ΔH is *large* and *positive* tend to be spontaneous only in the reverse direction.

Related Exercises: 5.47, 5.48

In many situations we will find it valuable to know the sign and magnitude of the enthalpy change associated with a given chemical process. As we see in the following sections, ΔH can be either determined directly by experiment or calculated from known enthalpy changes of other reactions.

5.5 | Calorimetry

The value of ΔH can be determined experimentally by measuring the heat flow accompanying a reaction at constant pressure. Typically, we can determine the magnitude of the heat flow by measuring the magnitude of the temperature change the heat flow produces. The measurement of heat flow is **calorimetry**; a device used to measure heat flow is a **calorimeter**.

Heat Capacity and Specific Heat

The more heat an object gains, the hotter it gets. All substances change temperature when they are heated, but the magnitude of the temperature change produced by a given quantity of heat varies from substance to substance. The temperature change experienced by an object when it absorbs a certain amount of heat is determined by its

GO FIGURE

Is the process shown in the figure endothermic or exothermic?

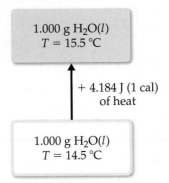

▲ Figure 5.16 **Specific heat of water.**

heat capacity, denoted C. The heat capacity of an object is the amount of heat required to raise its temperature by 1 K (or 1 °C). The greater the heat capacity, the greater the heat required to produce a given increase in temperature.

For pure substances the heat capacity is usually given for a specified amount of the substance. The heat capacity of one mole of a substance is called its **molar heat capacity**, C_m. The heat capacity of one gram of a substance is called its *specific heat capacity*, or merely its **specific heat**, C_s. The specific heat, C_s, of a substance can be determined experimentally by measuring the temperature change, ΔT, that a known mass m of the substance undergoes when it gains or loses a specific quantity of heat q:

$$\text{Specific heat} = \frac{(\text{quantity of heat transferred})}{(\text{grams of substance}) \times (\text{temperature change})}$$

$$C_s = \frac{q}{m \times \Delta T} \qquad [5.21]$$

For example, 209 J is required to increase the temperature of 50.0 g of water by 1.00 K. Thus, the specific heat of water is

$$C_s = \frac{209 \text{ J}}{(50.0 \text{ g})(1.00 \text{ K})} = 4.18 \text{ J/g-K}$$

Notice how the units combine in the calculation. A temperature change in kelvins is equal in magnitude to the temperature change in degrees Celsius: ΔT in K = ΔT in °C. ∞ (Section 1.4) Therefore, this specific heat for water can also be reported as 4.18 J/g-°C, where the unit is pronounced "Joules per gram-degree Celsius."

Because the specific heat values for a given substance can vary slightly with temperature, the temperature is often precisely specified. The 4.18 J/g-K value we use here for water, for instance, is for water initially at 14.5 °C (▲ Figure 5.16). Water's specific heat at this temperature is used to define the calorie at the value given in Section 5.1: 1 cal = 4.184 J exactly.

When a sample absorbs heat (positive q), its temperature increases (positive ΔT). Rearranging Equation 5.21, we get

$$q = C_s \times m \times \Delta T \qquad [5.22]$$

Thus, we can calculate the quantity of heat a substance gains or loses by using its specific heat together with its measured mass and temperature change.

▼ Table 5.2 lists the specific heats of several substances. Notice that the specific heat of liquid water is higher than those of the other substances listed. The high specific heat of water affects Earth's climate because it makes the temperatures of the oceans relatively resistant to change.

Give It Some Thought

Which substance in Table 5.2 undergoes the greatest temperature change when the same mass of each substance absorbs the same quantity of heat?

Table 5.2   Specific Heats of Some Substances at 298 K

Elements		Compounds	
Substance	Specific Heat (J/g-K)	Substance	Specific Heat (J/g-K)
$N_2(g)$	1.04	$H_2O(l)$	4.18
Al(s)	0.90	$CH_4(g)$	2.20
Fe(s)	0.45	$CO_2(g)$	0.84
Hg(l)	0.14	$CaCO_3(s)$	0.82

SAMPLE EXERCISE 5.6 Relating Heat, Temperature Change, and Heat Capacity

(a) How much heat is needed to warm 250 g of water (about 1 cup) from 22 °C (about room temperature) to 98 °C (near its boiling point)? (b) What is the molar heat capacity of water?

SOLUTION

Analyze In part (a) we must find the quantity of heat (q) needed to warm the water, given the mass of water (m), its temperature change (ΔT), and its specific heat (C_s). In part (b) we must calculate the molar heat capacity (heat capacity per mole, C_m) of water from its specific heat (heat capacity per gram).

Plan (a) Given C_s, m, and ΔT, we can calculate the quantity of heat, q, using Equation 5.22. (b) We can use the molar mass of water and dimensional analysis to convert from heat capacity per gram to heat capacity per mole.

Solve

(a) The water undergoes a temperature change of

$$\Delta T = 98\,°C - 22\,°C = 76\,°C = 76\,K$$

Using Equation 5.22, we have

$$q = C_s \times m \times \Delta T$$
$$= (4.18\,J/g\text{-}K)(250\,g)(76\,K) = 7.9 \times 10^4\,J$$

(b) The molar heat capacity is the heat capacity of one mole of substance. Using the atomic weights of hydrogen and oxygen, we have

$$1\,mol\,H_2O = 18.0\,g\,H_2O$$

From the specific heat given in part (a), we have

$$C_m = \left(4.18\,\frac{J}{g\text{-}K}\right)\left(\frac{18.0\,g}{1\,mol}\right) = 75.2\,J/mol\text{-}K$$

Practice Exercise 1

Suppose you have equal masses of two substances, A and B. When the same amount of heat is added to samples of each, the temperature of A increases by 14 °C whereas that of B increases by 22 °C. Which of the following statements is true? (a) The heat capacity of B is greater than that of A. (b) The specific heat of A is greater than that of B. (c) The molar heat capacity of B is greater than that of A. (d) The volume of A is greater than that of B. (e) The molar mass of A is greater than that of B.

Practice Exercise 2

(a) Large beds of rocks are used in some solar-heated homes to store heat. Assume that the specific heat of the rocks is 0.82 J/g-K. Calculate the quantity of heat absorbed by 50.0 kg of rocks if their temperature increases by 12.0 °C. (b) What temperature change would these rocks undergo if they emitted 450 kJ of heat?

Constant-Pressure Calorimetry

The techniques and equipment employed in calorimetry depend on the nature of the process being studied. For many reactions, such as those occurring in solution, it is easy to control pressure so that ΔH is measured directly. Although the calorimeters used for highly accurate work are precision instruments, a simple "coffee-cup" calorimeter (▶ **Figure 5.17**) is often used in general chemistry laboratories to illustrate the principles of calorimetry. Because the calorimeter is not sealed, the reaction occurs under the essentially constant pressure of the atmosphere.

Imagine adding two aqueous solutions, each containing a reactant, to a coffee-cup calorimeter. Once mixed, a reaction occurs. In this case there is no physical boundary between the system and the surroundings. The reactants and products of the reaction are the system, and the water in which they are dissolved is part of the surroundings. (The calorimeter apparatus is also part of the surroundings.) If we assume that the calorimeter is perfectly insulated, then any heat released or absorbed by the reaction will raise or lower the temperature of the water in the solution. Thus, we measure the temperature change of the solution and assume that any changes are due to heat transferred from the reaction to the water (for an exothermic process) or transferred from

▲ **Go Figure**

Propose a reason for why two Styrofoam® cups are often used instead of just one.

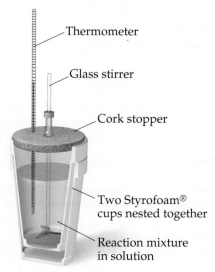

Thermometer

Glass stirrer

Cork stopper

Two Styrofoam® cups nested together

Reaction mixture in solution

▲ **Figure 5.17 Coffee-cup calorimeter.** This simple apparatus is used to measure temperature changes of reactions at constant pressure.

the water to the reaction (endothermic). In other words, by monitoring the temperature of the solution, we are seeing the flow of heat between the system (the reactants and products in the solution) and the surroundings (the water that forms the bulk of the solution).

For an exothermic reaction, heat is "lost" by the reaction and "gained" by the water in the solution, so the temperature of the solution rises. The opposite occurs for an endothermic reaction: Heat is gained by the reaction and lost by the water in the solution, and the temperature of the solution decreases. The heat gained or lost by the solution, q_{soln}, is therefore equal in magnitude but opposite in sign to the heat absorbed or released by the reaction, q_{rxn}: $q_{soln} = -q_{rxn}$. The value of q_{soln} is readily calculated from the mass of the solution, its specific heat, and the temperature change:

$$q_{soln} = (\text{specific heat of solution}) \times (\text{grams of solution}) \times \Delta T = -q_{rxn} \quad [5.23]$$

For dilute aqueous solutions we usually assume that the specific heat of the solution is the same as that of water, 4.18 J/g-K.

Equation 5.23 makes it possible to calculate q_{rxn} from the temperature change of the solution in which the reaction occurs. A temperature increase ($\Delta T > 0$) means the reaction is exothermic ($q_{rxn} < 0$).

SAMPLE EXERCISE 5.7 | Measuring ΔH Using a Coffee-Cup Calorimeter

When a student mixes 50 mL of 1.0 M HCl and 50 mL of 1.0 M NaOH in a coffee-cup calorimeter, the temperature of the resultant solution increases from 21.0 to 27.5 °C. Calculate the enthalpy change for the reaction in kJ/mol HCl, assuming that the calorimeter loses only a negligible quantity of heat, that the total volume of the solution is 100 mL, that its density is 1.0 g/mL, and that its specific heat is 4.18 J/g-K.

SOLUTION

Analyze Mixing solutions of HCl and NaOH results in an acid–base reaction:

$$HCl(aq) + NaOH(aq) \longrightarrow H_2O(l) + NaCl(aq)$$

We need to calculate the heat produced per mole of HCl, given the temperature increase of the solution, the number of moles of HCl and NaOH involved, and the density and specific heat of the solution.

Plan The total heat produced can be calculated using Equation 5.23. The number of moles of HCl consumed in the reaction must be calculated from the volume and molarity of this substance, and this amount is then used to determine the heat produced per mol HCl.

Solve

Because the total volume of the solution is 100 mL, its mass is	$(100 \text{ mL})(1.0 \text{ g/mL}) = 100 \text{ g}$
The temperature change is	$\Delta T = 27.5 \,°C - 21.0 \,°C = 6.5 \,°C = 6.5 \text{ K}$
Using Equation 5.23, we have	$q_{rxn} = -C_s \times m \times \Delta T$
	$= -(4.18 \text{ J/g} - \text{K})(100 \text{ g})(6.5 \text{ K}) = -2.7 \times 10^3 \text{ J} = -2.7 \text{ kJ}$
Because the process occurs at constant pressure,	$\Delta H = q_P = -2.7 \text{ kJ}$
To express the enthalpy change on a molar basis, we use the fact that the number of moles of HCl is given by the product of the volume (50 mL = 0.050 L) and concentration (1.0 M = 1.0 mol/L) of the HCl solution:	$(0.050 \text{ L})(1.0 \text{ mol/L}) = 0.050 \text{ mol}$
Thus, the enthalpy change per mole of HCl is	$\Delta H = -2.7 \text{ kJ}/0.050 \text{ mol} = -54 \text{ kJ/mol}$

Check ΔH is negative (exothermic), as evidenced by the observed increase in the temperature. The magnitude of the molar enthalpy change seems reasonable.

Practice Exercise 1

When 0.243 g of Mg metal is combined with enough HCl to make 100 mL of solution in a constant-pressure calorimeter, the following reaction occurs:

$$Mg(s) + 2 HCl(aq) \longrightarrow MgCl_2(aq) + H_2(g)$$

If the temperature of the solution increases from 23.0 to 34.1 °C as a result of this reaction, calculate ΔH in kJ/mol Mg. Assume that the solution has a specific heat of 4.18 J/g-°C.
(a) −19.1 kJ/mol **(b)** −111 kJ/mol **(c)** −191 kJ/mol **(d)** −464 kJ/mol **(e)** −961 kJ/mol

Bomb Calorimetry (Constant-Volume Calorimetry)

An important type of reaction studied using calorimetry is combustion, in which a compound reacts completely with excess oxygen. ∞ (Section 3.2) Combustion reactions are most accurately studied using a **bomb calorimeter** (▶ Figure 5.18). The substance to be studied is placed in a small cup within an insulated sealed vessel called a *bomb*. The bomb, which is designed to withstand high pressures, has an inlet valve for adding oxygen and electrical leads for initiating the reaction. After the sample has been placed in the bomb, the bomb is sealed and pressurized with oxygen. It is then placed in the calorimeter and covered with an accurately measured quantity of water. The combustion reaction is initiated by passing an electrical current through a fine wire in contact with the sample. When the wire becomes sufficiently hot, the sample ignites.

The heat released when combustion occurs is absorbed by the water and the various components of the calorimeter (which all together make up the surroundings), causing the water temperature to rise. The change in water temperature caused by the reaction is measured very precisely.

To calculate the heat of combustion from the measured temperature increase, we must know the total heat capacity of the calorimeter, C_{cal}. This quantity is determined by combusting a sample that releases a known quantity of heat and measuring the temperature change. For example, combustion of exactly 1 g of benzoic acid, C_6H_5COOH, in a bomb calorimeter produces 26.38 kJ of heat. Suppose 1.000 g of benzoic acid is combusted in a calorimeter, leading to a temperature increase of 4.857 °C. The heat capacity of the calorimeter is then $C_{cal} = 26.38 \text{ kJ}/4.857 \text{ °C} = 5.431 \text{ kJ/°C}$. Once we know C_{cal}, we can measure temperature changes produced by other reactions, and from these we can calculate the heat evolved in the reaction, q_{rxn}:

$$q_{rxn} = -C_{cal} \times \Delta T \qquad [5.24]$$

GO FIGURE

Why is a stirrer used in calorimeters?

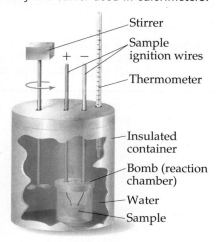

▲ **Figure 5.18 Bomb calorimeter.**

SAMPLE
EXERCISE 5.8 | Measuring q_{rxn} Using a Bomb Calorimeter

The combustion of methylhydrazine (CH_6N_2), a liquid rocket fuel, produces $N_2(g)$, $CO_2(g)$, and $H_2O(l)$:

$$2\,\text{CH}_6\text{N}_2(l) + 5\,\text{O}_2(g) \longrightarrow 2\,\text{N}_2(g) + 2\,\text{CO}_2(g) + 6\,\text{H}_2\text{O}(l)$$

When 4.00 g of methylhydrazine is combusted in a bomb calorimeter, the temperature of the calorimeter increases from 25.00 to 39.50 °C. In a separate experiment the heat capacity of the calorimeter is measured to be 7.794 kJ/°C. Calculate the heat of reaction for the combustion of a mole of CH_6N_2.

SOLUTION

Analyze We are given a temperature change and the total heat capacity of the calorimeter. We are also given the amount of reactant combusted. Our goal is to calculate the enthalpy change per mole for combustion of the reactant.

Plan We will first calculate the heat evolved for the combustion of the 4.00-g sample. We will then convert this heat to a molar quantity.

Solve

For combustion of the 4.00-g sample of methylhydrazine, the temperature change of the calorimeter is

$$\Delta T = (39.50\,°\text{C} - 25.00\,°\text{C}) = 14.50\,°\text{C}$$

We can use ΔT and the value for C_{cal} to calculate the heat of reaction (Equation 5.24):

$$q_{rxn} = -C_{cal} \times \Delta T = -(7.794\,\text{kJ}/°\text{C})(14.50\,°\text{C}) = -113.0\,\text{kJ}$$

We can readily convert this value to the heat of reaction for a mole of CH_6N_2:

$$\left(\frac{-113.0\,\text{kJ}}{4.00\,\text{g CH}_6\text{N}_2}\right) \times \left(\frac{46.1\,\text{g CH}_6\text{N}_2}{1\,\text{mol CH}_6\text{N}_2}\right) = -1.30 \times 10^3\,\text{kJ/mol CH}_6\text{N}_2$$

Check The units cancel properly, and the sign of the answer is negative as it should be for an exothermic reaction. The magnitude of the answer seems reasonable.

Practice Exercise 1

The combustion of exactly 1.000 g of benzoic acid in a bomb calorimeter releases 26.38 kJ of heat. If the combustion of 0.550 g of benzoic acid causes the temperature of the calorimeter to increase from 22.01 to 24.27 °C, calculate the heat capacity of the calorimeter. **(a)** 0.660 kJ/°C **(b)** 6.42 kJ/°C **(c)** 14.5 kJ/°C **(d)** 21.2 kJ/g-°C **(e)** 32.7 kJ/°C

Practice Exercise 2

A 0.5865-g sample of lactic acid $(HC_3H_5O_3)$ is burned in a calorimeter whose heat capacity is 4.812 kJ/°C. The temperature increases from 23.10 to 24.95 °C. Calculate the heat of combustion of lactic acid **(a)** per gram and **(b)** per mole.

Because reactions in a bomb calorimeter are carried out at constant volume, the heat transferred corresponds to the change in internal energy, ΔE, rather than the change in enthalpy, ΔH (Equation 5.14). For most reactions, however, the difference between ΔE and ΔH is very small. For the reaction discussed in Sample Exercise 5.8, for example, the difference between ΔE and ΔH is about 1 kJ/mol—a difference of less than 0.1%. It is possible to calculate ΔH from ΔE, but we need not concern ourselves with how these small corrections are made.

Chemistry and Life

The Regulation of Body Temperature

For most of us, being asked the question "Are you running a fever?" was one of our first introductions to medical diagnosis. Indeed, a deviation in body temperature of only a few degrees indicates something amiss. Maintaining a near-constant temperature is one of the primary physiological functions of the human body.

To understand how the body's heating and cooling mechanisms operate, we can view the body as a thermodynamic system. The body increases its internal energy content by ingesting foods from the surroundings. The foods, such as glucose $(C_6H_{12}O_6)$, are metabolized—a process that is essentially controlled oxidation to CO_2 and H_2O:

$$C_6H_{12}O_6(s) + 6\,O_2(g) \longrightarrow 6\,CO_2(g) + 6\,H_2O(l)$$

$$\Delta H = -2803\,\text{kJ}$$

Roughly 40% of the energy produced is ultimately used to do work in the form of muscle contractions and nerve cell activities. The remainder is released as heat, part of which is used to maintain body temperature. When the body produces too much heat, as in times of heavy physical exertion, it dissipates the excess to the surroundings.

Heat is transferred from the body to its surroundings primarily by *radiation*, *convection*, and *evaporation*. Radiation is the direct loss of heat from the body to cooler surroundings, much as a hot stovetop radiates heat to its surroundings. Convection is heat loss by virtue of heating air that is in contact with the body. The heated air rises and is replaced with cooler air, and the process continues.

Warm clothing decreases convective heat loss in cold weather. Evaporative cooling occurs when perspiration is generated at the skin surface by the sweat glands (▼ Figure 5.19). Heat is removed from the body as the perspiration evaporates. Perspiration is predominantly water, so the process is the endothermic conversion of liquid water into water vapor:

$$H_2O(l) \longrightarrow H_2O(g) \qquad \Delta H = +44.0\,\text{kJ}$$

The speed with which evaporative cooling occurs decreases as the atmospheric humidity increases, which is why we feel more sweaty and uncomfortable on hot, humid days.

When body temperature becomes too high, heat loss increases in two principal ways. First, blood flow near the skin surface increases, which allows for increased radiational and convective cooling. The reddish, "flushed" appearance of a hot individual is due to this increased blood flow. Second, we sweat, which increases evaporative

▲ Figure 5.19 **Perspiration.**

cooling. During extreme activity, the amount of perspiration can be as high as 2 to 4 liters per hour. As a result, the body's water supply must be replenished during these periods. If the body loses too much liquid through perspiration, it will no longer be able to cool itself and blood volume decreases, which can lead to either *heat exhaustion* or the more serious *heat stroke*. However, replenishing water without replenishing the electrolytes lost during perspiration can also lead to serious problems. If the normal blood sodium level drops too low, dizziness and confusion set in, and the condition can become critical.

Drinking a sport drink that contains some electrolytes helps to prevent this problem.

When body temperature drops too low, blood flow to the skin surface decreases, thereby decreasing heat loss. The lower temperature also triggers small involuntary contractions of the muscles (shivering); the biochemical reactions that generate the energy to do this work also generate heat for the body. If the body is unable to maintain a normal temperature, the very dangerous condition called *hypothermia* can result.

5.6 | Hess's Law

It is often possible to calculate the ΔH for a reaction from the tabulated ΔH values of other reactions. Thus, it is not necessary to make calorimetric measurements for all reactions.

Because enthalpy is a state function, the enthalpy change, ΔH, associated with any chemical process depends only on the amount of matter that undergoes change and on the nature of the initial state of the reactants and the final state of the products. This means that whether a particular reaction is carried out in one step or in a series of steps, the sum of the enthalpy changes associated with the individual steps must be the same as the enthalpy change associated with the one-step process. As an example, combustion of methane gas, $CH_4(g)$, to form $CO_2(g)$ and $H_2O(l)$ can be thought of as occurring in one step, as represented on the left in ▶ Figure 5.20, or in two steps, as represented on the right in Figure 5.20: (1) combustion of $CH_4(g)$ to form $CO_2(g)$ and $H_2O(g)$ and (2) condensation of $H_2O(g)$ to form $H_2O(l)$. The enthalpy change for the overall process is the sum of the enthalpy changes for these two steps:

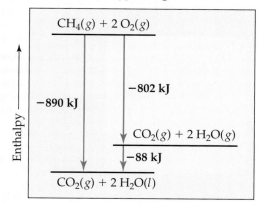

▲ GO FIGURE

What process corresponds to the −88 kJ enthalpy change?

▶ **Figure 5.20 Enthalpy diagram for combustion of 1 mol of methane.** The enthalpy change of the one-step reaction equals the sum of the enthalpy changes of the reaction run in two steps: −890 kJ = −802 kJ + (−88 kJ).

$$CH_4(g) + 2\,O_2(g) \longrightarrow CO_2(g) + 2\,H_2O(g) \qquad \Delta H = -802 \text{ kJ}$$
(Add) $\qquad\qquad 2\,H_2O(g) \longrightarrow 2\,H_2O(l) \qquad\qquad\qquad \Delta H = -88 \text{ kJ}$

$$CH_4(g) + 2\,O_2(g) + 2\,H_2O(g) \longrightarrow CO_2(g) + 2\,H_2O(l) + 2\,H_2O(g)$$

$$\Delta H = -890 \text{ kJ}$$

The net equation is

$$CH_4(g) + 2\,O_2(g) \longrightarrow CO_2(g) + 2\,H_2O(l) \quad \Delta H = -890 \text{ kJ}$$

Hess's law states that *if a reaction is carried out in a series of steps, ΔH for the overall reaction equals the sum of the enthalpy changes for the individual steps*. The overall enthalpy change for the process is independent of the number of steps and independent of the path by which the reaction is carried out. This law is a consequence of the fact that enthalpy is a state function. We can therefore calculate ΔH for any process as long as we find a route for which ΔH is known for each step. This means that a relatively small number of experimental measurements can be used to calculate ΔH for a vast number of reactions.

Hess's law provides a useful means of calculating energy changes that are difficult to measure directly. For instance, it is impossible to measure directly the enthalpy for the combustion of carbon to form carbon monoxide. Combustion of 1 mol of carbon with 0.5 mol of O_2 produces both CO and CO_2, leaving some carbon unreacted. However, solid carbon and carbon monoxide can both be completely burned in O_2 to produce CO_2. We can therefore use the enthalpy changes of these reactions to calculate the heat of combustion of carbon.

Give It Some Thought

What effect do these changes have on ΔH for a reaction:
(a) Reversing the reaction
(b) Multiplying the coefficients of the equation for the reaction by 2?

SAMPLE EXERCISE 5.9 Using Hess's Law to Calculate ΔH

The enthalpy of reaction for the combustion of C to CO_2 is -393.5 kJ/mol C, and the enthalpy for the combustion of CO to CO_2 is -283.0 kJ/mol CO:

(1) $\quad\quad C(s) + O_2(g) \longrightarrow CO_2(g) \quad\quad \Delta H = -393.5$ kJ

(2) $\quad\quad CO(g) + \frac{1}{2}O_2(g) \longrightarrow CO_2(g) \quad\quad \Delta H = -283.0$ kJ

Using these data, calculate the enthalpy for the combustion of C to CO:

(3) $\quad\quad C(s) + \frac{1}{2}O_2(g) \longrightarrow CO(g) \quad\quad \Delta H = ?$

SOLUTION

Analyze We are given two thermochemical equations, and our goal is to combine them in such a way as to obtain the third equation and its enthalpy change.

Plan We will use Hess's law. In doing so, we first note the numbers of moles of substances among the reactants and products in the target equation (3). We then manipulate equations (1) and (2) to give the same number of moles of these substances, so that when the resulting equations are added, we obtain the target equation. At the same time, we keep track of the enthalpy changes, which we add.

Solve To use equations (1) and (2), we arrange them so that C(s) is on the reactant side and CO(g) is on the product side of the arrow, as in the target reaction, equation (3). Because equation (1) has C(s) as a reactant, we can use that equation just as it is. We need to turn equation (2) around, however, so that CO(g) is a product. Remember that when reactions are turned around, the sign of ΔH is reversed. We arrange the two equations so that they can be added to give the desired equation:

$$C(s) + O_2(g) \longrightarrow CO_2(g) \quad\quad \Delta H = -393.5 \text{ kJ}$$
$$\underline{CO_2(g) \longrightarrow CO(g) + \tfrac{1}{2}O_2(g) \quad\quad -\Delta H = 283.0 \text{ kJ}}$$
$$C(s) + \tfrac{1}{2}O_2(g) \longrightarrow CO(g) \quad\quad \Delta H = -110.5 \text{ kJ}$$

When we add the two equations, $CO_2(g)$ appears on both sides of the arrow and therefore cancels out. Likewise, $\frac{1}{2}O_2(g)$ is eliminated from each side.

Practice Exercise 1

Calculate ΔH for $2NO(g) + O_2(g) \longrightarrow N_2O_4(g)$, using the following information:

$$N_2O_4(g) \longrightarrow 2NO_2(g) \quad \Delta H = +57.9 \text{ kJ}$$
$$2NO(g) + O_2(g) \longrightarrow 2NO_2(g) \quad \Delta H = -113.1 \text{ kJ}$$

(a) 2.7 kJ (b) −55.2 kJ (c) −85.5 kJ (d) −171.0 kJ (e) +55.2 kJ

Practice Exercise 2

Carbon occurs in two forms, graphite and diamond. The enthalpy of the combustion of graphite is −393.5 kJ/mol, and that of diamond is −395.4 kJ/mol:

$$C(graphite) + O_2(g) \longrightarrow CO_2(g) \quad \Delta H = -393.5 \text{ kJ}$$
$$C(diamond) + O_2(g) \longrightarrow CO_2(g) \quad \Delta H = -395.4 \text{ kJ}$$

Calculate ΔH for the conversion of graphite to diamond:

$$C(graphite) \longrightarrow C(diamond) \quad \Delta H = ?$$

SAMPLE EXERCISE 5.10 Using Three Equations with Hess's Law to Calculate ΔH

Calculate ΔH for the reaction

$$2\,C(s) + H_2(g) \longrightarrow C_2H_2(g)$$

given the following chemical equations and their respective enthalpy changes:

$$C_2H_2(g) + \tfrac{5}{2}O_2(g) \longrightarrow 2\,CO_2(g) + H_2O(l) \quad \Delta H = -1299.6 \text{ kJ}$$
$$C(s) + O_2(g) \longrightarrow CO_2(g) \quad \Delta H = -393.5 \text{ kJ}$$
$$H_2(g) + \tfrac{1}{2}O_2(g) \longrightarrow H_2O(l) \quad \Delta H = -285.8 \text{ kJ}$$

SOLUTION

Analyze We are given a chemical equation and asked to calculate its ΔH using three chemical equations and their associated enthalpy changes.

Plan We will use Hess's law, summing the three equations or their reverses and multiplying each by an appropriate coefficient so that they add to give the net equation for the reaction of interest. At the same time, we keep track of the ΔH values, reversing their signs if the reactions are reversed and multiplying them by whatever coefficient is employed in the equation.

Solve Because the target equation has C_2H_2 as a product, we turn the first equation around; the sign of ΔH is therefore changed. The desired equation has $2\ C(s)$ as a reactant, so we multiply the second equation and its ΔH by 2. Because the target equation has H_2 as a reactant, we keep the third equation as it is. We then add the three equations and their enthalpy changes in accordance with Hess's law:

$$2\ \cancel{CO_2(g)} + H_2O(l) \longrightarrow C_2H_2(g) + \cancel{\tfrac{5}{2}O_2(g)} \qquad \Delta H = 1299.6\ \text{kJ}$$
$$2\ C(s) + 2\ \cancel{O_2(g)} \longrightarrow 2\ \cancel{CO_2(g)} \qquad \Delta H = -787.0\ \text{kJ}$$
$$\underline{H_2(g) + \tfrac{1}{2}\cancel{O_2(g)} \longrightarrow H_2O(l) \qquad \Delta H = -285.8\ \text{kJ}}$$
$$2\ C(s) + H_2(g) \longrightarrow C_2H_2(g) \qquad \Delta H = 226.8\ \text{kJ}$$

When the equations are added, there are $2\ CO_2$, $\tfrac{5}{2}\ O_2$, and H_2O on both sides of the arrow. These are canceled in writing the net equation.

Check The procedure must be correct because we obtained the correct net equation. In cases like this you should go back over the numerical manipulations of the ΔH values to ensure that you did not make an inadvertent error with signs.

Practice Exercise 1

We can calculate ΔH for the reaction

$$C(s) + H_2O(g) \longrightarrow CO(g) + H_2(g)$$

using the following thermochemical equations:

$$C(s) + O_2(g) \longrightarrow CO_2(g) \qquad \Delta H_1 = -393.5\ \text{kJ}$$
$$2\ CO(g) + O_2(g) \longrightarrow 2\ CO_2(g) \qquad \Delta H_2 = -566.0\ \text{kJ}$$
$$2\ H_2(g) + O_2(g) \longrightarrow 2\ H_2O(g) \qquad \Delta H_3 = -483.6\ \text{kJ}$$

By what coefficient do you need to multiply ΔH_2 in determining ΔH for the target equation?
(a) $-1/2$ **(b)** -1 **(c)** -2 **(d)** $1/2$ **(e)** 2

Practice Exercise 2

Calculate ΔH for the reaction

$$NO(g) + O(g) \longrightarrow NO_2(g)$$

given the following information:

$$NO(g) + O_3(g) \longrightarrow NO_2(g) + O_2(g) \qquad \Delta H = -198.9\ \text{kJ}$$
$$O_3(g) \longrightarrow \tfrac{3}{2}O_2(g) \qquad \Delta H = -142.3\ \text{kJ}$$
$$O_2(g) \longrightarrow 2\ O(g) \qquad \Delta H = 495.0\ \text{kJ}$$

The key point of these examples is that H is a state function, so *for a particular set of reactants and products, ΔH is the same whether the reaction takes place in one step or in a series of steps.* We reinforce this point by giving one more example of an enthalpy diagram and Hess's law. Again we use combustion of methane to form CO_2 and H_2O, our reaction from Figure 5.20. This time we envision a different two-step path, with the initial formation of CO, which is then combusted to CO_2 (▶ Figure 5.21). Even though the two-step path is different from that in Figure 5.20, the overall reaction again has $\Delta H_1 = -890$ kJ. Because H is a state function, both paths *must* produce the same value of ΔH. In Figure 5.21, that means $\Delta H_1 = \Delta H_2 + \Delta H_3$. We will soon see that breaking up reactions in this way allows us to derive the enthalpy changes for reactions that are hard to carry out in the laboratory.

5.7 | Enthalpies of Formation

We can use the methods just discussed to calculate enthalpy changes for a great many reactions from tabulated ΔH values. For example, extensive tables exist of *enthalpies of vaporization* (ΔH for converting liquids to gases), *enthalpies of fusion* (ΔH for melting solids),

 GO FIGURE

Suppose the overall reaction were modified to produce 2 $H_2O(g)$ rather than 2 $H_2O(l)$. Would any of the values of ΔH in the diagram stay the same?

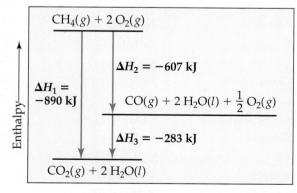

▲ **Figure 5.21 Enthalpy diagram illustrating Hess's law.** The net reaction is the same as in Figure 5.20, but here we imagine different reactions in our two-step version. As long as we can write a series of equations that add up to the equation we need, and as long as we know a value for ΔH for all intermediate reactions, we can calculate the overall ΔH.

enthalpies of combustion (ΔH for combusting a substance in oxygen), and so forth. A particularly important process used for tabulating thermochemical data is the formation of a compound from its constituent elements. The enthalpy change associated with this process is called the **enthalpy of formation** (or *heat of formation*), ΔH_f, where the subscript f indicates that the substance has been *formed* from its constituent elements.

The magnitude of any enthalpy change depends on the temperature, pressure, and state (gas, liquid, or solid crystalline form) of the reactants and products. To compare enthalpies of different reactions, we must define a set of conditions, called a **standard state**, at which most enthalpies are tabulated. The standard state of a substance is its pure form at atmospheric pressure (1 atm) and the temperature of interest, which we usually choose to be 298 K (25 °C).* The **standard enthalpy change** of a reaction is defined as the enthalpy change when all reactants and products are in their standard states. We denote a standard enthalpy change as $\Delta H°$, where the superscript ° indicates standard-state conditions.

The **standard enthalpy of formation** of a compound, $\Delta H_f°$, is the change in enthalpy for the reaction that forms one mole of the compound from its elements with all substances in their standard states:

If: elements (in standard state) $\longrightarrow$ compound (1 mol in standard state)
Then: $\Delta H = \Delta H_f°$

We usually report $\Delta H_f°$ values at 298 K. If an element exists in more than one form under standard conditions, the most stable form of the element is usually used for the formation reaction. For example, the standard enthalpy of formation for ethanol, C_2H_5OH, is the enthalpy change for the reaction

$$2\,C(graphite) + 3\,H_2(g) + \tfrac{1}{2}O_2(g) \longrightarrow C_2H_5OH(l) \qquad \Delta H_f° = -277.7\ \text{kJ} \quad [5.25]$$

The elemental source of oxygen is O_2, not O or O_3, because O_2 is the stable form of oxygen at 298 K and atmospheric pressure. Similarly, the elemental source of carbon is graphite and not diamond because graphite is the more stable (lower-energy) form at 298 K and atmospheric pressure. Likewise, the most stable form of hydrogen under standard conditions is $H_2(g)$, so this is used as the source of hydrogen in Equation 5.25.

The stoichiometry of formation reactions always indicates that one mole of the desired substance is produced, as in Equation 5.25. As a result, standard enthalpies of formation are reported in kJ/mol of the substance being formed. Some values are given in ▼ Table 5.3, and a more extensive table is provided in Appendix C.

Table 5.3 Standard Enthalpies of Formation, $\Delta H_f°$, at 298 K

Substance	Formula	$\Delta H_f°$ (kJ/mol)	Substance	Formula	$\Delta H_f°$ (kJ/mol)
Acetylene	$C_2H_2(g)$	226.7	Hydrogen chloride	$HCl(g)$	−92.30
Ammonia	$NH_3(g)$	−46.19	Hydrogen fluoride	$HF(g)$	−268.60
Benzene	$C_6H_6(l)$	49.0	Hydrogen iodide	$HI(g)$	25.9
Calcium carbonate	$CaCO_3(s)$	−1207.1	Methane	$CH_4(g)$	−74.80
Calcium oxide	$CaO(s)$	−635.5	Methanol	$CH_3OH(l)$	−238.6
Carbon dioxide	$CO_2(g)$	−393.5	Propane	$C_3H_8(g)$	−103.85
Carbon monoxide	$CO(g)$	−110.5	Silver chloride	$AgCl(s)$	−127.0
Diamond	$C(s)$	1.88	Sodium bicarbonate	$NaHCO_3(s)$	−947.7
Ethane	$C_2H_6(g)$	−84.68	Sodium carbonate	$Na_2CO_3(s)$	−1130.9
Ethanol	$C_2H_5OH(l)$	−277.7	Sodium chloride	$NaCl(s)$	−410.9
Ethylene	$C_2H_4(g)$	52.30	Sucrose	$C_{12}H_{22}O_{11}(s)$	−2221
Glucose	$C_6H_{12}O_6(s)$	−1273	Water	$H_2O(l)$	−285.8
Hydrogen bromide	$HBr(g)$	−36.23	Water vapor	$H_2O(g)$	−241.8

---

*The definition of the standard state for gases has been changed to 1 bar (1 atm = 1.013 bar), a slightly lower pressure than 1 atm. For most purposes, this change makes very little difference in the standard enthalpy changes.

By definition, *the standard enthalpy of formation of the most stable form of any element is zero* because there is no formation reaction needed when the element is already in its standard state. Thus, the values of ΔH_f° for C(graphite), $H_2(g)$, $O_2(g)$, and the standard states of other elements are zero by definition.

 Give It Some Thought

Ozone, $O_3(g)$, is a form of elemental oxygen produced during electrical discharge. Is ΔH_f° for $O_3(g)$ necessarily zero?

SAMPLE EXERCISE 5.11 Equations Associated with Enthalpies of Formation

For which of these reactions at 25 °C does the enthalpy change represent a standard enthalpy of formation? For each that does not, what changes are needed to make it an equation whose ΔH is an enthalpy of formation?

(a) $2\,\text{Na}(s) + \frac{1}{2}O_2(g) \longrightarrow \text{Na}_2\text{O}(s)$

(b) $2\,\text{K}(l) + \text{Cl}_2(g) \longrightarrow 2\,\text{KCl}(s)$

(c) $\text{C}_6\text{H}_{12}\text{O}_6(s) \longrightarrow 6\,\text{C}(diamond) + 6\,\text{H}_2(g) + 3\,\text{O}_2(g)$

SOLUTION

Analyze The standard enthalpy of formation is represented by a reaction in which each reactant is an element in its standard state and the product is one mole of the compound.

Plan We need to examine each equation to determine (1) whether the reaction is one in which one mole of substance is formed from the elements, and (2) whether the reactant elements are in their standard states.

Solve In (a) 1 mol Na_2O is formed from the elements sodium and oxygen in their proper states, solid Na and O_2 gas, respectively. Therefore, the enthalpy change for reaction (a) corresponds to a standard enthalpy of formation.

In (b) potassium is given as a liquid. It must be changed to the solid form, its standard state at room temperature. Furthermore, 2 mol KCl are formed, so the enthalpy change for the reaction as written is twice the standard enthalpy of formation of KCl(s). The equation for the formation reaction of 1 mol of KCl(s) is

$$\text{K}(s) + \tfrac{1}{2}\text{Cl}_2(g) \longrightarrow \text{KCl}(s)$$

Reaction (c) does not form a substance from its elements. Instead, a substance decomposes to its elements, so this reaction must be reversed. Next, the element carbon is given as diamond, whereas graphite is the standard state of carbon at room temperature and 1 atm pressure. The equation that correctly represents the enthalpy of formation of glucose from its elements is

$$6\,\text{C}(graphite) + 6\,\text{H}_2(g) + 3\,\text{O}_2(g) \longrightarrow \text{C}_6\text{H}_{12}\text{O}_6(s)$$

Practice Exercise 1

If the heat of formation of $H_2O(l)$ is -286 kJ/mol, which of the following thermochemical equations is correct?

(a) $2\text{H}(g) + \text{O}(g) \longrightarrow \text{H}_2\text{O}(l)$ $\Delta H = -286$ kJ

(b) $2\text{H}_2(g) + \text{O}_2(g) \longrightarrow 2\text{H}_2\text{O}(l)$ $\Delta H = -286$ kJ

(c) $\text{H}_2(g) + \frac{1}{2}\text{O}_2(g) \longrightarrow \text{H}_2\text{O}(l)$ $\Delta H = -286$ kJ

(d) $\text{H}_2(g) + \text{O}(g) \longrightarrow \text{H}_2\text{O}(g)$ $\Delta H = -286$ kJ

(e) $\text{H}_2\text{O}(l) \longrightarrow \text{H}_2(g) + \frac{1}{2}\text{O}_2(g)$ $\Delta H = -286$ kJ

Practice Exercise 2

Write the equation corresponding to the standard enthalpy of formation of liquid carbon tetrachloride (CCl_4) and look up ΔH_f° for this compound in Appendix C.

Using Enthalpies of Formation to Calculate Enthalpies of Reaction

We can use Hess's law and tabulations of ΔH_f° values, such as those in Table 5.3 and Appendix C, to calculate the standard enthalpy change for any reaction for which we know the ΔH_f° values for all reactants and products. For example, consider the combustion of propane under standard conditions:

$$C_3H_8(g) + 5\,O_2(g) \longrightarrow 3\,CO_2(g) + 4\,H_2O(l)$$

We can write this equation as the sum of three equations associated with standard enthalpies of formation:

$$C_3H_8(g) \longrightarrow 3\,C(s) + 4\,H_2(g) \qquad \Delta H_1 = -\Delta H_f^\circ[C_3H_8(g)] \quad [5.26]$$

$$3\,C(s) + 3\,O_2(g) \longrightarrow 3\,CO_2(g) \qquad \Delta H_2 = 3\Delta H_f^\circ[CO_2(g)] \quad [5.27]$$

$$4\,H_2(g) + 2\,O_2(g) \longrightarrow 4\,H_2O(l) \qquad \Delta H_3 = 4\Delta H_f^\circ[H_2O(l)] \quad [5.28]$$

$$\overline{C_3H_8(g) + 5\,O_2(g) \longrightarrow 3\,CO_2(g) + 4\,H_2O(l)\quad \Delta H_{rxn}^\circ = \Delta H_1 + \Delta H_2 + \Delta H_3 \quad [5.29]}$$

(Note that it is sometimes useful to add subscripts to the enthalpy changes, as we have done here, to keep track of the associations between reactions and their ΔH values.)

Notice that we have used Hess's law to write the standard enthalpy change for Equation 5.29 as the sum of the enthalpy changes for Equations 5.26 through 5.28. We can use values from Table 5.3 to calculate ΔH_{rxn}°:

$$\begin{aligned} \Delta H_{rxn}^\circ &= \Delta H_1 + \Delta H_2 + \Delta H_3 \\ &= -\Delta H_f^\circ[C_3H_8(g)] + 3\Delta H_f^\circ[CO_2(g)] + \Delta 4H_f^\circ[H_2O(l)] \\ &= -(-103.85\text{ kJ}) + 3(-393.5\text{ kJ}) + 4(-285.8\text{ kJ}) = -2220\text{ kJ} \qquad [5.30] \end{aligned}$$

The enthalpy diagram in ▼ **Figure 5.22** shows the components of this calculation. In Step **1** the reactants are decomposed into their constituent elements in their standard states. In Steps **2** and **3** the products are formed from the elements. Several

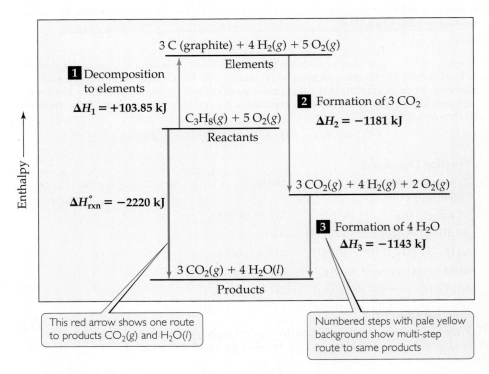

▲ Figure 5.22 **Enthalpy diagram for propane combustion.**

aspects of how we use enthalpy changes in this process depend on the guidelines we discussed in Section 5.4.

1 Decomposition. Equation 5.26 is the reverse of the formation reaction for $C_3H_8(g)$, so the enthalpy change for this decomposition reaction is the negative of the ΔH_f° value for the propane formation reaction: $-\Delta H_f^\circ[C_3H_8(g)]$.

2 Formation of CO₂. Equation 5.27 is the formation reaction for 3 mol of $CO_2(g)$. Because enthalpy is an extensive property, the enthalpy change for this step is $3\Delta H_f^\circ[CO_2(g)]$.

3 Formation of H₂O. The enthalpy change for Equation 5.28, formation of 4 mol of H_2O, is $4\Delta H_f^\circ[H_2O(l)]$. The reaction specifies that $H_2O(l)$ is produced, so be careful to use the value of ΔH_f° for $H_2O(l)$ and not the value for $H_2O(g)$.

Note that in this analysis we assume that the stoichiometric coefficients in the balanced equation represent the number of moles of each substance. For Equation 5.29, therefore, $\Delta H_{rxn}^\circ = -2220$ kJ represents the enthalpy change for the reaction of 1 mol C_3H_8 and 5 mol O_2 to form 3 mol CO_2 and 4 mol H_2O. The product of the number of moles and the enthalpy change in kJ/mol has the units kJ: (number of moles) $\times$ (ΔH_f° in kJ/mol) = kJ. We therefore report ΔH_{rxn}° in kJ.

We can break down any reaction into formation reactions as we have done here. When we do, we obtain the general result that the standard enthalpy change of a reaction is the sum of the standard enthalpies of formation of the products minus the standard enthalpies of formation of the reactants:

$$\Delta H_{rxn}^\circ = \Sigma n\Delta H_f^\circ(\text{products}) - \Sigma m\Delta H_f^\circ(\text{reactants}) \qquad [5.31]$$

The symbol Σ (sigma) means "the sum of," and n and m are the stoichiometric coefficients of the relevant chemical equation. The first term on the right in Equation 5.31 represents the formation reactions of the products, which are written in the "forward" direction in the chemical equation, that is, elements reacting to form products. This term is analogous to Equations 5.27 and 5.28. The second term on the right in Equation 5.31 represents the reverse of the formation reactions of the reactants, analogous to Equation 5.26, which is why this term is preceded by a minus sign.

SAMPLE EXERCISE 5.12 Calculating an Enthalpy of Reaction from Enthalpies of Formation

(a) Calculate the standard enthalpy change for the combustion of 1 mol of benzene, $C_6H_6(l)$, to $CO_2(g)$ and $H_2O(l)$. **(b)** Compare the quantity of heat produced by combustion of 1.00 g propane with that produced by 1.00 g benzene.

SOLUTION

Analyze (a) We are given a reaction [combustion of $C_6H_6(l)$ to form $CO_2(g)$ and $H_2O(l)$] and asked to calculate its standard enthalpy change, ΔH°. **(b)** We then need to compare the quantity of heat produced by combustion of 1.00 g C_6H_6 with that produced by 1.00 g C_3H_8, whose combustion was treated previously in the text. (See Equations 5.29 and 5.30.)

Plan (a) We first write the balanced equation for the combustion of C_6H_6. We then look up ΔH_f° values in Appendix C or in Table 5.3 and apply Equation 5.31 to calculate the enthalpy change for the reaction. **(b)** We use the molar mass of C_6H_6 to change the enthalpy change per mole to that per gram. We similarly use the molar mass of C_3H_8 and the enthalpy change per mole calculated in the text previously to calculate the enthalpy change per gram of that substance.

Solve

(a) We know that a combustion reaction involves $O_2(g)$ as a reactant. Thus, the balanced equation for the combustion reaction of 1 mol $C_6H_6(l)$ is

$$C_6H_6(l) + \tfrac{15}{2}O_2(g) \longrightarrow 6CO_2(g) + 3H_2O(l)$$

We can calculate ΔH° for this reaction by using Equation 5.31 and data in Table 5.3. Remember to multiply the ΔH_f° value for each substance in the reaction by that substance's stoichiometric coefficient. Recall also that $\Delta H_f^\circ = 0$ for any element in its most stable form under standard conditions, so $\Delta H_f^\circ[O_2(g)] = 0$.

$$\Delta H_{rxn}^\circ = [6\Delta H_f^\circ(CO_2) + 3\Delta H_f^\circ(H_2O)] - [\Delta H_f^\circ(C_6H_6) + \tfrac{15}{2}\Delta H_f^\circ(O_2)]$$
$$= [6(-393.5 \text{ kJ}) + 3(-285.8 \text{ kJ})] - [(49.0 \text{ kJ}) + \tfrac{15}{2}(0 \text{ kJ})]$$
$$= (-2361 - 857.4 - 49.0) \text{ kJ}$$
$$= -3267 \text{ kJ}$$

(b) From the example worked in the text, $\Delta H° = -2220$ kJ for the combustion of 1 mol of propane. In part (a) of this exercise we determined that $\Delta H° = -3267$ kJ for the combustion of 1 mol benzene. To determine the heat of combustion per gram of each substance, we use the molar masses to convert moles to grams:

$C_3H_8(g)$: $(-2220 \text{ kJ/mol})(1 \text{ mol}/44.1 \text{ g}) = -50.3$ kJ/g
$C_6H_6(l)$: $(-3267 \text{ kJ/mol})(1 \text{ mol}/78.1 \text{ g}) = -41.8$ kJ/g

Comment Both propane and benzene are hydrocarbons. As a rule, the energy obtained from the combustion of a gram of hydrocarbon is between 40 and 50 kJ.

Practice Exercise 1

Calculate the enthalpy change for the reaction

$2H_2O_2(l) \longrightarrow 2H_2O(l) + O_2(g)$

using enthalpies of formation:

$\Delta H_f°(H_2O_2) = -187.8$ kJ/mol $\Delta H_f°(H_2O) = -285.8$ kJ/mol

(a) −88.0 kJ, **(b)** −196.0 kJ, **(c)** +88.0 kJ, **(d)** +196.0 kJ, **(e)** more information needed

Practice Exercise 2

Use Table 5.3 to calculate the enthalpy change for the combustion of 1 mol of ethanol:

$C_2H_5OH(l) + 3 O_2(g) \longrightarrow 2 CO_2(g) + 3 H_2O(l)$

SAMPLE EXERCISE 5.13 Calculating an Enthalpy of Formation Using an Enthalpy of Reaction

The standard enthalpy change for the reaction $CaCO_3(s) \longrightarrow CaO(s) + CO_2(g)$ is 178.1 kJ. Use Table 5.3 to calculate the standard enthalpy of formation of $CaCO_3(s)$.

SOLUTION

Analyze Our goal is to obtain $\Delta H_f°(CaCO_3)$.

Plan We begin by writing the expression for the standard enthalpy change for the reaction:

$\Delta H_{rxn}° = \Delta H_f°CaO) + \Delta H_f°(CO_2) - \Delta H_f°1CaCO_3)$

Solve Inserting the given $\Delta H_{rxn}°$ and the known $\Delta H_f°$ values from Table 5.3 or Appendix C, we have

$178.1 = -635.5 \text{ kJ} - 393.5 \text{ kJ} - \Delta H_f°(CaCO_3)$

Solving for $\Delta H_f°(CaCO_3)$ gives

$\Delta H_f°(CaCO_3) = -1207.1$ kJ/mol

Check We expect the enthalpy of formation of a stable solid such as calcium carbonate to be negative, as obtained.

Practice Exercise 1

Given $2 SO_2(g) + O_2(g) \longrightarrow 2 SO_3(g)$, which of the following equations is correct?
(a) $\Delta H_f°(SO_3) = \Delta H_{rxn}° - \Delta H_f°(SO_2)$
(b) $\Delta H_f°(SO_3) = \Delta H_{rxn}° + \Delta H_f°(SO_2)$
(c) $2\Delta H_f°(SO_3) = \Delta H_{rxn}° + 2\Delta H_f°(SO_2)$
(d) $2\Delta H_f°(SO_3) = \Delta H_{rxn}° - 2\Delta H_f°(SO_2)$
(e) $2\Delta H_f°(SO_3) = 2\Delta H_f°(SO_2) - \Delta H_{rxn}°$

Practice Exercise 2

Given the following standard enthalpy change, use the standard enthalpies of formation in Table 5.3 to calculate the standard enthalpy of formation of $CuO(s)$:

$CuO(s) + H_2(g) \longrightarrow Cu(s) + H_2O(l)$ $\Delta H° = -129.7$ kJ

5.8 | Foods and Fuels

Most chemical reactions used for the production of heat are combustion reactions. The energy released when one gram of any substance is combusted is the **fuel value** of the substance. The fuel value of any food or fuel can be measured by calorimetry.

Foods

Most of the energy our bodies need comes from carbohydrates and fats. The carbohydrates known as starches are decomposed in the intestines into glucose, $C_6H_{12}O_6$.

Glucose is soluble in blood, and in the human body it is known as blood sugar. It is transported by the blood to cells where it reacts with O_2 in a series of steps, eventually producing $CO_2(g)$, $H_2O(l)$, and energy:

$$C_6H_{12}O_6(s) + 6\,O_2(g) \longrightarrow 6\,CO_2(g) + 6\,H_2O(l) \qquad \Delta H° = -2803 \text{ kJ}$$

Because carbohydrates break down rapidly, their energy is quickly supplied to the body. However, the body stores only a very small amount of carbohydrates. The average fuel value of carbohydrates is 17 kJ/g (4 kcal/g).*

Like carbohydrates, fats produce CO_2 and H_2O when metabolized. The reaction of tristearin, $C_{57}H_{110}O_6$, a typical fat, is

$$2\,C_{57}H_{110}O_6(s) + 163\,O_2(g) \longrightarrow 114\,CO_2(g) + 110\,H_2O(l) \qquad \Delta H° = 275,520 \text{ kJ}$$

The body uses the chemical energy from foods to maintain body temperature (see the "Chemistry and Life" box in Section 5.5), to contract muscles, and to construct and repair tissues. Any excess energy is stored as fats. Fats are well suited to serve as the body's energy reserve for at least two reasons: (1) They are insoluble in water, which facilitates storage in the body, and (2) they produce more energy per gram than either proteins or carbohydrates, which makes them efficient energy sources on a mass basis. The average fuel value of fats is 38 kJ/g (9 kcal/g).

The combustion of carbohydrates and fats in a bomb calorimeter gives the same products as when they are metabolized in the body. The metabolism of proteins produces less energy than combustion in a calorimeter because the products are different. Proteins contain nitrogen, which is released in the bomb calorimeter as N_2. In the body this nitrogen ends up mainly as urea, $(NH_2)_2CO$. Proteins are used by the body mainly as building materials for organ walls, skin, hair, muscle, and so forth. On average, the metabolism of proteins produces 17 kJ/g (4 kcal/g), the same as for carbohydrates.

Fuel values for some common foods are shown in ▼ Table 5.4. Labels on packaged foods show the amounts of carbohydrate, fat, and protein contained in an average serving, as well as the amount of energy supplied by a serving (▶ Figure 5.23).

The amount of energy our bodies require varies considerably, depending on such factors as weight, age, and muscular activity. About 100 kJ per kilogram of body mass per day is required to keep the body functioning at a minimal level. An average 70-kg

GO FIGURE

Which value would change most if this label were for skim milk instead of whole milk: grams of fat, grams of total carbohydrate, or grams of protein?

▲ Figure 5.23 **Nutrition label for whole milk.**

Table 5.4 Compositions and Fuel Values of Some Common Foods

	Approximate Composition (% by Mass)			Fuel Value	
	Carbohydrate	Fat	Protein	kJ/g	kcal/g (Cal/g)
Carbohydrate	100	—	—	17	4
Fat	—	100	—	38	9
Protein	—	—	100	17	4
Apples	13	0.5	0.4	2.5	0.59
Beer[a]	1.2	—	0.3	1.8	0.42
Bread	52	3	9	12	2.8
Cheese	4	37	28	20	4.7
Eggs	0.7	10	13	6.0	1.4
Fudge	81	11	2	18	4.4
Green beans	7.0	—	1.9	1.5	0.38
Hamburger	—	30	22	15	3.6
Milk (whole)	5.0	4.0	3.3	3.0	0.74
Peanuts	22	39	26	23	5.5

[a]Beer typically contains 3.5% ethanol, which has fuel value.

---

*Although fuel values represent the heat *released* in a combustion reaction, fuel values are reported as positive numbers.

Other Energy Sources

Nuclear energy is the energy released in either the fission (splitting) or the fusion (combining) of atomic nuclei. Nuclear power based on nuclear fission is currently used to produce about 21% of the electric power in the United States and makes up about 8.5% of the total U.S. energy production (Figure 5.24). Nuclear energy is, in principle, free of the polluting emissions that are a major problem with fossil fuels. However, nuclear power plants produce radioactive waste products, and their use has therefore been controversial. We will discuss issues related to the production of nuclear energy in Chapter 21.

Fossil fuels and nuclear energy are *nonrenewable* sources of energy—they are limited resources that we are consuming at a much greater rate than they can be regenerated. Eventually these fuels will be expended, although estimates vary greatly as to when this will occur. Because nonrenewable energy sources will eventually be used up, a great deal of research is being conducted on **renewable energy sources**, sources that are essentially inexhaustible. Renewable energy sources include *solar energy* from the Sun, *wind energy* harnessed by windmills, *geothermal energy* from the heat stored inside Earth, *hydroelectric energy* from flowing rivers, and *biomass energy* from crops and biological waste matter. Currently, renewable sources provide about 7.4% of the U.S. annual energy consumption, with hydroelectric and biomass sources the major contributors.

Fulfilling our future energy needs will depend on developing technology to harness solar energy with greater efficiency. Solar energy is the world's largest energy source. On a clear day about 1 kJ of solar energy reaches each square meter of Earth's surface every second. The average solar energy falling on only 0.1% of U.S. land area is equivalent to all the energy this nation currently uses. Harnessing this energy is difficult because it is dilute (that is, distributed over a wide area) and varies with time of day and weather conditions. The effective use of solar energy will depend on the development of some means of storing and distributing it. Any practical means for doing this will almost certainly involve an endothermic chemical process that can be later reversed to release heat. One such reaction is

$$CH_4(g) + H_2O(g) + heat \longrightarrow CO(g) + 3 H_2(g)$$

This reaction proceeds in the forward direction at high temperatures, which can be obtained in a solar furnace. The CO and H_2 formed in the reaction could then be stored and allowed to react later, with the heat released being put to useful work.

Chemistry Put to Work

The Scientific and Political Challenges of Biofuels*

One of the biggest challenges facing us in the twenty-first century is production of abundant sources of energy, both food and fuels. At the end of 2012, the global population was about 7.0 billion people, and it is growing at a rate of about 750 million per decade. A growing world population puts greater demands on the global food supply, especially in Asia and Africa, which together make up more than 75% of the world population.

A growing population also increases demands on the production of fuels for transportation, industry, electricity, heating, and cooling. As populous countries such as China and India have modernized, their per capita consumption of energy has increased significantly. In China, for instance, per capita energy consumption roughly doubled between 1990 and 2010, and in 2010 China passed the United States as the world's largest user of energy (although it is still less than 20% of U.S. per capita energy consumption).

Global fuel energy consumption in 2012 was more than 5×10^{17} kJ, a staggeringly large number. More than 80% of current energy requirements comes from combustion of nonrenewable fossil fuels, especially coal and petroleum. The exploration of new fossil-fuel sources often involves environmentally sensitive regions, making the search for new supplies of fossil fuels a major political and economic issue.

The global importance of petroleum is in large part because it provides liquid fuels, such as gasoline, that are critical to supplying transportation needs. One of the most promising—but controversial—alternatives to petroleum-based fuels is *biofuels*, liquid fuels derived from biological matter. The most common approach to producing biofuels is to transform plant sugars and other carbohydrates into combustible liquids.

The most commonly produced biofuel is *bioethanol*, which is ethanol (C_2H_5OH) made from fermentation of plant carbohydrates. The fuel value of ethanol is about two-thirds that of gasoline and is therefore comparable to that of coal (Table 5.5). The United States and Brazil dominate bioethanol production, together supplying 85% of the world's total.

In the United States, nearly all the bioethanol currently produced is made from yellow feed corn. Glucose ($C_6H_{12}O_6$) in the corn is converted to ethanol and CO_2:

$$C_6H_{12}O_6(s) \longrightarrow 2 C_2H_5OH(l) + 2 CO_2(g) \qquad \Delta H = 15.8 \text{ KJ}$$

---

*Data from the Annual Energy Outlook 2012, U.S. Energy Information Administration.

Glucose is soluble in blood, and in the human body it is known as blood sugar. It is transported by the blood to cells where it reacts with O_2 in a series of steps, eventually producing $CO_2(g)$, $H_2O(l)$, and energy:

$$C_6H_{12}O_6(s) + 6\,O_2(g) \longrightarrow 6\,CO_2(g) + 6\,H_2O(l) \qquad \Delta H^\circ = -2803 \text{ kJ}$$

Because carbohydrates break down rapidly, their energy is quickly supplied to the body. However, the body stores only a very small amount of carbohydrates. The average fuel value of carbohydrates is 17 kJ/g (4 kcal/g).*

Like carbohydrates, fats produce CO_2 and H_2O when metabolized. The reaction of tristearin, $C_{57}H_{110}O_6$, a typical fat, is

$$2\,C_{57}H_{110}O_6(s) + 163\,O_2(g) \longrightarrow 114\,CO_2(g) + 110\,H_2O(l) \qquad \Delta H^\circ = 275{,}520 \text{ kJ}$$

The body uses the chemical energy from foods to maintain body temperature (see the "Chemistry and Life" box in Section 5.5), to contract muscles, and to construct and repair tissues. Any excess energy is stored as fats. Fats are well suited to serve as the body's energy reserve for at least two reasons: (1) They are insoluble in water, which facilitates storage in the body, and (2) they produce more energy per gram than either proteins or carbohydrates, which makes them efficient energy sources on a mass basis. The average fuel value of fats is 38 kJ/g (9 kcal/g).

The combustion of carbohydrates and fats in a bomb calorimeter gives the same products as when they are metabolized in the body. The metabolism of proteins produces less energy than combustion in a calorimeter because the products are different. Proteins contain nitrogen, which is released in the bomb calorimeter as N_2. In the body this nitrogen ends up mainly as urea, $(NH_2)_2CO$. Proteins are used by the body mainly as building materials for organ walls, skin, hair, muscle, and so forth. On average, the metabolism of proteins produces 17 kJ/g (4 kcal/g), the same as for carbohydrates.

Fuel values for some common foods are shown in ▼ Table 5.4. Labels on packaged foods show the amounts of carbohydrate, fat, and protein contained in an average serving, as well as the amount of energy supplied by a serving (▶ Figure 5.23).

The amount of energy our bodies require varies considerably, depending on such factors as weight, age, and muscular activity. About 100 kJ per kilogram of body mass per day is required to keep the body functioning at a minimal level. An average 70-kg

▲ GO FIGURE

Which value would change most if this label were for skim milk instead of whole milk: grams of fat, grams of total carbohydrate, or grams of protein?

▲ Figure 5.23 **Nutrition label for whole milk.**

Table 5.4 Compositions and Fuel Values of Some Common Foods

	Approximate Composition (% by Mass)			Fuel Value	
	Carbohydrate	Fat	Protein	kJ/g	kcal/g (Cal/g)
Carbohydrate	100	—	—	17	4
Fat	—	100	—	38	9
Protein	—	—	100	17	4
Apples	13	0.5	0.4	2.5	0.59
Beer[a]	1.2	—	0.3	1.8	0.42
Bread	52	3	9	12	2.8
Cheese	4	37	28	20	4.7
Eggs	0.7	10	13	6.0	1.4
Fudge	81	11	2	18	4.4
Green beans	7.0	—	1.9	1.5	0.38
Hamburger	—	30	22	15	3.6
Milk (whole)	5.0	4.0	3.3	3.0	0.74
Peanuts	22	39	26	23	5.5

[a]Beer typically contains 3.5% ethanol, which has fuel value.

*Although fuel values represent the heat *released* in a combustion reaction, fuel values are reported as positive numbers.

(154-lb) person expends about 800 kJ/h when doing light work, and strenuous activity often requires 2000 kJ/h or more. When the fuel value, or caloric content, of the food we ingest exceeds the energy we expend, our body stores the surplus as fat.

 Give It Some Thought

Which releases the greatest amount of energy per gram when metabolized? Carbohydrates, proteins, or fats.

SAMPLE EXERCISE 5.14   Estimating the Fuel Value of a Food from Its Composition

(a) A 28-g (1-oz) serving of a popular breakfast cereal served with 120 mL of skim milk provides 8 g protein, 26 g carbohydrates, and 2 g fat. Using the average fuel values of these substances, estimate the fuel value (caloric content) of this serving. (b) A person of average weight uses about 100 Cal/mi when running or jogging. How many servings of this cereal provide the fuel value requirements to run 3 mi?

SOLUTION

Analyze The fuel value of the serving will be the sum of the fuel values of the protein, carbohydrates, and fat.

Plan We are given the masses of the protein, carbohydrates, and fat contained in a serving. We can use the data in Table 5.4 to convert these masses to their fuel values, which we can sum to get the total fuel value.

Solve

$$(8 \text{ g protein})\left(\frac{17 \text{ kJ}}{1 \text{ g protein}}\right) + (26 \text{ g carbohydrate})\left(\frac{17 \text{ kJ}}{1 \text{ g carbohydrate}}\right) +$$

$$(2 \text{ g fat})\left(\frac{38 \text{ kJ}}{1 \text{ g fat}}\right) = 650 \text{ kJ (to two significant figures)}$$

This corresponds to 160 kcal:

$$(650 \text{ kJ})\left(\frac{1 \text{ kcal}}{4.18 \text{ kJ}}\right) = 160 \text{ kcal}$$

Recall that the dietary Calorie is equivalent to 1 kcal. Thus, the serving provides 160 Cal.

Analyze Here we are faced with the reverse problem, calculating the quantity of food that provides a specific fuel value.

Plan The problem statement provides a conversion factor between Calories and miles. The answer to part (a) provides us with a conversion factor between servings and Calories.

Solve We can use these factors in a straightforward dimensional analysis to determine the number of servings needed, rounded to the nearest whole number:

$$\text{Servings} = (3 \text{ mi})\left(\frac{100 \text{ Cal}}{1 \text{ mi}}\right)\left(\frac{1 \text{ serving}}{160 \text{ Cal}}\right) = 2 \text{ servings}$$

Practice Exercise 1

A stalk of celery has a caloric content (fuel value) of 9.0 kcal. If 1.0 kcal is provided by fat and there is very little protein, estimate the number of grams of carbohydrate and fat in the celery. (a) 2 g carbohydrate and 0.1 g fat, (b) 2 g carbohydrate and 1 g fat, (c) 1 g carbohydrate and 2 g fat, (d) 2.2 g carbohydrate and 0.1 g fat, (e) 32 g carbohydrate and 10 g fat.

Practice Exercise 2

(a) Dry red beans contain 62% carbohydrate, 22% protein, and 1.5% fat. Estimate the fuel value of these beans. (b) During a very light activity, such as reading or watching television, the average adult expends about 7 kJ/min. How many minutes of such activity can be sustained by the energy provided by a serving of chicken noodle soup containing 13 g protein, 15 g carbohydrate, and 5 g fat?

Fuels

During the complete combustion of fuels, carbon is converted to CO_2 and hydrogen is converted to H_2O, both of which have large negative enthalpies of formation. Consequently, the greater the percentage of carbon and hydrogen in a fuel, the higher its fuel value. In ▼ Table 5.5, for example, compare the compositions and fuel values of bituminous coal and wood. The coal has a higher fuel value because of its greater carbon content.

In 2011 the United States consumed 1.03×10^{17} kJ of energy. This value corresponds to an average daily energy consumption per person of 9.3×10^5 kJ roughly 100 times greater than the per capita food-energy needs. ▶ Figure 5.24 illustrates the sources of this energy.

Coal, petroleum, and natural gas, which are the world's major sources of energy, are known as **fossil fuels**. All have formed over millions of years from the decomposition of plants and animals and are being depleted far more rapidly than they are being formed.

Natural gas consists of gaseous hydrocarbons, compounds of hydrogen and carbon. It contains primarily methane (CH_4), with small amounts of ethane (C_2H_6), propane (C_3H_8), and butane (C_4H_{10}). We determined the fuel value of propane in Sample Exercise 5.11. Natural gas burns with far fewer byproducts and produces less CO_2 than either petroleum or coal. **Petroleum** is a liquid composed of hundreds of compounds, most of which are hydrocarbons, with the remainder being chiefly organic compounds containing sulfur, nitrogen, or oxygen. **Coal**, which is solid, contains hydrocarbons of high molecular weight as well as compounds containing sulfur, oxygen, or nitrogen. Coal is the most abundant fossil fuel; current reserves are projected to last for well over 100 years at current consumption rates. However, the use of coal presents a number of problems.

Coal is a complex mixture of substances, and it contains components that cause air pollution. When coal is combusted, the sulfur it contains is converted mainly to sulfur dioxide, SO_2, a troublesome air pollutant. Because coal is a solid, recovery from its underground deposits is expensive and often dangerous. Furthermore, coal deposits are not always close to locations of high-energy use, so there are often substantial shipping costs.

Fossil fuels release energy in combustion reactions, which ideally produce only CO_2 and H_2O. The production of CO_2 has become a major issue that involves science and public policy because of concerns that increasing concentrations of atmospheric CO_2 are causing global climate changes. We will discuss the environmental aspects of atmospheric CO_2 in Chapter 18.

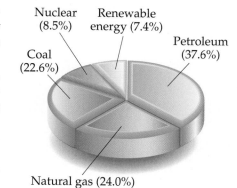

▲ **Figure 5.24 Energy consumption in the United States.*** In 2011 the United States consumed a total of 1.03×10^{17} kJ of energy.

Table 5.5 Fuel Values and Compositions of Some Common Fuels

	Approximate Elemental Composition (Mass %)			
	C	H	O	Fuel Value (kJ/g)
Wood (pine)	50	6	44	18
Anthracite coal (Pennsylvania)	82	1	2	31
Bituminous coal (Pennsylvania)	77	5	7	32
Charcoal	100	0	0	34
Crude oil (Texas)	85	12	0	45
Gasoline	85	15	0	48
Natural gas	70	23	0	49
Hydrogen	0	100	0	142

*Annual Energy Review 2011, U.S Energy Information Administration, U. S. Department of Energy.

Other Energy Sources

Nuclear energy is the energy released in either the fission (splitting) or the fusion (combining) of atomic nuclei. Nuclear power based on nuclear fission is currently used to produce about 21% of the electric power in the United States and makes up about 8.5% of the total U.S. energy production (Figure 5.24). Nuclear energy is, in principle, free of the polluting emissions that are a major problem with fossil fuels. However, nuclear power plants produce radioactive waste products, and their use has therefore been controversial. We will discuss issues related to the production of nuclear energy in Chapter 21.

Fossil fuels and nuclear energy are *nonrenewable* sources of energy—they are limited resources that we are consuming at a much greater rate than they can be regenerated. Eventually these fuels will be expended, although estimates vary greatly as to when this will occur. Because nonrenewable energy sources will eventually be used up, a great deal of research is being conducted on **renewable energy sources**, sources that are essentially inexhaustible. Renewable energy sources include *solar energy* from the Sun, *wind energy* harnessed by windmills, *geothermal energy* from the heat stored inside Earth, *hydroelectric energy* from flowing rivers, and *biomass energy* from crops and biological waste matter. Currently, renewable sources provide about 7.4% of the U.S. annual energy consumption, with hydroelectric and biomass sources the major contributors.

Fulfilling our future energy needs will depend on developing technology to harness solar energy with greater efficiency. Solar energy is the world's largest energy source. On a clear day about 1 kJ of solar energy reaches each square meter of Earth's surface every second. The average solar energy falling on only 0.1% of U.S. land area is equivalent to all the energy this nation currently uses. Harnessing this energy is difficult because it is dilute (that is, distributed over a wide area) and varies with time of day and weather conditions. The effective use of solar energy will depend on the development of some means of storing and distributing it. Any practical means for doing this will almost certainly involve an endothermic chemical process that can be later reversed to release heat. One such reaction is

$$CH_4(g) + H_2O(g) + \text{heat} \longrightarrow CO(g) + 3 H_2(g)$$

This reaction proceeds in the forward direction at high temperatures, which can be obtained in a solar furnace. The CO and H_2 formed in the reaction could then be stored and allowed to react later, with the heat released being put to useful work.

Chemistry Put to Work

The Scientific and Political Challenges of Biofuels*

One of the biggest challenges facing us in the twenty-first century is production of abundant sources of energy, both food and fuels. At the end of 2012, the global population was about 7.0 billion people, and it is growing at a rate of about 750 million per decade. A growing world population puts greater demands on the global food supply, especially in Asia and Africa, which together make up more than 75% of the world population.

A growing population also increases demands on the production of fuels for transportation, industry, electricity, heating, and cooling. As populous countries such as China and India have modernized, their per capita consumption of energy has increased significantly. In China, for instance, per capita energy consumption roughly doubled between 1990 and 2010, and in 2010 China passed the United States as the world's largest user of energy (although it is still less than 20% of U.S. per capita energy consumption).

Global fuel energy consumption in 2012 was more than 5×10^{17} kJ, a staggeringly large number. More than 80% of current energy requirements comes from combustion of nonrenewable fossil fuels, especially coal and petroleum. The exploration of new fossil-fuel sources often involves environmentally sensitive regions, making the search for new supplies of fossil fuels a major political and economic issue.

The global importance of petroleum is in large part because it provides liquid fuels, such as gasoline, that are critical to supplying transportation needs. One of the most promising—but controversial— alternatives to petroleum-based fuels is *biofuels*, liquid fuels derived from biological matter. The most common approach to producing biofuels is to transform plant sugars and other carbohydrates into combustible liquids.

The most commonly produced biofuel is *bioethanol*, which is ethanol (C_2H_5OH) made from fermentation of plant carbohydrates. The fuel value of ethanol is about two-thirds that of gasoline and is therefore comparable to that of coal (Table 5.5). The United States and Brazil dominate bioethanol production, together supplying 85% of the world's total.

In the United States, nearly all the bioethanol currently produced is made from yellow feed corn. Glucose ($C_6H_{12}O_6$) in the corn is converted to ethanol and CO_2:

$$C_6H_{12}O_6(s) \longrightarrow 2 C_2H_5OH(l) + 2 CO_2(g) \qquad \Delta H = 15.8 \text{ KJ}$$

---

*Data from the Annual Energy Outlook 2012, U.S. Energy Information Administration.

Notice that this reaction is *anaerobic*—it does not involve $O_2(g)$—and that the enthalpy change is positive and much smaller in magnitude than for most combustion reactions. Other carbohydrates can be converted to ethanol in similar fashion.

Producing bioethanol from corn is controversial for two main reasons. First, growing and transporting corn are both energy-intensive processes, and growing it requires the use of fertilizers. It is estimated that the *energy return* on corn-based bioethanol is only 34%—that is, for each 1.00 J of energy expended to produce the corn, 1.34 J of energy is produced in the form of bioethanol. Second, the use of corn as a starting material for making bioethanol competes with its use as an important component of the food chain (the so-called food versus fuel debate).

Much current research focuses on the formation of bioethanol from *cellulosic* plants, plants that contain the complex carbohydrate cellulose. Cellulose is not readily metabolized and so does not compete with the food supply. However, the chemistry for converting cellulose to ethanol is much more complex than that for converting corn. Cellulosic bioethanol could be produced from very fast-growing nonfood plants, such as prairie grasses and switchgrass, which readily renew themselves without the use of fertilizers.

The Brazilian bioethanol industry uses sugarcane as its feedstock (▶ Figure 5.25). Sugarcane grows much faster than corn and without the need for fertilizers or tending. Because of these differences, the energy return for sugarcane is much higher than the energy return for corn. It is estimated that for each 1.0 J of energy expended in growing and processing sugarcane, 8.0 J of energy is produced as bioethanol.

▲ Figure 5.25 **Sugarcane can be converted to a sustainable bioethanol product.**

Other biofuels that are also becoming a major part of the world economy include *biodiesel*, a substitute for petroleum-derived diesel fuel. Biodiesel is typically produced from crops that have a high oil content, such as soybeans and canola. It can also be produced from animal fats and waste vegetable oil from the food and restaurant industry.

Related Exercises: **5.89, 5.90, 5.111, 5.119**

Plants utilize solar energy in *photosynthesis*, the reaction in which the energy of sunlight is used to convert CO_2 and H_2O into carbohydrates and O_2:

$$6\,CO_2(g) + 6\,H_2O(l) + \text{sunlight} \longrightarrow C_6H_{12}O_6(s) + 6\,O_2(g) \qquad [5.32]$$

Photosynthesis is an important part of Earth's ecosystem because it replenishes atmospheric O_2, produces an energy-rich molecule that can be used as fuel, and consumes some atmospheric CO_2.

Perhaps the most direct way to use the Sun's energy is to convert it directly into electricity in photovoltaic devices, or *solar cells*, which we mentioned at the beginning of this chapter. The efficiencies of such devices have increased dramatically during the past few years. Technological advances have led to solar panels that last longer and produce electricity with greater efficiency at steadily decreasing unit cost. Indeed, the future of solar energy is, like the Sun itself, very bright.

SAMPLE INTEGRATIVE EXERCISE | Putting Concepts Together

Trinitroglycerin, $C_3H_5N_3O_9$ (usually referred to simply as nitroglycerin), has been widely used as an explosive. Alfred Nobel used it to make dynamite in 1866. Rather surprisingly, it also is used as a medication, to relieve angina (chest pains resulting from partially blocked arteries to the heart) by dilating the blood vessels. At 1 atm pressure and 25 °C, the enthalpy of decomposition of trinitroglycerin to form nitrogen gas, carbon dioxide gas, liquid water, and oxygen gas is −1541.4 kJ/mol.

(a) Write a balanced chemical equation for the decomposition of trinitroglycerin.
(b) Calculate the standard heat of formation of trinitroglycerin.
(c) A standard dose of trinitroglycerin for relief of angina is 0.60 mg. If the sample is eventually oxidized in the body (not explosively, though!) to nitrogen gas, carbon dioxide gas, and liquid water, what number of calories is released?
(d) One common form of trinitroglycerin melts at about 3 °C. From this information and the formula for the substance, would you expect it to be a molecular or ionic compound? Explain.
(e) Describe the various conversions of forms of energy when trinitroglycerin is used as an explosive to break rockfaces in highway construction.

SOLUTION

(a) The general form of the equation we must balance is

$$C_3H_5N_3O_9(l) \longrightarrow N_2(g) + CO_2(g) + H_2O(l) + O_2(g)$$

We go about balancing in the usual way. To obtain an even number of nitrogen atoms on the left, we multiply the formula for $C_3H_5N_3O_9$ by 2, which gives us 3 mol of N_2, 6 mol of CO_2 and 5 mol of H_2O. Everything is then balanced except for oxygen. We have an odd number of oxygen atoms on the right. We can balance the oxygen by using the coefficient $\frac{1}{2}$ for O_2 on the right:

$$2\,C_3H_5N_3O_9(l) \longrightarrow 3\,N_2(g) + 6\,CO_2(g) + 5\,H_2O(l) + \tfrac{1}{2}O_2(g)$$

We multiply through by 2 to convert all coefficients to whole numbers:

$$4\,C_3H_5N_3O_9(l) \longrightarrow 6\,N_2(g) + 12\,CO_2(g) + 10\,H_2O(l) + O_2(g)$$

(At the temperature of the explosion, water is a gas. The rapid expansion of the gaseous products creates the force of an explosion.)

(b) We can obtain the standard enthalpy of formation of nitroglycerin by using the heat of decomposition of trinitroglycerin together with the standard enthalpies of formation of the other substances in the decomposition equation:

$$4\,C_3H_5N_3O_9(l) \longrightarrow 6\,N_2(g) + 12\,CO_2(g) + 10\,H_2O(l) + O_2(g)$$

The enthalpy change for this decomposition is $4(-1541.4\text{ kJ}) = -6165.6\text{ kJ}$. [We need to multiply by 4 because there are 4 mol of $C_3H_5N_3O_9(l)$ in the balanced equation.]

This enthalpy change equals the sum of the heats of formation of the products minus the heats of formation of the reactants, each multiplied by its coefficient in the balanced equation:

$$-6165.6\text{ kJ} = 6\Delta H_f^{\circ}[N_2(g)] + 12\Delta H_f^{\circ}[CO_2(g)] + 10\Delta H_f^{\circ}[H_2O(l)] + \Delta H_f^{\circ}[O_2(g)]$$
$$-4\Delta H_f^{\circ}[C_3H_5N_3O_9(l)]$$

The ΔH_f° values for $N_2(g)$ and $O_2(g)$ are zero, by definition. Using the values for $H_2O(l)$ and $CO_2(g)$ from Table 5.3 or Appendix C, we have

$$-6165.6\text{ kJ} = 12(-393.5\text{ kJ}) + 10(-285.8\text{ kJ}) - 4\Delta H_f^{\circ}[C_3H_5N_3O_9(l)]$$
$$\Delta H_f^{\circ}[C_3H_5N_3O_9(l)] = -353.6\text{ kJ/mol}$$

(c) Converting 0.60 mg $C_3H_5N_3O_9(l)$ to moles and using the fact that the decomposition of 1 mol of $C_3H_5N_3O_9(l)$ yields 1541.4 kJ we have:

$$(0.60 \times 10^{-3}\text{ g }C_3H_5N_3O_9)\left(\frac{1\text{ mol }C_3H_5N_3O_9}{227\text{ g }C_3H_5N_3O_9}\right)\left(\frac{1541.4\text{ kJ}}{1\text{ mol }C_3H_5N_3O_9}\right) = 4.1 \times 10^{-3}\text{ kJ}$$
$$= 4.1\text{ J}$$

(d) Because trinitroglycerin melts below room temperature, we expect that it is a molecular compound. With few exceptions, ionic substances are generally hard, crystalline materials that melt at high temperatures. ∞ (Sections 2.6 and 2.7) Also, the molecular formula suggests that it is a molecular substance because all of its constituent elements are nonmetals.

(e) The energy stored in trinitroglycerin is chemical potential energy. When the substance reacts explosively, it forms carbon dioxide, water, and nitrogen gas, which are of lower potential energy. In the course of the chemical transformation, energy is released in the form of heat; the gaseous reaction products are very hot. This high heat energy is transferred to the surroundings. Work is done as the gases expand against the surroundings, moving the solid materials and imparting kinetic energy to them. For example, a chunk of rock might be impelled upward. It has been given kinetic energy by transfer of energy from the hot, expanding gases. As the rock rises, its kinetic energy is transformed into potential energy. Eventually, it again acquires kinetic energy as it falls to Earth. When it strikes Earth, its kinetic energy is converted largely to thermal energy, though some work may be done on the surroundings as well.

Chapter Summary and Key Terms

ENERGY (INTRODUCTION AND SECTION 5.1) **Thermodynamics** is the study of energy and its transformations. In this chapter we have focused on **thermochemistry**, the transformations of energy—especially heat—during chemical reactions.

An object can possess energy in two forms: (1) **kinetic energy**, which is the energy due to the motion of the object, and (2) **potential energy**,

which is the energy that an object possesses by virtue of its position relative to other objects. An electron in motion near a proton has kinetic energy because of its motion and potential energy because of its electrostatic attraction to the proton.

The SI unit of energy is the **joule** (J): $1\text{ J} = 1\text{ kg-m}^2/\text{s}^2$. Another common energy unit is the **calorie** (cal), which was originally defined as the quantity of energy necessary to increase the temperature of 1 g of water by 1 °C: 1 cal = 4.184 J.

When we study thermodynamic properties, we define a specific amount of matter as the **system**. Everything outside the system is the **surroundings**. When we study a chemical reaction, the system is generally the reactants and products. A closed system can exchange energy, but not matter, with the surroundings.

Energy can be transferred between the system and the surroundings as work or heat. **Work** is the energy expended to move an object against a **force**. **Heat** is the energy that is transferred from a hotter object to a colder one. **Energy** is the capacity to do work or to transfer heat.

THE FIRST LAW OF THERMODYNAMICS (SECTION 5.2) The

internal energy of a system is the sum of all the kinetic and potential energies of its component parts. The internal energy of a system can change because of energy transferred between the system and the surroundings.

According to the **first law of thermodynamics**, the change in the internal energy of a system, ΔE, is the sum of the heat, q, transferred into or out of the system and the work, w, done on or by the system: $\Delta E = q + w$. Both q and w have a sign that indicates the direction of energy transfer. When heat is transferred from the surroundings to the system, $q > 0$. Likewise, when the surroundings do work on the system, $w > 0$. In an **endothermic** process the system absorbs heat from the surroundings; in an **exothermic** process the system releases heat to the surroundings.

The internal energy, E, is a **state function**. The value of any state function depends only on the state or condition of the system and not on the details of how it came to be in that state. The heat, q, and the work, w, are not state functions; their values depend on the particular way by which a system changes its state.

ENTHALPY (SECTIONS 5.3 AND 5.4) When a gas is produced or

consumed in a chemical reaction occurring at constant pressure, the system may perform **pressure–volume (P–V) work** against the prevailing pressure of the surroundings. For this reason, we define a new state function called **enthalpy**, H, which is related to energy: $H = E + PV$. In systems where only pressure–volume work is involved, the change in the enthalpy of a system, ΔH, equals the heat gained or lost by the system at constant pressure: $\Delta H = q_p$ (the subscript P denotes constant pressure). For an endothermic process, $\Delta H > 0$; for an exothermic process, $\Delta H < 0$.

In a chemical process, the **enthalpy of reaction** is the enthalpy of the products minus the enthalpy of the reactants: $\Delta H_{rxn} = H(\text{products}) - H(\text{reactants})$. Enthalpies of reaction follow some simple rules: (1) The enthalpy of reaction is proportional to the amount of reactant that reacts. (2) Reversing a reaction changes the sign of ΔH. (3) The enthalpy of reaction depends on the physical states of the reactants and products.

CALORIMETRY (SECTION 5.5) The amount of heat transferred be-

tween the system and the surroundings is measured experimentally by **calorimetry**. A **calorimeter** measures the temperature change accompanying a process. The temperature change of a calorimeter depends on its **heat capacity**, the amount of heat required to raise its temperature by 1 K. The heat capacity for one mole of a pure substance is called its **molar heat capacity**; for one gram of the substance, we use the term **specific heat**. Water has a very high specific heat, 4.18 J/g-K. The amount of heat, q, absorbed by a substance is the product of its specific heat (C_s), its mass, and its temperature change: $q = C_s \times m \times \Delta T$.

If a calorimetry experiment is carried out under a constant pressure, the heat transferred provides a direct measure of the enthalpy change of the reaction. Constant-volume calorimetry is carried out in a vessel of fixed volume called a **bomb calorimeter**. The heat transferred under constant-volume conditions is equal to ΔE. Corrections can be applied to ΔE values to yield ΔH.

HESS'S LAW (SECTION 5.6) Because enthalpy is a state function,

ΔH depends only on the initial and final states of the system. Thus, the enthalpy change of a process is the same whether the process is carried out in one step or in a series of steps. **Hess's law** states that if a reaction is carried out in a series of steps, ΔH for the reaction will be equal to the sum of the enthalpy changes for the steps. We can therefore calculate ΔH for any process, as long as we can write the process as a series of steps for which ΔH is known.

ENTHALPIES OF FORMATION (SECTION 5.7) The **enthalpy of forma-**

tion, ΔH_f, of a substance is the enthalpy change for the reaction in which the substance is formed from its constituent elements. Usually enthalpies are tabulated for reactions where reactants and products are in their **standard states**. The standard state of a substance is its pure, most stable form at 1 atm and the temperature of interest (usually 298 K). Thus, the **standard enthalpy change** of a reaction, $\Delta H°$, is the enthalpy change when all reactants and products are in their standard states. The **standard enthalpy of formation**, $\Delta H_f°$, of a substance is the change in enthalpy for the reaction that forms one mole of the substance from its elements in their standard states. For any element in its standard state, $\Delta H_f° = 0$.

The standard enthalpy change for any reaction can be readily calculated from the standard enthalpies of formation of the reactants and products in the reaction:

$$\Delta H_{rxn}° = \Sigma n \Delta H_f°(\text{products}) - \Sigma m \Delta H_f°(\text{reactants})$$

FOODS AND FUELS (SECTION 5.8) The **fuel value** of a substance is

the heat released when one gram of the substance is combusted. Different types of foods have different fuel values and differing abilities to be stored in the body. The most common fuels are hydrocarbons that are found as **fossil fuels**, such as **natural gas**, **petroleum**, and **coal**. **Renewable energy sources** include solar energy, wind energy, biomass, and hydroelectric energy. Nuclear power does not utilize fossil fuels but does create controversial waste-disposal problems.

Learning Outcomes After studying this chapter, you should be able to:

- Interconvert energy units. (Section 5.1)
- Distinguish between the system and the surroundings in thermodynamics. (Section 5.1)
- Calculate internal energy from heat and work and state the sign conventions of these quantities. (Section 5.2)
- Explain the concept of a state function and give examples. (Section 5.2)

- Calculate ΔH from ΔE and $P\Delta V$ (Section 5.3)
- Relate q_p to ΔH and indicate how the signs of q and ΔH relate to whether a process is exothermic or endothermic. (Sections 5.2 and 5.3)
- Use thermochemical equations to relate the amount of heat energy transferred in reactions at constant pressure (ΔH) to the amount of substance involved in the reaction. (Section 5.4)

- Calculate the heat transferred in a process from temperature measurements together with heat capacities or specific heats (calorimetry). (Section 5.5)

- Use Hess's law to determine enthalpy changes for reactions. (Section 5.6)

- Use standard enthalpies of formation to calculate $\Delta H°$ for reactions. (Section 5.7)

Key Equations

- $E_k = \frac{1}{2}mv^2$ [5.1]   Kinetic energy

- $w = F \times d$ [5.3]   Relates work to force and distance

- $\Delta E = E_{\text{final}} - E_{\text{initial}}$ [5.4]   The change in internal energy

- $\Delta E = q + w$ [5.5]   Relates the change in internal energy to heat and work (the first law of thermodynamics)

- $H = E + PV$ [5.6]   Defines enthalpy

- $w = -P\,\Delta V$ [5.8]   The work done by an expanding gas at constant pressure

- $\Delta H = \Delta E + P\,\Delta V = q_P$ [5.10]   Enthalpy change at constant pressure

- $q = C_s \times m \times \Delta T$ [5.22]   Heat gained or lost based on specific heat, mass, and temperature change

- $q_{\text{rxn}} = -C_{\text{cal}} \times \Delta T$ [5.24]   Heat exchanged between a reaction and calorimeter

- $\Delta H°_{\text{rxn}} = \Sigma n \Delta H_f°(\text{products}) - \Sigma m \Delta H_f°(\text{reactants})$ [5.31]   Standard enthalpy change of a reaction

Exercises

Visualizing Concepts

5.1 Imagine a book that is falling from a shelf. At a particular moment during its fall, the book has a kinetic energy of 24 J and a potential energy with respect to the floor of 47 J. **(a)** How do the book's kinetic energy and its potential energy change as it continues to fall? **(b)** What was the initial potential energy of the book, and what is its total kinetic energy at the instant just before it strikes the floor? **(c)** If a heavier book fell from the same shelf, would it have the same kinetic energy when it strikes the floor? [Section 5.1]

5.2 The accompanying photo shows a pipevine swallowtail caterpillar climbing up a twig. **(a)** As the caterpillar climbs, its potential energy is increasing. What source of energy has been used to effect this change in potential energy? **(b)** If the caterpillar is the system, can you predict the sign of q as the caterpillar climbs? **(c)** Does the caterpillar do work in climbing the twig? Explain. **(d)** Does the amount of work done in climbing a 12-inch section of the twig depend on the speed of the caterpillar's climb? **(e)** Does the change

in potential energy depend on the caterpillar's speed of climb? [Section 5.1]

5.3 Consider the accompanying energy diagram. **(a)** Does this diagram represent an increase or decrease in the internal energy of the system? **(b)** What sign is given to ΔE for this process? **(c)** If there is no work associated with the process, is it exothermic or endothermic? [Section 5.2]

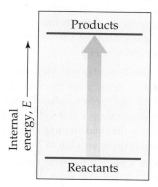

5.4 The contents of the closed box in each of the following illustrations represent a system, and the arrows show the changes to the system during some process. The lengths of the arrows represent the relative magnitudes of q and w. (a) Which of these processes is endothermic? (b) For which of these processes, if any, is $\Delta E < 0$? (c) For which process, if any, does the system experience a net gain in internal energy? [Section 5.2]

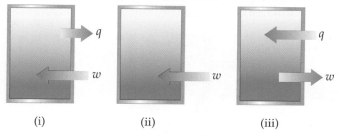

(i)   (ii)   (iii)

5.5 Imagine that you are climbing a mountain. (a) Is the distance you travel to the top a state function? Why or why not? (b) Is the change in elevation between your base camp and the peak a state function? Why or why not? [Section 5.2]

5.6 The diagram shows four states of a system, each with different internal energy, E. (a) Which of the states of the system has the greatest internal energy? (b) In terms of the ΔE values, write two expressions for the difference in internal energy between State A and State B. (c) Write an expression for the difference in energy between State C and State D. (d) Suppose there is another state of the system, State E, and its energy relative to State A is $\Delta E = \Delta E_1 + \Delta E_4$. Where would State E be on the diagram? [Section 5.2]

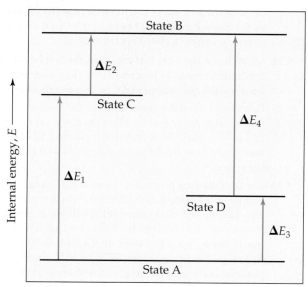

5.7 You may have noticed that when you compress the air in a bicycle pump, the body of the pump gets warmer. (a) Assuming the pump and the air in it comprise the system, what is the sign of w when you compress the air? (b) What is the sign of q for this process? (c) Based on your answers to parts (a) and (b), can you determine the sign of ΔE for compressing the air in the pump? If not, what would you expect for the sign of ΔE? What is your reasoning? [Section 5.2]

5.8 Imagine a container placed in a tub of water, as depicted in the accompanying diagram. (a) If the contents of the container are the system and heat is able to flow through the container walls, what qualitative changes will occur in the temperatures of the system and in its surroundings? What is the sign of q associated with each change? From the system's perspective, is the process exothermic or endothermic? (b) If neither the volume nor the pressure of the system changes during the process, how is the change in internal energy related to the change in enthalpy? [Sections 5.2 and 5.3]

5.9 In the accompanying cylinder diagram a chemical process occurs at constant temperature and pressure. (a) Is the sign of w indicated by this change positive or negative? (b) If the process is endothermic, does the internal energy of the system within the cylinder increase or decrease during the change and is ΔE positive or negative? [Sections 5.2 and 5.3]

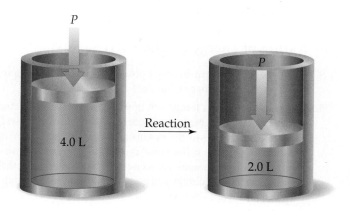

5.10 The gas-phase reaction shown, between N_2 and O_2, was run in an apparatus designed to maintain a constant pressure. (a) Write a balanced chemical equation for the reaction depicted and predict whether w is positive, negative, or zero. (b) Using data from Appendix C, determine ΔH for the formation of one mole of the product. Why is this enthalpy change called the enthalpy of formation of the involved product? [Sections 5.3 and 5.7]

5.11 Consider the two diagrams that follow. (a) Based on (i), write an equation showing how ΔH_A is related to ΔH_B and ΔH_C. How do both diagram (i) and your equation relate to the fact that enthalpy is a state function? (b) Based on (ii), write an equation relating ΔH_Z to the other enthalpy changes in the diagram. (c) How do these diagrams relate to Hess's law? [Section 5.6]

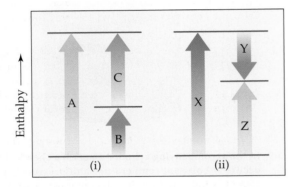

5.12 Consider the conversion of compound A into compound B: A ⟶ B. For both compounds A and B, $\Delta H_f^{\circ} > 0$. (a) Sketch an enthalpy diagram for the reaction that is analogous to Figure 5.22. (b) Suppose the overall reaction is exothermic. What can you conclude? [Section 5.7]

The Nature of Energy (Section 5.1)

5.13 In what two ways can an object possess energy? How do these two ways differ from one another?

5.14 Suppose you toss a tennis ball upward. (a) Does the kinetic energy of the ball increase or decrease as it moves higher? (b) What happens to the potential energy of the ball as it moves higher? (c) If the same amount of energy were imparted to a ball the same size as a tennis ball but of twice the mass, how high would the ball go in comparison to the tennis ball? Explain your answers.

5.15 (a) Calculate the kinetic energy, in joules, of a 1200-kg automobile moving at 18 m/s. (b) Convert this energy to calories. (c) What happens to this energy when the automobile brakes to a stop?

5.16 (a) A baseball weighs 5.13 oz. What is the kinetic energy, in joules, of this baseball when it is thrown by a major-league pitcher at 95.0 mi/h? (b) By what factor will the kinetic energy change if the speed of the baseball is decreased to 55.0 mi/h? (c) What happens to the kinetic energy when the baseball is caught by the catcher?

5.17 The use of the British thermal unit (Btu) is common in much engineering work. A Btu is the amount of heat required to raise the temperature of 1 lb of water by 1 °F. Calculate the number of joules in a Btu.

5.18 A watt is a measure of power (the rate of energy change) equal to 1 J/s. (a) Calculate the number of joules in a kilowatt-hour. (b) An adult person radiates heat to the surroundings at about the same rate as a 100-watt electric incandescent light-bulb. What is the total amount of energy in kcal radiated to the surroundings by an adult in 24 h?

5.19 (a) What is meant by the term *system* in thermodynamics? (b) What is a *closed system*? (c) What do we call the part of the universe that is not part of the system?

5.20 In a thermodynamic study a scientist focuses on the properties of a solution in an apparatus as illustrated. A solution is continuously flowing into the apparatus at the top and out at the bottom, such that the amount of solution in the apparatus is constant with time. (a) Is the solution in the apparatus a closed system, open system, or isolated system? Explain your choice. (b) If it is not a closed system, what could be done to make it a closed system?

5.21 Identify the force present and explain whether work is being performed in the following cases: (a) You lift a pencil off the top of a desk. (b) A spring is compressed to half its normal length.

5.22 Identify the force present and explain whether work is done when (a) a positively charged particle moves in a circle at a fixed distance from a negatively charged particle, (b) an iron nail is pulled off a magnet.

The First Law of Thermodynamics (Section 5.2)

5.23 (a) State the first law of thermodynamics. (b) What is meant by the *internal energy* of a system? (c) By what means can the internal energy of a closed system increase?

5.24 (a) Write an equation that expresses the first law of thermodynamics in terms of heat and work. (b) Under what conditions will the quantities q and w be negative numbers?

5.25 Calculate ΔE and determine whether the process is endothermic or exothermic for the following cases: (a) $q = 0.763$ kJ and $w = -840$ J. (b) A system releases 66.1 kJ of heat to its surroundings while the surroundings do 44.0 kJ of work on the system.

5.26 For the following processes, calculate the change in internal energy of the system and determine whether the process is endothermic or exothermic: (a) A balloon is cooled by removing 0.655 kJ of heat. It shrinks on cooling, and the atmosphere does 382 J of work on the balloon. (b) A 100.0-g bar of gold is heated from 25 °C to 50 °C during which it absorbs 322 J of heat. Assume the volume of the gold bar remains constant.

5.27 A gas is confined to a cylinder fitted with a piston and an electrical heater, as shown here:

Suppose that current is supplied to the heater so that 100 J of energy is added. Consider two different situations. In case (1) the piston is allowed to move as the energy is added. In case (2) the piston is fixed so that it cannot move. **(a)** In which case does the gas have the higher temperature after addition of the electrical energy? Explain. **(b)** What can you say about the values of q and w in each case? **(c)** What can you say about the relative values of ΔE for the system (the gas in the cylinder) in the two cases?

5.28 Consider a system consisting of two oppositely charged spheres hanging by strings and separated by a distance r_1, as shown in the accompanying illustration. Suppose they are separated to a larger distance r_2, by moving them apart along a track. **(a)** What change, if any, has occurred in the potential energy of the system? **(b)** What effect, if any, does this process have on the value of ΔE? **(c)** What can you say about q and w for this process?

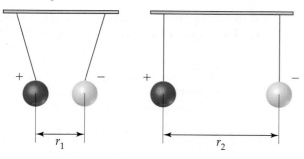

5.29 **(a)** What is meant by the term *state function*? **(b)** Give an example of a quantity that is a state function and one that is not. **(c)** Is the volume of a system a state function? Why or why not?

5.30 Indicate which of the following is independent of the path by which a change occurs: **(a)** the change in potential energy when a book is transferred from table to shelf, **(b)** the heat evolved when a cube of sugar is oxidized to $CO_2(g)$ and $H_2O(g)$, **(c)** the work accomplished in burning a gallon of gasoline.

Enthalpy (Sections 5.3 and 5.4)

5.31 During a normal breath, our lungs expand about 0.50 L against an external pressure of 1.0 atm. How much work is involved in this process (in J)?

5.32 How much work (in J) is involved in a chemical reaction if the volume decreases from 5.00 to 1.26 L against a constant pressure of 0.857 atm?

5.33 **(a)** Why is the change in enthalpy usually easier to measure than the change in internal energy? **(b)** H is a state function, but q is not a state function. Explain. **(c)** For a given process at constant pressure, ΔH is positive. Is the process endothermic or exothermic?

5.34 **(a)** Under what condition will the enthalpy change of a process equal the amount of heat transferred into or out of the system? **(b)** During a constant-pressure process, the system releases heat to the surroundings. Does the enthalpy of the system increase or decrease during the process? **(c)** In a constant-pressure process, $\Delta H = 0$. What can you conclude about ΔE, q, and w?

5.35 Assume that the following reaction occurs at constant pressure:

$$2\,Al(s) + 3\,Cl_2(g) \longrightarrow 2\,AlCl_3(s)$$

(a) If you are given ΔH for the reaction, what additional information do you need to determine ΔE for the process? **(b)** Which quantity is larger for this reaction? **(c)** Explain your answer to part **(b)**.

5.36 Suppose that the gas-phase reaction $2\,NO(g) + O_2(g) \longrightarrow 2\,NO_2(g)$ were carried out in a constant-volume container at constant temperature. **(a)** Would the measured heat change represent ΔH or ΔE? **(b)** If there is a difference, which quantity is larger for this reaction? **(c)** Explain your answer to part **(b)**.

5.37 A gas is confined to a cylinder under constant atmospheric pressure, as illustrated in Figure 5.4. When the gas undergoes a particular chemical reaction, it absorbs 824 J of heat from its surroundings and has 0.65 kJ of P–V work done on it by its surroundings. What are the values of ΔH and ΔE for this process?

5.38 A gas is confined to a cylinder under constant atmospheric pressure, as illustrated in Figure 5.4. When 0.49 kJ of heat is added to the gas, it expands and does 214 J of work on the surroundings. What are the values of ΔH and ΔE for this process?

5.39 The complete combustion of ethanol, $C_2H_5OH(l)$, to form $H_2O(g)$ and $CO_2(g)$ at constant pressure releases 1235 kJ of heat per mole of C_2H_5OH. **(a)** Write a balanced thermochemical equation for this reaction. **(b)** Draw an enthalpy diagram for the reaction.

5.40 The decomposition of $Ca(OH)_2(s)$ into $CaO(s)$ and $H_2O(g)$ at constant pressure requires the addition of 109 kJ of heat per mole of $Ca(OH)_2$. **(a)** Write a balanced thermochemical equation for the reaction. **(b)** Draw an enthalpy diagram for the reaction.

5.41 Ozone, $O_3(g)$, is a form of elemental oxygen that plays an important role in the absorption of ultraviolet radiation in the stratosphere. It decomposes to $O_2(g)$ at room temperature and pressure according to the following reaction:

$$2\,O_3(g) \longrightarrow 3\,O_2(g) \qquad \Delta H = -284.6\ kJ$$

(a) What is the enthalpy change for this reaction per mole of $O_3(g)$?

(b) Which has the higher enthalpy under these conditions, $2\,O_3(g)$ or $3\,O_2(g)$?

5.42 Without referring to tables, predict which of the following has the higher enthalpy in each case: (a) 1 mol $CO_2(s)$ or 1 mol $CO_2(g)$ at the same temperature, (b) 2 mol of hydrogen atoms or 1 mol of H_2, (c) 1 mol $H_2(g)$ and 0.5 mol $O_2(g)$ at 25 °C or 1 mol $H_2O(g)$ at 25 °C, (d) 1 mol $N_2(g)$ at 100 °C or 1 mol $N_2(g)$ at 300 °C.

5.43 Consider the following reaction:

$$2\,Mg(s) + O_2(g) \longrightarrow 2\,MgO(s) \quad \Delta H = -1204\,kJ$$

(a) Is this reaction exothermic or endothermic?

(b) Calculate the amount of heat transferred when 3.55 g of Mg(s) reacts at constant pressure.

(c) How many grams of MgO are produced during an enthalpy change of −234 kJ?

(d) How many kilojoules of heat are absorbed when 40.3 g of MgO(s) is decomposed into Mg(s) and $O_2(g)$ at constant pressure?

5.44 Consider the following reaction:

$$2\,CH_3OH(g) \longrightarrow 2\,CH_4(g) + O_2(g) \quad \Delta H = +252.8\,kJ$$

(a) Is this reaction exothermic or endothermic? (b) Calculate the amount of heat transferred when 24.0 g of $CH_3OH(g)$ is decomposed by this reaction at constant pressure. (c) For a given sample of CH_3OH, the enthalpy change during the reaction is 82.1 kJ. How many grams of methane gas are produced? (d) How many kilojoules of heat are released when 38.5 g of $CH_4(g)$ reacts completely with $O_2(g)$ to form $CH_3OH(g)$ at constant pressure?

5.45 When solutions containing silver ions and chloride ions are mixed, silver chloride precipitates:

$$Ag^+(aq) + Cl^-(aq) \longrightarrow AgCl(s) \quad \Delta H = -65.5\,kJ$$

(a) Calculate ΔH for the production of 0.450 mol of AgCl by this reaction. (b) Calculate ΔH for the production of 9.00 g of AgCl. (c) Calculate ΔH when 9.25×10^{-4} mol of AgCl dissolves in water.

5.46 At one time, a common means of forming small quantities of oxygen gas in the laboratory was to heat $KClO_3$:

$$2\,KClO_3(s) \longrightarrow 2\,KCl(s) + 3\,O_2(g) \quad \Delta H = -89.4\,kJ$$

For this reaction, calculate ΔH for the formation of (a) 1.36 mol of O_2 and (b) 10.4 g of KCl. (c) The decomposition of $KClO_3$ proceeds spontaneously when it is heated. Do you think that the reverse reaction, the formation of $KClO_3$ from KCl and O_2, is likely to be feasible under ordinary conditions? Explain your answer.

5.47 Consider the combustion of liquid methanol, $CH_3OH(l)$:

$$CH_3OH(l) + \tfrac{3}{2}O_2(g) \longrightarrow CO_2(g) + 2\,H_2O(l)$$
$$\Delta H = -726.5\,kJ$$

(a) What is the enthalpy change for the reverse reaction?
(b) Balance the forward reaction with whole-number coefficients. What is ΔH for the reaction represented by this equation? (c) Which is more likely to be thermodynamically favored, the forward reaction or the reverse reaction? (d) If the reaction were written to produce $H_2O(g)$ instead of $H_2O(l)$, would you expect the magnitude of ΔH to increase, decrease, or stay the same? Explain.

5.48 Consider the decomposition of liquid benzene, $C_6H_6(l)$, to gaseous acetylene, $C_2H_2(g)$:

$$C_6H_6(l) \longrightarrow 3\,C_2H_2(g) \quad \Delta H = +630\,kJ$$

(a) What is the enthalpy change for the reverse reaction?

(b) What is ΔH for the formation of 1 mol of acetylene?

(c) Which is more likely to be thermodynamically favored, the forward reaction or the reverse reaction?

(d) If $C_6H_6(g)$ were consumed instead of $C_6H_6(l)$, would you expect the magnitude of ΔH to increase, decrease, or stay the same? Explain.

Calorimetry (Section 5.5)

5.49 (a) What are the units of molar heat capacity? (b) What are the units of specific heat? (c) If you know the specific heat of copper, what additional information do you need to calculate the heat capacity of a particular piece of copper pipe?

5.50 Two solid objects, A and B, are placed in boiling water and allowed to come to the temperature of the water. Each is then lifted out and placed in separate beakers containing 1000 g water at 10.0 °C. Object A increases the water temperature by 3.50 °C; B increases the water temperature by 2.60 °C. (a) Which object has the larger heat capacity? (b) What can you say about the specific heats of A and B?

5.51 (a) What is the specific heat of liquid water? (b) What is the molar heat capacity of liquid water? (c) What is the heat capacity of 185 g of liquid water? (d) How many kJ of heat are needed to raise the temperature of 10.00 kg of liquid water from 24.6 to 46.2 °C?

5.52 (a) Which substance in Table 5.2 requires the smallest amount of energy to increase the temperature of 50.0 g of that substance by 10 K? (b) Calculate the energy needed for this temperature change.

5.53 The specific heat of octane, $C_8H_{18}(l)$, is 2.22 J/g-K. (a) How many J of heat are needed to raise the temperature of 80.0 g of octane from 10.0 to 25.0 °C? (b) Which will require more heat, increasing the temperature of 1 mol of $C_8H_{18}(l)$ by a certain amount or increasing the temperature of 1 mol of $H_2O(l)$ by the same amount?

5.54 Consider the data about gold metal in Exercise 5.26(b). (a) Based on the data, calculate the specific heat of Au(s). (b) Suppose that the same amount of heat is added to two 10.0-g blocks of metal, both initially at the same temperature. One block is gold metal, and one is iron metal. Which block will have the greater rise in temperature after the addition of the heat? (c) What is the molar heat capacity of Au(s)?

5.55 When a 6.50-g sample of solid sodium hydroxide dissolves in 100.0 g of water in a coffee-cup calorimeter (Figure 5.17), the temperature rises from 21.6 to 37.8 °C. (a) Calculate the quantity of heat (in kJ) released in the reaction. (b) Using your result from part (a), calculate ΔH (in kJ/mol NaOH) for the solution process. Assume that the specific heat of the solution is the same as that of pure water.

5.56 (a) When a 4.25-g sample of solid ammonium nitrate dissolves in 60.0 g of water in a coffee-cup calorimeter (Figure 5.17), the temperature drops from 22.0 to 16.9 °C. Calculate ΔH (in kJ/mol NH_4NO_3) for the solution process:

$$NH_4NO_3(s) \longrightarrow NH_4^+(aq) + NO_3^-(aq)$$

Assume that the specific heat of the solution is the same as that of pure water. (**b**) Is this process endothermic or exothermic?

5.57 A 2.200-g sample of quinone ($C_6H_4O_2$) is burned in a bomb calorimeter whose total heat capacity is 7.854 kJ/°C. The temperature of the calorimeter increases from 23.44 to 30.57 °C. What is the heat of combustion per gram of quinone? Per mole of quinone?

5.58 A 1.800-g sample of phenol (C_6H_5OH) was burned in a bomb calorimeter whose total heat capacity is 11.66 kJ/°C. The temperature of the calorimeter plus contents increased from 21.36 to 26.37 °C. (**a**) Write a balanced chemical equation for the bomb calorimeter reaction. (**b**) What is the heat of combustion per gram of phenol? Per mole of phenol?

5.59 Under constant-volume conditions, the heat of combustion of glucose ($C_6H_{12}O_6$) is 15.57 kJ/g. A 3.500-g sample of glucose is burned in a bomb calorimeter. The temperature of the calorimeter increases from 20.94 to 24.72 °C. (**a**) What is the total heat capacity of the calorimeter? (**b**) If the size of the glucose sample had been exactly twice as large, what would the temperature change of the calorimeter have been?

5.60 Under constant-volume conditions, the heat of combustion of benzoic acid (C_6H_5COOH) is 26.38 kJ/g. A 2.760-g sample of benzoic acid is burned in a bomb calorimeter. The temperature of the calorimeter increases from 21.60 to 29.93 °C. (**a**) What is the total heat capacity of the calorimeter? (**b**) A 1.440-g sample of a new organic substance is combusted in the same calorimeter. The temperature of the calorimeter increases from 22.14 to 27.09 °C. What is the heat of combustion per gram of the new substance? (**c**) Suppose that in changing samples, a portion of the water in the calorimeter were lost. In what way, if any, would this change the heat capacity of the calorimeter?

Hess's Law (Section 5.6)

5.61 What is the connection between Hess's law and the fact that *H* is a state function?

5.62 Consider the following hypothetical reactions:

$$A \longrightarrow B \quad \Delta H = +30 \text{ kJ}$$
$$B \longrightarrow C \quad \Delta H = +60 \text{ kJ}$$

(**a**) Use Hess's law to calculate the enthalpy change for the reaction A $\longrightarrow$ C.

(**b**) Construct an enthalpy diagram for substances A, B, and C, and show how Hess's law applies.

5.63 Calculate the enthalpy change for the reaction
$$P_4O_6(s) + 2 O_2(g) \longrightarrow P_4O_{10}(s)$$
given the following enthalpies of reaction:
$$P_4(s) + 3 O_2(g) \longrightarrow P_4O_6(s) \quad \Delta H = -1640.1 \text{ kJ}$$
$$P_4(s) + 5 O_2(g) \longrightarrow P_4O_{10}(s) \quad \Delta H = -2940.1 \text{ kJ}$$

5.64 From the enthalpies of reaction
$$2 C(s) + O_2(g) \longrightarrow 2 CO(g) \quad \Delta H = -221.0 \text{ kJ}$$
$$2 C(s) + O_2(g) + 4 H_2(g) \longrightarrow 2 CH_3OH(g) \quad \Delta H = -402.4 \text{ kJ}$$

calculate ΔH for the reaction
$$CO(g) + 2 H_2(g) \longrightarrow CH_3OH(g)$$

5.65 From the enthalpies of reaction
$$H_2(g) + F_2(g) \longrightarrow 2 HF(g) \quad \Delta H = -537 \text{ kJ}$$
$$C(s) + 2 F_2(g) \longrightarrow CF_4(g) \quad \Delta H = -680 \text{ kJ}$$
$$2 C(s) + 2 H_2(g) \longrightarrow C_2H_4(g) \quad \Delta H = +52.3 \text{ kJ}$$

calculate ΔH for the reaction of ethylene with F_2:
$$C_2H_4(g) + 6 F_2(g) \longrightarrow 2 CF_4(g) + 4 HF(g)$$

5.66 Given the data
$$N_2(g) + O_2(g) \longrightarrow 2 NO(g) \quad \Delta H = +180.7 \text{ kJ}$$
$$2 NO(g) + O_2(g) \longrightarrow 2 NO_2(g) \quad \Delta H = -113.1 \text{ kJ}$$
$$2 N_2O(g) \longrightarrow 2 N_2(g) + O_2(g) \quad \Delta H = -163.2 \text{ kJ}$$

use Hess's law to calculate ΔH for the reaction
$$N_2O(g) + NO_2(g) \longrightarrow 3 NO(g)$$

Enthalpies of Formation (Section 5.7)

5.67 (**a**) What is meant by the term *standard conditions* with reference to enthalpy changes? (**b**) What is meant by the term *enthalpy of formation*? (**c**) What is meant by the term *standard enthalpy of formation*?

5.68 (**a**) Why are tables of standard enthalpies of formation so useful? (**b**) What is the value of the standard enthalpy of formation of an element in its most stable form? (**c**) Write the chemical equation for the reaction whose enthalpy change is the standard enthalpy of formation of sucrose (table sugar), $C_{12}H_{22}O_{11}(s)$, $\Delta H_f°[C_{12}H_{22}O_{11}]$.

5.69 For each of the following compounds, write a balanced thermochemical equation depicting the formation of one mole of the compound from its elements in their standard states and then look up $\Delta H_f°$ for each substance in Appendix C. (**a**) $NO_2(g)$, (**b**) $SO_3(g)$, (**c**) $NaBr(s)$, (**d**) $Pb(NO_3)_2(s)$.

5.70 Write balanced equations that describe the formation of the following compounds from elements in their standard states, and then look up the standard enthalpy of formation for each substance in Appendix C: (**a**) $H_2O_2(g)$, (**b**) $CaCO_3(s)$, (**c**) $POCl_3(l)$, (**d**) $C_2H_5OH(l)$.

5.71 The following is known as the thermite reaction:
$$2 Al(s) + Fe_2O_3(s) \longrightarrow Al_2O_3(s) + 2 Fe(s)$$

This highly exothermic reaction is used for welding massive units, such as propellers for large ships. Using standard enthalpies of formation in Appendix C, calculate $\Delta H°$ for this reaction.

5.72 Many portable gas heaters and grills use propane, $C_3H_8(g)$, as a fuel. Using standard enthalpies of formation, calculate the quantity of heat produced when 10.0 g of propane is completely combusted in air under standard conditions.

5.73 Using values from Appendix C, calculate the standard enthalpy change for each of the following reactions:

(**a**) $2 SO_2(g) + O_2(g) \longrightarrow 2 SO_3(g)$

(**b**) $Mg(OH)_2(s) \longrightarrow MgO(s) + H_2O(l)$

(**c**) $N_2O_4(g) + 4 H_2(g) \longrightarrow N_2(g) + 4 H_2O(g)$

(**d**) $SiCl_4(l) + 2 H_2O(l) \longrightarrow SiO_2(s) + 4 HCl(g)$

5.74 Using values from Appendix C, calculate the value of $\Delta H°$ for each of the following reactions:

(a) $CaO(s) + 2\,HCl(g) \longrightarrow CaCl_2(s) + H_2O(g)$

(b) $4\,FeO(s) + O_2(g) \longrightarrow 2\,Fe_2O_3(s)$

(c) $2\,CuO(s) + NO(g) \longrightarrow Cu_2O(s) + NO_2(g)$

(d) $4\,NH_3(g) + O_2(g) \longrightarrow 2\,N_2H_4(g) + 2\,H_2O(l)$

5.75 Complete combustion of 1 mol of acetone (C_3H_6O) liberates 1790 kJ:

$$C_3H_6O(l) + 4\,O_2(g) \longrightarrow 3\,CO_2(g) + 3\,H_2O(l)$$
$$\Delta H° = -1790\text{ kJ}$$

Using this information together with the standard enthalpies of formation of $O_2(g)$, $CO_2(g)$, and $H_2O(l)$ from Appendix C, calculate the standard enthalpy of formation of acetone.

5.76 Calcium carbide (CaC_2) reacts with water to form acetylene (C_2H_2) and $Ca(OH)_2$. From the following enthalpy of reaction data and data in Appendix C, calculate $\Delta H_f°$ for $CaC_2(s)$:

$$CaC_2(s) + 2\,H_2O(l) \longrightarrow Ca(OH)_2(s) + C_2H_2(g)$$
$$\Delta H° = -127.2\text{ kJ}$$

5.77 Gasoline is composed primarily of hydrocarbons, including many with eight carbon atoms, called *octanes*. One of the cleanest-burning octanes is a compound called 2,3,4-trimethylpentane, which has the following structural formula:

$$
\begin{array}{ccccc}
 & CH_3 & CH_3 & CH_3 & \\
 & | & | & | & \\
H_3C- & CH- & CH- & CH- & CH_3
\end{array}
$$

The complete combustion of one mole of this compound to $CO_2(g)$ and $H_2O(g)$ leads to $\Delta H° = -5064.9$ kJ/mol. **(a)** Write a balanced equation for the combustion of 1 mol of $C_8H_{18}(l)$. **(b)** By using the information in this problem and data in Table 5.3, calculate $\Delta H_f°$ for 2,3,4-trimethylpentane.

5.78 Diethyl ether, $C_4H_{10}O(l)$, a flammable compound that has long been used as a surgical anesthetic, has the structure

$$H_3C-CH_2-O-CH_2-CH_3$$

The complete combustion of 1 mol of $C_4H_{10}O(l)$ to $CO_2(g)$ and $H_2O(l)$ yields $\Delta H° = -2723.7$ kJ. **(a)** Write a balanced equation for the combustion of 1 mol of $C_4H_{10}O(l)$. **(b)** By using the information in this problem and data in Table 5.3, calculate $\Delta H_f°$ for diethyl ether.

5.79 Ethanol (C_2H_5OH) is currently blended with gasoline as an automobile fuel. **(a)** Write a balanced equation for the combustion of liquid ethanol in air. **(b)** Calculate the standard enthalpy change for the reaction, assuming $H_2O(g)$ as a product. **(c)** Calculate the heat produced per liter of ethanol by combustion of ethanol under constant pressure. Ethanol has a density of 0.789 g/mL. **(d)** Calculate the mass of CO_2 produced per kJ of heat emitted.

5.80 Methanol (CH_3OH) is used as a fuel in race cars. **(a)** Write a balanced equation for the combustion of liquid methanol in air. **(b)** Calculate the standard enthalpy change for the reaction, assuming $H_2O(g)$ as a product. **(c)** Calculate the heat produced by combustion per liter of methanol. Methanol has a density of 0.791 g/mL. **(d)** Calculate the mass of CO_2 produced per kJ of heat emitted.

Foods and Fuels (Section 5.8)

5.81 **(a)** What is meant by the term *fuel value*? **(b)** Which is a greater source of energy as food, 5 g of fat or 9 g of carbohydrate? **(c)** The metabolism of glucose produces $CO_2(g)$ and $H_2O(l)$. How does the human body expel these reaction products?

5.82 **(a)** Why are fats well suited for energy storage in the human body? **(b)** A particular chip snack food is composed of 12% protein, 14% fat, and the rest carbohydrate. What percentage of the calorie content of this food is fat? **(c)** How many grams of protein provide the same fuel value as 25 g of fat?

5.83 **(a)** A serving of a particular ready-to-serve chicken noodle soup contains 2.5 g fat, 14 g carbohydrate, and 7 g protein. Estimate the number of Calories in a serving. **(b)** According to its nutrition label, the same soup also contains 690 mg of sodium. Do you think the sodium contributes to the caloric content of the soup?

5.84 A pound of plain M&M® candies contains 96 g fat, 320 g carbohydrate, and 21 g protein. What is the fuel value in kJ in a 42-g (about 1.5 oz) serving? How many Calories does it provide?

5.85 The heat of combustion of fructose, $C_6H_{12}O_6$, is -2812 kJ/mol. If a fresh golden delicious apple weighing 4.23 oz (120 g) contains 16.0 g of fructose, what caloric content does the fructose contribute to the apple?

5.86 The heat of combustion of ethanol, $C_2H_5OH(l)$, is -1367 kJ/mol. A batch of Sauvignon Blanc wine contains 10.6% ethanol by mass. Assuming the density of the wine to be 1.0 g/mL, what is the caloric content due to the alcohol (ethanol) in a 6-oz glass of wine (177 mL)?

5.87 The standard enthalpies of formation of gaseous propyne (C_3H_4), propylene (C_3H_6), and propane (C_3H_8) are $+185.4$, $+20.4$, and -103.8 kJ/mol, respectively. **(a)** Calculate the heat evolved per mole on combustion of each substance to yield $CO_2(g)$ and $H_2O(g)$. **(b)** Calculate the heat evolved on combustion of 1 kg of each substance. **(c)** Which is the most efficient fuel in terms of heat evolved per unit mass?

5.88 It is interesting to compare the "fuel value" of a hydrocarbon in a world where oxygen is the combustion agent. The enthalpy of formation of $CF_4(g)$ is -679.9 kJ/mol. Which of the following two reactions is the more exothermic?

$$CH_4(g) + 2\,O_2(g) \longrightarrow CO_2(g) + 2\,H_2O(g)$$
$$CH_4(g) + 4\,F_2(g) \longrightarrow CF_4(g) + 4\,HF(g)$$

5.89 At the end of 2012, global population was about 7.0 billion people. What mass of glucose in kg would be needed to provide 1500 cal/person/day of nourishment to the global population for one year? Assume that glucose is metabolized entirely to $CO_2(g)$ and $H_2O(l)$ according to the following thermochemical equation:

$$C_6H_{12}O_6(s) + 6\,O_2(g) \longrightarrow 6\,CO_2(g) + 6\,H_2O(l)$$
$$\Delta H° = -2803\text{ kJ}$$

5.90 The automobile fuel called E85 consists of 85% ethanol and 15% gasoline. E85 can be used in the so-called flex-fuel vehicles (FFVs), which can use gasoline, ethanol, or a mix as fuels. Assume that gasoline consists of a mixture of octanes (different isomers of C_8H_{18}), that the average heat of combustion of $C_8H_{18}(l)$ is 5400 kJ/mol, and that gasoline has an average density of 0.70 g/mL. The density of ethanol is 0.79 g/mL. (a) By using the information given as well as data in Appendix C, compare the energy produced by combustion of 1.0 L of gasoline and of 1.0 L of ethanol. (b) Assume that the density and heat of combustion of E85 can be obtained by using 85% of the values for ethanol and 15% of the values for gasoline. How much energy could be released by the combustion of 1.0 L of E85? (c) How many gallons of E85 would be needed to provide the same energy as 10 gal of gasoline? (d) If gasoline costs $3.88 per gallon in the United States, what is the break-even price per gallon of E85 if the same amount of energy is to be delivered?

Additional Exercises

5.91 At 20 °C (approximately room temperature) the average velocity of N_2 molecules in air is 1050 mph. (a) What is the average speed in m/s? (b) What is the kinetic energy (in J) of an N_2 molecule moving at this speed? (c) What is the total kinetic energy of 1 mol of N_2 molecules moving at this speed?

5.92 Suppose an Olympic diver who weighs 52.0 kg executes a straight dive from a 10-m platform. At the apex of the dive, the diver is 10.8 m above the surface of the water. (a) What is the potential energy of the diver at the apex of the dive, relative to the surface of the water? (b) Assuming that all the potential energy of the diver is converted into kinetic energy at the surface of the water, at what speed, in m/s, will the diver enter the water? (c) Does the diver do work on entering the water? Explain.

5.93 The air bags that provide protection in automobiles in the event of an accident expand because of a rapid chemical reaction. From the viewpoint of the chemical reactants as the system, what do you expect for the signs of q and w in this process?

5.94 An aluminum can of a soft drink is placed in a freezer. Later, you find that the can is split open and its contents frozen. Work was done on the can in splitting it open. Where did the energy for this work come from?

5.95 Consider a system consisting of the following apparatus, in which gas is confined in one flask and there is a vacuum in the other flask. The flasks are separated by a valve. Assume that the flasks are perfectly insulated and will not allow the flow of heat into or out of the flasks to the surroundings. When the valve is opened, gas flows from the filled flask to the evacuated one. (a) Is work performed during the expansion of the gas? (b) Why or why not? (c) Can you determine the value of ΔE for the process?

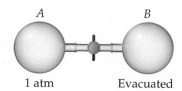

1 atm   Evacuated

5.96 A sample of gas is contained in a cylinder-and-piston arrangement. It undergoes the change in state shown in the drawing. (a) Assume first that the cylinder and piston are perfect thermal insulators that do not allow heat to be transferred. What is the value of q for the state change? What is the sign of w for the state change? What can be said about ΔE for the state change? (b) Now assume that the cylinder and piston

are made up of a thermal conductor such as a metal. During the state change, the cylinder gets warmer to the touch. What is the sign of q for the state change in this case? Describe the difference in the state of the system at the end of the process in the two cases. What can you say about the relative values of ΔE?

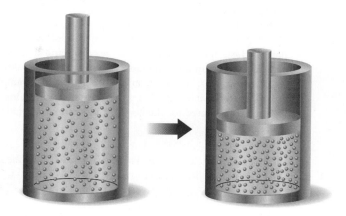

5.97 Limestone stalactites and stalagmites are formed in caves by the following reaction:

$$Ca^{2+}(aq) + 2\,HCO_3^-(aq) \longrightarrow CaCO_3(s) + CO_2(g) + H_2O(l)$$

If 1 mol of $CaCO_3$ forms at 298 K under 1 atm pressure, the reaction performs 2.47 kJ of P–V work, pushing back the atmosphere as the gaseous CO_2 forms. At the same time, 38.95 kJ of heat is absorbed from the environment. What are the values of ΔH and of ΔE for this reaction?

5.98 Consider the systems shown in Figure 5.10. In one case the battery becomes completely discharged by running the current through a heater and in the other case by running a fan. Both processes occur at constant pressure. In both cases the change in state of the system is the same: The battery goes from being fully charged to being fully discharged. Yet in one case the heat evolved is large, and in the other it is small. Is the enthalpy change the same in the two cases? If not, how can enthalpy be considered a state function? If it is, what can you say about the relationship between enthalpy change and q in this case, as compared with others that we have considered?

5.99 A house is designed to have passive solar energy features. Brickwork incorporated into the interior of the house acts as a heat absorber. Each brick weighs approximately 1.8 kg. The specific heat of the brick is 0.85 J/g-K. How many bricks must be incorporated into the interior of the house to provide the same total heat capacity as 1.7×10^3 gal of water?

5.100 A coffee-cup calorimeter of the type shown in Figure 5.17 contains 150.0 g of water at 25.1 °C. A 121.0-g block of copper metal is heated to 100.4 °C by putting it in a beaker of boiling water. The specific heat of $Cu(s)$ is 0.385 J/g-K. The Cu is added to the calorimeter, and after a time the contents of the cup reach a constant temperature of 30.1 °C. (a) Determine the amount of heat, in J, lost by the copper block. (b) Determine the amount of heat gained by the water. The specific heat of water is 4.18 J/g-K. (c) The difference between your answers for (a) and (b) is due to heat loss through the Styrofoam® cups and the heat necessary to raise the temperature of the inner wall of the apparatus. The heat capacity of the calorimeter is the amount of heat necessary to raise the temperature of the apparatus (the cups and the stopper) by 1 K. Calculate the heat capacity of the calorimeter in J/K. **(d)** What would be the final temperature of the system if all the heat lost by the copper block were absorbed by the water in the calorimeter?

5.101 **(a)** When a 0.235-g sample of benzoic acid is combusted in a bomb calorimeter (Figure 5.18), the temperature rises 1.642 °C. When a 0.265-g sample of caffeine, $C_8H_{10}O_2N_4$, is burned, the temperature rises 1.525 °C. Using the value 26.38 kJ/g for the heat of combustion of benzoic acid, calculate the heat of combustion per mole of caffeine at constant volume. **(b)** Assuming that there is an uncertainty of 0.002 °C in each temperature reading and that the masses of samples are measured to 0.001 g, what is the estimated uncertainty in the value calculated for the heat of combustion per mole of caffeine?

5.102 Meals-ready-to-eat (MREs) are military meals that can be heated on a flameless heater. The heat is produced by the following reaction:

$$Mg(s) + 2 H_2O(l) \longrightarrow Mg(OH)_2(s) + 2H_2(g)$$

(a) Calculate the standard enthalpy change for this reaction. **(b)** Calculate the number of grams of Mg needed for this reaction to release enough energy to increase the temperature of 75 mL of water from 21 to 79 °C.

5.103 Burning methane in oxygen can produce three different carbon-containing products: soot (very fine particles of graphite), $CO(g)$, and $CO_2(g)$. **(a)** Write three balanced equations for the reaction of methane gas with oxygen to produce these three products. In each case assume that $H_2O(l)$ is the only other product. **(b)** Determine the standard enthalpies for the reactions in part (a). **(c)** Why, when the oxygen supply is adequate, is $CO_2(g)$ the predominant carbon-containing product of the combustion of methane?

5.104 We can use Hess's law to calculate enthalpy changes that cannot be measured. One such reaction is the conversion of methane to ethylene:

$$2 CH_4(g) \longrightarrow C_2H_4(g) + H_2(g)$$

Calculate the $\Delta H°$ for this reaction using the following thermochemical data:

$$CH_4(g) + 2 O_2(g) \longrightarrow CO_2(g) + 2 H_2O(l) \quad \Delta H° = -890.3 \text{ kJ}$$
$$C_2H_4(g) + H_2(g) \longrightarrow C_2H_6(g) \quad \Delta H° = -136.3 \text{ kJ}$$
$$2 H_2(g) + O_2(g) \longrightarrow 2 H_2O(l) \quad \Delta H° = -571.6 \text{ kJ}$$
$$2 C_2H_6(g) + 7 O_2(g) \longrightarrow 4 CO_2(g) + 6 H_2O(l) \quad \Delta H° = -3120.8 \text{ kJ}$$

5.105 From the following data for three prospective fuels, calculate which could provide the most energy per unit volume:

Fuel	Density at 20 °C (g/cm³)	Molar Enthalpy of Combustion (kJ/mol)
Nitroethane, $C_2H_5NO_2(l)$	1.052	−1368
Ethanol, $C_2H_5OH(l)$	0.789	−1367
Methylhydrazine, $CH_6N_2(l)$	0.874	−1307

5.106 The hydrocarbons acetylene (C_2H_2) and benzene (C_6H_6) have the same empirical formula. Benzene is an "aromatic" hydrocarbon, one that is unusually stable because of its structure. **(a)** By using data in Appendix C, determine the standard enthalpy change for the reaction $3 C_2H_2(g) \longrightarrow C_6H_6(l)$.

(b) Which has greater enthalpy, 3 mol of acetylene gas or 1 mol of liquid benzene? **(c)** Determine the fuel value, in kJ/g, for acetylene and benzene.

5.107 Ammonia (NH_3) boils at −33 °C; at this temperature it has a density of 0.81 g/cm³. The enthalpy of formation of $NH_3(g)$ is −46.2 kJ/mol, and the enthalpy of vaporization of $NH_3(l)$ is 23.2 kJ/mol. Calculate the enthalpy change when 1 L of liquid NH_3 is burned in air to give $N_2(g)$ and $H_2O(g)$. How does this compare with ΔH for the complete combustion of 1 L of liquid methanol, $CH_3OH(l)$? For $CH_3OH(l)$, the density at 25 °C is 0.792 g/cm³, and $\Delta H_f° = -239$ kJ/mol.

5.108 Three common hydrocarbons that contain four carbons are listed here, along with their standard enthalpies of formation:

Hydrocarbon	Formula	$\Delta H_f°$ (kJ/mol)
1,3-Butadiene	$C_4H_6(g)$	111.9
1-Butene	$C_4H_8(g)$	1.2
n-Butane	$C_4H_{10}(g)$	−124.7

(a) For each of these substances, calculate the molar enthalpy of combustion to $CO_2(g)$ and $H_2O(l)$. **(b)** Calculate the fuel value, in kJ/g, for each of these compounds. **(c)** For each hydrocarbon, determine the percentage of hydrogen by mass. **(d)** By comparing your answers for parts (b) and (c), propose a relationship between hydrogen content and fuel value in hydrocarbons.

5.109 A 200-lb man decides to add to his exercise routine by walking up three flights of stairs (45 ft) 20 times per day. He figures that the work required to increase his potential energy in this way will permit him to eat an extra order of French fries, at 245 Cal, without adding to his weight. Is he correct in this assumption?

5.110 The Sun supplies about 1.0 kilowatt of energy for each square meter of surface area (1.0 kW/m², where a watt = 1 J/s). Plants produce the equivalent of about 0.20 g of sucrose ($C_{12}H_{22}O_{11}$) per hour per square meter. Assuming that the sucrose is produced as follows, calculate the percentage of sunlight used to produce sucrose.

$$12 CO_2(g) + 11 H_2O(l) \longrightarrow C_{12}H_{22}O_{11} + 12 O_2(g)$$
$$\Delta H = 5645 \text{ kJ}$$

5.111 It is estimated that the net amount of carbon dioxide fixed by photosynthesis on the landmass of Earth is 5.5×10^{16} g/yr of CO_2. Assume that all this carbon is converted into glucose. **(a)** Calculate the energy stored by photosynthesis on land per year, in kJ. **(b)** Calculate the average rate of conversion of solar energy into plant energy in megawatts, MW (1W = 1 J/s). A large nuclear power plant produces about 10^3 MW. The energy of how many such nuclear power plants is equivalent to the solar energy conversion?

Integrative Exercises

5.112 Consider the combustion of a single molecule of $CH_4(g)$, forming $H_2O(l)$ as a product. **(a)** How much energy, in J, is produced during this reaction? **(b)** A typical X-ray light source has an energy of 8 keV. How does the energy of combustion compare to the energy of the X-ray?

5.113 Consider the following unbalanced oxidation-reduction reactions in aqueous solution:

$$Ag^+(aq) + Li(s) \longrightarrow Ag(s) + Li^+(aq)$$
$$Fe(s) + Na^+(aq) \longrightarrow Fe^{2+}(aq) + Na(s)$$
$$K(s) + H_2O(l) \longrightarrow KOH(aq) + H_2(g)$$

(a) Balance each of the reactions. (b) By using data in Appendix C, calculate $\Delta H°$ for each of the reactions. (c) Based on the values you obtain for $\Delta H°$, which of the reactions would you expect to be thermodynamically favored? (d) Use the activity series to predict which of these reactions should occur. ∞ (Section 4.4) Are these results in accord with your conclusion in part (c) of this problem?

5.114 Consider the following acid-neutralization reactions involving the strong base NaOH(aq):

$$HNO_3(aq) + NaOH(aq) \longrightarrow NaNO_3(aq) + H_2O(l)$$
$$HCl(aq) + NaOH(aq) \longrightarrow NaCl(aq) + H_2O(l)$$
$$NH_4^+(aq) + NaOH(aq) \longrightarrow NH_3(aq) + Na^+(aq) + H_2O(l)$$

(a) By using data in Appendix C, calculate $\Delta H°$ for each of the reactions. (b) As we saw in Section 4.3, nitric acid and hydrochloric acid are strong acids. Write net ionic equations for the neutralization of these acids. (c) Compare the values of $\Delta H°$ for the first two reactions. What can you conclude? (d) In the third equation $NH_4^+(aq)$ is acting as an acid. Based on the value of $\Delta H°$ for this reaction, do you think it is a strong or a weak acid? Explain.

5.115 Consider two solutions, the first being 50.0 mL of 1.00 M CuSO₄ and the second 50.0 mL of 2.00 M KOH. When the two solutions are mixed in a constant-pressure calorimeter, a precipitate forms and the temperature of the mixture rises from 21.5 to 27.7 °C. (a) Before mixing, how many grams of Cu are present in the solution of CuSO₄? (b) Predict the identity of the precipitate in the reaction. (c) Write complete and net ionic equations for the reaction that occurs when the two solutions are mixed. (d) From the calorimetric data, calculate ΔH for the reaction that occurs on mixing. Assume that the calorimeter absorbs only a negligible quantity of heat, that the total volume of the solution is 100.0 mL, and that the specific heat and density of the solution after mixing are the same as those of pure water.

5.116 The precipitation reaction between AgNO₃(aq) and NaCl(aq) proceeds as follows:

$$AgNO_3(aq) + NaCl(aq) \longrightarrow NaNO_3(aq) + AgCl(s)$$

(a) By using data in Appendix C, calculate $\Delta H°$ for the net ionic equation of this reaction. (b) What would you expect for the value of $\Delta H°$ of the overall molecular equation compared to that for the net ionic equation? Explain. (c) Use the results from (a) and (b) along with data in Appendix C to determine the value of $\Delta H_f°$ for AgNO₃(aq).

5.117 A sample of a hydrocarbon is combusted completely in O₂(g) to produce 21.83 g CO₂(g), 4.47 g H₂O(g), and 311 kJ of heat. (a) What is the mass of the hydrocarbon sample that was combusted? (b) What is the empirical formula of the hydrocarbon? (c) Calculate the value of $\Delta H_f°$ per empirical-formula unit of the hydrocarbon. (d) Do you think that the hydrocarbon is one of those listed in Appendix C? Explain your answer.

5.118 The methane molecule, CH₄, has the geometry shown in Figure 2.17. Imagine a hypothetical process in which the methane molecule is "expanded," by simultaneously extending all four C—H bonds to infinity. We then have the process

$$CH_4(g) \longrightarrow C(g) + 4 H(g)$$

(a) Compare this process with the reverse of the reaction that represents the standard enthalpy of formation of CH₄(g). (b) Calculate the enthalpy change in each case. Which is the more endothermic process? What accounts for the difference in $\Delta H°$ values? (c) Suppose that 3.45 g CH₄(g) reacts with 1.22 g F₂(g), forming CH₄(g) and HF(g) as sole products. What is the limiting reagent in this reaction? If the reaction occurs at constant pressure, what amount of heat is evolved?

5.119 World energy supplies are often measured in the unit of quadrillion British thermal units (10^{12} Btu), generally called a "quad." In 2015, world energy consumption is projected to be 5.81 × 10^{17} kJ. (a) With reference to Exercise 5.17, how many quads of energy does this quantity represent? (b) Current annual energy consumption in the United States is 99.5 quads. Assume that all this energy is to be generated by burning CH₄(g) in the form of natural gas. If the combustion of the CH₄(g) were complete and 100% efficient, how many moles of CH₄(g) would need to be combusted to provide the U.S. energy demand? (c) How many kilograms of CO₂(g) would be generated in the combustion in part (b)? (d) Compare your answer to part (c) with information given in Exercise 5.111. Do you think that photosynthesis is an adequate means to maintain a stable level of CO₂ in the atmosphere?

Design an Experiment

One of the important ideas of thermodynamics is that energy can be transferred in the form of heat or work. Imagine that you lived 150 years ago when the relationships between heat and work were not well understood. You have formulated a hypothesis that work can be converted to heat with the same amount of work always generating the same amount of heat. To test this idea, you have designed an experiment using a device in which a falling weight is connected through pulleys to a shaft with an attached paddle wheel that is immersed in water. This is actually a classic experiment performed by James Joule in the 1840s. You can see various images of Joule's apparatus by Googling "Joule experiment images."

(a) Using this device, what measurements would you need to make to test your hypothesis? (b) What equations would you use in analyzing your experiment? (c) Do you think you could obtain a reasonable result from a single experiment? Why or why not? (d) In what way could the precision of your instruments affect the conclusions that you make? (e) List ways that you could modify the equipment to improve the data you obtain if you were performing this experiment today instead of 150 years ago. (f) Give an example of how you could demonstrate the relationship between heat and a form of energy other than mechanical work.

6
Electronic Structure of Atoms

The beginning of the twentieth century was truly one of the most revolutionary periods of scientific discovery. Two theoretical developments caused dramatic changes in our view of the universe. The first, Einstein's theory of relativity, forever changed our views of the relationships between space and time. The second—which will be the focus of this chapter—is the *quantum theory*, which explains much of the behavior of electrons in atoms.

The quantum theory has led to the explosion in technological developments in the twentieth century, including remarkable new light sources, such as light-emitting diodes (LEDs) that are now being used as high-quality, low-energy-consumption light sources in many applications, and lasers, which have revolutionized so many aspects of our lives. The quantum theory also led to the development of the solid-state electronics that has allowed computers, cellular telephones, and countless other electronic devices to transform our daily lives.

In this chapter we explore the quantum theory and its importance in chemistry. We begin by looking at the nature of light and how our description of light was changed by the quantum theory. We will explore some of the tools used in *quantum mechanics*, the "new" physics that had to be developed to describe atoms correctly. We will then use the quantum theory to describe the arrangements of electrons in atoms—what we call the **electronic structure** of atoms. The electronic structure of an atom refers to the number of electrons in the atom as well as their distribution around the nucleus and their energies. We will see that the quantum description of the electronic structure of atoms helps us to understand the arrangement of the elements in the periodic table—why, for example, helium and neon are both unreactive gases, whereas sodium and potassium are both soft, reactive metals.

▶ **A LASER LIGHTSHOW** at an entertainment event. Lasers produce light with very specific colors because of step-like energy transitions by electrons in the laser materials.

WHAT'S AHEAD ▶

6.1 THE WAVE NATURE OF LIGHT We learn that light (radiant energy, or *electromagnetic radiation*) has wave-like properties and is characterized by *wavelength*, *frequency*, and *speed*.

6.2 QUANTIZED ENERGY AND PHOTONS From studies of the radiation given off by hot objects and of the interaction of light with metal surfaces, we recognize that electromagnetic radiation also has particle-like properties and can be described as *photons*, "particles" of light.

6.3 LINE SPECTRA AND THE BOHR MODEL We examine the light emitted by electrically excited atoms (*line spectra*). Line spectra indicate that there are only certain energy levels that are allowed for electrons in atoms and that energy is involved when an electron jumps from one level to another. The Bohr model of the atom pictures the electrons moving only in certain allowed orbits around the nucleus.

6.4 THE WAVE BEHAVIOR OF MATTER We recognize that matter also has wave-like properties. As a result, it is impossible to determine simultaneously the exact position and the exact momentum of an electron in an atom (*Heisenberg's uncertainty principle*).

6.5 QUANTUM MECHANICS AND ATOMIC ORBITALS We can describe how the electron exists in the hydrogen atom by treating it as if it were a wave. The *wave functions* that mathematically describe the electron's position and energy in an atom are called *atomic orbitals*. Each orbital is characterized by a set of *quantum numbers*.

6.6 REPRESENTATIONS OF ORBITALS We consider the three-dimensional shapes of orbitals and how they can be represented by graphs of electron density.

6.7 MANY-ELECTRON ATOMS We learn that the energy levels of an atom having more than one electron are different from those of the hydrogen atom. In addition, we learn that each electron has an additional quantum-mechanical property called *spin*. The *Pauli*

exclusion principle states that no two electrons in an atom can have the same four quantum numbers (three for the orbital and one for the spin). Therefore, each orbital can hold a maximum of two electrons.

6.8 ELECTRON CONFIGURATIONS We learn how the orbitals of the hydrogen atom can be used to describe the arrangements of electrons in many-electron atoms. Using patterns in orbital energies as well as some fundamental characteristics of electrons described by *Hund's rule*, we determine how electrons are distributed among the orbitals (*electron configurations*).

6.9 ELECTRON CONFIGURATIONS AND THE PERIODIC TABLE We observe that the electron configuration of an atom is related to the location of the element in the periodic table.

▲ **Figure 6.1 Water waves.** The movement of a boat through the water forms waves. The regular variation of peaks and troughs enables us to sense the motion of the waves away from the boat.

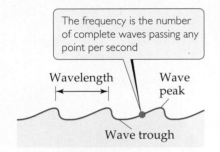

The frequency is the number of complete waves passing any point per second

Wavelength

Wave peak

Wave trough

▲ **Figure 6.2 Water waves.** The *wavelength* is the distance between two adjacent peaks or two adjacent troughs.

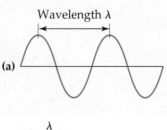

GO FIGURE

If wave (a) has a wavelength of 2.0 m and a frequency of 1.5×10^8 cycles/s, what are the wavelength and frequency of wave (b)?

Wavelength λ

(a)

λ

(b)

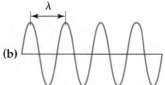

▲ **Figure 6.3 Electromagnetic waves.** Like water waves, electromagnetic radiation can be characterized by a wavelength. Notice that the shorter the wavelength, λ, the higher the frequency, ν. The wavelength in (b) is half as long as that in (a), and the frequency of the wave in (b) is therefore twice as great as that in (a).

6.1 | The Wave Nature of Light

Much of our present understanding of the electronic structure of atoms has come from analysis of the light either emitted or absorbed by substances. To understand electronic structure, therefore, we must first learn more about light. The light we see with our eyes, *visible light*, is one type of **electromagnetic radiation**. Because electromagnetic radiation carries energy through space, it is also known as *radiant energy*.

There are many types of electromagnetic radiation in addition to visible light. These different types—radio waves that carry music to our radios, infrared radiation (heat) from a glowing fireplace, X rays—may seem very different from one another, but they all share certain fundamental characteristics.

All types of electromagnetic radiation move through a vacuum at 2.998×10^8 m/s, the *speed of light*. All have wave-like characteristics similar to those of waves that move through water. Water waves are the result of energy imparted to the water, perhaps by the dropping of a stone or the movement of a boat on the water surface (◀ Figure 6.1). This energy is expressed as the up-and-down movements of the water.

A cross section of a water wave (◀ Figure 6.2) shows that it is *periodic*, which means that the pattern of peaks and troughs repeats itself at regular intervals. The distance between two adjacent peaks (or between two adjacent troughs) is called the **wavelength**. The number of complete wavelengths, or *cycles*, that pass a given point each second is the **frequency** of the wave.

Just as with water waves, we can assign a frequency and wavelength to electromagnetic waves, as illustrated in ◀ Figure 6.3. These and all other wave characteristics of electromagnetic radiation are due to the periodic oscillations in the intensities of the electric and magnetic fields associated with the radiation.

The speed of water waves can vary depending on how they are created—for example, the waves produced by a speedboat travel faster than those produced by a rowboat. In contrast, *all electromagnetic radiation moves at the same speed, namely, the speed of light*. As a result, the wavelength and frequency of electromagnetic radiation are always related in a straightforward way. If the wavelength is long, fewer cycles of the wave pass a given point per second, and so the frequency is low. Conversely, for a wave to have a high frequency, it must have a short wavelength. This inverse relationship between the frequency and wavelength of electromagnetic radiation is expressed by the equation

$$\lambda \nu = c \qquad [6.1]$$

where λ (lambda) is wavelength, ν (nu) is frequency, and c is the speed of light.

Why do different types of electromagnetic radiation have different properties? Their differences are due to their different wavelengths. ▶ Figure 6.4 shows the various types of electromagnetic radiation arranged in order of increasing wavelength, a display called the *electromagnetic spectrum*. Notice that the wavelengths span an enormous range. The wavelengths of gamma rays are comparable to the diameters of atomic nuclei, whereas the wavelengths of radio waves can be longer than a football field. Notice also that visible light, which corresponds to wavelengths of about 400 to 750 nm (4×10^{-7} to 7×10^{-7} m), is an extremely small portion of the electromagnetic spectrum. The unit of length chosen to express wavelength depends on the type of radiation, as shown in ▶ Table 6.1.

Frequency is expressed in cycles per second, a unit also called a *hertz* (Hz). Because it is understood that cycles are involved, the units of frequency are normally given simply as "per second," which is denoted by s^{-1} or /s. For example, a frequency of 698 megahertz (MHz), a typical frequency for a cellular telephone, could be written as 698 MHz, 698,000,000 Hz, 698,000,000 s^{-1}, or 698,000,000/s.

Give It Some Thought

Our bodies are penetrated by X rays but not by visible light. Is this because X rays travel faster than visible light?

▲ GO FIGURE

Is the wavelength of a microwave longer or shorter than the wavelength of visible light? By how many orders of magnitude do the two waves differ in wavelength?

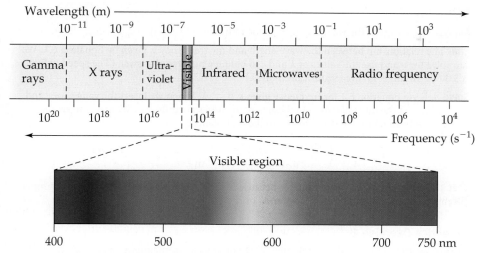

▲ **Figure 6.4 The electromagnetic spectrum.*** Wavelengths in the spectrum range from very short gamma rays to very long radio waves.

Table 6.1 Common Wavelength Units for Electromagnetic Radiation

Unit	Symbol	Length (m)	Type of Radiation
Angstrom	Å	10^{-10}	X ray
Nanometer	nm	10^{-9}	Ultraviolet, visible
Micrometer	μm	10^{-6}	Infrared
Millimeter	mm	10^{-3}	Microwave
Centimeter	cm	10^{-2}	Microwave
Meter	m	1	Television, radio
Kilometer	km	1000	Radio

SAMPLE EXERCISE 6.1 Concepts of Wavelength and Frequency

Two electromagnetic waves are represented in the margin. (**a**) Which wave has the higher frequency? (**b**) If one wave represents visible light and the other represents infrared radiation, which wave is which?

SOLUTION

(**a**) Wave 1 has a longer wavelength (greater distance between peaks). The longer the wavelength, the lower the frequency ($\nu = c/\lambda$). Thus, Wave 1 has the lower frequency, and Wave 2 has the higher frequency.

(**b**) The electromagnetic spectrum (Figure 6.4) indicates that infrared radiation has a longer wavelength than visible light. Thus, Wave 1 would be the infrared radiation.

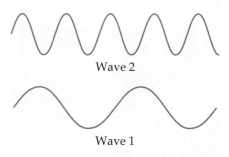

Wave 2

Wave 1

Practice Exercise 1

A source of electromagnetic radiation produces infrared light. Which of the following could be the wavelength of the light? (**a**) 3.0 nm (**b**) 4.7 cm (**c**) 66.8 m (**d**) 34.5 μm (**e**) 16.5 Å

Practice Exercise 2

If one of the waves in the margin represents blue light and the other red light, which wave is which?

*Based on B.A. Averill and P. Eldredge, *Chemistry: Principles, Patterns, and Applications* 1e, © 2007 Pearson Education, Inc.

SAMPLE
SAMPLE
EXERCISE 6.2 Calculating Frequency from Wavelength

The yellow light given off by a sodium vapor lamp used for public lighting has a wavelength of 589 nm. What is the frequency of this radiation?

SOLUTION

Analyze We are given the wavelength, λ, of the radiation and asked to calculate its frequency, ν.

Plan The relationship between the wavelength and the frequency is given by Equation 6.1. We can solve for ν and use the values of λ and c to obtain a numerical answer. (The speed of light, c, is 3.00×10^8 m/s to three significant figures.)

Solve Solving Equation 6.1 for frequency gives $\nu = c/\lambda$. When we insert the values for c and λ, we note that the units of length in these two quantities are different. We can convert the wavelength from nanometers to meters, so the units cancel:

$$\nu = \frac{c}{\lambda} = \left(\frac{3.00 \times 10^8 \text{ m/s}}{589 \text{ nm}}\right)\left(\frac{1 \text{ nm}}{10^{-9} \text{ m}}\right) = 5.09 \times 10^{14} \text{ s}^{-1}$$

Check The high frequency is reasonable because of the short wavelength. The units are proper because frequency has units of "per second," or s^{-1}.

Practice Exercise 1

Consider the following three statements: (i) For any electromagnetic radiation, the product of the wavelength and the frequency is a constant. (ii) If a source of light has a wavelength of 3.0 Å, its frequency is 1.0×10^{18} Hz. (iii) The speed of ultraviolet light is greater than the speed of microwave radiation. Which of these three statements is or are true? **(a)** Only one statement is true. **(b)** Statements (i) and (ii) are true. **(c)** Statements (i) and (iii) are true. **(d)** Statements (ii) and (iii) are true. **(e)** All three statements are true.

Practice Exercise 2

(a) A laser used in orthopedic spine surgery produces radiation with a wavelength of 2.10 μm. Calculate the frequency of this radiation. **(b)** An FM radio station broadcasts electromagnetic radiation at a frequency of 103.4 MHz (megahertz; 1 MHz = 10^6 s^{-1}). Calculate the wavelength of this radiation. The speed of light is 2.998×10^8 m/s to four significant figures.

6.2 | Quantized Energy and Photons

Although the wave model of light explains many aspects of the behavior of light, several observations cannot be resolved by this model. Three of these are particularly pertinent to our understanding of how electromagnetic radiation and atoms interact: (1) the emission of light from hot objects (referred to as *blackbody radiation* because the objects studied appear black before heating), (2) the emission of electrons from metal surfaces on which light shines (the *photoelectric effect*), and (3) the emission of light from electronically excited gas atoms (*emission spectra*). We examine the first two phenomena here and the third in Section 6.3.

Hot Objects and the Quantization of Energy

When solids are heated, they emit radiation, as seen in the red glow of an electric stove burner or the bright white light of a tungsten lightbulb. The wavelength distribution of the radiation depends on temperature; a red-hot object, for instance, is cooler than a yellowish or white-hot one (◄ Figure 6.5). During the late 1800s, a number of physicists studied this phenomenon, trying to understand the relationship between the temperature and the intensity and wavelength of the emitted radiation. The prevailing laws of physics could not account for the observations.

In 1900 a German physicist named Max Planck (1858–1947) solved the problem by making a daring assumption: He proposed that energy can be either released or absorbed by atoms only in discrete "chunks" of some minimum size.

GO FIGURE

Which area in the photograph corresponds to the highest temperature?

▲ Figure 6.5 **Color and temperature.** The color and intensity of the light emitted by a hot object, such as this pour of molten steel, depend on the temperature of the object.

Planck gave the name **quantum** (meaning "fixed amount") to the smallest quantity of energy that can be emitted or absorbed as electromagnetic radiation. He proposed that the energy, E, of a single quantum equals a constant times the frequency of the radiation:

$$E = h\nu \qquad [6.2]$$

The constant h is called **Planck constant** and has a value of 6.626×10^{-34} joule-second (J-s).

According to Planck's theory, matter can emit and absorb energy only in whole-number multiples of $h\nu$, such as $h\nu$, $2h\nu$, $3h\nu$, and so forth. If the quantity of energy emitted by an atom is $3h\nu$, for example, we say that three quanta of energy have been emitted (*quanta* being the plural of *quantum*). Because the energy can be released only in specific amounts, we say that the allowed energies are *quantized*—their values are restricted to certain quantities. Planck's revolutionary proposal that energy is quantized was proved correct, and he was awarded the 1918 Nobel Prize in Physics for his work on the quantum theory.

If the notion of quantized energies seems strange, it might be helpful to draw an analogy by comparing a ramp and a staircase (▶ Figure 6.6). As you walk up a ramp, your potential energy increases in a uniform, continuous manner. When you climb a staircase, you can step only *on* individual stairs, not *between* them, so that your potential energy is restricted to certain values and is therefore quantized.

If Planck's quantum theory is correct, why are its effects not obvious in our daily lives? Why do energy changes seem continuous rather than quantized, or "jagged"? Notice that Planck constant is an extremely small number. Thus, a quantum of energy, $h\nu$, is an extremely small amount. Planck's rules regarding the gain or loss of energy are always the same, whether we are concerned with objects on the scale of our ordinary experience or with microscopic objects. With everyday objects, however, the gain or loss of a single quantum of energy is so small that it goes completely unnoticed. In contrast, when dealing with matter at the atomic level, the impact of quantized energies is far more significant.

Potential energy of person walking up ramp increases in uniform, continuous manner

Potential energy of person walking up steps increases in stepwise, quantized manner

▲ Figure 6.6 **Quantized versus continuous change in energy.**

 Give It Some Thought

Consider the notes that can be played on a piano. In what way is a piano an example of a quantized system? In this analogy, would a violin be continuous or quantized?

The Photoelectric Effect and Photons

A few years after Planck presented his quantum theory, scientists began to see its applicability to many experimental observations. In 1905, Albert Einstein (1879–1955) used Planck's theory to explain the **photoelectric effect** (▶ Figure 6.7). Light shining on a clean metal surface causes electrons to be emitted from the surface. A minimum frequency of light, different for different metals, is required for the emission of electrons. For example, light with a frequency of $4.60 \times 10^{14}\ \text{s}^{-1}$ or greater causes cesium metal to emit electrons, but if the light has frequency less than that, no electrons are emitted.

To explain the photoelectric effect, Einstein assumed that the radiant energy striking the metal surface behaves like a stream of tiny energy packets. Each packet, which is like a "particle" of energy, is called a **photon**. Extending Planck's quantum theory, Einstein deduced that each photon must have an energy equal to Planck constant times the frequency of the light:

$$\text{Energy of photon} = E = h\nu \qquad [6.3]$$

Thus, radiant energy itself is quantized.

Under the right conditions, photons striking a metal surface can transfer their energy to electrons in the metal. A certain amount of energy—called the *work function* —is required for the electrons to overcome the attractive forces holding

 GO FIGURE

What is the source of the energy that causes electrons to be emitted from the surface?

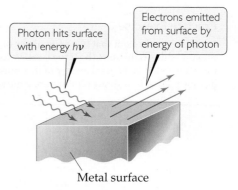

Photon hits surface with energy $h\nu$

Electrons emitted from surface by energy of photon

Metal surface

▲ Figure 6.7 **The photoelectric effect.**

them in the metal. If the photons striking the metal have less energy than the work function, the electrons do not acquire sufficient energy to escape from the metal, even if the light beam is intense. If the photons have energy greater than the work function of the particular metal, however, electrons are emitted; any excess energy of the photon is converted into kinetic energy of the emitted electron. The intensity (brightness) of the light is related to the number of photons striking the surface per unit time but not to the energy of each photon. Einstein won the Nobel Prize in Physics in 1921 primarily for his explanation of the photoelectric effect.

▲ Give It Some Thought

In Figure 6.7, will the kinetic energy of an emitted electron equal the energy of the photon that causes its emission?

To better understand what a photon is, imagine you have a light source that produces radiation of a single wavelength. Further suppose that you could switch the light on and off faster and faster to provide ever-smaller bursts of energy. Einstein's photon theory tells us that you would eventually come to the smallest energy burst, given by $E = h\nu$. This smallest burst consists of a single photon of light.

SAMPLE EXERCISE 6.3 | Energy of a Photon

Calculate the energy of one photon of yellow light that has a wavelength of 589 nm.

SOLUTION

Analyze Our task is to calculate the energy, E, of a photon, given $\lambda = 589$ nm.

Plan We can use Equation 6.1 to convert the wavelength to frequency:

$$\nu = c/\lambda$$

We can then use Equation 6.3 to calculate energy:

$$E = h\nu$$

Solve The frequency, ν, is calculated from the given wavelength, as shown in Sample Exercise 6.2:

$$\nu = c/\lambda = 5.09 \times 10^{14}\,\text{s}^{-1}$$

The value of Planck constant, h, is given both in the text and in the table of physical constants on the inside back cover of the text, and so we can easily calculate E:

$$E = (6.626 \times 10^{-34}\,\text{J-s})(5.09 \times 10^{14}\,\text{s}^{-1}) = 3.37 \times 10^{-19}\,\text{J}$$

Comment If one photon of radiant energy supplies 3.37×10^{-19} J, we calculate that one mole of these photons will supply:

$$(6.02 \times 10^{23}\,\text{photons/mol})(3.37 \times 10^{-19}\,\text{J/photon})$$
$$= 2.03 \times 10^{5}\,\text{J/mol}$$

Practice Exercise 1

Which of the following expressions correctly gives the energy of a mole of photons with wavelength λ?

(a) $E = \dfrac{h}{\lambda}$ (b) $E = N_A \dfrac{\lambda}{h}$ (c) $E = \dfrac{hc}{\lambda}$ (d) $E = N_A \dfrac{h}{\lambda}$ (e) $E = N_A \dfrac{hc}{\lambda}$

Practice Exercise 2

(a) A laser emits light that has a frequency of $4.69 \times 10^{14}\,\text{s}^{-1}$. What is the energy of one photon of this radiation? (b) If the laser emits a pulse containing 5.0×10^{17} photons of this radiation, what is the total energy of that pulse? (c) If the laser emits 1.3×10^{-2} J of energy during a pulse, how many photons are emitted?

The idea that the energy of light depends on its frequency helps us understand the diverse effects that different kinds of electromagnetic radiation have on matter. For example, because of the high frequency (short wavelength) of X rays (Figure 6.4), X-ray photons cause tissue damage and even cancer. Thus, signs are normally posted around X-ray equipment warning us of high-energy radiation.

Although Einstein's theory of light as a stream of photons rather than a wave explains the photoelectric effect and a great many other observations, it also poses a dilemma. Is light a wave, or does it consist of particles? The only way to resolve this dilemma is to adopt what might seem to be a bizarre position: We must consider that light possesses both wave-like and particle-like characteristics and, depending on the situation, will behave more like waves or more like particles. We will soon see that this dual wave-particle nature is also a characteristic trait of matter.

Give It Some Thought

Do you think that the formation of a rainbow is more a demonstration of the wave-like or particle-like behavior of light?

6.3 | Line Spectra and the Bohr Model

The work of Planck and Einstein paved the way for understanding how electrons are arranged in atoms. In 1913, the Danish physicist Niels Bohr (▶ Figure 6.8) offered a theoretical explanation of *line spectra*, another phenomenon that had puzzled scientists during the nineteenth century. We will see that Bohr used the ideas of Planck and Einstein to explain the line spectrum of hydrogen.

▲ Figure 6.8 **Quantum giants.** Niels Bohr (right) with Albert Einstein. Bohr (1885–1962) made major contributions to the quantum theory and was awarded the Nobel Prize in Physics in 1922.

Line Spectra

A particular source of radiant energy may emit a single wavelength, as in the light from a laser. Radiation composed of a single wavelength is *monochromatic*. However, most common radiation sources, including lightbulbs and stars, produce radiation containing many different wavelengths and is *polychromatic*. A **spectrum** is produced when radiation from such sources is separated into its component wavelengths, as shown in ▲ Figure 6.9. The resulting spectrum consists of a continuous range of colors—violet merges into indigo, indigo into blue, and so forth, with no (or very few) blank spots. This rainbow of colors, containing light of all wavelengths, is called a **continuous spectrum**. The most familiar example of a continuous spectrum is the rainbow produced when raindrops or mist acts as a prism for sunlight.

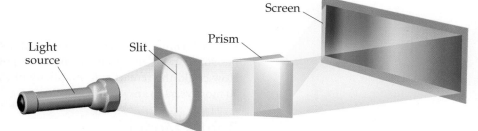

▲ Figure 6.9 **Creating a spectrum.** A continuous visible spectrum is produced when a narrow beam of white light is passed through a prism. The white light could be sunlight or light from an incandescent lamp.

Not all radiation sources produce a continuous spectrum. When a high voltage is applied to tubes that contain different gases under reduced pressure, the gases emit different colors of light (▶ Figure 6.10). The light emitted by neon gas is the familiar red-orange glow of many "neon" lights, whereas sodium vapor emits the yellow light characteristic of some modern streetlights. When light coming from such tubes is passed through a prism, only a few wavelengths are present in the resultant spectra (Figure 6.11). Each colored line in such spectra represents light of one wavelength. A spectrum containing radiation of only specific wavelengths is called a **line spectrum.**

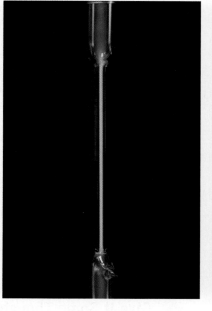

Neon (Ne)   Hydrogen (H)

▲ Figure 6.10 **Atomic emission of hydrogen and neon.** Different gases emit light of different characteristic colors when an electric current is passed through them.

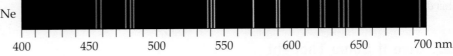

▲ Figure 6.11 **Line spectra of hydrogen and neon.** The colored lines occur at wavelengths present in the emission. The black regions are wavelengths for which no light is produced in the emission.

When scientists first detected the line spectrum of hydrogen in the mid-1800s, they were fascinated by its simplicity. At that time, only four lines at wavelengths of 410 nm (violet), 434 nm (blue), 486 nm (blue-green), and 656 nm (red) were observed (Figure 6.11). In 1885, a Swiss schoolteacher named Johann Balmer showed that the wavelengths of these four lines fit an intriguingly simple formula that relates the wavelengths to integers. Later, additional lines were found in the ultraviolet and infrared regions of hydrogen's line spectrum. Soon Balmer's equation was extended to a more general one, called the *Rydberg equation*, which allows us to calculate the wavelengths of all the spectral lines of hydrogen:

$$\frac{1}{\lambda} = (R_H)\left(\frac{1}{n_1^2} - \frac{1}{n_2^2}\right) \qquad [6.4]$$

In this formula λ is the wavelength of a spectral line, R_H is the *Rydberg constant* ($1.096776 \times 10^7 \text{ m}^{-1}$), and n_1 and n_2 are positive integers, with n_2 being larger than n_1. How could the remarkable simplicity of this equation be explained? It took nearly 30 more years to answer this question.

Bohr's Model

Rutherford's discovery of the nuclear atom ∞ (Section 2.2) suggested that an atom might be thought of as a "microscopic solar system" in which the electrons orbit the nucleus. To explain the line spectrum of hydrogen, Bohr assumed that electrons in hydrogen atoms move in circular orbits around the nucleus, but this assumption posed a problem. According to classical physics, a charged particle (such as an electron) moving in a circular path should continuously lose energy. As an electron loses energy, therefore, it should spiral into the positively charged nucleus. This behavior, however, does not happen—hydrogen atoms are stable. So how can we explain this apparent violation of the laws of physics? Bohr approached this problem in much the same way that Planck had approached the problem of the nature of the radiation emitted by hot objects: He assumed that the prevailing laws of physics were inadequate to describe all aspects of atoms. Furthermore, he adopted Planck's idea that energies are quantized.

Bohr based his model on three postulates:

1. Only orbits of certain radii, corresponding to certain specific energies, are permitted for the electron in a hydrogen atom.

2. An electron in a permitted orbit is in an "allowed" energy state. An electron in an allowed energy state does not radiate energy and, therefore, does not spiral into the nucleus.

3. Energy is emitted or absorbed by the electron only as the electron changes from one allowed energy state to another. This energy is emitted or absorbed as a photon that has energy $E = h\nu$.

 Give It Some Thought

With reference to Figure 6.6, in what way is the Bohr model for the H atom more like steps than a ramp?

The Energy States of the Hydrogen Atom

Starting with his three postulates and using classical equations for motion and for interacting electrical charges, Bohr calculated the energies corresponding to the allowed orbits for the electron in the hydrogen atom. Ultimately, the calculated energies fit the formula

$$E = (-hcR_H)\left(\frac{1}{n^2}\right) = (-2.18 \times 10^{-18}\,J)\left(\frac{1}{n^2}\right) \qquad [6.5]$$

where h, c, and R_H are the Planck constant, the speed of light, and the Rydberg constant, respectively. The integer n, which can have whole-number values of 1, 2, 3, ... ∞, is called the **principal quantum number**.

Each allowed orbit corresponds to a different value of n. The radius of the orbit gets larger as n increases. Thus, the first allowed orbit (the one closest to the nucleus) has $n = 1$, the next allowed orbit (the one second closest to the nucleus) has $n = 2$, and so forth. The electron in the hydrogen atom can be in any allowed orbit, and Equation 6.5 tells us the energy the electron has in each allowed orbit.

Note that the energies of the electron given by Equation 6.5 are negative for all values of n. The lower (more negative) the energy is, the more stable the atom is. The energy is lowest (most negative) for $n = 1$. As n gets larger, the energy becomes less negative and therefore increases. We can liken the situation to a ladder in which the rungs are numbered from the bottom. The higher one climbs (the greater the value of n), the higher the energy. The lowest-energy state ($n = 1$, analogous to the bottom rung) is called the **ground state** of the atom. When the electron is in a higher-energy state ($n = 2$ or higher), the atom is said to be in an **excited state**. ▶ Figure 6.12 shows the allowed energy levels for the hydrogen atom for several values of n.

Give It Some Thought

Why does it make sense that an orbit with a larger radius has a higher energy than one with a smaller radius?

What happens to the orbit radius and the energy as n becomes infinitely large? The radius increases as n^2, so when $n = \infty$ the electron is completely separated from the nucleus, and the energy of the electron is zero:

$$E = (-2.18 \times 10^{-18}\,J)\left(\frac{1}{\infty^2}\right) = 0$$

The state in which the electron is completely separated from the nucleus is called the *reference*, or zero-energy, state of the hydrogen atom.

In his third postulate, Bohr assumed that the electron can "jump" from one allowed orbit to another by either absorbing or emitting photons whose radiant energy corresponds exactly to the energy difference between the two orbits. The electron must absorb energy in order to move to a higher-energy state (higher value of n). Conversely, radiant energy is emitted when the electron jumps to a lower-energy state (lower value of n).

Let's consider a case in which the electron jumps from an initial state with principal quantum number n_i and energy E_i to a final state with principal quantum number n_f and energy E_f. Using Equation 6.5, we see that the change in energy for this transition is

$$\Delta E = E_f - E_i = (-2.18 \times 10^{-18}\,J)\left(\frac{1}{n_f^2} - \frac{1}{n_i^2}\right) \qquad [6.6]$$

What is the significance of the *sign* of ΔE? Notice that ΔE is positive when n_f is greater than n_i. That makes sense to us because that means the electron is jumping to a higher-energy orbit. Conversely, ΔE is negative when n_f is less than n_i; the electron is falling in energy to a lower-energy orbit.

GO FIGURE

If the transition of an electron from the $n = 3$ state to the $n = 2$ state results in emission of visible light, is the transition from the $n = 2$ state to the $n = 1$ state more likely to result in the emission of infrared or ultraviolet radiation?

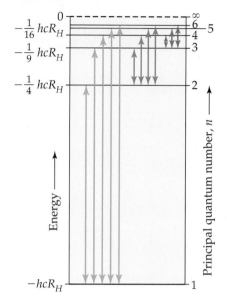

▲ **Figure 6.12 Energy levels in the hydrogen atom from the Bohr model.** The arrows refer to the transitions of the electron from one allowed energy state to another. The states shown are those for which $n = 1$ through $n = 6$ and the state for $n = \infty$ for which the energy, E, equals zero.

As noted above, transitions from one allowed state to another will involve a photon. *The energy of the photon (E_{photon}) must equal the difference in energy between the two states (ΔE).* When ΔE is positive, a photon must be *absorbed* as the electron jumps to a higher energy. When ΔE is negative, a photon is *emitted* as the electron falls to a lower energy level. In both cases, the energy of the photon must match the energy difference between the states. Because the frequency ν is always a positive number, the energy of the photon ($h\nu$) must always be positive. Thus, the sign of ΔE tells us whether the photon is absorbed or emitted:

$$\Delta E > 0 \ (n_f > n_i): \text{Photon } \textit{absorbed} \text{ with } E_{photon} = h\nu = \Delta E$$

$$\Delta E < 0 \ (n_f < n_i): \text{Photon } \textit{emitted} \text{ with } E_{photon} = h\nu = -\Delta E \qquad [6.7]$$

These two situations are summarized in ▼ Figure 6.13. We see that Bohr's model of the hydrogen atom leads to the conclusion that only the specific frequencies of light that satisfy Equation 6.7 can be absorbed or emitted by the atom.

Let's see how to apply these concepts by considering a transition in which the electron moves from $n_i = 3$ to $n_f = 1$. From Equation 6.6 we have

$$\Delta E = (-2.18 \times 10^{-18}\,\text{J})\left(\frac{1}{1^2} - \frac{1}{3^2}\right) = (-2.18 \times 10^{-18}\,\text{J})\left(\frac{8}{9}\right) = -1.94 \times 10^{-18}\,\text{J}$$

The value of ΔE is negative—that makes sense because the electron is falling from a higher-energy orbit ($n = 3$) to a lower-energy orbit ($n = 1$). A photon is *emitted* during this transition, and the energy of the photon is equal to $E_{photon} = h\nu = -\Delta E = +1.94 \times 10^{-18}\,\text{J}$.

Knowing the energy of the emitted photon, we can calculate either its frequency or its wavelength. For the wavelength, we recall that $\lambda = c/\nu = hc/E_{photon}$ and obtain

$$\lambda = \frac{c}{\nu} = \frac{hc}{E_{photon}} = \frac{hc}{-\Delta E} = \frac{(6.626 \times 10^{-34}\,\text{J-s})(2.998 \times 10^8\,\text{m/s})}{+1.94 \times 10^{-18}\,\text{J}} = 1.02 \times 10^{-7}\,\text{m}$$

Thus, a photon of wavelength 1.02×10^{-7} m (102 nm) is *emitted*.

 GO FIGURE

Which transition will lead to the emission of light with longer wavelength, $n = 3$ to $n = 2$, or $n = 4$ to $n = 3$?

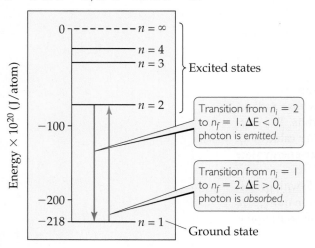

▲ **Figure 6.13 Change in energy states for absorption and emission.**

Figure labels:
- Energy × 10^{20} (J/atom)
- 0 ---- $n = \infty$
- $n = 4$
- $n = 3$ — Excited states
- $n = 2$
- -100
- Transition from $n_i = 2$ to $n_f = 1$. $\Delta E < 0$, photon is *emitted*.
- Transition from $n_i = 1$ to $n_f = 2$. $\Delta E > 0$, photon is *absorbed*.
- -200
- -218 — $n = 1$ — Ground state

 Give It Some Thought

What is the significance of the minus sign in front of ΔE in the above equation?

We are now in a position to understand the remarkable simplicity of the line spectra of hydrogen, first discovered by Balmer. We recognize that the line spectra are the result of emission, so $E_{photon} = h\nu = hc/\lambda = -\Delta E$ for these transitions. Combining Equations 6.5 and 6.6 we see that

$$E_{photon} = \frac{hc}{\lambda} = -\Delta E = hcR_H\left(\frac{1}{n_f^2} - \frac{1}{n_i^2}\right) \text{ (for emission)}$$

which gives us

$$\frac{1}{\lambda} = \frac{hcR_H}{hc}\left(\frac{1}{n_f^2} - \frac{1}{n_i^2}\right) = R_H\left(\frac{1}{n_f^2} - \frac{1}{n_i^2}\right), \text{ where } n_f < n_i$$

Thus, the existence of discrete spectral lines can be attributed to the quantized jumps of electrons between energy levels.

 Give It Some Thought

What is the relationship between $1/\lambda$ and ΔE for a transition of the electron from a lower value of n to a higher one?

SAMPLE
EXERCISE 6.4 Electronic Transitions in the Hydrogen Atom

Using Figure 6.12, predict which of these electronic transitions produces the spectral line having the longest wavelength: $n = 2$ to $n = 1$, $n = 3$ to $n = 2$, or $n = 4$ to $n = 3$.

SOLUTION
The wavelength increases as frequency decreases ($\lambda = c/\nu$). Hence, the longest wavelength will be associated with the lowest frequency. According to Planck's equation, $E = h\nu$, the lowest frequency is associated with the lowest energy. In Figure 6.12 the shortest vertical line represents the smallest energy change. Thus, the $n = 4$ to $n = 3$ transition produces the longest wavelength (lowest frequency) line.

Practice Exercise 1
In the top part of Figure 6.11, the four lines in the H atom spectrum are due to transitions from a level for which $n_i > 2$ to the $n_f = 2$ level. What is the value of n_i for the blue-green line in the spectrum? (a) 3 (b) 4 (c) 5 (d) 6 (e) 7

Practice Exercise 2
For each of the following transitions, give the sign of ΔE and indicate whether a photon is emitted or absorbed: (a) $n = 3$ to $n = 1$; (b) $n = 2$ to $n = 4$.

Limitations of the Bohr Model

Although the Bohr model explains the line spectrum of the hydrogen atom, it cannot explain the spectra of other atoms, except in a crude way. Bohr also avoided the problem of why the negatively charged electron would not just fall into the positively charged nucleus, by simply assuming it would not happen. Furthermore, there is a problem with describing an electron merely as a small particle circling the nucleus. As we will see in Section 6.4, the electron exhibits wave-like properties, a fact that any acceptable model of electronic structure must accommodate.

As it turns out, the Bohr model was only an important step along the way toward the development of a more comprehensive model. What is most significant about Bohr's model is that it introduces two important ideas that are also incorporated into our current model:

1. *Electrons exist only in certain discrete energy levels, which are described by quantum numbers.*
2. *Energy is involved in the transition of an electron from one level to another.*

We will now start to develop the successor to the Bohr model, which requires that we take a closer look at the behavior of matter.

6.4 | The Wave Behavior of Matter

In the years following the development of Bohr's model for the hydrogen atom, the dual nature of radiant energy became a familiar concept. Depending on the experimental circumstances, radiation appears to have either a wave-like or a particle-like (photon) character. Louis de Broglie (1892–1987), who was working on his Ph.D. thesis in physics at the Sorbonne in Paris, boldly extended this idea: If radiant energy could, under appropriate conditions, behave as though it were a stream of particles (photons), could matter, under appropriate conditions, possibly show the properties of a wave?

De Broglie suggested that an electron moving about the nucleus of an atom behaves like a wave and therefore has a wavelength. He proposed that the wavelength of the electron, or of any other particle, depends on its mass, m, and on its velocity, v:

$$\lambda = \frac{h}{mv} \tag{6.8}$$

where h is the Planck constant. The quantity mv for any object is called its **momentum**. De Broglie used the term **matter waves** to describe the wave characteristics of material particles.

Because de Broglie's hypothesis is applicable to all matter, any object of mass m and velocity v would give rise to a characteristic matter wave. However, Equation 6.8 indicates that the wavelength associated with an object of ordinary size, such as a golf ball, is so tiny as to be completely unobservable. This is not so for an electron because its mass is so small, as we see in Sample Exercise 6.5.

SAMPLE EXERCISE 6.5 | Matter Waves

What is the wavelength of an electron moving with a speed of 5.97×10^6 m/s? The mass of the electron is 9.11×10^{-31} kg.

SOLUTION

Analyze We are given the mass, m, and velocity, v, of the electron, and we must calculate its de Broglie wavelength, λ.

Plan The wavelength of a moving particle is given by Equation 6.8, so λ is calculated by inserting the known quantities h, m, and v. In doing so, however, we must pay attention to units.

Solve Using the value of the Planck constant: $\quad h = 6.626 \times 10^{-34}$ J-s

we have the following:

$$\lambda = \frac{h}{mv}$$

$$= \frac{(6.626 \times 10^{-34}\,\text{J-s})}{(9.11 \times 10^{-31}\,\text{kg})(5.97 \times 10^6\,\text{m/s})}\left(\frac{1\,\text{kg-m}^2/\text{s}^2}{1\,\text{J}}\right)$$

$$= 1.22 \times 10^{-10}\,\text{m} = 0.122\,\text{nm} = 1.22\,\text{Å}$$

Comment By comparing this value with the wavelengths of electromagnetic radiation shown in Figure 6.4, we see that the wavelength of this electron is about the same as that of X rays.

Practice Exercise 1

Consider the following three moving objects: (i) a golf ball with a mass of 45.9 g moving at a speed of 50.0 m/s, (ii) An electron moving at a speed of 3.50×10^5 m/s, (iii) A neutron moving at a speed of 2.3×10^2 m/s. List the three objects in order from shortest to longest de Broglie wavelength.
(a) i < iii < ii   (b) ii < iii < i   (c) iii < ii < i   (d) i < ii < iii   (e) iii < i < ii

Practice Exercise 2

Calculate the velocity of a neutron whose de Broglie wavelength is 505 pm. The mass of a neutron is given in the table inside the back cover of the text.

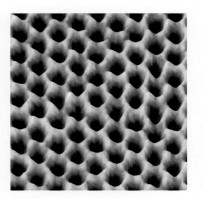

▲ **Figure 6.14 Electrons as waves.** Transmission electron micrograph of *graphene*, which has a hexagonal honeycomb arrangement of carbon atoms. Each of the bright yellow "mountains" indicates a carbon atom.

A few years after de Broglie published his theory, the wave properties of the electron were demonstrated experimentally. When X rays pass through a crystal, an interference pattern results that is characteristic of the wave-like properties of electromagnetic radiation, a phenomenon called *X-ray diffraction*. As electrons pass through a crystal, they are similarly diffracted. Thus, a stream of moving electrons exhibits the same kinds of wave behavior as X rays and all other types of electromagnetic radiation.

The technique of electron diffraction has been highly developed. In the electron microscope, for instance, the wave characteristics of electrons are used to obtain images at the atomic scale. This microscope is an important tool for studying surface phenomena at very high magnifications (◄ Figure 6.14). Electron microscopes can magnify objects by 3,000,000 times, far more than can be done with visible light (1000×), because the wavelength of the electrons is so much smaller than the wavelength of visible light.

▲ Give It Some Thought

A baseball pitcher throws a fastball that moves at 95 miles per hour. Does that moving baseball generate matter waves? If so, can we observe them?

The Uncertainty Principle

The discovery of the wave properties of matter raised some new and interesting questions. Consider, for example, a ball rolling down a ramp. Using the equations of classical physics, we can calculate, with great accuracy, the ball's position, direction of motion, and speed at any instant. Can we do the same for an electron, which exhibits wave properties? A wave extends in space and its location is not precisely defined. We might therefore anticipate that it is impossible to determine exactly where an electron is located at a specific instant.

The German physicist Werner Heisenberg (▶ Figure 6.15) proposed that the dual nature of matter places a fundamental limitation on how precisely we can know both the location and the momentum of an object at a given instant. The limitation becomes important only when we deal with matter at the subatomic level (that is, with masses as small as that of an electron). Heisenberg's principle is called the **uncertainty principle**. When applied to the electrons in an atom, this principle states that it is impossible for us to know simultaneously both the exact momentum of the electron and its exact location in space.

Heisenberg mathematically related the uncertainty in position, Δx, and the uncertainty in momentum, $\Delta(mv)$, to a quantity involving Planck constant:

$$\Delta x \cdot \Delta(mv) \geq \frac{h}{4\pi} \qquad [6.9]$$

▲ **Figure 6.15 Werner Heisenberg (1901–1976).** During his postdoctoral assistantship with Niels Bohr, Heisenberg formulated his famous uncertainty principle. At 32 he was one of the youngest scientists to receive a Nobel Prize.

A brief calculation illustrates the dramatic implications of the uncertainty principle. The electron has a mass of 9.11×10^{-31} kg and moves at an average speed of about 5×10^6 m/s in a hydrogen atom. Let's assume that we know the speed to an uncertainty of 1% [that is, an uncertainty of $(0.01)(5 \times 10^6 \text{ m/s}) = 5 \times 10^4$ m/s] and that this is the only important source of uncertainty in the momentum, so that $\Delta(mv) = m\,\Delta v$. We can use Equation 6.9 to calculate the uncertainty in the position of the electron:

$$\Delta x \geq \frac{h}{4\pi m \Delta v} = \left(\frac{6.626 \times 10^{-34} \text{ J-s}}{4\pi(9.11 \times 10^{-31} \text{ kg})(5 \times 10^4 \text{ m/s})} \right) = 1 \times 10^{-9} \text{ m}$$

Because the diameter of a hydrogen atom is about 1×10^{-10} m, the uncertainty in the position of the electron in the atom is an order of magnitude greater than the size of the atom. Thus, we have essentially no idea where the electron is located in the atom. On the other hand, if we were to repeat the calculation with an object of ordinary mass, such as a tennis ball, the uncertainty would be so small that it would be inconsequential. In that case, m is large and Δx is out of the realm of measurement and therefore of no practical consequence.

De Broglie's hypothesis and Heisenberg's uncertainty principle set the stage for a new and more broadly applicable theory of atomic structure. In this approach, any attempt to define precisely the instantaneous location and momentum of the electron is abandoned. The wave nature of the electron is recognized, and its behavior is described in terms appropriate to waves. The result is a model that precisely describes the energy of the electron while describing its location not precisely but rather in terms of probabilities.

 A Closer Look

Measurement and the Uncertainty Principle

Whenever any measurement is made, some uncertainty exists. Our experience with objects of ordinary dimensions, such as balls or trains or laboratory equipment, indicates that using more precise instruments can decrease the uncertainty of a measurement. In fact, we might expect that the uncertainty in a measurement can be made indefinitely small. However, the uncertainty principle states that there is an actual limit to the accuracy of measurements. This limit is not a restriction on how well instruments can be made; rather, it is inherent in nature. This limit has no practical consequences when dealing with ordinary-sized objects, but its implications are enormous when dealing with subatomic particles, such as electrons.

To measure an object, we must disturb it, at least a little, with our measuring device. Imagine using a flashlight to locate a large rubber ball in a dark room. You see the ball when the light from the flashlight bounces off the ball and strikes your eyes. When a beam of photons strikes an object of this size, it does not alter its position or momentum to any practical extent. Imagine, however, that you wish to locate an electron by similarly bouncing light off it into some detector. Objects can be located to an accuracy no greater than the wavelength of the radiation used. Thus, if we want an accurate position measurement for an electron, we must use a short wavelength. This means that photons of high energy must be employed. The more energy the photons have, the more momentum they impart to the electron when they strike it, which changes the electron's motion in an unpredictable way. The attempt to measure accurately the electron's position introduces considerable uncertainty in its momentum; the act of measuring the electron's position at one moment makes our knowledge of its future position inaccurate.

Suppose, then, that we use photons of longer wavelength. Because these photons have lower energy, the momentum of the electron is not so appreciably changed during measurement, but its position will be correspondingly less accurately known. This is the essence of the uncertainty principle: *There is an uncertainty in simultaneously knowing both the position and the momentum of the electron that cannot be reduced beyond a certain minimum level.* The more accurately one is known, the less accurately the other is known.

Although we can never know the exact position and momentum of the electron, we can talk about the probability of its being at certain locations in space. In Section 6.5 we introduce a model of the atom that provides the probability of finding electrons of specific energies at certain positions in atoms.

Related Exercises: 6.51, 6.52, 6.96, 6.97

 Give It Some Thought

What is the principal reason we must consider the uncertainty principle when discussing electrons and other subatomic particles but not when discussing our macroscopic world?

6.5 | Quantum Mechanics and Atomic Orbitals

In 1926 the Austrian physicist Erwin Schrödinger (1887–1961) proposed an equation, now known as *Schrödinger's wave equation*, that incorporates both the wave-like and particle-like behaviors of the electron. His work opened a new approach to dealing with subatomic particles, an approach known as *quantum mechanics* or *wave mechanics*. The application of Schrödinger's equation requires advanced calculus, and so we will not be concerned with its details. We will, however, qualitatively consider the results Schrödinger obtained because they give us a powerful new way to view electronic structure. Let's begin by examining the electronic structure of the simplest atom, hydrogen.

Schrödinger treated the electron in a hydrogen atom like the wave on a plucked guitar string (▼ **Figure 6.16**). Because such waves do not travel in space, they are called

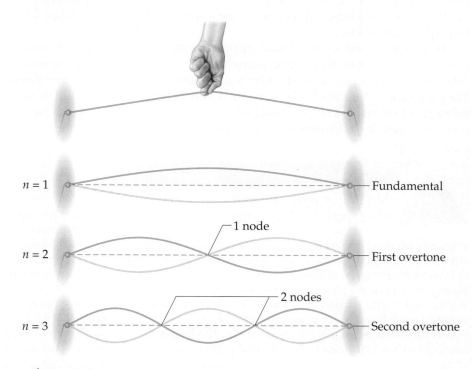

▲ Figure 6.16 **Standing waves in a vibrating string.**

standing waves. Just as the plucked guitar string produces a standing wave that has a fundamental frequency and higher overtones (harmonics), the electron exhibits a lowest-energy standing wave and higher-energy ones. Furthermore, just as the overtones of the guitar string have *nodes*, points where the magnitude of the wave is zero, so do the waves characteristic of the electron.

Solving Schrödinger's equation for the hydrogen atom leads to a series of mathematical functions called **wave functions** that describe the electron in the atom. These wave functions are usually represented by the symbol ψ (lowercase Greek letter *psi*). Although the wave function has no direct physical meaning, the square of the wave function, ψ^2, provides information about the electron's location when it is in an allowed energy state.

For the hydrogen atom, the allowed energies are the same as those predicted by the Bohr model. However, the Bohr model assumes that the electron is in a circular orbit of some particular radius about the nucleus. In the quantum mechanical model, the electron's location cannot be described so simply.

According to the uncertainty principle, if we know the momentum of the electron with high accuracy, our simultaneous knowledge of its location is very uncertain. Thus, we cannot hope to specify the exact location of an individual electron around the nucleus. Rather, we must be content with a kind of statistical knowledge. We therefore speak of the *probability* that the electron will be in a certain region of space at a given instant. As it turns out, the square of the wave function, ψ^2, at a given point in space represents the probability that the electron will be found at that location. For this reason, ψ^2 is called either the **probability density** or the **electron density**.

 Give It Some Thought

What is the difference between stating "The electron is located at a particular point in space" and "There is a high probability that the electron is located at a particular point in space"?

One way of representing the probability of finding the electron in various regions of an atom is shown in ▶ Figure 6.17, where the density of the dots represents the probability of finding the electron. The regions with a high density of dots correspond to relatively large values for ψ^2 and are therefore regions where there is a high probability of finding the electron. Based on this representation, we often describe atoms as consisting of a nucleus surrounded by an electron cloud.

 GO FIGURE

Where in the figure is the region of highest electron density?

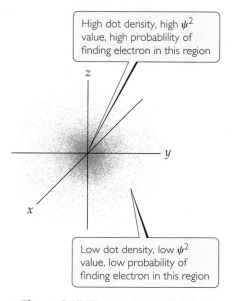

High dot density, high ψ^2 value, high probablility of finding electron in this region

Low dot density, low ψ^2 value, low probability of finding electron in this region

▲ **Figure 6.17 Electron-density distribution.** This rendering represents the probability, ψ^2, of finding the electron in a hydrogen atom in its ground state. The origin of the coordinate system is at the nucleus.

 A Closer Look

Thought Experiments and Schrödinger's Cat

The revolutions in scientific thinking caused by the theory of relativity and quantum theory not only changed science; it also caused deep changes in how we understand the world around us. Before relativity and quantum theory, the prevailing physical theories were inherently *deterministic*: Once the specific conditions of an object were given (position, velocity, forces acting on the object), we could determine exactly the position and motion of the object at any time in the future. These theories, from Newton's laws to Maxwell's theory of electromagnetism, successfully described physical phenomena such as motion of the planets, the trajectories of projectiles, and the diffraction of light.

Relativity and quantum theory both challenged the deterministic view of the universe, and did so in a way that caused a great deal of uneasiness among even the scientists who were developing the theories. One of the common methods scientists used to test these new theories was through so-called "thought experiments." Thought experiments are hypothetical scenarios that can lead to paradoxes within a given theory. Let's briefly discuss one of these thought experiments that was used to test ideas within quantum theory.

The quantum theory caused a great deal of discussion with respect to its nondeterministic description of matter. We have touched on two such areas in this chapter. First, we have seen that the descriptions of light and matter have become less distinct—light has particle-like properties and matter has wave-like properties. The description of matter that results—in which we can talk only about the probability of an electron being at a certain place as opposed to knowing exactly where it is—was very bothersome to many. Einstein, for example, famously said that "God doesn't play dice with the world"* about this probabilistic description. Heisenberg's uncertainty principle, which assures that we can't know the position and momentum of a particle exactly, also raised many philosophical questions—so many, in fact, that Heisenberg wrote a book entitled *Physics and Philosophy* in 1958.

One of the most famous thought experiments put forward in the early days of the quantum theory was formulated by Schrödinger and is now known as "Schrödinger's cat." This experiment called into question whether a system could have multiple acceptable wave functions prior to observation of the system. In other words, if we don't actually observe a system, can we know anything about the state it is in?

---

*Hermanns, William, *Einstein and the Poet: In Search of the Cosmic Man*, 1st edition, Branden Books, 1983.

In this paradox, a hypothetical cat is placed in a sealed box with an apparatus that will randomly trigger a lethal dose of poison to the cat (as morbid as that sounds). According to some interpretations of quantum theory, until the box is opened and the cat is observed, the cat must be considered simultaneously alive and dead.

Schrödinger posed this paradox to point out weaknesses in some interpretations of quantum results, but the paradox has led instead to a continuing and lively debate about the fate and meaning of Schrödinger's cat. In 2012, the Nobel Prize in physics was awarded to Serge Haroche of France and David Wineland of the United States for their ingenious methods for observing the quantum states of photons or particles without having the act of observation destroy the states. In so doing, they observed what is generally called the "cat state" of the systems, in which the photon or particle exists simultaneously in two different quantum states. A puzzling paradox, indeed, but that one might ultimately lead to new ways to harness the simultaneous states to create so-called quantum computers and more accurate clocks.

Related Exercise: 6.97

Orbitals and Quantum Numbers

The solution to Schrödinger's equation for the hydrogen atom yields a set of wave functions called **orbitals**. Each orbital has a characteristic shape and energy. For example, the lowest-energy orbital in the hydrogen atom has the spherical shape illustrated in Figure 6.17 and an energy of -2.18×10^{-18} J. Note that an *orbital* (quantum mechanical model, which describes electrons in terms of probabilities, visualized as "electron clouds") is not the same as an *orbit* (the Bohr model, which visualizes the electron moving in a physical orbit, like a planet around a star). The quantum mechanical model does not refer to orbits because the motion of the electron in an atom cannot be precisely determined (Heisenberg's uncertainty principle).

The Bohr model introduced a single quantum number, n, to describe an orbit. The quantum mechanical model uses three quantum numbers, n, l, and m_l, which result naturally from the mathematics used to describe an orbital.

1. The principal quantum number, n, can have positive integral values 1, 2, 3, ... As n increases, the orbital becomes larger, and the electron spends more time farther from the nucleus. An increase in n also means that the electron has a higher energy and is therefore less tightly bound to the nucleus. For the hydrogen atom, $E_n = -(2.18 \times 10^{-18} \text{ J})(1/n^2)$, as in the Bohr model.

2. The second quantum number—the **angular momentum quantum number**, l—can have integral values from 0 to $(n - 1)$ for each value of n. This quantum number defines the shape of the orbital. The value of l for a particular orbital is generally designated by the letters s, p, d, and f,* corresponding to l values of 0, 1, 2, and 3:

Value of l	0	1	2	3
Letter used	s	p	d	f

3. The **magnetic quantum number**, m_l, can have integral values between $-l$ and l, including zero. This quantum number describes the orientation of the orbital in space, as we discuss in Section 6.6.

Notice that because the value of n can be any positive integer, there is an infinite number of orbitals for the hydrogen atom. At any given instant, however, the electron in a hydrogen atom is described by only one of these orbitals—we say that the electron *occupies* a certain orbital. The remaining orbitals are *unoccupied* for that particular state of the hydrogen atom. We will focus mainly on orbitals that have small values of n.

 Give It Some Thought

What is the difference between an *orbit* in the Bohr model of the hydrogen atom and an *orbital* in the quantum mechanical model?

---

*The letters come from the words *sharp, principal, diffuse,* and *fundamental,* which were used to describe certain features of spectra before quantum mechanics was developed.

Table 6.2 Relationship among Values of n, l, and m_l through $n = 4$

n	Possible Values of l	Subshell Designation	Possible Values of m_l	Number of Orbitals in Subshell	Total Number of Orbitals in Shell
1	0	$1s$	0	1	1
2	0	$2s$	0	1	
	1	$2p$	$1, 0, -1$	3	4
3	0	$3s$	0	1	
	1	$3p$	$1, 0, -1$	3	
	2	$3d$	$2, 1, 0, -1, -2$	5	9
4	0	$4s$	0	1	
	1	$4p$	$1, 0, -1$	3	
	2	$4d$	$2, 1, 0, -1, -2$	5	
	3	$4f$	$3, 2, 1, 0, -1, -2, -3$	7	16

The collection of orbitals with the same value of n is called an **electron shell**. All the orbitals that have $n = 3$, for example, are said to be in the third shell. The set of orbitals that have the same n and l values is called a **subshell**. Each subshell is designated by a number (the value of n) and a letter (s, p, d, or f, corresponding to the value of l). For example, the orbitals that have $n = 3$ and $l = 2$ are called $3d$ orbitals and are in the $3d$ subshell.

▲ Table 6.2 summarizes the possible values of l and m_l for values of n through $n = 4$. The restrictions on possible values give rise to the following very important observations:

1. *The shell with principal quantum number* n *consists of exactly* n *subshells.* Each subshell corresponds to a different allowed value of l from 0 to $(n - 1)$. Thus, the first shell ($n = 1$) consists of only one subshell, the $1s$ ($l = 0$); the second shell ($n = 2$) consists of two subshells, the $2s$ ($l = 0$) and $2p$ ($l = 1$); the third shell consists of three subshells, $3s$, $3p$, and $3d$, and so forth.

2. *Each subshell consists of a specific number of orbitals.* Each orbital corresponds to a different allowed value of m_l. For a given value of l, there are $(2l + 1)$ allowed values of m_l, ranging from $-l$ to $+l$. Thus, each s ($l = 0$) subshell consists of one orbital; each p ($l = 1$) subshell consists of three orbitals; each d ($l = 2$) subshell consists of five orbitals, and so forth.

3. *The total number of orbitals in a shell is* n^2, *where* n *is the principal quantum number of the shell.* The resulting number of orbitals for the shells—1, 4, 9, 16—is related to a pattern seen in the periodic table: We see that the number of elements in the rows of the periodic table—2, 8, 18, and 32—equals twice these numbers. We will discuss this relationship further in Section 6.9.

▶ Figure 6.18 shows the relative energies of the hydrogen atom orbitals through $n = 3$. Each box represents an orbital, and orbitals of the same subshell, such as the three $2p$ orbitals, are grouped together. When the electron occupies the lowest-energy orbital ($1s$), the hydrogen atom is said to be in its *ground state*. When the electron occupies any other orbital, the atom is in an *excited state*. (The electron can be excited to a higher-energy orbital by absorption of a photon of appropriate energy.) At ordinary temperatures, essentially all hydrogen atoms are in the ground state.

▲ **GO FIGURE**

If the fourth shell (the $n = 4$ energy level) were shown, how many subshells would it contain? How would they be labeled?

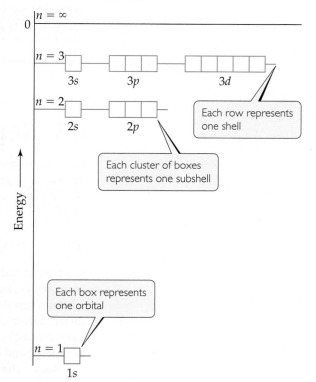

$n = 1$ shell has one orbital
$n = 2$ shell has two subshells composed of four orbitals
$n = 3$ shell has three subshells composed of nine orbitals

▲ Figure 6.18 **Energy levels in the hydrogen atom.**

Give It Some Thought

In Figure 6.18, why is the energy difference between the $n = 1$ and $n = 2$ levels much greater than the energy difference between the $n = 2$ and $n = 3$ levels?

SAMPLE EXERCISE 6.6 | Subshells of the Hydrogen Atom

(**a**) Without referring to Table 6.2, predict the number of subshells in the fourth shell, that is, for $n = 4$. (**b**) Give the label for each of these subshells. (**c**) How many orbitals are in each of these subshells?

SOLUTION

Analyze and Plan We are given the value of the principal quantum number, n. We need to determine the allowed values of l and m_l for this given value of n and then count the number of orbitals in each subshell.

Solve There are four subshells in the fourth shell, corresponding to the four possible values of l (0, 1, 2, and 3).

These subshells are labeled $4s$, $4p$, $4d$, and $4f$. The number given in the designation of a subshell is the principal quantum number, n; the letter designates the value of the angular momentum quantum number, l: for $l = 0$, s; for $l = 1$, p; for $l = 2$, d; for $l = 3$, f.

There is one $4s$ orbital (when $l = 0$, there is only one possible value of m_l: 0). There are three $4p$ orbitals (when $l = 1$, there are three possible values of m_l: 1, 0, −1). There are five $4d$ orbitals (when $l = 2$, there are five allowed values of m_l: 2, 1, 0, −1, −2). There are seven $4f$ orbitals (when $l = 3$, there are seven permitted values of m_l: 3, 2, 1, 0, −1, −2, −3).

Practice Exercise 1

An orbital has $n = 4$ and $m_l = -1$. What are the possible values of l for this orbital?
(**a**) 0, 1, 2, 3 (**b**) −3, −2, −1, 0, 1, 2, 3 (**c**) 1, 2, 3 (**d**) −3, −2 (**e**) 1, 2, 3, 4

Practice Exercise 2

(**a**) What is the designation for the subshell with $n = 5$ and $l = 1$? (**b**) How many orbitals are in this subshell? (**c**) Indicate the values of m_l for each of these orbitals.

6.6 | Representations of Orbitals

So far we have emphasized orbital energies, but the wave function also provides information about an electron's probable location in space. Let's examine the ways in which we can picture orbitals because their shapes help us visualize how the electron density is distributed around the nucleus.

The s Orbitals

We have already seen one representation of the lowest-energy orbital of the hydrogen atom, the 1s (Figure 6.17). The first thing we notice about the electron density for the 1s orbital is that it is *spherically symmetric*—in other words, the electron density at a given distance from the nucleus is the same regardless of the direction in which we proceed from the nucleus. All of the other s orbitals (2s, 3s, 4s, and so forth) are also spherically symmetric and centered on the nucleus.

Recall that the l quantum number for the s orbitals is 0; therefore, the m_l quantum number must be 0. Thus, for each value of n, there is only one s orbital. So how do s orbitals differ as the value of n changes? For example, how does the electron-density distribution of the hydrogen atom change when the electron is excited from the 1s orbital to the 2s orbital? To address this question, we will look at the *radial probability density*, which is the probability that the electron is at a specific distance from the nucleus.

▲ GO FIGURE

How many maxima would you expect to find in the radial probability function for the 4s orbital of the hydrogen atom? How many nodes would you expect in this function?

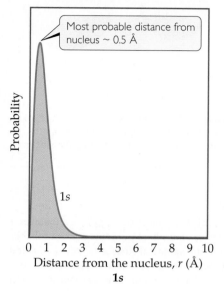

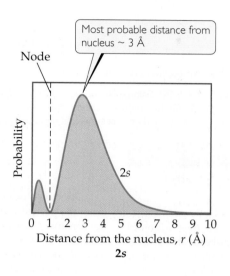

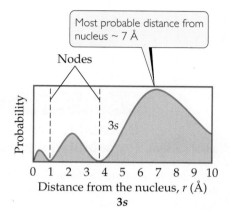

▲ **Figure 6.19 Radial probability functions for the 1s, 2s, and 3s orbitals of hydrogen.** These plots show the probability of finding the electron as a function of distance from the nucleus. As *n* increases, the most likely distance at which to find the electron (the highest peak) moves farther from the nucleus.

▲ Figure 6.19 shows the radial probability densities for the 1s, 2s, and 3s orbitals of hydrogen as a function of *r*, the distance from the nucleus—each resulting curve is the **radial probability function** for the orbital. Three features of these plots are noteworthy: the number of peaks, the number of points at which the probability function goes to zero (called **nodes**), and how spread out the distribution is, which gives a sense of the size of the orbital.

For the 1s orbital, we see that the probability rises rapidly as we move away from the nucleus, maximizing at about 0.5 Å. Thus, when the electron occupies the 1s orbital, it is *most likely* to be found this distance from the nucleus*—we still use the probabilistic description, consistent with the uncertainty principle. Notice also that in the 1s orbital the probability of finding the electron at a distance greater than about 3 Å from the nucleus is essentially zero.

Comparing the radial probability distributions for the 1s, 2s, and 3s orbitals reveals three trends:

1. *For an ns orbital, the number of peaks is equal to* n, *with the outermost peak being larger than inner ones.*

2. *For an ns orbital, the number of nodes is equal to* $n - 1$.

3. *As* n *increases, the electron density becomes more spread out, that is, there is a greater probability of finding the electron further from the nucleus.*

One widely used method of representing orbital *shape* is to draw a boundary surface that encloses some substantial portion, say 90%, of the electron density for the orbital. This type of drawing is called a *contour representation*, and the contour representations for the *s* orbitals are spheres (**Figure 6.20**). All the orbitals have the same shape, but they differ in size, becoming larger as *n* increases, reflecting the fact that the electron density becomes more spread out as *n* increases.

---

*In the quantum mechanical model, the most probable distance at which to find the electron in the 1s orbital is actually 0.529 Å, the same as the radius of the orbit predicted by Bohr for $n = 1$. The distance 0.529 Å is often called the Bohr radius.

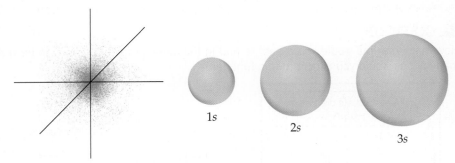

(a) An electron density model

(b) Contour models

▲ **Figure 6.20 Comparison of the 1s, 2s, and 3s orbitals. (a)** Electron-density distribution of a 1s orbital. **(b)** Contour representations of the 1s, 2s, and 3s orbitals. Each sphere is centered on the atom's nucleus and encloses the volume in which there is a 90% probability of finding the electron.

A Closer Look

Probability Density and Radial Probability Functions

According to quantum mechanics, we must describe the position of the electron in the hydrogen atom in terms of probabilities rather than exact locations. The information about the probability is contained in the wave functions, ψ, obtained from Schrödinger's equation. The square of the wave function, ψ^2, called either the probability density or the electron density, as noted earlier, gives the probability that the electron is at any *point* in space. Because *s* orbitals are spherically symmetric, the value of ψ for an *s* electron depends only on its distance from the nucleus, *r*. Thus, the probability density can be written as $[\psi(r)]^2$, where $\psi(r)$ is the value of ψ at *r*. This function $[\psi(r)]^2$ gives the probability density for any point located a distance *r* from the nucleus.

The radial probability function, which we used in Figure 6.19, differs from the probability density. The radial probability function equals the *total* probability of finding the electron at all the points at any distance *r* from the nucleus. In other words, to calculate this function, we need to "add up" the probability densities $[\psi(r)]^2$ over all points located a distance *r* from the nucleus. ▼ **Figure 6.21** compares the probability density at a point $[\psi(r)]^2$ with the radial probability function.

Let's examine the difference between probability density and radial probability function more closely. ▶ **Figure 6.22** shows plots of $[\psi(r)]^2$ as a function of *r* for the 1s, 2s, and 3s orbitals of the hydrogen atom. You will notice that these plots look distinctly different from the radial probability functions shown in Figure 6.19.

As shown in Figure 6.21, the collection of points a distance *r* from the nucleus is the surface of a sphere of radius *r*. The probability density *at each point* on that spherical surface is $[\psi(r)]^2$. To add up all the individual probability densities requires calculus and so is beyond the scope of this text. However, the result of that calculation tells us that the radial probability function is the probability density, $[\psi(r)]^2$, multiplied by the surface area of the sphere, $4\pi r^2$:

Radial probability function at distance $r = 4\pi r^2 [\psi(r)]^2$

Thus, the plots of radial probability function in Figure 6.19 are equal to the plots of $[\psi(r)]^2$ in Figure 6.22 multiplied by $4\pi r^2$. The fact that $4\pi r^2$ increases rapidly as we move away from the nucleus makes the two sets of plots look very different from each other. For example, the plot of $[\psi(r)]^2$ for the 3s orbital in Figure 6.22 shows that the function generally gets smaller the farther we go from the nucleus. But when we multiply by $4\pi r^2$, we see peaks that get larger and larger as we move away from the nucleus (Figure 6.19).

The radial probability functions in Figure 6.19 provide us with the more useful information because they tell us the probability of finding the electron at *all* points a distance *r* from the nucleus, not just one particular point.

Related Exercises: 6.54, 6.65, 6.66, 6.98

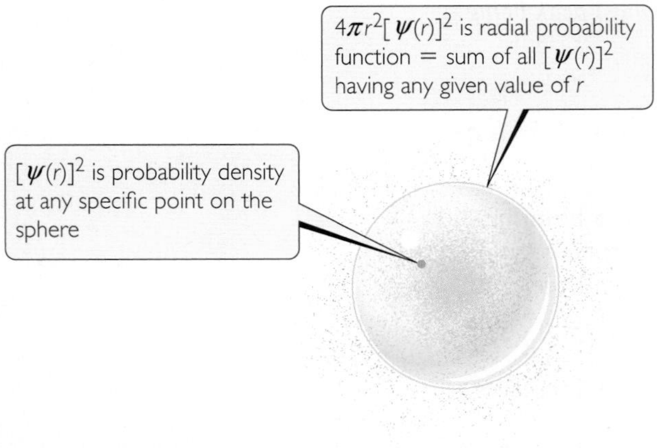

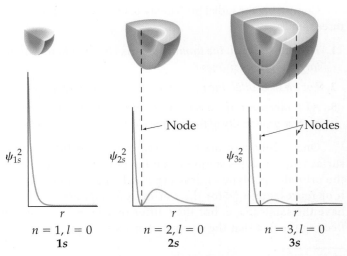

$4\pi r^2 [\psi(r)]^2$ is radial probability function = sum of all $[\psi(r)]^2$ having any given value of *r*

$[\psi(r)]^2$ is probability density at any specific point on the sphere

ψ_{1s}^2

ψ_{2s}^2 — Node

ψ_{3s}^2 — Nodes

r
n = 1, *l* = 0
1s

r
n = 2, *l* = 0
2s

r
n = 3, *l* = 0
3s

▲ **Figure 6.21 Comparing probability density $[\psi(r)]^2$ and radial probability function $4\pi r^2 [\psi(r)]^2$.**

▲ **Figure 6.22 Probability density $[\psi(r)]^2$ in the 1s, 2s, and 3s orbitals of hydrogen.**

Although the details of how electron density varies within a given contour representation are lost in these representations, this is not a serious disadvantage. For qualitative discussions, the most important features of orbitals are shape and relative size, which are adequately displayed by contour representations.

The *p* Orbitals

Recall that the orbitals for which $l = 1$ are the *p* orbitals. Each *p* subshell has three orbitals, corresponding to the three allowed values of m_l: $-1, 0,$ and 1. The distribution of electron density for a 2*p* orbital is shown in ▼ Figure 6.23(**a**). The electron density is not distributed spherically as in an *s* orbital. Instead, the density is concentrated in two regions on either side of the nucleus, separated by a node at the nucleus. We say that this dumbbell-shaped orbital has two *lobes*. Recall that we are making no statement of how the electron is moving within the orbital. Figure 6.23(a) portrays only the *averaged* distribution of the electron density in a 2*p* orbital.

Beginning with the $n = 2$ shell, each shell has three *p* orbitals (Table 6.2). Thus, there are three 2*p* orbitals, three 3*p* orbitals, and so forth. Each set of *p* orbitals has the dumbbell shapes shown in Figure 6.23(a) for the 2*p* orbitals. For each value of *n*, the three *p* orbitals have the same size and shape but differ from one another in spatial orientation. We usually represent *p* orbitals by drawing the shape and orientation of their wave functions, as shown in the contour representations in Figure 6.23(**b**). It is convenient to label these as $p_x, p_y,$ and p_z orbitals. The letter subscript indicates the Cartesian axis along which the orbital is oriented.* Like *s* orbitals, *p* orbitals increase in size as we move from 2*p* to 3*p* to 4*p*, and so forth.

The *d* and *f* Orbitals

When *n* is 3 or greater, we encounter the *d* orbitals (for which $l = 2$). There are five 3*d* orbitals, five 4*d* orbitals, and so forth, because in each shell there are five possible values for the m_l quantum number: $-2, -1, 0, 1,$ and 2. The different *d* orbitals in a given shell have different shapes and orientations in space, as shown in **Figure 6.24**. Four of the *d*-orbital contour representations have a "four-leaf clover" shape, with four lobes, and each lies primarily in a plane. The $d_{xy}, d_{xz},$ and d_{yz} orbitals lie in the *xy, xz,* and *yz* planes,

---

▲ GO FIGURE

(**a**) Note on the left that the color is deep pink in the interior of each lobe but fades to pale pink at the edges. What does this change in color represent? (**b**) What label is applied to the 2*p* orbital aligned along the *x* axis?

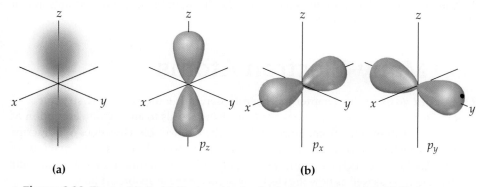

(a)   (b)

▲ **Figure 6.23 The *p* orbitals.** (**a**) Electron-density distribution of a 2*p* orbital. (**b**) Contour representations of the three *p* orbitals. The subscript on the orbital label indicates the axis along which the orbital lies.

---

*We cannot make a simple correspondence between the subscripts ($x, y,$ and z) and the allowed m_l values (1, 0, and -1). To explain why this is so is beyond the scope of an introductory text.

▲ **GO FIGURE**

Which of the *d* orbitals most resembles a p_z orbital?

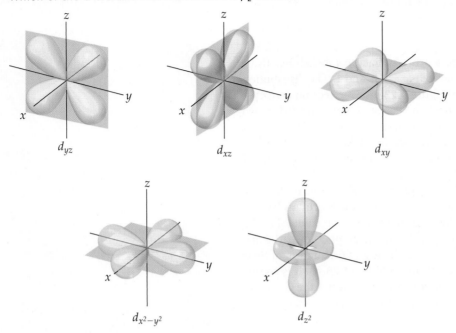

d_{yz} d_{xz} d_{xy}

$d_{x^2-y^2}$ d_{z^2}

▲ **Figure 6.24** **Contour representations of the five d orbitals.**

respectively, with the lobes oriented *between* the axes. The lobes of the $d_{x^2-y^2}$ orbital also lie in the *xy* plane, but the lobes lie *along* the *x* and *y* axes. The d_{z^2} orbital looks very different from the other four: It has two lobes along the *z* axis and a "doughnut" in the *xy* plane. Even though the d_{z^2} orbital looks different from the other *d* orbitals, it has the same energy as the other four *d* orbitals. The representations in Figure 6.24 are commonly used for all *d* orbitals, regardless of the principal quantum number.

When *n* is 4 or greater, there are seven equivalent *f* orbitals (for which $l = 3$). The shapes of the *f* orbitals are even more complicated than those of the *d* orbitals and are not presented here. As you will see in the next section, however, you must be aware of *f* orbitals as we consider the electronic structure of atoms in the lower part of the periodic table.

In many instances later in the text you will find that knowing the number and shapes of atomic orbitals will help you understand chemistry at the molecular level. You will therefore find it useful to memorize the shapes of the *s, p,* and *d* orbitals shown in Figures 6.20, 6.23, and 6.24.

6.7 | Many-Electron Atoms

One of our goals in this chapter has been to determine the electronic structures of atoms. So far, we have seen that quantum mechanics leads to an elegant description of the hydrogen atom. This atom, however, has only one electron. How does our description change when we consider an atom with two or more electrons (a *many-electron* atom)? To describe such an atom, we must consider the nature of orbitals and their relative energies as well as how the electrons populate the available orbitals.

Orbitals and Their Energies

We can describe the electronic structure of a many-electron atom by using the orbitals we described for the hydrogen atom in Table 6.2 (p. 229). Thus, the orbitals of a many-electron atom are designated 1*s*, 2p_x, and so forth (Table 6.2), and have the same general shapes as the corresponding hydrogen orbitals.

Although the shapes of the orbitals of a many-electron atom are the same as those for hydrogen, the presence of more than one electron greatly changes the energies of the orbitals. In hydrogen the energy of an orbital depends only on its principal quantum number, n (Figure 6.18). For instance, in a hydrogen atom the $3s$, $3p$, and $3d$ subshells all have the same energy. In a many-electron atom, however, the energies of the various subshells in a given shell are *different* because of electron–electron repulsions. To explain why this happens, we must consider the forces between the electrons and how these forces are affected by the shapes of the orbitals. We will, however, forgo this analysis until Chapter 7.

The important idea is this: *In a many-electron atom, for a given value of n, the energy of an orbital increases with increasing value of l,* as illustrated in ▶ Figure 6.25. For example, notice in Figure 6.25 that the $n = 3$ orbitals increase in energy in the order $3s < 3p < 3d$. Notice also that all orbitals of a given subshell (such as the five $3d$ orbitals) have the same energy, just as they do in the hydrogen atom. Orbitals with the same energy are said to be **degenerate**.

Figure 6.25 is a *qualitative* energy-level diagram; the exact energies of the orbitals and their spacings differ from one atom to another.

Give It Some Thought

In a many-electron atom, can we predict unambiguously whether the $4s$ orbital is lower or higher in energy than the $3d$ orbitals?

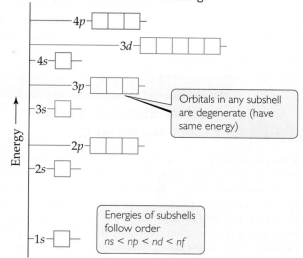

▲ **GO FIGURE**

Not all of the orbitals in the $n = 4$ shell are shown in this figure. Which subshells are missing?

Orbitals in any subshell are degenerate (have same energy)

Energies of subshells follow order
$ns < np < nd < nf$

▲ **Figure 6.25 General energy ordering of orbitals for a many-electron atom.**

Electron Spin and the Pauli Exclusion Principle

We have now seen that we can use hydrogen-like orbitals to describe many-electron atoms. What, however, determines which orbitals the electrons occupy? That is, how do the electrons of a many-electron atom populate the available orbitals? To answer this question, we must consider an additional property of the electron.

When scientists studied the line spectra of many-electron atoms in great detail, they noticed a very puzzling feature: Lines that were originally thought to be single were actually closely spaced pairs. This meant, in essence, that there were twice as many energy levels as there were "supposed" to be. In 1925 the Dutch physicists George Uhlenbeck and Samuel Goudsmit proposed a solution to this dilemma. They postulated that electrons have an intrinsic property, called **electron spin**, that causes each electron to behave as if it were a tiny sphere spinning on its own axis.

By now it may not surprise you to learn that electron spin is quantized. This observation led to the assignment of a new quantum number for the electron, in addition to n, l, and m_l, which we have already discussed. This new quantum number, the **spin magnetic quantum number**, is denoted m_s (the subscript s stands for *spin*). Two possible values are allowed for m_s, $+\frac{1}{2}$ or $-\frac{1}{2}$, which were first interpreted as indicating the two opposite directions in which the electron can spin. A spinning charge produces a magnetic field. The two opposite directions of spin therefore produce oppositely directed magnetic fields (▶ Figure 6.26).* These two opposite magnetic fields lead to the splitting of spectral lines into closely spaced pairs.

Electron spin is crucial for understanding the electronic structures of atoms. In 1925 the Austrian-born physicist Wolfgang Pauli (1900–1958) discovered the principle that governs the arrangement of electrons in many-electron atoms. The **Pauli exclusion principle** states that *no two electrons in an atom can have the same set of four*

▲ **GO FIGURE**

From this figure, why are there only two possible values for the spin quantum number?

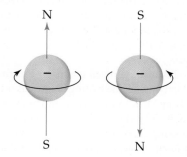

▲ **Figure 6.26 Electron spin.** The electron behaves as if it were spinning about an axis, thereby generating a magnetic field whose direction depends on the direction of spin. The two directions for the magnetic field correspond to the two possible values for the spin quantum number, m_s.

---

*As we discussed earlier, the electron has both particle-like and wave-like properties. Thus, the picture of an electron as a spinning charged sphere is, strictly speaking, just a useful pictorial representation that helps us understand the two directions of magnetic field that an electron can possess.

quantum numbers n, l, m$_l$, *and* m$_s$. For a given orbital, the values of *n*, *l*, and *m$_l$* are fixed. Thus, if we want to put more than one electron in an orbital *and* satisfy the Pauli exclusion principle, our only choice is to assign different *m$_s$* values to the electrons. Because there are only two such values, we conclude that *an orbital can hold a maximum of two electrons and they must have opposite spins*. This restriction allows us to index the electrons in an atom, giving their quantum numbers and thereby defining the region in space where each electron is most likely to be found. It also provides the key to understanding the remarkable structure of the periodic table of the elements.

Chemistry and Life

Nuclear Spin and Magnetic Resonance Imaging

A major challenge facing medical diagnosis is seeing inside the human body. Until recently, this was accomplished primarily by X-ray technology. X rays do not, however, give well-resolved images of overlapping physiological structures, and sometimes fail to discern diseased or injured tissue. Moreover, because X rays are high-energy radiation, they potentially can cause physiological harm, even in low doses. An imaging technique developed in the 1980s called *magnetic resonance imaging* (MRI) does not have these disadvantages.

The foundation of MRI is a phenomenon called *nuclear magnetic resonance* (NMR), which was discovered in the mid-1940s. Today NMR has become one of the most important spectroscopic methods used in chemistry. NMR is based on the observation that, like electrons, the nuclei of many elements possess an intrinsic spin. Like electron spin, nuclear spin is quantized. For example, the nucleus of ^{1}H has two possible magnetic nuclear spin quantum numbers, $+\frac{1}{2}$ and $-\frac{1}{2}$.

A spinning hydrogen nucleus acts like a tiny magnet. In the absence of external effects, the two spin states have the same energy. However, when the nuclei are placed in an external magnetic field, they can align either parallel or opposed (antiparallel) to the field, depending on their spin. The parallel alignment is lower in energy than the antiparallel one by a certain amount, ΔE (▶ Figure 6.27). If the nuclei are irradiated with photons having energy equal to ΔE, the spin of the nuclei can be "flipped," that is, excited from the parallel to the antiparallel alignment. Detection of the flipping of nuclei between the two spin states leads to an NMR spectrum. The radiation used in an NMR experiment is in the radiofrequency range, typically 100 to 900 MHz, which is far less energetic per photon than X rays.

Because hydrogen is a major constituent of aqueous body fluids and fatty tissue, the hydrogen nucleus is the most convenient one for study by MRI. In MRI a person's body is placed in a strong magnetic field. By irradiating the body with pulses of radiofrequency radiation and using sophisticated detection techniques, medical technicians can image tissue at specific depths in the body, giving pictures with spectacular detail (▶ Figure 6.28). The ability to sample at different depths allows the technicians to construct a three-dimensional picture of the body.

MRI has had such a profound influence on the modern practice of medicine that Paul Lauterbur, a chemist, and Peter Mansfield, a physicist, were awarded the 2003 Nobel Prize in Physiology or Medicine for their discoveries concerning MRI. The major drawback of this technique is expense: The current cost of a new standard MRI instrument for clinical applications is typically $1.5 million. In the 2000s, a new technique was developed, called *prepolarized MRI*, that requires

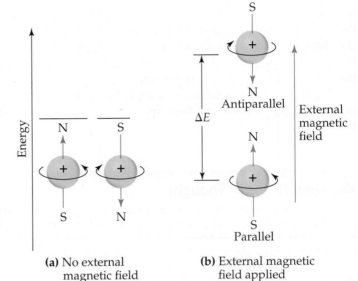

(a) No external magnetic field   **(b)** External magnetic field applied

▲ Figure 6.27 **Nuclear spin.** Like electron spin, nuclear spin generates a small magnetic field and has two allowed values. **(a)** In the absence of an external magnetic field, the two spin states have the same energy. **(b)** When an external magnetic field is applied, the spin state in which the spin direction is parallel to the direction of the external field is lower in energy than the spin state in which the spin direction is antiparallel to the field direction. The energy difference, ΔE, is in the radio frequency portion of the electromagnetic spectrum.

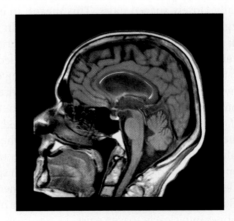

▲ Figure 6.28 **MRI image.** This image of a human head, obtained using magnetic resonance imaging, shows a normal brain, airways, and facial tissues.

much less expensive equipment and will lead to an even greater application of this important diagnostic tool.

Related Exercise: 6.100

6.8 | Electron Configurations

Armed with knowledge of the relative energies of orbitals and the Pauli exclusion principle, we are in a position to consider the arrangements of electrons in atoms. The way electrons are distributed among the various orbitals of an atom is called the **electron configuration** of the atom.

The most stable electron configuration—the ground state—is that in which the electrons are in the lowest possible energy states. If there were no restrictions on the possible values for the quantum numbers of the electrons, all the electrons would crowd into the $1s$ orbital because it is the lowest in energy (Figure 6.25). The Pauli exclusion principle tells us, however, that there can be at most two electrons in any single orbital. Thus, *the orbitals are filled in order of increasing energy, with no more than two electrons per orbital.* For example, consider the lithium atom, which has three electrons. (Recall that the number of electrons in a neutral atom equals its atomic number.) The $1s$ orbital can accommodate two of the electrons. The third one goes into the next lowest-energy orbital, the $2s$.

We can represent any electron configuration by writing the symbol for the occupied subshell and adding a superscript to indicate the number of electrons in that subshell. For example, for lithium we write $1s^2 2s^1$ (read "$1s$ two, $2s$ one"). We can also show the arrangement of the electrons as

Li   [↑↓] [↑]
      $1s$ $2s$

In this representation, which we call an **orbital diagram**, each orbital is denoted by a box and each electron by a half arrow. A half arrow pointing up (↑) represents an electron with a positive spin magnetic quantum number ($m_s = +\frac{1}{2}$), and a half arrow pointing down (↓) represents an electron with a negative spin magnetic quantum number ($m_s = -\frac{1}{2}$). This pictorial representation of electron spin, which corresponds to the directions of the magnetic fields in Figure 6.26, is quite convenient. Chemists refer to the two possible spin states as "spin-up" and "spin-down" corresponding to the directions of the half arrows.

Electrons having opposite spins are said to be *paired* when they are in the same orbital (↑↓). An *unpaired electron* is one not accompanied by a partner of opposite spin. In the lithium atom the two electrons in the $1s$ orbital are paired and the electron in the $2s$ orbital is unpaired.

Hund's Rule

Consider now how the electron configurations of the elements change as we move from element to element across the periodic table. Hydrogen has one electron, which occupies the $1s$ orbital in its ground state:

H   [↑] : $1s^1$
     $1s$

The choice of a spin-up electron here is arbitrary; we could equally well show the ground state with one spin-down electron. It is customary, however, to show unpaired electrons with their spins up.

The next element, helium, has two electrons. Because two electrons with opposite spins can occupy the same orbital, both of helium's electrons are in the $1s$ orbital:

He   [↑↓] : $1s^2$
      $1s$

The two electrons present in helium complete the filling of the first shell. This arrangement represents a very stable configuration, as is evidenced by the chemical inertness of helium.

The electron configurations of lithium and several elements that follow it in the periodic table are shown in ▼ Table 6.3. For the third electron of lithium, the change in the principal quantum number from $n = 1$ for the first two electrons to $n = 2$ for the third electron represents a large jump in energy and a corresponding jump in the average distance of the electron from the nucleus. In other words, it represents the start of a new shell occupied with electrons. As you can see by examining the periodic table, lithium starts a new row of the table. It is the first member of the alkali metals (group 1A).

The element that follows lithium is beryllium; its electron configuration is $1s^2 2s^2$ (Table 6.3). Boron, atomic number 5, has the electron configuration $1s^2 2s^2 2p^1$. The fifth electron must be placed in a $2p$ orbital because the $2s$ orbital is filled. Because all three of the $2p$ orbitals are of equal energy, it does not matter which $2p$ orbital we place this fifth electron in.

With the next element, carbon, we encounter a new situation. We know that the sixth electron must go into a $2p$ orbital. However, does this new electron go into the $2p$ orbital that already has one electron or into one of the other two $2p$ orbitals? This question is answered by **Hund's rule**, which states that *for degenerate orbitals, the lowest energy is attained when the number of electrons having the same spin is maximized.* This means that electrons occupy orbitals singly to the maximum extent possible and that these single electrons in a given subshell all have the same spin magnetic quantum number. Electrons arranged in this way are said to have *parallel spins*. For a carbon atom to achieve its lowest energy, therefore, the two $2p$ electrons must have the same spin. For this to happen, the electrons must be in different $2p$ orbitals, as shown in Table 6.3. Thus, a carbon atom in its ground state has two unpaired electrons.

Similarly, for nitrogen in its ground state, Hund's rule requires that the three $2p$ electrons singly occupy each of the three $2p$ orbitals. This is the only way that all three electrons can have the same spin. For oxygen and fluorine, we place four and five electrons, respectively, in the $2p$ orbitals. To achieve this, we pair up electrons in the $2p$ orbitals, as we will see in Sample Exercise 6.7.

Hund's rule is based in part on the fact that electrons repel one another. By occupying different orbitals, the electrons remain as far as possible from one another, thus minimizing electron–electron repulsions.

Table 6.3 Electron Configurations of Several Lighter Elements

Element	Total Electrons	Orbital Diagram				Electron Configuration
		1s	2s	2p	3s	
Li	3	⇅	↑	☐☐☐	☐	$1s^2 2s^1$
Be	4	⇅	⇅	☐☐☐	☐	$1s^2 2s^2$
B	5	⇅	⇅	↑ ☐☐	☐	$1s^2 2s^2 2p^1$
C	6	⇅	⇅	↑ ↑ ☐	☐	$1s^2 2s^2 2p^2$
N	7	⇅	⇅	↑ ↑ ↑	☐	$1s^2 2s^2 2p^3$
Ne	10	⇅	⇅	⇅ ⇅ ⇅	☐	$1s^2 2s^2 2p^6$
Na	11	⇅	⇅	⇅ ⇅ ⇅	↑	$1s^2 2s^2 2p^6 3s^1$

SAMPLE
EXERCISE 6.7 | Orbital Diagrams and Electron Configurations

Draw the orbital diagram for the electron configuration of oxygen, atomic number 8. How
many unpaired electrons does an oxygen atom possess?

SOLUTION

Analyze and Plan Because oxygen has an atomic number of 8, each oxygen atom has eight
electrons. Figure 6.25 shows the ordering of orbitals. The electrons (represented as half arrows)
are placed in the orbitals (represented as boxes) beginning with the lowest-energy orbital, the $1s$.
Each orbital can hold a maximum of two electrons (the Pauli exclusion principle). Because the
$2p$ orbitals are degenerate, we place one electron in each of these orbitals (spin-up) before pairing any electrons (Hund's rule).

Solve Two electrons each go into the $1s$ and $2s$ orbitals with their spins paired. This leaves four
electrons for the three degenerate $2p$ orbitals. Following Hund's rule, we put one electron into
each $2p$ orbital until all three orbitals have one electron each. The fourth electron is then paired
up with one of the three electrons already in a $2p$ orbital, so that the orbital diagram is

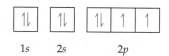

$1s$ $2s$ $2p$

The corresponding electron configuration is written $1s^2 2s^2 2p^4$. The atom has two unpaired
electrons.

Practice Exercise 1

How many of the elements in the second row of the periodic table (Li through Ne) will have
at least one unpaired electron in their electron configurations? **(a)** 3 **(b)** 4 **(c)** 5 **(d)** 6 **(e)** 7

Practice Exercise 2

(a) Write the electron configuration for silicon, element 14, in its ground state. **(b)** How many
unpaired electrons does a ground-state silicon atom possess?

Condensed Electron Configurations

The filling of the $2p$ subshell is complete at neon (Table 6.3), which has a stable configuration with eight electrons (an *octet*) in the outermost occupied shell. The next element, sodium, atomic number 11, marks the beginning of a new row of the periodic
table. Sodium has a single $3s$ electron beyond the stable configuration of neon. We can
therefore abbreviate the electron configuration of sodium as

Na: $[\text{Ne}]3s^1$

The symbol [Ne] represents the electron configuration of the ten electrons of neon,
$1s^2 2s^2 2p^6$. Writing the electron configuration as $[\text{Ne}]3s^1$ focuses attention on the outermost electron of the atom, which is the one largely responsible for how sodium behaves
chemically.

We can generalize what we have just done for the electron configuration of sodium.
In writing the *condensed electron configuration* of an element, the electron configuration of the nearest noble-gas element of lower atomic number is represented by its
chemical symbol in brackets. For lithium, for example, we can write

Li: $[\text{He}]2s^1$

We refer to the electrons represented by the bracketed symbol as the *noble-gas core*
of the atom. More usually, these inner-shell electrons are referred to as the **core electrons**. The electrons given after the noble-gas core are called the *outer-shell electrons*.
The outer-shell electrons include the electrons involved in chemical bonding, which
are called the **valence electrons**. For the elements with atomic number of 30 or less, all
of the outer-shell electrons are valence electrons. By comparing the condensed electron
configurations of lithium and sodium, we can appreciate why these two elements are so

1A

3
Li
[He]$2s^1$
11
Na
[Ne]$3s^1$
19
K
[Ar]$4s^1$
37
Rb
[Kr]$5s^1$
55
Cs
[Xe]$6s^1$
87
Fr
[Rn]$7s^1$

Alkali
metals

▲ Figure 6.29 **The condensed electron configurations of the alkali metals (group 1A in the periodic table).**

similar chemically. They have the same type of electron configuration in the outermost occupied shell. Indeed, all the members of the alkali metal group (1A) have a single *s* valence electron beyond a noble-gas configuration (◄ Figure 6.29).

Transition Metals

The noble-gas element argon ($1s^2 2s^2 2p^6 3s^2 3p^6$) marks the end of the row started by sodium. The element following argon in the periodic table is potassium (K), atomic number 19. In all its chemical properties, potassium is clearly a member of the alkali metal group. The experimental facts about the properties of potassium leave no doubt that the outermost electron of this element occupies an *s* orbital. But this means that the electron with the highest energy has *not* gone into a 3*d* orbital, which we might expect it to do. Because the 4*s* orbital is lower in energy than the 3*d* orbital (Figure 6.25), the condensed electron configuration of potassium is

$$K: \quad [Ar]4s^1$$

Following the complete filling of the 4*s* orbital (this occurs in the calcium atom), the next set of orbitals to be filled is the 3*d*. (You will find it helpful as we go along to refer often to the periodic table on the front-inside cover.) Beginning with scandium and extending through zinc, electrons are added to the five 3*d* orbitals until they are completely filled. Thus, the fourth row of the periodic table is ten elements wider than the two previous rows. These ten elements are known as either **transition elements** or **transition metals**. Note the position of these elements in the periodic table.

In writing the electron configurations of the transition elements, we fill orbitals in accordance with Hund's rule—we add them to the 3*d* orbitals singly until all five orbitals have one electron each and then place additional electrons in the 3*d* orbitals with spin pairing until the shell is completely filled. The condensed electron configurations and the corresponding orbital diagram representations of two transition elements are as follows:

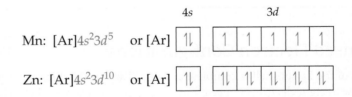

Mn: [Ar]$4s^2 3d^5$ or [Ar] ...

Zn: [Ar]$4s^2 3d^{10}$ or [Ar] ...

Once all the 3*d* orbitals have been filled with two electrons each, the 4*p* orbitals begin to be occupied until the completed octet of outer electrons ($4s^2 4p^6$) is reached with krypton (Kr), atomic number 36, another of the noble gases. Rubidium (Rb) marks the beginning of the fifth row. Refer again to the periodic table on the front-inside cover. Notice that this row is in every respect like the preceding one, except that the value for *n* is greater by 1.

 Give It Some Thought

Based on the structure of the periodic table, which becomes occupied first, the 6*s* orbital or the 5*d* orbitals?

The Lanthanides and Actinides

The sixth row of the periodic table begins with Cs and Ba, which have [Xe]$6s^1$ and [Xe]$6s^2$ configurations, respectively. Notice, however, that the periodic table then has a break, with elements 57–70 placed below the main portion of the table. This break point is where we begin to encounter a new set of orbitals, the 4*f*.

There are seven degenerate 4*f* orbitals, corresponding to the seven allowed values of m_l, ranging from 3 to −3. Thus, it takes 14 electrons to fill the 4*f* orbitals completely. The 14 elements corresponding to the filling of the 4*f* orbitals are known as either the **lanthanide elements** or the **rare earth elements**. These elements are set below the other elements to avoid making the periodic table unduly wide. The properties of the lanthanide elements are all quite similar, and these elements occur together in nature. For many years it was virtually impossible to separate them from one another.

Because the energies of the $4f$ and $5d$ orbitals are very close to each other, the electron configurations of some of the lanthanides involve $5d$ electrons. For example, the elements lanthanum (La), cerium (Ce), and praseodymium (Pr) have the following electron configurations:

$$[Xe]6s^2 5d^1 \qquad [Xe]6s^2 5d^1 4f^1 \qquad [Xe]6s^2 4f^3$$
Lanthanum　　　　Cerium　　　　Praseodymium

Because La has a single $5d$ electron, it is sometimes placed below yttrium (Y) as the first member of the third series of transition elements; Ce is then placed as the first member of the lanthanides. Based on its chemical properties, however, La can be considered the first element in the lanthanide series. Arranged this way, there are fewer apparent exceptions to the regular filling of the $4f$ orbitals among the subsequent members of the series.

After the lanthanide series, the third transition element series is completed by the filling of the $5d$ orbitals, followed by the filling of the $6p$ orbitals. This brings us to radon (Rn), heaviest of the known noble-gas elements.

The final row of the periodic table begins by filling the $7s$ orbitals. The **actinide elements**, of which uranium (U, element 92) and plutonium (Pu, element 94) are the best known, are then built up by completing the $5f$ orbitals. All of the actinide elements are radioactive, and most of them are not found in nature.

6.9 | Electron Configurations and the Periodic Table

We just saw that the electron configurations of the elements correspond to their locations in the periodic table. Thus, elements in the same column of the table have related outer-shell (valence) electron configurations. As ▶ **Table 6.4** shows, for example, all 2A elements have an ns^2 outer configuration, and all 3A elements have an $ns^2 np^1$ outer configuration, with the value of n increasing as we move down each column.

In Table 6.2 we saw that the total number of orbitals in each shell equals n^2: 1, 4, 9, or 16. Because we can place two electrons in each orbital, each shell accommodates up to $2n^2$ electrons: 2, 8, 18, or 32. We see that the overall structure of the periodic table reflects these electron numbers: Each row of the table has 2, 8, 18, or 32 elements in it. As shown in ▼ **Figure 6.30**, the periodic table can be further divided into four blocks based on the filling order of orbitals. On the left are *two* blue columns of elements. These elements, known as the alkali metals (group 1A) and alkaline earth metals (group 2A), are those in which the valence s orbitals are being filled. These two columns make up the s block of the periodic table.

Table 6.4 Electron Configurations of Group 2A and 3A Elements

Group 2A	
Be	$[He]2s^2$
Mg	$[Ne]3s^2$
Ca	$[Ar]4s^2$
Sr	$[Kr]5s^2$
Ba	$[Xe]6s^2$
Ra	$[Rn]7s^2$
Group 3A	
B	$[He]2s^2 2p^1$
Al	$[Ne]3s^2 3p^1$
Ga	$[Ar]3d^{10}4s^2 4p^1$
In	$[Kr]4d^{10}5s^2 5p^1$
Tl	$[Xe]4f^{14}5d^{10}6s^2 6p^1$

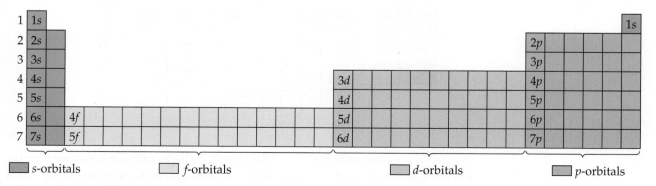

▲ **Figure 6.30 Regions of the periodic table.** The order in which electrons are added to orbitals is read left to right beginning in the top-left corner.

On the right is a block of *six* pink columns that comprises the *p* block, where the valence *p* orbitals are being filled. The *s* block and the *p* block elements together are the **representative elements**, sometimes called the **main-group elements**.

The orange block in Figure 6.30 has *ten* columns containing the transition metals. These are the elements in which the valence *d* orbitals are being filled and make up the *d* block.

The elements in the two tan rows containing *14* columns are the ones in which the valence *f* orbitals are being filled and make up the *f* block. Consequently, these elements are often referred to as the *f*-block metals. In most tables, the *f* block is positioned below the periodic table to save space:

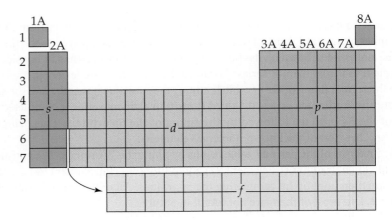

The number of columns in each block corresponds to the maximum number of electrons that can occupy each kind of subshell. Recall that 2, 6, 10, and 14 are the numbers of electrons that can fill the *s*, *p*, *d*, and *f* subshells, respectively. Thus, the *s* block has 2 columns, the *p* block has 6, the *d* block has 10, and the *f* block has 14. Recall also that 1*s* is the first *s* subshell, 2*p* is the first *p* subshell, 3*d* is the first *d* subshell, and 4*f* is the first *f* subshell, as Figure 6.30 shows. Using these facts, you can write the electron configuration of an element based merely on its position in the periodic table. Remember: *The periodic table is your best guide to the order in which orbitals are filled.*

Let's use the periodic table to write the electron configuration of selenium (Se, element 34). We first locate Se in the table and then move backward from it through the table, from element 34 to 33 to 32, and so forth, until we come to the noble gas that precedes Se. In this case, the noble gas is argon, Ar, element 18. Thus, the noble-gas core for Se is [Ar]. Our next step is to write symbols for the outer electrons. We do this by moving across period 4 from K, the element following Ar, to Se:

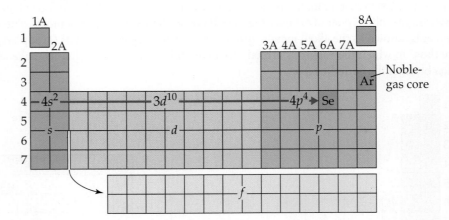

Because K is in the fourth period and the *s* block, we begin with the 4*s* electrons, meaning our first two outer electrons are written $4s^2$. We then move into the *d* block, which begins with the 3*d* electrons. (The principal quantum number in the *d* block is always one less than that of the preceding elements in the *s* block, as seen in Figure 6.30.)

Traversing the d block adds ten electrons, $3d^{10}$. Finally, we move into the p block, whose principal quantum number is always the same as that of the s block. Counting the squares as we move across the p block to Se tells us that we need four electrons, $4p^4$. The electron configuration for Se is therefore $[\text{Ar}]4s^2 3d^{10} 4p^4$. This configuration can also be written with the subshells arranged in order of increasing principal quantum number: $[\text{Ar}]3d^{10}4s^2 4p^4$.

As a check, we add the number of electrons in the [Ar] core, 18, to the number of electrons we added to the $4s$, $3d$, and $4p$ subshells. This sum should equal the atomic number of Se, 34: $18 + 2 + 10 + 4 = 34$.

SAMPLE EXERCISE 6.8 | Electron Configurations for a Group

What is the characteristic valence electron configuration of the group 7A elements, the halogens?

SOLUTION

Analyze and Plan We first locate the halogens in the periodic table, write the electron configurations for the first two elements, and then determine the general similarity between the configurations.

Solve The first member of the halogen group is fluorine (F, element 9). Moving backward from F, we find that the noble-gas core is [He]. Moving from He to the element of next higher atomic number brings us to Li, element 3. Because Li is in the second period of the s block, we add electrons to the $2s$ subshell. Moving across this block gives $2s^2$. Continuing to move to the right, we enter the p block. Counting the squares to F gives $2p^5$. Thus, the condensed electron configuration for fluorine is

$$F: \quad [\text{He}]2s^2 2p^5$$

The electron configuration for chlorine, the second halogen, is

$$Cl: \quad [\text{Ne}]3s^2 3p^5$$

From these two examples, we see that the characteristic valence electron configuration of a halogen is $ns^2 np^5$, where n ranges from 2 in the case of fluorine to 6 in the case of astatine.

Practice Exercise 1

A certain atom has an $ns^2 np^6$ electron configuration in its outermost occupied shell. Which of the following elements could it be? **(a)** Be **(b)** Si **(c)** I **(d)** Kr **(e)** Rb

Practice Exercise 2

Which family of elements is characterized by an $ns^2 np^2$ electron configuration in the outermost occupied shell?

SAMPLE EXERCISE 6.9 | Electron Configurations from the Periodic Table

(a) Based on its position in the periodic table, write the condensed electron configuration for bismuth, element 83. **(b)** How many unpaired electrons does a bismuth atom have?

SOLUTION

(a) Our first step is to write the noble-gas core. We do this by locating bismuth, element 83, in the periodic table. We then move backward to the nearest noble gas, which is Xe, element 54. Thus, the noble-gas core is [Xe].

Next, we trace the path in order of increasing atomic numbers from Xe to Bi. Moving from Xe to Cs, element 55, we find ourselves in period 6 of the s block. Knowing the block and the period identifies the subshell in which we begin placing outer electrons, $6s$. As we move through the s block, we add two electrons: $6s^2$.

As we move beyond the s block, from element 56 to element 57, the curved arrow below the periodic table reminds us that we are entering the f block. The first row of the f block corresponds to the $4f$ subshell. As we move across this block, we add 14 electrons: $4f^{14}$.

With element 71, we move into the third row of the d block. Because the first row of the d block is $3d$, the second row is $4d$ and the third row is $5d$. Thus, as we move through the ten elements of the d block, from element 71 to element 80, we fill the $5d$ subshell with ten electrons: $5d^{10}$.

Moving from element 80 to element 81 puts us into the p block in the $6p$ subshell. (Remember that the principal quantum number in the p block is the same as that in the s block.) Moving across to Bi requires three electrons: $6p^3$. The path we have taken is

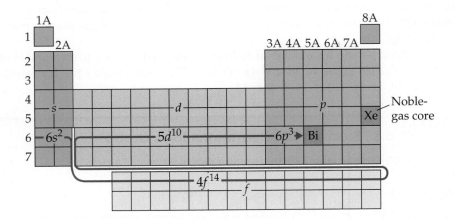

Putting the parts together, we obtain the condensed electron configuration: $[Xe]6s^2 4f^{14} 5d^{10} 6p^3$. This configuration can also be written with the subshells arranged in order of increasing principal quantum number: $[Xe]4f^{14} 5d^{10} 6s^2 6p^3$.

Finally, we check our result to see if the number of electrons equals the atomic number of Bi, 83: Because Xe has 54 electrons (its atomic number), we have $54 + 2 + 14 + 10 + 3 = 83$. (If we had 14 electrons too few, we would realize that we have missed the f block.)

(b) We see from the condensed electron configuration that the only partially occupied subshell is $6p$. The orbital diagram representation for this subshell is

In accordance with Hund's rule, the three $6p$ electrons occupy the three $6p$ orbitals singly, with their spins parallel. Thus, there are three unpaired electrons in the bismuth atom.

Practice Exercise 1

A certain atom has an $[\text{noble gas}]5s^2 4d^{10} 5p^4$ electron configuration. Which element is it?
(a) Cd **(b)** Te **(c)** Sm **(d)** Hg **(e)** More information is needed

Practice Exercise 2

Use the periodic table to write the condensed electron configuration for **(a)** Co (element 27), **(b)** In (element 49).

Figure 6.31 gives, for all the elements, the ground-state electron configurations for the outer-shell electrons. You can use this figure to check your answers as you practice writing electron configurations. We have written these configurations with orbitals listed in order of increasing principal quantum number. As we saw in Sample Exercise 6.9, the orbitals can also be listed in order of filling, as they would be read off of the periodic table.

Figure 6.31 allow us to reexamine the concept of *valence electrons*. Notice, for example, that as we proceed from Cl ($[Ne]3s^2 3p^5$) to Br ($[Ar]3d^{10} 4s^2 4p^5$) we add a complete subshell of $3d$ electrons to the electrons beyond the $[Ar]$ core. Although the $3d$ electrons are outer-shell electrons, they are not involved in chemical bonding and are therefore not considered valence electrons. Thus, we consider only the $4s$ and $4p$ electrons of Br to be valence electrons. Similarly, if we compare the electron configurations of Ag (element 47) and Au (element 79), we see that Au has a completely full $4f^{14}$ subshell beyond its noble-gas core, but those $4f$ electrons are not involved in bonding. In general, *for representative elements we do not consider the electrons in completely filled d or f subshells to be valence electrons*, and *for transition elements we do not consider the electrons in a completely filled f subshell to be valence electrons*.

 **GO FIGURE**

A friend tells you that her favorite element has an electron configuration of [noble gas]$6s^2 4f^{14} 5d^6$. Which element is it?

	1A 1																	8A 18
Core	1 **H** $1s^1$	2A 2											3A 13	4A 14	5A 15	6A 16	7A 17	2 **He** $1s^2$
[He]	3 **Li** $2s^1$	4 **Be** $2s^2$											5 **B** $2s^2 2p^1$	6 **C** $2s^2 2p^2$	7 **N** $2s^2 2p^3$	8 **O** $2s^2 2p^4$	9 **F** $2s^2 2p^5$	10 **Ne** $2s^2 2p^6$
[Ne]	11 **Na** $3s^1$	12 **Mg** $3s^2$	3B 3	4B 4	5B 5	6B 6	7B 7	8	8B 9	10	1B 11	2B 12	13 **Al** $3s^2 3p^1$	14 **Si** $3s^2 3p^2$	15 **P** $3s^2 3p^3$	16 **S** $3s^2 3p^4$	17 **Cl** $3s^2 3p^5$	18 **Ar** $3s^2 3p^6$
[Ar]	19 **K** $4s^1$	20 **Ca** $4s^2$	21 **Sc** $4s^2 3d^1$	22 **Ti** $4s^2 3d^2$	23 **V** $4s^2 3d^3$	24 **Cr** $4s^1 3d^5$	25 **Mn** $4s^2 3d^5$	26 **Fe** $4s^2 3d^6$	27 **Co** $4s^2 3d^7$	28 **Ni** $4s^2 3d^8$	29 **Cu** $4s^1 3d^{10}$	30 **Zn** $4s^2 3d^{10}$	31 **Ga** $4s^2 3d^{10}$ $4p^1$	32 **Ge** $4s^2 3d^{10}$ $4p^2$	33 **As** $4s^2 3d^{10}$ $4p^3$	34 **Se** $4s^2 3d^{10}$ $4p^4$	35 **Br** $4s^2 3d^{10}$ $4p^5$	36 **Kr** $4s^2 3d^{10}$ $4p^6$
[Kr]	37 **Rb** $5s^1$	38 **Sr** $5s^2$	39 **Y** $5s^2 4d^1$	40 **Zr** $5s^2 4d^2$	41 **Nb** $5s^1 4d^3$	42 **Mo** $5s^1 4d^5$	43 **Tc** $5s^2 4d^5$	44 **Ru** $5s^1 4d^7$	45 **Rh** $5s^1 4d^8$	46 **Pd** $4d^{10}$	47 **Ag** $5s^1 4d^{10}$	48 **Cd** $5s^2 4d^{10}$	49 **In** $5s^2 4d^{10}$ $5p^1$	50 **Sn** $5s^2 4d^{10}$ $5p^2$	51 **Sb** $5s^2 4d^{10}$ $5p^3$	52 **Te** $5s^2 4d^{10}$ $5p^4$	53 **I** $5s^2 4d^{10}$ $5p^5$	54 **Xe** $5s^2 4d^{10}$ $5p^6$
[Xe]	55 **Cs** $6s^1$	56 **Ba** $6s^2$	71 **Lu** $6s^2 4f^{14}$ $5d^1$	72 **Hf** $6s^2 4f^{14}$ $5d^2$	73 **Ta** $6s^2 4f^{14}$ $5d^3$	74 **W** $6s^2 4f^{14}$ $5d^4$	75 **Re** $6s^2 4f^{14}$ $5d^5$	76 **Os** $6s^2 4f^{14}$ $5d^6$	77 **Ir** $6s^2 4f^{14}$ $5d^7$	78 **Pt** $6s^1 4f^{14}$ $5d^9$	79 **Au** $6s^1 4f^{14}$ $5d^{10}$	80 **Hg** $6s^2 4f^{14}$ $5d^{10}$	81 **Tl** $6s^2 4f^{14}$ $5d^{10} 6p^1$	82 **Pb** $6s^2 4f^{14}$ $5d^{10} 6p^2$	83 **Bi** $6s^2 4f^{14}$ $5d^{10} 6p^3$	84 **Po** $6s^2 4f^{14}$ $5d^{10} 6p^4$	85 **At** $6s^2 4f^{14}$ $5d^{10} 6p^5$	86 **Rn** $6s^2 4f^{14}$ $5d^{10} 6p^6$
[Rn]	87 **Fr** $7s^1$	88 **Ra** $7s^2$	103 **Lr** $7s^2 5f^{14}$ $6d^1$	104 **Rf** $7s^2 5f^{14}$ $6d^2$	105 **Db** $7s^2 5f^{14}$ $6d^3$	106 **Sg** $7s^2 5f^{14}$ $6d^4$	107 **Bh** $7s^2 5f^{14}$ $6d^5$	108 **Hs** $7s^2 5f^{14}$ $6d^6$	109 **Mt** $7s^2 5f^{14}$ $6d^7$	110 **Ds** $7s^2 5f^{14}$ $6d^8$	111 **Rg** $7s^2 5f^{14}$ $6d^9$	112 **Cn** $7s^2 5f^{14}$ $6d^{10}$	113 **Fl** $7s^2 5f^{14}$ $6d^{10} 7p^1$	114 **Cn** $7s^2 5f^{14}$ $6d^{10} 7p^2$	115 $7s^2 5f^{14}$ $6d^{10} 7p^3$	116 **Lv** $7s^2 5f^{14}$ $6d^{10} 7p^4$	117 $7s^2 5f^{14}$ $6d^{10} 7p^5$	118 $7s^2 5f^{14}$ $6d^{10} 7p^6$

Lanthanide series [Xe]

57 **La** $6s^2 5d^1$	58 **Ce** $6s^2 4f^1$ $5d^1$	59 **Pr** $6s^2 4f^3$	60 **Nd** $6s^2 4f^4$	61 **Pm** $6s^2 4f^5$	62 **Sm** $6s^2 4f^6$	63 **Eu** $6s^2 4f^7$	64 **Gd** $6s^2 4f^7$ $5d^1$	65 **Tb** $6s^2 4f^9$	66 **Dy** $6s^2 4f^{10}$	67 **Ho** $6s^2 4f^{11}$	68 **Er** $6s^2 4f^{12}$	69 **Tm** $6s^2 4f^{13}$	70 **Yb** $6s^2 4f^{14}$

Actinide series [Rn]

89 **Ac** $7s^2 6d^1$	90 **Th** $7s^2 6d^2$	91 **Pa** $7s^2 5f^2$ $6d^1$	92 **U** $7s^2 5f^3$ $6d^1$	93 **Np** $7s^2 5f^4$ $6d^1$	94 **Pu** $7s^2 5f^6$	95 **Am** $7s^2 5f^7$	96 **Cm** $7s^2 5f^7$ $6d^1$	97 **Bk** $7s^2 5f^9$	98 **Cf** $7s^2 5f^{10}$	99 **Es** $7s^2 5f^{11}$	100 **Fm** $7s^2 5f^{12}$	101 **Md** $7s^2 5f^{13}$	102 **No** $7s^2 5f^{14}$

☐ Metals   ▨ Metalloids   ☐ Nonmetals

▲ **Figure 6.31 Outer-shell electron configurations of the elements.**

Anomalous Electron Configurations

The electron configurations of certain elements appear to violate the rules we have just discussed. For example, Figure 6.31 shows that the electron configuration of chromium (element 24) is $[Ar]3d^5 4s^1$ rather than the $[Ar]3d^4 4s^2$ configuration we might expect. Similarly, the configuration of copper (element 29) is $[Ar]3d^{10} 4s^1$ instead of $[Ar]3d^9 4s^2$.

This anomalous behavior is largely a consequence of the closeness of the $3d$ and $4s$ orbital energies. It frequently occurs when there are enough electrons to form precisely half-filled sets of degenerate orbitals (as in chromium) or a completely filled d subshell (as in copper). There are a few similar cases among the heavier transition metals (those with partially filled $4d$ or $5d$ orbitals) and among the f-block metals. Although these minor departures from the expected are interesting, they are not of great chemical significance.

 Give It Some Thought

The elements Ni, Pd, and Pt are all in the same group. By examining the electron configurations for these elements in Figure 6.31, what can you conclude about the relative energies of the nd and $(n + 1)s$ orbitals for this group?

SAMPLE INTEGRATIVE EXERCISE | Putting Concepts Together

Boron, atomic number 5, occurs naturally as two isotopes, ^{10}B and ^{11}B, with natural abundances of 19.9% and 80.1%, respectively. **(a)** In what ways do the two isotopes differ from each other? Does the electronic configuration of ^{10}B differ from that of ^{11}B? **(b)** Draw the orbital diagram for an atom of ^{11}B. Which electrons are the valence electrons? **(c)** Indicate three major ways in which the $1s$ electrons in boron differ from its $2s$ electrons. **(d)** Elemental boron reacts with fluorine to form BF_3, a gas. Write a balanced chemical equation for the reaction of solid boron with fluorine gas. **(e)** ΔH_f° for $BF_3(g)$ is -1135.6 kJ/mol. Calculate the standard enthalpy change in the reaction of boron with fluorine. **(f)** Will the mass percentage of F be the same in $^{10}BF_3$ and $^{11}BF_3$? If not, why is that the case?

SOLUTION

(a) The two isotopes of boron differ in the number of neutrons in the nucleus. ∞ (Sections 2.3 and 2.4) Each of the isotopes contains five protons, but ^{10}B contains five neutrons, whereas ^{11}B contains six neutrons. The two isotopes of boron have identical electron configurations, $1s^2 2s^2 2p^1$, because each has five electrons.

(b) The complete orbital diagram is

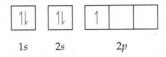

$$\begin{array}{ccc} 1s & 2s & 2p \end{array}$$

The valence electrons are the ones in the outermost occupied shell, the $2s^2$ and $2p^1$ electrons. The $1s^2$ electrons constitute the core electrons, which we represent as [He] when we write the condensed electron configuration, $[He]2s^2 2p^1$.

(c) The $1s$ and $2s$ orbitals are both spherical, but they differ in three important respects: First, the $1s$ orbital is lower in energy than the $2s$ orbital. Second, the average distance of the $2s$ electrons from the nucleus is greater than that of the $1s$ electrons, so the $1s$ orbital is smaller than the $2s$. Third, the $2s$ orbital has one node, whereas the $1s$ orbital has no nodes (Figure 6.19).

(d) The balanced chemical equation is

$$2\,B(s) + 3\,F_2(g) \longrightarrow 2\,BF_3(g)$$

(e) $\Delta H^\circ = 2(-1135.6) - [0 + 0] = -2271.2$ kJ. The reaction is strongly exothermic.

(f) As we saw in Equation 3.10 (Section 3.3), the mass percentage of an element in a substance depends on the formula weight of the substance. The formula weights of $^{10}BF_3$ and $^{11}BF_3$ are different because of the difference in the masses of the two isotopes (the isotope masses of ^{10}B and ^{11}B are 10.01294 and 11.00931 amu, respectively). The denominators in Equation 3.10 would therefore be different for the two isotopes, whereas the numerators would remain the same.

Chapter Summary and Key Terms

WAVELENGTHS AND FREQUENCIES OF LIGHT (INTRODUCTION AND SECTION 6.1) The **electronic structure** of an atom describes the energies and arrangement of electrons around the atom. Much of what is known about the electronic structure of atoms was obtained by observing the interaction of light with matter.

Visible light and other forms of **electromagnetic radiation** (also known as radiant energy) move through a vacuum at the speed of light, $c = 2.998 \times 10^8$ m/s. Electromagnetic radiation has both electric and magnetic components that vary periodically in wave-like fashion. The wave characteristics of radiant energy allow it to be described in terms of **wavelength**, λ, and **frequency**, ν, which are interrelated: $\lambda\nu = c$.

QUANTIZED ENERGY AND PHOTONS (SECTION 6.2) Planck proposed that the minimum amount of radiant energy that an object can gain or lose is related to the frequency of the radiation: $E = h\nu$. This smallest quantity is called a **quantum** of energy. The constant h is called **Planck constant**: $h = 6.626 \times 10^{-34}$ J-s.

In the quantum theory, energy is quantized, meaning that it can have only certain allowed values. Einstein used the quantum theory to explain the **photoelectric effect**, the emission of electrons from metal surfaces when exposed to light. He proposed that light behaves as if it consists of quantized energy packets called **photons**. Each photon carries energy, $E = h\nu$.

BOHR MODEL OF THE HYDROGEN ATOM (SECTION 6.3) Dispersion of radiation into its component wavelengths produces a **spectrum**. If the spectrum contains all wavelengths, it is called a **continuous spectrum**; if it contains only certain specific wavelengths, the spectrum is called a **line spectrum**. The radiation emitted by excited hydrogen atoms forms a line spectrum.

Bohr proposed a model of the hydrogen atom that explains its line spectrum. In this model the energy of the electron in the hydrogen atom depends on the value of a quantum number, n, called the

principal quantum number. The value of n must be a positive integer $(1, 2, 3, \ldots)$, and each value of n corresponds to a different specific energy, E_n. The energy of the atom increases as n increases. The lowest energy is achieved for $n = 1$; this is called the **ground state** of the hydrogen atom. Other values of n correspond to **excited states**. Light is emitted when the electron drops from a higher-energy state to a lower-energy state; light is absorbed to excite the electron from a lower energy state to a higher one. The frequency of light emitted or absorbed is such that $h\nu$ equals the difference in energy between two allowed states.

WAVE BEHAVIOR OF MATTER (SECTION 6.4) De Broglie proposed that matter, such as electrons, should exhibit wave-like properties. This hypothesis of **matter waves** was proved experimentally by observing the diffraction of electrons. An object has a characteristic wavelength that depends on its **momentum**, mv: $\lambda = h/mv$.

Discovery of the wave properties of the electron led to Heisenberg's **uncertainty principle**, which states that there is an inherent limit to the accuracy with which the position and momentum of a particle can be measured simultaneously.

QUANTUM MECHANICS AND ORBITALS (SECTION 6.5) In the quantum mechanical model of the hydrogen atom, the behavior of the electron is described by mathematical functions called **wave functions**, denoted with the Greek letter ψ. Each allowed wave function has a precisely known energy, but the location of the electron cannot be determined exactly; rather, the probability of it being at a particular point in space is given by the **probability density**, ψ^2. The **electron density** distribution is a map of the probability of finding the electron at all points in space.

The allowed wave functions of the hydrogen atom are called **orbitals**. An orbital is described by a combination of an integer and a letter, corresponding to values of three quantum numbers. The *principal quantum number*, n, is indicated by the integers $1, 2, 3, \ldots$. This quantum number relates most directly to the size and energy of the orbital. The **angular momentum quantum number**, l, is indicated by the letters s, p, d, f, and so on, corresponding to the values of $0, 1, 2, 3, \ldots$. The l quantum number defines the shape of the orbital. For a given value of n, l can have integer values ranging from 0 to $(n-1)$. The **magnetic quantum number**, m_l, relates to the orientation of the orbital in space. For a given value of l, m_l can have integral values ranging from $-l$ to l, including 0. Subscripts can be used to label the orientations of the orbitals. For example, the three $3p$ orbitals are designated $3p_x, 3p_y,$ and $3p_z$, with the subscripts indicating the axis along which the orbital is oriented.

An electron shell is the set of all orbitals with the same value of n, such as $3s, 3p,$ and $3d$. In the hydrogen atom all the orbitals in an electron shell have the same energy. A **subshell** is the set of one or more orbitals with the same n and l values; for example, $3s, 3p,$ and $3d$ are each subshells of the $n = 3$ shell. There is one orbital in an s subshell, three in a p subshell, five in a d subshell, and seven in an f subshell.

REPRESENTATIONS OF ORBITALS (SECTION 6.6) Contour representations are useful for visualizing the shapes of the orbitals. Represented this way, s orbitals appear as spheres that increase in size as n increases. The **radial probability function** tells us the probability that the electron will be found at a certain distance from the nucleus. The wave function for each p orbital has two lobes on opposite sides of the nucleus. They are oriented along the $x, y,$ and z axes. Four of the d orbitals appear as shapes with four lobes around the nucleus; the fifth one, the d_{z^2} orbital, is represented as two lobes along the z axis and a "doughnut" in the xy plane. Regions in which the wave function is zero are called **nodes**. There is zero probability that the electron will be found at a node.

MANY-ELECTRON ATOMS (SECTION 6.7) In many-electron atoms, different subshells of the same electron shell have different energies. For a given value of n, the energy of the subshells increases as the value of l increases: $ns < np < nd < nf$. Orbitals within the same subshell are **degenerate**, meaning they have the same energy.

Electrons have an intrinsic property called **electron spin**, which is quantized. The **spin magnetic quantum number**, m_s, can have two possible values, $+\frac{1}{2}$ and $-\frac{1}{2}$, which can be envisioned as the two directions of an electron spinning about an axis. The **Pauli exclusion principle** states that no two electrons in an atom can have the same values for $n, l, m_l,$ and m_s. This principle places a limit of two on the number of electrons that can occupy any one atomic orbital. These two electrons differ in their value of m_s.

ELECTRON CONFIGURATIONS AND THE PERIODIC TABLE (SECTIONS 6.8 AND 6.9) The **electron configuration** of an atom describes how the electrons are distributed among the orbitals of the atom. The ground-state electron configurations are generally obtained by placing the electrons in the atomic orbitals of lowest possible energy with the restriction that each orbital can hold no more than two electrons. We depict the arrangement of the electrons pictorially using an **orbital diagram**. When electrons occupy a subshell with more than one degenerate orbital, such as the $2p$ subshell, **Hund's rule** states that the lowest energy is attained by maximizing the number of electrons with the same electron spin. For example, in the ground-state electron configuration of carbon, the two $2p$ electrons have the same spin and must occupy two different $2p$ orbitals.

Elements in any given group in the periodic table have the same type of electron arrangements in their outermost shells. For example, the electron configurations of the halogens fluorine and chlorine are $[\text{He}]2s^22p^5$ and $[\text{Ne}]3s^23p^5$, respectively. The outer-shell electrons are those that lie outside the orbitals occupied in the next lowest noble-gas element. The outer-shell electrons that are involved in chemical bonding are the **valence electrons** of an atom; for the elements with atomic number 30 or less, all the outer-shell electrons are valence electrons. The electrons that are not valence electrons are called **core electrons**.

The periodic table is partitioned into different types of elements, based on their electron configurations. Those elements in which the outermost subshell is an s or p subshell are called the **representative** (or **main-group**) elements. The alkali metals (group 1A), halogens (group 7A), and noble gases (group 8A) are representative elements. Those elements in which a d subshell is being filled are called the **transition elements** (or **transition metals**). The elements in which the $4f$ subshell is being filled are called the **lanthanide** (or **rare earth**) elements. The actinide elements are those in which the $5f$ subshell is being filled. The lanthanide and **actinide elements** are collectively referred to as the **f-block metals**. These elements are shown as two rows of 14 elements below the main part of the periodic table. The structure of the periodic table, summarized in Figure 6.31, allows us to write the electron configuration of an element from its position in the periodic table.

Learning Outcomes   After studying this chapter, you should be able to:

- Calculate the wavelength of electromagnetic radiation given its frequency or its frequency given its wavelength. (Section 6.1)
- Order the common kinds of radiation in the electromagnetic spectrum according to their wavelengths or energy. (Section 6.1)
- Explain what photons are and be able to calculate their energies given either their frequency or wavelength. (Section 6.2)
- Explain how line spectra relate to the idea of quantized energy states of electrons in atoms. (Section 6.3)
- Calculate the wavelength of a moving object. (Section 6.4)
- Explain how the uncertainty principle limits how precisely we can specify the position and the momentum of subatomic particles such as electrons. (Section 6.4)

- Relate the quantum numbers to the number and type of orbitals and recognize the different orbital shapes. (Section 6.5)
- Interpret radial probability function graphs for the orbitals. (Section 6.6)
- Explain how and why the energies of the orbitals are different in a many-electron atom from those in the hydrogen atom (Section 6.7)
- Draw an energy-level diagram for the orbitals in a many-electron atom and describe how electrons populate the orbitals in the ground state of an atom, using the Pauli exclusion principle and Hund's rule. (Section 6.8)
- Use the periodic table to write condensed electron configurations and determine the number of unpaired electrons in an atom. (Section 6.9)

Key Equations

- $\lambda \nu = c$ [6.1]   light as a wave: λ = wavelength in meters, ν = frequency in s^{-1}, c = speed of light (2.998×10^8 m/s)

- $E = h\nu$ [6.2]   light as a particle (photon): E = energy of photon in joules, h = Planck constant (6.626×10^{-34} J-s), ν = frequency in s^{-1} (same frequency as previous formula)

- $E = (-hcR_H)\left(\dfrac{1}{n^2}\right) = (-2.18 \times 10^{-18}\text{ J})\left(\dfrac{1}{n^2}\right)$ [6.5]   energies of the allowed states of the hydrogen atom: $h =$ Planck constant; $c =$ speed of light; $R_H =$ Rydberg constant $(1.096776 \times 10^7\text{ m}^{-1})$; $n = 1, 2, 3, \ldots$ (any positive integer)

- $\lambda = h/mv$ [6.8]   matter as a wave: $\lambda =$ wavelength, $h =$ Planck constant, $m =$ mass of object in kg, $v =$ speed of object in m/s

- $\Delta x \cdot \Delta(mv) \geq \dfrac{h}{4\pi}$ [6.9]   Heisenberg's uncertainty principle. The uncertainty in position (Δx) and momentum $[\Delta(mv)]$ of an object cannot be zero; the smallest value of their product is $h/4\pi$

Exercises

Visualizing Concepts

6.1 Consider the water wave shown here. (a) How could you measure the speed of this wave? (b) How would you determine the wavelength of the wave? (c) Given the speed and wavelength of the wave, how could you determine the frequency of the wave? (d) Suggest an independent experiment to determine the frequency of the wave. [Section 6.1]

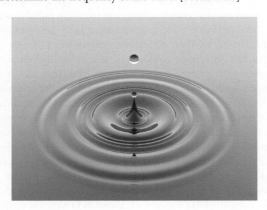

6.2 A popular kitchen appliance produces electromagnetic radiation with a frequency of 2450 MHz. With reference to Figure 6.4, answer the following: (a) Estimate the wavelength of this radiation. (b) Would the radiation produced by the appliance be visible to the human eye? (c) If the radiation is not visible, do photons of this radiation have more or less energy than photons of visible light? (d) Which of the following is the appliance likely to be? (i) A toaster oven, (ii) A microwave oven, or (iii) An electric hotplate. [Section 6.1]

6.3 The following diagrams represent two electromagnetic waves. Which wave corresponds to the higher-energy radiation? [Section 6.2]

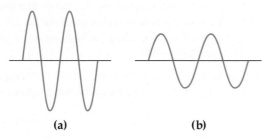

(a)   (b)

6.4 As shown in the accompanying photograph, an electric stove burner on its highest setting exhibits an orange glow. (a) When the burner setting is changed to low, the burner continues to produce heat but the orange glow disappears. How can this observation be explained with reference to one of the fundamental observations that led to the notion of quanta? (b) Suppose that the energy provided to the burner could be increased beyond the highest setting of the stove. What would we expect to observe with regard to visible light emitted by the burner? [Section 6.2]

6.5 Stars do not all have the same temperature. The color of light emitted by stars is characteristic of the light emitted by hot objects. Telescopic photos of three stars are shown below: (i) the Sun, which is classified as a *yellow* star, (ii) *Rigel*, in the constellation Orion, which is classified as a *blue-white* star, and (iii) *Betelgeuse*, also in Orion, which is classified as a *red* star. (a) Place these three stars in order of increasing temperature. (b) Which of the following principles is relevant to your choice of answer for part (a): The uncertainty principle, the photoelectric effect, blackbody radiation, or line spectra? [Section 6.2]

(i) Sun   (ii) Rigel   (iii) Betelgeuse

6.6 The familiar phenomenon of a rainbow results from the diffraction of sunlight through raindrops. (a) Does the wavelength of light increase or decrease as we proceed outward from the innermost band of the rainbow? (b) Does the frequency of light increase or decrease as we proceed outward? (c) Suppose that instead of sunlight, the visible light from a hydrogen discharge tube (Figure 6.10) was used as the light source. What do you think the resulting "hydrogen discharge rainbow" would look like? [Section 6.3]

6.7 A certain quantum mechanical system has the energy levels shown in the accompanying diagram. The energy levels are indexed by a single quantum number n that is an integer. (a) As drawn, which quantum numbers are involved in the transition that requires the most energy? (b) Which quantum numbers are involved in the transition that requires the least energy? (c) Based on the drawing, put the following in order of increasing wavelength of the light absorbed or emitted during the transition: (i) $n = 1$ to $n = 2$; (ii) $n = 3$ to $n = 2$; (iii) $n = 2$ to $n = 4$; (iv) $n = 3$ to $n = 1$. [Section 6.3]

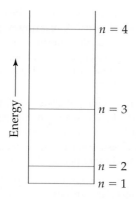

6.8 Consider a fictitious one-dimensional system with one electron. The wave function for the electron, drawn below, is $\psi(x) = \sin x$ from $x = 0$ to $x = 2\pi$. (a) Sketch the probability density, $\psi^2(x)$, from $x = 0$ to $x = 2\pi$. (b) At what value or values of x will there be the greatest probability of finding the electron? (c) What is the probability that the electron will be found at $x = \pi$? What is such a point in a wave function called? [Section 6.5]

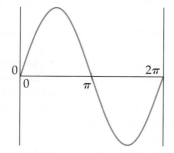

6.9 The contour representation of one of the orbitals for the $n = 3$ shell of a hydrogen atom is shown as follows. (a) What is the quantum number l for this orbital? (b) How do we label this orbital? (c) In which of the following ways would you modify this sketch to show the analogous orbital for the $n = 4$ shell: (i) It doesn't change, (ii) it would be drawn larger, (iii) another lobe would be added along the $+x$ axis, or (iv) the lobe on the $+y$ axis would be larger than the lobe on the $-y$ axis? [Section 6.6]

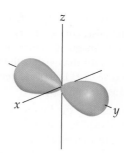

6.10 The accompanying drawing shows the shape of a d orbital. (a) Based on the shape, how many of the d orbitals could it be? (b) Which of the following would you need to determine which of the d orbitals it is: (i) the direction of the z-axis, (ii) the identity of the element, (iii) the number of electrons in the orbital, or (iv) the directions of two of the major axes? [Section 6.6]

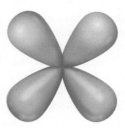

6.11 The accompanying drawing shows part of the orbital diagram for an element. (a) As drawn, the drawing is *incorrect*. Why? (b) How would you correct the drawing without changing the number of electrons? (c) To which group in the periodic table does the element belong? [Section 6.8]

11	1	1

6.12 State where in the periodic table these elements appear:
(a) elements with the valence-shell electron configuration ns^2np^5
(b) elements that have three unpaired p electrons
(c) an element whose valence electrons are $4s^24p^1$
(d) the d-block elements [Section 6.9]

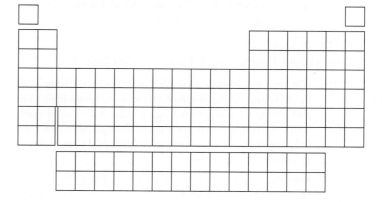

The Wave Nature of Light (Section 6.1)

6.13 What are the basic SI units for (a) the wavelength of light, (b) the frequency of light, (c) the speed of light?

6.14 (a) What is the relationship between the wavelength and the frequency of radiant energy? (b) Ozone in the upper atmosphere absorbs energy in the 210–230-nm range of the spectrum. In what region of the electromagnetic spectrum does this radiation occur?

6.15 Label each of the following statements as true or false. For those that are false, correct the statement. (a) Visible light is a form of electromagnetic radiation. (b) Ultraviolet light has longer wavelengths than visible light. (c) X rays travel faster than microwaves. (d) Electromagnetic radiation and sound waves travel at the same speed.

6.16 Determine which of the following statements are false and correct them. (a) The frequency of radiation increases as the wavelength increases. (b) Electromagnetic radiation travels through a vacuum at a constant speed, regardless of wavelength. (c) Infrared light has higher frequencies than visible light. (d) The glow from a fireplace, the energy within a microwave oven, and a foghorn blast are all forms of electromagnetic radiation.

6.17 Arrange the following kinds of electromagnetic radiation in order of increasing wavelength: infrared, green light, red light, radio waves, X rays, ultraviolet light.

6.18 List the following types of electromagnetic radiation in order of increasing wavelength: (a) the gamma rays produced by a radioactive nuclide used in medical imaging; (b) radiation from an FM radio station at 93.1 MHz on the dial; (c) a radio signal from an AM radio station at 680 kHz on the dial; (d) the yellow light from sodium vapor streetlights; (e) the red light of a light-emitting diode, such as in a calculator display.

6.19 (a) What is the frequency of radiation that has a wavelength of 10 μm, about the size of a bacterium? (b) What is the wavelength of radiation that has a frequency of 5.50×10^{14} s^{-1}? (c) Would the radiations in part (a) or part (b) be visible to the human eye? (d) What distance does electromagnetic radiation travel in 50.0 μs?

6.20 (a) What is the frequency of radiation whose wavelength is 0.86 nm? (b) What is the wavelength of radiation that has a frequency of 6.4×10^{11} s^{-1}? (c) Would the radiations in part (a) or part (b) be detected by an X-ray detector? (d) What distance does electromagnetic radiation travel in 0.38 ps?

6.21 A laser pointer used in a lecture hall emits light at 650 nm. What is the frequency of this radiation? Using Figure 6.4, predict the color associated with this wavelength.

6.22 It is possible to convert radiant energy into electrical energy using photovoltaic cells. Assuming equal efficiency of conversion, would infrared or ultraviolet radiation yield more electrical energy on a per-photon basis?

Quantized Energy and Photons (Section 6.2)

6.23 If human height were quantized in 1-foot increments, what would happen to the height of a child as she grows up: (i) The child's height would never change, (ii) the child's height would continuously get greater, (iii) the child's height would increase in "jumps" of 1 foot at a time, or (iv) the child's height would increase in jumps of 6 in?

6.24 Einstein's 1905 paper on the photoelectric effect was the first important application of Planck's quantum hypothesis. Describe Planck's original hypothesis, and explain how Einstein made use of it in his theory of the photoelectric effect.

6.25 (a) Calculate the energy of a photon of electromagnetic radiation whose frequency is 2.94×10^{14} s^{-1}. (b) Calculate the energy of a photon of radiation whose wavelength is 413 nm. (c) What wavelength of radiation has photons of energy 6.06×10^{-19} J?

6.26 (a) A green laser pointer emits light with a wavelength of 532 nm. What is the frequency of this light? (b) What is the

energy of one of these photons? (c) The laser pointer emits light because electrons in the material are excited (by a battery) from their ground state to an upper excited state. When the electrons return to the ground state, they lose the excess energy in the form of 532-nm photons. What is the energy gap between the ground state and excited state in the laser material?

6.27 (a) Calculate and compare the energy of a photon of wavelength 3.3 μm with that of wavelength 0.154 nm. (b) Use Figure 6.4 to identify the region of the electromagnetic spectrum to which each belongs.

6.28 An AM radio station broadcasts at 1010 kHz, and its FM partner broadcasts at 98.3 MHz. Calculate and compare the energy of the photons emitted by these two radio stations.

6.29 One type of sunburn occurs on exposure to UV light of wavelength in the vicinity of 325 nm. (a) What is the energy of a photon of this wavelength? (b) What is the energy of a mole of these photons? (c) How many photons are in a 1.00 mJ burst of this radiation? (d) These UV photons can break chemical bonds in your skin to cause sunburn—a form of radiation damage. If the 325-nm radiation provides exactly the energy to break an average chemical bond in the skin, estimate the average energy of these bonds in kJ/mol.

6.30 The energy from radiation can be used to cause the rupture of chemical bonds. A minimum energy of 242 kJ/mol is required to break the chlorine–chlorine bond in Cl_2. What is the longest wavelength of radiation that possesses the necessary energy to break the bond? What type of electromagnetic radiation is this?

6.31 A diode laser emits at a wavelength of 987 nm. (a) In what portion of the electromagnetic spectrum is this radiation found? (b) All of its output energy is absorbed in a detector that measures a total energy of 0.52 J over a period of 32 s. How many photons per second are being emitted by the laser?

6.32 A stellar object is emitting radiation at 3.55 mm. (a) What type of electromagnetic spectrum is this radiation? (b) If a detector is capturing 3.2×10^8 photons per second at this wavelength, what is the total energy of the photons detected in 1.0 hour?

6.33 Molybdenum metal must absorb radiation with a minimum frequency of 1.09×10^{15} s^{-1} before it can eject an electron from its surface via the photoelectric effect. (a) What is the minimum energy needed to eject an electron? (b) What wavelength of radiation will provide a photon of this energy? (c) If molybdenum is irradiated with light of wavelength of 120 nm, what is the maximum possible kinetic energy of the emitted electrons?

6.34 Titanium metal requires a photon with a minimum energy of 6.94×10^{-19} J to emit electrons. (a) What is the minimum frequency of light necessary to emit electrons from titanium via the photoelectric effect? (b) What is the wavelength of this light? (c) Is it possible to eject electrons from titanium metal using visible light? (d) If titanium is irradiated with light of wavelength 233 nm, what is the maximum possible kinetic energy of the emitted electrons? (d) What is the maximum number of electrons that can be freed by a burst of light whose total energy is 2.00 μJ?

Bohr's Model; Matter Waves (Sections 6.3 and 6.4)

6.35 Explain how the existence of line spectra is consistent with Bohr's theory of quantized energies for the electron in the hydrogen atom.

6.36 (a) Consider the following three statements: (i) A hydrogen atom in the $n = 3$ state can emit light at only two specific wavelengths, (ii) a hydrogen atom in the $n = 2$ state is at a lower energy than the $n = 1$ state, and (iii) the energy of an emitted photon equals the energy difference of the two states involved in the emission. Which of these statements is or are true? (b) Does a hydrogen atom "expand" or "contract" as it moves from its ground state to an excited state?

6.37 Is energy emitted or absorbed when the following electronic transitions occur in hydrogen? (a) from $n = 4$ to $n = 2$, (b) from an orbit of radius 2.12 Å to one of radius 8.46 Å, (c) an electron adds to the H^+ ion and ends up in the $n = 3$ shell?

6.38 Indicate whether energy is emitted or absorbed when the following electronic transitions occur in hydrogen: (a) from $n = 2$ to $n = 6$, (b) from an orbit of radius 4.76 Å to one of radius 0.529 Å, (c) from the $n = 6$ to the $n = 9$ state.

6.39 (a) Using Equation 6.5, calculate the energy of an electron in the hydrogen atom when $n = 2$ and when $n = 6$. Calculate the wavelength of the radiation released when an electron moves from $n = 6$ to $n = 2$. (b) Is this line in the visible region of the electromagnetic spectrum? If so, what color is it?

6.40 Consider a transition of the electron in the hydrogen atom from $n = 4$ to $n = 9$. (a) Is ΔE for this process positive or negative? (b) Determine the wavelength of light that is associated with this transition. Will the light be absorbed or emitted? (c) In which portion of the electromagnetic spectrum is the light in part (b)?

6.41 The visible emission lines observed by Balmer all involved $n_f = 2$. (a) Which of the following is the best explanation of why the lines with $n_f = 3$ are not observed in the visible portion of the spectrum: (i) Transitions to $n_f = 3$ are not allowed to happen, (ii) transitions to $n_f = 3$ emit photons in the infrared portion of the spectrum, (iii) transitions to $n_f = 3$ emit photons in the ultraviolet portion of the spectrum, or (iv) transitions to $n_f = 3$ emit photons that are at exactly the same wavelengths as those to $n_f = 2$. (b) Calculate the wavelengths of the first three lines in the Balmer series—those for which $n_i = 3, 4,$ and 5—and identify these lines in the emission spectrum shown in Figure 6.11.

6.42 The Lyman series of emission lines of the hydrogen atom are those for which $n_f = 1$. (a) Determine the region of the electromagnetic spectrum in which the lines of the Lyman series are observed. (b) Calculate the wavelengths of the first three lines in the Lyman series—those for which $n_i = 2, 3,$ and 4.

6.43 One of the emission lines of the hydrogen atom has a wavelength of 93.07 nm. (a) In what region of the electromagnetic spectrum is this emission found? (b) Determine the initial and final values of n associated with this emission.

6.44 The hydrogen atom can absorb light of wavelength 1094 nm. (a) In what region of the electromagnetic spectrum is this absorption found? (b) Determine the initial and final values of n associated with this absorption.

6.45 Order the following transitions in the hydrogen atom from smallest to largest frequency of light absorbed: $n = 3$ to $n = 6, n = 4$ to $n = 9, n = 2$ to $n = 3,$ and $n = 1$ to $n = 2$.

6.46 Place the following transitions of the hydrogen atom in order from shortest to longest wavelength of the photon emitted: $n = 5$ to $n = 3, n = 4$ to $n = 2, n = 7$ to $n = 4,$ and $n = 3$ to $n = 2$.

6.47 Use the de Broglie relationship to determine the wavelengths of the following objects: (a) an 85-kg person skiing at 50 km/hr, (b) a 10.0-g bullet fired at 250 m/s, (c) a lithium atom moving at 2.5×10^5 m/s, (d) an ozone (O_3) molecule in the upper atmosphere moving at 550 m/s.

6.48 Among the elementary subatomic particles of physics is the muon, which decays within a few nanoseconds after formation. The muon has a rest mass 206.8 times that of an electron. Calculate the de Broglie wavelength associated with a muon traveling at 8.85×10^5 cm/s.

6.49 Neutron diffraction is an important technique for determining the structures of molecules. Calculate the velocity of a neutron needed to achieve a wavelength of 1.25 Å. (Refer to the inside cover for the mass of the neutron.)

6.50 The electron microscope has been widely used to obtain highly magnified images of biological and other types of materials. When an electron is accelerated through a particular potential field, it attains a speed of 9.47×10^6 m/s. What is the characteristic wavelength of this electron? Is the wavelength comparable to the size of atoms?

6.51 Using Heisenberg's uncertainty principle, calculate the uncertainty in the position of (a) a 1.50-mg mosquito moving at a speed of 1.40 m/s if the speed is known to within ± 0.01 m/s; (b) a proton moving at a speed of $(5.00 \pm 0.01) \times 10^4$ m/s. (The mass of a proton is given in the table of fundamental constants in the inside cover of the text.)

6.52 Calculate the uncertainty in the position of (a) an electron moving at a speed of $(3.00 \pm 0.01) \times 10^5$ m/s, (b) a neutron moving at this same speed. (The masses of an electron and a neutron are given in the table of fundamental constants in the inside cover of the text.) (c) Based on your answers to parts (a) and (b), which can we know with greater precision, the position of the electron or of the neutron?

Quantum Mechanics and Atomic Orbitals (Sections 6.5 and 6.6)

6.53 (a) Why does the Bohr model of the hydrogen atom violate the uncertainty principle? (b) In what way is the description of the electron using a wave function consistent with de Broglie's hypothesis? (c) What is meant by the term *probability density*? Given the wave function, how do we find the probability density at a certain point in space?

6.54 (a) According to the Bohr model, an electron in the ground state of a hydrogen atom orbits the nucleus at a specific radius of 0.53 Å. In the quantum mechanical description of the hydrogen atom, the most probable distance of the electron from the nucleus is 0.53 Å. Why are these two statements different? (b) Why is the use of Schrödinger's wave equation to describe the location of a particle very different from the description obtained from classical physics? (c) In the quantum mechanical description of an electron, what is the physical significance of the square of the wave function, ψ^2?

6.55 (a) For $n = 4$, what are the possible values of l? (b) For $l = 2$, what are the possible values of m_l? (c) If m_l is 2, what are the possible values for l?

6.56 How many possible values for l and m_l are there when (a) $n = 3$, (b) $n = 5$?

6.57 Give the numerical values of n and l corresponding to each of the following orbital designations: (a) $3p$, (b) $2s$, (c) $4f$, (d) $5d$.

6.58 Give the values for n, l, and m_l for (a) each orbital in the $2p$ subshell, (b) each orbital in the $5d$ subshell.

6.59 A certain orbital of the hydrogen atom has $n = 4$ and $l = 2$. (a) What are the possible values of m_l for this orbital? (b) What are the possible values of m_s for the orbital?

6.60 A hydrogen atom orbital has $n = 5$ and $m_l = -2$. **(a)** What are the possible values of l for this orbital? **(b)** What are the possible values of m_s for the orbital?

6.61 Which of the following represent impossible combinations of n and l? **(a)** $1p$, **(b)** $4s$, **(c)** $5f$, **(d)** $2d$

6.62 For the table that follows, write which orbital goes with the quantum numbers. Don't worry about x, y, z subscripts. If the quantum numbers are not allowed, write "not allowed."

N	l	m_l	Orbital
2	1	−1	$2p$ (example)
1	0	0	
3	−3	2	
3	2	−2	
2	0	−1	
0	0	0	
4	2	1	
5	3	0	

6.63 Sketch the shape and orientation of the following types of orbitals: **(a)** s, **(b)** p_z, **(c)** d_{xy}.

6.64 Sketch the shape and orientation of the following types of orbitals: **(a)** p_x, **(b)** d_{z^2}, **(c)** $d_{x^2-y^2}$.

6.65 **(a)** What are the similarities of and differences between the $1s$ and $2s$ orbitals of the hydrogen atom? **(b)** In what sense does a $2p$ orbital have directional character? Compare the "directional" characteristics of the p_x and $d_{x^2-y^2}$ orbitals. (That is, in what direction or region of space is the electron density concentrated?) **(c)** What can you say about the average distance from the nucleus of an electron in a $2s$ orbital as compared with a $3s$ orbital? **(d)** For the hydrogen atom, list the following orbitals in order of increasing energy (that is, most stable ones first): $4f$, $6s$, $3d$, $1s$, $2p$.

6.66 **(a)** With reference to Figure 6.19, what is the relationship between the number of nodes in an s orbital and the value of the principal quantum number? **(b)** Identify the number of nodes; that is, identify places where the electron density is zero, in the $2p_x$ orbital; in the $3s$ orbital. **(c)** What information is obtained from the radial probability functions in Figure 6.19? **(d)** For the hydrogen atom, list the following orbitals in order of increasing energy: $3s$, $2s$, $2p$, $5s$, $4d$.

Many-Electron Atoms and Electron Configurations (Sections 6.7–6.9)

6.67 For a given value of the principal quantum number, n, how do the energies of the s, p, d, and f subshells vary for **(a)** hydrogen, **(b)** a many-electron atom?

6.68 **(a)** The average distance from the nucleus of a $3s$ electron in a chlorine atom is smaller than that for a $3p$ electron. In light of this fact, which orbital is higher in energy? **(b)** Would you expect it to require more or less energy to remove a $3s$ electron from the chlorine atom, as compared with a $2p$ electron?

6.69 **(a)** What experimental evidence is there for the electron having a "spin"? **(b)** Draw an energy-level diagram that shows the relative energetic positions of a $1s$ orbital and a $2s$ orbital. Put two electrons in the $1s$ orbital. **(c)** Draw an arrow showing the excitation of an electron from the $1s$ to the $2s$ orbital.

6.70 **(a)** State the Pauli exclusion principle in your own words. **(b)** The Pauli exclusion principle is, in an important sense, the key to understanding the periodic table. **(c)** Explain.

6.71 What is the maximum number of electrons that can occupy each of the following subshells? **(a)** $3p$, **(b)** $5d$, **(c)** $2s$, **(d)** $4f$.

6.72 What is the maximum number of electrons in an atom that can have the following quantum numbers? **(a)** $n = 3$, $m_l = -2$; **(b)** $n = 4$, $l = 3$; **(c)** $n = 5$, $l = 3$, $m_l = 2$, **(d)** $n = 4$, $l = 1$, $m_l = 0$.

6.73 **(a)** What are "valence electrons"? **(b)** What are "core electrons"? **(c)** What does each box in an orbital diagram represent? **(d)** What quantity is represented by the half arrows in an orbital diagram?

6.74 For each element, indicate the number of valence electrons, core electrons, and unpaired electrons in the ground state: **(a)** nitrogen, **(b)** silicon, **(c)** chlorine.

6.75 Write the condensed electron configurations for the following atoms, using the appropriate noble-gas core abbreviations: **(a)** Cs, **(b)** Ni, **(c)** Se, **(d)** Cd, **(e)** U, **(f)** Pb.

6.76 Write the condensed electron configurations for the following atoms and indicate how many unpaired electrons each has: **(a)** Mg, **(b)** Ge, **(c)** Br, **(d)** V, **(e)** Y, **(f)** Lu.

6.77 Identify the specific element that corresponds to each of the following electron configurations and indicate the number of unpaired electrons for each: **(a)** $1s^2 2s^2$, **(b)** $1s^2 2s^2 2p^4$, **(c)** $[\text{Ar}]4s^1 3d^5$, **(d)** $[\text{Kr}]5s^2 4d^{10} 5p^4$.

6.78 Identify the group of elements that corresponds to each of the following generalized electron configurations and indicate the number of unpaired electrons for each:

(a) [noble gas] $ns^2 np^5$

(b) [noble gas] $ns^2 (n-1)d^2$

(c) [noble gas] $ns^2 (n-1)d^{10} np^1$

(d) [noble gas] $ns^2 (n-2)f^6$

6.79 What is wrong with the following electron configurations for atoms in their ground states? **(a)** $1s^2 2s^2 3s^1$, **(b)** $[\text{Ne}]2s^2 2p^3$, **(c)** $[\text{Ne}]3s^2 3d^5$.

6.80 The following electron configurations represent excited states. Identify the element and write its ground-state condensed electron configuration. **(a)** $1s^2 2s^2 2p^4 3s^1$, **(b)** $[\text{Ar}]4s^1 3d^{10} 4p^2 5p^1$, **(c)** $[\text{Kr}]5s^2 4d^2 5p^1$.

Additional Exercises

6.81 Consider the two waves shown here, which we will consider to represent two electromagnetic radiations:

(a) What is the wavelength of wave A? Of wave B?

(b) What is the frequency of wave A? Of wave B?

(c) Identify the regions of the electromagnetic spectrum to which waves A and B belong.

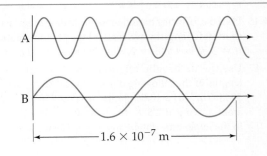

Additional Exercises **253**

6.82 If you put 120 volts of electricity through a pickle, the pickle will smoke and start glowing orange-yellow. The light is emitted because sodium ions in the pickle become excited; their return to the ground state results in light emission. (a) The wavelength of this emitted light is 589 nm. Calculate its frequency. (b) What is the energy of 0.10 mol of these photons? (c) Calculate the energy gap between the excited and ground states for the sodium ion. (d) If you soaked the pickle for a long time in a different salt solution, such as strontium chloride, would you still observe 589-nm light emission?

6.83 Certain elements emit light of a specific wavelength when they are burned. Historically, chemists used such emission wavelengths to determine whether specific elements were present in a sample. Characteristic wavelengths for some of the elements are given in the following table:

Ag	328.1 nm	Fe	372.0 nm
Au	267.6 nm	K	404.7 nm
Ba	455.4 nm	Mg	285.2 nm
Ca	422.7 nm	Na	589.6 nm
Cu	324.8 nm	Ni	341.5 nm

(a) Determine which elements emit radiation in the visible part of the spectrum. (b) Which element emits photons of highest energy? Of lowest energy? (c) When burned, a sample of an unknown substance is found to emit light of frequency $9.23 \times 10^{14} \text{ s}^{-1}$. Which of these elements is probably in the sample?

6.84 In August 2011, the Juno spacecraft was launched from Earth with the mission of orbiting Jupiter in 2016. The closest distance between Jupiter and Earth is 391 million miles. (a) If it takes 5.0 years for Juno to reach Jupiter, what is its average speed in mi/hr over this period? (b) Once Juno reaches Jupiter, what is the minimum amount of time it takes for the transmitted signals to travel from the spacecraft to Earth?

6.85 The rays of the Sun that cause tanning and burning are in the ultraviolet portion of the electromagnetic spectrum. These rays are categorized by wavelength. So-called UV-A radiation has wavelengths in the range of 320–380 nm, whereas UV-B radiation has wavelengths in the range of 290–320 nm. (a) Calculate the frequency of light that has a wavelength of 320 nm. (b) Calculate the energy of a mole of 320-nm photons. (c) Which are more energetic, photons of UV-A radiation or photons of UV-B radiation? (d) The UV-B radiation from the Sun is considered a greater cause of sunburn in humans than is UV-A radiation. Is this observation consistent with your answer to part (c)?

6.86 The watt is the derived SI unit of power, the measure of energy per unit time: $1 \text{ W} = 1 \text{ J/s}$. A semiconductor laser in a CD player has an output wavelength of 780 nm and a power level of 0.10 mW. How many photons strike the CD surface during the playing of a CD 69 minutes in length?

6.87 Carotenoids are yellow, orange, and red pigments synthesized by plants. The observed color of an object is not the color of light it absorbs but rather the complementary color, as described by a color wheel such as the one shown here. On this wheel, complementary colors are across from each other.

(a) Based on this wheel, what color is absorbed most strongly if a plant is orange? (b) If a particular carotenoid absorbs photons at 455 nm, what is the energy of the photon?

[6.88] In an experiment to study the photoelectric effect, a scientist measures the kinetic energy of ejected electrons as a function of the frequency of radiation hitting a metal surface. She obtains the following plot. The point labeled "ν_0" corresponds to light with a wavelength of 542 nm. (a) What is the value of ν_0 in s^{-1}? (b) What is the value of the work function of the metal in units of kJ/mol of ejected electrons? (c) Note that when the frequency of the light is greater than ν_0, the plot shows a straight line with a nonzero slope. What is the slope of this line segment?

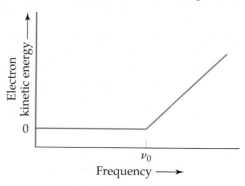

[6.89] Consider a transition in which the hydrogen atom is excited from $n = 1$ to $n = \infty$. (a) What is the end result of this transition? (b) What is the wavelength of light that must be absorbed to accomplish this process? (c) What will occur if light with a shorter wavelength than that in part (b) is used to excite the hydrogen atom? (d) How the results of parts (b) and (c) related to the plot shown in Exercise 6.88?

6.90 The human retina has three types of receptor cones, each sensitive to a different range of wavelengths of visible light, as shown in this figure (the colors are merely to differentiate the three curves from one another; they do not indicate the actual colors represented by each curve):

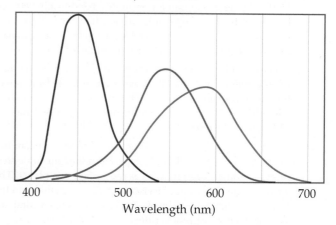

(a) Estimate the energies of photons with wavelengths at the maximum for each type of cone. (b) The color of the sky is due to scattering of solar light by the molecules of the atmosphere. Lord Rayleigh was one of the first to study scattering of this kind. He showed that the amount of scattering for very small particles such as molecules is inversely proportional to the fourth power of the wavelength. Estimate the ratio of the scattering efficiency of light at the wavelength of the maximum for the "blue" cones, as compared with that for the "green" cones. (c) Explain why the sky appears blue even though all wavelengths of solar light are scattered by the atmosphere.

6.91 The series of emission lines of the hydrogen atom for which $n_f = 3$ is called the *Paschen series*. (a) Determine the region of the electromagnetic spectrum in which the lines of the Paschen series are observed. (b) Calculate the wavelengths of the first three lines in the Paschen series—those for which $n_i = 4, 5,$ and 6.

6.92 When the spectrum of light from the Sun is examined in high resolution in an experiment similar to that illustrated in Figure 6.9, dark lines are evident. These are called Fraunhofer lines, after the scientist who studied them extensively in the early nineteenth century. Altogether, about 25,000 lines have been identified in the solar spectrum between 2950 Å and 10,000 Å. The Fraunhofer lines are attributed to absorption of certain wavelengths of the Sun's "white" light by gaseous elements in the Sun's atmosphere. (a) Describe the process that causes absorption of specific wavelengths of light from the solar spectrum. (b) To determine which Fraunhofer lines belong to a given element, say, neon, what experiments could a scientist conduct here on Earth?

6.93 Determine whether each of the following sets of quantum numbers for the hydrogen atom are valid. If a set is not valid, indicate which of the quantum numbers has a value that is not valid:

(a) $n = 4, l = 1, m_l = 2, m_s = -\frac{1}{2}$

(b) $n = 4, l = 3, m_l = -3, m_s = +\frac{1}{2}$

(c) $n = 3, l = 2, m_l = -1, m_s = +\frac{1}{2}$

(d) $n = 5, l = 0, m_l = 0, m_s = 0$

(e) $n = 2, l = 2, m_l = 1, m_s = +\frac{1}{2}$

[6.94] Bohr's model can be used for hydrogen-like ions—ions that have only one electron, such as He^+ and Li^{2+}. (a) Why is the Bohr model applicable to He^+ ions but not to neutral He atoms? (b) The ground-state energies of H, He^+, and Li^{2+} are tabulated as follows:

Atom or ion	H	He^+	Li^{2+}
Ground-state energy	-2.18×10^{-18} J	-8.72×10^{-18} J	-1.96×10^{-17} J

By examining these numbers, propose a relationship between the ground-state energy of hydrogen-like systems and the nuclear charge, Z. (c) Use the relationship you derive in part (b) to predict the ground-state energy of the C^{5+} ion.

[6.95] An electron is accelerated through an electric potential to a kinetic energy of 13.4 keV. What is its characteristic wavelength? [*Hint:* Recall that the kinetic energy of a moving object is $E = \frac{1}{2}mv^2$, where m is the mass of the object and v is the speed of the object.]

6.96 In the television series *Star Trek,* the transporter beam is a device used to "beam down" people from the *Starship Enterprise* to another location, such as the surface of a planet. The writers of the show put a "Heisenberg compensator" into the transporter beam mechanism. Explain why such a compensator (which is entirely fictional) would be necessary to get around Heisenberg's uncertainty principle.

[6.97] As discussed in the A Closer Look box on "Measurement and the Uncertainty Principle," the essence of the uncertainty principle is that we can't make a measurement without disturbing the system that we are measuring. (a) Why can't we measure the position of a subatomic particle without disturbing it? (b) How is this concept related to the paradox discussed in the Closer Look box on "Thought Experiments and Schrödinger's Cat"?

[6.98] Consider the discussion of radial probability functions in "A Closer Look" in Section 6.6. (a) What is the difference between the probability density as a function of r and the radial probability function as a function of r? (b) What is the significance of the term $4\pi r^2$ in the radial probability functions for the s orbitals? (c) Based on Figures 6.19 and 6.22, make sketches of what you think the probability density as a function of r and the radial probability function would look like for the 4s orbital of the hydrogen atom.

[6.99] For orbitals that are symmetric but not spherical, the contour representations (as in Figures 6.23 and 6.24) suggest where nodal planes exist (that is, where the electron density is zero). For example, the p_x orbital has a node wherever $x = 0$. This equation is satisfied by all points on the yz plane, so this plane is called a nodal plane of the p_x orbital. (a) Determine the nodal plane of the p_z orbital. (b) What are the two nodal planes of the d_{xy} orbital? (c) What are the two nodal planes of the $d_{x^2-y^2}$ orbital?

[6.100] The Chemistry and Life box in Section 6.7 described the techniques called NMR and MRI. (a) Instruments for obtaining MRI data are typically labeled with a frequency, such as 600 MHz. Why do you suppose this label is relevant to the experiment? (b) What is the value of ΔE in Figure 6.27 that would correspond to the absorption of a photon of radiation with frequency 450 MHz? (c) In general, the stronger the magnetic field, the greater the information obtained from an NMR or MRI experiment. Why do you suppose this is the case?

[6.101] Suppose that the spin quantum number, m_s, could have *three* allowed values instead of two. How would this affect the number of elements in the first four rows of the periodic table?

6.102 Using the periodic table as a guide, write the condensed electron configuration and determine the number of unpaired electrons for the ground state of (a) Br, (b) Ga, (c) Hf, (d) Sb, (e) Bi, (f) Sg.

6.103 Scientists have speculated that element 126 might have a moderate stability, allowing it to be synthesized and characterized. Predict what the condensed electron configuration of this element might be.

[6.104] In the experiment shown schematically below, a beam of neutral atoms is passed through a magnetic field. Atoms that have unpaired electrons are deflected in different directions in the magnetic field depending on the value of the electron spin quantum number. In the experiment illustrated, we envision that a beam of hydrogen atoms splits into two beams. (a) What is the significance of the observation that the single beam splits into two beams? (b) What do you think would happen if the strength of the magnet were increased? (c) What do you think would happen if the beam of hydrogen atoms were replaced with a beam of helium atoms? Why? (d) The relevant experiment was first performed by Otto Stern and Walter Gerlach in 1921. They used a beam of Ag atoms in the experiment. By considering the electron configuration of a silver atom, explain why the single beam splits into two beams.

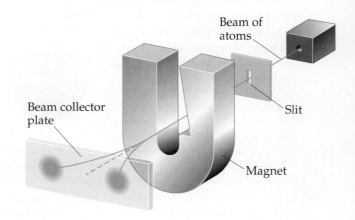

Beam of atoms

Beam collector plate

Slit

Magnet

Integrative Exercises

6.105 Microwave ovens use microwave radiation to heat food. The energy of the microwaves is absorbed by water molecules in food and then transferred to other components of the food. **(a)** Suppose that the microwave radiation has a wavelength of 11.2 cm. How many photons are required to heat 200 mL of coffee from 23 to 60 °C? **(b)** Suppose the microwave's power is 900 W (1 watt = 1 joule-second). How long would you have to heat the coffee in part (a)?

6.106 The stratospheric ozone (O_3) layer helps to protect us from harmful ultraviolet radiation. It does so by absorbing ultraviolet light and falling apart into an O_2 molecule and an oxygen atom, a process known as photodissociation.

$$O_3(g) \longrightarrow O_2(g) + O(g)$$

Use the data in Appendix C to calculate the enthalpy change for this reaction. What is the maximum wavelength a photon can have if it is to possess sufficient energy to cause this dissociation? In what portion of the spectrum does this wavelength occur?

6.107 The discovery of hafnium, element number 72, provided a controversial episode in chemistry. G. Urbain, a French chemist, claimed in 1911 to have isolated an element number 72 from a sample of rare earth (elements 58–71) compounds. However, Niels Bohr believed that hafnium was more likely to be found along with zirconium than with the rare earths. D. Coster and G. von Hevesy, working in Bohr's laboratory in Copenhagen, showed in 1922 that element 72 was present in a sample of Norwegian zircon, an ore of zirconium. (The name *hafnium* comes from the Latin name for Copenhagen, *Hafnia*). **(a)** How would you use electron configuration arguments to justify Bohr's prediction? **(b)** Zirconium, hafnium's neighbor in group 4B, can be produced as a metal by reduction of solid $ZrCl_4$ with molten sodium metal. Write a balanced chemical equation for the reaction. Is this an oxidation-reduction reaction? If yes, what is reduced and what is oxidized? **(c)** Solid zirconium dioxide, ZrO_2, reacts with chlorine gas in the presence of carbon. The products of the reaction are $ZrCl_4$ and two gases, CO_2 and CO in the ratio 1:2. Write a balanced chemical equation for the reaction. Starting with a 55.4-g sample of ZrO_2, calculate the mass of $ZrCl_4$ formed, assuming that ZrO_2 is the limiting reagent and assuming 100% yield. **(d)** Using their electron configurations, account for the fact that Zr and Hf form chlorides MCl_4 and oxides MO_2.

6.108 **(a)** Account for formation of the following series of oxides in terms of the electron configurations of the elements and the discussion of ionic compounds in Section 2.7: K_2O, CaO, Sc_2O_3, TiO_2, V_2O_5, CrO_3. **(b)** Name these oxides. **(c)** Consider the metal oxides whose enthalpies of formation (in kJ mol^{-1}) are listed here.

Oxide	$K_2O(s)$	$CaO(s)$	$TiO_2(s)$	$V_2O_5(s)$
ΔH_f°	−363.2	−635.1	−938.7	−1550.6

Calculate the enthalpy changes in the following general reaction for each case:

$$M_nO_m(s) + H_2(g) \longrightarrow nM(s) + mH_2O(g)$$

(You will need to write the balanced equation for each case and then compute ΔH°.) **(d)** Based on the data given, estimate a value of ΔH_f° for $Sc_2O_3(s)$.

6.109 The first 25 years of the twentieth century were momentous for the rapid pace of change in scientists' understanding of the nature of matter. **(a)** How did Rutherford's experiments on the scattering of α particles by a gold foil set the stage for Bohr's theory of the hydrogen atom? **(b)** In what ways is de Broglie's hypothesis, as it applies to electrons, consistent with J. J. Thomson's conclusion that the electron has mass? In what sense is it consistent with proposals preceding Thomson's work that the cathode rays are a wave phenomenon?

6.110 The two most common isotopes of uranium are ^{235}U and ^{238}U. **(a)** Compare the number of protons, the number of electrons, and the number of neutrons in atoms of these two isotopes. **(b)** Using the periodic table in the front-inside cover, write the electron configuration for a U atom. **(c)** Compare your answer to part (b) to the electron configuration given in Figure 6.30. How can you explain any differences between these two electron configurations? **(d)** ^{238}U undergoes radioactive decay to ^{234}Th. How many protons, electrons, and neutrons are gained or lost by the ^{238}U atom during this process? **(e)** Examine the electron configuration for Th in Figure 6.31. Are you surprised by what you find? Explain.

Design an Experiment

In this chapter, we have learned about the *photoelectric effect* and its impact on the formulation of light as photons. We have also seen that some anomalous electron configurations of the elements are particularly favorable if each atom has one or more half-filled shell, such as the case for the Cr atom with its $[Ar]4s^13d^5$ electron configuration. Let's suppose it is hypothesized that it requires more energy to remove an electron from a metal that has atoms with one or more half-filled shells than from those that do not. **(a)** Design a series of experiments involving the photoelectric effect that would test the hypothesis. **(b)** What experimental apparatus would be needed to test the hypothesis? It's not necessary that you name actual equipment but rather that you imagine how the apparatus would work—think in terms of the types of measurements that would be needed, and what capability you would need in your apparatus. **(c)** Describe the type of data you would collect and how you would analyze the data to see whether the hypothesis were correct. **(d)** Could your experiments be extended to test the hypothesis for other parts of the periodic table, such as the lanthanide or actinide elements?

7

Periodic Properties of the Elements

Why do some elements react more dramatically than others? If we drop a piece of gold metal into water, nothing happens. Dropping lithium metal into water initiates a slow reaction in which bubbles gradually form on the surface of the metal. By contrast, potassium metal reacts suddenly and violently with water, as shown here. Why do lithium and potassium react so differently with water, even though they are from the same family of the periodic table? To understand such differences we will examine how some key atomic properties change systematically as we traverse the periodic table.

As we saw in Chapter 6, the periodic nature of the periodic table arises from repeating patterns in the electron configurations of the elements. Elements in the same column contain the same number of electrons in their **valence orbitals**—the occupied orbitals that hold the electrons involved in bonding. For example, O ($[He]2s^22p^4$) and S ($[Ne]3s^23p^4$) are both in group 6A. The similarity of the electron distribution in their valence s and p orbitals leads to similarities in the properties of these two elements.

Electron configurations can be used to explain differences as well as similarities in the properties of elements. Despite similarities in their electron distributions, elemental oxygen and sulfur differ in fundamental ways. For example, at room temperature oxygen is a colorless gas but sulfur is a yellow solid. Might we explain these physical differences in noting that the outermost electrons of O are in the second shell, whereas those of S are in the third shell? We will see that even though elements share some similarities for being in the same *column* of the periodic table, there can be differences because elements are in different *rows* of the table.

▶ **POTASSIUM METAL REACTING WITH WATER.**

WHAT'S AHEAD

7.1 DEVELOPMENT OF THE PERIODIC TABLE
We begin our discussion with a brief history of the discovery of the periodic table.

7.2 EFFECTIVE NUCLEAR CHARGE We begin exploring many of the properties of atoms by examining the net attraction of the outer electrons to the nucleus and the average distance of those electrons from the nucleus. The net positive charge of the nucleus experienced by the outer electrons is called the *effective nuclear charge*.

7.3 SIZES OF ATOMS AND IONS We explore the relative sizes of atoms and ions, both of which follow trends that are related to their placement in the periodic table and the trends in effective nuclear charge.

7.4 IONIZATION ENERGY We next look at the trends in *ionization energy*, which is the energy required to remove one or more electrons from an atom. The periodic trends in ionization energy depend on variations in effective nuclear charge and atomic radii.

7.5 ELECTRON AFFINITY Next we examine periodic trends in the *electron affinity*, the energy released when an electron is added to an atom.

7.6 METALS, NONMETALS, AND METALLOIDS

We learn that the physical and chemical properties of metals are different from those of nonmetals. These properties arise from the fundamental characteristics of atoms, particularly ionization energy. Metalloids display properties that are intermediate between those of metals and those of nonmetals.

7.7 TRENDS FOR GROUP 1A AND GROUP 2A METALS

We examine some periodic trends in chemistry of group 1A and group 2A metals.

7.8 TRENDS FOR SELECTED NONMETALS

Finally, we examine some of the periodic trends in the chemistry of hydrogen and the elements in groups 6A, 7A, and 8A.

In this chapter we explore how some of the important properties of elements change as we move across a row or down a column of the periodic table. In many cases the trends in a row or column allow us to predict the physical and chemical properties of elements.

7.1 | Development of the Periodic Table

The discovery of chemical elements has been ongoing since ancient times (▼ Figure 7.1). Certain elements, such as gold (Au), appear in nature in elemental form and were thus discovered thousands of years ago. In contrast, some elements, such as technetium (Tc), are radioactive and intrinsically unstable. We know about them only because of technology developed during the twentieth century.

The majority of elements readily form compounds and, consequently, are not found in nature in their elemental form. For centuries, therefore, scientists were unaware of their existence. During the early nineteenth century, advances in chemistry made it easier to isolate elements from their compounds. As a result, the number of known elements more than doubled from 31 in 1800 to 63 by 1865.

As the number of known elements increased, scientists began classifying them. In 1869, Dmitri Mendeleev in Russia and Lothar Meyer in Germany published nearly identical classification schemes. Both noted that similar chemical and physical properties recur periodically when the elements are arranged in order of increasing atomic weight. Scientists at that time had no knowledge of atomic numbers. Atomic weights, however, generally increase with increasing atomic number, so both Mendeleev and Meyer fortuitously arranged the elements in nearly the proper sequence.

Although Mendeleev and Meyer came to essentially the same conclusion about the periodicity of elemental properties, Mendeleev is given credit for advancing his ideas more vigorously and stimulating new work. His insistence that elements with similar

GO FIGURE

Copper, silver, and gold have all been known since ancient times, whereas most of the other metals have not. Can you suggest an explanation?

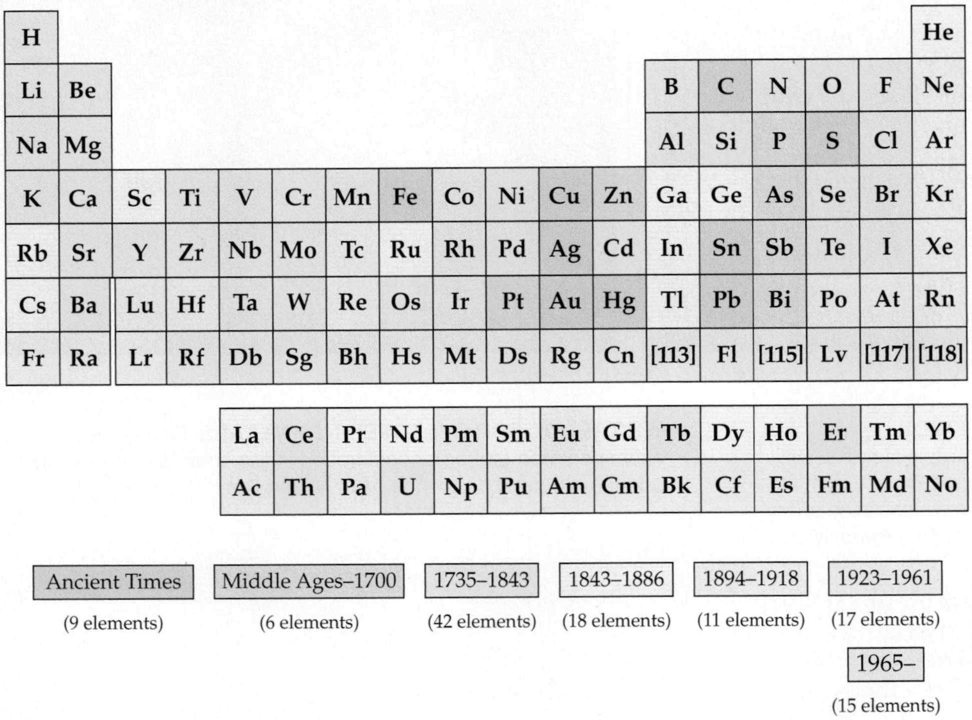

Ancient Times (9 elements)	Middle Ages–1700 (6 elements)	1735–1843 (42 elements)	1843–1886 (18 elements)	1894–1918 (11 elements)	1923–1961 (17 elements)

1965– (15 elements)

▲ Figure 7.1 **Discovering the elements.**

Table 7.1 Comparison of the Properties of Eka-Silicon Predicted by Mendeleev with the Observed Properties of Germanium

Property	Mendeleev's Predictions for Eka-Silicon (made in 1871)	Observed Properties of Germanium (discovered in 1886)
Atomic weight	72	72.59
Density (g/cm³)	5.5	5.35
Specific heat (J/g-K)	0.305	0.309
Melting point (°C)	High	947
Color	Dark gray	Grayish white
Formula of oxide	XO_2	GeO_2
Density of oxide (g/cm³)	4.7	4.70
Formula of chloride	XCl_4	$GeCl_4$
Boiling point of chloride (°C)	A little under 100	84

characteristics be listed in the same column forced him to leave blank spaces in his table. For example, both gallium (Ga) and germanium (Ge) were unknown to Mendeleev. He boldly predicted their existence and properties, referring to them as *eka-aluminum* ("under" aluminum) and *eka-silicon* ("under" silicon), respectively, after the elements under which they appeared in his table. When these elements were discovered, their properties closely matched those predicted by Mendeleev, as shown in ▲ Table 7.1.

In 1913, two years after Rutherford proposed the nuclear model of the atom ∞ (Section 2.2), English physicist Henry Moseley (1887–1915) developed the concept of atomic numbers. Bombarding different elements with high-energy electrons, Moseley found that each element produced X rays of a unique frequency and that the frequency generally increased as the atomic mass increased. He arranged the X-ray frequencies in order by assigning a unique whole number, called an *atomic number*, to each element. Moseley correctly identified the atomic number as the number of protons in the nucleus of the atom. ∞ (Section 2.3)

The concept of atomic number clarified some problems in the periodic table of Moseley's day, which was based on atomic weights. For example, the atomic weight of Ar (atomic number 18) is greater than that of K (atomic number 19), yet the chemical and physical properties of Ar are much more like those of Ne and Kr than like those of Na and Rb. When the elements are arranged in order of increasing atomic number, Ar and K appear in their correct places in the table. Moseley's studies also made it possible to identify "holes" in the periodic table, which led to the discovery of new elements.

Give It Some Thought

Looking at the periodic table on the front inside cover, can you find an example other than Ar and K where the order of the elements would be different if the elements were arranged in order of increasing atomic weight?

7.2 | Effective Nuclear Charge

Many properties of atoms depend on electron configuration and on how strongly the outer electrons in the atoms are attracted to the nucleus. Coulomb's law tells us that the strength of the interaction between two electrical charges depends on the magnitudes of the charges and on the distance between them. ∞ (Section 2.3) Thus, the attractive force between an electron and the nucleus depends on the magnitude of the nuclear charge and on the average distance between the nucleus and the electron. The force increases as the nuclear charge increases and decreases as the electron moves farther from the nucleus.

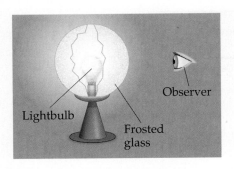

▲ Figure 7.2 **An analogy for effective nuclear charge.** We envision the nucleus as a light bulb, and a valence electron an observer. The amount of light seen by the observer depends on the screening by the frosted glass lampshade.

Understanding the attraction between the electron and the nucleus in a hydrogen atom is straightforward because we have only one electron and one proton. In a many-electron atom, however, the situation is more complicated. In addition to the attraction of each electron to the nucleus, each electron experiences the repulsion due to other electrons. These electron–electron repulsions cancel some of the attraction of the electron to the nucleus so that the electron experiences less attraction than it would if the other electrons weren't there. In essence, each electron in a many-electron atom is *screened* from the nucleus by the other electrons. It therefore experiences a net attraction that is less than it would in the absence of other electrons.

How can we account for the combination of nuclear attraction and electron repulsions for our electron of interest? The simplest way to do so is to imagine that the electron experiences a net attraction that is the result of the nuclear attraction decreased by the electron–electron repulsions. We call this partially screened nuclear charge the **effective nuclear charge**, Z_{eff}. Because the full attractive force of the nucleus has been decreased by the electron repulsions, we see that the effective nuclear charge is always less than the *actual* nuclear charge ($Z_{eff} < Z$). We can define the amount of screening of the nuclear charge quantitatively by using a *screening constant*, S, such that

$$Z_{eff} = Z - S \qquad [7.1]$$

where S is a positive number. For a valence electron, most of the shielding is due to the core electrons, which are much closer to the nucleus. As a result, for the valence electrons in an atom *the value of S is usually close to the number of core electrons in the atom.* (Electrons in the same valence shell do not screen one another very effectively, but they do affect the value of S slightly; see "A Closer Look: Effective Nuclear Charge.")

To understand better the notion of effective nuclear charge, we can use an analogy of a light bulb with a frosted glass shade (▲ Figure 7.2). The light bulb represents the nucleus, and the observer is the electron of interest, which is usually a valence electron. The amount of light that the electron "sees" is analogous to the amount of net nuclear attraction experienced by the electron. The other electrons in the atom, especially the core electrons, act like a frosted glass lampshade, decreasing the amount of light that gets to the observer. If the light bulb gets brighter while the lampshade stays the same, more light is observed. Likewise, if the lampshade gets thicker, less light is observed. We will find it helpful to keep this analogy in mind as we discuss trends in effective nuclear charge.

Let's consider what we would expect for the magnitude of Z_{eff} for the sodium atom. Sodium has the electron configuration [Ne]$3s^1$. The nuclear charge is $Z = 11+$, and there are 10 core electrons ($1s^2 2s^2 2p^6$), which serve as a "lampshade" to screen the nuclear charge "seen" by the $3s$ electron. Therefore, in the simplest approach, we expect S to equal 10 and the $3s$ electron to experience an effective nuclear charge of $Z_{eff} = 11 - 10 = 1+$ (◄ Figure 7.3). The situation is more complicated, however, because the $3s$ electron has a small probability of being closer to the nucleus, in the region occupied by the core electrons. ∞ (Section 6.6) Thus, this electron experiences a greater net attraction than our simple $S = 10$ model suggests: The actual value of Z_{eff} for the $3s$ electron in Na is $Z_{eff} = 2.5+$. In other words, because there is a small probability that the $3s$ electron is close to the nucleus, the value of S in Equation 7.1 changes from 10 to 8.5.

The notion of effective nuclear charge also explains an important effect we noted in Section 6.7: For a many-electron atom, the energies of orbitals with the same *n* value increase with increasing *l* value. For example, in the carbon atom, whose electron configuration is $1s^2 2s^2 2p^2$, the energy of the $2p$ orbital ($l = 1$) is higher than that of the $2s$ orbital ($l = 0$) even though both orbitals are in the $n = 2$ shell (Figure 6.25). This difference in energies is due to

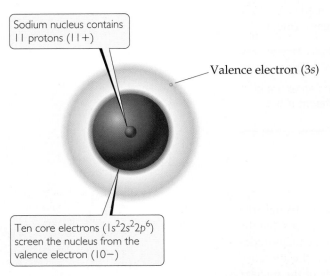

Sodium nucleus contains 11 protons (11+)

Valence electron (3s)

Ten core electrons ($1s^2 2s^2 2p^6$) screen the nucleus from the valence electron (10−)

▲ Figure 7.3 **Effective nuclear charge.** The effective nuclear charge experienced by the $3s$ electron in a sodium atom depends on the 11+ charge of the nucleus and the 10− charge of the core electrons.

▲ GO FIGURE

Based on this figure, is it possible for an electron in a 2s orbital to be closer to the nucleus than an electron in a 1s orbital?

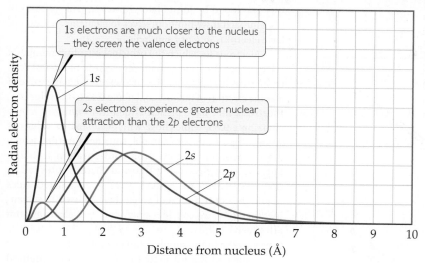

▲ **Figure 7.4 Comparison of 1s, 2s, and 2p radial probability functions.**

the radial probability functions for the orbitals (▲ Figure 7.4). We see first that the 1s electrons are much closer to the nucleus—they serve as an effective "lampshade" for the 2s and 2p electrons. Notice next that the 2s probability function has a small peak fairly close to the nucleus, whereas the 2p probability function does not. As a result, a 2s electron is not screened as much by the core orbitals as is a 2p electron. The greater attraction between the 2s electron and the nucleus leads to a lower energy for the 2s orbital than for the 2p orbital. The same reasoning explains the general trend in orbital energies ($ns < np < nd$) in many-electron atoms.

Finally, let's examine trends in valence-electron Z_{eff} values. *The effective nuclear charge increases from left to right across any period of the periodic table.* Although the number of core electrons stays the same across the period, the number of protons increases—in our analogy, we are increasing the brightness of the lightbulb while keeping the shade the same. The valence electrons added to counterbalance the increasing nuclear charge screen one another ineffectively. Thus, Z_{eff} increases steadily. For example, the core electrons of lithium ($1s^2 2s^1$) screen the 2s valence electron from the 3+ nucleus fairly efficiently. Consequently, the valence electron experiences an effective nuclear charge of roughly $3 - 2 = 1+$. For beryllium ($1s^2 2s^2$) the effective nuclear charge experienced by each valence electron is larger because here the 1s electrons screen a 4+ nucleus, and each 2s electron only partially screens the other. Consequently, the effective nuclear charge experienced by each 2s electron is about $4 - 2 = 2+$.

A Closer Look

Effective Nuclear Charge

To get a sense of how effective nuclear charge varies as both nuclear charge and number of electrons increase, consider **Figure 7.5**. Although the details of how the Z_{eff} values in the graph were calculated are beyond the scope of our discussion, the trends are instructive.

The effective nuclear charge felt by the outermost electrons is smaller than that felt by inner electrons because of screening by the inner electrons. In addition, the effective nuclear charge felt by the outermost electrons does not increase as steeply with increasing

atomic number because the valence electrons make a small but non-negligible contribution to the screening constant S. The most striking feature associated with the Z_{eff} value for the outermost electrons is the sharp drop between the last period 2 element (Ne) and the first period 3 element (Na). This drop reflects the fact that the core electrons are much more effective than the valence electrons at screening the nuclear charge.

Because Z_{eff} can be used to understand many physically measurable quantities, it is desirable to have a simple method for estimating it. The value of Z in Equation 7.1 is known exactly, so the

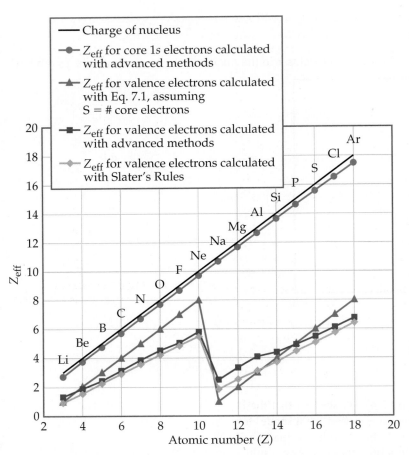

▲ **Figure 7.5 Variations in effective nuclear charge for period 2 and period 3 elements.** Moving from one element to the next in the periodic table, the increase in Z_{eff} felt by the innermost (1s) electrons (red circles) closely tracks the increase in nuclear charge Z (black line) because these electrons are not screened much. The results of several methods to calculate Z_{eff} for valence electrons are shown in other colors.

challenge boils down to estimating the value of S. In the text, we estimated S very simply by assuming that each core electron contributes 1.00 to S and the outer electrons contribute nothing. A more accurate approach was developed by John Slater, however, and we can use his approach if we limit ourselves to elements that do not have electrons in d or f subshells.

Electrons for which the principal quantum number n is larger than the value of n for the electron of interest contribute 0 to the value of S. Electrons with the same value of n as the electron of interest contribute 0.35 to the value of S. Electrons that have principal quantum number $n - 1$ contribute 0.85, while those with even smaller values of n contribute 1.00. For example, consider fluorine, which has the ground-state electron configuration $1s^2 2s^2 2p^5$. For a valence electron in fluorine, Slater's rules tell us that $S = (0.35 \times 6) + (0.85 \times 2) = 3.8$. (Slater's rules ignore the contribution of an electron to itself in screening; therefore, we consider only six $n = 2$ electrons, not all seven). Thus, $Z_{eff} = Z - S = 9 - 3.8 = 5.2+$, a little lower than the simple estimate of $9 - 2 = 7+$.

Values of Z_{eff} estimated using the simple method outlined in the text, as well as those estimated with Slater's rules, are plotted in Figure 7.5. While neither of these methods exactly replicate the values of Z_{eff} obtained from more sophisticated calculations, both methods effectively capture the periodic variation in Z_{eff}. While Slater's approach is more accurate, the method outlined in the text does a reasonably good job of estimating Z_{eff} despite its simplicity. For our purposes, therefore, we can assume that the screening constant S in Equation 7.1 is roughly equal to the number of core electrons.

Related Exercises: 7.13, 7.14, 7.15, 7.16, 7.31, 7.32, 7.80, 7.81, 7.110

Going down a column, the effective nuclear charge experienced by valence electrons changes far less than it does across a period. For example, using our simple estimate for S we would expect the effective nuclear charge experienced by the valence electrons in lithium and sodium to be about the same, roughly $3 - 2 = 1+$ for lithium and $11 - 10 = 1+$ for sodium. In fact, however, *effective nuclear charge increases slightly as we go down a column* because the more diffuse core electron cloud is less able to screen the valence electrons from the nuclear charge. In the case of the alkali metals, Z_{eff} increases from $1.3+$ for lithium, to $2.5+$ for sodium, to $3.5+$ for potassium.

 Give It Some Thought

Which would you expect to experience a greater effective nuclear charge, a $2p$ electron of a Ne atom or a $3s$ electron of a Na atom?

7.3 | Sizes of Atoms and Ions

It is tempting to think of atoms as hard, spherical objects. According to the quantum mechanical model, however, atoms do not have sharply defined boundaries at which the electron distribution becomes zero. ∞ (Section 6.5) Nevertheless, we can define atomic size in several ways, based on the distances between atoms in various situations.

Imagine a collection of argon atoms in the gas phase. When two of these atoms collide, they ricochet apart like colliding billiard balls. This ricocheting happens because the electron clouds of the colliding atoms cannot penetrate each other to any significant extent. The shortest distance separating the two nuclei during such collisions is twice the radii of the atoms. We call this radius the *nonbonding atomic radius* or the *van der Waals* radius (▶ Figure 7.6).

In molecules, the attractive interaction between any two adjacent atoms is what we recognize as a chemical bond. We discuss bonding in Chapters 8 and 9. For now, we need to realize that two bonded atoms are closer together than they would be in a nonbonding collision where the atoms ricochet apart. We can therefore define an atomic radius based on the distance between the nuclei when two atoms are bonded to each other, shown as distance *d* in Figure 7.6. The **bonding atomic radius** for any atom in a molecule is equal to half of the bond distance *d*. Note from Figure 7.6 that the bonding atomic radius (also known as the *covalent radius*) is smaller than the nonbonding atomic radius. Unless otherwise noted, we mean the bonding atomic radius when we speak of the "size" of an atom.

Scientists have developed a variety of techniques for measuring the distances separating nuclei in molecules. From observations of these distances in many molecules, each element can be assigned a bonding atomic radius. For example, in the I_2 molecule, the distance separating the nuclei is observed to be 2.66 Å, which means the bonding atomic radius of an iodine atom in I_2 is $(2.66$ Å$)/2 = 1.33$ Å.* Similarly, the distance separating adjacent carbon nuclei in diamond (a three-dimensional solid network of carbon atoms) is 1.54 Å; thus, the bonding atomic radius of carbon in diamond is 0.77 Å. By using structural information on more than 30,000 substances, a consistent set of bonding atomic radii of the elements can be defined (▼ Figure 7.7). Note that for helium and neon, the bonding atomic radii must be estimated because there are no known compounds of these elements.

The atomic radii in Figure 7.7 allows us to estimate bond lengths in molecules. For example, the bonding atomic radii for C and Cl are 1.02 Å and 0.76 Å, respectively. In CCl_4 the measured length of the C—Cl bond is 1.77 Å, very close to the sum $(1.02 + 0.76$ Å$)$ of the bonding atomic radii of Cl and C.

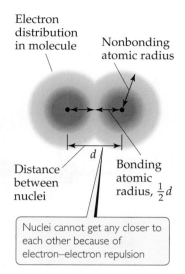

▲ Figure 7.6 **Distinction between nonbonding and bonding atomic radii within a molecule.**

GO FIGURE

Which part of the periodic table (top/bottom, left/right) has the elements with the largest atoms?

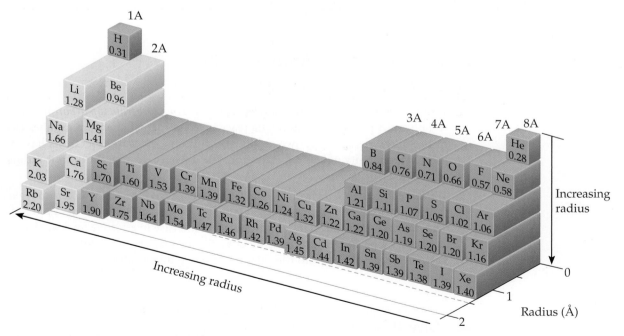

▲ Figure 7.7 **Trends in bonding atomic radii for periods 1 through 5.**

---

*Remember: The angstrom (1 Å $= 10^{-10}$ m) is a convenient metric unit for atomic measurements of length. It is not an SI unit. The most commonly used SI unit for atomic measurements is the picometer (1 pm $= 10^{-12}$ m; 1 Å $= 100$ pm).

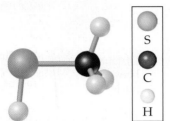

Methyl mercaptan

SAMPLE
EXERCISE 7.1 Bond Lengths in a Molecule

Natural gas used in home heating and cooking is odorless. Because natural gas leaks pose the danger of explosion or suffocation, various smelly substances are added to the gas to allow detection of a leak. One such substance is methyl mercaptan, CH_3SH. Use Figure 7.7 to predict the lengths of the C—S, C—H, and S—H bonds in this molecule.

SOLUTION

Analyze and Plan We are given three bonds and told to use Figure 7.7 for bonding atomic radii. We will assume that each bond length is the sum of the bonding atomic radii of the two atoms involved.

Solve

$$\text{C—S bond length} = \text{bonding atomic radius of C} + \text{bonding atomic radius of S}$$
$$= 0.76 \text{ Å} + 1.05 \text{ Å} = 1.81 \text{ Å}$$
$$\text{C—H bond length} = 0.76 \text{ Å} + 0.31 \text{ Å} = 1.07 \text{ Å}$$
$$\text{S—H bond length} = 1.05 \text{ Å} + 0.31 \text{ Å} = 1.36 \text{ Å}$$

Check The experimentally determined bond lengths are C—S = 1.82 Å, C—H = 1.10 Å, and S—H = 1.33 Å. (In general, the lengths of bonds involving hydrogen show larger deviations from the values predicted from bonding atomic radii than do bonds involving larger atoms.)

Comment Notice that our estimated bond lengths are close but not exact matches to the measured bond lengths. Bonding atomic radii must be used with some caution in estimating bond lengths.

> **Practice Exercise 1**
>
> Hypothetical elements X and Y form a molecule XY_2, in which both Y atoms are bonded to atom X (and not to one another). The X—X distance in the elemental form of X is 2.04 Å, and the Y—Y distance in elemental Y is 1.68 Å. What would you predict for the X—Y distance in the XY_2 molecule?
> **(a)** 0.84 Å   **(b)** 1.02 Å   **(c)** 1.86 Å   **(d)** 2.70 Å   **(e)** 3.72 Å
>
> **Practice Exercise 2**
>
> Using Figure 7.7, predict which is longer, the P—Br bond in PBr_3 or the As—Cl bond in $AsCl_3$.

Periodic Trends in Atomic Radii

Figure 7.7 shows two interesting trends:

1. *Within each group, bonding atomic radius tends to increase from top to bottom.* This trend results primarily from the increase in the principal quantum number (n) of the outer electrons. As we go down a column, the outer electrons have a greater probability of being farther from the nucleus, causing the atomic radius to increase.

2. *Within each period, bonding atomic radius tends to decrease from left to right* (although there are some minor exceptions, such as for Cl to Ar or As to Se). The major factor influencing this trend is the increase in effective nuclear charge Z_{eff} across a period. The increasing effective nuclear charge steadily draws the valence electrons closer to the nucleus, causing the bonding atomic radius to decrease.

 Give It Some Thought

In Section 7.2 we said that Z_{eff} generally increases when you move down a column of the periodic table, whereas in Chapter 6 we saw that the "size" of an orbital increases as the principal quantum number n increases. With respect to atomic radii, do these trends work together or against each other? Which effect is larger?

SAMPLE
EXERCISE 7.2 | Predicting Relative Sizes of Atomic Radii

Referring to the periodic table, arrange (as much as possible) the atoms B, C, Al, and Si in order of increasing size.

SOLUTION

Analyze and Plan We are given the chemical symbols for four elements and told to use their relative positions in the periodic table to predict the relative size of their atomic radii. We can use the two periodic trends just described to help with this problem.

Solve C and B are in the same period, with C to the right of B. Therefore, we expect the radius of C to be smaller than that of B because radii usually decrease as we move across a period. Likewise, the radius of Si is expected to be smaller than that of Al. Al is directly below B, and Si is directly below C. We expect, therefore, the radius of B to be smaller than that of Al and the radius of C to be smaller than that of Si. Thus, so far we can say C < B, B < Al, C < Si, and Si < Al. We can therefore conclude that C has the smallest radius and Al has the largest radius, and so can write C < ? < ? < Al.

Our two periodic trends in atomic size do not supply enough information to allow us to determine, however, whether B or Si (represented by the two question marks) has the larger radius. Going from B to Si in the periodic table, we move down (radius tends to increase) and to the right (radius tends to decrease). It is only because Figure 7.7 provides numerical values for each atomic radius that we know that the radius of Si is greater than that of B. If you examine the figure carefully, you will discover that for the *s*- and *p*-block elements the increase

in radius moving down a column tends to be the greater effect. There are exceptions, however.

Check From Figure 7.7, we have

$$C(0.76 \text{ Å}) < B(0.84 \text{ Å}) < Si(1.11 \text{ Å}) < Al(1.21 \text{ Å}).$$

Comment Note that the trends we have just discussed are for the *s*- and *p*-block elements. Figure 7.7 shows that the transition elements do not show a regular decrease moving across a period.

Practice Exercise 1
By referring to the periodic table, but not to Figure 7.7, place the following atoms in order of increasing bonding atomic radius: N, O, P, Ge.
(a) N < O < P < Ge
(b) P < N < O < Ge
(c) O < N < Ge < P
(d) O < N < P < Ge
(e) N < P < Ge < O

Practice Exercise 2
Arrange Be, C, K, and Ca in order of increasing atomic radius.

Periodic Trends in Ionic Radii

Just as bonding atomic radii can be determined from interatomic distances in molecules, ionic radii can be determined from interatomic distances in ionic compounds. Like the size of an atom, the size of an ion depends on its nuclear charge, the number of electrons it possesses, and the orbitals in which the valence electrons reside. When a cation is formed from a neutral atom, electrons are removed from the occupied atomic orbitals that are the most spatially extended from the nucleus. Also, when a cation is formed the number of electron–electron repulsions is reduced. Therefore, *cations are smaller than their parent atoms* (Figure 7.8). The opposite is true of anions. When electrons are added to an atom to form an anion, the increased electron–electron repulsions cause the electrons to spread out more in space. Thus, *anions are larger than their parent atoms.*

For ions carrying the same charge, ionic radius increases as we move down a column in the periodic table (Figure 7.8). In other words, as the principal quantum number of the outermost occupied orbital of an ion increases, the radius of the ion increases.

SAMPLE
EXERCISE 7.3 | Predicting Relative Sizes of Atomic and Ionic Radii

Arrange Mg^{2+}, Ca^{2+}, and Ca in order of decreasing radius.

SOLUTION

Cations are smaller than their parent atoms, and so Ca^{2+} < Ca. Because Ca is below Mg in group 2A, Ca^{2+} is larger than Mg^{2+}. Consequently, Ca > Ca^{2+} > Mg^{2+}.

Practice Exercise 1
Arrange the following atoms and ions in order of increasing ionic radius: F, S^{2-}, Cl, and Se^{2-}.
(a) F < S^{2-} < Cl < Se^{2-} **(b)** F < Cl < S^{2-} < Se^{2-} **(c)** F < S^{2-} < Se^{2-} < Cl
(d) Cl < F < Se^{2-} < S^{2-} **(e)** S^{2-} < F < Se^{2-} < Cl

Practice Exercise 2
Which of the following atoms and ions is largest: S^{2-}, S, O^{2-}?

▲ GO FIGURE

How do cations of the same charge change in radius as you move down a column in the periodic table?

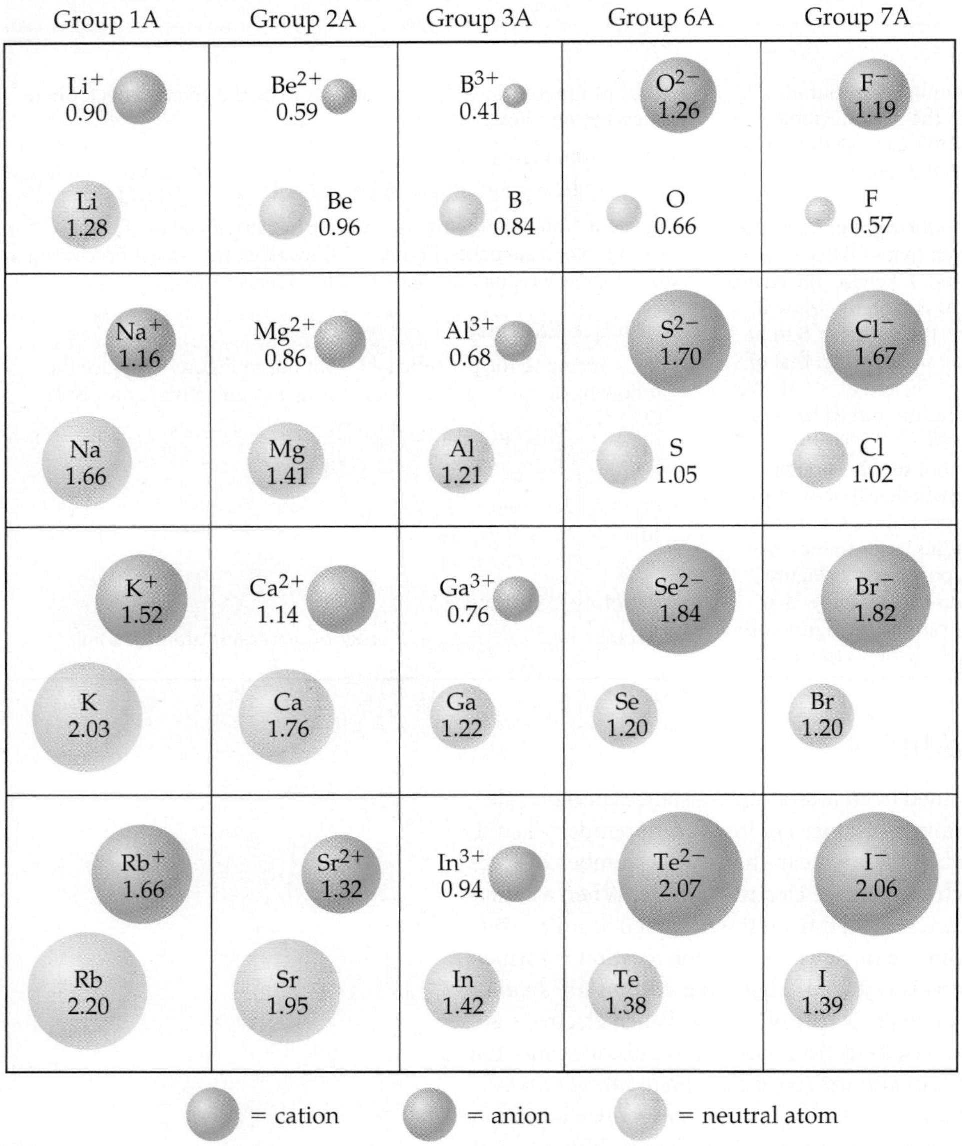

Group 1A	Group 2A	Group 3A	Group 6A	Group 7A
Li$^+$ 0.90	Be^{2+} 0.59	B^{3+} 0.41	O^{2-} 1.26	F$^-$ 1.19
Li 1.28	Be 0.96	B 0.84	O 0.66	F 0.57
Na$^+$ 1.16	Mg^{2+} 0.86	Al^{3+} 0.68	S^{2-} 1.70	Cl$^-$ 1.67
Na 1.66	Mg 1.41	Al 1.21	S 1.05	Cl 1.02
K$^+$ 1.52	Ca^{2+} 1.14	Ga^{3+} 0.76	Se^{2-} 1.84	Br$^-$ 1.82
K 2.03	Ca 1.76	Ga 1.22	Se 1.20	Br 1.20
Rb$^+$ 1.66	Sr^{2+} 1.32	In^{3+} 0.94	Te^{2-} 2.07	I$^-$ 2.06
Rb 2.20	Sr 1.95	In 1.42	Te 1.38	I 1.39

= cation   = anion   = neutral atom

▲ **Figure 7.8 Cation and anion size.** Radii, in angstroms, of atoms and their ions for five groups of representative elements.

An **isoelectronic series** is a group of ions all containing the same number of electrons. For example, each ion in the isoelectronic series O^{2-}, F^-, Na^+, Mg^{2+}, and Al^{3+} has 10 electrons. In any isoelectronic series we can list the members in order of increasing atomic number; therefore, the nuclear charge increases as we move through the series. Because the number of electrons remains constant, ionic radius decreases with increasing nuclear charge as the electrons are more strongly attracted to the nucleus:

—Increasing nuclear charge ⟶

O^{2-}	F^-	Na^+	Mg^{2+}	Al^{3+}
1.26 Å	1.19 Å	1.16 Å	0.86 Å	0.68 Å

—Decreasing ionic radius ⟶

Notice the positions and atomic numbers of these elements in the periodic table. The nonmetal anions precede the noble gas Ne in the table. The metal cations follow Ne. Oxygen, the largest ion in this isoelectronic series, has the lowest atomic number, 8. Aluminum, the smallest of these ions, has the highest atomic number, 13.

Chemistry Put to Work

Ionic Size and Lithium-Ion Batteries

Ionic size plays a major role in determining the properties of devices that rely on movement of ions. "Lithium-ion" batteries, which have become common energy sources for electronic devices such as cell phones, iPads, and laptop computers, rely in part on the small size of the lithium ion for their operation.

A fully charged battery spontaneously produces an electric current and, therefore, power when its positive and negative electrodes are connected to an electrical load, such as a device to be powered. The positive electrode is called the anode, and the negative electrode is called the cathode. The materials used for the electrodes in lithium-ion batteries are under intense development. Currently the anode material is graphite, a form of carbon, and the cathode is usually $LiCoO_2$, lithium cobalt oxide (▼ Figure 7.9). Between anode and cathode is a *separator*, a porous solid material that allows the passage of lithium ions but not electrons.

When the battery is being charged by an external source, lithium ions migrate through the separator from the cathode to the anode where they insert between the layers of carbon atoms. The ability of an ion to move through a solid increases as the size of the ion decreases and as the charge on the ion decreases. Lithium ions are smaller than most other cations, and they carry only a 1+ charge, which allows them to migrate more readily than other ions can. When the battery discharges, the lithium ions move from anode to cathode. To maintain charge balance, electrons simultaneously migrate from anode to cathode through an external circuit, thereby producing electricity.

At the cathode, lithium ions then insert in the oxide material. Again, the small size of lithium ions is an advantage. For every lithium ion that inserts into the lithium cobalt oxide cathode, a Co^{4+} ion is reduced to a Co^{3+} by an electron that has traveled through the external circuit.

The ion migration and the changes in structure that result when lithium ions enter and leave the electrode materials are complicated. Furthermore, the operation of all batteries generate heat because they are not perfectly efficient. In the case of Li-ion batteries, the heating of the separator material (typically a polymer) has led to problems as the size of the batteries has been scaled larger to increase energy capacity. In 2013, problems with overheating of large Li-ion batteries led to the temporary grounding of the new Boeing 787 Dreamliner aircraft.

Teams worldwide are trying to discover new cathode and anode materials that will easily accept and release lithium ions without falling apart over many repeated cycles. New separator materials that allow for faster passage of lithium ions with less heat generation are also under development. Some research groups are looking at using sodium ions instead of lithium ions because sodium is far more abundant than lithium, although the larger size of sodium ions poses additional challenges. In the next decade we expect great advances in battery technology based on alkali metal ions.

Related Exercise: 7.89

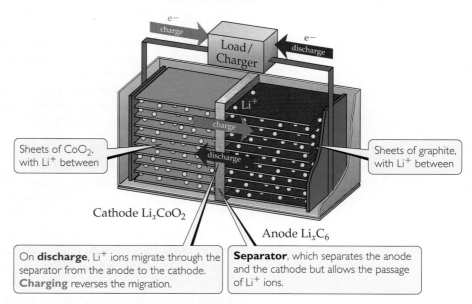

▲ Figure 7.9 **Schematic of a lithium-ion battery.**

SAMPLE EXERCISE 7.4 Ionic Radii in an Isoelectronic Series

Arrange the ions K^+, Cl^-, Ca^{2+}, and S^{2-} in order of decreasing size.

SOLUTION

This is an isoelectronic series, with all ions having 18 electrons. In such a series, size decreases as nuclear charge (atomic number) increases. The atomic numbers of the ions are S 16, Cl 17, K 19, Ca 20. Thus, the ions decrease in size in the order $S^{2-} > Cl^- > K^+ > Ca^{2+}$.

(c) $Rb^+ < Sr^{2+} < Se^{2-} < Te^{2-} < Br^-$
(d) $Rb^+ < Br^- < Sr^{2+} < Se^{2-} < Te^{2-}$
(e) $Sr^{2+} < Rb^+ < Br^- < Te^{2-} < Se^{2-}$

Practice Exercise 2

In the isoelectronic series Ca^{2+}, Cs^+, Y^{3+}, which ion is largest?

Practice Exercise 1

Arrange the following atoms and ions in order of increasing ionic radius: Br^-, Rb^+, Se^{2-}, Sr^{2+}, Te^{2-}.

(a) $Sr^{2+} < Rb^+ < Br^- < Se^{2-} < Te^{2-}$
(b) $Br^- < Sr^{2+} < Se^{2-} < Te^{2-} < Rb^+$

7.4 | Ionization Energy

The ease with which electrons can be removed from an atom or ion has a major impact on chemical behavior. The **ionization energy** of an atom or ion is the minimum energy required to remove an electron from the ground state of the isolated gaseous atom or ion. We first encountered ionization in our discussion of the Bohr model of the hydrogen atom. ∞ (Section 6.3) If the electron in an H atom is excited from $n = 1$ (the ground state) to $n = \infty$, the electron is completely removed from the atom; the atom is *ionized*.

In general, the *first ionization energy*, I_1, is the energy needed to remove the first electron from a neutral atom. For example, the first ionization energy for the sodium atom is the energy required for the process

$$\text{Na}(g) \longrightarrow \text{Na}^+(g) + e^- \qquad [7.2]$$

The *second ionization energy*, I_2, is the energy needed to remove the second electron, and so forth, for successive removals of additional electrons. Thus, I_2 for the sodium atom is the energy associated with the process

$$\text{Na}^+(g) \longrightarrow \text{Na}^{2+}(g) + e^- \qquad [7.3]$$

Variations in Successive Ionization Energies

The magnitude of the ionization energy tells us how much energy is required to remove an electron; the greater the ionization energy, the more difficult it is to remove an electron. Notice in ▼ Table 7.2 that ionization energies for a given element increase as successive electrons are removed: $I_1 < I_2 < I_3$, and so forth. This trend makes sense because with each successive removal, an electron is being pulled away from an increasingly more positive ion, requiring increasingly more energy.

 Give It Some Thought

Light can be used to ionize atoms and ions. Which of the two processes shown in Equations 7.2 and 7.3 requires shorter-wavelength radiation?

A second important feature shown in Table 7.2 is the sharp increase in ionization energy that occurs when an inner-shell electron is removed. For example, consider silicon, $1s^2 2s^2 2p^6 3s^2 3p^2$. The ionization energies increase steadily from 786 to 4356 kJ/mol for the four electrons in the $3s$ and $3p$ subshells. Removal of the fifth electron, which comes from the $2p$ subshell, requires a great deal more energy: 16,091 kJ/mol. The large increase occurs because the $2p$ electron is much more likely to be found close to the nucleus than are the four $n = 3$ electrons, and, therefore, the $2p$ electron experiences a much greater effective nuclear charge than do the $3s$ and $3p$ electrons.

Table 7.2 Successive Values of Ionization Energies, I, for the Elements Sodium through Argon (kJ/mol)

Element	I_1	I_2	I_3	I_4	I_5	I_6	I_7
Na	496	4562			(inner-shell electrons)		
Mg	738	1451	7733				
Al	578	1817	2745	11,577			
Si	786	1577	3232	4356	16,091		
P	1012	1907	2914	4964	6274	21,267	
S	1000	2252	3357	4556	7004	8496	27,107
Cl	1251	2298	3822	5159	6542	9362	11,018
Ar	1521	2666	3931	5771	7238	8781	11,995

Give It Some Thought

Which would you expect to be greater, I_1 for a boron atom or I_2 for a carbon atom?

Every element exhibits a large increase in ionization energy when the first of its inner-shell electrons is removed. This observation supports the idea that only the outermost electrons are involved in the sharing and transfer of electrons that give rise to chemical bonding and reactions. As we will see when we talk about chemical bonds in Chapters 8 and 9, the inner electrons are too tightly bound to the nucleus to be lost from the atom or even shared with another atom.

SAMPLE
EXERCISE 7.5 Trends in Ionization Energy

Three elements are indicated in the periodic table in the margin. Which one has the largest second ionization energy?

SOLUTION

Analyze and Plan The locations of the elements in the periodic table allow us to predict the electron configurations. The greatest ionization energies involve removal of core electrons. Thus, we should look first for an element with only one electron in the outermost occupied shell.

Solve The red box represents Na, which has one valence electron. The second ionization energy of this element is associated, therefore, with the removal of a core electron. The other elements indicated, S (green) and Ca (blue), have two or more valence electrons. Thus, Na should have the largest second ionization energy.

Check A chemistry handbook gives these I_2 values: Ca, 1145 kJ/mol; S, 2252 kJ/mol; Na, 4562 kJ/mol.

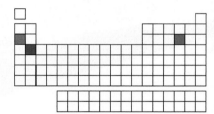

Practice Exercise 1

The third ionization energy of bromine is the energy required for which of the following processes?
(a) $Br(g) \longrightarrow Br^+(g) + e^-$ **(b)** $Br^+(g) \longrightarrow Br^{2+}(g) + e^-$ **(c)** $Br(g) \longrightarrow Br^{2+}(g) + 2e^-$
(d) $Br(g) \longrightarrow Br^{3+}(g) + 3e^-$ **(e)** $Br^{2+}(g) \longrightarrow Br^{3+}(g) + e^-$

Practice Exercise 2

Which has the greater third ionization energy, Ca or S?

Periodic Trends in First Ionization Energies

Figure 7.10 shows, for the first 54 elements, the trends we observe in first ionization energies as we move from one element to another in the periodic table. The important trends are as follows:

1. I_1 *generally increases as we move across a period.* The alkali metals show the lowest ionization energy in each period, and the noble gases show the highest. There are slight irregularities in this trend that we will discuss shortly.

2. I_1 *generally decreases as we move down any column in the periodic table.* For example, the ionization energies of the noble gases follow the order He > Ne > Ar > Kr > Xe.

3. *The s- and p-block elements show a larger range of I_1 values than do the transition-metal elements.* Generally, the ionization energies of the transition metals increase slowly from left to right in a period. The *f*-block metals (not shown in Figure 7.10) also show only a small variation in the values of I_1.

In general, smaller atoms have higher ionization energies. The same factors that influence atomic size also influence ionization energies. The energy needed to remove an electron from the outermost occupied shell depends on both the effective nuclear charge and the average distance of the electron from the nucleus. Either increasing the effective nuclear charge or decreasing the distance from the nucleus increases the attraction between the electron and the nucleus. As this attraction increases, it becomes more difficult to remove the electron, and, thus, the ionization energy increases.

▲ GO FIGURE

The value for astatine, At, is missing in this figure. To the nearest 100 kJ/mol, what estimate would you make for the first ionization energy of At?

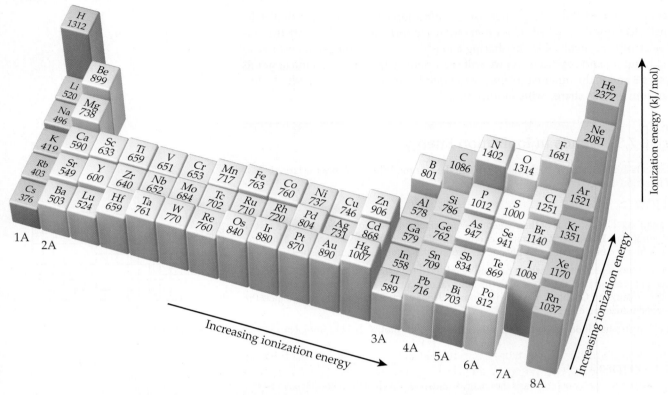

▲ **Figure 7.10 Trends in first ionization energies of the elements.**

GO FIGURE

Why is it easier to remove a 2p electron from an oxygen atom than from a nitrogen atom?

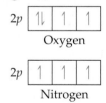

▲ **Figure 7.11 2p orbital filling in nitrogen and oxygen.**

As we move across a period, there is both an increase in effective nuclear charge and a decrease in atomic radius, causing the ionization energy to increase. As we move down a column, the atomic radius increases while the effective nuclear charge increases only gradually. The increase in radius dominates, so the attraction between the nucleus and the electron decreases, causing the ionization energy to decrease.

The irregularities in a given period are subtle but still readily explained. For example, the decrease in ionization energy from beryllium ($[\text{He}]2s^2$) to boron ($[\text{He}]2s^22p^1$), shown in Figure 7.10, occurs because the third valence electron of B must occupy the $2p$ subshell, which is empty for Be. Recall that the $2p$ subshell is at a higher energy than the $2s$ subshell (Figure 6.25). The slight decrease in ionization energy when moving from nitrogen ($[\text{He}]2s^22p^3$) to oxygen ($[\text{He}]2s^22p^4$) is the result of the repulsion of paired electrons in the p^4 configuration (◀ Figure 7.11). Remember that according to Hund's rule, each electron in the p^3 configuration resides in a different p orbital, which minimizes the electron–electron repulsion among the three $2p$ electrons. ∞ (Section 6.8)

SAMPLE EXERCISE 7.6 | Periodic Trends in Ionization Energy

Referring to the periodic table, arrange the atoms Ne, Na, P, Ar, K in order of increasing first ionization energy.

SOLUTION

Analyze and Plan We are given the chemical symbols for five elements. To rank them according to increasing first ionization energy, we need to locate each element in the periodic table. We can then use their relative positions and the trends in first ionization energies to predict their order.

Solve Ionization energy increases as we move left to right across a period and decreases as we move down a group. Because Na, P, and Ar are in the same period, we expect I_1 to vary in the order Na < P < Ar.

Because Ne is above Ar in group 8A, we expect Ar < Ne. Similarly, K is directly below Na in group 1A, and so we expect K < Na.

From these observations, we conclude that the ionization energies follow the order

$$\text{K} < \text{Na} < \text{P} < \text{Ar} < \text{Ne}$$

Check The values shown in Figure 7.10 confirm this prediction.

Electron Configurations of Ions

When electrons are removed from an atom to form a cation, they are always removed first from the occupied orbitals having the largest principal quantum number, n. For example, when one electron is removed from a lithium atom ($1s^2 2s^1$), it is the $2s^1$ electron:

$$\text{Li}(1s^2 2s^1) \Rightarrow \text{Li}^+(1s^2) + e^-$$

Likewise, when two electrons are removed from Fe ($[\text{Ar}]4s^2 3d^6$), the $4s^2$ electrons are the ones removed:

$$\text{Fe}([\text{Ar}]4s^2 3d^6) \Rightarrow \text{Fe}^{2+}([\text{Ar}]3d^6) + 2e^-$$

If an additional electron is removed, forming Fe^{3+}, it comes from a $3d$ orbital because all the orbitals with $n = 4$ are empty:

$$\text{Fe}^{2+}([\text{Ar}]3d^6) \Rightarrow \text{Fe}^{3+}([\text{Ar}]3d^5) + e^-$$

It may seem odd that $4s$ electrons are removed before $3d$ electrons in forming transition-metal cations. After all, in writing electron configurations, we added the $4s$ electrons before the $3d$ ones. In writing electron configurations for atoms, however, we are going through an imaginary process in which we move through the periodic table from one element to another. In doing so, we are adding both an electron to an orbital and a proton to the nucleus to change the identity of the element. In ionization, we do not reverse this process because no protons are being removed.

If there is more than one occupied subshell for a given value of n, the electrons are first removed from the orbital with the highest value of l. For example, a tin atom loses its $5p$ electrons before it loses its $5s$ electrons:

$$\text{Sn}([\text{Kr}]5s^2 4d^{10} 5p^2) \Rightarrow \text{Sn}^{2+}([\text{Kr}]5s^2 4d^{10}) + 2e^- \Rightarrow \text{Sn}^{4+}([\text{Kr}]4d^{10}) + 4e^-$$

Electrons added to an atom to form an anion are added to the empty or partially filled orbital having the lowest value of n. For example, an electron added to a fluorine atom to form the F^- ion goes into the one remaining vacancy in the $2p$ subshell:

$$\text{F}(1s^2 2s^2 2p^5) + e^- \Rightarrow \text{F}^-(1s^2 2s^2 2p^6)$$

 Give It Some Thought

Do Cr^{3+} and V^{2+} have the same or different electron configurations?

SAMPLE EXERCISE 7.7 Electron Configurations of Ions

Write the electron configurations for **(a)** Ca^{2+}, **(b)** Co^{3+}, and **(c)** S^{2-}.

SOLUTION

Analyze and Plan We are asked to write electron configurations for three ions. To do so, we first write the electron configuration of each parent atom, and then remove or add electrons to form the ions. Electrons are first removed from the orbitals having the highest value of n. They are added to the empty or partially filled orbitals having the lowest value of n.

Solve

(a) Calcium (atomic number 20) has the electron configuration $[\text{Ar}]4s^2$. To form a 2+ ion, the two outer $4s$ electrons must be removed, giving an ion that is isoelectronic with Ar:

$$\text{Ca}^{2+}: [\text{Ar}]$$

(b) Cobalt (atomic number 27) has the electron configuration $[Ar]4s^23d^7$. To form a 3+ ion, three electrons must be removed. As discussed in the text, the $4s$ electrons are removed before the $3d$ electrons. Consequently, we remove the two $4s$ electrons and one of the $3d$ electrons, and the electron configuration for Co^{3+} is

$$Co^{3+}: [Ar]3d^6$$

(c) Sulfur (atomic number 16) has the electron configuration $[Ne]3s^23p^4$. To form a 2− ion, two electrons must be added. There is room for two additional electrons in the $3p$ orbitals. Thus, the S^{2-} electron configuration is

$$S^{2-}: [Ne]3s^23p^6 = [Ar]$$

Comment Remember that many of the common ions of the s- and p-block elements, such as Ca^{2+} and S^{2-}, have the same number of electrons as the closest noble gas. ∞ (Section 2.7)

Practice Exercise 1

The ground electron configuration of a Tc atom is $[Kr]5s^24d^5$. What is the electron configuration of a Tc^{3+} ion?
(a) $[Kr]4d^4$ **(b)** $[Kr]5s^24d^2$ **(c)** $[Kr]5s^14d^3$ **(d)** $[Kr]5s^24d^8$
(e) $[Kr]4d^{10}$

Practice Exercise 2

Write the electron configurations for **(a)** Ga^{3+}, **(b)** Cr^{3+}, and **(c)** Br^-.

7.5 | Electron Affinity

The first ionization energy of an atom is a measure of the energy change associated with removing an electron from the atom to form a cation. For example, the first ionization energy of $Cl(g)$, 1251 kJ/mol, is the energy change associated with the process

$$\textit{Ionization energy:} \quad Cl(g) \longrightarrow Cl^+(g) + e^- \quad \Delta E = 1251 \text{ kJ/mol} \quad [7.4]$$
$$[Ne]3s^23p^5 \quad [Ne]3s^23p^4$$

The positive value of the ionization energy means that energy must be put into the atom to remove the electron. *All ionization energies for atoms are positive: Energy must be absorbed to remove an electron.*

Most atoms can also gain electrons to form anions. The energy change that occurs when an electron is added to a gaseous atom is called the **electron affinity** because it measures the attraction, or *affinity,* of the atom for the added electron. For most atoms, energy is *released* when an electron is added. For example, the addition of an electron to a chlorine atom is accompanied by an energy change of −349 kJ/mol, the negative sign indicating that energy is released during the process. We therefore say that the electron affinity of Cl is −349 kJ/mol.*

$$\textit{Electron affinity:} \quad Cl(g) + e^- \longrightarrow Cl^-(g) \quad \Delta E = -349 \text{ kJ/mol} \quad [7.5]$$
$$[Ne]3s^23p^5 \quad [Ne]3s^23p^6$$

It is important to understand the difference between ionization energy and electron affinity: Ionization energy measures the energy change when an atom *loses* an electron, whereas electron affinity measures the energy change when an atom *gains* an electron.

The greater the attraction between an atom and an added electron, the more negative the atom's electron affinity. For some elements, such as the noble gases, the electron affinity has a positive value, meaning that the anion is higher in energy than are the separated atom and electron:

$$Ar(g) + e^- \longrightarrow Ar^-(g) \quad \Delta E > 0 \quad [7.6]$$
$$[Ne]3s^23p^6 \quad [Ne]3s^23p^64s^1$$

The fact that the electron affinity is positive means that an electron will not attach itself to an Ar atom; the Ar^- ion is unstable and does not form.

◄ Figure 7.12 shows the electron affinities for the s- and p-block elements of the first five periods. Notice that the trends are not as evident as they are for ionization energy. The

▲ **GO FIGURE**

Why are the electron affinities of the Group 4A elements more negative than those of the Group 5A elements?

1A							8A
H −73	2A	3A	4A	5A	6A	7A	**He** >0
Li −60	**Be** >0	**B** −27	**C** −122	**N** >0	**O** −141	**F** −328	**Ne** >0
Na −53	**Mg** >0	**Al** −43	**Si** −134	**P** −72	**S** −200	**Cl** −349	**Ar** >0
K −48	**Ca** −2	**Ga** −30	**Ge** −119	**As** −78	**Se** −195	**Br** −325	**Kr** >0
Rb −47	**Sr** −5	**In** −30	**Sn** −107	**Sb** −103	**Te** −190	**I** −295	**Xe** >0

▲ **Figure 7.12 Electron affinity in kJ/mol for selected s- and p-block elements.**

*Two sign conventions are used for electron affinity. In most introductory texts, including this one, the thermodynamic sign convention is used: A negative sign indicates that addition of an electron is an exothermic process, as in the electron affinity for chlorine, −349 kJ/mol. Historically, however, electron affinity has been defined as the energy released when an electron is added to a gaseous atom or ion. Because 349 kJ/mol is released when an electron is added to Cl(g), the electron affinity by this convention would be +349 kJ/mol.

halogens, which are one electron shy of a filled p subshell, have the most negative electron affinities. By gaining an electron, a halogen atom forms a stable anion that has a noble-gas configuration (Equation 7.5). The addition of an electron to a noble gas, however, requires that the electron reside in a higher-energy subshell that is empty in the atom (Equation 7.6). Because occupying a higher-energy subshell is energetically unfavorable, the electron affinity is highly positive. The electron affinities of Be and Mg are positive for the same reason; the added electron would reside in a previously empty p subshell that is higher in energy.

The electron affinities of the group 5A elements are also interesting. Because these elements have half-filled p subshells, the added electron must be put in an orbital that is already occupied, resulting in larger electron–electron repulsions. Consequently, these elements have electron affinities that are either positive (N) or less negative than those of their neighbors to the left (P, As, Sb). Recall that in Section 7.4 we saw a discontinuity in the trends in first ionization energy for the same reason.

Electron affinities do not change greatly as we move down a group (Figure 7.12). For F, for instance, the added electron goes into a $2p$ orbital, for Cl a $3p$ orbital, for Br a $4p$ orbital, and so forth. As we proceed from F to I, therefore, the average distance between the added electron and the nucleus steadily increases, causing the electron–nucleus attraction to decrease. However, the orbital that holds the outermost electron is increasingly spread out, so that as we proceed from F to I, the electron–electron repulsions are also reduced. As a result, the reduction in the electron–nucleus attraction is counterbalanced by the reduction in electron–electron repulsions.

 Give It Some Thought

What is the relationship between the value for the first ionization energy of a $Cl^-(g)$ ion and the electron affinity of $Cl(g)$?

7.6 | Metals, Nonmetals, and Metalloids

Atomic radii, ionization energies, and electron affinities are properties of individual atoms. With the exception of the noble gases, however, none of the elements exist in nature as individual atoms. To get a broader understanding of the properties of elements, we must also examine periodic trends in properties that involve large collections of atoms.

The elements can be broadly grouped as metals, nonmetals, and metalloids (**Figure 7.13**). ∞ (Section 2.5) Some of the distinguishing properties of metals and nonmetals are summarized in ▼ **Table 7.3**.

In the following sections, we explore some common patterns of reactivity across the periodic table. We will examine reactivity for selected nonmetals and metals in more depth in later chapters.

The more an element exhibits the physical and chemical properties of metals, the greater its **metallic character**. As indicated in Figure 7.13, metallic character generally increases as we proceed down a group of the periodic table and decreases as we proceed right across a period. Let's now examine the close relationships that exist between electron configurations and the properties of metals, nonmetals, and metalloids.

Table 7.3   **Characteristic Properties of Metals and Nonmetals**

Metals	Nonmetals
Have a shiny luster; various colors, although most are silvery	Do not have a luster; various colors
Solids are malleable and ductile	Solids are usually brittle; some are hard, and some are soft
Good conductors of heat and electricity	Poor conductors of heat and electricity
Most metal oxides are ionic solids that are basic	Most nonmetal oxides are molecular substances that form acidic solutions
Tend to form cations in aqueous solution	Tend to form anions or oxyanions in aqueous solution

GO FIGURE

How do the periodic trends in metallic character compare to those for ionization energy?

Increasing metallic character ←

▲ Figure 7.13 Metals, metalloids, and nonmetals.

Metals

Most metallic elements exhibit the shiny luster we associate with metals (◀ Figure 7.14). Metals conduct heat and electricity. In general they are malleable (can be pounded into thin sheets) and ductile (can be drawn into wires). All are solids at room temperature except mercury (melting point = −39 °C), which is a liquid at room temperature. Two metals melt at slightly above room temperature, cesium at 28.4 °C and gallium at 29.8 °C. At the other extreme, many metals melt at very high temperatures. For example, tungsten, which is used for the filaments of incandescent light bulbs, melts at 3400 °C.

Metals tend to have low ionization energies (Figure 7.10) *and therefore tend to form cations relatively easily.* As a result, metals are oxidized (lose electrons) when they undergo chemical reactions. Among the fundamental atomic properties (radius, electron configuration, electron affinity, and so forth), first ionization energy is the best indicator of whether an element behaves as a metal or a nonmetal.

Figure 7.15 shows the oxidation states of representative ions of metals and nonmetals. As noted in Section 2.7, the charge on any alkali metal ion in a compound is always 1+, and that on any alkaline earth metal is always 2+. For atoms belonging to either of these groups, the outer s electrons are easily lost, yielding a noble-gas electron configuration. For metals belonging to groups with partially occupied p orbitals (groups 3A–7A), cations are formed either by losing only the outer p electrons (such as Sn^{2+}) or the outer s and p electrons (such as Sn^{4+}). The charge on transition-metal ions does not follow an obvious pattern. One characteristic of the transition metals is their ability to form more than one cation. For example, compounds of Fe^{2+} and Fe^{3+} are both very common.

▲ Figure 7.14 Metals are shiny, malleable, and ductile.

 Give It Some Thought

Arsenic forms binary compounds with Cl and with Mg. Will it be in the same oxidation state in these two compounds?

Compounds made up of a metal and a nonmetal tend to be ionic substances. For example, most metal oxides and halides are ionic solids. To illustrate, the reaction between nickel metal and oxygen produces nickel oxide, an ionic solid containing Ni^{2+} and O^{2-} ions:

$$2 \, Ni(s) \, + \, O_2(g) \longrightarrow 2 \, NiO(s) \qquad [7.7]$$

GO FIGURE

The red stepped line divides metals from nonmetals. How are common oxidation states divided by this line?

1A																7A	8A
H^+	2A											3A	4A	5A	6A	H^-	N
Li^+														N^{3-}	O^{2-}	F^-	O B
Na^+	Mg^{2+}				Transition metals							Al^{3+}		P^{3-}	S^{2-}	Cl^-	L E
K^+	Ca^{2+}	Sc^{3+}	Ti^{4+}	V^{5+} V^{4+}	Cr^{3+}	Mn^{2+} Mn^{4+}	Fe^{2+} Fe^{3+}	Co^{2+} Co^{3+}	Ni^{2+}	Cu^+ Cu^{2+}	Zn^{2+}				Se^{2-}	Br^-	G A S
Rb^+	Sr^{2+}							Pd^{2+}	Ag^+	Cd^{2+}		Sn^{2+} Sn^{4+}	Sb^{3+} Sb^{5+}	Te^{2-}	I^-		E S
Cs^+	Ba^{2+}							Pt^{2+}	Au^+ Au^{3+}	Hg_2^{2+} Hg^{2+}		Pb^{2+} Pb^{4+}	Bi^{3+} Bi^{5+}				

▲ **Figure 7.15 Representative oxidation states of the elements.** Note that hydrogen has both positive and negative oxidation numbers, +1 and −1.

The oxides are particularly important because of the great abundance of oxygen in our environment.

Most metal oxides are basic. Those that dissolve in water react to form metal hydroxides, as in the following examples:

$$\text{Metal oxide + water} \longrightarrow \text{metal hydroxide}$$
$$Na_2O(s) + H_2O(l) \longrightarrow 2\,NaOH(aq) \qquad [7.8]$$
$$CaO(s) + H_2O(l) \longrightarrow Ca(OH)_2(aq) \qquad [7.9]$$

The basicity of metal oxides is due to the oxide ion, which reacts with water:

$$O^{2-}(aq) + H_2O(l) \longrightarrow 2\,OH^-(aq) \qquad [7.10]$$

Even metal oxides that are insoluble in water demonstrate their basicity by reacting with acids to form a salt plus water, as illustrated in ▼ Figure 7.16:

$$\text{Metal oxide + acid} \longrightarrow \text{salt + water}$$
$$NiO(s) + 2\,HNO_3(aq) \longrightarrow Ni(NO_3)_2(aq) + H_2O(l) \qquad [7.11]$$

▲ **GO FIGURE**

Would you expect NiO to dissolve in an aqueous solution of $NaNO_3$?

Nickle oxide (NiO), nitric acid (HNO_3), and water

NiO is insoluble in water but reacts with HNO_3 to give a green solution of the salt $Ni(NO_3)_2$

▲ **Figure 7.16 Metal oxides react with acids.** NiO does not dissolve in water but does react with nitric acid (HNO_3) to give a green solution of $Ni(NO_3)_2$.

SAMPLE EXERCISE 7.8 | Properties of Metal Oxides

(a) Would you expect scandium oxide to be a solid, liquid, or gas at room temperature?
(b) Write the balanced chemical equation for the reaction of scandium oxide with nitric acid.

SOLUTION

Analyze and Plan We are asked about one physical property of scandium oxide—its state at room temperature—and one chemical property—how it reacts with nitric acid.

Solve

(a) Because scandium oxide is the oxide of a metal, we expect it to be an ionic solid. Indeed it is, with the very high melting point of 2485 °C.

(b) In compounds, scandium has a 3+ charge, Sc^{3+}, and the oxide ion is O^{2-}. Consequently, the formula of scandium oxide is Sc_2O_3. Metal oxides tend to be basic and, therefore, to react with acids to form a salt plus water. In this case the salt is scandium nitrate, $Sc(NO_3)_3$:

$$Sc_2O_3(s) + 6\ HNO_3(aq) \longrightarrow 2\ Sc(NO_3)_3(aq) + 3\ H_2O(l)$$

Practice Exercise 1

Suppose that a metal oxide of formula M_2O_3 were soluble in water. What would be the major product or products of dissolving the substance in water?
(a) $MH_3(aq) + O_2(g)$ (b) $M(s) + H_2(g) + O_2(g)$ (c) $M^{3+}(aq) + H_2O_2(aq)$
(d) $M(OH)_2(aq)$ (e) $M(OH)_3(aq)$

Practice Exercise 2

Write the balanced chemical equation for the reaction between copper(II) oxide and sulfuric acid.

▲ **GO FIGURE**
Do you think sulfur is malleable?

▲ **Figure 7.17** Sulfur, known to the medieval world as "brimstone," is a nonmetal.

Nonmetals

Nonmetals can be solid, liquid, or gas. They are not lustrous and generally are poor conductors of heat and electricity. Their melting points are generally lower than those of metals (although diamond, a form of carbon, is an exception and melts at 3570 °C). Under ordinary conditions, seven nonmetals exist as diatomic molecules. Five of these are gases (H_2, N_2, O_2, F_2, and Cl_2), one is a liquid (Br_2), and one is a volatile solid (I_2). Excluding the noble gases, the remaining nonmetals are solids that can be either hard, such as diamond, or soft, such as sulfur (◀ **Figure 7.17**).

Because of their relatively large, negative electron affinities, nonmetals tend to gain electrons when they react with metals. For example, the reaction of aluminum with bromine produces the ionic compound aluminum bromide:

$$2\ Al(s) + 3\ Br_2(l) \longrightarrow 2\ AlBr_3(s) \qquad [7.12]$$

A nonmetal will typically gain enough electrons to fill its outermost occupied p subshell, giving a noble-gas electron configuration. For example, the bromine atom gains one electron to fill its $4p$ subshell:

$$Br([Ar]4s^2 3d^{10} 4p^5) + e^- \Longrightarrow Br^-([Ar]4s^2 3d^{10} 4p^6)$$

Compounds composed entirely of nonmetals are typically molecular substances that tend to be gases, liquids, or low-melting solids at room temperature. Examples include the common hydrocarbons we use for fuel (methane, CH_4; propane, C_3H_8; octane, C_8H_{18}) and the gases HCl, NH_3, and H_2S. Many pharmaceuticals are molecules composed of C, H, N, O, and other nonmetals. For example, the molecular formula for the drug Celebrex® is $C_{17}H_{14}F_3N_3O_2S$.

Most nonmetal oxides are acidic, which means that those that dissolve in water form acids:

$$\text{Nonmetal oxide} + \text{water} \longrightarrow \text{acid}$$
$$CO_2(g) + H_2O(l) \longrightarrow H_2CO_3(aq) \qquad [7.13]$$
$$P_4O_{10}(s) + 6\ H_2O(l) \longrightarrow 4\ H_3PO_4(aq) \qquad [7.14]$$

The reaction of carbon dioxide with water (▶ Figure 7.18) accounts for the acidity of carbonated water and, to some extent, rainwater. Because sulfur is present in oil and coal, combustion of these common fuels produces sulfur dioxide and sulfur trioxide. These substances dissolve in water to produce *acid rain*, a major pollutant in many parts of the world. Like acids, most nonmetal oxides dissolve in basic solutions to form a salt plus water:

$$\text{Nonmetal oxide} + \text{base} \longrightarrow \text{salt} + \text{water}$$

$$CO_2(g) + 2\,NaOH(aq) \longrightarrow Na_2CO_3(aq) + H_2O(l) \quad [7.15]$$

 Give It Some Thought

A compound ACl_3 (A is an element) has a melting point of $-112\,°C$. Would you expect the compound to be molecular or ionic? If you were told that A is either scandium or phosphorus, which do you think is the more likely choice?

▲ **Figure 7.18 The reaction of CO_2 with water containing a bromthymol blue indicator.** Initially, the blue color tells us the water is slightly basic. When a piece of solid carbon dioxide ("dry ice") is added, the color changes to yellow, indicating an acidic solution. The mist is water droplets condensed from the air by the cold CO_2 gas.

SAMPLE EXERCISE 7.9 | Reactions of Nonmetal Oxides

Write a balanced chemical equation for the reaction of solid selenium dioxide, $SeO_2(s)$, with (a) water, (b) aqueous sodium hydroxide.

SOLUTION

Analyze and Plan We note that selenium is a nonmetal. We therefore need to write chemical equations for the reaction of a nonmetal oxide with water and with a base, NaOH. Nonmetal oxides are acidic, reacting with water to form an acid and with bases to form a salt and water.

Solve

(a) The reaction between selenium dioxide and water is like that between carbon dioxide and water (Equation 7.13):

$$SeO_2(s) + H_2O(l) \longrightarrow H_2SeO_3(aq)$$

(It does not matter that SeO_2 is a solid and CO_2 is a gas under ambient conditions; the point is that both are water-soluble nonmetal oxides.)

(b) The reaction with sodium hydroxide is like that in Equation 7.15:

$$SeO_2(s) + 2\,NaOH(aq) \longrightarrow Na_2SeO_3(aq) + H_2O(l)$$

Practice Exercise 1

Consider the following oxides: SO_2, Y_2O_3, MgO, Cl_2O, N_2O_5. How many are expected to form acidic solutions in water?
(a) 1   (b) 2   (c) 3   (d) 4   (e) 5

Practice Exercise 2

Write a balanced chemical equation for the reaction of solid tetraphosphorus hexoxide with water.

Metalloids

Metalloids have properties intermediate between those of metals and those of nonmetals. They may have some characteristic metallic properties but lack others. For example, the metalloid silicon *looks* like a metal (▶ Figure 7.19), but it is brittle rather than malleable and does not conduct heat or electricity nearly as well as metals do. Compounds of metalloids can have characteristics of the compounds of metals or nonmetals.

Several metalloids, most notably silicon, are electrical semiconductors and are the principal elements used in integrated circuits and computer chips. One of the reasons metalloids can be used for integrated circuits is that their electrical conductivity is intermediate between that of metals and that of nonmetals. Very pure silicon is an electrical insulator, but its conductivity can be dramatically increased with the addition of specific impurities called *dopants*. This modification provides a mechanism for controlling the electrical conductivity by controlling the chemical composition. We will return to this point in Chapter 12.

▲ **Figure 7.19 Elemental silicon.**

▲ Figure 7.20 **Sodium, like the other alkali metals, is soft enough to be cut with a knife.**

7.7 | Trends for Group 1A and Group 2A Metals

As we have seen, elements in a given group possess general similarities. However, trends also exist within each group. In this section we use the periodic table and our knowledge of electron configurations to examine the chemistry of the **alkali metals** and **alkaline earth metals**.

Group 1A: The Alkali Metals

The alkali metals are soft metallic solids (◀ Figure 7.20). All have characteristic metallic properties, such as a silvery, metallic luster, and high thermal and electrical conductivity. The name *alkali* comes from an Arabic word meaning "ashes." Many compounds of sodium and potassium, two alkali metals, were isolated from wood ashes by early chemists.

As ▼ Table 7.4 shows, the alkali metals have low densities and melting points, and these properties vary in a fairly regular way with increasing atomic number. We see the usual trends as we move down the group, such as increasing atomic radius and decreasing first ionization energy. The alkali metal of any given period has the lowest I_1 value in the period (Figure 7.10), which reflects the relative ease with which its outer *s* electron can be removed. As a result, the alkali metals are all very reactive, readily losing one electron to form ions carrying a 1+ charge. ∞ (Section 2.7)

The alkali metals exist in nature only as compounds. Sodium and potassium are relatively abundant in Earth's crust, in seawater, and in biological systems, always as the cations of ionic compounds. All alkali metals combine directly with most nonmetals. For example, they react with hydrogen to form hydrides and with sulfur to form sulfides:

$$2\,M(s) + H_2(g) \longrightarrow 2\,MH(s) \qquad [7.16]$$
$$2\,M(s) + S(s) \longrightarrow M_2S(s) \qquad [7.17]$$

where M represents any alkali metal. In hydrides of the alkali metals (LiH, NaH, and so forth), hydrogen is present as H^-, the **hydride ion**. A hydrogen atom that has *gained* an electron, this ion is distinct from the hydrogen ion, H^+, formed when a hydrogen atom *loses* its electron.

The alkali metals react vigorously with water, producing hydrogen gas and a solution of an alkali metal hydroxide:

$$2\,M(s) + 2\,H_2O(l) \longrightarrow 2\,MOH(aq) + H_2(g) \qquad [7.18]$$

These reactions are very exothermic (▶ Figure 7.21). In many cases enough heat is generated to ignite the H_2, producing a fire or sometimes even an explosion, as in the chapter-opening photograph of K reacting with water. The reaction is even more violent for Rb and, especially, Cs, because their ionization energies are even lower than that of K.

Recall that the most common ion of oxygen is the oxide ion, O^{2-}. We would therefore expect that the reaction of an alkali metal with oxygen would produce the corresponding metal oxide. Indeed, reaction of Li metal with oxygen does form lithium oxide:

$$4\,Li(s) + O_2(g) \longrightarrow 2\,\underset{\text{lithium oxide}}{Li_2O(s)} \qquad [7.19]$$

When dissolved in water, Li_2O and other soluble metal oxides form hydroxide ions from the reaction of O^{2-} ions with H_2O (Equation 7.10).

Table 7.4 Some Properties of the Alkali Metals

Element	Electron Configuration	Melting Point (°C)	Density (g/cm³)	Atomic Radius (Å)	I_1 (kJ/mol)
Lithium	$[He]2s^1$	181	0.53	1.28	520
Sodium	$[Ne]3s^1$	98	0.97	1.66	496
Potassium	$[Ar]4s^1$	63	0.86	2.03	419
Rubidium	$[Kr]5s^1$	39	1.53	2.20	403
Cesium	$[Xe]6s^1$	28	1.88	2.44	376

GO FIGURE

Would you expect rubidium metal to be more or less reactive with water than potassium metal?

Li   Na   K

▲ Figure 7.21 **The alkali metals react vigorously with water.**

The reactions of the other alkali metals with oxygen are more complex than we would anticipate. For example, when sodium reacts with oxygen, the main product is sodium *peroxide*, which contains the O_2^{2-} ion:

$$2\,Na(s)\ +\ O_2(g)\ \longrightarrow\ \underset{\text{sodium peroxide}}{Na_2O_2(s)} \qquad [7.20]$$

Potassium, rubidium, and cesium react with oxygen to form compounds that contain the O_2^- ion, which we call the *superoxide ion*. For example, potassium forms potassium superoxide, KO_2:

$$K(s)\ +\ O_2(g)\ \longrightarrow\ \underset{\text{potassium superoxide}}{KO_2(s)} \qquad [7.21]$$

Be aware that the reactions in Equations 7.20 and 7.21 are somewhat unexpected; in most cases, the reaction of oxygen with a metal forms the metal oxide.

As is evident from Equations 7.18 through 7.21, the alkali metals are extremely reactive toward water and oxygen. Because of this reactivity, the metals are usually stored submerged in a liquid hydrocarbon, such as mineral oil or kerosene.

Although alkali metal ions are colorless, each emits a characteristic color when placed in a flame (▼ Figure 7.22). The ions are reduced to gaseous metal atoms in

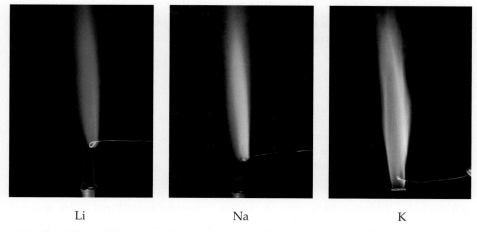

Li   Na   K

▲ Figure 7.22 **Placed in a flame, ions of each alkali metal emit light of a characteristic wavelength.**

GO FIGURE

If we had potassium vapor lamps, what color would they be?

Electrical energy used to excite the electron from the 3s to the 3p orbital

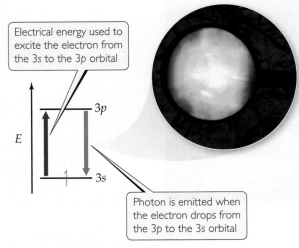

E

3p

3s

Photon is emitted when the electron drops from the 3p to the 3s orbital

▲ **Figure 7.23 The characteristic yellow light in a sodium lamp results from excited electrons in the high-energy 3p orbital falling back to the lower-energy 3s orbital.**

the flame. The high temperature excites the valence electron from the ground state to a higher-energy orbital, causing the atom to be in an excited state. The atom then emits energy in the form of visible light as the electron falls back into the lower-energy orbital and the atom returns to its ground state. The light emitted is at a specific wavelength for each element, just as we saw earlier for line spectra of hydrogen and sodium. ∞ (Section 6.3) The characteristic yellow emission of sodium at 589 nm is the basis for sodium vapor lamps (◀ Figure 7.23).

 Give It Some Thought

Cesium tends to be the most reactive of the stable alkali metals (francium, Fr, is radioactive and has not been extensively studied). What *atomic* property of Cs is most responsible for its high reactivity?

SAMPLE
EXERCISE 7.10 Reactions of an Alkali Metal

Write a balanced equation for the reaction of cesium metal with (**a**) $Cl_2(g)$, (**b**) $H_2O(l)$, (**c**) $H_2(g)$.

SOLUTION

Analyze and Plan Because cesium is an alkali metal, we expect its chemistry to be dominated by oxidation of the metal to Cs^+ ions. Further, we recognize that Cs is far down the periodic table, which means it is among the most active of all metals and probably reacts with all three substances.

Solve The reaction between Cs and Cl_2 is a simple combination reaction between a metal and a nonmetal, forming the ionic compound CsCl:

$$2\,Cs(s) + Cl_2(g) \longrightarrow 2\,CsCl(s)$$

From Equations 7.18 and 7.16, we predict the reactions of cesium with water and hydrogen to proceed as follows:

$$2\,Cs(s) + 2\,H_2O(l) \longrightarrow 2\,CsOH(aq) + H_2(g)$$

$$2\,Cs(s) + H_2(g) \longrightarrow 2\,CsH(s)$$

All three reactions are redox reactions where cesium forms a Cs^+ ion. The Cl^-, OH^-, and H^- are all 1− ions, which means the products have 1:1 stoichiometry with Cs^+.

Practice Exercise 1

Consider the following three statements about the reactivity of an alkali metal M with oxygen gas:
  (i) Based on their positions in the periodic table, the expected product is the ionic oxide M_2O.
  (ii) Some of the alkali metals produce metal peroxides or metal superoxides when they react with oxygen.
  (iii) When dissolved in water, an alkali metal oxide produces a basic solution.

Which of the statements (i), (ii), and (iii) is or are true?
(**a**) Only one of the statements is true.
(**b**) Statements (i) and (ii) are true.
(**c**) Statements (i) and (iii) are true.
(**d**) Statements (ii) and (iii) are true.
(**e**) All three statements are true.

Practice Exercise 2

Write a balanced equation for the expected reaction between potassium metal and elemental sulfur, S(s).

Chemistry and Life

The Improbable Development of Lithium Drugs

Alkali metal ions tend to play an unexciting role in most chemical reactions. As noted in Section 4.2, all salts of the alkali metal ions are soluble in water, and the ions are spectators in most aqueous reactions (except for those involving the alkali metals in their elemental form, such as those in Equations 7.16 through 7.21). However, these ions play an important role in human physiology. Sodium and potassium ions, for example, are major components of blood plasma and intracellular fluid, respectively, with average concentrations of 0.1 M. These electrolytes serve as vital charge carriers in normal cellular function. In contrast, the lithium ion has no known function in normal human physiology. Since the discovery of lithium in 1817, however, people have believed that salts of the element possessed almost mystical healing powers. There were even claims that lithium ions were an ingredient in ancient "fountain of youth" formulas. In 1927, C. L. Grigg began marketing a soft drink that contained lithium. The original unwieldy name of the beverage was "Bib-Label Lithiated Lemon-Lime Soda," which was soon changed to the simpler and more familiar name 7UP® (▶ Figure 7.24).

Because of concerns of the Food and Drug Administration, lithium was removed from 7UP® during the early 1950s. At nearly the same time, psychiatrists discovered that the lithium ion has a remarkable therapeutic effect on the mental condition called *bipolar disorder*. More than 5 million American adults each year suffer from this psychosis, undergoing severe mood swings from deep depression to a manic euphoria. The lithium ion

smoothes these mood swings, allowing the bipolar patient to function more effectively in daily life.

The antipsychotic action of Li$^+$ was discovered by accident in the 1940s by Australian psychiatrist John Cade as he was researching the use of uric acid—a component of urine—to treat manic-depressive illness. He administered the acid to manic laboratory animals in the form of its most soluble salt, lithium urate, and found that many of the manic symptoms seemed to disappear. Later studies showed that uric acid has no role in the therapeutic effects observed; rather, the Li$^+$ ions were responsible. Because lithium overdose can cause severe side effects in humans, including kidney failure and death, lithium salts were not approved as antipsychotic drugs for humans until 1970. Today Li$^+$ is usually administered orally in the form of Li$_2$CO$_3$, which is the active ingredient in prescription drugs such as Eskalith®. Lithium drugs are effective for about 70% of bipolar patients who take it.

In this age of sophisticated drug design and biotechnology, the simple lithium ion is still the most effective treatment of this destructive psychiatric disorder. Remarkably, in spite of intensive research, scientists still do not fully understand the biochemical action of lithium that leads to its therapeutic effects. Because of its similarity to Na$^+$, Li$^+$ is incorporated into blood plasma, where it can affect the behavior of nerve and muscle cells. Because Li$^+$ has a smaller radius than Na$^+$ (Figure 7.8), the way Li$^+$ interacts with molecules in human cells is different from the way Na$^+$ interacts with the molecules. Other studies indicate that Li$^+$ alters the function of certain neurotransmitters, which might lead to its effectiveness as an antipsychotic drug.

▲ **Figure 7.24 Lithium no more.** The soft drink 7UP® originally contained a lithium salt that was claimed to give the beverage healthful benefits, including "an abundance of energy, enthusiasm, a clear complexion, lustrous hair, and shining eyes!" The lithium was removed from the beverage in the early 1950s, about the time that the antipsychotic action of Li$^+$ was discovered.

Group 2A: The Alkaline Earth Metals

Like the alkali metals, the alkaline earth metals are all solids at room temperature and have typical metallic properties (▼ Table 7.5). Compared with the alkali metals, the alkaline earth metals are harder and denser, and melt at higher temperatures.

The first ionization energies of the alkaline earth metals are low but not as low as those of the alkali metals. Consequently, the alkaline earth metals are less reactive than their alkali metal neighbors. As noted in Section 7.4, the ease with which the elements lose electrons decreases as we move across a period and increases as we move down a

Table 7.5 Some Properties of the Alkaline Earth Metals

Element	Electron Configuration	Melting Point (°C)	Density (g/cm^3)	Atomic Radius (Å)	I$_1$ (kJ/mol)
Beryllium	[He]2s^2	1287	1.85	0.96	899
Magnesium	[Ne]3s^2	650	1.74	1.41	738
Calcium	[Ar]4s^2	842	1.55	1.76	590
Strontium	[Kr]5s^2	777	2.63	1.95	549
Barium	[Xe]6s^2	727	3.51	2.15	503

▲GO FIGURE

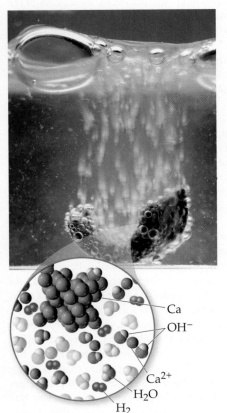

What is the cause of the bubbles that are formed? How could you test your answer?

▲ Figure 7.25 **Elemental calcium reacts with water.**

group. Thus, beryllium and magnesium, the lightest alkaline earth metals, are the least reactive.

The trend of increasing reactivity within the group is shown by the way the alkaline earth metals behave in the presence of water. Beryllium does not react with either water or steam, even when heated red-hot. Magnesium reacts slowly with liquid water and more readily with steam:

$$Mg(s) + H_2O(g) \longrightarrow MgO(s) + H_2(g) \qquad [7.22]$$

Calcium and the elements below it react readily with water at room temperature (although more slowly than the alkali metals adjacent to them in the periodic table). The reaction between calcium and water (◄ Figure 7.25), for example, is

$$Ca(s) + 2\,H_2O(l) \longrightarrow Ca(OH)_2(aq) + H_2(g) \qquad [7.23]$$

Equations 7.22 and 7.23 illustrate the dominant pattern in the reactivity of the alkaline earth elements: They tend to lose their two outer s electrons and form 2+ ions. For example, magnesium reacts with chlorine at room temperature to form $MgCl_2$ and burns with dazzling brilliance in air to give MgO:

$$Mg(s) + Cl_2(g) \longrightarrow MgCl_2(s) \qquad [7.24]$$

$$2\,Mg(s) + O_2(g) \longrightarrow 2\,MgO(s) \qquad [7.25]$$

In the presence of O_2, magnesium metal is protected by a thin coating of water-insoluble MgO. Thus, even though Mg is high in the activity series ∞ (Section 4.4), it can be incorporated into lightweight structural alloys used in, for example, automobile wheels. The heavier alkaline earth metals (Ca, Sr, and Ba) are even more reactive toward nonmetals than is magnesium.

The heavier alkaline earth ions give off characteristic colors when heated in a hot flame. Strontium salts produce the brilliant red color in fireworks, and barium salts produce the green color.

Like their neighbors sodium and potassium, magnesium and calcium are relatively abundant on Earth and in seawater and are essential for living organisms as cations in ionic compounds. Calcium is particularly important for growth and maintenance of bones and teeth.

▲ **Give It Some Thought**

Calcium carbonate, $CaCO_3$, is often used as a dietary calcium supplement for bone health. Although $CaCO_3(s)$ is insoluble in water (Table 4.1), it can be taken orally to allow for the delivery of $Ca^{2+}(aq)$ ions to the musculoskeletal system. Why is this the case? [*Hint:* Recall the reactions of metal carbonates discussed in Section 4.3.]

7.8 | Trends for Selected Nonmetals

Hydrogen

We have seen that the chemistry of the alkali metals is dominated by the loss of their outer ns^1 electron to form cations. The $1s^1$ electron configuration of hydrogen suggests that its chemistry should have some resemblance to that of the alkali metals. The chemistry of hydrogen is much richer and more complex than that of the alkali metals, however, mainly because the ionization energy of hydrogen, 1312 kJ/mol, is more than double that of any of the alkali metals. As a result, hydrogen is a nonmetal that occurs as a colorless diatomic gas, $H_2(g)$, under most conditions.

The reactivity of hydrogen with nonmetals reflects its much greater tendency to hold on to its electron relative to the alkali metals. Unlike the alkali metals, hydrogen reacts with most nonmetals to form molecular compounds in which its electron is shared with, rather than completely transferred to, the other nonmetal. For example, we have seen that sodium metal reacts vigorously with chlorine gas to produce the ionic compound sodium chloride, in which the outermost sodium electron is completely transferred to a chlorine atom (Figure 2.21):

$$Na(s) + \tfrac{1}{2}Cl_2(g) \longrightarrow \underset{\text{ionic}}{NaCl(s)} \qquad \Delta H° = -410.9 \text{ kJ} \qquad [7.26]$$

By contrast, molecular hydrogen reacts with chlorine gas to form hydrogen chloride gas, which consists of HCl molecules:

$$\tfrac{1}{2}H_2(g) + \tfrac{1}{2}Cl_2(g) \longrightarrow \underset{\text{molecular}}{HCl(g)} \qquad \Delta H° = -92.3 \text{ kJ} \qquad [7.27]$$

Hydrogen readily forms molecular compounds with other nonmetals, such as the formation of water, $H_2O(l)$; ammonia, $NH_3(g)$; and methane, $CH_4(g)$. The ability of hydrogen to form bonds with carbon is one of the most important aspects of organic chemistry, as will see in later chapters.

We have seen that, particularly in the presence of water, hydrogen does readily form H^+ ions in which the hydrogen atom has lost its electron. ∞ (Section 4.3) For example, $HCl(g)$ dissolves in H_2O to form a solution of hydrochloric acid, $HCl(aq)$, in which the electron of the hydrogen atom *is* transferred to the chlorine atom—a solution of hydrochloric acid consists largely of $H^+(aq)$ and $Cl^-(aq)$ ions stabilized by the H_2O solvent. Indeed, the ability of molecular compounds of hydrogen with nonmetals to form acids in water is one of the most important aspects of aqueous chemistry. We will discuss the chemistry of acids and bases in detail later in the text, particularly in Chapter 16.

Finally, as is typical for nonmetals, hydrogen also has the ability to gain an electron from a metal with a low ionization energy. For example, we saw in Equation 7.16 that hydrogen reacts with active metals to form solid metal hydrides that contain the hydride ion, H^-. The fact that hydrogen can gain an electron further illustrates that it behaves much more like a nonmetal than an alkali metal.

Group 6A: The Oxygen Group

As we proceed down group 6A, there is a change from nonmetallic to metallic character (Figure 7.13). Oxygen, sulfur, and selenium are typical nonmetals. Tellurium is a metalloid, and polonium, which is radioactive and quite rare, is a metal. Oxygen is a colorless gas at room temperature; all of the other members of group 6A are solids. Some of the physical properties of the group 6A elements are given in ▼ Table 7.6.

As we saw in Section 2.6, oxygen exists in two molecular forms, O_2 and O_3. Because O_2 is the more common form, people generally mean it when they say "oxygen," although the name *dioxygen* is more descriptive. The O_3 form is **ozone**. The two forms of oxygen are examples of *allotropes*, defined as different forms of the same element. About 21% of dry air consists of O_2 molecules. Ozone is present in very small amounts in the upper atmosphere and in polluted air. It is also formed from O_2 in electrical discharges, such as in lightning storms:

$$3\,O_2(g) \longrightarrow 2\,O_3(g) \qquad \Delta H° = 284.6 \text{ kJ} \qquad [7.28]$$

This reaction is strongly endothermic, telling us that O_3 is less stable than O_2.

Table 7.6 Some Properties of the Group 6A Elements

Element	Electron Configuration	Melting Point (°C)	Density	Atomic Radius (Å)	I_1 (kJ/mol)
Oxygen	$[He]2s^2 2p^4$	−218	1.43 g/L	0.66	1314
Sulfur	$[Ne]3s^2 3p^4$	115	1.96 g/cm³	1.05	1000
Selenium	$[Ar]3d^{10}4s^2 4p^4$	221	4.82 g/cm³	1.20	941
Tellurium	$[Kr]4d^{10}5s^2 5p^4$	450	6.24 g/cm³	1.38	869
Polonium	$[Xe]4f^{14}5d^{10}6s^2 6p^4$	254	9.20 g/cm³	1.40	812

GO FIGURE

Why is it okay to store water in a bottle with a normal, nonventing cap?

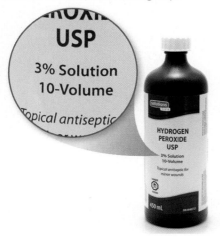

▲ Figure 7.26 **Hydrogen peroxide solution in a bottle with venting cap.**

GO FIGURE

Suppose it were possible to flatten the S_8 ring. What shape would you expect the flattened ring to have?

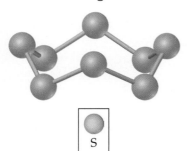

▲ Figure 7.27 **Elemental sulfur exists as the S_8 molecule.** At room temperature, this is the most common allotropic form of sulfur.

Although both O_2 and O_3 are colorless and therefore do not absorb visible light, O_3 absorbs certain wavelengths of ultraviolet light that O_2 does not. Because of this difference, the presence of ozone in the upper atmosphere is beneficial, filtering out harmful UV light. Ozone and oxygen also have different chemical properties. Ozone, which has a pungent odor, is a powerful oxidizing agent. Because of this property, ozone is sometimes added to water to kill bacteria or used in low levels to help to purify air. However, the reactivity of ozone also makes its presence in polluted air near Earth's surface detrimental to human health.

Oxygen has a great tendency to attract electrons from other elements (to *oxidize* them). Oxygen in combination with a metal is almost always present as the oxide ion, O^{2-}. This ion has a noble-gas configuration and is particularly stable. As shown in Figure 5.14, the formation of nonmetal oxides is also often very exothermic and thus energetically favorable.

In our discussion of the alkali metals, we noted two less common oxygen anions—the peroxide (O_2^{2-}) ion and the superoxide (O_2^{-}) ion. Compounds of these ions often react to produce an oxide and O_2:

$$2\,H_2O_2(aq) \longrightarrow 2\,H_2O(l) + O_2(g) \quad \Delta H° = -196.1\,kJ \qquad [7.29]$$

For this reason, bottles of aqueous hydrogen peroxide are topped with caps that are able to release the $O_2(g)$ produced before the pressure inside becomes too great (◀ Figure 7.26).

 Give It Some Thought

Hydrogen peroxide is light sensitive and so is stored in brown bottles because its O—O bond is relatively weak. If we assume that the brown bottle absorbs all visible wavelengths of light ∞ (Section 6.1), how might you estimate the energy of the O—O bond in hydrogen peroxide?

After oxygen, the most important member of group 6A is sulfur. This element exists in several allotropic forms, the most common and stable of which is the yellow solid having the molecular formula S_8. This molecule consists of an eight-membered ring of sulfur atoms (◀ Figure 7.27). Even though solid sulfur consists of S_8 rings, we usually write it simply as S(s) in chemical equations to simplify the stoichiometric coefficients.

Like oxygen, sulfur has a tendency to gain electrons from other elements to form sulfides, which contain the S^{2-} ion. In fact, most sulfur in nature is present as metal sulfides. Sulfur is below oxygen in the periodic table, and the tendency of sulfur to form sulfide anions is not as great as that of oxygen to form oxide ions. As a result, the chemistry of sulfur is more complex than that of oxygen. In fact, sulfur and its compounds (including those in coal and petroleum) can be burned in oxygen. The main product is sulfur dioxide, a major air pollutant:

$$S(s) + O_2(g) \longrightarrow SO_2(g) \qquad [7.30]$$

Below sulfur in group 6A is selenium, Se. This relatively rare element is essential for life in trace quantities, although it is toxic at high doses. There are many allotropes of Se, including several eight-membered ring structures that resemble the S_8 ring.

The next element in the group is tellurium, Te. Its elemental structure is even more complex than that of Se, consisting of long, twisted chains of Te—Te bonds. Both Se and Te favor the −2 oxidation state, as do O and S.

From O to S to Se to Te, the elements form larger and larger molecules and become increasingly metallic. The thermal stability of group 6A compounds with hydrogen decreases down the column: $H_2O > H_2S > H_2Se > H_2Te$, with H_2O, water, being the most stable of the series.

Group 7A: The Halogens

Some of the properties of the group 7A elements, the **halogens**, are given in ▶ Table 7.7. Astatine, which is both extremely rare and radioactive, is omitted because many of its properties are not yet known.

Table 7.7 Some Properties of the Halogens

Element	Electron Configuration	Melting Point (°C)	Density	Atomic Radius (Å)	I_1 (kJ/mol)
Fluorine	$[\text{He}]2s^2 2p^5$	−220	1.69 g/L	0.57	1681
Chlorine	$[\text{Ne}]3s^2 3p^5$	−102	3.12 g/L	1.02	1251
Bromine	$[\text{Ar}]4s^2 3d^{10} 4p^5$	−7.3	3.12 g/cm³	1.20	1140
Iodine	$[\text{Kr}]5s^2 4d^{10} 5p^5$	114	4.94 g/cm³	1.39	1008

Unlike the group 6A elements, all the halogens are typical nonmetals. Their melting and boiling points increase with increasing atomic number. Fluorine and chlorine are gases at room temperature, bromine is a liquid, and iodine is a solid. Each element consists of diatomic molecules: F_2, Cl_2, Br_2, and I_2 (▼ Figure 7.28).

 Give It Some Thought

The halogens do not exist as X_8 molecules like sulfur and selenium do. Can you speculate why?

The halogens have highly negative electron affinities (Figure 7.12). Thus, it is not surprising that the chemistry of the halogens is dominated by their tendency to gain electrons from other elements to form halide ions, X^-. (In many equations X is used to indicate any one of the halogen elements.) Fluorine and chlorine are more reactive than bromine and iodine. In fact, fluorine removes electrons from almost any substance with

 GO FIGURE

Why are more molecules of I_2 seen in the molecular view relative to the number of Cl_2 molecules?

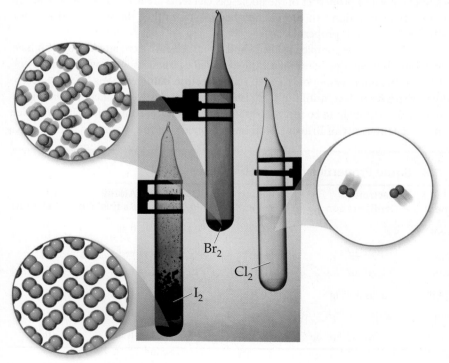

▲ Figure 7.28 The elemental halogens exist as diatomic molecules.

which it comes into contact, including water, and usually does so very exothermically, as in the following examples:

$$2\,H_2O(l) + 2\,F_2(g) \longrightarrow 4\,HF(aq) + O_2(g) \qquad \Delta H = -758.9\,kJ \qquad [7.31]$$

$$SiO_2(s) + 2\,F_2(g) \longrightarrow SiF_4(g) + O_2(g) \qquad \Delta H = -704.0\,kJ \qquad [7.32]$$

As a result, fluorine gas is difficult and dangerous to use in the laboratory, requiring specialized equipment.

Chlorine is the most industrially useful of the halogens. The annual production of chlorine is more than 20 billion pounds, making it one of the top ten most produced chemicals in the United States. ∞ (Section 1.1) Unlike fluorine, chlorine reacts slowly with water to form relatively stable aqueous solutions of HCl and HOCl (hypochlorous acid):

$$Cl_2(g) + H_2O(l) \longrightarrow HCl(aq) + HOCl(aq) \qquad [7.33]$$

Chlorine is often added to drinking water and swimming pools because the HOCl(aq) that is generated serves as a disinfectant.

The halogens react directly with most metals to form ionic halides. The halogens also react with hydrogen to form gaseous hydrogen halide compounds:

$$H_2(g) + X_2 \longrightarrow 2\,HX(g) \qquad [7.34]$$

These compounds are all very soluble in water and dissolve to form the hydrohalic acids. As we discussed in Section 4.3, HCl(aq), HBr(aq), and HI(aq) are strong acids, whereas HF(aq) is a weak acid.

 Give It Some Thought

Can you use data in Table 7.7 to estimate the atomic radius and first ionization energy of an astatine atom?

Group 8A: The Noble Gases

The group 8A elements, known as the **noble gases**, are all nonmetals that are gases at room temperature. They are all *monatomic* (that is, they consist of single atoms rather than molecules). Some physical properties of the noble-gas elements are listed in ▼ Table 7.8. The high radioactivity of radon (Rn, atomic number 86) has limited the study of its reaction chemistry and some of its properties.

The noble gases have completely filled *s* and *p* subshells. All elements of group 8A have large first ionization energies, and we see the expected decrease as we move down the column. Because the noble gases possess such stable electron configurations, they are exceptionally unreactive. In fact, until the early 1960s the elements were called the *inert gases* because they were thought to be incapable of forming chemical compounds. In 1962, Neil Bartlett at the University of British Columbia reasoned that the ionization energy of Xe

Table 7.8 Some Properties of the Noble Gases

Element	Electron Configuration	Boiling Point (K)	Density (g/L)	Atomic Radius* (Å)	I_1 (kJ/mol)
Helium	$1s^2$	4.2	0.18	0.28	2372
Neon	$[He]2s^22p^6$	27.1	0.90	0.58	2081
Argon	$[Ne]3s^23p^6$	87.3	1.78	1.06	1521
Krypton	$[Ar]4s^23d^{10}4p^6$	120	3.75	1.16	1351
Xenon	$[Kr]5s^24d^{10}5p^6$	165	5.90	1.40	1170
Radon	$[Xe]6s^24f^{14}5d^{10}6p^6$	211	9.73	1.50	1037

*Only the heaviest of the noble-gas elements form chemical compounds. Thus, the atomic radii for the lighter noble gas elements are estimated values.

might be low enough to allow it to form compounds. For this to happen, Xe would have to react with a substance with an extremely high ability to remove electrons from other substances, such as fluorine. Bartlett synthesized the first noble-gas compound by combining Xe with the fluorine-containing compound PtF_6. Xenon also reacts directly with $F_2(g)$ to form the molecular compounds XeF_2, XeF_4, and XeF_6. Krypton has a higher I_1 value than xenon and is therefore less reactive. In fact, only a single stable compound of krypton is known, KrF_2. In 2000, Finnish scientists reported the first neutral molecule that contains argon, the HArF molecule, which is stable only at low temperatures.

SAMPLE INTEGRATIVE EXERCISE   Putting Concepts Together

The element bismuth (Bi, atomic number 83) is the heaviest member of group 5A. A salt of the element, bismuth subsalicylate, is the active ingredient in Pepto-Bismol®, an over-the-counter medication for gastric distress.

(a) Based on values presented in Figure 7.7 and Tables 7.5 and 7.6, what might you expect for the bonding atomic radius of bismuth?

(b) What accounts for the general increase in atomic radius going down the group 5A elements?

(c) Another major use of bismuth has been as an ingredient in low-melting metal alloys, such as those used in fire sprinkler systems and in typesetting. The element itself is a brittle white crystalline solid. How do these characteristics fit with the fact that bismuth is in the same periodic group with such nonmetallic elements as nitrogen and phosphorus?

(d) Bi_2O_3 is a basic oxide. Write a balanced chemical equation for its reaction with dilute nitric acid. If 6.77 g of Bi_2O_3 is dissolved in dilute acidic solution to make 0.500 L of solution, what is the molarity of the solution of Bi^{3+} ion?

(e) ^{209}Bi is the heaviest stable isotope of any element. How many protons and neutrons are present in this nucleus?

(f) The density of Bi at 25 °C is 9.808 g/cm^3. How many Bi atoms are present in a cube of the element that is 5.00 cm on each edge? How many moles of the element are present?

SOLUTION

(a) Bismuth is directly below antimony, Sb, in group 5A. Based on the observation that atomic radii increase as we go down a column, we would expect the radius of Bi to be greater than that of Sb, which is 1.39 Å. We also know that atomic radii generally decrease as we proceed from left to right in a period. Tables 7.5 and 7.6 each give an element in the same period, namely Ba and Po. We would therefore expect that the radius of Bi is smaller than that of Ba (2.15 Å) and larger than that of Po (1.40 Å). We also see that in other periods, the difference in radius between the neighboring group 5A and group 6A elements is relatively small. We might therefore expect that the radius of Bi is slightly larger than that of Po—much closer to the radius of Po than to the radius of Ba. The tabulated value for the atomic radius on Bi is 1.48 Å, in accord with our expectations.

(b) The general increase in radius with increasing atomic number in the group 5A elements occurs because additional shells of electrons are being added, with corresponding increases in nuclear charge. The core electrons in each case largely screen the outermost electrons from the nucleus, so the effective nuclear charge does not vary greatly as we go to higher atomic numbers. However, the principal quantum number, n, of the outermost electrons steadily increases, with a corresponding increase in orbital radius.

(c) The contrast between the properties of bismuth and those of nitrogen and phosphorus illustrates the general rule that there is a trend toward increased metallic character as we move down in a given group. Bismuth, in fact, is a metal. The increased metallic character occurs because the outermost electrons are more readily lost in bonding, a trend that is consistent with its lower ionization energy.

(d) Following the procedures described in Section 4.2 for writing molecular and net ionic equations, we have the following:

Molecular equation: $Bi_2O_3(s) + 6\,HNO_3(aq) \longrightarrow 2\,Bi(NO_3)_3(aq) + 3\,H_2O(l)$

Net ionic equation: $Bi_2O_3(s) + 6\,H^+(aq) \longrightarrow 2\,Bi^{3+}(aq) + 3\,H_2O(l)$

In the net ionic equation, nitric acid is a strong acid and $Bi(NO_3)_3$ is a soluble salt, so we need to show only the reaction of the solid with the hydrogen ion forming the $Bi^{3+}(aq)$ ion and water. To calculate the concentration of the solution, we proceed as follows (Section 4.5):

$$\frac{6.77\text{ g }Bi_2O_3}{0.500\text{ L soln}} \times \frac{1\text{ mol }Bi_2O_3}{466.0\text{ g }Bi_2O_3} \times \frac{2\text{ mol }Bi^{3+}}{1\text{ mol }Bi_2O_3} = \frac{0.0581\text{ mol }Bi^{3+}}{\text{L soln}} = 0.0581\ M$$

(e) Recall that the atomic number of any element is the number of protons and electrons in a neutral atom of the element. ∞ (Section 2.3) Bismuth is element 83; there are therefore 83 protons in the nucleus. Because the atomic mass number is 209, there are $209 - 83 = 126$ neutrons in the nucleus.

(f) We can use the density and the atomic weight to determine the number of moles of Bi, and then use Avogadro's number to convert the result to the number of atoms. ∞ (Sections 1.4 and 3.4) The volume of the cube is $(5.00)^3 \, cm^3 = 125 \, cm^3$. Then we have

$$125 \, cm^3 \, Bi \times \frac{9.808 \, g \, Bi}{1 \, cm^3} \times \frac{1 \, mol \, Bi}{209.0 \, g \, Bi} = 5.87 \, mol \, Bi$$

$$5.87 \, mol \, Bi \times \frac{6.022 \times 10^{23} \, atom \, Bi}{1 \, mol \, Bi} = 3.53 \times 10^{24} \, atoms \, Bi$$

Chapter Summary and Key Terms

DEVELOPMENT OF THE PERIODIC TABLE (INTRODUCTION AND SECTION 7.1) The periodic table was first developed by Mendeleev and Meyer on the basis of the similarity in chemical and physical properties exhibited by certain elements. Moseley established that each element has a unique atomic number, which added more order to the periodic table.

We now recognize that elements in the same column of the periodic table have the same number of electrons in their **valence orbitals.** This similarity in valence electronic structure leads to the similarities among elements in the same group. The differences among elements in the same group arise because their valence orbitals are in different shells.

EFFECTIVE NUCLEAR CHARGE (SECTION 7.2) Many properties of atoms depend on the **effective nuclear charge**, which is the portion of the nuclear charge that an outer electron experiences after accounting for repulsions by other electrons in the atom. The core electrons are very effective in screening the outer electrons from the full charge of the nucleus, whereas electrons in the same shell do not screen each other as effectively. Because the actual nuclear charge increases as we progress through a period, the effective nuclear charge experienced by valence electrons increases as we move left to right across a period.

SIZES OF ATOMS AND IONS (SECTION 7.3) The size of an atom can be gauged by its **bonding atomic radius,** which is based on measurements of the distances separating atoms in their chemical compounds. In general, atomic radii increase as we go down a column in the periodic table and decrease as we proceed left to right across a row.

Cations are smaller than their parent atoms; anions are larger than their parent atoms. For ions of the same charge, size increases going down a column of the periodic table. An **isoelectronic series** is a series of ions that has the same number of electrons. For such a series, size decreases with increasing nuclear charge as the electrons are attracted more strongly to the nucleus.

IONIZATION ENERGY (SECTION 7.4) The first **ionization energy** of an atom is the minimum energy needed to remove an electron from the atom in the gas phase, forming a cation. The second ionization energy is the energy needed to remove a second electron, and so forth. Ionization energies show a sharp increase after all the valence electrons have been removed because of the much higher effective nuclear charge experienced by the core electrons. The first ionization energies of the elements show periodic trends that are opposite those seen for atomic radii, with smaller atoms having higher first ionization energies. Thus, first ionization energies decrease as we go down a column and increase as we proceed left to right across a row.

We can write electron configurations for ions by first writing the electron configuration of the neutral atom and then removing or adding the appropriate number of electrons. For cations, electrons are removed first from the orbitals of the neutral atom with the largest value of n. If there are two valence orbitals with the same value of n (such as $4s$ and $4p$), then the electrons are lost first from the orbital with a higher value of l (in this case, $4p$). For anions, electrons are added to orbitals in the reverse order.

ELECTRON AFFINITY (SECTION 7.5) The **electron affinity** of an element is the energy change upon adding an electron to an atom in the gas phase, forming an anion. A negative electron affinity means that energy is released when the electron is added; hence, when the electron affinity is negative the anion is stable. By contrast, a positive electron affinity means that the anion is not stable relative to the separated atom and electron, in which case its exact value cannot be measured. In general, electron affinities become more negative as we proceed from left to right across the periodic table. The halogens have the most negative electron affinities. The electron affinities of the noble gases are positive because the added electron would have to occupy a new, higher-energy subshell.

METALS, NONMETALS, AND METALLOIDS (SECTION 7.6) The elements can be categorized as metals, nonmetals, and metalloids. Most elements are metals; they occupy the left side and the middle of the periodic table. Nonmetals appear in the upper-right section of the table. Metalloids occupy a narrow band between the metals and nonmetals. The tendency of an element to exhibit the properties of metals, called the **metallic character,** increases as we proceed down a column and decreases as we proceed from left to right across a row.

Metals have a characteristic luster, and they are good conductors of heat and electricity. When metals react with nonmetals, the metal atoms are oxidized to cations and ionic substances are generally formed. Most metal oxides are basic; they react with acids to form salts and water.

Nonmetals lack metallic luster and are generally poor conductors of heat and electricity. Several are gases at room temperature. Compounds composed entirely of nonmetals are generally molecular. Nonmetals usually form anions in their reactions with metals. Nonmetal oxides are acidic; they react with bases to form salts and water. Metalloids have properties that are intermediate between those of metals and nonmetals.

TRENDS FOR GROUP 1A AND GROUP 2A METALS (SECTION 7.7) The periodic properties of the elements can help us understand the properties of groups of the representative elements. The **alkali metals**

(group 1A) are soft metals with low densities and low melting points. They have the lowest ionization energies of the elements. As a result, they are very reactive toward nonmetals, easily losing their outer *s* electron to form 1+ ions.

The **alkaline earth metals** (group 2A) are harder and denser, and have higher melting points than the alkali metals. They are also very reactive toward nonmetals, although not as reactive as the alkali metals. The alkaline earth metals readily lose their two outer *s* electrons to form 2+ ions. Both alkali and alkaline earth metals react with hydrogen to form ionic substances that contain the **hydride ion,** H^-.

TRENDS FOR SELECTED NONMETALS (SECTION 7.8) Hydrogen is a nonmetal with properties that are distinct from any of the groups of the periodic table. It forms molecular compounds with other non-metals, such as oxygen and the halogens.

Oxygen and sulfur are the most important elements in group 6A. Oxygen is usually found as a diatomic molecule, O_2. **Ozone,** O_3, is an important allotrope of oxygen. Oxygen has a strong tendency to gain electrons from other elements, thus oxidizing them. In combination with metals, oxygen is usually found as the oxide ion, O^{2-}, although salts of the peroxide ion, O_2^{2-}, and superoxide ion, O_2^-, are sometimes formed. Elemental sulfur is most commonly found as S_8 molecules. In combination with metals, it is most often found as the sulfide ion, S^{2-}.

The **halogens** (group 7A) exist as diatomic molecules. The halogens have the most negative electron affinities of the elements. Thus, their chemistry is dominated by a tendency to form 1− ions, especially in reactions with metals.

The **noble gases** (group 8A) exist as monatomic gases. They are very unreactive because they have completely filled *s* and *p* subshells. Only the heaviest noble gases are known to form compounds, and they do so only with very active nonmetals, such as fluorine.

Learning Outcomes   After studying this chapter, you should be able to:

- Explain the meaning of effective nuclear charge, Z_{eff}, and how Z_{eff} depends on nuclear charge and electron configuration. (Section 7.2)

- Predict the trends in atomic radii, ionic radii, ionization energy, and electron affinity by using the periodic table. (Sections 7.2, 7.3, 7.4, and 7.5)

- Explain how the radius of an atom changes upon losing electrons to form a cation or gaining electrons to form an anion. (Section 7.3)

- Write the electron configurations of ions. (Section 7.3)

- Explain how the ionization energy changes as we remove successive electrons, and the jump in ionization energy that occurs when the ionization corresponds to removing a core electron. (Section 7.4)

- Explain how irregularities in the periodic trends for electron affinity can be related to electron configuration. (Section 7.5)

- Explain the differences in chemical and physical properties of metals and nonmetals, including the basicity of metal oxides and the acidity of nonmetal oxides. (Section 7.6)

- Correlate atomic properties, such as ionization energy, with electron configuration, and explain how these relate to the chemical reactivity and physical properties of the alkali and alkaline earth metals (groups 1A and 2A). (Section 7.7)

- Write balanced equations for the reactions of the group 1A and 2A metals with water, oxygen, hydrogen, and the halogens. (Sections 7.7 and 7.8)

- List and explain the unique characteristics of hydrogen. (Section 7.7)

- Correlate the atomic properties (such as ionization energy, electron configuration, and electron affinity) of group 6A, 7A, and 8A elements with their chemical reactivity and physical properties. (Section 7.8)

Key Equations

- $Z_{eff} = Z - S$ [7.1]   Estimating effective nuclear charge

Exercises

Visualizing Concepts

7.1 As discussed in the text, we can draw an analogy between the attraction of an electron to a nucleus and the act of perceiving light from a lightbulb—in essence, the more nuclear charge the electron "sees," the greater the attraction. **(a)** Using this analogy, discuss how screening by core electrons is analogous to putting a frosted-glass lampshade between the lightbulb and your eyes, as shown in the illustration.

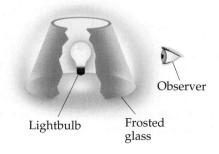

Observer

Lightbulb   Frosted glass

(b) Explain how we could mimic moving to the right in a row of the periodic table by changing the wattage of the lightbulb. (c) How would you change the wattage of the bulb and/or change the frosted glass to mimic the effect of moving down a column of the periodic table? [Section 7.2]

7.2 Which of these spheres represents F, which represents Br, and which represents Br⁻? [Section 7.3]

7.3 Consider the Mg^{2+}, Cl^-, K^+, and Se^{2-} ions. The four spheres below represent these four ions, scaled according to ionic size. (a) Without referring to Figure 7.8, match each ion to its appropriate sphere. (b) In terms of size, between which of the spheres would you find the (i) Ca^{2+} and (ii) S^{2-} ions? [Section 7.3]

7.4 In the following reaction

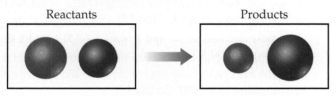

which sphere represents a metal and which represents a nonmetal? Explain your answer. [Section 7.3]

7.5 Consider the A_2X_4 molecule depicted here, where A and X are elements. The A—A bond length in this molecule is d_1, and the four A—X bond lengths are each d_2. (a) In terms of d_1 and d_2, how could you define the bonding atomic radii of atoms A and X? (b) In terms of d_1 and d_2, what would you predict for the X—X bond length of an X_2 molecule? [Section 7.3]

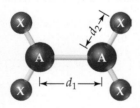

7.6 Shown below is a qualitative diagram of the atomic orbital energies for an Na atom. The number of orbitals in each subshell is not shown.

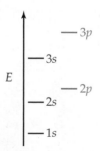

(a) Are all of the subshells for $n = 1$, $n = 2$, and $n = 3$ shown? If not, what is missing?

(b) The 2s and 2p energy levels are shown as different. Which of the following is the best explanation for why this is the case? (i) The 2s and 2p energy levels have different energies in the hydrogen atom, so of course they will have different energies in the sodium atom, (ii) The energy of the 2p orbital is higher than that of the 2s in all many-electron atoms. (iii) The 2s level in Na has electrons in it, whereas the 2p does not.

(c) Which of the energy levels holds the highest-energy electron in a sodium atom?

(d) A sodium vapor lamp (Figure 7.23) operates by using electricity to excite the highest-energy electron to the next highest-energy level. Light is produced when the excited electron drops back to the lower level. Which two energy levels are involved in this process for the Na atom? [Section 7.7]

7.7 (a) Which of the following charts below shows the general periodic trends for each of the following properties of the main-group elements (you can neglect small deviations going either across a row or down a column of the periodic table)? (1) Bonding atomic radius, (2) first ionization energy, and (3) metallic character. (b) Do any of the charts show the general periodic trends in the electron affinities of the main-group elements? [Sections 7.2–7.6]

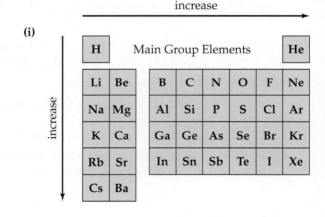

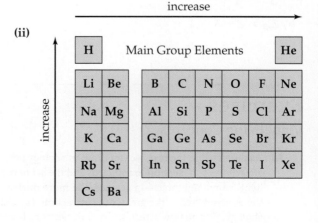

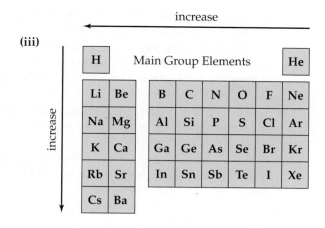

(iii)

increase

H   Main Group Elements   He

Li Be   B   C   N   O   F   Ne
Na Mg   Al Si P   S   Cl Ar
K   Ca   Ga Ge As Se Br Kr
Rb Sr   In Sn Sb Te I   Xe
Cs Ba

increase

(iv)

increase

H   Main Group Elements   He

Li Be   B   C   N   O   F   Ne
Na Mg   Al Si P   S   Cl Ar
K   Ca   Ga Ge As Se Br Kr
Rb Sr   In Sn Sb Te I   Xe
Cs Ba

7.8 An element X reacts with $F_2(g)$ to form the molecular product shown here. (**a**) Write a balanced equation for this reaction (do not worry about the phases for X and the product). (**b**) Do you think that X is a metal or nonmetal? Explain. [Section 7.6]

Periodic Table; Effective Nuclear Charge (Sections 7.1 and 7.2)

7.9 (**a**) Evaluate the expressions 2×1, $2 \times (1 + 3)$, $2 \times (1 + 3 + 5)$, and $2 \times (1 + 3 + 5 + 7)$. (**b**) How do the atomic numbers of the noble gases relate to the numbers from part (a)? (**c**) What topic discussed in Chapter 6 is the source of the number "2" in the expressions in part (a)?

7.10 The prefix *eka-* comes from the Sanskrit word for "one." Mendeleev used this prefix to indicate that the unknown element was one place away from the known element that followed the prefix. For example, *eka-silicon*, which we now call germanium, is one element below silicon. Mendeleev also predicted the existence of *eka-manganese*, which was not experimentally confirmed until 1937 because this element is radioactive and does not occur in nature. Based on the periodic table shown in Figure 7.1, what do we now call the element Mendeleev called *eka-manganese*.

7.11 You might have expected that the elements would have been discovered in order of their relative abundance in the Earth's crust (Figure 1.6), but this is not the case. Suggest a general reason.

7.12 (**a**) Moseley's experiments on X rays emitted from atoms led to the concept of atomic numbers. Where exactly do these X rays come from? Draw an energy-level diagram to explain. (**b**) Why are chemical and physical properties of the elements more closely related to atomic numbers than they are to atomic weights?

7.13 (**a**) What is meant by the term *effective nuclear charge*? (**b**) How does the effective nuclear charge experienced by the valence electrons of an atom vary going from left to right across a period of the periodic table?

7.14 Which of the following statements about effective nuclear charge for the outermost valence electron of an atom is *incorrect*? (**i**) The effective nuclear charge can be thought of as the true nuclear charge minus a screening constant due to the other electrons in the atom. (**ii**) Effective nuclear charge increases going left to right across a row of the periodic table. (**iii**) Valence electrons screen the nuclear charge more effectively than do core electrons. (**iv**) The effective nuclear charge shows a sudden decrease when we go from the end of one row to the beginning of the next row of the periodic table. (**v**) The change in effective nuclear charge going down a column of the periodic table is generally less than that going across a row of the periodic table.

7.15 Detailed calculations show that the value of Z_{eff} for the outermost electrons in Na and K atoms is 2.51+ and 3.49+, respectively. (**a**) What value do you estimate for Z_{eff} experienced by the outermost electron in both Na and K by assuming core electrons contribute 1.00 and valence electrons contribute 0.00 to the screening constant? (**b**) What values do you estimate for Z_{eff} using Slater's rules? (**c**) Which approach gives a more accurate estimate of Z_{eff}? (**d**) Does either method of approximation account for the gradual increase in Z_{eff} that occurs upon moving down a group? (**e**) Predict Z_{eff} for the outermost electrons in the Rb atom based on the calculations for Na and K.

7.16 Detailed calculations show that the value of Z_{eff} for the outermost electrons in Si and Cl atoms is 4.29+ and 6.12+, respectively. (**a**) What value do you estimate for Z_{eff} experienced by the outermost electron in both Si and Cl by assuming core electrons contribute 1.00 and valence electrons contribute 0.00 to the screening constant? (**b**) What values do you estimate for Z_{eff} using Slater's rules? (**c**) Which approach gives a more accurate estimate of Z_{eff}? (**d**) Which method of approximation more accurately accounts for the steady increase in Z_{eff} that occurs upon moving left to right across a period? (**e**) Predict Z_{eff} for a valence electron in P, phosphorus, based on the calculations for Si and Cl.

7.17 Which will experience the greater effective nuclear charge, the electrons in the $n = 3$ shell in Ar or the $n = 3$ shell in Kr? Which will be closer to the nucleus?

7.18 Arrange the following atoms in order of increasing effective nuclear charge experienced by the electrons in the $n = 3$ electron shell: K, Mg, P, Rh, Ti.

Atomic and Ionic Radii (Section 7.3)

7.19 (**a**) Because an exact outer boundary cannot be measured or even calculated for an atom, how are atomic radii determined? (**b**) What is the difference between a bonding radius and a nonbonding radius? (**c**) For a given element, which one is larger? (**d**) If a free atom reacts to become part of a molecule, would you say that the atom gets smaller or larger?

7.20 (a) Why does the quantum mechanical description of many-electron atoms make it difficult to define a precise atomic radius? (b) When nonbonded atoms come up against one another, what determines how closely the nuclear centers can approach?

7.21 Tungsten has the highest melting point of any metal in the periodic table: 3422 °C. The distance between W atoms in tungsten metal is 2.74 Å. (a) What is the atomic radius of a tungsten atom in this environment? (This radius is called the *metallic radius*.) (b) If you put tungsten metal under high pressure, predict what would happen to the distance between W atoms.

7.22 Which of the following statements about the bonding atomic radii in Figure 7.7 is *incorrect*? (i) For a given period, the radii of the representative elements generally decrease from left to right across period. (ii) The radii of the representative elements for the $n = 3$ period are all larger than those of the corresponding elements in the $n = 2$ period. (iii) For most of the representative elements, the change in radius from the $n = 2$ to the $n = 3$ period is greater than the change in radius from $n = 3$ to $n = 4$. (iv) The radii of the transition elements generally increase moving from left to right within a period. (v) The large radii of the Group 1A elements are due to their relatively small effective nuclear charges.

7.23 Estimate the As—I bond length from the data in Figure 7.7 and compare your value to the experimental As—I bond length in arsenic triiodide, AsI_3, 2.55 Å.

7.24 The experimental Bi—I bond length in bismuth triiodide, BiI_3, is 2.81 Å. Based on this value and data in Figure 7.7, predict the atomic radius of Bi.

7.25 Using only the periodic table, arrange each set of atoms in order from largest to smallest: (a) K, Li, Cs; (b) Pb, Sn, Si; (c) F, O, N.

7.26 Using only the periodic table, arrange each set of atoms in order of increasing radius: (a) Ba, Ca, Na; (b) In, Sn, As; (c) Al, Be, Si.

7.27 Identify each statement as true or false: (a) Cations are larger than their corresponding neutral atoms. (b) Li^+ is smaller than Li. (c) Cl^- is bigger than I^-.

7.28 Explain the following variations in atomic or ionic radii:
(a) $I^- > I > I^+$
(b) $Ca^{2+} > Mg^{2+} > Be^{2+}$
(c) $Fe > Fe^{2+} > Fe^{3+}$

7.29 Which neutral atom is isoelectronic with each of the following ions? Ga^{3+}, Zr^{4+}, Mn^{7+}, I^-, Pb^{2+}.

7.30 Some ions do not have a corresponding neutral atom that has the same electron configuration. For each of the following ions, identify the neutral atom that has the same number of electrons and determine if this atom has the same electron configuration. If such an atom does not exist, explain why. (a) Cl^-, (b) Sc^{3+}, (c) Fe^{2+}, (d) Zn^{2+}, (e) Sn^{4+}.

7.31 Consider the isoelectronic ions F^- and Na^+. (a) Which ion is smaller? (b) Using Equation 7.1 and assuming that core electrons contribute 1.00 and valence electrons contribute 0.00 to the screening constant, S, calculate Z_{eff} for the $2p$ electrons in both ions. (c) Repeat this calculation using Slater's rules to estimate the screening constant, S. (d) For isoelectronic ions, how are effective nuclear charge and ionic radius related?

7.32 Consider the isoelectronic ions Cl^- and K^+. (a) Which ion is smaller? (b) Using Equation 7.1 and assuming that core electrons contribute 1.00 and valence electrons contribute nothing to the screening constant, S, calculate Z_{eff} for these two ions. (c) Repeat this calculation using Slater's rules to estimate

the screening constant, S. (d) For isoelectronic ions, how are effective nuclear charge and ionic radius related?

7.33 Consider S, Cl, and K and their most common ions. (a) List the atoms in order of increasing size. (b) List the ions in order of increasing size. (c) Explain any differences in the orders of the atomic and ionic sizes.

7.34 Arrange each of the following sets of atoms and ions, in order of increasing size: (a) Se^{2-}, Te^{2-}, Se; (b) Co^{3+}, Fe^{2+}, Fe^{3+}; (c) Ca, Ti^{4+}, Sc^{3+}; (d) Be^{2+}, Na^+, Ne.

7.35 Provide a brief explanation for each of the following: (a) O^{2-} is larger than O. (b) S^{2-} is larger than O^{2-}. (c) S^{2-} is larger than K^+. (d) K^+ is larger than Ca^{2+}.

7.36 In the ionic compounds LiF, NaCl, KBr, and RbI, the measured cation–anion distances are 2.01 Å (Li–F), 2.82 Å (Na–Cl), 3.30 Å (K–Br), and 3.67 Å (Rb–I), respectively. (a) Predict the cation–anion distance using the values of ionic radii given in Figure 7.8. (b) Calculate the difference between the experimentally measured ion–ion distances and the ones predicted from Figure 7.8. Assuming we have an accuracy of 0.04 Å in the measurement, would you say that the two sets of ion–ion distances are the same or not? (c) What estimates of the cation–anion distance would you obtain for these four compounds using *bonding atomic radii*? Are these estimates as accurate as the estimates using ionic radii?

Ionization Energies; Electron Affinities (Sections 7.4 and 7.5)

7.37 Write equations that show the processes that describe the first, second, and third ionization energies of an aluminum atom. Which process would require the least amount of energy?

7.38 Write equations that show the process for (a) the first two ionization energies of lead and (b) the fourth ionization energy of zirconium.

7.39 (a) Why does Li have a larger first ionization energy than Na? (b) The difference between the third and fourth ionization energies of scandium is much larger than that of titanium. Why? (c) Why does Li have a much larger second ionization energy than Be?

7.40 Identify each statement as true or false: (a) Ionization energies are always negative quantities. (b) Oxygen has a larger first ionization energy than fluorine. (c) The second ionization energy of an atom is always greater than its first ionization energy. (d) The third ionization energy is the energy needed to ionize three electrons from a neutral atom.

7.41 (a) What is the general relationship between the size of an atom and its first ionization energy? (b) Which element in the periodic table has the largest ionization energy? Which has the smallest?

7.42 (a) What is the trend in first ionization energies as one proceeds down the group 7A elements? Explain how this trend relates to the variation in atomic radii. (b) What is the trend in first ionization energies as one moves across the fourth period from K to Kr? How does this trend compare with the trend in atomic radii?

7.43 Based on their positions in the periodic table, predict which atom of the following pairs will have the smaller first ionization energy: (a) Cl, Ar; (b) Be, Ca; (c) K, Co; (d) S, Ge; (e) Sn, Te.

7.44 For each of the following pairs, indicate which element has the smaller first ionization energy: (a) Ti, Ba; (b) Ag, Cu; (c) Ge, Cl; (d) Pb, Sb.

7.45 Write the electron configurations for the following ions, and determine which have noble-gas configurations: **(a)** Co^{2+}, **(b)** Sn^{2+}, **(c)** Zr^{4+}, **(d)** Ag^+, **(e)** S^{2-}.

7.46 Write the electron configurations for the following ions, and determine which have noble-gas configurations: **(a)** Ru^{3+}, **(b)** As^{3-}, **(c)** Y^{3+}, **(d)** Pd^{2+}, **(e)** Pb^{2+}, **(f)** Au^{3+}.

7.47 Find three examples of ions in the periodic table that have an electron configuration of nd^8 ($n = 3, 4, 5, \ldots$).

7.48 Find three atoms in the periodic table whose ions have an electron configuration of nd^6 ($n = 3, 4, 5, \ldots$).

7.49 The first ionization energy and electron affinity of Ar are both positive values. **(a)** What is the significance of the positive value in each case? **(b)** What are the units of electron affinity?

7.50 If the electron affinity for an element is a negative number, does it mean that the anion of the element is more stable than the neutral atom? Explain.

7.51 Although the electron affinity of bromine is a negative quantity, it is positive for Kr. Use the electron configurations of the two elements to explain the difference.

7.52 What is the relationship between the ionization energy of an anion with a 1− charge such as F^- and the electron affinity of the neutral atom, F?

7.53 Consider the first ionization energy of neon and the electron affinity of fluorine. **(a)** Write equations, including electron configurations, for each process. **(b)** These two quantities have opposite signs. Which will be positive, and which will be negative? **(c)** Would you expect the magnitudes of these two quantities to be equal? If not, which one would you expect to be larger?

7.54 Consider the following equation:

$$Ca^+(g) + e^- \longrightarrow Ca(g)$$

Which of the following statements are true? **(i)** The energy change for this process is the electron affinity of the Ca^+ ion. **(ii)** The energy change for this process is the negative of the first ionization energy of the Ca atom. **(iii)** The energy change for this process is the negative of the electron affinity of the Ca atom.

Properties of Metals and Nonmetals (Section 7.6)

7.55 **(a)** Does metallic character increase, decrease, or remain unchanged as one goes from left to right across a row of the periodic table? **(b)** Does metallic character increase, decrease, or remain unchanged as one goes down a column of the periodic table? **(c)** Are the periodic trends in (a) and (b) the same as or different from those for first ionization energy?

7.56 You read the following statement about two elements X and Y: One of the elements is a good conductor of electricity, and the other is a semiconductor. Experiments show that the first ionization energy of X is twice as great as that of Y. Which element has the greater metallic character?

7.57 Discussing this chapter, a classmate says, "An element that commonly forms a cation is a metal." Do you agree or disagree? Explain your answer.

7.58 Discussing this chapter, a classmate says, "Since elements that form cations are metals and elements that form anions are nonmetals, elements that do not form ions are metalloids." Do you agree or disagree? Explain your answer.

7.59 Predict whether each of the following oxides is ionic or molecular: SnO_2, Al_2O_3, CO_2, Li_2O, Fe_2O_3, H_2O.

7.60 Some metal oxides, such as Sc_2O_3, do not react with pure water, but they do react when the solution becomes either acidic or basic. Do you expect Sc_2O_3 to react when the solution becomes acidic or when it becomes basic? Write a balanced chemical equation to support your answer.

7.61 **(a)** What is meant by the terms *acidic oxide* and *basic oxide*? **(b)** How can we predict whether an oxide will be acidic or basic based on its composition?

7.62 Arrange the following oxides in order of increasing acidity: CO_2, CaO, Al_2O_3, SO_3, SiO_2, P_2O_5.

7.63 Chlorine reacts with oxygen to form Cl_2O_7. **(a)** What is the name of this product (see Table 2.6)? **(b)** Write a balanced equation for the formation of $Cl_2O_7(l)$ from the elements. **(c)** Under usual conditions, Cl_2O_7 is a colorless liquid with a boiling point of 81 °C. Is this boiling point expected or surprising? **(d)** Would you expect Cl_2O_7 to be more reactive toward $H^+(aq)$ or $OH^-(aq)$? **(e)** If the oxygen in Cl_2O_7 is considered to have the −2 oxidation state, what is the oxidation state of the Cl? What is the electron configuration of Cl in this oxidation state?

[7.64] An element X reacts with oxygen to form XO_2 and with chlorine to form XCl_4. XO_2 is a white solid that melts at high temperatures (above 1000 °C). Under usual conditions, XCl_4 is a colorless liquid with a boiling point of 58 °C. **(a)** XCl_4 reacts with water to form XO_2 and another product. What is the likely identity of the other product? **(b)** Do you think that element X is a metal, nonmetal, or metalloid? **(c)** By using a sourcebook such as the *CRC Handbook of Chemistry and Physics*, try to determine the identity of element X.

7.65 Write balanced equations for the following reactions: **(a)** barium oxide with water, **(b)** iron(II) oxide with perchloric acid, **(c)** sulfur trioxide with water, **(d)** carbon dioxide with aqueous sodium hydroxide.

7.66 Write balanced equations for the following reactions: **(a)** potassium oxide with water, **(b)** diphosphorus trioxide with water, **(c)** chromium(III) oxide with dilute hydrochloric acid, **(d)** selenium dioxide with aqueous potassium hydroxide.

Group Trends in Metals and Nonmetals (Sections 7.7 and 7.8)

7.67 **(a)** Why is calcium generally more reactive than magnesium? **(b)** Why is calcium generally less reactive than potassium?

7.68 Silver and rubidium both form +1 ions, but silver is far less reactive. Suggest an explanation, taking into account the ground-state electron configurations of these elements and their atomic radii.

7.69 Write a balanced equation for the reaction that occurs in each of the following cases: **(a)** Potassium metal is exposed to an atmosphere of chlorine gas. **(b)** Strontium oxide is added to water. **(c)** A fresh surface of lithium metal is exposed to oxygen gas. **(d)** Sodium metal reacts with molten sulfur.

7.70 Write a balanced equation for the reaction that occurs in each of the following cases: **(a)** Cesium is added to water. **(b)** Strontium is added to water. **(c)** Sodium reacts with oxygen. **(d)** Calcium reacts with iodine.

7.71 **(a)** As described in Section 7.7, the alkali metals react with hydrogen to form hydrides and react with halogens to form halides. Compare the roles of hydrogen and halogens in these reactions. How are the forms of hydrogen and halogens in the products alike? **(b)** Write balanced equations for the reaction of fluorine with calcium and for the reaction of hydrogen with calcium. What are the similarities among the products of these reactions?

7.72 Potassium and hydrogen react to form the ionic compound potassium hydride. (a) Write a balanced equation for this reaction. (b) Use data in Figures 7.10 and 7.12 to determine the energy change in kJ/mol for the following two reactions:

$$K(g) + H(g) \longrightarrow K^+(g) + H^-(g)$$
$$K(g) + H(g) \longrightarrow K^-(g) + H^+(g)$$

(c) Based on your calculated energy changes in (b), which of these reactions is energetically more favorable (or less unfavorable)? (d) Is your answer to (c) consistent with the description of potassium hydride as containing hydride ions?

7.73 Compare the elements bromine and chlorine with respect to the following properties: (a) electron configuration, (b) most common ionic charge, (c) first ionization energy, (d) reactivity toward water, (e) electron affinity, (f) atomic radius. Account for the differences between the two elements.

7.74 Little is known about the properties of astatine, At, because of its rarity and high radioactivity. Nevertheless, it is possible for us to make many predictions about its properties. (a) Do you expect the element to be a gas, liquid, or solid at room temperature? Explain. (b) Would you expect At to be a metal, nonmetal, or metalloid? Explain. (c) What is the chemical formula of the compound it forms with Na?

7.75 Until the early 1960s the group 8A elements were called the inert gases. (a) Why was the term *inert gases* dropped? (b) What discovery triggered this change in name? (c) What name is applied to the group now?

7.76 (a) Why does xenon react with fluorine, whereas neon does not? (b) Using appropriate reference sources, look up the bond lengths of Xe—F bonds in several molecules. How do these numbers compare to the bond lengths calculated from the atomic radii of the elements?

7.77 Write a balanced equation for the reaction that occurs in each of the following cases: (a) Ozone decomposes to dioxygen. (b) Xenon reacts with fluorine. (Write three different equations.) (c) Sulfur reacts with hydrogen gas. (d) Fluorine reacts with water.

7.78 Write a balanced equation for the reaction that occurs in each of the following cases: (a) Chlorine reacts with water. (b) Barium metal is heated in an atmosphere of hydrogen gas. (c) Lithium reacts with sulfur. (d) Fluorine reacts with magnesium metal.

Additional Exercises

7.79 Consider the stable elements through lead ($Z = 82$). In how many instances are the atomic weights of the elements out of order relative to the atomic numbers of the elements?

[7.80] Figure 7.4 shows the radial probability distribution functions for the 2s orbitals and 2p orbitals. (a) Which orbital, 2s or 2p, has more electron density close to the nucleus? (b) How would you modify Slater's rules to adjust for the difference in electronic penetration of the nucleus for the 2s and 2p orbitals?

7.81 (a) If the core electrons were totally effective at screening the valence electrons and the valence electrons provided no screening for each other, what would be the effective nuclear charge acting on the 3s and 3p valence electrons in P? (b) Repeat these calculations using Slater's rules. (c) Detailed calculations indicate that the effective nuclear charge is 5.6+ for the 3s electrons and 4.9+ for the 3p electrons. Why are the values for the 3s and 3p electrons different? (d) If you remove a single electron from a P atom, which orbital will it come from?

7.82 As we move across a period of the periodic table, why do the sizes of the transition elements change more gradually than those of the representative elements?

7.83 In the series of group 5A hydrides, of general formula MH_3, the measured bond distances are P—H, 1.419 Å; As—H, 1.519 Å; Sb—H, 1.707 Å. (a) Compare these values with those estimated by use of the atomic radii in Figure 7.7. (b) Explain the steady increase in M—H bond distance in this series in terms of the electron configurations of the M atoms.

7.84 In Table 7.8, the bonding atomic radius of neon is listed as 0.58 Å, whereas that for xenon is listed as 1.40 Å. A classmate of yours states that the value for Xe is more realistic than the one for Ne. Is she correct? If so, what is the basis for her statement?

7.85 The As—As bond length in elemental arsenic is 2.48 Å. The Cl—Cl bond length in Cl_2 is 1.99 Å. (a) Based on these data, what is the predicted As—Cl bond length in arsenic trichloride, $AsCl_3$, in which each of the three Cl atoms is bonded to the As atom? (b) What bond length is predicted for $AsCl_3$, using the atomic radii in Figure 7.7?

7.86 The following observations are made about two hypothetical elements A and B: The A—A and B—B bond lengths in elemental A and B are 2.36 and 1.94 Å, respectively. A and B react to form the binary compound AB_2, which has a *linear* structure (that is $\angle B - A - B = 180°$). Based on these statements, predict the separation between the two B nuclei in a molecule of AB_2.

7.87 Elements in group 7A in the periodic table are called the halogens; elements in group 6A are called the chalcogens. (a) What is the most common oxidation state of the chalcogens compared to the halogens? (b) For each of the following periodic properties, state whether the halogens or the chalcogens have larger values: atomic radii, ionic radii of the most common oxidation state, first ionization energy, second ionization energy.

7.88 Note from the following table that there is a significant increase in atomic radius upon moving from Y to La whereas the radii of Zr to Hf are the same. Suggest an explanation for this effect.

Atomic Radii (Å)			
Sc	1.70	Ti	1.60
Y	1.90	Zr	1.75
La	2.07	Hf	1.75

[7.89] (a) Which ion is smaller, Co^{3+} or Co^{4+}? (b) In a lithium-ion battery that is discharging to power a device, for every Li^+ that inserts into the lithium cobalt oxide electrode, a Co^{4+} ion must be reduced to a Co^{3+} ion to balance charge. Using the *CRC Handbook of Chemistry and Physics* or other standard reference, find the ionic radii of Li^+, Co^{3+}, and Co^{4+}. Order these ions from smallest to largest. (c) Will the lithium cobalt electrode expand or contract as lithium ions are inserted? (d) Lithium is not nearly as abundant as sodium. If sodium ion batteries were developed that function as lithium ion ones, do you think "sodium cobalt oxide" would still work as the electrode material? Explain. (e) If you don't think cobalt would work as the redox-active partner ion in the sodium version of the electrode, suggest an alternative metal ion and explain your reasoning.

[7.90] The ionic substance strontium oxide, SrO, forms from the reaction of strontium metal with molecular oxygen. The arrangement of the ions in solid SrO is analogous to that in solid NaCl:

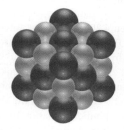

(a) Write a balanced equation for the formation of SrO(s) from its elements. (b) Based on the ionic radii in Figure 7.8, predict the length of the side of the cube in the figure (the distance from the center of an atom at one corner to the center of an atom at a neighboring corner). (c) The density of SrO is 5.10 g/cm^3. Given your answer to part (b), how many formula units of SrO are contained in the cube shown here?

7.91 Explain the variation in the ionization energies of carbon, as displayed in this graph:

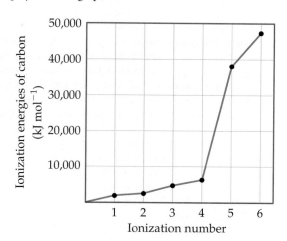

7.92 Group 4A elements have much more negative electron affinities than their neighbors in groups 3A and 5A (see Figure 7.12). Which of the following statements best explains this observation? (i) The group 4A elements have much higher first ionization energies than their neighbors in groups 3A and 5A. (ii) The addition of an electron to a group 4A element leads to a half-filled np^3 outer electron configuration. (iii) The group 4A elements have unusually large radii. (iv) The group 4A elements are easier to vaporize than are the group 3A and 5A elements.

7.93 In the chemical process called *electron transfer*, an electron is transferred from one atom or molecule to another. (We will talk about electron transfer extensively in Chapter 20.) A simple electron transfer reaction is

$$A(g) + A(g) \longrightarrow A^+(g) + A^-(g)$$

In terms of the ionization energy and electron affinity of atom A, what is the energy change for this reaction? For a representative nonmetal such as chlorine, is this process exothermic? For a representative metal such as sodium, is this process exothermic?

7.94 (a) Use orbital diagrams to illustrate what happens when an oxygen atom gains two electrons. (b) Why does O^{3-} not exist?

[7.95] Use electron configurations to explain the following observations: (a) The first ionization energy of phosphorus is greater than that of sulfur. (b) The electron affinity of nitrogen is lower (less negative) than those of both carbon and oxygen. (c) The second ionization energy of oxygen is greater than the first ionization energy of fluorine. (d) The third ionization energy of manganese is greater than those of both chromium and iron.

7.96 Identify two ions that have the following ground-state electron configurations: (a) [Ar], (b) [Ar]$3d^5$, (c) [Kr]$5s^2 4d^{10}$.

7.97 Which of the following chemical equations is connected to the definitions of (a) the first ionization energy of oxygen, (b) the second ionization energy of oxygen, and (c) the electron affinity of oxygen?

(i) $O(g) + e^- \longrightarrow O^-(g)$ (ii) $O(g) \longrightarrow O^+(g) + e^-$

(iii) $O(g) + 2e^- \longrightarrow O^{2-}(g)$ (iv) $O(g) \longrightarrow O^{2+}(g) + 2e^-$

(v) $O^+(g) \longrightarrow O^{2+}(g) + e^-$

7.98 The electron affinities, in kJ/mol, for the group 1B and group 2B metals are as follows:

Cu −119	Zn > 0
Ag −126	Cd > 0
Au −223	Hg > 0

(a) Why are the electron affinities of the group 2B elements greater than zero? (b) Why do the electron affinities of the group 1B elements become more negative as we move down the group? [*Hint:* Examine the trends in the electron affinities of other groups as we proceed down the periodic table.]

7.99 Hydrogen is an unusual element because it behaves in some ways like the alkali metal elements and in other ways like nonmetals. Its properties can be explained in part by its electron configuration and by the values for its ionization energy and electron affinity. (a) Explain why the electron affinity of hydrogen is much closer to the values for the alkali elements than for the halogens. (b) Is the following statement true? "Hydrogen has the smallest bonding atomic radius of any element that forms chemical compounds." If not, correct it. If it is, explain in terms of electron configurations. (c) Explain why the ionization energy of hydrogen is closer to the values for the halogens than for the alkali metals. (d) The hydride ion is H$^-$. Write out the process corresponding to the first ionization energy of the hydride ion. (e) How does the process in part (d) compare to the process for the electron affinity of a neutral hydrogen atom?

[7.100] The first ionization energy of the oxygen molecule is the energy required for the following process:

$$O_2(g) \longrightarrow O_2^+(g) + e^-$$

The energy needed for this process is 1175 kJ/mol, very similar to the first ionization energy of Xe. Would you expect O_2 to react with F_2? If so, suggest a product or products of this reaction.

[7.101] It is possible to define *metallic character* as we do in this book and base it on the reactivity of the element and the ease with which it loses electrons. Alternatively, one could measure how well electricity is conducted by each of the elements to determine how "metallic" the elements are. On the basis of conductivity, there is not much of a trend in the periodic table: Silver is the most conductive metal, and manganese the least. Look up the first ionization energies of silver and manganese; which of these two elements would you call more metallic based on the way we define it in this book?

7.102 Which of the following is the expected product of the reaction of K(s) and $H_2(g)$? (i) KH(s), (ii) $K_2H(s)$, (iii) $KH_2(s)$, (iv) $K_2H_2(s)$, or (v) K(s) and $H_2(g)$ will not react with one another.

7.103 Elemental cesium reacts more violently with water than does elemental sodium. Which of the following best explains this difference in reactivity? (i) Sodium has greater metallic character than does cesium. (ii) The first ionization energy of cesium is less than that of sodium. (iii) The electron affinity of sodium is smaller than that of cesium. (iv) The effective nuclear charge for cesium is less than that of sodium. (v) The atomic radius of cesium is smaller than that of sodium.

7.104 (a) One of the alkali metals reacts with oxygen to form a solid white substance. When this substance is dissolved in water, the solution gives a positive test for hydrogen peroxide, H_2O_2. When the solution is tested in a burner flame, a lilac-purple flame is produced. What is the likely identity of the metal? (b) Write a balanced chemical equation for the reaction of the white substance with water.

7.105 Zinc in its 2+ oxidation state is an essential metal ion for life. Zn^{2+} is found bound to many proteins that are involved in biological processes, but unfortunately Zn^{2+} is hard to detect by common chemical methods. Therefore, scientists who are interested in studying Zn^{2+}-containing proteins frequently substitute Cd^{2+} for Zn^{2+}, since Cd^{2+} is easier to detect. (a) On the basis of the properties of the elements and ions discussed in this chapter and their positions in the periodic table, describe the pros and cons of using Cd^{2+} as a Zn^{2+} substitute. (b) Proteins that speed up (catalyze) chemical reactions are called *enzymes*. Many enzymes are required for proper metabolic reactions in the body. One problem with using Cd^{2+} to replace Zn^{2+} in enzymes is that Cd^{2+} substitution can decrease or even eliminate enzymatic activity. Can you suggest a different metal ion that might replace Zn^{2+} in enzymes instead of Cd^{2+}? Justify your answer.

[7.106] A historian discovers a nineteenth-century notebook in which some observations, dated 1822, were recorded on a substance thought to be a new element. Here are some of the data recorded in the notebook: "Ductile, silver-white, metallic looking. Softer than lead. Unaffected by water. Stable in air. Melting point: 153 °C. Density: 7.3 g/cm³. Electrical conductivity: 20% that of copper. Hardness: About 1% as hard as iron. When 4.20 g of the unknown is heated in an excess of oxygen, 5.08 g of a white solid is formed. The solid could be sublimed by heating to over 800 °C." (a) Using information in the text and the *CRC Handbook of Chemistry and Physics*, and making allowances for possible variations in numbers from current values, identify the element reported. (b) Write a balanced chemical equation for the reaction with oxygen. (c) Judging from Figure 7.1, might this nineteenth-century investigator have been the first to discover a new element?

7.107 In April 2010, a research team reported that it had made Element 117. This discovery was confirmed in 2012 by additional experiments. Write the ground-state electron configuration for Element 117 and estimate values for its first ionization energy, electron affinity, atomic size, and common oxidation state based on its position in the periodic table.

7.108 We will see in Chapter 12 that semiconductors are materials that conduct electricity better than nonmetals but not as well as metals. The only two elements in the periodic table that are technologically useful semiconductors are silicon and germanium. Integrated circuits in computer chips today are based on silicon. Compound semiconductors are also used in the electronics industry. Examples are gallium arsenide, GaAs; gallium phosphide, GaP; cadmium sulfide, CdS; and cadmium selenide, CdSe. (a) What is the relationship between the compound semiconductors' compositions and the positions of their elements on the periodic table relative to Si and Ge? (b) Workers in the semiconductor industry refer to "II–VI" and "III–V" materials, using Roman numerals. Can you identify which compound semiconductors are II–VI and which are III–V? (c) Suggest other compositions of compound semiconductors based on the positions of their elements in the periodic table.

Integrative Exercises

[7.109] Moseley established the concept of atomic number by studying X rays emitted by the elements. The X rays emitted by some of the elements have the following wavelengths:

Element	Wavelength (Å)
Ne	14.610
Ca	3.358
Zn	1.435
Zr	0.786
Sn	0.491

(a) Calculate the frequency, ν, of the X rays emitted by each of the elements, in Hz. (b) Plot the square root of ν versus the atomic number of the element. What do you observe about the plot? (c) Explain how the plot in part (b) allowed Moseley to predict the existence of undiscovered elements. (d) Use the result from part (b) to predict the X-ray wavelength emitted by iron. (e) A particular element emits X rays with a wavelength of 0.980 Å. What element do you think it is?

[7.110] (a) Write the electron configuration for Li and estimate the effective nuclear charge experienced by the valence electron.

(b) The energy of an electron in a one-electron atom or ion equals $(-2.18 \times 10^{-18} \text{ J}) \left(\dfrac{Z^2}{n^2} \right)$, where Z is the nuclear charge and n is the principal quantum number of the electron. Estimate the first ionization energy of Li. (c) Compare the result of your calculation with the value reported in Table 7.4 and explain the difference. (d) What value of the effective nuclear charge gives the proper value for the ionization energy? Does this agree with your explanation in part (c)?

[7.111] One way to measure ionization energies is ultraviolet photoelectron spectroscopy (PES), a technique based on the photoelectric effect. ∞ (Section 6.2) In PES, monochromatic light is directed onto a sample, causing electrons to be emitted. The kinetic energy of the emitted electrons is measured. The difference between the energy of the photons and the kinetic energy of the electrons corresponds to the energy needed to remove the electrons (that is, the ionization energy). Suppose that a PES experiment is performed in which mercury vapor is irradiated with ultraviolet light of wavelength 58.4 nm. (a) What is the energy of a photon of this light, in eV? (b) Write an equation that shows the process corresponding to the first ionization energy of

Hg. **(c)** The kinetic energy of the emitted electrons is measured to be 10.75 eV. What is the first ionization energy of Hg, in kJ/mol? **(d)** Using Figure 7.10, determine which of the halogen elements has a first ionization energy closest to that of mercury.

7.112 Mercury in the environment can exist in oxidation states 0, +1, and +2. One major question in environmental chemistry research is how to best measure the oxidation state of mercury in natural systems; this is made more complicated by the fact that mercury can be reduced or oxidized on surfaces differently than it would be if it were free in solution. XPS, X-ray photoelectron spectroscopy, is a technique related to PES (see Exercise 7.111), but instead of using ultraviolet light to eject valence electrons, X rays are used to eject core electrons. The energies of the core electrons are different for different oxidation states of the element. In one set of experiments, researchers examined mercury contamination of minerals in water. They measured the XPS signals that corresponded to electrons ejected from mercury's $4f$ orbitals at 105 eV, from an X-ray source that provided 1253.6 eV of energy. The oxygen on the mineral surface gave emitted electron energies at 531 eV, corresponding to the $1s$ orbital of oxygen. Overall the researchers concluded that oxidation states were +2 for Hg and −2 for O. **(a)** Calculate the wavelength of the X rays used in this experiment. **(b)** Compare the energies of the $4f$ electrons in mercury and the $1s$ electrons in oxygen from these data to the first ionization energies of mercury and oxygen from the data in this chapter. **(c)** Write out the ground-state electron configurations for Hg^{2+} and O^{2-}; which electrons are the valence electrons in each case? **(d)** Use Slater's rules to estimate Z_{eff} for the $4f$ and valence electrons of Hg^{2+} and O^{2-}; assume for this purpose that all the inner electrons with $(n-3)$ or less screen a full +1.

[**7.113**] When magnesium metal is burned in air (Figure 3.6), two products are produced. One is magnesium oxide, MgO. The other is the product of the reaction of Mg with molecular nitrogen, magnesium nitride. When water is added to magnesium nitride, it reacts to form magnesium oxide and ammonia gas. **(a)** Based on the charge of the nitride ion (Table 2.5), predict the formula of magnesium nitride. **(b)** Write a balanced equation for the reaction of magnesium nitride with water. What is the driving force for this reaction? **(c)** In an experiment, a piece of magnesium ribbon is burned in air in a crucible. The mass of the

mixture of MgO and magnesium nitride after burning is 0.470 g. Water is added to the crucible, further reaction occurs, and the crucible is heated to dryness until the final product is 0.486 g of MgO. What was the mass percentage of magnesium nitride in the mixture obtained after the initial burning? **(d)** Magnesium nitride can also be formed by reaction of the metal with ammonia at high temperature. Write a balanced equation for this reaction. If a 6.3-g Mg ribbon reacts with 2.57 g $NH_3(g)$ and the reaction goes to completion, which component is the limiting reactant? What mass of $H_2(g)$ is formed in the reaction? **(e)** The standard enthalpy of formation of solid magnesium nitride is −461.08 kJ/mol. Calculate the standard enthalpy change for the reaction between magnesium metal and ammonia gas.

7.114 (a) The measured Bi—Br bond length in bismuth tribromide, $BiBr_3$, is 2.63 Å. Based on this value and the data in Figure 7.8, predict the atomic radius of Bi. **(b)** Bismuth tribromide is soluble in acidic solution. It is formed by treating solid bismuth(III) oxide with aqueous hydrobromic acid. Write a balanced chemical equation for this reaction. **(c)** While bismuth(III) oxide is soluble in acidic solutions, it is insoluble in basic solutions such as $NaOH(aq)$. Based on these properties, is bismuth characterized as a metallic, metalloid, or nonmetallic element? **(d)** Treating bismuth with fluorine gas forms BiF_5. Use the electron configuration of Bi to explain the formation of a compound with this formulation. **(e)** While it is possible to form BiF_5 in the manner just described, pentahalides of bismuth are not known for the other halogens. Explain why the pentahalide might form with fluorine but not with the other halogens. How does the behavior of bismuth relate to the fact that xenon reacts with fluorine to form compounds but not with the other halogens?

7.115 Potassium superoxide, KO_2, is often used in oxygen masks (such as those used by firefighters) because KO_2 reacts with CO_2 to release molecular oxygen. Experiments indicate that 2 mol of $KO_2(s)$ react with each mole of $CO_2(g)$. **(a)** The products of the reaction are $K_2CO_3(s)$ and $O_2(g)$. Write a balanced equation for the reaction between $KO_2(s)$ and $CO_2(g)$. **(b)** Indicate the oxidation number for each atom involved in the reaction in part (a). What elements are being oxidized and reduced? **(c)** What mass of $KO_2(s)$ is needed to consume 18.0 g $CO_2(g)$? What mass of $O_2(g)$ is produced during this reaction?

Design an Experiment

In this chapter we have seen that the reaction of potassium metal with oxygen leads to a product that we might not expect, namely, potassium superoxide, $KO_2(s)$. Let's design some experiments to learn more about this unusual product.

(a) One of your team members proposes that the capacity to form a superoxide such as KO_2 is related to a low value for the first ionization energy. How would you go about testing this hypothesis for the metals of group 1A? What other periodic property of the alkali metals might be considered as a factor favoring superoxide formation?

(b) $KO_2(s)$ is the active ingredient in many breathing masks used by firefighters because it can be used as a source of $O_2(g)$. In principle, $KO_2(s)$ can react with both major components of human breath, $H_2O(g)$ and $CO_2(g)$, to produce $O_2(g)$ and other products (all of which follow the expected patterns of reactivity we have seen). Predict the other products in these reactions and design experiments to determine whether $KO_2(s)$ does actually react with both $H_2O(g)$ and $CO_2(g)$.

(c) Propose an experiment to determine whether either of the reactions in part (b) is more important in the operation of a firefighter's breathing mask.

(d) The reaction of K(s) and $O_2(g)$ leads to a mixture of $KO_2(s)$ and $K_2O(s)$. Use ideas presented in this exercise to design an experiment to determine the percentages of $KO_2(s)$ and $K_2O(s)$ in the product mixture that results from the reaction of K(s) with excess $O_2(g)$.

Basic Concepts of Chemical Bonding

Whenever two atoms or ions are strongly held together, we say there is a **chemical bond** between them. There are three general types of chemical bonds: *ionic, covalent,* and *metallic*. We can get a glimpse of these three types of bonds by thinking about the simple act of using a stainless-steel

spoon to add table salt to a glass of water (Figure 8.1). Table salt is sodium chloride, NaCl, which consists of sodium ions, Na^+, and chloride ions, Cl^-. The structure is held together by **ionic bonds**, which are due to the electrostatic attractions between oppositely charged ions. The water consists mainly of H_2O molecules. The hydrogen and oxygen atoms are bonded to one another through **covalent bonds**, in which molecules are formed by the sharing of electrons between atoms. The spoon consists mainly of iron metal, in which Fe atoms are connected to one another via **metallic bonds**, which are formed by electrons that are relatively free to move through the metal. These different substances—NaCl, H_2O and Fe metal—behave as they do because of the ways in which their constituent atoms are connected to one another. For example, NaCl readily dissolves in water, but Fe metal does not.

What determines the type of bonding in any substance? How do the characteristics of these bonds give rise to different physical and chemical properties? The keys to answering the first question are found in the electronic structure of the atoms involved, discussed in Chapters 6 and 7. In this chapter and the next, we examine the relationship between the electronic structure of atoms and the ionic and covalent chemical bonds they form. We will discuss metallic bonding in Chapter 12.

▶ **CHEMICAL BONDING AS ART.** The *Atomium* is a 110-m-high steel sculpture commissioned for the 1958 World's Fair in Brussels. The nine spheres represent atoms, and the connecting rods evoke the chemical bonds holding them together. One sphere sits in the center of a cube formed by the other eight, a common arrangement of the atoms in metallic elements, such as iron.

WHAT'S AHEAD ▶

8.1 LEWIS SYMBOLS AND THE OCTET RULE
We begin with descriptions of the three main types of chemical bonds: *ionic, covalent,* and *metallic*. In evaluating bonding, *Lewis symbols* provide a useful shorthand for keeping track of valence electrons.

8.2 IONIC BONDING
We learn that in ionic substances, the atoms are held together by the electrostatic attractions between ions of opposite charge. We discuss the energetics of forming ionic substances and describe the *lattice energy* of these substances.

8.3 COVALENT BONDING
We examine the bonding in molecular substances, in which atoms bond by sharing one or more electron pairs. In general, the electrons are shared in such a way that each atom attains an *octet* of electrons.

8.4 BOND POLARITY AND ELECTRONEGATIVITY
We define *electronegativity* as the ability of an atom in a compound to attract electrons to itself. In general, electron pairs are shared unequally between atoms with different electronegativities, leading to *polar covalent bonds*.

8.5 DRAWING LEWIS STRUCTURES
We see that *Lewis structures* are a simple yet powerful way of predicting covalent

bonding patterns in molecules. In addition to the octet rule, we see that the concept of *formal charge* can be used to identify the dominant Lewis structure.

8.6 RESONANCE STRUCTURES We find that in some cases, more than one equivalent Lewis structure can be drawn for a molecule or polyatomic ion. The bonding description in such cases is a blend of two or more *resonance structures*.

8.7 EXCEPTIONS TO THE OCTET RULE We recognize that the octet rule is more of a guideline than an absolute rule. Exceptions to the rule include molecules with an odd number of electrons,

molecules where large differences in electronegativity prevent an atom from completing its octet, and molecules where an element from period 3 or below in the periodic table attains more than an octet of electrons.

8.8 STRENGTHS AND LENGTHS OF COVALENT BONDS We observe that bond strengths vary with the number of shared electron pairs as well as other factors. We use *average bond enthalpy* values to estimate the enthalpies of reactions in cases where thermodynamic data are unavailable.

GO FIGURE

If the white powder were sugar, $C_{12}H_{22}O_{11}$, how would we have to change this picture?

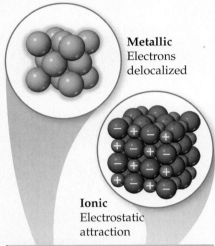

Metallic
Electrons delocalized

Ionic
Electrostatic attraction

Covalent
Electrons shared

▲ **Figure 8.1 Ionic, covalent, and metallic bonds.** The three different substances shown here are held together by different types of chemical bonds.

8.1 | Lewis Symbols and the Octet Rule

The electrons involved in chemical bonding are the *valence electrons*, which, for most atoms, are those in the outermost occupied shell. (Section 6.8) The American chemist G. N. Lewis (1875–1946) suggested a simple way of showing the valence electrons in an atom and tracking them during bond formation, using what are now known as either *Lewis electron-dot symbols* or simply Lewis symbols.

The **Lewis symbol** for an element consists of the element's chemical symbol plus a dot for each valence electron. Sulfur, for example, has the electron configuration $[Ne]3s^2 3p^4$ and therefore six valence electrons. Its Lewis symbol is

$$\cdot \ddot{\underset{\cdot}{S}} \cdot$$

The dots are placed on the four sides of the symbol—top, bottom, left, and right—and each side can accommodate up to two electrons. All four sides are equivalent, which means that the choice of on which sides to place two electrons rather than one electron is arbitrary. In general, we spread out the dots as much as possible. In the Lewis symbol for S, for instance, we prefer the dot arrangement shown rather than the arrangement having two electrons on three of the sides and none on the fourth.

The electron configurations and Lewis symbols for the main-group elements of periods 2 and 3 are shown in ▼ Table 8.1. Notice that the number of valence electrons in any representative element is the same as the element's group number. For example, the Lewis symbols for oxygen and sulfur, members of group 6A, both show six dots.

Give It Some Thought

Are all these Lewis symbols for Cl correct?

$$:\ddot{\underset{\cdot}{Cl}}\cdot \qquad :\ddot{Cl}: \qquad :\ddot{Cl}\cdot$$

The Octet Rule

Atoms often gain, lose, or share electrons to achieve the same number of electrons as the noble gas closest to them in the periodic table. The noble gases have very stable electron arrangements, as evidenced by their high ionization energies, low affinity for additional electrons, and general lack of chemical reactivity. (Section 7.8) Because all the noble gases except He have eight valence electrons, many atoms undergoing reactions end up with eight valence electrons. This observation has led to a guideline known as the **octet rule**: *Atoms tend to gain, lose, or share electrons until they are surrounded by eight valence electrons.*

Table 8.1   Lewis Symbols

Group	Element	Electron Configuration	Lewis Symbol	Element	Electron Configuration	Lewis Symbol
1A	Li	$[He]2s^1$	Li·	Na	$[Ne]3s^1$	Na·
2A	Be	$[He]2s^2$	·Be·	Mg	$[Ne]3s^2$	·Mg·
3A	B	$[He]2s^2 2p^1$	·Ḃ·	Al	$[Ne]3s^2 3p^1$	·Ȧl·
4A	C	$[He]2s^2 2p^2$	·Ċ·	Si	$[Ne]3s^2 3p^2$	·Ṡi·
5A	N	$[He]2s^2 2p^3$	·Ṅ:	P	$[Ne]3s^2 3p^3$	·Ṗ:
6A	O	$[He]2s^2 2p^4$	:Ö:	S	$[Ne]3s^2 3p^4$	:Ṡ:
7A	F	$[He]2s^2 2p^5$	·Ḟ:	Cl	$[Ne]3s^2 3p^5$	·Ċl:
8A	Ne	$[He]2s^2 2p^6$	:Ṅe:	Ar	$[Ne]3s^2 3p^6$	:Ȧr:

An octet of electrons consists of full *s* and *p* subshells in an atom. In a Lewis symbol, an octet is shown as four pairs of valence electrons arranged around the element symbol, as in the Lewis symbols for Ne and Ar in Table 8.1. There are exceptions to the octet rule, as we will see later in the chapter, but it provides a useful framework for introducing many important concepts of bonding.

8.2 | Ionic Bonding

Ionic substances generally result from the interaction of metals on the left side of the periodic table with nonmetals on the right side (excluding the noble gases, group 8A). For example, when sodium metal, Na(*s*), is brought into contact with chlorine gas, $Cl_2(g)$, a violent reaction ensues (▼ Figure 8.2). The product of this very exothermic reaction is sodium chloride, NaCl(*s*):

$$Na(s) + \tfrac{1}{2}Cl_2(g) \longrightarrow NaCl(s) \qquad \Delta H_f^\circ = -410.9 \, kJ \qquad [8.1]$$

Sodium chloride is composed of Na^+ and Cl^- ions arranged in a three-dimensional array (Figure 8.3).

The formation of Na^+ from Na and Cl^- from Cl_2 indicates that an electron has been lost by a sodium atom and gained by a chlorine atom—we say there has been an *electron transfer* from the Na atom to the Cl atom. Two of the atomic properties discussed in Chapter 7 give us an indication of how readily electron transfer occurs: ionization energy, which indicates how easily an electron can be removed from an atom; and electron affinity, which measures how much an atom wants to gain an electron.  (Sections 7.4 and 7.5) Electron transfer to form oppositely charged ions occurs when one atom readily gives up an electron (low ionization energy) and another atom readily gains an electron (high electron affinity). Thus, NaCl is a typical ionic compound because it consists of a metal of low ionization energy and a nonmetal of high electron affinity. Using Lewis electron-dot symbols (and showing a chlorine atom rather than the Cl_2 molecule), we can represent this reaction as

$$Na\cdot + \cdot\ddot{\underset{..}{Cl}}: \longrightarrow Na^+ + [:\ddot{\underset{..}{Cl}}:]^- \qquad [8.2]$$

▲ GO FIGURE

Do you expect a similar reaction between potassium metal and elemental bromine?

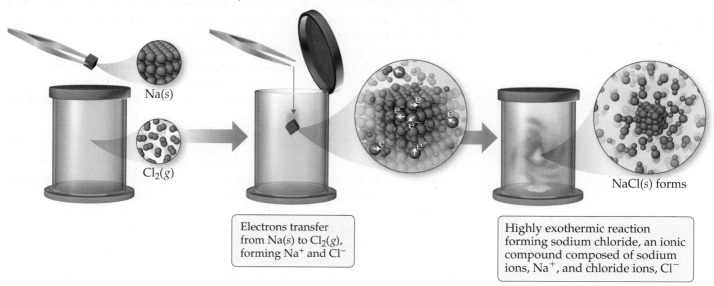

Na(*s*)

$Cl_2(g)$

Electrons transfer from Na(*s*) to $Cl_2(g)$, forming Na^+ and Cl^-

NaCl(*s*) forms

Highly exothermic reaction forming sodium chloride, an ionic compound composed of sodium ions, Na^+, and chloride ions, Cl^-

▲ Figure 8.2 **Reaction of sodium metal with chlorine gas to form the ionic compound sodium chloride.**

GO FIGURE

If no color key were provided, how would you know which color ball represented Na^+ and which represented Cl^-?

 $= Na^+$ $= Cl^-$

| Each Na^+ ion surrounded by six Cl^- ions | Each Cl^- ion surrounded by six Na^+ ions |

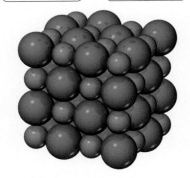

▲ **Figure 8.3 The crystal structure of sodium chloride.**

The arrow indicates the transfer of an electron from the Na atom to the Cl atom. Each ion has an octet of electrons, the Na^+ octet being the $2s^2 2p^6$ electrons that lie below the single $3s$ valence electron of the Na atom. We have put a bracket around the chloride ion to emphasize that all eight electrons are located on it.

Give It Some Thought

Describe the electron transfers that occur in the formation of calcium fluoride from elemental calcium and elemental fluorine.

Ionic substances possess several characteristic properties. They are usually brittle substances with high melting points. They are usually crystalline. Furthermore, ionic crystals often can be cleaved; that is, they break apart along smooth, flat surfaces. These characteristics result from electrostatic forces that maintain the ions in a rigid, well-defined, three-dimensional arrangement such as that shown in Figure 8.3.

Energetics of Ionic Bond Formation

The formation of sodium chloride from sodium and chlorine is *very* exothermic, as indicated by the large negative enthalpy of formation value given in Equation 8.1, $\Delta H_f^\circ = -410.9$ kJ. Appendix C shows that the heat of formation of other ionic substances is also quite negative. What factors make the formation of ionic compounds so exothermic?

In Equation 8.2 we represented the formation of NaCl as the transfer of an electron from Na to Cl. Recall from Section 7.4 that the loss of electrons from an atom is always an endothermic process. Removing an electron from $Na(g)$ to form $Na^+(g)$ for instance, requires 496 kJ/mol. Recall from Section 7.5 that when a nonmetal gains an electron, the process is generally exothermic, as seen from the negative electron affinities of the elements. Adding an electron to $Cl(g)$, for example, releases 349 kJ/mol. From the magnitudes of these energies, we can see that the transfer of an electron from a Na atom to a Cl atom would not be exothermic—the overall process would be an endothermic process that requires $496 - 349 = 147$ kJ/mol. This endothermic process corresponds to the formation of sodium and chloride ions that are infinitely far apart—in other words, the positive energy change assumes that the ions do not interact with each other, which is quite different from the situation in ionic solids.

The principal reason ionic compounds are stable is the attraction between ions of opposite charge. This attraction draws the ions together, releasing energy and causing the ions to form a solid array, or lattice, such as that shown in Figure 8.3. A measure of how much stabilization results from arranging oppositely charged ions in an ionic solid is given by the **lattice energy**, which is *the energy required to completely separate one mole of a solid ionic compound into its gaseous ions.*

To envision this process for NaCl, imagine that the structure in Figure 8.3 expands from within, so that the distances between the ions increase until the ions are very far apart. This process requires 788 kJ/mol, which is the value of the lattice energy:

$$NaCl(s) \longrightarrow Na^+(g) + Cl^-(g) \qquad \Delta H_{\text{lattice}} = +788 \text{ kJ/mol} \qquad [8.3]$$

Notice that this process is highly endothermic. The reverse process—the coming together of $Na^+(g)$ and $Cl^-(g)$ to form $NaCl(s)$—is therefore highly exothermic $(\Delta H = -788$ kJ/mol$)$.

▶ Table 8.2 lists the lattice energies for a number of ionic compounds. The large positive values indicate that the ions are strongly attracted to one another in ionic solids. The energy released by the attraction between ions of unlike charge more than makes up for the endothermic nature of ionization energies, making the formation of ionic compounds an exothermic process. The strong attractions also cause most ionic materials to be hard and brittle with high melting points—for example, NaCl melts at 801 °C.

Table 8.2 Lattice Energies for Some Ionic Compounds

Compound	Lattice Energy (kJ/mol)	Compound	Lattice Energy (kJ/mol)
LiF	1030	$MgCl_2$	2326
LiCl	834	$SrCl_2$	2127
LiI	730		
NaF	910	MgO	3795
NaCl	788	CaO	3414
NaBr	732	SrO	3217
NaI	682		
KF	808	ScN	7547
KCl	701		
KBr	671		
CsCl	657		
CsI	600		

 ## Give It Some Thought

If you were to perform the reaction $KCl(s) \longrightarrow K^+(g) + Cl^-(g)$, would energy be released?

The magnitude of the lattice energy of an ionic solid depends on the charges of the ions, their sizes, and their arrangement in the solid. We saw in Section 5.1 that the electrostatic potential energy of two interacting charged particles is given by

$$E_{el} = \frac{\kappa Q_1 Q_2}{d} \qquad [8.4]$$

In this equation Q_1 and Q_2 are the charges on the particles in Coulombs, with their signs; d is the distance between their centers in meters, and κ is a constant, 8.99×10^9 J-m/C^2. Equation 8.4 indicates that the attractive interaction between two oppositely charged ions increases as the magnitudes of their charges increase and as the distance between their centers decreases. Thus, *for a given arrangement of ions, the lattice energy increases as the charges on the ions increase and as their radii decrease.* The variation in the magnitude of lattice energies depends more on ionic charge than on ionic radius because ionic radii vary over only a limited range compared to charges.

SAMPLE EXERCISE 8.1 | Magnitudes of Lattice Energies

Without consulting Table 8.2, arrange the ionic compounds NaF, CsI, and CaO in order of increasing lattice energy.

SOLUTION

Analyze From the formulas for three ionic compounds, we must determine their relative lattice energies.

Plan We need to determine the charges and relative sizes of the ions in the compounds. We then use Equation 8.4 qualitatively to determine the relative energies, knowing that (a) the larger the ionic charges, the greater the energy and (b) the farther apart the ions are, the lower the energy.

Solve NaF consists of Na$^+$ and F$^-$ ions, CsI of Cs$^+$ and I$^-$ ions, and CaO of Ca^{2+} and O^{2-} ions. Because the product $Q_1 Q_2$ appears in the numerator of Equation 8.4, the lattice energy increases dramatically when the charges increase. Thus, we expect the lattice energy of CaO, which has 2+ and 2− ions, to be the greatest of the three.

The ionic charges are the same in NaF and CsI. The difference in their lattice energies thus depends on the difference in the distance between ions in the lattice. Because ionic size increases as we go down a group in the periodic table ∞∞(Section 7.3), we know that Cs⁺ is larger than Na⁺ and I⁻ is larger than F⁻. Therefore, the distance between Na⁺ and F⁻ ions in NaF is less than the distance between the Cs⁺ and I⁻ ions in CsI. As a result, the lattice energy of NaF should be greater than that of CsI. In order of increasing energy, therefore, we have CsI < NaF < CaO.

Check Table 8.2 confirms our predicted order is correct.

Practice Exercise 1

Without looking at Table 8.2, predict which one of the following orderings of lattice energy is correct for these ionic compounds. (a) NaCl > MgO > CsI > ScN, (b) ScN > MgO > NaCl > CsI, (c) NaCl > CsI > ScN > CaO, (d) MgO > NaCl > ScN > CsI, (e) ScN > CsI > NaCl > MgO.

Practice Exercise 2

Which substance do you expect to have the greatest lattice energy: MgF_2, CaF_2, or ZrO_2?

Because lattice energy decreases as distance between ions increases, lattice energies follow trends that parallel those in ionic radius shown in Figure 7.8. In particular, because ionic radius increases as we go down a group of the periodic table, we find that, for a given type of ionic compound, lattice energy decreases as we go down a group. ▼ Figure 8.4 illustrates this trend for the alkali chlorides MCl (M = Li, Na, K, Rb, Cs) and the sodium halides NaX (X = F, Cl, Br, I).

 GO FIGURE

Using this figure, find the most likely range of values for the lattice energy of KF.

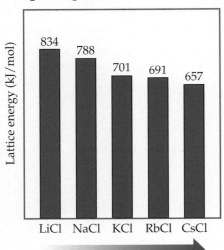

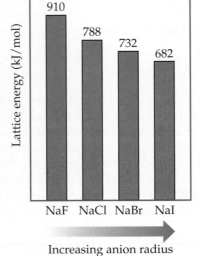

▲ Figure 8.4 **Periodic trends in lattice energy as a function of cation or anion radius.**

A Closer Look

Calculation of Lattice Energies: The Born–Haber Cycle

Lattice energies cannot be determined directly by experiment. They can, however, be calculated by envisioning the formation of an ionic compound as occurring in a series of well-defined steps. We can then use Hess's law ∞∞(Section 5.6) to combine the steps in a way that gives the lattice energy for the compound. By so doing, we construct a **Born–Haber cycle**, a thermochemical cycle named after the German scientists Max Born (1882–1970) and Fritz Haber (1868–1934), who introduced it to analyze the factors contributing to the stability of ionic compounds.

Let's use NaCl as an example. In Equation 8.3, which defines lattice energy, NaCl(s) is the reactant, and the gas-phase ions Na⁺(g) and Cl⁻(g) are the products. This equation is our target as we apply Hess's Law.

In seeking a set of other equations that can be added up to give our target equation, we can use the heat of formation for NaCl ∞∞(Section 5.7):

$$Na(s) + \tfrac{1}{2}Cl_2(g) \longrightarrow NaCl(s) \qquad \Delta H_f^\circ[NaCl(s)] = -411 \text{ kJ} \quad [8.5]$$

Of course, we will have to turn this equation around so we have $NaCl(s)$ as the reactant as we do in the equation for the lattice energy. We can use two other equations to arrive at our target, as shown below:

1. $NaCl(s) \longrightarrow Na(s) + \frac{1}{2}Cl_2(g)$ $\quad \Delta H_1 = -\Delta H_f^{\circ}[NaCl(s)]$
   $\quad\quad\quad\quad\quad\quad\quad\quad\quad\quad\quad\quad\quad\quad\quad\quad\quad\quad = +411 \text{ kJ}$

2. $Na(s) \longrightarrow Na^+(g)$ $\quad\quad\quad\quad \Delta H_2 = ??$

3. $\frac{1}{2}Cl_2(g) \longrightarrow Cl^-(g)$ $\quad\quad\quad \Delta H_3 = ???$

4. $NaCl(s) \longrightarrow Na^+(g) + Cl^-(g)$ $\quad \Delta H_4 = \Delta H_1 + \Delta H_2 + \Delta H_3$
   $\quad\quad\quad\quad\quad\quad\quad\quad\quad\quad\quad\quad\quad\quad\quad\quad\quad\quad = \Delta H_{\text{lattice}}$

Step 2 involves the formation of sodium ion from solid sodium, which is just the heat of formation for sodium gas and the first ionization energy for sodium (Appendix C and Figure 7.10 list numbers for these processes):

$Na(s) \longrightarrow Na(g)$ $\quad\quad\quad \Delta H = \Delta H_f^{\circ}[Na(g)] = 108 \text{ kJ}$ [8.6]

$Na(g) \longrightarrow Na^+(g) + e^-$ $\quad \Delta H = I_1(Na) = 496 \text{ kJ}$ [8.7]

The sum of these two processes gives us the required energy for Step 2 (above), which is 604 kJ.

Similarly, for Step 3, we have to create chlorine atoms, and then anions, from the Cl_2 molecule, in two steps. The enthalpy changes for these two steps are the sum of the enthalpy of formation of $Cl(g)$ and the electron affinity of chlorine, $E(Cl)$:

$\frac{1}{2}Cl_2(g) \longrightarrow Cl(g)$ $\quad \Delta H = \Delta H_f^{\circ}[Cl(g)] = 122 \text{ kJ}$ [8.8]

$e^- + Cl(g) \longrightarrow Cl^-(g)$ $\quad \Delta H = E(Cl) = -349 \text{ kJ}$ [8.9]

The sum of these two processes gives us the required energy for Step 3 (above), which is -227 kJ.

Finally, when we put it all together, we have:

1. $NaCl(s) \longrightarrow Na(s) + \frac{1}{2}Cl_2(g)$ $\quad \Delta H_1 = -\Delta H_f^{\circ}[NaCl(s)]$
   $\quad\quad\quad\quad\quad\quad\quad\quad\quad\quad\quad\quad\quad\quad\quad\quad\quad\quad = +411 \text{ kJ}$

2. $Na(s) \longrightarrow Na^+(g)$ $\quad\quad\quad\quad \Delta H_2 = 604 \text{ kJ}$

3. $\frac{1}{2}Cl_2(g) \longrightarrow Cl^-(g)$ $\quad\quad\quad\quad \Delta H_3 = -227 \text{ kJ}$

4. $NaCl(s) \longrightarrow Na^+(g) + Cl^-(g)$ $\quad \Delta H_4 = 788 \text{ kJ} = \Delta H_{\text{lattice}}$

This process is described as a "cycle" because it corresponds to the scheme in ▼ **Figure 8.5**, which shows how all the quantities we have just calculated are related. The sum of all the blue "up" arrow energies has to be equal to the sum of all the red "down" arrow energies in this cycle. Born and Haber recognized that if we know the value of every quantity in the cycle except the lattice energy, we can calculate it from this cycle.

Related Exercises: 8.28, 8.29, 8.30, 8.83, 8.102, 8.103

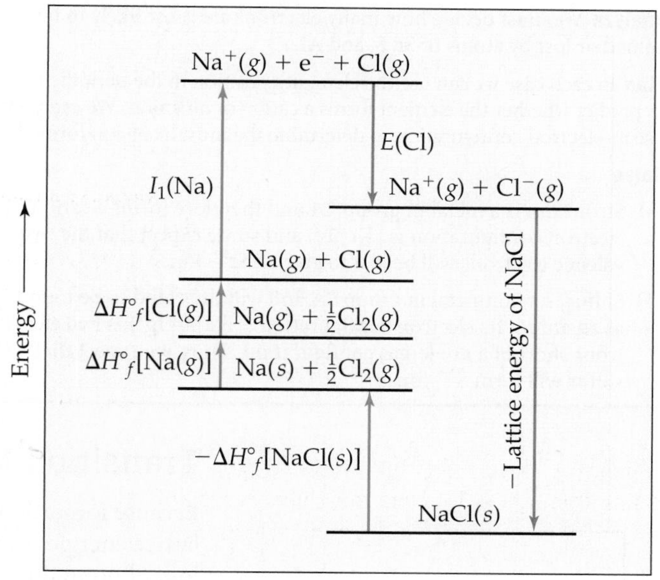

▲ **Figure 8.5 Born–Haber cycle for formation of NaCl.** This Hess's law representation shows the energetic relationships in the formation of the ionic solid from its elements.

Electron Configurations of Ions of the s- and p-Block Elements

The energetics of ionic bond formation helps explain why many ions tend to have noble-gas electron configurations. For example, sodium readily loses one electron to form Na^+, which has the same electron configuration as Ne:

$$Na \quad 1s^2 2s^2 2p^6 3s^1 = [Ne]3s^1$$

$$Na^+ \quad 1s^2 2s^2 2p^6 \quad = [Ne]$$

Even though lattice energy increases with increasing ionic charge, we never find ionic compounds that contain Na^{2+} ions. The second electron removed would have to come from an inner shell of the sodium atom, and removing electrons from an inner shell requires a very large amount of energy. ∞ (Section 7.4) The increase in lattice energy is not enough to compensate for the energy needed to remove an inner-shell electron. Thus, sodium and the other group 1A metals are found in ionic substances only as $1+$ ions.

Similarly, adding electrons to nonmetals is either exothermic or only slightly endothermic as long as the electrons are added to the valence shell. Thus, a Cl atom easily adds an electron to form Cl^-, which has the same electron configuration as Ar:

$$Cl \quad 1s^2 2s^2 2p^6 3s^2 3p^5 = [Ne]3s^2 3p^5$$

$$Cl^- \quad 1s^2 2s^2 2p^6 3s^2 3p^6 = [Ne]3s^2 3p^6 = [Ar]$$

To form a Cl^{2-} ion, the second electron would have to be added to the next higher shell of the Cl atom, an addition that is energetically very unfavorable. Therefore, we never observe Cl^{2-} ions in ionic compounds. We thus expect ionic compounds of the representative metals from groups 1A, 2A, and 3A to contain 1+, 2+ and 3+ cations, respectively, and usually expect ionic compounds of the representative nonmetals of groups 5A, 6A, and 7A to contain 3−, 2−, and 1− anions, respectively.

SAMPLE EXERCISE 8.2 | Charges on Ions

Predict the ion generally formed by (**a**) Sr, (**b**) S, (**c**) Al.

SOLUTION

Analyze We must decide how many electrons are most likely to be gained or lost by atoms of Sr, S, and Al.

Plan In each case we can use the element's position in the periodic table to predict whether the element forms a cation or an anion. We can then use its electron configuration to determine the most likely ion formed.

Solve

(**a**) Strontium is a metal in group 2A and therefore forms a cation. Its electron configuration is $[Kr]5s^2$, and so we expect that the two valence electrons will be lost to give an Sr^{2+} ion.

(**b**) Sulfur is a nonmetal in group 6A and will thus tend to be found as an anion. Its electron configuration ($[Ne]3s^23p^4$) is two electrons short of a noble-gas configuration. Thus, we expect that sulfur will form S^{2-} ions.

(**c**) Aluminum is a metal in group 3A. We therefore expect it to form Al^{3+} ions.

Check The ionic charges we predict here are confirmed in Tables 2.4 and 2.5.

Practice Exercise 1
Which of these elements is most likely to form ions with a 2+ charge?
(**a**) Li, (**b**) Ca, (**c**) O, (**d**) P, (**e**) Cl.

Practice Exercise 2
Predict the charges on the ions formed when magnesium reacts with nitrogen.

Transition Metal Ions

Because ionization energies increase rapidly for each successive electron removed, the lattice energies of ionic compounds are generally large enough to compensate for the loss of up to only three electrons from atoms. Thus, we find cations with charges of 1+, 2+, or 3+ in ionic compounds. Most transition metals, however, have more than three electrons beyond a noble-gas core. Silver, for example, has a $[Kr]4d^{10}5s^1$ electron configuration. Metals of group 1B (Cu, Ag, Au) often occur as 1+ ions (as in CuBr and AgCl). In forming Ag^+, the 5s electron is lost, leaving a completely filled 4d subshell. As in this example, transition metals generally do not form ions that have a noble-gas configuration. The octet rule, although useful, is clearly limited in scope.

Recall from Section 7.4 that when a positive ion forms from an atom, electrons are always lost first from the subshell having the largest value of *n*. Thus, *in forming ions, transition metals lose the valence-shell s electrons first, then as many d electrons as required to reach the charge of the ion.* For instance, in forming Fe^{2+} from Fe, which has the electron configuration $[Ar]3d^64s^2$, the two 4s electrons are lost, leading to an $[Ar]3d^6$ configuration. Removal of an additional electron gives Fe^{3+}, whose electron configuration is $[Ar]3d^5$.

 Give It Some Thought

Which element forms a 3+ ion that has the electron configuration $[Kr]4d^6$?

8.3 | Covalent Bonding

The vast majority of chemical substances do not have the characteristics of ionic materials. Most of the substances with which we come into daily contact—such as water—tend to be gases, liquids, or solids with low melting points. Many, such as gasoline, vaporize readily. Many are pliable in their solid forms—for example, plastic bags and wax.

For the very large class of substances that do not behave like ionic substances, we need a different model to describe the bonding between atoms. G. N. Lewis reasoned that atoms might acquire a noble-gas electron configuration by sharing electrons with other atoms. A chemical bond formed by sharing a pair of electrons is a *covalent bond*.

The hydrogen molecule, H_2, provides the simplest example of a covalent bond. When two hydrogen atoms are close to each other, the two positively charged nuclei repel each other, the two negatively charged electrons repel each other, and the nuclei and electrons attract each other, as shown in ▶ Figure 8.6(a). Because the molecule is stable, we know that the attractive forces must overcome the repulsive ones. Let's take a closer look at the attractive forces that hold this molecule together.

By using quantum mechanical methods analogous to those used for atoms in Section 6.5, we can calculate the distribution of electron density in molecules. Such a calculation for H_2 shows that the attractions between the nuclei and the electrons cause electron density to concentrate between the nuclei, as shown in Figure 8.6(**b**). As a result, the overall electrostatic interactions are attractive. Thus, the atoms in H_2 are held together principally because the two positive nuclei are attracted to the concentration of negative charge between them. In essence, the shared pair of electrons in any covalent bond acts as a kind of "glue" to bind atoms together.

 Give It Some Thought

Ionizing an H_2 molecule to H_2^+ changes the strength of the bond. Based on the description of covalent bonding given previously, do you expect the H—H bond in H_2^+ to be weaker or stronger than the H—H bond in H_2?

Lewis Structures

The formation of covalent bonds can be represented with Lewis symbols. The formation of the H_2 molecule from two H atoms, for example, can be represented as

$$H\cdot \ + \ \cdot H \longrightarrow \left(H{:}H \right)$$

In forming the covalent bond, each hydrogen atom acquires a second electron, achieving the stable, two-electron, noble-gas electron configuration of helium.

Formation of a covalent bond between two Cl atoms to give a Cl_2 molecule can be represented in a similar way:

$$:\ddot{\underset{..}{Cl}}\cdot \ + \ \cdot \ddot{\underset{..}{Cl}}: \longrightarrow \left(:\ddot{\underset{..}{Cl}}{:}\ddot{\underset{..}{Cl}}: \right)$$

By sharing the bonding electron pair, each chlorine atom has eight electrons (an octet) in its valence shell, thus achieving the noble-gas electron configuration of argon.

The structures shown here for H_2 and Cl_2 are **Lewis structures**, or *Lewis dot structures*. While these structures show circles to indicate electron sharing, the more common convention is to show each shared electron pair as a line and any unshared electron pairs (also called **lone pairs** or **nonbonding pairs**) as dots. Written this way, the Lewis structures for H_2 and Cl_2 are

$$H{-}H \qquad :\ddot{\underset{..}{Cl}}{-}\ddot{\underset{..}{Cl}}:$$

For nonmetals, the number of valence electrons in a neutral atom is the same as the group number. Therefore, one might predict that 7A elements, such as F, would form one covalent bond to achieve an octet; 6A elements, such as O, would form two covalent bonds; 5A elements, such as N, would form three; and 4A elements, such as C, would form four. These predictions are borne out in many compounds, as in, for example, the compounds with hydrogen of the nonmetals of the second row of the periodic table:

$$H{-}\ddot{\underset{..}{F}}: \qquad H{-}\underset{|}{\overset{..}{O}} \qquad H{-}\underset{|}{\overset{..}{N}}{-}H \qquad H{-}\underset{|}{\overset{H}{\underset{H}{C}}}{-}H$$

▶ Figure 8.6(a)

▲ **GO FIGURE**

What would happen to the magnitudes of the attractions and repulsions represented in (a) if the nuclei were farther apart?

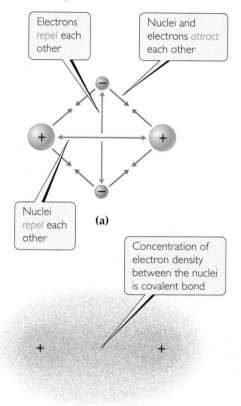

▲ **Figure 8.6 The covalent bond in H_2.** (**a**) The attractions and repulsions among electrons and nuclei in the hydrogen molecule. (**b**) Electron distribution in the H_2 molecule.

SAMPLE EXERCISE 8.3 | Lewis Structure of a Compound

Given the Lewis symbols for nitrogen and fluorine in Table 8.1, predict the formula of the stable binary compound (a compound composed of two elements) formed when nitrogen reacts with fluorine and draw its Lewis structure.

SOLUTION

Analyze The Lewis symbols for nitrogen and fluorine reveal that nitrogen has five valence electrons and fluorine has seven.

Plan We need to find a combination of the two elements that results in an octet of electrons around each atom. Nitrogen requires three additional electrons to complete its octet, and fluorine requires one. Sharing a pair of electrons between one N atom and one F atom will result in an octet of electrons for fluorine but not for nitrogen. We therefore need to figure out a way to get two more electrons for the N atom.

Solve Nitrogen must share a pair of electrons with three fluorine atoms to complete its octet. Thus, the binary compound these two elements form must be NF_3:

$$\cdot\ddot{N}\cdot \;+\; 3\;\cdot\ddot{\underset{\cdot\cdot}{F}}: \;\longrightarrow\; :\ddot{\underset{\cdot\cdot}{F}}:\ddot{N}:\ddot{\underset{\cdot\cdot}{F}}: \;\longrightarrow\; :\ddot{\underset{\cdot\cdot}{F}}-\overset{\displaystyle}{\underset{\displaystyle |}{\ddot{N}}}-\ddot{\underset{\cdot\cdot}{F}}:$$

Check The Lewis structure in the center shows that each atom is surrounded by an octet of electrons. Once you are accustomed to thinking of each line in a Lewis structure as representing *two* electrons, you can just as easily use the structure on the right to check for octets.

> **Practice Exercise 1**
> Which of these molecules has the same number of shared electron pairs as unshared electron pairs? (**a**) HCl, (**b**) H_2S, (**c**) PF_3, (**d**) CCl_2F_2 (**e**) Br_2.
>
> **Practice Exercise 2**
> Compare the Lewis symbol for neon with the Lewis structure for methane, CH_4. How many valence electrons are in each structure? How many bonding pairs and how many nonbonding pairs does each structure have?

Multiple Bonds

A shared electron pair constitutes a single covalent bond, generally referred to simply as a **single bond**. In many molecules, atoms attain complete octets by sharing more than one pair of electrons. When two electron pairs are shared by two atoms, two lines are drawn in the Lewis structure, representing a **double bond**. In carbon dioxide, for example, bonding occurs between carbon, with four valence electrons, and oxygen, with six:

$$:\ddot{O}: \;+\; \cdot\dot{C}\cdot \;+\; :\ddot{O}: \;\longrightarrow\; \ddot{O}::C::\ddot{O} \qquad (\text{or } \ddot{O}{=}C{=}\ddot{O})$$

As the diagram shows, each oxygen atom acquires an octet by sharing two electron pairs with carbon. In the case of CO_2, carbon acquires an octet by sharing two electron pairs with each of the two oxygen atoms; each double bond involves four electrons.

A **triple bond** corresponds to the sharing of three pairs of electrons, such as in the N_2 molecule:

$$:\dot{N}\cdot \;+\; \cdot\dot{N}: \;\longrightarrow\; :N:::N: \qquad (\text{or } :N{\equiv}N:)$$

Because each nitrogen atom has five valence electrons, three electron pairs must be shared to achieve the octet configuration.

The properties of N_2 are in complete accord with its Lewis structure. Nitrogen is a diatomic gas with exceptionally low reactivity that results from the very stable nitrogen–nitrogen bond. The nitrogen atoms are separated by only 1.10 Å. The short separation distance between the two N atoms is a result of the triple bond between the atoms. From studies of the structures of many different substances in which nitrogen atoms share one or two electron pairs, we have learned that the average distance between bonded nitrogen atoms varies with the number of shared electron pairs:

$$N-N \qquad N=N \qquad N\equiv N$$
$$1.47\ \text{Å} \qquad 1.24\ \text{Å} \qquad 1.10\ \text{Å}$$

As a general rule, the length of the bond between two atoms decreases as the number of shared electron pairs increases.

Give It Some Thought

The C—O bond length in carbon monoxide, CO, is 1.13 Å, whereas the C—O bond length in CO_2 is 1.24 Å. Without drawing a Lewis structure, do you think that CO contains a single, double, or triple bond?

8.4 | Bond Polarity and Electronegativity

When two identical atoms bond, as in Cl_2 or H_2, the electron pairs must be shared equally. When two atoms from opposite sides of the periodic table bond, such as NaCl, there is relatively little sharing of electrons, which means that NaCl is best described as an ionic compound composed of Na^+ and Cl^- ions. The $3s$ electron of the Na atom is, in effect, transferred completely to chlorine. The bonds that are found in most substances fall somewhere between these extremes.

Bond polarity is a measure of how equally or unequally the electrons in any covalent bond are shared. A **nonpolar covalent bond** is one in which the electrons are shared equally, as in Cl_2 and N_2. In a **polar covalent bond**, one of the atoms exerts a greater attraction for the bonding electrons than the other. If the difference in relative ability to attract electrons is large enough, an ionic bond is formed.

Electronegativity

We use a quantity called electronegativity to estimate whether a given bond is nonpolar covalent, polar covalent, or ionic. **Electronegativity** is defined as the ability of an atom *in a molecule* to attract electrons to itself. The greater an atom's electronegativity, the greater its ability to attract electrons to itself. The electronegativity of an atom in a molecule is related to the atom's ionization energy and electron affinity, which are properties of isolated atoms. An atom with a very negative electron affinity and a high ionization energy both attracts electrons from other atoms and resists having its electrons attracted away; therefore it is highly electronegative.

Electronegativity values can be based on a variety of properties, not just ionization energy and electron affinity. The American chemist Linus Pauling (1901–1994) developed the first and most widely used electronegativity scale, which is based on thermochemical data. As **Figure 8.7** shows, there is generally an increase in electronegativity from left to right across a period—that is, from the most metallic to the most nonmetallic elements. With some exceptions (especially in the transition metals), electronegativity decreases with increasing atomic number in a group. This is what we expect because we know that ionization energies decrease with increasing atomic number in a group and electron affinities do not change very much.

You do not need to memorize electronegativity values. Instead, you should know the periodic trends so that you can predict which of two elements is more electronegative.

Give It Some Thought

How does the *electronegativity* of an element differ from its *electron affinity*?

▲ **GO FIGURE**

For the group 6A elements, what is the trend in electronegativity with increasing atomic number?

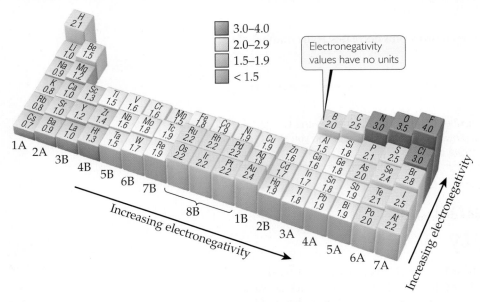

▲ Figure 8.7 **Electronegativity values based on Pauling's thermochemical data.**

Electronegativity and Bond Polarity

We can use the difference in electronegativity between two atoms to gauge the polarity of the bond the atoms form. Consider these three fluorine-containing compounds:

	F_2	HF	LiF
Electronegativity difference	$4.0 - 4.0 = 0$	$4.0 - 2.1 = 1.9$	$4.0 - 1.0 = 3.0$
Type of bond	Nonpolar covalent	Polar covalent	Ionic

In F_2 the electrons are shared equally between the fluorine atoms and, thus, the covalent bond is *nonpolar*. A nonpolar covalent bond results when the electronegativities of the bonded atoms are equal.

In HF the fluorine atom has a greater electronegativity than the hydrogen atom, with the result that the electrons are shared unequally—the bond is *polar*. In general, a polar covalent bond results when the atoms differ in electronegativity. In HF the more electronegative fluorine atom attracts electron density away from the less electronegative hydrogen atom, leaving a partial positive charge on the hydrogen atom and a partial negative charge on the fluorine atom. We can represent this charge distribution as

$$\overset{\delta+}{H}\!-\!\overset{\delta-}{F}$$

The $\delta+$ and $\delta-$ (read "delta plus" and "delta minus") symbolize the partial positive and negative charges, respectively. In a polar bond, these numbers are less than the full charges of the ions.

In LiF the electronegativity difference is very large, meaning that the electron density is shifted far toward F. The resultant bond is therefore most accurately described as *ionic*. Thus, if we considered the bond in LiF to be fully ionic, we could say $\delta+$ for Li is $1+$ and $\delta-$ for F is $1-$.

The shift of electron density toward the more electronegative atom in a bond can be seen from the results of calculations of electron-density distributions. For the three species in our example, the calculated electron-density distributions are shown in

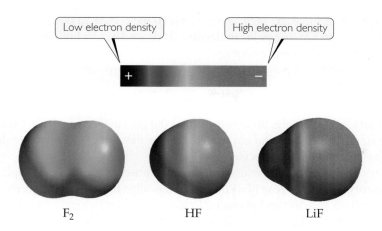

Low electron density   High electron density

+   −

F_2 HF   LiF

◀ **Figure 8.8 Electron-density distribution.** This computer-generated rendering shows the calculated electron-density distribution on the surface of the F_2, HF, and LiF molecules.

▲ **Figure 8.8.** You can see that in F_2 the distribution is symmetrical, in HF the electron density is clearly shifted toward fluorine, and in LiF the shift is even greater. These examples illustrate, therefore, that *the greater the difference in electronegativity between two atoms, the more polar their bond.*

 Give It Some Thought

Based on differences in electronegativity, how would you characterize the bonding in sulfur dioxide, SO_2? Do you expect the bonds between S and O to be nonpolar, polar covalent, or ionic?

SAMPLE EXERCISE 8.4 | Bond Polarity

In each case, which bond is more polar? (**a**) B—Cl or C—Cl, (**b**) P—F or P—Cl. Indicate in each case which atom has the partial negative charge.

SOLUTION

Analyze We are asked to determine relative bond polarities, given nothing but the atoms involved in the bonds.

Plan Because we are not asked for quantitative answers, we can use the periodic table and our knowledge of electronegativity trends to answer the question.

Solve

(**a**) The chlorine atom is common to both bonds. Therefore, we just need to compare the electronegativities of B and C. Because boron is to the left of carbon in the periodic table, we predict that boron has the lower electronegativity. Chlorine, being on the right side of the table, has high electronegativity. The more polar bond will be the one between the atoms with the biggest differences in electronegativity. Consequently, the B—Cl bond is more polar; the chlorine atom carries the partial negative charge because it has a higher electronegativity.

(**b**) In this example phosphorus is common to both bonds, and so we just need to compare the electronegativities of F and Cl. Because fluorine is above chlorine in the periodic table, it should be more electronegative and will form the more polar bond with P. The

higher electronegativity of fluorine means that it will carry the partial negative charge.

Check

(**a**) Using Figure 8.7: The difference in the electronegativities of chlorine and boron is $3.0 - 2.0 = 1.0$; the difference between the electronegativities of chlorine and carbon is $3.0 - 2.5 = 0.5$. Hence, the B—Cl bond is more polar, as we had predicted.

(**b**) Using Figure 8.7: The difference in the electronegativities of chlorine and phosphorus is $3.0 - 2.1 = 0.9$; the difference between the electronegativities of fluorine and phosphorus is $4.0 - 2.1 = 1.9$. Hence, the P—F bond is more polar, as we had predicted.

Practice Exercise 1

Which of the following bonds is the most polar? (**a**) H—F, (**b**) H—I, (**c**) Se—F, (**d**) N—P, (**e**) Ga—Cl.

Practice Exercise 2

Which of the following bonds is most polar: S—Cl, S—Br, Se—Cl, or Se—Br?

Dipole Moments

The difference in electronegativity between H and F leads to a polar covalent bond in the HF molecule. As a consequence, there is a concentration of negative charge on the more electronegative F atom, leaving the less electronegative H atom at the positive end

GO FIGURE

If the charged particles are moved closer together, does μ increase, decrease, or stay the same?

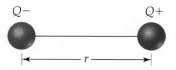

Dipole moment $\mu = Qr$

▲ **Figure 8.9 Dipole and dipole moment.** When charges of equal magnitude and opposite sign $Q+$ and $Q-$ are separated by a distance r, a dipole is produced.

of the molecule. A molecule such as HF, in which the centers of positive and negative charge do not coincide, is a **polar molecule**. Thus, we describe both bonds and entire molecules as being polar and nonpolar.

We can indicate the polarity of the HF molecule in two ways:

$$\overset{\delta+}{\text{H}} - \overset{\delta-}{\text{F}} \quad or \quad \overset{\longleftrightarrow}{\text{H} - \text{F}}$$

In the notation on the right, the arrow denotes the shift in electron density toward the fluorine atom. The crossed end of the arrow can be thought of as a plus sign designating the positive end of the molecule.

Polarity helps determine many properties we observe at the macroscopic level in the laboratory and in everyday life. Polar molecules align themselves with respect to one another, with the negative end of one molecule and the positive end of another attracting each other. Polar molecules are likewise attracted to ions. The negative end of a polar molecule is attracted to a positive ion, and the positive end is attracted to a negative ion. These interactions account for many properties of liquids, solids, and solutions, as you will see in Chapters 11, 12, and 13. Charge separation within molecules plays an important role in energy conversion processes such as photosynthesis (Chapter 23) and in organic solar cells (Chapter 12).

How can we quantify the polarity of a molecule? Whenever two electrical charges of equal magnitude but opposite sign are separated by a distance, a **dipole** is established. The quantitative measure of the magnitude of a dipole is called its **dipole moment**, denoted with the Greek letter mu, μ. If two equal and opposite charges $Q+$ and $Q-$ are separated by a distance r, as in ◀ **Figure 8.9**, the magnitude of the dipole moment is the product of Q and r:

$$\mu = Qr \qquad [8.10]$$

This expression tells us that dipole moment increases as the magnitude of Q increases and as r increases. The larger the dipole moment, the more polar the bond. For a nonpolar molecule, such as F_2, the dipole moment is zero because there is no charge separation.

Give It Some Thought

Chlorine monofluoride, ClF, and iodine monofluoride, IF, are *interhalogen* compounds—compounds that contain bonds between different halogen elements. Which of these molecules has the larger dipole moment?

Dipole moments are experimentally measurable and are usually reported in *debyes* (D), a unit that equals 3.34×10^{-30} coulomb-meters (C-m). For molecules, we usually measure charge in units of the electronic charge e, 1.60×10^{-19} C, and distance in angstroms. This means we need to convert units whenever we want to report a dipole moment in debyes. Suppose that two charges $1+$ and $1-$ (in units of e) are separated by 1.00 Å. The dipole moment produced is

$$\mu = Qr = (1.60 \times 10^{-19}\text{C})(1.00\text{Å})\left(\frac{10^{-10}\text{ m}}{1\text{Å}}\right)\left(\frac{1\text{ D}}{3.34 \times 10^{-30}\text{ C-m}}\right) = 4.79 \text{ D}$$

Measurement of the dipole moments can provide us with valuable information about the charge distributions in molecules, as illustrated in Sample Exercise 8.5.

SAMPLE EXERCISE 8.5 Dipole Moments of Diatomic Molecules

The bond length in the HCl molecule is 1.27 Å. **(a)** Calculate the dipole moment, in debyes, that results if the charges on the H and Cl atoms were $1+$ and $1-$ respectively. **(b)** The experimentally measured dipole moment of HCl(*g*) is 1.08 D. What magnitude of charge, in units of *e*, on the H and Cl atoms leads to this dipole moment?

SOLUTION

Analyze and Plan We are asked in (a) to calculate the dipole moment of HCl that would result if there were a full charge transferred from H to Cl. We can use Equation 8.10 to obtain this result. In (b), we are given the actual dipole moment for the molecule and will use that value to calculate the actual partial charges on the H and Cl atoms.

Solve

(a) The charge on each atom is the electronic charge, $e = 1.60 \times 10^{-19}$ C. The separation is 1.27 Å. The dipole moment is therefore

$$\mu = Qr = (1.60 \times 10^{-19} \text{ C})(1.27\text{Å})\left(\frac{10^{-10} \text{ m}}{1\text{Å}}\right)\left(\frac{1 \text{ D}}{3.34 \times 10^{-30} \text{ C-m}}\right) = 6.08 \text{ D}$$

(b) We know the value of μ, 1.08 D, and the value of r, 1.27 Å. We want to calculate the value of Q:

$$Q = \frac{\mu}{r} = \frac{(1.08 \text{ D})\left(\dfrac{3.34 \times 10^{-30} \text{ C-m}}{1 \text{ D}}\right)}{(1.27 \text{ Å})\left(\dfrac{10^{-10} \text{ m}}{1\text{Å}}\right)} = 2.84 \times 10^{-20} \text{ C}$$

We can readily convert this charge to units of e:

$$\text{Charge in } e = (2.84 \times 10^{-20} \text{ C})\left(\frac{1e}{1.60 \times 10^{-19} \text{ C}}\right) = 0.178e$$

Thus, the experimental dipole moment indicates that the charge separation in the HCl molecule is

$$\overset{0.178+}{\text{H}} \relbar \overset{0.178-}{\text{Cl}}$$

Because the experimental dipole moment is less than that calculated in part (a), the charges on the atoms are much less than a full electronic charge. We could have anticipated this because the H—Cl bond is polar covalent rather than ionic.

Practice Exercise 1

Calculate the dipole moment for HF (bond length 0.917 Å), assuming that the bond is completely ionic. (a) 0.917 D, (b) 1.91 D, (c) 2.75 D, (d) 4.39 D, (e) 7.37 D

Practice Exercise 2

The dipole moment of chlorine monofluoride, ClF(g), is 0.88 D. The bond length of the molecule is 1.63 Å. (a) Which atom is expected to have the partial negative charge? (b) What is the charge on that atom in units of e?

▼ Table 8.3 presents the bond lengths and dipole moments of the hydrogen halides. Notice that as we proceed from HF to HI, the electronegativity difference decreases and the bond length increases. The first effect decreases the amount of charge separated and causes the dipole moment to decrease from HF to HI, even though the bond length is increasing. Calculations identical to those used in Sample Exercise 8.5 show that the charges on the atoms decrease from 0.41+ and 0.41− in HF to 0.057+ and 0.057− in HI. We can visualize the varying degree of electronic charge shift in these substances from computer-generated renderings based on calculations of electron distribution, as shown in **Figure 8.10**. For these molecules, the change in the electronegativity difference has a greater effect on the dipole moment than does the change in bond length.

 Give It Some Thought

The bond between carbon and hydrogen is one of the most important types of bonds in chemistry. The length of an H—C bond is approximately 1.1 Å. Based on this distance and differences in electronegativity, do you expect the dipole moment of an individual H—C bond to be larger or smaller than that of an H—I bond?

Table 8.3 Bond Lengths, Electronegativity Differences, and Dipole Moments of the Hydrogen Halides

Compound	Bond Length (Å)	Electronegativity Difference	Dipole Moment (D)
HF	0.92	1.9	1.82
HCl	1.27	0.9	1.08
HBr	1.41	0.7	0.82
HI	1.61	0.4	0.44

▲ GO FIGURE

How do you interpret the fact that there is no red in the HBr and HI representations?

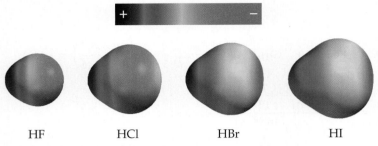

| HF | HCl | HBr | HI |

▲ **Figure 8.10 Charge separation in the hydrogen halides.** In HF, the strongly electronegative F pulls much of the electron density away from H. In HI, the I, being much less electronegative than F, does not attract the shared electrons as strongly, and, consequently, there is far less polarization of the bond.

Before leaving this section, let's return to the LiF molecule in Figure 8.8. Under standard conditions, LiF exists as an ionic solid with an arrangement of atoms analogous to the sodium chloride structure shown in Figure 8.3. However, it is possible to generate LiF *molecules* by vaporizing the ionic solid at high temperature. The molecules have a dipole moment of 6.28 D and a bond distance of 1.53 Å. From these values we can calculate the charge on lithium and fluorine to be 0.857+ and 0.857−, respectively. This bond is extremely polar, and the presence of such large charges strongly favors the formation of an extended ionic lattice in which each lithium ion is surrounded by fluoride ions and vice versa. But even here, the experimentally determined charges on the ions are still not 1+ and 1−. This tells us that even in ionic compounds, there is still some covalent contribution to the bonding.

Differentiating Ionic and Covalent Bonding

To understand the interactions responsible for chemical bonding, it is advantageous to treat ionic and covalent bonding separately. That is the approach taken in this chapter, as well as in most other undergraduate-level chemistry texts. In reality, however, there is a continuum between the extremes of ionic and covalent bonding. This lack of a well-defined separation between the two types of bonding may seem unsettling or confusing at first.

The simple models of ionic and covalent bonding presented in this chapter go a long way toward understanding and predicting the structures and properties of chemical compounds. When covalent bonding is dominant, we expect compounds to exist as molecules,* having all the properties we associate with molecular substances, such as relatively low melting and boiling points and nonelectrolyte behavior when dissolved in water. When ionic bonding is dominant, we expect the compounds to be brittle, high-melting solids with extended lattice structures, exhibiting strong electrolyte behavior when dissolved in water.

Naturally, there are exceptions to these general characterizations, some of which we examine later in the book. Nonetheless, the ability to quickly categorize the predominant bonding interactions in a substance as covalent or ionic imparts considerable insight into the properties of that substance. The question then becomes the best way to recognize which type of bonding dominates.

*There are some exceptions to this statement, such as network solids, including diamond, silicon, and germanium, where an extended structure is formed even though the bonding is clearly covalent. These examples are discussed in Section 12.7.

The simplest approach is to assume that the interaction between a metal and a nonmetal is ionic and that between two nonmetals is covalent. While this classification scheme is reasonably predictive, there are far too many exceptions to use it blindly. For example, tin is a metal and chlorine is a nonmetal, but $SnCl_4$ is a molecular substance that exists as a colorless liquid at room temperature. It freezes at $-33\,°C$ and boils at $114\,°C$. The characteristics of $SnCl_4$ are not those typical of an ionic substance. Is there a more predictable way of determining what type of bonding is prevalent in a compound? A more sophisticated approach is to use the difference in electronegativity as the main criterion for determining whether ionic or covalent bonding will be dominant. This approach correctly predicts the bonding in $SnCl_4$ to be polar covalent based on an electronegativity difference of 1.2 and at the same time correctly predicts the bonding in NaCl to be predominantly ionic based on an electronegativity difference of 2.1.

Evaluating bonding based on electronegativity difference is a useful system, but it has one shortcoming. The electronegativity values given in Figure 8.7 do not take into account changes in bonding that accompany changes in the oxidation state of the metal. For example, Figure 8.7 gives the electronegativity difference between manganese and oxygen as $3.5 - 1.5 = 2.0$, which falls in the range where the bonding is normally considered ionic (the electronegativity difference for NaCl is $3.0 - 0.9 = 2.1$). Therefore, it is not surprising to learn that manganese(II) oxide, MnO, is a green solid that melts at $1842\,°C$ and has the same crystal structure as NaCl.

However, the bonding between manganese and oxygen is not always ionic. Manganese(VII) oxide, Mn_2O_7, is a green liquid that freezes at $5.9\,°C$, an indication that covalent rather than ionic bonding dominates. The change in the oxidation state of manganese is responsible for the change in bonding. In general, as the oxidation state of a metal increases, so does the degree of covalent bonding. When the oxidation state of the metal is highly positive (roughly speaking, $+4$ or larger), we can expect significant covalency in the bonds it forms with nonmetals. Thus, metals in high oxidation states form molecular substances, such as Mn_2O_7, or polyatomic ions, such as MnO_4^- and CrO_4^{2-}, rather than ionic compounds.

Give It Some Thought

You have a yellow solid that melts at $41\,°C$ and boils at $131\,°C$ and a green solid that melts at $2320\,°C$. If you are told that one of them is Cr_2O_3 and the other is OsO_4, which one do you expect to be the yellow solid?

8.5 | Drawing Lewis Structures

Lewis structures can help us understand the bonding in many compounds and are frequently used when discussing the properties of molecules. For this reason, drawing Lewis structures is an important skill that you should practice. Here is the procedure you should follow in drawing Lewis structures:

1. **Sum the valence electrons from all atoms, taking into account overall charge.** Use the periodic table to help you determine the number of valence electrons in each atom. For an anion, add one electron to the total for each negative charge. For a cation, subtract one electron from the total for each positive charge. Do not worry about keeping track of which electrons come from which atoms. Only the total number is important.

2. **Write the symbols for the atoms, show which atoms are attached to which, and connect them with a single bond (*a line, representing* two *electrons*).** Chemical formulas are often written in the order in which the atoms are connected in the molecule or ion. The formula HCN, for example, tells you that the carbon atom is bonded to the H and to the N. In many polyatomic molecules and

ions, the central atom is usually written first, as in CO_3^{2-} and SF_4. Remember that the central atom is generally less electronegative than the atoms surrounding it. In other cases, you may need more information before you can draw the Lewis structure.

3. **Complete the octets around all the atoms bonded to the central atom.** Keep in mind that a hydrogen atom has only a single pair of electrons around it.

4. **Place any leftover electrons on the central atom,** even if doing so results in more than an octet of electrons around the atom.

5. **If there are not enough electrons to give the central atom an octet, try multiple bonds.** Use one or more of the unshared pairs of electrons on the atoms bonded to the central atom to form double or triple bonds.

The following examples of this procedure will help you put it into practice.

SAMPLE EXERCISE 8.6 | **Drawing a Lewis Structure**

Draw the Lewis structure for phosphorus trichloride, PCl_3.

SOLUTION

Analyze and Plan We are asked to draw a Lewis structure from a molecular formula. Our plan is to follow the five-step procedure just described.

Solve First, we sum the valence electrons. Phosphorus (group 5A) has five valence electrons, and each chlorine (group 7A) has seven. The total number of valence electrons is therefore

$$5 + (3 \times 7) = 26$$

Second, we arrange the atoms to show which atom is connected to which, and we draw a single bond between them. There are various ways the atoms might be arranged. It helps to know, though, that in binary compounds the first element in the chemical formula is generally surrounded by the remaining atoms. So we proceed to draw a skeleton structure in which a single bond connects the P atom to each Cl atom:

$$\text{Cl} \!-\! \text{P} \!-\! \text{Cl}$$
$$| $$
$$\text{Cl}$$

(It is not crucial that the Cl atoms be to the left of, right of, and below the P atom—any structure that shows each of the three Cl atoms bonded to P will work.)

Third, we add Lewis electron dots to complete the octets on the atoms bonded to the central atom. Completing the octets around each Cl

atom accounts for 24 electrons (remember, each line in our structure represents *two* electrons):

$$:\!\ddot{\text{Cl}} \!-\! \text{P} \!-\! \ddot{\text{Cl}}\!:$$
$$|$$
$$:\!\ddot{\text{Cl}}\!:$$

Fourth, recalling that our total number of electrons is 26, we place the remaining two electrons on the central atom, P, which completes its octet:

$$:\!\ddot{\text{Cl}} \!-\! \ddot{\text{P}} \!-\! \ddot{\text{Cl}}\!:$$
$$|$$
$$:\!\ddot{\text{Cl}}\!:$$

This structure gives each atom an octet, so we stop at this point. (In checking for octets, remember to count a single bond as two electrons.)

Practice Exercise 1

Which of these molecules has a Lewis structure with a central atom having no nonbonding electron pairs? **(a)** CO_2, **(b)** H_2S, **(c)** PF_3, **(d)** SiF_4, **(e)** more than one of a, b, c, d.

Practice Exercise 2

(a) How many valence electrons should appear in the Lewis structure for CH_2Cl_2?
(b) Draw the Lewis structure.

SAMPLE EXERCISE 8.7 | **Lewis Structure with a Multiple Bond**

Draw the Lewis structure for HCN.

SOLUTION

Hydrogen has one valence electron, carbon (group 4A) has four, and nitrogen (group 5A) has five. The total number of valence electrons is, therefore, $1 + 4 + 5 = 10$. In principle, there are different ways in which we might choose to arrange the atoms. Because hydrogen can accommodate only one electron pair, it always has only one single bond associated with it. Therefore, C—H—N is an impossible arrangement. The remaining two possibilities are H—C—N and H—N—C. The first is the arrangement found experimentally. You might have guessed this because (a) the formula is written with the

atoms in this order and (b) carbon is less electronegative than nitrogen. Thus, we begin with the skeleton structure

$$\text{H} \!-\! \text{C} \!-\! \text{N}$$

The two bonds account for four electrons. The H atom can have only two electrons associated with it, and so we will not add any more electrons to it. If we place the remaining six electrons around N to give it an octet, we do not achieve an octet on C:

$$\text{H} \!-\! \text{C} \!-\! \ddot{\text{N}}\!:$$

We therefore try a double bond between C and N, using one of the unshared pairs we placed on N. Again we end up with fewer than eight electrons on C, and so we next try a triple bond. This structure gives an octet around both C and N:

$$H—C \overset{\curvearrowleft}{\underset{\curvearrowright}{\ddot{N}}}: \longrightarrow H—C≡N:$$

The octet rule is satisfied for the C and N atoms, and the H atom has two electrons around it. This is a correct Lewis structure.

Practice Exercise 1

Draw the Lewis structure(s) for the molecule with the chemical formula C_2H_3N, where the N is connected to only one other atom. How many double bonds are there in the correct Lewis structure? **(a)** zero, **(b)** one, **(c)** two, **(d)** three, **(e)** four.

Practice Exercise 2

Draw the Lewis structure for **(a)** NO^+ ion, **(b)** C_2H_4.

SAMPLE EXERCISE 8.8 | Lewis Structure for a Polyatomic Ion

Draw the Lewis structure for the BrO_3^- ion.

SOLUTION

Bromine (group 7A) has seven valence electrons, and oxygen (group 6A) has six. We must add one more electron to our sum to account for the 1− charge of the ion. The total number of valence electrons is, therefore, $7 + (3 \times 6) + 1 = 26$. For oxyanions — SO_4^{2-}, NO_3^-, CO_3^{2-}, and so forth — the oxygen atoms surround the central nonmetal atom. After arranging the O atoms around the Br atom, drawing single bonds, and distributing the unshared electron pairs, we have

$$\left[:\ddot{O}—\ddot{Br}—\ddot{O}: \atop :\ddot{O}: \right]^-$$

Notice that the Lewis structure for an ion is written in brackets and the charge is shown outside the brackets at the upper right.

Practice Exercise 1

How many nonbonding electron pairs are there in the Lewis structure of the peroxide ion, O_2^{2-}? **(a)** 7, **(b)** 6, **(c)** 5, **(d)** 4, **(e)** 3.

Practice Exercise 2

Draw the Lewis structure for **(a)** ClO_2^-, **(b)** PO_4^{3-}.

Formal Charge and Alternative Lewis Structures

When we draw a Lewis structure, we are describing how the electrons are distributed in a molecule or polyatomic ion. In some instances we can draw two or more valid Lewis structures for a molecule that all obey the octet rule. All of these structures can be thought of as contributing to the *actual* arrangement of the electrons in the molecule, but not all of them will contribute to the same extent. How do we decide which one of several Lewis structures is the most important? One approach is to do some "bookkeeping" of the valence electrons to determine the *formal charge* of each atom in each Lewis structure. The **formal charge** of any atom in a molecule is the charge the atom would have if each bonding electron pair in the molecule were shared equally between its two atoms.

To calculate the formal charge on any atom in a Lewis structure, we assign electrons to the atom as follows:

1. *All* unshared (nonbonding) electrons are assigned to the atom on which they are found.

2. For any bond—single, double, or triple—*half* of the bonding electrons are assigned to each atom in the bond.

The formal charge of each atom is calculated by subtracting the number of electrons assigned to the atom from the number of valence electrons in the neutral atom:

Formal charge = valence electrons − $\frac{1}{2}$ (bonding electrons) − nonbonding electrons [8.11]

Let's practice by calculating the formal charges for the atoms in the cyanide ion, CN^-, which has the Lewis structure

$$[:C≡N:]^-$$

The neutral C atom has four valence electrons. There are six electrons in the cyanide triple bond and two nonbonding electrons on C. We calculate the formal charge on C as $4 − \frac{1}{2}(6) − 2 = −1$. For N, the valence electron count is five; there are six electrons in the cyanide triple bond, and two nonbonding electrons on the N.

The formal charge on N is $5 - \frac{1}{2}(6) - 2 = 0$. We can draw the whole ion with its formal charges as

$$\overset{-1 \quad 0}{[:C{\equiv}N:]^-}$$

Notice that the sum of the formal charges equals the overall charge on the ion, 1−. The formal charges on a neutral molecule must add to zero, whereas those on an ion add to give the charge on the ion.

If we can draw several Lewis structures for a molecule, the concept of formal charge can help us decide which is the most important, which we shall call the *dominant* Lewis structure. One Lewis structure for CO_2 for instance, has two double bonds. However, we can also satisfy the octet rule by drawing a Lewis structure having one single bond and one triple bond. Calculating formal charges in these structures, we have

	:Ö=C=Ö:			:Ö—C≡O:	
Valence electrons:	6	4	6	6	4	6
−(Electrons assigned to atom):	6	4	6	7	4	5
Formal charge:	0	0	0	−1	0	+1

Note that in both cases the formal charges add up to zero, as they must because CO_2 is a neutral molecule. So, which is the more correct structure? As a general rule, when more than one Lewis structure is possible, we will use the following guidelines to choose the dominant one:

1. The dominant Lewis structure is generally the one in which the atoms bear formal charges closest to zero.

2. A Lewis structure in which any negative charges reside on the more electronegative atoms is generally more dominant than one that has negative charges on the less electronegative atoms.

Thus, the first Lewis structure of CO_2 is the dominant one because the atoms carry no formal charges and so satisfy the first guideline. The other Lewis structure shown (and the similar one that has a triple bond to the left O and a single bond to the right O) contribute to the actual structure to a much smaller extent.

Although the concept of formal charge helps us to arrange alternative Lewis structures in order of importance, it is important that you remember that *formal charges do not represent real charges on atoms*. These charges are just a bookkeeping convention. The actual charge distributions in molecules and ions are determined not by formal charges but by a number of other factors, including electronegativity differences between atoms.

▲ Give It Some Thought

Suppose a Lewis structure for a neutral fluorine-containing molecule results in a formal charge of +1 on the fluorine atom. What conclusion would you draw?

SAMPLE EXERCISE 8.9 | Lewis Structures and Formal Charges

Three possible Lewis structures for the thiocyanate ion, NCS⁻, are

$$[:\ddot{N}-C{\equiv}S:]^- \qquad [\ddot{N}=C=\ddot{S}:]^- \qquad [:N{\equiv}C-\ddot{S}:]^-$$

(a) Determine the formal charges in each structure.

(b) Based on the formal charges, which Lewis structure is the dominant one?

SOLUTION

(a) Neutral N, C, and S atoms have five, four, and six valence electrons, respectively. We can determine the formal charges in the three structures by using the rules we just discussed:

$$\overset{-2 \quad 0 \quad +1}{[:\ddot{N}-C{\equiv}S:]^-} \qquad \overset{-1 \quad 0 \quad 0}{[\ddot{N}=C=\ddot{S}:]^-} \qquad \overset{0 \quad 0 \quad -1}{[:N{\equiv}C-\ddot{S}:]^-}$$

As they must, the formal charges in all three structures sum to $1-$, the overall charge of the ion.

(b) The dominant Lewis structure generally produces formal charges of the smallest magnitude (guideline 1). That eliminates the left structure as the dominant one. Further, as discussed in Section 8.4, N is more electronegative than C or S. Therefore, we expect any negative formal charge to reside on the N atom (guideline 2). For these two reasons, the middle Lewis structure is the dominant one for NCS^-.

Practice Exercise 1

Phosphorus oxychloride has the chemical formula $POCl_3$, with P as the central atom. To minimize formal charge, how many bonds does phosphorus make to the other atoms in the molecule? (Count each single bond as one, each double bond as two, and each triple bond as three.)
(a) 3, **(b)** 4, **(c)** 5, **(d)** 6, **(e)** 7

Practice Exercise 2

The cyanate ion, NCO^-, has three possible Lewis structures. **(a)** Draw these three structures and assign formal charges in each. **(b)** Which Lewis structure is dominant?

A Closer Look

Oxidation Numbers, Formal Charges, and Actual Partial Charges

In Chapter 4 we introduced the rules for assigning *oxidation numbers* to atoms. The concept of electronegativity is the basis of these numbers. An atom's oxidation number is the charge the atom would have if its bonds were completely ionic. That is, in determining oxidation number, all shared electrons are counted with the more electronegative atom. For example, consider the Lewis structure of HCl in ▼ Figure 8.11(**a**). To assign oxidation numbers, both electrons in the covalent bond between the atoms are assigned to the more electronegative Cl atom. This procedure gives Cl eight valence electrons, one more than in the neutral atom. Thus, its oxidation number is -1. Hydrogen has no valence electrons when they are counted this way, giving it an oxidation number of $+1$.

In assigning formal charges to the atoms in HCl [Figure 8.11(**b**)], we ignore electronegativity; the electrons in bonds are assigned equally to the two bonded atoms. In this case Cl has seven assigned electrons, the same as that of the neutral Cl atom, and H has one assigned electron. Thus, the formal charges of both Cl and H in this compound are 0.

Neither oxidation number nor formal charge gives an accurate depiction of the actual charges on atoms because oxidation numbers overstate the role of electronegativity and formal charges ignore it. It seems reasonable that electrons in covalent bonds should be apportioned according to the relative electronegativities of the bonded atoms. Figure 8.7

(page 310) shows that Cl has an electronegativity of 3.0, while the electronegativity value of H is 2.1. The more electronegative Cl atom might therefore be expected to have roughly $3.0/(3.0 + 2.1) = 0.59$ of the electrical charge in the bonding pair, whereas the H atom would have $2.1/(3.0 + 2.1) = 0.41$ of the charge. Because the bond consists of two electrons, the Cl atom's share is $0.59 \times 2e = 1.18e$, or $0.18e$ more than the neutral Cl atom. This gives rise to a partial negative charge of $0.18-$ on Cl and therefore a partial positive charge of $0.18+$ on H. (Notice again that we place the plus and minus signs *before* the magnitude in writing oxidation numbers and formal charges but *after* the magnitude in writing actual charges.)

The dipole moment of HCl gives an experimental measure of the partial charge on each atom. In Sample Exercise 8.5 we saw that the dipole moment of HCl corresponds to a partial charge of $0.178+$ on H and $0.178-$ on Cl, in remarkably good agreement with our simple approximation based on electronegativities. Although our approximation method provides ballpark numbers for the magnitude of charge on atoms, the relationship between electronegativities and charge separation is generally more complicated. As we have already seen, computer programs employing quantum mechanical principles have been developed to obtain more accurate estimates of the partial charges on atoms, even in complex molecules. A computer-graphical representation of the calculated charge distribution in HCl is shown in Figure 8.11(**c**).

Related Exercises: *8.8, 8.49, 8.50, 8.51, 8.52, 8.86, 8.87, 8.90, 8.91*

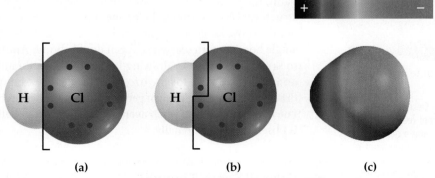

▲ **Figure 8.11 (a)** Oxidation number, **(b)** formal charge, and **(c)** electron-density distribution for the HCl molecule.

8.6 | Resonance Structures

GO FIGURE

What feature of this structure suggests that the two outer O atoms are in some way equivalent to each other?

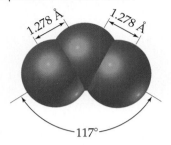

▲ Figure 8.12 **Molecular structure of ozone.**

Is the electron density consistent with equal contributions from the two resonance structures for O_3? Explain.

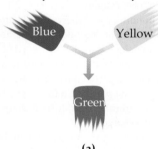

Primary color   Primary color

Blue   Yellow

Green

(a)

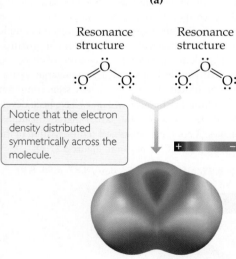

Resonance   Resonance
structure   structure

Notice that the electron density distributed symmetrically across the molecule.

Ozone molecule

(b)

▲ Figure 8.13 **Resonance.** Describing a molecule as a blend of different resonance structures is similar to describing a paint color as a blend of primary colors. **(a)** Green paint is a blend of blue and yellow. We cannot describe green as a single primary color. **(b)** The ozone molecule is a blend of two resonance structures. We cannot describe the ozone molecule in terms of a single Lewis structure.

We sometimes encounter molecules and ions in which the experimentally determined arrangement of atoms is not adequately described by a single dominant Lewis structure. Consider ozone, O_3, which is a bent molecule with two equal O—O bond lengths (◄ Figure 8.12). Because each oxygen atom contributes 6 valence electrons, the ozone molecule has 18 valence electrons. This means the Lewis structure must have one O—O single bond and one O=O double bond to attain an octet about each atom:

However, this single structure cannot by itself be dominant because it requires that one O—O bond be different from the other, contrary to the observed structure— we would expect the O=O double bond to be shorter than the O—O single bond. ∞ (Section 8.3) In drawing the Lewis structure, however, we could just as easily have put the O=O bond on the left:

There is no reason for one of these Lewis structures to be dominant because they are equally valid representations of the molecule. The placement of the atoms in these two alternative but completely equivalent Lewis structures is the same, but the placement of the electrons is different; we call Lewis structures of this sort **resonance structures**. To describe the structure of ozone properly, we write both resonance structures and use a double-headed arrow to indicate that the real molecule is described by an average of the two:

To understand why certain molecules require more than one resonance structure, we can draw an analogy to mixing paint (◄ Figure 8.13). Blue and yellow are both primary colors of paint pigment. An equal blend of blue and yellow pigments produces green pigment. We cannot describe green paint in terms of a single primary color, yet it still has its own identity. Green paint does not oscillate between its two primary colors: It is not blue part of the time and yellow the rest of the time. Similarly, molecules such as ozone cannot be described as oscillating between the two individual Lewis structures shown previously—there are two equivalent dominant Lewis structures that contribute equally to the actual structure of the molecule.

The actual arrangement of the electrons in molecules such as O_3 must be considered as a blend of two (or more) Lewis structures. By analogy to the green paint, the molecule has its own identity separate from the individual resonance structures. For example, the ozone molecule always has two equivalent O—O bonds whose lengths are intermediate between the lengths of an oxygen–oxygen single bond and an oxygen–oxygen double bond. Another way of looking at it is to say that the rules for drawing Lewis structures do not allow us to have a single dominant structure for the ozone molecule. For example, there are no rules for drawing half-bonds. We can get around this limitation by drawing two equivalent Lewis structures that, when averaged, amount to something very much like what is observed experimentally.

▲ **Give It Some Thought**

The O—O bonds in ozone are often described as "one-and-a-half" bonds. Is this description consistent with the idea of resonance?

As an additional example of resonance structures, consider the nitrate ion, NO_3^-, for which three equivalent Lewis structures can be drawn:

Notice that the arrangement of atoms is the same in each structure—only the placement of electrons differs. In writing resonance structures, the same atoms must be bonded to each other in all structures, so that the only differences are in the arrangements of electrons. All three NO_3^- Lewis structures are equally dominant and taken together adequately describe the ion, in which all three N—O bond lengths are the same.

▲ Give It Some Thought

In the same sense that we describe the O—O bonds in O_3 as "one-and-a-half" bonds, how would you describe the N—O bonds in NO_3^-?

There are some molecules or ions for which all possible Lewis structures may not be equivalent; in other words, one or more resonance structures are more dominant than others. We will encounter examples of this as we proceed.

SAMPLE EXERCISE 8.10 | Resonance Structures

Which is predicted to have the shorter sulfur–oxygen bonds, SO_3 or SO_3^{2-}?

SOLUTION

The sulfur atom has six valence electrons, as does oxygen. Thus, SO_3 contains 24 valence electrons. In writing the Lewis structure, we see that three equivalent resonance structures can be drawn:

As with NO_3^- the actual structure of SO_3 is an equal blend of all three. Thus, each S—O bond length should be about one-third of the way between the length of a single bond and the length of a double bond. That is, S—O should be shorter than single bonds but not as short as double bonds.

The SO_3^{2-} ion has 26 electrons, which leads to a dominant Lewis structure in which all the S—O bonds are single:

Our analysis of the Lewis structures thus far leads us to conclude that SO_3 should have the shorter S—O bonds and SO_3^{2-} the longer ones. This conclusion is correct: The experimentally measured S—O bond lengths are 1.42 Å in SO_3 and 1.51 Å in SO_3^{2-}.

Practice Exercise 1

Which of these statements about resonance is true?
(a) When you draw resonance structures, it is permissible to alter the way atoms are connected.
(b) The nitrate ion has one long N—O bond and two short N—O bonds.
(c) "Resonance" refers to the idea that molecules are resonating rapidly between different bonding patterns.
(d) The cyanide ion has only one dominant resonance structure.
(e) All of the above are true.

Practice Exercise 2

Draw two equivalent resonance structures for the formate ion, HCO_2^-.

What is the significance of the dashed bonds in this ball-and-stick model?

▲ **Figure 8.14 Benzene, an "aromatic" organic compound.** The benzene molecule is a regular hexagon of carbon atoms with a hydrogen atom bonded to each one. The dashed lines represent the blending of two equivalent resonance structures, leading to C—C bonds that are intermediate between single and double bonds.

Resonance in Benzene

Resonance is an important concept in describing the bonding in organic molecules, particularly *aromatic* organic molecules, a category that includes the hydrocarbon *benzene*, C_6H_6. The six C atoms are bonded in a hexagonal ring, and one H atom is bonded to each C atom. We can write two equivalent dominant Lewis structures for benzene, each of which satisfies the octet rule. These two structures are in resonance:

Note that the double bonds are in different places in the two structures. Each of these resonance structures shows three carbon–carbon single bonds and three carbon–carbon double bonds. However, experimental data show that all six C—C bonds are of equal length, 1.40 Å, intermediate between the typical bond lengths for a C—C single bond (1.54 Å) and a C=C double bond (1.34 Å). Each of the C—C bonds in benzene can be thought of as a blend of a single bond and a double bond (◀ Figure 8.14).

Benzene is commonly represented by omitting the hydrogen atoms and showing only the carbon–carbon framework with the vertices unlabeled. In this convention, the resonance in the molecule is represented either by two structures separated by a double-headed arrow or by a shorthand notation in which we draw a hexagon with a circle inside:

The shorthand notation reminds us that benzene is a blend of two resonance structures—it emphasizes that the C=C double bonds cannot be assigned to specific edges of the hexagon. Chemists use both representations of benzene interchangeably.

The bonding arrangement in benzene confers special stability to the molecule. As a result, millions of organic compounds contain the six-membered ring characteristic of benzene. Many of these compounds are important in biochemistry, in pharmaceuticals, and in the production of modern materials.

▲ **Give It Some Thought**

Each Lewis structure of benzene has three C=C double bonds. Another hydrocarbon containing three C=C double bonds is *hexatriene*, C_6H_8. A Lewis structure of hexatriene is

Do you expect hexatriene to have multiple resonance structures? If not, why is this molecule different from benzene with respect to resonance?

8.7 | Exceptions to the Octet Rule

The octet rule is so simple and useful in introducing the basic concepts of bonding that you might assume it is always obeyed. In Section 8.2, however, we noted its limitation in dealing with ionic compounds of the transition metals. The rule also fails in many situations involving covalent bonding. These exceptions to the octet rule are of three main types:

1. Molecules and polyatomic ions containing an odd number of electrons
2. Molecules and polyatomic ions in which an atom has fewer than an octet of valence electrons
3. Molecules and polyatomic ions in which an atom has more than an octet of valence electrons

Odd Number of Electrons

In the vast majority of molecules and polyatomic ions, the total number of valence electrons is even, and complete pairing of electrons occurs. However, in a few molecules and polyatomic ions, such as ClO_2, NO, NO_2 and O_2^-, the number of valence electrons is odd. Complete pairing of these electrons is impossible, and an octet around each atom cannot be achieved. For example, NO contains $5 + 6 = 11$ valence electrons. The two most important Lewis structures for this molecule are

$$\ddot{\text{N}}=\ddot{\text{O}} \quad \text{and} \quad \dot{\text{N}}=\ddot{\text{O}}$$

 Give It Some Thought

Which of the Lewis structures for NO is dominant based on analysis of the formal charges?

Less Than an Octet of Valence Electrons

A second type of exception occurs when there are fewer than eight valence electrons around an atom in a molecule or polyatomic ion. This situation is also relatively rare (with the exception of hydrogen and helium as we have already discussed), most often encountered in compounds of boron and beryllium. As an example, let's consider boron trifluoride, BF_3. If we follow the first steps of our procedure for drawing Lewis structures, we obtain the structure

which has only six electrons around the boron atom. The formal charge is zero on both B and F, and we could complete the octet around boron by forming a double bond (recall that if there are not enough electrons to give the central atom an octet, a multiple bond may be the answer). In so doing, we see that there are three equivalent resonance structures (the formal charges are shown in red):

Each of these structures forces a fluorine atom to share additional electrons with the boron atom, which is inconsistent with the high electronegativity of fluorine. In fact, the formal charges tell us that this is an unfavorable situation. In each structure, the F atom involved in the B=F double bond has a formal charge of +1 while the less electronegative B atom has a formal charge of −1. Thus, the resonance structures containing a B=F double bond are less important than the one in which there are fewer than an octet of valence electrons around boron:

Dominant   Less important

We usually represent BF_3 solely by the dominant resonance structure, in which there are only six valence electrons around boron. The chemical behavior of BF_3 is consistent with this representation. In particular, BF_3 reacts energetically with molecules having an unshared pair of electrons that can be used to form a bond with boron, as, for example, in the reaction

$$
\begin{array}{ccc}
& H & F \\
& | & | \\
H-N: & + & B-F \\
& | & | \\
& H & F
\end{array}
\longrightarrow
\begin{array}{cc}
H & F \\
| & | \\
H-N-B-F \\
| & | \\
H & F
\end{array}
$$

In the stable compound NH_3BF_3, boron has an octet of valence electrons.

More Than an Octet of Valence Electrons

The third and largest class of exceptions consists of molecules or polyatomic ions in which there are more than eight electrons in the valence shell of an atom. When we draw the Lewis structure for PF_5, for example, we are forced to place ten electrons around the central phosphorus atom:

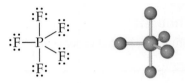

Molecules and ions with more than an octet of electrons around the central atom are often called *hypervalent*. Other examples of hypervalent species are SF_4, AsF_6^-, and ICl_4^-. The corresponding molecules with a second-period atom as the central atom, such as NCl_5 and OF_4, do *not* exist.

Hypervalent molecules are formed only for central atoms from period 3 and below in the periodic table. The principal reason for their formation is the relatively larger size of the central atom. For example, a P atom is large enough that five F (or even five Cl) atoms can be bonded to it without being too crowded. By contrast, an N atom is too small to accommodate five atoms bonded to it. Because size is a factor, hypervalent molecules occur most often when the central atom is bonded to the smallest and most electronegative atoms—F, Cl, and O.

The notion that a valence shell can contain more than eight electrons is also consistent with the presence of unfilled nd orbitals in atoms from period 3 and below. ∞ (Section 6.8) By comparison, in elements of the second period, only the $2s$ and $2p$ valence orbitals are available for bonding. However, theoretical work on the bonding in molecules such as PF_5 and SF_6 suggests that the presence of unfilled $3d$ orbitals in P and S has a relatively minor impact on the formation of hypervalent molecules. Most chemists now believe that the larger size of the atoms from periods 3 through 6 is more important to explain hypervalency than is the presence of unfilled d orbitals.

SAMPLE EXERCISE 8.11 | Lewis Structure for an Ion with More Than an Octet of Electrons

Draw the Lewis structure for ICl_4^-.

SOLUTION

Iodine (group 7A) has seven valence electrons. Each chlorine atom (group 7A) also has seven. An extra electron is added to account for the 1− charge of the ion. Therefore, the total number of valence electrons is $7 + (4 \times 7) + 1 = 36$.

The I atom is the central atom in the ion. Putting eight electrons around each Cl atom (including a pair of electrons between I and each Cl to represent the single bond between these atoms) requires $8 \times 4 = 32$ electrons.

We are thus left with $36 - 32 = 4$ electrons to be placed on the larger iodine:

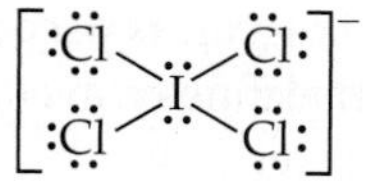

Iodine has 12 valence electrons around it, four more than needed for an octet.

> **Practice Exercise 1**
>
> In which of these molecules or ions is there only one lone pair of electrons on the central sulfur atom?
> (a) SF_4, (b) SF_6, (c) SOF_4, (d) SF_2, (e) SO_4^{2-}
>
> **Practice Exercise 2**
>
> (a) Which of the following atoms is never found with more than an octet of valence electrons around it? S, C, P, Br, I. (b) Draw the Lewis structure for XeF_2.

Finally, there are Lewis structures where you might have to choose between satisfying the octet rule and obtaining the most favorable formal charges by using more than an octet of electrons. For example, consider these Lewis structures for the phosphate ion, PO_4^{3-}:

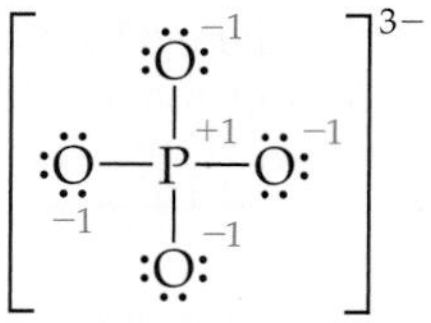

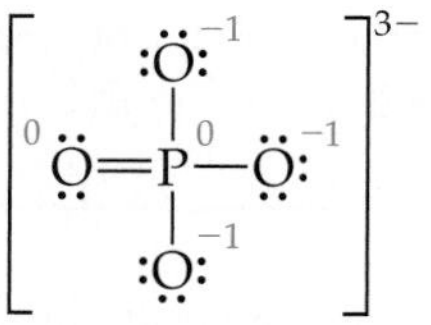

The formal charges on the atoms are shown in red. In the left structure, the P atom obeys the octet rule. In the right structure, however, the P atom has five electron pairs, leading to smaller formal charges on the atoms. (You should be able to see that there are three additional resonance structures for the Lewis structure on the right.)

Chemists are still debating which of these two structures is dominant for PO_4^{3-}. Recent theoretical calculations based on quantum mechanics suggest to some researchers that the left structure is the dominant one. Other researchers claim that the bond lengths in the ion are more consistent with the right structure being dominant. This disagreement is a convenient reminder that, in general, multiple Lewis structures can contribute to the actual electron distribution in an atom or molecule.

8.8 | Strengths and Lengths of Covalent Bonds

The stability of a molecule is related to the strengths of its covalent bonds. The strength of a covalent bond between two atoms is determined by the energy required to break the bond. It is easiest to relate bond strength to the enthalpy change in reactions in which bonds are broken. ∞ (Section 5.4) The **bond enthalpy** is the enthalpy change, ΔH, for the breaking of a particular bond in one mole of a gaseous substance. For example, the bond enthalpy for the bond in Cl_2 is the enthalpy change when 1 mol of $Cl_2(g)$ dissociates into chlorine atoms:

$$:\ddot{\underset{..}{C}}l-\ddot{\underset{..}{C}}l:(g) \longrightarrow 2 :\ddot{\underset{..}{C}}l\cdot(g)$$

We use the letter D followed by the bond in question to represent bond enthalpies. Thus, for example, $D(Cl-Cl)$ is the bond enthalpy for the Cl_2 bond, and $D(H-Br)$ is the bond enthalpy for the HBr bond.

It is relatively simple to assign bond enthalpies to the bond in a diatomic molecule because in these cases the bond enthalpy is just the energy required to break the molecule into its atoms. However, many important bonds, such as the C—H bond, exist only in polyatomic molecules. For these bonds, we usually use *average* bond enthalpies. For example, the enthalpy change for the following process in which a methane molecule is decomposed into its five atoms (a process called *atomization*) can be used to define an average bond enthalpy for the C—H bond,

$$H—\overset{\displaystyle H}{\underset{\displaystyle H}{\overset{|}{\underset{|}{C}}}}—H(g) \longrightarrow \cdot\dot{C}\cdot(g) \; + \; 4\,H\cdot(g) \qquad \Delta H = 1660 \text{ kJ}$$

Because there are four equivalent C—H bonds in methane, the enthalpy of atomization is equal to the sum of the bond enthalpies of the four C—H bonds. Therefore, the average C—H bond enthalpy for CH_4 is $D(C—H) = (1660/4) \text{kJ/mol} = 415 \text{ kJ/mol}$.

The bond enthalpy for a given pair of atoms, say C—H, depends on the rest of the molecule containing the atom pair. However, the variation from one molecule to another is generally small, which supports the idea that bonding electron pairs are localized between atoms. If we consider C—H bond enthalpies in many different compounds, we find that the average bond enthalpy is 413 kJ/mol, close to the 415 kJ/mol we just calculated from CH_4.

▲ Give It Some Thought

How can you use the enthalpy of atomization of the hydrocarbon ethane, $C_2H_6(g)$, along with the value $D(C—H) = 413 \text{ kJ/mol}$ to estimate the value for $D(C—C)$?

▼ Table 8.4 lists average bond enthalpies for a number of atom pairs. *The bond enthalpy is always a positive quantity; energy is always required to break chemical bonds.* Conversely, *energy is always released when a bond forms between two gaseous atoms or*

Table 8.4 Average Bond Enthalpies (kJ/mol)

Single Bonds

C—H	413	N—H	391	O—H	463	F—F	155
C—C	348	N—N	163	O—O	146		
C—N	293	N—O	201	O—F	190	Cl—F	253
C—O	358	N—F	272	O—Cl	203	Cl—Cl	242
C—F	485	N—Cl	200	O—I	234		
C—Cl	328	N—Br	243			Br—F	237
C—Br	276			S—H	339	Br—Cl	218
C—I	240	H—H	436	S—F	327	Br—Br	193
C—S	259	H—F	567	S—Cl	253		
		H—Cl	431	S—Br	218	I—Cl	208
Si—H	323	H—Br	366	S—S	266	I—Br	175
Si—Si	226	H—I	299			I—I	151
Si—C	301						
Si—O	368						
Si—Cl	464						

Multiple Bonds

C=C	614	N=N	418	O=O	495	
C≡C	839	N≡N	941			
C=N	615	N=O	607	S=O	523	
C≡N	891			S=S	418	
C=O	799					
C≡O	1072					

molecular fragments. The greater the bond enthalpy, the stronger the bond. Further, a molecule with strong chemical bonds generally has less tendency to undergo chemical change than does one with weak bonds. For example, N_2, which has a very strong $N\equiv N$ triple bond, is very unreactive, whereas hydrazine, N_2H_4, which has an $N-N$ single bond, is highly reactive.

 Give It Some Thought

Based on bond enthalpies, which do you expect to be more reactive, oxygen, O_2, or hydrogen peroxide, H_2O_2?

Bond Enthalpies and the Enthalpies of Reactions

We can use average bond enthalpies to estimate the enthalpies of reactions in which bonds are broken and new bonds are formed. This procedure allows us to estimate quickly whether a given reaction will be endothermic ($\Delta H > 0$) or exothermic ($\Delta H < 0$) even if we do not know ΔH_f° for all the species involved.

Our strategy for estimating reaction enthalpies is a straightforward application of Hess's law. ∞ (Section 5.6) We use the fact that breaking bonds is always endothermic and forming bonds is always exothermic. We therefore imagine that the reaction occurs in two steps:

1. We supply enough energy to break those bonds in the reactants that are not present in the products. The enthalpy of the system is increased by the sum of the bond enthalpies of the bonds that are broken.

2. We form the bonds in the products that were not present in the reactants. This step releases energy and therefore lowers the enthalpy of the system by the sum of the bond enthalpies of the bonds that are formed.

The enthalpy of the reaction, ΔH_{rxn}, is estimated as the sum of the bond enthalpies of the bonds broken minus the sum of the bond enthalpies of the bonds formed:

$$\Delta H_{rxn} = \sum (\text{bond enthalpies of bonds broken}) -$$
$$\sum (\text{bond enthalpies of bonds formed}) \quad [8.12]$$

Consider, for example, the gas-phase reaction between methane, CH_4, and chlorine to produce methyl chloride, CH_3Cl, and hydrogen chloride, HCl:

$$H-CH_3(g) + Cl-Cl(g) \longrightarrow Cl-CH_3(g) + H-Cl(g) \quad \Delta H_{rxn} = ? \quad [8.13]$$

Our two-step procedure is outlined in **Figure 8.15**. We note that the following bonds are broken and made:

Bonds broken: 1 mol $C-H$, 1 mol $Cl-Cl$

Bonds made: 1 mol $C-Cl$, 1 mol $H-Cl$

We first supply enough energy to break the $C-H$ and $Cl-Cl$ bonds, which raises the enthalpy of the system (indicated as $\Delta H_1 > 0$ in Figure 8.15). We then form the $C-Cl$ and $H-Cl$ bonds, which release energy and lower the enthalpy of the system ($\Delta H_2 < 0$). We then use Equation 8.12 and data from Table 8.4 to estimate the enthalpy of the reaction:

$$\Delta H_{rxn} = [D(C-H) + D(Cl-Cl)] - [D(C-Cl) + D(H-Cl)]$$
$$= (413\text{ kJ} + 242\text{ kJ}) - (328\text{ kJ} + 431\text{ kJ}) = -104\text{ kJ}$$

The reaction is exothermic because the bonds in the products (especially the $H-Cl$ bond) are stronger than the bonds in the reactants (especially the $Cl-Cl$ bond).

We usually use bond enthalpies to estimate ΔH_{rxn} only if we do not have the needed ΔH_f° values readily available. For the preceding reaction, we cannot calculate ΔH_{rxn} from ΔH_f° values and Hess's law because ΔH_f° for $CH_3Cl(g)$ is not given in Appendix C. If we obtain the value of ΔH_f° for $CH_3Cl(g)$ from another source and use Equation 5.31,

$$\Delta H_{rxn}^\circ = \sum n\Delta H_f^\circ (\text{products}) - \sum m\Delta H_f^\circ (\text{reactants})$$

GO FIGURE

Is this reaction exothermic or endothermic?

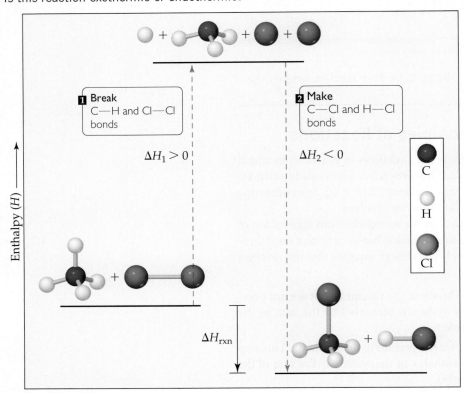

▲ **Figure 8.15 Using bond enthalpies to calculate ΔH_{rxn}.** Average bond enthalpies are used to estimate ΔH_{rxn} for the reaction of methane with chlorine to make methyl chloride and hydrogen chloride.

we find that $\Delta H_{rxn} = -99.8$ kJ for the reaction in Equation 8.13. Thus, the use of average bond enthalpies provides a reasonably accurate estimate of the actual reaction enthalpy change.

It is important to remember that bond enthalpies are derived for *gaseous* molecules and that they are often *averaged* values. Nonetheless, average bond enthalpies are useful for estimating reaction enthalpies quickly, especially for gas-phase reactions.

SAMPLE EXERCISE 8.12 Using Average Bond Enthalpies

Using data from Table 8.4, estimate ΔH for the combustion reaction

$$2\ \text{H}-\underset{\underset{\text{H}}{|}}{\overset{\overset{\text{H}}{|}}{\text{C}}}-\underset{\underset{\text{H}}{|}}{\overset{\overset{\text{H}}{|}}{\text{C}}}-\text{H}(g)\ +\ 7\ \text{O}_2(g)\ \longrightarrow\ 4\ \text{O}=\text{C}=\text{O}(g)\ +\ 6\ \text{H}-\text{O}-\text{H}(g)$$

SOLUTION

Analyze We are asked to estimate the enthalpy change for a chemical reaction by using average bond enthalpies for the bonds broken and formed.

Plan In the reactants, we must break twelve C—H bonds and two C—C bonds in the two molecules of C_2H_6 and seven O=O bonds in the seven O_2 molecules. In the products, we form eight C=O bonds (two in each CO_2) and twelve O—H bonds (two in each H_2O).

Solve Using Equation 8.12 and data from Table 8.4, we have

$$\Delta H = [12D(C-H) + 2D(C-C) + 7D(O=O)] - [8D(C=O) + 12D(O-H)]$$
$$= [12(413\,kJ) + 2(348\,kJ) + 7(495\,kJ)] - [8(799\,kJ) + 12(463\,kJ)]$$
$$= 9117\,kJ - 11948\,kJ$$
$$= -2831\,kJ$$

Check This estimate can be compared with the value of −2856 kJ calculated from more accurate thermochemical data; the agreement is good.

Practice Exercise 1

Using Table 8.4, estimate ΔH for the "water splitting reaction": $H_2O(g) \rightarrow H_2(g) + \frac{1}{2}O_2(g)$.
(a) 242 kJ, **(b)** 417 kJ, **(c)** 5 kJ, **(d)** −5 kJ, **(e)** −468 kJ

Practice Exercise 2

Using Table 8.4, estimate ΔH for the reaction

$$H-\underset{\underset{H}{|}}{N}-\underset{\underset{H}{|}}{N}-H(g) \longrightarrow N\equiv N(g) + 2\,H-H(g)$$

Bond Enthalpy and Bond Length

Just as we can define an average bond enthalpy, we can also define an average bond length for a number of common bonds (▼ Table 8.5). Of particular interest is the relationship, in any atom pair, among bond enthalpy, bond length, and number of bonds between the atoms. For example, we can use data in Tables 8.4 and 8.5 to compare the bond lengths and bond enthalpies of carbon–carbon single, double, and triple bonds:

C—C	C=C	C≡C
1.54 Å	1.34 Å	1.20 Å
348 kJ/mol	614 kJ/mol	839 kJ/mol

Table 8.5 Average Bond Lengths for Some Single, Double, and Triple Bonds

Bond	Bond Length (Å)	Bond	Bond Length (Å)
C—C	1.54	N—N	1.47
C=C	1.34	N=N	1.24
C≡C	1.20	N≡N	1.10
C—N	1.43	N—O	1.36
C=N	1.38	N=O	1.22
C≡N	1.16		
		O—O	1.48
C—O	1.43	O=O	1.21
C=O	1.23		
C≡O	1.13		

Chemistry Put to Work

Explosives and Alfred Nobel

Enormous amounts of energy can be released in chemical reactions. Perhaps the most graphic illustration of this fact is seen in certain molecular substances used as explosives. Our discussion of bond enthalpies allows us to examine more closely some of the properties of such explosive substances.

A useful explosive substance must (1) decompose very exothermically, (2) have gaseous products so that a tremendous gas pressure accompanies the decomposition, (3) decompose very rapidly, and (4) be stable enough so that it can be detonated predictably. The combination of the first three effects leads to the violent evolution of heat and gases.

To give the most exothermic reaction, an explosive should have weak chemical bonds and should decompose into molecules that have very strong bonds. Table 8.4 tells us that $N \equiv N$, $C \equiv O$ and $C = O$ bonds are among the strongest. Not surprisingly, explosives are usually designed to produce the gaseous products $N_2(g)$, $CO(g)$, and $CO_2(g)$. Water vapor is nearly always produced as well.

Many common explosives are organic molecules that contain nitro (NO_2) or nitrate (NO_3) groups attached to a carbon skeleton. The Lewis structures of two of the most familiar explosives, nitroglycerin and trinitrotoluene (TNT), are shown here (resonance structures are not shown for clarity). TNT contains the six-membered ring characteristic of benzene.

Nitroglycerin is a pale yellow, oily liquid. It is highly *shock-sensitive*: Merely shaking the liquid can cause its explosive decomposition into nitrogen, carbon dioxide, water, and oxygen gases:

$$4\,C_3H_5N_3O_9(l) \longrightarrow 6\,N_2(g) + 12\,CO_2(g) + 10\,H_2O(g) + O_2(g)$$

The large bond enthalpies of N_2 (941 kJ/mol), CO_2(2 × 799 kJ/mol), and H_2O(2 × 463 kJ/mol) make this reaction enormously exothermic. Nitroglycerin is an exceptionally unstable explosive because it is in nearly perfect *explosive balance*: With the exception of a small amount of $O_2(g)$ produced, the only products are N_2, CO_2, and H_2O. Note also that, unlike combustion reactions ∞ (Section 3.2), explosions are entirely self-contained. No other reagent, such as $O_2(g)$, is needed for the explosive decomposition.

Because nitroglycerin is so unstable, it is difficult to use it as a controllable explosive. The Swedish inventor Alfred Nobel (▼ Figure 8.16) found that mixing nitroglycerin with an absorbent solid material such as diatomaceous earth or cellulose gives a solid explosive (*dynamite*) that is much safer than liquid nitroglycerin.

Related Exercises: 8.98, 8.99

Nitroglycerin

Trinitrotoluene (TNT)

▲ Figure 8.16 Alfred Nobel (1833–1896), Swedish inventor of dynamite. By many accounts Nobel's discovery that nitroglycerin could be made more stable by absorbing it onto cellulose was an accident, but the discovery made Nobel a wealthy man. Although he had invented the most powerful military explosive to date, he strongly supported international peace movements. His will stated that his fortune be used to establish prizes awarding those who "have conferred the greatest benefit on mankind," including the promotion of peace and "fraternity between nations." The Nobel Prize is probably the most coveted award that a scientist, writer, or peace advocate can receive.

As the number of bonds between the carbon atoms increases, the bond length decreases and the bond enthalpy increases. That is, the carbon atoms are held more closely and more tightly together. In general, *as the number of bonds between two atoms increases, the bond grows shorter and stronger*. This trend is illustrated in ▶ Figure 8.17 for N—N single, double, and triple bonds.

▲ GO FIGURE

Predict the N—N bond enthalpy for an N—N bond that has resonance forms that include equal contributions from single and double N—N bonds.

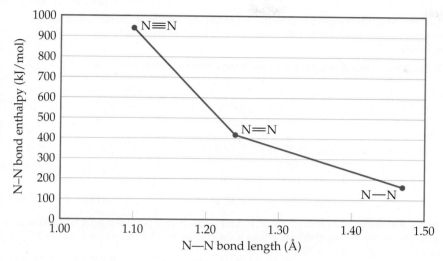

▲ **Figure 8.17 Bond strength versus bond length for N—N bonds.**

SAMPLE INTEGRATIVE EXERCISE | Putting Concepts Together

Phosgene, a substance used in poisonous gas warfare during World War I, is so named because it was first prepared by the action of sunlight on a mixture of carbon monoxide and chlorine gases. Its name comes from the Greek words *phos* (light) and *genes* (born of). Phosgene has the following elemental composition: 12.14% C, 16.17% O, and 71.69% Cl by mass. Its molar mass is 98.9 g/mol. **(a)** Determine the molecular formula of this compound. **(b)** Draw three Lewis structures for the molecule that satisfy the octet rule for each atom. (The Cl and O atoms bond to C.) **(c)** Using formal charges, determine which Lewis structure is the dominant one. **(d)** Using average bond enthalpies, estimate ΔH for the formation of gaseous phosgene from $CO(g)$ and $Cl_2(g)$.

SOLUTION

(a) The empirical formula of phosgene can be determined from its elemental composition.
∞ (Section 3.5) Assuming 100 g of the compound and calculating the number of moles of C, O, and Cl in this sample, we have

$$(12.14 \text{ g C})\left(\frac{1 \text{ mol C}}{12.01 \text{ g C}}\right) = 1.011 \text{ mol C}$$

$$(16.17 \text{ g O})\left(\frac{1 \text{ mol O}}{16.00 \text{ g O}}\right) = 1.011 \text{ mol O}$$

$$(71.69 \text{ g Cl})\left(\frac{1 \text{ mol Cl}}{35.45 \text{ g Cl}}\right) = 2.022 \text{ mol Cl}$$

The ratio of the number of moles of each element, obtained by dividing each number of moles by the smallest quantity, indicates that there is one C and one O for each two Cl in the empirical formula, $COCl_2$.

The molar mass of the empirical formula is $12.01 + 16.00 + 2(35.45) = 98.91$ g/mol, the same as the molar mass of the molecule. Thus, $COCl_2$ is the molecular formula.

(b) Carbon has four valence electrons, oxygen has six, and chlorine has seven, giving $4 + 6 + 2(7) = 24$ electrons for the Lewis structures. Drawing a Lewis structure with all single bonds does not give the central carbon atom an octet. Using multiple bonds, three structures satisfy the octet rule:

(c) Calculating the formal charges on each atom gives

$$
\overset{0}{:\!\ddot{O}\!:} \qquad\qquad \overset{-1}{:\!\ddot{O}\!:}^{\;-1} \qquad\qquad \overset{-1}{:\!\ddot{O}\!:}^{\;-1}
$$

$$
\overset{0}{:\!\ddot{C}l}\!-\!\overset{\;0}{\underset{\;\;\|}{C}}\!-\!\overset{0}{\ddot{C}l\!:} \quad\longleftrightarrow\quad \overset{+1}{:\!\ddot{C}l}\!=\!\overset{\;0}{C}\!-\!\overset{0}{\ddot{C}l\!:} \quad\longleftrightarrow\quad \overset{0}{:\!\ddot{C}l}\!-\!\overset{\;0}{C}\!=\!\overset{+1}{\ddot{C}l\!:}
$$

The first structure is expected to be the dominant one because it has the lowest formal charges on each atom. Indeed, the molecule is usually represented by this single Lewis structure.

(d) Writing the chemical equation in terms of the Lewis structures of the molecules, we have

$$
:\!C\!\equiv\!O\!: \quad + \quad :\!\ddot{C}l\!-\!\ddot{C}l\!: \quad\longrightarrow\quad :\!\ddot{C}l\!-\!\overset{:\ddot{O}:}{\underset{\|}{C}}\!-\!\ddot{C}l\!:
$$

Thus, the reaction involves breaking a $C\equiv O$ bond and a $Cl-Cl$ bond and forming a $C=O$ bond and two $C-Cl$ bonds. Using bond enthalpies from Table 8.4, we have

$$\Delta H = [D(C\equiv O) + D(Cl - Cl)] - [D(C=O) + 2D(C-Cl)]$$

$$= [1072\,kJ + 242\,kJ] - [799\,kJ + 2(328\,kJ)] = -141\,kJ$$

Notice that the reaction is exothermic. Nevertheless, energy is needed from sunlight or another source for the reaction to begin, as is the case for the combustion of $H_2(g)$ and $O_2(g)$ to form $H_2O(g)$ (Figure 5.14).

Chapter Summary and Key Terms

CHEMICAL BONDS, LEWIS SYMBOLS, AND THE OCTET RULE (INTRODUCTION AND SECTION 8.1) In this chapter we have focused on the interactions that lead to the formation of **chemical bonds**. We classify these bonds into three broad groups: **ionic bonds**, which result from the electrostatic forces that exist between ions of opposite charge; **covalent bonds**, which result from the sharing of electrons by two atoms; and **metallic bonds**, which result from a delocalized sharing of electrons in metals. The formation of bonds involves interactions of the outermost electrons of atoms, their valence electrons. The valence electrons of an atom can be represented by electron-dot symbols, called **Lewis symbols**. The tendencies of atoms to gain, lose, or share their valence electrons often follow the **octet rule**, which says that the atoms in molecules or ions (usually) have eight valence electrons.

IONIC BONDING (SECTION 8.2) Ionic bonding results from the transfer of electrons from one atom to another, leading to the formation of a three-dimensional lattice of charged particles. The stabilities of ionic substances result from the strong electrostatic attractions between an ion and the surrounding ions of opposite charge. The magnitude of these interactions is measured by the **lattice energy**, which is the energy needed to separate an ionic lattice into gaseous ions. Lattice energy increases with increasing charge on the ions and with decreasing distance between the ions. The **Born–Haber cycle** is a useful thermochemical cycle in which we use Hess's law to calculate the lattice energy as the sum of several steps in the formation of an ionic compound.

COVALENT BONDING (SECTION 8.3) A covalent bond results from the sharing of valence electrons between atoms. We can represent the electron distribution in molecules by means of **Lewis structures**, which indicate how many valence electrons are involved in forming bonds and how many remain as **nonbonding electron pairs** (or **lone pairs**). The octet rule helps determine how many bonds will be formed between two atoms. The sharing of one pair of electrons produces a **single bond**; the sharing of two or three pairs of electrons between two atoms produces **double** or **triple bonds**, respectively. Double and triple bonds are examples of multiple bonding between atoms. The bond length decreases as the number of bonds between the atoms increases.

BOND POLARITY AND ELECTRONEGATIVITY (SECTION 8.4) In covalent bonds, the electrons may not necessarily be shared equally between two atoms. **Bond polarity** helps describe unequal sharing of electrons in a bond. In a **nonpolar covalent bond** the electrons in the bond are shared equally by the two atoms; in a **polar covalent bond** one of the atoms exerts a greater attraction for the electrons than the other.

Electronegativity is a numerical measure of the ability of an atom to compete with other atoms for the electrons shared between them. Fluorine is the most electronegative element, meaning it has the greatest ability to attract electrons from other atoms. Electronegativity values range from 0.7 for Cs to 4.0 for F. Electronegativity generally increases from left to right in a row of the periodic table and decreases going down a column. The difference in the electronegativities of bonded atoms can be used to determine the polarity of a bond. The greater the electronegativity difference, the more polar the bond.

A **polar molecule** is one whose centers of positive and negative charge do not coincide. Thus, a polar molecule has a positive side and a negative side. This separation of charge produces a **dipole**, the magnitude of which is given by the **dipole moment**, which is measured in debyes (D). Dipole moments increase with increasing amount of charge separated and increasing distance of separation. Any diatomic molecule $X—Y$ in which X and Y have different electronegativities is a polar molecule.

Most bonding interactions lie between the extremes of covalent and ionic bonding. While it is generally true that the bonding between a metal and a nonmetal is predominantly ionic, exceptions to this guideline are not uncommon when the difference in electronegativity of the atoms is relatively small or when the oxidation state of the metal becomes large.

DRAWING LEWIS STRUCTURES AND RESONANCE STRUCTURES (SECTIONS 8.5 AND 8.6)

If we know which atoms are connected to one another, we can draw Lewis structures for molecules and ions by a simple procedure. Once we do so, we can determine the **formal charge** of each atom in a Lewis structure, which is the charge that the atom would have if all atoms had the same electronegativity. In general, the dominant Lewis structure will have low formal charges with any negative formal charges residing on more electronegative atoms.

Sometimes a single dominant Lewis structure is inadequate to represent a particular molecule (or ion). In such situations, we describe the molecule by using two or more **resonance structures** for the molecule. The molecule is envisioned as a blend of these multiple resonance structures. Resonance structures are important in describing the bonding in molecules such as ozone, O_3, and the organic molecule benzene, C_6H_6.

EXCEPTIONS TO THE OCTET RULE (SECTION 8.7)

The octet rule is not obeyed in all cases. Exceptions occur when (a) a molecule has an odd number of electrons, (b) it is not possible to complete an octet around an atom without forcing an unfavorable distribution of electrons, or (c) a large atom is surrounded by a sufficiently large number of small electronegative atoms that it has more than an octet of electrons around it. Lewis structures with more than an octet of electrons are observed for atoms in the third row and beyond in the periodic table.

STRENGTHS AND LENGTHS OF COVALENT BONDS (SECTION 8.8)

The strength of a covalent bond is measured by its **bond enthalpy**, which is the molar enthalpy change upon breaking a particular bond. Average bond enthalpies can be determined for a wide variety of covalent bonds. The strengths of covalent bonds increase with the number of electron pairs shared between two atoms. We can use bond enthalpies to estimate the enthalpy change during chemical reactions in which bonds are broken and new bonds formed. The average bond length between two atoms decreases as the number of bonds between the atoms increases, consistent with the bond being stronger as the number of bonds increases.

Learning Outcomes After studying this chapter, you should be able to:

- Write Lewis symbols for atoms and ions. (Section 8.1)
- Define lattice energy and be able to arrange compounds in order of increasing lattice energy based on the charges and sizes of the ions involved. (Section 8.2)
- Use atomic electron configurations and the octet rule to draw Lewis structures for molecules. (Section 8.3)
- Use electronegativity differences to identify nonpolar covalent, polar covalent, and ionic bonds. (Section 8.4)
- Calculate charge separation in diatomic molecules based on the experimentally measured dipole moment and bond length. (Section 8.4)

- Calculate formal charges from Lewis structures and use those formal charges to identify the dominant Lewis structure for a molecule or ion. (Section 8.5)
- Recognize molecules where resonance structures are needed to describe the bonding and draw the dominant resonance structures. (Section 8.6)
- Recognize exceptions to the octet rule and draw accurate Lewis structures even when the octet rule is not obeyed. (Section 8.7)
- Predict the relationship between bond type (single, double, and triple), bond strength (or enthalpy), and bond length. (Section 8.8)
- Use bond enthalpies to estimate enthalpy changes for reactions involving gas-phase reactants and products. (Section 8.8)

Key Equations

- $$E_{el} = \frac{\kappa Q_1 Q_2}{d}$$ [8.4]   The potential energy of two interacting charges

- $$\mu = Qr$$ [8.10]   The dipole moment of two charges of equal magnitude but opposite sign, separated by a distance r

- Formal charge = valence electrons − $\frac{1}{2}$ (bonding electrons) − nonbonding electrons   [8.11]   The definition of formal charge

- $\Delta H_{rxn} = \Sigma$ (bond enthalpies of bonds broken) − Σ (bond enthalpies of bonds formed)   [8.12]   The enthalpy change as a function of bond enthalpies for reactions involving gas-phase molecules

Exercises

Visualizing Concepts

8.1 For each of these Lewis symbols, indicate the group in the periodic table in which the element X belongs: [Section 8.1]

(a) $\cdot \dot{X} \cdot$ (b) $\cdot X \cdot$ (c) $:\dot{X}\cdot$

8.2 Illustrated are four ions — A, B, X, and Y— showing their relative ionic radii. The ions shown in red carry positive charges:

a 2+ charge for A and a 1+ charge for B. Ions shown in blue carry negative charges: a 1− charge for X and a 2− charge for Y. (a) Which combinations of these ions produce ionic compounds where there is a 1:1 ratio of cations and anions? (b) Among the combinations in part (a), which leads to the ionic compound having the largest lattice energy? [Section 8.2]

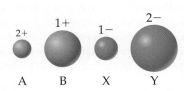

8.3 A portion of a two-dimensional "slab" of NaCl(s) is shown here (see Figure 8.3) in which the ions are numbered. **(a)** Which colored balls must represent sodium ions? **(b)** Which colored balls must represent chloride ions? **(c)** Consider ion 5. How many attractive electrostatic interactions are shown for it? **(d)** Consider ion 5. How many repulsive interactions are shown for it? **(e)** Is the sum of the attractive interactions in part (c) larger or smaller than the sum of the repulsive interactions in part (d)? **(f)** If this pattern of ions was extended indefinitely in two dimensions, would the lattice energy be positive or negative? [Section 8.2]

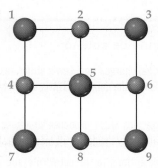

8.4 The orbital diagram that follows shows the valence electrons for a 2+ ion of an element. **(a)** What is the element? **(b)** What is the electron configuration of an atom of this element? [Section 8.2]

$$\boxed{1\!\downarrow}\ \boxed{1}\ \boxed{1}\ \boxed{1}\ \boxed{1}$$

$$4d$$

8.5 In the Lewis structure shown here, A, D, E, Q, X, and Z represent elements in the first two rows of the periodic table. Identify all six elements so that the formal charges of all atoms are zero. [Section 8.3]

8.6 Incomplete Lewis structures for the nitrous acid molecule, HNO_2, and the nitrite ion, NO_2^-, are shown here. **(a)** Complete each Lewis structure by adding electron pairs as needed. **(b)** Is the formal charge on N the same or different in these two species? **(c)** Would either HNO_2 or NO_2^- be expected to exhibit resonance? **(d)** Would you expect the $N\!=\!O$ bond in HNO_2 to be longer, shorter, or the same length as the $N\!-\!O$ bonds in NO_2^-? [Sections 8.5 and 8.6]

$$H-O-N\!=\!O \qquad O-N\!=\!O$$

8.7 The partial Lewis structure that follows is for a hydrocarbon molecule. In the full Lewis structure, each carbon atom satisfies the octet rule, and there are no unshared electron pairs in the molecule. The carbon–carbon bonds are labeled 1, 2, and 3. **(a)** How many hydrogen atoms are in the molecule? **(b)** Rank the carbon–carbon bonds in order of increasing bond length. **(c)** Rank the carbon–carbon bonds in order of increasing bond enthalpy. [Sections 8.3 and 8.8]

$$C\overset{1}{=\!\!=}C\overset{2}{-\!\!-}C\overset{3}{\equiv}C$$

8.8 Consider the Lewis structure for the polyatomic oxyanion shown here, where X is an element from the third period (Na − Ar). By changing the overall charge, n, from 1− to 2− to 3− we get three different polyatomic ions. For each of these ions **(a)** identify the central atom, X; **(b)** determine the formal charge of the central atom, X; **(c)** draw a Lewis structure that makes the formal charge on the central atom equal to zero. [Sections 8.5, 8.6, and 8.7]

$$\left[\ :\!\ddot{O}\!:\atop :\!\ddot{O}\!-\!X\!-\!\ddot{O}\!:\ \atop :\!\ddot{O}\!:\ \right]^{n-}$$

Lewis Symbols (Section 8.1)

8.9 **(a)** True or false: An element's number of valence electrons is the same as its atomic number. **(b)** How many valence electrons does a nitrogen atom possess? **(c)** An atom has the electron configuration $1s^2 2s^2 2p^6 3s^2 3p^2$. How many valence electrons does the atom have?

8.10 **(a)** True or false: The hydrogen atom is most stable when it has a full octet of electrons. **(b)** How many electrons must a sulfur atom gain to achieve an octet in its valence shell? **(c)** If an atom has the electron configuration $1s^2 2s^2 2p^3$, how many electrons must it gain to achieve an octet?

8.11 Consider the element silicon, Si. **(a)** Write its electron configuration. **(b)** How many valence electrons does a silicon atom have? **(c)** Which subshells hold the valence electrons?

8.12 **(a)** Write the electron configuration for the element titanium, Ti. How many valence electrons does this atom possess? **(b)** Hafnium, Hf, is also found in group 4B. Write the electron configuration for Hf. **(c)** Ti and Hf behave as though they possess the same number of valence electrons. Which of the subshells in the electron configuration of Hf behave as valence orbitals? Which behave as core orbitals?

8.13 Write the Lewis symbol for atoms of each of the following elements: **(a)** Al, **(b)** Br, **(c)** Ar, **(d)** Sr.

8.14 What is the Lewis symbol for each of the following atoms or ions? **(a)** K, **(b)** As, **(c)** Sn^{2+}, **(d)** N^{3-}.

Ionic Bonding (Section 8.2)

8.15 **(a)** Using Lewis symbols, diagram the reaction between magnesium and oxygen atoms to give the ionic substance MgO. **(b)** How many electrons are transferred? **(c)** Which atom loses electrons in the reaction?

8.16 **(a)** Use Lewis symbols to represent the reaction that occurs between Ca and F atoms. **(b)** What is the chemical formula of the most likely product? **(c)** How many electrons are transferred? **(d)** Which atom loses electrons in the reaction?

8.17 Predict the chemical formula of the ionic compound formed between the following pairs of elements: **(a)** Al and F, **(b)** K and S, **(c)** Y and O, **(d)** Mg and N.

8.18 Which ionic compound is expected to form from combining the following pairs of elements? (**a**) barium and fluorine, (**b**) cesium and chlorine, (**c**) lithium and nitrogen, (**d**) aluminum and oxygen.

8.19 Write the electron configuration for each of the following ions, and determine which ones possess noble-gas configurations: (**a**) Sr^{2+}, (**b**) Ti^{2+}, (**c**) Se^{2-}, (**d**) Ni^{2+}, (**e**) Br^-, (**f**) Mn^{3+}.

8.20 Write electron configurations for the following ions, and determine which have noble-gas configurations: (**a**) Cd^{2+}, (**b**) P^{3-}, (**c**) Zr^{4+}, (**d**) Ru^{3+}, (**e**) As^{3-}, (**f**) Ag^+.

8.21 (**a**) Is lattice energy usually endothermic or exothermic? (**b**) Write the chemical equation that represents the process of lattice energy for the case of NaCl. (**c**) Would you expect salts like NaCl, which have singly-charged ions, to have larger or smaller lattice energies compared to salts like CaO which are composed of doubly-charged ions?

8.22 NaCl and KF have the same crystal structure. The only difference between the two is the distance that separates cations and anions. (**a**) The lattice energies of NaCl and KF are given in Table 8.2. Based on the lattice energies, would you expect the Na—Cl or the K—F distance to be longer? (**b**) Use the ionic radii given in Figure 7.8 to estimate the Na—Cl and K—F distances.

8.23 The ionic substances NaF, CaO, and ScN are isoelectronic (they have the same number of electrons). Examine the lattice energies for these substances in Table 8.2. Make a graph of lattice energy on the vertical axis versus the charge on the cation on the horizontal axis. (**a**) What is the slope of the line? (**b**) Make a graph of lattice energy on the vertical axis versus the square of the cation charge on the horizontal axis. What is the slope of this line? (**c**) Compare how well the data points fall on a line for the graphs in (a) and (b). Which trend is more linear, lattice energy versus cation charge or lattice energy versus cation charge squared? (**d**) Predict the lattice energy for the compound TiC, if we consider the carbon to have a 4− charge.

8.24 (**a**) Does the lattice energy of an ionic solid increase or decrease (i) as the charges of the ions increase, (ii) as the sizes of the ions increase? (**b**) Arrange the following substances not listed in Table 8.2 according to their expected lattice energies, listing them from lowest lattice energy to the highest: MgS, KI, GaN, LiBr.

8.25 Consider the ionic compounds KF, NaCl, NaBr, and LiCl. (**a**) Use ionic radii (Figure 7.8) to estimate the cation–anion distance for each compound. (**b**) Based on your answer to part (a), arrange these four compounds in order of decreasing lattice energy. (**c**) Check your predictions in part (b) with the experimental values of lattice energy from Table 8.2. Are the predictions from ionic radii correct?

8.26 Which of the following trends in lattice energy is due to differences in ionic radii? (**a**) NaCl > RbBr > CsBr, (**b**) BaO > KF, (**c**) SrO > $SrCl_2$.

8.27 Energy is required to remove two electrons from Ca to form Ca^{2+}, and energy is required to add two electrons to O to form O^{2-}. Yet CaO is stable relative to the free elements. Which statement is the best explanation? (**a**) The lattice energy of CaO is large enough to overcome these processes. (**b**) CaO is a covalent compound, and these processes are irrelevant. (**c**) CaO has a higher molar mass than either Ca or O. (**d**) The enthalpy of formation of CaO is small. (**e**) CaO is stable to atmospheric conditions.

8.28 List the individual steps used in constructing a Born–Haber cycle for the formation of BaI_2 from the elements. Which of the steps would you expect to be exothermic?

8.29 Use data from Appendix C, Figure 7.10, and Figure 7.12 to calculate the lattice energy of RbCl.

8.30 (**a**) Based on the lattice energies of $MgCl_2$ and $SrCl_2$ given in Table 8.2, what is the range of values that you would expect for the lattice energy of $CaCl_2$? (**b**) Using data from Appendix C, Figure 7.11, Figure 7.13 and the value of the second ionization energy for Ca, 1145 kJ/mol, calculate the lattice energy of $CaCl_2$.

Covalent Bonding, Electronegativity, and Bond Polarity (Sections 8.3 and 8.4)

8.31 (**a**) State whether the bonding in each compound is likely to be covalent or not: (i) iron, (ii) sodium chloride, (iii) water, (iv) oxygen, (v) argon. (**b**) A substance XY, formed from two different elements, boils at −33 °C. Is XY likely to be a covalent or an ionic substance?

8.32 Which of these elements are unlikely to form covalent bonds? S, H, K, Ar, Si.

8.33 Using Lewis symbols and Lewis structures, diagram the formation of $SiCl_4$ from Si and Cl atoms, showing valence-shell electrons. (**a**) How many valence electrons does Si have initially? (**b**) How many valence electrons does each Cl have initially? (**c**) How many valence electrons surround the Si in the $SiCl_4$ molecule? (**d**) How many valence electrons surround each Cl in the $SiCl_4$ molecule? (**e**) How many bonding pairs of electrons are in the $SiCl_4$ molecule?

8.34 Use Lewis symbols and Lewis structures to diagram the formation of PF_3 from P and F atoms, showing valence-shell electrons. (**a**) How many valence electrons does P have initially? (**b**) How many valence electrons does each F have initially? (**c**) How many valence electrons surround the P in the PF_3 molecule? (**d**) How many valence electrons surround each F in the PF_3 molecule? (**e**) How many bonding pairs of electrons are in the PF_3 molecule?

8.35 (**a**) Construct a Lewis structure for O_2 in which each atom achieves an octet of electrons. (**b**) How many bonding electrons are in the structure? (**c**) Would you expect the O—O bond in O_2 to be shorter or longer than the O—O bond in compounds that contain an O—O single bond? Explain.

8.36 (**a**) Construct a Lewis structure for hydrogen peroxide, H_2O_2, in which each atom achieves an octet of electrons. (**b**) How many bonding electrons are between the two oxygen atoms? (**c**) Do you expect the O—O bond in H_2O_2 to be longer or shorter than the O—O bond in O_2? Explain.

8.37 Which of the following statements about electronegativity is false? (**a**) Electronegativity is the ability of an atom in a molecule to attract electron density toward itself. (**b**) Electronegativity is the same thing as electron affinity. (**c**) The numerical values for electronegativity have no units. (**d**) Fluorine is the most electronegative element. (**e**) Cesium is the least electronegative element.

8.38 (**a**) What is the trend in electronegativity going from left to right in a row of the periodic table? (**b**) How do electronegativity values generally vary going down a column in the periodic table? (**c**) True or false: The most easily ionizable elements are the most electronegative.

8.39 Using only the periodic table as your guide, select the most electronegative atom in each of the following sets: (a) Na, Mg, K, Ca; (b) P, S, As, Se; (c) Be, B, C, Si; (d) Zn, Ge, Ga, As.

8.40 By referring only to the periodic table, select (a) the most electronegative element in group 6A; (b) the least electronegative element in the group Al, Si, P; (c) the most electronegative element in the group Ga, P, Cl, Na; (d) the element in the group K, C, Zn, F that is most likely to form an ionic compound with Ba.

8.41 Which of the following bonds are polar? (a) B—F, (b) Cl—Cl, (c) Se—O, (d) H—I. Which is the more electronegative atom in each polar bond?

8.42 Arrange the bonds in each of the following sets in order of increasing polarity: (a) C—F, O—F, Be—F; (b) O—Cl, S—Br, C—P; (c) C—S, B—F, N—O.

8.43 (a) From the data in Table 8.3, calculate the effective charges on the H and Br atoms of the HBr molecule in units of the electronic charge, e. (b) If you were to put HBr under very high pressure, so its bond length decreased significantly, would its dipole moment increase, decrease, or stay the same, if you assume that the effective charges on the atoms do not change?

8.44 The iodine monobromide molecule, IBr, has a bond length of 2.49 Å and a dipole moment of 1.21 D. (a) Which atom of the molecule is expected to have a negative charge? (b) Calculate the effective charges on the I and Br atoms in IBr in units of the electronic charge, e.

8.45 In the following pairs of binary compounds determine which one is a molecular substance and which one is an ionic substance. Use the appropriate naming convention (for ionic or molecular substances) to assign a name to each compound: (a) SiF_4 and LaF_3, (b) $FeCl_2$ and $ReCl_6$, (c) $PbCl_4$ and RbCl.

8.46 In the following pairs of binary compounds determine which one is a molecular substance and which one is an ionic substance. Use the appropriate naming convention (for ionic or molecular substances) to assign a name to each compound: (a) $TiCl_4$ and CaF_2, (b) ClF_3 and VF_3, (c) $SbCl_5$ and AlF_3.

Lewis Structures; Resonance Structures (Sections 8.5 and 8.6)

8.47 Draw Lewis structures for the following: (a) SiH_4, (b) CO, (c) SF_2, (d) H_2SO_4 (H is bonded to O), (e) ClO_2^-, (f) NH_2OH.

8.48 Write Lewis structures for the following: (a) H_2CO (both H atoms are bonded to C), (b) H_2O_2, (c) C_2F_6 (contains a C—C bond), (d) AsO_3^{3-}, (e) H_2SO_3 (H is bonded to O), (f) NH_2Cl.

8.49 Which one of these statements about formal charge is true? (a) Formal charge is the same as oxidation number. (b) To draw the best Lewis structure, you should minimize formal charge. (c) Formal charge takes into account the different electronegativities of the atoms in a molecule. (d) Formal charge is most useful for ionic compounds. (e) Formal charge is used in calculating the dipole moment of a diatomic molecule.

8.50 (a) Draw the dominant Lewis structure for the phosphorus trifluoride molecule, PF_3. (b) Determine the oxidation numbers of the P and F atoms. (c) Determine the formal charges of the P and F atoms.

8.51 Write Lewis structures that obey the octet rule for each of the following, and assign oxidation numbers and formal charges to each atom: (a) OCS, (b) $SOCl_2$ (S is the central atom), (c) BrO_3^-, (d) $HClO_2$ (H is bonded to O).

8.52 For each of the following molecules or ions of sulfur and oxygen, write a single Lewis structure that obeys the octet rule, and calculate the oxidation numbers and formal charges on all the atoms: (a) SO_2, (b) SO_3, (c) SO_3^{2-}. (d) Arrange these molecules/ions in order of increasing S—O bond length.

8.53 (a) Draw the best Lewis structure(s) for the nitrite ion, NO_2^-. (b) With what allotrope of oxygen is it isoelectronic? (c) What would you predict for the lengths of the bonds in NO_2^- relative to N—O single bonds and double bonds?

8.54 Consider the formate ion, HCO_2^-, which is the anion formed when formic acid loses an H^+ ion. The H and the two O atoms are bonded to the central C atom. (a) Draw the best Lewis structure(s) for this ion. (b) Are resonance structures needed to describe the structure? (c) Would you predict that the C—O bond lengths in the formate ion would be longer or shorter relative to those in CO_2?

8.55 Predict the ordering, from shortest to longest, of the bond lengths in CO, CO_2, and CO_3^{2-}.

8.56 Based on Lewis structures, predict the ordering, from shortest to longest, of N—O bond lengths in NO^+, NO_2^-, and NO_3^-.

8.57 (a) Do the C—C bond lengths in benzene alternate short-long-short-long around the ring? Why or why not? (b) Are C—C bond lengths in benzene shorter than C—C single bonds? (c) Are C—C bond lengths in benzene shorter than C=C double bonds?

[8.58] Mothballs are composed of naphthalene, $C_{10}H_8$, a molecule that consists of two six-membered rings of carbon fused along an edge, as shown in this incomplete Lewis structure:

(a) Draw all of the resonance structures of naphthalene. How many are there? (b) Do you expect the C—C bond lengths in the molecule to be similar to those of C—C single bonds, C=C double bonds, or intermediate between C—C single and C=C double bonds? (c) Not all of the C—C bond lengths in naphthalene are equivalent. Based on your resonance structures, how many C—C bonds in the molecule do you expect to be shorter than the others?

Exceptions to the Octet Rule (Section 8.7)

8.59 Indicate whether each statement is true or false: (a) The octet rule is based on the fact that filling in all s and p valence electrons in a shell gives eight electrons. (b) The Si in SiH_4 does not follow the octet rule because hydrogen is in an unusual oxidation state. (c) Boron compounds are frequent exceptions to the octet rule because they have too few electrons surrounding the boron. (d) Compounds in which nitrogen is the central atom are frequent exceptions to the octet rule because they have too many electrons surrounding the nitrogen.

8.60 Fill in the blank with the appropriate numbers for both electrons and bonds (considering that single bonds are counted as one, double bonds as two, and triple bonds as three).

(a) Fluorine has ___ valence electrons and makes ___ bond(s) in compounds.

(b) Oxygen has ___ valence electrons and makes ___ bond(s) in compounds.

(c) Nitrogen has ___ valence electrons and makes ___ bond(s) in compounds.

(d) Carbon has ___ valence electrons and makes ___ bond(s) in compounds.

8.61 Draw the dominant Lewis structures for these chlorine–oxygen molecules/ions: ClO, ClO^-, ClO_2^-, ClO_3^-, ClO_4^-. Which of these do not obey the octet rule?

8.62 For elements in the third row of the periodic table and beyond, the octet rule is often not obeyed. A friend of yours says this is because these heavier elements are more likely to make double or triple bonds. Another friend of yours says that this is because the heavier elements are larger and can make bonds to more than four atoms at a time. Which friend is most correct?

8.63 Draw the Lewis structures for each of the following ions or molecules. Identify those in which the octet rule is not obeyed; state which atom in each compound does not follow the octet rule; and state, for those atoms, how many electrons surround these atoms: (a) PH_3, (b) AlH_3, (c) N_3^-, (d) CH_2Cl_2, (e) SnF_6.

8.64 Draw the Lewis structures for each of the following molecules or ions. Identify instances where the octet rule is not obeyed; state which atom in each compound does not follow the octet rule; and state how many electrons surround these atoms: (a) NO, (b) BF_3, (c) ICl_2^-, (d) $OPBr_3$ (the P is the central atom), (e) XeF_4.

8.65 In the vapor phase, $BeCl_2$ exists as a discrete molecule. (a) Draw the Lewis structure of this molecule, using only single bonds. Does this Lewis structure satisfy the octet rule? (b) What other resonance structures are possible that satisfy the octet rule? (c) On the basis of the formal charges, which Lewis structure is expected to be dominant for $BeCl_2$?

8.66 (a) Describe the molecule xenon trioxide, XeO_3, using four possible Lewis structures, one each with zero, one, two, or three Xe—O double bonds. (b) Do any of these resonance structures satisfy the octet rule for every atom in the molecule? (c) Do any of the four Lewis structures have multiple resonance structures? If so, how many resonance structures do you find? (d) Which of the Lewis structures in (a) yields the most favorable formal charges for the molecule?

8.67 Consider the following statement: "For some molecules and ions, a Lewis structure that satisfies the octet rule does not lead to the lowest formal charges, and a Lewis structure that leads to the lowest formal charges does not satisfy the octet rule." Illustrate this statement using the hydrogen sulfite ion, HSO_3^-, as an example (the H atom is bonded to one of the O atoms).

8.68 Some chemists believe that satisfaction of the octet rule should be the top criterion for choosing the dominant Lewis structure of a molecule or ion. Other chemists believe that achieving the best formal charges should be the top criterion. Consider the dihydrogen phosphate ion, $H_2PO_4^-$, in which the H atoms are bonded to O atoms. (a) What is the predicted dominant Lewis structure if satisfying the octet rule is the top criterion? (b) What is the predicted dominant Lewis structure if achieving the best formal charges is the top criterion?

Bond Enthalpies (Section 8.8)

8.69 Using Table 8.4, estimate ΔH for each of the following gas-phase reactions (note that lone pairs on atoms are not shown):

(a)

(b)

(c)

8.70 Using Table 8.4, estimate ΔH for the following gas-phase reactions:

(a)

(b)

(c)

8.71 Using Table 8.4, estimate ΔH for each of the following reactions:

(a) $2\,CH_4(g) + O_2(g) \longrightarrow 2\,CH_3OH(g)$

(b) $H_2(g) + Br_2(l) \longrightarrow 2\,HBr(g)$

(c) $2\,H_2O_2(g) \longrightarrow 2\,H_2O(g) + O_2(g)$

8.72 Use Table 8.4 to estimate the enthalpy change for each of the following reactions:

(a) $H_2C{=}O(g) + HCl(g) \longrightarrow H_3C{-}O{-}Cl(g)$

(b) $H_2O_2(g) + 2\,CO(g) \longrightarrow H_2(g) + 2\,CO_2(g)$

(c) $3\,H_2C{=}CH_2(g) \longrightarrow C_6H_{12}(g)$ (the six carbon atoms form a six-membered ring with two H atoms on each C atom)

8.73 Ammonia is produced directly from nitrogen and hydrogen by using the Haber process, which is perhaps the most widely used industrial chemical reaction on Earth. The chemical reaction is

$$N_2(g) + 3 H_2(g) \longrightarrow 2 NH_3(g)$$

(a) Use Table 8.4 to estimate the enthalpy change for the reaction. Is it exothermic or endothermic?
(b) Calculate the enthalpy change as obtained using ΔH_f° values.

8.74 (a) Use bond enthalpies to estimate the enthalpy change for the reaction of hydrogen with ethylene:

$$H_2(g) + C_2H_4(g) \longrightarrow C_2H_6(g)$$

(b) Calculate the standard enthalpy change for this reaction, using heat of formation.

8.75 Given the following bond-dissociation energies, calculate the average bond enthalpy for the Ti—Cl bond.

	$\Delta H(kJ/mol)$
$TiCl_4(g) \longrightarrow TiCl_3(g) + Cl(g)$	335
$TiCl_3(g) \longrightarrow TiCl_2(g) + Cl(g)$	423
$TiCl_2(g) \longrightarrow TiCl(g) + Cl(g)$	444
$TiCl(g) \longrightarrow Ti(g) + Cl(g)$	519

8.76 (a) Using average bond enthalpies, predict which of the following reactions will be most exothermic:

(i) $C(g) + 2 F_2(g) \longrightarrow CF_4(g)$

(ii) $CO(g) + 3 F_2(g) \longrightarrow CF_4(g) + OF_2(g)$

(iii) $CO_2(g) + 4 F_2(g) \longrightarrow CF_4(g) + 2 OF_2(g)$

(b) Make a graph that plots the reaction enthalpy you calculated on the vertical axis versus oxidation state of carbon on the horizontal axis. Draw the best-fit line through your data points. Is the slope of this line negative or positive? Does this mean that as the oxidation state of the carbon increases, the reaction with elemental fluorine becomes more or less exothermic?

(c) Predict the reaction enthalpy for the reaction of carbonate ion with fluorine from your graph, and compare to what you predict using average bond enthalpies.

Additional Exercises

8.77 How many elements in the periodic table are represented by a Lewis symbol with a single dot? Which groups are they in?

[8.78] From Equation 8.4 and the ionic radii given in Figure 7.8, calculate the potential energy of the following pairs of ions. Assume that the ions are separated by a distance equal to the sum of their ionic radii: (a) Na^+, Br^-; (b) Rb^+, Br^-; (c) Sr^{2+}, S^{2-}.

8.79 (a) Consider the lattice energies for the following compounds: BeH_2, 3205 kJ/mol; MgH_2, 2791 kJ/mol; CaH_2, 2410 kJ/mol; SrH_2, 2250 kJ/mol; BaH_2, 2121 kJ/mol. Plot lattice energy versus cation radius for these compounds. If you draw a line through your points, is the slope negative or positive? Explain. (b) The lattice energy of ZnH_2 is 2870 kJ/mol. Based on the data given in part (a), the radius of the Zn^{2+} ion is expected to be closest to that of which group 2A element?

8.80 Based on data in Table 8.2, estimate (within 30 kJ/mol) the lattice energy for (a) LiBr, (b) CsBr, (c) $CaCl_2$.

8.81 An ionic substance of formula MX has a lattice energy of 6×10^3 kJ/mol. Is the charge on the ion M likely to be 1+, 2+, or 3+? Explain.

[8.82] From the ionic radii given in Figure 7.8, calculate the potential energy of a Ca^{2+} and O^{2-} ion pair that is just touching (the magnitude of the electronic charge is given on the back-inside cover). Calculate the energy of a mole of such pairs. How does this value compare with the lattice energy of CaO (Table 8.2)? Explain.

8.83 Construct a Born–Haber cycle for the formation of the hypothetical compound $NaCl_2$, where the sodium ion has a 2+ charge (the second ionization energy for sodium is given in Table 7.2). (a) How large would the lattice energy need to be for the formation of $NaCl_2$ to be exothermic? (b) If we were to estimate the lattice energy of $NaCl_2$ to be roughly equal to that of $MgCl_2$ (2326 kJ/mol from Table 8.2), what value would you obtain for the standard enthalpy of formation, ΔH_f°, of $NaCl_2$?

8.84 A classmate of yours is convinced that he knows everything about electronegativity. (a) In the case of atoms X and Y having different electronegativities, he says, the diatomic molecule X—Y must be polar. Is your classmate correct? (b) Your classmate says that the farther the two atoms are apart in a bond, the larger the dipole moment will be. Is your classmate correct?

8.85 Consider the collection of nonmetallic elements O, P, Te, I and B. (a) Which two would form the most polar single bond? (b) Which two would form the longest single bond? (c) Which two would be likely to form a compound of formula XY_2? (d) Which combinations of elements would likely yield a compound of empirical formula X_2Y_3?

8.86 The substance chlorine monoxide, $ClO(g)$, is important in atmospheric processes that lead to depletion of the ozone layer. The ClO molecule has an experimental dipole moment of 1.24 D, and the Cl—O bond length is 1.60 Å. (a) Determine the magnitude of the charges on the Cl and O atoms in units of the electronic charge, e. (b) Based on the electronegativities of the elements, which atom would you expect to have a partial negative charge in the ClO molecule? (c) Using formal charges as a guide, propose the dominant Lewis structure for the molecule. (d) The anion ClO^- exists. What is the formal charge on the Cl for the best Lewis structure for ClO^-?

[8.87] (a) Using the electronegativities of Br and Cl, estimate the partial charges on the atoms in the Br—Cl molecule.

(**b**) Using these partial charges and the atomic radii given in Figure 7.8, estimate the dipole moment of the molecule. (**c**) The measured dipole moment of BrCl is 0.57 D. If you assume the bond length in BrCl is the sum of the atomic radii, what are the partial charges on the atoms in BrCl using the experimental dipole moment?

8.88 A major challenge in implementing the "hydrogen economy" is finding a safe, lightweight, and compact way of storing hydrogen for use as a fuel. The hydrides of light metals are attractive for hydrogen storage because they can store a high weight percentage of hydrogen in a small volume. For example, $NaAlH_4$ can release 5.6% of its mass as H_2 upon decomposing to $NaH(s)$, $Al(s)$, and $H_2(g)$. $NaAlH_4$ possesses both covalent bonds, which hold polyatomic anions together, and ionic bonds. (**a**) Write a balanced equation for the decomposition of $NaAlH_4$. (**b**) Which element in $NaAlH_4$ is the most electronegative? Which one is the least electronegative? (**c**) Based on electronegativity differences, predict the identity of the polyatomic anion. Draw a Lewis structure for this ion. (**d**) What is the formal charge on hydrogen in the polyatomic ion?

8.89 Although I_3^- is known, F_3^- is not. Which statement is the most correct explanation? (**a**) Iodine is more likely to be electron-deficient. (**b**) Fluorine is too small to accommodate three nonbonding electron pairs and two bonding electron pairs. (**c**) Fluorine is too electronegative to form anions. (**d**) I_2 is known but F_2 is not. (**e**) Iodine has a larger electron affinity than fluorine.

8.90 Calculate the formal charge on the indicated atom in each of the following molecules or ions: (**a**) the central oxygen atom in O_3, (**b**) phosphorus in PF_6^-, (**c**) nitrogen in NO_2, (**d**) iodine in ICl_3, (**e**) chlorine in $HClO_4$ (hydrogen is bonded to O).

8.91 (**a**) Determine the formal charge on the chlorine atom in the hypochlorite ion, ClO^-, and the perchlorate ion, ClO_4^-, using resonance structures where the Cl atom has an octet. (**b**) What are the oxidation numbers of chlorine in ClO^- and in ClO_4^-? (**c**) Perchlorate is a much stronger oxidizing agent than hypochlorite. Suggest an explanation.

8.92 The following three Lewis structures can be drawn for N_2O:

$$:N\equiv N-\ddot{\underset{..}{O}}: \longleftrightarrow :\ddot{\underset{.}{N}}-N\equiv O: \longleftrightarrow :\ddot{\underset{.}{N}}=N=\ddot{\underset{..}{O}}:$$

(**a**) Using formal charges, which of these three resonance forms is likely to be the most important? (**b**) The N—N bond length in N_2O is 1.12 Å, slightly longer than a typical N≡N bond; and the N—O bond length is 1.19 Å, slightly shorter than a typical bond (see Table 8.5). Based on these data, which resonance structure best represents N_2O?

[8.93] (**a**) Triazine, $C_3H_3N_3$, is like benzene except that in triazine every other C—H group is replaced by a nitrogen atom. Draw the Lewis structure(s) for the triazine molecule. (**b**) Estimate the carbon–nitrogen bond distances in the ring.

[8.94] Ortho-dichlorobenzene, $C_6H_4Cl_2$, is obtained when two of the adjacent hydrogen atoms in benzene are replaced with Cl atoms. A skeleton of the molecule is shown here. (**a**) Complete a Lewis structure for the molecule using bonds and electron pairs as needed. (**b**) Are there any resonance structures for the molecule? If so, sketch them. (**c**) Are the resonance structures in (a) and (b) equivalent to one another as they are in benzene?

8.95 Consider the hypothetical molecule B—A=B. Are the following statements true or false? (**a**) This molecule cannot exist. (**b**) If resonance was important, the molecule would have identical A–B bond lengths.

8.96 An important reaction for the conversion of natural gas to other useful hydrocarbons is the conversion of methane to ethane.

$$2\ CH_4(g) \longrightarrow C_2H_6(g) + H_2(g)$$

In practice, this reaction is carried out in the presence of oxygen, which converts the hydrogen produced into water.

$$2\ CH_4(g) + \tfrac{1}{2}O_2(g) \longrightarrow C_2H_6(g) + H_2O(g)$$

Use Table 8.4 to estimate ΔH for these two reactions. Why is the conversion of methane to ethane more favorable when oxygen is used?

8.97 Two compounds are isomers if they have the same chemical formula but different arrangements of atoms. Use Table 8.4 to estimate ΔH for each of the following gas-phase isomerization reactions and indicate which isomer has the lower enthalpy.

(**a**) Ethanol → Dimethyl ether

(**b**) Ethylene oxide → Acetaldehyde

(**c**) Cyclopentene → Pentadiene

(**d**) Methyl isocyanide → Acetonitrile

[8.98] With reference to the Chemistry Put to Work box on explosives, (**a**) use bond enthalpies to estimate the enthalpy change for the explosion of 1.00 g of nitroglycerin. (**b**) Write a balanced equation for the decomposition of TNT. Assume that, upon explosion, TNT decomposes into $N_2(g)$, $CO_2(g)$, $H_2O(g)$, and $C(s)$.

8.99 The "plastic" explosive C-4, often used in action movies, contains the molecule *cyclotrimethylenetrinitramine,* which is often called RDX (for Royal Demolition eXplosive):

Cyclotrimethylenetrinitramine (RDX)

(a) Complete the Lewis structure for the molecule by adding unshared electron pairs where they are needed. (b) Does the Lewis structure you drew in part (a) have any resonance structures? If so, how many? (c) The molecule causes an explosion by decomposing into $CO(g)$, $N_2(g)$, and $H_2O(g)$. Write a balanced equation for the decomposition reaction. (d) With reference to Table 8.4, which is the weakest type of bond in the molecule? (e) Use average bond enthalpies to estimate the enthalpy change when 5.0 g of RDX decomposes.

8.100 The bond lengths of carbon–carbon, carbon–nitrogen, carbon–oxygen, and nitrogen–nitrogen single, double, and triple bonds are listed in Table 8.5. Plot bond enthalpy (Table 8.4) versus bond length for these bonds (as in Figure 8.17). (a) Is this statement true: "The longer the bond, the stronger the bond"? (b) Order the relative strengths of C—C, C—N, C—O, and N—N bonds from weakest to strongest. (c) From your graph for the carbon–carbon bond in part (a), estimate the bond enthalpy of the hypothetical C≡C quadruple bond.

Integrative Exercises

8.101 The Ti^{2+} ion is isoelectronic with the Ca atom. (a) Write the electron configurations of Ti^{2+} and Ca. (b) Calculate the number of unpaired electrons for Ca and for Ti^{2+}. (c) What charge would Ti have to be isoelectronic with Ca^{2+}?

8.102 (a) Write the chemical equations that are used in calculating the lattice energy of $SrCl_2(s)$ via a Born–Haber cycle. (b) The second ionization energy of $Sr(g)$ is 1064 kJ/mol. Use this fact along with data in Appendix C, Figure 7.10, Figure 7.12, and Table 8.2 to calculate ΔH_f° for $SrCl_2(s)$.

8.103 The electron affinity of oxygen is −141 kJ/mol, corresponding to the reaction

$$O_2(g) + e^- \longrightarrow O^-(g)$$

The lattice energy of $K_2O(s)$ is 2238 kJ/mol. Use these data along with data in Appendix C and Figure 7.10 to calculate the "second electron affinity" of oxygen, corresponding to the reaction

$$O^-(g) + e^- \longrightarrow O^{2-}(g)$$

8.104 You and a partner are asked to complete a lab entitled "Oxides of Ruthenium" that is scheduled to extend over two lab periods. The first lab, which is to be completed by your partner, is devoted to carrying out compositional analysis. In the second lab, you are to determine melting points. Upon going to lab you find two unlabeled vials, one containing a soft yellow substance and the other a black powder. You also find the following notes in your partner's notebook—*Compound 1*: 76.0% Ru and 24.0% O (by mass), *Compound 2*: 61.2% Ru and 38.8% O (by mass).

(a) What is the empirical formula for Compound 1?

(b) What is the empirical formula for Compound 2?

Upon determining the melting points of these two compounds, you find that the yellow compound melts at 25 °C, while the black powder does not melt up to the maximum temperature of your apparatus, 1200 °C.

(c) What is the identity of the yellow compound?

(d) What is the identity of the black compound?

(e) Which compound is molecular?

(f) Which compound is ionic?

8.105 One scale for electronegativity is based on the concept that the electronegativity of any atom is proportional to the ionization energy of the atom minus its electron affinity: electronegativity = $k(I - EA)$, where k is a proportionality constant. (a) How does this definition explain why the electronegativity of F is greater than that of Cl even though Cl has the greater electron affinity? (b) Why are both ionization energy and electron affinity relevant to the notion of electronegativity? (c) By using data in Chapter 7, determine the value of k that would lead to an electronegativity of 4.0 for F under this definition. (d) Use your result from part (c) to determine the electronegativities of Cl and O using this scale. (e) Another scale for electronegativity defines electronegativity as the average of an atom's first ionization energy and its electron affinity. Using this scale, calculate the electronegativities for the halogens, and scale them so fluorine has an electronegativity of 4.0. On this scale, what is Br's electronegativity?

8.106 The compound chloral hydrate, known in detective stories as knockout drops, is composed of 14.52% C, 1.83% H, 64.30% Cl, and 13.35% O by mass, and has a molar mass of 165.4 g/mol. (a) What is the empirical formula of this substance? (b) What is the molecular formula of this substance? (c) Draw the Lewis structure of the molecule, assuming that the Cl atoms bond to a single C atom and that there are a C—C bond and two C—O bonds in the compound.

8.107 Barium azide is 62.04% Ba and 37.96% N. Each azide ion has a net charge of 1−. (a) Determine the chemical formula of the azide ion. (b) Write three resonance structures for the azide ion. (c) Which structure is most important? (d) Predict the bond lengths in the ion.

8.108 Acetylene (C_2H_2) and nitrogen (N_2) both contain a triple bond, but they differ greatly in their chemical properties. (a) Write the Lewis structures for the two substances. (b) By referring to Appendix C, look up the enthalpies of formation of acetylene and nitrogen. Which compound is more stable? (c) Write balanced chemical equations for the complete oxidation of N_2 to form $N_2O_5(g)$ and of acetylene to form $CO_2(g)$ and $H_2O(g)$. (d) Calculate the enthalpy of oxidation per mole for N_2 and for C_2H_2 (the enthalpy of formation of $N_2O_5(g)$ is 11.30 kJ/mol). (e) Both N_2 and C_2H_2 possess triple bonds with quite high bond enthalpies (Table 8.4). Calculate the enthalpy of hydrogenation per mole for both compounds: acetylene plus H_2 to make methane, CH_4; nitrogen plus H_2 to make ammonia, NH_3.

8.109 Under special conditions, sulfur reacts with anhydrous liquid ammonia to form a binary compound of sulfur and nitrogen. The compound is found to consist of 69.6% S and 30.4% N. Measurements of its molecular mass yield a value of 184.3 g/mol. The compound occasionally detonates on being struck or when heated rapidly. The sulfur and nitrogen atoms of the molecule are joined in a ring. All the bonds in the ring are of the same length. **(a)** Calculate the empirical and molecular formulas for the substance. **(b)** Write Lewis structures for the molecule, based on the information you are given. (*Hint:* You should find a relatively small number of dominant Lewis structures.) **(c)** Predict the bond distances between the atoms in the ring. (*Note:* The S—S distance in the S_8 ring is 2.05 Å.) **(d)** The enthalpy of formation of the compound is estimated to be 480 kJ/mol^{-1}. ΔH_f° of $S(g)$ is 222.8 kJ/mol^{-1}. Estimate the average bond enthalpy in the compound.

8.110 A common form of elemental phosphorus is the tetrahedral P_4 molecule, where all four phosphorus atoms are equivalent:

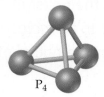

P_4

At room temperature phosphorus is a solid. **(a)** Are there any lone pairs of electrons in the P_4 molecule? **(b)** How many P—P bonds are there in the molecule? **(c)** Draw a Lewis structure for a linear P_4 molecule that satisfies the octet rule. Does this molecule have resonance structures? **(d)** On the basis of formal charges, which is more stable, the linear molecule or the tetrahedral molecule?

8.111 Consider benzene (C_6H_6) in the gas phase. **(a)** Write the reaction for breaking all the bonds in $C_6H_6(g)$, and use data in Appendix C to determine the enthalpy change for this reaction. **(b)** Write a reaction that corresponds to breaking all the carbon–carbon bonds in $C_6H_6(g)$. **(c)** By combining your answers to parts (a) and (b) and using the average bond enthalpy for C—H from Table 8.4, calculate the average bond enthalpy for the carbon–carbon bonds in $C_6H_6(g)$. **(d)** Compare your answer from part (c) to the values for C—C single bonds and C=C double bonds in Table 8.4. Is benzene's C—C bond enthalpy exactly halfway between them? If not, which bond type is more similar to that of benzene, CC single or CC double bonds?

8.112 Average bond enthalpies are generally defined for gas-phase molecules. Many substances are liquids in their standard state. ∞ (Section 5.7) By using appropriate thermochemical data from Appendix C, calculate average bond enthalpies in the liquid state for the following bonds, and compare these values to the gas-phase values given in Table 8.4: **(a)** Br—Br, from $Br_2(l)$; **(b)** C—Cl, from $CCl_4(l)$; **(c)** O—O, from $H_2O_2(l)$ (assume that the O—H bond enthalpy is the same as that in the gas phase). **(d)** Does the process of breaking bonds in the liquid as compared to the gas phase cost more energy? Explain the difference in the ΔH values between the two phases.

8.113 Silicon, the element, is the heart of integrated circuits and computer chips in almost all of our electronic devices. Si has the same structure as diamond; each atom is singly bonded to four neighbors. Unlike diamond, silicon has a tendency to oxidize (to SiO_2, another extended solid) if exposed to air. **(a)** Estimate the enthalpy of reaction for the conversion of 1 cm^3 of silicon into SiO_2. **(b)** Unlike carbon, silicon rarely forms multiple bonds. Estimate the bond enthalpy of the Si=Si bond, assuming that the ratio of the Si=Si double bond enthalpy to that of the Si—Si single bond is the same as that for carbon–carbon bonds.

Design an Experiment

You have learned that the resonance of benzene, C_6H_6, gives the compound special stability.

(a) By using data in Appendix C, compare the heat of combustion of 1.0 mol $C_6H_6(g)$ to the heat of combustion of 3.0 mol acetylene, $C_2H_2(g)$. Which has the greater fuel value, 1.0 mol $C_6H_6(g)$ or 3.0 mol $C_2H_2(g)$? Are your calculations consistent with benzene being especially stable? **(b)** Repeat part (a), with the appropriate molecules, for toluene ($C_6H_5CH_3$), a derivative of benzene that has a —CH_3 group in place of one H. **(c)** Another reaction you can use to compare molecules is *hydrogenation*, the reaction of a carbon-carbon double bond with H_2 to make a C—C single bond and two C—H single bonds. The experimental heat of hydrogenation of benzene to make cyclohexane (C_6H_{12}, a six-membered ring with 6 C—C single bonds with 12 C—H bonds) is 208 kJ/mol. The experimental heat of hydrogenation of cyclohexene (C_6H_{10}, a six-membered ring with one C=C double bond, 5 C—C single bonds, and 10 C—H bonds) to make cyclohexane is 120 kJ/mol. Show how these data can provide you with an estimate of the *resonance stabilization energy* of benzene. **(d)** Are the bond lengths or angles in benzene, compared to other hydrocarbons, sufficient to decide if benzene exhibits resonance and is especially stable? Discuss. **(e)** Consider cyclooctatetraene, C_8H_8, which has the octagonal structure shown below.

Cyclooctatetraene

What experiments or calculations could you perform to determine whether cyclooctatetraene exhibits resonance?

Molecular Geometry and Bonding Theories

We saw in Chapter 8 that Lewis structures help us understand the compositions of molecules and their covalent bonds. However, Lewis structures do not show one of the most important aspects of molecules—their overall shapes. The shape and size of molecules—sometimes referred to as *molecular architecture*—are defined by the angles and distances between the nuclei of the component atoms.

The shape and size of a molecule of a substance, together with the strength and polarity of its bonds, largely determine the properties of that substance. Some of the most dramatic examples of the important roles of molecular architecture are seen in biochemical reactions. For example, the chapter-opening photograph shows a molecular model of atorvastatin, better known as Lipitor®. In the body, Lipitor inhibits the action of a key enzyme, called *HMG-CoA reductase* (we will discuss enzymes in Section 14.7). HMG-CoA reductase is a large complex biomolecule that is critical in the biochemical sequence that synthesizes cholesterol in the liver, and inhibition of its action leads to reduced cholesterol production. The molecules of Lipitor have two properties that lead to their pharmaceutical effectiveness: First, the molecule has the correct overall *shape* to fit perfectly in an important cavity in the HMG-CoA reductase enzyme, thus blocking that site from the molecules involved in cholesterol synthesis. Second, the molecule has the right atoms and arrangements of electrons to form strong interactions within the cavity, assuring that the Lipitor molecule will "stick" where it should. Thus, the drug action of Lipitor is largely a consequence of the shape and size of the molecule as well as the charge distributions within it. Even a small modification to molecular shape or size alters the drug's effectiveness.

▶ **THE DRUG SHOWN HERE IS ATORVASTATIN,** better known by its trade name Lipitor®. It is a member of a class of pharmaceuticals called *statins*, which lower blood cholesterol levels, thereby reducing the risk of heart attacks and strokes. Lipitor was first synthesized in 1985 by Bruce Roth of Warner-Lambert/Parke Davis (now part of Pfizer) and was approved for use in 1996. It is the best-selling drug in pharmaceutical history, with sales of more than $125 billion from 1997 to 2011. It became available as a generic drug in 2011.

WHAT'S AHEAD ▶

9.1 MOLECULAR SHAPES We begin by discussing *molecular shapes* and examining some shapes commonly encountered in molecules.

9.2 THE VSEPR MODEL We see how molecular geometries can be predicted using the *valence-shell electron-pair repulsion*, or *VSEPR*, model, which is based on Lewis structures and the repulsions between regions of high electron density.

9.3 MOLECULAR SHAPE AND MOLECULAR POLARITY Once we know the geometry of a molecule and the types of bonds it contains, we can determine whether the molecule is *polar* or *nonpolar*.

9.4 COVALENT BONDING AND ORBITAL OVERLAP We explore how electrons are shared between atoms in a covalent bond. In *valence-bond theory*, the bonding electrons are visualized as originating in atomic orbitals on two atoms. A covalent bond is formed when these orbitals overlap.

9.5 HYBRID ORBITALS To account for molecular shape, we examine how the orbitals of one atom mix with one another, or *hybridize*, to create *hybrid orbitals*.

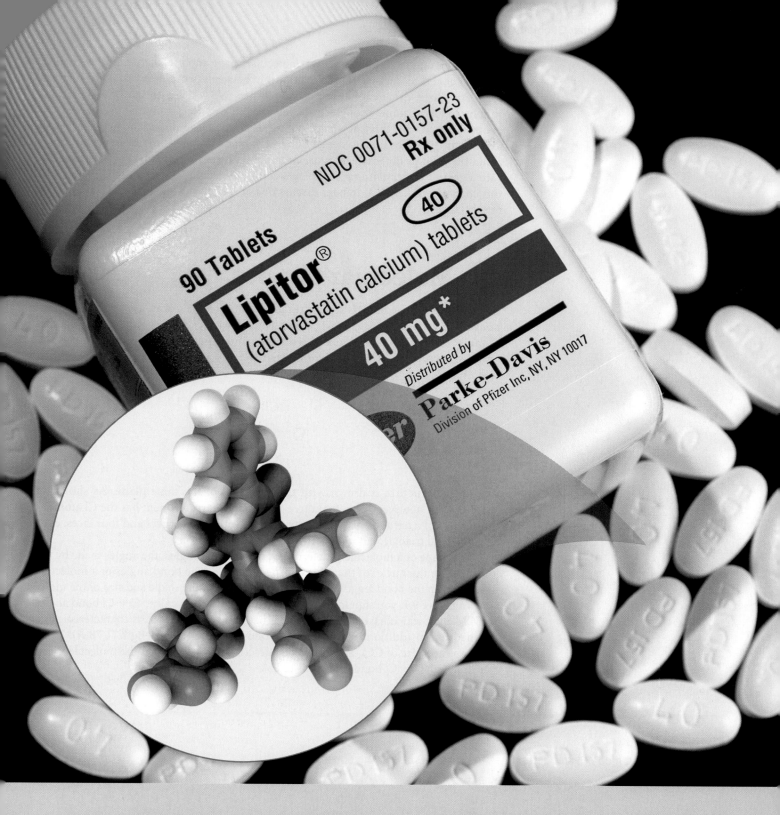

9.6 MULTIPLE BONDS Atomic orbitals that contribute to covalent bonding in a molecule can overlap in multiple ways to produce *sigma* and *pi* bonds between atoms. Single bonds consist of one sigma bond; multiple bonds involve one sigma and one or more pi bonds. We examine the geometric arrangements of these bonds and how they are exemplified in organic compounds.

9.7 MOLECULAR ORBITALS We examine a more sophisticated treatment of bonding called *molecular orbital theory*, which

introduces the concepts of *bonding* and *antibonding molecular orbitals*.

9.8 PERIOD 2 DIATOMIC MOLECULES We extend the concepts of molecular orbital theory to construct *energy-level diagrams* for second-row diatomic molecules.

As the example of Lipitor shows, molecular shape and size matter. In this chapter, our first goal is to understand the relationship between two-dimensional Lewis structures and three-dimensional molecular shapes. We will see the intimate relationship between the number of electrons involved in a molecule and the overall shape it adopts. Armed with this knowledge, we can examine more closely the nature of covalent bonds. The lines used to depict bonds in Lewis structures provide important clues about the orbitals that molecules use in bonding. By examining these orbitals, we can gain a greater understanding of the behavior of molecules. Mastering the material in this chapter will help you in later discussions of the physical and chemical properties of substances.

9.1 | Molecular Shapes

In Chapter 8 we used Lewis structures to account for the formulas of covalent compounds. ∞ (Section 8.5) Lewis structures, however, do not indicate the shapes of molecules; they simply show the number and types of bonds. For example, the Lewis structure of CCl_4 tells us only that four Cl atoms are bonded to a central C atom:

$$:\ddot{C}l:$$
$$|$$
$$:\ddot{C}l—C—\ddot{C}l:$$
$$|$$
$$:\ddot{C}l:$$

The Lewis structure is drawn with the atoms all in the same plane. As shown in ▼ Figure 9.1, however, the actual three-dimensional arrangement has the Cl atoms at the corners of a *tetrahedron*, a geometric object with four corners and four faces, each an equilateral triangle.

The shape of a molecule is determined by its **bond angles**, the angles made by the lines joining the nuclei of the atoms in the molecule. The bond angles of a molecule, together with the bond lengths ∞ (Section 8.8), define the shape and size of the molecule. In Figure 9.1, you should be able to see that there are six Cl—C—Cl bond angles in CCl_4, all of which have the same value. That bond angle, 109.5°, is characteristic of a tetrahedron. In addition, all four C—Cl bonds are of the same length (1.78 Å). Thus, the shape and size of CCl_4 are completely described by stating that the molecule is tetrahedral with C—Cl bonds of length 1.78 Å.

▲ GO FIGURE

In the space-filling model, what determines the relative sizes of the spheres?

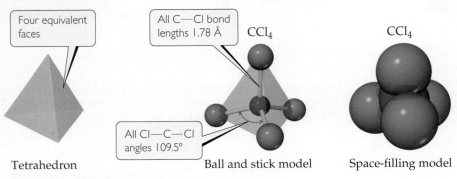

Four equivalent faces

All C—Cl bond lengths 1.78 Å CCl_4 CCl_4

All Cl—C—Cl angles 109.5°

Tetrahedron   Ball and stick model   Space-filling model

▲ **Figure 9.1 Tetrahedral shape of CCl_4.**

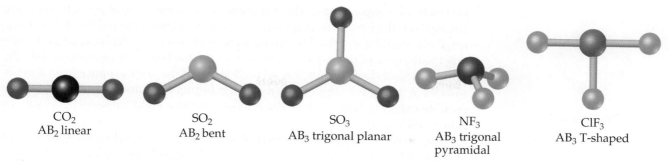

| CO_2 | SO_2 | SO_3 | NF_3 | ClF_3 |
| AB₂ linear | AB₂ bent | AB₃ trigonal planar | AB₃ trigonal pyramidal | AB₃ T-shaped |

▲ **Figure 9.2 Shapes of AB₂ and AB₃ molecules.**

We begin our discussion of molecular shapes with molecules (and ions) that, like CCl_4, have a single central atom bonded to two or more atoms of the same type. Such molecules have the general formula AB_n in which the central atom A is bonded to n B atoms. Both CO_2 and H_2O are AB_2 molecules, for example, whereas SO_3 and NH_3 are AB_3 molecules, and so on.

The number of shapes possible for AB_n molecules depends on the value of n. Those commonly found for AB_2 and AB_3 molecules are shown in ▲ **Figure 9.2.** An AB_2 molecule must be either *linear* (bond angle = 180°) or *bent* (bond angle ≠ 180°). For AB_3 molecules, the two most common shapes place the B atoms at the corners of an equilateral triangle. If the A atom lies in the same plane as the B atoms, the shape is called *trigonal planar*. If the A atom lies above the plane of the B atoms, the shape is called *trigonal pyramidal* (a pyramid with an equilateral triangle as its base). Some AB_3 molecules, such as ClF_3, are *T-shaped*, a relatively unusual shape shown in Figure 9.2. The atoms lie in one plane with two B—A—B angles of about 90°, and a third angle close to 180°.

Quite remarkably, the shapes of most AB_n molecules can be derived from just five basic geometric arrangements, shown in ▼ **Figure 9.3.** All of these are highly

▲ **GO FIGURE**

Which of these molecular shapes do you expect for the SF_6 molecule?

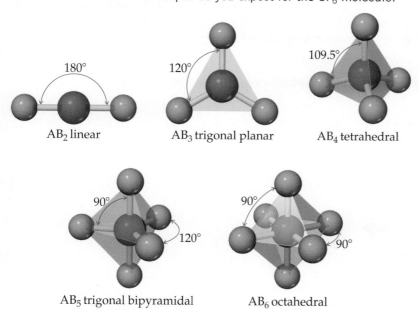

| AB₂ linear | AB₃ trigonal planar | AB₄ tetrahedral |
| AB₅ trigonal bipyramidal | AB₆ octahedral | |

▲ **Figure 9.3 Shapes allowing maximum distances between B atoms in AB$_n$ molecules.**

Octahedron

symmetric arrangements of the *n* B atoms around the central A atom. We have already seen the first three shapes: linear, trigonal planar, and tetrahedral. The trigonal bipyramid shape for AB_5 can be thought of as a trigonal planar AB_3 arrangement with two additional atoms, one above and one below the equilateral triangle. The octahedral shape for AB_6 has all six B atoms at the same distance from atom A with 90° B—A—B angles between all neighboring B atoms. Its symmetric shape (and its name) is derived from the *octahedron*, with eight faces, all of which are equilateral triangles.

You may have noticed that some of the shapes we have already discussed are *not* among the five shapes in Figure 9.3. For example, in Figure 9.2, neither the bent shape of the SO_2 molecule nor the trigonal pyramidal shape of the NF_3 molecule is among the shapes in Figure 9.3. However, as we soon will see, we can derive additional shapes, such as bent and trigonal pyramidal, by starting with one of our five basic arrangements. Starting with a tetrahedron, for example, we can remove atoms successively from the vertices, as shown in ▼ Figure 9.4. When an atom is removed from one vertex of a tetrahedron, the remaining AB_3 fragment has a trigonal-pyramidal geometry. When a second atom is removed, the remaining AB_2 fragment has a bent geometry.

Why do most AB_n molecules have shapes related to those shown in Figure 9.3? Can we predict these shapes? When A is a representative element (one from the *s* block or *p* block of the periodic table), we can answer these questions by using the **valence-shell electron-pair repulsion (VSEPR) model**. Although the name is rather imposing, the model is quite simple. It has useful predictive capabilities, as we will see in Section 9.2.

 Give It Some Thought

In addition to tetrahedral, another common shape for AB_4 molecules is *square planar*. All five atoms lie in the same plane, with the B atoms at the corners of a square and the A atom at the center of the square. Which shape in Figure 9.3 could lead to a square-planar shape upon removal of one or more atoms?

 GO FIGURE

In going from the tetrahedral shape to the bent shape, does it matter which two of the atoms we choose to remove?

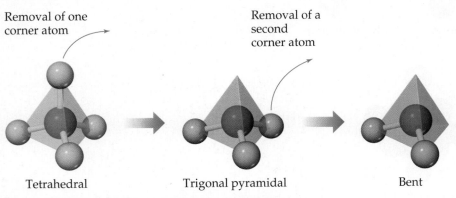

Removal of one corner atom

Removal of a second corner atom

Tetrahedral   Trigonal pyramidal   Bent

▲ Figure 9.4 **Derivatives of the tetrahedral molecular shape.**

9.2 | The VSEPR Model

Imagine tying two identical balloons together at their ends. As shown in ▶ Figure 9.5, the two balloons naturally orient themselves to point away from each other; that is, they try to "get out of each other's way" as much as possible. If we add a third balloon, the balloons orient themselves toward the vertices of an equilateral triangle, and if we add a fourth balloon, they adopt a tetrahedral shape. We see that an optimum geometry exists for each number of balloons.

In some ways, the electrons in molecules behave like these balloons. We have seen that a single covalent bond is formed between two atoms when a pair of electrons occupies the space between the atoms. ∞ (Section 8.3) A *bonding pair* of electrons thus defines a region in which the electrons are most likely to be found. We will refer to such a region as an **electron domain**. Likewise, a *nonbonding pair* (or *lone pair*) of electrons, which was also discussed in Section 8.3, defines an electron domain that is located principally on one atom. For example, the Lewis structure of NH_3 has four electron domains around the central nitrogen atom (three bonding pairs, represented as usual by short lines, and one nonbonding pair, represented by dots):

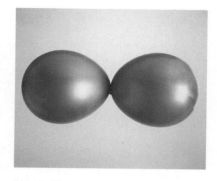

Two balloons
linear orientation

Nonbonding pair

$$H-\ddot{N}-H$$
$$|$$
$$H$$

Bonding pairs

Three balloons
trigonal-planar orientation

Each multiple bond in a molecule also constitutes a single electron domain. Thus, the following resonance structure for O_3 has three electron domains around the central oxygen atom (a single bond, a double bond, and a nonbonding pair of electrons):

$$:\ddot{O}-\ddot{O}=\ddot{O}:$$

In general, *each nonbonding pair, single bond, or multiple bond produces a single electron domain around the central atom in a molecule.*

 ### Give It Some Thought

Suppose a particular AB_3 molecule has the resonance structure

$$:\ddot{B}:$$
$$\|$$
$$:\ddot{B}-A-\ddot{B}:$$

Does this structure follow the octet rule? How many electron domains are there around the A atom?

Four balloons
tetrahedral orientation

▲ Figure 9.5 **A balloon analogy for electron domains.**

The VSEPR model is based on the idea that electron domains are negatively charged and therefore repel one another. Like the balloons in Figure 9.5, electron domains try to stay out of one another's way. *The best arrangement of a given number of electron domains is the one that minimizes the repulsions among them.* In fact, the analogy between electron domains and balloons is so close that the same preferred geometries are found in both cases. Like the balloons in Figure 9.5, two electron domains orient *linearly*, three domains orient in a *trigonal-planar* fashion, and four orient *tetrahedrally*. These arrangements, together with those for five- and six-electron domains, are summarized in **Table 9.1.** If you compare the geometries in Table 9.1 with those in Figure 9.3, you will see that they are the same. *The shapes of different AB_n molecules or ions depend on the number of electron domains surrounding the central atom.*

The arrangement of electron domains about the central atom of an AB_n molecule or ion is called its **electron-domain geometry**. In contrast, the **molecular geometry** is

Table 9.1 **Electron-Domain Geometries as a Function of Number of Electron Domains**

Number of Electron Domains	Arrangement of Electron Domains	Electron-Domain Geometry	Predicted Bond Angles
2	180°	Linear	180°
3	120°	Trigonal planar	120°
4	109.5°	Tetrahedral	109.5°
5	90° 120°	Trigonal bipyramidal	120° 90°
6	90° 90°	Octahedral	90°

the arrangement of *only the atoms* in a molecule or ion—any nonbonding pairs in the molecule are *not* part of the description of the molecular geometry.

In determining the shape of any molecule, we first use the VSEPR model to predict the electron-domain geometry. From knowing how many of the domains are due to nonbonding pairs, we can then predict the molecular geometry. When all the electron domains in a molecule arise from bonds, the molecular geometry is identical to the electron-domain geometry. When, however, one or more domains involve nonbonding pairs of electrons, we must remember that *the molecular geometry involves only electron domains due to bonds* even though the nonbonding pairs contribute to the electron-domain geometry.

We can generalize the steps we follow in using the VSEPR model to predict the shapes of molecules or ions:

1. Draw the Lewis structure of the molecule or ion ⚭ (Section 8.5), and count the number of electron domains around the central atom. Each nonbonding electron

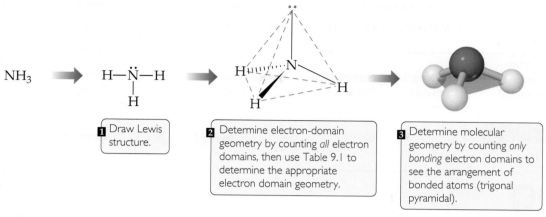

NH_3

$$H-\overset{\displaystyle\cdot\cdot}{N}-H$$
$$\overset{|}{H}$$

1 Draw Lewis structure.

2 Determine electron-domain geometry by counting *all* electron domains, then use Table 9.1 to determine the appropriate electron domain geometry.

3 Determine molecular geometry by counting *only* bonding electron domains to see the arrangement of bonded atoms (trigonal pyramidal).

▲ **Figure 9.6 Determining the molecular geometry of NH_3.**

pair, each single bond, each double bond, and each triple bond counts as one electron domain.

2. Determine the *electron-domain geometry* by arranging the electron domains about the central atom so that the repulsions among them are minimized, as shown in Table 9.1.

3. Use the arrangement of the bonded atoms to determine the *molecular geometry.*

▲ Figure 9.6 shows how these steps are applied to predict the geometry of the NH_3 molecule. The three bonds and one nonbonding pair in the Lewis structure tell us we have four electron domains. Thus, from Table 9.1, the electron-domain geometry of NH_3 is tetrahedral. We know from the Lewis structure that one electron domain is due to a nonbonding pair, which occupies one of the four vertices of the tetrahedron. In determining the molecular geometry, we consider only the three N—H bond domains, which leads to a trigonal pyramidal geometry. The situation is just like the middle drawing in Figure 9.4 in which removing one atom from a tetrahedral molecule results in a trigonal pyramidal molecule. Notice that the tetrahedral arrangement of the four electron domains leads us to predict the trigonal-pyramidal molecular geometry.

Because the trigonal-pyramidal molecular geometry is based on a tetrahedral electron-domain geometry, the *ideal bond angles* are 109.5°. As we will soon see, bond angles deviate from ideal values when the surrounding atoms and electron domains are not identical.

 Give It Some Thought

From the standpoint of the VSEPR model, what do nonbonding electron pairs, single bonds, and multiple bonds have in common?

As one more example, let's determine the shape of the CO_2 molecule. Its Lewis structure reveals two electron domains (each one a double bond) around the central carbon:

$$\overset{\displaystyle\cdot\cdot}{\underset{\displaystyle\cdot\cdot}{O}}=C=\overset{\displaystyle\cdot\cdot}{\underset{\displaystyle\cdot\cdot}{O}}$$

Two electron domains orient in a linear electron-domain geometry (Table 9.1). Because neither domain is a nonbonding pair of electrons, the molecular geometry is also linear, and the O—C—O bond angle is 180°.

Table 9.2 summarizes the possible molecular geometries when an AB_n molecule has four or fewer electron domains about A. These geometries are important because they include all the shapes usually seen in molecules or ions that obey the octet rule.

Table 9.2 Electron-Domain and Molecular Geometries for Two, Three, and Four Electron Domains around a Central Atom

Number of Electron Domains	Electron-Domain Geometry	Bonding Domains	Nonbonding Domains	Molecular Geometry	Example
2	Linear	2	0	Linear	$\ddot{O}{=}C{=}\ddot{O}$
3	Trigonal planar	3	0	Trigonal planar	
		2	1	Bent	
4	Tetrahedral	4	0	Tetrahedral	
		3	1	Trigonal pyramidal	
		2	2	Bent	

SAMPLE EXERCISE 9.1 | Using the VSEPR Model

Use the VSEPR model to predict the molecular geometry of (**a**) O_3, (**b**) $SnCl_3^-$.

SOLUTION

Analyze We are given the molecular formulas of a molecule and a polyatomic ion, both conforming to the general formula AB_n and both having a central atom from the *p* block of the periodic table. (Notice that for O_3, the A and B atoms are all oxygen atoms.)

Plan To predict the molecular geometries, we draw their Lewis structures and count electron domains around the central atom to get the electron-domain geometry. We then obtain the molecular geometry from the arrangement of the domains that are due to bonds.

Solve

(a) We can draw two resonance structures for O_3:

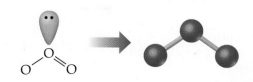

Because of resonance, the bonds between the central O atom and the outer O atoms are of equal length. In both resonance structures the central O atom is bonded to the two outer O atoms and has one non-bonding pair. Thus, there are three electron domains about the central O atoms. (Remember that a double bond counts as a single electron domain.) The arrangement of three electron domains is trigonal planar (Table 9.1). Two of the domains are from bonds, and one is due to a nonbonding pair. So, the molecular geometry is bent with an ideal bond angle of 120° (Table 9.2).

Comment As this example illustrates, when a molecule exhibits resonance, any one of the resonance structures can be used to predict the molecular geometry.

(b) The Lewis structure for $SnCl_3^-$ is

The central Sn atom is bonded to the three Cl atoms and has one nonbonding pair; thus, we have four electron domains, meaning a tetrahedral electron-domain geometry (Table 9.1) with one vertex occupied by a nonbonding pair of electrons. A tetrahedral electron-domain geometry with three bonding and one nonbonding domains leads to a trigonal-pyramidal molecular geometry (Table 9.2).

Practice Exercise 1

Consider the following AB_3 molecules and ions: PCl_3, SO_3, $AlCl_3$, SO_3^{2-}, and CH_3^+. How many of these molecules and ions do you predict to have a trigonal-planar molecular geometry?
(a) 1   **(b)** 2   **(c)** 3   **(d)** 4   **(e)** 5

Practice Exercise 2

Predict the electron-domain and molecular geometries for **(a)** $SeCl_2$, **(b)** CO_3^{2-}.

Effect of Nonbonding Electrons and Multiple Bonds on Bond Angles

We can refine the VSEPR model to explain slight distortions from the ideal geometries summarized in Table 9.2. For example, consider methane (CH_4), ammonia (NH_3), and water (H_2O). All three have a tetrahedral electron-domain geometry, but their bond angles differ slightly:

Notice that the bond angles decrease as the number of nonbonding electron pairs increases. A bonding pair of electrons is attracted by both nuclei of the bonded atoms, but a nonbonding pair is attracted primarily by only one nucleus. Because a nonbonding pair experiences less nuclear attraction, its electron domain is spread out more in space than is the electron domain for a bonding pair (**Figure 9.7**). Nonbonding electron pairs therefore take up more space than bonding pairs; in essence, they act as larger and fatter balloons in our analogy of Figure 9.5. As a result, *electron domains for nonbonding electron pairs exert greater repulsive forces on adjacent electron domains and tend to compress bond angles.*

Because multiple bonds contain a higher electronic-charge density than single bonds, multiple bonds also represent enlarged electron domains. Consider the Lewis structure of *phosgene*, Cl_2CO:

GO FIGURE

Why is the volume occupied by the nonbonding electron pair domain larger than the volume occupied by the bonding domain?

Bonding electron pair

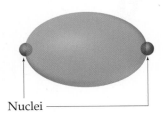

Nuclei

Nonbonding pair

Nucleus

▲ **Figure 9.7 Relative volumes occupied by bonding and nonbonding electron domains.**

 GO FIGURE

What is the bond angle formed by an axial atom, the central atom, and any equatorial atom?

Axial position

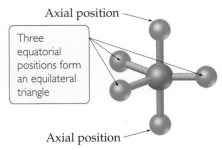

Three equatorial positions form an equilateral triangle

Axial position

▲ **Figure 9.8 In a trigonal-bipyramidal geometry, there are two types of positions for the outer atoms.**

Because three electron domains surround the central atom, we might expect a trigonal-planar geometry with 120° bond angles. The double bond, however, seems to act much like a nonbonding pair of electrons, reducing the Cl—C—Cl bond angle to 111.4°:

In general, *electron domains for multiple bonds exert a greater repulsive force on adjacent electron domains than do electron domains for single bonds.*

▲ **Give It Some Thought**

One resonance structure of the nitrate ion is

The bond angles in this ion are 120°. Is this observation consistent with the preceding discussion of the effect of multiple bonds on bond angles?

Molecules with Expanded Valence Shells

Atoms from period 3 and beyond may be surrounded by more than four electron pairs. ∞ (Section 8.7) Molecules with five or six electron domains around the central atom have molecular geometries based on either a *trigonal-bipyramidal* (five domains) or *octahedral* (six domains) electron-domain geometry (▶ Table 9.3).

The most stable electron-domain geometry for five electron domains is the trigonal bipyramid (two trigonal pyramids sharing a base). Unlike the other arrangements we have seen, the electron domains in a trigonal bipyramid can point toward two geometrically distinct types of positions. Two domains point toward *axial positions* and three point toward *equatorial positions* (◀ Figure 9.8). Each axial domain makes a 90° angle with any equatorial domain. Each equatorial domain makes a 120° angle with either of the other two equatorial domains and a 90° angle with either axial domain.

Suppose a molecule has five electron domains, and there are one or more nonbonding pairs. Will the domains from the nonbonding pairs occupy axial or equatorial positions? To answer this question, we must determine which location minimizes repulsion between domains. Repulsion between two domains is much greater when they are situated 90° from each other than when they are at 120°. An equatorial domain is 90° from only two other domains (the axial domains), but an axial domain is 90° from *three* other domains (the equatorial domains). Hence, an equatorial domain experiences less repulsion than an axial domain. Because the domains from nonbonding pairs exert larger repulsions than those from bonding pairs, nonbonding domains *always* occupy the equatorial positions in a trigonal bipyramid.

▲ **Give It Some Thought**

It might seem that a square-planar geometry of four electron domains around a central atom would be more favorable than a tetrahedron. Can you rationalize why the tetrahedron is preferred, based on angles between electron domains?

The most stable electron-domain geometry for six electron domains is the *octahedron.* An octahedron is a polyhedron with six vertices and eight faces, each an equilateral triangle. An atom with six electron domains around it can be visualized as being at the center of the octahedron with the electron domains pointing toward the six vertices, as

Table 9.3   Electron-Domain and Molecular Geometries for Five and Six Electron Domains around a Central Atom

Number of Electron Domains	Electron-Domain Geometry	Bonding Domains	Nonbonding Domains	Molecular Geometry	Example
5	Trigonal bipyramidal	5	0	Trigonal bipyramidal	PCl_5
		4	1	Seesaw	SF_4
		3	2	T-shaped	ClF_3
		2	3	Linear	XeF_2
6	Octahedral	6	0	Octahedral	SF_6
		5	1	Square pyramidal	BrF_5
		4	2	Square planar	XeF_4

shown in Table 9.3. All the bond angles are 90°, and all six vertices are equivalent. There-fore, if an atom has five bonding electron domains and one nonbonding domain, we can put the nonbonding domain at any of the six vertices of the octahedron. The result is always a *square-pyramidal* molecular geometry. When there are two nonbonding electron domains, however, their repulsions are minimized by pointing them toward opposite sides of the octahedron, producing a *square-planar* molecular geometry, as shown in Table 9.3.

SAMPLE EXERCISE 9.2 Molecular Geometries of Molecules with Expanded Valence Shells

Use the VSEPR model to predict the molecular geometry of (a) SF_4, (b) IF_5.

SOLUTION

Analyze The molecules are of the AB_n type with a central *p*-block atom.

Plan We first draw Lewis structures and then use the VSEPR model to determine the electron-domain geometry and molecular geometry.

Solve

(a) The Lewis structure for SF_4 is

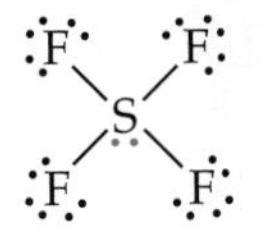

The sulfur has five electron domains around it: four from the S—F bonds and one from the nonbonding pair. Each domain points toward a vertex of a trigonal bipyramid. The domain from the nonbonding pair will point toward an equatorial position. The four bonds point toward the remaining four positions, resulting in a molecular geometry that is described as seesaw-shaped:

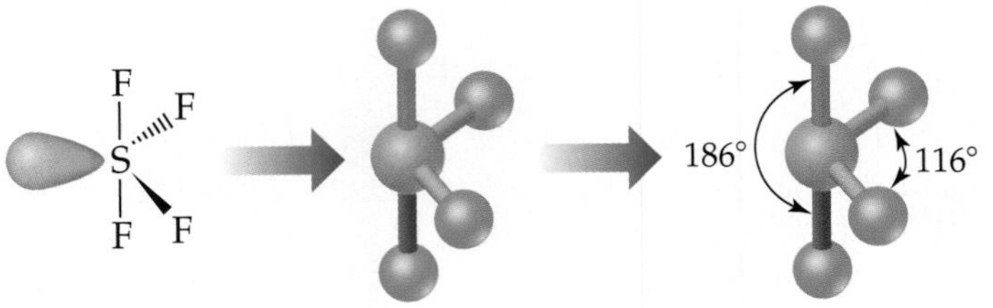

Comment The experimentally observed structure is shown on the right. We can infer that the nonbonding electron domain occupies an equatorial position, as predicted. The axial and equatorial S—F bonds are slightly bent away from the nonbonding domain, suggesting that the bonding domains are "pushed" by the nonbonding domain, which exerts a greater repulsion (Figure 9.7).

(b) The Lewis structure of IF_5 is

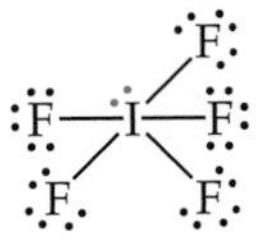

The iodine has six electron domains around it, one of which is non-bonding. The electron-domain geometry is therefore octahedral, with one position occupied by the nonbonding pair, and the molecular geometry is *square pyramidal* (Table 9.3):

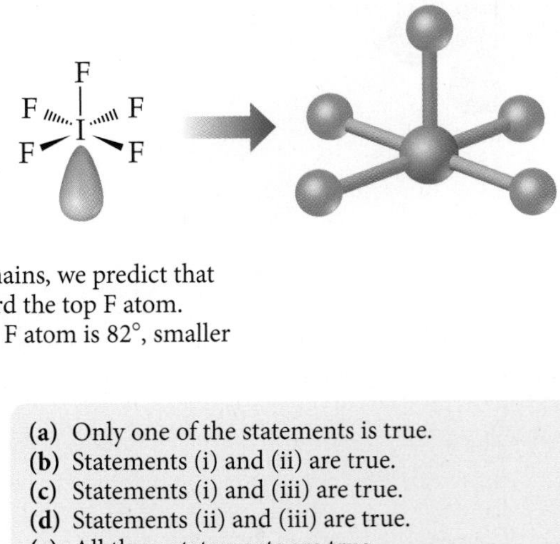

Comment Because the nonbonding domain is larger than the bonding domains, we predict that the four F atoms in the base of the pyramid will be tipped up slightly toward the top F atom. Experimentally, we find that the angle between the base atoms and the top F atom is 82°, smaller than the ideal 90° angle of an octahedron.

Practice Exercise 1

A certain AB_4 molecule has a square-planar molecular geometry. Which of the following statements about the molecule is or are true?:
   (i) The molecule has four electron domains about the central atom A.
   (ii) The B—A—B angles between neighboring B atoms is 90°.
   (iii) The molecule has two nonbonding pairs of electrons on atom A.

   (a) Only one of the statements is true.
   (b) Statements (i) and (ii) are true.
   (c) Statements (i) and (iii) are true.
   (d) Statements (ii) and (iii) are true.
   (e) All three statements are true.

Practice Exercise 2

Predict the electron-domain and molecular geometries of (a) BrF_3, (b) SF_5^+.

Shapes of Larger Molecules

Although the molecules and ions we have considered contain only a single central atom, the VSEPR model can be extended to more complex molecules. Consider the acetic acid molecule, for example:

$$
\begin{array}{ccc}
H & \ddot{O} & \\
| & || & \\
H-C-C-\ddot{O}-H & &
\end{array}
$$

We can use the VSEPR model to predict the geometry about each atom:

	H | H—C— | H	:O: || C—	Ö—H
Number of electron domains	4	3	4
Electron-domain geometry	Tetrahedral	Trigonal planar	Tetrahedral
Predicted bond angles	109.5°	120°	109.5°

The C on the left has four electron domains (all bonding), so the electron-domain and molecular geometries around that atom are both tetrahedral. The central C has three electron domains (counting the double bond as one domain), making both the electron-domain and the molecular geometries trigonal planar. The O on the right has four electron domains (two bonding, two nonbonding), so its electron-domain geometry is tetrahedral and its molecular geometry is bent. The bond angles about the central C atom and the O atom are expected to deviate slightly from the ideal values of 120° and 109.5° because of the spatial demands of multiple bonds and nonbonding electron pairs.

Our analysis of the acetic acid molecule is shown in ▶ **Figure 9.9.**

SAMPLE EXERCISE 9.3 | Predicting Bond Angles

Eyedrops for dry eyes usually contain a water-soluble polymer called *poly(vinyl alcohol)*, which is based on the unstable organic molecule *vinyl alcohol*:

$$
\begin{array}{cc}
H & H \\
| & | \\
H-\ddot{O}-C=C-H &
\end{array}
$$

Predict the approximate values for the H—O—C and O—C—C bond angles in vinyl alcohol.

SOLUTION

Analyze We are given a Lewis structure and asked to determine two bond angles.

Plan To predict a bond angle, we determine the number of electron domains surrounding the middle atom in the bond. The ideal angle corresponds to the electron-domain geometry around the atom. The angle will be compressed somewhat by nonbonding electrons or multiple bonds.

Solve In H—O—C, the O atom has four electron domains (two bonding, two nonbonding). The electron-domain geometry around O is therefore tetrahedral, which gives an ideal angle of 109.5°. The H—O—C angle is compressed somewhat by the nonbonding pairs, so we expect this angle to be slightly less than 109.5°.

To predict the O—C—C bond angle, we examine the middle atom in the angle. In the molecule, there are three atoms bonded to this C atom and no nonbonding pairs, and so it has three electron domains about it. The predicted electron-domain geometry is trigonal planar, resulting in an ideal bond angle of 120°. Because of the larger size of the C=C domain, the bond angle should be slightly greater than 120°.

▲ GO FIGURE

In the actual structure of acetic acid, which bond angle is expected to be the smallest?

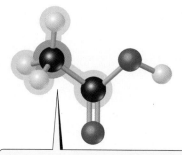

Electron-domain geometry tetrahedral, molecular geometry tetrahedral

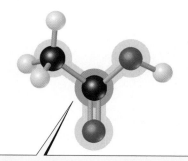

Electron-domain geometry trigonal planar, molecular geometry trigonal planar

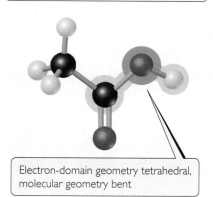

Electron-domain geometry tetrahedral, molecular geometry bent

▲ **Figure 9.9 The electron-domain and molecular geometries around the three central atoms of acetic acid, CH_3COOH.**

Practice Exercise 1

The atoms of the compound methylhydrazine, CH_6N_2, which is used as a rocket propellant, are connected as follows (note that lone pairs are not shown):

$$
\begin{array}{c}
\text{H} \\
| \\
\text{H}-\text{C}-\text{H} \\
| \\
\text{N}-\text{N}-\text{H} \\
| \quad | \\
\text{H} \quad \text{H}
\end{array}
$$

What do you predict for the ideal values of the C—N—N and H—N—H angles, respectively?
(a) 109.5° and 109.5°   **(b)** 109.5° and 120°   **(c)** 120° and 109.5°   **(d)** 120° and 120°
(e) None of the above

Practice Exercise 2

Predict the H—C—H and C—C—C bond angles in *propyne*:

$$
\begin{array}{c}
\text{H} \\
| \\
\text{H}-\text{C}-\text{C}\equiv\text{C}-\text{H} \\
| \\
\text{H}
\end{array}
$$

9.3 | Molecular Shape and Molecular Polarity

Now that we have a sense of the shapes that molecules adopt and why they do so, we will return to some topics that we first discussed in Section 8.4, namely, *bond polarity* and *dipole moments*. Recall that bond polarity is a measure of how equally the electrons in a bond are shared between the two atoms of the bond. As the difference in electronegativity between the two atoms increases, so does the bond polarity. ∞ (Section 8.4)

We saw that the dipole moment of a diatomic molecule is a measure of the amount of charge separation in the molecule.

For a molecule consisting of more than two atoms, *the dipole moment depends on both the polarities of the individual bonds and the geometry of the molecule.* For each bond in the molecule, we consider the **bond dipole**, which is the dipole moment due only to the two atoms in that bond. Consider the linear CO_2 molecule, for example. As shown in ◀ Figure 9.10, each C=O bond is polar, and because the C=O bonds are identical, the bond dipoles are equal in magnitude. A plot of the molecule's electron density clearly shows that the individual bonds are polar, but what can we say about the *overall* dipole moment of the molecule?

Bond dipoles and dipole moments are *vector quantities*; that is, they have both a magnitude and a direction. The dipole moment of a polyatomic molecule is the vector sum of its bond dipoles. Both the magnitudes *and* the directions of the bond dipoles must be considered when summing vectors. The two bond dipoles in CO_2, although equal in magnitude, are opposite in direction. Adding them is the same as adding two numbers that are equal in magnitude but opposite in sign, such as $100 + (-100)$. The bond dipoles, like the numbers, "cancel" each other. Therefore, the dipole moment of CO_2 is zero, even though the individual bonds are polar. The geometry of the molecule dictates that the overall dipole moment be zero, making CO_2 a *nonpolar* molecule.

Now let's consider H_2O, a bent molecule with two polar bonds (▶ Figure 9.11). Again, the two bonds are identical, and the bond dipoles are equal in magnitude. Because the molecule is bent, however, the bond dipoles do not directly oppose each other and therefore do not cancel. Hence, the H_2O molecule has an overall nonzero dipole moment ($\mu = 1.85\,\text{D}$) and is therefore a *polar* molecule. The oxygen atom carries a partial negative charge, and the hydrogen atoms each have a partial positive charge, as shown in the electron-density model.

GO FIGURE

What is the sum of the two red vectors at the top of the figure?

Equal and oppositely directed bond dipoles

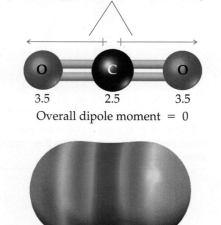

3.5   2.5   3.5

Overall dipole moment = 0

Low electron density   High electron density

+   −

▲ Figure 9.10 **CO₂, a nonpolar molecule.**
The numbers are electronegativity values for these two atoms.

Give It Some Thought

The molecule O=C=S is linear and has a Lewis structure analogous to that of CO_2. Would you expect this molecule to be nonpolar?

Figure 9.12 shows some examples of polar and nonpolar molecules, all with polar bonds. The molecules in which the central atom is symmetrically surrounded by identical atoms (BF_3 and CCl_4) are nonpolar. For AB_n molecules in which all the B atoms are the same, certain symmetrical shapes—linear (AB_2), trigonal planar (AB_3), tetrahedral and square planar (AB_4), trigonal bipyramidal (AB_5), and octahedral (AB_6)—must lead to nonpolar molecules even though the individual bonds might be polar.

SAMPLE
EXERCISE 9.4   Polarity of Molecules

Predict whether these molecules are polar or nonpolar: (a) BrCl, (b) SO_2, (c) SF_6.

SOLUTION

Analyze We are given three molecular formulas and asked to predict whether the molecules are polar.

Plan A molecule containing only two atoms is polar if the atoms differ in electronegativity. The polarity of a molecule containing three or more atoms depends on both the molecular geometry and the individual bond polarities. Thus, we must draw a Lewis structure for each molecule containing three or more atoms and determine its molecular geometry. We then use electronegativity values to determine the direction of the bond dipoles. Finally, we see whether the bond dipoles cancel to give a nonpolar molecule or reinforce each other to give a polar one.

Solve

(a) Chlorine is more electronegative than bromine. All diatomic molecules with polar bonds are polar molecules. Consequently, BrCl is polar, with chlorine carrying the partial negative charge:

$$\overset{\longrightarrow}{\text{Br—Cl}}$$

The measured dipole moment of BrCl is $\mu = 0.57$ D.

(b) Because oxygen is more electronegative than sulfur, SO_2 has polar bonds. Three resonance forms can be written:

$$:\ddot{\text{O}}\text{—}\ddot{\text{S}}\text{=}\ddot{\text{O}}: \longleftrightarrow :\ddot{\text{O}}\text{=}\ddot{\text{S}}\text{—}\ddot{\text{O}}: \longleftrightarrow :\ddot{\text{O}}\text{=}\ddot{\text{S}}\text{=}\ddot{\text{O}}:$$

For each of these, the VSEPR model predicts a bent molecular geometry. Because the molecule is bent, the bond dipoles do not cancel, and the molecule is polar:

Experimentally, the dipole moment of SO_2 is $\mu = 1.63$ D.

(c) Fluorine is more electronegative than sulfur, so the bond dipoles point toward fluorine. For clarity, only one S—F dipole is shown. The six S—F bonds are arranged octahedrally around the central sulfur:

$$\begin{array}{c} \text{F} \quad \text{F} \\ | \nearrow \\ \text{F—S}\rightleftharpoons\text{F} \\ \nearrow | \\ \text{F} \quad \text{F} \end{array}$$

Because the octahedral molecular geometry is symmetrical, the bond dipoles cancel, and the molecule is nonpolar, meaning that $\mu = 0$.

Practice Exercise 1

Consider an AB_3 molecule in which A and B differ in electronegativity. You are told that the molecule has an overall dipole moment of zero. Which of the following could be the molecular geometry of the molecule?   (a) Trigonal pyramidal   (b) Trigonal planar   (c) T-shaped   (d) Tetrahedral   (e) More than one of the above

Practice Exercise 2

Determine whether the following molecules are polar or nonpolar: (a) SF_4, (b) $SiCl_4$.

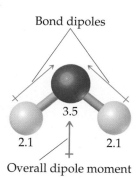

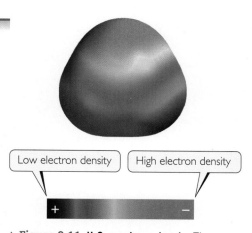

Bond dipoles

3.5

2.1   2.1

Overall dipole moment

Low electron density   High electron density

+   −

▲ **Figure 9.11 H_2O, a polar molecule.** The numbers are electronegativity values.

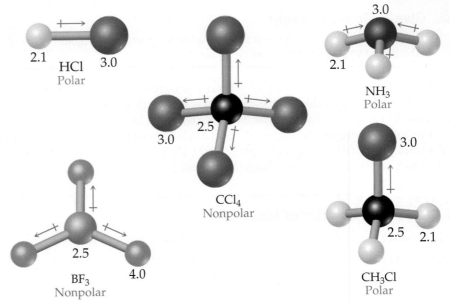

▲ **Figure 9.12 Polar and nonpolar molecules containing polar bonds.** The numbers are electronegativity values.

9.4 | Covalent Bonding and Orbital Overlap

The VSEPR model provides a simple means for predicting molecular geometries but does not explain why bonds exist between atoms. In developing theories of covalent bonding, chemists have approached the problem from another direction, using quantum mechanics. How can we use atomic orbitals to explain bonding and to account for molecular geometries? The marriage of Lewis's notion of electron-pair bonds and the idea of atomic orbitals leads to a model of chemical bonding, called **valence-bond theory**, in which bonding electron pairs are concentrated in the regions between atoms, and nonbonding electron pairs lie in directed regions of space. By extending this approach to include the ways in which atomic orbitals can mix with one another, we obtain an explanatory picture that corresponds to the VSEPR model.

In Lewis theory, covalent bonding occurs when atoms share electrons because the sharing concentrates electron density between the nuclei. In valence-bond theory, we visualize the buildup of electron density between two nuclei as occurring when a valence atomic orbital of one atom shares space, or *overlaps*, with a valence atomic orbital of another atom. The overlap of orbitals allows two electrons of opposite spin to share the space between the nuclei, forming a covalent bond.

◀ Figure 9.13 shows three examples of how valence-bond theory describes the coming together of two atoms to form a molecule. In the example of the formation of H_2, each hydrogen atom has a single electron in a $1s$ orbital. As the orbitals overlap, electron density is concentrated between the nuclei. Because the electrons in the overlap region are simultaneously attracted to both nuclei, they hold the atoms together, forming a covalent bond.

The idea of orbital overlap producing a covalent bond applies equally well to other molecules. In HCl, for example, chlorine has the electron configuration $[Ne]3s^23p^5$. All the valence orbitals of chlorine are full except one $3p$ orbital, which contains a single electron. This $3p$ electron pairs with the single $1s$ electron of H to form the covalent bond that holds HCl together (Figure 9.13). Because the other two chlorine $3p$ orbitals are already filled with a pair of electrons, they do not participate in the bonding to hydrogen. Likewise, we can explain the covalent bond in Cl_2 in terms of the overlap of the singly occupied $3p$ orbital of one Cl atom with the singly occupied $3p$ orbital of another.

GO FIGURE

How does the notion of overlap explain why the bond in HCl is longer than the bond in H_2?

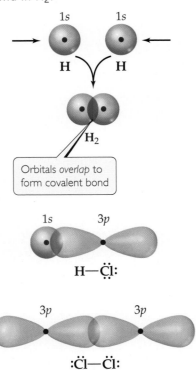

▲ **Figure 9.13 Covalent bonds in H_2, HCl, and Cl_2 result from overlap of atomic orbitals.**

⚠ GO FIGURE

On the left part of the curve the potential energy rises above zero. What causes this to happen?

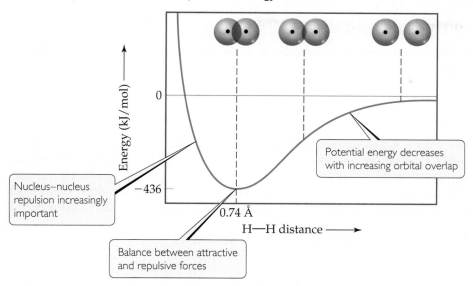

Nucleus–nucleus repulsion increasingly important

Balance between attractive and repulsive forces

Potential energy decreases with increasing orbital overlap

Energy (kJ/mol)

0

−436

0.74 Å

H—H distance

▲ **Figure 9.14 Formation of the H₂ molecule as atomic orbitals overlap.**

There is always an optimum distance between the two nuclei in any covalent bond. ▲ Figure 9.14 shows how the potential energy of a system consisting of two H atoms changes as the atoms come together to form an H_2 molecule. When the atoms are infinitely far apart, they do not "feel" each other and so the energy approaches zero. As the distance between the atoms decreases, the overlap between their $1s$ orbitals increases. Because of the resultant increase in electron density between the nuclei, the potential energy of the system decreases. That is, the strength of the bond increases, as shown by the decrease in the potential energy of the two-atom system. However, Figure 9.14 also shows that the energy increases sharply when the distance between the two hydrogen nuclei is less than 0.74 Å. The increase in potential energy of the system, which becomes significant at short internuclear distances, is due mainly to the electrostatic repulsion between the nuclei. The internuclear distance at the minimum of the potential-energy curve (in this example, at 0.74 Å) corresponds to the bond length of the molecule. The potential energy at this minimum corresponds to the bond strength. Thus, the observed bond length is the distance at which the attractive forces between unlike charges (electrons and nuclei) are balanced by the repulsive forces between like charges (electron–electron and nucleus–nucleus).

9.5 | Hybrid Orbitals

The VSEPR model, simple as it is, does a surprisingly good job at predicting molecular shape, despite the fact that it has no obvious relationship to the filling and shapes of atomic orbitals. For example, we would like to understand how to account for the tetrahedral arrangement of C—H bonds in methane in terms of the $2s$ and $2p$ orbitals of the central carbon atom, which are not directed toward the apices of a tetrahedron. How can we reconcile the notion that covalent bonds are formed from overlap of atomic orbitals with the molecular geometries that come from the VSEPR model?

To begin with, we recall that atomic orbitals are mathematical functions that come from the quantum mechanical model for atomic structure. ⚬⚬⚬ (Section 6.5) To explain molecular geometries, we often assume that the atomic orbitals on an atom (usually the central atom) mix to form new orbitals called **hybrid orbitals**. The shape of any hybrid orbital is different from the shapes of the original atomic orbitals. The process of mixing atomic orbitals is a mathematical operation called **hybridization**. The total number of atomic orbitals on an atom remains constant, so the number of hybrid orbitals on an atom equals the number of atomic orbitals that are mixed.

As we examine the common types of hybridization, notice the connection between the type of hybridization and some of the molecular geometries predicted by the VSEPR model: linear, bent, trigonal planar, and tetrahedral.

sp Hybrid Orbitals

To illustrate the process of hybridization, consider the BeF_2 molecule, which has the Lewis structure

$$:\ddot{F}—Be—\ddot{F}:$$

The VSEPR model correctly predicts that BeF_2 is linear with two identical Be—F bonds. How can we use valence-bond theory to describe the bonding? The electron configuration of F $(1s^2 2s^2 2p^5)$ indicates an unpaired electron in a $2p$ orbital. This electron can be paired with an unpaired Be electron to form a polar covalent bond. Which orbitals on the Be atom, however, overlap with those on the F atoms to form the Be—F bonds?

The orbital diagram for a ground-state Be atom is

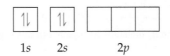

Because it has no unpaired electrons, the Be atom in its ground state cannot bond with the fluorine atoms. The Be atom could form two bonds, however, by envisioning that we "promote" one of the $2s$ electrons to a $2p$ orbital:

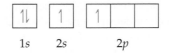

The Be atom now has two unpaired electrons and can therefore form two polar covalent bonds with F atoms. The two bonds would not be identical, however, because a Be $2s$ orbital would be used to form one of the bonds and a $2p$ orbital would be used to form the other. Therefore, although the promotion of an electron allows two Be—F bonds to form, we still have not explained the structure of BeF_2.

We can solve this dilemma by "mixing" the $2s$ orbital with one $2p$ orbital to generate two new orbitals, as shown in ▼ Figure 9.15. Like p orbitals, each new orbital has two lobes. Unlike p orbitals, however, one lobe is much larger than the other. The two new orbitals are identical in shape, but their large lobes point in opposite directions. These two new orbitals, which are shown in purple in Figure 9.15, are hybrid orbitals. Because we have hybridized one s and one p orbital, we call each hybrid an sp hybrid orbital. *According to the valence-bond model, a linear arrangement of electron domains implies* sp *hybridization.*

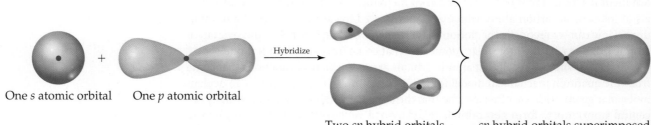

▲ Figure 9.15 **Formation of *sp* hybrid orbitals.**

For the Be atom of BeF_2, we write the orbital diagram for the formation of two sp hybrid orbitals as

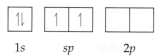

1s   sp   2p

The electrons in the sp hybrid orbitals can form bonds with the two fluorine atoms (▼ Figure 9.16). Because the sp hybrid orbitals are equivalent but point in opposite directions, BeF_2 has two identical bonds and a linear geometry. We have used one of the $2p$ orbitals in making the hybrids; the remaining two $2p$ atomic orbitals of Be remain unhybridized and are vacant. Remember also that each fluorine atom has two other valence p atomic orbitals, each containing one nonbonding electron pair. Those atomic orbitals are omitted from Figure 9.16 to keep the illustration simpler.

GO FIGURE

Why is it reasonable to take account of only the large lobes of the Be hybrid orbitals in considering the bonding to F?

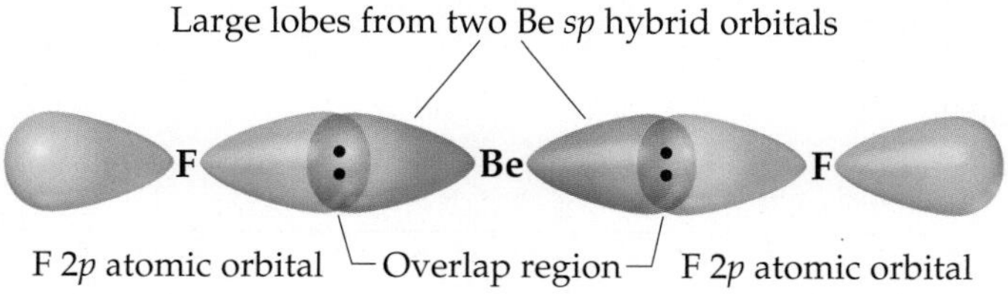

Large lobes from two Be sp hybrid orbitals

F 2p atomic orbital   └ Overlap region ┘   F 2p atomic orbital

▲ Figure 9.16 **Formation of two equivalent Be—F bonds in BeF₂.**

Give It Some Thought

What is the orientation of the two unhybridized p orbitals on Be with respect to the two Be—F bonds?

---

sp^2 and sp^3 Hybrid Orbitals

Whenever we mix a certain number of atomic orbitals, we get the same number of hybrid orbitals. Each hybrid orbital is equivalent to the others but points in a different direction. Thus, mixing one $2s$ and one $2p$ atomic orbital yields two equivalent sp hybrid orbitals that point in opposite directions (Figure 9.15). Other combinations of atomic orbitals can be hybridized to obtain different geometries. In BF_3, for example, mixing the $2s$ and two of the $2p$ atomic orbitals yields three equivalent sp^2 (pronounced "s-p-two") hybrid orbitals (Figure 9.17).

The three sp^2 hybrid orbitals lie in the same plane, 120° apart from one another. They are used to make three equivalent bonds with the three fluorine atoms, leading to the trigonal-planar molecular geometry of BF_3. Notice that an unfilled $2p$ atomic orbital remains unhybridized; it is oriented perpendicular to the plane defined by the three sp^2 hybrid orbitals, with one lobe above and one below the plane. This unhybridized orbital will be important when we discuss double bonds in Section 9.6.

An s atomic orbital can mix with all three p atomic orbitals in the same subshell. For example, the carbon atom in CH_4 forms four equivalent bonds with the four hydrogen atoms. We envision this process as resulting from the mixing of the $2s$ and all three $2p$ atomic orbitals of carbon to create four equivalent sp^3 (pronounced "s-p-three") hybrid orbitals. Each sp^3 hybrid orbital has a large lobe that

GO FIGURE

What is the angle formed between the large lobes of the three sp^2 hybrid orbitals?

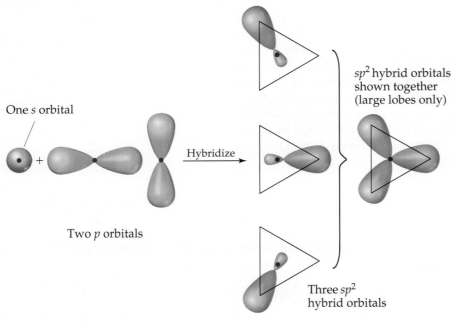

One *s* orbital

Two *p* orbitals

Hybridize

sp^2 hybrid orbitals shown together (large lobes only)

Three sp^2 hybrid orbitals

▲ Figure 9.17 **Formation of sp^2 hybrid orbitals.**

points toward one vertex of a tetrahedron (▶ Figure 9.18). These hybrid orbitals can be used to form two-electron bonds by overlap with the atomic orbitals of another atom, such as H. Using valence-bond theory, we can describe the bonding in CH_4 as the overlap of four equivalent sp^3 hybrid orbitals on C with the 1*s* orbitals of the four H atoms to form four equivalent bonds.

 Give It Some Thought

In an sp^2 hybridized atom, we saw that there was one unhybridized 2*p* orbital. How many unhybridized 2*p* orbitals remain on an atom that has sp^3 hybrid orbitals?

The idea of hybridization is also used to describe the bonding in molecules containing nonbonding pairs of electrons. In H_2O, for example, the electron-domain geometry around the central O atom is approximately tetrahedral (▶ Figure 9.19). Thus, the four electron pairs can be envisioned as occupying sp^3 hybrid orbitals. Two of the hybrid orbitals contain nonbonding pairs of electrons, and the other two form bonds with the hydrogen atoms.

Hypervalent Molecules

So far our discussion of hybridization has extended only to period 2 elements, specifically carbon, nitrogen, and oxygen. The elements of period 3 and beyond introduce a new consideration because in many of their compounds these elements are **hypervalent**—they have more than an octet of electrons around the central atom. ∞ (Section 8.7) We saw in Section 9.2 that the VSEPR model works well to predict the geometries of hypervalent molecules such as PCl_5, SF_6, or BrF_5. But can we extend the use of hybrid orbitals to describe the bonding in these molecules? In short, the answer to this question is that it is best *not* to use hybrid orbitals for hypervalent molecules, as we now briefly discuss.

The valence-bond model we developed for period 2 elements works well for compounds of period 3 elements so long as we have no more than an octet of electrons in

▲ GO FIGURE

Which of the p orbitals do you think contributes the most in the mixing that leads to the right-most sp^3 hybrid orbital in the second row of the figure?

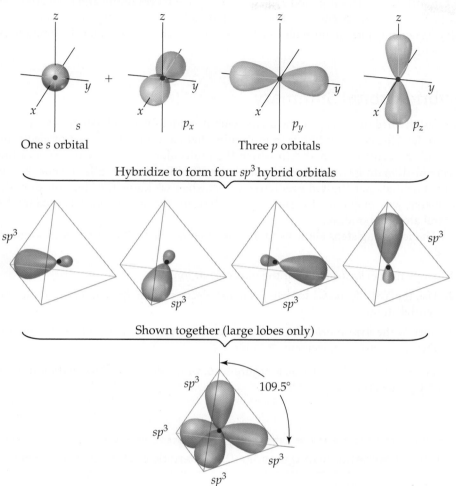

One s orbital   Three p orbitals

Hybridize to form four sp^3 hybrid orbitals

Shown together (large lobes only)

$109.5°$

▲ **Figure 9.18 Formation of sp^3 hybrid orbitals.**

the valence-shell orbitals. Thus, for example, it is appropriate to discuss the bonding in PF_3 or H_2Se in terms of hybridized s and p orbitals on the central atom.

For compounds with more than an octet, we could imagine increasing the number of hybrid orbitals formed by including valence-shell d orbitals. For example, for SF_6 we could envision mixing in two sulfur $3d$ orbitals in addition to the $3s$ and three $3p$ orbitals to make a total of six hybrid orbitals. However, the sulfur $3d$ orbitals are substantially higher in energy than the $3s$ and $3p$ orbitals, so the amount of energy needed to form the six hybrid orbitals is greater than the amount returned by forming bonds with the six fluorine atoms. Theoretical calculations suggest that the sulfur $3d$ orbitals do not participate to a significant degree in the bonding between sulfur and the six fluorine atoms, and that it would not be valid to describe the bonding in SF_6 in terms of six hybrid orbitals. The more detailed bonding model needed to discuss the bonding in SF_6 and other hypervalent molecules requires a treatment beyond the scope of a general chemistry text. Fortunately, the VSEPR model, which explains the geometrical properties of such molecules in terms of electrostatic repulsions, does a good job of predicting their geometries.

▲ GO FIGURE

Does it matter which of the two sp^3 hybrid orbitals are used to hold the two nonbonding electron pairs?

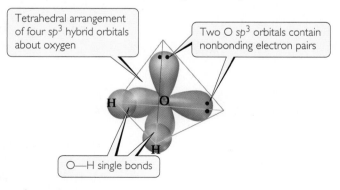

Tetrahedral arrangement of four sp^3 hybrid orbitals about oxygen

Two O sp^3 orbitals contain nonbonding electron pairs

O—H single bonds

▲ **Figure 9.19 Hybrid orbital description of H_2O.**

This discussion reminds us that models in science are not reality but rather are our attempts to describe aspects of reality that we have been able to measure, such as bond distances, bond energies, molecular geometries, and so on. A model may work well up to a certain point but not beyond it, as is the case for hybrid orbitals. The hybrid orbital model for period 2 elements has proven very useful and is an essential part of any modern discussion of bonding and molecular geometry in organic chemistry. When it comes to molecules such as SF_6, however, we encounter the limitations of the model.

Hybrid Orbital Summary

Overall, hybrid orbitals provide a convenient model for using valence-bond theory to describe covalent bonds in molecules that have an octet or less of electrons around the central atom and in which the molecular geometry conforms to the electron-domain geometry predicted by the VSEPR model. While the concept of hybrid orbitals has limited predictive value, when we know the electron-domain geometry, we can employ hybridization to describe the atomic orbitals used by the central atom in bonding.

The following steps allow us to describe the hybrid orbitals used by an atom in bonding:

1. Draw the *Lewis structure* for the molecule or ion.

2. Use the VSEPR model to determine the electron-domain geometry around the central atom.

3. Specify the *hybrid orbitals* needed to accommodate the electron pairs based on their geometric arrangement (▼ **Table 9.4**).

These steps are illustrated in ▶ **Figure 9.20**, which shows how the hybridization at N in NH_3 is determined.

Table 9.4   Geometric Arrangements Characteristic of Hybrid Orbital Sets

Atomic Orbital Set	Hybrid Orbital Set	Geometry	Examples
s, p	Two sp	180° Linear	BeF_2, $HgCl_2$
s, p, p	Three sp^2	120° Trigonal planar	BF_3, SO_3
s, p, p, p	Four sp^3	109.5° Tetrahedral	CH_4, NH_3, H_2O, NH_4^+

GO FIGURE

How would we modify the figure if we were looking at PH$_3$ rather than NH$_3$?

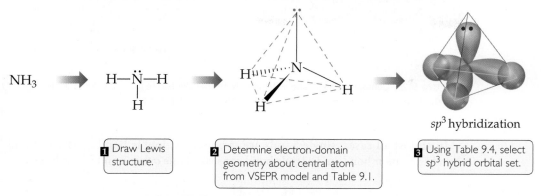

sp^3 hybridization

1 Draw Lewis structure.

2 Determine electron-domain geometry about central atom from VSEPR model and Table 9.1.

3 Using Table 9.4, select sp^3 hybrid orbital set.

▲ **Figure 9.20 Hybrid orbital description of bonding in NH$_3$.** Note the comparison with Figure 9.6. Here we focus on the hybrid orbitals used to make bonds and to hold nonbonding electron pairs.

SAMPLE EXERCISE 9.5 | Describing the Hybridization of a Central Atom

Describe the orbital hybridization around the central atom in NH$_2^-$.

SOLUTION

Analyze We are given the chemical formula for a polyatomic anion and asked to describe the type of hybrid orbitals surrounding the central atom.

Plan To determine the central atom hybrid orbitals, we must know the electron-domain geometry around the atom. Thus, we draw the Lewis structure to determine the number of electron domains around the central atom. The hybridization conforms to the number and geometry of electron domains around the central atom as predicted by the VSEPR model.

Solve The Lewis structure is

$$\left[\text{H} \!:\! \ddot{\text{N}} \!:\! \text{H} \right]^-$$

Because there are four electron domains around N, the electron-domain geometry is tetrahedral. The hybridization that gives a tetrahedral electron-domain geometry is sp^3 (Table 9.4). Two of the sp^3

hybrid orbitals contain nonbonding pairs of electrons, and the other two are used to make bonds with the hydrogen atoms.

Practice Exercise 1

For which of the following molecules or ions does the following description apply? "The bonding can be explained using a set of sp^2 hybrid orbitals on the central atom, with one of the hybrid orbitals holding a nonbonding pair of electrons."

(a) CO$_2$ **(b)** H$_2$S **(c)** O$_3$ **(d)** CO$_3^{2-}$
(e) More than one of the above

Practice Exercise 2

Predict the electron-domain geometry and hybridization of the central atom in SO$_3^{2-}$.

9.6 | Multiple Bonds

In the covalent bonds we have considered thus far, the electron density is concentrated along the line connecting the nuclei (the *internuclear axis*). The line joining the two nuclei passes through the middle of the overlap region, forming a type of covalent bond called a **sigma (σ) bond**. The overlap of two s orbitals in H$_2$, the overlap of an s and a p orbital in HCl, the overlap of two p orbitals in Cl$_2$ (all shown in Figure 9.13), and the overlap of a p orbital and an sp hybrid orbital in BeF$_2$ (Figure 9.16) are all σ bonds.

To describe multiple bonding, we must consider a second kind of bond, this one the result of overlap between two p orbitals oriented perpendicularly to the internuclear axis **(Figure 9.21)**. The sideways overlap of p orbitals produces what is called a **pi (π) bond**. A π bond is one in which the overlap regions lie above and below the internuclear axis. Unlike a σ bond, in a π bond the electron density is not concentrated on the internuclear axis. Although it is not evident in Figure 9.21, the sideways orientation of p orbitals in a π bond makes for weaker overlap. As a result, π bonds are generally weaker than σ bonds.

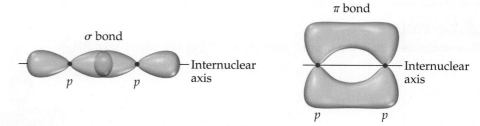

▲ **Figure 9.21 Comparison of σ and π bonds.** Note that the two regions of overlap in the π bond, above and below the internuclear axis, constitute a *single* π bond.

In almost all cases, single bonds are σ bonds. A double bond consists of one σ bond and one π bond, and a triple bond consists of one σ bond and two π bonds:

H—H   One σ bond

H_C=C_H ... One σ bond plus one π bond

:N≡N: One σ bond plus two π bonds

▲ **Figure 9.22 Trigonal-planar molecular geometry of ethylene.** The double bond is made up of one C—C σ bond and one C—C π bond.

To see how these ideas are used, consider ethylene (C_2H_4), which has a C=C double bond. As illustrated by the ball-and-stick model of ◄ **Figure 9.22,** the three bond angles about each carbon are all approximately 120°, suggesting that each carbon atom uses sp^2 hybrid orbitals (Figure 9.17) to form σ bonds with the other carbon and with two hydrogens. Because carbon has four valence electrons, after sp^2 hybridization one electron in each carbon remains in the *unhybridized* 2p orbital. Note that this unhybridized 2p orbital is directed perpendicular to the plane that contains the three sp^2 hybrid orbitals.

Let's go through the steps of building the bonds in the ethylene molecule. Each sp^2 hybrid orbital on a carbon atom contains one electron. ► **Figure 9.23** shows how we can first envision forming the C—C σ bond by the overlap of two sp^2 hybrid orbitals, one on each carbon atom. Two electrons are used in forming the C—C σ bond. Next, the C—H σ bonds are formed by overlap of the remaining sp^2 hybrid orbitals on the C atoms with the 1s orbitals on each H atom. We use eight more electrons to form these four C—H bonds. Thus, 10 of the 12 valence electrons in the C_2H_4 molecule are used to form five σ bonds.

The remaining two valence electrons reside in the unhybridized 2p orbitals, one electron on each carbon. These two orbitals can overlap sideways with each other, as shown in Figure 9.23. The resultant electron density is concentrated above and below the C—C bond axis: It is a π bond (Figure 9.21). Thus, the C=C double bond in ethylene consists of one σ bond and one π bond.

We often refer to the unhybridized 2p atomic orbital of an sp^2 hybridized atom as a p_π (**"pee-pie"**) **orbital** (because it is a p orbital that can be involved in forming a π bond). Thus, the two-electron π bond in ethylene is formed from the overlap of two p_π orbitals, one on each C atom and each holding one electron. Remember that the formation of the π bond involves the "sideways" overlap of the p_π orbitals as compared to the more direct "head on" overlap used in making the C—C and C—H σ bonds. As a result, as pointed out earlier, π bonds are generally weaker than σ bonds.

Although we cannot experimentally observe a π bond directly (all we can observe are the positions of the atoms), the structure of ethylene provides strong support for its presence. First, the C—C bond length in ethylene (1.34 Å) is much shorter than in compounds with C—C single bonds (1.54 Å), consistent with the presence of a stronger C=C double bond. Second, all six atoms in C_2H_4 lie in the same plane. The p_π orbitals on each C atom that make up the π bond can

GO FIGURE

Why is it important that the sp^2 hybrid orbitals of the two carbon atoms lie in the same plane?

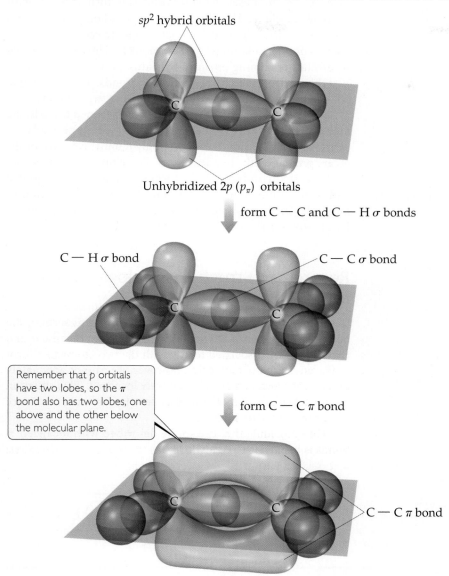

sp^2 hybrid orbitals

Unhybridized $2p$ (p_π) orbitals

form C — C and C — H σ bonds

C — H σ bond

C — C σ bond

Remember that p orbitals have two lobes, so the π bond also has two lobes, one above and the other below the molecular plane.

form C — C π bond

C — C π bond

▲ Figure 9.23 **The orbital structure of ethylene, C$_2$H$_4$.**

achieve a good overlap only when the two CH$_2$ fragments lie in the same plane. Because π bonds require that portions of a molecule be planar, they can introduce rigidity into molecules.

 Give It Some Thought

The molecule called *diazine* has the formula N$_2$H$_2$ and the Lewis structure

$$\text{H—}\ddot{\text{N}}\text{=}\ddot{\text{N}}\text{—H}$$

Do you expect diazine to be a linear molecule (all four atoms on the same line)? If not, do you expect the molecule to be planar (all four atoms in the same plane)?

GO FIGURE

Based on the models of bonding in ethylene and acetylene, which molecule should have the greater carbon–carbon bond energy?

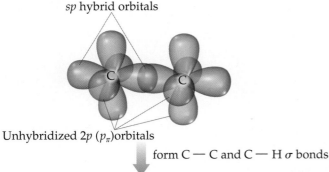

sp hybrid orbitals

Unhybridized 2p (p_π)orbitals

form C — C and C — H σ bonds

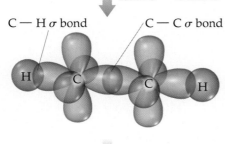

C — H σ bond   C — C σ bond

form C — C π bonds

C — C π bonds

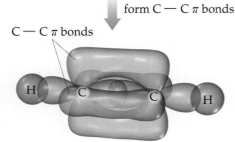

▲ **Figure 9.24 Formation of two π bonds in acetylene, C₂H₂.**

Triple bonds can also be explained using hybrid orbitals. Acetylene (C_2H_2), for example, is a linear molecule containing a triple bond: H—C≡C—H. The linear geometry suggests that each carbon atom uses *sp* hybrid orbitals to form σ bonds with the other carbon and one hydrogen. Each carbon atom thus has two unhybridized 2p orbitals at right angles to each other and to the axis of the *sp* hybrid set (◀ Figure 9.24). Thus, there are *two* p_π orbitals remaining on an *sp*-hybridized carbon atom. These p_π orbitals overlap to form a pair of π bonds. Thus, the triple bond in acetylene consists of one σ bond and two π bonds.

Although it is possible to make π bonds from *d* orbitals, the only π bonds we will consider are those formed by the overlap of *p* orbitals. These π bonds can form only if unhybridized *p* orbitals are present on the bonded atoms. Therefore, only atoms having *sp* or *sp²* hybridization can form π bonds. Further, double and triple bonds (and hence π bonds) are more common in molecules made up of period 2 atoms, especially C, N, and O. Larger atoms, such as S, P, and Si, form π bonds less readily.

Resonance Structures, Delocalization, and π Bonding

In the molecules we have discussed thus far in this section, the bonding electrons are *localized*. By this we mean that the σ and π electrons are associated totally with the two atoms that form the bond. In many molecules, however, we cannot adequately describe the bonding as being entirely localized. This situation arises particularly in molecules that have two or more resonance structures involving π bonds.

One molecule that cannot be described with localized π bonds is benzene (C_6H_6), which has two resonance structures: ∞ (Section 8.6)

SAMPLE
EXERCISE 9.6 Describing σ and π Bonds in a Molecule

Formaldehyde has the Lewis structure

$$\begin{array}{c} H \\ \\ H \end{array} \hspace{-0.3em} \Big\backslash \hspace{-0.2em} C = \ddot{O} :$$

Describe how the bonds in formaldehyde are formed in terms of overlaps of hybrid and unhybridized orbitals.

SOLUTION

Analyze We are asked to describe the bonding in formaldehyde in terms of hybrid orbitals.

Plan Single bonds are σ bonds, and double bonds consist of one σ bond and one π bond. The ways in which these bonds form can be deduced from the molecular geometry, which we predict using the VSEPR model.

Solve The C atom has three electron domains around it, which suggests a trigonal-planar geometry with bond angles of about 120°. This geometry implies *sp²* hybrid orbitals on C

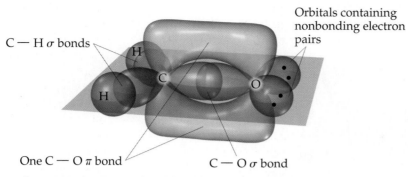

C—H σ bonds

Orbitals containing
nonbonding electron
pairs

One C—O π bond

C—O σ bond

▲ **Figure 9.25 Formation of σ and π bonds in formaldehyde, H₂CO.**

(Table 9.4). These hybrids are used to make the two C—H and one C—O σ bonds to C. There remains an unhybridized $2p$ orbital (a p_π orbital) on carbon, perpendicular to the plane of the three sp^2 hybrids.

The O atom also has three electron domains around it, and so we assume it has sp^2 hybridization as well. One of these hybrid orbitals participates in the C—O σ bond, while the other two hold the two nonbonding electron pairs of the O atom. Like the C atom, therefore, the O atom has a p_π orbital that is perpendicular to the plane of the molecule. The two p_π orbitals overlap to form a C—O π bond (▲ **Figure 9.25**).

Practice Exercise 1

We have just arrived at a bonding description for the formaldehyde molecule. Which of the following statements about the molecule is or are true?

(i) Two of the electrons in the molecule are used to make the π bond in the molecule.

(ii) Six of the electrons in the molecule are used to make the σ bonds in the molecule.

(iii) The C—O bond length in formaldehyde should be shorter than that in methanol, H₃COH.

(a) Only one of the statements is true.

(b) Statements (i) and (ii) are true.

(c) Statements (i) and (iii) are true.

(d) Statements (ii) and (iii) are true.

(e) All three statements are true.

Practice Exercise 2

(a) Predict the bond angles around each carbon atom in acetonitrile:

$$\underset{\overset{\displaystyle |}{H}}{\overset{\displaystyle H}{H-C-C\equiv N:}}$$

(b) Describe the hybridization at each carbon atom, and (c) determine the number of σ and π bonds in the molecule.

Benzene has a total of 30 valence electrons. To describe the bonding in benzene using hybrid orbitals, we first choose a hybridization scheme consistent with the geometry of the molecule. Because each carbon is surrounded by three atoms at 120° angles, the appropriate hybrid set is sp^2. Six localized C—C σ bonds and six localized C—H σ bonds are formed from the sp^2 hybrid orbitals, as shown in **Figure 9.26(a)**. Thus, 24 of the valence electrons are used to form the σ bonds in the molecule.

Because the hybridization at each C atom is sp^2, there is one p_π orbital on each C atom, each oriented perpendicular to the plane of the molecule. The situation is very much like that in ethylene except we now have six p_π orbitals arranged in a ring

▲GO FIGURE
What are the two kinds of σ bonds found in benzene?

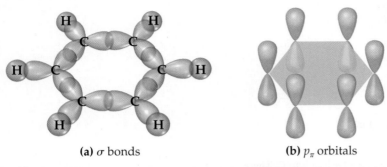

(a) σ bonds   **(b)** p_π orbitals

▲ **Figure 9.26** σ **and** π **bond networks in benzene, C_6H_6.** **(a)** The σ bond framework. **(b)** The π bonds are formed from overlap of the unhybridized $2p$ orbitals on the six carbon atoms.

[Figure 9.26(b)]. The remaining six valence electrons occupy these six p_π orbitals, one per orbital.

We could envision using the p_π orbitals to form three localized π bonds. As shown in ▼ Figure 9.27, there are two equivalent ways to make these localized bonds, each corresponding to one resonance structure. However, a representation that reflects *both* resonance structures has the six π electrons "smeared out" among all six carbon atoms, as shown on the right in Figure 9.27. Notice how this smeared representation corresponds to the circle-in-a-hexagon drawing we often use to represent benzene. This model leads us to predict that all the carbon–carbon bond lengths will be identical, with a bond length between that of a C—C single bond (1.54 Å) and that of a C=C double bond (1.34 Å). This prediction is consistent with the observed carbon–carbon bond length in benzene (1.40 Å).

Localized π bond   Delocalized π bond

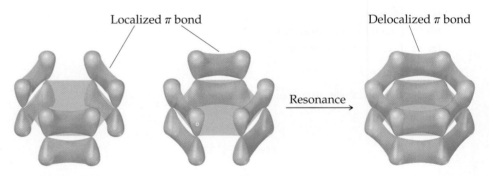

Resonance

▲ **Figure 9.27** Delocalized π bonds in benzene.

Because we cannot describe the π bonds in benzene as individual bonds between neighboring atoms, we say that benzene has a six-electron π system **delocalized** among the six carbon atoms. Delocalization of the electrons in its π bonds gives benzene a special stability. Electron delocalization in π bonds is also responsible for the color of many organic molecules. A final important point to remember about delocalized π bonds is the constraint they place on the geometry of a molecule. For optimal overlap of the p_π orbitals, all the atoms involved in a delocalized π bonding network should lie in the same plane. This restriction imparts a certain rigidity to the molecule that is absent in molecules containing only σ bonds (see the "Chemistry and Life" box on vision).

If you take a course in organic chemistry, you will see many examples of how electron delocalization influences the properties of organic molecules.

SAMPLE
EXERCISE 9.7   Delocalized Bonding

Describe the bonding in the nitrate ion, NO_3^-. Does this ion have delocalized π bonds?

SOLUTION

Analyze Given the chemical formula for a polyatomic anion, we are asked to describe the bonding and determine whether the ion has delocalized π bonds.

Plan Our first step is to draw Lewis structures. Multiple resonance structures involving the placement of the double bonds in different locations would suggest that the π component of the double bonds is delocalized.

Solve In Section 8.6 we saw that NO_3^- has three resonance structures:

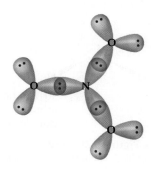

In each structure, the electron-domain geometry at nitrogen is trigonal planar, which implies sp^2 hybridization of the N atom. It is helpful when considering delocalized π bonding to consider atoms with lone pairs that are bonded to the central atom to be sp^2 hybridized as well. Thus, we can envision that each of the O atoms in the anion has three sp^2 hybrid orbitals in the plane of the ion. Each of the four atoms has an unhybridized p_π orbital oriented perpendicular to the plane of the ion.

The NO_3^- ion has 24 valence electrons. We can first use the sp^2 hybrid orbitals on the four atoms to construct the three N—O σ bonds. That uses all of the sp^2 hybrids on the N atom and one sp^2 hybrid on each O atom. Each of the two remaining sp^2 hybrids on each O atom is used to hold a nonbonding pair of electrons. Thus, for any of the resonance structures, we have the following arrangement in the plane of the ion:

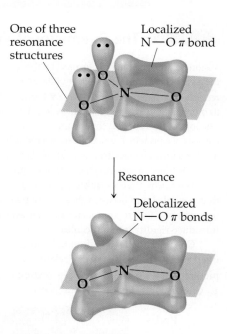

Notice that we have accounted for a total of 18 electrons — six in the three N—O σ bonds, and 12 as nonbonding pairs on the O atoms. The remaining six electrons will reside in the π system of the ion.

The four p_π orbitals — one on each of the four atoms — are used to build the π system. For any one of the three resonance structures shown, we might imagine a single localized N—O π bond formed by the overlap of the p_π orbital on N and a p_π orbital on one of the O atoms. The remaining two O atoms have nonbonding pairs in their p_π orbitals. Thus, for each of the resonance structures, we have the situation shown in ▶ **Figure 9.28.** Because each resonance structure contributes equally to the observed structure of NO_3^-, however, we represent the π bonding as delocalized over the three N—O bonds, as shown in the figure. We see that the NO_3^- ion has a six-electron π system delocalized among the four atoms in the ion.

▲ **Figure 9.28 Localized and delocalized representations of the six-electron π system in NO_3^-.**

Practice Exercise 1

How many electrons are in the π system of the ozone molecule, O_3?
(a) 2   **(b)** 4   **(c)** 6   **(d)** 14   **(e)** 18

Practice Exercise 2

Which of these species have delocalized bonding: SO_2, SO_3, SO_3^{2-}, H_2CO, NH_4^+?

General Conclusions about σ and π Bonding

On the basis of the examples we have seen, we can draw a few helpful conclusions for using hybrid orbitals to describe molecular structures:

1. Every pair of bonded atoms shares one or more pairs of electrons. Each bond line we draw in a Lewis structures represents two shared electrons. In every σ bond at least one pair of electrons is localized in the space between the atoms. The appropriate set of hybrid orbitals used to form the σ bonds between an atom and its neighbors is determined by the observed geometry of the molecule. The correlation between the set of hybrid orbitals and the geometry about an atom is given in Table 9.4.

2. Because the electrons in σ bonds are localized in the region between two bonded atoms, they do not make a significant contribution to the bonding between any other two atoms.

3. When atoms share more than one pair of electrons, one pair is used to form a σ bond; the additional pairs form π bonds. The centers of charge density in a π bond lie above and below the internuclear axis.

4. Molecules can have π systems that extend over more than two bonded atoms. Electrons in extended π systems are said to be "delocalized." We can determine the number of electrons in the π system of a molecule using the procedures we discussed in this section.

Chemistry and Life

The Chemistry of Vision

Vision begins when light is focused by the lens of the eye onto the retina, the layer of cells lining the interior of the eyeball. The retina contains *photoreceptor* cells called rods and cones (▼ Figure 9.29). The rods are sensitive to dim light and are used in night vision. The cones are sensitive to colors. The tops of the rods and cones contain a molecule called *rhodopsin*, which consists of a protein, *opsin*, bonded to a reddish purple pigment called *retinal*. Structural changes around a double bond in the retinal portion of the molecule trigger a series of chemical reactions that result in vision.

We know that a double bond between two atoms is stronger than a single bond between those same two atoms (Table 8.4). We are now in a position to appreciate another aspect of double bonds: They introduce rigidity into molecules.

Consider the C—C double bond in ethylene. Imagine rotating one —CH$_2$ group in ethylene relative to the other —CH$_2$ group, as pictured in ▶ Figure 9.30. This rotation destroys the overlap of the p_π orbitals, breaking the π bond, a process that requires considerable

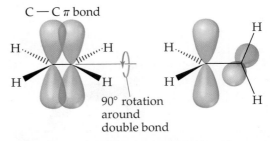

▲ Figure 9.30 **Rotation about the carbon–carbon double bond in ethylene breaks the π bond.**

energy. Thus, the presence of a double bond restricts bond rotation in a molecule. In contrast, molecules can rotate almost freely around the bond axis in single (σ) bonds because this motion has no effect on the orbital overlap for a σ bond. Rotation allows molecules with single bonds to twist and fold almost as if their atoms were attached by hinges.

Our vision depends on the rigidity of double bonds in retinal. In its normal form, retinal is held rigid by its double bonds. Light entering the eye is absorbed by rhodopsin, and it is the energy of that light which is used to break the π-bond portion of the double bond shown in red in ▶ Figure 9.31. Breaking the double bond allows rotation around the bond axis, changing the geometry of the retinal molecule. The retinal then separates from the opsin, triggering the reactions that produce a nerve impulse that the brain interprets as the sensation of vision. It takes as few as five closely spaced molecules reacting in this fashion to produce the sensation of vision. Thus, only five photons of light are necessary to stimulate the eye.

The retinal slowly reverts to its original form and reattaches to the opsin. The slowness of this process helps explain why intense bright light causes temporary blindness. The light causes all the retinal to separate from opsin, leaving no molecules to absorb light.

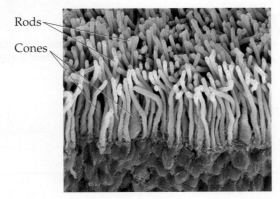

▲ Figure 9.29 **Inside the eye.** A color-enhanced scanning electron micrograph of the rods and cones in the retina of the human eye.

Related Exercises: 9.112, 9.116

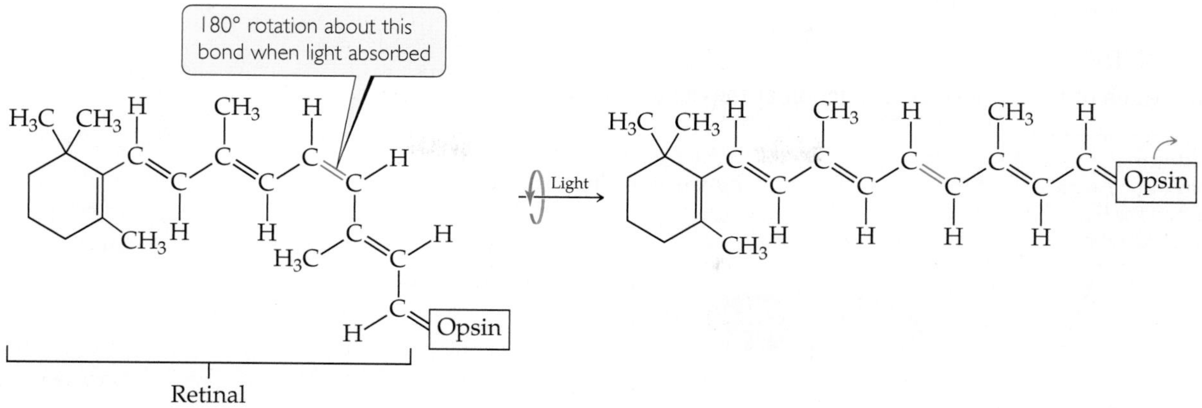

180° rotation about this bond when light absorbed

Retinal

▲ Figure 9.31 **The rhodopsin molecule, the chemical basis of vision.** When rhodopsin absorbs visible light, the π component of the double bond shown in red breaks, allowing rotation that produces a change in molecular geometry before the π bond re-forms.

Give It Some Thought

When two atoms are bonded by a triple bond, what is the hybridization of the orbitals that make up the σ-bond component of the bond?

9.7 | Molecular Orbitals

While valence-bond theory helps explain some of the relationships among Lewis structures, atomic orbitals, and molecular geometries, it does not explain all aspects of bonding. It is not successful, for example, in describing the excited states of molecules, which we must understand to explain how molecules absorb light, giving them color.

Some aspects of bonding are better explained by a more sophisticated model called **molecular orbital theory**. In Chapter 6 we saw that electrons in atoms can be described by wave functions, which we call atomic orbitals. In a similar way, molecular orbital theory describes the electrons in molecules by using specific wave functions, each of which is called a **molecular orbital (MO)**.

Molecular orbitals have many of the same characteristics as atomic orbitals. For example, an MO can hold a maximum of two electrons (with opposite spins), it has a definite energy, and we can visualize its electron-density distribution by using a contour representation, as we did with atomic orbitals. Unlike atomic orbitals, however, MOs are associated with an entire molecule, not with a single atom.

Molecular Orbitals of the Hydrogen Molecule

We begin our study of MO theory with the hydrogen molecule, H_2. We will use the two $1s$ atomic orbitals (one on each H atom) to construct molecular orbitals for H_2. *Whenever two atomic orbitals overlap, two molecular orbitals form.* Thus, the overlap of the $1s$ orbitals of two hydrogen atoms to form H_2 produces two MOs. The first MO, which is shown at the bottom right of **Figure 9.32**, is formed by adding the wave functions for the two $1s$ orbitals. We refer to this as *constructive combination*. The energy of the resulting MO is lower in energy than the two atomic orbitals from which it was made. It is called the **bonding molecular orbital**.

The second MO is formed by what is called *destructive combination:* combining the two atomic orbitals in a way that causes the electron density to be canceled in the central region where the two overlap. The process is discussed more fully in the "Closer Look" box later in the chapter. The energy of the resulting MO, referred to as the **antibonding molecular orbital**, is higher than the energy of the atomic orbitals. The antibonding MO of H_2 is shown at the top right in Figure 9.32.

▲ **GO FIGURE**
What is the value of the σ_{1s}^* MO wave function at the nodal plane?

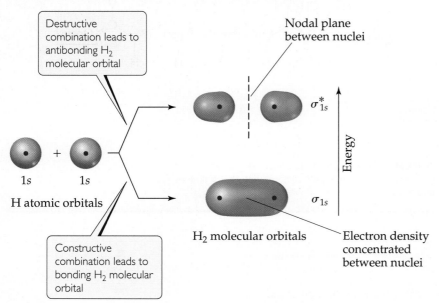

▲ **Figure 9.32** The two molecular orbitals of H₂, one a bonding MO and one an antibonding MO.

As illustrated in Figure 9.32, in the bonding MO electron density is concentrated in the region between the two nuclei. This sausage-shaped MO results from summing the two atomic orbitals so that the atomic orbital wave functions combine in the region between the two nuclei. Because an electron in this MO is attracted to both nuclei, the electron is more stable (it has lower energy) than it is in the 1s atomic orbital of an isolated hydrogen atom. Further, because this bonding MO concentrates electron density between the nuclei, it holds the atoms together in a covalent bond.

By contrast, the antibonding MO has very little electron density between the nuclei. Instead of combining in the region between the nuclei, the atomic orbital wave functions cancel each other in this region, leaving the greatest electron density on opposite sides of the two nuclei. Thus, an antibonding MO excludes electrons from the very region in which a bond must be formed. Antibonding orbitals invariably have a *plane* in the region between the nuclei where the electron density is zero. This plane is called a **nodal plane** of the MO. (The nodal plane is shown as a dashed line in Figure 9.32 and subsequent figures.) An electron in an antibonding MO is repelled from the bonding region and is therefore less stable (it has higher energy) than it is in the 1s atomic orbital of a hydrogen atom.

Notice from Figure 9.32 that the electron density in both the bonding MO and the antibonding MO of H₂ is centered about the internuclear axis. MOs of this type are called **sigma (σ) molecular orbitals** (by analogy to σ bonds). The bonding sigma MO of H₂ is labeled σ_{1s}; the subscript indicates that the MO is formed from two 1s orbitals. The antibonding sigma MO of H₂ is labeled σ_{1s}^* (read "sigma-star-one-s"); the asterisk denotes that the MO is antibonding.

The relative energies of two 1s atomic orbitals and the molecular orbitals formed from them are represented by an **energy-level diagram** (also called a **molecular orbital diagram**). Such diagrams show the interacting atomic orbitals on the left and right and the MOs in the middle, as shown in ▶ **Figure 9.33**. Like atomic orbitals, each MO can accommodate two electrons with their spins paired (Pauli exclusion principle). ∞ (Section 6.7)

As the MO diagram for H₂ in Figure 9.33 shows, each H atom has one electron, so there are two electrons in H₂. These two electrons occupy the lower-energy bonding

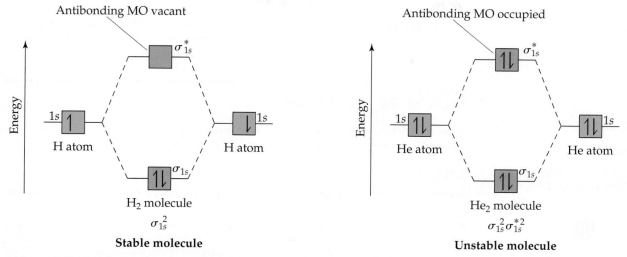

▲ **GO FIGURE**

What would happen to the energy of the σ_{1s} MO if the H atoms in H_2 were pulled apart to a distance twice as long as its normal bond length?

▲ Figure 9.33 **Energy-level diagrams and electron configurations for H₂ and He₂.**

(σ_{1s}) MO, and their spins are paired. Electrons occupying a bonding molecular orbital are called *bonding electrons*. Because the σ_{1s} MO is lower in energy than the H $1s$ atomic orbitals, the H_2 molecule is more stable than the two separate H atoms.

By analogy with atomic electron configurations, the electron configurations for molecules can be written with superscripts to indicate electron occupancy. The electron configuration for H_2, then, is σ_{1s}^2.

Figure 9.33 also shows the energy-level diagram for the hypothetical He_2 molecule, which requires four electrons to fill its molecular orbitals. Because only two electrons can go in the σ_{1s} MO, the other two electrons must go in the σ_{1s}^* MO. The electron configuration of He_2 is thus $\sigma_{1s}^2 \sigma_{1s}^{*2}$. The energy decrease realized in going from He atomic orbitals to the He bonding MO is offset by the energy increase realized in going from the atomic orbitals to the He antibonding MO.* Hence, He_2 is an unstable molecule. Molecular orbital theory correctly predicts that hydrogen forms diatomic molecules but helium does not.

Bond Order

In molecular orbital theory, the stability of a covalent bond is related to its **bond order**, defined as half the difference between the number of bonding electrons and the number of antibonding electrons:

$$\text{Bond order} = \tfrac{1}{2}(\text{no. of bonding electrons} - \text{no. of antibonding electrons}) \quad [9.1]$$

We take half the difference because we are used to thinking of bonds as pairs of electrons. *A bond order of 1 represents a single bond, a bond order of 2 represents a double bond, and a bond order of 3 represents a triple bond.* Because MO theory also treats molecules containing an odd number of electrons, bond orders of $1/2$, $3/2$, or $5/2$ are possible.

Let's now consider the bond order in H_2 and He_2, referring to Figure 9.33. H_2 has two bonding electrons and zero antibonding electrons, so it has a bond order of 1.

---

*Antibonding MOs are slightly more energetically unfavorable than bonding MOs are energetically favorable. Thus, whenever there is an equal number of electrons in bonding and antibonding orbitals, the energy of the molecule is slightly higher than that for the separated atoms. As a result, no bond is formed.

Because He_2 has two bonding electrons and two antibonding electrons, it has a bond order of 0. A bond order of 0 means that no bond exists.

 Give It Some Thought

Suppose one electron in H_2 is excited from the σ_{1s} MO to the σ_{1s}^* MO. Would you expect the H atoms to remain bonded to each other, or would the molecule fall apart?

SAMPLE EXERCISE 9.8 **Bond Order**

What is the bond order of the He_2^+ ion? Would you expect this ion to be stable relative to the separated He atom and He^+ ion?

SOLUTION

Analyze We will determine the bond order for the He_2^+ ion and use it to predict whether the ion is stable.

Plan To determine the bond order, we must determine the number of electrons in the molecule and how these electrons populate the available MOs. The valence electrons of He are in the $1s$ orbital, and the $1s$ orbitals combine to give an MO diagram like that for H_2 or He_2 (Figure 9.33). If the bond order is greater than 0, we expect a bond to exist, and the ion is stable.

Solve The energy-level diagram for the He_2^+ ion is shown in ◀ Figure 9.34. This ion has three electrons. Two are placed in the bonding orbital and the third in the antibonding orbital. Thus, the bond order is

$$\text{Bond order} = \tfrac{1}{2}(2 - 1) = \tfrac{1}{2}$$

Because the bond order is greater than 0, we predict the He_2^+ ion to be stable relative to the separated He and He^+. Formation of He_2^+ in the gas phase has been demonstrated in laboratory experiments.

Practice Exercise 1

How many of the following molecules and ions have a bond order of $\tfrac{1}{2}$: H_2, H_2^+, H_2^-, and He_2^{2+}?
(a) 0 **(b)** 1 **(c)** 2 **(d)** 3 **(e)** 4

Practice Exercise 2

What are the electron configuration and the bond order of the H_2^- ion?

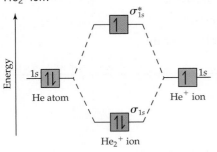

▲ **GO FIGURE**

Which electrons in this diagram contribute to the stability of the He_2^+ ion?

▲ Figure 9.34 Energy-level diagram for the He_2^+ ion

9.8 | Period 2 Diatomic Molecules

In considering the MO description of diatomic molecules other than H_2, we will initially restrict our discussion to *homonuclear* diatomic molecules (those composed of two identical atoms) of period 2 elements.

Period 2 atoms have valence $2s$ and $2p$ orbitals, and we need to consider how they interact to form MOs. The following rules summarize some of the guiding principles for the formation of MOs and for how they are populated by electrons:

1. The number of MOs formed equals the number of atomic orbitals combined.

2. Atomic orbitals combine most effectively with other atomic orbitals of similar energy.

3. The effectiveness with which two atomic orbitals combine is proportional to their overlap. That is, as the overlap increases, the energy of the bonding MO is lowered and the energy of the antibonding MO is raised.

4. Each MO can accommodate, at most, two electrons, with their spins paired (Pauli exclusion principle). ∞ (Section 6.7)

5. When MOs of the same energy are populated, one electron enters each orbital (with the same spin) before spin pairing occurs (Hund's rule). ∞ (Section 6.8)

Molecular Orbitals for Li$_2$ and Be$_2$

Lithium has the electron configuration $1s^2 2s^1$. When lithium metal is heated above its boiling point (1342 °C), Li$_2$ molecules are found in the vapor phase. The Lewis structure for Li$_2$ indicates a Li—Li single bond. We will now use MOs to describe the bonding in Li$_2$.

▶ **Figure 9.35** shows that the Li 1s and 2s atomic orbitals have substantially different energy levels. From this, we can assume that the 1s orbital on one Li atom interacts only with the 1s orbital on the other atom (rule 2). Likewise, the 2s orbitals interact only with each other. Notice that combining four atomic orbitals produces four MOs (rule 1).

The Li 1s orbitals combine to form σ_{1s} and σ_{1s}^* bonding and antibonding MOs, as they did for H$_2$. The 2s orbitals interact with one another in exactly the same way, producing bonding (σ_{2s}) and antibonding (σ_{2s}^*) MOs. In general, the separation between bonding and antibonding MOs depends on the extent to which the constituent atomic orbitals overlap. Because the Li 2s orbitals extend farther from the nucleus than the 1s orbitals do, the 2s orbitals overlap more effectively. As a result, the energy difference between the σ_{2s} and σ_{2s}^* orbitals is greater than the energy difference between the σ_{1s} and σ_{1s}^* orbitals. The 1s orbitals of Li are so much lower in energy than the 2s orbitals; however, the energy of the σ_{1s}^* antibonding MO is much lower than the energy of σ_{2s} bonding MO.

Each Li atom has three electrons, so six electrons must be placed in Li$_2$ MOs. As shown in Figure 9.35, these electrons occupy the σ_{1s}, σ_{1s}^*, and σ_{2s} MOs, each with two electrons. There are four electrons in bonding orbitals and two in antibonding orbitals, so the bond order is $\frac{1}{2}(4 - 2) = 1$. The molecule has a single bond, in agreement with its Lewis structure.

Because both the σ_{1s} and σ_{1s}^* MOs of Li$_2$ are completely filled, the 1s orbitals contribute almost nothing to the bonding. The single bond in Li$_2$ is due essentially to the interaction of the valence 2s orbitals on the Li atoms. This example illustrates the general rule that *core electrons usually do not contribute significantly to bonding in molecules*. The rule is equivalent to using only the valence electrons when drawing Lewis structures. Thus, we need not consider further the 1s orbitals while discussing the other period 2 diatomic molecules.

The MO description of Be$_2$ follows readily from the energy-level diagram for Li$_2$. Each Be atom has four electrons $(1s^2 2s^2)$, so we must place eight electrons in molecular orbitals. Therefore, we completely fill the σ_{1s}, σ_{1s}^*, σ_{2s}, and σ_{2s}^* MOs. With equal numbers of bonding and antibonding electrons, the bond order is zero; thus, Be$_2$ does not exist.

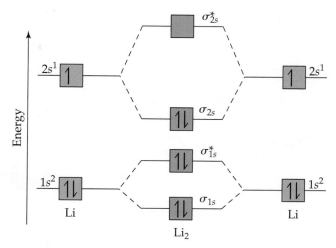

▲ GO FIGURE

Which of the MOs in the diagram will have nodal planes?

▲ **Figure 9.35** Energy-level diagram for the Li$_2$ molecule.

 Give It Some Thought

Would you expect Be$_2^+$ to be a stable ion?

Molecular Orbitals from 2p Atomic Orbitals

Before we can consider the remaining period 2 diatomic molecules, we must look at the MOs that result from combining 2p atomic orbitals. The interactions between p orbitals are shown in **Figure 9.36**, where we have arbitrarily chosen the internuclear axis to be the z-axis. The $2p_z$ orbitals face each other head to head. Just as with s orbitals, we can combine $2p_z$ orbitals in two ways. One combination concentrates electron density between the nuclei and is, therefore, a bonding molecular orbital. The other combination excludes electron density from the bonding region and so is an antibonding

▲**GO FIGURE**
In which MO is the overlap of atomic orbitals greater, a σ_{2p} or a π_{2p}?

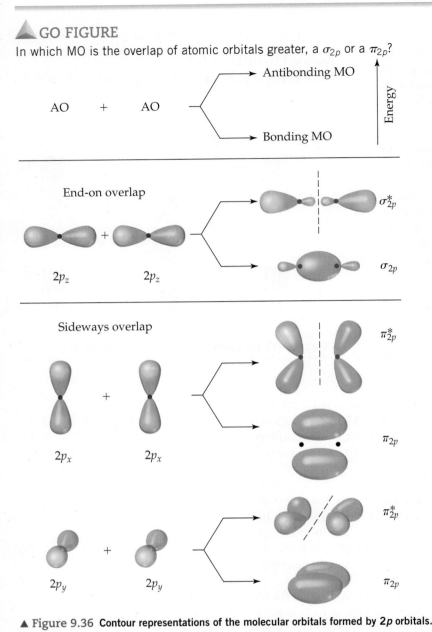

▲ **Figure 9.36 Contour representations of the molecular orbitals formed by 2p orbitals.**

molecular orbital. In both MOs, the electron density lies along the internuclear axis, so they are σ molecular orbitals: σ_{2p} and σ_{2p}^*.

The other $2p$ orbitals overlap sideways and thus concentrate electron density above and below the internuclear axis. MOs of this type are called **pi (π) molecular orbitals** by analogy to π bonds. We get one π bonding MO by combining the $2p_x$ atomic orbitals and another from the $2p_y$ atomic orbitals. These two π_{2p} molecular orbitals have the same energy; in other words, they are degenerate. Likewise, we get two degenerate π_{2p}^* antibonding MOs that are perpendicular to each other like the $2p$ orbitals from which they were made. These π_{2p}^* orbitals have four lobes, pointing away from the two nuclei, as shown in Figure 9.36.

The $2p_z$ orbitals on two atoms point directly at each other. Hence, the overlap of two $2p_z$ orbitals is greater than that of two $2p_x$ or $2p_y$ orbitals. From rule 3 we therefore expect the σ_{2p} MO to be lower in energy (more stable) than the π_{2p} MOs. Similarly, the σ_{2p}^* MO should be higher in energy (less stable) than the π_{2p}^* MOs.

A Closer Look

Phases in Atomic and Molecular Orbitals

Our discussion of atomic orbitals in Chapter 6 and molecular orbitals in this chapter highlights some of the most important applications of quantum mechanics in chemistry. In the quantum mechanical treatment of electrons in atoms and molecules, we are mainly interested in determining two characteristics of the electrons—their energies and their distribution in space. Recall that solving Schrödinger's wave equation yields the electron's energy, E, and wave function, ψ, but that ψ does not have a direct physical meaning. ∞ (Section 6.5) The contour representations of atomic and molecular orbitals we have presented thus far are based on ψ^2 (the *probability density*), which gives the probability of finding the electron at a given point in space.

Because probability densities are squares of functions, their values must be nonnegative (zero or positive) at all points in space. However, the functions themselves can have negative values. The situation is like that of the sine function plotted in ▶ Figure 9.37. In the top graph, the sine function is negative for x between 0 and $-\pi$ and positive for x between 0 and $+\pi$. We say that the *phase* of the sine function is negative between 0 and $-\pi$ and positive between 0 and $+\pi$. If we square the sine function (bottom graph), we get two peaks that are symmetrical about the origin. Both peaks are positive because squaring a negative number produces a positive number. In other words, *we lose the phase information of the function upon squaring it.*

Like the sine function, the more complicated wave functions for atomic orbitals can also have phases. Consider, for example, the representations of the $1s$ orbital in ▼ Figure 9.38. Note that here we plot this orbital a bit differently from what is shown in Section 6.6. The origin is the point where the nucleus resides, and the wave function for the $1s$ orbital extends from the origin out into space. The plot shows

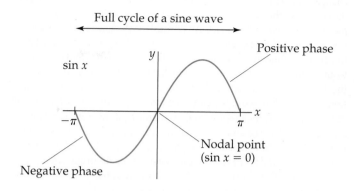

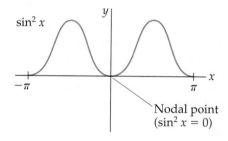

▲ **Figure 9.37 Graphs for a sine function and the same function squared.**

the value of ψ for a slice taken along the z-axis. Below the plot is a contour representation of the $1s$ orbital. Notice that the value of the $1s$ wave function is always a positive number (we show positive values in red in Figure 9.38). Thus, it has only one phase. Notice also that the wave function approaches zero only at a long distance from the nucleus. It therefore has no nodes, as we saw in Figure 6.22.

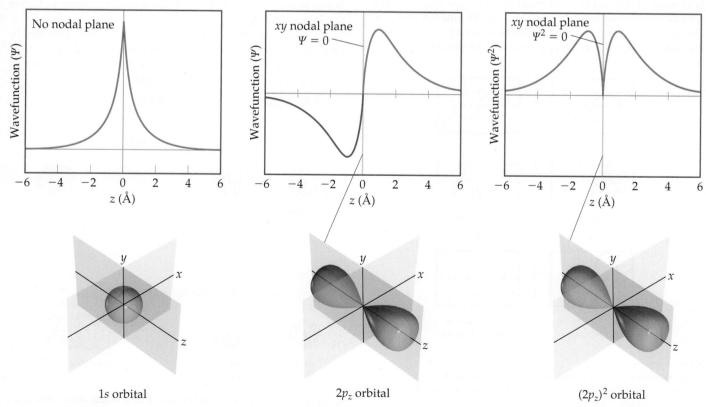

1s orbital $2p_z$ orbital $(2p_z)^2$ orbital

▲ **Figure 9.38 Phases in wave functions of *s* and *p* atomic orbitals.** Red shading means a positive value for the wave function and blue shading means a negative value.

In the graph for the $2p_z$ orbital in Figure 9.38, the wave function changes sign when it passes through $z = 0$. Notice that the two halves of the wave have the same shape except that one has positive (red) values and the other negative (blue) values. Analogously to the sine function, the wave function changes phase when it passes through the origin. Mathematically the $2p_z$ wave function is equal to zero whenever $z = 0$. This corresponds to any point on the xy plane, so we say that the xy plane is a *nodal plane* of the $2p_z$ orbital. The wave function for a p orbital is much like a sine function because it has two equal parts that have opposite phases. Figure 9.38 gives a typical representation used by chemists of the wave function for a p_z orbital.* The red and blue lobes indicate the different phases of the orbital. (Note: The colors do *not* represent charge, as they did in the plots in Figures 9.10 and 9.11.) As with the sine function, the origin is a node.

The third graph in Figure 9.38 shows that when we square the wave function of the $2p_z$ orbital, we get two peaks that are symmetrical about the origin. Both peaks are positive because squaring a negative number produces a positive number. Thus, *we lose the phase information of the function upon squaring it* just as we did for the sine function. When we square the wave function for the p_z orbital, we get the probability density for the orbital, which is given as a contour representation in Figure 9.38. This is what we saw in the earlier presentation of p orbitals. ⚬⚬⚬ (Section 6.6) For this squared wave function, both lobes have the same phase and therefore the same sign. We use this representation throughout most of this book because it has a simple physical interpretation: The square of the wave function at any point in space represents the electron density at that point.

The lobes of the wave functions for the d orbitals also have different phases. For example, the wave function for a d_{xy} orbital has four lobes, with the phase of each lobe opposite the phase of its nearest neighbors (▶ Figure 9.39). The wave functions for the other d orbitals likewise have lobes in which the phase in one lobe is opposite that in an adjacent lobe.

Why do we need to consider the complexity introduced by considering the phase of the wave function? While it is true that the phase is not necessary to visualize the shape of an atomic orbital in an isolated atom, it does become important when we consider overlap of orbitals in molecular orbital theory. Let's use the sine function as an example again. If you add two sine functions having the same phase, they add *constructively*, resulting in increased amplitude:

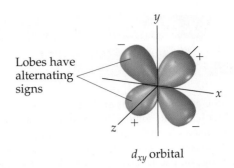

Lobes have alternating signs

d_{xy} orbital

▲ Figure 9.39 **Phases in *d* orbitals.**

but if you add two sine functions having opposite phases, they add *destructively* and cancel each other.

The idea of constructive and destructive interactions of wave functions is key to understanding the origin of bonding and antibonding molecular orbitals. For example, the wave function of the σ_{1s} MO of H_2 is generated by adding the wave function for the $1s$ orbital on one atom to the wave function for the $1s$ orbital on the other atom, with both orbitals having the same phase. The atomic wave functions overlap *constructively* in this case to increase the electron density between the two atoms (▼ Figure 9.40). The wave function of the σ_{1s}^* MO of H_2 is generated by subtracting the wave function for a $1s$ orbital on one atom from the wave function for a $1s$ orbital on the other atom. The result is that the atomic orbital wave functions overlap *destructively* to create a region of zero electron density between the two atoms—a node. Notice the similarity between this figure and Figure 9.32. In Figure 9.40, we use red and blue shading to denote positive and negative phases in the H atomic orbitals. However, chemists may alternatively draw contour representations in different colors, or with one phase shaded and one unshaded, to denote the two phases.

When we square the wave function of the σ_{1s}^* MO, we get the electron density representation which we saw earlier, in Figure 9.32. Notice once again that we lose the phase information when we look at the electron density.

The wave functions of atomic and molecular orbitals are used by chemists to understand many aspects of chemical bonding, spectroscopy, and reactivity. If you take a course in organic chemistry, you will probably see orbitals drawn to show the phases as in this box.

Related Exercises: 9.107, 9.119, 9.121

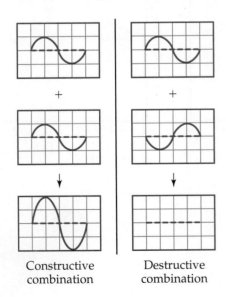

Constructive combination

Destructive combination

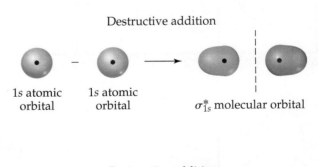

Destructive addition

$1s$ atomic orbital — $1s$ atomic orbital → σ_{1s}^* molecular orbital

Contructive addition

$1s$ atomic orbital + $1s$ atomic orbital → σ_{1s} molecular orbital

▲ Figure 9.40 **Molecular orbitals from atomic orbital wave functions.**

---

*The mathematical development of this three-dimensional function (and its square) is beyond the scope of this book, and, as is typically done by chemists, we have used lobes that are the same shape as in Figure 6.23.

Electron Configurations for B_2 through Ne_2

We can combine our analyses of MOs formed from *s* orbitals (Figure 9.32) and from *p* orbitals (Figure 9.36) to construct an energy-level diagram (▼ Figure 9.41) for homonuclear diatomic molecules of the elements boron through neon, all of which have valence 2*s* and 2*p* atomic orbitals. The following features of the diagram are notable:

1. The 2*s* atomic orbitals are substantially lower in energy than the 2*p* atomic orbitals. ∞∞(Section 6.7) Consequently, both MOs formed from the 2*s* orbitals are lower in energy than the lowest-energy MO derived from the 2*p* atomic orbitals.

2. The overlap of the two $2p_z$ orbitals is greater than that of the two $2p_x$ or $2p_y$ orbitals. As a result, the bonding σ_{2p} MO is lower in energy than the π_{2p} MOs, and the antibonding σ_{2p}^* MO is higher in energy than the π_{2p}^* MOs.

3. Both the π_{2p} and π_{2p}^* MOs are *doubly degenerate*; that is, there are two degenerate MOs of each type.

Before we can add electrons to Figure 9.41, we must consider one more effect. We have constructed the diagram assuming no interaction between the 2*s* orbital on one atom and the 2*p* orbitals on the other. In fact, such interactions can and do take place. Figure 9.42 shows the overlap of a 2*s* orbital on one of the atoms with a 2*p* orbital on the other. These interactions increase the energy difference between the σ_{2s} and σ_{2p} MOs, with the σ_{2s} energy decreasing and the σ_{2p} energy increasing (Figure 9.42). These 2*s*–2*p* interactions can be strong enough that the energetic ordering of the MOs can be altered: For B_2, C_2, and N_2, the σ_{2p} MO is above the π_{2p} MOs in energy. For O_2, F_2, and Ne_2, the σ_{2p} MO is below the π_{2p} MOs.

Given the energy ordering of the molecular orbitals, it is a simple matter to determine the electron configurations for the diatomic molecules B_2 through Ne_2. For example, a boron atom has three valence electrons. (Remember that we are ignoring the core 1*s* electrons.) Thus, for B_2 we must place six electrons in MOs. Four of them fill the σ_{2s} and σ_{2s}^* MOs, leading to no net bonding. The fifth electron goes in one π_{2p} MO, and the

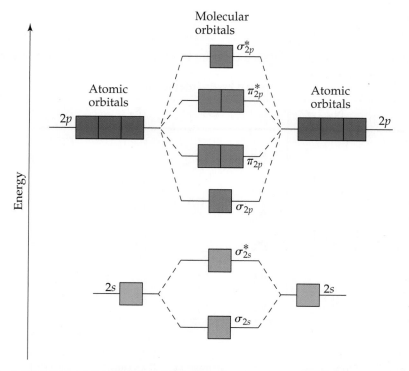

▲ **Figure 9.41 Energy-level diagram for MOs of period 2 homonuclear diatomic molecules.** The diagram assumes no interaction between the 2*s* atomic orbital on one atom and the 2*p* atomic orbitals on the other atom, and experiment shows that it fits only for O_2, F_2, and Ne_2.

▲ GO FIGURE

Which molecular orbitals have switched relative energy in the group on the right as compared with the group on the left?

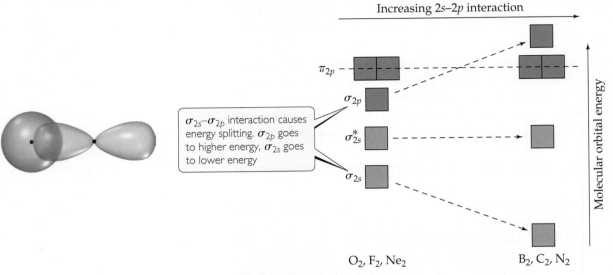

▲ **Figure 9.42** **The effect of interactions between 2s and 2p atomic orbitals.**

sixth goes in the other π_{2p} MO, with the two electrons having the same spin. Therefore, B_2 has a bond order of 1.

Each time we move one element to the right in period 2, two more electrons must be placed in the diagram of Figure 9.41. For example, on moving to C_2, we have two more electrons than in B_2, and these electrons are placed in the π_{2p} MOs, completely filling them. The electron configurations and bond orders for B_2 through Ne_2 are given in ▼ Figure 9.43.

▲ GO FIGURE

What difference in electron configuration accounts for most of the difference between the bond enthalpy of N_2 and that of F_2?

	Large 2s–2p interaction			**Small 2s–2p interaction**		
	B_2	**C_2**	**N_2**	**O_2**	**F_2**	**Ne_2**
σ^*_{2p}	☐	☐	☐	σ^*_{2p} ☐	☐	⇅
π^*_{2p}	☐ ☐	☐ ☐	☐ ☐	π^*_{2p} ↑ ↑	⇅ ⇅	⇅ ⇅
σ_{2p}	☐	☐	⇅	π_{2p} ⇅ ⇅	⇅ ⇅	⇅ ⇅
π_{2p}	↑ ↑	⇅ ⇅	⇅ ⇅	σ_{2p} ⇅	⇅	⇅
σ^*_{2s}	⇅	⇅	⇅	σ^*_{2s} ⇅	⇅	⇅
σ_{2s}	⇅	⇅	⇅	σ_{2s} ⇅	⇅	⇅
Bond order	1	2	3	2	1	0
Bond enthalpy (kJ/mol)	290	620	941	495	155	—
Bond length (Å)	1.59	1.31	1.10	1.21	1.43	—
Magnetic behavior	Paramagnetic	Diamagnetic	Diamagnetic	Paramagnetic	Diamagnetic	—

▲ **Figure 9.43** **Molecular orbital electron configurations and some experimental data for period 2 diatomic molecules.**

Electron Configurations and Molecular Properties

The way a substance behaves in a magnetic field can in some cases provide insight into the arrangements of its electrons. Molecules with one or more unpaired electrons are attracted to a magnetic field. The more unpaired electrons in a species, the stronger the attractive force. This type of magnetic behavior is called **paramagnetism**.

Substances with no unpaired electrons are weakly repelled by a magnetic field. This property is called **diamagnetism**. The distinction between paramagnetism and diamagnetism is nicely illustrated in an older method for measuring magnetic properties (▼ Figure 9.44). It involves weighing the substance in the presence and absence of a magnetic field. A paramagnetic substance appears to weigh more in the magnetic field; a diamagnetic substance appears to weigh less. The magnetic behaviors observed for the period 2 diatomic molecules agree with the electron configurations shown in Figure 9.43.

 Give It Some Thought

Figure 9.43 indicates that C_2 is diamagnetic. Would that be expected if the σ_{2p} MO were lower in energy than the π_{2p} MOs?

Electron configurations in molecules can also be related to bond distances and bond enthalpies. ⟳ (Section 8.8) As bond order increases, bond distances decrease and bond enthalpies increase. N_2, for example, whose bond order is 3, has a short bond distance and a large bond enthalpy. The N_2 molecule does not react readily with other substances to form nitrogen compounds. The high bond order of the molecule helps explain its exceptional stability. We should also note, however, that molecules with the same bond orders do *not* have the same bond distances and bond enthalpies. Bond order is only one factor influencing these properties. Other factors include nuclear charge and extent of orbital overlap.

Bonding in O_2 provides an interesting test case for molecular orbital theory. The Lewis structure for this molecule shows a double bond and complete pairing of electrons:

$$\ddot{O}=\ddot{O}$$

The short O—O bond distance (1.21 Å) and relatively high bond enthalpy (495 kJ/mol) are in agreement with the presence of a double bond. However, Figure 9.43 tells us that the molecule contains two unpaired electrons and should therefore be paramagnetic, a detail not discernible in the Lewis structure. The paramagnetism of O_2 is demonstrated in Figure 9.45, which confirms the prediction from MO theory. The MO description also correctly predicts a bond order of 2 as did the Lewis structure.

Going from O_2 to F_2, we add two electrons, completely filling the π_{2p}^* MOs. Thus, F_2 is expected to be diamagnetic and have an F—F single bond, in accord with its Lewis structure. Finally, the addition of two more electrons to make Ne_2 fills all the bonding and antibonding MOs. Therefore, the bond order of Ne_2 is zero, and the molecule is not expected to exist.

A diamagnetic sample appears to weigh less in magnetic field (weak effect)

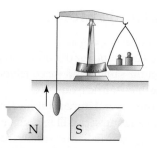

A paramagnetic sample appears to weigh more in magnetic field

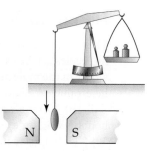

Weigh sample in absence of a magnetic field

Sample

▲ Figure 9.44 **Determining the magnetic properties of a sample.**

GO FIGURE

What would you expect to see if liquid nitrogen were poured between the poles of the magnet?

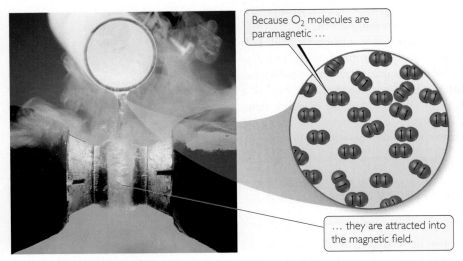

Because O_2 molecules are paramagnetic ...

... they are attracted into the magnetic field.

▲ Figure 9.45 **Paramagnetism of O_2.** When liquid oxygen is poured through a magnet, it "sticks" to the poles.

SAMPLE EXERCISE 9.9 | Molecular Orbitals of a Period 2 Diatomic Ion

For the O_2^+ ion predict (**a**) number of unpaired electrons, (**b**) bond order, (**c**) bond enthalpy and bond length.

SOLUTION

Analyze Our task is to predict several properties of the cation O_2^+.

Plan We will use the MO description of O_2^+ to determine the desired properties. We must first determine the number of electrons in O_2^+ and then draw its MO energy diagram. The unpaired electrons are those without a partner of opposite spin. The bond order is one-half the difference between the number of bonding and antibonding electrons. After calculating the bond order, we can use Figure 9.43 to estimate the bond enthalpy and bond length.

Solve

(**a**) The O_2^+ ion has 11 valence electrons, one fewer than O_2. The electron removed from O_2 to form O_2^+ is one of the two unpaired π_{2p}^* electrons (see Figure 9.43). Therefore, O_2^+ has one unpaired electron.

(**b**) The molecule has eight bonding electrons (the same as O_2) and three antibonding electrons (one fewer than O_2). Thus, its bond order is

$$\tfrac{1}{2}(8 - 3) = 2\tfrac{1}{2}$$

(**c**) The bond order of O_2^+ is between that for O_2 (bond order 2) and N_2 (bond order 3). Thus, the bond enthalpy and bond length should be about midway between those for O_2 and N_2, approximately 700 kJ/mol and 1.15 Å. (The experimentally measured values are 625 kJ/mol and 1.123 Å.)

Practice Exercise 1

Place the following molecular ions in order from smallest to largest bond order: C_2^{2+}, N_2^-, O_2^-, and F_2^-.
(**a**) $C_2^{2+} < N_2^- < O_2^- < F_2^-$ (**b**) $F_2^- < O_2^- < N_2^- < C_2^{2+}$
(**c**) $O_2^- < C_2^{2+} < F_2^- < N_2^-$ (**d**) $C_2^{2+} < F_2^- < O_2^- < N_2^-$
(**e**) $F_2^- < C_2^{2+} < O_2^- < N_2^-$

Practice Exercise 2

Predict the magnetic properties and bond orders of (**a**) the peroxide ion, O_2^{2-}; (**b**) the acetylide ion, C_2^{2-}.

Heteronuclear Diatomic Molecules

The principles we have used in developing an MO description of homonuclear diatomic molecules can be extended to *heteronuclear* diatomic molecules—those in which the two atoms in the molecule are not the same—and we conclude this section with a fascinating heteronuclear diatomic molecule—nitric oxide, NO.

The NO molecule controls several important human physiological functions. Our bodies use it, for example, to relax muscles, kill foreign cells, and reinforce memory. The 1998 Nobel Prize in Physiology or Medicine was awarded to three scientists for their research that uncovered the importance of NO as a "signaling" molecule in the cardiovascular system. NO also functions as a neurotransmitter and is implicated in many other biological pathways. That NO plays such an important role in human metabolism was unsuspected before 1987 because NO has an odd number of electrons and is highly reactive. The molecule has 11 valence electrons, and two possible Lewis structures can be drawn. The Lewis structure with the lower formal charges places the odd electron on the N atom:

$$\overset{0}{\ddot{N}}=\overset{0}{\ddot{O}} \longleftrightarrow \overset{-1}{\ddot{N}}=\overset{+1}{\ddot{O}}$$

Both structures indicate the presence of a double bond, but when compared with the molecules in Figure 9.43, the experimental bond length of NO (1.15 Å) suggests a bond order greater than 2. How do we treat NO using the MO model?

If the atoms in a heteronuclear diatomic molecule do not differ too greatly in electronegativities, their MOs resemble those in homonuclear diatomics, with one important modification: The energy of the atomic orbitals of the more electronegative atom is lower than that of the atomic orbitals of the less electronegative element. In ▶ Figure 9.46, you see that the 2s and 2p atomic orbitals of oxygen are slightly lower than those of nitrogen because oxygen is more electronegative than nitrogen. The MO energy-level diagram for NO is much like that of a homonuclear diatomic molecule—because the 2s and 2p orbitals on the two atoms interact, the same types of MOs are produced.

There is one other important difference in the MOs of heteronuclear molecules. The MOs are still a mix of atomic orbitals from both atoms, but in general *an MO in a heteronuclear diatomic molecule has a greater contribution from the atomic orbital to which it is closer in energy*. In the case of NO, for example, the σ_{2s} bonding MO is closer in energy to the O 2s atomic orbital than to the N 2s atomic orbital. As a result, the σ_{2s} MO has a slightly greater contribution from O than from N—the orbital is no longer an equal mixture of the two atoms, as was the case for the homonuclear diatomic molecules. Similarly, the σ_{2s}^* antibonding MO is weighted more heavily toward the N atom because that MO is closest in energy to the N 2s atomic orbital.

▲ GO FIGURE

How many valence-shell electrons are there in NO?

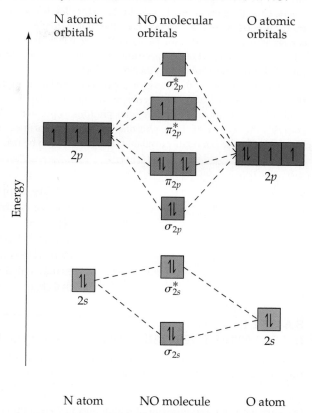

▲ **Figure 9.46 The energy-level diagram for atomic and molecular orbitals in NO.**

Chemistry Put to Work

Orbitals and Energy

Asked to identify the major technological challenge for the twenty-first century, you might say "energy," reasoning that development of sustainable energy sources is crucial to meet the needs of future generations of people on our planet. One of the most remarkable sources of clean energy is the Sun, which receives plenty of energy daily to power the world for millions of years. Our challenge is to capture enough of this energy in a form that allows us to use it as needed. *Photovoltaic solar cells* convert the light from the Sun into usable electricity, and the development of more efficient solar cells is one way to address Earth's future energy needs.

How does solar energy conversion work? Fundamentally, we need to be able to use photons from the Sun, especially from the visible portion of the spectrum, to excite electrons in molecules and materials to different energy levels. The brilliant colors around you—those of your clothes, the photographs in this book, the foods you eat—are due to the selective absorption of visible light by chemicals. It is helpful to think of this process in the context of molecular orbital theory: Light excites an electron from a filled molecular orbital to an empty one at higher energy. Because MOs have definite energies, only light of the proper wavelengths can excite electrons.

In discussing light absorption by molecules, we can focus on the two MOs shown in ▶ Figure 9.47. The *highest occupied molecular orbital* (HOMO) is the MO of highest energy that has electrons in it. The

lowest unoccupied molecular orbital (LUMO) is the MO of lowest energy that does not have electrons in it. In N_2, for example, the HOMO is the σ_{2p} MO and the LUMO is the π_{2p}^* MO (Figure 9.43).

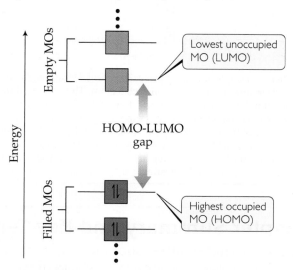

▲ **Figure 9.47 Definitions of the highest occupied and lowest unoccupied molecular orbitals.** The energy difference between these is the HOMO-LUMO gap.

The energy difference between the HOMO and the LUMO—known as the HOMO–LUMO gap—is related to the minimum energy needed to excite an electron in the molecule. Colorless or white substances usually have such a large HOMO–LUMO gap that visible light is not energetic enough to excite an electron to the higher level. The minimum energy needed to excite an electron from the HOMO to the LUMO in N_2 corresponds to light with a wavelength of less than 200 nm, which is far into the ultraviolet part of the spectrum. ∞ (Figure 6.4) As a result, N_2 cannot absorb visible light and is therefore colorless.

The magnitude of the energy gap between filled and empty electronic states is critical for solar energy conversion. Ideally, we want a substance that absorbs as many solar photons as possible and then converts the energy of those photons into a useful form of energy. Titanium dioxide is a readily available material that can be reasonably efficient at converting light directly into electricity. However, TiO_2 is white and absorbs only a small amount of the Sun's radiant energy. Scientists are working to make solar cells in which TiO_2 is mixed with highly colored molecules, whose HOMO–LUMO gaps correspond to visible and near-infrared light—molecules that can absorb more of the solar spectrum. If the HOMO of these molecules is higher in energy than the HOMO of TiO_2, the excited electrons will flow from the molecules into the TiO_2, thereby generating electricity when the device is illuminated with light and connected to an external circuit.

Efficient solar energy conversion promises to be one of the most interesting and important areas of both scientific and technological development in our future. Many of you may ultimately end up working in fields that have an impact on the world's energy portfolio.

Related Exercises: 9.109, 9.120, Design an Experiment

We complete the MO diagram for NO by filling the MOs in Figure 9.46 with the 11 valence electrons. Eight bonding and three antibonding electrons give a bond order of $\frac{1}{2}(8-3) = 2\frac{1}{2}$, which agrees better with experiment than the Lewis structures do. The unpaired electron resides in one of the π^*_{2p} MOs, which have a greater contribution from the N atom. (We could have placed this electron in either the left or right π^*_{2p} MO.) Thus, the Lewis structure that places the unpaired electron on nitrogen (the one preferred on the basis of formal charge) is the more accurate description of the true electron distribution in the molecule.

SAMPLE INTEGRATIVE EXERCISE | Putting Concepts Together

Elemental sulfur is a yellow solid that consists of S_8 molecules. The structure of the S_8 molecule is a puckered, eight-membered ring (see Figure 7.27). Heating elemental sulfur to high temperatures produces gaseous S_2 molecules:

$$S_8(s) \longrightarrow 4\,S_2(g)$$

(a) The electron configuration of which period 2 element is most similar to that of sulfur? **(b)** Use the VSEPR model to predict the S—S—S bond angles in S_8 and the hybridization at S in S_8. **(c)** Use MO theory to predict the sulfur–sulfur bond order in S_2. Do you expect this molecule to be diamagnetic or paramagnetic? **(d)** Use average bond enthalpies (Table 8.4) to estimate the enthalpy change for this reaction. Is the reaction exothermic or endothermic?

SOLUTION

(a) Sulfur is a group 6A element with an $[Ne]3s^2 3p^4$ electron configuration. It is expected to be most similar electronically to oxygen (electron configuration, $[He]2s^2 2p^4$), which is immediately above it in the periodic table.

(b) The Lewis structure of S_8 is

(Lewis structure of S_8 ring shown)

There is a single bond between each pair of S atoms and two nonbonding electron pairs on each S atom. Thus, we see four electron domains around each S atom and expect a tetrahedral electron-domain geometry corresponding to sp^3 hybridization. Because of the nonbonding pairs, we expect the S—S—S angles to be somewhat less than $109.5°$, the tetrahedral angle. Experimentally, the S—S—S angle in S_8 is $108°$, in good agreement with this prediction. Interestingly, if S_8 were a planar ring, it would have S—S—S angles of $135°$. Instead, the S_8 ring puckers to accommodate the smaller angles dictated by sp^3 hybridization.

(c) The MOs of S_2 are analogous to those of O_2, although the MOs for S_2 are constructed from the $3s$ and $3p$ atomic orbitals of sulfur. Further, S_2 has the same number of valence electrons as O_2. Thus, by analogy with O_2, we expect S_2 to have a bond order of 2 (a double bond) and to be paramagnetic with two unpaired electrons in the π^*_{3p} molecular orbitals of S_2.

(d) We are considering the reaction in which an S_8 molecule falls apart into four S_2 molecules. From parts (b) and (c), we see that S_8 has S—S single bonds and S_2 has S=S double bonds. During the reaction, therefore, we are breaking eight S—S single bonds and forming four S=S double bonds. We can estimate the enthalpy of the reaction by using Equation 8.12 and the average bond enthalpies in Table 8.4:

$$\Delta H_{rxn} = 8\,D(S{-}S) - 4\,D(S{=}S) = 8(266\text{ kJ}) - 4(418\text{ kJ}) = +456\text{ kJ}$$

Recall that $D(X{-}Y)$ represents the X—Y bond enthalpy. Because $\Delta H_{rxn} > 0$, the reaction is endothermic. (Section 5.4) The very positive value of ΔH_{rxn} suggests that high temperatures are required to cause the reaction to occur.

Chapter Summary and Key Terms

MOLECULAR SHAPES (INTRODUCTION AND SECTION 9.1) The three-dimensional shapes and sizes of molecules are determined by their **bond angles** and bond lengths. Molecules with a central atom A surrounded by n atoms B, denoted AB_n, adopt a number of different geometric shapes, depending on the value of n and on the particular atoms involved. In the overwhelming majority of cases, these geometries are related to five basic shapes (linear, trigonal pyramidal, tetrahedral, trigonal bipyramidal, and octahedral).

THE VSEPR MODEL (SECTION 9.2) The **valence-shell electron-pair repulsion (VSEPR) model** rationalizes molecular geometries based on the repulsions between **electron domains**, which are regions about a central atom in which electrons are likely to be found. **Bonding pairs** of electrons, which are those involved in making bonds, and **nonbonding pairs** of electrons, also called **lone pairs**, both create electron domains around an atom. According to the VSEPR model, electron domains orient themselves to minimize electrostatic repulsions; that is, they remain as far apart as possible.

Electron domains from nonbonding pairs exert slightly greater repulsions than those from bonding pairs, which leads to certain preferred positions for nonbonding pairs and to the departure of bond angles from idealized values. Electron domains from multiple bonds exert slightly greater repulsions than those from single bonds. The arrangement of electron domains around a central atom is called the **electron-domain geometry**; the arrangement of atoms is called the **molecular geometry**.

MOLECULAR POLARITY (SECTION 9.3) The dipole moment of a polyatomic molecule depends on the vector sum of the dipole moments associated with the individual bonds, called the **bond dipoles**. Certain molecular shapes, such as linear AB_2 and trigonal planar AB_3, lead to cancellation of the bond dipoles, producing a nonpolar molecule, which is one whose overall dipole moment is zero. In other shapes, such as bent AB_2 and trigonal pyramidal AB_3, the bond dipoles do not cancel and the molecule will be polar (that is, it will have a nonzero dipole moment).

COVALENT BONDING AND VALENCE–BOND THEORY (SECTION 9.4) **Valence-bond theory** is an extension of Lewis's notion of electron-pair bonds. In valence-bond theory, covalent bonds are formed when atomic orbitals on neighboring atoms overlap one another. The overlap region is one of greater stability for the two electrons because of their simultaneous attraction to two nuclei. The greater the overlap between two orbitals, the stronger the bond that is formed.

HYBRID ORBITALS (SECTION 9.5) To extend the ideas of valence-bond theory to polyatomic molecules, we must envision mixing s and p orbitals to form **hybrid orbitals**. The process of **hybridization** leads to hybrid atomic orbitals that have a large lobe directed to overlap with orbitals on another atom to make a bond. Hybrid orbitals can also accommodate nonbonding pairs. A particular mode of hybridization can be associated with each of three common electron-domain geometries (linear = sp; trigonal planar = sp^2; tetrahedral = sp^3). The bonding in **hypervalent** molecules—those with more than an octet of electrons—are not as readily discussed in terms of hybrid orbitals.

MULTIPLE BONDS (SECTION 9.6) Covalent bonds in which the electron density lies along the line connecting the atoms (the internuclear axis) are called **sigma (σ) bonds**. Bonds can also be formed from the sideways overlap of p orbitals. Such a bond is called a **pi (π) bond**. A double bond, such as that in C_2H_4, consists of one σ bond and one π bond; each carbon atom has an unhybridized p_π orbital, and these are the orbitals that overlap to form π bonds. A triple bond, such as that in C_2H_2, consists of one σ and two π bonds. The formation of a π bond requires that molecules adopt a specific orientation; the two CH_2 groups in C_2H_4, for example, must lie in the same plane. As a result, the presence of π bonds introduces rigidity into molecules. In molecules that have multiple bonds and more than one resonance structure, such as C_6H_6, the π bonds are **delocalized**; that is, the π bonds are spread among several atoms.

MOLECULAR ORBITALS (SECTION 9.7) **Molecular orbital theory** is another model used to describe the bonding in molecules. In this model the electrons exist in allowed energy states called **molecular orbitals (MOs)**. An MO can extend over all the atoms of a molecule. Like an atomic orbital, a molecular orbital has a definite energy and can hold two electrons of opposite spin. We can build molecular orbitals by combining atomic orbitals on different atomic centers. In the simplest case, the combination of two atomic orbitals leads to the formation of two MOs, one at lower energy and one at higher energy relative to the energy of the atomic orbitals. The lower-energy MO concentrates charge density in the region between the nuclei and is called a **bonding molecular orbital**. The higher-energy MO excludes electrons from the region between the nuclei and is called an **antibonding molecular orbital**. Antibonding MOs exclude electron density from the region between the nuclei and have a **nodal plane**—a place at which the electron density is zero—between the nuclei. Occupation of bonding MOs favors bond formation, whereas occupation of antibonding MOs is unfavorable. The bonding and antibonding MOs formed by the combination of s orbitals are **sigma (σ) molecular orbitals**; they lie on the internuclear axis.

The combination of atomic orbitals and the relative energies of the molecular orbitals are shown by an **energy-level** (or **molecular orbital**) **diagram**. When the appropriate number of electrons is put into the MOs, we can calculate the **bond order** of a bond, which is half the difference between the number of electrons in bonding MOs and the number of electrons in antibonding MOs. A bond order of 1 corresponds to a single bond, and so forth. Bond orders can be fractional numbers.

MOLECULAR ORBITALS OF PERIOD 2 DIATOMIC MOLECULES (SECTION 9.8) Electrons in core orbitals do not contribute to the bonding between atoms, so a molecular orbital description usually needs to consider only electrons in the outermost electron subshells. To describe the MOs of period 2 homonuclear diatomic molecules, we need to consider the MOs that can form by the combination of p orbitals. The p orbitals that point directly at one another can form σ bonding and σ^* antibonding MOs. The p orbitals that are oriented perpendicular to the internuclear axis combine to form **pi (π) molecular orbitals**. In diatomic molecules the π molecular orbitals occur as a pair of degenerate (same energy) bonding MOs and a pair of degenerate antibonding MOs. The σ_{2p} bonding MO is expected to be lower in energy than the π_{2p} bonding MOs because of larger orbital overlap of the p orbitals directed along the internuclear axis. However, this ordering is reversed in B_2, C_2, and N_2 because of interaction between the $2s$ and $2p$ atomic orbitals of different atoms.

The molecular orbital description of period 2 diatomic molecules leads to bond orders in accord with the Lewis structures of these molecules. Further, the model predicts correctly that O_2 should exhibit **paramagnetism**, which leads to attraction of a molecule into a magnetic field due to the influence of unpaired electrons. Molecules in which all the electrons are paired exhibit **diamagnetism**, which leads to weak repulsion from a magnetic field. The molecular orbitals of heteronuclear diatomic molecules are often closely related to those of homonuclear diatomic molecules.

Learning Outcomes   After studying this chapter, you should be able to:

- Predict the three-dimensional shapes of molecules using the VSEPR model. (Section 9.2)
- Determine whether a molecule is polar or nonpolar based on its geometry and the individual bond dipole moments. (Section 9.3)
- Explain the role of orbital overlap in the formation of covalent bonds. (Section 9.4)
- Determine the hybridization atoms in molecules based on observed molecular structures. (Section 9.5)
- Sketch how orbitals overlap to form sigma (σ) and pi (π) bonds. (Section 9.6)
- Explain the existence of delocalized π bonds in molecules such as benzene. (Section 9.6)

- Count the number of electrons in a delocalized π system. (Section 9.6)
- Explain the concept of bonding and antibonding molecular orbitals and draw examples of σ and π MOs. (Section 9.7)
- Draw molecular orbital energy-level diagrams and place electrons into them to obtain the bond orders and electron configurations of diatomic molecules using molecular orbital theory. (Sections 9.7 and 9.8)
- Correlate bond order, bond strength (bond enthalpy), bond length, and magnetic properties with molecular orbital descriptions of molecules. (Section 9.8)

Key Equations

- Bond order $= \frac{1}{2}$ (no. of bonding electrons − no. of antibonding electrons)   [9.1]

Exercises

Visualizing Concepts

9.1 A certain AB$_4$ molecule has a "seesaw" shape:

From which of the fundamental geometries shown in Figure 9.3 could you remove one or more atoms to create a molecule having this seesaw shape? [Section 9.1]

9.2 (a) If these three balloons are all the same size, what angle is formed between the red one and the green one? (b) If additional air is added to the blue balloon so that it gets larger, will the angle between the red and green balloons increase, decrease, or stay the same? (c) Which of the following aspects of the VSEPR model is illustrated by part (b): (i) The electron-domain geometry for four electron domains is tetrahedral. (ii) The electron domains for nonbonding pairs are larger than those for bonding pairs. (iii) The hybridization that corresponds to a trigonal planar electron-domain geometry is sp^2? [Section 9.2]

9.3 For each molecule (a)–(f), indicate how many different electron-domain geometries are consistent with the molecular geometry shown. [Section 9.2]

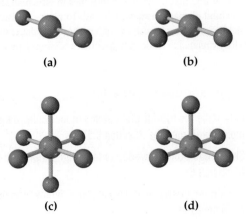

(a)   (b)

(c)   (d)

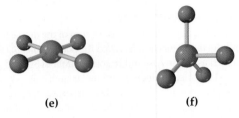

(e)   (f)

9.4 The molecule shown here is *difluoromethane* (CH_2F_2), which is used as a refrigerant called R-32. (a) Based on the structure, how many electron domains surround the C atom in this molecule? (b) Would the molecule have a nonzero dipole moment? (c) If the molecule is polar, which of the following describes the direction of the overall dipole moment vector in the molecule: (i) from the carbon atom toward a fluorine atom, (ii) from the carbon atom to a point midway between the fluorine atoms, (iii) from the carbon atom to a point midway between the hydrogen atoms, or (iv) From the carbon atom toward a hydrogen atom? [Sections 9.2 and 9.3]

9.5 The following plot shows the potential energy of two Cl atoms as a function of the distance between them. (a) To what does an energy of zero correspond in this diagram? (b) According to the valence-bond model, why does the energy decrease as the Cl atoms move from a large separation to a smaller one? (c) What is the significance of the Cl—Cl distance at the minimum point in the plot? (d) Why does the energy rise at Cl—Cl distances less than that at the minimum point in the plot? (e) How can you estimate the bond strength of the Cl—Cl bond from the plot? [Section 9.4]

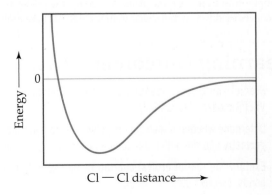

9.6 The orbital diagram that follows presents the final step in the formation of hybrid orbitals by a silicon atom. **(a)** Which of the following best describes what took place before the step pictured in the diagram: (i) Two $3p$ electrons became unpaired, (ii) An electron was promoted from the $2p$ orbital to the $3s$ orbital, or (iii) An electron was promoted from the $3s$ orbital to the $3p$ orbital? **(b)** What type of hybrid orbital is produced in this hybridization? [Section 9.5]

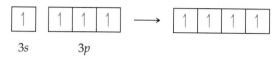

9.7 In the hydrocarbon

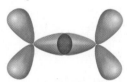

(a) What is the hybridization at each carbon atom in the molecule? **(b)** How many σ bonds are there in the molecule? **(c)** How many π bonds? **(d)** Identify all the 120° bond angles in the molecule. [Section 9.6]

9.8 The drawing below shows the overlap of two hybrid orbitals to form a bond in a hydrocarbon. **(a)** Which of the following types of bonds is being formed: (i) C—C σ, (ii) C—C π, or (iii) C—H σ? **(b)** Which of the following could be the identity of the hydrocarbon: (i) CH_4, (ii) C_2H_6, (iii) C_2H_4, or (iv) C_2H_2? [Section 9.6]

9.9 The molecule shown below is called *furan*. It is represented in typical shorthand way for organic molecules, with hydrogen atoms not shown.

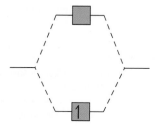

(a) What is the molecular formula for furan? **(b)** How many valence electrons are there in the molecule? **(c)** What is the hybridization at each of the carbon atoms? **(d)** How many electrons are in the π system of the molecule? **(e)** The C—C—C bond angles in furan are much smaller than those in benzene. The likely reason is which of the following: (i) The hybridization of the carbon atoms in furan is different from that in benzene, (ii) Furan does not have another resonance structure equivalent to the one above, or (iii) The atoms in a five-membered ring are forced to adopt smaller angles than in a six-membered ring. [Section 9.5]

9.10 The following is part of a molecular orbital energy-level diagram for MOs constructed from $1s$ atomic orbitals.

(a) What labels do we use for the two MOs shown? **(b)** For which of the following molecules or ions could this be the energy-level diagram:

$$H_2,\ He_2,\ H_2^+,\ He_2^+,\ \text{or}\ H_2^-?$$

(c) What is the bond order of the molecule or ion? **(d)** If an electron is added to the system, into which of the MOs will it be added? [Section 9.7]

9.11 For each of these contour representations of molecular orbitals, identify **(a)** the atomic orbitals (s or p) used to construct the MO **(b)** the type of MO (σ or π), **(c)** whether the MO is bonding or antibonding, and **(d)** the locations of nodal planes. [Sections 9.7 and 9.8]

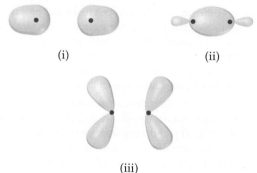

9.12 The diagram that follows shows the highest-energy occupied MOs of a neutral molecule CX, where element X is in the same row of the periodic table as C. **(a)** Based on the number of electrons, can you determine the identity of X? **(b)** Would the molecule be diamagnetic or paramagnetic? **(c)** Consider the π_{2p} MOs of the molecule. Would you expect them to have a greater atomic orbital contribution from C, have a greater atomic orbital contribution from X, or be an equal mixture of atomic orbitals from the two atoms? [Section 9.8]

σ_{2p} $\boxed{\uparrow}$

π_{2p} $\boxed{\uparrow\downarrow}$ $\boxed{\uparrow\downarrow}$

Molecular Shapes; the VSEPR Model (sections 9.1 and 9.2)

9.13 An AB_2 molecule is described as linear, and the A—B bond length is known. **(a)** Does this information completely describe the geometry of the molecule? **(b)** Can you tell how many nonbonding pairs of electrons are around the A atom from this information?

9.14 **(a)** Methane (CH_4) and the perchlorate ion (ClO_4^-) are both described as tetrahedral. What does this indicate about their bond angles? **(b)** The NH_3 molecule is trigonal pyramidal, while BF_3 is trigonal planar. Which of these molecules is flat?

9.15 How does a trigonal pyramid differ from a tetrahedron so far as molecular geometry is concerned?

9.16 Describe the bond angles to be found in each of the following molecular structures: **(a)** trigonal planar, **(b)** tetrahedral, **(c)** octahedral, **(d)** linear.

9.17 **(a)** How does one determine the number of electron domains in a molecule or ion? **(b)** What is the difference between a *bonding electron domain* and a *nonbonding electron domain*?

9.18 Would you expect the nonbonding electron-pair domain in NH_3 to be greater or less in size than for the corresponding one in PH_3?

9.19 In which of these molecules or ions does the presence of non-bonding electron pairs produce an effect on molecular shape? (**a**) SiH_4, (**b**) PF_3, (**c**) HBr, (**d**) HCN, (**e**) SO_2.

9.20 In which of the following molecules can you confidently predict the bond angles about the central atom, and for which would you be a bit uncertain? Explain in each case. (**a**) H_2S, (**b**) BCl_3, (**c**) CH_3I, (**d**) CBr_4, (**e**) $TeBr_4$.

9.21 How many nonbonding electron pairs are there in each of the following molecules: (**a**) $(CH_3)_2S$, (**b**) HCN, (**c**) H_2C_2, (**d**) CH_3F?

9.22 Describe the characteristic electron-domain geometry of each of the following numbers of electron domains about a central atom: (**a**) 3, (**b**) 4, (**c**) 5, (**d**) 6.

9.23 Give the electron-domain and molecular geometries of a molecule that has the following electron domains on its central atom: (**a**) four bonding domains and no nonbonding domains, (**b**) three bonding domains and two nonbonding domains, (**c**) five bonding domains and one nonbonding domain, (**d**) four bonding domains and two nonbonding domains.

9.24 What are the electron-domain and molecular geometries of a molecule that has the following electron domains on its central atom? (**a**) Three bonding domains and no nonbonding domains, (**b**) three bonding domains and one nonbonding domain, (**c**) two bonding domains and two nonbonding domains.

9.25 Give the electron-domain and molecular geometries for the following molecules and ions: (**a**) HCN, (**b**) SO_3^{2-}, (**c**) SF_4, (**d**) PF_6^-, (**e**) NH_3Cl^+, (**f**) N_3^-.

9.26 Draw the Lewis structure for each of the following molecules or ions, and predict their electron-domain and molecular geometries: (**a**) AsF_3, (**b**) CH_3^+, (**c**) BrF_3, (**d**) ClO_3^-, (**e**) XeF_2, (**f**) BrO_2^-.

9.27 The figure that follows shows ball-and-stick drawings of three possible shapes of an AF_3 molecule. (**a**) For each shape, give the electron-domain geometry on which the molecular geometry is based. (**b**) For each shape, how many nonbonding electron domains are there on atom A? (**c**) Which of the following elements will lead to an AF_3 molecule with the shape in (ii): Li, B, N, Al, P, Cl? (**d**) Name an element A that is expected to lead to the AF_3 structure shown in (iii). Explain your reasoning.

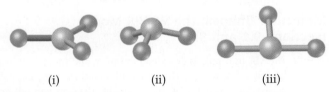

(i)   (ii)   (iii)

9.28 The figure that follows contains ball-and-stick drawings of three possible shapes of an AF_4 molecule. (**a**) For each shape, give the electron-domain geometry on which the molecular geometry is based. (**b**) For each shape, how many nonbonding electron domains are there on atom A? (**c**) Which of the following elements will lead to an AF_4 molecule with the shape in (iii): Be, C, S, Se, Si, Xe? (**d**) Name an element A that is expected to lead to the AF_4 structure shown in (i).

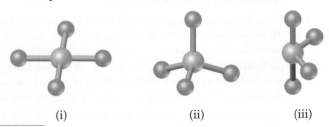

(i)   (ii)   (iii)

9.29 Give the approximate values for the indicated bond angles in the following molecules:

(**a**) H—Ö—Cl—Ö: (1, 2) with :Ö: below Cl

(**b**) H—C—Ö—H (with H above and below C; angles 3, 4)

(**c**) H—C≡C—H (angle 5)

(**d**) H—C—Ö—C—H (with :O: double-bonded to first C; angles 6, 7, 8)

9.30 Give approximate values for the indicated bond angles in the following molecules:

(**a**) H—Ö—N=Ö (angles 1, 2)

(**b**) H—C—C=Ö (with H atoms; angles 3, 4)

(**c**) H—N—Ö—H (with H atoms; angles 5, 6)

(**d**) H—C—C≡N: (with H atoms; angles 7, 8)

9.31 The three species NH_2^-, NH_3, and NH_4^+ have H—N—H bond angles of 105°, 107°, and 109°, respectively. Explain this variation in bond angles.

9.32 In which of the following AF_n molecules or ions is there more than one F—A—F bond angle: SiF_4, PF_5, SF_4, AsF_3?

9.33 (**a**) Explain why BrF_4^- is square planar, whereas BF_4^- is tetrahedral. (**b**) How would you expect the H—X—H bond angle to vary in the series H_2O, H_2S, H_2Se? Explain. (*Hint:* The size of an electron pair domain depends in part on the electronegativity of the central atom.)

9.34 (**a**) Explain why the following ions have different bond angles: ClO_2^- and NO_2^-. Predict the bond angle in each case. (**b**) Explain why the XeF_2 molecule is linear.

Shapes and Polarity of Polyatomic Molecules (section 9.3)

9.35 What is the distinction between a bond dipole and a molecular dipole moment?

9.36 Consider a molecule with formula AX_3. Supposing the A—X bond is polar, how would you expect the dipole moment of the AX_3 molecule to change as the X—A—X bond angle increases from 100° to 120°?

9.37 **(a)** Does SCl_2 have a dipole moment? If so, in which direction does the net dipole point? **(b)** Does $BeCl_2$ have a dipole moment? If so, in which direction does the net dipole point?

9.38 **(a)** The PH_3 molecule is polar. Does this offer experimental proof that the molecule cannot be planar? Explain. **(b)** It turns out that ozone, O_3, has a small dipole moment. How is this possible, given that all the atoms are the same?

9.39 **(a)** What conditions must be met if a molecule with polar bonds is nonpolar? **(b)** What geometries will signify nonpolar molecules for AB_2, AB_3, and AB_4 geometries?

9.40 **(a)** Consider the AF_3 molecules in Exercise 9.27. Which of these will have a nonzero dipole moment? **(b)** Which of the AF_4 molecules in Exercise 9.28 will have a zero dipole moment?

9.41 Predict whether each of the following molecules is polar or nonpolar: **(a)** IF, **(b)** CS_2, **(c)** SO_3, **(d)** PCl_3, **(e)** SF_6, **(f)** IF_5.

9.42 Predict whether each of the following molecules is polar or nonpolar: **(a)** CCl_4, **(b)** NH_3, **(c)** SF_4, **(d)** XeF_4, **(e)** CH_3Br, **(f)** GaH_3.

9.43 Dichloroethylene ($C_2H_2Cl_2$) has three forms (isomers), each of which is a different substance. **(a)** Draw Lewis structures of the three isomers, all of which have a carbon–carbon double bond. **(b)** Which of these isomers has a zero dipole moment? **(c)** How many isomeric forms can chloroethylene, C_2H_3Cl, have? Would they be expected to have dipole moments?

9.44 Dichlorobenzene, $C_6H_4Cl_2$, exists in three forms (isomers) called *ortho*, *meta*, and *para*:

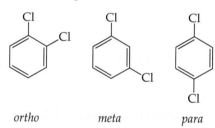

ortho   *meta*   *para*

Which of these has a nonzero dipole moment?

Orbital Overlap; Hybrid Orbitals (sections 9.4 and 9.5)

9.45 **(a)** What is meant by the term *orbital overlap*? **(b)** Describe what a chemical bond is in terms of electron density between two atoms.

9.46 Draw sketches illustrating the overlap between the following orbitals on two atoms: **(a)** the 2s orbital on each atom, **(b)** the $2p_z$ orbital on each atom (assume both atoms are on the z-axis), **(c)** the 2s orbital on one atom and the $2p_z$ orbital on the other atom.

9.47 Consider the bonding in an MgH_2 molecule. **(a)** Draw a Lewis structure for the molecule, and predict its molecular geometry. **(b)** What hybridization scheme is used in MgH_2? **(c)** Sketch one of the two-electron bonds between an Mg hybrid orbital and an H 1s atomic orbital.

9.48 How would you expect the extent of overlap of the bonding atomic orbitals to vary in the series IF, ICl, IBr, and I_2? Explain your answer.

9.49 **(a)** Starting with the orbital diagram of a boron atom, describe the steps needed to construct hybrid orbitals appropriate to describe the bonding in BF_3. **(b)** What is the name given to the hybrid orbitals constructed in (a)? **(c)** Sketch the large lobes of the hybrid orbitals constructed in part (a). **(d)** Are any valence atomic orbitals of B left unhybridized? If so, how are they oriented relative to the hybrid orbitals?

9.50 **(a)** Starting with the orbital diagram of a sulfur atom, describe the steps needed to construct hybrid orbitals appropriate to describe the bonding in SF_2. **(b)** What is the name given to the hybrid orbitals constructed in (a)? **(c)** Sketch the large lobes of these hybrid orbitals. **(d)** Would the hybridization scheme in part (a) be appropriate for SF_4? Explain.

9.51 Indicate the hybridization of the central atom in **(a)** BCl_3, **(b)** $AlCl_4^-$, **(c)** CS_2, **(d)** GeH_4.

9.52 What is the hybridization of the central atom in **(a)** $SiCl_4$, **(b)** HCN, **(c)** SO_3, **(d)** $TeCl_2$.

9.53 Shown here are three pairs of hybrid orbitals, with each set at a characteristic angle. For each pair, determine the type of hybridization, if any, that could lead to hybrid orbitals at the specified angle.

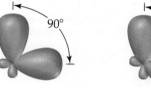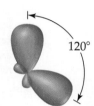

9.54 **(a)** Which geometry and central atom hybridization would you expect in the series BH_4^-, CH_4, NH_4^+? **(b)** What would you expect for the magnitude and direction of the bond dipoles in this series? **(c)** Write the formulas for the analogous species of the elements of period 3; would you expect them to have the same hybridization at the central atom?

Multiple Bonds (section 9.6)

9.55 **(a)** Draw a picture showing how two p orbitals on two different atoms can be combined to make a σ bond. **(b)** Sketch a π bond that is constructed from p orbitals. **(c)** Which is generally stronger, a σ bond or a π bond? Explain. **(d)** Can two s orbitals combine to form a π bond? Explain.

9.56 **(a)** If the valence atomic orbitals of an atom are sp hybridized, how many unhybridized p orbitals remain in the valence shell? How many π bonds can the atom form? **(b)** Imagine that you could hold two atoms that are bonded together, twist them, and not change the bond length. Would it be easier to twist (rotate) around a single σ bond or around a double (σ plus π) bond, or would they be the same?

9.57 **(a)** Draw Lewis structures for ethane (C_2H_6), ethylene (C_2H_4), and acetylene (C_2H_2). **(b)** What is the hybridization of the carbon atoms in each molecule? **(c)** Predict which molecules, if any, are planar. **(d)** How many σ and π bonds are there in each molecule?

9.58 The nitrogen atoms in N_2 participate in multiple bonding, whereas those in hydrazine, N_2H_4, do not. **(a)** Draw Lewis structures for both molecules. **(b)** What is the hybridization of the nitrogen atoms in each molecule? **(c)** Which molecule has the stronger N—N bond?

9.59 Propylene, C_3H_6, is a gas that is used to form the important polymer called polypropylene. Its Lewis structure is

$$H-\overset{\overset{\displaystyle H}{|}}{C}=\overset{\overset{\displaystyle H}{|}}{\underset{\underset{\displaystyle H}{|}}{C}}-\overset{\overset{\displaystyle H}{|}}{\underset{\underset{\displaystyle H}{|}}{C}}-H$$

(a) What is the total number of valence electrons in the propylene molecule? (b) How many valence electrons are used to make σ bonds in the molecule? (c) How many valence electrons are used to make π bonds in the molecule? (d) How many valence electrons remain in nonbonding pairs in the molecule? (e) What is the hybridization at each carbon atom in the molecule?

9.60 Ethyl acetate, $C_4H_8O_2$, is a fragrant substance used both as a solvent and as an aroma enhancer. Its Lewis structure is

$$H-\overset{\overset{\displaystyle H}{|}}{\underset{\underset{\displaystyle H}{|}}{C}}-\overset{\overset{\displaystyle :O:}{||}}{C}-\overset{..}{\underset{..}{O}}-\overset{\overset{\displaystyle H}{|}}{\underset{\underset{\displaystyle H}{|}}{C}}-\overset{\overset{\displaystyle H}{|}}{\underset{\underset{\displaystyle H}{|}}{C}}-H$$

(a) What is the hybridization at each of the carbon atoms of the molecule? (b) What is the total number of valence electrons in ethyl acetate? (c) How many of the valence electrons are used to make σ bonds in the molecule? (d) How many valence electrons are used to make π bonds? (e) How many valence electrons remain in nonbonding pairs in the molecule?

9.61 Consider the Lewis structure for glycine, the simplest amino acid:

$$H-\overset{..}{\underset{\underset{\displaystyle H}{|}}{N}}-\overset{\overset{\displaystyle H}{|}}{\underset{\underset{\displaystyle H}{|}}{C}}-\overset{\overset{\displaystyle :O:}{||}}{C}-\overset{..}{\underset{..}{O}}-H$$

(a) What are the approximate bond angles about each of the two carbon atoms, and what are the hybridizations of the orbitals on each of them? (b) What are the hybridizations of the orbitals on the two oxygens and the nitrogen atom, and what are the approximate bond angles at the nitrogen? (c) What is the total number of σ bonds in the entire molecule, and what is the total number of π bonds?

9.62 Acetylsalicylic acid, better known as aspirin, has the Lewis structure

(a) What are the approximate values of the bond angles labeled 1, 2, and 3? (b) What hybrid orbitals are used about the central atom of each of these angles? (c) How many σ bonds are in the molecule?

9.63 (a) What is the difference between a localized π bond and a delocalized one? (b) How can you determine whether a molecule or ion will exhibit delocalized π bonding? (c) Is the π bond in NO_2^- localized or delocalized?

9.64 (a) Write a single Lewis structure for SO_3, and determine the hybridization at the S atom. (b) Are there other equivalent Lewis structures for the molecule? (c) Would you expect SO_3 to exhibit delocalized π bonding?

9.65 In the formate ion, HCO_2^-, the carbon atom is the central atom with the other three atoms attached to it. (a) Draw a Lewis structure for the formate ion. (b) What hybridization is exhibited by the C atom? (c) Are there multiple equivalent resonance structures for the ion? (d) Which of the atoms in the ion have p_π orbitals? (e) How many electrons are in the π system of the ion?

9.66 Consider the Lewis structure shown below.

(a) Does the Lewis structure depict a neutral molecule or an ion? If it is an ion, what is the charge on the ion? (b) What hybridization is exhibited by each of the carbon atoms? (c) Are there multiple equivalent resonance structures for the species? (d) Which of the atoms in the species have p_π orbitals? (e) How many electrons are in the π system of the species?

9.67 Predict the molecular geometry of each of the following molecules:

(a) $H-C\equiv C-C\equiv C-C\equiv N$

(b) $H-O-\overset{\overset{\displaystyle O}{||}}{C}-\overset{\overset{\displaystyle O}{||}}{C}-O-H$

(c) $H-N=N-H$

9.68 What hybridization do you expect for the atom indicated in red in each of the following species?
(a) $CH_3CO_2^-$; (b) PH_4^+; (c) AlF_3; (d) $H_2C=CH-CH_2^+$

Molecular Orbitals and Period 2 Diatomic Molecules (sections 9.7 and 9.8)

9.69 (a) What is the difference between hybrid orbitals and molecular orbitals? (b) How many electrons can be placed into each MO of a molecule? (c) Can antibonding molecular orbitals have electrons in them?

9.70 (a) If you combine two atomic orbitals on two different atoms to make a new orbital, is this a hybrid orbital or a molecular orbital? (b) If you combine two atomic orbitals on *one* atom to make a new orbital, is this a hybrid orbital or a molecular orbital? (c) Does the Pauli exclusion principle (Section 6.7) apply to MOs? Explain.

9.71 Consider the H_2^+ ion. (a) Sketch the molecular orbitals of the ion and draw its energy-level diagram. (b) How many electrons are there in the H_2^+ ion? (c) Write the electron configuration of the ion in terms of its MOs. (d) What is the bond order in H_2^+? (e) Suppose that the ion is excited by light so that an electron moves from a lower-energy to a higher-energy MO. Would you expect the excited-state H_2^+ ion to be stable or to fall apart? (f) Which of the following statements about part (e) is correct: (i) The light excites an electron from a bonding orbital to an antibonding orbital, (ii) The light excites an electron from an antibonding orbital to a bonding orbital, or (iii) In the excited state there are more bonding electrons than antibonding electrons?

9.72 (a) Sketch the molecular orbitals of the H_2^- ion and draw its energy-level diagram. (b) Write the electron configuration of the ion in terms of its MOs. (c) Calculate the bond order in H_2^-. (d) Suppose that the ion is excited by light, so that an electron moves from a lower-energy to a higher-energy molecular orbital. Would you expect the excited-state H_2^- ion to be stable? (e) Which of the following statements about part (d) is correct: (i) The light excites an electron from a bonding orbital to an antibonding orbital, (ii) The light excites an electron from an antibonding orbital to a bonding orbital, or (iii) In the excited state there are more bonding electrons than antibonding electrons?

9.73 Draw a picture that shows all three $2p$ orbitals on one atom and all three $2p$ orbitals on another atom. (a) Imagine the atoms coming close together to bond. How many σ bonds can the two sets of $2p$ orbitals make with each other? (b) How many π bonds can the two sets of $2p$ orbitals make with each other? (c) How many antibonding orbitals, and of what type, can be made from the two sets of $2p$ orbitals?

9.74 (a) What is the probability of finding an electron on the internuclear axis if the electron occupies a π molecular orbital? (b) For a homonuclear diatomic molecule, what similarities and differences are there between the π_{2p} MO made from the $2p_x$ atomic orbitals and the π_{2p} MO made from the $2p_y$ atomic orbitals? (c) How do the π_{2p}^* MOs formed from the $2p_x$ and $2p_y$ atomic orbitals differ from the π_{2p} MOs in terms of energies and electron distributions?

9.75 (a) What are the relationships among bond order, bond length, and bond energy? (b) According to molecular orbital theory, would either Be_2 or Be_2^+ be expected to exist? Explain.

9.76 Explain the following: (a) The *peroxide* ion, O_2^{2-}, has a longer bond length than the *superoxide* ion, O_2^-. (b) The magnetic properties of B_2 are consistent with the π_{2p} MOs being lower in energy than the σ_{2p} MO. (c) The O_2^{2+} ion has a stronger O—O bond than O_2 itself.

9.77 (a) What does the term *diamagnetism* mean? (b) How does a diamagnetic substance respond to a magnetic field? (c) Which of the following ions would you expect to be diamagnetic: N_2^{2-}, O_2^{2-}, Be_2^{2+}, C_2^-?

9.78 (a) What does the term *paramagnetism* mean? (b) How can one determine experimentally whether a substance is paramagnetic? (c) Which of the following ions would you expect to be paramagnetic: O_2^+, N_2^{2-}, Li_2^+, O_2^{2-}? For those ions that are paramagnetic, determine the number of unpaired electrons.

9.79 Using Figures 9.35 and 9.43 as guides, draw the molecular orbital electron configuration for (a) B_2^+, (b) Li_2^+, (c) N_2^+, (d) Ne_2^{2+}. In each case indicate whether the addition of an electron to the ion would increase or decrease the bond order of the species.

9.80 If we assume that the energy-level diagrams for homonuclear diatomic molecules shown in Figure 9.43 can be applied to heteronuclear diatomic molecules and ions, predict the bond order and magnetic behavior of (a) CO^+, (b) NO^-, (c) OF^+, (d) NeF^+.

9.81 Determine the electron configurations for CN^+, CN, and CN^-. (a) Which species has the strongest C—N bond? (b) Which species, if any, has unpaired electrons?

9.82 (a) The nitric oxide molecule, NO, readily loses one electron to form the NO^+ ion. Which of the following is the best explanation of why this happens: (i) Oxygen is more electronegative than nitrogen, (ii) The highest energy electron in NO lies in a π_{2p}^* molecular orbital, or (iii) The π_{2p}^* MO in NO is completely filled. (b) Predict the order of the N—O bond strengths in NO, NO^+, and NO^-, and describe the magnetic properties of each. (c) With what neutral homonuclear diatomic molecules are the NO^+ and NO^- ions isoelectronic (same number of electrons)?

[9.83] Consider the molecular orbitals of the P_2 molecule. Assume that the MOs of diatomics from the third row of the periodic table are analogous to those from the second row. (a) Which valence atomic orbitals of P are used to construct the MOs of P_2? (b) The figure that follows shows a sketch of one of the MOs for P_2. What is the label for this MO? (c) For the P_2 molecule, how many electrons occupy the MO in the figure? (d) Is P_2 expected to be diamagnetic or paramagnetic?

[9.84] The iodine bromide molecule, IBr, is an *interhalogen compound*. Assume that the molecular orbitals of IBr are analogous to the homonuclear diatomic molecule F_2. (a) Which valence atomic orbitals of I and of Br are used to construct the MOs of IBr? (b) What is the bond order of the IBr molecule? (c) One of the valence MOs of IBr is sketched here. Why are the atomic orbital contributions to this MO different in size? (d) What is the label for the MO? (e) For the IBr molecule, how many electrons occupy the MO?

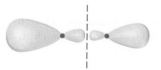

Additional Exercises

9.85 (a) What is the physical basis for the VSEPR model? (b) When applying the VSEPR model, we count a double or triple bond as a single electron domain. Why is this justified?

9.86 An AB_3 molecule is described as having a trigonal-bipyramidal electron-domain geometry. (a) How many nonbonding domains are on atom A? (b) Based on the information given, which of the following is the molecular geometry of the molecule: (i) trigonal planar, (ii) trigonal pyramidal, (iii) T-shaped, or (iv) tetrahedral?

9.87 Consider the following XF_4 ions: PF_4^-, BrF_4^-, ClF_4^+, and AlF_4^-. (a) Which of the ions have more than an octet of electrons around the central atom? (b) For which of the ions will the electron-domain and molecular geometries be the same? (c) Which of the ions will have an octahedral electron-domain geometry? (d) Which of the ions will exhibit a see-saw molecular geometry?

9.88 Consider the molecule PF_4Cl. (a) Draw a Lewis structure for the molecule, and predict its electron-domain geometry. (b) Which would you expect to take up more space, a P—F bond or a P—Cl bond? Explain. (c) Predict the molecular geometry of PF_4Cl. How did your answer for part (b) influence your answer here in part (c)? (d) Would you expect the molecule to distort from its ideal electron-domain geometry? If so, how would it distort?

[9.89] The vertices of a tetrahedron correspond to four alternating corners of a cube. By using analytical geometry, demonstrate that the angle made by connecting two of the vertices to a point at the center of the cube is 109.5°, the characteristic angle for tetrahedral molecules.

9.90 Fill in the blank spaces in the following chart. If the molecule column is blank, find an example that fulfills the conditions of the rest of the row.

Molecule	Electron-Domain Geometry	Hybridization of Central Atom	Dipole Moment? Yes or No
CO_2			
		sp^3	Yes
		sp^3	No
	Trigonal planar		No
SF_4			
	Octahedral		No
		sp^2	Yes
	Trigonal bipyramidal		No
XeF_2			

9.91 From their Lewis structures, determine the number of σ and π bonds in each of the following molecules or ions: (a) CO_2; (b) cyanogen, $(CN)_2$; (c) formaldehyde, H_2CO; (d) formic acid, HCOOH, which has one H and two O atoms attached to C.

9.92 The lactic acid molecule, $CH_3CH(OH)COOH$, gives sour milk its unpleasant, sour taste. (a) Draw the Lewis structure for the molecule, assuming that carbon always forms four bonds in its stable compounds. (b) How many π and how many σ bonds are in the molecule? (c) Which CO bond is shortest in the molecule? (d) What is the hybridization of atomic orbitals around the carbon atom associated with that short bond? (e) What are the approximate bond angles around each carbon atom in the molecule?

9.93 An AB_5 molecule adopts the geometry shown here. (a) What is the name of this geometry? (b) Do you think there are any nonbonding electron pairs on atom A? (c) Suppose the atoms B are halogen atoms. Of which group in the periodic table is atom A a member: (i) Group 5A, (ii) Group 6A, (iii) Group 7A, (iv) Group 8A, or (v) More information is needed?

9.94 There are two compounds of the formula $Pt(NH_3)_2Cl_2$:

The compound on the right, *cisplatin,* is used in cancer therapy. The compound on the left, *transplatin,* is ineffective for cancer therapy. Both compounds have a square-planar geometry. (a) Which compound has a nonzero dipole moment? (b) The reason cisplatin is a good anticancer drug is that it binds tightly to DNA. Cancer cells are rapidly dividing, producing a lot of DNA. Consequently, cisplatin kills cancer cells at a faster rate than normal cells. However, since normal cells also are making DNA, cisplatin also attacks healthy cells, which leads to unpleasant side effects. The way both molecules bind to DNA involves the Cl^- ions leaving the Pt ion, to be replaced by two nitrogens in DNA. Draw a picture in which a long vertical line represents a piece of DNA. Draw the $Pt(NH_3)_2$ fragments of cisplatin and transplatin with the proper shape. Also draw them attaching to your DNA line. Can you explain from your drawing why the shape of the cisplatin causes it to bind to DNA more effectively than transplatin?

[9.95] The O—H bond lengths in the water molecule (H_2O) are 0.96 Å, and the H—O—H angle is 104.5°. The dipole moment of the water molecule is 1.85 D. (a) In what directions do the bond dipoles of the O—H bonds point? In what direction does the dipole moment vector of the water molecule point? (b) Calculate the magnitude of the bond dipole of the O—H bonds. (*Note:* You will need to use vector addition to do this.) (c) Compare your answer from part (b) to the dipole moments of the hydrogen halides (Table 8.3). Is your answer in accord with the relative electronegativity of oxygen?

9.96 The reaction of three molecules of fluorine gas with a Xe atom produces the substance xenon hexafluoride, XeF_6:

$$Xe(g) + 3 F_2(g) \rightarrow XeF_6(s)$$

(a) Draw a Lewis structure for XeF_6. (b) If you try to use the VSEPR model to predict the molecular geometry of XeF_6, you run into a problem. What is it? (c) What could you do to resolve the difficulty in part (b)? (d) The molecule IF_7 has a pentagonal-bipyramidal structure (five equatorial fluorine atoms at the vertices of a regular pentagon and two axial fluorine atoms). Based on the structure of IF_7, suggest a structure for XeF_6.

9.97 Which of the following statements about hybrid orbitals is or are true? (i) After an atom undergoes sp hybridization there is one unhybridized p orbital on the atom, (ii) Under sp^2 hybridization, the large lobes point to the vertices of an equilateral triangle, and (iii) The angle between the large lobes of sp^3 hybrids is 109.5°.

[9.98] The Lewis structure for allene is

$$\underset{H}{\overset{H}{\diagdown}}C=C=C\underset{H}{\overset{H}{\diagup}}$$

Make a sketch of the structure of this molecule that is analogous to Figure 9.25. In addition, answer the following three questions: (a) Is the molecule planar? (b) Does it have a nonzero dipole moment? (c) Would the bonding in allene be described as delocalized? Explain.

9.99 Consider the molecule C_4H_5N, which has the connectivity shown below. (a) After the Lewis structure for the molecule is completed, how many σ and how many π bonds are there in this molecule? (b) How many atoms in the molecule exhibit (i) sp hybridization, (ii) sp^2 hybridization, and (iii) sp^3 hybridization?

$$H-\underset{\underset{H}{|}}{\overset{\overset{H}{|}}{C}}-C-C-N-\underset{}{\overset{\overset{H}{|}}{C}}-H$$

[9.100] The azide ion, N_3^-, is linear with two N—N bonds of equal length, 1.16 Å. **(a)** Draw a Lewis structure for the azide ion. **(b)** With reference to Table 8.5, is the observed N—N bond length consistent with your Lewis structure? **(c)** What hybridization scheme would you expect at each of the nitrogen atoms in N_3^-? **(d)** Show which hybridized and unhybridized orbitals are involved in the formation of σ and π bonds in N_3^-. **(e)** It is often observed that σ bonds that involve an sp hybrid orbital are shorter than those that involve only sp^2 or sp^3 hybrid orbitals. Can you propose a reason for this? Is this observation applicable to the observed bond lengths in N_3^-?

[9.101] In ozone, O_3, the two oxygen atoms on the ends of the molecule are equivalent to one another. **(a)** What is the best choice of hybridization scheme for the atoms of ozone? **(b)** For one of the resonance forms of ozone, which of the orbitals are used to make bonds and which are used to hold nonbonding pairs of electrons? **(c)** Which of the orbitals can be used to delocalize the π electrons? **(d)** How many electrons are delocalized in the π system of ozone?

9.102 Butadiene, C_4H_6, is a planar molecule that has the following carbon–carbon bond lengths:

$$\underset{1.34\,\text{Å}}{H_2C=\!\!=\!\!CH}\underset{1.48\,\text{Å}}{-\!\!-\!\!CH}\underset{1.34\,\text{Å}}{=\!\!=\!\!CH_2}$$

(a) Predict the bond angles around each of the carbon atoms and sketch the molecule.

(b) Compare the bond lengths to the average bond lengths listed in Table 8.5. Can you explain any differences?

9.103 The structure of *borazine*, $B_3N_3H_6$, is a six-membered ring of alternating B and N atoms. There is one H atom bonded to each B and to each N atom. The molecule is planar. **(a)** Write a Lewis structure for borazine in which the formal charges on every atom is zero. **(b)** Write a Lewis structure for borazine in which the octet rule is satisfied for every atom. **(c)** What are the formal charges on the atoms in the Lewis structure from part (b)? Given the electronegativities of B and N, do the formal charges seem favorable or unfavorable? **(d)** Do either of the Lewis structures in parts (a) and (b) have multiple resonance structures? **(e)** What are the hybridizations at the B and N atoms in the Lewis structures from parts (a) and (b)? Would you expect the molecule to be planar for both Lewis structures? **(f)** The six B—N bonds in the borazine molecule are all identical in length at 1.44 Å. Typical values for the bond lengths of B—N single and double bonds are 1.51 Å and 1.31 Å, respectively. Does the value of the B—N bond length seem to favor one Lewis structure over the other? **(g)** How many electrons are in the π system of borazine?

9.104 Suppose that silicon could form molecules that are precisely the analogs of ethane (C_2H_6), ethylene (C_2H_4), and acetylene (C_2H_2). How would you describe the bonding about Si in terms of hybrid orbitals? Silicon does not readily form some of the analogous compounds containing π bonds. Why might this be the case?

9.105 One of the molecular orbitals of the H_2^- ion is sketched below:

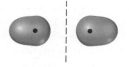

(a) Is the molecular orbital a σ or π MO? Is it bonding or antibonding? **(b)** In H_2^-, how many electrons occupy the MO shown above? **(c)** What is the bond order in the H_2^- ion? **(d)** Compared to the H—H bond in H_2, the H—H bond in H_2^- is expected to be which of the following: (i) Shorter and stronger, (ii) longer and stronger, (iii) shorter and weaker, (iv) longer and weaker, or (v) the same length and strength?

9.106 Place the following molecules and ions in order from smallest to largest bond order: H_2^+, B_2, N_2^+, F_2^+, and Ne_2.

[9.107] The following sketches show the atomic orbital wave functions (with phases) used to construct some of the MOs of a homonuclear diatomic molecule. For each sketch, determine the type of MO that will result from mixing the atomic orbital wave functions as drawn. Use the same labels for the MOs as in the "Closer Look" box on phases.

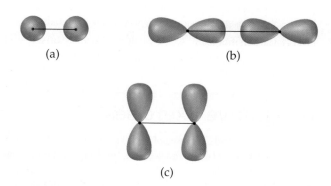

9.108 Write the electron configuration for the first excited state for N_2; that is, the state with the highest-energy electron moved to the next available energy level. **(a)** Is the nitrogen in its first excited state diamagnetic or paramagnetic? **(b)** Is the N—N bond strength in the first excited state stronger or weaker compared to that in the ground state?

9.109 *Azo dyes* are organic dyes that are used for many applications, such as the coloring of fabrics. Many azo dyes are derivatives of the organic substance *azobenzene*, $C_{12}H_{10}N_2$. A closely related substance is *hydrazobenzene*, $C_{12}H_{10}N_2$. The Lewis structures of these two substances are

Azobenzene   Hydrazobenzene

(Recall the shorthand notation used for benzene.) **(a)** What is the hybridization at the N atom in each of the substances? **(b)** How many unhybridized atomic orbitals are there on the N and the C atoms in each of the substances? **(c)** Predict the N—N—C angles in each of the substances. **(d)** Azobenzene is said to have greater delocalization of its π electrons than hydrazobenzene. Discuss this statement in light of your answers to (a) and (b). **(e)** All the atoms of azobenzene lie in one plane, whereas those of hydrazobenzene do not. Is this observation consistent with the statement in part (d)? **(f)** Azobenzene is an intense red-orange color, whereas hydrazobenzene is nearly colorless. Which molecule would be a better one to use in a solar energy conversion device? (See the "Chemistry Put to Work" box for more information about solar cells.)

[9.110] (a) Using only the valence atomic orbitals of a hydrogen atom and a fluorine atom, and following the model of Figure 9.46, how many MOs would you expect for the HF molecule? (b) How many of the MOs from part (a) would be occupied by electrons? (c) It turns out that the difference in energies between the valence atomic orbitals of H and F are sufficiently different that we can neglect the interaction of the $1s$ orbital of hydrogen with the $2s$ orbital of fluorine. The $1s$ orbital of hydrogen will mix only with one $2p$ orbital of fluorine. Draw pictures showing the proper orientation of all three $2p$ orbitals on F interacting with a $1s$ orbital on H. Which of the $2p$ orbitals can actually make a bond with a $1s$ orbital, assuming that the atoms lie on the z-axis? (d) In the most accepted picture of HF, all the other atomic orbitals on fluorine move over at the same energy into the molecular orbital energy-level diagram for HF. These are called "nonbonding orbitals." Sketch the energy-level diagram for HF using this information and calculate the bond order. (Nonbonding electrons do not contribute to bond order.) (e) Look at the Lewis structure for HF. Where are the nonbonding electrons?

[9.111] Carbon monoxide, CO, is isoelectronic to N_2. (a) Draw a Lewis structure for CO that satisfies the octet rule. (b) Assume that the diagram in Figure 9.46 can be used to describe the MOs of CO. What is the predicted bond order for CO? Is this answer in accord with the Lewis structure you drew in part (a)? (c) Experimentally, it is found that the highest-energy electrons in CO reside in a σ-type MO. Is that observation consistent with Figure 9.46? If not, what modification needs to be made to the diagram? How does this modification relate to Figure 9.43? (d) Would you expect the π_{2p} MOs of CO to have equal atomic orbital contributions from the C and O atoms? If not, which atom would have the greater contribution?

9.112 The energy-level diagram in Figure 9.36 shows that the sideways overlap of a pair of p orbitals produces two molecular orbitals, one bonding and one antibonding. In ethylene there is a pair of electrons in the bonding p orbital between the two carbons. Absorption of a photon of the appropriate wavelength can result in promotion of one of the bonding electrons from the π_{2p} to the π_{2p}^{*} molecular orbital. (a) What would you expect this electronic transition to do to the carbon–carbon bond order in ethylene? (b) How does this relate to the fact that absorption of a photon of appropriate wavelength can cause ready rotation about the carbon–carbon bond, as described in the "Chemistry and Life" box and shown in Figure 9.30?

Integrative Exercises

9.113 A compound composed of 2.1% H, 29.8% N, and 68.1% O has a molar mass of approximately 50 g/mol. (a) What is the molecular formula of the compound? (b) What is its Lewis structure if H is bonded to O? (c) What is the geometry of the molecule? (d) What is the hybridization of the orbitals around the N atom? (e) How many σ and how many π bonds are there in the molecule?

9.114 Sulfur tetrafluoride (SF_4) reacts slowly with O_2 to form sulfur tetrafluoride monoxide (OSF_4) according to the following unbalanced reaction:

$$SF_4(g) + O_2(g) \longrightarrow OSF_4(g)$$

The O atom and the four F atoms in OSF_4 are bonded to a central S atom. (a) Balance the equation. (b) Write a Lewis structure of OSF_4 in which the formal charges of all atoms are zero. (c) Use average bond enthalpies (Table 8.4) to estimate the enthalpy of the reaction. Is it endothermic or exothermic? (d) Determine the electron-domain geometry of OSF_4, and write two possible molecular geometries for the molecule based on this electron-domain geometry. (e) Which of the molecular geometries in part (d) is more likely to be observed for the molecule? Explain.

[9.115] The phosphorus trihalides (PX_3) show the following variation in the bond angle X—P—X: PF_3, 96.3°; PCl_3, 100.3°; PBr_3, 101.0°; PI_3, 102.0°. The trend is generally attributed to the change in the electronegativity of the halogen. (a) Assuming that all electron domains are the same size, what value of the X—P—X angle is predicted by the VSEPR model? (b) What is the general trend in the X—P—X angle as the halide electronegativity increases? (c) Using the VSEPR model, explain the observed trend in X—P—X angle as the electronegativity of X changes. (d) Based on your answer to part (c), predict the structure of $PBrCl_4$.

[9.116] The molecule 2-butene, C_4H_8, can undergo a geometric change called *cis-trans* isomerization:

cis-2-butene   *trans*-2-butene

As discussed in the "Chemistry and Life" box on the chemistry of vision, such transformations can be induced by light and are the key to human vision. (a) What is the hybridization at the two central carbon atoms of 2-butene? (b) The isomerization occurs by rotation about the central C—C bond. With reference to Figure 9.30, explain why the π bond between the two central carbon atoms is destroyed halfway through the rotation from *cis-* to *trans*-2-butene. (c) Based on average bond enthalpies (Table 8.4), how much energy per molecule must be supplied to break the C—C π bond? (d) What is the longest wavelength of light that will provide photons of sufficient energy to break the C—C π bond and cause the isomerization? (e) Is the wavelength in your answer to part (d) in the visible portion of the electromagnetic spectrum? Comment on the importance of this result for human vision.

9.117 (a) Compare the bond enthalpies (Table 8.4) of the carbon–carbon single, double, and triple bonds to deduce an average π-bond contribution to the enthalpy. What fraction of a single bond does this quantity represent? (b) Make a similar comparison of nitrogen–nitrogen bonds. What do you observe? (c) Write Lewis structures of N_2H_4, N_2H_2, and N_2, and determine the hybridization around nitrogen in each case. (d) Propose a reason for the large difference in your observations of parts (a) and (b).

9.118 Use average bond enthalpies (Table 8.4) to estimate ΔH for the atomization of benzene, C_6H_6:

$$C_6H_6(g) \longrightarrow 6 C(g) + 6 H(g)$$

Compare the value to that obtained by using ΔH_f° data given in Appendix C and Hess's law. To what do you attribute the large discrepancy in the two values?

[9.119] Many compounds of the transition-metal elements contain direct bonds between metal atoms. We will assume that the z-axis is defined as the metal–metal bond axis. (a) Which of the $3d$ orbitals (Figure 6.23) can be used to make a σ bond between metal atoms? (b) Sketch the σ_{3d} bonding and σ_{3d}^{*} antibonding MOs. (c) With reference to the "Closer Look" box on the phases of orbitals, explain why a node is generated in the σ^*_{3d} MO. (**d**) Sketch the energy-level diagram for the Sc_2 molecule, assuming that only the $3d$ orbital from part (a) is important. (**e**) What is the bond order in Sc_2?

[9.120] The organic molecules shown here are derivatives of benzene in which six-membered rings are "fused" at the edges of the hexagons.

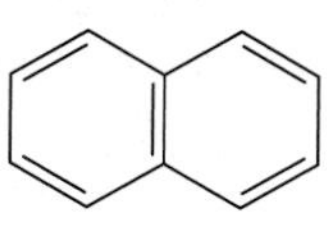

Naphthalene 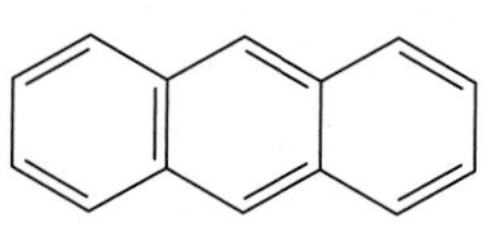   Anthracene

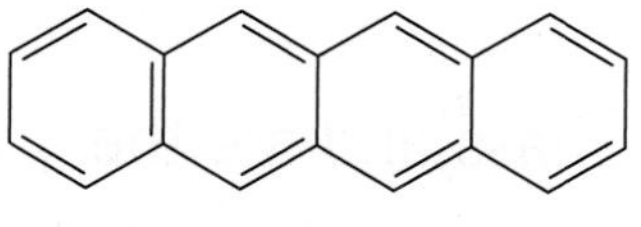

Tetracene

(**a**) Determine the empirical formula of benzene and of these three compounds. (**b**) Suppose you are given a sample of one of the compounds. Could combustion analysis be used to determine unambiguously which of the three it is? (**c**) Naphthalene, the active ingredient in mothballs, is a white solid. Write a balanced equation for the combustion of naphthalene to $CO_2(g)$ and $H_2O(g)$. (**d**) Using the Lewis structure for naphthalene and the average bond enthalpies in Table 8.4, estimate the heat of combustion of naphthalene in kJ/mol. (**e**) Would you expect naphthalene, anthracene, and tetracene to have multiple resonance structures? If so, draw the additional resonance structures for naphthalene. (**f**) Benzene, naphthalene, and anthracene are colorless, but tetracene is orange. What does this imply about the relative HOMO–LUMO energy gaps in these molecules? See the "Chemistry Put to Work" box on orbitals and energy.

[9.121] Antibonding molecular orbitals can be used to make bonds to other atoms in a molecule. For example, metal atoms can use appropriate d orbitals to overlap with the π^*_{2p} orbitals of the carbon monoxide molecule. This is called d-π backbonding. (**a**) Draw a coordinate axis system in which the y-axis is vertical in the plane of the paper and the x-axis horizontal. Write "M" at the origin to denote a metal atom. (**b**) Now, on the x-axis to the right of M, draw the Lewis structure of a CO molecule, with the carbon nearest the M. The CO bond axis should be on the x-axis. (**c**) Draw the CO π^*_{2p} orbital, with phases (see the "Closer Look" box on phases) in the plane of the paper. Two lobes should be pointing toward M. (**d**) Now draw the d_{xy} orbital of M, with phases. Can you see how they will overlap with the π^*_{2p} orbital of CO? (**e**) What kind of bond is being made with the orbitals between M and C, σ or π? (**f**) Predict what will happen to the strength of the CO bond in a metal–CO complex compared to CO alone.

9.122 Methyl isocyanate, CH_3NCO, was made infamous in 1984 when an accidental leakage of this compound from a storage tank in Bhopal, India, resulted in the deaths of about 3,800 people and severe and lasting injury to many thousands more. (**a**) Draw a Lewis structure for methyl isocyanate. (**b**) Draw a ball-and-stick model of the structure, including estimates of all the bond angles in the compound. (**c**) Predict all the bond distances in the molecule. (**d**) Do you predict that the molecule will have a dipole moment? Explain.

Design an Experiment

In this chapter we have seen a number of new concepts, including the delocalization of π systems of molecules and the molecular orbital description of molecular bonding. A connection between these concepts is provided by the field of *organic dyes*, molecules with delocalized π systems that have color. The color is due to the excitation of an electron from the *highest occupied molecular orbital* (HOMO) to the *lowest unoccupied molecular orbital* (LUMO). It is hypothesized that the energy gap between the HOMO and the LUMO depends on the length of the π system. Imagine that you are given samples of the following substances to test this hypothesis:

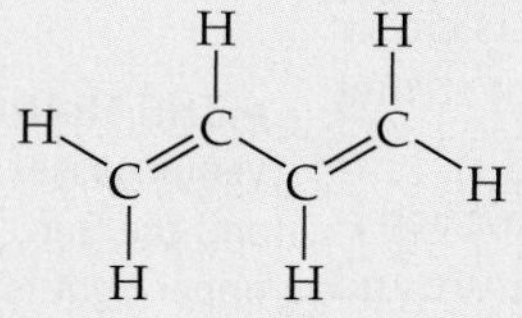

or, in shorthand notation for organic molecules

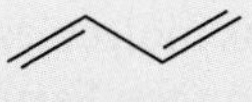

butadiene

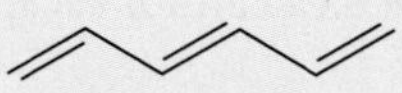

hexatriene

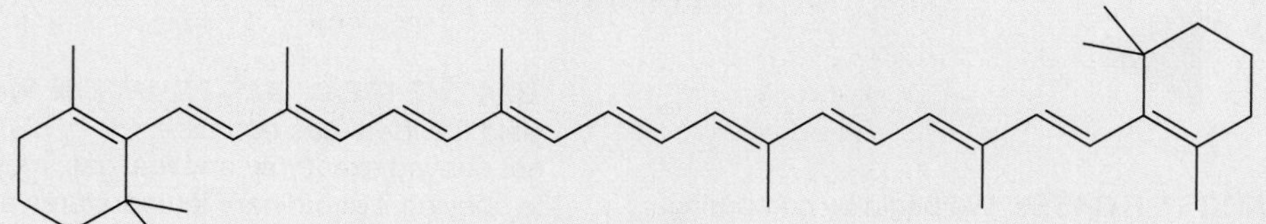

β-carotene

β-carotene is the substance chiefly responsible for the bright orange color of carrots. It is also an important nutrient for the body's production of retinal (see the "Chemistry and Life" box in Section 9.6). (**a**) What experiments could you design to determine the amount of energy needed to excite an electron from the HOMO to the LUMO in each of these molecules? (**b**) How might you graph your data to determine whether a relationship exists between the length of the π system and the excitation energy? (**c**) What additional molecules might you want to procure to further test the ideas developed here? (**d**) How could you design an experiment to determine whether the delocalized π systems and not some other molecular features, such as molecular length or the presence of π bonds, are important in making the excitations occur in the visible portion of the spectrum? (*Hint*: You might want to test some additional molecules not shown here.)

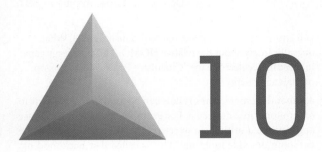

10

Gases

Hydrogen and helium make up over 98% of the Sun's mass and are the most abundant elements in the universe. The small black circle in the upper left is the planet Venus passing between the Sun and the Earth, a rare event that occurred in June of 2012 and will not occur again until 2117. The gases that make up the atmospheres of the inner planets like Venus have a very different composition from that of the Sun.

With the aid of sophisticated measuring instruments, astronomers have been able to discern significant differences in the atmospheres of the planets that make up our solar system. It is thought that the atmospheres of all of the planets initially consisted of largely hydrogen and helium, the two most abundant elements in the universe and the primary components of the Sun. The outer planets, those that are furthest from the Sun—Jupiter, Saturn, Uranus, and Neptune—can still reasonably be described as large balls of hydrogen and helium (which is why they are also known as the "gas giants"). The outer planets contain smaller amounts of other gases as well, including methane, which is responsible for the blue color of Uranus and Neptune.

The atmospheres of the inner planets—Mercury, Venus, Earth, and Mars—have changed dramatically and in quite different ways since the early days of the solar system. As the Sun matured, it became hotter, thereby warming the planetary atmospheres. As we will learn in this chapter, when the temperature of a gas increases its molecules move faster, an effect that becomes more pronounced for gases with low molecular weights, like H_2 and He. The heating effect was largest for the planets closer to the Sun, and the pull of gravity exerted by the smaller inner planets was not strong enough to keep these lightest of gas molecules from escaping into space.

The atmosphere that we find on the Earth today traces its origins to volcanic activity that released gases like CO_2, H_2O, and N_2 into the atmosphere. The temperature and gravity of the Earth are such that these heavier molecules cannot escape. Much of the

▶ **THE TRANSIT OF VENUS** occurs when Venus passes directly between the Earth and the Sun. The small black circle in the upper right is the shadow of Venus passing in front of the Sun.

WHAT'S AHEAD ▶

10.1 CHARACTERISTICS OF GASES We begin by comparing the distinguishing characteristics of gases with those of liquids and solids.

10.2 PRESSURE We study gas *pressure*, the units used to express it, and consider Earth's atmosphere and the pressure it exerts.

10.3 THE GAS LAWS We see that the state of a gas can be expressed in terms of its volume, pressure, temperature, and quantity and examine several *gas laws*, which are empirical relationships among these four variables.

10.4 THE IDEAL-GAS EQUATION We find that the gas laws yield the *ideal-gas equation*, $PV = nRT$. Although this equation is not obeyed exactly by any real gas, most gases come very close to obeying it at ordinary temperatures and pressures.

10.5 FURTHER APPLICATIONS OF THE IDEAL-GAS EQUATION We use the ideal-gas equation in many calculations, such as the calculation of the density or molar mass of a gas.

10.6 GAS MIXTURES AND PARTIAL PRESSURES We recognize that in a mixture of gases, each gas exerts a pressure that is part of the total pressure. This *partial pressure* is the pressure the gas would exert if it were by itself.

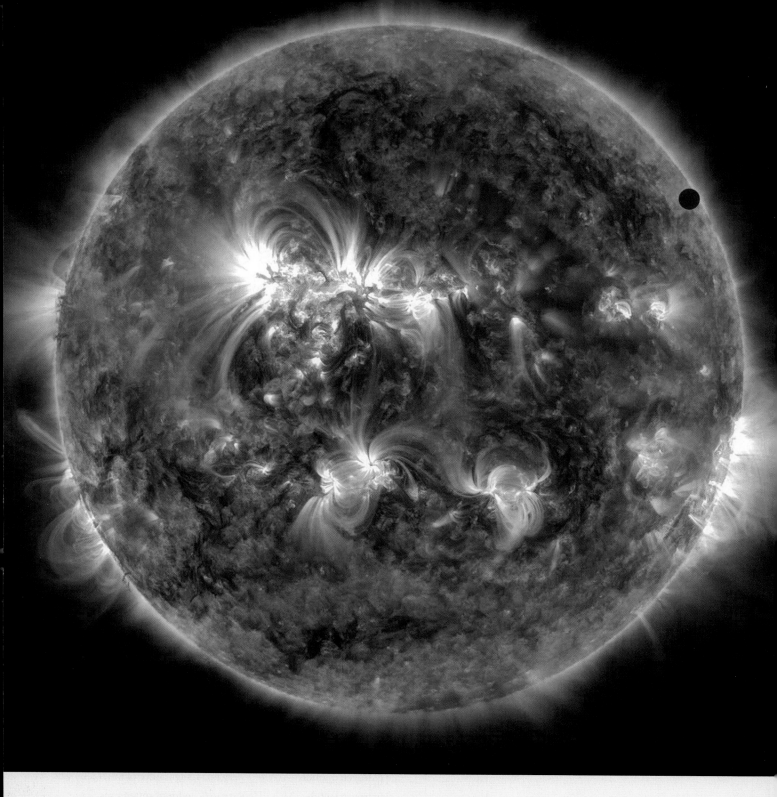

10.7 THE KINETIC-MOLECULAR THEORY OF GASES We see how the behavior of gases at the molecular level is responsible for their macroscopic properties. According to the theory, the atoms or molecules that make up a gas are in constant random motion and move with an average kinetic energy that is proportional to the temperature of the gas.

10.8 MOLECULAR EFFUSION AND DIFFUSION We observe that the kinetic-molecular theory helps account for such gas properties as *effusion* and *diffusion*.

10.9 REAL GASES: DEVIATIONS FROM IDEAL BEHAVIOR We learn that real gases deviate from ideal behavior because the gas molecules have finite volume and because attractive forces exist between molecules. The *van der Waals equation* accounts for real gas behavior at high pressures and low temperatures.

water vapor in early Earth's atmosphere condensed to form the oceans, and much of the CO_2 was dissolved in the ocean and went on to form carbonate minerals, leaving an atmosphere rich in N_2. Once life arose, plant photosynthesis produced O_2 so that now we have an atmosphere that is approximately 78% N_2 and 21% O_2. The hotter conditions on Venus led it down a very different path to a dense atmosphere that is 97% CO_2 with very little O_2 and inhospitable features like sulfuric acid clouds.

In this chapter, we will look more closely at the characteristics of gases as chemical systems. We will explore the macroscopic properties that we associate with gases—such as pressure—and see how these relate to the picture of a gas as a collection of molecules or atoms in random motion.

10.1 | Characteristics of Gases

In many ways, gases are the most easily understood form of matter. Even though different gaseous substances may have very different *chemical* properties, they behave quite similarly as far as their *physical* properties are concerned. For example, the N_2 and O_2 that account for approximately 99% of our atmosphere have very different chemical properties—O_2 supports human life but N_2 does not, to name just one difference—but these two components of air behave physically as one gaseous material because their physical properties are essentially identical. Of the few elements that exist as gases at ordinary temperatures and pressures, He, Ne, Ar, Kr, and Xe are monatomic and H_2, N_2, O_2, F_2, and Cl_2 are diatomic. Many molecular compounds are gases, and ▼ Table 10.1 lists a few of them. Notice that all of these gases are composed entirely of nonmetallic elements. Furthermore, all have simple molecular formulas and, therefore, low molar masses.

Substances that are liquids or solids under ordinary conditions can also exist in the gaseous state, where they are often referred to as **vapors**. The substance H_2O, for example, can exist as liquid water, solid ice, or water vapor.

Gases differ significantly from solids and liquids in several respects. ∞ (Section 1.2) For example, a gas expands spontaneously to fill its container. Consequently, the volume of a gas equals the volume of its container. Gases also are highly compressible: When pressure is applied to a gas, its volume readily decreases. Solids and liquids, on the other hand, do not expand to fill their containers and are not readily compressible.

Two or more gases form a homogeneous mixture regardless of the identities or relative proportions of the gases; the atmosphere serves as an excellent example. Two or more liquids or two or more solids may or may not form homogeneous mixtures, depending on their chemical nature. For example, when water and gasoline are mixed, the two liquids remain as separate layers. In contrast, the water vapor and gasoline vapors above the liquids form a homogeneous gas mixture.

Table 10.1 Some Common Compounds That Are Gases at Room Temperature

Formula	Name	Characteristics
HCN	Hydrogen cyanide	Very toxic, slight odor of bitter almonds
H_2S	Hydrogen sulfide	Very toxic, odor of rotten eggs
CO	Carbon monoxide	Toxic, colorless, odorless
CO_2	Carbon dioxide	Colorless, odorless
CH_4	Methane	Colorless, odorless, flammable
C_2H_4	Ethene (Ethylene)	Colorless, ripens fruit
C_3H_8	Propane	Colorless, odorless, bottled gas
N_2O	Nitrous oxide	Colorless, sweet odor, laughing gas
NO_2	Nitrogen dioxide	Toxic, red-brown, irritating odor
NH_3	Ammonia	Colorless, pungent odor
SO_2	Sulfur dioxide	Colorless, irritating odor

The characteristic properties of gases—expanding to fill a container, being highly compressible, forming homogeneous mixtures—arise because the molecules are relatively far apart. In any given volume of air, for example, the molecules take up only about 0.1% of the total volume with the rest being empty space. Thus, each molecule behaves largely as though the others were not present. As a result, different gases behave similarly even though they are made up of different molecules.

 Give It Some Thought

Xenon is the heaviest stable noble gas with a molar mass of 131 g/mol. Do any of the gases listed in Table 10.1 have molar masses larger than Xe?

10.2 | **Pressure**

In everyday terms, **pressure** conveys the idea of force, a push that tends to move something in a given direction. Pressure, P, is defined in science as the force, F, that acts on a given area, A.

$$P = \frac{F}{A} \qquad [10.1]$$

Gases exert a pressure on any surface with which they are in contact. The gas in an inflated balloon, for example, exerts a pressure on the inside surface of the balloon.

Atmospheric Pressure and the Barometer

People, coconuts, and nitrogen molecules all experience an attractive gravitational force that pulls them toward the center of the Earth. When a coconut comes loose from a tree, for example, this force causes the coconut to be accelerated toward Earth, its speed increasing as its potential energy is converted into kinetic energy. ∞ (Section 5.1) The gas atoms and molecules of the atmosphere also experience a gravitational acceleration. Because these particles have such tiny masses, however, their thermal energies of motion (their kinetic energies) override the gravitational forces, so the particles that make up the atmosphere don't pile up at the Earth's surface. Nevertheless, the gravitational force does operate, and it causes the atmosphere as a whole to press down on the Earth's surface, creating atmospheric pressure, defined as the force exerted by the atmosphere on a given surface area.

You can demonstrate the existence of atmospheric pressure with an empty plastic water bottle. If you suck on the mouth of the empty bottle, chances are you can cause the bottle to partially cave in. When you break the partial vacuum you have created, the bottle pops out to its original shape. The bottle caves in because, once you've sucked out some of the air molecules, the air molecules in the atmosphere exert a force on the outside of the bottle that is greater than the force exerted by the lesser number of air molecules inside the bottle. We calculate the magnitude of this atmospheric pressure as follows: The force, F, exerted by any object is the product of its mass, m, and its acceleration, a: $F = ma$. The acceleration given by Earth's gravitational force to any object located near the Earth's surface is 9.8 m/s^2. Now imagine a column of air, 1 m^2 in cross section, extending through the entire atmosphere (▶ Figure 10.1). That column has a mass of roughly 10,000 kg. The downward gravitational force exerted on this column is

$$F = (10{,}000 \text{ kg})(9.8 \text{ m/s}^2) = 1 \times 10^5 \text{ kg-m/s}^2 = 1 \times 10^5 \text{ N}$$

where N is the abbreviation for *newton*, the SI unit for force: $1 \text{ N} = 1 \text{ kg-m/s}^2$. The pressure exerted by the column is this force divided by the cross-sectional area, A, over which the force is applied. Because our air column has a cross-sectional area of 1 m^2, we have for the magnitude of atmospheric pressure at sea level

$$P = \frac{F}{A} = \frac{1 \times 10^5 \text{ N}}{1 \text{ m}^2} = 1 \times 10^5 \text{ N/m}^2 = 1 \times 10^5 \text{ Pa} = 1 \times 10^2 \text{ kPa}$$

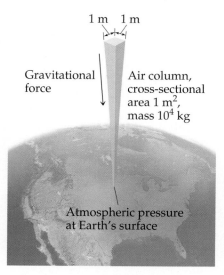

▲ Figure 10.1 **Calculating atmospheric pressure.**

The SI unit of pressure is the **pascal** (Pa), named for Blaise Pascal (1623–1662), a French scientist who studied pressure: $1\ \mathrm{Pa} = 1\ \mathrm{N/m^2}$. A related pressure unit is the **bar:** $1\ \mathrm{bar} = 10^5\ \mathrm{Pa} = 10^5\ \mathrm{N/m^2}$. Thus, the atmospheric pressure at sea level we just calculated, 100 kPa, can be reported as 1 bar. (The actual atmospheric pressure at any location depends on weather conditions and altitude.) Another pressure unit is pounds per square inch (psi, $\mathrm{lbs/in.^2}$). At sea level, atmospheric pressure is 14.7 psi.

 Give It Some Thought

Assume the top of your head has a surface area of 25 cm × 25 cm. How many pounds of air are you carrying on your head if you are at sea level?

In the seventeenth century, many scientists and philosophers believed that the atmosphere had no weight. Evangelista Torricelli (1608–1647), a student of Galileo's, proved this untrue. He invented the *barometer* (◀ Figure 10.2), which is made from a glass tube more than 760 mm long that is closed at one end, completely filled with mercury, and inverted into a dish of mercury. (Care must be taken so that no air gets into the tube.) When the tube is inverted into the dish, some of the mercury flows out of the tube, but a column of mercury remains in the tube. Torricelli argued that the mercury surface in the dish experiences the full force of Earth's atmosphere, which pushes the mercury up the tube until the pressure exerted by the mercury column downward, due to gravity, equals the atmospheric pressure at the base of the tube. Therefore the height, h, of the mercury column is a measure of atmospheric pressure and changes as atmospheric pressure changes.

Although Torricelli's explanation met with fierce opposition, it also had supporters. Blaise Pascal, for example, had one of Torricelli's barometers carried to the top of a mountain and compared its reading there with the reading on a duplicate barometer at the base of the mountain. As the barometer was carried up, the height of the mercury column diminished, as expected, because the amount of atmosphere pressing down on the mercury in the dish decreased as the instrument was carried higher. These and other experiments eventually prevailed, and the idea that the atmosphere has weight became accepted.

Standard atmospheric pressure, which corresponds to the typical pressure at sea level, is the pressure sufficient to support a column of mercury 760 mm high. In SI units, this pressure is 1.01325×10^5 Pa. Standard atmospheric pressure defines some common non-SI units used to express gas pressure, such as the **atmosphere** (atm) and the *millimeter of mercury* (mm Hg). The latter unit is also called the **torr**, after Torricelli: 1 torr = 1 mm Hg. Thus, we have

$$1\ \mathrm{atm} = 760.\ \mathrm{mm\ Hg} = 760.\ \mathrm{torr} = 1.01325 \times 10^5\ \mathrm{Pa} = 101.325\ \mathrm{kPa} = 1.01325\ \mathrm{bar}$$

We will express gas pressure in a variety of units throughout this chapter, so you should become comfortable converting pressures from one unit to another.

 Give It Some Thought

Convert a pressure of 745 torr to a pressure in (**a**) mm Hg, (**b**) atm, (**c**) kPa, (**d**) bar.

SAMPLE EXERCISE 10.1 Torricelli's Barometer

Torricelli used mercury in his barometer because it has a very high density, which makes it possible to make a more compact barometer than one based on a less dense fluid. Calculate the density of mercury, d_{Hg}, using the observation that the column of mercury is 760 mm high when the atmospheric pressure is 1.01×10^5 Pa. Assume the tube containing the mercury is a cylinder with a constant cross-sectional area.

SOLUTION

Analyze Torricelli's barometer is based on the principle that the pressure exerted by the atmosphere is equal to the pressure exerted by the mercury column. The latter quantity originates from the force of gravity acting on the mass of mercury in the column. By setting these two pressures equal to each other and cancelling variables, we can solve for the density of mercury, provided we use appropriate units.

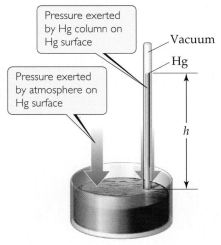

▲ **GO FIGURE**

What happens to h, the height of the mercury column, if the atmospheric pressure increases?

Pressure exerted by Hg column on Hg surface

Vacuum

Hg

Pressure exerted by atmosphere on Hg surface

h

▲ **Figure 10.2 A mercury barometer.**

Plan We will use Equation 10.1 to determine the pressure exerted by the mercury column and look to introduce density as a variable using the fact that $d = m/V$.

Solve We begin by calculating the pressure exerted by the mercury column, using Equation 10.1, and the knowledge that the force exerted by the mercury column is its mass multiplied by the acceleration of gravity near the Earth's surface, $(F = m \times g)$:

$$P_{Hg} = \frac{F}{A} = \frac{m_{Hg}g}{A}$$

The mass of the mercury is equal to its density times the volume of the mercury column. The expression can be simplified by recognizing that the volume of the column can be written in terms of its height and cross-sectional area $(V = h \times A)$, which allows us to cancel the cross-sectional area, which we do not know:

$$P_{Hg} = \frac{m_{Hg}g}{A} = \frac{d_{Hg}Vg}{A} = \frac{d_{Hg}(hA)g}{A} = d_{Hg}hg$$

We now set the pressure of the mercury column equal to atmospheric pressure

$$P_{atm} = P_{Hg} = d_{Hg}hg$$

Finally, we rearrange this expression to solve for d_{Hg} and substitute the appropriate values for the other variables. We can reduce the units of pressure down to base SI units $(Pa = N/m^2 = (kg\text{-}m/s^2)/m^2 = kg/m\text{-}s^2)$:

$$d_{Hg} = \frac{P_{atm}}{hg} = \frac{1.01 \times 10^5\,kg/m\text{-}s^2}{(0.760\,m)(9.81\,m/s^2)} = 1.35 \times 10^4\,kg/m^3$$

Check Water has a density of $1.00\,g/cm^3$, which can be converted to $1000\,kg/m^3$. Our estimate that the density of mercury is 14 times higher than water seems reasonable, given the fact that the molar mass of Hg is approximately 11 times larger than water.

Comment We see from this analysis that the height of the column does not depend upon its cross-sectional area, provided the area does not change along the height of the column.

Practice Exercise 1

What would be the height of the column if the external pressure was 101 kPa and water $(d = 1.00\,g/cm^3)$ was used in place of mercury? (a) 0.0558 m, (b) 0.760 m, (c) 1.03×10^4 m, (d) 10.3 m, (e) 0.103 m.

Practice Exercise 2

Gallium melts just above room temperature and is liquid over a very wide temperature range (30–2204 °C), which means it would be a suitable fluid for a high-temperature barometer. Given its density, $d_{Ga} = 6.0\,g/cm^3$, what would be the height of the column if gallium is used as the barometer fluid and the external pressure is 9.5×10^4 Pa?

We use various devices to measure the pressures of enclosed gases. Tire gauges, for example, measure the pressure of air in automobile and bicycle tires. In laboratories, we sometimes use a *manometer*, which operates on a principle similar to that of a barometer, as shown in Sample Exercise 10.2.

SAMPLE EXERCISE 10.2 Using a Manometer to Measure Gas Pressure

On a certain day, a laboratory barometer indicates that the atmospheric pressure is 764.7 torr. A sample of gas is placed in a flask attached to an open-end mercury manometer (▶ Figure 10.3), and a meter stick is used to measure the height of the mercury in the two arms of the U tube. The height of the mercury in the open ended arm is 136.4 mm, and the height in the arm in contact with the gas in the flask is 103.8 mm. What is the pressure of the gas in the flask (a) in atmospheres, (b) in kilopascals?

SOLUTION

Analyze We are given the atmospheric pressure (764.7 torr) and the mercury heights in the two arms of the manometer and asked to determine the gas pressure in the flask. Recall that millimeters of mercury is a pressure unit. We know that the gas pressure from the flask must be greater than atmospheric pressure because the mercury level in the arm on the flask side (103.8 mm) is lower than the level in the arm open to the atmosphere (136.4 mm). Therefore, the gas from the flask is pushing mercury from the arm in contact with the flask into the arm open to the atmosphere.

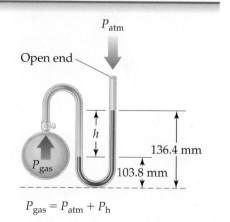

▲ Figure 10.3 **A mercury manometer.**

Plan We will use the difference in height between the two arms (h in Figure 10.3) to obtain the amount by which the pressure of the gas exceeds atmospheric pressure. Because an open-end mercury manometer is used, the height difference directly measures the pressure difference in mm Hg or torr between the gas and the atmosphere.

Solve

(a) The pressure of the gas equals the atmospheric pressure plus h:

$$P_{gas} = P_{atm} + h$$
$$= 764.7 \text{ torr} + (136.4 \text{ torr} - 103.8 \text{ torr})$$
$$= 797.3 \text{ torr}$$

We convert the pressure of the gas to atmospheres:

$$P_{gas} = (797.3 \text{ torr})\left(\frac{1 \text{ atm}}{760 \text{ torr}}\right) = 1.049 \text{ atm}$$

(b) To calculate the pressure in kPa, we employ the conversion factor between atmospheres and kPa:

$$1.049 \text{ atm}\left(\frac{101.3 \text{ kPa}}{1 \text{ atm}}\right) = 106.3 \text{ kPa}$$

Check The calculated pressure is a bit more than 1 atm, which is about 101 kPa. This makes sense because we anticipated that the pressure in the flask would be greater than the atmospheric pressure ($764.7 \text{ torr} = 1.01 \text{ atm}$) acting on the manometer.

Practice Exercise 1

If the gas inside the flask in the above exercise is cooled so that its pressure is reduced to a value of 715.7 torr, what will be the height of the mercury in the open ended arm? (Hint: The sum of the heights in both arms must remain constant regardless of the change in pressure.)
(a) 49.0 mm, (b) 95.6 mm, (c) 144.6 mm, (d) 120.1 mm.

Practice Exercise 2

If the pressure of the gas inside the flask were increased and the height of the column in the open ended arm went up by 5.0 mm, what would be the new pressure of the gas in the flask, in torr?

▲ **GO FIGURE**

Does atmospheric pressure increase or decrease as altitude increases? (Neglect changes in temperature.)

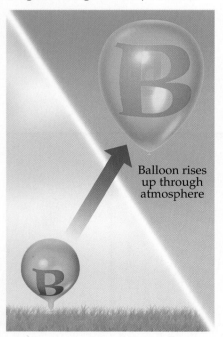

Balloon rises up through atmosphere

▲ **Figure 10.4 As a balloon rises in the atmosphere, its volume increases.**

10.3 | The Gas Laws

Four variables are needed to define the physical condition, or *state*, of a gas: temperature, pressure, volume, and amount of gas, usually expressed as number of moles. The equations that express the relationships among these four variables are known as the *gas laws*. Because volume is easily measured, the first gas laws to be studied expressed the effect of one of the variables on volume, with the remaining two variables held constant.

The Pressure–Volume Relationship: Boyle's Law

An inflated weather balloon released at the Earth's surface expands as it rises (◄ Figure 10.4) because the pressure of the atmosphere decreases with increasing elevation. Thus, for our first pressure–volume relationship we can use our experience with balloons to say that gas volume increases as the pressure exerted on the gas decreases.

British chemist Robert Boyle (1627–1691) first investigated the relationship between the pressure of a gas and its volume, using a J-shaped tube like that shown in Figure 10.5. In the tube on the left, a quantity of gas is trapped above a column of mercury. Boyle surmised that the pressure on the gas could be changed by adding mercury to the tube. He found that the volume of the gas decreased as the pressure increased. He learned, for example, that doubling the pressure will decrease gas volume to half its original value.

Boyle's law, which summarizes these observations, states that *the volume of a fixed quantity of gas maintained at constant temperature is inversely proportional to the pressure*. When two measurements are inversely proportional, one gets smaller as the other gets larger. Boyle's law can be expressed mathematically as

$$V = \text{constant} \times \frac{1}{P} \quad \text{or} \quad PV = \text{constant} \qquad [10.2]$$

The value of the constant depends on temperature and on the amount of gas in the sample.

GO FIGURE

▲ GO FIGURE

What is the total pressure on the gas after the 760 mm Hg has been added?

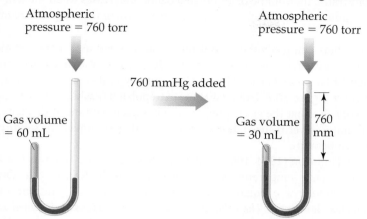

▲ **Figure 10.5 Boyle's experiment relating pressure and volume for a gas.**

The graph of V versus P in ▼ **Figure 10.6** shows the curve obtained for a given quantity of gas at a fixed temperature. A linear relationship is obtained when V is plotted versus $1/P$ as shown on the right in Figure 10.6.

Boyle's law occupies a special place in the history of science because Boyle was the first to carry out experiments in which one variable was systematically changed to determine the effect on another variable. The data from the experiments were then employed to establish an empirical relationship—a "law."

We apply Boyle's law every time we breathe. The rib cage, which can expand and contract, and the diaphragm, a muscle beneath the lungs, govern the volume of the lungs. Inhalation occurs when the rib cage expands and the diaphragm moves downward. Both actions increase the volume of the lungs, thus decreasing the gas pressure inside the lungs. Atmospheric pressure then forces air into the lungs until the pressure in the lungs equals atmospheric pressure. Exhalation reverses the process—the rib cage contracts and the diaphragm moves up, decreasing the volume of the lungs. Air is forced out of the lungs by the resulting increase in pressure.

 Give It Some Thought

What happens to the pressure of a gas in a closed container if you double its volume while its temperature is held constant?

 GO FIGURE

What would a plot of P versus $1/V$ look like for a fixed quantity of gas at a fixed temperature?

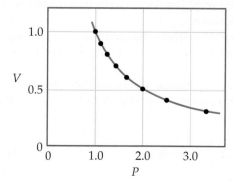
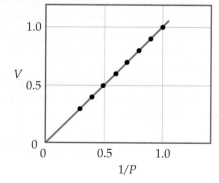

▲ **Figure 10.6 Boyle's Law.** For a fixed quantity of gas at constant temperature, the volume of the gas is inversely proportional to its pressure.

▲ Figure 10.7 **The effect of temperature on volume.**

The Temperature–Volume Relationship: Charles's Law

As ◄ **Figure 10.7** illustrates, the volume of an inflated balloon increases when the temperature of the gas inside the balloon increases and decreases when the temperature of the gas decreases.

The relationship between gas volume and temperature—volume increases as temperature increases and decreases as temperature decreases—was discovered in 1787 by French scientist Jacques Charles (1746–1823). Some typical volume–temperature data are shown in ▼ **Figure 10.8.** Notice that the extrapolated (dashed) line passes through −273 °C. Note also that the gas is predicted to have zero volume at this temperature. This condition is never realized, however, because all gases liquefy or solidify before reaching this temperature.

In 1848, William Thomson (1824–1907), a British physicist whose title was Lord Kelvin, proposed an absolute-temperature scale, now known as the Kelvin scale. On this scale, 0 K, called *absolute zero*, equals −273.15 °C. ∞ (Section 1.4) In terms of the Kelvin scale, **Charles's law** states: *The volume of a fixed amount of gas maintained at constant pressure is directly proportional to its absolute temperature.* Thus, doubling the absolute temperature causes the gas volume to double. Mathematically, Charles's law takes the form

$$V = \text{constant} \times T \quad \text{or} \quad \frac{V}{T} = \text{constant} \qquad [10.3]$$

with the value of the constant depending on the pressure and on the amount of gas.

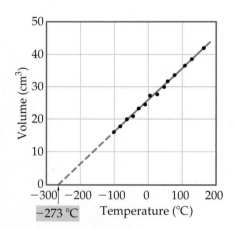

▲ Figure 10.8 **Charles's Law.** For a fixed quantity of gas at constant pressure, the volume of the gas is proportional to its temperature.

▲ **Give It Some Thought**

Does the volume of a fixed quantity of gas decrease to half its original value when the temperature is lowered from 100 °C to 50 °C?

The Quantity–Volume Relationship: Avogadro's Law

The relationship between the quantity of a gas and its volume follows from the work of Joseph Louis Gay-Lussac (1778–1823) and Amedeo Avogadro (1776–1856).

Gay-Lussac was one of those extraordinary figures in the history of science who could truly be called an adventurer. In 1804, he ascended to 23,000 ft in a hot-air balloon—an exploit that held the altitude record for several decades. To better control the balloon, Gay-Lussac studied the properties of gases. In 1808, he observed the *law of combining volumes*: At a given pressure and temperature, the volumes of gases that react with one another are in the ratios of small whole numbers. For example, two volumes of hydrogen gas react with one volume of oxygen gas to form two volumes of water vapor. ∞ (Section 3.1)

Three years later, Amedeo Avogadro interpreted Gay-Lussac's observation by proposing what is now known as **Avogadro's hypothesis**: *Equal volumes of gases at the same temperature and pressure contain equal numbers of molecules.* For example, 22.4 L of any gas at 0 °C and 1 atm contain 6.02×10^{23} gas molecules (that is, 1 mol), as depicted in ▶ Figure 10.9.

Avogadro's law follows from Avogadro's hypothesis: *The volume of a gas maintained at constant temperature and pressure is directly proportional to the number of moles of the gas.* That is,

$$V = \text{constant} \times n \quad \text{or} \quad \frac{V}{n} = \text{constant} \qquad [10.4]$$

where n is number of moles. Thus, for instance, doubling the number of moles of gas causes the volume to double if T and P remain constant.

GO FIGURE

How many moles of gas are in each vessel?

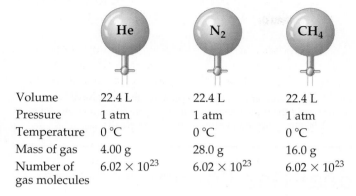

	He	N₂	CH₄
Volume	22.4 L	22.4 L	22.4 L
Pressure	1 atm	1 atm	1 atm
Temperature	0 °C	0 °C	0 °C
Mass of gas	4.00 g	28.0 g	16.0 g
Number of gas molecules	6.02×10^{23}	6.02×10^{23}	6.02×10^{23}

▲ **Figure 10.9 Avogadro's hypothesis.** At the same volume, pressure, and temperature, samples of different gases have the same number of molecules but different masses.

SAMPLE
EXERCISE 10.3 **Evaluating the Effects of Changes in P, V, n, and T on a Gas**

Suppose we have a gas confined to a cylinder with a movable piston that is sealed so there are no leaks. ∞ (Sections 5.2, 5.3) How will each of the following changes affect (i) the pressure of the gas, (ii) the number of moles of gas in the cylinder, (iii) the average distance between molecules: (**a**) Heating the gas while maintaining a constant pressure; (**b**) Reducing the volume while maintaining a constant temperature; (**c**) Injecting additional gas while keeping the temperature and volume constant.

SOLUTION

Analyze We need to think how each change affects (1) the pressure of the gas, (2) the number of moles of gas in the cylinder, and (3) the average distance between molecules.

Plan We can use the gas laws to evaluate the changes in pressure. The number of moles of gas in the cylinder will not change unless gas is either added or removed. Assessing the average distance between molecules is not quite as straightforward. For a given number of gas molecules, the average distance between molecules increases as the volume increases. Conversely, for constant volume, the average distance between molecules decreases as the number of moles increases. Thus the average distance between molecules will be proportional to V/n.

Solve

(**a**) Because it is stipulated that the pressure remains constant, pressure is not a variable in this problem, and the total number of moles of gas will also remain constant. We know from Charles's law, however, that heating the gas while maintaining constant pressure will cause the piston to move and the volume to increase. Thus, the distance between molecules will increase.

(**b**) The reduction in volume causes the pressure to increase (Boyle's law). Compressing the gas into a smaller volume does not change the total number of gas molecules; thus, the total number of moles remains the same. The average distance between molecules, however, must decrease because of the smaller volume.

(**c**) Injecting more gas into the cylinder means that more molecules are present and there will be an increase in the number of moles of gas in the cylinder. Because we have added more molecules while keeping the volume constant the average distance between molecules must decrease. Avogadro's law tells us that the volume of the cylinder should have increased when we added more gas, provided the pressure and temperature were held constant. Here the volume is held constant, as is the temperature, which means the pressure must change. Knowing from Boyle's law that there is an inverse relationship between volume and pressure ($PV =$ constant), we conclude that if the volume does not increase on injecting more gas the pressure must increase.

Practice Exercise 1

A helium balloon is filled to a volume of 5.60 liters at 25 °C. What will the volume of the balloon become if it is put into liquid nitrogen to lower the temperature of the helium to 77 K?
(**a**) 17 L
(**b**) 22 L
(**c**) 1.4 L
(**d**) 0.046 L
(**e**) 3.7 L

Practice Exercise 2

An oxygen cylinder used in a hospital contains 35.4 L of oxygen gas at a pressure of 149.6 atm. How much volume would the oxygen occupy if it were transferred to a container that maintained a pressure of 1.00 atm if the temperature remains constant?

10.4 | The Ideal-Gas Equation

All three laws we just examined were obtained by holding two of the four variables *P*, *V*, *T*, and *n* constant and seeing how the remaining two variables affect each other. We can express each law as a proportionality relationship. Using the symbol ∝ for "is proportional to," we have

$$\textit{Boyle's law:} \quad V \propto \frac{1}{P} \quad (\text{constant } n, T)$$

$$\textit{Charle's law:} \quad V \propto T \quad (\text{constant } n, P)$$

$$\textit{Avogadro's law:} \quad V \propto n \quad (\text{constant } P, T)$$

We can combine these relationships into a general gas law:

$$V \propto \frac{nT}{P}$$

and if we call the proportionality constant *R*, we obtain an equality:

$$V = R\left(\frac{nT}{P}\right)$$

which we can rearrange to

$$PV = nRT \qquad [10.5]$$

which is the **ideal-gas equation** (also called the **ideal-gas law**). An **ideal gas** is a hypothetical gas whose pressure, volume, and temperature relationships are described completely by the ideal-gas equation.

In deriving the ideal-gas equation, we assume (a) that the molecules of an ideal gas do not interact with one another and (b) that the combined volume of the molecules is much smaller than the volume the gas occupies; for this reason, we consider the molecules as taking up no space in the container. In many cases, the small error introduced by these assumptions is acceptable. If more accurate calculations are needed, we can correct for the assumptions if we know something about the attraction molecules have for one another and the size of the molecules.

The term *R* in the ideal-gas equation is the **gas constant**. The value and units of *R* depend on the units of *P*, *V*, *n*, and *T*. The value for *T* in the ideal-gas equation must *always* be the absolute temperature (in kelvins instead of degrees Celsius). The quantity of gas, *n*, is normally expressed in moles. The units chosen for pressure and volume are most often atmospheres and liters, respectively. However, other units can be used. In countries other than the United States, the pascal is the most commonly used unit for pressure. ◀ **Table 10.2** shows the numerical value for *R* in various units. In working with the ideal-gas equation, you must choose the form of *R* in which the units agree with the units of *P*, *V*, *n*, and *T* given in the problem. In this chapter we will most often use $R = 0.08206$ L-atm/mol-K because pressure is most often given in atmospheres.

Table 10.2 Numerical Values of the Gas Constant R in Various Units

Units	Numerical Value
L-atm/mol-K	0.08206
J/mol-K*	8.314
cal/mol-K	1.987
m^3-Pa/mol-K*	8.314
L-torr/mol-K	62.36

*SI unit

GO FIGURE

Which gas deviates most from ideal behavior?

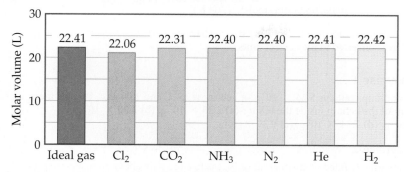

▲ Figure 10.10 **Comparison of molar volumes at STP.**

Suppose we have 1.000 mol of an ideal gas at 1.000 atm and 0.00 °C (273.15 K). According to the ideal-gas equation, the volume of the gas is

$$V = \frac{nRT}{P} = \frac{(1.000\ \text{mol})(0.08206\ \text{L-atm/mol-K})(273.15\ \text{K})}{1.000\ \text{atm}} = 22.41\ \text{L}$$

The conditions 0 °C and 1 atm are referred to as **standard temperature and pressure (STP)**. The volume occupied by 1 mol of ideal gas at STP, 22.41 L, is known as the *molar volume* of an ideal gas at STP.

 Give It Some Thought

If 1.00 mol of an ideal gas at STP were confined to a cube, what would be the length in cm of an edge of this cube?

The ideal-gas equation accounts adequately for the properties of most gases under a variety of circumstances. The equation is not exactly correct, however, for any real gas. Thus, the measured volume for given values of P, n, and T might differ from the volume calculated from $PV = nRT$ (▲ Figure 10.10). Although real gases do not always behave ideally, their behavior differs so little from ideal behavior that we can ignore any deviations for all but the most accurate work.

SAMPLE
EXERCISE 10.4 | Using the Ideal-Gas Equation

Calcium carbonate, $CaCO_3(s)$, the principal compound in limestone, decomposes upon heating to $CaO(s)$ and $CO_2(g)$. A sample of $CaCO_3$ is decomposed, and the carbon dioxide is collected in a 250-mL flask. After decomposition is complete, the gas has a pressure of 1.3 atm at a temperature of 31 °C. How many moles of CO_2 gas were generated?

SOLUTION

Analyze We are given the volume (250 mL), pressure (1.3 atm), and temperature (31 °C) of a sample of CO_2 gas and asked to calculate the number of moles of CO_2 in the sample.

Plan Because we are given V, P, and T, we can solve the ideal-gas equation for the unknown quantity, n.

Solve In analyzing and solving gas law problems, it is helpful to tabulate the information given in the problems and then to convert the values to units that are consistent with those for $R(0.08206\ \text{L-atm/mol-K})$. In this case, the given values are

$$V = 250\ \text{mL} = 0.250\ \text{L}$$
$$P = 1.3\ \text{atm}$$
$$T = 31\ °C = (31 + 273)\ \text{K} = 304\ \text{K}$$

Remember: *Absolute temperature must always be used when the ideal-gas equation is solved.*

We now rearrange the ideal-gas equation (Equation 10.5) to solve for n

$$n = \frac{PV}{RT}$$

$$n = \frac{(1.3\ \text{atm})(0.250\ \text{L})}{(0.08206\ \text{L-atm/mol-K})(304\ \text{K})} = 0.013\ \text{mol}\ CO_2$$

Check Appropriate units cancel, thus ensuring that we have properly rearranged the ideal-gas equation and have converted to the correct units.

Practice Exercise 1

The Goodyear blimp contains 5.74×10^6 L of helium at 25 °C and 1.00 atm. What is the mass in grams of the helium inside the blimp? **(a)** 2.30×10^7 g, **(b)** 2.80×10^6 g, **(c)** 1.12×10^7 g, **(d)** 2.34×10^5 g, **(e)** 9.39×10^5 g.

Practice Exercise 2

Tennis balls are usually filled with either air or N_2 gas to a pressure above atmospheric pressure to increase their bounce. If a tennis ball has a volume of 144 cm^3 and contains 0.33 g of N_2 gas, what is the pressure inside the ball at 24 °C?

Strategies in Chemistry

Calculations Involving Many Variables

In this chapter, we encounter a variety of problems based on the ideal-gas equation, which contains four variables—P, V, n, and T—and one constant, R. Depending on the type of problem, we might need to solve for any of the four variables.

To extract the necessary information from problems involving more than one variable, we suggest the following steps:

1. **Tabulate information.** Read the problems carefully to determine which variable is the unknown and which variables have numeric values given. Every time you encounter a numerical value, jot it down. In many cases, constructing a table of the given information will be useful.

2. **Convert to consistent units.** Make certain that quantities are converted to the proper units. In using the ideal-gas equation, for example, we usually use the value of R that has units of L-atm/mol-K. If you are given a pressure in torr, you will need to convert it to atmospheres before using this value of R in your calculations.

3. **If a single equation relates the variables, solve the equation for the unknown.** For the ideal-gas equation, these algebraic rearrangements will all be used at one time or another:

$$P = \frac{nRT}{V}, \quad V = \frac{nRT}{P}, \quad n = \frac{PV}{RT}, \quad T = \frac{PV}{nR}$$

4. **Use dimensional analysis.** Carry the units through your calculation. Using dimensional analysis enables you to check that you have solved an equation correctly. If the units in the equation cancel to give the units of the desired variable, you have probably used the equation correctly.

Sometimes you will not be given explicit values for several variables, making it look like a problem that cannot be solved. In these cases, however, you should look for information that can be used to determine the needed variables. For example, suppose you are using the ideal-gas equation to calculate a pressure in a problem that gives a value for T but not for n or V. However, the problem states that "the sample contains 0.15 mol of gas per liter." You can turn this statement into the expression

$$\frac{n}{V} = 0.15\ \text{mol/L}$$

Solving the ideal-gas equation for pressure yields

$$P = \frac{nRT}{V} = \left(\frac{n}{V}\right)RT$$

Thus, we can solve the equation even though we are not given values for n and V.

As we have continuously stressed, the most important thing you can do to become proficient at solving chemistry problems is to do the practice exercises and end-of-chapter exercises. By using systematic procedures, such as those described here, you should be able to minimize difficulties in solving problems involving many variables.

Relating the Ideal-Gas Equation and the Gas Laws

The gas laws we discussed in Section 10.3 are special cases of the ideal-gas equation. For example, when n and T are held constant, the product nRT contains three constants and so must itself be a constant:

$$PV = nRT = \text{constant} \quad \text{or} \quad PV = \text{constant} \qquad [10.6]$$

Note that this rearrangement gives Boyle's law. We see that if n and T are constant, the values of P and V can change, but the product PV must remain constant.

We can use Boyle's law to determine how the volume of a gas changes when its pressure changes. For example, if a cylinder fitted with a movable piston holds 50.0 L of O_2 gas at 18.5 atm and 21 °C, what volume will the gas occupy if the temperature is maintained at 21 °C while the pressure is reduced to 1.00 atm? Because the product PV is a constant when a gas is held at constant n and T, we know that

$$P_1V_1 = P_2V_2 \qquad [10.7]$$

where P_1 and V_1 are initial values and P_2 and V_2 are final values. Dividing both sides of this equation by P_2 gives the final volume, V_2:

$$V_2 = V_1 \times \frac{P_1}{P_2} = (50.0 \text{ L})\left(\frac{18.5 \text{ atm}}{1.00 \text{ atm}}\right) = 925 \text{ L}$$

The answer is reasonable because a gas expands as its pressure decreases.

In a similar way, we can start with the ideal-gas equation and derive relationships between any other two variables, V and T (Charles's law), n and V (Avogadro's law), or P and T.

SAMPLE EXERCISE 10.5 | Calculating the Effect of Temperature Changes on Pressure

The gas pressure in an aerosol can is 1.5 atm at 25 °C. Assuming that the gas obeys the ideal-gas equation, what is the pressure when the can is heated to 450 °C?

SOLUTION

Analyze We are given the initial pressure (1.5 atm) and temperature (25 °C) of the gas and asked for the pressure at a higher temperature (450 °C).

Plan The volume and number of moles of gas do not change, so we must use a relationship connecting pressure and temperature. Converting temperature to the Kelvin scale and tabulating the given information, we have

	P	T
Initial	1.5 atm	298 K
Final	P_2	723 K

Solve To determine how P and T are related, we start with the ideal-gas equation and isolate the quantities that do not change (n, V, and R) on one side and the variables (P and T) on the other side.

$$\frac{P}{T} = \frac{nR}{V} = \text{constant}$$

Because the quotient P/T is a constant, we can write

$$\frac{P_1}{T_1} = \frac{P_2}{T_2}$$

(where the subscripts 1 and 2 represent the initial and final states, respectively). Rearranging to solve for P_2 and substituting the given data give

$$P_2 = (1.5 \text{ atm})\left(\frac{723 \text{ K}}{298 \text{ K}}\right) = 3.6 \text{ atm}$$

Check This answer is intuitively reasonable—increasing the temperature of a gas increases its pressure.

Comment It is evident from this example why aerosol cans carry a warning not to incinerate.

Practice Exercise 1

If you fill your car tire to a pressure of 32 psi (pounds per square inch) on a hot summer day when the temperature is 35 °C (95 °F), what is the pressure (in psi) on a cold winter day when the temperature is −15 °C (5 °F)? Assume no gas leaks out between measurements and the volume of the tire does not change. **(a)** 38 psi, **(b)** 27 psi, **(c)** −13.7 psi, **(d)** 1.8 psi, **(e)** 13.7 psi.

Practice Exercise 2

The pressure in a natural-gas tank is maintained at 2.20 atm. On a day when the temperature is −15 °C, the volume of gas in the tank is $3.25 \times 10^3 \text{ m}^3$. What is the volume of the same quantity of gas on a day when the temperature is 31 °C?

We are often faced with the situation in which P, V, and T all change for a fixed number of moles of gas. Because n is constant in this situation, the ideal-gas equation gives

$$\frac{PV}{T} = nR = \text{constant}$$

If we represent the initial and final conditions by subscripts 1 and 2, respectively, we can write

$$\frac{P_1 V_1}{T_1} = \frac{P_2 V_2}{T_2} \qquad [10.8]$$

This equation is often called the *combined gas law*.

SAMPLE EXERCISE 10.6 | Using the Combined Gas Law

An inflated balloon has a volume of 6.0 L at sea level (1.0 atm) and is allowed to ascend until the pressure is 0.45 atm. During ascent, the temperature of the gas falls from 22 °C to −21 °C. Calculate the volume of the balloon at its final altitude.

SOLUTION

Analyze We need to determine a new volume for a gas sample when both pressure and temperature change.

Plan Let's again proceed by converting temperatures to kelvins and tabulating our information.

	P	V	T
Initial	1.0 atm	6.0 L	295 K
Final	0.45 atm	V_2	252 K

Because n is constant, we can use Equation 10.8.

Solve Rearranging Equation 10.8 to solve for V_2 gives

$$V_2 = V_1 \times \frac{P_1}{P_2} \times \frac{T_2}{T_1} = (6.0 \text{ L})\left(\frac{1.0 \text{ atm}}{0.45 \text{ atm}}\right)\left(\frac{252 \text{ K}}{295 \text{ K}}\right) = 11 \text{ L}$$

Check The result appears reasonable. Notice that the calculation involves multiplying the initial volume by a ratio of pressures and a ratio of temperatures. Intuitively, we expect decreasing pressure to cause the volume to increase, while decreasing the temperature should have the opposite effect. Because the change in pressure is more dramatic than the change in temperature, we expect the effect of the pressure change to predominate in determining the final volume, as it does.

> **Practice Exercise 1**
>
> A gas occupies a volume of 0.75 L at 20 °C at 720 torr. What volume would the gas occupy at 41 °C and 760 torr? (**a**) 1.45 L, (**b**) 0.85 L, (**c**) 0.76 L, (**d**) 0.66 L, (**e**) 0.35 L.
>
> **Practice Exercise 2**
>
> A 0.50-mol sample of oxygen gas is confined at 0 °C and 1.0 atm in a cylinder with a movable piston. The piston compresses the gas so that the final volume is half the initial volume and the final pressure is 2.2 atm. What is the final temperature of the gas in degrees Celsius?

10.5 | Further Applications of the Ideal-Gas Equation

In this section, we use the ideal-gas equation first to define the relationship between the density of a gas and its molar mass, and then to calculate the volumes of gases formed or consumed in chemical reactions.

Gas Densities and Molar Mass

Recall that density has units of mass per unit volume ($d = m/V$). ∞ (Section 1.4) We can arrange the ideal-gas equation to obtain similar units of moles per unit volume:

$$\frac{n}{V} = \frac{P}{RT}$$

If we multiply both sides of this equation by the molar mass, $\mathcal{M}$, which is the number of grams in 1 mol of a substance ∞ (Section 3.4), we obtain

$$\frac{n\mathcal{M}}{V} = \frac{P\mathcal{M}}{RT} \qquad [10.9]$$

The term on the left equals the density in grams per liter:

$$\frac{\text{moles}}{\text{liter}} \times \frac{\text{grams}}{\text{mole}} = \frac{\text{grams}}{\text{liter}}$$

Thus, the density of the gas is also given by the expression on the right in Equation 10.9:

$$d = \frac{n\mathcal{M}}{V} = \frac{P\mathcal{M}}{RT} \qquad [10.10]$$

This equation tells us that the density of a gas depends on its pressure, molar mass, and temperature. The higher the molar mass and pressure, the denser the gas. The higher the temperature, the less dense the gas. Although gases form homogeneous mixtures, a less dense gas will lie above a denser gas in the absence of mixing. For example, CO_2 has a higher molar mass than N_2 or O_2 and is therefore denser than air. For this reason, CO_2 released from a CO_2 fire extinguisher blankets a fire, preventing O_2 from reaching the combustible material. "Dry ice," which is solid CO_2, converts directly to CO_2 gas at room temperature, and the resulting "fog" (which is actually condensed water droplets cooled by the CO_2) flows downhill in air (▶ Figure 10.11).

When we have equal molar masses of two gases at the same pressure but different temperatures, the hotter gas is less dense than the cooler one, so the hotter gas rises. The difference between the densities of hot and cold air is responsible for the lift of hot-air balloons. It is also responsible for many phenomena in weather, such as the formation of large thunderheads during thunderstorms.

▲ **Figure 10.11** Carbon dioxide gas flows downward because it is denser than air.

 Give It Some Thought

Is water vapor more or less dense than N_2 under the same conditions of temperature and pressure?

SAMPLE EXERCISE 10.7 | Calculating Gas Density

What is the density of carbon tetrachloride vapor at 714 torr and 125 °C?

SOLUTION

Analyze We are asked to calculate the density of a gas given its name, its pressure, and its temperature. From the name, we can write the chemical formula of the substance and determine its molar mass.

Plan We can use Equation 10.10 to calculate the density. Before we can do that, however, we must convert the given quantities to the appropriate units, degrees Celsius to kelvins and pressure to atmospheres. We must also calculate the molar mass of CCl_4.

Solve The absolute temperature is $125 + 273 = 398$ K. The pressure is $(714 \text{ torr})(1 \text{ atm}/760 \text{ torr}) = 0.939$ atm. The molar mass of CCl_4 is $12.01 + (4)(35.45) = 153.8$ g/mol. Therefore,

$$d = \frac{(0.939 \text{ atm})(153.8 \text{ g/mol})}{(0.08206 \text{ L-atm/mol-K})(398 \text{ K})} = 4.42 \text{ g/L}$$

Check If we divide molar mass (g/mol) by density (g/L), we end up with L/mol. The numerical value is roughly $154/4.4 = 35$, which is in the right ballpark for the molar volume of a gas heated to 125 °C at near atmospheric pressure. We may thus conclude our answer is reasonable.

Practice Exercise 1

What is the density of methane, CH_4, in a vessel where the pressure is 910 torr and the temperature is 255 K? **(a)** 0.92 g/L, **(b)** 697 g/L, **(c)** 0.057 g/L, **(d)** 16 g/L, **(e)** 0.72 g/L.

Practice Exercise 2

The mean molar mass of the atmosphere at the surface of Titan, Saturn's largest moon, is 28.6 g/mol. The surface temperature is 95 K, and the pressure is 1.6 atm. Assuming ideal behavior, calculate the density of Titan's atmosphere.

Equation 10.10 can be rearranged to solve for the molar mass of a gas:

$$\mathcal{M} = \frac{dRT}{P} \qquad [10.11]$$

Thus, we can use the experimentally measured density of a gas to determine the molar mass of the gas molecules, as shown in Sample Exercise 10.8.

SAMPLE EXERCISE 10.8 Calculating the Molar Mass of a Gas

A large evacuated flask initially has a mass of 134.567 g. When the flask is filled with a gas of unknown molar mass to a pressure of 735 torr at 31 °C, its mass is 137.456 g. When the flask is evacuated again and then filled with water at 31 °C, its mass is 1067.9 g. (The density of water at this temperature is 0.997 g/mL.) Assuming the ideal-gas equation applies, calculate the molar mass of the gas.

SOLUTION

Analyze We are given the temperature (31 °C) and pressure (735 torr) for a gas, together with information to determine its volume and mass, and we are asked to calculate its molar mass.

Plan The data obtained when the flask is filled with water can be used to calculate the volume of the container. The mass of the empty flask and of the flask when filled with gas can be used to calculate the mass of the gas. From these quantities we calculate the gas density and then apply Equation 10.11 to calculate the molar mass of the gas.

Solve The gas volume equals the volume of water the flask can hold, calculated from the mass and density of the water. The mass of the water is the difference between the masses of the full and evacuated flask:

$$1067.9 \text{ g} - 134.567 \text{ g} = 933.3 \text{ g}$$

Rearranging the equation for density ($d = m/V$), we have

$$V = \frac{m}{d} = \frac{(933.3 \text{ g})}{(0.997 \text{ g/mL})} = 936 \text{ mL} = 0.936 \text{ L}$$

The gas mass is the difference between the mass of the flask filled with gas and the mass of the evacuated flask:

$$137.456 \text{ g} - 134.567 \text{ g} = 2.889 \text{ g}$$

Knowing the mass of the gas (2.889 g) and its volume (0.936 L), we can calculate the density of the gas:

$$d = 2.889 \text{ g}/0.936 \text{ L} = 3.09 \text{ g/L}$$

After converting pressure to atmospheres and temperature to kelvins, we can use Equation 10.11 to calculate the molar mass:

$$\mathcal{M} = \frac{dRT}{P}$$
$$= \frac{(3.09 \text{ g/L})(0.08206 \text{ L-atm/mol-K})(304 \text{ K})}{(0.967 \text{ atm})}$$
$$= 79.7 \text{ g/mol}$$

Check The units work out appropriately, and the value of molar mass obtained is reasonable for a substance that is gaseous near room temperature.

Practice Exercise 1

What is the molar mass of an unknown hydrocarbon whose density is measured to be 1.97 g/L at STP? **(a)** 4.04 g/mol, **(b)** 30.7 g/mol, **(c)** 44.1 g/mol, **(d)** 48.2 g/mol.

Practice Exercise 2

Calculate the average molar mass of dry air if it has a density of 1.17 g/L at 21 °C and 740.0 torr.

Volumes of Gases in Chemical Reactions

We are often concerned with knowing the identity and/or quantity of a gas involved in a chemical reaction. Thus, it is useful to be able to calculate the volumes of gases consumed or produced in reactions. Such calculations are based on the mole concept and balanced chemical equations. ∞ (Section 3.6) The coefficients in a balanced chemical equation tell us the relative amounts (in moles) of reactants and products in a reaction. The ideal-gas equation relates the number of moles of a gas to P, V, and T.

SAMPLE
EXERCISE 10.9 | Relating Gas Variables and Reaction Stoichiometry

Automobile air bags are inflated by nitrogen gas generated by the rapid decomposition of
sodium azide, NaN_3:

$$2\,NaN_3(s) \longrightarrow 2\,Na(s) + 3\,N_2(g)$$

If an air bag has a volume of 36 L and is to be filled with nitrogen gas at 1.15 atm and 26 °C, how
many grams of NaN_3 must be decomposed?

SOLUTION

Analyze This is a multistep problem. We are given the volume, pressure, and temperature of the
N_2 gas and the chemical equation for the reaction by which the N_2 is generated. We must use this
information to calculate the number of grams of NaN_3 needed to obtain the necessary N_2.

Plan We need to use the gas data (P, V, and T) and the ideal-gas equation to calculate the
number of moles of N_2 gas that should be formed for the air bag to operate correctly. We can
then use the balanced equation to determine the number of moles of NaN_3 needed. Finally, we
can convert moles of NaN_3 to grams.

Solve The number of moles of N_2 is determined
using the ideal-gas equation:

$$n = \frac{PV}{RT} = \frac{(1.15\ atm)(36\ L)}{(0.08206\ \text{L-atm/mol-K})(299\ K)} = 1.69\ mol\ N_2$$

We use the coefficients in the balanced equation to
calculate the number of moles of NaN_3:

$$(1.69\ mol\ N_2)\left(\frac{2\ mol\ NaN_3}{3\ mol\ N_2}\right) = 1.12\ mol\ NaN_3$$

Finally, using the molar mass of NaN_3, we convert
moles of NaN_3 to grams:

$$(1.12\ mol\ NaN_3)\left(\frac{65.0\ g\ NaN_3}{1\ mol\ NaN_3}\right) = 73\ g\ NaN_3$$

Check The units cancel properly at each step in the calculation, leaving us with the correct units in the
answer, g NaN_3. The number of significant figures is limited to be two because we only know the volume
to two significant figures. We keep more figures in the intermediate steps to avoid rounding errors.

Practice Exercise 1

When silver oxide is heated, it decomposes according to the reaction:

$$2\,Ag_2O(s) \xrightarrow{\Delta} 4\,Ag(s) + O_2(g)$$

If 5.76 g of Ag_2O is heated and the O_2 gas produced by the reaction is collected in an evacuated
flask, what is the pressure of the O_2 gas if the volume of the flask is 0.65 L and the gas temperature is 25 °C? **(a)** 0.94 atm, **(b)** 0.039 atm, **(c)** 0.012 atm, **(d)** 0.47 atm, **(e)** 3.2 atm.

Practice Exercise 2

In the first step of the industrial process for making nitric acid, ammonia reacts with oxygen in
the presence of a suitable catalyst to form nitric oxide and water vapor:

$$4\,NH_3(g) + 5\,O_2(g) \longrightarrow 4\,NO(g) + 6\,H_2O(g)$$

How many liters of $NH_3(g)$ at 850 °C and 5.00 atm are required to react with 1.00 mol of $O_2(g)$
in this reaction?

10.6 | Gas Mixtures and Partial Pressures

Thus far we have considered mainly pure gases—those that consist of only one
substance in the gaseous state. How do we deal with mixtures of two or more different gases? While studying the properties of air, John Dalton ∞ (Section 2.1) made
an important observation: *The total pressure of a mixture of gases equals the sum of the
pressures that each would exert if it were present alone.* The pressure exerted by a particular component of a mixture of gases is called the **partial pressure** of that component.
Dalton's observation is known as **Dalton's law of partial pressures**.

If we let P_t be the total pressure of a mixture of gases and P_1, P_2, P_3, and so forth be the partial pressures of the individual gases, we can write Dalton's law of partial pressures as

$$P_t = P_1 + P_2 + P_3 + \ldots \qquad [10.12]$$

This equation implies that each gas behaves independently of the others, as we can see by the following analysis. Let n_1, n_2, n_3, and so forth be the number of moles of each of the gases in the mixture and n_t be the total number of moles of gas. If each gas obeys the ideal-gas equation, we can write

$$P_1 = n_1\left(\frac{RT}{V}\right); \quad P_2 = n_2\left(\frac{RT}{V}\right); \quad P_3 = n_3\left(\frac{RT}{V}\right); \quad \text{and so forth}$$

All of the gases in a container must occupy the same volume and will come to the same temperature in a relatively short period of time. Using these facts to simplify Equation 10.12, we obtain

$$P_t = (n_1 + n_2 + n_3 + \ldots)\left(\frac{RT}{V}\right) = n_t\left(\frac{RT}{V}\right) \qquad [10.13]$$

That is, at constant temperature and constant volume the total pressure of a gas sample is determined by the total number of moles of gas present, whether that total represents just one gas or a mixture of gases.

 Give It Some Thought

How is the partial pressure exerted by N_2 gas affected when some O_2 is introduced into a container if the temperature and volume remain constant? How is the total pressure affected?

SAMPLE EXERCISE 10.10 | Applying Dalton's Law of Partial Pressures

A mixture of 6.00 g of $O_2(g)$ and 9.00 g of $CH_4(g)$ is placed in a 15.0-L vessel at 0 °C. What is the partial pressure of each gas, and what is the total pressure in the vessel?

SOLUTION

Analyze We need to calculate the pressure for two gases in the same volume and at the same temperature.

Plan Because each gas behaves independently, we can use the ideal-gas equation to calculate the pressure each would exert if the other were not present. Per Dalton's law, the total pressure is the sum of these two partial pressures.

Solve We first convert the mass of each gas to moles:

$$n_{O_2} = (6.00 \text{ g } O_2)\left(\frac{1 \text{ mol } O_2}{32.0 \text{ g } O_2}\right) = 0.188 \text{ mol } O_2$$

$$n_{CH_4} = (9.00 \text{ g } CH_4)\left(\frac{1 \text{ mol } CH_4}{16.0 \text{ g } CH_4}\right) = 0.563 \text{ mol } CH_4$$

We use the ideal-gas equation to calculate the partial pressure of each gas:

$$P_{O_2} = \frac{n_{O_2}RT}{V} = \frac{(0.188 \text{ mol})(0.08206 \text{ L-atm/mol-K})\,(273 \text{ K})}{15.0 \text{ L}} = 0.281 \text{ atm}$$

$$P_{CH_4} = \frac{n_{CH_4}RT}{V} = \frac{(0.563 \text{ mol})(0.08206 \text{ L-atm/mol-K})(273 \text{ K})}{15.0 \text{ L}} = 0.841 \text{ atm}$$

According to Dalton's law of partial pressures (Equation 10.12), the total pressure in the vessel is the sum of the partial pressures:

$$P_t = P_{O_2} + P_{CH_4} = 0.281 \text{ atm} + 0.841 \text{ atm} = 1.122 \text{ atm}$$

Check A pressure of roughly 1 atm seems right for a mixture of about 0.2 mol O_2 and a bit more than 0.5 mol CH_4, together in a 15-L volume, because 1 mol of an ideal gas at 1 atm pressure and 0 °C occupies about 22 L.

Practice Exercise 1

A 15-L cylinder contains 4.0 g of hydrogen and 28 g of nitrogen. If the temperature is 27 °C what is the total pressure of the mixture? (a) 0.44 atm, (b) 1.6 atm, (c) 3.3 atm, (d) 4.9 atm, (e) 9.8 atm.

Practice Exercise 2

What is the total pressure exerted by a mixture of 2.00 g of $H_2(g)$ and 8.00 g of $N_2(g)$ at 273 K in a 10.0-L vessel?

Partial Pressures and Mole Fractions

Because each gas in a mixture behaves independently, we can relate the amount of a given gas in a mixture to its partial pressure. For an ideal gas, we can write

$$\frac{P_1}{P_t} = \frac{n_1 RT/V}{n_t\, RT/V} = \frac{n_1}{n_t} \qquad [10.14]$$

The ratio n_1/n_t is called the *mole fraction of gas 1*, which we denote X_1. The **mole fraction**, X, is a dimensionless number that expresses the ratio of the number of moles of one component in a mixture to the total number of moles in the mixture. Thus, for gas 1 we have

$$X_1 = \frac{\text{Moles of compound 1}}{\text{Total moles}} = \frac{n_1}{n_t} \qquad [10.15]$$

We can combine Equations 10.14 and 10.15 to give

$$P_1 = \left(\frac{n_1}{n_t}\right) P_t = X_1 P_t \qquad [10.16]$$

The mole fraction of N_2 in air is 0.78—that is, 78% of the molecules in air are N_2. This means that if the barometric pressure is 760 torr, the partial pressure of N_2 is

$$P_{N_2} = (0.78)(760 \text{ torr}) = 590 \text{ torr}$$

This result makes intuitive sense: Because N_2 makes up 78% of the mixture, it contributes 78% of the total pressure.

SAMPLE EXERCISE 10.11 | **Relating Mole Fractions and Partial Pressures**

A study of the effects of certain gases on plant growth requires a synthetic atmosphere composed of 1.5 mol % CO_2, 18.0 mol % O_2, and 80.5 mol % Ar. (a) Calculate the partial pressure of O_2 in the mixture if the total pressure of the atmosphere is to be 745 torr. (b) If this atmosphere is to be held in a 121-L space at 295 K, how many moles of O_2 are needed?

SOLUTION

Analyze For (a) we need to calculate the partial pressure of O_2 given its mole percent and the total pressure of the mixture. For (b) we need to calculate the number of moles of O_2 in the mixture given its volume (121 L), temperature (745 torr), and partial pressure from part (a).

Plan We calculate the partial pressures using Equation 10.16, and then use P_{O_2}, V, and T in the ideal-gas equation to calculate the number of moles of O_2.

Solve

(a) The mole percent is the mole fraction times 100. Therefore, the mole fraction of O_2 is 0.180. Equation 10.16 gives

$$P_{O_2} = (0.180)(745 \text{ torr}) = 134 \text{ torr}$$

(b) Tabulating the given variables and converting to appropriate units, we have

$$P_{O_2} = (134\ \text{torr})\left(\frac{1\ \text{atm}}{760\ \text{torr}}\right) = 0.176\ \text{atm}$$

$$V = 121\ \text{L}$$

$$n_{O_2} = ?$$

$$R = 0.08206\ \frac{\text{L-atm}}{\text{mol-K}}$$

$$T = 295\ \text{K}$$

Solving the ideal-gas equation for n_{O_2}, we have

$$n_{O_2} = P_{O_2}\left(\frac{V}{RT}\right)$$

$$= (0.176\ \text{atm})\frac{121\ \text{L}}{(0.08206\ \text{L-atm}/\text{mol-K})(295\ \text{K})} = 0.880\ \text{mol}$$

Check The units check out, and the answer seems to be the right order of magnitude.

Practice Exercise 1

A 4.0-L vessel containing N_2 at STP and a 2.0-L vessel containing H_2 at STP are connected by a valve. If the valve is opened allowing the two gases to mix, what is the mole fraction of hydrogen in the mixture? **(a)** 0.034, **(b)** 0.33, **(c)** 0.50, **(d)** 0.67, **(e)** 0.96.

Practice Exercise 2

From data gathered by *Voyager 1*, scientists have estimated the composition of the atmosphere of Titan, Saturn's largest moon. The pressure on the surface of Titan is 1220 torr. The atmosphere consists of 82 mol % N_2, 12 mol % Ar, and 6.0 mol % CH_4. Calculate the partial pressure of each gas.

10.7 | The Kinetic-Molecular Theory of Gases

The ideal-gas equation describes *how* gases behave but not *why* they behave as they do. Why does a gas expand when heated at constant pressure? Or why does its pressure increase when the gas is compressed at constant temperature? To understand the physical properties of gases, we need a model that helps us picture what happens to gas particles when conditions such as pressure or temperature change. Such a model, known as the **kinetic-molecular theory of gases**, was developed over a period of about 100 years, culminating in 1857 when Rudolf Clausius (1822–1888) published a complete and satisfactory form of the theory.

The kinetic-molecular theory (the theory of moving molecules) is summarized by the following statements:

1. Gases consist of large numbers of molecules that are in continuous, random motion. (The word *molecule* is used here to designate the smallest particle of any gas even though some gases, such as the noble gases, consist of individual atoms. All we learn about gas behavior from the kinetic-molecular theory applies equally to atomic gases.)

2. The combined volume of all the molecules of the gas is negligible relative to the total volume in which the gas is contained.

3. Attractive and repulsive forces between gas molecules are negligible.

4. Energy can be transferred between molecules during collisions but, as long as temperature remains constant, the *average* kinetic energy of the molecules does not change with time.

5. The average kinetic energy of the molecules is proportional to the absolute temperature. At any given temperature, the molecules of all gases have the same average kinetic energy.

The kinetic-molecular theory explains both pressure and temperature at the molecular level. The pressure of a gas is caused by collisions of the molecules with the walls of the container (▶ Figure 10.12). The magnitude of the pressure is determined by how often and how forcefully the molecules strike the walls.

The absolute temperature of a gas is a measure of the *average* kinetic energy of its molecules. If two gases are at the same temperature, their molecules have the same average kinetic energy (statement 5 of the kinetic-molecular theory). If the absolute temperature of a gas is doubled, the average kinetic energy of its molecules doubles. Thus, molecular motion increases with increasing temperature.

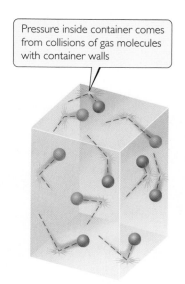

Pressure inside container comes from collisions of gas molecules with container walls

▲ **Figure 10.12 The molecular origin of gas pressure.**

Distributions of Molecular Speed

Although collectively the molecules in a sample of gas have an *average* kinetic energy and hence an average speed, the individual molecules are moving at different speeds. Each molecule collides frequently with other molecules. Momentum is conserved in each collision, but one of the colliding molecules might be deflected off at high speed while the other is nearly stopped. The result is that, at any instant, the molecules in the sample have a wide range of speeds. In ▼ Figure 10.13(a), which shows the distribution of molecular speeds for nitrogen gas at 0 °C and 100 °C, we see that a larger fraction of the 100 °C molecules moves at the higher speeds. This means that the 100 °C sample has the higher average kinetic energy.

In any graph of the distribution of molecular speeds in a gas sample, the peak of the curve represents the most probable speed, u_{mp} [Figure 10.13(**b**)]. The most probable speeds in Figure 10.13(a), for instance, are 4×10^2 m/s for the 0 °C sample and 5×10^2 m/s for the 100 °C sample. Figure 10.13(b) also shows the **root-mean-square (rms) speed**, u_{rms}, of the molecules. This is the speed of a molecule possessing a kinetic energy identical to the average kinetic energy of the sample. The rms speed is not quite the same as the average (mean) speed, u_{av}. The difference between the two is

▲ GO FIGURE

Estimate the fraction of molecules at 100 °C with speeds less than 300 m/s.

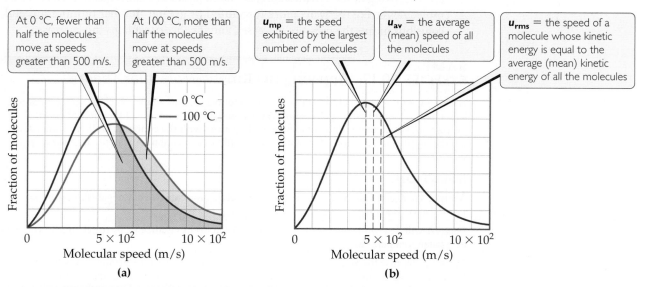

At 0 °C, fewer than half the molecules move at speeds greater than 500 m/s.

At 100 °C, more than half the molecules move at speeds greater than 500 m/s.

u_{mp} = the speed exhibited by the largest number of molecules

u_{av} = the average (mean) speed of all the molecules

u_{rms} = the speed of a molecule whose kinetic energy is equal to the average (mean) kinetic energy of all the molecules

▲ **Figure 10.13 Distribution of molecular speeds for nitrogen gas. (a)** The effect of temperature on molecular speed. The relative area under the curve for a range of speeds gives the relative fraction of molecules that have those speeds. **(b)** Position of most probable (u_{mp}), average (u_{av}), and root-mean-square (u_{rms}) speeds of gas molecules. The data shown here are for nitrogen gas at 0 °C.

small, however. In Figure 10.13(b), for example, the root-mean-square speed is about 4.9×10^2 m/s and the average speed is about 4.5×10^2 m/s.

If you calculate the rms speeds (as we will in Section 10.8), you will find that the rms speed is almost 6×10^2 m/s for the 100 °C sample but slightly less than 5×10^2 m/s for the 0 °C sample. Notice that the distribution curve broadens as we go to a higher temperature, which tells us that the range of molecular speeds increases with temperature.

The rms speed is important because the average kinetic energy of the gas molecules in a sample is equal to $\frac{1}{2} m(u_{rms})^2$. (Section 5.1) Because mass does not change with temperature, the increase in the average kinetic energy $\frac{1}{2} m(u_{rms})^2$ as the temperature increases implies that the rms speed of the molecules (as well as u_{av} and u_{mp}) increases as temperature increases.

▲ **Give It Some Thought**

Consider three gases all at 298 K: HCl, H_2, and O_2. List the gases in order of increasing average speed.

Application of Kinetic-Molecular Theory to the Gas Laws

The empirical observations of gas properties as expressed by the various gas laws are readily understood in terms of the kinetic-molecular theory. The following examples illustrate this point:

1. **An increase in volume at constant temperature causes pressure to decrease.** A constant temperature means that the average kinetic energy of the gas molecules remains unchanged. This means that the rms speed of the molecules remains unchanged. When the volume is increased, the molecules must move a longer distance between collisions. Consequently, there are fewer collisions per unit time with the container walls, which means the pressure decreases. Thus, kinetic-molecular theory explains Boyle's law.

2. **A temperature increase at constant volume causes pressure to increase.** An increase in temperature means an increase in the average kinetic energy of the molecules and in u_{rms}. Because there is no change in volume, the temperature increase causes more collisions with the walls per unit time because the molecules are all moving faster. Furthermore, the momentum in each collision increases (the molecules strike the walls more forcefully). A greater number of more forceful collisions means the pressure increases, and the theory explains this increase.

SAMPLE
EXERCISE 10.12 | Applying the Kinetic-Molecular Theory

A sample of O_2 gas initially at STP is compressed to a smaller volume at constant temperature. What effect does this change have on (**a**) the average kinetic energy of the molecules, (**b**) their average speed, (**c**) the number of collisions they make with the container walls per unit time, (**d**) the number of collisions they make with a unit area of container wall per unit time, (**e**) the pressure?

SOLUTION

Analyze We need to apply the concepts of the kinetic-molecular theory of gases to a gas compressed at constant temperature.

Plan We will determine how each of the quantities in (a)–(e) is affected by the change in volume at constant temperature.

Solve (**a**) Because the average kinetic energy of the O_2 molecules is determined only by temperature, this energy is unchanged by the compression. (**b**) Because the average kinetic energy of the molecules does not change, their average speed remains constant. (**c**) The number of collisions with the walls per unit time increases because the molecules are moving in a smaller volume but with the same average speed as before. Under these conditions they will strike the walls of the container more frequently. (**d**) The number of collisions with a unit area of wall per unit time increases because the total number of collisions with the walls per unit time increases and the area of the walls decreases. (**e**) Although the average force with which the molecules collide with the walls remains constant, the pressure increases because there are more collisions per unit area of wall per unit time.

Check In a conceptual exercise of this kind, there is no numerical answer to check. All we can check in such cases is our reasoning in the course of solving the problem. The increase in pressure seen in part (**e**) is consistent with Boyle's law.

Practice Exercise 1

Consider two gas cylinders of the same volume and temperature, one containing 1.0 mol of propane, C_3H_8, and the other 2.0 mol of methane, CH_4. Which of the following statements is true? (**a**) The C_3H_8 and CH_4 molecules have the same u_{rms}, (**b**) The C_3H_8 and CH_4 molecules have the same average kinetic energy, (**c**) The rate at which the molecules collide with the cylinder walls is the same for both cylinders, (**d**) The gas pressure is the same in both cylinders.

Practice Exercise 2

How is the rms speed of N_2 molecules in a gas sample changed by (**a**) an increase in temperature, (**b**) an increase in volume, (**c**) mixing with a sample of Ar at the same temperature?

A Closer Look

The Ideal-Gas Equation

The ideal-gas equation can be derived from the five statements given in the text for the kinetic-molecular theory. Rather than perform the derivation, however, let's consider in qualitative terms how the ideal-gas equation might follow from these statements. The total force of the molecular collisions on the walls and hence the pressure (force per unit area, Section 10.2) produced by these collisions depend both on how strongly the molecules strike the walls (impulse imparted per collision) and on the rate at which the collisions occur:

$$P \propto \text{impulse imparted per collision} \times \text{collision rate}$$

For a molecule traveling at the rms speed, the impulse imparted by a collision with a wall depends on the momentum of the molecule; that is, it depends on the product of the molecule's mass and speed: mu_{rms}. The collision rate is proportional to the number of molecules per unit volume, n/V, and to their speed, which is u_{rms} because we are talking about only molecules traveling at this speed. Thus, we have

$$P \propto mu_{rms} \times \frac{n}{V} \times u_{rms} \propto \frac{nm(u_{rms})^2}{V} \qquad [10.17]$$

Because the average kinetic energy, $\frac{1}{2}m(u_{rms})^2$, is proportional to temperature, we have $m(u_{rms})^2 \propto T$. Making this substitution in Equation 10.17 gives

$$P \propto \frac{nm(u_{rms})^2}{V} \propto \frac{nT}{V} \qquad [10.18]$$

If we put in a proportionality constant, calling it R, the gas constant, you can see that we obtain the ideal-gas equation:

$$P = \frac{nRT}{V} \qquad [10.19]$$

Swiss mathematician Daniel Bernoulli (1700–1782) conceived of a model for gases that was, for all practical purposes, the same as the model described by the kinetic-molecular theory of gases. From this model, Bernoulli derived Boyle's law and the ideal-gas equation. His was one of the first examples in science of developing a mathematical model from a set of assumptions, or hypothetical statements. However, Bernoulli's work on this subject was completely ignored, only to be rediscovered a hundred years later by Clausius and others. It was ignored because it conflicted with popular beliefs and was in conflict with Isaac Newton's incorrect model for gases. Those idols of the times had to fall before the way was clear for the kinetic-molecular theory. As this story illustrates, science is not a straight road running from here to the "truth." The road is built by humans, so it zigs and zags.

Related Exercises: 10.75, 10.76, 10.77, 10.78

10.8 | Molecular Effusion and Diffusion

According to the kinetic-molecular theory of gases, the average kinetic energy of *any* collection of gas molecules, $\frac{1}{2}m(u_{rms})^2$, has a specific value at a given temperature. Thus, for two gases at the same temperature a gas composed of low-mass particles, such as He, has the same average kinetic energy as one composed of more massive particles, such as Xe. The mass of the particles in the He sample is smaller than that in the

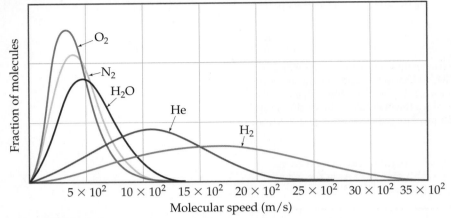

▲ **GO FIGURE**
Which of these gases has the largest molar mass? Which has the smallest?

▲ Figure 10.14 **The effect of molar mass on molecular speed at 25 °C.**

Xe sample. Consequently, the He particles must have a higher rms speed than the Xe particles. The equation that expresses this fact quantitatively is

$$u_{rms} = \sqrt{\frac{3RT}{\mathcal{M}}} \qquad [10.20]$$

where $\mathcal{M}$ is the molar mass of the particles, which can be derived from the kinetic-molecular theory. Because $\mathcal{M}$ appears in the denominator, the less massive the gas particles, the higher their rms speed.

▲ Figure 10.14 shows the distribution of molecular speeds for several gases at 25 °C. Notice how the distributions are shifted toward higher speeds for gases of lower molar masses.

The most probable speed of a gas molecule can also be derived:

$$u_{mp} = \sqrt{\frac{2RT}{\mathcal{M}}} \qquad [10.21]$$

▲ **Give It Some Thought**

What is the ratio of u_{rms} to u_{mp} for a sample of $O_2(g)$ at 300 K? Will this ratio change as the temperature changes? Will it be different for a different gas?

SAMPLE EXERCISE 10.13 | Calculating a Root-Mean-Square Speed

Calculate the rms speed of the molecules in a sample of N_2 gas at 25 °C.

SOLUTION

Analyze We are given the identity of a gas and the temperature, the two quantities we need to calculate the rms speed.

Plan We calculate the rms speed using Equation 10.20.

Solve We must convert each quantity in our equation to SI units. We will also use R in units of J/mol-K (Table 10.2) to make the units cancel correctly.

$$T = 25 + 273 = 298 \text{ K}$$
$$\mathcal{M} = 28.0 \text{ g/mol} = 28.0 \times 10^{-3} \text{ kg/mol}$$
$$R = 8.314 \text{ J/mol-K} = 8.314 \text{ kg-m}^2/s^2\text{-mol-K} \quad (\text{Since } 1 \text{ J} = 1 \text{ kg-m}^2/s^2)$$
$$u_{rms} = \sqrt{\frac{3(8.314 \text{ kg-m}^2/s^2\text{-mol-K})(298 \text{ K})}{28.0 \times 10^{-3} \text{ kg/mol}}} = 5.15 \times 10^2 \text{ m/s}$$

Comment This corresponds to a speed of 1150 mi/hr. Because the average molecular weight of air molecules is slightly greater than that of N_2, the rms speed of air molecules is a little smaller than that for N_2.

Practice Exercise 1

Fill in the blanks for the following statement: The rms speed of the molecules in a sample of H_2 gas at 300 K will be _____ times larger than the rms speed of O_2 molecules at the same temperature, and the ratio $u_{rms}(H_2)/u_{rms}(O_2)$ _____ with increasing temperature. **(a)** four, will not change, **(b)** four, will increase, **(c)** sixteen, will not change, **(d)** sixteen, will decrease **(e)** Not enough information is given to answer this question.

Practice Exercise 2

What is the rms speed of an atom in a sample of He gas at 25 °C?

The dependence of molecular speed on mass has two interesting consequences. The first is **effusion**, which is the escape of gas molecules through a tiny hole (▶ **Figure 10.15**). The second is **diffusion**, which is the spread of one substance throughout a space or throughout a second substance. For example, the molecules of a perfume diffuse throughout a room.

Graham's Law of Effusion

In 1846, Thomas Graham (1805–1869) discovered that the effusion rate of a gas is inversely proportional to the square root of its molar mass. Assume we have two gases at the same temperature and pressure in two containers with identical pinholes. If the rates of effusion of the two gases are r_1 and r_2 and their molar masses are M_1 and M_2, **Graham's law** states that

$$\frac{r_1}{r_2} = \sqrt{\frac{M_2}{M_1}} \qquad [10.22]$$

a relationship that indicates that the lighter gas has the higher effusion rate.

The only way for a molecule to escape from its container is for it to "hit" the hole in the partitioning wall of Figure 10.15. The faster the molecules are moving the more often they hit the partition wall and the greater the likelihood that a molecule will hit the hole and effuse. This implies that the rate of effusion is directly proportional to the rms speed of the molecules. Because R and T are constant, we have, from Equation 10.22

$$\frac{r_1}{r_2} = \frac{u_{rms1}}{u_{rms2}} = \sqrt{\frac{3RT/M_1}{3RT/M_2}} = \sqrt{\frac{M_2}{M_1}} \qquad [10.23]$$

As expected from Graham's law, helium escapes from containers through tiny pinhole leaks more rapidly than other gases of higher molecular weight (Figure 10.16).

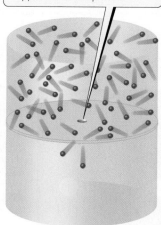

Gas molecules in top half effuse through pinhole only when they happen to hit the pinhole

▲ **Figure 10.15 Effusion.**

SAMPLE EXERCISE 10.14 **Applying Graham's Law**

An unknown gas composed of homonuclear diatomic molecules effuses at a rate that is 0.355 times the rate at which O_2 gas effuses at the same temperature. Calculate the molar mass of the unknown and identify it.

SOLUTION

Analyze We are given the rate of effusion of an unknown gas relative to that of O_2 and asked to find the molar mass and identity of the unknown. Thus, we need to connect relative rates of effusion to relative molar masses.

Plan We use Equation 10.22 to determine the molar mass of the unknown gas. If we let r_x and M_x represent the rate of effusion and molar mass of the gas, we can write

$$\frac{r_x}{r_{O_2}} = \sqrt{\frac{M_{O_2}}{M_x}}$$

Solve From the information given,

$$r_x = 0.355 \times r_{O_2}$$

Thus,

$$\frac{r_x}{r_{O_2}} = 0.355 = \sqrt{\frac{32.0 \text{ g/mol}}{\mathcal{M}_x}}$$

$$\frac{32.0 \text{ g/mol}}{\mathcal{M}_x} = (0.355)^2 = 0.126$$

$$\mathcal{M}_x = \frac{32.0 \text{ g/mol}}{0.126} = 254 \text{ g/mol}$$

Because we are told that the unknown gas is composed of homonuclear diatomic molecules, it must be an element. The molar mass must represent twice the atomic weight of the atoms in the unknown gas. We conclude that the unknown gas must have an atomic weight of 127 g/mol and therefore is I_2.

Practice Exercise 1

In a system for separating gases, a tank containing a mixture of hydrogen and carbon dioxide is connected to a much larger tank where the pressure is kept very low. The two tanks are separated by a porous membrane through which the molecules must effuse. If the initial partial pressure of each gas is 5.00 atm, what will be the mole fraction of hydrogen in the tank after the partial pressure of carbon dioxide has declined to 4.50 atm? **(a)** 52.1%, **(b)** 37.2%, **(c)** 32.1%, **(d)** 4.68%, **(e)** 27.4%.

Practice Exercise 2

Calculate the ratio of the effusion rates of N_2 gas and O_2 gas.

GO FIGURE

Because pressure and temperature are constant in this figure but volume changes, which other quantity in the ideal-gas equation must also change?

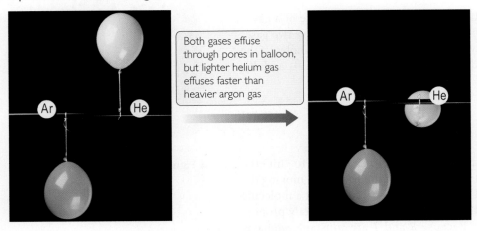

Both gases effuse through pores in balloon, but lighter helium gas effuses faster than heavier argon gas

▲ **Figure 10.16 An illustration of Graham's law of effusion.**

Diffusion and Mean Free Path

Although diffusion, like effusion, is faster for lower-mass molecules than for higher-mass ones, molecular collisions make diffusion more complicated than effusion.

Graham's law, Equation 10.22, approximates the ratio of the diffusion rates of two gases under identical conditions. We can see from the horizontal axis in Figure 10.14 that the speeds of molecules are quite high. For example, the rms speed of molecules of N_2 gas at room temperature is 515 m/s. In spite of this high speed, if someone opens a vial of perfume at one end of a room, some time elapses—perhaps a few minutes—before the scent is detected at the other end of the room. This tells us that the diffusion rate of gases throughout a volume of space is much

Chemistry Put to Work

Gas Separations

The fact that lighter molecules move at higher average speeds than more massive ones has many interesting applications. For example, developing the atomic bomb during World War II required scientists to separate the relatively low-abundance uranium isotope ^{235}U (0.7%) from the much more abundant ^{238}U (99.3%). This separation was accomplished by converting the uranium into a volatile compound, UF_6, that was then allowed to pass through a porous barrier (▼ **Figure 10.17**). Because of the pore diameters, this process is not simple effusion. Nevertheless, the way in which rate of passing through the pores depends on molar mass is essentially the same as that in effusion. The slight difference in molar mass between $^{235}UF_6$ and $^{238}UF_6$ caused the molecules to move at slightly different rates:

$$\frac{r_{235}}{r_{238}} = \sqrt{\frac{352.04}{349.03}} = 1.0043$$

Thus, the gas initially appearing on the opposite side of the barrier is very slightly enriched in ^{235}U. The process is repeated thousands of times, leading to a nearly complete separation of the two isotopes. Because of the large number of steps needed to adequately separate the isotopes, gaseous diffusion facilities are large-scale structures. The largest diffusion plant in the United States is located outside of Paducah, Kentucky. It contains approximately 400 miles of piping and the buildings where the separation takes place take up over 75 acres.

An increasingly popular method of separating uranium isotopes is by a technique that uses centrifuges. In this procedure, cylindrical rotors containing UF_6 vapor spin at high speed inside an evacuated casing. Molecules of $^{238}UF_6$ move closer to the spinning walls, whereas molecules of $^{235}UF_6$ remain in the middle of the cylinders. A stream of gas moves the $^{235}UF_6$ from the center of one centrifuge into another. Plants that use centrifuges consume less energy than those that use effusion and can be constructed in a more compact, modular fashion. Such plants are frequently in the news today as countries such as Iran and North Korea enrich uranium in the ^{235}U isotope for both nuclear power and nuclear weaponry.

Related Exercises: 10.87, 10.88

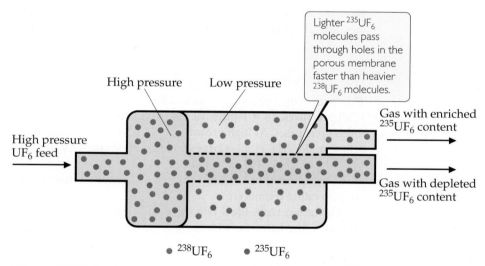

▲ **Figure 10.17 Uranium enrichment by gaseous diffusion.** The lighter $^{235}UF_6$ effuses through a porous barrier at a slightly faster rate than $^{238}UF_6$. The pressure difference across the membrane drives the effusion. The enrichment shown here for a single step is exaggerated for illustrative purposes.

slower than molecular speeds.* This difference is due to molecular collisions, which occur frequently for a gas at atmospheric pressure—about 10^{10} times per second for each molecule. Collisions occur because real gas molecules have finite volumes.

Because of molecular collisions, the direction of motion of a gas molecule is constantly changing. Therefore, the diffusion of a molecule from one point to another

---

*The rate at which the perfume moves across the room also depends on how well stirred the air is from temperature gradients and the movement of people. Nevertheless, even with the aid of these factors, it still takes much longer for the molecules to traverse the room than one would expect from their rms speed.

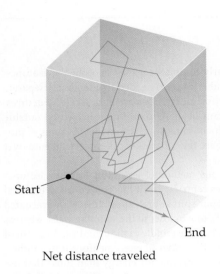

Start

End

Net distance traveled

▲ **Figure 10.18 Diffusion of a gas molecule.** For clarity, no other gas molecules in the container are shown.

consists of many short, straight-line segments as collisions buffet it around in random directions (◄ Figure 10.18).

The average distance traveled by a molecule between collisions, called the molecule's **mean free path**, varies with pressure as the following analogy illustrates. Imagine walking through a shopping mall. When the mall is crowded (high pressure), the average distance you can walk before bumping into someone is short (short mean free path). When the mall is empty (low pressure), you can walk a long way (long mean free path) before bumping into someone. The mean free path for air molecules at sea level is about 60 nm. At about 100 km in altitude, where the air density is much lower, the mean free path is about 10 cm, over 1 million times longer than at the Earth's surface.

> ### ▲ Give It Some Thought
>
> Will these changes increase, decrease, or have no effect on the mean free path of the molecules in a gas sample?
> **(a)** increasing pressure.
> **(b)** increasing temperature.

10.9 | Real Gases: Deviations from Ideal Behavior

The extent to which a real gas departs from ideal behavior can be seen by rearranging the ideal-gas equation to solve for n:

$$\frac{PV}{RT} = n \qquad [10.24]$$

This form of the equation tells us that for 1 mol of ideal gas, the quantity PV/RT equals 1 at all pressures. In ▼ Figure 10.19, PV/RT is plotted as a function of P for 1 mol of several real gases. At high pressures (generally above 10 atm), the deviation from ideal behavior ($PV/RT = 1$) is large and different for each gas. *Real gases, in other words, do not behave ideally at high pressure.* At lower pressures (usually below 10 atm), however, the deviation from ideal behavior is small, and we can use the ideal-gas equation without generating serious error.

Deviation from ideal behavior also depends on temperature. As temperature increases, the behavior of a real gas more nearly approaches that of the ideal gas

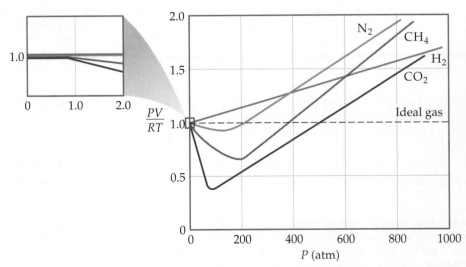

▲ **Figure 10.19 The effect of pressure on the behavior of several real gases.** Data for 1 mol of gas in all cases. Data for N_2, CH_4, and H_2 are at 300 K; for CO_2 data are at 313 K because under high pressure CO_2 liquefies at 300 K.

GO FIGURE

True or false: Nitrogen gas behaves more like an ideal gas as the temperature increases.

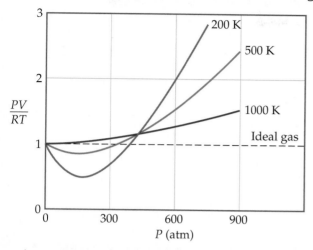

▲ Figure 10.20 **The effect of temperature and pressure on the behavior of nitrogen gas.**

(▲ Figure 10.20). In general, *the deviation from ideal behavior increases as temperature decreases*, becoming significant near the temperature at which the gas liquefies.

Give It Some Thought

Under which conditions do you expect helium gas to deviate most from ideal behavior? **(a)** 100 K and 1 atm, **(b)** 100 K and 5 atm, **(c)** 300 K and 2 atm.

The basic assumptions of the kinetic-molecular theory of gases give us insight into why real gases deviate from ideal behavior. The molecules of an ideal gas are assumed to occupy no space and have no attraction for one another. *Real molecules, however, do have finite volumes and do attract one another.* As ▶ Figure 10.21 shows, the unoccupied space in which real molecules can move is less than the container volume. At low pressures, the combined volume of the gas molecules is negligible relative to the container volume. Thus, the unoccupied volume available to the molecules is essentially the container volume. At high pressures, the combined volume of the gas molecules is *not* negligible relative to the container volume. Now the unoccupied volume available to the molecules is less than the container volume. At high pressures, therefore, gas volumes tend to be slightly greater than those predicted by the ideal-gas equation.

Another reason for nonideal behavior at high pressures is that the attractive forces between molecules come into play at the short intermolecular distances found when molecules are crowded together at high pressures. Because of these attractive forces, the impact of a given molecule with the container wall is lessened. If we could stop the motion in a gas, as illustrated in Figure 10.22, we would see that a molecule about to collide with the wall experiences the attractive forces of nearby molecules. These attractions lessen the force with which the molecule hits the wall. As a result, the gas pressure is less than that of an ideal gas. This effect decreases PV/RT to below its ideal value, as seen at the lower pressures in Figures 10.19 and 10.20. When the pressure is sufficiently high, however, the volume effects dominate and PV/RT increases to above the ideal value.

Gas molecules occupy a small fraction of the total volume.

Gas molecules occupy a larger fraction of the total volume.

Low pressure

High pressure

▲ Figure 10.21 **Gases behave more ideally at low pressure than at high pressure.**

▲ **GO FIGURE**

How would you expect the pressure of a gas to change if suddenly the intermolecular forces were repulsive rather than attractive?

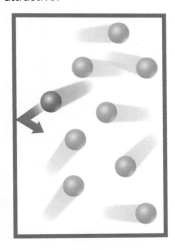

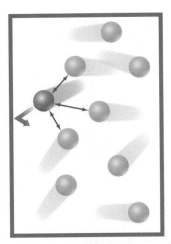

| Ideal gas | Real gas |

▲ Figure 10.22 **In any real gas, attractive intermolecular forces reduce pressure to values lower than in an ideal gas.**

Temperature determines how effective attractive forces between gas molecules are in causing deviations from ideal behavior at lower pressures. Figure 10.20 shows that, at pressures below about 400 atm, cooling increases the extent to which a gas deviates from ideal behavior. As the gas cools, the average kinetic energy of the molecules decreases. This drop in kinetic energy means the molecules do not have the energy needed to overcome intermolecular attraction, and the molecules will be more likely to stick to each other than bounce off each other.

As the temperature of a gas increases—as, say, from 200 to 1000 K in Figure 10.20—the negative deviation of PV/RT from the ideal value of 1 disappears. As noted earlier, the deviations seen at high temperatures stem mainly from the effect of the finite volumes of the molecules.

 Give It Some Thought

Explain the negative deviation from ideal gas behavior of N_2 below 300 atm in Figure 10.20.

The van der Waals Equation

Engineers and scientists who work with gases at high pressures often cannot use the ideal-gas equation because departures from ideal behavior are too large. One useful equation developed to predict the behavior of real gases was proposed by the Dutch scientist Johannes van der Waals (1837–1923).

As we have seen, a real gas has a lower pressure due to intermolecular forces, and a larger volume due to the finite volume of the molecules, relative to an ideal gas. Van der Waals recognized that it would be possible to retain the form of the ideal-gas equation, $PV = nRT$, if corrections were made to the pressure and the volume. He introduced two constants for these corrections: a, a measure of how strongly the gas molecules attract one another, and b, a measure of the finite volume occupied by the molecules. His description of gas behavior is known as the **van der Waals equation**:

$$\left(P + \frac{n^2a}{V^2}\right)(V - nb) = nRT \qquad [10.25]$$

The term n^2a/V^2 accounts for the attractive forces. The equation adjusts the pressure upward by adding n^2a/V^2 because attractive forces between molecules tend to reduce the pressure (Figure 10.22). The added term has the form n^2a/V^2 because the attractive force between pairs of molecules increases as the square of the number of molecules per unit volume, $(n/V)^2$.

The term nb accounts for the small but finite volume occupied by the gas molecules (Figure 10.21). The van der Waals equation subtracts nb to adjust the volume downward to give the volume that would be available to the molecules in the ideal case. The constants a and b, called *van der Waals constants*, are experimentally determined positive quantities that differ from one gas to another. Notice in ▼ Table 10.3 that a and b generally increase with increasing molecular mass. Larger, more massive molecules have larger volumes and tend to have greater intermolecular attractive forces.

Table 10.3   Van der Waals Constants for Gas Molecules

Substance	$a(L^2\text{-atm}/mol^2)$	$b(L/mol)$
He	0.0341	0.02370
Ne	0.211	0.0171
Ar	1.34	0.0322
Kr	2.32	0.0398
Xe	4.19	0.0510
H_2	0.244	0.0266
N_2	1.39	0.0391
O_2	1.36	0.0318
F_2	1.06	0.0290
Cl_2	6.49	0.0562
H_2O	5.46	0.0305
NH_3	4.17	0.0371
CH_4	2.25	0.0428
CO_2	3.59	0.0427
CCl_4	20.4	0.1383

SAMPLE EXERCISE 10.15 | Using the van der Waals Equation

If 10.00 mol of an ideal gas were confined to 22.41 L at 0.0 °C, it would exert a pressure of 10.00 atm. Use the van der Waals equation and Table 10.3 to estimate the pressure exerted by 1.000 mol of $Cl_2(g)$ in 22.41 L at 0.0 °C.

SOLUTION

Analyze We need to determine a pressure. Because we will use the van der Waals equation, we must identify the appropriate values for the constants in the equation.

Plan Rearrange Equation 10.25 to isolate P.

Solve Substituting $n = 10.00$ mol, $R = 0.08206$ L-atm/mol-K, $T = 273.2$ K, $V = 22.41$ L, $a = 6.49$ L^2-atm/mol^2, and $b = 0.0562$ L/mol:

$$P = \frac{(10.00 \text{ mol})(0.08206 \text{ L-atm}/\text{mol-K})(273.2 \text{ K})}{22.41 \text{ L} - (10.00 \text{ mol})(0.0562 \text{ L}/\text{mol})} - \frac{(10.00 \text{ mol})^2(6.49 \text{ L}^2\text{-atm}/\text{mol}^2)}{(22.41 \text{ L})^2}$$

$$= 10.26 \text{ atm} - 1.29 \text{ atm} = 8.97 \text{ atm}$$

Comment Notice that the term 10.26 atm is the pressure corrected for molecular volume. This value is higher than the ideal value, 10.00 atm, because the volume in which the molecules are free to move is smaller than the container volume, 22.41 L. Thus, the molecules collide more frequently with the container walls and the pressure is higher than that of a real gas. The term 1.29 atm makes a correction in the opposite direction for intermolecular forces. The correction for intermolecular forces is the larger of the two and thus the pressure 8.97 atm is smaller than would be observed for an ideal gas.

Practice Exercise 1

Calculate the pressure of a 2.975-mol sample of N_2 in a 0.7500-L flask at 300.0 °C using the van der Waals equation and then repeat the calculation using the ideal-gas equation. Within the limits of the significant figures justified by these parameters, will the ideal-gas equation overestimate or underestimate the pressure, and if so by how much?
(a) Underestimate by 17.92 atm, (b) Overestimate by 21.87 atm,
(c) Underestimate by 0.06 atm, (d) Overestimate by 0.06 atm.

Practice Exercise 2

A sample of 1.000 mol of $CO_2(g)$ is confined to a 3.000-L container at 0.000 °C. Calculate the pressure of the gas using (a) the ideal-gas equation and (b) the van der Waals equation.

SAMPLE INTEGRATIVE EXERCISE   Putting Concepts Together

Cyanogen, a highly toxic gas, is 46.2% C and 53.8% N by mass. At 25 °C and 751 torr, 1.05 g of cyanogen occupies 0.500 L. (a) What is the molecular formula of cyanogen? Predict (b) its molecular structure and (c) its polarity.

SOLUTION

Analyze We need to determine the molecular formula of a gas from elemental analysis data and data on its properties. Then we need to predict the structure of the molecule and from that, its polarity.

(a) **Plan** We can use the percentage composition of the compound to calculate its empirical formula. ∞ (Section 3.5) Then we can determine the molecular formula by comparing the mass of the empirical formula with the molar mass. ∞ (Section 3.5)

Solve To determine the empirical formula, we assume we have a 100-g sample and calculate the number of moles of each element in the sample:

$$\text{Moles C} = (46.2 \text{ g C})\left(\frac{1 \text{ mol C}}{12.01 \text{ g C}}\right) = 3.85 \text{ mol C}$$

$$\text{Moles N} = (53.8 \text{ g N})\left(\frac{1 \text{ mol N}}{14.01 \text{ g N}}\right) = 3.84 \text{ mol N}$$

Because the ratio of the moles of the two elements is essentially 1:1, the empirical formula is CN. To determine the molar mass, we use Equation 10.11.

$$\mathcal{M} = \frac{dRT}{P} = \frac{(1.05 \text{ g}/0.500 \text{ L})(0.08206 \text{ L-atm}/\text{mol-K})(298 \text{ K})}{(751/760)\text{atm}} = 52.0 \text{ g}/\text{mol}$$

The molar mass associated with the empirical formula CN is 12.0 + 14.0 = 26.0 g/mol. Dividing the molar mass by that of its empirical formula gives $(52.0 \text{ g}/\text{mol})/(26.0 \text{ g}/\text{mol}) = 2.00$. Thus, the molecule has twice as many atoms of each element as the empirical formula, giving the molecular formula C_2N_2.

(b) **Plan** To determine the molecular structure, we must determine the Lewis structure. ∞ (Section 8.5) We can then use the VSEPR model to predict the structure. ∞ (Section 9.2)

Solve The molecule has 2(4) + 2(5) = 18 valence-shell electrons. By trial and error, we seek a Lewis structure with 18 valence electrons in which each atom has an octet and the formal charges are as low as possible. The structure

$$:N≡C—C≡N:$$

meets these criteria. (This structure has zero formal charge on each atom.)

The Lewis structure shows that each atom has two electron domains. (Each nitrogen has a nonbonding pair of electrons and a triple bond, whereas each carbon has a triple bond and a single bond.) Thus, the electron-domain geometry around each atom is linear, causing the overall molecule to be linear.

(c) **Plan** To determine the polarity of the molecule, we must examine the polarity of the individual bonds and the overall geometry of the molecule.

Solve Because the molecule is linear, we expect the two dipoles created by the polarity in the carbon–nitrogen bond to cancel each other, leaving the molecule with no dipole moment.

Chapter Summary and Key Terms

CHARACTERISTICS OF GASES (SECTION 10.1) Substances that are gases at room temperature tend to be molecular substances with low molar masses. Air, a mixture composed mainly of N_2 and O_2, is the most common gas we encounter. Some liquids and solids can also exist in the gaseous state, where they are known as **vapors**. Gases are compressible; they mix in all proportions because their component molecules are far apart from each other.

PRESSURE (SECTION 10.2) To describe the state or condition of a gas, we must specify four variables: pressure (P), volume (V), temperature (T), and quantity (n). Volume is usually measured in liters, temperature in kelvins, and quantity of gas in moles. **Pressure** is the force per unit area and is expressed in SI units as **pascals**, Pa ($1 \text{ Pa} = 1 \text{ N/m}^2$). A related unit, the **bar**, equals 10^5 Pa. In chemistry, **standard atmospheric pressure** is used to define the **atmosphere** (atm) and the **torr** (also called the millimeter of mercury). One atmosphere of pressure equals 101.325 kPa, or 760 torr. A barometer is often used to measure the atmospheric pressure. A manometer can be used to measure the pressure of enclosed gases.

THE GAS LAWS (SECTION 10.3) Studies have revealed several simple gas laws: For a constant quantity of gas at constant temperature, the volume of the gas is inversely proportional to the pressure (**Boyle's law**). For a fixed quantity of gas at constant pressure, the volume is directly proportional to its absolute temperature (**Charles's law**). Equal volumes of gases at the same temperature and pressure contain equal numbers of molecules (**Avogadro's hypothesis**). For a gas at constant temperature and pressure, the volume of the gas is directly proportional to the number of moles of gas (**Avogadro's law**). Each of these gas laws is a special case of the ideal-gas equation.

THE IDEAL-GAS EQUATION (SECTIONS 10.4 AND 10.5) The **ideal-gas equation**, $PV = nRT$, is the equation of state for an **ideal gas**. The term R in this equation is the **gas constant**. We can use the ideal-gas equation to calculate variations in one variable when one or more of the others are changed. Most gases at pressures less than 10 atm and temperatures near 273 K and above obey the ideal-gas equation reasonably well. The conditions of 273 K (0 °C) and 1 atm are known as the **standard temperature and pressure (STP)**. In all applications of the ideal-gas equation we must remember to convert temperatures to the absolute-temperature scale (the kelvin scale).

Using the ideal-gas equation, we can relate the density of a gas to its molar mass: $\mathcal{M} = dRT/P$. We can also use the ideal-gas equation to solve problems involving gases as reactants or products in chemical reactions.

GAS MIXTURES AND PARTIAL PRESSURES (SECTION 10.6) In gas mixtures, the total pressure is the sum of the **partial pressures** that each gas would exert if it were present alone under the same conditions (**Dalton's law of partial pressures**). The partial pressure of a component of a mixture is equal to its mole fraction times the total pressure: $P_1 = X_1 P_t$. The **mole fraction** X is the ratio of the moles of one component of a mixture to the total moles of all components.

THE KINETIC-MOLECULAR THEORY OF GASES (SECTION 10.7) The **kinetic-molecular theory of gases** accounts for the properties of an ideal gas in terms of a set of statements about the nature of gases. Briefly, these statements are as follows: Molecules are in continuous chaotic motion. The volume of gas molecules is negligible compared to the volume of their container. The gas molecules neither attract nor repel each other. The average kinetic energy of the gas molecules is proportional to the absolute temperature and does not change if the temperature remains constant.

The individual molecules of a gas do not all have the same kinetic energy at a given instant. Their speeds are distributed over a wide range; the distribution varies with the molar mass of the gas and with temperature. The **root-mean-square (rms) speed**, u_{rms}, varies in proportion to the square root of the absolute temperature and inversely with the square root of the molar mass: $u_{rms} = \sqrt{3RT/\mathcal{M}}$. The most probable speed of a gas molecule is given by $u_{mp} = \sqrt{2RT/\mathcal{M}}$.

MOLECULAR EFFUSION AND DIFFUSION (SECTION 10.8) It follows from kinetic-molecular theory that the rate at which a gas undergoes **effusion** (escapes through a tiny hole) is inversely proportional to the square root of its molar mass (**Graham's law**). The **diffusion** of one gas through the space occupied by a second gas is another phenomenon related to the speeds at which molecules move. Because moving molecules undergo frequent collisions with one another, the **mean free path**—the mean distance traveled between collisions—is short. Collisions between molecules limit the rate at which a gas molecule can diffuse.

REAL GASES: DEVIATIONS FROM IDEAL BEHAVIOR (SECTION 10.9) Departures from ideal behavior increase in magnitude as pressure increases and as temperature decreases. Real gases depart from ideal behavior because (1) the molecules possess finite volume and (2) the molecules experience attractive forces for one another. These two effects make the volumes of real gases larger and their pressures smaller than those of an ideal gas. The **van der Waals equation** is an equation of state for gases, which modifies the ideal-gas equation to account for intrinsic molecular volume and intermolecular forces.

Learning Outcomes After studying this chapter, you should be able to:

- Convert between pressure units with an emphasis on torr and atmospheres. (Section 10.2)

- Calculate P, V, n, or T using the ideal-gas equation. (Section 10.4)

- Explain how the gas laws relate to the ideal-gas equation and apply the gas laws in calculations. (Sections 10.3 and 10.4)

- Calculate the density or molecular weight of a gas. (Section 10.5)

- Calculate the volume of gas consumed or formed in a chemical reaction. (Section 10.5)

- Calculate the total pressure of a gas mixture given its partial pressures or given information for calculating partial pressures. (Section 10.6)

- Describe the kinetic-molecular theory of gases and how it explains the pressure and temperature of a gas, the gas laws, and the rates of effusion and diffusion. (Sections 10.7 and 10.8)

- Explain why intermolecular attractions and molecular volumes cause real gases to deviate from ideal behavior at high pressure or low temperature. (Section 10.9)

Key Equations

- $PV = nRT$ [10.5]   Ideal-gas equation

- $\dfrac{P_1 V_1}{T_1} = \dfrac{P_2 V_2}{T_2}$ [10.8]   The combined gas law, showing how P, V, and T are related for a constant n

- $d = \dfrac{P\mathcal{M}}{RT}$ [10.10]   Calculating the density or molar mass of a gas

- $P_t = P_1 + P_2 + P_3 + \ldots$ [10.12]   Relating the total pressure of a gas mixture to the partial pressures of its components (Dalton's law of partial pressures)

- $P_1 = \left(\dfrac{n_1}{n_t}\right) P_t = X_1 P_t$ [10.16]   Relating partial pressure to mole fraction

- $u_{\text{rms}} = \sqrt{\dfrac{3RT}{\mathcal{M}}}$ [10.20]   Definition of the root-mean-square (rms) speed of gas molecules

- $\dfrac{r_1}{r_2} = \sqrt{\dfrac{\mathcal{M}_2}{\mathcal{M}_1}}$ [10.22]   Relating the relative rates of effusion of two gases to their molar masses

- $\left(P + \dfrac{n^2 a}{V^2}\right)(V - nb) = nRT$ [10.25]   The van der Waals equation

Exercises

Visualizing Concepts

10.1 Mars has an average atmospheric pressure of 0.007 atm. Would it be easier or harder to drink from a straw on Mars than on Earth? Explain. [Section 10.2]

10.2 You have a sample of gas in a container with a movable piston, such as the one in the drawing. (**a**) Redraw the container to show what it might look like if the temperature of the gas is increased from 300 to 500 K while the pressure is kept constant. (**b**) Redraw the container to show what it might look like if the external pressure on the piston is increased from 1.0 atm to 2.0 atm while the temperature is kept constant. (**c**) Redraw the container to show what it might look like if the temperature of the gas decreases from 300 to 200 K while the pressure is kept constant (assume the gas does not liquefy). [Section 10.3]

10.3 Consider the sample of gas depicted here. What would the drawing look like if the volume and temperature remained constant while you removed enough of the gas to decrease the pressure by a factor of 2? [Section 10.3]

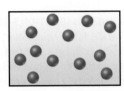

10.4 Imagine that the reaction $2\,CO(g) + O_2(g) \longrightarrow 2\,CO_2(g)$ occurs in a container that has a piston that moves to maintain a constant pressure when the reaction occurs at constant temperature. Which of the following statements describes how the volume of the container changes due to the reaction: (**a**) the volume increases by 50%, (**b**) the volume increases by 33%, (**c**) the volume remains constant, (**d**) the volume decreases by 33%, (**e**) the volume decreases by 50%. [Sections 10.3 and 10.4]

10.5 Suppose you have a fixed amount of an ideal gas at a constant volume. If the pressure of the gas is doubled while the volume is held constant, what happens to its temperature? [Section 10.4]

10.6 The apparatus shown here has two gas-filled containers and one empty container, all attached to a hollow horizontal tube. When the valves are opened and the gases are allowed to mix at constant temperature, what is the distribution of atoms in each container? Assume that the containers are of equal volume and ignore the volume of the connecting tube. Which gas has the greater partial pressure after the valves are opened? [Section 10.6]

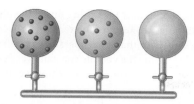

10.7 The accompanying drawing represents a mixture of three different gases. **(a)** Rank the three components in order of increasing partial pressure. **(b)** If the total pressure of the mixture is 1.40 atm, calculate the partial pressure of each gas. [Section 10.6]

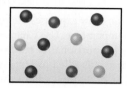

10.8 On a single plot, qualitatively sketch the distribution of molecular speeds for **(a)** $Kr(g)$ at $-50\ ^{\circ}C$, **(b)** $Kr(g)$ at $0\ ^{\circ}C$, **(c)** $Ar(g)$ at $0\ ^{\circ}C$. [Section 10.7]

10.9 Consider the following graph. **(a)** If curves A and B refer to two different gases, He and O_2, at the same temperature, which curve corresponds to He? **(b)** If A and B refer to the same gas at two different temperatures, which represents the higher temperature? **(c)** For each curve which speed is highest: the most probable speed, the root-mean-square speed, or the average speed? [Section 10.7]

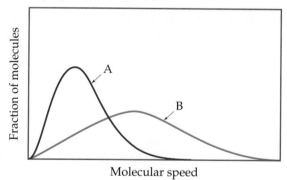

10.10 Consider the following samples of gases:

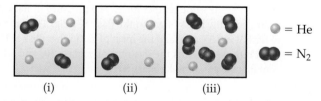

If the three samples are all at the same temperature, rank them with respect to **(a)** total pressure, **(b)** partial pressure of helium, **(c)** density, **(d)** average kinetic energy of particles. [Section 10.6 and 10.7]

10.11 A thin glass tube 1 m long is filled with Ar gas at 1 atm, and the ends are stoppered with cotton plugs:

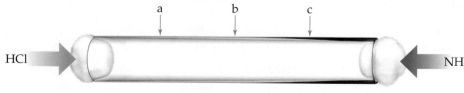

HCl gas is introduced at one end of the tube, and simultaneously NH_3 gas is introduced at the other end. When the two gases diffuse through the cotton plugs down the tube and meet, a white ring appears due to the formation of $NH_4Cl(s)$. At which location— a, b, or c—do you expect the ring to form? [Section 10.8]

10.12 The graph below shows the change in pressure as the temperature increases for a 1-mol sample of a gas confined to a 1-L container. The four plots correspond to an ideal gas and three real gases: CO_2, N_2, and Cl_2. **(a)** At room temperature, all three real gases have a pressure less than the ideal gas. Which van der Waals constant, a or b, accounts for the influence intermolecular forces have in lowering the pressure of a real gas? **(b)** Use the van der Waals constants in Table 10.3 to match the labels in the plot (A, B, and C) with the respective gases (CO_2, N_2, and Cl_2). [Section 10.9]

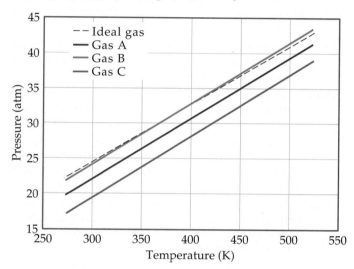

Gas Characteristics; Pressure (Sections 10.1 and 10.2)

10.13 How does a gas compare with a liquid for each of the following properties: **(a)** density, **(b)** compressibility, **(c)** ability to mix with other substances of the same phase to form homogeneous mixtures, **(d)** ability to conform to the shape of its container?

10.14 **(a)** A liquid and a gas are moved to larger containers. How does their behavior differ once they are in the larger containers? Explain the difference in molecular terms. **(b)** Although liquid water and carbon tetrachloride, $CCl_4(l)$, do not mix, their vapors form a homogeneous mixture. Explain. **(c)** Gas densities are generally reported in grams per liter, whereas liquid densities are reported in grams per milliliter. Explain the molecular basis for this difference.

10.15 Suppose that a woman weighing 130 lb and wearing high-heeled shoes momentarily places all her weight on the heel of one foot. If the area of the heel is 0.50 in.^{2}, calculate the pressure exerted on the underlying surface in **(a)** kilopascals, **(b)** atmospheres, and **(c)** pounds per square inch.

10.16 A set of bookshelves rests on a hard floor surface on four legs, each having a cross-sectional dimension of 3.0×4.1 cm in contact with the floor. The total mass of the shelves plus the books stacked on them is 262 kg. Calculate the pressure in pascals exerted by the shelf footings on the surface.

10.17 (a) How high in meters must a column of water be to exert a pressure equal to that of a 760-mm column of mercury? The density of water is 1.0 g/mL, whereas that of mercury is 13.6 g/mL. (b) What is the pressure, in atmospheres, on the body of a diver if he or she is 39 ft below the surface of the water when atmospheric pressure at the surface is 0.97 atm?

10.18 The compound 1-iodododecane is a nonvolatile liquid with a density of 1.20 g/mL. The density of mercury is 13.6 g/mL. What do you predict for the height of a barometer column based on 1-iodododecane, when the atmospheric pressure is 749 torr?

10.19 The typical atmospheric pressure on top of Mt. Everest (29,028 ft) is about 265 torr. Convert this pressure to (a) atm, (b) mm Hg, (c) pascals, (d) bars, (e) psi.

10.20 Perform the following conversions: (a) 0.912 atm to torr, (b) 0.685 bar to kilopascals, (c) 655 mm Hg to atmospheres, (d) 1.323×10^5 Pa to atmospheres, (e) 2.50 atm to psi.

10.21 In the United States, barometric pressures are generally reported in inches of mercury (in. Hg). On a beautiful summer day in Chicago, the barometric pressure is 30.45 in. Hg. (a) Convert this pressure to torr. (b) Convert this pressure to atm.

10.22 Hurricane Wilma of 2005 is the most intense hurricane on record in the Atlantic basin, with a low-pressure reading of 882 mbar (millibars). Convert this reading into (a) atmospheres, (b) torr, and (c) inches of Hg.

10.23 If the atmospheric pressure is 0.995 atm, what is the pressure of the enclosed gas in each of the three cases depicted in the drawing? Assume that the gray liquid is mercury.

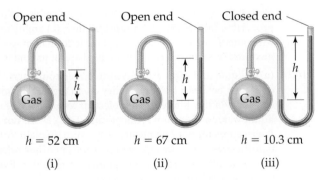

h = 52 cm   h = 67 cm   h = 10.3 cm

(i)   (ii)   (iii)

10.24 An open-end manometer containing mercury is connected to a container of gas, as depicted in Sample Exercise 10.2. What is the pressure of the enclosed gas in torr in each of the following situations? (a) The mercury in the arm attached to the gas is 15.4 mm higher than in the one open to the atmosphere; atmospheric pressure is 0.985 atm. (b) The mercury in the arm attached to the gas is 12.3 mm lower than in the one open to the atmosphere; atmospheric pressure is 0.99 atm.

The Gas Laws (Section 10.3)

10.25 You have a gas at 25 °C confined to a cylinder with a movable piston. Which of the following actions would double

the gas pressure? (a) Lifting up on the piston to double the volume while keeping the temperature constant; (b) Heating the gas so that its temperature rises from 25 °C to 50 °C, while keeping the volume constant; (c) Pushing down on the piston to halve the volume while keeping the temperature constant.

10.26 A fixed quantity of gas at 21 °C exhibits a pressure of 752 torr and occupies a volume of 5.12 L. (a) Calculate the volume the gas will occupy if the pressure is increased to 1.88 atm while the temperature is held constant. (b) Calculate the volume the gas will occupy if the temperature is increased to 175 °C while the pressure is held constant.

10.27 (a) Amonton's law expresses the relationship between pressure and temperature. Use Charles's law and Boyle's law to derive the proportionality relationship between P and T. (b) If a car tire is filled to a pressure of 32.0 lbs/in.^{2} (psi) measured at 75 °F, what will be the tire pressure if the tires heat up to 120 °F during driving?

10.28 Nitrogen and hydrogen gases react to form ammonia gas as follows:

$$N_2(g) + 3\,H_2(g) \longrightarrow 2\,NH_3(g)$$

At a certain temperature and pressure, 1.2 L of N_2 reacts with 3.6 L of H_2. If all the N_2 and H_2 are consumed, what volume of NH_3, at the same temperature and pressure, will be produced?

The Ideal-Gas Equation (Section 10.4)

10.29 (a) What conditions are represented by the abbreviation STP? (b) What is the molar volume of an ideal gas at STP? (c) Room temperature is often assumed to be 25 °C. Calculate the molar volume of an ideal gas at 25 °C and 1 atm pressure. (d) If you measure pressure in bars instead of atmospheres, calculate the corresponding value of R in L-bar/mol-K.

10.30 To derive the ideal-gas equation, we assume that the volume of the gas atoms/molecules can be neglected. Given the atomic radius of neon, 0.69 Å, and knowing that a sphere has a volume of $4\pi r^3/3$, calculate the fraction of space that Ne atoms occupy in a sample of neon at STP.

10.31 Suppose you are given two 1-L flasks and told that one contains a gas of molar mass 30, the other a gas of molar mass 60, both at the same temperature. The pressure in flask A is X atm, and the mass of gas in the flask is 1.2 g. The pressure in flask B is 0.5X atm, and the mass of gas in that flask is 1.2 g. Which flask contains the gas of molar mass 30, and which contains the gas of molar mass 60?

10.32 Suppose you are given two flasks at the same temperature, one of volume 2 L and the other of volume 3 L. The 2-L flask contains 4.8 g of gas, and the gas pressure is X atm. The 3-L flask contains 0.36 g of gas, and the gas pressure is 0.1X. Do the two gases have the same molar mass? If not, which contains the gas of higher molar mass?

10.33 Complete the following table for an ideal gas:

P	V	n	T
2.00 atm	1.00 L	0.500 mol	? K
0.300 atm	0.250 L	? mol	27 °C
650 torr	? L	0.333 mol	350 K
? atm	585 mL	0.250 mol	295 K

10.34 Calculate each of the following quantities for an ideal gas: (a) the volume of the gas, in liters, if 1.50 mol has a pressure of 1.25 atm at a temperature of $-6\,°C$; (b) the absolute temperature of the gas at which 3.33×10^{-3} mol occupies 478 mL at 750 torr; (c) the pressure, in atmospheres, if 0.00245 mol occupies 413 mL at $138\,°C$; (d) the quantity of gas, in moles, if 126.5 L at $54\,°C$ has a pressure of 11.25 kPa.

10.35 The Goodyear blimps, which frequently fly over sporting events, hold approximately 175,000 ft^3 of helium. If the gas is at $23\,°C$ and 1.0 atm, what mass of helium is in a blimp?

10.36 A neon sign is made of glass tubing whose inside diameter is 2.5 cm and whose length is 5.5 m. If the sign contains neon at a pressure of 1.78 torr at $35\,°C$, how many grams of neon are in the sign? (The volume of a cylinder is $\pi r^2 h$.)

10.37 (a) Calculate the number of molecules in a deep breath of air whose volume is 2.25 L at body temperature, $37\,°C$, and a pressure of 735 torr. (b) The adult blue whale has a lung capacity of 5.0×10^3 L. Calculate the mass of air (assume an average molar mass of 28.98 g/mol) contained in an adult blue whale's lungs at $0.0\,°C$ and 1.00 atm, assuming the air behaves ideally.

10.38 (a) If the pressure exerted by ozone, O_3, in the stratosphere is 3.0×10^{-3} atm and the temperature is 250 K, how many ozone molecules are in a liter? (b) Carbon dioxide makes up approximately 0.04% of Earth's atmosphere. If you collect a 2.0-L sample from the atmosphere at sea level (1.00 atm) on a warm day ($27\,°C$), how many CO_2 molecules are in your sample?

10.39 A scuba diver's tank contains 0.29 kg of O_2 compressed into a volume of 2.3 L. (a) Calculate the gas pressure inside the tank at $9\,°C$. (b) What volume would this oxygen occupy at $26\,°C$ and 0.95 atm?

10.40 An aerosol spray can with a volume of 250 mL contains 2.30 g of propane gas (C_3H_8) as a propellant. (a) If the can is at $23\,°C$, what is the pressure in the can? (b) What volume would the propane occupy at STP? (c) The can's label says that exposure to temperatures above $130\,°F$ may cause the can to burst. What is the pressure in the can at this temperature?

10.41 A 35.1 g sample of solid CO_2 (dry ice) is added to a container at a temperature of 100 K with a volume of 4.0 L. If the container is evacuated (all of the gas removed), sealed and then allowed to warm to room temperature ($T = 298$ K) so that all of the solid CO_2 is converted to a gas, what is the pressure inside the container?

10.42 A 334-mL cylinder for use in chemistry lectures contains 5.225 g of helium at $23\,°C$. How many grams of helium must be released to reduce the pressure to 75 atm assuming ideal gas behavior?

10.43 Chlorine is widely used to purify municipal water supplies and to treat swimming pool waters. Suppose that the volume of a particular sample of Cl_2 gas is 8.70 L at 895 torr and $24\,°C$. (a) How many grams of Cl_2 are in the sample? (b) What volume will the Cl_2 occupy at STP? (c) At what temperature will the volume be 15.00 L if the pressure is 8.76×10^2 torr? (d) At what pressure will the volume equal 5.00 L if the temperature is $58\,°C$?

10.44 Many gases are shipped in high-pressure containers. Consider a steel tank whose volume is 55.0 gallons that contains O_2 gas at a pressure of 16,500 kPa at $23\,°C$. (a) What mass of O_2 does the tank contain? (b) What volume would the gas occupy at STP? (c) At what temperature would the pressure in the tank equal 150.0 atm? (d) What would be the pressure of the gas, in kPa, if it were transferred to a container at $24\,°C$ whose volume is 55.0 L?

10.45 In an experiment reported in the scientific literature, male cockroaches were made to run at different speeds on a miniature treadmill while their oxygen consumption was measured. In 1 hr the average cockroach running at 0.08 km/hr consumed 0.8 mL of O_2 at 1 atm pressure and $24\,°C$ per gram of insect mass. (a) How many moles of O_2 would be consumed in 1 hr by a 5.2-g cockroach moving at this speed? (b) This same cockroach is caught by a child and placed in a 1-qt fruit jar with a tight lid. Assuming the same level of continuous activity as in the research, will the cockroach consume more than 20% of the available O_2 in a 48-hr period? (Air is 21 mol % O_2.)

10.46 The physical fitness of athletes is measured by "V_{O_2} max," which is the maximum volume of oxygen consumed by an individual during incremental exercise (for example, on a treadmill). An average male has a V_{O_2} max of 45 mL O_2/kg body mass/min, but a world-class male athlete can have a V_{O_2} max reading of 88.0 mL O_2/kg body mass/min. (a) Calculate the volume of oxygen, in mL, consumed in 1 hr by an average man who weighs 185 lbs and has a V_{O_2} max reading of 47.5 mL O_2/kg body mass/min. (b) If this man lost 20 lb, exercised, and increased his V_{O_2} max to 65.0 mL O_2/kg body mass/min, how many mL of oxygen would he consume in 1 hr?

Further Applications of the Ideal-Gas Equation (Section 10.5)

10.47 Which gas is most dense at 1.00 atm and 298 K: CO_2, N_2O, or Cl_2? Explain.

10.48 Rank the following gases from least dense to most dense at 1.00 atm and 298 K: SO_2, HBr, CO_2. Explain.

10.49 Which of the following statements best explains why a closed balloon filled with helium gas rises in air?

(a) Helium is a monatomic gas, whereas nearly all the molecules that make up air, such as nitrogen and oxygen, are diatomic.

(b) The average speed of helium atoms is greater than the average speed of air molecules, and the greater speed of collisions with the balloon walls propels the balloon upward.

(c) Because the helium atoms are of lower mass than the average air molecule, the helium gas is less dense than air. The mass of the balloon is thus less than the mass of the air displaced by its volume.

(d) Because helium has a lower molar mass than the average air molecule, the helium atoms are in faster motion. This means that the temperature of the helium is greater than the air temperature. Hot gases tend to rise.

10.50 Which of the following statements best explains why nitrogen gas at STP is less dense than Xe gas at STP?

(a) Because Xe is a noble gas, there is less tendency for the Xe atoms to repel one another, so they pack more densely in the gaseous state.

(b) Xe atoms have a higher mass than N_2 molecules. Because both gases at STP have the same number of molecules per unit volume, the Xe gas must be denser.

(c) The Xe atoms are larger than N_2 molecules and thus take up a larger fraction of the space occupied by the gas.

(d) Because the Xe atoms are much more massive than the N_2 molecules, they move more slowly and thus exert less upward force on the gas container and make the gas appear denser.

10.51 (a) Calculate the density of NO_2 gas at 0.970 atm and 35 °C. (b) Calculate the molar mass of a gas if 2.50 g occupies 0.875 L at 685 torr and 35 °C.

10.52 (a) Calculate the density of sulfur hexafluoride gas at 707 torr and 21 °C. (b) Calculate the molar mass of a vapor that has a density of 7.135 g/L at 12 °C and 743 torr.

10.53 In the Dumas-bulb technique for determining the molar mass of an unknown liquid, you vaporize the sample of a liquid that boils below 100 °C in a boiling-water bath and determine the mass of vapor required to fill the bulb. From the following data, calculate the molar mass of the unknown liquid: mass of unknown vapor, 1.012 g; volume of bulb, 354 cm³; pressure, 742 torr; temperature, 99 °C.

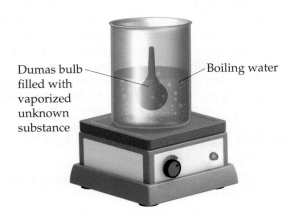

Dumas bulb filled with vaporized unknown substance

Boiling water

10.54 The molar mass of a volatile substance was determined by the Dumas-bulb method described in Exercise 10.53. The unknown vapor had a mass of 0.846 g; the volume of the bulb was 354 cm³, pressure 752 torr, and temperature 100 °C. Calculate the molar mass of the unknown vapor.

10.55 Magnesium can be used as a "getter" in evacuated enclosures to react with the last traces of oxygen. (The magnesium is usually heated by passing an electric current through a wire or ribbon of the metal.) If an enclosure of 0.452 L has a partial pressure of O_2 of 3.5×10^{-6} torr at 27 °C, what mass of magnesium will react according to the following equation?

$$2\,Mg(s) + O_2(g) \longrightarrow 2\,MgO(s)$$

10.56 Calcium hydride, CaH_2, reacts with water to form hydrogen gas:

$$CaH_2(s) + 2\,H_2O(l) \longrightarrow Ca(OH)_2(aq) + 2\,H_2(g)$$

This reaction is sometimes used to inflate life rafts, weather balloons, and the like, when a simple, compact means of generating H_2 is desired. How many grams of CaH_2 are needed to generate 145 L of H_2 gas if the pressure of H_2 is 825 torr at 21 °C?

10.57 The metabolic oxidation of glucose, $C_6H_{12}O_6$, in our bodies produces CO_2, which is expelled from our lungs as a gas:

$$C_6H_{12}O_6(aq) + 6\,O_2(g) \longrightarrow 6\,CO_2(g) + 6\,H_2O(l)$$

(a) Calculate the volume of dry CO_2 produced at body temperature (37 °C) and 0.970 atm when 24.5 g of glucose is consumed in this reaction. (b) Calculate the volume of oxygen you would need, at 1.00 atm and 298 K, to completely oxidize 50.0 g of glucose.

10.58 Both Jacques Charles and Joseph Louis Guy-Lussac were avid balloonists. In his original flight in 1783, Jacques Charles used a balloon that contained approximately 31,150 L of H_2. He generated the H_2 using the reaction between iron and hydrochloric acid:

$$Fe(s) + 2\,HCl(aq) \longrightarrow FeCl_2(aq) + H_2(g)$$

How many kilograms of iron were needed to produce this volume of H_2 if the temperature was 22 °C?

10.59 Hydrogen gas is produced when zinc reacts with sulfuric acid:

$$Zn(s) + H_2SO_4(aq) \longrightarrow ZnSO_4(aq) + H_2(g)$$

If 159 mL of wet H_2 is collected over water at 24 °C and a barometric pressure of 738 torr, how many grams of Zn have been consumed? (The vapor pressure of water is tabulated in Appendix B.)

10.60 Acetylene gas, $C_2H_2(g)$, can be prepared by the reaction of calcium carbide with water:

$$CaC_2(s) + 2\,H_2O(l) \longrightarrow Ca(OH)_2(aq) + C_2H_2(g)$$

Calculate the volume of C_2H_2 that is collected over water at 23 °C by reaction of 1.524 g of CaC_2 if the total pressure of the gas is 753 torr. (The vapor pressure of water is tabulated in Appendix B.)

Partial Pressures (Section 10.6)

10.61 Consider the apparatus shown in the following drawing. (a) When the valve between the two containers is opened and the gases allowed to mix, how does the volume occupied by the N_2 gas change? What is the partial pressure of N_2 after mixing? (b) How does the volume of the O_2 gas change when the gases mix? What is the partial pressure of O_2 in the mixture? (c) What is the total pressure in the container after the gases mix?

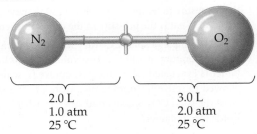

2.0 L
1.0 atm
25 °C

3.0 L
2.0 atm
25 °C

10.62 Consider a mixture of two gases, A and B, confined in a closed vessel. A quantity of a third gas, C, is added to the same vessel at the same temperature. How does the addition of gas C affect the following: (a) the partial pressure of gas A, (b) the total pressure in the vessel, (c) the mole fraction of gas B?

10.63 A mixture containing 0.765 mol He(g), 0.330 mol Ne(g), and 0.110 mol Ar(g) is confined in a 10.00-L vessel at 25 °C. (a) Calculate the partial pressure of each of the gases in the mixture. (b) Calculate the total pressure of the mixture.

10.64 A deep-sea diver uses a gas cylinder with a volume of 10.0 L and a content of 51.2 g of O_2 and 32.6 g of He. Calculate the partial pressure of each gas and the total pressure if the temperature of the gas is 19 °C.

10.65 The atmospheric concentration of CO_2 gas is presently 390 ppm (parts per million, by volume; that is, 390 L of every 10^6 L of the atmosphere are CO_2). What is the mole fraction of CO_2 in the atmosphere?

10.66 A plasma-screen TV contains thousands of tiny cells filled with a mixture of Xe, Ne, and He gases that emits light of specific wavelengths when a voltage is applied. A particular plasma cell, 0.900 mm × 0.300 mm × 10.0 mm, contains 4% Xe in a 1:1 Ne:He mixture at a total pressure of 500 torr. Calculate the number of Xe, Ne, and He atoms in the cell and state the assumptions you need to make in your calculation.

10.67 A piece of dry ice (solid carbon dioxide) with a mass of 5.50 g is placed in a 10.0-L vessel that already contains air at 705 torr and 24 °C. After the carbon dioxide has totally sublimed, what is the partial pressure of the resultant CO_2 gas, and the total pressure in the container at 24 °C?

10.68 A sample of 5.00 mL of diethylether ($C_2H_5OC_2H_5$, density = 0.7134 g/mL) is introduced into a 6.00-L vessel that already contains a mixture of N_2 and O_2, whose partial pressures are P_{N_2} = 0.751 atm and P_{O_2} = 0.208 atm. The temperature is held at 35.0 °C, and the diethylether totally evaporates. (a) Calculate the partial pressure of the diethylether. (b) Calculate the total pressure in the container.

10.69 A rigid vessel containing a 3:1 mol ratio of carbon dioxide and water vapor is held at 200 °C where it has a total pressure of 2.00 atm. If the vessel is cooled to 10 °C so that all of the water vapor condenses, what is the pressure of carbon dioxide? Neglect the volume of the liquid water that forms on cooling.

10.70 If 5.15 g of Ag_2O is sealed in a 75.0-mL tube filled with 760 torr of N_2 gas at 32 °C, and the tube is heated to 320 °C, the Ag_2O decomposes to form oxygen and silver. What is the total pressure inside the tube assuming the volume of the tube remains constant?

10.71 At an underwater depth of 250 ft, the pressure is 8.38 atm. What should the mole percent of oxygen be in the diving gas for the partial pressure of oxygen in the mixture to be 0.21 atm, the same as in air at 1 atm?

10.72 (a) What are the mole fractions of each component in a mixture of 15.08 g of O_2, 8.17 g of N_2, and 2.64 g of H_2? (b) What is the partial pressure in atm of each component of this mixture if it is held in a 15.50-L vessel at 15 °C?

10.73 A quantity of N_2 gas originally held at 5.25 atm pressure in a 1.00-L container at 26 °C is transferred to a 12.5-L container at 20 °C. A quantity of O_2 gas originally at 5.25 atm and 26 °C in a 5.00-L container is transferred to this same container. What is the total pressure in the new container?

10.74 A sample of 3.00 g of $SO_2(g)$ originally in a 5.00-L vessel at 21 °C is transferred to a 10.0-L vessel at 26 °C. A sample of 2.35 g of $N_2(g)$ originally in a 2.50-L vessel at 20 °C is transferred to this same 10.0-L vessel. (a) What is the partial pressure of $SO_2(g)$ in the larger container? (b) What is the partial pressure of $N_2(g)$ in this vessel? (c) What is the total pressure in the vessel?

Kinetic-Molecular Theory of Gases; Effusion and Diffusion (Sections 10.7 and 10.8)

10.75 Determine whether each of the following changes will increase, decrease, or not affect the rate with which gas molecules collide with the walls of their container: (a) increasing the volume of the container, (b) increasing the temperature, (c) increasing the molar mass of the gas.

10.76 Indicate which of the following statements regarding the kinetic-molecular theory of gases are correct. (a) The average kinetic energy of a collection of gas molecules at a given temperature is proportional to $m^{1/2}$. (b) The gas molecules are assumed to exert no forces on each other. (c) All the molecules of a gas at a given temperature have the same kinetic energy. (d) The volume of the gas molecules is negligible in comparison to the total volume in which the gas is contained. (e) All gas molecules move with the same speed if they are at the same temperature.

10.77 Which assumptions are common to both kinetic-molecular theory and the ideal-gas equation?

10.78 Newton had an incorrect theory of gases in which he assumed that all gas molecules repel one another and the walls of their container. Thus, the molecules of a gas are statically and uniformly distributed, trying to get as far apart as possible from one another and the vessel walls. This repulsion gives rise to pressure. Explain why Charles's law argues for the kinetic-molecular theory and against Newton's model.

10.79 WF_6 is one of the heaviest known gases. How much slower is the root-mean-square speed of WF_6 than He at 300 K?

[10.80] You have an evacuated container of fixed volume and known mass and introduce a known mass of a gas sample. Measuring the pressure at constant temperature over time, you are surprised to see it slowly dropping. You measure the mass of the gas-filled container and find that the mass is what it should be—gas plus container—and the mass does not change over time, so you do not have a leak. Suggest an explanation for your observations.

10.81 The temperature of a 5.00-L container of N_2 gas is increased from 20 °C to 250 °C. If the volume is held constant, predict qualitatively how this change affects the following: (a) the average kinetic energy of the molecules; (b) the root-mean-square speed of the molecules; (c) the strength of the impact of an average molecule with the container walls; (d) the total number of collisions of molecules with walls per second.

10.82 Suppose you have two 1-L flasks, one containing N_2 at STP, the other containing CH_4 at STP. How do these systems compare with respect to (a) number of molecules, (b) density, (c) average kinetic energy of the molecules, (d) rate of effusion through a pinhole leak?

10.83 (a) Place the following gases in order of increasing average molecular speed at 25 °C: Ne, HBr, SO_2, NF_3, CO. (b) Calculate the rms speed of NF_3 molecules at 25 °C. (c) Calculate the most probable speed of an ozone molecule in the stratosphere, where the temperature is 270 K.

10.84 (a) Place the following gases in order of increasing average molecular speed at 300 K: CO, SF_6, H_2S, Cl_2, HBr. (b) Calculate the rms speeds of CO and Cl_2 molecules at 300 K. (c) Calculate the most probable speeds of CO and Cl_2 molecules at 300 K.

10.85 Explain the difference between effusion and diffusion.

[10.86] At constant pressure, the mean free path (λ) of a gas molecule is directly proportional to temperature. At constant temperature, λ is inversely proportional to pressure. If you compare two different gas molecules at the same temperature and pressure, λ is inversely proportional to the square of the diameter of the gas molecules. Put these facts together to create a formula for the mean free path of a gas molecule with a proportionality constant (call it R_{mfp}, like the ideal-gas constant) and define units for R_{mfp}.

10.87 Hydrogen has two naturally occurring isotopes, ^1H and ^2H. Chlorine also has two naturally occurring isotopes, ^{35}Cl and ^{37}Cl. Thus, hydrogen chloride gas consists of four distinct types of molecules: $^1H^{35}Cl$, $^1H^{37}Cl$, $^2H^{35}Cl$, and $^2H^{37}Cl$. Place these four molecules in order of increasing rate of effusion.

10.88 As discussed in the "Chemistry Put to Work" box in Section 10.8, enriched uranium can be produced by effusion of gaseous UF_6 across a porous membrane. Suppose a process were developed to allow effusion of gaseous uranium atoms, $U(g)$. Calculate the ratio of effusion rates for ^{235}U and ^{238}U, and compare it to the ratio for UF_6 given in the essay.

10.89 Arsenic(III) sulfide sublimes readily, even below its melting point of 320 °C. The molecules of the vapor phase are found to effuse through a tiny hole at 0.28 times the rate of effusion of Ar atoms under the same conditions of temperature and pressure. What is the molecular formula of arsenic(III) sulfide in the gas phase?

10.90 A gas of unknown molecular mass was allowed to effuse through a small opening under constant-pressure conditions. It required 105 s for 1.0 L of the gas to effuse. Under identical experimental conditions it required 31 s for 1.0 L of O_2 gas to effuse. Calculate the molar mass of the unknown gas. (Remember that the faster the rate of effusion, the shorter the time required for effusion of 1.0 L; in other words, rate is the amount that diffuses over the time it takes to diffuse.)

Nonideal-Gas Behavior (Section 10.9)

10.91 (a) List two experimental conditions under which gases deviate from ideal behavior. (b) List two reasons why the gases deviate from ideal behavior.

10.92 The planet Jupiter has a surface temperature of 140 K and a mass 318 times that of Earth. Mercury (the planet) has a surface temperature between 600 K and 700 K and a mass 0.05 times that of Earth. On which planet is the atmosphere more likely to obey the ideal-gas law? Explain.

10.93 Based on their respective van der Waals constants (Table 10.3), is Ar or CO_2 expected to behave more nearly like an ideal gas at high pressures? Explain.

10.94 Briefly explain the significance of the constants a and b in the van der Waals equation.

10.95 In Sample Exercise 10.16, we found that one mole of Cl_2 confined to 22.41 L at 0 °C deviated slightly from ideal behavior. Calculate the pressure exerted by 1.00 mol Cl_2 confined to a smaller volume, 5.00 L, at 25 °C. (a) First use the ideal-gas equation and (b) then use the van der Waals equation for your calculation. (Values for the van der Waals constants are given in Table 10.3.) (c) Why is the difference between the result for an ideal gas and that calculated using the van der Waals equation greater when the gas is confined to 5.00 L compared to 22.4 L?

10.96 Calculate the pressure that CCl_4 will exert at 40 °C if 1.00 mol occupies 33.3 L, assuming that (a) CCl_4 obeys the ideal-gas equation; (b) CCl_4 obeys the van der Waals equation. (Values for the van der Waals constants are given in Table 10.3.) (c) Which would you expect to deviate more from ideal behavior under these conditions, Cl_2 or CCl_4? Explain.

[10.97] Table 10.3 shows that the van der Waals b parameter has units of L/mol. This implies that we can calculate the size of atoms or molecules from b. Using the value of b for Xe, calculate the radius of a Xe atom and compare it to the value found in Figure 7.7, that is, 1.40 Å. Recall that the volume of a sphere is $(4/3)\pi r^3$.

[10.98] Table 10.3 shows that the van der Waals b parameter has units of L/mol. This means that we can calculate the sizes of atoms or molecules from the b parameter. Refer back to the discussion in Section 7.3. Is the van der Waals radius we calculate from the b parameter of Table 10.3 more closely associated with the bonding or nonbonding atomic radius discussed there? Explain.

Additional Exercises

10.99 A gas bubble with a volume of 1.0 mm^3 originates at the bottom of a lake where the pressure is 3.0 atm. Calculate its volume when the bubble reaches the surface of the lake where the pressure is 730 torr, assuming that the temperature doesn't change.

10.100 A 15.0-L tank is filled with helium gas at a pressure of 1.00×10^2 atm. How many balloons (each 2.00 L) can be inflated to a pressure of 1.00 atm, assuming that the temperature remains constant and that the tank cannot be emptied below 1.00 atm?

10.101 To minimize the rate of evaporation of the tungsten filament, 1.4×10^{-5} mol of argon is placed in a 600-cm^3 lightbulb. What is the pressure of argon in the lightbulb at 23 °C?

10.102 Carbon dioxide, which is recognized as the major contributor to global warming as a "greenhouse gas," is formed when fossil fuels are combusted, as in electrical power plants fueled by coal, oil, or natural gas. One potential way to reduce the amount of CO_2 added to the atmosphere is to store it as a compressed gas in underground formations. Consider a 1000-megawatt coal-fired power plant that produces about 6×10^6 tons of CO_2 per year. (a) Assuming ideal-gas behavior, 1.00 atm, and 27 °C, calculate the volume of CO_2 produced by this power plant. (b) If the CO_2 is stored underground as a liquid at 10 °C and 120 atm and a density of 1.2 g/cm^3, what volume does it possess? (c) If it is stored underground as a gas at 36 °C and 90 atm, what volume does it occupy?

10.103 Propane, C_3H_8, liquefies under modest pressure, allowing a large amount to be stored in a container. (a) Calculate the number of moles of propane gas in a 110-L container at 3.00 atm and 27 °C. (b) Calculate the number of moles of liquid propane that can be stored in the same volume if the density of the liquid is 0.590 g/mL. (c) Calculate the ratio of the number of moles of liquid to moles of gas. Discuss this ratio in light of the kinetic-molecular theory of gases.

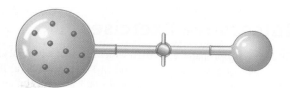

[**10.104**] Nickel carbonyl, $Ni(CO)_4$, is one of the most toxic substances known. The present maximum allowable concentration in laboratory air during an 8-hr workday is 1 ppb (parts per billion) by volume, which means that there is one mole of $Ni(CO)_4$ for every 10^9 moles of gas. Assume 24 °C and 1.00 atm pressure. What mass of $Ni(CO)_4$ is allowable in a laboratory room that is 12 ft × 20 ft × 9 ft?

10.105 When a large evacuated flask is filled with argon gas, its mass increases by 3.224 g. When the same flask is again evacuated and then filled with a gas of unknown molar mass, the mass increase is 8.102 g. (a) Based on the molar mass of argon, estimate the molar mass of the unknown gas. (b) What assumptions did you make in arriving at your answer?

10.106 Consider the arrangement of bulbs shown in the drawing. Each of the bulbs contains a gas at the pressure shown. What is the pressure of the system when all the stopcocks are opened, assuming that the temperature remains constant? (We can neglect the volume of the capillary tubing connecting the bulbs.)

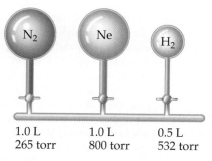

1.0 L 1.0 L 0.5 L
265 torr 800 torr 532 torr

10.107 Assume that a single cylinder of an automobile engine has a volume of 524 cm^3. (a) If the cylinder is full of air at 74 °C and 0.980 atm, how many moles of O_2 are present? (The mole fraction of O_2 in dry air is 0.2095.) (b) How many grams of C_8H_{18} could be combusted by this quantity of O_2, assuming complete combustion with formation of CO_2 and H_2O?

10.108 Assume that an exhaled breath of air consists of 74.8% N_2, 15.3% O_2, 3.7% CO_2, and 6.2% water vapor. (a) If the total pressure of the gases is 0.985 atm, calculate the partial pressure of each component of the mixture. (b) If the volume of the exhaled gas is 455 mL and its temperature is 37 °C, calculate the number of moles of CO_2 exhaled. (c) How many grams of glucose ($C_6H_{12}O_6$) would need to be metabolized to produce this quantity of CO_2? (The chemical reaction is the same as that for combustion of $C_6H_{12}O_6$. See Section 3.2 and Problem 10.57.)

10.109 A 1.42-g sample of helium and an unknown mass of O_2 are mixed in a flask at room temperature. The partial pressure of the helium is 42.5 torr, and that of the oxygen is 158 torr. What is the mass of the oxygen?

[**10.110**] An ideal gas at a pressure of 1.50 atm is contained in a bulb of unknown volume. A stopcock is used to connect this bulb with a previously evacuated bulb that has a volume of 0.800 L as shown here. When the stopcock is opened the gas expands into the empty bulb. If the temperature is held constant during this process and the final pressure is 695 torr, what is the volume of the bulb that was originally filled with gas?

10.111 The density of a gas of unknown molar mass was measured as a function of pressure at 0 °C, as in the table that follows. (a) Determine a precise molar mass for the gas. [*Hint:* Graph d/P versus P.] (b) Why is d/P not a constant as a function of pressure?

Pressure (atm)	1.00	0.666	0.500	0.333	0.250
Density (g/L)	2.3074	1.5263	1.1401	0.7571	0.5660

10.112 A glass vessel fitted with a stopcock valve has a mass of 337.428 g when evacuated. When filled with Ar, it has a mass of 339.854 g. When evacuated and refilled with a mixture of Ne and Ar, under the same conditions of temperature and pressure, it has a mass of 339.076 g. What is the mole percent of Ne in the gas mixture?

10.113 You have a sample of gas at −33 °C. You wish to increase the rms speed by a factor of 2. To what temperature should the gas be heated?

10.114 Consider the following gases, all at STP: Ne, SF_6, N_2, CH_4. (a) Which gas is most likely to depart from the assumption of the kinetic-molecular theory that says there are no attractive or repulsive forces between molecules? (b) Which one is closest to an ideal gas in its behavior? (c) Which one has the highest root-mean-square molecular speed at a given temperature? (d) Which one has the highest total molecular volume relative to the space occupied by the gas? (e) Which has the highest average kinetic-molecular energy? (f) Which one would effuse more rapidly than N_2? (g) Which one would have the largest van der Waals b parameter?

10.115 Does the effect of intermolecular attraction on the properties of a gas become more significant or less significant if (a) the gas is compressed to a smaller volume at constant temperature; (b) the temperature of the gas is increased at constant volume?

10.116 Which of the noble gases other than radon would you expect to depart most readily from ideal behavior? Use the density data in Table 7.8 to show evidence in support of your answer.

10.117 It turns out that the van der Waals constant b equals four times the total volume actually occupied by the molecules of a mole of gas. Using this figure, calculate the fraction of the volume in a container actually occupied by Ar atoms (a) at STP, (b) at 200 atm pressure and 0 °C. (Assume for simplicity that the ideal-gas equation still holds.)

[**10.118**] Large amounts of nitrogen gas are used in the manufacture of ammonia, principally for use in fertilizers. Suppose 120.00 kg of $N_2(g)$ is stored in a 1100.0-L metal cylinder at 280 °C. (a) Calculate the pressure of the gas, assuming ideal-gas behavior. (b) By using the data in Table 10.3, calculate the pressure of the gas according to the van der Waals equation. (c) Under the conditions of this problem, which correction dominates, the one for finite volume of gas molecules or the one for attractive interactions?

Integrative Exercises

10.119 Cyclopropane, a gas used with oxygen as a general anesthetic, is composed of 85.7% C and 14.3% H by mass. (a) If 1.56 g of cyclopropane has a volume of 1.00 L at 0.984 atm and 50.0 °C, what is the molecular formula of cyclopropane? (b) Judging from its molecular formula, would you expect cyclopropane to deviate more or less than Ar from ideal-gas behavior at moderately high pressures and room temperature? Explain. (c) Would cyclopropane effuse through a pinhole faster or more slowly than methane, CH_4?

[10.120] Consider the combustion reaction between 25.0 mL of liquid methanol (density = 0.850 g/mL) and 12.5 L of oxygen gas measured at STP. The products of the reaction are $CO_2(g)$ and $H_2O(g)$. Calculate the volume of liquid H_2O formed if the reaction goes to completion and you condense the water vapor.

10.121 An herbicide is found to contain only C, H, N, and Cl. The complete combustion of a 100.0-mg sample of the herbicide in excess oxygen produces 83.16 mL of CO_2 and 73.30 mL of H_2O vapor at STP. A separate analysis shows that the sample also contains 16.44 mg of Cl. (a) Determine the percentage of the composition of the substance. (b) Calculate its empirical formula. (c) What other information would you need to know about this compound to calculate its true molecular formula?

10.122 A 4.00-g sample of a mixture of CaO and BaO is placed in a 1.00-L vessel containing CO_2 gas at a pressure of 730 torr and a temperature of 25 °C. The CO_2 reacts with the CaO and BaO, forming $CaCO_3$ and $BaCO_3$. When the reaction is complete, the pressure of the remaining CO_2 is 150 torr. (a) Calculate the number of moles of CO_2 that have reacted. (b) Calculate the mass percentage of CaO in the mixture.

[10.123] Ammonia and hydrogen chloride react to form solid ammonium chloride:

$$NH_3(g) + HCl(g) \longrightarrow NH_4Cl(s)$$

Two 2.00-L flasks at 25 °C are connected by a valve, as shown in the drawing on the next page. One flask contains 5.00 g of $NH_3(g)$, and the other contains 5.00 g of $HCl(g)$. When the valve is opened, the gases react until one is completely consumed. (a) Which gas will remain in the system after the reaction is complete? (b) What will be the final pressure of the system after the reaction is complete? (Neglect the volume of the ammonium chloride formed.) (c) What mass of ammonium chloride will be formed?

NH₃(g)		HCl(g)
5.00 g		5.00 g
2.00 L		2.00 L
25 °C		25 °C

10.124 Gas pipelines are used to deliver natural gas (methane, CH_4) to the various regions of the United States. The total volume of natural gas that is delivered is on the order of 2.7×10^{12} L per day, measured at STP. Calculate the total enthalpy change for combustion of this quantity of methane.

(*Note*: Less than this amount of methane is actually combusted daily. Some of the delivered gas is passed through to other regions.)

10.125 Chlorine dioxide gas (ClO_2) is used as a commercial bleaching agent. It bleaches materials by oxidizing them. In the course of these reactions, the ClO_2 is itself reduced. (a) What is the Lewis structure for ClO_2? (b) Why do you think that ClO_2 is reduced so readily? (c) When a ClO_2 molecule gains an electron, the chlorite ion, ClO_2^-, forms. Draw the Lewis structure for ClO_2^-. (d) Predict the O—Cl—O bond angle in the ClO_2^- ion. (e) One method of preparing ClO_2 is by the reaction of chlorine and sodium chlorite:

$$Cl_2(g) + 2\,NaClO_2(s) \longrightarrow 2\,ClO_2(g) + 2\,NaCl(s)$$

If you allow 15.0 g of $NaClO_2$ to react with 2.00 L of chlorine gas at a pressure of 1.50 atm at 21 °C, how many grams of ClO_2 can be prepared?

10.126 Natural gas is very abundant in many Middle Eastern oil fields. However, the costs of shipping the gas to markets in other parts of the world are high because it is necessary to liquefy the gas, which is mainly methane and has a boiling point at atmospheric pressure of −164 °C. One possible strategy is to oxidize the methane to methanol, CH_3OH, which has a boiling point of 65 °C and can therefore be shipped more readily. Suppose that 10.7×10^9 ft³ of methane at atmospheric pressure and 25 °C is oxidized to methanol. (a) What volume of methanol is formed if the density of CH_3OH is 0.791 g/mL? (b) Write balanced chemical equations for the oxidations of methane and methanol to $CO_2(g)$ and $H_2O(l)$. Calculate the total enthalpy change for complete combustion of the 10.7×10^9 ft³ of methane just described and for complete combustion of the equivalent amount of methanol, as calculated in part (a). (c) Methane, when liquefied, has a density of 0.466 g/mL; the density of methanol at 25 °C is 0.791 g/mL. Compare the enthalpy change upon combustion of a unit volume of liquid methane and liquid methanol. From the standpoint of energy production, which substance has the higher enthalpy of combustion per unit volume?

[10.127] Gaseous iodine pentafluoride, IF_5, can be prepared by the reaction of solid iodine and gaseous fluorine:

$$I_2(s) + 5\,F_2(g) \longrightarrow 2\,IF_5(g)$$

A 5.00-L flask containing 10.0 g of I_2 is charged with 10.0 g of F_2, and the reaction proceeds until one of the reagents is completely consumed. After the reaction is complete, the temperature in the flask is 125 °C. (a) What is the partial pressure of IF_5 in the flask? (b) What is the mole fraction of IF_5 in the flask (c) Draw the Lewis structure of IF_5. (d) What is the total mass of reactants and products in the flask?

[10.128] A 6.53-g sample of a mixture of magnesium carbonate and calcium carbonate is treated with excess hydrochloric acid. The resulting reaction produces 1.72 L of carbon dioxide gas at 28 °C and 743 torr pressure. (a) Write balanced chemical equations for the reactions that occur between hydrochloric acid and each component of the mixture. (b) Calculate the total number of moles of carbon dioxide that forms from these reactions. (c) Assuming that the reactions are complete, calculate the percentage by mass of magnesium carbonate in the mixture.

Design an Experiment

You are given a cylinder of an unknown, non-radioactive, noble gas and tasked to determine its molar mass and use that value to identify the gas. The tools available to you are several empty mylar balloons that are about the size of a grapefruit when inflated (gases diffuse through mylar much more slowly than conventional latex balloons), an analytical balance, and three graduated glass beakers of different sizes (100 mL, 500 mL, and 2 L). (**a**) To how many significant figures would you need to determine the molar mass to identify the gas? (**b**) Propose an experiment or series of experiments that would allow you to determine the molar mass of the unknown gas. Describe the tools, calculations, and assumptions you would need to use. (**c**) If you had access to a broader range of analytical instruments, describe an alternative way you could identify the gas using any experimental methods that you have learned about in the earlier chapters.

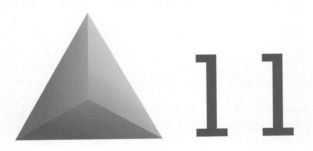

11

Liquids and Intermolecular Forces

The lotus plant grows in aquatic environments. To thrive in such an environment the surface of a lotus leaf is highly water repellent. Scientists call surfaces with this property "superhydrophobic." The superhydrophobic character of the lotus leaf not only allows it to float on water but also causes any water that falls on the leaf to bead up and roll off. The water drops collect dirt as they roll off, keeping the leaf clean, even in the muddy ponds and lakes where lotus plants tend to grow. Because of its self-cleaning properties, the lotus plant is considered a symbol of purity in many Eastern cultures.

What forces cause the lotus leaf to repel water so efficiently? Although this plant's self-cleaning nature has been known for millennia, the effect was not fully understood until the 1970s when scanning electron microscopy images revealed a rough, one might say, mountainous, leaf surface (**Figure 11.1**). The rough surface helps minimize contact between water and leaf.

Another important factor contributing to the plant's self-cleaning nature is the contrast between the molecular composition of the leaf and that of the water. The leaf is coated by hydrocarbon molecules, which are not attracted to water molecules. As a result the water molecules preferentially surround themselves with other water molecules, thereby minimizing their contact with the surface.

The lotus effect has inspired scientists to design superhydrophobic surfaces for applications such as self-cleaning windows and water-repellent clothing. To understand the lotus effect and other phenomena involving liquids and solids, we must understand **intermolecular forces**, the forces that exist *between* molecules. Only by understanding

▶ **BECAUSE THE LEAVES OF THE LOTUS PLANT** are highly water repellent, any water on a leaf beads up to minimize contact with the leaf surface.

WHAT'S AHEAD ▶

11.1 A MOLECULAR COMPARISON OF GASES, LIQUIDS, AND SOLIDS We begin with a comparison of solids, liquids, and gases from a molecular perspective. This comparison reveals the important roles that *temperature* and *intermolecular forces* play in determining the physical state of a substance.

11.2 INTERMOLECULAR FORCES We then examine four intermolecular forces: *dispersion forces, dipole–dipole forces, hydrogen bonds,* and *ion–dipole forces.*

11.3 SELECT PROPERTIES OF LIQUIDS We learn that the nature and strength of the intermolecular forces between molecules are largely responsible for many properties of liquids, including *viscosity* and *surface tension.*

11.4 PHASE CHANGES We explore *phase changes*—the transitions of matter between the gaseous, liquid, and solid states—and their associated energies.

11.5 VAPOR PRESSURE We examine the *dynamic equilibrium* that exists between a liquid and its gaseous state and introduce *vapor pressure.*

11.6 PHASE DIAGRAMS We learn how to read *phase diagrams*, which are graphic representations of the equilibria among the gaseous, liquid, and solid phases.

11.7 LIQUID CRYSTALS We learn about substances that pass into a liquid crystalline phase, which is an intermediate phase between the solid and liquid states. A substance in the liquid crystalline phase has some of the structural order of a solid and some of the freedom of motion of a liquid.

▲ **Figure 11.1 A microscopic view of a water droplet on the surface of a lotus leaf.**

the nature and strength of these forces can we understand how the composition and structure of a substance are related to its physical properties in the liquid or solid state.

11.1 | A Molecular Comparison of Gases, Liquids, and Solids

As we learned in Chapter 10, the molecules in a gas are widely separated and in a state of constant, chaotic motion. One of the key tenets of kinetic-molecular theory of gases is the assumption that we can neglect the interactions between molecules. ∞ (Section 10.7) The properties of liquids and solids are quite different from those of gases largely because the intermolecular forces in liquids and solids are much stronger. A comparison of the properties of gases, liquids, and solids is given in ▼ Table 11.1.

In liquids the intermolecular attractive forces are strong enough to hold particles close together. Thus, liquids are much denser and far less compressible than gases. Unlike gases, liquids have a definite volume, independent of the size and shape of their container. The attractive forces in liquids are not strong enough, however, to keep the particles from moving past one another. Thus, any liquid can be poured, and assumes the shape of the container it occupies.

In solids the intermolecular attractive forces are strong enough to hold particles close together and to lock them virtually in place. Solids, like liquids, are not very compressible because the particles have little free space between them. Because the particles in a solid or liquid are fairly close together compared with those of a gas, we often refer to solids and liquids as *condensed phases*. We will study solids in Chapter 12. For now it is sufficient to know that the particles of a solid are not free to undergo long-range movement, which makes solids rigid.*

▶ Figure 11.2 compares the three states of matter. *The state of a substance depends largely on the balance between the kinetic energies of the particles (atoms, molecules, or*

Table 11.1 Some Characteristic Properties of the States of Matter

Gas	Assumes both volume and shape of its container
	Expands to fill its container
	Is compressible
	Flows readily
	Diffusion within a gas occurs rapidly
Liquid	Assumes shape of portion of container it occupies
	Does not expand to fill its container
	Is virtually incompressible
	Flows readily
	Diffusion within a liquid occurs slowly
Solid	Retains own shape and volume
	Does not expand to fill its container
	Is virtually incompressible
	Does not flow
	Diffusion within a solid occurs extremely slowly

*The atoms in a solid are able to vibrate in place. As the temperature of the solid increases, the vibrational motion increases.

GO FIGURE

For a given substance, do you expect the density of the substance in its liquid state to be closer to the density in the gaseous state or in the solid state?

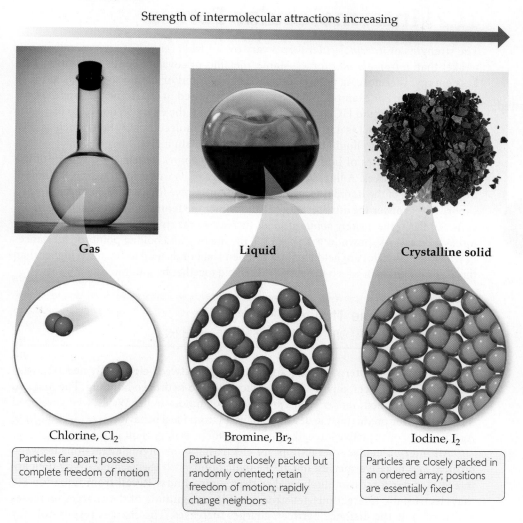

Strength of intermolecular attractions increasing

Gas

Liquid

Crystalline solid

Chlorine, Cl_2

Bromine, Br_2

Iodine, I_2

Particles far apart; possess complete freedom of motion

Particles are closely packed but randomly oriented; retain freedom of motion; rapidly change neighbors

Particles are closely packed in an ordered array; positions are essentially fixed

▲ Figure 11.2 **Gases, liquids, and solids.** Chlorine, bromine, and iodine are all made up of diatomic molecules as a result of covalent bonding. However, due to differences in the strength of the intermolecular forces, they exist in three different states at room temperature and standard pressure: Cl_2 gaseous, Br_2 liquid, I_2 solid.

ions) *and the interparticle energies of attraction,* as summarized in ▶ Table 11.2. The kinetic energies, which depend on temperature, tend to keep the particles apart and moving. The interparticle attractions tend to draw the particles together. Substances that are gases at room temperature have much weaker interparticle attractions than those that are liquids; substances that are liquids have weaker interparticle attractions than those that are solids. The different states of matter adopted by the halogens at room temperature—iodine is a solid, bromine is a liquid, and chlorine is a gas—are a direct consequence of a decrease in the strength of the intermolecular forces as we move from I_2 to Br_2 to Cl_2.

We can change a substance from one state to another by heating or cooling, which changes the average kinetic energy of the particles. NaCl, for example, a solid at room temperature, melts at 1074 K and boils at 1686 K under 1 atm pressure, and Cl_2, a gas at room temperature, liquefies at 239 K and solidifies at 172 K under 1 atm pressure. As the temperature of a gas decreases, the average kinetic energy of its particles decreases, allowing the attractions between the particles to first draw the particles close together, forming a liquid, and then to virtually lock them in place, forming a solid. Increasing the pressure on a gas can also drive transformations from gas to liquid to solid because the increased pressure

Table 11.2 Comparing Kinetic Energies and Energies of Attractions for States of Matter

Gas	Kinetic energies >> energies of attraction
Liquid	Comparable kinetic energies and energies of attraction
Solid	Energies of attraction >> kinetic energies

brings the molecules closer together, thus making intermolecular forces more effective. For example, propane (C_3H_8) is a gas at room temperature and 1 atm pressure, whereas liquefied propane (LP) is a liquid at room temperature because it is stored under much higher pressure.

11.2 | Intermolecular Forces

The strengths of *intermolecular* forces vary over a wide range but are generally much weaker than *intramolecular* forces—ionic, metallic, or covalent bonds (◀ Figure 11.3). Less energy, therefore, is required to vaporize a liquid or melt a solid than to break covalent bonds. For example, only 16 kJ/mol is required to overcome the intermolecular attractions in liquid HCl to vaporize it. In contrast, the energy required to break the covalent bond in HCl is 431 kJ/mol. Thus, when a molecular substance such as HCl changes from solid to liquid to gas, the molecules remain intact.

Many properties of liquids, including *boiling points,* reflect the strength of the intermolecular forces. A liquid boils when bubbles of its vapor form within the liquid. The molecules of the liquid must overcome their attractive forces to separate and form a vapor. The stronger the attractive forces, the higher the temperature at which the liquid boils. Similarly, the *melting points* of solids increase as the strengths of the intermolecular forces increase. As shown in ▼ Table 11.3, the melting and boiling points of substances in which the particles are held together by chemical bonds tend to be much higher than those of substances in which the particles are held together by intermolecular forces.

GO FIGURE

How would you expect the H—Cl distance represented by the red dotted line to compare with the H—Cl distance within the HCl molecule?

Strong intramolecular attraction (covalent bond)

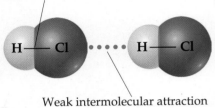

Weak intermolecular attraction

▲ **Figure 11.3 Intermolecular and intramolecular interactions.**

▲ Give It Some Thought

When water boils, what are the bubbles composed of?

Three types of intermolecular attractions exist between electrically neutral molecules: dispersion forces, dipole–dipole attractions, and hydrogen bonding. The first two are collectively called *van der Waals forces* after Johannes van der Waals, who developed the equation for predicting the deviation of gases from ideal behavior. ∞ (Section 10.9) Another kind of attractive force, the ion–dipole force, is important in solutions.

All intermolecular interactions are electrostatic, involving attractions between positive and negative species, much like ionic bonds. ∞ (Section 8.2) Why then are intermolecular forces so much weaker than ionic bonds? Recall from Equation 8.4 that electrostatic interactions get stronger as the magnitude of the charges increases and weaker as the distance between charges increases. The charges responsible for intermolecular forces are generally much smaller than the charges in ionic compounds. For example, from its dipole moment it is possible to estimate charges of +0.178 and −0.178 for the hydrogen and chlorine ends of the HCl molecule, respectively (see Sample Exercise 8.5). Furthermore, the distances between molecules are often larger than the distances between atoms held together by chemical bonds.

Table 11.3 Melting and Boiling Points of Representative Substances

Force Holding Particles Together	Substance	Melting Point (K)	Boiling Point (K)
Chemical bonds			
Ionic bonds	Lithium fluoride (LiF)	1118	1949
Metallic bonds	Beryllium (Be)	1560	2742
Covalent bonds	Diamond (C)	3800	4300
Intermolecular forces			
Dispersion force	Nitrogen (N_2)	63	77
Dipole–dipole force	Hydrogen chloride (HCl)	158	188
Hydrogen-bonding force	Hydrogen fluoride (HF)	190	293

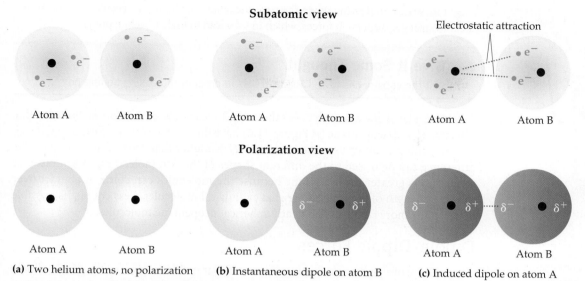

Subatomic view

Atom A   Atom B   Atom A   Atom B   Atom A   Atom B

Polarization view

Atom A   Atom B   Atom A   Atom B   Atom A   Atom B

(a) Two helium atoms, no polarization   **(b)** Instantaneous dipole on atom B   **(c)** Induced dipole on atom A

▲ **Figure 11.4 Dispersion forces.** "Snapshots" of the charge distribution for a pair of helium atoms at three instants.

Dispersion Forces

You might think there would be no electrostatic interactions between electrically neutral, nonpolar atoms and/or molecules. Yet some kind of attractive interactions must exist because nonpolar gases like helium, argon, and nitrogen can be liquefied. Fritz London, a German-American physicist, first proposed the origin of this attraction in 1930. London recognized that the motion of electrons in an atom or molecule can create an *instantaneous,* or momentary, dipole moment.

In a collection of helium atoms, for example, the *average* distribution of the electrons about each nucleus is spherically symmetrical as shown in ▲ **Figure 11.4(a)**. The atoms are nonpolar and so possess no permanent dipole moment. The *instantaneous* distribution of the electrons, however, can be different from the average distribution. If we could freeze the motion of the electrons at any given instant, both electrons could be on one side of the nucleus. At just that instant, the atom has an instantaneous dipole moment as shown in Figure 11.4(**b**). The motions of electrons in one atom influence the motions of electrons in its neighbors. The instantaneous dipole on one atom can induce an instantaneous dipole on an adjacent atom, causing the atoms to be attracted to each other as shown in Figure 11.4(**c**). This attractive interaction is called the **dispersion force** (or the *London dispersion force* in some texts). It is significant only when molecules are very close together.

The strength of the dispersion force depends on the ease with which the charge distribution in a molecule can be distorted to induce an instantaneous dipole. The ease with which the charge distribution is distorted is called the molecule's **polarizability**. We can think of the polarizability of a molecule as a measure of the "squashiness" of its electron cloud: The greater the polarizability, the more easily the electron cloud can be distorted to give an instantaneous dipole. Therefore, more polarizable molecules have larger dispersion forces.

In general, polarizability increases as the number of electrons in an atom or molecule increases. The strength of dispersion forces therefore tends to increase with increasing atomic or molecular size. Because molecular size and mass generally parallel each other, *dispersion forces tend to increase in strength with increasing molecular weight.* We can see this in the boiling points of the halogens and noble gases (▶ **Figure 11.5**), where dispersion forces are the only intermolecular forces at work. In both families the molecular

▲ **GO FIGURE**

Why is the boiling point of the halogen in each period greater than the noble gas?

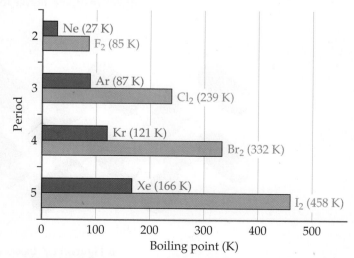

▲ **Figure 11.5 Boiling points of the halogens and noble gases.** This plot shows how the boiling points increase due to stronger dispersion forces as the molecular weight increases.

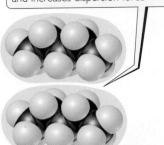

Linear molecule—larger surface area enhances intermolecular contact and increases dispersion force

n-Pentane (C_5H_{12})
bp = 309.4 K

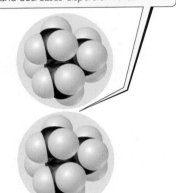

Spherical molecule—smaller surface area diminishes intermolecular contact and decreases dispersion force

Neopentane (C_5H_{12})
bp = 282.7 K

▲ **Figure 11.6 Molecular shape affects intermolecular attraction.** Molecules of *n*-pentane make more contact with each other than do neopentane molecules. Thus, *n*-pentane has stronger intermolecular attractive forces and therefore a higher boiling point.

weight increases on moving down the periodic table. The higher molecular weights translate into stronger dispersion forces, which in turn lead to higher boiling points.

 Give It Some Thought

List the substances CCl_4, CBr_4, and CH_4 in order of increasing boiling point.

Molecular shape also influences the magnitudes of dispersion forces. For example, *n*-pentane* and neopentane (◀ Figure 11.6) have the same molecular formula (C_5H_{12}), yet the boiling point of *n*-pentane is about 27 K higher than that of neopentane. The difference can be traced to the different shapes of the two molecules. Intermolecular attraction is greater for *n*-pentane because the molecules can come in contact over the entire length of the long, somewhat cylindrical molecules. Less contact is possible between the more compact and nearly spherical neopentane molecules.

Dipole–Dipole Forces

The presence of a permanent dipole moment in polar molecules gives rise to **dipole–dipole forces**. These forces originate from electrostatic attractions between the partially positive end of one molecule and the partially negative end of a neighboring molecule. Repulsions can also occur when the positive (or negative) ends of two molecules are in close proximity. Dipole–dipole forces are effective only when molecules are very close together.

To see the effect of dipole–dipole forces, we compare the boiling points of two compounds of similar molecular weight: acetonitrile (CH_3CN, MW 41 amu, bp 355 K) and propane ($CH_3CH_2CH_3$, MW 44 amu, bp 231 K). Acetonitrile is a polar molecule, with a dipole moment of 3.9 D, so dipole–dipole forces are present. However, propane is essentially nonpolar, which means that dipole–dipole forces are absent. Because acetonitrile and propane have similar molecular weights, dispersion forces are similar for these two molecules. Therefore, the higher boiling point of acetonitrile can be attributed to dipole–dipole forces.

To better understand these forces, consider how CH_3CN molecules pack together in the solid and liquid states. In the solid [▼ Figure 11.7(a)], the molecules are arranged with the negatively charged nitrogen end of each molecule close to the positively charged —CH_3 ends of its neighbors. In the liquid [Figure 11.7(b)], the molecules are free to move with respect to one another, and their arrangement becomes more disordered. This means that, at any given instant, both attractive and repulsive dipole–dipole interactions are present. However, not only are there more attractive interactions than repulsive ones, but also molecules that are attracting each other spend more time near each other than do molecules that are repelling each other. The overall effect is a net attraction strong enough to keep the molecules in liquid CH_3CN from moving apart to form a gas.

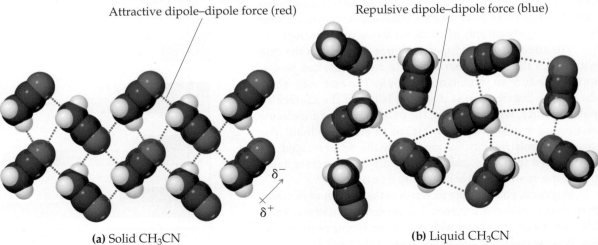

Attractive dipole–dipole force (red)

Repulsive dipole–dipole force (blue)

(a) Solid CH_3CN

(b) Liquid CH_3CN

▲ **Figure 11.7 Dipole–dipole interactions.** The dipole–dipole interactions in (a) crystalline CH_3CN and (b) liquid CH_3CN.

*The *n* in *n*-pentane is an abbreviation for the word *normal*. A normal hydrocarbon is one in which the carbon atoms are arranged in a straight chain. ☐☐☐ (Section 2.9)

GO FIGURE

Moving from left to right, do the dispersion forces get stronger, get weaker, or stay roughly the same in the molecules shown here?

Propane	Dimethyl ether	Acetaldehyde	Acetonitrile
$CH_3CH_2CH_3$	CH_3OCH_3	CH_3CHO	CH_3CN
MW = 44 amu	MW = 46 amu	MW = 44 amu	MW = 41 amu
μ = 0.1 D	μ = 1.3 D	μ = 2.7 D	μ = 3.9 D
bp = 231 K	bp = 248 K	bp = 294 K	bp = 355 K

Increasing polarity
Increasing strength of dipole–dipole forces

▲ Figure 11.8 **Molecular weights, dipole moments, and boiling points of several simple organic substances.**

For molecules of approximately equal mass and size, the strength of intermolecular attractions increases with increasing polarity, a trend we see in ▲ Figure 11.8. Notice how the boiling point increases as the dipole moment increases.

Hydrogen Bonding

▼ Figure 11.9 shows the boiling points of the binary compounds that form between hydrogen and the elements in groups 4A through 7A. The boiling points of the compounds containing group 4A elements (CH_4 through SnH_4, all nonpolar)

 GO FIGURE

Why is the boiling point of SiH_4 higher than that of CH_4?

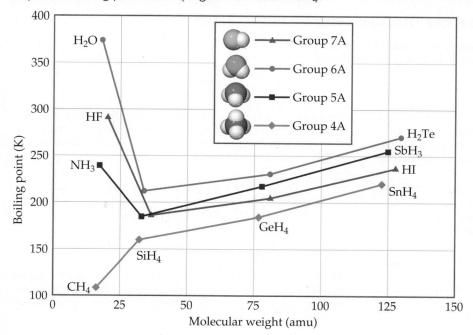

▲ Figure 11.9 **Boiling points of the covalent hydrides of the elements in groups 4A–7A as a function of molecular weight.**

GO FIGURE

To form a hydrogen bond, what must the non-hydrogen atom (N, O, or F) involved in the bond possess?

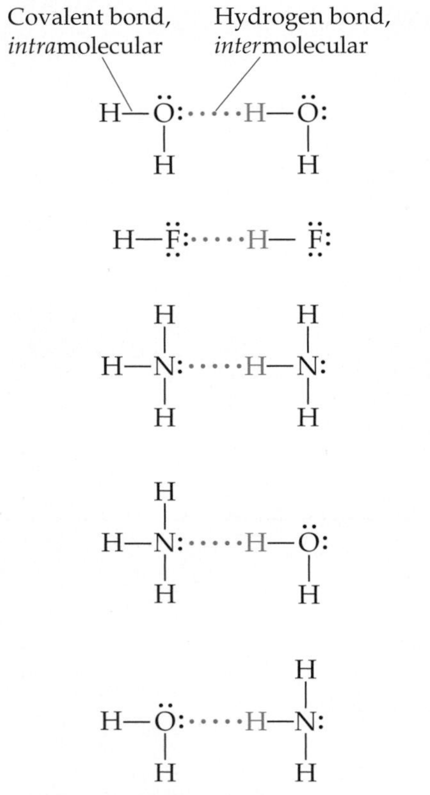

Covalent bond, *intra*molecular   Hydrogen bond, *inter*molecular

H—Ö:····H—Ö:
     | |
     H   H

H—F̈:·····H—F̈:

     H   H
     | |
H—N:·····H—N:
     | |
     H   H

     H   H
     | |
H—N:·····H—Ö:
     | |
     H   H

                    H
                    |
H—Ö:·····H—N:
     | |
     H   H

▲ **Figure 11.10 Hydrogen bonding.** Hydrogen bonding can occur when an H atom is bonded to an N, O, or F atom.

increase systematically moving down the group. This is the expected trend because polarizability and, hence, dispersion forces generally increase as molecular weight increases. The heavier members of groups 5A, 6A, and 7A follow the same trend, but NH_3, H_2O, and HF have boiling points that are much higher than expected. In fact, these three compounds also have many other characteristics that distinguish them from other substances of similar molecular weight and polarity. For example, water has a high melting point, a high specific heat, and a high heat of vaporization. Each of these properties indicates that the intermolecular forces are abnormally strong.

The strong intermolecular attractions in HF, H_2O, and NH_3 result from *hydrogen bonding. A **hydrogen bond** is an attraction between a hydrogen atom attached to a highly electronegative atom (usually F, O, or N) and a nearby small electronegative atom in another molecule or chemical group.* Thus, H—F, H—O, or H—N bonds in one molecule can form hydrogen bonds with an F, O, or N atom in another molecule. Several examples of hydrogen bonds are shown in ◄ Figure 11.10, including the hydrogen bond that exists between the H atom in an H_2O molecule and the O atom of an adjacent H_2O molecule. Notice in each case that the H atom in the hydrogen bond interacts with a nonbonding electron pair.

Hydrogen bonds can be considered a special type of dipole–dipole attraction. Because N, O, and F are so electronegative, a bond between hydrogen and any of these elements is quite polar, with hydrogen at the positive end (remember the + on the right-hand side of the dipole symbol represents the positive end of the dipole):

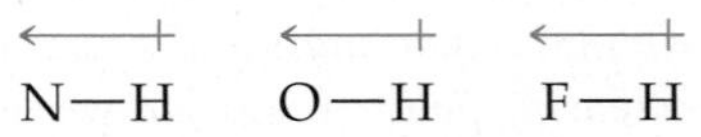

N—H   O—H   F—H

The hydrogen atom has no inner electrons. Thus, the positive side of the dipole has the concentrated charge of the nearly bare hydrogen nucleus. This positive charge is attracted to the negative charge of an electronegative atom in a nearby molecule. Because the electron-poor hydrogen is so small, it can approach an electronegative atom very closely and, thus, interact strongly with it.

SAMPLE EXERCISE 11.1 | Identifying Substances That Can Form Hydrogen Bonds

In which of these substances is hydrogen bonding likely to play an important role in determining physical properties: methane (CH_4), hydrazine (H_2NNH_2), methyl fluoride (CH_3F), hydrogen sulfide (H_2S)?

SOLUTION

Analyze We are given the chemical formulas of four compounds and asked to predict whether they can participate in hydrogen bonding. All the compounds contain H, but hydrogen bonding usually occurs only when the hydrogen is covalently bonded to N, O, or F.

Plan We analyze each formula to see if it contains N, O, or F directly bonded to H. There also needs to be a nonbonding pair of electrons on an electronegative atom (usually N, O, or F) in a nearby molecule, which can be revealed by drawing the Lewis structure for the molecule.

Solve The foregoing criteria eliminate CH_4 and H_2S, which do not contain H bonded to N, O, or F. They also eliminate CH_3F, whose Lewis structure shows a central C atom surrounded by three H atoms and an F atom. (Carbon always forms four bonds, whereas hydrogen and fluorine form one each.) Because the molecule contains a C—F

bond and not an H—F bond, it does not form hydrogen bonds. In H_2NNH_2, however, we find N—H bonds, and the Lewis structure shows a nonbonding pair of electrons on each N atom, telling us hydrogen bonds can exist between the molecules:

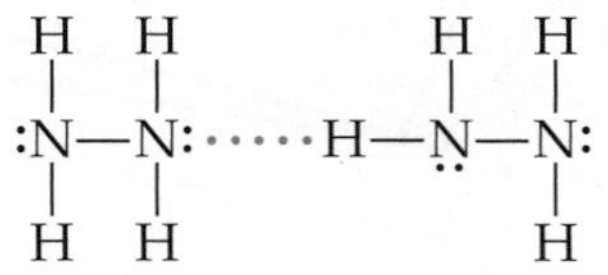

        H H   H H
        | | | |
    :N—N:·····H—N—N:
        | | ·· |
        H H   H

Check Although we can generally identify substances that participate in hydrogen bonding based on their containing N, O, or F covalently bonded to H, drawing the Lewis structure for the interaction provides a way to check the prediction.

Practice Exercise 1

Which of the following substances is most likely to be a liquid at room temperature?
(a) Formaldehyde, H_2CO; (b) fluoromethane, CH_3F; (c) hydrogen cyanide, HCN; (d) hydrogen peroxide, H_2O_2; (e) hydrogen sulfide, H_2S.

Practice Exercise 2

In which of these substances is significant hydrogen bonding possible: methylene chloride (CH_2Cl_2), phosphine (PH_3), chloramine (NH_2Cl), acetone (CH_3COCH_3)?

The energies of hydrogen bonds vary from about 5 to 25 kJ/mol, although there are isolated examples of hydrogen bond energies close to 100 kJ/mol. Thus, hydrogen bonds are typically much weaker than covalent bonds, which have bond enthalpies of 150–1100 kJ/mol (see Table 8.4). Nevertheless, because hydrogen bonds are generally stronger than dipole–dipole or dispersion forces, they play important roles in many chemical systems, including those of biological significance. For example, hydrogen bonds help stabilize the structures of proteins and are also responsible for the way that DNA is able to carry genetic information.

One remarkable consequence of hydrogen bonding is seen in the densities of ice and liquid water. In most substances the molecules in the solid are more densely packed than those in the liquid, making the solid phase denser than the liquid phase. By contrast, the density of ice at 0 °C (0.917 g/mL) is less than that of liquid water at 0 °C (1.00 g/mL), so ice floats on liquid water.

The lower density of ice can be understood in terms of hydrogen bonding. In ice, the H_2O molecules assume the ordered, open arrangement shown in ▼ Figure 11.11. This arrangement optimizes hydrogen bonding between molecules, with each H_2O molecule forming hydrogen bonds to four neighboring H_2O molecules. These hydrogen bonds, however, create the cavities seen in the middle image of Figure 11.11. When ice melts, the motions of the molecules cause the structure to collapse. The hydrogen bonding in the liquid is more random than that in the solid but is strong enough to hold the molecules close together. Consequently, liquid water has a denser structure than ice, meaning that a given mass of water occupies a smaller volume than the same mass of ice.

▲ GO FIGURE

What is the approximate H—O ··· H bond angle in ice, where H—O is the covalent bond and O ··· H is the hydrogen bond?

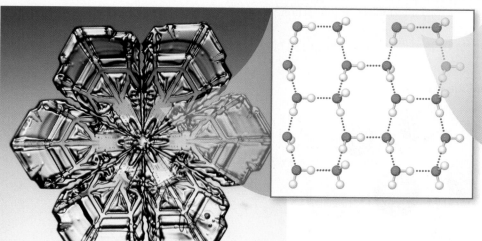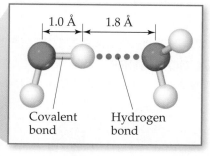

▲ **Figure 11.11 Hydrogen bonding in ice.** The empty channels in the structure of ice make water less dense as a solid than as a liquid.

▲ Figure 11.12 **Expansion of water upon freezing.**

The expansion of water upon freezing (◄ Figure 11.12) is responsible for many phenomena we take for granted. It causes icebergs to float and water pipes to burst in cold weather. The lower density of ice compared to liquid water also profoundly affects life on Earth. Because ice floats, it covers the top of the water when a lake freezes, thereby insulating the water. If ice were denser than water, ice forming at the top of a lake would sink to the bottom, and the lake could freeze solid. Most aquatic life could not survive under these conditions.

 Give It Some Thought

What major type of attractive interaction must be overcome for water to evaporate?

Ion–Dipole Forces

An **ion–dipole force** exists between an ion and a polar molecule (▼ Figure 11.13). Cations are attracted to the negative end of a dipole, and anions are attracted to the positive end. The magnitude of the attraction increases as either the ionic charge or the magnitude of the dipole moment increases. Ion–dipole forces are especially important for solutions of ionic substances in polar liquids, such as a solution of NaCl in water. ∞∞(Section 4.1)

 Give It Some Thought

In which mixture do you expect to find ion–dipole forces between solute and solvent? CH_3OH in water or $Ca(NO_3)_2$ in water.

Comparing Intermolecular Forces

We can identify the intermolecular forces operative in a substance by considering its composition and structure. *Dispersion forces are found in all substances.* The strength of these attractive forces increases with increasing molecular weight and depends on molecular shapes. With polar molecules dipole–dipole forces are also operative, but these forces often make a smaller contribution to the total intermolecular attraction than do dispersion forces. For example, in liquid HCl, dispersion forces are estimated to account for more than 80% of the total attraction between molecules, while dipole–dipole attractions account for the rest. Hydrogen bonds, when present, make an important contribution to the total intermolecular interaction.

 GO FIGURE

Why does the O side of H_2O point toward the Na^+ ion?

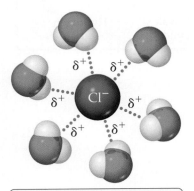

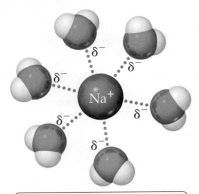

Positive ends of polar molecules are oriented toward negatively charged anion

Negative ends of polar molecules are oriented toward positively charged cation

▲ Figure 11.13 **Ion–dipole forces.**

In general, the energies associated with dispersion forces are 0.1–30 kJ/mol. The wide range reflects the wide range in polarizabilities of molecules. By comparison, the energies associated with dipole–dipole forces and hydrogen bonds are approximately 2–15 kJ/mol and 10–40 kJ/mol, respectively. Ion–dipole forces tend to be stronger than the aforementioned intermolecular forces, with energies typically exceeding 50 kJ/mol. All these interactions are considerably weaker than covalent and ionic bonds, which have energies that are hundreds of kilojoules per mole.

When comparing the relative strengths of intermolecular attractions, consider these generalizations:

1. **When the molecules of two substances have comparable molecular weights and shapes, dispersion forces are approximately equal in the two substances.** Differences in the magnitudes of the intermolecular forces are due to differences in the strengths of dipole–dipole attractions. The intermolecular forces get stronger as molecule polarity increases, with those molecules capable of hydrogen bonding having the strongest interactions.

2. **When the molecules of two substances differ widely in molecular weights, and there is no hydrogen bonding, dispersion forces tend to determine which substance has the stronger intermolecular attractions.** Intermolecular attractive forces are generally higher in the substance with higher molecular weight.

▼ Figure 11.14 presents a systematic way of identifying the intermolecular forces in a particular system.

It is important to realize that the effects of all these attractions are additive. For example, acetic acid, CH_3COOH, and 1-propanol, $CH_3CH_2CH_2OH$, have the same molecular weight, 60 amu, and both are capable of forming hydrogen bonds. However, a pair of acetic

▲ GO FIGURE

At which point in this flowchart would a distinction be made between SiH_4 and SiH_2Br_2?

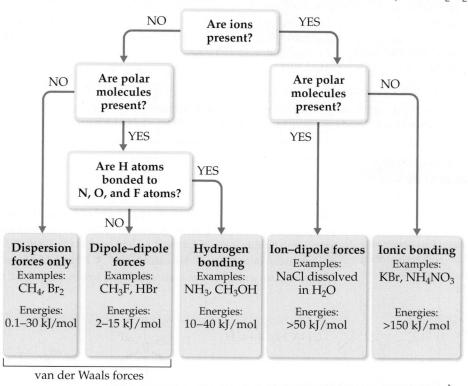

▲ **Figure 11.14 Flowchart for determining intermolecular forces.** Multiple types of intermolecular forces can be at work in a given substance or mixture. In particular, dispersion forces occur in all substances.

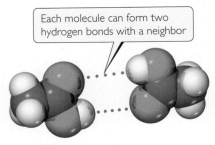

Each molecule can form two hydrogen bonds with a neighbor

Acetic acid, CH₃COOH
MW = 60 amu
bp = 391 K

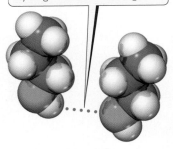

Each molecule can form one hydrogen bond with a neighbor

1-Propanol, CH₃CH₂CH₂OH
MW = 60 amu
bp = 370 K

▲ Figure 11.15 Hydrogen bonding in acetic acid and 1-propanol. The greater the number of hydrogen bonds possible, the more tightly the molecules are held together and, therefore, the higher the boiling point.

acid molecules can form two hydrogen bonds, whereas a pair of 1-propanol molecules can form only one (◄ Figure 11.15). Hence, the boiling point of acetic acid is higher. These effects can be important, especially for very large polar molecules such as proteins, which have multiple dipoles over their surfaces. These molecules can be held together in solution to a surprisingly high degree due to the presence of multiple dipole–dipole attractions.

SAMPLE EXERCISE 11.2 Predicting Types and Relative Strengths of Intermolecular Attractions

List the substances $BaCl_2$, H_2, CO, HF, and Ne in order of increasing boiling point.

SOLUTION

Analyze We need to assess the intermolecular forces in these substances and use that information to determine the relative boiling points.

Plan The boiling point depends in part on the attractive forces in each substance. We need to order these substances according to the relative strengths of the different kinds of intermolecular attractions.

Solve The attractive forces are stronger for ionic substances than for molecular ones, so $BaCl_2$ should have the highest boiling point. The intermolecular forces of the remaining substances depend on molecular weight, polarity, and hydrogen bonding. The molecular weights are H_2, 2; CO, 28; HF, 20; and Ne, 20. The boiling point of H_2 should be the lowest because it is nonpolar and has the lowest molecular weight. The molecular weights of CO, HF, and Ne are similar. Because HF can hydrogen-bond, it should have the highest boiling point of the three. Next is CO, which is slightly polar and has the highest molecular weight. Finally, Ne, which is nonpolar, should have the lowest boiling point of these three. The predicted order of boiling points is, therefore,

$$H_2 < Ne < CO < HF < BaCl_2$$

Check The boiling points reported in the literature are H_2, 20 K; Ne, 27 K; CO, 83 K; HF, 293 K; and $BaCl_2$, 1813 K—in agreement with our predictions.

Practice Exercise 1

List the substances Ar, Cl_2, CH_4, and CH_3COOH in order of increasing strength of intermolecular attractions.
(a) $CH_4 < Ar < CH_3COOH < Cl_2$
(b) $Cl_2 < CH_3COOH < Ar < CH_4$
(c) $CH_4 < Ar < Cl_2 < CH_3COOH$
(d) $CH_3COOH < Cl_2 < Ar < CH_4$
(e) $Ar < Cl_2 < CH_4 < CH_3COOH$

Practice Exercise 2

(a) Identify the intermolecular attractions present in the following substances and (b) select the substance with the highest boiling point: CH_3CH_3, CH_3OH, and CH_3CH_2OH.

Chemistry Put to Work

Ionic Liquids

The strong electrostatic attractions between cations and anions explain why most ionic compounds are solids at room temperature, with high melting and boiling points. However, the melting point of an ionic compound can be low if the ionic charges are not too high and the cation–anion distance is sufficiently large. For example, the melting point of NH_4NO_3, where both cation and anion are larger polyatomic ions, is 170 °C. If the ammonium cation is replaced by the even larger ethyl-ammonium cation, $CH_3CH_2NH_3^+$, the melting point drops to 12 °C, making ethylammonium nitrate a liquid at room temperature. Ethylammonium nitrate is an example of an *ionic liquid*: a salt that is a liquid at room temperature.

Not only is $CH_3CH_2NH_3^+$ larger than NH_4^+, but also it is less symmetric. In general, the larger

and more irregularly shaped the ions in an ionic substance, the better the chances of forming an ionic liquid. Among the cations that form ionic liquids, one of the most widely used is the 1-butyl-3-methylimidazolium cation (abbreviated bmim⁺, ▼ Figure 11.16 and ► Table 11.4), which has two arms of different lengths coming off a

$$H_3C-N \overset{\overset{H}{\underset{\|}{C}}}{\underset{}{}}N-CH_2CH_2CH_2CH_3$$

1-Butyl-3-methylimidazolium (bmim⁺) cation

PF_6^- anion

BF_4^- anion

▲ Figure 11.16 Representative ions found in ionic liquids.

Table 11.4 Melting Point and Decomposition Temperature of Four 1-Butyl-3-Methylimidazolium ($bmim^+$) Salts

Cation	Anion	Melting Point (°C)	Decomposition Temperature (°C)
$bmim^+$	Cl^-	41	254
$bmim^+$	I^-	−72	265
$bmim^+$	PF_6^-	10	349
$bmim^+$	BF_4^-	−81	403

five-atom central ring. This feature gives $bmim^+$ an irregular shape, which makes it difficult for the molecules to pack together in a solid.

Common anions found in ionic liquids include the PF_6^-, BF_4^-, and halide ions.

Ionic liquids have many useful properties. Unlike most molecular liquids, they are nonvolatile (that is, they don't evaporate readily) and nonflammable. They tend to remain in the liquid state at temperatures up to about 400 °C. Most molecular substances are liquids only at much lower temperatures, 100 °C or less in most cases (see Table 11.3). Because ionic liquids are good solvents for a wide range of substances, ionic liquids can be used for a variety of reactions and separations. These properties make them attractive replacements for volatile organic solvents in many industrial processes. Relative to traditional organic solvents, ionic liquids offer the promise of reduced volumes, safer handling, and easier reuse, thereby reducing the environmental impact of industrial chemical processes.

Related Exercises: 11.31, 11.32, 11.82

11.3 | Select Properties of Liquids

The intermolecular attractions we have just discussed can help us understand many familiar properties of liquids. In this section we examine three: viscosity, surface tension, and capillary action.

Viscosity

Some liquids, such as molasses and motor oil, flow very slowly; others, such as water and gasoline, flow easily. The resistance of a liquid to flow is called **viscosity**. The greater a liquid's viscosity, the more slowly it flows. Viscosity can be measured by timing how long it takes a certain amount of the liquid to flow through a thin vertical tube (▶ Figure 11.17). Viscosity can also be determined by measuring the rate at which steel balls fall through the liquid. The balls fall more slowly as the viscosity increases.

Viscosity is related to the ease with which the molecules of the liquid can move relative to one another. It depends on the attractive forces between molecules and on whether the shapes and flexibility of the molecules are such that they tend to become entangled (for example, long molecules can become tangled like spaghetti). For a series of related compounds, viscosity increases with molecular weight, as illustrated in ▼ Table 11.5. The SI unit for viscosity is kg/m-s.

SAE 40
higher number
higher viscosity
slower pouring

SAE 10
lower number
lower viscosity
faster pouring

▲ **Figure 11.17 Comparing viscosities.** The Society of Automotive Engineers (SAE) has established a numeric scale to indicate motor-oil viscosity.

Table 11.5 Viscosities of a Series of Hydrocarbons at 20 °C

Substance	Formula	Viscosity (kg/m-s)
Hexane	$CH_3CH_2CH_2CH_2CH_2CH_3$	3.26×10^{-4}
Heptane	$CH_3CH_2CH_2CH_2CH_2CH_2CH_3$	4.09×10^{-4}
Octane	$CH_3CH_2CH_2CH_2CH_2CH_2CH_2CH_3$	5.42×10^{-4}
Nonane	$CH_3CH_2CH_2CH_2CH_2CH_2CH_2CH_2CH_3$	7.11×10^{-4}
Decane	$CH_3CH_2CH_2CH_2CH_2CH_2CH_2CH_2CH_2CH_3$	1.42×10^{-3}

For any given substance, viscosity decreases with increasing temperature. Octane, for example, has a viscosity of 7.06×10^{-4} kg/m-s at 0 °C and 4.33×10^{-4} kg/m-s at 40 °C. At higher temperatures the greater average kinetic energy of the molecules overcomes the attractive forces between molecules.

Surface Tension

The surface of water behaves almost as if it had an elastic skin, as evidenced by the ability of certain insects to "walk" on water. This behavior is due to an imbalance of intermolecular forces at the surface of the liquid. As shown in ◀ Figure 11.18, molecules in the interior are attracted equally in all directions, but those at the surface experience a net inward force. This net force tends to pull surface molecules toward the interior, thereby reducing the surface area and making the molecules at the surface pack closely together.

Because spheres have the smallest surface area for their volume, water droplets assume an almost spherical shape. This explains the tendency of water to "bead up" when it contacts a surface made of nonpolar molecules, like a lotus leaf (chapter-opening photo) or a newly waxed car.

A measure of the net inward force that must be overcome to expand the surface area of a liquid is given by its surface tension. **Surface tension** is the energy required to increase the surface area of a liquid by a unit amount. For example, the surface tension of water at 20 °C is 7.29×10^{-2} J/m^2, which means that an energy of 7.29×10^{-2} J must be supplied to increase the surface area of a given amount of water by 1 m^2. Water has a high surface tension because of its strong hydrogen bonds. The surface tension of mercury is even higher (4.6×10^{-1} J/m^2) because of even stronger metallic bonds between the atoms of mercury.

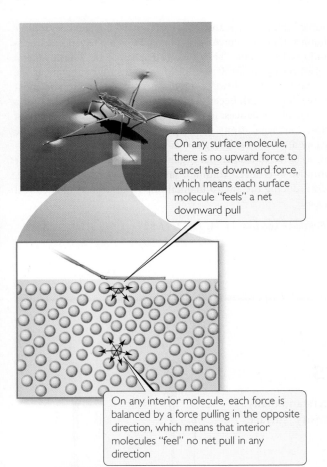

On any surface molecule, there is no upward force to cancel the downward force, which means each surface molecule "feels" a net downward pull

On any interior molecule, each force is balanced by a force pulling in the opposite direction, which means that interior molecules "feel" no net pull in any direction

▲ Figure 11.18 **Molecular-level view of surface tension.** The high surface tension of water keeps the water strider from sinking.

 Give It Some Thought

How do viscosity and surface tension change?
(a) as temperature increases
(b) as intermolecular forces of attraction become stronger

Capillary Action

Intermolecular forces that bind similar molecules to one another, such as the hydrogen bonding in water, are called *cohesive forces*. Intermolecular forces that bind a substance to a surface are called *adhesive forces*. Water placed in a glass tube adheres to the glass because the adhesive forces between the water and the glass are greater than the cohesive forces between water molecules. The curved surface, or *meniscus*, of the water is therefore U-shaped (▶ Figure 11.19). For mercury, however, the situation is different. Mercury atoms can form bonds with one another but not with the glass. As a result the cohesive forces are much greater than the adhesive forces and the meniscus is shaped like an inverted U.

When a small-diameter glass tube, or capillary, is placed in water, water rises in the tube. The rise of liquids up very narrow tubes is called **capillary action**. The adhesive forces between the liquid and the walls of the tube tend to increase the surface area of the liquid. The surface tension of the liquid tends to reduce the area, thereby pulling the liquid up the tube. The liquid climbs until the force of gravity on the liquid balances the adhesive and cohesive forces. Capillary action is widespread. For example, towels absorb liquid and "stay-dry" synthetic fabrics move sweat away from the skin by capillary action. Capillary action also plays a role in moving water and dissolved nutrients upward through plants.

▲ GO FIGURE

If the inside surface of each tube were coated with wax, would the general shape of the water meniscus change? Would the general shape of the mercury meniscus change?

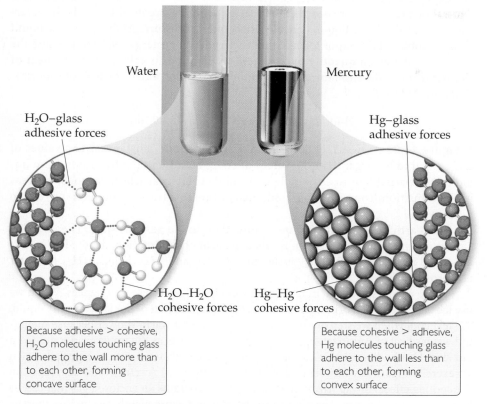

H₂O–glass
adhesive forces

Hg–glass
adhesive forces

H₂O–H₂O
cohesive forces

Hg–Hg
cohesive forces

Because adhesive > cohesive, H₂O molecules touching glass adhere to the wall more than to each other, forming concave surface

Because cohesive > adhesive, Hg molecules touching glass adhere to the wall less than to each other, forming convex surface

▲ Figure 11.19 **Meniscus shapes for water and mercury in glass tubes.**

11.4 | Phase Changes

Liquid water left uncovered in a glass eventually evaporates. An ice cube left in a warm room quickly melts. Solid CO_2 (sold as a product called dry ice) *sublimes* at room temperature; that is, it changes directly from solid to gas. In general, each state of matter—solid, liquid, gas—can transform into either of the other two states. ▶ Figure 11.20 shows the names associated with these transformations, which are called either **phase changes** or *changes of state*.

Energy Changes Accompanying Phase Changes

Every phase change is accompanied by a change in the energy of the system. In a solid, for example, the particles (molecules, ions, or atoms) are in more or less fixed positions with respect to one another and closely arranged to minimize the energy of the system. As the temperature of the solid increases, the particles vibrate about their equilibrium positions with increasing energetic motion. When the solid melts, the particles are freed to move relative to one another, which means their average kinetic energy increases.

Melting is called (somewhat confusingly) *fusion*. The increased freedom of motion of the particles requires energy, measured by the **heat of fusion** or *enthalpy of fusion*, ΔH_{fus}. The heat of fusion of ice, for example, is 6.01 kJ/mol:

$$H_2O(s) \rightarrow H_2O(l) \qquad \Delta H = 6.01 \text{ kJ}$$

▲ GO FIGURE

How is energy evolved in deposition related to those for condensation and freezing?

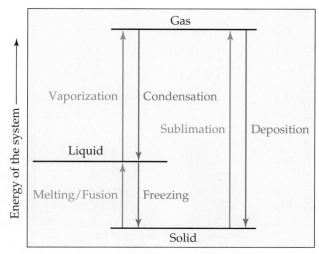

—— Endothermic process (energy added to substance)
—— Exothermic process (energy released from substance)

▲ Figure 11.20 **Phase changes and the names associated with them.**

As the temperature of the liquid increases, the particles move about more vigorously. The increased motion allows some particles to escape into the gas phase. As a result, the concentration of gas-phase particles above the liquid surface increases with temperature. These gas-phase particles exert a pressure called *vapor pressure*. We explore vapor pressure in Section 11.5. For now we just need to understand that vapor pressure increases with increasing temperature until it equals the external pressure above the liquid, typically atmospheric pressure. At this point the liquid boils—bubbles of the vapor form within the liquid. The energy required to cause the transition of a given quantity of the liquid to the vapor is called either the **heat of vaporization** or the *enthalpy of vaporization,* ΔH_{vap}. For water, the heat of vaporization is 40.7 kJ/mol.

$$\text{H}_2\text{O}(l) \longrightarrow \text{H}_2\text{O}(g) \qquad \Delta H = 40.7 \text{ kJ}$$

▼ Figure 11.21 shows ΔH_{fus} and ΔH_{vap} values for four substances. The values of ΔH_{vap} tend to be larger than the values of ΔH_{fus} because in the transition from liquid to gas, particles must essentially sever all their interparticle attractions, whereas in the transition from solid to liquid, many of these attractive interactions remain operative.

The particles of a solid can move directly into the gaseous state. The enthalpy change required for this transition is called the **heat of sublimation**, denoted ΔH_{sub}. As illustrated in Figure 11.20, ΔH_{sub} is the sum of ΔH_{fus} and ΔH_{vap}. Thus, ΔH_{sub} for water is approximately 47 kJ/mol.

Phase changes show up in important ways in our everyday experiences. When we use ice cubes to cool a drink, for instance, the heat of fusion of the ice cools the liquid. We feel cool when we step out of a swimming pool or a warm shower because the liquid water's heat of vaporization is drawn from our bodies as the water evaporates from our skin. Our bodies use this mechanism to regulate body temperature, especially when we exercise vigorously in warm weather. ∞ (Section 5.5) A refrigerator also relies on the cooling effects of vaporization. Its mechanism contains an enclosed gas that can be liquefied under pressure. The liquid absorbs heat as it subsequently evaporates, thereby cooling the interior of the refrigerator.

What happens to the heat absorbed when the liquid refrigerant vaporizes? According to the first law of thermodynamics ∞ (Section 5.2), this absorbed heat

---

▲ **GO FIGURE**

Is it possible to calculate the heat of sublimation for a substance given its heats of vaporization and fusion? If so, what is the relationship?

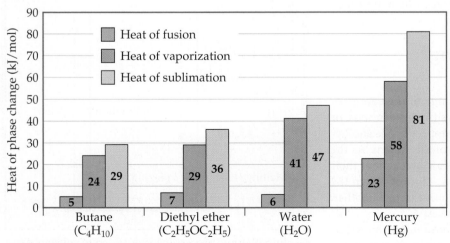

▲ **Figure 11.21 Heats of fusion, vaporization, and sublimation.**

must be released when the gas condenses to liquid. As this phase change occurs, the heat released is dissipated through cooling coils in the back of the refrigerator. Just as for a given substance the heat of condensation is equal in magnitude to the heat of vaporization and has the opposite sign, so too the *heat of deposition* for a given substance is exothermic to the same degree that the heat of sublimation is endothermic; the *heat of freezing* is exothermic to the same degree that the heat of fusion is endothermic (see Figure 11.20).

 Give It Some Thought

What is the name of the phase change that occurs when ice left at room temperature changes to liquid water? Is this change exothermic or endothermic?

Heating Curves

When we heat an ice cube initially at $-25\,°C$ and 1 atm pressure, the temperature of the ice increases. As long as the temperature is below $0\,°C$, the ice cube remains in the solid state. When the temperature reaches $0\,°C$, the ice begins to melt. Because melting is an endothermic process, the heat we add at $0\,°C$ is used to convert ice to liquid water, and *the temperature remains constant until all the ice has melted*. Once all the ice has melted, adding more heat causes the temperature of the liquid water to increase.

A graph of temperature versus amount of heat added is called a *heating curve*. ▶ Figure 11.22 shows the heating curve for transforming ice, $H_2O(s)$, initially at $-25\,°C$ to steam, $H_2O(g)$, at $125\,°C$. Heating the $H_2O(s)$ from -25 to $0\,°C$ is represented by the line segment AB, and converting the $H_2O(s)$ at $0\,°C$ to $H_2O(l)$ at $0\,°C$ is the horizontal segment BC. Additional heat increases the temperature of the $H_2O(l)$ until the temperature reaches $100\,°C$ (segment CD). The heat is then used to convert $H_2O(l)$ to $H_2O(g)$ at a constant temperature of $100\,°C$ (segment DE) as the water boils. Once all the $H_2O(l)$ has been converted to $H_2O(g)$, adding heat increases the temperature of the $H_2O(g)$ (segment EF).

We can calculate the enthalpy change of the system for each segment of the heating curve. In segments AB, CD, and EF we are heating a single phase from one temperature to another. As we saw in Section 5.5, the amount of heat needed to raise the temperature of a substance is given by the product of the specific heat, mass, and temperature change (Equation 5.22). The greater the specific heat of a substance, the more heat we must add to accomplish a certain temperature increase. Because the specific heat of water is greater than that of ice, the slope of segment CD is less than that of segment AB. This lesser slope means the amount of heat we must add to a given mass of liquid water to achieve a $1\,°C$ temperature change is greater than the amount we must add to achieve a $1\,°C$ temperature change in the same mass of ice.

In segments BC and DE we are converting one phase to another at a constant temperature. The temperature remains constant during these phase changes because the added energy is used to overcome the attractive forces between molecules rather than to increase their average kinetic energy. For segment BC, the enthalpy change can be calculated by using ΔH_{fus}, and for segment DE we can use ΔH_{vap}.

If we start with 1 mole of steam at $125\,°C$ and cool it, we move right to left across Figure 11.22. We first lower the temperature of the $H_2O(g)$ ($F \longrightarrow E$), then condense it ($E \longrightarrow D$) to $H_2O(l)$, and so forth.

Sometimes as we remove heat from a liquid, we can temporarily cool it below its freezing point without forming a solid. This phenomenon, called *supercooling*, occurs when the heat is removed so rapidly that the molecules have no time to assume the

 GO FIGURE

What process is occurring between points C and D?

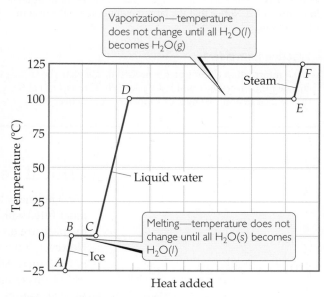

▲ **Figure 11.22 Heating curve for water.** Changes that occur when 1.00 mol of H_2O is heated from $H_2O(s)$ at $-25\,°C$ to $H_2O(g)$ at $125\,°C$ at a constant pressure of 1 atm. Even though heat is being added continuously, the system temperature does not change during the two phase changes (red lines).

ordered structure of a solid. A supercooled liquid is unstable; particles of dust entering the solution or gentle stirring is often sufficient to cause the substance to solidify quickly.

SAMPLE EXERCISE 11.3 | Calculating ΔH for Temperature and Phase Changes

Calculate the enthalpy change upon converting 1.00 mol of ice at $-25\,°C$ to steam at $125\,°C$ under a constant pressure of 1 atm. The specific heats of ice, liquid water, and steam are 2.03, 4.18, and 1.84 J/g-K, respectively. For H_2O, $\Delta H_{fus} = 6.01$ kJ/mol and $\Delta H_{vap} = 40.67$ kJ/mol.

SOLUTION

Analyze Our goal is to calculate the total heat required to convert 1 mol of ice at $-25\,°C$ to steam at $125\,°C$.

Plan We can calculate the enthalpy change for each segment and then sum them to get the total enthalpy change (Hess's law, Section 5.6).

Solve For segment AB in Figure 11.22, we are adding enough heat to ice to increase its temperature by 25 °C. A temperature change of 25 °C is the same as a temperature change of 25 K, so we can use the specific heat of ice to calculate the enthalpy change during this process:

$$AB: \Delta H = (1.00 \text{ mol})(18.0 \text{ g/mol})(2.03 \text{ J/g-K})(25 \text{ K}) = 914 \text{ J} = 0.91 \text{ kJ}$$

For segment BC in Figure 11.22, in which we convert ice to water at 0 °C, we can use the molar enthalpy of fusion directly:

$$BC: \Delta H = (1.00 \text{ mol})(6.01 \text{ kJ/mol}) = 6.01 \text{ kJ}$$

The enthalpy changes for segments CD, DE, and EF can be calculated in similar fashion:

$$CD: \Delta H = (1.00 \text{ mol})(18.0 \text{ g/mol})(4.18 \text{ J/g-K})(100. \text{ K}) = 7520 \text{ J} = 7.52 \text{ kJ}$$
$$DE: \Delta H = (1.00 \text{ mol})(40.67 \text{ kJ/mol}) = 40.7 \text{ kJ}$$
$$EF: \Delta H = (1.00 \text{ mol})(18.0 \text{ g/mol})(1.84 \text{ J/g-K})(25 \text{ K}) = 830 \text{ J} = 0.83 \text{ kJ}$$

The total enthalpy change is the sum of the changes of the individual steps:

$$\Delta H = 0.91 \text{ kJ} + 6.01 \text{ kJ} + 7.52 \text{ kJ} + 40.7 \text{ kJ} + 0.83 \text{ kJ} = 56.0 \text{ kJ}$$

Check The components of the total enthalpy change are reasonable relative to the horizontal lengths (heat added) of the segments in Figure 11.22. Notice that the largest component is the heat of vaporization.

Practice Exercise 1

What information about water is needed to calculate the enthalpy change for converting 1 mol $H_2O(g)$ at 100 °C to $H_2O(l)$ at 80 °C? (a) Heat of fusion, (b) heat of vaporization, (c) heat of vaporization and specific heat of $H_2O(g)$, (d) heat of vaporization and specific heat of $H_2O(l)$, (e) heat of fusion and specific heat of $H_2O(l)$.

Practice Exercise 2

What is the enthalpy change during the process in which 100.0 g of water at 50.0 °C is cooled to ice at $-30.0\,°C$? (Use the specific heats and enthalpies for phase changes given in Sample Exercise 11.3.)

Critical Temperature and Pressure

A gas normally liquefies at some point when pressure is applied. Suppose we have a cylinder fitted with a piston and the cylinder contains water vapor at 100 °C. If we increase the pressure on the water vapor, liquid water will form when the pressure is 760 torr. However, if the temperature is 110 °C, the liquid phase does not form until the pressure is 1075 torr. At 374 °C the liquid phase forms only at 1.655×10^5 torr (217.7 atm). Above this temperature no amount of pressure causes a distinct liquid phase to form. Instead, as pressure increases, the gas becomes steadily more compressed. The highest temperature at which a distinct liquid phase can form is called the **critical temperature**. The **critical pressure** is the pressure required to bring about liquefaction at this critical temperature.

The critical temperature is the highest temperature at which a liquid can exist. Above the critical temperature, the kinetic energies of the molecules are greater than the attractive forces that lead to the liquid state regardless of how much the substance

Table 11.6   Critical Temperatures and Pressures of Selected Substances

Substance	Critical Temperature (K)	Critical Pressure (atm)
Nitrogen, N_2	126.1	33.5
Argon, Ar	150.9	48.0
Oxygen, O_2	154.4	49.7
Methane, CH_4	190.0	45.4
Carbon dioxide, CO_2	304.3	73.0
Phosphine, PH_3	324.4	64.5
Propane, $CH_3CH_2CH_3$	370.0	42.0
Hydrogen sulfide, H_2S	373.5	88.9
Ammonia, NH_3	405.6	111.5
Water, H_2O	647.6	217.7

is compressed to bring the molecules closer together. *The greater the intermolecular forces, the higher the critical temperature of a substance.*

Several critical temperatures and pressures are listed in ▲ Table 11.6. Notice that nonpolar, low-molecular-weight substances, which have weak intermolecular attractions, have lower critical temperatures and pressures than substances that are polar or of higher molecular weight. Notice also that water and ammonia have exceptionally high critical temperatures and pressures as a consequence of strong intermolecular hydrogen-bonding forces.

 Give It Some Thought

Why are the critical temperature and pressure for H_2O so much higher than those for H_2S, a related substance (Table 11.6)?

Because they provide information about the conditions under which gases liquefy, critical temperatures and pressures are often of considerable importance to engineers and other people working with gases. Sometimes we want to liquefy a gas; other times we want to avoid liquefying it. It is useless to try to liquefy a gas by applying pressure if the gas is above its critical temperature. For example, O_2 has a critical temperature of 154.4 K. It must be cooled below this temperature before it can be liquefied by pressure. In contrast, ammonia has a critical temperature of 405.6 K. Thus, it can be liquefied at room temperature (approximately 295 K) by applying sufficient pressure.

When the temperature exceeds the critical temperature and the pressure exceeds the critical pressure, the liquid and gas phases are indistinguishable from each other, and the substance is in a state called a **supercritical fluid**. A supercritical fluid expands to fill its container (like a gas), but the molecules are still quite closely spaced (like a liquid).

Like liquids, supercritical fluids can behave as solvents dissolving a wide range of substances. Using *supercritical fluid extraction*, the components of mixtures can be separated from one another. Supercritical fluid extraction has been successfully used to separate complex mixtures in the chemical, food, pharmaceutical, and energy industries. Supercritical CO_2 is a popular choice because it is relatively inexpensive and there are no problems associated with disposing of solvent, nor are there toxic residues resulting from the process.

11.5 | Vapor Pressure

Molecules can escape from the surface of a liquid into the gas phase by evaporation. Suppose we place a quantity of ethanol (CH_3CH_2OH) in an evacuated, closed container, as in **Figure 11.23**. The ethanol quickly begins to evaporate. As a result, the pressure exerted by the vapor in the space above the liquid increases. After a short time the pressure of the vapor attains a constant value, which we call the **vapor pressure**.

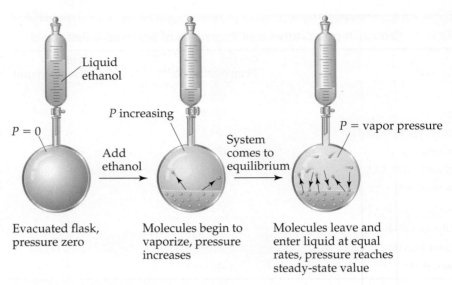

Liquid ethanol

$P = 0$

Add ethanol

P increasing

System comes to equilibrium

P = vapor pressure

Evacuated flask, pressure zero

Molecules begin to vaporize, pressure increases

Molecules leave and enter liquid at equal rates, pressure reaches steady-state value

▲ Figure 11.23 **Vapor pressure over a liquid.**

At any instant, some of the ethanol molecules at the liquid surface possess sufficient kinetic energy to overcome the attractive forces of their neighbors and, therefore, escape into the gas phase. At any particular temperature, the movement of molecules from liquid phase to gas phase goes on continuously. As the number of gas-phase molecules increases, however, the probability increases that a molecule in the gas phase will strike the liquid surface and be recaptured by the liquid, as shown in the flask on the right in Figure 11.23. Eventually, the rate at which molecules return to the liquid equals the rate at which they escape. The number of molecules in the gas phase then reaches a steady value, and the pressure exerted by the vapor becomes constant.

The condition in which two opposing processes occur simultaneously at equal rates is called **dynamic equilibrium** (or simply *equilibrium*). Chemical equilibrium, which we encountered in Section 4.1, is a kind of dynamic equilibrium in which the opposing processes are chemical reactions.

A liquid and its vapor are in dynamic equilibrium when evaporation and condensation occur at equal rates. It may appear that nothing is occurring at equilibrium because there is no net change in the system. In fact, though, a great deal is happening as molecules continuously pass from liquid state to gas state and from gas state to liquid state. *The vapor pressure of a liquid is the pressure exerted by its vapor when the liquid and vapor are in dynamic equilibrium.*

Volatility, Vapor Pressure, and Temperature

When vaporization occurs in an open container, as when water evaporates from a bowl, the vapor spreads away from the liquid. Little, if any, is recaptured at the surface of the liquid. Equilibrium never occurs, and the vapor continues to form until the liquid evaporates to dryness. Substances with high vapor pressure (such as gasoline) evaporate more quickly than substances with low vapor pressure (such as motor oil). Liquids that evaporate readily are said to be **volatile**.

Hot water evaporates more quickly than cold water because vapor pressure increases with increasing temperature. To see why this statement is true, we begin with the fact that the molecules of a liquid move at various speeds. ◄ Figure 11.24 shows the distribution of kinetic energies of the molecules at the surface of a liquid at two temperatures. (The curves are like those shown for gases in Section 10.7.) As the temperature is increased, the molecules move more energetically and more of them can break free from their neighbors and enter the gas phase, increasing the vapor pressure.

GO FIGURE

As the temperature increases, does the rate of molecules escaping into the gas phase increase or decrease?

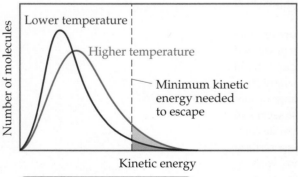

Number of molecules

Lower temperature

Higher temperature

Minimum kinetic energy needed to escape

Kinetic energy

Blue area = number of molecules having enough energy to evaporate at lower temperature

Red + blue areas = number of molecules having enough energy to evaporate at higher temperature

▲ Figure 11.24 **The effect of temperature on the distribution of kinetic energies in a liquid.**

▶ Figure 11.25 depicts the variation in vapor pressure with temperature for four common substances that differ greatly in volatility. Note that the vapor pressure in all cases increases nonlinearly with increasing temperature. The weaker the intermolecular forces in the liquid, the more easily molecules can escape and, therefore, the higher the vapor pressure at a given temperature.

Give It Some Thought

Which compound do you think is more volatile at 25 °C: CCl_4 or CBr_4?

Vapor Pressure and Boiling Point

The **boiling point** of a liquid is the temperature at which its vapor pressure equals the external pressure, acting on the liquid surface. At this temperature, the thermal energy of the molecules is great enough for the molecules in the interior of the liquid to break free from their neighbors and enter the gas phase. As a result, bubbles of vapor form within the liquid. The boiling point increases as the external pressure increases. The boiling point of a liquid at 1 atm (760 torr) pressure is called its **normal boiling point**. From Figure 11.25 we see that the normal boiling point of water is 100 °C.

The time required to cook food in boiling water depends on the water temperature. In an open container, that temperature is 100 °C, but it is possible to boil at higher temperatures. Pressure cookers work by allowing steam to escape only when it exceeds a predetermined pressure; the pressure above the water can therefore increase above atmospheric pressure. The higher pressure causes the water to boil at a higher temperature, thereby allowing the food to get hotter and to cook more rapidly.

The effect of pressure on boiling point also explains why it takes longer to cook food at high elevations than it does at sea level. The atmospheric pressure is lower at higher altitudes, so water boils at a temperature lower than 100 °C, and foods generally take longer to cook.

GO FIGURE

What is the vapor pressure of ethylene glycol at its normal boiling point?

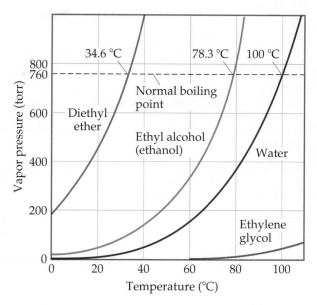

▲ Figure 11.25 **Vapor pressure for four liquids as a function of temperature.**

A Closer Look

The Clausius–Clapeyron Equation

Notice that the plots in Figure 11.25 have a distinct shape: For each substance, the vapor–pressure curves sharply upward with increasing temperature. The relationship between vapor pressure and temperature is given by the *Clausius–Clapeyron equation*:

$$\ln P = \frac{-\Delta H_{vap}}{RT} + C \qquad [11.1]$$

where P is the vapor pressure, T is the absolute temperature, R is the gas constant (8.314 J/mol-K), ΔH_{vap} is the molar enthalpy of vaporization, and C is a constant. This equation predicts that a graph of $\ln P$ versus $1/T$ should give a straight line with a slope equal to $\Delta H_{vap}/R$. Using this plot we can determine the enthalpy of vaporization of a substance:

$$\Delta H_{vap} = -slope \times R$$

As an example of how we use the Clausius–Clapeyron equation, the vapor-pressure data for ethanol shown in Figure 11.25 are graphed as $\ln P$ versus $1/T$ in ▶ Figure 11.26. The data lie on a straight line with a negative slope. We can use the slope to determine ΔH_{vap} for ethanol, 38.56 kJ/mol. We can also extrapolate the line to obtain the vapor pressure of ethanol at temperatures above and below the temperature range for which we have data.

Related Exercises: 11.84, 11.85, 11.86

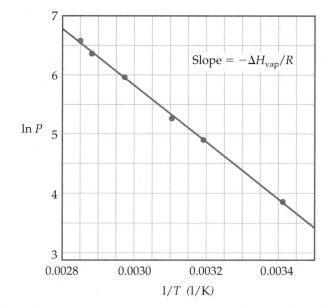

▲ Figure 11.26 **The natural logarithm of vapor pressure versus $1/T$ for ethanol.**

SAMPLE
EXERCISE 11.4 Relating Boiling Point to Vapor Pressure

Use Figure 11.25 to estimate the boiling point of diethyl ether under an external pressure of 0.80 atm.

SOLUTION

Analyze We are asked to read a graph of vapor pressure versus temperature to determine the boiling point of a substance at a particular pressure. The boiling point is the temperature at which the vapor pressure is equal to the external pressure.

Plan We need to convert 0.80 atm to torr because that is the pressure scale on the graph. We estimate the location of that pressure on the graph, move horizontally to the vapor–pressure curve, and then drop vertically from the curve to estimate the temperature.

Solve The pressure equals $(0.80 \text{ atm})(760 \text{ torr/atm}) = 610$ torr. From Figure 11.25 we see that the boiling point at this pressure is about 27 °C, which is close to room temperature.

Comment We can make a flask of diethyl ether boil at room temperature by using a vacuum pump to lower the pressure above the liquid to about 0.8 atm.

Practice Exercise 1

In the mountains, water in an open container will boil when

(a) its critical temperature exceeds room temperature
(b) its vapor pressure equals atmospheric pressure
(c) its temperature is 100 °C
(d) enough energy is supplied to break covalent bonds
(e) none of these is correct

Practice Exercise 2

Use Figure 11.25 to determine the external pressure if ethanol boils at 60 °C.

11.6 | Phase Diagrams

The equilibrium between a liquid and its vapor is not the only dynamic equilibrium that can exist between states of matter. Under appropriate conditions, a solid can be in equilibrium with its liquid or even with its vapor. The temperature at which solid and liquid phases coexist at equilibrium is the *melting point* of the solid or the *freezing point* of the liquid. Solids can also undergo evaporation and therefore possess a vapor pressure.

A **phase diagram** is a graphic way to summarize the conditions under which equilibria exist between the different states of matter. Such a diagram also allows us to predict which phase of a substance is present at any given temperature and pressure.

The phase diagram for any substance that can exist in all three phases of matter is shown in ▶ Figure 11.27. The diagram contains three important curves, each of which represents the temperature and pressure at which the various phases can coexist at equilibrium. The only substance present in the system is the one whose phase diagram is under consideration. The pressure shown in the diagram is either the pressure applied to the system or the pressure generated by the substance. The curves may be described as follows:

1. The red curve is the *vapor-pressure curve* of the liquid, representing equilibrium between the liquid and gas phases. The point on this curve where the vapor pressure is 1 atm is the normal boiling point of the substance. The vapor-pressure curve ends at the **critical point** (*C*), which corresponds to the critical temperature and critical pressure of the substance. At temperatures and pressures beyond the critical point, the liquid and gas phases are indistinguishable from each other, and the substance is a *supercritical fluid*.

2. The green curve, the *sublimation curve*, separates the solid phase from the gas phase and represents the change in the vapor pressure of the solid as it sublimes at different temperatures. Each point on this curve is a condition of equilibrium between the solid and the gas.

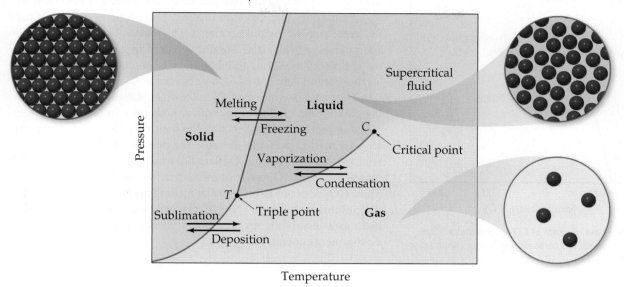

▲ **GO FIGURE**

Imagine that the pressure on the solid phase in the figure is decreased at constant temperature. If the solid eventually sublimes, what must be true about the temperature?

▲ Figure 11.27 **Generic phase diagram for a pure substance.** The green line is the sublimation curve, the blue line is the melting curve, and the red line is the vapor–pressure curve.

3. The blue curve, the *melting curve*, separates the solid phase from the liquid phase and represents the change in melting point of the solid with increasing pressure. Each point on this curve is an equilibrium between the solid and the liquid. This curve usually slopes slightly to the right as pressure increases because for most substances the solid form is denser than the liquid form. An increase in pressure usually favors the more compact solid phase; thus, higher temperatures are required to melt the solid at higher pressures. The melting point at 1 atm is the **normal melting point**.

Point *T*, where the three curves intersect, is the **triple point,** and here all three phases are in equilibrium. Any other point on any of the three curves represents equilibrium between two phases. Any point on the diagram that does not fall on one of the curves corresponds to conditions under which only one phase is present. The gas phase, for example, is stable at low pressures and high temperatures, whereas the solid phase is stable at low temperatures and high pressures. Liquids are stable in the region between the other two.

The Phase Diagrams of H_2O and CO_2

▶ Figure 11.28 shows the phase diagram of H_2O. Because of the large range of pressures covered in the diagram, a logarithmic scale is used to represent pressure. The melting curve (blue line) of H_2O is atypical, slanting slightly to the left with increasing pressure, indicating that for water the melting point *decreases* with increasing pressure. This unusual behavior occurs because water is among the very few substances whose liquid form is more compact than its solid form, as we learned in Section 11.2.

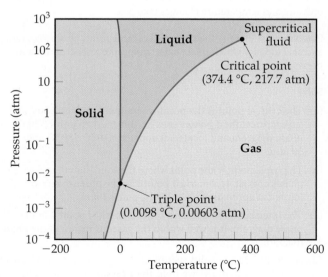

▲ Figure 11.28 **Phase diagram of** H_2O. Note that a linear scale is used to represent temperature and a logarithmic scale to represent pressure.

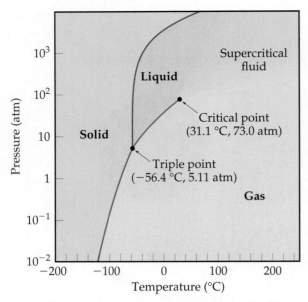

▲ **Figure 11.29 Phase diagram of CO₂.** Note that a linear scale is used to represent temperature and a logarithmic scale to represent pressure.

If the pressure is held constant at 1 atm, it is possible to move from the solid to liquid to gaseous regions of the phase diagram by changing the temperature, as we expect from our everyday encounters with water. The triple point of H_2O falls at a relatively low pressure, 0.00603 atm. Below this pressure, liquid water is not stable and ice sublimes to water vapor on heating. This property of water is used to "freeze-dry" foods and beverages. The food or beverage is frozen to a temperature below 0 °C. Next it is placed in a low-pressure chamber (below 0.00603 atm) and then warmed so that the water sublimes, leaving behind dehydrated food or beverage.

The phase diagram for CO_2 is shown in ◄ Figure 11.29. The melting curve (blue line) behaves typically, slanting to the right with increasing pressure, telling us that the melting point of CO_2 increases with increasing pressure. Because the pressure at the triple point is relatively high, 5.11 atm, CO_2 does not exist as a liquid at 1 atm, which means that solid CO_2 does not melt when heated, but instead sublimes. Thus, CO_2 does not have a normal melting point; instead, it has a normal sublimation point, −78.5 °C. Because CO_2 sublimes rather than melts as it absorbs energy at ordinary pressures, solid CO_2 (dry ice) is a convenient coolant.

SAMPLE EXERCISE 11.5 | Interpreting a Phase Diagram

Use the phase diagram for methane, CH_4, shown in ▼ Figure 11.30 to answer the following questions. (**a**) What are the approximate temperature and pressure of the critical point? (**b**) What are the approximate temperature and pressure of the triple point? (**c**) Is methane a solid, liquid, or gas at 1 atm and 0 °C? (**d**) If solid methane at 1 atm is heated while the pressure is held constant, will it melt or sublime? (**e**) If methane at 1 atm and 0 °C is compressed until a phase change occurs, in which state is the methane when the compression is complete?

SOLUTION

Analyze We are asked to identify key features of the phase diagram and to use it to deduce what phase changes occur when specific pressure and temperature changes take place.

Plan We must identify the triple and critical points on the diagram and also identify which phase exists at specific temperatures and pressures.

Solve

(**a**) The critical point is the point where the liquid, gaseous, and supercritical fluid phases coexist. It is marked point 3 in the phase diagram and located at approximately −80 °C and 50 atm.

(**b**) The triple point is the point where the solid, liquid, and gaseous phases coexist. It is marked point 1 in the phase diagram and located at approximately −180 °C and 0.1 atm.

(**c**) The intersection of 0 °C and 1 atm is marked point 2 in the phase diagram. It is well within the gaseous region of the phase diagram.

(**d**) If we start in the solid region at $P = 1$ atm and move horizontally (this means we hold the pressure constant), we cross first into the liquid region, at $T \approx -180$ °C, and then into the gaseous region, at $T \approx -160$ °C. Therefore, solid methane melts when the pressure is 1 atm. (For methane to sublime, the pressure must be below the triple point pressure.)

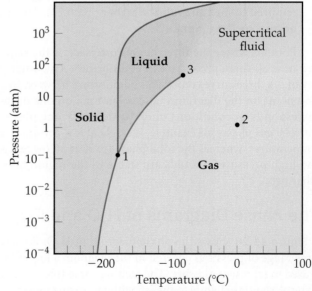

▲ **Figure 11.30 Phase diagram of CH₄.** Note that a linear scale is used to represent temperature and a logarithmic scale to represent pressure.

(**e**) Moving vertically up from point 2, which is 1 atm and 0 °C, the first phase change we come to is from gas to supercritical fluid. This phase change happens when we exceed the critical pressure (∼50 atm).

Check The pressure and temperature at the critical point are higher than those at the triple point, which is expected. Methane is the principal component of natural gas. So it seems reasonable that it exists as a gas at 1 atm and 0 °C.

(a) It sublimes at about −200 °C. (b) It melts at about −200 °C. (c) It boils at about −200 °C. (d) It condenses at about −200 °C. (e) It reaches the triple point at about −200 °C.

Practice Exercise 1

Based on the phase diagram for methane (Figure 11.30), what happens to methane as it is heated from −250 to 0 °C at a pressure of 10^{-2} atm?

Practice Exercise 2

Use the phase diagram of methane to answer the following questions. (a) What is the normal boiling point of methane? (b) Over what pressure range does solid methane sublime? (c) Above what temperature does liquid methane not exist?

11.7 | Liquid Crystals

In 1888 Frederick Reinitzer, an Austrian botanist, discovered that the organic compound cholesteryl benzoate has an interesting and unusual property, shown in ▼ Figure 11.31. Solid cholesteryl benzoate melts at 145 °C, forming a viscous milky liquid; then at 179 °C the milky liquid becomes clear and remains that way at temperatures above 179 °C. When cooled, the clear liquid turns viscous and milky at 179 °C, and the milky liquid solidifies at 145 °C.

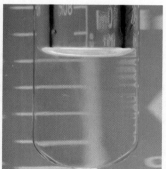

145 °C < *T* < 179 °C
Liquid crystalline phase

T > 179 °C
Liquid phase

▲ Figure 11.31 **Cholesteryl benzoate in its liquid and liquid crystalline states.**

Reinitzer's work represents the first systematic report of what we call a **liquid crystal**, the term we use today for the viscous, milky state that some substances exhibit between the liquid and solid states.

This intermediate phase has some of the structure of solids and some of the freedom of motion of liquids. Because of the partial ordering, liquid crystals may be viscous and possess properties intermediate between those of solids and those of liquids. The region in which they exhibit these properties is marked by sharp transition temperatures, as in Reinitzer's sample.

Today liquid crystals are used as pressure and temperature sensors and as liquid crystals displays (LCDs) in such devices as digital watches, televisions, and computers. They can be used for these applications because the weak intermolecular forces that hold the molecules together in the liquid crystalline phase are easily affected by changes in temperature, pressure, and electric fields.

Types of Liquid Crystals

Substances that form liquid crystals are often composed of rod-shaped molecules that are somewhat rigid. In the liquid phase, these molecules are oriented randomly. In the liquid crystalline phase, by contrast, the molecules are arranged in specific patterns as

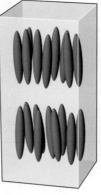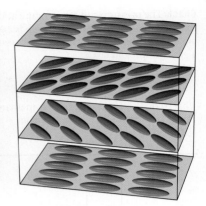

Liquid phase

Molecules arranged randomly

Nematic liquid crystalline phase

Long axes of molecules aligned, but ends are not aligned

Smectic A liquid crystalline phase

Molecules aligned in layers, long axes of molecules perpendicular to layer planes

Smectic C liquid crystalline phase

Molecules aligned in layers, long axes of molecules inclined with respect to layer planes

Cholesteric liquid crystalline phase

Molecules pack into layers, long axes of molecules in one layer rotated relative to the long axes in the layer above it

▲ Figure 11.32 **Molecular order in nematic, smectic, and cholesteric liquid crystals.** In the liquid phase of any substance, the molecules are arranged randomly, whereas in the liquid crystalline phases the molecules are arranged in a partially ordered way.

illustrated in ▲ Figure 11.32. Depending on the nature of the ordering, liquid crystals are classified as nematic, smectic A, smectic C, or cholesteric.

In a **nematic liquid crystal**, the molecules are aligned so that their long axes tend to point in the same direction but the ends are not aligned with one another. In **smectic A** and **smectic C liquid crystals**, the molecules maintain the long-axis alignment seen in nematic crystals, but in addition they pack into layers.

Two molecules that exhibit liquid crystalline phases are shown in ▼ Figure 11.33. The lengths of these molecules are much greater than their widths. The double bonds, including those in the benzene rings, add rigidity to the molecules, and the rings, because they are flat, help the molecules stack with one another. The polar —CH₃O and —COOH groups give rise to dipole–dipole interactions and promote alignment of the molecules. Thus, the molecules order themselves quite naturally along their long axes. They can, however, rotate around their axes and slide parallel to one another. In smectic liquid crystals, the intermolecular forces (dispersion forces, dipole–dipole attractions, and hydrogen bonding) limit the ability of the molecules to slide past one another.

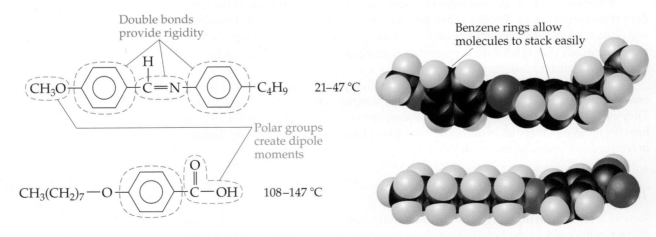

▲ Figure 11.33 **Molecular structure and liquid crystal temperature range for two typical liquid crystalline materials.**

In a **cholesteric liquid crystal**, the molecules are arranged in layers, with their long axes parallel to the other molecules within the same layer.* Upon moving from one layer to the next, the orientation of the molecules rotates by a fixed angle, resulting in a spiral pattern. These liquid crystals are so named because many derivatives of cholesterol adopt this structure.

The molecular arrangement in cholesteric liquid crystals produces unusual coloring patterns with visible light. Changes in temperature and pressure change the order and, hence, the color. Cholesteric liquid crystals are used to monitor temperature changes in situations where conventional methods are not feasible. For example, they can detect hot spots in microelectronic circuits, which may signal the presence of flaws. They can also be fashioned into thermometers for measuring the skin temperature of infants. Because cholesteric liquid crystal displays can be built that draw very little power, they are also being investigated for use in electronic paper (▶ Figure 11.34). Such applications are possible because an applied electrical field changes the orientation of liquid crystal molecules and thus affects the optical properties of the device.

▲ Figure 11.34 **Electronic paper (e-paper) based on cholesteric liquid crystal technology.** Electronic paper mimics the appearance of ordinary ink on paper (like a page of a book or magazine). It has numerous potential uses including thin displays on walls, electronic labels, and electronic book readers.

SAMPLE EXERCISE 11.6 | Properties of Liquid Crystals

Which of these substances is most likely to exhibit liquid crystalline behavior?

$$CH_3-CH_2-\underset{\underset{CH_3}{|}}{\overset{\overset{CH_3}{|}}{C}}-CH_2-CH_3$$

(i)

$$CH_3CH_2-\text{⟨○⟩}-N{=}N-\text{⟨○⟩}-\overset{\overset{O}{\|}}{C}-OCH_3$$

(ii)

$$\text{⟨○⟩}-CH_2-\overset{\overset{O}{\|}}{C}-O^-Na^+$$

(iii)

SOLUTION

Analyze We have three molecules with different structures, and we are asked to determine which one is most likely to be a liquid crystalline substance.

Plan We need to identify all structural features that might induce liquid crystalline behavior.

Solve Molecule (i) is not likely to be liquid crystalline because the absence of double and/or triple bonds makes this molecule flexible rather than rigid. Molecule (iii) is ionic and the generally high melting points of ionic materials make it unlikely that this substance is liquid crystalline. Molecule (ii) possesses the characteristic long axis and the kinds of structural features often seen in liquid crystals: The molecule has a rod-like shape, the double bonds and benzene rings provide rigidity, and the polar $COOCH_3$ group creates a dipole moment.

Practice Exercise 1

Liquid crystalline phases are produced by which of the following?
(**a**) Short, flexible molecules, (**b**) complete lack of order among molecules, (**c**) three-dimensional order among molecules, (**d**) highly branched molecules, (**e**) rod-shaped molecules.

Practice Exercise 2

Suggest a reason why decane ($CH_3CH_2CH_2CH_2CH_2CH_2CH_2CH_2CH_2CH_3$) does not exhibit liquid crystalline behavior.

*Cholesteric liquid crystals are sometimes called chiral nematic phases because the molecules within each plane adopt an arrangement similar to a nematic liquid crystal.

SAMPLE
INTEGRATIVE EXERCISE Putting Concepts Together

The substance CS_2 has a melting point of $-110.8\,°C$ and a boiling point of $46.3\,°C$. Its density at $20\,°C$ is $1.26\ g/cm^3$. It is highly flammable. **(a)** What is the name of this compound? **(b)** List the intermolecular forces that CS_2 molecules exert on one another. **(c)** Write a balanced equation for the combustion of this compound in air. (You will have to decide on the most likely oxidation products.) **(d)** The critical temperature and pressure for CS_2 are 552 K and 78 atm, respectively. Compare these values with those for CO_2 in Table 11.6 and discuss the possible origins of the differences.

SOLUTION

(a) The compound is named carbon disulfide, in analogy with the naming of other binary molecular compounds such as carbon dioxide. ∞ (Section 2.8)

(b) Because there is no H atom, there can be no hydrogen bonding. If we draw the Lewis structure, we see that carbon forms double bonds with each sulfur:

$$\ddot{\underset{\cdot\cdot}{S}}\!\!=\!\!C\!\!=\!\!\ddot{\underset{\cdot\cdot}{S}}$$

Using the VSEPR model ∞ (Section 9.2), we conclude that the molecule is linear and therefore has no dipole moment. ∞ (Section 9.3) Thus, there are no dipole–dipole forces. Only dispersion forces operate between the CS_2 molecules.

(c) The most likely products of the combustion will be CO_2 and SO_2. ∞ (Section 3.2) Under some conditions SO_3 might be formed, but this would be the less likely outcome. Thus, we have the following equation for combustion:

$$CS_2(l)\ +\ 3\,O_2(g) \rightarrow CO_2(g)\ +\ 2\,SO_2(g)$$

(d) The critical temperature and pressure of CS_2 (552 K and 78 atm, respectively) are both higher than those given for CO_2 in Table 11.6 (304 K and 73 atm, respectively). The difference in critical temperatures is especially notable. The higher values for CS_2 arise from the greater dispersion attractions between the CS_2 molecules compared with CO_2. These greater attractions are due to the larger size of the sulfur compared to oxygen and, therefore, its greater polarizability.

Chapter Summary and Key Terms

A MOLECULAR COMPARISON OF GASES, LIQUIDS, AND SOLIDS (INTRODUCTION AND SECTION 11.1) Substances that are gases or liquids at room temperature are usually composed of molecules. In gases the intermolecular attractive forces are negligible compared to the kinetic energies of the molecules; thus, the molecules are widely separated and undergo constant, chaotic motion. In liquids the **intermolecular forces** are strong enough to keep the molecules in close proximity; nevertheless, the molecules are free to move with respect to one another. In solids the intermolecular attractive forces are strong enough to restrain molecular motion and to force the particles to occupy specific locations in a three-dimensional arrangement.

INTERMOLECULAR FORCES (SECTION 11.2) Three types of intermolecular forces exist between neutral molecules: **dispersion forces, dipole–dipole forces,** and **hydrogen bonding.** Dispersion forces operate between all molecules (and atoms, for atomic substances such as He, Ne, Ar, and so forth). As molecular weight increases, the **polarizability** of a molecule increases, which results in stronger dispersion forces. Molecular shape is also an important factor. Dipole–dipole forces increase in strength as the polarity of the molecule increases. Hydrogen bonding occurs in compounds containing O—H, N—H, and F—H bonds. Hydrogen bonds are generally stronger than dipole–dipole or dispersion forces. **Ion–dipole forces** are important in solutions in which ionic compounds are dissolved in polar solvents.

SELECT PROPERTIES OF LIQUIDS (SECTION 11.3) The stronger the intermolecular forces, the greater is the **viscosity,** or resistance to flow, of a liquid. The surface tension of a liquid also increases as intermolecular forces increase in strength. **Surface tension** is a measure of the tendency of a liquid to maintain a minimum surface area. The adhesion of a liquid to the walls of a narrow tube and the cohesion of the liquid account for **capillary action.**

PHASE CHANGES (SECTION 11.4) A substance may exist in more than one state of matter, or phase. **Phase changes** are transformations from one phase to another. Changes of a solid to liquid (melting), solid to gas (sublimation), and liquid to gas (vaporization) are all endothermic processes. Thus, the **heat of fusion** (melting), the **heat of sublimation,** and the **heat of vaporization** are all positive quantities. The reverse processes (freezing, deposition, and condensation) are exothermic.

A gas cannot be liquefied by application of pressure if the temperature is above its **critical temperature.** The pressure required to liquefy a gas at its critical temperature is called the **critical pressure.**

When the temperature exceeds the critical temperature and the pressure exceeds the critical pressure, the liquid and gas phases coalesce to form a **supercritical fluid**.

VAPOR PRESSURE (SECTION 11.5) The **vapor pressure** of a liquid is the partial pressure of the vapor when it is in **dynamic equilibrium** with the liquid. At equilibrium the rate of transfer of molecules from the liquid to the vapor equals the rate of transfer from the vapor to the liquid. The higher the vapor pressure of a liquid, the more readily it evaporates and the more **volatile** it is. Vapor pressure increases with temperature. Boiling occurs when the vapor pressure equals the external pressure. Thus, the **boiling point** of a liquid depends on pressure. The **normal boiling point** is the temperature at which the vapor pressure equals 1 atm.

PHASE DIAGRAMS (SECTION 11.6) The equilibria between the solid, liquid, and gas phases of a substance as a function of temperature and pressure are displayed on a **phase diagram**. A line indicates equilibria between any two phases. The line through the melting point usually slopes slightly to the right as pressure increases, because the solid is usually more dense than the liquid. The melting point at 1 atm is the **normal melting point**. The point on the diagram at which all three phases coexist in equilibrium is called the **triple point**. The **critical point** corresponds to the critical temperature and critical pressure. Beyond the critical point, the substance is a supercritical fluid.

LIQUID CRYSTALS (SECTION 11.7) A **liquid crystal** is a substance that exhibits one or more ordered phases at a temperature above the melting point of the solid. In a **nematic liquid crystal** the molecules are aligned along a common direction, but the ends of the molecules are not lined up. In a **smectic liquid crystal** the ends of the molecules are lined up so that the molecules form layers. In **smectic A liquid crystals** the long axes of the molecules line up perpendicular to the layers. In **smectic C liquid crystals** the long axes of molecules are inclined with respect to the layers. A **cholesteric liquid crystal** is composed of molecules that align parallel to each other within a layer, as they do in nematic liquid crystalline phases, but the direction along which the long axes of the molecules align rotates from one layer to the next to form a helical structure. Substances that form liquid crystals are generally composed of molecules with fairly rigid, elongated shapes, as well as polar groups to help align molecules through dipole–dipole interactions.

Learning Outcomes After studying this chapter, you should be able to:

- Identify the intermolecular attractive interactions (dispersion, dipole–dipole, hydrogen bonding, ion–dipole) that exist between molecules or ions based on their composition and molecular structure and compare the relative strengths of these intermolecular forces. (Section 11.2)

- Explain the concept of polarizability and how it relates to dispersion forces. (Section 11.2)

- Explain the concepts of viscosity, surface tension, and capillary action. (Section 11.3)

- List the names of the various changes of state for a pure substance and indicate which are endothermic and which are exothermic. (Section 11.4)

- Interpret heating curves and calculate the enthalpy changes related to temperature changes and phase changes. (Section 11.4)

- Define critical pressure, critical temperature, vapor pressure, normal boiling point, normal melting point, critical point, and triple point. (Sections 11.5 and 11.6)

- Interpret and sketch phase diagrams. Explain how water's phase diagram differs from most other substances and why. (Section 11.6)

- Describe how the molecular arrangements characteristic of nematic, smectic, and cholesteric liquid crystals differ from ordinary liquids and from each other. Recognize the features of molecules that favor formation of liquid crystalline phases. (Section 11.7)

Exercises

Visualizing Concepts

11.1 **(a)** Does the diagram best describe a crystalline solid, a liquid, or a gas? **(b)** Explain. [Section 11.1]

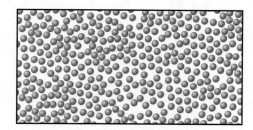

11.2 **(a)** Which kind of intermolecular attractive force is shown in each case here?

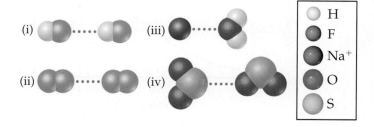

(b) Predict which of the four interactions is the weakest. [Section 11.2]

11.3 **(a)** Do you expect the viscosity of glycerol, $C_3H_5(OH)_3$, to be larger or smaller than that of 1-propanol, C_3H_7OH? **(b)** Explain. [Section 11.3]

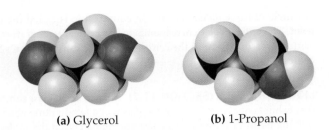

(a) Glycerol   **(b)** 1-Propanol

11.4 If 42.0 kJ of heat is added to a 32.0-g sample of liquid methane under 1 atm of pressure at a temperature of −170 °C, what are the final state and temperature of the methane once the system equilibrates? Assume no heat is lost to the surroundings. The normal boiling point of methane is −161.5 °C. The specific heats of liquid and gaseous methane are 3.48 and 2.22 J/g-K, respectively. [Section 11.4]

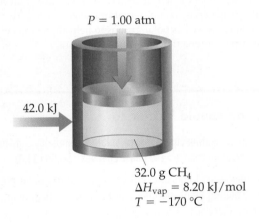

$P = 1.00$ atm

42.0 kJ

32.0 g CH_4
$\Delta H_{vap} = 8.20$ kJ/mol
$T = -170$ °C

11.5 Using this graph of CS_2 data,

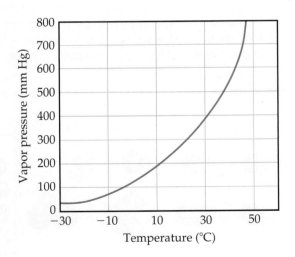

determine **(a)** the approximate vapor pressure of CS_2 at 30 °C, **(b)** the temperature at which the vapor pressure equals 300 torr, **(c)** the normal boiling point of CS_2. [Section 11.5]

11.6 The molecules

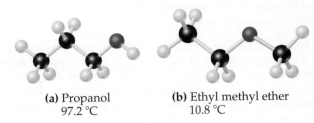

(a) Propanol   **(b)** Ethyl methyl ether
97.2 °C   10.8 °C

have the same molecular formula (C_3H_8O) but different normal boiling points, as shown. Rationalize the difference in boiling points. [Sections 11.2 and 11.5]

11.7 The phase diagram of a hypothetical substance is

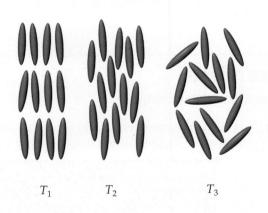

(a) Estimate the normal boiling point and freezing point of the substance.

(b) What is the physical state of the substance under the following conditions? (i) $T = 150$ K, $P = 0.2$ atm; (ii) $T = 100$ K, $P = 0.8$ atm; (iii) $T = 300$ K, $P = 1.0$ atm.

(c) What is the triple point of the substance? [Section 11.6]

11.8 At three different temperatures, T_1, T_2, and T_3, the molecules in a liquid crystal align in these ways:

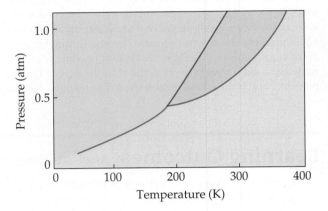

T_1 T_2 T_3

(a) At which temperature or temperatures is the substance in a liquid crystalline state? At those temperatures, which type of liquid crystalline phase is depicted?

(b) Which is the highest of these three temperatures? [Section 11.7]

Molecular Comparisons of Gases, Liquids, and Solids (Section 11.1)

11.9 List the three states of matter in order of (**a**) increasing molecular disorder and (**b**) increasing intermolecular attraction. (**c**) Which state of matter is most easily compressed?

11.10 (**a**) How does the average kinetic energy of molecules compare with the average energy of attraction between molecules in solids, liquids, and gases? (**b**) Why does increasing the temperature cause a solid substance to change in succession from a solid to a liquid to a gas? (**c**) What happens to a vapor if you put it under extremely high pressure?

11.11 As a metal such as lead melts, what happens to (**a**) the average kinetic energy of the atoms, (**b**) the average distance between the atoms?

11.12 At room temperature, Si is a solid, CCl_4 is a liquid, and Ar is a gas. List these substances in order of (**a**) increasing intermolecular energy of attraction, (**b**) increasing boiling point.

11.13 At standard temperature and pressure the molar volumes of Cl_2 and NH_3 gases are 22.06 and 22.40 L, respectively. (**a**) Given the different molecular weights, dipole moments, and molecular shapes, why are their molar volumes nearly the same? (**b**) On cooling to 160 K, both substances form crystalline solids. Do you expect the molar volumes to decrease or increase on cooling the gases to 160 K? (**c**) The densities of crystalline Cl_2 and NH_3 at 160 K are 2.02 and 0.84 g/cm^3, respectively. Calculate their molar volumes. (**d**) Are the molar volumes in the solid state as similar as they are in the gaseous state? Explain. (**e**) Would you expect the molar volumes in the liquid state to be closer to those in the solid or gaseous state?

11.14 Benzoic acid, C_6H_5COOH, melts at 122 °C. The density in the liquid state at 130 °C is 1.08 g/cm^3. The density of solid benzoic acid at 15 °C is 1.266 g/cm^3. (**a**) In which of these two states is the average distance between molecules greater? (**b**) Explain the difference in densities at the two temperatures in terms of the relative kinetic energies of the molecules.

Intermolecular Forces (Section 11.2)

11.15 (**a**) Which type of intermolecular attractive force operates between all molecules? (**b**) Which type of intermolecular attractive force operates only between polar molecules? (**c**) Which type of intermolecular attractive force operates only between the hydrogen atom of a polar bond and a nearby small electronegative atom?

11.16 (**a**) Which is generally stronger, intermolecular interactions or intramolecular interactions? (**b**) Which of these kinds of interactions are broken when a liquid is converted to a gas?

11.17 Describe the intermolecular forces that must be overcome to convert these substances from a liquid to a gas: (**a**) SO_2, (**b**) CH_3COOH, (**c**) H_2S.

11.18 Which type of intermolecular force accounts for each of these differences? (**a**) CH_3OH boils at 65 °C; CH_3SH boils at 6 °C. (**b**) Xe is a liquid at atmospheric pressure and 120 K, whereas Ar is a gas under the same conditions. (**c**) Kr, atomic weight 84, boils at 120.9 K, whereas Cl_2, molecular weight about 71, boils at 238 K. (**d**) Acetone boils at 56 °C, whereas 2-methylpropane boils at −12 °C.

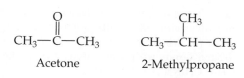

Acetone   2-Methylpropane

11.19 (**a**) What is meant by the term *polarizability*? (**b**) Which of the following atoms would you expect to be most polarizable: N, P, As, Sb? Explain. (**c**) List the following molecules in order of increasing polarizability: $GeCl_4$, CH_4, $SiCl_4$, SiH_4, and $GeBr_4$. (**d**) Predict the order of boiling points of the substances in part (**c**).

11.20 True or false:

(**a**) For molecules with similar molecular weights, the dispersion forces become stronger as the molecules become more polarizable.

(**b**) For the noble gases the dispersion forces decrease while the boiling points increase as you go down the column in the periodic table.

(**c**) In terms of the total attractive forces for a given substance, dipole–dipole interactions, when present, are always greater than dispersion forces.

(**d**) All other factors being the same, dispersion forces between linear molecules are greater than those between molecules whose shapes are nearly spherical.

11.21 Which member in each pair has the greater dispersion forces? (**a**) H_2O or H_2S, (**b**) CO_2 or CO, (**c**) SiH_4 or GeH_4.

11.22 Which member in each pair has the stronger intermolecular dispersion forces? (**a**) Br_2 or O_2, (**b**) $CH_3CH_2CH_2CH_2SH$ or $CH_3CH_2CH_2CH_2CH_2SH$, (**c**) $CH_3CH_2CH_2Cl$ or $(CH_3)_2CHCl$.

11.23 Butane and 2-methylpropane, whose space-filling models are shown here, are both nonpolar and have the same molecular formula, C_4H_{10}, yet butane has the higher boiling point (−0.5 °C compared to −11.7 °C). Explain.

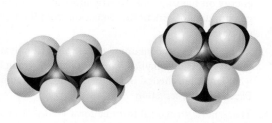

(**a**) Butane   (**b**) 2-Methylpropane

11.24 Propyl alcohol ($CH_3CH_2CH_2OH$) and isopropyl alcohol [$(CH_3)_2CHOH$], whose space-filling models are shown, have boiling points of 97.2 and 82.5 °C, respectively. Explain why the boiling point of propyl alcohol is higher, even though both have the molecular formula, C_3H_8O.

(**a**) Propyl alcohol   (**b**) Isopropyl alcohol

11.25 (a) What atoms must a molecule contain to participate in hydrogen bonding with other molecules of the same kind? (b) Which of the following molecules can form hydrogen bonds with other molecules of the same kind: CH_3F, CH_3NH_2, CH_3OH, CH_3Br?

11.26 Rationalize the difference in boiling points in each pair: (a) HF (20 °C) and HCl (−85 °C), (b) $CHCl_3$ (61 °C) and $CHBr_3$ (150 °C), (c) Br_2 (59 °C) and ICl (97 °C).

11.27 Ethylene glycol ($HOCH_2CH_2OH$), the major substance in antifreeze, has a normal boiling point of 198 °C. By comparison, ethyl alcohol (CH_3CH_2OH) boils at 78 °C at atmospheric pressure. Ethylene glycol dimethyl ether ($CH_3OCH_2CH_2OCH_3$) has a normal boiling point of 83 °C, and ethyl methyl ether ($CH_3CH_2OCH_3$) has a normal boiling point of 11 °C. (a) Explain why replacement of a hydrogen on the oxygen by a CH_3 group generally results in a lower boiling point. (b) What are the major factors responsible for the difference in boiling points of the two ethers?

11.28 Based on the type or types of intermolecular forces, predict the substance in each pair that has the higher boiling point: (a) propane (C_3H_8) or *n*-butane (C_4H_{10}), (b) diethyl ether ($CH_3CH_2OCH_2CH_3$) or 1-butanol ($CH_3CH_2CH_2CH_2OH$), (c) sulfur dioxide (SO_2) or sulfur trioxide (SO_3), (d) phosgene (Cl_2CO) or formaldehyde (H_2CO).

11.29 Look up and compare the normal boiling points and normal melting points of H_2O and H_2S. Based on these physical properties, which substance has stronger intermolecular forces? What kinds of intermolecular forces exist for each molecule?

11.30 The following quote about ammonia (NH_3) is from a textbook of inorganic chemistry: "It is estimated that 26% of the hydrogen bonding in NH_3 breaks down on melting, 7% on warming from the melting to the boiling point, and the final 67% on transfer to the gas phase at the boiling point." From the standpoint of the kinetic energy of the molecules, explain (a) why there is a decrease of hydrogen-bonding energy on melting and (b) why most of the loss in hydrogen bonding occurs in the transition from the liquid to the vapor state.

11.31 A number of salts containing the tetrahedral polyatomic anion, BF_4^-, are ionic liquids, whereas salts containing the somewhat larger tetrahedral ion SO_4^{2-} do not form ionic liquids. Explain this observation.

11.32 The generic structural formula for a 1-alkyl-3-methylimidazolium cation is

where R is a $-CH_2(CH_2)_nCH_3$ alkyl group. The melting points of the salts that form between the 1-alkyl-3-methylimidazolium cation and the PF_6^- anion are as follows: R = CH_2CH_3 (m.p. = 60 °C), R = $CH_2CH_2CH_3$ (m.p. = 40 °C), R = $CH_2CH_2CH_2CH_3$ (m.p. = 10 °C), and R = $CH_2CH_2CH_2CH_2CH_2CH_3$ (m.p. = −61 °C). Why does the melting point decrease as the length of alkyl group increases?

Select Properties of Liquids (Section 11.3)

11.33 (a) What is the relationship between surface tension and temperature? (b) What is the relationship between viscosity and temperature? (c) Why do substances with high surface tension also tend to have high viscosities?

11.34 Based on their composition and structure, list CH_2Cl_2, $CH_3CH_2CH_3$, and CH_3CH_2OH in order of (a) increasing intermolecular forces, (b) increasing viscosity, (c) increasing surface tension.

11.35 Explain the following observations: (a) The surface tension of $CHBr_3$ is greater than that of $CHCl_3$. (b) As temperature increases, oil flows faster through a narrow tube. (c) Raindrops that collect on a waxed automobile hood take on a nearly spherical shape. (d) Oil droplets that collect on a waxed automobile hood take on a flat shape.

11.36 Hydrazine (H_2NNH_2), hydrogen peroxide (HOOH), and water (H_2O) all have exceptionally high surface tensions compared with other substances of comparable molecular weights. (a) Draw the Lewis structures for these three compounds. (b) What structural property do these substances have in common, and how might that account for the high surface tensions?

11.37 The boiling points, surface tensions, and viscosities of water and several alcohols are as follows:

	Boiling Point (°C)	Surface Tension (J/m^2)	Viscosity (kg/m-s)
Water, H_2O	100	7.3×10^{-2}	0.9×10^{-3}
Ethanol, CH_3CH_2OH	78	2.3×10^{-2}	1.1×10^{-3}
Propanol, $CH_3CH_2CH_2OH$	97	2.4×10^{-2}	2.2×10^{-3}
n-Butanol, $CH_3CH_2CH_2CH_2OH$	117	2.6×10^{-2}	2.6×10^{-3}
Ethylene glycol, $HOCH_2CH_2OH$	197	4.8×10^{-2}	26×10^{-3}

(a) For ethanol, propanol, and *n*-butanol the boiling points, surface tensions, and viscosities all increase. What is the reason for this increase? (b) How do you explain the fact that propanol and ethylene glycol have similar molecular weights (60 versus 62 amu), yet the viscosity of ethylene glycol is more than 10 times larger than propanol? (c) How do you explain the fact that water has the highest surface tension but the lowest viscosity?

11.38 (a) Would you expect the viscosity of *n*-pentane, $CH_3CH_2CH_2CH_2CH_3$, to be larger or smaller than the viscosity of *n*-hexane, $CH_3CH_2CH_2CH_2CH_2CH_3$? (b) Would you expect the viscosity of neopentane, $(CH_3)_4C$, to be smaller or larger than the viscosity of *n*-pentane? (See Figure 11.6 to see the shapes of these molecules.)

Phase Changes (Section 11.4)

11.39 Name the phase transition in each of the following situations and indicate whether it is exothermic or endothermic: (a) When ice is heated, it turns to water. (b) Wet clothes dry on a warm summer day. (c) Frost appears on a window on a cold winter day. (d) Droplets of water appear on a cold glass of beer.

11.40 Name the phase transition in each of the following situations and indicate whether it is exothermic or endothermic: (a) Bromine vapor turns to bromine liquid as it is cooled. (b) Crystals of iodine disappear from an evaporating dish as they stand in a fume hood. (c) Rubbing alcohol in an open container slowly disappears. (d) Molten lava from a volcano turns into solid rock.

11.41 Explain why any substance's heat of fusion is generally lower than its heat of vaporization.

11.42 Ethyl chloride (C_2H_5Cl) boils at 12 °C. When liquid C_2H_5Cl under pressure is sprayed on a room-temperature (25 °C) surface in air, the surface is cooled considerably. (a) What does this observation tell us about the specific heat of $C_2H_5Cl(g)$ as compared with that of $C_2H_5Cl(l)$? (b) Assume that the heat lost by the surface is gained by ethyl chloride. What enthalpies must you consider if you were to calculate the final temperature of the surface?

11.43 For many years drinking water has been cooled in hot climates by evaporating it from the surfaces of canvas bags or porous clay pots. How many grams of water can be cooled from 35 to 20 °C by the evaporation of 60 g of water? (The heat of vaporization of water in this temperature range is 2.4 kJ/g. The specific heat of water is 4.18 J/g-K.)

11.44 Compounds like CCl_2F_2 are known as chlorofluorocarbons, or CFCs. These compounds were once widely used as refrigerants but are now being replaced by compounds that are believed to be less harmful to the environment. The heat of vaporization of CCl_2F_2 is 289 J/g. What mass of this substance must evaporate to freeze 200 g of water initially at 15 °C? (The heat of fusion of water is 334 J/g; the specific heat of water is 4.18 J/g-K.)

11.45 Ethanol (C_2H_5OH) melts at −114 °C and boils at 78 °C. The enthalpy of fusion of ethanol is 5.02 kJ/mol, and its enthalpy of vaporization is 38.56 kJ/mol. The specific heats of solid and liquid ethanol are 0.97 and 2.3 J/g-K, respectively. (a) How much heat is required to convert 42.0 g of ethanol at 35 °C to the vapor phase at 78 °C? (b) How much heat is required to convert the same amount of ethanol at −155 °C to the vapor phase at 78 °C?

11.46 The fluorocarbon compound $C_2Cl_3F_3$ has a normal boiling point of 47.6 °C. The specific heats of $C_2Cl_3F_3(l)$ and $C_2Cl_3F_3(g)$ are 0.91 and 0.67 J/g-K, respectively. The heat of vaporization for the compound is 27.49 kJ/mol. Calculate the heat required to convert 35.0 g of $C_2Cl_3F_3$ from a liquid at 10.00 °C to a gas at 105.00 °C.

11.47 (a) What is the significance of the critical pressure of a substance? (b) What happens to the critical temperature of a series of compounds as the force of attraction between molecules increases? (c) Which of the substances listed in Table 11.6 can be liquefied at the temperature of liquid nitrogen (−196 °C)?

11.48 The critical temperatures (K) and pressures (atm) of a series of halogenated methanes are as follows:

Compound	CCl_3F	CCl_2F_2	$CClF_3$	CF_4
Critical temperature	471	385	302	227
Critical pressure	43.5	40.6	38.2	37.0

(a) List the intermolecular forces that occur for each compound. (b) Predict the order of increasing intermolecular attraction, from least to most, for this series of compounds. (c) Predict the critical temperature and pressure for CCl_4 based on the trends in this table. Look up the experimentally determined critical temperatures and pressures for CCl_4, using a source such as the *CRC Handbook of Chemistry and Physics*, and suggest a reason for any discrepancies.

Vapor Pressure (Section 11.5)

11.49 Which of the following affects the vapor pressure of a liquid? (a) Volume of the liquid, (b) surface area, (c) intermolecular attractive forces, (d) temperature, (e) density of the liquid.

11.50 Acetone (H_3CCOCH_3) has a boiling point of 56 °C. Based on the data given in Figure 11.25, would you expect acetone to have a higher or lower vapor pressure than ethanol at 25 °C?

11.51 (a) Place the following substances in order of increasing volatility: CH_4, CBr_4, CH_2Cl_2, CH_3Cl, $CHBr_3$, and CH_2Br_2. (b) How do the boiling points vary through this series? (c) Explain your answer to part (b) in terms of intermolecular forces.

11.52 True or false:
(a) CBr_4 is more volatile than CCl_4.
(b) CBr_4 has a higher boiling point than CCl_4.
(c) CBr_4 has weaker intermolecular forces than CCl_4.
(d) CBr_4 has a higher vapor pressure at the same temperature than CCl_4.

11.53 (a) Two pans of water are on different burners of a stove. One pan of water is boiling vigorously, while the other is boiling gently. What can be said about the temperature of the water in the two pans? (b) A large container of water and a small one are at the same temperature. What can be said about the relative vapor pressures of the water in the two containers?

11.54 Explain the following observations: (a) Water evaporates more quickly on a hot, dry day than on a hot, humid day. (b) It takes longer to cook an egg in boiling water at high altitudes than it does at lower altitudes.

11.55 Using the vapor-pressure curves in Figure 11.25, (a) estimate the boiling point of ethanol at an external pressure of 200 torr, (b) estimate the external pressure at which ethanol will boil at 60 °C, (c) estimate the boiling point of diethyl ether at 400 torr, (d) estimate the external pressure at which diethyl ether will boil at 40 °C.

11.56 Title of Appendix B lists the vapor pressure of water at various external pressures.

(a) Plot the data in Title of Appendix B, vapor pressure (torr) versus temperature (°C). From your plot, estimate the vapor pressure of water at body temperature, 37 °C.

(b) Explain the significance of the data point at 760.0 torr, 100 °C.

(c) A city at an altitude of 5000 ft above sea level has a barometric pressure of 633 torr. To what temperature would you have to heat water to boil it in this city?

(d) A city at an altitude of 500 ft below sea level would have a barometric pressure of 774 torr. To what temperature would you have to heat water to boil it in this city?

(e) For the two cities in parts (c) and (d), compare the average kinetic energies of the water molecules at their boiling points. Are the kinetic energies the same or different? Explain.

Phase Diagrams (Section 11.6)

11.57 (a) What is the significance of the critical point in a phase diagram? (b) Why does the line that separates the gas and liquid phases end at the critical point?

11.58 (a) What is the significance of the triple point in a phase diagram? (b) Could you measure the triple point of water by measuring the temperature in a vessel in which water vapor, liquid water, and ice are in equilibrium under 1 atm of air? Explain.

11.59 Referring to Figure 11.28, describe all the phase changes that would occur in each of the following cases: (a) Water vapor originally at 0.005 atm and −0.5 °C is slowly compressed at constant temperature until the final pressure is 20 atm. (b) Water originally at 100.0 °C and 0.50 atm is cooled at constant pressure until the temperature is −10 °C.

11.60 Referring to Figure 11.29, describe the phase changes (and the temperatures at which they occur) when CO_2 is heated from −80 to −20 °C at (a) a constant pressure of 3 atm, (b) a constant pressure of 6 atm.

11.61 The phase diagram for neon is

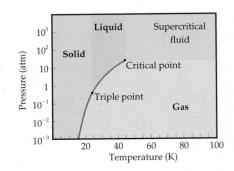

Use the phase diagram to answer the following questions. (a) What is the approximate value of the normal melting point? (b) Over what pressure range will solid neon sublime? (c) At room temperature ($T = 25$ °C) can neon be liquefied by compressing it?

11.62 Use the phase diagram of neon to answer the following questions. (a) What is the approximate value of the normal boiling point? (b) What can you say about the strength of the intermolecular forces in neon and argon based on the critical points of Ne and Ar (see Table 11.6.)?

11.63 The fact that water on Earth can readily be found in all three states (solid, liquid, and gas) is in part a consequence of the fact that the triple point of water ($T = 0.01$ °C, $P = 0.006$ atm) falls within a range of temperatures and pressures found on Earth. Saturn's largest moon Titan has a considerable amount of methane in its atmosphere. The conditions on the surface of Titan are estimated to be $P = 1.6$ atm and $T = -178$ °C. As seen from the phase diagram of methane (Figure 11.30), these conditions are not far from the triple point of methane, raising the tantalizing possibility that solid, liquid, and gaseous methane can be found on Titan. (a) In what state would you expect to find methane on the surface of Titan? (b) On

moving upward through the atmosphere the pressure will decrease. If we assume that the temperature does not change, what phase change would you expect to see as we move away from the surface?

11.64 At 25 °C gallium is a solid with a density of 5.91 g/cm³. Its melting point, 29.8 °C, is low enough that you can melt it by holding it in your hand. The density of liquid gallium just above the melting point is 6.1 g/cm³. Based on this information, what unusual feature would you expect to find in the phase diagram of gallium?

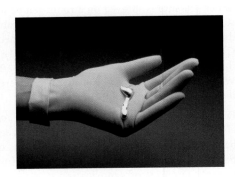

Liquid Crystals (Section 11.7)

11.65 In terms of the arrangement and freedom of motion of the molecules, how are the nematic liquid crystalline phase and an ordinary liquid phase similar? How are they different?

11.66 What observations made by Reinitzer on cholesteryl benzoate suggested that this substance possesses a liquid crystalline phase?

11.67 The molecules shown in Figure 11.33 possess polar groups (that is, groupings of atoms that give rise to sizable dipole moments within the molecules). How might the presence of polar groups enhance the tendency toward liquid crystal formation?

11.68 One of the more effective liquid crystalline substances employed in LCDs is the molecule

$$CH_3(CH_2)_2CH=CH-CH \underset{CH_2-CH_2}{\overset{CH_2-CH_2}{\diagup\diagdown}} CH-CH \underset{CH_2-CH_2}{\overset{CH_2-CH_2}{\diagup\diagdown}} CH-C\equiv N$$

(a) How many double bonds are there in this molecule?
(b) Describe the features of the molecule that make it prone to show liquid crystalline behavior.

11.69 For a given substance, the liquid crystalline phase tends to be more viscous than the liquid phase. Why?

11.70 Describe how a cholesteric liquid crystal phase differs from a nematic phase.

11.71 It often happens that a substance possessing a smectic liquid crystalline phase just above the melting point passes into a nematic liquid crystalline phase at a higher temperature. Account for this type of behavior.

11.72 The smectic liquid crystalline phase can be said to be more highly ordered than the nematic phase. In what sense is this true?

Additional Exercises

11.73 As the intermolecular attractive forces between molecules increase in magnitude, do you expect each of the following to increase or decrease in magnitude? (**a**) Vapor pressure, (**b**) heat of vaporization, (**c**) boiling point, (**d**) freezing point, (**e**) viscosity, (**f**) surface tension, (**g**) critical temperature.

11.74 The table below lists the density of O_2 at various temperatures and at 1 atm. Graph the data and predict the substance's normal boiling point.

Temperature (K)	Density (mol/L)
60	40.1
70	38.6
80	37.2
90	35.6
100	0.123
120	0.102
140	0.087

11.75 Suppose you have two colorless molecular liquids, one boiling at $-84\,°C$, the other at $34\,°C$, and both at atmospheric pressure. Which of the following statements is correct? For each statement that is not correct, modify the statement so that it is correct. (**a**) The higher-boiling liquid has greater total intermolecular forces than the lower-boiling liquid. (**b**) The lower-boiling liquid must consist of nonpolar molecules. (**c**) The lower-boiling liquid has a lower molecular weight than the higher-boiling liquid. (**d**) The two liquids have identical vapor pressures at their normal boiling points. (**e**) At $-84\,°C$ both liquids have vapor pressures of 760 mm Hg.

11.76 Two isomers of the planar compound 1,2-dichloroethylene are shown here.

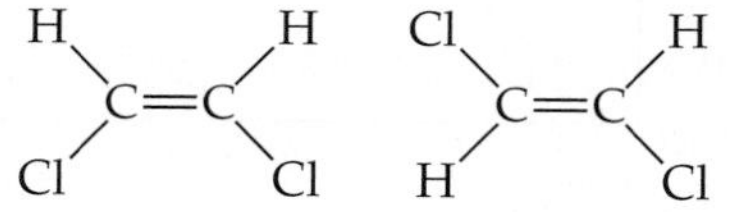

cis isomer   trans isomer

(**a**) Which of the two isomers will have the stronger dipole–dipole forces? (**b**) One isomer has a boiling point of $60.3\,°C$ and the other $47.5\,°C$. Which isomer has which boiling point?

11.77 In dichloromethane, CH_2Cl_2 ($\mu = 1.60$ D), the dispersion force contribution to the intermolecular attractive forces is about five times larger than the dipole–dipole contribution. Compared to CH_2Cl_2, would you expect the relative importance of the dipole–dipole contribution to increase or decrease (**a**) in dibromomethane ($\mu = 1.43$ D), (**b**) in difluoromethane ($\mu = 1.93$ D)? (**c**) Explain.

11.78 When an atom or a group of atoms is substituted for an H atom in benzene (C_6H_6), the boiling point changes. Explain the order of the following boiling points: C_6H_6 ($80\,°C$), C_6H_5Cl ($132\,°C$), C_6H_5Br ($156\,°C$), C_6H_5OH ($182\,°C$).

11.79 The DNA double helix (Figure 24.30) at the atomic level looks like a twisted ladder, where the "rungs" of the ladder consist of molecules that are hydrogen-bonded together. Sugar and phosphate groups make up the sides of the ladder. Shown are the structures of the adenine–thymine (AT) "base pair" and the guanine–cytosine (GC) base pair:

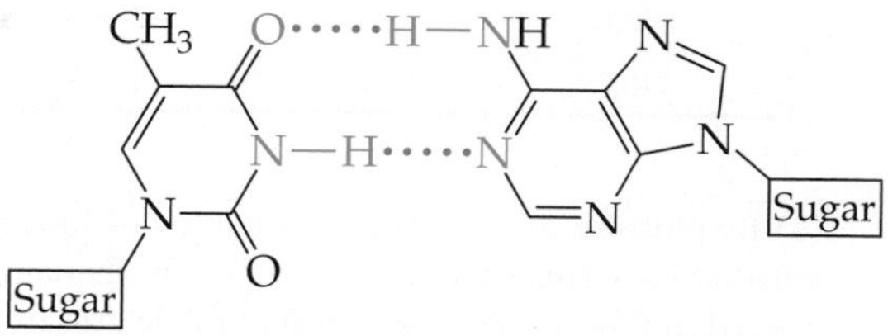

Thymine   Adenine

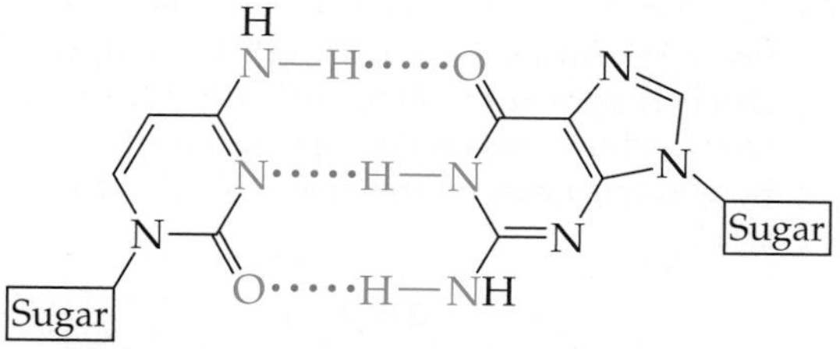

Cytosine   Guanine

You can see that AT base pairs are held together by two hydrogen bonds and the GC base pairs are held together by three hydrogen bonds. Which base pair is more stable to heating? Why?

11.80 Ethylene glycol ($HOCH_2CH_2OH$) is the major component of antifreeze. It is a slightly viscous liquid, not very volatile at room temperature, with a boiling point of $198\,°C$. Pentane (C_5H_{12}), which has about the same molecular weight, is a nonviscous liquid that is highly volatile at room temperature and whose boiling point is $36.1\,°C$. Explain the differences in the physical properties of the two substances.

11.81 Use the normal boiling points

propane (C_3H_8) $-42.1\,°C$
butane (C_4H_{10}) $-0.5\,°C$
pentane (C_5H_{12}) $36.1\,°C$
hexane (C_6H_{14}) $68.7\,°C$
heptane (C_7H_{16}) $98.4\,°C$

to estimate the normal boiling point of octane (C_8H_{18}). Explain the trend in the boiling points.

11.82 One of the attractive features of ionic liquids is their low vapor pressure, which in turn tends to make them nonflammable. Why do you think ionic liquids have lower vapor pressures than most room-temperature molecular liquids?

11.83 (**a**) When you exercise vigorously, you sweat. How does this help your body cool? (**b**) A flask of water is connected to a vacuum pump. A few moments after the pump is turned on, the water begins to boil. After a few minutes, the water begins to freeze. Explain why these processes occur.

11.84 The following table gives the vapor pressure of hexafluoro-benzene (C_6F_6) as a function of temperature:

Temperature (K)	Vapor Pressure (torr)
280.0	32.42
300.0	92.47
320.0	225.1
330.0	334.4
340.0	482.9

(a) By plotting these data in a suitable fashion, determine whether the Clausius–Clapeyron equation (Equation 11.1) is obeyed. If it is obeyed, use your plot to determine ΔH_{vap} for C_6F_6. **(b)** Use these data to determine the boiling point of the compound.

11.85 Suppose the vapor pressure of a substance is measured at two different temperatures. **(a)** By using the Clausius–Clapeyron equation (Equation 11.1) derive the following relationship between the vapor pressures, P_1 and P_2, and the absolute temperatures at which they were measured, T_1 and T_2:

$$\ln \frac{P_1}{P_2} = -\frac{\Delta H_{vap}}{R}\left(\frac{1}{T_1} - \frac{1}{T_2}\right)$$

(b) Gasoline is a mixture of hydrocarbons, a major component of which is octane ($CH_3CH_2CH_2CH_2CH_2CH_2CH_2CH_3$). Octane has a vapor pressure of 13.95 torr at 25 °C and a vapor pressure of 144.78 torr at 75 °C. Use these data and the equation in part (a) to calculate the heat of vaporization of octane. **(c)** By using the equation in part (a) and the data given in part (b), calculate the normal boiling point of octane. Compare your answer to the one you obtained from Exercise 11.80. **(d)** Calculate the vapor pressure of octane at −30 °C.

11.86 The following data present the temperatures at which certain vapor pressures are achieved for dichloromethane (CH_2Cl_2) and methyl iodide (CH_3I):

Vapor Pressure (torr)	10.0	40.0	100.0	400.0
T for CH_2Cl_2 (°C)	−43.3	−22.3	−6.3	24.1
T for CH_3I (°C)	−45.8	−24.2	−7.0	25.3

(a) Which of the two substances is expected to have the greater dipole–dipole forces? Which is expected to have the greater dispersion forces? Based on your answers, explain why it is difficult to predict which compound would be more volatile. **(b)** Which compound would you expect to have the higher boiling point? Check your answer in a reference book such as the *CRC Handbook of Chemistry and Physics*. **(c)** The order of volatility of these two substances changes as the temperature is increased. What quantity must be different for the two substances for this phenomenon to occur? **(d)** Substantiate your answer for part (c) by drawing an appropriate graph.

11.87 Naphalene, ($C_{10}H_8$) is the main ingredient in traditional moth-balls. Its normal melting point is 81 °C, its normal boiling point is 218 °C, and its triple point is 80 °C at 1000 Pa. Using the data, construct a phase diagram for naphthalene, labeling all the regions of your diagram.

11.88 A watch with a liquid crystal display (LCD) does not function properly when it is exposed to low temperatures during a trip to Antarctica. Explain why the LCD might not function well at low temperature.

11.89 A particular liquid crystalline substance has the phase diagram shown in the figure. By analogy with the phase diagram for a non-liquid crystalline substance, identify the phase present in each area.

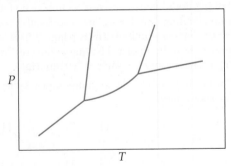

Integrative Exercises

11.90 **(a)** At the molecular level, what factor is responsible for the steady increase in viscosity with increasing molecular weight in the hydrocarbon series shown in Table 11.5? **(b)** Although the viscosity varies over a factor of more than two in the series from hexane to nonane, the surface tension at 25 °C increases by only about 20% in the same series. How do you account for this? **(c)** *n*-octyl alcohol [$CH_3(CH_2)_7OH$] has a viscosity of 1.01×10^{-2} kg/m-s, much higher than nonane, which has about the same molecular weight. What accounts for this difference? How does your answer relate to the difference in normal boiling points for these two substances?

11.91 Acetone [$(CH_3)_2CO$] is widely used as an industrial solvent. **(a)** Draw the Lewis structure for the acetone molecule and predict the geometry around each carbon atom. **(b)** Is the acetone molecule polar or nonpolar? **(c)** What kinds of intermolecular attractive forces exist between acetone molecules? **(d)** 1-Propanol ($CH_3CH_2CH_2OH$) has a molecular weight that is very similar to that of acetone, yet acetone boils at 56.5 °C and 1-propanol boils at 97.2 °C. Explain the difference.

11.92 The table shown here lists the molar heats of vaporization for several organic compounds. Use specific examples from this list to illustrate how the heat of vaporization varies with **(a)** molar mass, **(b)** molecular shape, **(c)** molecular polarity, **(d)** hydrogen-bonding interactions. Explain these comparisons in terms of the nature of the intermolecular forces at work. (You may find it helpful to draw out the structural formula for each compound.)

Compound	Heat of Vaporization (kJ/mol)
$CH_3CH_2CH_3$	19.0
$CH_3CH_2CH_2CH_2CH_3$	27.6
$CH_3CHBrCH_3$	31.8
CH_3COCH_3	32.0
$CH_3CH_2CH_2Br$	33.6
$CH_3CH_2CH_2OH$	47.3

11.93 The vapor pressure of ethanol (C_2H_5OH) at 19 °C is 40.0 torr. A 1.00-g sample of ethanol is placed in a 2.00 L container at 19 °C. If the container is closed and the ethanol is allowed to reach equilibrium with its vapor, how many grams of liquid ethanol remain?

11.94 Liquid butane (C_4H_{10}) is stored in cylinders to be used as a fuel. The normal boiling point of butane is listed as −0.5 °C. (**a**) Suppose the tank is standing in the Sun and reaches a temperature of 35 °C. Would you expect the pressure in the tank to be greater or less than atmospheric pressure? How does the pressure within the tank depend on how much liquid butane is in it? (**b**) Suppose the valve to the tank is opened and a few liters of butane are allowed to escape rapidly. What do you expect would happen to the temperature of the remaining liquid butane in the tank? Explain. (**c**) How much heat must be added to vaporize 250 g of butane if its heat of vaporization is 21.3 kJ/mol? What volume does this much butane occupy at 755 torr and 35 °C?

11.95 Using information in Titles of Appendices B and C, calculate the minimum grams of propane, $C_3H_8(g)$, that must be combusted to provide the energy necessary to convert 5.50 kg of ice at −20 °C to liquid water at 75 °C.

11.96 The vapor pressure of a volatile liquid can be determined by slowly bubbling a known volume of gas through it at a known temperature and pressure. In an experiment, 5.00 L of N_2 gas is passed through 7.2146 g of liquid benzene (C_6H_6) at 26.0 °C. The liquid remaining after the experiment weighs 5.1493 g. Assuming that the gas becomes saturated with benzene vapor and that the total gas volume and temperature remain constant, what is the vapor pressure of the benzene in torr?

11.97 The relative humidity of air equals the ratio of the partial pressure of water in the air to the equilibrium vapor pressure of water at the same temperature times 100%. If the relative humidity of the air is 58% and its temperature is 68 °F, how many molecules of water are present in a room measuring 12 ft × 10 ft × 8 ft?

Design an Experiment

Hydraulic fluids are used to transfer power in hydraulic machinery such as aircraft flight controls, excavating equipment, and hydraulic brakes. The power is transferred by the fluid, which is distributed through hoses and tubes by means of various pumps and valves. Imagine that an organic liquid is needed to serve as a hydraulic fluid over a range of temperatures. The three isomers of hexanol are hypothesized to be potentially suitable. These compounds all have a six-carbon backbone with —OH groups on the first, second, or third carbon of the chain. (See Section 2.9 for nomenclature of simple organic substances.) These three substances will have similar yet differing key properties. (**a**) What properties would be important in this application? (**b**) Once you have identified at least two properties required for the application, describe the experiments you would carry out to determine which of the three alcohols would be best suited for use as a hydraulic fluid. (**c**) The class of compounds called *hexanediols* is related to hexanols except that they have two —OH groups attached to the six-carbon backbone. Hypothesize how the key properties you discuss here would change from a hexanol to a hexanediol. Could your experimental procedures be used to explore these properties for hexanediols?

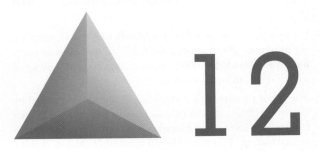

12

Solids and Modern Materials

Modern devices like computers and cell phones are built from solids with very specific physical properties. For example, the integrated circuit that is at the heart of many electronic devices is built from semiconductors like silicon, metals like copper, and insulators like hafnium oxide.

Scientists and engineers turn almost exclusively to solids for materials used in many other technologies: *alloys* for magnets and airplane turbines, *semiconductors* for solar cells and light-emitting diodes, and *polymers* for packaging and biomedical applications. Chemists have contributed to the discovery and development of new materials either by inventing new substances or by developing the means to process natural materials to form substances that have specific electrical, magnetic, optical, or mechanical properties. In this chapter, we explore the structures and properties of solids. As we do so, we will examine some of the solid materials used in modern technology.

12.1 | Classification of Solids

Solids can be as hard as diamond or as soft as wax. Some readily conduct electricity, whereas others do not. The shapes of some solids can easily be manipulated, while others are brittle and resistant to any change in shape. The physical properties as well as

▶ **THE TOUCH SCREEN OF A SMART PHONE** contains materials that are electrically connected to a power source and emit light. Each pixel shown here contains light-emitting diodes (LEDs), which are made of materials that produce specific colors of light in response to an applied voltage. Each pixel must switch on or off in less than a millisecond in response to touch. Devices such as this would not be possible without advanced solid-state and polymer materials.

WHAT'S AHEAD ▶

12.1 CLASSIFICATION OF SOLIDS We see that solids can be classified according to the types of bonding interactions that hold the atoms together. This classification helps us make general predictions about the properties of solids.

12.2 STRUCTURES OF SOLIDS We learn that in *crystalline solids* the atoms are arranged in an orderly, repeating pattern but in *amorphous solids* this order is missing. We learn about *lattices* and *unit cells*, which define the repeating patterns that characterize crystalline solids.

12.3 METALLIC SOLIDS We examine the properties and structures of metals. We learn that many metals have structures in which the atoms pack together as closely as possible. We examine various types of *alloys*, materials that contain more than one element and display the characteristic properties of a metal.

12.4 METALLIC BONDING We take a closer look at metallic bonding and how it is responsible for the properties of metals, in terms of two models—the *electron-sea model* and the molecular–orbital model. We learn how overlap of atomic orbitals gives rise to *bands* in metals.

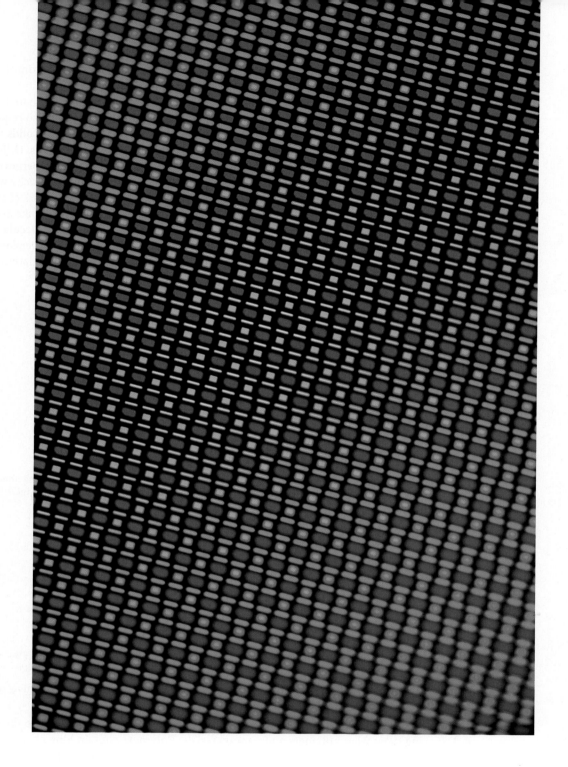

12.5 IONIC SOLIDS We examine the structures and properties of solids held together by the mutual attractions between cations and anions. We learn how the structures of ionic solids depend on the relative sizes of the ions and their stoichiometry.

12.6 MOLECULAR SOLIDS We take a brief look at the solids that form when molecules are held together by weak intermolecular forces.

12.7 COVALENT-NETWORK SOLIDS We learn about solids in which the atoms are held together by extended networks of covalent bonds. We learn how the electronic structure and properties of *semiconductors* differ from those of metals.

12.8 POLYMERS We investigate *polymers*—long chain-like molecules in which the motif of a small molecule is repeated many times over. We see how both molecular shape and interactions between polymer chains affect the physical properties of polymers.

12.9 NANOMATERIALS We learn how the physical and chemical properties of materials change when their crystals become very small. These effects begin to occur when materials have sizes on the order of 1–100 nm. We explore lower-dimensional forms of carbon—fullerenes, carbon nanotubes, and graphene.

the structures of solids are dictated by the types of bonds that hold the atoms in place. We can classify solids according to those bonds (▼ Figure 12.1).

Metallic solids are held together by a delocalized "sea" of collectively shared valence electrons. This form of bonding allows metals to conduct electricity. It is also responsible for the fact that most metals are relatively strong without being brittle. **Ionic solids** are held together by the mutual electrostatic attraction between cations and anions. Differences between ionic and metallic bonding make the electrical and mechanical properties of ionic solids very different from those of metals: Ionic solids do not conduct electricity well and are brittle. **Covalent-network solids** are held together by an extended network of covalent bonds. This type of bonding can result in materials that are extremely hard, like diamond, and it is also responsible for the unique properties of semiconductors. **Molecular solids** are held together by the intermolecular forces we studied in Chapter 11: dispersion forces, dipole–dipole interactions, and hydrogen bonds. Because these forces are relatively weak, molecular solids tend to be soft and have low melting points.

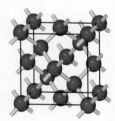

Metallic solids

Extended networks of atoms held together by metallic bonding (Cu, Fe)

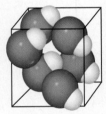

Ionic solids

Extended networks of ions held together by ion–ion interactions (NaCl, MgO)

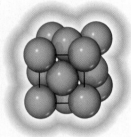

Covalent-network solids

Extended networks of atoms held together by covalent bonds (C, Si)

Molecular solids

Discrete molecules held together by intermolecular forces (HBr, H_2O)

▲ Figure 12.1 Classification and examples of solids according to predominant bonding type.

We will also consider two classes of solids that do not fall neatly into the preceding categories: polymers and nanomaterials. **Polymers** contain long chains of atoms (usually carbon), where the atoms within a given chain are connected by covalent bonds and adjacent chains are held to one another largely by weaker intermolecular forces. Polymers are normally stronger and have higher melting points than molecular solids, and they are more flexible than metallic, ionic, or covalent-network solids. **Nanomaterials** are solids in which the dimensions of individual crystals have been reduced to the order of 1–100 nm. As we will see, the properties of conventional materials change when their crystals become this small.

12.2 | Structures of Solids

Crystalline and Amorphous Solids

Solids contain large numbers of atoms. For example, a 1-carat diamond has a volume of 57 mm^3 and contains 1.0×10^{22} carbon atoms. How can we hope to describe such a large collection of atoms? Fortunately, the structures of many solids have patterns that repeat over and over in three dimensions. We can visualize the solid as being formed by stacking a large number of small, identical structural units, much like a wall can be built by stacking identical bricks.

Solids in which atoms are arranged in an orderly repeating pattern are called **crystalline solids**. These solids usually have flat surfaces, or *faces*, that make definite angles with one another. The orderly arrangements of atoms that produce these faces also cause the solids to have highly regular shapes (▶ Figure 12.2). Examples of crystalline solids include sodium chloride, quartz, and diamond.

Amorphous solids (from the Greek words for "without form") lack the order found in crystalline solids. At the atomic level the structures of amorphous solids are similar to the structures of liquids, but the molecules, atoms, and/or ions lack the freedom of motion they have in liquids. Amorphous solids do not have the well-defined faces and shapes of a crystal. Familiar amorphous solids are rubber, glass, and obsidian (volcanic glass).

Unit Cells and Crystal Lattices

In a crystalline solid there is a relatively small repeating unit, called a **unit cell**, that is made up of a unique arrangement of atoms and embodies the structure of the solid. The structure of the crystal can be built by stacking this unit over and over in all three dimensions. Thus, the structure of a crystalline solid is defined by (a) the size and shape of the unit cell and (b) the locations of atoms within the unit cell.

The geometrical pattern of points on which the unit cells are arranged is called a **crystal lattice**. The crystal lattice is, in effect, an abstract (that is, not real) scaffolding for the crystal structure. We can imagine forming the entire crystal structure by first building the scaffolding and then filling in each unit cell with the same atom or group of atoms.

Before describing the structures of solids, we need to understand the properties of crystal lattices. It is useful to begin with two-dimensional lattices because they are simpler to visualize than three-dimensional ones. ▶ Figure 12.3 shows a two-dimensional array of **lattice points**. Each lattice point has an identical environment. The positions of the lattice points are defined by the **lattice vectors** *a* and *b*. Beginning from any lattice point it is possible to move to any other lattice point by adding together whole-number multiples of the two lattice vectors.*

The parallelogram formed by the lattice vectors, the shaded region in Figure 12.3, defines the unit cell. In two dimensions the unit cells must *tile*, or fit together in space, in such a way that they completely cover the area of the lattice with no gaps. In three dimensions the unit cells must stack together to fill all space.

In a two-dimensional lattice, the unit cells can take only one of the five shapes shown in Figure 12.4. The most general type of lattice is the *oblique lattice*. In this lattice, the lattice vectors are of different lengths and the angle γ between them is of arbitrary size, which makes the unit cell an arbitrarily shaped parallelogram. A *square lattice* results when the lattice vectors are equal in length and the angle between them is 90°. A *rectangular lattice* is formed when the angle between the lattice vectors is 90° but the vectors have different lengths. The fourth type of two-dimensional lattice, where *a* and *b* are of the same length and γ is 120°, is a *hexagonal lattice*.† If *a* and *b* are of the same length but the angle between them is any value other than 90° or 120° a *rhombic lattice* results. For a rhombic lattice an alternative unit cell can be drawn, a rectangle with lattice points on its corners *and* its center (shown in green in Figure 12.4). Because of this the rhombic lattice is commonly referred to as a *centered rectangular lattice*. The lattices in Figure 12.4 represent five basic shapes: squares, rectangles, hexagons, rhombuses (diamonds), and arbitrary parallelograms. Other polygons, such as pentagons, cannot cover space without leaving gaps, as Figure 12.5 shows.

Iron pyrite (FeS_2), a crystaline solid

Obsidian (typically $KAlSi_3O_8$), an amorphous solid

▲ Figure 12.2 **Examples of crystalline and amorphous solids.** The atoms in crystalline solids repeat in an orderly, periodic fashion that leads to well-defined faces at the macroscopic level. This order is lacking in amorphous solids like obsidian (volcanic glass).

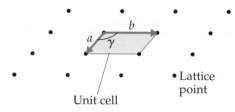

▲ Figure 12.3 **A crystalline lattice in two dimensions.** An infinite array of lattice points is generated by adding together the lattice vectors *a* and *b*. The unit cell is a parallelogram defined by the lattice vectors.

---

*A vector is a quantity involving both a direction and a magnitude. The magnitudes of the vectors in Figure 12.3 are indicated by their lengths, and their directions are indicated by the arrowheads.
†You may wonder why the hexagonal unit cell is not shaped like a hexagon. Remember that the unit cell is by definition a *parallelogram* whose size and shape are defined by the lattice vectors *a* and *b*.

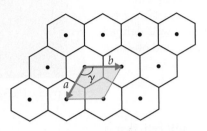

Oblique lattice ($a \neq b$, γ = arbitrary)

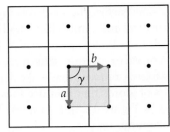

Square lattice ($a = b$, $\gamma = 90°$)

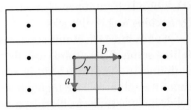

Rectangular lattice ($a \neq b$, $\gamma = 90°$)

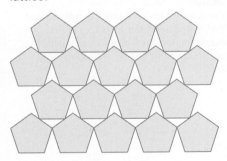

Hexagonal lattice ($a = b$, $\gamma = 120°$)

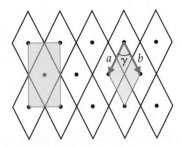

Rhombic lattice ($a = b$, γ = arbitrary)
Centered rectangular lattice

▲ **Figure 12.4 The five two-dimensional lattices.** The primitive unit cell for each lattice is shaded in blue. For the rhombic lattice the centered rectangular unit cell is shaded in green. Unlike the primitive rhombic unit cell, the centered cell has two lattice points per unit cell.

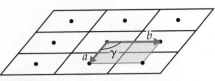

GO FIGURE

Why is there a centered rectangular lattice but not a centered square lattice?

▲ **Figure 12.5 Not all shapes tile space.** Tiling means covering a surface entirely, which is impossible for some geometric shapes, as shown here for pentagons.

To understand real crystals, we must consider three dimensions. A three-dimensional lattice is defined by *three* lattice vectors *a*, *b*, and *c* (▼ Figure 12.6). These lattice vectors define a unit cell that is a parallelepiped (a six-sided figure whose faces are all parallelograms) and is described by the lengths *a*, *b*, *c* of the cell edges and the angles α, β, γ between these edges. There are seven possible shapes for a three-dimensional unit cell, as shown in Figure 12.6.

Give It Some Thought

Imagine you generate a three-dimensional lattice by taking *a* and *b* vectors that form a two-dimensional square lattice. Then add a third vector, *c*, that is of different length and perpendicular to the first two. Which of the seven three-dimensional lattices results?

If we place a lattice point at each corner of a unit cell, we get a **primitive lattice**. All seven lattices in Figure 12.6 are primitive lattices. It is also possible to generate

Cubic
$a = b = c$
$\alpha = \beta = \gamma = 90°$

Tetragonal
$a = b \neq c$
$\alpha = \beta = \gamma = 90°$

Orthorhombic
$a \neq b \neq c$
$\alpha = \beta = \gamma = 90°$

Rhombohedral
$a = b = c$
$\alpha = \beta = \gamma \neq 90°$

Hexagonal
$a = b \neq c$
$\alpha = \beta = 90°, \gamma = 120°$

Monoclinic
$a \neq b \neq c$
$\alpha = \gamma = 90°, \beta \neq 90°$

Triclinic
$a \neq b \neq c$
$\alpha \neq \beta \neq \gamma$

▲ **Figure 12.6 The seven three-dimensional primitive lattices.**

what are called *centered lattices* by placing additional lattice points in specific locations in the unit cell. This is illustrated for a cubic lattice in ▶ Figure 12.7. A **body-centered cubic lattice** has one lattice point at the center of the unit cell in addition to the lattice points at the eight corners. A **face-centered cubic lattice** has one lattice point at the center of each of the six faces of the unit cell in addition to the lattice points at the eight corners. Centered lattices exist for other types of unit cells as well. For the crystals discussed in this chapter we need consider only the lattices shown in Figures 12.6 and 12.7.

Filling the Unit Cell

The lattice by itself does not define a crystal structure. To generate a crystal structure, we need to associate an atom or group of atoms with each lattice point. In the simplest case, the crystal structure consists of identical atoms, and each atom lies directly on a lattice point. When this happens, the crystal structure and the lattice points have identical patterns. Many metallic elements adopt such structures, as we will see in Section 12.3. Only for solids in which all the atoms are identical can this occur; in other words, *only elements* can form structures of this type. For compounds, even if we were to put an atom on every lattice point, the points would not be identical because the atoms are not all the same.

In most crystals, the atoms are not exactly coincident with the lattice points. Instead, a group of atoms, called a **motif**, is associated with each lattice point. The unit cell contains a specific motif of atoms, and the crystal structure is built up by repeating the unit cell over and over. This process is illustrated in ▼ Figure 12.8 for a two-dimensional crystal based on a hexagonal unit cell and a two-carbon-atom motif. The resulting infinite two-dimensional honeycomb structure is a two-dimensional crystal called *graphene*, a material that has so many interesting properties that its modern discoverers won the Nobel Prize in Physics in 2010. Each carbon atom is covalently bonded to three neighboring carbon atoms in what amounts to an infinite sheet of interconnected hexagonal rings.

The crystal structure of graphene illustrates two important characteristics of crystals. First, we see that no atoms lie on the lattice points. While most of the structures we discuss in this chapter do have atoms on the lattice points, there are many examples, like graphene, where this is not the case. Thus, to build up a structure you must know the location and orientation of the atoms in the motif with respect to the lattice points. Second, we see that bonds can be formed between atoms in neighboring unit cells and the bonds between atoms need not be parallel to the lattice vectors.

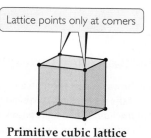

Lattice points only at corners

Primitive cubic lattice

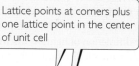

Lattice points at corners plus one lattice point in the center of unit cell

Body-centered cubic lattice

Lattice points at corners plus one lattice point at the center of each face

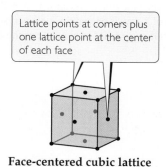

Face-centered cubic lattice

▲ **Figure 12.7 The three types of cubic lattices.**

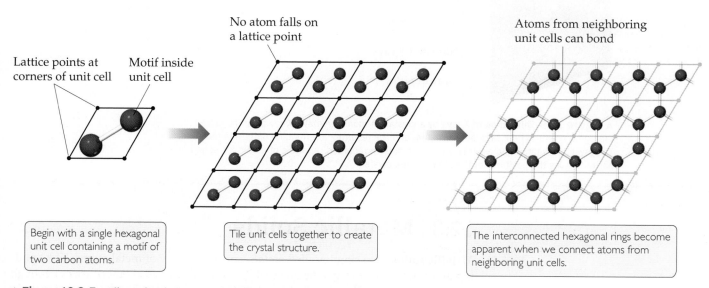

No atom falls on a lattice point

Atoms from neighboring unit cells can bond

Lattice points at corners of unit cell

Motif inside unit cell

Begin with a single hexagonal unit cell containing a motif of two carbon atoms.

Tile unit cells together to create the crystal structure.

The interconnected hexagonal rings become apparent when we connect atoms from neighboring unit cells.

▲ Figure 12.8 **Two-dimensional structure of graphene built up from a single unit cell.**

A Closer Look

X-ray Diffraction

When light waves pass through a narrow slit, they are scattered in such a way that the wave seems to spread out. This physical phenomenon is called *diffraction*. When light passes through many evenly spaced narrow slits (a *diffraction grating*), the scattered waves interact to form a series of bright and dark bands, known as a diffraction pattern. The bright bands correspond to constructive overlapping of the light waves, and the dark bands correspond to destructive overlapping of the light waves. ∞ (Section 9.8, "Phases in Atomic and Molecular Orbitals") The most effective diffraction of light occurs when the wavelength of the light and the width of the slits are similar in magnitude.

The spacing of the layers of atoms in solid crystals is usually about 2–20 Å. The wavelengths of X rays are also in this range. Thus, a crystal can serve as an effective diffraction grating for X rays. X-ray diffraction results from the scattering of X rays by a regular arrangement of atoms, molecules, or ions. Much of what we know about crystal structures has been obtained by looking at the diffraction patterns that result when X rays pass through a crystal, a technique known as *X-ray crystallography*. As shown in ▼ Figure 12.9 a monochromatic beam of X rays is passed through a crystal. The diffraction pattern that results is recorded. For many years the diffracted X rays were detected by photographic film. Today, crystallographers use an *array detector*, a device analogous to that used in digital cameras, to capture and measure the intensities of the diffracted rays.

The pattern of spots on the detector in Figure 12.9 depends on the particular arrangement of atoms in the crystal. The spacing and symmetry of the bright spots, where constructive interference occurs, provide information about the size and shape of the unit cell. The intensities of the spots provide information that can be used to determine the locations of the atoms within the unit cell. When combined, these two pieces of information give the atomic structure that defines the crystal.

X-ray crystallography is used extensively to determine the structures of molecules in crystals. The instruments used to measure X-ray diffraction, known as *X-ray diffractometers*, are now computer controlled, making the collection of diffraction data highly automated. The diffraction pattern of a crystal can be determined very accurately and quickly (sometimes in a matter of hours), even though thousands of diffraction spots are measured. Computer programs are then used to analyze the diffraction data and determine the arrangement and structure of the molecules in the crystal. X-ray diffraction is an important technique in industries ranging from steel and cement manufacture to pharmaceuticals.

Related Exercises: 12.113, 12.114, 12.115

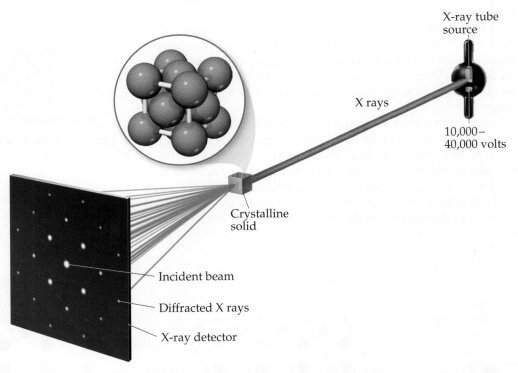

▲ Figure 12.9 **Diffraction of X rays by a crystal.** A monochromatic X-ray beam is passed through a crystal. The X rays are diffracted, and the resulting interference pattern is recorded. The crystal is rotated and another diffraction pattern recorded. Analysis of many diffraction patterns gives the positions of the atoms in the crystal.

12.3 | Metallic Solids

Metallic solids, also simply called *metals*, consist entirely of metal atoms. The bonding in metals is too strong to be due to dispersion forces, and yet there are not enough valence electrons to form covalent bonds between atoms. The bonding, called *metallic* *bonding*, happens because the valence electrons are *delocalized* throughout the entire solid. That is, the valence electrons are not associated with specific atoms or bonds but are spread throughout the solid. In fact, we can visualize a metal as an array of positive ions immersed in a "sea" of delocalized valence electrons.

You have probably held a length of copper wire or an iron bolt. Perhaps you have even seen the surface of a freshly cut piece of sodium metal. These substances, although distinct from one another, share certain similarities that enable us to classify them as metallic. A clean metal surface has a characteristic luster. Metals have a characteristic cold feeling when you touch them, related to their high thermal conductivity (ability to conduct heat). Metals also have high electrical conductivity, which means that electrically charged particles flow easily through them. The thermal conductivity of a metal usually parallels its electrical conductivity. Silver and copper, for example, which possess the highest electrical conductivities among the elements, also possess the highest thermal conductivities.

Most metals are *malleable*, which means that they can be hammered into thin sheets, and *ductile*, which means that they can be drawn into wires (▶ Figure 12.10). These properties indicate that the atoms are capable of slipping past one another. Ionic and covalent-network solids do not exhibit such behavior; they are typically brittle.

▲ **Figure 12.10 Malleability and ductility.** Gold leaf demonstrates the characteristic malleability of metals, and copper wire demonstrates their ductility.

 Give It Some Thought

Atoms in metals easily slip past one another as mechanical force is applied; can you think of why this would not be true for ionic solids?

The Structures of Metallic Solids

The crystal structures of many metals are simple enough that we can generate the structure by placing a single atom on each lattice point. The structures corresponding to the three cubic lattices are shown in ▼ Figure 12.11. Metals with a primitive cubic structure are rare, one of the few examples being the radioactive element polonium. Body-centered cubic metals include iron, chromium, sodium, and tungsten. Examples of face-centered cubic metals include aluminum, lead, copper, silver, and gold.

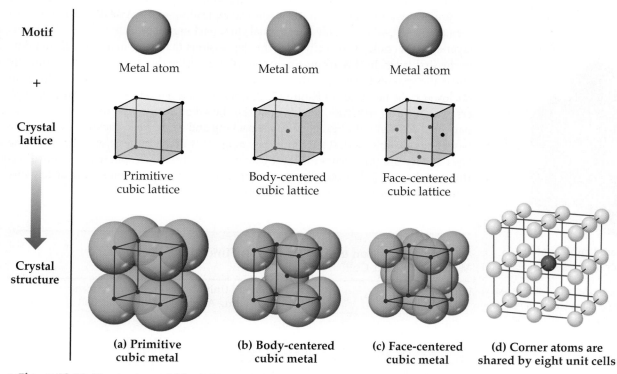

▲ **Figure 12.11 The structures of (a) primitive cubic, (b) body-centered cubic, and (c) face-centered cubic metals.** Each structure can be generated by the combination of a single-atom motif and the appropriate lattice. (**d**) Corner atoms (one shown in red) are shared among eight neighboring cubic unit cells.

GO FIGURE

Which one of these unit cells would you expect to represent the densest packing of spheres?

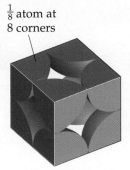

$\frac{1}{8}$ atom at 8 corners

(a) Primitive cubic metal
1 atom per unit cell

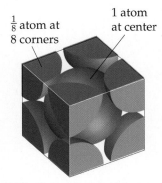

$\frac{1}{8}$ atom at 8 corners

1 atom at center

(b) Body-centered cubic metal
2 atoms per unit cell

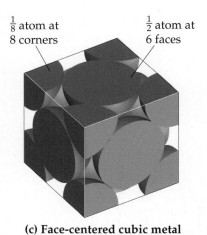

$\frac{1}{8}$ atom at 8 corners

$\frac{1}{2}$ atom at 6 faces

(c) Face-centered cubic metal
4 atoms per unit cell

▲ **Figure 12.12 A space-filling view of unit cells for metals with a cubic structure.** Only the portion of each atom that falls within the unit cell is shown.

Notice in the bottom row of Figure 12.11 that the atoms on the corners and faces of a unit cell do not lie wholly within the unit cell. These corner and face atoms are shared by neighboring unit cells. An atom that sits at the corner of a unit cell is shared among eight unit cells, and only 1/8 of the atom is in one particular unit cell. Because a cube has eight corners, each primitive cubic unit cell contains $(1/8) \times 8 = 1$ atom, as shown in ◀ Figure 12.12(**a**). Similarly, each body-centered cubic unit cell [Figure 12.12(**b**)] contains two atoms, $(1/8) \times 8 = 1$ from the corners and 1 at the center of the unit cell. Atoms that lie on the face of a unit cell, as they do in a face-centered cubic metal, are shared by two unit cells so that only one-half of the atom belongs to each unit cell. Therefore, a face-centered cubic unit cell [Figure 12.12(**c**)] contains four atoms, $(1/8) \times 8 = 1$ atom from the corners and $(1/2) \times 6 = 3$ atoms from the faces.

▼ Table 12.1 summarizes how the fractional part of each atom that resides within a unit cell depends on the atom's location within the cell.

Close Packing

The shortage of valence electrons and the fact that they are collectively shared make it favorable for the atoms in a metal to pack together closely. Because we can treat atoms as spherical objects, we can understand the structures of metals by considering how spheres pack. The most efficient way to pack one layer of equal-sized spheres is to surround each sphere by six neighbors, as shown at the top of ▶ Figure 12.13. To form a three-dimensional structure, we need to stack additional layers on top of this base layer. To maximize packing efficiency the second layer of spheres must sit in the depressions formed by the spheres in the first layer. We can either put the next layer of atoms into the depressions marked by the yellow dot or the depressions marked by the red dot (realizing that the spheres are too large to simultaneously fill both sets of depressions). For the sake of discussion we arbitrarily put the second layer in the yellow depressions.

For the third layer, we have two choices for where to place the spheres. One possibility is to put the third layer in the depressions that lie directly over the spheres in the first layer. This is done on the left-hand side of Figure 12.13, as shown by the dashed red lines in the side view. Continuing with this pattern, the fourth layer would lie directly over the spheres in the second layer, leading to the ABAB stacking pattern seen on the left, which is called **hexagonal close packing** (hcp). Alternatively, the third-layer spheres could lie directly over the depressions that were marked with red dots in the first layer. In this arrangement the spheres in the third layer do not sit directly above the spheres in either of the first two layers, as shown by the dashed red lines on the lower right-hand side of Figure 12.13. If this sequence is repeated in subsequent layers, we derive an ABCABC stacking pattern shown on the right known as **cubic close packing** (ccp). In both hexagonal close packing and cubic close packing, each sphere has 12 equidistant nearest neighbors: six neighbors in the same layer, three from the layer above, and three from the layer below. We say that each sphere has a **coordination number** of 12. The coordination number is the number of atoms immediately surrounding a given atom in a crystal structure.

Table 12.1 Fraction of Any Atom as a Function of Location Within the Unit Cell*

Atom Location	Number of Unit Cells Sharing Atom	Fraction of Atom Within Unit Cell
Corner	8	1/8 or 12.5%
Edge	4	1/4 or 25%
Face	2	1/2 or 50%
Anywhere else	1	1 or 100%

*It is only the position of the center of the atom that matters. Atoms that reside near the boundary of the unit cell but not on a corner, edge, or face are counted as residing 100% within the unit cell.

▲GO FIGURE

What type of two-dimensional lattice describes the structure of a single layer of close-packed atoms?

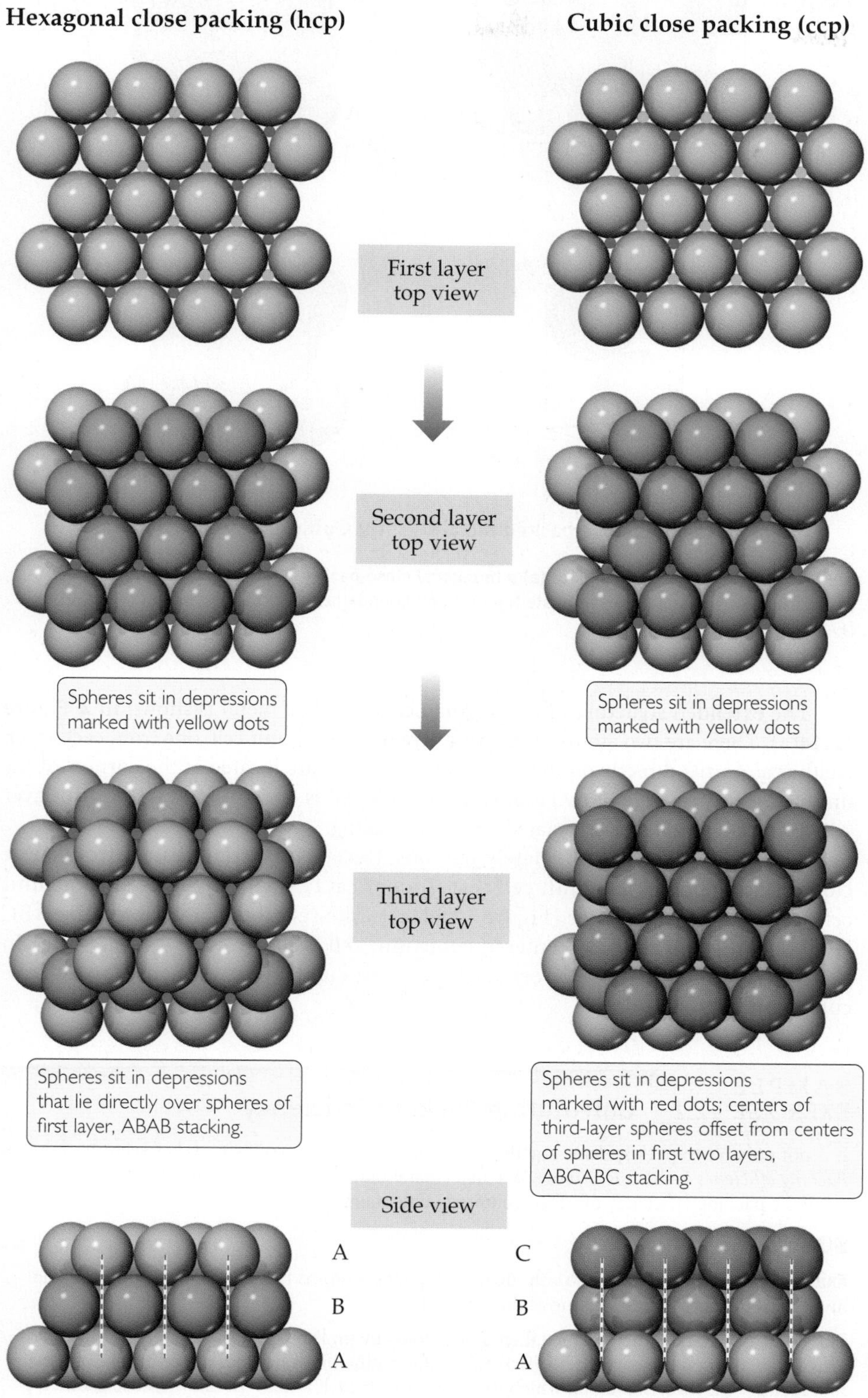

Hexagonal close packing (hcp)

Cubic close packing (ccp)

First layer top view

Spheres sit in depressions marked with yellow dots

Second layer top view

Spheres sit in depressions marked with yellow dots

Spheres sit in depressions that lie directly over spheres of first layer, ABAB stacking.

Third layer top view

Spheres sit in depressions marked with red dots; centers of third-layer spheres offset from centers of spheres in first two layers, ABCABC stacking.

Side view

A
B
A

C
B
A

▲ **Figure 12.13 Close packing of equal-sized spheres.** Hexagonal (left) close packing and cubic (right) close packing are equally efficient ways of packing spheres. The red and yellow dots indicate the positions of depressions between atoms.

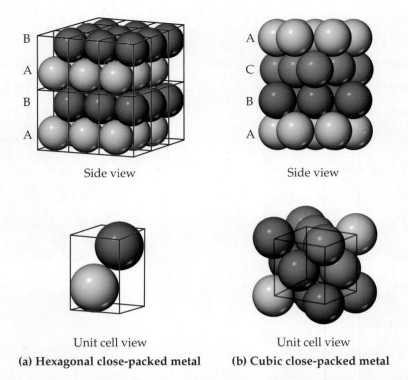

Side view   Side view

Unit cell view   Unit cell view

(a) Hexagonal close-packed metal   **(b) Cubic close-packed metal**

▲ Figure 12.14 **The unit cells for (a) a hexagonal close-packed metal and (b) a cubic close-packed metal.** The solid lines indicate the unit cell boundaries. Colors are used to distinguish one layer of atoms from another.

The extended structure of a hexagonal close-packed metal is shown in ▲ Figure 12.14(**a**). There are two atoms in the primitive hexagonal unit cell, one from each layer. Neither atom sits directly on the lattice points, which are located at the corners of the unit cell. The presence of two atoms in the unit cell is consistent with the two-layer ABAB stacking sequence associated with hcp packing.

Although it is not immediately obvious, the structure that results from cubic close packing possesses a unit cell that is identical to the face-centered cubic unit cell we encountered earlier [Figure 12.11(**c**)]. The relationship between the ABC layer stacking and the face-centered cubic unit cell is shown in Figure 12.14(**b**). In this figure we see that the layers stack perpendicular to the body diagonal of the cubic unit cell.

SAMPLE EXERCISE 12.1   Calculating Packing Efficiency

It is not possible to pack spheres together without leaving some void spaces between the spheres. *Packing efficiency* is the fraction of space in a crystal that is actually occupied by atoms. Determine the packing efficiency of a face-centered cubic metal.

SOLUTION

Analyze We must determine the volume taken up by the atoms that reside in the unit cell and divide this number by the volume of the unit cell.

Plan We can calculate the volume taken up by atoms by multiplying the number of atoms per unit cell by the volume of a sphere, $4\pi r^3/3$. To determine the volume of the unit cell, we must first identify the direction along which the atoms touch each other. We can then use geometry to express the length of the cubic unit cell edge, a, in terms of the radius of the atoms. Once we know the edge length, the cell volume is simply a^3.

Solve As shown in Figure 12.12, a face-centered cubic metal has four atoms per unit cell. Therefore, the volume occupied by the atoms is

$$\text{Occupied volume} = 4 \times \left(\frac{4\pi r^3}{3}\right) = \frac{16\pi r^3}{3}$$

For a face-centered cubic metal the atoms touch along the diagonal of a face of the unit cell:

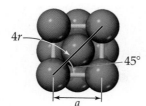

Therefore, a diagonal across a face of the unit cell is equal to four times the atomic radius, r. Using simple trigonometry, and the identity $\cos(45°) = \sqrt{2}/2$, we can show that

$$a = 4r \cos(45°) = 4r(\sqrt{2}/2) = (2\sqrt{2})r$$

Finally, we calculate the packing efficiency by dividing the volume occupied by atoms by the volume of the cubic unit cell, a^3:

$$\text{Packing efficiency} = \frac{\text{volume of atoms}}{\text{volume of unit cell}} = \frac{\left(\frac{16}{3}\right)\pi r^3}{(2\sqrt{2})^3 r^3} = 0.74 \text{ or } 74\%$$

Practice Exercise 1

Consider the two-dimensional square lattice of Figure 12.4. The "packing efficiency" for a two-dimensional structure would be the area of the atoms divided by the area of the unit cell, times 100%. What is the packing efficiency for a square lattice for atoms of radius $a/2$ that are centered at the lattice points? **(a)** 3.14% **(b)** 15.7% **(c)** 31.8% **(d)** 74.0% **(e)** 78.5%

Practice Exercise 2

Determine the packing efficiency by calculating the fraction of space occupied by atoms in a body-centered cubic metal.

 Give It Some Thought

For metallic structures, does the packing efficiency (see Sample Exercise 12.1) increase or decrease as the number of nearest neighbors (the coordination number) decreases?

Alloys

An **alloy** is a material that contains more than one element and has the characteristic properties of a metal. The alloying of metals is of great importance because it is one of the primary ways of modifying the properties of pure metallic elements. Nearly all the common uses of iron, for example, involve alloy compositions (for example, stainless steel). Bronze is formed by alloying copper and tin, while brass is an alloy of copper and zinc. Pure gold is too soft to be used in jewelry, but alloys of gold are much harder (see "Chemistry Put to Work: Alloys of Gold"). Other common alloys are described in ▼ Table 12.2.

Table 12.2   Some Common Alloys

Name	Primary Element	Typical Composition (by Mass)	Properties	Uses
Wood's metal	Bismuth	50% Bi, 25% Pb, 12.5% Sn, 12.5% Cd	Low melting point (70 °C)	Fuse plugs, automatic sprinklers
Yellow brass	Copper	67% Cu, 33% Zn	Ductile, takes polish	Hardware items
Bronze	Copper	88% Cu, 12% Sn	Tough and chemically stable in dry air	Important alloy for early civilizations
Stainless steel	Iron	80.6% Fe, 0.4% C, 18% Cr, 1% Ni	Resists corrosion	Cookware, surgical instruments
Plumber's solder	Lead	67% Pb, 33% Sn	Low melting point (275 °C)	Soldering joints
Sterling silver	Silver	92.5% Ag, 7.5% Cu	Bright surface	Tableware
Dental amalgam	Silver	70% Ag, 18% Sn, 10% Cu, 2% Hg	Easily worked	Dental fillings
Pewter	Tin	92% Sn, 6% Sb, 2% Cu	Low melting point (230 °C)	Dishes, jewelry

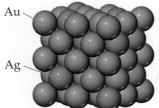
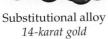

▲ **GO FIGURE**

What determines which species in a solid solution is the solute and which is the solvent?

Substitutional alloy
14-karat gold

Interstitial alloy
Steel

▲ **Figure 12.15 The distribution of solute and solvent atoms in a substitutional alloy and an interstitial alloy.** Both types of alloys are solid solutions and, therefore, homogeneous mixtures.

Alloys can be divided into four categories: substitutional alloys, interstitial alloys, heterogeneous alloys, and intermetallic compounds. Substitutional and interstitial alloys are both homogeneous mixtures in which components are dispersed randomly and uniformly (▲ Figure 12.15). ∞ (Section 1.2) Solids that form homogeneous mixtures are called solid solutions. When atoms of the solute in a solid solution occupy positions normally occupied by a solvent atom, we have a **substitutional alloy**. When the solute atoms occupy interstitial positions in the "holes" between solvent atoms, we have an **interstitial alloy** (Figure 12.15).

Substitutional alloys are formed when the two metallic components have similar atomic radii and chemical-bonding characteristics. For example, silver and gold form such an alloy over the entire range of possible compositions. When two metals differ in radii by more than about 15%, solubility is generally more limited.

For an interstitial alloy to form, the solute atoms must have a much smaller bonding atomic radius than the solvent atoms. Typically, the interstitial element is a nonmetal that makes covalent bonds to the neighboring metal atoms. The presence of the extra bonds provided by the interstitial component causes the metal lattice to become harder, stronger, and less ductile. For example, steel, which is much harder and stronger than pure iron, is an alloy of iron that contains up to 3% carbon. Other elements may be added to form *alloy steels*. Vanadium and chromium may be added to impart strength, for instance, and to increase resistance to fatigue and corrosion.

▲ **Give It Some Thought**

Would you expect the alloy $PdB_{0.15}$ to be a substitutional alloy or an interstitial alloy?

One of the most important iron alloys is stainless steel, which contains about 0.4% carbon, 18% chromium, and 1% nickel. The chromium is obtained by carbon reduction of chromite ($FeCr_2O_4$) in an electric furnace. The product of the reduction is *ferrochrome* ($FeCr_2$), which is added in the appropriate amount to molten iron to achieve the desired steel composition. The ratio of elements present in the steel may vary over a wide range, imparting a variety of specific physical and chemical properties to the materials.

In a **heterogeneous alloy** the components are not dispersed uniformly. For example, the heterogeneous alloy pearlite contains two phases (▶ Figure 12.16). One phase is essentially pure body-centered cubic iron, and the other is the compound Fe_3C,

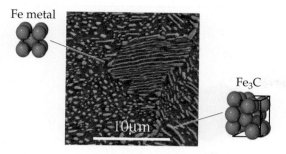

Fe metal

Fe₃C

▲ **Figure 12.16 Microscopic view of the structure of the heterogeneous alloy pearlite.**
The dark regions are body-centered cubic iron metal, and the lighter regions are cementite, Fe₃C.

known as cementite. In general, the properties of heterogeneous alloys depend on both the composition and the manner in which the solid is formed from the molten mixture. The properties of a heterogeneous alloy formed by rapid cooling of a molten mixture, for example, are distinctly different from the properties of an alloy formed by slow cooling of the same mixture.

Intermetallic compounds are compounds rather than mixtures. Because they are compounds, they have definite properties and their composition cannot be varied. Furthermore, the different types of atoms in an intermetallic compound are ordered rather than randomly distributed. The ordering of atoms in an intermetallic compound generally leads to better structural stability and higher melting points than what is observed in the constituent metals. These features can be attractive for high-temperature applications. On the negative side, intermetallic compounds are often more brittle than substitutional alloys.

Intermetallic compounds play many important roles in modern society. The intermetallic compound Ni₃Al is a major component of jet aircraft engines because of its strength at high temperature and its low density. Razor blades are often coated with Cr₃Pt, which adds hardness, allowing the blade to stay sharp longer. Both compounds have the structure shown on the left-hand side of ▼ **Figure 12.17.** The compound Nb₃Sn, also shown in Figure 12.17, is a superconductor, a substance that, when cooled below a critical temperature, conducts electricity with no resistance. In the case of Nb₃Sn superconductivity is observed only when the temperature falls below 18 K. Superconductors are used in the magnets in MRI scanners widely employed for medical imaging. ∞ (Section 6.7, "Nuclear Spin and Magnetic Resonance Imaging") The need to keep the magnets cooled to such a low temperature is part of the reason why MRI devices are expensive to operate. The hexagonal intermetallic compound SmCo₅, shown on the right-hand side of Figure 12.17, is used to make the permanent magnets found in lightweight headsets and high-fidelity speakers. A related compound with the same structure, LaNi₅, is used as the anode in nickel-metal hydride batteries.

▲ GO FIGURE

In the unit cell drawing on the right, why do we see eight Sm atoms and nine Co atoms if the empirical formula is SmCo₅?

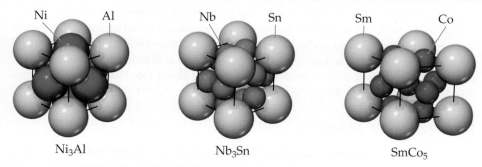

Ni   Al   Nb   Sn   Sm   Co

Ni₃Al   Nb₃Sn   SmCo₅

▲ **Figure 12.17 Three examples of intermetallic compounds.**

Chemistry Put to Work

Alloys of Gold

Gold has long been a metal of choice for decorative objects, jewelry, and coins. The popularity of gold is driven by its unusual color (for a metal), its resistance to many chemical reactions, and the fact that it can easily be worked. However, pure gold is too soft for many applications, including jewelry. To increase its strength and hardness, as well as to modify its color, gold is often alloyed with other metals. In the jewelry trade pure gold is termed 24 karat. The karat number decreases as the mass percent of gold decreases. The most common alloys used in jewelry are 14 karat, which is $(14/24) \times 100 = 58\%$ gold, and 18 karat, which is $(18/24) \times 100 = 75\%$ gold.

The color of gold varies depending on the metals it is alloyed with. Gold is typically alloyed with silver and/or copper. All three elements crystallize with a face-centered cubic structure. The fact that all three elements have similar radii (Au and Ag are nearly the same size; Cu is roughly 11% smaller) and crystallize with the same type of structure make it possible to form substitutional alloys with nearly any composition. The variations in color of these alloys as a function of composition are shown in ▶ Figure 12.18. Gold alloyed with equal parts silver and copper takes on the golden yellow color we associate with gold jewelry. Red or rose gold is a copper-rich alloy. Silver-rich alloys take on a greenish hue, eventually giving way to silvery-white colors as silver becomes the majority constituent.

Related Exercises: 12.43, 12.44, 12.117

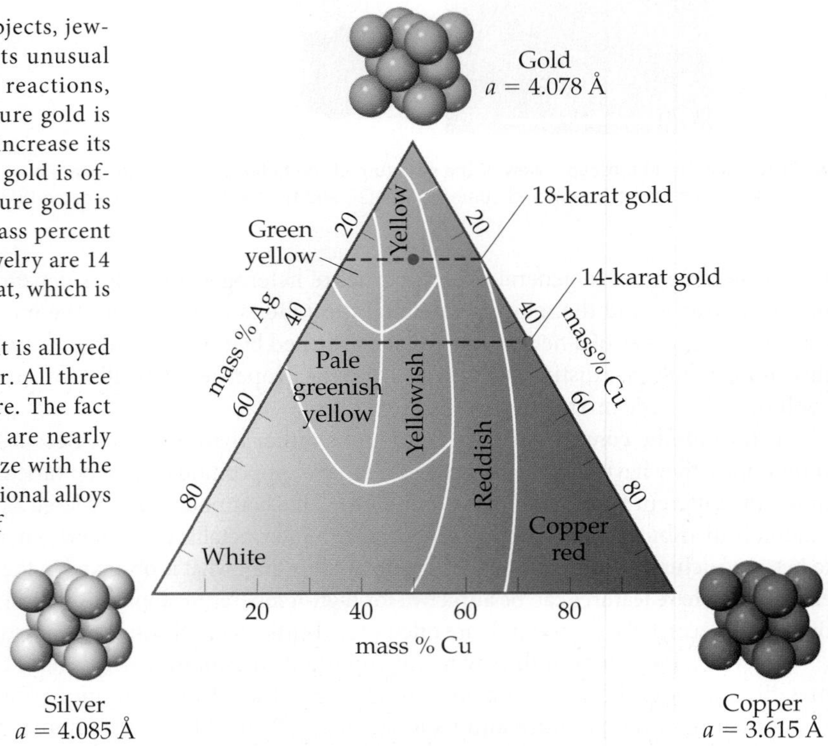

▲ Figure 12.18 **Colors of Au–Ag–Cu alloys as a function of composition.**

12.4 | Metallic Bonding

Consider the elements of the third period of the periodic table (Na–Ar). Argon with eight valence electrons has a complete octet; as a result it does not form any bonds. Chlorine, sulfur, and phosphorus form molecules (Cl_2, S_8, and P_4) in which the atoms make one, two, and three bonds, respectively (▶ Figure 12.19). Silicon forms an extended network solid in which each atom is bonded to four equidistant neighbors. Each of these elements forms $8 - N$ bonds, where N is the number of valence electrons. This behavior can easily be understood through the application of the octet rule.

If the $8 - N$ trend continued as we move left across the periodic table, we would expect aluminum (three valence electrons) to form five bonds. Like many other metals, however, aluminum adopts a close-packed structure with 12 nearest neighbors. Magnesium and sodium also adopt metallic structures. What is responsible for this abrupt change in the preferred bonding mechanism? The answer, as noted earlier, is that metals do not have enough valence-shell electrons to satisfy their bonding requirements by forming localized electron-pair bonds. In response to this deficiency, the valence electrons are collectively shared. A structure in which the atoms are close-packed facilitates this delocalized sharing of electrons.

Electron-Sea Model

A simple model that accounts for some of the most important characteristics of metals is the **electron-sea model**, which pictures the metal as an array of metal cations in a "sea" of valence electrons (▶ Figure 12.20). The electrons are confined to the metal by electrostatic attractions to the cations, and they are uniformly distributed throughout the structure.

▲ GO FIGURE

Which of these drawings represent molecules?

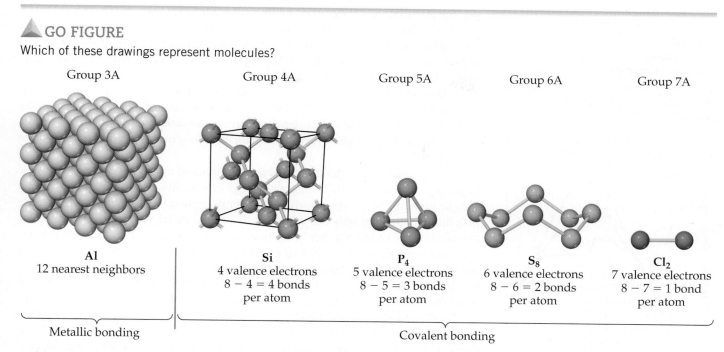

| Group 3A | Group 4A | Group 5A | Group 6A | Group 7A |

Al
12 nearest neighbors

Si
4 valence electrons
8 − 4 = 4 bonds
per atom

P₄
5 valence electrons
8 − 5 = 3 bonds
per atom

S₈
6 valence electrons
8 − 6 = 2 bonds
per atom

Cl₂
7 valence electrons
8 − 7 = 1 bond
per atom

Metallic bonding   Covalent bonding

▲ **Figure 12.19 Bonding in period 3 elements.**

The electrons are mobile, however, and no individual electron is confined to any particular metal ion. When a voltage is applied to a metal wire, the electrons, being negatively charged, flow through the metal toward the positively charged end of the wire.

The high thermal conductivity of metals is also accounted for by the presence of mobile electrons. The movement of electrons in response to temperature gradients permits ready transfer of kinetic energy throughout the solid.

The ability of metals to deform (their malleability and ductility) can be explained by the fact that metal atoms form bonds to many neighbors. Changes in the positions of the atoms brought about in reshaping the metal are partly accommodated by a redistribution of electrons.

Molecular–Orbital Model

Although the electron-sea model works surprisingly well given its simplicity, it does not adequately explain many properties of metals. According to the model, for example, the strength of bonding between metal atoms should steadily increase as the number of valence electrons increases, resulting in a corresponding increase in the melting points. However, elements near the middle of the transition metal series, rather than those at the end, have the highest melting points in their respective periods (**Figure 12.21**). This trend implies that the strength of metallic bonding first increases with increasing number of electrons and then decreases. Similar trends are seen in other physical properties of the metals, such as the boiling point, heat of fusion, and hardness.

To obtain a more accurate picture of the bonding in metals, we must turn to molecular orbital theory. In Sections 9.7 and 9.8 we learned how molecular orbitals are created from the overlap of atomic orbitals. Let's briefly review some of the rules of molecular orbital theory:

1. Atomic orbitals combine to make molecular orbitals that can extend over the entire molecule.

2. A molecular orbital can contain zero, one, or two electrons.

3. The number of molecular orbitals in a molecule equals the number of atomic orbitals that combine to form molecular orbitals.

4. Adding electrons to a bonding molecular orbital strengthens bonding, while adding electrons to antibonding molecular orbitals weakens bonding.

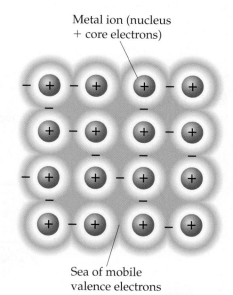

Metal ion (nucleus + core electrons)

Sea of mobile valence electrons

▲ **Figure 12.20 Electron-sea model of metallic bonding.** The valence electrons delocalize to form a sea of mobile electrons that surrounds and binds together an extended array of metal ions.

Which element in each period has the highest melting point? In each case, is the element you named at the beginning, middle, or end of its period?

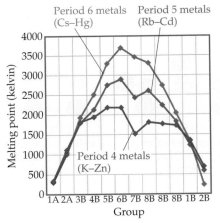

▲ Figure 12.21 The melting points of metals from periods 4, 5, and 6.

The electronic structures of crystalline solids and small molecules have similarities as well as differences. To illustrate, consider how the molecular-orbital diagram for a chain of lithium atoms changes as we increase the length of the chain (▼ Figure 12.22). Each lithium atom contains a half-filled 2s orbital in its valence shell. The molecular-orbital diagram for Li_2 is analogous to that of an H_2 molecule: one filled bonding molecular orbital and one empty antibonding molecular orbital with a nodal plane between the atoms. ∞∞(Section 9.7) For Li_4, there are four molecular orbitals, ranging from the lowest-energy orbital, where the orbital interactions are completely bonding (zero nodal planes), to the highest-energy orbital, where all interactions are antibonding (three nodal planes).

As the length of the chain increases, the number of molecular orbitals increases. Regardless of chain length, the lowest-energy orbitals are always the most bonding and the highest-energy orbitals always the most antibonding. Furthermore, because each lithium atom has only one valence shell atomic orbital, the number of molecular orbitals is equal to the number of lithium atoms in the chain. Because each lithium atom has one valence electron, half of the molecular orbitals are fully occupied and the other half are empty, regardless of chain length.*

If the chain becomes very long, there are so many molecular orbitals that the energy separation between them becomes vanishingly small. As the chain length goes to infinity, the allowed energy states become a continuous **band**. For a crystal large enough to see with the eye (or even an optical microscope), the number of atoms is extremely large. Consequently, the electronic structure of the crystal is like that of the infinite chain, consisting of bands, as shown on the right-hand side of Figure 12.22.

The electronic structures of most metals are more complicated than those shown in Figure 12.22 because we have to consider more than one type of atomic orbital on each atom. Because each type of orbital can give rise to its own band, the electronic

How does the energy spacing between molecular orbitals change as the number of atoms in the chain increases?

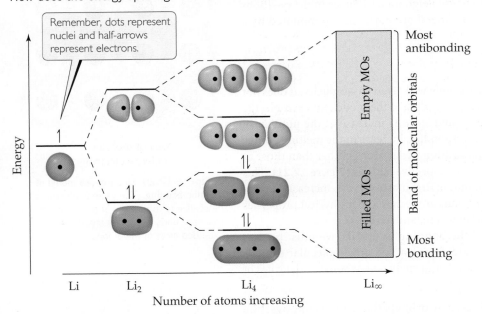

▲ Figure 12.22 Discrete energy levels in individual molecules become continuous energy bands in a solid. Occupied orbitals are shaded blue, and empty orbitals pink.

*This is strictly true only for chains with an even number of atoms.

▲GO FIGURE

If the metal were potassium rather than nickel, which bands—4*s*, 4*p*, and/or 3*d*—would be partially occupied?

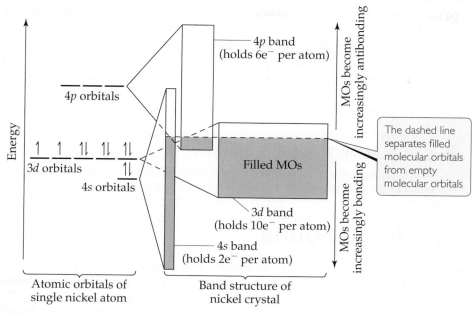

▲ **Figure 12.23** The electronic band structure of nickel.

structure of a solid usually consists of a series of bands. The electronic structure of a bulk solid is referred to as a **band structure**.

The band structure of a typical metal is shown schematically in ▲ **Figure 12.23**. The electron filling depicted corresponds to nickel metal, but the basic features of other metals are similar. The electron configuration of a nickel atom is $[Ar]4s^23d^8$, as shown on the left side of the figure. The energy bands that form from each of these orbitals are shown on the right side. The 4*s*, 4*p*, and 3*d* orbitals are treated independently, each giving rise to a band of molecular orbitals. In practice, these overlapping bands are not completely independent of each other, but for our purposes this simplification is reasonable.

The 4*s*, 4*p*, and 3*d* bands differ from one another in the energy range they span (represented by the heights of the rectangles on the right side of Figure 12.23) and in the number of electrons they can hold (represented by the area of the rectangles). The 4*s*, 4*p*, and 3*d* bands can hold 2, 6, and 10 electrons per atom, respectively, corresponding to two per orbital, as dictated by the Pauli exclusion principle. ∞ (Section 6.7) The energy range spanned by the 3*d* band is smaller than the range spanned by the 4*s* and 4*p* bands because the 3*d* orbitals are smaller and, therefore, overlap with orbitals on neighboring atoms less effectively.

Many properties of metals can be understood from Figure 12.23. We can think of the energy band as a partially filled container for electrons. The incomplete filling of the energy band gives rise to characteristic metallic properties. The electrons in orbitals near the top of the occupied levels require very little energy input to be "promoted" to higher-energy orbitals that are unoccupied. Under the influence of any source of excitation, such as an applied electrical potential or an input of thermal energy, electrons move into previously vacant levels and are thus freed to move through the lattice, giving rise to electrical and thermal conductivity.

Without the overlap of energy bands, the periodic properties of metals could not be explained. In the absence of the *d*- and *p*-bands, we would expect the *s*-band to be half-filled for the alkali metals (group 1A) and completely filled for the alkaline-earth metals (group 2A). If that were true, metals like magnesium, calcium, and strontium

▲ GO FIGURE

Why don't metals cleave in the way depicted here for ionic substances?

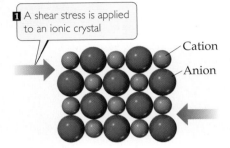

1 A shear stress is applied to an ionic crystal

Cation
Anion

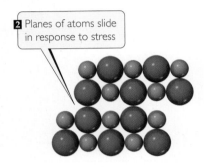

2 Planes of atoms slide in response to stress

3 Repulsive interactions between ions of like charge lead to separation of the layers

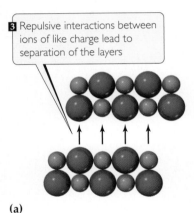

(a)

(b)

▲ **Figure 12.24 Brittleness and faceting in ionic crystals. (a)** When a shear stress (blue arrows) is applied to an ionic solid, the crystal separates along a plane of atoms as shown. **(b)** This property of ionic crystals is used to facet gemstones, such as rubies.

would not be good electrical and thermal conductors, in disagreement with experimental observations.

While the conductivity of metals can be qualitatively understood using either the electron–sea model or the molecular–orbital model, many physical properties of transition metals, such as the melting points plotted in Figure 12.21, can be explained only with the latter model. The molecular–orbital model predicts that bonding first becomes stronger as the number of valence electrons increases and the bonding orbitals are increasingly populated. Upon moving past the middle elements of the transition metal series, the bonds grow weaker as electrons populate antibonding orbitals. Strong bonds between atoms lead to metals with higher melting and boiling points, higher heats of fusion, higher hardness, and so forth.

▲ Give It Some Thought

Which element, W or Au, has the greater number of electrons in antibonding orbitals? Which one would you expect to have the higher melting point?

12.5 | Ionic Solids

Ionic solids are held together by the electrostatic attraction between cations and anions: ionic bonds. ∞(Section 8.2) The high melting and boiling points of ionic compounds are a testament to the strength of the ionic bonds. The strength of an ionic bond depends on the charges and sizes of the ions. As discussed in Chapters 8 and 11, the attractions between cations and anions increase as the charges of the ions increase. Thus NaCl, where the ions have charges of 1+ and 1−, melts at 801 °C, whereas MgO, where the ions have charges of 2+ and 2−, melts at 2852 °C. The interactions between cations and anions also increase as the ions get smaller, as we see from the melting points of the alkali metal halides in ▼ Table 12.3. These trends mirror the trends in lattice energy discussed in Section 8.2.

Although ionic and metallic solids both have high melting and boiling points, the differences between ionic and metallic bonding are responsible for important differences in their properties. Because the valence electrons in ionic compounds are confined to the anions, rather than being delocalized, ionic compounds are typically electrical insulators. They tend to be brittle, a property explained by repulsive interactions between ions of like charge. When stress is applied to an ionic solid, as in ◀ Figure 12.24, the planes of atoms, which before the stress were arranged with cations next to anions, shift so that the alignment becomes cation–cation, anion–anion. The resulting repulsive interaction causes the planes to split away from each other, a property that lends itself to the carving of certain gemstones (such as ruby, composed principally of Al_2O_3).

Structures of Ionic Solids

Like metallic solids, ionic solids tend to adopt structures with symmetric, close-packed arrangements of atoms. However, important differences arise because we now have to pack together spheres that have different radii and opposite charges. Because cations are often considerably smaller than anions ∞(Section 7.3), the coordination numbers in ionic compounds are smaller than those in close-packed metals. Even if the

Table 12.3 Properties of the Alkali Metal Halides

Compound	Cation–Anion Distance (Å)	Lattice Energy (kJ/mol)	Melting Point (°C)
LiF	2.01	1030	845
NaCl	2.83	788	801
KBr	3.30	671	734
RbI	3.67	632	674

anions and cations were the same size, the close-packed arrangements seen in metals cannot be replicated without letting ions of like charge come in contact with each other. The repulsions between ions of the same type make such arrangements unfavorable. The most favorable structures are those where the cation–anion distances are as close as those permitted by ionic radii, but the anion–anion and cation–cation distances are maximized.

Give It Some Thought

Is it possible for all atoms in an ionic compound to lie on the lattice points as they do in the metallic structures shown in Figure 12.11?

Three common ionic structure types are shown in ▼ Figure 12.25. The cesium chloride (CsCl) structure is based on a primitive cubic lattice. Anions sit on the lattice points at the corners of the unit cell, and a cation sits at the center of each cell. (Remember, there is no lattice point inside a primitive unit cell.) With this arrangement, both cations and anions are surrounded by a cube of eight ions of the opposite type.

The sodium chloride (NaCl; also called the rock salt structure) and zinc blende (ZnS) structures are based on a face-centered cubic lattice. In both structures the anions sit on the lattice points that lie on the corners and faces of the unit cell, but the two-atom motif is slightly different for the two structures. In NaCl the Na^+ ions are displaced from the Cl^- ions along the edge of the unit cell, whereas in ZnS the Zn^{2+} ions are displaced from the S^{2-} ions along the body diagonal of the unit cell. This difference leads to different coordination numbers. In sodium chloride, each cation and each anion are surrounded by six ions of the opposite type, leading to an octahedral coordination environment. In zinc blende, each cation and

GO FIGURE

Do the anions touch each other in any of these three structures? If not, which ions do touch each other?

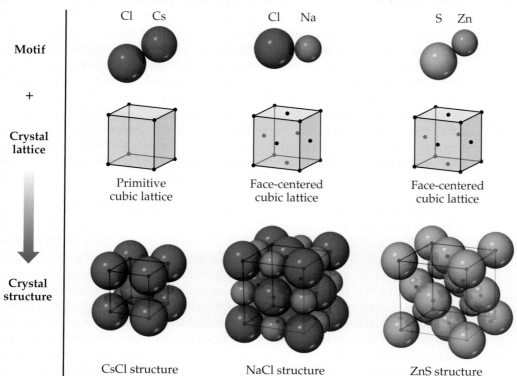

▲ **Figure 12.25 The structures of CsCl, NaCl, and ZnS.** Each structure type can be generated by the combination of a two-atom motif and the appropriate lattice.

Decreasing r_+/r_-

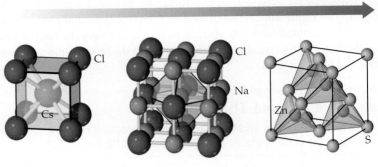

	CsCl	**NaCl**	**ZnS**
Cation radius, r_+ (Å)	1.81	1.16	0.88
Anion radius, r_- (Å)	1.67	1.67	1.70
r_+/r_-	1.08	0.69	0.52
Cation coordination number	8	6	4
Anion coordination number	8	6	4

▲ **Figure 12.26 Coordination environments in CsCl, NaCl, and ZnS.** The sizes of the ions have been reduced to show the coordination environments clearly.

each anion are surrounded by four ions of the opposite type, leading to a tetrahedral coordination geometry. The cation coordination environments can be seen in ▲ **Figure 12.26**.

For a given ionic compound, we might ask which type of structure is most favorable. There are a number of factors that come into play, but two of the most important are the relative sizes of the ions and the stoichiometry. Consider first ion size. Notice in Figure 12.26 that the coordination number changes from 8 to 6 to 4 on moving from CsCl to NaCl to ZnS. This trend is driven in part by the fact that for these three compounds the ionic radius of the cation gets smaller while the ionic radius of the anion changes very little. When the cation and anion are similar in size, a large coordination number is favored and the CsCl structure is often realized. As the relative size of the cation gets smaller, eventually it is no longer possible to maintain the cation–anion contacts and simultaneously keep the anions from touching each other. When this occurs, the coordination number drops from 8 to 6, and the sodium chloride structure becomes more favorable. As the cation size decreases further, eventually the coordination number must be reduced again, this time from 6 to 4, and the zinc blende structure becomes favored. Remember that, in ionic crystals, ions of opposite charge touch each other but ions of the same charge should not touch.

The relative number of cations and anions also helps determine the most stable structure type. All the structures in Figure 12.26 have equal numbers of cations and anions. These structure types (cesium chloride, sodium chloride, zinc blende) can be realized only for ionic compounds in which the number of cations and anions is equal. When this is not the case, other crystal structures must result. As an example, consider NaF, MgF_2, and ScF_3 (▶ **Figure 12.27**). Sodium fluoride has the sodium chloride structure with a coordination number of 6 for both cation and anion, as you might expect since NaF and NaCl are quite similar. Magnesium fluoride, however, has two anions for every cation, resulting in a tetragonal crystal structure called the *rutile structure*. The cation coordination number is still 6, but the fluoride coordination number is now only 3. In the scandium fluoride structure, there are three anions for every cation; the cation coordination number is still 6, but the fluoride coordination number has dropped to 2. As the cation/anion ratio goes down, there are fewer cations to surround each anion,

GO FIGURE

How many cations are there per unit cell for each of these structures? How many anions per unit cell?

Increasing anion-to-cation ratio

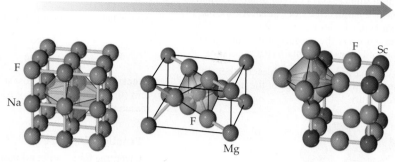

	NaF	MgF$_2$	ScF$_3$
Cation coordination number	6	6	6
Cation coordination geometry	Octahedral	Octahedral	Octahedral
Anion coordination number	6	3	2
Anion coordination geometry	Octahedral	Trigonal planar	Linear

▲ **Figure 12.27 Coordination numbers depend on stoichiometry.** The sizes of the ions have been reduced to show the coordination environments clearly.

and so the anion coordination number must decrease. The empirical formula of an ionic compound can be described quantitatively by the relationship

$$\frac{\text{Number of cations per formula unit}}{\text{Number of anions per formula unit}} = \frac{\text{anion coordination number}}{\text{cation coordination number}} \quad [12.1]$$

 Give It Some Thought

In the crystal structure of potassium oxide, the oxide ions are coordinated by eight potassium ions. What is the coordination number of potassium?

SAMPLE EXERCISE 12.2 Calculating the Density of an Ionic Solid

Rubidium iodide crystallizes with the same structure as sodium chloride. (**a**) How many iodide ions are there per unit cell? (**b**) How many rubidium ions are there per unit cell? (**c**) Use the ionic radii and molar masses of Rb$^+$ (1.66 Å, 85.47 g/mol) and I$^-$ (2.06 Å, 126.90 g/mol) to estimate the density of rubidium iodide in g/cm^3.

SOLUTION

Analyze and Plan

(**a**) We need to count the number of anions in the unit cell of the sodium chloride structure, remembering that ions on the corners, edges, and faces of the unit cell are only partially inside the unit cell.

(**b**) We can apply the same approach to determine the number of cations in the unit cell. We can double-check our answer by writing the empirical formula to make sure the charges of the cations and anions are balanced.

(**c**) Because density is an intensive property, the density of the unit cell is the same as the density of a bulk crystal. To calculate the density we must divide the mass of the atoms per unit cell by the

volume of the unit cell. To determine the volume of the unit cell we need to estimate the length of the unit cell edge by first identifying the direction along which the ions touch and then using ionic radii to estimate the length. Once we have the length of the unit cell edge we can cube it to determine its volume.

Solve

(**a**) The crystal structure of rubidium iodide looks just like NaCl with Rb$^+$ ions replacing Na$^+$ and I$^-$ ions replacing Cl$^-$. From the views of the NaCl structure in Figures 12.25 and 12.26 we see that there is an anion at each corner of the unit cell and at the center of each face. From Table 12.1 we see that the ions sitting on the corners are equally shared by eight unit cells (1/8 ion per unit cell), while those ions sitting on the faces are equally shared by two unit cells

(1/2 ion per unit cell). A cube has eight corners and six faces, so the total number of I^- ions is $8 (1/8)+6 (1/2) = 4$ per unit cell.

(b) Using the same approach for the rubidium cations we see that there is a rubidium ion on each edge and one at the center of the unit cell. Using Table 12.1 again we see that the ions sitting on the edges are equally shared by four unit cells (1/4 ion per unit cell), whereas the cation at the center of the unit cell is not shared. A cube has 12 edges, so the total number of rubidium ions is $12 (1/4)+1 = 4$. This answer makes sense because the number of Rb^+ ions must be the same as the number of I^- ions to maintain charge balance.

(c) In ionic compounds cations and anions touch each other. In RbI the cations and anions touch along the edge of the unit cell as shown in the following figure.

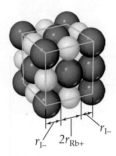

The distance the unit cell edge is equal to $r(I^-) + 2r(Rb^+) + r(I^-) = 2r(I^-) + 2r(Rb^+)$. Plugging in the ionic radii we get $2(2.06 \text{ Å}) + 2(1.66 \text{ Å}) = 7.44 \text{ Å}$. The volume of a cubic unit cell is just the edge length cubed. Converting from Å to cm and cubing we get

$$\text{Volume} = (7.44 \times 10^{-8} \text{ cm})^3 = 4.12 \times 10^{-22} \text{ cm}^3.$$

From parts (a) and (b) we know that there are four rubidium and four iodide ions per unit cell. Using this result and the molar masses we can calculate the mass per unit cell

$$\text{Mass} = \frac{4(85.47 \text{ g/mol}) + 4(126.90 \text{ g/mol})}{6.022 \times 10^{23}\text{mol}^{-1}} = 1.411 \times 10^{-21} \text{ g}$$

The density is the mass per unit cell divided by the volume of a unit cell

$$\text{Density} = \frac{\text{mass}}{\text{volume}} = \frac{1.411 \times 10^{-21} \text{ g}}{4.12 \times 10^{-22} \text{ cm}^3} = 3.43 \text{ g/cm}^3$$

Check The densities of most solids fall between the density of lithium (0.5 g/cm^3) and that of iridium (22.6 g/cm^3), so this value is reasonable.

Practice Exercise 1

Given the ionic radii and molar masses of Sc^{3+} (0.88 Å, 45.0 g/mol) and F^- (1.19 Å, 19.0 g/mol) what value do you estimate for the density of ScF_3, whose structure is shown in Figure 12.27? (**a**) 5.99 g/cm^3, (**b**) $1.44 \times 10^{24} \text{ g/mol}$, (**c**) 19.1 g/cm^3, (**d**) 2.39 g/cm^3, (**e**) 5.72 g/cm^3.

Practice Exercise 2

Estimate the length of the cubic unit cell edge and the density of CsCl (Figure 12.25) from the ionic radii of cesium, 1.81 Å, and chloride, 1.67 Å. (*Hint*: Ions in CsCl touch along the body diagonal, a vector running from one corner of a cube through the body center to the opposite corner. Using trigonometry it can be shown that the body diagonal of a cube is $\sqrt{3}$ times longer than the edge.)

12.6 | Molecular Solids

Molecular solids consist of atoms or neutral molecules held together by dipole–dipole forces, dispersion forces, and/or hydrogen bonds. Because these intermolecular forces are weak, molecular solids are soft and have relatively low melting points (usually below 200 °C). Most substances that are gases or liquids at room temperature form molecular solids at low temperature. Examples include Ar, H_2O, and CO_2.

The properties of molecular solids depend in large part on the strengths of the forces between molecules. Consider, for example, the properties of sucrose (table sugar, $C_{12}H_{22}O_{11}$). Each sucrose molecule has eight —OH groups, which allow for the formation of multiple hydrogen bonds. Consequently, sucrose exists as a crystalline solid at room temperature, and its melting point, 184 °C, is relatively high for a molecular solid.

Molecular shape is also important because it dictates how efficiently molecules pack together in three dimensions. Benzene (C_6H_6), for example, is a highly symmetrical planar molecule. ∞ (Section 8.6) It has a higher melting point than toluene, a compound in which one of the hydrogen atoms of benzene has been replaced by a CH_3 group (◄ Figure 12.28). The lower symmetry of toluene molecules prevents them from packing in a crystal as efficiently as benzene molecules. As a result, the intermolecular forces that depend on close contact are not as effective and the melting point is lower. In contrast, the boiling point of toluene is *higher* than that of benzene, indicating that the intermolecular attractive forces are larger in liquid toluene than in liquid benzene. The melting and boiling points of phenol, another substituted benzene shown in Figure 12.28, are higher than those of benzene because the OH group of phenol can form hydrogen bonds.

 GO FIGURE

In which substance, benzene or toluene, are the intermolecular forces stronger? In which substance do the molecules pack more efficiently?

	Benzene	Toluene	Phenol
Melting point (°C)	5	−95	43
Boiling point (°C)	80	111	182

▲ Figure 12.28 Melting and boiling points for benzene, toluene, and phenol.

12.7 | Covalent-Network Solids

Covalent-network solids consist of atoms held together in large networks by covalent bonds. Because covalent bonds are much stronger than intermolecular forces, these solids are much harder and have higher melting points than molecular solids. Diamond and graphite, two allotropes of carbon, are two of the most familiar covalent-network solids. Other examples are silicon, germanium, quartz (SiO_2), silicon carbide (SiC), and boron nitride (BN). In all cases, the bonding between atoms is either completely covalent or more covalent than ionic.

In diamond, each carbon atom is bonded tetrahedrally to four other carbon atoms (▼ Figure 12.29). The structure of diamond can be derived from the zinc blende structure (Figure 12.26) if carbon atoms replace both the zinc and sulfide ions. The carbon atoms are sp^3-hybridized and held together by strong carbon–carbon single covalent bonds. The strength and directionality of these bonds make diamond the hardest known material. For this reason, industrial-grade diamonds are employed in saw blades used for the most demanding cutting jobs. The stiff, interconnected bond network also explains why diamond is one of the best-known thermal conductors, yet is not electrically conductive. Diamond has a high melting point, 3550 °C.

In graphite, [Figure 12.29(b)], the carbon atoms form covalently bonded layers that are held together by intermolecular forces. The layers in graphite are the same as those in the graphene sheet shown in Figure 12.8. Graphite has a hexagonal unit cell containing two layers offset so that the carbon atoms in a given layer sit over the middle of the hexagons of the layer below. Each carbon is covalently bonded to three other carbons in the same layer to form interconnected hexagonal rings. The distance between adjacent carbon atoms in the plane, 1.42 Å, is very close to the C—C distance in benzene, 1.395 Å. In fact, the bonding resembles that of benzene, with delocalized π bonds extending over the layers. ∞ (Section 9.6) Electrons move freely through the delocalized orbitals, making graphite a good electrical conductor along the layers. (In fact, graphite is used as a conducting electrode in batteries.) These sp^2-hybridized sheets of carbon atoms are separated by 3.35 Å from one another, and the sheets are held together only by dispersion forces. Thus, the layers readily slide past one another when rubbed, giving graphite a greasy feel. This tendency is enhanced when impurity atoms are trapped between the layers, as is typically the case in commercial forms of the material.

Graphite is used as a lubricant and as the "lead" in pencils. The enormous differences in physical properties of graphite and diamond—both of which are pure carbon—arise from differences in their three-dimensional structure and bonding.

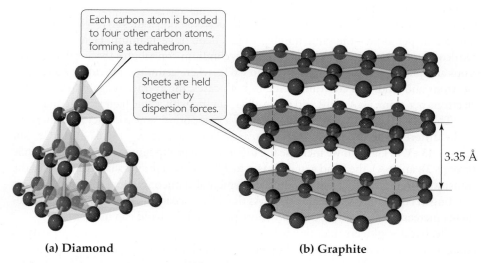

Each carbon atom is bonded to four other carbon atoms, forming a tedrahedron.

Sheets are held together by dispersion forces.

3.35 Å

(a) Diamond   **(b) Graphite**

▲ **Figure 12.29** The structures of **(a)** diamond and **(b)** graphite.

▲ GO FIGURE

If you draw a second diagram next to this one to represent an insulator, what aspect of the second diagram would be different?

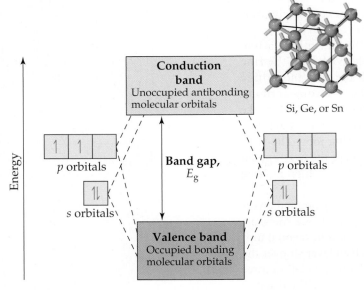

▲ **Figure 12.30 The electronic band structure of semiconductors that have the diamond crystal structure.**

Semiconductors

Metals conduct electricity extremely well. Many solids, however, conduct electricity somewhat, but nowhere near as well as metals, which is why such materials are called **semiconductors**. Two examples of semiconductors are silicon and germanium, which lie immediately below carbon in the periodic table. Like carbon, each of these elements has four valence electrons, just the right number to satisfy the octet rule by forming single covalent bonds with four neighbors. Hence, silicon and germanium, as well as the gray form of tin, crystallize with the same infinite network of covalent bonds as diamond.

When atomic *s* and *p* orbitals overlap, they form bonding molecular orbitals and antibonding molecular orbitals. Each pair of *s* orbitals overlaps to give one bonding and one antibonding molecular orbital, whereas the *p* orbitals overlap to give three bonding and three antibonding molecular orbitals. ∞ (Section 9.8) The extended network of bonds leads to the formation of the same type of bands we saw for metals in Section 12.4. However, unlike metals, in semiconductors an energy gap develops between the filled and empty states, much like the energy gap between bonding and antibonding orbitals. ∞ (Section 9.6) The band that forms from bonding molecular orbitals is called the **valence band**, and the band that forms the antibonding orbitals is called the **conduction band** (◄ **Figure 12.30**). In a semiconductor, the valence band is filled with electrons and the conduction band is empty. These two bands are separated by the energy **band gap** E_g. In the semiconductor community, energies are given in electron volts (eV); $1 \text{ eV} = 1.602 \times 10^{-19}$J. Band gaps greater than ~3.5 eV are so large that the material is not a semiconductor; it is an **insulator** and does not conduct electricity at all.

Semiconductors can be divided into two classes, elemental semiconductors, which contain only one type of atom, and compound semiconductors, which contain two or more elements. The elemental semiconductors all come from group 4A. As we move down the periodic table, bond distances increase, which decreases orbital overlap. This decrease in overlap reduces the energy difference between the top of the valence band and the bottom of the conduction band. As a result, the band gap decreases on going from diamond (5.5 eV, an insulator) to silicon (1.11 eV) to germanium (0.67 eV) to gray tin (0.08 eV). In the heaviest group 4A element, lead, the band gap collapses altogether. As a result, lead has the structure and properties of a metal.

Compound semiconductors maintain the same *average* valence electron count as elemental semiconductors—four per atom. For example, in gallium arsenide, GaAs, each Ga atom contributes three electrons and each As atom contributes five, which averages out to four per atom—the same number as in silicon or germanium. Hence, GaAs is a semiconductor. Other examples are InP, where indium contributes three valence electrons and phosphorus contributes five, and CdTe, where cadmium provides two valence electrons and tellurium contributes six. In both cases, the average is again four valence electrons per atom. GaAs, InP, and CdTe all crystallize with a zinc blende structure.

There is a tendency for the band gap of a compound semiconductor to increase as the difference in group numbers increases. For example, $E_g = 0.67$ eV in Ge, but $E_g = 1.43$ eV in GaAs. If we increase the difference in group number to four, as in ZnSe (groups 2B and 6A), the band gap increases to 2.70 eV. This progression is a result of the transition from pure covalent bonding in elemental semiconductors to polar covalent bonding in compound semiconductors. As the difference in electronegativity of the elements increases, the bonding becomes more polar and the band gap increases.

Electrical engineers manipulate both the orbital overlap and the bond polarity to control the band gaps of compound semiconductors for use in a wide range of electrical and optical devices. The band gaps of several elemental and compound semiconductors are given in ▶ Table 12.4.

Table 12.4 Band Gaps of Select Elemental and Compound Semiconductors

Material	Structure Type	E_g, eV[†]
Si	Diamond	1.11
AlP	Zinc blende	2.43
Ge	Diamond	0.67
GaAs	Zinc blende	1.43
ZnSe	Zinc blende	2.58
CuBr	Zinc blende	3.05
Sn[‡]	Diamond	0.08
InSb	Zinc blende	0.18
CdTe	Zinc blende	1.50

13 Al	14 Si	15 P

30 Zn	31 Ga	32 Ge	33 As	34 Se
48 Cd	49 In	50 Sn	51 Sb	52 Te

[†] Band gap energies are room temperature values, 1 eV = 1.602×10^{-19} J.
[‡] These data are for gray tin, the semiconducting allotrope of tin. The other allotrope, white tin, is a metal.

SAMPLE EXERCISE 12.3 Qualitative Comparison of Semiconductor Band Gaps

Will GaP have a larger or smaller band gap than ZnS? Will it have a larger or smaller band gap than GaN?

SOLUTION

Analyze The size of the band gap depends on the vertical and horizontal positions of the elements in the periodic table. The band gap will increase when either of the following conditions is met: (1) The elements are located higher up in the periodic table, where enhanced orbital overlap leads to a larger splitting between bonding and antibonding orbital energies, or (2) the horizontal separation between the elements increases, which leads to an increase in the electronegativity difference and bond polarity.

Plan We must look at the periodic table and compare the relative positions of the elements in each case.

Solve Gallium is in the fourth period and group 3A. Phosphorus is in the third period and group 5A. Zinc and sulfur are in the same periods as gallium and phosphorus, respectively. However, zinc, in group 2B, is one element to the left of gallium; sulfur in group 6A is one element to the right of phosphorus. Thus, we would expect the electronegativity difference to be larger for ZnS, which should result in ZnS having a larger band gap than GaP.

For both GaP and GaN the more electropositive element is gallium. So we need only compare the positions of the more electronegative elements, P and N. Nitrogen is located above phosphorus in group 5A. Therefore, based on increased orbital overlap, we would expect GaN to have a larger band gap than GaP.

Check External references show that the band gap of GaP is 2.26 eV, ZnS is 3.6 eV, and GaN is 3.4 eV.

Practice Exercise 1

Which of these statements is false?
(a) As you go down column 4A in the periodic table, the elemental solids become more electrically conducting. (b) As you go down column 4A in the periodic table, the band gaps of the elemental solids decrease. (c) The valence electron count for a compound semiconductor averages out to four per atom. (d) Band gap energies of semiconductors range from ~0.1 to 3.5 eV. (e) In general, the more polar the bonds are in compound semiconductors, the smaller the band gap.

Practice Exercise 2

Will ZnSe have a larger or smaller band gap than ZnS?

▲ GO FIGURE

Predict what would happen in panel (**b**) if you doubled the amount of doping shown in the n-type semiconductor.

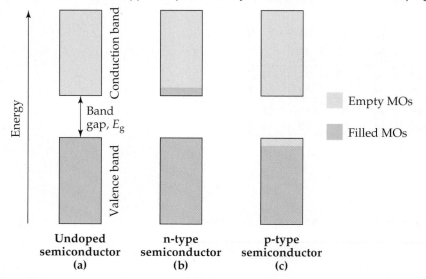

▲ **Figure 12.31** The addition of small amounts of impurities (doping) to a semiconductor changes the electronic properties of the material.

Semiconductor Doping

The electrical conductivity of a semiconductor is influenced by the presence of small numbers of impurity atoms. The process of adding controlled amounts of impurity atoms to a material is known as **doping**. Consider what happens when a few phosphorus atoms (known as dopants) replace silicon atoms in a silicon crystal. In pure Si all of the valence-band molecular orbitals are filled and all of the conduction-band molecular orbitals are empty, as ▲ **Figure 12.31(a)** shows. Because phosphorus has five valence electrons but silicon has only four, the "extra" electrons that come with the dopant phosphorus atoms are forced to occupy the conduction band [Figure 12.31(**b**)]. The doped material is called an *n-type* semiconductor, *n* signifying that the number of *n*egatively charged electrons in the conduction band has increased. These extra electrons can move very easily in the conduction band. Thus, just a few parts per million (ppm) of phosphorus in silicon can increase silicon's intrinsic conductivity by a factor of a million!

The dramatic change in conductivity in response to the addition of a trace amount of a dopant means that extreme care must be taken to control the impurities in semiconductors. The semiconductor industry uses "nine-nines" silicon to make integrated circuits; what this means is that Si must be 99.999999999% pure (nine nines after the decimal place) to be technologically useful! Doping provides an opportunity for controlling the electrical conductivity through precise control of the type and concentration of dopants.

It is also possible to dope semiconductors with atoms that have fewer valence electrons than the host material. Consider what happens when a few aluminum atoms replace silicon atoms in a silicon crystal. Aluminum has only three valence electrons compared to silicon's four. Thus, there are electron vacancies, known as **holes**, in the valence band when silicon is doped with aluminum [Figure 12.31(**c**)]. Since the negatively charged electron is not there, the hole can be thought of as having a positive charge. Any adjacent electron that jumps into the hole leaves behind a new hole. Thus, the positive hole moves about in the lattice like a particle.* A material like this is called

---

*This movement is analogous to watching people changing seats in a classroom; you can watch the people (electrons) move about the seats (atoms), or you can watch the empty seats (holes) "move."

a *p-type* semiconductor, *p* signifying that the number of *positive* holes in the material has increased.

As with n-type conductivity, p-type dopant levels of only parts per million can lead to a millionfold increase in conductivity—but in this case, the holes in the valence band are doing the conduction [Figure 12.31(c)].

The junction of an n-type semiconductor with a p-type semiconductor forms the basis for diodes, transistors, solar cells, and other devices.

SAMPLE EXERCISE 12.4 Identifying Types of Semiconductors

Which of the following elements, if doped into silicon, would yield an n-type semiconductor: Ga, As, or C?

SOLUTION

Analyze An n-type semiconductor means that the dopant atoms must have more valence electrons than the host material. Silicon is the host material in this case.

Plan We must look at the periodic table and determine the number of valence electrons associated with Si, Ga, As, and C. The elements with more valence electrons than silicon are the ones that will produce an n-type material upon doping.

Solve Si is in column 4A, and so has four valence electrons. Ga is in column 3A, and so has three valence electrons. As is in column 5A, and so has five valence electrons; C is in column 4A, and so has four valence electrons. Therefore, As, if doped into silicon, would yield an n-type semiconductor.

> **Practice Exercise 1**
> Which of these doped semiconductors would yield a p-type material? (These choices are written as host atom: dopant atom.)
> **(a)** Ge:P **(b)** Si:Ge **(c)** Si:Al **(d)** Ge:S **(e)** Si:N
>
> **Practice Exercise 2**
> Compound semiconductors can be doped to make n-type and p-type materials, but the scientist has to make sure that the proper atoms are substituted. For example, if Ge were doped into GaAs, Ge could substitute for Ga, making an n-type semiconductor; but if Ge substituted for As, the material would be p-type. Suggest a way to dope CdSe to create a p-type material.

12.8 | Polymers

In nature we find many substances of very high molecular weight, running up to millions of amu, that make up much of the structure of living organisms and tissues. Some examples are starch and cellulose, which abound in plants, as well as proteins, which are found in both plants and animals. In 1827 Jons Jakob Berzelius coined the word **polymer** (from the Greek *polys*, "many," and *meros*, "parts") to denote molecular substances of high molecular weight formed by the *polymerization* (joining together) of **monomers**, molecules with low molecular weight.

Historically natural polymers, such as wool, leather, silk, and natural rubber, were processed into usable materials. During the past 70 years or so, chemists have learned to form synthetic polymers by polymerizing monomers through controlled chemical reactions. A great many of these synthetic polymers have a backbone of carbon–carbon bonds because carbon atoms have an exceptional ability to form strong stable bonds with one another.

Chemistry Put to Work

Solid-State Lighting

Artificial lighting is so widespread we take it for granted. Major savings in energy would be realized if incandescent lights can be replaced by light-emitting diodes (LEDs). Because LEDs are made from semiconductors, this is an appropriate place to take a closer look at the operation of an LED.

The heart of an LED is a p–n diode, which is formed by bringing an n-type semiconductor into contact with a p-type semiconductor. In the junction where they meet there are very few electrons or holes to carry the charge across the interface between them, and the conductivity decreases. When an appropriate voltage is applied, electrons are driven from the conduction band of the n-doped side into the junction, where they meet holes that have been driven from the valence band of the p-doped side. The electrons fall into the empty holes, and their energy is converted into light whose photons have energy equal to the band gap (▼ **Figure 12.32**). In this way electrical energy is converted into optical energy.

Because the wavelength of light that is emitted depends on the band gap of the semiconductor, the color of light produced by the LED can be controlled by appropriate choice of semiconductor. Most red LEDs are made of a mixture of GaP and GaAs. The band gap of GaP is 2.26 eV (3.62×10^{-19} J), which corresponds to a green photon with a wavelength of 549 nm, while GaAs has a band gap of 1.43 eV (2.29×10^{-19} J), which corresponds to an infrared photon with a wavelength of 867 nm. ∞ (Sections 6.1 and 6.2) By forming solid solutions of these two compounds, with stoichiometries of $GaP_{1-x}As_x$, the band gap can be adjusted to any intermediate value. Thus, $GaP_{1-x}As_x$ is the solid solution of choice for red, orange, and yellow LEDs. Green LEDs are made from mixtures of GaP and AlP ($E_g = 2.43$ eV, $\lambda = 510$ nm).

Red LEDs have been in the market for decades, but to make white light, an efficient blue LED was needed. The first prototype bright blue LED was demonstrated in a Japanese laboratory in 1993. In 2010, less than 20 years later, over $10 billion worth of blue LEDs were sold worldwide. The blue LEDs are based on combinations of GaN ($E_g = 3.4$ eV, $\lambda = 365$ nm) and InN ($E_g = 2.4$ eV, $\lambda = 517$ nm). Many colors of LEDs are now available and are used in everything from barcode scanners to traffic lights (▼ **Figure 12.33**). Because the light emission results from semiconductor structures that can be made extremely small and because they emit little heat, LEDs are replacing standard incandescent and fluorescent light bulbs in many applications.

Related Exercises: 12.73, 12.74, 12.75, 12.76

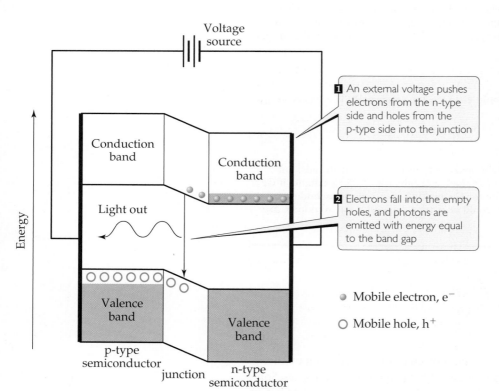

An external voltage pushes electrons from the n-type side and holes from the p-type side into the junction

Electrons fall into the empty holes, and photons are emitted with energy equal to the band gap

- Mobile electron, e^-
- Mobile hole, h^+

▲ Figure 12.32 **Light-emitting diodes.** The heart of a light-emitting diode is a p–n junction in which an applied voltage drives electrons and holes together, where they combine and give off light.

▲ Figure 12.33 **LEDs are all around us.**

Plastics are polymeric solids that can be formed into various shapes, usually by the application of heat and pressure. There are several types of plastics. **Thermoplastics** can be reshaped. For example, plastic milk containers are made from the thermoplastic polymer *polyethylene*. These containers can be melted down and the polymer recycled for some other use. In contrast, a **thermosetting plastic** (also called a *thermoset*) is shaped through irreversible chemical processes and, therefore, cannot be reshaped readily. Another type of plastic is the **elastomer**, which is a material that exhibits rubbery or elastic behavior. When subjected to stretching or bending, an elastomer regains its original shape upon removal of the distorting force, if it has not been distorted beyond some elastic limit. Rubber is the most familiar example of an elastomer.

Some polymers, such as nylon and polyesters, both of which are thermosetting plastics, can be formed into fibers that, like hair, are very long relative to their cross-sectional area. These fibers can be woven into fabrics and cords and fashioned into clothing, tire cord, and other useful objects.

Making Polymers

A good example of a polymerization reaction is the formation of polyethylene from ethylene molecules (▼ Figure 12.34). In this reaction, the double bond in each ethylene molecule "opens up," and two of the electrons originally in this bond are used to form new C—C single bonds with two other ethylene molecules. This type of polymerization, in which monomers are coupled through their multiple bonds, is called **addition polymerization**.

We can write the equation for the polymerization reaction as follows:

$$n\ CH_2{=}CH_2 \longrightarrow \left[\begin{array}{cc} H & H \\ | & | \\ C - C \\ | & | \\ H & H \end{array}\right]_n$$

Here n represents the large number—ranging from hundreds to many thousands—of monomer molecules (ethylene in this case) that react to form one polymer molecule. Within the polymer, a repeat unit (the unit shown in brackets in the previous equation) appears over and over along the entire chain. The ends of the chain are capped by carbon–hydrogen bonds or by some other bond, so that the end carbons have four bonds.

Polyethylene is an important material; its annual production exceeds 170 billion pounds each year. Although its composition is simple, the polymer is not easy to make. The right manufacturing conditions were identified only after many years of research. Today many forms of polyethylene, varying widely in physical properties, are known.

Polymers of other chemical compositions provide still greater variety in physical and chemical properties. Table 12.5 lists several other common polymers obtained by addition polymerization.

A second general reaction used to synthesize commercially important polymers is **condensation polymerization**. In a condensation reaction two molecules are joined to form a larger molecule by elimination of a small molecule, such as H_2O. For example, an amine (a compound containing —NH_2) reacts with a carboxylic acid

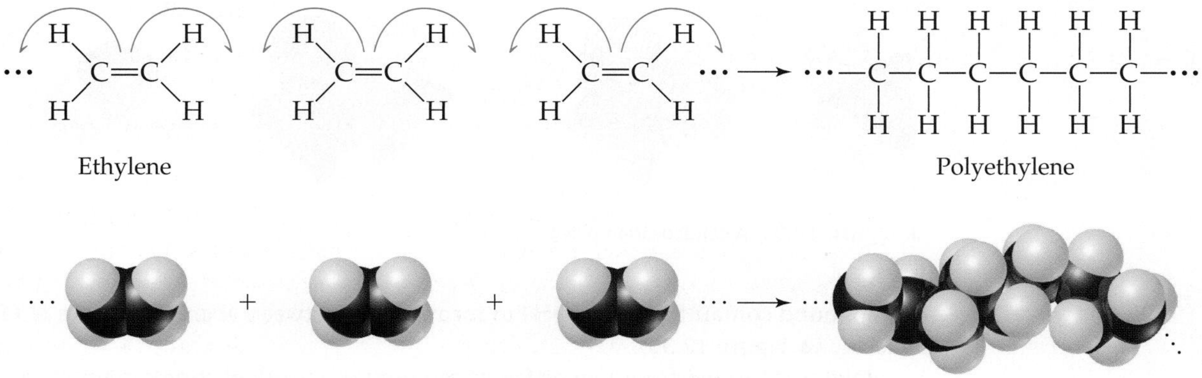

▲ Figure 12.34 **The polymerization of ethylene monomers to make the polymer polyethylene.**

Table 12.5 Polymers of Commercial Importance

Polymer	Structure	Uses
Addition Polymers		
Polyethylene	$-(CH_2-CH_2)_n$	Films, packaging, bottles
Polypropylene	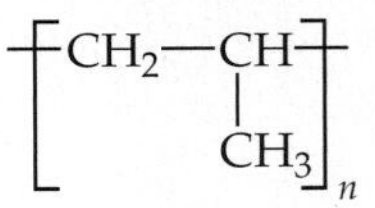 $\begin{bmatrix} CH_2-CH \\ \qquad\quad CH_3 \end{bmatrix}_n$	Kitchenware, fibers, appliances
Polystyrene	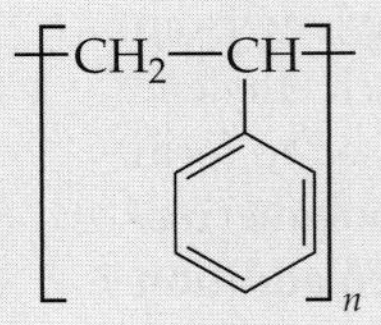 $\begin{bmatrix} CH_2-CH \\ \qquad\quad C_6H_5 \end{bmatrix}_n$	Packaging, disposable food containers, insulation
Polyvinyl chloride (PVC)	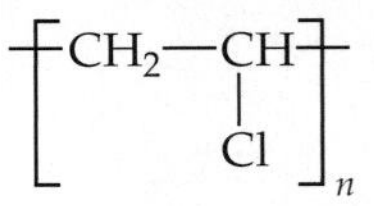 $\begin{bmatrix} CH_2-CH \\ \qquad\quad Cl \end{bmatrix}_n$	Pipe fittings, plumbing
Condensation Polymers		
Polyurethane	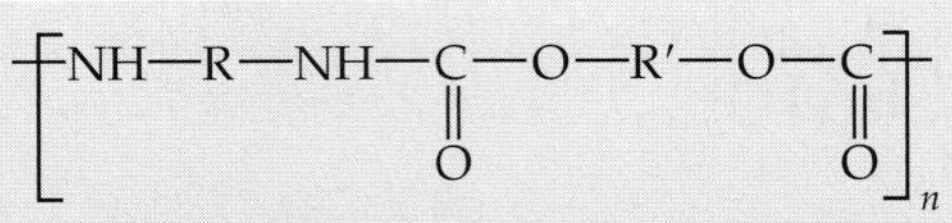 $\begin{bmatrix} NH-R-NH-\overset{O}{\underset{\|}{C}}-O-R'-O-\overset{O}{\underset{\|}{C}} \end{bmatrix}_n$ R, R' $= -CH_2-CH_2-$ (for example)	"Foam" furniture stuffing, spray-on insulation, automotive parts, footwear, water-protective coatings
Polyethylene terephthalate (a polyester)	$\begin{bmatrix} O-CH_2-CH_2-O-\overset{O}{\underset{\|}{C}}-\bigcirc-\overset{O}{\underset{\|}{C}} \end{bmatrix}_n$	Tire cord, magnetic tape, apparel, soft-drink bottles
Nylon 6,6	$\begin{bmatrix} NH-(CH_2)_6-NH-\overset{O}{\underset{\|}{C}}-(CH_2)_4-\overset{O}{\underset{\|}{C}} \end{bmatrix}_n$	Home furnishings, apparel, carpet, fishing line, toothbrush bristles
Polycarbonate	$\begin{bmatrix} O-\bigcirc-\overset{CH_3}{\underset{CH_3}{C}}-\bigcirc-O-\overset{O}{\underset{\|}{C}} \end{bmatrix}_n$	Shatterproof eyeglass lenses, CDs, DVDs, bulletproof windows, greenhouses

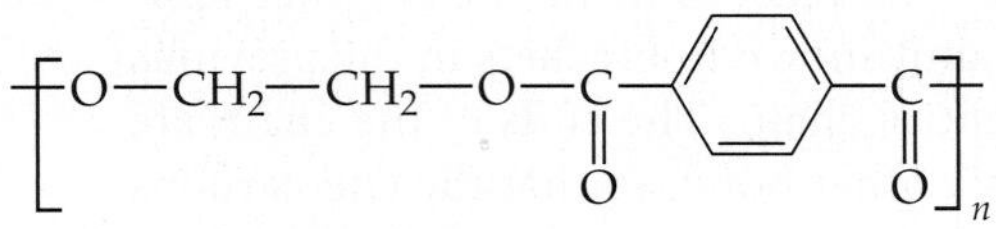

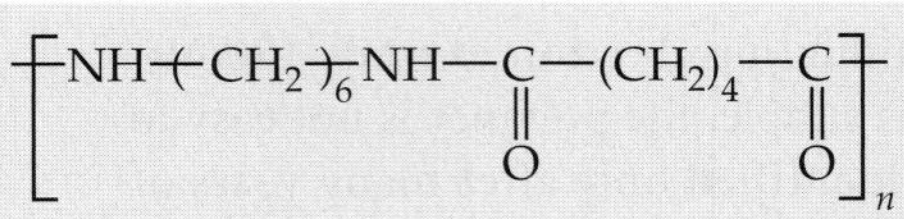

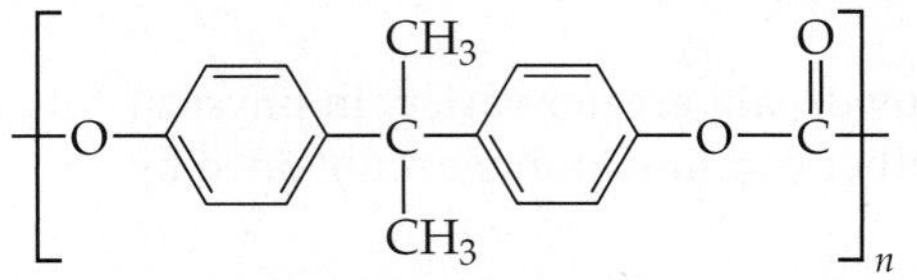

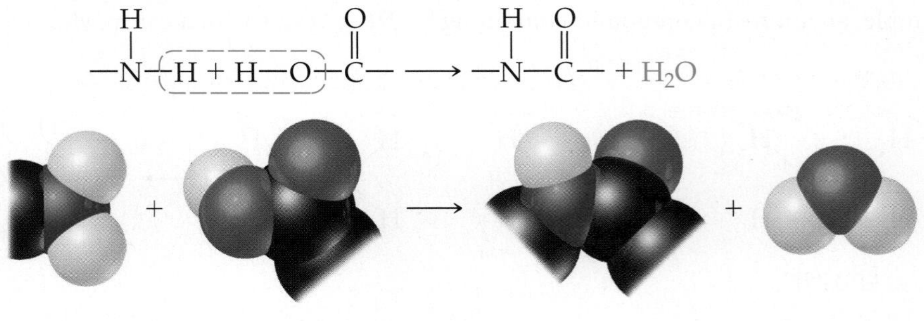

▲ Figure 12.35 A condensation polymerization.

(a compound containing —COOH) to form a bond between N and C plus an H_2O molecule (▲ Figure 12.35).

Polymers formed from two different monomers are called **copolymers**. In the formation of many nylons, a *diamine*, a compound with the —NH_2 group at each

Chemistry Put to Work

Recycling Plastics

If you look at the bottom of a plastic container, you are likely to see a recycle symbol containing a number, as shown in ▼ Figure 12.36. The number and the letter abbreviation below it indicate the kind of polymer from which the container is made, as summarized in ▼ Table 12.6. (The chemical structures of these polymers are shown in Table 12.5.) These symbols make it possible to sort containers by composition. In general, the lower the number, the greater the ease with which the material can be recycled.

▶ **Figure 12.36 Recycling symbols.** Most plastic containers manufactured today carry a recycling symbol indicating the type of polymer used to make the container and the polymer's suitability for recycling.

Table 12.6 Categories Used for Recycling Polymeric Materials in the United States

Number	Abbreviation	Polymer
1	PET or PETE	Polyethylene terephthalate
2	HDPE	High-density polyethylene
3	V or PVC	Polyvinyl chloride (PVC)
4	LDPE	Low-density polyethylene
5	PP	Polypropylene
6	P	Polystyrene
7	None	Other

end, is reacted with a *diacid*, a compound with the $-COOH$ group at each end. For example, the copolymer nylon 6,6 is formed when a diamine that has six carbon atoms and an amino group on each end is reacted with adipic acid, which also has six carbon atoms (▼ Figure 12.37). A condensation reaction occurs on each end of the diamine and the acid. Water is released, and $N-C$ bonds are formed between molecules.

Table 12.5 lists nylon 6,6 and some other common polymers obtained by condensation polymerization. Notice that these polymers have backbones containing N or O atoms as well as C atoms.

$$n\ H-\overset{\overset{H}{|}}{N}\!\!\leftarrow\!\!CH_2\!\!\xrightarrow{}_6\!\overset{\overset{H}{|}}{N}-H\ +\ n\ HO\overset{\overset{O}{\|}}{C}\!\!\leftarrow\!\!CH_2\!\!\xrightarrow{}_4\!\overset{\overset{O}{\|}}{C}OH\ \longrightarrow\ \left[\overset{\overset{H}{|}}{N}(CH_2)_6\overset{\overset{H}{|}}{N}-\overset{\overset{O}{\|}}{C}(CH_2)_4\overset{\overset{O}{\|}}{C}\right]_n\ +\ 2n\ H_2O$$

Diamine   Adipic acid   Nylon 6,6

▲ **Figure 12.37 The formation of the copolymer nylon 6,6.**

Give It Some Thought

Is this molecule a better starting material for an addition polymer or a condensation polymer?

$$H_2N-\!\!\bigcirc\!\!-\overset{\overset{O}{\|}}{C}-O-H$$

Structure and Physical Properties of Polymers

The simple structural formulas given for polyethylene and other polymers are deceptive. Because four bonds surround each carbon atom in polyethylene, the atoms are arranged in a tetrahedral fashion, so that the chain is not straight as we have depicted it. Furthermore, the atoms are relatively free to rotate around the $C-C$ single bonds. Rather than being straight and rigid, therefore, the chains are flexible, folding readily

▲ **Figure 12.38 A segment of a polyethylene chain.** This segment consists of 28 carbon atoms. In commercial polyethylenes, the chain lengths range from about 10^3 to 10^5 CH$_2$ units.

(◄ **Figure 12.38**). The flexibility in the molecular chains causes any material made of this polymer to be very flexible.

Both synthetic and natural polymers commonly consist of a collection of *macromolecules* (large molecules) of different molecular weights. Depending on the conditions of formation, the molecular weights may be distributed over a wide range or may be closely clustered around an average value. In part because of this distribution in molecular weights, polymers are largely amorphous (noncrystalline) materials. Rather than exhibiting a well-defined crystalline phase with a sharp melting point, polymers soften over a range of temperatures. They may, however, possess short-range order in some regions of the solid, with chains lined up in regular arrays as shown in ◄ **Figure 12.39**. The extent of such ordering is indicated by the degree of **crystallinity** of the polymer. Mechanical stretching or pulling to align the chains as the molten polymer is drawn through small holes can frequently enhance the crystallinity of a polymer. Intermolecular forces between the polymer chains hold the chains together in the ordered crystalline regions, making the polymer denser, harder, less soluble, and more resistant to heat. ▼ **Table 12.7** shows how the properties of polyethylene change as the degree of crystallinity increases.

The linear structure of polyethylene is conducive to intermolecular interactions that lead to crystallinity. However, the degree of crystallinity in polyethylene strongly depends on the average molecular weight. Polymerization results in a mixture of macromolecules with varying values of *n* (numbers of monomer molecules) and, hence, varying molecular weights. Low-density polyethylene (LDPE), used in forming films and sheets, has an average molecular weight in the range of 10^4 amu, has a density of less than 0.94 g/cm^3, and has substantial chain branching. That is, there are side chains off the main chain of the polymer. These side chains inhibit the formation of crystalline regions, reducing the density of the material. High-density polyethylene (HDPE), used to form bottles, drums, and pipes, has an average molecular weight in the range of 10^6 amu and has a density of 0.94 g/cm^3 or higher. This form has fewer side chains and thus a higher degree of crystallinity.

Polymers can be made stiffer by introducing chemical bonds between chains. Forming bonds between chains is called **cross-linking** (◄ **Figure 12.40**). The greater

Ordered regions

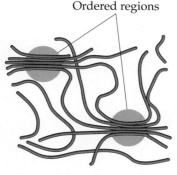

▲ **Figure 12.39 Interactions between polymer chains.** In the circled regions, the forces that operate between adjacent segments of the chains lead to ordering analogous to the ordering in crystals, though less regular.

▲ **Figure 12.40 Cross-linking of polymer chains.** The cross-linking groups (red) constrain the relative motions of the polymer chains, making the material harder and less flexible than when the cross-links are not present.

Give It Some Thought

In copolymers made of ethylene and vinyl acetate monomers, melting point and degree of crystallinity decrease as the percentage of vinyl acetate increases. Suggest an explanation.

Ethylene   Vinyl acetate

Table 12.7 **Properties of Polyethylene as a Function of Crystallinity**

Properties	Crystallinity				
	55%	62%	70%	77%	85%
Melting point (°C)	109	116	125	130	133
Density (g/cm^3)	0.92	0.93	0.94	0.95	0.96
Stiffness*	25	47	75	120	165
Yield stress*	1700	2500	3300	4200	5100

*These test results show that the mechanical strength of the polymer increases with increased crystallinity. The physical units for the stiffness test are psi × 10^{-3} (psi = pounds per square inch); those for the yield stress test are psi. Discussion of the exact meaning and significance of these tests is beyond the scope of this text.

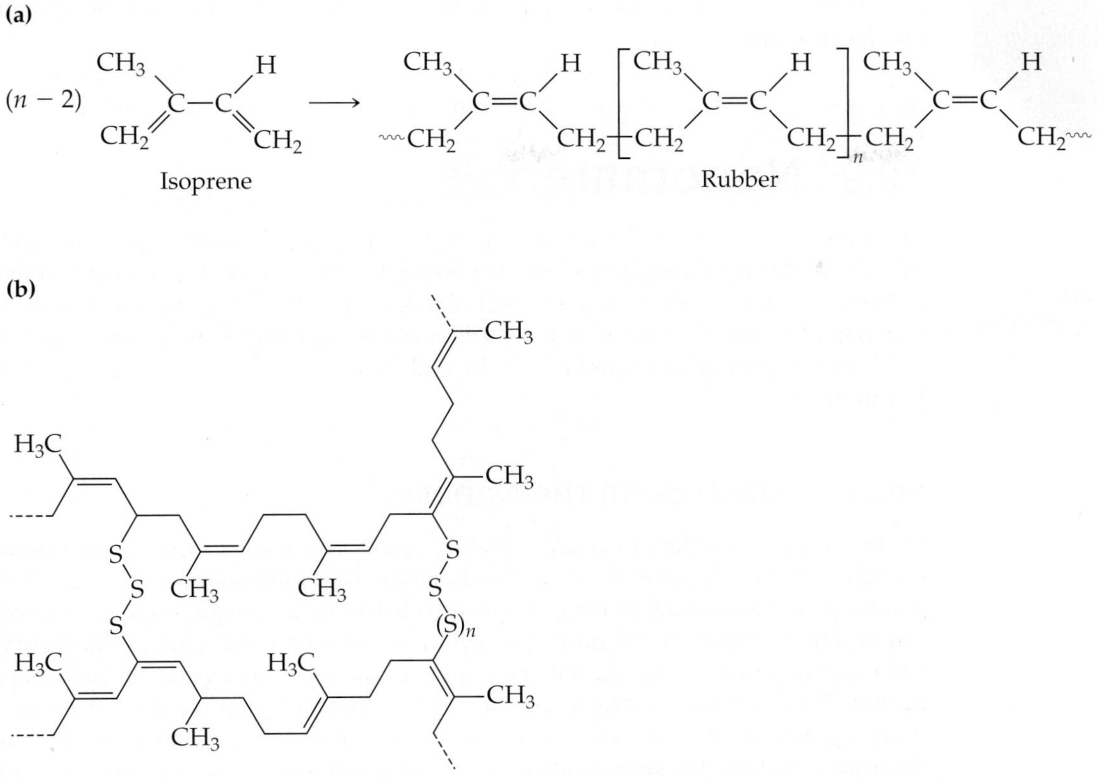

(a)

Isoprene   Rubber

(b)

▲ **Figure 12.41 Vulcanization of natural rubber.** (a) Formation of polymeric natural rubber from the monomer isoprene. (b) Adding sulfur to rubber creates carbon–sulfur bonds, and sulfur–atom links between chains.

the number of cross-links, the more rigid the polymer. Whereas thermoplastic materials consist of independent polymer chains, thermosetting plastics become cross-linked when heated; the cross-links allow them to hold their shapes.

An important example of cross-linking is the **vulcanization** of natural rubber, a process discovered by Charles Goodyear in 1839. Natural rubber is formed from a liquid resin derived from the inner bark of the *Hevea brasiliensis* tree. Chemically, it is a polymer of isoprene, C_5H_8 (▲ **Figure 12.41**). Because rotation about the carbon–carbon double bond does not readily occur, the orientation of the groups bound to the carbons is rigid. In natural rubber, the chain extensions are on the same side of the double bond, as shown in Figure 12.41(**a**).

Natural rubber is not a useful polymer because it is too soft and too chemically reactive. Goodyear accidentally discovered that adding sulfur and then heating the mixture makes the rubber harder and reduces its susceptibility to oxidation and other chemical degradation reactions. The sulfur changes rubber into a thermosetting polymer by cross-linking the polymer chains through reactions at some of the double bonds, as shown schematically in Figure 12.41(**b**). Cross-linking of about 5% of the double bonds creates a flexible, resilient rubber. When the rubber is stretched, the cross-links help prevent the chains from slipping, so that the rubber retains its elasticity. Because heating was an important step in his process, Goodyear named it after Vulcan, the Roman god of fire.

Most polymers contain sp^3-hybridized carbon atoms lacking delocalized π electrons ∞ (Section 9.6), so they are usually electrical insulators and are colorless (which implies a large band gap). However, if the backbone of the polymer has resonance ∞ (Sections 8.6 and 9.6), the electrons can become delocalized over long distances, which can lead to semiconducting behavior in the polymer. Such "plastic electronics" are of great current interest for lightweight and flexible organic solar cells, organic transistors, organic light-emitting diodes, and other

▲ **Figure 12.42 Plastic electronics.** Flexible organic solar cells are made from conducting polymers.

devices that are based on carbon rather than inorganic semiconductors like silicon (◀ Figure 12.42).

12.9 | Nanomaterials

The prefix *nano* means 10^{-9}. ∞(Section 1.4) When people speak of "nanotechnology," they usually mean making devices that are on the 1–100-nm scale. It turns out that the properties of semiconductors and metals change in this size range. **Nanomaterials**— materials that have dimensions on the 1–100-nm scale—are under intense investigation in research laboratories around the world, and chemistry plays a central role in this investigation.

Semiconductors on the Nanoscale

Figure 12.22 shows that, in small molecules, electrons occupy discrete molecular orbitals whereas in macroscale solids the electrons occupy delocalized bands. At what point does a molecule get so large that it starts behaving as though it has delocalized bands rather than localized molecular orbitals? For semiconductors, both theory and experiment tell us that the answer is roughly at 1 to 10 nm (about 10–100 atoms across). The exact number depends on the specific semiconductor material. The equations of quantum mechanics that were used for electrons in atoms can be applied to electrons (and holes) in semiconductors to estimate the size where materials undergo a crossover from molecular orbitals to bands. Because these effects become important at 1 to 10 nm, semiconductor particles with diameters in this size range are called *quantum dots*.

One of the most spectacular effects of reducing the size of a semiconductor crystal is that the band gap changes substantially with size in the 1–10-nm range. As the particle gets smaller, the band gap gets larger, an effect observable by the naked eye, as shown in ▼ Figure 12.43. On the macro level, the semiconductor cadmium phosphide looks black because its band gap is small ($E_g = 0.5$ eV), and it absorbs all wavelengths of visible light. As the crystals are made smaller, the material progressively changes color until it looks white! It looks white because now no visible light is absorbed. The band gap is so large that only high-energy ultraviolet light can excite electrons into the conduction band ($E_g > 3.0$ eV).

Making quantum dots is most easily accomplished using chemical reactions in solution. For example, to make CdS, you can mix $Cd(NO_3)_2$ and Na_2S in water. If you

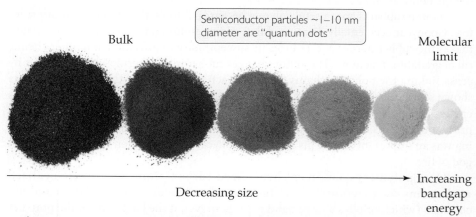

Bulk | Semiconductor particles ~1–10 nm diameter are "quantum dots" | Molecular limit

Decreasing size → Increasing bandgap energy

▲ **Figure 12.43 Cd₃P₂ powders with different particle sizes.** The arrow indicates decreasing particle size and a corresponding increase in the band gap energy, resulting in different colors.

do not do anything else, you will precipitate large crystals of CdS. How-ever, if you first add a negatively charged polymer to the water (such as polyphosphate, $-(OPO_2)_n-$), the Cd^{2+} associates with the polymer, like tiny "meatballs" in the polymer "spaghetti." When sulfide is added, CdS particles grow, but the polymer keeps them from forming large crys-tals. A great deal of fine-tuning of reaction conditions is necessary to pro-duce nanocrystals that are of uniform size and shape.

As we learned in Section 12.7, some semiconductor devices can emit light when a voltage is applied. Another way to make semicon-ductors emit light is to illuminate them with light whose photons have energies larger than the energy of the band gap of the semiconductor, a process called *photoluminescence*. A valence-band electron absorbs a photon and is promoted to the conduction band. If the excited electron then falls back down into the hole it left in the valence band, it emits a photon having energy equal to the band gap energy. In the case of quan-tum dots, the band gap is tunable with the crystal size, and thus all the colors of the rainbow can be obtained from just one material, as shown for CdSe in ▶ Figure 12.44.

Give It Some Thought

Large crystals of ZnS can show photoluminescence, emitting ultraviolet photons with energies equal to the band gap energy and a wavelength of 340 nm. Is it possible to shift the luminescence so that the emitted photons are in the visible region of the spectrum by making appropriately sized nanocrystals?

GO FIGURE

As the size of the quantum dots decreases, does the wavelength of the emitted light increase or decrease?

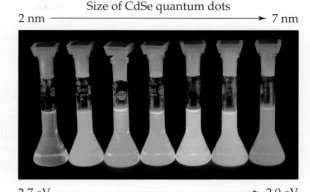

Size of CdSe quantum dots

2 nm ⟶ 7 nm

2.7 eV ⟶ 2.0 eV

Band gap energy, E_g

▲ **Figure 12.44 Photoluminescence depends on particle size at the nanoscale.** When illuminated with ultraviolet light, these solutions, each containing nanoparticles of the semiconductor CdSe, emit light that corresponds to their respective band gap energies. The wavelength of the light emitted depends on the size of the CdSe nanoparticles.

Quantum dots are being explored for applications ranging from electronics to lasers to medical imaging because they are very bright, very stable, and small enough to be taken up by living cells even after being coated with a biocompatible surface layer.

Semiconductors do not have to be shrunk to the nanoscale in all three dimensions to show new properties. They can be laid down in relatively large two-dimensional areas on a substrate but be only a few nanometers thick to make *quantum wells*. *Quantum wires*, in which the semiconductor wire diameter is only a few nanometers but its length is very long, have also been made by various chemical routes. In both quantum wells and quantum wires, measurements along the nanoscale dimension(s) show quan-tum behavior, but in the long dimension, the properties seem to be just like those of the bulk material.

Metals on the Nanoscale

Metals also have unusual properties on the 1–100-nm-length scale. Fun-damentally, this is because the mean free path (∞ Section 10.8) of an electron in a metal at room temperature is typically about 1–100 nm. So when the particle size of a metal is 100 nm or less, one might expect un-usual effects because the "sea of electrons" encounter a "shore" (the sur-face of the particle).

Although it was not fully understood, people have known for hun-dreds of years that metals are different when they are very finely divided. Dating back to the middle ages, the makers of stained-glass windows knew that gold dispersed in molten glass made the glass a beautiful deep red (▶ Figure 12.45). Much later, in 1857, Michael Faraday reported that disper-sions of small gold particles could be made stable and were deeply colored—some of the original colloidal solutions that he made are still in the Royal Institution of Great Britain's Faraday Museum in London (**Figure 12.46**).

▲ **Figure 12.45 Stained glass window from the Chartres Cathedral in France.** Gold nanoparticles are responsible for the red color in this window, which dates back to the twelfth century.

▲ **Figure 12.46 The solutions of colloidal gold nanoparticles made by Michael Faraday in the 1850s.** These are on display in the Faraday Museum, London.

▲ **GO FIGURE**

How many bonds does each carbon atom in C_{60} make? Based on this observation would you expect the bonding in C_{60} to be more like that in diamond or that in graphite?

▲ **Figure 12.47 Buckminsterfullerene, C_{60}.** The molecule has a highly symmetric structure in which the 60 carbon atoms sit at the vertices of a truncated icosahedron. The bottom view shows only the bonds between carbon atoms.

Other physical and chemical properties of metallic nanoparticles are also different from the properties of the bulk materials. Gold particles less than 20 nm in diameter melt at a far lower temperature than bulk gold, for instance, and when the particles are between 2 and 3 nm in diameter, gold is no longer a "noble," unreactive metal; in this size range it becomes chemically reactive.

At nanoscale dimensions, silver has properties analogous to those of gold in its beautiful colors, although it is more reactive than gold. Currently, there is great interest in research laboratories around the world in taking advantage of the unusual optical properties of metal nanoparticles for applications in biomedical imaging and chemical detection.

Carbon on the Nanoscale

We have seen that elemental carbon is quite versatile. In its bulk sp^3-hybridized solid-state form, it is diamond; in its bulk sp^2-hybridized solid-state form, it is graphite. Over the past three decades, scientists have discovered that sp^2-hybridized carbon can also form discrete molecules, one-dimensional nanoscale tubes, and two-dimensional nanoscale sheets. Each of these forms of carbon shows very interesting properties.

Until the mid-1980s, pure solid carbon was thought to exist in only two forms: the covalent-network solids diamond and graphite. In 1985, however, a group of researchers led by Richard Smalley and Robert Curl of Rice University and Harry Kroto of the University of Sussex, England, vaporized a sample of graphite with an intense pulse of laser light and used a stream of helium gas to carry the vaporized carbon into a mass spectrometer. ∞ (Section 2.4, "The Mass Spectrometer") The mass spectrum showed peaks corresponding to clusters of carbon atoms, with a particularly strong peak corresponding to molecules composed of 60 carbon atoms, C_{60}.

Because C_{60} clusters were so preferentially formed, the group proposed a radically different form of carbon, namely, nearly spherical C_{60} *molecules*. They proposed that the carbon atoms of C_{60} form a "ball" with 32 faces, 12 of them pentagons and 20 hexagons (◀ Figure 12.47), exactly like a soccer ball. The shape of this molecule is reminiscent of the geodesic dome invented by the U.S. engineer and philosopher R. Buckminster Fuller, so C_{60} was whimsically named "buckminsterfullerene," or "buckyball" for short. Since the discovery of C_{60}, other related molecules made of pure carbon have been discovered. These molecules are now known as fullerenes.

Appreciable amounts of buckyball can be prepared by electrically evaporating graphite in an atmosphere of helium gas. About 14% of the resulting soot consists of C_{60} and a related molecule, C_{70}, which has a more elongated structure. The carbon-rich gases from which C_{60} and C_{70} condense also contain other fullerenes, mostly containing more carbon atoms, such as C_{76} and C_{84}. The smallest possible fullerene, C_{20}, was first detected in 2000. This small, ball-shaped molecule is much more reactive than the larger fullerenes. Because fullerenes are molecules, they dissolve in various organic solvents, whereas diamond and graphite do not. This solubility permits fullerenes to be separated from the other components of soot and even from one another. It also allows the study of their reactions in solution.

Soon after the discovery of C_{60}, chemists discovered carbon nanotubes (▶ Figure 12.48). You can think of these as sheets of graphite rolled up and capped at one or both ends by half of a C_{60} molecule. Carbon nanotubes are made in a manner similar to that used to make C_{60}. They can be made in either *multiwall* or *single-walled* forms. Multiwall carbon nanotubes consist of tubes within tubes, nested together, whereas single-walled carbon nanotubes consist of single tubes. Single-walled carbon nanotubes can be 1000 nm long or even longer but are only about 1 nm in diameter. Depending on the diameter of the graphite sheet and how it is rolled up, carbon nanotubes can behave as either semiconductors or metals.

The fact that carbon nanotubes can be made either semiconducting or metallic without any doping is unique among solid-state materials, and laboratories worldwide are making and testing carbon-based electronic devices. Carbon nanotubes are also being explored for their mechanical properties. The carbon–carbon bonded framework of the nanotubes means that the imperfections that might appear in a metal nanowire of similar dimensions are nearly absent. Experiments on individual carbon nanotubes suggest that they are stronger than steel, if steel were the dimensions of a carbon nanotube. Carbon nanotubes have been spun into fibers with polymers, adding great strength and toughness to the composite material.

The two-dimensional form of carbon, graphene, is the most recent low-dimensional form of carbon to be experimentally isolated and studied. Although its properties had been the subject of theoretical predictions for over 60 years, it was not until 2004 that researchers at the University of Manchester in England isolated and identified individual sheets of carbon atoms with the honeycomb structure shown in ▶ Figure 12.49. Amazingly, the technique they used to isolate single-layer graphene was to successively peel away thin layers of graphite using adhesive tape. Individual layers of graphene were then transferred to a silicon wafer having a precisely defined overcoat of SiO_2. When a single layer of graphene is left on the wafer, an interference-like contrast pattern results that can be seen with an optical microscope. If not for this simple yet effective way to scan for individual graphene crystals, they would probably still remain undiscovered. Subsequently, it has been shown that graphene can be deposited on clean surfaces of other types of crystals. The scientists who led the effort at the University of Manchester, Andre Geim and Konstantin Novoselov, were awarded the 2010 Nobel Prize in Physics for their work.

The properties of graphene are remarkable. It is very strong and has a record thermal conductivity, topping carbon nanotubes in both categories. Graphene is a semimetal, which means its electronic structure is like that of a semiconductor in which the energy gap is exactly zero. The combination of graphene's two-dimensional character and the fact that it is a semimetal allows the electrons to travel very long distances, up to 0.3 μm, without scattering from another electron, atom, or impurity. Graphene can sustain electrical current densities six orders of magnitude higher than those sustainable in copper. Even though it is only one atom thick, graphene can absorb 2.3% of sunlight that strikes it. Scientists are currently exploring ways to incorporate graphene in various technologies including electronics, sensors, batteries, and solar cells.

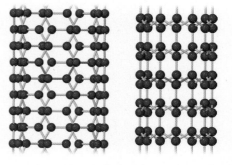

▲ Figure 12.48 **Atomic models of carbon nanotubes.** Left: "Armchair" nanotube, which shows metallic behavior. Right: "Zigzag" nanotube, which can be either semiconducting or metallic, depending on tube diameter.

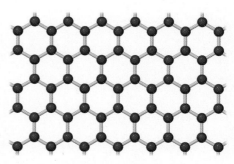

▲ Figure 12.49 **A portion of a two-dimensional graphene sheet.**

SAMPLE INTEGRATIVE EXERCISE | Putting Concepts Together

Polymers that can conduct electricity are called *conducting polymers*. Some polymers can be made semiconducting; others can be nearly metallic. Polyacetylene is an example of a polymer that is a semiconductor. It can also be doped to increase its conductivity.

Polyacetylene is made from acetylene in a reaction that looks simple but is actually tricky to do:

$$\mathrm{H-C\equiv C-H} \qquad \mathrm{+CH=CH+}_n$$

Acetylene   Polyacetylene

(a) What is the hybridization of the carbon atoms, and the geometry around those atoms, in acetylene and in polyacetylene?

(b) Write a balanced equation to make polyacetylene from acetylene.

(c) Acetylene is a gas at room temperature and pressure (298 K, 1.00 atm). How many grams of polyacetylene can you make from a 5.00-L vessel of acetylene gas at room temperature and room pressure? Assume acetylene behaves ideally, and that the polymerization reaction occurs with 100% yield.

(d) Using the average bond enthalpies in Table 8.4, predict whether the formation of polyacetylene from acetylene is endothermic or exothermic.

(e) A sample of polyacetylene absorbs light from 300 nm down to 650 nm. What is its band gap, in electron volts?

SOLUTION

Analyze For part (a), we need to recall what we have learned about sp, sp^2, and sp^3 hybridization and geometry. ∞∞ (Section 9.5) For part (b), we need to write a balanced equation. For part (c), we need to use the ideal-gas equation. ∞∞ (Section 10.4) For part (d), we need to recall the definitions of endothermic and exothermic and how bond enthalpies can be used to predict overall reaction enthalpies. ∞∞ (Section 8.8) For part (e), we need to relate the absorption of light to the differences in energy levels between filled and empty states in a material. ∞∞ (Section 6.3)

Plan For part (a), we should draw out the chemical structures of the reactant and product. For part (b), we need to make sure the equation is properly balanced. For part (c), we need to convert from liters of gas to moles of gas, using the ideal-gas equation ($PV = nRT$); then we need to convert from moles of acetylene gas to moles of polyacetylene using the answer from part (b); finally, we can convert to grams of polyacetylene. For part (d), we need to recall that $\Delta H_{rxn} = \Sigma$(bond enthalpies of bonds broken) $- \Sigma$(bond enthalpies of bonds formed). ∞∞ (Section 8.8) For part (e), we need to realize that the lowest energy absorbed by a material will tell us its band gap E_g (for a semiconductor or insulator) and combine $E = h\nu$ and $c = \lambda\nu$ together ($E = hc/\lambda$) to solve for E_g.

Solve

(a) Carbon always forms four bonds. Thus, each C atom must have a single bond to H and a triple bond to the other C atom in acetylene. As a result, each C atom has two electron domains and must be sp hybridized. This sp hybridization also means that the H—C—C angles in acetylene are 180° and the molecule is linear.

We can write out the partial structure of polyacetylene as follows:

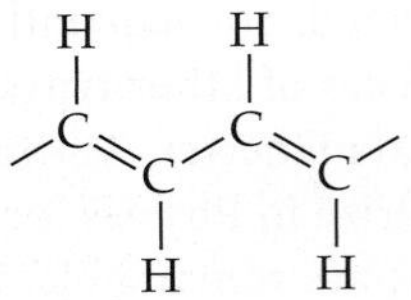

Each carbon is identical but now has three bonding electron domains that surround it. Therefore, the hybridization of each carbon atom is sp^2, and each carbon has local trigonal planar geometry with 120° angles.

(b) We can write:

$$n\,C_2H_2(g) \longrightarrow -[CH = CH]_n-$$

Note that all atoms originally present in acetylene end up in the polyacetylene product.

(c) We can use the ideal-gas equation as follows:

$$PV = nRT$$

$$(1.00\text{ atm})(5.00\text{ L}) = n(0.08206\text{ L-atm/K-mol})(298\text{ K})$$

$$n = 0.204\text{ mol}$$

Acetylene has a molar mass of 26.0 g/mol; therefore, the mass of 0.204 mol is

$$(0.204\text{ mol})(26.0\text{ g/mol}) = 5.32\text{ g acetylene}$$

Note that from the answer to part (b), all the atoms in acetylene go into polyacetylene. Due to conservation of mass, then, the mass of polyacetylene produced must also be 5.32 g, if we assume 100% yield.

(d) Let's consider the case for $n = 1$. We note that the reactant side of the equation in part (b) has one C≡C triple bond and two C—H single bonds. The product side of the equation in part (b) has one C=C double bond, one C—C single bond (to link to the adjacent monomer), and two C—H single bonds. Therefore, we are breaking one C≡C triple bond and are forming one C=C double bond and one C—C single bond. Accordingly, the enthalpy change for polyacetylene formation is

$$\Delta H_{rxn} = (\text{C≡C triple bond enthalpy}) - (\text{C=C double bond enthalpy}) - (\text{C—C single bond enthalpy})$$

$$\Delta H_{rxn} = (839\text{ kJ/mol}) - (614\text{ kJ/mol}) - (348\text{ kJ/mol})$$

$$= -123\text{ kJ/mol}$$

Because ΔH is a negative number, the reaction releases heat and is exothermic.

(e) The sample of polyacetylene absorbs many wavelengths of light, but the one we care about is the longest one, which corresponds to the lowest energy.

$$E = hc/\lambda$$

$$E = (6.626 \times 10^{-34}\text{ J s})(3.00 \times 10^8\text{ m s}^{-1})/(650 \times 10^{-9}\text{ m})$$

$$E = 3.06 \times 10^{-19}\text{ J}$$

We recognize that this energy corresponds to the energy difference between the bottom of the conduction band and the top of the valence band, and so is equivalent to the band gap E_g. Now we have to convert the number to electron volts. Since $1.602 \times 10^{-19}\text{ J} = 1\text{ eV}$, we find that

$$E_g = 1.91\text{ eV}$$

Chapter Summary and Key Terms

CLASSIFICATION OF SOLIDS (INTRODUCTION AND SECTION 12.1) The structures and properties of solids can be classified according to the forces that hold the atoms together. **Metallic solids** are held together by a delocalized sea of collectively shared valence electrons. **Ionic solids** are held together by the mutual attraction between cations and anions. **Covalent-network solids** are held together by an extended network of covalent bonds. **Molecular solids** are held together by weak intermolecular forces. **Polymers** contain very long chains of atoms held together by covalent bonds. These chains are usually held to one another by weaker intermolecular forces. **Nanomaterials** are solids where the dimensions of individual crystals are on the order of 1–100 nm.

STRUCTURES OF SOLIDS (SECTION 12.2) In **crystalline solids**, particles are arranged in a regularly repeating pattern. In **amorphous solids**, however, particles show no long-range order. In a crystalline solid the smallest repeating unit is called a **unit cell**. All unit cells in a crystal contain an identical arrangement of atoms. The geometrical pattern of points on which the unit cells are arranged is called a **crystal lattice**. To generate a crystal structure a **motif**, which is an atom or group of atoms, is associated with each and every **lattice point**.

In two dimensions the unit cell is a parallelogram whose size and shape are defined by two **lattice vectors** (*a* and *b*). There are five **primitive lattices**, lattices in which the lattice points are located only at the corners of the unit cell: square, hexagonal, rectangular, rhombic and oblique. In three dimensions the unit cell is a parallelepiped whose size and shape are defined by three lattice vectors (*a*, *b* and *c*), and there are seven primitive lattices: cubic, tetragonal, hexagonal, rhombohedral, orthorhombic, monoclinic, and triclinic. Placing an additional lattice point at the center of a cubic unit cell leads to a **body-centered cubic lattice**, while placing an additional point at the center of each face of the unit cell leads to a **face-centered cubic lattice**.

METALLIC SOLIDS (SECTION 12.3) **Metallic solids** are typically good conductors of electricity and heat, *malleable*, which means that they can be hammered into thin sheets, and *ductile*, which means that they can be drawn into wires. Metals tend to form structures where the atoms are closely packed. Two related forms of close packing, **cubic close packing** and **hexagonal close packing**, are possible. In both, each atom has a **coordination number** of 12.

Alloys are materials that possess characteristic metallic properties and are composed of more than one element. The elements in an alloy can be distributed either homogeneously or heterogeneously. Alloys which contain homogeneous mixtures of elements can either be substitutional or interstitial alloys. In a **substitutional alloy** the atoms of the minority element(s) occupy positions normally occupied by atoms of the majority element. In an **interstitial alloy** atoms of the minority element(s), often smaller nonmetallic atoms, occupy interstitial positions that lie in the "holes" between atoms of the majority element. In a **heterogeneous alloy** the elements are not distributed uniformly; instead, two or more distinct phases with characteristic compositions are present. **Intermetallic compounds** are alloys that have a fixed composition and definite properties.

METALLIC BONDING (SECTION 12.4) The properties of metals can be accounted for in a qualitative way by the **electron-sea model**, in which the electrons are visualized as being free to move throughout the metal. In the molecular-orbital model the valence atomic orbitals of the metal atoms interact to form energy **bands** that are incompletely filled by valence electrons. Consequently, the electronic structure of a bulk solid is referred to as a **band structure**. The orbitals that constitute the energy band are delocalized over the atoms of the metal, and their energies are closely spaced. In a metal the valence shell *s*, *p*, and *d* orbitals form bands, and these bands overlap resulting in one or more partially filled bands. Because the energy differences between orbitals *within a band* are extremely small, promoting electrons to higher-energy orbitals requires very little energy. This gives rise to high electrical and thermal conductivity, as well as other characteristic metallic properties.

IONIC SOLIDS (SECTION 12.5) **Ionic solids** consist of cations and anions held together by electrostatic attractions. Because these interactions are quite strong, ionic compounds tend to have high melting points. The attractions become stronger as the charges of the ions increase and/or the sizes of the ions decrease. The presence of both attractive (cation–anion) and repulsive (cation–cation and anion–anion) interactions helps to explain why ionic compounds are brittle. Like metals, the structures of ionic compounds tend to be symmetric, but to avoid direct contact between ions of like charge, the coordination numbers (typically 4 to 8) are necessarily smaller than those seen in close-packed metals. The exact structure depends on the relative sizes of the ions and the cation-to-anion ratio in the empirical formula.

MOLECULAR SOLIDS (SECTION 12.6) **Molecular solids** consist of atoms or molecules held together by intermolecular forces. Because these forces are relatively weak, molecular solids tend to be soft and possess low melting points. The melting point depends on the strength of the intermolecular forces, as well as the efficiency with which the molecules can pack together.

COVALENT-NETWORK SOLIDS (SECTION 12.7) **Covalent-network solids** consist of atoms held together in large networks by covalent bonds. These solids are much harder and have higher melting points than molecular solids. Important examples include diamond, where the carbons are tetrahedrally coordinated to each other, and graphite, where the sp^2-hybridized carbon atoms form hexagonal layers. **Semiconductors** are solids that do conduct electricity, but to a far lesser extent than metals. **Insulators** do not conduct electricity at all.

Elemental semiconductors, like Si and Ge, as well as compound semiconductors, like GaAs, InP, and CdTe, are important examples of covalent-network solids. In a semiconductor the filled bonding molecular orbitals make up the **valence band**, while the empty antibonding molecular orbitals make up the **conduction band**. The valence and conduction bands are separated by an energy that is referred to as the **band gap**, E_g. The size of the band gap increases as the bond length decreases, and as the difference in electronegativity between the two elements increases.

Doping semiconductors changes their ability to conduct electricity by orders of magnitude. An n-type semiconductor is one that is doped so that there are excess electrons in the conduction band; a p-type semiconductor is one that is doped so that there are missing electrons, which are called **holes**, in the valence band.

POLYMERS (SECTION 12.8) **Polymers** are molecules of high molecular weight formed by joining large numbers of small molecules called **monomers**. **Plastics** are materials that can be formed into various shapes, usually by the application of heat and pressure. **Thermoplastic** polymers can be reshaped, typically through heating, in contrast to **thermosetting plastics**, which are formed into objects through an irreversible chemical process and cannot readily be reshaped. An **elastomer** is a material that exhibits elastic behavior; that is, it returns to its original shape following stretching or bending.

In an **addition polymerization** reaction, the molecules form new linkages by opening existing π bonds. Polyethylene forms, for example, when the carbon–carbon double bonds of ethylene open up. In a **condensation polymerization** reaction, the monomers are joined by eliminating a small molecule between them. The various kinds of nylon are formed, for example, by removing a water molecule between an amine and a carboxylic acid. A polymer formed from two different monomers is called a **copolymer**.

Polymers are largely amorphous, but some materials possess a degree of **crystallinity**. For a given chemical composition, the crystallinity depends on the molecular weight and the degree of branching along the main polymer chain. Polymer properties are also strongly affected by **cross-linking**, in which short chains of atoms connect the long polymer chains. Rubber is cross-linked by short chains of sulfur atoms in a process called **vulcanization**.

NANOMATERIALS (SECTION 12.9) When one or more dimensions of a material become sufficiently small, generally smaller than 100 nm, the properties of the material change. Materials with dimensions on this length scale are called **nanomaterials**. Quantum dots are semiconductor particles with diameters of 1–10 nm. In this size range the material's band gap energy becomes size-dependent. Metal nanoparticles have different chemical and physical properties in the 1–100-nm size range. Nanoparticles of gold, for example, are more reactive than bulk gold and no longer have a golden color. Nanoscience has produced a number of previously unknown forms of sp^2-hybridized carbon. Fullerenes, like C_{60}, are large molecules containing only carbon atoms. Carbon nanotubes are sheets of graphite rolled up. They can behave as either semiconductors or metals depending on how the sheet was rolled. Graphene, which is an isolated layer from graphite, is a two-dimensional form of carbon. These nanomaterials are being developed now for many applications in electronics, batteries, solar cells, and medicine.

Learning Outcomes   After studying this chapter, you should be able to:

- Classify solids based on their bonding/intermolecular forces and understand how differences in bonding relate to physical properties. (Section 12.1)

- Describe the difference between crystalline and amorphous solids. Define and describe the relationships between unit cells, crystal lattice, lattice vectors, and lattice points. (Section 12.2)

- Explain why there are a limited number of lattices. Recognize the five two-dimensional and the seven three-dimensional primitive lattices. Describe the locations of lattice points for body-centered and face-centered lattices. (Section 12.2)

- State the characteristics and properties of metals. (Section 12.3)

- Calculate the empirical formula and density of ionic and metallic solids from a picture of the unit cell. Estimate the length of a cubic unit cell from the radii of the atoms/ions present. (Sections 12.3 and 12.5)

- Explain how homogeneous and heterogeneous alloys differ. Describe the differences between substitutional alloys, interstitial alloys, and intermetallic compounds. (Section 12.3)

- Explain the electron-sea model of metallic bonding. (Section 12.4)

- Use the molecular-orbital model of metallic bonding to generate the electronic band structures of metals and qualitatively predict the trends in melting point, boiling point, and hardness of metals. (Section 12.4)

- Predict the structures of ionic solids from their ionic radii and empirical formula. (Section 12.5)

- Interpret melting point and boiling point data of molecular solids in terms of intermolecular forces and crystalline packing. (Section 12.6)

- Define the terms *valence band, conduction band, band gap, holes* (the chemical meaning), *semiconductor*, and *insulator*. (Section 12.7)

- Account for the relative band gap energies of semiconductors in terms of periodic trends. (Section 12.7)

- Predict how n-type and p-type doping can be used to control the conductivity of semiconductors. (Section 12.7)

- Define the terms *plastic, thermoplastic, thermosetting plastic, elastomer, copolymers,* and *cross-linking*. (Section 12.8)

- Describe how polymers are formed from monomers and recognize the features of a molecule that allow it to react to form a polymer. Explain the differences between addition polymerization and condensation polymerization. (Section 12.8)

- Explain how the interactions between polymer chains impact the physical properties of polymers. (Section 12.8)

- Describe how the properties of bulk semiconductors and metals change as the size of the crystals decreases to the nanometer-length scale. (Section 12.9)

- Describe the structures and unique properties of fullerenes, carbon nanotubes, and graphene. (Section 12.9)

Key Equation

$$\frac{\text{Number of cations per formula unit}}{\text{Number of anions per formula unit}} = \frac{\text{anion coordination number}}{\text{cation coordination number}}$$

[12.1]   Relationship between cation and anion coordination numbers and the empirical formula of an ionic compound

Exercises

Visualizing Concepts

12.1 Two solids are shown below. One is a semiconductor and one is an insulator. Which one is which? Explain your reasoning. [Sections 12.1, 12.7]

12.2 For each of the two-dimensional structures shown here (**a**) draw the unit cell, (**b**) determine the type of two-dimensional lattice (from Figure 12.4), and (**c**) determine how many of each type of circle (white or black) there are per unit cell. [Section 12.2]

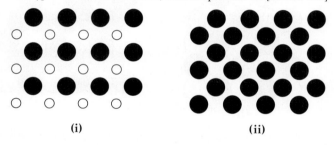

    (i)   **(ii)**

12.3 (**a**) What kind of packing arrangement is seen in the accompanying photo? (**b**) What is the coordination number of each cannonball in the interior of the stack? (**c**) What are the coordination numbers for the numbered cannonballs on the visible side of the stack? [Section 12.3]

12.4 Which arrangement of cations (yellow) and anions (blue) in a lattice is the more stable? Explain your reasoning. [Section 12.5]

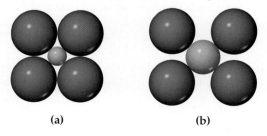

    (a)   **(b)**

12.5 Which of these molecular fragments would you expect to be more likely to give rise to electrical conductivity? Explain your reasoning. [Sections 12.6, 12.8]

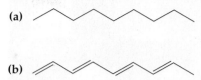

12.6 The electronic structure of a doped semiconductor is shown here. (**a**) Which band, A or B, is the valence band? (**b**) Which band is the conduction band? (**c**) Which band consists of bonding molecular orbitals? (**d**) Is this an example of an n-type or p-type semiconductor? (**e**) If the semiconductor is germanium, which of the following elements could be the dopant: Ga, Si, or P? [Section 12.7]

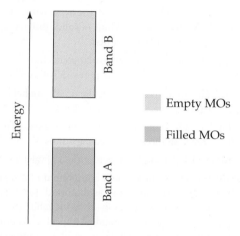

12.7 Shown here are cartoons of two different polymers. Which of these polymers would you expect to be more crystalline? Which one would have the higher melting point? [Section 12.8]

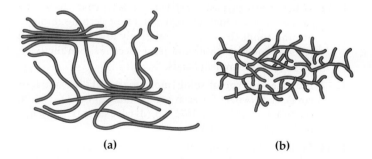

    (a)   **(b)**

12.8 The accompanying image shows photoluminescence from four different samples of CdTe nanocrystals, each embedded in a polymer matrix. The photoluminescence occurs because the samples are being irradiated by a UV light source. The nanocrystals in each vial have different average sizes. The sizes are 4.0, 3.5, 3.2, and 2.8 nm. (**a**) Which vial contains the 4.0-nm nanocrystals? (**b**) Which vial contains the 2.8-nm nanocrystals? (**c**) Crystals of CdTe that have sizes that are larger than approximately 100 nm have a band gap of 1.5 eV. What would be the wavelength and frequency of light emitted from these crystals? What type of light is this? [Sections 12.7 and 12.9]

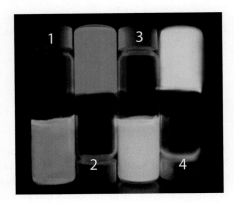

Classification of Solids (Section 12.1)

12.9 Covalent bonding occurs in both molecular and covalent-network solids. Which of the following statements best explains why these two kinds of solids differ so greatly in their hardness and melting points?

(a) The molecules in molecular solids have stronger covalent bonding than covalent-network solids do.

(b) The molecules in molecular solids are held together by weak intermolecular interactions.

(c) The atoms in covalent-network solids are more polarizable than those in molecular solids.

(d) Molecular solids are denser than covalent-network solids.

12.10 Silicon is the fundamental component of integrated circuits. Si has the same structure as diamond. **(a)** Is Si a molecular, metallic, ionic, or covalent-network solid? **(b)** Silicon readily reacts to form silicon dioxide, SiO_2, which is quite hard and is insoluble in water. Is SiO_2 most likely a molecular, metallic, ionic, or covalent-network solid?

12.11 What kinds of attractive forces exist between particles (atoms, molecules, or ions) in **(a)** molecular crystals, **(b)** covalent-network crystals, **(c)** ionic crystals, **(d)** and metallic crystals?

12.12 Which type (or types) of crystalline solid is characterized by each of the following? **(a)** High mobility of electrons throughout the solid; **(b)** softness, relatively low melting point; **(c)** high melting point and poor electrical conductivity; **(d)** network of covalent bonds.

12.13 Indicate the type of solid (molecular, metallic, ionic, or covalent-network) for each compound: **(a)** $CaCO_3$, **(b)** Pt, **(c)** ZrO_2 (melting point, 2677 °C), **(d)** table sugar ($C_{12}H_{22}O_{11}$), **(e)** benzene (C_6H_6), **(f)** I_2.

12.14 Indicate the type of solid (molecular, metallic, ionic, or covalent-network) for each compound: **(a)** InAs, **(b)** MgO, **(c)** HgS, **(d)** In, **(e)** HBr.

12.15 A white substance melts with some decomposition at 730 °C. As a solid, it does not conduct electricity, but it dissolves in water to form a conducting solution. Which type of solid (molecular, metallic, covalent-network, or ionic) might the substance be?

12.16 You are given a white substance that sublimes at 3000 °C; the solid is a nonconductor of electricity and is insoluble in water. Which type of solid (molecular, metallic, covalent-network, or ionic) might this substance be?

Structures of Solids (Section 12.2)

12.17 **(a)** Draw a picture that represents a crystalline solid at the atomic level. **(b)** Now draw a picture that represents an amorphous solid at the atomic level.

12.18 Amorphous silica, SiO_2, has a density of about 2.2 g/cm^3, whereas the density of crystalline quartz, another form of SiO_2, is 2.65 g/cm^3. Which of the following statements is the best explanation for the difference in density?

(a) Amorphous silica is a network-covalent solid, but quartz is metallic.

(b) Amorphous silica crystallizes in a primitive cubic lattice.

(c) Quartz is harder than amorphous silica.

(d) Quartz must have a larger unit cell than amorphous silica.

(e) The atoms in amorphous silica do not pack as efficiently in three dimensions as compared to the atoms in quartz.

12.19 Two patterns of packing for two different spheres are shown here. For each structure **(a)** draw the two-dimensional unit cell; **(b)** determine the angle between the lattice vectors, γ, and determine whether the lattice vectors are of the same length or of different lengths; and **(c)** determine the type of two-dimensional lattice (from Figure 12.4).

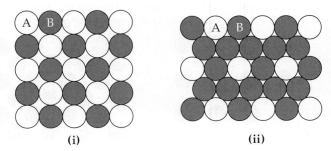

(i)   (ii)

12.20 Two patterns of packing two different spheres are shown here. For each structure **(a)** draw the two-dimensional unit cell; **(b)** determine the angle between the lattice vectors, γ, and determine whether the lattice vectors are of the same length or of different lengths; **(c)** determine the type of two-dimensional lattice (from Figure 12.4).

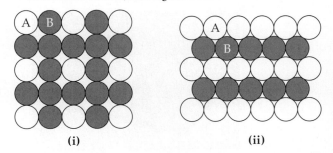

(i)   (ii)

12.21 Imagine the primitive cubic lattice. Now imagine grabbing the top of it and stretching it straight up. All angles remain 90°. What kind of primitive lattice have you made?

12.22 Imagine the primitive cubic lattice. Now imagine pushing on top of it, straight down. All angles remain 90°. What kind of primitive lattice have you made?

12.23 Which of the three-dimensional primitive lattices has a unit cell where none of the internal angles is 90°? (**a**) Orthorhombic, (**b**) hexagonal, (**c**) rhombohedral, (**d**) triclinic, (**e**) both rhombohedral and triclinic.

12.24 Besides the cubic unit cell, which other unit cell(s) has edge lengths that are all equal to each other? (**a**) Orthorhombic, (**b**) hexagonal, (**c**) rhombohedral, (**d**) triclinic, (**e**) both rhombohedral and triclinic.

12.25 What is the minimum number of atoms that could be contained in the unit cell of an element with a body-centered cubic lattice? (**a**) 1, (**b**) 2, (**c**) 3, (**d**) 4, (**e**) 5.

12.26 What is the minimum number of atoms that could be contained in the unit cell of an element with a face-centered cubic lattice? (**a**) 1, (**b**) 2, (**c**) 3, (**d**) 4, (**e**) 5.

12.27 The unit cell of nickel arsenide is shown here. (**a**) What type of lattice does this crystal possess? (**b**) What is the empirical formula?

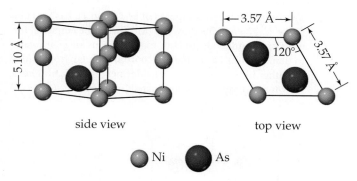

side view   top view

○ Ni   ● As

12.28 The unit cell of a compound containing strontium, iron, and oxygen is shown here. (**a**) What type of lattice does this crystal possess (all three lattice vectors are mutually perpendicular)? (**b**) What is the empirical formula?

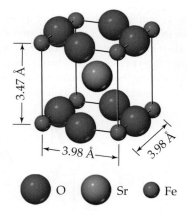

● O   ○ Sr   ● Fe

Metallic Solids (Section 12.3)

12.29 The densities of the elements K, Ca, Sc, and Ti are 0.86, 1.5, 3.2, and 4.5 g/cm³, respectively. One of these elements crystallizes in a body-centered cubic structure; the other three crystallize in a face-centered cubic structure. Which one crystallizes in the body-centered cubic structure? Justify your answer.

12.30 For each of these solids, state whether you would expect it to possess metallic properties: (**a**) $TiCl_4$, (**b**) NiCo alloy, (**c**) W, (**d**) Ge, (**e**) ScN.

12.31 Consider the unit cells shown here for three different structures that are commonly observed for metallic elements. (**a**) Which structure(s) corresponds to the densest packing of atoms? (**b**) Which structure(s) corresponds to the least dense packing of atoms?

Structure type A   Structure type B   Structure type C

12.32 Sodium metal (atomic weight 22.99 g/mol) adopts a body-centered cubic structure with a density of 0.97 g/cm³. (**a**) Use this information and Avogadro's number ($N_A = 6.022 \times 10^{23}$/mol) to estimate the atomic radius of sodium. (**b**) If sodium didn't react so vigorously, it could float on water. Use the answer from part (**a**) to estimate the density of Na if its structure were that of a cubic close-packed metal. Would it still float on water?

12.33 Iridium crystallizes in a face-centered cubic unit cell that has an edge length of 3.833 Å. (**a**) Calculate the atomic radius of an iridium atom. (**b**) Calculate the density of iridium metal.

12.34 Calcium crystallizes in a body-centered cubic structure. (**a**) How many Ca atoms are contained in each unit cell? (**b**) How many nearest neighbors does each Ca atom possess? (**c**) Estimate the length of the unit cell edge, a, from the atomic radius of calcium (1.97 Å). (**d**) Estimate the density of Ca metal.

12.35 Aluminum metal crystallizes in a face-centered cubic unit cell. (**a**) How many aluminum atoms are in a unit cell? (**b**) What is the coordination number of each aluminum atom? (**c**) Estimate the length of the unit cell edge, a, from the atomic radius of aluminum (1.43 Å). (**d**) Calculate the density of aluminum metal.

12.36 An element crystallizes in a body-centered cubic lattice. The edge of the unit cell is 2.86 Å, and the density of the crystal is 7.92 g/cm³. Calculate the atomic weight of the element.

12.37 Which of these statements about alloys and intermetallic compounds is false? (**a**) Bronze is an example of an alloy. (**b**) "Alloy" is just another word for "a chemical compound of fixed composition that is made of two or more metals." (**c**) Intermetallics are compounds of two or more metals that have a definite composition and are not considered alloys. (**d**) If you mix two metals together and, at the atomic level, they separate into two or more different compositional phases, you have created a heterogeneous alloy. (**e**) Alloys can be formed even if the atoms that comprise them are rather different in size.

12.38 Determine if each statement is true or false: (**a**) Substitutional alloys are solid solutions, but interstitial alloys are heterogenous alloys. (**b**) Substitutional alloys have "solute" atoms that replace "solvent" atoms in a lattice, but interstitial alloys have "solute" atoms that are in between the "solvent" atoms in a lattice. (**c**) The atomic radii of the atoms in a substitutional alloy are similar to each other, but in an interstitial alloy, the interstitial atoms are a lot smaller than the host lattice atoms.

12.39 For each of the following alloy compositions, indicate whether you would expect it to be a substitutional alloy, an interstitial alloy, or an intermetallic compound:

(a) $Fe_{0.97}Si_{0.03}$, (b) $Fe_{0.60}Ni_{0.40}$, (c) $SmCo_5$.

12.40 For each of the following alloy compositions, indicate whether you would expect it to be a substitutional alloy, an interstitial alloy, or an intermetallic compound:

(a) $Cu_{0.66}Zn_{0.34}$, (b) Ag_3Sn, (c) $Ti_{0.99}O_{0.01}$.

12.41 Indicate whether each statement is true or false:

(a) Substitutional alloys tend to be more ductile than interstitial alloys.

(b) Interstitial alloys tend to form between elements with similar ionic radii.

(c) Nonmetallic elements are never found in alloys.

12.42 Indicate whether each statement is true or false:

(a) Intermetallic compounds have a fixed composition.

(b) Copper is the majority component in both brass and bronze.

(c) In stainless steel, the chromium atoms occupy interstitial positions.

12.43 Which element or elements are alloyed with gold to make the following types of "colored gold" used in the jewelry industry? For each type, also indicate what type of alloy is formed: (a) white gold, (b) rose gold, (c) green gold.

12.44 Look up the chemical composition of purple gold. Is the composition variable? Why don't jewelers use purple gold to make rings or necklaces?

Metallic Bonding (Section 12.4)

12.45 State whether each sentence is true or false:

(a) Metals have high electrical conductivities because the electrons in the metal are delocalized.

(b) Metals have high electrical conductivities because they are denser than other solids.

(c) Metals have large thermal conductivities because they expand when heated.

(d) Metals have small thermal conductivities because the delocalized electrons cannot easily transfer the kinetic energy imparted to the metal from heat.

12.46 Imagine that you have a metal bar sitting half in the Sun and half in the dark. On a sunny day, the part of the metal that has been sitting in the Sun feels hot. If you touch the part of the metal bar that has been sitting in the dark, will it feel hot or cold? Justify your answer in terms of thermal conductivity.

12.47 The molecular-orbital diagrams for two- and four-atom linear chains of lithium atoms are shown in Figure 12.22. Construct a molecular-orbital diagram for a chain containing six lithium atoms and use it to answer the following questions: (a) How many molecular orbitals are there in the diagram? (b) How many nodes are in the lowest-energy molecular orbital? (c) How many nodes are in the highest-energy molecular orbital? (d) How many nodes are in the highest-energy occupied molecular orbital (HOMO)? (e) How many nodes are in the lowest-energy unoccupied molecular orbital (LUMO)? (f) How does the

HOMO–LUMO energy gap for this case compare to that of the four-atom case?

12.48 Repeat Exercise 12.47 for a linear chain of eight lithium atoms.

12.49 Which would you expect to be the more ductile element, (a) Ag or Mo, (b) Zn or Si? In each case explain your reasoning.

12.50 Which of the following statements does not follow from the fact that the alkali metals have relatively weak metal–metal bonding?

(a) The alkali metals are less dense than other metals.

(b) The alkali metals are soft enough to be cut with a knife.

(c) The alkali metals are more reactive than other metals.

(d) The alkali metals have higher melting points than other metals.

(e) The alkali metals have low ionization energies.

12.51 Explain this trend in melting points: Y 1522 °C, Zr 1852 °C, Nb 2468 °C, Mo 2617 °C.

12.52 For each of the following groups, which metal would you expect to have the highest melting point: (a) gold, rhenium, or cesium; (b) rubidium, molybdenum, or indium; (c) ruthenium, strontium, or cadmium?

Ionic and Molecular Solids (Sections 12.5 and 12.6)

12.53 Tausonite, a mineral composed of Sr, O, and Ti, has the cubic unit cell shown in the drawing. (a) What is the empirical formula of this mineral? (b) How many oxygens are coordinated to titanium? (c) To see the full coordination environment of the other ions, we have to consider neighboring unit cells. How many oxygens are coordinated to strontium?

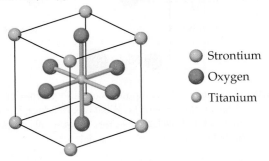

- Strontium
- Oxygen
- Titanium

12.54 Rutile is a mineral composed of Ti and O. Its unit cell, shown in the drawing, contains Ti atoms at each corner and a Ti atom at the center of the cell. Four O atoms are on the opposite faces of the cell, and two are entirely within the cell. (a) What is the chemical formula of this mineral? (b) What is the coordination number of each atom?

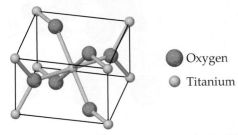

- Oxygen
- Titanium

12.55 NaF has the same structure as NaCl. (a) Use ionic radii from Chapter 7 to estimate the length of the unit cell edge for NaF. (b) Use the unit cell size calculated in part (a) to estimate the density of NaF.

12.56 Clausthalite is a mineral composed of lead selenide (PbSe). The mineral adopts the rock salt structure. The density of PbSe at 25 °C is 8.27 g/cm³. Calculate the length of an edge of the PbSe unit cell.

12.57 A particular form of cinnabar (HgS) adopts the zinc blende structure. The length of the unit cell edge is 5.852 Å. (a) Calculate the density of HgS in this form. (b) The mineral tiemmanite (HgSe) also forms a solid phase with the zinc blende structure. The length of the unit cell edge in this mineral is 6.085 Å. What accounts for the larger unit cell length in tiemmanite? (c) Which of the two substances has the higher density? How do you account for the difference in densities?

12.58 At room temperature and pressure RbI crystallizes with the NaCl-type structure. (a) Use ionic radii to predict the length of the cubic unit cell edge. (b) Use this value to estimate the density. (c) At high pressure the structure transforms to one with a CsCl-type structure. (c) Use ionic radii to predict the length of the cubic unit cell edge for the high-pressure form of RbI. (d) Use this value to estimate the density. How does this density compare with the density you calculated in part (b)?

12.59 CuI, CsI, and NaI each adopt a different type of structure. The three different structures are those shown in Figure 12.26. (a) Use ionic radii, Cs^+ ($r = 1.81$ Å), Na^+ ($r = 1.16$ Å), Cu^+ ($r = 0.74$ Å), and, I^- ($r = 2.06$ Å), to predict which compound will crystallize with which structure. (b) What is the coordination number of iodide in each of these structures?

12.60 The rutile and fluorite structures, shown here (anions are colored green), are two of the most common structure types of ionic compounds where the cation to anion ratio is 1:2. (a) For CaF_2 and ZnF_2 use ionic radii, Ca^{2+} ($r = 1.14$ Å), Zn^{2+} ($r = 0.88$ Å), and F^- ($r = 1.19$ Å), to predict which compound is more likely to crystallize with the fluorite structure and which with the rutile structure. (b) What are the coordination numbers of the cations and anions in each of these structures?

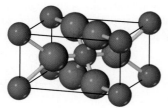

Rutile

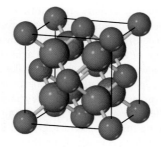

Fluorite

12.61 The coordination number for Mg^{2+} ion is usually six. Assuming this assumption holds, determine the anion coordination number in the following compounds: (a) MgS, (b) MgF_2, (c) MgO.

12.62 The coordination number for the Al^{3+} ion is typically between four and six. Use the anion coordination number to determine the Al^{3+} coordination number in the following compounds: (a) AlF_3 where the fluoride ions are two coordinate, (b) Al_2O_3 where the oxygen ions are six coordinate, (c) AlN where the nitride ions are four coordinate.

12.63 Classify each of the following statements as true or false:

(a) Although both molecular solids and covalent-network solids have covalent bonds, the melting points of molecular solids are much lower because their covalent bonds are much weaker.

(b) Other factors being equal, highly symmetric molecules tend to form solids with higher melting points than asymmetrically shaped molecules.

12.64 Classify each of the following statements as true or false:

(a) For molecular solids, the melting point generally increases as the strengths of the covalent bonds increase.

(b) For molecular solids, the melting point generally increases as the strengths of the intermolecular forces increase.

Covalent-Network Solids (Section 12.7)

12.65 Both covalent-network solids and ionic solids can have melting points well in excess of room temperature, and both can be poor conductors of electricity in their pure form. However, in other ways their properties are quite different.

(a) Which type of solid is more likely to dissolve in water?

(b) Which type of solid can become an electrical conductor via chemical substitution?

12.66 Which of the following properties are typical characteristics of a covalent-network solid, a metallic solid, or both: (a) ductility, (b) hardness, (c) high melting point?

12.67 For each of the following pairs of semiconductors, which one will have the larger band gap: (a) CdS or CdTe, (b) GaN or InP, (c) GaAs or InAs?

12.68 For each of the following pairs of semiconductors, which one will have the larger band gap: (a) InP or InAs, (b) Ge or AlP, (c) AgI or CdTe?

12.69 If you want to dope GaAs to make an n-type semiconductor with an element to replace Ga, which element(s) would you pick?

12.70 If you want to dope GaAs to make a p-type semiconductor with an element to replace As, which element(s) would you pick?

12.71 Silicon has a band gap of 1.1 eV at room temperature. (a) What wavelength of light would a photon of this energy correspond to? (b) Draw a vertical line at this wavelength in the figure shown, which shows the light output of the Sun as a function of wavelength. Does silicon absorb all, none, or a portion of the visible light that comes from the Sun? (c) You can estimate the portion of the overall solar spectrum that silicon absorbs by considering the area under the curve. If you call the area under the entire curve "100%," what approximate percentage of the area under the curve is absorbed by silicon?

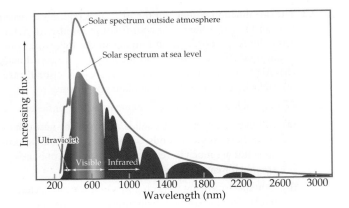

12.72 Cadmium telluride is an important material for solar cells. **(a)** What is the band gap of CdTe? **(b)** What wavelength of light would a photon of this energy correspond to? **(c)** Draw a vertical line at this wavelength in the figure shown in Exercise 12.71, which shows the light output of the sun as a function of wavelength. **(d)** With respect to silicon, does CdTe absorb a larger or smaller portion of the solar spectrum?

12.73 The semiconductor GaP has a band gap of 2.26 eV. What wavelength of light would be emitted from an LED made from GaP? What color is this?

12.74 The first LEDs were made from GaAs, which has a band gap of 1.43 eV. What wavelength of light would be emitted from an LED made from GaAs? What region of the electromagnetic spectrum does this light correspond to: ultraviolet, visible, or infrared?

12.75 GaAs and GaP make solid solutions that have the same crystal structure as the parent materials, with As and P randomly distributed throughout the crystal. GaP_xAs_{1-x} exists for any value of x. If we assume that the band gap varies linearly with composition between $x = 0$ and $x = 1$, estimate the band gap for $GaP_{0.5}As_{0.5}$. (See the previous two exercises for GaAs and GaP band gaps.) What wavelength of light does this correspond to?

12.76 Red light-emitting diodes are made from GaAs and GaP solid solutions, GaP_xAs_{1-x} (see Exercise 12.75). The original red LEDs emitted light with a wavelength of 660 nm. If we assume that the band gap varies linearly with composition between $x = 0$ and $x = 1$, estimate the composition (the value of x) that is used in these LEDs.

Polymers (Section 12.8)

12.77 **(a)** What is a monomer? **(b)** Which of these molecules can be used as a monomer: benzene, ethene (also called ethylene), methane?

12.78 The molecular formula of n-decane is $CH_3(CH_2)_8CH_3$. Decane is not considered a polymer, whereas polyethylene is. What is the distinction?

12.79 State whether each of these numbers is a reasonable value for a polymer's molecular weight: 100 amu, 10,000 amu, 100,000 amu, 1,000,000 amu?

12.80 Indicate whether the following statement is true or false: For an addition polymerization, there are no by-products of the reaction (assuming 100% yield).

12.81 An ester is a compound formed by a condensation reaction between a carboxylic acid and an alcohol that eliminates a water molecule. Read the discussion of esters in Section 24.4

and then give an example of a reaction forming an ester. How might this kind of reaction be extended to form a polymer (a polyester)?

12.82 Write a balanced chemical equation for the formation of a polymer via a condensation reaction from the monomers succinic acid ($HOOCCH_2CH_2COOH$) and ethylenediamine ($H_2NCH_2CH_2NH_2$).

12.83 Draw the structure of the monomer(s) that are used to make each of the polymers shown in Table 12.5: **(a)** polyvinyl chloride, **(b)** nylon 6,6, **(c)** polyethylene terephthalate.

12.84 Write the chemical equation that represents the formation of **(a)** polychloroprene from chloroprene (polychloroprene is used in highway-pavement seals, expansion joints, conveyor belts, and wire and cable jackets)

$$CH_2=CH-\underset{\underset{Cl}{|}}{C}=CH_2$$

Chloroprene

(b) polyacrylonitrile from acrylonitrile (polyacrylonitrile is used in home furnishings, craft yarns, clothing, and many other items).

$$CH_2=\underset{\underset{CN}{|}}{CH}$$

Acrylonitrile

12.85 The nylon Nomex, a condensation polymer, has the following structure:

$$\left[\begin{array}{c}\overset{O}{\overset{||}{C}}\cdots\overset{O}{\overset{||}{C}}-NH-\cdots-NH\end{array}\right]_n$$

Draw the structures of the two monomers that yield Nomex.

12.86 Proteins are naturally occurring polymers formed by condensation reactions of amino acids, which have the general structure

$$H-\overset{\overset{H}{|}}{N}-\overset{\overset{R}{|}}{\underset{\underset{H}{|}}{C}}-\overset{\overset{O}{||}}{C}-O-H$$

In this structure, $-R$ represents $-H$, $-CH_3$, or another group of atoms; there are 20 different natural amino acids, and each has one of 20 different R groups. **(a)** Draw the general structure of a protein formed by condensation polymerization of the generic amino acid shown here. **(b)** When only a few amino acids react to make a chain, the product is called a "peptide" rather than a protein; only when there are 50 amino acids or more in the chain would the molecule be called a protein. For three amino acids (distinguished by having three different R groups, R1, R2, and R3), draw the peptide that results from their condensation reactions. **(c)** The order in which the R groups exist in a peptide or protein has a huge influence on its biological activity. To distinguish different peptides and proteins, chemists call the first amino acid the one at the "N terminus" and the last one the one at the "C terminus." From your drawing in part **(b)** you should be able to figure out what "N terminus" and "C terminus" mean. How many different peptides can be made from your three different amino acids?

12.87 (a) What molecular features make a polymer flexible? (b) If you cross-link a polymer, is it more flexible or less flexible than it was before?

12.88 What molecular structural features cause high-density polyethylene to be denser than low-density polyethylene?

12.89 If you want to make a polymer for plastic wrap, should you strive to make a polymer that has a high or low degree of crystallinity?

12.90 Indicate whether each statement is true or false:

(a) Elastomers are rubbery solids.

(b) Thermosets cannot be reshaped.

(c) Thermoplastic polymers can be recycled.

Nanomaterials (Section 12.9)

12.91 Explain why "bands" may not be the most accurate description of bonding in a solid when the solid has nanoscale dimensions.

12.92 CdS has a band gap of 2.4 eV. If large crystals of CdS are illuminated with ultraviolet light, they emit light equal to the band gap energy. (a) What color is the emitted light? (b) Would appropriately sized CdS quantum dots be able to emit blue light? (c) What about red light?

12.93 Indicate whether each statement is true or false:

(a) The band gap of a semiconductor decreases as the particle size decreases in the 1–10-nm range.

(b) The light that is emitted from a semiconductor, upon external stimulation, becomes longer in wavelength as the particle size of the semiconductor decreases.

12.94 Indicate whether this statement is true or false:

If you want a semiconductor that emits blue light, you could either use a material that has a band gap corresponding to the energy of a blue photon or you could use a material that has a smaller band gap but make an appropriately sized nanoparticle of the same material.

12.95 Gold adopts a face-centered cubic structure with a unit cell edge of 4.08 Å (Figure 12.11). How many gold atoms are there in a sphere that is 20 nm in diameter? Recall that the volume of a sphere is $\frac{4}{3}\pi r^3$.

12.96 Cadmium telluride, CdTe, adopts the zinc blende structure with a unit cell edge length of 6.49 Å. There are four cadmium atoms and four tellurium atoms per unit cell. How many of each type of atom are there in a cubic crystal with an edge length of 5.00 nm?

12.97 Which statement correctly describes a difference between graphene and graphite?

(a) Graphene is a molecule but graphite is not. (b) Graphene is a single sheet of carbon atoms and graphite contains many, and larger, sheets of carbon atoms. (c) Graphene is an insulator but graphite is a metal. (d) Graphite is pure carbon but graphene is not. (e) The carbons are sp^2 hybridized in graphene but sp^3 hybridized in graphite.

12.98 What evidence supports the notion that buckyballs are actual molecules and not extended materials?

(a) Buckyballs are made of carbon.

(b) Buckyballs have a well-defined atomic structure and molecular weight.

(c) Buckyballs have a well-defined melting point.

(d) Buckyballs are semiconductors.

(e) More than one of the previous choices.

Additional Exercises

[12.99] A face-centered tetragonal lattice is not one of the 14 three-dimensional lattices. Show that a face-centered tetragonal unit cell can be redefined as a body-centered tetragonal lattice with a smaller unit cell.

12.100 Pure iron crystallizes in a body-centered cubic structure, but small amounts of impurities can stabilize a face-centered cubic structure. Which form of iron has a higher density?

[12.101] Introduction of carbon into a metallic lattice generally results in a harder, less ductile substance with lower electrical and thermal conductivities. Explain why this might be so.

12.102 Ni_3Al is used in the turbines of aircraft engines because of its strength and low density. Nickel metal has a cubic close-packed structure with a face-centered cubic unit cell, while Ni_3Al has the ordered cubic structure shown in Figure 12.17. The length of the cubic unit cell edge is 3.53 Å for nickel and 3.56 Å for Ni_3Al. Use these data to calculate and compare the densities of these two materials.

12.103 For each of the intermetallic compounds shown in Figure 12.17 determine the number of each type of atom in the unit cell. Do your answers correspond to the ratios expected from the empirical formulas: Ni_3Al, Nb_3Sn, and $SmCo_5$?

12.104 What type of lattice—primitive cubic, body-centered cubic, or face-centered cubic—does each of the following structure types possess: (a) CsCl, (b) Au, (c) NaCl, (d) Po, (e) ZnS?

12.105 Tin exists in two allotropic forms: Gray tin has the diamond structure, and white tin has a close-packed structure. One of these allotropic forms is a semiconductor with a small band gap, while the other is a metal. (a) Which one is which? (b) Which form would you expect to have the longer Sn—Sn bond distance?

[12.106] The electrical conductivity of aluminum is approximately 10^9 times greater than that of its neighbor in the periodic table, silicon. Aluminum has a face-centered cubic structure, and silicon has the diamond structure. A classmate of yours tells you that density is the reason aluminum is a metal but silicon is not; therefore, if you were to put silicon under high pressure, it too would act like a metal. Discuss this idea with your classmates, looking up data about Al and Si as needed.

12.107 Silicon carbide, SiC, has the three-dimensional structure shown in the figure.

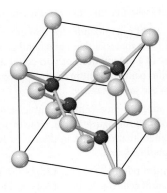

(a) Name another compound that has the same structure. (b) Would you expect the bonding in SiC to be predominantly ionic, metallic, or covalent? (c) How do the bonding and structure of SiC lead to its high thermal stability (to 2700 °C) and exceptional hardness?

[12.108] Unlike metals, semiconductors increase their conductivity as you heat them (up to a point). Suggest an explanation.

12.109 Rhenium oxide crystallizes with a structure that has a primitive cubic lattice, as shown here. In the image on the left, the sizes of the ions have been reduced to show the entire unit cell. (a) How many atoms of each type are there per unit cell? (b) Use the ionic radii of rhenium (0.70 Å) and oxygen (1.26 Å) to estimate the length of the edge of the unit cell. (c) Use your answers to parts (a) and (b) to estimate the density of this compound.

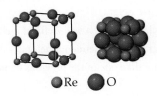

● Re   ● O

12.110 Teflon is a polymer formed by the polymerization of $F_2C = CF_2$. (a) Draw the structure of a section of this polymer.

(b) What type of polymerization reaction is required to form Teflon?

12.111 Hydrogen bonding between polyamide chains plays an important role in determining the properties of a nylon such as nylon 6,6 (Table 12.5). Draw the structural formulas for two adjacent chains of nylon 6,6 and show where hydrogen-bonding interactions could occur between them.

12.112 Explain why X rays can be used to measure atomic distances in crystals but visible light cannot be used for this purpose.

12.113 In their study of X-ray diffraction, William and Lawrence Bragg determined that the relationship among the wavelength of the radiation (λ), the angle at which the radiation is diffracted (θ), and the distance between planes of atoms in the crystal that cause the diffraction (d) is given by $n\lambda = 2d \sin \theta$. X rays from a copper X-ray tube that have a wavelength of 1.54 Å are diffracted at an angle of 14.22 degrees by crystalline silicon. Using the Bragg equation, calculate the distance between the planes of atoms responsible for diffraction in this crystal, assuming $n = 1$ (first-order diffraction).

12.114 Germanium has the same structure as silicon, but the unit cell size is different because Ge and Si atoms are not the same size. If you were to repeat the experiment described in the previous problem but replace the Si crystal with a Ge crystal, would you expect the X rays to be diffracted at a larger or smaller angle θ?

[12.115] (a) The density of diamond is 3.5 g/cm^3, and that of graphite is 2.3 g/cm^3. Based on the structure of buckminsterfullerene, what would you expect its density to be relative to these other forms of carbon? (b) X-ray diffraction studies of buckminsterfullerene show that it has a face-centered cubic lattice of C$_{60}$ molecules. The length of an edge of the unit cell is 14.2 Å. Calculate the density of buckminsterfullerene.

12.116 When you shine light of band gap energy or higher on a semiconductor and promote electrons from the valence band to the conduction band, do you expect the conductivity of the semiconductor to (a) remain unchanged, (b) increase, or (c) decrease?

Integrative Exercises

12.117 The karat scale used to describe gold alloys is based on mass percentages. (a) If an alloy is formed that is 50 mol% silver and 50 mol% gold, what is the karat number of the alloy? Use Figure 12.18 to estimate the color of this alloy. (b) If an alloy is formed that is 50 mol% copper and 50 mol% gold, what is the karat number of the alloy? What is the color of this alloy?

12.118 Spinel is a mineral that contains 37.9% Al, 17.1% Mg, and 45.0% O, by mass, and has a density of 3.57 g/cm^3. The unit cell is cubic with an edge length of 8.09 Å. How many atoms of each type are in the unit cell?

12.119 (a) What are the C—C—C bond angles in diamond? (b) What are they in graphite (in one sheet)? (c) What atomic orbitals are involved in the stacking of graphite sheets with each other?

[12.120] Employing the bond enthalpy values listed in Table 8.4, estimate the molar enthalpy change occurring upon (a) polymerization of ethylene, (b) formation of nylon 6,6, (c) formation of polyethylene terephthalate (PET).

[12.121] Although polyethylene can twist and turn in random ways, the most stable form is a linear one with the carbon backbone oriented as shown in the following figure:

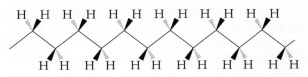

The solid wedges in the figure indicate bonds from carbon that come out of the plane of the page; the dashed wedges indicate bonds that lie behind the plane of the page.

(a) What is the hybridization of orbitals at each carbon atom? What angles do you expect between the bonds?

(b) Now imagine that the polymer is polypropylene rather than polyethylene. Draw structures for polypropylene in which (i) the CH$_3$ groups all lie on the same side of the plane of the paper (this form is called isotactic polypropylene), (ii) the CH$_3$ groups lie on alternating sides of the plane (syndiotactic polypropylene), or (iii) the CH$_3$ groups are randomly distributed on either side (atactic polypropylene). Which of these forms would you expect to have the highest crystallinity and melting point, and which the lowest? Explain in terms of intermolecular interactions and molecular shapes.

(c) Polypropylene fibers have been employed in athletic wear. The product is said to be superior to cotton or polyester clothing in wicking moisture away from the body through the fabric to the outside. Explain the difference between polypropylene and polyester or cotton (which has many —OH groups along the molecular chain) in terms of intermolecular interactions with water.

12.122 (a) In polyvinyl chloride shown in Table 12.5, which bonds have the lowest average bond enthalpy? (b) When subjected to high pressure and heated, polyvinyl chloride converts to diamond. During this transformation which bonds are most likely to break first? (c) Employing the values of average bond enthalpy in Table 8.4, estimate the overall enthalpy change for converting PVC to diamond.

[12.123] Silicon has the diamond structure with a unit cell edge length of 5.43 Å and eight atoms per unit cell. (a) How many silicon atoms are there in 1 cm^3 of material? (b) Suppose you dope that 1 cm^3 sample of silicon with 1 ppm of phosphorus that will increase the conductivity by a factor of a million. How many milligrams of phosphorus are required?

12.124 KCl has the same structure as NaCl. The length of the unit cell is 6.28 Å. The density of KCl is 1.984 g/cm^3, and its formula mass is 74.55 amu. Using this information, calculate Avogadro's number.

12.125 Look up the diameter of a silicon atom, in Å. The latest semiconductor chips have fabricated lines as small as 22 nm. How many silicon atoms does this correspond to?

Design an Experiment

Polymers were commercially made by the DuPont™ Company starting in the late 1920s. At that time, some chemists still could not believe that polymers were molecules; they thought covalent bonding would not "last" for millions of atoms, and that polymers were really clumps of molecules held together by weak intermolecular forces. Design an experiment to demonstrate that polymers really are large molecules and not little clumps of small molecules that are held together by weak intermolecular forces.

13

Properties of Solutions

In Chapters 10, 11, and 12, we explored the properties of pure gases, liquids, and solids. However, the matter that we encounter in our daily lives, such as soda, air, and glass, are frequently mixtures. In this chapter, we examine homogeneous mixtures.

As we noted in the earlier chapters, homogeneous mixtures are called *solutions*. ∞ (Sections 1.2 and 4.1)

When we think of solutions, we usually think of liquids, like those in this chapter's opening photograph. Solutions, however, can also be solids or gases. For example, sterling silver is a homogeneous mixture of about 7% copper in silver and so is a solid solution. The air we breathe is a homogeneous mixture of several gases, making air a gaseous solution. However, because liquid solutions are the most common, we focus our attention on them in this chapter.

Each substance in a solution is a *component* of the solution. As we saw in Chapter 4, the *solvent* is normally the component present in the greatest amount, and all the other components are called *solutes*. In this chapter, we compare the physical properties of solutions with the properties of the components in their pure form. We will be particularly concerned with aqueous solutions, which contain water as the solvent and either a gas, liquid, or solid as a solute.

13.1 | The Solution Process

A solution is formed when one substance disperses uniformly throughout another. The ability of substances to form solutions depends on two factors: (1) the natural tendency of substances to mix and spread into larger volumes when not restrained in some way and (2) the types of intermolecular interactions involved in the solution process.

▶ **A COLORED DYE DISSOLVES IN WATER.** The processes by which molecules mix with and spread into water are important for many events, including drug dissolution in your bloodstream and for nutrient cycling in the ocean.

WHAT'S AHEAD ▶

13.1 THE SOLUTION PROCESS We begin by considering what happens at the molecular level when one substance dissolves into another, paying particular attention to the role of intermolecular forces. Two important aspects of the solution process are the natural tendency of particles to mix and their accompanying changes in energy.

13.2 SATURATED SOLUTIONS AND SOLUBILITY We learn that when a *saturated solution* is in contact with an undissolved solute, the dissolved and undissolved solutes are in *equilibrium*. The amount of solute in a saturated solution defines the solubility of the solute, the extent to which a particular solute dissolves in a particular solvent.

13.3 FACTORS AFFECTING SOLUBILITY We next consider the major factors affecting solubility. The nature of the solute and solvent determines the kinds of intermolecular forces among solute and solvent particles and strongly influences solubility. Temperature also affects solubility: Most solids are more soluble in water at higher temperatures, whereas gases are less soluble in water at higher temperatures. The solubility of gases increases with increasing pressure.

13.4 EXPRESSING SOLUTION CONCENTRATION We examine several common ways of expressing concentration, including *mole fraction*, *molarity*, and *molality*.

13.5 COLLIGATIVE PROPERTIES We observe that some physical properties of solutions depend only on concentration and not on the identity of the solute. These *colligative properties* include the extent to which the solute lowers the vapor pressure, increases the boiling point, and decreases the freezing point of the solvent. The *osmotic pressure* of a solution is also a colligative property.

13.6 COLLOIDS We close the chapter by investigating *colloids*, mixtures that are not true solutions but consist of a solute-like phase (the dispersed phase) and a solvent-like phase (the dispersion medium). The dispersed phase consists of particles larger than typical molecular sizes.

The Natural Tendency toward Mixing

Suppose we have $O_2(g)$ and $Ar(g)$ separated by a barrier, as in ▼ Figure 13.1. If the barrier is removed, the gases mix to form a solution. The molecules experience very little in the way of intermolecular interactions and behave like ideal gas particles. As a result, their molecular motion causes them to spread through the larger volume, and a gaseous solution is formed.

The mixing of gases is a *spontaneous* process, meaning it occurs of its own accord without any input of energy from outside the system. When the molecules mix and become more randomly distributed, there is an increase in a thermodynamic quantity called *entropy*. We will examine spontaneous processes and entropy in more depth in Chapter 19. For now, it is sufficient to recognize that this mixing leads to an increase in the entropy of the system. Furthermore, the balance between increasing the entropy and decreasing the enthalpy of a system is what determines whether a process is spontaneous. Thus, the *formation of solutions is favored by the increase in entropy that accompanies mixing.*

When molecules of different types are brought together, mixing occurs spontaneously unless the molecules are restrained either by sufficiently strong intermolecular forces or by physical barriers. Thus, gases spontaneously mix unless restrained by their containers because with gases intermolecular forces are too weak to restrain the molecules. However, when the solvent or solute is a solid or liquid, intermolecular forces become important in determining whether a solution forms. For example, although ionic bonds hold sodium and chloride ions together in solid sodium chloride (Section 8.2), the solid dissolves in water because of the compensating strength of the attractive forces between the ions and water molecules. Sodium chloride does not dissolve in gasoline, however, because the intermolecular forces between the ions and the gasoline molecules are too weak.

▲ Give It Some Thought

In the chapter-opening photograph of dye dispersing in water, is entropy increasing or decreasing?

The Effect of Intermolecular Forces on Solution Formation

Any of the intermolecular forces discussed in Chapter 11 can operate between solute and solvent particles in a solution. These forces are summarized in ▶ Figure 13.2. Dispersion forces, for example, dominate when one nonpolar substance, such as C_7H_{16}, dissolves in another, such as C_5H_{12}, and ion–dipole forces dominate in solutions of ionic substances in water.

GO FIGURE

What aspect of the kinetic theory of gases tells us that the gases will mix?

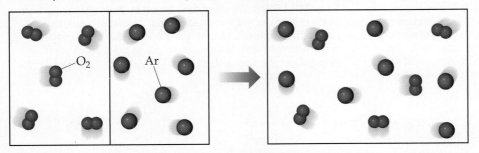

▲ Figure 13.1 **Spontaneous mixing of two gases to form a homogeneous mixture (a solution).**

▲ GO FIGURE

Why does the oxygen atom in H_2O point toward Na^+ in the ion–dipole interaction?

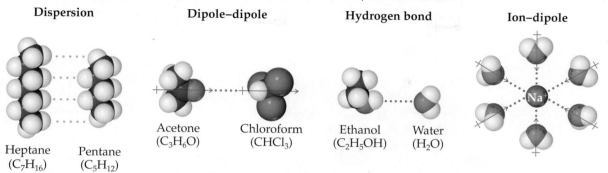

Dispersion   **Dipole–dipole**   **Hydrogen bond**   **Ion–dipole**

Heptane
(C_7H_{16})   Pentane
(C_5H_{12})

Acetone
(C_3H_6O)   Chloroform
($CHCl_3$)

Ethanol
(C_2H_5OH)   Water
(H_2O)

▲ **Figure 13.2 Intermolecular interactions involved in solutions.**

Three kinds of intermolecular interactions are involved in solution formation:

1. *Solute–solute* interactions between solute particles must be overcome to disperse the solute particles through the solvent.

2. *Solvent–solvent* interactions between solvent particles must be overcome to make room for the solute particles in the solvent.

3. *Solvent–solute* interactions between the solvent and solute particles occur as the particles mix.

The extent to which one substance is able to dissolve in another depends on the relative magnitudes of these three types of interactions. Solutions form when the magnitudes of the solvent–solute interactions are either comparable to or greater than the solute–solute and solvent–solvent interactions. For example, heptane (C_7H_{16}) and pentane (C_5H_{12}) dissolve in each other in all proportions. For this discussion, we can arbitrarily call heptane the solvent and pentane the solute. Both substances are nonpolar, and the magnitudes of the solvent–solute interactions (attractive dispersion forces) are comparable to the solute–solute and solvent–solvent interactions. Thus, no forces impede mixing, and the tendency to mix (increase entropy) causes the solution to form spontaneously.

Solid NaCl dissolves readily in water because the attractive solvent–solute interactions between the polar H_2O molecules and the ions are strong enough to overcome the attractive solute–solute interactions between ions in the NaCl(s) and the attractive solvent–solvent interactions between H_2O molecules. When NaCl is added to water (**Figure 13.3**), the water molecules orient themselves on the surface of the NaCl crystals with the positive end of the water dipole oriented toward Cl^- ions and the negative end oriented toward Na^+ ions. These ion–dipole attractions are strong enough to pull the surface ions away from the solid, thus overcoming the solute–solute interactions. For the solid to dissolve, some solvent–solvent interactions must also be overcome to create room for the ions to "fit" among all the water molecules.

Once separated from the solid, the Na^+ and Cl^- ions are surrounded by water molecules. Interactions such as this between solute and solvent molecules are known as **solvation**. When the solvent is water, the interactions are referred to as **hydration**.

 Give It Some Thought

Why doesn't NaCl dissolve in nonpolar solvents such as hexane, C_6H_{14}?

Energetics of Solution Formation

Solution processes are typically accompanied by changes in enthalpy. For example, when NaCl dissolves in water, the process is slightly endothermic, $\Delta H_{soln} = 3.9$ kJ/mol. We can use Hess's law to analyze how the solute–solute, solvent–solvent, and solute–solvent interactions influence the enthalpy of solution. ∞ (Section 5.6)

GO FIGURE

How does the orientation of H_2O molecules around Na^+ differ from that around Cl^-?

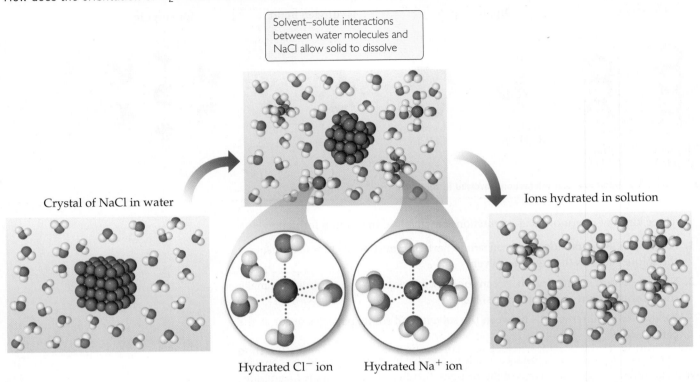

Crystal of NaCl in water

Solvent–solute interactions between water molecules and NaCl allow solid to dissolve

Ions hydrated in solution

Hydrated Cl^- ion   Hydrated Na^+ ion

▲ Figure 13.3 Dissolution of the ionic solid NaCl in water.

We can imagine the solution process as having three components, each with an associated enthalpy change: A cluster of n solute particles must separate from one another (ΔH_{solute}), a cluster of m solvent particles separate from one another ($\Delta H_{solvent}$), and these solute and solvent particles mix (ΔH_{mix}).

1. $(solute)_n \rightleftharpoons n$ solute ΔH_{solute}
2. $(solvent)_m \rightleftharpoons m$ solvent $\Delta H_{solvent}$
3. n solute $+ m$ solvent $\rightleftharpoons$ solution ΔH_{mix}

4. $(solute)_n + (solvent)_m \rightleftharpoons$ solution $\Delta H_{soln} = \Delta H_{solute} + \Delta H_{solvent} + \Delta H_{mix}$

As seen above, the overall enthalpy change, ΔH_{soln}, is the sum of the three steps:

$$\Delta H_{soln} = \Delta H_{solute} + \Delta H_{solvent} + \Delta H_{mix} \qquad [13.1]$$

Separation of the solute particles from one another always requires an input of energy to overcome their attractive interactions. The process is therefore endothermic ($\Delta H_{solute} > 0$). Likewise, separation of solvent molecules to accommodate the solute always requires energy ($\Delta H_{solvent} > 0$). The third component, which arises from the attractive interactions between solute particles and solvent particles, is always exothermic ($\Delta H_{mix} < 0$).

The three enthalpy terms in Equation 13.1 can be added together to give either a negative or a positive sum, depending on the actual numbers for the system being considered (▶ Figure 13.4). Thus, the formation of a solution can be either exothermic or endothermic. For example, when magnesium sulfate ($MgSO_4$) is added to water, the solution process is exothermic: $\Delta H_{soln} = -91.2$ kJ/mol. In contrast, the dissolution of ammonium nitrate (NH_4NO_3) is endothermic: $\Delta H_{soln} = 26.4$ kJ/mol. These particular salts are the main components in the instant heat packs and ice packs used to treat

GO FIGURE

How does the magnitude of ΔH_{mix} compare with the magnitude of $\Delta H_{solvent} + \Delta H_{solute}$ for exothermic solution processes?

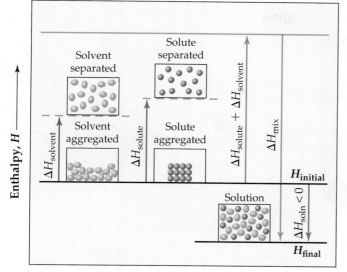

▲ **Figure 13.4 Enthalpy changes accompanying the solution process.**

athletic injuries (▶ **Figure 13.5**). The packs consist of a pouch of water and the solid salt sealed off from the water—$MgSO_4(s)$ for hot packs and $NH_4NO_3(s)$ for cold packs. When the pack is squeezed, the seal separating the solid from the water is broken and a solution forms, either increasing or decreasing the temperature.

The enthalpy change for a process can provide insight into the extent to which the process occurs. ∞ (Section 5.4) Exothermic processes tend to proceed spontaneously. On the other hand, if ΔH_{soln} is too endothermic, the solute might not dissolve to any significant extent in the chosen solvent. Thus, for solutions to form, the solvent–solute interaction must be strong enough to make ΔH_{mix} comparable in magnitude to $\Delta H_{solute} + \Delta H_{solvent}$. This fact further explains why ionic solutes do not dissolve in nonpolar solvents. The nonpolar solvent molecules experience only weak attractive interactions with the ions, and these interactions do not compensate for the energies required to separate the ions from one another.

By similar reasoning, a polar liquid solute, such as water, does not dissolve in a nonpolar liquid solvent, such as octane (C_8H_{18}). The water molecules experience strong hydrogen-bonding interactions with one another ∞ (Section 11.2)—attractive forces that must be overcome if the water molecules are to be dispersed throughout the octane solvent. The energy required to separate the H_2O molecules from one another is not recovered in the form of attractive interactions between the H_2O and C_8H_{18} molecules.

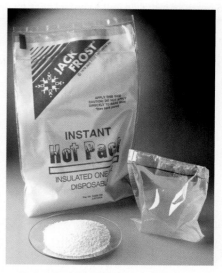

▲ **Figure 13.5 Magnesium sulfate instant hot pack.**

 Give It Some Thought

Label the following processes as exothermic or endothermic:
(a) breaking solvent–solvent interactions to form separated particles
(b) forming solvent–solute interactions from separated particles

Solution Formation and Chemical Reactions

In discussing solutions, we must be careful to distinguish the physical process of solution formation from chemical reactions that lead to a solution. For example, nickel metal dissolves on contact with an aqueous hydrochloric acid solution because the following reaction occurs:

$$Ni(s) + 2\,HCl(aq) \longrightarrow NiCl_2(aq) + H_2(g) \qquad [13.2]$$

GO FIGURE

What is the molar mass of nickel chloride hexahydrate, $NiCl_2 \cdot 6\,H_2O(s)$?

| Nickel metal and hydrochloric acid | Nickel reacts with hydrochloric acid, forming $NiCl_2(aq)$ and $H_2(g)$. The solution is of $NiCl_2$, not Ni metal | $NiCl_2 \cdot 6\,H_2O(s)$ remains when solvent evaporated |

▲ **Figure 13.6 The reaction between nickel metal and hydrochloric acid is *not* a simple dissolution.** The product is $NiCl_2 \cdot 6\,H_2O(s)$, nickel(II) chloride hexahydrate, which has exactly 6 waters of hydration in the crystal lattice for every nickel ion.

In this instance, one of the resulting solutes is not Ni metal but rather its salt $NiCl_2$. If the solution is evaporated to dryness, $NiCl_2 \cdot 6\,H_2O(s)$ is recovered (▲ **Figure 13.6**). Compounds such as $NiCl_2 \cdot 6\,H_2O(s)$ that have a defined number of water molecules in the crystal lattice are known as *hydrates*. When $NaCl(s)$ is dissolved in water, on the other hand, no chemical reaction occurs. If the solution is evaporated to dryness, NaCl is recovered. Our focus throughout this chapter is on solutions from which the solute can be recovered unchanged from the solution.

13.2 | Saturated Solutions and Solubility

As a solid solute begins to dissolve in a solvent, the concentration of solute particles in solution increases, increasing the chances that some solute particles will collide with the surface of the solid and reattach. This process, which is the opposite of the solution process, is called **crystallization**. Thus, two opposing processes occur in a solution in contact with undissolved solute. This situation is represented in this chemical equation:

$$\text{Solute} + \text{solvent} \underset{\text{crystallize}}{\overset{\text{dissolve}}{\rightleftharpoons}} \text{solution} \qquad \qquad [13.3]$$

When the rates of these opposing processes become equal, a *dynamic equilibrium* is established, and there is no further increase in the amount of solute in solution. ∞ (Section 4.1)

A solution that is in equilibrium with undissolved solute is **saturated**. Additional solute will not dissolve if added to a saturated solution. The amount of solute needed to form a saturated solution in a given quantity of solvent is known as the **solubility** of that solute. That is, *the solubility of a particular solute in a particular solvent is the maximum amount of the solute that can dissolve in a given amount of the solvent at a specified temperature, assuming that excess solute is present.* For example, the solubility of NaCl in water at 0 °C is 35.7 g per 100 mL of water. This is the maximum amount of NaCl that can be dissolved in water to give a stable equilibrium solution at that temperature.

If we dissolve less solute than the amount needed to form a saturated solution, the solution is **unsaturated**. Thus, a solution containing 10.0 g of NaCl per 100 mL of water at 0 °C is unsaturated because it has the capacity to dissolve more solute.

Under suitable conditions, it is possible to form solutions that contain a greater amount of solute than needed to form a saturated solution. Such solutions are **super-saturated**. For example, when a saturated solution of sodium acetate is made at a high temperature and then slowly cooled, all of the solute may remain dissolved even though its solubility decreases as the temperature decreases. Because the solute in a supersaturated solution is present in a concentration higher than the equilibrium concentration, supersaturated solutions are unstable. For crystallization to occur, however, the solute particles must arrange themselves properly to form crystals. The addition of a small crystal of the solute (a seed crystal) provides a template for crystallization of the excess solute, leading to a saturated solution in contact with excess solid (▼ Figure 13.7).

 Give It Some Thought

What happens if a solute is added to a saturated solution?

GO FIGURE

What is the evidence that the solution in the left photograph is supersaturated?

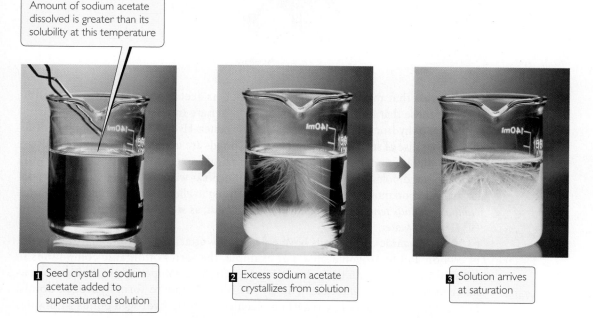

Amount of sodium acetate dissolved is greater than its solubility at this temperature

1 Seed crystal of sodium acetate added to supersaturated solution

2 Excess sodium acetate crystallizes from solution

3 Solution arrives at saturation

▲ **Figure 13.7 Precipitation from a supersaturated sodium acetate solution.** The solution on the left was formed by dissolving about 170 g of the salt in 100 mL of water at 100 °C and then slowly cooling it to 20 °C. Because the solubility of sodium acetate in water at 20 °C is 46 g per 100 mL of water, the solution is supersaturated. Addition of a sodium acetate crystal causes the excess solute to crystallize from solution.

Table 13.1 Solubilities of Gases in Water at 20 °C, with 1 atm Gas Pressure

Gas	Molar Mass (g/mol)	Solubility (M)
N_2	28.0	0.69×10^{-3}
O_2	32.0	1.38×10^{-3}
Ar	39.9	1.50×10^{-3}
Kr	83.8	2.79×10^{-3}

13.3 | Factors Affecting Solubility

The extent to which one substance dissolves in another depends on the nature of both substances. ⚬⚬ (Section 13.1) It also depends on temperature and, at least for gases, on pressure.

Solute–Solvent Interactions

The natural tendency of substances to mix and the various interactions among solute and solvent particles are all involved in determining solubilities. Nevertheless, insight into variations in solubility can often be gained by focusing on the interaction between the solute and solvent. The data in ◀ Table 13.1 show that the solubilities of various gases in water increase with increasing molecular mass. The attractive forces between the gas molecules and solvent molecules are mainly dispersion forces, which increase with increasing size and molecular mass. ⚬⚬ (Section 11.2) Thus, the data indicate that the solubilities of gases in water increase as the attraction between solute (gas) and solvent (water) increases. In general, when other factors are comparable, *the stronger the attractions between solute and solvent molecules, the greater the solubility of the solute in that solvent.*

Because of favorable dipole–dipole attractions between solvent molecules and solute molecules, *polar liquids tend to dissolve in polar solvents.* Water is both polar and able to form hydrogen bonds. ⚬⚬ (Section 11.2) Thus, polar molecules, especially those that can form hydrogen bonds with water molecules, tend to be soluble in water. For example, acetone, a polar molecule with the structural formula shown below, mixes in all proportions with water. Acetone has a strongly polar C=O bond and pairs of nonbonding electrons on the O atom that can form hydrogen bonds with water.

$$:\overset{\displaystyle :O:}{\underset{}{\parallel}}$$
$$CH_3CCH_3$$

Acetone

Liquids that mix in all proportions, such as acetone and water, are **miscible**, whereas those that do not dissolve in one another are **immiscible**. Gasoline, which is a mixture of hydrocarbons, is immiscible with water. Hydrocarbons are nonpolar substances because of several factors: The C—C bonds are nonpolar, the C—H bonds are nearly nonpolar, and the molecules are symmetrical enough to cancel much of the weak C—H bond dipoles. The attraction between the polar water molecules and the nonpolar hydrocarbon molecules is not sufficiently strong to allow the formation of a solution. *Nonpolar liquids tend to be insoluble in polar liquids,* as ◀ Figure 13.8 shows for hexane (C_6H_{14}) and water.

Many organic compounds have polar groups attached to a nonpolar framework of carbon and hydrogen atoms. For example, the series of organic compounds in ▶ Table 13.2 all contain the polar OH group. Organic compounds with this molecular feature are called *alcohols.* The O—H bond is able to form hydrogen bonds. For example, ethanol (CH_3CH_2OH) molecules can form hydrogen bonds with water molecules as well as with each other (▶ Figure 13.9). As a result, the solute–solute, solvent–solvent, and solute–solvent forces are not greatly different in a mixture of CH_3CH_2OH and H_2O. No major change occurs in the environments of the molecules as they are mixed. Therefore, the increased entropy when the components mix plays a significant role in solution formation, and ethanol is completely miscible with water.

Hexane

Water

▲ Figure 13.8 **Hexane is immiscible with water.** Hexane is the top layer because it is less dense than water.

Table 13.2 **Solubilities of Some Alcohols in Water and in Hexane***

Alcohol	Solubility in H_2O	Solubility in C_6H_{14}
CH_3OH (methanol)	∞	0.12
CH_3CH_2OH (ethanol)	∞	∞
$CH_3CH_2CH_2OH$ (propanol)	∞	∞
$CH_3CH_2CH_2CH_2OH$ (butanol)	0.11	∞
$CH_3CH_2CH_2CH_2CH_2OH$ (pentanol)	0.030	∞
$CH_3CH_2CH_2CH_2CH_2CH_2OH$ (hexanol)	0.0058	∞

*Expressed in mol alcohol/100 g solvent at 20 °C. The infinity symbol (∞) indicates that the alcohol is completely miscible with the solvent.

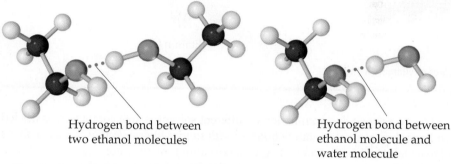

Hydrogen bond between two ethanol molecules

Hydrogen bond between ethanol molecule and water molecule

▲ Figure 13.9 **Hydrogen bonding involving OH groups.**

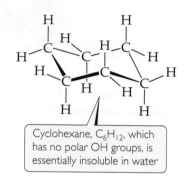

Cyclohexane, C_6H_{12}, which has no polar OH groups, is essentially insoluble in water

Notice in Table 13.2 that the number of carbon atoms in an alcohol affects its solubility in water. As this number increases, the polar OH group becomes an even smaller part of the molecule, and the molecule behaves more like a hydrocarbon. The solubility of the alcohol in water decreases correspondingly. On the other hand, the solubility of alcohols in a nonpolar solvent like hexane (C_6H_{14}) increases as the nonpolar hydrocarbon chain lengthens.

One way to enhance the solubility of a substance in water is to increase the number of polar groups the substance contains. For example, increasing the number of OH groups in a solute increases the extent of hydrogen bonding between that solute and water, thereby increasing solubility. Glucose ($C_6H_{12}O_6$, ▶ Figure 13.10) has five OH groups on a six-carbon framework, which makes the molecule very soluble in water: 830 g dissolves in 1.00 L of water at 17.5 °C. In contrast, cyclohexane (C_6H_{12}), which has a similar structure to glucose but with all of the OH groups replaced by H, is essentially insoluble in water (only 55 mg of cyclohexane can dissolve in 1.00 L of water at 25 °C).

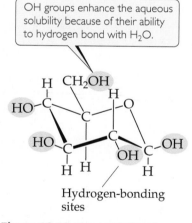

OH groups enhance the aqueous solubility because of their ability to hydrogen bond with H_2O.

Hydrogen-bonding sites

▲ Figure 13.10 **The correlation of molecular structure with solubility.**

Chemistry and Life

Fat-Soluble and Water-Soluble Vitamins

Vitamins have unique chemical structures that affect their solubilities in different parts of the human body. Vitamin C and the B vitamins are soluble in water, for example, whereas vitamins A, D, E, and K are soluble in nonpolar solvents and in fatty tissue (which is nonpolar). Because of their water solubility, vitamins B and C are not stored to any appreciable extent in the body, and so foods containing these vitamins should be included in the daily diet. In contrast, the fat-soluble vitamins are stored in sufficient quantities to keep vitamin-deficiency

diseases from appearing even after a person has subsisted for a long period on a vitamin-deficient diet.

That some vitamins are soluble in water and others are not can be explained in terms of their structures. Notice in **Figure 13.11** that vitamin A (retinol) is an alcohol with a very long carbon chain. Because the OH group is such a small part of the molecule, the molecule resembles the long-chain alcohols listed in Table 13.2. This vitamin is nearly nonpolar. In contrast, the vitamin C molecule is smaller and has several OH groups that can form hydrogen bonds with water, somewhat like glucose.

Related Exercises: 13.7, 13.48

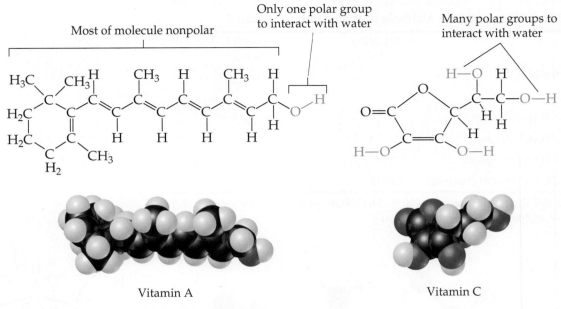

Most of molecule nonpolar

Only one polar group to interact with water

Many polar groups to interact with water

Vitamin A

Vitamin C

▲ Figure 13.11 **The molecular structures of vitamins A and C.**

Over years of study, examination of different solvent–solute combinations has led to an important generalization: *Substances with similar intermolecular attractive forces tend to be soluble in one another.* This generalization is often simply stated as *"like dissolves like."* Nonpolar substances are more likely to be soluble in nonpolar solvents; ionic and polar solutes are more likely to be soluble in polar solvents. Network solids such as diamond and quartz are not soluble in either polar or nonpolar solvents because of the strong bonding within the solid.

 Give It Some Thought

Suppose the hydrogens on the OH groups in glucose (Figure 13.10) were replaced with methyl groups, CH_3. Would you expect the water solubility of the resulting molecule to be higher than, lower than, or about the same as glucose?

SAMPLE EXERCISE 13.1 Predicting Solubility Patterns

Predict whether each of the following substances is more likely to dissolve in the nonpolar solvent carbon tetrachloride (CCl_4) or in water: C_7H_{16}, Na_2SO_4, HCl, and I_2.

SOLUTION

Analyze We are given two solvents, one that is nonpolar (CCl_4) and the other that is polar (H_2O), and asked to determine which will be the better solvent for each solute listed.

Plan By examining the formulas of the solutes, we can predict whether they are ionic or molecular. For those that are molecular, we can predict whether they are polar or nonpolar. We can then apply the idea that the nonpolar solvent will be better for the nonpolar solutes, whereas the polar solvent will be better for the ionic and polar solutes.

Solve C_7H_{16} is a hydrocarbon, so it is molecular and nonpolar. Na_2SO_4, a compound containing a metal and nonmetals, is ionic. HCl, a diatomic molecule containing two nonmetals that differ in electronegativity, is polar. I_2, a diatomic molecule with atoms of equal electronegativity, is nonpolar. We would therefore predict that C_7H_{16} and I_2 (the nonpolar solutes) would be more soluble in the nonpolar CCl_4 than in polar H_2O, whereas water would be the better solvent for Na_2SO_4 and HCl (the ionic and polar covalent solutes).

Practice Exercise 1

Which of the following solvents will best dissolve wax, which is a complex mixture of compounds that mostly are $CH_3-CH_2-CH_2-CH_2-CH_2-$?

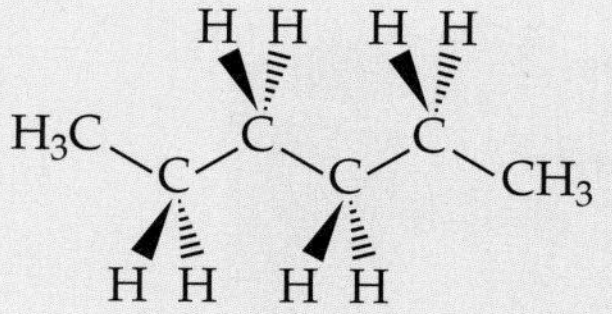

(a) Hexane

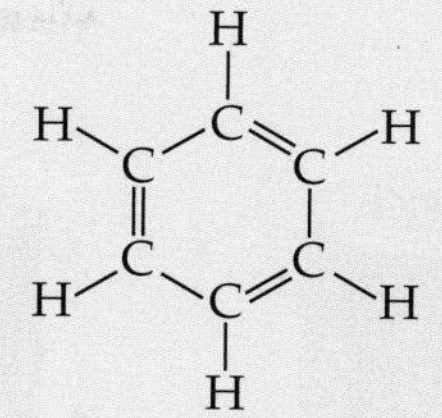

(b) Benzene

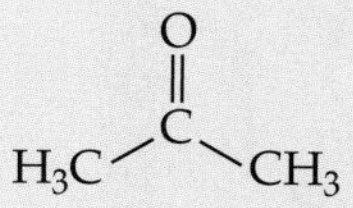

(c) Acetone

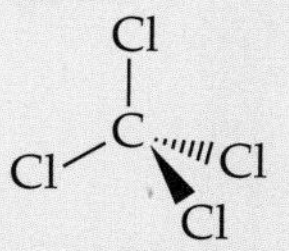

(d) Carbon tetrachloride

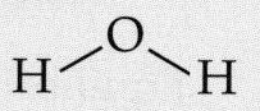

(e) Water

Practice Exercise 2

Arrange the following substances in order of increasing solubility in water:

(a)

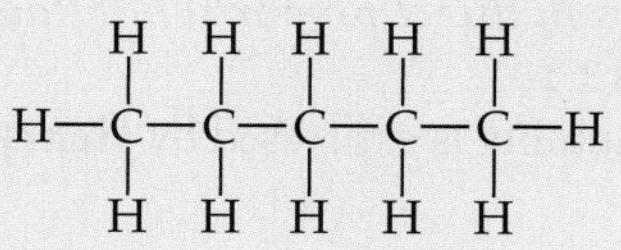

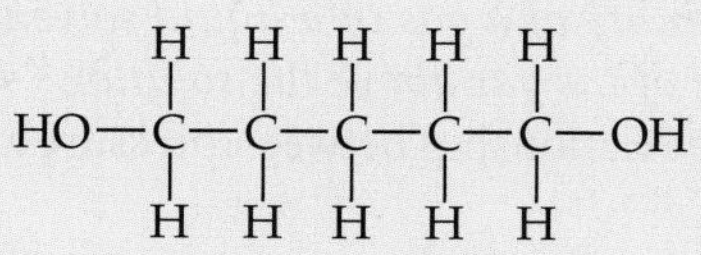

(b)

(c)

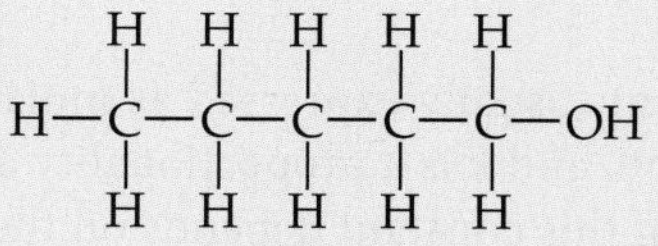

(d)

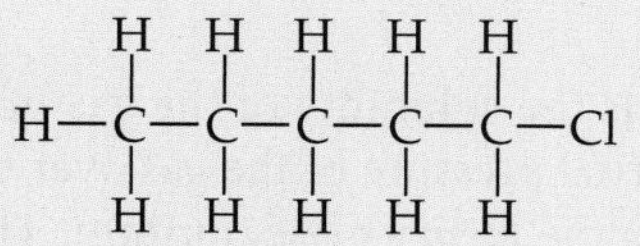

Pressure Effects

The solubilities of solids and liquids are not appreciably affected by pressure, whereas *the solubility of a gas in any solvent is increased as the partial pressure of the gas above the solvent increases*. We can understand the effect of pressure on gas solubility by considering **Figure 13.12**, which shows carbon dioxide gas distributed between the gas and solution phases. When equilibrium is established, the rate at which gas molecules enter the solution equals the rate at which solute molecules escape from the solution to enter the gas phase. The equal number of up and down arrows in the left container in Figure 13.12 represent these opposing processes.

Now suppose we exert greater pressure on the piston and compress the gas above the solution, as shown in the middle container in Figure 13.12. If we reduce the gas volume to half its original value, the pressure of the gas increases to about twice its original value. As a result of this pressure increase, the rate at which gas molecules strike the liquid surface and enter the solution phase increases. Thus, the solubility of the gas in the solution increases until equilibrium is again established; that is, solubility increases until the rate at which gas molecules enter the solution equals the rate at which they escape from the solution. Thus,

▲GO FIGURE

If the partial pressure of a gas over a solution is doubled, how has the concentration of gas in the solution changed after equilibrium is restored?

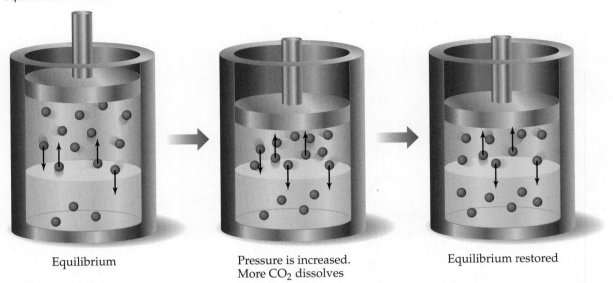

Equilibrium   Pressure is increased.   Equilibrium restored
                     More CO_2 dissolves

▲ **Figure 13.12 Effect of pressure on gas solubility.**

▲GO FIGURE

How do the slopes of the lines vary with the molecular weight of the gas? Explain the trend.

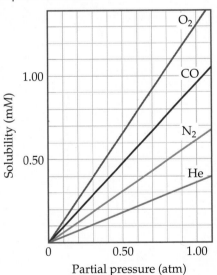

▲ **Figure 13.13 The solubility of a gas in water is directly proportional to the partial pressure of the gas.** The solubilities are in millimoles of gas per liter of solution.

---

the solubility of a gas in a liquid solvent increases in direct proportion to the partial pressure of the gas above the solution (◄ Figure 13.13).

The relationship between pressure and gas solubility is expressed by **Henry's law**:

$$S_g = kP_g \qquad [13.4]$$

Here, S_g is the solubility of the gas in the solvent (usually expressed as molarity), P_g is the partial pressure of the gas over the solution, and k is a proportionality constant known as the *Henry's law constant*. The value of this constant depends on the solute, solvent, and temperature. As an example, the solubility of N_2 gas in water at 25 °C and 0.78 atm pressure is $4.75 \times 10^{-4}\ M$. The Henry's law constant for N_2 in 25 °C water is thus $(4.75 \times 10^{-4}\ \text{mol/L})/0.78\ \text{atm} = 6.1 \times 10^{-4}\ \text{mol/L-atm}$. If the partial pressure of N_2 is doubled, Henry's law predicts that the solubility in water at 25 °C also doubles to $9.50 \times 10^{-4}\ M$.

Bottlers use the effect of pressure on solubility in producing carbonated beverages, which are bottled under a carbon dioxide pressure greater than 1 atm. When the bottles are opened to the air, the partial pressure of CO_2 above the solution decreases. Hence, the solubility of CO_2 decreases, and $CO_2(g)$ escapes from the solution as bubbles (► Figure 13.14).

SAMPLE EXERCISE 13.2 | A Henry's Law Calculation

Calculate the concentration of CO_2 in a soft drink that is bottled with a partial pressure of CO_2 of 4.0 atm over the liquid at 25 °C. The Henry's law constant for CO_2 in water at this temperature is $3.4 \times 10^{-2}\ \text{mol/L-atm}$.

SOLUTION

Analyze We are given the partial pressure of CO_2, P_{CO_2}, and the Henry's law constant, k, and asked to calculate the concentration of CO_2 in the solution.

Plan With the information given, we can use Henry's law, Equation 13.4, to calculate the solubility, S_{CO_2}.

Solve $S_{CO_2} = kP_{CO_2} = (3.4 \times 10^{-2}\,\text{mol/L-atm})(4.0\,\text{atm}) = 0.14\,\text{mol/L} = 0.14\,M$

Check The units are correct for solubility, and the answer has two significant figures consistent with both the partial pressure of CO_2 and the value of Henry's constant.

> **Practice Exercise 1**
>
> You double the partial pressure of a gas over a liquid at constant temperature. Which of these statements is then true?
> **(a)** The Henry's law constant is doubled.
> **(b)** The Henry's law constant is decreased by half.
> **(c)** There are half as many gas molecules in the liquid.
> **(d)** There are twice as many gas molecules in the liquid.
> **(e)** There is no change in the number of gas molecules in the liquid.
>
> **Practice Exercise 2**
>
> Calculate the concentration of CO_2 in a soft drink after the bottle is opened and the solution equilibrates at 25 °C under a CO_2 partial pressure of 3.0×10^{-4} atm.

Temperature Effects

The solubility of most solid solutes in water increases as the solution temperature increases, as ▼ Figure 13.15 shows. There are exceptions to this rule, however, as seen for $Ce_2(SO_4)_3$, whose solubility curve slopes downward with increasing temperature.

In contrast to solid solutes, *the solubility of gases in water decreases with increasing temperature* (▼ Figure 13.16). If a glass of cold tap water is warmed, you can see bubbles on the inside of the glass because some of the dissolved air comes out of solution.

Similarly, as carbonated beverages are allowed to warm, the solubility of CO_2 decreases, and $CO_2(g)$ escapes from the solution.

▲ **Figure 13.14 Gas solubility decreases as pressure decreases.** CO_2 bubbles out of solution when a carbonated beverage is opened because the CO_2 partial pressure above the solution is reduced.

 Give It Some Thought

Why do bubbles form on the inside wall of a cooking pot when water is heated on the stove, even though the water temperature is well below the boiling point of water?

GO FIGURE

How does the solubility of KCl at 80 °C compare with that of NaCl at the same temperature?

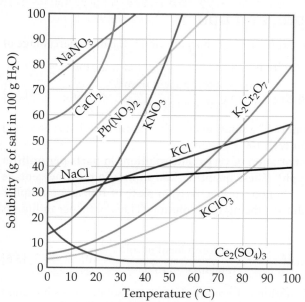

▲ **Figure 13.15 Solubilities of some ionic compounds in water as a function of temperature.**

GO FIGURE

Where would you expect N_2 to fit on this graph?

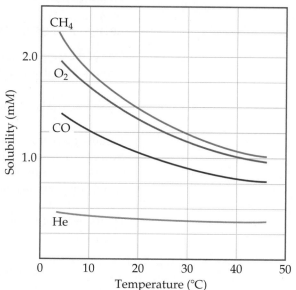

▲ **Figure 13.16 Solubilities of four gases in water as a function of temperature.** The solubilities are in millimoles per liter of solution, for a constant total pressure of 1 atm in the gas phase.

Chemistry and Life

Blood Gases and Deep-Sea Diving

Because gas solubility increases with increasing pressure, divers who breathe compressed air (▶ Figure 13.17) must be concerned about the solubility of gases in their blood. Although the gases are not very soluble at sea level, their solubilities can be appreciable at deep levels where their partial pressures are greater. Thus, divers must ascend slowly to prevent dissolved gases from being released rapidly from solution and forming bubbles in the blood and other fluids in the body. These bubbles affect nerve impulses and cause decompression sickness, or "the bends," which is a painful and potentially fatal condition. Nitrogen is the main problem because it is the most abundant gas in air and because it can be removed from the body only through the respiratory system. Oxygen, in contrast, is consumed in metabolism.

Deep-sea divers sometimes substitute helium for nitrogen in the air they breathe because helium has a much lower solubility in biological fluids than N_2. For example, divers working at a depth of 100 ft experience a pressure of about 4 atm. At this pressure, a mixture of 95% helium and 5% oxygen gives an oxygen partial pressure of about 0.2 atm, which is the partial pressure of oxygen in normal air at 1 atm.

▲ Figure 13.17 **Gas solubility increases as pressure increases.** Divers who use compressed gases must be concerned about the solubility of the gases in their blood.

If the oxygen partial pressure becomes too great, the urge to breathe is reduced, CO_2 is not removed from the body, and CO_2 poisoning occurs. At excessive concentrations in the body, carbon dioxide acts as a neurotoxin, interfering with nerve conduction and transmission.

Related Exercises: 13.59, 13.60, 13.107

13.4 | Expressing Solution Concentration

The concentration of a solution can be expressed either qualitatively or quantitatively. The terms *dilute* and *concentrated* are used to describe a solution qualitatively. A solution with a relatively small concentration of solute is said to be dilute; one with a large concentration is said to be concentrated. Chemists use various ways to express concentration quantitatively, and we examine several of these next.

Mass Percentage, ppm, and ppb

One of the simplest quantitative expressions of concentration is the **mass percentage** of a component in a solution, given by

$$\text{Mass \% of component} = \frac{\text{mass of component in soln}}{\text{total mass of soln}} \times 100 \qquad [13.5]$$

Because *percent* means "per hundred," a solution of hydrochloric acid that is 36% HCl by mass contains 36 g of HCl for each 100 g of solution.

We often express the concentration of very dilute solutions in **parts per million (ppm)** or **parts per billion (ppb)**. These quantities are similar to mass percentage but use 10^6 (a million) or 10^9 (a billion), respectively, in place of 100, as a multiplier for the ratio of the mass of solute to the mass of solution. Thus, parts per million is defined as

$$\text{ppm of component} = \frac{\text{mass of component in soln}}{\text{total mass of soln}} \times 10^6 \qquad [13.6]$$

A solution whose solute concentration is 1 ppm contains 1 g of solute for each million (10^6) grams of solution or, equivalently, 1 mg of solute per kilogram of solution. Because the density of water is 1 g/mL, 1 kg of a dilute aqueous solution has a volume very close to 1 L. Thus, 1 ppm also corresponds to 1 mg of solute per liter of aqueous solution.

The acceptable maximum concentrations of toxic or carcinogenic substances in the environment are often expressed in ppm or ppb. For example, the maximum allowable concentration of arsenic in drinking water in the United States is 0.010 ppm; that is, 0.010 mg of arsenic per liter of water. This concentration corresponds to 10 ppb.

 Give It Some Thought

A solution of SO_2 in water contains 0.00023 g of SO_2 per liter of solution. What is the concentration of SO_2 in ppm? In ppb?

SAMPLE
EXERCISE 13.3 | Calculation of Mass-Related Concentrations

(a) A solution is made by dissolving 13.5 g of glucose $(C_6H_{12}O_6)$ in 0.100 kg of water. What is the mass percentage of solute in this solution?

(b) A 2.5-g sample of groundwater was found to contain 5.4 μg of Zn^{2+}. What is the concentration of Zn^{2+} in parts per million?

SOLUTION

(a) Analyze We are given the number of grams of solute (13.5 g) and the number of grams of solvent (0.100 kg = 100 g). From this, we must calculate the mass percentage of solute.

Plan We can calculate the mass percentage by using Equation 13.5. The mass of the solution is the sum of the mass of solute (glucose) and the mass of solvent (water).

Solve Mass % of glucose $= \dfrac{\text{mass glucose}}{\text{mass soln}} \times 100 = \dfrac{13.5 \text{ g}}{13.5 \text{ g} + 100 \text{ g}} \times 100 = 11.9\%$

Comment The mass percentage of water in this solution is $(100 - 11.9)\% = 88.1\%$.

(b) Analyze In this case we are given the number of micrograms of solute. Because 1 μg is 1×10^{-6} g, 5.4μg $= 5.4 \times 10^{-6}$ g.

Plan We calculate the parts per million using Equation 13.6.

Solve ppm $= \dfrac{\text{mass of solute}}{\text{mass of soln}} \times 10^6 = \dfrac{5.4 \times 10^{-6} \text{ g}}{2.5 \text{ g}} \times 10^6 = 2.2$ ppm

Practice Exercise 1

Calculate the mass percentage of NaCl in a solution containing 1.50 g of NaCl in 50.0 g of water. **(a)** 0.0291%, **(b)** 0.0300%, **(c)** 0.0513%, **(d)** 2.91%, **(e)** 3.00%.

Practice Exercise 2

A commercial bleaching solution contains 3.62% by mass of sodium hypochlorite, NaOCl. What is the mass of NaOCl in a bottle containing 2.50 kg of bleaching solution?

Mole Fraction, Molarity, and Molality

Concentration expressions are often based on the number of moles of one or more components of the solution. Recall from Section 10.6 that the *mole fraction* of a component of a solution is given by

$$\text{Mole fraction of component} = \frac{\text{moles of component}}{\text{total moles of all components}} \qquad [13.7]$$

The symbol X is commonly used for mole fraction, with a subscript to indicate the component of interest. For example, the mole fraction of HCl in a hydrochloric acid solution is represented as X_{HCl}. Thus, if a solution contains 1.00 mol of HCl (36.5 g) and 8.00 mol of water (144 g), the mole fraction of HCl is $X_{HCl} = (1.00 \text{ mol})/(1.00 \text{ mol} + 8.00 \text{ mol}) = 0.111$.

Mole fractions have no units because the units in the numerator and the denominator cancel. The sum of the mole fractions of all components of a solution must equal 1. Thus, in the aqueous HCl solution, $X_{H_2O} = 1.000 - 0.111 = 0.889$. Mole fractions are very useful when dealing with gases, as we saw in Section 10.6, but have limited use when dealing with liquid solutions.

Recall from Section 4.5 that the *molarity* (*M*) of a solute in a solution is defined as

$$\text{Molarity} = \frac{\text{moles of solute}}{\text{liters of soln}} \qquad [13.8]$$

For example, if you dissolve 0.500 mol of Na_2CO_3 in enough water to form 0.250 L of solution, the molarity of Na_2CO_3 in the solution is $(0.500 \text{ mol})/(0.250 \text{ L}) = 2.00 \text{ } M$. Molarity is especially useful for relating the volume of a solution to the quantity of solute contained in that volume, as we saw in our discussions of titrations. ∞ (Section 4.6)

The **molality** of a solution, denoted *m*, is a concentration unit that is also based on moles of solute. Molality equals the number of moles of solute per kilogram of solvent:

$$\text{Molality} = \frac{\text{moles of solute}}{\text{kilograms of solvent}} \qquad [13.9]$$

Thus, if you form a solution by mixing 0.200 mol of NaOH (8.00 g) and 0.500 kg of water (500 g), the concentration of the solution is $(0.200 \text{ mol})/(0.500 \text{ kg}) = 0.400 \text{ } m$ (that is, 0.400 molal) in NaOH.

The definitions of molarity and molality are similar enough that they can be easily confused. Molarity depends on the *volume* of *solution*, whereas molality depends on the *mass* of *solvent*. When water is the solvent, the molality and molarity of dilute solutions are numerically about the same because 1 kg of solvent is nearly the same as 1 kg of solution, and 1 kg of the solution has a volume of about 1 L.

The molality of a given solution does not vary with temperature because masses do not vary with temperature. The molarity of the solution does change with temperature, however, because the volume of the solution expands or contracts with temperature. Thus, molality is often the concentration unit of choice when a solution is to be used over a range of temperatures.

 Give It Some Thought

If an aqueous solution is very dilute, will its molality be greater than its molarity, nearly the same as its molarity, or smaller than its molarity?

SAMPLE EXERCISE 13.4 Calculation of Molality

A solution is made by dissolving 4.35 g of glucose ($C_6H_{12}O_6$) in 25.0 mL of water at 25 °C. Calculate the molality of glucose in the solution. Water has a density of 1.00 g/mL.

SOLUTION

Analyze We are asked to calculate a solution concentration in units of molality. To do this, we must determine the number of moles of solute (glucose) and the number of kilograms of solvent (water).

Plan We use the molar mass of $C_6H_{12}O_6$ to convert grams of glucose to moles. We use the density of water to convert milliliters of water to kilograms. The molality equals the number of moles of solute (glucose) divided by the number of kilograms of solvent (water).

Solve Use the molar mass of glucose, 180.2 g/mol, to convert grams to moles:

$$\text{Mol } C_6H_{12}O_6 = (4.35 \text{ g } C_6H_{12}O_6)\left(\frac{1 \text{ mol } C_6H_{12}O_6}{180.2 \text{ g } C_6H_{12}O_6}\right) = 0.0241 \text{ mol } C_6H_{12}O_6$$

Because water has a density of 1.00 g/mL, the mass of the solvent is

$$(25.0 \text{ mL})(1.00 \text{ g/mL}) = 25.0 \text{ g} = 0.0250 \text{ kg}$$

Finally, use Equation 13.9 to obtain the molality:

$$\text{Molality of } C_6H_{12}O_6 = \frac{0.0241 \text{ mol } C_6H_{12}O_6}{0.0250 \text{ kg } H_2O} = 0.964 \text{ } m$$

Practice Exercise 1

Suppose you take a solution and add more solvent, so that the original mass of solvent is doubled. You take this new solution and add more solute, so that the original mass of the solute is doubled. What happens to the molality of the final solution, compared to the original molality?
(a) It is doubled.
(b) It is decreased by half.

(c) It is unchanged.
(d) It will increase or decrease depending on the molar mass of the solute.
(e) There is no way to tell without knowing the molar mass of the solute.

Practice Exercise 2

What is the molality of a solution made by dissolving 36.5 g of naphthalene $(C_{10}H_8)$ in 425 g of toluene (C_7H_8)?

Converting Concentration Units

If you follow the dimensional analysis techniques you learned in Chapter 1, you can convert between concentration units, as shown in Sample Exercise 13.5. To convert between molality and molarity, the density of the solution will be needed, as in Sample Exercise 13.6.

SAMPLE EXERCISE 13.5 Calculation of Mole Fraction and Molality

An aqueous solution of hydrochloric acid contains 36% HCl by mass. (a) Calculate the mole fraction of HCl in the solution. (b) Calculate the molality of HCl in the solution.

SOLUTION

Analyze We are asked to calculate the concentration of the solute, HCl, in two related concentration units, given only the percentage by mass of the solute in the solution.

Plan In converting concentration units based on the mass or moles of solute and solvent (mass percentage, mole fraction, and molality), it is useful to assume a certain total mass of solution. Let's assume that there is exactly 100 g of solution. Because the solution is 36% HCl, it contains 36 g of HCl and $(100 - 36)$ g = 64 g of H_2O. We must convert grams of solute (HCl) to moles to calculate either mole fraction or molality. We must convert grams of solvent (H_2O) to moles to calculate mole fractions and to kilograms to calculate molality.

Solve (a) To calculate the mole fraction of HCl, we convert the masses of HCl and H_2O to moles and then use Equation 13.7:

$$\text{Moles HCl} = (36 \text{ g HCl})\left(\frac{1 \text{ mol HCl}}{36.5 \text{ g HCl}}\right) = 0.99 \text{ mol HCl}$$

$$\text{Moles H}_2\text{O} = (64 \text{ g H}_2\text{O})\left(\frac{1 \text{ mol H}_2\text{O}}{18 \text{ g H}_2\text{O}}\right) = 3.6 \text{ mol H}_2\text{O}$$

$$X_{HCl} = \frac{\text{moles HCl}}{\text{moles H}_2\text{O} + \text{moles HCl}} = \frac{0.99}{3.6 + 0.99} = \frac{0.99}{4.6} = 0.22$$

(b) To calculate the molality of HCl in the solution, we use Equation 13.9. We calculated the number of moles of HCl in part (a), and the mass of solvent is 64 g = 0.064 kg:

$$\text{Molality of HCl} = \frac{0.99 \text{ mol HCl}}{0.064 \text{ kg H}_2\text{O}} = 15 \, m$$

Notice that we can't readily calculate the molarity of the solution because we don't know the volume of the 100 g of solution.

Practice Exercise 1

The solubility of oxygen gas in water at 40 °C is 1.0 mmol per liter of solution. What is this concentration in units of mole fraction?
(a) 1.00×10^{-6}, (b) 1.80×10^{-5}, (c) 1.00×10^{-2},
(d) 1.80×10^{-2}, (e) 5.55×10^{-2}.

Practice Exercise 2

A commercial bleach solution contains 3.62% by mass of NaOCl in water. Calculate (a) the mole fraction and (b) the molality of NaOCl in the solution.

Calculation of Molarity Using the Density of the Solution

A solution with a density of 0.876 g/mL contains 5.0 g of toluene (C_7H_8) and 225 g of benzene.
Calculate the molarity of the solution.

SOLUTION

Analyze Our goal is to calculate the molarity of a solution, given the masses of solute (5.0 g) and solvent (225 g) and the density of the solution (0.876 g/mL).

Plan The molarity of a solution is the number of moles of solute divided by the number of liters of solution (Equation 13.8). The number of moles of solute (C_7H_8) is calculated from the number of grams of solute and its molar mass. The volume of the solution is obtained from the mass of the solution (mass of solution = mass of solute + mass of solvent = 5.0 g + 225 g = 230 g) and its density.

Solve The number of moles of solute is:

$$\text{Moles } C_7H_8 = (5.0 \text{ g } C_7H_8)\left(\frac{1 \text{ mol } C_7H_8}{92 \text{ g } C_7H_8}\right) = 0.054 \text{ mol}$$

The density of the solution is used to convert the mass of the solution to its volume:

$$\text{Milliliters soln} = (230 \text{ g})\left(\frac{1 \text{ mL}}{0.876 \text{ g}}\right) = 263 \text{ mL}$$

Molarity is moles of solute per liter of solution:

$$\text{Molarity} = \left(\frac{\text{moles } C_7H_8}{\text{liter soln}}\right) = \left(\frac{0.054 \text{ mol } C_7H_8}{263 \text{ mL soln}}\right)\left(\frac{1000 \text{ mL soln}}{1 \text{ L soln}}\right) = 0.21 \text{ M}$$

Check The magnitude of our answer is reasonable. Rounding moles to 0.05 and liters to 0.25 gives a molarity of $(0.05 \text{ mol})/(0.25 \text{ L}) = 0.2 \text{ M}$.

The units for our answer (mol/L) are correct, and the answer, 0.21, has two significant figures, corresponding to the number of significant figures in the mass of solute (2).

Comment Because the mass of the solvent (0.225 kg) and the volume of the solution (0.263) are similar in magnitude, the molarity and molality are also similar in magnitude:
$(0.054 \text{ mol } C_7H_8)/(0.225 \text{ kg solvent}) = 0.24 \text{ m}$.

Practice Exercise 1

Maple syrup has a density of 1.325 g/mL, and 100.00 g of maple syrup contains 67 mg of calcium in the form of Ca^{2+} ions. What is the molarity of calcium in maple syrup?
(a) 0.017 M, **(b)** 0.022 M, **(c)** 0.89 M, **(d)** 12.6 M, **(e)** 45.4 M

Practice Exercise 2

A solution containing equal masses of glycerol ($C_3H_8O_3$) and water has a density of 1.10 g/mL. Calculate **(a)** the molality of glycerol, **(b)** the mole fraction of glycerol, **(c)** the molarity of glycerol in the solution.

13.5 | Colligative Properties

Some physical properties of solutions differ in important ways from those of the pure solvent. For example, pure water freezes at 0 °C, but aqueous solutions freeze at lower temperatures. We apply this behavior when we add ethylene glycol antifreeze to a car's radiator to lower the freezing point of the solution. The added solute also raises the boiling point of the solution above that of pure water, making it possible to operate the engine at a higher temperature.

Lowering of the freezing point and raising of the boiling point are physical properties of solutions that depend on the *quantity* (concentration) but not on the *kind* or *identity* of the solute particles. Such properties are called **colligative properties**. (*Colligative* means "depending on the collection"; colligative properties depend on the collective effect of the number of solute particles.)

In addition to freezing-point lowering and boiling-point raising, vapor-pressure lowering and osmotic pressure are also colligative properties. As we examine each one, notice how solute concentration quantitatively affects the property.

Vapor-Pressure Lowering

A liquid in a closed container establishes equilibrium with its vapor. ∞ (Section 11.5) The *vapor pressure* is the pressure exerted by the vapor when it is at equilibrium with the liquid (that is, when the rate of vaporization equals the rate of condensation). A substance that has no measurable vapor pressure is *nonvolatile,* whereas one that exhibits a vapor pressure is *volatile.*

A solution consisting of a *volatile* liquid solvent and a *nonvolatile* solute forms spontaneously because of the increase in entropy that accompanies their mixing. In

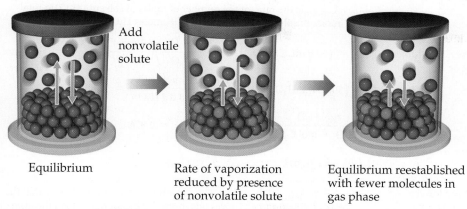

● Volatile solvent particles

● Nonvolatile solute particles

Equilibrium | Rate of vaporization reduced by presence of nonvolatile solute | Equilibrium reestablished with fewer molecules in gas phase

▲ **Figure 13.18 Vapor-pressure lowering.** The presence of nonvolatile solute particles in a liquid solvent results in a reduction of the vapor pressure above the liquid.

effect, the solvent molecules are stabilized in their liquid state by this process and thus have a lower tendency to escape into the vapor state. Therefore, when a *nonvolatile* solute is present, the vapor pressure of the solvent is lower than the vapor pressure of the pure solvent, as illustrated in ▲ **Figure 13.18.**

Ideally, the vapor pressure of a *volatile* solvent above a solution containing a *nonvolatile* solute is proportional to the solvent's concentration in the solution. This relationship is expressed quantitatively by **Raoult's law**, which states that the partial pressure exerted by solvent vapor above the solution, $P_{solution}$, equals the product of the mole fraction of the solvent, $X_{solvent}$, times the vapor pressure of the pure solvent, $P^{\circ}_{solvent}$:

$$P_{solution} = X_{solvent} P^{\circ}_{solvent} \qquad [13.10]$$

For example, the vapor pressure of pure water at 20 °C is $P^{\circ}_{H_2O} = 17.5$ torr. Imagine holding the temperature constant while adding glucose $(C_6H_{12}O_6)$ to the water so that the mole fractions in the resulting solution are $X_{H_2O} = 0.800$ and $X_{C_6H_{12}O_6} = 0.200$. According to Equation 13.10, the vapor pressure of the water above this solution is 80.0% of that of pure water:

$$P_{solution} = (0.800)(17.5 \text{ torr}) = 14.0 \text{ torr}$$

The presence of the *nonvolatile* solute lowers the vapor pressure of the *volatile* solvent by 17.5 torr − 14.0 torr = 3.5 torr.

The vapor-pressure lowering, ΔP, is directly proportional to the mole fraction of the solute, X_{solute}:

$$\Delta P = X_{solute} \, P^{\circ}_{solvent} \qquad [13.11]$$

Thus, for the example of the solution of glucose in water, we have

$$\Delta P = X_{C_6H_{12}O_6} \, P^{\circ}_{H_2O} = (0.200)(17.5 \text{ torr}) = 3.50 \text{ torr}$$

The vapor-pressure lowering caused by adding a *nonvolatile* solute depends on the total concentration of solute particles, regardless of whether they are molecules or ions. Remember that vapor-pressure lowering is a colligative property, so its value for any solution depends on the concentration of solute particles and not on their kind or identity.

 Give It Some Thought

Adding 1 mol of NaCl to 1 kg of water lowers the vapor pressure of water more than adding 1 mol of $C_6H_{12}O_6$. Explain.

SAMPLE EXERCISE 13.7 Calculation of Vapor Pressure of a Solution

Glycerin $(C_3H_8O_3)$ is a nonvolatile nonelectrolyte with a density of 1.26 g/mL at 25 °C. Calculate the vapor pressure at 25 °C of a solution made by adding 50.0 mL of glycerin to 500.0 mL of water. The vapor pressure of pure water at 25 °C is 23.8 torr (Appendix B), and its density is 1.00 g/mL.

SOLUTION

Analyze Our goal is to calculate the vapor pressure of a solution, given the volumes of solute and solvent and the density of the solute.

Plan We can use Raoult's law (Equation 13.10) to calculate the vapor pressure of a solution. The mole fraction of the solvent in the solution, $X_{solvent}$, is the ratio of the number of moles of solvent (H_2O) to total moles of solution (moles $C_3H_8O_3$ + moles H_2O).

Solve To calculate the mole fraction of water in the solution, we must determine the number of moles of $C_3H_8O_3$ and H_2O:

$$\text{Moles } C_3H_8O_3 = (50.0 \text{ mL } C_3H_8O_3)\left(\frac{1.26 \text{ g } C_3H_8O_3}{1 \text{ mL } C_3H_8O_3}\right)\left(\frac{1 \text{ mol } C_3H_8O_3}{92.1 \text{ g } C_3H_8O_3}\right) = 0.684 \text{ mol}$$

$$\text{Moles } H_2O = (500.0 \text{ mL } H_2O)\left(\frac{1.00 \text{ g } H_2O}{1 \text{ mL } H_2O}\right)\left(\frac{1 \text{ mol } H_2O}{18.0 \text{ g } H_2O}\right) = 27.8 \text{ mol}$$

$$X_{H_2O} = \frac{\text{mol } H_2O}{\text{mol } H_2O + \text{mol } C_3H_8O_3} = \frac{27.8}{27.8 + 0.684} = 0.976$$

We now use Raoult's law to calculate the vapor pressure of water for the solution:

$$P_{H_2O} = X_{H_2O} P^\circ_{H_2O} = (0.976)(23.8 \text{ torr}) = 23.2 \text{ torr}$$

Comment The vapor pressure of the solution has been lowered by 23.8 torr − 23.2 torr = 0.6 torr relative to that of pure water. The vapor-pressure lowering can be calculated directly using Equation 13.11 together with the mole fraction of the solute, $C_3H_8O_3$:

$\Delta P = X_{C_3H_8O_3} P^\circ_{H_2O} = (0.024)(23.8 \text{ torr}) = 0.57 \text{ torr}$. Notice that the use of Equation 13.11 gives one more significant figure than the number obtained by subtracting the vapor pressure of the solution from that of the pure solvent.

Practice Exercise 1

The vapor pressure of benzene, C_6H_6, is 100.0 torr at 26.1 °C. Assuming Raoult's law is obeyed, how many moles of a nonvolatile solute must be added to 100.0 mL of benzene to decrease its vapor pressure by 10.0% at 26.1 °C? The density of benzene is 0.8765 g/cm³.
(a) 0.011237, (b) 0.11237, (c) 0.1248, (d) 0.1282, (e) 8.765.

Practice Exercise 2

The vapor pressure of pure water at 110 °C is 1070 torr. A solution of ethylene glycol and water has a vapor pressure of 1.00 atm at 110 °C. Assuming that Raoult's law is obeyed, what is the mole fraction of ethylene glycol in the solution?

A Closer Look

Ideal Solutions with Two or More Volatile Components

Solutions sometimes have two or more volatile components. Gasoline, for example, is a solution of several volatile liquids. To gain some understanding of such mixtures, consider an ideal solution of two volatile liquids, A and B. (For our purposes here, it does not matter which we call the solute and which the solvent.) The partial pressures above the solution are given by Raoult's law:

$$P_A = X_A P^\circ_A \text{ and } P_B = X_B P^\circ_B$$

and the total vapor pressure above the solution is

$$P_{total} = P_A + P_B = X_A P^\circ_A + X_B P^\circ_B$$

▲ **Figure 13.19** The volatile components of organic mixtures can be separated on an industrial scale in these distillation towers.

Consider a mixture of 1.0 mol of benzene (C_6H_6) and 2.0 mol of toluene (C_7H_8) (X_{ben} = 0.33, X_{tol} = 0.67). At 20 °C, the vapor pressures of the pure substances are P°_{ben} = 75 torr and P°_{tol} = 22 torr. Thus, the partial pressures above the solution are

$$P_{ben} = (0.33)(75 \text{ torr}) = 25 \text{ torr}$$
$$P_{tol} = (0.67)(22 \text{ torr}) = 15 \text{ torr}$$

and the total vapor pressure above the liquid is

$$P_{total} = P_{ben} + P_{tol} = 25 \text{ torr} + 15 \text{ torr} = 40 \text{ torr}$$

Note that the vapor is richer in benzene, the more volatile component.

The mole fraction of benzene in the vapor is given by the ratio of its vapor pressure to the total pressure (Equations 10.14 and 10.15):

$$X_{ben} \text{ in vapor} = \frac{P_{ben}}{P_{tol}} = \frac{25 \text{ torr}}{40 \text{ torr}} = 0.63$$

Although benzene constitutes only 33% of the molecules in the solution, it makes up 63% of the molecules in the vapor.

When an ideal liquid solution containing two volatile components is in equilibrium with its vapor, the more volatile component will be relatively richer in the vapor. This fact forms the basis of *distillation*, a technique used to separate (or partially separate) mixtures containing volatile components. ∞ (Section 1.3) Distillation is a way of purifying liquids, and is the procedure by which petrochemical plants achieve the separation of crude petroleum into gasoline, diesel fuel, lubricating oil, and other products (◄ Figure 13.19). Distillation is also used routinely on a small scale in the laboratory.

Related Exercises: 13.67, 13.68

An ideal gas is defined as one that obeys the ideal-gas equation ⚬⚬ (Section 10.4), and an **ideal solution** is defined as one that obeys Raoult's law. Whereas ideality for a gas arises from a complete lack of intermolecular interaction, ideality for a solution implies total uniformity of interaction. The molecules in an ideal solution all influence one another in the same way—in other words, solute–solute, solvent–solvent, and solute–solvent interactions are indistinguishable from one another. Real solutions best approximate ideal behavior when the solute concentration is low and solute and solvent have similar molecular sizes and take part in similar types of intermolecular attractions.

Many solutions do not obey Raoult's law exactly and so are not ideal. If, for instance, the solvent–solute interactions in a solution are weaker than either the solvent–solvent or solute–solute interactions, the vapor pressure tends to be greater than that predicted by Raoult's law. When the solute–solvent interactions in a solution are exceptionally strong, as might be the case when hydrogen bonding exists, the vapor pressure is lower than that predicted by Raoult's law. Although you should be aware that these departures from ideality occur, we will ignore them for the remainder of this chapter.

Boiling-Point Elevation

In Sections 11.5 and 11.6, we examined the vapor pressures of pure substances and how to use them to construct phase diagrams. How does the phase diagram of a solution and, hence, its boiling and freezing points differ from that of the pure solvent? The addition of a nonvolatile solute lowers the vapor pressure of the solution. Thus, in ▼ **Figure 13.20** the vapor-pressure curve of the solution is shifted downward relative to the vapor-pressure curve of the pure solvent.

Recall from Section 11.5 that the normal boiling point of a liquid is the temperature at which its vapor pressure equals 1 atm. Because the solution has a lower vapor pressure than the pure solvent, a higher temperature is required for the solution to achieve a vapor pressure of 1 atm. As a result, *the boiling point of the solution is higher than that of the pure solvent.* This effect is seen in Figure 13.20. We find the normal boiling point of the pure solvent on the graph by locating the point where the 1-atm pressure horizontal line intersects the black vapor-pressure curve and then tracing this point down to the temperature axis. For the solution, the 1-atm line intersects the blue vapor-pressure curve at a higher temperature, indicating that the solution has a higher boiling point than the pure solvent.

The increase in the boiling point of a solution, relative to the pure solvent, depends on the molality of the solute. But it is important to remember that boiling-point

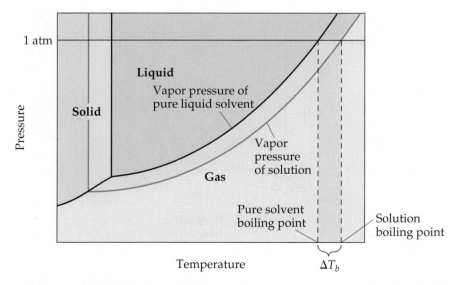

▲ Figure 13.20 **Phase diagram illustrating boiling-point elevation.** The black lines show the pure solvent's phase equilibria curves, and the blue lines show the solution's phase equilibria curves.

elevation is proportional to the *total* concentration of solute particles, regardless of whether the particles are molecules or ions. When NaCl dissolves in water, 2 mol of solute particles (1 mol of Na$^+$ and 1 mol of Cl$^-$) are formed for each mole of NaCl that dissolves. We take this into account by defining i, the **van't Hoff factor**, as the number of fragments that a solute breaks up into for a particular solvent. The change in boiling point for a solution compared to the pure solvent is:

$$\Delta T_b = T_b(\text{solution}) - T_b(\text{solvent}) = iK_b m \qquad [13.12]$$

In this equation, $T_b(\text{solution})$ is the boiling point of the solution, $T_b(\text{solvent})$ is the boiling point of the pure solvent, m is the molality of the solute, K_b is the **molal boiling-point-elevation constant** for the solvent (which is a proportionality constant that is experimentally determined for each solvent), and i is the van't Hoff factor. For a non-electrolyte, we can always assume $i = 1$; for an electrolyte, i will depend on how the substance ionizes in that solvent. For instance, $i = 2$ for NaCl in water, assuming complete dissociation of ions. As a result, we expect the boiling-point elevation of a 1 m aqueous solution of NaCl to be twice as large as the boiling-point elevation of a 1 m solution of a nonelectrolyte such as sucrose. Thus, to properly predict the effect of a particular solute on boiling-point elevation (or any other colligative property), it is important to know whether the solute is an electrolyte or a nonelectrolyte. ∞∞ (Sections 4.1 and 4.3)

 Give It Some Thought

A solute dissolved in water causes the boiling point to increase by 0.51 °C. Does this necessarily mean that the concentration of the solute is 1.0 m (▶ **Table 13.3**)?

Freezing-Point Depression

The vapor-pressure curves for the liquid and solid phases meet at the triple point. ∞∞ (Section 11.6) In ▼ **Figure 13.21** we see that the triple-point temperature of the solution is lower than the triple-point temperature of pure liquid because the solution has a lower vapor pressure than the pure liquid.

The freezing point of a solution is the temperature at which the first crystals of pure solvent form in equilibrium with the solution. Recall from Section 11.6 that the line representing the solid–liquid equilibrium rises nearly vertically from the triple point. It is easy to see in Figure 13.21 that the triple-point temperature of the solution is lower than that of the pure liquid, but it is also true for all points along the solid–liquid equilibrium curve: *the freezing point of the solution is lower than that of the pure liquid.*

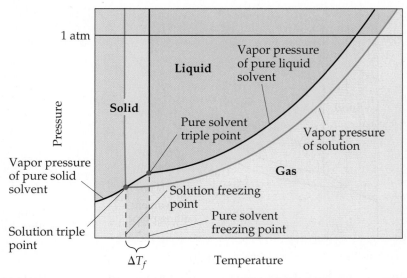

▲ **Figure 13.21 Phase diagram illustrating freezing-point depression.** The black lines show the pure solvent's phase equilibria curves, and the blue lines show the solution's phase equilibria curves.

Table 13.3 Molal Boiling-Point-Elevation and Freezing-Point-Depression Constants

Solvent	Normal Boiling Point (°C)	K_b(°C/m)	Normal Freezing Point (°C)	K_f(°C/m)
Water, H_2O	100.0	0.51	0.0	1.86
Benzene, C_6H_6	80.1	2.53	5.5	5.12
Ethanol, C_2H_5OH	78.4	1.22	−114.6	1.99
Carbon tetrachloride, CCl_4	76.8	5.02	−22.3	29.8
Chloroform, $CHCl_3$	61.2	3.63	−63.5	4.68

Like the boiling-point elevation, the change in freezing point ΔT_f is directly proportional to solute molality, taking into account the van't Hoff factor i:

$$\Delta T_f = T_f(\text{solution}) - T_f(\text{solvent}) = -iK_f m \qquad [13.13]$$

The proportionality constant K_f is the **molal freezing-point-depression constant**, analogous to K_b for boiling point elevation. Note that because the solution freezes at a *lower* temperature than does the pure solvent, the value of ΔT_f is *negative*.

Some typical values of K_b and K_f for several common solvents are given in ▲ Table 13.3. For water, the table shows $K_b = 0.51\,°C/m$, which means that the boiling point of any aqueous solution that is 1 m in nonvolatile solute particles is $0.51\,°C$ higher than the boiling point of pure water. Because solutions generally do not behave ideally, the constants listed in Table 13.3 serve well only for solutions that are rather dilute.

For water, K_f is $1.86\,°C/m$. Therefore, any aqueous solution that is 1 m in nonvolatile solute particles (such as 1 m $C_6H_{12}O_6$ or 0.5 m NaCl) freezes at the temperature that is $1.86\,°C$ lower than the freezing point of pure water.

The freezing-point depression caused by solutes has useful applications: it is why antifreeze works in car cooling systems, and why calcium chloride ($CaCl_2$) promotes the melting of ice on roads during winter.

SAMPLE EXERCISE 13.8 Calculation of Boiling-Point Elevation and Freezing-Point Depression

Automotive antifreeze contains ethylene glycol, $CH_2(OH)CH_2(OH)$, a nonvolatile nonelectrolyte, in water. Calculate the boiling point and freezing point of a 25.0% by mass solution of ethylene glycol in water.

SOLUTION

Analyze We are given that a solution contains 25.0% by mass of a nonvolatile, nonelectrolyte solute and asked to calculate the boiling and freezing points of the solution. To do this, we need to calculate the boiling-point elevation and freezing-point depression.

Plan To calculate the boiling-point elevation and the freezing-point depression using Equations 13.12 and 13.13, we must express the concentration of the solution as molality. Let's assume for convenience that we have 1000 g of solution. Because the solution is 25.0% by mass ethylene glycol, the masses of ethylene glycol and water in the solution are 250 and 750 g, respectively. Using these quantities, we can calculate the molality of the solution, which we use with the molal boiling-point-elevation and freezing-point-depression constants (Table 13.3) to calculate ΔT_b and ΔT_f. We add ΔT_b to the boiling point and ΔT_f to the freezing point of the solvent to obtain the boiling point and freezing point of the solution.

Solve The molality of the solution is calculated as follows:

$$\text{Molality} = \frac{\text{moles } C_2H_6O_2}{\text{kilograms } H_2O} = \left(\frac{250 \text{ g } C_2H_6O_2}{750 \text{ g } H_2O}\right)\left(\frac{1 \text{ mol } C_2H_6O_2}{62.1 \text{ g } C_2H_6O_2}\right)\left(\frac{1000 \text{ g } H_2O}{1 \text{ kg } H_2O}\right)$$

$$= 5.37 \ m$$

We can now use Equations 13.12 and 13.13 to calculate the changes in the boiling and freezing points:

$$\Delta T_b = iK_b m = (1)(0.51\,°C/m)(5.37\,m) = 2.7\,°C$$
$$\Delta T_f = -iK_f m = -(1)(1.86\,°C/m)(5.37\,m) = -10.0\,°C$$

Hence, the boiling and freezing points of the solution are readily calculated:

$$\Delta T_b = T_b(\text{solution}) - T_b(\text{solvent})$$
$$2.7\,°C = T_b(\text{solution}) - 100.0\,°C$$
$$T_b(\text{solution}) = 102.7\,°C$$
$$\Delta T_f = T_f(\text{solution}) - T_f(\text{solvent})$$
$$-10.0\,°C = T_f(\text{solution}) - 0.0\,°C$$
$$T_f(\text{solution}) = -10.0\,°C$$

Comment Notice that the solution is a liquid over a larger temperature range than the pure solvent.

Practice Exercise 1

Which aqueous solution will have the lowest freezing point? (a) 0.050 m $CaCl_2$, (b) 0.15 m NaCl, (c) 0.10 m HCl, (d) 0.050 m CH_3COOH, (e) 0.20 m $C_{12}H_{22}O_{11}$.

Practice Exercise 2

Referring to Table 13.3, calculate the freezing point of a solution containing 0.600 kg of $CHCl_3$ and 42.0 g of eucalyptol ($C_{10}H_{18}O$), a fragrant substance found in the leaves of eucalyptus trees.

Osmosis

Certain materials, including many membranes in biological systems and synthetic substances such as cellophane, are *semipermeable*. When in contact with a solution, these materials allow only ions or small molecules—water molecules, for instance—to pass through their network of tiny pores.

Consider a situation in which only solvent molecules are able to pass through a semipermeable membrane placed between two solutions of different concentrations. The rate at which the solvent molecules pass from the less concentrated solution (lower solute concentration but higher solvent concentration) to the more concentrated solution (higher solute concentration but lower solvent concentration) is greater than the rate in the opposite direction. Thus, there is a net movement of solvent molecules from the solution with a lower solute concentration into the one with a higher solute concentration. In this process, called **osmosis**, *the net movement of solvent is always toward the solution with the lower solvent (higher solute) concentration*, as if the solutions were driven to attain equal concentrations.

▶ Figure 13.22 shows the osmosis that occurs between an aqueous solution and pure water, separated by a semipermeable membrane. The U-tube contains water on the left and an aqueous solution on the right. Initially, there is a net movement of water through the membrane from left to right, leading to unequal liquid levels in the two arms of the U-tube. Eventually, at equilibrium (middle panel of Figure 13.22), the pressure difference resulting from the unequal liquid heights becomes so large that the net flow of water ceases. This pressure, which stops osmosis, is the **osmotic pressure, Π**, of the solution. If an external pressure equal to the osmotic pressure is applied to the solution, the liquid levels in the two arms can be equalized, as shown in the right panel of Figure 13.22.

The osmotic pressure obeys a law similar in form to the ideal-gas law, $\Pi V = inRT$ where Π is the osmotic pressure, V is the volume of the solution, i is the van't Hoff factor, n is the number of moles of solute, R is the ideal-gas constant, and T is the absolute temperature. From this equation, we can write

$$\Pi = i\left(\frac{n}{V}\right)RT = iMRT \qquad [13.14]$$

▲ GO FIGURE

If the pure water in the left arm of the U-tube is replaced by a solution more concentrated than the one in the right arm, what will happen?

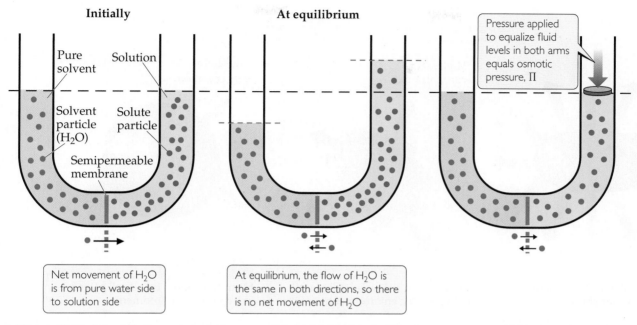

▲ **Figure 13.22 Osmosis is the process of a solvent moving from one compartment to another, across a semipermeable membrane, toward higher solute concentration.** Osmotic pressure is generated at equilibrium due to the different heights of liquid on either side of the membrane and is equivalent to the pressure needed to equalize the fluid levels across the membrane.

where M is the molarity of the solution. Because the osmotic pressure for any solution depends on the solution concentration, osmotic pressure is a colligative property.

If two solutions of identical osmotic pressure are separated by a semipermeable membrane, no osmosis will occur. The two solutions are *isotonic* with respect to each other. If one solution is of lower osmotic pressure, it is *hypotonic* with respect to the more concentrated solution. The more concentrated solution is *hypertonic* with respect to the dilute solution.

Give It Some Thought

Of two KBr solutions, one 0.50 *m* and the other 0.20 *m,* which is hypotonic with respect to the other?

Osmosis plays an important role in living systems. The membranes of red blood cells, for example, are semipermeable. Placing a red blood cell in a solution that is *hyper*tonic relative to the intracellular solution (the solution inside the cells) causes water to move out of the cell (**Figure 13.23**). This causes the cell to shrivel, a process called *crenation*. Placing the cell in a solution that is *hypo*tonic relative to the intracellular fluid causes water to move into the cell, which may cause the cell to rupture, a process called *hemolysis*. People who need body fluids or nutrients replaced but cannot be fed orally are given solutions by intravenous (IV) infusion, which feeds nutrients directly into the veins. To prevent crenation or hemolysis of red blood cells, the IV solutions must be isotonic with the intracellular fluids of the blood cells.

GO FIGURE

If the fluid surrounding a patient's red blood cells is depleted in electrolytes, is crenation or hemolysis more likely to occur?

> The arrows represent the net movement of water molecules.

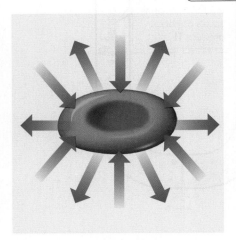

High solute
concentration

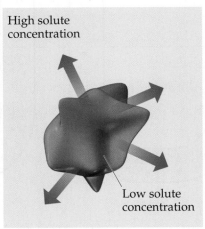

Low solute
concentration

Low solute
concentration

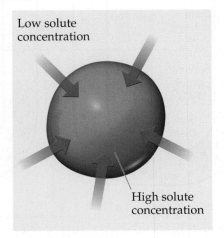

High solute
concentration

Red blood cell in isotonic medium
neither swells nor shrinks.

Crenation of red blood cell placed
in hypertonic environment

Hemolysis of red blood cell placed
in hypotonic environment

▲ **Figure 13.23 Osmosis through red blood cell walls.** If water moves out of the red blood cell, it shrivels (crenation); if water moves into the red blood cell, it will swell and may burst (hemolysis).

**SAMPLE
EXERCISE 13.9** Osmotic Pressure Calculations

The average osmotic pressure of blood is 7.7 atm at 25 °C. What molarity of glucose $(C_6H_{12}O_6)$ will be isotonic with blood?

SOLUTION

Analyze We are asked to calculate the concentration of glucose in water that would be isotonic with blood, given that the osmotic pressure of blood at 25 °C is 7.7 atm.

Plan Because we are given the osmotic pressure and temperature, we can solve for the concentration, using Equation 13.14. Because glucose is a nonelectrolyte, $i = 1$.

Solve

$$\Pi = iMRT$$

$$M = \frac{\Pi}{iRT} = \frac{(7.7 \text{ atm})}{(1)\left(0.0821\dfrac{\text{L-atm}}{\text{mol-K}}\right)(298 \text{ K})} = 0.31 \, M$$

Comment In clinical situations, the concentrations of solutions are generally expressed as mass percentages. The mass percentage of a 0.31 M solution of glucose is 5.3%. The concentration of NaCl that is isotonic with blood is 0.16 M, because $i = 2$ for NaCl in water (a 0.155 M solution of NaCl is 0.310 M in particles). A 0.16 M solution of NaCl is 0.9% mass in NaCl. This kind of solution is known as a physiological saline solution.

Practice Exercise 1

Which of the following actions will raise the osmotic pressure of a solution? (a) decreasing the solute concentration, (b) decreasing the temperature, (c) adding more solvent, (d) increasing the temperature, (e) none of the above.

Practice Exercise 2

What is the osmotic pressure, in atm, of a 0.0020 M sucrose $(C_{12}H_{22}O_{11})$ solution at 20.0 °C?

There are many interesting biological examples of osmosis. A cucumber placed in concentrated brine loses water via osmosis and shrivels into a pickle. People who eat a lot of salty food retain water in tissue cells and intercellular space because of osmosis. The resultant swelling or puffiness is called *edema*. Water moves from soil into plant roots partly because of osmosis. Bacteria on salted meat or candied fruit lose water through osmosis, shrivel, and die—thus preserving the food.

 Give It Some Thought

Is the osmotic pressure of a 0.10 *M* solution of NaCl greater than, less than, or equal to that of a 0.10 *M* solution of KBr?

Determination of Molar Mass from Colligative Properties

The colligative properties of solutions provide a useful means of determining the molar mass of solutes. Any of the four colligative properties can be used, as shown in Sample Exercises 13.10 and 13.11.

SAMPLE EXERCISE 13.10 | Molar Mass from Freezing-Point Depression

A solution of an unknown nonvolatile nonelectrolyte was prepared by dissolving 0.250 g of the substance in 40.0 g of CCl_4. The boiling point of the resultant solution was 0.357 °C higher than that of the pure solvent. Calculate the molar mass of the solute.

SOLUTION

Analyze Our goal is to calculate the molar mass of a solute based on knowledge of the boiling-point elevation of its solution in CCl_4, $\Delta T_b = 0.357$ °C, and the masses of solute and solvent. Table 13.3 gives K_b for the solvent (CCl_4), $K_b = 5.02$ °C/*m*.

Plan We can use Equation 13.12, $\Delta T_b = iK_b m$, to calculate the molality of the solution. Because the solute is a nonelectrolyte, $i = 1$. Then we can use molality and the quantity of solvent (40.0 g CCl_4) to calculate the number of moles of solute. Finally, the molar mass of the solute equals the number of grams per mole, so we divide the number of grams of solute (0.250 g) by the number of moles we have just calculated.

Solve From Equation 13.12, we have

$$\text{Molality} = \frac{\Delta T_b}{iK_b} = \frac{0.357\,°C}{(1)5.02\,°C/m} = 0.0711\ m$$

Thus, the solution contains 0.0711 mol of solute per kilogram of solvent. The solution was prepared using 40.0 g = 0.0400 kg of solvent (CCl_4). The number of moles of solute in the solution is therefore

$$(0.0400\ \text{kg}\ CCl_4)\left(0.0711\ \frac{\text{mol solute}}{\text{kg}\ CCl_4}\right) = 2.84 \times 10^{-3}\ \text{mol solute}$$

The molar mass of the solute is the number of grams per mole of the substance:

$$\text{Molar mass} = \frac{0.250\ \text{g}}{2.84 \times 10^{-3}\ \text{mol}} = 88.0\ \text{g/mol}$$

Practice Exercise 1

A mysterious white powder could be powdered sugar ($C_{12}H_{22}O_{11}$), cocaine ($C_{17}H_{21}NO_4$), codeine ($C_{18}H_{21}NO_3$), norfenefrine ($C_8H_{11}NO_2$), or fructose ($C_6H_{12}O_6$). When 80 mg of the powder is dissolved in 1.50 mL of ethanol ($d = 0.789$ g/cm^3, normal freezing point −114.6 °C, $K_f = 1.99$ °C/*m*), the freezing point is lowered to −115.5 °C. What is the identity of the white powder? **(a)** powdered sugar, **(b)** cocaine, **(c)** codeine, **(d)** norfenefrine, **(e)** fructose.

Practice Exercise 2

Camphor ($C_{10}H_{16}O$) melts at 179.8 °C, and it has a particularly large freezing-point-depression constant, $K_f = 40.0$ °C/*m*. When 0.186 g of an organic substance of unknown molar mass is dissolved in 22.01 g of liquid camphor, the freezing point of the mixture is found to be 176.7 °C. What is the molar mass of the solute?

SAMPLE EXERCISE 13.11 | Molar Mass from Osmotic Pressure

The osmotic pressure of an aqueous solution of a certain protein was measured to determine the protein's molar mass. The solution contained 3.50 mg of protein dissolved in sufficient water to form 5.00 mL of solution. The osmotic pressure of the solution at 25 °C was found to be 1.54 torr. Treating the protein as a nonelectrolyte, calculate its molar mass.

SOLUTION

Analyze Our goal is to calculate the molar mass of a high-molecular-mass protein, based on its osmotic pressure and a knowledge of the mass of protein and solution volume. Since the protein will be considered as a nonelectrolyte, $i = 1$.

Plan The temperature ($T = 25$ °C) and osmotic pressure ($\Pi = 1.54$ torr) are given, and we know the value of R so we can use Equation 13.14 to calculate the molarity of the solution, M. In doing so, we must convert temperature from °C to K and the osmotic pressure from torr to atm. We then use the molarity and the volume of the solution (5.00 mL) to determine the number of moles of solute. Finally, we obtain the molar mass by dividing the mass of the solute (3.50 mg) by the number of moles of solute.

Solve Solving Equation 13.14 for molarity gives

$$\text{Molarity} = \frac{\Pi}{iRT} = \frac{(1.54 \text{ torr})\left(\dfrac{1 \text{ atm}}{760 \text{ torr}}\right)}{(1)\left(0.0821 \dfrac{\text{L-atm}}{\text{mol-K}}\right)(298 \text{ K})} = 8.28 \times 10^{-5} \frac{\text{mol}}{\text{L}}$$

Because the volume of the solution is 5.00 mL = 5.00×10^{-3} L, the number of moles of protein must be

$$\text{Moles} = (8.28 \times 10^{-5} \text{ mol/L})(5.00 \times 10^{-3} \text{ L}) = 4.14 \times 10^{-7} \text{ mol}$$

The molar mass is the number of grams per mole of the substance. Because we know the sample has a mass of 3.50 mg = 3.50×10^{-3} g, we can calculate the molar mass by dividing the number of grams in the sample by the number of moles we just calculated:

$$\text{Molar mass} = \frac{\text{grams}}{\text{moles}} = \frac{3.50 \times 10^{-3} \text{ g}}{4.14 \times 10^{-7} \text{ mol}} = 8.45 \times 10^3 \text{ g/mol}$$

Comment Because small pressures can be measured easily and accurately, osmotic pressure measurements provide a useful way to determine the molar masses of large molecules.

Practice Exercise 1

Proteins frequently form complexes in which 2, 3, 4, or even more individual proteins ("monomers") interact specifically with each other via hydrogen bonds or electrostatic interactions. The entire assembly of proteins can act as one unit in solution, and this assembly is called the "quaternary structure" of the protein. Suppose you discover a new protein whose monomer molar mass is 25,000 g/mol. You measure an osmotic pressure of 0.0916 atm at 37 °C for 7.20 g of the protein in 10.00 mL of an aqueous solution. How many protein monomers form the quaternary protein structure in solution? Treat the protein as a nonelectrolyte. (a) 1, (b) 2, (c) 3, (d) 4 , (e) 8.

Practice Exercise 2

A sample of 2.05 g of polystyrene of uniform polymer chain length was dissolved in enough toluene to form 0.100 L of solution. The osmotic pressure of this solution was found to be 1.21 kPa at 25 °C. Calculate the molar mass of the polystyrene.

A Closer Look

The van't Hoff Factor

The colligative properties of solutions depend on the *total* concentration of solute particles, regardless of whether the particles are ions or molecules. Thus, we expect a 0.100 *m* solution of NaCl to have a freezing-point depression of $(2)(0.100 \text{ } m)(1.86 \text{ °C}/m) = 0.372 \text{ °C}$ because it is 0.100 *m* in $\text{Na}^+(aq)$ and 0.100 *m* in $\text{Cl}^-(aq)$. The measured freezing-point depression is only 0.348 °C, however, and the situation is similar for other strong electrolytes. A 0.100 *m* solution of KCl, for example, freezes at −0.344 °C.

The difference between expected and observed colligative properties for strong electrolytes is due to electrostatic attractions between ions. As the ions move about in solution, ions of opposite charge collide and "stick together" for brief moments. While they are together, they behave as a single particle called an *ion pair* (▶ Figure 13.24). The number of independent particles is thereby reduced, causing a reduction in the freezing-point depression (as well as in boiling-point elevation, vapor-pressure reduction, and osmotic pressure).

We have been assuming that the van't Hoff factor, *i*, is equal to the number of ions per formula unit of the electrolyte. The true (measured) value of this factor, however, is given by the ratio of the measured value of a colligative property to the value calculated when the substance is assumed to be a nonelectrolyte. Using the freezing-point depression, for example, we have

$$i = \frac{\Delta T_f(\text{measured})}{\Delta T_f(\text{calculated for nonelectrolyte})} \qquad [13.15]$$

The limiting value of *i* can be determined for a salt from the number of ions per formula unit. For NaCl, for example, the limiting van't Hoff factor is 2 because NaCl consists of one Na^+ and one Cl^- per formula unit; for K_2SO_4 it is 3 because K_2SO_4 consists of two K^+ and

one SO_4^{2-} per formula unit. In the absence of any information about the actual value of *i* for a solution, we will use the limiting value in calculations.

Two trends are evident in ▶ **Table 13.4**, which gives measured van't Hoff factors for several substances at different

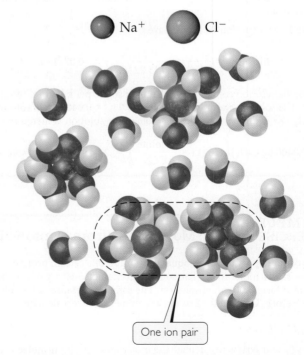

▲ **Figure 13.24 Ion pairing and colligative properties.** A solution of NaCl contains not only separated $\text{Na}^+(aq)$ and $\text{Cl}^-(aq)$ ions but ion pairs as well.

dilutions. First, dilution affects the value of *i* for electrolytes; the more dilute the solution, the more closely *i* approaches the expected value based on the number of ions in the formula unit. Thus, we conclude that the extent of ion pairing in electrolyte solutions decreases upon dilution. Second, the lower the charges on the ions, the less *i* departs from the expected value because the extent of ion pairing decreases as the ionic charges decrease. Both trends are consistent with simple electrostatics: The force of interaction between charged particles decreases as their separation increases and as their charges decrease.

Related Exercises: 13.83, 13.84, 13.103, 13.105

Table 13.4 Measured and Expected van't Hoff Factors for Several Substances at 25 °C

Compound	Concentration			Expected Value
	0.100 *m*	0.0100 *m*	0.00100 *m*	
Sucrose	1.00	1.00	1.00	1.00
NaCl	1.87	1.94	1.97	2.00
K_2SO_4	2.32	2.70	2.84	3.00
$MgSO_4$	1.21	1.53	1.82	2.00

13.6 | Colloids

Some substances appear to initially dissolve in a solvent, but over time, the substance separates from the pure solvent. For example, finely divided clay particles dispersed in water eventually settle out because of gravity. Gravity affects the clay particles because they are much larger than most molecules, consisting of thousands or even millions of atoms. In contrast, the dispersed particles in a true solution (ions in a salt solution or glucose molecules in a sugar solution) are small. Between these extremes lie dispersed particles that are larger than typical molecules but not so large that the components of the mixture separate under the influence of gravity. These intermediate types of dispersions are called either **colloidal dispersions** or simply **colloids**. Colloids form the dividing line between solutions and heterogeneous mixtures. Like solutions, colloids can be gases, liquids, or solids. Examples of each are listed in ▼ Table 13.5.

Particle size can be used to classify a mixture as colloid or solution. Colloid particles range in diameter from 5 to 1000 nm; solute particles are smaller than 5 nm in diameter. The nanomaterials we saw in Chapter 12 ∞ (Section 12.9), when dispersed in a liquid, are colloids. A colloid particle may even consist of a single giant molecule. The hemoglobin molecule, for example, which carries oxygen in your blood, has molecular dimensions of $6.5 \times 5.5 \times 5.0$ nm and a molar mass of 64,500 g/mol.

Although colloid particles may be so small that the dispersion appears uniform even under a microscope, they are large enough to scatter light. Consequently, most colloids appear cloudy or opaque unless they are very dilute. (For example, homogenized milk is a colloid of fat and protein molecules dispersed in water.) Furthermore, because they scatter light, a light beam can be seen as it passes through a colloidal

Table 13.5 Types of Colloids

Phase of Colloid	Dispersing (solvent-like) Substance	Dispersed (solute-like) Substance	Colloid Type	Example
Gas	Gas	Gas	—	None (all are solutions)
Gas	Gas	Liquid	Aerosol	Fog
Gas	Gas	Solid	Aerosol	Smoke
Liquid	Liquid	Gas	Foam	Whipped cream
Liquid	Liquid	Liquid	Emulsion	Milk
Liquid	Liquid	Solid	Sol	Paint
Solid	Solid	Gas	Solid foam	Marshmallow
Solid	Solid	Liquid	Solid emulsion	Butter
Solid	Solid	Solid	Solid sol	Ruby glass

▲ Figure 13.25 **Tyndall effect in the laboratory.** The glass on the right contains a colloidal dispersion; that on the left contains a solution.

dispersion (▲ Figure 13.25). This scattering of light by colloidal particles, known as the **Tyndall effect**, makes it possible to see the light beam of an automobile on a dusty dirt road, or the sunlight streaming through trees or clouds. Not all wavelengths are scattered to the same extent. Colors at the blue end of the visible spectrum are scattered more than those at the red end by the molecules and small dust particles in the atmosphere. As a result, our sky appears blue. At sunset, light from the sun travels through more of the atmosphere; blue light is scattered even more, allowing the reds and yellows to pass through and be seen.

Hydrophilic and Hydrophobic Colloids

The most important colloids are those in which the dispersing medium is water. These colloids may be **hydrophilic** ("water loving") or **hydrophobic** ("water fearing"). Hydrophilic colloids are most like the solutions that we have previously examined. In the human body, the extremely large protein molecules such as enzymes and antibodies are kept in suspension by interaction with surrounding water molecules. A hydrophilic molecule folds in such a way that its hydrophobic groups are away from the water molecules, on the inside of the folded molecule, while its hydrophilic, polar groups are on the surface, interacting with the water molecules. The hydrophilic groups generally contain oxygen or nitrogen and often carry a charge (◀ Figure 13.26).

▲ **GO FIGURE**

What is the chemical composition of the groups that carry a negative charge?

Hydrophilic polar and charged groups on molecule surface help molecule remain dispersed in water and other polar solvents

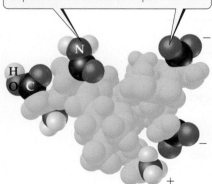

▲ Figure 13.26 **Hydrophilic colloidal particle.** Examples of the hydrophilic groups that help to keep a giant molecule (macromolecule) suspended in water.

▲ **Give it Some Thought**

Some proteins reside in the hydrophobic lipid bilayer of the cell membrane. Would hydrophilic groups of these proteins still be facing the lipid "solvent"?

Hydrophobic colloids can be dispersed in water only if they are stabilized in some way. Otherwise, their natural lack of affinity for water causes them to separate from the water. One method of stabilization involves adsorbing ions on the surface of the hydrophobic particles (▶ Figure 13.27). (*Adsorption* means to adhere to a surface. It differs from *absorption*, which means to pass into the interior, as when a

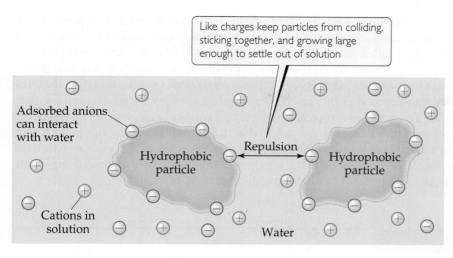

▲ Figure 13.27 **Hydrophobic colloids stabilized in water by adsorbed anions.**

sponge absorbs water.) The adsorbed ions can interact with water, thereby stabiliz-
ing the colloid. At the same time, the electrostatic repulsion between adsorbed ions
on neighboring colloid particles keeps the particles from sticking together rather
than dispersing in the water.

Hydrophobic colloids can also be stabilized by hydrophilic groups on their surfaces. Oil drops are hydrophobic, for example, and they do not remain suspended in
water. Instead, they aggregate, forming an oil slick on the water surface. Sodium stearate (▼ Figure 13.28), or any similar substance having one end that is hydrophilic
(either polar or charged) and one end that is hydrophobic (nonpolar), will stabilize a
suspension of oil in water. Stabilization results from the interaction of the hydrophobic
ends of the stearate ions with the oil drops and the hydrophilic ends with the water.

 Give It Some Thought

Why don't oil drops stabilized by sodium stearate coagulate to form larger oil drops?

GO FIGURE

Which kind of intermolecular force attracts the stearate ion to the oil drop?

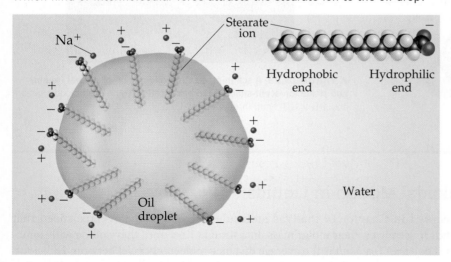

▲ Figure 13.28 **Stabilization of an emulsion of oil in water by stearate ions.**

Colloid stabilization has an interesting application in the human digestive system. When fats in our diet reach the small intestine, a hormone causes the gallbladder to excrete a fluid called bile. Among the components of bile are compounds that have chemical structures similar to sodium stearate; that is, they have a hydrophilic (polar) end and a hydrophobic (nonpolar) end. These compounds emulsify the fats in the intestine and thus permit digestion and absorption of fat-soluble vitamins through the intestinal wall. The term *emulsify* means "to form an emulsion," a suspension of one liquid in another, with milk being one example (Table 13.5). A substance that aids in the formation of an emulsion is called an emulsifying agent. If you read the labels on foods and other materials, you will find that a variety of chemicals are used as emulsifying agents. These chemicals typically have a hydrophilic end and a hydrophobic end.

Chemistry and Life

Sickle-Cell Anemia

Our blood contains the complex protein hemoglobin, which carries oxygen from the lungs to other parts of the body. In the genetic disease sickle-cell anemia, hemoglobin molecules are abnormal and have a lower solubility in water, especially in their unoxygenated form. Consequently, as much as 85% of the hemoglobin in red blood cells crystallizes out of solution.

The cause of the insolubility is a structural change in one part of an amino acid. Normal hemoglobin molecules contain an amino acid that has a —CH_2CH_2COOH group:

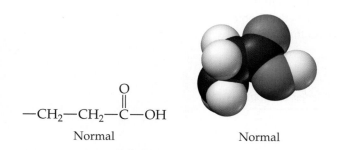

Normal Normal

The polarity of the —COOH group contributes to the solubility of the hemoglobin molecule in water. In the hemoglobin molecules of sickle-cell anemia patients, the —CH_2CH_2COOH chain is absent and in its place is the nonpolar (hydrophobic) —$CH(CH_3)_2$ group:

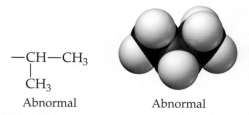

Abnormal Abnormal

This change leads to the aggregation of the defective form of hemoglobin into particles too large to remain suspended in biological fluids. It also causes the cells to distort into the sickle shape shown in ▼ Figure 13.29. The sickled cells tend to clog capillaries, causing severe pain, weakness, and the gradual deterioration of vital organs. The disease is hereditary, and if both parents carry the defective genes, it is likely that their children will possess only abnormal hemoglobin.

You might wonder how it is that a life-threatening disease such as sickle-cell anemia has persisted in humans through evolutionary time. The answer in part is that people with the disease are far less susceptible to malaria. Thus, in tropical climates rife with malaria, those with sickle-cell disease have lower incidence of this debilitating disease.

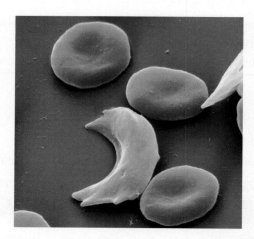

▲ Figure 13.29 **A scanning electron micrograph of normal (round) and sickle (crescent-shaped) red blood cells.** Normal red blood cells are about 6×10^{-3} mm in diameter.

Colloidal Motion in Liquids

We learned in Chapter 10 that gas molecules move at some average speed that depends inversely on their molar mass, in a straight line, until they collide with something. The *mean free path* is the average distance molecules travel between collisions. ∞∞(Section 10.8) Recall also that the kinetic-molecular theory of gases assumes that gas molecules are in continuous, random motion. ∞∞(Section 10.7)

Colloidal particles in a solution undergo random motion as a result of collisions with solvent molecules. Because the colloidal particles are massive in comparison with solvent molecules, their movements from any one collision are very tiny. However, there are many such collisions, and they cause a random motion of the entire colloidal particle, called **Brownian motion**. In 1905, Einstein developed an equation for the average square of the displacemet of a colloidal particle, a historically very important development. As you might expect, the larger the colloidal particle, the shorter its mean free path in a given liquid (▼ Table 13.6). Today, the understanding of Brownian motion is applied to diverse problems in everything from cheese-making to medical imaging.

Table 13.6 Calculated Mean Free Path, after One Hour, for Uncharged Colloidal Spheres in Water at 20 °C

Radius of sphere, nm	Mean Free Path, mm
1	1.23
10	0.390
100	0.123
1000	0.039

SAMPLE INTEGRATIVE EXERCISE   Putting Concepts Together

A 0.100-L solution is made by dissolving 0.441 g of $CaCl_2(s)$ in water. (a) Calculate the osmotic pressure of this solution at 27 °C, assuming that it is completely dissociated into its component ions. (b) The measured osmotic pressure of this solution is 2.56 atm at 27 °C. Explain why it is less than the value calculated in (a), and calculate the van't Hoff factor, i, for the solute in this solution. (c) The enthalpy of solution for $CaCl_2$ is $\Delta H = -81.3$ kJ/mol. If the final temperature of the solution is 27 °C, what was its initial temperature? (Assume that the density of the solution is 1.00 g/mL, that its specific heat is 4.18 J/g-K, and that the solution loses no heat to its surroundings.)

SOLUTION

(a) The osmotic pressure is given by Equation 13.14, $\Pi = iMRT$. We know the temperature, $T = 27\,°C = 300$ K, and the gas constant, $R = 0.0821$ L-atm/mol-K. We can calculate the molarity of the solution from the mass of $CaCl_2$ and the volume of the solution:

$$\text{Molarity} = \left(\frac{0.441 \text{ g CaCl}_2}{0.100 \text{ L}} \right)\left(\frac{1 \text{ mol CaCl}_2}{110 \text{ g CaCl}_2} \right) = 0.0397 \text{ mol CaCl}_2/\text{L}$$

Soluble ionic compounds are strong electrolytes. ∞ (Sections 4.1 and 4.3) Thus, $CaCl_2$ consists of metal cations (Ca^{2+}) and nonmetal anions (Cl^-). When completely dissociated, each $CaCl_2$ unit forms three ions (one Ca^{2+} and two Cl^-). Hence, the calculated osmotic pressure is

$$\Pi = iMRT = (3)(0.0397 \text{ mol/L})(0.0821 \text{ L-atm/mol-K})(300 \text{ K}) = 2.93 \text{ atm}$$

(b) The actual values of colligative properties of electrolytes are less than those calculated because the electrostatic interactions between ions limit their independent movements. In this case, the van't Hoff factor, which measures the extent to which electrolytes actually dissociate into ions, is given by

$$i = \frac{\Pi(\text{measured})}{\Pi(\text{calculated for nonelectrolyte})}$$

$$= \frac{2.56 \text{ atm}}{(0.0397 \text{ mol/L})(0.0821 \text{ L-atm/mol-K})(300 \text{ K})} = 2.62$$

Thus, the solution behaves as if the $CaCl_2$ has dissociated into 2.62 particles instead of the ideal 3.

(c) If the solution is 0.0397 M in $CaCl_2$ and has a total volume of 0.100 L, the number of moles of solute is $(0.100 \text{ L})(0.0397 \text{ mol/L}) = 0.00397$ mol. Hence, the quantity of heat generated in forming the solution is $(0.00397 \text{ mol})(-81.3 \text{ kJ/mol}) = -0.323$ kJ. The solution absorbs this heat, causing its temperature to increase. The relationship between temperature change and heat is given by Equation 5.22:

$$q = (\text{specific heat})(\text{grams})(\Delta T)$$

The heat absorbed by the solution is $q = +0.323 \text{ kJ} = 323 \text{ J}$. The mass of the 0.100 L of solution is $(100 \text{ mL})(1.00 \text{ g/mL}) = 100 \text{ g}$ (to three significant figures). Thus, the temperature change is

$$\Delta T = \frac{q}{(\text{specific heat of solution})(\text{grams of solution})}$$

$$= \frac{323 \text{ J}}{(4.18 \text{ J/g-K})(100 \text{ g})} = 0.773 \text{ K}$$

A kelvin has the same size as a degree Celsius. ∞ (Section 1.4) Because the solution temperature increases by 0.773 °C, the initial temperature was 27.0 °C − 0.773 °C = 26.2 °C.

Chapter Summary and Key Terms

THE SOLUTION PROCESS (SECTION 13.1) Solutions form when one substance disperses uniformly throughout another. The attractive interaction of solvent molecules with solute is called **solvation**. When the solvent is water, the interaction is called **hydration**. The dissolution of ionic substances in water is promoted by hydration of the separated ions by the polar water molecules. The overall enthalpy change upon solution formation may be either positive or negative. Solution formation is favored both by a positive entropy change, corresponding to an increased dispersal of the components of the solution, and by a negative enthalpy change, indicating an exothermic process.

SATURATED SOLUTIONS AND SOLUBILITY (SECTION 13.2) The equilibrium between a saturated solution and undissolved solute is dynamic; the process of solution and the reverse process, **crystallization**, occur simultaneously. In a solution in equilibrium with undissolved solute, the two processes occur at equal rates, giving a **saturated** solution. If there is less solute present than is needed to saturate the solution, the solution is **unsaturated**. When solute concentration is greater than the equilibrium concentration value, the solution is **supersaturated**. This is an unstable condition, and separation of some solute from the solution will occur if the process is initiated with a solute seed crystal. The amount of solute needed to form a saturated solution at any particular temperature is the **solubility** of that solute at that temperature.

FACTORS AFFECTING SOLUBILITY (SECTION 13.3) The solubility of one substance in another depends on the tendency of systems to become more random, by becoming more dispersed in space, and on the relative intermolecular solute–solute and solvent–solvent energies compared with solute–solvent interactions. Polar and ionic solutes tend to dissolve in polar solvents, and nonpolar solutes tend to dissolve in nonpolar solvents ("like dissolves like"). Liquids that mix in all proportions are **miscible**; those that do not dissolve significantly in one another are **immiscible**. Hydrogen-bonding interactions between solute and solvent often play an important role in determining solubility; for example, ethanol and water, whose molecules form hydrogen bonds with each other, are miscible. The solubilities of gases in a liquid are generally proportional to the pressure of the gas over the solution, as expressed by **Henry's law**: $S_g = kP_g$. The solubilities of most solid solutes in water increase as the temperature of the solution increases. In contrast, the solubilities of gases in water generally decrease with increasing temperature.

EXPRESSING SOLUTION CONCENTRATIONS (SECTION 13.4) Concentrations of solutions can be expressed quantitatively by several different measures, including **mass percentage** [(mass solute/mass solution) × 100] **parts per million (ppm)**, **parts per billion (ppb)**, and mole fraction. Molarity, M, is defined as moles of solute per liter of solution; **molality**, m, is defined as moles of solute per kilogram of solvent. Molarity can be converted to these other concentration units if the density of the solution is known.

COLLIGATIVE PROPERTIES (SECTION 13.5) A physical property of a solution that depends on the concentration of solute particles present, regardless of the nature of the solute, is a **colligative property**. Colligative properties include vapor-pressure lowering, freezing-point lowering, boiling-point elevation, and osmotic pressure. **Raoult's law** expresses the lowering of vapor pressure. An **ideal solution** obeys Raoult's law. Differences in solvent–solute as compared with solvent–solvent and solute–solute intermolecular forces cause many solutions to depart from ideal behavior.

A solution containing a nonvolatile solute possesses a higher boiling point than the pure solvent. The **molal boiling-point-elevation constant**, K_b, represents the increase in boiling point for a 1 m solution of solute particles as compared with the pure solvent. Similarly, the **molal freezing-point-depression constant**, K_f, measures the lowering of the freezing point of a solution for a 1 m solution of solute particles. The temperature changes are given by the equations $\Delta T_b = iK_b m$ and $\Delta T_f = -iK_f m$ where i is the **van't Hoff factor**, which represents how many particles the solute breaks up into in the solvent. When NaCl dissolves in water, two moles of solute particles are formed for each mole of dissolved salt. The boiling point or freezing point is thus elevated or depressed, respectively, approximately twice as much as that of a nonelectrolyte solution of the same concentration. Similar considerations apply to other strong electrolytes.

Osmosis is the movement of solvent molecules through a semipermeable membrane from a less concentrated to a more concentrated solution. This net movement of solvent generates an **osmotic pressure**, Π, which can be measured in units of gas pressure, such as atm. The osmotic pressure of a solution is proportional to the solution molarity: $\Pi = iMRT$. Osmosis is a very important process in living systems, in which cell walls act as semipermeable membranes, permitting the passage of water but restricting the passage of ionic and macromolecular components.

COLLOIDS (SECTION 13.6) Particles that are large on the molecular scale but still small enough to remain suspended indefinitely in a solvent system form **colloids**, or **colloidal dispersions**. Colloids, which are intermediate between solutions and heterogeneous mixtures, have many practical applications. One useful physical property of colloids, the scattering of visible light, is referred to as the **Tyndall effect**. Aqueous colloids are classified as **hydrophilic** or **hydrophobic**. Hydrophilic colloids are common in living organisms, in which large molecular aggregates (enzymes, antibodies) remain suspended because they have many polar, or charged, atomic groups on their surfaces that interact with water. Hydrophobic colloids, such as small droplets of oil, may remain in suspension through adsorption of charged particles on their surfaces.

Colloids undergo **Brownian motion** in liquids, analogous to the random three-dimensional motion of gas molecules.

Learning Outcomes After studying this chapter, you should be able to:

- Describe how enthalpy and entropy changes affect solution formation. (Section 13.1)

- Describe the relationship between intermolecular forces and solubility, including use of the "like dissolves like" rule. (Sections 13.1 and 13.3)

- Describe the role of equilibrium in the solution process and its relationship to the solubility of a solute. (Section 13.2)

- Describe the effect of temperature on the solubility of solids and gases in liquids. (Section 13.3)

- Describe the relationship between the partial pressure of a gas and its solubility. (Section 13.3)

- Calculate the concentration of a solution in terms of molarity, molality, mole fraction, percent composition, and parts per million and be able to interconvert between them. (Section 13.4)

- Describe what a colligative property is and explain the difference between the effects of nonelectrolytes and electrolytes on colligative properties. (Section 13.5)

- Calculate the vapor pressure of a solvent over a solution. (Section 13.5)

- Calculate the boiling-point elevation and freezing-point depression of a solution. (Section 13.5)

- Calculate the osmotic pressure of a solution. (Section 13.5)

- Explain the difference between a solution and a colloid. (Section 13.6)

- Describe the similarities between the motions of gas molecules and the motions of colloids in a liquid. (Section 13.6)

Key Equations

- $S_g = kP_g$ [13.4]   Henry's law, which relates gas solubility to partial pressure

- Mass % of component $= \dfrac{\text{mass of component in soln}}{\text{total mass of soln}} \times 100$ [13.5]   Concentration in terms of mass percent

- ppm of component $= \dfrac{\text{mass of component in soln}}{\text{total mass of soln}} \times 10^6$ [13.6]   Concentration in terms of parts per million (ppm)

- Mole fraction of component $= \dfrac{\text{moles of component}}{\text{total moles of all components}}$ [13.7]   Concentration in terms of mole fraction

- Molarity $= \dfrac{\text{moles of solute}}{\text{liters of soln}}$ [13.8]   Concentration in terms of molarity

- Molality $= \dfrac{\text{moles of solute}}{\text{kilograms of solvent}}$ [13.9]   Concentration in terms of molality

- $P_{\text{solution}} = X_{\text{solvent}} P^\circ_{\text{solvent}}$ [13.10]   Raoult's law, calculating vapor pressure of solvent above a solution

- $\Delta T_b = iK_b m$ [13.12]   Calculating the boiling-point elevation of a solution

- $\Delta T_f = -iK_f m$ [13.13]   Calculating the freezing-point depression of a solution

- $\Pi = i\left(\dfrac{n}{V}\right)RT = iMRT$ [13.14]   Calculating the osmotic pressure of a solution

Exercises

Visualizing Concepts

13.1 Rank the contents of the following containers in order of increasing entropy: [Section 13.1]

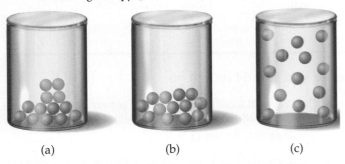

(a) (b) (c)

13.2 This figure shows the interaction of a cation with surrounding water molecules.

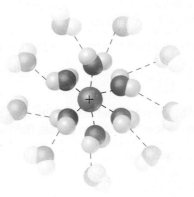

(a) Which atom of water is associated with the cation? Explain.

(b) Which of the following explanations accounts for the fact that the ion-solvent interaction is greater for Li$^+$ than for K$^+$?

a. Li$^+$ is of lower mass than K$^+$.

b. The ionization energy of Li is higher than that for K.

c. Li$^+$ has a smaller ionic radius than K$^+$.

d. Li has a lower density than K.

e. Li reacts with water more slowly than K. [Section 13.1]

13.3 Consider two ionic solids, both composed of singly-charged ions, that have different lattice energies. (a) Will the solids have the same solubility in water? (b) If not, which solid will be more soluble in water, the one with the larger lattice energy or the one with the smaller lattice energy? Assume that solute-solvent interactions are the same for both solids. [Section 13.1]

13.4 Are gases always miscible with each other? Explain. [Section 13.1]

13.5 Which of the following is the best representation of a saturated solution? Explain your reasoning. [Section 13.2]

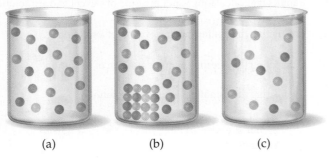

(a) (b) (c)

13.6 The solubility of Xe in water at 1 atm pressure and 20 °C is approximately 5×10^{-3} M. (a) Compare this with the solubilities of Ar and Kr in water (Table 13.1). (b) What properties of the rare gas atoms account for the variation in solubility? [Section 13.3]

13.7 The structures of vitamins E and B$_6$ are shown below. Predict which is more water soluble and which is more fat soluble. Explain. [Section 13.3]

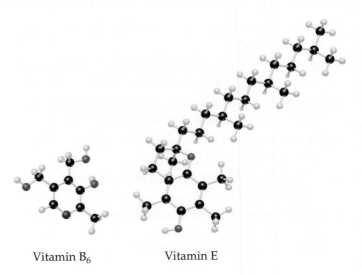

Vitamin B$_6$ Vitamin E

13.8 You take a sample of water that is at room temperature and in contact with air and put it under a vacuum. Right away, you see bubbles leave the water, but after a little while, the bubbles stop. As you keep applying the vacuum, more bubbles appear. A friend tells you that the first bubbles were water vapor, and the low pressure had reduced the boiling point of water, causing the water to boil. Another friend tells you that the first bubbles were gas molecules from the air (oxygen, nitrogen, and so forth) that were dissolved in the water. Which friend is mostly likely to be correct? What, then, is responsible for the second batch of bubbles? [Section 13.4]

13.9 The figure shows two identical volumetric flasks containing the same solution at two temperatures.

(a) Does the molarity of the solution change with the change in temperature? Explain.

(b) Does the molality of the solution change with the change in temperature? Explain. [Section 13.4]

25 °C   55 °C

13.10 This portion of a phase diagram shows the vapor-pressure curves of a volatile solvent and of a solution of that solvent containing a nonvolatile solute. **(a)** Which line represents the solution? **(b)** What are the normal boiling points of the solvent and the solution? [Section 13.5]

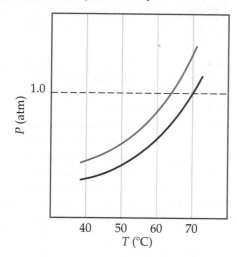

13.11 Suppose you had a balloon made of some highly flexible semipermeable membrane. The balloon is filled completely with a 0.2 *M* solution of some solute and is submerged in a 0.1 *M* solution of the same solute:

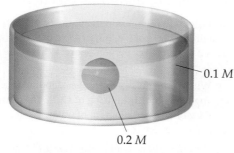

0.1 *M*

0.2 *M*

Initially, the volume of solution in the balloon is 0.25 L. Assuming the volume outside the semipermeable membrane is large, as the illustration shows, what would you expect for the solution volume inside the balloon once the system has come to equilibrium through osmosis? [Section 13.5]

13.12 The molecule *n*-octylglucoside, shown here, is widely used in biochemical research as a nonionic detergent for "solubilizing" large hydrophobic protein molecules. What characteristics of this molecule are important for its use in this way? [Section 13.6]

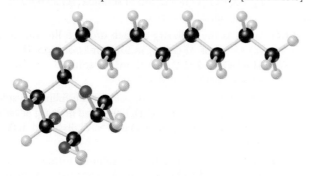

The Solution Process (Section 13.1)

13.13 Indicate whether each statement is true or false:

    (a) A solute will dissolve in a solvent if solute–solute interactions are stronger than solute-solvent interactions.

    (b) In making a solution, the enthalpy of mixing is always a positive number.

    (c) An increase in entropy favors mixing.

13.14 Indicate whether each statement is true or false: **(a)** NaCl dissolves in water but not in benzene (C_6H_6) because benzene is denser than water. **(b)** NaCl dissolves in water but not in benzene because water has a large dipole moment and benzene has zero dipole moment. **(c)** NaCl dissolves in water but not in benzene because the water–ion interactions are stronger than benzene–ion interactions.

13.15 Indicate the type of solute–solvent interaction (Section 11.2) that should be most important in each of the following solutions: **(a)** CCl_4 in benzene (C_6H_6), **(b)** methanol (CH_3OH) in water, **(c)** KBr in water, **(d)** HCl in acetonitrile (CH_3CN).

13.16 Indicate the principal type of solute–solvent interaction in each of the following solutions and rank the solutions from weakest to strongest solute–solvent interaction: **(a)** KCl in water, **(b)** CH_2Cl_2 in benzene (C_6H_6), **(c)** methanol (CH_3OH) in water.

13.17 An ionic compound has a very negative ΔH_{soln} in water. **(a)** Would you expect it to be very soluble or nearly insoluble in water? **(b)** Which term would you expect to be the largest negative number: $\Delta H_{solvent}$, ΔH_{solute}, or ΔH_{mix}?

13.18 When ammonium chloride dissolves in water, the solution becomes colder. **(a)** Is the solution process exothermic or endothermic? **(b)** Why does the solution form?

13.19 **(a)** In Equation 13.1, which of the enthalpy terms for dissolving an ionic solid would correspond to the lattice energy? **(b)** Which energy term in this equation is always exothermic?

13.20 For the dissolution of LiCl in water, $\Delta H_{soln} = -37 \text{ kJ/mol}$. Which term would you expect to be the largest negative number: $\Delta H_{solvent}$, ΔH_{solute}, or ΔH_{mix}?

13.21 Two nonpolar organic liquids, hexane (C_6H_{14}) and heptane (C_7H_{16}), are mixed. **(a)** Do you expect ΔH_{soln} to be a large positive number, a large negative number, or close to zero? Explain. **(b)** Hexane and heptane are miscible with each other in all proportions. In making a solution of them, is the entropy of the system increased, decreased, or close to zero, compared to the separate pure liquids?

13.22 The enthalpy of solution of KBr in water is about $+198 \text{ kJ/mol}$. Nevertheless, the solubility of KBr in water is relatively high. Why does the solution process occur even though it is endothermic?

Saturated Solutions; Factors Affecting Solubility (Sections 13.2 and 13.3)

13.23 The solubility of $Cr(NO_3)_3 \cdot 9 H_2O$ in water is 208 g per 100 g of water at 15 °C. A solution of $Cr(NO_3)_3 \cdot 9 H_2O$ in water at 35 °C is formed by dissolving 324 g in 100 g of water. When this solution is slowly cooled to 15 °C, no precipitate forms. **(a)** What term describes this solution? **(b)** What action might you take to initiate crystallization? Use molecular-level processes to explain how your suggested procedure works.

13.24 The solubility of $MnSO_4 \cdot H_2O$ in water at 20 °C is 70 g per 100 mL of water. **(a)** Is a 1.22 *M* solution of $MnSO_4 \cdot H_2O$ in water at 20 °C saturated, supersaturated, or unsaturated? **(b)** Given a solution of $MnSO_4 \cdot H_2O$ of unknown concentration, what experiment could you perform to determine whether the new solution is saturated, supersaturated, or unsaturated?

13.25 By referring to Figure 13.15, determine whether the addition of 40.0 g of each of the following ionic solids to 100 g of water at 40 °C will lead to a saturated solution: (a) $NaNO_3$, (b) KCl, (c) $K_2Cr_2O_7$, (d) $Pb(NO_3)_2$.

13.26 By referring to Figure 13.15, determine the mass of each of the following salts required to form a saturated solution in 250 g of water at 30 °C: (a) $KClO_3$, (b) $Pb(NO_3)_2$, (c) $Ce_2(SO_4)_3$.

13.27 Consider water and glycerol, $CH_2(OH)CH(OH)CH_2OH$. (a) Would you expect them to be miscible in all proportions? Explain. (b) List the intermolecular attractions that occur between a water molecule and a glycerol molecule.

13.28 Oil and water are immiscible. Which is the most likely reason? (a) Oil molecules are denser than water. (b) Oil molecules are composed mostly of carbon and hydrogen. (c) Oil molecules have higher molar masses than water. (d) Oil molecules have higher vapor pressures than water. (e) Oil molecules have higher boiling points than water.

13.29 Common laboratory solvents include acetone (CH_3COCH_3), methanol (CH_3OH), toluene ($C_6H_5CH_3$), and water. Which of these is the best solvent for nonpolar solutes?

13.30 Would you expect alanine (an amino acid) to be more soluble in water or in hexane? Explain.

Alanine

13.31 (a) Would you expect stearic acid, $CH_3(CH_2)_{16}COOH$, to be more soluble in water or in carbon tetrachloride? Explain.

(b) Which would you expect to be more soluble in water, cyclohexane or dioxane? Explain.

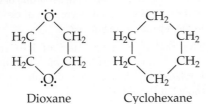

Dioxane   Cyclohexane

13.32 Ibuprofen, widely used as a pain reliever, has a limited solubility in water, less than 1 mg/mL. Which part of the molecule's structure (gray, white, red) contributes to its water solubility? Which part of the molecule (gray, white, red) contributes to its water insolubility?

Ibuprofen

13.33 Which of the following in each pair is likely to be more soluble in hexane, C_6H_{14}: (a) CCl_4 or $CaCl_2$, (b) benzene (C_6H_6) or glycerol, $CH_2(OH)CH(OH)CH_2OH$, (c) octanoic acid, $CH_3CH_2CH_2CH_2CH_2CH_2CH_2COOH$, or acetic acid, CH_3COOH? Explain your answer in each case.

13.34 Which of the following in each pair is likely to be more soluble in water: (a) cyclohexane (C_6H_{12}) or glucose ($C_6H_{12}O_6$), (b) propionic acid (CH_3CH_2COOH) or sodium propionate (CH_3CH_2COONa), (c) HCl or ethyl chloride (CH_3CH_2Cl)? Explain in each case.

13.35 (a) Explain why carbonated beverages must be stored in sealed containers. (b) Once the beverage has been opened, why does it maintain more carbonation when refrigerated than at room temperature?

13.36 Explain why pressure substantially affects the solubility of O_2 in water but has little effect on the solubility of NaCl in water.

13.37 The Henry's law constant for helium gas in water at 30 °C is $3.7 \times 10^{-4}\ M/\text{atm}$ and the constant for N_2 at 30 °C is $6.0 \times 10^{-4}\ M/\text{atm}$. If the two gases are each present at 1.5 atm pressure, calculate the solubility of each gas.

13.38 The partial pressure of O_2 in air at sea level is 0.21 atm. Using the data in Table 13.1, together with Henry's law, calculate the molar concentration of O_2 in the surface water of a mountain lake saturated with air at 20 °C and an atmospheric pressure of 650 torr.

Concentrations of Solutions (Section 13.4)

13.39 (a) Calculate the mass percentage of Na_2SO_4 in a solution containing 10.6 g of Na_2SO_4 in 483 g of water. (b) An ore contains 2.86 g of silver per ton of ore. What is the concentration of silver in ppm?

13.40 (a) What is the mass percentage of iodine in a solution containing 0.035 mol I_2 in 125 g of CCl_4? (b) Seawater contains 0.0079 g of Sr^{2+} per kilogram of water. What is the concentration of Sr^{2+} in ppm?

13.41 A solution is made containing 14.6 g of CH_3OH in 184 g of H_2O. Calculate (a) the mole fraction of CH_3OH, (b) the mass percent of CH_3OH, (c) the molality of CH_3OH.

13.42 A solution is made containing 20.8 g of phenol (C_6H_5OH) in 425 g of ethanol (CH_3CH_2OH). Calculate (a) the mole fraction of phenol, (b) the mass percent of phenol, (c) the molality of phenol.

13.43 Calculate the molarity of the following aqueous solutions: (a) 0.540 g of $Mg(NO_3)_2$ in 250.0 mL of solution, (b) 22.4 g of $LiClO_4 \cdot 3\ H_2O$ in 125 mL of solution, (c) 25.0 mL of 3.50 M HNO_3 diluted to 0.250 L.

13.44 What is the molarity of each of the following solutions: (a) 15.0 g of $Al_2(SO_4)_3$ in 0.250 mL solution, (b) 5.25 g of $Mn(NO_3)_2 \cdot 2\ H_2O$ in 175 mL of solution, (c) 35.0 mL of 9.00 M H_2SO_4 diluted to 0.500 L?

13.45 Calculate the molality of each of the following solutions: (a) 8.66 g of benzene (C_6H_6) dissolved in 23.6 g of carbon tetrachloride (CCl_4), (b) 4.80 g of NaCl dissolved in 0.350 L of water.

13.46 (a) What is the molality of a solution formed by dissolving 1.12 mol of KCl in 16.0 mol of water? (b) How many grams of sulfur (S_8) must be dissolved in 100.0 g of naphthalene ($C_{10}H_8$) to make a 0.12 m solution?

13.47 A sulfuric acid solution containing 571.6 g of H_2SO_4 per liter of solution has a density of 1.329 g/cm^3. Calculate

(a) the mass percentage, (b) the mole fraction, (c) the molality, (d) the molarity of H_2SO_4 in this solution.

13.48 Ascorbic acid (vitamin C, $C_6H_8O_6$) is a water-soluble vitamin. A solution containing 80.5 g of ascorbic acid dissolved in 210 g of water has a density of 1.22 g/mL at 55 °C. Calculate (a) the mass percentage, (b) the mole fraction, (c) the molality, (d) the molarity of ascorbic acid in this solution.

13.49 The density of acetonitrile (CH_3CN) is 0.786 g/mL and the density of methanol (CH_3OH) is 0.791 g/mL. A solution is made by dissolving 22.5 mL of CH_3OH in 98.7 mL of CH_3CN. (a) What is the mole fraction of methanol in the solution? (b) What is the molality of the solution? (c) Assuming that the volumes are additive, what is the molarity of CH_3OH in the solution?

13.50 The density of toluene (C_7H_8) is 0.867 g/mL, and the density of thiophene (C_4H_4S) is 1.065 g/mL. A solution is made by dissolving 8.10 g of thiophene in 250.0 mL of toluene. (a) Calculate the mole fraction of thiophene in the solution. (b) Calculate the molality of thiophene in the solution. (c) Assuming that the volumes of the solute and solvent are additive, what is the molarity of thiophene in the solution?

13.51 Calculate the number of moles of solute present in each of the following aqueous solutions: (a) 600 mL of 0.250 M $SrBr_2$, (b) 86.4 g of 0.180 m KCl, (c) 124.0 g of a solution that is 6.45% glucose ($C_6H_{12}O_6$) by mass.

13.52 Calculate the number of moles of solute present in each of the following solutions: (a) 255 mL of 1.50 M $HNO_3(aq)$, (b) 50.0 mg of an aqueous solution that is 1.50 m NaCl, (c) 75.0 g of an aqueous solution that is 1.50% sucrose ($C_{12}H_{22}O_{11}$) by mass.

13.53 Describe how you would prepare each of the following aqueous solutions, starting with solid KBr: (a) 0.75 L of 1.5×10^{-2} M KBr, (b) 125 g of 0.180 m KBr, (c) 1.85 L of a solution that is 12.0% KBr by mass (the density of the solution is 1.10 g/mL), (d) a 0.150 M solution of KBr that contains just enough KBr to precipitate 16.0 g of AgBr from a solution containing 0.480 mol of $AgNO_3$.

13.54 Describe how you would prepare each of the following aqueous solutions: (a) 1.50 L of 0.110 M $(NH_4)_2SO_4$ solution, starting with solid $(NH_4)_2SO_4$; (b) 225 g of a solution that is 0.65 m in Na_2CO_3, starting with the solid solute; (c) 1.20 L of a solution that is 15.0% $Pb(NO_3)_2$ by mass (the density of the solution is 1.16 g/mL), starting with solid solute; (d) a 0.50 M solution of HCl that would just neutralize 5.5 g of $Ba(OH)_2$ starting with 6.0 M HCl.

13.55 Commercial aqueous nitric acid has a density of 1.42 g/mL and is 16 M. Calculate the percent HNO_3 by mass in the solution.

13.56 Commercial concentrated aqueous ammonia is 28% NH_3 by mass and has a density of 0.90 g/mL. What is the molarity of this solution?

13.57 Brass is a substitutional alloy consisting of a solution of copper and zinc. A particular sample of red brass consisting of 80.0% Cu and 20.0% Zn by mass has a density of 8750 kg/m³. (a) What is the molality of Zn in the solid solution? (b) What is the molarity of Zn in the solution?

13.58 Caffeine ($C_8H_{10}N_4O_2$) is a stimulant found in coffee and tea. If a solution of caffeine in the solvent chloroform ($CHCl_3$) has a concentration of 0.0500 m, calculate (a) the percentage of caffeine by mass, (b) the mole fraction of caffeine in the solution.

Caffeine

13.59 During a person's typical breathing cycle, the CO_2 concentration in the expired air rises to a peak of 4.6% by volume. (a) Calculate the partial pressure of the CO_2 in the expired air at its peak, assuming 1 atm pressure and a body temperature of 37 °C. (b) What is the molarity of the CO_2 in the expired air at its peak, assuming a body temperature of 37 °C?

13.60 Breathing air that contains 4.0% by volume CO_2 over time causes rapid breathing, throbbing headache, and nausea, among other symptoms. What is the concentration of CO_2 in such air in terms of (a) mol percentage, (b) molarity, assuming 1 atm pressure and a body temperature of 37 °C?

Colligative Properties (Section 13.5)

13.61 You make a solution of a nonvolatile solute with a liquid solvent. Indicate whether each of the following statements is true or false. (a) The freezing point of the solution is higher than that of the pure solvent. (b) The freezing point of the solution is lower than that of the pure solvent. (c) The boiling point of the solution is higher than that of the pure solvent. (d) The boiling point of the solution is lower than that of the pure solvent.

13.62 You make a solution of a nonvolatile solute with a liquid solvent. Indicate if each of the following statements is true or false. (a) The freezing point of the solution is unchanged by addition of the solvent. (b) The solid that forms as the solution freezes is nearly pure solute. (c) The freezing point of the solution is independent of the concentration of the solute. (d) The boiling point of the solution increases in proportion to the concentration of the solute. (e) At any temperature, the vapor pressure of the solvent over the solution is lower than what it would be for the pure solvent.

13.63 Consider two solutions, one formed by adding 10 g of glucose ($C_6H_{12}O_6$) to 1 L of water and the other formed by adding 10 g of sucrose ($C_{12}H_{22}O_{11}$) to 1 L of water. Calculate the vapor pressure for each solution at 20 °C; the vapor pressure of pure water at this temperature is 17.5 torr.

13.64 (a) What is an *ideal solution*? (b) The vapor pressure of pure water at 60 °C is 149 torr. The vapor pressure of water over a solution at 60 °C containing equal numbers of moles of water and ethylene glycol (a nonvolatile solute) is 67 torr. Is the solution ideal according to Raoult's law? Explain.

13.65 (a) Calculate the vapor pressure of water above a solution prepared by adding 22.5 g of lactose ($C_{12}H_{22}O_{11}$) to 200.0 g of water at 338 K. (Vapor-pressure data for water are given in Appendix B.) (b) Calculate the mass of propylene glycol ($C_3H_8O_2$) that must be added to 0.340 kg of water to reduce the vapor pressure by 2.88 torr at 40 °C.

13.66 **(a)** Calculate the vapor pressure of water above a solution prepared by dissolving 28.5 g of glycerin ($C_3H_8O_3$) in 125 g of water at 343 K. (The vapor pressure of water is given in Appendix B.) **(b)** Calculate the mass of ethylene glycol ($C_2H_6O_2$) that must be added to 1.00 kg of ethanol (C_2H_5OH) to reduce its vapor pressure by 10.0 torr at 35 °C. The vapor pressure of pure ethanol at 35 °C is 1.00×10^2 torr.

[13.67] At 63.5 °C, the vapor pressure of H_2O is 175 torr, and that of ethanol (C_2H_5OH) is 400 torr. A solution is made by mixing equal masses of H_2O and C_2H_5OH. **(a)** What is the mole fraction of ethanol in the solution? **(b)** Assuming ideal-solution behavior, what is the vapor pressure of the solution at 63.5 °C? **(c)** What is the mole fraction of ethanol in the vapor above the solution?

[13.68] At 20 °C, the vapor pressure of benzene (C_6H_6) is 75 torr, and that of toluene (C_7H_8) is 22 torr. Assume that benzene and toluene form an ideal solution. **(a)** What is the composition in mole fraction of a solution that has a vapor pressure of 35 torr at 20 °C? **(b)** What is the mole fraction of benzene in the vapor above the solution described in part (a)?

13.69 **(a)** Does a 0.10 m aqueous solution of NaCl have a higher boiling point, a lower boiling point, or the same boiling point as a 0.10 m aqueous solution of $C_6H_{12}O_6$? **(b)** The experimental boiling point of the NaCl solution is lower than that calculated assuming that NaCl is completely dissociated in solution. Why is this the case?

13.70 Arrange the following aqueous solutions, each 10% by mass in solute, in order of increasing boiling point: glucose ($C_6H_{12}O_6$), sucrose ($C_{12}H_{22}O_{11}$), sodium nitrate ($NaNO_3$).

13.71 List the following aqueous solutions in order of increasing boiling point: 0.120 m glucose, 0.050 m LiBr, 0.050 m $Zn(NO_3)_2$.

13.72 List the following aqueous solutions in order of decreasing freezing point: 0.040 m glycerin ($C_3H_8O_3$), 0.020 m KBr, 0.030 m phenol (C_6H_5OH).

13.73 Using data from Table 13.3, calculate the freezing and boiling points of each of the following solutions: **(a)** 0.22 m glycerol ($C_3H_8O_3$) in ethanol, **(b)** 0.240 mol of naphthalene ($C_{10}H_8$) in 2.45 mol of chloroform, **(c)** 1.50 g NaCl in 0.250 kg of water, **(d)** 2.04 g KBr and 4.82 g glucose ($C_6H_{12}O_6$) in 188 g of water.

13.74 Using data from Table 13.3, calculate the freezing and boiling points of each of the following solutions: **(a)** 0.25 m glucose in ethanol; **(b)** 20.0 g of decane, $C_{10}H_{22}$, in 50.0 g $CHCl_3$; **(c)** 3.50 g NaOH in 175 g of water, **(d)** 0.45 mol ethylene glycol and 0.15 mol KBr in 150 g H_2O.

13.75 How many grams of ethylene glycol ($C_2H_6O_2$) must be added to 1.00 kg of water to produce a solution that freezes at −5.00 °C?

13.76 What is the freezing point of an aqueous solution that boils at 105.0 °C?

13.77 What is the osmotic pressure formed by dissolving 44.2 mg of aspirin ($C_9H_8O_4$) in 0.358 L of water at 25 °C?

13.78 Seawater contains 3.4 g of salts for every liter of solution. Assuming that the solute consists entirely of NaCl (in fact, over 90% of the salt is indeed NaCl), calculate the osmotic pressure of seawater at 20 °C.

13.79 Adrenaline is the hormone that triggers the release of extra glucose molecules in times of stress or emergency. A solution of 0.64 g of adrenaline in 36.0 g of CCl_4 elevates the boiling point by 0.49 °C. Calculate the approximate molar mass of adrenaline from this data.

Adrenaline

13.80 Lauryl alcohol is obtained from coconut oil and is used to make detergents. A solution of 5.00 g of lauryl alcohol in 0.100 kg of benzene freezes at 4.1 °C. What is the molar mass of lauryl alcohol from this data?

13.81 Lysozyme is an enzyme that breaks bacterial cell walls. A solution containing 0.150 g of this enzyme in 210 mL of solution has an osmotic pressure of 0.953 torr at 25 °C. What is the molar mass of lysozyme?

13.82 A dilute aqueous solution of an organic compound soluble in water is formed by dissolving 2.35 g of the compound in water to form 0.250 L of solution. The resulting solution has an osmotic pressure of 0.605 atm at 25 °C. Assuming that the organic compound is a nonelectrolyte, what is its molar mass?

[13.83] The osmotic pressure of a 0.010 M aqueous solution of $CaCl_2$ is found to be 0.674 atm at 25 °C. **(a)** Calculate the van't Hoff factor, i, for the solution. **(b)** How would you expect the value of i to change as the solution becomes more concentrated? Explain.

[13.84] Based on the data given in Table 13.4, which solution would give the larger freezing-point lowering, a 0.030 m solution of NaCl or a 0.020 m solution of K_2SO_4? How do you explain the departure from ideal behavior and the differences observed between the two salts?

Colloids (Section 13.6)

13.85 **(a)** Do colloids made only of gases exist? Why or why not? **(b)** In the 1850's, Michael Faraday prepared ruby-red colloids of gold nanoparticles in water that are still stable today. These brightly colored colloids look like solutions. What experiment(s) could you do to determine whether a given colored preparation is a solution or colloid?

13.86 Choose the best answer: A colloidal dispersion of one liquid in another is called **(a)** a gel, **(b)** an emulsion, **(c)** a foam, **(d)** an aerosol.

13.87 An "emulsifying agent" is a compound that helps stabilize a hydrophobic colloid in a hydrophilic solvent (or a hydrophilic colloid in a hydrophobic solvent). Which of the following choices is the best emulsifying agent? **(a)** CH_3COOH, **(b)** $CH_3CH_2CH_2COOH$, **(c)** $CH_3(CH_2)_{11}COOH$, **(d)** $CH_3(CH_2)_{11}COONa$.

13.88 Aerosols are important components of the atmosphere. Does the presence of aerosols in the atmosphere increase or decrease the amount of sunlight that arrives at the Earth's surface, compared to an "aerosol-free" atmosphere? Explain your reasoning.

[13.89] Proteins can be precipitated out of aqueous solution by the addition of an electrolyte; this process is called "salting out" the protein. (a) Do you think that all proteins would be precipitated out to the same extent by the same concentration of the same electrolyte? (b) If a protein has been salted out, are the protein–protein interactions stronger or weaker than they were before the electrolyte was added? (c) A friend of yours who is taking a biochemistry class says that salting out works because the waters of hydration that surround the protein prefer to surround the electrolyte as the electrolyte is added; therefore, the protein's hydration shell is stripped away, leading to protein precipitation. Another friend of yours in the same biochemistry class says that salting out works because the incoming ions adsorb tightly to the protein, making ion pairs on the protein surface, which end up giving the protein a zero net charge in water and therefore leading to precipitation. Discuss these two hypotheses. What kind of measurements would you need to make to distinguish between these two hypotheses?

13.90 Explain how (a) a soap such as sodium stearate stabilizes a colloidal dispersion of oil droplets in water; (b) milk curdles upon addition of an acid.

Additional Exercises

13.91 Butylated hydroxytoluene (BHT) has the following molecular structure:

BHT

It is widely used as a preservative in a variety of foods, including dried cereals. Based on its structure, would you expect BHT to be more soluble in water or in hexane (C_6H_{14})? Explain.

13.92 A saturated solution of sucrose ($C_{12}H_{22}O_{11}$) is made by dissolving excess table sugar in a flask of water. There are 50 g of undissolved sucrose crystals at the bottom of the flask in contact with the saturated solution. The flask is stoppered and set aside. A year later a single large crystal of mass 50 g is at the bottom of the flask. Explain how this experiment provides evidence for a dynamic equilibrium between the saturated solution and the undissolved solute.

13.93 Most fish need at least 4 ppm dissolved O_2 in water for survival. (a) What is this concentration in mol/L? (b) What partial pressure of O_2 above water is needed to obtain 4 ppm O_2 in water at 10 °C? (The Henry's law constant for O_2 at this temperature is 1.71×10^{-3} mol/L-atm.)

13.94 The presence of the radioactive gas radon (Rn) in well water presents a possible health hazard in parts of the United States. (a) Assuming that the solubility of radon in water with 1 atm pressure of the gas over the water at 30 °C is $7.27 \times 10^{-3}\ M$, what is the Henry's law constant for radon in water at this temperature? (b) A sample consisting of various gases contains 3.5×10^{-6} mole fraction of radon. This gas at a total pressure of 32 atm is shaken with water at 30 °C. Calculate the molar concentration of radon in the water.

13.95 Glucose makes up about 0.10% by mass of human blood. Calculate this concentration in (a) ppm, (b) molality. (c) What further information would you need to determine the molarity of the solution?

13.96 The concentration of gold in seawater has been reported to be between 5 ppt (parts per trillion) and 50 ppt. Assuming that seawater contains 13 ppt of gold, calculate the number of grams of gold contained in 1.0×10^3 gal of seawater.

13.97 The maximum allowable concentration of lead in drinking water is 9.0 ppb. (a) Calculate the molarity of lead in a 9.0-ppb solution. (b) How many grams of lead are in a swimming pool containing 9.0 ppb lead in 60 m³ of water?

13.98 Acetonitrile (CH_3CN) is a polar organic solvent that dissolves a wide range of solutes, including many salts. The density of a 1.80 M LiBr solution in acetonitrile is 0.826 g/cm³. Calculate the concentration of the solution in (a) molality, (b) mole fraction of LiBr, (c) mass percentage of CH_3CN.

13.99 A "canned heat" product used to warm buffet dishes consists of a homogeneous mixture of ethanol (C_2H_5OH) and paraffin, which has an average formula of $C_{24}H_{50}$. What mass of C_2H_5OH should be added to 620 kg of the paraffin to produce 8 torr of ethanol vapor pressure at 35 °C? The vapor pressure of pure ethanol at 35 °C is 100 torr.

13.100 A solution contains 0.115 mol H_2O and an unknown number of moles of sodium chloride. The vapor pressure of the solution at 30 °C is 25.7 torr. The vapor pressure of pure water at this temperature is 31.8 torr. Calculate the number of grams of sodium chloride in the solution. (*Hint:* Remember that sodium chloride is a strong electrolyte.)

[13.101] Two beakers are placed in a sealed box at 25 °C. One beaker contains 30.0 mL of a 0.050 M aqueous solution of a nonvolatile nonelectrolyte. The other beaker contains 30.0 mL of a 0.035 M aqueous solution of NaCl. The water vapor from the two solutions reaches equilibrium. (a) In which beaker does the solution level rise, and in which one does it fall? (b) What are the volumes in the two beakers when equilibrium is attained, assuming ideal behavior?

13.102 A car owner who knows no chemistry has to put antifreeze in his car's radiator. The instructions recommend a mixture of 30% ethylene glycol and 70% water. Thinking he will improve his protection he uses pure ethylene glycol, which is a liquid at room temperature. He is saddened to find that the solution does not provide as much protection as he hoped. The pure ethylene glycol freezes solid in his radiator on a very cold day, while his neighbor, who did use the 30/70 mixture, has no problem. Suggest an explanation.

13.103 Calculate the freezing point of a 0.100 m aqueous solution of K_2SO_4, (a) ignoring interionic attractions, and (b) taking interionic attractions into consideration by using the van't Hoff factor (Table 13.4).

13.104 Carbon disulfide (CS_2) boils at 46.30 °C and has a density of 1.261 g/mL. (a) When 0.250 mol of a nondissociating solute is dissolved in 400.0 mL of CS_2, the solution boils at 47.46 °C. What is the molal boiling-point-elevation constant for CS_2? **(b)** When 5.39 g of a nondissociating unknown is dissolved in 50.0 mL of CS_2, the solution boils at 47.08 °C. What is the molecular weight of the unknown?

[13.105] A lithium salt used in lubricating grease has the formula $LiC_nH_{2n+1}O_2$. The salt is soluble in water to the extent of 0.036 g per 100 g of water at 25 °C. The osmotic pressure of this solution is found to be 57.1 torr. Assuming that molality and molarity in such a dilute solution are the same and that the lithium salt is completely dissociated in the solution, determine an appropriate value of n in the formula for the salt.

Integrative Exercises

13.106 Fluorocarbons (compounds that contain both carbon and fluorine) were, until recently, used as refrigerants. The compounds listed in the following table are all gases at 25 °C, and their solubilities in water at 25 °C and 1 atm fluorocarbon pressure are given as mass percentages. **(a)** For each fluorocarbon, calculate the molality of a saturated solution. **(b)** Explain why the molarity of each of the solutions should be very close numerically to the molality. **(c)** Based on their molecular structures, account for the differences in solubility of the four fluorocarbons. **(d)** Calculate the Henry's law constant at 25 °C for $CHClF_2$, and compare its magnitude to that for N_2 (6.8×10^{-4} mol/L-atm). Suggest a reason for the difference in magnitude.

Fluorocarbon	Solubility (mass %)
CF_4	0.0015
$CClF_3$	0.009
CCl_2F_2	0.028
$CHClF_2$	0.30

[13.107] At ordinary body temperature (37 °C), the solubility of N_2 in water at ordinary atmospheric pressure (1.0 atm) is 0.015 g/L. Air is approximately 78 mol % N_2. **(a)** Calculate the number of moles of N_2 dissolved per liter of blood, assuming blood is a simple aqueous solution. **(b)** At a depth of 100 ft in water, the external pressure is 4.0 atm. What is the solubility of N_2 from air in blood at this pressure? **(c)** If a scuba diver suddenly surfaces from this depth, how many milliliters of N_2 gas, in the form of tiny bubbles, are released into the bloodstream from each liter of blood?

[13.108] Consider the following values for enthalpy of vaporization (kJ/mol) of several organic substances:

$$CH_3\overset{\displaystyle O}{\overset{\|}{C}}\!-\!H \quad 30.4$$
Acetaldehyde

$$H_2C\overset{\displaystyle O}{\underset{}{\triangle}}CH_2 \quad 28.5$$
Ethylene oxide

$$CH_3\overset{\displaystyle O}{\overset{\|}{C}}CH_3 \quad 32.0$$
Acetone

$$H_2C\overset{CH_2}{\underset{}{\triangle}}CH_2 \quad 24.7$$
Cyclopropane

(a) Account for the variations in heats of vaporization for these substances, considering their relative intermolecular forces. **(b)** How would you expect the solubilities of these substances to vary in hexane as solvent? In ethanol? Use intermolecular forces, including hydrogen-bonding interactions where applicable, to explain your responses.

[13.109] A textbook on chemical thermodynamics states, "The heat of solution represents the difference between the lattice energy of the crystalline solid and the solvation energy of the gaseous ions." **(a)** Draw a simple energy diagram to illustrate this statement. **(b)** A salt such as NaBr is insoluble in most polar nonaqueous solvents such as acetonitrile (CH_3CN) or nitromethane (CH_3NO_2), but salts of large cations, such as tetramethylammonium bromide $[(CH_3)_4NBr]$, are generally more soluble. Use the thermochemical cycle you drew in part (a) and the factors that determine the lattice energy (Section 8.2) to explain this fact.

13.110 **(a)** A sample of hydrogen gas is generated in a closed container by reacting 2.050 g of zinc metal with 15.0 mL of 1.00 M sulfuric acid. Write the balanced equation for the reaction, and calculate the number of moles of hydrogen formed, assuming that the reaction is complete. **(b)** The volume over the solution in the container is 122 mL. Calculate the partial pressure of the hydrogen gas in this volume at 25 °C, ignoring any solubility of the gas in the solution. **(c)** The Henry's law constant for hydrogen in water at 25 °C is 7.8×10^{-4} mol/L-atm. Estimate the number of moles of hydrogen gas that remain dissolved in the solution. What fraction of the gas molecules in the system is dissolved in the solution? Was it reasonable to ignore any dissolved hydrogen in part (b)?

[13.111] The following table presents the solubilities of several gases in water at 25 °C under a total pressure of gas and water vapor of 1 atm. **(a)** What volume of $CH_4(g)$ under standard conditions of temperature and pressure is contained in 4.0 L of a saturated solution at 25 °C? **(b)** Explain the variation in solubility among the hydrocarbons listed (the first three compounds), based on their molecular structures and intermolecular forces. **(c)** Compare the solubilities of O_2, N_2, and NO, and account for the variations based on molecular structures and intermolecular forces. **(d)** Account for the much larger values observed for H_2S and SO_2 as compared with the other gases listed. **(e)** Find several pairs of substances with the same or nearly the same molecular masses (for example, C_2H_4 and N_2), and use intermolecular interactions to explain the differences in their solubilities.

Gas	Solubility (mM)
CH_4 (methane)	1.3
C_2H_6 (ethane)	1.8
C_2H_4 (ethylene)	4.7
N_2	0.6
O_2	1.2
NO	1.9
H_2S	99
SO_2	1476

13.112 A small cube of lithium (density = 0.535 g/cm³) measuring 1.0 mm on each edge is added to 0.500 L of water. The following reaction occurs:

$$2\,Li(s) + 2\,H_2O(l) \longrightarrow 2\,LiOH(aq) + H_2(g)$$

What is the freezing point of the resultant solution, assuming that the reaction goes to completion?

[13.113] At 35 °C the vapor pressure of acetone, $(CH_3)_2CO$, is 360 torr, and that of chloroform, $CHCl_3$, is 300 torr. Acetone and chloroform can form very weak hydrogen bonds between one another; the chlorines on the carbon give the carbon a sufficient partial positive charge to enable this behavior:

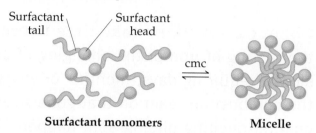

A solution composed of an equal number of moles of acetone and chloroform has a vapor pressure of 250 torr at 35 °C. (a) What would be the vapor pressure of the solution if it exhibited ideal behavior? (b) Use the existence of hydrogen bonds between acetone and chloroform molecules to explain the deviation from ideal behavior. (c) Based on the behavior of the solution, predict whether the mixing of acetone and chloroform is an exothermic ($\Delta H_{soln} < 0$) or endothermic ($\Delta H_{soln} > 0$) process. (d) Would you expect the same vapor-pressure behavior for acetone and chloromethane (CH_3Cl)? Explain.

13.114 Compounds like sodium stearate, called "surfactants" in general, can form structures known as micelles in water, once the solution concentration reaches the value known as the critical micelle concentration (cmc). Micelles contain dozens to hundreds of molecules. The cmc depends on the substance, the solvent, and the temperature.

Surfactant tail   Surfactant head

cmc

Surfactant monomers   **Micelle**

At and above the cmc, the properties of the solution vary drastically.

(a) The turbidity (the amount of light scattering) of solutions increases dramatically at the cmc. Suggest an explanation. (b) The ionic conductivity of the solution dramatically changes at the cmc. Suggest an explanation. (c) Chemists have developed fluorescent dyes that glow brightly only when the dye molecules are in a hydrophobic environment. Predict how the intensity of such fluorescence would relate to the concentration of sodium stearate as the sodium stearate concentration approaches and then increases past the cmc.

Design an Experiment

Based on Figure 13.18, you might think that the reason volatile solvent molecules in a solution are less likely to escape to the gas phase, compared to the pure solvent, is because the solute molecules are physically blocking the solvent molecules from leaving at the surface. This is a common misconception. Design an experiment to test the hypothesis that solute blocking of solvent vaporization is not the reason that solutions have lower vapor pressures than pure solvents.

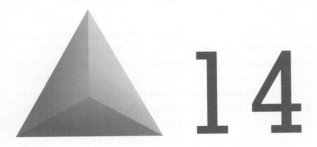

14

Chemical Kinetics

Chemical reactions take time to occur. Some reactions, such as the rusting of iron or the changing of color in leaves, occur relatively slowly, requiring days, months, or years to complete. Others, such as the combustion reaction that generates the thrust for a rocket, as in the chapter-opening photograph, happen much more rapidly. As chemists, we need to be concerned about the *speed* with which chemical reactions occur as well as the products of the reactions. For example,

the chemical reactions that govern the metabolism of food, the transport of essential nutrients, and your body's ability to adjust to temperature changes (see the Chemistry and Life box on the regulation of body temperature in Section 5.5) all require that reactions occur with the appropriate speed. Indeed, considerations of the speeds of reactions are among the most important aspects of designing new chemistry and chemical processes. The area of chemistry concerned with the speeds, or rates, of reactions is **chemical kinetics**.

So far, we have focused on the beginning and end of chemical reactions: We start with certain reactants and see what products they yield. This view is useful but does not tell us what happens in the middle—that is, which chemical bonds are broken, which are formed, and in what order these events occur. The speed at which a chemical reaction occurs is called the **reaction rate**. Reaction rates can occur over very different time

▶ **LAUNCH OF THE JUNO SPACECRAFT** at Cape Canaveral, Florida, in August 2011. The spacecraft is mounted to an Atlas V rocket, which at launch uses the very rapid combustion of kerosene and liquid oxygen to generate its thrust.

WHAT'S AHEAD

14.1 FACTORS THAT AFFECT REACTION RATES We see that four variables affect reaction rates: concentration, physical states of reactants, temperature, and presence of catalysts. These factors can be understood in terms of the collisions among reactant molecules that lead to reaction.

14.2 REACTION RATES We examine how to express *reaction rates* and how reactant disappearance rates and product appearance rates are related to the reaction stoichiometry.

14.3 CONCENTRATION AND RATE LAWS We show that the effect of concentration on rate is expressed quantitatively by *rate laws* and show how rate laws and *rate constants* are determined experimentally.

14.4 THE CHANGE OF CONCENTRATION WITH TIME We learn that rate equations can be written to express how concentrations change with time and look at several classifications of rate equations: *zero-order*, *first-order*, and *second-order* reactions.

14.5 TEMPERATURE AND RATE We explore the effect of temperature on rate. In order to occur, most reactions require a minimum input of energy called the *activation energy*.

14.6 REACTION MECHANISMS We look more closely at *reaction mechanisms*, the step-by-step molecular pathways leading from reactants to products.

14.7 CATALYSIS We end the chapter with a discussion of how *catalysts* increase reaction rates, including a discussion of biological catalysts called *enzymes*.

10^{-15} s   1 s 10^9 s (30 years) 10^{15} s (30 million years)

Time scale

▲ **Figure 14.1 Reaction rates span an enormous range of time scales.** The absorption of light by an atom or a molecule is complete within one femtosecond; explosions occur within seconds; corrosion can occur over years; and the weathering of rocks can occur over millions of years.

▲ **GO FIGURE**

If a heated steel nail were placed in pure O_2, would you expect it to burn as readily as the steel wool does?

Steel wool heated in air (about 20% O_2) glows red-hot but oxidizes to Fe_2O_3 slowly

Red-hot steel wool in 100% O_2 burns vigorously, forming Fe_2O_3 quickly

▲ **Figure 14.2 Effect of concentration on reaction rate.** The difference in behavior is due to the different concentrations of O_2 in the two environments.

scales (▲ **Figure 14.1**). To investigate how reactions happen, we must examine the reaction rates and the factors that influence them. Experimental information on the rate of a given reaction provides important evidence that helps us formulate a *reaction mechanism*, which is a step-by-step, molecular-level view of the pathway from reactants to products.

Our goal in this chapter is to understand how to determine reaction rates and to consider the factors that control these rates. What factors determine how rapidly food spoils, for instance? How does one design an automotive airbag that fills extremely rapidly following a car crash? What determines the rate at which steel rusts? How can we remove hazardous pollutants in automobile exhaust before the exhaust leaves the tailpipe? Although we will not address these specific questions, we will see that the rates of all chemical reactions are subject to the same principles.

14.1 | Factors that Affect Reaction Rates

Four factors affect the rate at which any particular reaction occurs:

1. *Physical state of the reactants.* Reactants must come together to react. The more readily reactant molecules collide with one another, the more rapidly they react. Reactions may broadly classified as being either *homogeneous*, involving either all gases or all liquids, or as *heterogeneous*, in which reactants are in different phases. Under heterogeneous conditions, a reaction is limited by the area of contact of the reactants. Thus, heterogeneous reactions that involve solids tend to proceed more rapidly if the surface area of the solid is increased. For example, a medicine in the form of a fine powder dissolves in the stomach and enters the blood more quickly than the same medicine in the form of a tablet.

2. *Reactant concentrations.* Most chemical reactions proceed more quickly if the concentration of one or more reactants is increased. For example, steel wool burns only slowly in air, which contains 20% O_2, but bursts into flame in pure oxygen (◄ **Figure 14.2**). As reactant concentration increases, the frequency with which the reactant molecules collide increases, leading to increased rates.

3. *Reaction temperature.* Reaction rates generally increase as temperature is increased. The bacterial reactions that spoil milk, for instance, proceed more rapidly at room temperature than at the lower temperature of a refrigerator. Increasing temperature increases the kinetic energies of molecules. ∞ (Section 10.7) As molecules move more rapidly, they collide more frequently and with higher energy, leading to increased reaction rates.

4. *The presence of a catalyst. Catalysts* are agents that increase reaction rates without themselves being used up. They affect the kinds of collisions (and therefore alter the mechanism) that lead to reaction. Catalysts play many crucial roles in living organisms, including ourselves.

On a molecular level, reaction rates depend on the frequency of collisions between molecules. *The greater the frequency of collisions, the higher the reaction rate.* For a collision to lead to a reaction, however, it must occur with sufficient energy to break bonds and with suitable orientation for new bonds to form in the proper locations. We will consider these factors as we proceed through this chapter.

Give It Some Thought

In a reaction involving reactants in the gas state, how does increasing the partial pressures of the gases affect the reaction rate?

14.2 | Reaction Rates

The *speed* of an event is defined as the *change* that occurs in a given *time* interval, which means that whenever we talk about speed, we necessarily bring in the notion of time. For example, the speed of a car is expressed as the change in the car's position over a certain time interval. In the United States, the speed of cars is usually measured in units of miles per hour—that is, the quantity that is changing (position measured in miles) divided by a time interval (measured in hours).

Similarly, the speed of a chemical reaction—its reaction rate—is the change in the concentration of reactants or products per unit of time. The units for reaction rate are usually molarity per second (M/s)—that is, the change in concentration measured in molarity divided by a time interval measured in seconds.

Let's consider the hypothetical reaction A $\longrightarrow$ B, depicted in ▼ Figure 14.3. Each red sphere represents 0.01 mol of A, each blue sphere represents 0.01 mol of B, and the container has a volume of 1.00 L. At the beginning of the reaction, there is 1.00 mol A, so the concentration is 1.00 $mol/L = 1.00\ M$. After 20 s, the concentration of A has fallen to 0.54 M and the concentration of B has risen to 0.46 M. The sum of the concentrations is still 1.00 M because 1 mol of B is produced for each mole of A that reacts. After 40 s, the concentration of A is 0.30 M and that of B is 0.70 M.

GO FIGURE

Estimate the number of moles of A in the mixture after 30 s.

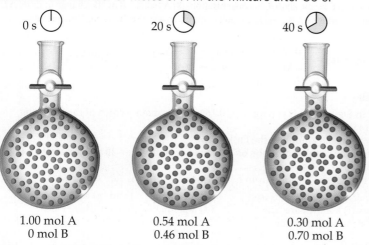

| 1.00 mol A | 0.54 mol A | 0.30 mol A |
| 0 mol B | 0.46 mol B | 0.70 mol B |

▲ Figure 14.3 **Progress of a hypothetical reaction A $\longrightarrow$ B.** The volume of the flask is 1.0 L.

The rate of this reaction can be expressed either as the rate of disappearance of reactant A or as the rate of appearance of product B. The *average* rate of appearance of B over a particular time interval is given by the change in concentration of B divided by the change in time:

$$\text{Average rate of appearance of B} = \frac{\text{change in concentration of B}}{\text{change in time}}$$

$$= \frac{[B] \text{ at } t_2 - [B] \text{ at } t_1}{t_2 - t_1} = \frac{\Delta[B]}{\Delta t} \quad [14.1]$$

We use brackets around a chemical formula, as in [B], to indicate molarity. The Greek letter delta, Δ, is read "change in" and is always equal to a final value minus an initial value. ∞ (Equation 5.4, Section 5.2) The average rate of appearance of B over the 20-s interval from the beginning of the reaction ($t_1 = 0$ s to $t_2 = 20$ s) is

$$\text{Average rate} = \frac{0.46\ M - 0.00\ M}{20\ s - 0\ s} = 2.3 \times 10^{-2}\ M/s$$

We could equally well express the reaction rate in term of the reactant, A. In this case, we would be describing the rate of disappearance of A, which we express as

$$\text{Average rate of disappearance of A} = -\frac{\text{change in concentration of A}}{\text{change in time}}$$

$$= -\frac{\Delta[A]}{\Delta t} \quad [14.2]$$

Notice the minus sign in this equation, which we use to indicate that the concentration of A decreases. By convention, *rates are always expressed as positive quantities.* Because [A] decreases, $\Delta[A]$ is a negative number. The minus sign we put in the equation converts the negative $\Delta[A]$ to a positive rate of disappearance.

Because one molecule of A is consumed for every molecule of B that forms, the average rate of disappearance of A equals the average rate of appearance of B:

$$\text{Average rate} = -\frac{\Delta[A]}{\Delta t} = -\frac{0.54\ M - 1.00\ M}{20\ s - 0\ s} = 2.3 \times 10^{-2}\ M/s$$

SAMPLE EXERCISE 14.1 | Calculating an Average Rate of Reaction

From the data in Figure 14.3, calculate the average rate at which A disappears over the time interval from 20 s to 40 s.

SOLUTION

Analyze We are given the concentration of A at 20 s ($0.54\ M$) and at 40 s ($0.30\ M$) and asked to calculate the average rate of reaction over this time interval.

Plan The average rate is given by the change in concentration, $\Delta[A]$, divided by the change in time, Δt. Because A is a reactant, a minus sign is used in the calculation to make the rate a positive quantity.

Solve

$$\text{Average rate} = -\frac{\Delta[A]}{\Delta t} = -\frac{0.30\ M - 0.54\ M}{40\ s - 20\ s} = 1.2 \times 10^{-2}\ M/s$$

Practice Exercise 1

If the experiment in Figure 14.3 is run for 60 s, 0.16 mol A remain. Which of the following statements is or are true?
   (i) After 60 s there are 0.84 mol B in the flask.
   (ii) The decrease in the number of moles of A from $t_1 = 0$ s to $t_2 = 20$ s is greater than that from $t_1 = 40$ to $t_2 = 60$ s.
   (iii) The average rate for the reaction from $t_1 = 40$ s to $t_2 = 60$ s is $7.0 \times 10^{-3}\ M/s$.

   (a) Only one of the statements is true.   **(b)** Statements (i) and (ii) are true.
   (c) Statements (i) and (iii) are true.   **(d)** Statements (ii) and (iii) are true.
   (e) All three statements are true.

Practice Exercise 2

Use the data in Figure 14.3 to calculate the average rate of appearance of B over the time interval from 0 s to 40 s.

Table 14.1 Rate Data for Reaction of C_4H_9Cl with Water

Time, t (s)	$[C_4H_9Cl]$ (M)		Average Rate (M/s)
0.0	0.1000		1.9×10^{-4}
50.0	0.0905		1.7×10^{-4}
100.0	0.0820		1.6×10^{-4}
150.0	0.0741		1.4×10^{-4}
200.0	0.0671		1.22×10^{-4}
300.0	0.0549		1.01×10^{-4}
400.0	0.0448		0.80×10^{-4}
500.0	0.0368		0.560×10^{-4}
800.0	0.0200		
10,000	0		

Change of Rate with Time

Now let's consider the reaction between butyl chloride (C_4H_9Cl) and water to form butyl alcohol (C_4H_9OH) and hydrochloric acid:

$$C_4H_9Cl(aq) + H_2O(l) \longrightarrow C_4H_9OH(aq) + HCl(aq) \qquad [14.3]$$

Suppose we prepare a 0.1000-M aqueous solution of C_4H_9Cl and then measure the concentration of C_4H_9Cl at various times after time zero (which is the instant at which the reactants are mixed, thereby initiating the reaction). We can use the resulting data, shown in the first two columns of ▲ Table 14.1, to calculate the average rate of disappearance of C_4H_9Cl over various time intervals; these rates are given in the third column. Notice that the average rate decreases over each 50-s interval for the first several measurements and continues to decrease over even larger intervals through the remaining measurements. *It is typical for rates to decrease as a reaction proceeds because the concentration of reactants decreases.* The change in rate as the reaction proceeds is also seen in a graph of $[C_4H_9Cl]$ versus time (▶ Figure 14.4). Notice how the steepness of the curve decreases with time, indicating a decreasing reaction rate.

Instantaneous Rate

Graphs such as Figure 14.4 that show how the concentration of a reactant or product changes with time allow us to evaluate the **instantaneous rate** of a reaction, which is the rate at a particular instant during the reaction. The instantaneous rate is determined from the slope of the curve at a particular point in time. We have drawn two tangent lines in Figure 14.4, a dashed line running through the point at $t = 0$ s and a solid line running through the point at $t = 600$ s. The slopes of these tangent lines give the instantaneous rates at these two time points.* To determine the instantaneous rate at 600 s, for example, we construct horizontal and vertical lines to form the blue right triangle in Figure 14.4. The slope of the tangent line is the ratio of the height of the vertical side to the length of the horizontal side:

$$\text{Instantaneous rate} = -\frac{\Delta[C_4H_9Cl]}{\Delta t} = -\frac{(0.017 - 0.042)\,M}{(800 - 400)\,s}$$

$$= 6.3 \times 10^{-5}\,M/s$$

▲ **GO FIGURE**

How does the instantaneous rate of reaction change as the reaction proceeds?

$$C_4H_9Cl(aq) + H_2O(l) \longrightarrow C_4H_9OH(aq) + HCl(aq)$$

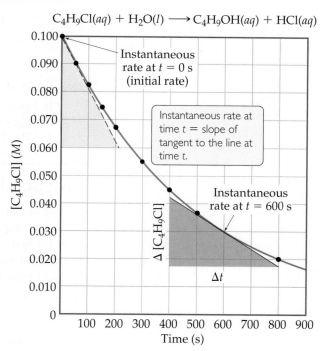

▲ Figure 14.4 **Concentration of butyl chloride (C_4H_9Cl) as a function of time.**

---

*You may wish to review graphical determination of slopes in Appendix A. If you are familiar with calculus, you may recognize that the average rate approaches the instantaneous rate as the time interval approaches zero. This limit, in the notation of calculus, is the negative of the derivative of the curve at time t, $-d[C_4H_9Cl]/dt$.

In discussions that follow, the term *rate* means instantaneous rate unless indicated otherwise. The instantaneous rate at $t = 0$ is called the *initial rate* of the reaction. To understand the difference between average and instantaneous rates, imagine you have just driven 98 mi in 2.0 h. Your average speed for the trip is 49 mi/hr, but your instantaneous speed at any moment during the trip is the speedometer reading at that moment.

SAMPLE EXERCISE 14.2 | Calculating an Instantaneous Rate of Reaction

Using Figure 14.4, calculate the instantaneous rate of disappearance of C_4H_9Cl at $t = 0$ s (the initial rate).

SOLUTION

Analyze We are asked to determine an instantaneous rate from a graph of reactant concentration versus time.

Plan To obtain the instantaneous rate at $t = 0$ s, we must determine the slope of the curve at $t = 0$. The tangent is drawn on the graph as the hypotenuse of the tan triangle. The slope of this straight line equals the change in the vertical axis divided by the corresponding change in the horizontal axis (which, in the case of this example, is the change in molarity over change in time).

Solve The tangent line falls from $[C_4H_9Cl] = 0.100\ M$ to $0.060\ M$ in the time change from 0 s to 210 s. Thus, the initial rate is

$$\text{Rate} = -\frac{\Delta[C_4H_9Cl]}{\Delta t} = -\frac{(0.060 - 0.100)\ M}{(210 - 0)\ s} = 1.9 \times 10^{-4}\ M/s$$

Practice Exercise 1

Which of the following could be the instantaneous rate of the reaction in Figure 14.4 at $t = 1000$ s?
(a) $1.2 \times 10^{-4}\ M/s$, (b) $8.8 \times 10^{-5}\ M/s$, (c) $6.3 \times 10^{-5}\ M/s$, (d) $2.7 \times 10^{-5}\ M/s$, (e) More than one of these.

Practice Exercise 2

Using Figure 14.4, determine the instantaneous rate of disappearance of C_4H_9Cl at $t = 300$ s.

 Give It Some Thought

In Figure 14.4, order the following three rates from fastest to slowest: (i) The average rate of the reaction between 0 s and 600 s, (ii) the instantaneous rate at $t = 0$ s, and (iii) the instantaneous rate at $t = 600$ s. You should not have to do any calculations.

Reaction Rates and Stoichiometry

During our discussion of the hypothetical reaction A $\longrightarrow$ B, we saw that the stoichiometry requires that the rate of disappearance of A equal the rate of appearance of B. Likewise, the stoichiometry of Equation 14.3 indicates that 1 mol of C_4H_9OH is produced for each mole of C_4H_9Cl consumed. Therefore, the rate of appearance of C_4H_9OH equals the rate of disappearance of C_4H_9Cl:

$$\text{Rate} = -\frac{\Delta[C_4H_9Cl]}{\Delta t} = \frac{\Delta[C_4H_9OH]}{\Delta t}$$

What happens when the stoichiometric relationships are not one-to-one? For example, consider the reaction $2\,HI(g) \longrightarrow H_2(g) + I_2(g)$. We can measure either the rate of disappearance of HI or the rate of appearance of either H_2 or I_2. Because 2 mol of HI disappears for each mole of H_2 or I_2 that forms, the rate of disappearance of HI is *twice* the rate of appearance of either H_2 or I_2. How do we decide which number to use for the rate of the reaction? Depending on whether we monitor HI, I_2, or H_2, the rates can differ by a factor of 2. To fix this problem, we need to take into account the reaction stoichiometry. To arrive at a number for the reaction rate that does not depend on which component we measured, we must divide the rate of disappearance of HI by 2 (its coefficient in the balanced chemical equation):

$$\text{Rate} = -\frac{1}{2}\frac{\Delta[\text{HI}]}{\Delta t} = \frac{\Delta[\text{H}_2]}{\Delta t} = \frac{\Delta[\text{I}_2]}{\Delta t}$$

In general, for the reaction

$$a\,\text{A} + b\,\text{B} \longrightarrow c\,\text{C} + d\,\text{D}$$

the rate is given by

$$\text{Rate} = -\frac{1}{a}\frac{\Delta[\text{A}]}{\Delta t} = -\frac{1}{b}\frac{\Delta[\text{B}]}{\Delta t} = \frac{1}{c}\frac{\Delta[\text{C}]}{\Delta t} = \frac{1}{d}\frac{\Delta[\text{D}]}{\Delta t} \qquad [14.4]$$

When we speak of the rate of a reaction without specifying a particular reactant or product, we utilize the definition in Equation 14.4.*

SAMPLE EXERCISE 14.3 | Relating Rates at Which Products Appear and Reactants Disappear

(a) How is the rate at which ozone disappears related to the rate at which oxygen appears in the reaction $2\,\text{O}_3(g) \longrightarrow 3\,\text{O}_2(g)$?

(b) If the rate at which O_2 appears, $\Delta[\text{O}_2]/\Delta t$, is $6.0 \times 10^{-5}\,M/s$ at a particular instant, at what rate is O_3 disappearing at this same time, $-\Delta[\text{O}_3]/\Delta t$?

SOLUTION

Analyze We are given a balanced chemical equation and asked to relate the rate of appearance of the product to the rate of disappearance of the reactant.

Plan We can use the coefficients in the chemical equation as shown in Equation 14.4 to express the relative rates of reactions.

Solve

(a) Using the coefficients in the balanced equation and the relationship given by Equation 14.4, we have:

$$\text{Rate} = -\frac{1}{2}\frac{\Delta[\text{O}_3]}{\Delta t} = \frac{1}{3}\frac{\Delta[\text{O}_2]}{\Delta t}$$

(b) Solving the equation from part (a) for the rate at which O_3 disappears, $-\Delta[\text{O}_3]/\Delta t$, we have:

$$-\frac{\Delta[\text{O}_3]}{\Delta t} = \frac{2}{3}\frac{\Delta[\text{O}_2]}{\Delta t} = \frac{2}{3}(6.0 \times 10^{-5}M/s) = 4.0 \times 10^{-5}M/s$$

Check We can apply a stoichiometric factor to convert the O_2 formation rate to the O_3 disappearance rate:

$$-\frac{\Delta[\text{O}_3]}{\Delta t} = \left(6.0 \times 10^{-5}\,\frac{\text{mol O}_2/\text{L}}{\text{s}}\right)\left(\frac{2\text{ mol O}_3}{3\text{ mol O}_2}\right) = 4.0 \times 10^{-5}\,\frac{\text{mol O}_3/\text{L}}{\text{s}}$$
$$= 4.0 \times 10^{-5}M/s$$

Practice Exercise 1

At a certain time in a reaction, substance A is disappearing at a rate of $4.0 \times 10^{-2}\,M/s$, substance B is appearing at a rate of $2.0 \times 10^{-2}\,M/s$, and substance C is appearing at a rate of $6.0 \times 10^{-2}\,M/s$. Which of the following could be the stoichiometry for the reaction being studied?
(a) $2\text{A} + \text{B} \longrightarrow 3\text{C}$ (b) $\text{A} \longrightarrow 2\text{B} + 3\text{C}$
(c) $2\text{A} \longrightarrow \text{B} + 3\text{C}$ (d) $4\text{A} \longrightarrow 2\text{B} + 3\text{C}$
(e) $\text{A} + 2\text{B} \longrightarrow 3\text{C}$

Practice Exercise 2

If the rate of decomposition of N_2O_5 in the reaction $2\,\text{N}_2\text{O}_5(g) \longrightarrow 4\,\text{NO}_2(g) + \text{O}_2(g)$ at a particular instant is $4.2 \times 10^{-7}M/s$, what is the rate of appearance of (a) NO_2 and (b) O_2 at that instant?

14.3 | Concentration and Rate Laws

One way of studying the effect of concentration on reaction rate is to determine the way in which the initial rate of a reaction depends on the initial concentrations. For example, we might study the rate of the reaction

$$\text{NH}_4^+(aq) + \text{NO}_2^-(aq) \longrightarrow \text{N}_2(g) + 2\,\text{H}_2\text{O}(l)$$

---

*Equation 14.4 does not hold true if substances other than C and D are formed in significant amounts. For example, sometimes intermediate substances build in concentration before forming the final products. In that case, the relationship between the rate of disappearance of reactants and the rate of appearance of products is not given by Equation 14.4. All reactions whose rates we consider in this chapter obey Equation 14.4.

Table 14.2 Rate Data for the Reaction of Ammonium and Nitrite Ions in Water at 25 °C

Experiment Number	Initial NH_4^+ Concentration (M)	Initial NO_2^- Concentration (M)	Observed Initial Rate (M/s)
1	0.0100	0.200	5.4×10^{-7}
2	0.0200	0.200	10.8×10^{-7}
3	0.0400	0.200	21.5×10^{-7}
4	0.200	0.0202	10.8×10^{-7}
5	0.200	0.0404	21.6×10^{-7}
6	0.200	0.0808	43.3×10^{-7}

by measuring the concentration of NH_4^+ or NO_2^- as a function of time or by measuring the volume of N_2 collected as a function of time. Because the stoichiometric coefficients on NH_4^+, NO_2^-, and N_2 are the same, all of these rates are the same.

▲ Table 14.2 shows that changing the initial concentration of either reactant changes the initial reaction rate. If we double $[NH_4^+]$ while holding $[NO_2^-]$ constant, the rate doubles (compare experiments 1 and 2). If we increase $[NH_4^+]$ by a factor of 4 but leave $[NO_2^-]$ unchanged (experiments 1 and 3), the rate changes by a factor of 4, and so forth. These results indicate that the initial reaction rate is proportional to $[NH_4^+]$. When $[NO_2^-]$ is similarly varied while $[NH_4^+]$ is held constant, the rate is affected in the same manner. Thus, the rate is also directly proportional to the concentration of $[NO_2^-]$.

A Closer Look

Using Spectroscopic Methods to Measure Reaction Rates: Beer's Law

A variety of techniques can be used to monitor reactant and product concentration during a reaction, including spectroscopic methods, which rely on the ability of substances to absorb (or emit) light. Spectroscopic kinetic studies are often performed with the reaction mixture in the sample compartment of a *spectrometer*, an instrument that measures the amount of light transmitted or absorbed by a sample at different wavelengths. For kinetic studies, the spectrometer is set to measure the light absorbed at a wavelength characteristic of one of the reactants or products. In the decomposition of $HI(g)$ into $H_2(g)$ and $I_2(g)$, for example, both HI and H_2 are colorless, whereas I_2 is violet. During the reaction, the violet color of the reaction mixture gets more intense as I_2 forms. Thus, visible light of appropriate wavelength can be used to monitor the reaction (▶ Figure 14.5).

▶ Figure 14.6 shows the components of a spectrometer. The spectrometer measures the amount of light absorbed by the sample by comparing the intensity of the light emitted from the light source with the intensity of the light transmitted through the sample, for various wavelengths. As the concentration of I_2 increases and its color becomes more intense, the amount of light absorbed by the reaction mixture increases, as Figure 14.5 shows, causing less light to reach the detector.

How can we relate the amount of light detected by the spectrometer to the concentration of a species? A relationship called *Beer's law* gives us a direct route to the information we seek. Beer's law connects the amount of light absorbed to the concentration of the absorbing substance:

$$A = \varepsilon bc \qquad [14.5]$$

In this equation, A is the measured absorbance, ε is the extinction coefficient (a characteristic of the substance being monitored at a given wavelength of light), b is the path length through which the light

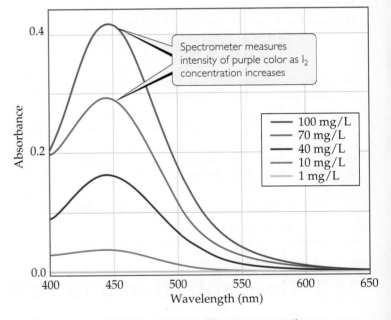

▲ Figure 14.5 **Visible spectra of I_2 at different concentrations.**

passes, and c is the molar concentration of the absorbing substance. Thus, the concentration is directly proportional to absorbance. Many chemical and pharmaceutical companies routinely use Beer's law to calculate the concentration of purified solutions of the compounds that they make. In the laboratory portion of your course, you may very well perform one or more experiments in which you use Beer's law to relate absorption of light to concentration.

Related Exercises: 14.101, 14.102, Design an Experiment

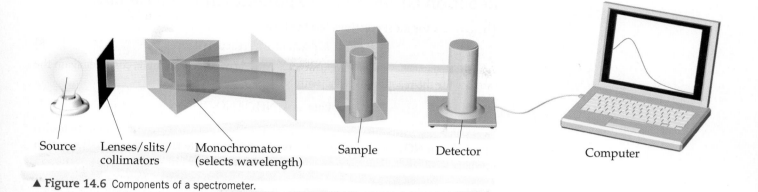

▲ **Figure 14.6** Components of a spectrometer.

We express the way in which the rate depends on the reactant concentrations by the equation

$$\text{Rate} = k[\text{NH}_4^+][\text{NO}_2^-] \qquad [14.6]$$

An equation such as Equation 14.6, which shows how the rate depends on reactant concentrations, is called a **rate law**. For the general reaction

$$a\,\text{A} + b\,\text{B} \longrightarrow c\,\text{C} + d\,\text{D}$$

the rate law generally has the form

$$\text{Rate} = k[\text{A}]^m[\text{B}]^n \qquad [14.7]$$

Notice that only the concentrations of the reactants generally appear in the rate law. The constant k is called the **rate constant**. The magnitude of k changes with temperature and therefore determines how temperature affects rate, as we will see in Section 14.5. The exponents m and n are typically small whole numbers. As we will learn shortly, if we know m and n for a reaction, we can gain great insight into the individual steps that occur during the reaction.

Give It Some Thought

How do reaction rate, rate law, and rate constant differ?

Once we know the rate law for a reaction and the reaction rate for a set of reactant concentrations, we can calculate the value of k. For example, using the values for experiment 1 in Table 14.2, we can substitute into Equation 14.6:

$$5.4 \times 10^{-7}\,M/s = k(0.0100\,M)(0.200\,M)$$

$$k = \frac{5.4 \times 10^{-7}\,M/s}{(0.0100\,M)(0.200\,M)} = 2.7 \times 10^{-4}\,M^{-1}s^{-1}$$

You should verify that this same value of k is obtained using any of the other experimental results in Table 14.2.

Once we have both the rate law and the k value for a reaction, we can calculate the reaction rate for any set of concentrations. For example, using Equation 14.7 with $k = 2.7 \times 10^{-4}\,M^{-1}\,s^{-1}$, $m = 1$, and $n = 1$, we can calculate the rate for $[\text{NH}_4^+] = 0.100\,M$ and $[\text{NO}_2^-] = 0.100\,M$:

$$\text{Rate} = (2.7 \times 10^{-4}M^{-1}\,s^{-1})(0.100\,M)(0.100\,M) = 2.7 \times 10^{-6}M/s$$

Give It Some Thought

Does the rate constant have the same units as the rate?

Reaction Orders: The Exponents in the Rate Law

The rate law for most reactions has the form

$$\text{Rate} = k[\text{reactant 1}]^m[\text{reactant 2}]^n \ldots \qquad [14.8]$$

The exponents m and n are called **reaction orders**. For example, consider again the rate law for the reaction of NH_4^+ with NO_2^-:

$$\text{Rate} = k[NH_4^+][NO_2^-]$$

Because the exponent of $[NH_4^+]$ is 1, the rate is *first order* in NH_4^+. The rate is also first order in NO_2^-. (The exponent 1 is not shown in rate laws.) The **overall reaction order** is the sum of the orders with respect to each reactant represented in the rate law. Thus, for the $NH_4^+ - NO_2^-$ reaction, the rate law has an overall reaction order of $1 + 1 = 2$, and the reaction is *second order overall*.

The exponents in a rate law indicate how the rate is affected by each reactant concentration. Because the rate at which NH_4^+ reacts with NO_2^- depends on $[NH_4^+]$ raised to the first power, the rate doubles when $[NH_4^+]$ doubles, triples when $[NH_4^+]$ triples, and so forth. Doubling or tripling $[NO_2^-]$ likewise doubles or triples the rate. If a rate law is second order with respect to a reactant, $[A]^2$, then doubling the concentration of that substance causes the reaction rate to quadruple because $[2]^2 = 4$, whereas tripling the concentration causes the rate to increase ninefold: $[3]^2 = 9$.

The following are some additional examples of experimentally determined rate laws:

$$2\,N_2O_5(g) \longrightarrow 4\,NO_2(g) + O_2(g) \quad \text{Rate} = k[N_2O_5] \qquad [14.9]$$

$$H_2(g) + I_2(g) \longrightarrow 2\,HI(g) \qquad\qquad \text{Rate} = k[H_2][I_2] \qquad [14.10]$$

$$CHCl_3(g) + Cl_2(g) \longrightarrow CCl_4(g) + HCl(g) \quad \text{Rate} = k[CHCl_3][Cl_2]^{1/2} \quad [14.11]$$

Although the exponents in a rate law are sometimes the same as the coefficients in the balanced equation, this is not necessarily the case, as Equations 14.9 and 14.11 show. For any reaction, *the rate law must be determined experimentally*. In most rate laws, reaction orders are 0, 1, or 2. However, we also occasionally encounter rate laws in which the reaction order is fractional (as is the case with Equation 14.11) or even negative.

 Give It Some Thought

The experimentally determined rate law for the reaction
$2\,NO(g) + 2\,H_2(g) \longrightarrow N_2(g) + 2\,H_2O(g)$ is rate $= k[NO]^2[H_2]$.
(a) What are the reaction orders in this rate law?
(b) Would the reaction rate increase more if we doubled the concentration of NO or the concentration of H_2?

SAMPLE EXERCISE 14.4 Relating a Rate Law to the Effect of Concentration on Rate

Consider a reaction $A + B \longrightarrow C$ for which rate $= k[A][B]^2$. Each of the following boxes represents a reaction mixture in which A is shown as red spheres and B as purple ones. Rank these mixtures in order of increasing rate of reaction.

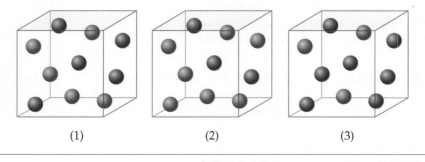

(1)   (2)   (3)

SOLUTION

Analyze We are given three boxes containing different numbers of spheres representing mixtures containing different reactant concentrations. We are asked to use the given rate law and the compositions of the boxes to rank the mixtures in order of increasing reaction rates.

Plan Because all three boxes have the same volume, we can put the number of spheres of each kind into the rate law and calculate the rate for each box.

Solve Box 1 contains 5 red spheres and 5 purple spheres, giving the following rate:

$$\text{Box 1: Rate} = k(5)(5)^2 = 125k$$

Box 2 contains 7 red spheres and 3 purple spheres:

$$\text{Box 2: Rate} = k(7)(3)^2 = 63k$$

Box 3 contains 3 red spheres and 7 purple spheres:

$$\text{Box 3: Rate} = k(3)(7)^2 = 147k$$

The slowest rate is $63k$ (Box 2), and the highest is $147k$ (Box 3). Thus, the rates vary in the order $2 < 1 < 3$.

Check Each box contains 10 spheres. The rate law indicates that in this case [B] has a greater influence on rate than [A] because B has a larger reaction order. Hence, the mixture with the highest concentration of B (most purple spheres) should react fastest. This analysis confirms the order $2 < 1 < 3$.

> **Practice Exercise 1**
>
> Suppose the rate law for the reaction in this Sample Exercise were rate $= k[A]^2[B]$. What would be the ordering of the rates for the three mixtures shown above, from slowest to fastest?
> **(a)** $1 < 2 < 3$ **(b)** $1 < 3 < 2$ **(c)** $3 < 2 < 1$
> **(d)** $2 < 1 < 3$ **(e)** $3 < 1 < 2$
>
> **Practice Exercise 2**
>
> Assuming that rate $= k[A][B]$, rank the mixtures represented in this Sample Exercise in order of increasing rate.

Magnitudes and Units of Rate Constants

If chemists want to compare reactions to evaluate which ones are relatively fast and which ones are relatively slow, the quantity of interest is the rate constant. A good general rule is that a large value of k ($\sim 10^9$ or higher) means a fast reaction and a small value of k (10 or lower) means a slow reaction.

 Give It Some Thought

Suppose the reactions A ⟶ B and X ⟶ Y have the same value of k. When [A] = [X], will the two reactions necessarily have the same rate?

The units of the rate constant depend on the overall reaction order of the rate law. In a reaction that is second order overall, for example, the units of the rate constant must satisfy the equation:

$$\text{Units of rate} = (\text{units of rate constant})(\text{units of concentration})^2$$

Hence, in our usual units of molarity for concentration and seconds for time, we have

$$\text{Units of rate constant} = \frac{\text{units of rate}}{(\text{units of concentration})^2} = \frac{M/s}{M^2} = M^{-1}\,s^{-1}$$

SAMPLE EXERCISE 14.5 | Determining Reaction Orders and Units for Rate Constants

(a) What are the overall reaction orders for the reactions described in Equations 14.9 and 14.11?
(b) What are the units of the rate constant for the rate law in Equation 14.9?

SOLUTION

Analyze We are given two rate laws and asked to express **(a)** the overall reaction order for each and **(b)** the units for the rate constant for the first reaction.

Plan The overall reaction order is the sum of the exponents in the rate law. The units for the rate constant, k, are found by using the normal units for rate (M/s) and concentration (M) in the rate law and applying algebra to solve for k.

Solve

(a) The rate of the reaction in Equation 14.9 is first order in N_2O_5 and first order overall. The reaction in Equation 14.11 is first

order in $CHCl_3$ and one-half order in Cl_2. The overall reaction order is three halves.

(b) For the rate law for Equation 14.9, we have

$$\text{Units of rate} = (\text{units of rate constant})(\text{units of concentration})$$

so

$$\text{Units of rate constant} = \frac{\text{units of rate}}{\text{units of concentration}} = \frac{M/s}{M} = s^{-1}$$

Notice that the units of the rate constant change as the overall order of the reaction changes.

Using Initial Rates to Determine Rate Laws

We have seen that the rate law for most reactions has the general form

$$\text{Rate} = k[\text{reactant 1}]^m[\text{reactant 2}]^n \ldots$$

Thus, the task of determining the rate law becomes one of determining the reaction orders, m and n. In most reactions, the reaction orders are 0, 1, or 2. As noted earlier in this section, we can use the response of the reaction rate to a change in initial concentration to determine the reaction order.

In working with rate laws, it is important to realize that the *rate* of a reaction depends on concentration but the *rate constant* does not. As we will see later in this chapter, the rate constants (and hence the reaction rate) are affected by temperature and by the presence of a catalyst.

SAMPLE EXERCISE 14.6 Determining a Rate Law from Initial Rate Data

The initial rate of a reaction A + B ⟶ C was measured for several different starting concentrations of A and B, and the results are as follows:

Experiment Number	[A] (M)	[B] (M)	Initial Rate (M/s)
1	0.100	0.100	4.0×10^{-5}
2	0.100	0.200	4.0×10^{-5}
3	0.200	0.100	16.0×10^{-5}

Using these data, determine **(a)** the rate law for the reaction, **(b)** the rate constant, **(c)** the rate of the reaction when $[A] = 0.050M$ and $[B] = 0.100\ M$.

SOLUTION

Analyze We are given a table of data that relates concentrations of reactants with initial rates of reaction and asked to determine **(a)** the rate law, **(b)** the rate constant, and **(c)** the rate of reaction for a set of concentrations not listed in the table.

Plan (a) We assume that the rate law has the following form: Rate $= k[A]^m[B]^n$. We will use the given data to deduce the reaction orders m and n by determining how changes in the concentration change the rate. **(b)** Once we know m and n, we can use the rate law and one of the sets of data to determine the rate constant k. **(c)** Upon determining both the rate constant and the reaction orders, we can use the rate law with the given concentrations to calculate rate.

Solve

(a) If we compare experiments 1 and 2, we see that [A] is held constant and [B] is doubled. Thus, this pair of experiments shows how [B] affects the rate, allowing us to deduce the order of the rate law with respect to B. Because the rate remains the same when [B] is doubled, the concentration of B has no effect on the reaction rate. The rate law is therefore zero order in B (that is, $n = 0$).

In experiments 1 and 3, [B] is held constant, so these data show how [A] affects rate. Holding [B] constant while doubling [A] increases the rate fourfold. This result indicates that rate is proportional to $[A]^2$ (that is, the reaction is second order in A). Hence, the rate law is

$$\text{Rate} = k[A]^2[B]^0 = k[A]^2$$

(b) Using the rate law and the data from experiment 1, we have

$$k = \frac{\text{rate}}{[A]^2} = \frac{4.0 \times 10^{-5}\ M/s}{(0.100\ M)^2} = 4.0 \times 10^{-3}\ M^{-1}\ s^{-1}$$

(c) Using the rate law from part (a) and the rate constant from part (b), we have

$$\text{Rate} = k[A]^2 = (4.0 \times 10^{-3}\ M^{-1}s^{-1})(0.050\ M)^2 = 1.0 \times 10^{-5}\ M/s$$

Because [B] is not part of the rate law, it is irrelevant to the rate if there is at least some B present to react with A.

Check A good way to check our rate law is to use the concentrations in experiment 2 or 3 and see if we can correctly calculate the rate. Using data from experiment 3, we have

$$\text{Rate} = k[A]^2 = (4.0 \times 10^{-3}\, M^{-1}\, s^{-1})(0.200\, M)^2 = 1.6 \times 10^{-4}\, M/s$$

Thus, the rate law correctly reproduces the data, giving both the correct number and the correct units for the rate.

Practice Exercise 1

A certain reaction X + Y $\longrightarrow$ Z is described as being first order in [X] and third order overall. Which of the following statements is or are true?:
(i) The rate law for the reaction is: Rate = $k[X][Y]^2$.
(ii) If the concentration of X is increased by a factor of 1.5, the rate will increase by a factor of 2.25.
(iii) If the concentration of Y is increased by a factor of 1.5, the rate will increase by a factor of 2.25.

(a) Only one of the statements is true.   (b) Statements (i) and (ii) are true.
(c) Statements (i) and (iii) are true.   (d) Statements (ii) and (iii) are true.
(e) All three statements are true.

Practice Exercise 2

The following data were measured for the reaction of nitric oxide with hydrogen:

$$2\,NO(g) + 2\,H_2(g) \longrightarrow N_2(g) + 2\,H_2O(g)$$

Experiment Number	[NO] (M)	[H$_2$] (M)	Initial Rate (M/s)
1	0.10	0.10	1.23×10^{-3}
2	0.10	0.20	2.46×10^{-3}
3	0.20	0.10	4.92×10^{-3}

(a) Determine the rate law for this reaction.   (b) Calculate the rate constant.
(c) Calculate the rate when $[NO] = 0.050\, M$ and $[H_2] = 0.150\, M$.

14.4 | The Change of Concentration with Time

The rate laws we have examined so far enable us to calculate the rate of a reaction from the rate constant and reactant concentrations. In this section, we will show that rate laws can also be converted into equations that show the relationship between concentrations of reactants or products and time. The mathematics required to accomplish this conversion involves calculus. We do not expect you to be able to perform the calculus operations, but you should be able to use the resulting equations. We will apply this conversion to three of the simplest rate laws: those that are first order overall, those that are second order overall, and those that are zero order overall.

First-Order Reactions

A **first-order reaction** is one whose rate depends on the concentration of a single reactant raised to the first power. If a reaction of the type A $\longrightarrow$ products is first order, the rate law is:

$$\text{Rate} = -\frac{\Delta[A]}{\Delta t} = k[A]$$

This form of a rate law, which expresses how rate depends on concentration, is called the *differential rate law*. Using the operation from calculus called integration, this relationship can be transformed into an equation known as the *integrated rate law* for a first-order reaction that relates the initial concentration of A, $[A]_0$, to its concentration at any other time t, $[A]_t$:

$$\ln[A]_t - \ln[A]_0 = -kt \quad \text{or} \quad \ln\frac{[A]_t}{[A]_0} = -kt \qquad [14.12]$$

The function "ln" in Equation 14.12 is the natural logarithm (Appendix A.2). Equation 14.12 can also be rearranged to

$$\ln[A]_t = -kt + \ln[A]_0 \qquad [14.13]$$

Equations 14.12 and 14.13 can be used with any concentration units as long as the units are the same for both $[A]_t$ and $[A]_0$.

For a first-order reaction, Equation 14.12 or 14.13 can be used in several ways. Given any three of the following quantities, we can solve for the fourth: k, t, $[A]_0$, and $[A]_t$. Thus, you can use these equations to determine (1) the concentration of a reactant remaining at any time after the reaction has started, (2) the time interval required for a given fraction of a sample to react, or (3) the time interval required for a reactant concentration to fall to a certain level.

SAMPLE EXERCISE 14.7 Using the Integrated First-Order Rate Law

The decomposition of a certain insecticide in water at 12 °C follows first-order kinetics with a rate constant of 1.45 yr^{-1}. A quantity of this insecticide is washed into a lake on June 1, leading to a concentration of 5.0×10^{-7} g/cm^3. Assume that the temperature of the lake is constant (so that there are no effects of temperature variation on the rate). **(a)** What is the concentration of the insecticide on June 1 of the following year? **(b)** How long will it take for the insecticide concentration to decrease to 3.0×10^{-7} g/cm^3?

SOLUTION

Analyze We are given the rate constant for a reaction that obeys first-order kinetics, as well as information about concentrations and times, and asked to calculate how much reactant (insecticide) remains after 1 yr. We must also determine the time interval needed to reach a particular insecticide concentration. Because the exercise gives time in (a) and asks for time in (b), we will find it most useful to use the integrated rate law, Equation 14.13.

Plan

(a) We are given $k = 1.45$ yr^{-1}, $t = 1.00$ yr, and $[\text{insecticide}]_0 = 5.0 \times 10^{-7}$ g/cm^3, and so Equation 14.13 can be solved for $[\text{insecticide}]_t$.

(b) We have $k = 1.45$ yr^{-1}, $[\text{insecticide}]_0 = 5.0 \times 10^{-7}$ g/cm^3, and $[\text{insecticide}]_t = 3.0 \times 10^{-7}$ g/cm^3, and so we can solve Equation 14.13 for time, t.

Solve

(a) Substituting the known quantities into Equation 14.13, we have

$$\ln[\text{insecticide}]_{t=1\,\text{yr}} = -(1.45\,\text{yr}^{-1})(1.00\,\text{yr}) + \ln(5.0 \times 10^{-7})$$

We use the ln function on a calculator to evaluate the second term on the right [that is, $\ln(5.0 \times 10^{-7})$], giving

$$\ln[\text{insecticide}]_{t=1\,\text{yr}} = -1.45 + (-14.51) = -15.96$$

To obtain $[\text{insecticide}]_{t=1\,\text{yr}}$, we use the inverse natural logarithm, or e^x, function on the calculator:

$$[\text{insecticide}]_{t=1\,\text{yr}} = e^{-15.96} = 1.2 \times 10^{-7}\,\text{g/cm}^3$$

Note that the concentration units for $[A]_t$ and $[A]_0$ must be the same.

(b) Again substituting into Equation 14.13, with $[\text{insecticide}]_t = 3.0 \times 10^{-7}$ g/cm^3, gives

$$\ln(3.0 \times 10^{-7}) = -(1.45\,\text{yr}^{-1})(t) + \ln(5.0 \times 10^{-7})$$

Solving for t gives

$$t = -[\ln(3.0 \times 10^{-7}) - \ln(5.0 \times 10^{-7})]/1.45\,\text{yr}^{-1}$$
$$= -(-15.02 + 14.51)/1.45\,\text{yr}^{-1} = 0.35\,\text{yr}$$

Check In part (a) the concentration remaining after 1.00 yr (that is, 1.2×10^{-7} g/cm^3) is less than the original concentration (5.0×10^{-7} g/cm^3), as it should be. In (b) the given concentration (3.0×10^{-7} g/cm^3) is greater than that remaining after 1.00 yr, indicating that the time must be less than a year. Thus, $t = 0.35$ yr is a reasonable answer.

Practice Exercise 1

At 25 °C, the decomposition of dinitrogen pentoxide, $N_2O_5(g)$, into $NO_2(g)$ and $O_2(g)$ follows first-order kinetics with $k = 3.4 \times 10^{-5}$ s^{-1}. A sample of N_2O_5 with an initial pressure of 760 torr decomposes at 25 °C until its partial pressure is 650 torr. How much time (in seconds) has elapsed? **(a)** 5.3×10^{-6} **(b)** 2000 **(c)** 4600 **(d)** 34,000 **(e)** 190,000

Practice Exercise 2

The decomposition of dimethyl ether, $(CH_3)_2O$, at 510 °C is a first-order process with a rate constant of 6.8×10^{-4} s^{-1}:

$$(CH_3)_2O(g) \longrightarrow CH_4(g) + H_2(g) + CO(g)$$

If the initial pressure of $(CH_3)_2O$ is 135 torr, what is its pressure after 1420 s?

Equation 14.13 can be used to verify whether a reaction is first order and to determine its rate constant. This equation has the form of the general equation for a straight line, $y = mx + b$, in which m is the slope and b is the y-intercept of the line (Appendix A.4):

$$\ln[A]_t = -kt + \ln[A]_0$$

$$y = mx + b$$

For a first-order reaction, therefore, a graph of $\ln[A]_t$ versus time gives a straight line with a slope of $-k$ and a y-intercept of $\ln[A]_0$. A reaction that is not first order will not yield a straight line.

As an example, consider the conversion of methyl isonitrile (CH_3NC) to its isomer acetonitrile (CH_3CN) (▶ **Figure 14.7**). Because experiments show that the reaction is first order, we can write the rate equation:

$$\ln[CH_3NC]_t = -kt + \ln[CH_3NC]_0$$

We run the reaction at a temperature at which methyl isonitrile is a gas (199 °C), and ▼ **Figure 14.8(a)** shows how the pressure of this gas varies with time. We can use pressure as a unit of concentration for a gas because we know from the ideal-gas law the pressure is directly proportional to the number of moles per unit volume. **Figure 14.8(b)** shows that a plot of the natural logarithm of the pressure versus time is a straight line. The slope of this line is $-5.1 \times 10^{-5}\,s^{-1}$. (You should verify this for yourself, remembering that your result may vary slightly from ours because of inaccuracies associated with reading the graph.) Because the slope of the line equals $-k$, the rate constant for this reaction equals $5.1 \times 10^{-5}\,s^{-1}$.

Second-Order Reactions

A **second-order reaction** is one for which the rate depends either on a reactant concentration raised to the second power or on the concentrations of two reactants each raised to the first power. For simplicity, let's consider reactions of the type A $\longrightarrow$ products or A + B $\longrightarrow$ products that are second order in just one reactant, A:

$$\text{Rate} = -\frac{\Delta[A]}{\Delta t} = k[A]^2$$

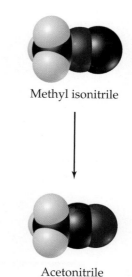

Methyl isonitrile

Acetonitrile

▲ **Figure 14.7 The first-order reaction of CH_3NC conversion into CH_3CN.**

▲ GO FIGURE

What can you conclude given that the plot of ln P versus t is linear?

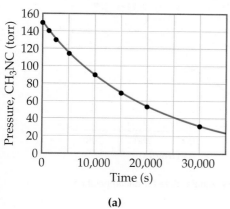

(a)

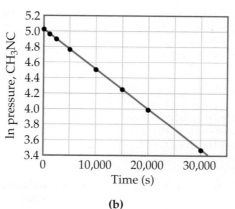

(b)

▲ **Figure 14.8 Kinetic data for conversion of methyl isonitrile into acetonitrile.**

With the use of calculus, this differential rate law can be used to derive the integrated rate law for second-order reactions:

$$\frac{1}{[A]_t} = kt + \frac{1}{[A]_0} \qquad [14.14]$$

This equation, like Equation 14.13, has four variables, k, t, $[A]_0$, and $[A]_t$, and any one of these can be calculated knowing the other three. Equation 14.14 also has the form of a straight line ($y = mx + b$). If the reaction is second order, a plot of $1/[A]_t$ versus t yields a straight line with slope k and y-intercept $1/[A]_0$. One way to distinguish between first- and second-order rate laws is to graph both $\ln[A]_t$ and $1/[A]_t$ against t. If the $\ln[A]_t$ plot is linear, the reaction is first order; if the $1/[A]_t$ plot is linear, the reaction is second order.

SAMPLE EXERCISE 14.8 Determining Reaction Order from the Integrated Rate Law

The following data were obtained for the gas-phase decomposition of nitrogen dioxide at 300 °C, $NO_2(g) \longrightarrow NO(g) + \frac{1}{2}O_2(g)$. Is the reaction first or second order in NO_2?

Time (s)	[NO₂] (M)
0.0	0.01000
50.0	0.00787
100.0	0.00649
200.0	0.00481
300.0	0.00380

SOLUTION

Analyze We are given the concentrations of a reactant at various times during a reaction and asked to determine whether the reaction is first or second order.

Plan We can plot $\ln[NO_2]$ and $1/[NO_2]$ against time. If one plot or the other is linear, we will know the reaction is either first or second order.

Solve To graph $\ln[NO_2]$ and $1/[NO_2]$ against time, we first make the following calculations from the data given:

Time (s)	[NO₂] (M)	ln [NO₂]	1/[NO₂] (1/M)
0.0	0.01000	−4.605	100
50.0	0.00787	−4.845	127
100.0	0.00649	−5.037	154
200.0	0.00481	−5.337	208
300.0	0.00380	−5.573	263

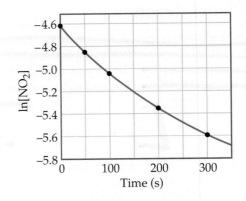

As ▶ **Figure 14.9** shows, only the plot of $1/[NO_2]$ versus time is linear. Thus, the reaction obeys a second-order rate law: Rate $= k[NO_2]^2$. From the slope of this straight-line graph, we determine that $k = 0.543\ M^{-1}\,s^{-1}$ for the disappearance of NO_2.

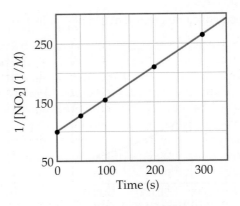

▲ **Figure 14.9 Kinetic data for decomposition of NO₂.**

Practice Exercise 1

For a certain reaction A $\longrightarrow$ products, a plot of $\ln[A]$ versus time produces a straight line with a slope of $-3.0 \times 10^{-2}\,s^{-1}$. Which of the following statements is or are true?:
  (i) The reaction follows first-order kinetics.
  (ii) The rate constant for the reaction is $3.0 \times 10^{-2}\,s^{-1}$.
  (iii) The initial concentration of $[A]$ was $1.0\ M$.

(a) Only one of the statements is true.
(b) Statements (i) and (ii) are true.
(c) Statements (i) and (iii) are true.
(d) Statements (ii) and (iii) are true.
(e) All three statements are true.

Practice Exercise 2

The decomposition of NO_2 discussed in the Sample Exercise is second order in NO_2 with $k = 0.543\ M^{-1}\ s^{-1}$. If the initial concentration of NO_2 in a closed vessel is 0.0500 M, what is the concentration of this reactant after 0.500 h?

Zero-Order Reactions

We have seen that in a first-order reaction the concentration of a reactant A decreases nonlinearly, as shown by the red curve in ▶ Figure 14.10. As [A] declines, the *rate* at which it disappears declines in proportion. A **zero-order reaction** is one in which the rate of disappearance of A is *independent* of [A]. The rate law for a zero-order reaction is

$$\text{Rate} = \frac{-\Delta[A]}{\Delta t} = k$$

The integrated rate law for a zero-order reaction is

$$[A]_t = -kt + [A]_0$$

where $[A]_t$ is the concentration of A at time t and $[A]_0$ is the initial concentration. This is the equation for a straight line with vertical intercept $[A]_0$ and slope $-kt$, as shown in the blue curve in Figure 14.10.

The most common type of zero-order reaction occurs when a gas undergoes decomposition on the surface of a solid. If the surface is completely covered by decomposing molecules, the rate of reaction is constant because the number of reacting surface molecules is constant, so long as there is some gas-phase substance left.

Half-Life

The **half-life** of a reaction, $t_{1/2}$, is the time required for the concentration of a reactant to reach half its initial value, $[A]_{t_{1/2}} = \frac{1}{2}[A]_0$. Half-life is a convenient way to describe how fast a reaction occurs, especially if it is a first-order process. A fast reaction has a short half-life.

We can determine the half-life of a first-order reaction by substituting $[A]_{t_{1/2}} = \frac{1}{2}[A]_0$ for $[A]_t$ and $t_{1/2}$ for t in Equation 14.12:

$$\ln\frac{\frac{1}{2}[A]_0}{[A]_0} = -kt_{1/2}$$

$$\ln\frac{1}{2} = -kt_{1/2}$$

$$t_{1/2} = -\frac{\ln\frac{1}{2}}{k} = \frac{0.693}{k} \qquad [14.15]$$

From Equation 14.15, we see that $t_{1/2}$ for a first-order rate law does not depend on the initial concentration of any reactant. Consequently, the half-life remains constant throughout the reaction. If, for example, the concentration of a reactant is 0.120 M at some instant in the reaction, it will be $\frac{1}{2}(0.120\ M) = 0.060\ M$ after one half-life. After one more half-life passes, the concentration will drop to 0.030 M, and so on. Equation 14.15 also indicates that, for a first-order reaction, we can calculate $t_{1/2}$ if we know k and calculate k if we know $t_{1/2}$.

The change in concentration over time for the first-order rearrangement of gaseous methyl isonitrile at 199 °C is graphed in ▶ Figure 14.11. Because the concentration of this gas is directly proportional to its pressure during the reaction, we have chosen to plot pressure rather than concentration in this graph. The first half-life occurs at 13,600 s (3.78 h). At a time 13,600 s later, the methyl isonitrile pressure (and therefore, concentration) has decreased to half of one-half, or one-fourth, of the initial value. *In a first-order reaction, the concentration of the reactant decreases by one-half in each of a series of regularly spaced time intervals, each interval equal to $t_{1/2}$.*

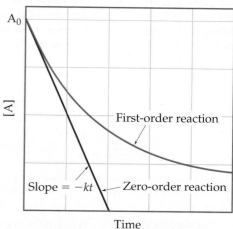

▲ GO FIGURE

At which times during the reaction would you have trouble distinguishing a zero-order reaction from a first-order reaction?

▲ Figure 14.10 Comparison of first-order and zero-order reactions for the disappearance of reactant A with time.

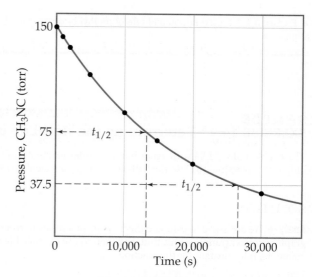

▲ Figure 14.11 Kinetic data for the rearrangement of methyl isonitrile to acetonitrile at 199 °C, showing the half-life of the reaction.

Give It Some Thought

If a solution containing 10.0 g of a substance reacts by first-order kinetics, how many grams remain after three half-lives?

Chemistry Put to Work

Methyl Bromide in the Atmosphere

The compounds known as chlorofluorocarbons (CFCs) are well-known agents responsible for the destruction of Earth's protective ozone layer. Another simple molecule that has the potential to destroy the stratospheric ozone layer is methyl bromide, CH₃Br (▼ **Figure 14.12**). Because this substance has a wide range of uses, including antifungal treatment of plant seeds, it has been produced in large quantities in the past (about 150 million pounds per year worldwide in 1997, at the height of its production). In the stratosphere, the C—Br bond is

broken through absorption of short-wavelength radiation. The resultant Br atoms then catalyze decomposition of O₃.

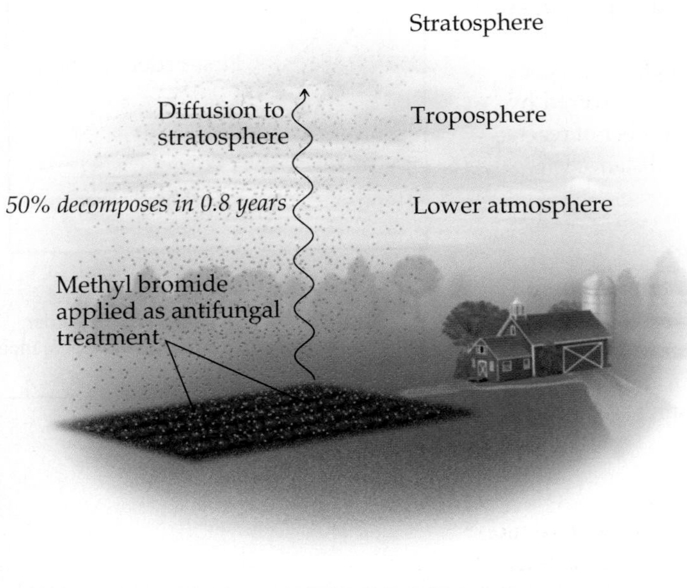

Methyl bromide is removed from the lower atmosphere by a variety of mechanisms, including a slow reaction with ocean water:

$$CH_3Br(g) + H_2O(l) \longrightarrow CH_3OH(aq) + HBr(aq) \quad [14.16]$$

To determine the potential importance of CH₃Br in destruction of the ozone layer, it is important to know how rapidly the reaction in Equation 14.16 and all other reactions remove CH₃Br from the lower atmosphere before it can diffuse into the stratosphere.

The average lifetime of CH₃Br in Earth's lower atmosphere is difficult to measure because the conditions that exist in the atmosphere are too complex to be simulated in the laboratory. Instead, scientists analyzed nearly 4000 atmospheric samples collected above the Pacific Ocean for the presence of several trace organic substances, including methyl bromide. From these measurements, it was possible to estimate the *atmospheric residence time* for CH₃Br.

The atmospheric residence time is related to the half-life for CH₃Br in the lower atmosphere, assuming CH₃Br decomposes by a first-order process. From the experimental data, the half-life for methyl bromide in the lower atmosphere is estimated to be 0.8 ± 0.1 yr. That is, a collection of CH₃Br molecules present at any given time will, on average, be 50% decomposed after 0.8 yr, 75% decomposed after 1.6 yr, and so on. A half-life of 0.8 yr, while comparatively short, is still sufficiently long so that CH₃Br contributes significantly to the destruction of the ozone layer.

In 1997 an international agreement was reached to phase out use of methyl bromide in developed countries by 2005. However, in recent years exemptions for critical agricultural use have been requested and granted. Nevertheless, authorized worldwide production was down to 26 million pounds in 2012, three-fourths of which is used in the United States.

Related Exercise: 14.122

▲ **Figure 14.12** Distribution and fate of methyl bromide in Earth's atmosphere.

SAMPLE EXERCISE 14.9 | Determining the Half-Life of a First-Order Reaction

The reaction of C₄H₉Cl with water is a first-order reaction. (a) Use Figure 14.4 to estimate the half-life for this reaction. (b) Use the half-life from (a) to calculate the rate constant.

SOLUTION

Analyze We are asked to estimate the half-life of a reaction from a graph of concentration versus time and then to use the half-life to calculate the rate constant for the reaction.

Plan

(a) To estimate a half-life, we can select a concentration and then determine the time required for the concentration to decrease to half of that value.

(b) Equation 14.15 is used to calculate the rate constant from the half-life.

Solve

(a) From the graph, we see that the initial value of [C₄H₉Cl] is 0.100 M. The half-life for this first-order reaction is the time required for [C₄H₉Cl] to decrease to 0.050 M, which we can read off the graph. This point occurs at approximately 340 s.

(b) Solving Equation 14.15 for k, we have

$$k = \frac{0.693}{t_{1/2}} = \frac{0.693}{340\ s} = 2.0 \times 10^{-3}\ s^{-1}$$

Check At the end of the second half-life, which should occur at 680 s, the concentration should have decreased by yet another factor of 2, to 0.025 M. Inspection of the graph shows that this is indeed the case.

Practice Exercise 1

We noted in an earlier Practice Exercise that at 25 °C the decomposition of $N_2O_5(g)$ into $NO_2(g)$ and $O_2(g)$ follows first-order

kinetics with $k = 3.4 \times 10^{-5}\ s^{-1}$. How long will it take for a sample originally containing 2.0 atm of N_2O_5 to reach a partial pressure of 380 torr?
(a) 5.7 h **(b)** 8.2 h **(c)** 11 h **(d)** 16 h **(e)** 32 h

Practice Exercise 2

(a) Using Equation 14.15, calculate $t_{1/2}$ for the decomposition of the insecticide described in Sample Exercise 14.7.
(b) How long does it take for the concentration of the insecticide to reach one-quarter of the initial value?

The half-life for second-order and other reactions depends on reactant concentrations and therefore changes as the reaction progresses. We obtained Equation 14.15 for the half-life for a first-order reaction by substituting $[A]_{t_{1/2}} = \frac{1}{2}[A]_0$ for $[A]_t$ and $t_{1/2}$ for t in Equation 14.12. We find the half-life of a second-order reaction by making the same substitutions into Equation 14.14:

$$\frac{1}{\frac{1}{2}[A]_0} = kt_{1/2} + \frac{1}{[A]_0}$$

$$\frac{2}{[A]_0} - \frac{1}{[A]_0} = kt_{1/2}$$

$$t_{1/2} = \frac{1}{k[A]_0} \qquad [14.17]$$

In this case, the half-life depends on the initial concentration of reactant—the lower the initial concentration, the longer the half-life.

 Give It Some Thought

Why can we report the half-life for a first-order reaction without knowing the initial concentration, but not for a second-order reaction?

14.5 | Temperature and Rate

The rates of most chemical reactions increase as the temperature rises. For example, dough rises faster at room temperature than when refrigerated, and plants grow more rapidly in warm weather than in cold. We can see the effect of temperature on reaction rate by observing a chemiluminescence reaction (one that produces light), such as that in Cyalume® light sticks (▶ Figure 14.13).

How is this experimentally observed temperature effect reflected in the rate law? The faster rate at higher temperature is due to an increase in the rate constant with increasing temperature. For example, let's reconsider the first-order reaction we saw in Figure 14.7, namely $CH_3NC \longrightarrow CH_3CN$. **Figure 14.14** shows the rate constant for this reaction as a function of temperature. The rate constant and, hence, the rate of the reaction increase rapidly with temperature, approximately doubling for each 10 °C rise.

The Collision Model

Reaction rates are affected both by reactant concentrations and by temperature. The **collision model**, based on the kinetic-molecular theory ∞ (Section 10.7), accounts for both of these effects at the molecular level. The central idea of the collision model is that molecules must collide to react. The greater the number of collisions per second, the greater the reaction rate. As reactant concentration increases, therefore, the number of collisions increases, leading to an increase in reaction rate. According to the kinetic-molecular theory of gases, increasing the temperature increases molecular speeds. As molecules move faster, they collide more forcefully (with more energy) and more frequently, increasing reaction rates.

▲ GO FIGURE

Why does the light stick glow with less light in cold water than in hot water?

Hot water   Cold water

▲ Figure 14.13 **Temperature affects the rate of the chemiluminescence reaction in light sticks:** The chemiluminescent reaction occurs more rapidly in hot water, and more light is produced.

▲ **GO FIGURE**
Would you expect this curve to
eventually go back down to lower
values? Why or why not?

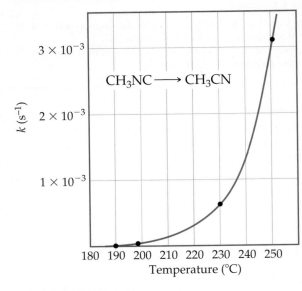

▲ **Figure 14.14 Temperature dependence
of the rate constant for methyl isonitrile
conversion to acetonitrile.** The four points
indicated are used in Sample Exercise 14.11.

For a reaction to occur, though, more is required than simply a collision—it must be the right kind of collision. For most reactions, in fact, only a tiny fraction of collisions leads to a reaction. For example, in a mixture of H_2 and I_2 at ordinary temperatures and pressures, each molecule undergoes about 10^{10} collisions per second. If every collision between H_2 and I_2 resulted in the formation of HI, the reaction would be over in much less than a second. Instead, at room temperature the reaction proceeds very slowly because only about one in every 10^{13} collisions produces a reaction. What keeps the reaction from occurring more rapidly?

The Orientation Factor

In most reactions, collisions between molecules result in a chemical reaction only if the molecules are oriented in a certain way when they collide. The relative orientations of the molecules during collision determine whether the atoms are suitably positioned to form new bonds. For example, consider the reaction

$$Cl + NOCl \longrightarrow NO + Cl_2$$

which takes place if the collision brings Cl atoms together to form Cl_2, as shown in the top panel of ▼ Figure 14.15. In contrast, in the collision shown in the lower panel, the two Cl atoms are not colliding directly with one another, and no products are formed.

Activation Energy

Molecular orientation is not the only factor influencing whether a molecular collision will produce a reaction. In 1888 the Swedish chemist Svante Arrhenius suggested that molecules must possess a certain minimum amount of energy to react. According to the collision model, this energy comes from the kinetic energies of the colliding molecules. Upon collision, the kinetic energy of the molecules can be used to stretch, bend, and ultimately break bonds, leading to chemical reactions. That is, the kinetic energy is used to change the potential energy of the molecule. If molecules are moving too slowly—in other words, with too little kinetic energy—they merely bounce off one another without changing. The minimum energy required to initiate a chemical reaction is called the **activation energy**, E_a, and its value varies from reaction to reaction.

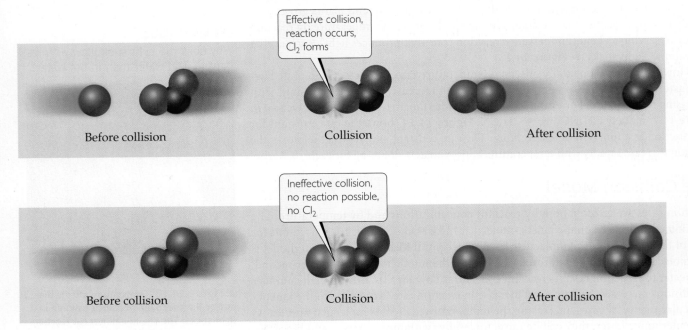

▲ **Figure 14.15 Molecular collisions may or may not lead to a chemical reaction between Cl and NOCl.**

▲ GO FIGURE

If the barrier were lower than as shown in the figure, would the golfer have to hit the ball as hard?

▲ Figure 14.16 **Energy is needed to overcome a barrier between initial and final states.**

The situation during reactions is analogous to that shown in ▲ **Figure 14.16.** The golfer hits the ball to make it move over the hill in the direction of the cup. The hill is a *barrier* between ball and cup. To reach the cup, the player must impart enough kinetic energy with the putter to move the ball to the top of the barrier. If he does not impart enough energy, the ball will roll partway up the hill and then back down toward him. In the same way, molecules require a certain minimum energy to break existing bonds during a chemical reaction. We can think of this minimum energy as an *energy barrier*. In the rearrangement of methyl isonitrile to acetonitrile, for example, we might imagine the reaction passing through an intermediate state in which the N≡C portion of the methyl isonitrile molecule is sideways:

$$H_3C-N\equiv C: \longrightarrow \left[H_3C\cdots\overset{\overset{\ddot{C}}{|||}}{\underset{\ddot{N}}{}} \right] \longrightarrow H_3C-C\equiv N:$$

▶ **Figure 14.17** shows that energy must be supplied to stretch the bond between the H_3C group and the N≡C group to allow the N≡C group to rotate. After the N≡C group has twisted sufficiently, the C—C bond begins to form, and the energy of the molecule drops. Thus, the barrier to formation of acetonitrile represents the energy necessary to force the molecule through the relatively unstable intermediate state, analogous to forcing the ball in Figure 14.16 over the hill. The difference between the energy of the starting molecule and the highest energy along the reaction pathway is the activation energy, E_a. The molecule having the arrangement of atoms shown at the top of the barrier is called either the **activated complex** or the **transition state**.

The conversion of $H_3C-N\equiv C$ to $H_3C-C\equiv N$ is exothermic. Figure 14.17 therefore shows the product as having a lower energy than the reactant. The energy change for the reaction, ΔE, has no effect on reaction rate, however. *The rate depends on the magnitude of E_a; generally, the lower the value of E_a is, the faster the reaction.*

Notice that the reverse reaction is endothermic. The activation energy for the reverse reaction is equal to the energy that must be overcome if approaching the barrier from the right: $\Delta E + E_a$. Thus, to reach the activated complex for the reverse reaction requires more energy than for the forward reaction—for this reaction, there is a larger barrier to overcome going from right to left than from left to right.

▲ GO FIGURE

How does the energy needed to overcome the energy barrier compare with the overall change in energy for this reaction?

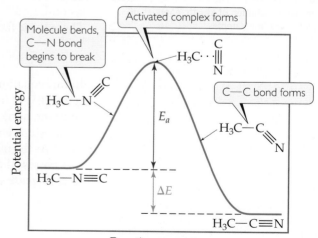

▲ Figure 14.17 **Energy profile for conversion of methyl isonitrile (H₃CNC) to its isomer acetonitrile (H₃CCN).**

Give It Some Thought

Suppose you could measure the rates for both the forward and reverse reactions of the process in Figure 14.17. In which direction would the rate be larger? Why?

Any particular methyl isonitrile molecule acquires sufficient energy to overcome the energy barrier through collisions with other molecules. Recall from the kinetic-molecular theory of gases that, at any instant, gas molecules are distributed in energy over a wide range. ⚮⚮ (Section 10.7) ▼ **Figure 14.18** shows the distribution of kinetic energies for two temperatures, comparing them with the minimum energy needed for reaction, E_a. At the higher temperature a much greater fraction of the molecules have kinetic energy greater than E_a, which leads to a greater rate of reaction.

Give It Some Thought

Suppose we have two reactions, A ⟶ B and B ⟶ C. You can isolate B, and it is stable. Is B the transition state for the reaction A ⟶ C?

▲ **GO FIGURE**

What would the curve look like for a temperature higher than that for the red curve in the figure?

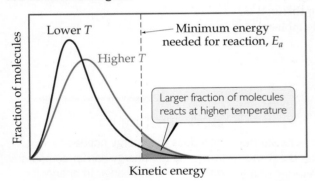

▲ **Figure 14.18 The effect of temperature on the distribution of kinetic energies of molecules in a sample.**

For a collection of molecules in the gas phase, the fraction of molecules that have kinetic energy equal to or greater than E_a is given by the expression

$$f = e^{-E_a/RT} \qquad [14.18]$$

In this equation, R is the gas constant (8.314 J/mol-K) and T is the absolute temperature. To get an idea of the magnitude of f, let's suppose that E_a is 100 kJ/mol, a value typical of many reactions, and that T is 300 K. The calculated value of f is 3.9×10^{-18}, an extremely small number! At 320 K, $f = 4.7 \times 10^{-17}$. Thus, only a 20° increase in temperature produces a more than tenfold increase in the fraction of molecules possessing at least 100 kJ/mol of energy.

The Arrhenius Equation

Arrhenius noted that for most reactions the increase in rate with increasing temperature is nonlinear (Figure 14.14). He found that most reaction-rate data obeyed an equation based on (a) the fraction of molecules possessing energy E_a or greater, (b) the number of collisions per second, and (c) the fraction of collisions that have the appropriate orientation. These three factors are incorporated into the **Arrhenius equation**:

$$k = Ae^{-E_a/RT} \qquad [14.19]$$

In this equation, k is the rate constant, E_a is the activation energy, R is the gas constant (8.314 J/mol-K), and T is the absolute temperature. The **frequency factor**, A, is constant, or nearly so, as temperature is varied. This factor is related to the frequency of collisions and the probability that the collisions are favorably oriented for reaction.* As the magnitude of E_a increases, k decreases because the fraction of molecules that possess the required energy is smaller. Thus, at fixed values of T and A, *reaction rates decrease as E_a increases.*

---

*Because collision frequency increases with temperature, A also has some temperature dependence, but this dependence is much smaller than the exponential term. Therefore, A is considered approximately constant.

SAMPLE
EXERCISE 14.10   Activation Energies and Speeds of Reaction

Consider a series of reactions having these energy profiles:

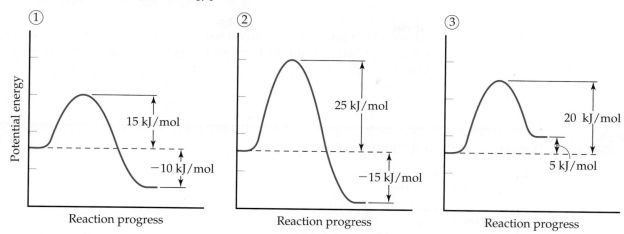

Rank the reactions from slowest to fastest assuming that they have nearly the same value for the frequency factor A.

SOLUTION

The lower the activation energy, the faster the reaction. The value of ΔE does not affect the rate. Hence, the order from slowest reaction to fastest is $2 < 3 < 1$.

Practice Exercise 1

Which of the following statements is or are true?
 (i) The activation energies for the forward and reverse directions of a reaction can be different.
(ii) Assuming that A is constant, if both E_a and T increase, then k will increase.

(iii) For two different reactions, the one with the smaller value of E_a will necessarily have the larger value for k.

(a) Only one of the statements is true.
(b) Statements (i) and (ii) are true.
(c) Statements (i) and (iii) are true.
(d) Statements (ii) and (iii) are true.
(e) All three statements are true.

Practice Exercise 2

Rank the reverse reactions from slowest to fastest.

Determining the Activation Energy

We can calculate the activation energy for a reaction by manipulating the Arrhenius equation. Taking the natural log of both sides of Equation 14.19, we obtain

$$\ln k = -\frac{E_a}{RT} + \ln A$$

$$\underset{\displaystyle y}{\Big\downarrow} \quad = \quad \underset{\displaystyle mx}{\Big\downarrow} \quad + \quad \underset{\displaystyle b}{\Big\downarrow}$$

[14.20]

which has the form of the equation for a straight line. A graph of $\ln k$ versus $1/T$ is a line with a slope equal to $-E_a/R$ and a y-intercept equal to $\ln A$. Thus, the activation energy can be determined by measuring k at a series of temperatures, graphing $\ln k$ versus $1/T$, and calculating E_a from the slope of the resultant line.

We can also use Equation 14.20 to evaluate E_a in a nongraphical way if we know the rate constant of a reaction at two or more temperatures. For example, suppose that at two different temperatures T_1 and T_2 a reaction has rate constants k_1 and k_2. For each condition, we have

$$\ln k_1 = -\frac{E_a}{RT_1} + \ln A \quad \text{and} \quad \ln k_2 = -\frac{E_a}{RT_2} + \ln A$$

Subtracting $\ln k_2$ from $\ln k_1$ gives

$$\ln k_1 - \ln k_2 = \left(-\frac{E_a}{RT_1} + \ln A\right) - \left(-\frac{E_a}{RT_2} + \ln A\right)$$

Simplifying this equation and rearranging gives

$$\ln \frac{k_1}{k_2} = \frac{E_a}{R}\left(\frac{1}{T_2} - \frac{1}{T_1}\right)$$

[14.21]

Equation 14.21 provides a convenient way to calculate a rate constant k_1 at some temperature T_1 when we know the activation energy and the rate constant k_2 at some other temperature T_2.

SAMPLE EXERCISE 14.11 | Determining the Activation Energy

The following table shows the rate constants for the rearrangement of methyl isonitrile at various temperatures (these are the data points in Figure 14.14):

Temperature (°C)	$k\,(s^{-1})$
189.7	2.52×10^{-5}
198.9	5.25×10^{-5}
230.3	6.30×10^{-4}
251.2	3.16×10^{-3}

(a) From these data, calculate the activation energy for the reaction. (b) What is the value of the rate constant at 430.0 K?

SOLUTION

Analyze We are given rate constants, k, measured at several temperatures and asked to determine the activation energy, E_a, and the rate constant, k, at a particular temperature.

Plan We can obtain E_a from the slope of a graph of $\ln k$ versus $1/T$. Once we know E_a, we can use Equation 14.21 together with the given rate data to calculate the rate constant at 430.0 K.

Solve

(a) We must first convert the temperatures from degrees Celsius to kelvins. We then take the inverse of each temperature, $1/T$, and the natural log of each rate constant, $\ln k$. This gives us the table shown at the right:

A graph of $\ln k$ versus $1/T$ is a straight line (▼ **Figure 14.19**).

T (K)	$1/T\,(K^{-1})$	$\ln k$
462.9	2.160×10^{-3}	-10.589
472.1	2.118×10^{-3}	-9.855
503.5	1.986×10^{-3}	-7.370
524.4	1.907×10^{-3}	-5.757

The slope of the line is obtained by choosing any two well-separated points and using the coordinates of each:

$$\text{Slope} = \frac{\Delta y}{\Delta x} = \frac{-6.6 - (-10.4)}{0.00195 - 0.00215} = -1.9 \times 10^4$$

Because logarithms have no units, the numerator in this equation is dimensionless. The denominator has the units of $1/T$, namely, K^{-1}. Thus, the overall units for the slope are K. The slope equals $-E_a/R$. We use the value for the gas constant R in units of J/mol-K (Table 10.2). We thus obtain

$$\text{Slope} = -\frac{E_a}{R}$$

$$E_a = -(\text{slope})(R) = -(-1.9 \times 10^4\,K)\left(8.314\frac{J}{\text{mol-K}}\right)\left(\frac{1\,kJ}{1000\,J}\right)$$

$$= 1.6 \times 10^2\,kJ/mol = 160\,kJ/mol$$

We report the activation energy to only two significant figures because we are limited by the precision with which we can read the graph in Figure 14.19.

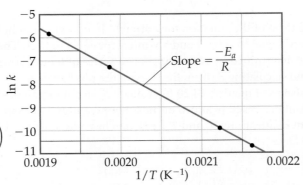

▲ **Figure 14.19 Graphical determination of activation energy E_a.**

(b) To determine the rate constant, k_1, at $T_1 = 430.0$ K, we can use Equation 14.21 with $E_a = 160$ kJ/mol and one of the rate constants and temperatures from the given data, such as $k_2 = 2.52 \times 10^{-5}$ s^{-1} and $T_2 = 462.9$ K:

$$\ln\left(\frac{k_1}{2.52 \times 10^{-5}\,\text{s}^{-1}}\right) =$$

$$\left(\frac{160\,\text{kJ/mol}}{8.314\,\text{J/mol-K}}\right)\left(\frac{1}{462.9\,\text{K}} - \frac{1}{430.0\,\text{K}}\right)\left(\frac{1000\,\text{J}}{1\,\text{kJ}}\right) = -3.18$$

Thus,

$$\frac{k_1}{2.52 \times 10^{-5}\,\text{s}^{-1}} = e^{-3.18} = 4.15 \times 10^{-2}$$

$$k_1 = (4.15 \times 10^{-2})(2.52 \times 10^{-5}\,\text{s}^{-1}) = 1.0 \times 10^{-6}\,\text{s}^{-1}$$

Note that the units of k_1 are the same as those of k_2.

Practice Exercise 1

Using the data in Sample Exercise 14.11, which of the following is the rate constant for the rearrangement of methyl isonitrile at 320 °C?
(a) 8.1×10^{-15} s^{-1} **(b)** 2.2×10^{-13} s^{-1} **(c)** 2.7×10^{-9} s^{-1}
(d) 2.3×10^{-1} s^{-1} **(e)** 9.2×10^{3} s^{-1}

Practice Exercise 2

To one significant figure, what is the value for the frequency factor A for the data presented in Sample Exercise 14.11.

14.6 | Reaction Mechanisms

A balanced equation for a chemical reaction indicates the substances present at the start of the reaction and those present at the end of the reaction. It provides no information, however, about the detailed steps that occur at the molecular level as the reactants are turned into products. The steps by which a reaction occurs is called the **reaction mechanism**. At the most sophisticated level, a reaction mechanism describes the order in which bonds are broken and formed and the changes in relative positions of the atoms in the course of the reaction.

Elementary Reactions

We have seen that reactions take place because of collisions between reacting molecules. For example, the collisions between molecules of methyl isonitrile (CH_3NC) can provide the energy to allow the CH_3NC to rearrange to acetonitrile:

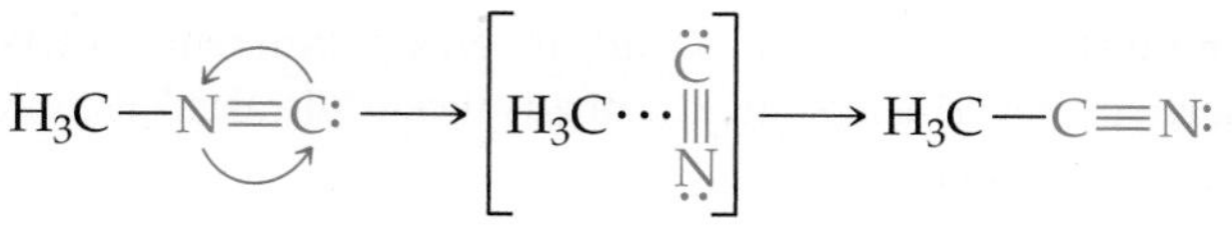

Similarly, the reaction of NO and O_3 to form NO_2 and O_2 appears to occur as a result of a single collision involving suitably oriented and sufficiently energetic NO and O_3 molecules:

$$NO(g) + O_3(g) \longrightarrow NO_2(g) + O_2(g) \qquad [14.22]$$

Both reactions occur in a single event or step and are called **elementary reactions**.

The number of molecules that participate as reactants in an elementary reaction defines the **molecularity** of the reaction. If a single molecule is involved, the reaction is **unimolecular**. The rearrangement of methyl isonitrile is a unimolecular process. Elementary reactions involving the collision of two reactant molecules are **bimolecular**. The reaction between NO and O_3 is bimolecular. Elementary reactions involving the simultaneous collision of three molecules are **termolecular**. Termolecular reactions are far less probable than unimolecular or bimolecular processes and are extremely rare. The chance that four or more molecules will collide simultaneously with any regularity is even more remote; consequently, such collisions are never proposed as part of a reaction mechanism. Thus, nearly all reaction mechanisms contain only unimolecular and bimolecular elementary reactions.

Give It Some Thought

What is the molecularity of the elementary reaction?

$$NO(g) + Cl_2(g) \longrightarrow NOCl(g) + Cl(g)$$

Multistep Mechanisms

The net change represented by a balanced chemical equation often occurs by a *multistep mechanism* consisting of a sequence of elementary reactions. For example, below 225 °C, the reaction

$$NO_2(g) + CO(g) \longrightarrow NO(g) + CO_2(g) \qquad [14.23]$$

appears to proceed in two elementary reactions (or two *elementary steps*), each of which is bimolecular. First, two NO_2 molecules collide, and an oxygen atom is transferred from one to the other. The resultant NO_3 then collides with a CO molecule and transfers an oxygen atom to it:

$$NO_2(g) + NO_2(g) \longrightarrow NO_3(g) + NO(g)$$

$$NO_3(g) + CO(g) \longrightarrow NO_2(g) + CO_2(g)$$

Thus, we say that the reaction occurs by a two-step mechanism.

The chemical equations for the elementary reactions in a multistep mechanism must always add to give the chemical equation of the overall process. In the present example, the sum of the two elementary reactions is

$$2\,NO_2(g) + NO_3(g) + CO(g) \longrightarrow NO_2(g) + NO_3(g) + NO(g) + CO_2(g)$$

Simplifying this equation by eliminating substances that appear on both sides gives Equation 14.23, the net equation for the process.

Because NO_3 is neither a reactant nor a product of the reaction—it is formed in one elementary reaction and consumed in the next—it is called an **intermediate**. Multistep mechanisms involve one or more intermediates. Intermediates are not the same as transition states, as shown in ◀ Figure 14.20. Intermediates can be stable and can therefore sometimes be identified and even isolated. Transition states, on the other hand, are always inherently unstable and as such can never be isolated. Nevertheless, the use of advanced "ultrafast" techniques sometimes allows us to characterize them.

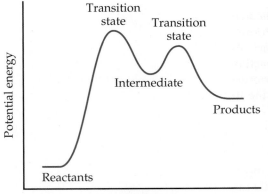 **GO FIGURE**

For this profile, is it easier for a molecule of the intermediate to convert to reactants or products?

▲ **Figure 14.20** The energy profile of a reaction, showing transition states and an intermediate.

SAMPLE EXERCISE 14.12 Determining Molecularity and Identifying Intermediates

It has been proposed that the conversion of ozone into O_2 proceeds by a two-step mechanism:

$$O_3(g) \longrightarrow O_2(g) + O(g)$$

$$O_3(g) + O(g) \longrightarrow 2\,O_2(g)$$

(a) Describe the molecularity of each elementary reaction in this mechanism.
(b) Write the equation for the overall reaction.
(c) Identify the intermediate(s).

SOLUTION

Analyze We are given a two-step mechanism and asked for **(a)** the molecularities of each of the two elementary reactions, **(b)** the equation for the overall process, and **(c)** the intermediate.

Plan The molecularity of each elementary reaction depends on the number of reactant molecules in the equation for that reaction. The overall equation is the sum of the equations for the elementary reactions. The intermediate is a substance formed in one step of the

mechanism and used in another and therefore not part of the equation for the overall reaction.

Solve

(a) The first elementary reaction involves a single reactant and is consequently unimolecular. The second reaction, which involves two reactant molecules, is bimolecular.

(b) Adding the two elementary reactions gives

$$2\,O_3(g) + O(g) \longrightarrow 3\,O_2(g) + O(g)$$

Because $O(g)$ appears in equal amounts on both sides of the equation, it can be eliminated to give the net equation for the chemical process:

$$2\,O_3(g) \longrightarrow 3\,O_2(g)$$

(c) The intermediate is $O(g)$. It is neither an original reactant nor a final product but is formed in the first step of the mechanism and consumed in the second.

Practice Exercise 1

Consider the following two-step reaction mechanism:

$$A(g) + B(g) \longrightarrow X(g) + Y(g)$$
$$X(g) + C(g) \longrightarrow Y(g) + Z(g)$$

Which of the following statements about this mechanism is or are true?
(i) Both of the steps in this mechanism are bimolecular.
(ii) The overall reaction is $A(g) + B(g) + C(g) \longrightarrow Y(g) + Z(g)$.

(iii) The substance $X(g)$ is an intermediate in this mechanism.

(a) Only one of the statements is true.
(b) Statements (i) and (ii) are true.
(c) Statements (i) and (iii) are true.
(d) Statements (ii) and (iii) are true.
(e) All three statements are true.

Practice Exercise 2

For the reaction

$$Mo(CO)_6 + P(CH_3)_3 \longrightarrow Mo(CO)_5P(CH_3)_3 + CO$$

the proposed mechanism is

$$Mo(CO)_6 \longrightarrow Mo(CO)_5 + CO$$
$$Mo(CO)_5 + P(CH_3)_3 \longrightarrow Mo(CO)_5P(CH_3)_3$$

(a) Is the proposed mechanism consistent with the equation for the overall reaction? **(b)** What is the molecularity of each step of the mechanism? **(c)** Identify the intermediate(s).

Rate Laws for Elementary Reactions

In Section 14.3, we stressed that rate laws must be determined experimentally; they cannot be predicted from the coefficients of balanced chemical equations. We are now in a position to understand why this is so. Every reaction is made up of a series of one or more elementary steps, and the rate laws and relative speeds of these steps dictate the overall rate law for the reaction. Indeed, the rate law for a reaction can be determined from its mechanism, as we will see shortly, and compared with the experimental rate law. Thus, our next challenge in kinetics is to arrive at reaction mechanisms that lead to rate laws consistent with those observed experimentally. We start by examining the rate laws of elementary reactions.

Elementary reactions are significant in a very important way: *If a reaction is elementary, its rate law is based directly on its molecularity.* For example, consider the unimolecular reaction

$$A \longrightarrow \text{products}$$

As the number of A molecules increases, the number that reacts in a given time interval increases proportionally. Thus, the rate of a unimolecular process is first order:

$$\text{Rate} = k[A]$$

For bimolecular elementary steps, the rate law is second order, as in the reaction

$$A + B \longrightarrow \text{products} \qquad \text{Rate} = k[A][B]$$

The second-order rate law follows directly from collision theory. If we double the concentration of A, the number of collisions between the molecules of A and B doubles; likewise, if we double [B], the number of collisions between A and B doubles. Therefore, the rate law is first order in both [A] and [B] and second order overall.

The rate laws for all feasible elementary reactions are given in ▼ **Table 14.3**. Notice how each rate law follows directly from the molecularity of the reaction. It is important

Table 14.3   Elementary Reactions and Their Rate Laws

Molecularity	Elementary Reaction	Rate Law
Unimolecular	$A \longrightarrow$ products	Rate $= k[A]$
Bimolecular	$A + A \longrightarrow$ products	Rate $= k[A]^2$
Bimolecular	$A + B \longrightarrow$ products	Rate $= k[A][B]$
Termolecular	$A + A + A \longrightarrow$ products	Rate $= k[A]^3$
Termolecular	$A + A + B \longrightarrow$ products	Rate $= k[A]^2[B]$
Termolecular	$A + B + C \longrightarrow$ products	Rate $= k[A][B][C]$

to remember, however, that we cannot tell by merely looking at a balanced, overall chemical equation whether the reaction involves one or several elementary steps.

Predicting the Rate Law for an Elementary Reaction

If the following reaction occurs in a single elementary reaction, predict its rate law:

$$H_2(g) + Br_2(g) \longrightarrow 2\,HBr(g)$$

SOLUTION

Analyze We are given the equation and asked for its rate law, assuming that it is an elementary process.

Plan Because we are assuming that the reaction occurs as a single elementary reaction, we are able to write the rate law using the coefficients for the reactants in the equation as the reaction orders.

Solve The reaction is bimolecular, involving one molecule of H_2 and one molecule of Br_2. Thus, the rate law is first order in each reactant and second order overall:

$$\text{Rate} = k[H_2][Br_2]$$

Comment Experimental studies of this reaction show that the reaction actually has a very different rate law:

$$\text{Rate} = k[H_2][Br_2]^{1/2}$$

Because the experimental rate law differs from the one obtained by assuming a single elementary reaction, we can conclude that the mechanism cannot occur by a single elementary step. It must, therefore, involve two or more elementary steps.

Practice Exercise 1
Consider the following reaction: $2\,A + B \longrightarrow X + 2\,Y$. You are told that the first step in the mechanism of this reaction has the following rate law: Rate $= k[A][B]$. Which of the following could be the first step in the reaction mechanism (note that substance Z is an intermediate)?
(a) $A + A \longrightarrow Y + Z$
(b) $A \longrightarrow X + Z$
(c) $A + A + B \longrightarrow X + Y + Y$
(d) $B \longrightarrow X + Y$
(e) $A + B \longrightarrow X + Z$

Practice Exercise 2
Consider the following reaction: $2\,NO(g) + Br_2(g) \longrightarrow 2\,NOBr(g)$.
(a) Write the rate law for the reaction, assuming it involves a single elementary reaction. (b) Is a single-step mechanism likely for this reaction?

The Rate-Determining Step for a Multistep Mechanism

As with the reaction in Sample Exercise 14.13, most reactions occur by mechanisms that involve two or more elementary reactions. Each step of the mechanism has its own rate constant and activation energy. Often one step is much slower than the others, and the overall rate of a reaction cannot exceed the rate of the slowest elementary step. Because the slow step limits the overall reaction rate, it is called the **rate-determining step** (or *rate-limiting step*).

To understand the concept of the rate-determining step for a reaction, consider a toll road with two toll plazas (▶ Figure 14.21). Cars enter the toll road at point 1 and pass through toll plaza A. They then pass an intermediate point 2 before passing through toll plaza B and arriving at point 3. We can envision this trip along the toll road as occurring in two elementary steps:

Step 1:	Point 1 $\longrightarrow$ Point 2	(through toll plaza A)
Step 2:	Point 2 $\longrightarrow$ Point 3	(through toll plaza B)
Overall:	Point 1 $\longrightarrow$ Point 3	(through both toll plazas)

Now suppose that one or more gates at toll plaza A are malfunctioning, so that traffic backs up behind the gates, as depicted in Figure 14.21(a). The rate at which cars can get to point 3 is limited by the rate at which they can get through the traffic jam at plaza A. Thus, step 1 is the rate-determining step of the journey along the toll road. If, however, all gates at A are functioning but one or more at B are not, traffic flows quickly through A but gets backed up at B, as depicted in Figure 14.21(b). In this case, step 2 is the rate-determining step.

In the same way, *the slowest step in a multistep reaction determines the overall rate.* By analogy to Figure 14.21(a), the rate of a fast step following the rate-determining step does not speed up the overall rate. If the slow step is not the first one, as is the case in

GO FIGURE

For which of the two scenarios in the figure will one get from point 1 to point 3 most rapidly?

(a) Cars slowed at toll plaza A, rate-determining step is passage through A

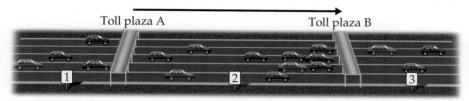

(b) Cars slowed at toll plaza B, rate-determining step is passage through B

▲ Figure 14.21 **Rate-determining steps in traffic flow on a toll road.**

Figure 14.21(b), the faster preceding steps produce intermediate products that accumulate before being consumed in the slow step. In either case, *the rate-determining step governs the rate law for the overall reaction.*

 Give It Some Thought

Why can't the rate law for a reaction generally be deduced from the balanced equation for the reaction?

Mechanisms with a Slow Initial Step

We can most easily see the relationship between the slow step in a mechanism and the rate law for the overall reaction by considering an example in which the first step in a multistep mechanism is the rate-determining step. Consider the reaction of NO_2 and CO to produce NO and CO_2 (Equation 14.23). Below 225 °C, it is found experimentally that the rate law for this reaction is second order in NO_2 and zero order in CO: Rate = $k[NO_2]^2$. Can we propose a reaction mechanism consistent with this rate law? Consider the two-step mechanism:*

$$\text{Step 1:} \quad NO_2(g) + NO_2(g) \xrightarrow{k_1} NO_3(g) + NO(g) \quad \text{(slow)}$$

$$\text{Step 2:} \quad NO_3(g) + CO(g) \xrightarrow{k_2} NO_2(g) + CO_2(g) \quad \text{(fast)}$$

$$\text{Overall:} \; NO_2(g) + CO(g) \longrightarrow NO(g) + CO_2(g)$$

Step 2 is much faster than step 1; that is, $k_2 \gg k_1$, telling us that the intermediate $NO_3(g)$ is slowly produced in step 1 and immediately consumed in step 2.

Because step 1 is slow and step 2 is fast, step 1 is the rate-determining step. Thus, the rate of the overall reaction depends on the rate of step 1, and the rate law of the

---

*Note the rate constants k_1 and k_2 written above the reaction arrows. The subscript on each rate constant identifies the elementary step involved. Thus, k_1 is the rate constant for step 1, and k_2 is the rate constant for step 2. A negative subscript refers to the rate constant for the reverse of an elementary step. For example, k_{-1} is the rate constant for the reverse of the first step.

overall reaction equals the rate law of step 1. Step 1 is a bimolecular process that has the rate law

$$\text{Rate} = k_1[NO_2]^2$$

Thus, the rate law predicted by this mechanism agrees with the one observed experimentally. The reactant CO is absent from the rate law because it reacts in a step that follows the rate-determining step.

A scientist would not, at this point, say that we have "proved" that this mechanism is correct. All we can say is that the rate law predicted by the mechanism is *consistent with experiment*. We can often envision a different sequence of steps that leads to the same rate law. If, however, the predicted rate law of the proposed mechanism disagrees with experiment, we know for certain that the mechanism cannot be correct.

SAMPLE EXERCISE 14.14 | Determining the Rate Law for a Multistep Mechanism

The decomposition of nitrous oxide, N_2O, is believed to occur by a two-step mechanism:

$$N_2O(g) \longrightarrow N_2(g) + O(g) \quad \text{(slow)}$$

$$N_2O(g) + O(g) \longrightarrow N_2(g) + O_2(g) \quad \text{(fast)}$$

(a) Write the equation for the overall reaction. (b) Write the rate law for the overall reaction.

SOLUTION

Analyze Given a multistep mechanism with the relative speeds of the steps, we are asked to write the overall reaction and the rate law for that overall reaction.

Plan (a) Find the overall reaction by adding the elementary steps and eliminating the intermediates. (b) The rate law for the overall reaction will be that of the slow, rate-determining step.

Solve

(a) Adding the two elementary reactions gives

$$2\,N_2O(g) + O(g) \longrightarrow 2\,N_2(g) + 2\,O_2(g) + O(g)$$

Omitting the intermediate, $O(g)$, which occurs on both sides of the equation, gives the overall reaction:

$$2\,N_2O(g) \longrightarrow 2\,N_2(g) + O_2(g)$$

(b) The rate law for the overall reaction is just the rate law for the slow, rate-determining elementary reaction. Because that slow step is a unimolecular elementary reaction, the rate law is first order:

$$\text{Rate} = k[N_2O]$$

Practice Exercise 1

Let's consider a hypothetical reaction similar to that in Practice Exercise 1 of Sample Exercise 14.13: $2\,C + D \longrightarrow J + 2\,K$. You are told that the rate of this reaction is second order overall and second order in [C]. Could any of the following be a rate-determining first step in a reaction mechanism that is consistent with the observed rate law for the reaction (note that substance Z is an intermediate)?
(a) $C + C \longrightarrow K + Z$ (b) $C \longrightarrow J + Z$ (c) $C + D \longrightarrow J + Z$
(d) $D \longrightarrow J + K$ (e) None of these are consistent with the observed rate law.

Practice Exercise 2

Ozone reacts with nitrogen dioxide to produce dinitrogen pentoxide and oxygen:

$$O_3(g) + 2\,NO_2(g) \longrightarrow N_2O_5(g) + O_2(g)$$

The reaction is believed to occur in two steps:

$$O_3(g) + NO_2(g) \longrightarrow NO_3(g) + O_2(g)$$

$$NO_3(g) + NO_2(g) \longrightarrow N_2O_5(g)$$

The experimental rate law is rate $= k[O_3][NO_2]$. What can you say about the relative rates of the two steps of the mechanism?

Mechanisms with a Fast Initial Step

It is possible, but not particularly straightforward, to derive the rate law for a mechanism in which an intermediate is a reactant in the rate-determining step. This situation arises in multistep mechanisms when the first step is fast and therefore *not* the rate-determining step. Let's consider one example: the gas-phase reaction of nitric oxide (NO) with bromine (Br_2):

$$2\,NO(g) + Br_2(g) \longrightarrow 2\,NOBr(g) \qquad [14.24]$$

The experimentally determined rate law for this reaction is second order in NO and first order in Br_2:

$$Rate = k[NO]^2[Br_2] \qquad [14.25]$$

We seek a reaction mechanism that is consistent with this rate law. One possibility is that the reaction occurs in a single termolecular step:

$$NO(g) + NO(g) + Br_2(g) \longrightarrow 2\,NOBr(g) \quad Rate = k[NO]^2[Br_2] \quad [14.26]$$

As noted in Practice Exercise 2 of Exercise 14.13, this does not seem likely because termolecular processes are so rare.

 Give It Some Thought

Why are termolecular elementary steps rare in gas-phase reactions?

Let's consider an alternative mechanism that does not involve a termolecular step:

Step 1: $\qquad NO(g) + Br_2(g) \underset{k_{-1}}{\overset{k_1}{\rightleftharpoons}} NOBr_2(g) \quad$ (fast)

Step 2: $\quad NOBr_2(g) + NO(g) \xrightarrow{k_2} 2\,NOBr(g) \quad$ (slow) $\qquad$ [14.27]

In this mechanism, step 1 involves two processes: a forward reaction and its reverse.

Because step 2 is the rate-determining step, the rate law for that step governs the rate of the overall reaction:

$$Rate = k_2[NOBr_2][NO] \qquad [14.28]$$

Note that $NOBr_2$ is an intermediate generated in the forward reaction of step 1. Intermediates are usually unstable and have a low, unknown concentration. Thus, the rate law of Equation 14.28 depends on the unknown concentration of an intermediate, which isn't desirable. We want instead to express the rate law for a reaction in terms of the reactants, or the products if necessary, of the reaction.

With the aid of some assumptions, we can express the concentration of the intermediate $NOBr_2$ in terms of the concentrations of the starting reactants NO and Br_2. We first assume that $NOBr_2$ is unstable and does not accumulate to any significant extent in the reaction mixture. Once formed, $NOBr_2$ can be consumed either by reacting with NO to form NOBr or by falling back apart into NO and Br_2. The first of these possibilities is step 2 of our alternative mechanism, a slow process. The second is the reverse of step 1, a unimolecular process:

$$NOBr_2(g) \xrightarrow{k_{-1}} NO(g) + Br_2(g)$$

Because step 2 is slow, we assume that most of the $NOBr_2$ falls apart according to this reaction. Thus, we have both the forward and reverse reactions of step 1 occurring much faster than step 2. Because they occur rapidly relative to step 2, the forward and reverse reactions of step 1 establish an equilibrium. As in any other dynamic equilibrium, the rate of the forward reaction equals that of the reverse reaction:

$$\underset{\text{Rate of forward reaction}}{k[NO][Br_2]} = \underset{\text{Rate of reverse reaction}}{k_{-1}[NOBr_2]}$$

Solving for $[NOBr_2]$, we have

$$[NOBr_2] = \frac{k_1}{k_{-1}}[NO][Br_2]$$

Substituting this relationship into Equation 14.28, we have

$$Rate = k_2\frac{k_1}{k_{-1}}[NO][Br_2][NO] = k[NO]^2[Br_2]$$

where the experimental rate constant k equals k_2k_1/k_{-1}. This expression is consistent with the experimental rate law (Equation 14.25). Thus, our alternative mechanism

(Equation 14.27), which involves two steps but only unimolecular and bimolecular processes, is far more probable than the single-step termolecular mechanism of Equation 14.26.

In general, *whenever a fast step precedes a slow one, we can solve for the concentration of an intermediate by assuming that an equilibrium is established in the fast step.*

SAMPLE EXERCISE 14.15 | Deriving the Rate Law for a Mechanism with a Fast Initial Step

Show that the following mechanism for Equation 14.24 also produces a rate law consistent with the experimentally observed one:

Step 1: $NO(g) + NO(g) \underset{k_{-1}}{\overset{k_1}{\rightleftharpoons}} N_2O_2(g)$ (fast, equilibrium)

Step 2: $N_2O_2(g) + Br_2(g) \overset{k_2}{\longrightarrow} 2\,NOBr(g)$ (slow)

SOLUTION

Analyze We are given a mechanism with a fast initial step and asked to write the rate law for the overall reaction.

Plan The rate law of the slow elementary step in a mechanism determines the rate law for the overall reaction. Thus, we first write the rate law based on the molecularity of the slow step. In this case, the slow step involves the intermediate N_2O_2 as a reactant. Experimental rate laws, however, do not contain the concentrations of intermediates; instead they are expressed in terms of the concentrations of starting substances. Thus, we must relate the concentration of N_2O_2 to the concentration of NO by assuming that an equilibrium is established in the first step.

Solve The second step is rate determining, so the overall rate is

$$Rate = k_2[N_2O_2][Br_2]$$

We solve for the concentration of the intermediate N_2O_2 by assuming that an equilibrium is established in step 1; thus, the rates of the forward and reverse reactions in step 1 are equal:

$$k_1[NO]^2 = k_{-1}[N_2O_2]$$

Solving for the concentration of the intermediate, N_2O_2, gives

$$[N_2O_2] = \frac{k_1}{k_{-1}}[NO]^2$$

Substituting this expression into the rate expression gives

$$Rate = k_2\frac{k_1}{k_{-1}}[NO]^2[Br_2] = k[NO]^2[Br_2]$$

Thus, this mechanism also yields a rate law consistent with the experimental one. Remember: There may be more than one mechanism that leads to an observed experimental rate law!

Practice Exercise 1

Consider the following hypothetical reaction:
$2\,P + Q \longrightarrow 2\,R + S$. The following mechanism is proposed for this reaction:

$$P + P \rightleftharpoons T \quad (fast)$$

$$Q + T \longrightarrow R + U \quad (slow)$$

$$U \longrightarrow R + S \quad (fast)$$

Substances T and U are unstable intermediates. What rate law is predicted by this mechanism?
(a) Rate $= k[P]^2$ **(b)** Rate $= k[P][Q]$ **(c)** Rate $= k[P]^2[Q]$
(d) Rate $= k[P][Q]^2$ **(e)** Rate $= k[U]$

Practice Exercise 2

The first step of a mechanism involving the reaction of bromine is

$$Br_2(g) \underset{k_{-1}}{\overset{k_1}{\rightleftharpoons}} 2\,Br(g) \quad (fast, equilibrium)$$

What is the expression relating the concentration of $Br(g)$ to that of $Br_2(g)$?

So far we have considered only three reaction mechanisms: one for a reaction that occurs in a single elementary step and two for simple multistep reactions where there is one rate-determining step. There are other more complex mechanisms, however. If you take a biochemistry class, for example, you will learn about cases in which the concentration of an intermediate cannot be neglected in deriving the rate law. Furthermore, some mechanisms require a large number of steps, sometimes 35 or more, to arrive at a rate law that agrees with experimental data!

14.7 | Catalysis

A **catalyst** is a substance that changes the speed of a chemical reaction without undergoing a permanent chemical change itself. Most reactions in the body, the atmosphere, and the oceans occur with the help of catalysts. Much industrial chemical research is

devoted to the search for more effective catalysts for reactions of commercial importance. Extensive research efforts also are devoted to finding means of inhibiting or removing certain catalysts that promote undesirable reactions, such as those that corrode
metals, age our bodies, and cause tooth decay.

Homogeneous Catalysis

A catalyst that is present in the same phase as the reactants in a reaction mixture is
called a **homogeneous catalyst**. Examples abound both in solution and in the gas phase.
Consider, for example, the decomposition of aqueous hydrogen peroxide, $H_2O_2(aq)$,
into water and oxygen:

$$2\,H_2O_2(aq) \longrightarrow 2\,H_2O(l) + O_2(g) \qquad [14.29]$$

In the absence of a catalyst, this reaction occurs extremely slowly. Many substances are capable of catalyzing the reaction, however, including bromide ion, which reacts with hydrogen
peroxide in acidic solution, forming aqueous bromine and water (▼ **Figure 14.22**).

$$2\,Br^-(aq) + H_2O_2(aq) + 2\,H^+ \longrightarrow Br_2(aq) + 2\,H_2O(l) \qquad [14.30]$$

 GO FIGURE

Why does the solution in the middle cylinder have a brownish color?

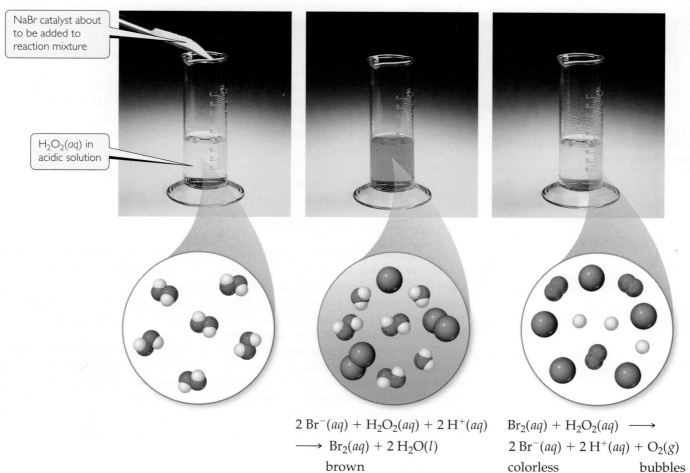

NaBr catalyst about to be added to reaction mixture

$H_2O_2(aq)$ in acidic solution

$$2\,Br^-(aq) + H_2O_2(aq) + 2\,H^+(aq)$$
$$\longrightarrow Br_2(aq) + 2\,H_2O(l)$$
brown

$$Br_2(aq) + H_2O_2(aq) \longrightarrow$$
$$2\,Br^-(aq) + 2\,H^+(aq) + O_2(g)$$
colorless   bubbles

▲ **Figure 14.22 Homogeneous catalysis.** Effect of catalyst on the speed of hydrogen peroxide
decomposition to water and oxygen gas.

If this were the complete reaction, bromide ion would not be a catalyst because it undergoes chemical change during the reaction. However, hydrogen peroxide also reacts with the $Br_2(aq)$ generated in Equation 14.30:

$$Br_2(aq) + H_2O_2(aq) \longrightarrow 2\,Br^-(aq) + 2\,H^+(aq) + O_2(g) \qquad [14.31]$$

The sum of Equations 14.30 and 14.31 is just Equation 14.29, a result which you can check for yourself.

When the H_2O_2 has been completely decomposed, we are left with a colorless solution of $Br^-(aq)$, which means that this ion is indeed a catalyst of the reaction because it speeds up the reaction without itself undergoing any net change. In contrast, Br_2 is an intermediate because it is first formed (Equation 14.30) and then consumed (Equation 14.31). Neither the catalyst nor the intermediate appears in the equation for the overall reaction. Notice, however, that *the catalyst is there at the start of the reaction, whereas the intermediate is formed during the course of the reaction.*

How does a catalyst work? If we think about the general form of rate laws (Equation 14.7, rate $= k[A]^m[B]^n$), we must conclude that the catalyst must affect the numerical value of k, the rate constant. On the basis of the Arrhenius equation (Equation 14.19, $k = Ae^{-E_a/RT}$), k is determined by the activation energy (E_a) and the frequency factor (A). A catalyst may affect the rate of reaction by altering the value of either E_a or A. We can envision this happening in two ways: The catalyst could provide a new mechanism for the reaction that has an E_a value lower than the E_a value for the uncatalyzed reaction, or the catalyst could assist in the orientation of reactants and so increase A. The most dramatic catalytic effects come from lowering E_a. As a general rule, *a catalyst lowers the overall activation energy for a chemical reaction.*

A catalyst can lower the activation energy for a reaction by providing a different mechanism for the reaction. In the decomposition of hydrogen peroxide, for example, two successive reactions of H_2O_2, first with bromide and then with bromine, take place. Because these two reactions together serve as a catalytic pathway for hydrogen peroxide decomposition, *both* of them must have significantly lower activation energies than the uncatalyzed decomposition (◄ **Figure 14.23**).

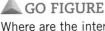

GO FIGURE

Where are the intermediates and transition states in this diagram?

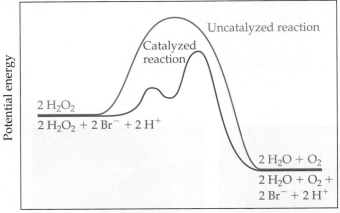

▲ **Figure 14.23 Energy profiles for the uncatalyzed and bromide-catalyzed decomposition of H₂O₂.**

Heterogeneous Catalysis

A **heterogeneous catalyst** is one that exists in a phase different from the phase of the reactant molecules, usually as a solid in contact with either gaseous reactants or reactants in a liquid solution. Many industrially important reactions are catalyzed by the surfaces of solids. For example, raw petroleum is transformed into smaller hydrocarbon molecules by using what are called "cracking" catalysts. Heterogeneous catalysts are often composed of metals or metal oxides.

The initial step in heterogeneous catalysis is usually **adsorption** of reactants. *Adsorption* refers to the binding of molecules to a surface, whereas *absorption* refers to the uptake of molecules into the interior of a substance. ∞ (Section 13.6) Adsorption occurs because the atoms or ions at the surface of a solid are extremely reactive. Because the catalyzed reaction occurs on the surface, special methods are often used to prepare catalysts so that they have very large surface areas. Unlike their counterparts in the interior of the substance, surface atoms and ions have unused bonding capacity that can be used to bond molecules from the gas or solution phase to the surface of the solid.

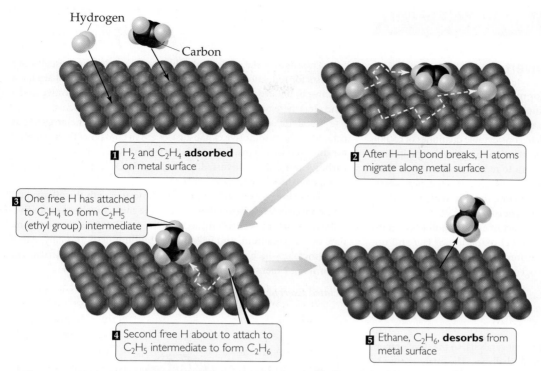

1 H_2 and C_2H_4 **adsorbed** on metal surface

2 After H—H bond breaks, H atoms migrate along metal surface

3 One free H has attached to C_2H_4 to form C_2H_5 (ethyl group) intermediate

4 Second free H about to attach to C_2H_5 intermediate to form C_2H_6

5 Ethane, C_2H_6, **desorbs** from metal surface

▲ **Figure 14.24 Heterogeneous catalysis.** Mechanism for reaction of ethylene with hydrogen on a catalytic surface.

The reaction of hydrogen gas with ethylene gas to form ethane gas provides an example of heterogeneous catalysis:

$$C_2H_4(g) + H_2(g) \longrightarrow C_2H_6(g) \qquad \Delta H° = -137 \, kJ/mol \qquad [14.32]$$

Ethylene   Ethane

Even though this reaction is exothermic, it occurs very slowly in the absence of a catalyst. In the presence of a finely powdered metal, however, such as nickel, palladium, or platinum, the reaction occurs easily at room temperature via the mechanism diagrammed in ▲ **Figure 14.24.** Both ethylene and hydrogen are adsorbed on the metal surface. Upon adsorption, the H—H bond of H_2 breaks, leaving two H atoms initially bonded to the metal surface but relatively free to move. When a hydrogen encounters an adsorbed ethylene molecule, it can form a σ bond to one of the carbon atoms, effectively destroying the C—C π bond and leaving an *ethyl group* (C_2H_5) bonded to the surface via a metal-to-carbon σ bond. This σ bond is relatively weak, so when the other carbon atom also encounters a hydrogen atom, a sixth C—H σ bond is readily formed, and an ethane molecule (C_2H_6) is released from the metal surface.

 Give It Some Thought

How does a homogeneous catalyst compare with a heterogeneous one regarding the ease of recovery of the catalyst from the reaction mixture?

Enzymes

The human body is characterized by an extremely complex system of interrelated chemical reactions, all of which must occur at carefully controlled rates to maintain life. A large number of marvelously efficient biological catalysts known as **enzymes** are necessary for many of these reactions to occur at suitable rates. Most enzymes

Chemistry Put to Work

Catalytic Converters

Heterogeneous catalysis plays a major role in the fight against urban air pollution. Two components of automobile exhausts that help form photochemical smog are nitrogen oxides and unburned hydrocarbons. In addition, automobile exhaust may contain considerable quantities of carbon monoxide. Even with the most careful attention to engine design, it is impossible under normal driving conditions to reduce the quantity of these pollutants to an acceptable level in the exhaust gases. It is therefore necessary to remove them from the exhaust before they are vented to the air. This removal is accomplished in the *catalytic converter*.

The catalytic converter, which is part of an automobile's exhaust system, must perform two functions: (1) oxidation of CO and unburned hydrocarbons (C_xH_y) to carbon dioxide and water, and (2) reduction of nitrogen oxides to nitrogen gas:

$$CO, C_xH_y \xrightarrow{O_2} CO_2 + H_2O$$

$$NO, NO_2 \longrightarrow N_2$$

These two functions require different catalysts, so the development of a successful catalyst system is a difficult challenge. The catalysts must be effective over a wide range of operating temperatures. They must continue to be active despite the fact that various components of the exhaust can block the active sites of the catalyst. And the catalysts must be sufficiently rugged to withstand exhaust gas turbulence and the mechanical shocks of driving under various conditions for thousands of miles.

Catalysts that promote the combustion of CO and hydrocarbons are, in general, the transition-metal oxides and the noble metals. These materials are supported on a structure (▶ Figure 14.25) that allows the best possible contact between the flowing exhaust gas and the catalyst surface. A honeycomb structure made from alumina (Al_2O_3) and impregnated with the catalyst is employed. Such catalysts operate by first adsorbing oxygen gas present in the exhaust gas. This adsorption weakens the O—O bond in O_2, so that oxygen atoms are available for reaction with adsorbed CO to form CO_2. Hydrocarbon oxidation probably proceeds somewhat similarly, with the hydrocarbons first being adsorbed followed by rupture of a C—H bond.

Transition-metal oxides and noble metals are also the most effective catalysts for reduction of NO to N_2 and O_2. The catalysts that are most effective in one reaction, however, are usually much less effective in the other. It is therefore necessary to have two catalytic components.

Catalytic converters contain remarkably efficient heterogeneous catalysts. The automotive exhaust gases are in contact with the catalyst for only 100 to 400 ms, but in this very short time, 96% of the hydrocarbons and CO is converted to CO_2 and H_2O, and the emission of nitrogen oxides is reduced by 76%.

There are costs as well as benefits associated with the use of catalytic converters, one being that some of the metals are very expensive. Catalytic converters currently account for about 35% of the platinum, 65% of the palladium, and 95% of the rhodium used annually. All of these metals, which come mainly from Russia and South Africa, can be far more expensive than gold.

Related Exercises: 14.62, 14.81, 14.82, 14.124

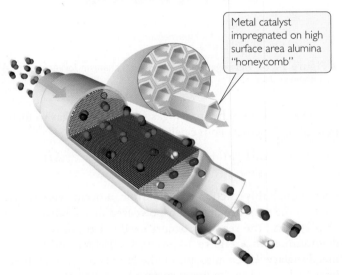

Metal catalyst impregnated on high surface area alumina "honeycomb"

▲ Figure 14.25 **Cross section of a catalytic converter.**

 GO FIGURE

Why is the reaction faster when the liver is ground up?

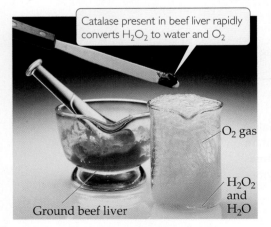

Catalase present in beef liver rapidly converts H_2O_2 to water and O_2

O_2 gas

H_2O_2 and H_2O

Ground beef liver

▲ Figure 14.26 **Enzymes speed up reactions.**

are large protein molecules with molecular weights ranging from about 10,000 to about 1 million amu. They are very selective in the reactions they catalyze, and some are absolutely specific, operating for only one substance in only one reaction. The decomposition of hydrogen peroxide, for example, is an important biological process. Because hydrogen peroxide is strongly oxidizing, it can be physiologically harmful. For this reason, the blood and liver of mammals contain an enzyme, *catalase*, that catalyzes the decomposition of hydrogen peroxide into water and oxygen (Equation 14.29). ◀ Figure 14.26 shows the dramatic acceleration of this chemical reaction by the catalase in beef liver.

The reaction any given enzyme catalyzes takes place at a specific location in the enzyme called the **active site**. The substances that react at this site are called **substrates**. The **lock-and-key model** provides a simple explanation for the specificity of an enzyme (▶ Figure 14.27). The substrate is pictured as fitting neatly into the active site, much like a key fits into a lock.

Lysozyme is an enzyme that is important to the functioning of our immune system because it accelerates reactions that damage (or "lyse") bacterial cell walls. ▶ Figure 14.28 shows a model of the enzyme lysozyme without and with a bound substrate molecule.

▲GO FIGURE

Which molecules must bind more tightly to the active site, substrates or products?

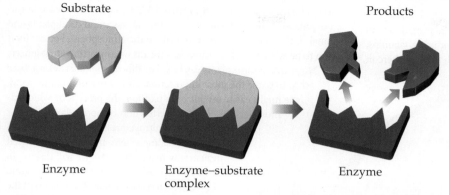

▲ Figure 14.27 **Lock-and-key model for enzyme action.**

The combination of enzyme and substrate is called the *enzyme–substrate complex.* Although Figure 14.27 shows both the active site and its substrate as having a fixed shape, the active site is often fairly flexible and so may change shape as it binds the substrate. The binding between substrate and active site involves dipole–dipole attractions, hydrogen bonds, and dispersion forces. ∞ (Section 11.2)

As substrate molecules enter the active site, they are somehow activated so that they are capable of reacting rapidly. This activation process may occur, for example, by the withdrawal or donation of electron density from a particular bond or group of atoms in the enzyme's active site. In addition, the substrate may become distorted in the process of fitting into the active site and made more reactive. Once the reaction occurs, the products depart from the active site, allowing another substrate molecule to enter.

The activity of an enzyme is destroyed if some molecule other than the substrate specific to that enzyme binds to the active site and blocks entry of the substrate. Such substances are called *enzyme inhibitors.* Nerve poisons and certain toxic metal ions, such as lead and mercury, are believed to act in this way to inhibit enzyme activity. Some other poisons act by attaching elsewhere on the enzyme, thereby distorting the active site so that the substrate no longer fits.

Enzymes are enormously more efficient than nonbiochemical catalysts. The number of individual catalyzed reaction events occurring at a particular active site, called the *turnover number*, is generally in the range of 10^3 to 10^7 per second. Such large turnover numbers correspond to very low activation energies. Compared with a simple chemical catalyst, enzymes can increase the rate constant for a given reaction by a millionfold or more.

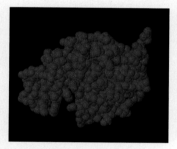

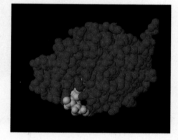

▲ Figure 14.28 **Lysozyme was one of the first enzymes for which a structure–function relationship was described.** This model shows how the substrate (yellow) "fits" into the active site of the enzyme.

 Give It Some Thought

Is it reasonable to say that enzymes lower the energy of the transition state for a reaction?

Chemistry and Life

Nitrogen Fixation and Nitrogenase

Nitrogen is one of the most essential elements in living organisms, found in many compounds vital to life, including proteins, nucleic acids, vitamins, and hormones. Nitrogen is continually cycling through the biosphere in various forms, as shown in ◀ **Figure 14.29**. For example, certain microorganisms convert the nitrogen in animal waste and dead plants and animals into $N_2(g)$, which then returns to the atmosphere. For the food chain to be sustained, there must be a means of converting atmospheric $N_2(g)$ into a form plants can use. For this reason, if a chemist were asked to name the most important chemical reaction in the world, she might easily say *nitrogen fixation*, the process by which atmospheric $N_2(g)$ is converted into compounds suitable for plant use. Some fixed nitrogen results from the action of lightning on the atmosphere, and some is produced industrially using a process we will discuss in Chapter 15. About 60% of fixed nitrogen, however, is a consequence of the action of the remarkable and complex enzyme *nitrogenase*. This enzyme is *not* present in humans or other animals; rather, it is found in bacteria that live in the root nodules of certain plants, such as the legumes clover and alfalfa.

Nitrogenase converts N_2 into NH_3, a process that, in the absence of a catalyst, has a very large activation energy. This process is a *reduction* reaction in which the oxidation state of N is reduced from 0 in N_2 to -3 in NH_3. The mechanism by which nitrogenase reduces N_2 is not fully understood. Like many other enzymes, including catalase, the active site of nitrogenase contains transition-metal atoms; such enzymes are called *metalloenzymes*. Because transition metals can readily change oxidation state, metalloenzymes are especially useful for effecting transformations in which substrates are either oxidized or reduced.

It has been known for nearly 40 years that a portion of nitrogenase contains iron and molybdenum atoms. This portion, called the *FeMo-cofactor*, is thought to serve as the active site of the enzyme. The FeMo-cofactor of nitrogenase is a cluster of seven Fe atoms and one Mo atom, all linked by sulfur atoms (▼ **Figure 14.30**).

It is one of the wonders of life that simple bacteria can contain beautifully complex and vitally important enzymes such as nitrogenase. Because of this enzyme, nitrogen is continually cycled between its comparatively inert role in the atmosphere and its critical role in living organisms. Without nitrogenase, life as we know it could not exist on Earth.

Related Exercises: 14.86, 14.115, 14.116

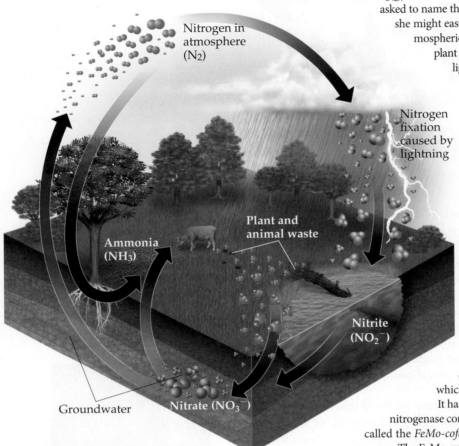

▲ Figure 14.29 Simplified picture of the nitrogen cycle.

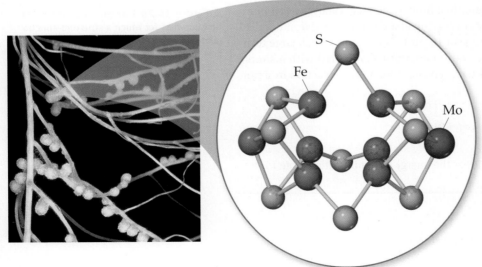

◀ Figure 14.30 The FeMo-cofactor of nitrogenase. Nitrogenase is found in nodules in the roots of certain plants, such as the white clover roots shown at the left. The cofactor, which is thought to be the active site of the enzyme, contains seven Fe atoms and one Mo atom, linked by sulfur atoms. The molecules on the outside of the cofactor connect it to the rest of the protein.

SAMPLE INTEGRATIVE EXERCISE   Putting Concepts Together

Formic acid (HCOOH) decomposes in the gas phase at elevated temperatures as follows:

$$HCOOH(g) \longrightarrow CO_2(g) + H_2(g)$$

The uncatalyzed decomposition reaction is determined to be first order. A graph of the partial pressure of HCOOH versus time for decomposition at 838 K is shown as the red curve in ▶ Figure 14.31. When a small amount of solid ZnO is added to the reaction chamber, the partial pressure of acid versus time varies as shown by the blue curve in Figure 14.31.

(a) Estimate the half-life and first-order rate constant for formic acid decomposition.

(b) What can you conclude from the effect of added ZnO on the decomposition of formic acid?

(c) The progress of the reaction was followed by measuring the partial pressure of formic acid vapor at selected times. Suppose that, instead, we had plotted the concentration of formic acid in units of mol/L. What effect would this have had on the calculated value of k?

(d) The pressure of formic acid vapor at the start of the reaction is 3.00×10^2 torr. Assuming constant temperature and ideal-gas behavior, what is the pressure in the system at the end of the reaction? If the volume of the reaction chamber is 436 cm³, how many moles of gas occupy the reaction chamber at the end of the reaction?

(e) The standard heat of formation of formic acid vapor is $\Delta H_f^\circ = -378.6$ kJ/mol. Calculate ΔH° for the overall reaction. If the activation energy (E_a) for the reaction is 184 kJ/mol, sketch an approximate energy profile for the reaction, and label E_a, ΔH°, and the transition state.

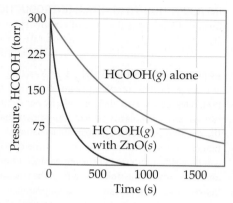

▲ **Figure 14.31** Variation in pressure of HCOOH(g) as a function of time at 838 K.

SOLUTION

(a) The initial pressure of HCOOH is 3.00×10^2 torr. On the graph we move to the level at which the partial pressure of HCOOH is 1.50×10^2 torr, half the initial value. This corresponds to a time of about 6.60×10^2 s, which is therefore the half-life. The first-order rate constant is given by Equation 14.15: $k = 0.693/t_{1/2} = 0.693/660$ s $= 1.05 \times 10^{-3}$ s^{-1}.

(b) The reaction proceeds much more rapidly in the presence of solid ZnO, so the surface of the oxide must be acting as a catalyst for the decomposition of the acid. This is an example of heterogeneous catalysis.

(c) If we had graphed the concentration of formic acid in units of moles per liter, we would still have determined that the half-life for decomposition is 660 s, and we would have computed the same value for k. Because the units for k are s^{-1}, the value for k is independent of the units used for concentration.

(d) According to the stoichiometry of the reaction, two moles of product are formed for each mole of reactant. When reaction is completed, therefore, the pressure will be 600 torr, just twice the initial pressure, assuming ideal-gas behavior. (Because we are working at quite high temperature and fairly low gas pressure, assuming ideal-gas behavior is reasonable.) The number of moles of gas present can be calculated using the ideal-gas equation ∞∞ (Section 10.4):

$$n = \frac{PV}{RT} = \frac{(600/760 \text{ atm})(0.436 \text{ L})}{(0.08206 \text{ L-atm/mol-K})(838 \text{ K})} = 5.00 \times 10^{-3} \text{ mol}$$

(e) We first calculate the overall change in energy, ΔH° ∞∞ (Section 5.7 and Appendix C), as in

$$\Delta H^\circ = \Delta H_f^\circ(CO_2(g)) + \Delta H_f^\circ(H_2(g)) - \Delta H_f^\circ(HCOOH(g))$$
$$= -393.5 \text{ kJ/mol} + 0 - (-378.6 \text{ kJ/mol})$$
$$= -14.9 \text{ kJ/mol}$$

From this and the given value for E_a, we can draw an approximate energy profile for the reaction, in analogy to Figure 14.17.

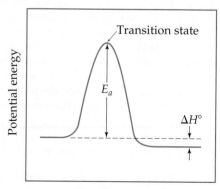

Chapter Summary and Key Terms

INTRODUCTION TO KINETICS (INTRODUCTION AND SECTION 14.1) **Chemical kinetics** is the area of chemistry in which **reaction rates** are studied. Factors that affect reaction rate are the physical state of the reactants; concentration; temperature; and the presence of catalysts.

REACTION RATES (SECTION 14.2) Reaction rates are usually expressed as changes in concentration per unit time: Typically, for reactions in solution, rates are given in units of molarity per second (M/s). For most reactions, a plot of molarity versus time shows that the rate slows down as the reaction proceeds. The **instantaneous rate** is the slope of a line drawn tangent to the concentration-versus-time curve at a specific time. Rates can be written in terms of the appearance of products or the disappearance of reactants; the stoichiometry of the reaction dictates the relationship between rates of appearance and disappearance.

RATES AND CONCENTRATION (SECTION 14.3) The quantitative relationship between rate and concentration is expressed by a **rate law**, which usually has the following form:

$$\text{Rate} = k[\text{reactant 1}]^m[\text{reactant 2}]^n \ldots$$

The constant k in the rate law is called the **rate constant**; the exponents m, n, and so forth are called **reaction orders** for the reactants. The sum of the reaction orders gives the **overall reaction order**. Reaction orders must be determined experimentally. The units of the rate constant depend on the overall reaction order. For a reaction in which the overall reaction order is 1, k has units of s^{-1}; for one in which the overall reaction order is 2, k has units of $M^{-1}\,s^{-1}$.

Spectroscopy is one technique that can be used to monitor the course of a reaction. According to Beer's law, the absorption of electromagnetic radiation by a substance at a particular wavelength is directly proportional to its concentration.

CONCENTRATION AND TIME (SECTION 14.4) Rate laws can be used to determine the concentrations of reactants or products at any time during a reaction. In a **first-order reaction** the rate is proportional to the concentration of a single reactant raised to the first power: Rate = $k[A]$. In such cases the integrated form of the rate law is $\ln[A]_t = -kt + \ln[A]_0$, where $[A]_t$ is the concentration of reactant A at time t, k is the rate constant, and $[A]_0$ is the initial concentration of A. Thus, for a first-order reaction, a graph of $\ln[A]$ versus time yields a straight line of slope $-k$.

A **second-order reaction** is one for which the overall reaction order is 2. If a second-order rate law depends on the concentration of only one reactant, then rate = $k[A]^2$, and the time dependence of $[A]$ is given by the integrated form of the rate law: $1/[A]_t = 1/[A]_0 + kt$. In this case, a graph of $1/[A]_t$ versus time yields a straight line. A **zero-order reaction** is one for which the overall reaction order is 0. Rate = k if the reaction is zero order.

The **half-life** of a reaction, $t_{1/2}$, is the time required for the concentration of a reactant to drop to one-half of its original value. For a first-order reaction, the half-life depends only on the rate constant and not on the initial concentration: $t_{1/2} = 0.693/k$. The half-life of a second-order reaction depends on both the rate constant and the initial concentration of A: $t_{1/2} = 1/k[A]_0$.

THE EFFECT OF TEMPERATURE ON RATES (SECTION 14.5) The **collision model**, which assumes that reactions occur as a result of collisions between molecules, helps explain why the magnitudes of rate constants increase with increasing temperature. The greater the kinetic energy of the colliding molecules, the greater is the energy of collision. The minimum energy required for a reaction to occur is called the **activation energy**, E_a. A collision with energy E_a or greater can cause the atoms of the colliding molecules to reach the **activated complex** (or **transition state**), which is the highest energy arrangement in the pathway from reactants to products. Even if a collision is energetic enough, it may not lead to reaction; the reactants must also be correctly oriented relative to one another in order for a collision to be effective.

Because the kinetic energy of molecules depends on temperature, the rate constant of a reaction is very dependent on temperature. The relationship between k and temperature is given by the **Arrhenius equation**: $k = Ae^{-E_a/RT}$. The term A is called the **frequency factor**; it relates to the number of collisions that are favorably oriented for reaction. The Arrhenius equation is often used in logarithmic form: $\ln k = \ln A - E_a/RT$. Thus, a graph of $\ln k$ versus $1/T$ yields a straight line with slope $-E_a/R$.

REACTION MECHANISMS (SECTION 14.6) A **reaction mechanism** details the individual steps that occur in the course of a reaction. Each of these steps, called **elementary reactions**, has a well-defined rate law that depends on the number of molecules (the **molecularity**) of the step. Elementary reactions are defined as either **unimolecular**, **bimolecular**, or **termolecular**, depending on whether one, two, or three reactant molecules are involved, respectively. Termolecular elementary reactions are very rare. Unimolecular, bimolecular, and termolecular reactions follow rate laws that are first order overall, second order overall, and third order overall, respectively.

Many reactions occur by a multistep mechanism, involving two or more elementary reactions, or steps. An **intermediate** is produced in one elementary step, is consumed in a later elementary step, and therefore does not appear in the overall equation for the reaction. When a mechanism has several elementary steps, the overall rate is limited by the slowest elementary step, called the **rate-determining step**. A fast elementary step that follows the rate-determining step will have no effect on the rate law of the reaction. A fast step that precedes the rate-determining step often creates an equilibrium that involves an intermediate. For a mechanism to be valid, the rate law predicted by the mechanism must be the same as that observed experimentally.

CATALYSTS (SECTION 14.7) A **catalyst** is a substance that increases the rate of a reaction without undergoing a net chemical change itself. It does so by providing a different mechanism for the reaction, one that has a lower activation energy. A **homogeneous catalyst** is one that is in the same phase as the reactants. A **heterogeneous catalyst** has a different phase from the reactants. Finely divided metals are often used as heterogeneous catalysts for solution- and gas-phase reactions. Reacting molecules can undergo binding, or **adsorption**, at the surface of the catalyst. The adsorption of a reactant at specific sites on the surface makes bond breaking easier, lowering the activation energy. Catalysis in living organisms is achieved by **enzymes**, large protein molecules that usually catalyze a very specific reaction. The specific reactant molecules involved in an enzymatic reaction are called **substrates**. The site of the enzyme where the catalysis occurs is called the **active site**. In the **lock-and-key model** for enzyme catalysis, substrate molecules bind very specifically to the active site of the enzyme, after which they can undergo reaction.

Learning Outcomes   After studying this chapter, you should be able to:

- List the factors that affect the rate of chemical reactions. (Section 14.1)

- Determine the rate of a reaction given time and concentration. (Section 14.2)

- Relate the rate of formation of products and the rate of disappearance of reactants given the balanced chemical equation for the reaction. (Section 14.2)

- Explain the form and meaning of a rate law, including the ideas of reaction order and rate constant. (Section 14.3)
- Determine the rate law and rate constant for a reaction from a series of experiments given the measured rates for various concentrations of reactants. (Section 14.3)
- Apply the integrated form of a rate law to determine the concentration of a reactant at a given time. (Section 14.4)

- Apply the relationship between the rate constant of a first-order reaction and its half-life. (Section 14.4)
- Explain how the activation energy affects a rate and be able to use the Arrhenius equation. (Section 14.5)
- Predict a rate law for a reaction having a multistep mechanism given the individual steps in the mechanism. (Section 14.6)
- Explain the principles underlying catalysis. (Section 14.7)

Key Equations

$$\text{Rate} = -\frac{1}{a}\frac{\Delta[A]}{\Delta t} = -\frac{1}{b}\frac{\Delta[B]}{\Delta t} = \frac{1}{c}\frac{\Delta[C]}{\Delta t} = \frac{1}{d}\frac{\Delta[D]}{\Delta t} \qquad [14.4]$$

Definition of reaction rate in terms of the components of the balanced chemical equation $a\,A + b\,B \longrightarrow c\,C + d\,D$

$$\text{Rate} = k[A]^m[B]^n \qquad [14.7]$$

General form of a rate law for the reaction $A + B \longrightarrow$ products

$$\ln[A]_t - \ln[A]_0 = -kt \quad \text{or} \quad \ln\frac{[A]_t}{[A]_0} = -kt \qquad [14.12]$$

The integrated form of a first-order rate law for the reaction $A \longrightarrow$ products

$$\frac{1}{[A]_t} = kt + \frac{1}{[A]_0} \qquad [14.14]$$

The integrated form of the second-order rate law for the reaction $A \longrightarrow$ products

$$t_{1/2} = \frac{0.693}{k} \qquad [14.15]$$

Relating the half-life and rate constant for a first-order reaction

$$k = Ae^{-E_a/RT} \qquad [14.19]$$

The Arrhenius equation, which expresses how the rate constant depends on temperature

$$\ln k = -\frac{E_a}{RT} + \ln A \qquad [14.20]$$

Logarithmic form of the Arrhenius equation

Exercises

Visualizing Concepts

14.1 An automotive fuel injector dispenses a fine spray of gasoline into the automobile cylinder, as shown in the bottom drawing here. When an injector gets clogged, as shown in the top drawing, the spray is not as fine or even and the performance of the car declines. How is this observation related to chemical kinetics? [Section 14.1]

14.2 Consider the following graph of the concentration of a substance X over time. Is each of the following statements true or false? (a) X is a product of the reaction. (b) The rate of the reaction remains the same as time progresses. (c) The average rate between points 1 and 2 is greater than the average rate between points 1 and 3. (d) As time progresses, the curve will eventually turn downward toward the x-axis. [Section 14.2]

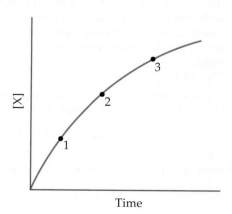

14.3 You study the rate of a reaction, measuring both the concentration of the reactant and the concentration of the product as a function of time, and obtain the following results:

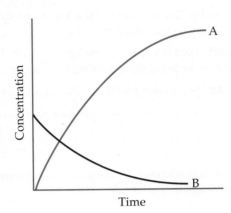

(a) Which chemical equation is consistent with these data: (i) A ⟶ B, (ii) B ⟶ A, (iii) A ⟶ 2 B, (iv) B ⟶ 2 A?

(b) Write equivalent expressions for the rate of the reaction in terms of the appearance or disappearance of the two substances. [Section 14.2]

14.4 Suppose that for the reaction K + L ⟶ M, you monitor the production of M over time, and then plot the following graph from your data:

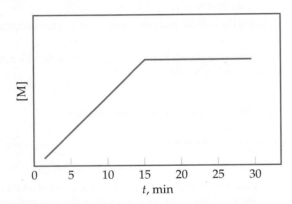

(a) Is the reaction occurring at a constant rate from $t = 0$ to $t = 15$ min? (b) Is the reaction completed at $t = 15$ min? (c) Suppose the reaction as plotted here were started with 0.20 mol K and 0.40 mol L. After 30 min, an additional 0.20 mol K are added to the reaction mixture. Which of the following correctly describes how the plot would look from $t = 30$ min to $t = 60$ min? (i) [M] would remain at the same constant value it has at $t = 30$ min, (ii) [M] would increase with the same slope as $t = 0$ to 15 min, until $t = 45$ min at which point the plot becomes horizontal again, or (iii) [M] decreases and reaches 0 at $t = 45$ min. [Section 14.2]

14.5 The following diagrams represent mixtures of $NO(g)$ and $O_2(g)$. These two substances react as follows:

$$2\,NO(g) + O_2(g) \longrightarrow 2\,NO_2(g)$$

It has been determined experimentally that the rate is second order in NO and first order in O_2. Based on this fact, which of the following mixtures will have the fastest initial rate? [Section 14.3]

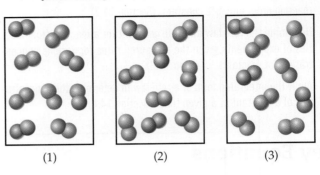

(1)   (2)   (3)

14.6 A friend studies a first-order reaction and obtains the following three graphs for experiments done at two different temperatures. (a) Which two graphs represent experiments done at the same temperature? What accounts for the difference in these two graphs? In what way are they the same? (b) Which two graphs represent experiments done with the same starting concentration but at different temperatures? Which graph probably represents the lower temperature? How do you know? [Section 14.4]

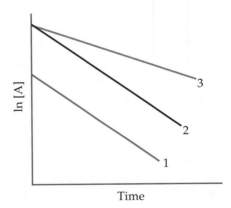

14.7 (a) Given the following diagrams at $t = 0$ min and $t = 30$ min, what is the half-life of the reaction if it follows first-order kinetics?

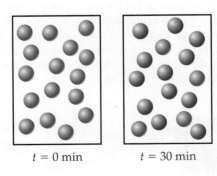

$t = 0$ min $t = 30$ min

(b) After four half-life periods for a first-order reaction, what fraction of reactant remains? [Section 14.4]

14.8 Which of the following linear plots do you expect for a reaction A ⟶ products if the kinetics are (a) zero order, (b) first order, or (c) second order? [Section 14.4]

(i) 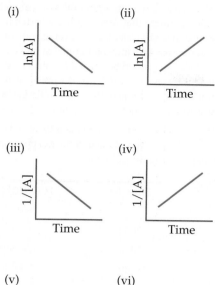 (ii)

(iii) 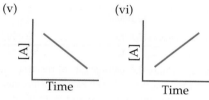 (iv)

(v) 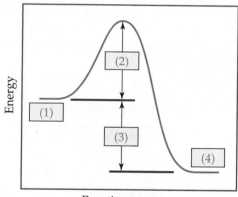 (vi)

14.9 The following diagram shows a reaction profile. Label the components indicated by the boxes. [Section 14.5]

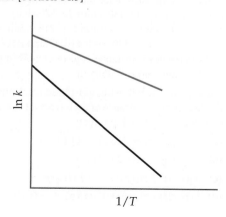

14.10 The accompanying graph shows plots of ln k versus $1/T$ for two different reactions. The plots have been extrapolated to the y-intercepts. Which reaction (red or blue) has (**a**) the larger value for E_a, and (**b**) the larger value for the frequency factor, A? [Section 14.5]

14.11 The following graph shows two different reaction pathways for the same overall reaction at the same temperature. Is each of the following statements true or false? (**a**) The rate is faster for the red path than for the blue path. (**b**) For both paths, the rate of the reverse reaction is slower than the rate of the forward reaction. (**c**) The energy change ΔE is the same for both paths. [Section 14.6]

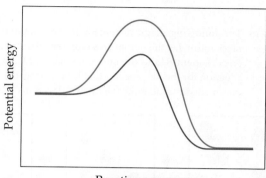

14.12 Consider the diagram that follows, which represents two steps in an overall reaction. The red spheres are oxygen, the blue ones nitrogen, and the green ones fluorine. (**a**) Write the chemical equation for each step in the reaction. (**b**) Write the equation for the overall reaction. (**c**) Identify the intermediate in the mechanism. (**d**) Write the rate law for the overall reaction if the first step is the slow, rate-determining step. [Section 14.6]

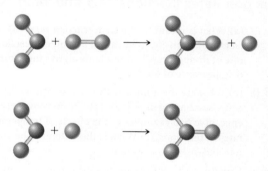

14.13 Based on the following reaction profile, how many intermediates are formed in the reaction A $\longrightarrow$ C? How many transition states are there? Which step, A $\longrightarrow$ B or B $\longrightarrow$ C, is the faster? For the reaction A $\longrightarrow$ C, is ΔE positive, negative, or zero? [Section 14.6]

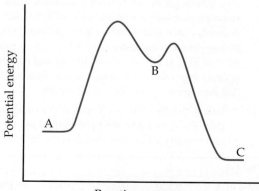

14.14 Draw a possible transition state for the bimolecular reaction depicted here. (The blue spheres are nitrogen atoms, and the red ones are oxygen atoms.) Use dashed lines to represent the bonds that are in the process of being broken or made in the transition state. [Section 14.6]

14.15 The following diagram represents an imaginary two-step mechanism. Let the red spheres represent element A, the green ones element B, and the blue ones element C. (**a**) Write the equation for the net reaction that is occurring. (**b**) Identify the intermediate. (**c**) Identify the catalyst. [Sections 14.6 and 14.7]

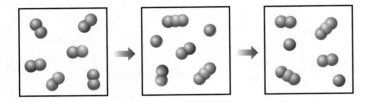

14.16 Draw a graph showing the reaction pathway for an overall exothermic reaction with two intermediates that are produced at different rates. On your graph indicate the reactants, products, intermediates, transition states, and activation energies. [Sections 14.6 and 14.7]

Reaction Rates (Sections 14.1 and 14.2)

14.17 (**a**) What is meant by the term *reaction rate*? (**b**) Name three factors that can affect the rate of a chemical reaction. (**c**) Is the rate of disappearance of reactants always the same as the rate of appearance of products?

14.18 (**a**) What are the units usually used to express the rates of reactions occurring in solution? (**b**) From your everyday experience, give two examples of the effects of temperature on the rates of reactions. (**c**) What is the difference between average rate and instantaneous rate?

14.19 Consider the following hypothetical aqueous reaction: $A(aq) \rightarrow B(aq)$. A flask is charged with 0.065 mol of A in a total volume of 100.0 mL. The following data are collected:

Time (min)	0	10	20	30	40
Moles of A	0.065	0.051	0.042	0.036	0.031

(**a**) Calculate the number of moles of B at each time in the table, assuming that there are no molecules of B at time zero, and that A cleanly converts to B with no intermediates. (**b**) Calculate the average rate of disappearance of A for each 10-min interval in units of M/s. (**c**) Between $t = 10$ min and $t = 30$ min, what is the average rate of appearance of B in units of M/s? Assume that the volume of the solution is constant.

14.20 A flask is charged with 0.100 mol of A and allowed to react to form B according to the hypothetical gas-phase reaction $A(g) \longrightarrow B(g)$. The following data are collected:

Time (s)	0	40	80	120	160
Moles of A	0.100	0.067	0.045	0.030	0.020

(**a**) Calculate the number of moles of B at each time in the table, assuming that A is cleanly converted to B with no intermediates. (**b**) Calculate the average rate of disappearance of A for each 40-s interval in units of mol/s. (**c**) Which of the following would be needed to calculate the rate in units of concentration per time: (**i**) the pressure of the gas at each time, (**ii**) the volume of the reaction flask, (**iii**) the temperature, or (**iv**) the molecular weight of A?

14.21 The isomerization of methyl isonitrile (CH_3NC) to acetonitrile (CH_3CN) was studied in the gas phase at 215 °C, and the following data were obtained:

Time (s)	$[CH_3NC]$ (M)
0	0.0165
2000	0.0110
5000	0.00591
8000	0.00314
12,000	0.00137
15,000	0.00074

(**a**) Calculate the average rate of reaction, in M/s, for the time interval between each measurement. (**b**) Calculate the average rate of reaction over the entire time of the data from $t = 0$ to $t = 15,000$ s. (**c**) Which is greater, the average rate between $t = 2000$ and $t = 12,000$ s, or between $t = 8000$ and $t = 15,000$ s? (**d**) Graph $[CH_3NC]$ versus time and determine the instantaneous rates in M/s at $t = 5000$ s and $t = 8000$ s.

14.22 The rate of disappearance of HCl was measured for the following reaction:

$$CH_3OH(aq) + HCl(aq) \longrightarrow CH_3Cl(aq) + H_2O(l)$$

The following data were collected:

Time (min)	[HCl] (M)
0.0	1.85
54.0	1.58
107.0	1.36
215.0	1.02
430.0	0.580

(**a**) Calculate the average rate of reaction, in M/s, for the time interval between each measurement. (**b**) Calculate the average rate of reaction for the entire time for the data from $t = 0.0$ min to $t = 430.0$ min. (**c**) Which is greater, the average rate between $t = 54.0$ and $t = 215.0$ min, or between $t = 107.0$ and $t = 430.0$ min? (**d**) Graph [HCl] versus time and determine the instantaneous rates in M/min and M/s at $t = 75.0$ min and $t = 250$ min.

14.23 For each of the following gas-phase reactions, indicate how the rate of disappearance of each reactant is related to the rate of appearance of each product:

(**a**) $H_2O_2(g) \longrightarrow H_2(g) + O_2(g)$

(**b**) $2 N_2O(g) \longrightarrow 2 N_2(g) + O_2(g)$

(**c**) $N_2(g) + 3 H_2(g) \longrightarrow 2 NH_3(g)$

(**d**) $C_2H_5NH_2(g) \longrightarrow C_2H_4(g) + NH_3(g)$

14.24 For each of the following gas-phase reactions, write the rate expression in terms of the appearance of each product and disappearance of each reactant:

(a) $2\,H_2O(g) \longrightarrow 2\,H_2(g) + O_2(g)$

(b) $2\,SO_2(g) + O_2(g) \longrightarrow 2\,SO_3(g)$

(c) $2\,NO(g) + 2\,H_2(g) \longrightarrow N_2(g) + 2\,H_2O(g)$

(d) $N_2(g) + 2\,H_2(g) \longrightarrow N_2H_4(g)$

14.25 (a) Consider the combustion of $H_2(g)$: $2\,H_2(g) + O_2(g) \longrightarrow 2\,H_2O(g)$. If hydrogen is burning at the rate of 0.48 mol/s, what is the rate of consumption of oxygen? What is the rate of formation of water vapor? (b) The reaction $2\,NO(g) + Cl_2(g) \longrightarrow 2\,NOCl(g)$ is carried out in a closed vessel. If the partial pressure of NO is decreasing at the rate of 56 torr/min, what is the rate of change of the total pressure of the vessel?

14.26 (a) Consider the combustion of ethylene, $C_2H_4(g) + 3\,O_2(g) \longrightarrow 2\,CO_2(g) + 2\,H_2O(g)$. If the concentration of C_2H_4 is decreasing at the rate of 0.036 M/s, what are the rates of change in the concentrations of CO_2 and H_2O? (b) The rate of decrease in N_2H_4 partial pressure in a closed reaction vessel from the reaction $N_2H_4(g) + H_2(g) \longrightarrow 2\,NH_3(g)$ is 74 torr per hour. What are the rates of change of NH_3 partial pressure and total pressure in the vessel?

Rate Laws (Section 14.3)

14.27 A reaction $A + B \longrightarrow C$ obeys the following rate law: Rate $= k[B]^2$. (a) If [A] is doubled, how will the rate change? Will the rate constant change? (b) What are the reaction orders for A and B? What is the overall reaction order? (c) What are the units of the rate constant?

14.28 Consider a hypothetical reaction between A, B, and C that is first order in A, zero order in B, and second order in C. (a) Write the rate law for the reaction. (b) How does the rate change when [A] is doubled and the other reactant concentrations are held constant? (c) How does the rate change when [B] is tripled and the other reactant concentrations are held constant? (d) How does the rate change when [C] is tripled and the other reactant concentrations are held constant? (e) By what factor does the rate change when the concentrations of all three reactants are tripled? (f) By what factor does the rate change when the concentrations of all three reactants are cut in half?

14.29 The decomposition reaction of N_2O_5 in carbon tetrachloride is $2\,N_2O_5 \longrightarrow 4\,NO_2 + O_2$. The rate law is first order in N_2O_5. At 64 °C the rate constant is $4.82 \times 10^{-3}\,s^{-1}$. (a) Write the rate law for the reaction. (b) What is the rate of reaction when $[N_2O_5] = 0.0240\,M$? (c) What happens to the rate when the concentration of N_2O_5 is doubled to 0.0480 M? (d) What happens to the rate when the concentration of N_2O_5 is halved to 0.0120 M?

14.30 Consider the following reaction:

$$2\,NO(g) + 2\,H_2(g) \longrightarrow N_2(g) + 2\,H_2O(g)$$

(a) The rate law for this reaction is first order in H_2 and second order in NO. Write the rate law. (b) If the rate constant for this reaction at 1000 K is $6.0 \times 10^4\,M^{-2}s^{-1}$, what is the reaction rate when $[NO] = 0.035\,M$ and $[H_2] = 0.015\,M$? (c) What is the reaction rate at 1000 K when the concentration of NO is increased to 0.10 M, while the concentration of H_2 is 0.010 M? (d) What is the reaction rate at 1000 K if [NO] is decreased to 0.010 M and $[H_2]$ is increased to 0.030 M?

14.31 Consider the following reaction:

$$CH_3Br(aq) + OH^-(aq) \longrightarrow CH_3OH(aq) + Br^-(aq)$$

The rate law for this reaction is first order in CH_3Br and first order in OH^-. When $[CH_3Br]$ is $5.0 \times 10^{-3}\,M$ and $[OH^-]$ is 0.050 M, the reaction rate at 298 K is 0.0432 M/s. (a) What is the value of the rate constant? (b) What are the units of the rate constant? (c) What would happen to the rate if the concentration of OH^- were tripled? (d) What would happen to the rate if the concentration of both reactants were tripled?

14.32 The reaction between ethyl bromide (C_2H_5Br) and hydroxide ion in ethyl alcohol at 330 K, $C_2H_5Br(alc) + OH^-(alc) \longrightarrow C_2H_5OH(l) + Br^-(alc)$, is first order each in ethyl bromide and hydroxide ion. When $[C_2H_5Br]$ is 0.0477 M and $[OH^-]$ is 0.100 M, the rate of disappearance of ethyl bromide is $1.7 \times 10^{-7}\,M$/s. (a) What is the value of the rate constant? (b) What are the units of the rate constant? (c) How would the rate of disappearance of ethyl bromide change if the solution were diluted by adding an equal volume of pure ethyl alcohol to the solution?

14.33 The iodide ion reacts with hypochlorite ion (the active ingredient in chlorine bleaches) in the following way: $OCl^- + I^- \longrightarrow OI^- + Cl^-$. This rapid reaction gives the following rate data:

$[OCl^-]\,(M)$	$[I^-]\,(M)$	Initial Rate (M/s)
1.5×10^{-3}	1.5×10^{-3}	1.36×10^{-4}
3.0×10^{-3}	1.5×10^{-3}	2.72×10^{-4}
1.5×10^{-3}	3.0×10^{-3}	2.72×10^{-4}

(a) Write the rate law for this reaction. (b) Calculate the rate constant with proper units. (c) Calculate the rate when $[OCl^-] = 2.0 \times 10^{-3}\,M$ and $[I^-] = 5.0 \times 10^{-4}\,M$.

14.34 The reaction $2\,ClO_2(aq) + 2\,OH^-(aq) \longrightarrow ClO_3^-(aq) + ClO_2^-(aq) + H_2O(l)$ was studied with the following results:

Experiment	$(ClO_2)\,(M)$	$[OH^-]\,(M)$	Initial Rate (M/s)
1	0.060	0.030	0.0248
2	0.020	0.030	0.00276
3	0.020	0.090	0.00828

(a) Determine the rate law for the reaction. (b) Calculate the rate constant with proper units. (c) Calculate the rate when $[ClO_2] = 0.100\,M$ and $[OH^-] = 0.050\,M$.

14.35 The following data were measured for the reaction $BF_3(g) + NH_3(g) \longrightarrow F_3BNH_3(g)$:

Experiment	$[BF_3]\,(M)$	$[NH_3]\,(M)$	Initial Rate (M/s)
1	0.250	0.250	0.2130
2	0.250	0.125	0.1065
3	0.200	0.100	0.0682
4	0.350	0.100	0.1193
5	0.175	0.100	0.0596

(a) What is the rate law for the reaction? (b) What is the overall order of the reaction? (c) Calculate the rate constant with proper units? (d) What is the rate when $[BF_3] = 0.100\,M$ and $[NH_3] = 0.500\,M$?

14.36 The following data were collected for the rate of disappearance of NO in the reaction $2 NO(g) + O_2(g) \longrightarrow 2 NO_2(g)$:

Experiment	[NO](M)	[O₂] (M)	Initial Rate (M/s)
1	0.0126	0.0125	1.41×10^{-2}
2	0.0252	0.0125	5.64×10^{-2}
3	0.0252	0.0250	1.13×10^{-1}

(a) What is the rate law for the reaction? (b) What are the units of the rate constant? (c) What is the average value of the rate constant calculated from the three data sets? (d) What is the rate of disappearance of NO when $[NO] = 0.0750\,M$ and $[O_2] = 0.0100\,M$? (e) What is the rate of disappearance of O_2 at the concentrations given in part (d)?

[**14.37**] Consider the gas-phase reaction between nitric oxide and bromine at 273 °C: $2 NO(g) + Br_2(g) \longrightarrow 2 NOBr(g)$. The following data for the initial rate of appearance of NOBr were obtained:

Experiment	[NO] (M)	[Br₂] (M)	Initial Rate (M/s)
1	0.10	0.20	24
2	0.25	0.20	150
3	0.10	0.50	60
4	0.35	0.50	735

(a) Determine the rate law. (b) Calculate the average value of the rate constant for the appearance of NOBr from the four data sets. (c) How is the rate of appearance of NOBr related to the rate of disappearance of Br_2? (d) What is the rate of disappearance of Br_2 when $[NO] = 0.075\,M$ and $[Br_2] = 0.25\,M$?

[**14.38**] Consider the reaction of peroxydisulfate ion $(S_2O_8^{2-})$ with iodide ion (I^-) in aqueous solution:

$$S_2O_8^{2-}(aq) + 3 I^-(aq) \longrightarrow 2 SO_4^{2-}(aq) + I_3^-(aq)$$

At a particular temperature the initial rate of disappearance of $S_2O_8^{2-}$ varies with reactant concentrations in the following manner:

Experiment	[S₂O₈²⁻] (M)	[I⁻] (M)	Initial Rate (M/s)
1	0.018	0.036	2.6×10^{-6}
2	0.027	0.036	3.9×10^{-6}
3	0.036	0.054	7.8×10^{-6}
4	0.050	0.072	1.4×10^{-5}

(a) Determine the rate law for the reaction and state the units of the rate constant. (b) What is the average value of the rate constant for the disappearance of $S_2O_8^{2-}$ based on the four sets of data? (c) How is the rate of disappearance of $S_2O_8^{2-}$ related to the rate of disappearance of I^-? (d) What is the rate of disappearance of I^- when $[S_2O_8^{2-}] = 0.025\,M$ and $[I^-] = 0.050\,M$?

Change of Concentration with Time (Section 14.4)

14.39 (a) Define the following symbols that are encountered in rate equations for the generic reaction $A \longrightarrow B$: $[A]_0, t_{1/2}, [A]_t, k$. (b) What quantity, when graphed versus time, will yield a

straight line for a first-order reaction? (c) How can you calculate the rate constant for a first-order reaction from the graph you made in part (b)?

14.40 (a) For a generic second-order reaction $A \longrightarrow B$, what quantity, when graphed versus time, will yield a straight line? (b) What is the slope of the straight line from part (a)? (c) How do the half-lives of first-order and second-order reactions differ?

14.41 (a) The gas-phase decomposition of SO_2Cl_2, $SO_2Cl_2(g) \longrightarrow SO_2(g) + Cl_2(g)$, is first order in SO_2Cl_2. At 600 K the half-life for this process is 2.3×10^5 s. What is the rate constant at this temperature? (b) At 320 °C the rate constant is $2.2 \times 10^{-5}\ s^{-1}$. What is the half-life at this temperature?

14.42 Molecular iodine, $I_2(g)$, dissociates into iodine atoms at 625 K with a first-order rate constant of $0.271\ s^{-1}$. (a) What is the half-life for this reaction? (b) If you start with $0.050\,M\ I_2$ at this temperature, how much will remain after 5.12 s assuming that the iodine atoms do not recombine to form I_2?

14.43 As described in Exercise 14.41, the decomposition of sulfuryl chloride (SO_2Cl_2) is a first-order process. The rate constant for the decomposition at 660 K is $4.5 \times 10^{-2}\ s^{-1}$. (a) If we begin with an initial SO_2Cl_2 pressure of 450 torr, what is the partial pressure of this substance after 60 s? (b) At what time will the partial pressure of SO_2Cl_2 decline to one-tenth its initial value?

14.44 The first-order rate constant for the decomposition of N_2O_5, $2 N_2O_5(g) \longrightarrow 4 NO_2(g) + O_2(g)$, at 70 °C is $6.82 \times 10^{-3}\ s^{-1}$. Suppose we start with 0.0250 mol of $N_2O_5(g)$ in a volume of 2.0 L. (a) How many moles of N_2O_5 will remain after 5.0 min? (b) How many minutes will it take for the quantity of N_2O_5 to drop to 0.010 mol? (c) What is the half-life of N_2O_5 at 70 °C?

14.45 The reaction $SO_2Cl_2(g) \longrightarrow SO_2(g) + Cl_2(g)$ is first order in SO_2Cl_2. Using the following kinetic data, determine the magnitude and units of the first-order rate constant:

Time (s)	Pressure SO₂Cl₂ (atm)
0	1.000
2500	0.947
5000	0.895
7500	0.848
10,000	0.803

14.46 From the following data for the first-order gas-phase isomerization of CH_3NC at 215 °C, calculate the first-order rate constant and half-life for the reaction:

Time (s)	Pressure CH₃NC (torr)
0	502
2000	335
5000	180
8000	95.5
12,000	41.7
15,000	22.4

14.47 Consider the data presented in Exercise 14.19. (a) By using appropriate graphs, determine whether the reaction is first order or second order. (b) What is the rate constant for the reaction? (c) What is the half-life for the reaction?

14.48 Consider the data presented in Exercise 14.20. **(a)** Determine whether the reaction is first order or second order. **(b)** What is the rate constant? **(c)** What is the half-life?

14.49 The gas-phase decomposition of NO_2, $2 NO_2(g) \longrightarrow 2 NO(g) + O_2(g)$, is studied at 383 °C, giving the following data:

Time (s)	[NO₂] (M)
0.0	0.100
5.0	0.017
10.0	0.0090
15.0	0.0062
20.0	0.0047

(a) Is the reaction first order or second order with respect to the concentration of NO_2? **(b)** What is the rate constant? **(c)** Predict the reaction rates at the beginning of the reaction for initial concentrations of 0.200 M, 0.100 M, and 0.050 M NO_2.

14.50 Sucrose ($C_{12}H_{22}O_{11}$), commonly known as table sugar, reacts in dilute acid solutions to form two simpler sugars, glucose and fructose, both of which have the formula $C_6H_{22}O_6$. At 23 °C and in 0.5 M HCl, the following data were obtained for the disappearance of sucrose:

Time (min)	[C₁₂H₂₂O₁₁] (M)
0	0.316
39	0.274
80	0.238
140	0.190
210	0.146

(a) Is the reaction first order or second order with respect to $[C_{12}H_{22}O_{11}]$? **(b)** What is the rate constant? **(c)** Using this rate constant, calculate the concentration of sucrose at 39, 80, 140, and 210 min if the initial sucrose concentration was 0.316 M and the reaction were zero order in sucrose.

Temperature and Rate (Section 14.5)

14.51 **(a)** What factors determine whether a collision between two molecules will lead to a chemical reaction? **(b)** According to the collision model, why does temperature affect the value of the rate constant? **(c)** Does the rate constant for a reaction generally increase or decrease with an increase in reaction temperature?

14.52 **(a)** In which of the following reactions would you expect the orientation factor to be least important in leading to reaction: $NO + O \longrightarrow NO_2$ or $H + Cl \longrightarrow HCl$? **(b)** How does the kinetic-molecular theory help us understand the temperature dependence of chemical reactions?

14.53 Calculate the fraction of atoms in a sample of argon gas at 400 K that has an energy of 10.0 kJ or greater.

14.54 **(a)** The activation energy for the isomerization of methyl isonitrile (Figure 14.7) is 160 kJ/mol. Calculate the fraction of methyl isonitrile molecules that has an energy of 160.0 kJ or greater at 500 K. **(b)** Calculate this fraction for a temperature of 520 K. What is the ratio of the fraction at 520 K to that at 500 K?

14.55 The gas-phase reaction $Cl(g) + HBr(g) \longrightarrow HCl(g) + Br(g)$ has an overall energy change of −66 kJ. The activation energy for the reaction is 7 kJ. **(a)** Sketch the energy profile for the reaction, and label E_a and ΔE. **(b)** What is the activation energy for the reverse reaction?

14.56 For the elementary process $N_2O_5(g) \longrightarrow NO_2(g) + NO_3(g)$ the activation energy (E_a) and overall ΔE are 154 kJ/mol and 136 kJ/mol, respectively. **(a)** Sketch the energy profile for this reaction, and label E_a and ΔE. **(b)** What is the activation energy for the reverse reaction?

14.57 Indicate whether each statement is true or false.

(a) If you compare two reactions with similar collision factors, the one with the larger activation energy will be faster.

(b) A reaction that has a small rate constant must have a small frequency factor.

(c) Increasing the reaction temperature increases the fraction of successful collisions between reactants.

14.58 Indicate whether each statement is true or false.

(a) If you measure the rate constant for a reaction at different temperatures, you can calculate the overall enthalpy change for the reaction.

(b) Exothermic reactions are faster than endothermic reactions.

(c) If you double the temperature for a reaction, you cut the activation energy in half.

14.59 Based on their activation energies and energy changes and assuming that all collision factors are the same, which of the following reactions would be fastest and which would be slowest?

(a) $E_a = 45$ kJ/mol; $\Delta E = -25$ kJ/mol

(b) $E_a = 35$ kJ/mol; $\Delta E = -10$ kJ/mol

(c) $E_a = 55$ kJ/mol; $\Delta E = 10$ kJ/mol

14.60 Which of the reactions in Exercise 14.61 will be fastest in the reverse direction? Which will be slowest?

14.61 **(a)** A certain first-order reaction has a rate constant of 2.75×10^{-2} s^{-1} at 20 °C. What is the value of k at 60 °C if $E_a = 75.5$ kJ/mol? **(b)** Another first-order reaction also has a rate constant of 2.75×10^{-2} s^{-1} at 20 °C What is the value of k at 60 °C if $E_a = 125$ kJ/mol? **(c)** What assumptions do you need to make in order to calculate answers for parts (a) and (b)?

14.62 Understanding the high-temperature behavior of nitrogen oxides is essential for controlling pollution generated in automobile engines. The decomposition of nitric oxide (NO) to N_2 and O_2 is second order with a rate constant of 0.0796 $M^{-1}s^{-1}$ at 737 °C and 0.0815 $M^{-1}s^{-1}$ at 947 °C. Calculate the activation energy for the reaction.

14.63 The rate of the reaction

$$CH_3COOC_2H_5(aq) + OH^-(aq) \longrightarrow CH_3COO^-(aq) + C_2H_5OH(aq)$$

was measured at several temperatures, and the following data were collected:

Temperature (°C)	k (M⁻¹s⁻¹)
15	0.0521
25	0.101
35	0.184
45	0.332

Calculate the value of E_a by constructing an appropriate graph.

14.64 The temperature dependence of the rate constant for a reaction is tabulated as follows:

Temperature (K)	k ($M^{-1}s^{-1}$)
600	0.028
650	0.22
700	1.3
750	6.0
800	23

Calculate E_a and A.

Reaction Mechanisms (Section 14.6)

14.65 **(a)** What is meant by the term *elementary reaction*? **(b)** What is the difference between a *unimolecular* and a *bimolecular* elementary reaction? **(c)** What is a *reaction mechanism*? **(d)** What is meant by the term *rate-determining step*?

14.66 **(a)** What is meant by the term *molecularity*? **(b)** Why are termolecular elementary reactions so rare? **(c)** What is an *intermediate* in a mechanism? **(d)** What are the differences between an intermediate and a transition state?

14.67 What is the molecularity of each of the following elementary reactions? Write the rate law for each.

(a) $Cl_2(g) \longrightarrow 2\,Cl(g)$

(b) $OCl^-(aq) + H_2O(l) \longrightarrow HOCl(aq) + OH^-(aq)$

(c) $NO(g) + Cl_2(g) \longrightarrow NOCl_2(g)$

14.68 What is the molecularity of each of the following elementary reactions? Write the rate law for each.

(a) $2\,NO(g) \longrightarrow N_2O_2(g)$

(b) $H_2C \overset{CH_2}{\overset{\wedge}{\frown}} CH_2(g) \longrightarrow CH_2{=}CH{-}CH_3(g)$

(c) $SO_3(g) \longrightarrow SO_2(g) + O(g)$

14.69 **(a)** Based on the following reaction profile, how many intermediates are formed in the reaction A $\longrightarrow$ D? **(b)** How many transition states are there? **(c)** Which step is the fastest? **(d)** For the reaction A $\longrightarrow$ D, is ΔE positive, negative, or zero?

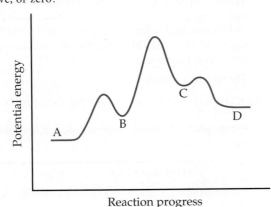

14.70 Consider the following energy profile.

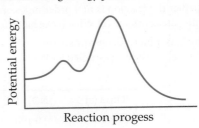

(a) How many elementary reactions are in the reaction mechanism? **(b)** How many intermediates are formed in the reaction? **(c)** Which step is rate limiting? **(d)** For the overall reaction, is ΔE positive, negative, or zero?

14.71 The following mechanism has been proposed for the gas-phase reaction of H_2 with ICl:

$$H_2(g) + ICl(g) \longrightarrow HI(g) + HCl(g)$$
$$HI(g) + ICl(g) \longrightarrow I_2(g) + HCl(g)$$

(a) Write the balanced equation for the overall reaction. **(b)** Identify any intermediates in the mechanism. **(c)** If the first step is slow and the second one is fast, which rate law do you expect to be observed for the overall reaction?

14.72 The decomposition of hydrogen peroxide is catalyzed by iodide ion. The catalyzed reaction is thought to proceed by a two-step mechanism:

$$H_2O_2(aq) + I^-(aq) \longrightarrow H_2O(l) + IO^-(aq) \quad (slow)$$
$$IO^-(aq) + H_2O_2(aq) \longrightarrow H_2O(l) + O_2(g) + I^-(aq) \quad (fast)$$

(a) Write the chemical equation for the overall process. **(b)** Identify the intermediate, if any, in the mechanism. **(c)** Assuming that the first step of the mechanism is rate determining, predict the rate law for the overall process.

[14.73] The reaction $2\,NO(g) + Cl_2(g) \longrightarrow 2\,NOCl(g)$ was performed and the following data obtained under conditions of constant $[Cl_2]$:

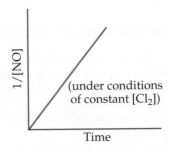

(a) Is the following mechanism consistent with the data?

$$NO(g) + Cl_2(g) \longrightarrow NOCl_2(g) \quad (fast)$$
$$NOCl_2(g) + NO(g) \longrightarrow 2\,NOCl(g) \quad (slow)$$

(b) Does the linear plot guarantee that the overall rate law is second order?

14.74 You have studied the gas-phase oxidation of HBr by O_2:

$$4\,HBr(g) + O_2(g) \longrightarrow 2\,H_2O(g) + 2\,Br_2(g)$$

You find the reaction to be first order with respect to HBr and first order with respect to O_2. You propose the following mechanism:

$$HBr(g) + O_2(g) \longrightarrow HOOBr(g)$$
$$HOOBr(g) + HBr(g) \longrightarrow 2\,HOBr(g)$$
$$HOBr(g) + HBr(g) \longrightarrow H_2O(g) + Br_2(g)$$

(a) Confirm that the elementary reactions add to give the overall reaction. (b) Based on the experimentally determined rate law, which step is rate determining? (c) What are the intermediates in this mechanism? (d) If you are unable to detect HOBr or HOOBr among the products, does this disprove your mechanism?

Catalysis (Section 14.7)

14.75 (a) What is a catalyst? (b) What is the difference between a homogeneous and a heterogeneous catalyst? (c) Do catalysts affect the overall enthalpy change for a reaction, the activation energy, or both?

14.76 (a) Most commercial heterogeneous catalysts are extremely finely divided solid materials. Why is particle size important? (b) What role does adsorption play in the action of a heterogeneous catalyst?

14.77 In Figure 14.22, we saw that $Br^-(aq)$ catalyzes the decomposition of $H_2O_2(aq)$ into $H_2O(l)$ and $O_2(g)$. Suppose that some KBr(s) is added to an aqueous solution of hydrogen peroxide. Make a sketch of $[Br^-(aq)]$ versus time from the addition of the solid to the end of the reaction.

14.78 In solution, chemical species as simple as H^+ and OH^- can serve as catalysts for reactions. Imagine you could measure the $[H^+]$ of a solution containing an acid-catalyzed reaction as it occurs. Assume the reactants and products themselves are neither acids nor bases. Sketch the $[H^+]$ concentration profile you would measure as a function of time for the reaction, assuming $t = 0$ is when you add a drop of acid to the reaction.

14.79 The oxidation of SO_2 to SO_3 is accelerated by NO_2. The reaction proceeds according to:

$$NO_2(g) + SO_2(g) \longrightarrow NO(g) + SO_3(g)$$
$$2\,NO(g) + O_2(g) \longrightarrow 2\,NO_2(g)$$

(a) Show that, with appropriate coefficients, the two reactions can be summed to give the overall oxidation of SO_2 by O_2 to give SO_3. (b) Do we consider NO_2 a catalyst or an intermediate in this reaction? (c) Is this an example of homogeneous catalysis or heterogeneous catalysis?

14.80 The addition of NO accelerates the decomposition of N_2O, possibly by the following mechanism:

$$NO(g) + N_2O(g) \longrightarrow N_2(g) + NO_2(g)$$
$$2\,NO_2(g) \longrightarrow 2\,NO(g) + O_2(g)$$

(a) What is the chemical equation for the overall reaction? Show how the two steps can be added to give the overall equation. (b) Is NO serving as a catalyst or an intermediate in this reaction? (c) If experiments show that during the decomposition of N_2O, NO_2 does not accumulate in measurable quantities, does this rule out the proposed mechanism?

14.81 Many metallic catalysts, particularly the precious-metal ones, are often deposited as very thin films on a substance of high surface area per unit mass, such as alumina (Al_2O_3) or silica (SiO_2). (a) Why is this an effective way of utilizing the catalyst material compared to having powdered metals? (b) How does the surface area affect the rate of reaction?

14.82 (a) If you were going to build a system to check the effectiveness of automobile catalytic converters on cars, what substances would you want to look for in the car exhaust? (b) Automobile catalytic converters have to work at high temperatures, as hot exhaust gases stream through them. In what ways could this be an advantage? In what ways a disadvantage? (c) Why is the rate of flow of exhaust gases over a catalytic converter important?

14.83 When D_2 reacts with ethylene (C_2H_4) in the presence of a finely divided catalyst, ethane with two deuteriums, $CH_2D—CH_2D$, is formed. (Deuterium, D, is an isotope of hydrogen of mass 2). Very little ethane forms in which two deuteriums are bound to one carbon (for example, $CH_3—CHD_2$). Use the sequence of steps involved in the reaction (Figure 14.24) to explain why this is so.

14.84 Heterogeneous catalysts that perform hydrogenation reactions, as illustrated in Figure 14.24, are subject to "poisoning," which shuts down their catalytic ability. Compounds of sulfur are often poisons. Suggest a mechanism by which such compounds might act as poisons.

[14.85] The enzyme carbonic anhydrase catalyzes the reaction $CO_2(g) + H_2O(l) \longrightarrow HCO_3^-(aq) + H^+(aq)$. In water, without the enzyme, the reaction proceeds with a rate constant of $0.039\,s^{-1}$ at 25 °C. In the presence of the enzyme in water, the reaction proceeds with a rate constant of $1.0 \times 10^6\,s^{-1}$ at 25 °C. Assuming the collision factor is the same for both situations, calculate the difference in activation energies for the uncatalyzed versus enzyme-catalyzed reaction.

[14.86] The enzyme urease catalyzes the reaction of urea, (NH_2CONH_2), with water to produce carbon dioxide and ammonia. In water, without the enzyme, the reaction proceeds with a first-order rate constant of $4.15 \times 10^{-5}\,s^{-1}$ at 100 °C. In the presence of the enzyme in water, the reaction proceeds with a rate constant of $3.4 \times 10^4\,s^{-1}$ at 21 °C. (a) Write out the balanced equation for the reaction catalyzed by urease. (b) If the rate of the catalyzed reaction were the same at 100 °C as it is at 21 °C, what would be the difference in the activation energy between the catalyzed and uncatalyzed reactions? (c) In actuality, what would you expect for the rate of the catalyzed reaction at 100 °C as compared to that at 21 °C? (d) On the basis of parts (c) and (d), what can you conclude about the difference in activation energies for the catalyzed and uncatalyzed reactions?

[14.87] The activation energy of an uncatalyzed reaction is 95 kJ/mol. The addition of a catalyst lowers the activation energy to 55 kJ/mol. Assuming that the collision factor remains the same, by what factor will the catalyst increase the rate of the reaction at (a) 25 °C, (b) 125 °C?

[14.88] Suppose that a certain biologically important reaction is quite slow at physiological temperature (37 °C) in the absence of a catalyst. Assuming that the collision factor remains the same, by how much must an enzyme lower the activation energy of the reaction to achieve a 1×10^5-fold increase in the reaction rate?

Additional Exercises

14.89 Consider the reaction A + B $\longrightarrow$ C + D. Is each of the following statements true or false? (a) The rate law for the reaction must be Rate = $k[A][B]$. (b) If the reaction is an elementary reaction, the rate law is second order. (c) If the reaction is an elementary reaction, the rate law of the reverse reaction is first order. (d) The activation energy for the reverse reaction must be greater than that for the forward reaction.

14.90 Hydrogen sulfide (H_2S) is a common and troublesome pollutant in industrial wastewaters. One way to remove H_2S is to treat the water with chlorine, in which case the following reaction occurs:

$$H_2S(aq) + Cl_2(aq) \longrightarrow S(s) + 2\,H^+(aq) + 2\,Cl^-(aq)$$

The rate of this reaction is first order in each reactant. The rate constant for the disappearance of H_2S at 28 °C is $3.5 \times 10^{-2}\,M^{-1}s^{-1}$. If at a given time the concentration of H_2S is $2.0 \times 10^{-4}\,M$ and that of Cl_2 s $0.025\,M$, what is the rate of formation of Cl^-?

14.91 The reaction $2\,NO(g) + O_2(g) \longrightarrow 2\,NO_2(g)$ is second order in NO and first order in O_2. When $[NO] = 0.040\,M$, and $[O_2] = 0.035\,M$, the observed rate of disappearance of NO is $9.3 \times 10^{-5}\,M/s$. (a) What is the rate of disappearance of O_2 at this moment? (b) What is the value of the rate constant? (c) What are the units of the rate constant? (d) What would happen to the rate if the concentration of NO were increased by a factor of 1.8?

14.92 You perform a series of experiments for the reaction A $\longrightarrow$ B + C and find that the rate law has the form rate = $k[A]^x$. Determine the value of x in each of the following cases: (a) There is no rate change when $[A]_0$ is tripled. (b) The rate increases by a factor of 9 when $[A]_0$ is tripled. (c) When $[A]_0$ is doubled, the rate increases by a factor of 8.

14.93 Consider the following reaction between mercury(II) chloride and oxalate ion:

$$2\,HgCl_2(aq) + C_2O_4{}^{2-}(aq) \longrightarrow 2\,Cl^-(aq) + 2\,CO_2(g) + Hg_2Cl_2(s)$$

The initial rate of this reaction was determined for several concentrations of $HgCl_2$ and $C_2O_4{}^{2-}$, and the following rate data were obtained for the rate of disappearance of $C_2O_4{}^{2-}$:

Experiment	[HgCl$_2$] (M)	[C$_2$O$_4$$^{2-}$] (M)	Rate (M/s)
1	0.164	0.15	3.2×10^{-5}
2	0.164	0.45	2.9×10^{-4}
3	0.082	0.45	1.4×10^{-4}
4	0.246	0.15	4.8×10^{-5}

(a) What is the rate law for this reaction? (b) What is the value of the rate constant with proper units? (c) What is the reaction rate when the initial concentration of $HgCl_2$ is $0.100\,M$ and that of $(C_2O_4{}^{2-})$ is $0.25\,M$ if the temperature is the same as that used to obtain the data shown?

14.94 The following kinetic data are collected for the initial rates of a reaction $2\,X + Z \longrightarrow$ products:

Experiment	[X]$_0$(M)	[Z]$_0$(M)	Rate (M/s)
1	0.25	0.25	4.0×10^1
2	0.50	0.50	3.2×10^2
3	0.50	0.75	7.2×10^2

(a) What is the rate law for this reaction? (b) What is the value of the rate constant with proper units? (c) What is the reaction rate when the initial concentration of X is $0.75\,M$ and that of Z is $1.25\,M$?

14.95 The reaction $2\,NO_2 \longrightarrow 2\,NO + O_2$ has the rate constant $k = 0.63\,M^{-1}s^{-1}$. (a) Based on the units for k, is the reaction first or second order in NO_2? (b) If the initial concentration of NO_2 is $0.100\,M$, how would you determine how long it would take for the concentration to decrease to $0.025\,M$?

14.96 Consider two reactions. Reaction (1) has a constant half-life, whereas reaction (2) has a half-life that gets longer as the reaction proceeds. What can you conclude about the rate laws of these reactions from these observations?

[14.97] A first-order reaction A $\longrightarrow$ B has the rate constant $k = 3.2 \times 10^{-3}\,s^{-1}$. If the initial concentration of A is $2.5 \times 10^{-2}\,M$, what is the rate of the reaction at $t = 660$ s?

14.98 (a) The reaction $H_2O_2(aq) \longrightarrow H_2O(l) + \frac{1}{2}O_2(g)$ is first order. Near room temperature, the rate constant equals $7.0 \times 10^{-4}\,s^{-1}$. Calculate the half-life at this temperature. (b) At 415 °C, $(CH_2)_2O$ decomposes in the gas phase, $(CH_2)_2O(g) \longrightarrow CH_4(g) + CO(g)$. If the reaction is first order with a half-life of 56.3 min at this temperature, calculate the rate constant in s^{-1}.

14.99 Americium-241 is used in smoke detectors. It has a first–order rate constant for radioactive decay of $k = 1.6 \times 10^{-3}\,yr^{-1}$. By contrast, iodine-125, which is used to test for thyroid functioning, has a rate constant for radioactive decay of $k = 0.011\,day^{-1}$. (a) What are the half-lives of these two isotopes? (b) Which one decays at a faster rate? (c) How much of a 1.00-mg sample of each isotope remains after 3 half-lives? (d) How much of a 1.00-mg sample of each isotope remains after 4 days?

14.100 Urea (NH_2CONH_2) is the end product in protein metabolism in animals. The decomposition of urea in 0.1 M HCl occurs according to the reaction

$$NH_2CONH_2(aq) + H^+(aq) + 2\,H_2O(l) \longrightarrow 2\,NH_4{}^+(aq) + HCO_3{}^-(aq)$$

The reaction is first order in urea and first order overall. When $[NH_2CONH_2] = 0.200\,M$, the rate at 61.05 °C is $8.56 \times 10^{-5}\,M/s$. (a) What is the rate constant, k? (b) What is the concentration of urea in this solution after 4.00×10^3 s if the starting concentration is $0.500\,M$? (c) What is the half-life for this reaction at 61.05 °C?

[14.101] The rate of a first-order reaction is followed by spectroscopy, monitoring the absorbance of a colored reactant at 520 nm. The reaction occurs in a 1.00-cm sample cell, and the only colored species in the reaction has an extinction coefficient of $5.60 \times 10^3\,M^{-1}\,cm^{-1}$ at 520 nm. (a) Calculate the initial concentration of the colored reactant if the absorbance is 0.605 at the beginning of the reaction. (b) The absorbance falls to 0.250 at 30.0 min. Calculate the rate constant in units of s^{-1}. (c) Calculate the half-life of the reaction. (d) How long does it take for the absorbance to fall to 0.100?

[14.102] A colored dye compound decomposes to give a colorless product. The original dye absorbs at 608 nm and has an extinction coefficient of $4.7 \times 10^4\,M^{-1}cm^{-1}$ at that wavelength. You perform the decomposition reaction in a 1-cm cuvette in a spectrometer and obtain the following data:

Time (min)	Absorbance at 608 nm
0	1.254
30	0.941
60	0.752
90	0.672
120	0.545

From these data, determine the rate law for the reaction "dye $\longrightarrow$ product" and determine the rate constant.

14.103 Cyclopentadiene (C_5H_6) reacts with itself to form dicyclopentadiene ($C_{10}H_{12}$). A 0.0400 M solution of C_5H_6 was monitored as a function of time as the reaction $2\,C_5H_6 \longrightarrow C_{10}H_{12}$ proceeded. The following data were collected:

Time (s)	$[C_5H_6]$ (M)
0.0	0.0400
50.0	0.0300
100.0	0.0240
150.0	0.0240
200.0	0.0174

Plot $[C_5H_6]$ versus time, $\ln[C_5H_6]$ versus time, and $1/[C_5H_6]$ versus time. **(a)** What is the order of the reaction? **(b)** What is the value of the rate constant?

14.104 The first-order rate constant for reaction of a particular organic compound with water varies with temperature as follows:

Temperature (K)	Rate Constant (s^{-1})
300	3.2×10^{-11}
320	1.0×10^{-9}
340	3.0×10^{-8}
355	2.4×10^{-7}

From these data, calculate the activation energy in units of kJ/mol.

14.105 At 28 °C, raw milk sours in 4.0 h but takes 48 h to sour in a refrigerator at 5 °C. Estimate the activation energy in kJ/mol for the reaction that leads to the souring of milk?

[14.106] The following is a quote from an article in the August 18, 1998, issue of *The New York Times* about the breakdown of cellulose and starch: "A drop of 18 degrees Fahrenheit [from 77 °F to 59 °F] lowers the reaction rate six times; a 36-degree drop [from 77 °F to 41°F] produces a fortyfold decrease in the rate." **(a)** Calculate activation energies for the breakdown process based on the two estimates of the effect of temperature on rate. Are the values consistent? **(b)** Assuming the value of E_a calculated from the 36° drop and that the rate of breakdown is first order with a half-life at 25 °C of 2.7 yr, calculate the half-life for breakdown at a temperature of −15 °C.

14.107 The following mechanism has been proposed for the reaction of NO with H_2 to form N_2O and H_2O:

$$NO(g) + NO(g) \longrightarrow N_2O_2(g)$$

$$N_2O_2(g) + H_2(g) \longrightarrow N_2O(g) + H_2O(g)$$

(a) Show that the elementary reactions of the proposed mechanism add to provide a balanced equation for the reaction. **(b)** Write a rate law for each elementary reaction in the mechanism. **(c)** Identify any intermediates in the mechanism. **(d)** The observed rate law is rate $= k[NO]^2[H_2]$. If the proposed mechanism is correct, what can we conclude about the relative speeds of the first and second reactions?

14.108 Ozone in the upper atmosphere can be destroyed by the following two-step mechanism:

$$Cl(g) + O_3(g) \longrightarrow ClO(g) + O_2(g)$$

$$ClO(g) + O(g) \longrightarrow Cl(g) + O_2(g)$$

(a) What is the overall equation for this process? **(b)** What is the catalyst in the reaction? **(c)** What is the intermediate in the reaction?

14.109 Using Figure 14.23 as your basis, draw the energy profile for the bromide-catalyzed decomposition of hydrogen peroxide. **(a)** Label the curve with the activation energies for reactions [14.30] and [14.31]. **(b)** Notice from Figure 14.22 that when $Br^-(aq)$ is first added, Br_2 accumulates to some extent during the reaction and the solution turns brown. What does this tell us about the relative rates of the reactions represented by Equations 14.30 and 14.31?

[14.110] The following mechanism has been proposed for the gas-phase reaction of chloroform ($CHCl_3$) and chlorine:

Step 1: $Cl_2(g) \underset{k_{-1}}{\overset{k_1}{\rightleftharpoons}} 2\,Cl(g)$ (fast)

Step 2: $Cl(g) + CHCl_3(g) \xrightarrow{k_2} HCl(g) + CCl_3(g)$ (slow)

Step 3: $Cl(g) + CCl_3(g) \xrightarrow{k_3} CCl_4$ (fast)

(a) What is the overall reaction? **(b)** What are the intermediates in the mechanism? **(c)** What is the molecularity of each of the elementary reactions? **(d)** What is the rate-determining step? **(e)** What is the rate law predicted by this mechanism? (*Hint:* The overall reaction order is not an integer.)

[14.111] Consider the hypothetical reaction $2\,A + B \longrightarrow 2\,C + D$. The following two-step mechanism is proposed for the reaction:

Step 1: $A + B \longrightarrow C + X$

Step 2: $A + X \longrightarrow C + D$

X is an unstable intermediate. **(a)** What is the predicted rate law expression if Step 1 is rate determining? **(b)** What is the predicted rate law expression if Step 2 is rate determining? **(c)** Your result for part (b) might be considered surprising for which of the following reasons: **(i)** The concentration of a product is in the rate law. **(ii)** There is a negative reaction order in the rate law. **(iii)** Both reasons (i) and (ii). **(iv)** Neither reasons (i) nor (ii).

[14.112] In a hydrocarbon solution, the gold compound $(CH_3)_3AuPH_3$ decomposes into ethane (C_2H_6) and a different gold compound, $(CH_3)AuPH_3$. The following mechanism has been proposed for the decomposition of $(CH_3)_3AuPH_3$:

Step 1: $(CH_3)_3AuPH_3 \underset{k_{-1}}{\overset{k_1}{\rightleftharpoons}} (CH_3)_3Au + PH_3$ (fast)

Step 2: $(CH_3)_3Au \xrightarrow{k_2} C_2H_6 + (CH_3)Au$ (slow)

Step 3: $(CH_3)Au + PH_3 \xrightarrow{k_3} (CH_3)AuPH_3$ (fast)

(a) What is the overall reaction? (b) What are the intermediates in the mechanism? (c) What is the molecularity of each of the elementary steps? (d) What is the rate-determining step? (e) What is the rate law predicted by this mechanism? (f) What would be the effect on the reaction rate of adding PH_3 to the solution of $(CH_3)_3AuPH_3$?

[14.113] Platinum nanoparticles of diameter ~2 nm are important catalysts in carbon monoxide oxidation to carbon dioxide. Platinum crystallizes in a face-centered cubic arrangement with an edge length of 3.924 Å. (a) Estimate how many platinum atoms would fit into a 2.0-nm sphere; the volume of a sphere is $(4/3)\pi r^3$. Recall that $1 \text{ Å} = 1 \times 10^{-10}$ m and $1 \text{ nm} = 1 \times 10^{-9}$ m. (b) Estimate how many platinum atoms are on the surface of a 2.0-nm Pt sphere, using the surface area of a sphere $(4\pi r^2)$ and assuming that the "footprint" of one Pt atom can be estimated from its atomic diameter of 2.8 Å. (c) Using your results from (a) and (b), calculate the percentage of Pt atoms that are on the surface of a 2.0-nm nanoparticle. (d) Repeat these calculations for a 5.0-nm platinum nanoparticle. (e) Which size of nanoparticle would you expect to be more catalytically active and why?

14.114 One of the many remarkable enzymes in the human body is carbonic anhydrase, which catalyzes the interconversion of carbon dioxide and water with bicarbonate ion and protons. If it were not for this enzyme, the body could not rid itself rapidly enough of the CO_2 accumulated by cell metabolism. The enzyme catalyzes the dehydration (release to air) of up to 10^7 CO_2 molecules per second. Which components of this description correspond to the terms *enzyme*, *substrate*, and *turnover number*?

14.115 Suppose that, in the absence of a catalyst, a certain biochemical reaction occurs x times per second at normal body temperature (37 °C). In order to be physiologically useful, the reaction needs to occur 5000 times faster than when it is uncatalyzed. By how many kJ/mol must an enzyme lower the activation energy of the reaction to make it useful?

14.116 Enzymes are often described as following the two-step mechanism:

$$E + S \rightleftharpoons ES \quad \text{(fast)}$$
$$ES \longrightarrow E + P \quad \text{(slow)}$$

where E = enzyme, S = substrate,

ES = enzyme–substrate complex, and P = product.

(a) If an enzyme follows this mechanism, what rate law is expected for the reaction? (b) Molecules that can bind to the active site of an enzyme but are not converted into product are called *enzyme inhibitors*. Write an additional elementary step to add into the preceding mechanism to account for the reaction of E with I, an inhibitor.

Integrative Exercises

[14.117] Dinitrogen pentoxide (N_2O_5) decomposes in chloroform as a solvent to yield NO_2 and O_2. The decomposition is first order with a rate constant at 45 °C of 1.0×10^{-5} s^{-1}. Calculate the partial pressure of O_2 produced from 1.00 L of 0.600 M N_2O_5 solution at 45 °C over a period of 20.0 h if the gas is collected in a 10.0-L container. (Assume that the products do not dissolve in chloroform.)

[14.118] The reaction between ethyl iodide and hydroxide ion in ethanol (C_2H_5OH) solution, $C_2H_5I(alc) + OH^-(alc) \longrightarrow C_2H_5OH(l) + I^-(alc)$, has an activation energy of 86.8 kJ/mol and a frequency factor of 2.10×10^{11} M^{-1} s^{-1}. (a) Predict the rate constant for the reaction at 35 °C. (b) A solution of KOH in ethanol is made up by dissolving 0.335 g KOH in ethanol to form 250.0 mL of solution. Similarly, 1.453 g of C_2H_5I is dissolved in ethanol to form 250.0 mL of solution. Equal volumes of the two solutions are mixed. Assuming the reaction is first order in each reactant, what is the initial rate at 35 °C? (c) Which reagent in the reaction is limiting, assuming the reaction proceeds to completion? (d) Assuming the frequency factor and activation energy do not change as a function of temperature, calculate the rate constant for the reaction at 50 °C.

[14.119] You obtain kinetic data for a reaction at a set of different temperatures. You plot ln k versus $1/T$ and obtain the following graph:

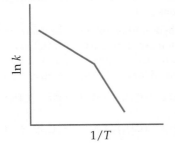

Suggest a molecular-level interpretation of these unusual data.

14.120 The gas-phase reaction of NO with F_2 to form NOF and F has an activation energy of $E_a = 6.3$ kJ/mol. and a frequency factor of $A = 6.0 \times 10^8$ M^{-1} s^{-1}. The reaction is believed to be bimolecular:

$$NO(g) + F_2(g) \longrightarrow NOF(g) + F(g)$$

(a) Calculate the rate constant at 100 °C. (b) Draw the Lewis structures for the NO and the NOF molecules, given that the chemical formula for NOF is misleading because the nitrogen atom is actually the central atom in the molecule. (c) Predict the shape for the NOF molecule. (d) Draw a possible transition state for the formation of NOF, using dashed lines to indicate the weak bonds that are beginning to form. (e) Suggest a reason for the low activation energy for the reaction.

14.121 The mechanism for the oxidation of HBr by O_2 to form 2 H_2O and Br_2 is shown in Exercise 14.74. (a) Calculate the overall standard enthalpy change for the reaction process. (b) HBr does not react with O_2 at a measurable rate at room temperature under ordinary conditions. What can you infer from this about the magnitude of the activation energy for the rate-determining step? (c) Draw a plausible Lewis structure for the intermediate HOOBr. To what familiar compound of hydrogen and oxygen does it appear similar?

[14.122] The rates of many atmospheric reactions are accelerated by the absorption of light by one of the reactants. For example, consider the reaction between methane and chlorine to produce methyl chloride and hydrogen chloride:

Reaction 1: $CH_4(g) + Cl_2(g) \longrightarrow CH_3Cl(g) + HCl(g)$

This reaction is very slow in the absence of light. However, $Cl_2(g)$ can absorb light to form Cl atoms:

$$\text{Reaction 2: } Cl_2(g) + h\nu \longrightarrow 2\,Cl(g)$$

Once the Cl atoms are generated, they can catalyze the reaction of CH_4 and Cl_2, according to the following proposed mechanism:

$$\text{Reaction 3: } CH_4(g) + Cl(g) \longrightarrow CH_3(g) + HCl(g)$$

$$\text{Reaction 4: } CH_3(g) + Cl_2(g) \longrightarrow CH_3Cl(g) + Cl(g)$$

The enthalpy changes and activation energies for these two reactions are tabulated as follows:

Reaction	$\Delta H°$ (kJ/mol)	E_a (kJ/mol)
3	+4	17
4	−109	4

(a) By using the bond enthalpy for Cl_2 (Table 8.4), determine the longest wavelength of light that is energetic enough to cause reaction 2 to occur. In which portion of the electromagnetic spectrum is this light found? (b) By using the data tabulated here, sketch a quantitative energy profile for the catalyzed reaction represented by reactions 3 and 4. (c) By using bond enthalpies, estimate where the reactants, $CH_4(g) + Cl_2(g)$, should be placed on your diagram in part (b). Use this result to estimate the value of E_a for the reaction $CH_4(g) + Cl_2(g) \longrightarrow CH_3(g) + HCl(g) + Cl(g)$. (d) The species $Cl(g)$ and $CH_3(g)$ in reactions 3 and 4 are radicals, that is, atoms or molecules with unpaired electrons. Draw a Lewis structure of CH_3, and verify that it is a radical. (e) The sequence of reactions 3 and 4 comprises a radical chain mechanism. Why do you think this is called a "chain reaction"? Propose a reaction that will terminate the chain reaction.

[14.123] Many primary amines, RNH_2, where R is a carbon-containing fragment such as CH_3, CH_3CH_2, and so on, undergo reactions where the transition state is tetrahedral. (a) Draw a hybrid orbital picture to visualize the bonding at the nitrogen in a primary amine (just use a C atom for "R"). (b) What kind of reactant with a primary amine can produce a tetrahedral intermediate?

[14.124] The NO_x waste stream from automobile exhaust includes species such as NO and NO_2. Catalysts that convert these species to N_2 are desirable to reduce air pollution. (a) Draw the Lewis dot and VSEPR structures of NO, NO_2, and N_2. (b) Using a resource such as Table 8.4, look up the energies of the bonds in these molecules. In what region of the electromagnetic spectrum are these energies? (c) Design a spectroscopic experiment to monitor the conversion of NO_x into N_2, describing what wavelengths of light need to be monitored as a function of time.

[14.125] As shown in Figure 14.24, the first step in the heterogeneous hydrogenation of ethylene is adsorption of the ethylene molecule on a metal surface. One proposed explanation for the "sticking" of ethylene to a metal surface is the interaction of the electrons in the C—C π bond with vacant orbitals on the metal surface. (a) If this notion is correct, would ethane be expected to adsorb to a metal surface, and, if so, how strongly would ethane bind compared to ethylene? (b) Based on its Lewis structure, would you expect ammonia to adsorb to a metal surface using a similar explanation as for ethylene?

Design an Experiment

Let's explore the chemical kinetics of our favorite hypothetical reaction: $a\,A + b\,B \longrightarrow c\,C + d\,D$. We shall assume that all the substances are soluble in water and that we carry out the reaction in aqueous solution. Substances A and C both absorb visible light, and the absorption maxima are 510 nm for A and for 640 nm for C. Substances B and D are colorless. You are provided with pure samples of all four substances and you know their chemical formulas. You are also provided appropriate instrumentation to obtain visible absorption spectra (see the Closer Look box on using spectroscopic methods in Section 14.3). Let's design an experiment to ascertain the kinetics of our reaction. (a) What experiments could you design to determine the rate law and the rate constant for the reaction at room temperature? Would you need to know the values of the stoichiometric constants a and c in order to find the rate law? (b) Design an experiment to determine the activation energy for the reaction. What challenges might you face in actually carrying out this experiment? (c) You now want to test whether a particular water-soluble substance Q is a homogeneous catalyst for the reaction. What experiments can you carry out to test this notion? (d) If Q does indeed catalyze the reaction, what follow-up experiments might you undertake to learn more about the reaction profile for the reaction?

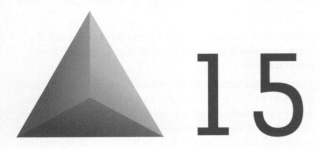

15
Chemical Equilibrium

To be in equilibrium is to be in a state of balance. A tug of war, in which the two sides pull with equal force so that the rope does not move, is an example of a *static* equilibrium, one in which an object is at rest. Equilibria can also be *dynamic*, whereby a forward process and the reverse process take place at the same rate so that no net change occurs.

A saturated solution ∞ (Section 13.2) of an ionic compound in contact with undissolved crystals of the same compound is a good example of dynamic equilibrium. The rate at which ions leave the surface of the crystals and enter the solution (*dissolution*) is equal to the rate at which ions leave the solution and become part of the solid (*crystallization*). Hence the concentration of ions in solution and the amount of undissolved solid do not change with time.

If some of the solvent is lost due to evaporation, the solution becomes more concentrated, which increases the rate of crystallization. This change leads to a net migration of ions from the solution into the solid until the solution concentration is reduced so that the rate of crystallization and dissolution are once again equal and equilibrium is reestablished. A striking example of this effect is the formation of intricate salt formations in the Dead Sea, as illustrated in the chapter-opening photograph. The Dead Sea, which borders Jordan and Israel, is the lowest point on the surface of the Earth, and has a salt concentration that is almost nine times higher than the ocean. An extended period of hot weather leads to extensive evaporation of the Dead Sea, resulting in an increase in the salt concentration and subsequent crystallization and growth of intricate salt formations.

A saturated solution is one of many instances of dynamic equilibrium that we have already encountered. Vapor pressure ∞ (Section 11.5) is another example of dynamic equilibrium. The pressure of a vapor above a liquid in a closed container reaches equilibrium with the liquid phase, and therefore stops changing when the rate at which molecules escape from the liquid into the gas phase equals the rate at which molecules

▶ **SALT PILLARS** in the Dead Sea. These pillars are formed in shallow bays where evaporation can lead to salt concentrations that exceed equilibrium values.

WHAT'S AHEAD ▶

15.6 APPLICATIONS OF EQUILIBRIUM CONSTANTS We learn that equilibrium constants can be used to predict equilibrium concentrations of reactants and products and to determine the direction in which a reaction mixture must proceed to achieve equilibrium.

15.7 LE CHÂTELIER'S PRINCIPLE We discuss *Le Châtelier's principle*, which predicts how a system at equilibrium responds to changes in concentration, volume, pressure, and temperature.

in the gas phase become part of the liquid. Henry's law ∞ (Section 13.3), which governs the solubility of gases in a solvent, is yet another example of dynamic equilibrium.

In this chapter, we consider dynamic equilibria in chemical reactions. ***Chemical equilibrium** occurs when opposing reactions proceed at equal rates*: The rate at which the products form from the reactants equals the rate at which the reactants form from the products. As a result, concentrations cease to change, making the reaction appear to be stopped. In this and the next two chapters, we will explore chemical equilibrium in some detail. Later, in Chapter 19, we will learn how to relate chemical equilibria to thermodynamics. Here, we learn how to express the equilibrium state of a reaction in quantitative terms and study the factors that determine the relative concentrations of reactants and products in equilibrium mixtures.

15.1 | The Concept of Equilibrium

Let's examine a simple chemical reaction to see how it reaches an *equilibrium state*—a mixture of reactants and products whose concentrations no longer change with time. We begin with N_2O_4, a colorless substance that dissociates to form brown NO_2. ▼ Figure 15.1 shows a sample of frozen N_2O_4 inside a sealed tube. The solid N_2O_4 becomes a gas as it is warmed above its boiling point (21.2 °C), and the gas turns darker as the colorless N_2O_4 gas dissociates into brown NO_2 gas. Eventually, even though there is still N_2O_4 in the tube, the color stops getting darker because the system reaches equilibrium. We are left with an

GO FIGURE

If you were to let the tube on the right sit overnight and then take another picture would the brown color look darker, lighter, or the same?

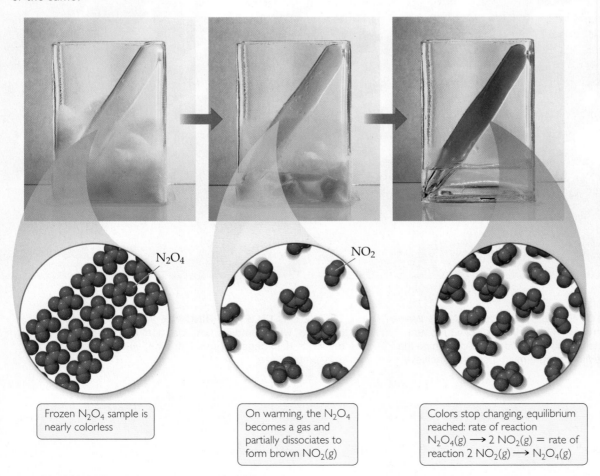

Frozen N_2O_4 sample is nearly colorless

On warming, the N_2O_4 becomes a gas and partially dissociates to form brown $NO_2(g)$

Colors stop changing, equilibrium reached: rate of reaction $N_2O_4(g) \longrightarrow 2\,NO_2(g)$ = rate of reaction $2\,NO_2(g) \longrightarrow N_2O_4(g)$

▲ Figure 15.1 The equilibrium between N_2O_4 and NO_2.

equilibrium mixture of N_2O_4 and NO_2 in which the concentrations of the gases no longer change as time passes. Because the reaction is in a closed system, where no gases can escape, equilibrium will eventually be reached.

The equilibrium mixture results because the reaction is *reversible*: N_2O_4 can form NO_2, and NO_2 can form N_2O_4. Dynamic equilibrium is represented by writing the equation for the reaction with two half arrows pointing in opposite directions ∞ (Section 4.1):

$$N_2O_4(g) \rightleftharpoons 2\,NO_2(g) \qquad [15.1]$$
$$\underset{\text{Colorless}}{} \qquad\qquad \underset{\text{Brown}}{}$$

We can analyze this equilibrium using our knowledge of kinetics. Let's call the decomposition of N_2O_4 the forward reaction and the formation of N_2O_4 the reverse reaction. In this case, both the forward reaction and the reverse reaction are *elementary reactions*. As we learned in Section 14.6, the rate laws for elementary reactions can be written from their chemical equations:

Forward reaction: $N_2O_4(g) \longrightarrow 2\,NO_2(g)$ $\text{Rate}_f = k_f[N_2O_4]$ [15.2]

Reverse reaction: $2\,NO_2(g) \longrightarrow N_2O_4(g)$ $\text{Rate}_r = k_r[NO_2]^2$ [15.3]

At equilibrium, the rate at which NO_2 forms in the forward reaction equals the rate at which N_2O_4 forms in the reverse reaction:

$$\underset{\text{Forward reaction}}{k_f[N_2O_4]} = \underset{\text{Reverse reaction}}{k_r[NO_2]^2} \qquad [15.4]$$

Rearranging this equation gives

$$\frac{[NO_2]^2}{[N_2O_4]} = \frac{k_f}{k_r} = \text{a constant} \qquad [15.5]$$

From Equation 15.5, we see that the quotient of two rate constants is another constant. We also see that, at equilibrium, the ratio of the concentration terms equals this same constant. (We consider this constant, called the equilibrium constant, in Section 15.2.) It makes no difference whether we start with N_2O_4 or with NO_2, or even with some mixture of the two. At equilibrium, at a given temperature, the ratio equals a specific value. Thus, there is an important constraint on the proportions of N_2O_4 and NO_2 at equilibrium.

Once equilibrium is established, the concentrations of N_2O_4 and NO_2 no longer change, as shown in ▼ **Figure 15.2(a)**. However, the fact that the composition of the equilibrium mixture remains constant with time does not mean that N_2O_4 and NO_2 stop reacting. On the contrary, the equilibrium is *dynamic*—which means some N_2O_4 is always converting to NO_2 and some NO_2 is always converting to N_2O_4. At equilibrium, however, the two processes occur at the same rate, as shown in Figure 15.2(b).

 GO FIGURE

At equilibrium, is the ratio $[NO_2]/[N_2O_4]$ less than, greater to, or equal to 1?

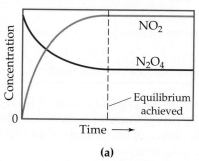

(a)

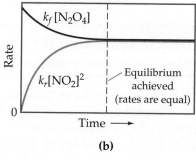

(b)

▲ **Figure 15.2 Achieving chemical equilibrium in the $N_2O_4(g) \rightleftharpoons 2\,NO_2(g)$ reaction.** Equilibrium occurs when the rate of the forward reaction equals the rate of the reverse reaction.

We learn several important lessons about equilibrium from this example:

- At equilibrium, the concentrations of reactants and products no longer change with time.
- For equilibrium to occur, neither reactants nor products can escape from the system.
- At equilibrium, a particular ratio of concentration terms equals a constant.

 Give It Some Thought

(a) Which quantities are equal in a dynamic equilibrium?

(b) If the rate constant for the forward reaction in Equation 15.1 is larger than the rate constant for the reverse reaction, will the constant in Equation 15.5 be greater than 1 or smaller than 1?

15.2 | The Equilibrium Constant

A reaction in which reactants convert to products and products convert to reactants in the same reaction vessel naturally leads to an equilibrium, regardless of how complicated the reaction is and regardless of the nature of the kinetic processes for the forward and reverse reactions. Consider the synthesis of ammonia from nitrogen and hydrogen:

$$N_2(g) + 3 H_2(g) \rightleftharpoons 2 NH_3(g) \qquad [15.6]$$

This reaction is the basis for the **Haber process**, which is critical for the production of fertilizers and therefore critical to the world's food supply. In the Haber process, N_2 and H_2 react at high pressure and temperature in the presence of a catalyst to form ammonia. In a closed system, however, the reaction does not lead to complete consumption of the N_2 and H_2. Rather, at some point the reaction appears to stop with all three components of the reaction mixture present at the same time.

How the concentrations of H_2, N_2, and NH_3 vary with time is shown in ▼ **Figure 15.3**. Notice that an equilibrium mixture is obtained regardless of whether we begin with N_2 and H_2 or with NH_3. *The equilibrium condition is reached from either direction.*

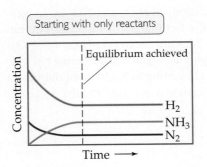
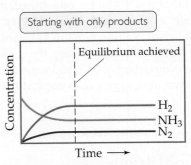

▲ Figure 15.3 **The same equilibrium is reached whether we start with only reactants (N_2 and H_2) or with only product (NH_3).**

 Give It Some Thought

How do we know when equilibrium has been reached in a chemical reaction?

An expression similar to Equation 15.5 governs the concentrations of N_2, H_2, and NH_3 at equilibrium. If we were to systematically change the relative amounts of the three gases in the starting mixture and then analyze each equilibrium mixture, we could determine the relationship among the equilibrium concentrations.

Chemists carried out studies of this kind on other chemical systems in the nineteenth century before Haber's work. In 1864, Cato Maximilian Guldberg (1836–1902) and Peter Waage (1833–1900) postulated their **law of mass action**, which expresses, for any reaction, the relationship between the concentrations of the reactants and products present at equilibrium. Suppose we have the general equilibrium equation

$$a\,A + b\,B \rightleftharpoons d\,D + e\,E \qquad [15.7]$$

where A, B, D, and E are the chemical species involved and a, b, d, and e are their coefficients in the balanced chemical equation. According to the law of mass action, the equilibrium condition is described by the expression

$$K_c = \frac{[\text{D}]^d[\text{E}]^e}{[\text{A}]^a[\text{B}]^b} \quad \begin{array}{l}\longleftarrow \text{ products} \\ \longleftarrow \text{ reactants}\end{array} \qquad [15.8]$$

We call this relationship the **equilibrium-constant expression** (or merely the *equilibrium expression*) for the reaction. The constant K_c, the **equilibrium constant**, is the numerical value obtained when we substitute molar equilibrium concentrations into the equilibrium-constant expression. The subscript c on the K indicates that concentrations expressed in molarity are used to evaluate the constant.

Chemistry Put to Work

The Haber Process

The quantity of food required to feed the ever-increasing human population far exceeds that provided by nitrogen-fixing plants. ∞ (Section 14.7, "Nitrogen Fixation and Nitrogenase") Therefore, human agriculture requires substantial amounts of ammonia-based fertilizers for croplands. Of all the chemical reactions that humans have learned to control for their own purposes, the synthesis of ammonia from hydrogen and atmospheric nitrogen is one of the most important.

In 1912, the German chemist Fritz Haber (1868–1934) developed the Haber process (Equation 15.6). The process is sometimes also called the *Haber–Bosch process* to honor Karl Bosch, the engineer who developed the industrial process on a large scale (▶ **Figure 15.4**). The engineering needed to implement the Haber process requires the use of temperatures and pressures (approximately 500 °C and 200 to 600 atm) that were difficult to achieve at that time.

The Haber process provides a historically interesting example of the complex impact of chemistry on our lives. At the start of World War I, in 1914, Germany depended on nitrate deposits in Chile for the nitrogen-containing compounds needed to manufacture explosives. During the war, the Allied naval blockade of South America cut off this supply. However, by using the Haber reaction to fix nitrogen from air, Germany was able to continue to produce explosives. Experts have estimated that World War I would have ended before 1918 had it not been for the Haber process.

From these unhappy beginnings as a major factor in international warfare, the Haber process has become the world's principal source of fixed nitrogen. The same process that prolonged World War I has enabled the manufacture of fertilizers that have increased crop yields, thereby saving millions of people from starvation. About 40 billion pounds of ammonia are manufactured annually in the United States, mostly by the Haber process. The ammonia can be applied directly to the soil, or it can be converted into ammonium salts that are also used as fertilizers.

Haber was a patriotic German who gave enthusiastic support to his nation's war effort. He served as chief of Germany's Chemical Warfare Service during World War I and developed the use of chlorine as a poison-gas weapon. Consequently, the decision to award him the Nobel Prize in Chemistry in 1918 was the subject of considerable controversy and criticism. The ultimate irony, however, came in 1933 when Haber was expelled from Germany because he was Jewish.

Related Exercises: 15.44, 15.76, 15.92

▲ **Figure 15.4** A high-pressure steel reactor used in the Haber process is on display at Karlsruhe Institute of Technology in Germany where the Haber process was developed.

The numerator of the equilibrium-constant expression is the product of the concentrations of all substances on the product side of the equilibrium equation, each raised to a power equal to its coefficient in the balanced equation. The denominator is similarly derived from the reactant side of the equilibrium equation. Thus, for the Haber process, $N_2(g) + 3\,H_2(g) \rightleftharpoons 2\,NH_3(g)$, the equilibrium-constant expression is

$$K_c = \frac{[NH_3]^2}{[N_2][H_2]^3} \tag{15.9}$$

Once we know the balanced chemical equation for a reaction that reaches equilibrium, we can write the equilibrium-constant expression even if we do not know the reaction mechanism. *The equilibrium-constant expression depends only on the stoichiometry of the reaction, not on its mechanism.*

The value of the equilibrium constant at any given temperature does not depend on the initial amounts of reactants and products. It also does not matter whether other substances are present, as long as they do not react with a reactant or a product. The value of K_c depends only on the particular reaction and on the temperature.

SAMPLE EXERCISE 15.1 | Writing Equilibrium-Constant Expressions

Write the equilibrium expression for K_c for the following reactions:

(a) $2\,O_3(g) \rightleftharpoons 3\,O_2(g)$

(b) $2\,NO(g) + Cl_2(g) \rightleftharpoons 2\,NOCl(g)$

(c) $Ag^+(aq) + 2\,NH_3(aq) \rightleftharpoons Ag(NH_3)_2{}^+(aq)$

SOLUTION

Analyze We are given three equations and are asked to write an equilibrium-constant expression for each.

Plan Using the law of mass action, we write each expression as a quotient having the product concentration terms in the numerator and the reactant concentration terms in the denominator. Each concentration term is raised to the power of its coefficient in the balanced chemical equation.

Solve

(a) $K_c = \dfrac{[O_2]^3}{[O_3]^2}$ (b) $K_c = \dfrac{[NOCl]^2}{[NO]^2[Cl_2]}$ (c) $K_c = \dfrac{[Ag(NH_3)_2{}^+]}{[Ag^+][NH_3]^2}$

Practice Exercise 1

For the reaction $2\,SO_2(g) + O_2(g) \rightleftharpoons 2\,SO_3(g)$ which of the following is the correct equilibrium-constant expression?

(a) $K_C = \dfrac{[SO_2]^2[O_2]}{[SO_3]^2}$ (b) $K_C = \dfrac{2[SO_2][O_2]}{2[SO_3]}$

(c) $K_C = \dfrac{[SO_3]^2}{[SO_2]^2[O_2]}$ (d) $K_C = \dfrac{2[SO_3]}{2[SO_2][O_2]}$

Practice Exercise 2

Write the equilibrium-constant expression K_c for

(a) $H_2(g) + I_2(g) \rightleftharpoons 2\,HI(g)$,

(b) $Cd^{2+}(aq) + 4\,Br^-(aq) \rightleftharpoons CdBr_4{}^{2-}(aq)$.

Evaluating K_c

We can illustrate how the law of mass action was discovered empirically and demonstrate that the equilibrium constant is independent of starting concentrations by examining a series of experiments involving dinitrogen tetroxide and nitrogen dioxide:

$$N_2O_4(g) \rightleftharpoons 2\,NO_2(g) \qquad K_c = \frac{[NO_2]^2}{[N_2O_4]} \tag{15.10}$$

We start with several sealed tubes containing different concentrations of NO_2 and N_2O_4. The tubes are kept at 100 °C until equilibrium is reached. We then analyze the mixtures and determine the equilibrium concentrations of NO_2 and N_2O_4, which are shown in ▶ Table 15.1.

To evaluate K_c, we insert the equilibrium concentrations into the equilibrium-constant expression. For example, using Experiment 1 data, $[NO_2] = 0.0172\,M$ and $[N_2O_4] = 0.00140\,M$, we find

$$K_c = \frac{[NO_2]^2}{[N_2O_4]} = \frac{[0.0172]^2}{0.00140} = 0.211$$

Table 15.1 Initial and Equilibrium Concentrations of $N_2O_4(g)$ and $NO_2(g)$ at 100 °C

Experiment	Initial [N_2O_4] (M)	Initial [NO_2] (M)	Equilibrium [N_2O_4] (M)	Equilibrium [NO_2] (M)	K_c
1	0.0	0.0200	0.00140	0.0172	0.211
2	0.0	0.0300	0.00280	0.0243	0.211
3	0.0	0.0400	0.00452	0.0310	0.213
4	0.0200	0.0	0.00452	0.0310	0.213

Proceeding in the same way, the values of K_c for the other samples are calculated. Note from Table 15.1 that the value for K_c is constant (within the limits of experimental error) even though the initial concentrations vary. Furthermore, Experiment 4 shows that equilibrium can be achieved beginning with N_2O_4 rather than with NO_2. That is, equilibrium can be approached from either direction. ▶ Figure 15.5 shows how Experiments 3 and 4 result in the same equilibrium mixture even though the two experiments start with very different NO_2 concentrations.

Notice that no units are given for K_c either in Table 15.1 or in the calculation we just did using Experiment 1 data. It is common practice to write equilibrium constants without units for reasons that we address later in this section.

 Give It Some Thought

How does the value of K_c in Equation 15.10 depend on the starting concentrations of NO_2 and N_2O_4?

Equilibrium Constants in Terms of Pressure, K_p

When the reactants and products in a chemical reaction are gases, we can formulate the equilibrium-constant expression in terms of partial pressures. When partial pressures in atmospheres are used in the expression, we denote the equilibrium constant K_p (where the subscript p stands for pressure). For the general reaction in Equation 15.7, we have

$$K_p = \frac{(P_D)^d (P_E)^e}{(P_A)^a (P_B)^b} \qquad [15.11]$$

where P_A is the partial pressure of A in atmospheres, P_B is the partial pressure of B in atmospheres, and so forth. For example, for our N_2O_4/NO_2 reaction, we have

$$K_p = \frac{(P_{NO_2})^2}{P_{N_2O_4}}$$

 Give It Some Thought

What is the difference between the equilibrium constant K_c and the equilibrium constant K_p?

For a given reaction, the numerical value of K_c is generally different from the numerical value of K_p. We must therefore take care to indicate, via subscript c or p, which constant we are using. It is possible, however, to calculate one from the other using the ideal-gas equation ∞ (Section 10.4):

$$PV = nRT, \text{ so } P = \frac{n}{V}RT \qquad [15.12]$$

The usual units for n/V are mol/L, which equals molarity, M. For substance A in our generic reaction, we therefore see that

$$P_A = \frac{n_A}{V}RT = [A]RT \qquad [15.13]$$

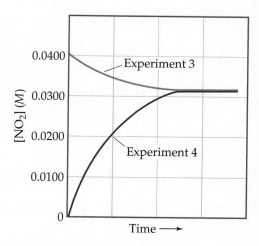

▲ Figure 15.5 **The same equilibrium mixture is produced regardless of the initial NO_2 concentration.** The concentration of NO_2 either increases or decreases until equilibrium is reached.

When we substitute Equation 15.13 and like expressions for the other gaseous components of the reaction into Equation 15.11, we obtain a general expression relating K_p and K_c:

$$K_p = K_c(RT)^{\Delta n} \qquad [15.14]$$

The quantity Δn is the change in the number of moles of gas in the balanced chemical equation. It equals the sum of the coefficients of the gaseous products minus the sum of the coefficients of the gaseous reactants:

$$\Delta n = (\text{moles of gaseous product}) - (\text{moles of gaseous reactant}) \qquad [15.15]$$

For example, in the $N_2O_4(g) \rightleftharpoons 2\,NO_2(g)$ reaction, there are two moles of product NO_2 and one mole of reactant N_2O_4. Therefore, $\Delta n = 2 - 1 = 1$, and $K_p = K_c(RT)$, for this reaction.

 Give It Some Thought

Is it possible to have a reaction where $K_c = K_p$? If so, under what conditions would this relationship hold?

SAMPLE EXERCISE 15.2 | **Converting between K_c and K_p**

For the Haber process,

$$N_2(g) + 3\,H_2(g) \rightleftharpoons 2\,NH_3(g)$$

$K_c = 9.60$ at 300 °C. Calculate K_p for this reaction at this temperature.

SOLUTION

Analyze We are given K_c for a reaction and asked to calculate K_p.

Plan The relationship between K_c and K_p is given by Equation 15.14. To apply that equation, we must determine Δn by comparing the number of moles of product with the number of moles of reactants (Equation 15.15).

Solve With 2 mol of gaseous products ($2\,NH_3$) and 4 mol of gaseous reactants ($1\,N_2 + 3\,H_2$), $\Delta n = 2 - 4 = -2$. (Remember that Δ functions are always based on *products minus reactants*.) The temperature is $273 + 300 = 573$ K. The value for the ideal-gas constant, R, is 0.08206 L-atm/mol-K. Using $K_c = 9.60$, we therefore have

$$K_p = K_c(RT)^{\Delta n} = (9.60)(0.08206 \times 573)^{-2} = \frac{(9.60)}{(0.08206 \times 573)^2} = 4.34 \times 10^{-3}$$

Practice Exercise 1

For which of the following reactions is the ratio K_p/K_c largest at 300 K?
(a) $N_2(g) + O_2(g) \rightleftharpoons 2\,NO(g)$ (b) $CaCO_3(s) \rightleftharpoons CaO(s) + CO_2(g)$
(c) $Ni(CO)_4(g) \rightleftharpoons Ni(s) + 4\,CO(g)$ (d) $C(s) + 2\,H_2(g) \rightleftharpoons CH_4(g)$

Practice Exercise 2

For the equilibrium $2\,SO_3(g) \rightleftharpoons 2\,SO_2(g) + O_2(g)$, K_c is 4.08×10^{-3} at 1000 K. Calculate the value for K_p.

Equilibrium Constants and Units

You may wonder why equilibrium constants are reported without units. The equilibrium constant is related to the kinetics of a reaction as well as to the thermodynamics. (We explore this latter connection in Chapter 19.) Equilibrium constants derived from thermodynamic measurements are defined in terms of *activities* rather than concentrations or partial pressures.

The activity of any substance in an *ideal* mixture is the ratio of the concentration or pressure of the substance either to a reference concentration (1 *M*) or to a reference pressure (1 atm). For example, if the concentration of a substance in an equilibrium mixture is 0.010 *M*, its activity is 0.010 *M*/1 *M* = 0.010. The units of such ratios always cancel and, consequently, activities have no units. Furthermore, the numerical value of the activity equals the concentration. For pure solids and pure liquids, the situation is even simpler because the activities then merely equal 1 (again with no units).

In real systems, activities are also ratios that have no units. Even though these activities may not be exactly numerically equal to concentrations, we will ignore the differences. All we need to know at this point is that activities have no units. As a result, the *thermodynamic equilibrium constants* derived from them also have no units. It is therefore common practice to write all types of equilibrium constants without units, a practice that we adhere to in this text. In more advanced chemistry courses, you may make more rigorous distinctions between concentrations and activities.

 Give It Some Thought

If the concentration of N_2O_4 in an equilibrium mixture is 0.00140 *M*, what is its activity? (Assume the solution is ideal.)

15.3 | Understanding and Working with Equilibrium Constants

Before doing calculations with equilibrium constants, it is valuable to understand what the magnitude of an equilibrium constant can tell us about the relative concentrations of reactants and products in an equilibrium mixture. It is also useful to consider how the magnitude of any equilibrium constant depends on how the chemical equation is expressed.

The Magnitude of Equilibrium Constants

The magnitude of the equilibrium constant for a reaction gives us important information about the composition of the equilibrium mixture. For example, consider the experimental data for the reaction of carbon monoxide gas and chlorine gas at 100 °C to form phosgene ($COCl_2$), a toxic gas used in the manufacture of certain polymers and insecticides:

$$CO(g) + Cl_2(g) \rightleftharpoons COCl_2(g) \qquad K_c = \frac{[COCl_2]}{[CO][Cl_2]} = 4.56 \times 10^9$$

For the equilibrium constant to be so large, the numerator of the equilibrium-constant expression must be approximately a billion (10^9) times larger than the denominator. Thus, the equilibrium concentration of $COCl_2$ must be much greater than that of CO or Cl_2, and in fact, this is just what we find experimentally. We say that this equilibrium *lies to the right* (that is, toward the product side). Likewise, a very small equilibrium constant indicates that the equilibrium mixture contains mostly reactants. We then say that the equilibrium *lies to the left*. In general,

If K ≫ 1 *(large K)*: Equilibrium lies to right, products predominate

If K ≪ 1 *(small K)*: Equilibrium lies to left, reactants predominate

These situations are summarized in ▶ **Figure 15.6**. Remember, it is forward and reverse reaction rates, not reactant and product concentrations, that are equal at equilibrium.

 GO FIGURE

What would this figure look like for a reaction in which $K \approx 1$?

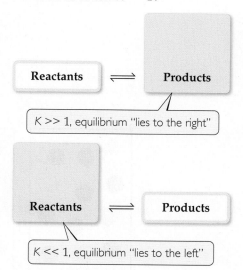

▲ Figure 15.6 **Relationship between magnitude of K and composition of an equilibrium mixture.**

SAMPLE EXERCISE 15.3 Interpreting the Magnitude of an Equilibrium Constant

The following diagrams represent three systems at equilibrium, all in the same-size containers.
(a) Without doing any calculations, rank the systems in order of increasing K_c. **(b)** If the volume of the containers is 1.0 L and each sphere represents 0.10 mol, calculate K_c for each system.

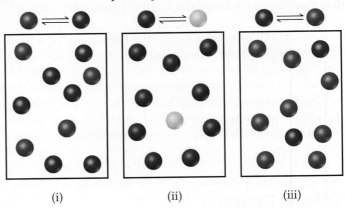

(i)　　　　　(ii)　　　　　(iii)

SOLUTION

Analyze We are asked to judge the relative magnitudes of three equilibrium constants and then to calculate them.

Plan (a) The more product present at equilibrium, relative to reactant, the larger the equilibrium constant. **(b)** The equilibrium constant is given by Equation 15.8.

Solve

(a) Each box contains 10 spheres. The amount of product in each varies as follows: (i) 6, (ii) 1, (iii) 8. Therefore, the equilibrium constant varies in the order (ii) < (i) < (iii), from smallest (most reactant) to largest (most products).

(b) In (i), we have 0.60 mol/L product and 0.40 mol/L reactant, giving $K_c = 0.60/0.40 = 1.5$. (You will get the same result by merely dividing the number of spheres of each kind: 6 spheres/4 spheres = 1.5.) In (ii), we have 0.10 mol/L product and 0.90 mol/L reactant, giving $K_c = 0.10/0.90 = 0.11$ (or 1 sphere/9 spheres = 0.11). In (iii), we have 0.80 mol/L product and 0.20 mol/L reactant, giving $K_c = 0.80/0.20 = 4.0$ (or 8 spheres/2 spheres = 4.0). These calculations verify the order in (a).

Comment Imagine a drawing that represents a reaction with a very small or very large value of K_c. For example, what would the drawing look like if $K_c = 1 \times 10^{-5}$? In that case there would need to be 100,000 reactant molecules for only 1 product molecule. But then, that would be impractical to draw.

Practice Exercise 1

The equilibrium constant for the reaction $N_2O_4(g) \rightleftharpoons 2\, NO_2(g)$ at 2 °C is $K_c = 2.0$. If each yellow sphere represents 1 mol of N_2O_4 and each brown sphere 1 mol of NO_2 which of the following 1.0 L containers represents the equilibrium mixture at 2 °C?

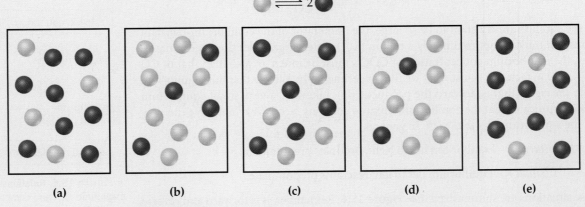

(a)　　　　(b)　　　　(c)　　　　(d)　　　　(e)

Practice Exercise 2

For the reaction $H_2(g) + I_2(g) \rightleftharpoons 2\, HI(g)$, $K_p = 794$ at 298 K and $K_p = 55$ at 700 K. Is the formation of HI favored more at the higher or lower temperature?

The Direction of the Chemical Equation and K

We have seen that we can represent the N_2O_4/NO_2 equilibrium as

$$N_2O_4(g) \rightleftharpoons 2\,NO_2(g) \quad K_c = \frac{[NO_2]^2}{[N_2O_4]} = 0.212 \quad \text{(at 100 °C)} \quad [15.16]$$

We could equally well consider this equilibrium in terms of the reverse reaction:

$$2\,NO_2(g) \rightleftharpoons N_2O_4(g)$$

The equilibrium expression is then

$$K_c = \frac{[N_2O_4]}{[NO_2]^2} = \frac{1}{0.212} = 4.72 \quad \text{(at 100 °C)} \quad [15.17]$$

Equation 15.17 is the reciprocal of the expression in Equation 15.16. *The equilibrium-constant expression for a reaction written in one direction is the reciprocal of the expression for the reaction written in the reverse direction.* Consequently, the numerical value of the equilibrium constant for the reaction written in one direction is the reciprocal of that for the reverse reaction. Both expressions are equally valid, but it is meaningless to say that the equilibrium constant for the equilibrium between NO_2 and N_2O_4 is "0.212" or "4.72" unless we indicate how the equilibrium reaction is written and specify the temperature. Therefore, whenever you are using an equilibrium constant, you should always write the associated balanced chemical equation.

Give It Some Thought

For the reaction $PCl_5(g) \rightleftharpoons PCl_3(g) + Cl_2(g)$, the equilibrium constant $K_c = 1.1 \times 10^{-2}$ at 400 K. What is the equilibrium constant for the reaction $PCl_3(g) + Cl_2(g) \rightleftharpoons PCl_5(g)$ at 400 K?

Relating Chemical Equation Stoichiometry and Equilibrium Constants

There are many ways to write a balanced chemical equation for a given reaction. For example, if we multiply Equation 15.1, $N_2O_4(g) \rightleftharpoons 2\,NO_2(g)$ by 2, we have

$$2\,N_2O_4(g) \rightleftharpoons 4\,NO_2(g)$$

This chemical equation is balanced and might be written this way in some contexts. Therefore, the equilibrium-constant expression for this equation is

$$K_c = \frac{[NO_2]^4}{[N_2O_4]^2}$$

which is the square of the equilibrium-constant expression given in Equation 15.10 for the reaction as written in Equation 15.1: $[NO_2]^2/[N_2O_4]$. Because the new equilibrium-constant expression equals the original expression squared, the new equilibrium constant K_c equals the original constant squared: $0.212^2 = 0.0449$ (at 100 °C). Once again, it is important to remember that you must relate each equilibrium constant you work with to a *specific* balanced chemical equation. The concentrations of the substances in the equilibrium mixture will be the same no matter how you write the chemical equation, but the value of K_c you calculate depends completely on how you write the reaction.

Give It Some Thought

How does the magnitude of K_p for the reaction $2\,HI(g) \rightleftharpoons H_2(g) + I_2(g)$ change if the equilibrium is written $6\,HI(g) \rightleftharpoons 3\,H_2(g) + 3\,I_2(g)$?

It is also possible to calculate the equilibrium constant for a reaction if we know the equilibrium constants for other reactions that add up to give us the one we want, similar to Hess's law. ∞ (Section 5.6) For example, consider the following two reactions, their equilibrium-constant expressions, and their equilibrium constants at 100 °C:

1. $2\,NOBr(g) \rightleftharpoons 2\,NO(g) + Br_2(g)$ $K_{c1} = \dfrac{[NO]^2[Br_2]}{[NOBr]^2} = 0.014$

2. $Br_2(g) + Cl_2(g) \rightleftharpoons 2\,BrCl(g)$ $K_{c2} = \dfrac{[BrCl]^2}{[Br_2][Cl_2]} = 7.2$

The net sum of these two equations is:

3. $2\,NOBr(g) + Cl_2(g) \rightleftharpoons 2\,NO(g) + 2\,BrCl(g)$

You can prove algebraically that the equilibrium-constant expression for reaction 3 is the product of the expressions for reactions 1 and 2:

$$K_{c3} = \frac{[NO]^2[BrCl]^2}{[NOBr]^2[Cl_2]} = \frac{[NO]^2[Br_2]}{[NOBr]^2} \times \frac{[BrCl]^2}{[Br_2][Cl_2]}$$

Thus,

$$K_{c3} = (K_{c1})(K_{c2}) = (0.014)(7.2) = 0.10$$

To summarize:

1. The equilibrium constant of a reaction in the *reverse* direction is the *inverse* (or *reciprocal*) of the equilibrium constant of the reaction in the forward direction:

$$A + B \rightleftharpoons C + D \quad K_1$$
$$C + D \rightleftharpoons A + B \quad K = 1/K_1$$

2. The equilibrium constant of a reaction that has been *multiplied* by a number is equal to the original equilibrium constant raised to a *power* equal to that number.

$$A + B \rightleftharpoons C + D \qquad K_1$$
$$nA + nB \rightleftharpoons nC + nD \quad K = K_1{}^n$$

3. The equilibrium constant for a net reaction made up of *two or more reactions* is the *product* of the equilibrium constants for the individual reactions:

$$A + B \rightleftharpoons C + D \quad K_1$$
$$C + F \rightleftharpoons G + A \quad K_2$$
$$\overline{B + F \rightleftharpoons D + G \quad K_3 = (K_1)(K_2)}$$

SAMPLE EXERCISE 15.4 Combining Equilibrium Expressions

Given the reactions

$$HF(aq) \rightleftharpoons H^+(aq) + F^-(aq) \qquad K_c = 6.8 \times 10^{-4}$$
$$H_2C_2O_4(aq) \rightleftharpoons 2\,H^+(aq) + C_2O_4{}^{2-}(aq) \quad K_c = 3.8 \times 10^{-6}$$

determine the value of K_c for the reaction

$$2\,HF(aq) + C_2O_4{}^{2-}(aq) \rightleftharpoons 2\,F^-(aq) + H_2C_2O_4(aq)$$

SOLUTION

Analyze We are given two equilibrium equations and the corresponding equilibrium constants and are asked to determine the equilibrium constant for a third equation, which is related to the first two.

Plan We cannot simply add the first two equations to get the third. Instead, we need to determine how to manipulate these equations to come up with equations that we can add to give us the desired equation.

Solve If we multiply the first equation by 2 and make the corresponding change to its equilibrium constant (raising to the power 2), we get

$$2\,HF(aq) \rightleftharpoons 2\,H^+(aq) + 2\,F^-(aq) \qquad K_c = (6.8 \times 10^{-4})^2 = 4.6 \times 10^{-7}$$

Reversing the second equation and again making the corresponding change to its equilibrium constant (taking the reciprocal) gives

$$2\,H^+(aq) + C_2O_4^{2-}(aq) \rightleftharpoons H_2C_2O_4(aq) \qquad K_c = \dfrac{1}{3.8 \times 10^{-6}} = 2.6 \times 10^5$$

Now, we have two equations that sum to give the net equation, and we can multiply the individual K_c values to get the desired equilibrium constant.

$$2\,HF(aq) \rightleftharpoons 2\,H^+(aq) + 2\,F^-(aq) \qquad K_c = 4.6 \times 10^{-7}$$
$$\underline{2\,H^+(aq) + C_2O_4^{2-}(aq) \rightleftharpoons H_2C_2O_4(aq) \qquad K_c = 2.5 \times 10^5}$$
$$2\,HF(aq) + C_2O_4^{2-}(aq) \rightleftharpoons 2\,F^-(aq) + H_2C_2O_4(aq) \qquad K_c = (4.6 \times 10^{-7})(2.6 \times 10^5) = 0.12$$

Practice Exercise 1

Given the equilibrium constants for the following two reactions in aqueous solution at 25 °C,

$$HNO_2(aq) \rightleftharpoons H^+(aq) + NO_2^-(aq) \qquad K_c = 4.5 \times 10^{-4}$$
$$H_2SO_3(aq) \rightleftharpoons 2\,H^+(aq) + SO_3^-(aq) \qquad K_c = 1.1 \times 10^{-9}$$

what is the value of K_c for the reaction?

$$2\,HNO_2(aq) + SO_3^{2-}(aq) \rightleftharpoons H_2SO_3(aq) + 2\,NO_2^-(aq)$$

(a) 4.9×10^{-13}, **(b)** 4.1×10^5, **(c)** 8.2×10^5, **(d)** 1.8×10^2, **(e)** 5.4×10^{-3}.

Practice Exercise 2

Given that, at 700 K, $K_p = 54.0$ for the reaction $H_2(g) + I_2(g) \rightleftharpoons 2\,HI(g)$ and $K_p = 1.04 \times 10^{-4}$ for the reaction $N_2(g) + 3\,H_2(g) \rightleftharpoons 2\,NH_3(g)$, determine the value of K_p for the reaction $2\,NH_3(g) + 3\,I_2(g) \rightleftharpoons 6\,HI(g) + N_2(g)$ at 700 K.

15.4 | Heterogeneous Equilibria

Many equilibria involve substances that are all in the same phase. Such equilibria are called **homogeneous equilibria**. In some cases, however, the substances in equilibrium are in different phases, giving rise to **heterogeneous equilibria**. As an example of the latter, consider the equilibrium that occurs when solid lead(II) chloride dissolves in water to form a saturated solution:

$$PbCl_2(s) \rightleftharpoons Pb^{2+}(aq) + 2\,Cl^-(aq) \qquad [15.18]$$

This system consists of a solid in equilibrium with two aqueous species. If we want to write the equilibrium-constant expression for this process, we encounter a problem we have not encountered previously: How do we express the concentration of a solid? If we were to carry out experiments starting with varying amounts of products and reactants we would find that the equilibrium-constant expression for the reaction of Equation 15.18 is

$$K_c = [Pb^{2+}][Cl^-]^2 \qquad [15.19]$$

Thus, our problem of how to express the concentration of a solid is not relevant in the end, because $PbCl_2(s)$ does not show up in the equilibrium-constant expression. More generally, we can state that *whenever a pure solid or a pure liquid is involved in a heterogeneous equilibrium, its concentration is not included in the equilibrium-constant expression.*

The fact that pure solids and pure liquids are excluded from equilibrium-constant expressions can be explained in two ways. First, the concentration of a pure solid or liquid has a constant value. If the mass of a solid is doubled, its volume also doubles. Thus, its concentration, which relates to the ratio of mass to volume, stays the same. Because equilibrium-constant expressions include terms only for reactants and products whose concentrations can change during a chemical reaction, the concentrations of pure solids and pure liquids are omitted.

The omission can also be rationalized in a second way. Recall from Section 15.2 that what is substituted into a thermodynamic equilibrium expression is the activity of each substance, which is a ratio of the concentration to a reference value. For a pure substance, the reference value is the concentration of the pure substance, so that the activity of any pure solid or liquid is always 1.

Give It Some Thought

Write the equilibrium-constant expression for the evaporation of water, $H_2O(l) \rightleftharpoons H_2O(g)$, in terms of partial pressures.

Decomposition of calcium carbonate is another example of a heterogeneous reaction:

$$CaCO_3(s) \rightleftharpoons CaO(s) + CO_2(g)$$

Omitting the concentrations of the solids from the equilibrium-constant expression gives

$$K_c = [CO_2] \quad \text{and} \quad K_p = P_{CO_2}$$

These equations tell us that at a given temperature, an equilibrium among $CaCO_3$, CaO, and CO_2 always leads to the same CO_2 partial pressure as long as all three components are present. As shown in ▼ Figure 15.7, we have the same CO_2 pressure regardless of the relative amounts of CaO and $CaCO_3$.

GO FIGURE

If some of the $CO_2(g)$ were released from the upper bell jar and the seal then restored and the system allowed to return to equilibrium, would the amount of $CaCO_3(s)$ increase, decrease or remain the same?

$$CaCO_3(s) \rightleftharpoons CaO(s) + CO_2(g)$$

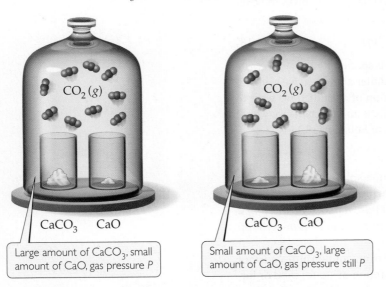

▲ Figure 15.7 **At a given temperature, the equilibrium pressure of CO$_2$ in the bell jars is the same no matter how much of each solid is present.**

SAMPLE EXERCISE 15.5 | Writing Equilibrium-Constant Expressions for Heterogeneous Reactions

Write the equilibrium-constant expression K_c for

(a) $CO_2(g) + H_2(g) \rightleftharpoons CO(g) + H_2O(l)$

(b) $SnO_2(s) + 2\,CO(g) \rightleftharpoons Sn(s) + 2\,CO_2(g)$

SOLUTION

Analyze We are given two chemical equations, both for heterogeneous equilibria, and asked to write the corresponding equilibrium-constant expressions.

Plan We use the law of mass action, remembering to omit any pure solids and pure liquids from the expressions.

Solve

(a) The equilibrium-constant expression is

$$K_c = \frac{[CO]}{[CO_2][H_2]}$$

Because H_2O appears in the reaction as a liquid, its concentration does not appear in the equilibrium-constant expression.

(b) The equilibrium-constant expression is

$$K_c = \frac{[CO_2]^2}{[CO]^2}$$

Because SnO_2 and Sn are pure solids, their concentrations do not appear in the equilibrium-constant expression.

Practice Exercise 1

Consider the equilibrium that is established in a saturated solution of silver chloride, $Ag^+(aq) + Cl^-(aq) \rightleftharpoons AgCl(s)$. If solid AgCl is added to this solution, what will happen to the concentration of Ag^+ and Cl^- ions in solution? (a) $[Ag^+]$ and $[Cl^-]$ will both increase, (b) $[Ag^+]$ and $[Cl^-]$ will both decrease, (c) $[Ag^+]$ will increase and $[Cl^-]$ will decrease, (d) $[Ag^+]$ will decrease and $[Cl^-]$ will increase, (e) neither $[Ag^+]$ nor $[Cl^-]$ will change.

Practice Exercise 2

Write the following equilibrium-constant expressions:
(a) K_c for $Cr(s) + 3\,Ag^+(aq) \rightleftharpoons Cr^{3+}(aq) + 3\,Ag(s)$,
(b) K_p for $3\,Fe(s) + 4\,H_2O(g) \rightleftharpoons Fe_3O_4(s) + 4\,H_2(g)$.

SAMPLE EXERCISE 15.6 | Analyzing a Heterogeneous Equilibrium

Each of these mixtures was placed in a closed container and allowed to stand:

(a) $CaCO_3(s)$

(b) $CaO(s)$ and $CO_2(g)$ at a pressure greater than the value of K_p

(c) $CaCO_3(s)$ and $CO_2(g)$ at a pressure greater than the value of K_p

(d) $CaCO_3(s)$ and $CaO(s)$

Determine whether or not each mixture can attain the equilibrium

$$CaCO_3(s) \rightleftharpoons CaO(s) + CO_2(g)$$

SOLUTION

Analyze We are asked which of several combinations of species can establish an equilibrium between calcium carbonate and its decomposition products, calcium oxide and carbon dioxide.

Plan For equilibrium to be achieved, it must be possible for both the forward process and the reverse process to occur. For the forward process to occur, some calcium carbonate must be present. For the reverse process to occur, both calcium oxide and carbon dioxide must be present. In both cases, either the necessary compounds may be present initially or they may be formed by reaction of the other species.

Solve Equilibrium can be reached in all cases except (c) as long as sufficient quantities of solids are present. (a) $CaCO_3$ simply decomposes, forming $CaO(s)$ and $CO_2(g)$ until the equilibrium pressure of CO_2 is attained. There must be enough $CaCO_3$, however, to allow the CO_2 pressure to reach equilibrium. (b) CO_2 continues to combine with CaO until the partial pressure of the CO_2 decreases to the equilibrium value. (c) Because there is no CaO present, equilibrium cannot be attained; there is no way the CO_2 pressure can decrease to its equilibrium value (which would require some CO_2 to react with CaO). (d) The situation is essentially the same as in (a): $CaCO_3$ decomposes

until equilibrium is attained. The presence of CaO initially makes no difference.

Practice Exercise 1

If 8.0 g of $NH_4HS(s)$ is placed in a sealed vessel with a volume of 1.0 L and heated to 200 °C the reaction $NH_4HS(s) \rightleftharpoons NH_3(g) + H_2S(g)$ will occur. When the system comes to equilibrium, some $NH_4HS(s)$ is still present. Which of the following changes will lead to a reduction in the amount of $NH_4HS(s)$ that is present, assuming in all cases that equilibrium is re-established following the change? (a) Adding more $NH_3(g)$ to the vessel, (b) Adding more $H_2S(g)$ to the vessel, (c) Adding more $NH_4HS(s)$ to the vessel, (d) Increasing the volume of the vessel, (e) decreasing the volume of the vessel.

Practice Exercise 2

When added to $Fe_3O_4(s)$ in a closed container, which one of the following substances—$H_2(g)$, $H_2O(g)$, $O_2(g)$—allows equilibrium to be established in the reaction $3\,Fe(s) + 4\,H_2O(g) \rightleftharpoons Fe_3O_4(s) + 4\,H_2(g)$?

When a solvent is a reactant or product in an equilibrium, its concentration is omitted from the equilibrium-constant expression, provided the concentrations of reactants and products are low, so that the solvent is essentially a pure substance. Applying this guideline to an equilibrium involving water as a solvent,

$$H_2O(l) + CO_3^{2-}(aq) \rightleftharpoons OH^-(aq) + HCO_3^-(aq) \qquad [15.20]$$

gives an equilibrium-constant expression that does not contain $[H_2O]$:

$$K_c = \frac{[OH^-][HCO_3^-]}{[CO_3^{2-}]} \qquad [15.21]$$

 Give It Some Thought

Write the equilibrium-constant expression for the reaction

$$NH_3(aq) + H_2O(l) \rightleftharpoons NH_4^+(aq) + OH^-(aq)$$

15.5 | Calculating Equilibrium Constants

If we can measure the equilibrium concentrations of all the reactants and products in a chemical reaction, as we did with the data in Table 15.1, calculating the value of the equilibrium constant is straightforward. We simply insert all the equilibrium concentrations into the equilibrium-constant expression for the reaction.

SAMPLE EXERCISE 15.7 | **Calculating K When All Equilibrium Concentrations Are Known**

After a mixture of hydrogen and nitrogen gases in a reaction vessel is allowed to attain equilibrium at 472 °C, it is found to contain 7.38 atm H_2, 2.46 atm N_2, and 0.166 atm NH_3. From these data, calculate the equilibrium constant K_p for the reaction

$$N_2(g) + 3 H_2(g) \rightleftharpoons 2 NH_3(g)$$

SOLUTION

Analyze We are given a balanced equation and equilibrium partial pressures and are asked to calculate the value of the equilibrium constant.

Plan Using the balanced equation, we write the equilibrium-constant expression. We then substitute the equilibrium partial pressures into the expression and solve for K_p.

Solve

$$K_p = \frac{(P_{NH_3})^2}{P_{N_2}(P_{H_2})^3} = \frac{(0.166)^2}{(2.46)(7.38)^3} = 2.79 \times 10^{-5}$$

Practice Exercise 1

A mixture of gaseous sulfur dioxide and oxygen are added to a reaction vessel and heated to 1000 K where they react to form $SO_3(g)$. If the vessel contains 0.669 atm $SO_2(g)$, 0.395 atm $O_2(g)$, and 0.0851 atm $SO_3(g)$ after the system has reached

equilibrium, what is the equilibrium constant K_p for the reaction $2 SO_2(g) + O_2(g) \rightleftharpoons 2 SO_3(g)$? **(a)** 0.0410, **(b)** 0.322, **(c)** 24.4, **(d)** 3.36, **(e)** 3.11.

Practice Exercise 2

An aqueous solution of acetic acid is found to have the following equilibrium concentrations at 25 °C: $[CH_3COOH] = 1.65 \times 10^{-2}\,M$; $[H^+] = 5.44 \times 10^{-4}\,M$; and $[CH_3COO^-] = 5.44 \times 10^{-4}\,M$. Calculate the equilibrium constant K_c for the ionization of acetic acid at 25 °C. The reaction is

$$CH_3COOH(aq) \rightleftharpoons H^+(aq) + CH_3COO^-(aq)$$

Often, we do not know the equilibrium concentrations of all species in an equilibrium mixture. If we know the initial concentrations and the equilibrium concentration of at least one species, however, we can generally use the stoichiometry of the reaction to deduce the equilibrium concentrations of the others. The following steps outline the procedure:

1. Tabulate all known initial and equilibrium concentrations of the species that appear in the equilibrium-constant expression.

2. For those species for which initial and equilibrium concentrations are known, calculate the change in concentration that occurs as the system reaches equilibrium.

3. Use the stoichiometry of the reaction (that is, the coefficients in the balanced chemical equation) to calculate the changes in concentration for all other species in the equilibrium-constant expression.

4. Use initial concentrations from step 1 and changes in concentration from step 3 to calculate any equilibrium concentrations not tabulated in step 1.

5. Determine the value of the equilibrium constant.

The best way to illustrate how to do this type of calculation is by example, as we do in Sample Exercise 15.8.

SAMPLE EXERCISE 15.8 | Calculating K from Initial and Equilibrium Concentrations

A closed system initially containing $1.000 \times 10^{-3}\ M\ H_2$ and $2.000 \times 10^{-3}\ M\ I_2$ at 448 °C is allowed to reach equilibrium, and at equilibrium the HI concentration is $1.87 \times 10^{-3}\ M$. Calculate K_c at 448 °C for the reaction taking place, which is

$$H_2(g) + I_2(g) \rightleftharpoons 2\ HI(g)$$

SOLUTION

Analyze We are given the initial concentrations of H_2 and I_2 and the equilibrium concentration of HI. We are asked to calculate the equilibrium constant K_c for $H_2(g) + I_2(g) \rightleftharpoons 2\ HI(g)$.

Plan We construct a table to find equilibrium concentrations of all species and then use the equilibrium concentrations to calculate the equilibrium constant.

Solve

(1) We tabulate the initial and equilibrium concentrations of as many species as we can. We also provide space in our table for listing the changes in concentrations. As shown, it is convenient to use the chemical equation as the heading for the table.

	$H_2(g)$	+	$I_2(g)$	$\rightleftharpoons$	$2\ HI(g)$
Initial concentration (M)	1.000×10^{-3}		2.000×10^{-3}		0
Change in concentration (M)					
Equilibrium concentration (M)					1.87×10^{-3}

(2) We calculate the change in HI concentration, which is the difference between the equilibrium and initial values:

Change in $[HI] = 1.87 \times 10^{-3}\ M - 0 = 1.87 \times 10^{-3}\ M$

(3) We use the coefficients in the balanced equation to relate the change in [HI] to the changes in $[H_2]$ and $[I_2]$:

$$\left(1.87 \times 10^{-3}\ \frac{\text{mol HI}}{\text{L}}\right)\left(\frac{1\ \text{mol}\ H_2}{2\ \text{mol HI}}\right) = 0.935 \times 10^{-3}\ \frac{\text{mol}\ H_2}{\text{L}}$$

$$\left(1.87 \times 10^{-3}\ \frac{\text{mol HI}}{\text{L}}\right)\left(\frac{1\ \text{mol}\ I_2}{2\ \text{mol HI}}\right) = 0.935 \times 10^{-3}\ \frac{\text{mol}\ I_2}{\text{L}}$$

(4) We calculate the equilibrium concentrations of H_2 and I_2, using initial concentrations and changes in concentration. The equilibrium concentration equals the initial concentration minus that consumed:

$[H_2] = 1.000 \times 10^{-3}\ M - 0.935 \times 10^{-3}\ M = 0.065 \times 10^{-3}\ M$

$[I_2] = 2.000 \times 10^{-3}\ M - 0.935 \times 10^{-3}\ M = 1.065 \times 10^{-3}\ M$

(5) Our table now is complete (with equilibrium concentrations in blue for emphasis):

	$H_2(g)$	+	$I_2(g)$	$\rightleftharpoons$	$2\ HI(g)$
Initial concentration (M)	1.000×10^{-3}		2.000×10^{-3}		0
Change in concentration (M)	-0.935×10^{-3}		-0.935×10^{-3}		$+1.87 \times 10^{-3}$
Equilibrium concentration (M)	0.065×10^{-3}		1.065×10^{-3}		1.87×10^{-3}

Notice that the entries for the changes are negative when a reactant is consumed and positive when a product is formed. Finally, we use the equilibrium-constant expression to calculate the equilibrium constant:

$$K_c = \frac{[HI]^2}{[H_2][I_2]} = \frac{(1.87 \times 10^{-3})^2}{(0.065 \times 10^{-3})(1.065 \times 10^{-3})} = 51$$

Comment The same method can be applied to gaseous equilibrium problems to calculate K_p, in which case partial pressures are used as table entries in place of molar concentrations. Your instructor may refer to this kind of table as an ICE chart, where ICE stands for *I*nitial – *C*hange – *E*quilibrium.

Practice Exercise 1

In Section 15.1, we discussed the equilibrium between $N_2O_4(g)$ and $NO_2(g)$. Let's return to that equation in a quantitative example. When 9.2 g of frozen N_2O_4 is added to a 0.50 L reaction vessel and the vessel is heated to 400 K and allowed to come to equilibrium, the concentration of N_2O_4 is determined to be 0.057 M. Given this information, what is the value of K_c for the reaction $N_2O_4(g) \longrightarrow 2\,NO_2(g)$ at 400 K? **(a)** 0.23, **(b)** 0.36, **(c)** 0.13, **(d)** 1.4, **(e)** 2.5.

Practice Exercise 2

The gaseous compound BrCl decomposes at high temperature in a sealed container: $2\,BrCl(g) \rightleftharpoons Br_2(g) + Cl_2(g)$. Initially, the vessel is charged at 500 K with $BrCl(g)$ at a partial pressure of 0.500 atm. At equilibrium, the $BrCl(g)$ partial pressure is 0.040 atm. Calculate the value of K_p at 500 K.

15.6 | Applications of Equilibrium Constants

We have seen that the magnitude of K indicates the extent to which a reaction proceeds. If K is very large, the equilibrium mixture contains mostly substances on the product side of the equation for the reaction. That is, the reaction proceeds far to the right. If K is very small (that is, much less than 1), the equilibrium mixture contains mainly substances on the reactant side of the equation. The equilibrium constant also allows us to (1) predict the direction in which a reaction mixture achieves equilibrium and (2) calculate equilibrium concentrations of reactants and products.

Predicting the Direction of Reaction

For the formation of NH_3 from N_2 and H_2 (Equation 15.6), $K_c = 0.105$ at 472 °C. Suppose we place 2.00 mol of H_2, 1.00 mol of N_2, and 2.00 mol of NH_3 in a 1.00-L container at 472 °C. How will the mixture react to reach equilibrium? Will N_2 and H_2 react to form more NH_3, or will NH_3 decompose to N_2 and H_2?

To answer this question, we substitute the starting concentrations of N_2, H_2, and NH_3 into the equilibrium-constant expression and compare its value to the equilibrium constant:

$$\frac{[NH_3]^2}{[N_2][H_2]^3} = \frac{(2.00)^2}{(1.00)(2.00)^3} = 0.500 \quad \text{whereas} \quad K_c = 0.105 \qquad [15.22]$$

To reach equilibrium, the quotient $[NH_3]^2/[N_2][H_2]^3$ must decrease from the starting value of 0.500 to the equilibrium value of 0.105. Because the system is closed, this change can happen only if $[NH_3]$ decreases and $[N_2]$ and $[H_2]$ increase. Thus, the reaction proceeds toward equilibrium by forming N_2 and H_2 from NH_3; that is, the reaction as written in Equation 15.6 proceeds from right to left.

This approach can be formalized by defining a quantity called the reaction quotient. The **reaction quotient**, *Q*, is *a number obtained by substituting reactant and product concentrations or partial pressures at any point during a reaction into an equilibrium-constant expression.* Therefore, for the general reaction

$$a\,A + b\,B \rightleftharpoons d\,D + e\,E$$

the reaction quotient in terms of molar concentrations is

$$Q_c = \frac{[\text{D}]^d[\text{E}]^e}{[\text{A}]^a[\text{B}]^b}$$ [15.23]

(A related quantity Q_p can be written for any reaction that involves gases by using partial pressures instead of concentrations.)

Although we use what looks like the equilibrium-constant expression to calculate the reaction quotient, the concentrations we use may or may not be the equilibrium concentrations. For example, when we substituted the starting concentrations into the equilibrium-constant expression of Equation 15.22, we obtained $Q_c = 0.500$ whereas $K_c = 0.105$. The equilibrium constant has only one value at each temperature. The reaction quotient, however, varies as the reaction proceeds.

Of what use is Q? One practical thing we can do with Q is tell whether our reaction really is at equilibrium, which is an especially valuable option when a reaction is very slow. We can take samples of our reaction mixture as the reaction proceeds, separate the components, and measure their concentrations. Then we insert these numbers into Equation 15.23 for our reaction. To determine whether we are at equilibrium, or in which direction the reaction proceeds to achieve equilibrium, we compare the values of Q_c and K_c or Q_p and K_p. Three possible situations arise:

- $Q < K$: The concentration of products is too small and that of reactants too large. The reaction achieves equilibrium by forming more products; it proceeds from left to right.

- $Q = K$: The reaction quotient equals the equilibrium constant only if the system is at equilibrium.

- $Q > K$: The concentration of products is too large and that of reactants too small. The reaction achieves equilibrium by forming more reactants; it proceeds from right to left.

These relationships are summarized in ▶ Figure 15.8.

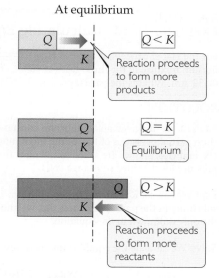

▲ Figure 15.8 **Predicting the direction of a reaction by comparing Q and K at a given temperature.**

SAMPLE EXERCISE 15.9 | Predicting the Direction of Approach to Equilibrium

At 448 °C, the equilibrium constant K_c for the reaction

$$\text{H}_2(g) + \text{I}_2(g) \rightleftharpoons 2\,\text{HI}(g)$$

is 50.5. Predict in which direction the reaction proceeds to reach equilibrium if we start with 2.0×10^{-2} mol of HI, 1.0×10^{-2} mol of H_2, and 3.0×10^{-2} mol of I_2 in a 2.00-L container.

SOLUTION

Analyze We are given a volume and initial molar amounts of the species in a reaction and asked to determine in which direction the reaction must proceed to achieve equilibrium.

Plan We can determine the starting concentration of each species in the reaction mixture. We can then substitute the starting concentrations into the equilibrium-constant expression to calculate the reaction quotient, Q_c. Comparing the magnitudes of the equilibrium constant, which is given, and the reaction quotient will tell us in which direction the reaction will proceed.

Solve The initial concentrations are

$$[\text{HI}] = 2.0 \times 10^{-2}\,\text{mol}/2.00\,\text{L} = 1.0 \times 10^{-2}\,M$$
$$[\text{H}_2] = 1.0 \times 10^{-2}\,\text{mol}/2.00\,\text{L} = 5.0 \times 10^{-3}\,M$$
$$[\text{I}_2] = 3.0 \times 10^{-2}\,\text{mol}/2.00\,\text{L} = 1.5 \times 10^{-2}\,M$$

The reaction quotient is therefore

$$Q_c = \frac{[\text{HI}]^2}{[\text{H}_2][\text{I}_2]} = \frac{(1.0 \times 10^{-2})^2}{(5.0 \times 10^{-3})(1.5 \times 10^{-2})} = 1.3$$

Because $Q_c < K_c$, the concentration of HI must increase and the concentrations of H_2 and I_2 must decrease to reach equilibrium; the reaction as written proceeds left to right to attain equilibrium.

Practice Exercise 1

Which of the following statements accurately describes what would happen to the direction of the reaction described in the sample exercise above, if the size of the container were different from 2.00 L? **(a)** The reaction would proceed in the opposite direction (from right to left) if the container volume were reduced sufficiently. **(b)** The reaction would proceed in the opposite direction if the container volume were expanded sufficiently. **(c)** The direction of this reaction does not depend on the volume of the container.

Practice Exercise 2

At 1000 K, the value of K_p for the reaction $2\,\text{SO}_3(g) \rightleftharpoons 2\,\text{SO}_2(g) + \text{O}_2(g)$ is 0.338. Calculate the value for Q_p, and predict the direction in which the reaction proceeds toward equilibrium if the initial partial pressures are $P_{\text{SO}_3} = 0.16$ atm; $P_{\text{SO}_2} = 0.41$ atm; $P_{\text{O}_2} = 2.5$ atm.

Calculating Equilibrium Concentrations

Chemists frequently need to calculate the amounts of reactants and products present at equilibrium in a reaction for which they know the equilibrium constant. The approach in solving problems of this type is similar to the one we used for evaluating equilibrium constants: We tabulate initial concentrations or partial pressures, changes in those concentrations or pressures, and final equilibrium concentrations or partial pressures. Usually, we end up using the equilibrium-constant expression to derive an equation that must be solved for an unknown quantity, as demonstrated in Sample Exercise 15.10.

SAMPLE EXERCISE 15.10 | Calculating Equilibrium Concentrations

For the Haber process, $N_2(g) + 3H_2(g) \rightleftharpoons 2NH_3(g)$, $K_p = 1.45 \times 10^{-5}$, at 500 °C. In an equilibrium mixture of the three gases at 500 °C, the partial pressure of H_2 is 0.928 atm and that of N_2 is 0.432 atm. What is the partial pressure of NH_3 in this equilibrium mixture?

SOLUTION

Analyze We are given an equilibrium constant, K_p, and the equilibrium partial pressures of two of the three substances in the equation (N_2 and H_2), and we are asked to calculate the equilibrium partial pressure for the third substance (NH_3).

Plan We can set K_p equal to the equilibrium-constant expression and substitute in the partial pressures that we know. Then we can solve for the only unknown in the equation.

Solve We tabulate the equilibrium pressures:

$$N_2(g) + 3H_2(g) \rightleftharpoons 2NH_3(g)$$

Equilibrium pressure (atm)	0.432	0.928	x

Because we do not know the equilibrium pressure of NH_3, we represent it with x. At equilibrium, the pressures must satisfy the equilibrium-constant expression:

$$K_p = \frac{(P_{NH_3})^2}{P_{N_2}(P_{H_2})^3} = \frac{x^2}{(0.432)(0.928)^3} = 1.45 \times 10^{-5}$$

We now rearrange the equation to solve for x:

$$x^2 = (1.45 \times 10^{-5})(0.432)(0.928)^3 = 5.01 \times 10^{-6}$$

$$x = \sqrt{5.01 \times 10^{-6}} = 2.24 \times 10^{-3} \text{ atm} = P_{NH_3}$$

Check We can always check our answer by using it to recalculate the value of the equilibrium constant:

$$K_p = \frac{(2.24 \times 10^{-3})^2}{(0.432)(0.928)^3} = 1.45 \times 10^{-5}$$

Practice Exercise 1

At 500 K, the reaction $2NO(g) + Cl_2(g) \rightleftharpoons 2NOCl(g)$ has $K_p = 51$. In an equilibrium mixture at 500 K, the partial pressure of NO is 0.125 atm and Cl_2 is 0.165 atm. What is the partial pressure of NOCl in the equilibrium mixture? **(a)** 0.13 atm, **(b)** 0.36 atm, **(c)** 1.0 atm, **(d)** 5.1×10^{-5} atm, **(e)** 0.125 atm.

Practice Exercise 2

At 500 K, the reaction $PCl_5(g) \rightleftharpoons PCl_3(g) + Cl_2(g)$ has $K_p = 0.497$. In an equilibrium mixture at 500 K, the partial pressure of PCl_5 is 0.860 atm and that of PCl_3 is 0.350 atm. What is the partial pressure of Cl_2 in the equilibrium mixture?

In many situations, we know the value of the equilibrium constant and the initial amounts of all species. We must then solve for the equilibrium amounts. Solving this type of problem usually entails treating the change in concentration as a variable. The stoichiometry of the reaction gives us the relationship between the changes in the amounts of all the reactants and products, as illustrated in Sample Exercise 15.11. The calculations frequently involve the quadratic formula, as you will see in this exercise.

SAMPLE EXERCISE 15.11 | Calculating Equilibrium Concentrations from Initial Concentrations

A 1.000-L flask is filled with 1.000 mol of $H_2(g)$ and 2.000 mol of $I_2(g)$ at 448 °C. The value of the equilibrium constant K_c for the reaction

$$H_2(g) + I_2(g) \rightleftharpoons 2HI(g)$$

at 448 °C is 50.5. What are the equilibrium concentrations of H_2, I_2, and HI in moles per liter?

SOLUTION

Analyze We are given the volume of a container, an equilibrium constant, and starting amounts of reactants in the container and are asked to calculate the equilibrium concentrations of all species.

Plan In this case, we are not given any of the equilibrium concentrations. We must develop some relationships that relate the initial concentrations to those at equilibrium. The procedure is similar in many regards to that outlined in Sample Exercise 15.8, where we calculated an equilibrium constant using initial concentrations.

Solve

(1) We note the initial concentrations of H_2 and I_2:

$$[H_2] = 1.000\ M \quad \text{and} \quad [I_2] = 2.000\ M$$

(2) We construct a table in which we tabulate the initial concentrations:

	$H_2(g)$ +	$I_2(g)$ $\rightleftharpoons$	2 HI(g)
Initial concentration (M)	1.000	2.000	0
Change in concentration (M)			
Equilibrium concentration (M)			

(3) We use the stoichiometry of the reaction to determine the changes in concentration that occur as the reaction proceeds to equilibrium. The H_2 and I_2 concentrations will decrease as equilibrium is established and that of HI will increase. Let's represent the change in concentration of H_2 by x. The balanced chemical equation tells us the relationship between the changes in the concentrations of the three gases. For each x mol of H_2 that reacts, x mol of I_2 are consumed and $2x$ mol of HI are produced:

	$H_2(g)$ +	$I_2(g)$ $\rightleftharpoons$	2 HI(g)
Initial concentration (M)	1.000	2.000	0
Change in concentration (M)	$-x$	$-x$	$+2x$
Equilibrium concentration (M)			

(4) We use initial concentrations and changes in concentrations, as dictated by stoichiometry, to express the equilibrium concentrations. With all our entries, our table now looks like this:

	$H_2(g)$ +	$I_2(g)$ $\rightleftharpoons$	2 HI(g)
Initial concentration (M)	1.000	2.000	0
Change in concentration (M)	$-x$	$-x$	$+2x$
Equilibrium concentration (M)	$1.000 - x$	$2.000 - x$	$2x$

(5) We substitute the equilibrium concentrations into the equilibrium-constant expression and solve for x:

$$K_c = \frac{[HI]^2}{[H_2][I_2]} = \frac{(2x)^2}{(1.000 - x)(2.000 - x)} = 50.5$$

If you have an equation-solving calculator, you can solve this equation directly for x. If not, expand this expression to obtain a quadratic equation in x:

$$4x^2 = 50.5(x^2 - 3.000x + 2.000)$$
$$46.5x^2 - 151.5x + 101.0 = 0$$

Solving the quadratic equation (Appendix A.3) leads to two solutions for x:

$$x = \frac{-(-151.5) \pm \sqrt{(-151.5)^2 - 4(46.5)(101.0)}}{2(46.5)} = 2.323 \text{ or } 0.935$$

When we substitute $x = 2.323$ into the expressions for the equilibrium concentrations, we find *negative* concentrations of H_2 and I_2. Because a negative concentration is not chemically meaningful, we reject this solution. We then use $x = 0.935$ to find the equilibrium concentrations:

$$[H_2] = 1.000 - x = 0.065\ M$$
$$[I_2] = 2.000 - x = 1.065\ M$$
$$[HI] = 2x = 1.87\ M$$

Check We can check our solution by putting these numbers into the equilibrium-constant expression to assure that we correctly calculate the equilibrium constant:

$$K_c = \frac{[HI]^2}{[H_2][I_2]} = \frac{(1.87)^2}{(0.065)(1.065)} = 51$$

Comment Whenever you use a quadratic equation to solve an equilibrium problem, one of the solutions to the equation will give you a value that leads to negative concentrations and thus is not chemically meaningful. Reject this solution to the quadratic equation.

15.7 | Le Châtelier's Principle

Many of the products we use in everyday life are obtained from the chemical industry. Chemists and chemical engineers in industry spend a great deal of time and
effort to maximize the yield of valuable products and minimize waste. For example,
when Haber developed his process for making ammonia from N_2 and H_2, he examined how reaction conditions might be varied to increase yield. Using the values
of the equilibrium constant at various temperatures, he calculated the equilibrium
amounts of NH_3 formed under a variety of conditions. Some of Haber's results are
shown in ▼ Figure 15.9.

Notice that the percent of NH_3 present at equilibrium decreases with increasing
temperature and increases with increasing pressure.

We can understand these effects in terms of a principle first put forward by
Henri-Louis Le Châtelier* (1850–1936), a French industrial chemist: *If a system at
equilibrium is disturbed by a change in temperature, pressure, or a component concentration, the system will shift its equilibrium position so as to counteract the effect of the
disturbance.*

▲ GO FIGURE

At what combination of pressure and temperature should you run the reaction to maximize NH_3 yield?

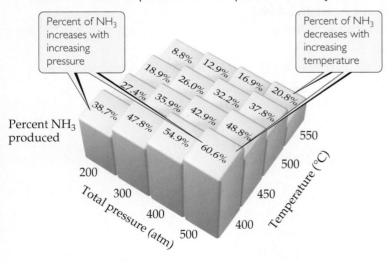

▲ Figure 15.9 **Effect of temperature and pressure on NH₃ yield in the Haber process.** Each
mixture was produced by starting with a 3 : 1 molar mixture of H_2 and N_2.

---

*Pronounced "le-SHOT-lee-ay."

Le Châtelier's Principle

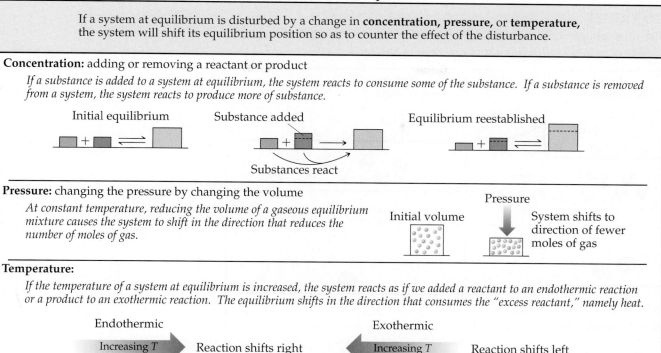

If a system at equilibrium is disturbed by a change in **concentration**, **pressure**, or **temperature**, the system will shift its equilibrium position so as to counter the effect of the disturbance.

Concentration: adding or removing a reactant or product

If a substance is added to a system at equilibrium, the system reacts to consume some of the substance. If a substance is removed from a system, the system reacts to produce more of substance.

Initial equilibrium   Substance added   Equilibrium reestablished

Substances react

Pressure: changing the pressure by changing the volume

At constant temperature, reducing the volume of a gaseous equilibrium mixture causes the system to shift in the direction that reduces the number of moles of gas.

Pressure

Initial volume   System shifts to direction of fewer moles of gas

Temperature:

If the temperature of a system at equilibrium is increased, the system reacts as if we added a reactant to an endothermic reaction or a product to an exothermic reaction. The equilibrium shifts in the direction that consumes the "excess reactant," namely heat.

Endothermic   Exothermic

Increasing T Reaction shifts right   Increasing T Reaction shifts left

Decreasing T Reaction shifts left   Decreasing T Reaction shifts right

In this section, we use Le Châtelier's principle to make qualitative predictions about how a system at equilibrium responds to various changes in external conditions. We consider three ways in which a chemical equilibrium can be disturbed: (1) adding or removing a reactant or product, (2) changing the pressure by changing the volume, and (3) changing the temperature.

Change in Reactant or Product Concentration

A system at dynamic equilibrium is in a state of balance. When the concentrations of species in the reaction are altered, the equilibrium shifts until a new state of balance is attained. What does *shift* mean? It means that reactant and product concentrations change over time to accommodate the new situation. *Shift* does *not* mean that the equilibrium constant itself is altered; the equilibrium constant remains the same. Le Châtelier's principle states that the shift is in the direction that minimizes or reduces the effect of the change. Therefore, *if a chemical system is already at equilibrium and the concentration of any substance in the mixture is increased (either reactant or product), the system reacts to consume some of that substance. Conversely, if the concentration of a substance is decreased, the system reacts to produce some of that substance.*

There is no change in the equilibrium constant when we change the concentrations of reactants or products. As an example, consider our familiar equilibrium mixture of N_2, H_2, and NH_3:

$$N_2(g) + 3H_2(g) \rightleftharpoons 2NH_3(g)$$

▲ GO FIGURE

Why does the nitrogen concentration decrease after hydrogen is added?

$$N_2(g) + 3\,H_2(g) \rightleftharpoons 2\,NH_3(g)$$

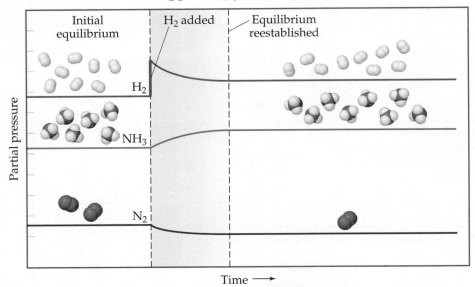

▲ **Figure 15.10 Effect of adding H_2 to an equilibrium mixture of N_2, H_2, and NH_3.** Adding H_2 causes the reaction as written to shift to the right, consuming some N_2 to produce more NH_3.

Adding H_2 causes the system to shift so as to reduce the increased concentration of H_2 (▲ **Figure 15.10**). This change can occur only if the reaction consumes H_2 and simultaneously consumes N_2 to form more NH_3. Adding N_2 to the equilibrium mixture likewise causes the reaction to shift toward forming more NH_3. Removing NH_3 also causes a shift toward producing more NH_3, whereas *adding* NH_3 to the system at equilibrium causes the reaction to shift in the direction that reduces the increased NH_3 concentration: Some of the added ammonia decomposes to form N_2 and H_2. All of these "shifts" are entirely consistent with predictions that we would make by comparing the reaction quotient Q with the equilibrium constant K.

In the Haber reaction, therefore, removing NH_3 from an equilibrium mixture of N_2, H_2, and NH_3 causes the reaction to shift right to form more NH_3. If the NH_3 can be removed continuously as it is produced, the yield can be increased dramatically. In the industrial production of ammonia, the NH_3 is continuously removed by selectively liquefying it (▶ **Figure 15.11**). (The boiling point of NH_3, $-33\,°C$, is much higher than those of N_2, $-196\,°C$, and H_2, $-253\,°C$.) The liquid NH_3 is removed, and the N_2 and H_2 are recycled to form more NH_3. As a result of the product being continuously removed, the reaction is driven essentially to completion.

Give It Some Thought

Does the equilibrium $2\,NO(g) + O_2(g) \rightleftharpoons 2\,NO_2(g)$ shift to the right (more products) or left (more reactants) if

(a) O_2 is added to the system?

(b) NO is removed?

Effects of Volume and Pressure Changes

If a system containing one or more gases is at equilibrium and its volume is decreased, thereby increasing its total pressure, Le Châtelier's principle indicates that the system responds by shifting its equilibrium position to reduce the pressure. A system can reduce its pressure by reducing the total number of gas molecules (fewer molecules of gas

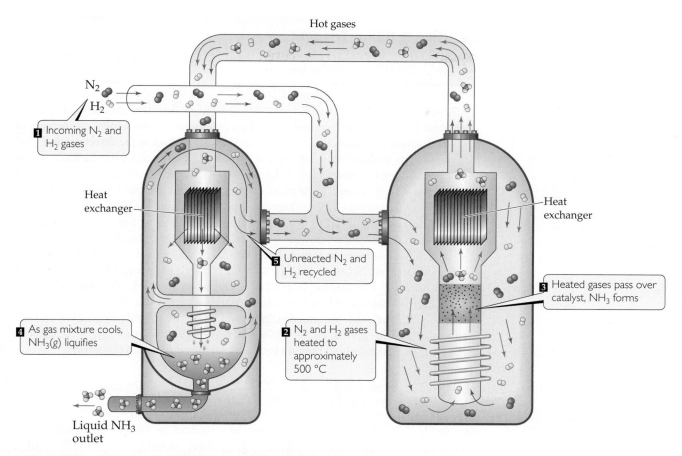

▲ **Figure 15.11 Diagram of the industrial production of ammonia.** Incoming $N_2(g)$ and $H_2(g)$ are heated to approximately 500 °C and passed over a catalyst. When the resultant N_2, H_2, and NH_3 mixture is cooled, the NH_3 liquefies and is removed from the mixture, shifting the reaction to produce more NH_3.

exert a lower pressure). Thus, at constant temperature, *reducing the volume of a gaseous equilibrium mixture causes the system to shift in the direction that reduces the number of moles of gas.* Increasing the volume causes a shift in the direction that produces more gas molecules (▼ Figure 15.12).

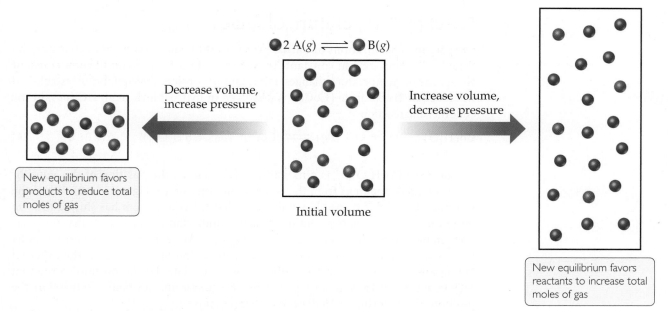

▲ **Figure 15.12 Pressure and Le Châtelier's principle.**

> ### ▲ Give It Some Thought
>
> What happens to the equilibrium $2\,SO_2(g) + O_2(g) \rightleftharpoons 2\,SO_3(g)$, if the volume of the system is increased?

In the reaction $N_2(g) + 3\,H_2(g) \rightleftharpoons 2\,NH_3(g)$, four molecules of reactant are consumed for every two molecules of product produced. Consequently, an increase in pressure (caused by a decrease in volume) shifts the reaction in the direction that produces fewer gas molecules, which leads to the formation of more NH_3, as indicated in Figure 15.9. In the reaction $H_2(g) + I_2(g) \rightleftharpoons 2\,HI(g)$, the number of molecules of gaseous products (two) equals the number of molecules of gaseous reactants; therefore, changing the pressure does not influence the position of equilibrium.

Keep in mind that, as long as temperature remains constant, pressure–volume changes do *not* change the value of K. Rather, these changes alter the partial pressures of the gaseous substances. In Sample Exercise 15.7, we calculated $K_p = 2.79 \times 10^{-5}$ for the Haber reaction, $N_2(g) + 3\,H_2(g) \rightleftharpoons 2\,NH_3(g)$, in an equilibrium mixture at 472 °C containing 7.38 atm H_2, 2.46 atm N_2, and 0.166 atm NH_3. Consider what happens when we suddenly reduce the volume of the system by one-half. If there were no shift in equilibrium, this volume change would cause the partial pressures of all substances to double, giving $P_{H_2} = 14.76$ atm, $P_{N_2} = 4.92$ atm, and $P_{NH_3} = 0.332$ atm. The reaction quotient would then no longer equal the equilibrium constant:

$$Q_p = \frac{(P_{NH_3})^2}{P_{N_2}(P_{H_2})^3} = \frac{(0.332)^2}{(4.92)(14.76)^3} = 6.97 \times 10^{-6} \neq K_p$$

Because $Q_p < K_p$, the system would no longer be at equilibrium. Equilibrium would be reestablished by increasing P_{NH_3} and/or decreasing P_{N_2} and P_{H_2} until $Q_p = K_p = 2.79 \times 10^{-5}$. Therefore, the equilibrium shifts to the right in the reaction as written, as Le Châtelier's principle predicts.

It is possible to change the pressure of a system in which a chemical reaction is running without changing its volume. For example, pressure increases if additional amounts of any reacting components are added to the system. We have already seen how to deal with a change in concentration of a reactant or product. However, the *total* pressure in the reaction vessel might also be increased by adding a gas that is not involved in the equilibrium. For example, argon might be added to the ammonia equilibrium system. The argon would not alter the *partial* pressures of any of the reacting components and therefore would not cause a shift in equilibrium.

Effect of Temperature Changes

Changes in concentrations or partial pressures shift equilibria without changing the value of the equilibrium constant. In contrast, almost every equilibrium constant changes as the temperature changes. For example, consider the equilibrium established when cobalt(II) chloride $(CoCl_2)$ is dissolved in hydrochloric acid, $HCl(aq)$, in the endothermic reaction

$$\underset{\text{Pale pink}}{Co(H_2O)_6^{2+}(aq)} + 4\,Cl^-(aq) \rightleftharpoons \underset{\text{Deep blue}}{CoCl_4^{2-}(aq)} + 6\,H_2O(l) \qquad \Delta H > 0 \qquad [15.24]$$

Because $Co(H_2O)_6^{2+}$ is pink and $CoCl_4^{2-}$ is blue, the position of this equilibrium is readily apparent from the color of the solution (▶ Figure 15.13). When the solution is heated it turns blue, indicating that the equilibrium has shifted to form more $CoCl_4^{2-}$. Cooling the solution leads to a pink solution, indicating that the equilibrium has shifted to produce more $Co(H_2O)_6^{2+}$. We can monitor this reaction by spectroscopic methods, measuring the concentration of all species at the different temperatures. ∞ (Section 14.2) We can then calculate the equilibrium constant at each temperature. How do we explain why the equilibrium constants and therefore the position of equilibrium both depend on temperature?

$\Delta H > 0$, endothermic reaction

$$\text{Heat} + \text{Co(H}_2\text{O)}_6{}^{2+}(aq) + 4\,\text{Cl}^-(aq) \rightleftharpoons \text{CoCl}_4{}^{2-}(aq) + 6\,\text{H}_2\text{O}(l)$$

Pink   Blue

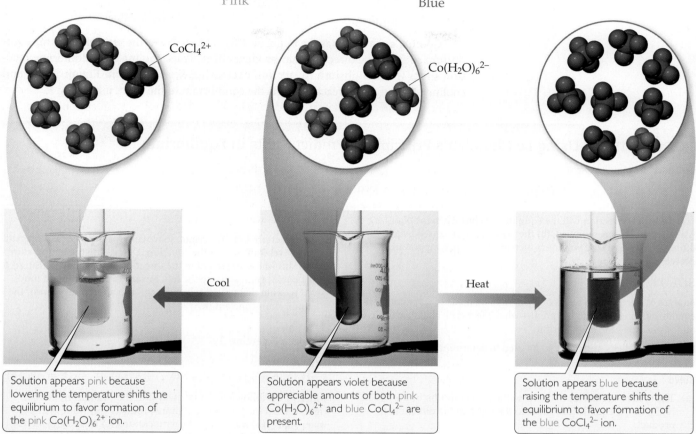

Cool   Heat

Solution appears pink because lowering the temperature shifts the equilibrium to favor formation of the pink $\text{Co(H}_2\text{O)}_6{}^{2+}$ ion.

Solution appears violet because appreciable amounts of both pink $\text{Co(H}_2\text{O)}_6{}^{2+}$ and blue $\text{CoCl}_4{}^{2-}$ are present.

Solution appears blue because raising the temperature shifts the equilibrium to favor formation of the blue $\text{CoCl}_4{}^{2-}$ ion.

▲ **Figure 15.13 Temperature and Le Châtelier's principle.** In the molecular level views, only the $\text{CoCl}_4{}^{2-}$ and $\text{Co(H}_2\text{O)}_6{}^{2+}$ ions are shown for clarity.

We can deduce the rules for the relationship between K and temperature from Le Châtelier's principle. We do this by treating heat as a chemical reagent. In an *endothermic* (heat-absorbing) reaction, we consider heat a *reactant*, and in an *exothermic* (heat-releasing) reaction, we consider heat a *product*:

    Endothermic: Reactants + *heat* $\rightleftharpoons$ products

    Exothermic: Reactants $\rightleftharpoons$ products + *heat*

When the temperature of a system at equilibrium is increased, the system reacts as if we added a reactant to an endothermic reaction or a product to an exothermic reaction. The equilibrium shifts in the direction that consumes the excess reactant (or product), namely heat.

 Give It Some Thought

Use Le Châtelier's principle to explain why the equilibrium vapor pressure of a liquid increases with increasing temperature.

In an endothermic reaction, such as Equation 15.24, heat is absorbed as reactants are converted to products. Thus, increasing the temperature causes the equilibrium to shift to the right, in the direction of making more products, and K increases. In an exothermic reaction, the opposite occurs: Heat is produced as reactants are converted to products. Thus, increasing the temperature in this case causes the equilibrium to shift to the left, in the direction of making more reactants, and K decreases.

Endothermic: Increasing T results in higher K value

Exothermic: Increasing T results in lower K value

Cooling a reaction has the opposite effect. As we lower the temperature, the equilibrium shifts in the direction that produces heat. Thus, cooling an endothermic reaction shifts the equilibrium to the left, decreasing K, as shown in Figure 15.13, and cooling an exothermic reaction shifts the equilibrium to the right, increasing K.

SAMPLE EXERCISE 15.12 Using Le Châtelier's Principle to Predict Shifts in Equilibrium

Consider the equilibrium

$$N_2O_4(g) \rightleftharpoons 2 NO_2(g) \qquad \Delta H^\circ = 58.0 \text{ kJ}$$

In which direction will the equilibrium shift when **(a)** N_2O_4 is added, **(b)** NO_2 is removed, **(c)** the pressure is increased by addition of $N_2(g)$, **(d)** the volume is increased, **(e)** the temperature is decreased?

SOLUTION

Analyze We are given a series of changes to be made to a system at equilibrium and are asked to predict what effect each change will have on the position of the equilibrium.

Plan Le Châtelier's principle can be used to determine the effects of each of these changes.

Solve

(a) The system will adjust to decrease the concentration of the added N_2O_4, so the equilibrium shifts to the right, in the direction of product.

(b) The system will adjust to the removal of NO_2 by shifting to the side that produces more NO_2; thus, the equilibrium shifts to the right.

(c) Adding N_2 will increase the total pressure of the system, but N_2 is not involved in the reaction. The partial pressures of NO_2 and N_2O_4 are therefore unchanged, and there is no shift in the position of the equilibrium.

(d) If the volume is increased, the system will shift in the direction that occupies a larger volume (more gas molecules); thus, the equilibrium shifts to the right.

(e) The reaction is endothermic, so we can imagine heat as a reagent on the reactant side of the equation. Decreasing the temperature will shift the equilibrium in the direction that produces heat, so the equilibrium shifts to the left, toward the formation of more N_2O_4. Note that only this last change also affects the value of the equilibrium constant, K.

Practice Exercise 1

For the reaction

$$4 NH_3(g) + 5 O_2(g) \rightleftharpoons 4 NO(g) + 6 H_2O(g) \qquad \Delta H^\circ = -904 \text{ kJ}$$

which of the following changes will shift the equilibrium to the right, toward the formation of more products? **(a)** Adding more water vapor, **(b)** Increasing the temperature, **(c)** Increasing the volume of the reaction vessel, **(d)** Removing $O_2(g)$, **(e)** Adding 1 atm of $Ne(g)$ to the reaction vessel.

Practice Exercise 2

For the reaction

$$PCl_5(g) \rightleftharpoons PCl_3(g) + Cl_2(g) \qquad \Delta H^\circ = 87.9 \text{ kJ}$$

in which direction will the equilibrium shift when **(a)** $Cl_2(g)$ is removed, **(b)** the temperature is decreased, **(c)** the volume of the reaction system is increased, **(d)** $PCl_3(g)$ is added?

SAMPLE EXERCISE 15.13 Predicting the Effect of Temperature on K

(a) Using the standard heat of formation data in Appendix C, determine the standard enthalpy change for the reaction

$$N_2(g) + 3 H_2(g) \rightleftharpoons 2 NH_3(g)$$

(b) Determine how the equilibrium constant for this reaction should change with temperature.

SOLUTION

Analyze We are asked to determine the standard enthalpy change of a reaction and how the equilibrium constant for the reaction varies with temperature.

Plan

(a) We can use standard enthalpies of formation to calculate ΔH° for the reaction.

(b) We can then use Le Châtelier's principle to determine what effect temperature will have on the equilibrium constant.

Solve

(a) Recall that the standard enthalpy change for a reaction is given by the sum of the standard molar enthalpies of formation of the products, each multiplied by its coefficient in the balanced chemical equation, minus the same quantities for the reactants. ⚭ (Section 5.7) At 25 °C, ΔH_f° for $NH_3(g)$ is -46.19 kJ/mol. The ΔH_f° values for $H_2(g)$ and $N_2(g)$ are zero by definition because the enthalpies of formation of the elements in their normal states at 25 °C are defined as zero. ∞ (Section 5.7) Because 2 mol of NH_3 is formed, the total enthalpy change is

$$(2\ mol)(-46.19\ kJ/mol) - 0 = -92.38\ kJ$$

(b) Because the reaction in the forward direction is exothermic, we can consider heat a product of the reaction. An increase in temperature causes the reaction to shift in the direction of less NH_3 and more N_2 and H_2. This effect is seen in the values for K_p presented in ▼ Table 15.2. Notice that K_p changes markedly with changes in temperature and that it is larger at lower temperatures.

Comment The fact that K_p for the formation of NH_3 from N_2 and H_2 decreases with increasing temperature is a matter of great practical importance. To form NH_3 at a reasonable rate requires higher temperatures. At higher temperatures, however, the equilibrium constant is smaller, and so the percentage conversion to NH_3 is smaller. To compensate for this, higher pressures are needed because high pressure favors NH_3 formation.

Table 15.2 **Variation in K_p with Temperature for $N_2 + 3H_2 \rightleftharpoons 2\,NH_3$**

Temperature (°C)	K_p
300	4.34×10^{-3}
400	1.64×10^{-4}
450	4.51×10^{-5}
500	1.45×10^{-5}
550	5.38×10^{-6}
600	2.25×10^{-6}

Practice Exercise 1

The standard enthalpy of formation of $HCl(g)$ is -92.3 kJ/mol. Given only this information, in which direction would you expect the equilibrium for the reaction $H_2(g) + Cl_2(g) \rightleftharpoons 2\,HCl(g)$ to shift as the temperature increases: **(a)** to the left, **(b)** to the right, **(c)** no shift in equilibrium?

Practice Exercise 2

Using the thermodynamic data in Appendix C, determine the enthalpy change for the reaction

$$2\,POCl_3(g) \rightleftharpoons 2\,PCl_3(g) + O_2(g)$$

Use this result to determine how the equilibrium constant for the reaction should change with temperature.

The Effect of Catalysts

What happens if we add a catalyst to a chemical system that is at equilibrium? As shown in ▼ Figure 15.14, a catalyst lowers the activation barrier between reactants and products. ∞ (Section 14.7) The activation energies for both the forward and reverse reactions are lowered. The catalyst thereby increases the rates of both forward and reverse reactions. Since K is the ratio of the forward and reverse rate constants for a reaction, you can predict, correctly, that the presence of a catalyst, even though it changes the reaction *rate*, does not affect the numeric value of K (Figure 15.14). As a result, *a catalyst increases the rate at which equilibrium is achieved but does not change the composition of the equilibrium mixture.*

The rate at which a reaction approaches equilibrium is an important practical consideration. As an example, let's again consider the synthesis of ammonia from

 GO FIGURE

What quantity dictates the speed of a reaction: **(a)** the energy difference between the initial state and the transition state or **(b)** the energy difference between the initial state and the final state?

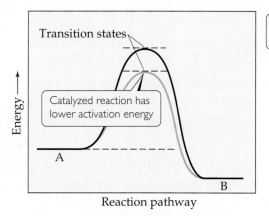 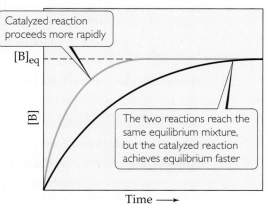

▲ **Figure 15.14** An energy profile for the reaction **A $\rightleftharpoons$ B (left), and the change in concentration of B as a function of time (right), with and without a catalyst.** Green curves show the reaction with a catalyst; black curves show the reaction without a catalyst.

N_2 and H_2. In designing his process, Haber had to deal with a rapid decrease in the equilibrium constant with increasing temperature (Table 15.2). At temperatures sufficiently high to give a satisfactory reaction rate, the amount of ammonia formed was too small. The solution to this dilemma was to develop a catalyst that would produce a reasonably rapid approach to equilibrium at a sufficiently low temperature, so that the equilibrium constant remained reasonably large. The development of a suitable catalyst thus became the focus of Haber's research efforts.

After trying different substances to see which would be most effective, Carl Bosch settled on iron mixed with metal oxides, and variants of this catalyst formulation are still used today. ∞ (Section 15.2, "The Haber Process") These catalysts make it possible to obtain a reasonably rapid approach to equilibrium at around 400 to 500 °C and 200 to 600 atm. The high pressures are needed to obtain a satisfactory equilibrium amount of NH_3. If a catalyst could be found that leads to sufficiently rapid reaction at temperatures lower than 400 °C, it would be possible to obtain the same extent of equilibrium conversion at pressures much lower than 200 to 600 atm. This would result in great savings in both the cost of the high-pressure equipment and the energy consumed in the production of ammonia. It is estimated that the Haber process consumes approximately 1% of the energy generated in the world each year. Not surprisingly chemists and chemical engineers are actively searching for improved catalysts for the Haber process. A breakthrough in this field would not only increase the supply of ammonia for fertilizers, it would also reduce the global consumption of fossil fuels in a significant way.

▲ Give It Some Thought

Can a catalyst be used to increase the amount of product produced for a reaction that reaches equilibrium quickly in the absence of a catalyst?

SAMPLE
INTEGRATIVE EXERCISE | **Putting Concepts Together**

At temperatures near 800 °C, steam passed over hot coke (a form of carbon obtained from coal) reacts to form CO and H_2:

$$C(s) + H_2O(g) \rightleftharpoons CO(g) + H_2(g)$$

The mixture of gases that results is an important industrial fuel called *water gas*. **(a)** At 800 °C the equilibrium constant for this reaction is $K_p = 14.1$. What are the equilibrium partial pressures of H_2O, CO, and H_2 in the equilibrium mixture at this temperature if we start with solid carbon and 0.100 mol of H_2O in a 1.00-L vessel? **(b)** What is the minimum amount of carbon required to achieve equilibrium under these conditions? **(c)** What is the total pressure in the vessel at equilibrium? **(d)** At 25 °C the value of K_p for this reaction is 1.7×10^{-21}. Is the reaction exothermic or endothermic? **(e)** To produce the maximum amount of CO and H_2 at equilibrium, should the pressure of the system be increased or decreased?

SOLUTION

(a) To determine the equilibrium partial pressures, we use the ideal-gas equation, first determining the starting partial pressure of water.

$$P_{H_2O} = \frac{n_{H_2O}RT}{V} = \frac{(0.100 \text{ mol})(0.08206 \text{ L-atm/mol-K})(1073 \text{ K})}{1.00 \text{ L}} = 8.81 \text{ atm}$$

We then construct a table of initial partial pressures and their changes as equilibrium is achieved:

	C(s)	+	$H_2O(g)$	$\rightleftharpoons$	CO(g)	+	$H_2(g)$
Initial partial pressure (atm)			8.81		0		0
Change in partial pressure (atm)			$-x$		$+x$		$+x$
Equilibrium partial pressure (atm)			$8.81 - x$		x		x

There are no entries in the table under C(s) because the reactant, being a solid, does not appear in the equilibrium-constant expression. Substituting the equilibrium partial pressures of the other species into the equilibrium-constant expression for the reaction gives

$$K_p = \frac{P_{CO}P_{H_2}}{P_{H_2O}} = \frac{(x)(x)}{(8.81 - x)} = 14.1$$

Multiplying through by the denominator gives a quadratic equation in x:

$$x^2 = (14.1)(8.81 - x)$$
$$x^2 + 14.1x - 124.22 = 0$$

Solving this equation for x using the quadratic formula yields $x = 6.14$ atm. Hence, the equilibrium partial pressures are $P_{CO} = x = 6.14$ atm, $P_{H_2} = x = 6.14$ atm, and $P_{H_2O} = (8.81 - x) = 2.67$ atm.

(b) Part (a) shows that $x = 6.14$ atm of H_2O must react for the system to achieve equilibrium. We can use the ideal-gas equation to convert this partial pressure into a mole amount.

$$n = \frac{PV}{RT} = \frac{(6.14 \text{ atm})(1.00 \text{ L})}{(0.08206 \text{ L-atm/mol-K})(1073 \text{ K})} = 0.0697 \text{ mol}$$

Thus, 0.0697 mol of H_2O and the same amount of C must react to achieve equilibrium. As a result, there must be at least 0.0697 mol of C (0.836 g C) present among the reactants at the start of the reaction.

(c) The total pressure in the vessel at equilibrium is simply the sum of the equilibrium partial pressures:

$$P_{total} = P_{H_2O} + P_{CO} + P_{H_2} = 2.67 \text{ atm} + 6.14 \text{ atm} + 6.14 \text{ atm} = 14.95 \text{ atm}$$

(d) In discussing Le Châtelier's principle, we saw that endothermic reactions exhibit an increase in K_p with increasing temperature. Because the equilibrium constant for this reaction increases as temperature increases, the reaction must be endothermic. From the enthalpies of formation given in Appendix C, we can verify our prediction by calculating the enthalpy change for the reaction,

$$\Delta H^\circ = \Delta H_f^\circ(CO(g)) + \Delta H_f^\circ(H_2(g)) - \Delta H_f^\circ(C(s, \text{graphite})) - \Delta H_f^\circ(H_2O(g)) = +131.3 \text{ kJ}$$

The positive sign for ΔH° indicates that the reaction is endothermic.

(e) According to Le Châtelier's principle, a decrease in the pressure causes a gaseous equilibrium to shift toward the side of the equation with the greater number of moles of gas. In this case, there are 2 mol of gas on the product side and only one on the reactant side. Therefore, the pressure should be decreased to maximize the yield of the CO and H_2.

Chemistry Put to Work

Controlling Nitric Oxide Emissions

The formation of NO from N_2 and O_2,

$$\tfrac{1}{2} N_2(g) + \tfrac{1}{2} O_2(g) \rightleftharpoons NO(g) \qquad \Delta H^\circ = 90.4 \text{ kJ} \qquad [15.25]$$

provides an interesting example of the practical importance of the fact that equilibrium constants and reaction rates change with temperature. By applying Le Châtelier's principle to this endothermic reaction and treating heat as a reactant, we deduce that an increase in temperature shifts the equilibrium in the direction of more NO. The equilibrium constant K_p for formation of 1 mol of NO from its elements at 300 K is only about 1×10^{-15} (▶ Figure 15.15). At 2400 K, however, the equilibrium constant is about 0.05, which is 10^{13} times larger than the 300 K value.

Figure 15.15 helps explain why NO is a pollution problem. In the cylinder of a modern high-compression automobile engine, the temperature during the fuel-burning part of the cycle is approximately 2400 K. Also, there is a fairly large excess of air in the cylinder. These conditions favor the formation of NO. After combustion, however, the gases cool quickly. As the temperature drops, the equilibrium in Equation 15.25 shifts to the left (because the reactant heat is being removed). However, the lower temperature also means that the reaction rate decreases, so the NO formed at 2400 K is essentially "trapped" in that form as the gas cools.

The gases exhausting from the cylinder are still quite hot, perhaps 1200 K. At this temperature, as shown in Figure 15.15, the equilibrium constant for formation of NO is about 5×10^{-4}, much smaller than the

 GO FIGURE

Estimate the value of K_p at 1200 K, the exhaust gas temperature.

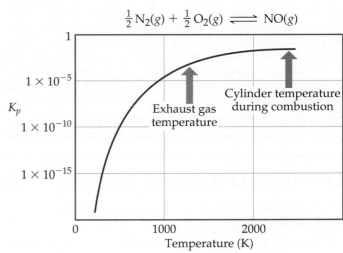

▲ Figure 15.15 **Equilibrium and temperature.** The equilibrium constant increases with increasing temperature because the reaction is endothermic. It is necessary to use a log scale for K_p because the values vary over such a large range.

value at 2400 K. However, the rate of conversion of NO to N_2 and O_2 is too slow to permit much loss of NO before the gases are cooled further.

As discussed in the "Chemistry Put to Work" box in Section 14.7, one of the goals of automotive catalytic converters is to achieve rapid conversion of NO to N_2 and O_2 at the temperature of the exhaust gas.

Some catalysts developed for this reaction are reasonably effective under the grueling conditions in automotive exhaust systems. Nevertheless, scientists and engineers are continuously searching for new materials that provide even more effective catalysis of the decomposition of nitrogen oxides.

Chapter Summary and Key Terms

THE CONCEPT OF EQUILIBRIUM (SECTION 15.1) A chemical reaction can achieve a state in which the forward and reverse processes are occurring at the same rate. This condition is called **chemical equilibrium**, and it results in the formation of an equilibrium mixture of the reactants and products of the reaction. The composition of an equilibrium mixture does not change with time if temperature is held constant.

THE EQUILIBRIUM CONSTANT (SECTION 15.2) An equilibrium that is used throughout this chapter is the reaction $N_2(g)$ + $3 H_2(g) \rightleftharpoons 2 NH_3(g)$. This reaction is the basis of the **Haber process** for the production of ammonia. The relationship between the concentrations of the reactants and products of a system at equilibrium is given by the **law of mass action**. For an equilibrium equation of the form $a A + b B \rightleftharpoons d D + e E$, the **equilibrium-constant expression** is written as

$$K_c = \frac{[D]^d[E]^e}{[A]^a[B]^b}$$

where K_c is a dimensionless constant called the **equilibrium constant**. When the equilibrium system of interest consists of gases, it is often convenient to express the concentrations of reactants and products in terms of gas pressures:

$$K_p = \frac{(P_D)^d(P_E)^e}{(P_A)^a(P_B)^b}$$

K_c and K_p are related by the expression $K_p = K_c(RT)^{\Delta n}$. To do this conversion properly, use $R = 0.08206$ L-atm/mol-K and temperature in kelvins.

UNDERSTANDING AND WORKING WITH EQUILIBRIUM CONSTANTS (SECTION 15.3) The value of the equilibrium constant changes with temperature. A large value of K_c indicates that the equilibrium mixture contains more products than reactants and therefore lies toward the product side of the equation. A small value for the equilibrium constant means that the equilibrium mixture contains less products than reactants and therefore lies toward the reactant side. The equilibrium-constant expression and the equilibrium constant of the reverse of a reaction are the reciprocals of those of the forward reaction. If a reaction is the sum of two or more reactions, its equilibrium constant will be the product of the equilibrium constants for the individual reactions.

HETEROGENEOUS EQUILIBRIA (SECTION 15.4) Equilibria for which all substances are in the same phase are called **homogeneous**

equilibria; in **heterogeneous equilibria**, two or more phases are present. Because their activities are exactly 1 the concentrations of pure solids and liquids are left out of the equilibrium-constant expression for a heterogeneous equilibrium.

CALCULATING EQUILIBRIUM CONSTANTS (SECTION 15.5) If the concentrations of all species in an equilibrium are known, the equilibrium-constant expression can be used to calculate the equilibrium constant. The changes in the concentrations of reactants and products on the way to achieving equilibrium are governed by the stoichiometry of the reaction.

APPLICATIONS OF EQUILIBRIUM CONSTANTS (SECTION 15.6) The **reaction quotient**, Q, is found by substituting reactant and product concentrations or partial pressures at any point during a reaction into the equilibrium-constant expression. If the system is at equilibrium, $Q = K$. If $Q \neq K$, however, the system is not at equilibrium. When $Q < K$, the reaction will move toward equilibrium by forming more products (the reaction proceeds from left to right); when $Q > K$, the reaction will move toward equilibrium by forming more reactants (the reaction proceeds from right to left). Knowing the value of K makes it possible to calculate the equilibrium amounts of reactants and products, often by the solution of an equation in which the unknown is the change in a partial pressure or concentration.

LE CHÂTELIER'S PRINCIPLE (SECTION 15.7) Le Châtelier's principle states that if a system at equilibrium is disturbed, the equilibrium will shift to minimize the disturbing influence. Therefore, if a reactant or product is added to a system at equilibrium, the equilibrium will shift to consume the added substance. The effects of removing reactants or products and of changing the pressure or volume of a reaction can be similarly deduced. For example, if the volume of the system is reduced, the equilibrium will shift in the direction that decreases the number of gas molecules. While changes in concentration or pressure lead to shifts in the equilibrium concentrations they do not change the value of the equilibrium constant, K.

Changes in temperature affect both the equilibrium concentrations and the equilibrium constant. We can use the enthalpy change for a reaction to determine how an increase in temperature affects the equilibrium: For an endothermic reaction, an increase in temperature shifts the equilibrium to the right; for an exothermic reaction, a temperature increase shifts the equilibrium to the left. Catalysts affect the speed at which equilibrium is reached but do not affect the magnitude of K.

Learning Outcomes   After studying this chapter, you should be able to:

- Explain what is meant by chemical equilibrium and how it relates to reaction rates. (Section 15.1)

- Write the equilibrium-constant expression for any reaction. (Section 15.2)

- Given the value of K_c convert to K_p and vice versa. (Section 15.2)

- Relate the magnitude of an equilibrium constant to the relative amounts of reactants and products present in an equilibrium mixture. (Section 15.3)

- Manipulate the equilibrium constant to reflect changes in the chemical equation. (Section 15.3)

- Write the equilibrium-constant expression for a heterogeneous reaction. (Section 15.4)
- Calculate an equilibrium constant from concentration measurements. (Section 15.5)
- Predict the direction of a reaction given the equilibrium constant and the concentrations of reactants and products. (Section 15.6)

- Calculate equilibrium concentrations given the equilibrium constant and all but one equilibrium concentration. (Section 15.6)
- Calculate equilibrium concentrations, given the equilibrium constant and the starting concentrations. (Section 15.6)
- Use Le Châtelier's principle to predict how changing the concentrations, volume, or temperature of a system at equilibrium affects the equilibrium position. (Section 15.7)

Key Equations

- $K_c = \dfrac{[D]^d[E]^e}{[A]^a[B]^b}$ [15.8]

  The equilibrium-constant expression for a general reaction of the type $a\,A + b\,B \rightleftharpoons d\,D + e\,E$, the concentrations are equilibrium concentrations only

- $K_p = \dfrac{(P_D)^d(P_E)^e}{(P_A)^a(P_B)^b}$ [15.11]

  The equilibrium-constant expression in terms of equilibrium partial pressures

- $K_p = K_c(RT)^{\Delta n}$ [15.14]

  Relating the equilibrium constant based on pressures to the equilibrium constant based on concentration

- $Q_c = \dfrac{[D]^d[E]^e}{[A]^a[B]^b}$ [15.23]

  The reaction quotient. The concentrations are for any time during a reaction. If the concentrations are equilibrium concentrations, then $Q_c = K_c$.

Exercises

Visualizing Concepts

15.1 **(a)** Based on the following energy profile, predict whether $k_f > k_r$ or $k_f < k_r$. **(b)** Using Equation 15.5, predict whether the equilibrium constant for the process is greater than 1 or less than 1. [Section 15.1]

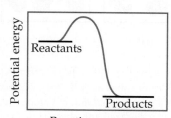

15.2 The following diagrams represent a hypothetical reaction A $\longrightarrow$ B, with A represented by red spheres and B represented by blue spheres. The sequence from left to right represents the system as time passes. Does the system reach equilibrium? If so, in which diagram is the system in equilibrium? [Sections 15.1 and 15.2]

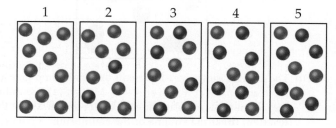

15.3 The following diagram represents an equilibrium mixture produced for a reaction of the type A + X $\rightleftharpoons$ AX. Is K greater or smaller than 1 if the volume is 1 L and each atom/molecule in the diagram represents 1 mol? [Section 15.2]

15.4 The following diagram represents a reaction shown going to completion. Each molecule in the diagram represents 0.1 mol and the volume of the box is 1.0 L. **(a)** Letting A = red spheres and B = blue spheres, write a balanced equation for the reaction. **(b)** Write the equilibrium-constant expression for the reaction. **(c)** Calculate the value of K_c. **(d)** Assuming that all of the molecules are in the gas phase, calculate Δn, the change in the number of gas molecules that accompanies the reaction. **(e)** Calculate the value of K_p. [Section 15.2]

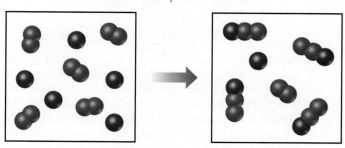

15.5 Snapshots of two hypothetical reactions, A(g) + B(g) $\rightleftharpoons$ AB(g) and X(g) + Y(g) $\rightleftharpoons$ XY(g) at five different times are shown here. Which reaction has a larger equilibrium constant? [Sections 15.1 and 15.2]

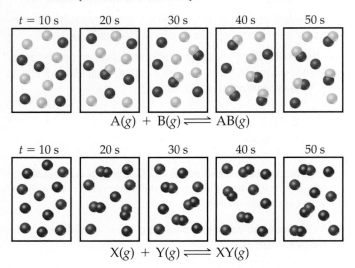

$$A(g) + B(g) \rightleftharpoons AB(g)$$

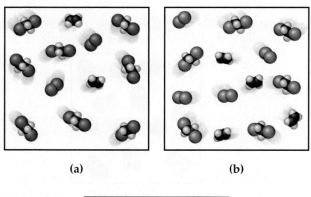

$$X(g) + Y(g) \rightleftharpoons XY(g)$$

15.6 Ethene (C_2H_4) reacts with halogens (X_2) by the following reaction:

$$C_2H_4(g) + X_2(g) \rightleftharpoons C_2H_4X_2(g)$$

The following figures represent the concentrations at equilibrium at the same temperature when X_2 is Cl_2 (green), Br_2 (brown), and I_2 (purple). List the equilibria from smallest to largest equilibrium constant. [Section 15.3]

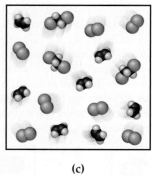

(a)   **(b)**

(c)

15.7 When lead (IV) oxide is heated above 300 °C it decomposes according to the following reaction $PbO_2(s) \rightleftharpoons PbO(s) + O_2(g)$. Consider the two sealed vessels of PbO_2 shown here. If both vessels are heated to 400 °C and allowed to come to equilibitum which of the following statements is true? **(a)** There will be less PbO_2 remaining in vessel A, **(b)** There will be less PbO_2 remaining in vessel B, **(c)** The amount of PbO_2 remaining in each vessel will be the same. [Section 15.4]

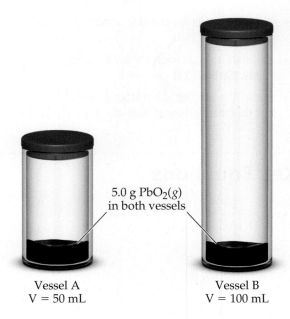

5.0 g $PbO_2(g)$ in both vessels

Vessel A   Vessel B
V = 50 mL   V = 100 mL

15.8 The reaction $A_2 + B_2 \rightleftharpoons 2\,AB$ has an equilibrium constant $K_c = 1.5$. The following diagrams represent reaction mixtures containing A_2 molecules (red), B_2 molecules (blue), and AB molecules. **(a)** Which reaction mixture is at equilibrium? **(b)** For those mixtures that are not at equilibrium, how will the reaction proceed to reach equilibrium? [Sections 15.5 and 15.6]

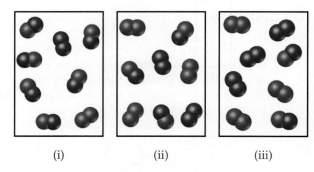

(i)   (ii)   (iii)

15.9 The reaction $A_2(g) + B(g) \rightleftharpoons A(g) + AB(g)$ has an equilibrium constant of $K_p = 2$. The accompanying diagram shows a mixture containing A atoms (red), A_2 molecules, and AB molecules (red and blue). How many B atoms should be added to the diagram to illustrate an equilibrium mixture? [Section 15.6]

15.10 The diagram shown here represents the equilibrium state for the reaction $A_2(g) + 2\,B(g) \rightleftharpoons 2\,AB(g)$. **(a)** Assuming the volume is 2 L, calculate the equilibrium constant K_c for the reaction. **(b)** If the volume of the equilibrium mixture is decreased, will the number of AB molecules increase or decrease? [Sections 15.5 and 15.7]

15.11 The following diagrams represent equilibrium mixtures for the reaction $A_2 + B \rightleftharpoons A + AB$ at 300 K and 500 K. The A atoms are red, and the B atoms are blue. Is the reaction exothermic or endothermic? [Section 15.7]

300 K   500 K

15.12 The following graph represents the yield of the compound AB at equilibrium in the reaction $A(g) + B(g) \longrightarrow AB(g)$ at two different pressures, x and y, as a function of temperature.

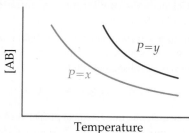

(a) Is this reaction exothermic or endothermic? (b) Is $P = x$ greater or smaller than $P = y$? [Section 15.7]

Equilibrium; The Equilibrium Constant (Sections 15.1–15.4)

15.13 Suppose that the gas-phase reactions $A \longrightarrow B$ and $B \longrightarrow A$ are both elementary processes with rate constants of $4.7 \times 10^{-3}\,s^{-1}$ and $5.8 \times 10^{-1}\,s^{-1}$, respectively. (a) What is the value of the equilibrium constant for the equilibrium $A(g) \rightleftharpoons B(g)$? (b) Which is greater at equilibrium, the partial pressure of A or the partial pressure of B?

15.14 Consider the reaction $A + B \rightleftharpoons C + D$. Assume that both the forward reaction and the reverse reaction are elementary processes and that the value of the equilibrium constant is very large. (a) Which species predominate at equilibrium, reactants or products? (b) Which reaction has the larger rate constant, the forward or the reverse?

15.15 Write the expression for K_c for the following reactions. In each case indicate whether the reaction is homogeneous or heterogeneous.

(a) $3\,NO(g) \rightleftharpoons N_2O(g) + NO_2(g)$
(b) $CH_4(g) + 2\,H_2S(g) \rightleftharpoons CS_2(g) + 4\,H_2(g)$
(c) $Ni(CO)_4(g) \rightleftharpoons Ni(s) + 4\,CO(g)$
(d) $HF(aq) \rightleftharpoons H^+(aq) + F^-(aq)$
(e) $2\,Ag(s) + Zn^{2+}(aq) \rightleftharpoons 2\,Ag^+(aq) + Zn(s)$
(f) $H_2O(l) \rightleftharpoons H^+(aq) + OH^-(aq)$
(g) $2\,H_2O(l) \rightleftharpoons 2\,H^+(aq) + 2\,OH^-(aq)$

15.16 Write the expressions for K_c for the following reactions. In each case indicate whether the reaction is homogeneous or heterogeneous.

(a) $2\,O_3(g) \rightleftharpoons 3\,O_2(g)$
(b) $Ti(s) + 2\,Cl_2(g) \rightleftharpoons TiCl_4(l)$
(c) $2\,C_2H_4(g) + 2\,H_2O(g) \rightleftharpoons 2\,C_2H_6(g) + O_2(g)$
(d) $C(s) + 2\,H_2(g) \rightleftharpoons CH_4(g)$
(e) $4\,HCl(aq) + O_2(g) \rightleftharpoons 2\,H_2O(l) + 2\,Cl_2(g)$
(f) $2\,C_8H_{18}(l) + 25\,O_2(g) \rightleftharpoons 16\,CO_2(g) + 18\,H_2O(g)$
(g) $2\,C_8H_{18}(l) + 25\,O_2(g) \rightleftharpoons 16\,CO_2(g) + 18\,H_2O(l)$

15.17 When the following reactions come to equilibrium, does the equilibrium mixture contain mostly reactants or mostly products?

(a) $N_2(g) + O_2(g) \rightleftharpoons 2\,NO(g);\, K_c = 1.5 \times 10^{-10}$
(b) $2\,SO_2(g) + O_2(g) \rightleftharpoons 2\,SO_3(g);\, K_p = 2.5 \times 10^9$

15.18 Which of the following reactions lies to the right, favoring the formation of products, and which lies to the left, favoring formation of reactants?

(a) $2\,NO(g) + O_2(g) \rightleftharpoons 2\,NO_2(g);\, K_p = 5.0 \times 10^{12}$
(b) $2\,HBr(g) \rightleftharpoons H_2(g) + Br_2(g);\, K_c = 5.8 \times 10^{-18}$

15.19 Which of the following statements are true and which are false? (a) The equilibrium constant can never be a negative number. (b) In reactions that we draw with a single-headed arrow, the equilibrium constant has a value that is very close to zero. (c) As the value of the equilibrium constant increases the speed at which a reaction reaches equilibrium increases.

15.20 Which of the following statements are true and which are false? (a) For the reaction $2\,A(g) + B(g) \rightleftharpoons A_2B(g)$ K_c and K_p are numerically the same. (b) It is possible to distinguish K_c from K_p by comparing the units used to express the equilibrium constant. (c) For the equilibrium in (a), the value of K_c increases with increasing pressure.

15.21 If $K_c = 0.042$ for $PCl_3(g) + Cl_2(g) \rightleftharpoons PCl_5(g)$ at 500 K, what is the value of K_p for this reaction at this temperature?

15.22 Calculate K_c at 303 K for $SO_2(g) + Cl_2(g) \rightleftharpoons SO_2Cl_2(g)$ if $K_p = 34.5$ at this temperature.

15.23 The equilibrium constant for the reaction

$$2\,NO(g) + Br_2(g) \rightleftharpoons 2\,NOBr(g)$$

is $K_c = 1.3 \times 10^{-2}$ at 1000 K. (a) At this temperature does the equilibrium favor NO and Br_2, or does it favor NOBr? (b) Calculate K_c for $2\,NOBr(g) \rightleftharpoons 2\,NO(g) + Br_2(g)$. (c) Calculate K_c for $NOBr(g) \longrightarrow NO(g) + \frac{1}{2}\,Br_2(g)$.

15.24 Consider the following equilibrium:

$$2\,H_2(g) + S_2(g) \rightleftharpoons 2\,H_2S(g) \quad K_c = 1.08 \times 10^7 \text{ at } 700\,°C$$

(a) Calculate K_p. **(b)** Does the equilibrium mixture contain mostly H_2 and S_2 or mostly H_2S? **(c)** Calculate the value of K_c if you rewrote the equation $H_2(g) + \frac{1}{2}S_2(g) \rightleftharpoons H_2S(g)$.

15.25 At 1000 K, $K_p = 1.85$ for the reaction

$$SO_2(g) + \tfrac{1}{2}O_2(g) \rightleftharpoons SO_3(g)$$

(a) What is the value of K_p for the reaction $SO_3(g) \rightleftharpoons SO_2(g) + \frac{1}{2}O_2(g)$? **(b)** What is the value of K_p for the reaction $2 SO_2(g) + O_2(g) \rightleftharpoons 2 SO_3(g)$? **(c)** What is the value of K_c for the reaction in part **(b)**?

15.26 Consider the following equilibrium, for which $K_p = 0.0752$ at 480 °C:

$$2 Cl_2(g) + 2 H_2O(g) \rightleftharpoons 4 HCl(g) + O_2(g)$$

(a) What is the value of K_p for the reaction
$4 HCl(g) + O_2(g) \rightleftharpoons 2 Cl_2(g) + 2 H_2O(g)$?
(b) What is the value of K_p for the reaction
$Cl_2(g) + H_2O(g) \rightleftharpoons 2 HCl(g) + \frac{1}{2}O_2(g)$?
(c) What is the value of K_c for the reaction in part (b)?

15.27 The following equilibria were attained at 823 K:

$$CoO(s) + H_2(g) \rightleftharpoons Co(s) + H_2O(g) \quad K_c = 67$$

$$CoO(s) + CO(g) \rightleftharpoons Co(s) + CO_2(g) \quad K_c = 490$$

Based on these equilibria, calculate the equilibrium constant for $H_2(g) + CO_2(g) \rightleftharpoons CO(g) + H_2O(g)$ at 823 K.

15.28 Consider the equilibrium

$$N_2(g) + O_2(g) + Br_2(g) \rightleftharpoons 2 NOBr(g)$$

Calculate the equilibrium constant K_p for this reaction, given the following information (at 298 K):

$$2 NO(g) + Br_2(g) \rightleftharpoons 2 NOBr(g) \quad K_c = 2.0$$

$$2 NO(g) \rightleftharpoons N_2(g) + O_2(g) \quad K_c = 2.1 \times 10^{30}$$

15.29 Mercury(I) oxide decomposes into elemental mercury and elemental oxygen: $2 Hg_2O(s) \rightleftharpoons 4 Hg(l) + O_2(g)$. **(a)** Write the equilibrium-constant expression for this reaction in terms of partial pressures. **(b)** Suppose you run this reaction in a solvent that dissolves elemental mercury and elemental oxygen. Rewrite the equilibrium-constant expression in terms of molarities for the reaction, using (solv) to indicate solvation.

15.30 Consider the equilibrium $Na_2O(s) + SO_2(g) \rightleftharpoons Na_2SO_3(s)$. **(a)** Write the equilibrium-constant expression for this reaction in terms of partial pressures. **(b)** All the compounds in this reaction are soluble in water. Rewrite the equilibrium-constant expression in terms of molarities for the aqueous reaction.

Calculating Equilibrium Constants
(Section 15.5)

15.31 Methanol (CH_3OH) is produced commercially by the catalyzed reaction of carbon monoxide and hydrogen: $CO(g) + 2 H_2(g) \rightleftharpoons CH_3OH(g)$. An equilibrium mixture in a 2.00-L vessel is found to contain 0.0406 mol CH_3OH, 0.170 mol CO, and 0.302 mol H_2 at 500 K. Calculate K_c at this temperature.

15.32 Gaseous hydrogen iodide is placed in a closed container at 425 °C, where it partially decomposes to hydrogen and iodine: $2 HI(g) \rightleftharpoons H_2(g) + I_2(g)$. At equilibrium it is found that $[HI] = 3.53 \times 10^{-3} M, [H_2] = 4.79 \times 10^{-4} M$, and $[I_2] = 4.79 \times 10^{-4} M$. What is the value of K_c at this temperature?

15.33 The equilibrium $2 NO(g) + Cl_2(g) \rightleftharpoons 2 NOCl(g)$ is established at 500 K. An equilibrium mixture of the three gases has partial pressures of 0.095 atm, 0.171 atm, and 0.28 atm for NO, Cl_2, and NOCl, respectively. **(a)** Calculate K_p for this reaction at 500.0 K. **(b)** If the vessel has a volume of 5.00 L, calculate K_c at this temperature.

15.34 Phosphorus trichloride gas and chlorine gas react to form phosphorus pentachloride gas: $PCl_3(g) + Cl_2(g) \rightleftharpoons PCl_5(g)$. A 7.5-L gas vessel is charged with a mixture of $PCl_3(g)$ and $Cl_2(g)$, which is allowed to equilibrate at 450 K. At equilibrium the partial pressures of the three gases are $P_{PCl_3} = 0.124$ atm, $P_{Cl_2} = 0.157$ atm, and $P_{PCl_5} = 1.30$ atm. **(a)** What is the value of K_p at this temperature? **(b)** Does the equilibrium favor reactants or products? **(c)** Calculate K_c for this reaction at 450 K.

15.35 A mixture of 0.10 mol of NO, 0.050 mol of H_2, and 0.10 mol of H_2O is placed in a 1.0-L vessel at 300 K. The following equilibrium is established:

$$2 NO(g) + 2 H_2(g) \rightleftharpoons N_2(g) + 2 H_2O(g)$$

At equilibrium $[NO] = 0.062 M$. **(a)** Calculate the equilibrium concentrations of H_2, N_2, and H_2O. **(b)** Calculate K_c.

15.36 A mixture of 1.374 g of H_2 and 70.31 g of Br_2 is heated in a 2.00-L vessel at 700 K. These substances react according to

$$H_2(g) + Br_2(g) \rightleftharpoons 2 HBr(g)$$

At equilibrium, the vessel is found to contain 0.566 g of H_2. **(a)** Calculate the equilibrium concentrations of H_2, Br_2, and HBr. **(b)** Calculate K_c.

15.37 A mixture of 0.2000 mol of CO_2, 0.1000 mol of H_2, and 0.1600 mol of H_2O is placed in a 2.000-L vessel. The following equilibrium is established at 500 K:

$$CO_2(g) + H_2(g) \rightleftharpoons CO(g) + H_2O(g)$$

(a) Calculate the initial partial pressures of CO_2, H_2, and H_2O. **(b)** At equilibrium $P_{H_2O} = 3.51$ atm. Calculate the equilibrium partial pressures of CO_2, H_2, and CO. **(c)** Calculate K_p for the reaction. **(d)** Calculate K_c for the reaction.

15.38 A flask is charged with 1.500 atm of $N_2O_4(g)$ and 1.00 atm $NO_2(g)$ at 25 °C, and the following equilibrium is achieved:

$$N_2O_4(g) \rightleftharpoons 2 NO_2(g)$$

After equilibrium is reached, the partial pressure of NO_2 is 0.512 atm. **(a)** What is the equilibrium partial pressure of N_2O_4? **(b)** Calculate the value of K_p for the reaction. **(c)** Calculate K_c for the reaction.

15.39 Two different proteins X and Y are dissolved in aqueous solution at 37 °C. The proteins bind in a 1:1 ratio to form XY. A solution that is initially 1.00 mM in each protein is allowed to reach equilibrium. At equilibrium, 0.20 mM of free X and 0.20 mM of free Y remain. What is K_c for the reaction?

15.40 A chemist at a pharmaceutical company is measuring equilibrium constants for reactions in which drug candidate molecules bind to a protein involved in cancer. The drug molecules bind the protein in a 1:1 ratio to form a drug–protein complex. The protein concentration in aqueous solution at 25 °C is $1.50 \times 10^{-6} M$. Drug A is introduced into the protein solution at an initial concentration of $2.00 \times 10^{-6} M$. Drug B is introduced into a separate, identical protein solution at an initial concentration of $2.00 \times 10^{-6} M$. At equilibrium, the drug A-protein solution has an A-protein complex concentration of $1.00 \times 10^{-6} M$, and the drug B solution has a B-protein complex concentration of $1.40 \times 10^{-6} M$. Calculate the K_c value for the A-protein binding reaction and for the B-protein binding reaction. Assuming that the drug that binds more strongly will be more effective, which drug is the better choice for further research?

Applications of Equilibrium Constants (Section 15.6)

15.41 (a) If $Q_c < K_c$, in which direction will a reaction proceed in order to reach equilibrium? (b) What condition must be satisfied so that $Q_c = K_c$?

15.42 (a) If $Q_c > K_c$, how must the reaction proceed to reach equilibrium? (b) At the start of a certain reaction, only reactants are present; no products have been formed. What is the value of Q_c at this point in the reaction?

15.43 At 100 °C, the equilibrium constant for the reaction $COCl_2(g) \rightleftharpoons CO(g) + Cl_2(g)$ has the value $K_c = 2.19 \times 10^{-10}$. Are the following mixtures of $COCl_2$, CO, and Cl_2 at 100 °C at equilibrium? If not, indicate the direction that the reaction must proceed to achieve equilibrium.

(a) $[COCl_2] = 2.00 \times 10^{-3} M$, $[CO] = 3.3 \times 10^{-6} M$, $[Cl_2] = 6.62 \times 10^{-6} M$

(b) $[COCl_2] = 4.50 \times 10^{-2} M$, $[CO] = 1.1 \times 10^{-7} M$, $[Cl_2] = 2.25 \times 10^{-6} M$

(c) $[COCl_2] = 0.0100 M$, $[CO] = [Cl_2] = 1.48 \times 10^{-6} M$

15.44 As shown in Table 15.2, K_p for the equilibrium

$$N_2(g) + 3 H_2(g) \rightleftharpoons 2 NH_3(g)$$

is 4.51×10^{-5} at 450 °C. For each of the mixtures listed here, indicate whether the mixture is at equilibrium at 450 °C. If it is not at equilibrium, indicate the direction (toward product or toward reactants) in which the mixture must shift to achieve equilibrium.

(a) 98 atm NH_3, 45 atm N_2, 55 atm H_2

(b) 57 atm NH_3, 143 atm N_2, no H_2

(c) 13 atm NH_3, 27 atm N_2, 82 atm H_2

15.45 At 100 °C, $K_c = 0.078$ for the reaction

$$SO_2Cl_2(g) \rightleftharpoons SO_2(g) + Cl_2(g)$$

In an equilibrium mixture of the three gases, the concentrations of SO_2Cl_2 and SO_2 are 0.108 M and 0.052 M, respectively. What is the partial pressure of Cl_2 in the equilibrium mixture?

15.46 At 900 K, the following reaction has $K_p = 0.345$:

$$2 SO_2(g) + O_2(g) \rightleftharpoons 2 SO_3(g)$$

In an equilibrium mixture the partial pressures of SO_2 and O_2 are 0.135 atm and 0.455 atm, respectively. What is the equilibrium partial pressure of SO_3 in the mixture?

15.47 At 1285 °C, the equilibrium constant for the reaction $Br_2(g) \rightleftharpoons 2 Br(g)$ is $K_c = 1.04 \times 10^{-3}$. A 0.200-L vessel containing an equilibrium mixture of the gases has 0.245 g $Br_2(g)$ in it. What is the mass of $Br(g)$ in the vessel?

15.48 For the reaction $H_2(g) + I_2(g) \rightleftharpoons 2 HI(g)$, $K_c = 55.3$ at 700 K. In a 2.00-L flask containing an equilibrium mixture of the three gases, there are 0.056 g H_2 and 4.36 g I_2. What is the mass of HI in the flask?

15.49 At 800 K, the equilibrium constant for $I_2(g) \rightleftharpoons 2 I(g)$ is $K_c = 3.1 \times 10^{-5}$. If an equilibrium mixture in a 10.0-L vessel contains 2.67×10^{-2} g of $I(g)$, how many grams of I_2 are in the mixture?

15.50 For $2 SO_2(g) + O_2(g) \rightleftharpoons 2 SO_3(g)$, $K_p = 3.0 \times 10^4$ at 700 K. In a 2.00-L vessel, the equilibrium mixture contains 1.17 g of SO_3 and 0.105 g of O_2. How many grams of SO_2 are in the vessel?

15.51 At 2000 °C, the equilibrium constant for the reaction

$$2 NO(g) \rightleftharpoons N_2(g) + O_2(g)$$

is $K_c = 2.4 \times 10^3$. If the initial concentration of NO is 0.175 M, what are the equilibrium concentrations of NO, N_2, and O_2?

15.52 For the equilibrium

$$Br_2(g) + Cl_2(g) \rightleftharpoons 2 BrCl(g)$$

at 400 K, $K_c = 7.0$. If 0.25 mol of Br_2 and 0.55 mol of Cl_2 are introduced into a 3.0-L container at 400 K, what will be the equilibrium concentrations of Br_2, Cl_2, and BrCl?

15.53 At 373 K, $K_p = 0.416$ for the equilibrium

$$2 NOBr(g) \rightleftharpoons 2 NO(g) + Br_2(g)$$

If the pressures of $NOBr(g)$ and $NO(g)$ are equal, what is the equilibrium pressure of $Br_2(g)$?

15.54 At 218 °C, $K_c = 1.2 \times 10^{-4}$ for the equilibrium

$$NH_4SH(s) \rightleftharpoons NH_3(g) + H_2S(g)$$

Calculate the equilibrium concentrations of NH_3 and H_2S if a sample of solid NH_4SH is placed in a closed vessel at 218 °C and decomposes until equilibrium is reached.

15.55 Consider the reaction

$$CaSO_4(s) \rightleftharpoons Ca^{2+}(aq) + SO_4^{2-}(aq)$$

At 25 °C, the equilibrium constant is $K_c = 2.4 \times 10^{-5}$ for this reaction. (a) If excess $CaSO_4(s)$ is mixed with water at 25 °C to produce a saturated solution of $CaSO_4$, what are the equilibrium concentrations of Ca^{2+} and SO_4^{2-}? (b) If the resulting solution has a volume of 1.4 L, what is the minimum mass of $CaSO_4(s)$ needed to achieve equilibrium?

15.56 At 80 °C, $K_c = 1.87 \times 10^{-3}$ for the reaction

$$PH_3BCl_3(s) \rightleftharpoons PH_3(g) + BCl_3(g)$$

(a) Calculate the equilibrium concentrations of PH_3 and BCl_3 if a solid sample of PH_3BCl_3 is placed in a closed vessel at 80 °C and decomposes until equilibrium is reached. (b) If the flask has a volume of 0.250 L, what is the minimum mass of $PH_3BCl_3(s)$ that must be added to the flask to achieve equilibrium?

15.57 For the reaction $I_2 + Br_2(g) \rightleftharpoons 2 IBr(g)$, $K_c = 280$ at 150 °C. Suppose that 0.500 mol IBr in a 2.00-L flask is allowed to reach equilibrium at 150 °C. What are the equilibrium concentrations of IBr, I_2, and Br_2?

15.58 At 25 °C, the reaction

$$CaCrO_4(s) \rightleftharpoons Ca^{2+}(aq) + CrO_4{}^{2-}(aq)$$

has an equilibrium constant $K_c = 7.1 \times 10^{-4}$. What are the equilibrium concentrations of Ca^{2+} and $CrO_4{}^{2-}$ in a saturated solution of $CaCrO_4$?

15.59 Methane, CH_4, reacts with I_2 according to the reaction $CH_4(g) + I_2(g) \rightleftharpoons CH_3I(g) + HI(g)$. At 630 K, K_p for this reaction is 2.26×10^{-4}. A reaction was set up at 630 K with initial partial pressures of methane of 105.1 torr and of 7.96 torr for I_2. Calculate the pressures, in torr, of all reactants and products at equilibrium.

15.60 The reaction of an organic acid with an alcohol, in organic solvent, to produce an ester and water is commonly done in the pharmaceutical industry. This reaction is catalyzed by strong acid (usually H_2SO_4). A simple example is the reaction of acetic acid with ethyl alcohol to produce ethyl acetate and water:

$$CH_3COOH(solv) + CH_3CH_2OH(solv) \rightleftharpoons$$
$$CH_3COOCH_2CH_3(solv) + H_2O(solv)$$

where "(solv)" indicates that all reactants and products are in solution but not an aqueous solution. The equilibrium constant for this reaction at 55 °C is 6.68. A pharmaceutical chemist makes up 15.0 L of a solution that is initially 0.275 M in acetic acid and 3.85 M in ethanol. At equilibrium, how many grams of ethyl acetate are formed?

Le Châtelier's Principle (Section 15.7)

15.61 Consider the following equilibrium for which $\Delta H < 0$

$$2 SO_2(g) + O_2(g) \rightleftharpoons 2 SO_3(g)$$

How will each of the following changes affect an equilibrium mixture of the three gases: (a) $O_2(g)$ is added to the system; (b) the reaction mixture is heated; (c) the volume of the reaction vessel is doubled; (d) a catalyst is added to the mixture; (e) the total pressure of the system is increased by adding a noble gas; (f) $SO_3(g)$ is removed from the system?

15.62 Consider the reaction

$$4 NH_3(g) + 5 O_2(g) \rightleftharpoons$$
$$4 NO(g) + 6 H_2O(g), \Delta H = -904.4 \text{ kJ}$$

Does each of the following increase, decrease, or leave unchanged the yield of NO at equilibrium? (a) increase $[NH_3]$; (b) increase $[H_2O]$; (c) decrease $[O_2]$;

(d) decrease the volume of the container in which the reaction occurs; (e) add a catalyst; (f) increase temperature.

15.63 How do the following changes affect the value of the equilibrium constant for a gas-phase exothermic reaction: (a) removal of a reactant, (b) removal of a product, (c) decrease in the volume, (d) decrease in the temperature, (e) addition of a catalyst?

15.64 For a certain gas-phase reaction, the fraction of products in an equilibrium mixture is increased by either increasing the temperature or by increasing the volume of the reaction vessel. (a) Is the reaction exothermic or endothermic? (b) Does the balanced chemical equation have more molecules on the reactant side or product side?

15.65 Consider the following equilibrium between oxides of nitrogen

$$3 NO(g) \rightleftharpoons NO_2(g) + N_2O(g)$$

(a) Use data in Appendix C to calculate $\Delta H°$ for this reaction. (b) Will the equilibrium constant for the reaction increase or decrease with increasing temperature? (c) At constant temperature, would a change in the volume of the container affect the fraction of products in the equilibrium mixture?

15.66 Methanol (CH_3OH) can be made by the reaction of CO with H_2:

$$CO(g) + 2 H_2(g) \rightleftharpoons CH_3OH(g)$$

(a) Use thermochemical data in Appendix C to calculate $\Delta H°$ for this reaction. (b) To maximize the equilibrium yield of methanol, would you use a high or low temperature? (c) To maximize the equilibrium yield of methanol, would you use a high or low pressure?

15.67 Ozone, O_3, decomposes to molecular oxygen in the stratosphere according to the reaction $2 O_3(g) \longrightarrow 3 O_2(g)$. Would an increase in pressure favor the formation of ozone or of oxygen?

15.68 The water–gas shift reaction $CO(g) + H_2O(g) \rightleftharpoons CO_2(g) + H_2(g)$ is used industrially to produce hydrogen. The reaction enthalpy is $\Delta H° = -41$ kJ. (a) To increase the equilibrium yield of hydrogen would you use high or low temperature? (b) Could you increase the equilibrium yield of hydrogen by controlling the pressure of this reaction? If so would high or low pressure favor formation of $H_2(g)$?

Additional Exercises

15.69 Both the forward reaction and the reverse reaction in the following equilibrium are believed to be elementary steps:

$$CO(g) + Cl_2(g) \rightleftharpoons COCl(g) + Cl(g)$$

At 25 °C, the rate constants for the forward and reverse reactions are $1.4 \times 10^{-28} M^{-1} s^{-1}$ and $9.3 \times 10^{10} M^{-1} s^{-1}$, respectively. (a) What is the value for the equilibrium constant at 25 °C? (b) Are reactants or products more plentiful at equilibrium?

15.70 If $K_c = 1$ for the equilibrium $2 A(g) \rightleftharpoons B(g)$, what is the relationship between [A] and [B] at equilibrium?

15.71 A mixture of CH_4 and H_2O is passed over a nickel catalyst at 1000 K. The emerging gas is collected in a 5.00-L flask and is found to contain 8.62 g of CO, 2.60 g of H_2, 43.0 g of CH_4, and 48.4 g of H_2O. Assuming that equilibrium has been reached, calculate K_c and K_p for the reaction $CH_4(g) + H_2O(g) \rightleftharpoons CO(g) + 3 H_2(g)$.

15.72 When 2.00 mol of SO_2Cl_2 is placed in a 2.00-L flask at 303 K, 56% of the SO_2Cl_2 decomposes to SO_2 and Cl_2:

$$SO_2Cl_2(g) \rightleftharpoons SO_2(g) + Cl_2(g)$$

(a) Calculate K_c for this reaction at this temperature. (b) Calculate K_p for this reaction at 303 K. (c) According to Le Châtelier's principle, would the percent of SO_2Cl_2 that decomposes increase, decrease or stay the same if the mixture were transferred to a 15.00-L vessel? **(d)** Use the equilibrium constant you calculated above to determine the percentage of SO_2Cl_2 that decomposes when 2.00 mol of SO_2Cl_2 is placed in a 15.00-L vessel at 303 K.

15.73 A mixture of H_2, S, and H_2S is held in a 1.0-L vessel at 90 °C and reacts according to the equation:

$$H_2(g) + S(s) \rightleftharpoons H_2S(g)$$

At equilibrium, the mixture contains 0.46 g of H_2S and 0.40 g H_2. **(a)** Write the equilibrium-constant expression for this reaction. **(b)** What is the value of K_c for the reaction at this temperature?

15.74 A sample of nitrosyl bromide (NOBr) decomposes according to the equation:

$$2\,NOBr(g) \rightleftharpoons 2\,NO(g) + Br_2(g)$$

An equilibrium mixture in a 5.00-L vessel at 100 °C contains 3.22 g of NOBr, 3.08 g of NO, and 4.19 g of Br_2. **(a)** Calculate K_c. **(b)** What is the total pressure exerted by the mixture of gases? **(c)** What was the mass of the original sample of NOBr?

15.75 Consider the hypothetical reaction $A(g) \rightleftharpoons 2\,B(g)$. A flask is charged with 0.75 atm of pure A, after which it is allowed to reach equilibrium at 0 °C. At equilibrium, the partial pressure of A is 0.36 atm. **(a)** What is the total pressure in the flask at equilibrium? **(b)** What is the value of K_p? **(c)** What could we do to maximize the yield of B?

15.76 As shown in Table 15.2, the equilibrium constant for the reaction $N_2(g) + 3\,H_2(g) \rightleftharpoons 2\,NH_3(g)$ is $K_p = 4.34 \times 10^{-3}$ at 300 °C. Pure NH_3 is placed in a 1.00-L flask and allowed to reach equilibrium at this temperature. There are 1.05 g NH_3 in the equilibrium mixture. **(a)** What are the masses of N_2 and H_2 in the equilibrium mixture? **(b)** What was the initial mass of ammonia placed in the vessel? **(c)** What is the total pressure in the vessel?

15.77 For the equilibrium

$$2\,IBr(g) \rightleftharpoons I_2(g) + Br_2(g)$$

$K_p = 8.5 \times 10^{-3}$ at 150 °C. If 0.025 atm of IBr is placed in a 2.0-L container, what is the partial pressure of all substances after equilibrium is reached?

15.78 For the equilibrium

$$PH_3BCl_3(s) \rightleftharpoons PH_3(g) + BCl_3(g)$$

$K_p = 0.052$ at 60 °C. **(a)** Calculate K_c. **(b)** After 3.00 g of solid PH_3BCl_3 is added to a closed 1.500-L vessel at 60 °C, the vessel is charged with 0.0500 g of $BCl_3(g)$. What is the equilibrium concentration of PH_3?

[15.79] Solid NH_4SH is introduced into an evacuated flask at 24 °C. The following reaction takes place:

$$NH_4SH(s) \rightleftharpoons NH_3(g) + H_2S(g)$$

At equilibrium, the total pressure (for NH_3 and H_2S taken together) is 0.614 atm. What is K_p for this equilibrium at 24 °C?

[15.80] A 0.831-g sample of SO_3 is placed in a 1.00-L container and heated to 1100 K. The SO_3 decomposes to SO_2 and O_2:

$$2\,SO_3(g) \rightleftharpoons 2\,SO_2(g) + O_2(g)$$

At equilibrium, the total pressure in the container is 1.300 atm. Find the values of K_p and K_c for this reaction at 1100 K.

15.81 Nitric oxide (NO) reacts readily with chlorine gas as follows:

$$2\,NO(g) + Cl_2(g) \rightleftharpoons 2\,NOCl(g)$$

At 700 K, the equilibrium constant K_p for this reaction is 0.26. Predict the behavior of each of the following mixtures at this temperature and indicate whether or not the mixtures are at equilibrium. If not, state whether the mixture will need to produce more products or reactants to reach equilibrium.

(a) $P_{NO} = 0.15$ atm, $P_{Cl_2} = 0.31$ atm, $P_{NOCl} = 0.11$ atm

(b) $P_{NO} = 0.12$ atm, $P_{Cl_2} = 0.10$ atm, $P_{NOCl} = 0.050$ atm

(c) $P_{NO} = 0.15$ atm, $P_{Cl_2} = 0.20$ atm, $P_{NOCl} = 5.10 \times 10^{-3}$ atm

15.82 At 900 °C, $K_c = 0.0108$ for the reaction

$$CaCO_3(s) \rightleftharpoons CaO(s) + CO_2(g)$$

A mixture of $CaCO_3$, CaO, and CO_2 is placed in a 10.0-L vessel at 900 °C. For the following mixtures, will the amount of $CaCO_3$ increase, decrease, or remain the same as the system approaches equilibrium?

(a) 15.0 g $CaCO_3$, 15.0 g CaO, and 4.25 g CO_2

(b) 2.50 g $CaCO_3$, 25.0 g CaO, and 5.66 g CO_2

(c) 30.5 g $CaCO_3$, 25.5 g CaO, and 6.48 g CO_2

15.83 When 1.50 mol CO_2 and 1.50 mol H_2 are placed in a 3.00-L container at 395 °C, the following reaction occurs: $CO_2(g) + H_2(g) \rightleftharpoons CO(g) + H_2O(g)$. If $K_c = 0.802$, what are the concentrations of each substance in the equilibrium mixture?

15.84 The equilibrium constant K_c for $C(s) + CO_2(g) \rightleftharpoons 2\,CO(g)$ is 1.9 at 1000 K and 0.133 at 298 K. **(a)** If excess C is allowed to react with 25.0 g of CO_2 in a 3.00-L vessel at 1000 K, how many grams of CO are produced? **(b)** How many grams of C are consumed? **(c)** If a smaller vessel is used for the reaction, will the yield of CO be greater or smaller? **(d)** Is the reaction endothermic or exothermic?

15.85 NiO is to be reduced to nickel metal in an industrial process by use of the reaction

$$NiO(s) + CO(g) \rightleftharpoons Ni(s) + CO_2(g)$$

At 1600 K, the equilibrium constant for the reaction is $K_p = 6.0 \times 10^2$. If a CO pressure of 150 torr is to be employed in the furnace and total pressure never exceeds 760 torr, will reduction occur?

15.86 Le Châtelier noted that many industrial processes of his time could be improved by an understanding of chemical equilibria. For example, the reaction of iron oxide with carbon monoxide was used to produce elemental iron and CO_2 according to the reaction

$$Fe_2O_3(s) + 3\,CO(g) \rightleftharpoons 2\,Fe(s) + 3\,CO_2(g)$$

Even in Le Châtelier's time, it was noted that a great deal of CO was wasted, expelled through the chimneys over the furnaces. Le Châtelier wrote, "Because this incomplete reaction was thought to be due to an insufficiently prolonged contact between carbon monoxide and the iron ore [oxide], the dimensions of the furnaces have been increased. In England, they have been made as high as 30 m. But the proportion of carbon monoxide escaping has not diminished, thus demonstrating, by an experiment costing several hundred thousand francs, that the reduction of iron oxide by carbon monoxide is a limited reaction. Acquaintance with the laws of chemical equilibrium would have permitted the same conclusion to be reached more rapidly and far more economically." What does this anecdote tell us about the equilibrium constant for this reaction?

[15.87] At 700 K, the equilibrium constant for the reaction

$$CCl_4(g) \rightleftharpoons C(s) + 2\,Cl_2(g)$$

is $K_p = 0.76$. A flask is charged with 2.00 atm of CCl_4, which then reaches equilibrium at 700 K. **(a)** What fraction of the CCl_4 is converted into C and Cl_2? **(b)** What are the partial pressures of CCl_4 and Cl_2 at equilibrium?

[15.88] The reaction $PCl_3(g) + Cl_2(g) \rightleftharpoons PCl_5(g)$ has $K_p = 0.0870$ at 300 °C. A flask is charged with 0.50 atm PCl_3, 0.50 atm Cl_2, and 0.20 atm PCl_5 at this temperature. **(a)** Use the reaction quotient to determine the direction the reaction must proceed to reach equilibrium. **(b)** Calculate the equilibrium partial pressures of the gases. **(c)** What effect will increasing the volume of the system have on the mole fraction of Cl_2 in the equilibrium mixture? **(d)** The reaction is exothermic. What effect will increasing the temperature of the system have on the mole fraction of Cl_2 in the equilibrium mixture?

[15.89] An equilibrium mixture of H_2, I_2, and HI at 458 °C contains 0.112 mol H_2, 0.112 mol I_2, and 0.775 mol HI in a 5.00-L vessel. What are the equilibrium partial pressures when equilibrium is reestablished following the addition of 0.200 mol of HI?

[15.90] Consider the hypothetical reaction $A(g) + 2\,B(g) \rightleftharpoons 2\,C(g)$, for which $K_c = 0.25$ at a certain temperature. A 1.00-L reaction vessel is loaded with 1.00 mol of compound C, which is allowed to reach equilibrium. Let the variable x represent the number of mol/L of compound A present at equilibrium. **(a)** In terms of x, what are the equilibrium concentrations of compounds B and C? **(b)** What limits must be placed on the value of x so that all concentrations are positive? **(c)** By putting the equilibrium concentrations (in terms of x) into the equilibrium-constant expression, derive an equation that can be solved for x. **(d)** The equation from part (c) is a cubic equation (one that has the form $ax^3 + bx^2 + cx + d = 0$). In general, cubic equations cannot be solved in closed form. However, you can estimate the solution by plotting the cubic equation in the allowed range of x that you specified in part (b). The point at which the cubic equation crosses the x-axis is the solution. **(e)** From the plot in part (d), estimate the equilibrium concentrations of A, B, and C. (*Hint:* You can check the accuracy of your answer by substituting these concentrations into the equilibrium expression.)

15.91 At 1200 K, the approximate temperature of automobile exhaust gases (Figure 15.15), K_p for the reaction

$$2\,CO_2(g) \rightleftharpoons 2\,CO(g) + O_2(g)$$

is about 1×10^{-13}. Assuming that the exhaust gas (total pressure 1 atm) contains 0.2% CO, 12% CO_2, and 3% O_2 by volume, is the system at equilibrium with respect to the CO_2 reaction? Based on your conclusion, would the CO concentration in the exhaust be decreased or increased by a catalyst that speeds up the CO_2 reaction? Recall that at a fixed pressure and temperature, volume % = mol %.

15.92 Suppose that you worked at the U.S. Patent Office and a patent application came across your desk claiming that a newly developed catalyst was much superior to the Haber catalyst for ammonia synthesis because the catalyst led to much greater equilibrium conversion of N_2 and H_2 into NH_3 than the Haber catalyst under the same conditions. What would be your response?

Integrative Exercises

15.93 Consider the reaction $IO_4^-(aq) + 2\,H_2O(l) \rightleftharpoons H_4IO_6^-(aq)$; $K_c = 3.5 \times 10^{-2}$. If you start with 25.0 mL of a 0.905 M solution of $NaIO_4$, and then dilute it with water to 500.0 mL, what is the concentration of $H_4IO_6^-$ at equilibrium?

[15.94] Silver chloride, AgCl(s), is an "insoluble" strong electrolyte. **(a)** Write the equation for the dissolution of AgCl(s) in $H_2O(l)$. **(b)** Write the expression for K_c for the reaction in part (a). **(c)** Based on the thermochemical data in Appendix C and Le Châtelier's principle, predict whether the solubility of AgCl in H_2O increases or decreases with increasing temperature. **(d)** The equilibrium constant for the dissolution of AgCl in water is 1.6×10^{-10} at 25 °C. In addition, $Ag^+(aq)$ can react with $Cl^-(aq)$ according to the reaction

$$Ag^+(aq) + 2\,Cl^-(aq) \rightleftharpoons AgCl_2^-(aq)$$

where $K_c = 1.8 \times 10^5$ at 25 °C. Although AgCl is "not soluble" in water, the complex $AgCl_2^-$ is soluble. At 25 °C, is the solubility of AgCl in a 0.100 M NaCl solution *greater* than the solubility of AgCl in pure water, due to the formation of soluble $AgCl_2^-$ ions? Or is the AgCl solubility in 0.100 M NaCl *less* than in pure water because of a Le Châtelier-type argument? Justify your answer with calculations. (*Hint:* Any form in which silver is in solution counts as "solubility.")

[15.95] Consider the equilibrium $A \rightleftharpoons B$ in which both the forward and reverse reactions are elementary (single-step) reactions. Assume that the only effect of a catalyst on the reaction is to lower the activation energies of the forward and reverse reactions, as shown in Figure 15.14. Using the Arrhenius equation (Section 14.5), prove that the equilibrium constant is the same for the catalyzed reaction as for the uncatalyzed one.

[15.96] The phase diagram for SO_2 is shown here. **(a)** What does this diagram tell you about the enthalpy change in the reaction $SO_2(l) \longrightarrow SO_2(g)$? **(b)** Calculate the equilibrium constant for this reaction at 100 °C and at 0 °C. **(c)** Why is it not possible to calculate an equilibrium constant between the gas and liquid phases in the supercritical region? **(d)** At which of the three points marked in red does $SO_2(g)$ most closely approach ideal-gas behavior? **(e)** At which of the three red points does $SO_2(g)$ behave least ideally?

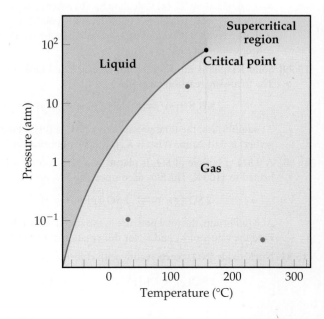

[15.97] Write the equilibrium-constant expression for the equilibrium

$$C(s) + CO_2(g) \rightleftharpoons 2\,CO(g)$$

The table that follows shows the relative mole percentages of $CO_2(g)$ and $CO(g)$ at a total pressure of 1 atm for several temperatures. Calculate the value of K_p at each temperature. Is the reaction exothermic or endothermic?

Temperature (°C)	CO_2 (mol %)	CO (mol %)
850	6.23	93.77
950	1.32	98.68
1050	0.37	99.63
1200	0.06	99.94

15.98 In Section 11.5, we defined the vapor pressure of a liquid in terms of an equilibrium. (a) Write the equation representing the equilibrium between liquid water and water vapor and the corresponding expression for K_p. (b) By using data in Appendix B, give the value of K_p for this reaction at 30 °C. (c) What is the value of K_p for any liquid in equilibrium with its vapor at the normal boiling point of the liquid?

15.99 Water molecules in the atmosphere can form hydrogen-bonded dimers, $(H_2O)_2$. The presence of these dimers is thought to be important in the nucleation of ice crystals in the atmosphere and in the formation of acid rain. (a) Using VSEPR theory, draw the structure of a water dimer, using dashed lines to indicate intermolecular interactions. (b) What kind of intermolecular forces are involved in water dimer formation? (c) The K_p for water dimer formation in the gas phase is 0.050 at 300 K and 0.020 at 350 K. Is water dimer formation endothermic or exothermic?

15.100 The protein hemoglobin (Hb) transports O_2 in mammalian blood. Each Hb can bind 4 O_2 molecules. The equilibrium constant for the O_2 binding reaction is higher in fetal hemoglobin than in adult hemoglobin. In discussing protein oxygen-binding capacity, biochemists use a measure called the *P50 value*, defined as the partial pressure of oxygen at which 50% of the protein is saturated. Fetal hemoglobin has a P50 value of 19 torr, and adult hemoglobin has a P50 value of 26.8 torr. Use these data to estimate how much larger K_c is for the aqueous reaction $4\,O_2(g) + Hb(aq) \longrightarrow [Hb(O_2)_4(aq)]$ in the fetal bloodstream.

Design an Experiment

The reaction between hydrogen and iodine to form hydrogen iodide was used to illustrate Beer's law in Chapter 14 (Figure 14.5). The reaction can be monitored using visible-light spectroscopy because I_2 has a violet color while H_2 and HI are colorless. At 300 K, the equilibrium constant for the reaction $H_2(g) + I_2(g) \rightleftharpoons 2\,HI(g)$ is $K_c = 794$. To answer the following questions assume you have access to hydrogen, iodine, hydrogen iodide, a transparent reaction vessel, a visible-light spectrometer, and a means for changing the temperature. (a) Which gas (or gases) concentration could you readily monitor with the spectrometer? (b) To use Beer's law (Equation 14.5) you need to determine the extinction coefficient, ε, for the substance in question. How would you determine ε? (c) Describe an experiment for determining the equilibrium constant at 600 K. (d) Use the bond enthalpies in Table 8.4 to estimate the enthalpy of this reaction. (e) Based on your answer to part (d), would you expect K_c at 600 K to be larger or smaller than at 300 K?

16

Acid–Base Equilibria

The acids and bases that you have used so far in the laboratory are probably solutions of relatively simple inorganic substances, such as hydrochloric acid, sulfuric acid, sodium hydroxide, and the like. But acids and bases are important even when we are not in the lab.

They are ubiquitous, including in the foods we eat. The characteristic flavor of the grapes shown in the opening photograph is largely due to tartaric acid ($H_2C_4H_4O_6$) and malic acid ($H_2C_4H_4O_5$) (Figure 16.1), two closely related (they differ by only one O atom) organic acids that are found in biological systems. Fermentation of the sugars in the grapes ultimately forms vinegar, the tangy, sour flavor of which is due to acetic acid (CH_3COOH), a substance we discussed in Section 4.3. The sour taste of oranges, lemons, and other citrus fruits is due to citric acid ($H_3C_6H_5O_7$), and, to a lesser extent, ascorbic acid ($H_2C_6H_6O_6$), better known as Vitamin C.

Acids and bases are among the most important substances in chemistry, and they affect our daily lives in innumerable ways. Not only are they present in our foods, but acids and bases are also crucial components of living systems, such as the amino acids that are used to synthesize proteins and the nucleic acids that code genetic information. Both citric and malic acids are among several acids involved in the Krebs cycle (also called the citric acid cycle) that is used to generate energy in aerobic organisms. The application of acid–base chemistry has also had critical roles in shaping modern society, including such human-driven activities as industrial manufacturing, the creation of advanced pharmaceuticals, and many aspects of the environment.

The impact of acids and bases depends not only on the type of acid or base, but also on how much is present. The time required for a metal object immersed in water to corrode, the ability of an aquatic environment to support fish and plant life, the fate of pollutants washed out of the air by rain, and even the rates of reactions that maintain

▶ **CLUSTERS OF GRAPES AND BALSAMIC VINEGAR**. Grapes contain several acids that contribute to their characteristic flavor. The distinctive flavor of all vinegars is due to acetic acid. Balsamic vinegar is obtained by fermenting grapes.

WHAT'S AHEAD ▶

16.1 ACIDS AND BASES: A BRIEF REVIEW We begin by reviewing the *Arrhenius* definition of acids and bases.

16.2 BRØNSTED–LOWRY ACIDS AND BASES We learn that a Brønsted–Lowry acid is a *proton donor* and a Brønsted–Lowry base is a *proton acceptor*. Two species that differ by the presence or absence of a proton are known as a *conjugate acid–base pair*.

16.3 THE AUTOIONIZATION OF WATER We see that the *autoionization* of water produces small quantities of H_3O^+ and OH^- ions. The *equilibrium constant* for autoionization,

$K_w = [H_3O^+][OH^-]$ defines the relationship between H_3O^+ and OH^- concentrations in aqueous solutions.

16.4 THE PH SCALE We use the pH scale to describe the acidity or basicity of an aqueous solution. Neutral solutions have a pH = 7, acidic solutions have pH below 7, and basic solutions have pH above 7.

16.5 STRONG ACIDS AND BASES We categorize acids and bases as being either strong or weak electrolytes. *Strong* acids and bases are strong electrolytes, ionizing or dissociating completely in aqueous solution. *Weak* acids and bases are weak electrolytes and ionize only partially.

16.6 WEAK ACIDS We learn that the ionization of a weak acid in water is an equilibrium process with an equilibrium constant K_a that can be used to calculate the pH of a weak acid solution.

16.7 WEAK BASES We learn that the ionization of a weak base in water is an equilibrium process with equilibrium constant K_b that can be used to calculate the pH of a weak base solution.

16.8 RELATIONSHIP BETWEEN K_a AND K_b We see that K_a and K_b are related by the relationship $K_a \times K_b = K_w$. Hence, the stronger an acid, the weaker its conjugate base.

16.9 ACID–BASE PROPERTIES OF SALT SOLUTIONS We learn that the ions of a soluble ionic compound can serve as Brønsted–Lowry acids or bases.

16.10 ACID–BASE BEHAVIOR AND CHEMICAL STRUCTURE We explore the relationship between chemical structure and acid–base behavior.

16.11 LEWIS ACIDS AND BASES Finally, we see the most general definition of acids and bases, namely the Lewis acid–base definition. A Lewis acid is an *electron-pair* acceptor and a Lewis base is an *electron-pair donor*.

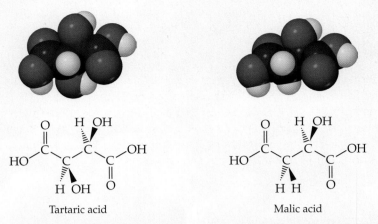

Tartaric acid   Malic acid

▲ **Figure 16.1 Two organic acids: Tartaric acid, $H_2C_4H_4O_6$, and malic acid, $H_2C_4H_4O_5$.**

our lives all critically depend on the acidity or basicity of solutions. We will thus explore in this chapter how we measure acidity and how the chemical reactions of acids and bases depend on their concentrations.

We first encountered acids and bases in Sections 2.8 and 4.3, in which we discussed the naming of acids and some simple acid–base reactions, respectively. In this chapter we take a closer look at how acids and bases are identified and characterized. In doing so, we consider their behavior both in terms of their structure and bonding and in terms of the chemical equilibria in which they participate.

16.1 | Acids and Bases: A Brief Review

From the earliest days of experimental chemistry, scientists have recognized acids and bases by their characteristic properties. Acids have a sour taste and cause certain dyes to change color, whereas bases have a bitter taste and feel slippery (soap is a good example). Use of the term *base* comes from the old English meaning of the word, "to bring low." (We still use the word *debase* in this sense, meaning to lower the value of something.) When a base is added to an acid, the base "lowers" the amount of acid. Indeed, when acids and bases are mixed in the right proportions, their characteristic properties seem to disappear altogether. ∞ (Section 4.3)

By 1830 it was evident that all acids contain hydrogen but not all hydrogen-containing substances are acids. During the 1880s, the Swedish chemist Svante Arrhenius (1859–1927) defined acids as substances that produce H^+ ions in water and bases as substances that produce OH^- ions in water. Over time the Arrhenius concept of acids and bases came to be stated in the following way:

- An *acid* is a substance that, when dissolved in water, increases the concentration of H^+ ions.

- A *base* is a substance that, when dissolved in water, increases the concentration of OH^- ions.

Hydrogen chloride gas, which is highly soluble in water, is an example of an Arrhenius acid. When it dissolves in water, $HCl(g)$ produces hydrated H^+ and Cl^- ions:

$$HCl(g) \xrightarrow{H_2O} H^+(aq) + Cl^-(aq) \qquad [16.1]$$

The aqueous solution of HCl is known as *hydrochloric acid*. Concentrated hydrochloric acid is about 37% HCl by mass and is 12 *M* in HCl.

Sodium hydroxide is an Arrhenius base. Because NaOH is an ionic compound, it dissociates into Na^+ and OH^- ions when it dissolves in water, thereby increasing the concentration of OH^- ions in the solution.

 Give It Some Thought

Which two ions are central to the Arrhenius definitions of acids and bases?

16.2 | Brønsted–Lowry Acids and Bases

The Arrhenius concept of acids and bases, while useful, is rather limited. For one thing, it is restricted to aqueous solutions. In 1923 the Danish chemist Johannes Brønsted (1879–1947) and the English chemist Thomas Lowry (1874–1936) independently proposed a more general definition of acids and bases. Their concept is based on the fact that *acid–base reactions involve the transfer of H^+ ions from one substance to another.* To understand this definition better, we need to examine the behavior of the H^+ ion in water more closely.

The H^+ Ion in Water

We might at first imagine that ionization of HCl in water produces just H^+ and Cl^-. A hydrogen ion is no more than a bare proton—a very small particle with a positive charge. As such, an H^+ ion interacts strongly with any source of electron density, such as the nonbonding electron pairs on the oxygen atoms of water molecules. For example, the interaction of a proton with water forms the **hydronium ion**, $H_3O^+(aq)$:

$$H^+ + \ddot{O}-H \longrightarrow \left[H-\ddot{O}-H \right]^+ \qquad [16.2]$$

The behavior of H^+ ions in liquid water is complex because hydronium ions interact with additional water molecules via the formation of hydrogen bonds. ∞ (Section 11.2) For example, the H_3O^+ ion bonds to additional H_2O molecules to generate such ions as $H_5O_2^+$ and $H_9O_4^+$ (▶ Figure 16.2).

Chemists use the notations $H^+(aq)$ and $H_3O^+(aq)$ interchangeably to represent the hydrated proton responsible for the characteristic properties of aqueous solutions of acids. We often use the notation $H^+(aq)$ for simplicity and convenience, as we did in Chapter 4 and Equation 16.1. The notation $H_3O^+(aq)$, however, more closely represents reality.

Proton-Transfer Reactions

In the reaction that occurs when HCl dissolves in water, the HCl molecule transfers an H^+ ion (a proton) to a water molecule. Thus, we can represent the reaction as occurring between an HCl molecule and a water molecule to form hydronium and chloride ions:

$$HCl(g) + H_2O(l) \longrightarrow Cl^-(aq) + H_3O^+(aq)$$

$$:\ddot{C}l-H + \ddot{O}-H \longrightarrow :\ddot{C}l:^- + \left[H-\ddot{O}-H \right]^+ \qquad [16.3]$$

       Acid   Base

Notice that the reaction in Equation 16.3 involves a *proton donor* (HCl) and a *proton acceptor* (H_2O). The notion of transfer from a proton donor to a proton acceptor is the key idea in the Brønsted–Lowry definition of acids and bases:

- An *acid* is a substance (molecule or ion) that *donates* a proton to another substance.
- A *base* is a substance that *accepts* a proton.

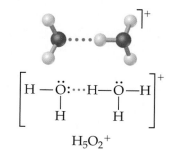

▲ GO FIGURE

Which type of intermolecular force do the dotted lines in this figure represent?

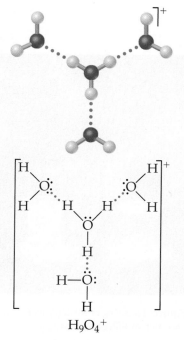

▲ Figure 16.2 **Ball-and-stick models and Lewis structures for two hydrated hydronium ions.**

Thus, when HCl dissolves in water (Equation 16.3), HCl acts as a **Brønsted–Lowry acid** (it donates a proton to H_2O), and H_2O acts as a **Brønsted–Lowry base** (it accepts a proton from HCl). We see that the H_2O molecule serves as a proton acceptor by using one of the nonbonding pairs of electrons on the O atom to "attach" the proton.

Because the emphasis in the Brønsted–Lowry concept is on proton transfer, the concept also applies to reactions that do not occur in aqueous solution. In the reaction between gas phase HCl and NH_3, for example, a proton is transferred from the acid HCl to the base NH_3:

$$[16.4]$$

Acid   Base

The hazy film that forms on the windows of general chemistry laboratories and on glassware in the laboratory (◀ Figure 16.3) is largely solid NH_4Cl formed by the gas-phase reaction between HCl and NH_3.

Let's consider another example that compares the relationship between the Arrhenius and Brønsted–Lowry definitions of acids and bases—an aqueous solution of ammonia, in which we have the equilibrium:

$$NH_3(aq) + H_2O(l) \rightleftharpoons NH_4^+(aq) + OH^-(aq) \qquad [16.5]$$
Base   Acid

Ammonia is a Brønsted–Lowry base because it accepts a proton from H_2O. Ammonia is also an Arrhenius base because adding it to water leads to an increase in the concentration of $OH^-(aq)$.

The transfer of a proton always involves both an acid (donor) and a base (acceptor). In other words, a substance can function as an acid only if another substance simultaneously behaves as a base. To be a Brønsted–Lowry acid, a molecule or ion must have a hydrogen atom it can lose as an H^+ ion. To be a Brønsted–Lowry base, a molecule or ion must have a nonbonding pair of electrons it can use to bind the H^+ ion.

Some substances can act as an acid in one reaction and as a base in another. For example, H_2O is a Brønsted–Lowry base in Equation 16.3 and a Brønsted–Lowry acid in Equation 16.5. A substance capable of acting as either an acid or a base is called **amphiprotic**. An amphiprotic substance acts as a base when combined with something more strongly acidic than itself and as an acid when combined with something more strongly basic than itself.

▲ Figure 16.3 **Fog of $NH_4Cl(s)$ caused by the reaction of $HCl(g)$ and $NH_3(g)$.**

 Give It Some Thought

In the forward reaction of this equilibrium, which substance acts as the Brønsted–Lowry base?

$$H_2S(aq) + CH_3NH_2(aq) \rightleftharpoons HS^-(aq) + CH_3NH_3^+(aq)$$

Conjugate Acid–Base Pairs

In any acid–base equilibrium, both the forward reaction (to the right) and the reverse reaction (to the left) involve proton transfer. For example, consider the reaction of an acid HA with water:

$$HA(aq) + H_2O(l) \rightleftharpoons A^-(aq) + H_3O^+(aq) \qquad [16.6]$$

In the forward reaction, HA donates a proton to H_2O. Therefore, HA is the Brønsted–Lowry acid and H_2O is the Brønsted–Lowry base. In the reverse reaction, the H_3O^+ ion donates a proton to the A^- ion, so H_3O^+ is the acid and A^- is the base. When the acid HA donates a proton, it leaves behind a substance, A^-, that can act as a base. Likewise, when H_2O acts as a base, it generates H_3O^+, which can act as an acid.

An acid and a base such as HA and A^- that differ only in the presence or absence of a proton are called a **conjugate acid–base pair.*** Every acid has a **conjugate base**, formed by removing a proton from the acid. For example, OH^- is the conjugate base of H_2O, and A^- is the conjugate base of HA. Every base has a **conjugate acid**, formed by adding a proton to the base. Thus, H_3O^+ is the conjugate acid of H_2O, and HA is the conjugate acid of A^-.

In any acid–base (proton-transfer) reaction, we can identify two sets of conjugate acid–base pairs. For example, consider the reaction between nitrous acid and water:

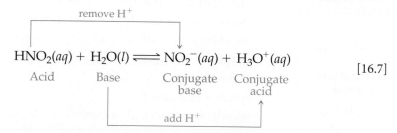

$$HNO_2(aq) + H_2O(l) \rightleftharpoons NO_2^-(aq) + H_3O^+(aq) \qquad [16.7]$$

Acid · Base · Conjugate base · Conjugate acid

Likewise, for the reaction between NH_3 and H_2O (Equation 16.5), we have

$$NH_3(aq) + H_2O(l) \rightleftharpoons NH_4^+(aq) + OH^-(aq) \qquad [16.8]$$

Base · Acid · Conjugate acid · Conjugate base

SAMPLE EXERCISE 16.1 Identifying Conjugate Acids and Bases

(a) What is the conjugate base of $HClO_4$, H_2S, PH_4^+, HCO_3^-?

(b) What is the conjugate acid of CN^-, SO_4^{2-}, H_2O, HCO_3^-?

SOLUTION

Analyze We are asked to give the conjugate base for several acids and the conjugate acid for several bases.

Plan The conjugate base of a substance is simply the parent substance minus one proton, and the conjugate acid of a substance is the parent substance plus one proton.

Solve

(a) If we remove a proton from $HClO_4$, we obtain ClO_4^-, which is its conjugate base. The other conjugate bases are HS^-, PH_3, and CO_3^{2-}.

(b) If we add a proton to CN^-, we get HCN, its conjugate acid. The other conjugate acids are HSO_4^-, H_3O^+, and H_2CO_3. Notice that the hydrogen carbonate ion (HCO_3^-) is amphiprotic. It can act as either an acid or a base.

Practice Exercise 1

Consider the following equilibrium reaction:

$$HSO_4^-(aq) + OH^-(aq) \rightleftharpoons SO_4^{2-}(aq) + H_2O(l)$$

Which substances are acting as acids in the reaction?
(a) HSO_4^- and OH^- **(b)** HSO_4^- and H_2O
(c) OH^- and SO_4^{2-} **(d)** SO_4^{2-} and H_2O
(e) None of the substances are acting as acids in this reaction.

Practice Exercise 2

Write the formula for the conjugate acid of each of the following: HSO_3^-, F^-, PO_4^{3-}, CO.

Once you become proficient at identifying conjugate acid–base pairs it is not difficult to write equations for reactions involving Brønsted–Lowry acids and bases (proton-transfer reactions).

---

*The word *conjugate* means "joined together as a pair."

SAMPLE EXERCISE 16.2 | Writing Equations for Proton-Transfer Reactions

The hydrogen sulfite ion (HSO_3^-) is amphiprotic. Write an equation for the reaction of HSO_3^- with water **(a)** in which the ion acts as an acid and **(b)** in which the ion acts as a base. In both cases identify the conjugate acid–base pairs.

SOLUTION

Analyze and Plan We are asked to write two equations representing reactions between HSO_3^- and water, one in which HSO_3^- should donate a proton to water, thereby acting as a Brønsted–Lowry acid, and one in which HSO_3^- should accept a proton from water, thereby acting as a base. We are also asked to identify the conjugate pairs in each equation.

Solve

(a) $\quad HSO_3^-(aq) + H_2O(l) \rightleftharpoons SO_3^{2-}(aq) + H_3O^+(aq)$

The conjugate pairs in this equation are HSO_3^- (acid) and SO_3^{2-} (conjugate base), and H_2O (base) and H_3O^+ (conjugate acid).

(b) $\quad HSO_3^-(aq) + H_2O(l) \rightleftharpoons H_2SO_3(aq) + OH^-(aq)$

The conjugate pairs in this equation are H_2O (acid) and OH^- (conjugate base), and HSO_3^- (base) and H_2SO_3 (conjugate acid).

Practice Exercise 1

The dihydrogen phosphate ion, $H_2PO_4^-$, is amphiprotic. In which of the following reactions is this ion serving as a base?

(i) $H_3O^+(aq) + H_2PO_4^-(aq) \rightleftharpoons H_3PO_4(aq) + H_2O(l)$

(ii) $H_3O^+(aq) + HPO_4^{2-}(aq) \rightleftharpoons H_2PO_4^-(aq) + H_2O(l)$

(iii) $H_3PO_4(aq) + HPO_4^{2-}(aq) \rightleftharpoons 2\,H_2PO_4^-(aq)$

(a) i only   **(b)** i and ii   **(c)** i and iii   **(d)** ii and iii
(e) i, ii, and iii

Practice Exercise 2

When lithium oxide (Li_2O) is dissolved in water, the solution turns basic from the reaction of the oxide ion (O^{2-}) with water. Write the equation for this reaction and identify the conjugate acid–base pairs.

Relative Strengths of Acids and Bases

Some acids are better proton donors than others, and some bases are better proton acceptors than others. If we arrange acids in order of their ability to donate a proton, we find that the more easily a substance gives up a proton, the less easily its conjugate base accepts a proton. Similarly, the more easily a base accepts a proton, the less easily its conjugate acid gives up a proton. In other words, *the stronger an acid, the weaker its conjugate base*, and *the stronger a base, the weaker its conjugate acid*. Thus, if we know how readily an acid donates protons, we also know something about how readily its conjugate base accepts protons.

The inverse relationship between the strengths of acids and their conjugate bases is illustrated in ▼ Figure 16.4. Here we have grouped acids and bases into three broad categories based on their behavior in water:

 GO FIGURE

If O^{2-} ions are added to water, what reaction, if any, occurs?

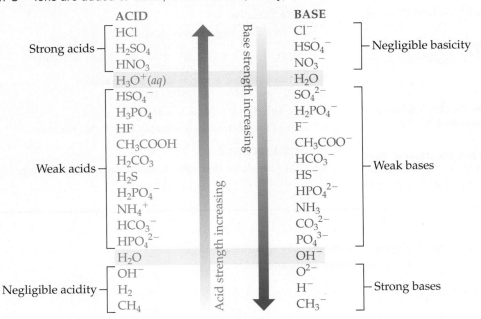

▲ Figure 16.4 **Relative strengths of select conjugate acid–base pairs.** The two members of each pair are listed opposite each other in the two columns.

1. A *strong acid* completely transfers its protons to water, leaving essentially no undissociated molecules in solution. ∞ (Section 4.3) Its conjugate base has a negligible tendency to accept protons in aqueous solution. (*The conjugate base of a strong acid shows negligible basicity.*)

2. A *weak acid* only partially dissociates in aqueous solution and therefore exists in the solution as a mixture of the undissociated acid and its conjugate base. The conjugate base of a weak acid shows a slight ability to remove protons from water. (*The conjugate base of a weak acid is a weak base.*)

3. A substance with *negligible acidity* contains hydrogen but does not demonstrate any acidic behavior in water. Its conjugate base is a strong base, reacting completely with water, to form OH^- ions. (*The conjugate base of a substance with negligible acidity is a strong base.*)

The ions $H_3O^+(aq)$ and $OH^-(aq)$ are, respectively, the strongest possible acid and strongest possible base that can exist at equilibrium in aqueous solution. Stronger acids react with water to produce $H_3O^+(aq)$ ions, and stronger bases react with water to produce $OH^-(aq)$ ions, a phenomenon known as the *leveling effect*.

 Give It Some Thought

Given that $HClO_4$ is a strong acid, how would you classify the basicity of ClO_4^-?

We can think of proton-transfer reactions as being governed by the relative abilities of two bases to abstract protons. For example, consider the proton transfer that occurs when an acid HA dissolves in water:

$$HA(aq) + H_2O(l) \rightleftharpoons H_3O^+(aq) + A^-(aq) \qquad [16.9]$$

If H_2O (the base in the forward reaction) is a stronger base than A^- (the conjugate base of HA), it is favorable to transfer the proton from HA to H_2O, producing H_3O^+ and A^-. As a result, the equilibrium lies to the right. This describes the behavior of a strong acid in water. For example, when HCl dissolves in water, the solution consists almost entirely of H_3O^+ and Cl^- ions with a negligible concentration of HCl molecules:

$$HCl(g) + H_2O(l) \longrightarrow H_3O^+(aq) + Cl^-(aq) \qquad [16.10]$$

H_2O is a stronger base than Cl^- (Figure 16.4), so H_2O acquires the proton to become the hydronium ion. Because the reaction lies completely to the right, we write Equation 16.10 with only an arrow to the right rather than using the double arrows for an equilibrium.

When A^- is a stronger base than H_2O, the equilibrium lies to the left. This situation occurs when HA is a weak acid. For example, an aqueous solution of acetic acid consists mainly of CH_3COOH molecules with only a relatively few H_3O^+ and CH_3COO^- ions:

$$CH_3COOH(aq) + H_2O(l) \rightleftharpoons H_3O^+(aq) + CH_3COO^-(aq) \qquad [16.11]$$

The CH_3COO^- ion is a stronger base than H_2O (Figure 16.4) and therefore the reverse reaction is favored more than the forward reaction.

From these examples, we conclude that *in every acid–base reaction, equilibrium favors transfer of the proton from the stronger acid to the stronger base to form the weaker acid and the weaker base.*

SAMPLE
EXERCISE 16.3   Predicting the Position of a Proton-Transfer
Equilibrium

For the following proton-transfer reaction use Figure 16.4 to predict whether the equilibrium lies to the left ($K_c < 1$) or to the right ($K_c > 1$):

$$HSO_4^-(aq) + CO_3^{2-}(aq) \rightleftharpoons SO_4^{2-}(aq) + HCO_3^-(aq)$$

SOLUTION

Analyze We are asked to predict whether an equilibrium lies to the right, favoring products, or to the left, favoring reactants.

Plan This is a proton-transfer reaction, and the position of the equilibrium will favor the proton going to the stronger of two bases. The two bases in the equation are CO_3^{2-}, the base in the forward reaction, and SO_4^{2-}, the conjugate base of HSO_4^-. We can find the relative positions of these two bases in Figure 16.4 to determine which is the stronger base.

Solve The CO_3^{2-} ion appears lower in the right-hand column in Figure 16.4 and is therefore a stronger base than SO_4^{2-}. Therefore, CO_3^{2-} will get the proton preferentially to become HCO_3^-, while SO_4^{2-} will remain mostly unprotonated. The resulting equilibrium lies to the right, favoring products (that is, $K_c > 1$):

$$\underset{\text{Acid}}{HSO_4^-(aq)} + \underset{\text{Base}}{CO_3^{2-}(aq)} \rightleftharpoons \underset{\text{Conjugate base}}{SO_4^{2-}(aq)} + \underset{\text{Conjugate acid}}{HCO_3^-(aq)} \quad K_c > 1$$

Comment Of the two acids HSO_4^- and HCO_3^-, the stronger one (HSO_4^-) gives up a proton more readily, and the weaker one (HCO_3^-) tends to retain its proton. Thus, the equilibrium favors the direction in which the proton moves from the stronger acid and becomes bonded to the stronger base.

Practice Exercise 1

Based on information in Figure 16.4, place the following equilibria in order from smallest to largest value of K_c:

(i) $CH_3COOH(aq) + HS^-(aq) \rightleftharpoons CH_3COO^-(aq) + H_2S(aq)$

(ii) $F^-(aq) + NH_4^+(aq) \rightleftharpoons HF(aq) + NH_3(aq)$

(iii) $H_2CO_3(aq) + Cl^-(aq) \rightleftharpoons HCO_3^-(aq) + HCl(aq)$

(a) i < ii < iii   **(b)** ii < i < iii   **(c)** iii < i < ii   **(d)** ii < iii < i

(e) iii < ii < i

Practice Exercise 2

For each reaction, use Figure 16.4 to predict whether the equilibrium lies to the left or to the right:

(a) $HPO_4^{2-}(aq) + H_2O(l) \rightleftharpoons H_2PO_4^-(aq) + OH^-(aq)$

(b) $NH_4^+(aq) + OH^-(aq) \rightleftharpoons NH_3(aq) + H_2O(l)$

16.3 | The Autoionization of Water

One of the most important chemical properties of water is its ability to act as either a Brønsted–Lowry acid or a Brønsted–Lowry base. In the presence of an acid, it acts as a proton acceptor; in the presence of a base, it acts as a proton donor. In fact, one water molecule can donate a proton to another water molecule:

$$H_2O(l) + H_2O(l) \rightleftharpoons OH^-(aq) + H_3O^+(aq)$$

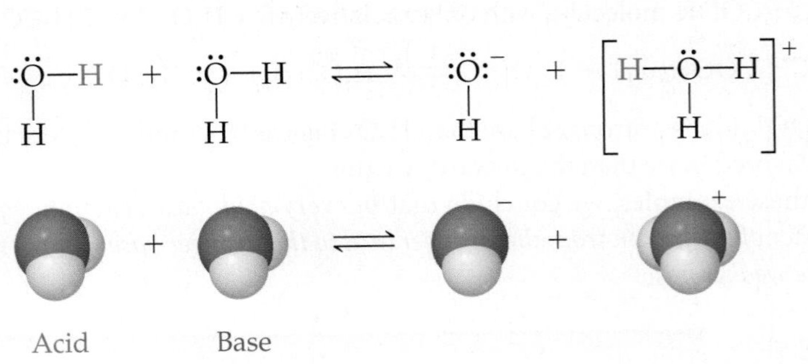

$$:\overset{\,}{\underset{H}{\ddot{O}}}-H + :\overset{\,}{\underset{H}{\ddot{O}}}-H \rightleftharpoons :\overset{\,}{\underset{H}{\ddot{O}}}:^- + \left[H-\overset{\,}{\underset{H}{\ddot{O}}}-H\right]^+ \qquad [16.12]$$

Acid   Base

We call this process the **autoionization** of water.

Because the forward and reverse reactions in Equation 16.12 are extremely rapid, no water molecule remains ionized for long. At room temperature only about two out of every 10^9 water molecules are ionized at any given instant. Thus, pure water consists almost entirely of H_2O molecules and is an extremely poor conductor of electricity. Nevertheless, the autoionization of water is very important, as we will soon see.

The Ion Product of Water

The equilibrium-constant expression for the autoionization of water is

$$K_c = [H_3O^+][OH^-] \qquad [16.13]$$

The term $[H_2O]$ is excluded from the equilibrium-constant expression because we exclude the concentrations of pure solids and liquids. ∞ (Section 15.4) Because this expression refers specifically to the autoionization of water, we use the symbol K_w to denote the equilibrium constant, which we call the **ion-product constant** for water. At 25 °C, K_w equals 1.0×10^{-14}. Thus, we have

$$K_w = [H_3O^+][OH^-] = 1.0 \times 10^{-14} \quad (\text{at } 25\,°C) \qquad [16.14]$$

Because we use $H^+(aq)$ and $H_3O^+(aq)$ interchangeably to represent the hydrated proton, the autoionization reaction for water can also be written as

$$H_2O(l) \rightleftharpoons H^+(aq) + OH^-(aq) \qquad [16.15]$$

Likewise, the expression for K_w can be written in terms of either H_3O^+ or H^+, and K_w has the same value in either case:

$$K_w = [H_3O^+][OH^-] = [H^+][OH^-] = 1.0 \times 10^{-14} \quad (\text{at } 25\,°C) \qquad [16.16]$$

This equilibrium-constant expression and the value of K_w at 25 °C are extremely important, and you should commit them to memory.

A solution in which $[H^+] = [OH^-]$ is said to be *neutral*. In most solutions, however, the H^+ and OH^- concentrations are not equal. As the concentration of one of these ions increases, the concentration of the other must decrease, so that the product of their concentrations always equals 1.0×10^{-14} (▼ **Figure 16.5**).

 GO FIGURE

Suppose that equal volumes of the middle and right samples in the figure were mixed. Would the resultant solution be acidic, neutral, or basic?

Hydrochloric acid HCl(aq)	Water H₂O	Sodium hydroxide NaOH(aq)
Acidic solution	**Neutral solution**	**Basic solution**
$[H^+] > [OH^-]$	$[H^+] = [OH^-]$	$[H^+] < [OH^-]$
$[H^+][OH^-] = 1.0 \times 10^{-14}$	$[H^+][OH^-] = 1.0 \times 10^{-14}$	$[H^+][OH^-] = 1.0 \times 10^{-14}$

▲ **Figure 16.5 Relative concentrations of H⁺ and OH⁻ in aqueous solutions at 25 °C.**

SAMPLE EXERCISE 16.4 Calculating $[H^+]$ for Pure Water

Calculate the values of $[H^+]$ and $[OH^-]$ in a neutral aqueous solution at 25 °C.

SOLUTION

Analyze We are asked to determine the concentrations of H^+ and OH^- ions in a neutral solution at 25 °C.

Plan We will use Equation 16.16 and the fact that, by definition, $[H^+] = [OH^-]$ in a neutral solution.

Solve We will represent the concentration of H^+ and OH^- in neutral solution with x. This gives

$$[H^+][OH^-] = (x)(x) = 1.0 \times 10^{-14}$$
$$x^2 = 1.0 \times 10^{-14}$$
$$x = 1.0 \times 10^{-7}M = [H^+] = [OH^-]$$

In an acid solution $[H^+]$ is greater than $1.0 \times 10^{-7}\,M$; in a basic solution $[H^+]$ is less than $1.0 \times 10^{-7}\,M$.

Practice Exercise 1

In a certain acidic solution at 25 °C, $[H^+]$ is 100 times greater than $[OH^-]$. What is the value for $[OH^-]$ for the solution?

(a) $1.0 \times 10^{-8}\ M$ **(b)** $1.0 \times 10^{-7}\ M$ **(c)** $1.0 \times 10^{-6}\ M$
(d) $1.0 \times 10^{-2}\ M$ **(e)** $1.0 \times 10^{-9}\ M$

Practice Exercise 2

Indicate whether solutions with each of the following ion concentrations are neutral, acidic, or basic: **(a)** $[H^+] = 4 \times 10^{-9}\ M$; **(b)** $[OH^-] = 1 \times 10^{-7}\ M$; **(c)** $[OH^-] = 1 \times 10^{-13}\ M$.

What makes Equation 16.16 particularly useful is that it is applicable both to pure water and to any aqueous solution. Although the equilibrium between $H^+(aq)$ and $OH^-(aq)$ as well as other ionic equilibria are affected somewhat by the presence of additional ions in solution, it is customary to ignore these ionic effects except in work requiring exceptional accuracy. Thus, Equation 16.16 is taken to be valid for any dilute aqueous solution and can be used to calculate either $[H^+]$ (if $[OH^-]$ is known) or $[OH^-]$ (if $[H^+]$ is known).

SAMPLE EXERCISE 16.5 Calculating $[H^+]$ from $[OH^-]$

Calculate the concentration of $H^+(aq)$ in **(a)** a solution in which $[OH^-]$ is 0.010 *M*, **(b)** a solution in which $[OH^-]$ is $1.8 \times 10^{-9} M$. *Note:* In this problem and all that follow, we assume, unless stated otherwise, that the temperature is 25 °C.

SOLUTION

Analyze We are asked to calculate the $[H^+]$ concentration in an aqueous solution where the hydroxide concentration is known.

Plan We can use the equilibrium-constant expression for the autoionization of water and the value of K_w to solve for each unknown concentration.

Solve

(a) Using Equation 16.16, we have

$$[H^+][OH^-] = 1.0 \times 10^{-14}$$

$$[H^+] = \frac{(1.0 \times 10^{-14})}{[OH^-]} = \frac{1.0 \times 10^{-14}}{0.010} = 1.0 \times 10^{-12}\ M$$

This solution is basic because

$$[OH^-] > [H^+]$$

(b) In this instance

$$[H^+] = \frac{(1.0 \times 10^{-14})}{[OH^-]} = \frac{1.0 \times 10^{-14}}{1.8 \times 10^{-9}} = 5.6 \times 10^{-6}\ M$$

This solution is acidic because

$$[H^+] > [OH^-]$$

Practice Exercise 1

A solution has $[OH^-] = 4.0 \times 10^{-8}$. What is the value of $[H^+]$ for the solution?

(a) $2.5 \times 10^{-8}\ M$ **(b)** $4.0 \times 10^{-8}\ M$ **(c)** $2.5 \times 10^{-7}\ M$
(d) $2.5 \times 10^{-6}\ M$ **(e)** $4.0 \times 10^{-6}\ M$

Practice Exercise 2

Calculate the concentration of $OH^-(aq)$ in a solution in which **(a)** $[H^+] = 2 \times 10^{-6}\ M$; **(b)** $[H^+] = [OH^-]$; **(c)** $[H^+] = 200 \times [OH^-]$.

16.4 | The pH Scale

The molar concentration of $H^+(aq)$ in an aqueous solution is usually very small. For convenience, we therefore usually express $[H^+]$ in terms of **pH**, which is the negative logarithm in base 10 of $[H^+]$:*

$$pH = -\log[H^+] \qquad [16.17]$$

If you need to review the use of logarithms, see Appendix A.

In Sample Exercise 16.4, we saw that $[H^+] = 1.0 \times 10^{-7}\ M$ for a neutral aqueous solution at 25 °C. We can now use Equation 16.17 to calculate the pH of a neutral solution at 25 °C:

$$pH = -\log(1.0 \times 10^{-7}) = -(-7.00) = 7.00$$

*Because $[H^+]$ and $[H_3O^+]$ are used interchangeably, you might see pH defined as $-\log[H_3O^+]$.

Table 16.1 Relationships among [H⁺], [OH⁻], and pH at 25 °C

Solution Type	$[H^+](M)$	$[OH^-](M)$	pH
Acidic	$>1.0 \times 10^{-7}$	$<1.0 \times 10^{-7}$	<7.00
Neutral	1.0×10^{-7}	1.0×10^{-7}	7.00
Basic	$<1.0 \times 10^{-7}$	$>1.0 \times 10^{-7}$	>7.00

Notice that the pH is reported with two decimal places. We do so because *only the numbers to the right of the decimal point are the significant figures in a logarithm.* Because our original value for the concentration ($1.0 \times 10^{-7} M$) has two significant figures, the corresponding pH has two decimal places (7.00).

What happens to the pH of a solution as we make the solution more acidic, so that $[H^+]$ increases? Because of the negative sign in the logarithm term of Equation 16.17, *the pH decreases as $[H^+]$ increases.* For example, when we add sufficient acid to make $[H^+] = 1.0 \times 10^{-3} M$ the pH is

$$pH = -\log(1.0 \times 10^{-3}) = -(-3.00) = 3.00$$

At 25 °C the pH of an acidic solution is less than 7.00.

We can also calculate the pH of a basic solution, one in which $[OH^-] > 1.0 \times 10^{-7} M$. Suppose $[OH^-] = 2.0 \times 10^{-3} M$. We can use Equation 16.16 to calculate $[H^+]$ for this solution and Equation 16.17 to calculate the pH:

$$[H^+] = \frac{K_w}{[OH^-]} = \frac{1.0 \times 10^{-14}}{2.0 \times 10^{-3}} = 5.0 \times 10^{-12} M$$

$$pH = -\log(5.0 \times 10^{-12}) = 11.30$$

At 25 °C the pH of a basic solution is greater than 7.00. The relationships among $[H^+]$, $[OH^-]$, and pH are summarized in ▲ Table 16.1.

Give It Some Thought

Is it possible for a solution to have a negative pH? If so, would that pH signify a basic or acidic solution?

One might think that when $[H^+]$ is very small, as is often the case, it would be unimportant. That reasoning is quite incorrect! Remember that many chemical processes depend on the ratio of changes in concentration. For example, if a kinetic rate law is first order in $[H^+]$, doubling the H^+ concentration doubles the rate even if the change is merely from $1 \times 10^{-7} M$ to $2 \times 10^{-7} M$. ∞ (Section 14.3) In biological systems, many reactions involve proton transfers and have rates that depend on $[H^+]$. Because the speeds of these reactions are crucial, the pH of biological fluids must be maintained within narrow limits. For example, human blood has a normal pH range of 7.35 to 7.45. Illness and even death can result if the pH varies much from this narrow range.

SAMPLE EXERCISE 16.6 | Calculating pH from [H⁺]

Calculate the pH values for the two solutions of Sample Exercise 16.5.

SOLUTION

Analyze We are asked to determine the pH of aqueous solutions for which we have already calculated $[H^+]$.

Plan We can calculate pH using its defining equation, Equation 16.17.

Solve

(a) In the first instance we found $[H^+]$ to be $1.0 \times 10^{-12} M$, so that

$$pH = -\log(1.0 \times 10^{-12}) = -(-12.00) = 12.00$$

Because 1.0×10^{-12} has two significant figures, the pH has two decimal places, 12.00.

(b) For the second solution, $[H^+] = 5.6 \times 10^{-6} M$. Before performing the calculation, it is helpful to estimate the pH. To do so, we note that $[H^+]$ lies between 1×10^{-6} and 1×10^{-5}. Thus, we expect the pH to lie between 6.0 and 5.0. We use Equation 16.17 to calculate the pH:

$$pH = -\log(5.6 \times 10^{-6}) = 5.25$$

Check After calculating a pH, it is useful to compare it to your estimate. In this case the pH, as we predicted, falls between 6 and 5. Had the calculated pH and the estimate not agreed, we should have reconsidered our calculation or estimate or both.

pOH and Other "p" Scales

The negative logarithm is a convenient way of expressing the magnitudes of other small quantities. We use the convention that the negative logarithm of a quantity is labeled "p" (quantity). Thus, we can express the concentration of OH^- as pOH:

$$pOH = -\log[OH^-] \qquad [16.18]$$

Likewise, pK_w equals $-\log K_w$.

By taking the negative logarithm of both sides of the equilibrium-constant expression for water, $K_w = [H^+][OH^-]$, we obtain

$$-\log[H^+] + (-\log[OH^-]) = -\log K_w \qquad [16.19]$$

from which we obtain the useful expression

$$pH + pOH = 14.00 \quad (at\ 25\,°C) \qquad [16.20]$$

The pH and pOH values characteristic of a number of familiar solutions are shown in (▼ Figure 16.6). Notice that a change in $[H^+]$ by a factor of 10 causes the pH to change by 1. Thus, the concentration of $H^+(aq)$ in a solution of pH 5 is 10 times the $H^+(aq)$ concentration in a solution of pH 6.

▲ GO FIGURE

Which is more acidic, black coffee or lemon juice?

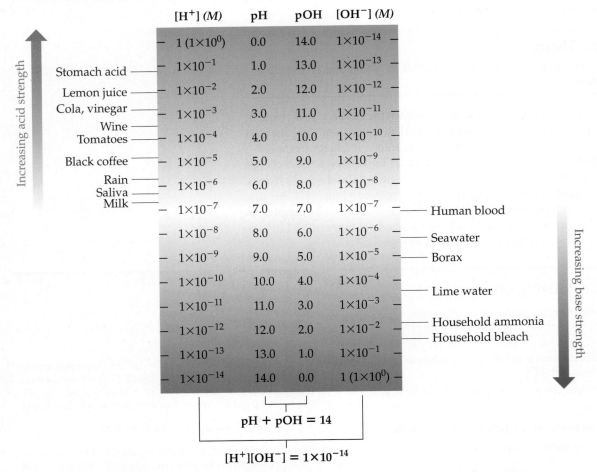

▲ Figure 16.6 **Concentrations of H$^+$ and OH$^-$, and pH and pOH values of some common substances at 25 °C.**

Give It Some Thought

If the pOH for a solution is 3.00, what is the pH? Is the solution acidic or basic?

SAMPLE EXERCISE 16.7 Calculating [H⁺] from pOH

A sample of freshly pressed apple juice has a pOH of 10.24. Calculate $[H^+]$.

SOLUTION

Analyze We need to calculate $[H^+]$ from pOH.

Plan We will first use Equation 16.20, pH + pOH = 14.00, to calculate pH from pOH. Then we will use Equation 16.17 to determine the concentration of H^+.

Solve From Equation 16.20, we have

$$pH = 14.00 - pOH$$
$$pH = 14.00 - 10.24 = 3.76$$

Next we use Equation 16.17:

$$pH = -\log[H^+] = 3.76$$

Thus,

$$\log[H^+] = -3.76$$

To find $[H^+]$, we need to determine the *antilogarithm* of -3.76. Your calculator will show this command as 10^x or INV log (these functions are usually above the log key). We use this function to perform the calculation:

$$[H^+] = \text{antilog}\,(-3.76) = 10^{-3.76} = 1.7 \times 10^{-4}\,M$$

Comment The number of significant figures in $[H^+]$ is two because the number of decimal places in the pH is two.

Check Because the pH is between 3.0 and 4.0, we know that $[H^+]$ will be between $1.0 \times 10^{-3}\,M$ and $1.0 \times 10^{-4}\,M$. Our calculated $[H^+]$ falls within this estimated range.

Practice Exercise 1

A solution at 25 °C has pOH = 10.53. Which of the following statements is or are true?
  (i) The solution is acidic.
 (ii) The pH of the solution is 14.00 − 10.53.
(iii) For this solution, $[OH^-] = 10^{-10.53}\,M$.

(a) Only one of the statements is true.
(b) Statements (i) and (ii) are true.
(c) Statements (i) and (iii) are true.
(d) Statements (ii) and (iii) are true.
(e) All three statements are true.

Practice Exercise 2

A solution formed by dissolving an antacid tablet has a pOH of 4.82. Calculate $[H^+]$.

Measuring pH

The pH of a solution can be measured with a *pH meter* (▶ Figure 16.7). A complete understanding of how this important device works requires a knowledge of electrochemistry, a subject we take up in Chapter 20. In brief, a pH meter consists of a pair of electrodes connected to a meter capable of measuring small voltages, on the order of millivolts. A voltage, which varies with pH, is generated when the electrodes are placed in a solution. This voltage is read by the meter, which is calibrated to give pH.

Although less precise, acid–base indicators can be used to measure pH. An acid–base indicator is a colored substance that can exist in either an acid or a base form. The two forms have different colors. Thus, the indicator has one color at lower pH and another at higher pH. If you know the pH at which the indicator turns from one form to the other, you can determine whether a solution has a higher or lower pH than this value. Litmus, for example, changes color in the vicinity of pH 7. The color change, however, is not very sharp. Red litmus indicates a pH of about 5 or lower, and blue litmus indicates a pH of about 8 or higher.

Some common indicators are listed in **Figure 16.8**. The chart tells us, for instance, that methyl red changes color over the pH interval from about 4.5 to 6.0. Below pH 4.5 it is in the acid form, which is red. In the interval between 4.5 and 6.0, it is gradually converted to its basic form, which is yellow. Once the pH rises above 6 the conversion

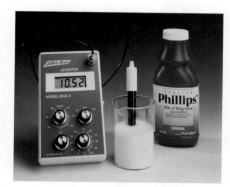

▲ **Figure 16.7 A digital pH meter.** The device is a millivoltmeter, and the electrodes immersed in a solution produce a voltage that depends on the pH of the solution.

▲ GO FIGURE

Which of these indicators is best suited to distinguish between a solution that is slightly acidic and one that is slightly basic?

Methyl red

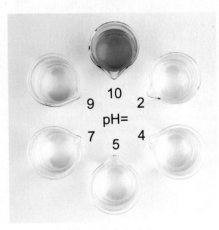

Bromthymol blue

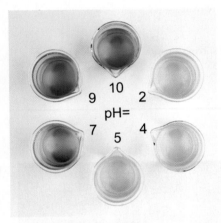

Phenolphthalein

▲ Figure 16.9 **Solutions containing three common acid–base indicators at various pH values.**

▲ GO FIGURE

If a colorless solution turns pink when we add phenolphthalein, what can we conclude about the pH of the solution?

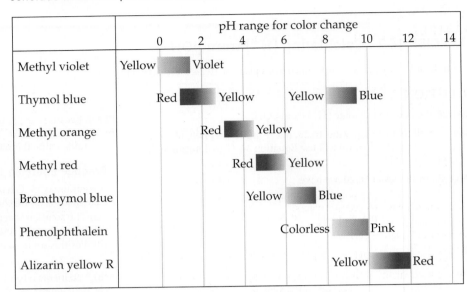

▲ Figure 16.8 **pH ranges for common acid–base indicators.** Most indicators have a useful range of about 2 pH units.

is complete, and the solution is yellow. This color change, along with that of the indicators bromthymol blue and phenolphthalein, is shown in ◄ Figure 16.9. Paper tape impregnated with several indicators is widely used for determining approximate pH values.

16.5 | Strong Acids and Bases

The chemistry of an aqueous solution often depends critically on pH. It is therefore important to examine how pH relates to acid and base concentrations. The simplest cases are those involving strong acids and strong bases. Strong acids and bases are *strong electrolytes*, existing in aqueous solution entirely as ions. There are relatively few common strong acids and bases (see Table 4.2).

Strong Acids

The seven most common strong acids include six monoprotic acids (HCl, HBr, HI, HNO_3, $HClO_3$, and $HClO_4$), and one diprotic acid (H_2SO_4). Nitric acid (HNO_3) exemplifies the behavior of the monoprotic strong acids. For all practical purposes, an aqueous solution of HNO_3 consists entirely of H_3O^+ and NO_3^- ions:

$$HNO_3(aq) + H_2O(l) \longrightarrow H_3O^+(aq) + NO_3^-(aq) \quad \text{(complete ionization)} \quad [16.21]$$

We have not used equilibrium arrows for this equation because the reaction lies entirely to the right. ∞ (Section 4.1) As noted in Section 16.3, we use $H_3O^+(aq)$ and $H^+(aq)$ interchangeably to represent the hydrated proton in water. Thus, we can simplify this acid ionization equation to

$$HNO_3(aq) \longrightarrow H^+(aq) + NO_3^-(aq)$$

In an aqueous solution of a strong acid, the acid is normally the only significant source of H^+ ions.* As a result, calculating the pH of a solution of a strong monoprotic

---

*If the concentration of the acid is $10^{-6}\,M$ or less, we also need to consider H^+ ions that result from H_2O autoionization. Normally, the concentration of H^+ from H_2O is so small that it can be neglected.

acid is straightforward because $[H^+]$ equals the original concentration of acid. In a 0.20 M solution of $HNO_3(aq)$, for example, $[H^+] = [NO_3^-] = 0.20\ M$. The situation with the diprotic acid H_2SO_4 is somewhat more complex, as we will see in Section 16.6.

SAMPLE
EXERCISE 16.8 | Calculating the pH of a Strong Acid

What is the pH of a 0.040 M solution of $HClO_4$?

SOLUTION

Analyze and Plan Because $HClO_4$ is a strong acid, it is completely ionized, giving $[H^+] = [ClO_4^-] = 0.040\ M$.

Solve

$$pH = -\log(0.040) = 1.40$$

Check Because $[H^+]$ lies between 1×10^{-2} and 1×10^{-1}, the pH will be between 2.0 and 1.0. Our calculated pH falls within the estimated range. Furthermore, because the concentration has two significant figures, the pH has two decimal places.

Practice Exercise 1
Order the following three solutions from smallest to largest pH:
(i) 0.20 M $HClO_3$ (ii) 0.0030 M HNO_3 (iii) 1.50 M HCl
(a) i < ii < iii **(b)** ii < i < iii **(c)** iii < i < ii
(d) ii < iii < i **(e)** iii < ii < i

Practice Exercise 2
An aqueous solution of HNO_3 has a pH of 2.34. What is the concentration of the acid?

Strong Bases

The most common soluble strong bases are the ionic hydroxides of the alkali metals, such as NaOH, KOH, and the ionic hydroxides heavier alkaline earth metals, such as $Sr(OH)_2$. These compounds completely dissociate into ions in aqueous solution. Thus, a solution labeled 0.30 M NaOH consists of 0.30 M $Na^+(aq)$ and 0.30 M $OH^-(aq)$; there is essentially no undissociated NaOH.

 Give It Some Thought

Which solution has the higher pH, a 0.001 M solution of NaOH or a 0.001 M solution of $Ba(OH)_2$?

SAMPLE
EXERCISE 16.9 | Calculating the pH of a Strong Base

What is the pH of **(a)** a 0.028 M solution of NaOH, **(b)** a 0.0011 M solution of $Ca(OH)_2$?

SOLUTION

Analyze We are asked to calculate the pH of two solutions of strong bases.

Plan We can calculate each pH by either of two equivalent methods. First, we could use Equation 16.16 to calculate $[H^+]$ and then use Equation 16.17 to calculate the pH. Alternatively, we could use $[OH^-]$ to calculate pOH and then use Equation 16.20 to calculate the pH.

Solve

(a) NaOH dissociates in water to give one OH^- ion per formula unit. Therefore, the OH^- concentration for the solution in (a) equals the stated concentration of NaOH, namely 0.028 M.

Method 1:

$$[H^+] = \frac{1.0 \times 10^{-14}}{0.028} = 3.57 \times 10^{-13}\ M \qquad pH = -\log(3.57 \times 10^{-13}) = 12.45$$

Method 2:

$$pOH = -\log(0.028) = 1.55 \qquad pH = 14.00 - pOH = 12.45$$

(b) $Ca(OH)_2$ is a strong base that dissociates in water to give *two* OH^- ions per formula unit. Thus, the concentration of $OH^-(aq)$ for the solution in part (b) is $2 \times (0.0011\ M) = 0.0022\ M$.

Method 1:

$$[H^+] = \frac{1.0 \times 10^{-14}}{0.0022} = 4.55 \times 10^{-12}\ M \qquad pH = -\log(4.55 \times 10^{-12}) = 11.34$$

Method 2:

$$pOH = -\log(0.0022) = 2.66 \qquad pH = 14.00 - pOH = 11.34$$

Although all of the alkali metal hydroxides are strong electrolytes, LiOH, RbOH, and CsOH are not commonly encountered in the laboratory. The hydroxides of the heavier alkaline earth metals—Ca(OH)$_2$, Sr(OH)$_2$, and Ba(OH)$_2$—are also strong electrolytes. They have limited solubility, however, so they are used only when high solubility is not critical.

Strongly basic solutions are also created by certain substances that react with water to form OH$^-$(aq). The most common of these contain the oxide ion. Ionic metal oxides, especially Na$_2$O and CaO, are often used in industry when a strong base is needed. The O^{2-} reacts very exothermically with water to form OH$^-$, leaving virtually no O^{2-} in the solution:

$$O^{2-}(aq) + H_2O(l) \longrightarrow 2\,OH^-(aq) \qquad [16.22]$$

Thus, a solution formed by dissolving 0.010 mol of Na$_2$O(s) in enough water to form 1.0 L of solution has [OH$^-$] = 0.020 M and a pH of 12.30.

 Give It Some Thought

The CH$_3^-$ ion is the conjugate base of CH$_4$, and CH$_4$ shows no evidence of being an acid in water. Write a balanced equation for the reaction of CH$_3^-$ and water.

16.6 | Weak Acids

Most acidic substances are weak acids and therefore only partially ionized in aqueous solution (▶ Figure 16.10). We can use the equilibrium constant for the ionization reaction to express the extent to which a weak acid ionizes. If we represent a general weak acid as HA, we can write the equation for its ionization in either of the following ways, depending on whether the hydrated proton is represented as H$_3$O$^+$(aq) or H$^+$(aq):

$$HA(aq) + H_2O(l) \rightleftharpoons H_3O^+(aq) + A^-(aq) \qquad [16.23]$$

or

$$HA(aq) \rightleftharpoons H^+(aq) + A^-(aq) \qquad [16.24]$$

These equilibria are in aqueous solution, so we will use equilibrium-constant expressions based on concentrations. Because H$_2$O is the solvent, it is omitted from the equilibrium-constant expression. ∞ (Section 15.4) Further, we add a subscript a on the equilibrium constant to indicate that it is an equilibrium constant for the ionization of an *acid*. Thus, we can write the equilibrium-constant expression as either:

$$K_a = \frac{[H_3O^+][A^-]}{[HA]} \quad \text{or} \quad K_a = \frac{[H^+][A^-]}{[HA]} \qquad [16.25]$$

K_a is called the **acid-dissociation constant** for acid HA.

▶ Table 16.2 shows the structural formulas, conjugate bases, and K_a values for a number of weak acids. Appendix D provides a more complete list. Many weak acids are organic compounds composed entirely of carbon, hydrogen, and oxygen. These compounds usually contain some hydrogen atoms bonded to carbon atoms and some bonded to oxygen atoms. In almost all cases, the hydrogen atoms bonded to carbon do not ionize in water; instead, the acidic behavior of these compounds is due to the hydrogen atoms attached to oxygen atoms.

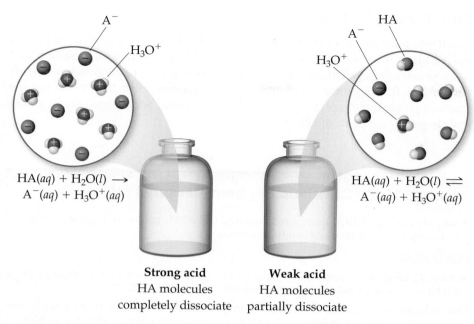

$$HA(aq) + H_2O(l) \longrightarrow \\ A^-(aq) + H_3O^+(aq)$$

$$HA(aq) + H_2O(l) \rightleftharpoons \\ A^-(aq) + H_3O^+(aq)$$

Strong acid
HA molecules
completely dissociate

Weak acid
HA molecules
partially dissociate

▲ **Figure 16.10 Species present in a solution of a strong acid and a weak acid.**

Table 16.2   Some Weak Acids in Water at 25 °C

Acid	Structural Formula*	Conjugate Base	K_a
Chlorous ($HClO_2$)	H—O—Cl—O	ClO_2^-	1.0×10^{-2}
Hydrofluoric (HF)	H—F	F^-	6.8×10^{-4}
Nitrous (HNO_2)	H—O—N=O	NO_2^-	4.5×10^{-4}
Benzoic (C_6H_5COOH)	H—O—C(=O)—⬡	$C_6H_5COO^-$	6.3×10^{-5}
Acetic (CH_3COOH)	H—O—C(=O)—C(H)(H)—H	CH_3COO^-	1.8×10^{-5}
Hypochlorous (HOCl)	H—O—Cl	OCl^-	3.0×10^{-8}
Hydrocyanic (HCN)	H—C≡N	CN^-	4.9×10^{-10}
Phenol (HOC_6H_5)	H—O—⬡	$C_6H_5O^-$	1.3×10^{-10}

*The proton that ionizes is shown in red.

The magnitude of K_a indicates the tendency of the acid to ionize in water: *The larger the value of K_a, the stronger the acid.* Chlorous acid ($HClO_2$), for example, is the strongest acid in Table 16.2, and phenol (HOC_6H_5) is the weakest. For most weak acids K_a values range from 10^{-2} to 10^{-10}.

Give It Some Thought

Based on the entries in Table 16.2, which element is most commonly bonded to the acidic hydrogen?

Calculating K_a from pH

In order to calculate either the K_a value for a weak acid or the pH of its solutions, we will use many of the skills for solving equilibrium problems developed in Section 15.5. In many cases the small magnitude of K_a allows us to use approximations to simplify the problem. In doing these calculations, it is important to realize that proton-transfer reactions are generally very rapid. As a result, the measured or calculated pH for a weak acid always represents an equilibrium condition.

SAMPLE EXERCISE 16.10 Calculating K_a from Measured pH

A student prepared a 0.10 M solution of formic acid (HCOOH) and found its pH at 25 °C to be 2.38. Calculate K_a for formic acid at this temperature.

SOLUTION

Analyze We are given the molar concentration of an aqueous solution of weak acid and the pH of the solution, and we are asked to determine the value of K_a for the acid.

Plan Although we are dealing specifically with the ionization of a weak acid, this problem is very similar to the equilibrium problems we encountered in Chapter 15. We can solve this problem using the method first outlined in Sample Exercise 15.8, starting with the chemical reaction and a tabulation of initial and equilibrium concentrations.

Solve The first step in solving any equilibrium problem is to write the equation for the equilibrium reaction. The ionization of formic acid can be written as

$$HCOOH(aq) \rightleftharpoons H^+(aq) + HCOO^-(aq)$$

The equilibrium-constant expression is

$$K_a = \frac{[H^+][HCOO^-]}{[HCOOH]}$$

From the measured pH, we can calculate $[H^+]$:

$$pH = -\log[H^+] = 2.38$$
$$\log[H^+] = -2.38$$
$$[H^+] = 10^{-2.38} = 4.2 \times 10^{-3} M$$

To determine the concentrations of the species involved in the equilibrium, we imagine that the solution is initially 0.10 M in HCOOH molecules. We then consider the ionization of the acid into H^+ and $HCOO^-$. For each HCOOH molecule that ionizes, one H^+ ion and one $HCOO^-$ ion are produced in solution. Because the pH measurement indicates that $[H^+] = 4.2 \times 10^{-3} M$ at equilibrium, we can construct the following table:

	$HCOOH(aq)$	$\rightleftharpoons$ $H^+(aq)$	$+$ $HCOO^-(aq)$
Initial concentration (M)	0.10	0	0
Change in concentration (M)	-4.2×10^{-3}	$+4.2 \times 10^{-3}$	$+4.2 \times 10^{-3}$
Equilibrium concentration (M)	$(0.10 - 4.2 \times 10^{-3})$	4.2×10^{-3}	4.2×10^{-3}

Notice that we have neglected the very small concentration of $H^+(aq)$ due to H_2O autoionization. Notice also that the amount of HCOOH that ionizes is very small compared with the initial concentration of the acid. To the number of significant figures we are using, the subtraction yields 0.10 M:

$$(0.10 - 4.2 \times 10^{-3}) M \approx 0.10 M$$

We can now insert the equilibrium concentrations into the expression for K_a:

$$K_a = \frac{(4.2 \times 10^{-3})(4.2 \times 10^{-3})}{0.10} = 1.8 \times 10^{-4}$$

Check The magnitude of our answer is reasonable because K_a for a weak acid is usually between 10^{-2} and 10^{-10}.

Percent Ionization

We have seen that the magnitude of K_a indicates the strength of a weak acid. Another measure of acid strength is **percent ionization**, defined as

$$\text{Percent ionization} = \frac{\text{concentration of ionized HA}}{\text{original concentration of HA}} \times 100\% \qquad [16.26]$$

The stronger the acid, the greater the percent ionization.

If we assume that the autoionization of H_2O is negligible, the concentration of acid that ionizes equals the concentration of $H^+(aq)$ that forms. Thus, the percent ionization for an acid HA can be expressed as

$$\text{Percent ionization} = \frac{[H^+]_{\text{equilibrium}}}{[HA]_{\text{initial}}} \times 100\% \qquad [16.27]$$

For example, a 0.035 M solution of HNO_2 contains $3.7 \times 10^{-3}\ M\ H^+(aq)$ and its percent ionization is

$$\text{Percent ionization} = \frac{[H^+]_{\text{equilibrium}}}{[HNO_2]_{\text{initial}}} \times 100\% = \frac{3.7 \times 10^{-3}\ M}{0.035\ M} \times 100\% = 11\%$$

SAMPLE EXERCISE 16.11 | Calculating Percent Ionization

As calculated in Sample Exercise 16.10, a 0.10 M solution of formic acid (HCOOH) contains $4.2 \times 10^{-3}\ M\ H^+(aq)$. Calculate the percentage of the acid that is ionized.

SOLUTION

Analyze We are given the molar concentration of an aqueous solution of weak acid and the equilibrium concentration of $H^+(aq)$ and asked to determine the percent ionization of the acid.

Plan The percent ionization is given by Equation 16.27.

Solve

$$\text{Percent ionization} = \frac{[H^+]_{\text{equilibrium}}}{[HCOOH]_{\text{initial}}} \times 100\% = \frac{4.2 \times 10^{-3}\ M}{0.10\ M} \times 100\% = 4.2$$

Using K_a to Calculate pH

Knowing the value of K_a and the initial concentration of a weak acid, we can calculate the concentration of $H^+(aq)$ in a solution of the acid. Let's calculate the pH at 25 °C of a 0.30 M solution of acetic acid (CH_3COOH), the weak acid responsible for the characteristic odor and acidity of vinegar.

1. Our first step is to write the ionization equilibrium:

$$CH_3COOH(aq) \rightleftharpoons H^+(aq) + CH_3COO^-(aq) \qquad [16.28]$$

Notice that the hydrogen that ionizes is the one attached to an oxygen atom.

2. The second step is to write the equilibrium-constant expression and the value for the equilibrium constant. Taking $K_a = 1.8 \times 10^{-5}$ from Table 16.2, we write

$$K_a = \frac{[H^+][CH_3COO^-]}{[CH_3COOH]} = 1.8 \times 10^{-5} \qquad [16.29]$$

3. The third step is to express the concentrations involved in the equilibrium reaction. This can be done with a little accounting, as described in Sample Exercise 16.10. Because we want to find the equilibrium value for $[H^+]$, let's call this quantity x. The concentration of acetic acid before any of it ionizes is 0.30 M. The chemical equation tells us that for each molecule of CH_3COOH that ionizes, one $H^+(aq)$ and one $CH_3COO^-(aq)$ are formed. Consequently, if x moles per liter of $H^+(aq)$ form at equilibrium, x moles per liter of $CH_3COO^-(aq)$ must also form and x moles per liter of CH_3COOH must be ionized:

$$CH_3COOH(aq) \rightleftharpoons H^+(aq) \quad + \quad CH_3COO^-(aq)$$

	CH_3COOH	H^+	CH_3COO^-
Initial concentration (M)	0.30	0	0
Change in concentration (M)	$-x$	$+x$	$+x$
Equilibrium concentration (M)	$(0.30 - x)$	x	x

4. The fourth step is to substitute the equilibrium concentrations into the equilibrium-constant expression and solve for x:

$$K_a = \frac{[H^+][CH_3COO^-]}{[CH_3COOH]} = \frac{(x)(x)}{0.30 - x} = 1.8 \times 10^{-5} \qquad [16.30]$$

This expression leads to a quadratic equation in x, which we can solve by using either an equation-solving calculator or the quadratic formula. We can simplify the problem, however, by noting that the value of K_a is quite small. As a result, we anticipate that the equilibrium lies far to the left and that x is much smaller than the initial concentration of acetic acid. Thus, we *assume* that x is negligible relative to 0.30, so that $0.30 - x$ is essentially equal to 0.30. We can (and should!) check the validity of this assumption when we finish the problem. By using this assumption, Equation 16.30 becomes

$$K_a = \frac{x^2}{0.30} = 1.8 \times 10^{-5}$$

Solving for x, we have

$$x^2 = (0.30)(1.8 \times 10^{-5}) = 5.4 \times 10^{-6}$$

$$x = \sqrt{5.4 \times 10^{-6}} = 2.3 \times 10^{-3}$$

$$[H^+] = x = 2.3 \times 10^{-3}\, M$$

$$pH = -\log(2.3 \times 10^{-3}) = 2.64$$

Now we check the validity of our simplifying assumption that $0.30 - x \simeq 0.30$. The value of x we determined is so small that, for this number of significant figures, the assumption is entirely valid. We are thus satisfied that the assumption was a reasonable one to make. Because x represents the moles per liter of acetic acid that ionize, we see that, in this particular case, less than 1% of the acetic acid molecules ionize:

$$\text{Percent ionization of CH}_3\text{COOH} = \frac{0.0023 \ M}{0.30 \ M} \times 100\% = 0.77\%$$

As a general rule, if x is more than about 5% of the initial concentration value, it is better to use the quadratic formula. You should always check the validity of any simplifying assumptions after you have finished solving a problem.

We have also made one other assumption, namely that all of the H^+ in the solution comes from ionization of CH_3COOH. Are we justified in neglecting the autoionization of H_2O? The answer is yes—the additional $[H^+]$ due to water, which would be on the order of $10^{-7}\ M$, is negligible compared to the $[H^+]$ from the acid (which in this case is on the order of $10^{-3}\ M$). In extremely precise work, or in cases involving very dilute solutions of acids, we would need to consider the autoionization of water more fully.

 Give It Some Thought

Would a $1.0 \times 10^{-8}M$ solution of HCl have pH < 7, pH $= 7$, or pH > 7?

Finally, we can compare the pH value of this weak acid with the pH of a solution of a strong acid of the same concentration. The pH of the 0.30 M acetic acid is 2.64, but the pH of a 0.30 M solution of a strong acid such as HCl is $-\log(0.30) = 0.52$. As expected, the pH of a solution of a weak acid is higher than that of a solution of a strong acid of the same molarity. (Remember, the higher the pH value, the *less* acidic the solution.)

SAMPLE EXERCISE 16.12 Using K_a to Calculate pH

Calculate the pH of a 0.20 M solution of HCN. (Refer to Table 16.2 or Appendix D for the value of K_a.)

SOLUTION

Analyze We are given the molarity of a weak acid and are asked for the pH. From Table 16.2, K_a for HCN is 4.9×10^{-10}.

Plan We proceed as in the example just worked in the text, writing the chemical equation and constructing a table of initial and equilibrium concentrations in which the equilibrium concentration of H^+ is our unknown.

Solve Writing both the chemical equation for the ionization reaction that forms $H^+(aq)$ and the equilibrium-constant (K_a) expression for the reaction:

$$HCN(aq) \rightleftharpoons H^+(aq) + CN^-(aq)$$

$$K_a = \frac{[H^+][CN^-]}{[HCN]} = 4.9 \times 10^{-10}$$

Next, we tabulate the concentrations of the species involved in the equilibrium reaction, letting $x = [H^+]$ at equilibrium:

	$HCN(aq)$ $\rightleftharpoons$	$H^+(aq)$ +	$CN^-(aq)$
Initial concentration (M)	0.20	0	0
Change in concentration (M)	$-x$	$+x$	$+x$
Equilibrium concentration (M)	$(0.20 - x)$	x	x

Substituting the equilibrium concentrations into the equilibrium-constant expression yields

$$K_a = \frac{(x)(x)}{0.20 - x} = 4.9 \times 10^{-10}$$

We next make the simplifying approximation that x, the amount of acid that dissociates, is small compared with the initial concentration of acid, $0.20 - x \simeq 0.20$. Thus,

$$\frac{x^2}{0.20} = 4.9 \times 10^{-10}$$

Solving for x, we have

$$x^2 = (0.20)(4.9 \times 10^{-10}) = 0.98 \times 10^{-10}$$
$$x = \sqrt{0.98 \times 10^{-10}} = 9.9 \times 10^{-6}\,M = [\text{H}^+]$$

A concentration of $9.9 \times 10^{-6}\,M$ is much smaller than 5% of 0.20, the initial HCN concentration. Our simplifying approximation is therefore appropriate. We now calculate the pH of the solution:

$$\text{pH} = -\log[\text{H}^+] = -\log(9.9 \times 10^{-6}) = 5.00$$

Practice Exercise 1

What is the pH of a 0.40 M solution of benzoic acid, C_6H_5COOH? (The K_a value for benzoic acid is given in Table 16.2.)

(a) 2.30   (b) 2.10   (c) 1.90   (d) 4.20   (e) 4.60

Practice Exercise 2

The K_a for niacin (Practice Exercise 16.10) is 1.5×10^{-5}. What is the pH of a 0.010 M solution of niacin?

The properties of an acid solution that relate directly to the concentration of $\text{H}^+(aq)$, such as electrical conductivity and rate of reaction with an active metal, are less evident for a solution of a weak acid than for a solution of a strong acid of the same concentration. ▼ Figure 16.11 presents an experiment that demonstrates this difference with 1 M CH_3COOH and 1 M HCl. The concentration of $\text{H}^+(aq)$ in 1 M CH_3COOH is only 0.004 M, whereas the 1 M HCl solution contains 1 M $\text{H}^+(aq)$. As a result, the reaction rate with the metal is much faster in the HCl solution.

As the concentration of a weak acid increases, the equilibrium concentration of $\text{H}^+(aq)$ increases, as expected. However, as shown in ▶ Figure 16.12, *the percent ionization decreases as the concentration increases*. Thus, the concentration of $\text{H}^+(aq)$ is not directly proportional to the concentration of the weak acid. For example, doubling the concentration of a weak acid does not double the concentration of $\text{H}^+(aq)$.

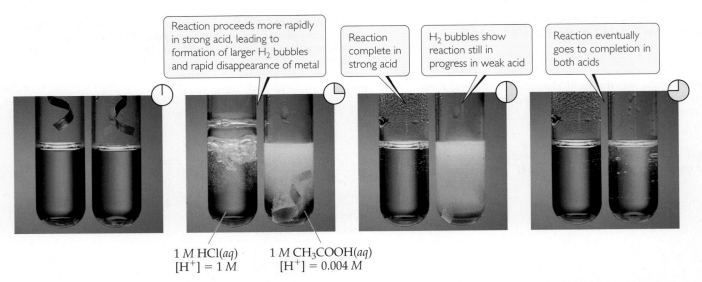

Reaction proceeds more rapidly in strong acid, leading to formation of larger H_2 bubbles and rapid disappearance of metal

Reaction complete in strong acid

H_2 bubbles show reaction still in progress in weak acid

Reaction eventually goes to completion in both acids

1 M HCl(aq)
[H^+] = 1 M

1 M CH_3COOH(aq)
[H^+] = 0.004 M

▲ **Figure 16.11 Rates of the same reaction run in a weak acid and a strong acid.** The bubbles are H_2 gas, which along with metal cations, is produced when a metal is oxidized by an acid. ∞(Section 4.4)

▲ GO FIGURE

Is the trend observed in this graph consistent with Le Châtelier's principle? Explain.

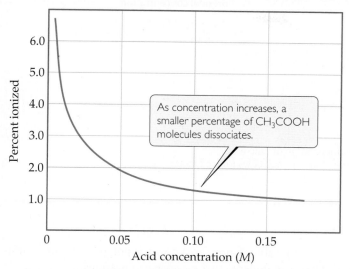

As concentration increases, a smaller percentage of CH_3COOH molecules dissociates.

▲ **Figure 16.12 Effect of concentration on percent ionization in an acetic acid solution.**

SAMPLE EXERCISE 16.13 | Using the Quadratic Equation to Calculate pH and Percent Ionization

Calculate the pH and percentage of HF molecules ionized in a 0.10 M HF solution.

SOLUTION

Analyze We are asked to calculate the percent ionization of a solution of HF. From Appendix D, we find $K_a = 6.8 \times 10^{-4}$.

Plan We approach this problem as for previous equilibrium problems: We write the chemical equation for the equilibrium and tabulate the known and unknown concentrations of all species. We then substitute the equilibrium concentrations into the equilibrium-constant expression and solve for the unknown concentration of H^+.

Solve The equilibrium reaction and equilibrium concentrations are as follows:

	$HF(aq)$	$\rightleftharpoons$	$H^+(aq)$	$+$	$F^-(aq)$
Initial concentration (M)	0.10		0		0
Change in concentration (M)	$-x$		$+x$		$+x$
Equilibrium concentration (M)	$(0.10 - x)$		x		x

The equilibrium-constant expression is

$$K_a = \frac{[H^+][F^-]}{[HF]} = \frac{(x)(x)}{0.10 - x} = 6.8 \times 10^{-4}$$

When we try solving this equation using the approximation $0.10 - x \simeq 0.10$ (that is, by neglecting the concentration of acid that ionizes), we obtain

$$x = 8.2 \times 10^{-3} M$$

Because this approximation is greater than 5% of 0.10 M, however, we should work the problem in standard quadratic form. Rearranging, we have

$$x^2 = (0.10 - x)(6.8 \times 10^{-4})$$
$$= 6.8 \times 10^{-5} - (6.8 \times 10^{-4})x$$
$$x^2 + (6.8 \times 10^{-4})x - 6.8 \times 10^{-5} = 0$$

Substituting these values in the standard quadratic formula gives

$$x = \frac{-6.8 \times 10^{-4} \pm \sqrt{(6.8 \times 10^{-4})^2 - 4(-6.8 \times 10^{-5})}}{2}$$
$$= \frac{-6.8 \times 10^{-4} \pm 1.6 \times 10^{-2}}{2}$$

Of the two solutions, only the positive value for x is chemically reasonable. From that value, we can determine $[H^+]$ and hence the pH

$$x = [H^+] = [F^-] = 7.9 \times 10^{-3}\,M, \text{ so pH} = -\log[H^+] = 2.10$$

From our result, we can calculate the percent of molecules ionized:

$$\text{Percent ionization of HF} = \frac{\text{concentration ionized}}{\text{original concentration}} \times 100\%$$

$$= \frac{7.9 \times 10^{-3}\,M}{0.10\,M} \times 100\% = 7.9\%$$

Practice Exercise 1

What is the pH of a 0.010 M solution of HF?
(a) 1.58   **(b)** 2.10   **(c)** 2.30   **(d)** 2.58   **(e)** 2.64

Practice Exercise 2

In Practice Exercise 2 for Sample Exercise 16.11, we found that the percent ionization of niacin ($K_a = 1.5 \times 10^{-5}$) in a 0.020 M solution is 2.7%. Calculate the percentage of niacin molecules ionized in a solution that is **(a)** 0.010 M, **(b)** $1.0 \times 10^{-3}\,M$.

Polyprotic Acids

Acids that have more than one ionizable H atom are known as **polyprotic acids**. Sulfurous acid (H_2SO_3), for example, can undergo two successive ionizations:

$$H_2SO_3(aq) \rightleftharpoons H^+(aq) + HSO_3^-(aq) \qquad K_{a1} = 1.7 \times 10^{-2} \qquad [16.31]$$

$$HSO_3^-(aq) \rightleftharpoons H^+(aq) + SO_3^{2-}(aq) \qquad K_{a2} = 6.4 \times 10^{-8} \qquad [16.32]$$

Note that the acid-dissociation constants are labeled K_{a1} and K_{a2}. The numbers on the constants refer to the particular proton of the acid that is ionizing. Thus, K_{a2} always refers to the equilibrium involving removal of the second proton of a polyprotic acid.

We see that K_{a2} for sulfurous acid is much smaller than K_{a1}. Because of electrostatic attractions, we would expect a positively charged proton to be lost more readily from the neutral H_2SO_3 molecule than from the negatively charged HSO_3^- ion. This observation is general: *It is always easier to remove the first proton from a polyprotic acid than to remove the second.* Similarly, for an acid with three ionizable protons, it is easier to remove the second proton than the third. Thus, the K_a values become successively smaller as successive protons are removed.

Give It Some Thought

What is the equilibrium associated with K_{a3} for H_3PO_4?

The acid-dissociation constants for common polyprotic acids are listed in ▼ Table 16.3, and Appendix D provides a more complete list. The structure of citric acid illustrates the presence of multiple ionizable protons ◀ Figure 16.13.

GO FIGURE

Citric acid has four hydrogen atoms bonded to oxygen. How does the hydrogen atom that is not an acidic proton differ from the other three?

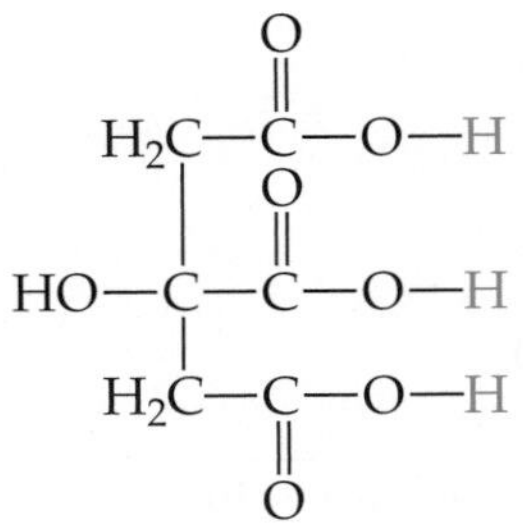

Citric acid

▲ **Figure 16.13 The structure of the polyprotic acid, citric acid.**

Table 16.3   Acid-Dissociation Constants of Some Common Polyprotic Acids

Name	Formula	K_{a1}	K_{a2}	K_{a3}
Ascorbic	$H_2C_6H_6O_6$	8.0×10^{-5}	1.6×10^{-12}	
Carbonic	H_2CO_3	4.3×10^{-7}	5.6×10^{-11}	
Citric	$H_3C_6H_5O_7$	7.4×10^{-4}	1.7×10^{-5}	4.0×10^{-7}
Oxalic	$HOOC—COOH$	5.9×10^{-2}	6.4×10^{-5}	
Phosphoric	H_3PO_4	7.5×10^{-3}	6.2×10^{-8}	4.2×10^{-13}
Sulfurous	H_2SO_3	1.7×10^{-2}	6.4×10^{-8}	
Sulfuric	H_2SO_4	Large	1.2×10^{-2}	
Tartaric	$C_2H_2O_2(COOH)_2$	1.0×10^{-3}	4.6×10^{-5}	

Notice in Table 16.3 that in most cases the K_a values for successive losses of protons differ by a factor of at least 10^3. Notice also that the value of K_{a1} for sulfuric acid is listed simply as "large." Sulfuric acid is a strong acid with respect to the removal of the first proton. Thus, the reaction for the first ionization step lies completely to the right:

$$H_2SO_4(aq) \longrightarrow H^+(aq) + HSO_4^-(aq) \quad (\text{complete ionization})$$

However, HSO_4^- is a weak acid for which $K_{a2} = 1.2 \times 10^{-2}$.

For many polyprotic acids K_{a1} is much larger than subsequent dissociation constants, in which case the $H^+(aq)$ in the solution comes almost entirely from the first ionization reaction. As long as successive K_a values differ by a factor of 10^3 or more, it is usually possible to obtain a satisfactory estimate of the pH of polyprotic acid solutions by treating the acids as if they were monoprotic, considering only K_{a1}.

SAMPLE EXERCISE 16.14 | Calculating the pH of a Solution of a Polyprotic Acid

The solubility of CO_2 in water at 25 °C and 0.1 atm is 0.0037 M. The common practice is to assume that all the dissolved CO_2 is in the form of carbonic acid (H_2CO_3), which is produced in the reaction

$$CO_2(aq) + H_2O(l) \rightleftharpoons H_2CO_3(aq)$$

What is the pH of a 0.0037 M solution of H_2CO_3?

SOLUTION

Analyze We are asked to determine the pH of a 0.0037 M solution of a polyprotic acid.

Plan H_2CO_3 is a diprotic acid; the two acid-dissociation constants, K_{a1} and K_{a2} (Table 16.3), differ by more than a factor of 10^3. Consequently, the pH can be determined by considering only K_{a1}, thereby treating the acid as if it were a monoprotic acid.

Solve Proceeding as in Sample Exercises 16.12 and 16.13, we can write the equilibrium reaction and equilibrium concentrations as

$$H_2CO_3(aq) \rightleftharpoons H^+(aq) + HCO_3^-(aq)$$

Initial concentration (M)	0.0037	0	0
Change in concentration (M)	$-x$	$+x$	$+x$
Equilibrium concentration (M)	$(0.0037 - x)$	x	x

The equilibrium-constant expression is

$$K_{a1} = \frac{[H^+][HCO_3^-]}{[H_2CO_3]} = \frac{(x)(x)}{0.0037 - x} = 4.3 \times 10^{-7}$$

Solving this quadratic equation, we get

$$x = 4.0 \times 10^{-5} M$$

Alternatively, because K_{a1} is small, we can make the simplifying approximation that x is small, so that

$$0.0037 - x \approx 0.0037$$

Thus,

$$\frac{(x)(x)}{0.0037} = 4.3 \times 10^{-7}$$

Solving for x, we have

$$x^2 = (0.0037)(4.3 \times 10^{-7}) = 1.6 \times 10^{-9}$$
$$x = [H^+] = [HCO_3^-] = \sqrt{1.6 \times 10^{-9}} = 4.0 \times 10^{-5} M$$

Because we get the same value (to 2 significant figures) our simplifying assumption was justified. The pH is therefore

$$pH = -\log[H^+] = -\log(4.0 \times 10^{-5}) = 4.40$$

Comment If we were asked for $[CO_3^{2-}]$ we would need to use K_{a2}. Let's illustrate that calculation. Using our calculated values of $[HCO_3^-]$ and $[H^+]$ and setting $[CO_3^{2-}] = y$, we have

$$HCO_3^-(aq) \rightleftharpoons H^+(aq) + CO_3^{2-}(aq)$$

	HCO_3^-	H^+	CO_3^{2-}
Initial concentration (M)	4.0×10^{-5}	4.0×10^{-5}	0
Change in concentration (M)	$-y$	$+y$	$+y$
Equilibrium concentration (M)	$(4.0 \times 10^{-5} - y)$	$(4.0 \times 10^{-5} + y)$	y

Assuming that y is small relative to 4.0×10^{-5}, we have

$$K_{a2} = \frac{[H^+][CO_3^{2-}]}{[HCO_3^-]} = \frac{(4.0 \times 10^{-5})(y)}{4.0 \times 10^{-5}} = 5.6 \times 10^{-11}$$

$$y = 5.6 \times 10^{-11}\, M = [CO_3^{2-}]$$

We see that the value for y is indeed very small compared with 4.0×10^{-5}, showing that our assumption was justified. It also shows that the ionization of HCO_3^- is negligible relative to that of H_2CO_3, as far as production of H^+ is concerned. However, it is the *only* source of CO_3^{2-}, which has a very low concentration in the solution.

Our calculations thus tell us that in a solution of carbon dioxide in water, most of the CO_2 is in the form of CO_2 or H_2CO_3, only a small fraction ionizes to form H^+ and HCO_3^-, and an even smaller fraction ionizes to give CO_3^{2-}. Notice also that $[CO_3^{2-}]$ is numerically equal to K_{a2}.

Practice Exercise 1

What is the pH of a 0.28 M solution of ascorbic acid (Vitamin C)? (See Table 16.3 for K_{a1} and K_{a2}.)
(a) 2.04 (b) 2.32 (c) 2.82 (d) 4.65 (e) 6.17

Practice Exercise 2

(a) Calculate the pH of a 0.020 M solution of oxalic acid ($H_2C_2O_4$). (See Table 16.3 for K_{a1} and K_{a2}.)
(b) Calculate the concentration of oxalate ion, $[C_2O_4^{2-}]$, in this solution.

16.7 | Weak Bases

Many substances behave as weak bases in water. Weak bases react with water, abstracting protons from H_2O, thereby forming the conjugate acid of the base and OH^- ions:

$$B(aq) + H_2O(l) \rightleftharpoons HB^+(aq) + OH^-(aq) \qquad [16.33]$$

The equilibrium-constant expression for this reaction can be written as

$$K_b = \frac{[BH^+][OH^-]}{[B]} \qquad [16.34]$$

Water is the solvent, so it is omitted from the equilibrium-constant expression. One of the most commonly encountered weak bases is ammonia, NH_3:

$$NH_3(aq) + H_2O(l) \rightleftharpoons NH_4^+(aq) + OH^-(aq) \quad K_b = \frac{[NH_4^+][OH^-]}{[NH_3]} \quad [16.35]$$

As with K_w and K_a, the subscript b in K_b denotes that the equilibrium constant refers to a particular type of reaction, namely the ionization of a weak base in water. The constant K_b, the **base-dissociation constant**, *always refers to the equilibrium in which a base reacts with H_2O to form the corresponding conjugate acid and OH^-.*

▶ Table 16.4 lists the Lewis structures, conjugate acids, and K_b values for a number of weak bases in water. Appendix D includes a more extensive list. These bases contain one or more lone pairs of electrons because a lone pair is necessary to form the bond with H^+. Notice that in the neutral molecules in Table 16.4, the lone pairs are on nitrogen atoms. The other bases listed are anions derived from weak acids.

Table 16.4 Some Weak Bases in Water at 25 °C

Base	Structural Formula*	Conjugate Acid	K_b

Ammonia (NH_3)	H—N̈—H
	    |
	    H
		NH_4^+	1.8×10^{-5}

Pyridine (C_5H_5N)	⬡N:	$C_5H_5NH^+$	1.7×10^{-9}

Hydroxylamine ($HONH_2$)	H—N̈—Ö̈H
	    |
	    H
		$HONH_3^+$	1.1×10^{-8}

Methylamine (CH_3NH_2)	H—N̈—CH_3
	    |
	    H
		$CH_3NH_3^+$	4.4×10^{-4}

Hydrosulfide ion (HS^-)	$[H-\ddot{\underset{..}{S}}:]^-$	H_2S	1.8×10^{-7}
Carbonate ion (CO_3^{2-})	$\left[\begin{array}{c} :\ddot{O}: \\ \| \\ ..\ddot{O}—C=\ddot{O}.. \end{array} \right]^{2-}$	HCO_3^-	1.8×10^{-4}
Hypochlorite ion (ClO^-)	$[:\ddot{\underset{..}{Cl}}—\ddot{\underset{..}{O}}:]^-$	$HClO$	3.3×10^{-7}

*The atom that accepts the proton is shown in blue.

SAMPLE EXERCISE 16.15 Using K_b to Calculate OH^-

Calculate the concentration of OH^- in a 0.15 M solution of NH_3.

SOLUTION

Analyze We are given the concentration of a weak base and asked to determine the concentration of OH^-.

Plan We will use essentially the same procedure here as used in solving problems involving the ionization of weak acids—that is, write the chemical equation and tabulate initial and equilibrium concentrations.

Solve The ionization reaction and equilibrium-constant expression are

$$NH_3(aq) + H_2O(l) \rightleftharpoons NH_4^+(aq) + OH^-(aq)$$

$$K_b = \frac{[NH_4^+][OH^-]}{[NH_3]} = 1.8 \times 10^{-5}$$

Ignoring the concentration of H_2O because it is not involved in the equilibrium-constant expression, the equilibrium concentrations are

	$NH_3(aq)$	+	$H_2O(l)$	$\rightleftharpoons$	$NH_4^+(aq)$	+	OH^-
Initial concentration (M)	0.15		—		0		0
Change in concentration (M)	$-x$		—		$+x$		$+x$
Equilibrium concentration (M)	$(0.15 - x)$		—		x		x

Inserting these quantities into the equilibrium-constant expression gives

$$K_b = \frac{[NH_4^+][OH^-]}{[NH_3]} = \frac{(x)(x)}{0.15 - x} = 1.8 \times 10^{-5}$$

Because K_b is small, the amount of NH_3 that reacts with water is much smaller than the NH_3 concentration, and so we can neglect x relative to 0.15 M. Then we have

$$\frac{x^2}{0.15} = 1.8 \times 10^{-5}$$

$$x^2 = (0.15)(1.8 \times 10^{-5}) = 2.7 \times 10^{-6}$$

$$x = [NH_4^+] = [OH^-] = \sqrt{2.7 \times 10^{-6}} = 1.6 \times 10^{-3} M$$

Check The value obtained for x is only about 1% of the NH_3 concentration, 0.15 M. Therefore, neglecting x relative to 0.15 was justified.

Comment You may be asked to find the pH of a solution of a weak base. Once you have found $[OH^-]$, you can proceed as in Sample Exercise 16.9, where we calculated the pH of a strong base. In the present sample exercise, we have seen that the 0.15 M solution of NH_3 contains $[OH^-]$ = 1.6×10^{-3} M. Thus, pOH = $-\log(1.6 \times 10^{-3})$ = 2.80, and pH = 14.00 − 2.80 = 11.20. The pH of the solution is above 7 because we are dealing with a solution of a base.

Practice Exercise 1

What is the pH of a 0.65 M solution of pyridine, C_5H_5N? (See Table 16.4 for K_b.)
(a) 4.48 **(b)** 8.96 **(c)** 9.52 **(d)** 9.62 **(e)** 9.71

Practice Exercise 2

Which of the following compounds should produce the highest pH as a 0.05 M solution: pyridine, methylamine, or nitrous acid?

GO FIGURE

When hydroxylamine acts as a base, which atom accepts the proton?

Ammonia
NH_3

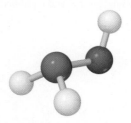

Methylamine
CH_3NH_2

Hydroxylamine
NH_2OH

▲ **Figure 16.14 Structures of ammonia and two simple amines.**

Types of Weak Bases

Weak bases fall into two general categories. The first category is neutral substances that have an atom with a nonbonding pair of electrons that can accept a proton. Most of these bases, including all uncharged bases in Table 16.4, contain a nitrogen atom. These substances include ammonia and a related class of compounds called **amines** (◀ Figure 16.14). In organic amines, at least one N—H bond in NH_3 is replaced with an N—C bond. Like NH_3, amines can abstract a proton from a water molecule by forming an N—H bond, as shown here for methylamine:

$$H-\ddot{N}-CH_3(aq) + H_2O(l) \rightleftharpoons \left[H-N-CH_3 \right]^+ (aq) + OH^-(aq) \qquad [16.36]$$

Anions of weak acids make up the second general category of weak bases. In an aqueous solution of sodium hypochlorite (NaClO), for example, NaClO dissociates to Na^+ and ClO^- ions. The Na^+ ion is always a spectator ion in acid–base reactions. ⚬⚬ (Section 4.3) The ClO^- ion, however, is the conjugate base of a weak acid, hypochlorous acid. Consequently, the ClO^- ion acts as a weak base in water:

$$ClO^-(aq) + H_2O(l) \rightleftharpoons HClO(aq) + OH^-(aq) \qquad K_b = 3.3 \times 10^{-7} \qquad [16.37]$$

In Figure 16.6 we saw that bleach is quite basic (pH values of 12–13). Common chlorine bleach is typically a 5% NaOCl solution.

SAMPLE EXERCISE 16.16 | **Using pH to Determine the Concentration of a Salt**

A solution made by adding solid sodium hypochlorite (NaClO) to enough water to make 2.00 L of solution has a pH of 10.50. Using the information in Equation 16.37, calculate the number of moles of NaClO added to the water.

SOLUTION

Analyze NaClO is an ionic compound consisting of Na^+ and ClO^- ion. As such, it is a strong electrolyte that completely dissociates in solution into Na^+, a spectator ion, and ClO^- ion, a weak base with $K_b = 3.3 \times 10^{-7}$ (Equation 16.37). Given this information we must calculate the number of moles of NaClO needed to increase the pH of 2.00-L of water to 10.50.

Plan From the pH, we can determine the equilibrium concentration of OH^-. We can then construct a table of initial and equilibrium concentrations in which the initial concentration of ClO^- is our unknown. We can calculate $[ClO^-]$ using the expression for K_b.

Solve We can calculate $[OH^-]$ by using either Equation 16.16 or Equation 16.20; we will use the latter method here:

$$pOH = 14.00 - pH = 14.00 - 10.50 = 3.50$$
$$[OH^-] = 10^{-3.50} = 3.2 \times 10^{-4}\,M$$

This concentration is high enough that we can assume that Equation 16.37 is the only source of OH^-; that is, we can neglect any OH^- produced by the autoionization of H_2O. We now assume a value of x for the initial concentration of ClO^- and solve the equilibrium problem in the usual way.

	$ClO^-(aq)$	$+$ $H_2O(l)$	$\rightleftharpoons$ $HClO(aq)$	$+$ $OH^-(aq)$
Initial concentration (M)	x	—	0	0
Change in concentration (M)	-3.2×10^{-4}	—	$+3.2 \times 10^{-4}$	$+3.2 \times 10^{-4}$
Equilibrium concentration (M)	$(x - 3.2 \times 10^{-4})$	—	3.2×10^{-4}	3.2×10^{-4}

We now use the expression for the base-dissociation constant to solve for x:

$$K_b = \frac{[HClO][OH^-]}{[ClO^-]} = \frac{(3.2 \times 10^{-4})^2}{x - 3.2 \times 10^{-4}} = 3.3 \times 10^{-7}$$

$$x = \frac{(3.2 \times 10^{-4})^2}{3.3 \times 10^{-7}} + (3.2 \times 10^{-4}) = 0.31\,M$$

We say that the solution is $0.31\,M$ in NaClO even though some of the ClO^- ions have reacted with water. Because the solution is $0.31\,M$ in NaClO and the total volume of solution is 2.00 L, 0.62 mol of NaClO is the amount of the salt that was added to the water.

Practice Exercise 1

The benzoate ion, $C_6H_5COO^-$, is a weak base with $K_b = 1.6 \times 10^{-10}$. How many moles of sodium benzoate are present in 0.50 L of a solution of NaC_6H_5COO if the pH is 9.04?
(a) 0.38 **(b)** 0.66 **(c)** 0.76 **(d)** 1.5 **(e)** 2.9

Practice Exercise 2

What is the molarity of an aqueous NH_3 solution that has a pH of 11.17?

16.8 | Relationship Between K_a and K_b

We have seen in a qualitative way that the stronger an acid, the weaker its conjugate base. To see if we can find a corresponding *quantitative* relationship, let's consider the NH_4^+ and NH_3 conjugate acid–base pair. Each species reacts with water. For the acid, NH_4^+, the equilibrium is

$$NH_4^+(aq) + H_2O(l) \rightleftharpoons NH_3(aq) + H_3O^+(aq)$$

or written in its simpler form:

$$NH_4^+(aq) \rightleftharpoons NH_3(aq) + H^+(aq) \qquad [16.38]$$

For the base, NH_3, the equilibrium is

$$NH_3(aq) + H_2O(l) \rightleftharpoons NH_4^+(aq) + OH^-(aq) \qquad [16.39]$$

Each equilibrium is expressed by a dissociation constant:

$$K_a = \frac{[NH_3][H^+]}{[NH_4^+]} \qquad K_b = \frac{[NH_4^+][OH^-]}{[NH_3]}$$

When we add Equations 16.38 and 16.39, the NH_4^+ and NH_3 species cancel and we are left with the autoionization of water:

$$NH_4^+(aq) \rightleftharpoons NH_3(aq) + H^+(aq)$$
$$\underline{NH_3(aq) + H_2O(l) \rightleftharpoons NH_4^+(aq) + OH^-(aq)}$$
$$H_2O(l) \rightleftharpoons H^+(aq) + OH^-(aq)$$

Recall that when two equations are added to give a third, the equilibrium constant associated with the third equation equals the product of the equilibrium constants of the first two equations. ∞ (Section 15.3)

Applying this rule to our present example, we see that when we multiply K_a and K_b, we obtain

$$K_a \times K_b = \left(\frac{[NH_3][H^+]}{[NH_4^+]} \right) \left(\frac{[NH_4^+][OH^-]}{[NH_3]} \right)$$

$$= [H^+][OH^-] = K_w$$

Thus, the product of K_a and K_b is the ion-product constant for water, K_w (Equation 16.16). We expect this result because adding Equations 16.38 and 16.39 gave us the autoionization equilibrium for water, for which the equilibrium constant is K_w.

The above result holds for any conjugate acid–base pair. In general, *the product of the acid-dissociation constant for an acid and the base-dissociation constant for its conjugate base equals the ion-product constant for water:*

$$K_a \times K_b = K_w \quad \text{(for a conjugate acid–base pair)} \qquad [16.40]$$

As the strength of an acid increases (K_a gets larger), the strength of its conjugate base must decrease (K_b gets smaller) so that the product $K_a \times K_b$ remains 1.0×10^{-14} at 25 °C. ▼ Table 16.5 demonstrates this relationship. Remember, this important relationship applies *only* to conjugate acid–base pairs.

By using Equation 16.40, we can calculate K_b for any weak base if we know K_a for its conjugate acid. Similarly, we can calculate K_a for a weak acid if we know K_b for its conjugate base. As a practical consequence, ionization constants are often listed for only one member of a conjugate acid–base pair. For example, Appendix D does not contain K_b values for the anions of weak acids because they can be readily calculated from the tabulated K_a values for their conjugate acids.

Recall that we often express $[H^+]$ as pH: $pH = -\log[H^+]$. ∞ (Section 16.4) This "p" nomenclature is often used for other very small numbers. For example, if you look up the values for acid- or base-dissociation constants in a chemistry handbook, you may find them expressed as pK_a or pK_b:

$$pK_a = -\log K_a \quad \text{and} \quad pK_b = -\log K_b \qquad [16.41]$$

Using this nomenclature, Equation 16.40 can be written in terms of pK_a and pK_b by taking the negative logarithm of both sides:

$$pK_a + pK_b = pK_w = 14.00 \quad \text{at 25 °C (conjugate acid–base pair)} \qquad [16.42]$$

 Give It Some Thought

K_a for acetic acid is 1.8×10^{-5}. What is the first digit of the pK_a value for acetic acid?

Table 16.5 Some Conjugate Acid–Base Pairs

Acid	K_a	Base	K_b
HNO_3	(Strong acid)	NO_3^-	(Negligible basicity)
HF	6.8×10^{-4}	F^-	1.5×10^{-11}
CH_3COOH	1.8×10^{-5}	CH_3COO^-	5.6×10^{-10}
H_2CO_3	4.3×10^{-7}	HCO_3^-	2.3×10^{-8}
NH_4^+	5.6×10^{-10}	NH_3	1.8×10^{-5}
HCO_3^-	5.6×10^{-11}	CO_3^{2-}	1.8×10^{-4}
OH^-	(Negligible acidity)	O^{2-}	(Strong base)

Chemistry Put to Work

Amines and Amine Hydrochlorides

Many low-molecular-weight amines have a fishy odor. Amines and NH_3 are produced by the anaerobic (absence of O_2) decomposition of dead animal or plant matter. Two such amines with very disagreeable aromas are $H_2N(CH_2)_4NH_2$, *putrescine*, and $H_2N(CH_2)_5NH_2$, *cadaverine*. The names of these substances are testaments to their repugnant odors!

Many drugs, including quinine, codeine, caffeine, and amphetamine, are amines. Like other amines, these substances are weak bases; the amine nitrogen is readily protonated upon treatment with an acid. The resulting products are called *acid salts*. If we use A as the abbreviation for an amine, the acid salt formed by reaction with hydrochloric acid can be written AH^+Cl^-. It can also be written as $A \cdot HCl$ and referred to as a hydrochloride. Amphetamine hydrochloride, for example, is the acid salt formed by treating amphetamine with HCl:

$$\langle\!\!\bigcirc\!\!\rangle - CH_2 - \underset{\underset{CH_3}{|}}{CH} - \ddot{N}H_2(aq) + HCl(aq) \longrightarrow$$

Amphetamine

$$\langle\!\!\bigcirc\!\!\rangle - CH_2 - \underset{\underset{CH_3}{|}}{CH} - NH_3^+Cl^-(aq)$$

Amphetamine hydrochloride

Acid salts are much less volatile, more stable, and generally more water soluble than the corresponding amines. For this reason, many drugs that are amines are sold and administered as acid salts. Some examples of over-the-counter medications that contain amine hydrochlorides as active ingredients are shown in ▼ Figure 16.15.

Related Exercises: 16.9, 16.73, 16.74, 16.101, 16.114, 16.124

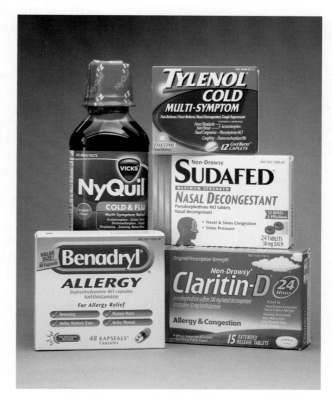

▲ Figure 16.15 **Some over-the-counter medications in which an amine hydrochloride is a major active ingredient.**

SAMPLE EXERCISE 16.17 | Calculating K_a or K_b for a Conjugate Acid–Base Pair

Calculate **(a)** K_b for the fluoride ion, **(b)** K_a for the ammonium ion.

SOLUTION

Analyze We are asked to determine dissociation constants for F^-, the conjugate base of HF, and NH_4^+, the conjugate acid of NH_3.

Plan We can use the tabulated K values for HF and NH_3 and the relationship between K_a and K_b to calculate the dissociation constants for their conjugates, F^- and NH_4^+.

Solve

(a) For the weak acid HF, Table 16.2 and Appendix D give $K_a = 6.8 \times 10^{-4}$. We can use Equation 16.40 to calculate K_b for the conjugate base, F^-:

$$K_b = \frac{K_w}{K_a} = \frac{1.0 \times 10^{-14}}{6.8 \times 10^{-4}} = 1.5 \times 10^{-11}$$

(b) For NH_3, Table 16.4 and in Appendix D give $K_b = 1.8 \times 10^{-5}$, and this value in Equation 16.40 gives us K_a for the conjugate acid, NH_4^+:

$$K_a = \frac{K_w}{K_b} = \frac{1.0 \times 10^{-14}}{1.8 \times 10^{-5}} = 5.6 \times 10^{-10}$$

Check The respective K values for F^- and NH_4^+ are listed in Table 16.5, where we see that the values calculated here agree with those in Table 16.5.

Practice Exercise 1

By using information from Appendix D, put the following three substances in order of weakest to strongest base: (i) $(CH_3)_3N$, (ii) $HCOO^-$, (iii) BrO^-.

(a) i < ii < iii **(b)** ii < i < iii **(c)** iii < i < ii
(d) ii < iii < i **(e)** iii < ii < i.

Practice Exercise 2

(a) Based on information in Appendix D, which of these anions has the largest base-dissociation constant: NO_2^-, PO_4^{3-}, or N_3^-?
(b) The base quinoline has the structure

Its conjugate acid is listed in handbooks as having a pK_a of 4.90. What is the base-dissociation constant for quinoline?

16.9 | Acid–Base Properties of Salt Solutions

Even before you began this chapter, you were undoubtedly aware of many substances that are acidic, such as HNO_3, HCl, and H_2SO_4, and others that are basic, such as $NaOH$ and NH_3. However, our discussion up to this point in the chapter has indicated that ions can also exhibit acidic or basic properties. For example, we calculated K_a for NH_4^+ and K_b for F^- in Sample Exercise 16.17. Such behavior implies that salt solutions can be acidic or basic. Before proceeding with further discussions of acids and bases, let's examine the way dissolved salts can affect pH.

Because nearly all salts are strong electrolytes, we can assume that any salt dissolved in water is completely dissociated. Consequently, the acid–base properties of salt solutions are due to the behavior of the cations and anions. Many ions react with water to generate $H^+(aq)$ or $OH^-(aq)$ ions. This type of reaction is often called **hydrolysis**. The pH of an aqueous salt solution can be predicted qualitatively by considering the salt's cations and anions.

An Anion's Ability to React with Water

In general, an anion A^- in solution can be considered the conjugate base of an acid. For example, Cl^- is the conjugate base of HCl, and CH_3COO^- is the conjugate base of CH_3COOH. Whether an anion reacts with water to produce hydroxide ions depends on the strength of the anion's conjugate acid. To identify the acid and assess its strength, we add a proton to the anion's formula. If the acid HA determined in this way is one of the seven strong acids listed at the beginning of Section 16.5, the anion has a negligible tendency to produce OH^- ions from water and does not affect the pH of the solution. The presence of Cl^- in an aqueous solution, for example, does not result in the production of any OH^- and does not affect the pH. Thus, Cl^- is always a spectator ion in acid–base chemistry.

If HA is *not* one of the seven common strong acids, it is a weak acid. In this case, the conjugate base A^- is a weak base and it reacts to a small extent with water to produce the weak acid and hydroxide ions:

$$A^-(aq) + H_2O(l) \rightleftharpoons HA(aq) + OH^-(aq) \qquad [16.43]$$

The OH^- ion generated in this way increases the pH of the solution, making it basic. Acetate ion, for example, being the conjugate base of a weak acid, reacts with water to produce acetic acid and hydroxide ions, thereby increasing the pH of the solution:

$$CH_3COO^-(aq) + H_2O(l) \rightleftharpoons CH_3COOH(aq) + OH^-(aq) \qquad [16.44]$$

 Give It Some Thought

Will NO_3^- ions affect the pH of a solution? What about CO_3^{2-} ions?

The situation is more complicated for salts containing anions that have ionizable protons, such as HSO_3^-. These salts are amphiprotic (Section 16.2), and how they behave in water is determined by the relative magnitudes of K_a and K_b for the ion, as shown in Sample Exercise 16.19. If $K_a > K_b$, the ion causes the solution to be acidic. If $K_b > K_a$, the solution is made basic by the ion.

A Cation's Ability to React with Water

Polyatomic cations containing one or more protons can be considered the conjugate acids of weak bases. The NH_4^+ ion, for example, is the conjugate acid of the weak base NH_3. Thus, NH_4^+ is a weak acid and will donate a proton to water, producing hydronium ions and thereby lowering the pH:

$$NH_4^+(aq) + H_2O(l) \rightleftharpoons NH_3(aq) + H_3O^+(aq) \qquad [16.45]$$

Many metal ions react with water to decrease the pH of an aqueous solution. This effect is most pronounced for small, highly charged cations like Fe^{3+} and Al^{3+}, as illustrated by the K_a values for metal cations in ▶ Table 16.6. A comparison of Fe^{2+} and Fe^{3+} values in the table illustrates how acidity increases as ionic charge increases.

Notice that K_a values for the 3+ ions in Table 16.6 are comparable to the values for familiar weak acids, such as acetic acid ($K_a = 1.8 \times 10^{-5}$). In contrast, the ions of alkali and alkaline earth metals, being relatively large and not highly charged, do not react with water and therefore do not affect pH. Note that these are the same cations found in the strong bases (Section 16.5). The different tendencies of four cations to lower the pH of a solution are illustrated in ▼ Figure 16.16.

The mechanism by which metal ions produce acidic solutions is shown in ▼ Figure 16.17. Because metal ions are positively charged, they attract the unshared

Table 16.6 Acid-Dissociation Constants for Metal Cations in Aqueous Solution at 25 °C

Cation	K_a
Fe^{2+}	3.2×10^{-10}
Zn^{2+}	2.5×10^{-10}
Ni^{2+}	2.5×10^{-11}
Fe^{3+}	6.3×10^{-3}
Cr^{3+}	1.6×10^{-4}
Al^{3+}	1.4×10^{-5}

▲ GO FIGURE

Why do we need to use two different acid-base indicators in this figure?

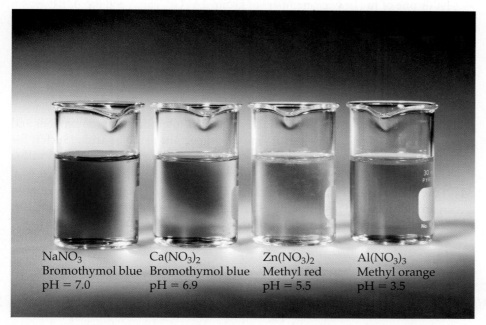

$NaNO_3$	$Ca(NO_3)_2$	$Zn(NO_3)_2$	$Al(NO_3)_3$
Bromothymol blue	Bromothymol blue	Methyl red	Methyl orange
pH = 7.0	pH = 6.9	pH = 5.5	pH = 3.5

▲ **Figure 16.16 Effect of cations on solution pH.** The pH values of 1.0 *M* solutions of four nitrate salts are estimated using acid–base indicators.

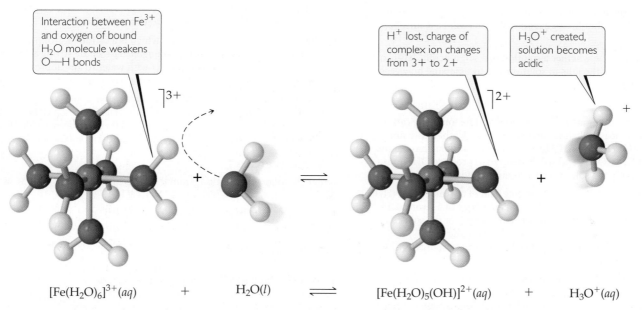

Interaction between Fe^{3+} and oxygen of bound H_2O molecule weakens O—H bonds

H^+ lost, charge of complex ion changes from 3+ to 2+

H_3O^+ created, solution becomes acidic

$$[Fe(H_2O)_6]^{3+}(aq) \quad + \quad H_2O(l) \quad \rightleftharpoons \quad [Fe(H_2O)_5(OH)]^{2+}(aq) \quad + \quad H_3O^+(aq)$$

▲ **Figure 16.17 A hydrated Fe^{3+} ion acts as an acid by donating an H^+ to a free H_2O molecule, forming H_3O^+.**

electron pairs of water molecules and become hydrated. ⮐ (Section 13.1) The larger the charge on the metal ion, the stronger the interaction between the ion and the oxygen of its hydrating water molecules. As the strength of this interaction increases, the O—H bonds in the hydrating water molecules become weaker. This facilitates transfer of protons from the hydration water molecules to solvent water molecules.

Combined Effect of Cation and Anion in Solution

To determine whether a salt forms an acidic, a basic, or a neutral solution when dissolved in water, we must consider the action of both cation and anion. There are four possible combinations.

1. If the salt contains an anion that does not react with water and a cation that does not react with water, we expect the pH to be neutral. Such is the case when the anion is a conjugate base of a strong acid and the cation is either from group 1A or one of the heavier members of group 2A (Ca^{2+}, Sr^{2+}, and Ba^{2+}). *Examples*: $NaCl$, $Ba(NO_3)_2$, $RbClO_4$.

2. If the salt contains an anion that reacts with water to produce hydroxide ions and a cation that does not react with water, we expect the pH to be basic. Such is the case when the anion is the conjugate base of a weak acid and the cation is either from group 1A or one of the heavier members of group 2A (Ca^{2+}, Sr^{2+}, and Ba^{2+}). *Examples*: $NaClO$, RbF, $BaSO_3$.

3. If the salt contains a cation that reacts with water to produce hydronium ions and an anion that does not react with water, we expect the pH to be acidic. Such is the case when the cation is a conjugate acid of a weak base or a small cation with a charge of 2+ or greater. *Examples*: NH_4NO_3, $AlCl_3$, $Fe(NO_3)_3$.

4. If the salt contains an anion and a cation *both* capable of reacting with water, both hydroxide ions and hydronium ions are produced. Whether the solution is basic, neutral, or acidic depends on the relative abilities of the ions to react with water. *Examples*: NH_4ClO, $Al(CH_3COO)_3$, CrF_3.

SAMPLE EXERCISE 16.18 | Determining Whether Salt Solutions Are Acidic, Basic, or Neutral

Determine whether aqueous solutions of each of these salts are acidic, basic, or neutral:
(a) $Ba(CH_3COO)_2$, (b) NH_4Cl, (c) CH_3NH_3Br, (d) KNO_3, (e) $Al(ClO_4)_3$.

SOLUTION

Analyze We are given the chemical formulas of five ionic compounds (salts) and asked whether their aqueous solutions will be acidic, basic, or neutral.

Plan We can determine whether a solution of a salt is acidic, basic, or neutral by identifying the ions in solution and by assessing how each ion will affect the pH.

Solve

(a) This solution contains barium ions and acetate ions. The cation is an ion of a heavy alkaline earth metal and will therefore not affect the pH. The anion, CH_3COO^-, is the conjugate base of the weak acid CH_3COOH and will hydrolyze to produce OH^- ions, thereby making the solution basic (combination 2).

(b) In this solution, NH_4^+ is the conjugate acid of a weak base (NH_3) and is therefore acidic. Cl^- is the conjugate base of a strong acid (HCl) and therefore has no influence on the pH of the solution. Because the solution contains an ion that is acidic (NH_4^+) and one that has no influence on pH (Cl^-), the solution of NH_4Cl will be acidic (combination 3).

(c) Here $CH_3NH_3^+$ is the conjugate acid of a weak base (CH_3NH_2, an amine) and is therefore acidic, and Br^- is the conjugate base of a strong acid (HBr) and therefore pH neutral. Because the solution

contains one ion that is acidic and one that has no influence on pH, the solution of CH_3NH_3Br will be acidic (combination 3).

(d) This solution contains the K^+ ion, which is a cation of group 1A, and the NO_3^- ion, which is the conjugate base of the strong acid HNO_3. Neither of the ions will react with water to any appreciable extent, making the solution neutral (combination 1).

(e) This solution contains Al^{3+} and ClO_4^- ions. Cations, such as Al^{3+}, that have a charge of 3+ or higher are acidic. The ClO_4^- ion is the conjugate base of a strong acid ($HClO_4$) and therefore does not affect pH. Thus, the solution of $Al(ClO_4)_3$ will be acidic (combination 3).

Practice Exercise 1

Order the following solutions from lowest to highest pH: (i) 0.10 M $NaClO$, (ii) 0.10 M KBr, (iii) 0.10 M NH_4ClO_4.

(a) i < ii < iii (b) ii < i < iii (c) iii < i < ii
(d) ii < iii < i (e) iii < ii < i

Practice Exercise 2

Indicate which salt in each of the following pairs forms the more acidic (or less basic) 0.010 M solution: (a) $NaNO_3$ or $Fe(NO_3)_3$, (b) KBr or $KBrO$, (c) CH_3NH_3Cl or $BaCl_2$, (d) NH_4NO_2 or NH_4NO_3.

SAMPLE
EXERCISE 16.19   Predicting Whether the Solution of
an Amphiprotic Anion Is Acidic or Basic

Predict whether the salt Na_2HPO_4 forms an acidic solution or a basic solution when dissolved in water.

SOLUTION

Analyze We are asked to predict whether a solution of Na_2HPO_4 is acidic or basic. This substance is an ionic compound composed of Na^+ and HPO_4^{2-} ions.

Plan We need to evaluate each ion, predicting whether it is acidic or basic. Because Na^+ is a cation of group 1A, it has no influence on pH. Thus, our analysis of whether the solution is acidic or basic must focus on the behavior of the HPO_4^{2-} ion. We need to consider that HPO_4^{2-} can act as either an acid or a base:

$$As\ acid \quad HPO_4^{2-}(aq) \rightleftharpoons H^+(aq) + PO_4^{3-}(aq) \qquad [16.46]$$

$$As\ base \quad HPO_4^{2-}(aq) + H_2O \rightleftharpoons H_2PO_4^-(aq) + OH^-(aq) \qquad [16.47]$$

Of these two reactions, the one with the larger equilibrium constant determines whether the solution is acidic or basic.

Solve The value of K_a for Equation 16.46 is K_{a3} for H_3PO_4: 4.2×10^{-13} (Table 16.3). For Equation 16.47, we must calculate K_b for the base HPO_4^{2-} from the value of K_a for its conjugate acid, $H_2PO_4^-$, and the relationship $K_a \times K_b = K_w$ (Equation 16.40). The relevant value of K_a for $H_2PO_4^-$ is K_{a2} for H_3PO_4: 6.2×10^{-8} (from Table 16.3). We therefore have

$$K_b(HPO_4^{2-}) \times K_a(H_2PO_4^-) = K_w = 1.0 \times 10^{-14}$$

$$K_b(HPO_4^{2-}) = \frac{1.0 \times 10^{-14}}{6.2 \times 10^{-8}} = 1.6 \times 10^{-7}.$$

This K_b value is more than 10^5 times larger than K_a for HPO_4^{2-}; thus, the reaction in Equation 16.47 predominates over that in Equation 16.46, and the solution is basic.

Practice Exercise 1

How many of the following salts are expected to produce acidic solutions (see Table 16.3 for data): $NaHSO_4$, $NaHC_2O_4$, NaH_2PO_4, and $NaHCO_3$?
(a) 0   **(b)** 1   **(c)** 2   **(d)** 3   **(e)** 4

Practice Exercise 2

Predict whether the dipotassium salt of citric acid ($K_2HC_6H_5O_7$) forms an acidic or basic solution in water (see Table 16.3 for data).

16.10 | Acid–Base Behavior and Chemical Structure

When a substance is dissolved in water, it may behave as an acid, behave as a base, or exhibit no acid–base properties. How does the chemical structure of a substance determine which of these behaviors is exhibited by the substance? For example, why do some substances that contain OH groups behave as bases, releasing OH^- ions into solution, whereas others behave as acids, ionizing to release H^+ ions? In this section we discuss briefly the effects of chemical structure on acid–base behavior.

Factors That Affect Acid Strength

A molecule containing H will act as a proton donor (an acid) only if the H—A bond is polarized such that the H atom has a partial positive charge. ∞ (Section 8.4) Recall that we indicate such polarization in this way:

$$\overset{\longmapsto}{H-A}$$

In ionic hydrides, such as NaH, the bond is polarized in the opposite way: the H atom possesses a negative charge and behaves as a proton acceptor (a base). Nonpolar H—A bonds, such as the H—C bond in CH_4, produce neither acidic nor basic aqueous solutions.

A second factor that helps determine whether a molecule containing an H—A bond will donate a proton is the strength of the bond. ∞ (Section 8.8) Very strong bonds are less easily broken than weaker ones. This factor is important, for example, in the hydrogen halides. The H—F bond is the most polar H—A bond. You therefore might expect HF to be a very strong acid if bond polarity were all that mattered. However, the H—A bond strength increases as you move up the group: 299 kJ/mol in HI, 366 kJ/mol in HBr, 431 kJ/mol in HCl, and 567 kJ/mol in HF. Because HF has the highest bond strength among the hydrogen halides, it is a weak acid, whereas all the other hydrogen halides are strong acids in water.

A third factor that affects the ease with which a hydrogen atom ionizes from HA is the stability of the conjugate base, A^-. In general, the greater the stability of the conjugate base, the stronger the acid.

The strength of an acid is often a combination of all three factors.

Binary Acids

For a series of binary acids HA in which A represents members of the same *group* in the periodic table, the strength of the H—A bond is generally the most important factor determining acid strength. The strength of an H—A bond tends to decrease as the element A increases in size. As a result, the bond strength decreases and acidity increases down a group. Thus, HCl is a stronger acid than HF, and H_2S is a stronger acid than H_2O.

Bond polarity is the major factor determining acidity for binary acids HA when A represents members of the same *period*. Thus, acidity increases as the electronegativity of the element A increases, as it generally does moving from left to right across a period. ∞ (Section 8.4) For example, the difference in acidity of the period 2 elements is $CH_4 < NH_3 \ll H_2O < HF$. Because the C—H bond is essentially nonpolar, CH_4 shows no tendency to form H^+ and CH_3^- ions. Although the N—H bond is polar, NH_3 has a nonbonding pair of electrons on the nitrogen atom that dominates its chemistry, so NH_3 acts as a base rather than an acid.

The periodic trends in the acid strengths of binary compounds of hydrogen and the nonmetals of periods 2 and 3 are summarized in ▼ Figure 16.18.

GO FIGURE

Are the acid properties of HI what you would expect from this figure?

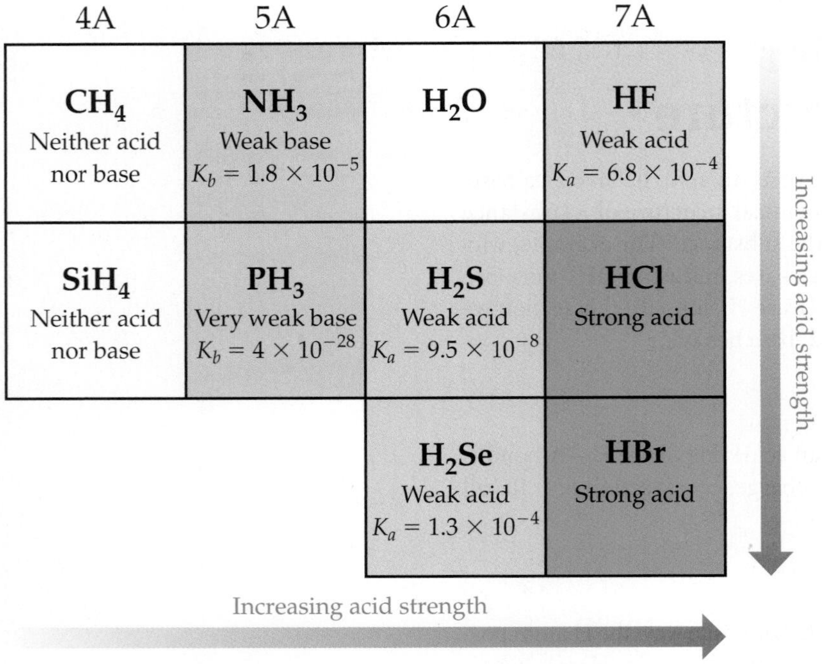

4A	5A	6A	7A
CH_4 Neither acid nor base	**NH_3** Weak base $K_b = 1.8 \times 10^{-5}$	**H_2O**	**HF** Weak acid $K_a = 6.8 \times 10^{-4}$
SiH_4 Neither acid nor base	**PH_3** Very weak base $K_b = 4 \times 10^{-28}$	**H_2S** Weak acid $K_a = 9.5 \times 10^{-8}$	**HCl** Strong acid
		H_2Se Weak acid $K_a = 1.3 \times 10^{-4}$	**HBr** Strong acid

Increasing acid strength →

Increasing acid strength (vertical)

▲ Figure 16.18 **Trends in acid strength for the binary hydrides of periods 2–4.**

Oxyacids

Many common acids, such as sulfuric acid, contain one or more O—H bonds:

$$\text{H}-\overset{..}{\underset{..}{\text{O}}}-\overset{\overset{\displaystyle :\overset{..}{\text{O}}:}{|}}{\underset{\underset{\displaystyle :\underset{..}{\text{O}}:}{|}}{\text{S}}}-\overset{..}{\underset{..}{\text{O}}}-\text{H}$$

Acids in which OH groups and possibly additional oxygen atoms are bound to a central atom are called **oxyacids**. At first it may seem confusing that the OH group, which we know behaves as a base, is also present in some acids. Let's take a closer look at what factors determine whether a given OH group behaves as a base or as an acid.

Consider an OH group bound to some atom Y, which might in turn have other groups attached to it:

$$\overset{\diagdown}{\underset{\diagup}{\text{Y}}}\!\!-\text{O}-\text{H}$$

At one extreme, Y might be a metal, such as Na or Mg. Because of the low electronegativity of metals, the pair of electrons shared between Y and O is completely transferred to oxygen, and an ionic compound containing OH^- is formed. Such compounds are therefore sources of OH^- ions and behave as bases, as in NaOH and Mg(OH)_2.

When Y is a nonmetal, the bond to O is covalent and the substance does not readily lose OH^-. Instead, these compounds are either acidic or neutral. *Generally, as the electronegativity of Y increases, so does the acidity of the substance.* This happens for two reasons: First, as electron density is drawn toward Y, the O—H bond becomes weaker and more polar, thereby favoring loss of H^+. Second, because the conjugate base of any acid YOH is usually an anion, its stability generally increases as the electronegativity of Y increases. This trend is illustrated by the K_a values of the hypohalous acids (YOH acids where Y is a halide ion), which decrease as the electronegativity of the halogen atom decreases (▼ Figure 16.19).

▲ GO FIGURE

At equilibrium, which of the two species with a halogen atom (green) is present in greater concentration?

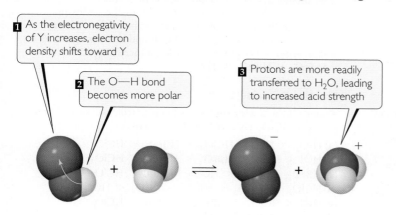

1 As the electronegativity of Y increases, electron density shifts toward Y

2 The O—H bond becomes more polar

3 Protons are more readily transferred to H_2O, leading to increased acid strength

Substance	Y—OH	Electronegativity of Y	Dissociation constant
Hypochlorous acid	Cl—OH	3.0	$K_a = 3.0 \times 10^{-8}$
Hypobromous acid	Br—OH	2.8	$K_a = 2.5 \times 10^{-9}$
Hypoiodous acid	I—OH	2.5	$K_a = 2.3 \times 10^{-11}$
Water	H—OH	2.1	$K_w = 1.0 \times 10^{-14}$

▲ **Figure 16.19 Acidity of the hypohalous oxyacids (YOH) as a function of electronegativity of Y.**

Many oxyacids contain additional oxygen atoms bonded to the central atom Y. These atoms pull electron density from the O—H bond, further increasing its polarity. Increasing the number of oxygen atoms also helps stabilize the conjugate base by increasing its ability to "spread out" its negative charge. Thus, *the strength of an acid increases as additional electronegative atoms bond to the central atom Y.* For example, the strength of the chlorine oxyacids (Y = Cl) steadily increases as O atoms are added:

| Hypochlorous | Chlorous | Chloric | Perchloric |

| $K_a = 3.0 \times 10^{-8}$ | $K_a = 1.1 \times 10^{-2}$ | Strong acid | Strong acid |

Increasing acid strength

Because the oxidation number of Y increases as the number of attached O atoms increases, this correlation can be stated in an equivalent way: In a series of oxyacids, the acidity increases as the oxidation number of the central atom increases.

Give It Some Thought

Which acid has the larger acid-dissociation constant, HIO_2 or $HBrO_3$?

SAMPLE EXERCISE 16.20 Predicting Relative Acidities from Composition and Structure

Arrange the compounds in each series in order of increasing acid strength:
(a) AsH_3, HBr, KH, H_2Se; (b) H_2SO_4, H_2SeO_3, H_2SeO_4.

SOLUTION

Analyze We are asked to arrange two sets of compounds in order from weakest acid to strongest acid. In (a), the substances are binary compounds containing H, and in (b) the substances are oxyacids.

Plan For the binary compounds, we will consider the electronegativities of As, Br, K, and Se relative to the electronegativity of H. The higher the electronegativity of these atoms, the higher the partial positive charge on H and so the more acidic the compound.

For the oxyacids, we will consider both the electronegativities of the central atom and the number of oxygen atoms bonded to the central atom.

Solve

(a) Because K is on the left side of the periodic table, it has a very low electronegativity (0.8, from Figure 8.7, p. 310). As a result, the hydrogen in KH carries a negative charge. Thus, KH should be the least acidic (most basic) compound in the series.

Arsenic and hydrogen have similar electronegativities, 2.0 and 2.1, respectively. This means that the As—H bond is nonpolar, and so AsH_3 has little tendency to donate a proton in aqueous solution.

The electronegativity of Se is 2.4, and that of Br is 2.8. Consequently, the H—Br bond is more polar than the H—Se bond, giving HBr the greater tendency to donate a proton. (This expectation is confirmed by Figure 16.18, where we see that H_2Se is a weak acid and HBr a strong acid.) Thus, the order of increasing acidity is KH < AsH_3 < H_2Se < HBr.

(b) The acids H_2SO_4 and H_2SeO_4 have the same number of O atoms and the same number of OH groups. In such cases, the acid strength increases with increasing electronegativity of the central atom. Because S is slightly more electronegative than Se (2.5 vs 2.4), we predict that H_2SO_4 is more acidic than H_2SeO_4.

For acids with the same central atom, the acidity increases as the number of oxygen atoms bonded to the central atom increases. Thus, H_2SeO_4 should be a stronger acid than H_2SeO_3. We predict the order of increasing acidity to be $H_2SeO_3 < H_2SeO_4 < H_2SO_4$.

Practice Exercise 1

Arrange the following substances in order from weakest to strongest acid: $HClO_3$, HOI, $HBrO_2$, $HClO_2$, HIO_2

(a) $HIO_2 < HOI < HClO_3 < HBrO_2 < HClO_2$
(b) $HOI < HIO_2 < HBrO_2 < HClO_2 < HClO_3$
(c) $HBrO_2 < HIO_2 < HClO_2 < HOI < HClO_3$
(d) $HClO_3 < HClO_2 < HBrO_2 < HIO_2 < HOI$
(e) $HOI < HClO_2 < HBrO_2 < HIO_2 < HClO_3$

Practice Exercise

In each pair, choose the compound that gives the more acidic (or less basic) solution: (a) HBr, HF; (b) PH_3, H_2S; (c) HNO_2, HNO_3; (d) H_2SO_3, H_2SeO_3.

Carboxylic Acids

Another large group of acids is illustrated by acetic acid, a weak acid ($K_a = 1.8 \times 10^{-5}$):

The portion of the structure shown in red is called the *carboxyl group*, which is often written COOH. Thus, the chemical formula of acetic acid is written as CH_3COOH, where only the hydrogen atom in the carboxyl group can be ionized. Acids that contain a carboxyl group are called **carboxylic acids**, and they form the largest category of organic acids. Formic acid and benzoic acid are further examples of this large and important category of acids:

Formic acid   Benzoic acid

Two factors contribute to the acidic behavior of carboxylic acids. First, the additional oxygen atom attached to the carbon of the carboxyl group draws electron density from the O—H bond, increasing its polarity and helping to stabilize the conjugate base. Second, the conjugate base of a carboxylic acid (a *carboxylate anion*) can exhibit resonance ∞∞ (Section 8.6), which contributes to the stability of the anion by spreading the negative charge over several atoms:

Give It Some Thought

What group of atoms is present in all carboxylic acids?

Chemistry and Life

The Amphiprotic Behavior of Amino Acids

As we will discuss in greater detail in Chapter 24, *amino acids* are the building blocks of proteins. The general structure of amino acids is

Amine group   Carboxyl group
(basic)   (acidic)

where different amino acids have different R groups attached to the central carbon atom. For example, in *glycine*, the simplest amino acid, R is a hydrogen atom, and in *alanine* R is a CH_3 group:

Glycine   Alanine

Amino acids contain a carboxyl group and can therefore serve as acids. They also contain an NH_2 group, characteristic of amines

(Section 16.7), and thus they can also act as bases. Amino acids, therefore, are amphiprotic. For glycine, we might expect the acid and base reactions with water to be

Acid: $H_2N-CH_2-COOH(aq) + H_2O(l) \rightleftharpoons$
$$H_2N-CH_2-COO^-(aq) + H_3O^+(aq) \quad [16.48]$$

Base: $H_2N-CH_2-COOH(aq) + H_2O(l) \rightleftharpoons$
$$^+H_3N-CH_2-COOH(aq) + OH^-(aq) \quad [16.49]$$

The pH of a solution of glycine in water is about 6.0, indicating that it is a slightly stronger acid than base.

The acid–base chemistry of amino acids is more complicated than shown in Equations 16.48 and 16.49, however. Because the COOH group can act as an acid and the NH_2 group can act as a base, amino acids undergo a "self-contained" Brønsted–Lowry acid–base reaction in which the proton of the carboxyl group is transferred to the basic nitrogen atom:

proton transfer
Neutral molecule   Zwitterion

Although the form of the amino acid on the right in this equation is electrically neutral overall, it has a positively charged end and a negatively charged end. A molecule of this type is called a *zwitterion* (German for "hybrid ion").

Do amino acids exhibit any properties indicating that they behave as zwitterions? If so, their behavior should be similar to that of ionic substances. ∞ (Section 8.2) Crystalline amino acids have relatively high melting points, usually above 200 °C, which is characteristic of ionic solids. Amino acids are far more soluble in water than in nonpolar solvents. In addition, the dipole moments of amino acids are large, consistent with a large separation of charge in the molecule. Thus, the ability of amino acids to act simultaneously as acids and bases has important effects on their properties.

Related Exercise: 16.105, 16.114

16.11 | Lewis Acids and Bases

For a substance to be a proton acceptor (a Brønsted–Lowry base), it must have an unshared pair of electrons for binding the proton, as, for example, in NH_3. Using Lewis structures, we can write the reaction between H^+ and NH_3 as

G. N. Lewis was the first to notice this aspect of acid–base reactions. He proposed a more general definition of acids and bases that emphasizes the shared electron pair: A **Lewis acid** is an electron-pair acceptor, and a **Lewis base** is an electron-pair donor.

Every base that we have discussed thus far—whether OH^-, H_2O, an amine, or an anion—is an electron-pair donor. Everything that is a base in the Brønsted–Lowry sense (a proton acceptor) is also a base in the Lewis sense (an electron-pair donor). In the Lewis theory, however, a base can donate its electron pair to something other than H^+. The Lewis definition therefore greatly increases the number of species that can be considered acids; in other words, H^+ is a Lewis acid but not the only one. For example, the reaction between NH_3 and BF_3 occurs because BF_3 has a vacant orbital in its valence shell. ∞ (Section 8.7) It therefore acts as an electron-pair acceptor (a Lewis acid) toward NH_3, which donates the electron pair:

▲ **Give It Some Thought**

What feature must any molecule or ion have in order to act as a Lewis acid?

Our emphasis throughout this chapter has been on water as the solvent and on the proton as the source of acidic properties. In such cases we find the Brønsted–Lowry definition of acids and bases to be the most useful. In fact, when we speak of a substance as being acidic or basic, we are usually thinking of aqueous solutions and using these terms in the Arrhenius or Brønsted–Lowry sense. The advantage of the Lewis definitions of acid and base is that they allow us to treat a wider variety of reactions, including those that do not involve proton transfer, as acid–base reactions. To avoid confusion, a substance such as BF_3 is rarely called an acid unless it is clear from the context that we are using the term in the sense of the Lewis definition. Instead, substances that function as electron-pair acceptors are referred to explicitly as "Lewis acids."

Lewis acids include molecules that, like BF_3, have an incomplete octet of electrons. In addition, many simple cations can function as Lewis acids. For example, Fe^{3+} interacts strongly with cyanide ions to form the ferricyanide ion:

$$Fe^{3+} + 6[:C\equiv N:]^- \longrightarrow [Fe(C\equiv N:)_6]^{3-}$$

The Fe^{3+} ion has vacant orbitals that accept the electron pairs donated by the cyanide ions. (We will learn more in Chapter 23 about just which orbitals are used by the Fe^{3+} ion.) The metal ion is highly charged, too, which contributes to the interaction with CN^- ions.

Some compounds containing multiple bonds can behave as Lewis acids. For example, the reaction of carbon dioxide with water to form carbonic acid (H_2CO_3) can be pictured as an attack by a water molecule on CO_2, in which the water acts as an electron-pair donor and the CO_2 as an electron-pair acceptor:

One electron pair of one of the carbon–oxygen double bonds is moved onto the oxygen, leaving a vacant orbital on the carbon, which means the carbon can accept an electron pair donated by H_2O. The initial acid–base product rearranges by transferring a proton from the water oxygen to a carbon dioxide oxygen, forming carbonic acid.

The hydrated cations we encountered in Section 16.9, such as $[Fe(H_2O)_6]^{3+}$ in Figure 16.17, form through the reaction between the cation acting as a Lewis acid and the water molecules acting as Lewis bases. When a water molecule interacts with the positively charged metal ion, electron density is drawn from the oxygen (▼ Figure 16.20). This flow of electron density causes the O—H bond to become more polarized; as a result, water molecules bound to the metal ion are more acidic than

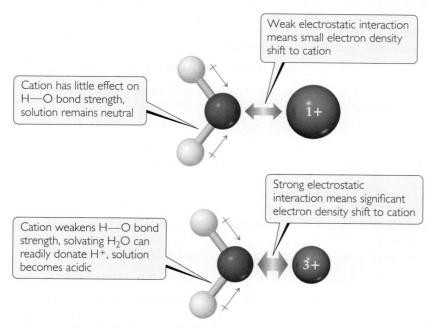

Weak electrostatic interaction means small electron density shift to cation

Cation has little effect on H—O bond strength, solution remains neutral

1+

Strong electrostatic interaction means significant electron density shift to cation

Cation weakens H—O bond strength, solvating H_2O can readily donate H^+, solution becomes acidic

3+

▲ Figure 16.20 **The acidity of a hydrated cation depends on cation charge.**

those in the bulk solvent. This effect becomes more pronounced as the charge of the cation increases, which explains why 3+ cations are much more acidic than cations with smaller charges.

The Lewis acid–base concept allows many ideas developed in this chapter to be used more broadly in chemistry, including rections in solvents other than water. If you take a course in organic chemistry, you will see a number of important rections that require the presence of a Lewis acid in order to proceed. The interaction of lone pairs on one molecule or ion with vacant orbitals on another molecule or ion is one of the most important concepts in chemistry, as you will see throughout your studies.

SAMPLE INTEGRATIVE EXERCISE | Putting Concepts Together

Phosphorous acid (H_3PO_3) has the Lewis structure

$$
\begin{array}{c}
\text{H} \\
| \\
\ddot{\text{O}}\!-\!\text{P}\!-\!\ddot{\text{O}}\!-\!\text{H} \\
| \\
\ddot{\text{O}}\!-\!\text{H}
\end{array}
$$

(a) Explain why H_3PO_3 is diprotic and not triprotic. **(b)** A 25.0-mL sample of an H_3PO_3 solution titrated with 0.102 M NaOH requires 23.3 mL of NaOH to neutralize both acidic protons. What is the molarity of the H_3PO_3 solution? **(c)** The original solution from part (b) has a pH of 1.59. Calculate the percent ionization and K_{a1} for H_3PO_3, assuming that $K_{a1} \gg K_{a2}$. **(d)** How does the osmotic pressure of a 0.050 M solution of HCl compare qualitatively with that of a 0.050 M solution of H_3PO_3? Explain.

SOLUTION

We will use what we have learned about molecular structure and its impact on acidic behavior to answer part (a). We will then use stoichiometry and the relationship between pH and $[H^+]$ to answer parts (b) and (c). Finally, we will consider percent ionization in order to compare the osmotic pressure of the two solutions in part (d).

(a) Acids have polar H—X bonds. From Figure 8.7 (p. 310) we see that the electronegativity of H is 2.1 and that of P is also 2.1. Because the two elements have the same electronegativity, the H—P bond is nonpolar. ∞ (Section 8.4) Thus, this H cannot be acidic. The other two H atoms, however, are bonded to O, which has an electronegativity of 3.5. The H—O bonds are, therefore, polar with H having a partial positive charge. These two H atoms are consequently acidic.

(b) The chemical equation for the neutralization reaction is

$$H_3PO_3(aq) + 2NaOH(aq) \longrightarrow Na_2HPO_3(aq) + 2H_2O(l)$$

From the definition of molarity, M = mol/L, we see that moles = $M \times$ L. ∞ (Section 4.5) Thus, the number of moles of NaOH added to the solution is

$$(0.0233\ \text{L})(0.102\ \text{mol/L}) = 2.38 \times 10^{-3}\ \text{mol NaOH}$$

The balanced equation indicates that 2 mol of NaOH is consumed for each mole of H_3PO_3. Thus, the number of moles of H_3PO_3 in the sample is

$$(2.38 \times 10^{-3}\ \text{mol NaOH})\left(\frac{1\ \text{mol}\,H_3PO_3}{2\ \text{mol}\,\text{NaOH}}\right) = 1.19 \times 10^{-3}\ \text{mol}\,H_3PO_3$$

The concentration of the H_3PO_3 solution, therefore, equals $(1.19 \times 10^{-3}\ \text{mol})/(0.0250\ \text{L}) = 0.0476\ M$.

(c) From the pH of the solution, 1.59, we can calculate $[H^+]$ at equilibrium:

$$[H^+] = \text{antilog}(-1.59) = 10^{-1.59} = 0.026\ M\ \text{(two significant figures)}$$

Because $K_{a1} \gg K_{a2}$, the vast majority of the ions in solution are from the first ionization step of the acid.

$$H_3PO_3(aq) \rightleftharpoons H^+(aq) + H_2PO_3^-(aq)$$

Because one $H_2PO_3^-$ ion forms for each H^+ ion formed, the equilibrium concentrations of H^+ and $H_2PO_3^-$ are equal: $[H^+] = [H_2PO_3^-] = 0.026\ M$. The equilibrium concentration of H_3PO_3 equals the initial concentration minus the amount that ionizes to form H^+

and $H_2PO_3^-$: $[H_3PO_3] = 0.0476\,M - 0.026\,M = 0.022\,M$ (two significant figures). These results can be tabulated as follows:

	$H_3PO_3(aq)$	$\rightleftharpoons$ $H^+(aq)$	$+$ $H_2PO_3^-(aq)$
Initial concentration (M)	0.0476	0	0
Change in concentration (M)	-0.026	$+0.026$	$+0.026$
Equilibrium concentration (M)	0.022	0.026	0.026

The percent ionization is

$$\text{percent ionization} = \frac{[H^+]_{\text{equilibrium}}}{[H_3PO_3]_{\text{initial}}} \times 100\% = \frac{0.026\,M}{0.0476\,M} \times 100\% = 55\%$$

The first acid-dissociation constant is

$$K_{a1} = \frac{[H^+][H_2PO_3^-]}{[H_3PO_3]} = \frac{(0.026)(0.026)}{0.022} = 0.031$$

(d) Osmotic pressure is a colligative property and depends on the total concentration of particles in solution. ∞ (Section 13.5) Because HCl is a strong acid, a 0.050 M solution will contain 0.050 M $H^+(aq)$ and 0.050 M $Cl^-(aq)$, or a total of 0.100 mol/L of particles. Because H_3PO_3 is a weak acid, it ionizes to a lesser extent than HCl and, hence, there are fewer particles in the H_3PO_3 solution. As a result, the H_3PO_3 solution will have the lower osmotic pressure.

Chapter Summary and Key Terms

INTRODUCTION TO ACIDS AND BASES (SECTION 16.1) Acids and bases were first recognized by the properties of their aqueous solutions. For example, acids turn litmus red, whereas bases turn litmus blue. Arrhenius recognized that the properties of acidic solutions are due to $H^+(aq)$ ions and those of basic solutions are due to $OH^-(aq)$ ions.

BRØNSTED–LOWRY ACIDS AND BASES (SECTION 16.2) The Brønsted–Lowry concept of acids and bases is more general than the Arrhenius concept and emphasizes the transfer of a proton (H^+) from an acid to a base. The H^+ ion, which is merely a proton with no surrounding valence electrons, is strongly bound to water. For this reason, the **hydronium ion**, $H_3O^+(aq)$, is often used to represent the predominant form of H^+ in water instead of the simpler $H^+(aq)$.

A **Brønsted–Lowry acid** is a substance that donates a proton to another substance; a **Brønsted–Lowry base** is a substance that accepts a proton from another substance. Water is an **amphiprotic** substance, one that can function as either a Brønsted–Lowry acid or base, depending on the substance with which it reacts.

The **conjugate base** of a Brønsted–Lowry acid is the species that remains when a proton is removed from the acid. The **conjugate acid** of a Brønsted–Lowry base is the species formed by adding a proton to the base. Together, an acid and its conjugate base (or a base and its conjugate acid) are called a **conjugate acid–base pair**.

The acid–base strengths of conjugate acid–base pairs are related: The stronger an acid, the weaker is its conjugate base; the weaker an acid, the stronger is its conjugate base. In every acid–base reaction, the position of the equilibrium favors the transfer of the proton from the stronger acid to the stronger base.

AUTOIONIZATION OF WATER (SECTION 16.3) Water ionizes to a slight degree, forming $H^+(aq)$ and $OH^-(aq)$. The extent of this **autoionization** is expressed by the **ion-product constant** for water: $K_w = [H^+][OH^-] = 1.0 \times 10^{-14}$ (25 °C). This relationship holds for both pure water and aqueous solutions. The K_w expression indicates that the product of $[H^+]$ and $[OH^-]$ is a constant. Thus, as $[H^+]$ increases, $[OH^-]$ decreases. Acidic solutions are those that contain more $H^+(aq)$ than $OH^-(aq)$, whereas basic solutions contain more $OH^-(aq)$ than $H^+(aq)$. When $[H^+] = [OH^-]$, the solution is neutral.

THE pH SCALE (SECTION 16.4) The concentration of $H^+(aq)$ can be expressed in terms of **pH**: $pH = -\log[H^+]$. At 25 °C the pH of a neutral solution is 7.00, whereas the pH of an acidic solution is below 7.00, and the pH of a basic solution is above 7.00. This p notation is also used to represent the negative logarithm of other small quantities, as in pOH and pK_w. The pH of a solution can be measured using a pH meter, or it can be estimated using acid–base indicators.

STRONG ACIDS AND BASES (SECTION 16.5) Strong acids are strong electrolytes, ionizing completely in aqueous solution. The common strong acids are HCl, HBr, HI, HNO_3, $HClO_3$, $HClO_4$, and H_2SO_4. The conjugate bases of strong acids have negligible basicity.

Common strong bases are the ionic hydroxides of the alkali metals and the heavy alkaline earth metals.

WEAK ACIDS AND BASES (SECTIONS 16.6 AND 16.7) Weak acids are weak electrolytes; only a small fraction of the molecules exist in solution in ionized form. The extent of ionization is expressed by the **acid-dissociation constant**, K_a, which is the equilibrium constant for the reaction $HA(aq) \rightleftharpoons H^+(aq) + A^-(aq)$, which can also be written as $HA(aq) + H_2O(l) \rightleftharpoons H_3O^+(aq) + A^-(aq)$. The larger the value of K_a, the stronger is the acid. For solutions of the same concentration, a stronger acid also has a larger **percent ionization**. The concentration of a weak acid and its K_a value can be used to calculate the pH of a solution.

Polyprotic acids, such as H_3PO_4, have more than one ionizable proton. These acids have acid-dissociation constants that decrease in magnitude in the order $K_{a1} > K_{a2} > K_{a3}$. Because nearly all the

$H^+(aq)$ in a polyprotic acid solution comes from the first dissociation step, the pH can usually be estimated satisfactorily by considering only K_{a1}. Weak bases include NH_3, **amines**, and the anions of weak acids. The extent to which a weak base reacts with water to generate the corresponding conjugate acid and OH^- is measured by the **base-dissociation constant**, K_b. K_b is the equilibrium constant for the reaction $B(aq) + H_2O(l) \rightleftharpoons HB^+(aq) + OH^-(aq)$, where B is the base.

RELATIONSHIP BETWEEN K_a AND K_b (SECTION 16.8) The relationship between the strength of an acid and the strength of its conjugate base is expressed quantitatively by the equation $K_a \times K_b = K_w$, where K_a and K_b are dissociation constants for conjugate acid–base pairs. This equation explains the inverse relationship between the strength of an acid and the strength of its conjugate base.

ACID–BASE PROPERTIES OF SALTS (SECTION 16.9) The acid–base properties of salts can be ascribed to the behavior of their respective cations and anions. The reaction of ions with water, with a resultant change in pH, is called **hydrolysis**. The cations of the alkali metals and the alkaline earth metals as well as the anions of strong acids, such as Cl^-, Br^-, I^-, and NO_3^-, do not undergo hydrolysis. They are always spectator ions in acid–base chemistry. A cation that is the conjugate acid of a weak base produces H^+ upon hydrolysis.

An anion that is the conjugate base of a weak acid produces OH^- upon hydrolysis.

ACID–BASE BEHAVIOR AND CHEMICAL STRUCTURE (SECTION 16.10) The tendency of a substance to show acidic or basic characteristics in water can be correlated with its chemical structure. Acid character requires the presence of a highly polar $H—X$ bond. Acidity is also favored when the $H—X$ bond is weak and when the X^- ion is very stable.

For **oxyacids** with the same number of OH groups and the same number of O atoms, acid strength increases with increasing electronegativity of the central atom. For oxyacids with the same central atom, acid strength increases as the number of oxygen atoms attached to the central atom increases. **Carboxylic acids**, which are organic acids containing the COOH group, are the most important class of organic acids. The presence of delocalized π bonding in the conjugate base is a major factor responsible for the acidity of these compounds.

LEWIS ACIDS AND BASES (SECTION 16.11) The Lewis concept of acids and bases emphasizes the shared electron pair rather than the proton. A **Lewis acid** is an electron-pair acceptor, and a **Lewis base** is an electron-pair donor. The Lewis concept is more general than the Brønsted–Lowry concept because it can apply to cases in which the acid is some substance other than H^+, and to solvents other than water.

Learning Outcomes   After studying this chapter, you should be able to:

- Define and identify Arrhenius acids and bases. (Section 16.1)

- Describe the nature of the hydrated proton, represented as either $H^+(aq)$ or $H_3O^+(aq)$. (Section 16.2)

- Define and identify Brønsted–Lowry acids and bases and identify conjugate acid–base pairs. (Section 16.2)

- Correlate the strength of an acid to the strength of its conjugate base. (Section 16.2)

- Explain how the equilibrium position of a proton-transfer reaction relates to the strengths of the acids and bases involved. (Section 16.3)

- Describe the autoionization of water and explain how $[H_3O^+]$ and $[OH^-]$ are related via K_w. (Section 16.3)

- Calculate the pH of a solution given $[H_3O^+]$ or $[OH^-]$. (Section 16.4)

- Calculate the pH of a strong acid or strong base given its concentration. (Section 16.5)

- Calculate K_a or K_b for a weak acid or weak base given its concentration and the pH of the solution, and vice versa. (Sections 16.6 and 16.7)

- Calculate the pH of a weak acid or weak base or its percent ionization given its concentration and K_a or K_b. (Sections 16.6 and 16.7)

- Calculate K_b for a weak base given K_a of its conjugate acid, and similarly calculate K_a from K_b. (Section 16.8)

- Predict whether an aqueous solution of a salt will be acidic, basic, or neutral. (Section 16.9)

- Predict the relative strength of a series of acids from their molecular structures. (Section 16.10)

- Define and identify Lewis acids and bases. (Section 16.11)

Key Equations

- $K_w = [H_3O^+][OH^-] = [H^+][OH^-] = 1.0 \times 10^{-14}$ [16.16]   Ion product of water at 25 °C

- $pH = -\log[H^+]$ [16.17]   Definition of pH

- $pOH = -\log[OH^-]$ [16.18]   Definition of pOH

- $pH + pOH = 14.00$ [16.20]   Relationship between pH and pOH

- $K_a = \dfrac{[H_3O^+][A^-]}{[HA]}$ or $K_a = \dfrac{[H^+][A^-]}{[HA]}$ [16.25]   Acid-dissociation constant for a weak acid, HA

- Percent ionization $= \dfrac{[H^+]_{equilibrium}}{[HA]_{initial}} \times 100\%$ [16.27]   Percent ionization of a weak acid

- $K_b = \dfrac{[BH^+][OH^-]}{[B]}$ [16.34]   Base-dissociation constant for a weak base, B

- $K_a \times K_b = K_w$ [16.40]   Relationship between acid- and base-dissociation constants of a conjugate acid–base pair

- $pK_a = -\log K_a$ and $pK_b = -\log K_b$ [16.41]   Definitions of pK_a and pK_b

Exercises

Visualizing Concepts

16.1 **(a)** Identify the Brønsted–Lowry acid and base in the reaction

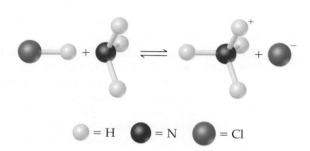

= H   = N   = Cl

(b) Identify the Lewis acid and base in the reaction. [Sections 16.2 and 16.11]

16.2 The following diagrams represent aqueous solutions of two monoprotic acids, HA (A = X or Y). The water molecules have been omitted for clarity. **(a)** Which is the stronger acid, HX or HY? **(b)** Which is the stronger base, X⁻ or Y⁻? **(c)** If you mix equal concentrations of HX and NaY, will the equilibrium

$$HX(aq) + Y^-(aq) \rightleftharpoons HY(aq) + X^-(aq)$$

lie mostly to the right ($K_c > 1$) or to the left ($K_c < 1$)? [Section 16.2]

= HA   = H_3O^+ = A^-

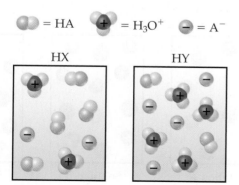

HX   HY

16.3 The indicator methyl orange has been added to both of the following solutions. Based on the colors, classify each statement as true or false:

(a) The pH of solution A is definitely less than 7.00.

(b) The pH of solution B is definitely greater than 7.00.

(c) The pH of solution B is greater than that of solution A. [Section 16.4]

Solution A   Solution B

16.4 The probe of the pH meter shown here is sitting in a beaker that contains a clear liquid. **(a)** You are told the liquid is pure water, a solution of HCl(aq), or a solution of KOH(aq). Which one is it? **(b)** If the liquid is one of the solutions, what is its molarity? **(c)** Why is the temperature given on the pH meter? [Sections 16.4 and 16.5]

16.5 The following diagrams represent aqueous solutions of three acids, HX, HY, and HZ. The water molecules have been omitted for clarity, and the hydrated proton is represented as H⁺ rather than H_3O^+. **(a)** Which of the acids is a strong acid? Explain. **(b)** Which acid would have the smallest acid-dissociation constant, K_a? **(c)** Which solution would have the highest pH? [Sections 16.5 and 16.6]

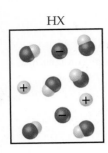
HX

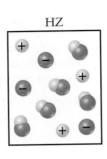
HY

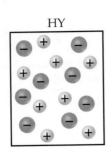
HZ

16.6 The graph given below shows $[H^+]$ vs. concentration for an aqueous solution of an unknown substance. **(a)** Is the substance a strong acid, a weak acid, a strong base, or a weak base? **(b)** Based on your answer to (a), can you determine the value of the pH of the solution when the concentration is 0.18 *M*? **(c)** Would the line go exactly through the origin of the plot? [Sections 16.5 and 16.6]

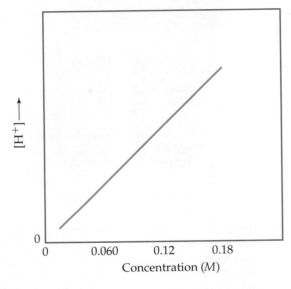

16.7 **(a)** Which of these three lines represents the effect of concentration on the percent ionization of a weak acid? **(b)** Explain in qualitative terms why the curve you chose has the shape it does. [Section 16.6]

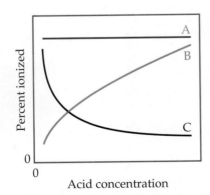

16.8 Each of the three molecules shown here contains an OH group, but one molecule acts as a base, one as an acid, and the third is neither acid nor base. **(a)** Which one acts as a base? Why does only this molecule act as a base? **(b)** Which one acts as an acid? **(c)** Why is the remaining molecule neither acidic nor basic? [Sections 16.6 and 16.7]

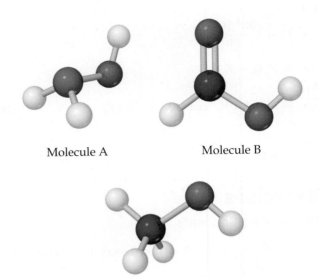

Molecule A   Molecule B

Molecule C

16.9 *Phenylephrine*, an organic substance with molecular formula $C_9H_{13}NO_2$, is used as a nasal decongenstant in over-the-counter medications. The molecular structure of phenylephrine is shown below using the usual shortcut organic nomenclature. **(a)** Would you expect a solution of phenylephrine to be acidic, neutral, or basic? **(b)** One of the active ingredients in Alka-Seltzer PLUS® cold medication is "phenylephrine hydrochloride." How does this ingredient differ from the structure shown below? **(c)** Would you expect a solution of phenylephrine hydrochloride to be acidic, neutral, or basic? [Sections 16.8 and 16.9]

16.10 Which of the following diagrams best represents an aqueous solution of NaF? (For clarity, the water molecules are not shown.) Will this solution be acidic, neutral, or basic? [Section 16.9]

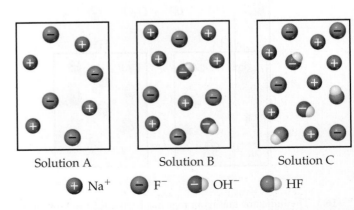

Solution A   Solution B   Solution C

● Na⁺   ● F⁻   ● OH⁻   ● HF

16.11 Consider the molecular models shown here, where X represents a halogen atom. **(a)** If X is the same atom in both molecules, which molecule will be more acidic? **(b)** Does the acidity of each molecule increase or decrease as the electronegativity of the atom X increases? [Section 16.10]

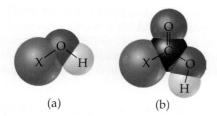

(a)   (b)

16.12 (a) The following diagram represents the reaction of PCl_4^+ with Cl^-. Draw the Lewis structures for the reactants and products, and identify the Lewis acid and the Lewis base in the reaction.

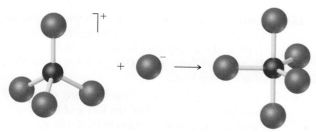

(b) The following reaction represents a hydrated cation losing a proton. How does the equilibrium constant for the reaction change as the charge of the cation increases? [Sections 16.9 and 16.11]

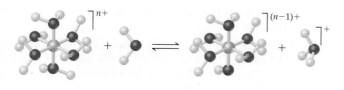

Arrhenius and Brønsted–Lowry Acids and Bases (Sections 16.1 and 16.2)

16.13 (a) What is the difference between the Arrhenius and the Brønsted–Lowry definitions of an acid? **(b)** $NH_3(g)$ and $HCl(g)$ react to form the ionic solid $NH_4Cl(s)$. Which substance is the Brønsted–Lowry acid in this reaction? Which is the Brønsted–Lowry base?

16.14 (a) What is the difference between the Arrhenius and the Brønsted–Lowry definitions of a base? **(b)** Can a substance behave as an Arrhenius base if it does not contain an OH group? Explain.

16.15 (a) Give the conjugate base of the following Brønsted–Lowry acids: **(i)** HIO_3, **(ii)** NH_4^+. **(b)** Give the conjugate acid of the following Brønsted–Lowry bases: **(i)** O^{2-}, **(ii)** $H_2PO_4^-$.

16.16 (a) Give the conjugate base of the following Brønsted–Lowry acids: **(i)** $HCOOH$, **(ii)** HPO_4^{2-}. **(b)** Give the conjugate acid of the following Brønsted–Lowry bases: **(i)** SO_4^{2-}, **(ii)** CH_3NH_2.

16.17 Designate the Brønsted–Lowry acid and the Brønsted–Lowry base on the left side of each of the following equations, and also designate the conjugate acid and conjugate base of each on the right side:

(a) $NH_4^+(aq) + CN^-(aq) \rightleftharpoons HCN(aq) + NH_3(aq)$

(b) $(CH_3)_3N(aq) + H_2O(l) \rightleftharpoons$
$(CH_3)_3NH^+(aq) + OH^-(aq)$

(c) $HCOOH(aq) + PO_4^{3-}(aq) \rightleftharpoons$
$HCOO^-(aq) + HPO_4^{2-}(aq)$

16.18 Designate the Brønsted–Lowry acid and the Brønsted–Lowry base on the left side of each equation, and also designate the conjugate acid and conjugate base of each on the right side.

(a) $HBrO(aq) + H_2O(l) \rightleftharpoons H_3O^+(aq) + BrO^-(aq)$

(b) $HSO_4^-(aq) + HCO_3^-(aq) \rightleftharpoons$
$SO_4^{2-}(aq) + H_2CO_3(aq)$

(c) $HSO_3^-(aq) + H_3O^+(aq) \rightleftharpoons H_2SO_3(aq) + H_2O(l)$

16.19 (a) The hydrogen sulfite ion (HSO_3^-) is amphiprotic. Write a balanced chemical equation showing how it acts as an acid toward water and another equation showing how it acts as a base toward water. **(b)** What is the conjugate acid of HSO_3^-? What is its conjugate base?

16.20 (a) Write an equation for the reaction in which $H_2C_6H_7O_5^-(aq)$ acts as a base in $H_2O(l)$. **(b)** Write an equation for the reaction in which $H_2C_6H_7O_5^-(aq)$ acts as an acid in $H_2O(l)$. **(c)** What is the conjugate acid of $H_2C_6H_7O_5^-(aq)$? What is its conjugate base?

16.21 Label each of the following as being a strong base, a weak base, or a species with negligible basicity. In each case write the formula of its conjugate acid, and indicate whether the conjugate acid is a strong acid, a weak acid, or a species with negligible acidity: **(a)** CH_3COO^-, **(b)** HCO_3^-, **(c)** O^{2-}, **(d)** Cl^-, **(e)** NH_3.

16.22 Label each of the following as being a strong acid, a weak acid, or a species with negligible acidity. In each case write the formula of its conjugate base, and indicate whether the conjugate base is a strong base, a weak base, or a species with negligible basicity: **(a)** $HCOOH$, **(b)** H_2, **(c)** CH_4, **(d)** HF, **(e)** NH_4^+.

16.23 (a) Which of the following is the stronger Brønsted–Lowry acid, HBrO or HBr? **(b)** Which is the stronger Brønsted–Lowry base, F^- or Cl^-?

16.24 (a) Which of the following is the stronger Brønsted–Lowry acid, $HClO_3$ or $HClO_2$? **(b)** Which is the stronger Brønsted–Lowry base, HS^- or HSO_4^-?

16.25 Predict the products of the following acid–base reactions, and predict whether the equilibrium lies to the left or to the right of the equation:

(a) $O^{2-}(aq) + H_2O(l) \rightleftharpoons$

(b) $CH_3COOH(aq) + HS^-(aq) \rightleftharpoons$

(c) $NO_2^-(aq) + H_2O(l) \rightleftharpoons$

16.26 Predict the products of the following acid–base reactions, and predict whether the equilibrium lies to the left or to the right of the equation:

(a) $NH_4^+(aq) + OH^-(aq) \rightleftharpoons$

(b) $CH_3COO^-(aq) + H_3O^+(aq) \rightleftharpoons$

(c) $HCO_3^-(aq) + F^-(aq) \rightleftharpoons$

Autoionization of Water (Section 16.3)

16.27 If a neutral solution of water, with pH = 7.00, is cooled to 10 °C, the pH rises to 7.27. Which of the following three statements is correct for the cooled water: **(i)** $[H^+] > [OH^-]$, **(ii)** $[H^+] = [OH^-]$, or **(iii)** $[H^+] < [OH^-]$?

16.28 (a) Write a chemical equation that illustrates the autoionization of water. **(b)** Write the expression for the ion-product constant for water, K_w. **(c)** If a solution is described as basic, which of the following is true: **(i)** $[H^+] > [OH^-]$, **(ii)** $[H^+] = [OH^-]$, or **(iii)** $[H^+] < [OH^-]$?

16.29 Calculate $[H^+]$ for each of the following solutions, and indicate whether the solution is acidic, basic, or neutral: **(a)** $[OH^-] = 0.00045\,M$; **(b)** $[OH^-] = 8.8 \times 10^{-9}\,M$; **(c)** a solution in which $[OH^-]$ is 100 times greater than $[H^+]$.

16.30 Calculate $[OH^-]$ for each of the following solutions, and indicate whether the solution is acidic, basic, or neutral: **(a)** $[H^+] = 0.0505\,M$; **(b)** $[H^+] = 2.5 \times 10^{-10}\,M$; **(c)** a solution in which $[H^+]$ is 1000 times greater than $[OH^-]$.

16.31 At the freezing point of water $(0\,°C)$, $K_w = 1.2 \times 10^{-15}$. Calculate $[H^+]$ and $[OH^-]$ for a neutral solution at this temperature.

16.32 Deuterium oxide (D_2O, where D is deuterium, the hydrogen-2 isotope) has an ion-product constant, K_w, of 8.9×10^{-16} at $20\,°C$. Calculate $[D^+]$ and $[OD^-]$ for pure (neutral) D_2O at this temperature.

The pH Scale (Section 16.4)

16.33 By what factor does $[H^+]$ change for a pH change of **(a)** 2.00 units, **(b)** 0.50 units?

16.34 Consider two solutions, solution A and solution B. $[H^+]$ in solution A is 250 times greater than that in solution B. What is the difference in the pH values of the two solutions?

16.35 Complete the following table by calculating the missing entries and indicating whether the solution is acidic or basic.

$[H^+]$	$OH^-\,(aq)$	pH	pOH	Acidic or Basic?
$7.5 \times 10^{-3}\,M$				
	$3.6 \times 10^{-10}\,M$			
		8.25		
			5.70	

16.36 Complete the following table by calculating the missing entries. In each case indicate whether the solution is acidic or basic.

pH	pOH	$[H^+]$	$[OH^-]$	Acidic or Basic?
5.25				
	2.02			
		$4.4 \times 10^{-10}\,M$		
			$8.5 \times 10^{-2}\,M$	

16.37 The average pH of normal arterial blood is 7.40. At normal body temperature $(37\,°C)$, $K_w = 2.4 \times 10^{-14}$. Calculate $[H^+]$, $[OH^-]$, and pOH for blood at this temperature.

16.38 Carbon dioxide in the atmosphere dissolves in raindrops to produce carbonic acid (H_2CO_3), causing the pH of clean, unpolluted rain to range from about 5.2 to 5.6. What are the ranges of $[H^+]$ and $[OH^-]$ in the raindrops?

16.39 Addition of the indicator methyl orange to an unknown solution leads to a yellow color. The addition of bromthymol blue to the same solution also leads to a yellow color. **(a)** Is the solution acidic, neutral, or basic? **(b)** What is the range (in whole numbers) of possible pH values for the solution? **(c)** Is there another indicator you could use to narrow the range of possible pH values for the solution?

16.40 Addition of phenolphthalein to an unknown colorless solution does not cause a color change. The addition of bromthymol blue to the same solution leads to a yellow color. **(a)** Is the solution acidic, neutral, or basic? **(b)** Which of the following can you establish about the solution: (i) A minimum pH, (ii) A maximum pH, or (iii) A specific range of pH values? **(c)** What other indicator or indicators would you want to use to determine the pH of the solution more precisely?

Strong Acids and Bases (Section 16.5)

16.41 Is each of the following statements true or false? **(a)** All strong acids contain one or more H atoms. **(b)** A strong acid is a strong electrolyte. **(c)** A 1.0-M solution of a strong acid will have pH $= 1.0$.

16.42 Determine whether each of the following is true or false: **(a)** All strong bases are salts of the hydroxide ion. **(b)** The addition of a strong base to water produces a solution of pH > 7.0. **(c)** Because $Mg(OH)_2$ is not very soluble, it cannot be a strong base.

16.43 Calculate the pH of each of the following strong acid solutions: **(a)** $8.5 \times 10^{-3}\,M$ HBr, **(b)** 1.52 g of HNO_3 in 575 mL of solution, **(c)** 5.00 mL of 0.250 M $HClO_4$ diluted to 50.0 mL, **(d)** a solution formed by mixing 10.0 mL of 0.100 M HBr with 20.0 mL of 0.200 M HCl.

16.44 Calculate the pH of each of the following strong acid solutions: **(a)** 0.0167 M HNO_3, **(b)** 0.225 g of $HClO_3$ in 2.00 L of solution, **(c)** 15.00 mL of 1.00 M HCl diluted to 0.500 L, **(d)** a mixture formed by adding 50.0 mL of 0.020 M HCl to 125 mL of 0.010 M HI.

16.45 Calculate $[OH^-]$ and pH for **(a)** $1.5 \times 10^{-3}\,M$ $Sr(OH)_2$, **(b)** 2.250 g of LiOH in 250.0 mL of solution, **(c)** 1.00 mL of 0.175 M NaOH diluted to 2.00 L, **(d)** a solution formed by adding 5.00 mL of 0.105 M KOH to 15.0 mL of $9.5 \times 10^{-2}\,M$ $Ca(OH)_2$.

16.46 Calculate $[OH^-]$ and pH for each of the following strong base solutions: **(a)** 0.182 M KOH, **(b)** 3.165 g of KOH in 500.0 mL of solution, **(c)** 10.0 mL of 0.0105 M $Ca(OH)_2$ diluted to 500.0 mL, **(d)** a solution formed by mixing 20.0 mL of 0.015 M $Ba(OH)_2$ with 40.0 mL of $8.2 \times 10^{-3}\,M$ NaOH.

16.47 Calculate the concentration of an aqueous solution of NaOH that has a pH of 11.50.

16.48 Calculate the concentration of an aqueous solution of $Ca(OH)_2$ that has a pH of 10.05.

Weak Acids (Section 16.6)

16.49 Write the chemical equation and the K_a expression for the ionization of each of the following acids in aqueous solution. First show the reaction with $H^+(aq)$ as a product and then with the hydronium ion: **(a)** $HBrO_2$, **(b)** C_2H_5COOH.

16.50 Write the chemical equation and the K_a expression for the acid dissociation of each of the following acids in aqueous solution. First show the reaction with $H^+(aq)$ as a product and then with the hydronium ion: **(a)** C_6H_5COOH, **(b)** HCO_3^-.

16.51 Lactic acid ($CH_3CH(OH)COOH$) has one acidic hydrogen. A 0.10 M solution of lactic acid has a pH of 2.44. Calculate K_a.

16.52 Phenylacetic acid ($C_6H_5CH_2COOH$) is one of the substances that accumulates in the blood of people with phenylketonuria, an inherited disorder that can cause mental retardation or even death. A 0.085 M solution of $C_6H_5CH_2COOH$ has a pH of 2.68. Calculate the K_a value for this acid.

16.53 A 0.100 M solution of chloroacetic acid $(ClCH_2COOH)$ is 11.0% ionized. Using this information, calculate $[ClCH_2COO^-]$, $[H^+]$, $[ClCH_2COOH]$, and K_a for chloroacetic acid.

16.54 A 0.100 M solution of bromoacetic acid $(BrCH_2COOH)$ is 13.2% ionized. Calculate $[H^+]$, $[BrCH_2COO^-]$, $[BrCH_2COOH]$ and K_a for bromoacetic acid.

16.55 A particular sample of vinegar has a pH of 2.90. If acetic acid is the only acid that vinegar contains $(K_a = 1.8 \times 10^{-5})$, calculate the concentration of acetic acid in the vinegar.

16.56 If a solution of HF $(K_a = 6.8 \times 10^{-4})$ has a pH of 3.65, calculate the concentration of hydrofluoric acid.

16.57 The acid-dissociation constant for benzoic acid (C_6H_5COOH) is 6.3×10^{-5}. Calculate the equilibrium concentrations of H_3O^+, $C_6H_5COO^-$, and C_6H_5COOH in the solution if the initial concentration of C_6H_5COOH is 0.050 M.

16.58 The acid-dissociation constant for chlorous acid $(HClO_2)$ is 1.1×10^{-2}. Calculate the concentrations of H_3O^+, ClO_2^-, and $HClO_2$ at equilibrium if the initial concentration of $HClO_2$ is 0.0125 M.

16.59 Calculate the pH of each of the following solutions (K_a and K_b values are given in Appendix D): **(a)** 0.095 M propionic acid (C_2H_5COOH), **(b)** 0.100 M hydrogen chromate ion $(HCrO_4^-)$, **(c)** 0.120 M pyridine (C_5H_5N).

16.60 Determine the pH of each of the following solutions (K_a and K_b values are given in Appendix D): **(a)** 0.095 M hypochlorous acid, **(b)** 0.0085 M hydrazine, **(c)** 0.165 M hydroxylamine.

16.61 Saccharin, a sugar substitute, is a weak acid with $pK_a = 2.32$ at 25 °C. It ionizes in aqueous solution as follows:

$$HNC_7H_4SO_3(aq) \rightleftharpoons H^+(aq) + NC_7H_4SO_3^-(aq)$$

What is the pH of a 0.10 M solution of this substance?

16.62 The active ingredient in aspirin is acetylsalicylic acid $(HC_9H_7O_4)$, a monoprotic acid with $K_a = 3.3 \times 10^{-4}$ at 25 °C. What is the pH of a solution obtained by dissolving two extra-strength aspirin tablets, containing 500 mg of acetylsalicylic acid each, in 250 mL of water?

16.63 Calculate the percent ionization of hydrazoic acid (HN_3) in solutions of each of the following concentrations (K_a is given in Appendix D): **(a)** 0.400 M, **(b)** 0.100 M, **(c)** 0.0400 M.

16.64 Calculate the percent ionization of propionic acid (C_2H_5COOH) in solutions of each of the following concentrations (K_a is given in Appendix D): **(a)** 0.250 M, **(b)** 0.0800 M, **(c)** 0.0200 M.

16.65 Citric acid, which is present in citrus fruits, is a triprotic acid (Table 16.3). **(a)** Calculate the pH of a 0.040 M solution of citric acid. **(b)** Did you have to make any approximations or assumptions in completing your calculations? **(c)** Is the concentration of citrate ion $(C_6H_5O_7^{3-})$ equal to, less than, or greater than the H^+ ion concentration?

16.66 Tartaric acid is found in many fruits, including grapes, and is partially responsible for the dry texture of certain wines. Calculate the pH and the tartrate ion $(C_4H_4O_6^{2-})$ concentration for a 0.250 M solution of tartaric acid, for which the acid-dissociation constants are listed in Table 16.3. Did you have to make any approximations or assumptions in your calculation?

Weak Bases (Section 16.7)

16.67 Consider the base hydroxylamine, NH_2OH. **(a)** What is the conjugate acid of hydroxylamine? **(b)** When it acts as a base, which atom in hydroxylamine accepts a proton? **(c)** There are two atoms in hydroxylamine that have nonbonding electron pairs that could act as proton acceptors. Use Lewis structures and formal charges ∞ (Section 8.5) to rationalize why one of these two atoms is a much better proton acceptor than the other.

16.68 The hypochlorite ion, ClO^-, acts as a weak base. **(a)** Is ClO^- a stronger or weaker base than hydroxylamine? **(b)** When ClO^- acts as a base, which atom, Cl or O, acts as the proton acceptor? **(c)** Can you use formal charges to rationalize your answer to part (b)?

16.69 Write the chemical equation and the K_b expression for the reaction of each of the following bases with water: **(a)** dimethylamine, $(CH_3)_2NH$; **(b)** carbonate ion, CO_3^{2-}; **(c)** formate ion, CHO_2^-.

16.70 Write the chemical equation and the K_b expression for the reaction of each of the following bases with water: **(a)** propylamine, $C_3H_7NH_2$; **(b)** monohydrogen phosphate ion, HPO_4^{2-}; **(c)** benzoate ion, $C_6H_5CO_2^-$.

16.71 Calculate the molar concentration of OH^- in a 0.075 M solution of ethylamine $(C_2H_5NH_2; K_b = 6.4 \times 10^{-4})$. Calculate the pH of this solution.

16.72 Calculate the molar concentration of OH^- in a 0.724 M solution of hypobromite ion $(BrO^-; K_b = 4.0 \times 10^{-6})$. What is the pH of this solution?

16.73 Ephedrine, a central nervous system stimulant, is used in nasal sprays as a decongestant. This compound is a weak organic base:

$$C_{10}H_{15}ON(aq) + H_2O(l) \rightleftharpoons C_{10}H_{15}ONH^+(aq) + OH^-(aq)$$

A 0.035 M solution of ephedrine has a pH of 11.33. **(a)** What are the equilibrium concentrations of $C_{10}H_{15}ON$, $C_{10}H_{15}ONH^+$, and OH^-? **(b)** Calculate K_b for ephedrine.

16.74 Codeine $(C_{18}H_{21}NO_3)$ is a weak organic base. A $5.0 \times 10^{-3} M$ solution of codeine has a pH of 9.95. Calculate the value of K_b for this substance. What is the pK_b for this base?

The $K_a - K_b$ Relationship; Acid–Base Properties of Salts (Sections 16.8 and 16.9)

16.75 Although the acid-dissociation constant for phenol (C_6H_5OH) is listed in Appendix D, the base-dissociation constant for the phenolate ion $(C_6H_5O^-)$ is not. **(a)** Explain why it is not necessary to list both K_a for phenol and K_b for the phenolate ion. **(b)** Calculate K_b for the phenolate ion. **(c)** Is the phenolate ion a weaker or stronger base than ammonia?

16.76 Use the acid-dissociation constants in Table 16.3 to arrange these oxyanions from strongest base to weakest: SO_4^{2-}, CO_3^{2-}, SO_3^{2-}, and PO_4^{3-}.

16.77 **(a)** Given that K_a for acetic acid is 1.8×10^{-5} and that for hypochlorous acid is 3.0×10^{-8}, which is the stronger acid? **(b)** Which is the stronger base, the acetate ion or the hypochlorite ion? **(c)** Calculate K_b values for CH_3COO^- and ClO^-.

16.78 **(a)** Given that K_b for ammonia is 1.8×10^{-5} and that for hydroxylamine is 1.1×10^{-8}, which is the stronger base?

(b) Which is the stronger acid, the ammonium ion or the hydroxylammonium ion? **(c)** Calculate K_a values for NH_4^+ and H_3NOH^+.

16.79 Using data from Appendix D, calculate $[OH^-]$ and pH for each of the following solutions: **(a)** 0.10 M $NaBrO$, **(b)** 0.080 M $NaHS$, **(c)** a mixture that is 0.10 M in $NaNO_2$ and 0.20 M in $Ca(NO_2)_2$.

16.80 Using data from Appendix D, calculate $[OH^-]$ and pH for each of the following solutions: **(a)** 0.105 M NaF, **(b)** 0.035 M Na_2S, **(c)** a mixture that is 0.045 M in $NaCH_3COO$ and 0.055 M in $Ba(CH_3COO)_2$.

16.81 A solution of sodium acetate ($NaCH_3COO$) has a pH of 9.70. What is the molarity of the solution?

16.82 Pyridinium bromide (C_5H_5NHBr) is a strong electrolyte that dissociates completely into $C_5H_5NH^+$ and Br^-. A solution of pyridinium bromide has a pH of 2.95. What is the molarity of the solution?

16.83 Predict whether aqueous solutions of the following compounds are acidic, basic, or neutral: **(a)** NH_4Br, **(b)** $FeCl_3$, **(c)** Na_2CO_3, **(d)** $KClO_4$, **(e)** $NaHC_2O_4$.

16.84 Predict whether aqueous solutions of the following substances are acidic, basic, or neutral: **(a)** $AlCl_3$, **(b)** $NaBr$, **(c)** $NaClO$, **(d)** $[CH_3NH_3]NO_3$, **(e)** Na_2SO_3.

16.85 An unknown salt is either NaF, $NaCl$, or $NaOCl$. When 0.050 mol of the salt is dissolved in water to form 0.500 L of solution, the pH of the solution is 8.08. What is the identity of the salt?

16.86 An unknown salt is either KBr, NH_4Cl, KCN, or K_2CO_3. If a 0.100 M solution of the salt is neutral, what is the identity of the salt?

Acid–Base Character and Chemical Structure (Section 16.10)

16.87 Explain the following observations: **(a)** HNO_3 is a stronger acid than HNO_2; **(b)** H_2S is a stronger acid than H_2O; **(c)** H_2SO_4 is a stronger acid than HSO_4^-; **(d)** H_2SO_4 is a stronger acid than H_2SeO_4; **(e)** CCl_3COOH is a stronger acid than CCl_3COOH.

16.88 Explain the following observations: **(a)** HCl is a stronger acid than H_2S; **(b)** H_3PO_4 is a stronger acid than H_3AsO_4; **(c)** $HBrO_3$ is a stronger acid than $HBrO_2$; **(d)** $H_2C_2O_4$ is a stronger acid than $HC_2O_4^-$; **(e)** benzoic acid (C_6H_5COOH) is a stronger acid than phenol (C_6H_5OH).

16.89 Based on their compositions and structures and on conjugate acid–base relationships, select the stronger base in each of the following pairs: **(a)** BrO^- or ClO^-, **(b)** BrO^- or BrO_2^-, **(c)** HPO_4^{2-} or $H_2PO_4^-$.

16.90 Based on their compositions and structures and on conjugate acid–base relationships, select the stronger base in each of the following pairs: **(a)** NO_3^- or NO_2^-, **(b)** PO_4^{3-} or AsO_4^{3-}, **(c)** HCO_3^- or CO_3^{2-}.

16.91 Indicate whether each of the following statements is true or false. For each statement that is false, correct the statement to make it true. **(a)** In general, the acidity of binary acids increases from left to right in a given row of the periodic table. **(b)** In a series of acids that have the same central atom, acid strength increases with the number of hydrogen atoms bonded to the central atom. **(c)** Hydrotelluric acid (H_2Te) is a stronger acid than H_2S because Te is more electronegative than S.

16.92 Indicate whether each of the following statements is true or false. For each statement that is false, correct the statement to make it true. **(a)** Acid strength in a series of H—A molecules increases with increasing size of A. **(b)** For acids of the same general structure but differing electronegativities of the central atoms, acid strength decreases with increasing electronegativity of the central atom. **(c)** The strongest acid known is HF because fluorine is the most electronegative element.

Lewis Acids and Bases (Section 16.11)

16.93 If a substance is an Arrhenius base, is it necessarily a Brønsted–Lowry base? Is it necessarily a Lewis base?

16.94 If a substance is a Lewis acid, is it necessarily a Brønsted–Lowry acid? Is it necessarily an Arrhenius acid?

16.95 Identify the Lewis acid and Lewis base among the reactants in each of the following reactions:

(a) $Fe(ClO_4)_3(s) + 6 H_2O(l) \rightleftharpoons$
$$Fe(H_2O)_6^{3+}(aq) + 3 ClO_4^-(aq)$$

(b) $CN^-(aq) + H_2O(l) \rightleftharpoons HCN(aq) + OH^-(aq)$

(c) $(CH_3)_3N(g) + BF_3(g) \rightleftharpoons (CH_3)_3NBF_3(s)$

(d) $HIO(lq) + NH_2^-(lq) \rightleftharpoons NH_3(lq) + IO^-(lq)$
(*lq* denotes liquid ammonia as solvent)

16.96 Identify the Lewis acid and Lewis base in each of the following reactions:

(a) $HNO_2(aq) + OH^-(aq) \rightleftharpoons NO_2^-(aq) + H_2O(l)$

(b) $FeBr_3(s) + Br^-(aq) \rightleftharpoons FeBr_4^-(aq)$

(c) $Zn^{2+}(aq) + 4 NH_3(aq) \rightleftharpoons Zn(NH_3)_4^{2+}(aq)$

(d) $SO_2(g) + H_2O(l) \rightleftharpoons H_2SO_3(aq)$

16.97 Predict which member of each pair produces the more acidic aqueous solution: **(a)** K^+ or Cu^{2+}, **(b)** Fe^{2+} or Fe^{3+}, **(c)** Al^{3+} or Ga^{3+}.

16.98 Which member of each pair produces the more acidic aqueous solution: **(a)** $ZnBr_2$ or $CdCl_2$, **(b)** $CuCl$ or $Cu(NO_3)_2$, **(c)** $Ca(NO_3)_2$ or $NiBr_2$?

Additional Exercises

16.99 Indicate whether each of the following statements is correct or incorrect.

(a) Every Brønsted–Lowry acid is also a Lewis acid.

(b) Every Lewis acid is also a Brønsted–Lowry acid.

(c) Conjugate acids of weak bases produce more acidic solutions than conjugate acids of strong bases.

(d) K^+ ion is acidic in water because it causes hydrating water molecules to become more acidic.

(e) The percent ionization of a weak acid in water increases as the concentration of acid decreases.

16.100 A solution is made by adding 0.300 g $Ca(OH)_2(s)$, 50.0 mL of 1.40 M HNO_3, and enough water to make a final volume of 75.0 mL Assuming that all of the solid dissolves, what is the pH of the final solution?

16.101 The odor of fish is due primarily to amines, especially methylamine (CH_3NH_2). Fish is often served with a wedge of lemon, which contains citric acid. The amine and the acid react forming a product with no odor, thereby making the less-than-fresh fish more appetizing. Using data from Appendix D, calculate the equilibrium constant for the reaction of citric acid with methylamine, if only the first proton of the citric acid (K_{a1}) is important in the neutralization reaction.

[16.102] For solutions of a weak acid, a graph of pH versus the logarithm of the initial acid concentration should be a straight line. What is the magnitude of the slope of that line?

16.103 Hemoglobin plays a part in a series of equilibria involving protonation-deprotonation and oxygenation-deoxygenation. The overall reaction is approximately as follows:

$$HbH^+(aq) + O_2(aq) \rightleftharpoons HbO_2(aq) + H^+(aq)$$

where Hb stands for hemoglobin and HbO_2 for oxyhemoglobin. (a) The concentration of O_2 is higher in the lungs and lower in the tissues. What effect does high $[O_2]$ have on the position of this equilibrium? (b) The normal pH of blood is 7.4. Is the blood acidic, basic, or neutral? (c) If the blood pH is lowered by the presence of large amounts of acidic metabolism products, a condition known as acidosis results. What effect does lowering blood pH have on the ability of hemoglobin to transport O_2?

[16.104] Calculate the pH of a solution made by adding 2.50 g of lithium oxide (Li_2O) to enough water to make 1.500 L of solution.

16.105 Benzoic acid (C_6H_5COOH) and aniline ($C_6H_5NH_2$) are both derivatives of benzene. Benzoic acid is an acid with $K_a = 6.3 \times 10^{-5}$ and aniline is a base with $K_a = 4.3 \times 10^{-10}$.

Benzoic acid   Aniline

(a) What are the conjugate base of benzoic acid and the conjugate acid of aniline? (b) Anilinium chloride ($C_6H_5NH_3Cl$) is a strong electrolyte that dissociates into anilinium ions ($C_6H_5NH_3^+$) and chloride ions. Which will be more acidic, a 0.10 M solution of benzoic acid or a 0.10 M solution of anilinium chloride? (c) What is the value of the equilibrium constant for the following equilibrium?

$$C_6H_5COOH(aq) + C_6H_5NH_2(aq) \rightleftharpoons$$
$$C_6H_5COO^-(aq) + C_6H_5NH_3^+(aq)$$

16.106 What is the pH of a solution that is 2.5×10^{-9} M in NaOH? Does your answer make sense? What assumption do we normally make that is not valid in this case?

16.107 Oxalic acid ($H_2C_2O_4$) is a diprotic acid. By using data in Appendix D as needed, determine whether each of the following statements is true: (a) $H_2C_2O_4$ can serve as both a Brønsted–Lowry acid and a Brønsted–Lowry base. (b) $C_2O_4^{2-}$ is the conjugate base of $HC_2O_4^-$. (c) An aqueous solution of the strong electrolyte KHC_2O_4 will have pH < 7.

16.108 Succinic acid ($H_2C_4H_6O_4$), which we will denote H_2Suc, is a biologically relevant diprotic acid with the structure shown below. It is closely related to tartaric acid and malic acid (Figure 16.1). At 25 °C, the acid-dissociation constants for succinic acid are $K_{a1} = 6.9 \times 10^{-5}$ and $K_{a2} = 2.5 \times 10^{-6}$. (a) Determine the pH of a 0.32 M solution of H_2Suc at 25 °C, assuming that only the first dissociation is relevant. (b) Determine the molar concentration of Suc^{2-} in the solution in part (a). (c) Is the assumption you made in part (a) justified by the result from part (b)? (d) Will a solution of the salt $NaHSuc$ be acidic, neutral, or basic?

16.109 Butyric acid is responsible for the foul smell of rancid butter. The pK_a of butyric acid is 4.84. (a) Calculate the pK_b for the butyrate ion. (b) Calculate the pH of a 0.050 M solution of butyric acid. (c) Calculate the pH of a 0.050 M solution of sodium butyrate.

16.110 Arrange the following 0.10 M solutions in order of increasing acidity (decreasing pH): (i) NH_4NO_3, (ii) $NaNO_3$, (iii) CH_3COONH_4, (iv) NaF, (v) CH_3COONa.

16.111 A 0.25 M solution of a salt NaA has pH = 9.29. What is the value of K_a for the parent acid HA?

[16.112] The following observations are made about a diprotic acid H_2A: (i) A 0.10 M solution of H_2A has pH = 3.30. (ii) A 0.10 M solution of the salt NaHA is acidic. Which of the following could be the value of pK_{a2} for H_2A: (i) 3.22, (ii) 5.30, (iii) 7.47, or (iv) 9.82?

16.113 Many moderately large organic molecules containing basic nitrogen atoms are not very soluble in water as neutral molecules, but they are frequently much more soluble as their acid salts. Assuming that pH in the stomach is 2.5, indicate whether each of the following compounds would be present in the stomach as the neutral base or in the protonated form: nicotine, $K_b = 7 \times 10^{-7}$; caffeine, $K_b = 4 \times 10^{-14}$; strychnine, $K_b = 1 \times 10^{-6}$; quinine, $K_b = 1.1 \times 10^{-6}$.

16.114 The amino acid glycine (H_2N-CH_2-COOH) can participate in the following equilibria in water:

$$H_2N-CH_2-COOH + H_2O \rightleftharpoons$$
$$H_2N-CH_2-COO^- + H_3O^+ \quad K_a = 4.3 \times 10^{-3}$$

$$H_2N-CH_2-COOH + H_2O \rightleftharpoons$$
$${}^+H_3N-CH_2-COOH + OH^- \quad K_b = 6.0 \times 10^{-5}$$

(a) Use the values of K_a and K_b to estimate the equilibrium constant for the intramolecular proton transfer to form a zwitterion:

$$H_2N-CH_2-COOH \rightleftharpoons {}^+H_3N-CH_2-COO^-$$

(b) What is the pH of a 0.050 M aqueous solution of glycine?

(c) What would be the predominant form of glycine in a solution with pH 13? With pH 1?

16.115 The structural formula for acetic acid is shown in Table 16.2. Replacing hydrogen atoms on the carbon with chlorine atoms causes an increase in acidity, as follows:

Acid	Formula	$K_a(25\,°C)$
Acetic	CH_3COOH	1.8×10^{-5}
Chloroacetic	$CH_2ClCOOH$	1.4×10^{-3}
Dichloroacetic	$CHCl_2COOH$	3.3×10^{-2}
Trichloroacetic	CCl_3COOH	2×10^{-1}

Using Lewis structures as the basis of your discussion, explain the observed trend in acidities in the series. Calculate the pH of a 0.010 M solution of each acid.

Integrative Exercises

16.116 Calculate the number of $H^+(aq)$ ions in 1.0 mL of pure water at 25 °C.

16.117 How many milliliters of concentrated hydrochloric acid solution (36.0% HCl by mass, density $= 1.18$ g/mL) are required to produce 10.0 L of a solution that has a pH of 2.05?

16.118 The volume of an adult's stomach ranges from about 50 mL when empty to 1 L when full. If the stomach volume is 400 mL and its contents have a pH of 2, how many moles of H^+ does the stomach contain? Assuming that all the H^+ comes from HCl, how many grams of sodium hydrogen carbonate will totally neutralize the stomach acid?

16.119 Atmospheric CO_2 levels have risen by nearly 20% over the past 40 years from 315 ppm to 380 ppm. **(a)** Given that the average pH of clean, unpolluted rain today is 5.4, determine the pH of unpolluted rain 40 years ago. Assume that carbonic acid (H_2CO_3) formed by the reaction of CO_2 and water is the only factor influencing pH.

$$CO_2(g) + H_2O(l) \rightleftharpoons H_2CO_3(aq)$$

(b) What volume of CO_2 at 25 °C and 1.0 atm is dissolved in a 20.0-L bucket of today's rainwater?

16.120 At 50 °C, the ion-product constant for H_2O has the value $K_w = 5.48 \times 10^{-14}$. **(a)** What is the pH of pure water at 50 °C? **(b)** Based on the change in K_w with temperature, predict whether ΔH is positive, negative, or zero for the autoionization reaction of water:

$$2\,H_2O(l) \rightleftharpoons H_3O^+(aq) + OH^-(aq)$$

16.121 In many reactions the addition of $AlCl_3$ produces the same effect as the addition of H^+. **(a)** Draw a Lewis structure for $AlCl_3$ in which no atoms carry formal charges, and determine its structure using the VSEPR method. **(b)** What characteristic is notable about the structure in part (a) that helps us understand the acidic character of $AlCl_3$? **(c)** Predict the result of the reaction between $AlCl_3$ and NH_3 in a solvent that does not participate as a reactant. **(d)** Which acid–base theory is most suitable for discussing the similarities between $AlCl_3$ and H^+?

16.122 What is the boiling point of a 0.10 M solution of $NaHSO_4$ if the solution has a density of 1.002 g/mL?

16.123 Use average bond enthalpies from Table 8.4 to estimate the enthalpies of the following gas-phase reactions:

Reaction 1: $HF(g) + H_2O(g) \rightleftharpoons F^-(g) + H_3O^+(g)$

Reaction 2: $HCl(g) + H_2O(g) \rightleftharpoons Cl^-(g) + H_3O^+(g)$

Are both reactions exothermic? How do these values relate to the different strengths of hydrofluoric and hydrochloric acid?

16.124 Cocaine is a weak organic base whose molecular formula is $C_{17}H_{21}NO_4$. An aqueous solution of cocaine was found to have a pH of 8.53 and an osmotic pressure of 52.7 torr at 15 °C. Calculate K_b for cocaine.

[16.125] The iodate ion is reduced by sulfite according to the following reaction:

$$IO_3^-(aq) + 3\,SO_3^{2-}(aq) \longrightarrow I^-(aq) + 3\,SO_4^{2-}(aq)$$

The rate of this reaction is found to be first order in IO_3^-, first order in SO_3^{2-}, and first order in H^+. **(a)** Write the rate law for the reaction. **(b)** By what factor will the rate of the reaction change if the pH is lowered from 5.00 to 3.50? Does the reaction proceed more quickly or more slowly at the lower pH? **(c)** By using the concepts discussed in Section 14.6, explain how the reaction can be pH-dependent even though H^+ does not appear in the overall reaction.

16.126 **(a)** Using dissociation constants from Appendix D, determine the value for the equilibrium constant for each of the following reactions.

(i) $HCO_3^-(aq) + OH^-(aq) \rightleftharpoons CO_3^{2-}(aq) + H_2O(l)$

(ii) $NH_4^+(aq) + CO_3^{2-}(aq) \rightleftharpoons NH_3(aq) + HCO_3^-(aq)$

(b) We usually use single arrows for reactions when the forward reaction is appreciable (K much greater than 1) or when products escape from the system, so that equilibrium is never established. If we follow this convention, which of these equilibria might be written with a single arrow?

Design an Experiment

Your professor gives you a bottle that contains a clear liquid. You are told that the liquid is a pure substance that is volatile, soluble in water, and might be an acid or a base. Design experiments to elucidate the following about this unknown sample. **(a)** Determine whether the substance in the sample is an acid or a base. **(b)** Suppose the substance is an acid. How would you determine whether it is a strong acid or a weak acid? **(c)** If the substance is a weak acid, how would you determine the value of K_a for the substance? **(d)** Suppose the substance were a weak acid and that you are also given a solution of NaOH(aq) of known molarity. What procedure would you use to isolate a pure sample of the sodium salt of the substance? **(e)** Now suppose that the substance were a base rather than an acid. How would you adjust the procedures in parts (b) and (c) to determine if the substance were a strong or weak base, and, if weak, the value of K_b?

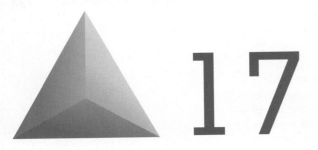

17
Additional Aspects of Aqueous Equilibria

Water, the most common and most important solvent on Earth, occupies its position of importance because of its abundance and its exceptional ability to dissolve a wide variety of substances. Coral reefs are a striking example of aqueous chemistry at work in nature. Coral reefs are built by tiny animals called stony corals, which secrete a hard calcium carbonate exoskeleton. Over time, the stony corals build up large networks of calcium carbonate upon which a reef is built. The size of such structures can be immense, as illustrated by the Great Barrier Reef.

Stony corals make their exoskeletons from dissolved Ca^{2+} and CO_3^{2-} ions. This process is aided by the fact that the CO_3^{2-} concentration is supersaturated in most parts of the ocean. However, well documented increases in the amount of CO_2 in the atmosphere threaten to upset the aqueous chemistry that stony corals depend on. As atmospheric CO_2 levels increase, the amount of CO_2 dissolved in the ocean also increases. This lowers the pH of the ocean and leads to a decrease in the CO_3^{2-} concentration. As a result it becomes more difficult for stony corals and other important ocean creatures to maintain their exoskeletons. We will take a closer look at the consequences of ocean acidification later in the chapter.

To understand the chemistry that underlies coral reef formation and other processes in the ocean and in aqueous systems such as living cells, we must understand the concepts of aqueous equilibria. In this chapter, we take a step toward understanding such complex solutions by looking first at further applications of acid–base equilibria. The idea is to consider not only solutions in which there is a single solute but

▶ A MICROSCOPIC VIEW OF CORAL AND SAND. This image shows coral and sand particles, magnified by 100x.

WHAT'S AHEAD ▶

17.1 THE COMMON-ION EFFECT We begin by considering a specific example of Le Châtelier's principle known as the common-ion effect.

17.2 BUFFERS We consider the composition of buffered solutions and learn how they resist pH change when small amounts of a strong acid or strong base are added to them.

17.3 ACID–BASE TITRATIONS We examine acid–base titrations and explore how to determine pH at any point in an acid–base titration.

17.4 SOLUBILITY EQUILIBRIA We learn how to use *solubility-product constants* to determine to what extent a sparingly soluble salt dissolves in water.

17.5 FACTORS THAT AFFECT SOLUBILITY We investigate some of the factors that affect solubility, including the common-ion effect and the effect of acids.

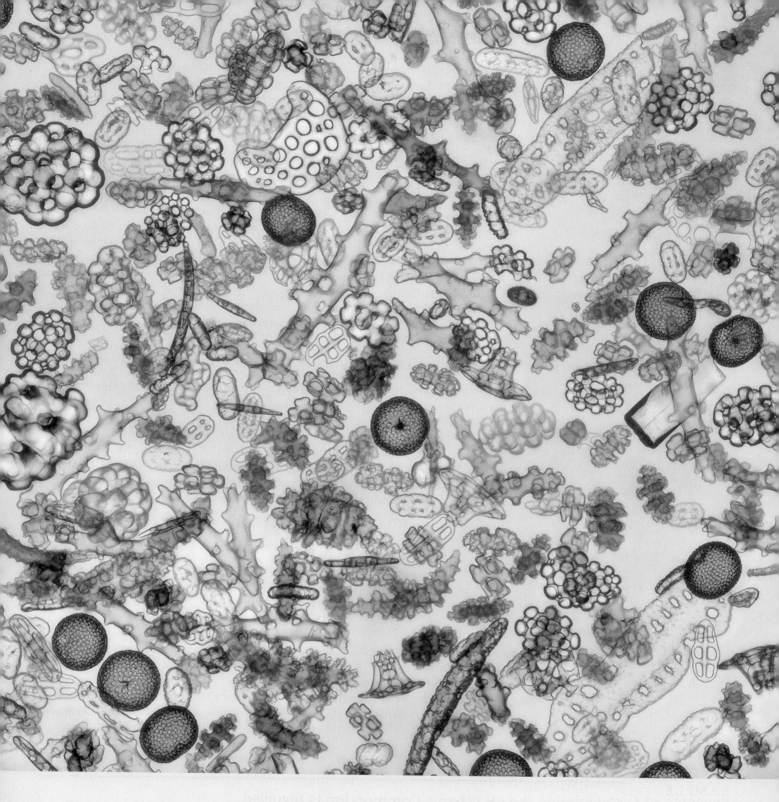

17.6 PRECIPITATION AND SEPARATION OF IONS We learn how differences in solubility can be used to separate ions through selective precipitation.

17.7 QUALITATIVE ANALYSIS FOR METALLIC ELEMENTS We explain how the principles of solubility and complexation equilibria can be used to identify ions in solution.

also those containing a mixture of solutes. We then broaden our discussion to include two additional types of aqueous equilibria: those involving slightly soluble salts and those involving the formation of metal complexes in solution. For the most part, the discussions and calculations in this chapter are extensions of those in Chapters 15 and 16.

17.1 | The Common-Ion Effect

In Chapter 16, we examined the equilibrium concentrations of ions in solutions containing a weak acid or a weak base. We now consider solutions that contain a weak acid, such as acetic acid (CH_3COOH), and a soluble salt of that acid, such as sodium acetate (CH_3COONa). Notice that these solutions contain two substances that share a *common ion*, CH_3COO^-. It is instructive to view these solutions from the perspective of Le Châtelier's principle. ∞ (Section 15.7)

Sodium acetate is a soluble ionic compound and therefore a strong electrolyte. ∞ (Section 4.1) Consequently, it dissociates completely in aqueous solution to form Na^+ and CH_3COO^- ions:

$$CH_3COONa(aq) \longrightarrow Na^+(aq) + CH_3COO^-(aq)$$

In contrast, CH_3COOH is a weak electrolyte that ionizes only partially, represented by the dynamic equilibrium

$$CH_3COOH(aq) \rightleftharpoons H^+(aq) + CH_3COO^-(aq) \qquad [17.1]$$

The equilibrium constant for Equation 17.1 is $K_a = 1.8 \times 10^{-5}$ at 25 °C (Table 16.2). If we add sodium acetate to a solution of acetic acid in water, the CH_3COO^- from CH_3COONa causes the equilibrium concentrations of the substances in Equation 17.1 to shift to the left, thereby decreasing the equilibrium concentration of $H^+(aq)$:

$$CH_3COOH(aq) \rightleftharpoons H^+(aq) + CH_3COO^-(aq)$$

Addition of CH_3COO^- shifts equilibrium concentrations, lowering $[H^+]$

In other words, the presence of the added acetate ion causes the acetic acid to ionize less than it normally would. *The equilibrium constant itself does not change; it is the relative concentrations of products and reactants in the equilibrium expression that change.*

Whenever a weak electrolyte and a strong electrolyte containing a common ion are together in solution, the weak electrolyte ionizes less than it would if it were alone in solution. We call this observation the **common-ion effect**.

SAMPLE EXERCISE 17.1 Calculating the pH When a Common Ion Is Involved

What is the pH of a solution made by adding 0.30 mol of acetic acid and 0.30 mol of sodium acetate to enough water to make 1.0 L of solution?

SOLUTION

Analyze We are asked to determine the pH of a solution of a weak electrolyte (CH_3COOH) and a strong electrolyte (CH_3COONa) that share a common ion, CH_3COO^-.

Plan In any problem in which we must determine the pH of a solution containing a mixture of solutes, it is helpful to proceed by a series of logical steps:

(1) Consider which solutes are strong electrolytes and which are weak electrolytes, and identify the major species in solution.

(2) Identify the important equilibrium reaction that is the source of H^+ and therefore determines pH.

(3) Tabulate the concentrations of ions involved in the equilibrium.

(4) Use the equilibrium-constant expression to calculate $[H^+]$ and then pH.

Solve First, because CH_3COOH is a weak electrolyte and CH_3COONa is a strong electrolyte, the major species in the solution are CH_3COOH (a weak acid), Na^+ (which is neither acidic nor basic and is therefore a spectator in the acid–base chemistry), and CH_3COO^- (which is the conjugate base of CH_3COOH).

Second, $[H^+]$ and, therefore, the pH are controlled by the dissociation equilibrium of CH_3COOH:

$$CH_3COOH(aq) \rightleftharpoons H^+(aq) + CH_3COO^-(aq)$$

(We have written the equilibrium using $H^+(aq)$ rather than $H_3O^+(aq)$, but both representations of the hydrated hydrogen ion are equally valid.)

Third, we tabulate the initial and equilibrium concentrations as we did in solving other equilibrium problems in Chapters 15 and 16:

	$CH_3COOH(aq)$	$\rightleftharpoons$	$H^+(aq)$	+	$CH_3COO^-(aq)$
Initial (M)	0.30		0		0.30
Change (M)	$-x$		$+x$		$+x$
Equilibrium (M)	$(0.30 - x)$		x		$(0.30 + x)$

The equilibrium concentration of CH_3COO^- (the common ion) is the initial concentration that is due to CH_3COONa (0.30 M) plus the change in concentration (x) that is due to the ionization of CH_3COOH.

Now we can use the equilibrium-constant expression:

$$K_a = 1.8 \times 10^{-5} = \frac{[H^+][CH_3COO^-]}{[CH_3COOH]}$$

The dissociation constant for CH_3COOH at 25 °C is from Table 16.2, or Appendix D; addition of CH_3COONa does not change the value of this constant. Substituting the equilibrium-constant concentrations from our table into the equilibrium expression gives:

$$K_a = 1.8 \times 10^{-5} = \frac{x(0.30 + x)}{0.30 - x}$$

Because K_a is small, we assume that x is small compared to the original concentrations of CH_3COOH and CH_3COO^- (0.30 M each). Thus, we can ignore the very small x relative to 0.30 M, giving

$$K_a = 1.8 \times 10^{-5} = \frac{x(0.30)}{0.30}$$

The resulting value of x is indeed small relative to 0.30, justifying the approximation made in simplifying the problem.

$$x = 1.8 \times 10^{-5} M = [H^+]$$

Finally, we calculate the pH from the equilibrium concentration of $H^+(aq)$:

$$pH = -\log(1.8 \times 10^{-5}) = 4.74$$

Comment In Section 16.6, we calculated that a 0.30 M solution of CH_3COOH has a pH of 2.64, corresponding to $[H^+] = 2.3 \times 10^{-3} M$. Thus, the addition of CH_3COONa has substantially decreased $[H^+]$, as we expect from Le Châtelier's principle.

Practice Exercise 1

For the generic equilibrium $HA\ (aq) \rightleftharpoons H^+\ (aq) + A^-(aq)$, which of these statements is true?
(a) The equilibrium constant for this reaction changes as the pH changes.
(b) If you add the soluble salt KA to a solution of HA that is at equilibrium, the concentration of HA would decrease.
(c) If you add the soluble salt KA to a solution of HA that is at equilibrium, the concentration of A^- would decrease.
(d) If you add the soluble salt KA to a solution of HA that is at equilibrium, the pH would increase.

Practice Exercise 2

Calculate the pH of a solution containing 0.085 M nitrous acid (HNO_2, $K_a = 4.5 \times 10^{-4}$) and 0.10 M potassium nitrite (KNO_2).

EXERCISE 17.2 Calculating Ion Concentrations When a Common Ion Is Involved

Calculate the fluoride ion concentration and pH of a solution that is 0.20 M in HF and 0.10 M in HCl.

SOLUTION

Analyze We are asked to determine the concentration of F^- and the pH in a solution containing the weak acid HF and the strong acid HCl. In this case the common ion is H^+.

Plan We can again use the four steps outlined in Sample Exercise 17.1.

Solve Because HF is a weak acid and HCl is a strong acid, the major species in solution are HF, H^+, and Cl^-. The Cl^-, which is the conjugate base of a strong acid, is merely a spectator ion in any acid–base chemistry. The problem asks for $[F^-]$, which is formed by ionization of HF. Thus, the important equilibrium is

$$HF(aq) \rightleftharpoons H^+(aq) + F^-(aq)$$

The common ion in this problem is the hydrogen (or hydronium) ion. Now we can tabulate the initial and equilibrium concentrations of each species involved in this equilibrium:

	$HF(aq)$	$\rightleftharpoons$ $H^+(aq)$	$+$ $F^-(aq)$
Initial (M)	0.20	0.10	0
Change (M)	$-x$	$+x$	$+x$
Equilibrium (M)	$(0.20 - x)$	$(0.10 + x)$	x

The equilibrium constant for the ionization of HF, from Appendix D, is 6.8×10^{-4}. Substituting the equilibrium-constant concentrations into the equilibrium expression gives

$$K_a = 6.8 \times 10^{-4} = \frac{[H^+][F^-]}{[HF]} = \frac{(0.10 + x)(x)}{0.20 - x}$$

If we assume that x is small relative to 0.10 or 0.20 M, this expression simplifies to

$$\frac{(0.10)(x)}{0.20} = 6.8 \times 10^{-4}$$

$$x = \frac{0.20}{0.10}(6.8 \times 10^{-4}) = 1.4 \times 10^{-3} M = [F^-]$$

This F^- concentration is substantially smaller than it would be in a 0.20 M solution of HF with no added HCl. The common ion, H^+, suppresses the ionization of HF. The concentration of $H^+(aq)$ is

$$[H^+] = (0.10 + x) M \simeq 0.10 M$$

Thus,

$$pH = 1.00$$

Comment Notice that for all practical purposes, the hydrogen ion concentration is due entirely to the HCl; the HF makes a negligible contribution by comparison.

Practice Exercise 1

Calculate the concentration of the lactate ion in a solution that is 0.100 M in lactic acid ($CH_3CH(OH)COOH$, $pK_a = 3.86$) and 0.080 M in HCl.
(a) 4.83 M, **(b)** 0.0800 M, **(c)** $7.3 \times 10^{-3} M$, **(d)** $3.65 \times 10^{-3} M$, **(e)** $1.73 \times 10^{-4} M$.

Practice Exercise 2

Calculate the formate ion concentration and pH of a solution that is 0.050 M in formic acid ($HCOOH$, $K_a = 1.8 \times 10^{-4}$) and 0.10 M in HNO_3.

Sample Exercises 17.1 and 17.2 both involve weak acids. The ionization of a weak base is also decreased by the addition of a common ion. For example, the addition of NH_4^+ (as from the strong electrolyte NH_4Cl) causes the equilibrium concentrations of the reagents in Equation 17.2 to shift to the left, decreasing the equilibrium concentration of OH^- and lowering the pH:

$$NH_3(aq) + H_2O(l) \rightleftharpoons NH_4^+(aq) + OH^-(aq) \qquad [17.2]$$

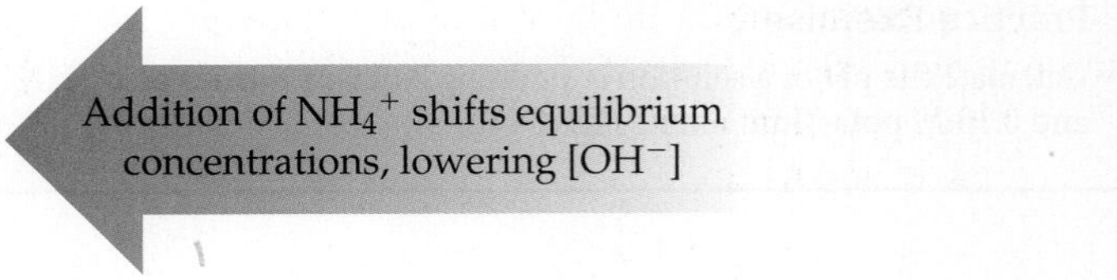

Addition of NH_4^+ shifts equilibrium concentrations, lowering $[OH^-]$

Give It Some Thought

If solutions of $NH_4Cl(aq)$ and $NH_3(aq)$ are mixed, which ions in the resulting solution are spectator ions in any acid–base chemistry occurring in the solution? What equilibrium reaction determines $[OH^-]$ and, therefore, the pH of the solution?

17.2 | Buffers

Solutions that contain high concentrations ($10^{-3}\ M$ or more) of a weak conjugate acid–base pair and that resist drastic changes in pH when small amounts of strong acid or strong base are added to them are called **buffered solutions** (or merely **buffers**). Human blood, for example, is a complex buffered solution that maintains the blood pH at about 7.4. (Section 17.2, "Blood as a Buffered Solution") Much of the chemical behavior of seawater is determined by its pH, buffered at about 8.1 to 8.3 near the surface. (Section 15.5, "Ocean Acidification") Buffers find many important applications in the laboratory and in medicine (▶ Figure 17.1). Many biological reactions occur at the optimal rates only when properly buffered. If you ever work in a biochemistry lab, you will very likely to have prepare specific buffers in which to run your biochemical reactions.

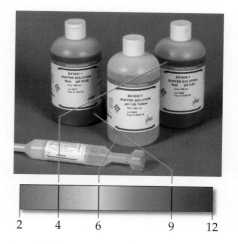

▲ Figure 17.1 **Standard buffers.** For laboratory work, prepackaged buffers at specific pH values can be purchased.

Composition and Action of Buffers

A buffer resists changes in pH because it contains both an acid to neutralize added OH^- ions and a base to neutralize added H^+ ions. The acid and base that make up the buffer, however, must not consume each other through a neutralization reaction. ∞ (Section 4.3) These requirements are fulfilled by a weak acid–base conjugate pair, such as CH_3COOH/CH_3COO^- or NH_4^+/NH_3. The key is to have roughly equal concentrations of both the weak acid and its conjugate base.

There are two ways to make a buffer:

1. Mix a weak acid or a weak base with a salt of that acid or base. For example, the CH_3COOH/CH_3COO^- buffer can be prepared by adding CH_3COONa to a solution of CH_3COOH. Similarly, the NH_4^+/NH_3 buffer can be prepared by adding NH_4Cl to a solution of NH_3.

2. Make the conjugate acid or base from a solution of weak base or acid by the addition of strong acid or base. For example, to make the CH_3COOH/CH_3COO^- buffer, you could start with a solution of CH_3COOH and add some NaOH to the solution—enough to neutralize about half of CH_3COOH according to the reaction.

$$CH_3COOH + OH^- \longrightarrow CH_3COO^- + H_2O$$

∞ (Section 4.3) Neutralization reactions have very large equilibrium constants, and so the amount of acetate formed will only be limited by the relative amounts of the acid and strong base that are mixed. The resulting solution is the same as if you added sodium acetate to the acetic acid solution: You will have comparable quantities of both acetic acid and its conjugate base in solution.

By choosing appropriate components and adjusting their relative concentrations, we can buffer a solution at virtually any pH.

 Give It Some Thought

Which of these conjugate acid–base pairs will *not* function as a buffer: C_2H_5COOH and $C_2H_5COO^-$, HCO_3^- and CO_3^{2-}, or HNO_3 and NO_3^-? Explain.

To understand how a buffer works, let's consider one composed of a weak acid HA and one of its salts MA, where M^+ could be Na^+, K^+, or any other cation that does not react with water. The acid-dissociation equilibrium in this buffered solution involves both the acid and its conjugate base:

$$HA(aq) \rightleftharpoons H^+(aq) + A^-(aq) \qquad [17.3]$$

The corresponding acid-dissociation-constant expression is

$$K_a = \frac{[H^+][A^-]}{[HA]}$$ [17.4]

Solving this expression for $[H^+]$, we have

$$[H^+] = K_a\frac{[HA]}{[A^-]}$$ [17.5]

We see from this expression that $[H^+]$ and, thus, the pH are determined by two factors: the value of K_a for the weak-acid component of the buffer and the ratio of the concentrations of the conjugate acid–base pair, $[HA]/[A^-]$.

If OH^- ions are added to this buffered solution, they react with the buffer acid component to produce water and A^-:

$$OH^-(aq) + HA(aq) \longrightarrow H_2O(l) + A^-(aq)$$ [17.6]
added base

This neutralization reaction causes [HA] to decrease and $[A^-]$ to increase. As long as the amounts of HA and A^- in the buffer are large relative to the amount of OH^- added, the ratio $[HA]/[A^-]$ does not change much and, thus, the change in pH is small.

If H^+ ions are added, they react with the base component of the buffer:

$$H^+(aq) + A^-(aq) \longrightarrow HA(aq)$$ [17.7]
added acid

This reaction can also be represented using H_3O^+:

$$H_3O^+(aq) + A^-(aq) \longrightarrow HA(aq) + H_2O(l)$$

Using either equation, we see that this reaction causes $[A^-]$ to decrease and [HA] to increase. As long as the change in the ratio $[HA]/[A^-]$ is small, the change in pH will be small.

▼ Figure 17.2 shows an HA/A^- buffer consisting of equal concentrations of hydrofluoric acid and fluoride ion (center). The addition of OH^- reduces [HF] and increases $[F^-]$, whereas the addition of $[H^+]$ reduces $[F^-]$ and increases [HF].

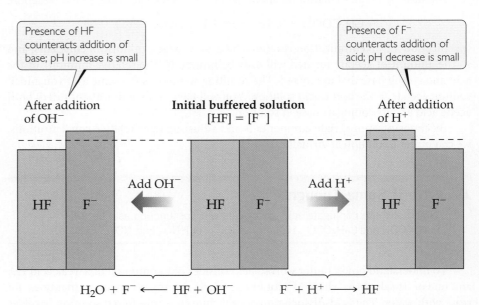

▲ Figure 17.2 Buffer action. The pH of an HF/F^- buffered solution changes by only a small amount in response to addition of an acid or base.

It is possible to overwhelm a buffer by adding too much strong acid or strong base. We will examine this in more detail a little later in this chapter.

 Give It Some Thought

(a) What happens when NaOH is added to a buffer composed of CH_3COOH and CH_3COO^-?

(b) What happens when HCl is added to this buffer?

Calculating the pH of a Buffer

Because conjugate acid–base pairs share a common ion, we can use the same procedures to calculate the pH of a buffer that we used to treat the common-ion effect in Sample Exercise 17.1. Alternatively, we can take an approach based on an equation derived from Equation 17.5. Taking the negative logarithm of both sides of Equation 17.5, we have

$$-\log[H^+] = -\log\left(K_a\frac{[HA]}{[A^-]}\right) = -\log K_a - \log\frac{[HA]}{[A^-]}$$

Because $-\log[H^+] = pH$ and $-\log K_a = pK_a$, we have

$$pH = pK_a - \log\frac{[HA]}{[A^-]} = pK_a + \log\frac{[A^-]}{[HA]} \qquad [17.8]$$

(Remember the logarithm rules in Appendix A.2, if you are not sure how this calculation works.)

In general,

$$pH = pK_a + \log\frac{[base]}{[acid]} \qquad [17.9]$$

where [acid] and [base] refer to the equilibrium concentrations of the *conjugate acid–base pair*. Note that when [base] = [acid], we have pH = pK_a.

Equation 17.9 is known as the **Henderson–Hasselbalch equation**. Biologists, biochemists, and others who work frequently with buffers often use this equation to calculate the pH of buffers. In doing equilibrium calculations, we have seen that we can normally neglect the amounts of the acid and base of the buffer that ionize. Therefore, we can usually use the initial concentrations of the acid and base components of the buffer directly in Equation 17.9, as seen in Sample Exercise 17.3. However, the assumption that the initial concentrations of the acid and base components in the buffer are equal to the equilibrium concentrations is just that: an assumption. There may be times when you will need to be more careful, as seen in Sample Exercise 17.4.

SAMPLE EXERCISE 17.3 | Calculating the pH of a Buffer

What is the pH of a buffer that is 0.12 M in lactic acid $[CH_3CH(OH)COOH$, or $HC_3H_5O_3]$ and 0.10 M in sodium lactate $[CH_3CH(OH)COONa$ or $NaC_3H_5O_3]$? For lactic acid, $K_a = 1.4 \times 10^{-4}$.

SOLUTION

Analyze We are asked to calculate the pH of a buffer containing lactic acid $(HC_3H_5O_3)$ and its conjugate base, the lactate ion $(C_3H_5O_3^-)$.

Plan We will first determine the pH using the method described in Section 17.1. Because $HC_3H_5O_3$ is a weak electrolyte and $NaC_3H_5O_3$ is a strong electrolyte, the major species in solution are $HC_3H_5O_3$, Na^+, and $C_3H_5O_3^-$. The Na^+ ion is a spectator ion. The $HC_3H_5O_3/C_3H_5O_3^-$ conjugate acid–base pair determines $[H^+]$ and, thus, pH; $[H^+]$ can be determined using the acid-dissociation equilibrium of lactic acid.

Solve The initial and equilibrium concentrations of the species involved in this equilibrium are

$$CH_3CH(OH)COOH\ (aq) \rightleftharpoons H^+\ (aq) + CH_3CH(OH)COO^-(aq)$$

	CH₃CH(OH)COOH	H⁺	CH₃CH(OH)COO⁻
Initial (M)	0.12	0	0.10
Change (M)	−x	+x	+x
Equilibrium (M)	(0.12 − x)	x	(0.10 + x)

The equilibrium concentrations are governed by the equilibrium expression:

$$K_a = 1.4 \times 10^{-4} = \frac{[H^+][C_3H_5O_3^-]}{[HC_3H_5O_3]} = \frac{x(0.10 + x)}{0.12 - x}$$

Because K_a is small and a common ion is present, we expect x to be small relative to either 0.12 or 0.10 M. Thus, our equation can be simplified to give

$$K_a = 1.4 \times 10^{-4} = \frac{x(0.10)}{0.12}$$

Solving for x gives a value that justifies our approximation:

$$[H^+] = x = \left(\frac{0.12}{0.10}\right)(1.4 \times 10^{-4}) = 1.7 \times 10^{-4}\ M$$

Then, we can solve for pH:

$$pH = -\log(1.7 \times 10^{-4}) = 3.77$$

Alternatively, we can use the Henderson–Hasselbalch equation with the initial concentrations of acid and base to calculate pH directly:

$$pH = pK_a + \log\frac{[base]}{[acid]} = 3.85 + \log\left(\frac{0.10}{0.12}\right)$$

$$= 3.85 + (-0.08) = 3.77$$

Practice Exercise 1

If the pH of a buffer solution is equal to the pK_a of the acid in the buffer, what does this tell you about the relative concentrations of the acid and conjugate base forms of the buffer components?
(a) The acid concentration must be zero. (b) The base concentration must be zero.
(c) The acid and base concentrations must be equal. (d) The acid and base concentrations must be equal to the K_a. (e) The base concentration must be 2.3 times as large as the acid concentration.

Practice Exercise 2

Calculate the pH of a buffer composed of 0.12 M benzoic acid and 0.20 M sodium benzoate. (Refer to Appendix D.)

SAMPLE EXERCISE 17.4 Calculating pH When the Henderson–Hasselbalch Equation May Not Be Accurate

Calculate the pH of a buffer that initially contains $1.00 \times 10^{-3}\ M$ CH₃COOH and $1.00 \times 10^{-4}\ M$ CH₃COONa in the following two ways: (i) using the Henderson–Hasselbalch equation and (ii) making no assumptions about quantities (which means you will need to use the quadratic equation). The K_a of CH₃COOH is 1.80×10^{-5}.

SOLUTION

Analyze We are asked to calculate the pH of a buffer two different ways. We know the initial concentrations of the weak acid and its conjugate base, and the K_a of the weak acid.

Plan We will first use the Henderson–Hasselbalch equation, which relates pK_a and ratio of acid–base concentrations to the pH. This will be straightforward. Then, we will redo the calculation making no assumptions about any quantities, which means we will need to write out the initial/change/equilibrium concentrations, as we have done before. In addition, we will need to solve for quantities using the quadratic equation (since we cannot make assumptions about unknowns being small).

Solve

(i) The Henderson–Hasselbalch equation is

$$pH = pK_a + \log\frac{[base]}{[acid]}$$

We know the K_a of the acid (1.8×10^{-5}), so we know pK_a ($pK_a = -\log K_a = 4.74$). We know the initial concentrations of the base, sodium acetate, and the acid, acetic acid, which we will assume are the same as the equilibrium concentrations.

Therefore, we have

$$pH = 4.74 + \log\frac{(1.00 \times 10^{-4})}{(1.00 \times 10^{-3})}$$

Therefore,

$$pH = 4.74 - 1.00 = 3.74$$

(ii) Now we will redo the calculation, without making any assumptions at all. We will solve for x, which represents the H^+ concentration at equilibrium, in order to calculate pH.

	$CH_3COOH\ (aq)$	$\rightleftharpoons$ $CH_3COO^-(aq)$	$+$ $H^+(aq)$
Initial (M)	1.00×10^{-3}	1.00×10^{-4}	0
Change (M)	$-x$	$+x$	$+x$
Equilibrium (M)	$(1.00 \times 10^{-3} - x)$	$(1.00 \times 10^{-4} + x)$	x

$$\frac{[CH_3COO^-][H^+]}{[CH_3COOH]} = K_a$$

$$\frac{(1.00 \times 10^{-4} + x)(x)}{(1.00 \times 10^{-3} - x)} = 1.8 \times 10^{-5}$$

$$1.00 \times 10^{-4}x + x^2 = 1.8 \times 10^{-5}(1.00 \times 10^{-3} - x)$$

$$x^2 + 1.00 \times 10^{-4}x = 1.8 \times 10^{-8} - 1.8 \times 10^{-5}x$$

$$x^2 + 1.18 \times 10^{-4}x - 1.8 \times 10^{-8} = 0$$

$$x = \frac{-1.18 \times 10^{-4} \pm \sqrt{(1.18 \times 10^{-4})^2 - 4(1)(-1.8 \times 10^{-8})}}{2(1)}$$

$$= \frac{-1.18 \times 10^{-4} \pm \sqrt{8.5924 \times 10^{-8}}}{2}$$

$$= 8.76 \times 10^{-5} = [H^+]$$

$$pH = 4.06$$

Comment In Sample Exercise 17.3, the calculated pH is the same whether we solve exactly using the quadratic equation or make the simplifying assumption that the equilibrium concentrations of acid and base are equal to their initial concentrations. The simplifying assumption works because the concentrations of the acid–base conjugate pair are both a thousand times larger than K_a. In this Sample Exercise, the acid–base conjugate pair concentrations are only 10–100 as large as K_a. Therefore, we cannot assume that x is small compared to the initial concentrations (that is, that the initial concentrations are essentially equal to the equilibrium concentrations). The best answer to this Sample Exercise is pH = 4.06, obtained without assuming x is small. Thus we see that the assumptions behind the Henderson–Hasselbalch equation become less accurate as the acid/base becomes stronger and/or its concentration becomes smaller.

Practice Exercise 1

A buffer is made with sodium acetate (CH_3COONa) and acetic acid (CH_3COOH); the K_a for acetic acid is 1.80×10^{-5}. The pH of the buffer is 3.98. What is the ratio of the equilibrium concentration of sodium acetate to that of acetic acid? **(a)** -0.760, **(b)** 0.174, **(c)** 0.840, **(d)** 5.75, **(e)** Not enough information is given to answer this question.

Practice Exercise 2

Calculate the final, equilibrium pH of a buffer that initially contains $6.50 \times 10^{-4}\ M$ HOCl and $7.50 \times 10^{-4}\ M$ NaOCl. The K_a of HOCl is 3.0×10^{-5}.

In Sample Exercise 17.3 we calculated the pH of a buffered solution. Often we will need to work in the opposite direction by calculating the amounts of the acid and its conjugate base needed to achieve a specific pH. This calculation is illustrated in Sample Exercise 17.5.

SAMPLE EXERCISE 17.5 Preparing a Buffer

How many moles of NH_4Cl must be added to 2.0 L of 0.10 M NH_3 to form a buffer whose pH is 9.00? (Assume that the addition of NH_4Cl does not change the volume of the solution.)

SOLUTION

Analyze We are asked to determine the amount of NH_4^+ ion required to prepare a buffer of a specific pH.

Plan The major species in the solution will be NH_4^+, Cl^-, and NH_3. Of these, the Cl^- ion is a spectator (it is the conjugate base of a strong acid). Thus, the NH_4^+/NH_3 conjugate acid–base pair will determine the pH of the buffer. The equilibrium relationship between NH_4^+ and NH_3 is given by the base-dissociation reaction for NH_3:

$$NH_3(aq) + H_2O(l) \rightleftharpoons NH_4^+(aq) + OH^-(aq) \quad K_b = \frac{[NH_4^+][OH^-]}{[NH_3]} = 1.8 \times 10^{-5}$$

The key to this exercise is to use this K_b expression to calculate $[NH_4^+]$.

Solve We obtain $[OH^-]$ from the given pH:

$$pOH = 14.00 - pH = 14.00 - 9.00 = 5.00$$

and so

$$[OH^-] = 1.0 \times 10^{-5}\,M$$

Because K_b is small and the common ion $[NH_4^+]$ is present, the equilibrium concentration of NH_3 essentially equals its initial concentration:

$$[NH_3] = 0.10\,M$$

We now use the expression for K_b to calculate $[NH_4^+]$:

$$[NH_4^+] = K_b\frac{[NH_3]}{[OH^-]} = (1.8 \times 10^{-5})\frac{(0.10)}{(1.0 \times 10^{-5})} = 0.18\,M$$

Thus, for the solution to have pH = 9.00, $[NH_4^+]$ must equal 0.18 M. The number of moles of NH_4Cl needed to produce this concentration is given by the product of the volume of the solution and its molarity:

$$(2.0\,L)(0.18\,mol\,NH_4Cl/L) = 0.36\,mol\,NH_4Cl$$

Comment Because NH_4^+ and NH_3 are a conjugate acid–base pair, we could use the Henderson–Hasselbalch equation (Equation 17.9) to solve this problem. To do so requires first using Equation 16.41 to calculate pK_a for NH_4^+ from the value of pK_b for NH_3. We suggest you try this approach to convince yourself that you can use the Henderson–Hasselbalch equation for buffers for which you are given K_b for the conjugate base rather than K_a for the conjugate acid.

Practice Exercise 1

Calculate the number of grams of ammonium chloride that must be added to 2.00 L of a 0.500 M ammonia solution to obtain a buffer of pH = 9.20. Assume the volume of the solution does not change as the solid is added. K_b for ammonia is 1.80×10^{-5}.
(a) 60.7 g, (b) 30.4 g, (c) 1.52 g, (d) 0.568 g, (e) 1.59×10^{-5} g.

Practice Exercise 2

Calculate the concentration of sodium benzoate that must be present in a 0.20 M solution of benzoic acid (C_6H_5COOH) to produce a pH of 4.00. Refer to Appendix D.

Buffer Capacity and pH Range

Two important characteristics of a buffer are its capacity and its effective pH range. **Buffer capacity** is the amount of acid or base the buffer can neutralize before the pH begins to change to an appreciable degree. The buffer capacity depends on the amount of acid and base used to prepare the buffer. According to Equation 17.5, for example, the pH of a 1-L solution that is 1 M in CH_3COOH and 1 M in CH_3COONa is the same as the pH of a 1-L solution that is 0.1 M in CH_3COOH and 0.1 M in CH_3COONa. The first solution has a greater buffering capacity, however, because it contains more CH_3COOH and CH_3COO^-.

The **pH range** of any buffer is the pH range over which the buffer acts effectively. Buffers most effectively resist a change in pH in *either* direction when the concentrations of weak acid and conjugate base are about the same. From Equation 17.9 we see that when the concentrations of weak acid and conjugate base are equal, pH = pK_a. This relationship gives the optimal pH of any buffer. Thus, we usually try to select a buffer whose acid form has a pK_a close to the desired pH. In practice, we find that if the concentration of one component of the buffer is more than 10 times the concentration of the other component, the buffering action is poor. Because log 10 = 1, *buffers usually have a usable range within ±1 pH unit of* pK_a (that is, a range of pH = $pK_a \pm 1$).

Give It Some Thought

The K_a values for nitrous acid (HNO_2) and hypochlorous ($HClO$) acid are 4.5×10^{-4} and 3.0×10^{-8}, respectively. Which one would be more suitable for use in a solution buffered at pH = 7.0? What other substances would be needed to make the buffer?

Addition of Strong Acids or Bases to Buffers

Let's now consider in a more quantitative way how a buffered solution responds to addition of a strong acid or base. In this discussion, it is important to understand that *neutralization reactions between strong acids and weak bases proceed essentially to completion, as do those between strong bases and weak acids.* This is because water is a produce of the reaction, and you have an equilibrium constant of $1/K_w = 10^{14}$ in your favor when making water. ∞ (Section 16.3) Thus, as long as we do not exceed the buffering capacity of the buffer, we can assume that the strong acid or strong base is completely consumed by reaction with the buffer.

Consider a buffer that contains a weak acid HA and its conjugate base A⁻. When a strong acid is added to this buffer, the added H^+ is consumed by A⁻ to produce HA; thus, [HA] increases and [A⁻] decreases. (See Equation 17.7.) Upon addition of a strong base, the added OH^- is consumed by HA to produce A⁻; in this case [HA] decreases and [A⁻] increases. (See Equation 17.6.) These two situations are summarized in Figure 17.2.

To calculate how the pH of the buffer responds to the addition of a strong acid or a strong base, we follow the strategy outlined in ▼ Figure 17.3:

1. Consider the acid–base neutralization reaction and determine its effect on [HA] and [A⁻]. This step is a *limiting reactant stoichiometry calculation.* ∞ (Section 3.6 and 3.7)

2. Use the calculated values of [HA] and [A⁻] along with K_a to calculate [H⁺]. This step is an *equilibrium calculation* and is most easily done using the Henderson–Hasselbalch equation (if the concentrations of the weak acid–base pair are very large compared to K_a for the acid).

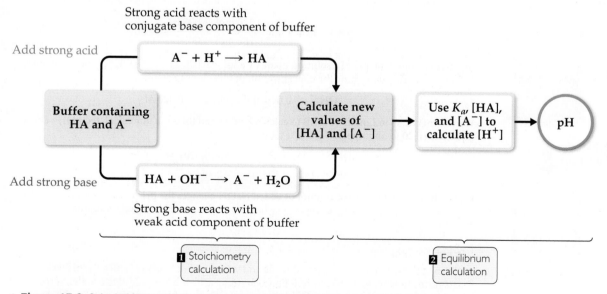

▲ Figure 17.3 **Calculating the pH of a buffer after addition of a strong acid or strong base.**

SAMPLE EXERCISE 17.6 Calculating pH Changes in Buffers

A buffer is made by adding 0.300 mol CH_3COOH and 0.300 mol CH_3COONa to enough water to make 1.000 L of solution. The pH of the buffer is 4.74 (Sample Exercise 17.1). **(a)** Calculate the pH of this solution after 5.0 mL of 4.0 M $NaOH(aq)$ solution is added. **(b)** For comparison, calculate the pH of a solution made by adding 5.0 mL of 4.0 M $NaOH(aq)$ solution to 1.000 L of pure water.

SOLUTION

Analyze We are asked to determine the pH of a buffer after addition of a small amount of strong base and to compare the pH change with the pH that would result if we were to add the same amount of strong base to pure water.

Plan Solving this problem involves the two steps outlined in Figure 17.3. First we do a stoichiometry calculation to determine how the added OH^- affects the buffer composition. Then we use the resultant buffer composition and either the Henderson–Hasselbalch equation or the equilibrium-constant expression for the buffer to determine the pH.

Solve

1.000 L buffer
0.300 M CH_3COOH
0.300 M CH_3COO^- **pH = 4.74**

Add 5.0 mL of
4.0 M $NaOH(aq)$

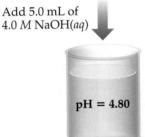

pH = 4.80

**pH increases by
0.06 pH units**

(a) *Stoichiometry Calculation*: The OH^- provided by NaOH reacts with CH_3COOH, the weak acid component of the buffer. Since volumes are changing, it is prudent to figure out how many moles of reactants and products would be produced, then divide by the final volume later to obtain concentrations. Prior to this neutralization reaction, there are 0.300 mol each of CH_3COOH and CH_3COO^-. The amount of base added is 0.0050 L × 4.0 mol/L = 0.020 mol. Neutralizing the 0.020 mol OH^- requires 0.020 mol of CH_3COOH. Consequently, the amount of CH_3COOH *decreases* by 0.020 mol, and the amount of the product of the neutralization, CH_3COO^-, *increases* by 0.020 mol. We can create a table to see how the composition of the buffer changes as a result of its reaction with OH^-:

$$CH_3COOH(aq) + OH^-(aq) \longrightarrow H_2O(l) + CH_3COO^-(aq)$$

	CH_3COOH	OH^-	H_2O	CH_3COO^-
Before reaction (mol)	0.300	0.020	—	0.300
Change (limiting reactant) (mol)	−0.020	−0.020	—	+0.020
After reaction (mol)	0.280	0	—	0.320

Equilibrium Calculation: We now turn our attention to the equilibrium for the ionization of acetic acid, the relationship that determines the buffer pH:

$$CH_3COOH(aq) \rightleftharpoons H^+(aq) + CH_3COO^-(aq)$$

Using the quantities of CH_3COOH and CH_3COO^- remaining in the buffer after the reaction with strong base, we determine the pH using the Henderson–Hasselbalch equation. The volume of the solution is now 1.000 L + 0.0050 L = 1.005 L due to addition of the NaOH solution:

$$pH = 4.74 + \log \frac{0.320 \text{ mol}/1.005 \text{ L}}{0.280 \text{ mol}/1.005 \text{ L}} = 4.80$$

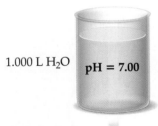

1.000 L H_2O **pH = 7.00**

Add 5.0 mL of
4.0 M $NaOH(aq)$

pH = 12.30

**pH increases by
5.30 pH units**

▲ **Figure 17.4 Effect of adding a strong base to a buffered solution and to water.**

(b) To determine the pH of a solution made by adding 0.020 mol of NaOH to 1.000 L of pure water, we first determine the concentration of OH^- ions in solution,

$$[OH^-] = 0.020 \text{ mol}/1.005 \text{ L} = 0.020 \, M$$

We use this value in Equation 16.18 to calculate pOH and then use our calculated pOH value in Equation 16.20 to obtain pH:

$$pOH = -\log[OH^-] = -\log(0.020) = +1.70$$
$$pH = 14 - (+1.70) = 12.30$$

Comment Note that the small amount of added NaOH changes the pH of water significantly. In contrast, the pH of the buffer changes very little when the NaOH is added, as summarized in ◀ Figure 17.4.

Practice Exercise 1

Which of these statements is true? **(a)** If you add strong acid or base to a buffer, the pH will never change. **(b)** In order to do calculations in which strong acid or base is added to a buffer, you only need to use the Henderson–Hasselbalch equation. **(c)** Strong bases react with strong acids, but not weak acids. **(d)** If you add a strong acid or base to a buffer, the buffer's pK_a or pK_b will change. **(e)** In order to do calculations in which strong acid or base is added to a buffer, you need to calculate the amounts of substances from the neutralization reaction and then equilibrate.

Practice Exercise 2

Determine **(a)** the pH of the original buffer described in Sample Exercise 17.6 after the addition of 0.020 mol HCl and **(b)** the pH of the solution that would result from the addition of 0.020 mol HCl to 1.000 L of pure water.

Chemistry and Life

Blood as a Buffered Solution

Chemical reactions that occur in living systems are often extremely sensitive to pH. Many of the enzymes that catalyze important biochemical reactions are effective only within a narrow pH range. For this reason, the human body maintains a remarkably intricate system of buffers, both within cells and in the fluids that transport cells. Blood, the fluid that transports oxygen to all parts of the body, is one of the most prominent examples of the importance of buffers in living beings.

Human blood has a normal pH of 7.35 to 7.45. Any deviation from this range can have extremely disruptive effects on the stability of cell membranes, the structures of proteins, and the activities of enzymes. Death may result if the blood pH falls below 6.8 or rises above 7.8. When the pH falls below 7.35, the condition is called *acidosis*; when it rises above 7.45, the condition is called *alkalosis*. Acidosis is the more common tendency because metabolism generates several acids in the body.

The major buffer system used to control blood pH is the *carbonic acid–bicarbonate buffer system*. Carbonic acid (H_2CO_3) and bicarbonate ion (HCO_3^-) are a conjugate acid–base pair. In addition, carbonic acid decomposes into carbon dioxide gas and water. The important equilibria in this buffer system are

$$H^+(aq) + HCO_3^-(aq) \rightleftharpoons H_2CO_3(aq) \rightleftharpoons H_2O(l) + CO_2(g)$$
$$[17.10]$$

Several aspects of these equilibria are notable. First, although carbonic acid is diprotic, the carbonate ion (CO_3^{2-}) is unimportant in this system. Second, one component of this equilibrium, CO_2, is a gas, which provides a mechanism for the body to adjust the equilibria. Removal of CO_2 via exhalation shifts the equilibria to the right, consuming H^+ ions. Third, the buffer system in blood operates at pH 7.4, which is fairly far removed from the pK_{a1} value of H_2CO_3 (6.1 at physiological temperatures). For the buffer to have a pH of 7.4, the ratio [base]/[acid] must be about 20. In normal blood plasma, the concentrations of HCO_3^- and H_2CO_3 are about 0.024 M and 0.0012 M, respectively. Consequently, the buffer has a high capacity to neutralize additional acid but only a low capacity to neutralize additional base.

The principal organs that regulate the pH of the carbonic acid–bicarbonate buffer system are the lungs and kidneys. When the concentration of CO_2 rises, the equilibrium concentrations in Equation 17.10 shift to the left, which leads to the formation of more H^+ and a drop in pH. This change is detected by receptors in the brain that trigger a reflex to breathe faster and deeper, increasing the rate at which CO_2 is expelled from the lungs and thereby shifting the equilibrium concentrations back to the right. When the blood pH becomes too high, the kidneys remove HCO_3^- from the blood. This shifts the equilibrium concentrations to the left, increasing the concentration of H^+. As a result, the pH decreases.

Regulation of blood pH relates directly to the effective transport of O_2 throughout the body. The protein hemoglobin, found in red blood cells (▶ **Figure 17.5**), carries oxygen. Hemoglobin (Hb)

reversibly binds both O_2 and H^+. These two substances compete for the Hb, which can be represented approximately by the equilibrium

$$HbH^+ + O_2 \rightleftharpoons HbO_2 + H^+ \qquad [17.11]$$

Oxygen enters the blood through the lungs, where it passes into the red blood cells and binds to Hb. When the blood reaches tissue in which the concentration of O_2 is low, the equilibrium concentrations in Equation 17.11 shift to the left and O_2 is released.

During periods of strenuous exertion, three factors work together to ensure delivery of O_2 to active tissues. The role of each factor can be understood by applying Le Châtelier's principle to Equation 17.11:

1. O_2 is consumed, causing the equilibrium concentrations to shift to the left, releasing more O_2.
2. Large amounts of CO_2 are produced by metabolism, which increases $[H^+]$ and causes the equilibrium concentrations to shift to the left, releasing O_2.
3. Body temperature rises. Because Equation 17.11 is exothermic, the increase in temperature shifts the equilibrium concentrations to the left, releasing O_2.

In addition to the factors causing release of O_2 to tissues, the decrease in pH stimulates an increase in breathing rate, which furnishes more O_2 and eliminates CO_2. Without this elaborate series of equilibrium shifts and pH changes, the O_2 in tissues would be rapidly depleted, making further activity impossible. Under such conditions, the buffering capacity of the blood and the exhalation of CO_2 through the lungs are essential to keep the pH from dropping too low, thereby triggering acidosis.

Related Exercises: 17.29, 17.97

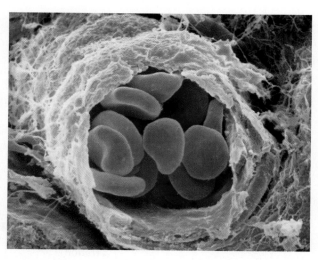

▲ **Figure 17.5 Red blood cells.** A scanning electron micrograph of red blood cells traveling through a small branch of an artery. The red blood cells are approximately 0.010 mm in diameter.

▲ GO FIGURE

In which direction do you expect the pH to change as NaOH is added to the HCl solution?

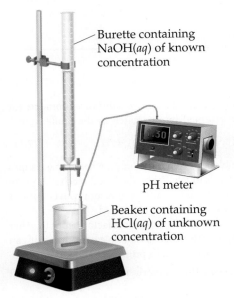

- Burette containing NaOH(aq) of known concentration

- pH meter

- Beaker containing HCl(aq) of unknown concentration

▲ Figure 17.6 Measuring pH during a titration.

17.3 | Acid–Base Titrations

Titrations are procedures in which one reactant is slowly added into a solution of another reactant, while equilibrium concentrations along the way are monitored. ∞ (Section 4.6) There are two main reasons to do titrations: (1) you want to know the concentration of one of the reactants or (2) you want to know the equilibrium constant for the reaction.

In an acid–base titration, a solution containing a known concentration of base is slowly added to an acid (or the acid is added to the base). ∞ (Section 4.6) Acid–base indicators can be used to signal the *equivalence point* of a titration (the point at which stoichiometrically equivalent quantities of acid and base have been brought together). Alternatively, a pH meter can be used to monitor the progress of the reaction (◄ Figure 17.6), producing a **pH titration curve**, a graph of the pH as a function of the volume of titrant added. The shape of the titration curve makes it possible to determine the equivalence point. The curve can also be used to select suitable indicators and to determine the K_a of the weak acid or the K_b of the weak base being titrated.

To understand why titration curves have certain characteristic shapes, we will examine the curves for three kinds of titrations: (1) strong acid–strong base, (2) weak acid–strong base, and (3) polyprotic acid–strong base. We will also briefly consider how these curves relate to those involving weak bases.

Strong Acid–Strong Base Titrations

The titration curve produced when a strong base is added to a strong acid has the general shape shown in ▼ Figure 17.7, which depicts the pH change that occurs as 0.100 *M*

▲ GO FIGURE

What volume of NaOH(aq) would be needed to reach the equivalence point if the concentration of the added base were 0.200 *M*?

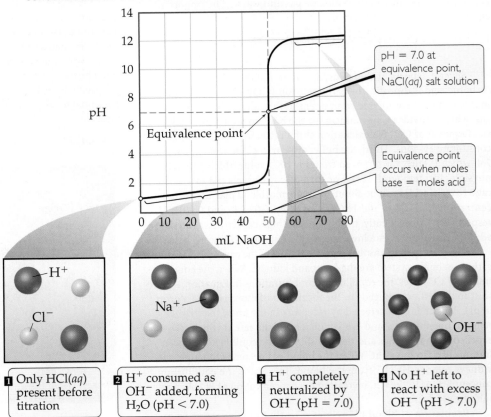

pH = 7.0 at equivalence point, NaCl(aq) salt solution

Equivalence point occurs when moles base = moles acid

1 Only HCl(aq) present before titration

2 H⁺ consumed as OH⁻ added, forming H₂O (pH < 7.0)

3 H⁺ completely neutralized by OH⁻ (pH = 7.0)

4 No H⁺ left to react with excess OH⁻ (pH > 7.0)

▲ Figure 17.7 Titration of a strong acid with a strong base. The pH curve for titration of 50.0 mL of a 0.100 *M* solution of hydrochloric acid with a 0.100 *M* solution of NaOH(aq). For clarity, water molecules have been omitted from the molecular art.

NaOH is added to 50.0 mL of 0.100 M HCl. The pH can be calculated at various stages of the titration. To help understand these calculations, we can divide the curve into four regions:

1. **Initial pH:** The pH of the solution before the addition of any base is determined by the initial concentration of the strong acid. For a solution of 0.100 M HCl, $[H^+] = 0.100\ M$ and pH $= -\log(0.100) = 1.000$. Thus, the initial pH is low.

2. **Between initial pH and equivalence point:** As NaOH is added, the pH increases slowly at first and then rapidly in the vicinity of the equivalence point. The pH before the equivalence point is determined by the concentration of acid not yet neutralized. This calculation is illustrated in Sample Exercise 17.7(a).

3. **Equivalence point:** At the equivalence point an equal number of moles of NaOH and HCl have reacted, leaving only a solution of their salt, NaCl. The pH of the solution is 7.00 because the cation of a strong base (in this case Na^+) and the anion of a strong acid (in this case Cl^-) are neither acids nor bases and, therefore, have no appreciable effect on pH. ⚬⚬⚬ (Section 16.9)

4. **After equivalence point:** The pH of the solution after the equivalence point is determined by the concentration of excess NaOH in the solution. This calculation is illustrated in Sample Exercise 17.7(b).

SAMPLE EXERCISE 17.7 | Calculations for a Strong Acid–Strong Base Titration

Calculate the pH when (a) 49.0 mL and (b) 51.0 mL of 0.100 M NaOH solution have been added to 50.0 mL of 0.100 M HCl solution.

SOLUTION

Analyze We are asked to calculate the pH at two points in the titration of a strong acid with a strong base. The first point is just before the equivalence point, so we expect the pH to be determined by the small amount of strong acid that has not yet been neutralized. The second point is just after the equivalence point, so we expect this pH to be determined by the small amount of excess strong base.

Plan (a) As the NaOH solution is added to the HCl solution, $H^+(aq)$ reacts with $OH^-(aq)$ to form H_2O. Both Na^+ and Cl^- are spectator ions, having negligible effect on the pH. To determine the pH of the solution, we must first determine how many moles of H^+ were originally present and how many moles of OH^- were added. We can then calculate how many moles of each ion remain after the neutralization reaction. To calculate $[H^+]$, and hence pH, we must also remember that the volume of the solution increases as we add titrant, thus diluting the concentration of all solutes present. Therefore, it is best to deal with moles first, and then convert to molarities using total solution volumes (volume of acid plus volume of base).

Solve The number of moles of H^+ in the original HCl solution is given by the product of the volume of the solution and its molarity:

$$(0.0500\ \text{L soln})\left(\frac{0.100\ \text{mol } H^+}{1\ \text{L soln}}\right) = 5.00 \times 10^{-3}\ \text{mol } H^+$$

Likewise, the number of moles of OH^-, in 49.0 mL of 0.100 M NaOH is

$$(0.0490\ \text{L soln})\left(\frac{0.100\ \text{mol } OH^-}{1\ \text{L soln}}\right) = 4.90 \times 10^{-3}\ \text{mol } OH^-$$

Because we have not reached the equivalence point, there are more moles of H^+ present than OH^-. Therefore, OH^- is the limiting reactant. Each mole of OH^- reacts with 1 mol of H^+. Using the convention introduced in Sample Exercise 17.6, we have

	$H^+(aq)$	$+$ $OH^-(aq)$	$\longrightarrow$ $H_2O(l)$
Before reaction (mol)	5.00×10^{-3}	4.90×10^{-3}	—
Change (limiting reactant) (mol)	-4.90×10^{-3}	-4.90×10^{-3}	—
After reaction (mol)	0.10×10^{-3}	0	—

The volume of the reaction mixture increases as the NaOH solution is added to the HCl solution. Thus, at this point in the titration, the volume in the titration flask is

$$50.0\ \text{mL} + 49.0\ \text{mL} = 99.0\ \text{mL} = 0.0990\ \text{L}$$

Thus, the concentration of $H^+(aq)$ in the flask is

$$[H^+] = \frac{\text{moles } H^+(aq)}{\text{liters soln}} = \frac{0.10 \times 10^{-3}\ \text{mol}}{0.09900\ \text{L}} = 1.0 \times 10^{-3}\ M$$

The corresponding pH is

$$-\log(1.0 \times 10^{-3}) = 3.00$$

Plan (b) We proceed in the same way as we did in part (a) except we are now past the equivalence point and have more OH^- in the solution than H^+. As before, the initial number of moles of each reactant is determined from their volumes and concentrations. The reactant present in smaller stoichiometric amount (the limiting reactant) is consumed completely, leaving an excess of hydroxide ion.

Solve

	$H^+(aq)$	$+$	$OH^-(aq)$	$\longrightarrow$	$H_2O(l)$
Before reaction (mol)	5.00×10^{-3}		5.10×10^{-3}		—
Change (limiting reactant) (mol)	-5.00×10^{-3}		-5.00×10^{-3}		—
After reaction (mol)	0		0.10×10^{-3}		—

In this case, the volume in the titration flask is

$$50.0 \text{ mL} + 51.0 \text{ mL} = 101.0 \text{ mL} = 0.1010 \text{ L}$$

Hence, the concentration of $OH^-(aq)$ in the flask is

$$[OH^-] = \frac{\text{moles } OH^-(aq)}{\text{liters soln}} = \frac{0.10 \times 10^{-3} \text{ mol}}{0.1010 \text{ L}} = 1.0 \times 10^{-3} \text{ M}$$

and we have

$$pOH = -\log(1.0 \times 10^{-3}) = 3.00$$
$$pH = 14.00 - pOH = 14.00 - 3.00 = 11.00$$

Comment Note that the pH increased by only two pH units, from 1.00 (Figure 17.7) to 3.00, after the first 49.0 mL of NaOH solution was added, but jumped by eight pH units, from 3.00 to 11.00, as 2.0 mL of base solution was added near the equivalence point. Such a rapid rise in pH near the equivalence point is a characteristic of titrations involving strong acids and strong bases.

Practice Exercise 1

An acid–base titration is performed: 250.0 mL of an unknown concentration of HCl (*aq*) is titrated to the equivalence point with 36.7 mL of a 0.1000 M aqueous solution of NaOH. Which of the following statements is *not* true of this titration?
(a) The HCl solution is less concentrated than the NaOH solution. **(b)** The pH is less than 7 after adding 25 mL of NaOH solution. **(c)** The pH at the equivalence point is 7.00. **(d)** If an additional 1.00 mL of NaOH solution is added beyond the equivalence point, the pH of the solution is more than 7.00. **(e)** At the equivalence point, the OH^- concentration in the solution is 3.67×10^{-3} M.

Practice Exercise 2

Calculate the pH when **(a)** 24.9 mL and **(b)** 25.1 mL of 0.100 M HNO_3 have been added to 25.0 mL of 0.100 M KOH solution.

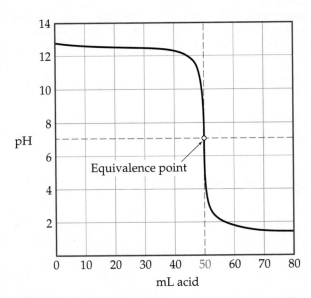

▲ **Figure 17.8 Titration of a strong base with a strong acid.** The pH curve for titration of 50.0 mL of a 0.100 M solution of a strong base with a 0.100 M solution of a strong acid.

Titration of a solution of a strong base with a solution of a strong acid yields an analogous curve of pH versus added acid. In this case, however, the pH is high at the outset of the titration and low at its completion (◀ Figure 17.8). The pH at the equivalence point is still 7.0 (at 25 °C), just like the strong acid–strong base titration.

▲ **Give It Some Thought**

What is the pH at the equivalence point when 0.10 M HNO_3 is used to titrate a volume of solution containing 0.30 g of KOH?

Weak Acid–Strong Base Titrations

The curve for titration of a weak acid by a strong base is similar to the curve in Figure 17.7. Consider, for example, the curve for titration of 50.0 mL of 0.100 M acetic acid with 0.100 M NaOH shown in ▶ Figure 17.9. We can calculate the pH at points along this curve, using principles we discussed earlier, which means again dividing the curve into four regions:

1. **Initial pH:** We use K_a to calculate this pH, as shown in Section 16.6. The calculated pH of 0.100 M CH_3COOH is 2.89.

▲ **GO FIGURE**

If the acetic acid being titrated here were replaced by hydrochloric acid, would the amount of base needed to reach the equivalence point change? Would the pH at the equivalence point change?

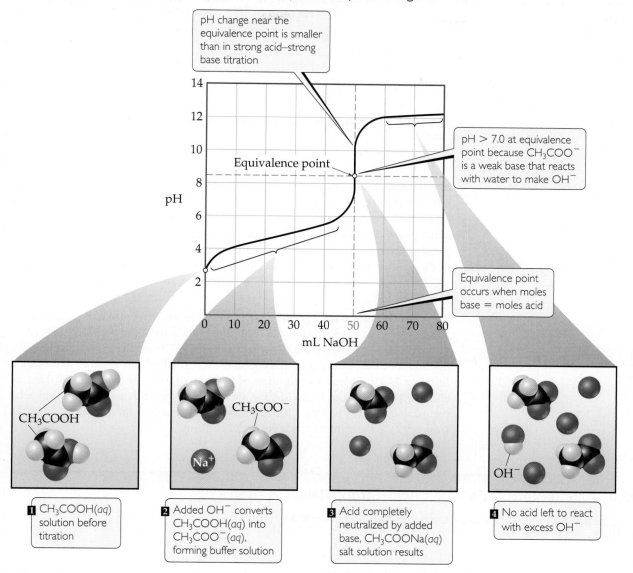

① CH$_3$COOH(aq) solution before titration

② Added OH$^-$ converts CH$_3$COOH(aq) into CH$_3$COO$^-$(aq), forming buffer solution

③ Acid completely neutralized by added base, CH$_3$COONa(aq) salt solution results

④ No acid left to react with excess OH$^-$

▲ **Figure 17.9 Titration of a weak acid with a strong base.** The pH curve for titration of 50.0 mL of a 0.100 M solution of acetic acid with a 0.100 M solution of NaOH(aq). For clarity, water molecules have been omitted from the molecular art.

2. **Between initial pH and equivalence point:** Prior to reaching the equivalence point, the acid is being neutralized, and its conjugate base is being formed:

$$CH_3COOH(aq) + OH^-(aq) \longrightarrow CH_3COO^-(aq) + H_2O(l) \quad [17.12]$$

Thus, the solution contains a mixture of CH$_3$COOH and CH$_3$COO$^-$. Calculating the pH in this region involves two steps. First, we consider the neutralization reaction between CH$_3$COOH and OH$^-$ to determine [CH$_3$COOH] and [CH$_3$COO$^-$]. Next, we calculate the pH of this buffer pair using procedures developed in Sections 17.1 and 17.2. The general procedure is diagrammed in **Figure 17.10** and illustrated in Sample Exercise 17.7.

3. **Equivalence point:** The equivalence point is reached when 50.0 mL of 0.100 M NaOH has been added to the 50.0 mL of 0.100 M CH$_3$COOH. At this point, the 5.00×10^{-3} mol of NaOH completely reacts with the 5.00×10^{-3} mol

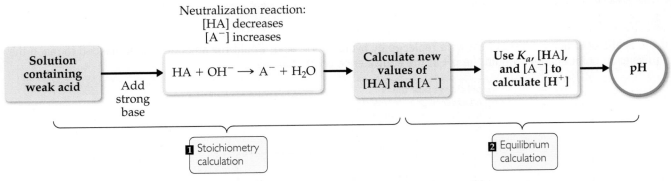

▲ Figure 17.10 **Procedure for calculating pH when a weak acid is partially neutralized by a strong base.**

of CH_3COOH to form 5.00×10^{-3} mol of CH_3COONa. The Na^+ ion of this salt has no significant effect on the pH. The CH_3COO^- ion, however, is a weak base whose reaction with water cannot be neglected, and the pH at the equivalence point is therefore greater than 7. In general, the pH at the equivalence point is always above 7 in a weak acid–strong base titration because the anion of the salt formed is a weak base. The procedure for calculating the pH of the solution of a weak base is described in Section 16.7 and is shown in Sample Exercise 17.8.

4. **After equivalence point (excess base):** In this region, $[OH^-]$ from the reaction of CH_3COO^- with water is negligible relative to $[OH^-]$ from the excess NaOH. Thus, the pH is determined by the concentration of OH^- from the excess NaOH. The method for calculating pH in this region is therefore like that illustrated in Sample Exercise 17.7(b). Thus, the addition of 51.0 mL of 0.100 M NaOH to 50.0 mL of either 0.100 M HCl or 0.100 M CH_3COOH yields the same pH, 11.00. Notice by comparing Figures 17.7 and 17.9 that the titration curves for a strong acid and a weak acid are the same after the equivalence point.

SAMPLE EXERCISE 17.8 Calculations for a Weak Acid–Strong Base Titration

Calculate the pH of the solution formed when 45.0 mL of 0.100 M NaOH is added to 50.0 mL of 0.100 M CH_3COOH ($K_a = 1.8 \times 10^{-5}$).

SOLUTION

Analyze We are asked to calculate the pH before the equivalence point of the titration of a weak acid with a strong base.

Plan We first must determine the number of moles of CH_3COOH and CH_3COO^- present after the neutralization reaction (the stoichiometry calculation). We then calculate pH using K_a, $[CH_3COOH]$, and $[CH_3COO^-]$ (the equilibrium calculation).

Solve *Stoichiometry Calculation:* The product of the volume and concentration of each solution gives the number of moles of each reactant present before the neutralization:

$$(0.0500 \text{ L soln})\left(\frac{0.100 \text{ mol } CH_3COOH}{1 \text{ L soln}}\right) = 5.00 \times 10^{-3} \text{ mol } CH_3COOH$$

$$(0.0450 \text{ L soln})\left(\frac{0.100 \text{ mol NaOH}}{1 \text{ L soln}}\right) = 4.50 \times 10^{-3} \text{ mol NaOH}$$

The 4.50×10^{-3} mol of NaOH consumes 4.50×10^{-3} mol of CH_3COOH:

$$CH_3COOH(aq) + OH^-(aq) \longrightarrow CH_3COO^-(aq) + H_2O(l)$$

	CH_3COOH	OH^-	CH_3COO^-	
Before reaction (mol)	5.00×10^{-3}	4.50×10^{-3}	0	—
Change (limiting reactant) (mol)	-4.50×10^{-3}	-4.50×10^{-3}	$+4.50 \times 10^{-3}$	
After reaction (mol)	0.50×10^{-3}	0	4.50×10^{-3}	—

The total volume of the solution is   45.0 mL + 50.0 mL = 95.0 mL = 0.0950 L

The resulting molarities of CH_3COOH and CH_3COO^- after the reaction are therefore

$$[CH_3COOH] = \frac{0.50 \times 10^{-3}\,mol}{0.0950\,L} = 0.0053\,M$$

$$[CH_3COO^-] = \frac{4.50 \times 10^{-3}\,mol}{0.0950\,L} = 0.0474\,M$$

Equilibrium Calculation: The equilibrium between CH_3COOH and CH_3COO^- must obey the equilibrium-constant expression for CH_3COOH:

$$K_a = \frac{[H^+][CH_3COO^-]}{[CH_3COOH]} = 1.8 \times 10^{-5}$$

Solving for $[H^+]$ gives

$$[H^+] = K_a \times \frac{[CH_3COOH]}{[CH_3COO^-]} = (1.8 \times 10^{-5}) \times \left(\frac{0.0053}{0.0474}\right) = 2.0 \times 10^{-6}\,M$$

$$pH = -\log(2.0 \times 10^{-6}) = 5.70$$

Comment We could have solved for pH equally well using the Henderson–Hasselbalch equation in the last step.

Practice Exercise 1

If you think carefully about what happens during the course of a weak acid–strong base titration, you can learn some very interesting things. For example, let's look back at Figure 17.9 and pretend you did not know that acetic acid was the acid being titrated. You can figure out the pK_a of a weak acid just by thinking about the definition of K_a and looking at the right place on the titration curve! Which of the following choices is the best way to do this?
(a) At the equivalence point, $pH = pK_a$. **(b)** Halfway to the equivalence point, $pH = pK_a$.
(c) Before any base is added, $pH = pK_a$. **(d)** At the top of the graph with excess base added, $pH = pK_a$.

Practice Exercise 2

(a) Calculate the pH in the solution formed by adding 10.0 mL of 0.050 M NaOH to 40.0 mL of 0.0250 M benzoic acid $(C_6H_5COOH, K_a = 6.3 \times 10^{-5})$. **(b)** Calculate the pH in the solution formed by adding 10.0 mL of 0.100 M HCl to 20.0 mL of 0.100 M NH$_3$.

In order to further monitor the evolution of pH as a function of added base, we can calculate the pH at the equivalence point.

SAMPLE EXERCISE 17.9 | Calculating the pH at the Equivalence Point

Calculate the pH at the equivalence point in the titration of 50.0 mL of 0.100 M CH_3COOH with 0.100 M NaOH.

SOLUTION

Analyze We are asked to determine the pH at the equivalence point of the titration of a weak acid with a strong base. Because the neutralization of a weak acid produces its anion, a conjugate base that can react with water, we expect the pH at the equivalence point to be greater than 7.

Plan The initial number of moles of acetic acid equals the number of moles of acetate ion at the equivalence point. We use the volume of the solution at the equivalence point to calculate the concentration of acetate ion. Because the acetate ion is a weak base, we can calculate the pH using K_b and $[CH_3COO^-]$.

Solve The number of moles of acetic acid in the initial solution is obtained from the volume and molarity of the solution:

$$Moles = M \times L = (0.100\,mol/L)(0.0500\,L) = 5.00 \times 10^{-3}\,mol\ CH_3COOH$$

Hence, 5.00×10^{-3} mol of CH_3COO^- is formed. It will take 50.0 mL of NaOH to reach the equivalence point (Figure 17.9). The volume of this salt solution at the equivalence point is the sum of the volumes of the acid and base, 50.0 mL + 50.0 mL = 100.0 mL = 0.1000 L. Thus, the concentration of CH_3COO^- is

$$[CH_3COO^-] = \frac{5.00 \times 10^{-3}\,mol}{0.1000\,L} = 0.0500\,M$$

The CH_3COO^- ion is a weak base:

$$CH_3COO^-(aq) + H_2O(l) \rightleftharpoons CH_3COOH(aq) + OH^-(aq)$$

The K_b for CH_3COO^- can be calculated from the K_a value of its conjugate acid, $K_b = K_w/K_a = (1.0 \times 10^{-14})/(1.8 \times 10^{-5}) = 5.6 \times 10^{-10}$. Using the K_b expression, we have

$$K_b = \frac{[CH_3COOH][OH^-]}{[CH_3COO^-]} = \frac{(x)(x)}{0.0500 - x} = 5.6 \times 10^{-10}$$

Making the approximation that $0.0500 - x \approx 0.0500$, and then solving for x, we have $x = [OH^-] = 5.3 \times 10^{-6} M$, which gives pOH = 5.28 and pH = 8.72.

Check The pH is above 7, as expected for the salt of a weak acid and strong base.

Practice Exercise 1

Why is pH at the equivalence point larger than 7 when you titrate a weak acid with a strong base? (a) There is excess strong base at the equivalence point. (b) There is excess weak acid at the equivalence point. (c) The conjugate base that is formed at the equivalence point is a strong base. (d) The conjugate base that is formed at the equivalence point reacts with water. (e) This statement is false: the pH is always 7 at an equivalence point in a pH titration.

Practice Exercise 2

Calculate the pH at the equivalence point when (a) 40.0 mL of 0.025 M benzoic acid (C_6H_5COOH, $K_a = 6.3 \times 10^{-5}$) is titrated with 0.050 M NaOH and (b) 40.0 mL of 0.100 M NH_3 is titrated with 0.100 M HCl.

The titration curve for a weak acid–strong base titration (Figure 17.9) differs from the curve for a strong acid–strong base titration (Figure 17.7) in three noteworthy ways:

1. The solution of the weak acid has a higher initial pH than a solution of a strong acid of the same concentration.

2. The pH change in the rapid-rise portion of the curve near the equivalence point is smaller for the weak acid than for the strong acid.

3. The pH at the equivalence point is above 7.00 for the weak acid titration.

GO FIGURE

How does the pH at the equivalence point change as the acid being titrated becomes weaker? How does the volume of NaOH(*aq*) needed to reach the equivalence point change?

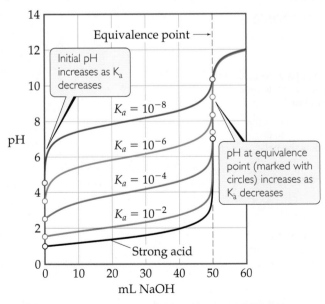

▲ **Figure 17.11 A set of curves showing the effect of acid strength on the characteristics of the titration curve when a weak acid is titrated by a strong base.** Each curve represents titration of 50.0 mL of 0.10 M acid with 0.10 M NaOH.

Give It Some Thought

Describe the reasons why statement 3 above is true.

The weaker the acid, the more pronounced these differences become. To illustrate this consider the family of titration curves shown in ◀ Figure 17.11. Notice that as the acid becomes weaker (that is, as K_a becomes smaller), the initial pH increases and the pH change near the equivalence point becomes less marked. Furthermore, the pH at the equivalence point steadily increases as K_a decreases, because the strength of the conjugate base of the weak acid increases. It is virtually impossible to determine the equivalence point when pK_a is 10 or higher because the pH change is too small and gradual.

Titrating with an Acid–Base Indicator

Oftentimes in an acid–base titration, an indicator is used rather than a pH meter. An indicator is a compound that changes color in solution over a specific pH range. Optimally, an indicator should change color at the equivalence point in a titration. In practice, however, an indicator need not precisely mark the equivalence point. The pH changes very rapidly near the equivalence point, and in this region one drop of titrant can change the pH by several units. Thus, an indicator beginning and ending its color change anywhere on the rapid-rise portion of the titration curve gives a sufficiently accurate measure of the titrant volume needed to reach the equivalence point. The point in a titration where the indicator changes color is called the *end point* to distinguish it from the equivalence point that it closely approximates.

▶ **Figure 17.12** shows the curve for titration of a strong base (NaOH) with a strong acid (HCl). We see from the vertical part of the curve that the pH changes rapidly from roughly 11 to 3 near the equivalence point. Consequently, an indicator for this titration can change color anywhere in this range. Most strong acid–strong base titrations are carried out using phenolphthalein as an indicator because it changes color in this range (see Figure 16.8, page 684). Several other indicators would also be satisfactory, including methyl red, which, as the lower color band in Figure 17.12 shows, changes color in the pH range from about 4.2 to 6.0 (see Figure 16.8, page 684).

As noted in our discussion of Figure 17.11, because the pH change near the equivalence point becomes smaller as K_a decreases, the choice of indicator for a weak acid–strong base titration is more critical than it is for titrations where both acid and base are strong. When 0.100 M CH$_3$COOH ($K_a = 1.8 \times 10^{-5}$) is titrated with 0.100 M NaOH, for example, the pH increases rapidly only over the pH range from about 7 to 11 (▼ **Figure 17.13**). Phenolphthalein is therefore an ideal indicator because it changes color from pH 8.3 to 10.0, close to the pH at the equivalence point. Methyl red is a poor choice, however, because its color change, from 4.2 to 6.0, begins well before the equivalence point is reached.

Titration of a weak base (such as 0.100 M NH$_3$) with a strong acid solution (such as 0.100 M HCl) leads to the titration curve shown in **Figure 17.14**. In this example, the equivalence point occurs at pH 5.28. Thus, methyl red is an ideal indicator but phenolphthalein would be a poor choice.

▲ **GO FIGURE**

Is methyl red a suitable indicator when you are titrating a strong acid with a strong base? Explain your answer.

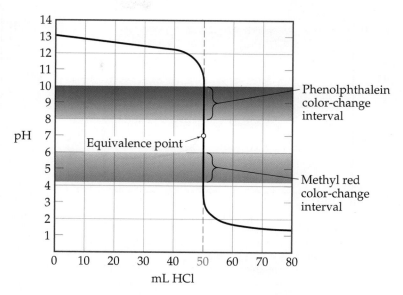

▲ **Figure 17.12 Using color indicators for titration of a strong base with a strong acid.** Both phenolphthalein and methyl red change color in the rapid-rise portion of the titration curve.

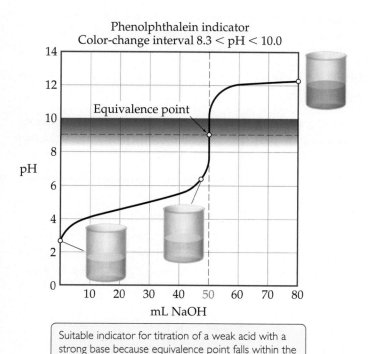

Suitable indicator for titration of a weak acid with a strong base because equivalence point falls within the color-change interval

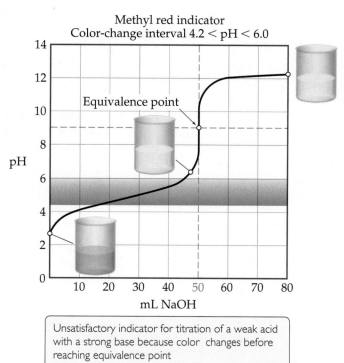

Unsatisfactory indicator for titration of a weak acid with a strong base because color changes before reaching equivalence point

▲ **Figure 17.13 Good and poor indicators for titration of a weak acid with a strong base.**

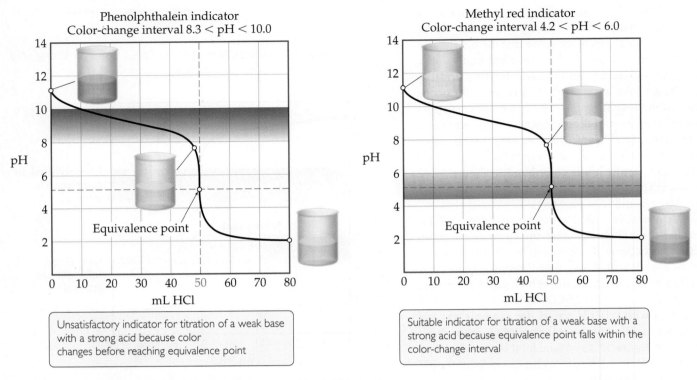

Phenolphthalein indicator
Color-change interval 8.3 < pH < 10.0

Equivalence point

Unsatisfactory indicator for titration of a weak base with a strong acid because color changes before reaching equivalence point

Methyl red indicator
Color-change interval 4.2 < pH < 6.0

Equivalence point

Suitable indicator for titration of a weak base with a strong acid because equivalence point falls within the color-change interval

▲ Figure 17.14 **Good and poor indicators for titration of a weak base with a strong acid.**

▲ Give It Some Thought

Why is the choice of indicator more crucial for a weak acid–strong base titration than for a strong acid–strong base titration?

Titrations of Polyprotic Acids

When weak acids contain more than one ionizable H atom, the reaction with OH^- occurs in a series of steps. Neutralization of H_3PO_3, for example, proceeds in two steps (the third H is bonded to the P and does not ionize):

$$H_3PO_3(aq) + OH^-(aq) \longrightarrow H_2PO_3^-(aq) + H_2O(l) \qquad [17.13]$$

$$H_2PO_3^-(aq) + OH^-(aq) \longrightarrow HPO_3^{2-}(aq) + H_2O(l) \qquad [17.14]$$

When the neutralization steps of a polyprotic acid or polybasic base are sufficiently separated, the titration has multiple equivalence points. ▶ **Figure 17.15** shows the two equivalence points corresponding to Equations 17.13 and 17.14.

▲ Give It Some Thought

Sketch an approximate titration curve for the titration of Na_2CO_3 with HCl.

You can use titration data such as that shown in Figure 17.15 to figure out the pK_as of the weak polyprotic acid. For example, let's write the K_{a1} and K_{a2} reactions for phosphorous acid:

$$H_3PO_3(aq) \rightleftharpoons H_2PO_3^-(aq) + H^+(aq) \qquad K_{a1} = \frac{[H_2PO_3^-][H^+]}{H_3PO_3}$$

$$H_2PO_3^-(aq) \rightleftharpoons HPO_3^{2-}(aq) + H^+(aq) \qquad K_{a2} = \frac{[HPO_3^{2-}][H^+]}{H_2PO_3^-}$$

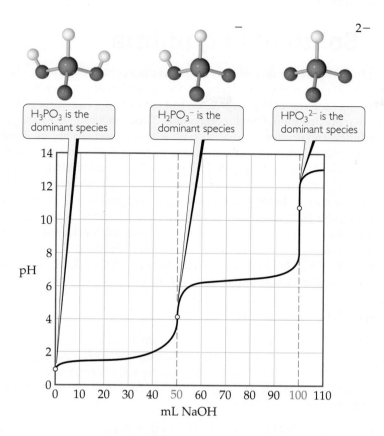

▲ **Figure 17.15 Titration curve for a diprotic acid.** The curve shows the pH change when 50.0 mL of 0.10 M H_3PO_3 is titrated with 0.10 M NaOH.

If we rearrange these equilibrium expressions, you see that we obtain Henderson–Hasselbalch equations:

$$pH = pK_{a1} + \log\frac{[H_2PO_3^-]}{[H_3PO_3]}$$

$$pH = pK_{a2} + \log\frac{[HPO_3^{2-}]}{[H_2PO_3^-]}$$

Therefore, if the concentrations of the each acid and base conjugate pairs were identical for each equilibrium, $\log(1) = 0$ and so $pH = pK_a$. When does this happen during the titration? At the beginning of the titration, the acid is H_3PO_3 initially; at the first equivalence point, it is all converted to $H_2PO_3^-$. Therefore, halfway to the first equivalence point, half of the H_3PO_3 is converted to $H_2PO_3^-$. Thus, halfway to the equivalence point, the concentration of H_3PO_3 is equal to that of $H_2PO_3^-$, and at that point, $pH = pK_{a1}$. Similar logic is true for the second equilibrium reaction: halfway toward its equivalence point, $pH = pK_{a2}$.

We can then just look at titration data and estimate the pK_as for the polyprotic acid directly from the titration curve. This procedure is especially useful if you are trying to identify an unknown polyprotic acid. In Figure 17.15, for instance, the first equivalence point occurs for 50 mL NaOH added. Halfway to the equivalence point corresponds to 25 mL NaOH. Because the pH at 25 mL NaOH is about 1.5 we may estimate $pK_{a1} = 1.5$ for phosphorous acid. The second equivalence point occurs at 100 mL NaOH added; halfway there (from the first equivalence point) is at 75 mL NaOH added. The graph indicates the pH at 75 mL NaOH added is about 6.5, and we therefore estimate that pK_{a2} for phosphorous acid is 6.5. The actual values for the two pK_as are $pK_{a1} = 1.3$ and $pK_{a2} = 6.7$ (close to our estimates).

17.4 | Solubility Equilibria

The equilibria we have considered thus far in this chapter have involved acids and bases. Furthermore, they have been homogeneous; that is, all the species have been in the same phase. Through the rest of the chapter, we will consider the equilibria involved in the dissolution or precipitation of ionic compounds. These reactions are heterogeneous.

Dissolution and precipitation occur both within us and around us. Tooth enamel dissolves in acidic solutions, for example, causing tooth decay, and the precipitation of certain salts in our kidneys produces kidney stones. The waters of Earth contain salts dissolved as water passes over and through the ground. Coral reefs are principally made of $CaCO_3$, as we saw in the beginning of this chapter. Precipitation of $CaCO_3$ from groundwater is responsible for the formation of stalactites and stalagmites within limestone caves.

In our earlier discussion of precipitation reactions, we considered general rules for predicting the solubility of common salts in water. ∞ (Section 4.2) These rules give us a qualitative sense of whether a compound has a low or high solubility in water. By considering solubility equilibria, however, we can make quantitative predictions about solubility.

The Solubility-Product Constant, K_{sp}

Recall that a *saturated solution* is one in which the solution is in contact with undissolved solute. ∞ (Section 13.2) Consider, for example, a saturated aqueous solution of $BaSO_4$ in contact with solid $BaSO_4$. Because the solid is an ionic compound, it is a strong electrolyte and yields $Ba^{2+}(aq)$ and $SO_4^{2-}(aq)$ ions when dissolved in water, readily establishing the equilibrium

$$BaSO_4(s) \rightleftharpoons Ba^{2+}(aq) + SO_4^{2-}(aq) \qquad [17.15]$$

As with any other equilibrium, the extent to which this dissolution reaction occurs is expressed by the magnitude of the equilibrium constant. Because this equilibrium equation describes the dissolution of a solid, the equilibrium constant indicates how soluble the solid is in water and is referred to as the **solubility-product constant** (or simply the **solubility product**). It is denoted K_{sp}, where *sp* stands for solubility product.

The equilibrium-constant expression for the equilibrium between a solid and an aqueous solution of its component ions (K_{sp}) is written according to the rules that apply to any other equilibrium-constant expression. Remember, however, that solids do not appear in the equilibrium-constant expressions for heterogeneous equilibrium. ∞ (Section 15.4)

Thus, the solubility-product expression for $BaSO_4$, which is based on Equation 17.15, is

$$K_{sp} = [Ba^{2+}][SO_4^{2-}] \qquad [17.16]$$

In general, *the solubility product K_{sp} of a compound equals the product of the concentration of the ions involved in the equilibrium, each raised to the power of its coefficient in the equilibrium equation.* The coefficient for each ion in the equilibrium equation also equals its subscript in the compound's chemical formula.

The values of K_{sp} at 25 °C for many ionic solids are tabulated in Appendix D. The value of K_{sp} for $BaSO_4$ is 1.1×10^{-10}, a very small number indicating that only a very small amount of the solid dissolves in 25 °C water.

SAMPLE EXERCISE 17.10 Writing Solubility-Product (K_{sp}) Expressions

Write the expression for the solubility-product constant for CaF_2, and look up the corresponding K_{sp} value in Appendix D.

SOLUTION

Analyze We are asked to write an equilibrium-constant expression for the process by which CaF_2 dissolves in water.

Plan We apply the general rules for writing an equilibrium-constant expression, excluding the solid reactant from the expression. We assume that the compound dissociates into its component ions:

$$CaF_2(s) \rightleftharpoons Ca^{2+}(aq) + 2 F^-(aq)$$

Solve The expression for K_{sp} is

$$K_{sp} = [Ca^{2+}][F^-]^2$$

Appendix D gives 3.9×10^{-11} for this K_{sp}.

Practice Exercise 1

Which of these expressions correctly expresses the solubility-product constant for Ag_3PO_4 in water? (a) $[Ag][PO_4]$, (b) $[Ag^+][PO_4^{3-}]$, (c) $[Ag^+]^3[PO_4^{3-}]$, (d) $[Ag^+][PO_4^{3-}]^3$, (e) $[Ag^+]^3[PO_4^{3-}]^3$.

Practice Exercise 2

Give the solubility-product-constant expressions and K_{sp} values (from Appendix D) for (a) barium carbonate and (b) silver sulfate.

Solubility and K_{sp}

It is important to distinguish carefully between solubility and the solubility-product constant. The solubility of a substance is the quantity that dissolves to form a saturated solution. ∞ (Section 13.2) Solubility is often expressed as grams of solute per liter of solution (g/L). *Molar solubility* is the number of moles of solute that dissolve in forming 1 L of saturated solution of the solute (mol/L). The solubility-product constant (K_{sp}) is the equilibrium constant for the equilibrium between an ionic solid and its saturated solution and is a unitless number. Thus, the magnitude of K_{sp} is a measure of how much of the solid dissolves to form a saturated solution.

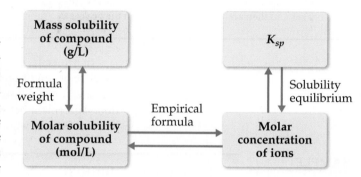

▲ Figure 17.16 **Procedure for converting between solubility and K_{sp}.** Starting from the mass solubility, follow the green arrows to determine K_{sp}. Starting from K_{sp}, follow the red arrows to determine either molar solubility or mass solubility.

 Give It Some Thought

Without doing a calculation, predict which of these compounds has the greatest molar solubility in water: AgCl ($K_{sp} = 1.8 \times 10^{-10}$), AgBr ($K_{sp} = 5.0 \times 10^{-13}$), or AgI ($K_{sp} = 8.3 \times 10^{-17}$).

The solubility of a substance can change considerably in response to a number of factors. For example, the solubilities of hydroxide salts, like $Mg(OH)_2$, are dependent upon the pH of the solution. The solubility is also affected by concentrations of other ions in solution, especially common ions. In other words, the numeric value of the solubility of a given solute does change as the other species in solution change. In contrast, the solubility-product constant, K_{sp}, has only one value for a given solute at any specific temperature.* ▲ Figure 17.16 summarizes the relationships among various expressions of solubility and K_{sp}.

SAMPLE EXERCISE 17.11 | Calculating K_{sp} from Solubility

Solid silver chromate is added to pure water at 25 °C, and some of the solid remains undissolved. The mixture is stirred for several days to ensure that equilibrium is achieved between the undissolved $Ag_2CrO_4(s)$ and the solution. Analysis of the equilibrated solution shows that its silver ion concentration is 1.3×10^{-4} M. Assuming that the Ag_2CrO_4 solution is saturated and that there are no other important equilibria involving Ag^+ or CrO_4^{2-} ions in the solution, calculate K_{sp} for this compound.

SOLUTION

Analyze We are given the equilibrium concentration of Ag^+ in a saturated solution of Ag_2CrO_4 and asked to determine the value of K_{sp} for Ag_2CrO_4.

Plan The equilibrium equation and the expression for K_{sp} are

$$Ag_2CrO_4(s) \rightleftharpoons 2\,Ag^+(aq) + CrO_4^{2-}(aq)$$
$$K_{sp} = [Ag^+]^2[CrO_4^{2-}]$$

To calculate K_{sp}, we need the equilibrium concentrations of Ag^+ and CrO_4^{2-}. We know that at equilibrium $[Ag^+] = 1.3 \times 10^{-4}$ M. All the Ag^+ and CrO_4^{2-} ions in the solution come from the Ag_2CrO_4 that dissolves. Thus, we can use $[Ag^+]$ to calculate $[CrO_4^{2-}]$.

---

*This is strictly true only for very dilute solutions, for K_{sp} values change somewhat when the concentration of ionic substances in water is increased. However, we will ignore these effects, which are taken into consideration only for work that requires exceptional accuracy.

Solve From the chemical formula of silver chromate, we know that there must be two Ag^+ ions in solution for each CrO_4^{2-} ion in solution. Consequently, the concentration of CrO_4^{2-} is half the concentration of Ag^+:

$$[CrO_4^{2-}] = \left(\frac{1.3 \times 10^{-4}\ mol\ Ag^+}{L}\right)\left(\frac{1\ mol\ CrO_4^{2-}}{2\ mol\ Ag^+}\right) = 6.5 \times 10^{-5}\ M$$

and K_{sp} is

$$K_{sp} = [Ag^+]^2[CrO_4^{2-}] = (1.3 \times 10^{-4})^2(6.5 \times 10^{-5}) = 1.1 \times 10^{-12}$$

Check We obtain a small value, as expected for a slightly soluble salt. Furthermore, the calculated value agrees well with the one given in Appendix D, 1.2×10^{-12}.

Practice Exercise 1

You add 10.0 grams of solid copper(II) phosphate, $Cu_3(PO_4)_2$, to a beaker and then add 100.0 mL of water to the beaker at

$T = 298$ K. The solid does not appear to dissolve. You wait a long time, with occasional stirring and eventually measure the equilibrium concentration of $Cu^{2+}(aq)$ in the water to be 5.01×10^{-8} M. What is the K_{sp} of copper(II) phosphate?

(a) 5.01×10^{-8} (b) 2.50×10^{-15} (c) 4.20×10^{-15}
(d) 3.16×10^{-37} (e) 1.40×10^{-37}

Practice Exercise 2

A saturated solution of $Mg(OH)_2$ in contact with undissolved $Mg(OH)_2(s)$ is prepared at 25 °C. The pH of the solution is found to be 10.17. Assuming that there are no other simultaneous equilibria involving the Mg^{2+} or OH^- ions, calculate K_{sp} for this compound.

In principle, it is possible to use the K_{sp} value of a salt to calculate solubility under a variety of conditions. In practice, great care must be taken in doing so for the reasons indicated in "A Closer Look: Limitations of Solubility Products" at the end of this section. Agreement between the measured solubility and that calculated from K_{sp} is usually best for salts whose ions have low charges (1+ and 1−) and do not react with water.

SAMPLE EXERCISE 17.12 | Calculating Solubility from K_{sp}

The K_{sp} for CaF_2 is 3.9×10^{-11} at 25 °C. Assuming equilibrium is established between solid and dissolved CaF_2, and that there are no other important equilibria affecting its solubility, calculate the solubility of CaF_2 in grams per liter.

SOLUTION

Analyze We are given K_{sp} for CaF_2 and asked to determine solubility. Recall that the solubility of a substance is the quantity that can dissolve in solvent, whereas the solubility-product constant, K_{sp}, is an equilibrium constant.

Plan To go from K_{sp} to solubility, we follow the steps indicated by the red arrows in Figure 17.16. We first write the chemical equation for the dissolution and set up a table of initial and equilibrium concentrations. We then use the equilibrium-constant expression. In this case we know K_{sp}, and so we solve for the concentrations of the ions in solution. Once we know these concentrations, we use the formula weight to determine solubility in g/L.

Solve Assume that initially no salt has dissolved, and then allow x mol/L of CaF_2 to dissociate completely when equilibrium is achieved:

	$CaF_2(s)$	$\rightleftharpoons$	$Ca^{2+}(aq)$	+	$2\ F^-(aq)$
Initial concentration (*M*)	—		0		0
Change (*M*)	—		$+x$		$+2x$
Equilibrium concentration (*M*)	—		x		$2x$

The stoichiometry of the equilibrium dictates that $2x$ mol/L of F^- are produced for each x mol/L of CaF_2 that dissolve. We now use the expression for K_{sp} and substitute the equilibrium concentrations to solve for the value of x:

$$K_{sp} = [Ca^{2+}][F^-]^2 = (x)(2x)^2 = 4x^3 = 3.9 \times 10^{-11}$$

(Remember that $\sqrt[3]{y} = y^{1/3}$.) Thus, the molar solubility of CaF_2 is 2.1×10^{-4} mol/L.

$$x = \sqrt[3]{\frac{3.9 \times 10^{-11}}{4}} = 2.1 \times 10^{-4}$$

The mass of CaF_2 that dissolves in water to form 1 L of solution is

$$\left(\frac{2.1 \times 10^{-4}\ mol\ CaF_2}{1\ L\ soln}\right)\left(\frac{78.1\ g\ CaF_2}{1\ mol\ CaF_2}\right) = 1.6 \times 10^{-2}\ g\ CaF_2/L\ soln$$

Check We expect a small number for the solubility of a slightly soluble salt. If we reverse the calculation, we should be able to recalculate the solubility product: $K_{sp} = (2.1 \times 10^{-4})(4.2 \times 10^{-4})^2 = 3.7 \times 10^{-11}$, close to the value given in the problem statement, 3.9×10^{-11}.

Comment Because F^- is the anion of a weak acid, you might expect hydrolysis of the ion to affect the solubility of CaF_2. The basicity of F^- is so small ($K_b = 1.5 \times 10^{-11}$), however, that the hydrolysis occurs to only a slight extent and does not significantly influence the solubility. The reported solubility is 0.017 g/L at 25 °C, in good agreement with our calculation.

Practice Exercise 1

Of the five salts listed below, which has the highest concentration of its cation in water? Assume that all salt solutions are saturated and that the ions do not undergo any additional reactions in water.

(a) lead (II) chromate, $K_{sp} = 2.8 \times 10^{-13}$ **(b)** cobalt(II) hydroxide, $K_{sp} = 1.3 \times 10^{-15}$

(c) cobalt(II) sulfide, $K_{sp} = 5 \times 10^{-22}$ **(d)** chromium(III) hydroxide, $K_{sp} = 1.6 \times 10^{-30}$

(e) silver sulfide, $K_{sp} = 6 \times 10^{-51}$

Practice Exercise 2

The K_{sp} for LaF_3 is 2×10^{-19}. What is the solubility of LaF_3 in water in moles per liter?

A Closer Look

Limitations of Solubility Products

Ion concentrations calculated from K_{sp} values sometimes deviate appreciably from those found experimentally. In part, these deviations are due to electrostatic interactions between ions in solution, which can lead to ion pairs. ∞ (Section 13.5, "The van't Hoff Factor") These interactions increase in magnitude both as the concentrations of the ions increase and as their charges increase. The solubility calculated from K_{sp} tends to be low unless corrected to account for these interactions.

As an example of the effect of these interactions, consider $CaCO_3$ (calcite), whose solubility product, 4.5×10^{-9}, gives a calculated solubility of 6.7×10^{-5} mol/L; correcting for ionic interactions in the solution yields 7.3×10^{-5} mol/L. The reported solubility, however, is 1.4×10^{-4} mol/L, indicating that there must be additional factors involved.

Another common source of error in calculating ion concentrations from K_{sp} is ignoring other equilibria that occur simultaneously in the solution. It is possible, for example, that acid–base equilibria take place simultaneously with solubility equilibria. In particular, both basic anions and cations with high charge-to-size ratios undergo hydrolysis reactions that can measurably increase the solubilities of their salts. For example, $CaCO_3$ contains the basic carbonate ion $(K_b = 1.8 \times 10^{-4})$, which reacts with water:

$$CO_3^{2-}(aq) + H_2O(l) \rightleftharpoons HCO_3^-(aq) + OH^-(aq)$$

If we consider the effect of ion–ion interactions as well as simultaneous solubility and K_b equilibria, we calculate a solubility of 1.4×10^{-4} mol/L, in agreement with the measured value for calcite.

Finally, we generally assume that ionic compounds dissociate completely when they dissolve, but this assumption is not always valid. When MgF_2 dissolves, for example, it yields not only Mg^{2+} and F^- ions but also MgF^+ ions.

17.5 | Factors That Affect Solubility

Solubility is affected by temperature and by the presence of other solutes. The presence of an acid, for example, can have a major influence on the solubility of a substance. In Section 17.4, we considered the dissolving of ionic compounds in pure water. In this section, we examine three factors that affect the solubility of ionic compounds: (1) presence of common ions, (2) solution pH, and (3) presence of complexing agents. We will also examine the phenomenon of *amphoterism*, which is related to the effects of both pH and complexing agents.

Common-Ion Effect

The presence of either $Ca^{2+}(aq)$ or $F^-(aq)$ in a solution reduces the solubility of CaF_2, shifting the equilibrium concentrations to the left:

$$CaF_2(s) \rightleftharpoons Ca^{2+}(aq) + 2\,F^-(aq)$$

⬅ Addition of Ca^{2+} or F^- shifts equilibrium concentrations, reducing solubility

This reduction in solubility is another manifestation of the common-ion effect we looked at in Section 17.1. In general, *the solubility of a slightly soluble salt is decreased by the presence of a second solute that furnishes a common ion*, as ▶ **Figure 17.17** shows for CaF_2.

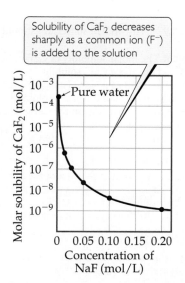

▲ **Figure 17.17 Common-ion effect.** Notice that the CaF_2 solubility is on a logarithmic scale.

SAMPLE EXERCISE 17.13 | Calculating the Effect of a Common Ion on Solubility

Calculate the molar solubility of CaF_2 at 25 °C in a solution that is **(a)** 0.010 M in $Ca(NO_3)_2$ and **(b)** 0.010 M in NaF.

SOLUTION

Analyze We are asked to determine the solubility of CaF_2 in the presence of two strong electrolytes, each containing an ion common to CaF_2. In **(a)** the common ion is Ca^{2+}, and NO_3^- is a spectator ion. In **(b)** the common ion is F^-, and Na^+ is a spectator ion.

Plan Because the slightly soluble compound is CaF_2, we need to use K_{sp} for this compound, which Appendix D gives as 3.9×10^{-11}. *The value of K_{sp} is unchanged by the presence of additional solutes.* Because of the common-ion effect, however, the solubility of the salt decreases in the presence of common ions. We use our standard equilibrium techniques of starting with the equation for CaF_2 dissolution, setting up a table of initial and equilibrium concentrations, and using the K_{sp} expression to determine the concentration of the ion that comes only from CaF_2.

Solve

(a) The initial concentration of Ca^{2+} is 0.010 M because of the dissolved $Ca(NO_3)_2$:

	$CaF_2(s)$ $\rightleftharpoons$	$Ca^{2+}(aq)$ +	$2 F^-(aq)$
Initial Concentration (M)	—	0.010	0
Change (M)	—	$+x$	$+2x$
Equilibrium Concentration (M)	—	$(0.010 + x)$	$2x$

Substituting into the solubility-product expression gives

$$K_{sp} = 3.9 \times 10^{-11} = [Ca^{2+}][F^-]^2 = (0.010 + x)(2x)^2$$

If we assume that x is small compared to 0.010, we have

$$3.9 \times 10^{-11} = (0.010)(2x)^2$$

$$x^2 = \frac{3.9 \times 10^{-11}}{4(0.010)} = 9.8 \times 10^{-10}$$

$$x = \sqrt{9.8 \times 10^{-10}} = 3.1 \times 10^{-5} M$$

This very small value for x validates the simplifying assumption we made. Our calculation indicates that 3.1×10^{-5} mol of solid CaF_2 dissolves per liter of 0.010 M $Ca(NO_3)_2$ solution.

(b) The common ion is F^-, and at equilibrium we have

$$[Ca^{2+}] = x \quad \text{and} \quad [F^-] = 0.010 + 2x$$

Assuming that $2x$ is much smaller than 0.010 M (that is, $0.010 + 2x \simeq 0.010$), we have

$$3.9 \times 10^{-11} = (x)(0.010 + 2x)^2 \simeq x(0.010)^2$$

$$x = \frac{3.9 \times 10^{-11}}{(0.010)^2} = 3.9 \times 10^{-7} M$$

Thus, 3.9×10^{-7} mol of solid CaF_2 should dissolve per liter of 0.010 M NaF solution.

Comment The molar solubility of CaF_2 in water is $2.1 \times 10^{-4} M$ (Sample Exercise 17.12). By comparison, our calculations here give a CaF_2 solubility of $3.1 \times 10^{-5} M$ in the presence of 0.010 M Ca^{2+} and $3.9 \times 10^{-7} M$ in the presence of 0.010 M F^- ion. Thus, the addition of either Ca^{2+} or F^- to a solution of CaF_2 decreases the solubility. However, the effect of F^- on the solubility is more pronounced than that of Ca^{2+} because $[F^-]$ appears to the second power in the K_{sp} expression for CaF_2, whereas $[Ca^{2+}]$ appears to the first power.

Practice Exercise 1

Consider a saturated solution of the salt MA_3, in which M is a metal cation with a 3+ charge and A is an anion with a 1− charge, in water at 298 K. Which of the following will affect the K_{sp} of MA_3 in water? **(a)** The addition of more M^{3+} to the solution. **(b)** The addition of more A^- to the solution. **(c)** Diluting the solution. **(d)** Raising the temperature of the solution. **(e)** More than one of the above factors.

Practice Exercise 2

For manganese(II) hydroxide, $Mn(OH)_2$, $K_{sp} = 1.6 \times 10^{-13}$. Calculate the molar solubility of $Mn(OH)_2$ in a solution that contains 0.020 M NaOH.

Solubility and pH

The pH of a solution affects the solubility of any substance whose anion is basic. Consider $Mg(OH)_2$, for which the solubility equilibrium is

$$Mg(OH)_2(s) \rightleftharpoons Mg^{2+}(aq) + 2\,OH^-(aq) \quad K_{sp} = 1.8 \times 10^{-11} \quad [17.17]$$

A saturated solution of $Mg(OH)_2$ has a calculated pH of 10.52 and its Mg^{2+} concentration is $1.7 \times 10^{-4}\,M$. Now suppose that solid $Mg(OH)_2$ is equilibrated with a solution buffered at pH 9.0. The pOH, therefore, is 5.0, so $[OH^-] = 1.0 \times 10^{-5}$. Inserting this value for $[OH^-]$ into the solubility-product expression, we have

$$K_{sp} = [Mg^{2+}][OH^-]^2 = 1.8 \times 10^{-11}$$

$$[Mg^{2+}](1.0 \times 10^{-5})^2 = 1.8 \times 10^{-11}$$

$$[Mg^{2+}] = \frac{1.8 \times 10^{-11}}{(1.0 \times 10^{-5})^2} = 0.18\,M$$

Thus, the $Mg(OH)_2$ dissolves until $[Mg^{2+}] = 0.18\,M$. It is apparent that $Mg(OH)_2$ is much more soluble in this solution.

Chemistry and Life

Ocean Acidification

Seawater is a weakly basic solution, with pH values typically between 8.0 and 8.3. This pH range is maintained through a carbonic acid buffer system similar to the one in blood (see Equation 17.10). Because the pH of seawater is higher than that of blood (7.35–7.45), the second dissociation of carbonic acid cannot be neglected and CO_3^{2-} becomes an important aqueous species.

The availability of carbonate ions plays an important role in shell formation for a number of marine organisms, including stony corals (▶ Figure 17.18). These organisms, which are referred to as marine *calcifiers* and play an important role in the food chains of nearly all oceanic ecosystems, depend on dissolved Ca^{2+} and CO_3^{2-} ions to form their shells and exoskeletons. The relatively low solubility-product constant of $CaCO_3$,

$$CaCO_3(s) \rightleftharpoons Ca^{2+}(aq) + CO_3^{2-}(aq) \quad K_{sp} = 4.5 \times 10^{-9}$$

and the fact that the ocean contains saturated concentrations of Ca^{2+} and CO_3^{2-} mean that $CaCO_3$ is usually quite stable once formed. In fact, calcium carbonate skeletons of creatures that died millions of years ago are not uncommon in the fossil record.

Just as in our bodies, the carbonic acid buffer system can be perturbed by removing or adding $CO_2(g)$. The concentration of dissolved CO_2 in the ocean is sensitive to changes in atmospheric CO_2 levels. As discussed in Chapter 18, the atmospheric CO_2 concentration has risen by approximately 30% over the past three centuries to the present level of 400 ppm. Human activity has played a prominent role in this increase. Scientists estimate that one-third to one-half of the CO_2 emissions resulting from human activity have been absorbed by Earth's oceans. While this absorption helps mitigate the greenhouse gas effects of CO_2, the extra CO_2 in the ocean produces carbonic acid, which lowers the pH. Because CO_3^{2-} is the conjugate base of the weak

acid HCO_3^-, the carbonate ion readily combines with the hydrogen ion:

$$CO_3^{2-}(aq) + H^+(aq) \longrightarrow HCO_3^-(aq)$$

This consumption of carbonate ion shifts the $CaCO_3$ dissolution equilibrium to the right, increasing the solubility of $CaCO_3$, which can lead to partial dissolution of calcium carbonate shells and exoskeletons. If the amount of atmospheric CO_2 continues to increase at the present rate, scientists estimate that seawater pH will fall to 7.9 sometime over the next 50 years. While this change might sound small, it has dramatic ramifications for oceanic ecosystems.

Related Exercises: 17.99

▲ **Figure 17.18 Marine calcifiers.** Many sea-dwelling organisms use $CaCO_3$ for their shells and exoskeletons. Examples include stony coral, crustaceans, some phytoplankton, and echinoderms, such as sea urchins and starfish.

If $[OH^-]$ were reduced further by making the solution even more acidic, the Mg^{2+} concentration would have to increase to maintain the equilibrium condition. Thus, a sample of $Mg(OH)_2(s)$ dissolves completely if sufficient acid is added, as we saw in Figure 4.9 (page 137).

The solubility of almost any ionic compound is affected if the solution is made sufficiently acidic or basic. The effects are noticeable, however, only when one (or both) ions in the compound are at least moderately acidic or basic. The metal hydroxides, such as $Mg(OH)_2$, are examples of compounds containing a strongly basic ion, the hydroxide ion.

In general, *the solubility of a compound containing a basic anion (that is, the anion of a weak acid) increases as the solution becomes more acidic.* As we have seen, the solubility of $Mg(OH)_2$ greatly increases as the acidity of the solution increases. The solubility of PbF_2 increases as the solution becomes more acidic, too, because F^- is a base (it is the conjugate base of the weak acid HF). As a result, the solubility equilibrium of PbF_2 is shifted to the right as the concentration of F^- is reduced by protonation to form HF. Thus, the solution process can be understood in terms of two consecutive reactions:

$$PbF_2(s) \rightleftharpoons Pb^{2+}(aq) + 2\,F^-(aq) \qquad [17.18]$$

$$F^-(aq) + H^+(aq) \rightleftharpoons HF(aq) \qquad [17.19]$$

The equation for the overall process is

$$PbF_2(s) + 2\,H^+(aq) \rightleftharpoons Pb^{2+}(aq) + 2\,HF(aq) \qquad [17.20]$$

The processes responsible for the increase in solubility of PbF_2 in acidic solution are illustrated in ▼ Figure 17.19(a).

Other salts that contain basic anions, such as CO_3^{2-}, PO_4^{3-}, CN^-, or S^{2-}, behave similarly. These examples illustrate a general rule: *The solubility of slightly soluble salts containing basic anions increases as $[H^+]$ increases (as pH is lowered).* The more basic the anion, the more the solubility is influenced by pH. The solubility of salts with anions of negligible basicity (the anions of strong acids), such as Cl^-, Br^-, I^-, and NO_3^-, is unaffected by pH changes, as shown in Figure 17.19(b).

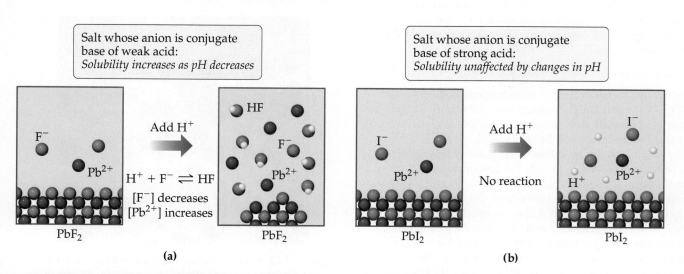

▲ Figure 17.19 **Response of two ionic compounds to addition of a strong acid.** (a) The solubility of PbF_2 increases upon addition of acid. (b) The solubility of PbI_2 is not affected by addition of acid. The water molecules and the anion of the strong acid have been omitted for clarity.

SAMPLE
EXERCISE 17.14 | Predicting the Effect of Acid on Solubility

Which of these substances are more soluble in acidic solution than in basic solution:
(a) $Ni(OH)_2(s)$, **(b)** $CaCO_3(s)$, **(c)** $BaF_2(s)$, **(d)** $AgCl(s)$?

SOLUTION

Analyze The problem lists four sparingly soluble salts, and we are asked to determine which are
more soluble at low pH than at high pH.

Plan We will identify ionic compounds that dissociate to produce a basic anion, as these are
especially soluble in acid solution.

Solve

(a) $Ni(OH)_2(s)$ is more soluble in acidic solution because of the basicity of OH^-; the H^+ reacts
with the OH^- ion, forming water:

$$Ni(OH)_2(s) \rightleftharpoons Ni^{2+}(aq) + 2\,OH^-(aq)$$
$$\underline{2\,OH^-(aq) + 2\,H^+(aq) \longrightarrow 2\,H_2O(l)}$$
$$\text{Overall:}\quad Ni(OH)_2(s) + 2\,H^+(aq) \rightleftharpoons Ni^{2+}(aq) + 2\,H_2O(l)$$

(b) Similarly, $CaCO_3(s)$ dissolves in acid solutions because CO_3^{2-} is a basic anion:

$$CaCO_3(s) \rightleftharpoons Ca^{2+}(aq) + CO_3^{2-}(aq)$$
$$CO_3^{2-}(aq) + 2\,H^+(aq) \rightleftharpoons H_2CO_3(aq)$$
$$\underline{H_2CO_3(aq) \rightleftharpoons CO_2(g) + H_2O(l)}$$
$$\text{Overall:}\quad CaCO_3(s) + 2\,H^+(aq) \rightleftharpoons Ca^{2+}(aq) + CO_2(g) + H_2O(l)$$

The reaction between CO_3^{2-} and H^+ occurs in steps, with HCO_3^- forming first and
H_2CO_3 forming in appreciable amounts only when $[H^+]$ is sufficiently high.

(c) The solubility of BaF_2 is enhanced by lowering the pH because F^- is a basic anion:

$$BaF_2(s) \rightleftharpoons Ba^{2+}(aq) + 2\,F^-(aq)$$
$$\underline{2\,F^-(aq) + 2\,H^+(aq) \rightleftharpoons 2\,HF(aq)}$$
$$\text{Overall:}\quad BaF_2(s) + 2\,H^+(aq) \rightleftharpoons Ba^{2+}(aq) + 2\,HF(aq)$$

(d) The solubility of $AgCl$ is unaffected by changes in pH because Cl^- is the anion of a strong
acid and therefore has negligible basicity.

Practice Exercise 1

Which of the following actions will increase the solubility of $AgBr$ in water? **(a)** increasing the
pH, **(b)** decreasing the pH, **(c)** adding $NaBr$, **(d)** adding $NaNO_3$, **(e)** none of the above.

Practice Exercise 2

Write the net ionic equation for the reaction between a strong acid and **(a)** CuS, **(b)** $Cu(N_3)_2$.

Chemistry and Life

Tooth Decay and Fluoridation

Tooth enamel consists mainly of the mineral hydroxyapatite,
$Ca_{10}(PO_4)_6(OH)_2$, the hardest substance in the body. Tooth cavities
form when acids dissolve tooth enamel:

$$Ca_{10}(PO_4)_6(OH)_2(s) + 8\,H^+(aq) \longrightarrow$$
$$10\,Ca^{2+}(aq) + 6\,HPO_4^{2-}(aq) + 2\,H_2O(l)$$

The Ca^{2+} and HPO_4^{2-} ions diffuse out of the enamel and are washed
away by saliva. The acids that attack the hydroxyapatite are formed by
the action of bacteria on sugars and other carbohydrates present in the
plaque adhering to the teeth.

Fluoride ion, which is added to municipal water systems and
toothpastes, can react with hydroxyapatite to form fluoroapatite,
$Ca_{10}(PO_4)_6F_2$. This mineral, in which F^- has replaced OH^-, is much
more resistant to attack by acids because the fluoride ion is a much
weaker Brønsted–Lowry base than the hydroxide ion.

The usual concentration of F^- in municipal water systems is
1 mg/L (1 ppm). The compound added may be NaF or Na_2SiF_6. The
silicon-fluorine anion reacts with water to release fluoride ions:

$$SiF_6^{2-}(aq) + 2\,H_2O(l) \longrightarrow 6\,F^-(aq) + 4\,H^+(aq) + SiO_2(s)$$

About 80% of all toothpastes now sold in the United States contain fluoride compounds, usually at the level of 0.1% fluoride by mass.
The most common compounds in toothpastes are sodium fluoride
(NaF), sodium monofluorophosphate (Na_2PO_3F), and stannous fluoride (SnF_2).

Related Exercises: 17.100, 17.118

Formation of Complex Ions

A characteristic property of metal ions is their ability to act as Lewis acids toward water molecules, which act as Lewis bases. ∞ (Section 16.11) Lewis bases other than water can also interact with metal ions, particularly transition-metal ions. Such interactions can dramatically affect the solubility of a metal salt. For example, AgCl ($K_{sp} = 1.8 \times 10^{-10}$) dissolves in the presence of aqueous ammonia because Ag^+ interacts with the Lewis base NH_3, as shown in ▼ Figure 17.20. This process can be viewed as the sum of two reactions:

$$AgCl(s) \rightleftharpoons Ag^+(aq) + Cl^-(aq) \qquad [17.21]$$

$$Ag^+(aq) + 2\,NH_3(aq) \rightleftharpoons Ag(NH_3)_2^+(aq) \qquad [17.22]$$

$$\text{Overall:} \quad AgCl(s) + 2\,NH_3(aq) \rightleftharpoons Ag(NH_3)_2^+(aq) + Cl^-(aq) \quad [17.23]$$

The presence of NH_3 drives the reaction, the dissolution of AgCl, to the right as $Ag^+(aq)$ is consumed to form $Ag(NH_3)_2^+$, which is a very soluble species.

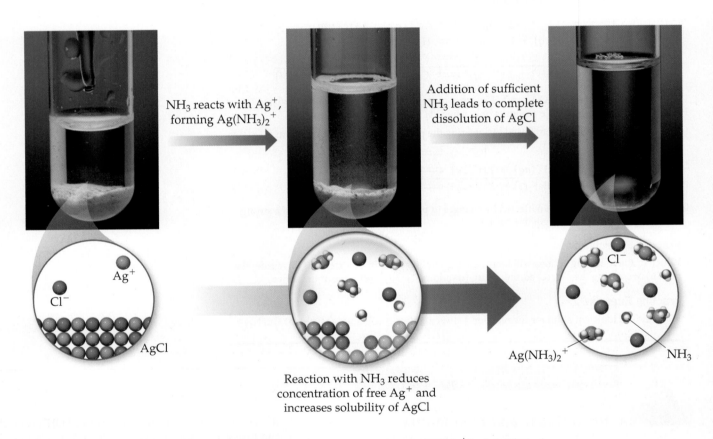

NH_3 reacts with Ag^+, forming $Ag(NH_3)_2^+$

Addition of sufficient NH_3 leads to complete dissolution of AgCl

Reaction with NH_3 reduces concentration of free Ag^+ and increases solubility of AgCl

$$AgCl(s) + 2\,NH_3(aq) \rightleftharpoons Ag(NH_3)_2^+(aq) + Cl^-(aq)$$

▲ Figure 17.20 **Concentrated NH₃(aq) dissolves AgCl(s), which otherwise has very low solubility in water.**

For a Lewis base such as NH_3 to increase the solubility of a metal salt, the base must be able to interact more strongly with the metal ion than water does. In other words, the NH_3 must displace solvating H_2O molecules ∞ (Sections 13.1 and 16.11) in order to form $[Ag(NH_3)_2]^+$:

$$Ag^+(aq) + 2\,NH_3(aq) \rightleftharpoons Ag(NH_3)_2^+(aq) \qquad [17.24]$$

Table 17.1 Formation Constants for Some Metal Complex Ions in Water at 25 °C

Complex Ion	K_f	Equilibrium Equation
$Ag(NH_3)_2^+$	1.7×10^7	$Ag^+(aq) + 2\,NH_3(aq) \rightleftharpoons Ag(NH_3)_2^+(aq)$
$Ag(CN)_2^-$	1×10^{21}	$Ag^+(aq) + 2\,CN^-(aq) \rightleftharpoons Ag(CN)_2^-(aq)$
$Ag(S_2O_3)_2^{3-}$	2.9×10^{13}	$Ag^+(aq) + 2\,S_2O_3^{2-}(aq) \rightleftharpoons Ag(S_2O_3)_2^{3-}(aq)$
$CdBr_4^{2-}$	5×10^3	$Cd^{2+}(aq) + 4\,Br^-(aq) \rightleftharpoons CdBr_4^{2-}(aq)$
$Cr(OH)_4^-$	8×10^{29}	$Cr^{3+}(aq) + 4\,OH^-(aq) \rightleftharpoons Cr(OH)_4^-(aq)$
$Co(SCN)_4^{2-}$	1×10^3	$Co^{2+}(aq) + 4\,SCN^-(aq) \rightleftharpoons Co(SCN)_4^{2-}(aq)$
$Cu(NH_3)_4^{2+}$	5×10^{12}	$Cu^{2+}(aq) + 4\,NH_3(aq) \rightleftharpoons Cu(NH_3)_4^{2+}(aq)$
$Cu(CN)_4^{2-}$	1×10^{25}	$Cu^{2+}(aq) + 4\,CN^-(aq) \rightleftharpoons Cu(CN)_4^{2+}(aq)$
$Ni(NH_3)_6^{2+}$	1.2×10^9	$Ni^{2+}(aq) + 6\,NH_3(aq) \rightleftharpoons Ni(NH_3)_6^{2+}(aq)$
$Fe(CN)_6^{4-}$	1×10^{35}	$Fe^{2+}(aq) + 6\,CN^-(aq) \rightleftharpoons Fe(CN)_3^{4-}(aq)$
$Fe(CN)_6^{3-}$	1×10^{42}	$Fe^{3+}(aq) + 6\,CN^-(aq) \rightleftharpoons Fe(CN)_6^{3-}(aq)$

An assembly of a metal ion and the Lewis bases bonded to it, such as $Ag(NH_3)_2^+$, is called a **complex ion.** Complex ions are very soluble in water. The stability of a complex ion in aqueous solution can be judged by the size of the equilibrium constant for its formation from the hydrated metal ion. For example, the equilibrium constant for Equation 17.24 is

$$K_f = \frac{[Ag(NH_3)_2^+]}{[Ag^+][NH_3]^2} = 1.7 \times 10^7 \qquad [17.25]$$

Note that the equilibrium constant for this kind of reaction is called a **formation constant**, K_f. The formation constants for several complex ions are listed in ▲ Table 17.1.

SAMPLE EXERCISE 17.15 Evaluating an Equilibrium Involving a Complex Ion

Calculate the concentration of Ag^+ present in solution at equilibrium when concentrated ammonia is added to a 0.010 M solution of $AgNO_3$ to give an equilibrium concentration of $[NH_3] = 0.20$ M. Neglect the small volume change that occurs when NH_3 is added.

SOLUTION

Analyze Addition of $NH_3(aq)$ to $Ag^+(aq)$ forms $Ag(NH_3)_2^+$, as shown in Equation 17.22. We are asked to determine what concentration of $Ag^+(aq)$ remains uncombined when the NH_3 concentration is brought to 0.20 M in a solution originally 0.010 M in $AgNO_3$.

Plan We assume that the $AgNO_3$ is completely dissociated, giving 0.010 M Ag^+. Because K_f for the formation of $Ag(NH_3)_2^+$ is quite large, we assume that essentially all the Ag^+ is converted to $Ag(NH_3)_2^+$ and approach the problem as though we are concerned with the dissociation of $Ag(NH_3)_2^+$ rather than its formation. To facilitate this approach, we need to reverse Equation 17.22 and make the corresponding change to the equilibrium constant:

$$Ag(NH_3)_2^+(aq) \rightleftharpoons Ag^+(aq) + 2\,NH_3(aq)$$
$$\frac{1}{K_f} = \frac{1}{1.7 \times 10^7} = 5.9 \times 10^{-8}$$

Solve If $[Ag^+]$ is 0.010 M initially, $[Ag(NH_3)_2^+]$ will be 0.010 M following addition of the NH_3. We construct a table to solve this equilibrium problem. Note that the NH_3 concentration given in the problem is an equilibrium concentration rather than an initial concentration.

	$Ag(NH_3)_2^+(aq)$ $\rightleftharpoons$	$Ag^+(aq)$ +	$2\,NH_3(aq)$
Initial (M)	0.010	0	–
Change (M)	$-x$	$+x$	–
Equilibrium (M)	$(0.010 - x)$	x	0.20

Because $[Ag^+]$ is very small, we can assume x is small compared to 0.010. Substituting these values into the equilibrium-constant expression for the dissociation of $Ag(NH_3)_2^+$, we obtain

$$\frac{[Ag^+][NH_3]^2}{[Ag(NH_3)_2^+]} = \frac{(x)(0.20)^2}{0.010} = 5.9 \times 10^{-8}$$

$$x = 1.5 \times 10^{-8}\,M = [Ag^+]$$

Formation of the $Ag(NH_3)_2^+$ complex drastically reduces the concentration of free Ag^+ ion in solution.

Practice Exercise 1

You have an aqueous solution of chromium(III) nitrate that you titrate with an aqueous solution of sodium hydroxide. After a certain amount of titrant has been added, you observe a precipitate forming. You add more sodium hydroxide solution and the precipitate dissolves, leaving a solution again. What has happened? **(a)** The precipitate was sodium hydroxide, which redissolved in the larger volume. **(b)** The precipitate was chromium hydroxide, which dissolved once more solution was added, forming $Cr^{3+}(aq)$. **(c)** The precipitate was chromium hydroxide, which then reacted with more hydroxide to produce a soluble complex ion, $Cr(OH)_4^-$. **(d)** The precipitate was sodium nitrate, which reacted with more nitrate to produce the soluble complex ion $Na(NO_3)_2^{2-}$.

Practice Exercise 2

Calculate $[Cr^{3+}]$ in equilibrium with $Cr(OH)_4^-$ when 0.010 mol of $Cr(NO_3)_3$ is dissolved in 1 L of solution buffered at pH 10.0.

The general rule is that the solubility of metal salts increases in the presence of suitable Lewis bases, such as NH_3, CN^-, or OH^-, provided the metal forms a complex with the base. The ability of metal ions to form complexes is an extremely important aspect of their chemistry.

Amphoterism

Some metal oxides and hydroxides that are relatively insoluble in water dissolve in strongly acidic and strongly basic solutions. These substances, called **amphoteric oxides** and **amphoteric hydroxides,*** are soluble in strong acids and bases because they themselves are capable of behaving as either an acid or base. Examples of amphoteric substances include the oxides and hydroxides of Al^{3+}, Cr^{3+}, Zn^{2+}, and Sn^{2+}.

Like other metal oxides and hydroxides, amphoteric species dissolve in acidic solutions because their anions, O^{2-} or OH^-, react with acids. What makes amphoteric oxides and hydroxides special, though, is that they also dissolve in strongly basic solutions. This behavior results from the formation of complex anions containing several (typically four) hydroxides bound to the metal ion (▼ **Figure 17.21**):

$$Al(OH)_3(s) + OH^-(aq) \rightleftharpoons Al(OH)_4^-(aq) \qquad [17.26]$$

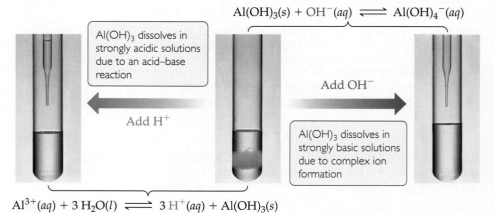

$$Al^{3+}(aq) + 3\,H_2O(l) \rightleftharpoons 3\,H^+(aq) + Al(OH)_3(s)$$

▲ **Figure 17.21 Amphoterism.** Some metal oxides and hydroxides, such as $Al(OH)_3$, are amphoteric, which means they dissolve in both strongly acidic and strongly basic solutions.

---

*Notice that the term *amphoteric* is applied to the behavior of insoluble oxides and hydroxides that dissolve in acidic or basic solutions. The similar term *amphiprotic* (Section 16.2) relates more generally to any molecule or ion that can either gain or lose a proton.

The extent to which an insoluble metal hydroxide reacts with either acid or base varies with the particular metal ion involved. Many metal hydroxides—such as $Ca(OH)_2$, $Fe(OH)_2$, and $Fe(OH)_3$—are capable of dissolving in acidic solution but do not react with excess base. These hydroxides are not amphoteric.

The purification of aluminum ore in the manufacture of aluminum metal provides an interesting application of amphoterism. As we have seen, $Al(OH)_3$ is amphoteric, whereas $Fe(OH)_3$ is not. Aluminum occurs in large quantities as the ore *bauxite*, which is essentially hydrated Al_2O_3 contaminated with Fe_2O_3. When bauxite is added to a strongly basic solution, the Al_2O_3 dissolves because the aluminum forms complex ions, such as $Al(OH)_4^-$. The Fe_2O_3 impurity, however, is not amphoteric and remains as a solid. The solution is filtered, getting rid of the iron impurity. Aluminum hydroxide is then precipitated by addition of acid. The purified hydroxide receives further treatment and eventually yields aluminum metal.

 Give It Some Thought

What is the difference between an amphoteric substance and an amphiprotic substance?

17.6 | Precipitation and Separation of Ions

Equilibrium can be achieved starting with the substances on either side of a chemical equation. For example, the equilibrium that exists between $BaSO_4(s)$, $Ba^{2+}(aq)$, and $SO_4^{2-}(aq)$ (Equation 17.15), can be achieved either by starting with $BaSO_4(s)$ or by starting with solutions containing Ba^{2+} and SO_4^{2-}. If we mix, say, a $BaCl_2$ aqueous solution with a Na_2SO_4 aqueous solution, $BaSO_4$ might precipitate out. How can we predict whether a precipitate will form under various conditions?

Recall that we used the reaction quotient Q in Section 15.6 to determine the direction in which a reaction must proceed to reach equilibrium. The form of Q is the same as the equilibrium-constant expression for a reaction, but instead of only equilibrium concentrations, you can use whatever concentrations are being considered. The direction in which a reaction proceeds to reach equilibrium depends on the relationship between Q and K for the reaction. If $Q < K$, the product concentrations are too low and reactant concentrations are too high relative to the equilibrium concentrations, and so the reaction will proceed to the right (toward products) to achieve equilibrium. If $Q > K$, product concentrations are too high and reactant concentrations are too low, and so the reaction will proceed to the left to achieve equilibrium. If $Q = K$, the reaction is at equilibrium.

For solubility-product equilibria, the relationship between Q and K_{sp} is exactly like that for other equilibria. For K_{sp} reactions, products are always the soluble ions, and the reactant is always the solid.

Therefore, for solubility equilibria,

- If $Q = K_{sp}$, the system is at equilibrium, which means the solution is saturated; this is the highest concentration the solution can have without precipitating.

- If $Q < K_{sp}$, the reaction will proceed to the right, towards the soluble ions; no precipitate will form.

- If $Q > K_{sp}$, the reaction will proceed to the left, towards the solid; precipitate will form.

For the case of the barium sulfate solution, then we would calculate $Q = [Ba^{2+}][SO_4^{2-}]$, and compare this quantity to the K_{sp} for barium sulfate.

SAMPLE EXERCISE 17.16 | Predicting Whether a Precipitate Forms

Does a precipitate form when 0.10 L of $8.0 \times 10^{-3}\,M\,Pb(NO_3)_2$ is added to 0.40 L of $5.0 \times 10^{-3}\,M\,Na_2SO_4$?

SOLUTION

Analyze The problem asks us to determine whether a precipitate forms when two salt solutions are combined.

Plan We should determine the concentrations of all ions just after the solutions are mixed and compare the value of Q with K_{sp} for any potentially insoluble product. The possible metathesis products are $PbSO_4$ and $NaNO_3$. Like all sodium salts $NaNO_3$ is soluble, but $PbSO_4$ has a K_{sp} of 6.3×10^{-7} (Appendix D) and will precipitate if the Pb^{2+} and SO_4^{2-} concentrations are high enough for Q to exceed K_{sp}.

Solve When the two solutions are mixed, the volume is $0.10\,L + 0.40\,L = 0.50\,L$. The number of moles of Pb^{2+} in 0.10 L of $8.0 \times 10^{-3}\,M\,Pb(NO_3)_2$ is

$$(0.10\,L)\left(\frac{8.0 \times 10^{-3}\,mol}{L}\right) = 8.0 \times 10^{-4}\,mol$$

The concentration of Pb^{2+} in the 0.50-L mixture is therefore

$$[Pb^{2+}] = \frac{8.0 \times 10^{-4}\,mol}{0.50\,L} = 1.6 \times 10^{-3}\,M$$

The number of moles of SO_4^{2-} in 0.40 L of $5.0 \times 10^{-3}\,M\,Na_2SO_4$ is

$$(0.40\,L)\left(\frac{5.0 \times 10^{-3}\,mol}{L}\right) = 2.0 \times 10^{-3}\,mol$$

Therefore

$$[SO_4{}^{2-}] = \frac{2.0 \times 10^{-3}\,mol}{0.50\,L} = 4.0 \times 10^{-3}\,M$$

and

$$Q = [Pb^{2+}][SO_4{}^{2-}] = (1.6 \times 10^{-3})(4.0 \times 10^{-3}) = 6.4 \times 10^{-6}$$

Because $Q > K_{sp}$, $PbSO_4$ precipitates.

Practice Exercise 1

An insoluble salt MA has a K_{sp} of 1.0×10^{-16}. Two solutions, MNO_3 and NaA are mixed, to yield a final solution that is $1.0 \times 10^{-8}\,M$ in $M^+(aq)$ and $1.00 \times 10^{-7}\,M$ in $A^-(aq)$. Will a precipitate form?

(a) Yes.   (b) No.

Practice Exercise 2

Does a precipitate form when 0.050 L of $2.0 \times 10^{-2}\,M$ NaF is mixed with 0.010 L of $1.0 \times 10^{-2}\,M\,Ca(NO_3)_2$?

Selective Precipitation of Ions

Ions can be separated from each other based on the solubilities of their salts. Consider a solution containing both Ag^+ and Cu^{2+}. If HCl is added to the solution, AgCl $(K_{sp} = 1.8 \times 10^{-10})$ precipitates, while Cu^{2+} remains in solution because $CuCl_2$ is soluble. Separation of ions in an aqueous solution by using a reagent that forms a precipitate with one or more (but not all) of the ions is called *selective precipitation*.

SAMPLE EXERCISE 17.17 | Selective Precipitation

A solution contains $1.0 \times 10^{-2}\,M\,Ag^+$ and $2.0 \times 10^{-2}\,M\,Pb^{2+}$. When Cl^- is added, both AgCl $(K_{sp} = 1.8 \times 10^{-10})$ and $PbCl_2(K_{sp} = 1.7 \times 10^{-5})$ can precipitate. What concentration of Cl^- is necessary to begin the precipitation of each salt? Which salt precipitates first?

SOLUTION

Analyze We are asked to determine the concentration of Cl^- necessary to begin the precipitation from a solution containing Ag^+ and Pb^{2+} ions, and to predict which metal chloride will begin to precipitate first.

Plan We are given K_{sp} values for the two precipitates. Using these and the metal ion concentrations, we can calculate what Cl^- concentration is necessary to precipitate each salt. The salt requiring the lower Cl^- ion concentration precipitates first.

Solve For AgCl we have $K_{sp} = [Ag^+][Cl^-] = 1.8 \times 10^{-10}$.

Because $[Ag^+] = 1.0 \times 10^{-2}\, M$, the greatest concentration of Cl^- that can be present without causing precipitation of AgCl can be calculated from the K_{sp} expression:

$$K_{sp} = (1.0 \times 10^{-2})[Cl^-] = 1.8 \times 10^{-10}$$

$$[Cl^-] = \frac{1.8 \times 10^{-10}}{1.0 \times 10^{-2}} = 1.8 \times 10^{-8}\, M$$

Any Cl^- in excess of this very small concentration will cause AgCl to precipitate from solution. Proceeding similarly for $PbCl_2$, we have

$$K_{sp} = [Pb^{2+}][Cl^-]^2 = 1.7 \times 10^{-5}$$

$$(2.0 \times 10^{-2})[Cl^-]^2 = 1.7 \times 10^{-5}$$

$$[Cl^-]^2 = \frac{1.7 \times 10^{-5}}{2.0 \times 10^{-2}} = 8.5 \times 10^{-4}$$

$$[Cl^-] = \sqrt{8.5 \times 10^{-4}} = 2.9 \times 10^{-2}\, M$$

Thus, a concentration of Cl^- in excess of $2.9 \times 10^{-2}\, M$ causes $PbCl_2$ to precipitate.

Comparing the Cl^- concentration required to precipitate each salt, we see that as Cl^- is added, AgCl precipitates first because it requires a much smaller concentration of Cl^-. Thus, Ag^+ can be separated from Pb^{2+} by slowly adding Cl^- so that the chloride ion concentration remains between $1.8 \times 10^{-8}\, M$ and $2.9 \times 10^{-2}\, M$.

Comment Precipitation of AgCl will keep the Cl^- concentration low until the number of moles of Cl^- added exceeds the number of moles of Ag^+ in the solution. Once past this point, $[Cl^-]$ rises sharply and $PbCl_2$ will soon begin to precipitate.

Practice Exercise 1

Under what conditions does an ionic compound precipitate from a solution of the constituent ions?
(a) always, (b) when $Q = K_{sp}$, (c) when Q exceeds K_{sp}, (d) when Q is less than K_{sp}, (e) never, if it is very soluble.

Practice Exercise 2

A solution consists of $0.050\, M$ Mg^{2+} and Cu^{2+}. Which ion precipitates first as OH^- is added? What concentration of OH^- is necessary to begin the precipitation of each cation? [$K_{sp} = 1.8 \times 10^{-11}$ for $Mg(OH)_2$, and $K_{sp} = 4.8 \times 10^{-20}$ for $Cu(OH)_2$.]

Sulfide ion is often used to separate metal ions because the solubilities of sulfide salts span a wide range and depend greatly on solution pH. For example, Cu^{2+} and Zn^{2+} can be separated by bubbling H_2S gas through an acidified solution containing these two cations. Because CuS ($K_{sp} = 6 \times 10^{-37}$) is less soluble than ZnS ($K_{sp} = 2 \times 10^{-25}$), CuS precipitates from an acidified solution pH $\approx$ 1 while ZnS does not (▼ **Figure 17.22**):

$$Cu^{2+}(aq) + H_2S(aq) \rightleftharpoons CuS(s) + 2\,H^+(aq) \qquad [17.27]$$

▲ **GO FIGURE**

What would happen if the pH were raised to 8 first and then H_2S were added?

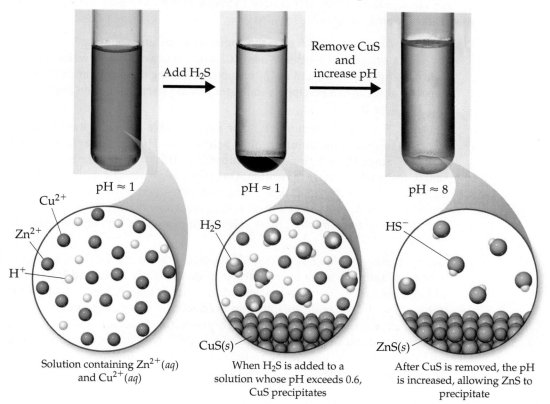

▲ **Figure 17.22 Selective precipitation.** In this example, Cu^{2+} ions are separated from Zn^{2+} ions.

The CuS can be separated from the Zn^{2+} solution by filtration. The separated CuS can then be dissolved by raising the concentration of H^+ even further, shifting the equilibrium concentrations of the compounds in Equation 17.27 to the left.

17.7 | Qualitative Analysis for Metallic Elements

In this final section, we look at how solubility equilibria and complex-ion formation can be used to detect the presence of particular metal ions in solution. Before the development of modern analytical instrumentation, it was necessary to analyze mixtures of metals in a sample by what were called *wet chemical methods*. For example, an ore sample that might contain several metallic elements was dissolved in a concentrated acid solution that was then tested in a systematic way for the presence of various metal ions.

Qualitative analysis determines only the presence or absence of a particular metal ion relative to some threshold, whereas **quantitative analysis** determines how much of a given substance is present. Even though wet methods of qualitative analysis have become less important in the chemical industry, they are frequently used in general chemistry laboratory programs to illustrate equilibria, to teach the properties of common metal ions in solution, and to develop laboratory skills. Typically, such analyses proceed in three stages: (1) The ions are separated into broad groups on the basis of solubility properties. (2) The ions in each group are separated by selectively dissolving members in the group. (3) The ions are identified by means of specific tests.

A scheme in general use divides the common cations into five groups (▶ Figure 17.23). The order in which reagents are added is important in this scheme. The most selective separations—those that involve the smallest number of ions—are carried out first. The reactions used must proceed so far toward completion that any concentration of cations remaining in the solution is too small to interfere with subsequent tests.

Let's look at each of these five groups of cations, briefly examining the logic used in this qualitative analysis scheme.

Group 1. *Insoluble chlorides:* Of the common metal ions, only Ag^+, Hg_2^{2+}, and Pb^{2+} form insoluble chlorides. When HCl is added to a mixture of cations, therefore, only AgCl, Hg_2Cl_2, and $PbCl_2$ precipitate, leaving the other cations in solution. The absence of a precipitate indicates that the starting solution contains no Ag^+, Hg_2^{2+}, or Pb^{2+}.

Group 2. *Acid-insoluble sulfides:* After any insoluble chlorides have been removed, the remaining solution, now acidic from HCl treatment, is treated with H_2S. Since H_2S is a weak acid compared to HCl, its role here is to act as a source for small amounts of sulfide. Only the most insoluble metal sulfides—CuS, Bi_2S_3, CdS, PbS, HgS, As_2S_3, Sb_2S_3, and SnS_2—precipitate. (Note the very small values of K_{sp} for some of these sulfides in Appendix D.) Those metal ions whose sulfides are somewhat more soluble—for example, ZnS or NiS—remain in solution.

Group 3. *Base-insoluble sulfides and hydroxides:* After the solution is filtered to remove any acid-insoluble sulfides, it is made slightly basic, and $(NH_4)_2S$ is added. In basic solutions the concentration of S^{2-} is higher than in acidic solutions. Under these conditions, the ion products for many of the more soluble sulfides exceed their K_{sp} values and thus precipitation occurs. The metal ions precipitated at this stage are Al^{3+}, Cr^{3+}, Fe^{3+}, Zn^{2+}, Ni^{2+}, Co^{2+}, and Mn^{2+}. (The Al^{3+}, Fe^{3+}, and Cr^{3+} ions do not form insoluble sulfides; instead they precipitate as insoluble hydroxides, as Figure 17.23 shows.)

GO FIGURE

If a solution contained a mixture of Cu^{2+} and Zn^{2+} ions, would this separation scheme work? After which step would the first precipitate be observed?

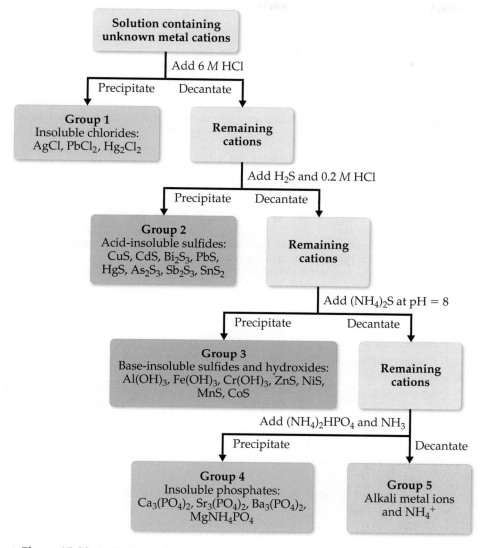

▲ **Figure 17.23 Qualitative analysis.** A flowchart showing a common scheme for identifying cations.

Group 4. *Insoluble phosphates:* At this point, the solution contains only metal ions from groups 1A and 2A of the periodic table. Adding $(NH_4)_2HPO_4$ to a basic solution precipitates the group 2A elements Mg^{2+}, Ca^{2+}, Sr^{2+}, and Ba^{2+} because these metals form insoluble phosphates.

Group 5. *The alkali metal ions and NH_4^+:* The ions that remain after removing the insoluble phosphates are tested for individually. A flame test can be used to determine the presence of K^+, for example, because the flame turns a characteristic violet color if K^+ is present (Figure 7.22).

 Give It Some Thought

If a precipitate forms when HCl is added to an aqueous solution, what conclusions can you draw about the contents of the solution?

SAMPLE INTEGRATIVE EXERCISE   Putting Concepts Together

A sample of 1.25 L of HCl gas at 21 °C and 0.950 atm is bubbled through 0.500 L of 0.150 M NH_3 solution. Calculate the pH of the resulting solution assuming that all the HCl dissolves and that the volume of the solution remains 0.500 L.

SOLUTION

The number of moles of HCl gas is calculated from the ideal-gas law,

$$n = \frac{PV}{RT} = \frac{(0.950 \text{ atm})(1.25 \text{ L})}{(0.0821 \text{ L-atm/mol-K})(294 \text{ K})} = 0.0492 \text{ mol HCl}$$

The number of moles of NH_3 in the solution is given by the product of the volume of the solution and its concentration,

$$\text{Moles } NH_3 = (0.500 \text{ L})(0.150 \text{ mol } NH_3/\text{L}) = 0.0750 \text{ mol } NH_3$$

The acid HCl and base NH_3 react, transferring a proton from HCl to NH_3, producing NH_4^+ and Cl^- ions,

$$HCl(g) + NH_3(aq) \longrightarrow NH_4^+(aq) + Cl^-(aq)$$

To determine the pH of the solution, we first calculate the amount of each reactant and each product present at the completion of the reaction. Because you can assume this neutralization reaction proceeds as far toward the product side as possible, this is a limiting reactant problem.

	$HCl(g)$ +	$NH_3(aq)$ $\longrightarrow$	$NH_4^+(aq)$ +	$Cl^-(aq)$
Before reaction (mol)	0.0492	0.0750	0	0
Change (limiting reactant) (mol)	−0.0492	−0.0492	+0.0492	+0.0492
After reaction (mol)	0	0.0258	0.0492	0.0492

Thus, the reaction produces a solution containing a mixture of NH_3, NH_4^+, and Cl^-. The NH_3 is a weak base $(K_b = 1.8 \times 10^{-5})$, NH_4^+ is its conjugate acid, and Cl^- is neither acidic nor basic. Consequently, the pH depends on $[NH_3]$ and $[NH_4^+]$,

$$[NH_3] = \frac{0.0258 \text{ mol } NH_3}{0.500 \text{ L soln}} = 0.0516 \, M$$

$$[NH_4^+] = \frac{0.0492 \text{ mol } NH_4^+}{0.500 \text{ L soln}} = 0.0984 \, M$$

We can calculate the pH using either K_b for NH_3 or K_a for NH_4^+. Using the K_b expression, we have

	$NH_3(aq)$ +	$H_2O(l)$ $\rightleftharpoons$	$NH_4^+(aq)$ +	$OH^-(aq)$
Initial (M)	0.0516	—	0.0984	0
Change (M)	−x	—	+x	+x
Equilibrium (M)	(0.0516 − x)	—	(0.0984 + x)	x

$$K_b = \frac{[NH_4^+][OH^-]}{[NH_3]} = \frac{(0.0984 + x)(x)}{(0.0516 - x)} \cong \frac{(0.0984)x}{0.0516} = 1.8 \times 10^{-5}$$

$$x = [OH^-] = \frac{(0.0516)(1.8 \times 10^{-5})}{0.0984} = 9.4 \times 10^{-6} \, M$$

Hence, pOH $= -\log(9.4 \times 10^{-6}) = 5.03$ and pH $= 14.00 - \text{pOH} = 14.00 - 5.03 = 8.97$.

Chapter Summary and Key Terms

THE COMMON ION EFFECT (SECTION 17.1) In this chapter, we have considered several types of important equilibria that occur in aqueous solution. Our primary emphasis has been on acid–base equilibria in solutions containing two or more solutes and on solubility equilibria. The dissociation of a weak acid or weak base is repressed by the presence of a strong electrolyte that provides an ion common to the equilibrium (the **common-ion effect**).

BUFFERS (SECTION 17.2) A particularly important type of acid–base mixture is that of a weak conjugate acid–base pair that functions as a **buffered solution (buffer)**. Addition of small amounts of a strong acid or a strong base to a buffered solution causes only small changes in pH because the buffer reacts with the added acid or base. (Strong acid–strong base, strong acid–weak base, and weak acid–strong base reactions proceed essentially to completion.) Buffered solutions are usually prepared from a weak acid and a salt of that acid or from a weak base and a salt of that base. Two important characteristics of a buffered solution are its **buffer capacity** and its **pH range**. The optimal pH of a buffer is equal to pK_a (or pK_b) of the acid (or base) used to prepare the buffer. The relationship between pH, pK_a, and the concentrations of an acid and its conjugate base can be expressed by the **Henderson–Hasselbalch equation**. It is important to realize that the Henderson–Hasselbalch equation is an approximation, and more detailed calculations may need to be performed to obtain equilibrium concentrations.

ACID–BASE TITRATIONS (SECTION 17.3) The plot of the pH of an acid (or base) as a function of the volume of added base (or acid) is called a **pH titration curve**. The titration curve of a strong acid–strong base titration exhibits a large change in pH in the immediate vicinity of the equivalence point; at the equivalence point for such a titration pH = 7. For strong acid–weak base or weak acid–strong base titrations, the pH change in the vicinity of the equivalence point is not as large as the strong acid–strong base titration, nor will the pH equal 7 at the equivalence point in these cases. Instead, what determines the pH at the equivalence point is the conjugate base or acid salt solution that results from the neutralization reaction. For this reason, it is important to choose an indicator whose color change is near the pH at the equivalence point for titrations involving either weak acids or weak bases. It is possible to calculate the pH at any point of the titration curve by first considering the effects of the acid–base reaction on solution concentrations and then examining equilibria involving the remaining solute species.

SOLUBILITY EQUILIBRIA (SECTION 17.4) The equilibrium between a solid compound and its ions in solution provides an example of heterogeneous equilibrium. The **solubility-product constant** (or simply the **solubility product**), K_{sp}, is an equilibrium constant that expresses quantitatively the extent to which the compound dissolves. The K_{sp} can be used to calculate the solubility of an ionic compound, and the solubility can be used to calculate K_{sp}.

FACTORS THAT AFFECT SOLUBILITY (SECTION 17.5) Several experimental factors, including temperature, affect the solubilities of ionic compounds in water. The solubility of a slightly soluble ionic compound is decreased by the presence of a second solute that furnishes a common ion (the common-ion effect). The solubility of compounds containing basic anions increases as the solution is made more acidic (as pH decreases). Salts with anions of negligible basicity (the anions of strong acids) are unaffected by pH changes.

The solubility of metal salts is also affected by the presence of certain Lewis bases that react with metal ions to form stable **complex ions**. Complex-ion formation in aqueous solution involves the displacement by Lewis bases (such as NH_3 and CN^-) of water molecules attached to the metal ion. The extent to which such complex formation occurs is expressed quantitatively by the **formation constant** for the complex ion. **Amphoteric oxides** and **hydroxides** are those that are only slightly soluble in water but dissolve on addition of either acid or base.

PRECIPITATION AND SEPARATION OF IONS (SECTION 17.6) Comparison of the reaction quotient, Q, with the value of K_{sp} can be used to judge whether a precipitate will form when solutions are mixed or whether a slightly soluble salt will dissolve under various conditions. Precipitates form when $Q > K_{sp}$. If two salts have sufficiently different solubilities, selective precipitation can be used to precipitate one ion while leaving the other in solution, effectively separating the two ions.

QUALITATIVE ANALYSIS FOR METALLIC ELEMENTS (SECTION 17.7) Metallic elements vary a great deal in the solubilities of their salts, in their acid–base behavior, and in their tendencies to form complex ions. These differences can be used to separate and detect the presence of metal ions in mixtures. **Qualitative analysis** determines the presence or absence of species in a sample, whereas **quantitative analysis** determines how much of each species is present. The qualitative analysis of metal ions in solution can be carried out by separating the ions into groups on the basis of precipitation reactions and then analyzing each group for individual metal ions.

Learning Outcomes   After studying this chapter, you should be able to:

- Describe the common-ion effect. (Section 17.1)

- Explain how a buffer functions. (Section 17.2)

- Calculate the pH of a buffered solution. (Section 17.2)

- Calculate the pH of a buffer after the addition of small amounts of a strong acid or a strong base. (Section 17.2)

- Calculate the appropriate quantities of compounds to make a buffer at a given pH. (Section 17.2)

- Calculate the pH at any point in a strong acid–strong base titration. (Section 17.3)

- Calculate the pH at any point in a weak acid–strong base or weak base–strong acid titration. (Section 17.3)

- Describe the differences between the titration curves for a strong acid–strong base titration and those when either the acid or base is weak. (Section 17.3)

- Estimate the pK_a for monoprotic or polyprotic acids from titration curves. (Section 17.3)

- Given either K_{sp}, molar solubility or mass solubility for a substance, calculate the other two quantities. (Section 17.4)

- Calculate molar solubility in the presence of a common ion. (Section 17.5)

- Predict the effect of pH on solubility. (Section 17.5)

- Predict whether a precipitate will form when solutions are mixed, by comparing Q and K_{sp}. (Section 17.6)

- Calculate the ion concentrations required to begin precipitation. (Section 17.6)

- Explain the effect of complex-ion formation on solubility. (Section 17.6)

- Explain the logic of identification of metal ions in aqueous solution by a series of reactions. (Section 17.7)

Key Equations

- $pH = pK_a + \log\dfrac{[base]}{[acid]}$ [17.9]   The Henderson–Hasselbalch equation, used to estimate the pH of a buffer from the concentrations of a conjugate acid–base pair

Exercises

Visualizing Concepts

17.1 The following boxes represent aqueous solutions containing a weak acid, HA and its conjugate base, A⁻. Water molecules, hydronium ions, and cations are not shown. Which solution has the highest pH? Explain. [Section 17.1]

 = HA   = A⁻

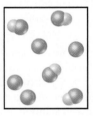 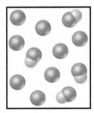 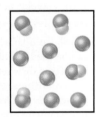

17.2 The beaker on the right contains 0.1 M acetic acid solution with methyl orange as an indicator. The beaker on the left contains a mixture of 0.1 M acetic acid and 0.1 M sodium acetate with methyl orange. **(a)** Using Figure 16.7, which solution has a higher pH? **(b)** Which solution is better able to maintain its pH when small amounts of NaOH are added? Explain. [Sections 17.1 and 17.2]

17.3 A buffer contains a weak acid, HA, and its conjugate base. The weak acid has a pK_a of 4.5, and the buffer has a pH of 4.3. Without doing a calculation, state which of these possibilities are correct. **(a)** [HA] = [A⁻], **(b)** [HA] > [A⁻], or **(c)** [HA] < [A⁻]. [Section 17.2]

17.4 The following diagram represents a buffer composed of equal concentrations of a weak acid, HA, and its conjugate base, A⁻. The heights of the columns are proportional to the concentrations of the components of the buffer.

(a) Which of the three drawings, (1), (2), or (3), represents the buffer after the addition of a strong acid? **(b)** Which of the

three represents the buffer after the addition of a strong base? **(c)** Which of the three represents a situation that cannot arise from the addition of either an acid or a base? [Section 17.2]

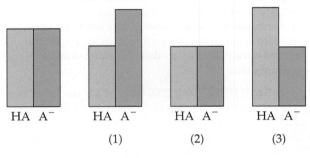

17.5 The following figure represents solutions at various stages of the titration of a weak acid, HA, with NaOH. (The Na⁺ ions and water molecules have been omitted for clarity.) To which of the following regions of the titration curve does each drawing correspond: **(a)** before addition of NaOH, **(b)** after addition of NaOH but before the equivalence point, **(c)** at the equivalence point, **(d)** after the equivalence point? [Section 17.3]

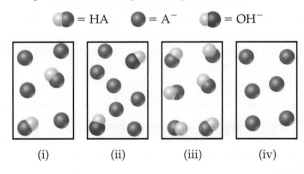

17.6 Match the following descriptions of titration curves with the diagrams: **(a)** strong acid added to strong base, **(b)** strong base added to weak acid, **(c)** strong base added to strong acid, **(d)** strong base added to polyprotic acid. [Section 17.3]

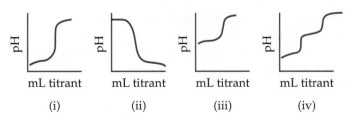

17.7 Equal volumes of two acids are titrated with 0.10 M NaOH resulting in the two titration curves shown in the following figure. (**a**) Which curve corresponds to the more concentrated acid solution? (**b**) Which corresponds to the acid with the larger K_a? [Section 17.3]

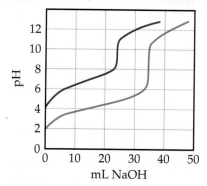

17.8 A saturated solution of $Cd(OH)_2$ is shown in the middle beaker. If hydrochloric acid solution is added, the solubility of

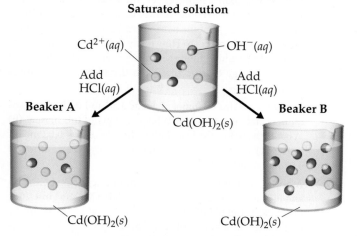

$Cd(OH)_2$ will increase, causing additional solid to dissolve. Which of the two choices, Beaker A or Beaker B, accurately represents the solution after equilibrium is reestablished? (The water molecules and Cl^- ions are omitted for clarity.) [Sections 17.4 and 17.5]

17.9 The following graphs represent the behavior of $BaCO_3$ under different circumstances. In each case, the vertical axis indicates the solubility of the $BaCO_3$ and the horizontal axis represents the concentration of some other reagent. (**a**) Which graph represents what happens to the solubility of $BaCO_3$ as HNO_3 is added? (**b**) Which graph represents what happens to the $BaCO_3$ solubility as Na_2CO_3 is added? (**c**) Which represents what happens to the $BaCO_3$ solubility as $NaNO_3$ is added? [Section 17.5]

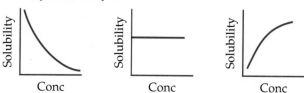

17.10 $Ca(OH)_2$ has a K_{sp} of 6.5×10^{-6}. (**a**) If 0.370 g of $Ca(OH)_2$ is added to 500 mL of water and the mixture is allowed to come to equilibrium, will the solution be saturated? (**b**) If 50 mL of the solution from part (a) is added to each of the beakers shown here, in which beakers, if any, will a precipitate form? In those cases where a precipitate forms, what is its identity? [Section 17.6]

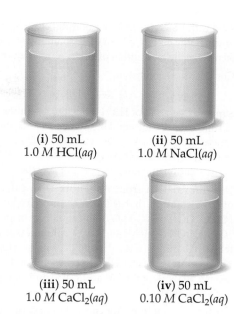

(**i**) 50 mL
1.0 M HCl(*aq*)

(**ii**) 50 mL
1.0 M NaCl(*aq*)

(**iii**) 50 mL
1.0 M CaCl$_2$(*aq*)

(**iv**) 50 mL
0.10 M CaCl$_2$(*aq*)

17.11 The graph below shows the solubility of a salt as a function of pH. Which of the following choices explain the shape of this graph? (**a**) None; this behavior is not possible. (**b**) A soluble salt reacts with acid to form a precipitate, and additional acid reacts with this product to dissolve it. (**c**) A soluble salt forms an insoluble hydroxide, then additional base reacts with this product to dissolve it. (**d**) The solubility of the salt increases with pH then decreases because of the heat generated from the neutralization reactions. [Section 17.5]

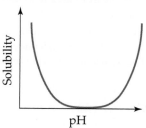

17.12 Three cations, Ni^{2+}, Cu^{2+}, and Ag^+, are separated using two different precipitating agents. Based on Figure 17.23, what two precipitating agents could be used? Using these agents, indicate which of the cations is A, which is B, and which is C. [Section 17.7]

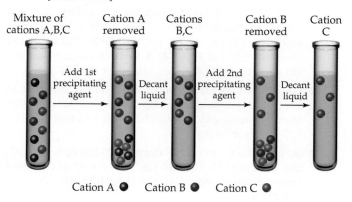

The Common-Ion Effect (Section 17.1)

17.13 Which of these statements about the common-ion effect is most correct? (**a**) The solubility of a salt MA is decreased in a solution that already contains either M^+ or A^-. (**b**) Common ions alter the equilibrium constant for the reaction of an ionic solid with water. **(c)** The common-ion effect does not apply to unusual ions like SO_3^{2-}. **(d)** The solubility of a salt MA is affected equally by addition of either A^- or a non-common ion.

17.14 Consider the equilibrium

$$B(aq) + H_2O(l) \rightleftharpoons HB^+(aq) + OH^-(aq).$$

Suppose that a salt of HB^+ is added to a solution of B at equilibrium. **(a)** Will the equilibrium constant for the reaction increase, decrease, or stay the same? **(b)** Will the concentration of B(aq) increase, decrease, or stay the same? **(c)** Will the pH of the solution increase, decrease, or stay the same?

17.15 Use information from Appendix D to calculate the pH of **(a)** a solution that is 0.060 M in potassium propionate (C_2H_5COOK or $KC_3H_5O_2$) and 0.085 M in propionic acid (C_2H_5COOH or $HC_3H_5O_2$); **(b)** a solution that is 0.075 M in trimethylamine, $(CH_3)_3N$, and 0.10 M in trimethylammonium chloride, $(CH_3)_3NHCl$; **(c)** a solution that is made by mixing 50.0 mL of 0.15 M acetic acid and 50.0 mL of 0.20 M sodium acetate.

17.16 Use information from Appendix D to calculate the pH of **(a)** a solution that is 0.250 M in sodium formate (HCOONa) and 0.100 M in formic acid (HCOOH), **(b)** a solution that is 0.510 M in pyridine (C_5H_5N) and 0.450 M in pyridinium chloride (C_5H_5NHCl), **(c)** a solution that is made by combining 55 mL of 0.050 M hydrofluoric acid with 125 mL of 0.10 M sodium fluoride.

17.17 **(a)** Calculate the percent ionization of 0.0075 M butanoic acid ($K_a = 1.5 \times 10^{-5}$). **(b)** Calculate the percent ionization of 0.0075 M butanoic acid in a solution containing 0.085 M sodium butanoate.

17.18 **(a)** Calculate the percent ionization of 0.125 M lactic acid ($K_a = 1.4 \times 10^{-4}$). **(b)** Calculate the percent ionization of 0.125 M lactic acid in a solution containing 0.0075 M sodium lactate.

Buffers (Section 17.2)

17.19 Which of the following solutions is a buffer? **(a)** 0.10 M CH_3COOH and 0.10 $M CH_3COONa$, **(b)** 0.10 M CH_3COOH, **(c)** 0.10 M HCl and 0.10 M NaCl, **(d)** both a and c, **(e)** all of a, b, and c.

17.20 Which of the following solutions is a buffer? **(a)** A solution made by mixing 100 mL of 0.100 M CH_3COOH and 50 mL of 0.100 M NaOH, **(b)** a solution made by mixing 100 mL of 0.100 M CH_3COOH and 500 mL of 0.100 M NaOH, **(c)** A solution made by mixing 100 mL of 0.100 M CH_3COOH and 50 mL of 0.100 M HCl, **(d)** A solution made by mixing 100 mL of 0.100 M CH_3COOK and 50 mL of 0.100 M KCl.

17.21 **(a)** Calculate the pH of a buffer that is 0.12 M in lactic acid and 0.11 M in sodium lactate. **(b)** Calculate the pH of a buffer formed by mixing 85 mL of 0.13 M lactic acid with 95 mL of 0.15 M sodium lactate.

17.22 **(a)** Calculate the pH of a buffer that is 0.105 M in $NaHCO_3$ and 0.125 M in Na_2CO_3. **(b)** Calculate the pH of a solution formed by mixing 65 mL of 0.20 M $NaHCO_3$ with 75 mL of 0.15 M Na_2CO_3.

17.23 A buffer is prepared by adding 20.0 g of sodium acetate (CH_3COONa) to 500 mL of a 0.150 M acetic acid (CH_3COOH) solution. **(a)** Determine the pH of the buffer. **(b)** Write the complete ionic equation for the reaction that occurs when a few drops of hydrochloric acid are added to the

buffer. **(c)** Write the complete ionic equation for the reaction that occurs when a few drops of sodium hydroxide solution are added to the buffer.

17.24 A buffer is prepared by adding 10.0 g of ammonium chloride (NH_4Cl) to 250 mL of 1.00 M NH_3 solution. **(a)** What is the pH of this buffer? **(b)** Write the complete ionic equation for the reaction that occurs when a few drops of nitric acid are added to the buffer. **(c)** Write the complete ionic equation for the reaction that occurs when a few drops of potassium hydroxide solution are added to the buffer.

17.25 You are asked to prepare a pH = 3.00 buffer solution starting from 1.25 L of a 1.00 M solution of hydrofluoric acid (HF) and any amount you need of sodium fluoride (NaF). **(a)** What is the pH of the hydrofluoric acid solution prior to adding sodium fluoride? **(b)** How many grams of sodium fluoride should be added to prepare the buffer solution? Neglect the small volume change that occurs when the sodium fluoride is added.

17.26 You are asked to prepare a pH = 4.00 buffer starting from 1.50 L of 0.0200 M solution of benzoic acid (C_6H_5COOH) and any amount you need of sodium benzoate (C_6H_5COONa). **(a)** What is the pH of the benzoic acid solution prior to adding sodium benzoate? **(b)** How many grams of sodium benzoate should be added to prepare the buffer? Neglect the small volume change that occurs when the sodium benzoate is added.

17.27 A buffer contains 0.10 mol of acetic acid and 0.13 mol of sodium acetate in 1.00 L. **(a)** What is the pH of this buffer? **(b)** What is the pH of the buffer after the addition of 0.02 mol of KOH? **(c)** What is the pH of the buffer after the addition of 0.02 mol of HNO_3?

17.28 A buffer contains 0.15 mol of propionic acid (C_2H_5COOH) and 0.10 mol of sodium propionate (C_2H_5COONa) in 1.20 L. **(a)** What is the pH of this buffer? **(b)** What is the pH of the buffer after the addition of 0.01 mol of NaOH? **(c)** What is the pH of the buffer after the addition of 0.01 mol of HI?

17.29 **(a)** What is the ratio of HCO_3^- to H_2CO_3 in blood of pH 7.4? **(b)** What is the ratio of HCO_3^- to H_2CO_3 in an exhausted marathon runner whose blood pH is 7.1?

17.30 A buffer, consisting of $H_2PO_4^-$ and HPO_4^{2-}, helps control the pH of physiological fluids. Many carbonated soft drinks also use this buffer system. What is the pH of a soft drink in which the major buffer ingredients are 6.5 g of NaH_2PO_4 and 8.0 g of Na_2HPO_4 per 355 mL of solution?

17.31 You have to prepare a pH 3.50 buffer, and you have the following 0.10 M solutions available: HCOOH, CH_3COOH, H_3PO_4, HCOONa, CH_3COONa, and NaH_2PO_4. Which solutions would you use? How many milliliters of each solution would you use to make approximately 1 L of the buffer?

17.32 You have to prepare a pH 5.00 buffer, and you have the following 0.10 M solutions available: HCOOH, HCOONa, CH_3COOH, CH_3COONa, HCN, and NaCN. Which solutions would you use? How many milliliters of each solution would you use to make approximately 1 L of the buffer?

Acid–Base Titrations (Section 17.3)

17.33 The accompanying graph shows the titration curves for two monoprotic acids. **(a)** Which curve is that of a strong acid? **(b)** What is the approximate pH at the equivalence point of each titration? (**c**) 40.0 mL of each acid was titrated with 0.100 *M* base. Which acid is more concentrated? (**d**) Estimate the pK_a of the weak acid.

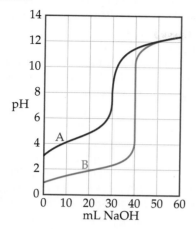

17.34 Compare the titration of a strong, monoprotic acid with a strong base to the titration of a weak, monoprotic acid with a strong base. Assume the strong and weak acid solutions initially have the same concentrations. Indicate whether the following statements are true or false. (**a**) More base is required to reach the equivalence point for the strong acid than the weak acid. (**b**) The pH at the beginning of the titration is lower for the weak acid than the strong acid. (**c**) The pH at the equivalence point is 7 no matter which acid is titrated.

17.35 The samples of nitric and acetic acids shown here are both titrated with a 0.100 *M* solution of NaOH(*aq*).

25.0 mL of 1.0 *M* HNO$_3$(*aq*)   25.0 mL of 1.0 *M* CH$_3$COOH(*aq*)

Determine whether each of the following statements concerning these titrations is true or false.

(**a**) A larger volume of NaOH(*aq*) is needed to reach the equivalence point in the titration of HNO$_3$.

(**b**) The pH at the equivalence point in the HNO$_3$ titration will be lower than the pH at the equivalence point in the CH$_3$COOH titration.

(**c**) Phenolphthalein would be a suitable indicator for both titrations.

17.36 Determine whether each of the following statements concerning the titrations in Problem 17.35 is true or false.

(**a**) The pH at the beginning of the two titrations will be the same.

(**b**) The titration curves will both be essentially the same after passing the equivalence point.

(**c**) Methyl red would be a suitable indicator for both titrations.

17.37 Predict whether the equivalence point of each of the following titrations is below, above, or at pH 7: (**a**) NaHCO$_3$ titrated

with NaOH, (**b**) NH$_3$ titrated with HCl, (**c**) KOH titrated with HBr.

17.38 Predict whether the equivalence point of each of the following titrations is below, above, or at pH 7: (**a**) formic acid titrated with NaOH, (**b**) calcium hydroxide titrated with perchloric acid, (**c**) pyridine titrated with nitric acid.

17.39 As shown in Figure 16.8, the indicator thymol blue has two color changes. Which color change will generally be more suitable for the titration of a weak acid with a strong base?

17.40 Assume that 30.0 mL of a 0.10 *M* solution of a weak base B that accepts one proton is titrated with a 0.10 *M* solution of the monoprotic strong acid HA. (**a**) How many moles of HA have been added at the equivalence point? (**b**) What is the predominant form of B at the equivalence point? (**c**) Is the pH 7, less than 7, or more than 7 at the equivalence point? (**d**) Which indicator, phenolphthalein or methyl red, is likely to be the better choice for this titration?

17.41 How many milliliters of 0.0850 *M* NaOH are required to titrate each of the following solutions to the equivalence point: (**a**) 40.0 mL of 0.0900 *M* HNO$_3$, (**b**) 35.0 mL of 0.0850 *M* CH$_3$COOH, (**c**) 50.0 mL of a solution that contains 1.85 g of HCl per liter?

17.42 How many milliliters of 0.105 *M* HCl are needed to titrate each of the following solutions to the equivalence point: (**a**) 45.0 mL of 0.0950 *M* NaOH, (**b**) 22.5 mL of 0.118 *M* NH$_3$, (**c**) 125.0 mL of a solution that contains 1.35 g of NaOH per liter?

17.43 A 20.0-mL sample of 0.200 *M* HBr solution is titrated with 0.200 *M* NaOH solution. Calculate the pH of the solution after the following volumes of base have been added: (**a**) 15.0 mL, (**b**) 19.9 mL, (**c**) 20.0 mL, (**d**) 20.1 mL, (**e**) 35.0 mL.

17.44 A 20.0-mL sample of 0.150 *M* KOH is titrated with 0.125 *M* HClO$_4$ solution. Calculate the pH after the following volumes of acid have been added: (**a**) 20.0 mL, (**b**) 23.0 mL, (**c**) 24.0 mL, (**d**) 25.0 mL, (**e**) 30.0 mL.

17.45 A 35.0-mL sample of 0.150 *M* acetic acid (CH$_3$COOH) is titrated with 0.150 *M* NaOH solution. Calculate the pH after the following volumes of base have been added: (**a**) 0 mL, (**b**) 17.5 mL, (**c**) 34.5 mL, (**d**) 35.0 mL, (**e**) 35.5 mL, (**f**) 50.0 mL.

17.46 Consider the titration of 30.0 mL of 0.050 *M* NH$_3$ with 0.025 *M* HCl. Calculate the pH after the following volumes of titrant have been added: (**a**) 0 mL, (**b**) 20.0 mL, (**c**) 59.0 mL, (**d**) 60.0 mL, (**e**) 61.0 mL, (**f**) 65.0 mL.

17.47 Calculate the pH at the equivalence point for titrating 0.200 *M* solutions of each of the following bases with 0.200 *M* HBr: (**a**) sodium hydroxide (NaOH), (**b**) hydroxylamine (NH$_2$OH), (**c**) aniline (C$_6$H$_5$NH$_2$).

17.48 Calculate the pH at the equivalence point in titrating 0.100 *M* solutions of each of the following with 0.080 *M* NaOH: (**a**) hydrobromic acid (HBr), (**b**) chlorous acid (HClO$_2$), (**c**) benzoic acid (C$_6$H$_5$COOH).

Solubility Equilibria and Factors Affecting Solubility (Sections 17.4 and 17.5)

17.49 For each statement, indicate whether it is true or false.

(**a**) The solubility of a slightly soluble salt can be expressed in units of moles per liter.

(**b**) The solubility product of a slightly soluble salt is simply the square of the solubility.

(c) The solubility of a slightly soluble salt is independent of the presence of a common ion.

(d) The solubility product of a slightly soluble salt is independent of the presence of a common ion.

17.50 The solubility of two slightly soluble salts of M^{2+}, MA and MZ_2, are the same, 4×10^{-4} mol/L. (a) Which has the larger numerical value for the solubility product constant? (b) In a saturated solution of each salt in water, which has the higher concentration of M^{2+}? (c) If you added an equal volume of a solution saturated in MA to one saturated in MZ_2, what would be the equilibrium concentration of the cation, M^{2+}?

17.51 (a) Why is the concentration of undissolved solid not explicitly included in the expression for the solubility-product constant? (b) Write the expression for the solubility-product constant for each of the following strong electrolytes: AgI, $SrSO_4$, $Fe(OH)_2$, and Hg_2Br_2.

17.52 (a) True or false: "solubility" and "solubility-product constant" are the same number for a given compound. (b) Write the expression for the solubility-product constant for each of the following ionic compounds: $MnCO_3$, $Hg(OH)_2$, and $Cu_3(PO_4)_2$.

17.53 (a) If the molar solubility of CaF_2 at 35 °C is 1.24×10^{-3} mol/L, what is K_{sp} at this temperature? (b) It is found that 1.1×10^{-2} g SrF_2 dissolves per 100 mL of aqueous solution at 25 °C. Calculate the solubility product for SrF_2. (c) The K_{sp} of $Ba(IO_3)_2$ at 25 °C is 6.0×10^{-10}. What is the molar solubility of $Ba(IO_3)_2$?

17.54 (a) The molar solubility of $PbBr_2$ at 25 °C is 1.0×10^{-2} mol/L. Calculate K_{sp}. (b) If 0.0490 g of $AgIO_3$ dissolves per liter of solution, calculate the solubility-product constant. (c) Using the appropriate K_{sp} value from Appendix D, calculate the pH of a saturated solution of $Ca(OH)_2$.

17.55 A 1.00-L solution saturated at 25 °C with calcium oxalate (CaC_2O_4) contains 0.0061 g of CaC_2O_4. Calculate the solubility-product constant for this salt at 25 °C.

17.56 A 1.00-L solution saturated at 25 °C with lead(II) iodide contains 0.54 g of PbI_2. Calculate the solubility-product constant for this salt at 25 °C.

17.57 Using Appendix D, calculate the molar solubility of AgBr in (a) pure water, (b) 3.0×10^{-2} M $AgNO_3$ solution, (c) 0.10 M NaBr solution.

17.58 Calculate the solubility of LaF_3 in grams per liter in (a) pure water, (b) 0.010 M KF solution, (c) 0.050 M $LaCl_3$ solution.

17.59 Consider a beaker containing a saturated solution of CaF_2 in equilibrium with undissolved $CaF_2(s)$. Solid $CaCl_2$ is then added to the solution. (a) Will the amount of solid CaF_2 at the bottom of the beaker increase, decrease, or remain the same? (b) Will the concentration of Ca^{2+} ions in solution increase or decrease? (c) Will the concentration of F^- ions in solution increase or decrease?

17.60 Consider a beaker containing a saturated solution of PbI_2 in equilibrium with undissolved $PbI_2(s)$. Now solid KI is added to this solution. (a) Will the amount of solid PbI_2 at the bottom of the beaker increase, decrease, or remain the same? (b) Will the concentration of Pb^{2+} ions in solution increase or decrease? (c) Will the concentration of I^- ions in solution increase or decrease?

17.61 Calculate the solubility of $Mn(OH)_2$ in grams per liter when buffered at pH (a) 7.0, (b) 9.5, (c) 11.8.

17.62 Calculate the molar solubility of $Ni(OH)_2$ when buffered at pH (a) 8.0, (b) 10.0, (c) 12.0.

17.63 Which of the following salts will be substantially more soluble in acidic solution than in pure water: (a) $ZnCO_3$, (b) ZnS, (c) BiI_3, (d) AgCN, (e) $Ba_3(PO_4)_2$?

17.64 For each of the following slightly soluble salts, write the net ionic equation, if any, for reaction with a strong acid: (a) MnS, (b) PbF_2, (c) $AuCl_3$, (d) $Hg_2C_2O_4$, (e) CuBr.

17.65 From the value of K_f listed in Table 17.1, calculate the concentration of Ni^{2+} in 1.0 L of a solution that contains a total of 1×10^{-3} mol of nickel(II) ion and that is 0.20 M in NH_3.

17.66 To what final concentration of NH_3 must a solution be adjusted to just dissolve 0.020 mol of NiC_2O_4 ($K_{sp} = 4 \times 10^{-10}$) in 1.0 L of solution? (*Hint:* You can neglect the hydrolysis of $C_2O_4{}^{2-}$ because the solution will be quite basic.)

17.67 Use values of K_{sp} for AgI and K_f for $Ag(CN)_2{}^-$ to (a) calculate the molar solubility of AgI in pure water, (b) calculate the equilibrium constant for the reaction $AgI(s) + 2\,CN^-(aq) \rightleftharpoons Ag(CN)_2{}^-(aq) + I^-(aq)$, (c) determine the molar solubility of AgI in a 0.100 M NaCN solution.

17.68 Using the value of K_{sp} for Ag_2S, K_{a1} and K_{a2} for H_2S, and $K_f = 1.1 \times 10^5$ for $AgCl_2{}^-$, calculate the equilibrium constant for the following reaction:

$$Ag_2S(s) + 4\,Cl^-(aq) + 2\,H^+(aq) \rightleftharpoons 2\,AgCl_2{}^-(aq) + H_2S(aq)$$

Precipitation and Separation of Ions (Section 17.6)

17.69 (a) Will $Ca(OH)_2$ precipitate from solution if the pH of a 0.050 M solution of $CaCl_2$ is adjusted to 8.0? (b) Will Ag_2SO_4 precipitate when 100 mL of 0.050 M $AgNO_3$ is mixed with 10 mL of 5.0×10^{-2} M Na_2SO_4 solution?

17.70 (a) Will $Co(OH)_2$ precipitate from solution if the pH of a 0.020 M solution of $Co(NO_3)_2$ is adjusted to 8.5? (b) Will $AgIO_3$ precipitate when 20 mL of 0.010 M $AgIO_3$ is mixed with 10 mL of 0.015 M $NaIO_3$? (K_{sp} of $AgIO_3$ is 3.1×10^{-8}).

17.71 Calculate the minimum pH needed to precipitate $Mn(OH)_2$ so completely that the concentration of Mn^{2+} is less than 1 μg per liter [1 part per billion (ppb)].

17.72 Suppose that a 10-mL sample of a solution is to be tested for I^- ion by addition of 1 drop (0.2 mL) of 0.10 M $Pb(NO_3)_2$. What is the minimum number of grams of I^- that must be present for $PbI_2(s)$ to form?

17.73 A solution contains 2.0×10^{-4} M Ag^+ and 1.5×10^{-3} M Pb^{2+}. If NaI is added, will AgI ($K_{sp} = 8.3 \times 10^{-17}$) or PbI_2 ($K_{sp} = 7.9 \times 10^{-9}$) precipitate first? Specify the concentration of I^- needed to begin precipitation.

17.74 A solution of Na_2SO_4 is added dropwise to a solution that is 0.010 M in Ba^{2+} and 0.010 M in Sr^{2+}. (a) What concentration of $SO_4{}^{2-}$ is necessary to begin precipitation? (Neglect volume changes. $BaSO_4$: $K_{sp} = 1.1 \times 10^{-10}$; $SrSO_4$: $K_{sp} = 3.2 \times 10^{-7}$.) (b) Which cation precipitates first? (c) What is the concentration of $SO_4{}^{2-}$ when the second cation begins to precipitate?

17.75 A solution contains three anions with the following concentrations: 0.20 M $CrO_4{}^{2-}$, 0.10 M $CO_3{}^{2-}$, and 0.010 M Cl^-. If a dilute

$AgNO_3$ solution is slowly added to the solution, what is the first compound to precipitate: Ag_2CrO_4 ($K_{sp} = 1.2 \times 10^{-12}$), Ag_2CO_3 ($K_{sp} = 8.1 \times 10^{-12}$), or $AgCl$ ($K_{sp} = 1.8 \times 10^{-10}$)?

17.76 A 1.0 M Na_2SO_4 solution is slowly added to 10.0 mL of a solution that is 0.20 M in Ca^{2+} and 0.30 M in Ag^+. **(a)** Which compound will precipitate first: $CaSO_4$($K_{sp} = 2.4 \times 10^{-5}$) or Ag_2SO_4 ($K_{sp} = 1.5 \times 10^{-5}$)? **(b)** How much Na_2SO_4 solution must be added to initiate the precipitation?

Qualitative Analysis for Metallic Elements (Section 17.7)

17.77 A solution containing an unknown number of metal ions is treated with dilute HCl; no precipitate forms. The pH is adjusted to about 1, and H_2S is bubbled through. Again, no precipitate forms. The pH of the solution is then adjusted to about 8. Again, H_2S is bubbled through. This time a precipitate forms. The filtrate from this solution is treated with $(NH_4)_2HPO_4$. No precipitate forms. Which metal ions discussed in Section 17.7 are possibly present? Which are definitely absent within the limits of these tests?

17.78 An unknown solid is entirely soluble in water. On addition of dilute HCl, a precipitate forms. After the precipitate is filtered off, the pH is adjusted to about 1 and H_2S is bubbled in; a precipitate again forms. After filtering off this precipitate, the pH is adjusted to 8 and H_2S is again added; no precipitate forms.

No precipitate forms upon addition of $(NH_4)_2HPO_4$. (See Figure 7.23.) The remaining solution shows a yellow color in a flame test (see Figure 7.22). Based on these observations, which of the following compounds might be present, which are definitely present, and which are definitely absent: CdS, $Pb(NO_3)_2$, HgO, $ZnSO_4$, $Cd(NO_3)_2$, and Na_2SO_4?

17.79 In the course of various qualitative analysis procedures, the following mixtures are encountered: **(a)** Zn^{2+} and Cd^{2+}, **(b)** $Cr(OH)_3$ and $Fe(OH)_3$, **(c)** Mg^{2+} and K^+, **(d)** Ag^+ and Mn^{2+}. Suggest how each mixture might be separated.

17.80 Suggest how the cations in each of the following solution mixtures can be separated: **(a)** Na^+ and Cd^{2+}, **(b)** Cu^{2+} and Mg^{2+}, **(c)** Pb^{2+} and Al^{3+}, **(d)** Ag^+ and Hg^{2+}.

17.81 **(a)** Precipitation of the group 4 cations of Figure 17.23 requires a basic medium. Why is this so? **(b)** What is the most significant difference between the sulfides precipitated in group 2 and those precipitated in group 3? **(c)** Suggest a procedure that would serve to redissolve the group 3 cations following their precipitation.

17.82 A student who is in a great hurry to finish his laboratory work decides that his qualitative analysis unknown contains a metal ion from group 4 of Figure 17.23. He therefore tests his sample directly with $(NH_4)_2HPO_4$, skipping earlier tests for the metal ions in groups 1, 2, and 3. He observes a precipitate and concludes that a metal ion from group 4 is indeed present. Why is this possibly an erroneous conclusion?

Additional Exercises

17.83 Derive an equation similar to the Henderson–Hasselbalch equation relating the pOH of a buffer to the pK_b of its base component.

[17.84] Rainwater is acidic because $CO_2(g)$ dissolves in the water, creating carbonic acid, H_2CO_3. If the rainwater is too acidic, it will react with limestone and seashells (which are principally made of calcium carbonate, $CaCO_3$). Calculate the concentrations of carbonic acid, bicarbonate ion (HCO_3^-) and carbonate ion (CO_3^{2-}) that are in a raindrop that has a pH of 5.60, assuming that the sum of all three species in the raindrop is 1.0×10^{-5} M.

17.85 Furoic acid ($HC_5H_3O_3$) has a K_a value of 6.76×10^{-4} at 25 °C. Calculate the pH at 25 °C of **(a)** a solution formed by adding 25.0 g of furoic acid and 30.0 g of sodium furoate ($NaC_5H_3O_3$) to enough water to form 0.250 L of solution, **(b)** a solution formed by mixing 30.0 mL of 0.250 M $HC_5H_3O_3$ and 20.0 mL of 0.22 M $NaC_5H_3O_3$ and diluting the total volume to 125 mL, **(c)** a solution prepared by adding 50.0 mL of 1.65 M NaOH solution to 0.500 L of 0.0850 M $HC_5H_3O_3$.

17.86 The acid–base indicator bromcresol green is a weak acid. The yellow acid and blue base forms of the indicator are present in equal concentrations in a solution when the pH is 4.68. What is the pK_a for bromcresol green?

17.87 Equal quantities of 0.010 M solutions of an acid HA and a base B are mixed. The pH of the resulting solution is 9.2. **(a)** Write the chemical equation and equilibrium-constant expression for the reaction between HA and B. **(b)** If K_a for HA is 8.0×10^{-5}, what is the value of the equilibrium constant for the reaction between HA and B? **(c)** What is the value of K_b for B?

17.88 Two buffers are prepared by adding an equal number of moles of formic acid (HCOOH) and sodium formate (HCOONa) to enough water to make 1.00 L of solution. Buffer A is prepared using 1.00 mol each of formic acid and sodium formate. Buffer B is prepared by using 0.010 mol of each. **(a)** Calculate the pH of each buffer. **(b)** Which buffer will have the greater buffer capacity? **(c)** Calculate the change in pH for each buffer upon the addition of 1.0 mL of 1.00 M HCl. **(d)** Calculate the change in pH for each buffer upon the addition of 10 mL of 1.00 M HCl.

17.89 A biochemist needs 750 mL of an acetic acid–sodium acetate buffer with pH 4.50. Solid sodium acetate (CH_3COONa) and glacial acetic acid (CH_3COOH) are available. Glacial acetic acid is 99% CH_3COOH by mass and has a density of 1.05 g/mL. If the buffer is to be 0.15 M in CH_3COOH, how many grams of CH_3COONa and how many milliliters of glacial acetic acid must be used?

17.90 A sample of 0.2140 g of an unknown monoprotic acid was dissolved in 25.0 mL of water and titrated with 0.0950 M NaOH. The acid required 27.4 mL of base to reach the equivalence point. **(a)** What is the molar mass of the acid? **(b)** After 15.0 mL of base had been added in the titration, the pH was found to be 6.50. What is the K_a for the unknown acid?

17.91 A sample of 0.1687 g of an unknown monoprotic acid was dissolved in 25.0 mL of water and titrated with 0.1150 M NaOH. The acid required 15.5 mL of base to reach the equivalence point. **(a)** What is the molecular weight of the acid? **(b)** After 7.25 mL of base had been added in the titration, the pH was found to be 2.85. What is the K_a for the unknown acid?

17.92 Mathematically prove that the pH at the halfway point of a titration of a weak acid with a strong base (where the volume of added base is half of that needed to reach the equivalence point) is equal to pK_a for the acid.

17.93 A weak monoprotic acid is titrated with 0.100 M NaOH. It requires 50.0 mL of the NaOH solution to reach the equivalence point. After 25.0 mL of base is added, the pH of the solution is 3.62. Estimate the pKa of the weak acid.

17.94 What is the pH of a solution made by mixing 0.30 mol NaOH, 0.25 mol Na_2HPO_4, and 0.20 mol H_3PO_4 with water and diluting to 1.00 L?

17.95 Suppose you want to do a physiological experiment that calls for a pH 6.50 buffer. You find that the organism with which you are working is not sensitive to the weak acid H_2A ($K_{a1} = 2 \times 10^{-2}$; $K_{a2} = 5.0 \times 10^{-7}$) or its sodium salts. You have available a 1.0 M solution of this acid and a 1.0 M solution of NaOH. How much of the NaOH solution should be added to 1.0 L of the acid to give a buffer at pH 6.50? (Ignore any volume change.)

17.96 How many microliters of 1.000 M NaOH solution must be added to 25.00 mL of a 0.1000 M solution of lactic acid $[CH_3CH(OH)COOH$ or $HC_3H_5O_3]$ to produce a buffer with pH = 3.75?

17.97 A person suffering from anxiety begins breathing rapidly and as a result suffers alkalosis, an increase in blood pH. (a) Using Equation 17.10, explain how rapid breathing can cause the pH of blood to increase. (b) One cure for this problem is breathing in a paper bag. Why does this procedure lower blood pH?

17.98 For each pair of compounds, use K_{sp} values to determine which has the greater molar solubility: (a) CdS or CuS, (b) $PbCO_3$ or $BaCrO_4$, (c) $Ni(OH)_2$ or $NiCO_3$, (d) AgI or Ag_2SO_4.

17.99 The solubility of $CaCO_3$ is pH dependent. (a) Calculate the molar solubility of $CaCO_3$ ($K_{sp} = 4.5 \times 10^{-9}$) neglecting the acid–base character of the carbonate ion. (b) Use the K_b expression for the CO_3^{2-} ion to determine the equilibrium constant for the reaction

$$CaCO_3(s) + H_2O(l) \rightleftharpoons$$
$$Ca^{2+}(aq) + HCO_3^-(aq) + OH^-(aq)$$

(c) If we assume that the only sources of Ca^{2+}, HCO_3^-, and OH^- ions are from the dissolution of $CaCO_3$, what is the molar solubility of $CaCO_3$ using the equilibrium expression from part (b)? (d) What is the molar solubility of $CaCO_3$ at the pH of the ocean (8.3)? (e) If the pH is buffered at 7.5, what is the molar solubility of $CaCO_3$?

17.100 Tooth enamel is composed of hydroxyapatite, whose simplest formula is $Ca_5(PO_4)_3OH$, and whose corresponding $K_{sp} = 6.8 \times 10^{-27}$. As discussed in the "Chemistry and Life" box on page 755, fluoride in fluorinated water or in toothpaste reacts with hydroxyapatite to form fluoroapatite, $Ca_5(PO_4)_3F$, whose $K_{sp} = 1.0 \times 10^{-60}$. (a) Write the expression for the solubility-constant for hydroxyapatite and for fluoroapatite. (b) Calculate the molar solubility of each of these compounds.

17.101 Use the solubility-product constant for $Cr(OH)_3$ ($K_{sp} = 6.7 \times 10^{-31}$) and the formation constant for $Cr(OH)_4^-$ from Table 17.1 to determine the concentration of $Cr(OH)_4^-$ in a solution that is buffered at pH = 10.0 and is in equilibrium with solid $Cr(OH)_3$.

17.102 Calculate the solubility of $Mg(OH)_2$ in 0.50 M NH_4Cl.

17.103 The solubility-product constant for barium permanganate, $Ba(MnO_4)_2$, is 2.5×10^{-10}. Assume that solid $Ba(MnO_4)_2$ is in equilibrium with a solution of $KMnO_4$. What concentration of $KMnO_4$ is required to establish a concentration of 2.0×10^{-8} M for the Ba^{2+} ion in solution?

17.104 Calculate the ratio of $[Ca^{2+}]$ to $[Fe^{2+}]$ in a lake in which the water is in equilibrium with deposits of both $CaCO_3$ and $FeCO_3$. Assume that the water is slightly basic and that the hydrolysis of the carbonate ion can therefore be ignored.

17.105 The solubility product constants of $PbSO_4$ and $SrSO_4$ are 6.3×10^{-7} and 3.2×10^{-7}, respectively. What are the values of $[SO_4^{2-}]$, $[Pb^{2+}]$, and $[Sr^{2+}]$ in a solution at equilibrium with both substances?

17.106 A buffer of what pH is needed to give a Mg^{2+} concentration of 3.0×10^{-2} M in equilibrium with solid magnesium oxalate?

17.107 The value of K_{sp} for $Mg_3(AsO_4)_2$ is 2.1×10^{-20}. The AsO_4^{3-} ion is derived from the weak acid H_3AsO_4 ($pK_{a1} = 2.22$; $pK_{a2} = 6.98$; $pK_{a3} = 11.50$). When asked to calculate the molar solubility of $Mg_3(AsO_4)_2$ in water, a student used the K_{sp} expression and assumed that $[Mg^{2+}] = 1.5[AsO_4^{3-}]$. Why was this a mistake?

17.108 The solubility product for $Zn(OH)_2$ is 3.0×10^{-16}. The formation constant for the hydroxo complex, $Zn(OH)_4^{2-}$, is 4.6×10^{17}. What concentration of OH^- is required to dissolve 0.015 mol of $Zn(OH)_2$ in a liter of solution?

17.109 The value of K_{sp} for $Cd(OH)_2$ is 2.5×10^{-14}. (a) What is the molar solubility of $Cd(OH)_2$? (b) The solubility of $Cd(OH)_2$ can be increased through formation of the complex ion $CdBr_4^{2-}$ ($K_f = 5 \times 10^3$). If solid $Cd(OH)_2$ is added to a NaBr solution, what is the initial concentration of NaBr needed to increase the molar solubility of $Cd(OH)_2$ to 1.0×10^{-3} mol/L?

Integrative Exercises

17.110 (a) Write the net ionic equation for the reaction that occurs when a solution of hydrochloric acid (HCl) is mixed with a solution of sodium formate ($NaCHO_2$). (b) Calculate the equilibrium constant for this reaction. (c) Calculate the equilibrium concentrations of Na^+, Cl^-, H^+, CHO_2^-, and $HCHO_2$ when 50.0 mL of 0.15 M HCl is mixed with 50.0 mL of 0.15 M $NaCHO_2$.

17.111 (a) A 0.1044-g sample of an unknown monoprotic acid requires 22.10 mL of 0.0500 M NaOH to reach the end point. What is the molecular weight of the unknown? (b) As the acid is titrated, the pH of the solution after the addition of 11.05 mL of the base is 4.89. What is the K_a for the acid? (c) Using Appendix D, suggest the identity of the acid.

17.112 A sample of 7.5 L of NH_3 gas at 22 °C and 735 torr is bubbled into a 0.50-L solution of 0.40 M HCl. Assuming that all the NH_3 dissolves and that the volume of the solution remains 0.50 L, calculate the pH of the resulting solution.

17.113 Aspirin has the structural formula

At body temperature (37 °C), K_a for aspirin equals 3×10^{-5}. If two aspirin tablets, each having a mass of 325 mg, are dissolved in a full stomach whose volume is 1 L and whose pH is 2, what percent of the aspirin is in the form of neutral molecules?

17.114 What is the pH at 25 °C of water saturated with CO_2 at a partial pressure of 1.10 atm? The Henry's law constant for CO_2 at 25 °C is 3.1×10^{-2} mol/L-atm.

17.115 Excess $Ca(OH)_2$ is shaken with water to produce a saturated solution. The solution is filtered, and a 50.00-mL sample titrated with HCl requires 11.23 mL of 0.0983 M HCl to reach the end point. Calculate K_{sp} for $Ca(OH)_2$. Compare your result with that in Appendix D. 25 °C. Suggest a reason for any differences you find between your value and the one in Appendix D.

17.116 The osmotic pressure of a saturated solution of strontium sulfate at 25 °C is 21 torr. What is the solubility product of this salt at 25 °C?

17.117 A concentration of 10–100 parts per billion (by mass) of Ag^+ is an effective disinfectant in swimming pools. However, if the concentration exceeds this range, the Ag^+ can cause adverse health effects. One way to maintain an appropriate concentration of Ag^+ is to add a slightly soluble salt to the pool. Using K_{sp} values from Appendix D, calculate the equilibrium concentration of Ag^+ in parts per billion that would exist in equilibrium with **(a)** AgCl, **(b)** AgBr, **(c)** AgI.

17.118 Fluoridation of drinking water is employed in many places to aid in the prevention of tooth decay. Typically the F^- ion concentration is adjusted to about 1 ppb. Some water supplies are also "hard"; that is, they contain certain cations such as Ca^{2+} that interfere with the action of soap. Consider a case where the concentration of Ca^{2+} is 8 ppb. Could a precipitate of CaF_2 form under these conditions? (Make any necessary approximations.)

17.119 Baking soda (sodium bicarbonate, $NaHCO_3$) reacts with acids in foods to form carbonic acid (H_2CO_3), which in turn decomposes to water and carbon dioxide gas. In a cake batter, the $CO_2(g)$ forms bubbles and causes the cake to rise. **(a)** A rule of thumb in baking is that 1/2 teaspoon of baking soda is neutralized by one cup of sour milk. The acid component in sour milk is lactic acid, $CH_3CH(OH)COOH$. Write the chemical equation for this neutralization reaction. **(b)** The density of baking soda is 2.16 g/cm^3. Calculate the concentration of lactic acid in one cup of sour milk (assuming the rule of thumb applies), in units of mol/L. (One cup = 236.6 mL = 48 teaspoons). **(c)** If 1/2 teaspoon of baking soda is indeed completely neutralized by the lactic acid in sour milk, calculate the volume of carbon dioxide gas that would be produced at 1 atm pressure, in an oven set to 350 °F.

17.120 In nonaqueous solvents, it is possible to react HF to create H_2F^+. Which of these statements follows from this observation? **(a)** HF can act like a strong acid in nonaqueous solvents, **(b)** HF can act like a base in nonaqueous solvents, **(c)** HF is thermodynamically unstable, **(d)** There is an acid in the nonaqueous medium that is a stronger acid than HF.

Design an Experiment

The pK_a of acetic acid is 4.74. The pK_a of chloroacetic acid, $CH_2ClCOOH$, is 2.86. The pK_a of trichloroacetic acid, CCl_3COOH, is 0.66. **(a)** Why might this be the case? For example, one hypothesis is that the O—H bond of trichloroacetic acid is significantly more polar than the O—H bond in acetic acid due to chlorine being more electron-withdrawing than hydrogen, making the O—H bond in trichloroacetic acid weak. Another hypothesis is that the chlorines thermodynamically stabilize the conjugate base forms of these acids, and the more chlorines there are, the more stable the conjugate bases. Design a set of experiments or calculations to test these hypotheses. **(b)** Would you predict that the differences in pK_as of these acids would lead to differences in aqueous solubilities of their sodium salts? Design an experiment to test this hypothesis.

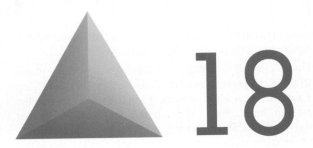

18

Chemistry of the Environment

The richness of life on Earth, represented in the chapter-opening photograph, is made possible by our planet's supportive atmosphere, the energy received from the Sun, and an abundance of water. These are the signature environmental features believed to be necessary for life.

As technology has advanced and the world human population has increased, humans have put new and greater stresses on the environment. Paradoxically, the very technology that can cause pollution also provides the tools to help understand and manage the environment in a beneficial way. Chemistry is often at the heart of environmental issues. The economic growth of both developed and developing nations depends critically on chemical processes that range from treatment of water supplies to industrial processes. Some of these processes produce products or by-products that are harmful to the environment.

We are now in a position to apply the principles we have learned in preceding chapters to an understanding of how our environment operates and how human activities affect it. To understand and protect the environment in which we live, we must understand how human-made and natural chemical compounds interact on land and in the sea and sky. Our daily actions as consumers turn on the same choices made by leading experts and governmental leaders: Each decision should reflect the costs versus the benefits of our choices. Unfortunately, the environmental impacts of our decisions are often subtle and not immediately evident.

▶ **TAHQUOMENON FALLS IN THE UPPER PENINSULA OF MICHIGAN,** one of the largest waterfalls east of the Mississippi in the United States.

18.4 HUMAN ACTIVITIES AND WATER QUALITY We consider how Earth's water is connected to the global climate and examine one measure of water quality: dissolved oxygen concentration. Water for drinking and for irrigation must be free of salts and pollutants.

18.5 GREEN CHEMISTRY We conclude by examining *green chemistry*, an international initiative to make industrial products, processes, and chemical reactions compatible with a sustainable society and environment.

18.1 | Earth's Atmosphere

Because most of us have never been very far from Earth's surface, we often take for granted the many ways in which the atmosphere determines the environment in which we live. In this section we examine some of the important characteristics of our planet's atmosphere.

The temperature of the atmosphere varies with altitude (▼ Figure 18.1), and the atmosphere is divided into four regions based on this temperature profile. Just above the surface, in the **troposphere**, the temperature normally decreases with increasing altitude, reaching a minimum of about 215 K at about 10 km. Nearly all of us live our entire lives in the troposphere. Howling winds and soft breezes, rain, and sunny skies—all that we normally think of as "weather"—occur in this region. Commercial jet aircraft typically fly about 10 km (33,000 ft) above Earth, an altitude that defines the upper limit of the troposphere, which we call the *tropopause*.

Above the tropopause, air temperature increases with altitude, reaching a maximum of about 275 K at about 50 km. The region from 10 km to 50 km is the **stratosphere**, and above it are the *mesosphere* and *thermosphere*. Notice in Figure 18.1 that the temperature extremes that form the boundaries between adjacent regions are denoted by the suffix *-pause*. The boundaries are important because gases mix across them relatively slowly. For example, pollutant gases generated in the troposphere pass through the tropopause and find their way into the stratosphere only very slowly.

Atmospheric pressure decreases with increasing elevation (Figure 18.1), declining much more rapidly at lower elevations than at higher ones because of the atmosphere's compressibility. Thus, the pressure decreases from an average value of 760 torr (101 kPa) at sea level to 2.3×10^{-3} torr (3.1×10^{-4} kPa) at 100 km, to only 1.0×10^{-6} torr (1.3×10^{-7} kPa) at 200 km.

The troposphere and stratosphere together account for 99.9% of the mass of the atmosphere, 75% of which is the mass in the troposphere. Nevertheless the thin upper atmosphere plays many important roles in determining the conditions of life at the surface.

▲ **GO FIGURE**

At what altitude is the atmospheric temperature lowest?

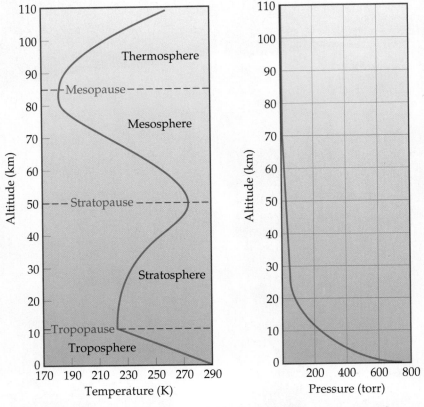

▲ **Figure 18.1 Temperature and pressure in the atmosphere vary as a function of altitude above sea level.**

Composition of the Atmosphere

Earth's atmosphere is constantly bombarded by radiation and energetic particles from the Sun. This barrage of energy has profound chemical and physical effects, especially in the upper regions of the atmosphere, above about 80 km (▶ Figure 18.2). In addition, because of Earth's gravitational field, heavier atoms and molecules tend to sink in the atmosphere, leaving lighter atoms and molecules at the top of the atmosphere. (This is why, as just noted, 75% of the atmosphere's mass is in the troposphere.) Because of all these factors, the composition of the atmosphere is not uniform.

▶ Table 18.1 shows the composition of dry air near sea level. Note that although traces of many substances are present, N_2 and O_2 make up about 99% of sea-level air. The noble gases and CO_2 make up most of the remainder.

Table 18.1   The Major Components of Dry Air near Sea Level

Component*	Content (mole fraction)	Molar Mass (g/mol)
Nitrogen	0.78084	28.013
Oxygen	0.20948	31.998
Argon	0.00934	39.948
Carbon dioxide	0.000400	44.0099
Neon	0.00001818	20.183
Helium	0.00000524	4.003
Methane	0.000002	16.043
Krypton	0.00000114	83.80
Hydrogen	0.0000005	2.0159
Nitrous oxide	0.0000005	44.0128
Xenon	0.000000087	131.30

*Ozone, sulfur dioxide, nitrogen dioxide, ammonia, and carbon monoxide are present as trace gases in variable amounts.

Give It Some Thought

How would you expect the ratio of atmospheric helium to argon to differ at 100 km elevation as compared with sea level?

High-energy solar particles create excited N and O atoms; visible light results as electrons in these atoms fall from upper-level states to lower-level states

▲ **Figure 18.2 The aurora borealis (northern lights).**

When applied to substances in aqueous solution, the concentration unit *parts per million* (ppm) refers to grams of substance per million grams of solution. ∞ (Section 13.4) When dealing with gases, however, 1 ppm means one part by *volume* in 1 million volumes of the whole. Because volume is proportional to number of moles of gas via the ideal-gas equation ($PV = nRT$), volume fraction and mole fraction are the same. Thus, 1 ppm of a trace constituent of the atmosphere amounts to 1 mol of that constituent in 1 million moles of air; that is, the concentration in parts per million is equal to the mole fraction times 10^6. For example, Table 18.1 lists the mole fraction of CO_2 in the atmosphere as 0.000400, which means its concentration in parts per million is $0.000400 \times 10^6 = 400$ ppm.

Other minor constituents of the troposphere, in addition to CO_2, are listed in ▼ Table 18.2.

Before we consider the chemical processes that occur in the atmosphere, let's review some of the properties of the two major components, N_2 and O_2. Recall that the N_2 molecule possesses a triple bond between the nitrogen atoms. ∞ (Section 8.3) This very strong bond (bond energy 941 kJ/mol) is largely responsible for the very low reactivity of N_2. The bond energy in O_2 is only 495 kJ/mol, making O_2 much more reactive than N_2. For example, oxygen reacts with many substances to form oxides. The oxides of nonmetals,

Table 18.2   Sources and Typical Concentrations of Some Minor Atmospheric Constituents

Constituent	Sources	Typical Concentration
Carbon dioxide, CO_2	Decomposition of organic matter, release from oceans, fossil-fuel combustion	400 ppm throughout troposphere
Carbon monoxide, CO	Decomposition of organic matter, industrial processes, fossil-fuel combustion	0.05 ppm in unpolluted air; 1–50 ppm in urban areas
Methane, CH_4	Decomposition of organic matter, natural-gas seepage, livestock emissions	1.82 ppm throughout troposphere
Nitric oxide, NO	Atmospheric electrical discharges, internal combustion engines, combustion of organic matter	0.01 ppm in unpolluted air; 0.2 ppm in smog
Ozone, O_3	Atmospheric electrical discharges, diffusion from the stratosphere, photochemical smog	0–0.01 ppm in unpolluted air; 0.5 ppm in photochemical smog
Sulfur dioxide, SO_2	Volcanic gases, forest fires, bacterial action, fossil-fuel combustion, industrial processes	0–0.01 ppm in unpolluted air; 0.1–2 ppm in polluted urban areas

SAMPLE
EXERCISE 18.1 | Calculating Concentration from Partial Pressure

What is the concentration, in parts per million, of water vapor in a sample of air if the partial pressure of the water is 0.80 torr and the total pressure of the air is 735 torr?

SOLUTION

Analyze We are given the partial pressure of water vapor and the total pressure of an air sample and asked to determine the water vapor concentration.

Plan Recall that the partial pressure of a component in a mixture of gases is given by the product of its mole fraction and the total pressure of the mixture ⚬⚬⚬ (Section 10.6):

$$P_{H_2O} = X_{H_2O}P_t$$

Solve Solving for the mole fraction of water vapor in the mixture, X_{H_2O}, gives

$$X_{H_2O} = \frac{P_{H_2O}}{P_t} = \frac{0.80 \text{ torr}}{735 \text{ torr}} = 0.0011$$

The concentration in ppm is the mole fraction times 10^6:

$$0.0011 \times 10^6 = 1100 \text{ ppm}$$

Practice Exercise 1

From the data in Table 18.1, the partial pressure of argon in dry air at an atmospheric pressure of 668 mm Hg is **(a)** 3.12 mm Hg, **(b)** 7.09 mm Hg, **(c)** 6.24 mm Hg, **(d)** 9.34 mm Hg, **(e)** 39.9 mm Hg.

Practice Exercise 2

The concentration of CO in a sample of air is 4.3 ppm. What is the partial pressure of the CO if the total air pressure is 695 torr?

such as SO_2, usually form acidic solutions when dissolved in water. The oxides of active metals, such as CaO, form basic solutions when dissolved in water. ⚬⚬⚬ (Section 7.7)

Photochemical Reactions in the Atmosphere

Although the atmosphere beyond the stratosphere contains only a small fraction of the atmospheric mass, it forms the outer defense against the hail of radiation and high-energy particles that continuously bombard Earth. As the bombarding radiation passes through the upper atmosphere, it causes two kinds of chemical changes: *photodissociation* and *photoionization*. These processes protect us from high-energy radiation by absorbing most of the radiation before it reaches the troposphere. If it were not for these photochemical processes, plant and animal life as we know it could not exist on Earth.

The Sun emits radiant energy over a wide range of wavelengths (▼ Figure 18.3). To understand the connection between the wavelength of radiation and its effect on

▲ GO FIGURE

Why does not the solar spectrum at sea level perfectly match the solar spectrum outside the atmosphere?

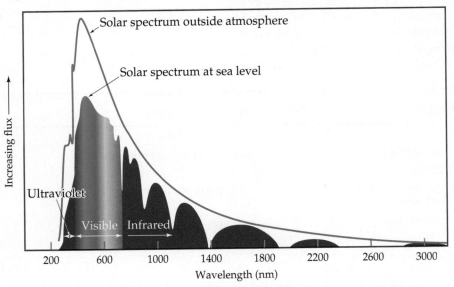

▲ Figure 18.3 **The solar spectrum above Earth's atmosphere compared to that at sea level.** The more structured curve at sea level is due to gases in the atmosphere absorbing specific wavelengths of light. "Flux," the unit on the vertical axis, is light energy per area per unit of time.

atoms and molecules, recall that electromagnetic radiation can be pictured as a stream of photons. ∞ (Section 6.2) The energy of each photon is given by $E = h\nu$, where h is Planck constant and ν is the radiation frequency. For a chemical change to occur when radiation strikes atoms or molecules, two conditions must be met. First, the incoming photons must have sufficient energy to break a chemical bond or remove an electron from the atom or molecule. Second, the atoms or molecules being bombarded must absorb these photons. When these requirements are met, the energy of the photons is used to do the work associated with some chemical change.

The rupture of a chemical bond resulting from absorption of a photon by a molecule is called **photodissociation**. No ions are formed when the bond between two atoms is cleaved by photodissociation. Instead, half the bonding electrons stay with one atom and half stay with the other atom. The result is two electrically neutral particles.

One of the most important processes occurring above an altitude of about 120 km is photodissociation of the oxygen molecule:

$$\ddot{O}{=}\ddot{O} + h\nu \longrightarrow :\ddot{O} + \ddot{O}: \qquad [18.1]$$

The minimum energy required to cause this change is determined by the bond energy (or *dissociation energy*) of O_2, 495 kJ/mol.

SAMPLE EXERCISE 18.2 Calculating the Wavelength Required to Break a Bond

What is the maximum wavelength of light, in nanometers, that has enough energy per photon to dissociate the O_2 molecule?

SOLUTION

Analyze We are asked to determine the wavelength of a photon that has just enough energy to break the O=O double bond in O_2.

Plan We first need to calculate the energy required to break the O=O double bond in one molecule and then find the wavelength of a photon of this energy.

Solve The dissociation energy of O_2 is 495 kJ/mol. Using this value and Avogadro's number, we can calculate the amount of energy needed to break the bond in a single O_2 molecule:

$$\left(495 \times 10^3 \frac{J}{mol}\right)\left(\frac{1\ mol}{6.022 \times 10^{23}\ molecules}\right) = 8.22 \times 10^{-19} \frac{J}{molecule}$$

We next use the Planck relationship, $E = h\nu$, ∞ (Equation 6.2) to calculate the frequency ν of a photon that has this amount of energy:

$$\nu = \frac{E}{h} = \frac{8.22 \times 10^{-19}\ J}{6.626 \times 10^{-34}\ J\text{-}s} = 1.24 \times 10^{15}\ s^{-1}$$

Finally, we use the relationship between frequency and wavelength ∞ (Section 6.1) to calculate the wavelength of the light:

$$\lambda = \frac{c}{\nu} = \left(\frac{3.00 \times 10^8\ m/s}{1.24 \times 10^{15}/s}\right)\left(\frac{10^9\ nm}{1\ m}\right) = 242\ nm$$

Thus, light of wavelength 242 nm, which is in the ultraviolet region of the electromagnetic spectrum, has sufficient energy per photon to photodissociate an O_2 molecule. Because photon energy increases as wavelength *decreases*, any photon of wavelength *shorter* than 242 nm will have sufficient energy to dissociate O_2.

Practice Exercise 1

The bond dissociation energy of the Br—Br bond is 193 kJ/mol. What wavelength of light has just sufficient energy to cause Br—Br bond dissociation?
(a) 620 nm (b) 310 nm (c) 148 nm (d) 6200 nm
(e) 563 nm

Practice Exercise 2

The bond energy in N_2 is 941 kJ/mol. What is the longest wavelength a photon can have and still have sufficient energy to dissociate N_2?

Fortunately for us, O_2 absorbs much of the high-energy, short-wavelength radiation from the solar spectrum before that radiation reaches the lower atmosphere. As it does, atomic oxygen, O, is formed. The dissociation of O_2 is very extensive at higher elevations. At 400 km, for example, only 1% of the oxygen is in the form of O_2; 99% is atomic oxygen. At 130 km, O_2 and atomic oxygen are just about equally abundant. Below 130 km, O_2 is more abundant than atomic oxygen because most of the solar energy has been absorbed in the upper atmosphere.

The dissociation energy of N_2 is very high, 941 kJ/mol. As you should have seen in working out Practice Exercise 2 of Sample Exercise 18.2, only photons having a wavelength shorter than 127 nm possess sufficient energy to dissociate N_2. Furthermore, N_2 does not readily absorb photons, even when they possess sufficient energy. As a result, very little atomic nitrogen is formed in the upper atmosphere by photodissociation of N_2.

Other photochemical processes besides photodissociation occur in the upper atmosphere, although their discovery has taken many twists and turns. In 1901 Guglielmo Marconi received a radio signal in St. John's, Newfoundland, that had been transmitted from Land's End, England, 2900 km away. Because people at the time thought radio waves traveled in straight lines, they assumed that the curvature of Earth's surface would make radio communication over large distances impossible. Marconi's successful experiment suggested that Earth's atmosphere in some way substantially affects radio-wave propagation. His discovery led to intensive study of the upper atmosphere. In about 1924, the existence of electrons in the upper atmosphere was established by experimental studies.

The electrons in the upper atmosphere result mainly from **photoionization**, which occurs when a molecule in the upper atmosphere absorbs solar radiation and the absorbed energy causes an electron to be ejected from the molecule. The molecule then becomes a positively charged ion. For photoionization to occur, therefore, a molecule must absorb a photon, and the photon must have enough energy to remove an electron. ∞ (Section 7.4) Notice that this is a very different process from photodissociation.

Four important photoionization processes occurring in the atmosphere above about 90 km are shown in ▼ Table 18.3. Photons of any wavelength shorter than the maximum lengths given in the table have enough energy to cause photoionization. A look back at Figure 18.3 shows you that virtually all of these high-energy photons are filtered out of the radiation reaching Earth because they are absorbed by the upper atmosphere.

Give It Some Thought

Explain the difference between photoionization and photodissociation.

Ozone in the Stratosphere

Although N_2, O_2, and atomic oxygen absorb photons having wavelengths shorter than 240 nm, ozone, O_3, is the key absorber of photons having wavelengths ranging from 240 to 310 nm, in the ultraviolet region of the electromagnetic spectrum. Ozone in the upper atmosphere protects us from these harmful high-energy photons, which would otherwise penetrate to Earth's surface. Let's consider how ozone forms in the upper atmosphere and how it absorbs photons.

By the time radiation from the Sun reaches an altitude of 90 km above Earth's surface, most of the short-wavelength radiation capable of photoionization has been absorbed. At this altitude, however, radiation capable of dissociating the O_2 molecule is sufficiently intense for photodissociation of O_2 (Equation 18.1) to remain important down to an altitude of 30 km. In the region between 30 and 90 km, however, the concentration of O_2 is much greater than the concentration of atomic oxygen. From this

Table 18.3 Photoionization Reactions for Four Components of the Atmosphere

Process	Ionization Energy (kJ/mol)	λ_{max}(nm)
$N_2 + h\nu \longrightarrow N_2^+ + e^-$	1495	80.1
$O_2 + h\nu \longrightarrow O_2^+ + e^-$	1205	99.3
$O + h\nu \longrightarrow O^+ + e^-$	1313	91.2
$NO + h\nu \longrightarrow NO^+ + e^-$	890	134.5

finding, we conclude that the oxygen atoms formed by photodissociation of O_2 in this region frequently collide with O_2 molecules and form ozone:

$$\ddot{\ddot{O}} + O_2 \longrightarrow O_3{}^* \qquad\qquad [18.2]$$

The asterisk on O_3 denotes that the product contains an excess of energy, because the reaction is exothermic. The 105 kJ/mol that is released must be transferred away from the $O_3{}^*$ molecule quickly or else the molecule will fly apart into O_2 and atomic O—a decomposition that is the reverse of the reaction by which $O_3{}^*$ is formed.

An energy-rich $O_3{}^*$ molecule can release its excess energy by colliding with another atom or molecule and transferring some of the excess energy to it. Let's use M to represent the atom or molecule with which $O_3{}^*$ collides. (Usually M is N_2 or O_2 because these are the most abundant molecules in the atmosphere.) The formation of $O_3{}^*$ and the transfer of excess energy to M are summarized by the equations

$$O(g) + O_2(g) \rightleftharpoons O_3{}^*(g) \qquad\qquad [18.3]$$
$$\underline{O_3{}^*(g) + M(g) \longrightarrow O_3(g) + M^*(g)} \qquad\qquad [18.4]$$
$$O(g) + O_2(g) + M(g) \longrightarrow O_3(g) + M^*(g) \qquad\qquad [18.5]$$

The rate at which the reactions of Equations 18.3 and 18.4 proceed depends on two factors that vary in opposite directions with increasing altitude. First, the Equation 18.3 reaction depends on the presence of O atoms. At low altitudes, most of the radiation energetic enough to dissociate O_2 into O atoms has been absorbed; thus, O atoms are more plentiful at higher altitudes. Second, Equations 18.3 and 18.4 both depend on molecular collisions. ∞ (Section 14.5) The concentration of molecules is greater at low altitudes, and so the rates of both reactions are greater at lower altitudes. Because these two reactions vary with altitude in opposite directions, the highest rate of O_3 formation occurs in a band at an altitude of about 50 km, near the stratopause (Figure 18.1). Overall, roughly 90% of Earth's ozone is found in the stratosphere.

 Give It Some Thought

Why do O_2 and N_2 molecules fail to filter out ultraviolet light with wavelengths between 240 and 310 nm?

The photodissociation of ozone reverses the reaction that forms it. We thus have a cycle of ozone formation and decomposition, summarized as follows:

$$O_2(g) + h\nu \longrightarrow O(g) + O(g)$$
$$O(g) + O_2(g) + M(g) \longrightarrow O_3(g) + M^*(g) \quad \text{(heat released)}$$
$$O_3(g) + h\nu \longrightarrow O_2(g) + O(g)$$
$$O(g) + O(g) + M(g) \longrightarrow O_2(g) + M^*(g) \quad \text{(heat released)}$$

The first and third processes are photochemical; they use a solar photon to initiate a chemical reaction. The second and fourth are exothermic chemical reactions. The net result of the four reactions is a cycle in which solar radiant energy is converted into thermal energy. The ozone cycle in the stratosphere is responsible for the rise in temperature that reaches its maximum at the stratopause (Figure 18.1).

The reactions of the ozone cycle account for some, but not all, of the facts about the ozone layer. Many chemical reactions occur that involve substances other than oxygen. We must also consider the effects of turbulence and winds that mix up the stratosphere. A complicated picture results. The overall result of ozone formation and removal reactions, coupled with atmospheric turbulence and other factors, is to produce the upper-atmosphere ozone profile shown in ▶ Figure 18.4, with a maximum ozone concentration occurring at an altitude of about 25 km. This band of relatively high ozone concentration is referred to as the "ozone layer" or the "ozone shield."

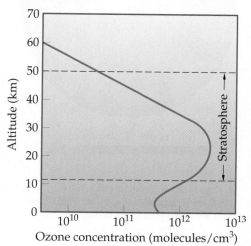

GO FIGURE

Estimate the ozone concentration in moles per liter for the peak value in this graph.

▲ Figure 18.4 **Variation in ozone concentration in the atmosphere as a function of altitude.**

Photons with wavelengths shorter than about 300 nm are energetic enough to break many kinds of single chemical bonds. Thus, the "ozone shield" is essential for our continued well-being. The ozone molecules that form this essential shield against high-energy radiation represent only a tiny fraction of the oxygen atoms present in the stratosphere, however, because these molecules are continually destroyed even as they are formed.

18.2 | Human Activities and Earth's Atmosphere

Both natural and *anthropogenic* (human-caused) events can modify Earth's atmosphere. One impressive natural event was the eruption of Mount Pinatubo in June 1991 (◀ Figure 18.5). The volcano ejected approximately $10 \, km^3$ of material into the stratosphere, causing a 10% drop in the amount of sunlight reaching Earth's surface during the next 2 years. That drop in sunlight led to a temporary 0.5 °C drop in Earth's surface temperature. The volcanic particles that made it to the stratosphere remained there for approximately 3 years, *raising* the temperature of the stratosphere by several degrees due to light absorption. Measurements of the stratospheric ozone concentration showed significantly increased ozone decomposition in this 3-year period.

▲ Figure 18.5 **Mount Pinatubo erupts, June 1991.**

The Ozone Layer and Its Depletion

The ozone layer protects Earth's surface from the damaging ultraviolet (UV) radiation. Therefore, if the concentration of ozone in the stratosphere decreases substantially, more UV radiation will reach Earth's surface, causing unwanted photochemical reactions, including reactions correlated with skin cancer. Satellite monitoring of ozone, which began in 1978, has revealed a depletion of ozone in the stratosphere that is particularly severe over Antarctica, a phenomenon known as the *ozone hole* (◀ Figure 18.6). The first scientific paper on this phenomenon appeared in 1985, and the National Aeronautics and Space Administration (NASA) maintains an "Ozone Hole Watch" website with daily updates and data from 1999 to the present.

In 1995 the Nobel Prize in Chemistry was awarded to F. Sherwood Rowland, Mario Molina, and Paul Crutzen for their studies of ozone depletion. In 1970 Crutzen showed that naturally occurring nitrogen oxides catalytically destroy ozone. Rowland and Molina recognized in 1974 that chlorine from **chlorofluorocarbons** (CFCs) may deplete the ozone layer. These substances, principally $CFCl_3$ and CF_2Cl_2, do not occur in nature and have been widely used as propellants in spray cans, as refrigerant and air-conditioner gases, and as foaming agents for plastics. They are virtually unreactive in the lower atmosphere. Furthermore, they are relatively insoluble in water and are therefore not removed from the atmosphere by rainfall or by dissolution in the oceans. Unfortunately, the lack of reactivity that makes them commercially useful also allows them to survive in the atmosphere and to diffuse into the stratosphere. It is estimated that several million tons of chlorofluorocarbons are now present in the atmosphere.

As CFCs diffuse into the stratosphere, they are exposed to high-energy radiation, which can cause photodissociation. Because C—Cl bonds are considerably weaker than C—F bonds, free chlorine atoms are formed readily in the presence of light with wavelengths in the range from 190 to 225 nm, as shown in this typical reaction:

$$CF_2Cl_2(g) + h\nu \longrightarrow CF_2Cl(g) + Cl(g) \qquad [18.6]$$

Calculations suggest that chlorine atom formation occurs at the greatest rate at an altitude of about 30 km, the altitude at which ozone is at its highest concentration.

Atomic chlorine reacts rapidly with ozone to form chlorine monoxide and molecular oxygen:

$$Cl(g) + O_3(g) \longrightarrow ClO(g) + O_2(g) \qquad [18.7]$$

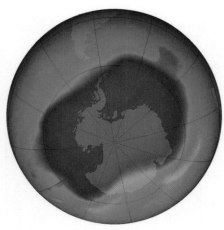

Total ozone (Dobson units)

110   220   330   440   550

▲ Figure 18.6 **Ozone present in the Southern Hemisphere, Sept. 24, 2006.** The data were taken from an orbiting satellite. This day had the lowest stratospheric ozone concentration yet recorded. One "Dobson unit" corresponds to 2.69×10^{16} ozone molecules in a 1 cm^2 column of atmosphere.

This reaction follows a second-order rate law with a very large rate constant:

$$\text{Rate} = k[\text{Cl}][\text{O}_3] \quad k = 7.2 \times 10^9 \, M^{-1} \, s^{-1} \text{ at 298 K} \qquad [18.8]$$

Under certain conditions, the ClO generated in Equation 18.7 can react to regenerate free Cl atoms. One way that this can happen is by photodissociation of ClO:

$$\text{ClO}(g) + h\nu \longrightarrow \text{Cl}(g) + \text{O}(g) \qquad [18.9]$$

The Cl atoms generated in Equations 18.6 and 18.9 can react with more O_3, according to Equation 18.7. The result is a sequence of reactions that accomplishes the Cl-catalyzed decomposition of O_3 to O_2:

$$2\,\text{Cl}(g) + 2\,\text{O}_3(g) \longrightarrow 2\,\text{ClO}(g) + 2\,\text{O}_2(g)$$
$$2\,\text{ClO}(g) + h\nu \longrightarrow 2\,\text{Cl}(g) + 2\,\text{O}(g)$$
$$\underline{\text{O}(g) + \text{O}(g) \longrightarrow \text{O}_2(g)}$$
$$2\,\text{Cl}(g) + 2\,\text{O}_3(g) + 2\,\text{ClO}(g) + 2\,\text{O}(g) \longrightarrow 2\,\text{Cl}(g) + 2\,\text{ClO}(g) + 3\,\text{O}_2(g) + 2\,\text{O}(g)$$

The equation can be simplified by eliminating like species from each side to give

$$2\,\text{O}_3(g) \xrightarrow{\text{Cl}} 3\,\text{O}_2(g) \qquad [18.10]$$

Because the rate of Equation 18.7 increases linearly with [Cl], the rate at which ozone is destroyed increases as the quantity of Cl atoms increases. Thus, the greater the amount of CFCs that diffuse into the stratosphere, the faster the destruction of the ozone layer. Even though troposphere-to-stratosphere diffusion rates are slow, a substantial thinning of the ozone layer over the South Pole has been observed, particularly during September and October (Figure 18.6).

Give It Some Thought

Since the rate of ozone destruction depends on [Cl], can Cl be considered a catalyst for the reaction of Equation 18.10?

Because of the environmental problems associated with CFCs, steps have been taken to limit their manufacture and use. A major step was the signing in 1987 of the Montreal Protocol on Substances That Deplete the Ozone Layer, in which participating nations agreed to reduce CFC production. More stringent limits were set in 1992, when representatives of approximately 100 nations agreed to ban the production and use of CFCs by 1996, with some exceptions for "essential uses." Since then, the production of CFCs has indeed dropped precipitously. Images such as that shown in Figure 18.6 taken annually reveal that the depth and size of the ozone hole has begun to decline. Nevertheless, because CFCs are unreactive and because they diffuse so slowly into the stratosphere, scientists estimate that ozone depletion will continue for many years to come. What substances have replaced CFCs? At this time, the main alternatives are hydrofluorocarbons (HFCs), compounds in which C—H bonds replace the C—Cl bonds of CFCs. One such compound in current use is CH_2FCF_3, known as HFC-134a. While the HFCs are a big improvement over the CFCs because they contain no C—Cl bonds, it turns out that they are potent greenhouse warming gases, with which we will deal shortly.

There are no naturally occurring CFCs, but some natural sources contribute chlorine and bromine to the atmosphere, and, just like halogens from CFC, these naturally occurring Cl and Br atoms can participate in ozone-depleting reactions. The principal natural sources are methyl bromide and methyl chloride, which are emitted from the oceans. It is estimated that these molecules contribute less than a third of the total Cl and Br in the atmosphere; the remaining two-thirds is a result of human activities.

Volcanoes are a source of HCl, but generally the HCl they release reacts with water in the troposphere and does not make it to the upper atmosphere.

Sulfur Compounds and Acid Rain

Sulfur-containing compounds are present to some extent in the natural, unpolluted atmosphere. They originate in the bacterial decay of organic matter, in volcanic gases, and from other sources listed in Table 18.2. The amount of these compounds released into the atmosphere worldwide from natural sources is about 24×10^{12} g per year, less than the amount from human activities, about 80×10^{12} g per year (principally related to combustion of fuels).

Sulfur compounds, chiefly sulfur dioxide, SO_2, are among the most unpleasant and harmful of the common pollutant gases. ◀ Table 18.4 shows the concentrations of several pollutant gases in a *typical* urban environment (where by *typical* we mean one that is not particularly affected by smog). According to these data, the level of sulfur dioxide is 0.08 ppm or higher about half the time. This concentration is considerably lower than that of other pollutants, notably carbon monoxide. Nevertheless, SO_2 is regarded as the most serious health hazard among the pollutants shown, especially for people with respiratory difficulties.

Combustion of coal accounts for about 65% of the SO_2 released annually in the United States, and combustion of oil accounts for another 20%. The majority of this amount is from coal-burning electrical power plants, which generate about 50% of our electricity. The extent to which SO_2 emissions are a problem when coal is burned depends on the amount of sulfur in the coal. Because of concern about SO_2 pollution, low-sulfur coal is in greater demand and is thus more expensive. Much of the coal from east of the Mississippi is relatively high in sulfur content, up to 6% by mass. Much of the coal from the western states has a lower sulfur content, but also a lower heat content per unit mass, so the difference in sulfur content per unit of heat produced is not as large as is often assumed.

In 2010, the U.S. Environmental Protection Agency set new standards to reduce SO_2 emissions, the first change in nearly 40 years. The old standard of 140 parts per billion, measured over 24 h, has been replaced by a standard of 75 parts per billion, measured over 1 h. The impact of SO_2 emissions is not restricted to the United States, however. China, which generates about 70% of its energy from coal, is the world's largest generator of SO_2, producing about 35×10^{12} g annually from coal and other sources. As a result, that nation has a major problem with SO_2 pollution and has set targets to reduce emissions with some success. India, which is projected to surpass China as the largest importer of coal by 2014, is also concerned about increased SO_2 emissions. Nations will need to work together to address what has truly become a global issue.

Sulfur dioxide is harmful to both human health and property; furthermore, atmospheric SO_2 can be oxidized to SO_3 by several pathways (such as reaction with O_2 or O_3). When SO_3 dissolves in water, it produces sulfuric acid:

$$SO_3(g) + H_2O(l) \longrightarrow H_2SO_4(aq)$$

Many of the environmental effects ascribed to SO_2 are actually due to H_2SO_4.

The presence of SO_2 in the atmosphere and the sulfuric acid it produces result in the phenomenon of **acid rain**. (Nitrogen oxides, which form nitric acid, are also major contributors to acid rain.) Uncontaminated rainwater generally has a pH value of about 5.6. The primary source of this natural acidity is CO_2, which reacts with water to form carbonic acid, H_2CO_3. Acid rain typically has a pH value of about 4. This shift toward greater acidity has affected many lakes in northern Europe, the northern United States, and Canada, reducing fish populations and affecting other parts of the ecological network in the lakes and surrounding forests.

The pH of most natural waters containing living organisms is between 6.5 and 8.5, but as ▶ Figure 18.7 shows, freshwater pH values are far below 6.5 in many parts of the continental United States. At pH levels below 4.0, all vertebrates, most invertebrates, and many microorganisms are destroyed. The lakes most susceptible to damage are those with low concentrations of basic ions, such as HCO_3^-, that would act as a buffer to minimize changes in pH. Some of these lakes are recovering as sulfur emissions from fossil

Table 18.4 Median Concentrations of Atmospheric Pollutants in a Typical Urban Atmosphere

Pollutant	Concentration (ppm)
Carbon monoxide	10
Hydrocarbons	3
Sulfur dioxide	0.08
Nitrogen oxides	0.05
Total oxidants (ozone and others)	0.02

▲ GO FIGURE

Why is the pH found in freshwater sources in the Eastern half of the United States dramatically lower than found in the western half?

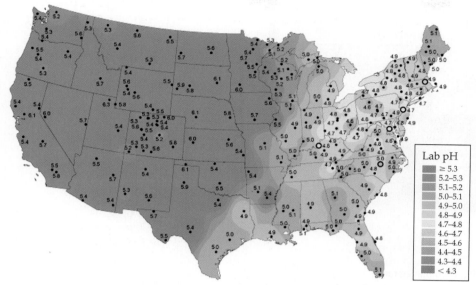

Lab pH	
	≥ 5.3
	5.2–5.3
	5.1–5.2
	5.0–5.1
	4.9–5.0
	4.8–4.9
	4.7–4.8
	4.6–4.7
	4.5–4.6
	4.4–4.5
	4.3–4.4
	< 4.3

▲ **Figure 18.7 Water pH values from freshwater sites across the United States, 2008.**
The numbered dots indicate the locations of monitoring stations.

(a)

(b)

▲ Figure 18.8 **Damage from acid rain.** The right photograph, recently taken, shows how the statue has lost detail in its carvings.

fuel combustion decrease, in part because of the Clean Air Act. In the period 1990–2010 the average ambient air concentration of SO_2 nationwide has declined by 75%.

Because acids react with metals and with carbonates, acid rain is corrosive both to metals and to stone building materials. Marble and limestone, for example, whose major constituent is $CaCO_3$, are readily attacked by acid rain (▲ Figure 18.8). Billions of dollars each year are lost because of corrosion due to SO_2 pollution.

GO FIGURE

What is the major solid product resulting from removal of SO_2 from furnace gas?

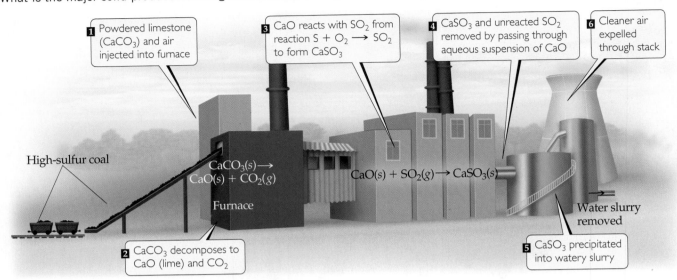

1 Powdered limestone ($CaCO_3$) and air injected into furnace

3 CaO reacts with SO_2 from reaction $S + O_2 \longrightarrow SO_2$ to form $CaSO_3$

4 $CaSO_3$ and unreacted SO_2 removed by passing through aqueous suspension of CaO

6 Cleaner air expelled through stack

High-sulfur coal

$$CaCO_3(s) \longrightarrow CaO(s) + CO_2(g)$$

Furnace

$$CaO(s) + SO_2(g) \longrightarrow CaSO_3(s)$$

Water slurry removed

2 $CaCO_3$ decomposes to CaO (lime) and CO_2

5 $CaSO_3$ precipitated into watery slurry

▲ Figure 18.9 **One method for removing SO_2 from combusted fuel.**

One way to reduce the quantity of SO_2 released into the environment is to remove sulfur from coal and oil before these fuels are burned. Although difficult and expensive, several methods have been developed. Powdered limestone ($CaCO_3$), for example, can be injected into the furnace of a power plant, where it decomposes into lime (CaO) and carbon dioxide:

$$CaCO_3(s) \longrightarrow CaO(s) + CO_2(g)$$

The CaO then reacts with SO_2 to form calcium sulfite:

$$CaO(s) + SO_2(g) \longrightarrow CaSO_3(s)$$

The solid particles of $CaSO_3$, as well as much of the unreacted SO_2, can be removed from the furnace gas by passing it through an aqueous suspension of CaO (▲ Figure 18.9). Not all the SO_2 is removed, however, and given the enormous quantities of coal and oil burned worldwide, pollution by SO_2 will probably remain a problem for some time.

▲ **Give It Some Thought**

What chemical behavior associated with sulfur oxides gives rise to acid rain?

Nitrogen Oxides and Photochemical Smog

Nitrogen oxides are primary components of smog, a phenomenon with which city dwellers are all too familiar. The term *smog* refers to the pollution condition that occurs in certain urban environments when weather conditions produce a relatively stagnant air mass. The smog made famous by Los Angeles, but now common in many other urban areas as well, is more accurately described as **photochemical smog** because photochemical processes play a major role in its formation (◀ Figure 18.10).

The majority of nitrogen oxide emissions (about 50%) comes from cars, buses, and other forms of transportation. Nitric oxide, NO, forms in small quantities in the cylinders of internal combustion engines in the reaction

$$N_2(g) + O_2(g) \rightleftharpoons 2\,NO(g) \quad \Delta H = 180.8\ kJ \qquad [18.11]$$

▲ Figure 18.10 **Photochemical smog is produced largely by the action of sunlight on vehicle exhaust gases.**

As noted in the "Chemistry Put to Work" box in Section 15.7, the equilibrium constant for this reaction increases from about 10^{-15} at 300 K to about 0.05 at 2400 K (approximate temperature in the cylinder of an engine during combustion). Thus, the reaction is more favorable at higher temperatures. In fact, some NO is formed in any high-temperature combustion. As a result, electrical power plants are also major contributors to nitrogen oxide pollution.

Before the installation of pollution-control devices on automobiles, typical emission levels of NO_x were 4 g/mi. (The x is either 1 or 2 because both NO and NO_2 are formed, although NO predominates.) Starting in 2004, the auto emission standards for NO_x called for a phased-in reduction to 0.07 g/mi by 2009, which was achieved.

In air, nitric oxide is rapidly oxidized to nitrogen dioxide:

$$2\,NO(g) + O_2(g) \rightleftharpoons 2\,NO_2(g) \quad \Delta H = -113.1\ kJ \qquad [18.12]$$

The equilibrium constant for this reaction decreases from about 10^{12} at 300 K to about 10^{-5} at 2400 K.

The photodissociation of NO_2 initiates the reactions associated with photochemical smog. Dissociation of NO_2 requires 304 kJ/mol, which corresponds to a photon wavelength of 393 nm. In sunlight, therefore, NO_2 dissociates to NO and O:

$$NO_2(g) + h\nu \longrightarrow NO(g) + O(g) \qquad [18.13]$$

The atomic oxygen formed undergoes several reactions, one of which gives ozone, as described earlier:

$$O(g) + O_2 + M(g) \longrightarrow O_3(g) + M^*(g) \qquad [18.14]$$

Although it is an essential UV screen in the upper atmosphere, ozone is an undesirable pollutant in the troposphere. It is extremely reactive and toxic, and breathing air that contains appreciable amounts of ozone can be especially dangerous for asthma sufferers, exercisers, and the elderly. We therefore have two ozone problems: excessive amounts in many urban environments, where it is harmful, and depletion in the stratosphere, where its presence is vital.

In addition to nitrogen oxides and carbon monoxide, an automobile engine also emits unburned *hydrocarbons* as pollutants. These organic compounds are the principal components of gasoline and of many compounds we use as fuel (propane, C_3H_8, and butane, C_4H_{10}; for example), but are major components of smog. A typical engine without effective emission controls emits about 10 to 15 g of hydrocarbons per mile. Current standards require that hydrocarbon emissions be less than 0.075 g/mi. Hydrocarbons are also emitted naturally from living organisms (see "A Closer Look" box later in this section).

Reduction or elimination of smog requires that the ingredients essential to its formation be removed from automobile exhaust. Catalytic converters reduce the levels of NO_x and hydrocarbons, two of the major components of smog. (See the "Chemistry Put to Work: Catalytic Converters" in Section 14.7.)

 Give It Some Thought

What photochemical reaction involving nitrogen oxides initiates the formation of photochemical smog?

Greenhouse Gases: Water Vapor, Carbon Dioxide, and Climate

In addition to screening out harmful short-wavelength radiation, the atmosphere is essential in maintaining a reasonably uniform and moderate temperature on Earth's surface. Earth is in overall thermal balance with its surroundings. This means that the planet radiates energy into space at a rate equal to the rate at which it absorbs energy from the Sun. **Figure 18.11** shows the distribution of radiation to and from Earth's surface, and **Figure 18.12** shows which portion of the infrared radiation leaving the surface is absorbed by atmospheric water vapor and carbon dioxide. In absorbing this radiation, these two atmospheric gases help maintain a livable uniform temperature at the surface by holding in, as it were, the infrared radiation, which we feel as heat.

▲GO FIGURE

What fraction of the incoming solar radiation is absorbed by Earth's surface?

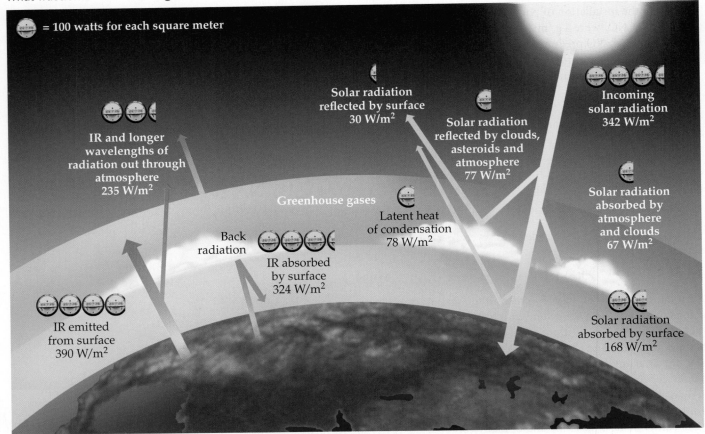

▲ **Figure 18.11 Earth's thermal balance.** The amount of radiation reaching the surface of the planet is approximately equal to the amount radiated back into space.

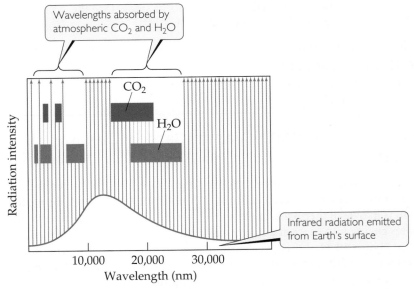

▲ **Figure 18.12 Portions of the infrared radiation emitted by Earth's surface that are absorbed by atmospheric CO₂ and H₂O.**

The influence of H_2O, CO_2, and certain other atmospheric gases on Earth's temperature is called the *greenhouse effect* because in trapping infrared radiation these gases act much like the glass of a greenhouse. The gases themselves are called **greenhouse gases**.

Water vapor makes the largest contribution to the greenhouse effect. The partial pressure of water vapor in the atmosphere varies greatly from place to place and time to time but is generally highest near Earth's surface and drops off with increasing elevation. Because water vapor absorbs infrared radiation so strongly, it plays the major role in maintaining the atmospheric temperature at night, when the surface is emitting radiation into space and not receiving energy from the Sun. In very dry desert climates, where the water-vapor concentration is low, it may be extremely hot during the day but very cold at night. In the absence of a layer of water vapor to absorb and then radiate part of the infrared radiation back to Earth, the surface loses this radiation into space and cools off very rapidly.

Carbon dioxide plays a secondary but very important role in maintaining the surface temperature. The worldwide combustion of fossil fuels, principally coal and oil, on a prodigious scale in the modern era has sharply increased carbon dioxide levels in the atmosphere. To get a sense of the amount of CO_2 produced—for example, by the combustion of hydrocarbons and other carbon-containing substances, which are the components of fossil fuels—consider the combustion of butane, C_4H_{10}. Combustion of 1.00 g of C_4H_{10} produces 3.03 g of CO_2. ∞ (Section 3.6) Similarly, a gallon (3.78 L) of gasoline (density 0.7 g/mL, approximate composition C_8H_{18}) produces about 8 kg of CO_2. Combustion of fossil fuels releases about 2.2×10^{16} g (24 billion tons) of CO_2 into the atmosphere annually, with the largest quantity coming from transportation vehicles.

Much CO_2 is absorbed into oceans or used by plants. Nevertheless, we are now generating CO_2 much faster than it is being absorbed or used. Analysis of air trapped in ice cores taken from Antarctica and Greenland makes it possible to determine the atmospheric levels of CO_2 during the past 160,000 years. These measurements reveal that the level of CO_2 remained fairly constant from the last Ice Age, some 10,000 years ago, until roughly the beginning of the Industrial Revolution, about 300 years ago. Since that time, the concentration of CO_2 has increased by about 30% to a current high of about 400 ppm (▼ Figure 18.13). Climate scientists believe that the CO_2 level has not been this high since 3 to 5 million years ago.

▲ GO FIGURE

What is the source of the slight but steady increase in *slope* of this curve over time?

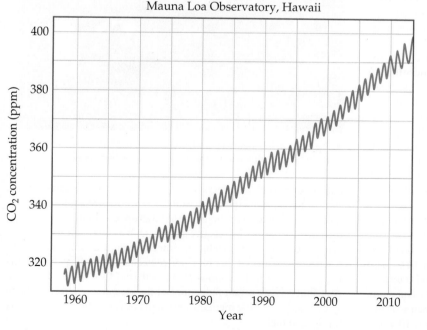

▲ **Figure 18.13 Rising CO₂ levels.** The sawtooth shape of the graph is due to regular seasonal variations in CO_2 concentration for each year.

The consensus among climate scientists is that the increase in atmospheric CO_2 is perturbing Earth's climate and is very likely playing a role in the observed increase in the average global air temperature of 0.3–0.6 °C over the past century. Scientists often use the term *climate change* instead of *global warming* to refer to this effect because as the Earth's temperature increases, winds and ocean currents are affected in ways that cool some areas and warm others.

On the basis of present and expected future rates of fossil-fuel use, the atmospheric CO_2 level is expected to double from its present level sometime between 2050 and 2100. Computer models predict that this increase will result in an average global temperature increase of 1 °C to 3 °C. Because so many factors go into determining climate, we cannot predict with certainty what changes will occur because of this warming. Clearly, however, humanity has acquired the potential, by changing the concentrations of CO_2 and other heat-trapping gases in the atmosphere, to substantially alter the climate of the planet.

The climate change threat posed by atmospheric CO_2 has sparked considerable research into ways of capturing the gas at its largest combustion sources and storing it under ground or under the seafloor. There is also much interest in developing new ways to use CO_2 as a chemical feedstock. However, the approximately 115 million tons of CO_2 used annually by the global chemical industry is but a small fraction of the approximately 24 billion tons of annual CO_2 emissions. The use of CO_2 as a raw material will probably never be great enough to significantly reduce its atmospheric concentration.

 Give It Some Thought

Explain why nighttime temperatures remain higher in locations where there is higher humidity.

 A Closer Look

Other Greenhouse Gases

Although CO_2 receives most of the attention, other gases contribute to the greenhouse effect, including methane, CH_4, hydrofluorocarbons (HFCs), and chlorofluorocarbons (CFCs).

HFCs have replaced CFCs in a host of applications, including refrigerants and air-conditioner gases. Although they do not contribute to the depletion of the ozone layer, HFCs are nevertheless potent greenhouse gases. For example, one of the byproduct molecules from production of HFCs that are used in commerce is HCF_3, which is estimated to have a global warming potential, gram for gram, more than 14,000 times that of CO_2. The total concentration of HFCs in the atmosphere has been increasing about 10% per year. Thus, these substances are becoming increasingly important contributors to the greenhouse effect. Methane already makes a significant contribution to the greenhouse effect. Studies of atmospheric gas trapped long ago in the Greenland and Antarctic ice sheets show that the atmospheric methane concentration has increased from preindustrial values of 0.3 to 0.7 ppm to the present value of about 1.8 ppm. The major sources of methane are associated with agriculture and fossil-fuel use.

Methane is formed in biological processes that occur in low-oxygen environments. Anaerobic bacteria, which flourish in swamps and landfills, near the roots of rice plants, and in the digestive systems of cows and other ruminant animals, produce methane (▶ Figure 18.14). It also leaks into the atmosphere during natural-gas extraction and transport. It is estimated that about two-thirds of present-day methane emissions, which are increasing by about 1% per year, are related to human activities.

Methane has a half-life in the atmosphere of about 10 years, whereas CO_2 is much longer-lived. This might seem a good thing,

▲ Figure 18.14 **Methane production.** Ruminant animals, such as cows and sheep, produce methane in their digestive systems.

but there are indirect effects to consider. Methane is oxidized in the stratosphere, producing water vapor, a powerful greenhouse gas that is otherwise virtually absent from the stratosphere. In the troposphere, methane is attacked by reactive species such as OH radicals or nitrogen oxides, eventually producing other greenhouse gases, such as O_3. It has been estimated that on a per-molecule level, the global warming potential of CH_4 is about 21 times that of CO_2. Given this large contribution, important reductions of the greenhouse effect could be achieved by reducing methane emissions or capturing the emissions for use as a fuel.

Related Exercises: 18.67, 18.69

18.3 | Earth's Water

Water covers 72% of Earth's surface and is essential to life. Our bodies are about 65% water by mass. Because of extensive hydrogen bonding, water has unusually high melting and boiling points and a high heat capacity. ∞ (Section 11.2) Water's highly polar character is responsible for its exceptional ability to dissolve a wide range of ionic and polar-covalent substances. Many reactions occur in water, including reactions in which H_2O itself is a reactant. Recall, for example, that H_2O can participate in acid–base reactions as either a proton donor or a proton acceptor. ∞ (Section 16.3) All these properties play a role in our environment.

The Global Water Cycle

All the water on Earth is connected in a global water cycle (▼ Figure 18.15). Most of the processes depicted here rely on the phase changes of water. For instance, warmed by the Sun, liquid water in the oceans evaporates into the atmosphere as water vapor and condenses into liquid water droplets that we see as clouds. Water droplets in the clouds can crystallize to ice, which can precipitate as hail or snow. Once on the ground, the hail or snow melts to liquid water, which soaks into the ground. If conditions are right, it is also possible for ice on the ground to sublime to water vapor in the atmosphere.

Give It Some Thought

Consider the phase diagram for water shown in Figure 11.28 (page 465). In what pressure range and in what temperature range must H_2O exist in order for $H_2O(s)$ to sublime to $H_2O(g)$?

GO FIGURE

Which processes shown in this figure involve the phase transition $H_2O(l) \longrightarrow H_2O(g)$?

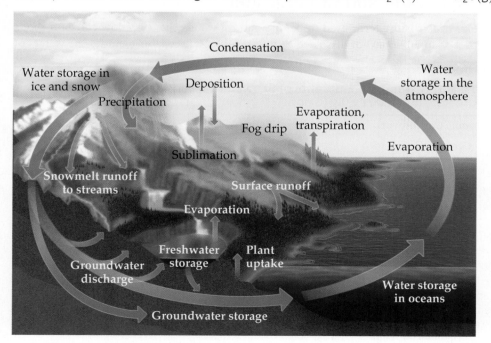

▲ Figure 18.15 **The global water cycle.**

Salt Water: Earth's Oceans and Seas

The vast layer of salty water that covers so much of the planet is in actuality one large connected body and is generally constant in composition. For this reason, oceanographers speak of a *world ocean* rather than of the separate oceans we learn about in geography books.

The world ocean is huge, having a volume of 1.35×10^9 km^3 and containing 97.2% of all the water on Earth. Of the remaining 2.8%, 2.1% is in the form of ice caps and glaciers. All the freshwater—in lakes, in rivers, and in the ground—amounts to only 0.6%. Most of the remaining 0.1% is in brackish (salty) water, such as that in the Great Salt Lake in Utah.

Seawater is often referred to as saline water. The **salinity** of seawater is the mass in grams of dry salts present in 1 kg of seawater. In the world ocean, salinity averages about 35. To put it another way, seawater contains about 3.5% dissolved salts by mass. The list of elements present in seawater is very long. Most, however, are present only in very low concentrations. ▼ Table 18.5 lists the 11 ionic species most abundant in seawater.

Seawater temperature varies as a function of depth (▶ Figure 18.16), as does salinity and density. Sunlight penetrates well only 200 m into the water; the region between 200 m and 1000 m deep is the "twilight zone," where visible light is faint. Below 1000 m, the ocean is pitch-black and cold, about 4 °C. The transport of heat, salt, and other chemicals throughout the ocean is influenced by these changes in the physical properties of seawater, and in turn the changes in the way heat and substances are transported affects ocean currents and the global climate.

The sea is so vast that if the concentration of a substance in seawater is 1 part per billion (1×10^{-6} g/kg of water), there is 1×10^{12} kg of the substance in the world ocean. Nevertheless, because of high extracting costs, only three substances are obtained from seawater in commercially important amounts: sodium chloride, bromine (from bromide salts), and magnesium (from its salts).

Absorption of CO_2 by the ocean plays a large role in global climate. Because carbon dioxide and water form carbonic acid, the H_2CO_3 concentration in the ocean increases as the water absorbs atmospheric CO_2. Most of the carbon in the ocean, however, is in the form of HCO_3^- and CO_3^{2-} ions, which form a buffer system that maintains the ocean's pH between 8.0 and 8.3. The pH of the ocean is predicted to decrease as the concentration of CO_2 in the atmosphere increases, as discussed in the "Chemistry and Life" box on ocean acidification in Section 17.5.

Freshwater and Groundwater

Freshwater is the term used to denote natural waters that have low concentrations (less than 500 ppm) of dissolved salts and solids. Freshwater includes the waters of

Table 18.5 Ionic Constituents of Seawater Present in Concentrations Greater than 0.001 g/kg (1 ppm)

Ionic Constituent	Salinity	Concentration (M)
Chloride, Cl$^-$	19.35	0.55
Sodium, Na$^+$	10.76	0.47
Sulfate, SO$_4^{2-}$	2.71	0.028
Magnesium, Mg^{2+}	1.29	0.054
Calcium, Ca^{2+}	0.412	0.010
Potassium, K$^+$	0.40	0.010
Carbon dioxide*	0.106	2.3×10^{-3}
Bromide, Br$^-$	0.067	8.3×10^{-4}
Boric acid, H$_3$BO$_3$	0.027	4.3×10^{-4}
Strontium, Sr^{2+}	0.0079	9.1×10^{-5}
Fluoride, F$^-$	0.0013	7.0×10^{-5}

*CO_2 is present in seawater as HCO_3^- and CO_3^{2-}.

▲ GO FIGURE

How would you expect the temperature variation to affect the density of seawater in the range 0 to 100 m depth?

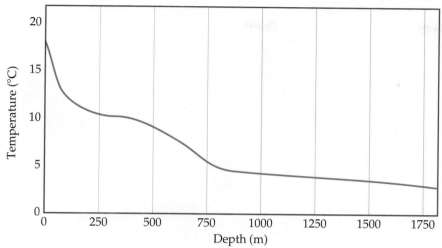

▲ **Figure 18.16 Typical average temperature of mid-latitude seawater as a function of depth.**

lakes, rivers, ponds, and streams. The United States is fortunate in its abundance of freshwater—1.7×10^{15} L (660 trillion gallons) is the estimated reserve, which is renewed by rainfall. An estimated 9×10^{11} L of freshwater is used every day in the United States. Most of this is used for agriculture (41%) and hydroelectric power (39%), with small amounts for industry (6%), household needs (6%), and drinking water (1%). An adult drinks about 2 L of water per day. In the United States, our daily use of water per person far exceeds this subsistence level, amounting to an average of about 300 L/ day for personal consumption and hygiene. We use about 8 L/person for cooking and drinking, about 120 L/person for cleaning (bathing, laundering, and housecleaning), 80 L/person for flushing toilets, and 80 L/person for watering lawns.

The total amount of freshwater on Earth is not a very large fraction of the total water present. Indeed, freshwater is one of our most precious resources. It forms by evaporation from the oceans and the land. The water vapor that accumulates in the atmosphere is transported by global atmospheric circulation, eventually returning to Earth as rain, snow, and other forms of precipitation (Figure 18.15).

As water runs off the land on its way to the oceans, it dissolves a variety of cations (mainly Na^+, K^+, Mg^{2+}, Ca^{2+}, and Fe^{2+}), anions (mainly Cl^-, SO_4^{2-}, and HCO_3^-), and gases (principally O_2, N_2, and CO_2). As we use water, it becomes laden with additional dissolved material, including the wastes of human society. As our population and output of environmental pollutants increase, ever-increasing amounts of money and resources must be spent to guarantee a supply of freshwater.

The availability and cost of fresh water that is clean enough to sustain daily life varies greatly among nations. To illustrate, daily fresh water use in the United States approaches 600 L/person, whereas in the relatively underdeveloped nations of sub-Sahara Africa it is only about 30 L. To make matters worse, for many people, water is not only scarce, it is so contaminated that it is a continuing source of diseases.

Approximately 20% of the world's freshwater is under the soil, in the form of *groundwater*. Groundwater resides in *aquifers*, which are layers of porous rock that hold water. The water in aquifers can be very pure, and accessible for human consumption if near the surface. Dense underground formations that do not allow water to readily penetrate can hold groundwater for years or even millennia. When their water is removed by drilling and pumping, such aquifers are slow to recharge via the diffusion of surface water.

A Closer Look

The Ogallala Aquifer— A shrinking resource

The Ogallala Aquifer, also referred to as the High Plains Aquifer, is an enormous underground body of water lying beneath the Great Plains of the United States. One of the world's largest aquifers, it covers an area of approximately 450,000 km² (170,000 mi²) encompassing portions of the eight states of South Dakota, Nebraska, Wyoming, Colorado, Kansas, Oklahoma, New Mexico, and Texas. (▶ Figure 18.17). The depth of the Ogallala formation that gives rise to the aquifer ranges from about 120 m, to more than 300 m, particularly in the northern portion. Before large-scale pumping in the modern era, the depth of water in the aquifer ranged up to more than 120 m in the northern portion.

Anyone who has flown over the Great Plains is familiar with the view of huge circles made by the center pivot irrigators nearly covering the land. The center post irrigation system, developed in the post–World War II era, permitted application of water onto large areas. As a result, the Great Plains became one of the most productive agricultural areas in the world. Unfortunately, the premise that the aquifer is an inexhaustible source of fresh water proved false. Recharge of the aquifer from surface water is slow, taking hundreds, perhaps thousands of years. Recently, water levels in many regions of the Ogallala have declined to the point where the costs of bringing water to the surface have become prohibitive. As the aquifer levels continue to drop, less water will be available for the needs of cities, residences and businesses.

Related Exercises: 18.41, 18.42

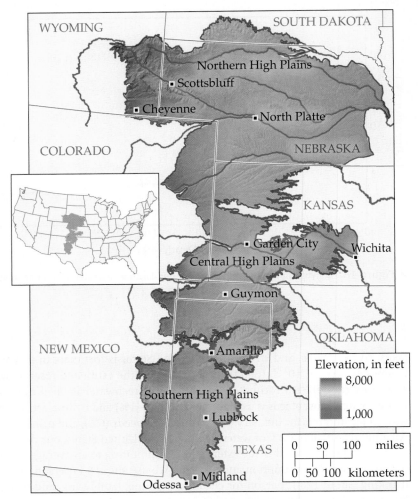

▲ **Figure 18.17** A map showing the extent of the Ogallala (High Plains) aquifer. Note that the elevation of the land varies greatly. The aquifer follows the topography of the formations that underlie the area.

The nature of the rock that contains the groundwater has a large influence on the water's chemical composition. If minerals in the rock are water soluble to some extent, ions can leach out of the rock and remain dissolved in the groundwater. Arsenic in the form of $HAsO_4^{2-}$, $H_2AsO_4^{-}$, and H_3AsO_3 is found in many groundwater sources across the world, most infamously in Bangladesh, at concentrations poisonous to humans.

18.4 | Human Activities and Water Quality

All life on Earth depends on the availability of suitable water. Many human activities entail waste disposal into natural waters without any treatment. These practices result in contaminated water that is detrimental to both plant and animal aquatic life. Unfortunately people in many parts of the world do not have access to water that has been treated to remove harmful contaminants, including disease-bearing bacteria.

Dissolved Oxygen and Water Quality

The amount of O_2 dissolved in water is an important indicator of water quality. Water fully saturated with air at 1 atm and 20 °C contains about 9 ppm of O_2. Oxygen is necessary for fish and most other aquatic life. Cold-water fish require water containing at least 5 ppm of dissolved oxygen for survival. Aerobic bacteria consume dissolved oxygen to oxidize organic materials for energy. The organic material the bacteria are able to oxidize is said to be **biodegradable**.

Excessive quantities of biodegradable organic materials in water are detrimental because they remove the oxygen necessary to sustain normal animal life. Typical sources of these biodegradable materials, which are called *oxygen-demanding wastes*, include sewage, industrial wastes from food-processing plants and paper mills, and liquid waste from meatpacking plants.

In the presence of oxygen, the carbon, hydrogen, nitrogen, sulfur, and phosphorus in biodegradable material end up mainly as CO_2, HCO_3^-, H_2O, NO_3^-, SO_4^{2-}, and phosphates. The formation of these oxidation products sometimes reduces the amount of dissolved oxygen to the point where aerobic bacteria can no longer survive. Anaerobic bacteria then take over the decomposition process, forming CH_4, NH_3, H_2S, PH_3, and other products, several of which contribute to the offensive odors of some polluted waters.

Plant nutrients, particularly nitrogen and phosphorus, contribute to water pollution by stimulating excessive growth of aquatic plants. The most visible results of excessive plant growth are floating algae and murky water. What is more significant, however, is that as plant growth becomes excessive, the amount of dead and decaying plant matter increases rapidly, a process called *eutrophication* (▶ Figure 18.18). The processes by which plants decay consumes O_2, and without sufficient oxygen, the water cannot sustain animal life.

The most significant sources of nitrogen and phosphorus compounds in water are domestic sewage (phosphate-containing detergents and nitrogen-containing body wastes), runoff from agricultural land (fertilizers contain both nitrogen and phosphorus), and runoff from livestock areas (animal wastes contain nitrogen).

▲ Figure 18.18 **Eutrophication.** This rapid accumulation of dead and decaying plant matter in a body of water uses up the water's oxygen supply, making the water unsuitable for aquatic animals.

 Give It Some Thought

If a test on a sample of polluted water shows a considerable decrease in dissolved oxygen over a five-day period, what can we conclude about the nature of the pollutants present?

Water Purification: Desalination

Because of its high salt content, seawater is unfit for human consumption and for most of the uses to which we put water. In the United States, the salt content of municipal water supplies is restricted by health codes to no more than about 0.05% by mass. This amount is much lower than the 3.5% dissolved salts present in seawater and the 0.5% or so present in brackish water found underground in some regions. The removal of salts from seawater or brackish water to make the water usable is called **desalination**.

Water can be separated from dissolved salts by *distillation* because water is a volatile substance and the salts are nonvolatile. (Section 1.3, "Separation of Mixtures") The principle of distillation is simple enough, but carrying out the process on a large scale presents many problems. As water is distilled from seawater, for example, the salts become more and more concentrated and eventually precipitate out. Distillation is also an energy-intensive process.

Seawater can also be desalinated using **reverse osmosis**. Recall that osmosis is the net movement of solvent molecules, but not solute molecules, through a semipermeable membrane. (Section 13.5) In osmosis, the solvent passes from the more dilute solution into the more concentrated one. However, if sufficient external pressure is applied, osmosis can be stopped and, at still higher pressures, reversed. When reverse osmosis occurs, solvent passes from the more concentrated into the more dilute solution. In a modern reverse-osmosis facility, hollow fibers are used as the semipermeable membrane (▶ Figure 18.19). Saline water (water containing significant salts) is introduced under pressure into the fibers, and desalinated water is recovered.

The world's largest desalination plant, in Jubail, Saudi Arabia, provides 50% of that country's drinking water by using reverse osmosis to desalinate seawater from the Persian Gulf. An even larger plant, which will produce 600 million L/day (160 million gallons) of drinking water, is scheduled for completion in Saudi Arabia in 2018. Such plants are becoming increasingly common in the United States. The largest, near

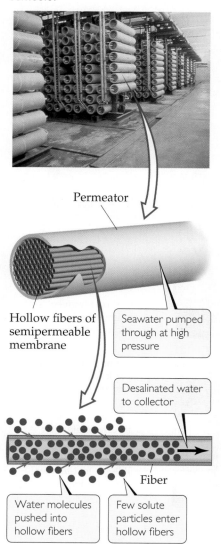

▲ **GO FIGURE**

What feature of this process is responsible for its being called *reverse osmosis*?

Permeator

Hollow fibers of semipermeable membrane

Seawater pumped through at high pressure

Desalinated water to collector

Fiber

Water molecules pushed into hollow fibers

Few solute particles enter hollow fibers

▲ Figure 18.19 **Reverse osmosis.**

▲ GO FIGURE

What is the primary function of the aeration step in water treatment?

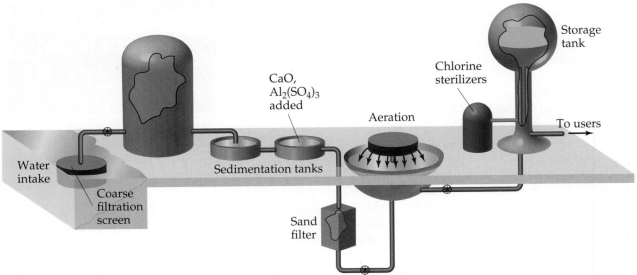

▲ Figure 18.20 **Common steps in treating water for a public water system.**

— Carbon removes iodine smells and parasites

— Iodine-impregnated beads kill bacteria, viruses, and parasites

— 15-μm textile filter removes debris

— 100-μm textile filter removes debris

▲ Figure 18.21 **A LifeStraw purifies water as it is drunk.**

Tampa Bay, Florida, has been operating since 2007 and produces 35 million gallons of drinking water a day by reverse osmosis. Small-scale, manually operated reverse-osmosis desalinators are used in camping, traveling, and at sea.

Water Purification: Municipal Treatment

The water needed for domestic, agricultural, and industrial use is taken either from lakes, rivers, and underground sources or from reservoirs. Much of the water that finds its way into municipal water systems is "used" water, meaning it has already passed through one or more sewage systems or industrial plants. Consequently, this water must be treated before it is distributed to our faucets.

Municipal water treatment usually involves five steps (▲ Figure 18.20). After coarse filtration through a screen, the water is allowed to stand in large sedimentation tanks where sand and other minute particles settle out. To aid in removing very small particles, the water may first be made slightly basic with CaO. Then $Al_2(SO_4)_3$ is added and reacts with OH^- ions to form a spongy, gelatinous precipitate of $Al(OH)_3$ ($K_{sp} = 1.3 \times 10^{-33}$). This precipitate settles slowly, carrying suspended particles down with it, thereby removing nearly all finely divided matter and most bacteria. The water is then filtered through a sand bed. Following filtration, the water may be sprayed into the air (aeration) to hasten oxidation of dissolved inorganic ions of iron and manganese, reduce concentrations of any H_2S or NH_3 that may be present and reduce bacterial concentrations.

The final step normally involves treating the water with a chemical agent to ensure the destruction of bacteria. Ozone is more effective, but chlorine is less expensive. Liquefied Cl_2 is dispensed from tanks through a metering device directly into the water supply. The amount used depends on the presence of other substances with which the chlorine might react and on the concentrations of bacteria and viruses to be removed. The sterilizing action of chlorine is probably due not to Cl_2 itself but to hypochlorous acid, which forms when chlorine reacts with water:

$$Cl_2(aq) + H_2O(l) \longrightarrow HClO(aq) + H^+(aq) + Cl^-(aq) \qquad [18.15]$$

It is estimated that about 800 million people worldwide lack access to clean water. According to the United Nations, 95% of the world's cities still dump raw sewage into their water supplies. Thus, it should come as no surprise that 80% of all the health maladies in developing countries can be traced to waterborne diseases associated with unsanitary water.

One promising development is a device called the LifeStraw (◄ Figure 18.21). When a person sucks water through the straw, the water first encounters a textile

A Closer Look

Fracking and Water Quality

In recent years **fracking**, short for *hydraulic fracturing,* has become widely used to greatly increase the availability of petroleum reserves. In fracking, a large volume of water, typically two million gallons or more, mixed with various additives, is injected at high pressure into wellbores extended horizontally into rock formations (▶ Figure 18.22). The water is laden with sand, ceramic materials and other additives, including gels, foams, and compressed gases, that serve to increase the yield in the process. The high pressure fluid finds its way into tiny faults in geological formations, releasing petroleum and natural gas. Fracking has greatly increased petroleum reserves, particularly of natural gas, in many parts of the world. The technique has been so productive that more than 20,000 new wells are being drilled annually in the United States alone, in all areas of the country.

Unfortunately, the potential for environmental damage from fracking is significant. The large volume of fracking fluid required to create a well must be returned to the surface. Without purification the fluid is rendered unfit for other uses, and

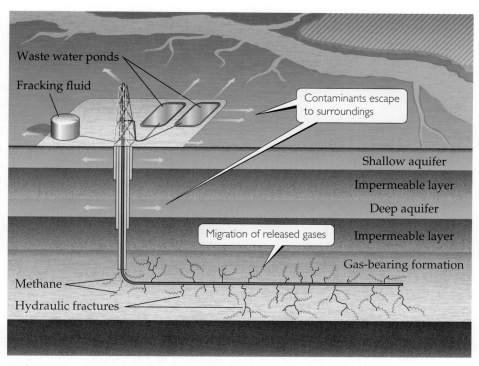

▲ Figure 18.22 **A schematic of a well site employing fracking.** The yellow arrows indicate the avenues through which contaminants enter the environment.

becomes a large-scale environmental problem. Often the waste water is allowed to sit in open waste water pits. The 2005 Energy Policy act and other federal legislation exempts hydraulic fracturing operations from certain provisions of the Safe Drinking Water act and other regulations. Some areas of the country that are already facing water shortages thus have one more large demand for a limited supply. Because fracturing of rock formations increases the pathways for flows of petroleum and various gases, bodies of underground water that have been serving as municipal water supplies or wells for individual homes in some locales have become contaminated with petroleum, hydrogen sulfide and other toxic substances. The escape of

a variety of gases, including methane and other hydrocarbons, from the wellheads contributes to air pollution. In a study published in 2013, methane emissions to the atmosphere during hydraulic fracking operations in Utah were estimated to be in the range of 6–12% of the amount of methane produced. As related in the Closer Look box on page 790, methane is a potent greenhouse gas.

The many environmental issues surrounding the practice of fracking have generated widespread concern and adverse public reaction. Fracking represents yet one more instance of the conflict between those who advocate the availability of low cost energy and those who are more focused on sustaining long term the quality of the environment.

filter with a mesh opening of 100 µm followed by a second textile filter with a mesh opening of 15 µm. These filters remove debris and even clusters of bacteria. The water next encounters a chamber of iodine-impregnated beads, where bacteria, viruses, and parasites are killed. Finally, the water passes through granulated active carbon, which removes the smell of iodine as well as the parasites that have not been taken by the filters or killed by the iodine. At present the Lifestraw is too costly to permit widespread use in underdeveloped countries, but there is hope that its cost can be greatly reduced.

Water disinfection is one of the greatest public health innovations in human history. It has dramatically decreased the incidences of waterborne bacterial diseases such as cholera and typhus. However, this great benefit comes at a price.

In 1974 scientists in Europe and the United States discovered that chlorination of water produces a group of by-products previously undetected. These by-products are called *trihalomethanes* (THMs) because all have a single carbon atom and three halogen atoms: $CHCl_3$, $CHCl_2Br$, $CHClBr_2$, and $CHBr_3$. These and many other chlorine- and bromine-containing organic substances are produced by the reaction of dissolved chlorine with the organic materials present in nearly all natural waters, as well as with substances that are by-products of human activity. Recall that chlorine dissolves in water to form the oxidizing agent HClO:

$$Cl_2(g) + H_2O(l) \longrightarrow HClO(aq) + H^+(aq) + Cl^-(aq) \qquad [18.16]$$

The HClO in turn reacts with organic substances to form THMs. Bromine enters the reaction sequence through the reaction of HClO with dissolved bromide ion:

$$HClO(aq) + Br^-(aq) \longrightarrow HBrO(aq) + Cl^-(aq) \qquad [18.17]$$

Then both HBrO(aq) and HClO(aq) can halogenate organic substances to form the THMs.

Some THMs and other halogenated organic substances are suspected carcinogens; others interfere with the body's endocrine system. As a result, the World Health Organization and the U.S. Environmental Protection Agency have placed concentration limits of 80 μg/L (80 ppb) on the total quantity of THMs in drinking water. The goal is to reduce the levels of THMs and other disinfection by-products in the drinking water supply while preserving the antibacterial effectiveness of the water treatment. In some cases, lowering the concentration of chlorine may provide adequate disinfection while reducing the concentrations of THMs formed. Alternative oxidizing agents, such as ozone or chlorine dioxide, produce less of the halogenated substances but have their own disadvantages. For example, each is capable of oxidizing dissolved bromide, as shown here for ozone:

$$O_3(aq) + Br^-(aq) + H_2O(l) \longrightarrow HBrO(aq) + O_2(aq) + OH^-(aq) \qquad [18.18]$$

$$HBrO(aq) + 2\,O_3(aq) \longrightarrow BrO_3^-(aq) + 2\,O_2(aq) + H^+(aq) \qquad [18.19]$$

Bromate ion, BrO_3^-, has been shown to cause cancer in animal tests.

At present, there seem to be no completely satisfactory alternatives to chlorination or ozonation, and we are faced with a consideration of benefit versus risk. In this case, the risks of cancer from THMs and related substances in municipal water are very low relative to the risks of cholera, typhus, and gastrointestinal disorders from untreated water. When the water supply is cleaner to begin with, less disinfectant is needed and thus the risk of THMs is lowered. Once THMs form, their concentrations in the water supply can be reduced by aeration because the THMs are more volatile than water. Alternatively, they can be removed by adsorption onto activated charcoal or other adsorbents.

18.5 | Green Chemistry

The planet on which we live is, to a large extent, a *closed system*, one that exchanges energy but not matter with its surroundings. If humankind is to thrive in the future, all the processes we carry out should be in balance with Earth's natural processes and physical resources. This goal requires that no toxic materials be released to the environment, that our needs be met with renewable resources, and that we consume the least possible amount of energy. Although the chemical industry is but a small part of human activity, chemical processes are involved in nearly all aspects of modern life. Chemistry is therefore at the heart of efforts to accomplish these goals.

Green chemistry is an initiative that promotes the design and application of chemical products and processes that are compatible with human health and that preserve the environment. Green chemistry rests on a set of 12 principles:

1. **Prevention** It is better to prevent waste than to clean it up after it has been created.

2. **Atom Economy** Methods to make chemical compounds should be designed to maximize the incorporation of all starting atoms into the final product.

3. **Less Hazardous Chemical Syntheses** Wherever practical, synthetic methods should be designed to use and generate substances that possess little or no toxicity to human health and the environment.

4. **Design of Safer Chemicals** Chemical products should be designed to minimize toxicity and yet maintain their desired function.

5. **Safer Solvents and Auxiliaries** Auxiliary substances (for example, solvents, separation agents, etc.) should be used as little as possible. Those that are used should be as nontoxic as possible.

6. **Design for Energy Efficiency** Energy requirements of chemical processes should be recognized for their environmental and economic impacts and should be minimized. If possible, chemical reactions should be conducted at room temperature and pressure.

7. **Use of Renewable Feedstocks** A raw material or feedstock should be renewable whenever technically and economically practical.

8. **Reduction of Derivatives** Unnecessary derivatization (intermediate compound formation, temporary modification of physical/chemical processes) should be minimized or avoided if possible, because such steps require additional reagents and can generate waste.

9. **Catalysis** Catalytic reagents (as selective as possible) improve product yields within a given time and with a lower energy cost compared to noncatalytic processes and are, therefore, preferred to noncatalytic alternatives.

10. **Design for Degradation** The end products of chemical processing should break down at the end of their useful lives into innocuous degradation products that do not persist in the environment.

11. **Real-Time Analysis for Pollution Prevention** Analytical methods need to be developed that allow for real-time, in-process monitoring and control prior to the formation of hazardous substances.

12. **Inherently Safer Chemistry for Accident Prevention** Reagents and solvents used in a chemical process should be chosen to minimize the potential for chemical accidents, including releases, explosions, and fires.*

 Give It Some Thought

Explain how a chemical reaction that uses a catalyst can be "greener" than the same reaction run without a catalyst.

To illustrate how green chemistry works, consider the manufacture of styrene, an important building block for many polymers, including the expanded polystyrene packages used to pack eggs and restaurant takeout food. The global demand for styrene is more than 25 billion kg per year. For many years, styrene has been produced in a two-step process: Benzene and ethylene react to form ethyl benzene, followed by the ethyl benzene being mixed with high-temperature steam and passed over an iron oxide catalyst to form styrene:

$$
\text{Benzene} + H_2C{=}CH_2 \xrightarrow[\text{catalyst}]{\text{Acid}} \text{Ethyl benzene} \xrightarrow[-H_2]{\text{Iron oxide catalyst}} \text{Styrene}
$$

Benzene   Ethylene   Ethyl benzene   Styrene

This process has several shortcomings. One is that both benzene, which is formed from crude oil, and ethylene, formed from natural gas, are high-priced starting materials for a product that should be a low-priced commodity. Another is that benzene is a known carcinogen. In a recently-developed process that bypasses some of these

*Adapted from P. T. Anastas and J. C. Warner, *Green Chemistry: Theory and Practice.* New York: Oxford University Press 1998, p. 30. See also, Mike Lancaster, *Green Chemistry: An Introductory Text.* Cambridge, UK: RSC Publishing, 2010, Second Edition, Chapter 1.

shortcomings, the two-step process is replaced by a one-step process in which toluene is reacted with methanol at 425 °C over a special catalyst:

$$\text{Toluene} \quad + \quad CH_3OH \quad \xrightarrow[-H_2, -H_2O]{\text{Base catalyst}} \quad \text{Styrene}$$

Toluene   Methanol   Styrene

The one-step process saves money both because toluene and methanol are less expensive than benzene and ethylene, and because the reaction requires less energy input. Additional benefits are that the methanol could be produced from biomass and that benzene is replaced by less-toxic toluene. The hydrogen formed in the reaction can be recycled as a source of energy. (This example demonstrates how finding the right catalyst is often key in discovering a new process.)

Let's consider some other examples in which green chemistry can operate to improve environmental quality.

Supercritical Solvents

A major area of concern in chemical processes is the use of volatile organic compounds as solvents. Generally, the solvent in which a reaction is run is not consumed in the reaction, and there are unavoidable releases of solvent into the atmosphere even in the most carefully controlled processes. Further, the solvent may be toxic or may decompose to some extent during the reaction, thus creating waste products.

The use of supercritical fluids represents a way to replace conventional solvents. Recall that a supercritical fluid is an unusual state of matter that has properties of both a gas and a liquid. ∞ (Section 11.4) Water and carbon dioxide are the two most popular choices as supercritical fluid solvents. One recently developed industrial process, for example, replaces chlorofluorocarbon solvents with liquid or supercritical CO_2 in the production of polytetrafluoroethylene ($[CF_2CF_2]n$, sold as Teflon®). Though CO_2 is a greenhouse gas, no new CO_2 need be manufactured for use as a supercritical fluid solvent.

As a further example, *para*-xylene is oxidized to form terephthalic acid, which is used to make polyethylene terephthalate (PET) plastic and polyester fiber ∞ (Section 12.8, Table 12.5):

$$CH_3\text{-}\langle\bigcirc\rangle\text{-}CH_3 + 3\,O_2 \xrightarrow[\text{Catalyst}]{190\,°C,\ 20\ atm} HO\text{-}\overset{O}{\underset{\|}{C}}\text{-}\langle\bigcirc\rangle\text{-}\overset{O}{\underset{\|}{C}}\text{-}OH + 2\,H_2O$$

para-Xylene   Terephthalic acid

This commercial process requires pressurization and a relatively high temperature. Oxygen is the oxidizing agent, and acetic acid (CH_3COOH) is the solvent. An alternative route employs supercritical water as the solvent and hydrogen peroxide as the oxidant. This alternative process has several potential advantages, most particularly the elimination of acetic acid as solvent.

Give It Some Thought

We noted earlier that increasing carbon dioxide levels contribute to global climate change, which seems like a bad thing, but now we are saying that using carbon dioxide in industrial processes is a good thing for the environment. Explain this seeming contradiction.

Greener Reagents and Processes

Let us examine two more examples of green chemistry in action.

Hydroquinone, $HO-C_6H_4-OH$, is a common intermediate used to make polymers. The standard industrial route to hydroquinone, used until recently, yields many by-products that are treated as waste:

Using the principles of green chemistry, researchers have improved this process. The new process for hydroquinone production uses a new starting material. Two of the by-products of the new reaction (shown in green) can be isolated and used to make the new starting material.

By-products recycled to make starting material

The new process is an example of "atom economy," a phrase that means that a high percentage of the atoms from the starting materials end up in the product.

Give It Some Thought

Where might there be room to make changes in this process that would make hydroquinone production even greener?

Another example of atom economy is a reaction in which, at room temperature and in the presence of a copper(I) catalyst, an organic *azide* and an *alkyne* form one product molecule:

Azide   Alkyne

This reaction is informally called a *click reaction*. The yield—actual, not just theoretical—is close to 100%, and there are no by-products. Depending on the type of azide and type of alkyne we start with, this very efficient click reaction can be used to create any number of valuable product molecules.

 Give It Some Thought

What are the hybridizations of the two alkyne C atoms before and after the click reaction?

SAMPLE
INTEGRATIVE EXERCISE | Putting Concepts Together

(a) Acid rain is no threat to lakes in areas where the rock is limestone (calcium carbonate), which can neutralize the acid. Where the rock is granite, however, no neutralization occurs. How does limestone neutralize acid? **(b)** Acidic water can be treated with basic substances to increase the pH, although such a procedure is usually only a temporary cure. Calculate the minimum mass of lime, CaO, needed to adjust the pH of a small lake ($V = 4 \times 10^9$ L) from 5.0 to 6.5. Why might more lime be needed?

SOLUTION

Analyze We need to remember what a neutralization reaction is and calculate the amount of a substance needed to effect a certain change in pH.

Plan For (a), we need to think about how acid can react with calcium carbonate, a reaction that evidently does not happen with acid and granite. For (b), we need to think about what reaction between an acid and CaO is possible and do stoichiometric calculations. From the proposed change in pH, we can calculate the change in proton concentration needed and then figure out how much CaO is needed.

Solve

(a) The carbonate ion, which is the anion of a weak acid, is basic ∞ (Sections 16.2 and 16.7) and so reacts with $H^+(aq)$. If the concentration of $H^+(aq)$ is low, the major product is the bicarbonate ion, HCO_3^-. If the concentration of $H^+(aq)$ is high, H_2CO_3 forms and decomposes to CO_2 and H_2O. ∞ (Section 4.3)

(b) The initial and final concentrations of $H^+(aq)$ in the lake are obtained from their pH values:

$$[H^+]_{initial} = 10^{-5.0} = 1 \times 10^{-5}\, M \quad \text{and} \quad [H^+]_{final} = 10^{-6.5} = 3 \times 10^{-7}\, M$$

Using the lake volume, we can calculate the number of moles of $H^+(aq)$ at both pH values:

$$(1 \times 10^{-5}\, \text{mol/L})(4.0 \times 10^9\, \text{L}) = 4 \times 10^4\, \text{mol}$$
$$(3 \times 10^{-7}\, \text{mol/L})(4.0 \times 10^9\, \text{L}) = 1 \times 10^3\, \text{mol}$$

Hence, the change in the amount of $H^+(aq)$ is 4×10^4 mol $- 1 \times 10^3$ mol $\approx 4 \times 10^4$ mol.

Let's assume that all the acid in the lake is completely ionized, so that only the free $H^+(aq)$ contributing to the pH needs to be neutralized. We need to neutralize at least that much acid, although there may be a great deal more than that amount in the lake.

The oxide ion of CaO is very basic. ∞ (Section 16.5) In the neutralization reaction, 1 mol of O^{2-} reacts with 2 mol of H^+ to form H_2O. Thus, 4×10^4 mol of H^+ requires

$$(4 \times 10^4\, \text{mol H}^+)\left(\frac{1\, \text{mol CaO}}{2\, \text{mol H}^+}\right)\left(\frac{56.1\, \text{g CaO}}{1\, \text{mol CaO}}\right) = 1 \times 10^6\, \text{g CaO}$$

This is slightly more than a ton of CaO. That would not be very costly because CaO is inexpensive, selling for less than $100 per ton when purchased in large quantities. This amount of CaO is the minimum amount needed, however, because there are likely to be weak acids in the water that must also be neutralized.

This liming procedure has been used to bring the pH of some small lakes into the range necessary for fish to live. The lake in our example would be about a half mile long and a half mile wide and have an average depth of 20 ft.

Chapter Summary and Key Terms

EARTH'S ATMOSPHERE (SECTION 18.1) In this section we examined the physical and chemical properties of Earth's atmosphere. The complex temperature variations in the atmosphere give rise to four regions, each with characteristic properties. The lowest of these regions, the **troposphere**, extends from Earth's surface up to an altitude of about 12 km. Above the troposphere, in order of increasing altitude, are the **stratosphere**, mesosphere, and thermosphere. In the upper reaches of the atmosphere, only the simplest chemical species can survive the bombardment of highly energetic particles and radiation from the Sun. The average molecular weight of the atmosphere at high elevations is lower than that at Earth's surface because the lightest atoms and molecules diffuse upward and also because of **photodissociation**, which is the breaking of bonds in molecules because of the absorption of light. Absorption of radiation may also lead to the formation of ions via **photoionization**.

HUMAN ACTIVITIES AND EARTH'S ATMOSPHERE (SECTION 18.2) Ozone is produced in the upper atmosphere from the reaction of atomic oxygen with O_2. Ozone is itself decomposed by absorption of a photon or by reaction with an active species such as Cl. **Chlorofluorocarbons** can undergo photodissociation in the stratosphere, introducing atomic chlorine, which is capable of catalytically destroying ozone. A marked reduction in the ozone level in the upper atmosphere would have serious adverse consequences because the ozone layer filters out certain wavelengths of harmful ultraviolet light that are not removed by any other atmospheric component. In the troposphere the chemistry of trace atmospheric components is of major importance. Many of these minor components are pollutants. Sulfur dioxide is one of the more noxious and prevalent examples. It is oxidized in air to form sulfur trioxide, which, upon dissolving in water, forms sulfuric acid. The oxides of sulfur are major contributors to **acid rain**. One method of preventing the escape of SO_2 from industrial operations is to react it with CaO to form calcium sulfite $CaSO_3$.

Photochemical smog is a complex mixture in which both nitrogen oxides and ozone play important roles. Smog components are generated mainly in automobile engines, and smog control consists largely of controlling auto emissions.

Carbon dioxide and water vapor are the major components of the atmosphere that strongly absorb infrared radiation. CO_2 and H_2O are therefore critical in maintaining Earth's surface temperature. The concentrations of CO_2 and other so-called **greenhouse gases** in the atmosphere are thus important in determining worldwide climate.

Because of the extensive combustion of fossil fuels (coal, oil, and natural gas), the concentration of carbon dioxide in the atmosphere is steadily increasing.

EARTH'S WATER (SECTION 18.3) Earth's water is largely in the oceans and seas; only a small fraction is freshwater. Seawater contains about 3.5% by mass of dissolved salts and is described as having a **salinity** (grams of dry salts per 1 kg seawater) of 35. Seawater's density and salinity vary with depth. Because most of the world's water is in the oceans, humans may eventually need to recover freshwater from seawater. The global water cycle involves continuous phase changes of water.

HUMAN ACTIVITIES AND WATER QUALITY (SECTION 18.4) Freshwater contains many dissolved substances including dissolved oxygen, which is necessary for fish and other aquatic life. Substances that are decomposed by bacteria are said to be **biodegradable**. Because the oxidation of biodegradable substances by aerobic bacteria consumes dissolved oxygen, these substances are called oxygen-demanding wastes. The presence of an excess amount of oxygen-demanding wastes in water can sufficiently deplete the dissolved oxygen to kill fish and produce offensive odors. Plant nutrients can contribute to the problem by stimulating the growth of plants that become oxygen-demanding wastes when they die. **Desalination** is the removal of dissolved salts from seawater or brackish water to make it fit for human consumption. Desalination may be accomplished by distillation or by **reverse osmosis**.

The water available from freshwater sources may require treatment before it can be used domestically. The several steps generally used in municipal water treatment include coarse filtration, sedimentation, sand filtration, aeration, sterilization, and sometimes water softening.

Water supplies may be impacted by the practice of **fracking**, in which water laden with sand and a variety of chemicals is pumped at high pressure into rock formations to release natural gas and other petroleum materials.

GREEN CHEMISTRY (SECTION 18.5) The **green chemistry** initiative promotes the design and application of chemical products and processes that are compatible with human health and that preserve the environment. The areas in which the principles of green chemistry can operate to improve environmental quality include choices of solvents and reagents for chemical reactions, development of alternative processes, and improvements in existing systems and practices.

Learning Outcomes After studying this chapter, you should be able to:

- Describe the regions of Earth's atmosphere in terms of how temperature varies with altitude. (Section 18.1)

- Describe the composition of the atmosphere in terms of the major components in dry air at sea level. (Section 18.1)

- Calculate concentrations of gases in parts per million (ppm). (Section 18.1)

- Describe the processes of photodissociation and photoionization and their role in the upper atmosphere. (Section 18.1)

- Use bond energies and ionization energies to calculate the minimum frequency or maximum wavelength needed to cause photodissociation or photoionization. (Section 18.1)

- Explain how ozone in the upper atmosphere functions to filter short wavelength solar radiation. (Section 18.1)

- Explain how chlorofluorocarbons (CFCs) cause depletion of the ozone layer. (Section 18.2)

- Describe the origins and behavior of sulfur oxides and nitrogen oxides as air pollutants, including the generation of acid rain and photochemical smog. (Section 18.2)

- Describe how water and carbon dioxide cause an increase in atmospheric temperature near Earth's surface. (Section 18.2)

- Describe the global water cycle. (Section 18.3)

- Explain what is meant by the salinity of water and describe the process of reverse osmosis as a means of desalination. (Section 18.4)

- List the major cations, anions, and gases present in natural waters and describe the relationship between dissolved oxygen and water quality. (Section 18.4)

- List the main steps involved in treating water for domestic uses. (Section 18.4)

- Describe the process of fracking and name its potential adverse environmental effects. (Section 18.4)

- Describe the main goals of green chemistry. (Section 18.5)

- Compare reactions and decide which reaction is greener. (Section 18.5)

Exercises

Visualizing Concepts

18.1 At 273 K and 1 atm pressure, 1 mol of an ideal gas occupies 22.4 L. (Section 10.4) **(a)** Looking at Figure 18.1 predict whether a 1 mol sample of the atmosphere in the middle of the stratosphere would occupy a greater or smaller volume than 22.4 L **(b)** Looking at Figure 18.1, we see that the temperature is lower at 85 km altitude than at 50 km. Does this mean that one mole of an ideal gas would occupy less volume at 85 km than at 50 km? Explain. **(c)** In which parts of the atmosphere would you expect gases to behave most ideally (ignoring any photochemical reactions)? [Section 18.1]

18.2 Molecules in the upper atmosphere tend to contain double and triple bonds rather than single bonds. Suggest an explanation. [Section 18.1]

18.3 The figure shows the three lowest regions of Earth's atmosphere. **(a)** Name each and indicate the approximate elevations at which the boundaries occur. **(b)** In which region is ozone a pollutant? In which region does it filter UV solar radiation? **(c)** In which region is infrared radiation from Earth's surface most strongly reflected back? **(d)** An aurora borealis is due to excitation of atoms and molecules in the atmosphere 55–95 km above Earth's surface. Which regions in the figure are involved in an aurora borealis? **(e)** Compare the changes in relative concentrations of water vapor and carbon dioxide with increasing elevation in these three regions.

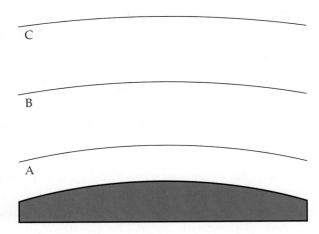

18.4 You are working with an artist who has been commissioned to make a sculpture for a big city in the eastern United States. The artist is wondering what material to use to make her sculpture because she has heard that acid rain in the eastern United States might destroy it over time. You take samples of granite, marble, bronze, and other materials, and place them outdoors for a long time in the big city. You periodically examine the appearance and measure the mass of the samples. **(a)** What observations would lead you to conclude that one or more of the materials are well-suited for the sculpture? **(b)** What chemical process (or processes) is (are) the most likely responsible for any observed changes in the materials? [Section 18.2]

18.5 Where does the energy come from to evaporate the estimated 425,000 km^3 of water that annually leaves the oceans, as illustrated here? [Section 18.3]

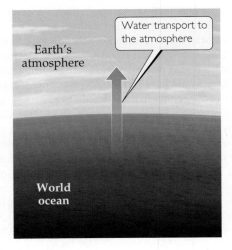

18.6 Describe the properties that most clearly distinguish among salt water, freshwater, and groundwater. [Section 18.3]

18.7 Describe what changes occur when atmospheric CO_2 interacts with the world ocean as illustrated here. [Section 18.3]

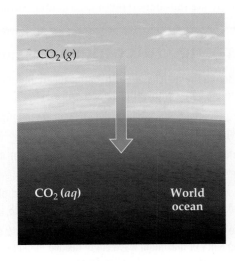

18.8 The following picture represents an ion-exchange column, in which water containing "hard" ions, such as Ca^{2+}, is added to the top of the column, and water containing "soft" ions, such as Na^+, comes out the bottom. Explain what is happening in the column. [Section 18.4]

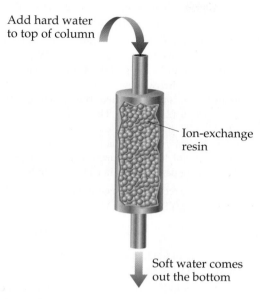

Add hard water to top of column

Ion-exchange resin

Soft water comes out the bottom

18.9 From study of Figure 18.22 describe the various ways in which operation of a fracking well site could lead to environmental contamination.

18.10 One mystery in environmental science is the imbalance in the "carbon dioxide budget." Considering only human activities, scientists have estimated that 1.6 billion metric tons of CO_2 is added to the atmosphere every year because of deforestation (plants use CO_2, and fewer plants will leave more CO_2 in the atmosphere). Another 5.5 billion tons per year is put into the atmosphere because of burning fossil fuels. It is further estimated (again, considering only human activities) that the atmosphere actually takes up about 3.3 billion tons of this CO_2 per year, while the oceans take up 2 billion tons per year, leaving about 1.8 billion tons of CO_2 per year unaccounted for. This "missing" CO_2 is assumed to be taken up by the "land." What do you think might be happening? [Sections 18.1–18.3]

Earth's Atmosphere (Section 18.1)

18.11 **(a)** What is the primary basis for the division of the atmosphere into different regions? **(b)** Name the regions of the atmosphere, indicating the altitude interval for each one.

18.12 **(a)** How are the boundaries between the regions of the atmosphere determined? **(b)** Explain why the stratosphere, which is about 35 km thick, has a smaller total mass than the troposphere, which is about 12 km thick.

18.13 Air pollution in the Mexico City metropolitan area is among the worst in the world. The concentration of ozone in Mexico City has been measured at 441 ppb (0.441 ppm). Mexico City sits at an altitude of 7400 feet, which means its atmospheric pressure is only 0.67 atm. **(a)** Calculate the partial pressure of ozone at 441 ppb if the atmospheric pressure is 0.67 atm. **(b)** How many ozone molecules are in 1.0 L of air in Mexico City? Assume $T = 25\,°C$.

18.14 From the data in Table 18.1, calculate the partial pressures of carbon dioxide and argon when the total atmospheric pressure is 1.05 bar.

18.15 The average concentration of carbon monoxide in air in an Ohio city in 2006 was 3.5 ppm. Calculate the number of CO molecules in 1.0 L of this air at a pressure of 759 torr and a temperature of 22 °C.

18.16 **(a)** From the data in Table 18.1, what is the concentration of neon in the atmosphere in ppm? **(b)** What is the concentration of neon in the atmosphere in molecules per liter, assuming an atmospheric pressure of 730 torr and a temperature of 296 K?

18.17 The dissociation energy of a carbon–bromine bond is typically about 210 kJ/mol. **(a)** What is the maximum wavelength of photons that can cause C—Br bond dissociation? **(b)** Which kind of electromagnetic radiation—ultraviolet, visible, or infrared—does the wavelength you calculated in part (a) correspond to?

18.18 In CF_3Cl the C—Cl bond-dissociation energy is 339 kJ/mol. In CCl_4 the C—Cl bond-dissociation energy is 293 kJ/mol. What is the range of wavelengths of photons that can cause C—Cl bond rupture in one molecule but not in the other?

18.19 **(a)** Distinguish between *photodissociation* and *photoionization*. **(b)** Use the energy requirements of these two processes to explain why photodissociation of oxygen is more important than photoionization of oxygen at altitudes below about 90 km.

18.20 Why is the photodissociation of N_2 in the atmosphere relatively unimportant compared with the photodissociation of O_2?

Human Activities and Earth's Atmosphere (Section 18.2)

18.21 Do the reactions involved in ozone depletion involve changes in oxidation state of the O atoms? Explain.

18.22 Which of the following reactions in the stratosphere cause an increase in temperature in this region?

(a) $O(g) + O_2(g) \longrightarrow O_3^*(g)$

(b) $O_3^*(g) + M(g) \longrightarrow O_3(g) + M^*(g)$

(c) $O_2(g) + h\nu \longrightarrow 2\,O(g)$

(d) $O(g) + N_2(g) \longrightarrow NO(g) + N(g)$

(e) All of the above

18.23 **(a)** What is the difference between chlorofluorocarbons and hydrofluorocarbons? **(b)** Why are hydrofluorocarbons potentially less harmful to the ozone layer than CFCs?

18.24 Draw the Lewis structure for the chlorofluorocarbon CFC-11, $CFCl_3$. What chemical characteristics of this substance allow it to effectively deplete stratospheric ozone?

18.25 **(a)** Why is the fluorine present in chlorofluorocarbons not a major contributor to depletion of the ozone layer? **(b)** What are the chemical forms in which chlorine exists in the stratosphere following cleavage of the carbon–chlorine bond?

18.26 Would you expect the substance $CFBr_3$ to be effective in depleting the ozone layer, assuming that it is present in the stratosphere? Explain.

18.27 For each of the following gases, make a list of known or possible naturally occurring sources: (a) CH_4, (b) SO_2, (c) NO.

18.28 Why is rainwater naturally acidic, even in the absence of polluting gases such as SO_2?

18.29 (a) Write a chemical equation that describes the attack of acid rain on limestone, $CaCO_3$. (b) If a limestone sculpture were treated to form a surface layer of calcium sulfate, would this help to slow down the effects of acid rain? Explain.

18.30 The first stage in corrosion of iron upon exposure to air is oxidation to Fe^{2+}. (a) Write a balanced chemical equation to show the reaction of iron with oxygen and protons from acid rain. (b) Would you expect the same sort of reaction to occur with a silver surface? Explain.

18.31 Alcohol-based fuels for automobiles lead to the production of formaldehyde (CH_2O) in exhaust gases. Formaldehyde undergoes photodissociation, which contributes to photochemical smog:

$$CH_2O + h\nu \longrightarrow CHO + H$$

The maximum wavelength of light that can cause this reaction is 335 nm. (a) In what part of the electromagnetic spectrum is light with this wavelength found? (b) What is the maximum strength of a bond, in kJ/mol, that can be broken by absorption of a photon of 335-nm light? (c) Compare your answer from part (b) to the appropriate value from Table 8.4. What do you conclude about C—H bond energy in formaldehyde? (d) Write out the formaldehyde photodissociation reaction, showing Lewis-dot structures.

18.32 An important reaction in the formation of photochemical smog is the photodissociation of NO_2:

$$NO_2 + h\nu \longrightarrow NO(g) + O(g)$$

The maximum wavelength of light that can cause this reaction is 420 nm. (a) In what part of the electromagnetic spectrum is light with this wavelength found? (b) What is the maximum strength of a bond, in kJ/mol, that can be broken by absorption of a photon of 420-nm light? (c) Write out the photodissociation reaction showing Lewis-dot structures.

18.33 Explain why an increasing concentration of CO_2 in the atmosphere affects the quantity of energy leaving Earth but does not affect the quantity of energy entering from the Sun.

18.34 (a) With respect to absorption of radiant energy, what distinguishes a greenhouse gas from a non-greenhouse gas? (b) CH_4 is a greenhouse gas, but N_2 is not. How might the molecular structure of CH_4 explain why it is a greenhouse gas?

Earth's Water (Section 18.3)

18.35 What is the molarity of Na^+ in a solution of NaCl whose salinity is 5.6 if the solution has a density of 1.03 g/mol?

18.36 Phosphorus is present in seawater to the extent of 0.07 ppm by mass. Assuming that the phosphorus is present as dihydrogenphosphate, $H_2PO_4^-$, calculate the corresponding molar concentration of phosphate in seawater.

18.37 The enthalpy of evaporation of water is 40.67 kJ/mol. Sunlight striking Earth's surface supplies 168 W per square meter (1 W = 1 watt = 1 J/s). (a) Assuming that evaporation of water is due only to energy input from the Sun, calculate how many grams of water could be evaporated from a 1.00 square meter patch of ocean over a 12-h day. (b) The specific heat capacity of liquid water is 4.184 J/g °C. If the initial surface temperature of a 1.00 square meter patch of ocean is 26 °C, what is its final temperature after being in sunlight for 12 h, assuming no phase changes and assuming that sunlight penetrates uniformly to depth of 10.0 cm?

[18.38] The enthalpy of fusion of water is 6.01 kJ/mol. Sunlight striking Earth's surface supplies 168 W per square meter (1 W = 1 watt = 1 J/s). (a) Assuming that melting of ice is due only to energy input from the Sun, calculate how many grams of ice could be melted from a 1.00 square meter patch of ice over a 12-h day. (b) The specific heat capacity of ice is 2.032 J/g °C. If the initial temperature of a 1.00 square meter patch of ice is −5.0 °C, what is its final temperature after being in sunlight for 12 h, assuming no phase changes and assuming that sunlight penetrates uniformly to a depth of 1.00 cm?

18.39 A first-stage recovery of magnesium from seawater is precipitation of $Mg(OH)_2$ with CaO:

$$Mg^{2+}(aq) + CaO(s) + H_2O(l) \longrightarrow Mg(OH)_2(s) + Ca^{2+}(aq)$$

What mass of CaO, in grams, is needed to precipitate 1000 lb of $Mg(OH)_2$?

18.40 Gold is found in seawater at very low levels, about 0.05 ppb by mass. Assuming that gold is worth about $1300 per troy ounce, how many liters of seawater would you have to process to obtain $1,000,000 worth of gold? Assume the density of seawater is 1.03 g/mL and that your gold recovery process is 50% efficient.

18.41 (a) What is *groundwater*? (b) What is an *aquifer*?

18.42 The Ogallala aquifer described in the Closer Look box in Section 18.3, provides 82% of the drinking water for the people who live in the region, although more than 75% of the water that is pumped from it is for irrigation. Irrigation withdrawals are approximately 18 billion gallons per day. (a) Assuming that 2% of the rainfall that falls on an area of 600,000 km^2 recharges the aquifer, what average annual rainfall would be required to replace the water removed for irrigation? (b) What process or processes accounts for the presence of arsenic in well water?

Human Activities and Water Quality (Section 18.4)

18.43 Suppose that one wishes to use reverse osmosis to reduce the salt content of brackish water containing 0.22 M total salt concentration to a value of 0.01 M, thus rendering it usable for human consumption. What is the minimum pressure that needs to be applied in the permeators (Figure 18.19) to achieve this goal, assuming that the operation occurs at 298 K? (*Hint:* Refer to Section 13.5.)

18.44 Assume that a portable reverse-osmosis apparatus operates on seawater, whose concentrations of constituent ions are listed in Table 18.5, and that the desalinated water output has an effective molarity of about 0.02 M. What minimum pressure must be applied by hand pumping at 297 K to cause reverse osmosis to occur? (*Hint:* Refer to Section 13.5.)

18.45 List the common products formed when an organic material containing the elements carbon, hydrogen, oxygen, sulfur, and nitrogen decomposes **(a)** under aerobic conditions, **(b)** under anaerobic conditions.

18.46 **(a)** Explain why the concentration of dissolved oxygen in freshwater is an important indicator of the quality of the water. **(b)** Find graphical data in the text that show variations of gas solubility with temperature, and estimate to two significant figures the percent solubility of O_2 in water at 30 °C as compared with 20 °C. How do these data relate to the quality of natural waters?

18.47 The organic anion

is found in most detergents. Assume that the anion undergoes aerobic decomposition in the following manner:

$$2\ C_{18}H_{29}SO_3^-(aq) + 51\ O_2(aq) \longrightarrow$$
$$36\ CO_2(aq) + 28\ H_2O(l) + 2\ H^+(aq) + 2\ SO_4^{2-}(aq)$$

What is the total mass of O_2 required to biodegrade 10.0 g of this substance?

18.48 The average daily mass of O_2 taken up by sewage discharged in the United States is 59 g per person. How many liters of water at 9 ppm O_2 are 50 % depleted of oxygen in 1 day by a population of 1,200,000 people?

18.49 Magnesium ions are removed in water treatment by the addition of slaked lime, $Ca(OH)_2$. Write a balanced chemical equation to describe what occurs in this process.

18.50 **(a)** Which of the following ionic species could be responsible for hardness in a water supply: Ca^{2+}, K^+, Mg^{2+}, Fe^{2+}, Na^+? **(b)** What properties of an ion determine whether it will contribute to water hardness?

18.51 In the lime soda process, at one time used in large scale municipal water softening, calcium hydroxide prepared from lime and sodium carbonate are added to precipitate Ca^{2+} as $CaCO_3(s)$ and Mg^{2+} as $Mg(OH)_2\ (s)$:

$$Ca^{2+}(aq) + CO_3^{2-}(aq) \longrightarrow CaCO_3(s)$$
$$Mg^{2+}(aq) + 2OH^-(aq) \longrightarrow MgOH_2(aq)$$

How many moles of $Ca(OH)_2$ and Na_2CO_3 should be added to soften 1200 L of water in which

$$[Ca^{2+}] = 5.0 \times 10^{-4}\ M \quad \text{and}$$
$$[Mg^{2+}] = 7.0 \times 10^{-4}\ M?$$

18.52 The concentration of Ca^{2+} in a particular water supply is $5.7 \times 10^{-3}\ M$. The concentration of bicarbonate ion, HCO_3^-, in the same water is $1.7 \times 10^{-3}\ M$. What masses of $Ca(OH)_2$ and Na_2CO_3 must be added to 5.0×10^7 L of this water to reduce the level of Ca^{2+} to 20% of its original level?

18.53 Ferrous sulfate $(FeSO_4)$ is often used as a coagulant in water purification. The iron(II) salt is dissolved in the water to be purified, then oxidized to the iron(III) state by dissolved oxygen, at which time gelatinous $Fe(OH)_3$ forms, assuming the pH is above approximately 6. Write balanced chemical

equations for the oxidation of Fe^{2+} to Fe^{3+} by dissolved oxygen and for the formation of $Fe(OH)_3(s)$ by reaction of $Fe^{3+}(aq)$ with $HCO_3^-(aq)$.

18.54 What properties make a substance a good coagulant for water purification?

18.55 **(a)** What are *trihalomethanes* (THMs)? **(b)** Draw the Lewis structures of two example THMs.

18.56 **(a)** Suppose that tests of a municipal water system reveal the presence of bromate ion, BrO_3^-. What are the likely origins of this ion? **(b)** Is bromate ion an oxidizing or reducing agent? Write a chemical equation for the reaction of bromate ion with hyponitrite ion.

Green Chemistry (Section 18.5)

18.57 One of the principles of green chemistry is that it is better to use as few steps as possible in making new chemicals. In what ways does following this rule advance the goals of green chemistry? How does this principle relate to energy efficiency?

18.58 Discuss how catalysts can make processes more energy efficient.

18.59 A reaction for converting ketones to lactones, called the Baeyer–Villiger reaction,

Ketone   3-Chloroperbenzoic acid

Lactone   3-Chlorobenzoic acid

is used in the manufacture of plastics and pharmaceuticals. 3-Chloroperbenzoic acid is shock-sensitive, however, and prone to explode. Also, 3-chlorobenzoic acid is a waste product. An alternative process being developed uses hydrogen peroxide and a catalyst consisting of tin deposited within a solid support. The catalyst is readily recovered from the reaction mixture. **(a)** What would you expect to be the other product of oxidation of the ketone to lactone by hydrogen peroxide? **(b)** What principles of green chemistry are addressed by use of the proposed process?

18.60 The hydrogenation reaction shown here was performed with an iridium catalyst, both in supercritical CO_2 (scCO_2) and in the chlorinated solvent CH_2Cl_2. The kinetic data for the reaction in both solvents are plotted in the graph. In what respects is the use of scCO_2 a good example of a green chemical reaction?

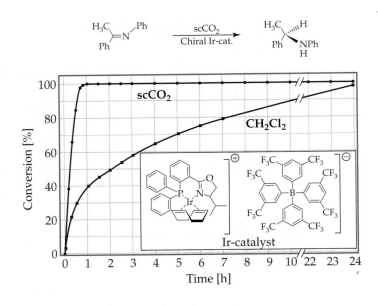

18.61 In the following three instances which choice is greener in each situation? Explain. **(a)** Benzene as a solvent or water as a solvent. **(b)** The reaction temperature is 500 K, or 1000 K. **(c)** Sodium chloride as a by-product or chloroform ($CHCl_3$) as a by-product.

18.62 In the following three instances which choice is greener in a chemical process? Explain. **(a)** A reaction that can be run at 350 K for 12 h without a catalyst or one that can be run at 300 K for 1 h with a reusable catalyst. **(b)** A reagent for the reaction that can be obtained from corn husks or one that is obtained from petroleum. **(c)** A process that produces no by-products or one in which the by-products are recycled for another process.

Additional Exercises

18.63 A friend of yours has seen each of the following items in newspaper articles and would like an explanation: **(a)** acid rain, **(b)** greenhouse gas, **(c)** photochemical smog, **(d)** ozone depletion. Give a brief explanation of each term and identify one or two of the chemicals associated with each.

18.64 Suppose that on another planet the atmosphere consists of 17% Kr, 38% CH_4, and 45% O_2. What is the average molar mass at the surface? What is the average molar mass at an altitude at which all the O_2 is photodissociated?

18.65 If an average O_3 molecule "lives" only 100–200 seconds in the stratosphere before undergoing dissociation, how can O_3 offer any protection from ultraviolet radiation?

18.66 Show how Equations 18.7 and 18.9 can be added to give Equation 18.10.

18.67 What properties of CFCs make them ideal for various commercial applications but also make them a long-term problem in the stratosphere?

18.68 *Halons* are fluorocarbons that contain bromine, such as $CBrF_3$. They are used extensively as foaming agents for fighting fires. Like CFCs, halons are very unreactive and ultimately can diffuse into the stratosphere. **(a)** Based on the data in Table 8.4, would you expect photodissociation of Br atoms to occur in the stratosphere? **(b)** Propose a mechanism by which the presence of halons in the stratosphere could lead to the depletion of stratospheric ozone.

18.69 **(a)** What is the difference between a CFC and an HFC? **(b)** It is estimated that the lifetime for HFCs in the stratosphere is 2–7 years. Why is this number significant? **(c)** Why have HFCs been used to replace CFCs? **(d)** What is the major disadvantage of HFCs as replacements for CFCs?

18.70 Explain, using Le Châtelier's principle, why the equilibrium constant for the formation of NO from N_2 and O_2 increases with increasing temperature, whereas the equilibrium constant for the formation of NO_2 from NO and O_2 decreases with increasing temperature.

18.71 Natural gas consists primarily of methane, $CH_4(g)$. **(a)** Write a balanced chemical equation for the complete combustion of methane to produce $CO_2(g)$ as the only carbon-containing product. **(b)** Write a balanced chemical equation for the incomplete combustion of methane to produce $CO(g)$ as the only carbon-containing product. **(c)** At 25 °C and 1.0 atm pressure, what is the minimum quantity of dry air needed to combust 1.0 L of $CH_4(g)$ completely to $CO_2(g)$?

18.72 It was estimated that the eruption of the Mount Pinatubo volcano resulted in the injection of 20 million metric tons of SO_2 into the atmosphere. Most of this SO_2 underwent oxidation to SO_3, which reacts with atmospheric water to form an aerosol. **(a)** Write chemical equations for the processes leading to formation of the aerosol. **(b)** The aerosols caused a 0.5–0.6 °C drop in surface temperature in the northern hemisphere. What is the mechanism by which this occurs? **(c)** The sulfate aerosols, as they are called, also cause loss of ozone from the stratosphere. How might this occur?

18.73 One of the possible consequences of climate change is an increase in the temperature of ocean water. The oceans serve as a "sink" for CO_2 by dissolving large amounts of it. **(a)** The figure below shows the solubility of CO_2 in water as a function of temperature. Does CO_2 behave more or less similarly to other gases in this respect?

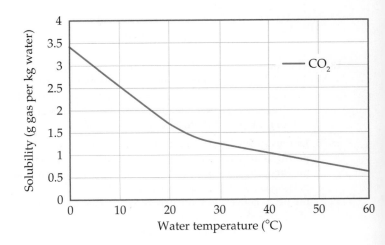

(b) What are the implications of this figure for the problem of climate change?

18.74 The rate of solar energy striking Earth averages 168 watts per square meter. The rate of energy radiated from Earth's surface averages 390 watts per square meter. Comparing these numbers, one might expect that the planet would cool quickly, yet it does not. Why not?

18.75 The solar power striking Earth every day averages 168 watts per square meter. The peak electrical power usage in New York City is 12,000 MW. Considering that present technology for solar energy conversion is about 10% efficient, from how many square meters of land must sunlight be collected in order to provide this peak power? (For comparison, the total area of New York city is 830 km^2.)

18.76 Write balanced chemical equations for each of the following reactions: **(a)** The nitric oxide molecule undergoes photodissociation in the upper atmosphere. **(b)** The nitric oxide molecule undergoes photoionization in the upper atmosphere. **(c)** Nitric oxide undergoes oxidation by ozone in the stratosphere.

(d) Nitrogen dioxide dissolves in water to form nitric acid and nitric oxide.

18.77 (a) Explain why $Mg(OH)_2$ precipitates when CO_3^{2-} ion is added to a solution containing Mg^{2+}. **(b)** Will $Mg(OH)_2$ precipitate when 4.0 g of Na_2CO_3 is added to 1.00 L of a solution containing 125 ppm of Mg^{2+}?

[18.78] It has been pointed out that there may be increased amounts of NO in the troposphere as compared with the past because of massive use of nitrogen-containing compounds in fertilizers. Assuming that NO can eventually diffuse into the stratosphere, how might it affect the conditions of life on Earth? Using the index to this text, look up the chemistry of nitrogen oxides. What chemical pathways might NO in the troposphere follow?

[18.79] As of the writing of this text, EPA standards limit atmospheric ozone levels in urban environments to 84 ppb. How many moles of ozone would there be in the air above Los Angeles County (area about 4000 square miles; consider a height of 100 m above the ground) if ozone was at this concentration?

Integrative Exercises

18.80 The estimated average concentration of NO_2 in air in the United States in 2006 was 0.016 ppm. **(a)** Calculate the partial pressure of the NO_2 in a sample of this air when the atmospheric pressure is 755 torr (99.1 kPa). **(b)** How many molecules of NO_2 are present under these conditions at 20 °C in a room that measures 15 × 14 × 8 ft?

[18.81] In 1986 an electrical power plant in Taylorsville, Georgia, burned 8,376,726 tons of coal, a national record at that time. **(a)** Assuming that the coal was 83% carbon and 2.5% sulfur and that combustion was complete, calculate the number of tons of carbon dioxide and sulfur dioxide produced by the plant during the year. **(b)** If 55% of the SO_2 could be removed by reaction with powdered CaO to form $CaSO_3$, how many tons of $CaSO_3$ would be produced?

18.82 The water supply for a midwestern city contains the following impurities: coarse sand, finely divided particulates, nitrate ion, trihalomethanes, dissolved phosphorus in the form of phosphates, potentially harmful bacterial strains, dissolved organic substances. Which of the following processes or agents, if any, is effective in removing each of these impurities: coarse sand filtration, activated carbon filtration, aeration, ozonization, precipitation with aluminum hydroxide?

[18.83] The *hydroxyl radical*, OH, is formed at low altitudes via the reaction of excited oxygen atoms with water:

$$O^*(g) + H_2O(g) \longrightarrow 2\,OH(g)$$

(a) Write the Lewis structure for the hydroxyl radical. (*Hint:* It has one unpaired electron.)

Once produced, the hydroxyl radical is very reactive. Explain the significance of each of the following reactions or series of reactions with respect to pollution in the troposphere:

(b) $OH + NO_2 \longrightarrow HNO_3$

(c) $OH + CO + O_2 \longrightarrow CO_2 + OOH$
$OOH + NO \longrightarrow OH + NO_2$

(d) $OH + CH_4 \longrightarrow H_2O + CH_3$
$CH_3 + O_2 \longrightarrow OOCH_3$
$OOCH_3 + NO \longrightarrow OCH_3 + NO_2$

(e) The concentration of hydroxyl radicals in the troposphere is approximately 2×10^6 radicals per cm^3. This estimate is based on a method called long path absorption spectroscopy (LPAS), similar in principle to the Beer's law measurement discussed in the *Closer Look* essay on p. 582, except that the length of the light path in the LPAS measurement is 20 km. Why must the path length be so large?

(f) The reactions shown in (d) also illustrate a second characteristic of the hydroxyl radical: its ability to cleanse the atmosphere of certain pollutants. Which of the reactions in (d) illustrate this?

18.84 An impurity in water has an extinction coefficient of $3.45 \times 10^3\,M^{-1}\,cm^{-1}$ at 280 nm, its absorption maximum (A Closer Look, p. 582). Below 50 ppb, the impurity is not a problem for human health. Given that most spectrometers cannot detect absorbances less than 0.0001 with good reliability, is measuring the absorbance of a water sample at 280 nm a good way to detect concentrations of the impurity above the 50-ppb threshold?

18.85 The concentration of H_2O in the stratosphere is about 5 ppm. It undergoes photodissociation according to:

$$H_2O(g) \longrightarrow H(g) + OH(g)$$

(a) Write out the Lewis-dot structures for both products and reactant.

(b) Using Table 8.4, calculate the wavelength required to cause this dissociation.

(c) The hydroxyl radicals, OH, can react with ozone, giving the following reactions:

$$OH\,(g) + O_3(g) \longrightarrow HO_2(g) + O_2(g)$$
$$HO_2(g) + O(g) \longrightarrow OH(g) + O_2(g)$$

What overall reaction results from these two elementary reactions? What is the catalyst in the overall reaction? Explain.

18.86 Bioremediation is the process by which bacteria repair their environment in response, for example, to an oil spill. The efficiency of bacteria for "eating" hydrocarbons depends on the amount of oxygen in the system, pH, temperature, and many other factors. In a certain oil spill, hydrocarbons from the oil disappeared with a first-order rate constant of $2 \times 10^{-6}\,s^{-1}$. At that rate, how many days would it take for the hydrocarbons to decrease to 10% of their initial value?

18.87 The standard enthalpies of formation of ClO and ClO_2 are 101 and 102 kJ/mol, respectively. Using these data and the thermodynamic data in Appendix C, calculate the overall enthalpy change for each step in the following catalytic cycle:

$$ClO(g) + O_3(g) \longrightarrow ClO_2(g) + O_2(g)$$
$$ClO_2(g) + O(g) \longrightarrow ClO(g) + O_2(g)$$

What is the enthalpy change for the overall reaction that results from these two steps?

18.88 The main reason that distillation is a costly method for purifying water is the high energy required to heat and vaporize water. **(a)** Using the density, specific heat, and heat of vaporization of water from Appendix B, calculate the amount of energy required to vaporize 1.00 gal of water beginning with water at 20 °C. **(b)** If the energy is provided by electricity costing \$0.085/kWh, calculate its cost. **(c)** If distilled water sells in a grocery store for \$1.26 per gal, what percentage of the sales price is represented by the cost of the energy?

[18.89] A reaction that contributes to the depletion of ozone in the stratosphere is the direct reaction of oxygen atoms with ozone:

$$O(g) + O_3(g) \longrightarrow 2\,O_2(g)$$

At 298 K the rate constant for this reaction is $4.8 \times 10^5\,M^{-1}\,s^{-1}$. **(a)** Based on the units of the rate constant, write the likely rate law for this reaction. **(b)** Would you expect this reaction to occur via a single elementary process? Explain why or why not **(c)** Use ΔH_f° values from Appendix C to estimate the enthalpy change for this reaction. Would this reaction raise or lower the temperature of the stratosphere?

18.90 The following data were collected for the destruction of O_3 by H ($O_3 + H \longrightarrow O_2 + OH$) at very low concentrations:

Trial	$[O_3]$ (M)	[H] (M)	Initial Rate (M/s)
1	5.17×10^{-33}	3.22×10^{-26}	1.88×10^{-14}
2	2.59×10^{-33}	3.25×10^{-26}	9.44×10^{-15}
3	5.19×10^{-33}	6.46×10^{-26}	3.77×10^{-14}

(a) Write the rate law for the reaction.

(b) Calculate the rate constant.

18.91 The degradation of CF_3CH_2F (an HFC) by OH radicals in the troposphere is first order in each reactant and has a rate constant of $k = 1.6 \times 10^8\,M^{-1}s^{-1}$ at 4 °C. If the tropospheric concentrations of OH and CF_3CH_2F are 8.1×10^5 and 6.3×10^8 molecules/cm³, respectively, what is the rate of reaction at this temperature in M/s?

18.92 The Henry's law constant for CO_2 in water at 25 °C is $3.1 \times 10^{-2}\,M\,atm^{-1}$. **(a)** What is the solubility of CO_2 in water at this temperature if the solution is in contact with air at normal atmospheric pressure? **(b)** Assume that all of this CO_2 is in the form of H_2CO_3 produced by the reaction between CO_2 and H_2O:

$$CO_2(aq) + H_2O(l) \longrightarrow H_2CO_3(aq)$$

What is the pH of this solution?

18.93 The precipitation of $Al(OH)_3$ ($K_{sp} = 1.3 \times 10^{-33}$) is sometimes used to purify water. **(a)** Estimate the pH at which precipitation of $Al(OH)_3$ will begin if 5.0 lb of $Al_2(SO_4)_3$ is added to 2000 gal of water. **(b)** Approximately how many pounds of CaO must be added to the water to achieve this pH?

[18.94] The valuable polymer polyurethane is made by a condensation reaction of alcohols (ROH) with compounds that contain an isocyanate group (RNCO). Two reactions that can generate a urethane monomer are shown here:

(i) $RNH_2 + CO_2 \longrightarrow R-N=C=O + 2\,H_2O$

(ii) $RNH_2 + $ $\longrightarrow R-N=C=O + 2\,HCl$

(a) Which process, i or ii, is greener? Explain.

(b) What are the hybridization and geometry of the carbon atoms in each C-containing compound in each reaction?

(c) If you wanted to promote the formation of the isocyanate intermediate in each reaction, what could you do, using Le Châtelier's principle?

18.95 The pH of a particular raindrop is 5.6. **(a)** Assuming the major species in the raindrop are $H_2CO_3(aq)$, $HCO_3^-(aq)$, and $CO_3^{2-}(aq)$, calculate the concentrations of these species in the raindrop, assuming the total carbonate concentration is $1.0 \times 10^{-5}\,M$. The appropriate K_a values are given in Table 16.3. **(b)** What experiments could you do to test the hypothesis that the rain also contains sulfur-containing species that contribute to its pH? Assume you have a large sample of rain to test.

Design an Experiment

Considerable fracking of petroleum/gas wells (see Closer Look box in Section 18.4) has occurred in recent years in a particular rural area. The residents have complained that the water in the residential wells serving their domestic water needs has become contaminated with chemicals associated with the fracking operations. The well operators respond that the chemicals about which complaints are lodged occur naturally, and are not the result of well-drilling activities.

Describe experiments that you could conduct on the waters from residential wells to help determine whether and to what extent well contaminants are due to fracking operations. Among the chemicals that might be expected to be employed in fracking operations are hydrochloric acid, sodium chloride, ethylene glycol, borate salts, water-soluble gelling agents such as guar gum, citric acid, methanol, and other alcohols such as isopropanol, and methane. Assume that you have available the techniques to make measurements of the concentrations of these substances in the residential wells. What experiments would you conduct and what analyses of the results would you carry out in an attempt to settle the question of whether fracking operations have led to contamination of the well water? Would simply measuring the concentrations of some or all of these substances in the well waters be sufficient to settle the issue?

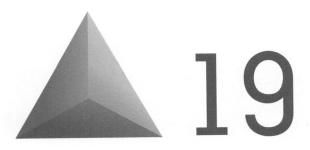

19
Chemical Thermodynamics

The amazing organization of living systems, from complex molecular structures such as the nucleosome, to cells, to tissues, and finally to whole plants and animals, is an unending source of wonder and delight to the chemists, biochemists, physicists, and biologists who study them. Energy must be spent, somehow, to keep all of these organized systems in good working order. But we have not yet learned enough about energy to understand how the underlying chemical and physical processes of life are governed.

Understanding biochemical processes such as DNA replication, photosynthesis, or metabolism requires us to ask increasingly more sophisticated questions, and anticipate increasingly sophisticated answers to them. Fortunately, the general laws that govern chemical reactions—from those that we perform in the laboratory to those that occur in biology—can help us understand the most complicated of processes. Two of the most important questions chemists ask when designing and using chemical reactions are "How fast is the reaction?" and "How far does it proceed?" The first question is addressed by chemical kinetics, which we discussed in Chapter 14. The second question involves the equilibrium constant, the focus of Chapter 15. Let's briefly review how these concepts are related.

In Chapter 14 we learned that the rate of any chemical reaction is controlled largely by a factor related to energy, namely, the activation energy of the reaction. ∞ (Section 14.5) In general, the lower the activation energy, the faster a reaction proceeds. In Chapter 15 we saw that chemical equilibrium is reached when a given reaction and its reverse reaction occur at the same rate. ∞ (Section 15.1)

Because reaction rates are closely tied to energy, it is logical that equilibrium also depends in some way on energy. In this chapter we explore the connection between

▶ **THE NUCLEOSOME** Inside the nucleus of a living cell, DNA (the outer double-helical ribbon) surrounds eight protein molecules (the colored spiral ribbons). This overall DNA/protein structure, called the nucleosome, is the basic unit of chromosomes in the nuclei of our cells. These structures are highly ordered, yet also must be unraveled in order for gene expression to take place. Both packaging and unpackaging of DNA in the nucleosome involve changes in the energy of the system.

WHAT'S AHEAD

19.1 SPONTANEOUS PROCESSES We see that changes that occur in nature have a directional character. They move *spontaneously* in one direction but not in the reverse direction.

19.2 ENTROPY AND THE SECOND LAW OF THERMODYNAMICS We discuss *entropy*, a thermodynamic state function that is important in determining whether a process is spontaneous. The *second law of thermodynamics* tells us that in any spontaneous process, the entropy of the universe (system plus surroundings) increases.

19.3 THE MOLECULAR INTERPRETATION OF ENTROPY AND THE THIRD LAW OF THERMODYNAMICS On the molecular level, we learn that the entropy of a system is related to the number of accessible *microstates*. The entropy of the system increases as the randomness of the system increases. The *third law of thermodynamics* states that, at 0 K, the entropy of a perfect crystalline solid is zero.

19.4 ENTROPY CHANGES IN CHEMICAL REACTIONS Using tabulated *standard molar entropies*, we can calculate the standard entropy changes for systems undergoing reaction.

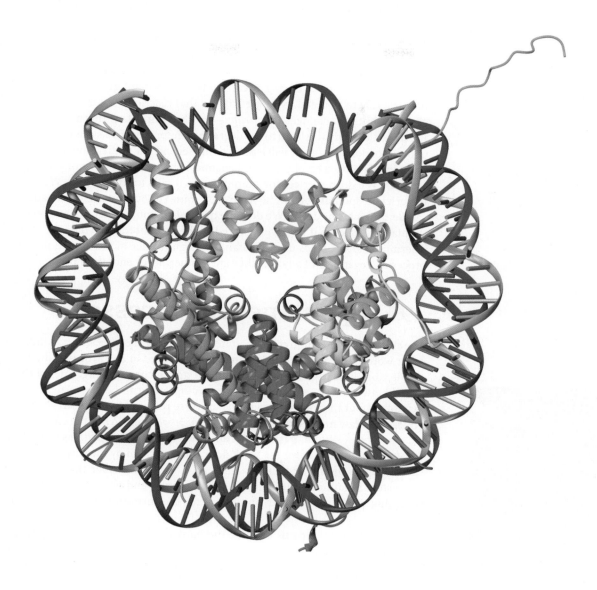

19.5 GIBBS FREE ENERGY We encounter another thermodynamic state function, *free energy* (or *Gibbs free energy*), a measure of how far removed a system is from equilibrium. The change in free energy measures the maximum amount of useful work obtainable from a process and tells us the direction in which a chemical reaction is spontaneous.

19.6 FREE ENERGY AND TEMPERATURE We consider how the relationship among free-energy change, enthalpy change, and entropy change provides insight into how temperature affects the spontaneity of a process.

19.7 FREE ENERGY AND THE EQUILIBRIUM CONSTANT Finally, we consider how the standard free-energy change for a chemical reaction can be used to calculate the equilibrium constant for the reaction.

energy and the extent of a reaction. Doing so requires a deeper look at *chemical thermodynamics*, the area of chemistry that deals with energy relationships. We first encountered thermodynamics in Chapter 5, where we discussed the nature of energy, the first law of thermodynamics, and the concept of enthalpy. Recall that the enthalpy change for any system is the heat transferred between the system and its surroundings during a constant-pressure process. ∞ (Equation 5.10)

In the "Strategies in Chemistry" box in Section 5.4, we pointed out that the enthalpy change that takes place during a reaction is an important guide as to whether the reaction is likely to proceed. Now we will see that reactions involve not only changes in enthalpy but also changes in *entropy*—another important thermodynamic quantity. Recall that entropy is related to the degree of randomness in a system. ∞ (Section 13.1) Our discussion of entropy will lead us to the second law of thermodynamics, which provides insight into why physical and chemical changes tend to favor one direction over another. We drop a brick, for example, and it falls to the ground. We do not expect the brick to spontaneously rise from the ground to our outstretched hand. We light a candle, and it burns down. We do not expect a half-consumed candle to regenerate itself spontaneously, even if we have captured all the gases produced when the candle burned. Thermodynamics helps us understand the significance of this directional character of processes, regardless of whether they are exothermic or endothermic.

19.1 | Spontaneous Processes

The first law of thermodynamics states that *energy is conserved.* ∞ (Section 5.2) In other words, energy is neither created nor destroyed in any process, whether that process is a brick falling, a candle burning, or an ice cube melting. Energy can be transferred between a system and the surroundings and can be converted from one form to another, but the total energy of the universe remains constant. In Chapter 5, we expressed this law mathematically as $\Delta E = q + w$, where ΔE is the change in the internal energy of a system, q is the heat absorbed (or released) by the system from (or to) the surroundings, and w is the work done on the system by the surroundings, or on the surroundings by the system. ∞ (Equation 5.5) Remember that $q > 0$ means that the system is gaining heat from the surroundings, and $w > 0$ means that the system is gaining work from the surroundings (i.e., the surroundings are doing work on the system).

The first law helps us balance the books, so to speak, on the heat transferred between a system and its surroundings and the work done by or on a system. However, because energy is conserved, we cannot simply use the value of ΔE to tell us whether a process is favored to occur because anything we do to lower the energy of the system raises the energy of the surroundings, and vice versa. Nevertheless, experience tells us that certain processes *always* occur, even though, so far as we can observe in our study of chemical and physical processes, energy is conserved. Water placed in a freezer turns into ice, for instance, and if you touch a hot object, heat is transferred to your hand. The first law guarantees that energy is conserved in these processes, but it says nothing about the preferred direction for the process. Furthermore, the situations we have just mentioned occur without any outside intervention. We say they are *spontaneous*. A **spontaneous process** is one that proceeds on its own without any outside assistance.

A spontaneous process occurs in one direction only, and the reverse of any spontaneous process is always *nonspontaneous*. Drop an egg above a hard surface, for example, and it breaks on impact (◀ Figure 19.1). Now, imagine seeing a video clip in which a broken egg rises from the floor, reassembles itself, and ends up in someone's hand. You would conclude that the video is running in reverse because you know that broken eggs simply do not magically rise and reassemble themselves! An egg falling and breaking is spontaneous. The reverse process is *nonspontaneous*, even though energy is conserved in both processes. ∞ (Strategies in Chemistry Box, Section 5.4)

We have touched upon several spontaneous and nonspontaneous processes in our study of chemistry thus far. For example, a gas spontaneously expands into a vacuum (▶ Figure 19.2), but the reverse process, in which the gas moves back entirely into one

▲ **GO FIGURE**

Does the potential energy of the eggs change during this process?

Spontaneous   Not spontaneous

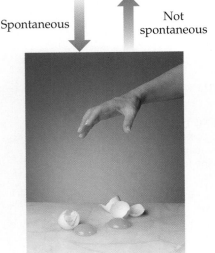

▲ **Figure 19.1 A spontaneous process!**

of the flasks, does not happen. In other words, expansion of the gas is spontaneous, but the reverse process is nonspontaneous. In general, *processes that are spontaneous in one direction are nonspontaneous in the opposite direction.*

Experimental conditions, such as temperature and pressure, are often important in determining whether a process is spontaneous. We are all familiar with situations in which a forward process is spontaneous at one temperature but the reverse process is spontaneous at a different temperature. Consider, for example, ice melting. At atmospheric pressure, when the temperature of the surroundings is above 0 °C, ice melts spontaneously, and the reverse process—liquid water turning into ice—is not spontaneous. However, when the temperature of the surroundings is below 0 °C, the opposite is true—liquid water turns to ice spontaneously, but the reverse process is *not* spontaneous (▼ Figure 19.3).

What happens at $T = 0$ °C, the normal melting point of water, when the flask of Figure 19.3 contains both water and ice? At the normal melting point of a substance, the solid and liquid phases are in equilibrium. ∞(Section 11.6) At this temperature, the two phases are interconverting at the same rate and there is no preferred direction for the process.

Just because a process is spontaneous does not necessarily mean that it will occur at an observable rate. A chemical reaction is spontaneous if it occurs on its own accord, regardless of its speed. A spontaneous reaction can be very fast, as in the case of acid–base neutralization, or very slow, as in the rusting of iron. Thermodynamics tells us the *direction* and *extent* of a reaction but nothing about the *rate*; rate is the domain of kinetics rather than of thermodynamics.

 Give It Some Thought

If a process is nonspontaneous, does that mean the process cannot occur under any circumstances?

**GO FIGURE**

In which direction is this process exothermic?

Spontaneous for $T > 0$ °C

Spontaneous for $T < 0$ °C

▲ Figure 19.3 **Spontaneity can depend on temperature.** At $T > 0$ °C, ice melts spontaneously to liquid water. At $T < 0$ °C, the reverse process, water freezing to ice, is spontaneous. At $T = 0$ °C the two states are in equilibrium.

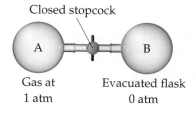▲**GO FIGURE**

If flask B were smaller than flask A, would the final pressure after the stopcock is opened be greater than, equal to, or less than 0.5 atm?

Closed stopcock

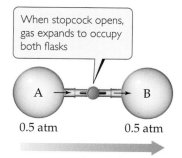

Gas at 1 atm

Evacuated flask 0 atm

When stopcock opens, gas expands to occupy both flasks

A 0.5 atm 0.5 atm B

This process is spontaneous

All gas molecules move back into flask A

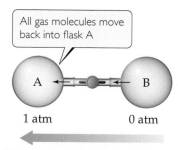

A 1 atm 0 atm B

This process is not spontaneous

▲ Figure 19.2 **Expansion of a gas into an evacuated space is a spontaneous process.** The reverse process—gas molecules initially distributed evenly in two flasks all moving into one flask—is not spontaneous.

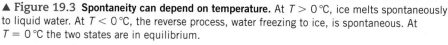
SAMPLE
EXERCISE 19.1 **Identifying Spontaneous Processes**

Predict whether each process is spontaneous as described, spontaneous in the reverse direction, or at equilibrium: **(a)** Water at 40 °C gets hotter when a piece of metal heated to 150 °C is added. **(b)** Water at room temperature decomposes into $H_2(g)$ and $O_2(g)$. **(c)** Benzene vapor, $C_6H_6(g)$, at a pressure of 1 atm condenses to liquid benzene at the normal boiling point of benzene, 80.1 °C.

SOLUTION

Analyze We are asked to judge whether each process is spontaneous in the direction indicated, in the reverse direction, or in neither direction.

Plan We need to think about whether each process is consistent with our experience about the natural direction of events or whether we expect the reverse process to occur.

Solve

(a) This process is spontaneous. Whenever two objects at different temperatures are brought into contact, heat is transferred from the hotter object to the colder one. ∞ (Section 5.1) Thus, heat is transferred from the hot metal to the cooler water. The final temperature, after the metal and water achieve the same temperature (thermal equilibrium), will be somewhere between the initial temperatures of the metal and the water. **(b)** Experience tells us that this process is not spontaneous—we certainly have never seen hydrogen and oxygen gases spontaneously bubbling up out of water! Rather, the *reverse* process—the reaction of H_2 and O_2 to form H_2O—is spontaneous. **(c)** The normal boiling point is the temperature at which a vapor at 1 atm is in equilibrium with its liquid. Thus, this is an equilibrium situation. If the temperature were below 80.1 °C, condensation would be spontaneous.

> **Practice Exercise 1**
>
> The process of iron being oxidized to make iron(III) oxide (rust) is spontaneous. Which of these statements about this process is/are true? **(a)** The reduction of iron(III) oxide to iron is also spontaneous. **(b)** Because the process is spontaneous, the oxidation of iron must be fast. **(c)** The oxidation of iron is endothermic. **(d)** Equilibrium is achieved in a closed system when the rate of iron oxidation is equal to the rate of iron(III) oxide reduction. **(e)** The energy of the universe is decreased when iron is oxidized to rust.
>
> **Practice Exercise 2**
>
> At 1 atm pressure, $CO_2(s)$ sublimes at −78 °C. Is this process spontaneous at −100 °C and 1 atm pressure?

Seeking a Criterion for Spontaneity

A marble rolling down an incline or a brick falling from your hand loses potential energy. The loss of some form of energy is a common feature of spontaneous change in mechanical systems. In the 1870s Marcellin Bertholet (1827–1907) suggested that the direction of spontaneous changes in chemical systems is determined by the loss of energy. He proposed that all spontaneous chemical and physical changes are exothermic. It takes only a few moments, however, to find exceptions to this generalization. For example, the melting of ice at room temperature is spontaneous and endothermic. Similarly, many spontaneous dissolution processes, such as the dissolving of NH_4NO_3, are endothermic, as we discovered in Section 13.1. We conclude that although the majority of spontaneous reactions are exothermic, there are spontaneous endothermic ones as well. Clearly, some other factor must be at work in determining the natural direction of processes.

To understand why certain processes are spontaneous, we need to consider more closely the ways in which the state of a system can change. Recall from Section 5.2 that quantities such as temperature, internal energy, and enthalpy are *state functions*, properties that define a state and do not depend on how we reach that state. The heat transferred between a system and its surroundings, q, and the work done by or on the system, w, are *not* state functions—their values depend on the specific path taken between states. One key to understanding spontaneity is understanding differences in the paths between states.

Reversible and Irreversible Processes

In 1824, a 28-year-old French engineer named Sadi Carnot (1796–1832) published an analysis of the factors that determine how efficiently a steam engine can convert heat to work. Carnot considered what an *ideal engine*, one with the highest possible efficiency, would be like. He observed that it is impossible to convert the energy content of a fuel completely to work because a significant amount of heat is always lost to the surroundings. Carnot's analysis gave insight into how to build better, more efficient engines, and it was one of the earliest studies in what has developed into the discipline of thermodynamics.

An ideal engine operates under an ideal set of conditions in which all the processes are reversible. A **reversible process** is a specific way in which a system changes its state. In a reversible process, the change occurs in such a way that the system and surroundings can be restored to their original states by *exactly* reversing the change. In other words, we can restore the system to its original condition with no net change to either the system or its surroundings. An **irreversible process** is one that cannot simply be reversed to restore the system and its surroundings to their original states. What Carnot discovered is that the amount of work we can extract from any process depends on the manner in which the process is carried out. He concluded that *a reversible change produces the maximum amount of work that can be done by a system on its surroundings.*

GO FIGURE

If the flow of heat into or out of the system is to be reversible, what must be true of δT?

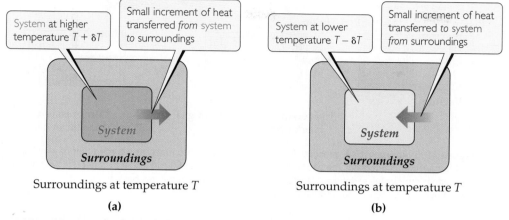

Surroundings at temperature T

(a)

Surroundings at temperature T

(b)

▲ **Figure 19.4 Reversible flow of heat.** Heat can flow reversibly between a system and its surroundings only if the two have an infinitesimally small difference in temperature δT. **(a)** Increasing the temperature of the system by δT causes heat to flow from the hotter system to the colder surroundings. **(b)** Decreasing the temperature of the system by δT causes heat to flow from the hotter surroundings to the colder system.

 Give It Some Thought

Suppose you have a system made up of water only, with the container and everything beyond being the surroundings. Consider a process in which the water is first evaporated and then condensed back into its original container. Is this two-step process necessarily reversible?

Let's next examine some aspects of reversible and irreversible processes, first with respect to the transfer of heat. When two objects at different temperatures are in contact, heat flows spontaneously from the hotter object to the colder one. Because it is impossible to make heat flow in the opposite direction, from colder object to hotter one, the flow of heat is an irreversible process. Given these facts, can we imagine any conditions under which heat transfer can be made reversible?

To answer this question, we must consider temperature differences that are infinitesimally small, as opposed to the discrete temperature differences with which we are most familiar. For example, consider a system and its surroundings at essentially the same temperature, with just an infinitesimal temperature difference δT, between them (▲ **Figure 19.4**). If the surroundings are at temperature T and the system is at the infinitesimally higher temperature $T + \delta T$, then an infinitesimal amount of heat flows from system to surroundings. We can reverse the direction of heat flow by making an infinitesimal change of temperature in the opposite direction, lowering the system temperature to $T - \delta T$. Now the direction of heat flow is from surroundings to system. *Reversible processes are those that reverse direction whenever an infinitesimal change is made in some property of the system.* *

Now let's consider another example, the expansion of an ideal gas at constant temperature (referred to as an **isothermal** process). In the cylinder-piston arrangement of **Figure 19.5**, when the partition is removed, the gas expands spontaneously to fill the evacuated space. Can we determine whether this particular isothermal expansion is reversible or irreversible? Because the gas expands into a vacuum with no external

---

*For a process to be truly reversible, the amounts of heat must be infinitesimally small and the transfer of heat must occur infinitely slowly; thus, no process that we can observe is truly reversible. The notion of infinitesimal amounts is related to the infinitesimals that you may have studied in a calculus course.

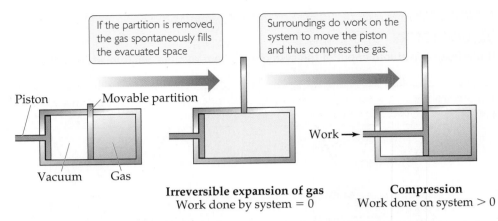

▲ **Figure 19.5 An irreversible process.** Initially an ideal gas is confined to the right half of a cylinder. When the partition is removed, the gas spontaneously expands to fill the whole cylinder. No work is done by the system during this expansion. Using the piston to compress the gas back to its original state requires the surroundings to do work on the system.

pressure, it does no $P-V$ work on the surroundings. ∞ (Section 5.3) Thus, for the expansion, $w = 0$. We can use the piston to compress the gas back to its original state, but doing so requires that the surroundings do work on the system, meaning that $w > 0$ for the compression. In other words, the path that restores the system to its original state requires a different value of w (and, by the first law, a different value of q) than the path by which the system was first changed. The fact that the same path cannot be followed to restore the system to its original state indicates that the process is irreversible.

What might a *reversible* isothermal expansion of an ideal gas be? This process will occur only if initially, when the gas is confined to half the cylinder, the external pressure acting on the piston exactly balances the pressure exerted by the gas on the piston. If the external pressure is reduced infinitely slowly, the piston will move outward, allowing the pressure of the confined gas to readjust to maintain the pressure balance. This infinitely slow process in which the external pressure and internal pressure are always in equilibrium is reversible. If we reverse the process and compress the gas in the same infinitely slow manner, we can return the gas to its original volume. The complete cycle of expansion and compression in this hypothetical process, moreover, is accomplished without any net change to the surroundings.

Because real processes can at best only approximate the infinitely slow change associated with reversible processes, all real processes are irreversible. Further, the reverse of any spontaneous process is a nonspontaneous process. A nonspontaneous process can occur only if the surroundings do work on the system. Thus, *any spontaneous process is irreversible*. Even if we return the system to the original condition, the surroundings will have changed.

19.2 | Entropy and the Second Law of Thermodynamics

Knowing that any spontaneous process is irreversible, can we make predictions about the spontaneity of an unfamiliar process? Understanding spontaneity requires us to examine the thermodynamic quantity called **entropy**, which was first mentioned in Section 13.1. In general, entropy is associated either with the extent of *randomness* in a system or with the extent to which energy is distributed among the various motions of the molecules of the system. In this section we consider how entropy changes are related to heat transfer and temperature. Our analysis will bring us to a profound statement about spontaneity known as the second law of thermodynamics.

The Relationship between Entropy and Heat

The entropy, S, of a system is a state function just like internal energy, E, and enthalpy, H. As with these other quantities, the value of S is a characteristic of the state of a system. ∞ (Section 5.2) Thus, the change in entropy, ΔS, in a system depends only on the initial and final states of the system and not on the path taken from one state to the other:

$$\Delta S = S_{final} - S_{initial} \qquad [19.1]$$

For the special case of an isothermal process, ΔS is equal to the heat that would be transferred if the process were reversible, q_{rev}, divided by the absolute temperature at which the process occurs:

$$\Delta S = \frac{q_{rev}}{T} \quad \text{(constant } T \text{)} \qquad [19.2]$$

Although many possible paths can take the system from one state to another, only one path is associated with a reversible process. Thus, the value of q_{rev} is uniquely defined for any two states of the system. Because S is a state function, we can use Equation 19.2 to calculate ΔS for *any* isothermal process between states, not just the reversible one.

▲ **Give It Some Thought**

How can S be a state function when ΔS depends on q, which is not a state function?

ΔS for Phase Changes

The melting of a substance at its melting point and the vaporization of a substance at its boiling point are isothermal processes. (Section 11.4) Consider the melting of ice. At 1 atm pressure, ice and liquid water are in equilibrium at 0 °C. Imagine melting 1 mol of ice at 0 °C, 1 atm to form 1 mol of liquid water at 0 °C, 1 atm. We can achieve this change by adding heat to the system from the surroundings: $q = \Delta H_{fusion}$, where ΔH_{fusion} is the heat of melting. Imagine adding the heat infinitely slowly, raising the temperature of the surroundings only infinitesimally above 0 °C. When we make the change in this fashion, the process is reversible because we can reverse it by infinitely slowly removing the same amount of heat, ΔH_{fusion}, from the system, using immediate surroundings that are infinitesimally below 0 °C. Thus, $q_{rev} = \Delta H_{fusion}$ for the melting of ice at $T = 0$ °C $= 273$ K.

The enthalpy of fusion for H_2O is $\Delta H_{fusion} = 6.01$ kJ/mol (a positive value because melting is an endothermic process). Thus, we can use Equation 19.2 to calculate ΔS_{fusion} for melting 1 mol of ice at 273 K:

$$\Delta S_{fusion} = \frac{q_{rev}}{T} = \frac{\Delta H_{fusion}}{T} = \frac{(1 \text{ mol})(6.01 \times 10^3 \text{ J/mol})}{273 \text{ K}} = 22.0 \text{ J/K}$$

Notice that (1) we must use the absolute temperature in Equation 19.2, and (2) the units for ΔS, J/K, are energy divided by absolute temperature, as we expect from Equation 19.2.

SAMPLE EXERCISE 19.2 Calculating ΔS for a Phase Change

Elemental mercury is a silver liquid at room temperature. Its normal freezing point is -38.9 °C, and its molar enthalpy of fusion is $\Delta H_{fusion} = 2.29$ kJ/mol. What is the entropy change of the system when 50.0 g of Hg(l) freezes at the normal freezing point?

SOLUTION

Analyze We first recognize that freezing is an *exothermic* process, which means heat is transferred from the system to the surroundings and $q < 0$. Because freezing is the reverse of melting, the enthalpy change that accompanies the freezing of 1 mol of Hg is $-\Delta H_{fusion} = -2.29$ kJ/mol.

Plan We can use $-\Delta H_{fusion}$ and the atomic weight of Hg to calculate q for freezing 50.0 g of Hg. Then we use this value of q as q_{rev} in Equation 19.2 to determine ΔS for the system.

Solve For q we have

$$q = (50.0 \text{ g Hg})\left(\frac{1 \text{ mol Hg}}{200.59 \text{ g Hg}}\right)\left(\frac{-2.29 \text{ kJ}}{1 \text{ mol Hg}}\right)\left(\frac{1000 \text{ J}}{1 \text{ kJ}}\right) = -571 \text{ J}$$

Before using Equation 19.2, we must first convert the given Celsius temperature to kelvins:

$$-38.9 \text{ °C} = (-38.9 + 273.15) \text{ K} = 234.3 \text{ K}$$

We can now calculate ΔS_{sys}:

$$\Delta S_{sys} = \frac{q_{rev}}{T} = \frac{-571\ J}{234.3\ K} = -2.44\ J/K$$

Check The entropy change is negative because our q_{rev} value is negative, which it must be because heat flows out of the system in this exothermic process.

Comment This procedure can be used to calculate ΔS for other isothermal phase changes, such as the vaporization of a liquid at its boiling point.

Practice Exercise 1

Do all exothermic phase changes have a negative value for the entropy change of the system? **(a)** Yes, because the heat transferred from the system has a negative sign. **(b)** Yes, because the temperature decreases during the phase transition. **(c)** No, because the entropy change depends on the sign of the heat transferred to or from the system. **(d)** No, because the heat transferred to the system has a positive sign. **(e)** More than one of the previous answers is correct.

Practice Exercise 2

The normal boiling point of ethanol, C_2H_5OH, is 78.3 °C, and its molar enthalpy of vaporization is 38.56 kJ/mol. What is the change in entropy in the system when 68.3 g of $C_2H_5OH(g)$ at 1 atm condenses to liquid at the normal boiling point?

A Closer Look

The Entropy Change When a Gas Expands Isothermally

In general, the entropy of any system increases as the system becomes more random or more spread out. Thus, we expect the spontaneous expansion of a gas to result in an increase in entropy. To see how this entropy increase can be calculated, consider the expansion of an ideal gas that is initially constrained by a piston, as in the rightmost part of Figure 19.5. Imagine that we allow the gas to undergo a reversible isothermal expansion by infinitesimally decreasing the external pressure on the piston. The work done on the surroundings by the reversible expansion of the system against the piston can be calculated with the aid of calculus (we do not show the derivation):

$$w_{rev} = -nRT \ln\frac{V_2}{V_1}$$

In this equation, n is the number of moles of gas, R is the ideal-gas constant ∞ (Section 10.4), T is the absolute temperature, V_1 is the initial volume, and V_2 is the final volume. Notice that if $V_2 > V_1$, as it must be in our expansion, then $w_{rev} < 0$, meaning that the expanding gas does work on the surroundings.

One characteristic of an ideal gas is that its internal energy depends only on temperature, not on pressure. Thus, when an ideal gas expands isothermally, $\Delta E = 0$. Because $\Delta E = q_{rev} + w_{rev} = 0$, we

see that $q_{rev} = -w_{rev} = nRT \ln(V_2/V_1)$. Then, using Equation 19.2, we can calculate the entropy change in the system:

$$\Delta S_{sys} = \frac{q_{rev}}{T} = \frac{nRT \ln\frac{V_2}{V_1}}{T} = nR \ln\frac{V_2}{V_1} \qquad [19.3]$$

Let's calculate the entropy change for 1.00 L of an ideal gas at 1.00 atm pressure, 0 °C temperature, expanding to 2.00 L. From the ideal-gas equation, we can calculate the number of moles in 1.00 L of an ideal gas at 1.00 atm and 0 °C as we did in Chapter 10:

$$n = \frac{PV}{RT} = \frac{(1.00\ atm)(1.00\ L)}{(0.08206\ L\text{-}atm/mol\text{-}K)(273\ K)} = 4.46 \times 10^{-2}\ mol$$

The gas constant, R, can also be expressed as 8.314 J/mol-K (Table 10.2), and this is the value we must use in Equation 19.3 because we want our answer to be expressed in terms of J rather than in L-atm. Thus, for the expansion of the gas from 1.00 L to 2.00 L, we have

$$\Delta S_{sys} = (4.46 \times 10^{-2}\ mol)\left(8.314\ \frac{J}{mol\text{-}K}\right)\left(\ln\frac{2.00\ L}{1.00\ L}\right)$$
$$= 0.26\ J/K$$

In Section 19.3 we will see that this increase in entropy is a measure of the increased randomness of the molecules due to the expansion.
Related Exercises: 19.29, 19.30, 19.106

The Second Law of Thermodynamics

The key idea of the first law of thermodynamics is that energy is conserved in any process. ∞ (Section 5.2) Is entropy also conserved in a spontaneous process in the same way that energy is conserved?

Let's try to answer this question by calculating the entropy change of a system and the entropy change of its surroundings when our system is 1 mol of ice (a piece roughly the size of an ice cube) melting in the palm of your hand, which is part of the surroundings. The process is not reversible because the system and surroundings are at different temperatures. Nevertheless, because ΔS is a state function, its value is the same regardless of whether the process is reversible or irreversible. We calculated the entropy change of the system just before Sample Exercise 19.2:

$$\Delta S_{sys} = \frac{q_{rev}}{T} = \frac{(1\ mol)(6.01 \times 10^3\ J/mol)}{273\ K} = 22.0\ J/K$$

The surroundings immediately in contact with the ice are your hand, which we assume is at body temperature, 37 °C = 310 K. The quantity of heat lost by your hand is -6.01×10^3 J/mol, which is equal in magnitude to the quantity of heat gained by the ice but has the opposite sign. Hence, the entropy change of the surroundings is

$$\Delta S_{surr} = \frac{q_{rev}}{T} = \frac{(1 \text{ mol})(-6.01 \times 10^3 \text{ J/mol})}{310 \text{ K}} = -19.4 \text{ J/K}$$

We can consider that everything in the universe is either the system of interest, or its surroundings. Therefore $\Delta S_{univ} = \Delta S_{sys} + \Delta S_{surr}$. Thus, the overall entropy change of the universe is positive in our example:

$$\Delta S_{univ} = \Delta S_{sys} + \Delta S_{surr} = (22.0 \text{ J/K}) + (-19.4 \text{ J/K}) = 2.6 \text{ J/K}$$

If the temperature of the surroundings were not 310 K but rather some temperature infinitesimally above 273 K, the melting would be reversible instead of irreversible. In that case the entropy change of the surroundings would equal -22.0 J/K and ΔS_{univ} would be zero.

In general, any irreversible process results in an increase in the entropy of the universe, whereas any reversible process results in no change in the entropy of the universe, a statement known as the **second law of thermodynamics**. The second law of thermodynamics can be expressed in terms of either of the following two equations:

Reversible Process: $\qquad \Delta S_{univ} = \Delta S_{sys} + \Delta S_{surr} = 0$

Irreversible Process: $\qquad \Delta S_{univ} = \Delta S_{sys} + \Delta S_{surr} > 0 \qquad$ [19.4]

Because spontaneous processes are irreversible, we can say that *the entropy of the universe increases in any spontaneous process.* This profound generalization is yet another way of expressing the second law of thermodynamics.

 Give It Some Thought

The rusting of iron is spontaneous and is accompanied by a decrease in the entropy of the system (the iron and oxygen). What can we conclude about the entropy change of the surroundings?

The second law of thermodynamics tells us the essential character of any spontaneous change—it is always accompanied by an increase in the entropy of the universe. We can use this criterion to predict whether a given process is spontaneous or not. Before seeing how this is done, however, we will find it useful to explore entropy from a molecular perspective.

A word on notation before we proceed: Throughout most of the remainder of this chapter, we will focus on systems rather than surroundings. To simplify the notation, we will usually refer to the entropy change of the system as ΔS rather than explicitly indicating ΔS_{sys}.

19.3 | The Molecular Interpretation of Entropy and the Third Law of Thermodynamics

As chemists, we are interested in molecules. What does entropy have to do with them and with their transformations? What molecular property does entropy reflect? Ludwig Boltzmann (1844–1906) gave another conceptual meaning to the notion of entropy, and to understand his contribution, we need to examine the ways in which we can interpret entropy at the molecular level.

Expansion of a Gas at the Molecular Level

In discussing Figure 19.2, we talked about the expansion of a gas into a vacuum as a spontaneous process. We now understand that it is an irreversible process and that the entropy of the universe increases during the expansion. How can we explain

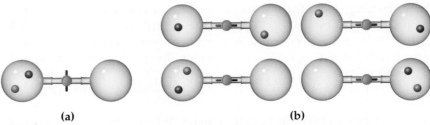

(a)

The two molecules are colored red and blue to keep track of them.

(b)

Four possible arrangements (microstates) once the stopcock is opened.

▲ **Figure 19.6 Possible arrangements of two gas molecules in two flasks. (a)** Before the stopcock is opened, both molecules are in the left flask. **(b)** After the stopcock is opened, there are four possible arrangements of the two molecules.

the spontaneity of this process at the molecular level? We can get a sense of what makes this expansion spontaneous by envisioning the gas as a collection of particles in constant motion, as we did in discussing the kinetic-molecular theory of gases. ∞ (Section 10.7) When the stopcock in Figure 19.2 is opened, we can view the expansion of the gas as the ultimate result of the gas molecules moving randomly throughout the larger volume.

Let's look at this idea more closely by tracking two of the gas molecules as they move around. Before the stopcock is opened, both molecules are confined to the left flask, as shown in ▲ **Figure 19.6(a)**. After the stopcock is opened, the molecules travel randomly throughout the entire apparatus. As Figure 19.6(b) shows, there are four possible arrangements for the two molecules once both flasks are available to them. Because the molecular motion is random, all four arrangements are equally likely. Note that now only one arrangement corresponds to the situation before the stopcock was opened: both molecules in the left flask.

Figure 19.6(b) shows that with both flasks available to the molecules, the probability of the red molecule being in the left flask is two in four (top right and bottom left arrangements), and the probability of the blue molecule being in the left flask is the same (top left and bottom left arrangements). Because the probability is $^2/_4 = ^1/_2$ that each molecule is in the left flask, the probability that *both* are there is $(^1/_2)^2 = ^1/_4$. If we apply the same analysis to *three* gas molecules, we find that the probability that all three are in the left flask at the same time is $(^1/_2)^3 = ^1/_8$.

Now let's consider a *mole* of gas. The probability that all the molecules are in the left flask at the same time is $(^1/_2)^N$, where $N = 6.02 \times 10^{23}$. This is a vanishingly small number! Thus, there is essentially zero likelihood that all the gas molecules will be in the left flask at the same time. This analysis of the microscopic behavior of the gas molecules leads to the expected macroscopic behavior: The gas spontaneously expands to fill both the left and right flasks, and it does not spontaneously all go back in the left flask.

This molecular view of gas expansion shows the tendency of the molecules to "spread out" among the different arrangements they can take. Before the stopcock is opened, there is only one possible arrangement: all molecules in the left flask. When the stopcock is opened, the arrangement in which all the molecules are in the left flask is but one of an extremely large number of possible arrangements. The most probable arrangements by far are those in which there are essentially equal numbers of molecules in the two flasks. When the gas spreads throughout the apparatus, any given molecule can be in either flask rather than confined to the left flask. We say that with the stopcock opened, the arrangement of gas molecules is more random or disordered than when the molecules are all confined in the left flask.

We will see this notion of increasing randomness helps us understand entropy at the molecular level.

Boltzmann's Equation and Microstates

The science of thermodynamics developed as a means of describing the properties of matter in our macroscopic world without regard to microscopic structure. In fact, thermodynamics was a well-developed field before the modern view of atomic and molecular structure was even known. The thermodynamic properties of water, for example, addressed the behavior of bulk water (or ice or water vapor) as a substance without considering any specific properties of individual H_2O molecules.

To connect the microscopic and macroscopic descriptions of matter, scientists have developed the field of *statistical thermodynamics*, which uses the tools of statistics and probability to link the microscopic and macroscopic worlds. Here we show how entropy, which is a property of bulk matter, can be connected to the behavior of atoms and molecules. Because the mathematics of statistical thermodynamics is complex, our discussion will be largely conceptual.

In our discussion of two gas molecules in the two-flask system in Figure 19.6, we saw that the number of possible arrangements helped explain why the gas expands.

Suppose we now consider one mole of an ideal gas in a particular thermodynamic state, which we can define by specifying the temperature, T, and volume, V, of the gas. What is happening to this gas at the microscopic level, and how does what is going on at the microscopic level relate to the entropy of the gas?

Imagine taking a snapshot of the positions and speeds of all the molecules at a given instant. The speed of each molecule tells us its kinetic energy. That particular set of 6×10^{23} positions and kinetic energies of the individual gas molecules is what we call a *microstate* of the system. A **microstate** is a single possible arrangement of the positions and kinetic energies of the molecules when the molecules are in a specific thermodynamic state. We could envision continuing to take snapshots of our system to see other possible microstates.

As you no doubt see, there would be such a staggeringly large number of microstates that taking individual snapshots of all of them is not feasible. Because we are examining such a large number of particles, however, we can use the tools of statistics and probability to determine the total number of microstates for the thermodynamic state. (That is where the *statistical* part of the name *statistical thermodynamics* comes in.) Each thermodynamic state has a characteristic number of microstates associated with it, and we will use the symbol W for that number.

Students sometimes have difficulty distinguishing between the state of a system and the microstates associated with the state. The difference is that *state* is used to describe the macroscopic view of our system as characterized, for example, by the pressure or temperature of a sample of gas. A *microstate* is a particular microscopic arrangement of the atoms or molecules of the system that corresponds to the given state of the system. Each of the snapshots we described is a microstate—the positions and kinetic energies of individual gas molecules will change from snapshot to snapshot, but each one is a possible arrangement of the collection of molecules corresponding to a single state. For macroscopically sized systems, such as a mole of gas, there is a very large number of microstates for each state—that is, W is generally an extremely large number.

The connection between the number of microstates of a system, W, and the entropy of the system, S, is expressed in a beautifully simple equation developed by Boltzmann and engraved on his tombstone (▶ Figure 19.7):

$$S = k \ln W \qquad [19.5]$$

In this equation, k is the Boltzmann constant, 1.38×10^{-23} J/K. Thus, *entropy is a measure of how many microstates are associated with a particular macroscopic state.*

▲ **Figure 19.7 Ludwig Boltzmann's gravestone.** Boltzmann's gravestone in Vienna is inscribed with his famous relationship between the entropy of a state, S, and the number of available microstates, W. (In Boltzmann's time, "log" was used to represent the natural logarithm.)

 Give It Some Thought

What is the entropy of a system that has only a single microstate?

From Equation 19.5, we see that the entropy change accompanying any process is

$$\Delta S = k \ln W_{\text{final}} - k \ln W_{\text{initial}} = k \ln \frac{W_{\text{final}}}{W_{\text{initial}}} \qquad [19.6]$$

Any change in the system that leads to an increase in the number of microstates ($W_{\text{final}} > W_{\text{initial}}$) leads to a positive value of ΔS: *Entropy increases with the number of microstates of the system.*

Let's consider two modifications to our ideal-gas sample and see how the entropy changes in each case. First, suppose we increase the volume of the system, which is analogous to allowing the gas to expand isothermally. A greater volume means a greater number of positions available to the gas atoms and therefore a greater number of microstates. The entropy therefore increases as the volume increases, as we saw in the "A Closer Look" box in Section 19.2.

Second, suppose we keep the volume fixed but increase the temperature. How does this change affect the entropy of the system? Recall the distribution of molecular speeds presented in Figure 10.13(a). An increase in temperature increases the most probable speed of the molecules and also broadens the distribution of speeds. Hence, the molecules have a greater number of possible kinetic energies, and the number of microstates increases. Thus, the entropy of the system increases with increasing temperature.

Molecular Motions and Energy

When a substance is heated, the motion of its molecules increases. In Section 10.7, we found that the average kinetic energy of the molecules of an ideal gas is directly proportional to the absolute temperature of the gas. That means the higher the temperature, the faster the molecules move and the more kinetic energy they possess. Moreover, hotter systems have a *broader distribution* of molecular speeds, as Figure 10.13(a) shows.

The particles of an ideal gas are idealized points with no volume and no bonds, however, points that we visualize as flitting around through space. Any real molecule can undergo three kinds of more complex motion. The entire molecule can move in one direction, which is the simple motion we visualize for an ideal particle and see in a macroscopic object, such as a thrown baseball. We call such movement **translational motion**. The molecules in a gas have more freedom of translational motion than those in a liquid, which have more freedom of translational motion than the molecules of a solid.

A real molecule can also undergo **vibrational motion**, in which the atoms in the molecule move periodically toward and away from one another, and **rotational motion**, in which the molecule spins about an axis. ▼ Figure 19.8 shows the vibrational motions and one of the rotational motions possible for the water molecule. These different forms of motion are ways in which a molecule can store energy.

 Give It Some Thought

Can an argon atom undergo vibrational motion?

The vibrational and rotational motions possible in real molecules lead to arrangements that a single atom cannot have. A collection of real molecules therefore has a

 GO FIGURE

Describe another possible rotational motion for this molecule.

 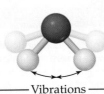 

├─────────── Vibrations ───────────┤   └─ Rotation ─┘

▲ **Figure 19.8 Vibrational and rotational motions in a water molecule.**

greater number of possible microstates than does the same number of ideal-gas particles. In general, *the number of microstates possible for a system increases with an increase in volume, an increase in temperature, or an increase in the number of molecules because any of these changes increases the possible positions and kinetic energies of the molecules making up the system.* We will also see that the number of microstates increases as the complexity of the molecule increases because there are more vibrational motions available.

Chemists have several ways of describing an increase in the number of microstates possible for a system and therefore an increase in the entropy for the system. Each way seeks to capture a sense of the increased freedom of motion that causes molecules to spread out when not restrained by physical barriers or chemical bonds.

The most common way of describing an increase in entropy is the increase in the *randomness*, or *disorder*, of the system. Another way likens an entropy increase to an increased *dispersion (spreading out) of energy* because there is an increase in the number of ways the positions and energies of the molecules can be distributed throughout the system. Each description (randomness or energy dispersal) is conceptually helpful if applied correctly.

Making Qualitative Predictions about ΔS

It is usually not difficult to estimate qualitatively how the entropy of a system changes during a simple process. As noted earlier, an increase in either the temperature or the volume of a system leads to an increase in the number of microstates, and hence an increase in the entropy. One more factor that correlates with number of microstates is the number of independently moving particles.

We can usually make qualitative predictions about entropy changes by focusing on these factors. For example, when water vaporizes, the molecules spread out into a larger volume. Because they occupy a larger volume, there is an increase in their freedom of motion, giving rise to a greater number of possible microstates, and hence an increase in entropy.

Now consider the phases of water. In ice, hydrogen bonding leads to the rigid structure shown in ▼ Figure 19.9. Each molecule in the ice is free to vibrate, but its translational and rotational motions are much more restricted than in liquid water. Although there are hydrogen bonds in liquid water, the molecules can more readily move about relative to

---

▲ GO FIGURE

In which phase are water molecules least able to have rotational motion?

Increasing entropy

| Ice | Liquid water | Water vapor |

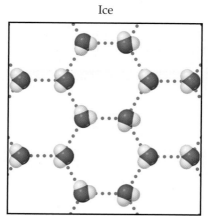 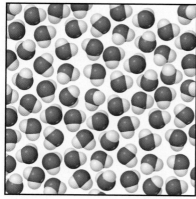 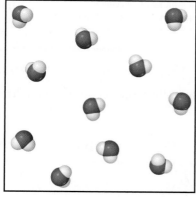

Rigid, crystalline structure

Motion restricted to **vibration** only

Smallest number of microstates

Increased freedom with respect to **translation**

Free to **vibrate** and **rotate**

Larger number of microstates

Molecules spread out, essentially independent of one another

Complete freedom for **translation**, **vibration**, and **rotation**

Largest number of microstates

▲ **Figure 19.9 Entropy and the phases of water.** The larger the number of possible microstates, the higher the entropy of the system.

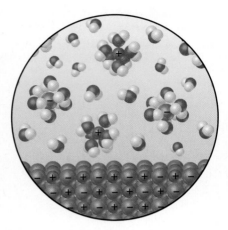

▲ **Figure 19.10 Entropy changes when an ionic solid dissolves in water.** The ions become more spread out and disordered, but the water molecules that hydrate the ions become less disordered.

▲ GO FIGURE

What major factor leads to a decrease in entropy as this reaction takes place?

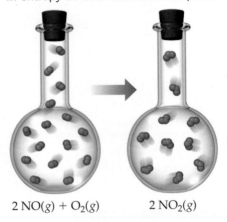

$$2\,NO(g) + O_2(g) \qquad 2\,NO_2(g)$$

▲ **Figure 19.11 Entropy decreases when NO(g) is oxidized by O₂(g) to NO₂(g).** A decrease in the number of gaseous molecules leads to a decrease in the entropy of the system.

one another (translation) and tumble around (rotation). During melting, therefore, the number of possible microstates increases and so does the entropy. In water vapor, the molecules are essentially independent of one another and have their full range of translational, vibrational, and rotational motions. Thus, water vapor has an even greater number of possible microstates and therefore a higher entropy than liquid water or ice.

When an ionic solid dissolves in water, a mixture of water and ions replaces the pure solid and pure water, as shown for KCl in ◄ Figure 19.10. The ions in the liquid move in a volume that is larger than the volume in which they were able to move in the crystal lattice and so undergo more motion. This increased motion might lead us to conclude that the entropy of the system has increased. We have to be careful, however, because some of the water molecules have lost some freedom of motion because they are now held around the ions as water of hydration. ∞ (Section 13.1) These water molecules are in a *more* ordered state than before because they are now confined to the immediate environment of the ions. Therefore, the dissolving of a salt involves both a disordering process (the ions become less confined) and an ordering process (some water molecules become more confined). The disordering processes are usually dominant, and so the overall effect is an increase in the randomness of the system when most salts dissolve in water.

Now, imagine arranging biomolecules into a highly organized biochemical system, such as the nucleosome in the chapter-opening figure. We might expect that the creation of this well-ordered structure would lead to a decrease in the entropy of the system. But this is frequently not the case. Waters of hydration and counterions can be expelled from the interface as two large biomolecules interact, and so the entropy of the system can actually increase—if you consider the water and counterions to be part of the system.

The same ideas apply to chemical reactions. Consider the reaction between nitric oxide gas and oxygen gas to form nitrogen dioxide gas:

$$2\,NO(g) + O_2(g) \longrightarrow 2\,NO_2(g) \qquad [19.7]$$

which results in a decrease in the number of molecules—three molecules of gaseous reactants form two molecules of gaseous products (◄ Figure 19.11). The formation of new N—O bonds reduces the motions of the atoms in the system. The formation of new bonds decreases the *number of degrees of freedom*, or forms of motion, available to the atoms. That is, the atoms are less free to move in random fashion because of the formation of new bonds. The decrease in the number of molecules and the resultant decrease in motion result in fewer possible microstates and therefore a decrease in the entropy of the system.

In summary, we generally expect the entropy of a system to increase for processes in which

1. Gases form from either solids or liquids.
2. Liquids or solutions form from solids.
3. The number of gas molecules increases during a chemical reaction.

SAMPLE
EXERCISE 19.3 | Predicting the Sign of ΔS

Predict whether ΔS is positive or negative for each process, assuming each occurs at constant temperature:

(a) $H_2O(l) \longrightarrow H_2O(g)$

(b) $Ag^+(aq) + Cl^-(aq) \longrightarrow AgCl(s)$

(c) $4\,Fe(s) + 3\,O_2(g) \longrightarrow 2\,Fe_2O_3(s)$

(d) $N_2(g) + O_2(g) \longrightarrow 2\,NO(g)$

SOLUTION

Analyze We are given four reactions and asked to predict the sign of ΔS for each.

Plan We expect ΔS to be positive if there is an increase in temperature, increase in volume, or increase in number of gas particles. The question states that the temperature is constant, and so we need to concern ourselves only with volume and number of particles.

Solve

(a) Evaporation involves a large increase in volume as liquid changes to gas. One mole of water (18 g) occupies about 18 mL as a liquid and if it could exist as a gas at STP it would occupy 22.4 L. Because the molecules are distributed throughout a much larger volume in the gaseous state, an increase in motional freedom accompanies vaporization and ΔS is positive.

(b) In this process, ions, which are free to move throughout the volume of the solution, form a solid, in which they are confined to a smaller volume and restricted to more highly constrained positions. Thus, ΔS is negative.

(c) The particles of a solid are confined to specific locations and have fewer ways to move (fewer microstates) than do the molecules of a gas. Because O_2 gas is converted into part of the solid product Fe_2O_3, ΔS is negative.

(d) The number of moles of reactant gases is the same as the number of moles of product gases, and so the entropy change is expected to be small. The sign of ΔS is impossible to predict based on our discussions thus far, but we can predict that ΔS will be close to zero.

Practice Exercise 1

Indicate whether each process produces an increase or decrease in the entropy of the system:

(a) $CO_2(s) \longrightarrow CO_2(g)$
(b) $CaO(s) + CO_2(g) \longrightarrow CaCO_3(s)$
(c) $HCl(g) + NH_3(g) \longrightarrow NH_4Cl(s)$
(d) $2 SO_2(g) + O_2(g) \longrightarrow 2 SO_3(g)$

Practice Exercise 2

Since the entropy of the universe increases for spontaneous processes, does it mean that the entropy of the universe decreases for nonspontaneous processes?

SAMPLE EXERCISE 19.4 | **Predicting Relative Entropies**

In each pair, choose the system that has greater entropy and explain your choice: **(a)** 1 mol of NaCl(s) or 1 mol of HCl(g) at 25 °C, **(b)** 2 mol of HCl(g) or 1 mol of HCl(g) at 25 °C, **(c)** 1 mol of HCl(g) or 1 mol of Ar(g) at 298 K.

SOLUTION

Analyze We need to select the system in each pair that has the greater entropy.

Plan We examine the state of each system and the complexity of the molecules it contains.

Solve **(a)** HCl(g) has the higher entropy because the particles in gases are more disordered and have more freedom of motion than the particles in solids. **(b)** When these two systems are at the same pressure, the sample containing 2 mol of HCl has twice the number of molecules as the sample containing 1 mol. Thus, the 2-mol sample has twice the number of microstates and twice the entropy. **(c)** The HCl system has the higher entropy because the number of ways in which an HCl molecule can store energy is greater than the number of ways

in which an Ar atom can store energy. (Molecules can rotate and vibrate; atoms cannot.)

Practice Exercise 1

Which system has the greatest entropy? **(a)** 1 mol of $H_2(g)$ at STP, **(b)** 1 mol of $H_2(g)$ at 100 °C and 0.5 atm, **(c)** 1 mol of $H_2O(s)$ at 0 °C, **(d)** 1 mol of $H_2O(l)$ at 25 °C.

Practice Exercise 2

Choose the system with the greater entropy in each case: **(a)** 1 mol of $H_2(g)$ at STP or 1 mol of $SO_2(g)$ at STP, **(b)** 1 mol of $N_2O_4(g)$ at STP or 2 mol of $NO_2(g)$ at STP.

The Third Law of Thermodynamics

If we decrease the thermal energy of a system by lowering the temperature, the energy stored in translational, vibrational, and rotational motion decreases. As less energy is stored, the entropy of the system decreases because it has fewer and fewer microstates available. If we keep lowering the temperature, do we reach a state in which these motions are essentially shut down, a point described by a single microstate? This question is addressed by the **third law of thermodynamics**, which states that *the entropy of a pure, perfect crystalline substance at absolute zero is zero*: $S(0\text{ K}) = 0$.

Consider a pure, perfect crystalline solid. At absolute zero, the individual atoms or molecules in the lattice would be perfectly ordered in position. Because none of them would have thermal motion, there is only one possible microstate. As a result, Equation 19.5 becomes $S = k \ln W = k \ln 1 = 0$. As the temperature is increased from absolute zero, the atoms or molecules in the crystal gain energy in the form of vibrational motion about their lattice positions. This means that the degrees of freedom and the entropy both increase. What happens to the entropy, however, as we continue to heat the crystal? We consider this important question in the next section.

 Give It Some Thought

If you are told that the entropy of a system is zero, what do you know about the system?

Chemistry and Life

Entropy and Human Society

Any living organism is a complex, highly organized, well-ordered system, even at the molecular level like the nucleosome we saw at the beginning of this chapter. Our entropy content is much lower than it would be if we were completely decomposed into carbon dioxide, water, and several other simple chemicals. Does this mean that life is a violation of the second law? No, because the thousands of chemical reactions necessary to produce and maintain life have caused a large increase in the entropy of the rest of the universe. Thus, as the second law requires, the overall entropy change during the lifetime of a human, or any other living system, is positive.

The second law of thermodynamics applies also to the way we humans order our surroundings. In addition to being complex living systems ourselves, we humans are masters of producing order in the world around us. We manipulate and order matter at the nanoscale level in order to produce the technological breakthroughs that have become so commonplace in the twenty-first century. We use tremendous quantities of raw materials to produce highly ordered materials. In so doing, we expend a great deal of energy to, in essence, "fight" the second law of thermodynamics.

For every bit of order we produce, however, we produce an even greater amount of disorder. Petroleum, coal, and natural gas are burned to provide the energy necessary for us to achieve highly ordered structures, but their combustion increases the entropy of the universe by releasing $CO_2(g)$, $H_2O(g)$, and heat. Oxide and sulfide ores release $CO_2(g)$ and $SO_2(g)$ that spread throughout our atmosphere. Thus, even as we strive to create more impressive discoveries and greater order in our society, we drive the entropy of the universe higher, as the second law says we must.

We humans are, in effect, using up our storehouse of energy-rich materials to create order and advance technology. As noted in Chapter 5, we must learn to harness new energy sources, such as solar energy, before we exhaust the supplies of readily available energy of other kinds.

19.4 | Entropy Changes in Chemical Reactions

In Section 5.5 we discussed how calorimetry can be used to measure ΔH for chemical reactions. No comparable method exists for measuring ΔS for a reaction. However, because the third law establishes a zero point for entropy, we can use experimental measurements to determine the *absolute value of the entropy, S*. To see schematically how this is done, let's review in greater detail the variation in the entropy of a substance with temperature.

We know that the entropy of a pure, perfect crystalline solid at 0 K is zero and that the entropy increases as the temperature of the crystal is increased. ▼ Figure 19.12 shows that the entropy of the solid increases steadily with increasing temperature up to the melting point of the solid. When the solid melts, the atoms or molecules are free

 GO FIGURE

Why does the plot show vertical jumps at the melting and boiling points?

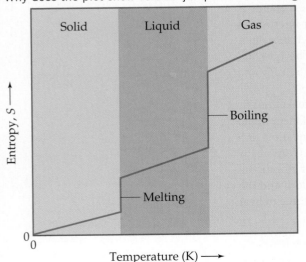

▲ Figure 19.12 **Entropy increases with increasing temperature.**

to move about the entire volume of the sample. The added degrees of freedom increase the randomness of the substance, thereby increasing its entropy. We therefore see a sharp increase in the entropy at the melting point. After all the solid has melted, the temperature again increases and with it, the entropy.

At the boiling point of the liquid, another abrupt increase in entropy occurs. We can understand this increase as resulting from the increased volume available to the atoms or molecules as they enter the gaseous state. When the gas is heated further, the entropy increases steadily as more energy is stored in the translational motion of the gas atoms or molecules.

Another change occurring at higher temperatures is the skewing of molecular speeds toward higher values (Figure 10.13(a)). The expansion of the range of speeds leads to increased kinetic energy and increased disorder and, hence, increased entropy. The conclusions we reach in examining Figure 10.13 are consistent with what we noted earlier: Entropy generally increases with increasing temperature because the increased motional energy leads to a greater number of possible microstates.

Entropy versus temperature graphs such as Figure 19.12 can be obtained by carefully measuring how the heat capacity of a substance ∞ (Section 5.5) varies with temperature, and we can use the data to obtain the absolute entropies at different temperatures. (The theory and methods used for these measurements and calculations are beyond the scope of this text.) Entropies are usually tabulated as molar quantities, in units of joules per mole-kelvin (J/mol-K).

Molar entropies for substances in their standard states are known as **standard molar entropies** and denoted $S°$. The standard state for any substance is defined as the pure substance at 1 atm pressure.* ▶ Table 19.1 lists the values of $S°$ for a number of substances at 298 K; Appendix C gives a more extensive list.

We can make several observations about the $S°$ values in Table 19.1:

1. Unlike enthalpies of formation, standard molar entropies of elements at the reference temperature of 298 K are *not* zero.

2. The standard molar entropies of gases are greater than those of liquids and solids, consistent with our interpretation of experimental observations, as represented in Figure 19.12.

3. Standard molar entropies generally increase with increasing molar mass.

4. Standard molar entropies generally increase with an increasing number of atoms in the formula of a substance.

Point 4 is related to the molecular motion discussed in Section 19.3. In general, the number of degrees of freedom for a molecule increases with increasing number of atoms, and thus the number of possible microstates also increases. **Figure 19.13** compares the standard molar entropies of three hydrocarbons in the gas phase. Notice how the entropy increases as the number of atoms in the molecule increases.

The entropy change in a chemical reaction equals the sum of the entropies of the products minus the sum of the entropies of the reactants:

$$\Delta S° = \sum n S°(\text{products}) - \sum m S°(\text{reactants}) \qquad [19.8]$$

As in Equation 5.31, the coefficients n and m are the coefficients in the balanced chemical equation for the reaction.

Table 19.1 Standard Molar Entropies of Selected Substances at 298 K

Substance	$S°$ (J/mol-K)
$H_2(g)$	130.6
$N_2(g)$	191.5
$O_2(g)$	205.0
$H_2O(g)$	188.8
$NH_3(g)$	192.5
$CH_3OH(g)$	237.6
$C_6H_6(g)$	269.2
$H_2O(l)$	69.9
$CH_3OH(l)$	126.8
$C_6H_6(l)$	172.8
$Li(s)$	29.1
$Na(s)$	51.4
$K(s)$	64.7
$Fe(s)$	27.23
$FeCl_3(s)$	142.3
$NaCl(s)$	72.3

SAMPLE EXERCISE 19.5 Calculating $\Delta S°$ from Tabulated Entropies

Calculate the change in the standard entropy of the system, $\Delta S°$, for the synthesis of ammonia from $N_2(g)$ and $H_2(g)$ at 298 K:

$$N_2(g) + 3\,H_2(g) \longrightarrow 2\,NH_3(g)$$

---

*The standard pressure used in thermodynamics is no longer 1 atm but rather is based on the SI unit for pressure, the pascal (Pa). The standard pressure is 10^5 Pa, a quantity known as a *bar*: 1 bar = 10^5 Pa = 0.987 atm. Because 1 bar differs from 1 atm by only 1.3%, we will continue to refer to the standard pressure as 1 atm.

SOLUTION

Analyze We are asked to calculate the standard entropy change for the synthesis of $NH_3(g)$ from its constituent elements.

Plan We can make this calculation using Equation 19.8 and the standard molar entropy values in Table 19.1 and Appendix C.

Solve Using Equation 19.8, we have

Substituting the appropriate $S°$ values from Table 19.1 yields

$$\Delta S° = 2S°(NH_3) - [S°(N_2) + 3S°(H_2)]$$
$$\Delta S° = (2 \text{ mol})(192.5 \text{ J/mol-K}) - [(1 \text{ mol})(191.5 \text{ J/mol-K}) + (3 \text{ mol})(130.6 \text{ J/mol-K})]$$
$$= -198.3 \text{ J/K}$$

Check: The value for $\Delta S°$ is negative, in agreement with our qualitative prediction based on the decrease in the number of molecules of gas during the reaction.

> ### Practice Exercise 1
>
> Using the standard molar entropies in Appendix C, calculate the standard entropy change, $\Delta S°$, for the "water-splitting" reaction at 298 K:
>
> $$2 H_2O(l) \longrightarrow 2 H_2(g) + O_2(g)$$
>
> **(a)** 326.3 J/K **(b)** 265.7 J/K **(c)** 163.2 J/K **(d)** 88.5 J/K **(e)** −326.3 J/K.
>
> ### Practice Exercise 2
>
> Using the standard molar entropies in Appendix C, calculate the standard entropy change, $\Delta S°$, for the following reaction at 298 K:
>
> $$Al_2O_3(s) + 3 H_2(g) \longrightarrow 2 Al(s) + 3 H_2O(g)$$

▲ **GO FIGURE**

What might you expect for the value of $S°$ for butane, C_4H_{10}?

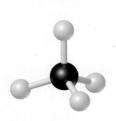

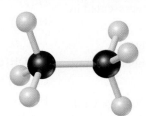

Methane, CH_4
$S° = 186.3 \text{ J/mol-K}$

Ethane, C_2H_6
$S° = 229.6 \text{ J/mol-K}$

Propane, C_3H_8
$S° = 270.3 \text{ J/mol-K}$

▲ **Figure 19.13 Entropy increases with increasing molecular complexity.**

Entropy Changes in the Surroundings

We can use tabulated absolute entropy values to calculate the standard entropy change in a system, such as a chemical reaction, as just described. But what about the entropy change in the surroundings? We encountered this situation in Section 19.2, but it is good to revisit it now that we are examining chemical reactions.

We should recognize that the surroundings for any system serve essentially as a large, constant-temperature heat source (or heat sink if the heat flows from the system to the surroundings). The change in entropy of the surroundings depends on how much heat is absorbed or given off by the system.

For an isothermal process, the entropy change of the surroundings is given by

$$\Delta S_{surr} = \frac{-q_{sys}}{T}$$

Because in a constant-pressure process, q_{sys} is simply the enthalpy change for the reaction, ΔH, we can write

$$\Delta S_{surr} = \frac{-\Delta H_{sys}}{T} \quad [\text{at constant P}] \qquad [19.9]$$

For the ammonia synthesis reaction in Sample Exercise 19.5, q_{sys} is the enthalpy change for the reaction under standard conditions, $\Delta H°$, so the changes in entropy will be standard entropy changes, $\Delta S°$. Therefore, using the procedures described in Section 5.7, we have

$$\Delta H°_{rxn} = 2\,\Delta H°_f[NH_3(g)] - 3\,\Delta H°_f[H_2(g)] - \Delta H°_f[N_2(g)]$$
$$= 2(-46.19\,\text{kJ}) - 3(0\,\text{kJ}) - (0\,\text{kJ}) = -92.38\,\text{kJ}$$

The negative value tells us that at 298 K the formation of ammonia from $H_2(g)$ and $N_2(g)$ is exothermic. The surroundings absorb the heat given off by the system, which means an increase in the entropy of the surroundings:

$$\Delta S°_{surr} = \frac{92.38\,\text{kJ}}{298\,\text{K}} = 0.310\,\text{kJ/K} = 310\,\text{J/K}$$

Notice that the magnitude of the entropy gained by the surroundings is greater than that lost by the system, calculated as $-198.3\,\text{J/K}$ in Sample Exercise 19.5.

The overall entropy change for the reaction is

$$\Delta S°_{univ} = \Delta S°_{sys} + \Delta S°_{surr} = -198.3\,\text{J/K} + 310\,\text{J/K} = 112\,\text{J/K}$$

Because $\Delta S°_{univ}$ is positive for any spontaneous reaction, this calculation indicates that when $NH_3(g)$, $H_2(g)$, and $N_2(g)$ are together at 298 K in their standard states (each at 1 atm pressure), the reaction moves spontaneously toward formation of $NH_3(g)$.

Keep in mind that while the thermodynamic calculations indicate that formation of ammonia is spontaneous, they do not tell us anything about the rate at which ammonia is formed. Establishing equilibrium in this system within a reasonable period requires a catalyst, as discussed in Section 15.7.

 Give It Some Thought

If a process is exothermic, does the entropy of the surroundings (a) always increase, (b) always decrease, or (c) sometimes increase and sometimes decrease, depending on the process?

19.5 | Gibbs Free Energy

We have seen examples of endothermic processes that are spontaneous, such as the dissolution of ammonium nitrate in water. ∞ (Section 13.1) We learned in our discussion of the solution process that a spontaneous process that is endothermic must be accompanied by an increase in the entropy of the system. However, we have also encountered processes that are spontaneous and yet proceed with a *decrease* in the entropy of the system, such as the highly exothermic formation of sodium chloride from its constituent elements. ∞ (Section 8.2) Spontaneous processes that result in a decrease in the system's entropy are always exothermic. Thus, the spontaneity of a reaction seems to involve two thermodynamic concepts, enthalpy and entropy.

How can we use ΔH and ΔS to predict whether a given reaction occurring at constant temperature and pressure will be spontaneous? The means for doing so was first developed by the American mathematician J. Willard Gibbs (1839–1903). Gibbs proposed a new state function, now called the **Gibbs free energy** (or just **free energy**), G, and defined as

$$G = H - TS \qquad [19.10]$$

where T is the absolute temperature. For an isothermal process, the change in the free energy of the system, ΔG, is

$$\Delta G = \Delta H - T\Delta S \qquad [19.11]$$

Under standard conditions, this equation becomes

$$\Delta G^\circ = \Delta H^\circ - T\Delta S^\circ \qquad [19.12]$$

To see how the state function G relates to reaction spontaneity, recall that for a reaction occurring at constant temperature and pressure

$$\Delta S_{univ} = \Delta S_{sys} + \Delta S_{surr} = \Delta S_{sys} + \left(\frac{-\Delta H_{sys}}{T} \right)$$

where Equation 19.9 substitutes for ΔS_{surr}. Multiplying both sides by $-T$ gives

$$-T\Delta S_{univ} = \Delta H_{sys} - T\Delta S_{sys} \qquad [19.13]$$

Comparing Equations 19.11 and 19.13, we see that in a process occurring at constant temperature and pressure, the free-energy change, ΔG, is equal to $-T\Delta S_{univ}$. We know that for spontaneous processes, ΔS_{univ} is always positive and, therefore, $-T\Delta S_{univ}$ is always negative. Thus, the sign of ΔG provides us with extremely valuable information about the spontaneity of processes that occur at constant temperature and pressure. If both T and P are constant, the relationship between the sign of ΔG and the spontaneity of a reaction is:

1. If $\Delta G < 0$, the reaction is spontaneous in the forward direction.
2. If $\Delta G = 0$, the reaction is at equilibrium.
3. If $\Delta G > 0$, the reaction in the forward direction is nonspontaneous (work must be done to make it occur) but the reverse reaction is spontaneous.

It is more convenient to use ΔG as a criterion for spontaneity than to use ΔS_{univ} because ΔG relates to the system alone and avoids the complication of having to examine the surroundings.

An analogy is often drawn between the free-energy change during a spontaneous reaction and the potential-energy change when a boulder rolls down a hill (◄ **Figure 19.14**). Potential energy in a gravitational field "drives" the boulder until it reaches a state of minimum potential energy in the valley. Similarly, the free energy of a chemical system decreases until it reaches a minimum value. When this minimum is reached, a state of equilibrium exists. *In any spontaneous process carried out at constant temperature and pressure, the free energy always decreases.*

To illustrate these ideas, let's return to the Haber process for the synthesis of ammonia from nitrogen and hydrogen, which we discussed extensively in Chapter 15:

$$N_2(g) + 3\,H_2(g) \rightleftharpoons 2\,NH_3(g)$$

Imagine that we have a reaction vessel that allows us to maintain a constant temperature and pressure and that we have a catalyst that allows the reaction to proceed at a reasonable rate. What happens when we load the vessel with a certain number of moles of N_2 and three times that number of moles of H_2? As we saw in Figure 15.3 (p. 632), the N_2 and H_2 react spontaneously to form NH_3 until equilibrium is achieved. Similarly, Figure 15.3 shows that if we load the vessel with pure NH_3, it decomposes spontaneously to N_2 and H_2 until equilibrium is reached. In each case the free energy of the system gets progressively lower and lower as the reaction moves toward equilibrium, which represents a minimum in the free energy. We illustrate these cases in ▶ **Figure 19.15**.

 GO FIGURE

Are the processes that move a system toward equilibrium spontaneous or nonspontaneous?

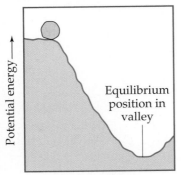

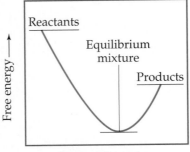

▲ Figure 19.14 **Potential energy and free energy.** An analogy is shown between the gravitational potential-energy change of a boulder rolling down a hill and the free-energy change in a spontaneous reaction. Free energy always decreases in a spontaneous process when pressure and temperature are held constant.

 Give It Some Thought

What are the criteria for spontaneity
(a) in terms of entropy and
(b) in terms of free energy?

GO FIGURE

Why are the spontaneous processes shown sometimes said to be "downhill" in free energy?

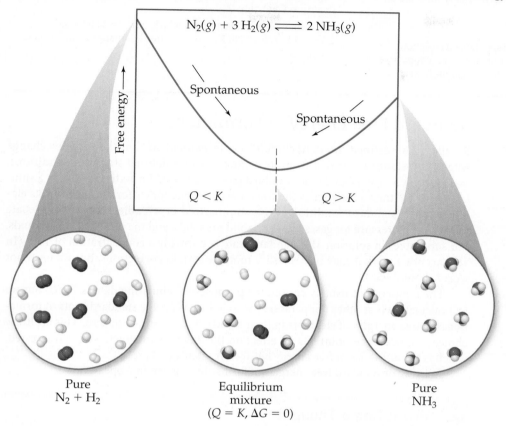

Pure
$N_2 + H_2$

Equilibrium
mixture
$(Q = K, \Delta G = 0)$

Pure
NH_3

▲ **Figure 19.15 Free energy and approaching equilibrium.** In the reaction
$N_2(g) + 3\,H_2(g) \rightleftharpoons 2\,NH_3(g)$, if the reaction mixture has too much N_2 and H_2 relative to
NH_3 (left), Q < K and NH_3 forms spontaneously. If there is more NH_3 in the mixture relative to
the reactants N_2 and H_2 (right), Q > K and the NH_3 decomposes spontaneously into N_2 and H_2.

This is a good time to remind ourselves of the significance of the reaction quotient, Q,
for a system that is not at equilibrium. ∞ (Section 15.6) Recall that when $Q < K$, there
is an excess of reactants relative to products and the reaction proceeds spontaneously in
the forward direction to reach equilibrium, as noted in Figure 19.15. When $Q > K$, the
reaction proceeds spontaneously in the reverse direction. At equilibrium $Q = K$.

**SAMPLE
EXERCISE 19.6** | Calculating Free-Energy Change from $\Delta H°$, T, and $\Delta S°$

Calculate the standard free-energy change for the formation of $NO(g)$ from $N_2(g)$ and $O_2(g)$ at 298 K:

$$N_2(g) + O_2(g) \longrightarrow 2\,NO(g)$$

given that $\Delta H° = 180.7$ kJ and $\Delta S° = 24.7$ J/K. Is the reaction spontaneous under these
conditions?

SOLUTION

Analyze We are asked to calculate $\Delta G°$ for the indicated reaction
(given $\Delta H°$, $\Delta S°$, and T) and to predict whether the reaction is spontaneous under standard conditions at 298 K.

Plan To calculate $\Delta G°$, we use Equation 19.12, $\Delta G° = \Delta H° - T\Delta S°$.
To determine whether the reaction is spontaneous under standard
conditions, we look at the sign of $\Delta G°$.

Solve

$$\Delta G° = \Delta H° - T\Delta S°$$
$$= 180.7 \text{ kJ} - (298 \text{ K})(24.7 \text{ J/K})\left(\frac{1 \text{ kJ}}{10^3 \text{ J}}\right)$$
$$= 180.7 \text{ kJ} - 7.4 \text{ kJ}$$
$$= 173.3 \text{ kJ}$$

Because $\Delta G°$ is positive, the reaction is not spontaneous under standard conditions at 298 K.

Comment Notice that we had to convert the units of the $T\Delta S°$ term to kJ so that they could be added to the $\Delta H°$ term, whose units are kJ.

Practice Exercise 1

Which of these statements is true? (**a**) All spontaneous reactions have a negative enthalpy change, (**b**) All spontaneous reactions have a positive entropy change, (**c**) All spontaneous reactions have a

positive free-energy change, (**d**) All spontaneous reactions have a negative free-energy change, (**e**) All spontaneous reactions have a negative entropy change.

Practice Exercise 2

Calculate $\Delta G°$ for a reaction for which $\Delta H° = 24.6$ kJ and $\Delta S° = 132$ J/K at 298 K. Is the reaction spontaneous under these conditions?

Standard Free Energy of Formation

Table 19.2 Conventions Used in Establishing Standard Free Energies

State of Matter	Standard State
Solid	Pure solid
Liquid	Pure liquid
Gas	1 atm pressure
Solution	1 M concentration
Element	$\Delta G_f° = 0$ for element in standard state

Recall that we defined *standard enthalpies of formation*, $\Delta H_f°$, as the enthalpy change when a substance is formed from its elements under defined standard conditions. ∞ (Section 5.7) We can define **standard free energies of formation**, $\Delta G_f°$, in a similar way: $\Delta G_f°$ for a substance is the free-energy change for its formation from its elements under standard conditions. As is summarized in ◄ Table 19.2, standard state means 1 atm pressure for gases, the pure solid for solids, and the pure liquid for liquids. For substances in solution, the standard state is normally a concentration of 1 M. (In very accurate work it may be necessary to make certain corrections, but we need not worry about these.)

The temperature usually chosen for purposes of tabulating data is 25 °C, but we will calculate ΔG at other temperatures as well. Just as for the standard heats of formation, the free energies of elements in their standard states are set to zero. This arbitrary choice of a reference point has no effect on the quantity in which we are interested, which is the *difference* in free energy between reactants and products.

A listing of standard free energies of formation is given in Appendix C.

 Give It Some Thought

What does the superscript ° indicate when associated with a thermodynamic quantity, as in $\Delta H°$, $\Delta S°$, or $\Delta G°$?

Standard free energies of formation are useful in calculating the *standard free-energy change* for chemical processes. The procedure is analogous to the calculation of $\Delta H°$ (Equation 5.31) and $\Delta S°$ (Equation 19.8):

$$\Delta G° = \sum n\Delta G_f°(\text{products}) - \sum m\Delta G_f°(\text{reactants}) \qquad [19.14]$$

SAMPLE EXERCISE 19.7 Calculating Standard Free-Energy Change from Free Energies of Formation

(**a**) Use data from Appendix C to calculate the standard free-energy change for the reaction $P_4(g) + 6\,Cl_2(g) \longrightarrow 4\,PCl_3(g)$ at 298 K. (**b**) What is $\Delta G°$ for the reverse of this reaction?

SOLUTION

Analyze We are asked to calculate the free-energy change for a reaction and then to determine the free-energy change for the reverse reaction.

Plan We look up the free-energy values for the products and reactants and use Equation 19.14: We multiply the molar quantities by the coefficients in the balanced equation and subtract the total for the reactants from that for the products.

Solve

(**a**) $Cl_2(g)$ is in its standard state, so $\Delta G_f°$ is zero for this reactant. $P_4(g)$, however, is not in its standard state, so $\Delta G_f°$ is not zero for

this reactant. From the balanced equation and values from Appendix C, we have

$$\Delta G_{rxn}° = 4\,\Delta G_f°[PCl_3(g)] - \Delta G_f°[P_4(g)] - 6\,\Delta G_f°[Cl_2(g)]$$
$$= (4\text{ mol})(-269.6\text{ kJ/mol}) - (1\text{ mol})(24.4\text{ kJ/mol}) - 0$$
$$= -1102.8\text{ kJ}$$

That $\Delta G°$ is negative tells us that a mixture of $P_4(g)$, $Cl_2(g)$, and $PCl_3(g)$ at 25 °C, each present at a partial pressure of 1 atm, would react spontaneously in the forward direction to form more PCl_3. Remember, however, that the value of $\Delta G°$ tells us nothing about the rate at which the reaction occurs.

(b) When we reverse the reaction, we reverse the roles of the reactants and products. Thus, reversing the reaction changes the sign of ΔG in Equation 19.14, just as reversing the reaction changes the sign of ΔH. ∞∞ (Section 5.4) Hence, using the result from part (a), we have

$$4\,PCl_3(g) \longrightarrow P_4(g) + 6\,Cl_2(g) \quad \Delta G° = +1102.8\,kJ$$

Practice Exercise 1

The following chemical equations describe the same chemical reaction. How do the free energies of these two chemical equations compare?

(1) $2\,H_2O(l) \longrightarrow 2\,H_2(g) + O_2(g)$
(2) $H_2O(l) \longrightarrow H_2(g) + \frac{1}{2}\,O_2(g)$
(a) $\Delta G_1° = \Delta G_2°$, **(b)** $\Delta G_1° = 2\,\Delta G_2°$, **(c)** $2\Delta G_1° = \Delta G_2°$,
(d) None of the above.

Practice Exercise 2

Use data from Appendix C to calculate $\Delta G°$ at 298 K for the combustion of methane:
$$CH_4(g) + 2\,O_2(g) \longrightarrow CO_2(g) + 2\,H_2O(g).$$

SAMPLE EXERCISE 19.8 Predicting and Calculating $\Delta G°$

In Section 5.7 we used Hess's law to calculate $\Delta H°$ for the combustion of propane gas at 298 K:

$$C_3H_8(g) + 5\,O_2(g) \longrightarrow 3\,CO_2(g) + 4\,H_2O(l) \quad \Delta H° = -2220\,kJ$$

(a) *Without using data from Appendix C*, predict whether $\Delta G°$ for this reaction is more negative or less negative than $\Delta H°$. **(b)** Use data from Appendix C to calculate $\Delta G°$ for the reaction at 298 K. Is your prediction from part (a) correct?

SOLUTION

Analyze In part (a) we must predict the value for $\Delta G°$ relative to that for $\Delta H°$ on the basis of the balanced equation for the reaction. In part (b) we must calculate the value for $\Delta G°$ and compare this value with our qualitative prediction.

Plan The free-energy change incorporates both the change in enthalpy and the change in entropy for the reaction (Equation 19.11), so under standard conditions

$$\Delta G° = \Delta H° - T\Delta S°$$

To determine whether $\Delta G°$ is more negative or less negative than $\Delta H°$, we need to determine the sign of the term $T\Delta S°$. Because T is the absolute temperature, 298 K, it is always a positive number. We can predict the sign of $\Delta S°$ by looking at the reaction.

Solve

(a) The reactants are six molecules of gas, and the products are three molecules of gas and four molecules of liquid. Thus, the number of molecules of gas has decreased significantly during the reaction. By using the general rules discussed in Section 19.3, we expect a decrease in the number of gas molecules to lead to a decrease in the entropy of the system—the products have fewer possible microstates than the reactants. We therefore expect $\Delta S°$ and $T\Delta S°$ to be negative. Because we are subtracting $T\Delta S°$, which is a negative number, we predict that $\Delta G°$ is *less negative* than $\Delta H°$.

(b) Using Equation 19.14 and values from Appendix C, we have

$$\Delta G° = 3\,\Delta G_f°[CO_2(g)] + 4\,\Delta G_f°[H_2O(l)] - \Delta G_f°[C_3H_8(g)] - 5\,\Delta G_f°[O_2(g)]$$
$$= 3\,mol(-394.4\,kJ/mol) + 4\,mol(-237.13\,kJ/mol) -$$
$$1\,mol(-23.47\,kJ/mol) - 5\,mol(0\,kJ/mol) = -2108\,kJ$$

Notice that we have been careful to use the value of $\Delta G_f°$ for $H_2O(l)$. As in calculating ΔH values, the phases of the reactants and products are important. As we predicted, $\Delta G°$ is less negative than $\Delta H°$ because of the decrease in entropy during the reaction.

Practice Exercise 1

If a reaction is exothermic and its entropy change is positive, which statement is true?
(a) The reaction is spontaneous at all temperatures, **(b)** The reaction is nonspontaneous at all temperatures, **(c)** The reaction is spontaneous only at higher temperatures, **(d)** The reaction is spontaneous only at lower temperatures.

Practice Exercise 2

For the combustion of propane at 298 K, $C_3H_8(g) + 5\,O_2(g) \longrightarrow 3\,CO_2(g) + 4\,H_2O(g)$, do you expect $\Delta G°$ to be more negative or less negative than $\Delta H°$?

A Closer Look

What's "Free" about Free Energy?

The Gibbs free energy is a remarkable thermodynamic quantity. Because so many chemical reactions are carried out under conditions of near-constant pressure and temperature, chemists, biochemists, and engineers consider the sign and magnitude of ΔG as exceptionally useful tools in the design of chemical and biochemical reactions. We will see examples of the usefulness of ΔG throughout the remainder of this chapter and this text.

When first learning about ΔG, two common questions often arise: Why is the sign of ΔG an indicator of the spontaneity of reactions? And what is "free" about free energy?

In Section 19.2 we saw that the second law of thermodynamics governs the spontaneity of processes. In order to apply the second law (Equation 19.4), however, we must determine ΔS_{univ}, which is often difficult to evaluate. When T and P are constant, however, we can relate ΔS_{univ} to the changes in entropy and enthalpy of just the *system* by substituting the Equation 19.9 expression for ΔS_{surr} in Equation 19.4:

$$\Delta S_{univ} = \Delta S_{sys} + \Delta S_{surr} = \Delta S_{sys} + \left(\frac{-\Delta H_{sys}}{T}\right) \quad \text{(constant } T, P)$$

[19.15]

Thus, at constant temperature and pressure, the second law becomes

Reversible process: $\quad \Delta S_{univ} = \Delta S_{sys} - \dfrac{\Delta H_{sys}}{T} = 0$

Irreversible process: $\quad \Delta S_{univ} = \Delta S_{sys} - \dfrac{\Delta H_{sys}}{T} > 0$ [19.16]

$\text{(constant } T, P)$

Now we can see the relationship between ΔG_{sys} (which we call simply ΔG) and the second law. From Equation 19.11 we know that $\Delta G = \Delta H_{sys} - T\Delta S_{sys}$. If we multiply Equations 19.16 by $-T$ and rearrange, we reach the following conclusion:

Reversible process: $\quad \Delta G = \Delta H_{sys} - T\,\Delta S_{sys} = 0$

Irreversible process: $\quad \Delta G = \Delta H_{sys} - T\,\Delta S_{sys} < 0$ [19.17]

$\text{(constant } T, P)$

Equations 19.17 allow us to use the sign of ΔG to conclude whether a reaction is spontaneous, nonspontaneous, or at equilibrium. When $\Delta G < 0$, a process is irreversible and, therefore, spontaneous. When $\Delta G = 0$, the process is reversible and, therefore, at equilibrium. If a process has $\Delta G > 0$, then the reverse process will have $\Delta G < 0$; thus, the process as written is nonspontaneous but its reverse reaction will be irreversible and spontaneous.

The magnitude of ΔG is also significant. A reaction for which ΔG is large and negative, such as the burning of gasoline, is much more capable of doing work on the surroundings than is a reaction for which ΔG is small and negative, such as ice melting at room temperature. In fact, thermodynamics tells us that *the change in free energy for a process, ΔG, equals the maximum useful work that can be done by the system on its surroundings in a spontaneous process occurring at constant temperature and pressure:*

$$\Delta G = -w_{max}$$

[19.18]

(Remember our sign convention from Table 5.1: Work done *by* a system is negative.) In other words, ΔG gives the theoretical limit to how much work can be done by a process.

The relationship in Equation 19.18 explains why ΔG is called *free energy*—it is the portion of the energy change of a spontaneous reaction that is free to do useful work. The remainder of the energy enters the environment as heat. For example, the theoretical maximum work obtained for the combustion of gasoline is given by the value of ΔG for the combustion reaction. On average, standard internal combustion engines are inefficient in utilizing this potential work—more than 60% of the potential work is lost (primarily as heat) in converting the chemical energy of the gasoline to mechanical energy to move the vehicle. When other losses are considered—idling time, braking, aerodynamic drag, and so forth—only about 15% of the potential work from the gasoline is used to move the car. Advances in automobile design—such as hybrid technology, efficient diesel engines, and new lightweight materials—have the potential to increase the percentage of useful work obtained from the gasoline.

For nonspontaneous processes ($\Delta G > 0$), the free-energy change is a measure of the *minimum* amount of work that must be done to cause the process to occur. In actuality, we always need to do more than this theoretical minimum amount because of the inefficiencies in the way the changes occur.

19.6 | Free Energy and Temperature

Tabulations of ΔG_f°, such as those in Appendix C, make it possible to calculate ΔG° for reactions at the standard temperature of 25 °C, but we are often interested in examining reactions at other temperatures. To see how ΔG is affected by temperature, let's look again at Equation 19.11:

$$\Delta G = \Delta H - T\Delta S = \Delta H + (-T\Delta S)$$

$$\underset{\text{term}}{\text{Enthalpy}} \qquad \underset{\text{term}}{\text{Entropy}}$$

Notice that we have written the expression for ΔG as a sum of two contributions, an enthalpy term, ΔH, and an entropy term, $-T\Delta S$. Because the value of $-T\Delta S$ depends directly on the absolute temperature T, ΔG varies with temperature. We know that the enthalpy term, ΔH, can be either positive or negative and that T is positive at all temperatures other than absolute zero. The entropy term, $-T\Delta S$, can also be positive or

Table 19.3 **How Signs of ΔH and ΔS Affect Reaction Spontaneity**

ΔH	ΔS	$-T\Delta S$	$\Delta G = \Delta H - T\Delta S$	Reaction Characteristics	Example
$-$	$+$	$-$	$-$	Spontaneous at all temperatures	$2\,O_3(g) \longrightarrow 3\,O_2(g)$
$+$	$-$	$+$	$+$	Nonspontaneous at all temperatures	$3\,O_2(g) \longrightarrow 2\,O_3(g)$
$-$	$-$	$+$	$+$ or $-$	Spontaneous at low T; nonspontaneous at high T	$H_2O(l) \longrightarrow H_2O(s)$
$+$	$+$	$-$	$+$ or $-$	Spontaneous at high T; nonspontaneous at low T	$H_2O(s) \longrightarrow H_2O(l)$

negative. When ΔS is positive, which means the final state has greater randomness (a greater number of microstates) than the initial state, the term $-T\Delta S$ is negative. When ΔS is negative, $-T\Delta S$ is positive.

The sign of ΔG, which tells us whether a process is spontaneous, depends on the signs and magnitudes of ΔH and $-T\Delta S$. The various combinations of ΔH and $-T\Delta S$ signs are given in ▲ Table 19.3.

Note in Table 19.3 that when ΔH and $-T\Delta S$ have opposite signs, the sign of ΔG depends on the magnitudes of these two terms. In these instances temperature is an important consideration. Generally, ΔH and ΔS change very little with temperature. However, the value of T directly affects the magnitude of $-T\Delta S$. As the temperature increases, the magnitude of $-T\Delta S$ increases, and this term becomes relatively more important in determining the sign and magnitude of ΔG.

As an example, let's consider once more the melting of ice to liquid water at 1 atm:

$$H_2O(s) \longrightarrow H_2O(l) \quad \Delta H > 0, \Delta S > 0$$

This process is endothermic, which means that ΔH is positive. Because the entropy increases during the process, ΔS is positive, which makes $-T\Delta S$ negative. At temperatures below 0 °C (273 K), the magnitude of ΔH is greater than that of $-T\Delta S$. Hence, the positive enthalpy term dominates, and ΔG is positive. This positive value of ΔG means that ice melting is not spontaneous at $T < 0\,°C$, just as our everyday experience tells us; rather, the reverse process, the freezing of liquid water into ice, is spontaneous at these temperatures.

What happens at temperatures greater than 0 °C? As T increases, so does the magnitude of $-T\Delta S$. When $T > 0\,°C$, the magnitude of $-T\Delta S$ is greater than the magnitude of ΔH, which means that the $-T\Delta S$ term dominates and ΔG is negative. The negative value of ΔG tells us that ice melting is spontaneous at $T > 0\,°C$.

At the normal melting point of water, $T = 0\,°C$, the two phases are in equilibrium. Recall that $\Delta G = 0$ at equilibrium; at $T = 0\,°C$, ΔH and $-T\Delta S$ are equal in magnitude and opposite in sign, so they cancel and give $\Delta G = 0$.

 Give It Some Thought

The normal boiling point of benzene is 80 °C. At 100 °C and 1 atm, which term is greater in magnitude for the vaporization of benzene, ΔH or $T\Delta S$?

Our discussion of the temperature dependence of ΔG is also relevant to standard free-energy changes. We can calculate the values of $\Delta H°$ and $\Delta S°$ at 298 K from the data in Appendix C. If we assume that these values do not change with temperature, we can then use Equation 19.12 to estimate ΔG at temperatures other than 298 K.

SAMPLE EXERCISE 19.9 Determining the Effect of Temperature on Spontaneity

The Haber process for the production of ammonia involves the equilibrium

$$N_2(g) + 3\,H_2(g) \rightleftharpoons 2\,NH_3(g)$$

Assume that $\Delta H°$ and $\Delta S°$ for this reaction do not change with temperature. **(a)** Predict the direction in which ΔG for the reaction changes with increasing temperature. **(b)** Calculate ΔG at 25 °C and at 500 °C.

SOLUTION

Analyze In part (a) we are asked to predict the direction in which ΔG changes as temperature increases. In part (b) we need to determine ΔG for the reaction at two temperatures.

Plan We can answer part (a) by determining the sign of ΔS for the reaction and then using that information to analyze Equation 19.12. In part (b) we first calculate $\Delta H°$ and $\Delta S°$ for the reaction using data in Appendix C and then use Equation 19.12 to calculate ΔG.

Solve

(a) The temperature dependence of ΔG comes from the entropy term in Equation 19.12, $\Delta G = \Delta H - T\Delta S$. We expect ΔS for this reaction to be negative because the number of molecules of gas is smaller in the products. Because ΔS is negative, $-T\Delta S$ is positive and increases with increasing temperature. As a result, ΔG becomes less negative (or more positive) with increasing temperature. Thus, the driving force for the production of NH_3 becomes smaller with increasing temperature.

(b) We calculated $\Delta H°$ for this reaction in Sample Exercise 15.14 and $\Delta S°$ in Sample Exercise 19.5: $\Delta H° = -92.38$ kJ and $\Delta S° = -198.3$ J/K. If we assume that these values do not change with temperature, we can calculate ΔG at any temperature by using Equation 19.12. At $T = 25\,°C = 298$ K, we have

$$\Delta G° = -92.38 \text{ kJ} - (298 \text{ K})(-198.3 \text{ J/K})\left(\frac{1 \text{ kJ}}{1000 \text{ J}}\right)$$

$$= -92.38 \text{ kJ} + 59.1 \text{ kJ} = -33.3 \text{ kJ}$$

At $T = 500\,°C = 773$ K, we have

$$\Delta G = -92.38 \text{ kJ} - (773 \text{ K})(-198.3 \text{ J/K})\left(\frac{1 \text{ kJ}}{1000 \text{ J}}\right)$$

$$= -92.38 \text{ kJ} + 153 \text{ kJ} = 61 \text{ kJ}$$

Notice that we had to convert the units of $-T\Delta S°$ to kJ in both calculations so that this term can be added to the $\Delta H°$ term, which has units of kJ.

Comment Increasing the temperature from 298 K to 773 K changes ΔG from -33.3 kJ to $+61$ kJ. Of course, the result at 773 K assumes that $\Delta H°$ and $\Delta S°$ do not change with temperature. Although these values do change slightly with temperature, the result at 773 K should be a reasonable approximation.

The positive increase in ΔG with increasing T agrees with our prediction in part (a). Our result indicates that in a mixture of $N_2(g)$, $H_2(g)$, and $NH_3(g)$, each present at a partial pressure of 1 atm, the $N_2(g)$ and $H_2(g)$ react spontaneously at 298 K to form more $NH_3(g)$. At 773 K, the positive value of ΔG tells us that the reverse reaction is spontaneous. Thus, when the mixture of these gases, each at a partial pressure of 1 atm, is heated to 773 K, some of the $NH_3(g)$ spontaneously decomposes into $N_2(g)$ and $H_2(g)$.

Practice Exercise 1

What is the temperature above which the Haber ammonia process becomes nonspontaneous?
(a) 25 °C, (b) 47 °C, (c) 61 °C, (d) 193 °C, (e) 500 °C.

Practice Exercise 2

(a) Using standard enthalpies of formation and standard entropies in Appendix C, calculate $\Delta H°$ and $\Delta S°$ at 298 K for the reaction $2 SO_2(g) + O_2(g) \longrightarrow 2 SO_3(g)$. (b) Use your values from part (a) to estimate ΔG at 400 K.

19.7 | Free Energy and the Equilibrium Constant

In Section 19.5 we saw a special relationship between ΔG and equilibrium: For a system at equilibrium, $\Delta G = 0$. We have also seen how to use tabulated thermodynamic data to calculate values of the standard free-energy change, $\Delta G°$. In this final section, we learn two more ways in which we can use free energy to analyze chemical reactions: using $\Delta G°$ to calculate ΔG under *nonstandard* conditions and relating the values of $\Delta G°$ and K for a reaction.

Free Energy under Nonstandard Conditions

The set of standard conditions for which $\Delta G°$ values pertain is given in Table 19.2. Most chemical reactions occur under nonstandard conditions. For any chemical process, the relationship between the free-energy change under standard conditions, $\Delta G°$, and the free-energy change under any other conditions, ΔG, is given by

$$\Delta G = \Delta G° + RT \ln Q \qquad [19.19]$$

In this equation R is the ideal-gas constant, 8.314 J/mol-K; T is the absolute temperature; and Q is the reaction quotient for the reaction mixture of interest. ∞ (Section 15.6) Recall that the reaction quotient Q is calculated like an equilibrium constant, except that you use the concentrations at any point of interest in the reaction; if $Q = K$, then the reaction is at equilibrium. Under standard conditions, the concentrations of all the reactants and products are equal to 1 M. Thus, under standard conditions $Q = 1$, $\ln Q = 0$, and Equation 19.19 reduces to $\Delta G = \Delta G°$ under standard conditions, as it should.

SAMPLE
EXERCISE 19.10 Relating ΔG to a Phase Change at Equilibrium

(a) Write the chemical equation that defines the normal boiling point of liquid carbon tetrachloride, $CCl_4(l)$. **(b)** What is the value of $\Delta G°$ for the equilibrium in part (a)? **(c)** Use data from Appendix C and Equation 19.12 to estimate the normal boiling point of CCl_4.

SOLUTION

Analyze (a) We must write a chemical equation that describes the physical equilibrium between liquid and gaseous CCl_4 at the normal boiling point. **(b)** We must determine the value of $\Delta G°$ for CCl_4, in equilibrium with its vapor at the normal boiling point. **(c)** We must estimate the normal boiling point of CCl_4, based on available thermodynamic data.

Plan (a) The chemical equation is the change of state from liquid to gas. For **(b)**, we need to analyze Equation 19.19 at equilibrium ($\Delta G = 0$), and for **(c)** we can use Equation 19.12 to calculate T when $\Delta G = 0$.

Solve

(a) The normal boiling point is the temperature at which a pure liquid is in equilibrium with its vapor at a pressure of 1 atm:

$$CCl_4(l) \rightleftharpoons CCl_4(g) \quad P = 1 \text{ atm}$$

(b) At equilibrium, $\Delta G = 0$. In any normal boiling-point equilibrium, both liquid and vapor are in their standard state of pure liquid and vapor at 1 atm (Table 19.2). Consequently, $Q = 1$, $\ln Q = 0$, and $\Delta G = \Delta G°$ for this process. We conclude that $\Delta G° = 0$ for the equilibrium representing the normal boiling point of any liquid. (We would also find that $\Delta G° = 0$ for the equilibria relevant to normal melting points and normal sublimation points.)

(c) Combining Equation 19.12 with the result from part (b), we see that the equality at the normal boiling point, T_b, of $CCl_4(l)$ (or any other pure liquid) is

$$\Delta G° = \Delta H° - T_b \Delta S° = 0$$

Solving the equation for T_b, we obtain

$$T_b = \Delta H°/\Delta S°$$

Strictly speaking, we need the values of $\Delta H°$ and $\Delta S°$ for the $CCl_4(l)/CCl_4(g)$ equilibrium at the normal boiling point to do this calculation. However, we can *estimate* the boiling point by using the values of $\Delta H°$ and $\Delta S°$ for the phases of CCl_4 at 298 K, which we obtain from Appendix C and Equations 5.31 and 19.8:

$$\Delta H° = (1 \text{ mol})(-106.7 \text{ kJ/mol}) - (1 \text{ mol})(-139.3 \text{ kJ/mol}) = +32.6 \text{ kJ}$$
$$\Delta S° = (1 \text{ mol})(309.4 \text{ J/mol-K}) - (1 \text{ mol})(214.4 \text{ J/mol-K}) = +95.0 \text{ J/K}$$

As expected, the process is endothermic ($\Delta H > 0$) and produces a gas, thus increasing the entropy ($\Delta S > 0$). We now use these values to estimate T_b for $CCl_4(l)$:

$$T_b = \frac{\Delta H°}{\Delta S°} = \left(\frac{32.6 \text{ kJ}}{95.0 \text{ J/K}}\right)\left(\frac{1000 \text{ J}}{1 \text{ kJ}}\right) = 343 \text{ K} = 70 \,°\text{C}$$

Note that we have used the conversion factor between joules and kilojoules to make the units of $\Delta H°$ and $\Delta S°$ match.

Check The experimental normal boiling point of $CCl_4(l)$ is 76.5 °C. The small deviation of our estimate from the experimental value is due to the assumption that $\Delta H°$ and $\Delta S°$ do not change with temperature.

Practice Exercise 1

If the normal boiling point of a liquid is 67 °C, and the standard molar entropy change for the boiling process is $+100$ J/K, estimate the standard molar enthalpy change for the boiling process.
(a) $+6700$ J, **(b)** -6700 J, **(c)** $+34,000$ J, **(d)** $-34,000$ J.

Practice Exercise 2

Use data in Appendix C to estimate the normal boiling point, in K, for elemental bromine, $Br_2(l)$. (The experimental value is given in Figure 11.5.)

When the concentrations of reactants and products are nonstandard, we must calculate Q in order to determine ΔG. We illustrate how this is done in Sample Exercise 19.11. At this stage in our discussion, therefore, it becomes important to note the units used to calculate Q when using Equation 19.19. The convention used for standard states is used when applying this equation: In determining the value of Q, the concentrations of gases are always expressed as partial pressures in atmospheres and solutes are expressed as their concentrations in molarities.

SAMPLE
EXERCISE 19.11 | Calculating the Free-Energy Change under
Nonstandard Conditions

Calculate ΔG at 298 K for a mixture of 1.0 atm N_2, 3.0 atm H_2, and 0.50 atm NH_3 being used in the Haber process:

$$N_2(g) + 3\,H_2(g) \rightleftharpoons 2\,NH_3(g)$$

SOLUTION

Analyze We are asked to calculate ΔG under nonstandard conditions.

Plan We can use Equation 19.19 to calculate ΔG. Doing so requires that we calculate the value of the reaction quotient Q for the specified partial pressures, for which we use the partial-pressures form of Equation 15.23: $Q = [D]^d[E]^e/[A]^a[B]^b$. We then use a table of standard free energies of formation to evaluate $\Delta G°$.

Solve The partial-pressures form of Equation 15.23 gives

$$Q = \frac{P_{NH_3}{}^2}{P_{N_2}\,P_{H_2}{}^3} = \frac{(0.50)^2}{(1.0)(3.0)^3} = 9.3 \times 10^{-3}$$

In Sample Exercise 19.9 we calculated $\Delta G° = -33.3$ kJ for this reaction. We will have to change the units of this quantity in applying Equation 19.19, however. For the units in Equation 19.19 to work out, we will use kJ/mol as our units for $\Delta G°$, where "per mole" means "per mole of the reaction as written." Thus, $\Delta G° = -33.3$ kJ/mol implies per 1 mol of N_2, per 3 mol of H_2, and per 2 mol of NH_3.

We now use Equation 19.19 to calculate ΔG for these nonstandard conditions:

$$\begin{aligned}
\Delta G &= \Delta G° + RT \ln Q \\
&= (-33.3\text{ kJ/mol}) + (8.314\text{ J/mol-K})(298\text{ K})(1\text{ kJ}/1000\text{ J})\ln(9.3 \times 10^{-3}) \\
&= (-33.3\text{ kJ/mol}) + (-11.6\text{ kJ/mol}) = -44.9\text{ kJ/mol}
\end{aligned}$$

Comment We see that ΔG becomes more negative as the pressures of N_2, H_2, and NH_3 are changed from 1.0 atm (standard conditions, $\Delta G°$) to 1.0 atm, 3.0 atm, and 0.50 atm, respectively. The larger negative value for ΔG indicates a larger "driving force" to produce NH_3.

We would make the same prediction based on Le Châtelier's principle. ∞∞ (Section 15.7) Relative to standard conditions, we have increased the pressure of a reactant (H_2) and decreased the pressure of the product (NH_3). Le Châtelier's principle predicts that both changes shift the reaction to the product side, thereby forming more NH_3.

Practice Exercise 1

Which of the following statements is true? (**a**) The larger the Q, the larger the $\Delta G°$. (**b**) If $Q = 0$, the system is at equilibrium. (**c**) If a reaction is spontaneous under standard conditions, it is spontaneous under all conditions. (**d**) The free-energy change for a reaction is independent of temperature. (**e**) If $Q > 1$, $\Delta G > \Delta G°$.

Practice Exercise 2

Calculate ΔG at 298 K for the Haber reaction if the reaction mixture consists of 0.50 atm N_2, 0.75 atm H_2, and 2.0 atm NH_3.

Relationship between $\Delta G°$ and K

We can now use Equation 19.19 to derive the relationship between $\Delta G°$ and the equilibrium constant, K. At equilibrium, $\Delta G = 0$ and $Q = K$. Thus, at equilibrium, Equation 19.19 transforms as follows:

$$\begin{aligned}
\Delta G &= \Delta G° + RT \ln Q \\
0 &= \Delta G° + RT \ln K \\
\Delta G° &= -RT \ln K \qquad\qquad\qquad\qquad\qquad [19.20]
\end{aligned}$$

Equation 19.20 is a very important one, with broad significance in chemistry. By relating K to $\Delta G°$, we can also relate K to entropy and enthalpy changes for a reaction.

We can also solve Equation 19.20 for K, to yield an expression that allows us to calculate K if we know the value of $\Delta G°$:

$$\ln K = \frac{\Delta G°}{-RT}$$

$$K = e^{-\Delta G°/RT} \hspace{3cm} [19.21]$$

As usual, we must be careful in our choice of units. In Equations 19.20 and 19.21 we again express $\Delta G°$ in kJ/mol. In the equilibrium-constant expression, we use atmospheres for gas pressures, molarities for solutions; and solids, liquids, and solvents do not appear in the expression. ∞ (Section 15.4) Thus, the equilibrium constant is K_p for gas-phase reactions and K_c for reactions in solution. ∞ (Section 15.2)

From Equation 19.20 we see that if $\Delta G°$ is negative, $\ln K$ must be positive, which means $K > 1$. Therefore, the more negative $\Delta G°$ is, the larger K is. Conversely, if $\Delta G°$ is positive, $\ln K$ is negative, which means $K < 1$. Finally, if $\Delta G°$ is zero, $K = 1$.

 Give It Some Thought

Can $K = 0$?

SAMPLE EXERCISE 19.12 | Calculating an Equilibrium Constant from $\Delta G°$

The standard free-energy change for the Haber process at 25 °C was obtained in Sample Exercise 19.9 for the Haber reaction:

$$N_2(g) + 3\,H_2(g) \rightleftharpoons 2\,NH_3(g) \hspace{1cm} \Delta G° = -33.3 \text{ kJ/mol} = -33,300 \text{ J/mol}$$

Use this value of $\Delta G°$ to calculate the equilibrium constant for the process at 25 °C.

SOLUTION

Analyze We are asked to calculate K for a reaction, given $\Delta G°$.

Plan We can use Equation 19.21 to calculate K.

Solve Remembering to use the absolute temperature for T in Equation 19.21 and the form of R that matches our units, we have

$$K = e^{-\Delta G°/RT} = e^{-(-33,300 \text{ J/mol})/(8.314 \text{ J/mol-K})(298 \text{ K})} = e^{13.4} = 7 \times 10^5$$

Comment This is a large equilibrium constant, which indicates that the product, NH_3, is greatly favored in the equilibrium mixture at 25 °C. The equilibrium constants for the Haber reaction at temperatures in the range 300 °C to 600 °C, given in Table 15.2, are much smaller than the value at 25 °C. Clearly, a low-temperature equilibrium favors the production of ammonia more than a high-temperature one. Nevertheless, the Haber process is carried out at high temperatures because the reaction is extremely slow at room temperature.

Remember Thermodynamics can tell us the direction and extent of a reaction but tells us nothing about the rate at which it will occur. If a catalyst were found that would permit the reaction to proceed at a rapid rate at room temperature, high pressures would not be needed to force the equilibrium toward NH_3.

Practice Exercise 1
The K_{sp} for a very insoluble salt is 4.2×10^{-47} at 298 K. What is $\Delta G°$ for the dissolution of the salt in water?
(a) -265 kJ/mol,   (b) -115 kJ/mol,   (c) -2.61 kJ/mol,   (d) $+115$ kJ/mol,
(e) $+265$ kJ/mol.

Practice Exercise 2
Use data from Appendix C to calculate $\Delta G°$ and K at 298 K for the reaction
$H_2(g) + Br_2(l) \rightleftharpoons 2\,HBr(g)$.

Chemistry and Life

Driving Nonspontaneous Reactions: Coupling Reactions

Many desirable chemical reactions, including a large number that are central to living systems, are nonspontaneous as written. For example, consider the extraction of copper metal from the mineral *chalcocite*, which contains Cu_2S. The decomposition of Cu_2S to its elements is nonspontaneous:

$$Cu_2S(s) \longrightarrow 2\,Cu(s) + S(s) \qquad \Delta G° = +86.2\ kJ$$

Because $\Delta G°$ is very positive, we cannot obtain $Cu(s)$ directly via this reaction. Instead, we must find some way to "do work" on the reaction to force it to occur as we wish. We can do this by coupling the reaction to another one so that the overall reaction *is* spontaneous. For example, we can envision the $S(s)$ reacting with $O_2(g)$ to form $SO_2(g)$:

$$S(s) + O_2(g) \longrightarrow SO_2(g) \qquad \Delta G° = -300.4\ kJ$$

By coupling (adding together) these reactions, we can extract much of the copper metal via a spontaneous reaction:

$$Cu_2S(s) + O_2(g) \longrightarrow 2\,Cu(s) + SO_2(g)$$
$$\Delta G° = (+86.2\ kJ) + (-300.4\ kJ) = -214.2\ kJ$$

In essence, we have used the spontaneous reaction of $S(s)$ with $O_2(g)$ to provide the free energy needed to extract the copper metal from the mineral.

Biological systems employ the same principle of using spontaneous reactions to drive nonspontaneous ones. Many of the biochemical reactions that are essential for the formation and maintenance of highly ordered biological structures are not spontaneous. These necessary reactions are made to occur by coupling them with spontaneous reactions that release energy. The metabolism of food is the usual source of the free energy needed to do the work of maintaining biological systems. For example, complete oxidation of the sugar glucose, $C_6H_{12}O_6$, to CO_2 and H_2O yields substantial free energy:

$$C_6H_{12}O_6(s) + 6\,O_2(g) \longrightarrow 6\,CO_2(g) + 6\,H_2O(l) \quad \Delta G° = -2880\ kJ$$

This energy can be used to drive nonspontaneous reactions in the body. However, a means is necessary to transport the energy released by glucose metabolism to the reactions that require energy. One way, shown in ▼ Figure 19.16, involves the interconversion of adenosine triphosphate (ATP) and adenosine diphosphate (ADP), molecules that are related to the building blocks of nucleic acids. The conversion of ATP to ADP releases free energy ($\Delta G° = -30.5\ kJ$) that can be used to drive other reactions.

In the human body the metabolism of glucose occurs via a complex series of reactions, most of which release free energy. The free energy released during these steps is used in part to reconvert lower-energy ADP back to higher-energy ATP. Thus, the ATP–ADP interconversions are used to store energy during metabolism and to release it as needed to drive nonspontaneous reactions in the body. If you take a course in general biology or biochemistry, you will have the opportunity to learn more about the remarkable sequence of reactions used to transport free energy throughout the human body.

Related Exercises: 19.102, 19.103

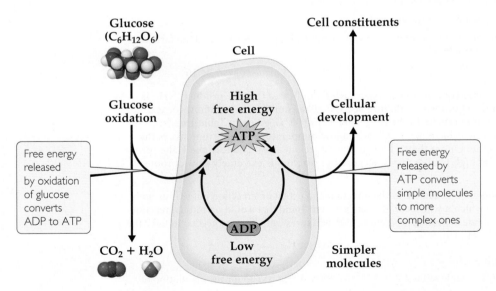

▲ **Figure 19.16 Schematic representation of free-energy changes during cell metabolism.** The oxidation of glucose to CO_2 and H_2O produces free energy that is then used to convert ADP into the more energetic ATP. The ATP is then used, as needed, as an energy source to drive nonspontaneous reactions, such as the conversion of simple molecules into more complex cell constituents.

SAMPLE
INTEGRATIVE EXERCISE Putting Concepts Together

Consider the simple salts NaCl(s) and AgCl(s). We will examine the equilibria in which these salts dissolve in water to form aqueous solutions of ions:

$$NaCl(s) \rightleftharpoons Na^+(aq) + Cl^-(aq)$$

$$AgCl(s) \rightleftharpoons Ag^+(aq) + Cl^-(aq)$$

(a) Calculate the value of ΔG° at 298 K for each of the preceding reactions. **(b)** The two values from part (a) are very different. Is this difference primarily due to the enthalpy term or the entropy term of the standard free-energy change? **(c)** Use the values of ΔG° to calculate the K_{sp} values for the two salts at 298 K. **(d)** Sodium chloride is considered a soluble salt, whereas silver chloride is considered insoluble. Are these descriptions consistent with the answers to part (c)? **(e)** How will ΔG° for the solution process of these salts change with increasing T? What effect should this change have on the solubility of the salts?

SOLUTION

(a) We will use Equation 19.14 along with ΔG_f° values from Appendix C to calculate the ΔG_{soln}° values for each equilibrium. (As we did in Section 13.1, we use the subscript "soln" to indicate that these are thermodynamic quantities for the formation of a solution.) We find

$$\Delta G_{soln}^\circ(NaCl) = (-261.9 \text{ kJ/mol}) + (-131.2 \text{ kJ/mol}) - (-384.0 \text{ kJ/mol})$$

$$= -9.1 \text{ kJ/mol}$$

$$\Delta G_{soln}^\circ(AgCl) = (+77.11 \text{ kJ/mol}) + (-131.2 \text{ kJ/mol}) - (-109.70 \text{ kJ/mol})$$

$$= +55.6 \text{ kJ/mol}$$

(b) We can write ΔG_{soln}° as the sum of an enthalpy term, ΔH_{soln}°, and an entropy term, $-T\Delta S_{soln}^\circ$: $\Delta G_{soln}^\circ = \Delta H_{soln}^\circ + (-T\Delta S_{soln}^\circ)$. We can calculate the values of ΔH_{soln}° and ΔS_{soln}° by using Equations 5.31 and 19.8. We can then calculate $-T\Delta S_{soln}^\circ$ at $T = 298$ K. All these calculations are now familiar to us. The results are summarized in the following table:

Salt	ΔH_{soln}°	ΔS_{soln}°	$T\Delta S_{soln}^\circ$
NaCl	+3.6 kJ/mol	+43.2 kJ/mol-K	−12.9 kJ/mol
AgCl	+65.7 kJ/mol	+34.3 kJ/mol-K	−10.2 kJ/mol

The entropy terms for the solution of the two salts are very similar. That seems sensible because each solution process should lead to a similar increase in randomness as the salt dissolves, forming hydrated ions. ∞ (Section 13.1) In contrast, we see a very large difference in the enthalpy term for the solution of the two salts. The difference in the values of ΔG_{soln}° is dominated by the difference in the values of ΔH_{soln}°.

(c) The solubility product, K_{sp}, is the equilibrium constant for the solution process. ∞ (Section 17.4) As such, we can relate K_{sp} directly to ΔG_{soln}° by using Equation 19.21:

$$K_{sp} = e^{-\Delta G_{soln}^\circ/RT}$$

We can calculate the K_{sp} values in the same way we applied Equation 19.21 in Sample Exercise 19.12. We use the ΔG_{soln}° values we obtained in part (a), remembering to convert them from kJ/mol to J/mol:

$$\text{NaCl: } K_{sp} = [Na^+(aq)][Cl^-(aq)] = e^{-(-9100)/[(8.314)(298)]} = e^{+3.7} = 40$$

$$\text{AgCl: } K_{sp} = [Ag^+(aq)][Cl^-(aq)] = e^{-(+55,600)/[(8.314)(298)]} = e^{-22.4} = 1.9 \times 10^{-10}$$

The value calculated for the K_{sp} of AgCl is very close to that listed in Appendix D.

(d) A soluble salt is one that dissolves appreciably in water. ∞ (Section 4.2) The K_{sp} value for NaCl is greater than 1, indicating that NaCl dissolves to a great extent. The K_{sp} value for AgCl is very small, indicating that very little dissolves in water. Silver chloride should indeed be considered an insoluble salt.

(e) As we expect, the solution process has a positive value of ΔS for both salts (see the table in part b). As such, the entropy term of the free-energy change, $-T\Delta S_{soln}^\circ$, is negative. If we assume that ΔH_{soln}° and ΔS_{soln}° do not change much with temperature, then an increase in T will serve to make ΔG_{soln}° more negative. Thus, the driving force for dissolution of the salts will increase with increasing T, and we therefore expect the solubility of the salts to increase with increasing T. In Figure 13.18 we see that the solubility of NaCl (and that of nearly any other salt) increases with increasing temperature. ∞ (Section 13.3)

Chapter Summary and Key Terms

SPONTANEOUS PROCESSES (SECTION 19.1) Most reactions and chemical processes have an inherent directionality: They are **spontaneous** in one direction and nonspontaneous in the reverse direction. The spontaneity of a process is related to the thermodynamic path the system takes from the initial state to the final state. In a **reversible process**, both the system and its surroundings can be restored to their original state by exactly reversing the change. In an **irreversible process** the system cannot return to its original state without a permanent change in the surroundings. Any spontaneous process is irreversible. A process that occurs at a constant temperature is said to be **isothermal**.

ENTROPY AND THE SECOND LAW OF THERMODYNAMICS (SECTION 19.2) The spontaneous nature of processes is related to a thermodynamic state function called **entropy**, denoted S. For a process that occurs at constant temperature, the entropy change of the system is given by the heat absorbed by the system along a reversible path, divided by the temperature: $\Delta S = q_{rev}/T$. For any process, the entropy change of the universe equals the entropy change of the system plus the entropy change of the surroundings: $\Delta S_{univ} = \Delta S_{sys} + \Delta S_{surr}$. The way entropy controls the spontaneity of processes is given by the **second law of thermodynamics**, which states that in a reversible process $\Delta S_{univ} = 0$; in an irreversible (spontaneous) process $\Delta S_{univ} > 0$. Entropy values are usually expressed in units of joules per kelvin, J/K.

THE MOLECULAR INTERPRETATION OF ENTROPY AND THE THIRD LAW OF THERMODYNAMICS (SECTION 19.3) A particular combination of motions and locations of the atoms and molecules of a system at a particular instant is called a **microstate**. The entropy of a system is a measure of its randomness or disorder. The entropy is related to the number of microstates, W, corresponding to the state of the system: $S = k \ln W$. Molecules can undergo three kinds of motion: In **translational motion** the entire molecule moves in space. Molecules can also undergo **vibrational motion**, in which the atoms of the molecule move toward and away from one another in periodic fashion, and **rotational motion**, in which the entire molecule spins like a top. The number of available microstates, and therefore the entropy, increases with an increase in volume, temperature, or motion of molecules because any of these changes increases the possible motions and locations of the molecules. As a result, entropy generally increases when liquids or solutions are formed from solids, gases are formed from either solids or liquids, or the number of molecules of gas increases during a chemical reaction. The **third law of thermodynamics** states that the entropy of a pure crystalline solid at 0 K is zero.

ENTROPY CHANGES IN CHEMICAL REACTIONS (SECTION 19.4) The third law allows us to assign entropy values for substances at different temperatures. Under standard conditions the entropy of a mole of a substance is called its **standard molar entropy**, denoted $S°$. From tabulated values of $S°$, we can calculate the entropy change for any process under standard conditions. For an isothermal process, the entropy change in the surroundings is equal to $-\Delta H/T$.

GIBBS FREE ENERGY (SECTION 19.5) The **Gibbs free energy** (or just **free energy**), G, is a thermodynamic state function that combines the two state functions enthalpy and entropy: $G = H - TS$. For processes that occur at constant temperature, $\Delta G = \Delta H - T\Delta S$. For a process occurring at constant temperature and pressure, the sign of ΔG relates to the spontaneity of the process. When ΔG is negative, the process is spontaneous. When ΔG is positive, the process is nonspontaneous but the reverse process is spontaneous. At equilibrium the process is reversible and ΔG is zero. The free energy is also a measure of the maximum useful work that can be performed by a system in a spontaneous process. The standard free-energy change, $\Delta G°$, for any process can be calculated from tabulations of **standard free energies of formation**, $\Delta G_f°$, which are defined in a fashion analogous to standard enthalpies of formation, $\Delta G_f°$. The value of $\Delta G_f°$ for a pure element in its standard state is defined to be zero.

FREE ENERGY, TEMPERATURE, AND THE EQUILIBRIUM CONSTANT (SECTIONS 19.6 AND 19.7) The values of ΔH and ΔS for a chemical process generally do not vary much with temperature. Therefore, the dependence of ΔG with temperature is governed mainly by the value of T in the expression $\Delta G = \Delta H - T\Delta S$. The entropy term $-T\Delta S$ has the greater effect on the temperature dependence of ΔG and, hence, on the spontaneity of the process. For example, a process for which $\Delta H > 0$ and $\Delta S > 0$, such as the melting of ice, can be nonspontaneous ($\Delta G > 0$) at low temperatures and spontaneous ($\Delta G > 0$) at higher temperatures. Under nonstandard conditions ΔG is related to $\Delta G°$ and the value of the reaction quotient, Q: $\Delta G = \Delta G° + RT \ln Q$. At equilibrium ($\Delta G = 0, Q = K$), $\Delta G° = -RT \ln K$. Thus, the standard free-energy change is directly related to the equilibrium constant for the reaction. This relationship expresses the temperature dependence of equilibrium constants.

Learning Outcomes   After studying this chapter, you should be able to:

- Explain the meaning of spontaneous process, reversible process, irreversible process, and isothermal process. (Section 19.1)

- Define entropy and state the second law of thermodynamics. (Section 19.2)

- Explain how the entropy of a system is related to the number of possible microstates. (Section 19.3)

- Describe the kinds of molecular motion that a molecule can possess. (Section 19.3)

- Predict the sign of ΔS for physical and chemical processes. (Section 19.3)

- State the third law of thermodynamics. (Section 19.3)

- Calculate standard entropy changes for a system from standard molar entropies. (Section 19.4)

- Calculate entropy changes in the surroundings for isothermal processes. (Section 19.4)

- Calculate the Gibbs free energy from the enthalpy change and entropy change at a given temperature. (Section 19.5)

- Use free-energy changes to predict whether reactions are spontaneous. (Section 19.5)

- Calculate standard free-energy changes using standard free energies of formation. (Section 19.5)

- Predict the effect of temperature on spontaneity given ΔH and ΔS. (Section 19.6)

- Calculate ΔG under nonstandard conditions. (Section 19.7)

- Relate $\Delta G°$ and equilibrium constant. (Section 19.7)

Key Equations

- $\Delta S = \dfrac{q_{rev}}{T}$ (constant T)   [19.2]   Relating entropy change to the heat absorbed or released in a reversible process

- *Reversible process:* $\Delta S_{univ} = \Delta S_{sys} + \Delta S_{surr} = 0$ ⎫
  Irreversible process: $\Delta S_{univ} = \Delta S_{sys} + \Delta S_{surr} > 0$ ⎭   [19.4]   The second law of thermodynamics

- $S = k \ln W$ [19.5]   Relating entropy to the number of microstates

- $\Delta S° = \sum nS°(\text{products}) - \sum mS°(\text{reactants})$ [19.8]   Calculating the standard entropy change from standard molar entropies

- $\Delta S_{surr} = \dfrac{-\Delta H_{sys}}{T}$ [19.9]   The entropy change of the surroundings for a process at constant temperature and pressure

- $\Delta G = \Delta H - T\Delta S$ [19.11]   Calculating the Gibbs free-energy change from enthalpy and entropy changes at constant temperature

- $\Delta G° = \sum n\Delta G_f°(\text{products}) - \sum m\Delta G_f°(\text{reactants})$ [19.14]   Calculating the standard free-energy change from standard free energies of formation

- *Reversible process:* $\Delta G = \Delta H_{sys} - T\Delta S_{sys} = 0$ ⎫
  Irreversible process: $\Delta G = \Delta H_{sys} - T\Delta S_{sys} < 0$ ⎭   [19.17]   Relating the free-energy change to the reversibility of a process at constant temperature and pressure

- $\Delta G = -w_{max}$ [19.18]   Relating the free-energy change to the maximum work a system can perform

- $\Delta G = \Delta G° + RT \ln Q$ [19.19]   Calculating free-energy change under nonstandard conditions

- $\Delta G° = -RT \ln K$ [19.20]   Relating the standard free-energy change and the equilibrium constant

Exercises

Visualizing Concepts

19.1 Two different gases occupy the two bulbs shown here. Consider the process that occurs when the stopcock is opened, assuming the gases behave ideally. **(a)** Draw the final (equilibrium) state. **(b)** Predict the signs of ΔH and ΔS for the process. **(c)** Is the process that occurs when the stopcock is opened a reversible one? **(d)** How does the process affect the entropy of the surroundings? [Sections 19.1 and 19.2]

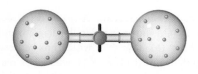

19.2 As shown here, one type of computer keyboard cleaner contains liquefied 1,1-difluoroethane ($C_2H_4F_2$), which is a gas at atmospheric pressure. When the nozzle is squeezed, the 1,1-difluoroethane vaporizes out of the nozzle at high pressure, blowing dust out of objects. **(a)** Based on your experience, is the vaporization a spontaneous process at room temperature? **(b)** Defining the 1,1-difluoroethane as the system, do you expect q_{sys} for the process to be positive or negative? **(c)** Predict whether ΔS is positive or negative for this process. **(d)** Given your answers to (a), (b), and (c), do you think the operation of this product depends more on enthalpy or entropy? [Sections 19.1 and 19.2]

Vaporized $C_2H_4F_2$
Liquefied $C_2H_4F_2$

19.3 **(a)** What are the signs of ΔS and ΔH for the process depicted here? **(b)** If energy can flow in and out of the system to maintain a constant temperature during the process, what can you say about the entropy change of the surroundings as a result of this process? [Sections 19.2 and 19.5]

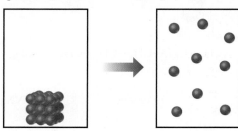

19.4 Predict the signs of ΔH and ΔS for this reaction. Explain your choice. [Section 19.3]

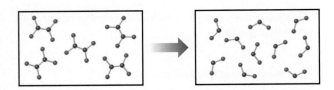

19.5 The accompanying diagram shows how entropy varies with temperature for a substance that is a gas at the highest temperature shown. **(a)** What processes correspond to the entropy increases along the vertical lines labeled 1 and 2? **(b)** Why is the entropy change for 2 larger than that for 1? **(c)** If this substance is a perfect crystal at $T = 0$ K, what is the value of S at this temperature? [Section 19.3]

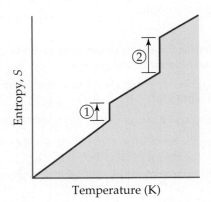

19.6 *Isomers* are molecules that have the same chemical formula but different arrangements of atoms, as shown here for two isomers of pentane, C_5H_{12}. **(a)** Do you expect a significant difference in the enthalpy of combustion of the two isomers? Explain. **(b)** Which isomer do you expect to have the higher standard molar entropy? Explain. [Section 19.4]

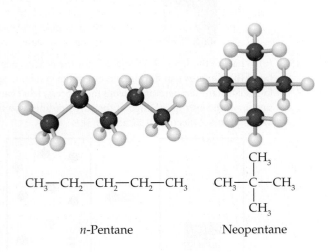

$CH_3-CH_2-CH_2-CH_2-CH_3$

$$CH_3-\underset{\underset{CH_3}{|}}{\overset{\overset{CH_3}{|}}{C}}-CH_3$$

n-Pentane   Neopentane

19.7 The accompanying diagram shows how ΔH (red line) and $T\Delta S$ (blue line) change with temperature for a hypothetical reaction. **(a)** What is the significance of the point at 300 K, where ΔH and $T\Delta S$ are equal? **(b)** In what temperature range is this reaction spontaneous? [Section 19.6]

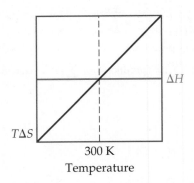

19.8 The accompanying diagram shows how ΔG for a hypothetical reaction changes as temperature changes. **(a)** At what temperature is the system at equilibrium? **(b)** In what temperature range is the reaction spontaneous? **(c)** Is ΔH positive or negative? **(d)** Is ΔS positive or negative? [Sections 19.5 and 19.6]

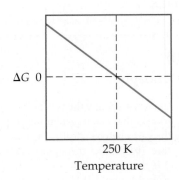

19.9 Consider a reaction $A_2(g) + B_2(g) \rightleftharpoons 2\,AB(g)$, with atoms of A shown in red in the diagram and atoms of B shown in blue. **(a)** If $K_c = 1$, which box represents the system at equilibrium? **(b)** If $K_c = 1$, which box represents the system at $Q < K_c$? **(c)** Rank the boxes in order of increasing magnitude of ΔG for the reaction. [Sections 19.5 and 19.7]

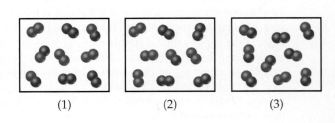

(1)   (2)   (3)

19.10 The accompanying diagram shows how the free energy, G, changes during a hypothetical reaction $A(g) + B(g) \longrightarrow C(g)$. On the left are pure reactants A and B, each at 1 atm, and on the right is the pure product, C, also at 1 atm. Indicate whether each of the following statements is true or false. **(a)** The minimum of the graph corresponds to the equilibrium mixture of reactants and products for this reaction. **(b)** At equilibrium, all of A and B have reacted to give pure C. **(c)** The entropy change for this reaction is positive. **(d)** The "x" on the graph corresponds to ΔG for the reaction. **(e)** ΔG for the reaction corresponds to the difference between the top left of the curve and the bottom of the curve. [Section 19.7]

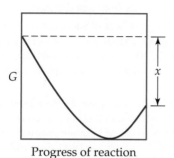

Progress of reaction

Spontaneous Processes (Section 19.1)

19.11 Which of the following processes are spontaneous and which are nonspontaneous: **(a)** the ripening of a banana, **(b)** dissolution of sugar in a cup of hot coffee, **(c)** the reaction of nitrogen atoms to form N_2 molecules at 25 °C and 1 atm, **(d)** lightning, **(e)** formation of CH_4 and O_2 molecules from CO_2 and H_2O at room temperature and 1 atm of pressure?

19.12 Which of the following processes are spontaneous: **(a)** the melting of ice cubes at −10 °C and 1 atm pressure; **(b)** separating a mixture of N_2 and O_2 into two separate samples, one that is pure N_2 and one that is pure O_2; **(c)** alignment of iron filings in a magnetic field; **(d)** the reaction of hydrogen gas with oxygen gas to form water vapor at room temperature; **(e)** the dissolution of $HCl(g)$ in water to form concentrated hydrochloric acid?

19.13 **(a)** Can endothermic chemical reactions be spontaneous? **(b)** Can a process that is spontaneous at one temperature be nonspontaneous at a different temperature?

19.14 The crystalline hydrate $Cd(NO_3)_2 \cdot 4 H_2O(s)$ loses water when placed in a large, closed, dry vessel at room temperature:

$$Cd(NO_3)_2 \cdot 4 H_2O(s) \longrightarrow Cd(NO_3)_2(s) + 4 H_2O(g)$$

This process is spontaneous and $\Delta H°$ is positive at room temperature. **(a)** What is the sign of $\Delta S°$ at room temperature? **(b)** If the hydrated compound is placed in a large, closed vessel that already contains a large amount of water vapor, does $\Delta S°$ change for this reaction at room temperature?

19.15 Consider the vaporization of liquid water to steam at a pressure of 1 atm. **(a)** Is this process endothermic or exothermic? **(b)** In what temperature range is it a spontaneous process? **(c)** In what temperature range is it a nonspontaneous process? **(d)** At what temperature are the two phases in equilibrium?

19.16 The normal freezing point of n-octane (C_8H_{18}) is −57 °C. **(a)** Is the freezing of n-octane an endothermic or exothermic process? **(b)** In what temperature range is the freezing of n-octane a spontaneous process? **(c)** In what temperature range is it a nonspontaneous process? **(d)** Is there any temperature at which liquid n-octane and solid n-octane are in equilibrium? Explain.

19.17 Indicate whether each statement is true or false. **(a)** If a system undergoes a reversible process, the entropy of the universe increases. **(b)** If a system undergoes a reversible process, the change in entropy of the system is exactly matched by an equal and opposite change in the entropy of the surroundings. **(c)** If a system undergoes a reversible process, the entropy change of the system must be zero. **(d)** Most spontaneous processes in nature are reversible.

19.18 Indicate whether each statement is true or false. **(a)** All spontaneous processes are irreversible. **(b)** The entropy of the universe increases for spontaneous processes. **(c)** The change in entropy of the surroundings is equal in magnitude and opposite in sign for the change in entropy of the system, for an irreversible process. **(d)** The maximum amount of work can be gotten out of a system that undergoes an irreversible process, as compared to a reversible process.

19.19 Consider a process in which an ideal gas changes from state 1 to state 2 in such a way that its temperature changes from 300 K to 200 K. **(a)** Does the temperature change depend on whether the process is reversible or irreversible? **(b)** Is this process isothermal? **(c)** Does the change in the internal energy, ΔE, depend on the particular pathway taken to carry out this change of state?

19.20 A system goes from state 1 to state 2 and back to state 1. **(a)** Is ΔE the same in magnitude for both the forward and reverse processes? **(b)** Without further information, can you conclude that the amount of heat transferred to the system as it goes from state 1 to state 2 is the same or different as compared to that upon going from state 2 back to state 1? **(c)** Suppose the changes in state are reversible processes. Is the work done by the system upon going from state 1 to state 2 the same or different as compared to that upon going from state 2 back to state 1?

19.21 Consider a system consisting of an ice cube. **(a)** Under what conditions can the ice cube melt reversibly? **(b)** If the ice cube melts reversibly, is ΔE zero for the process?

19.22 Consider what happens when a sample of the explosive TNT (Section 8.8: "Chemistry Put to Work: Explosives and Alfred Nobel") is detonated under atmospheric pressure. **(a)** Is the detonation a spontaneous process? **(b)** What is the sign of q for this process? **(c)** Is it possible to tell whether w is positive, negative, or zero for the process? Explain. **(d)** Can you determine the sign of ΔE for the process? Explain.

Entropy and the Second Law of Thermodynamics (Section 19.2)

19.23 Indicate whether each statement is true or false. (a) ΔS for an isothermal process depends on both the temperature and the amount of heat reversibly transferred. (b) ΔS is a state function. (c) The second law of thermodynamics says that the entropy of the system increases for all spontaneous processes.

19.24 Suppose we vaporize a mole of liquid water at 25 °C and another mole of water at 100 °C. (a) Assuming that the enthalpy of vaporization of water does not change much between 25 °C and 100 °C, which process involves the larger change in entropy? (b) Does the entropy change in either process depend on whether we carry out the process reversibly or not? Explain.

19.25 The normal boiling point of $Br_2(l)$ is 58.8 °C, and its molar enthalpy of vaporization is $\Delta H_{vap} = 29.6$ kJ/mol. (a) When $Br_2(l)$ boils at its normal boiling point, does its entropy increase or decrease? (b) Calculate the value of ΔS when 1.00 mol of $Br_2(l)$ is vaporized at 58.8 °C.

19.26 The element gallium (Ga) freezes at 29.8 °C, and its molar enthalpy of fusion is $\Delta H_{fus} = 5.59$ kJ/mol. (a) When molten gallium solidifies to Ga(s) at its normal melting point, is ΔS positive or negative? (b) Calculate the value of ΔS when 60.0 g of Ga(l) solidifies at 29.8 °C.

19.27 Indicate whether each statement is true or false. (a) The second law of thermodynamics says that entropy is conserved. (b) If the entropy of the system increases during a reversible process, the entropy change of the surroundings must decrease by the same amount. (c) In a certain spontaneous process the system undergoes an entropy change of 4.2 J/K; therefore, the entropy change of the surroundings must be −4.2 J/K.

19.28 (a) Does the entropy of the surroundings increase for spontaneous processes? (b) In a particular spontaneous process the entropy of the system decreases. What can you conclude about the sign and magnitude of ΔS_{surr}? (c) During a certain reversible process, the surroundings undergo an entropy change, $\Delta S_{surr} = -78$ J/K. What is the entropy change of the system for this process?

19.29 (a) What sign for ΔS do you expect when the volume of 0.200 mol of an ideal gas at 27 °C is increased isothermally from an initial volume of 10.0 L? (b) If the final volume is 18.5 L, calculate the entropy change for the process. (c) Do you need to specify the temperature to calculate the entropy change? Explain.

19.30 (a) What sign for ΔS do you expect when the pressure on 0.600 mol of an ideal gas at 350 K is increased isothermally from an initial pressure of 0.750 atm? (b) If the final pressure on the gas is 1.20 atm, calculate the entropy change for the process. (c) Do you need to specify the temperature to calculate the entropy change? Explain.

The Molecular Interpretation of Entropy and the Third Law of Thermodynamics (Section 19.3)

19.31 For the isothermal expansion of a gas into a vacuum, $\Delta E = 0, q = 0,$ and $w = 0$. (a) Is this a spontaneous

process? (b) Explain why no work is done by the system during this process. (c) What is the "driving force" for the expansion of the gas: enthalpy or entropy?

19.32 (a) What is the difference between a *state* and a *microstate* of a system? (b) As a system goes from state A to state B, its entropy decreases. What can you say about the number of microstates corresponding to each state? (c) In a particular spontaneous process, the number of microstates available to the system decreases. What can you conclude about the sign of ΔS_{surr}?

19.33 Would each of the following changes increase, decrease, or have no effect on the number of microstates available to a system: (a) increase in temperature, (b) decrease in volume, (c) change of state from liquid to gas?

19.34 (a) Using the heat of vaporization in Appendix B, calculate the entropy change for the vaporization of water at 25 °C and at 100 °C. (b) From your knowledge of microstates and the structure of liquid water, explain the difference in these two values.

19.35 (a) What do you expect for the sign of ΔS in a chemical reaction in which two moles of gaseous reactants are converted to three moles of gaseous products? (b) For which of the processes in Exercise 19.11 does the entropy of the system increase?

19.36 (a) In a chemical reaction two gases combine to form a solid. What do you expect for the sign of ΔS? (b) How does the entropy of the system change in the processes described in Exercise 19.12?

19.37 Does the entropy of the system increase, decrease, or stay the same when (a) a solid melts, (b) a gas liquefies, (c) a solid sublimes?

19.38 Does the entropy of the system increase, decrease, or stay the same when (a) the temperature of the system increases, (b) the volume of a gas increases, (c) equal volumes of ethanol and water are mixed to form a solution?

19.39 Indicate whether each statement is true or false. (a) The third law of thermodynamics says that the entropy of a perfect, pure crystal at absolute zero increases with the mass of the crystal. (b) "Translational motion" of molecules refers to their change in spatial location as a function of time. (c) "Rotational" and "vibrational" motions contribute to the entropy in atomic gases like He and Xe. (d) The larger the number of atoms in a molecule, the more degrees of freedom of rotational and vibrational motion it likely has.

19.40 Indicate whether each statement is true or false. (a) Unlike enthalpy, where we can only ever know changes in H, we can know absolute values of S. (b) If you heat a gas such as CO_2, you will increase its degrees of translational, rotational and vibrational motions. (c) $CO_2(g)$ and Ar(g) have nearly the same molar mass. At a given temperature, they will have the same number of microstates.

19.41 For each of the following pairs, choose the substance with the higher entropy per mole at a given temperature: (a) Ar(l) or Ar(g), (b) He(g) at 3 atm pressure or He(g) at 1.5 atm pressure, (c) 1 mol of Ne(g) in 15.0 L or 1 mol of Ne(g) in 1.50 L, (d) $CO_2(g)$ or $CO_2(s)$.

19.42 For each of the following pairs, indicate which substance possesses the larger standard entropy: (a) 1 mol of $P_4(g)$ at 300 °C, 0.01 atm, or 1 mol of $As_4(g)$ at 300 °C, 0.01 atm; (b) 1 mol of $H_2O(g)$ at 100 °C, 1 atm, or 1 mol of $H_2O(l)$ at

100 °C, 1 atm; **(c)** 0.5 mol of $N_2(g)$ at 298 K, 20-L volume, or 0.5 mol $CH_4(g)$ at 298 K, 20-L volume; **(d)** 100 g $Na_2SO_4(s)$ at 30 °C or 100 g $Na_2SO_4(aq)$ at 30 °C.

19.43 Predict the sign of the entropy change of the system for each of the following reactions:

(a) $N_2(g) + 3 H_2(g) \longrightarrow 2 NH_3(g)$

(b) $CaCO_3(s) \longrightarrow CaO(s) + CO_2(g)$

(c) $3 C_2H_2(g) \longrightarrow C_6H_6(g)$

(d) $Al_2O_3(s) + 3 H_2(g) \longrightarrow 2 Al(s) + 3 H_2O(g)$

19.44 Predict the sign of ΔS_{sys} for each of the following processes: **(a)** Molten gold solidifies. **(b)** Gaseous Cl_2 dissociates in the stratosphere to form gaseous Cl atoms. **(c)** Gaseous CO reacts with gaseous H_2 to form liquid methanol, CH_3OH. **(d)** Calcium phosphate precipitates upon mixing $Ca(NO_3)_2(aq)$ and $(NH_4)_3PO_4(aq)$.

Entropy Changes in Chemical Reactions (Section 19.4)

19.45 **(a)** Using Figure 19.12 as a model, sketch how the entropy of water changes as it is heated from -50 °C to 110 °C at sea level. Show the temperatures at which there are vertical increases in entropy. **(b)** Which process has the larger entropy change: melting ice or boiling water? Explain.

19.46 Propanol (C_3H_7OH) melts at -126.5 °C and boils at 97.4 °C. Draw a qualitative sketch of how the entropy changes as propanol vapor at 150 °C and 1 atm is cooled to solid propanol at -150 °C and 1 atm.

19.47 In each of the following pairs, which compound would you expect to have the higher standard molar entropy: **(a)** $C_2H_2(g)$ or $C_2H_6(g)$, **(b)** $CO_2(g)$ or $CO(g)$?

19.48 Cyclopropane and propylene are isomers that both have the formula C_3H_6. Based on the molecular structures shown, which of these isomers would you expect to have the higher standard molar entropy at 25 °C?

Cyclopropane   Propylene

19.49 Use Appendix C to compare the standard entropies at 25 °C for the following pairs of substances: **(a)** $Sc(s)$ and $Sc(g)$, **(b)** $NH_3(g)$ and $NH_3(aq)$, **(c)** 1 mol $P_4(g)$ and 2 mol $P_2(g)$, **(d)** C(graphite) and C(diamond).

19.50 Using Appendix C, compare the standard entropies at 25 °C for the following pairs of substances: **(a)** $CuO(s)$ and $Cu_2O(s)$, **(b)** 1 mol $N_2O_4(g)$ and 2 mol $NO_2(g)$, **(c)** $SiO_2(s)$ and $CO_2(g)$, **(d)** $CO(g)$ and $CO_2(g)$.

[19.51] The standard entropies at 298 K for certain group 4A elements are: C(s, diamond) = 2.43 J/mol-K, Si(s) = 18.81 J/mol-K, Ge(s) = 31.09 J/mol-K, and Sn(s) = 51.818 J/mol-K. All but Sn have the same (diamond) structure. How do you account for the trend in the $S°$ values?

[19.52] Three of the forms of elemental carbon are graphite, diamond, and buckminsterfullerene. The entropies at 298 K for graphite and diamond are listed in Appendix C. **(a)** Account for the difference in the $S°$ values of graphite and diamond in light of their structures (Figure 12.29, p. 503). **(b)** What would you expect for the $S°$ value of buckminsterfullerene (Figure 12.47, p. 516) relative to the values for graphite and diamond? Explain.

19.53 Using $S°$ values from Appendix C, calculate $\Delta S°$ values for the following reactions. In each case account for the sign of $\Delta S°$.

(a) $C_2H_4(g) + H_2(g) \longrightarrow C_2H_6(g)$

(b) $N_2O_4(g) \longrightarrow 2 NO_2(g)$

(c) $Be(OH)_2(s) \longrightarrow BeO(s) + H_2O(g)$

(d) $2 CH_3OH(g) + 3 O_2(g) \longrightarrow 2 CO_2(g) + 4 H_2O(g)$

19.54 Calculate $\Delta S°$ values for the following reactions by using tabulated $S°$ values from Appendix C. In each case explain the sign of $\Delta S°$.

(a) $HNO_3(g) + NH_3(g) \longrightarrow NH_4NO_3(s)$

(b) $2 Fe_2O_3(s) \longrightarrow 4 Fe(s) + 3 O_2(g)$

(c) $CaCO_3(s, calcite) + 2HCl(g) \longrightarrow CaCl_2(s) + CO_2(g) + H_2O(l)$

(d) $3 C_2H_6(g) \longrightarrow C_6H_6(l) + 6 H_2(g)$

Gibbs Free Energy (Sections 19.5 and 19.6)

19.55 **(a)** For a process that occurs at constant temperature, does the change in Gibbs free energy depend on changes in the enthalpy and entropy of the system? **(b)** For a certain process that occurs at constant T and P, the value of ΔG is positive. Is the process spontaneous? **(c)** If ΔG for a process is large, is the rate at which it occurs fast?

19.56 **(a)** Is the standard free-energy change, $\Delta G°$, always larger than ΔG? **(b)** For any process that occurs at constant temperature and pressure, what is the significance of $\Delta G = 0$? **(c)** For a certain process, ΔG is large and negative. Does this mean that the process necessarily has a low activation barrier?

19.57 For a certain chemical reaction, $\Delta H° = -35.4$ kJ and $\Delta S° = -85.5$ J/K. **(a)** Is the reaction exothermic or endothermic? **(b)** Does the reaction lead to an increase or decrease in the randomness or disorder of the system? **(c)** Calculate $\Delta G°$ for the reaction at 298 K. **(d)** Is the reaction spontaneous at 298 K under standard conditions?

19.58 A certain reaction has $\Delta H° = +23.7$ kJ and $\Delta S° = +52.4$ J/K. **(a)** Is the reaction exothermic or endothermic? **(b)** Does the reaction lead to an increase or decrease in the randomness or disorder of the system? **(c)** Calculate $\Delta G°$ for the reaction at 298 K. **(d)** Is the reaction spontaneous at 298 K under standard conditions?

19.59 Using data in Appendix C, calculate $\Delta H°$, $\Delta S°$, and $\Delta G°$ at 298 K for each of the following reactions.

(a) $H_2(g) + F_2(g) \longrightarrow 2 HF(g)$

(b) $C(s, graphite) + 2 Cl_2(g) \longrightarrow CCl_4(g)$

(c) $2 PCl_3(g) + O_2(g) \longrightarrow 2 POCl_3(g)$

(d) $2 CH_3OH(g) + H_2(g) \longrightarrow C_2H_6(g) + 2 H_2O(g)$

19.60 Use data in Appendix C to calculate $\Delta H°$, $\Delta S°$, and $\Delta G°$ at 25 °C for each of the following reactions.

(a) $4\,Cr(s) + 3\,O_2(g) \longrightarrow 2\,Cr_2O_3(s)$

(b) $BaCO_3(s) \longrightarrow BaO(s) + CO_2(g)$

(c) $2\,P(s) + 10\,HF(g) \longrightarrow 2\,PF_5(g) + 5\,H_2(g)$

(d) $K(s) + O_2(g) \longrightarrow KO_2(s)$

19.61 Using data from Appendix C, calculate $\Delta G°$ for the following reactions. Indicate whether each reaction is spontaneous at 298 K under standard conditions.

(a) $2\,SO_2(g) + O_2(g) \longrightarrow 2\,SO_3(g)$

(b) $NO_2(g) + N_2O(g) \longrightarrow 3\,NO(g)$

(c) $6\,Cl_2(g) + 2\,Fe_2O_3(s) \longrightarrow 4\,FeCl_3(s) + 3\,O_2(g)$

(d) $SO_2(g) + 2\,H_2(g) \longrightarrow S(s) + 2\,H_2O(g)$

19.62 Using data from Appendix C, calculate the change in Gibbs free energy for each of the following reactions. In each case indicate whether the reaction is spontaneous at 298 K under standard conditions.

(a) $2\,Ag(s) + Cl_2(g) \longrightarrow 2\,AgCl(s)$

(b) $P_4O_{10}(s) + 16\,H_2(g) \longrightarrow 4\,PH_3(g) + 10\,H_2O(g)$

(c) $CH_4(g) + 4\,F_2(g) \longrightarrow CF_4(g) + 4\,HF(g)$

(d) $2\,H_2O_2(l) \longrightarrow 2\,H_2O(l) + O_2(g)$

19.63 Octane (C_8H_{18}) is a liquid hydrocarbon at room temperature that is the primary constituent of gasoline. (a) Write a balanced equation for the combustion of $C_8H_{18}(l)$ to form $CO_2(g)$ and $H_2O(l)$. (b) Without using thermochemical data, predict whether $\Delta G°$ for this reaction is more negative or less negative than $\Delta H°$.

19.64 Sulfur dioxide reacts with strontium oxide as follows:

$$SO_2(g) + SrO(g) \longrightarrow SrSO_3(s)$$

(a) Without using thermochemical data, predict whether $\Delta G°$ for this reaction is more negative or less negative than $\Delta H°$. (b) If you had only standard enthalpy data for this reaction, estimate of the value of $\Delta G°$ at 298 K, using data from Appendix C on other substances.

19.65 Classify each of the following reactions as one of the four possible types summarized in Table 19.3: (i) spontanous at all temperatures; (ii) not spontaneous at any temperature; (iii) spontaneous at low T but not spontaneous at high T; (iv) spontaneous at high T but not spontaneous at low T.

(a) $N_2(g) + 3\,F_2(g) \longrightarrow 2\,NF_3(g)$
$\Delta H° = -249\,kJ;\ \Delta S° = -278\,J/K$

(b) $N_2(g) + 3\,Cl_2(g) \longrightarrow 2\,NCl_3(g)$
$\Delta H° = 460\,kJ;\ \Delta S° = -275\,J/K$

(c) $N_2F_4(g) \longrightarrow 2\,NF_2(g)$
$\Delta H° = 85\,kJ;\ \Delta S° = 198\,J/K$

19.66 From the values given for $\Delta H°$ and $\Delta S°$, calculate $\Delta G°$ for each of the following reactions at 298 K. If the reaction is not spontaneous under standard conditions at 298 K, at what temperature (if any) would the reaction become spontaneous?

(a) $2\,PbS(s) + 3\,O_2(g) \longrightarrow 2\,PbO(s) + 2\,SO_2(g)$
$\Delta H° = -844\,kJ;\ \Delta S° = -165\,J/K$

(b) $2\,POCl_3(g) \longrightarrow 2\,PCl_3(g) + O_2(g)$
$\Delta H° = 572\,kJ;\ \Delta S° = 179\,J/K$

19.67 A particular constant-pressure reaction is barely spontaneous at 390 K. The enthalpy change for the reaction is +23.7 kJ. Estimate ΔS for the reaction.

19.68 A certain constant-pressure reaction is barely nonspontaneous at 45 °C. The entropy change for the reaction is 72 J/K. Estimate ΔH.

19.69 For a particular reaction, $\Delta H = -32$ kJ and $\Delta S = -98$ J/K. Assume that ΔH and ΔS do not vary with temperature. (a) At what temperature will the reaction have $\Delta G = 0$? (b) If T is increased from that in part (a), will the reaction be spontaneous or nonspontaneous?

19.70 Reactions in which a substance decomposes by losing CO are called *decarbonylation* reactions. The decarbonylation of acetic acid proceeds according to:

$$CH_3COOH(l) \longrightarrow CH_3OH(g) + CO(g)$$

By using data from Appendix C, calculate the minimum temperature at which this process will be spontaneous under standard conditions. Assume that $\Delta H°$ and $\Delta S°$ do not vary with temperature.

19.71 Consider the following reaction between oxides of nitrogen:

$$NO_2(g) + N_2O(g) \longrightarrow 3\,NO(g)$$

(a) Use data in Appendix C to predict how ΔG for the reaction varies with increasing temperature. (b) Calculate ΔG at 800 K, assuming that $\Delta H°$ and $\Delta S°$ do not change with temperature. Under standard conditions is the reaction spontaneous at 800 K? (c) Calculate ΔG at 1000 K. Is the reaction spontaneous under standard conditions at this temperature?

19.72 Methanol (CH_3OH) can be made by the controlled oxidation of methane:

$$CH_4(g) + \tfrac{1}{2}O_2(g) \longrightarrow CH_3OH(g)$$

(a) Use data in Appendix C to calculate $\Delta H°$ and $\Delta S°$ for this reaction. (b) Will ΔG for the reaction increase, decrease, or stay unchanged with increasing temperature? (c) Calculate $\Delta G°$ at 298 K. Under standard conditions, is the reaction spontaneous at this temperature? (d) Is there a temperature at which the reaction would be at equilibrium under standard conditions and that is low enough so that the compounds involved are likely to be stable?

19.73 (a) Use data in Appendix C to estimate the boiling point of benzene, $C_6H_6(l)$. (b) Use a reference source, such as the *CRC Handbook of Chemistry and Physics*, to find the experimental boiling point of benzene. How do you explain any deviation between your answer in part (a) and the experimental value?

19.74 (a) Using data in Appendix C, estimate the temperature at which the free-energy change for the transformation from $I_2(s)$ to $I_2(g)$ is zero. What assumptions must you make in arriving at this estimate? (b) Use a reference source, such as Web Elements (www.webelements.com), to find the experimental melting and boiling points of I_2. (c) Which of the values in part (b) is closer to the value you obtained in part (a)? Can you explain why this is so?

19.75 Acetylene gas, $C_2H_2(g)$, is used in welding. (a) Write a balanced equation for the combustion of acetylene gas to $CO_2(g)$ and $H_2O(l)$. (b) How much heat is produced in burning 1 mol of C_2H_2 under standard conditions if both reactants and products are brought to 298 K? (c) What is the maximum amount of useful work that can be accomplished under standard conditions by this reaction?

19.76 The fuel in high-efficiency natural gas vehicles consists primarily of methane (CH_4). **(a)** How much heat is produced in burning 1 mol of $CH_4(g)$ under standard conditions if reactants and products are brought to 298 K and $H_2O(l)$ is formed? **(b)** What is the maximum amount of useful work that can be accomplished under standard conditions by this system?

Free Energy and Equilibrium (Section 19.7)

19.77 Indicate whether ΔG increases, decreases, or stays the same for each of the following reactions as the partial pressure of O_2 is increased:

(a) $2\,CO(g) + O_2(g) \longrightarrow 2\,CO_2(g)$

(b) $2\,H_2O_2(l) \longrightarrow 2\,H_2O(l) + O_2(g)$

(c) $2\,KClO_3(s) \longrightarrow 2\,KCl(s) + 3\,O_2(g)$

19.78 Indicate whether ΔG increases, decreases, or does not change when the partial pressure of H_2 is increased in each of the following reactions:

(a) $N_2(g) + 3\,H_2(g) \longrightarrow 2\,NH_3(g)$

(b) $2\,HBr(g) \longrightarrow H_2(g) + Br_2(g)$

(c) $2\,H_2(g) + C_2H_2(g) \longrightarrow C_2H_6(g)$

19.79 Consider the reaction $2\,NO_2(g) \longrightarrow N_2O_4(g)$. **(a)** Using data from Appendix C, calculate $\Delta G°$ at 298 K. **(b)** Calculate ΔG at 298 K if the partial pressures of NO_2 and N_2O_4 are 0.40 atm and 1.60 atm, respectively.

19.80 Consider the reaction $3\,CH_4(g) \longrightarrow C_3H_8(g) + 2\,H_2(g)$. **(a)** Using data from Appendix C, calculate $\Delta G°$ at 298 K. **(b)** Calculate ΔG at 298 K if the reaction mixture consists of 40.0 atm of CH_4, 0.0100 atm of $C_3H_8(g)$, and 0.0180 atm of H_2.

19.81 Use data from Appendix C to calculate the equilibrium constant, K, and $\Delta G°$ at 298 K for each of the following reactions:

(a) $H_2(g) + I_2(g) \rightleftharpoons 2\,HI(g)$

(b) $C_2H_5OH(g) \rightleftharpoons C_2H_4(g) + H_2O(g)$

(c) $3\,C_2H_2(g) \rightleftharpoons C_6H_6(g)$

19.82 Using data from Appendix C, write the equilibrium-constant expression and calculate the value of the equilibrium constant and the free-energy change for these reactions at 298 K:

(a) $NaHCO_3(s) \rightleftharpoons NaOH(s) + CO_2(g)$

(b) $2\,HBr(g) + Cl_2(g) \rightleftharpoons 2\,HCl(g) + Br_2(g)$

(c) $2\,SO_2(g) + O_2(g) \rightleftharpoons 2\,SO_3(g)$

19.83 Consider the decomposition of barium carbonate:

$$BaCO_3(s) \rightleftharpoons BaO(s) + CO_2(g)$$

Using data from Appendix C, calculate the equilibrium pressure of CO_2 at **(a)** 298 K and **(b)** 1100 K.

19.84 Consider the reaction

$$PbCO_3(s) \rightleftharpoons PbO(s) + CO_2(g)$$

Using data in Appendix C, calculate the equilibrium pressure of CO_2 in the system at **(a)** 400 °C and **(b)** 180 °C.

19.85 The value of K_a for nitrous acid (HNO_2) at 25 °C is given in Appendix D. **(a)** Write the chemical equation for the equilibrium that corresponds to K_a. **(b)** By using the value of K_a, calculate $\Delta G°$ for the dissociation of nitrous acid in aqueous solution. **(c)** What is the value of ΔG at equilibrium? **(d)** What is the value of ΔG when $[H^+] = 5.0 \times 10^{-2}\,M$, $[NO_2{}^-] = 6.0 \times 10^{-4}\,M$, and $[HNO_2] = 0.20\,M$?

19.86 The K_b for methylamine (CH_3NH_2) at 25 °C is given in Appendix D. **(a)** Write the chemical equation for the equilibrium that corresponds to K_b. **(b)** By using the value of K_b, calculate $\Delta G°$ for the equilibrium in part (a). **(c)** What is the value of ΔG at equilibrium? **(d)** What is the value of ΔG when $[H^+] = 6.7 \times 10^{-9}\,M$, $[CH_3NH_3{}^+] = 2.4 \times 10^{-3}\,M$, and $[CH_3NH_2] = 0.098\,M$?

Additional Exercises

19.87 **(a)** Which of the thermodynamic quantities T, E, q, w, and S are state functions? **(b)** Which depend on the path taken from one state to another? **(c)** How many *reversible* paths are there between two states of a system? **(d)** For a reversible isothermal process, write an expression for ΔE in terms of q and w and an expression for ΔS in terms of q and T.

19.88 Indicate whether each of the following statements is true or false. If it is false, correct it. **(a)** The feasibility of manufacturing NH_3 from N_2 and H_2 depends entirely on the value of ΔH for the process $N_2(g) + 3\,H_2(g) \longrightarrow 2\,NH_3(g)$. **(b)** The reaction of $Na(s)$ with $Cl_2(g)$ to form $NaCl(s)$ is a spontaneous process. **(c)** A spontaneous process can in principle be conducted reversibly. **(d)** Spontaneous processes in general require that work be done to force them to proceed. **(e)** Spontaneous processes are those that are exothermic and that lead to a higher degree of order in the system.

19.89 For each of the following processes, indicate whether the signs of ΔS and ΔH are expected to be positive, negative, or about zero. **(a)** A solid sublimes. **(b)** The temperature of a sample of Co(s) is lowered from 60 °C to 25 °C. **(c)** Ethyl alcohol evaporates from a beaker. **(d)** A diatomic molecule dissociates into atoms. **(e)** A piece of charcoal is combusted to form $CO_2(g)$ and $H_2O(g)$.

19.90 The reaction $2\,Mg(s) + O_2(g) \longrightarrow 2\,MgO(s)$ is highly spontaneous. A classmate calculates the entropy change for this reaction and obtains a large negative value for $\Delta S°$. Did your classmate make a mistake in the calculation? Explain.

[19.91] Suppose four gas molecules are placed in the left flask in Figure 19.6(a). Initially, the right flask is evacuated and the stopcock is closed. **(a)** After the stopcock is opened, how many different arrangements of the molecules are possible? **(b)** How many of the arrangements from part (a) have all the molecules in the left flask? **(c)** How does the answer to part (b) explain the spontaneous expansion of the gas?

[19.92] Consider a system that consists of two standard playing dice, with the state of the system defined by the sum of the values shown on the top faces. **(a)** The two arrangements of top faces shown here can be viewed as two possible microstates of the system. Explain. **(b)** To which state does each microstate correspond? **(c)** How many possible states are there for the system? **(d)** Which state or states have the highest entropy? Explain. **(e)** Which state or states have the lowest entropy? Explain. **(f)** Calculate the absolute entropy of the two-dice system.

19.93 Ammonium nitrate dissolves spontaneously and endothermally in water at room temperature. What can you deduce about the sign of ΔS for this solution process?

[19.94] A standard air conditioner involves a *refrigerant* that is typically now a fluorinated hydrocarbon, such as CH_2F_2. An air-conditioner refrigerant has the property that it readily vaporizes at atmospheric pressure and is easily compressed to its liquid phase under increased pressure. The operation of an air conditioner can be thought of as a closed system made up of the refrigerant going through the two stages shown here (the air circulation is not shown in this diagram).

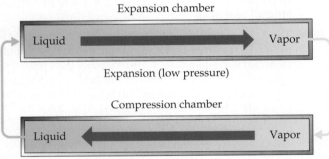

Expansion chamber

Expansion (low pressure)

Compression chamber

Compression (high pressure)

During *expansion*, the liquid refrigerant is released into an expansion chamber at low pressure, where it vaporizes. The vapor then undergoes *compression* at high pressure back to its liquid phase in a compression chamber. **(a)** What is the sign of q for the expansion? **(b)** What is the sign of q for the compression? **(c)** In a central air-conditioning system, one chamber is inside the home and the other is outside. Which chamber is where, and why? **(d)** Imagine that a sample of liquid refrigerant undergoes expansion followed by compression, so that it is back to its original state. Would you expect that to be a reversible process? **(e)** Suppose that a house and its exterior are both initially at 31 °C. Some time after the air conditioner is turned on, the house is cooled to 24 °C. Is this process spontaneous or nonspontaneous?

[19.95] *Trouton's rule* states that for many liquids at their normal boiling points, the standard molar entropy of vaporization is about 88 J/mol-K. **(a)** Estimate the normal boiling point of bromine, Br_2, by determining ΔH°_{vap} for Br_2 using data from Appendix C. Assume that ΔH°_{vap} remains constant with temperature and that Trouton's rule holds. **(b)** Look

up the normal boiling point of Br_2 in a chemistry handbook or at the WebElements Web site (www.webelements.com) and compare it to your calculation. What are the possible sources of error, or incorrect assumptions, in the calculation?

[19.96] For the majority of the compounds listed in Appendix C, the value of ΔG°_f is more positive (or less negative) than the value of ΔH°_f. **(a)** Explain this observation, using $NH_3(g)$, $CCl_4(l)$, and $KNO_3(s)$ as examples. **(b)** An exception to this observation is $CO(g)$. Explain the trend in the ΔH°_f and ΔG°_f values for this molecule.

19.97 Consider the following three reactions:
(i) $Ti(s) + 2 Cl_2(g) \longrightarrow TiCl_4(g)$
(ii) $C_2H_6(g) + 7 Cl_2(g) \longrightarrow 2 CCl_4(g) + 6 HCl(g)$
(iii) $BaO(s) + CO_2(g) \longrightarrow BaCO_3(s)$

(a) For each of the reactions, use data in Appendix C to calculate ΔH°, ΔG°, K, and ΔS° at 25 °C. **(b)** Which of these reactions are spontaneous under standard conditions at 25 °C? **(c)** For each of the reactions, predict the manner in which the change in free energy varies with an increase in temperature.

19.98 Using the data in Appendix C and given the pressures listed, calculate K_p and ΔG for each of the following reactions:
(a) $N_2(g) + 3 H_2(g) \longrightarrow 2 NH_3(g)$
$P_{N_2} = 2.6$ atm, $P_{H_2} = 5.9$ atm, $P_{NH_3} = 1.2$ atm
(b) $2 N_2H_4(g) + 2 NO_2(g) \longrightarrow 3 N_2(g) + 4 H_2O(g)$
$P_{N_2H_4} = P_{NO_2} = 5.0 \times 10^{-2}$ atm,
$P_{N_2} = 0.5$ atm, $P_{H_2O} = 0.3$ atm
(c) $N_2H_4(g) \longrightarrow N_2(g) + 2 H_2(g)$
$P_{N_2H_4} = 0.5$ atm, $P_{N_2} = 1.5$ atm, $P_{H_2} = 2.5$ atm

19.99 **(a)** For each of the following reactions, predict the sign of ΔH° and ΔS° without doing any calculations. **(b)** Based on your general chemical knowledge, predict which of these reactions will have $K > 1$. **(c)** In each case indicate whether K should increase or decrease with increasing temperature.
(i) $2 Mg(s) + O_2(g) \rightleftharpoons 2 MgO(s)$
(ii) $2 KI(s) \rightleftharpoons 2 K(g) + I_2(g)$
(iii) $Na_2(g) \rightleftharpoons 2 Na(g)$
(iv) $2 V_2O_5(s) \rightleftharpoons 4 V(s) + 5 O_2(g)$

19.100 Acetic acid can be manufactured by combining methanol with carbon monoxide, an example of a *carbonylation* reaction:

$$CH_3OH(l) + CO(g) \longrightarrow CH_3COOH(l)$$

(a) Calculate the equilibrium constant for the reaction at 25 °C. **(b)** Industrially, this reaction is run at temperatures above 25 °C. Will an increase in temperature produce an increase or decrease in the mole fraction of acetic acid at equilibrium? Why are elevated temperatures used? **(c)** At what temperature will this reaction have an equilibrium constant equal to 1? (You may assume that ΔH° and ΔS° are temperature-independent, and you may ignore any phase changes that might occur.)

19.101 The oxidation of glucose ($C_6H_{12}O_6$) in body tissue produces CO_2 and H_2O. In contrast, anaerobic decomposition, which occurs during fermentation, produces ethanol (C_2H_5OH)

and CO_2. **(a)** Using data given in Appendix C, compare the equilibrium constants for the following reactions:

$$C_6H_{12}O_6(s) + 6\,O_2(g) \rightleftharpoons 6\,CO_2(g) + 6\,H_2O(l)$$

$$C_6H_{12}O_6(s) \rightleftharpoons 2\,C_2H_5OH(l) + 2\,CO_2(g)$$

(b) Compare the maximum work that can be obtained from these processes under standard conditions.

[**19.102**] The conversion of natural gas, which is mostly methane, into products that contain two or more carbon atoms, such as ethane (C_2H_6), is a very important industrial chemical process. In principle, methane can be converted into ethane and hydrogen:

$$2\,CH_4(g) \longrightarrow C_2H_6(g) + H_2(g)$$

In practice, this reaction is carried out in the presence of oxygen:

$$2\,CH_4(g) + \tfrac{1}{2}O_2(g) \longrightarrow C_2H_6(g) + H_2O(g)$$

(a) Using the data in Appendix C, calculate K for these reactions at 25 °C and 500 °C. **(b)** Is the difference in $\Delta G°$ for the two reactions due primarily to the enthalpy term (ΔH) or the entropy term ($-T\Delta S$)? **(c)** Explain how the preceding reactions are an example of driving a nonspontaneous reaction, as discussed in the "Chemistry and Life" box in Section 19.7. **(d)** The reaction of CH_4 and O_2 to form C_2H_6 and H_2O must be carried out carefully to avoid a competing reaction. What is the most likely competing reaction?

[**19.103**] Cells use the hydrolysis of adenosine triphosphate (ATP) as a source of energy (Figure 19.16). The conversion of ATP to ADP has a standard free-energy change of -30.5 kJ/mol. If all the free energy from the metabolism of glucose,

$$C_6H_{12}O_6(s) + 6\,O_2(g) \longrightarrow 6\,CO_2(g) + 6\,H_2O(l)$$

goes into the conversion of ADP to ATP, how many moles of ATP can be produced for each mole of glucose?

[**19.104**] The potassium-ion concentration in blood plasma is about 5.0×10^{-3} M, whereas the concentration in muscle-cell fluid is much greater (0.15 M). The plasma and intracellular fluid are separated by the cell membrane, which we assume is permeable only to K^+. **(a)** What is ΔG for the transfer of 1 mol of K^+ from blood plasma to the cellular fluid at body temperature 37 °C? **(b)** What is the minimum amount of work that must be used to transfer this K^+?

[**19.105**] One way to derive Equation 19.3 depends on the observation that at constant T the number of ways, W, of arranging m ideal-gas particles in a volume V is proportional to the volume raised to the m power:

$$W \propto V^m$$

Use this relationship and Boltzmann's relationship between entropy and number of arrangements (Equation 19.5) to derive the equation for the entropy change for the isothermal expansion or compression of n moles of an ideal gas.

[**19.106**] About 86% of the world's electrical energy is produced by using steam turbines, a form of heat engine. In his analysis of an ideal heat engine, Sadi Carnot concluded that the maximum possible efficiency is defined by the total work that could be done by the engine, divided by the quantity of heat available to do the work (for example, from hot steam produced by combustion of a fuel such as coal or methane). This efficiency is given by the ratio $(T_{high} - T_{low})/T_{high}$, where T_{high} is the temperature of the heat going into the engine and T_{low} is that of the heat leaving the engine. **(a)** What is the maximum possible efficiency of a heat engine operating between an input temperature of 700 K and an exit temperature of 288 K? **(b)** Why is it important that electrical power plants be located near bodies of relatively cool water? **(c)** Under what conditions could a heat engine operate at or near 100% efficiency? **(d)** It is often said that if the energy of combustion of a fuel such as methane were captured in an electrical fuel cell instead of by burning the fuel in a heat engine, a greater fraction of the energy could be put to useful work. Make a qualitative drawing like that in Figure 5.10 (p. 175) that illustrates the fact that in principle the fuel cell route will produce more useful work than the heat engine route from combustion of methane.

Integrative Exercises

19.107 Most liquids follow Trouton's rule (see Exercise 19.95), which states that the molar entropy of vaporization is approximately 88 ± 5 J/mol-K. The normal boiling points and enthalpies of vaporization of several organic liquids are as follows:

Substance	Normal Boiling Point (°C)	ΔH_{vap} (kJ/mol)
Acetone, $(CH_3)_2CO$	56.1	29.1
Dimethyl ether, $(CH_3)_2O$	-24.8	21.5
Ethanol, C_2H_5OH	78.4	38.6
Octane, C_8H_{18}	125.6	34.4
Pyridine, C_5H_5N	115.3	35.1

(a) Calculate ΔS_{vap} for each of the liquids. Do all the liquids obey Trouton's rule? **(b)** With reference to intermolecular forces (Section 11.2), can you explain any exceptions to the rule? **(c)** Would you expect water to obey Trouton's rule? By using data in Appendix B, check the accuracy of your conclusion. **(d)** Chlorobenzene (C_6H_5Cl) boils at 131.8 °C. Use Trouton's rule to estimate ΔH_{vap} for this substance.

19.108 In chemical kinetics the *entropy of activation* is the entropy change for the process in which the reactants reach the activated complex. The entropy of activation for bimolecular processes is usually negative. Explain this observation with reference to Figure 14.15.

19.109 At what temperatures is the following reaction, the reduction of magnetite by graphite to elemental iron, spontaneous?

$$Fe_3O_4(s) + 2\,C\,(s, graphite) \longrightarrow 2\,CO_2(g) + 3\,Fe\,(s)$$

19.110 The following processes were all discussed in Chapter 18, "Chemistry of the Environment." Estimate whether the entropy of the system increases or decreases during each process: **(a)** photodissociation of $O_2(g)$, **(b)** formation of ozone from oxygen molecules and oxygen atoms, **(c)** diffusion of CFCs into the stratosphere, **(d)** desalination of water by reverse osmosis.

[19.111] An ice cube with a mass of 20 g at $-20\,°C$ (typical freezer temperature) is dropped into a cup that holds 500 mL of hot water, initially at $83\,°C$. What is the final temperature in the cup? The density of liquid water is $1.00\,g/mL$; the specific heat capacity of ice is $2.03\,J/g\text{-}C$; the specific heat capacity of liquid water is $4.184\,J/g\text{-}C$; the enthalpy of fusion of water is $6.01\,kJ/mol$.

19.112 Carbon disulfide (CS_2) is a toxic, highly flammable substance. The following thermodynamic data are available for $CS_2(l)$ and $CS_2(g)$ at 298 K:

	$\Delta H_f^\circ(kJ/mol)$	$\Delta G_f^\circ(kJ/mol)$
$CS_2(l)$	89.7	65.3
$CS_2(g)$	117.4	67.2

(a) Draw the Lewis structure of the molecule. What do you predict for the bond order of the C—S bonds? **(b)** Use the VSEPR method to predict the structure of the CS_2 molecule. **(c)** Liquid CS_2 burns in O_2 with a blue flame, forming $CO_2(g)$ and $SO_2(g)$. Write a balanced equation for this reaction. **(d)** Using the data in the preceding table and in Appendix C, calculate ΔH° and ΔG° for the reaction in part (c). Is the reaction exothermic? Is it spontaneous at 298 K? **(e)** Use the data in the table to calculate ΔS° at 298 K for the vaporization of $CS_2(l)$. Is the sign of ΔS° as you would expect for a vaporization? **(f)** Using data in the table and your answer to part (e), estimate the boiling point of $CS_2(l)$. Do you predict that the substance will be a liquid or a gas at 298 K and 1 atm?

[19.113] The following data compare the standard enthalpies and free energies of formation of some crystalline ionic substances and aqueous solutions of the substances:

Substance	$\Delta H_f^\circ(kJ/mol)$	$\Delta G_f^\circ(kJ/mol)$
$AgNO_3(s)$	-124.4	-33.4
$AgNO_3(aq)$	-101.7	-34.2
$MgSO_4(s)$	-1283.7	-1169.6
$MgSO_4(aq)$	-1374.8	-1198.4

Write the formation reaction for $AgNO_3(s)$. Based on this reaction, do you expect the entropy of the system to increase or decrease upon the formation of $AgNO_3(s)$? **(b)** Use ΔH_f° and ΔG_f° of $AgNO_3(s)$ to determine the entropy change upon formation of the substance. Is your answer consistent with your reasoning in part (a)? **(c)** Is dissolving $AgNO_3$ in water an exothermic or endothermic process? What about dissolving $MgSO_4$ in water? **(d)** For both $AgNO_3$ and $MgSO_4$, use the data to calculate the entropy change when the solid is dissolved in water. **(e)** Discuss the results from part (d) with reference to material presented in this chapter and in the "A Closer Look" box on page 820.

[19.114] Consider the following equilibrium:

$$N_2O_4(g) \rightleftharpoons 2\,NO_2(g)$$

Thermodynamic data on these gases are given in Appendix C. You may assume that ΔH° and ΔS° do not vary with temperature. **(a)** At what temperature will an equilibrium mixture contain equal amounts of the two gases? **(b)** At what temperature will an equilibrium mixture of 1 atm total pressure contain twice as much NO_2 as N_2O_4? **(c)** At what temperature will an equilibrium mixture of 10 atm total pressure contain twice as much NO_2 as N_2O_4? **(d)** Rationalize the results from parts (b) and (c) by using Le Châtelier's principle. [Section 15.7]

[19.115] The reaction

$$SO_2(g) + 2\,H_2S(g) \rightleftharpoons 3\,S(s) + 2\,H_2O(g)$$

is the basis of a suggested method for removal of SO_2 from power-plant stack gases. The standard free energy of each substance is given in Appendix C. **(a)** What is the equilibrium constant for the reaction at 298 K? **(b)** In principle, is this reaction a feasible method of removing SO_2? **(c)** If $P_{SO_2} = P_{H_2S}$ and the vapor pressure of water is 25 torr, calculate the equilibrium SO_2 pressure in the system at 298 K. **(d)** Would you expect the process to be more or less effective at higher temperatures?

19.116 When most elastomeric polymers (e.g., a rubber band) are stretched, the molecules become more ordered, as illustrated here:

Suppose you stretch a rubber band. **(a)** Do you expect the entropy of the system to increase or decrease? **(b)** If the rubber band were stretched isothermally, would heat need to be absorbed or emitted to maintain constant temperature? **(c)** Try this experiment: Stretch a rubber band and wait a moment. Then place the stretched rubber band on your upper lip, and let it return suddenly to its unstretched state (remember to keep holding on!). What do you observe? Are your observations consistent with your answer to part (b)?

Design an Experiment

You are measuring the equilibrium constant for a drug candidate binding to its DNA target over a series of different temperatures. You chose your drug candidate based on computer-aided molecular modeling, which indicates that the drug molecule likely would make many hydrogen bonds and favorable dipole–dipole interactions with the DNA site. You perform a set of experiments in buffer solution for the drug–DNA complex and generate a table of K's at different T's. (**a**) Derive an equation that relates equilibrium constant to standard enthalpy and entropy changes. (*Hint: equilibrium constant, enthalpy and entropy are all related to free energy*). (**b**) Show how you can graph your K and T data to calculate the standard entropy and enthalpy changes for the drug candidate + DNA binding interaction. (**c**) You are surprised to learn that the enthalpy change for the binding reaction is close to zero, and the entropy change is large and positive. Suggest an explanation and design an experiment to test it. (*Hint: think about water and ions*). (**d**) You try another drug candidate with the DNA target and find that this drug candidate has a large negative enthalpy change upon DNA binding, and the entropy change is small and positive. Suggest an explanation, at the molecular level, and design an experiment to test your hypothesis.

20 Electrochemistry

The electricity that powers much of modern society has many favorable characteristics, but it has a serious shortcoming: it cannot easily be stored. The electricity that flows into power company lines is consumed as it is generated, but for many other applications, stored electrical energy is needed.

In such cases electrical energy is converted into chemical energy, which can be stored and is portable, and then converted back to electricity when needed. Batteries are the most familiar devices for converting between electrical and chemical energies. Objects like laptop computers, cell phones, pacemakers, portable music players, cordless power tools, wristwatches, and countless other devices rely on batteries to provide the electricity needed for operation. A considerable amount of effort is currently being focused on research and development of new batteries, such as the one shown at right, particularly for powering electric vehicles. For that application new batteries are needed that are lighter, can charge faster, deliver more power, and have longer lifetimes. Cost and toxicity of the materials used in the battery are also important. At the heart of such developments are the oxidation–reduction reactions that power batteries.

As discussed in Chapter 4, *oxidation* is the loss of electrons in a chemical reaction, and *reduction* is the gain of electrons. ∞ (Section 4.4) Thus, oxidation–reduction (redox) reactions occur when electrons are transferred from an atom that is oxidized to an atom that is reduced. Redox reactions are involved not only in the operation of batteries but also in a wide variety of important natural processes, including the rusting of iron, the browning of foods, and the respiration of animals. **Electrochemistry** is the study of the relationships between electricity and chemical reactions. It includes the study of both spontaneous and nonspontaneous processes.

▶ **AN ADVANCED LI-ION BATTERY** pack built for use in Mercedes S-class hybrid automobiles.

WHAT'S AHEAD ▶

20.1 OXIDATION STATES AND OXIDATION–REDUCTION REACTIONS We review oxidation states and *oxidation–reduction (redox) reactions*.

20.2 BALANCING REDOX EQUATIONS We learn how to balance redox equations using the method of *half-reactions*.

20.3 VOLTAIC CELLS We consider *voltaic cells*, which produce electricity from spontaneous redox reactions. Solid electrodes serve as the surfaces at which oxidation and reduction take place. The electrode where oxidation occurs is the *anode*, and the electrode where reduction occurs is the *cathode*.

20.4 CELL POTENTIALS UNDER STANDARD CONDITIONS We see that an important characteristic of a voltaic cell is its *cell potential*, which is the difference in the electrical potentials at the two electrodes and is measured in units of volts. Half-cell potentials are tabulated for reduction half-reactions under standard conditions (*standard reduction potentials*).

20.5 FREE ENERGY AND REDOX REACTIONS We relate the Gibbs free energy, $\Delta G°$, to cell potential.

20.6 CELL POTENTIALS UNDER NONSTANDARD CONDITIONS We calculate cell potentials under nonstandard conditions using standard cell potentials and the Nernst equation.

20.7 BATTERIES AND FUEL CELLS We describe batteries and fuel cells, which are commercially important energy sources that use electrochemical reactions.

20.8 CORROSION We discuss *corrosion*, a spontaneous electrochemical process involving metals.

20.9 ELECTROLYSIS Finally, we focus on nonspontaneous redox reactions, examining *electrolytic cells*, which use electricity to perform chemical reactions.

20.1 | Oxidation States and Oxidation–Reduction Reactions

We determine whether a given chemical reaction is an oxidation–reduction reaction by keeping track of the *oxidation numbers* (*oxidation states*) of the elements involved in the reaction. ∞ (Section 4.4) This procedure identifies whether the oxidation number changes for any elements involved in the reaction. For example, consider the reaction that occurs spontaneously when zinc metal is added to a strong acid (▼ Figure 20.1):

$$Zn(s) + 2H^+(aq) \longrightarrow Zn^{2+}(aq) + H_2(g) \qquad [20.1]$$

The chemical equation for this reaction can be written as

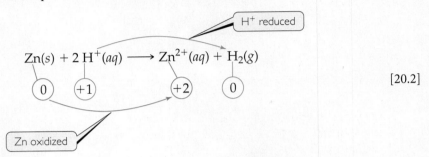

$$[20.2]$$

The oxidation numbers below the equation show that the oxidation number of Zn changes from 0 to +2 while that of H changes from +1 to 0. Thus, this is an oxidation–reduction reaction. Electrons are transferred from zinc atoms to hydrogen ions; Zn is oxidized and H^+ is reduced.

▲ GO FIGURE

Explain (a) the vigorous bubbling in the beaker on the right and (b) the formation of steam above that beaker.

▲ Figure 20.1 **Oxidation of zinc by hydrochloric acid.**

In a reaction such as Equation 20.2, a clear transfer of electrons occurs. In some reactions, however, the oxidation numbers change, but we cannot say that any substance literally gains or loses electrons. For example, in the combustion of hydrogen gas,

$$2\,H_2(g) + O_2(g) \longrightarrow 2\,H_2O(g)$$

[20.3]

with oxidation states 0, 0, +1, −2

hydrogen is oxidized from the 0 to the +1 oxidation state and oxygen is reduced from the 0 to the −2 oxidation state, indicating that Equation 20.3 is an oxidation–reduction reaction. Water is not an ionic substance, therefore, a complete transfer of electrons from hydrogen to oxygen does not occur as water is formed. So, while keeping track of oxidation states offers a convenient form of "bookkeeping," you should not generally equate the oxidation state of an atom with its actual charge in a chemical compound. ∞ (Section 8.5, "Oxidation Numbers, Formal Charges, and Actual Partial Charges")

Give It Some Thought

What are the oxidation numbers of the elements in the nitrite ion, NO_2^-?

In any redox reaction, both oxidation and reduction must occur. If one substance is oxidized, another must be reduced. The substance that makes it possible for another substance to be oxidized is called either the **oxidizing agent** or the **oxidant**. The oxidizing agent acquires electrons from the other substance and so is itself reduced. A **reducing agent**, or **reductant**, is a substance that gives up electrons, thereby causing another substance to be reduced. The reducing agent is therefore oxidized in the process. In Equation 20.2, $H^+(aq)$, the species that is reduced, is the oxidizing agent and $Zn(s)$, the species that is oxidized, is the reducing agent.

SAMPLE EXERCISE 20.1 | Identifying Oxidizing and Reducing Agents

The nickel–cadmium (nicad) battery uses the following redox reaction to generate electricity:

$$Cd(s) + NiO_2(s) + 2H_2O(l) \longrightarrow Cd(OH)_2(s) + Ni(OH)_2(s)$$

Identify the substances that are oxidized and reduced, and indicate which is the oxidizing agent and which is the reducing agent.

SOLUTION

Analyze We are given a redox equation and asked to identify the substance oxidized and the substance reduced and to label the oxidizing agent and the reducing agent.

Plan First, we use the rules outlined earlier ∞ (Section 4.4) to assign oxidation states, or numbers, to all the atoms and determine which elements change oxidation state. Second, we apply the definitions of oxidation and reduction.

Solve $Cd(s) + NiO_2(s) + 2H_2O(l) \longrightarrow Cd(OH)_2(s) + Ni(OH)_2(s)$

with oxidation states: Cd 0; Ni +4, O −2; H +1, O −2; Cd +2, O −2, H +1; Ni +2, O −2, H +1

The oxidation state of Cd increases from 0 to +2, and that of Ni decreases from +4 to +2. Thus, the Cd atom is oxidized (loses electrons) and is the reducing agent. The oxidation state of Ni decreases as NiO_2 is converted into $Ni(OH)_2$. Thus, NiO_2 is reduced (gains electrons) and is the oxidizing agent.

Comment A common mnemonic for remembering oxidation and reduction is "LEO the lion says GER": *l*osing *e*lectrons is *o*xidation; *g*aining *e*lectrons is *r*eduction.

Practice Exercise 1

What is the reducing agent in the following reaction?

$$2\,Br^-(aq) + H_2O_2(aq) + 2\,H^+(aq) \longrightarrow Br_2(aq) + 2\,H_2O(l)$$

(a) $Br^-(aq)$ **(b)** $H_2O_2(aq)$ **(c)** $H^+(aq)$ **(d)** $Br_2(aq)$ **(e)** $Na^+(aq)$

Practice Exercise 2

Identify the oxidizing and reducing agents in the reaction

$$2\,H_2O(l) + Al(s) + MnO_4^-(aq) \longrightarrow Al(OH)_4^-(aq) + MnO_2(s)$$

20.2 | Balancing Redox Equations

Whenever we balance a chemical equation, we must obey the law of conservation of mass: The amount of each element must be the same on both sides of the equation. (Atoms are neither created nor destroyed in any chemical reaction.) As we balance oxidation–reduction reactions, there is an additional requirement: The gains and losses of electrons must be balanced. If a substance loses a certain number of electrons during a reaction, another substance must gain that same number of electrons. (Electrons are neither created nor destroyed in any chemical reaction.)

In many simple chemical equations, such as Equation 20.2, balancing the electrons is handled "automatically"—that is, we balance the equation without explicitly accounting for the transfer of electrons. Many redox equations are more complex than Equation 20.2, however, and cannot be balanced easily without taking into account the number of electrons lost and gained. In this section, we examine the *method of half-reactions*, a systematic procedure for balancing redox equations.

Half-Reactions

Although oxidation and reduction must take place simultaneously, it is often convenient to consider them as separate processes. For example, the oxidation of Sn^{2+} by Fe^{3+},

$$Sn^{2+}(aq) + 2\ Fe^{3+}(aq) \longrightarrow Sn^{4+}(aq) + 2\ Fe^{2+}(aq)$$

can be considered as consisting of two processes: oxidation of Sn^{2+} and reduction of Fe^{3+}:

$$\text{\textit{Oxidation}:} \qquad Sn^{2+}(aq) \longrightarrow Sn^{4+}(aq) + 2e^{-} \qquad [20.4]$$

$$\text{\textit{Reduction}:} \quad 2\ Fe^{3+}(aq) + 2e^{-} \longrightarrow 2\ Fe^{2+}(aq) \qquad [20.5]$$

Notice that electrons are shown as products in the oxidation process and as reactants in the reduction process.

Equations that show either oxidation or reduction alone, such as Equations 20.4 and 20.5, are called **half-reactions**. In the overall redox reaction, the number of electrons lost in the oxidation half-reaction must equal the number of electrons gained in the reduction half-reaction. When this condition is met and each half-reaction is balanced, the electrons on the two sides cancel when the two half-reactions are added to give the balanced oxidation–reduction equation.

Balancing Equations by the Method of Half-Reactions

In the half-reaction method, we usually begin with a "skeleton" ionic equation showing only the substances undergoing oxidation and reduction. In such cases, we assign oxidation numbers only when we are unsure whether the reaction involves oxidation–reduction. We will find that H^{+} (for acidic solutions), OH^{-} (for basic solutions), and H_2O are often involved as reactants or products in redox reactions. Unless H^{+}, OH^{-}, or H_2O is being oxidized or reduced, these species do not appear in the skeleton equation. Their presence, however, can be deduced as we balance the equation.

For balancing a redox reaction that occurs *in acidic aqueous* solution, the procedure is as follows:

1. Divide the equation into one oxidation half-reaction and one reduction half-reaction.

2. Balance each half-reaction.

   (a) First, balance elements other than H and O.

   (b) Next, balance O atoms by adding H_2O as needed.

   (c) Then balance H atoms by adding H^{+} as needed.

   (d) Finally, balance charge by adding e^{-} as needed.

This specific sequence (a)–(d) is important, and it is summarized in the diagram in the margin. At this point, you can check whether the number of electrons in each half-reaction corresponds to the changes in oxidation state.

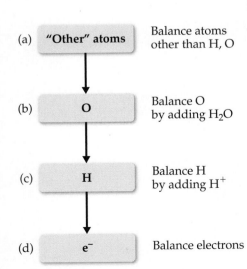

(a) "Other" atoms — Balance atoms other than H, O

(b) O — Balance O by adding H_2O

(c) H — Balance H by adding H^{+}

(d) e^{-} — Balance electrons

3. Multiply half-reactions by integers as needed to make the number of electrons lost in the oxidation half-reaction equal the number of electrons gained in the reduction half-reaction.

4. Add half-reactions and, if possible, simplify by canceling species appearing on both sides of the combined equation.

5. Check to make sure that atoms and charges are balanced.

As an example, let's consider the reaction between permanganate ion (MnO_4^-) and oxalate ion $(C_2O_4^{2-})$ in acidic aqueous solution (▼ Figure 20.2). When MnO_4^- is added to an acidified solution of $C_2O_4^{2-}$, the deep purple color of the MnO_4^- ion fades, bubbles of CO_2 form, and the solution takes on the pale pink color of Mn^{2+}. We can write the skeleton equation as

$$MnO_4^-(aq) + C_2O_4^{2-}(aq) \longrightarrow Mn^{2+}(aq) + CO_2(aq) \qquad [20.6]$$

Experiments show that H^+ is consumed and H_2O is produced in the reaction. We will see that their involvement in the reaction is deduced in the course of balancing the equation.

To complete and balance Equation 20.6, we first write the two half-reactions (step 1). One half-reaction must have Mn on both sides of the arrow, and the other must have C on both sides of the arrow:

$$MnO_4^-(aq) \longrightarrow Mn^{2+}(aq)$$
$$C_2O_4^{2-}(aq) \longrightarrow CO_2(g)$$

We next complete and balance each half-reaction. First, we balance all the atoms except H and O (step 2a). In the permanganate half-reaction, we have one manganese atom on each side of the equation and so need to do nothing. In the oxalate half-reaction, we add a coefficient 2 on the right to balance the two carbons on the left:

$$MnO_4^-(aq) \longrightarrow Mn^{2+}(aq)$$
$$C_2O_4^{2-}(aq) \longrightarrow 2\,CO_2(g)$$

Next we balance O (step 2b). The permanganate half-reaction has four oxygens on the left and none on the right; to balance these four oxygen atoms, we add four H_2O molecules on the right:

$$MnO_4^-(aq) \longrightarrow Mn^{2+}(aq) + 4\,H_2O(l)$$

▲ **GO FIGURE**

Which species is reduced in this reaction? Which species is the reducing agent?

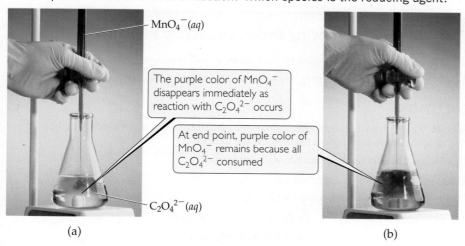

(a)   (b)

▲ **Figure 20.2** Titration of an acidic solution of $Na_2C_2O_4$ with $KMnO_4(aq)$.

The eight hydrogen atoms now in the products must be balanced by adding 8 H^+ to the reactants (step 2c):

$$8\,H^+(aq) + MnO_4^-(aq) \longrightarrow Mn^{2+}(aq) + 4\,H_2O(l)$$

Now there are equal numbers of each type of atom on the two sides of the equation, but the charge still needs to be balanced. The charge of the reactants is $8(1+) + 1(1-) = 7+$, and that of the products is $1(2+) + 4(0) = 2+$. To balance the charge, we add five electrons to the reactant side (step 2d):

$$5\,e^- + 8\,H^+(aq) + MnO_4^-(aq) \longrightarrow Mn^{2+}(aq) + 4\,H_2O(l)$$

We can use oxidation states to check our result. In this half-reaction Mn goes from the +7 oxidation state in MnO_4^- to the +2 oxidation state of Mn^{2+}. Therefore, each Mn atom gains five electrons, in agreement with our balanced half-reaction.

In the oxalate half-reaction, we have C and O balanced (step 2a). We balance the charge (step 2d) by adding two electrons to the products:

$$C_2O_4^{2-}(aq) \longrightarrow 2\,CO_2(g) + 2\,e^-$$

We can check this result using oxidation states. Carbon goes from the +3 oxidation state in $C_2O_4^{2-}$ to the +4 oxidation state in CO_2. Thus, each C atom loses one electron; therefore, the two C atoms in $C_2O_4^{2-}$ lose two electrons, in agreement with our balanced half-reaction.

Now we multiply each half-reaction by an appropriate integer so that the number of electrons gained in one half-reaction equals the number of electrons lost in the other (step 3). We multiply the MnO_4^- half-reaction by 2 and the $C_2O_4^{2-}$ half-reaction by 5:

$$10\,e^- + 16\,H^+(aq) + 2\,MnO_4^-(aq) \longrightarrow 2\,Mn^{2+}(aq) + 8\,H_2O(l)$$
$$5\,C_2O_4^{2-}(aq) \longrightarrow 10\,CO_2(g) + 10\,e^-$$

$$16\,H^+(aq) + 2\,MnO_4^-(aq) + 5\,C_2O_4^{2-}(aq) \longrightarrow$$
$$2\,Mn^{2+}(aq) + 8\,H_2O(l) + 10\,CO_2(g)$$

The balanced equation is the sum of the balanced half-reactions (step 4). Note that the electrons on the reactant and product sides of the equation cancel each other.

We check the balanced equation by counting atoms and charges (step 5). There are 16 H, 2 Mn, 28 O, 10 C, and a net charge of 4+ on each side of the equation, confirming that the equation is correctly balanced.

 Give It Some Thought

Do free electrons appear anywhere in the balanced equation for a redox reaction?

SAMPLE
EXERCISE 20.2 Balancing Redox Equations in Acidic Solution

Complete and balance this equation by the method of half-reactions:

$$Cr_2O_7^{2-}(aq) + Cl^-(aq) \longrightarrow Cr^{3+}(aq) + Cl_2(g) \quad \text{(acidic solution)}$$

SOLUTION

Analyze We are given an incomplete, unbalanced (skeleton) equation for a redox reaction occurring in acidic solution and asked to complete and balance it.

Plan We use the half-reaction procedure we just learned.

Solve

Step 1: We divide the equation into two half-reactions:

$$Cr_2O_7^{2-}(aq) \longrightarrow Cr^{3+}(aq)$$
$$Cl^-(aq) \longrightarrow Cl_2(g)$$

Step 2: We balance each half-reaction. In the first half-reaction the presence of one $Cr_2O_7^{2-}$ among the reactants requires two Cr^{3+} among the products. The seven oxygen atoms in $Cr_2O_7^{2-}$ are balanced by adding seven H_2O to the products. The 14 hydrogen atoms in 7 H_2O are then balanced by adding 14 H^+ to the reactants:

$$14\,H^+(aq) + Cr_2O_7^{2-}(aq) \longrightarrow 2\,Cr^{3+}(aq) + 7\,H_2O(l)$$

We then balance the charge by adding electrons to the left side of the equation so that the total charge is the same on the two sides:

$$6\,e^- + 14\,H^+(aq) + Cr_2O_7^{2-}(aq) \longrightarrow 2\,Cr^{3+}(aq) + 7\,H_2O(l)$$

We can check this result by looking at the oxidation state changes. Each chromium atom goes from +6 to +3, gaining three electrons; therefore, the two Cr atoms in $Cr_2O_7^{2-}$ gain six electrons, in agreement with our half-reaction.

In the second half-reaction, two Cl^- are required to balance one Cl_2:

$$2\,Cl^-(aq) \longrightarrow Cl_2(g)$$

We add two electrons to the right side to attain charge balance:

$$2\,Cl^-(aq) \longrightarrow Cl_2(g) + 2\,e^-$$

This result agrees with the oxidation state changes. Each chlorine atom goes from −1 to 0, losing one electron; therefore, the two chlorine atoms lose two electrons.

Step 3: We equalize the number of electrons transferred in the two half-reactions. To do so, we multiply the Cl half-reaction by 3 so that the number of electrons gained in the Cr half-reaction (6) equals the number lost in the Cl half-reaction, allowing the electrons to cancel when the half-reactions are added:

$$6\,Cl^-(aq) \longrightarrow 3\,Cl_2(g) + 6\,e^-$$

Step 4: The equations are added to give the balanced equation:

$$14\,H^+(aq) + Cr_2O_7^{2-}(aq) + 6\,Cl^-(aq) \longrightarrow 2\,Cr^{3+}(aq) + 7\,H_2O(l) + 3\,Cl_2(g)$$

Step 5: There are equal numbers of atoms of each kind on the two sides of the equation (14 H, 2 Cr, 7 O, 6 Cl). In addition, the charge is the same on the two sides (6+). Thus, the equation is balanced.

Practice Exercise 1

If you complete and balance the following equation in acidic solution

$$Mn^{2+}(aq) + NaBiO_3(s) \longrightarrow Bi^{3+}(aq) + MnO_4^-(aq) + Na^+(aq)$$

how many water molecules are there in the balanced equation (for the reaction balanced with the smallest whole-number coefficients)? **(a)** Four on the reactant side, **(b)** Three on the product side, **(c)** One on the reactant side, **(d)** Seven on the product side, **(e)** Two on the product side.

Practice Exercise 2

Complete and balance the following equation in acidic solution using the method of half-reactions.

$$Cu(s) + NO_3^-(aq) \longrightarrow Cu^{2+}(aq) + NO_2(g)$$

Balancing Equations for Reactions Occurring in Basic Solution

If a redox reaction occurs in basic solution, the equation must be balanced by using OH^- and H_2O rather than H^+ and H_2O. Because the water molecule and the hydroxide ion both contain hydrogen, this approach can take more moving back and forth from one side of the equation to the other to arrive at the appropriate half-reaction. An alternate approach is to first balance the half-reactions as if they occurred in acidic solution, count the number of H^+ in each half-reaction, and then add the same number of OH^- *to each side* of the half-reaction. This way, the reaction is mass-balanced because you are adding the same thing to both sides. In essence, what you are doing is "neutralizing" the protons to form water ($H^+ + OH^- \longrightarrow H_2O$) on the side containing H^+, and the other side ends up with the OH^-. The resulting water molecules can be canceled as needed.

SAMPLE EXERCISE 20.3 | **Balancing Redox Equations in Basic Solution**

Complete and balance this equation for a redox reaction that takes place in basic solution:

$$CN^-(aq) + MnO_4^-(aq) \longrightarrow CNO^-(aq) + MnO_2(s) \quad \text{(basic solution)}$$

SOLUTION

Analyze We are given an incomplete equation for a basic redox reaction and asked to balance it.

Plan We go through the first steps of our procedure as if the reaction were occurring in acidic solution. We then add the appropriate number of OH^- to each side of the equation, combining H^+ and OH^- to form H_2O. We complete the process by simplifying the equation.

Solve

Step 1: We write the incomplete, unbalanced half-reactions:

$$CN^-(aq) \longrightarrow CNO^-(aq)$$
$$MnO_4^-(aq) \longrightarrow MnO_2(s)$$

Step 2: We balance each half-reaction as if it took place in acidic solution:

$$CN^-(aq) + H_2O(l) \longrightarrow CNO^-(aq) + 2\,H^+(aq) + 2\,e^-$$
$$3\,e^- + 4\,H^+(aq) + MnO_4^-(aq) \longrightarrow MnO_2(s) + 2\,H_2O(l)$$

Now we must take into account that the reaction occurs in basic solution, adding OH^- to both sides of both half-reactions to neutralize H^+:

$$CN^-(aq) + H_2O(l) + 2\,OH^-(aq) \longrightarrow CNO^-(aq) + 2\,H^+(aq) + 2\,e^- + 2\,OH^-(aq)$$
$$3\,e^- + 4\,H^+(aq) + MnO_4^-(aq) + 4\,OH^-(aq) \longrightarrow MnO_2(s) + 2\,H_2O(l) + 4\,OH^-(aq)$$

We "neutralize" H^+ and OH^- by forming H_2O when they are on the same side of either half-reaction:

$$CN^-(aq) + H_2O(l) + 2\,OH^-(aq) \longrightarrow CNO^-(aq) + 2\,H_2O(l) + 2\,e^-$$
$$3\,e^- + 4\,H_2O(l) + MnO_4^-(aq) \longrightarrow MnO_2(s) + 2\,H_2O(l) + 4\,OH^-(aq)$$

Next, we cancel water molecules that appear as both reactants and products:

$$CN^-(aq) + 2\,OH^-(aq) \longrightarrow CNO^-(aq) + H_2O(l) + 2\,e^-$$
$$3\,e^- + 2\,H_2O(l) + MnO_4^-(aq) \longrightarrow MnO_2(s) + 4\,OH^-(aq)$$

Both half-reactions are now balanced. You can check the atoms and the overall charge.

Step 3: We multiply the cyanide half-reaction by 3, which gives 6 electrons on the product side, and multiply the permanganate half-reaction by 2, which gives 6 electrons on the reactant side:

$$3\,CN^-(aq) + 6\,OH^-(aq) \longrightarrow 3\,CNO^-(aq) + 3\,H_2O(l) + 6\,e^-$$
$$6\,e^- + 4\,H_2O(l) + 2\,MnO_4^-(aq) \longrightarrow 2\,MnO_2(s) + 8\,OH^-(aq)$$

Step 4: We add the two half-reactions together and simplify by canceling species that appear as both reactants and products:

$$3\,CN^-(aq) + H_2O(l) + 2\,MnO_4^-(aq) \longrightarrow 3\,CNO^-(aq) + 2\,MnO_2(s) + 2\,OH^-(aq)$$

Step 5: Check that the atoms and charges are balanced.

There are 3 C, 3 N, 2 H, 9 O, 2 Mn, and a charge of 5– on both sides of the equation.

Comment It is important to remember that this procedure does not imply that H^+ ions are involved in the chemical reaction. Recall that in aqueous solutions at 20 °C, $K_w = [H^+][OH^-] = 1.0 \times 10^{-14}$. Thus, $[H^+]$ is very small in this basic solution. ∞ (Section 16.3)

Practice Exercise 1

If you complete and balance the following oxidation–reduction reaction in basic solution

$$NO_2^-(aq) + Al(s) \longrightarrow NH_3(aq) + Al(OH)_4^-(aq)$$

how many hydroxide ions are there in the balanced equation (for the reaction balanced with the smallest whole-number coefficients)? **(a)** One on the reactant side, **(b)** One on the product side, **(c)** Four on the reactant side, **(d)** Seven on the product side, **(e)** None.

Practice Exercise 2

Complete and balance the following oxidation–reduction reaction in basic solution:

$$Cr(OH)_3(s) + ClO^-(aq) \longrightarrow CrO_4^{2-}(aq) + Cl_2(g)$$

GO FIGURE

Why does the intensity of the blue solution color lessen as the reaction proceeds?

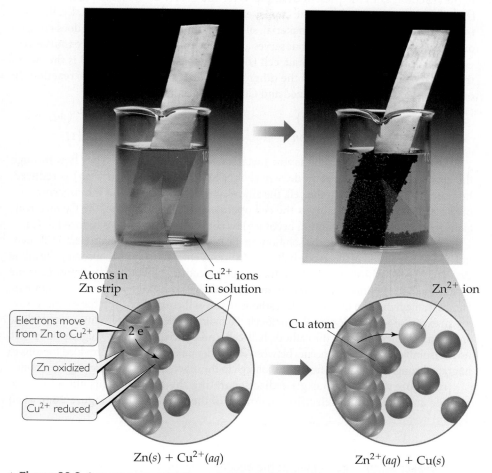

Zn(s) + Cu²⁺(aq)   Zn²⁺(aq) + Cu(s)

▲ **Figure 20.3 A spontaneous oxidation–reduction reaction involving zinc and copper.**

20.3 | Voltaic Cells

The energy released in a spontaneous redox reaction can be used to perform electrical work. This task is accomplished through a **voltaic** (or **galvanic**) **cell**, a device in which the transfer of electrons takes place through an external pathway rather than directly between reactants present in the same reaction vessel.

One such spontaneous reaction occurs when a strip of zinc is placed in contact with a solution containing Cu^{2+}. As the reaction proceeds, the blue color of $Cu^{2+}(aq)$ ions fades and copper metal deposits on the zinc. At the same time, the zinc begins to dissolve. These transformations, shown in ▲ **Figure 20.3,** are summarized by the equation

$$Zn(s) + Cu^{2+}(aq) \longrightarrow Zn^{2+}(aq) + Cu(s) \qquad [20.7]$$

▶ Figure 20.4 shows a voltaic cell that uses the redox reaction given in Equation 20.7. Although the setup in Figure 20.4 is more complex than that in Figure 20.3, the reaction is the same in both cases. The significant difference is that in the voltaic cell the Zn metal and $Cu^{2+}(aq)$ are not in direct contact with each other. Instead, Zn metal is in contact with $Zn^{2+}(aq)$ in one compartment, and Cu metal is in contact with $Cu^{2+}(aq)$ in the other compartment. Consequently, $Cu^{2+}(aq)$ reduction can occur only by the flow of electrons through an external circuit, namely, a wire connecting the Zn and Cu strips. Electrons flowing through a wire and ions moving in solution both constitute an *electrical current*. This flow of electrical charge can be used to accomplish electrical work.

GO FIGURE

Which metal, Cu or Zn, is oxidized in this voltaic cell?

Zn electrode in
1 M ZnSO₄ solution

Cu electrode in
1 M CuSO₄ solution

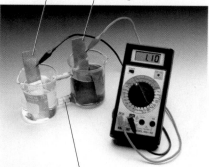

Solutions in contact with each other through porous glass disc

▲ **Figure 20.4 A Cu–Zn voltaic cell based on the reaction in Equation 20.7.**

The two solid metals connected by the external circuit are called *electrodes*. By definition, the electrode at which oxidation occurs is the **anode** and the electrode at which reduction occurs is the **cathode**.* The electrodes can be made of materials that participate in the reaction, as in the present example. Over the course of the reaction, the Zn electrode gradually disappears and the copper electrode gains mass. More typically, the electrodes are made of a conducting material, such as platinum or graphite, that does not gain or lose mass during the reaction but serves as a surface at which electrons are transferred.

Each compartment of a voltaic cell is called a *half-cell*. One half-cell is the site of the oxidation half-reaction, and the other is the site of the reduction half-reaction. In our present example, Zn is oxidized and Cu^{2+} is reduced:

Anode (*oxidation half-reaction*)	$Zn(s) \longrightarrow Zn^{2+}(aq) + 2\,e^-$
Cathode (*reduction half-reaction*)	$Cu^{2+}(aq) + 2\,e^- \longrightarrow Cu(s)$

Electrons become available as zinc metal is oxidized at the anode. They flow through the external circuit to the cathode, where they are consumed as $Cu^{2+}(aq)$ is reduced. Because $Zn(s)$ is oxidized in the cell, the zinc electrode loses mass, and the concentration of the Zn^{2+} solution increases as the cell operates. At the same time, the Cu electrode gains mass, and the Cu^{2+} solution becomes less concentrated as Cu^{2+} is reduced to $Cu(s)$.

For a voltaic cell to work, the solutions in the two half-cells must remain electrically neutral. As Zn is oxidized in the anode half-cell, Zn^{2+} ions enter the solution, upsetting the initial Zn^{2+}/SO_4^{2-} charge balance. To keep the solution electrically neutral, there must be some means for Zn^{2+} cations to migrate out of the anode half-cell and for anions to migrate in. Similarly, the reduction of Cu^{2+} at the cathode removes these cations from the solution, leaving an excess of SO_4^{2-} anions in that half-cell. To maintain electrical neutrality, some of these anions must migrate out of the cathode half-cell, and positive ions must migrate in. In fact, no measurable electron flow occurs between electrodes unless a means is provided for ions to migrate through the solution from one half-cell to the other, thereby completing the circuit.

In Figure 20.4, a porous glass disc separating the two half-cells allows ions to migrate and maintain the electrical neutrality of the solutions. In ▼ Figure 20.5, a *salt*

GO FIGURE

How is electrical balance maintained in the left beaker as Zn^{2+} are formed at the anode?

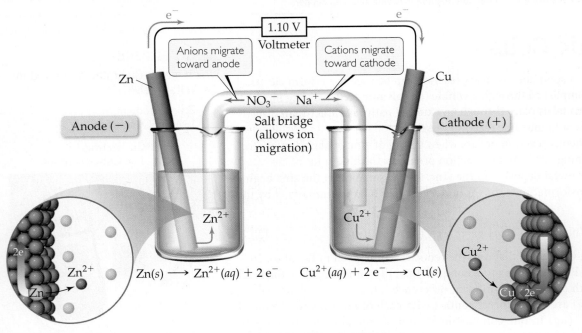

▲ **Figure 20.5 A voltaic cell that uses a salt bridge to complete the electrical circuit.**

*To help remember these definitions, note that *anode* and *oxidation* both begin with a vowel, and *cathode* and *reduction* both begin with a consonant.

bridge serves this purpose. The salt bridge consists of a U-shaped tube containing an electrolyte solution, such as $NaNO_3(aq)$, whose ions will not react with other ions in the voltaic cell or with the electrodes. The electrolyte is often incorporated into a paste or gel so that the electrolyte solution does not pour out when the U-tube is inverted. As oxidation and reduction proceed at the electrodes, ions from the salt bridge migrate into the two half-cells—cations migrating to the cathode half-cell and anions migrating to the anode half-cell—to neutralize charge in the half-cell solutions. Whichever device is used to allow ions to migrate between half-cells, *anions always migrate toward the anode and cations toward the cathode.*

▶ Figure 20.6 summarizes the various relationships in a voltaic cell. Notice in particular that *electrons flow from the anode through the external circuit to the cathode.* Because of this directional flow, the anode in a voltaic cell is labeled with a negative sign and the cathode is labeled with a positive sign. We can envision the electrons as being attracted to the positive cathode from the negative anode through the external circuit.

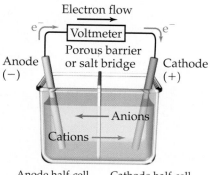

▲ **Figure 20.6 Summary of reactions occurring in a voltaic cell.** The half-cells can be separated by either a porous glass disc (as in Figure 20.4) or by a salt bridge (as in Figure 20.5).

SAMPLE EXERCISE 20.4 | Describing a Voltaic Cell

The oxidation–reduction reaction

$$Cr_2O_7^{2-}(aq) + 14\,H^+(aq) + 6\,I^-(aq) \longrightarrow 2\,Cr^{3+}(aq) + 3\,I_2(s) + 7\,H_2O(l)$$

is spontaneous. A solution containing $K_2Cr_2O_7$ and H_2SO_4 is poured into one beaker, and a solution of KI is poured into another. A salt bridge is used to join the beakers. A metallic conductor that will not react with either solution (such as platinum foil) is suspended in each solution, and the two conductors are connected with wires through a voltmeter or some other device to detect an electric current. The resultant voltaic cell generates an electric current. Indicate the reaction occurring at the anode, the reaction at the cathode, the direction of electron migration, the direction of ion migration, and the signs of the electrodes.

SOLUTION

Analyze We are given the equation for a spontaneous reaction that takes place in a voltaic cell and a description of how the cell is constructed. We are asked to write the half-reactions occurring at the anode and at the cathode, as well as the directions of electron and ion movements and the signs assigned to the electrodes.

Plan Our first step is to divide the chemical equation into half-reactions so that we can identify the oxidation and the reduction processes. We then use the definitions of anode and cathode and the other terminologies summarized in Figure 20.6.

Solve In one half-reaction, $Cr_2O_7^{2-}(aq)$ is converted into $Cr^{3+}(aq)$. Starting with these ions and then completing and balancing the half-reaction, we have

$$Cr_2O_7^{2-}(aq) + 14\,H^+(aq) + 6\,e^- \longrightarrow 2\,Cr^{3+}(aq) + 7\,H_2O(l)$$

In the other half-reaction, $I^-(aq)$ is converted to $I_2(s)$:

$$6\,I^-(aq) \longrightarrow 3\,I_2(s) + 6\,e^-$$

Now we can use the summary in Figure 20.6 to help us describe the voltaic cell. The first half-reaction is the reduction process (electrons on the reactant side of the equation). By definition, the reduction process occurs at the cathode. The second half-reaction is the oxidation process (electrons on the product side of the equation), which occurs at the anode.

The I^- ions are the source of electrons, and the $Cr_2O_7^{2-}$ ions accept the electrons. Hence, the electrons flow through the external circuit from the electrode immersed in the KI solution (the anode) to the electrode immersed in the $K_2Cr_2O_7$–H_2SO_4 solution (the cathode). The electrodes themselves do not react in any way; they merely provide a means of transferring electrons from or to the solutions. The cations move through the solutions toward the cathode, and the anions move toward the anode. The anode (from which the electrons move) is the negative electrode, and the cathode (toward which the electrons move) is the positive electrode.

Practice Exercise 1

The following two half-reactions occur in a voltaic cell:

$$Ni(s) \longrightarrow Ni^{2+}(aq) + 2\,e^- \qquad \text{(electrode = Ni)}$$
$$Cu^{2+}(aq) + 2\,e^- \longrightarrow Cu(s) \qquad \text{(electrode = Cu)}$$

Which one of the following descriptions most accurately describes what is occurring in the half-cell containing the Cu electrode and $Cu^{2+}(aq)$ solution?
(a) The electrode is losing mass and cations from the salt bridge are flowing into the half-cell.
(b) The electrode is gaining mass and cations from the salt bridge are flowing into the half-cell.

(c) The electrode is losing mass and anions from the salt bridge are flowing into the half-cell.
(d) The electrode is gaining mass and anions from the salt bridge are flowing into the half-cell.

Practice Exercise 2

The two half-reactions in a voltaic cell are

$$Zn(s) \longrightarrow Zn^{2+}(aq) + 2\,e^- \qquad (\text{electrode} = Zn)$$
$$ClO_3^-(aq) + 6\,H^+(aq) + 6\,e^- \longrightarrow Cl^-(aq) + 3\,H_2O(l) \quad (\text{electrode} = Pt)$$

(a) Indicate which reaction occurs at the anode and which at the cathode. (b) Does the zinc electrode gain, lose, or retain the same mass as the reaction proceeds? (c) Does the platinum electrode gain, lose, or retain the same mass as the reaction proceeds? (d) Which electrode is positive?

20.4 | Cell Potentials Under Standard Conditions

Why do electrons transfer spontaneously from a Zn atom to a Cu^{2+} ion, either directly as in Figure 20.3 or through an external circuit as in Figure 20.4? In a simple sense, we can compare the electron flow to the flow of water in a waterfall (▼ Figure 20.7). Water flows spontaneously over a waterfall because of a difference in potential energy between the top of the falls and the bottom. ∞ (Section 5.1) In a similar fashion, electrons flow spontaneously through an external circuit from the anode of a voltaic cell to the cathode because of a difference in potential energy. The potential energy of electrons is higher in the anode than in the cathode. Thus, electrons flow spontaneously toward the electrode with the more positive electrical potential.

The difference in potential energy per electrical charge (the *potential difference*) between two electrodes is measured in volts. One volt (V) is the potential difference required to impart 1 joule (J) of energy to a charge of 1 coulomb (C):

$$1\,V = 1\,\frac{J}{C}$$

Recall that one electron has a charge of 1.60×10^{-19} C. ∞ (Section 2.2)

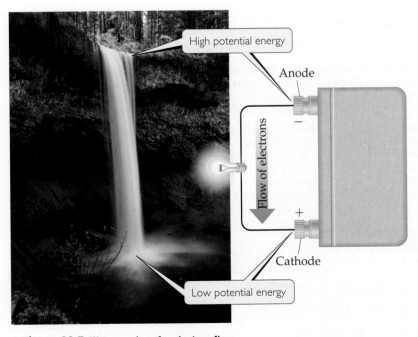

▲ **Figure 20.7 Water analogy for electron flow.**

The potential difference between the two electrodes of a voltaic cell is called the **cell potential**, denoted E_{cell}. Because the potential difference provides the driving force that pushes electrons through the external circuit, we also call it the **electromotive** ("causing electron motion") **force**, or **emf**. Because E_{cell} is measured in volts, it is also commonly called the *voltage* of the cell.

The cell potential of any voltaic cell is positive. The magnitude of the cell potential depends on the reactions that occur at the cathode and anode, the concentrations of reactants and products, and the temperature, which we will assume to be 25 °C unless otherwise noted. In this section, we focus on cells that are operated at 25 °C under *standard conditions*. Recall from Table 19.2 that standard conditions include 1 *M* concentration for reactants and products in solution and 1 atm pressure for gaseous reactants and products. The cell potential under standard conditions is called either the **standard cell potential** or **standard emf** and is denoted $E°_{cell}$. For the Zn–Cu voltaic cell in Figure 20.5, for example, the standard cell potential at 25 °C is +1.10 V:

$$Zn(s) + Cu^{2+}(aq, 1\ M) \longrightarrow Zn^{2+}(aq, 1\ M) + Cu(s) \qquad E°_{cell} = +1.10\ V$$

Recall that the superscript ° indicates standard-state conditions. ∞ (Section 5.7)

Give It Some Thought

If a standard cell potential is $E°_{cell} = +0.85$ V at 25 °C, is the redox reaction of the cell spontaneous?

Standard Reduction Potentials

The standard cell potential of a voltaic cell, $E°_{cell}$, depends on the particular cathode and anode half-cells. We could, in principle, tabulate the standard cell potentials for all possible cathode–anode combinations. However, it is not necessary to undertake this arduous task. Rather, we can assign a standard potential to each half-cell and then use these half-cell potentials to determine $E°_{cell}$. The cell potential is the difference between two half-cell potentials. By convention, the potential associated with each electrode is chosen to be the potential for *reduction* at that electrode. Thus, standard half-cell potentials are tabulated for reduction reactions, which means they are **standard reduction potentials**, denoted $E°_{red}$. The standard cell potential, $E°_{cell}$, is the standard reduction potential of the cathode reaction, $E°_{red}$ (cathode), *minus* the standard reduction potential of the anode reaction, $E°_{red}$ (anode):

$$E°_{cell} = E°_{red}\ (\text{cathode}) - E°_{red}\ (\text{anode}) \qquad [20.8]$$

It is not possible to measure the standard reduction potential of a half-reaction directly. If we assign a standard reduction potential to a certain reference half-reaction, however, we can then determine the standard reduction potentials of other half-reactions relative to that reference value. The reference half-reaction is the reduction of $H^+(aq)$ to $H_2(g)$ under standard conditions, which is assigned a standard reduction potential of exactly 0 V:

$$2\ H^+(aq, 1\ M) + 2\ e^- \longrightarrow H_2(g, 1\ atm) \quad E°_{red} = 0\ V \qquad [20.9]$$

An electrode designed to produce this half-reaction is called a **standard hydrogen electrode** (SHE). An SHE consists of a platinum wire connected to a piece of platinum foil covered with finely divided platinum that serves as an inert surface for the reaction (**Figure 20.8**). The SHE allows the platinum to be in contact with both 1 *M* $H^+(aq)$ and a stream of hydrogen gas at 1 atm. The SHE can operate as either the anode or cathode of a cell, depending on the nature of the other electrode.

Figure 20.9 shows a voltaic cell using an SHE. The spontaneous reaction is the one shown in Figure 20.1, namely, oxidation of Zn and reduction of H^+:

$$Zn(s) + 2\ H^+(aq) \longrightarrow Zn^{2+}(aq) + H_2(g)$$

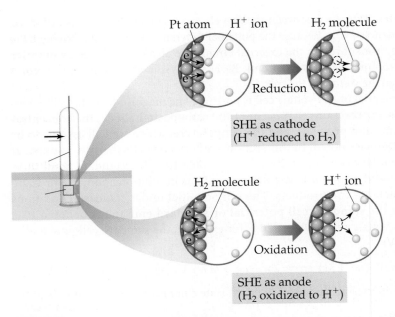

Pt atom   H⁺ ion   H₂ molecule

Reduction

SHE as cathode
(H⁺ reduced to H₂)

H₂ molecule   H⁺ ion

Oxidation

SHE as anode
(H₂ oxidized to H⁺)

▲ Figure 20.8 **The standard hydrogen electrode (SHE) is used as a reference electrode.**

▲ GO FIGURE

Why do Na⁺ ions migrate into the cathode half-cell as the cell reaction proceeds?

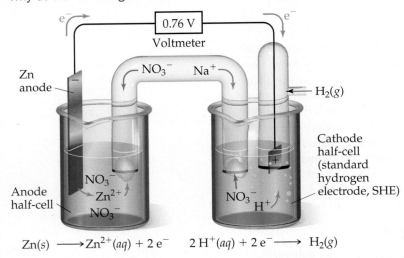

$$Zn(s) \longrightarrow Zn^{2+}(aq) + 2\,e^- \qquad 2\,H^+(aq) + 2\,e^- \longrightarrow H_2(g)$$

▲ Figure 20.9 **A voltaic cell using a standard hydrogen electrode (SHE).** The anode half-cell is Zn metal in a $Zn(NO_3)_2(aq)$ solution, and the cathode half-cell is the SHE in a $HNO_3(aq)$ solution.

When the cell is operated under standard conditions, the cell potential is +0.76 V. By using the standard cell potential ($E^\circ_{cell} = 0.76$ V), the defined standard reduction potential of H⁺($E^\circ_{red} = 0$ V) and Equation 20.8, we can determine the standard reduction potential for the Zn^{2+}/Zn half-reaction:

$$E^\circ_{cell} = E^\circ_{red}\,(\text{cathode}) - E^\circ_{red}\,(\text{anode})$$

$$+0.76\text{ V} = 0\text{ V} - E^\circ_{red}\,(\text{anode})$$

$$E^\circ_{red}\,(\text{anode}) = -0.76\text{ V}$$

Thus, a standard reduction potential of −0.76 V can be assigned to the reduction of Zn^{2+} to Zn:

$$Zn^{2+}(aq, 1\,M) + 2\,e^- \longrightarrow Zn(s) \quad E^\circ_{red} = -0.76\text{ V}$$

We write the reaction as a reduction even though the Zn reaction in Figure 20.9 is an oxidation. *Whenever we assign an electrical potential to a half-reaction, we write the reaction as a reduction.* Half-reactions, however, are reversible, being able to operate as either reductions or oxidations. Consequently, half-reactions are sometimes written using two arrows ($\rightleftharpoons$) between reactants and products, as in equilibrium reactions.

The standard reduction potentials for other half-reactions can be determined in a fashion analogous to that used for the Zn^{2+}/Zn half-reaction. ▼ **Table 20.1** lists some standard reduction potentials; a more complete list is found in Appendix E. These standard reduction potentials, often called *half-cell potentials*, can be combined to calculate E°_{cell} values for a large variety of voltaic cells.

 Give It Some Thought

For the half-reaction $Cl_2(g) + 2\,e^- \longrightarrow 2\,Cl^-(aq)$, what are the standard conditions for the reactant and product?

Because electrical potential measures potential energy per electrical charge, standard reduction potentials are intensive properties. ∞ (Section 1.3) In other words, if we increase the amount of substances in a redox reaction, we increase both the energy and the charges involved, but the ratio of energy (joules) to electrical charge (coulombs) remains constant ($V = J/C$). Thus, *changing the stoichiometric coefficient in a half-reaction does not affect the value of the standard reduction potential.*

Table 20.1 Standard Reduction Potentials in Water at 25 °C

$E^\circ_{red}(V)$	Reduction Half-Reaction
+2.87	$F_2(g) + 2\,e^- \longrightarrow 2\,F^-(aq)$
+1.51	$MnO_4^-(aq) + 8\,H^+(aq) + 5\,e^- \longrightarrow Mn^{2+}(aq) + 4\,H_2O(l)$
+1.36	$Cl_2(g) + 2\,e^- \longrightarrow 2\,Cl^-(aq)$
+1.33	$Cr_2O_7^{2-}(aq) + 14\,H^+(aq) + 6\,e^- \longrightarrow 2\,Cr^{3+}(aq) + 7\,H_2O(l)$
+1.23	$O_2(g) + 4\,H^+(aq) + 4\,e^- \longrightarrow 2\,H_2O(l)$
+1.06	$Br_2(l) + 2\,e^- \longrightarrow 2\,Br^-(aq)$
+0.96	$NO_3^-(aq) + 4\,H^+(aq) + 3\,e^- \longrightarrow NO(g) + 2\,H_2O(l)$
+0.80	$Ag^+(aq) + e^- \longrightarrow Ag(s)$
+0.77	$Fe^{3+}(aq) + e^- \longrightarrow Fe^{2+}(aq)$
+0.68	$O_2(g) + 2\,H^+(aq) + 2\,e^- \longrightarrow H_2O_2(aq)$
+0.59	$MnO_4^-(aq) + 2\,H_2O(l) + 3\,e^- \longrightarrow MnO_2(s) + 4\,OH^-(aq)$
+0.54	$I_2(s) + 2\,e^- \longrightarrow 2\,I^-(aq)$
+0.40	$O_2(g) + 2\,H_2O(l) + 4\,e^- \longrightarrow 4\,OH^-(aq)$
+0.34	$Cu^{2+}(aq) + 2\,e^- \longrightarrow Cu(s)$
0 [defined]	$2\,H^+(aq) + 2\,e^- \longrightarrow H_2(g)$
−0.28	$Ni^{2+}(aq) + 2\,e^- \longrightarrow Ni(s)$
−0.44	$Fe^{2+}(aq) + 2\,e^- \longrightarrow Fe(s)$
−0.76	$Zn^{2+}(aq) + 2\,e^- \longrightarrow Zn(s)$
−0.83	$2\,H_2O(l) + 2\,e^- \longrightarrow H_2(g) + 2\,OH^-(aq)$
−1.66	$Al^{3+}(aq) + 3\,e^- \longrightarrow Al(s)$
−2.71	$Na^+(aq) + e^- \longrightarrow Na(s)$
−3.05	$Li^+(aq) + e^- \longrightarrow Li(s)$

For example, $E°_{red}$ for the reduction of 10 mol Zn^{2+} is the same as that for the reduction of 1 mol Zn^{2+}:

$$10\ Zn^{2+}(aq, 1\ M) + 20\ e^- \longrightarrow 10\ Zn(s) \qquad E°_{red} = -0.76\ V$$

SAMPLE EXERCISE 20.5 Calculating $E°_{red}$ from $E°_{cell}$

For the Zn–Cu^{2+} voltaic cell shown in Figure 20.5, we have

$$Zn(s) + Cu^{2+}(aq, 1\ M) \longrightarrow Zn^{2+}(aq, 1\ M) + Cu(s) \qquad E°_{cell} = 1.10\ V$$

Given that the standard reduction potential of Zn^{2+} to Zn(s) is −0.76 V, calculate the $E°_{red}$ for the reduction of Cu^{2+} to Cu:

$$Cu^{2+}(aq, 1\ M) + 2\ e^- \longrightarrow Cu(s)$$

SOLUTION

Analyze We are given $E°_{cell}$ and $E°_{red}$ for Zn^{2+} and asked to calculate $E°_{red}$ for Cu^{2+}.

Plan In the voltaic cell, Zn is oxidized and is therefore the anode. Thus, the given $E°_{red}$ for Zn^{2+} is $E°_{red}$ (anode). Because Cu^{2+} is reduced, it is in the cathode half-cell. Thus, the unknown reduction potential for Cu^{2+} is $E°_{red}$ (cathode). Knowing $E°_{cell}$ and $E°_{red}$ (anode), we can use Equation 20.8 to solve for $E°_{red}$ (cathode).

Solve

$$E°_{cell} = E°_{red}\ (\text{cathode}) - E°_{red}\ (\text{anode})$$

$$1.10\ V = E°_{red}\ (\text{cathode}) - (-0.76\ V)$$

$$E°_{red}\ (\text{cathode}) = 1.10\ V - 0.76\ V = 0.34\ V$$

Check This standard reduction potential agrees with the one listed in Table 20.1.

Comment The standard reduction potential for Cu^{2+} can be represented as $E°_{Cu^{2+}} = 0.34\ V$ and that for Zn^{2+} as $E°_{Zn^{2+}} = -0.76\ V$. The subscript identifies the ion that is reduced in the reduction half-reaction.

Practice Exercise 1

A voltaic cell based on the reaction
$$2\ Eu^{2+}(aq) + Ni^{2+}(aq) \longrightarrow 2\ Eu^{3+}(aq) + Ni(s)$$ generates
$E°_{cell} = 0.07\ V$. Given the standard reduction potential of Ni^{2+} given in Table 20.1 what is the standard reduction potential for the reaction $Eu^{3+}(aq) + e^- \longrightarrow Eu^{2+}(aq)$? (a) −0.35 V, (b) 0.35 V, (c) −0.21 V, (d) 0.21 V, (e) 0.07 V.

Practice Exercise 2

The standard cell potential is 1.46 V for a voltaic cell based on the following half-reactions:

$$In^+(aq) \longrightarrow In^{3+}(aq) + 2\ e^-$$

$$Br_2(l) + 2\ e^- \longrightarrow 2\ Br^-(aq)$$

Using Table 20.1, calculate $E°_{red}$ for the reduction of In^{3+} to In^+.

SAMPLE EXERCISE 20.6 Calculating $E°_{cell}$ from $E°_{red}$

Use Table 20.1 to calculate $E°_{cell}$ for the voltaic cell described in Sample Exercise 20.4, which is based on the reaction

$$Cr_2O_7^{2-}(aq) + 14\ H^+(aq) + 6\ I^-(aq) \longrightarrow 2\ Cr^{3+}(aq) + 3\ I_2(s) + 7\ H_2O(l)$$

SOLUTION

Analyze We are given the equation for a redox reaction and asked to use data in Table 20.1 to calculate the standard cell potential for the associated voltaic cell.

Plan Our first step is to identify the half-reactions that occur at the cathode and anode, which we did in Sample Exercise 20.4. Then we use Table 20.1 and Equation 20.8 to calculate the standard cell potential.

Solve The half-reactions are

Cathode: $Cr_2O_7^{2-}(aq) + 14\ H^+(aq) + 6\ e^- \longrightarrow 2\ Cr^{3+}(aq) + 7\ H_2O(l)$

Anode: $6\ I^-(aq) \longrightarrow 3\ I_2(s) + 6\ e^-$

According to Table 20.1, the standard reduction potential for the reduction of $Cr_2O_7^{2-}$ to Cr^{3+} is +1.33 V and the standard reduction potential for the reduction of I_2 to I^- (the reverse of the oxidation half-reaction) is +0.54 V. We use these values in Equation 20.8:

$$E°_{cell} = E°_{red}\ (\text{cathode}) - E°_{red}\ (\text{anode}) = 1.33\ V - 0.54\ V = 0.79\ V$$

Although we must multiply the iodide half-reaction by 3 to obtain a balanced equation, we do *not* multiply the E_{red}° value by 3. As we have noted, the standard reduction potential is an intensive property and so is independent of the stoichiometric coefficients.

Check The cell potential, 0.79 V, is a positive number. As noted earlier, a voltaic cell must have a positive potential.

> **Practice Exercise 1**
>
> Using the data in Table 20.1 what value would you calculate for the standard emf (E_{cell}°) for a voltaic cell that employs the overall cell reaction $2\,Ag^+(aq) + Ni(s) \longrightarrow 2\,Ag(s) + Ni^{2+}(aq)$?
> **(a)** +0.52 V,   **(b)** −0.52 V,   **(c)** +1.08 V,   **(d)** −1.08 V,   **(e)** +0.80 V.
>
> **Practice Exercise 2**
>
> Using data in Table 20.1, calculate the standard emf for a cell that employs the overall cell reaction $2\,Al(s) + 3\,I_2(s) \longrightarrow 2\,Al^{3+}(aq) + 6\,I^-(aq)$.

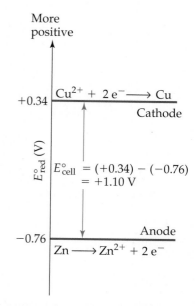

▲ **Figure 20.10 Half-cell potentials and standard cell potential for the Zn–Cu voltaic cell.**

For each half-cell in a voltaic cell, the standard reduction potential provides a measure of the tendency for reduction to occur: *The more positive the value of E_{red}°, the greater the tendency for reduction under standard conditions.* In any voltaic cell operating under standard conditions, the E_{red}° value for the reaction at the cathode is more positive than the E_{red}° value for the reaction at the anode. Thus, electrons flow spontaneously through the external circuit from the electrode with the more negative value of E_{red}° to the electrode with the more positive value of E_{red}°. ▶ Figure 20.10 graphically illustrates the relationship between the standard reduction potentials for the two half-reactions in the Zn–Cu voltaic cell of Figure 20.5.

▲ Give It Some Thought

The standard reduction potential of $Ni^{2+}(aq)$ is $E_{red}^\circ = -0.28$ V and that of $Fe^{2+}(aq)$ is $E_{red}^\circ = -0.44$ V. In a Ni–Fe voltaic cell which electrode is the cathode, Ni or Fe?

SAMPLE EXERCISE 20.7 | **Determining Half-Reactions at Electrodes and Calculating Cell Potentials**

A voltaic cell is based on the two standard half-reactions

$$Cd^{2+}(aq) + 2\,e^- \longrightarrow Cd(s)$$

$$Sn^{2+}(aq) + 2\,e^- \longrightarrow Sn(s)$$

Use data in Appendix E to determine **(a)** which half-reaction occurs at the cathode and which occurs at the anode and **(b)** the standard cell potential.

SOLUTION

Analyze We have to look up E_{red}° for two half-reactions. We then use these values first to determine the cathode and the anode and then to calculate the standard cell potential, E_{cell}°.

Plan The cathode will have the reduction with the more positive E_{red}° value, and the anode will have the less positive E_{red}°. To write the half-reaction at the anode, we reverse the half-reaction written for the reduction, so that the half-reaction is written as an oxidation.

Solve

(a) According to Appendix E, $E_{red}^\circ(Cd^{2+}/Cd) = -0.403$ V and $E_{red}^\circ(Sn^{2+}/Sn) = -0.136$ V. The standard reduction potential for Sn^{2+} is more positive (less negative) than that for Cd^{2+}. Hence, the reduction of Sn^{2+} is the reaction that occurs at the cathode:

$$Cathode:\quad Sn^{2+}(aq) + 2\,e^- \longrightarrow Sn(s)$$

The anode reaction, therefore, is the loss of electrons by Cd:

$$Anode:\quad Cd(s) \longrightarrow Cd^{2+}(aq) + 2\,e^-$$

(b) The cell potential is given by the difference in the standard reduction potentials at the cathode and anode (Equation 20.8):

$$E_{cell}^\circ = E_{red}^\circ(\text{cathode}) - E_{red}^\circ(\text{anode}) = (-0.136\ V) - (-0.403\ V) = 0.267\ V$$

Notice that it is unimportant that the E°_{red} values of both half-reactions are negative; the negative values merely indicate how these reductions compare to the reference reaction, the reduction of $H^+(aq)$.

Check The cell potential is positive, as it must be for a voltaic cell.

Practice Exercise 1

Consider three voltaic cells, each similar to the one shown in Figure 20.5. In each voltaic cell, one half-cell contains a $1.0\ M\ Fe(NO_3)_2(aq)$ solution with an Fe electrode. The contents of the other half-cells are as follows:

> Cell 1: a $1.0\ M\ CuCl_2(aq)$ solution with a Cu electrode
>
> Cell 2: a $1.0\ M\ NiCl_2(aq)$ solution with a Ni electrode
>
> Cell 3: a $1.0\ M\ ZnCl_2(aq)$ solution with a Zn electrode

In which voltaic cell(s) does iron act as the anode? **(a)** Cell 1, **(b)** Cell 2, **(c)** Cell 3, **(d)** Cells 1 and 2, **(e)** All three cells.

Practice Exercise 2

A voltaic cell is based on a Co^{2+}/Co half-cell and a $AgCl/Ag$ half-cell.
(a) What half-reaction occurs at the anode? **(b)** What is the standard cell potential?

Strengths of Oxidizing and Reducing Agents

Table 20.1 lists half-reactions in the order of decreasing tendency to undergo reduction. For example, F_2 is located at the top of the table, having the most positive value for E°_{red}. Thus, F_2 is the most easily reduced species in Table 20.1 and therefore the strongest oxidizing agent listed.

Among the most frequently used oxidizing agents are the halogens, O_2, and oxyanions such as MnO_4^-, $Cr_2O_7^{2-}$, and NO_3^-, whose central atoms have high positive oxidation states. As seen in Table 20.1, all these species have large positive values of E°_{red} and therefore easily undergo reduction.

The lower the tendency for a half-reaction to occur in one direction, the greater the tendency for it to occur in the opposite direction. Thus, *the half-reaction with the most negative reduction potential in Table 20.1 is the one most easily reversed and run as an oxidation.* Being at the bottom of Table 20.1, $Li^+(aq)$ is the most difficult species in the list to reduce and is therefore the poorest oxidizing agent listed. Although $Li^+(aq)$ has little tendency to gain electrons, the reverse reaction, oxidation of $Li(s)$ to $Li^+(aq)$, is highly favorable. Thus, Li is the strongest reducing agent among the substances listed in Table 20.1. (Note that, because Table 20.1 lists half-reactions as reductions, only the substances on the reactant side of these equations can serve as oxidizing agents; only those on the product side can serve as reducing agents.)

Commonly used reducing agents include H_2 and the active metals, such as the alkali metals and the alkaline earth metals. Other metals whose cations have negative E°_{red} values—Zn and Fe, for example—are also used as reducing agents. Solutions of reducing agents are difficult to store for extended periods because of the ubiquitous presence of O_2, a good oxidizing agent.

The information contained in Table 20.1 is summarized graphically in ▶ Figure 20.11. For the half-reactions at the top of Table 20.1 the substances on the reactant side of the equation are the most readily reduced species in the table and are therefore the strongest oxidizing agents. Substances on the product side of these reactions are the most difficult to oxidize and so are the weakest reducing agents in the table. Thus, Figure 20.11 shows $F_2(g)$ as the strongest oxidizing agent and $F^-(aq)$ as the weakest reducing agent. Conversely, the reactants in half-reactions at the bottom of Table 20.1, such as $Li^+(aq)$ are the most difficult to reduce and so are the weakest oxidizing agents, while the products of these reactions, such as $Li(s)$, are the most readily oxidized species in the table and so are the strongest reducing agents.

GO FIGURE

Can an acidic solution oxidize a piece of aluminum?

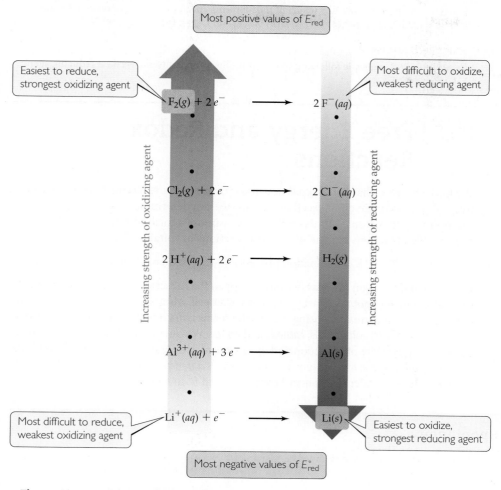

▲ **Figure 20.11 Relative strengths of oxidizing and reducing agents.**

This inverse relationship between oxidizing and reducing strength is similar to the inverse relationship between the strengths of conjugate acids and bases. ∞ (Section 16.2 and Figure 16.3)

SAMPLE
EXERCISE 20.8   Determining Relative Strengths of Oxidizing Agents

Using Table 20.1, rank the following ions in the order of increasing strength as oxidizing agents: $NO_3^-(aq), Ag^+(aq), Cr_2O_7^{2-}$.

SOLUTION

Analyze We are asked to rank the abilities of several ions to act as oxidizing agents.

Plan The more readily an ion is reduced (the more positive its $E°_{red}$ value), the stronger it is as an oxidizing agent.

Solve From Table 20.1, we have

$$NO_3^-(aq) + 4\,H^+(aq) + 3\,e^- \longrightarrow NO(g) + 2\,H_2O(l) \qquad E°_{red} = +0.96\ V$$

$$Ag^+(aq) + e^- \longrightarrow Ag(s) \qquad E°_{red} = +0.80\ V$$

$$Cr_2O_7^{2-}(aq) + 14\,H^+(aq) + 6\,e^- \longrightarrow 2\,Cr^{3+}(aq) + 7\,H_2O(l) \qquad E°_{red} = +1.33\ V$$

Because the standard reduction potential of $Cr_2O_7^{2-}$ is the most positive, $Cr_2O_7^{2-}$ is the strongest oxidizing agent of the three. The rank order is $Ag^+ < NO_3^- < Cr_2O_7^{2-}$.

> **Practice Exercise 1**
>
> Based on the data in Table 20.1 which of the following species would you expect to be the strongest oxidizing agent? **(a)** $Cl^-(aq)$, **(b)** $Cl_2(g)$, **(c)** $O_2(g)$, **(d)** $H^+(aq)$, **(e)** $Na^+(aq)$.
>
> **Practice Exercise 2**
>
> Using Table 20.1, rank the following species from the strongest to the weakest reducing agent: $I^-(aq)$, $Fe(s)$, $Al(s)$.

20.5 | Free Energy and Redox Reactions

We have observed that voltaic cells use spontaneous redox reactions to produce a positive cell potential. Given half-cell potentials, we can determine whether a given redox reaction is spontaneous. In this endeavor, we can use a form of Equation 20.8 that describes redox reactions in general, not just reactions in voltaic cells:

$$E° = E°_{red} \text{ (reduction process)} - E°_{red} \text{ (oxidation process)} \qquad [20.10]$$

In writing the equation this way, we have dropped the subscript "cell" to indicate that the calculated emf does not necessarily refer to a voltaic cell. Also, we have generalized the standard reduction potentials by using the general terms *reduction* and *oxidation* rather than the terms specific to voltaic cells, *cathode* and *anode*. We can now make a general statement about the spontaneity of a reaction and its associated emf, E: *A positive value of* E *indicates a spontaneous process; a negative value of* E *indicates a nonspontaneous process.* We use E to represent the emf under nonstandard conditions and $E°$ to indicate the standard emf.

SAMPLE EXERCISE 20.9 | Determining Spontaneity

Use Table 20.1 to determine whether the following reactions are spontaneous under standard conditions.

(a) $Cu(s) + 2 H^+(aq) \longrightarrow Cu^{2+}(aq) + H_2(g)$

(b) $Cl_2(g) + 2 I^-(aq) \longrightarrow 2 Cl^-(aq) + I_2(s)$

SOLUTION

Analyze We are given two reactions and must determine whether each is spontaneous.

Plan To determine whether a redox reaction is spontaneous under standard conditions, we first need to write its reduction and oxidation half-reactions. We can then use the standard reduction potentials and Equation 20.10 to calculate the standard emf, $E°$, for the reaction. If a reaction is spontaneous, its standard emf must be a positive number.

Solve

(a) We first must identify the oxidation and reduction half-reactions that when combined give the overall reaction.

Reduction: $2 H^+(aq) + 2 e^- \longrightarrow H_2(g)$

Oxidation: $Cu(s) \longrightarrow Cu^{2+}(aq) + 2 e^-$

We look up standard reduction potentials for both half-reactions and use them to calculate $E°$ using Equation 20.10:

$E° = E°_{red} \text{ (reduction process)} - E°_{red} \text{ (oxidation process)}$
$= (0 \text{ V}) - (0.34 \text{ V}) = -0.34 \text{ V}$

Because $E°$ is negative, the reaction is not spontaneous in the direction written. Copper metal does not react with acids as written in Equation (a). The reverse reaction, however, *is* spontaneous and has a positive $E°$ value:

$Cu^{2+}(aq) + H_2(g) \longrightarrow Cu(s) + 2 H^+(aq) \quad E° = +0.34 \text{ V}$

Thus, Cu^{2+} can be reduced by H_2.

(b) We follow a procedure analogous to that in (a):

Reduction: $Cl_2(g) + 2 e^- \longrightarrow 2 Cl^-(aq)$

Oxidation: $2 I^-(aq) \longrightarrow I_2(s) + 2 e^-$

In this case

$$E° = (1.36 \text{ V}) - (0.54 \text{ V}) = +0.82 \text{ V}$$

Because the value of $E°$ is positive, this reaction is spontaneous and could be used to build a voltaic cell.

Practice Exercise 1

Which of the following elements is capable of oxidizing $Fe^{2+}(aq)$ ions to $Fe^{3+}(aq)$ ions: chlorine, bromine, iodine? **(a)** I_2, **(b)** Cl_2, **(c)** Cl_2 and I_2, **(d)** Cl_2 and Br_2, **(e)** all three elements.

Practice Exercise 2

Using the standard reduction potentials listed in Appendix E, determine which of the following reactions are spontaneous under standard conditions:
(a) $I_2(s) + 5 \text{ Cu}^{2+}(aq) + 6 \text{ H}_2\text{O}(l) \longrightarrow 2 \text{ IO}_3^-(aq) + 5 \text{ Cu}(s) + 12 \text{ H}^+(aq)$
(b) $Hg^{2+}(aq) + 2 \text{ I}^-(aq) \longrightarrow Hg(l) + I_2(s)$
(c) $H_2SO_3(aq) + 2 \text{ Mn}(s) + 4 \text{ H}^+(aq) \longrightarrow S(s) + 2 \text{ Mn}^{2+}(aq) + 3 \text{ H}_2\text{O}(l)$

We can use standard reduction potentials to understand the activity series of metals. ∞ (Section 4.4) Recall that any metal in the activity series (Table 4.5) is oxidized by the ions of any metal below it. We can now recognize the origin of this rule based on standard reduction potentials. The activity series is based on the oxidation reactions of the metals, ordered from strongest reducing agent at the top to weakest reducing agent at the bottom. (Thus, the ordering is inverted relative to that in Table 20.1.) For example, nickel lies above silver in the activity series, making nickel the stronger reducing agent. Because a reducing agent is oxidized in any redox reaction, nickel is more easily oxidized than silver. In a mixture of nickel metal and silver cations, therefore, we expect a displacement reaction in which the silver ions are displaced in the solution by nickel ions:

$$Ni(s) + 2 \text{ Ag}^+(aq) \longrightarrow Ni^{2+}(aq) + 2 \text{ Ag}(s)$$

In this reaction Ni is oxidized and Ag^+ is reduced. Therefore, the standard emf for the reaction is

$$E° = E°_{red}(Ag^+/Ag) - E°_{red}(Ni^{2+}/Ni)$$
$$= (+0.80 \text{ V}) - (-0.28 \text{ V}) = +1.08 \text{ V}$$

The positive value of $E°$ indicates that the displacement of silver by nickel resulting from oxidation of Ni metal and reduction of Ag^+ is a spontaneous process. Remember that although we multiply the silver half-reaction by 2, the reduction potential is not multiplied.

Give It Some Thought

Based on their relative positions in Table 4.5, which will have a more positive standard reduction potential, Sn^{2+} or Ni^{2+}?

Emf, Free Energy, and the Equilibrium Constant

The change in the Gibbs free energy, ΔG, is a measure of the spontaneity of a process that occurs at constant temperature and pressure. ∞ (Section 19.5) The emf, E, of a redox reaction also indicates whether the reaction is spontaneous. The relationship between emf and the free-energy change is

$$\Delta G = -nFE \qquad [20.11]$$

In this equation, n is a positive number without units that represents the number of moles of electrons transferred according to the balanced equation for the reaction, and F is the **Faraday constant**, named after Michael Faraday (▶ Figure 20.12):

$$F = 96,485 \text{ C/mol} = 96,485 \text{ J/V-mol}$$

The Faraday constant is the quantity of electrical charge on 1 mol of electrons.

▲ **Figure 20.12 Michael Faraday.** Faraday (1791–1867) was born in England, a child of a poor blacksmith. At the age of 14 he was apprenticed to a bookbinder who gave him time to read and to attend lectures. In 1812 he became an assistant in Humphry Davy's laboratory at the Royal Institution. He succeeded Davy as the most famous and influential scientist in England, making an amazing number of important discoveries, including his formulation of the quantitative relationships between electrical current and the extent of chemical reaction in electrochemical cells.

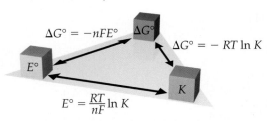

GO FIGURE

What does the variable n represent in the $\Delta G°$ and $E°$ equations?

$$\Delta G° = -nFE°$$

$$\Delta G° = -RT \ln K$$

$$E° = \frac{RT}{nF} \ln K$$

▲ **Figure 20.13 Relationships of $E°$, $\Delta G°$, and K.** Any one of these important parameters can be used to calculate the other two. The signs of $E°$ and $\Delta G°$ determine the direction in which the reaction proceeds under standard conditions. The magnitude of K determines the relative amounts of reactants and products in an equilibrium mixture.

The units of ΔG calculated with Equation 20.11 are J/mol. As in Equation 19.19, we use "per mole" to mean per mole of reaction as indicated by the coefficients in the balanced equation. ⟳ (Section 19.7)

Because both n and F are positive numbers, a positive value of E in Equation 20.11 leads to a negative value of ΔG. Remember: *A positive value of* E *and a negative value of* ΔG *both indicate a spontaneous reaction.* When the reactants and products are all in their standard states, Equation 20.11 can be modified to relate $\Delta G°$ and $E°$.

$$\Delta G° = -nFE° \qquad [20.12]$$

Because $\Delta G°$ is related to the equilibrium constant, K, for a reaction by the expression $\Delta G° = -RT \ln K$ (Equation 19.20), we can relate $E°$ to K by solving Equation 20.12 for $E°$ and then substituting the Equation 19.20 expression for $\Delta G°$.

$$E° = \frac{\Delta G°}{-nF} = \frac{-RT \ln K}{-nF} = \frac{RT}{nF} \ln K \qquad [20.13]$$

◀ Figure 20.13 summarizes the relationships among $E°$, $\Delta G°$, and K.

SAMPLE EXERCISE 20.10 | Using Standard Reduction Potentials to Calculate $\Delta G°$ and K

(a) Use the standard reduction potentials in Table 20.1 to calculate the standard free-energy change, $\Delta G°$, and the equilibrium constant, K, at 298 K for the reaction

$$4 \, Ag(s) + O_2(g) + 4 \, H^+(aq) \longrightarrow 4 \, Ag^+(aq) + 2 \, H_2O(l)$$

(b) Suppose the reaction in part (a) is written

$$2 \, Ag(s) + \frac{1}{2} O_2(g) + 2 \, H^+(aq) \longrightarrow 2 \, Ag^+(aq) + H_2O(l)$$

What are the values of $E°$, $\Delta G°$, and K when the reaction is written in this way?

SOLUTION

Analyze We are asked to determine $\Delta G°$ and K for a redox reaction, using standard reduction potentials.

Plan We use the data in Table 20.1 and Equation 20.10 to determine $E°$ for the reaction and then use $E°$ in Equation 20.12 to calculate $\Delta G°$. We can then use either Equation 19.20 or Equation 20.13 to calculate K.

Solve

(a) We first calculate $E°$ by breaking the equation into two half-reactions and obtaining $E°_{red}$ values from Table 20.1 (or Appendix E):

Reduction:	$O_2(g) + 4 \, H^+(aq) + 4 \, e^- \longrightarrow 2 \, H_2O(l)$	$E°_{red} = +1.23 \, V$
Oxidation:	$4 \, Ag(s) \longrightarrow 4 \, Ag^+(aq) + 4 \, e^-$	$E°_{red} = +0.80 \, V$

Even though the second half-reaction has 4 Ag, we use the $E°_{red}$ value directly from Table 20.1 because emf is an intensive property.

Using Equation 20.10, we have

$$E° = (1.23 \, V) - (0.80 \, V) = 0.43 \, V$$

The half-reactions show the transfer of four electrons. Thus, for this reaction $n = 4$. We now use Equation 20.12 to calculate $\Delta G°$:

$$\Delta G° = -nFE°$$
$$= -(4)(96{,}485 \, J/V\text{-}mol)(+0.43 \, V)$$
$$= -1.7 \times 10^5 \, J/mol = -170 \, kJ/mol$$

Now we need to calculate the equilibrium constant, K, using $\Delta G° = RT \ln K$. Because $\Delta G°$ is a large negative number, which means the reaction is thermodynamically very favorable, we expect K to be large.

$$\Delta G° = -RT \ln K$$
$$-1.7 \times 10^5 \, J/mol = -(8.314 \, J/K \, mol)(298 \, K) \ln K$$

$$\ln K = \frac{-1.7 \times 10^5 J/mol}{-(8.314 \, J/K \, mol)(298 \, K)}$$

$$\ln K = 69$$
$$K = 9 \times 10^{29}$$

(b) The overall equation is the same as that in part (a), multiplied by $\frac{1}{2}$. The half-reactions are

Reduction: $\frac{1}{2}O_2(g) + 2\,H^+(aq) + 2\,e^- \longrightarrow H_2O(l)$ $E^\circ_{red} = +1.23\text{ V}$

Oxidation: $2\,Ag(s) \longrightarrow 2\,Ag^+(aq) + 2\,e^-$ $E^\circ_{red} = +0.80\text{ V}$

The values of E°_{red} are the same as they were in part (a); they are not changed by multiplying the half-reactions by $\frac{1}{2}$. Thus, E° has the same value as in part (a): $E^\circ = +0.43$ V. Notice, though, that the value of n has changed to $n = 2$, which is one-half the value in part (a). Thus, ΔG° is half as large as in part (a):

$$\Delta G^\circ = -(2)(96{,}485\text{ J/V-mol})(+0.43\text{ V}) = -83\text{ kJ/mol}$$

The value of ΔG° is half that in part (a) because the coefficients in the chemical equation are half those in (a).

Now we can calculate K as before:

$$-8.3 \times 10^4\text{ J/mol} = -(8.314\text{ J/K mol})(298\text{ K})\ln K$$

$$K = 4 \times 10^{14}$$

Comment E° is an *intensive* quantity, so multiplying a chemical equation by a certain factor will not affect the value of E°. Multiplying an equation will change the value of n, however, and hence the value of ΔG°. The change in free energy, in units of J/mol of reaction as written, is an *extensive* quantity. The equilibrium constant is also an extensive quantity.

Practice Exercise 1

For the reaction

$$3\,Ni^{2+}(aq) + 2\,Cr(OH)_3(s) + 10\,OH^-(aq) \longrightarrow 3\,Ni(s) + 2\,CrO_4^{2-}(aq) + 8\,H_2O(l)$$

$\Delta G^\circ = +87\text{ kJ/mol}$. Given the standard reduction potential of $Ni^{2+}(aq)$ in Table 20.1, what value do you calculate for the standard reduction potential of the half-reaction

$$CrO_4^{2-}(aq) + 4\,H_2O(l) + 3\,e^- \longrightarrow Cr(OH)_3(s) + 5\,OH^-(aq)?$$

(a) -0.43 V   **(b)** -0.28 V   **(c)** 0.02 V   **(d)** -0.13 V   **(e)** -0.15 V

Practice Exercise 2

Consider the reaction $2\,Ag^+(aq) + H_2(g) \longrightarrow 2\,Ag(s) + 2\,H^+(aq)$. Calculate ΔG°_f for the $Ag^+(aq)$ ion from the standard reduction potentials in Table 20.1 and the fact that ΔG°_f for $H_2(g)$, $Ag(s)$, and $H^+(aq)$ are all zero. Compare your answer with the value given in Appendix C.

A Closer Look

Electrical Work

For any spontaneous process, ΔG is a measure of the maximum useful work, w_{max}, that can be extracted from the process: $\Delta G = w_{max}$. ∞ (Section 19.5) Because $\Delta G = -nFE$, the maximum useful electrical work obtainable from a voltaic cell is

$$w_{max} = -nFE_{cell} \qquad \text{[20.14]}$$

Because cell emf, E_{cell}, is always positive for a voltaic cell, w_{max} is negative, indicating that work is done *by* a system *on* its surroundings, as we expect for a voltaic cell. ∞ (Section 5.2)

As Equation 20.14 shows, the more charge a voltaic cell moves through a circuit (that is, the larger nF is) and the larger the emf pushing the electrons through the circuit (that is, the larger E°_{cell} is), the more work the cell can accomplish. In Sample Exercise 20.10, we calculated $\Delta G^\circ = -170\text{ kJ/mol}$ for the reaction $4\,Ag(s) + O_2(g) + 4\,H^+(aq) \longrightarrow 4\,Ag^+(aq) + 2\,H_2O(l)$. Thus, a voltaic cell utilizing this reaction could perform a maximum of 170 kJ of work in consuming 4 mol Ag, 1 mol O_2, and 4 mol H^+.

If a reaction is not spontaneous, ΔG is positive and E is negative. To force a nonspontaneous reaction to occur in an electrochemical cell, we need to apply an external potential, E_{ext}, that exceeds $|E_{cell}|$. For example, if a nonspontaneous process has $E = -0.9$ V, then the external potential E_{ext} must be greater than $+0.9$ V in order for the process to occur. We will examine such nonspontaneous processes in Section 20.9.

Electrical work can be expressed in energy units of watts times time. The *watt* (W) is a unit of electrical power (that is, rate of energy expenditure):

$$1\text{ W} = 1\text{ J/s}$$

Thus, a watt-second is a joule. The unit employed by electric utilities is the kilowatt-hour (kWh), which equals 3.6×10^6 J:

$$1\text{ kWh} = (1000\text{ W})(1\text{ h})\left(\frac{3600\text{ s}}{1\text{ h}}\right)\left(\frac{1\text{ J/s}}{1\text{ W}}\right) = 3.6 \times 10^6\text{ J}$$

Related Exercises: 20.59, 20.60

20.6 | Cell Potentials Under Nonstandard Conditions

We have seen how to calculate the emf of a cell when the reactants and products are under standard conditions. As a voltaic cell is discharged, however, reactants are consumed and products are generated, so concentrations change. The emf progressively drops until $E = 0$, at which point we say the cell is "dead." In this section, we examine how the emf generated under nonstandard conditions can be calculated by using an equation first derived by Walther Nernst (1864–1941), a German chemist who established many of the theoretical foundations of electrochemistry.

The Nernst Equation

The effect of concentration on cell emf can be obtained from the effect of concentration on free-energy change. ∞ (Section 19.7) Recall that the free-energy change for any chemical reaction, ΔG, is related to the standard free-energy change for the reaction, $\Delta G°$:

$$\Delta G = \Delta G° + RT \ln Q \qquad [20.15]$$

The quantity Q is the reaction quotient, which has the form of the equilibrium-constant expression except that the concentrations are those that exist in the reaction mixture at a given moment. ∞ (Section 15.6)

Substituting $\Delta G = -nFE$ (Equation 20.11) into Equation 20.15 gives

$$-nFE = -nFE° + RT \ln Q$$

Solving this equation for E gives the **Nernst equation**:

$$E = E° - \frac{RT}{nF} \ln Q \qquad [20.16]$$

This equation is customarily expressed in terms of the base-10 logarithm:

$$E = E° - \frac{2.303 \, RT}{nF} \log Q \qquad [20.17]$$

At $T = 298$ K, the quantity $2.303 \, RT/F$ equals 0.0592, with units of volts, and so the Nernst equation simplifies to

$$E = E° - \frac{0.0592 \text{ V}}{n} \log Q \qquad (T = 298 \text{ K}) \qquad [20.18]$$

We can use this equation to find the emf E produced by a cell under nonstandard conditions or to determine the concentration of a reactant or product by measuring E for the cell. For example, consider the following reaction:

$$\text{Zn}(s) + \text{Cu}^{2+}(aq) \longrightarrow \text{Zn}^{2+}(aq) + \text{Cu}(s)$$

In this case $n = 2$ (two electrons are transferred from Zn to Cu^{2+}), and the standard emf is +1.10 V. ∞ (Section 20.4) Thus, at 298 K the Nernst equation gives

$$E = 1.10 \text{ V} - \frac{0.0592 \text{ V}}{2} \log \frac{[\text{Zn}^{2+}]}{[\text{Cu}^{2+}]} \qquad [20.19]$$

Recall that pure solids are excluded from the expression for Q. ∞ (Section 15.6) According to Equation 20.19, the emf increases as $[\text{Cu}^{2+}]$ increases and as $[\text{Zn}^{2+}]$ decreases. For example, when $[\text{Cu}^{2+}]$ is 5.0 M and $[\text{Zn}^{2+}]$ is 0.050 M, we have

$$E = 1.10 \text{ V} - \frac{0.0592 \text{ V}}{2} \log \left(\frac{0.050}{5.0} \right)$$

$$= 1.10 \text{ V} - \frac{0.0592 \text{ V}}{2} (-2.00) = 1.16 \text{ V}$$

Thus, increasing the concentration of reactant Cu^{2+} and decreasing the concentration of product Zn^{2+} relative to standard conditions increases the emf of the cell relative to standard conditions ($E° = +1.10$ V).

The Nernst equation helps us understand why the emf of a voltaic cell drops as the cell discharges. As reactants are converted to products, the value of Q increases, so the value of E decreases, eventually reaching $E = 0$. Because $\Delta G = -nFE$ (Equation 20.11), it follows that $\Delta G = 0$ when $E = 0$. Recall that a system is at equilibrium when $\Delta G = 0$. ∞ (Section 19.7) Thus, when $E = 0$, the cell reaction has reached equilibrium, and no net reaction occurs.

In general, increasing the concentration of reactants or decreasing the concentration of products increases the driving force for the reaction, resulting in a higher emf. Conversely, decreasing the concentration of reactants or increasing the concentration of products causes the emf to decrease from its value under standard conditions.

SAMPLE EXERCISE 20.11 | Cell Potential under Nonstandard Conditions

Calculate the emf at 298 K generated by a voltaic cell in which the reaction is

$$Cr_2O_7^{2-}(aq) + 14\,H^+(aq) + 6\,I^-(aq) \longrightarrow 2\,Cr^{3+}(aq) + 3\,I_2(s) + 7\,H_2O(l)$$

when

$$[Cr_2O_7^{2-}] = 2.0\,M, [H^+] = 1.0\,M, [I^-] = 1.0\,M, \text{ and } [Cr^{3+}] = 1.0 \times 10^{-5}\,M$$

SOLUTION

Analyze We are given a chemical equation for a voltaic cell and the concentrations of reactants and products under which it operates. We are asked to calculate the emf of the cell under these nonstandard conditions.

Plan To calculate the emf of a cell under nonstandard conditions, we use the Nernst equation in the form of Equation 20.18.

Solve We calculate $E°$ for the cell from standard reduction potentials (Table 20.1 or Appendix E). The standard emf for this reaction was calculated in Sample Exercise 20.6: $E° = 0.79$ V. As that exercise shows, six electrons are transferred from reducing agent to oxidizing agent, so $n = 6$. The reaction quotient, Q, is

$$Q = \frac{[Cr^{3+}]^2}{[Cr_2O_7^{2-}][H^+]^{14}[I^-]^6} = \frac{(1.0 \times 10^{-5})^2}{(2.0)(1.0)^{14}(1.0)^6} = 5.0 \times 10^{-11}$$

Using Equation 20.18, we have

$$E = 0.79\,V - \left(\frac{0.0592\,V}{6}\right)\log(5.0 \times 10^{-11})$$

$$= 0.79\,V - \left(\frac{0.0592\,V}{6}\right)(-10.30)$$

$$= 0.79\,V + 0.10\,V = 0.89\,V$$

Check This result is qualitatively what we expect: Because the concentration of $Cr_2O_7^{2-}$ (a reactant) is greater than 1 M and the concentration of Cr^{3+} (a product) is less than 1 M, the emf is greater than $E°$. Because Q is about 10^{-10}, $\log Q$ is about -10. Thus, the correction to $E°$ is about $0.06 \times 10/6$, which is 0.1, in agreement with the more detailed calculation.

Practice Exercise 1

Consider a voltaic cell whose overall reaction is $Pb^{2+}(aq) + Zn(s) \longrightarrow Pb(s) + Zn^{2+}(aq)$. What is the emf generated by this voltaic cell when the ion concentrations are $[Pb^{2+}] = 1.5 \times 10^{-3}\,M$ and $[Zn^{2+}] = 0.55\,M$? **(a)** 0.71 V, **(b)** 0.56 V, **(c)** 0.49 V, **(d)** 0.79 V, **(e)** 0.64 V.

Practice Exercise 2

For the Zn–Cu voltaic cell depicted in Figure 20.5 would the emf increase, decrease, or stay the same, if you increased the $Cu^{2+}(aq)$ concentration by adding $CuSO_4 \cdot 5H_2O$ to the cathode compartment?

SAMPLE EXERCISE 20.12 | Calculating Concentrations in a Voltaic Cell

If the potential of a Zn–H^+ cell (like that in Figure 20.9) is 0.45 V at 25 °C when $[Zn^{2+}] = 1.0\ M$ and $P_{H_2} = 1.0$ atm, what is the pH of the cathode solution?

SOLUTION

Analyze We are given a description of a voltaic cell, its emf, the concentration of Zn^{2+}, and the partial pressure of H_2 (both products in the cell reaction). We are asked to calculate the pH of the cathode solution, which we can calculate from the concentration of H^+, a reactant.

Plan We write the equation for the cell reaction and use standard reduction potentials to calculate $E°$ for the reaction. After determining the value of n from our reaction equation, we solve the Nernst equation, Equation 20.18, for Q. We use the equation for the cell reaction to write an expression for Q that contains $[H^+]$ to determine $[H^+]$. Finally, we use $[H^+]$ to calculate pH.

Solve The cell reaction is

$$Zn(s) + 2\,H^+(aq) \longrightarrow Zn^{2+}(aq) + H_2(g)$$

The standard emf is

$$E° = E°_{red}\,(\text{reduction}) - E°_{red}\,(\text{oxidation})$$
$$= 0\,V - (-0.76\,V) = +0.76\,V$$

Because each Zn atom loses two electrons,

$$n = 2$$

Using Equation 20.18, we can solve for Q:

$$0.45\,V = 0.76\,V - \frac{0.0592\,V}{2} \log Q$$
$$Q = 10^{10.5} = 3 \times 10^{10}$$

Q has the form of the equilibrium constant for the reaction:

$$Q = \frac{[Zn^{2+}]P_{H_2}}{[H^+]^2} = \frac{(1.0)(1.0)}{[H^+]^2} = 3 \times 10^{10}$$

Solving for $[H^+]$, we have

$$[H^+]^2 = \frac{1.0}{3 \times 10^{10}} = 3 \times 10^{-11}$$
$$[H^+] = \sqrt{3 \times 10^{-11}} = 6 \times 10^{-6}\,M$$

Finally, we use $[H^+]$ to calculate the pH of the cathode solution.

$$pH = \log[H^+] = -\log(6 \times 10^{-6}) = 5.2$$

Comment A voltaic cell whose cell reaction involves H^+ can be used to measure $[H^+]$ or pH. A pH meter is a specially designed voltaic cell with a voltmeter calibrated to read pH directly. ∞ (Section 16.4)

Practice Exercise 1

Consider a voltaic cell where the anode half-reaction is $Zn(s) \longrightarrow Zn^{2+}(aq) + 2\,e^-$ and the cathode half-reaction is $Sn^{2+}(aq) + 2\,e^- \longrightarrow Sn(s)$. What is the concentration of Sn^{2+} if Zn^{2+} is $2.5 \times 10^{-3}\ M$ and the cell emf is 0.660 V? Use the reduction potentials in Appendix E that are reported to three significant figures. **(a)** $3.3 \times 10^{-2}\,M$, **(b)** $1.9 \times 10^{-4}\,M$, **(c)** $9.0 \times 10^{-3}\,M$, **(d)** $6.9 \times 10^{-4}\,M$, **(e)** $7.6 \times 10^{-3}\,M$.

Practice Exercise 2

What is the pH of the solution in the cathode half-cell in Figure 20.9 when $P_{H_2} = 1.0$ atm, $[Zn^{2+}]$ in the anode half-cell is 0.10 M, and the cell emf is 0.542 V?

Concentration Cells

In the voltaic cells we have looked at thus far, the reactive species at the anode has been different from the reactive species at the cathode. Cell emf depends on concentration, however, so a voltaic cell can be constructed using the *same* species in both half-cells as long as the concentrations are different. A cell based solely on the emf generated because of a difference in a concentration is called a **concentration cell**.

An example of a concentration cell is diagrammed in ▶ Figure 20.14(a). One half-cell consists of a strip of nickel metal immersed in a $1.00 \times 10^{-3}\ M$ solution of $Ni^{2+}(aq)$. The other half-cell also has an Ni(s) electrode, but it is immersed in a 1.00 M solution of $Ni^{2+}(aq)$. The two half-cells are connected by a salt bridge and by an

GO FIGURE

Assuming that the solutions are made from $Ni(NO_3)_2$, how do the ions migrate as the cell operates?

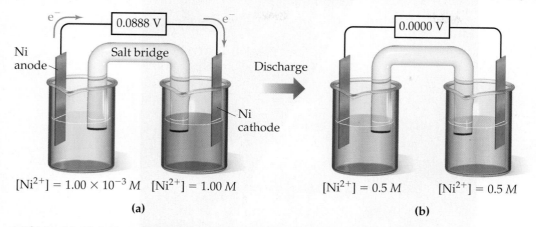

$[Ni^{2+}] = 1.00 \times 10^{-3}\ M$ $[Ni^{2+}] = 1.00\ M$ $[Ni^{2+}] = 0.5\ M$ $[Ni^{2+}] = 0.5\ M$

(a)   **(b)**

▲ **Figure 20.14 Concentration cell based on Ni^{2+}–Ni cell reaction.** (a) Concentrations of $Ni^{2+}(aq)$ in the two half-cells are unequal, and the cell generates an electrical current. (b) The cell operates until $[Ni^{2+}(aq)]$ is the same in the two half-cells, at which point the cell has reached equilibrium and the emf goes to zero.

external wire running through a voltmeter. The half-cell reactions are the reverse of each other:

Anode:	$Ni(s) \longrightarrow Ni^{2+}(aq) + 2\ e^-$	$E^\circ_{red} = -0.28\ V$
Cathode:	$Ni^{2+}(aq) + 2\ e^- \longrightarrow Ni(s)$	$E^\circ_{red} = -0.28\ V$

Although the *standard* emf for this cell is zero,

$$E^\circ_{cell} = E^\circ_{red}\,(\text{cathode}) - E^\circ_{red}\,(\text{anode}) = (-0.28\ V) - (-0.28\ V) = 0\ V$$

the cell operates under *nonstandard* conditions because the concentration of $Ni^{2+}(aq)$ is not 1 M in both half-cells. In fact, the cell operates until $[Ni^{2+}]_{anode} = [Ni^{2+}]_{cathode}$. Oxidation of $Ni(s)$ occurs in the half-cell containing the more dilute solution, which means this is the anode of the cell. Reduction of $Ni^{2+}(aq)$ occurs in the half-cell containing the more concentrated solution, making it the cathode. The *overall* cell reaction is therefore

Anode:	$Ni(s) \longrightarrow Ni^{2+}(aq, \text{dilute}) + 2\ e^-$
Cathode:	$Ni^{2+}(aq, \text{concentrated}) + 2\ e^- \longrightarrow Ni(s)$
Overall:	$Ni^{2+}(aq, \text{concentrated}) \longrightarrow Ni^{2+}(aq, \text{dilute})$

We can calculate the emf of a concentration cell by using the Nernst equation. For this particular cell, we see that $n = 2$. The expression for the reaction quotient for the overall reaction is $Q = [Ni^{2+}]_{dilute}/[Ni^{2+}]_{concentrated}$. Thus, the emf at 298 K is

$$E = E^\circ - \frac{0.0592\ V}{n} \log Q$$

$$= 0 - \frac{0.0592\ V}{2} \log \frac{[Ni^{2+}]_{dilute}}{[Ni^{2+}]_{concentrated}} = -\frac{0.0592\ V}{2} \log \frac{1.00 \times 10^{-3}\ M}{1.00\ M}$$

$$= +0.0888\ V$$

This concentration cell generates an emf of nearly 0.09 V even though $E^\circ = 0$. The difference in concentration provides the driving force for the cell. When the concentrations in the two half-cells become the same, $Q = 1$ and $E = 0$.

The idea of generating a potential by a concentration difference is the basis for the operation of pH meters. It is also a critical aspect in biology. For example, nerve cells in the

▲ Figure 20.15 **An electric eel.** Differences in ion concentrations, mainly Na⁺ and K⁺, in special cells called electrocytes produce an emf on the order of 0.1 V. By connecting thousands of these cells in series these South American fish are able to generate short electric pulses as high as 500 V.

brain generate a potential across the cell membrane by having different concentrations of ions on the two sides of the membrane. Electric eels use cells called electrocytes that are based on a similar principle to generate short, but intense pulses of electricity to stun prey and dissuade predators. (▲ Figure 20.15). The regulation of the heartbeat in mammals, as discussed in the following "Chemistry and Life" box, is another example of the importance of electrochemistry to living organisms.

Chemistry and Life

Heartbeats and Electrocardiography

The human heart is a marvel of efficiency and dependability. In a typical day an adult's heart pumps more than 7000 L of blood through the circulatory system, usually with no maintenance required beyond a sensible diet and lifestyle. We generally think of the heart as a mechanical device, a muscle that circulates blood via regularly spaced muscular contractions. However, more than two centuries ago, two pioneers in electricity, Luigi Galvani (1729–1787) and Alessandro Volta (1745–1827), discovered that the contractions of the heart are controlled by electrical phenomena, as are nerve impulses throughout the body. The pulses of electricity that cause the heart to beat result from a remarkable combination of electrochemistry and the properties of semipermeable membranes. ∞ (Section 13.5)

Cell walls are membranes with variable permeability with respect to a number of physiologically important ions (especially Na⁺, K⁺, and Ca²⁺). The concentrations of these ions are different for the fluids inside the cells (the *intracellular fluid*, or ICF) and outside the cells (the

extracellular fluid, or ECF). In cardiac muscle cells, for example, the concentrations of K⁺ in the ICF and ECF are typically about 135 millimolar (mM) and 4 mM, respectively. For Na⁺, however, the concentration difference between the ICF and ECF is opposite to that for K⁺; typically, $[Na^+]_{ICF} = 10$ mM and $[Na^+]_{ECF} = 145$ mM.

The cell membrane is initially permeable to K⁺ ions but is much less so to Na⁺ and Ca²⁺. The difference in concentration of K⁺ ions between the ICF and ECF generates a concentration cell. Even though the same ions are present on both sides of the membrane, there is a potential difference between the two fluids that we can calculate using the Nernst equation with $E° = 0$. At physiological temperature (37 °C) the potential in millivolts for moving K⁺ from the ECF to the ICF is

$$E = E° - \frac{2.30\,RT}{nF} \log \frac{[K^+]_{ICF}}{[K^+]_{ECF}}$$

$$= 0 - (61.5\text{ mV}) \log \left(\frac{135\text{ m}M}{4\text{ m}M} \right) = -94\text{ mV}$$

In essence, the interior of the cell and the ECF together serve as a voltaic cell. The negative sign for the potential indicates that work is required to move K^+ into the ICF.

Changes in the relative concentrations of the ions in the ECF and ICF lead to changes in the emf of the voltaic cell. The cells of the heart that govern the rate of heart contraction are called the *pacemaker cells*. The membranes of the cells regulate the concentrations of ions in the ICF, allowing them to change in a systematic way. The concentration changes cause the emf to change in a cyclic fashion, as shown in ▼ Figure 20.16. The emf cycle determines the rate at which the heart beats. If the pacemaker cells malfunction because of disease or injury, an artificial pacemaker can be surgically implanted. The artificial

pacemaker contains a small battery that generates the electrical pulses needed to trigger the contractions of the heart.

During the late 1800s, scientists discovered that the electrical impulses that cause the contraction of the heart muscle are strong enough to be detected at the surface of the body. This observation formed the basis for *electrocardiography*, noninvasive monitoring of the heart by using a complex array of electrodes on the skin to measure voltage changes during heartbeats. A typical electrocardiogram is shown in ▼ Figure 20.17. It is quite striking that, although the heart's major function is the *mechanical* pumping of blood, it is most easily monitored by using the *electrical* impulses generated by tiny voltaic cells.

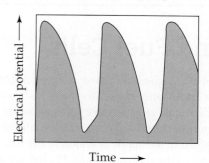

▲ **Figure 20.16 Changes in electrical potential in the human heart.** Variation of the electrical potential caused by changes of ion concentrations in the pacemaker cells of the heart.

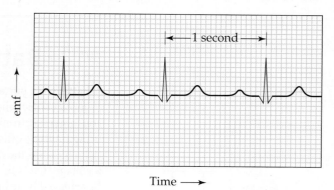

▲ **Figure 20.17 A typical electrocardiogram.** The printout records the electrical events monitored by electrodes attached to the body surface.

SAMPLE EXERCISE 20.13 Determining pH Using a Concentration Cell

A voltaic cell is constructed with two hydrogen electrodes. Electrode 1 has $P_{H_2} = 1.00$ atm and an unknown concentration of $H^+(aq)$. Electrode 2 is a standard hydrogen electrode ($P_{H_2} = 1.00$ atm, $[H^+] = 1.00\ M$). At 298 K the measured cell potential is 0.211 V, and the electrical current is observed to flow from electrode 1 through the external circuit to electrode 2. Calculate $[H^+]$ for the solution at electrode 1. What is the pH of the solution?

SOLUTION

Analyze We are given the potential of a concentration cell and the direction in which the current flows. We also have the concentrations or partial pressures of all reactants and products except for $[H^+]$ in half-cell 1, which is our unknown.

Plan We can use the Nernst equation to determine Q and then use Q to calculate the unknown concentration. Because this is a concentration cell, $E^{\circ}_{cell} = 0$ V.

Solve Using the Nernst equation, we have

$$0.211\ \text{V} = 0 - \frac{0.0592\ \text{V}}{2}\log Q$$

$$\log Q = -(0.211\ \text{V})\left(\frac{2}{0.0592\ \text{V}}\right) = -7.13$$

$$Q = 10^{-7.13} = 7.4 \times 10^{-8}$$

Because electrons flow from electrode 1 to electrode 2, electrode 1 is the anode of the cell and electrode 2 is the cathode. The electrode reactions are therefore as follows, with the concentration of $H^+(aq)$ in electrode 1 represented with the unknown x:

Electrode 1:	$H_2(g, 1.00\ \text{atm}) \longrightarrow 2\,H^+(aq, x\,M) + 2\,e^- \quad E^{\circ}_{red} = 0$
Electrode 2:	$2\,H^+(aq;\ 1.00\ M) + 2\,e^- \longrightarrow H_2(g, 1.00\ \text{atm}) \quad\quad E^{\circ}_{red} = 0$
Overall:	$2\,H^+(aq;\ 1.00\ M) \longrightarrow 2\,H^+(aq, x\,M)$

Thus,

$$Q = \frac{[H^+(aq, x\,M)]^2}{[H^+(aq, 1.00\,M)]^2}$$

$$= \frac{x^2}{(1.00)^2} = x^2 = 7.4 \times 10^{-8}$$

$$x = [H^+] = \sqrt{7.4 \times 10^{-8}} = 2.7 \times 10^{-4}$$

At electrode 1, therefore, the pH of the solution is

$$\text{pH} = -\log[H^+] = -\log(2.7 \times 10^{-4}) = 3.57$$

Comment The concentration of H^+ at electrode 1 is lower than that in electrode 2, which is why electrode 1 is the anode of the cell: The oxidation of H_2 to $H^+(aq)$ increases $[H^+]$ at electrode 1.

Practice Exercise 1

A concentration cell constructed from two hydrogen electrodes, both with $P_{H_2} = 1.00$. One electrode is immersed in pure H_2O and the other in 6.0 M hydrochloric acid. What is the emf generated by the cell and what is the identity of the electrode that is immersed in hydrochloric acid? **(a)** −0.23 V, cathode, **(b)** 0.46 V, anode, **(c)** 0.023 V, anode, **(d)** 0.23 V, cathode, **(e)** 0.23 V, anode.

Practice Exercise 2

A concentration cell is constructed with two $Zn(s)-Zn^{2+}(aq)$ half-cells. In one half-cell $[Zn^{2+}] = 1.35$ M, and in the other $[Zn^{2+}] = 3.75 \times 10^{-4}$ M. **(a)** Which half-cell is the anode? **(b)** What is the emf of the cell?

20.7 | Batteries and Fuel Cells

A **battery** is a portable, self-contained electrochemical power source that consists of one or more voltaic cells. For example, the 1.5-V batteries used to power flashlights and many consumer electronic devices are single voltaic cells. Greater voltages can be achieved by using multiple cells, as in 12-V automotive batteries. When cells are connected in series (which means the cathode of one attached to the anode of another), the battery produces a voltage that is the sum of the voltages of the individual cells. Higher voltages can also be achieved by using multiple batteries in series (◀ Figure 20.18). Battery electrodes are marked following the convention of Figure 20.6—plus for cathode and minus for anode.

Although any spontaneous redox reaction can serve as the basis for a voltaic cell, making a commercial battery that has specific performance characteristics requires considerable ingenuity. The substances oxidized at the anode and reduced by the cathode determine the voltage, and the usable life of the battery depends on the quantities of these substances packaged in the battery. Usually a barrier analogous to the porous barrier of Figure 20.6 separates the anode and cathode half-cells.

Different applications require batteries with different properties. The battery required to start a car, for example, must be capable of delivering a large electrical current for a short time period, whereas the battery that powers a heart pacemaker must be very small and capable of delivering a small but steady current over an extended time period. Some batteries are **primary cells**, meaning they cannot be recharged and must be either discarded or recycled after the voltage drops to zero. A **secondary cell** can be recharged from an external power source after its voltage has dropped.

As we consider some common batteries, notice how the principles we have discussed so far help us understand these important sources of portable electrical energy.

▲ **Figure 20.18 Combining batteries.** When batteries are connected in series, as in most flashlights, the total voltage is the sum of the individual voltages.

Lead–Acid Battery

A 12-V lead–acid automotive battery consists of six voltaic cells in series, each producing 2 V. The cathode of each cell is lead dioxide (PbO_2) packed on a lead grid (▶ Figure 20.19). The anode of each cell is lead. Both electrodes are immersed in sulfuric acid.

The reactions that occur during discharge are

Cathode: $\quad PbO_2(s) + HSO_4^-(aq) + 3\,H^+(aq) + 2\,e^- \longrightarrow PbSO_4(s) + 2\,H_2O(l)$

Anode: $\quad\quad\quad\quad\quad Pb(s) + HSO_4^-(aq) \longrightarrow PbSO_4(s) + H^+(aq) + 2\,e^-$

Overall: $\quad PbO_2(s) + Pb(s) + 2\,HSO_4^-(aq) + 2\,H^+(aq) \longrightarrow 2\,PbSO_4(s) + 2\,H_2O(l)$ $\quad\quad$ [20.20]

The standard cell potential can be obtained from the standard reduction potentials in Appendix E:

$$E°_{cell} = E°_{red}(\text{cathode}) - E°_{red}(\text{anode}) = (+1.685\text{ V}) - (-0.356\text{ V}) = +2.041\text{ V}$$

The reactants Pb and PbO_2 are the electrodes. Because these reactants are solids, there is no need to separate the cell into half-cells; the Pb and PbO_2 cannot come into contact with each other unless one electrode touches another. To keep the electrodes from touching, wood or glass-fiber spacers are placed between them (Figure 20.19). Using a reaction whose reactants and products are solids has another benefit. Because solids are excluded from the reaction quotient Q, the relative amounts of $Pb(s)$, $PbO_2(s)$, and $PbSO_4(s)$ have no effect on the voltage of the lead storage battery, helping the battery maintain a relatively constant voltage during discharge. The voltage does vary somewhat with use because the concentration of H_2SO_4 varies with the extent of discharge. As Equation 20.20 indicates, H_2SO_4 is consumed during the discharge.

A major advantage of the lead–acid battery is that it can be recharged. During recharging, an external source of energy is used to reverse the direction of the cell reaction, regenerating $Pb(s)$ and $PbO_2(s)$:

$$2\,PbSO_4(s) + 2\,H_2O(l) \longrightarrow PbO_2(s) + Pb(s) + 2\,HSO_4^-(aq) + 2\,H^+(aq)$$

In an automobile, the alternator provides the energy necessary for recharging the battery. Recharging is possible because $PbSO_4$ formed during discharge adheres to the electrodes. As the external source forces electrons from one electrode to the other, the $PbSO_4$ is converted to Pb at one electrode and to PbO_2 at the other.

Alkaline Battery

The most common primary (nonrechargeable) battery is the alkaline battery (▶ Figure 20.20). The anode is powdered zinc metal immobilized in a gel in contact with a concentrated solution of KOH (hence, the name *alkaline* battery). The cathode is a mixture of $MnO_2(s)$ and graphite, separated from the anode by a porous fabric. The battery is sealed in a steel can to reduce the risk of any of the concentrated KOH escaping.

The cell reactions are complex but can be approximately represented as follows:

Cathode: $\quad 2\,MnO_2(s) + 2\,H_2O(l) + 2\,e^- \longrightarrow 2\,MnO(OH)(s) + 2\,OH^-(aq)$

Anode: $\qquad\quad Zn(s) + 2\,OH^-(aq) \longrightarrow Zn(OH)_2(s) + 2\,e^-$

Nickel–Cadmium and Nickel–Metal Hydride Batteries

The tremendous growth in high-power-demand portable electronic devices in the last decade has increased the demand for lightweight, readily rechargable batteries. One relatively common rechargeable battery is the nickel–cadmium (nicad) battery. During discharge, cadmium metal is oxidized at the anode while nickel oxyhydroxide $[NiO(OH)(s)]$ is reduced at the cathode:

Cathode: $\quad 2\,NiO(OH)(s) + 2\,H_2O(l) + 2\,e^- \longrightarrow 2\,Ni(OH)_2(s) + 2\,OH^-(aq)$

Anode: $\qquad\qquad Cd(s) + 2\,OH^-(aq) \longrightarrow Cd(OH)_2(s) + 2\,e^-$

As in the lead–acid battery, the solid reaction products adhere to the electrodes, which permits the electrode reactions to be reversed during charging. A single nicad voltaic cell has a voltage of 1.30 V. Nicad battery packs typically contain three or more cells in series to produce the higher voltages needed by most electronic devices.

Although nickel–cadmium batteries have a number of attractive characteristics the use of cadmium as the anode introduces significant limitations. Because cadmium is toxic, these batteries must be recycled. The toxicity of cadmium has led to a decline in their popularity from a peak annual production level of approximately 1.5 billion batteries in the early 2000s. Cadmium also has a relatively high density, which increases battery weight, an undesirable characteristic for use in portable devices and electric vehicles. These shortcomings have fueled the development of the nickel–metal hydride (NiMH) battery. The cathode reaction is the same as that for nickel–cadmium batteries, but the anode reaction is very different. The anode consists of a metal alloy, typically with AM_5 stoichiometry, where A is lanthanum (La) or a mixture of metals from the lanthanide series, and M is mostly nickel alloyed with smaller amounts of other transition metals. On charging, water is reduced at the anode to form hydroxide ions and hydrogen atoms that are absorbed into the AM_5 alloy. When the battery is operating (discharging) the hydrogen atoms are oxidized and the resultant H^+ ions react with OH^- ions to form H_2O.

Lithium-Ion Batteries

Currently most portable electronic devices, including cell phones and laptop computers, are powered by rechargeable lithium-ion (Li-ion) batteries. Because lithium

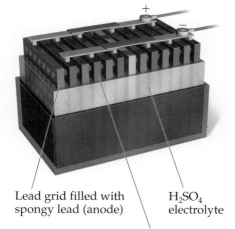

△ GO FIGURE

What is the oxidation state of lead in the cathode of this battery?

Lead grid filled with spongy lead (anode)

H_2SO_4 electrolyte

Lead grid filled with PbO_2 (cathode)

▲ **Figure 20.19 A 12-V automotive lead–acid battery.** Each anode/cathode pair in this schematic cutaway produces a voltage of about 2 V. Six pairs of electrodes are connected in series, producing 12 V.

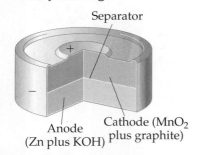

△ GO FIGURE

What substance is oxidized as the battery discharges?

Separator

Anode (Zn plus KOH)

Cathode (MnO_2 plus graphite)

▲ **Figure 20.20 Cutaway view of a miniature alkaline battery.**

is a very light element, Li-ion batteries achieve a greater *specific energy density*—the amount of energy stored per unit mass—than nickel-based batteries. Because Li^+ has a very large negative standard reduction potential (Table 20.1). Li-ion batteries produce a higher voltage per cell than other batteries. A Li-ion battery produces a maximum voltage of 3.7 V per cell, nearly three times higher than the 1.3 V per cell that nickel–cadmium and nickel–metal hydride batteries generate. As a result, a Li-ion battery can deliver more power than other batteries of comparable size, which leads to a higher *volumetric energy density*–the amount of energy stored per unit volume.

The technology of Li-ion batteries is based on the ability of Li^+ ions to be inserted into and removed from certain layered solids. In most commercial cells, the anode is made of graphite, which contains layers of sp^2 bonded carbon atoms (Figure 12.29(b)). The cathode is made of a transition metal oxide that also has a layered structure, typically lithium cobalt oxide ($LiCoO_2$). The two electrodes are separated by an electrolyte, which functions like a salt bridge by allowing Li^+ ions to pass through it. When the cell is being charged, cobalt ions are oxidized and Li^+ ions migrate out of $LiCoO_2$ and into the graphite. During discharge, when the battery is producing electricity for use, the Li^+ ions spontaneously migrate from the graphite anode through the electrolyte to the cathode, enabling electrons to flow through the external circuit. (▼ Figure 20.21)

◢ GO FIGURE

When a Li-ion battery is fully discharged the cathode has an empirical formula of $LiCoO_2$. What is the oxidation number of cobalt in this state? Does the oxidation number of the cobalt increase or decrease as the battery charges?

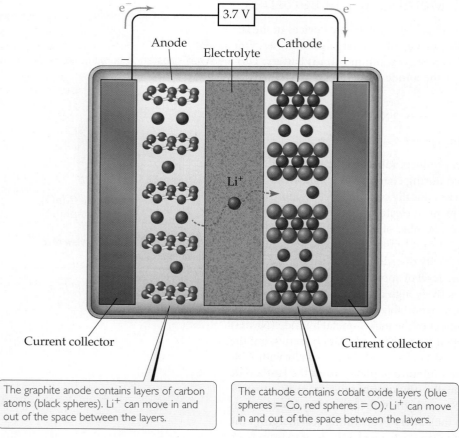

Current collector

Current collector

The graphite anode contains layers of carbon atoms (black spheres). Li^+ can move in and out of the space between the layers.

The cathode contains cobalt oxide layers (blue spheres = Co, red spheres = O). Li^+ can move in and out of the space between the layers.

▲ **Figure 20.21 Schematic of a Li-ion battery.** When the battery is discharging (operating) Li^+ ions move out of the anode and migrate through the electrolyte where they enter the spaces between the cobalt oxide layers, reducing the cobalt ions. To recharge the battery, electrical energy is used to drive the Li^+ back to the anode, oxidizing the cobalt ions in the cathode.

Chemistry Put to Work

Batteries for Hybrid and Electric Vehicles

There has been a tremendous growth over the last two decades in the development of electric vehicles. This growth has been driven by a desire to reduce the use of fossil fuels and lower emissions. Today both hybrid electric vehicles and fully electric vehicles are commercially available. Hybrid electric vehicles can be powered either by electricity from batteries or by a conventional combustion engine, while fully electric vehicles are powered exclusively by the batteries (▼ Figure 20.22). Hybrid electric vehicles can be further divided into plug-in hybrids which require the owner to charge the battery by plugging it into a conventional outlet, or regular hybrids which use regenerative braking and power from the combustion engine to charge the batteries.

Among the many technological advances needed to make electric vehicles practical, none is more important than advances in battery technology. Batteries for electric vehicles must have a high specific energy density, to reduce the weight of the car, as well as a high volumetric energy density, to minimize the space needed for the battery pack. A plot of energy densities for various types of rechargeable batteries is shown in ▶ Figure 20.23. The lead–acid batteries used in gasoline-powered automobiles are reliable and inexpensive, but their energy densities are far too low for practical use in an electric vehicle. Nickel–metal hydride batteries offer roughly three times higher energy density and until recently were the batteries of choice for commercial hybrid vehicles, such as the Toyota Prius.

Fully electric vehicles and plug-in hybrids use Li-ion batteries because they offer the highest energy density of all commercially

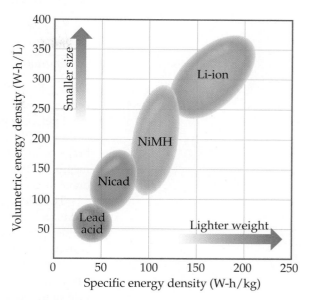

▲ Figure 20.23 **Energy densities of various types of batteries.** The higher the volumetric energy density, the smaller the amount of space needed for the batteries. The higher the specific energy density, the smaller the mass of the batteries. A Watt-hour (W-h) is equivalent to 3.6×10^3 Joules.

available batteries. As Li-ion battery technology has advanced, these batteries have started to displace the nickel–metal hydride batteries used in hybrid electric cars. Concern over safety is one factor that has delayed implementation of Li-ion batteries in commercial automobiles. In rare cases overheating and/or overcharging can cause Li-ion batteries to combust (the most high profile cases occurring in Boeing's 787 Dreamliner airplanes, ⬭ Section 7.3). Most electric vehicles now use Li-ion batteries where the $LiCoO_2$ cathode has been replaced by a cathode made from lithium manganese spinel ($LiMn_2O_4$). Batteries made with $LiMn_2O_4$ cathodes have several advantages. They are not prone to thermal runaway events that can lead to combustion, they tend to have longer lifetimes, and manganese is less expensive and more environmentally friendly than cobalt. However, they do have one important shortcoming—the capacity of batteries made from $LiMn_2O_4$ is only about two-third of that of batteries with $LiCoO_2$ cathodes. Scientists and engineers are intensively looking for new materials that will lead to further improvements in the energy density, cost, lifetime, and safety of batteries.

Related Exercises: 20.10, 20.79, 20.80

▲ Figure 20.22 **Electric automobile.** The Tesla Roadster is a fully electric vehicle powered by Li-ion batteries that can go over 200 miles per charge.

Hydrogen Fuel Cells

The thermal energy released by burning fuels can be converted to electrical energy. The thermal energy may convert water to steam, for instance, which drives a turbine that in turn drives an electrical generator. Typically, a maximum of only 40% of the energy from combustion is converted to electricity in this manner; the remainder is lost as heat. The direct production of electricity from fuels by a voltaic cell could, in principle, yield a higher rate of conversion of chemical energy to electrical energy. Voltaic cells that perform this conversion using conventional fuels, such as H_2 and CH_4, are called **fuel cells**. Fuel cells are *not* batteries because they are not self-contained systems—the fuel must be continuously supplied to generate electricity.

The most common fuel-cell systems involve the reaction of $H_2(g)$ and $O_2(g)$ to form $H_2O(l)$. These cells can generate electricity twice as efficiently as the best internal combustion engine. Under acidic conditions, the reactions are

Cathode:	$O_2(g) + 4\,H^+ + 4\,e^- \longrightarrow 2\,H_2O(l)$
Anode:	$2\,H_2(g) \longrightarrow 4\,H^+ + 4\,e^-$
Overall:	$2\,H_2(g) + O_2(g) \longrightarrow 2\,H_2O(l)$

These cells employ hydrogen gas as the fuel and oxygen gas from air as the oxidant and generate about 1 V.

Fuel cells are often named for either the fuel or the electrolyte used. In the hydrogen-PEM fuel cell (the acronym PEM stands for either proton-exchange membrane or polymer-electrolyte membrane), the anode and cathode are separated by a membrane that is permeable to protons but not to electrons (▼ Figure 20.24). The membrane therefore acts as the salt bridge. The electrodes are typically made from graphite.

The hydrogen-PEM cell operates at around 80 °C. At this temperature the electrochemical reactions would normally occur very slowly, and so small islands of platinum are deposited on each electrode to catalyze the reactions. The high cost and relative scarcity of platinum is one factor that limits wider use of hydrogen-PEM fuel cells.

In order to power a vehicle, multiple cells must be assembled into a fuel cell *stack*. The amount of power generated by a stack depends on the number and size of the fuel cells in the stack and on the surface area of the PEM.

Much fuel cell research today is directed toward improving electrolytes and catalysts and toward developing cells that use fuels such as hydrocarbons and alcohols, which are less difficult to handle and distribute than hydrogen gas.

▲ GO FIGURE

What half-reaction occurs at the cathode?

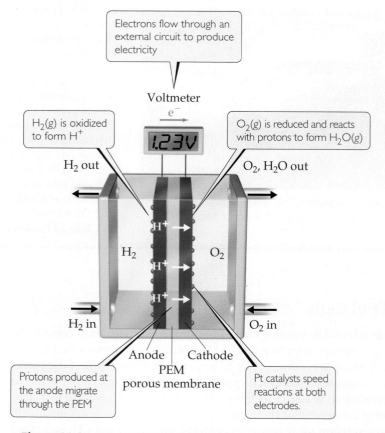

▲ **Figure 20.24 A hydrogen-PEM fuel cell.** The proton-exchange membrane (PEM) allows H^+ ions generated by H_2 oxidation at the anode to migrate to the cathode, where H_2O is formed.

20.8 | Corrosion

In this section, we examine the undesirable redox reactions that lead to **corrosion** of metals. Corrosion reactions are spontaneous redox reactions in which a metal is attacked by some substance in its environment and converted to an unwanted compound.

For nearly all metals, oxidation is thermodynamically favorable in air at room temperature. When oxidation of a metal object is not inhibited, it can destroy the object. Oxidation can form an insulating protective oxide layer, however, that prevents further reaction of the underlying metal. Based on the standard reduction potential for Al^{3+}, for example, we expect aluminum metal to be readily oxidized. The many aluminum soft-drink and beer cans that litter the environment are ample evidence, however, that aluminum undergoes only very slow chemical corrosion. The exceptional stability of this active metal in air is due to the formation of a thin protective coat of oxide—a hydrated form of Al_2O_3—on the metal surface. The oxide coat is impermeable to O_2 or H_2O and so protects the underlying metal from further corrosion.

Magnesium metal is similarly protected, and some metal alloys, such as stainless steel, likewise form protective impervious oxide coats.

Corrosion of Iron (Rusting)

The rusting of iron is a familiar corrosion process that carries a significant economic impact. Up to 20% of the iron produced annually in the United States is used to replace iron objects that have been discarded because of rust damage.

Rusting of iron requires both oxygen and water, and the process can be accelerated by other factors such as pH, presence of salts, contact with metals more difficult to oxidize than iron, and stress on the iron. The corrosion process involves oxidation and reduction, and the metal conducts electricity. Thus, electrons can move through the metal from a region where oxidation occurs to a region where reduction occurs, as in voltaic cells. Because the standard reduction potential for reduction of $Fe^{2+}(aq)$ is less positive than that for reduction of O_2, $Fe(s)$ can be oxidized by $O_2(g)$:

Cathode: $\quad O_2(g) + 4\,H^+(aq) + 4\,e^- \longrightarrow 2\,H_2O(l) \qquad E^\circ_{red} = 1.23$ V

Anode: $\qquad\qquad\qquad\qquad Fe(s) \longrightarrow Fe^{2+}(aq) + 2\,e^- \quad E^\circ_{red} = -0.44$ V

A portion of the iron, often associated with a dent or region of strain, can serve as an anode at which Fe is oxidized to Fe^{2+} (▼ **Figure 20.25**). The electrons produced in the

▲ GO FIGURE

What is the oxidizing agent in this corrosion reaction?

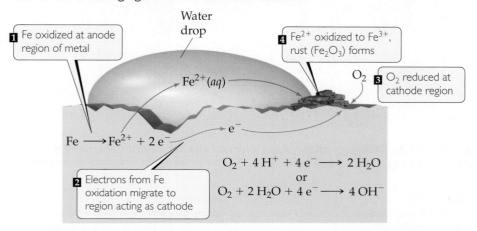

▲ **Figure 20.25 Corrosion of iron in contact with water.** One region of the iron acts as the cathode and another region acts as the anode.

oxidation migrate through the metal from this anodic region to another portion of the surface, which serves as the cathode where O$_2$ is reduced. The reduction of O$_2$ requires H$^+$, so lowering the concentration of H$^+$ (increasing the pH) makes O$_2$ reduction less favorable. Iron in contact with a solution whose pH is greater than 9 does not corrode.

The Fe^{2+} formed at the anode is eventually oxidized to Fe^{3+}, which forms the hydrated iron(III) oxide known as rust:*

$$4\,Fe^{2+}(aq) + O_2(g) + 4\,H_2O(l) + 2x\,H_2O(l) \longrightarrow 2\,Fe_2O_3 \cdot x\,H_2O(s) + 8\,H^+(aq)$$

Because the cathode is generally the area having the largest supply of O$_2$, rust often deposits there. If you look closely at a shovel after it has stood outside in the moist air with wet dirt adhered to its blade, you may notice that pitting has occurred under the dirt but that rust has formed elsewhere, where O$_2$ is more readily available. The enhanced corrosion caused by the presence of salts is usually evident on autos in areas where roads are heavily salted during winter. Like a salt bridge in a voltaic cell, the ions of the salt provide the electrolyte necessary to complete the electrical circuit.

Preventing Corrosion of Iron

Objects made of iron are often covered with a coat of paint or another metal to protect against corrosion. Covering the surface with paint prevents oxygen and water from reaching the iron surface. If the coating is broken, however, and the iron exposed to oxygen and water, corrosion begins as the iron is oxidized.

With *galvanized iron*, which is iron coated with a thin layer of zinc, the iron is protected from corrosion even after the surface coat is broken. The standard reduction potentials are

$$Fe^{2+}(aq) + 2\,e^- \longrightarrow Fe(s) \quad E^\circ_{red} = -0.44\ V$$
$$Zn^{2+}(aq) + 2\,e^- \longrightarrow Zn(s) \quad E^\circ_{red} = -0.76\ V$$

Because E°_{red} for Fe^{2+} is less negative (more positive) than E°_{red} for Zn^{2+}, Zn(s) is more readily oxidized than Fe(s). Thus, even if the zinc coating is broken and the galvanized iron is exposed to oxygen and water, as in ▼ Figure 20.26, the zinc serves as the anode and is corroded (oxidized) instead of the iron. The iron serves as the cathode at which O$_2$ is reduced.

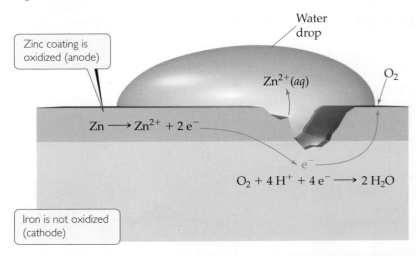

▲ **Figure 20.26 Cathodic protection of iron in contact with zinc.** The standard reduction potentials are $E^\circ_{red,\ Fe^{2+}} = -0.440\ V$, $E^\circ_{red,\ Zn^{2+}} = -0.763\ V$, making the zinc more readily oxidized.

---

*Frequently, metal compounds obtained from aqueous solution have water associated with them. For example, copper(II) sulfate crystallizes from water with 5 mol of water per mole of CuSO$_4$. We represent this substance by the formula CuSO$_4 \cdot$ 5H$_2$O. Such compounds are called hydrates. ∞ (Section 13.1) Rust is a hydrate of iron(III) oxide with a variable amount of water of hydration. We represent this variable water content by writing the formula Fe$_2$O$_3 \cdot x$H$_2$O.

Protecting a metal from corrosion by making it the cathode in an electrochemical cell is known as **cathodic protection**. The metal that is oxidized while protecting the cathode is called the *sacrificial anode*. Underground pipelines and storage tanks made of iron are often protected against corrosion by making the iron the cathode of a voltaic cell. For example, pieces of a metal that is more easily oxidized than iron, such as magnesium ($E_{red}^\circ = -2.37\,\text{V}$), are buried near the pipe or storage tank and connected to it by wire (▶ Figure 20.27). In moist soil, where corrosion can occur, the sacrificial metal serves as the anode, and the pipe or tank experiences cathodic protection.

 Give It Some Thought

Based on the values in Table 20.1, which of these metals could provide cathodic protection to iron: Al, Cu, Ni, Zn?

20.9 | Electrolysis

Voltaic cells are based on spontaneous redox reactions. It is also possible for *nonspontaneous* redox reactions to occur, however, by using electrical energy to drive them. For example, electricity can be used to decompose molten sodium chloride into its component elements Na and Cl_2. Such processes driven by an outside source of electrical energy are called **electrolysis reactions** and take place in **electrolytic cells**.

An electrolytic cell consists of two electrodes immersed either in a molten salt or in a solution. A battery or some other source of electrical energy acts as an electron pump, pushing electrons into one electrode and pulling them from the other. Just as in voltaic cells, the electrode at which reduction occurs is called the cathode, and the electrode at which oxidation occurs is called the anode.

In the electrolysis of molten NaCl, Na^+ ions pick up electrons and are reduced to Na at the cathode, ▶ Figure 20.28. As Na^+ ions near the cathode are depleted, additional Na^+ ions migrate in. Similarly, there is net movement of Cl^- ions to the anode where they are oxidized. The electrode reactions for the electrolysis are

Cathode: $\quad 2\,Na^+(l) + 2\,e^- \longrightarrow 2\,Na(l)$

Anode: $\quad\quad\quad 2\,Cl^-(l) \longrightarrow Cl_2(g) + 2\,e^-$

Overall: $\quad 2\,Na^+(l) + 2\,Cl^-(l) \longrightarrow 2\,Na(l) + Cl_2(g)$

Notice how the energy source is connected to the electrodes in Figure 20.28. The positive terminal is connected to the anode and the negative terminal is connected to the cathode, which forces electrons to move from the anode to the cathode.

Because of the high melting points of ionic substances, the electrolysis of molten salts requires very high temperatures. Do we obtain the same products if we electrolyze the aqueous solution of a salt instead of the molten salt? Frequently the answer is no because water itself might be oxidized to form O_2 or reduced to form H_2 rather than the ions of the salt.

In our examples of the electrolysis of NaCl, the electrodes are *inert*; they do not react but merely serve as the surface where oxidation and reduction occur. Several practical applications of electrochemistry, however, are based on *active* electrodes—electrodes that participate in the electrolysis process. *Electroplating*, for example, uses electrolysis to deposit a thin layer of one metal on another metal to improve beauty or resistance to corrosion. Examples include electroplating nickel or chromium onto steel and electroplating a precious metal like silver onto a less expensive one.

Figure 20.29 illustrates an electrolytic cell for electroplating nickel onto a piece of steel. The anode is a strip of nickel metal, and the cathode is the steel. The electrodes are

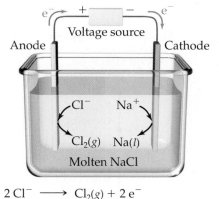

▲ Figure 20.27 **Cathodic protection of an iron pipe.** A mixture of gypsum, sodium sulfate, and clay surrounds the sacrificial magnesium anode to promote conductivity of ions.

$2\,Cl^- \longrightarrow Cl_2(g) + 2\,e^-$

$2\,Na^+ + 2\,e^- \longrightarrow 2\,Na(l)$

▲ Figure 20.28 **Electrolysis of molten sodium chloride.** Pure NaCl melts at 801 °C.

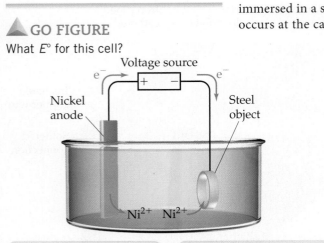

▲GO FIGURE
What $E°$ for this cell?

Anode	Cathode
$Ni(s) \longrightarrow Ni^{2+}(aq) + 2\,e^-$	$Ni^{2+}(aq) + 2\,e^- \longrightarrow Ni(s)$

▲ Figure 20.29 **Electrolytic cell with an active metal electrode.** Nickel dissolves from the anode to form $Ni^{2+}(aq)$. At the cathode $Ni^{2+}(aq)$ is reduced and forms a nickel "plate" on the steel cathode.

immersed in a solution of $NiSO_4(aq)$. When an external voltage is applied, reduction occurs at the cathode. The standard reduction potential of Ni^{2+} ($E°_{red} = -0.28$ V) is less negative than that of H_2O ($E°_{red} = -0.83$ V), so Ni^{2+} is preferentially reduced, depositing a layer of nickel metal on the steel cathode.

At the anode, the nickel metal is oxidized. To explain this behavior, we need to compare the substances in contact with the anode, H_2O and $NiSO_4(aq)$, with the anode material, Ni. For the $NiSO_4(aq)$ solution, neither Ni^{2+} nor SO_4^{2-} can be oxidized (because both already have their elements in their highest possible oxidation state). Both, the H_2O solvent and the Ni atoms in the anode, however, can undergo oxidation:

$$2\,H_2O(l) \longrightarrow O_2(g) + 4\,H^+(aq) + 4\,e^- \qquad E°_{red} = +1.23 \text{ V}$$
$$Ni(s) \longrightarrow Ni^{2+}(aq) + 2\,e^- \qquad E°_{red} = -0.28 \text{ V}$$

We saw in Section 20.4 that the half-reaction with the more negative $E°_{red}$ undergoes oxidation more readily. (Remember Figure 20.11: The strongest reducing agents, which are the substances oxidized most readily, have the most negative $E°_{red}$ values.) Thus, it is the $Ni(s)$, with its $E°_{red} = -0.28$ V, that is oxidized at the anode rather than the H_2O. If we look at the overall reaction, it appears as if nothing has been accomplished. However, this is not true because Ni atoms are transferred from the Ni anode to the steel cathode, plating the steel with a thin layer of nickel atoms.

The standard emf for the overall reaction is

$$E°_{cell} = E°_{red}(\text{cathode}) - E°_{red}(\text{anode}) = (-0.28 \text{ V}) - (-0.28 \text{ V}) = 0$$

Because the standard emf is zero, only a small emf is needed to cause the transfer of nickel atoms from one electrode to the other.

Quantitative Aspects of Electrolysis

The stoichiometry of a half-reaction shows how many electrons are needed to achieve an electrolytic process. For example, the reduction of Na^+ to Na is a one-electron process:

$$Na^+ + e^- \longrightarrow Na$$

Thus, 1 mol of electrons plates out 1 mol of Na metal, 2 mol of electrons plates out 2 mol of Na metal, and so forth. Similarly, 2 mol of electrons are required to produce 1 mol of Cu from Cu^{2+}, and 3 mol of electrons are required to produce 1 mol of Al from Al^{3+}:

$$Cu^{2+} + 2\,e^- \longrightarrow Cu$$
$$Al^{3+} + 3\,e^- \longrightarrow Al$$

For any half-reaction, the amount of substance reduced or oxidized in an electrolytic cell is directly proportional to the number of electrons passed into the cell.

The quantity of charge passing through an electrical circuit, such as that in an electrolytic cell, is generally measured in *coulombs*. As noted in Section 20.5, the charge on 1 mol of electrons is 96,485 C. A coulomb is the quantity of charge passing a point in a circuit in 1 s when the current is 1 ampere (A). Therefore, the number of coulombs passing through a cell can be obtained by multiplying the current in amperes by the elapsed time in seconds.

$$\text{coulombs} = \text{amperes} \times \text{seconds} \qquad [20.21]$$

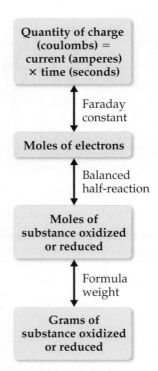

▲ Figure 20.30 **Relationship between charge and amount of reactant and product in electrolysis reactions.**

◀ Figure 20.30 shows how the quantities of substances produced or consumed in electrolysis are related to the quantity of electrical charge used. The same relationships can also be applied to voltaic cells. In other words, electrons can be thought of as "reagents" in electrolysis reactions.

SAMPLE
EXERCISE 20.14 | Relating Electrical Charge and Quantity of Electrolysis

Calculate the number of grams of aluminum produced in 1.00 h by the electrolysis of molten $AlCl_3$ if the electrical current is 10.0 A.

SOLUTION

Analyze We are told that $AlCl_3$ is electrolyzed to form Al and asked to calculate the number of grams of Al produced in 1.00 h with 10.0 A.

Plan Figure 20.30 provides a roadmap for this problem. Using the current, time, a balanced half-reaction, and the atomic weight of aluminum we can calculate the mass of Al produced.

Solve First, we calculate the coulombs of electrical charge passed into the electrolytic cell:

$$\text{Coulombs} = \text{amperes} \times \text{seconds} = (10.0\ \text{A})(1.00\ \text{h})\left(\frac{3600\ \text{s}}{\text{h}}\right) = 3.60 \times 10^4\ \text{C}$$

Second, we calculate the number of moles of electrons that pass into the cell:

$$\text{Moles e}^- = (3.60 \times 10^4\ \text{C})\left(\frac{1\ \text{mol e}^-}{96{,}485\ \text{C}}\right) = 0.373\ \text{mol e}^-$$

Third, we relate number of moles of electrons to number of moles of aluminum formed, using the half-reaction for the reduction of Al^{3+}:

$$Al^{3+} + 3\ e^- \longrightarrow Al$$

Thus, 3 mol of electrons are required to form 1 mol of Al:

$$\text{Moles Al} = (0.373\ \text{mol e}^-)\left(\frac{1\ \text{mol Al}}{3\ \text{mol e}^-}\right) = 0.124\ \text{mol Al}$$

Finally, we convert moles to grams:

$$\text{Grams Al} = (0.124\ \text{mol Al})\left(\frac{27.0\ \text{g Al}}{1\ \text{mol Al}}\right) = 3.36\ \text{g Al}$$

Because each step involves multiplication by a new factor, we can combine all the steps:

$$\text{Grams Al} = (3.60 \times 10^4\ \text{C})\left(\frac{1\ \text{mole e}^-}{96{,}485\ \text{C}}\right)\left(\frac{1\ \text{mol Al}}{3\ \text{mol e}^-}\right)\left(\frac{27.0\ \text{g Al}}{1\ \text{mol Al}}\right) = 3.36\ \text{g Al}$$

Practice Exercise 1

How much time is needed to deposit 1.0 g of chromium metal from an aqueous solution of $CrCl_3$ using a current of 1.5 A? **(a)** 3.8×10^{-2} s, **(b)** 21 min, **(c)** 62 min, **(d)** 139 min, **(e)** 3.2×10^3 min.

Practice Exercise 2

(a) The half-reaction for formation of magnesium metal upon electrolysis of molten $MgCl_2$ is $Mg^{2+} + 2\ e^- \longrightarrow Mg$. Calculate the mass of magnesium formed upon passage of a current of 60.0 A for a period of 4.00×10^3 s. **(b)** How many seconds would be required to produce 50.0 g of Mg from $MgCl_2$ if the current is 100.0 A?

Chemistry Put to Work

Electrometallurgy of Aluminum

Many processes used to produce or refine metals are based on electrolysis. Collectively these processes are referred to as *electrometallurgy*. Electrometallurgical procedures can be broadly differentiated according to whether they involve electrolysis of a molten salt or of an aqueous solution.

Electrolytic methods using molten salts are important for obtaining the more active metals, such as sodium, magnesium, and aluminum. These metals cannot be obtained from aqueous solution because water is more easily reduced than the metal ions. The standard reduction potentials of water under both acidic ($E^\circ_{red} = 0.00$ V) and basic ($E^\circ_{red} = -0.83$ V) conditions are more positive than

those of $Na^+ (E^\circ_{red} = -2.71$ V$)$, $Mg^{2+} (E^\circ_{red} = -2.37$ V$)$, and Al^{3+} $(E^\circ_{red} = -1.66$ V$)$.

Historically, obtaining aluminum metal has been a challenge. It is obtained from bauxite ore, which is chemically treated to concentrate aluminum oxide (Al_2O_3). The melting point of aluminum oxide is above 2000 °C, which is too high to permit its use as a molten medium for electrolysis.

The electrolytic process used commercially to produce aluminum is the *Hall–Héroult process*, named after its inventors, Charles M. Hall and Paul Héroult. Hall (**Figure 20.31**) began working on the problem of reducing aluminum in about 1885 after he had learned from a professor of the difficulty of reducing ores of very active metals. Before the development of an electrolytic process, aluminum was

▲ **Figure 20.31 Charles M. Hall (1863–1914) as a young man.**

obtained by a chemical reduction using sodium or potassium as the reducing agent, a costly procedure that made aluminum metal expensive. As late as 1852, the cost of aluminum was $545 per pound, far greater than the cost of gold. During the Paris Exposition in 1855, aluminum was exhibited as a rare metal, even though it is the third most abundant element in Earth's crust.

Hall, who was 21 years old when he began his research, utilized handmade and borrowed equipment in his studies and used a woodshed near his Ohio home as his laboratory. In about a year's time, he developed an electrolytic procedure using an ionic compound that melts to form a conducting medium that dissolves Al_2O_3 but does not interfere with the electrolysis reactions. The ionic compound he selected was the relatively rare mineral cryolite (Na_3AlF_6). Héroult, who was the same age as Hall, independently made the same discovery in France at about the same time. Because of the research of these two unknown young scientists, large-scale production of aluminum became commercially feasible, and aluminum became a common and familiar metal. Indeed, the factory that Hall subsequently built to produce aluminum evolved into Alcoa Corporation.

In the Hall–Héroult process, Al_2O_3 is dissolved in molten cryolite, which melts at 1012 °C and is an effective electrical conductor (▼ **Figure 20.32**). Graphite rods are employed as anodes and are consumed in the electrolysis:

Anode: $C(s) + 2\,O^{2-}(l) \longrightarrow CO_2(g) + 4\,e^-$

Cathode: $3e^- + Al^{3+}(l) \longrightarrow Al(l)$

A large amount of electrical energy is needed in the Hall–Héroult process with the result that the aluminum industry consumes about 2% of the electrical energy generated in the United States. Because recycled aluminum requires only 5% of the energy needed to produce "new" aluminum, considerable energy savings can be realized by increasing the amount of aluminum recycled. Approximately 65% of aluminum beverage containers are recycled in the United States.

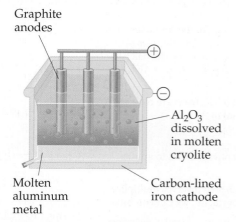

▲ **Figure 20.32 The Hall–Héroult process.** Because molten aluminum is denser than the mixture of cryolite (Na_3AlF_6) and Al_2O_3, the metal collects at the bottom of the cell.

SAMPLE INTEGRATIVE EXERCISE Putting Concepts Together

The K_{sp} at 298 K for iron(II) fluoride is 2.4×10^{-6}. **(a)** Write a half-reaction that gives the likely products of the two-electron reduction of $FeF_2(s)$ in water. **(b)** Use the K_{sp} value and the standard reduction potential of $Fe^{2+}(aq)$ to calculate the standard reduction potential for the half-reaction in part (a). **(c)** Rationalize the difference between the reduction potential in part (a) and the reduction potential for $Fe^{2+}(aq)$.

SOLUTION

Analyze We are going to combine what we know about equilibrium constants and electrochemistry to obtain reduction potentials.

Plan For (a) we need to determine which ion, Fe^{2+} or F^-, is more likely to be reduced by two electrons and complete the overall reaction $FeF_2 + 2\,e^- \longrightarrow$?. For (b) we need to write the chemical equation associated with the K_{sp} and see how it relates to $E°$ for the reduction half-reaction in (a). For (c) we need to compare $E°$ from (b) with the value for the reduction of Fe^{2+}.

Solve

(a) Iron(II) fluoride is an ionic substance that consists of Fe^{2+} and F^- ions. We are asked to predict where two electrons could be added to FeF_2. We cannot envision adding the electrons to the F^- ions to form F^{2-}, so it seems likely that we could reduce the Fe^{2+} ions to $Fe(s)$. We therefore predict the half-reaction

$$FeF_2(s) + 2\,e^- \longrightarrow Fe(s) + 2\,F^-(aq)$$

(b) The K_{sp} for FeF_2 refers to the following equilibrium: ∞ (Section 17.4)

$$FeF_2(s) \rightleftharpoons Fe^{2+}(aq) + 2\,F^-(aq) \qquad K_{sp} = [Fe^{2+}][F^-]^2 = 2.4 \times 10^{-6}$$

We were also asked to use the standard reduction potential of Fe^{2+}, whose half-reaction and standard reduction potentials are listed in Appendix E:

$$Fe^{2+}(aq) + 2\,e^- \longrightarrow Fe(s) \qquad E = -0.440\ V$$

According to Hess's law, if we can add chemical equations to get a desired equation, then we can add their associated thermodynamic state functions, like ΔH or ΔG, to determine the thermodynamic quantity for the desired reaction. ∞ (Section 5.6) So we need to consider whether the three equations we are working with can be combined in a similar fashion. Notice that if we add the K_{sp} reaction to the standard reduction half-reaction for Fe^{2+}, we get the half-reaction we want:

1. $\qquad FeF_2(s) \longrightarrow Fe^{2+}(aq) + 2\,F^-(aq)$

2. $\underline{Fe^{2+}(aq) + 2\,e^- \longrightarrow Fe(s)}$

Overall: 3. $\quad FeF_2(s) + 2\,e^- \longrightarrow Fe(s) + 2\,F^-(aq)$

Reaction 3 is still a half-reaction, so we do see the free electrons.

If we knew $\Delta G°$ for reactions 1 and 2, we could add them to get $\Delta G°$ for reaction 3. We can relate $\Delta G°$ to $E°$ by $\Delta G° = -nFE°$ (Equation 20.12) and to K by $\Delta G° = -RT\ln K$ (Equation 19.20, see also Figure 20.13). Furthermore, we know that K for reaction 1 is the K_{sp} of FeF_2, and we know $E°$ for reaction 2. Therefore, we can calculate $\Delta G°$ for reactions 1 and 2:

Reaction 1:

$$\Delta G° = -RT\ln K = -(8.314\ J/K\ mol)(298\ K)\ln(2.4 \times 10^{-6}) = 3.21 \times 10^4\ J/mol$$

Reaction 2:

$$\Delta G° = -nFE° = -(2)(96{,}485\ C/mol)(-0.440\ J/C) = 8.49 \times 10^4\ J/mol$$

(Recall that 1 volt is 1 joule per coulomb.)

Then $\Delta G°$ for reaction 3, the one we want, is the sum of the $\Delta G°$ values for reactions 1 and 2:

$$3.21 \times 10^4\ J/mol + 8.49 \times 10^4\ J/mol = 1.17 \times 10^5\ J/mol$$

We can convert this to $E°$ from the relationship $\Delta G° = -nFE°$:

$$1.17 \times 10^5\ J/mol = -(2)(96{,}485\ C/mol)E°$$

$$E° = \frac{1.17 \times 10^5\ J/mol}{-(2)(96{,}485\ C/mol)} = -0.606\ J/C = -0.606\ V$$

(c) The standard reduction potential for $FeF_2 (-0.606\ V)$ is more negative than that for $Fe^{2+} (-0.440\ V)$, telling us that the reduction of FeF_2 is the less favorable process. When FeF_2 is reduced, we not only reduce the Fe^{2+} but also break up the ionic solid. Because this additional energy must be overcome, the reduction of FeF_2 is less favorable than the reduction of Fe^{2+}.

Chapter Summary and Key Terms

OXIDATION STATES AND OXIDATION REDUCTION REACTIONS (INTRODUCTION AND SECTION 20.1) In this chapter, we have focused on **electrochemistry**, the branch of chemistry that relates electricity and chemical reactions. Electrochemistry involves oxidation–reduction reactions, also called redox reactions. These reactions involve a change in the oxidation state of one or more elements. In every oxidation–reduction reaction one substance is oxidized (its oxidation state, or number, increases) and one substance is reduced (its oxidation state, or number, decreases). The substance that is oxidized is referred to as a **reducing agent**, or **reductant**, because it causes the reduction of some other substance. Similarly, the substance that is reduced is referred to as an **oxidizing agent**, or **oxidant**, because it causes the oxidation of some other substance.

BALANCING REDOX EQUATIONS (SECTION 20.2) An oxidation–reduction reaction can be balanced by dividing the reaction into two **half-reactions**, one for oxidation and one for reduction. A half-reaction is a balanced chemical equation that includes electrons. In oxidation half-reactions the electrons are on the product (right) side of the equation. In reduction half-reactions the electrons are on the reactant (left)

side of the equation. Each half-reaction is balanced separately, and the two are brought together with proper coefficients to balance the electrons on each side of the equation, so the electrons cancel when the half-reactions are added.

VOLTAIC CELLS (SECTION 20.3) A **voltaic** (or **galvanic**) **cell** uses a spontaneous oxidation–reduction reaction to generate electricity. In a voltaic cell the oxidation and reduction half-reactions often occur in separate half-cells. Each half-cell has a solid surface called an electrode, where the half-reaction occurs. The electrode where oxidation occurs is called the **anode**, and the electrode where reduction occurs is called the **cathode**. The electrons released at the anode flow through an external circuit (where they do electrical work) to the cathode. Electrical neutrality in the solution is maintained by the migration of ions between the two half-cells through a device such as a salt bridge.

CELL POTENTIALS UNDER STANDARD CONDITIONS (SECTION 20.4) A voltaic cell generates an **electromotive force (emf)** that moves the electrons from the anode to the cathode through the external circuit. The origin of emf is a difference in the electrical potential energy of the two electrodes in the cell. The emf of a cell is called its **cell potential**, E_{cell}, and is measured in volts ($1 \text{ V} = 1 \text{ J/C}$). The cell potential under standard conditions is called the **standard emf**, or the **standard cell potential**, and is denoted E°_{cell}.

A **standard reduction potential**, E°_{red}, can be assigned for an individual half-reaction. This is achieved by comparing the potential of the half-reaction to that of the **standard hydrogen electrode** (SHE), which is defined to have $E^{\circ}_{red} = 0$ V and is based on the following half-reaction:

$$2 \text{ H}^+(aq, 1 \text{ M}) + 2 \text{ e}^- \longrightarrow \text{H}_2(g, 1 \text{ atm}) \quad E^{\circ}_{red} = 0 \text{ V}$$

The standard cell potential of a voltaic cell is the difference between the standard reduction potentials of the half-reactions that occur at the cathode and the anode:

$$E^{\circ}_{cell} = E^{\circ}_{red} (\text{cathode}) - E^{\circ}_{red} (\text{anode}).$$

The value of E°_{cell} is positive for a voltaic cell.

For a reduction half-reaction, E°_{red} is a measure of the tendency of the reduction to occur; the more positive the value for E°_{red}, the greater the tendency of the substance to be reduced. Substances that are easily reduced act as strong oxidizing agents; thus, E°_{red} provides a measure of the oxidizing strength of a substance. Substances that are strong oxidizing agents produce products that are weak reducing agents and vice versa.

FREE ENERGY AND REDOX REACTIONS (SECTION 20.5)

The emf, E, is related to the change in the Gibbs free energy, $\Delta G = -nFE$, where n is the number of moles of electrons transferred during the redox process and F is the **Faraday constant**, defined as the quantity of electrical charge on one mole of electrons: $F = 96,485 \text{ C/mol}$. Because E is related to ΔG, the sign of E indicates whether a redox process is spontaneous: $E > 0$ indicates a spontaneous process, and $E < 0$ indicates a nonspontaneous one. Because ΔG is also related to the equilibrium constant for a reaction ($\Delta G^{\circ} = -RT \ln K$), we can relate E to K as well.

The maximum amount of electrical work produced by a voltaic cell is given by the product of the total charge delivered, nF, and the emf, E: $w_{max} = -nFE$. The watt is a unit of power: $1 \text{ W} = 1 \text{ J/s}$. Electrical work is often measured in kilowatt-hours.

CELL POTENTIALS UNDER NONSTANDARD CONDITIONS (SECTION 20.6)

The emf of a redox reaction varies with temperature and with the concentrations of reactants and products. The **Nernst equation** relates the emf under nonstandard conditions to the standard emf and the reaction quotient Q:

$$E = E^{\circ} - (RT/nF) \ln Q = E^{\circ} - (0.0592/n) \log Q$$

The factor 0.0592 is valid when $T = 298$ K. A **concentration cell** is a voltaic cell in which the same half-reaction occurs at both the anode and the cathode but with different concentrations of reactants in each half-cell. At equilibrium, $Q = K$ and $E = 0$.

BATTERIES AND FUEL CELLS (SECTION 20.7)

A **battery** is a self-contained electrochemical power source that contains one or more voltaic cells. Batteries are based on a variety of different redox reactions. Batteries that cannot be recharged are called **primary cells**, while those that can be recharged are called **secondary cells**. The common alkaline dry cell battery is an example of a primary cell battery. Lead–acid, nickel–cadmium, nickel–metal hydride, and lithium-ion batteries are examples of secondary cells. **Fuel cells** are voltaic cells that utilize redox reactions in which reactants such as H_2 have to be continuously supplied to the cell to generate voltage.

CORROSION (SECTION 20.8)

Electrochemical principles help us understand **corrosion**, undesirable redox reactions in which a metal is attacked by some substance in its environment. The corrosion of iron into rust is caused by the presence of water and oxygen, and it is accelerated by the presence of electrolytes, such as road salt. The protection of a metal by putting it in contact with another metal that more readily undergoes oxidation is called **cathodic protection**. Galvanized iron, for example, is coated with a thin layer of zinc; because zinc is oxidized more readily than iron, the zinc serves as a sacrificial anode in the redox reaction.

ELECTROLYSIS (SECTION 20.9)

An **electrolysis reaction**, which is carried out in an **electrolytic cell**, employs an external source of electricity to drive a nonspontaneous electrochemical reaction. The current-carrying medium within an electrolytic cell may be either a molten salt or an electrolyte solution. The products of electrolysis can generally be predicted by comparing the reduction potentials associated with possible oxidation and reduction processes. The electrodes in an electrolytic cell can be inert or active, meaning that the electrode can be involved in the electrolysis reaction. Active electrodes are important in electroplating and in metallurgical processes.

The quantity of substances formed during electrolysis can be calculated by considering the number of electrons involved in the redox reaction and the amount of electrical charge that passes into the cell. The amount of electrical charge is measured in coulombs and is related to the magnitude of the current and the time it flows ($1 \text{ C} = 1 \text{ A-s}$).

Learning Outcomes
After studying this chapter, you should be able to:

- Identify oxidation, reduction, oxidizing agent, and reducing agent in a chemical equation. (Section 20.1)
- Complete and balance redox equations using the method of half-reactions. (Section 20.2)
- Sketch a voltaic cell and identify its cathode, anode, and the directions in which electrons and ions move. (Section 20.3)
- Calculate standard emfs (cell potentials), E°_{cell}, from standard reduction potentials. (Section 20.4)
- Use reduction potentials to predict whether a redox reaction is spontaneous. (Section 20.4)
- Relate E°_{cell}, to ΔG° and equilibrium constants. (Section 20.5)

- Calculate emf under nonstandard conditions. (Section 20.6)
- Identify the components of common batteries. (Section 20.7)
- Describe the construction of a lithium-ion battery and explain how it works (Section 20.7)
- Describe the construction of a fuel cell and explain how it generates electrical energy. (Section 20.7)
- Explain how corrosion occurs and how it is prevented by cathodic protection. (Section 20.8)
- Describe the reactions in electrolytic cells. (Section 20.9)
- Relate amounts of products and reactants in redox reactions to electrical charge. (Section 20.9)

Key Equations

- $E°_{cell} = E°_{red}(\text{cathode}) - E°_{red}(\text{anode})$ [20.8]   Relating standard emf to standard reduction potentials of the reduction (cathode) and oxidation (anode) half-reactions

- $\Delta G = -nFE$ [20.11]   Relating free-energy change and emf

- $E = E° - \dfrac{0.0592\text{ V}}{n}\log Q$ (at 298 K)   [20.18]   The Nernst equation, expressing the effect of concentration on cell potential

Exercises

Visualizing Concepts

20.1 In the Brønsted–Lowry concept of acids and bases, acid–base reactions are viewed as proton-transfer reactions. The stronger the acid, the weaker is its conjugate base. If we were to think of redox reactions in a similar way, what particle would be analogous to the proton? Would strong oxidizing agents be analogous to strong acids or strong bases? [Sections 20.1 and 20.2]

20.2 You may have heard that "antioxidants" are good for your health. Based on what you have learned in this chapter, what do you deduce an "antioxidant" is? [Sections 20.1 and 20.2]

20.3 The diagram that follows represents a molecular view of a process occurring at an electrode in a voltaic cell.

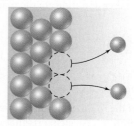

(a) Does the process represent oxidation or reduction? (b) Is the electrode the anode or cathode? (c) Why are the atoms in the electrode represented by larger spheres than the ions in the solution? [Section 20.3]

20.4 Assume that you want to construct a voltaic cell that uses the following half-reactions:

$$A^{2+}(aq) + 2\,e^- \longrightarrow A(s) \qquad E°_{red} = -0.10\text{ V}$$
$$B^{2+}(aq) + 2\,e^- \longrightarrow B(s) \qquad E°_{red} = -1.10\text{ V}$$

You begin with the incomplete cell pictured here in which the electrodes are immersed in water.

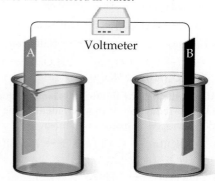

(a) What additions must you make to the cell for it to generate a standard emf? (b) Which electrode functions as the cathode? (c) Which direction do electrons move through the

external circuit? (d) What voltage will the cell generate under standard conditions? [Sections 20.3 and 20.4]

20.5 For a spontaneous reaction $A(aq) + B(aq) \longrightarrow A^-(aq) + B^+(aq)$, answer the following questions:

(a) If you made a voltaic cell out of this reaction, what half-reaction would be occurring at the cathode, and what half-reaction would be occurring at the anode?

(b) Which half-reaction from (a) is higher in potential energy?

(c) What is the sign of $E°_{cell}$? [Section 20.3]

20.6 Consider the following table of standard electrode potentials for a series of hypothetical reactions in aqueous solution:

Reduction Half-Reaction	E°(V)
$A^+(aq) + e^- \longrightarrow A(s)$	1.33
$B^{2+}(aq) + 2\,e^- \longrightarrow B(s)$	0.87
$C^{3+}(aq) + e^- \longrightarrow C^{2+}(aq)$	-0.12
$D^{3+}(aq) + 3\,e^- \longrightarrow D(s)$	-1.59

(a) Which substance is the strongest oxidizing agent? Which is weakest?

(b) Which substance is the strongest reducing agent? Which is weakest?

(c) Which substance(s) can oxidize C^{2+}? [Sections 20.4 and 20.5]

20.7 Consider a redox reaction for which $E°$ is a negative number.

(a) What is the sign of $\Delta G°$ for the reaction?

(b) Will the equilibrium constant for the reaction be larger or smaller than 1?

(c) Can an electrochemical cell based on this reaction accomplish work on its surroundings? [Section 20.5]

20.8 Consider the following voltaic cell:

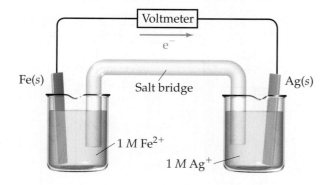

(a) Which electrode is the cathode?

(b) What is the standard emf generated by this cell?

(c) What is the change in the cell voltage when the ion concentrations in the cathode half-cell are increased by a factor of 10?

(d) What is the change in the cell voltage when the ion concentrations in the anode half-cell are increased by a factor of 10? [Sections 20.4 and 20.6]

20.9 Consider the half-reaction $Ag^+(aq) + e^- \longrightarrow Ag(s)$. **(a)** Which of the lines in the following diagram indicates how the reduction potential varies with the concentration of Ag^+? **(b)** What is the value of $E°_{red}$ when $\log[Ag^+] = 0$? [Section 20.6]

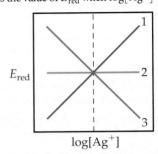

20.10 The electrodes in a silver oxide battery are silver oxide (Ag_2O) and zinc. **(a)** Which electrode acts as the anode? **(b)** Which battery do you think has an energy density most similar to the silver oxide battery: a Li-ion battery, a nickel–cadmium battery, or a lead–acid battery? [Section 20.7]

20.11 Bars of iron are put into each of the three beakers as shown here. In which beaker—A, B, or C—would you expect the iron to show the most corrosion ? [Section 20.8]

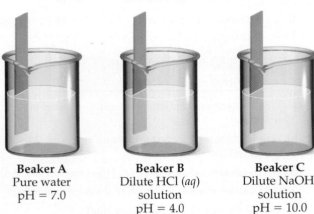

Beaker A	**Beaker B**	**Beaker C**
Pure water	Dilute HCl (aq)	Dilute NaOH
pH = 7.0	solution	solution
	pH = 4.0	pH = 10.0

20.12 Magnesium is produced commercially by electrolysis from a molten salt using a cell similar to the one shown here. **(a)** What salt is used as the electrolyte? **(b)** Which electrode is the anode, and which one is the cathode? **(c)** Write the overall cell reaction and individual half-reactions. **(d)** What precautions would need to be taken with respect to the magnesium formed? [Section 20.9]

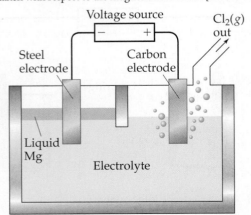

Oxidation–Reduction Reactions (Section 20.1)

20.13 **(a)** What is meant by the term *oxidation*? **(b)** On which side of an oxidation half-reaction do the electrons appear? **(c)** What is meant by the term *oxidant*? **(d)** What is meant by the term *oxidizing agent*?

20.14 **(a)** What is meant by the term *reduction*? **(b)** On which side of a reduction half-reaction do the electrons appear? **(c)** What is meant by the term *reductant*? **(d)** What is meant by the term *reducing agent*?

20.15 Indicate whether each of the following statements is true or false:

(a) If something is oxidized, it is formally losing electrons.

(b) For the reaction $Fe^{3+}(aq) + Co^{2+}(aq) \longrightarrow Fe^{2+}(aq) + Co^{3+}(aq)$, $Fe^{3+}(aq)$ is the reducing agent and $Co^{2+}(aq)$ is the oxidizing agent.

(c) If there are no changes in the oxidation state of the reactants or products of a particular reaction, that reaction is not a redox reaction.

20.16 Indicate whether each of the following statements is true or false:

(a) If something is reduced, it is formally losing electrons.

(b) A reducing agent gets oxidized as it reacts.

(c) An oxidizing agent is needed to convert CO into CO_2.

20.17 In each of the following balanced oxidation–reduction equations, identify those elements that undergo changes in oxidation number and indicate the magnitude of the change in each case.

(a) $I_2O_5(s) + 5\,CO(g) \longrightarrow I_2(s) + 5\,CO_2(g)$

(b) $2\,Hg^{2+}(aq) + N_2H_4(aq) \longrightarrow 2\,Hg(l) + N_2(g) + 4\,H^+(aq)$

(c) $3\,H_2S(aq) + 2\,H^+(aq) + 2\,NO_3^-(aq) \longrightarrow 3\,S(s) + 2\,NO(g) + 4\,H_2O(l)$

20.18 In each of the following balanced oxidation–reduction equations, identify those elements that undergo changes in oxidation number and indicate the magnitude of the change in each case.

(a) $2\,MnO_4^-(aq) + 3\,S^{2-}(aq) + 4\,H_2O(l) \longrightarrow 3\,S(s) + 2\,MnO_2(s) + 8\,OH^-(aq)$

(b) $4\,H_2O_2(aq) + Cl_2O_7(g) + 2\,OH^-(aq) \longrightarrow 2\,ClO_2^-(aq) + 5\,H_2O(l) + 4\,O_2(g)$

(c) $Ba^{2+}(aq) + 2\,OH^-(aq) + H_2O_2(aq) + 2\,ClO_2(aq) \longrightarrow Ba(ClO_2)_2(s) + 2\,H_2O(l) + O_2(g)$

20.19 Indicate whether the following balanced equations involve oxidation–reduction. If they do, identify the elements that undergo changes in oxidation number.

(a) $PBr_3(l) + 3\,H_2O(l) \longrightarrow H_3PO_3(aq) + 3\,HBr(aq)$

(b) $NaI(aq) + 3\,HOCl(aq) \longrightarrow NaIO_3(aq) + 3\,HCl(aq)$

(c) $3\,SO_2(g) + 2\,HNO_3(aq) + 2\,H_2O(l) \longrightarrow 3\,H_2SO_4(aq) + 2\,NO(g)$

20.20 Indicate whether the following balanced equations involve oxidation–reduction. If they do, identify the elements that undergo changes in oxidation number.

(a) $2\,AgNO_3(aq) + CoCl_2(aq) \longrightarrow 2\,AgCl(s) + Co(NO_3)_2(aq)$

(b) $2\,PbO_2(s) \longrightarrow 2\,PbO(s) + O_2(g)$

(c) $2\,H_2SO_4(aq) + 2\,NaBr(s) \longrightarrow Br_2(l) + SO_2(g) + Na_2SO_4(aq) + 2\,H_2O(l)$

Balancing Oxidation–Reduction Reactions (Section 20.2)

20.21 At 900 °C titanium tetrachloride vapor reacts with molten magnesium metal to form solid titanium metal and molten magnesium chloride. **(a)** Write a balanced equation for this reaction. **(b)** What is being oxidized, and what is being reduced? **(c)** Which substance is the reductant, and which is the oxidant?

20.22 Hydrazine (N_2H_4) and dinitrogen tetroxide (N_2O_4) form a self-igniting mixture that has been used as a rocket propellant. The reaction products are N_2 and H_2O. **(a)** Write a balanced chemical equation for this reaction. **(b)** What is being oxidized, and what is being reduced? **(c)** Which substance serves as the reducing agent and which as the oxidizing agent?

20.23 Complete and balance the following half-reactions. In each case indicate whether the half-reaction is an oxidation or a reduction.

(a) $Sn^{2+}(aq) \longrightarrow Sn^{4+}(aq)$ (acidic solution)

(b) $TiO_2(s) \longrightarrow Ti^{2+}(aq)$ (acidic solution)

(c) $ClO_3^-(aq) \longrightarrow Cl^-(aq)$ (acidic solution)

(d) $N_2(g) \longrightarrow NH_4^+(aq)$ (acidic solution)

(e) $OH^-(aq) \longrightarrow O_2(g)$ (basic solution)

(f) $SO_3^{2-}(aq) \longrightarrow SO_4^{2-}(aq)$ (basic solution)

(g) $N_2(g) \longrightarrow NH_3(g)$ (basic solution)

20.24 Complete and balance the following half-reactions. In each case indicate whether the half-reaction is an oxidation or a reduction.

(a) $Mo^{3+}(aq) \longrightarrow Mo(s)$ (acidic solution)

(b) $H_2SO_3(aq) \longrightarrow SO_4^{2-}(aq)$ (acidic solution)

(c) $NO_3^-(aq) \longrightarrow NO(g)$ (acidic solution)

(d) $O_2(g) \longrightarrow H_2O(l)$ (acidic solution)

(e) $O_2(g) \longrightarrow H_2O(l)$ (basic solution)

(f) $Mn^{2+}(aq) \longrightarrow MnO_2(s)$ (basic solution)

(g) $Cr(OH)_3(s) \longrightarrow CrO_4^{2-}(aq)$ (basic solution)

20.25 Complete and balance the following equations, and identify the oxidizing and reducing agents:

(a) $Cr_2O_7^{2-}(aq) + I^-(aq) \longrightarrow Cr^{3+}(aq) + IO_3^-(aq)$ (acidic solution)

(b) $MnO_4^-(aq) + CH_3OH(aq) \longrightarrow Mn^{2+}(aq) + HCO_2H(aq)$ (acidic solution)

(c) $I_2(s) + OCl^-(aq) \longrightarrow IO_3^-(aq) + Cl^-(aq)$ (acidic solution)

(d) $As_2O_3(s) + NO_3^-(aq) \longrightarrow H_3AsO_4(aq) + N_2O_3(aq)$ (acidic solution)

(e) $MnO_4^-(aq) + Br^-(aq) \longrightarrow MnO_2(s) + BrO_3^-(aq)$ (basic solution)

(f) $Pb(OH)_4^{2-}(aq) + ClO^-(aq) \longrightarrow PbO_2(s) + Cl^-(aq)$ (basic solution)

20.26 Complete and balance the following equations, and identify the oxidizing and reducing agents. (Recall that the O atoms in hydrogen peroxide, H_2O_2, have an atypical oxidation state.)

(a) $NO_2^-(aq) + Cr_2O_7^{2-}(aq) \longrightarrow Cr^{3+}(aq) + NO_3^-(aq)$ (acidic solution)

(b) $S(s) + HNO_3(aq) \longrightarrow H_2SO_3(aq) + N_2O(g)$ (acidic solution)

(c) $Cr_2O_7^{2-}(aq) + CH_3OH(aq) \longrightarrow HCO_2H(aq) + Cr^{3+}(aq)$ (acidic solution)

(d) $BrO_3^-(aq) + N_2H_4(g) \longrightarrow Br^-(aq) + N_2(g)$ (acidic solution)

(e) $NO_2^-(aq) + Al(s) \longrightarrow NH_4^+(aq) + AlO_2^-(aq)$ (basic solution)

(f) $H_2O_2(aq) + ClO_2(aq) \longrightarrow ClO_2^-(aq) + O_2(g)$ (basic solution)

Voltaic Cells (Section 20.3)

20.27 **(a)** What are the similarities and differences between Figure 20.3 and Figure 20.4? **(b)** Why are Na^+ ions drawn into the cathode half-cell as the voltaic cell shown in Figure 20.5 operates?

20.28 **(a)** What is the role of the porous glass disc shown in Figure 20.4? **(b)** Why do NO_3^- ions migrate into the anode half-cell as the voltaic cell shown in Figure 20.5 operates?

20.29 A voltaic cell similar to that shown in Figure 20.5 is constructed. One electrode half-cell consists of a silver strip placed in a solution of $AgNO_3$, and the other has an iron strip placed in a solution of $FeCl_2$. The overall cell reaction is

$$Fe(s) + 2\,Ag^+(aq) \longrightarrow Fe^{2+}(aq) + 2\,Ag(s)$$

(a) What is being oxidized, and what is being reduced? **(b)** Write the half-reactions that occur in the two half-cells. **(c)** Which electrode is the anode, and which is the cathode? **(d)** Indicate the signs of the electrodes. **(e)** Do electrons flow from the silver electrode to the iron electrode or from the iron to the silver? **(f)** In which directions do the cations and anions migrate through the solution?

20.30 A voltaic cell similar to that shown in Figure 20.5 is constructed. One half-cell consists of an aluminum strip placed in a solution of $Al(NO_3)_3$, and the other has a nickel strip placed in a solution of $NiSO_4$. The overall cell reaction is

$$2\,Al(s) + 3\,Ni^{2+}(aq) \longrightarrow 2\,Al^{3+}(aq) + 3\,Ni(s)$$

(a) What is being oxidized, and what is being reduced? **(b)** Write the half-reactions that occur in the two half-cells. **(c)** Which electrode is the anode, and which is the cathode? **(d)** Indicate the signs of the electrodes. **(e)** Do electrons flow from the aluminum electrode to the nickel electrode or from the nickel to the aluminum? **(f)** In which directions do the cations and anions migrate through the solution? Assume the Al is not coated with its oxide.

Cell Potentials under Standard Conditions (Section 20.4)

20.31 **(a)** What does the term *electromotive force* mean? **(b)** What is the definition of the *volt*? **(c)** What does the term *cell potential* mean?

20.32 **(a)** Which electrode of a voltaic cell, the cathode or the anode, corresponds to the higher potential energy for the electrons? **(b)** What are the units for electrical potential? How does this unit relate to energy expressed in joules?

20.33 **(a)** Write the half-reaction that occurs at a hydrogen electrode in acidic aqueous solution when it serves as the cathode of a voltaic cell. **(b)** Write the half-reaction that occurs at a hydrogen electrode in acidic aqueous solution when it serves as the anode of a voltaic cell. **(c)** What is *standard* about the standard hydrogen electrode?

20.34 (a) What conditions must be met for a reduction potential to be a *standard reduction potential*? (b) What is the standard reduction potential of the standard hydrogen electrode? (c) Why is it impossible to measure the standard reduction potential of a single half-reaction?

20.35 A voltaic cell that uses the reaction

$$Tl^{3+}(aq) + 2\,Cr^{2+}(aq) \longrightarrow Tl^+(aq) + 2\,Cr^{3+}(aq)$$

has a measured standard cell potential of +1.19 V. (a) Write the two half-cell reactions. (b) By using data from Appendix E, determine E°_{red} for the reduction of $Tl^{3+}(aq)$ to $Tl^+(aq)$. (c) Sketch the voltaic cell, label the anode and cathode, and indicate the direction of electron flow.

20.36 A voltaic cell that uses the reaction

$$PdCl_4^{2-}(aq) + Cd(s) \longrightarrow Pd(s) + 4\,Cl^-(aq) + Cd^{2+}(aq)$$

has a measured standard cell potential of +1.03 V. (a) Write the two half-cell reactions. (b) By using data from Appendix E, determine E°_{red} for the reaction involving Pd. (c) Sketch the voltaic cell, label the anode and cathode, and indicate the direction of electron flow.

20.37 Using standard reduction potentials (Appendix E), calculate the standard emf for each of the following reactions:

(a) $Cl_2(g) + 2\,I^-(aq) \longrightarrow 2\,Cl^-(aq) + I_2(s)$

(b) $Ni(s) + 2\,Ce^{4+}(aq) \longrightarrow Ni^{2+}(aq) + 2\,Ce^{3+}(aq)$

(c) $Fe(s) + 2\,Fe^{3+}(aq) \longrightarrow 3\,Fe^{2+}(aq)$

(d) $2\,NO_3^-(aq) + 8\,H^+(aq) + 3\,Cu(s) \longrightarrow 2\,NO(g) + 4\,H_2O(l) + 3\,Cu^{2+}(aq)$

20.38 Using data in Appendix E, calculate the standard emf for each of the following reactions:

(a) $H_2(g) + F_2(g) \longrightarrow 2\,H^+(aq) + 2\,F^-(aq)$

(b) $Cu^{2+}(aq) + Ca(s) \longrightarrow Cu(s) + Ca^{2+}(aq)$

(c) $3\,Fe^{2+}(aq) \longrightarrow Fe(s) + 2\,Fe^{3+}(aq)$

(d) $2\,ClO_3^-(aq) + 10\,Br^-(aq) + 12\,H^+(aq) \longrightarrow Cl_2(g) + 5\,Br_2(l) + 6\,H_2O(l)$

20.39 The standard reduction potentials of the following half-reactions are given in Appendix E:

$$Ag^+(aq) + e^- \longrightarrow Ag(s)$$
$$Cu^{2+}(aq) + 2\,e^- \longrightarrow Cu(s)$$
$$Ni^{2+}(aq) + 2\,e^- \longrightarrow Ni(s)$$
$$Cr^{3+}(aq) + 3\,e^- \longrightarrow Cr(s)$$

(a) Determine which combination of these half-cell reactions leads to the cell reaction with the largest positive cell potential and calculate the value. (b) Determine which combination of these half-cell reactions leads to the cell reaction with the smallest positive cell potential and calculate the value.

20.40 Given the following half-reactions and associated standard reduction potentials:

$$AuBr_4^-(aq) + 3\,e^- \longrightarrow Au(s) + 4\,Br^-(aq)$$
$$E^\circ_{red} = -0.858\ \text{V}$$
$$Eu^{3+}(aq) + e^- \longrightarrow Eu^{2+}(aq)$$
$$E^\circ_{red} = -0.43\ \text{V}$$
$$IO^-(aq) + H_2O(l) + 2\,e^- \longrightarrow I^-(aq) + 2\,OH^-(aq)$$
$$E^\circ_{red} = +0.49\ \text{V}$$

(a) Write the equation for the combination of these half-cell reactions that leads to the largest positive emf and calculate the value. (b) Write the equation for the combination of half-cell reactions that leads to the smallest positive emf and calculate that value.

20.41 A 1 *M* solution of $Cu(NO_3)_2$ is placed in a beaker with a strip of Cu metal. A 1 *M* solution of $SnSO_4$ is placed in a second beaker with a strip of Sn metal. A salt bridge connects the two beakers, and wires to a voltmeter link the two metal electrodes. (a) Which electrode serves as the anode and which as the cathode? (b) Which electrode gains mass and which loses mass as the cell reaction proceeds? (c) Write the equation for the overall cell reaction. (d) What is the emf generated by the cell under standard conditions?

20.42 A voltaic cell consists of a strip of cadmium metal in a solution of $Cd(NO_3)_2$ in one beaker, and in the other beaker a platinum electrode is immersed in a NaCl solution, with Cl_2 gas bubbled around the electrode. A salt bridge connects the two beakers. (a) Which electrode serves as the anode and which as the cathode? (b) Does the Cd electrode gain or lose mass as the cell reaction proceeds? (c) Write the equation for the overall cell reaction. (d) What is the emf generated by the cell under standard conditions?

Strengths of Oxidizing and Reducing Agents (Section 20.4)

20.43 From each of the following pairs of substances, use data in Appendix E to choose the one that is the stronger reducing agent:

(a) Fe(s) or Mg(s)

(b) Ca(s) or Al(s)

(c) H_2 (g, acidic solution) or $H_2S(g)$

(d) $BrO_3^-(aq)$ or $IO_3^-(aq)$

20.44 From each of the following pairs of substances, use data in Appendix E to choose the one that is the stronger oxidizing agent:

(a) $Cl_2(g)$ or $Br_2(l)$

(b) $Zn^{2+}(aq)$ or $Cd^{2+}(aq)$

(c) $Cl^-(aq)$ or $ClO_3^-(aq)$

(d) $H_2O_2(aq)$ or $O_3(g)$

20.45 By using the data in Appendix E, determine whether each of the following substances is likely to serve as an oxidant or a reductant: (a) $Cl_2(g)$, (b) MnO_4^- (aq, acidic solution), (c) Ba(s), (d) Zn(s).

20.46 By using the data in Appendix E, determine whether each of the following substances is likely to serve as an oxidant or a reductant: (a) $Ce^{3+}(aq)$, (b) Ca(s), (c) $ClO_3^-(aq)$, (d) $N_2O_5(g)$?

20.47 (a) Assuming standard conditions, arrange the following in order of increasing strength as oxidizing agents in acidic solution: $Cr_2O_7^{2-}$, H_2O_2, Cu^{2+}, Cl_2, O_2. (b) Arrange the following in order of increasing strength as reducing agents in acidic solution: Zn, I^-, Sn^{2+}, H_2O_2, Al.

20.48 Based on the data in Appendix E, (a) which of the following is the strongest oxidizing agent and which is the weakest in acidic solution: Br_2, H_2O_2, Zn, $Cr_2O_7^{2-}$? (b) Which of the following is the strongest reducing agent, and which is the weakest in acidic solution: F^-, Zn, $N_2H_5^+$, I_2, NO?

20.49 The standard reduction potential for the reduction of $Eu^{3+}(aq)$ to $Eu^{2+}(aq)$ is -0.43 V. Using Appendix E, which of the following substances is capable of reducing $Eu^{3+}(aq)$ to $Eu^{2+}(aq)$ under standard conditions: Al, Co, H_2O_2, $N_2H_5^+$, $H_2C_2O_4$?

20.50 The standard reduction potential for the reduction of $RuO_4^-(aq)$ to $RuO_4^{2-}(aq)$ is $+0.59$ V. By using Appendix E, which of the following substances can oxidize $RuO_4^{2-}(aq)$ to $RuO_4^-(aq)$ under standard conditions: $Br_2(l)$, $BrO_3^-(aq)$, $Mn^{2+}(aq)$, $O_2(g)$, $Sn^{2+}(aq)$?

Free Energy and Redox Reactions (Section 20.5)

20.51 Given the following reduction half-reactions:

$$Fe^{3+}(aq) + e^- \longrightarrow Fe^{2+}(aq) \quad E^\circ_{red} = +0.77 \text{ V}$$
$$S_2O_6^{2-}(aq) + 4\,H^+(aq) + 2\,e^- \longrightarrow 2\,H_2SO_3(aq) \quad E^\circ_{red} = +0.60 \text{ V}$$
$$N_2O(g) + 2\,H^+(aq) + 2\,e^- \longrightarrow N_2(g) + H_2O(l) \quad E^\circ_{red} = -1.77 \text{ V}$$
$$VO_2^+(aq) + 2\,H^+(aq) + e^- \longrightarrow VO^{2+} + H_2O(l) \quad E^\circ_{red} = +1.00 \text{ V}$$

(a) Write balanced chemical equations for the oxidation of $Fe^{2+}(aq)$ by $S_2O_6^{2-}(aq)$, by $N_2O(aq)$, and by $VO_2^+(aq)$. (b) Calculate ΔG° for each reaction at 298 K. (c) Calculate the equilibrium constant K for each reaction at 298 K.

20.52 For each of the following reactions, write a balanced equation, calculate the standard emf, calculate ΔG° at 298 K, and calculate the equilibrium constant K at 298 K. (a) Aqueous iodide ion is oxidized to $I_2(s)$ by $Hg_2^{2+}(aq)$. (b) In acidic solution, copper(I) ion is oxidized to copper(II) ion by nitrate ion. (c) In basic solution, $Cr(OH)_3(s)$ is oxidized to $CrO_4^{2-}(aq)$ by $ClO^-(aq)$.

20.53 If the equilibrium constant for a two-electron redox reaction at 298 K is 1.5×10^{-4}, calculate the corresponding ΔG° and E°_{red}.

20.54 If the equilibrium constant for a one-electron redox reaction at 298 K is 8.7×10^4, calculate the corresponding ΔG° and E°_{red}.

20.55 Using the standard reduction potentials listed in Appendix E, calculate the equilibrium constant for each of the following reactions at 298 K:

(a) $Fe(s) + Ni^{2+}(aq) \longrightarrow Fe^{2+}(aq) + Ni(s)$

(b) $Co(s) + 2\,H^+(aq) \longrightarrow Co^{2+}(aq) + H_2(g)$

(c) $10\,Br^-(aq) + 2\,MnO_4^-(aq) + 16\,H^+(aq) \longrightarrow$ $2\,Mn^{2+}(aq) + 8\,H_2O(l) + 5\,Br_2(l)$

20.56 Using the standard reduction potentials listed in Appendix E, calculate the equilibrium constant for each of the following reactions at 298 K:

(a) $Cu(s) + 2\,Ag^+(aq) \longrightarrow Cu^{2+}(aq) + 2\,Ag(s)$

(b) $3\,Ce^{4+}(aq) + Bi(s) + H_2O(l) \longrightarrow 3\,Ce^{3+}(aq) +$ $BiO^+(aq) + 2H^+(aq)$

(c) $N_2H_5^+(aq) + 4\,Fe(CN)_6^{3-}(aq) \longrightarrow N_2(g) +$ $5\,H^+(aq) + 4\,Fe(CN)_6^{4-}(aq)$

20.57 A cell has a standard cell potential of $+0.177$ V at 298 K. What is the value of the equilibrium constant for the reaction (a) if $n = 1$? (b) if $n = 2$? (c) if $n = 3$?

20.58 At 298 K a cell reaction has a standard cell potential of $+0.17$ V. The equilibrium constant for the reaction is 5.5×10^5. What is the value of n for the reaction?

20.59 A voltaic cell is based on the reaction

$$Sn(s) + I_2(s) \longrightarrow Sn^{2+}(aq) + 2I^-(aq)$$

Under standard conditions, what is the maximum electrical work, in joules, that the cell can accomplish if 75.0 g of Sn is consumed?

20.60 Consider the voltaic cell illustrated in Figure 20.5, which is based on the cell reaction

$$Zn(s) + Cu^{2+}(aq) \longrightarrow Zn^{2+}(aq) + Cu(s)$$

Under standard conditions, what is the maximum electrical work, in joules, that the cell can accomplish if 50.0 g of copper is formed?

Cell EMF under Nonstandard Conditions (Section 20.6)

20.61 (a) In the Nernst equation what is the numerical value of the reaction quotient, Q, under standard conditions? (b) Can the Nernst equation be used at temperatures other than room temperature?

20.62 (a) A voltaic cell is constructed with all reactants and products in their standard states. Will the concentration of the reactants increase, decrease, or remain the same as the cell operates? (b) What happens to the emf of a cell if the concentrations of the products are increased?

20.63 What is the effect on the emf of the cell shown in Figure 20.9, which has the overall reaction $Zn(s) + 2\,H^+(aq) \longrightarrow Zn^{2+}(aq) + H_2(g)$, for each of the following changes? (a) The pressure of the H_2 gas is increased in the cathode half-cell. (b) Zinc nitrate is added to the anode half-cell. (c) Sodium hydroxide is added to the cathode half-cell, decreasing $[H^+]$. (d) The surface area of the anode is doubled.

20.64 A voltaic cell utilizes the following reaction:

$$Al(s) + 3\,Ag^+(aq) \longrightarrow Al^{3+}(aq) + 3\,Ag(s)$$

What is the effect on the cell emf of each of the following changes? (a) Water is added to the anode half-cell, diluting the solution. (b) The size of the aluminum electrode is increased. (c) A solution of $AgNO_3$ is added to the cathode half-cell, increasing the quantity of Ag^+ but not changing its concentration. (d) HCl is added to the $AgNO_3$ solution, precipitating some of the Ag^+ as $AgCl$.

20.65 A voltaic cell is constructed that uses the following reaction and operates at 298 K:

$$Zn(s) + Ni^{2+}(aq) \longrightarrow Zn^{2+}(aq) + Ni(s)$$

(a) What is the emf of this cell under standard conditions? (b) What is the emf of this cell when $[Ni^{2+}] = 3.00\ M$? and $[Zn^{2+}] = 0.100\ M$? (c) What is the emf of the cell when $[Ni^{2+}] = 0.200\ M$ and $[Zn^{2+}] = 0.900\ M$?

20.66 A voltaic cell utilizes the following reaction and operates at 298 K:

$$3\,Ce^{4+}(aq) + Cr(s) \longrightarrow 3\,Ce^{3+}(aq) + Cr^{3+}(aq)$$

(a) What is the emf of this cell under standard conditions? (b) What is the emf of this cell when $[Ce^{4+}] = 3.0\ M$, $[Ce^{3+}] = 0.10\ M$, and $[Cr^{3+}] = 0.010\ M$? (c) What is the emf of the cell when $[Ce^{4+}] = 0.010\ M$, $[Ce^{3+}] = 2.0\ M$, and $[Cr^{3+}] = 1.5\ M$?

20.67 A voltaic cell utilizes the following reaction:

$$4\,Fe^{2+}(aq) + O_2(g) + 4\,H^+(aq) \longrightarrow 4\,Fe^{3+}(aq) + 2\,H_2O(l)$$

(a) What is the emf of this cell under standard conditions?
(b) What is the emf of this cell when $[Fe^{2+}] = 1.3\,M$, $[Fe^{3+}] = 0.010\,M$, $P_{O_2} = 0.50$ atm, and the pH of the solution in the cathode half-cell is 3.50?

20.68 A voltaic cell utilizes the following reaction:

$$2\,Fe^{3+}(aq) + H_2(g) \longrightarrow 2\,Fe^{2+}(aq) + 2\,H^+(aq)$$

(a) What is the emf of this cell under standard conditions?
(b) What is the emf for this cell when $[Fe^{3+}] = 3.50\,M$, $P_{H_2} = 0.95$ atm, $[Fe^{2+}] = 0.0010\,M$, and the pH in both half-cells is 4.00?

20.69 A voltaic cell is constructed with two $Zn^{2+} - Zn$ electrodes. The two half-cells have $[Zn^{2+}] = 1.8\,M$ and $[Zn^{2+}] = 1.00 \times 10^{-2}\,M$, respectively. **(a)** Which electrode is the anode of the cell? **(b)** What is the standard emf of the cell? **(c)** What is the cell emf for the concentrations given? **(d)** For each electrode, predict whether $[Zn^{2+}]$ will increase, decrease, or stay the same as the cell operates.

20.70 A voltaic cell is constructed with two silver–silver chloride electrodes, each of which is based on the following half-reaction:

$$AgCl(s) + e^- \longrightarrow Ag(s) + Cl^-(aq)$$

The two half-cells have $[Cl^-] = 0.0150\,M$ and $[Cl^-] = 2.55\,M$, respectively. **(a)** Which electrode is the cathode of the cell? **(b)** What is the standard emf of the cell? **(c)** What is the cell emf for the concentrations given? **(d)** For each electrode, predict whether $[Cl^-]$ will increase, decrease, or stay the same as the cell operates.

20.71 The cell in Figure 20.9 could be used to provide a measure of the pH in the cathode half-cell. Calculate the pH of the cathode half-cell solution if the cell emf at 298 K is measured to be $+0.684$ V when $[Zn^{2+}] = 0.30\,M$ and $P_{H_2} = 0.90$ atm.

20.72 A voltaic cell is constructed that is based on the following reaction:

$$Sn^{2+}(aq) + Pb(s) \longrightarrow Sn(s) + Pb^{2+}(aq)$$

(a) If the concentration of Sn^{2+} in the cathode half-cell is $1.00\,M$ and the cell generates an emf of $+0.22$ V, what is the concentration of Pb^{2+} in the anode half-cell? **(b)** If the anode half-cell contains $[SO_4^{2-}] = 1.00\,M$ in equilibrium with $PbSO_4(s)$, what is the K_{sp} of $PbSO_4$?

Batteries and Fuel Cells (Section 20.7)

20.73 During a period of discharge of a lead–acid battery, 402 g of Pb from the anode is converted into $PbSO_4(s)$. **(a)** What mass of $PbO_2(s)$ is reduced at the cathode during this same period? **(b)** How many coulombs of electrical charge are transferred from Pb to PbO_2?

20.74 During the discharge of an alkaline battery, 4.50 g of Zn is consumed at the anode of the battery. **(a)** What mass of MnO_2 is reduced at the cathode during this discharge? **(b)** How many coulombs of electrical charge are transferred from Zn to MnO_2?

20.75 Heart pacemakers are often powered by lithium–silver chromate "button" batteries. The overall cell reaction is

$$2\,Li(s) + Ag_2CrO_4(s) \longrightarrow Li_2CrO_4(s) + 2\,Ag(s)$$

(a) Lithium metal is the reactant at one of the electrodes of the battery. Is it the anode or the cathode? **(b)** Choose the two half-reactions from Appendix E that *most closely approximate* the reactions that occur in the battery. What standard emf would be generated by a voltaic cell based on these half-reactions? **(c)** The battery generates an emf of $+3.5$ V. How close is this value to the one calculated in part (b)? **(d)** Calculate the emf that would be generated at body temperature, 37 °C. How does this compare to the emf you calculated in part (b)?

20.76 Mercuric oxide dry-cell batteries are often used where a flat discharge voltage and long life are required, such as in watches and cameras. The two half-cell reactions that occur in the battery are

$$HgO(s) + H_2O(l) + 2\,e^- \longrightarrow Hg(l) + 2\,OH^-(aq)$$
$$Zn(s) + 2\,OH^-(aq) \longrightarrow ZnO(s) + H_2O(l) + 2\,e^-$$

(a) Write the overall cell reaction. **(b)** The value of E°_{red} for the cathode reaction is $+0.098$ V. The overall cell potential is $+1.35$ V. Assuming that both half-cells operate under standard conditions, what is the standard reduction potential for the anode reaction? **(c)** Why is the potential of the anode reaction different than would be expected if the reaction occurred in an acidic medium?

20.77 **(a)** Suppose that an alkaline battery was manufactured using cadmium metal rather than zinc. What effect would this have on the cell emf? **(b)** What environmental advantage is provided by the use of nickel–metal hydride batteries over nickel–cadmium batteries?

20.78 In some applications nickel–cadmium batteries have been replaced by nickel–zinc batteries. The overall cell reaction for this relatively new battery is:

$$2\,H_2O(l) + 2\,NiO(OH)(s) + Zn(s) \longrightarrow 2\,Ni(OH)_2(s) + Zn(OH)_2(s)$$

(a) What is the cathode half-reaction? **(b)** What is the anode half-reaction? **(c)** A single nickel–cadmium cell has a voltage of 1.30 V. Based on the difference in the standard reduction potentials of Cd^{2+} and Zn^{2+}, what voltage would you estimate a nickel–zinc battery will produce? **(d)** Would you expect the specific energy density of a nickel–zinc battery to be higher or lower than that of a nickel–cadmium battery?

20.79 In a Li-ion battery the composition of the cathode is $LiCoO_2$ when completely discharged. On charging approximately 50% of the Li^+ ions can be extracted from the cathode and transported to the graphite anode where they intercalate between the layers. **(a)** What is the composition of the cathode when the battery is fully charged? **(b)** If the $LiCoO_2$ cathode has a mass of 10 g (when fully discharged), how many coulombs of electricity can be delivered on completely discharging a fully charged battery?

20.80 Li-ion batteries used in automobiles typically use a $LiMn_2O_4$ cathode in place of the $LiCoO_2$ cathode found in most Li-ion batteries. **(a)** Calculate the mass percent lithium in each electrode material? **(b)** Which material has a higher percentage of lithium? Does this help to explain why batteries made with

LiMn$_2$O$_4$ cathodes deliver less power on discharging? **(c)** In a battery that uses a LiCoO$_2$ cathode approximately 50% of the lithium migrates from the cathode to the anode on charging. In a battery that uses a LiMn$_2$O$_4$ cathode what fraction of the lithium in LiMn$_2$O$_4$ would need to migrate out of the cathode to deliver the same amount of lithium to the graphite anode?

20.81 The hydrogen–oxygen fuel cell has a standard emf of 1.23 V. What advantages and disadvantages are there to using this device as a source of power compared to a 1.55-V alkaline battery?

20.82 (a) What is the difference between a battery and a fuel cell? **(b)** Can the "fuel" of a fuel cell be a solid? Explain.

Corrosion (Section 20.8)

20.83 (a) Write the anode and cathode reactions that cause the corrosion of iron metal to aqueous iron(II). **(b)** Write the balanced half-reactions involved in the air oxidation of Fe^{2+}(aq) to Fe$_2$O$_3 \cdot$ 3 H$_2$O.

20.84 (a) Based on standard reduction potentials, would you expect copper metal to oxidize under standard conditions in the presence of oxygen and hydrogen ions? **(b)** When the Statue of Liberty was refurbished, Teflon® spacers were placed between the iron skeleton and the copper metal on the surface of the statue. What role do these spacers play?

20.85 (a) Magnesium metal is used as a sacrificial anode to protect underground pipes from corrosion. Why is the magnesium referred to as a "sacrificial anode"? **(b)** Looking in Appendix E, suggest what metal the underground pipes could be made from in order for magnesium to be successful as a sacrificial anode.

20.86 An iron object is plated with a coating of cobalt to protect against corrosion. Does the cobalt protect iron by cathodic protection? Explain.

20.87 A plumber's handbook states that you should not connect a brass pipe directly to a galvanized steel pipe because electrochemical reactions between the two metals will cause corrosion. The handbook recommends you use instead an insulating fitting to connect them. Brass is a mixture of copper and zinc. What spontaneous redox reaction(s) might cause the corrosion? Justify your answer with standard emf calculations.

20.88 A plumber's handbook states that you should not connect a copper pipe directly to a steel pipe because electrochemical reactions between the two metals will cause corrosion. The handbook recommends you use instead an insulating fitting to connect them. What spontaneous redox reaction(s) might cause the corrosion? Justify your answer with standard emf calculations.

Electrolysis (Section 20.9)

20.89 (a) What is *electrolysis*? **(b)** Are electrolysis reactions thermodynamically spontaneous? Explain. **(c)** What process occurs at the anode in the electrolysis of molten NaCl? **(d)** Why is sodium metal not obtained when an aqueous solution of NaCl undergoes electrolysis?

20.90 (a) What is an *electrolytic cell*? **(b)** The negative terminal of a voltage source is connected to an electrode of an electrolytic cell. Is the electrode the anode or the cathode of the cell? Explain. **(c)** The electrolysis of water is often done with a small amount of sulfuric acid added to the water. What is the role of the sulfuric acid? **(d)** Why are active metals such as Al obtained by electrolysis using molten salts rather than aqueous solutions?

20.91 (a) A Cr^{3+}(aq) solution is electrolyzed, using a current of 7.60 A. What mass of Cr(s) is plated out after 2.00 days? **(b)** What amperage is required to plate out 0.250 mol Cr from a Cr^{3+} solution in a period of 8.00 h?

20.92 Metallic magnesium can be made by the electrolysis of molten MgCl$_2$. **(a)** What mass of Mg is formed by passing a current of 4.55 A through molten MgCl$_2$, for 4.50 days? **(b)** How many minutes are needed to plate out 25.00 g Mg from molten MgCl$_2$ using 3.50 A of current?

20.93 (a) Calculate the mass of Li formed by electrolysis of molten LiCl by a current of 7.5 $\times$ 10^4 A flowing for a period of 24 h. Assume the electrolytic cell is 85% efficient. **(b)** What is the minimum voltage required to drive the reaction?

20.94 Elemental calcium is produced by the electrolysis of molten CaCl$_2$. **(a)** What mass of calcium can be produced by this process if a current of 7.5 $\times$ 10^3 A is applied for 48 h? Assume that the electrolytic cell is 68% efficient. **(b)** What is the minimum voltage needed to cause the electrolysis?

20.95 Metallic gold is collected from below the anode when crude copper metal is refined by electrolysis. Explain this behavior.

20.96 The crude copper that is subjected to electrorefining contains tellurium as an impurity. The standard reduction potential between tellurium and its lowest common oxidation state, Te^{4+}, is

$$Te^{4+}(aq) + 4\,e^- \longrightarrow Te(s) \quad E°_{red} = 0.57\ V$$

Given this information, describe the probable fate of tellurium impurities during electrorefining. Do the impurities fall to the bottom of the refining bath, unchanged, as copper is oxidized, or do they go into solution as ions? If they go into solution, do they plate out on the cathode?

Additional Exercises

20.97 A *disproportionation* reaction is an oxidation–reduction reaction in which the same substance is oxidized and reduced. Complete and balance the following disproportionation reactions:

(a) Ni$^+$(aq) $\longrightarrow$ Ni^{2+}(aq) + Ni(s) (acidic solution)

(b) MnO$_4^{2-}$(aq) $\longrightarrow$ MnO$_4^-$(aq) + MnO$_2$(s) (acidic solution)

(c) H$_2$SO$_3$(aq) $\longrightarrow$ S(s) + HSO$_4^-$(aq) (acidic solution)

(d) Cl$_2$(aq) $\longrightarrow$ Cl$^-$(aq) + ClO$^-$(aq) (basic solution)

20.98 A common shorthand way to represent a voltaic cell is

anode | anode solution || cathode solution | cathode

A double vertical line represents a salt bridge or a porous barrier. A single vertical line represents a change in phase, such as from solid to solution. **(a)** Write the half-reactions and overall cell reaction represented by $Fe|Fe^{2+}||Ag^+|Ag$; sketch the cell. **(b)** Write the half-reactions and overall cell reaction represented by $Zn|Zn^{2+}||H^+|H_2$; sketch the cell. **(c)** Using the notation just described, represent a cell based on the following reaction:

$$ClO_3^-(aq) + 3\,Cu(s) + 6\,H^+(aq)$$
$$\longrightarrow Cl^-(aq) + 3\,Cu^{2+}(aq) + 3\,H_2O(l)$$

Pt is used as an inert electrode in contact with the ClO_3^- and Cl^-. Sketch the cell.

20.99 Predict whether the following reactions will be spontaneous in acidic solution under standard conditions: **(a)** oxidation of Sn to Sn^{2+} by I_2 (to form I^-), **(b)** reduction of Ni^{2+} to Ni by I^- (to form I_2), **(c)** reduction of Ce^{4+} to Ce^{3+} by H_2O_2, **(d)** reduction of Cu^{2+} to Cu by Sn^{2+} (to form Sn^{4+}).

[20.100] Gold exists in two common positive oxidation states, +1 and +3. The standard reduction potentials for these oxidation states are

$$Au^+(aq) + e^- \longrightarrow Au(s) \quad E^\circ_{red} = +1.69\text{ V}$$
$$Au^{3+}(aq) + 3\,e^- \longrightarrow Au(s) \quad E^\circ_{red} = +1.50\text{ V}$$

(a) Can you use these data to explain why gold does not tarnish in the air? **(b)** Suggest several substances that should be strong enough oxidizing agents to oxidize gold metal. **(c)** Miners obtain gold by soaking gold-containing ores in an aqueous solution of sodium cyanide. A very soluble complex ion of gold forms in the aqueous solution because of the redox reaction

$$4\,Au(s) + 8\,NaCN(aq) + 2\,H_2O(l) + O_2(g)$$
$$\longrightarrow 4\,Na[Au(CN)_2](aq) + 4\,NaOH(aq)$$

What is being oxidized and what is being reduced in this reaction? **(d)** Gold miners then react the basic aqueous product solution from part (c) with Zn dust to get gold metal. Write a balanced redox reaction for this process. What is being oxidized, and what is being reduced?

20.101 A voltaic cell is constructed from an $Ni^{2+}(aq) - Ni(s)$ half-cell and an $Ag^+(aq) - Ag(s)$ half-cell. The initial concentration of $Ni^{2+}(aq)$ in the $Ni^{2+} - Ni$ half-cell is $[Ni^{2+}] = 0.0100\,M$. The initial cell voltage is +1.12 V. **(a)** By using data in Table 20.1, calculate the standard emf of this voltaic cell. **(b)** Will the concentration of $Ni^{2+}(aq)$ increase or decrease as the cell operates? **(c)** What is the initial concentration of $Ag^+(aq)$ in the Ag^+-Ag half-cell?

[20.102] A voltaic cell is constructed that uses the following half-cell reactions:

$$Cu^+(aq) + e^- \longrightarrow Cu(s)$$
$$I_2(s) + 2\,e^- \longrightarrow 2\,I^-(aq)$$

The cell is operated at 298 K with $[Cu^+] = 0.25\,M$ and $[I^-] = 3.5\,M$. **(a)** Determine E for the cell at these concentrations. **(b)** Which electrode is the anode of the cell? **(c)** Is the answer to part (b) the same as it would be if the cell were operated under standard conditions? **(d)** If $[Cu^+]$ were equal to 0.15 M, at what concentration of I^- would the cell have zero potential?

20.103 Using data from Appendix E, calculate the equilibrium constant for the disproportionation of the copper(I) ion at room temperature:

$$2\,Cu^+(aq) \longrightarrow Cu^{2+}(aq) + Cu(s).$$

20.104 **(a)** Write the reactions for the discharge and charge of a nickel–cadmium (nicad) rechargeable battery. **(b)** Given the following reduction potentials, calculate the standard emf of the cell:

$$Cd(OH)_2(s) + 2\,e^- \longrightarrow Cd(s) + 2\,OH^-(aq)$$
$$E^\circ_{red} = -0.76\text{ V}$$
$$NiO(OH)(s) + H_2O(l) + e^- \longrightarrow Ni(OH)_2(s) + OH^-(aq)$$
$$E^\circ_{red} = +0.49\text{ V}$$

(c) A typical nicad voltaic cell generates an emf of +1.30 V. Why is there a difference between this value and the one you calculated in part (b)? **(d)** Calculate the equilibrium constant for the overall nicad reaction based on this typical emf value.

20.105 The capacity of batteries such as the typical AA alkaline battery is expressed in units of milliamp-hours (mAh). An AA alkaline battery yields a nominal capacity of 2850 mAh. **(a)** What quantity of interest to the consumer is being expressed by the units of mAh? **(b)** The starting voltage of a fresh alkaline battery is 1.55 V. The voltage decreases during discharge and is 0.80 V when the battery has delivered its rated capacity. If we assume that the voltage declines linearly as current is withdrawn, estimate the total maximum electrical work the battery could perform during discharge.

20.106 If you were going to apply a small potential to a steel ship resting in the water as a means of inhibiting corrosion, would you apply a negative or a positive charge? Explain.

[20.107] **(a)** How many coulombs are required to plate a layer of chromium metal 0.25 mm thick on an auto bumper with a total area of 0.32 m^2 from a solution containing CrO_4^{2-}? The density of chromium metal is 7.20 g/cm^3. **(b)** What current flow is required for this electroplating if the bumper is to be plated in 10.0 s? **(c)** If the external source has an emf of +6.0 V and the electrolytic cell is 65% efficient, how much electrical power is expended to electroplate the bumper?

20.108 Magnesium is obtained by electrolysis of molten $MgCl_2$. **(a)** Why is an aqueous solution of $MgCl_2$ not used in the electrolysis? **(b)** Several cells are connected in parallel by very large copper bars that convey current to the cells. Assuming that the cells are 96% efficient in producing the desired products in electrolysis, what mass of Mg is formed by passing a current of 97,000 A for a period of 24 h?

20.109 Calculate the number of kilowatt-hours of electricity required to produce 1.0×10^3 kg (1 metric ton) of aluminum by electrolysis of Al^{3+} if the applied voltage is 4.50 V and the process is 45% efficient.

20.110 Some years ago a unique proposal was made to raise the *Titanic*. The plan involved placing pontoons within the ship using a surface-controlled submarine-type vessel. The pontoons would contain cathodes and would be filled with hydrogen gas formed by the electrolysis of water. It has been estimated that it would require about 7×10^8 mol of H_2 to provide the buoyancy to lift the ship (*J. Chem. Educ.*, 1973, Vol. 50, 61). **(a)** How many coulombs of electrical charge would be required? **(b)** What is the minimum voltage required to generate H_2 and O_2 if the pressure on the gases at the depth of the wreckage (2 mi) is 300 atm? **(c)** What is the minimum electrical energy required to raise the *Titanic* by electrolysis? **(d)** What is the minimum cost of the electrical energy required to generate the necessary H_2 if the electricity costs 85 cents per kilowatt-hour to generate at the site?

Integrative Exercises

20.111 The Haber process is the principal industrial route for converting nitrogen into ammonia:

$$N_2(g) + 3 H_2(g) \longrightarrow 2 NH_3(g)$$

(a) What is being oxidized, and what is being reduced? **(b)** Using the thermodynamic data in Appendix C, calculate the equilibrium constant for the process at room temperature. **(c)** Calculate the standard emf of the Haber process at room temperature.

[20.112] In a galvanic cell the cathode is an $Ag^+(1.00\ M)/Ag(s)$ half-cell. The anode is a standard hydrogen electrode immersed in a buffer solution containing 0.10 M benzoic acid (C_6H_5COOH) and 0.050 M sodium benzoate $(C_6H_5COO^-Na^+)$. The measured cell voltage is 1.030 V. What is the pK_a of benzoic acid?

20.113 Consider the general oxidation of a species A in solution: $A \longrightarrow A^+ + e^-$. The term *oxidation potential* is sometimes used to describe the ease with which species A is oxidized—the easier a species is to oxidize, the greater its oxidation potential. **(a)** What is the relationship between the standard oxidation potential of A and the standard reduction potential of A^+? **(b)** Which of the metals listed in Table 4.5 has the highest standard oxidation potential? Which has the lowest? **(c)** For a series of substances, the trend in oxidation potential is often related to the trend in the first ionization energy. Explain why this relationship makes sense.

20.114 A voltaic cell is based on $Ag^+(aq)/Ag(s)$ and $Fe^{3+}(aq)/Fe^{2+}(aq)$ half-cells. **(a)** What is the standard emf of the cell? **(b)** Which reaction occurs at the cathode and which at the anode of the cell? **(c)** Use $S°$ values in Appendix C and the relationship between cell potential and free-energy change to predict whether the standard cell potential increases or decreases when the temperature is raised above 25 °C.

20.115 Hydrogen gas has the potential for use as a clean fuel in reaction with oxygen. The relevant reaction is

$$2 H_2(g) + O_2(g) \longrightarrow 2 H_2O(l)$$

Consider two possible ways of utilizing this reaction as an electrical energy source: (i) Hydrogen and oxygen gases are combusted and used to drive a generator, much as coal is currently used in the electric power industry; (ii) hydrogen and oxygen gases are used to generate electricity directly by using fuel cells that operate at 85 °C. **(a)** Use data in Appendix C to calculate $\Delta H°$ and $\Delta S°$ for the reaction. We will assume that these values do not change appreciably with temperature. **(b)** Based on the values from part (a), what trend would you expect for the magnitude of ΔG for the reaction as the temperature increases? **(c)** What is the significance of the change in the magnitude of ΔG with temperature with respect to the utility of hydrogen as a fuel? **(d)** Based on the analysis here, would it be more efficient to use the combustion method or the fuel-cell method to generate electrical energy from hydrogen?

20.116 Cytochrome, a complicated molecule that we will represent as $CyFe^{2+}$, reacts with the air we breathe to supply energy required to synthesize adenosine triphosphate (ATP). The body uses ATP as an energy source to drive other reactions. (Section 19.7) At pH 7.0 the following reduction potentials pertain to this oxidation of $CyFe^{2+}$:

$$O_2(g) + 4 H^+(aq) + 4 e^- \longrightarrow 2 H_2O(l) \qquad E°_{red} = +0.82\ V$$

$$CyFe^{3+}(aq) + e^- \longrightarrow CyFe^{2+}(aq) \qquad E°_{red} = +0.22\ V$$

(a) What is ΔG for the oxidation of $CyFe^{2+}$ by air? **(b)** If the synthesis of 1.00 mol of ATP from adenosine diphosphate (ADP) requires a ΔG of 37.7 kJ, how many moles of ATP are synthesized per mole of O_2?

[20.117] The standard potential for the reduction of AgSCN(s) is +0.0895 V.

$$AgSCN(s) + e^- \longrightarrow Ag(s) + SCN^-(aq)$$

Using this value and the electrode potential for $Ag^+(aq)$, calculate the K_{sp} for AgSCN.

[20.118] The K_{sp} value for PbS(s) is 8.0×10^{-28}. By using this value together with an electrode potential from Appendix E, determine the value of the standard reduction potential for the reaction

$$PbS(s) + 2 e^- \longrightarrow Pb(s) + S^{2-}(aq)$$

20.119 A student designs an ammeter (a device that measures electrical current) that is based on the electrolysis of water into hydrogen and oxygen gases. When electrical current of unknown magnitude is run through the device for 2.00 min, 12.3 mL of water-saturated $H_2(g)$ is collected. The temperature of the system is 25.5 °C, and the atmospheric pressure is 768 torr. What is the magnitude of the current in amperes?

Design an Experiment

You are asked to construct a voltaic cell that would simulate an alkaline battery by providing an electrical output of 1.50 V at the beginning of its discharge. After you complete it your voltaic cell will be used to power an external device that draws a constant current of 0.50 amperes for 2.0 hours. You are given the following supplies: electrodes of each transition metal from manganese to zinc, the chloride salts of the +2 transition metal ions from Mn^{2+} to Zn^{2+} ($MnCl_2$, $FeCl_2$, $CoCl_2$, $NiCl_2$, $CuCl_2$ and $ZnCl_2$), two 100 mL beakers, a salt bridge, a voltmeter, and wires to make electrical connections between the electrodes and the voltmeter. **(a)** Sketch out your voltaic cell labeling the metal used for each electrode and the type and concentration of the solutions in which each electrode is immersed. Be sure to describe how many grams of the salt are dissolved and the total volume of solution in each beaker. **(b)** What will be the concentrations of the transition metal ion in each solution at the end of the 2-h discharge? **(c)** What voltage will the cell register at the end of the discharge? **(d)** How long would your cell run before it died because the reactant in one of the half-cells was completely consumed? Assume the current stays constant throughout the discharge.

21

Nuclear Chemistry

The chemical energy we have discussed thus far originates in the making and breaking of chemical bonds, which result from the interaction of electrons between atoms. When chemical bonds are made or broken, the manner in which atoms are connected changes, but the number of atoms of each type is the same on both sides of the chemical equation. Indeed, for all the chemical reactions we have studied thus far, atoms are neither created nor destroyed. In this chapter, however, we examine a very different type of chemical transformation: reactions in which the nuclei of atoms undergo change, thereby changing the very identities of the atoms involved.

Transformations of atomic nuclei, which are called **nuclear reactions**, can involve enormous changes in energy—far greater than that involved in bond making and breaking. The energy of stars, such as our Sun, and the energy created by nuclear power plants offer examples of the tremendous energy released in nuclear reactions. Our chapter-opening photo shows a pellet of the dioxide of plutonium-238, an unstable isotope that undergoes a spontaneous process called *nuclear decay*. The pellet's glow is a consequence of the significant amount of heat generated by the nuclear decay of plutonium-238. This heat is utilized to generate electricity in space vehicles using a device called a *radioisotope thermoelectric generator* (RTG). For example, all of the electricity used in the *Curiosity* rover vehicle currently exploring the surface of Mars is generated using an RTG that contains 4.8 kg of plutonium-238 dioxide. After its nuclear decay, the atoms of plutonium have changed into a different element—because the number of protons and neutrons in the nucleus generally changes during a nuclear decay, the identity of the atom changes.

▶ **A PELLET OF PLUTONIUM-238 DIOXIDE,** which generates heat and light from its radioactive decay. Pellets such as these are used in radioisotope thermoelectric generators (RTGs) to generate electricity in space vehicles.

WHAT'S AHEAD ▶

21.1 RADIOACTIVITY AND NUCLEAR EQUATIONS We begin by learning how to describe nuclear reactions using equations analogous to chemical equations, in which the nuclear charges and masses of reactants and products are in balance. We see that radioactive nuclei most commonly decay by emission of *alpha*, *beta*, or *gamma* radiation.

21.2 PATTERNS OF NUCLEAR STABILITY We see that nuclear stability is determined largely by the *neutron-to-proton ratio*. For stable nuclei, this ratio increases with increasing atomic number. All nuclei with 84 or more protons are radioactive. Heavy nuclei gain stability by a series of nuclear disintegrations leading to stable nuclei.

21.3 NUCLEAR TRANSMUTATIONS We explore *nuclear transmutations*, which are nuclear reactions induced by bombardment of a nucleus by a neutron or an accelerated charged particle.

21.4 RATES OF RADIOACTIVE DECAY We learn that radioisotope decays are first-order kinetic processes with characteristic half-lives. Decay rates can be used to determine the age of ancient artifacts and geological formations.

21.5 DETECTION OF RADIOACTIVITY We see that the radiation emitted by a radioactive substance can be detected by a variety of devices, including dosimeters, Geiger counters, and scintillation counters.

21.6 ENERGY CHANGES IN NUCLEAR REACTIONS We learn
that energy changes in nuclear reactions are related to mass
changes via Einstein's equation, $E = mc^2$. The *nuclear binding
energy* of a nucleus is the difference between the mass of the
nucleus and the sum of the masses of its nucleons.

21.7 NUCLEAR POWER: FISSION We explore *nuclear fission*,
in which a heavy nucleus splits to form two or more product
nuclei. Nuclear fission is the energy source for nuclear power
plants, and we look at the operating principles of these plants.

21.8 NUCLEAR POWER: FUSION We learn that in *nuclear
fusion* two light nuclei are fused together to form a more stable,
heavier nucleus.

**21.9 RADIATION IN THE ENVIRONMENT AND LIVING
SYSTEMS** We discover that naturally occurring radioisotopes
bathe our planet—and us—with low levels of radiation. The
radiation emitted in nuclear reactions can damage living cells but
also has diagnostic and therapeutic applications.

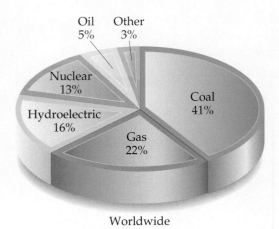

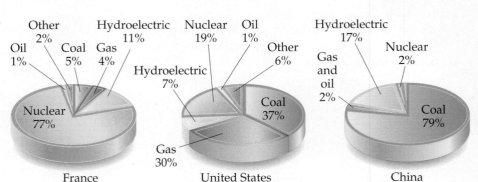

▲ **Figure 21.1 Sources of electricity generation, worldwide and for select countries** (**Sources**: U.S. Energy Information Administration and the International Energy Agency, 2010–2012 data).

Nuclear reactions are used to generate electricity on Earth as well as in space. Roughly 13% of the electricity generated worldwide comes from nuclear power plants, though the percentage varies from one country to the next, as ▲ **Figure 21.1** shows.

The use of nuclear energy for power generation and the disposal of nuclear wastes from power plants, as well as concerns about nuclear weaponry, are controversial social and political issues. It is imperative, therefore, that as a citizen with a stake in these matters, you have some understanding of nuclear reactions and the uses of radioactive substances.

Nuclear chemistry is the study of nuclear reactions, with an emphasis on their uses and their effects on biological systems. Nuclear chemistry affects our lives in many ways, particularly in energy and medical applications. Radioactivity is also used to help determine the mechanisms of chemical reactions, to trace the movement of atoms in biological systems and the environment, and to date historical artifacts. Different isotopes of the same element can undergo very different nuclear reactions, and one of our goals in this chapter is to gain a deeper appreciation for the differences among different radioactive isotopes and the ways in which they undergo decay and other nuclear transformations.

21.1 | Radioactivity and Nuclear Equations

To understand nuclear reactions, we must review and develop some ideas introduced in Section 2.3. First, recall that two types of subatomic particles reside in the nucleus: *protons* and *neutrons*. We will refer to these particles as **nucleons**. Recall also that all atoms of a given element have the same number of protons; this number is the element's *atomic number*. The atoms of a given element can have different numbers of neutrons, however, so they can have different *mass numbers*; the mass number is the total number of nucleons in the nucleus. Atoms with the same atomic number but different mass numbers are known as *isotopes*.

The different isotopes of an element are distinguished by their mass numbers. For example, the three naturally occurring isotopes of uranium are uranium-234, uranium-235, and uranium-238, where the numerical suffixes represent the mass numbers. These isotopes are also written $^{234}_{92}U$, $^{235}_{92}U$, and $^{238}_{92}U$, where the superscript is the mass number and the subscript is the atomic number.*

Different isotopes of an element have different natural abundances. For example, 99.3% of naturally occurring uranium is uranium-238, 0.7% is uranium-235, and only a trace is uranium-234. Different isotopes of an element also have different stabilities. Indeed, the nuclear properties of any given isotope depend on the number of protons and neutrons in its nucleus.

A *nuclide* is a nucleus containing a specified number of protons and neutrons. Nuclides that are radioactive are called **radionuclides**, and atoms containing these nuclei are called **radioisotopes**.

Nuclear Equations

Most nuclei in nature are stable and remain intact indefinitely. Radionuclides, however, are unstable and spontaneously emit particles and electromagnetic radiation. Emission of radiation is one of the ways in which an unstable nucleus is transformed into a more stable one that has less energy. The emitted radiation is the carrier of the excess energy. Uranium-238, for example, is radioactive, undergoing a nuclear reaction emitting helium-4 nuclei. The helium-4 particles are known as **alpha (α) particles**, and a stream of them is called *alpha radiation*. When a $^{238}_{92}U$ nucleus loses an alpha particle, the remaining fragment has an atomic number of 90 and a mass number of 234. The element with atomic number 90 is Th, thorium. Therefore, the products of uranium-238 decomposition are an alpha particle and a thorium-234 nucleus. We represent this reaction by the *nuclear equation*

$$^{238}_{92}U \longrightarrow {}^{234}_{90}Th + {}^{4}_{2}He \qquad [21.1]$$

When a nucleus spontaneously decomposes in this way, it is said either to have decayed or to have undergone *radioactive decay*. Because an alpha particle is involved in this reaction, scientists also describe the process as **alpha decay**.

 Give It Some Thought

What change in the mass number of a nucleus occurs when the nucleus emits an alpha particle?

In Equation 21.1 the sum of the mass numbers is the same on both sides of the equation (238 = 234 + 4). Likewise, the sum of the atomic numbers on both sides of the equation is equal (92 = 90 + 2). Mass numbers and atomic numbers must be balanced in all nuclear equations.

The radioactive properties of the nucleus in an atom are independent of the chemical state of the atom. In writing nuclear equations, therefore, we are not concerned with the chemical form (element or compound) of the atom in which the nucleus resides.

---

*As noted in Section 2.3, we often do not explicitly write the atomic number of an isotope because the element symbol is specific to the atomic number. In studying nuclear chemistry, however, it is often useful to include the atomic number in order to help us keep track of changes in the nuclei.

SAMPLE EXERCISE 21.1 Predicting the Product of a Nuclear Reaction

What product is formed when radium-226 undergoes alpha decay?

SOLUTION

Analyze We are asked to determine the nucleus that results when radium-226 loses an alpha particle.

Plan We can best do this by writing a balanced nuclear reaction for the process.

Solve The periodic table shows that radium has an atomic number of 88. The complete chemical symbol for radium-226 is therefore $^{226}_{88}\text{Ra}$. An alpha particle is a helium-4 nucleus, and so its symbol is ^4_2He. The alpha particle is a product of the nuclear reaction, and so the equation is of the form

$$^{226}_{88}\text{Ra} \longrightarrow {}^A_Z\text{X} + {}^4_2\text{He}$$

where A is the mass number of the product nucleus and Z is its atomic number. Mass numbers and atomic numbers must balance, so

$$226 = A + 4$$

and

$$88 = Z + 2$$

Hence,

$$A = 222 \quad \text{and} \quad Z = 86$$

Again, from the periodic table, the element with $Z = 86$ is radon (Rn). The product, therefore, is $^{222}_{86}\text{Rn}$, and the nuclear equation is

$$^{226}_{88}\text{Ra} \longrightarrow {}^{222}_{86}\text{Rn} + {}^4_2\text{He}$$

Practice Exercise 1

The plutonium-238 that is shown in the chapter-opening photograph undergoes alpha decay. What product forms when this radionuclide decays?
(a) Plutonium-234   (b) Uranium-234   (c) Uranium-238
(d) Thorium-236   (e) Neptunium-237

Practice Exercise 2

Which element undergoes alpha decay to form lead-208?

Types of Radioactive Decay

The three most common kinds of radiation given off when a radionuclide decays are alpha (α), beta (β), and gamma (γ) radiation. ∞ (Section 2.2) ▼ Table 21.1 summarizes some of the important properties of these types of radiation. As just described, alpha radiation consists of a stream of helium-4 nuclei known as alpha particles, which we denote as ^4_2He or simply α.

Beta radiation consists of streams of **beta (β) particles**, which are high-speed electrons emitted by an unstable nucleus. Beta particles are represented in nuclear equations by $^0_{-1}\text{e}$ or more commonly by β^-. The superscript 0 indicates that the mass of the electron is exceedingly small relative to the mass of a nucleon. The subscript -1 represents the negative charge of the beta particle, which is opposite that of the proton.

Iodine-131 is an isotope that undergoes decay by **beta emission**:

$$^{131}_{53}\text{I} \longrightarrow {}^{131}_{54}\text{Xe} + {}^0_{-1}\text{e} \qquad [21.2]$$

You can see from this equation that beta decay causes the atomic number of the reactant to increase from 53 to 54, which means a proton was created. Therefore, beta emission is equivalent to the conversion of a neutron (^1_0n or simply n) to a proton (^1_1H or simply p):

$$^1_0\text{n} \longrightarrow {}^1_1\text{H} + {}^0_{-1}\text{e} \quad \text{or} \quad \text{n} \longrightarrow \text{p} + \beta^- \qquad [21.3]$$

Just because an electron is emitted from a nucleus in beta decay, we should not think that the nucleus is composed of these particles any more than we consider a match to

Table 21.1 Properties of Alpha, Beta, and Gamma Radiation

Property	Type of Radiation		
	α	β	γ
Charge	2+	1−	0
Mass	6.64×10^{-24} g	9.11×10^{-28} g	0
Relative penetrating power	1	100	10,000
Nature of radiation	^4_2He nuclei	Electrons	High-energy photons

be composed of sparks simply because it gives them off when struck. The beta-particle electron comes into being only when the nucleus undergoes a nuclear reaction. Furthermore, the speed of the beta particle is sufficiently high that it does not end up in an orbital of the decaying atom.

Gamma (γ) radiation (or gamma rays) consists of high-energy photons (that is, electromagnetic radiation of very short wavelength). It changes neither the atomic number nor the mass number of a nucleus and is represented as either $_{0}^{0}\gamma$ or simply γ. Gamma radiation usually accompanies other radioactive emission because it represents the energy lost when the nucleons in a nuclear reaction reorganize into more stable arrangements. Often gamma rays are not explicitly shown when writing nuclear equations.

Two other types of radioactive decay are positron emission and electron capture. A **positron**, $_{+1}^{0}e$, or, often simply as β^{+}, is a particle that has the same mass as an electron (thus, we use the letter e and superscript 0 for the mass) but the opposite charge (represented by the +1 subscript).*

The isotope carbon-11 decays by **positron emission**:

$$_{6}^{11}C \longrightarrow {}_{5}^{11}B + {}_{+1}^{0}e \qquad [21.4]$$

Positron emission causes the atomic number of the reactant in this equation to decrease from 6 to 5. In general, positron emission has the effect of converting a proton to a neutron, thereby decreasing the atomic number of the nucleus by 1 while not changing the mass number:

$$_{1}^{1}p \longrightarrow {}_{0}^{1}n + {}_{+1}^{0}e \quad \text{or} \quad p \longrightarrow n + \beta^{+} \qquad [21.5]$$

Electron capture is the capture by the nucleus of an electron from the electron cloud surrounding the nucleus, as in this rubidium-81 decay:

$$_{37}^{81}Rb + {}_{-1}^{0}e \text{ (orbital electron)} \longrightarrow {}_{36}^{81}Kr \qquad [21.6]$$

Because the electron is consumed rather than formed in the process, it is shown on the reactant side of the equation. Electron capture, like positron emission, has the effect of converting a proton to a neutron:

$$_{1}^{1}p + {}_{-1}^{0}e \longrightarrow {}_{0}^{1}n \qquad [21.7]$$

▶ Table 21.2 summarizes the symbols used to represent the particles commonly encountered in nuclear reactions. The various types of radioactive decay are summarized in ▼ Table 21.3.

Table 21.2 Particles Found in Nuclear Reactions

Particle	Symbol
Neutron	$_{0}^{1}n$ or n
Proton	$_{1}^{1}H$ or p
Electron	$_{-1}^{0}e$
Alpha particle	$_{2}^{4}He$ or α
Beta particle	$_{-1}^{0}e$ or β^{-}
Positron	$_{+1}^{0}e$ or β^{+}

 Give It Some Thought

Which particles in Table 21.2 result in no change in nuclear charge when emitted in nuclear decay?

Table 21.3 Types of Radioactive Decay

Type	Nuclear Equation	Change in Atomic Number	Change in Mass Number
Alpha decay	$_{Z}^{A}X \longrightarrow {}_{Z-2}^{A-4}Y + {}_{2}^{4}He$	-2	-4
Beta emission	$_{Z}^{A}X \longrightarrow {}_{Z+1}^{A}Y + {}_{-1}^{0}e$	$+1$	Unchanged
Positron emission	$_{Z}^{A}X \longrightarrow {}_{Z-1}^{A}Y + {}_{+1}^{0}e$	-1	Unchanged
Electron capture*	$_{Z}^{A}X + {}_{-1}^{0}e \longrightarrow {}_{Z-1}^{A}Y$	-1	Unchanged

*The electron captured comes from the electron cloud surrounding the nucleus.

---

*The positron has a very short life because it is annihilated when it collides with an electron, producing gamma rays: $_{+1}^{0}e + {}_{-1}^{0}e \longrightarrow 2{}_{0}^{0}\gamma$.

SAMPLE EXERCISE 21.2 | Writing Nuclear Equations

Write nuclear equations for **(a)** mercury-201 undergoing electron capture; **(b)** thorium-231 decaying to protactinium-231.

SOLUTION

Analyze We must write balanced nuclear equations in which the masses and charges of reactants and products are equal.

Plan We can begin by writing the complete chemical symbols for the nuclei and decay particles that are given in the problem.

Solve

(a) The information given in the question can be summarized as

$$^{201}_{80}\text{Hg} + ^{0}_{-1}\text{e} \longrightarrow ^{A}_{Z}\text{X}$$

The mass numbers must have the same sum on both sides of the equation:

$$201 + 0 = A$$

Thus, the product nucleus must have a mass number of 201. Similarly, balancing the atomic numbers gives

$$80 - 1 = Z$$

Thus, the atomic number of the product nucleus must be 79, which identifies it as gold (Au):

$$^{201}_{80}\text{Hg} + ^{0}_{-1}\text{e} \longrightarrow ^{201}_{79}\text{Au}$$

(b) In this case we must determine what type of particle is emitted in the course of the radioactive decay:

$$^{231}_{90}\text{Th} \longrightarrow ^{231}_{91}\text{Pa} + ^{A}_{Z}\text{X}$$

From $231 = 231 + A$ and $90 = 91 + Z$, we deduce $A = 0$ and $Z = -1$. According to Table 21.2, the particle with these characteristics is the beta particle (electron). We therefore write

$$^{231}_{90}\text{Th} \longrightarrow ^{231}_{91}\text{Pa} + ^{0}_{-1}\text{e} \quad \text{or} \quad ^{231}_{90}\text{Th} \longrightarrow ^{231}_{91}\text{Pa} + \beta^-$$

Practice Exercise 1

The radioactive decay of thorium-232 occurs in multiple steps, called a *radioactive decay chain*. The second product produced in this chain is actinium-228. Which of the following processes could lead to this product starting with thorium-232?
(a) Alpha decay followed by beta emission
(b) Beta emission followed by electron capture
(c) Positron emission followed by alpha decay
(d) Electron capture followed by positron emission
(e) More than one of the above is consistent with the observed transformation

Practice Exercise 2

Write a balanced nuclear equation for the reaction in which oxygen-15 undergoes positron emission.

21.2 | Patterns of Nuclear Stability

Some nuclides, such as $^{12}_{6}\text{C}$ and $^{13}_{6}\text{C}$ are stable, whereas others, such as $^{14}_{6}\text{C}$ are unstable and undergo fission. Why is it that some nuclides are stable but others that may have only one more or one fewer neutron are not? No single rule allows us to predict whether a particular nucleus is radioactive and, if it is, how it might decay. However, several empirical observations can help us predict the stability of a nucleus.

Neutron-to-Proton Ratio

Because like charges repel each other, it may seem surprising that a large number of protons can reside within the small volume of the nucleus. At close distances, however, a strong force of attraction, called the *strong nuclear force*, exists between nucleons. Neutrons are intimately involved in this attractive force. All nuclei other than $^{1}_{1}\text{H}$ contain neutrons. As the number of protons in a nucleus increases, there is an ever greater need for neutrons to counteract the proton–proton repulsions. Stable nuclei with atomic numbers up to about 20 have approximately equal numbers of neutrons and protons. For nuclei with atomic number above 20, the number of neutrons exceeds the number of protons. Indeed, the number of neutrons necessary to create a stable nucleus increases more rapidly than the number of protons. Thus, the neutron-to-proton ratios of stable nuclei increase with increasing atomic number, as illustrated by the most common isotopes of carbon, $^{12}_{6}\text{C}$ (n/p = 1), manganese, $^{55}_{25}\text{Mn}$ (n/p = 1.20), and gold, $^{197}_{79}\text{Au}$ (n/p = 1.49).

▶ Figure 21.2 shows all known isotopes of the elements through $Z = 100$ plotted according to their numbers of protons and neutrons. Notice how the plot goes above the line for 1:1 neutron-to-proton for heavier elements. The dark blue dots in the figure represent stable (nonradioactive) isotopes. The region of the graph covered by these dark blue dots is known as the *belt of stability*. The belt of stability ends at element 83 (bismuth), which means that *all nuclei with 84 or more protons are radioactive*. For example, all isotopes of uranium, $Z = 92$, are radioactive.

▲ GO FIGURE

Estimate the optimal number of neutrons for a nucleus containing 70 protons.

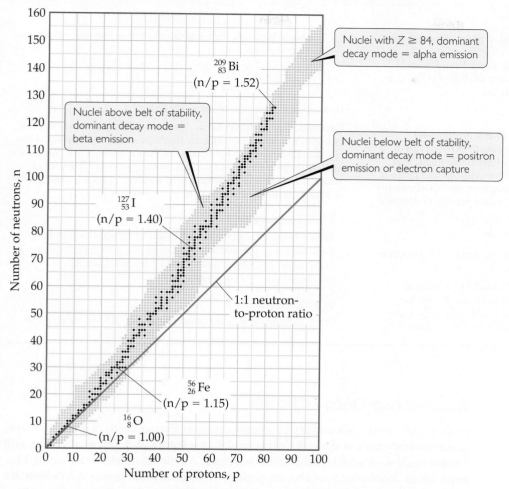

▲ Figure 21.2 **Stable and radioactive isotopes as a function of numbers of neutrons and protons in a nucleus.** The stable nuclei (dark blue dots) define a region known as the belt of stability.

The type of radioactive decay that a particular radionuclide undergoes depends largely on how its neutron-to-proton ratio compares with those of nearby nuclei that lie within the belt of stability. We can envision three general situations:

1. **Nuclei above the belt of stability (high neutron-to-proton ratios).** These neutron-rich nuclei can lower their ratio and thereby move toward the belt of stability by emitting a beta particle because beta emission decreases the number of neutrons and increases the number of protons (Equation 21.3).

2. **Nuclei below the belt of stability (low neutron-to-proton ratios).** These proton-rich nuclei can increase their ratio and so move closer to the belt of stability by either positron emission or electron capture because both decays increase the number of neutrons and decrease the number of protons (Equations 21.5 and 21.7). Positron emission is more common among lighter nuclei. Electron capture becomes increasingly common as the nuclear charge increases.

3. **Nuclei with atomic numbers ≥84.** These heavy nuclei tend to undergo alpha emission, which decreases both the number of neutrons and the number of protons by two, moving the nucleus diagonally toward the belt of stability.

SAMPLE EXERCISE 21.3 Predicting Modes of Nuclear Decay

Predict the mode of decay of (a) carbon-14, (b) xenon-118.

SOLUTION

Analyze We are asked to predict the modes of decay of two nuclei.

Plan To do this, we must locate the respective nuclei in Figure 21.2 and determine their positions with respect to the belt of stability in order to predict the most likely mode of decay.

Solve

(a) Carbon is element 6. Thus, carbon-14 has 6 protons and $14 - 6 = 8$ neutrons, giving it a neutron-to-proton ratio of 1.25. Elements with $Z < 20$ normally have stable nuclei that contain approximately equal numbers of neutrons and protons ($n/p = 1$). Thus, carbon-14 is located above the belt of stability and we expect it to decay by emitting a beta particle to decrease the n/p ratio:

$$^{14}_{6}\text{C} \longrightarrow {}^{14}_{7}\text{N} + {}^{0}_{-1}\text{e}$$

This is indeed the mode of decay observed for carbon-14, a reaction that lowers the n/p ratio from 1.25 to 1.0.

(b) Xenon is element 54. Thus, xenon-118 has 54 protons and $118 - 54 = 64$ neutrons, giving it an n/p ratio of 1.18. According to Figure 21.2, stable nuclei in this region of the belt of stability have higher neutron-to-proton ratios than xenon-118.

The nucleus can increase this ratio by either positron emission or electron capture:

$$^{118}_{54}\text{Xe} \longrightarrow {}^{118}_{53}\text{I} + {}^{0}_{+1}\text{e}$$

$$^{118}_{54}\text{Xe} + {}^{0}_{-1}\text{e} \longrightarrow {}^{118}_{53}\text{I}$$

In this case both modes of decay are observed.

Comment Keep in mind that our guidelines do not always work. For example, thorium-233, which we might expect to undergo alpha decay, actually undergoes beta emission. Furthermore, a few radioactive nuclei lie within the belt of stability. Both $^{146}_{60}\text{Nd}$ and $^{148}_{60}\text{Nd}$, for example, are stable and lie in the belt of stability. $^{147}_{60}\text{Nd}$, however, which lies between them, is radioactive.

Practice Exercise 1

Which of the following radioactive nuclei is most likely to decay via emission of a β^- particle?
(a) nitrogen-13 (b) magnesium-23 (c) rubidium-83
(d) iodine-131 (e) neptunium-237

Practice Exercise 2

Predict the mode of decay of (a) plutonium-239, (b) indium-120.

Radioactive Decay Chains

Some nuclei cannot gain stability by a single emission. Consequently, a series of successive emissions occurs as shown for uranium-238 in ▶ Figure 21.3. Decay continues until a stable nucleus—lead-206 in this case—is formed. A series of nuclear reactions that begins with an unstable nucleus and terminates with a stable one is known as a **radioactive decay chain** or a **nuclear disintegration series**. Three such series occur in nature: uranium-238 to lead-206, uranium-235 to lead-207, and thorium-232 to lead-208. All of the decay processes in these series are either alpha emissions or beta emissions.

Further Observations

Two further observations can help us to predict stable nuclei:

- Nuclei with the **magic numbers** of 2, 8, 20, 28, 50, or 82 protons or 2, 8, 20, 28, 50, 82, or 126 neutrons are generally more stable than nuclei that do not contain these numbers of nucleons.

- Nuclei with even numbers of protons, neutrons, or both are more likely to be stable than those with odd numbers of protons and/or neutrons. Approximately 60% of stable nuclei have an even number of both protons and neutrons, whereas less than 2% have odd numbers of both (◀ Table 21.4).

These observations can be understood in terms of the *shell model of the nucleus*, in which nucleons are described as residing in shells analogous to the shell structure for electrons in atoms. Just as certain numbers of electrons correspond to stable filled-shell electron configurations, so also the magic numbers of nucleons represent filled shells in nuclei.

There are several examples of the stability of nuclei with magic numbers of nucleons. For example, the radioactive series in Figure 21.3 ends with the stable $^{203}_{82}\text{Pb}$ nucleus, which has a magic number of protons (82). Another example is the observation that tin, which has a magic number of protons (50), has ten stable isotopes, more than any other element.

Table 21.4 Number of Stable Isotopes with Even and Odd Numbers of Protons and Neutrons

Number of Stable Isotopes	Proton Number	Neutron Number
157	Even	Even
53	Even	Odd
50	Odd	Even
5	Odd	Odd

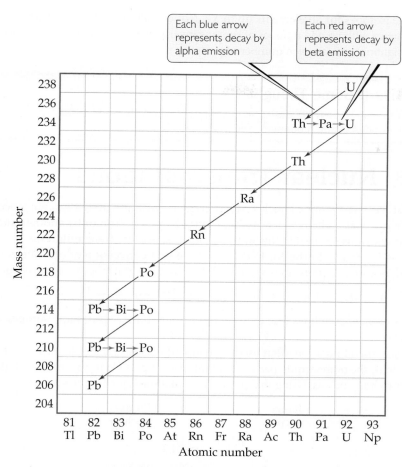

▲ Figure 21.3 **Nuclear decay chain for uranium-238.** The decay continues until the stable nucleus ^{206}Pb is formed.

Evidence also suggests that pairs of protons and pairs of neutrons have a special stability, analogous to the pairs of electrons in molecules. This evidence accounts for the second observation noted earlier, that stable nuclei with an even number of protons and/or neutrons are far more numerous than those with odd numbers. The preference for even numbers of protons is illustrated in ▼ **Figure 21.4**, which shows the number

 GO FIGURE

Among the elements shown here, how many have an even number of protons and fewer than three stable isotopes? How many have an odd number of protons and more than two stable isotopes?

1 H (2)																	2 He (2)
3 Li (2)	4 Be (1)											5 B (2)	6 C (2)	7 N (2)	8 O (3)	9 F (1)	10 Ne (3)
11 Na (1)	12 Mg (3)											13 Al (1)	14 Si (3)	15 P (1)	16 S (4)	17 Cl (2)	18 Ar (3)
19 K (2)	20 Ca (5)	21 Sc (1)	22 Ti (5)	23 V (2)	24 Cr (4)	25 Mn (1)	26 Fe (4)	27 Co (1)	28 Ni (5)	29 Cu (2)	30 Zn (5)	31 Ga (2)	32 Ge (4)	33 As (1)	34 Se (5)	35 Br (2)	36 Kr (6)
37 Rb (1)	38 Sr (3)	39 Y (1)	40 Zr (4)	41 Nb (1)	42 Mo (6)	43 Tc (0)	44 Ru (7)	45 Rh (1)	46 Pd (6)	47 Ag (2)	48 Cd (6)	49 In (1)	50 Sn (10)	51 Sb (2)	52 Te (6)	53 I (1)	54 Xe (9)

Number of stable isotopes

Elements with two or fewer stable isotopes

Elements with three or more stable isotopes

▲ Figure 21.4 **Number of stable isotopes for elements 1–54.**

of stable isotopes for all elements up to Xe. Notice that once we move past nitrogen, the elements with an odd number of protons invariably have fewer stable isotopes than their neighbors with an even number of protons.

 Give It Some Thought

What can you say about the number of neutrons in the stable isotopes of fluorine, sodium, aluminum, and phosphorus?

21.3 | Nuclear Transmutations

Thus far we have examined nuclear reactions in which a nucleus decays spontaneously. A nucleus can also change identity if it is struck by a neutron or by another nucleus. Nuclear reactions induced in this way are known as **nuclear transmutations**.

In 1919, Ernest Rutherford performed the first conversion of one nucleus into another, using alpha particles emitted by radium to convert nitrogen-14 into oxygen-17:

$$^{14}_{7}\text{N} + {}^{4}_{2}\text{He} \longrightarrow {}^{17}_{8}\text{O} + {}^{1}_{1}\text{H} \quad \text{or} \quad {}^{14}_{7}\text{N} + \alpha \longrightarrow {}^{17}_{8}\text{O} + \text{p} \qquad [21.8]$$

Such reactions have allowed scientists to synthesize hundreds of radioisotopes in the laboratory.

A shorthand notation often used to represent nuclear transmutations lists the target nucleus, the bombarding particle and the ejected particle in parentheses, followed by the product nucleus. Using this condensed notation, Equation 21.8 becomes

Target nucleus →   Product nucleus

$$^{14}_{7}\text{N} (\alpha, \text{p}) {}^{17}_{8}\text{O}$$

Bombarding particle   Ejected particle

SAMPLE EXERCISE 21.4 Writing a Balanced Nuclear Equation

Write the balanced nuclear equation for the process summarized as $^{27}_{13}\text{Al}(\text{n}, \alpha)^{24}_{11}\text{Na}$.

SOLUTION

Analyze We must go from the condensed descriptive form of the reaction to the balanced nuclear equation.

Plan We arrive at the balanced equation by writing n and α, each with its associated subscripts and superscripts.

Solve The n is the abbreviation for a neutron ($^{1}_{0}\text{n}$) and α represents an alpha particle ($^{4}_{2}\text{He}$). The neutron is the bombarding particle, and the alpha particle is a product. Therefore, the nuclear equation is

$$^{27}_{13}\text{Al} + {}^{1}_{0}\text{n} \longrightarrow {}^{24}_{11}\text{Na} + {}^{4}_{2}\text{He} \quad \text{or} \quad {}^{27}_{13}\text{Al} + \text{n} \longrightarrow {}^{24}_{11}\text{Na} + \alpha$$

Practice Exercise 1

Consider the following nuclear transmutation: $^{238}_{92}\text{U}(\text{n}, \beta^{-})\text{X}$. What is the identity of nucleus X?
(a) $^{238}_{93}\text{Np}$ **(b)** $^{239}_{92}\text{U}$ **(c)** $^{239}_{92}\text{U}^{+}$ **(d)** $^{235}_{90}\text{Th}$ **(e)** $^{239}_{93}\text{Np}$

Practice Exercise 2

Write the condensed version of the nuclear reaction

$$^{16}_{8}\text{O} + {}^{1}_{1}\text{H} \longrightarrow {}^{13}_{7}\text{N} + {}^{4}_{2}\text{He}$$

Accelerating Charged Particles

Alpha particles and other positively charged particles must move very fast to overcome the electrostatic repulsion between them and the target nucleus. The higher the nuclear charge on either the bombarding particle or the target nucleus, the faster the bombarding particle must move to bring about a nuclear reaction. Many methods have been devised to accelerate charged particles, using strong magnetic and electrostatic fields. These **particle accelerators**, popularly called "atom smashers," bear such names as *cyclotron* and *synchrotron*.

4 Finally the ions are transferred to RHIC, which has a circumference of 3.8 km. Ions moving in opposite directions can collide at one of six points on the ring, marked with white rectangles

3 The booster synchrotron and alternate gradient synchrotron (AGS) further accelerate the ions to 99.7% of the speed of light

1 Gold atoms are ionized, creating ions that are accelerated in a Tandem van de Graaff accelerator

2 If needed, beams of H^+ ions can be generated in the Linac

▲ **Figure 21.5 The Relativistic Heavy Ion Collider.** This particle accelerator is located at Brookhaven National Laboratory on Long Island, New York.

A common theme in all particle accelerators is the need to create charged particles so that they can be manipulated by electric and magnetic fields. The tubes through which the particles move must be kept at high vacuum so that the particles do not inadvertently collide with any gas-phase molecules.

▲ **Figure 21.5** shows the Relativistic Heavy Ion Collider (RHIC) located at Brookhaven National Laboratory. This facility and the Large Hadron Collider (LHC) at CERN (Conseil Européen pour la Recherche Nucléaire) near Geneva are two of the largest particle accelerators in the world. Both LHC and RHIC are capable of accelerating protons, as well as heavy ions such as gold and lead, to speeds approaching the speed of light. Scientists study the outcomes of collisions involving these ultra-high-energy particles. These experiments are used to investigate the fundamental structure of matter and ultimately answer questions about the beginning of the universe. In 2013, the existence of an important fundamental particle in particle physics, called the *Higgs boson*, was experimentally confirmed at the LHC.

Reactions Involving Neutrons

Most synthetic isotopes used in medicine and scientific research are made using neutrons as the bombarding particles. Because neutrons are neutral, they are not repelled by the nucleus. Consequently, they do not need to be accelerated to cause nuclear reactions. The neutrons are produced in nuclear reactors. For example, cobalt-60, which is used in cancer radiation therapy, is produced by neutron capture. Iron-58 is placed in a nuclear reactor and bombarded by neutrons to trigger the following sequence of reactions:

$$^{58}_{26}\text{Fe} + ^{1}_{0}\text{n} \longrightarrow ^{59}_{26}\text{Fe} \qquad [21.9]$$

$$^{59}_{26}\text{Fe} \longrightarrow ^{59}_{27}\text{Co} + ^{0}_{-1}\text{e} \qquad [21.10]$$

$$^{59}_{27}\text{Co} + ^{1}_{0}\text{n} \longrightarrow ^{60}_{27}\text{Co} \qquad [21.11]$$

▲ **Give It Some Thought**

Can an electrostatic or magnetic field be used to accelerate neutrons in a particle accelerator? Why or why not?

Transuranium Elements

Nuclear transmutations have been used to produce the elements with atomic number above 92, collectively known as the **transuranium elements** because they follow uranium in the periodic table. Elements 93 (neptunium, Np) and 94 (plutonium, Pu) were produced in 1940 by bombarding uranium-238 with neutrons:

$$^{238}_{92}\text{U} + ^{1}_{0}\text{n} \longrightarrow ^{239}_{92}\text{U} \longrightarrow ^{239}_{93}\text{Np} + ^{0}_{-1}\text{e} \qquad [21.12]$$

$$^{239}_{93}\text{Np} \longrightarrow ^{239}_{94}\text{Pu} + ^{0}_{-1}\text{e} \qquad [21.13]$$

Elements with still larger atomic numbers are normally formed in small quantities in particle accelerators. Curium-242, for example, is formed when a plutonium-239 target is bombarded with accelerated alpha particles:

$$^{239}_{94}\text{Pu} + ^{4}_{2}\text{He} \longrightarrow ^{242}_{96}\text{Cm} + ^{1}_{0}\text{n} \qquad [21.14]$$

New advances in the detection of the decay patterns of single atoms have led to recent additions to the periodic table. Between 1994 and 2010, elements 110 through 118 were discovered via the nuclear reactions that occur when nuclei of much lighter elements collide with high energy. For example, in 1996 a team of European scientists based in Germany synthesized element 112, copernicium, Cn, by bombarding a lead target continuously for three weeks with a beam of zinc atoms:

$$^{208}_{82}\text{Pb} + ^{70}_{30}\text{Zn} \longrightarrow ^{277}_{112}\text{Cn} + ^{1}_{0}\text{n} \qquad [21.15]$$

Amazingly, their discovery was based on the detection of only one atom of the new element, which decays after roughly 100 μs by alpha decay to form darmstadtium-273 (element 110). Within one minute, another five alpha decays take place producing fermium-253 (element 100). The finding has been verified in both Japan and Russia.

Because experiments to create new elements are very complicated and produce only a very small number of atoms of the new elements, they need to be carefully evaluated and reproduced before the new element is made an official part of the periodic table. The International Union for Pure and Applied Chemistry (IUPAC) is the international body that authorizes names of new elements after their experimental discovery and confirmation. In 2012, IUPAC officially approved names for the two latest elements added to the periodic table, which are *flerovium* (element 114) and *livermorium* (element 116).

21.4 | Rates of Radioactive Decay

Some radioisotopes, such as uranium-238, are found in nature even though they are not stable. Other radioisotopes do not exist in nature but can be synthesized in nuclear reactions. To understand this distinction, we must realize that different nuclei undergo radioactive decay at different rates. Many radioisotopes decay essentially completely in fractions of a second, so we do not find them in nature. Uranium-238, on the other hand, decays very slowly. Therefore, despite its instability, we can still observe what remains from its formation in the early history of the universe.

Radioactive decay is a first-order kinetic process. Recall that a first-order process has a characteristic **half-life**, which is the time required for half of any given quantity of a substance to react. ∞ (Section 14.4) Nuclear decay rates are commonly expressed in terms of half-lives. Each radioisotope has its own characteristic half-life. For example, the half-life of strontium-90 is 28.8 yr (◀ **Figure 21.6**). If we start with 10.0 g of strontium-90, only

▲ **GO FIGURE**

If we start with a 50.0-g sample, how much of it remains after three half-lives have passed?

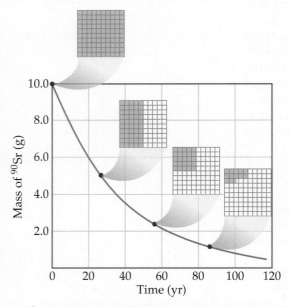

▲ **Figure 21.6 Decay of a 10.0-g sample of strontium-90** ($t_{1/2}$ = **28.8 yr**). The 10 × 10 grids show how much of the radioactive isotope remains after various amounts of time.

Table 21.5   The Half-Lives and Type of Decay for Several Radioisotopes

	Isotope	Half-Life (yr)	Type of Decay
Natural radioisotopes	$^{238}_{92}U$	4.5×10^9	Alpha
	$^{235}_{92}U$	7.0×10^8	Alpha
	$^{232}_{90}Th$	1.4×10^{10}	Alpha
	$^{40}_{19}K$	1.3×10^9	Beta
	$^{14}_{6}C$	5700	Beta
Synthetic radioisotopes	$^{239}_{94}Pu$	24,000	Alpha
	$^{137}_{55}Cs$	30.2	Beta
	$^{90}_{38}Sr$	28.8	Beta
	$^{131}_{53}I$	0.022	Beta

5.0 g of that isotope remains after 28.8 yr, 2.5 g remains after another 28.8 yr, and so on. Strontium-90 decays to yttrium-90 via beta emission:

$$^{90}_{38}Sr \longrightarrow {}^{90}_{39}Y + {}^{0}_{-1}e \qquad\qquad [21.16]$$

Half-lives as short as millionths of a second and as long as billions of years are known. The half-lives of some radioisotopes are listed in ▲ Table 21.5. One important feature of half-lives for nuclear decay is that they are unaffected by external conditions such as temperature, pressure, or state of chemical combination. Unlike toxic chemicals, therefore, radioactive atoms cannot be rendered harmless by chemical reaction or by any other practical treatment. At this point, we can do nothing but allow these nuclei to lose radioactivity at their characteristic rates. In the meantime, we must take precautions to prevent radioisotopes, such as those produced in nuclear power plants (Section 21.7), from entering the environment because of the damage radiation can cause.

SAMPLE EXERCISE 21.5   Calculation Involving Half-Lives

The half-life of cobalt-60 is 5.27 yr. How much of a 1.000-mg sample of cobalt-60 is left after 15.81 yr?

SOLUTION

Analyze We are given the half-life for cobalt-60 and asked to calculate the amount of cobalt-60 remaining from an initial 1.000-mg sample after 15.81 yr.

Plan We will use the fact that the amount of a radioactive substance decreases by 50% for every half-life that passes.

Solve Because $5.27 \times 3 = 15.81$, 15.81 yr is three half-lives for cobalt-60. At the end of one half-life, 0.500 mg of cobalt-60 remains, 0.250 mg at the end of two half-lives, and 0.125 mg at the end of three half-lives.

Practice Exercise 1

A radioisotope of technetium is useful in medical imaging techniques. A sample initially contains 80.0 mg of this isotope. After 24.0 h, only 5.0 mg of the technetium isotope remains. What is the half-life of the isotope?
(a) 3.0 h   **(b)** 6.0 h   **(c)** 12.0 h   **(d)** 16.0 h   **(e)** 24.0 h

Practice Exercise 2

Carbon-11, used in medical imaging, has a half-life of 20.4 min. The carbon-11 nuclides are formed, and the carbon atoms are then incorporated into an appropriate compound. The resulting sample is injected into a patient, and the medical image is obtained. If the entire process takes five half-lives, what percentage of the original carbon-11 remains at this time?

Radiometric Dating

Because the half-life of any particular nuclide is constant, the half-life can serve as a "nuclear clock" to determine the age of objects. The method of dating objects based on their isotopes and isotope abundances is called *radiometric dating*.

When carbon-14 is used in radiometric dating, the technique is known as *radiocarbon dating*. The procedure is based on the formation of carbon-14 as neutrons created by

▲ **GO FIGURE**

How does $^{14}CO_2$ become incorporated into the mammalian food chain?

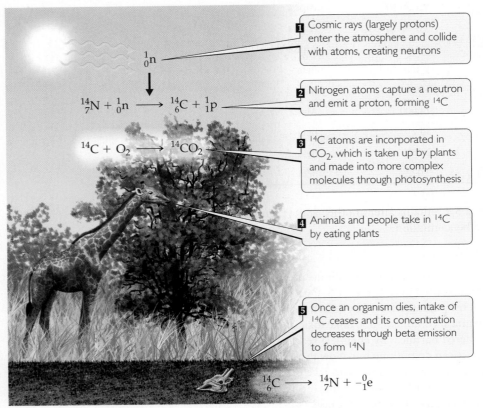

▲ **Figure 21.7 Creation and distribution of carbon-14.** The ratio of carbon-14 to carbon-12 in a dead animal or plant is related to the time since death occurred.

cosmic rays in the upper atmosphere convert nitrogen-14 into carbon-14 (▲ **Figure 21.7**). The ^{14}C reacts with oxygen to form $^{14}CO_2$ in the atmosphere, and this "labeled" CO_2 is taken up by plants and introduced into the food chain through photosynthesis. This process provides a small but reasonably constant source of carbon-14, which is radioactive and undergoes beta decay with a half-life of 5700 yr (to two significant figures):

$$^{14}_{6}C \longrightarrow ^{14}_{7}N + ^{0}_{-1}e \qquad [21.17]$$

Because a living plant or animal has a constant intake of carbon compounds, it is able to maintain a ratio of carbon-14 to carbon-12 that is nearly identical with that of the atmosphere. Once the organism dies, however, it no longer ingests carbon compounds to replenish the carbon-14 lost through radioactive decay. The ratio of carbon-14 to carbon-12 therefore decreases. By measuring this ratio and comparing it with that of the atmosphere, we can estimate the age of an object. For example, if the ratio diminishes to half that of the atmosphere, we can conclude that the object is one half-life, or 5700 yr old.

This method cannot be used to date objects older than about 50,000 yr because after this length of time the radioactivity is too low to be measured accurately.

In radiocarbon dating, a reasonable assumption is that the ratio of carbon-14 to carbon-12 in the atmosphere has been relatively constant for the past 50,000 yr. However, because variations in solar activity control the amount of carbon-14 produced in the atmosphere, that ratio can fluctuate. We can correct for this effect by using other kinds of data. Recently scientists have compared carbon-14 data with data from tree rings, corals, lake sediments, ice cores, and other natural sources to correct variations in the carbon-14 "clock" back to 26,000 yr.

Other isotopes can be similarly used to date other types of objects. For example, it takes 4.5×10^9 yr for half of a sample of uranium-238 to decay to lead-206. The age of rocks containing uranium can therefore be determined by measuring the ratio of lead-206 to uranium-238. If the lead-206 had somehow become incorporated into the rock by normal chemical processes instead of by radioactive decay, the rock would also contain large amounts of the more abundant isotope lead-208. In the absence of large amounts of this "geonormal" isotope of lead, it is assumed that all of the lead-206 was at one time uranium-238.

The oldest rocks found on Earth are approximately 3×10^9 yr old. This age indicates that Earth's crust has been solid for at least this length of time. Scientists estimate that it required 1×10^9 to 1.5×10^9 yr for Earth to cool and its surface to become solid, making the age of Earth 4.0 to 4.5×10^9 yr.

Calculations Based on Half-Life

So far, our discussion has been mainly qualitative. We now consider the topic of half-lives from a more quantitative point of view. This approach enables us to determine the half-life of a radioisotope or the age of an object.

As noted earlier, radioactive decay is a first-order kinetic process. Its rate, therefore, is proportional to the number of radioactive nuclei N in a sample:

$$\text{Rate} = kN \qquad [21.18]$$

The first-order rate constant, k, is called the *decay constant*.

The rate at which a sample decays is called its **activity**, and it is often expressed as number of disintegrations per unit time. The **becquerel** (Bq) is the SI unit for expressing activity. A becquerel is defined as one nuclear disintegration per second. An older, but still widely used, unit of activity is the **curie** (Ci), defined as 3.7×10^{10} disintegrations per second, which is the rate of decay of 1 g of radium. Thus, a 4.0-mCi sample of cobalt-60 undergoes

$$4.0 \times 10^{-3} \, \cancel{\text{Ci}} \times \frac{3.7 \times 10^{10} \text{ disintegrations/s}}{1 \, \cancel{\text{Ci}}} = 1.5 \times 10^8 \text{ disintegrations/s}$$

and so has an activity of 1.5×10^8 Bq.

As a radioactive sample decays, the amount of radiation emanating from the sample decays as well. For example, the half-life of cobalt-60 is 5.27 yr. The 4.0-mCi sample of cobalt-60 would, after 5.27 yr, have a radiation activity of 2.0 mCi, or 7.5×10^7 Bq.

 Give It Some Thought

If the size of a radioactive sample is doubled, what happens to the activity of the sample in Bq?

As we saw in Section 14.4, a first-order rate law can be transformed into the equation

$$\ln \frac{N_t}{N_0} = -kt \qquad [21.19]$$

In this equation t is the time interval of decay, k is the decay constant, N_0 is the initial number of nuclei (at time zero), and N_t is the number remaining after the time interval. Both the mass of a particular radioisotope and its activity are proportional to the number of radioactive nuclei. Thus, either the ratio of the mass at any time t to the mass at time $t = 0$ or the ratio of the activities at time t and $t = 0$ can be substituted for N_t/N_0 in Equation 21.19.

From Equation 21.19 we can obtain the relationship between the decay constant, k, and half-life, $t_{1/2}$: ∞ (Section 14.4)

$$k = \frac{0.693}{t_{1/2}} \qquad [21.20]$$

where we have used the value $\ln(N_t/N_0) = \ln(0.5) = -0.693$ for one half-life. Thus, if we know the value of either the decay constant or the half-life, we can calculate the value of the other.

 Give It Some Thought

Would doubling the mass of a radioactive sample change the half-life for the radioactive decay?

SAMPLE EXERCISE 21.6 Calculating the Age of Objects Using Radioactive Decay

A rock contains 0.257 mg of lead-206 for every milligram of uranium-238. The half-life for the decay of uranium-238 to lead-206 is 4.5×10^9 yr. How old is the rock?

SOLUTION

Analyze We are told that a rock sample has a certain amount of lead-206 for every unit mass of uranium-238 and asked to estimate the age of the rock.

Plan Lead-206 is the product of the radioactive decay of uranium-238. We will assume that the only source of lead-206 in the rock is from the decay of uranium-238, with a known half-life. To apply first-order kinetics expressions (Equations 21.19 and 21.20) to calculate the time elapsed since the rock was formed, we first need to calculate how much initial uranium-238 there was for every 1 mg that remains today.

Solve Let's assume that the rock currently contains 1.000 mg of uranium-238 and therefore 0.257 mg of lead-206. The amount of uranium-238 in the rock when it was first formed therefore equals 1.000 mg plus the quantity that has decayed to lead-206. Because the mass of lead atoms is not the same as the mass of uranium atoms, we cannot just add 1.000 mg and 0.257 mg. We have to multiply the present mass of lead-206 (0.257 mg) by the ratio of the mass number of uranium to that of lead, into which it has decayed. Therefore, the original mass of $^{238}_{92}\text{U}$ was

$$\text{Original } ^{238}_{92}\text{U} = 1.000 \text{ mg} + \frac{238}{206}(0.257 \text{ mg})$$

$$= 1.297 \text{ mg}$$

Using Equation 21.20, we can calculate the decay constant for the process from its half-life:

$$k = \frac{0.693}{4.5 \times 10^9 \text{ yr}} = 1.5 \times 10^{-10} \text{ yr}^{-1}$$

Rearranging Equation 21.19 to solve for time, t, and substituting known quantities gives

$$t = -\frac{1}{k}\ln\frac{N_t}{N_0} = -\frac{1}{1.5 \times 10^{-10} \text{ yr}^{-1}}\ln\frac{1.000}{1.297} = 1.7 \times 10^9 \text{ yr}$$

Comment To check this result, you could use the fact that the decay of uranium-235 to lead-207 has a half-life of 7×10^8 yr and measure the relative amounts of uranium-235 and lead-207 in the rock.

Practice Exercise 1

Cesium-137, which has a half-life of 30.2 yr, is a component of the radioactive waste from nuclear power plants. If the activity due to cesium-137 in a sample of radioactive waste has decreased to 35.2% of its initial value, how old is the sample?
(a) 1.04 yr **(b)** 15.4 yr **(c)** 31.5 yr **(d)** 45.5 yr **(e)** 156 yr

Practice Exercise 2

A wooden object from an archeological site is subjected to radiocarbon dating. The activity due to ^{14}C is measured to be 11.6 disintegrations per second. The activity of a carbon sample of equal mass from fresh wood is 15.2 disintegrations per second. The half-life of ^{14}C is 5700 yr. What is the age of the archeological sample?

SAMPLE EXERCISE 21.7 Calculations Involving Radioactive Decay and Time

If we start with 1.000 g of strontium-90, 0.953 g will remain after 2.00 yr. **(a)** What is the half-life of strontium-90? **(b)** How much strontium-90 will remain after 5.00 yr? **(c)** What is the initial activity of the sample in becquerels and curies?

SOLUTION

(a) Analyze We are asked to calculate a half-life, $t_{1/2}$, based on data that tell us how much of a radioactive nucleus has decayed in a time interval $t = 2.00$ yr and the information $N_0 = 1.000$ g, $N_t = 0.953$ g.

Plan We first calculate the rate constant for the decay, k, and then use that to compute $t_{1/2}$.

Solve Equation 21.19 is solved for the decay constant, k, and then Equation 21.20 is used to calculate half-life, $t_{1/2}$:

$$k = -\frac{1}{t} \ln \frac{N_t}{N_0} = -\frac{1}{2.00 \text{ yr}} \ln \frac{0.953 \text{ g}}{1.000 \text{ g}}$$

$$= -\frac{1}{2.00 \text{ yr}} (-0.0481) = 0.0241 \text{ yr}^{-1}$$

$$t_{1/2} = \frac{0.693}{k} = \frac{0.693}{0.0241 \text{ yr}^{-1}} = 28.8 \text{ yr}$$

(b) Analyze We are asked to calculate the amount of a radionuclide remaining after a given period of time.

Plan We need to calculate N_t, the amount of strontium present at time t, using the initial quantity, N_0, and the rate constant for decay, k, calculated in part (a).

Solve Again using Equation 21.19, with $k = 0.0241 \text{ yr}^{-1}$, we have

$$\ln \frac{N_t}{N_0} = -kt = -(0.0241 \text{ yr}^{-1})(5.00 \text{ yr}) = -0.120$$

N_t/N_0 is calculated from $\ln(N_t/N_0) = -0.120$ using the e^x or INV LN function of a calculator:

$$\frac{N_t}{N_0} = e^{-0.120} = 0.887$$

Because $N_0 = 1.000$ g, we have

$$N_t = (0.887)N_0 = (0.887)(1.000 \text{ g}) = 0.887 \text{ g}$$

(c) Analyze We are asked to calculate the activity of the sample in becquerels and curies.

Plan We must calculate the number of disintegrations per atom per second and then multiply by the number of atoms in the sample.

Solve The number of disintegrations per atom per second is given by the decay constant, k:

$$k = \left(\frac{0.0241}{\text{yr}}\right)\left(\frac{1 \text{ yr}}{365 \text{ days}}\right)\left(\frac{1 \text{ day}}{24 \text{ h}}\right)\left(\frac{1 \text{ h}}{3600 \text{ s}}\right) = 7.64 \times 10^{-10} \text{ s}^{-1}$$

To obtain the total number of disintegrations per second, we calculate the number of atoms in the sample. We multiply this quantity by k, where we express k as the number of disintegrations per atom per second, to obtain the number of disintegrations per second:

$$(1.000 \text{ g } ^{90}\text{Sr})\left(\frac{1 \text{ mol } ^{90}\text{Sr}}{90 \text{ g } ^{90}\text{Sr}}\right)\left(\frac{6.022 \times 10^{23} \text{ atoms Sr}}{1 \text{ mol } ^{90}\text{Sr}}\right) = 6.7 \times 10^{21} \text{ atoms } ^{90}\text{Sr}$$

$$\text{Total disintegrations/s} = \left(\frac{7.64 \times 10^{-10} \text{ disintegrations}}{\text{atom} \cdot \text{s}}\right)(6.7 \times 10^{21} \text{ atoms})$$

$$= 5.1 \times 10^{12} \text{ disintegrations/s}$$

Because 1 Bq is one disintegration per second, the activity is 5.1×10^{12} Bq. The activity in curies is given by

$$(5.1 \times 10^{12} \text{ disintegrations/s})\left(\frac{1 \text{ Ci}}{3.7 \times 10^{10} \text{ disintegrations/s}}\right) = 1.4 \times 10^2 \text{ Ci}$$

We have used only two significant figures in products of these calculations because we do not know the atomic weight of ^{90}Sr to more than two significant figures without looking it up in a special source.

Practice Exercise 1

As mentioned in the previous Practice Exercise 1, cesium-137, a component of radioactive waste, has a half-life of 30.2 yr. If a sample of waste has an initial activity of 15.0 Ci due to cesium-137, how long will it take for the activity due to cesium-137 to drop to 0.250 Ci?
(a) 0.728 yr   **(b)** 60.4 yr   **(c)** 78.2 yr   **(d)** 124 yr   **(e)** 178 yr

Practice Exercise 2

A sample to be used for medical imaging is labeled with ^{18}F, which has a half-life of 110 min. What percentage of the original activity in the sample remains after 300 min?

21.5 | Detection of Radioactivity

A variety of methods have been devised to detect emissions from radioactive substances. Henri Becquerel discovered radioactivity because radiation caused fogging of photographic plates, and since that time photographic plates and film have been used to detect radioactivity. The radiation affects photographic film in much the same way as X-rays do. The greater the extent of exposure to radiation, the darker the area of the developed negative. People who work with radioactive substances carry film badges to record the extent of their exposure to radiation (▼ Figure 21.8).

Radioactivity can also be detected and measured by a Geiger counter. The operation of this device is based on the fact that radiation is able to ionize matter. The ions and electrons produced by the ionizing radiation permit conduction of an electrical current. The basic design of a Geiger counter is shown in ▶ Figure 21.9. A current pulse between the anode and the metal cylinder occurs whenever entering radiation produces ions. Each pulse is counted in order to estimate the amount of radiation.

> ▲ **Give It Some Thought**
> Which type of radiation—alpha, beta, or gamma—is most likely to be stopped by the window of a Geiger counter?

Substances that are electronically excited by radiation can also be used to detect and measure radiation. For example, some substances excited by radiation give off light as electrons return to their lower-energy states. These substances are called *phosphors*. Different substances respond to different particles. Zinc sulfide, for example, responds to alpha particles. An instrument called a scintillation counter is used to detect and measure radiation, based on the tiny flashes of light produced when radiation strikes a suitable phosphor. The flashes of light are magnified electronically and counted to measure the amount of radiation.

▲ GO FIGURE

Which type of radiation—alpha, beta, or gamma—is likely to fog a film that is sensitive to X rays?

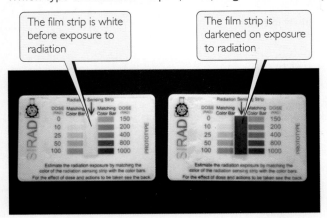

The film strip is white before exposure to radiation

The film strip is darkened on exposure to radiation

▲ **Figure 21.8 Badge dosimeters monitor the extent to which the individual has been exposed to high-energy radiation.** The radiation dose is determined from the extent of darkening of the film in the dosimeter.

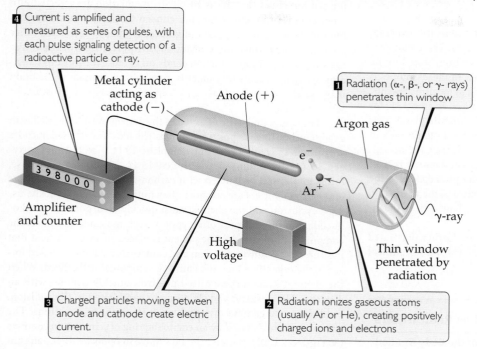

 GO FIGURE

Which property of the atoms of gas inside a Geiger counter is most relevant to the operation of the device?

4 Current is amplified and measured as series of pulses, with each pulse signaling detection of a radioactive particle or ray.

Metal cylinder acting as cathode (−)

Anode (+)

1 Radiation (α-, β-, or γ- rays) penetrates thin window

Argon gas

e⁻

Ar⁺

398000

Amplifier and counter

γ-ray

High voltage

Thin window penetrated by radiation

3 Charged particles moving between anode and cathode create electric current.

2 Radiation ionizes gaseous atoms (usually Ar or He), creating positively charged ions and electrons

▲ **Figure 21.9 Schematic drawing of a Geiger counter.**

Radiotracers

Because radioisotopes can be detected readily, they can be used to follow an element through its chemical reactions. The incorporation of carbon atoms from CO_2 into glucose during photosynthesis, for example, has been studied using CO_2 enriched in carbon-14:

$$6 \, {}^{14}CO_2 + 6 \, H_2O \xrightarrow[\text{Chlorophyll}]{\text{Sunlight}} {}^{14}C_6H_{12}O_6 + 6 \, O_2 \qquad [21.21]$$

The use of the carbon-14 label provides direct experimental evidence that carbon dioxide in the environment is chemically converted to glucose in plants. Analogous labeling experiments using oxygen-18 show that the O_2 produced during photosynthesis comes from water, not carbon dioxide. When it is possible to isolate and purify intermediates and products from reactions, detection devices such as scintillation counters can be used to "follow" the radioisotope as it moves from starting material through intermediates to final product. These types of experiments are useful for identifying elementary steps in a reaction mechanism. ∞ (Section 14.6)

The use of radioisotopes is possible because all isotopes of an element have essentially identical chemical properties. When a small quantity of a radioisotope is mixed with the naturally occurring stable isotopes of the same element, all the isotopes go through the same reactions together. The element's path is revealed by the radioactivity of the radioisotope. Because the radioisotope can be used to trace the path of the element, it is called a **radiotracer**.

▲ Give It Some Thought

Can you think of a process not involving radioactive decay for which $^{14}CO_2$ would behave *differently* from $^{12}CO_2$?

Chemistry and Life

Medical Applications of Radiotracers

Radiotracers have found wide use as diagnostic tools in medicine. ▼ Table 21.6 lists some radiotracers and their uses. These radioisotopes are incorporated into a compound that is administered to the patient, usually intravenously. The diagnostic use of these isotopes is based on the ability of the radioactive compound to localize and concentrate in the organ or tissue under investigation. Iodine-131, for example, has been used to test the activity of the thyroid gland. This gland is the only place in which iodine is incorporated significantly in the body. The patient drinks a solution of NaI containing iodine-131. Only a very small amount is used so that the patient does not receive a harmful dose of radioactivity. A Geiger counter placed close to the thyroid, in the neck region, determines the ability of the thyroid to take up the iodine. A normal thyroid will absorb about 12% of the iodine within a few hours.

The medical applications of radiotracers are further illustrated by positron emission tomography (PET). PET is used for clinical diagnosis of many diseases. In this method, compounds containing radionuclides that decay by positron emission are injected into a patient. These compounds are chosen to enable researchers to monitor blood flow, oxygen and glucose metabolic rates, and other biological functions. Some of the most interesting work involves the study of the brain, which depends on glucose for most of its energy. Changes in how this sugar is metabolized or used by the brain may signal a disease such as cancer, epilepsy, Parkinson's disease, or schizophrenia.

The compound to be detected in the patient must be labeled with a radionuclide that is a positron emitter. The most widely used nuclides are carbon-11 ($t_{1/2} = 20.4$ min), fluorine-18 ($t_{1/2} = 110$ min), oxygen-15 ($t_{1/2} = 2$ min), and nitrogen-13 ($t_{1/2} = 10$ min). Glucose, for example, can be labeled with carbon-11. Because the half-lives of positron emitters are so short, they must be generated on site using a cyclotron and the chemist must quickly incorporate the radionuclide into the sugar (or other appropriate) molecule and inject the compound immediately. The patient is placed in an instrument that measures the positron emission and constructs a computer-based image of the organ in which the emitting compound is localized. When the element decays, the emitted positron quickly collides with an electron. The positron and electron are annihilated in the collision, producing two gamma rays that move in opposite directions. The gamma rays are detected by an encircling ring of scintillation counters (▼ Figure 21.10). Because the rays move in opposite directions but were created in the same place at the same time, it is possible to accurately locate the point in the body where the radioactive isotope decayed. The nature of this image provides clues to the presence of disease or other abnormality and helps medical researchers understand how a particular disease affects the functioning of the brain. For example, the images shown in ▼ Figure 21.11 reveal that levels of activity in brains of patients with Alzheimer's disease are different from the levels in those without the disease.

Related Exercises: 21.55, 21.56, 21.82, 21.83

Table 21.6   Some Radionuclides Used as Radiotracers

Nuclide	Half-Life	Area of the Body Studied
Iodine-131	8.04 days	Thyroid
Iron-59	44.5 days	Red blood cells
Phosphorus-32	14.3 days	Eyes, liver, tumors
Technetium-99[a]	6.0 hours	Heart, bones, liver, and lungs
Thallium-201	73 hours	Heart, arteries
Sodium-24	14.8 hours	Circulatory system

[a]The isotope of technetium is actually a special isotope of Tc-99 called Tc-99*m*, where the *m* indicates a so-called *metastable* isotope.

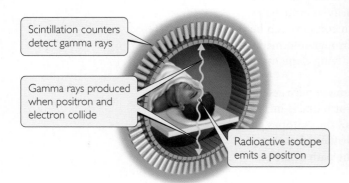

▲ Figure 21.10 Schematic representation of a positron emission tomography (PET) scanner.

Scintillation counters detect gamma rays

Gamma rays produced when positron and electron collide

Radioactive isotope emits a positron

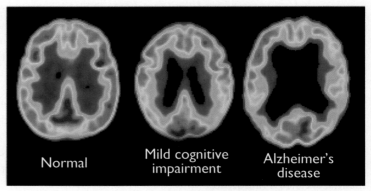

Normal

Mild cognitive impairment

Alzheimer's disease

▲ Figure 21.11 Positron emission tomography (PET) scans showing glucose metabolism levels in the brain. Red and yellow colors show higher levels of glucose metabolism.

21.6 | Energy Changes in Nuclear Reactions

Why are the energies associated with nuclear reactions so large, in many cases orders of magnitude larger than those associated with nonnuclear chemical reactions? The answer to this question begins with Einstein's celebrated equation from the theory of relativity that relates mass and energy:

$$E = mc^2 \qquad\qquad [21.22]$$

In this equation E stands for energy, m for mass, and c for the speed of light, 2.9979×10^8 m/s. This equation states that mass and energy are equivalent and can be converted into one another. If a system loses mass, it loses energy; if it gains mass, it gains energy. Because the proportionality constant between energy and mass, c^2, is such a large number, even small changes in mass are accompanied by large changes in energy.

The mass changes in chemical reactions are too small to detect. For example, the mass change associated with the combustion of 1 mol of CH_4 (an exothermic process) is -9.9×10^{-9} g. Because the mass change is so small, it is possible to treat chemical reactions as though mass is conserved. ∞ (Section 2.1)

The mass changes and the associated energy changes in nuclear reactions are much greater than those in chemical reactions. The mass change accompanying the radioactive decay of 1 mol of uranium-238, for example, is 50,000 times greater than that for the combustion of 1 mol of CH_4. Let's examine the energy change for the nuclear reaction

$$^{238}_{92}U \longrightarrow \ ^{234}_{90}Th + \ ^{4}_{2}He$$

The masses of the nuclei are $^{238}_{92}U$, 238.0003 amu; $^{234}_{90}Th$, 233.9942 amu; and $^{4}_{2}He$, 4.0015 amu. The mass change, Δm, is the total mass of the products minus the total mass of the reactants. The mass change for the decay of 1 mol of uranium-238 can then be expressed in grams:

$$233.9942 \text{ g} + 4.0015 \text{ g} - 238.0003 \text{ g} = -0.0046 \text{ g}$$

The fact that the system has lost mass indicates that the process is exothermic. All spontaneous nuclear reactions are exothermic.

The energy change per mole associated with this reaction is

$$\Delta E = \Delta(mc^2) = c^2 \Delta m$$

$$= (2.9979 \times 10^8 \text{ m/s})^2 (-0.0046 \text{ g})\left(\frac{1 \text{ kg}}{1000 \text{ g}}\right)$$

$$= -4.1 \times 10^{11} \frac{\text{kg-m}^2}{\text{s}^2} = -4.1 \times 10^{11} \text{J}$$

Notice that Δm must be converted to kilograms, the SI unit of mass, to obtain ΔE in joules, the SI unit of energy. The negative sign for the energy change indicates that energy is released in the reaction—in this case, over 400 billion joules per mole of uranium!

SAMPLE EXERCISE 21.8 | Calculating Mass Change in a Nuclear Reaction

How much energy is lost or gained when 1 mol of cobalt-60 undergoes beta decay, $^{60}_{27}Co \longrightarrow \ ^{60}_{28}Ni + \ ^{0}_{-1}e$? The mass of a $^{60}_{27}Co$ atom is 59.933819 amu, and that of a $^{60}_{28}Ni$ atom is 59.930788 amu.

SOLUTION

Analyze We are asked to calculate the energy change in a nuclear reaction.

Plan We must first calculate the mass change in the process. We are given atomic masses, but we need the masses of the nuclei in the reaction. We calculate these by taking account of the masses of the electrons that contribute to the atomic masses.

Solve A $^{60}_{27}\text{Co}$ atom has 27 electrons. The mass of an electron is 5.4858×10^{-4} amu. (See the list of fundamental constants in the back inside cover.) We subtract the mass of the 27 electrons from the mass of the $^{60}_{27}\text{Co}$ *atom* to find the mass of the $^{60}_{27}\text{Co}$ *nucleus*:

$$59.933819 \text{ amu} - (27)(5.4858 \times 10^{-4} \text{ amu}) = 59.919007 \text{ amu (or 59.919007 g/mol)}$$

Likewise, for $^{60}_{28}\text{Ni}$, the mass of the nucleus is

$$59.930788 \text{ amu} - (28)(5.4858 \times 10^{-4} \text{ amu}) = 59.915428 \text{ amu (or 59.915428 g/mol)}$$

The mass change in the nuclear reaction is the total mass of the products minus the mass of the reactant:

$$\Delta m = \text{mass of electron} + \text{mass } ^{60}_{28}\text{Ni nucleus} - \text{mass of } ^{60}_{27}\text{Co nucleus}$$

$$= 0.00054858 \text{ amu} + 59.915428 \text{ amu} - 59.919007 \text{ amu}$$

$$= -0.003030 \text{ amu}$$

Thus, when a mole of cobalt-60 decays,

$$\Delta m = -0.003030 \text{ g}$$

Because the mass decreases ($\Delta m < 0$), energy is released ($\Delta E < 0$). The quantity of energy released *per mole* of cobalt-60 is calculated using Equation 21.22:

$$\Delta E = c^2 \, \Delta m$$

$$= (2.9979 \times 10^8 \text{ m/s})^2(-0.003030 \text{ g})\left(\frac{1 \text{ kg}}{1000 \text{ g}}\right)$$

$$= -2.723 \times 10^{11} \frac{\text{kg-m}^2}{\text{s}^2} = -2.723 \times 10^{11} \text{ J}$$

Practice Exercise 1

The nuclear reaction that powers the radioisotope thermoelectric generator shown in the chapter-opening photograph is $^{238}_{94}\text{Pu} \longrightarrow {}^{234}_{92}\text{U} + {}^{4}_{2}\text{He}$. The atomic masses of plutonium-238 and uranium-234 are 238.049554 amu and 234.040946 amu, respectively. The mass of an alpha particle is 4.001506 amu. How much energy in kJ is released when 1.00 g of plutonium-238 decays to uranium-234?
(a) 2.27×10^6 kJ   **(b)** 2.68×10^6 kJ   **(c)** 3.10×10^6 kJ   **(d)** 3.15×10^6 kJ
(e) 7.37×10^8 kJ

Practice Exercise 2

Positron emission from ^{11}C, $^{11}_{6}\text{C} \longrightarrow {}^{11}_{5}\text{B} + {}^{0}_{+1}\text{e}$, occurs with release of 2.87×10^{11} J per mole of ^{11}C. What is the mass change per mole of ^{11}C in this nuclear reaction? The masses of ^{11}B and ^{11}C are 11.009305 and 11.011434 amu, respectively.

Nuclear Binding Energies

Scientists discovered in the 1930s that the masses of nuclei are always less than the masses of the individual nucleons of which they are composed. For example, the helium-4 nucleus (an alpha particle) has a mass of 4.00150 amu. The mass of a proton is 1.00728 amu and that of a neutron is 1.00866 amu. Consequently, two protons and two neutrons have a total mass of 4.03188 amu:

$$\text{Mass of two protons} = 2(1.00728 \text{ amu}) = 2.01456 \text{ amu}$$

$$\text{Mass of two neutrons} = 2(1.00866 \text{ amu}) = \underline{2.01732 \text{ amu}}$$

$$\text{Total mass} = 4.03188 \text{ amu}$$

The mass of the individual nucleons is 0.03038 amu greater than that of the helium-4 nucleus:

$$\text{Mass of two protons and two neutrons} = 4.03188 \text{ amu}$$

$$\text{Mass of } {}^{4}_{2}\text{He nucleus} = \underline{4.00150 \text{ amu}}$$

$$\text{Mass difference } \Delta m = 0.03038 \text{ amu}$$

The mass difference between a nucleus and its constituent nucleons is called the **mass defect**. The origin of the mass defect is readily understood if we consider that energy must be added to a nucleus to break it into separated protons and neutrons:

$$\text{Energy} + {}^{4}_{2}\text{He} \longrightarrow 2 {}^{1}_{1}\text{H} + 2 {}^{1}_{0}\text{n} \qquad [21.23]$$

Table 21.7   **Mass Defects and Binding Energies for Three Nuclei**

Nucleus	Mass of Nucleus (amu)	Mass of Individual Nucleons (amu)	Mass Defect (amu)	Binding Energy (J)	Binding Energy per Nucleon (J)
$^{4}_{2}\text{He}$	4.00150	4.03188	0.03038	4.53×10^{-12}	1.13×10^{-12}
$^{56}_{26}\text{Fe}$	55.92068	56.44914	0.52846	7.90×10^{-11}	1.41×10^{-12}
$^{238}_{92}\text{U}$	238.00031	239.93451	1.93420	2.89×10^{-10}	1.21×10^{-12}

By Einstein's relation, the addition of energy to a system must be accompanied by a proportional increase in mass. The mass change we just calculated for the conversion of helium-4 into separated nucleons is $\Delta m = 0.03038$ amu. Therefore, the energy required for this process is

$$\Delta E = c^2 \, \Delta m$$

$$= (2.9979 \times 10^8 \text{ m/s})^2 (0.03038 \text{ amu}) \left(\frac{1 \text{ g}}{6.022 \times 10^{23} \text{ amu}} \right) \left(\frac{1 \text{ kg}}{1000 \text{ g}} \right)$$

$$= 4.534 \times 10^{-12} \text{ J}$$

The energy required to separate a nucleus into its individual nucleons is called the **nuclear binding energy**. The mass defect and nuclear binding energy for three elements are compared in ▲ Table 21.7.

Give It Some Thought

The mass of a single atom of iron-56 is 55.93494 amu. Why is this number different from the mass of the nucleus given in Table 21.7?

Values of binding energies per nucleon can be used to compare the stabilities of different combinations of nucleons (such as two protons and two neutrons arranged either as $^{4}_{2}\text{He}$ or as $2\,^{2}_{1}\text{H}$). ▶ Figure 21.12 shows average binding energy per nucleon plotted against mass number. Binding energy per nucleon at first increases in magnitude as mass number increases, reaching about 1.4×10^{-12} J for nuclei whose mass numbers are in the vicinity of iron-56. It then decreases slowly to about 1.2×10^{-12} J for very heavy nuclei. This trend indicates that nuclei of intermediate mass numbers are more tightly bound (and therefore more stable) than those with either smaller or larger mass numbers.

This trend has two significant consequences: First, heavy nuclei gain stability and therefore give off energy if they are fragmented into two midsized nuclei. This process, known as **fission**, is used to generate energy in nuclear power plants. Second, because of the sharp increase in the graph for small mass numbers, even greater amounts of energy are released if very light nuclei are combined, or fused together, to give more massive nuclei. This **fusion** process is the essential energy-producing process in the Sun and other stars.

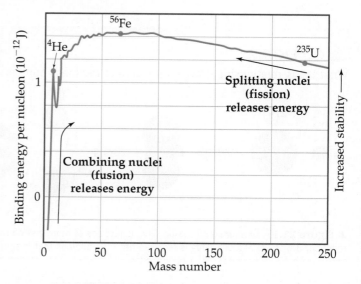

▲ **Figure 21.12 Nuclear binding energies.** The average binding energy per nucleon increases initially as the mass number increases and then decreases slowly. Because of these trends, fusion of light nuclei and fission of heavy nuclei are exothermic processes.

Give It Some Thought

Would fusing two stable nuclei that have mass numbers in the vicinity of 100 be an energy-releasing process?

21.7 | Nuclear Power: Fission

Commercial nuclear power plants and most forms of nuclear weaponry depend on nuclear fission for their operation. The first nuclear fission reaction to be discovered was that of uranium-235. This nucleus, as well as those of uranium-233 and plutonium-239, undergoes fission when struck by a slow-moving neutron (▼ **Figure 21.13**).*

A heavy nucleus can split in many ways. Two ways that the uranium-235 nucleus splits, for instance, are

$$\ ^{1}_{0}n + \ ^{235}_{92}U \left\{ \begin{array}{ll} \longrightarrow \ ^{137}_{52}Te + \ ^{97}_{40}Zr + 2\ ^{1}_{0}n & \text{[21.24]} \\ \longrightarrow \ ^{142}_{56}Ba + \ ^{91}_{36}Kr + 3\ ^{1}_{0}n & \text{[21.25]} \end{array} \right.$$

The nuclei produced in equations 21.24 and 21.25—called the *fission products*—are themselves radioactive and undergo further nuclear decay. More than 200 isotopes of 35 elements have been found among the fission products of uranium-235. Most of them are radioactive.

Slow-moving neutrons are required in fission because the process involves initial absorption of the neutron by the nucleus. The resulting more massive nucleus is often unstable and spontaneously undergoes fission. Fast neutrons tend to bounce off the nucleus, and little fission occurs.

Note that the coefficients of the neutrons produced in Equations 21.24 and 21.25 are 2 and 3. On average, 2.4 neutrons are produced by every fission of a uranium-235 nucleus. If one fission produces two neutrons, the two neutrons can cause two additional fissions, each producing two neutrons. The four neutrons thereby released can produce four fissions, and so forth, as shown in ▶ **Figure 21.14**. The number of fissions and the energy released quickly escalate, and if the process is unchecked, the result is a violent explosion. Reactions that multiply in this fashion are called **chain reactions**.

For a fission chain reaction to occur, the sample of fissionable material must have a certain minimum mass. Otherwise, neutrons escape from the sample before they have the opportunity to strike other nuclei and cause additional fission. The amount of fissionable material large enough to maintain a chain reaction with a constant rate of fission is called the **critical mass**. When a critical mass of material is present, one neutron on average from each fission is subsequently effective in producing another fission and

▲ GO FIGURE

What is the relationship between the sum of the mass numbers on the two sides of this reaction?

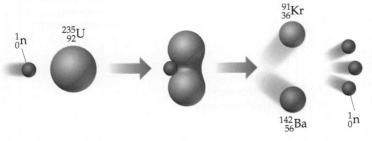

▲ **Figure 21.13 Uranium-235 fission.** This is just one of many fission patterns. In this reaction, 3.5×10^{-11} J of energy is released per ^{235}U nucleus that is split.

*Other heavy nuclei can be induced to undergo fission. However, these three are the only ones of practical importance.

the fission continues at a constant, controllable rate. The critical mass of uranium-235 is about 50 kg for a bare sphere of the metal.*

If more than a critical mass of fissionable material is present, very few neutrons escape. The chain reaction thus multiplies the number of fissions, which can lead to a nuclear explosion. A mass in excess of a critical mass is referred to as a **supercritical mass**. The effect of mass on a fission reaction is illustrated in ▼ Figure 21.15.

▼ Figure 21.16 shows a schematic diagram of the first atomic bomb used in warfare, the bomb, code-named "Little Boy," that was dropped on Hiroshima, Japan, on August 6, 1945. The bomb contained about 64 kg of uranium-235, which had been separated from the nonfissionable uranium-238 primarily by gaseous diffusion of uranium hexafluoride, UF_6. ∞ (Section 10.8) To trigger the fission reaction, two subcritical masses of uranium-235 were slammed together using chemical explosives. The combined masses of the uranium formed a supercritical mass, which led to a rapid, uncontrolled chain reaction and, ultimately, a nuclear explosion. The energy released by the bomb dropped on Hiroshima was equivalent to that of 16,000 tons of TNT (it therefore is called a *16-kiloton* bomb). Unfortunately, the basic design of a fission-based atomic bomb is quite simple, and the fissionable materials are potentially available to any nation with a nuclear reactor. The combination of design simplicity and materials availability has generated international concerns about the proliferation of atomic weapons.

▲ **GO FIGURE**

If this figure were extended one more "generation" down, how many neutrons would be produced?

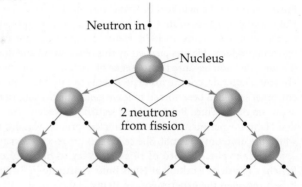

Neutron in

Nucleus

2 neutrons from fission

▲ Figure 21.14 **Fission chain reaction.**

▲ **GO FIGURE**

Which of these criticality scenarios—subcritical, critical, or supercritical—is desirable in a nuclear power plant that generates electricity?

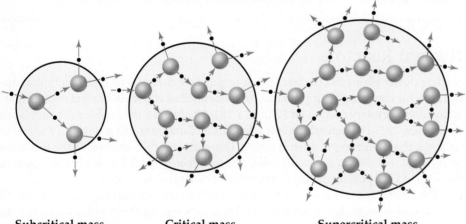

Subcritical mass	**Critical mass**	**Supercritical mass**
Rate of neutron loss > rate of neutron creation by fission	Rate of neutron loss = rate of neutron creation by fission	Rate of neutron loss < rate of neutron creation by fission

▲ Figure 21.15 **Subcritical, critical, and supercritical nuclear fission.**

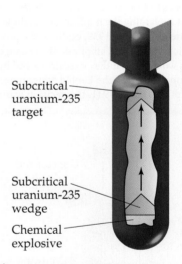

Subcritical uranium-235 target

Subcritical uranium-235 wedge

Chemical explosive

▲ Figure 21.16 **Schematic drawing of an atomic bomb.** A conventional explosive is used to bring two subcritical masses together to form a supercritical mass.

*The exact value of the critical mass depends on the shape of the radioactive substance. The critical mass can be reduced if the radioisotope is surrounded by a material that reflects some neutrons.

A Closer Look

The Dawning of the Nuclear Age

Uranium-235 fission was first achieved during the late 1930s by Enrico Fermi and coworkers in Rome and shortly thereafter by Otto Hahn and coworkers in Berlin. Both groups were trying to produce transuranium elements. In 1938, Hahn identified barium among his reaction products. He was puzzled by this observation and questioned the identification because the presence of barium was so unexpected. He sent a letter describing his experiments to Lise Meitner, a former coworker who had been forced to leave Germany because of the anti-Semitism of the Third Reich and had settled in Sweden. She surmised that Hahn's experiment indicated a nuclear process was occurring in which the uranium-235 split. She called this process *nuclear fission*.

Meitner passed word of this discovery to her nephew, Otto Frisch, a physicist working at Niels Bohr's institute in Copenhagen. Frisch repeated the experiment, verifying Hahn's observations, and found that tremendous energies were involved. In January 1939, Meitner and Frisch published a short article describing the reaction. In March 1939, Leo Szilard and Walter Zinn at Columbia University discovered that more neutrons are produced than are used in each fission. As we have seen, this result allows a chain reaction to occur.

News of these discoveries and an awareness of their potential use in explosive devices spread rapidly within the scientific community. Several scientists finally persuaded Albert Einstein, the most famous physicist of the time, to write a letter to President Franklin D. Roosevelt explaining the implications of these discoveries. Einstein's letter, written in August 1939, outlined the possible military applications of nuclear fission and emphasized the danger that weapons based on fission would pose if they were developed by the Nazis. Roosevelt judged it imperative that the United States investigate the possibility of such weapons. Late in 1941, the decision was made to build a bomb based on the fission reaction. An enormous research project, known as the Manhattan Project, began.

On December 2, 1942, the first artificial self-sustaining nuclear fission chain reaction was achieved in an abandoned squash court at the University of Chicago. This accomplishment led to the development of the first atomic bomb, at Los Alamos National Laboratory in New Mexico in July 1945 (▼ **Figure 21.17**). In August 1945 the United States dropped atomic bombs on two Japanese cities, Hiroshima and Nagasaki. The nuclear age had arrived, albeit in a sadly destructive fashion. Humanity has struggled with the conflict between the positive potential of nuclear energy and its terrifying potential as a weapon ever since.

▲ Figure 21.17 **The Trinity test for the atom bomb developed during World War II.** The first human-made nuclear explosion took place on July 16, 1945, on the Alamogordo test range in New Mexico.

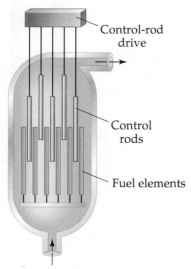

Water acts as both moderator and coolant

▲ Figure 21.18 **Schematic diagram of a pressurized water reactor core.**

Nuclear Reactors

Nuclear power plants use nuclear fission to generate energy. The core of a typical nuclear reactor consists of four principal components: fuel elements, control rods, a moderator, and a primary coolant (◀ **Figure 21.18**). The fuel is a fissionable substance, such as uranium-235. The natural isotopic abundance of uranium-235 is only 0.7, too low to sustain a chain reaction in most reactors. Therefore, the ^{235}U content of the fuel must be enriched to 3–5% for use in a reactor. The *fuel elements* contain enriched uranium in the form of UO_2 pellets encased in zirconium or stainless steel tubes.

The *control rods* are composed of materials that absorb neutrons, such as boron-10 or an alloy of silver, indium, and cadmium. These rods regulate the flux of neutrons to keep the reaction chain self-sustaining and also prevent the reactor core from overheating.*

The probability that a neutron will trigger fission of a ^{235}U nucleus depends on the speed of the neutron. The neutrons produced by fission have high speeds (typically in excess of 10,000 km/s). The function of the *moderator* is to slow down the neutrons (to speeds of a few kilometers per second) so that they can be captured more readily by the fissionable nuclei. The moderator is typically either water or graphite.

The *primary coolant* is a substance that transports the heat generated by the nuclear chain reaction away from the reactor core. In a *pressurized water reactor*, which is the most

---

*The reactor core cannot reach supercritical levels and explode with the violence of an atomic bomb because the concentration of uranium-235 is too low. However, if the core overheats, sufficient damage can lead to release of radioactive materials into the environment.

GO FIGURE

Why are nuclear power plants usually located near a large body of water?

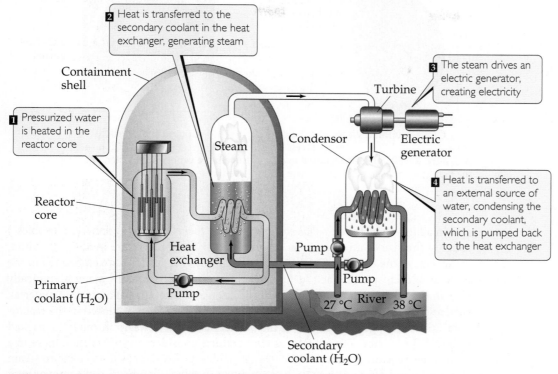

▲ Figure 21.19 **Basic design of a pressurized water reactor nuclear power plant.**

common commercial reactor design, water acts as both the moderator and the primary coolant.

The design of a nuclear power plant is basically the same as that of a power plant that burns fossil fuel (except that the burner is replaced by a reactor core). The nuclear power plant design shown in ▲ **Figure 21.19**, a pressurized water reactor, is currently the most popular. The primary coolant passes through the core in a closed system, which lessens the chance that radioactive products could escape the core. As an added safety precaution, the reactor is surrounded by a reinforced concrete *containment shell* to shield personnel and nearby residents from radiation and to protect the reactor from external forces. After passing through the reactor core, the very hot primary coolant passes through a heat exchanger where much of its heat is transferred to a *secondary coolant*, converting the latter to high-pressure steam that is used to drive a turbine. The secondary coolant is then condensed by transferring heat to an external source of water, such as a river or lake.

Approximately two-thirds of all commercial reactors are pressurized water reactors, but there are several variations on this basic design, each with advantages and disadvantages. A *boiling water reactor* generates steam by boiling the primary coolant; thus, no secondary coolant is needed. Pressurized water reactors and boiling water reactors are collectively referred to as *light water reactors* because they use H_2O as moderator and primary coolant. A *heavy water reactor* uses D_2O (D = deuterium, ^2H) as moderator and primary coolant, and a *gas-cooled reactor* uses a gas, typically CO_2, as primary coolant and graphite as the moderator. Use of either D_2O or graphite as the moderator has the advantage that both substances absorb fewer neutrons than H_2O. Consequently, the uranium fuel does not need to be as enriched.

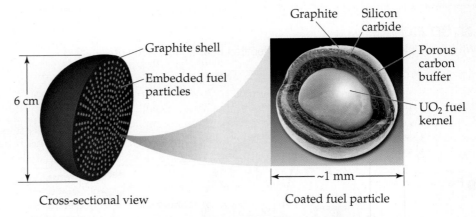

▲ **Figure 21.20 Fuel spheres used in a high-temperature pebble-bed reactor.** The image on the right is an optical microscope image of a fuel particle.

In a *high-temperature pebble-bed reactor*, the fuel elements are spheres ("pebbles") roughly the size of an orange (▲ **Figure 21.20**). The spheres are made of graphite, which acts as the moderator, and thousands of tiny fuel particles are embedded in the interior of each sphere. Each fuel particle is a kernel of fissionable material, typically ^{235}U in the form of UO_2, surrounded by carbon and a coating of a ceramic material, such as SiC. Hundreds of thousands of these spheres are loosely packed in the reactor core, and helium gas, which acts as the primary coolant, flows up through the packed spheres. The reactor core operates at temperatures considerably higher than those in a light water reactor, approaching 950 °C. A pebble-bed reactor is not subject to steam explosions and does not need to be shut down to refuel. Engineers can remove spent spheres from the bottom of the reactor core and add fresh ones to the top. This design is relatively new and is not yet in commercial use.

Nuclear Waste

The fission products that accumulate as a reactor operates decrease the efficiency of the reactor by capturing neutrons. For this reason, commercial reactors must be stopped periodically to either replace or reprocess the nuclear fuel. When the fuel elements are removed from the reactor, they are initially very radioactive. It was originally intended that they be stored for several months in pools at the reactor site to allow decay of short-lived radioactive nuclei. They were then to be transported in shielded containers to reprocessing plants where the unspent fuel would be separated from the fission products. Reprocessing plants have been plagued with operational difficulties, however, and there is intense opposition in the United States to the transport of nuclear wastes on the nation's roads and rails.

Even if the transportation difficulties could be overcome, the high level of radioactivity of the spent fuel makes reprocessing a hazardous operation. At present in the United States spent fuel elements are kept in storage at reactor sites. Spent fuel is reprocessed, however, in France, Russia, the United Kingdom, India, and Japan.

Storage of spent nuclear fuel poses a major problem because the fission products are extremely radioactive. It is estimated that 10 half-lives are required for their radioactivity to reach levels acceptable for biological exposure. Based on the 28.8-yr half-life of strontium-90, one of the longer-lived and most dangerous of the products, the wastes must be stored for nearly 300 yr. Plutonium-239 is one of the by-products present in spent fuel elements. It is formed by absorption of a neutron by uranium-238, followed by two successive beta emissions. (Remember that most of the uranium in the fuel elements is uranium-238.) If the elements are reprocessed, the plutonium-239 is largely recovered because it can be used as a nuclear fuel. However, if the plutonium is not removed, spent elements must be stored for a very long time because plutonium-239 has a half-life of 24,000 yr.

A *fast breeder reactor* offers one approach to getting more power out of existing uranium sources and potentially reducing radioactive waste. This type of reactor is so named because it creates ("breeds") more fissionable material than it consumes. The reactor operates without a moderator, which means the neutrons used are not slowed down. In order to capture the fast neutrons, the fuel must be highly enriched with both uranium-235 and plutonium-239. Water cannot be used as a primary coolant because it would moderate the neutrons, and so a liquid metal, usually sodium, is used. The core is surrounded by a blanket of uranium-238 that captures neutrons that escape the core, producing plutonium-239 in the process. The plutonium can later be separated by reprocessing and used as fuel in a future cycle.

Because fast neutrons are more effective at decaying many radioactive nuclides, the material separated from the uranium and plutonium during reprocessing is less radioactive than waste from other reactors. However, generation of relatively high levels of plutonium coupled with the need for reprocessing is problematic in terms of nuclear nonproliferation. Thus, political factors coupled with increased safety concerns and higher operational costs make fast breeder reactors quite rare.

A considerable amount of research is being devoted to disposal of radioactive wastes. At present, the most attractive possibilities appear to be formation of glass, ceramic, or synthetic rock from the wastes, as a means of immobilizing them. These solid materials would then be placed in containers of high corrosion resistance and durability and buried deep underground. The U.S. Department of Energy (DOE) had designated Yucca Mountain in Nevada as a disposal site, and extensive construction has been done there. However, in 2010 this project was discontinued because of technological and political concerns. The long-term solution to nuclear waste storage in the United States remains unclear. Whatever solution is finally decided on, there must be assurances that the solids and their containers will not crack from the heat generated by nuclear decay, allowing radioactivity to find its way into underground water supplies.

In spite of all these difficulties, nuclear power is making a modest comeback as an energy source. Concerns about climate change caused by escalating atmospheric CO_2 levels ⟶ (Section 18.2) have increased support for nuclear power as a major energy source in the future. Increasing demand for power in rapidly developing countries, particularly China, has sparked a rise in construction of new nuclear power plants in those parts of the world.

21.8 | Nuclear Power: Fusion

Energy is produced when light nuclei fuse into heavier ones. Reactions of this type are responsible for the energy produced by the Sun. Spectroscopic studies indicate that the mass composition of the Sun is 73% H, 26% He, and only 1% all other elements. The following reactions are among the numerous fusion processes believed to occur in the Sun:

$$^1_1H + {}^1_1H \longrightarrow {}^2_1H + {}^0_{+1}e \qquad [21.26]$$

$$^1_1H + {}^2_1H \longrightarrow {}^3_2He \qquad [21.27]$$

$$^3_2He + {}^3_2He \longrightarrow {}^4_2He + 2\,{}^1_1H \qquad [21.28]$$

$$^3_2He + {}^1_1H \longrightarrow {}^4_2He + {}^0_{+1}e \qquad [21.29]$$

Fusion is appealing as an energy source because of the availability of light isotopes on Earth and because fusion products are generally not radioactive. Despite this fact, fusion is not presently used to generate energy. The problem is that, in order for two nuclei to fuse, extremely high temperatures and pressures are needed to overcome the electrostatic repulsion between nuclei in order to fuse them. Fusion reactions are therefore also known as **thermonuclear reactions**. The lowest temperature required for any fusion is about 40,000,000 K, the temperature needed to fuse deuterium and tritium:

$$^2_1H + {}^3_1H \longrightarrow {}^4_2He + {}^1_0n \qquad [21.30]$$

Such high temperatures have been achieved by using an atomic bomb to initiate fusion. This is the operating principle behind a thermonuclear, or hydrogen, bomb. This approach is obviously unacceptable, however, for a power generation plant.*

Numerous problems must be overcome before fusion becomes a practical energy source. In addition to the high temperatures necessary to initiate the reaction, there is the problem of confining the reaction. No known structural material is able to withstand the enormous temperatures necessary for fusion. Research has centered on the use of an apparatus called a *tokamak*, which uses strong magnetic fields to contain and to heat the reaction. Temperatures of over 100,000,000 K have been achieved in a tokamak. Unfortunately, scientists have not yet been able to generate more power than is consumed over a sustained period of time.

21.9 | Radiation in the Environment and Living Systems

We are continuously bombarded by radiation from both natural and artificial sources. We are exposed to infrared, ultraviolet, and visible radiation from the Sun; radio waves from radio and television stations; microwaves from microwave ovens; X rays from medical procedures; and radioactivity from natural materials (▼ Table 21.8). Understanding the different energies of these various kinds of radiation is necessary in order to understand their different effects on matter.

When matter absorbs radiation, the radiation energy can cause atoms in the matter to be either excited or ionized. In general, radiation that causes ionization, called **ionizing radiation**, is far more harmful to biological systems than radiation that does not cause ionization. The latter, called **nonionizing radiation**, is generally of lower energy, such as radiofrequency electromagnetic radiation ∞ (Section 6.7) or slow-moving neutrons.

Most living tissue contains at least 70% water by mass. When living tissue is irradiated, water molecules absorb most of the energy of the radiation. Thus, it is common to define ionizing radiation as radiation that can ionize water, a process requiring a minimum energy of 1216 kJ/mol. Alpha, beta, and gamma rays (as well as X rays and higher-energy ultraviolet radiation) possess energies in excess of this quantity and are therefore forms of ionizing radiation.

Table 21.8 Average Abundances and Activities of Natural Radionuclides[†]

	Potassium-40	Rubidium-87	Thorium-232	Uranium-238
Land elemental abundance (ppm)	28,000	112	10.7	2.8
Land activity (Bq/kg)	870	102	43	35
Ocean elemental concentration (mg/L)	339	0.12	1×10^{-7}	0.0032
Ocean activity (Bq/L)	12	0.11	4×10^{-7}	0.040
Ocean sediments elemental abundance (ppm)	17,000	—	5.0	1.0
Ocean sediments activity (Bq/kg)	500	—	20	12
Human body activity (Bq)	4000	600	0.08	0.4[‡]

[†]Data from "Ionizing Radiation Exposure of the Population of the United States," Report 93, 1987, and Report 160, 2009, National Council on Radiation Protection.

[‡]Includes lead-210 and polonium-210, daughter nuclei of uranium-238.

*Historically a nuclear weapon that relies solely on a fission process to release energy is called an atomic bomb, whereas one that also releases energy via a fusion reaction is called a hydrogen bomb.

A Closer Look

Nuclear Synthesis of the Elements

The lightest elements—hydrogen and helium along with very small amounts of lithium and beryllium—were formed as the universe expanded in the moments following the Big Bang. All the heavier elements owe their existence to nuclear reactions that occur in stars. These heavier elements are not all created equally, however. In our solar system, for example, carbon and oxygen are a million times more abundant than lithium and boron, for instance, and over 100 million times more abundant that beryllium (▼ **Figure 21.21**)! In fact, of the elements heavier than helium, carbon and oxygen are the most abundant. This is more than an academic curiosity given the fact that these elements, together with hydrogen, are the most important elements for

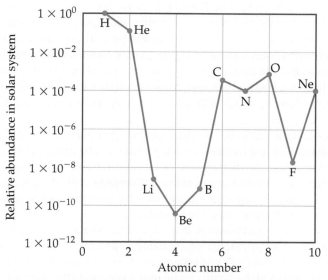

▲ **Figure 21.21 Relative abundance of elements 1–10 in the solar system.** Note the logarithmic scale used for the *y*-axis.

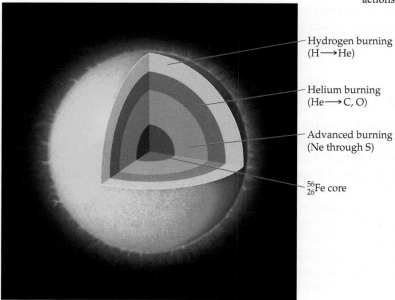

▲ Figure 21.22 **Fusion processes going on in a red giant just prior to a supernova explosion.**

life on Earth. Let's look at the factors responsible for the relatively high abundance of carbon and oxygen in the universe.

A star is born from a cloud of gas and dust called a *nebula*. When conditions are right, gravitational forces collapse the cloud, and its core density and temperature rise until nuclear fusion commences. Hydrogen nuclei fuse to form deuterium, $_1^2\text{H}$, and eventually $_2^4\text{He}$ through the reactions shown in Equations 21.26 through 21.29. Because $_2^4\text{He}$ has a larger binding energy than any of its immediate neighbors (Figure 21.12), these reactions release an enormous amount of energy. This process, called *hydrogen burning*, is the dominant process for most of a star's lifetime.

Once a star's supply of hydrogen is nearly exhausted, several important changes occur as the star enters next phase of its life, and is transformed into a *red giant*. The decrease in nuclear fusion causes the core to contract, triggering an increase in core temperature and pressure. At the same time, the outer regions expand and cool enough to make the star emit red light (thus, the name *red giant*). The star now must use $_2^4\text{He}$ nuclei as its fuel. The simplest reaction that can occur in the He-rich core, fusion of two alpha particles to form a $_4^8\text{Be}$ nucleus, does occur. The binding energy per nucleon for $_4^8\text{Be}$ is very slightly smaller than that for $_2^4\text{He}$, so this fusion process is very slightly endothermic. The $_4^8\text{Be}$ nucleus is highly unstable (half-life of 7×10^{-17} s) and so falls apart almost immediately. In a tiny fraction of cases, however, a third $_2^4\text{He}$ collides with a $_4^8\text{Be}$ nucleus before it decays, forming carbon-12 through the *triple-alpha process*:

$$_2^4\text{He} + {}_2^4\text{He} \longrightarrow {}_4^8\text{Be}$$

$$_4^8\text{Be} + {}_2^4\text{He} \longrightarrow {}_6^{12}\text{C}$$

Some of the $_6^{12}\text{C}$ nuclei go on to react with alpha particles to form oxygen-16:

$$_6^{12}\text{C} + {}_2^4\text{He} \longrightarrow {}_8^{16}\text{O}$$

This stage of nuclear fusion is called *helium burning*. Notice that carbon, element 6, is formed without prior formation of elements 3, 4, and 5, explaining in part their unusually low abundance. Nitrogen is relatively abundant because it can be produced from carbon through a series of reactions involving proton capture and positron emission.

Most stars gradually cool and dim as the helium is converted to carbon and oxygen, ending their lives as *white dwarfs*, a phase in which stars become incredibly dense—generally about one million times denser than the Sun. The extreme density of white dwarfs is accompanied by much higher temperatures and pressures at the core, where a variety of fusion processes lead to synthesis of the elements from neon to sulfur. These fusion reactions are collectively called *advanced burning*.

Eventually progressively heavier elements form at the core until it becomes predominantly ^{56}Fe as shown in ◄ **Figure 21.22.** Because this is such a stable nucleus, further fusion to heavier nuclei consumes energy rather than releasing it. When this happens, the fusion reactions that power the star diminish, and immense gravitational forces lead to a dramatic collapse called a supernova *explosion*. Neutron capture coupled with subsequent radioactive decays in the dying moments of such a star are responsible for the presence of all elements heavier than iron and nickel.

Without these dramatic supernova events in the past history of the universe, heavier elements that are so familiar to us, such as silver, gold, iodine, lead, and uranium, would not exist.

Related Exercises: 21.73, 21.75

When ionizing radiation passes through living tissue, electrons are removed from water molecules, forming highly reactive H_2O^+ ions. An H_2O^+ ion can react with another water molecule to form an H_3O^+ ion and a neutral OH molecule:

$$H_2O^+ + H_2O \longrightarrow H_3O^+ + OH \qquad [21.31]$$

The unstable and highly reactive OH molecule is a **free radical**, a substance with one or more unpaired electrons, as seen in the Lewis structure shown in the margin. The OH molecule is also called the *hydroxyl radical*, and the presence of the unpaired electron is often emphasized by writing the species with a single dot, ·OH. In cells and tissues, hydroxyl radicals can attack biomolecules to produce new free radicals, which in turn attack yet other biomolecules. Thus, the formation of a single hydroxyl radical via Equation 21.31 can initiate a large number of chemical reactions that are ultimately able to disrupt the normal operations of cells.

The damage produced by radiation depends on the activity and energy of the radiation, the length of exposure, and whether the source is inside or outside the body. Gamma rays are particularly harmful outside the body because they penetrate human tissue very effectively, just as X rays do. Consequently, their damage is not limited to the skin. In contrast, most alpha rays are stopped by skin, and beta rays are able to penetrate only about 1 cm beyond the skin surface (◀ Figure 21.23). Neither alpha rays nor beta rays are as dangerous as gamma rays, therefore, *unless* the radiation source somehow enters the body. Within the body, alpha rays are particularly dangerous because they transfer their energy efficiently to the surrounding tissue, causing considerable damage.

In general, the tissues damaged most by radiation are those that reproduce rapidly, such as bone marrow, blood-forming tissues, and lymph nodes. The principal effect of extended exposure to low doses of radiation is to cause cancer. Cancer is caused by damage to the growth-regulation mechanism of cells, inducing the cells to reproduce uncontrollably. Leukemia, which is characterized by excessive growth of white blood cells, is probably the major type of radiation-caused cancer.

In light of the biological effects of radiation, it is important to determine whether any levels of exposure are safe. Unfortunately, we are hampered in our attempts to set realistic standards because we do not fully understand the effects of long-term exposure. Scientists concerned with setting health standards have used the hypothesis that the effects of radiation are proportional to exposure. *Any* amount of radiation is assumed to cause some finite risk of injury, and the effects of high dosage rates are extrapolated to those of lower ones. Other scientists believe, however, that there is a threshold below which there are no radiation risks. Until scientific evidence enables us to settle the matter with some confidence, it is safer to assume that even low levels of radiation present some danger.

GO FIGURE

Why are alpha rays much more dangerous when the source of radiation is located inside the body?

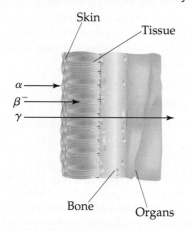

▲ Figure 21.23 **Relative penetrating abilities of alpha, beta, and gamma radiation.**

Radiation Doses

Two units are commonly used to measure exposure to radiation. The **gray** (Gy), the SI unit of absorbed dose, corresponds to the absorption of 1 J of energy per kilogram of tissue. The **rad** (radiation *a*bsorbed *d*ose) corresponds to the absorption of 1×10^{-2} J of energy per kilogram of tissue. Thus, 1 Gy = 100 rad. The rad is the unit most often used in medicine.

Not all forms of radiation harm biological materials to the same extent even at the same level of exposure. For example, 1 rad of alpha radiation can produce more damage than 1 rad of beta radiation. To correct for these differences, the radiation dose is multiplied by a factor that measures the relative damage caused by the radiation. This multiplication factor is known as the *relative biological effectiveness*, *RBE*. The RBE is approximately 1 for gamma and beta radiation, and 10 for alpha radiation.

The exact value of the RBE varies with dose rate, total dose, and type of tissue affected. The product of the radiation dose in rads and the RBE of the radiation give the *effective dosage* in **rem** (roentgen equivalent for *man*):

$$\text{Number of rem} = (\text{number of rad})(\text{RBE}) \qquad [21.32]$$

The SI unit for effective dose is the *sievert* (Sv), obtained by multiplying the RBE times the SI unit for radiation dose, the gray; because a gray is 100 times larger than a rad, 1 Sv = 100 rem. The rem is the unit of radiation damage usually used in medicine.

Give It Some Thought

If a 50-kg person is uniformly irradiated by 0.10-J alpha radiation, what is the absorbed dosage in rad and the effective dosage in rem?

The effects of short-term exposure to radiation appear in ▼ Table 21.9. An exposure of 600 rem is fatal to most humans. To put this number in perspective, a typical dental X-ray entails an exposure of about 0.5 mrem. The average exposure for a person in 1 yr due to all natural sources of ionizing radiation (called *background radiation*) is about 360 mrem (▼ Figure 21.24).

Table 21.9   Effects of Short-Term Exposures to Radiation

Dose (rem)	Effect
0–25	No detectable clinical effects
25–50	Slight, temporary decrease in white blood cell counts
100–200	Nausea; marked decrease in white blood cell counts
500	Death of half the exposed population within 30 days

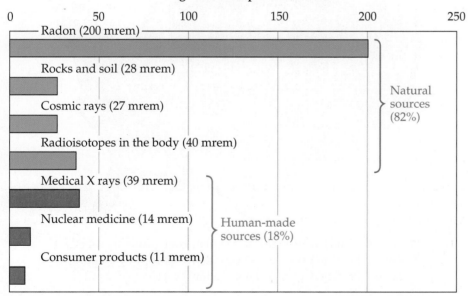

▲ **Figure 21.24 Sources of U.S. average annual exposure to high-energy radiation.** The total average annual exposure is 360 mrem.
Data from "Ionizing Radiation Exposure of the Population of the United States," Report 93, 1987 and Report 160, 2009, National Council on Radiation Protection.

Radon

Radon-222 is a product of the nuclear disintegration series of uranium-238 (Figure 21.3) and is continuously generated as uranium in rocks and soil decays. As Figure 21.24 indicates, radon exposure is estimated to account for more than half the 360-mrem average annual exposure to ionizing radiation.

The interplay between the chemical and nuclear properties of radon makes it a health hazard. Because radon is a noble gas, it is extremely unreactive and is therefore free to escape from the ground without chemically reacting along the way. It is readily inhaled and exhaled with no direct chemical effects. Its half-life, however, is only 3.82 days. It decays, by losing an alpha particle, into a radioisotope of polonium:

$$\begin{array}{ccc} ^{222}_{86}\text{Rn} & \longrightarrow & ^{218}_{84}\text{Po} + {}^{4}_{2}\text{He} \end{array} \qquad [21.33]$$

Because radon has such a short half-life and because alpha particles have a high RBE, inhaled radon is considered a probable cause of lung cancer. Even worse than the radon, however, is the decay product because polonium-218 is an alpha-emitting chemically active element that has an even shorter half-life (3.11 min) than radon-222:

$$\begin{array}{ccc} ^{218}_{84}\text{Po} & \longrightarrow & ^{214}_{82}\text{Pb} + {}^{4}_{2}\text{He} \end{array} \qquad [21.34]$$

When a person inhales radon, therefore, atoms of polonium-218 can become trapped in the lungs, where they bathe the delicate tissue with harmful alpha radiation. The resulting damage is estimated to contribute to 10% of all lung cancer deaths in the United States.

The U.S. Environmental Protection Agency (EPA) has recommended that radon-222 levels not exceed 4 pCi per liter of air in homes. Homes located in areas where the natural uranium content of the soil is high often have levels much greater than that (▼ Figure 21.25). Because of public awareness, radon-testing kits are readily available in many parts of the country.

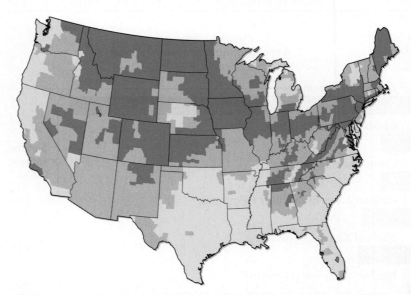

■ **Zone 1** Predicted average indoor radon level greater than 4 pCi/L
■ **Zone 2** Predicted average indoor radon level between 2 and 4 pCi/L
□ **Zone 3** Predicted average indoor radon level less than 2 pCi/L

▲ Figure 21.25 **EPA map of radon zones in the United States.*** The color coding shows average indoor radon levels as a function of geographic location.

---

*Data from "Ionizing Radiation Exposure of the Population of the United States," Report 93, 1987, National Council on Radiation Protection and Measurements.

Chemistry and Life

Radiation Therapy

Healthy cells are either destroyed or damaged by high-energy radiation, leading to physiological disorders. This radiation can also destroy *unhealthy* cells, however, including cancerous cells. All cancers are characterized by runaway cell growth that can produce *malignant tumors*. These tumors can be caused by the exposure of healthy cells to high-energy radiation. Paradoxically, however, they can be destroyed by the same radiation that caused them because the rapidly reproducing cells of the tumors are very susceptible to radiation damage. Thus, cancerous cells are more susceptible to destruction by radiation than healthy ones, allowing radiation to be used effectively in the treatment of cancer. As early as 1904, physicians used the radiation emitted by radioactive substances to treat tumors by destroying the mass of unhealthy tissue. The treatment of disease by high-energy radiation is called *radiation therapy*.

Many radionuclides are currently used in radiation therapy. Some of the more commonly used ones are listed in ▼ **Table 21.10**. Most of them have short half-lives, meaning that they emit a great deal of radiation in a short period of time.

The radiation source used in radiation therapy may be inside or outside the body. In almost all cases, radiation therapy uses gamma radiation emitted by radioisotopes. Any alpha or beta radiation that is emitted concurrently can be blocked by appropriate packaging. For example, ^{192}Ir is often administered as "seeds" consisting of a core of radioactive isotope coated with 0.1 mm of platinum metal. The platinum coating stops the alpha and beta rays, but the gamma rays penetrate it readily. The radioactive seeds can be surgically implanted in a tumor.

In some cases, human physiology allows a radioisotope to be ingested. For example, most of the iodine in the human body ends up in the thyroid gland, so thyroid cancer can be treated by using large doses of ^{131}I. Radiation therapy on deep organs, where a surgical implant is impractical, often uses a ^{60}Co "gun" outside the body to shoot a beam of gamma rays at the tumor. Particle accelerators are also used as an external source of high-energy radiation for radiation therapy.

Because gamma radiation is so strongly penetrating, it is nearly impossible to avoid damaging healthy cells during radiation therapy. Many cancer patients undergoing radiation treatment experience unpleasant and dangerous side effects such as fatigue, nausea, hair loss, a weakened immune system, and occasionally even death. However, if other treatments such as *chemotherapy* (the use of drugs to combat cancer) fail, radiation therapy can be a good option.

Much current research in radiation therapy is engaged in developing new drugs that specifically target tumors using a method called *neutron capture therapy*. In this technique, a nonradioactive isotope, usually boron-10, is concentrated in the tumor by using specific tumor-seeking reagents. The boron-10 is then irradiated with neutrons, where it undergoes the following nuclear reaction:

$$^{10}_{5}\text{B} + ^{1}_{0}\text{n} \longrightarrow ^{7}_{3}\text{Li} + ^{4}_{2}\text{He}$$

Tumor cells are killed or damaged by exposure to the alpha particles. Healthy tissue farther away from the tumor is unaffected because of the short-range penetrating power of alpha particles. Thus, neutron-capture therapy has the promise to be a "silver bullet" that specifically targets unhealthy cells for exposure to radiation.

Related Exercises: 21.37, 21.55, 21.56

Table 21.10   Some Radioisotopes Used in Radiation Therapy

Isotope	Half-Life	Isotope	Half-Life
^{32}P	14.3 days	^{137}Cs	30 yr
^{60}Co	5.27 yr	^{192}Ir	74.2 days
^{90}Sr	28.8 yr	^{198}Au	2.7 days
^{125}I	60.25 days	^{222}Rn	3.82 days
^{131}I	8.04 days	^{226}Ra	1600 yr

SAMPLE INTEGRATIVE EXERCISE   Putting Concepts Together

Potassium ion is present in foods and is an essential nutrient in the human body. One of the naturally occurring isotopes of potassium, potassium-40, is radioactive. Potassium-40 has a natural abundance of 0.0117% and a half-life $t_{1/2} = 1.28 \times 10^9$ yr. It undergoes radioactive decay in three ways: 98.2% is by electron capture, 1.35% is by beta emission, and 0.49% is by positron emission. **(a)** Why should we expect ^{40}K to be radioactive? **(b)** Write the nuclear equations for the three modes by which ^{40}K decays. **(c)** How many ^{40}K$^+$ ions are present in 1.00 g of KCl? **(d)** How long does it take for 1.00% of the ^{40}K in a sample to undergo radioactive decay?

SOLUTION

(a) The ^{40}K nucleus contains 19 protons and 21 neutrons. There are very few stable nuclei with odd numbers of both protons and neutrons (Section 21.2).

(b) Electron capture is capture of an inner-shell electron by the nucleus:

$$^{40}_{19}\text{K} + ^{0}_{-1}\text{e} \longrightarrow ^{40}_{18}\text{Ar}$$

Beta emission is loss of a beta particle ($_{-1}^{0}$e) by the nucleus:

$$_{19}^{40}\text{K} \longrightarrow {}_{20}^{40}\text{Ca} + {}_{-1}^{0}\text{e}$$

Positron emission is loss of a positron ($_{+1}^{0}$e) by the nucleus:

$$_{19}^{40}\text{K} \longrightarrow {}_{18}^{40}\text{Ar} + {}_{+1}^{0}\text{e}$$

(c) The total number of K^+ ions in the sample is

$$(1.00 \text{ g KCl})\left(\frac{1 \text{ mol KCl}}{74.55 \text{ g KCl}}\right)\left(\frac{1 \text{ mol K}^+}{1 \text{ mol KCl}}\right)\left(\frac{6.022 \times 10^{23} \text{ K}^+}{1 \text{ mol K}^+}\right) = 8.08 \times 10^{21} \text{ K}^+ \text{ ions}$$

Of these, 0.0117% are $^{40}K^+$ ions:

$$(8.08 \times 10^{21} \text{ K}^+ \text{ ions})\left(\frac{0.0117 \ ^{40} \text{ K}^+ \text{ ions}}{100^+ \text{ ions}}\right) = 9.45 \times 10^{17} \text{ potassium-40 ions}$$

(d) The decay constant (the rate constant) for the radioactive decay can be calculated from the half-life, using Equation 21.20:

$$k = \frac{0.693}{t_{1/2}} = \frac{0.693}{1.28 \times 10^9 \text{ yr}} = (5.41 \times 10^{-10})/\text{yr}$$

The rate equation, Equation 21.19, then allows us to calculate the time required:

$$\ln\frac{N_t}{N_0} = -kt$$

$$\ln\frac{99}{100} = -[(5.41 \times 10^{-10})/\text{yr}]t$$

$$-0.01005 = -[(5.41 \times 10^{-10})/\text{yr}]t$$

$$t = \frac{-0.01005}{(-5.41 \times 10^{-10})/\text{yr}} = 1.86 \times 10^7 \text{ yr}$$

That is, it would take 18.6 million years for just 1.00% of the ^{40}K in a sample to decay.

Chapter Summary and Key Terms

INTRODUCTION TO RADIOACTIVITY AND NUCLEAR EQUATIONS (SECTION 21.1) The nucleus of an atom contains protons and neutrons, both of which are called **nucleons**. Reactions that involve changes in atomic nuclei are called **nuclear reactions**. Nuclei that spontaneously change by emitting radiation are said to be radioactive. Radioactive nuclei are called **radionuclides**, and the atoms containing them are called **radioisotopes**. Radionuclides spontaneously change through a process called radioactive decay. The three most important types of radiation given off as a result of radioactive decay are **alpha** (α) **particles** ($_{2}^{4}$He or α), **beta** (β) **particles** ($_{-1}^{0}$e or β^-), and **gamma** (γ) **radiation** ($_{0}^{0}\gamma$ or γ). **Positrons** ($_{+1}^{0}$e or β^+), which are particles with the same mass as an electron but the opposite charge, can also be produced when a radioisotope decays.

In nuclear equations, reactant and product nuclei are represented by giving their mass numbers and atomic numbers, as well as their chemical symbol. The totals of the mass numbers on both sides of the equation are equal; the totals of the atomic numbers on both sides are also equal. There are four common modes of radioactive decay: **alpha decay**, which reduces the atomic number by 2 and the mass number by 4, **beta emission**, which increases the atomic number by 1 and leaves the mass number unchanged, **positron emission** and **electron capture**, both of which reduce the atomic number by 1 and leave the mass number unchanged.

PATTERNS OF NUCLEAR STABILITY (SECTION 21.2) The neutron-to-proton ratio is an important factor determining nuclear stability. By comparing a nuclide's neutron-to-proton ratio with those of stable nuclei, we can predict the mode of radioactive decay. In general, neutron-rich nuclei tend to emit beta particles; proton-rich nuclei tend to either emit positrons or undergo electron capture; and heavy nuclei tend to emit alpha particles. The presence of **magic numbers** of nucleons and an even number of protons and neutrons also help determine the stability of a nucleus. A nuclide may undergo a series of decay steps before a stable nuclide forms. This series of steps is called a **radioactive decay chain** or a **nuclear disintegration series**.

NUCLEAR TRANSMUTATIONS (SECTION 21.3) **Nuclear transmutations**, induced conversions of one nucleus into another, can be brought about by bombarding nuclei with either charged particles or neutrons. **Particle accelerators** increase the kinetic energies of positively charged particles, allowing these particles to overcome their electrostatic repulsion by the nucleus. Nuclear transmutations are used to produce the **transuranium elements**, those elements with atomic numbers greater than that of uranium.

RADIOACTIVE DECAY RATES AND DETECTION OF RADIOACTIVITY (SECTIONS 21.4 AND 21.5) The SI unit for the activity of a radioactive source is the **becquerel** (Bq), defined as one nuclear disintegration per second. A related unit, the **curie** (Ci), corresponds to 3.7×10^{10} disintegrations per second. Nuclear decay is a first-order process. The decay rate (**activity**) is therefore directly proportional to the number of radioactive nuclei. The **half-life** of a radionuclide, which is a constant independent of temperature, is the time needed for one-half of the nuclei to decay. Some radioisotopes can be used to date objects; ^{14}C, for example, is used to date organic objects. Geiger counters and scintillation counters count the emissions from radioactive samples. The ease of detection of radioisotopes also permits their use as **radiotracers** to follow elements through reactions.

ENERGY CHANGES IN NUCLEAR REACTIONS (SECTION 21.6) The energy produced in nuclear reactions is accompanied by measurable changes of mass in accordance with Einstein's relationship, $\Delta E = c^2 \, \Delta m$. The difference in mass between nuclei and the nucleons of which they are composed is known as the **mass defect**. The mass defect of a nuclide makes it possible to calculate its **nuclear binding energy**, the energy required to separate the nucleus into individual nucleons. Because of trends in the nuclear binding energy with atomic number, energy is produced when heavy nuclei split (**fission**) and when light nuclei fuse (**fusion**).

NUCLEAR FISSION AND FUSION (SECTIONS 21.7 AND 21.8) Uranium-235, uranium-233, and plutonium-239 undergo fission when they capture a neutron, splitting into lighter nuclei and releasing more neutrons. The neutrons produced in one fission can cause further fission reactions, which can lead to a nuclear **chain reaction**. A reaction that maintains a constant rate is said to be critical, and the mass necessary to maintain this constant rate is called a **critical mass**. A mass in excess of the critical mass is termed a **supercritical mass**.

In nuclear reactors the fission rate is controlled to generate a constant power. The reactor core consists of fuel elements containing fissionable nuclei, control rods, a moderator, and a primary coolant. A nuclear power plant resembles a conventional power plant except that the reactor core replaces the fuel burner. There is concern about the disposal of highly radioactive nuclear wastes that are generated in nuclear power plants.

Nuclear fusion requires high temperatures because nuclei must have large kinetic energies to overcome their mutual repulsions. Fusion reactions are therefore called **thermonuclear reactions**. It is not yet possible to generate power on Earth through a controlled fusion process.

NUCLEAR CHEMISTRY AND LIVING SYSTEMS (SECTION 21.9) **Ionizing radiation** is energetic enough to remove an electron from a water molecule; radiation with less energy is called **nonionizing radiation**. Ionizing radiation generates **free radicals**, reactive substances with one or more unpaired electrons. The effects of long-term exposure to low levels of radiation are not completely understood, but there is evidence that the extent of biological damage varies in direct proportion to the level of exposure.

The amount of energy deposited in biological tissue by radiation is called the radiation dose and is measured in units of gray or rad. One **gray** (Gy) corresponds to a dose of 1 J/kg of tissue. It is the SI unit of radiation dose. The **rad** is a smaller unit; 100 rad = 1 Gy. The effective dose, which measures the biological damage created by the deposited energy, is measured in units of rem or sievert (Sv). The **rem** is obtained by multiplying the number of rad by the relative biological effectiveness (RBE); 100 rem = 1 Sv.

Learning Outcomes   After studying this chapter, you should be able to:

- Write balanced nuclear equations. (Section 21.1)

- Predict nuclear stability and expected type of nuclear decay from the neutron-to-proton ratio of an isotope. (Section 21.2)

- Write balanced nuclear equations for nuclear transmutations. (Section 21.3)

- Calculate ages of objects and/or the amount of a radionuclide remaining after a given period of time using the half-life of the radionuclide in question. (Section 21.4)

- Calculate mass and energy changes for nuclear reactions. (Section 21.6)

- Calculate the binding energies for nuclei. (Section 21.6)

- Describe the difference between fission and fusion. (Sections 21.7 and 21.8)

- Explain how a nuclear power plant operates and know the differences among various types of nuclear power plants. (Section 21.7)

- Compare different measurements and units of radiation dosage. (Section 21.9)

- Describe the biological effects of different kinds of radiation. (Section 21.9)

Key Equations

- $\ln\dfrac{N_t}{N_0} = -kt$ 　　　　　　　[21.19]　　　First-order rate law for nuclear decay

- $k = \dfrac{0.693}{t_{1/2}}$ 　　　　　　　[21.20]　　　Relationship between nuclear decay constant and half-life; this is derived from the previous equation at $N_t = \frac{1}{2}N_0$

- $E = mc^2$ 　　　　　　　[21.22]　　　Einstein's equation that relates mass and energy

Exercises

Visualizing Concepts

21.1 Indicate whether each of the following nuclides lies within the belt of stability in Figure 21.2: **(a)** neon-24, **(b)** chlorine-32, **(c)** tin-108, **(d)** polonium-216. For any that do not, describe a nuclear decay process that would alter the neutron-to-proton ratio in the direction of increased stability. [Section 21.2]

21.2 Write the balanced nuclear equation for the reaction represented by the diagram shown here. [Section 21.2]

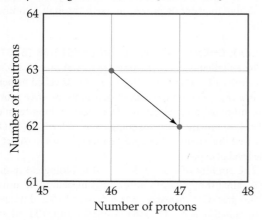

21.3 Draw a diagram similar to that shown in Exercise 21.2 that illustrates the nuclear reaction $^{211}_{83}Bi \longrightarrow \, ^{4}_{2}He + \, ^{207}_{81}Tl$. [Section 21.2]

21.4 In the sketch below, the red spheres represent protons and the gray spheres represent neutrons. **(a)** What are the identities of the four particles involved in the reaction depicted? **(b)** Write the transformation represented below using condensed notation. **(c)** Based on its atomic number and mass number, do you think the product nucleus is stable or radioactive? [Section 21.3]

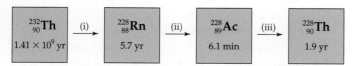

21.5 The steps below show three of the steps in the radioactive decay chain for $^{232}_{90}Th$. The half-life of each isotope is shown below the symbol of the isotope. **(a)** Identify the type of radioactive decay for each of the steps (i), (ii), and (iii). **(b)** Which of the isotopes shown has the highest activity? **(c)** Which of the isotopes shown has the lowest activity? **(d)** The next step in the decay chain is an alpha emission. What is the next isotope in the chain? [Sections 21.2 and 21.4]

21.6 The accompanying graph illustrates the decay of $^{88}_{42}Mo$, which decays via positron emission. **(a)** What is the half-life of the decay? **(b)** What is the rate constant for the decay? **(c)** What fraction of the original sample of $^{88}_{42}Mo$ remains after 12 min? **(d)** What is the product of the decay process? [Section 21.4]

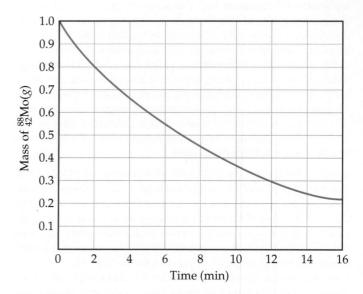

21.7 All the stable isotopes of boron, carbon, nitrogen, oxygen, and fluorine are shown in the accompanying chart (in red), along with their radioactive isotopes with $t_{1/2} > 1$ min (in blue). **(a)** Write the chemical symbols, including mass and atomic numbers, for all of the stable isotopes. **(b)** Which radioactive isotopes are most likely to decay by beta emission? **(c)** Some of the isotopes shown are used in positron emission tomography. Which ones would you expect to be most useful for this application? **(d)** Which isotope would decay to 12.5% of its original concentration after 1 hour? [Sections 21.2, 21.4, and 21.5]

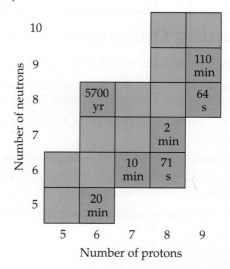

21.8 The diagram shown here illustrates a fission process. **(a)** What is the unidentified product of the fission? **(b)** Use Figure 21.2 to predict whether the nuclear products of this fission reaction are stable. [Section 21.7]

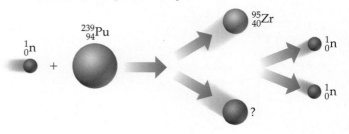

Radioactivity and Nuclear Equations (Section 21.1)

21.9 Indicate the number of protons and neutrons in the following nuclei: (a) $^{56}_{24}Cr$, (b) ^{193}Tl, (c) argon-38.

21.10 Indicate the number of protons and neutrons in the following nuclei: (a) $^{129}_{53}I$, (b) ^{138}Ba, (c) neptunium-237.

21.11 Give the symbol for (a) a neutron, (b) an alpha particle, (c) gamma radiation.

21.12 Give the symbol for (a) a proton, (b) a beta particle, (c) a positron.

21.13 Write balanced nuclear equations for the following processes: (a) rubidium-90 undergoes beta emission; (b) selenium-72 undergoes electron capture; (c) krypton-76 undergoes positron emission; (d) radium-226 emits alpha radiation.

21.14 Write balanced nuclear equations for the following transformations: (a) bismuth-213 undergoes alpha decay; (b) nitrogen-13 undergoes electron capture; (c) technicium-98 undergoes electron capture; (d) gold-188 decays by positron emission.

21.15 Decay of which nucleus will lead to the following products: (a) bismuth-211 by beta decay; (b) chromium-50 by positron emission; (c) tantalum-179 by electron capture; (d) radium-226 by alpha decay?

21.16 What particle is produced during the following decay processes: (a) sodium-24 decays to magnesium-24; (b) mercury-188 decays to gold-188; (c) iodine-122 decays to xenon-122; (d) plutonium-242 decays to uranium-238?

21.17 The naturally occurring radioactive decay series that begins with $^{235}_{92}U$ stops with formation of the stable $^{207}_{82}Pb$ nucleus. The decays proceed through a series of alpha-particle and beta-particle emissions. How many of each type of emission are involved in this series?

21.18 A radioactive decay series that begins with $^{232}_{90}Th$ ends with formation of the stable nuclide $^{208}_{82}Pb$. How many alpha-particle emissions and how many beta-particle emissions are involved in the sequence of radioactive decays?

Patterns of Nuclear Stability (Section 21.2)

21.19 Predict the type of radioactive decay process for the following radionuclides: (a) $^{8}_{5}B$, (b) $^{68}_{29}Cu$, (c) phosphorus-32, (d) chlorine-39.

21.20 Each of the following nuclei undergoes either beta decay or positron emission. Predict the type of emission for each: (a) tritium, $^{3}_{1}H$, (b) $^{89}_{38}Sr$, (c) iodine-120, (d) silver-102.

21.21 One of the nuclides in each of the following pairs is radioactive. Predict which is radioactive and which is stable: (a) $^{39}_{19}K$ and $^{40}_{19}K$, (b) ^{209}Bi and ^{208}Bi, (c) nickel-58 and nickel-65.

21.22 One nuclide in each of these pairs is radioactive. Predict which is radioactive and which is stable: (a) $^{40}_{20}Ca$ and $^{45}_{20}Ca$, (b) ^{12}C and ^{14}C, (c) lead-206 and thorium-230. Explain your choice in each case.

21.23 Which of the following nuclides have magic numbers of both protons and neutrons: (a) helium-4, (b) oxygen-18, (c) calcium-40, (d) zinc-66, (e) lead-208?

21.24 Despite the similarities in the chemical reactivity of elements in the lanthanide series, their abundances in Earth's crust vary by two orders of magnitude. This graph shows the relative abundance as a function of atomic number. How do you explain the sawtooth variation across the series?

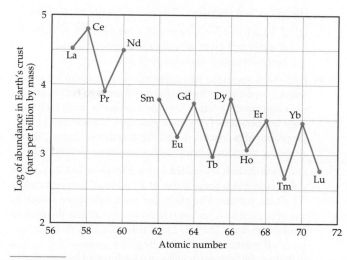

21.25 Using the concept of magic numbers, explain why alpha emission is relatively common, but proton emission is nonexistent.

21.26 Which of the following nuclides would you expect to be radioactive: $^{58}_{26}Fe$, $^{60}_{27}Co$, $^{92}_{41}Nb$, mercury-202, radium-226? Justify your choices.

Nuclear Transmutations (Section 21.3)

21.27 Why are nuclear transmutations involving neutrons generally easier to accomplish than those involving protons or alpha particles?

21.28 In 1930 the American physicist Ernest Lawrence designed the first cyclotron in Berkeley, California. In 1937 Lawrence bombarded a molybdenum target with deuterium ions, producing for the first time an element not found in nature. What was this element? Starting with molybdenum-96 as your reactant, write a nuclear equation to represent this process.

21.29 Complete and balance the following nuclear equations by supplying the missing particle:

(a) $^{252}_{98}Cf + ^{10}_{5}B \longrightarrow 3\,^{1}_{0}n + ?$

(b) $^{2}_{1}H + ^{3}_{2}He \longrightarrow ^{4}_{2}He + ?$

(c) $^{1}_{1}H + ^{11}_{5}B \longrightarrow 3?$

(d) $^{122}_{53}I \longrightarrow ^{122}_{54}Xe + ?$

(e) $^{59}_{26}Fe \longrightarrow ^{0}_{-1}e + ?$

21.30 Complete and balance the following nuclear equations by supplying the missing particle:

(a) $^{14}_{7}N + ^{4}_{2}He \longrightarrow ? + ^{1}_{1}H$

(b) $^{40}_{19}K + ^{0}_{-1}e$ (orbital electron) $\longrightarrow ?$

(c) $? + ^{4}_{2}He \longrightarrow ^{30}_{14}Si + ^{1}_{1}H$

(d) $^{58}_{26}Fe + 2\,^{1}_{0}n \longrightarrow ^{60}_{27}Co + ?$

(e) $^{235}_{92}U + ^{1}_{0}n \longrightarrow ^{135}_{54}Xe + 2\,^{1}_{0}n + ?$

21.31 Write balanced equations for (a) $^{238}_{92}U(\alpha, n)^{241}_{94}Pu$, (b) $^{14}_{7}N(\alpha, p)^{17}_{8}O$, (c) $^{56}_{26}Fe(\alpha, \beta^-)^{60}_{29}Cu$.

21.32 Write balanced equations for each of the following nuclear reactions: (a) $^{238}_{92}U(n, \gamma)^{239}_{92}U$, (b) $^{16}_{8}O(p, \alpha)^{13}_{7}N$, (c) $^{18}_{8}O(n, \beta^-)^{19}_{9}F$.

Rates of Radioactive Decay (Section 21.4)

21.33 Each statement that follows refers to a comparison between two radioisotopes, A and X. Indicate whether each of the following statements is true or false, and why.

(a) If the half-life for A is shorter than the half-life for X, A has a larger decay rate constant.

(b) If X is "not radioactive," its half-life is essentially zero.

(c) If A has a half-life of 10 yr, and X has a half-life of 10,000 yr, A would be a more suitable radioisotope to measure processes occurring on the 40-yr time scale.

21.34 It has been suggested that strontium-90 (generated by nuclear testing) deposited in the hot desert will undergo radioactive decay more rapidly because it will be exposed to much higher average temperatures. **(a)** Is this a reasonable suggestion? **(b)** Does the process of radioactive decay have an activation energy, like the Arrhenius behavior of many chemical reactions (Section 14.5)?

21.35 Some watch dials are coated with a phosphor, like ZnS, and a polymer in which some of the ^{1}H atoms have been replaced by ^{3}H atoms, tritium. The phosphor emits light when struck by the beta particle from the tritium decay, causing the dials to glow in the dark. The half-life of tritium is 12.3 yr. If the light given off is assumed to be directly proportional to the amount of tritium, by how much will a dial be dimmed in a watch that is 50 yr old?

21.36 It takes 4 h 39 min for a 2.00-mg sample of radium-230 to decay to 0.25 mg. What is the half-life of radium-230?

21.37 Cobalt-60 is a strong gamma emitter that has a half-life of 5.26 yr. The cobalt-60 in a radiotherapy unit must be replaced when its radioactivity falls to 75% of the original sample. If an original sample was purchased in June 2013, when will it be necessary to replace the cobalt-60?

21.38 How much time is required for a 6.25-mg sample of ^{51}Cr to decay to 0.75 mg if it has a half-life of 27.8 days?

[21.39] Radium-226, which undergoes alpha decay, has a half-life of 1600 yr. **(a)** How many alpha particles are emitted in 5.0 min by a 10.0-mg sample of ^{226}Ra? **(b)** What is the activity of the sample in mCi?

21.40 Cobalt-60, which undergoes beta decay, has a half-life of 5.26 yr. **(a)** How many beta particles are emitted in 600 s by a 3.75-mg sample of ^{60}Co? **(b)** What is the activity of the sample in Bq?

21.41 The cloth shroud from around a mummy is found to have a ^{14}C activity of 9.7 disintegrations per minute per gram of carbon as compared with living organisms that undergo 16.3 disintegrations per minute per gram of carbon. From the half-life for ^{14}C decay, 5715 yr, calculate the age of the shroud.

21.42 A wooden artifact from a Chinese temple has a ^{14}C activity of 38.0 counts per minute as compared with an activity of 58.2 counts per minute for a standard of zero age. From the half-life for ^{14}C decay, 5715 yr, determine the age of the artifact.

21.43 Potassium-40 decays to argon-40 with a half-life of 1.27×10^9 yr. What is the age of a rock in which the mass ratio of ^{40}Ar to ^{40}K is 4.2?

21.44 The half-life for the process ^{238}U $\longrightarrow$ ^{206}Pb is 4.5×10^9 yr. A mineral sample contains 75.0 mg of ^{238}U and 18.0 mg of ^{206}Pb. What is the age of the mineral?

Energy Changes in Nuclear Reactions (Section 21.6)

21.45 The thermite reaction, $Fe_2O_3(s) + 2\,Al(s) \longrightarrow 2\,Fe(s) + Al_2O_3(s)$, $\Delta H° = -851.5$ kJ/mol, is one of the most exothermic reactions known. Because the heat released is sufficient to melt the iron product, the reaction is used to weld metal under the ocean. How much heat is released per mole of Fe_2O_3 produced? How does this amount of thermal energy compare with the energy released when 2 mol of protons and 2 mol of neutrons combine to form 1 mol of alpha particles?

21.46 An analytical laboratory balance typically measures mass to the nearest 0.1 mg. What energy change would accompany the loss of 0.1 mg in mass?

21.47 How much energy must be supplied to break a single aluminum-27 nucleus into separated protons and neutrons if an aluminum-27 atom has a mass of 26.9815386 amu? How much energy is required for 100.0 grams of aluminum-27? (The mass of an electron is given on the inside back cover.)

21.48 How much energy must be supplied to break a single ^{21}Ne nucleus into separated protons and neutrons if the nucleus has a mass of 20.98846 amu? What is the nuclear binding energy for 1 mol of ^{21}Ne?

21.49 The atomic masses of hydrogen-2 (deuterium), helium-4, and lithium-6 are 2.014102 amu, 4.002602 amu, and 6.0151228 amu, respectively. For each isotope, calculate **(a)** the nuclear mass, **(b)** the nuclear binding energy, **(c)** the nuclear binding energy per nucleon. **(d)** Which of these three isotopes has the largest nuclear binding energy per nucleon? Does this agree with the trends plotted in Figure 21.12?

21.50 The atomic masses of nitrogen-14, titanium-48, and xenon-129 are 13.999234 amu, 47.935878 amu, and 128.904779 amu, respectively. For each isotope, calculate **(a)** the nuclear mass, **(b)** the nuclear binding energy, **(c)** the nuclear binding energy per nucleon.

21.51 The energy from solar radiation falling on Earth is 1.07×10^{16} kJ/min. **(a)** How much loss of mass from the Sun occurs in one day from just the energy falling on Earth? **(b)** If the energy released in the reaction

$$^{235}\text{U} + {}^{1}_{0}\text{n} \longrightarrow {}^{141}_{56}\text{Ba} + {}^{92}_{36}\text{Kr} + 3\,{}^{1}_{0}\text{n}$$

(^{235}U nuclear mass, 234.9935 amu; ^{141}Ba nuclear mass, 140.8833 amu; ^{92}Kr nuclear mass, 91.9021 amu) is taken as typical of that occurring in a nuclear reactor, what mass of uranium-235 is required to equal 0.10% of the solar energy that falls on Earth in 1.0 day?

21.52 Based on the following atomic mass values—^{1}H, 1.00782 amu; ^{2}H, 2.01410 amu; ^{3}H, 3.01605 amu; ^{3}He, 3.01603 amu; ^{4}He, 4.00260 amu—and the mass of the neutron given in the text, calculate the energy released per mole in each of the following nuclear reactions, all of which are possibilities for a controlled fusion process:

(a) $^{2}_{1}\text{H} + {}^{3}_{1}\text{H} \longrightarrow {}^{4}_{2}\text{He} + {}^{1}_{0}\text{n}$

(b) $^{2}_{1}\text{H} + {}^{2}_{1}\text{H} \longrightarrow {}^{3}_{2}\text{He} + {}^{1}_{0}\text{n}$

(c) $^{2}_{1}\text{H} + {}^{3}_{2}\text{He} \longrightarrow {}^{4}_{2}\text{He} + {}^{1}_{1}\text{H}$

21.53 Which of the following nuclei is likely to have the largest mass defect per nucleon: **(a)** ^{59}Co, **(b)** ^{11}B, **(c)** ^{118}Sn, **(d)** ^{243}Cm? Explain your answer.

21.54 The isotope $^{62}_{28}$Ni has the largest binding energy per nucleon of any isotope. Calculate this value from the atomic mass of nickel-62 (61.928345 amu) and compare it with the value given for iron-56 in Table 21.7.

Nuclear Power and Radioisotopes (Sections 21.7–21.9)

21.55 Iodine-131 is a convenient radioisotope to monitor thyroid activity in humans. It is a beta emitter with a half-life of 8.02 days. The thyroid is the only gland in the body that uses iodine. A person undergoing a test of thyroid activity drinks a solution of NaI, in which only a small fraction of the iodide is radioactive.

(a) Why is NaI a good choice for the source of iodine? (b) If a Geiger counter is placed near the person's thyroid (which is near the neck) right after the sodium iodide solution is taken, what will the data look like as a function of time? (c) A normal thyroid will take up about 12% of the ingested iodide in a few hours. How long will it take for the radioactive iodide taken up and held by the thyroid to decay to 0.01% of the original amount?

21.56 Why is it important that radioisotopes used as diagnostic tools in nuclear medicine produce gamma radiation when they decay? Why are alpha emitters not used as diagnostic tools?

21.57 (a) Which of the following are required characteristics of an isotope to be used as a fuel in a nuclear power reactor? (i) It must emit gamma radiation. (ii) On decay, it must release two or more neutrons. (iii) It must have a half-life less than one hour. (iv) It must undergo fission upon the absorption of a neutron. (b) What is the most common fissionable isotope in a commercial nuclear power reactor?

21.58 (a) Which of the following statements about the uranium used in nuclear reactors is or are true? (i) Natural uranium has too little ^{235}U to be used as a fuel. (ii) ^{238}U cannot be used as a fuel because it forms a supercritical mass too easily. (iii) To be used as fuel, uranium must be enriched so that it is more than 50% ^{235}U in composition. (iv) The neutron-induced fission of ^{235}U releases more neutrons per nucleus than fission of ^{238}U. (b) Which of the following statements about the plutonium shown in the chapter-opening photograph explains why it cannot be used for nuclear power plants or nuclear weapons? (i) None of the isotopes of Pu possess the characteristics needed to support nuclear fission chain reactions. (ii) The orange glow indicates that the only radioactive decay products are heat and visible light. (iii) The particular isotope of plutonium used for RTGs is incapable of sustaining a chain reaction. (iv) Plutonium can be used as a fuel, but only after it decays to uranium.

21.59 What is the function of the control rods in a nuclear reactor? What substances are used to construct control rods? Why are these substances chosen?

21.60 (a) What is the function of the moderator in a nuclear reactor? (b) What substance acts as the moderator in a pressurized water generator? (c) What other substances are used as a moderator in nuclear reactor designs?

21.61 Complete and balance the nuclear equations for the following fission or fusion reactions:

(a) $^{2}_{1}H + ^{2}_{1}H \longrightarrow ^{3}_{2}He + \_\_$

(b) $^{239}_{92}U + ^{1}_{0}n \longrightarrow ^{133}_{51}Sb + ^{98}_{41}Nb + \_\_ ^{1}_{0}n$

21.62 Complete and balance the nuclear equations for the following fission reactions:

(a) $^{235}_{92}U + ^{1}_{0}n \longrightarrow ^{160}_{62}Sm + ^{72}_{30}Zn + \_\_ ^{1}_{0}n$

(b) $^{239}_{94}Pu + ^{1}_{0}n \longrightarrow ^{144}_{58}Ce + \_\_ + 2 ^{1}_{0}n$

21.63 A portion of the Sun's energy comes from the reaction

$$4 ^{1}_{1}H \longrightarrow ^{4}_{2}He + 2 ^{0}_{1}e$$

which requires a temperature of 10^6 to 10^7 K. (a) Use the mass of the helium-4 nucleus given in Table 21.7 to determine how much energy is released when the reaction is run with 1 mol of hydrogen atoms. (b) Why is such a high temperature required?

21.64 The spent fuel elements from a fission reactor are much more intensely radioactive than the original fuel elements. (a) What does this tell you about the products of the fission process in relationship to the belt of stability, Figure 21.2? (b) Given that only two or three neutrons are released per fission event and knowing that the nucleus undergoing fission has a neutron-to-proton ratio characteristic of a heavy nucleus, what sorts of decay would you expect to be dominant among the fission products?

21.65 Which type or types of nuclear reactors have these characteristics?

(a) Does not use a secondary coolant
(b) Creates more fissionable material than it consumes
(c) Uses a gas, such as He or CO_2, as the primary coolant

21.66 Which type or types of nuclear reactors have these characteristics?

(a) Can use natural uranium as a fuel
(b) Does not use a moderator
(c) Can be refueled without shutting down

21.67 Hydroxyl radicals can pluck hydrogen atoms from molecules ("hydrogen abstraction"), and hydroxide ions can pluck protons from molecules ("deprotonation"). Write the reaction equations and Lewis dot structures for the hydrogen abstraction and deprotonation reactions for the generic carboxylic acid R—COOH with hydroxyl radical and hydroxide ion, respectively. Why is hydroxyl radical more toxic to living systems than hydroxide ion?

21.68 Which are classified as ionizing radiation: X rays, alpha particles, microwaves from a cell phone, and gamma rays?

21.69 A laboratory rat is exposed to an alpha-radiation source whose activity is 14.3 mCi. (a) What is the activity of the radiation in disintegrations per second? In becquerels? (b) The rat has a mass of 385 g and is exposed to the radiation for 14.0 s, absorbing 35% of the emitted alpha particles, each having an energy of 9.12×10^{-13} J. Calculate the absorbed dose in millirads and grays. (c) If the RBE of the radiation is 9.5, calculate the effective absorbed dose in mrem and Sv.

21.70 A 65-kg person is accidentally exposed for 240 s to a 15-mCi source of beta radiation coming from a sample of ^{90}Sr. (a) What is the activity of the radiation source in disintegrations per second? In becquerels? (b) Each beta particle has an energy of 8.75×10^{-14} J. and 7.5% of the radiation is absorbed by the person. Assuming that the absorbed radiation is spread over the person's entire body, calculate the absorbed dose in rads and in grays. (c) If the RBE of the beta particles is 1.0, what is the effective dose in mrem and in sieverts? (d) Is the radiation dose equal to, greater than, or less than that for a typical mammogram (300 mrem)?

Additional Exercises

21.71 The table to the right gives the number of protons (p) and neutrons (n) for four isotopes. (a) Write the symbol for each of the isotopes. (b) Which of the isotopes is most likely to be unstable? (c) Which of the isotopes involves a magic number of protons and/or neutrons? (d) Which isotope will yield potassium-39 following positron emission?

	(i)	(ii)	(iii)	(iv)
p	19	19	20	20
n	19	21	19	20

21.72 Radon-222 decays to a stable nucleus by a series of three alpha emissions and two beta emissions. What is the stable nucleus that is formed?

21.73 Equation 21.28 is the nuclear reaction responsible for much of the helium-4 production in our Sun. How much energy is released in this reaction?

21.74 Chlorine has two stable nuclides, ^{35}Cl and ^{37}Cl. In contrast, ^{36}Cl is a radioactive nuclide that decays by beta emission. **(a)** What is the product of decay of ^{36}Cl? **(b)** Based on the empirical rules about nuclear stability, explain why the nucleus of ^{36}Cl is less stable than either ^{35}Cl or ^{37}Cl.

21.75 When two protons fuse in a star, the product is ^2H plus a positron (Equation 21.26). Why do you think the more obvious product of the reaction, ^2He, is unstable?

21.76 Nuclear scientists have synthesized approximately 1600 nuclei not known in nature. More might be discovered with heavy-ion bombardment using high-energy particle accelerators. Complete and balance the following reactions, which involve heavy-ion bombardments:

(a) $^6_3Li + ^{56}_{28}Ni \longrightarrow$?

(b) $^{40}_{20}Ca + ^{248}_{96}Cm \longrightarrow ^{147}_{62}Sm + $?

(c) $^{88}_{38}Sr + ^{84}_{36}Kr \longrightarrow ^{116}_{46}Pd + $?

(d) $^{40}_{20}Ca + ^{238}_{92}U \longrightarrow ^{70}_{30}Zn + 4\,^1_0n + 2$?

21.77 In 2010, a team of scientists from Russia and the U.S. reported creation of the first atom of element 117, which is not yet named and is denoted [117]. The synthesis involved the collision of a target of $^{249}_{97}Bk$ with accelerated ions of an isotope which we will denote Q. The product atom, which we will call Z, immediately releases neutrons and forms $^{294}_{117}[117]$:

$$^{249}_{97}Bk + Q \longrightarrow Z \longrightarrow ^{294}_{117}[117] + 3\,^1_0n$$

(a) What are the identities of isotopes Q and Z? **(b)** Isotope Q is unusual in that it is very long-lived (its half-life is on the order of 10^{19} yr) in spite of having an unfavorable neutron-to-proton ratio (Figure 21.2). Can you propose a reason for its unusual stability? **(c)** Collision of ions of isotope Q with a target was also used to produce the first atoms of livermorium, Lv. The initial product of this collision was $^{296}_{116}Lv$. What was the target isotope with which Q collided in this experiment?

21.78 The synthetic radioisotope technetium-99, which decays by beta emission, is the most widely used isotope in nuclear medicine. The following data were collected on a sample of ^{99}Tc:

Disintegrations per Minute	Time (h)
180	0
130	2.5
104	5.0
77	7.5
59	10.0
46	12.5
24	17.5

Using these data, make an appropriate graph and curve fit to determine the half-life.

[21.79] According to current regulations, the maximum permissible dose of strontium-90 in the body of an adult is 1 μCi(1×10^{-6} Ci). Using the relationship rate $= kN$, calculate the number of atoms of strontium-90 to which this dose corresponds. To what mass of strontium-90 does this correspond? The half-life for strontium-90 is 28.8 yr.

[21.80] Suppose you had a detection device that could count every decay event from a radioactive sample of plutonium-239 ($t_{1/2}$ is 24,000 yr). How many counts per second would you obtain from a sample containing 0.385 g of plutonium-239?

21.81 Methyl acetate (CH_3COOCH_3) is formed by the reaction of acetic acid with methyl alcohol. If the methyl alcohol is labeled with oxygen-18, the oxygen-18 ends up in the methyl acetate:

$$CH_3\overset{O}{\overset{\|}{C}}OH + H^{18}OCH_3 \longrightarrow CH_3\overset{O}{\overset{\|}{C}}^{18}OCH_3 + H_2O$$

(a) Do the C—OH bond of the acid and the O—H bond of the alcohol break in the reaction, or do the O—H bond of the acid and the C—OH bond of the alcohol break? **(b)** Imagine a similar experiment using the radioisotope ^3H, which is called *tritium* and is usually denoted T. Would the reaction between CH_3COOH and $TOCH_3$ provide the same information about which bond is broken as does the above experiment with $H^{18}OCH_3$?

21.82 An experiment was designed to determine whether an aquatic plant absorbed iodide ion from water. Iodine-131 ($t_{1/2} = 8.02$ days) was added as a tracer, in the form of iodide ion, to a tank containing the plants. The initial activity of a 1.00-μL sample of the water was 214 counts per minute. After 30 days the level of activity in a 1.00-μL sample was 15.7 counts per minute. Did the plants absorb iodide from the water?

21.83 Each of the following transmutations produces a radionuclide used in positron emission tomography (PET). **(a)** In equations (i) and (ii), identify the species signified as "X." **(b)** In equation (iii), one of the species is indicated as "d." What do you think it represents?

(i) $^{14}N(p, \alpha)X$ **(ii)** $^{18}O(p, X)^{18}F$ **(iii)** $^{14}N(d, n)^{15}O$

21.84 The nuclear masses of ^7Be, ^9Be, and ^{10}Be are 7.0147, 9.0100, and 10.0113 amu, respectively. Which of these nuclei has the largest binding energy per nucleon?

21.85 A 26.00-g sample of water containing tritium, ^3_1H, emits 1.50×10^3 beta particles per second. Tritium is a weak beta emitter with a half-life of 12.3 yr. What fraction of all the hydrogen in the water sample is tritium?

21.86 The Sun radiates energy into space at the rate of 3.9×10^{26} J/s. **(a)** Calculate the rate of mass loss from the Sun in kg/s. **(b)** How does this mass loss arise? **(c)** It is estimated that the Sun contains 9×10^{56} free protons. How many protons per second are consumed in nuclear reactions in the Sun?

21.87 The average energy released in the fission of a single uranium-235 nucleus is about 3×10^{-11} J. If the conversion of this energy to electricity in a nuclear power plant is 40% efficient, what mass of uranium-235 undergoes fission in a year in a plant that produces 1000 megawatts? Recall that a watt is 1 J/s.

21.88 Tests on human subjects in Boston in 1965 and 1966, following the era of atomic bomb testing, revealed average quantities of about 2 pCi of plutonium radioactivity in the average person. How many disintegrations per second does this level of activity imply? If each alpha particle deposits 8×10^{-13} J of energy and if the average person weighs 75 kg, calculate the number of rads and rems of radiation in 1 yr from such a level of plutonium.

Integrative Exercises

21.89 A 53.8-mg sample of sodium perchlorate contains radioactive chlorine-36 (whose atomic mass is 36.0 amu). If 29.6% of the chlorine atoms in the sample are chlorine-36 and the remainder are naturally occurring nonradioactive chlorine atoms, how many disintegrations per second are produced by this sample? The half-life of chlorine-36 is 3.0×10^5 yr.

21.90 Calculate the mass of octane, $C_8H_{18}(l)$, that must be burned in air to evolve the same quantity of energy as produced by the fusion of 1.0 g of hydrogen in the following fusion reaction:

$$4\,{}^1_1\text{H} \longrightarrow {}^4_2\text{He} + 2\,{}^0_1\text{e}$$

Assume that all the products of the combustion of octane are in their gas phases. Use data from Exercise 21.50, Appendix C, and the inside covers of the text. The standard enthalpy of formation of octane is −250.1 kJ/mol.

21.91 Naturally found uranium consists of 99.274% ^{238}U, 0.720% ^{235}U, and 0.006% ^{233}U. As we have seen, ^{235}U is the isotope that can undergo a nuclear chain reaction. Most of the ^{235}U used in the first atomic bomb was obtained by gaseous diffusion of uranium hexafluoride, $UF_6(g)$. **(a)** What is the mass of UF_6 in a 30.0-L vessel of UF_6 at a pressure of 695 torr at 350 K? **(b)** What is the mass of ^{235}U in the sample described in part (a)? **(c)** Now suppose that the UF_6 is diffused through a porous barrier and that the change in the ratio of ^{238}U and ^{235}U in the diffused gas can be described by Equation 10.23. What is the mass of ^{235}U in a sample of the diffused gas analogous to that in part (a)? **(d)** After one more cycle of gaseous diffusion, what is the percentage of ^{235}UF$_6$ in the sample?

21.92 A sample of an alpha emitter having an activity of 0.18 Ci is stored in a 25.0-mL sealed container at 22 °C for 245 days. **(a)** How many alpha particles are formed during this time? **(b)** Assuming that each alpha particle is converted to a helium atom, what is the partial pressure of helium gas in the container after this 245-day period?

[21.93] Charcoal samples from Stonehenge in England were burned in O_2, and the resultant CO_2 gas bubbled into a solution of $Ca(OH)_2$ (limewater), resulting in the precipitation of $CaCO_3$. The $CaCO_3$ was removed by filtration and dried. A 788-mg sample of the $CaCO_3$ had a radioactivity of 1.5×10^{-2} Bq due to carbon-14. By comparison, living organisms undergo 15.3 disintegrations per minute per gram of carbon. Using the half-life of carbon-14, 5700 yr, calculate the age of the charcoal sample.

21.94 A 25.0-mL sample of 0.050 *M* barium nitrate solution was mixed with 25.0 mL of 0.050 *M* sodium sulfate solution labeled with radioactive sulfur-35. The activity of the initial sodium sulfate solution was 1.22×10^6 Bq/mL. After the resultant precipitate was removed by filtration, the remaining filtrate was found to have an activity of 250 Bq/mL. **(a)** Write a balanced chemical equation for the reaction that occurred. **(b)** Calculate the K_{sp} for the precipitate under the conditions of the experiment.

Design an Experiment

This chapter has focused on the properties of elements that exhibit radioactivity. Because radioactivity can have harmful effects on human health, very stringent experimental procedures and precautions are required when undertaking experiments on radioactive materials. As such, we typically do not have experiments involving radioactive substances in general chemistry laboratories. We can nevertheless ponder the design of some hypothetical experiments that would allow us to explore some of the properties of radium, which was discovered by Marie and Pierre Curie in 1898.

(a) A key aspect of the discovery of radium was Marie Curie's observation that *pitchblende*, a natural ore of uranium, had greater radioactivity than pure uranium metal. Design an experiment to reproduce this observation and to obtain a ratio of the activity of pitchblende relative to that of pure uranium.

(b) Radium was first isolated as halide salts. Suppose you had pure samples of radium metal and radium bromide. The sample sizes are on the order of milligrams and are not amenable to the usual forms of elemental analysis. Could you use a device that measures radioactivity quantitatively to determine the empirical formula of radium bromide? What information must you use that the Curies may not have had at the time of their discovery?

(c) Suppose you had a 1-yr time period in order to measure the half-life of radium and related elements. You have some pure samples and a device that measures radioactivity quantitatively. Could you determine the half-life of the elements in the samples? Would you have different experimental constraints depending on whether the half-life were 10 yr or 1000 yr?

(d) Before its negative health effects were better understood, small amounts of radium salts were used in "glow in the dark" watches, such as the one shown here. The glow is not due to the radioactivity of radium directly; rather, the radium is combined with a luminescent substance, such as zinc sulfide, which glows when it is exposed to radiation. Suppose you had pure samples of radium and zinc sulfide. How could you determine whether the glow of zinc sulfide is due to alpha, beta, or gamma radiation? What type of device could you design to use the glow as a quantitative measure of the amount of radioactivity in a sample?

22
Chemistry of the Nonmetals

Everything we see in the chapter-opening photo is composed of nonmetals. The water, of course, is H_2O, and the sand is mostly SiO_2. Although we cannot see it, the air contains principally N_2 and O_2 with much lesser amounts of other nonmetallic substances. The palm tree is also composed mostly of nonmetallic elements.

In this chapter, we take a panoramic view of the descriptive chemistry of the nonmetallic elements, starting with hydrogen and progressing group by group across the periodic table. We will consider how the elements occur in nature, how they are isolated from their sources, and how they are used. We will emphasize hydrogen, oxygen, nitrogen, and carbon because these four nonmetals form many commercially important compounds and account for 99% of the atoms required by living cells.

As you study this *descriptive chemistry*, it is important to look for trends rather than trying to memorize all the facts presented. The periodic table is your most valuable tool in this task.

22.1 | Periodic Trends and Chemical Reactions

Recall that we can classify elements as metals, metalloids, and nonmetals. ∞ (Section 7.6) Except for hydrogen, which is a special case, the nonmetals occupy the upper right portion of the periodic table. This division of elements relates nicely

▶ A TROPICAL BEACH.

22.6 THE OTHER GROUP 6A ELEMENTS: S, Se, Te, AND Po
We study the other members of group 6A (S, Se, Te, and Po), of which sulfur is the most important.

22.7 NITROGEN We next consider nitrogen, a key component of our atmosphere. It forms compounds in which its oxidation number ranges from -3 to $+5$, including such important compounds as NH_3 and HNO_3.

22.8 THE OTHER GROUP 5A ELEMENTS: P, As, Sb, AND Bi
Of the other members of group 5A (P, As, Sb, and Bi), we take a closer look at phosphorus—the most commercially important one and the only one that plays an important and beneficial role in biological systems.

22.9 CARBON We next focus on the inorganic compounds of carbon.

22.10 THE OTHER GROUP 4A ELEMENTS: Si, Ge, Sn, AND Pb
We then consider silicon, the element most abundant and significant of the heavier members of group 4A.

22.11 BORON Finally, we examine boron—the sole nonmetallic element of group 3A.

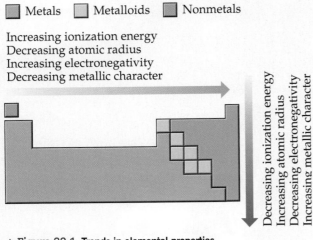

☐ Metals ☐ Metalloids ☐ Nonmetals

Increasing ionization energy
Decreasing atomic radius
Increasing electronegativity
Decreasing metallic character

Decreasing ionization energy
Increasing atomic radius
Decreasing electronegativity
Increasing metallic character

▲ Figure 22.1 **Trends in elemental properties.**

to trends in the properties of the elements as summarized in ◀ Figure 22.1. Electronegativity, for example, increases as we move left to right across a period and decreases as we move down a group. The nonmetals thus have higher electronegativities than the metals. This difference leads to the formation of ionic solids in reactions between metals and nonmetals. ∞ (Sections 7.6, 8.2, and 8.4) In contrast, compounds formed between two or more nonmetals are usually molecular substances. ∞ (Sections 7.8 and 8.4)

The chemistry exhibited by the first member of a nonmetal group can differ from that of subsequent members in important ways. Two differences are particularly notable: (1) The first member is able to accommodate fewer bonded neighbors. ∞ (Section 8.7) For example, nitrogen is able to bond to a maximum of three Cl atoms, NCl_3, whereas phosphorus can bond to five, PCl_5. The small size of nitrogen is largely responsible for this difference. (2) The first member can more readily form π bonds. This trend is also due, in part, to size because small atoms are able to approach each other more closely. As a result, the overlap of p orbitals, which results in the formation of π bonds, is more effective for the first element in each group (▼ Figure 22.2). More effective overlap means stronger π bonds, reflected in bond enthalpies. ∞ (Section 8.8) For example, the difference between the enthalpies of the C—C bond and the C=C bond is about 270 kJ/mol ∞ (Table 8.4); this large value reflects the "strength" of a carbon–carbon π bond. On the other hand, the difference between Si—Si and Si=Si bonds is only about 100 kJ/mol, significantly lower than that for carbon, reflecting much weaker π bonding.

As we shall see, π bonds are particularly important in the chemistry of carbon, nitrogen, and oxygen, each the first member in its group. The heavier elements in these groups have a tendency to form only single bonds.

SAMPLE EXERCISE 22.1 | Identifying Elemental Properties

Of the elements Li, K, N, P, and Ne, which (**a**) is the most electronegative, (**b**) has the greatest metallic character, (**c**) can bond to more than four atoms in a molecule, and (**d**) forms π bonds most readily?

SOLUTION

Analyze We are given a list of elements and asked to predict several properties that can be related to periodic trends.

Plan We can use Figures 22.1 and 22.2 to guide us to the answers.

Solve

(**a**) Electronegativity increases as we proceed toward the upper right portion of the periodic table, excluding the noble gases. Thus, N is the most electronegative element of our choices.

(**b**) Metallic character correlates inversely with electronegativity—the less electronegative an element, the greater its metallic character. The element with the greatest metallic character is therefore K, which is closest to the lower left corner of the periodic table.

(**c**) Nonmetals tend to form molecular compounds, so we can narrow our choice to the three nonmetals on the list: N, P, and Ne. To form more than four bonds, an element must be able to expand its valence shell to allow more than an octet of electrons around it. Valence-shell expansion occurs for period 3 elements and below; N and Ne are both in period 2 and do not undergo valence-shell expansion. Thus, the answer is P.

(**d**) Period 2 nonmetals form π bonds more readily than elements in period 3 and below. There are no compounds known that contain covalent bonds to Ne. Thus, N is the element from the list that forms π bonds most readily.

C—C   Si—Si

Smaller nucleus-to-nucleus distance, more orbital overlap, stronger π bond

Larger nucleus-to-nucleus distance, less orbital overlap, weaker π bond

▲ Figure 22.2 **π Bonds in period 2 and period 3 elements.**

Practice Exercise 1

Which description correctly describes a difference between the chemistry of oxygen and sulfur?
(a) Oxygen is a nonmetal and sulfur is a metalloid. (b) Oxygen can form more than four bonds whereas sulfur cannot. (c) Sulfur has a higher electronegativity than oxygen. (d) Oxygen is better able to form π bonds than sulfur.

Practice Exercise 2

Of the elements Be, C, Cl, Sb, and Cs, which (a) has the lowest electronegativity, (b) has the greatest nonmetallic character, (c) is most likely to participate in extensive π bonding, (d) is most likely to be a metalloid?

The ready ability of period 2 elements to form π bonds is an important factor in determining the elemental forms of these elements. Compare, for example, carbon and silicon. Carbon has five major crystalline allotropes: diamond, graphite, buckminster-fullerene, graphene, and carbon nanotubes. ∞ (Sections 12.7 and 12.9) Diamond is a covalent-network solid that has C—C σ bonds but no π bonds. Graphite, buckminster-fullerene, graphene, and carbon nanotubes have π bonds that result from the sideways overlap of p orbitals. Elemental silicon, however, exists only as a diamond-like covalent-network solid with σ bonds; it has no forms analogous to graphite, buckminsterfuller-ene, graphene, or carbon nanotubes, apparently because Si—Si π bonds are too weak.

We likewise see significant differences in the dioxides of carbon and silicon as a result of their relative abilities to form π bonds (▶ Figure 22.3). CO_2 is a molecular substance containing C=O double bonds, whereas SiO_2 is a covalent-network solid in which four oxygen atoms are bonded to each silicon atom by single bonds, forming an extended structure that has the empirical formula SiO_2.

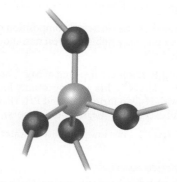

Fragment of extended SiO_2 lattice; Si forms only single bonds

CO_2; C forms double bonds

▲ Figure 22.3 **Comparison of the bonds in SiO_2 and CO_2.**

 Give It Some Thought

Nitrogen is found in nature as $N_2(g)$. Would you expect phosphorus to be found in nature as $P_2(g)$? Explain.

Chemical Reactions

Because O_2 and H_2O are abundant in our environment, it is particularly important to consider how these substances react with other compounds. About one-third of the reactions discussed in this chapter involve either O_2 (oxidation or combustion reactions) or H_2O (especially proton-transfer reactions).

In combustion reactions ∞ (Section 3.2), hydrogen-containing compounds produce H_2O. Carbon-containing ones produce CO_2 (unless the amount of O_2 is insufficient, in which case CO or even C can form). Nitrogen-containing compounds tend to form N_2, although NO can form in special cases or in small amounts. A reaction illustrating these points is:

$$4\,CH_3NH_2(g) + 9\,O_2(g) \longrightarrow 4\,CO_2(g) + 10\,H_2O(g) + 2\,N_2(g) \quad [22.1]$$

The formation of H_2O, CO_2, and N_2 reflects the high thermodynamic stability of these substances, indicated by the large bond energies for the O—H, C=O, and N≡N bonds (463, 799, and 941 kJ/mol, respectively). ∞ (Section 8.8)

When dealing with proton-transfer reactions, remember that the weaker a Brønsted–Lowry acid, the stronger its conjugate base. ∞ (Section 16.2) For example, H_2, OH^-, NH_3, and CH_4 are exceedingly weak proton donors that have *no* tendency to act as acids in water. Thus, the species formed by removing one or more protons from them are extremely strong bases. All of them react readily with water, removing protons from H_2O to form OH^-. Two representative reactions are:

$$CH_3^-(aq) + H_2O(l) \longrightarrow CH_4(g) + OH^-(aq) \quad [22.2]$$

$$N^{3-}(aq) + 3\,H_2O(l) \longrightarrow NH_3(aq) + 3\,OH^-(aq) \quad [22.3]$$

SAMPLE EXERCISE 22.2 | Predicting the Products of Chemical Reactions

Predict the products formed in each of the following reactions, and write a balanced equation:

(a) $CH_3NHNH_2(g) + O_2(g) \longrightarrow$?   (b) $Mg_3P_2(s) + H_2O(l) \longrightarrow$?

SOLUTION

Analyze We are given the reactants for two chemical equations and asked to predict the products and then balance the equations.

Plan We need to examine the reactants to see if we might recognize a reaction type. In (a) the carbon compound is reacting with O_2, which suggests a combustion reaction. In (b) water reacts with an ionic compound. The anion P^{3-} is a strong base and H_2O is able to act as an acid, so the reactants suggest an acid–base (proton-transfer) reaction.

Solve

(a) Based on the elemental composition of the carbon compound, this combustion reaction should produce CO_2, H_2O, and N_2:

$$2\,CH_3NHNH_2(g) + 5\,O_2(g) \longrightarrow 2\,CO_2(g) + 6\,H_2O(g) + 2\,N_2(g)$$

(b) Mg_3P_2 is ionic, consisting of Mg^{2+} and P^{3-} ions. The P^{3-} ion, like N^{3-}, has a strong affinity for protons and reacts with H_2O to form OH^- and PH_3 (PH^{2-}, PH_2^-, and PH_3 are all exceedingly weak proton donors).

$$Mg_3P_2(s) + 6\,H_2O(l) \longrightarrow 2\,PH_3(g) + 3\,Mg(OH)_2(s)$$

$Mg(OH)_2$ has low solubility in water and will precipitate.

Practice Exercise 1

When CaC_2 reacts with water, what carbon-containing compound forms?

(a) CO, (b) CO_2, (c) CH_4, (d) C_2H_2, (e) H_2CO_3.

Practice Exercise 2

Write a balanced equation for the reaction of solid sodium hydride with water.

22.2 | Hydrogen

The English chemist Henry Cavendish (1731–1810) was the first to isolate hydrogen. Because the element produces water when burned in air, the French chemist Antoine Lavoisier ∞ (Figure 3.1) gave it the name *hydrogen*, which means "water producer" (Greek: *hydro*, water; *gennao*, to produce).

Hydrogen is the most abundant element in the universe. It is the nuclear fuel consumed by our Sun and other stars to produce energy. ∞ (Section 21.8) Although about 75% of the known mass of the universe is hydrogen, it constitutes only 0.87% of Earth's mass. Most of the hydrogen on our planet is found associated with oxygen. Water, which is 11% hydrogen by mass, is the most abundant hydrogen compound.

Isotopes of Hydrogen

The most common isotope of hydrogen, ^1_1H, has a nucleus consisting of a single proton. This isotope, sometimes referred to as **protium**,* makes up 99.9844% of naturally occurring hydrogen.

Two other isotopes are known: ^2_1H, whose nucleus contains a proton and a neutron, and ^3_1H, whose nucleus contains a proton and two neutrons. The ^2_1H isotope, **deuterium**, makes up 0.0156% of naturally occurring hydrogen. It is not radioactive and is often given the symbol D in chemical formulas, as in D_2O (deuterium oxide), which is known as *heavy water*.

Because an atom of deuterium is about twice as massive as an atom of protium, the properties of deuterium-containing substances vary somewhat from those

---

*Giving unique names to isotopes is limited to hydrogen. Because of the proportionally large differences in their masses, the isotopes of H show appreciably more differences in their properties than isotopes of heavier elements.

of the protium-containing analogs. For example, the normal melting and boiling points of D_2O are 3.81 °C and 101.42 °C, respectively, versus 0.00 °C and 100.00 °C for H_2O. Not surprisingly, the density of D_2O at 25 °C (1.104 g/mL) is greater than that of H_2O (0.997 g/mL). Replacing protium with deuterium (a process called *deuteration*) can also have a profound effect on reaction rates, a phenomenon called a *kinetic-isotope effect*. For example, heavy water can be obtained from the electrolysis $[2 H_2O(l) \longrightarrow 2 H_2(g) + O_2(g)]$ of ordinary water because the small amount of naturally occurring D_2O in the sample undergoes electrolysis more slowly than H_2O and, therefore, becomes concentrated during the reaction.

The third isotope, ^3_1H, **tritium**, is radioactive, with a half-life of 12.3 yr:

$$^3_1H \longrightarrow \, ^3_2He + \, ^{\,\,0}_{-1}e \quad t_{1/2} = 12.3 \text{ yr} \qquad [22.4]$$

Because of its short half-life, only trace quantities of tritium exist naturally. The isotope can be synthesized in nuclear reactors by neutron bombardment of lithium-6:

$$^6_3Li + \, ^1_0n \longrightarrow \, ^3_1H + \, ^4_2He \qquad [22.5]$$

Deuterium and tritium are useful in studying reactions of compounds containing hydrogen. A compound is "labeled" by replacing one or more ordinary hydrogen atoms with deuterium or tritium at specific locations in a molecule. By comparing the locations of the label atoms in reactants and products, the reaction mechanism can often be inferred. When methyl alcohol (CH_3OH) is placed in D_2O, for example, the H atom of the O—H bond exchanges rapidly with the D atoms, forming CH_3OD. The H atoms of the CH_3 group do not exchange. This experiment demonstrates the kinetic stability of C—H bonds and reveals the speed at which the O—H bond in the molecule breaks and re-forms.

Properties of Hydrogen

Hydrogen is the only element that is not a member of any family in the periodic table. Because of its ls^1 electron configuration, it is generally placed above lithium in the table. However, it is definitely *not* an alkali metal. It forms a positive ion much less readily than any alkali metal. The ionization energy of the hydrogen atom is 1312 kJ/mol, whereas that of lithium is 520 kJ/mol.

Hydrogen is sometimes placed above the halogens in the periodic table because the hydrogen atom can pick up one electron to form the *hydride ion*, H^-, which has the same electron configuration as helium. However, the electron affinity of hydrogen, $E = -73$ kJ/mol, is not as large as that of any halogen. In general, hydrogen shows no closer resemblance to the halogens than it does to the alkali metals.

Elemental hydrogen exists at room temperature as a colorless, odorless, tasteless gas composed of diatomic molecules. We can call H_2 *dihydrogen*, but it is more commonly referred to as either *molecular hydrogen* or simply hydrogen. Because H_2 is nonpolar and has only two electrons, attractive forces between molecules are extremely weak. As a result, its melting point (-259 °C) and boiling point (-253 °C) are very low.

The H—H bond enthalpy (436 kJ/mol) is high for a single bond. ∞ (Table 8.4) By comparison, the Cl—Cl bond enthalpy is only 242 kJ/mol. Because H_2 has a strong bond, most reactions involving H_2 are slow at room temperature. However, the molecule is readily activated by heat, irradiation, or catalysis. The activation generally produces hydrogen atoms, which are very reactive. Once H_2 is activated, it reacts rapidly and exothermically with a wide variety of substances.

 Give It Some Thought

If H_2 is activated to produce H^+, what must the other product be?

Hydrogen forms strong covalent bonds with many other elements, including oxygen; the O—H bond enthalpy is 463 kJ/mol. The formation of the strong O—H bond makes hydrogen an effective reducing agent for many metal oxides. When H_2 is passed over heated CuO, for example, copper is produced:

$$CuO(s) + H_2(g) \longrightarrow Cu(s) + H_2O(g) \qquad [22.6]$$

When H_2 is ignited in air, a vigorous reaction occurs, forming H_2O. Air containing as little as 4% H_2 by volume is potentially explosive. Combustion of hydrogen–oxygen mixtures is used in liquid-fuel rocket engines such as those of the Space Shuttle. The hydrogen and oxygen are stored at low temperatures in liquid form.

Production of Hydrogen

When a small quantity of H_2 is needed in the laboratory, it is usually obtained by the reaction between an active metal such as zinc and a dilute strong acid such as HCl or H_2SO_4:

$$Zn(s) + 2\,H^+(aq) \longrightarrow Zn^{2+}(aq) + H_2(g) \qquad [22.7]$$

Large quantities of H_2 are produced by reacting methane with steam at 1100 °C. We can view this process as involving two reactions:

$$CH_4(g) + H_2O(g) \longrightarrow CO(g) + 3\,H_2(g) \qquad [22.8]$$

$$CO(g) + H_2O(g) \longrightarrow CO_2(g) + H_2(g) \qquad [22.9]$$

Carbon heated with water to about 1000 °C is another source of H_2:

$$C(s) + H_2O(g) \longrightarrow H_2(g) + CO(g) \qquad [22.10]$$

This mixture, known as *water gas*, is used as an industrial fuel.

Electrolysis of water consumes too much energy and is consequently too costly to be used commercially to produce H_2. However, H_2 is produced as a by-product in the electrolysis of brine (NaCl) solutions in the course of commercial Cl_2 and NaOH manufacture:

$$2\,NaCl(aq) + 2\,H_2O(l) \xrightarrow{\text{electrolysis}} H_2(g) + Cl_2(g) + 2\,NaOH(aq) \qquad [22.11]$$

Give It Some Thought

What are the oxidation states of the H atoms in Equations 22.7–22.11?

Closer Look

The Hydrogen Economy

The reaction of hydrogen with oxygen is highly exothermic:

$$2\,H_2(g) + O_2(g) \longrightarrow 2\,H_2O(g) \quad \Delta H = -483.6\ kJ \qquad [22.12]$$

Because H_2 has a low molar mass and a high enthalpy of combustion, it has a high energy density by mass. (That is, its combustion produces high energy per gram.) Furthermore, the only product of the reaction is water vapor, which means that hydrogen is environmentally cleaner than fossil fuels. Thus, the prospect of using hydrogen widely as a fuel is attractive.

The term "hydrogen economy" is used to describe the concept of delivering and using hydrogen as a fuel in place of fossil fuels. In order to develop a hydrogen economy, it would be necessary to generate elemental hydrogen on a large scale and arrange for its transport and storage. These matters provide significant technical challenges.

▶ **Figure 22.4** illustrates various sources and uses of H_2 fuel. The generation of H_2 through electrolysis of water is in principle the cleanest route, because this process—the reverse of Equation 22.11—produces only hydrogen and oxygen. ∞∞ (Figure 1.7 and Section 20.9) However, the energy required to electrolyze water must come from somewhere. If we burn fossil fuels to generate this energy, we have not advanced very far toward a true hydrogen economy. If the energy for electrolysis came instead from a hydroelectric or nuclear power plant, solar cells, or wind generators, consumption of nonrenewable energy sources and undesired production of CO_2 could be avoided.

The storage of hydrogen is another technical obstacle that must be overcome in developing a hydrogen economy. Although $H_2(g)$ has a high energy density by mass, it has a low energy density by volume. Thus, storing hydrogen as a gas requires a large volume compared to the energy it delivers. There are also safety issues associated with handling and storing the gas because its combustion can be explosive. Storing hydrogen in the form of various hydride compounds such as $LiAlH_4$ is being investigated as a means of reducing the volume and increasing the safety. One problem with this approach, however, is that such compounds have high energy density by volume but low energy density by mass.

Related Exercises: 22.29, 22.30, 22.94

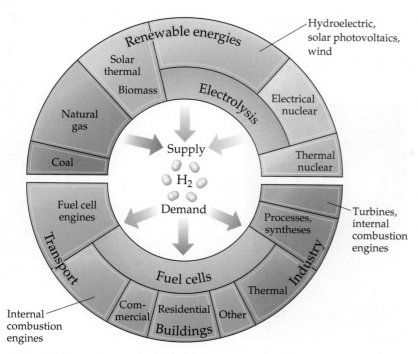

▲ **Figure 22.4** The "hydrogen economy" would require hydrogen to be produced from various sources and would use hydrogen in energy-related applications.

Uses of Hydrogen

Hydrogen is commercially important. About 5.0×10^{10} kg (50 million metric tons) is produced annually across the world. About half of the H_2 produced is used to synthesize ammonia by the Haber process. ∞∞ (Section 15.2) Much of the remaining hydrogen is used to convert high-molecular-weight hydrocarbons from petroleum into lower-molecular-weight hydrocarbons suitable for fuel (gasoline, diesel, and others) in a process known as *cracking*. Hydrogen is also used to manufacture methanol via the catalytic reaction of CO and H_2 at high pressure and temperature:

$$CO(g) + 2\,H_2(g) \longrightarrow CH_3OH(g) \qquad [22.13]$$

Binary Hydrogen Compounds

Hydrogen reacts with other elements to form three types of compounds: (1) ionic hydrides, (2) metallic hydrides, and (3) molecular hydrides.

The **ionic hydrides** are formed by the alkali metals and by the heavier alkaline earths (Ca, Sr, and Ba). These active metals are much less electronegative than hydrogen. Consequently, hydrogen acquires electrons from them to form hydride ions (H^-):

$$Ca(s) + H_2(g) \longrightarrow CaH_2(s) \qquad [22.14]$$

The hydride ion is very basic and reacts readily with compounds having even weakly acidic protons to form H_2:

$$H^-(aq) + H_2O(l) \longrightarrow H_2(g) + OH^-(aq) \qquad [22.15]$$

Ionic hydrides can therefore be used as convenient (although expensive) sources of H_2.

Calcium hydride (CaH_2) is used to inflate life rafts, weather balloons, and the like where a simple, compact means of generating H_2 is desired (**Figure 22.5**).

GO FIGURE

This reaction is exothermic. Is the beaker on the right warmer or colder than the beaker on the left?

CaH₂

H₂O with pH indicator

H₂ gas

Color change indicates presence of OH⁻

▲ Figure 22.5 The reaction of CaH₂ with water.

GO FIGURE

Which is the most thermodynamically stable hydride? Which is the least thermodynamically stable?

4A	5A	6A	7A
$CH_4(g)$	$NH_3(g)$	$H_2O(l)$	$HF(g)$
−50.8	−16.7	−237	−271
$SiH_4(g)$	$PH_3(g)$	$H_2S(g)$	$HCl(g)$
+56.9	+18.2	−33.0	−95.3
$GeH_4(g)$	$AsH_3(g)$	$H_2Se(g)$	$HBr(g)$
+117	+111	+71	−53.2
	$SbH_3(g)$	$H_2Te(g)$	$HI(g)$
	+187	+138	+1.30

▲ Figure 22.6 Standard free energies of formation of molecular hydrides. All values are kilojoules per mole of hydride.

The reaction between H^- and H_2O (Equation 22.15) is an acid–base reaction *and* a redox reaction. The H^- ion, therefore, is a good base *and* a good reducing agent. In fact, hydrides are able to reduce O_2 to OH^-:

$$2\,NaH(s) + O_2(g) \longrightarrow 2\,NaOH(s) \qquad [22.16]$$

For this reason, hydrides are normally stored in an environment that is free of both moisture and air.

Metallic hydrides are formed when hydrogen reacts with transition metals. These compounds are so named because they retain their metallic properties. In many metallic hydrides, the ratio of metal atoms to hydrogen atoms is not fixed or in small whole numbers. The composition can vary within a range, depending on reaction conditions. TiH_2 can be produced, for example, but preparations usually yield $TiH_{1.8}$. These nonstoichiometric metallic hydrides are sometimes called *interstitial hydrides*. Because hydrogen atoms are small enough to fit between the sites occupied by the metal atoms, many metal hydrides behave like interstitial alloys. ∞ (Section 12.3)

The **molecular hydrides,** formed by nonmetals and metalloids, are either gases or liquids under standard conditions. The simple molecular hydrides are listed in ◀ Figure 22.6, together with their standard free energies of formation, ΔG_f°. ∞ (Section 19.5) In each family, the thermal stability (*measured as* ΔG_f°) decreases as we move down the family. (Recall that the more stable a compound is with respect to its elements under standard conditions, the more negative ΔG_f° is.)

22.3 | Group 8A: The Noble Gases

The elements of group 8A are chemically unreactive. Indeed, most of our references to these elements have been in relation to their physical properties, as when we discussed intermolecular forces. ∞ (Section 11.2) The relative inertness of these elements is due to the presence of a completed octet of valence-shell electrons (except He, which only has a filled 1s shell). The stability of such an arrangement is reflected in the high ionization energies of the group 8A elements. ∞ (Section 7.4)

The group 8A elements are all gases at room temperature. They are components of Earth's atmosphere, except for radon, which exists only as a short-lived radioisotope. ∞ (Section 21.9) Only argon is relatively abundant. ∞ (Table 18.1)

Neon, argon, krypton, and xenon are used in lighting, display, and laser applications in which the atoms are excited electrically and electrons that are in a higher energy state emit light as they fall to the ground state. ∞ (Section 6.2) Argon is used as a blanketing atmosphere in electric lightbulbs. The gas conducts heat away from the filament but does not react with it. Argon is also used as a protective atmosphere to prevent oxidation in welding and certain high-temperature metallurgical processes.

Helium is in many ways the most important noble gas. Liquid helium is used as a coolant to conduct experiments at very low temperatures. Helium boils at 4.2 K and 1 atm, the lowest boiling point of any substance. It is found in relatively high concentrations in many natural-gas wells from which it is isolated.

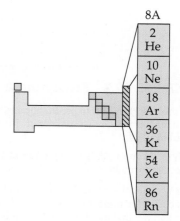

Noble-Gas Compounds

Because the noble gases are exceedingly stable, they react only under rigorous conditions. We expect the heavier ones to be most likely to form compounds because their ionization energies are lower. ∞ (Figure 7.9) A lower ionization energy suggests the possibility of sharing an electron with another atom, leading to a chemical bond. In addition, because the group 8A elements (except helium) already contain eight electrons in their valence shell, formation of covalent bonds will require an expanded valence shell. Valence-shell expansion occurs most readily with larger atoms. ∞ (Section 8.7)

The first noble-gas compound was reported in 1962. This discovery caused a sensation because it undercut the belief that the noble-gas elements were inert. The initial study involved xenon in combination with fluorine, the element we would expect to be most reactive in pulling electron density from another atom. Since that time chemists have prepared several xenon compounds of fluorine and oxygen (▼ Table 22.1). The fluorides XeF_2, XeF_4, and XeF_6 are made by direct reaction of the elements. By varying the ratio of reactants and altering reaction conditions, one of the three compounds can be obtained. The oxygen-containing compounds are formed when the fluorides react with water as, for example,

$$XeF_6(s) + 3\,H_2O(l) \longrightarrow XeO_3(aq) + 6\,HF(aq) \qquad [22.17]$$

The other noble-gas elements form compounds much less readily than xenon. For many years, only one binary krypton compound, KrF_2, was known with certainty, and

Table 22.1 Properties of Xenon Compounds

Compound	Oxidation State of Xe	Melting Point (°C)	ΔH_f° (kJ/mol)[a]
XeF_2	+2	129	−109(g)
XeF_4	+4	117	−218(g)
XeF_6	+6	49	−298(g)
$XeOF_4$	+6	−41 to −28	+146(l)
XeO_3	+6	—[b]	+402(s)
XeO_2F_2	+6	31	+145(s)
XeO_4	+8	—[c]	—

[a]At 25 °C, for the compound in the state indicated.

[b]A solid; decomposes at 40 °C.

[c]A solid; decomposes at −40 °C.

it decomposes to its elements at $-10\,°C$. Other compounds of krypton have been isolated at very low temperatures (40 K).

SAMPLE EXERCISE 22.3 Predicting a Molecular Structure

Use the VSEPR model to predict the structure of XeF_4.

SOLUTION

Analyze We must predict the geometrical structure given only the molecular formula.

Plan We must first write the Lewis structure for the molecule. We then count the number of electron pairs (domains) around the Xe atom and use that number and the number of bonds to predict the geometry.

Solve There are 36 valence-shell electrons (8 from xenon and 7 from each fluorine). If we make four single Xe—F bonds, each fluorine has its octet satisfied. Xe then has 12 electrons in its valence shell, so we expect an octahedral arrangement of six electron pairs. Two of these are nonbonded pairs. Because nonbonded pairs require more volume than bonded pairs ∞ (Section 9.2), it is reasonable to expect these nonbonded pairs to be opposite each other. The expected structure is square planar, as shown in ▶ Figure 22.7.

Comment The experimentally determined structure agrees with this prediction.

▲ Figure 22.7 **Xenon tetrafluoride.**

Practice Exercise 1

Compounds containing the XeF_3^+ ion have been characterized. Describe the electron-domain geometry and molecular geometry of this ion.
(a) trigonal planar, trigonal planar; **(b)** tetrahedral, trigonal pyramidal; **(c)** trigonal bipyramidal, T shaped; **(d)** tetrahedral, tetrahedral; **(e)** octahedral, square planar.

Practice Exercise 2

Describe the electron-domain geometry and molecular geometry of KrF_2.

22.4 | Group 7A: The Halogens

7A
9 F
17 Cl
35 Br
53 I
85 At

The elements of group 7A, the halogens, have the outer-electron configuration ns^2np^5, where n ranges from 2 through 6. The halogens have large negative electron affinities ∞ (Section 7.5), and they most often achieve a noble-gas configuration by gaining an electron, which results in a -1 oxidation state. Fluorine, being the most electronegative element, exists in compounds only in the -1 state. The other halogens exhibit positive oxidation states up to $+7$ in combination with more electronegative atoms such as O. In the positive oxidation states, the halogens tend to be good oxidizing agents, readily accepting electrons.

Chlorine, bromine, and iodine are found as the halides in seawater and in salt deposits. Fluorine occurs in the minerals fluorspar (CaF_2), cryolite (Na_3AlF_6), and fluorapatite $[Ca_5(PO_4)_3F]$.* Only fluorspar is an important commercial source of fluorine.

All isotopes of astatine are radioactive. The longest-lived isotope is astatine-210, which has a half-life of 8.1 h and decays mainly by electron capture. Because astatine is so unstable, very little is known about its chemistry.

Properties and Production of the Halogens

Most properties of the halogens vary in a regular fashion as we go from fluorine to iodine (▶ Table 22.2).

Under ordinary conditions the halogens exist as diatomic molecules. The molecules are held together in the solid and liquid states by dispersion forces. ∞ (Section 11.2) Because I_2 is the largest and most polarizable halogen molecule, the intermolecular forces between I_2 molecules are the strongest. Thus, I_2 has the highest melting point and boiling point. At room temperature and 1 atm, I_2 is a purple solid, Br_2 is a red-brown liquid, and Cl_2 and F_2 are gases. ∞ (Figure 7.27) Chlorine readily liquefies upon compression

---

*Minerals are solid substances that occur in nature. They are usually known by their common names rather than by their chemical names. What we know as rock is merely an aggregate of different minerals.

Table 22.2 Some Properties of the Halogens

Property	F	Cl	Br	I
Atomic radius (Å)	0.57	1.02	1.20	1.39
Ionic radius, X⁻ (Å)	1.33	1.81	1.96	2.20
First ionization energy (kJ/mol)	1681	1251	1140	1008
Electron affinity (kJ/mol)	−328	−349	−325	−295
Electronegativity	4.0	3.0	2.8	2.5
X—X single-bond enthalpy (kJ/mol)	155	242	193	151
Reduction potential (V):				
$\frac{1}{2}X_2(aq) + e^- \longrightarrow X^-(aq)$	2.87	1.36	1.07	0.54

at room temperature and is normally stored and handled in liquid form under pressure in steel containers.

The comparatively low bond enthalpy of F_2 (155 kJ/mol) accounts in part for the extreme reactivity of elemental fluorine. Because of its high reactivity, F_2 is difficult to work with. Certain metals, such as copper and nickel, can be used to contain F_2 because their surfaces form a protective coating of metal fluoride. Chlorine and the heavier halogens are also reactive, although less so than fluorine.

Because of their high electronegativities, the halogens tend to gain electrons from other substances and thereby serve as oxidizing agents. The oxidizing ability of the halogens, indicated by their standard reduction potentials, decreases going down the group. As a result, a given halogen is able to oxidize the halide anions below it. For example, Cl_2 oxidizes Br^- and I^- but not F^-, as seen in ▶ Figure 22.8.

SAMPLE EXERCISE 22.4 Predicting Chemical Reactions among the Halogens

Write the balanced equation for the reaction, if any, between (a) $I^-(aq)$ and $Br_2(l)$, (b) $Cl^-(aq)$ and $I_2(s)$.

SOLUTION

Analyze We are asked to determine whether a reaction occurs when a particular halide and halogen are combined.

Plan A given halogen is able to oxidize anions of the halogens below it in the periodic table. Thus, in each pair the halogen having the smaller atomic number ends up as the halide ion. If the halogen with the smaller atomic number is already the halide ion, there is no reaction. Thus, the key to determining whether a reaction occurs is locating the elements in the periodic table.

Solve

(a) Br_2 can oxidize (remove electrons from) the anions of the halogens below it in the periodic table. Thus, it oxidizes I^-:

$$2\,I^-(aq) + Br_2(aq) \longrightarrow I_2(s) + 2\,Br^-(aq)$$

(b) Cl^- is the anion of a halogen above iodine in the periodic table. Thus, I_2 cannot oxidize Cl^-; there is no reaction.

Practice Exercise 1

Which is (are) able to oxidize Cl^-?

(a) F_2

(b) F^-

(c) Both Br_2 and I_2

(d) Both Br^- and I^-

Practice Exercise 2

Write the balanced chemical equation for the reaction between $Br^-(aq)$ and $Cl_2(aq)$.

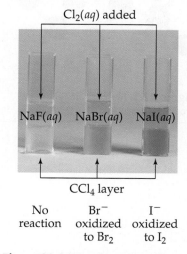

▲GO FIGURE

Do Br_2 and I_2 appear to be more or less soluble in CCl_4 than in H_2O?

▲ **Figure 22.8 Reaction of Cl_2 with aqueous solutions of NaF, NaBr, and NaI in the presence of carbon tetrachloride.** The top liquid layer in each vial is water; the bottom liquid layer is carbon tetrachloride. The $Cl_2(aq)$, which has been added to each vial, is colorless. The brown color in the carbon tetrachloride layer indicates the presence of Br_2, whereas purple indicates the presence of I_2.

Notice in Table 22.2 that the standard reduction potential of F_2 is exceptionally high. As a result, fluorine gas readily oxidizes water:

$$F_2(aq) + H_2O(l) \longrightarrow 2\,HF(aq) + \tfrac{1}{2}O_2(g) \quad E° = 1.80\text{ V} \quad [22.18]$$

Fluorine cannot be prepared by electrolytic oxidation of aqueous solutions of fluoride salts because water is oxidized more readily than F^-. ∞ (Section 20.9) In practice, the element is formed by electrolytic oxidation of a solution of KF in anhydrous HF.

Chlorine is produced mainly by electrolysis of either molten or aqueous sodium chloride. Both bromine and iodine are obtained commercially from brines containing the halide ions; the reaction used is oxidation with Cl_2.

Uses of the Halogens

Fluorine is used to prepare fluorocarbons—very stable carbon–fluorine compounds used as refrigerants, lubricants, and plastics. Teflon® (◀ **Figure 22.9**) is a polymeric fluorocarbon noted for its high thermal stability and lack of chemical reactivity.

Chlorine is by far the most commercially important halogen. About 1×10^{10} kg (10 million tons) of Cl_2 is produced annually in the United States. In addition, hydrogen chloride production is about 4.0×10^9 kg (4.4 million tons) annually. About half of this chlorine finds its way eventually into the manufacture of chlorine-containing organic compounds, such as the vinyl chloride (C_2H_3Cl) used in making polyvinyl chloride (PVC) plastics. ∞ (Section 12.8) Much of the remainder is used as a bleaching agent in the paper and textile industries.

When Cl_2 dissolves in cold dilute base, it converts into Cl^- and hypochlorite, ClO^-:

$$Cl_2(aq) + 2\,OH^-(aq) \rightleftharpoons Cl^-(aq) + ClO^-(aq) + H_2O(l) \quad [22.19]$$

Sodium hypochlorite (NaClO) is the active ingredient in many liquid bleaches. Chlorine is also used in water treatment to oxidize and thereby destroy bacteria. ∞ (Section 18.4)

▲ Figure 22.9 Structure of Teflon®, a fluorocarbon polymer.

 Give It Some Thought

What is the oxidation state of Cl in each Cl species in Equation 22.19?

A common use of iodine is as KI in table salt. Iodized salt provides the small amount of iodine necessary in our diets; it is essential for the formation of thyroxin, a hormone secreted by the thyroid gland. Lack of iodine in the diet results in an enlarged thyroid gland, a condition called *goiter*.

The Hydrogen Halides

All the halogens form stable diatomic molecules with hydrogen. Aqueous solutions of HCl, HBr, and HI are strong acids. The hydrogen halides can be formed by direct reaction of the elements. The most important means of preparing HF and HCl, however, is by reacting a salt of the halide with a strong nonvolatile acid, as in the reaction

$$CaF_2(s) + H_2SO_4(l) \xrightarrow{\Delta} 2\,HF(g) + CaSO_4(s) \quad [22.20]$$

Neither HBr nor HI can be prepared in this way, however, because H_2SO_4 oxidizes Br^- and I^- (▶ **Figure 22.10**). This difference in reactivity reflects the greater ease of oxidation of Br^- and I^- relative to F^- and Cl^-. These undesirable oxidations are avoided by using a nonvolatile acid, such as H_3PO_4, that is a weaker oxidizing agent than H_2SO_4.

GO FIGURE

Are these reactions acid–base reactions or oxidation–reduction reactions?

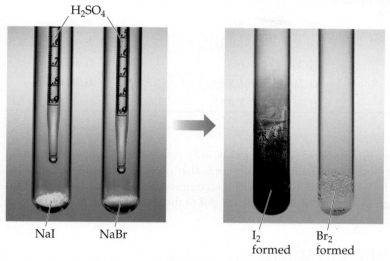

H$_2$SO$_4$

NaI   NaBr   I$_2$ formed   Br$_2$ formed

▲ Figure 22.10 Reaction of H$_2$SO$_4$ with NaI and NaBr.

SAMPLE EXERCISE 22.5 Writing a Balanced Chemical Equation

Write a balanced equation for the formation of hydrogen bromide gas from the reaction of solid sodium bromide with phosphoric acid.

SOLUTION

Analyze We are asked to write a balanced equation for the reaction between NaBr and H$_3$PO$_4$ to form HBr and another product.

Plan As in Equation 22.20, a metathesis reaction takes place. ∞ (Section 4.2) Let's assume that only one H in H$_3$PO$_4$ reacts. (The actual number depends on the reaction conditions.) The H$_2$PO$_4^-$ and Na$^+$ will form NaH$_2$PO$_4$ as one product.

Solve The balanced equation is

$$NaBr(s) + H_3PO_4(l) \longrightarrow NaH_2PO_4(s) + HBr(g)$$

Practice Exercise 1

Which of the following are oxidized by H$_2$SO$_4$?
(a) Cl$^-$, **(b)** Cl$^-$ and Br$^-$, **(c)** Br$^-$ and I$^-$, **(d)** Cl$_2$, **(e)** Br$_2$ and I$_2$.

Practice Exercise 2

Write the balanced equation for the preparation of HI from NaI and H$_3$PO$_4$.

The hydrogen halides form hydrohalic acid solutions when dissolved in water. These solutions have the characteristic properties of acids, such as reactions with active metals to produce hydrogen gas. ∞ (Section 4.4) Hydrofluoric acid also reacts readily with **silica** (SiO$_2$) and with silicates to form hexafluorosilicic acid (H$_2$SiF$_6$):

$$SiO_2(s) + 6\,HF(aq) \longrightarrow H_2SiF_6(aq) + 2\,H_2O(l) \qquad [22.21]$$

Interhalogen Compounds

Because the halogens exist as diatomic molecules, diatomic molecules made up of two different halogen atoms exist. These compounds are the simplest examples of **interhalogens**, compounds, such as ClF and IF$_5$, formed between two halogen elements.

The vast majority of the higher interhalogen compounds have a central Cl, Br, or I atom surrounded by fluorine atoms. The large size of the iodine atom allows the formation of IF$_3$, IF$_5$, and IF$_7$, in which the oxidation state of I is +3, +5, and +7, respectively.

Table 22.3 The Stable Oxyacids of the Halogens

Oxidation State of Halogen	Formula of Acid			Acid Name
	Cl	Br	I	
+1	HClO	HBrO	HIO	*Hypohalous* acid
+3	$HClO_2$	—	—	*Halous* acid
+5	$HClO_3$	$HBrO_3$	HIO_3	Hal*ic* acid
+7	$HClO_4$	$HBrO_4$	HIO_4	*Perhalic* acid

With the smaller bromine and chlorine atoms, only compounds with 3 or 5 fluorines form. The only higher interhalogen compounds that do not have outer F atoms are ICl_3 and ICl_5; the large size of the I atom can accommodate 5 Cl atoms, whereas Br is not large enough to allow even $BrCl_3$ to form. All of the interhalogen compounds are powerful oxidizing agents.

Oxyacids and Oxyanions

▲ Table 22.3 summarizes the formulas of the known oxyacids of the halogens and the way they are named*. ∞ (Section 2.8) The acid strengths of the oxyacids increase with increasing oxidation state of the central halogen atom. ∞ (Section 16.10) All the oxyacids are strong oxidizing agents. The oxyanions, formed on removal of H^+ from the oxyacids, are generally more stable than the oxyacids. Hypochlorite salts are used as bleaches and disinfectants because of the powerful oxidizing capabilities of the ClO^- ion. Chlorate salts are similarly very reactive. For example, potassium chlorate is used to make matches and fireworks.

 Give It Some Thought

Which do you expect to be the stronger oxidizing agent, $NaBrO_3$ or $NaClO_3$?

Perchloric acid and its salts are the most stable oxyacids and oxyanions. Dilute solutions of perchloric acid are quite safe, and many perchlorate salts are stable except when heated with organic materials. When heated, however, perchlorates can become vigorous, even violent, oxidizers. Considerable caution should be exercised, therefore, when handling these substances, and it is crucial to avoid contact between perchlorates and readily oxidized material. The use of ammonium perchlorate (NH_4ClO_4) as the oxidizer in the solid booster rockets for the Space Shuttle demonstrates the oxidizing power of perchlorates. The solid propellant contains a mixture of NH_4ClO_4 and powdered aluminum, the reducing agent. Each shuttle launch requires about 6×10^5 kg (700 tons) of NH_4ClO_4 (◀ Figure 22.11).

22.5 | Oxygen

By the middle of the seventeenth century, scientists recognized that air contained a component associated with burning and breathing. That component was not isolated until 1774, however, when English scientist Joseph Priestley discovered oxygen. Lavoisier subsequently named the element *oxygen*, meaning "acid former."

Oxygen is found in combination with other elements in a great variety of compounds—water (H_2O), silica (SiO_2), alumina (Al_2O_3), and the iron oxides

$$10\,Al(s) + 6\,NH_4ClO_4(s) \longrightarrow$$
$$4\,Al_2O_3(s) + 2\,AlCl_3(s)$$
$$+ 12\,H_2O(g) + 3\,N_2(g)$$

The large volume of gases produced provides thrust for the booster rockets

▲ **Figure 22.11 Launch of the Space Shuttle *Columbia* from the Kennedy Space Center.**

---

*Fluorine forms one oxyacid, HOF. Because the electronegativity of fluorine is greater than that of oxygen, we must consider fluorine to be in a −1 oxidation state and oxygen to be in the 0 oxidation state in this compound.

(Fe_2O_3, Fe_3O_4) are obvious examples. Indeed, oxygen is the most abundant element by mass both in Earth's crust and in the human body. ∞ (Section 1.2) It is the oxidizing agent for the metabolism of our foods and is crucial to human life.

Properties of Oxygen

Oxygen has two allotropes, O_2 and O_3. When we speak of molecular oxygen or simply oxygen, it is usually understood that we are speaking of *dioxygen* (O_2), the normal form of the element; O_3 is ozone.

At room temperature, dioxygen is a colorless, odorless gas. It condenses to a liquid at $-183\,°C$ and freezes at $-218\,°C$. It is only slightly soluble in water (0.04 g/L, or 0.001 M at 25 °C), but its presence in water is essential to marine life.

The electron configuration of the oxygen atom is $[He]2s^2 2p^4$. Thus, oxygen can complete its octet of valence electrons either by adding two electrons to form the oxide ion (O^{2-}) or by sharing two electrons. In its covalent compounds, it tends to form either two single bonds, as in H_2O, or a double bond, as in formaldehyde $(H_2C{=}O)$. The O_2 molecule contains a double bond. The bond in O_2 is very strong (bond enthalpy 495 kJ/mol). Oxygen also forms strong bonds with many other elements. Consequently, many oxygen-containing compounds are thermodynamically more stable than O_2. In the absence of a catalyst, however, most reactions of O_2 have high activation energies and thus require high temperatures to proceed at a suitable rate. Once a sufficiently exothermic reaction begins, it may accelerate rapidly, producing a reaction of explosive violence.

Production of Oxygen

Nearly all commercial oxygen is obtained from air. The normal boiling point of O_2 is $-183\,°C$, whereas that of N_2, the other principal component of air, is $-196\,°C$. Thus, when air is liquefied and then allowed to warm, the N_2 boils off, leaving liquid O_2 contaminated mainly by small amounts of N_2 and Ar.

In the laboratory, O_2 can be obtained by heating either aqueous hydrogen peroxide or solid potassium chlorate $(KClO_3)$:

$$2\,KClO_3(s) \longrightarrow 2\,KCl(s) + 3\,O_2(g) \qquad [22.22]$$

Manganese dioxide (MnO_2) catalyzes both reactions.

Much of the O_2 in the atmosphere is replenished through photosynthesis, in which green plants use the energy of sunlight to generate O_2 (along with glucose, $C_6H_{12}O_6$) from atmospheric CO_2:

$$6\,CO_2(g) + 6\,H_2O(l) \longrightarrow C_6H_{12}O_6(aq) + 6\,O_2(g)$$

Uses of Oxygen

In industrial use, oxygen ranks behind only sulfuric acid (H_2SO_4) and nitrogen (N_2). About 3×10^{10} kg (30 million tons) of O_2 is used annually in the United States. It is shipped and stored either as a liquid or in steel containers as a compressed gas. About 70% of the O_2 output, however, is generated where it is needed.

Oxygen is by far the most widely used oxidizing agent in industry. Over half of the O_2 produced is used in the steel industry, mainly to remove impurities from steel. It is also used to bleach pulp and paper. (Oxidation of colored compounds often gives colorless products.) Oxygen is used together with acetylene (C_2H_2) in oxyacetylene welding (▶ **Figure 22.12**). The reaction between C_2H_2 and O_2 is highly exothermic, producing temperatures in excess of 3000 °C.

Ozone

Ozone is a pale blue, poisonous gas with a sharp, irritating odor. Many people can detect as little as 0.01 ppm in air. Exposure to 0.1 to 1 ppm produces headaches, burning eyes, and irritation to the respiratory passages.

$2\,C_2H_2(g) + 5\,O_2(g) \longrightarrow$
$4\,CO_2(g) + 2\,H_2O(g);$
$\Delta H° = -2510\ kJ$

▲ **Figure 22.12 Welding with an oxyacetylene torch.**

The O_3 molecule possesses π electrons that are delocalized over the three oxygen atoms. ∞∞ (Section 8.6) The molecule dissociates readily, forming reactive oxygen atoms:

$$O_3(g) \longrightarrow O_2(g) + O(g) \quad \Delta H° = 105 \text{ kJ} \qquad [22.23]$$

Ozone is a stronger oxidizing agent than dioxygen. Ozone forms oxides with many elements under conditions where O_2 will not react; indeed, it oxidizes all the common metals except gold and platinum.

Ozone can be prepared by passing electricity through dry O_2 in a flow-through apparatus. The electrical discharge causes the O_2 bond to break, resulting in reactions like those described in Section 18.1. During thunderstorms, ozone is generated (and can be smelled, if you are too close) from lightning strikes:

$$3 O_2(g) \xrightarrow{\text{electricity}} 2 O_3(g) \quad \Delta H° = 285 \text{ kJ} \qquad [22.24]$$

Ozone is sometimes used to treat drinking water. Like Cl_2, ozone kills bacteria and oxidizes organic compounds. The largest use of ozone, however, is in the preparation of pharmaceuticals, synthetic lubricants, and other commercially useful organic compounds, where O_3 is used to sever carbon–carbon double bonds.

Ozone is an important component of the upper atmosphere, where it screens out ultraviolet radiation and so protects us from the effects of these high-energy rays. For this reason, depletion of stratospheric ozone is a major scientific concern. ∞∞ (Section 18.2) In the lower atmosphere, ozone is considered an air pollutant and is a major constituent of smog. ∞∞ (Section 18.2) Because of its oxidizing power, ozone damages living systems and structural materials, especially rubber.

Oxides

The electronegativity of oxygen is second only to that of fluorine. As a result, oxygen has negative oxidation states in all compounds except OF_2 and O_2F_2. The -2 oxidation state is by far the most common. Compounds that contain oxygen in this oxidation state are called *oxides*.

Nonmetals form covalent oxides, most of which are simple molecules with low melting and boiling points. Both SiO_2 and B_2O_3, however, have extended structures. Most nonmetal oxides combine with water to give oxyacids. Sulfur dioxide (SO_2), for example, dissolves in water to give sulfurous acid (H_2SO_3):

$$SO_2(g) + H_2O(l) \longrightarrow H_2SO_3(aq) \qquad [22.25]$$

This reaction and that of SO_3 with H_2O to form H_2SO_4 are largely responsible for acid rain. ∞∞ (Section 18.2) The analogous reaction of CO_2 with H_2O to form carbonic acid (H_2CO_3) causes the acidity of carbonated water.

Oxides that form acids when they react with water are called either **acidic anhydrides** (anhydride means "without water") or **acidic oxides**. A few nonmetal oxides, especially ones with the nonmetal in a low oxidation state—such as N_2O, NO, and CO—do not react with water and are not acidic anhydrides.

 Give It Some Thought

What acid is produced by the reaction of I_2O_5 with water?

Most metal oxides are ionic compounds. The ionic oxides that dissolve in water form hydroxides and, consequently, are called either **basic anhydrides** or **basic oxides**. Barium oxide, for example, reacts with water to form barium hydroxide (▶ Figure 22.13). These kinds of reactions are due to the high basicity of the O^{2-} ion and its virtually complete hydrolysis in water:

$$O^{2-}(aq) + H_2O(l) \longrightarrow 2 OH^-(aq) \qquad [22.26]$$

Even those ionic oxides that are insoluble in water tend to dissolve in strong acids. Iron(III) oxide, for example, dissolves in acids:

$$Fe_2O_3(s) + 6 H^+(aq) \longrightarrow 2 Fe^{3+}(aq) + 3 H_2O(l) \qquad [22.27]$$

▲GO FIGURE

Is this reaction a redox reaction?

H₂O
with
indicator

Pink color
indicates
basic
solution

$$BaO(s) \ + \ H_2O(l) \ \longrightarrow \ Ba(OH)_2(aq)$$

▲ Figure 22.13 **Reaction of a basic oxide with water.**

This reaction is used to remove rust $(Fe_2O_3 \cdot nH_2O)$ from iron or steel before a protective coat of zinc or tin is applied.

Oxides that can exhibit both acidic and basic characters are said to be *amphoteric*. ∞ (Section 17.5) If a metal forms more than one oxide, the basic character of the oxide decreases as the oxidation state of the metal increases (▼ Table 22.4).

Table 22.4   Acid–Base Character of Chromium Oxides

Oxide	Oxidation State of Cr	Nature of Oxide
CrO	+2	Basic
Cr_2O_3	+3	Amphoteric
CrO_3	+6	Acidic

Peroxides and Superoxides

Compounds containing O—O bonds and oxygen in the −1 oxidation state are *peroxides*. Oxygen has an oxidation state of $-\frac{1}{2}$ in O_2^-, which is called the *superoxide* ion. The most active (easily oxidized) metals (K, Rb, and Cs) react with O_2 to give superoxides $(KO_2, RbO_2, \text{and } CsO_2)$. Their active neighbors in the periodic table (Na, Ca, Sr, and Ba) react with O_2, producing peroxides $(Na_2O_2, CaO_2, SrO_2, \text{and } BaO_2)$. Less active metals and nonmetals produce normal oxides. ∞ (Section 7.6)

When superoxides dissolve in water, O_2 is produced:

$$4\,KO_2(s) \ + \ 2\,H_2O(l) \ \longrightarrow \ 4\,K^+(aq) \ + \ 4\,OH^-(aq) \ + \ 3\,O_2(g) \qquad [22.28]$$

Because of this reaction, potassium superoxide is used as an oxygen source in masks worn by rescue workers (▶ Figure 22.14). For proper breathing in toxic environments, oxygen must be generated in the mask and exhaled carbon dioxide in the mask must be eliminated. Moisture in the breath causes the KO_2 to decompose to O_2 and KOH, and the KOH removes CO_2 from the exhaled breath:

$$2\,OH^-(aq) \ + \ CO_2(g) \ \longrightarrow \ H_2O(l) \ + \ CO_3^{2-}(aq) \qquad [22.29]$$

$$4\,KO_2(s) + 2\,H_2O(l, \text{from breath}) \longrightarrow$$
$$4\,K^+(aq) + 4\,OH^-(aq) + 3\,O_2(g)$$

$$2\,OH^-(aq) + CO_2(g, \text{from breath}) \longrightarrow$$
$$H_2O(l) + CO_3^{2-}(aq)$$

▲ Figure 22.14 **A self-contained breathing apparatus.**

Does H_2O_2 have a dipole moment?

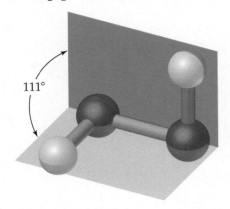

111°

▲ **Figure 22.15 Molecular structure of hydrogen peroxide.** The repulsive interaction of the O—H bonds with the lone-pairs of electrons on each O atom restricts the free rotation around the O—O single bond.

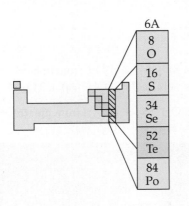

6A
8 O
16 S
34 Se
52 Te
84 Po

▲ **Figure 22.16 Massive amounts of sulfur are extracted every year from the earth.**

Hydrogen peroxide (◄ **Figure 22.15**) is the most familiar and commercially important peroxide. Pure hydrogen peroxide is a clear, syrupy liquid that melts at −0.4 °C. Concentrated hydrogen peroxide is dangerously reactive because the decomposition to water and oxygen is very exothermic:

$$2\,H_2O_2(l) \longrightarrow 2\,H_2O(l) + O_2(g) \quad \Delta H° = -196.1\ kJ \qquad [22.30]$$

This is another example of a **disproportionation** reaction, in which an element is simultaneously oxidized and reduced. The oxidation number of oxygen changes from −1 to −2 and 0.

Hydrogen peroxide is marketed as a chemical reagent in aqueous solutions of up to about 30% by mass. A solution containing about 3% H_2O_2 by mass is sold in drugstores and used as a mild antiseptic. Somewhat more concentrated solutions are used to bleach fabrics.

The peroxide ion is a by-product of metabolism that results from the reduction of O_2.

The body disposes of this reactive ion with enzymes such as peroxidase and catalase.

22.6 | The Other Group 6A Elements: S, Se, Te, and Po

The other group 6A elements are sulfur, selenium, tellurium, and polonium. In this section, we will survey the properties of the group as a whole and then examine the chemistry of sulfur, selenium, and tellurium. We will not discuss polonium, which has no stable isotopes and is found only in minute quantities in radium-containing minerals.

General Characteristics of the Group 6A Elements

The group 6A elements possess the general outer-electron configuration ns^2np^4 with n ranging from 2 to 6. Thus, these elements attain a noble-gas electron configuration by adding two electrons, which results in a −2 oxidation state. Except for oxygen, the group 6A elements are also commonly found in positive oxidation states up to +6, and they can have expanded valence shells. Thus, we have such compounds as SF_6, SeF_6, and TeF_6 with the central atom in the +6 oxidation state.

▼ **Table 22.5** summarizes some properties of the group 6A elements.

Occurrence and Production of S, Se, and Te

Sulfur, selenium, and tellurium can all be mined from the earth. Large underground deposits are the principal source of elemental sulfur (◄ **Figure 22.16**). Sulfur also occurs widely as sulfide (S^{2-}) and sulfate (SO_4^{2-}) minerals. Its presence as a minor component of coal and petroleum poses a major problem. Combustion of these "unclean"

Table 22.5 Some Properties of the Group 6A Elements

Property	O	S	Se	Te
Atomic radius (Å)	0.66	1.05	1.21	1.38
X^{2-} ionic radius (Å)	1.40	1.84	1.98	2.21
First ionization energy (kJ/mol)	1314	1000	941	869
Electron affinity (kJ/mol)	−141	−200	−195	−190
Electronegativity	3.5	2.5	2.4	2.1
X—X single-bond enthalpy (kJ/mol)	146*	266	172	126
Reduction potential to H_2X in acidic solution (V)	1.23	0.14	−0.40	−0.72

*Based on O—O bond energy in H_2O_2.

fuels leads to serious pollution by sulfur oxides. ∞∞ (Section 18.2) Much effort has been directed at removing this sulfur, and these efforts have increased the availability of sulfur.

Selenium and tellurium occur in rare minerals, such as Cu_2Se, PbSe, Cu_2Te, and PbTe, and as minor constituents in sulfide ores of copper, iron, nickel, and lead.

Properties and Uses of Sulfur, Selenium, and Tellurium

Elemental sulfur is yellow, tasteless, and nearly odorless. It is insoluble in water and exists in several allotropic forms. The thermodynamically stable form at room temperature is rhombic sulfur, which consists of puckered S_8 rings with each sulfur atom forming two bonds (Figure 7.26). Rhombic sulfur melts at 113 °C.

Most of the approximately 1×10^{10} kg (10 million tons) of sulfur produced in the United States each year is used to manufacture sulfuric acid. Sulfur is also used to vulcanize rubber, a process that toughens rubber by introducing cross-linking between polymer chains. ∞∞ (Section 12.8)

Selenium and tellurium do not form eight-membered rings in their elemental forms. ∞∞ (Section 7.8) The most stable allotropes of these elements are crystalline substances containing helical chains of atoms (▶ Figure 22.17). In all the allotropes each atom forms two bonds to its neighbors. Each atom is close to atoms in adjacent chains, and it appears that some sharing of electron pairs between these atoms occurs.

The electrical conductivity of elemental selenium is low in the dark but increases greatly upon exposure to light. This property is exploited in photoelectric cells and light meters. Photocopiers also depend on the photoconductivity of selenium. Photocopy machines contain a belt or drum coated with a film of selenium. This drum is electrostatically charged and then exposed to light reflected from the image being photocopied. The charge drains from the regions where the selenium film has been made conductive by exposure to light. A black powder (the toner) sticks only to the areas that remain charged. The photocopy is made when the toner is transferred to a sheet of plain paper.

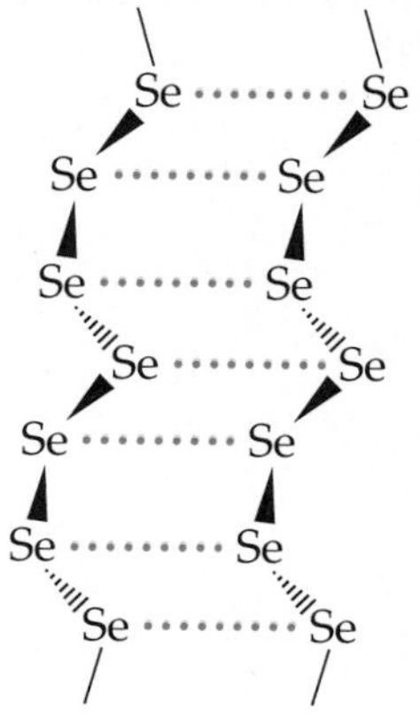

▲ **Figure 22.17 Portion of helical chains making up the structure of crystalline selenium.**

Sulfides

When an element is less electronegative than sulfur, *sulfides* that contain S^{2-} form. Many metallic elements are found in the form of sulfide ores, such as PbS (galena) and HgS (cinnabar). A series of related ores containing the disulfide ion, S_2^{2-} (analogous to the peroxide ion), are known as *pyrites*. Iron pyrite, FeS_2, occurs as golden yellow cubic crystals (▶ Figure 22.18). Because it has been occasionally mistaken for gold by miners, iron pyrite is often called fool's gold.

One of the most important sulfides is hydrogen sulfide (H_2S). This substance is normally prepared by action of dilute acid on iron(II) sulfide:

$$FeS(s) + 2\,H^+(aq) \longrightarrow H_2S(aq) + Fe^{2+}(aq) \qquad [22.31]$$

▲ **Figure 22.18 Iron pyrite (FeS_2, on the right) with gold for comparison.**

One of hydrogen sulfide's most readily recognized properties is its odor, which is most frequently encountered as the offensive odor of rotten eggs. Hydrogen sulfide is toxic but our noses can detect H_2S in extremely low, nontoxic concentrations. A sulfur-containing organic molecule, such as dimethyl sulfide, $(CH_3)_2S$, which is similarly odoriferous and can be detected by smell at a level of one part per trillion, is added to natural gas as a safety factor to give it a detectable odor.

Oxides, Oxyacids, and Oxyanions of Sulfur

Sulfur dioxide, formed when sulfur burns in air, has a choking odor and is poisonous. The gas is particularly toxic to lower organisms, such as fungi, so it is used to sterilize dried fruit and wine. At 1 atm and room temperature, SO_2 dissolves in water to produce a 1.6 M solution. The SO_2 solution is acidic, and we describe it as sulfurous acid (H_2SO_3).

▲ **Figure 22.19 Food label warning of sulfites.**

Salts of SO_3^{2-} (sulfites) and HSO_3^- (hydrogen sulfites or bisulfites) are well known. Small quantities of Na_2SO_3 or $NaHSO_3$ are used as food additives to prevent bacterial spoilage. However, they are known to increase asthma symptoms in approximately 5% of asthmatics. Thus, all food products with sulfites must now carry a warning label disclosing their presence (◀ Figure 22.19).

Although combustion of sulfur in air produces mainly SO_2, small amounts of SO_3 are also formed. The reaction produces chiefly SO_2 because the activation-energy barrier for oxidation to SO_3 is very high unless the reaction is catalyzed. Interestingly, the SO_3 by-product is used industrially to make H_2SO_4, which is the ultimate product of the reaction between SO_3 and water. In the manufacture of sulfuric acid, SO_2 is obtained by burning sulfur and then oxidized to SO_3, using a catalyst such as V_2O_5 or platinum. The SO_3 is dissolved in H_2SO_4 because it does not dissolve quickly in water, and then the $H_2S_2O_7$ formed in this reaction, called pyrosulfuric acid, is added to water to form H_2SO_4:

$$SO_3(g) + H_2SO_4(l) \longrightarrow H_2S_2O_7(l) \qquad [22.32]$$

$$H_2S_2O_7(l) + H_2O(l) \longrightarrow 2\,H_2SO_4(l) \qquad [22.33]$$

 Give It Some Thought

What is the net reaction of Equations 22.32 and 22.33?

Commercial sulfuric acid is 98% H_2SO_4. It is a dense, colorless, oily liquid that boils at 340 °C. It is a strong acid, a good dehydrating agent (▼ Figure 22.20), and a moderately good oxidizing agent.

Year after year, the production of sulfuric acid is the largest of any chemical produced in the United States. About 4×10^{10} kg (40 million tons) is produced annually in this country. Sulfuric acid is employed in some way in almost all manufacturing.

Sulfuric acid is a strong acid, but only the first hydrogen is completely ionized in aqueous solution:

$$H_2SO_4(aq) \longrightarrow H^+(aq) + HSO_4^-(aq) \qquad [22.34]$$

$$HSO_4^-(aq) \rightleftharpoons H^+(aq) + SO_4^{2-}(aq) \quad K_a = 1.1 \times 10^{-2} \qquad [22.35]$$

▲ **GO FIGURE**

In this reaction, what has happened to the H and O atoms in the sucrose?

H_2SO_4

Table sugar (sucrose), $C_{12}H_{22}O_{11}$

Pure carbon, C

▲ **Figure 22.20 Sulfuric acid dehydrates table sugar to produce elemental carbon.**

Consequently, sulfuric acid forms both sulfates (SO_4^{2-} salts) and bisulfates (or hydrogen sulfates, HSO_4^- salts). Bisulfate salts are common components of the "dry acids" used for adjusting the pH of swimming pools and hot tubs; they are also components of many toilet bowl cleaners.

The term *thio* indicates substitution of sulfur for oxygen, and the thiosulfate ion ($S_2O_3^{2-}$) is formed by boiling an alkaline solution of SO_3^{2-} with elemental sulfur:

$$8\,SO_3^{2-}(aq) + S_8(s) \longrightarrow 8\,S_2O_3^{2-}(aq) \qquad [22.36]$$

The structures of the sulfate and thiosulfate ions are compared in ▶ Figure 22.21.

▲ **GO FIGURE**

What are the oxidation states of the sulfur atoms in the $S_2O_3^{2-}$ ion?

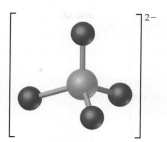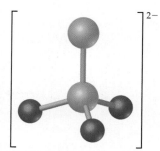

▲ **Figure 22.21 Structures of the sulfate (left) and thiosulfate (right) ions.**

22.7 | Nitrogen

Nitrogen constitutes 78% by volume of Earth's atmosphere, where it occurs as N_2 molecules. Although nitrogen is a key element in living organisms, compounds of nitrogen are not abundant in Earth's crust. The major natural deposits of nitrogen compounds are those of KNO_3 (saltpeter) in India and $NaNO_3$ (Chile saltpeter) in Chile and other desert regions of South America.

Properties of Nitrogen

Nitrogen is a colorless, odorless, tasteless gas composed of N_2 molecules. Its melting point is $-210\,°C$, and its normal boiling point is $-196\,°C$.

The N_2 molecule is very unreactive because of the strong triple bond between nitrogen atoms (the $N{\equiv}N$ bond enthalpy is 941 kJ/mol, nearly twice that for the bond in O_2. ∞ (Table 8.4) When substances burn in air, they normally react with O_2 but not with N_2.

The electron configuration of the nitrogen atom is $[He]2s^2 2p^3$. The element exhibits all formal oxidation states from +5 to −3 (▶ **Table 22.6**). The +5, 0, and −3 oxidation states are the most common and generally the most stable of these. Because nitrogen is more electronegative than all other elements except fluorine, oxygen, and chlorine, it exhibits positive oxidation states only in combination with these three elements.

Table 22.6 Oxidation States of Nitrogen

Oxidation State	Examples
+5	N_2O_5, HNO_3, NO_3^-
+4	NO_2, N_2O_4
+3	HNO_2, NO_2^-, NF_3
+2	NO
+1	N_2O, $H_2N_2O_2$, $N_2O_2^{2-}$, HNF_2
0	N_2
−1	NH_2OH, NH_2F
−2	N_2H_4
−3	NH_3, NH_4^+, NH_2^-

Production and Uses of Nitrogen

Elemental nitrogen is obtained in commercial quantities by fractional distillation of liquid air. About 4×10^{10} kg (40 million tons) of N_2 is produced annually in the United States.

Because of its low reactivity, large quantities of N_2 are used as an inert gaseous blanket to exclude O_2 in food processing, manufacture of chemicals, metal fabrication, and production of electronic devices. Liquid N_2 is employed as a coolant to freeze foods rapidly.

The largest use of N_2 is in the manufacture of nitrogen-containing fertilizers, which provide a source of *fixed* nitrogen. We have previously discussed nitrogen fixation in the "Chemistry and Life" box in Section 14.7 and in the "Chemistry Put to Work" box in Section 15.2. Our starting point in fixing nitrogen is the manufacture of ammonia via the Haber process. ∞ (Section 15.2) The ammonia can then be converted into a variety of useful, simple nitrogen-containing species (**Figure 22.22**).

Hydrogen Compounds of Nitrogen

Ammonia is one of the most important compounds of nitrogen. It is a colorless, toxic gas that has a characteristic irritating odor. As noted in previous discussions, the NH_3 molecule is basic ($K_b = 1.8 \times 10^{-5}$). ∞ (Section 16.7)

GO FIGURE

In which of these species is the oxidation number of nitrogen +3?

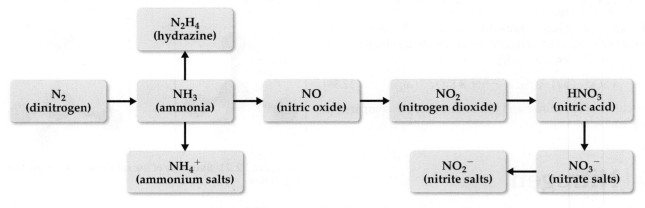

▲ Figure 22.22 **Sequence of conversion of N_2 into common nitrogen compounds.**

 GO FIGURE

Is the N—N bond length in these molecules shorter or longer than the N—N bond length in N_2?

▲ Figure 22.23 **Hydrazine (top, N_2H_4) and methylhydrazine (bottom, CH_3NHNH_2).**

In the laboratory, NH_3 can be prepared by the action of NaOH on an ammonium salt. The NH_4^+ ion, which is the conjugate acid of NH_3, transfers a proton to OH^-. The resultant NH_3 is volatile and is driven from the solution by mild heating:

$$NH_4Cl(aq) + NaOH(aq) \longrightarrow NH_3(g) + H_2O(l) + NaCl(aq) \quad [22.37]$$

Commercial production of NH_3 is achieved by the Haber process:

$$N_2(g) + 3\,H_2(g) \longrightarrow 2\,NH_3(g) \quad [22.38]$$

About 1×10^{10} kg (10 million tons) of ammonia is produced annually in the United States. About 75% is used for fertilizer.

Hydrazine (N_2H_4) is another important hydride of nitrogen. The hydrazine molecule contains an N—N single bond (◀ Figure 22.23). Hydrazine is quite poisonous. It can be prepared by the reaction of ammonia with hypochlorite ion (OCl^-) in aqueous solution:

$$2\,NH_3(aq) + OCl^-(aq) \longrightarrow N_2H_4(aq) + Cl^-(aq) + H_2O(l) \quad [22.39]$$

The reaction involves several intermediates, including chloramine (NH_2Cl). The poisonous NH_2Cl bubbles out of solution when household ammonia and chlorine bleach (which contains OCl^-) are mixed. This reaction is one reason for the frequently cited warning not to mix bleach and household ammonia.

Pure hydrazine is a strong and versatile reducing agent. The major use of hydrazine and compounds related to it, such as methylhydrazine (Figure 22.23), is as rocket fuel.

SAMPLE
EXERCISE 22.6 **Writing a Balanced Equation**

Hydroxylamine (NH_2OH) reduces copper(II) to the free metal in acid solutions. Write a balanced equation for the reaction, assuming that N_2 is the oxidation product.

SOLUTION

Analyze We are asked to write a balanced oxidation–reduction equation in which NH_2OH is converted to N_2 and Cu^{2+} is converted to Cu.

Plan Because this is a redox reaction, the equation can be balanced by the method of half-reactions discussed in Section 20.2. Thus, we begin with two half-reactions, one involving the NH_2OH and N_2 and the other involving Cu^{2+} and Cu.

Solve The unbalanced and incomplete half-reactions are

$$Cu^{2+}(aq) \longrightarrow Cu(s)$$
$$NH_2OH(aq) \longrightarrow N_2(g)$$
$$Cu^{2+}(aq) + 2\,e^- \longrightarrow Cu(s)$$

Balancing these equations as described in Section 20.2 gives

$$2\,NH_2OH(aq) \longrightarrow N_2(g) + 2\,H_2O(l) + 2\,H^+(aq) + 2\,e^-$$

Adding these half-reactions gives the balanced equation:

$$Cu^{2+}(aq) + 2\,NH_2OH(aq) \longrightarrow Cu(s) + N_2(g) + 2\,H_2O(l) + 2\,H^+(aq)$$

Practice Exercise 1

In power plants, hydrazine is used to prevent corrosion of the metal parts of steam boilers by the O_2 dissolved in the water. The hydrazine reacts with O_2 in water to give N_2 and H_2O. Write a balanced equation for this reaction.

Practice Exercise 2

Methylhydrazine, $N_2H_3CH_3(l)$, is used with the oxidizer dinitrogen tetroxide, $N_2O_4(l)$, to power the steering rockets of the Space Shuttle orbiter. The reaction of these two substances produces N_2, CO_2, and H_2O. Write a balanced equation for this reaction.

Oxides and Oxyacids of Nitrogen

Nitrogen forms three common oxides: N_2O (nitrous oxide), NO (nitric oxide), and NO_2 (nitrogen dioxide). It also forms two unstable oxides that we will not discuss, N_2O_3 (dinitrogen trioxide) and N_2O_5 (dinitrogen pentoxide).

Nitrous oxide (N_2O) is also known as laughing gas because a person becomes giddy after inhaling a small amount. This colorless gas was the first substance used as a general anesthetic. It is used as the compressed gas propellant in several aerosols and foams, such as in whipped cream. It can be prepared in the laboratory by carefully heating ammonium nitrate to about 200 °C:

$$NH_4NO_3(s) \xrightarrow{\Delta} N_2O(g) + 2\,H_2O(g) \qquad [22.40]$$

Nitric oxide (NO) is also a colorless gas but, unlike N_2O, it is slightly toxic. It can be prepared in the laboratory by reduction of dilute nitric acid, using copper or iron as a reducing agent:

$$3\,Cu(s) + 2\,NO_3^-(aq) + 8\,H^+(aq) \longrightarrow 3\,Cu^{2+}(aq) + 2\,NO(g) + 4\,H_2O(l) \quad [22.41]$$

Nitric oxide is also produced by direct reaction of N_2 and O_2 at high temperatures. This reaction is a significant source of nitrogen oxide air pollutants. ∞ (Section 18.2) The direct combination of N_2 and O_2 is not used for commercial production of NO, however, because the yield is low, the equilibrium constant K_p at 2400 K being only 0.05. ∞ (Section 15.7, "Chemistry Put to Work: Controlling Nitric Oxide Emissions")

The commercial route to NO (and hence to other oxygen-containing compounds of nitrogen) is via the catalytic oxidation of NH_3:

$$4\,NH_3(g) + 5\,O_2(g) \xrightarrow[850\,°C]{Pt\ catalyst} 4\,NO(g) + 6\,H_2O(g) \qquad [22.42]$$

This reaction is the first step in the **Ostwald process**, by which NH_3 is converted commercially into nitric acid (HNO_3).

When exposed to air, nitric oxide reacts readily with O_2 (▼ Figure 22.24):

$$2\,NO(g) + O_2(g) \longrightarrow 2\,NO_2(g) \qquad [22.43]$$

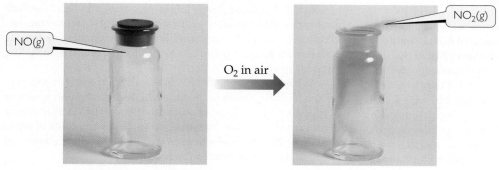

▲ Figure 22.24 **Formation of NO₂(g) as NO(g) combines with O₂(g) in the air.**

GO FIGURE

Which is the shortest NO bond in these two molecules?

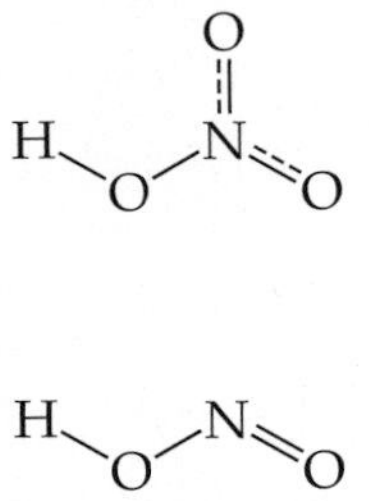

▲ Figure 22.25 **Structures of nitric acid (top) and nitrous acid (bottom).**

When dissolved in water, NO_2 forms nitric acid:

$$3\,NO_2(g) + H_2O(l) \longrightarrow 2\,H^+(aq) + 2\,NO_3^-(aq) + NO(g) \qquad [22.44]$$

Nitrogen is both oxidized and reduced in this reaction, which means it disproportionates. The NO can be converted back into NO_2 by exposure to air (Equation 22.43) and thereafter dissolved in water to prepare more HNO_3.

NO is an important neurotransmitter in the human body. It causes the muscles that line blood vessels to relax, thus allowing an increased passage of blood (see the "Chemistry and Life").

Nitrogen dioxide (NO_2) is a yellow-brown gas (Figure 22.24). Like NO, it is a major constituent of smog. ∞ (Section 18.2) It is poisonous and has a choking odor. As discussed in Section 15.1, NO_2 and N_2O_4 exist in equilibrium:

$$2\,NO_2(g) \rightleftharpoons N_2O_4(g) \qquad \Delta H^\circ = -58\text{ kJ} \qquad [22.45]$$

The two common oxyacids of nitrogen are nitric acid (HNO_3) and nitrous acid (HNO_2) (◀ Figure 22.25). *Nitric acid* is a strong acid. It is also a powerful oxidizing agent, as indicated by the standard reduction potential in the reaction

$$NO_3^-(aq) + 4\,H^+(aq) + 3\,e^- \longrightarrow NO(g) + 2\,H_2O(l) \qquad E^\circ = +0.96\text{ V} \qquad [22.46]$$

Concentrated nitric acid attacks and oxidizes most metals except Au, Pt, Rh, and Ir.

About 7×10^9 kg (8 million tons) of nitric acid is produced annually in the United States. Its largest use is in the manufacture of NH_4NO_3 for fertilizers. It is also used in the production of plastics, drugs, and explosives. Among the explosives made from nitric acid are nitroglycerin, trinitrotoluene (TNT), and nitrocellulose. The following reaction occurs when nitroglycerin explodes:

$$4\,C_3H_5N_3O_9(l) \longrightarrow 6\,N_2(g) + 12\,CO_2(g) + 10\,H_2O(g) + O_2(g) \qquad [22.47]$$

All the products of this reaction contain very strong bonds and are gases. As a result, the reaction is very exothermic, and the volume of the products is far larger than the volume occupied by the reactant. Thus, the expansion resulting from the heat generated by the reaction produces the explosion. ∞ (Section 8.8, "Explosives and Alfred Nobel")

Nitrous acid is considerably less stable than HNO_3 and tends to disproportionate into NO and HNO_3. It is normally made by action of a strong acid, such as H_2SO_4, on a cold solution of a nitrite salt, such as $NaNO_2$. Nitrous acid is a weak acid ($K_a = 4.5 \times 10^{-4}$).

Give It Some Thought

What are the oxidation numbers of the nitrogen atoms in
(a) nitric acid
(b) nitrous acid

Chemistry and Life

Nitroglycerin, Nitric Oxide, and Heart Disease

During the 1870s, an interesting observation was made in Alfred Nobel's dynamite factories. Workers who suffered from heart disease that caused chest pains when they exerted themselves found relief from the pains during the workweek. It quickly became apparent that nitroglycerin, present in the air of the factory, acted to enlarge blood vessels. Thus, this powerfully explosive chemical became a standard treatment for angina pectoris, the chest pains accompanying heart failure. It took more than 100 yrs to discover that nitroglycerin was converted in the vascular smooth muscle into NO, which is the chemical agent actually causing dilation of the blood vessels.

In 1998, the Nobel Prize in Physiology or Medicine was awarded to Robert F. Furchgott, Louis J. Ignarro, and Ferid Murad for their discoveries of the detailed pathways by which NO acts in the cardiovascular system. It was a sensation that this simple, common air pollutant could exert important functions in mammals, including humans.

As useful as nitroglycerin is to this day in treating angina pectoris, it has a limitation in that prolonged administration results in development of tolerance, or desensitization, of the vascular muscle to further vasorelaxation by nitroglycerin. The bioactivation of nitroglycerin is the subject of active research in the hope that a means of circumventing desensitization can be found.

22.8 | The Other Group 5A Elements: P, As, Sb, and Bi

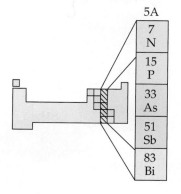

Of the other group 5A elements—phosphorus, arsenic, antimony, and bismuth—phosphorus has a central role in several aspects of biochemistry and environmental chemistry.

General Characteristics of the Group 5A Elements

The group 5A elements have the outer-shell electron configuration ns^2np^3, with n ranging from 2 to 6. A noble-gas configuration is achieved by adding three electrons to form the -3 oxidation state. Ionic compounds containing X^{3-} ions are not common, however. More commonly, the group 5A element acquires an octet of electrons via covalent bonding and oxidation numbers ranging from -3 to $+5$.

Because of its lower electronegativity, phosphorus is found more frequently in positive oxidation states than is nitrogen. Furthermore, compounds in which phosphorus has the $+5$ oxidation state are not as strongly oxidizing as the corresponding compounds of nitrogen. Compounds in which phosphorus has a -3 oxidation state are much stronger reducing agents than are the corresponding nitrogen compounds.

Some properties of the group 5A elements are listed in ▼ **Table 22.7**. The general pattern is similar to what we saw with other groups: Size and metallic character increase as atomic number increases in the group.

The variation in properties among group 5A elements is more striking than that seen in groups 6A and 7A. Nitrogen at the one extreme exists as a gaseous diatomic molecule, clearly nonmetallic. At the other extreme, bismuth is a reddish white, metallic-looking substance that has most of the characteristics of a metal.

The values listed for X—X single-bond enthalpies are not reliable because it is difficult to obtain such data from thermochemical experiments. However, there is no doubt about the general trend: a low value for the N—N single bond, an increase at phosphorus, and then a gradual decline to arsenic and antimony. From observations of the elements in the gas phase, it is possible to estimate the X≡X triple-bond enthalpies. Here, we see a trend that is different from that for the X—X single bond. Nitrogen forms a much stronger triple bond than the other elements, and there is a steady decline in the triple-bond enthalpy down through the group. These data help us to appreciate why nitrogen alone of the group 5A elements exists as a diatomic molecule in its stable state at 25 °C. All the other elements exist in structural forms with single bonds between the atoms.

Occurrence, Isolation, and Properties of Phosphorus

Phosphorus occurs mainly in the form of phosphate minerals. The principal source of phosphorus is phosphate rock, which contains phosphate principally as $Ca_3(PO_4)_2$. The element is produced commercially by the reduction of calcium phosphate with carbon in the presence of SiO_2:

$$2\,Ca_3(PO_4)_2(s) + 6\,SiO_2(s) + 10\,C(s) \xrightarrow{1500\,°C} P_4(g) + 6\,CaSiO_3(l) + 10\,CO(g)$$

[22.48]

Table 22.7 Properties of the Group 5A Elements

Property	N	P	As	Sb	Bi
Atomic radius (Å)	0.71	1.07	1.19	1.39	1.48
First ionization energy (kJ/mol)	1402	1012	947	834	703
Electron affinity (kJ/mol)	> 0	−72	−78	−103	−91
Electronegativity	3.0	2.1	2.0	1.9	1.9
X—X single-bond enthalpy (kJ/mol)*	163	200	150	120	—
X≡X triple-bond enthalpy (kJ/mol)	941	490	380	295	192

*Approximate values only.

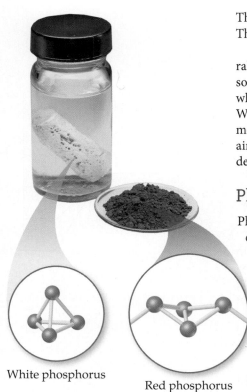

White phosphorus

Red phosphorus

▲ **Figure 22.26 White and red phosphorus.**
Despite the fact that both contain only
phosphorus atoms, these two forms of
phosphorus differ greatly in reactivity. The
white allotrope, which reacts violently with
oxygen, must be stored under water so that it
is not exposed to air. The much less reactive
red form does not need to be stored this way.

The phosphorus produced in this fashion is the allotrope known as white phosphorus. This form distills from the reaction mixture as the reaction proceeds.

Phosphorus exists in several allotropic forms. White phosphorus consists of P_4 tetrahedra (◄ Figure 22.26). The bond angles in this molecule, 60°, are unusually small, so there is much strain in the bonding, which is consistent with the high reactivity of white phosphorus. This allotrope bursts spontaneously into flames if exposed to air. When heated in the absence of air to about 400 °C, white phosphorus is converted to a more stable allotrope known as red phosphorus, which does not ignite on contact with air. Red phosphorus is also considerably less poisonous than the white form. We will denote elemental phosphorus as simply $P(s)$.

Phosphorus Halides

Phosphorus forms a wide range of compounds with the halogens, the most important of which are the trihalides and pentahalides. Phosphorus trichloride (PCl_3) is commercially the most significant of these compounds and is used to prepare a wide variety of products, including soaps, detergents, plastics, and insecticides.

Phosphorus chlorides, bromides, and iodides can be made by direct oxidation of elemental phosphorus with the elemental halogen. PCl_3, for example, which is a liquid at room temperature, is made by passing a stream of dry chlorine gas over white or red phosphorus:

$$2 P(s) + 3 Cl_2(g) \longrightarrow 2 PCl_3(l) \qquad [22.49]$$

If excess chlorine gas is present, an equilibrium is established between PCl_3 and PCl_5.

$$PCl_3(l) + Cl_2(g) \rightleftharpoons PCl_5(s) \qquad [22.50]$$

The phosphorus halides react readily with water, and most fume in air because of reaction with water vapor. In the presence of excess water, the products are the corresponding phosphorus oxyacid and hydrogen halide:

$$PBr_3(l) + 3 H_2O(l) \longrightarrow H_3PO_3(aq) + 3 HBr(aq) \qquad [22.51]$$

$$PCl_5(l) + 4 H_2O(l) \longrightarrow H_3PO_4(aq) + 5 HCl(aq) \qquad [22.52]$$

 Give It Some Thought

Which oxyacid is produced when PF_3 reacts with water?

Oxy Compounds of Phosphorus

Probably the most significant phosphorus compounds are those in which the element is combined with oxygen. Phosphorus(III) oxide (P_4O_6) is obtained by allowing white phosphorus to oxidize in a limited supply of oxygen. When oxidation takes place in the presence of excess oxygen, phosphorus(V) oxide (P_4O_{10}) forms. This compound is also readily formed by oxidation of P_4O_6. These two oxides represent the two most common oxidation states for phosphorus, +3 and +5. The structural relationship between P_4O_6 and P_4O_{10} is shown in ▶ Figure 22.27. Notice the resemblance these molecules have to the P_4 molecule (Figure 22.27); all three substances have a P_4 core.

Phosphorus(V) oxide is the anhydride of phosphoric acid (H_3PO_4), a weak triprotic acid. In fact, P_4O_{10} has a very high affinity for water and is consequently used as a drying agent. Phosphorus(III) oxide is the anhydride of phosphorous acid (H_3PO_3), a weak diprotic acid (▶ Figure 22.28).*

---

*Note that the element phosphorus (*FOS · for · us*) has a *-us* suffix, whereas the first word in the name phosphorous (*fos · FOR · us*) acid has an *-ous* suffix.

One characteristic of phosphoric and phosphorous acids is their tendency to undergo condensation reactions when heated. ∞ (Section 12.8) For example, two H_3PO_4 molecules are joined by the elimination of one H_2O molecule to form $H_4P_2O_7$:

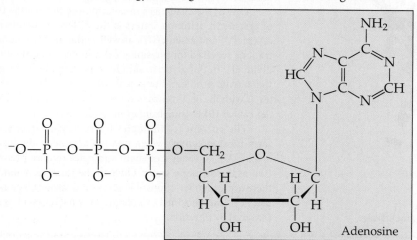

These atoms are eliminated as H_2O [22.53]

Phosphoric acid and its salts find their most important uses in detergents and fertilizers. The phosphates in detergents are often in the form of sodium tripolyphosphate ($Na_5P_3O_{10}$).

The phosphate ions "soften" water by binding their oxygen groups to the metal ions that contribute to the hardness of water. This keeps the metal ions from interfering with the action of the detergent. The phosphates also keep the pH above 7 and thus prevent the detergent molecules from becoming protonated.

Most mined phosphate rock is converted to fertilizers. The $Ca_3(PO_4)_2$ in phosphate rock is insoluble ($K_{sp} = 2.0 \times 10^{-29}$). It is converted to a soluble form for use in fertilizers by treatment with sulfuric or phosphoric acid. The reaction with phosphoric acid yields $Ca(H_2PO_4)_2$:

$$Ca_3(PO_4)_2(s) + 4\,H_3PO_4(aq) \longrightarrow 3\,Ca^{2+}(aq) + 6\,H_2PO_4^-(aq) \quad [22.54]$$

Although the solubility of $Ca(H_2PO_4)_2$ allows it to be assimilated by plants, it also allows it to be washed from the soil and into bodies of water, thereby contributing to water pollution. ∞ (Section 18.4)

Phosphorus compounds are important in biological systems. The element occurs in phosphate groups in RNA and DNA, the molecules responsible for the control of protein biosynthesis and transmission of genetic information. It also occurs in adenosine triphosphate (ATP), which stores energy in biological cells and has the following structure:

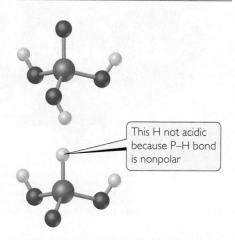

The P—O—P bond of the end phosphate group is broken by hydrolysis with water, forming adenosine diphosphate (ADP):

$$\text{ATP} + H_2O \longrightarrow$$

[22.55]

$$\text{ADP}$$

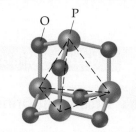

▲ GO FIGURE
How do the electron domains about P in P_4O_6 differ from those about P in P_4O_{10}?

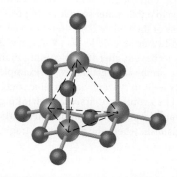

▲ Figure 22.27 Structures of P_4O_6 (top) and P_4O_{10} (bottom).

This H not acidic because P–H bond is nonpolar

▲ Figure 22.28 Structures of H_3PO_4 (top) and H_3PO_3 (bottom).

This reaction releases 33 kJ of energy under standard conditions, but in the living cell, the Gibbs free energy change for the reaction is closer to -57 kJ/mol. The concentration of ATP inside a living cell is in the range of 1–10 mM, which means a typical human metabolizes her or his body mass of ATP in one day! ATP is continually made from ADP and continually converted back to ADP, releasing energy that can be harnessed by other cellular reactions.

Chemistry and Life

Arsenic in Drinking Water

Arsenic, in the form of its oxides, has been known as a poison for centuries. The current Environmental Protection Agency (EPA) standard for arsenic in public water supplies is 10 ppb (equivalent to 10 μg/L). Most regions of the United States tend to have low to moderate (2–10 ppb) groundwater arsenic levels (▼ **Figure 22.29**). The western region tends to have higher levels, coming mainly from natural geological sources in the area. Estimates, for example, indicate that 35% of water-supply wells in Arizona have arsenic concentrations above 10 ppb.

The problem of arsenic in drinking water in the United States is dwarfed by the problem in other parts of the world—especially in Bangladesh, where the problem is tragic. Historically, surface water

sources in that country have been contaminated with micro-organisms, causing significant health problems. During the 1970s, international agencies, headed by the United Nations Children's Fund (UNICEF), began investing millions of dollars of aid money in Bangladesh for wells to provide "clean" drinking water. Unfortunately, no one tested the well water for the presence of arsenic; the problem was not discovered until the 1980s. The result has been the biggest outbreak of mass poisoning in history. Up to half of the country's estimated 10 million wells have arsenic concentrations above 50 ppb.

In water, the most common forms of arsenic are the arsenate ion and its protonated hydrogen anions (AsO_4^{3-}, $HAsO_4^{2-}$, and $H_2AsO_4^-$) and the arsenite ion and its protonated forms (AsO_3^{3-}, $HAsO_3^{2-}$, $H_2AsO_3^-$, and H_3AsO_3). These species are collectively referred to by the oxidation number of the arsenic as arsenic(V) and arsenic(III), respectively. Arsenic(V) is more prevalent in oxygen-rich (aerobic) surface waters, whereas arsenic(III) is more likely to occur in oxygen-poor (anaerobic) groundwaters.

One of the challenges in determining the health effects of arsenic in drinking waters is the different chemistry of arsenic(V) and arsenic(III), as well as the different concentrations required for physiological responses in different individuals. In Bangladesh, skin lesions were the first sign of the arsenic problem. Statistical studies correlating arsenic levels with the occurrence of disease indicate a lung and bladder cancer risk arising from even low levels of arsenic.

The current technologies for removing arsenic perform most effectively when treating arsenic in the form of arsenic(V), so water treatment strategies require preoxidation of the drinking water. Once in the form of arsenic(V), there are a number of possible removal strategies. For example, Fe^{3+} can be added to precipitate $FeAsO_4$, which is then removed by filtration.

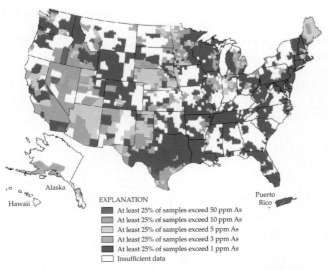

EXPLANATION
- At least 25% of samples exceed 50 ppm As
- At least 25% of samples exceed 10 ppm As
- At least 25% of samples exceed 5 ppm As
- At least 25% of samples exceed 3 ppm As
- At least 25% of samples exceed 1 ppm As
- Insufficient data

▲ Figure 22.29 **Geographic distribution of arsenic in groundwater.**

22.9 | Carbon

Carbon constitutes only 0.027% of Earth's crust. Although some carbon occurs in elemental form as graphite and diamond, most is found in combined form. Over half occurs in carbonate compounds, and carbon is also found in coal, petroleum, and natural gas. The importance of the element stems in large part from its occurrence in all living organisms: Life as we know it is based on carbon compounds.

Elemental Forms of Carbon

We have seen that carbon exists in several allotropic crystalline forms: graphite, diamond, fullerenes, carbon nanotubes, and graphene. Fullerenes, nanotubes, and graphene are discussed in Chapter 12; here we focus on graphite and diamond.

Graphite is a soft, black, slippery solid that has a metallic luster and conducts electricity. It consists of parallel sheets of sp^2-hybridized carbon atoms held together by dispersion forces. ∞ (Section 12.7) *Diamond* is a clear, hard solid in which the carbon atoms form an sp^3-hybridized covalent network. ∞ (Section 12.7) Diamond is denser than graphite (d = 2.25 g/cm^3 for graphite; d = 3.51 g/cm^3 for diamond). At approximately 100,000 atm at 3000 °C, graphite converts to diamond. In fact, almost any carbon-containing substance, if put under sufficiently high pressure, forms diamonds; scientists at General Electric in the 1950s used peanut butter to make diamonds. About 3×10^4 kg of industrial-grade diamonds are synthesized each year, mainly for use in cutting, grinding, and polishing tools.

Graphite has a well-defined crystalline structure, but it also exists in two common amorphous forms: **carbon black** and **charcoal**. Carbon black is formed when hydrocarbons are heated in a very limited supply of oxygen, such as in this methane reaction:

$$CH_4(g) + O_2(g) \longrightarrow C(s) + 2 H_2O(g) \qquad [22.56]$$

Carbon black is used as a pigment in black inks; large amounts are also used in making automobile tires.

Charcoal is formed when wood is heated strongly in the absence of air. Charcoal has an open structure, giving it an enormous surface area per unit mass. "Activated charcoal," a pulverized form of charcoal whose surface is cleaned by heating with steam, is widely used to adsorb molecules. It is used in filters to remove offensive odors from air and colored or bad-tasting impurities from water.

Oxides of Carbon

Carbon forms two principal oxides: carbon monoxide (CO) and carbon dioxide (CO_2). *Carbon monoxide* is formed when carbon or hydrocarbons are burned in a limited supply of oxygen:

$$2 C(s) + O_2(g) \longrightarrow 2 CO(g) \qquad [22.57]$$

CO is a colorless, odorless, tasteless gas that is toxic because it binds to hemoglobin in the blood and thus interferes with oxygen transport. Low-level poisoning results in headaches and drowsiness; high-level poisoning can cause death.

Carbon monoxide is unusual in that it has a nonbonding pair of electrons on carbon: :C≡O:. It is isoelectronic with N_2, so you might expect CO to be equally unreactive. Moreover, both substances have high bond energies (1072 kJ/mol for C≡O and 941 kJ/mol for N≡N). Because of the lower nuclear charge on carbon (compared with either N or O), however, the carbon nonbonding pair is not held as strongly as that on N or O. Consequently, CO is better able to function as a Lewis base than is N_2; for example, CO can coordinate its nonbonding pair to the iron of hemoglobin, displacing O_2, but N_2 cannot. In addition, CO forms a variety of covalent compounds, known as metal carbonyls, with transition metals. $Ni(CO)_4$, for example, is a volatile, toxic compound formed by warming metallic nickel in the presence of CO. The formation of metal carbonyls is the first step in the transition-metal catalysis of a variety of reactions of CO.

Carbon monoxide has several commercial uses. Because it burns readily, forming CO_2, it is employed as a fuel:

$$2 CO(g) + O_2(g) \longrightarrow 2 CO_2(g) \qquad \Delta H° = -566 \text{ kJ} \qquad [22.58]$$

Carbon monoxide is an important reducing agent, widely used in metallurgical operations to reduce metal oxides, such as the iron oxides:

$$Fe_3O_4(s) + 4 CO(g) \longrightarrow 3 Fe(s) + 4 CO_2(g) \qquad [22.59]$$

Carbon dioxide is produced when carbon-containing substances are burned in excess oxygen, such as in this reaction involving ethanol:

$$C_2H_5OH(l) + 3 O_2(g) \longrightarrow 2 CO_2(g) + 3 H_2O(g) \qquad [22.60]$$

Chemistry Put to Work

Carbon Fibers and Composites

The properties of graphite are anisotropic; that is, they differ in different directions through the solid. Along the carbon planes, graphite possesses great strength because of the number and strength of the carbon–carbon bonds in this direction. The bonds between planes are relatively weak, however, making graphite weak in that direction.

Fibers of graphite can be prepared in which the carbon planes are aligned to varying extents parallel to the fiber axis. These fibers are lightweight (density of about 2 g/cm^3) and chemically quite unreactive. The oriented fibers are made by first slowly pyrolyzing (decomposing by action of heat) organic fibers at about 150 to 300 °C. These fibers are then heated to about 2500 °C to graphitize them (convert amorphous carbon to graphite). Stretching the fiber during pyrolysis helps orient the graphite planes parallel to the fiber axis. More amorphous carbon fibers are formed by pyrolysis of organic fibers at lower temperatures (1200 to 400 °C). These amorphous materials, commonly called *carbon fibers*, are the type most often used in commercial materials.

Composite materials that take advantage of the strength, stability, and low density of carbon fibers are widely used. Composites are combinations of two or more materials. These materials are present as separate phases and are combined to form structures that take advantage of certain desirable properties of each component. In carbon composites, the graphite fibers are often woven into a fabric that is embedded in a matrix that binds them into a solid structure.

The fibers transmit loads evenly throughout the matrix. The finished composite is thus stronger than any one of its components.

Carbon composite materials are used widely in a number of applications, including high-performance graphite sports equipment such as tennis racquets, golf clubs, and bicycle wheels (▼ Figure 22.30). Heat-resistant composites are required for many aerospace applications, where carbon composites now find wide use.

▲ Figure 22.30 **Carbon composites in commercial products.**

▲ Figure 22.31 **CO$_2$ formation from the reaction between an acid and calcium carbonate in rock.**

It is also produced when many carbonates are heated:

$$CaCO_3(s) \xrightarrow{\Delta} CaO(s) + CO_2(g) \qquad [22.61]$$

In the laboratory, CO$_2$ can be produced by the action of acids on carbonates (◀ Figure 22.31):

$$CO_3{}^{2-}(aq) + 2\,H^+(aq) \longrightarrow CO_2(g) + H_2O(l) \qquad [22.62]$$

Carbon dioxide is a colorless, odorless gas. It is a minor component of Earth's atmosphere but a major contributor to the greenhouse effect. ∞ (Section 18.2) Although it is not toxic, high concentrations of CO$_2$ increase respiration rate and can cause suffocation. It is readily liquefied by compression. When cooled at atmospheric pressure, however, CO$_2$ forms a solid rather than liquefying. The solid sublimes at atmospheric pressure at −78 °C. This property makes solid CO$_2$, known as *dry ice*, valuable as a refrigerant. About half of the CO$_2$ consumed annually is used for refrigeration. The other major use of CO$_2$ is in the production of carbonated beverages. Large quantities are also used to manufacture *washing soda* (Na$_2$CO$_3$ · 10 H$_2$O), used to precipitate metal ions that interfere with the cleansing action of soap, and *baking soda* (NaHCO$_3$). Baking soda is so named because this reaction occurs during baking:

$$NaHCO_3(s) + H^+(aq) \longrightarrow Na^+(aq) + CO_2(g) + H_2O(l) \qquad [22.63]$$

The H$^+$(aq) is provided by vinegar, sour milk, or the hydrolysis of certain salts. The bubbles of CO$_2$ that form are trapped in the baking dough, causing it to rise.

Give It Some Thought

Yeast are living organisms that make bread rise in the absence of baking soda and acid. What must the yeast be producing to make bread rise?

Carbonic Acid and Carbonates

Carbon dioxide is moderately soluble in H_2O at atmospheric pressure. The resulting solution is moderately acidic because of the formation of carbonic acid (H_2CO_3):

$$CO_2(aq) + H_2O(l) \rightleftharpoons H_2CO_3(aq) \qquad [22.64]$$

Carbonic acid is a weak diprotic acid. Its acidic character causes carbonated beverages to have a sharp, slightly acidic taste.

Although carbonic acid cannot be isolated, hydrogen carbonates (bicarbonates) and carbonates can be obtained by neutralizing carbonic acid solutions. Partial neutralization produces HCO_3^-, and complete neutralization gives CO_3^{2-}. The HCO_3^- ion is a stronger base than acid ($K_b = 2.3 \times 10^{-8}$; $K_a = 5.6 \times 10^{-11}$). The carbonate ion is much more strongly basic ($K_b = 1.8 \times 10^{-4}$).

The principal carbonate minerals are calcite ($CaCO_3$), magnesite ($MgCO_3$), dolomite [$MgCa(CO_3)_2$], and siderite ($FeCO_3$). Calcite is the principal mineral in limestone and the main constituent of marble, chalk, pearls, coral reefs, and the shells of marine animals such as clams and oysters. Although $CaCO_3$ has low solubility in pure water, it dissolves readily in acidic solutions with evolution of CO_2:

$$CaCO_3(s) + 2\,H^+(aq) \rightleftharpoons Ca^{2+}(aq) + H_2O(l) + CO_2(g) \qquad [22.65]$$

Because water containing CO_2 is slightly acidic (Equation 22.64), $CaCO_3$ dissolves slowly in this medium:

$$CaCO_3(s) + H_2O(l) + CO_2(g) \longrightarrow Ca^{2+}(aq) + 2\,HCO_3^-(aq) \qquad [22.66]$$

This reaction occurs when surface waters move underground through limestone deposits. It is the principal way Ca^{2+} enters groundwater, producing "hard water." If the limestone deposit is deep enough underground, dissolution of the limestone produces a cave.

One of the most important reactions of $CaCO_3$ is its decomposition into CaO and CO_2 at elevated temperatures (Equation 22.61). About 2×10^{10} kg (20 million tons) of calcium oxide, known as lime or quicklime, is produced in the United States annually. Because calcium oxide reacts with water to form $Ca(OH)_2$, it is an important commercial base. It is also important in making mortar, the mixture of sand, water, and CaO used to bind bricks, blocks, or rocks together. Calcium oxide reacts with water and CO_2 to form $CaCO_3$, which binds the sand in the mortar:

$$CaO(s) + H_2O(l) \longrightarrow Ca^{2+}(aq) + 2\,OH^-(aq) \qquad [22.67]$$

$$Ca^{2+}(aq) + 2\,OH^-(aq) + CO_2(aq) \longrightarrow CaCO_3(s) + H_2O(l) \qquad [22.68]$$

Carbides

The binary compounds of carbon with metals, metalloids, and certain nonmetals are called **carbides**. The more active metals form *ionic carbides*, and the most common of these contain the *acetylide* ion (C_2^{2-}). This ion is isoelectronic with N_2, and its Lewis structure, $[:C \equiv C:]^{2-}$, has a carbon–carbon triple bond. The most important ionic carbide is calcium carbide (CaC_2), produced by the reduction of CaO with carbon at high temperature:

$$2\,CaO(s) + 5\,C(s) \longrightarrow 2\,CaC_2(s) + CO_2(g) \qquad [22.69]$$

The carbide ion is a very strong base that reacts with water to form acetylene ($H - C \equiv C - H$):

$$CaC_2(s) + 2\,H_2O(l) \longrightarrow Ca(OH)_2(aq) + C_2H_2(g) \qquad [22.70]$$

Calcium carbide is therefore a convenient solid source of acetylene, which is used in welding (Figure 22.13).

Interstitial carbides are formed by many transition metals. The carbon atoms occupy open spaces (interstices) between the metal atoms in a manner analogous to the interstitial hydrides. ∞ (Section 22.2) This process generally makes the metal harder.

Tungsten carbide (WC), for example, is very hard and very heat-resistant and, thus, used to make cutting tools.

Covalent carbides are formed by boron and silicon. Silicon carbide (SiC), known as Carborundum®, is used as an abrasive and in cutting tools. Almost as hard as diamond, SiC has a diamond-like structure with alternating Si and C atoms.

22.10 | The Other Group 4A Elements: Si, Ge, Sn, and Pb

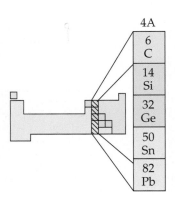

The trend from nonmetallic to metallic character as we go down a family is strikingly evident in group 4A. Carbon is a nonmetal; silicon and germanium are metalloids; tin and lead are metals. In this section, we consider a few general characteristics of group 4A and then look more thoroughly at silicon.

General Characteristics of the Group 4A Elements

The group 4A elements possess the outer-shell electron configuration ns^2np^2. The electronegativities of the elements are generally low (▼ Table 22.8); carbides that formally contain C^{4-} ions are observed only in the case of a few compounds of carbon with very active metals. Formation of 4+ ions by electron loss is not observed for any of these elements; the ionization energies are too high. The +4 oxidation state is common, however, and is found in the vast majority of the compounds of the group 4A elements. The +2 oxidation state is found in the chemistry of germanium, tin, and lead, however, and it is the principal oxidation state for lead. Carbon, except in highly unusual examples, forms a maximum of four bonds. The other members of the family are able to form more than four bonds. ∞ (Section 8.7)

Table 22.8 shows that the strength of a bond between two atoms of a given element decreases as we go down group 4A. Carbon–carbon bonds are quite strong. Carbon, therefore, has a striking ability to form compounds in which carbon atoms are bonded to one another in extended chains and rings, which accounts for the large number of organic compounds that exist. Other elements can form chains and rings, but these bonds are far less important in the chemistries of these other elements. The Si—Si bond strength (226 kJ/mol), for example, is much lower than the Si—O bond strength (386 kJ/mol). As a result, the chemistry of silicon is dominated by the formation of Si—O bonds, and Si—Si bonds play a minor role.

Occurrence and Preparation of Silicon

Silicon is the second most abundant element, after oxygen, in Earth's crust. It occurs in SiO_2 and in an enormous variety of silicate minerals. The element is obtained by the reduction of molten silicon dioxide with carbon at high temperature:

$$SiO_2(l) + 2\,C(s) \longrightarrow Si(l) + 2\,CO(g) \qquad [22.71]$$

Elemental silicon has a diamond-like structure. Crystalline silicon is a gray metallic-looking solid that melts at 1410 °C. The element is a semiconductor, as we saw in Chapters 7 and 12, and is used to make solar cells and transistors for computer chips. To be used as a semiconductor, it must be extremely pure, possessing less than $10^{-7}\%$

Table 22.8 Some Properties of the Group 4A Elements

Property	C	Si	Ge	Sn	Pb
Atomic radius (Å)	0.76	1.11	1.20	1.39	1.46
First ionization energy (kJ/mol)	1086	786	762	709	716
Electronegativity	2.5	1.8	1.8	1.8	1.9
X—X single-bond enthalpy (kJ/mol)	348	226	188	151	—

(1 ppb) impurities. One method of purification is to treat the element with Cl_2 to form $SiCl_4$, a volatile liquid that is purified by fractional distillation and then converted back to elemental silicon by reduction with H_2:

$$SiCl_4(g) + 2 H_2(g) \longrightarrow Si(s) + 4 HCl(g) \qquad [22.72]$$

The process known as *zone refining* can further purify the element (▶ Figure 22.30). As a heated coil is passed slowly along a silicon rod, a narrow band of the element is melted. As the molten section is swept slowly along the length of the rod, the impurities concentrate in this section, following it to the end of the rod. The purified top portion of the rod crystallizes as 99.999999999% pure silicon.

Silicates

Silicon dioxide and other compounds that contain silicon and oxygen make up over 90% of Earth's crust. In **silicates,** a silicon atom is surrounded by four oxygens and silicon is found in its most common oxidation state, +4. The orthosilicate ion, $SiO_4{}^{4-}$, is found in very few silicate minerals, but we can view it as the "building block" for many mineral structures. As ▼ Figure 22.33 shows, adjacent tetrahedra are linked by a common oxygen atom. Two tetrahedra joined in this way, called the *disilicate* ion, contain two Si atoms and seven O atoms. Silicon and oxygen are in the +4 and −2 oxidation states, respectively, in all silicates, so the overall charge of any silicate ion must be consistent with these oxidation states. For example, the charge on Si_2O_7 is $(2)(+4) + (7)(-2) = -6$; it is the $Si_2O_7{}^{6-}$ ion.

In most silicate minerals, silicate tetrahedra are linked together to form chains, sheets, or three-dimensional structures. We can connect two vertices of each tetrahedron to two other tetrahedra, for example, leading to an infinite chain with an $\cdots$ O—Si—O—Si $\cdots$ backbone as shown in Figure 22.33(b). Notice that each silicon in this structure has two unshared (terminal) oxygens and two shared (bridging) oxygens. The stoichiometry is then $2(1) + 2(1/2) = 3$ oxygens per silicon. Thus, the formula unit for this chain is $SiO_3{}^{2-}$. The mineral *enstatite* $(MgSiO_3)$ has this kind of structure, consisting of rows of single-strand silicate chains with Mg^{2+} ions between the strands to balance charge.

In Figure 22.33(c), each silicate tetrahedron is linked to three others, forming an infinite sheet structure. Each silicon in this structure has one unshared oxygen and three shared oxygens. The stoichiometry is then $1(1) + 3(\frac{1}{2}) = 2\frac{1}{2}$ oxygens per silicon. The simplest formula of this sheet is $Si_2O_5{}^{2-}$. The mineral *talc*, also known as talcum powder, has the formula $Mg_3(Si_2O_5)_2(OH)_2$ and is based on this sheet

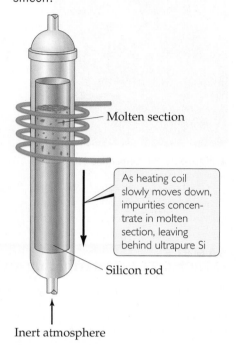

▲ **GO FIGURE**

What limits the range of temperatures you can use for zone refining of silicon?

Molten section

As heating coil slowly moves down, impurities concentrate in molten section, leaving behind ultrapure Si

Silicon rod

Inert atmosphere

▲ **Figure 22.32 Zone-refining apparatus for the production of ultrapure silicon.**

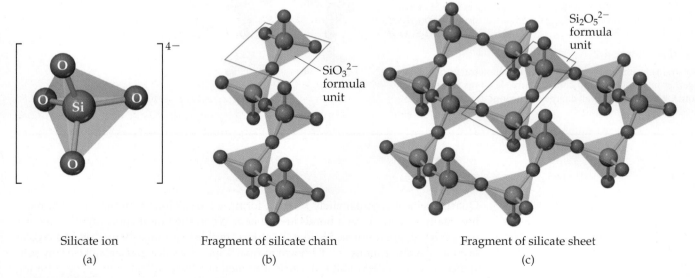

Silicate ion

(a)

Fragment of silicate chain

(b)

$SiO_3{}^{2-}$ formula unit

Fragment of silicate sheet

(c)

$Si_2O_5{}^{2-}$ formula unit

▲ **Figure 22.33 Silicate chains and sheets.**

▲ **Figure 22.34 Serpentine asbestos.**

structure. The Mg^{2+} and OH^- ions lie between the silicate sheets. The slippery feel of talcum powder is due to the silicate sheets sliding relative to one another.

Numerous minerals are based on silicates, and many are useful as clays, ceramics, and other materials. A few silicates have harmful effects on human health, the best-known example being *asbestos*, a general term applied to a group of fibrous silicate minerals. The structure of these minerals is either chains of silicate tetrahedra or sheets formed into rolls. The result is that the minerals have a fibrous character (◀ **Figure 22.34**). Asbestos minerals were once widely used as thermal insulation, especially in high-temperature applications, because of the great chemical stability of the silicate structure. In addition, the fibers can be woven into asbestos cloth, which was used for fireproof curtains and other applications. However, the fibrous structure of asbestos minerals poses a health risk because the fibers readily penetrate soft tissues, such as the lungs, where they can cause diseases, including cancer. The use of asbestos as a common building material has therefore been discontinued.

When all four vertices of each SiO_4 tetrahedron are linked to other tetrahedra, the structure extends in three dimensions. This linking of the tetrahedra forms quartz (SiO_2). Because the structure is locked together in a three-dimensional array much like diamond ∞ (Section 12.7), quartz is harder than strand- or sheet-type silicates.

SAMPLE EXERCISE 22.7 Determining an Empirical Formula

The mineral *chrysotile* is a noncarcinogenic asbestos mineral that is based on the sheet structure shown in Figure 22.33(c). In addition to silicate tetrahedra, the mineral contains Mg^{2+} and OH^- ions. Analysis of the mineral shows that there are 1.5 Mg atoms per Si atom. What is the empirical formula for chrysotile?

SOLUTION

Analyze A mineral is described that has a sheet silicate structure with Mg^{2+} and OH^- ions to balance charge and 1.5 Mg for each 1 Si. We are asked to write the empirical formula for the mineral.

Plan As shown in Figure 22.33(c), the silicate sheet structure has the simplest formula $Si_2O_5{}^{2-}$. We first add Mg^{2+} to give the proper Mg:Si ratio. We then add OH^- ions to obtain a neutral compound.

Solve The observation that the Mg:Si ratio equals 1.5 is consistent with three Mg^{2+} ions per $Si_2O_5{}^{2-}$ unit. The addition of three Mg^{2+} ions would make $Mg_3(Si_2O_5)^{4+}$. In order to achieve charge balance in the mineral, there must be four OH^- per $Si_2O_5{}^{2-}$. Thus, the formula of chrysotile is $Mg_3(Si_2O_5)(OH)_4$. Since this is not reducible to a simpler formula, this is the empirical formula.

Practice Exercise 1

In the mineral beryl, six silicate tetrahedra are connected to form a ring as shown here. The negative charge of this polyanion is balanced by Be^{2+} and Al^{3+} cations. If elemental analysis gives a Be:Si ratio of 1:2 and an Al:Si ratio of 1:3 what is the empirical formula of beryl: (a) $Be_2Al_3Si_6O_{19}$, (b) $Be_3Al_2(SiO_4)_6$, (c) $Be_3Al_2Si_6O_{18}$, (d) $BeAl_2Si_6O_{15}$?

Practice Exercise 2

The cyclosilicate ion consists of three silicate tetrahedra linked together in a ring. The ion contains three Si atoms and nine O atoms. What is the overall charge on the ion?

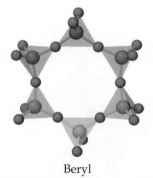

Beryl

Glass

Quartz melts at approximately 1600 °C, forming a tacky liquid. In the course of melting, many silicon–oxygen bonds are broken. When the liquid cools rapidly, silicon–oxygen bonds are re-formed before the atoms are able to arrange themselves in a regular fashion. An amorphous solid, known as quartz glass or silica glass, results. Many substances can be added to SiO_2 to cause it to melt at a lower temperature. The common **glass** used in windows and bottles, known as soda-lime glass, contains CaO and Na_2O

in addition to SiO_2 from sand. The CaO and Na_2O are produced by heating two inexpensive chemicals, limestone ($CaCO_3$) and soda ash (Na_2CO_3), which decompose at high temperatures:

$$CaCO_3(s) \longrightarrow CaO(s) + CO_2(g) \qquad [22.73]$$

$$Na_2CO_3(s) \longrightarrow Na_2O(s) + CO_2(g) \qquad [22.74]$$

Other substances can be added to soda-lime glass to produce color or to change the properties of the glass in various ways. The addition of CoO, for example, produces the deep blue color of "cobalt glass." Replacing Na_2O with K_2O results in a harder glass that has a higher melting point. Replacing CaO with PbO results in a denser "lead crystal" glass with a higher refractive index. Lead crystal is used for decorative glassware; the higher refractive index gives this glass a particularly sparkling appearance. Addition of nonmetal oxides, such as B_2O_3 and P_4O_{10}, which form network structures related to the silicates, also changes the properties of the glass. Adding B_2O_3 creates a "borosilicate" glass with a higher melting point and a greater ability to withstand temperature changes. Such glasses, sold commercially under trade names such as Pyrex® and Kimax®, are used where resistance to thermal shock is important, such as in laboratory glassware or coffeemakers.

Silicones

Silicones consist of O—Si—O chains in which the two remaining bonding positions on each silicon are occupied by organic groups such as CH_3:

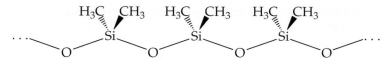

Depending on chain length and degree of cross-linking, silicones can be either oils or rubber-like materials. Silicones are nontoxic and have good stability toward heat, light, oxygen, and water. They are used commercially in a wide variety of products, including lubricants, car polishes, sealants, and gaskets. They are also used for waterproofing fabrics. When applied to a fabric, the oxygen atoms form hydrogen bonds with the molecules on the surface of the fabric. The hydrophobic (water-repelling) organic groups of the silicone are then left pointing away from the surface as a barrier.

 Give It Some Thought

Distinguish among the substances silicon, silica, and silicone.

22.11 | Boron

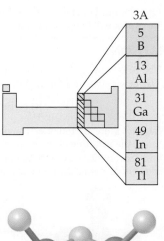

Boron is the only group 3A element that can be considered nonmetallic and thus is our final element in this chapter. The element has an extended network structure with a melting point (2300 °C) that is intermediate between the melting points of carbon (3550 °C) and silicon (1410 °C). The electron configuration of boron is $[He]2s^2 2p^1$.

In the family of compounds called **boranes,** the molecules contain only boron and hydrogen. The simplest borane is BH_3. This molecule contains only six valence electrons and is therefore an exception to the octet rule. As a result, BH_3 reacts with itself to form *diborane* (B_2H_6). This reaction can be viewed as a Lewis acid–base reaction in which one B—H bonding pair of electrons in each BH_3 molecule is donated to the other. As a result, diborane is an unusual molecule in which hydrogen atoms form a bridge between two B atoms (▶ **Figure 22.35**). Such hydrogens, called *bridging hydrogens*, exhibit interesting chemical reactivity, which you may learn about in a more advanced chemistry course.

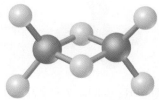

▲ **Figure 22.35 The structure of diborane (B_2H_6).**

Sharing hydrogen atoms between the two boron atoms compensates somewhat for the deficiency in valence electrons around each boron. Nevertheless, diborane is an extremely reactive molecule, spontaneously flammable in air in a highly exothermic reaction:

$$B_2H_6(g) + 3\,O_2(g) \longrightarrow B_2O_3(s) + 3\,H_2O(g) \qquad \Delta H° = -2030\ kJ \quad [22.75]$$

Boron and hydrogen form a series of anions called *borane anions*. Salts of the borohydride ion (BH_4^-) are widely used as reducing agents. For example, sodium borohydride $(NaBH_4)$ is a commonly used reducing agent for certain organic compounds.

 Give It Some Thought

Recall that the hydride ion is H^-. What is the oxidation state of boron in sodium borohydride?

The only important oxide of boron is boric oxide (B_2O_3). This substance is the anhydride of boric acid, which we may write as H_3BO_3 or $B(OH)_3$. Boric acid is so weak an acid $(K_a = 5.8 \times 10^{-10})$ that solutions of H_3BO_3 are used as an eyewash. Upon heating, boric acid loses water by a condensation reaction similar to that described for phosphorus in Section 22.8:

$$4\,H_3BO_3(s) \longrightarrow H_2B_4O_7(s) + 5\,H_2O(g) \qquad [22.76]$$

The diprotic acid $H_2B_4O_7$ is tetraboric acid. The hydrated sodium salt $Na_2B_4O_7 \cdot 10\,H_2O$, called borax, occurs in dry lake deposits in California and can also be prepared from other borate minerals. Solutions of borax are alkaline, and the substance is used in various laundry and cleaning products.

SAMPLE INTEGRATIVE EXERCISE / Putting Concepts Together

The interhalogen compound BrF_3 is a volatile, straw-colored liquid. The compound exhibits appreciable electrical conductivity because of autoionization ("solv" refers to BrF_3 as the solvent):

$$2\,BrF_3(l) \rightleftharpoons BrF_2^+(solv) + BrF_4^-(solv)$$

(a) What are the molecular structures of the BrF_2^+ and BrF_4^- ions?

(b) The electrical conductivity of BrF_3 decreases with increasing temperature. Is the autoionization process exothermic or endothermic?

(c) One chemical characteristic of BrF_3 is that it acts as a Lewis acid toward fluoride ions. What do we expect will happen when KBr is dissolved in BrF_3?

SOLUTION

(a) The BrF_2^+ ion has $7 + 2(7) - 1 = 20$ valence-shell electrons. The Lewis structure for the ion is

$$\left[:\ddot{F}—\ddot{Br}—\ddot{F}:\right]^+$$

Because there are four electron domains around the central Br atom, the resulting electron domain geometry is tetrahedral ∞ (Section 9.2). Because bonding pairs of electrons occupy two of these domains, the molecular geometry is bent:

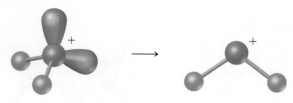

The BrF_4^- ion has $7 + 4(7) + 1 = 36$ electrons, leading to the Lewis structure

Because there are six electron domains around the central Br atom in this ion, the electron-domain geometry is octahedral. The two nonbonding pairs of electrons are located opposite each other on the octahedron, leading to a square-planar molecular geometry:

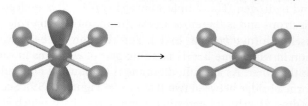

(b) The observation that conductivity decreases as temperature increases indicates that there are fewer ions present in the solution at the higher temperature. Thus, increasing the temperature causes the equilibrium to shift to the left. According to Le Châtelier's principle, this shift indicates that the reaction is exothermic as it proceeds from left to right. ∞ (Section 15.7)

(c) A Lewis acid is an electron-pair acceptor. ∞ (Section 16.11) The fluoride ion has four valence-shell electron pairs and can act as a Lewis base (an electron-pair donor). Thus, we can envision the following reaction occurring:

$$F^- + BrF_3 \longrightarrow BrF_4^-$$

Chapter Summary and Key Terms

THE PERIODIC TABLE AND CHEMICAL REACTIONS (SECTION 22.1) The periodic table is useful for organizing and remembering the descriptive chemistry of the elements. Among elements of a given group, size increases with increasing atomic number, and electronegativity and ionization energy decrease. Nonmetallic character parallels electronegativity, so the most nonmetallic elements are found in the upper right portion of the periodic table.

Among the nonmetallic elements, the first member of each group differs dramatically from the other members; it forms a maximum of four bonds to other atoms and exhibits a much greater tendency to form π bonds than the heavier elements in its group.

Because O_2 and H_2O are abundant in our world, we focus on two important and general reaction types as we discuss the nonmetals: oxidation by O_2 and proton-transfer reactions involving H_2O or aqueous solutions.

HYDROGEN (SECTION 22.2) Hydrogen has three isotopes: **protium** ($_1^1H$), **deuterium** ($_1^2H$), and **tritium** ($_1^3H$). Hydrogen is not a member of any particular periodic group, although it is usually placed above lithium. The hydrogen atom can either lose an electron, forming H^+, or gain one, forming H^- (the hydride ion). Because the H—H bond is relatively strong, H_2 is fairly unreactive unless activated by heat or a catalyst. Hydrogen forms a very strong bond to oxygen, so the reactions of H_2 with oxygen-containing compounds usually lead to the formation of H_2O. Because the bonds in CO and CO_2 are even stronger than the O—H bond, the reaction of H_2O with carbon or certain organic compounds leads to the formation of H_2. The $H^+(aq)$ ion is able to oxidize many metals, forming $H_2(g)$. The electrolysis of water also forms $H_2(g)$.

The binary compounds of hydrogen are of three general types: **ionic hydrides** (formed by active metals), **metallic hydrides** (formed by transition metals), and **molecular hydrides** (formed by nonmetals). The ionic hydrides contain the H^- ion; because this ion is extremely basic, ionic hydrides react with H_2O to form H_2 and OH^-.

GROUP 8A: THE NOBLE GASES AND GROUP 7A: HALOGENS (SECTIONS 22.3 AND 22.4) The noble gases (group 8A) exhibit a very limited chemical reactivity because of the exceptional stability of their electron configurations. The xenon fluorides and oxides and KrF_2 are the best-established compounds of the noble gases.

The halogens (group 7A) occur as diatomic molecules. All except fluorine exhibit oxidation states varying from -1 to $+7$. Fluorine is the most electronegative element, so it is restricted to the oxidation states 0 and -1. The oxidizing power of the element (the tendency to form the -1 oxidation state) decreases as we proceed down the group.

The hydrogen halides are among the most useful compounds of these elements; these gases dissolve in water to form the hydrohalic acids, such as HCl(aq). Hydrofluoric acid reacts with **silica**. The **interhalogens** are compounds formed between two different halogen elements. Chlorine, bromine, and iodine form a series of oxyacids, in which the halogen atom is in a positive oxidation state. These compounds and their associated oxyanions are strong oxidizing agents.

OXYGEN AND THE OTHER GROUP 6A ELEMENTS (SECTIONS 22.5 AND 22.6) Oxygen has two allotropes, O_2 and O_3 (ozone). Ozone is unstable compared to O_2, and it is a stronger oxidizing agent than O_2. Most reactions of O_2 lead to oxides, compounds in which oxygen is in the -2 oxidation state. The soluble oxides of nonmetals generally produce acidic aqueous solutions; they are called **acidic anhydrides** or **acidic oxides**. In contrast, soluble metal oxides produce basic solutions and are called **basic anhydrides** or **basic oxides**. Many metal oxides that are insoluble in water dissolve in acid, accompanied by the formation of H_2O. Peroxides contain O—O bonds and oxygen in the -1 oxidation state. Peroxides are unstable, decomposing to O_2 and oxides. In such reactions peroxides are simultaneously oxidized and reduced, a process called **disproportionation**. Superoxides contain the O_2^- ion in which oxygen is in the $-\frac{1}{2}$ oxidation state.

Sulfur is the most important of the other group 6A elements. It has several allotropic forms; the most stable one at room temperature consists of S_8 rings. Sulfur forms two oxides, SO_2 and SO_3, and both are important atmospheric pollutants. Sulfur trioxide is the anhydride of sulfuric acid, the most important sulfur compound and the most-produced industrial chemical. Sulfuric acid is a strong acid and a good dehydrating agent. Sulfur forms several oxyanions as well, including the SO_3^{2-} (sulfite), SO_4^{2-} (sulfate), and $S_2O_3^{2-}$ (thiosulfate) ions. Sulfur is found combined with many metals as a sulfide, in which sulfur is in the -2 oxidation state. These compounds often react with acids to form hydrogen sulfide (H_2S), which smells like rotten eggs.

NITROGEN AND THE OTHER GROUP 5A ELEMENTS (SECTIONS 22.7 AND 22.8) Nitrogen is found in the atmosphere as N_2 molecules. Molecular nitrogen is chemically very stable because of the strong N≡N bond. Molecular nitrogen can be converted into ammonia via the Haber process. Once the ammonia is made, it can be converted into a variety of different compounds that exhibit nitrogen oxidation states ranging from -3 to $+5$. The most important industrial conversion of ammonia is the **Ostwald process**, in which ammonia is oxidized to nitric acid (HNO_3).

Nitrogen has three important oxides: nitrous oxide (N_2O), nitric oxide (NO), and nitrogen dioxide (NO_2). Nitrous acid (HNO_2) is a weak acid; its conjugate base is the nitrite ion (NO_2^-). Another important nitrogen compound is hydrazine (N_2H_4).

Phosphorus is the most important of the remaining group 5A elements. It occurs in nature as phosphate minerals. Phosphorus has several allotropes, including white phosphorus, which consists of P_4 tetrahedra. In reaction with the halogens, phosphorus forms trihalides PX_3 and pentahalides PX_5. These compounds undergo hydrolysis to produce an oxyacid of phosphorus and HX.

Phosphorus forms two oxides, P_4O_6 and P_4O_{10}. Their corresponding acids, phosphorous acid and phosphoric acid, undergo condensation reactions when heated. Phosphorus compounds are important in biochemistry and as fertilizers.

CARBON AND THE OTHER GROUP 4A ELEMENTS (SECTIONS 22.9 AND 22.10) The allotropes of carbon include diamond, graphite, fullerenes, carbon nanotubes, and graphene. Amorphous forms of graphite include **charcoal** and **carbon black**. Carbon forms two common oxides, CO and CO_2. Aqueous solutions of CO_2 produce the weak diprotic acid carbonic acid (H_2CO_3), which is the parent acid of hydrogen carbonate and carbonate salts. Binary compounds of carbon are called **carbides**. Carbides may be ionic, interstitial, or covalent. Calcium carbide (CaC_2) contains the strongly basic acetylide ion (C_2^{2-}), which reacts with water to form acetylene.

The other group 4A elements show great diversity in physical and chemical properties. Silicon, the second most abundant element, is a semiconductor. It reacts with Cl_2 to form $SiCl_4$, a liquid at room temperature, a reaction that is used to help purify silicon from its native minerals. Silicon forms strong Si—O bonds and therefore occurs in a variety of silicate minerals. Silica is SiO_2; **silicates** consist of SiO_4 tetrahedra, linked together at their vertices to form chains, sheets, or three-dimensional structures. The most common three-dimensional silicate is quartz (SiO_2). **Glass** is an amorphous (noncrystalline) form of SiO_2. Silicones contain O—Si—O chains with organic groups bonded to the Si atoms. Like silicon, germanium is a metalloid; tin and lead are metallic.

BORON (SECTION 22.11) Boron is the only group 3A element that is a nonmetal. It forms a variety of compounds with hydrogen called boron hydrides, or **boranes**. Diborane (B_2H_6) has an unusual structure with two hydrogen atoms that bridge between the two boron atoms. Boranes react with oxygen to form boric oxide (B_2O_3), in which boron is in the +3 oxidation state. Boric oxide is the anhydride of boric acid (H_3BO_3). Boric acid readily undergoes condensation reactions.

Learning Outcomes After studying this chapter, you should be able to:

- Use periodic trends to explain the basic differences between the elements of a group or period. (Section 22.1)

- Explain two ways in which the first element in a group differs from subsequent elements in the group. (Section 22.1)

- Be able to determine electron configurations, oxidation numbers, and molecular shapes of elements and compounds. (Sections 22.2–22.11)

- Know the sources of the common nonmetals, how they are obtained, and how they are used. (Sections 22.2–22.11)

- Understand how phosphoric and phosphorous acids undergo condensation reactions. (Section 22.8)

- Explain how the bonding and structures of silicates relate to their chemical formulas and properties. (Section 22.10)

Exercises

Visualizing Concepts

22.1 **(a)** One of these structures is a stable compound; the other is not. Identify the stable compound, and explain why it is stable. Explain why the other compound is not stable. **(b)** What is the geometry around the central atoms of the stable compound? [Section 22.1]

22.2 **(a)** Identify the *type* of chemical reaction represented by the following diagram. **(b)** Place appropriate charges on the species on both sides of the equation. **(c)** Write the chemical equation for the reaction. [Section 22.1]

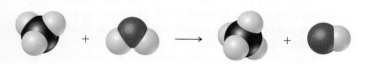

22.3 Which of the following species (there may be more than one) is/are likely to have the structure shown here: **(a)** XeF_4, **(b)** BrF_4^+, **(c)** SiF_4, **(d)** $TeCl_4$, **(e)** $HClO_4$? The colors do not reflect atom identities.) [Sections 22.3, 22.4, 22.6, and 22.10]

22.4 You have two glass bottles, one containing oxygen and one filled with nitrogen. How could you determine which one is which? [Sections 22.5 and 22.7]

22.5 Write the molecular formula and Lewis structure for each of the following oxides of nitrogen: [Section 22.7]

22.6 Which property of the group 6A elements might be the one depicted in the graph shown here: (**a**) electronegativity, (**b**) first ionization energy, (**c**) density, (**d**) X — X single-bond enthalpy, (**e**) electron affinity? Explain your answer. [Sections 22.5 and 22.6]

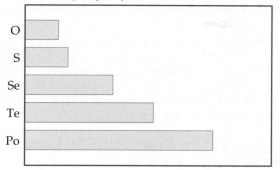

22.7 The atomic and ionic radii of the first three group 6A elements are

	Atomic radius (Å)	Ionic radius (Å)

O O^{2-}

0.66   1.40

S S^{2-}

1.05   1.84

Se Se^{2-}

1.20   1.98

(**a**) Explain why the atomic radius increases in moving downward in the group. (**b**) Explain why the ionic radii are larger than the atomic radii. (**c**) Which of the three anions would you expect to be the strongest base in water? Explain. [Sections 22.5 and 22.6]

22.8 Which property of the third-row nonmetallic elements might be the one depicted below: (**a**) first ionization energy, (**b**) atomic radius, (**c**) electronegativity, (**d**) melting point, (**e**) X — X single-bond enthalpy? Explain both your choice and why the other choices would not be correct. [Sections 22.3, 22.4, 22.6, 22.8, and 22.10]

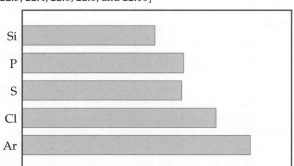

22.9 Which of the following compounds would you expect to be the most generally reactive, and why? (Each corner in these structures represents a CH_2 group.) [Section 22.8]

22.10 (**a**) Draw the Lewis structures for at least four species that have the general formula

$$\left[:X \equiv Y:\right]^n$$

where X and Y may be the same or different, and n may have a value from +1 to −2. (**b**) Which of the compounds is likely to be the strongest Brønsted base? Explain. [Sections 22.1, 22.7, and 22.9]

Periodic Trends and Chemical Reactions (Section 22.1)

22.11 Identify each of the following elements as a metal, nonmetal, or metalloid: (**a**) phosphorus, (**b**) strontium, (**c**) manganese, (**d**) selenium, (**e**) sodium, (**f**) krypton.

22.12 Identify each of the following elements as a metal, nonmetal, or metalloid: (**a**) gallium, (**b**) molybdenum, (**c**) tellurium, (**d**) arsenic, (**e**) xenon, (**f**) ruthenium.

22.13 Consider the elements O, Ba, Co, Be, Br, and Se. From this list, select the element that (**a**) is most electronegative, (**b**) exhibits a maximum oxidation state of +7, (**c**) loses an electron most readily, (**d**) forms π bonds most readily, (**e**) is a transition metal, (**f**) is a liquid at room temperature and pressure.

22.14 Consider the elements Li, K, Cl, C, Ne, and Ar. From this list, select the element that (**a**) is most electronegative, (**b**) has the greatest metallic character, (**c**) most readily forms a positive ion, (**d**) has the smallest atomic radius, (**e**) forms π bonds most readily, (**f**) has multiple allotropes.

22.15 Explain the following observations: (**a**) The highest fluoride compound formed by nitrogen is NF_3, whereas phosphorus readily forms PF_5. (**b**) Although CO is a well-known compound, SiO does not exist under ordinary conditions. (**c**) AsH_3 is a stronger reducing agent than NH_3.

22.16 Explain the following observations: (**a**) HNO_3 is a stronger oxidizing agent than H_3PO_4. **b**) Silicon can form an ion with six fluorine atoms, SiF_6^{2-}, whereas carbon is able to bond to a maximum of four, CF_4. (**c**) There are three compounds formed by carbon and hydrogen that contain two carbon atoms each (C_2H_2, C_2H_4, and C_2H_6), whereas silicon forms only one analogous compound (Si_2H_6).

22.17 Complete and balance the following equations:

(**a**) $NaOCH_3(s) + H_2O(l) \longrightarrow$

(**b**) $CuO(s) + HNO_3(aq) \longrightarrow$

(**c**) $WO_3(s) + H_2(g) \xrightarrow{\Delta}$

(**d**) $NH_2OH(l) + O_2(g) \longrightarrow$

(**e**) $Al_4C_3(s) + H_2O(l) \longrightarrow$

22.18 Complete and balance the following equations:

(**a**) $Mg_3N_2(s) + H_2O(l) \longrightarrow$

(**b**) $C_3H_7OH(l) + O_2(g) \longrightarrow$

(**c**) $MnO_2(s) + C(s) \xrightarrow{\Delta}$

(**d**) $AlP(s) + H_2O(l) \longrightarrow$

(**e**) $Na_2S(s) + HCl(aq) \longrightarrow$

Hydrogen, the Noble Gases, and the Halogens (Sections 22.2, 22.3, and 22.4)

22.19 (a) Give the names and chemical symbols for the three isotopes of hydrogen. (b) List the isotopes in order of decreasing natural abundance. (c) Which hydrogen isotope is radioactive? (d) Write the nuclear equation for the radioactive decay of this isotope.

22.20 Are the physical properties of H_2O different from D_2O? Explain.

22.21 Give a reason why hydrogen might be placed along with the group 1A elements of the periodic table.

22.22 What does hydrogen have in common with the halogens? Explain.

22.23 Write a balanced equation for the preparation of H_2 using (a) Mg and an acid, (b) carbon and steam, (c) methane and steam.

22.24 List (a) three commercial means of producing H_2, (b) three industrial uses of H_2.

22.25 Complete and balance the following equations:

(a) $NaH(s) + H_2O(l) \longrightarrow$

(b) $Fe(s) + H_2SO_4(aq) \longrightarrow$

(c) $H_2(g) + Br_2(g) \longrightarrow$

(d) $Na(l) + H_2(g) \longrightarrow$

(e) $PbO(s) + H_2(g) \longrightarrow$

22.26 Write balanced equations for each of the following reactions (some of these are analogous to reactions shown in the chapter). (a) Aluminum metal reacts with acids to form hydrogen gas. (b) Steam reacts with magnesium metal to give magnesium oxide and hydrogen. (c) Manganese(IV) oxide is reduced to manganese(II) oxide by hydrogen gas. (d) Calcium hydride reacts with water to generate hydrogen gas.

22.27 Identify the following hydrides as ionic, metallic, or molecular: (a) BaH_2, (b) H_2Te, (c) $TiH_{1.7}$.

22.28 Identify the following hydrides as ionic, metallic, or molecular: (a) B_2H_6, (b) RbH, (c) $Th_4H_{1.5}$.

22.29 Describe two characteristics of hydrogen that are favorable for its use as a general energy source in vehicles.

22.30 The H_2/O_2 fuel cell converts elemental hydrogen and oxygen into water, producing, theoretically, 1.23 V. What is the most sustainable way to obtain hydrogen to run a large number of fuel cells? Explain.

22.31 Why does xenon form stable compounds with fluorine, whereas argon does not?

22.32 A friend tells you that the "neon" in neon signs is a compound of neon and aluminum. Can your friend be correct? Explain.

22.33 Write the chemical formula for each of the following, and indicate the oxidation state of the halogen or noble-gas atom in each: (a) calcium hypobromite, (b) bromic acid, (c) xenon trioxide, (d) perchlorate ion, (e) iodous acid, (f) iodine pentafluoride.

22.34 Write the chemical formula for each of the following compounds, and indicate the oxidation state of the halogen or noble-gas atom in each: (a) chlorate ion, (b) hydroiodic acid, (c) iodine trichloride, (d) sodium hypochlorite, (e) perchloric acid, (f) xenon tetrafluoride.

22.35 Name the following compounds and assign oxidation states to the halogens in them: (a) $Fe(ClO_3)_3$, (b) $HClO_2$, (c) XeF_6, (d) BrF_5, (e) $XeOF_4$, (f) HIO_3.

22.36 Name the following compounds and assign oxidation states to the halogens in them: (a) $KClO_3$, (b) $Ca(IO_3)_2$, (c) $AlCl_3$, (d) $HBrO_3$, (e) H_5IO_6, (f) XeF_4.

22.37 Explain each of the following observations: (a) At room temperature I_2 is a solid, Br_2 is a liquid, and Cl_2 and F_2 are both gases. (b) F_2 cannot be prepared by electrolytic oxidation of aqueous F^- solutions. (c) The boiling point of HF is much higher than those of the other hydrogen halides. (d) The halogens decrease in oxidizing power in the order $F_2 > Cl_2 > Br_2 > I_2$.

22.38 Explain the following observations: (a) For a given oxidation state, the acid strength of the oxyacid in aqueous solution decreases in the order chlorine > bromine > iodine. (b) Hydrofluoric acid cannot be stored in glass bottles. (c) HI cannot be prepared by treating NaI with sulfuric acid. (d) The interhalogen ICl_3 is known, but $BrCl_3$ is not.

Oxygen and the Other Group 6A Elements (Sections 22.5 and 22.6)

22.39 Write balanced equations for each of the following reactions. (a) When mercury(II) oxide is heated, it decomposes to form O_2 and mercury metal. (b) When copper(II) nitrate is heated strongly, it decomposes to form copper(II) oxide, nitrogen dioxide, and oxygen. (c) Lead(II) sulfide, PbS(s), reacts with ozone to form $PbSO_4(s)$ and $O_2(g)$. (d) When heated in air, ZnS(s) is converted to ZnO. (e) Potassium peroxide reacts with $CO_2(g)$ to give potassium carbonate and O_2. (f) Oxygen is converted to ozone in the upper atmosphere.

22.40 Complete and balance the following equations:

(a) $CaO(s) + H_2O(l) \longrightarrow$

(b) $Al_2O_3(s) + H^+(aq) \longrightarrow$

(c) $Na_2O_2(s) + H_2O(l) \longrightarrow$

(d) $N_2O_3(g) + H_2O(l) \longrightarrow$

(e) $KO_2(s) + H_2O(l) \longrightarrow$

(f) $NO(g) + O_3(g) \longrightarrow$

22.41 Predict whether each of the following oxides is acidic, basic, amphoteric, or neutral: (a) NO_2, (b) CO_2, (c) Al_2O_3, (d) CaO.

22.42 Select the more acidic member of each of the following pairs: (a) Mn_2O_7 and MnO_2, (b) SnO and SnO_2, (c) SO_2 and SO_3, (d) SiO_2 and SO_2, (e) Ga_2O_3 and In_2O_3, (f) SO_2 and SeO_2.

22.43 Write the chemical formula for each of the following compounds, and indicate the oxidation state of the group 6A element in each: (a) selenous acid, (b) potassium hydrogen sulfite, (c) hydrogen telluride, (d) carbon disulfide, (e) calcium sulfate, (f) cadmium sulfide, (g) zinc telluride.

22.44 Write the chemical formula for each of the following compounds, and indicate the oxidation state of the group 6A element in each: (a) sulfur tetrachloride, (b) selenium trioxide, (c) sodium thiosulfate, (d) hydrogen sulfide, (e) sulfuric acid, (f) sulfur dioxide, (g) mercury telluride.

22.45 In aqueous solution, hydrogen sulfide reduces (a) Fe^{3+} to Fe^{2+}, (b) Br_2 to Br^-, (c) MnO_4^- to Mn^{2+}, (d) HNO_3 to NO_2. In all cases, under appropriate conditions, the product is elemental sulfur. Write a balanced net ionic equation for each reaction.

22.46 An aqueous solution of SO_2 reduces (a) aqueous $KMnO_4$ to $MnSO_4(aq)$, (b) acidic aqueous $K_2Cr_2O_7$ to aqueous Cr^{3+}, (c) aqueous $Hg_2(NO_3)_2$ to mercury metal. Write balanced equations for these reactions.

22.47 Write the Lewis structure for each of the following species, and indicate the structure of each: (a) SeO_3^{2-}; (b) S_2Cl_2; (c) chlorosulfonic acid, HSO_3Cl (chlorine is bonded to sulfur).

22.48 The SF_5^- ion is formed when $SF_4(g)$ reacts with fluoride salts containing large cations, such as CsF(s). Draw the Lewis structures for SF_4 and SF_5^-, and predict the molecular structure of each.

22.49 Write a balanced equation for each of the following reactions: **(a)** Sulfur dioxide reacts with water. **(b)** Solid zinc sulfide reacts with hydrochloric acid. **(c)** Elemental sulfur reacts with sulfite ion to form thiosulfate. **(d)** Sulfur trioxide is dissolved in sulfuric acid.

22.50 Write a balanced equation for each of the following reactions. (You may have to guess at one or more of the reaction products, but you should be able to make a reasonable guess, based on your study of this chapter.) **(a)** Hydrogen selenide can be prepared by reaction of an aqueous acid solution on aluminum selenide. **(b)** Sodium thiosulfate is used to remove excess Cl_2 from chlorine-bleached fabrics. The thiosulfate ion forms SO_4^{2-} and elemental sulfur, while Cl_2 is reduced to Cl^-.

Nitrogen and the Other Group 5A Elements (Sections 22.7 and 22.8)

22.51 Write the chemical formula for each of the following compounds, and indicate the oxidation state of nitrogen in each: **(a)** sodium nitrite, **(b)** ammonia, **(c)** nitrous oxide, **(d)** sodium cyanide, **(e)** nitric acid, **(f)** nitrogen dioxide, **(g)** nitrogen, **(h)** boron nitride.

22.52 Write the chemical formula for each of the following compounds, and indicate the oxidation state of nitrogen in each: **(a)** nitric oxide, **(b)** hydrazine, **(c)** potassium cyanide, **(d)** sodium nitrite, **(e)** ammonium chloride, **(f)** lithium nitride.

22.53 Write the Lewis structure for each of the following species, describe its geometry, and indicate the oxidation state of the nitrogen: **(a)** HNO_2, **(b)** N_3^-, **(c)** $N_2H_5^+$, **(d)** NO_3^-.

22.54 Write the Lewis structure for each of the following species, describe its geometry, and indicate the oxidation state of the nitrogen: **(a)** NH_4^+, **(b)** NO_2^-, **(c)** N_2O, **(d)** NO_2.

22.55 Complete and balance the following equations:
(a) $Mg_3N_2(s) + H_2O(l) \longrightarrow$
(b) $NO(g) + O_2(g) \longrightarrow$
(c) $N_2O_5(g) + H_2O(l) \longrightarrow$
(d) $NH_3(aq) + H^+(aq) \longrightarrow$
(e) $N_2H_4(l) + O_2(g) \longrightarrow$

Which ones of these are redox reactions?

22.56 Write a balanced net ionic equation for each of the following reactions: **(a)** Dilute nitric acid reacts with zinc metal with formation of nitrous oxide. **(b)** Concentrated nitric acid reacts with sulfur with formation of nitrogen dioxide. **(c)** Concentrated nitric acid oxidizes sulfur dioxide with formation of nitric oxide. **(d)** Hydrazine is burned in excess fluorine gas, forming NF_3. **(e)** Hydrazine reduces CrO_4^{2-} to $Cr(OH)_4^-$ in base (hydrazine is oxidized to N_2).

22.57 Write complete balanced half-reactions for **(a)** oxidation of nitrous acid to nitrate ion in acidic solution, **(b)** oxidation of N_2 to N_2O in acidic solution.

22.58 Write complete balanced half-reactions for **(a)** reduction of nitrate ion to NO in acidic solution, **(b)** oxidation of HNO_2 to NO_2 in acidic solution.

22.59 Write a molecular formula for each compound, and indicate the oxidation state of the group 5A element in each formula: **(a)** phosphorous acid, **(b)** pyrophosphoric acid, **(c)** antimony trichloride, **(d)** magnesium arsenide, **(e)** diphosphorus pentoxide, **(f)** sodium phosphate.

22.60 Write a chemical formula for each compound or ion, and indicate the oxidation state of the group 5A element in each formula: **(a)** phosphate ion, **(b)** arsenous acid, **(c)** antimony(III) sulfide, **(d)** calcium dihydrogen phosphate, **(e)** potassium phosphide, **(f)** gallium arsenide.

22.61 Account for the following observations: **(a)** Phosphorus forms a pentachloride, but nitrogen does not. **(b)** H_3PO_2 is a monoprotic acid. **(c)** Phosphonium salts, such as PH_4Cl, can be formed under anhydrous conditions, but they cannot be made in aqueous solution. **(d)** White phosphorus is more reactive than red phosphorus.

22.62 Account for the following observations: **(a)** H_3PO_3 is a diprotic acid. **(b)** Nitric acid is a strong acid, whereas phosphoric acid is weak. **(c)** Phosphate rock is ineffective as a phosphate fertilizer. **(d)** Phosphorus does not exist at room temperature as diatomic molecules, but nitrogen does. **(e)** Solutions of Na_3PO_4 are quite basic.

22.63 Write a balanced equation for each of the following reactions: **(a)** preparation of white phosphorus from calcium phosphate, **(b)** hydrolysis of PBr_3, **(c)** reduction of PBr_3 to P_4 in the gas phase, using H_2.

22.64 Write a balanced equation for each of the following reactions: **(a)** hydrolysis of PCl_5, **(b)** dehydration of phosphoric acid (also called orthophosphoric acid) to form pyrophosphoric acid, **(c)** reaction of P_4O_{10} with water.

Carbon, the Other Group 4A Elements, and Boron (Sections 22.9, 22.10, and 22.11)

22.65 Give the chemical formula for **(a)** hydrocyanic acid, **(b)** nickel tetracarbonyl, **(c)** barium bicarbonate, **(d)** calcium acetylide, **(e)** potassium carbonate.

22.66 Give the chemical formula for **(a)** carbonic acid, **(b)** sodium cyanide, **(c)** potassium hydrogen carbonate, **(d)** acetylene, **(e)** iron pentacarbonyl.

22.67 Complete and balance the following equations:
(a) $ZnCO_3(s) \xrightarrow{\Delta}$
(b) $BaC_2(s) + H_2O(l) \longrightarrow$
(c) $C_2H_2(g) + O_2(g) \longrightarrow$
(d) $CS_2(g) + O_2(g) \longrightarrow$
(e) $Ca(CN)_2(s) + HBr(aq) \longrightarrow$

22.68 Complete and balance the following equations:
(a) $CO_2(g) + OH^-(aq) \longrightarrow$
(b) $NaHCO_3(s) + H^+(aq) \longrightarrow$
(c) $CaO(s) + C(s) \xrightarrow{\Delta}$
(d) $C(s) + H_2O(g) \xrightarrow{\Delta}$
(e) $CuO(s) + CO(g) \longrightarrow$

22.69 Write a balanced equation for each of the following reactions: **(a)** Hydrogen cyanide is formed commercially by passing a mixture of methane, ammonia, and air over a catalyst at 800 °C. Water is a by-product of the reaction. **(b)** Baking soda reacts with acids to produce carbon dioxide gas. **(c)** When barium carbonate reacts in air with sulfur dioxide, barium sulfate and carbon dioxide form.

22.70 Write a balanced equation for each of the following reactions: **(a)** Burning magnesium metal in a carbon dioxide atmosphere reduces the CO_2 to carbon. **(b)** In photosynthesis, solar energy is used to produce glucose ($C_6H_{12}O_6$) and O_2 from carbon dioxide and water. **(c)** When carbonate salts dissolve in water, they produce basic solutions.

22.71 Write the formulas for the following compounds, and indicate the oxidation state of the group 4A element or of boron in each: **(a)** boric acid, **(b)** silicon tetrabromide, **(c)** lead(II) chloride, **(d)** sodium tetraborate decahydrate (borax), **(e)** boric oxide, **(f)** germanium dioxide.

22.72 Write the formulas for the following compounds, and indicate the oxidation state of the group 4A element or of boron in each: (a) silicon dioxide, (b) germanium tetrachloride, (c) sodium borohydride, (d) stannous chloride, (e) diborane, (f) boron trichloride.

22.73 Select the member of group 4A that best fits each description: (a) has the lowest first ionization energy, (b) is found in oxidation states ranging from −4 to +4, (c) is most abundant in Earth's crust.

22.74 Select the member of group 4A that best fits each description: (a) forms chains to the greatest extent, (b) forms the most basic oxide, (c) is a metalloid that can form 2+ ions.

22.75 (a) What is the characteristic geometry about silicon in all silicate minerals? (b) Metasilicic acid has the empirical formula H_2SiO_3. Which of the structures shown in Figure 22.34 would you expect metasilicic acid to have?

22.76 Speculate as to why carbon forms carbonate rather than silicate analogs.

22.77 (a) Determine the number of calcium ions in the chemical formula of the mineral hardystonite, $Ca_xZn(Si_2O_7)$. (b) Determine the number of hydroxide ions in the chemical formula of the mineral pyrophyllite, $Al_2(Si_2O_5)_2(OH)_x$.

22.78 (a) Determine the number of sodium ions in the chemical formula of albite, $Na_xAlSi_3O_8$. (b) Determine the number of hydroxide ions in the chemical formula of tremolite, $Ca_2Mg_5(Si_4O_{11})_2(OH)_x$.

22.79 (a) How does the structure of diborane (B_2H_6) differ from that of ethane (C_2H_6)? (b) Explain why diborane adopts the geometry that it does. (c) What is the significance of the statement that the hydrogen atoms in diborane are described as "hydridic"?

22.80 Write a balanced equation for each of the following reactions: (a) Diborane reacts with water to form boric acid and molecular hydrogen. (b) Upon heating, boric acid undergoes a condensation reaction to form tetraboric acid. (c) Boron oxide dissolves in water to give a solution of boric acid.

Additional Exercises

22.81 Indicate whether each of the following statements is true or false (a) $H_2(g)$ and $D_2(g)$ are allotropic forms of hydrogen. (b) ClF_3 is an interhalogen compound. (c) $MgO(s)$ is an acidic anhydride. (d) $SO_2(g)$ is an acidic anhydride. (e) $2 H_3PO_4(l) \rightarrow H_4P_2O_7(l) + H_2O(g)$ is an example of a condensation reaction. (f) Tritium is an isotope of the element hydrogen. (g) $2SO_2(g) + O_2(g) \rightarrow 2SO_3(g)$ is an example of a disproportionation reaction.

22.82 Although the ClO_4^- and IO_4^- ions have been known for a long time, BrO_4^- was not synthesized until 1965. The ion was synthesized by oxidizing the bromate ion with xenon difluoride, producing xenon, hydrofluoric acid, and the perbromate ion. (a) Write the balanced equation for this reaction. (b) What are the oxidation states of Br in the Br-containing species in this reaction?

22.83 Write a balanced equation for the reaction of each of the following compounds with water: (a) $SO_2(g)$, (b) $Cl_2O_7(g)$, (c) $Na_2O_2(s)$, (d) $BaC_2(s)$, (e) $RbO_2(s)$ (f) $Mg_3N_2(s)$, (g) $NaH(s)$.

22.84 What is the anhydride for each of the following acids: (a) H_2SO_4, (b) $HClO_3$, (c) HNO_2, (d) H_2CO_3, (e) H_3PO_4?

22.85 Hydrogen peroxide is capable of oxidizing (a) hydrazine to N_2 and H_2O, (b) SO_2 to SO_4^{2-}, (c) NO_2^- to NO_3^-, (d) $H_2S(g)$ to $S(s)$, (e) Fe^{2+} to Fe^{3+}. Write a balanced net ionic equation for each of these redox reactions.

22.86 Explain why SO_2 can be used as a reducing agent but SO_3 cannot.

22.87 A sulfuric acid plant produces a considerable amount of heat. This heat is used to generate electricity, which helps reduce operating costs. The synthesis of H_2SO_4 consists of three main chemical processes: (a) oxidation of S to SO_2, (b) oxidation of SO_2 to SO_3, (c) the dissolving of SO_3 in H_2SO_4 and its reaction with water to form H_2SO_4. If the third process produces 130 kJ/mol, how much heat is produced in preparing a mole of H_2SO_4 from a mole of S? How much heat is produced in preparing 5000 pounds of H_2SO_4?

22.88 (a) What is the oxidation state of P in PO_4^{3-} and of N in NO_3^-? (b) Why doesn't N form a stable NO_4^{3-}. ion analogous to P?

22.89 (a) The P_4, P_4O_6 and P_4O_{10} molecules have a common structural feature of four P atoms arranged in a tetrahedron (Figures 22.27 and 22.28). Does this mean that the bonding between the P atoms is the same in all these cases? Explain. (b) Sodium trimetaphosphate ($Na_3P_3O_9$) and sodium tetrametaphosphate ($Na_4P_4O_{12}$) are used as water-softening agents. They contain cyclic $P_3O_9^{3-}$ and $P_4O_{12}^{4-}$ ions, respectively. Propose reasonable structures for these ions.

22.90 Ultrapure germanium, like silicon, is used in semiconductors. Germanium of "ordinary" purity is prepared by the high-temperature reduction of GeO_2 with carbon. The Ge is converted to $GeCl_4$ by treatment with Cl_2 and then purified by distillation; $GeCl_4$ is then hydrolyzed in water to GeO_2 and reduced to the elemental form with H_2. The element is then zone refined. Write a balanced chemical equation for each of the chemical transformations in the course of forming ultrapure Ge from GeO_2.

22.91 (a) Determine the charge of the aluminosilicate ion whose composition is $AlSi_3O_{10}$. (b) Using Figure 22.33, propose a reasonable description of the structure of this aluminosilicate.

Integrative Exercises

[22.92] (a) How many grams of H_2 can be stored in 100.0 kg of the alloy FeTi if the hydride $FeTiH_2$ is formed? (b) What volume does this quantity of H_2 occupy at STP? (c) If this quantity of hydrogen was combusted in air to produce liquid water, how much energy could be produced?

[22.93] Using the thermochemical data in Table 22.1 and Appendix C, calculate the average Xe—F bond enthalpies in XeF_2, XeF_4, and XeF_6, respectively. What is the significance of the trend in these quantities?

22.94 Hydrogen gas has a higher fuel value than natural gas on a mass basis but not on a volume basis. Thus, hydrogen is not competitive with natural gas as a fuel transported long distances through pipelines. Calculate the heats of combustion of H_2 and CH_4 (the principal component of natural gas) (a) per mole of each, (b) per gram of each, (c) per cubic meter of each at STP. Assume $H_2O(l)$ as a product.

22.95 Using ΔG_f° for ozone from Appendix C, calculate the equilibrium constant for Equation 22.24 at 298.0 K, assuming no electrical input.

22.96 The solubility of Cl_2 in 100 g of water at STP is 310 cm³. Assume that this quantity of Cl_2 is dissolved and equilibrated as follows:

$$Cl_2(aq) + H_2O \rightleftharpoons Cl^-(aq) + HClO(aq) + H^+(aq)$$

(a) If the equilibrium constant for this reaction is 4.7×10^{-4}, calculate the equilibrium concentration of HClO formed. (b) What is the pH of the final solution?

[22.97] When ammonium perchlorate decomposes thermally, the products of the reaction are $N_2(g)$, $O_2(g)$, $H_2O(g)$, and $HCl(g)$. (a) Write a balanced equation for the reaction. (*Hint:* You might find it easier to use fractional coefficients for the products.) (b) Calculate the enthalpy change in the reaction per mole of NH_4ClO_4. The standard enthalpy of formation of $NH_4ClO_4(s)$ is −295.8 kJ. (c) When $NH_4ClO_4(s)$ is employed in solid-fuel booster rockets, it is packed with powdered aluminum. Given the high temperature needed for $NH_4ClO_4(s)$ decomposition and what the products of the reaction are, what role does the aluminum play? (d) Calculate the volume of all the gases that would be produced at STP, assuming complete reaction of one pound of ammonium perchlorate.

22.98 The dissolved oxygen present in any highly pressurized, high-temperature steam boiler can be extremely corrosive to its metal parts. Hydrazine, which is completely miscible with water, can be added to remove oxygen by reacting with it to form nitrogen and water. (a) Write the balanced equation for the reaction between gaseous hydrazine and oxygen. (b) Calculate the enthalpy change accompanying this reaction. (c) Oxygen in air dissolves in water to the extent of 9.1 ppm at 20 °C at sea level. How many grams of hydrazine are required to react with all the oxygen in 3.0×10^4 L (the volume of a small swimming pool) under these conditions?

22.99 One method proposed for removing SO_2 from the flue gases of power plants involves reaction with aqueous H_2S. Elemental sulfur is the product. (a) Write a balanced chemical equation for the reaction. (b) What volume of H_2S at 27 °C and 760 torr would be required to remove the SO_2 formed by burning 2.0 tons of coal containing 3.5% S by mass? (c) What mass of elemental sulfur is produced? Assume that all reactions are 100% efficient.

22.100 The maximum allowable concentration of $H_2S(g)$ in air is 20 mg per kilogram of air (20 ppm by mass). How many grams of FeS would be required to react with hydrochloric acid to produce this concentration at 1.00 atm and 25 °C in an average room measuring 12 ft × 20 ft × 8 ft? (Under these conditions, the average molar mass of air is 29.0 g/mol.)

22.101 The standard heats of formation of $H_2O(g)$, $H_2S(g)$, $H_2Se(g)$, and $H_2Te(g)$ are −241.8, −20.17, +29.7, and +99.6 kJ/mol, respectively. The enthalpies necessary to convert the elements in their standard states to one mole of gaseous atoms are 248, 277, 227, and 197 kJ/mol of atoms for O, S, Se,

and Te, respectively. The enthalpy for dissociation of H_2 is 436 kJ/mol. Calculate the average H—O, H—S, H—Se, and H—Te bond enthalpies, and comment on their trend.

22.102 Manganese silicide has the empirical formula MnSi and melts at 1280 °C. It is insoluble in water but does dissolve in aqueous HF. (a) What type of compound do you expect MnSi to be: metallic, molecular, covalent-network, or ionic? (b) Write a likely balanced chemical equation for the reaction of MnSi with concentrated aqueous HF.

[22.103] Chemists tried for a long time to make molecular compounds containing silicon–silicon double bonds; they finally succeeded in 1981. The trick is having large, bulky R groups on the silicon atoms to make $R_2Si=SiR_2$ compounds. What experiments could you do to prove that a new compound has a silicon–silicon double bond rather than a silicon–silicon single bond?

22.104 Hydrazine has been employed as a reducing agent for metals. Using standard reduction potentials, predict whether the following metals can be reduced to the metallic state by hydrazine under standard conditions in acidic solution: (a) Fe^{2+}, (b) Sn^{2+}, (c) Cu^{2+}, (d) Ag^+, (e) Cr^{3+}, (f) Co^{3+}.

22.105 Both dimethylhydrazine, $(CH_3)_2NNH_2$, and methylhydrazine, CH_3NHNH_2, have been used as rocket fuels. When dinitrogen tetroxide (N_2O_4) is used as the oxidizer, the products are H_2O, CO_2, and N_2. If the thrust of the rocket depends on the volume of the products produced, which of the substituted hydrazines produces a greater thrust per gram total mass of oxidizer plus fuel? (Assume that both fuels generate the same temperature and that $H_2O(g)$ is formed.)

22.106 Carbon forms an unusual unstable oxide of formula C_3O_2, called carbon suboxide. Carbon suboxide is made by using P_2O_5 to dehydrate the dicarboxylic acid called malonic acid, which has the formula HOOC—CH_2—COOH. (a) Write a balanced reaction for the production of carbon suboxide from malonic acid. (b) How many grams of carbon suboxide could be made from 20.00 g of malonic acid? (c) Suggest a Lewis structure for C_3O_2. (*Hint:* The Lewis structure of malonic acid suggests which atoms are connected to which.) (d) By using the information in Table 8.5, predict the C—C and C—O bond lengths in C_3O_2. (e) Sketch the Lewis structure of a product that could result by the addition of 2 mol of H_2 to 1 mol of C_3O_2.

22.107 Borazine, $(BH)_3(NH)_3$, is an analog of C_6H_6, benzene. It can be prepared from the reaction of diborane with ammonia, with hydrogen as another product; or from lithium borohydride and ammonium chloride, with lithium chloride and hydrogen as the other products. (a) Write balanced chemical equations for the production of borazine using both synthetic methods. (b) Draw the Lewis dot structure of borazine. (c) How many grams of borazine can be prepared from 2.00 L of ammonia at STP, assuming diborane is in excess?

Designing an Experiment

You are given samples of five substances. At room temperature, three are colorless gases, one is a colorless liquid, and one is a white solid. You are told that the substances are NF_3, PF_3, PCl_3, PF_5, and PCl_5. Let's design experiments to determine which substance is which, using concepts from this and earlier chapters.

(a) Assuming that you don't have access to either the internet or to a handbook of chemistry (as is the case during your exams!), design experiments that would allow you to identify the substances. (b) How might you proceed differently if you had access to data from the internet? (c) Which of the substances could undergo reaction to add more atoms around the central atom? What types of reactions might you choose to test this hypothesis? (d) Based on what you know about intermolecular forces, which of the substances is likely the solid?

23

Transition Metals and Coordination Chemistry

The colors of our world are beautiful, but to a chemist they are also informative—providing insights into the structure and bonding of matter. Compounds of the transition metals constitute an important group of colored substances. Some of them are used in pigments; others produce the colors in glass and precious gems. The use of vivid green, yellow, and

blue colors in the paintings of impressionists like Monet, Cezanne, and van Gogh was made possible by the development of synthetic pigments in the 1800s. Three such pigments used extensively by the impressionists were cobalt blue, $CoAl_2O_4$, chrome yellow, $PbCrO_4$, and emerald green, $Cu_4(CH_3COO)_2(AsO_2)_6$. In each case, the presence of a transition metal ion is directly responsible for the color of the pigment—Co^{2+} in cobalt blue, Cr^{6+} in chrome yellow, and Cu^{2+} in emerald green.

The color of a given transition-metal compound depends upon not only the transition-metal ion but also upon the identity and geometry of the ions and/or molecules that surround it. To appreciate the importance of the transition-metal ion surroundings, consider the colors of the minerals azurite, $Cu_3(CO_3)_2(OH)_2$, and malachite, $Cu_2(CO_3)(OH)_2$ **(Figure 23.1)**. Both minerals contain Cu^{2+} ions surrounded by carbonate and hydroxide ions, but subtle differences in the local surroundings of the Cu^{2+} ion lead to the contrasting colors of these two minerals. In this chapter, we will learn why transition-metal compounds are colored and how changes in the identity and environment of the transition-metal ion lead to changes in color.

▶ **WATER LILLIES AND THE JAPANESE BRIDGE**, was painted by Claude Monet in 1899.

WHAT'S AHEAD

23.1 THE TRANSITION METALS We examine the physical properties, electron configurations, oxidation states, and magnetic properties of the *transition metals*.

23.2 TRANSITION-METAL COMPLEXES We introduce the concepts of *metal complexes* and *ligands* and provide a brief history of the development of *coordination chemistry*.

23.3 COMMON LIGANDS IN COORDINATION CHEMISTRY We examine some common geometries found in coordination complexes and how the geometries relate to *coordination numbers*.

23.4 NOMENCLATURE AND ISOMERISM IN COORDINATION CHEMISTRY We introduce the nomenclature used for coordination compounds. We see that coordination compounds exhibit *isomerism*, in which two compounds have the same composition but different structures, and then look at two types: *structural isomers* and *stereoisomers*.

23.5 COLOR AND MAGNETISM IN COORDINATION CHEMISTRY We discuss color and magnetism in coordination compounds, emphasizing the visible portion of the electromagnetic spectrum and the notion of *complementary colors*. We then see that many transition-metal complexes are paramagnetic because they contain unpaired electrons.

23.6 CRYSTAL-FIELD THEORY We explore how *crystal-field theory* allows us to explain some of the interesting spectral and magnetic properties of coordination compounds.

▲ **Figure 23.1 Crystals of blue azurite and green malachite.** These semiprecious stones were ground up and used as pigments in the Middle Ages and Renaissance, but they were eventually replaced with blue and green pigments that have superior chemical stability.

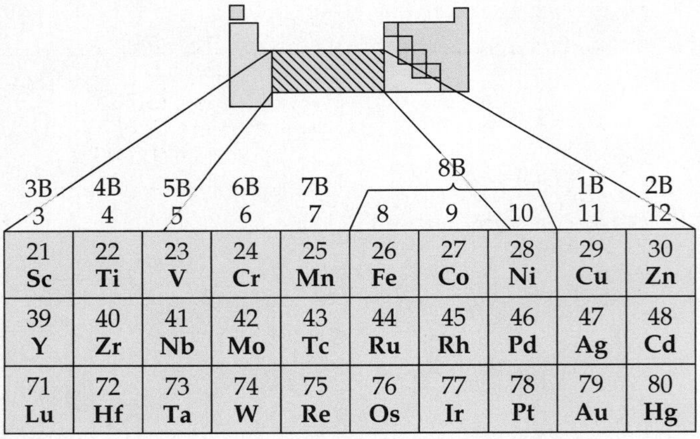

Transition metals and their compounds are important for much more than their color. For example, they are used as catalysts and as magnets, and they play an important role in biology.

In earlier chapters, we saw that metal ions can function as Lewis acids, forming covalent bonds with molecules and ions functioning as Lewis bases. ∞∞ (Section 16.11) We have encountered many ions and compounds that result from such interactions, such as $[Ag(NH_3)_2]^+$ in ∞∞ Section 17.5 and hemoglobin in ∞∞ Section 13.6. In this chapter, we focus on the rich and important chemistry associated with such complex assemblies of metal ions surrounded by molecules and ions. Metal compounds of this kind are called *coordination compounds*, and the branch of chemistry that focuses on them is called *coordination chemistry*.

23.1 | The Transition Metals

The part of the periodic table in which the *d* orbitals are being filled as we move left to right across a row is the home of the transition metals (◀ Figure 23.2). ∞∞ (Section 6.8)

With some exceptions (e.g., platinum, gold), metallic elements are found in nature as solid inorganic compounds called **minerals**. Notice from ▼ Table 23.1 that minerals are identified by common names rather than chemical names.

Most transition metals in minerals have oxidation states ranging from +1 to +4. To obtain a pure metal from its mineral, various chemical processes must be performed to reduce the metal to the 0 oxidation state. **Metallurgy** is the science and technology of extracting metals from their natural sources and preparing them for practical use. It usually involves several steps: (1) mining, that is, removing the relevant *ore* (a mixture of minerals) from the ground, (2) concentrating the ore or otherwise preparing it for further treatment, (3) reducing the ore to obtain the free metal, (4) purifying the metal, and (5) mixing it with other elements to modify its properties. This last process produces an *alloy*, a metallic material composed of two or more elements. ∞∞ (Section 12.3)

3B	4B	5B	6B	7B		8B		1B	2B
3	4	5	6	7	8	9	10	11	12
21 Sc	22 Ti	23 V	24 Cr	25 Mn	26 Fe	27 Co	28 Ni	29 Cu	30 Zn
39 Y	40 Zr	41 Nb	42 Mo	43 Tc	44 Ru	45 Rh	46 Pd	47 Ag	48 Cd
71 Lu	72 Hf	73 Ta	74 W	75 Re	76 Os	77 Ir	78 Pt	79 Au	80 Hg

▲ **Figure 23.2 The position of the transition metals in the periodic table.** They are the B groups in periods 4, 5, and 6. The short lived, radioactive transition metals from period 7 are not shown.

Physical Properties

Some physical properties of the period 4 (also known as "first-row") transition metals are listed in ▶ Table 23.2. The properties of the heavier transition metals vary similarly across periods 5 and 6.

Table 23.1 Principal Mineral Sources of Some Transition Metals

Metal	Mineral	Mineral Composition
Chromium	Chromite	$FeCr_2O_4$
Cobalt	Cobaltite	$CoAsS$
Copper	Chalcocite	Cu_2S
	Chalcopyrite	$CuFeS_2$
	Malachite	$Cu_2CO_3(OH)_2$
Iron	Hematite	Fe_2O_3
	Magnetite	Fe_3O_4
Manganese	Pyrolusite	MnO_2
Mercury	Cinnabar	HgS
Molybdenum	Molybdenite	MoS_2
Titanium	Rutile	TiO_2
	Ilmenite	$FeTiO_3$
Zinc	Sphalerite	ZnS

Table 23.2 Properties of the Period 4 Transition Metals

Group	3B	4B	5B	6B	7B	8B			1B	2B
Element:	Sc	Ti	V	Cr	Mn	Fe	Co	Ni	Cu	Zn
Ground state electron configuration	$3d^14s^2$	$3d^24s^2$	$3d^34s^2$	$3d^54s^1$	$3d^54s^2$	$3d^64s^2$	$3d^74s^2$	$3d^84s^2$	$3d^{10}4s^1$	$3d^{10}4s^2$
First ionization energy (kJ/mol)	631	658	650	653	717	759	758	737	745	906
Metallic radius (Å)	1.64	1.47	1.35	1.29	1.37	1.26	1.25	1.25	1.28	1.37
Density (g/cm³)	3.0	4.5	6.1	7.9	7.2	7.9	8.7	8.9	8.9	7.1
Melting point (°C)	1541	1660	1917	1857	1244	1537	1494	1455	1084	420
Crystal structure*	hcp	hcp	bcc	bcc	**	bcc	hcp	fcc	fcc	hcp

*Abbreviations for crystal structures are hcp = hexagonal close packed, fcc = face centered cubic, bcc = body centered cubic. ∞(Section 12.3)

**Manganese has a more complex crystal structure.

▶ **Figure 23.3** shows the atomic radius observed in close-packed metallic structures as a function of group number.* The trends seen in the graph are a result of two competing forces. On the one hand, increasing effective nuclear charge favors a decrease in radius as we move left to right across each period. ∞(Section 7.2) On the other hand, the metallic bonding strength increases until we reach the middle of each period and then decreases as we fill antibonding orbitals. ∞(Section 12.4) As a general rule, a bond shortens as it becomes stronger. ∞(Section 8.8) For groups 3B through 6B, these two effects work cooperatively and the result is a marked decrease in radius. In elements to the right of group 6B, the two effects counteract each other, reducing the decrease and eventually leading to an increase in radius.

Give It Some Thought

Which element has the largest bonding atomic radius: Sc, Fe, or Au?

In general, atomic radii increase as we move down in a given group in the periodic table because of the increasing principal quantum number of the outer-shell electrons. (∞ Section 7.3) Note in Figure 23.2, however, that once we move beyond the group 3B elements, the period 5 and period 6 transition elements in a given group have virtually the same radii. In group 5B, for example, tantalum in period 6 has virtually the same radius as niobium in period 5. This interesting and important effect has its origin in the lanthanide series, elements 57 through 70. The filling of 4f orbitals through the lanthanide elements ∞(Figure 6.31) causes a steady increase in the effective nuclear charge, producing a size decrease, called the **lanthanide contraction**, that just offsets the increase we expect as we go from period 5 transition metals to period 6. Thus, the period 5 and period 6 transition metals in each group have about the same radii and similar chemical properties. For example, the chemical properties of the group 4B metals zirconium (period 5) and hafnium (period 6) are remarkably similar. These two metals always occur together in nature, and they are very difficult to separate.

 GO FIGURE

Does the variation in radius of the transition metals follow the same trend as the effective nuclear charge on moving from left to right across the periodic table?

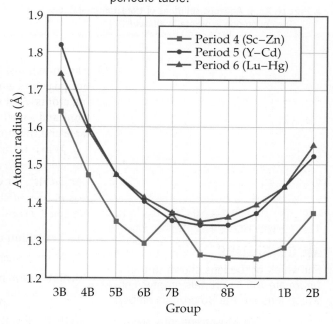

▲ **Figure 23.3 Radii of transition metals as a function of group number.**

Electron Configurations and Oxidation States

Transition metals owe their location in the periodic table to the filling of the *d* subshells, as you saw in Figure 6.31. Many of the chemical and physical properties of transition metals result from the unique characteristics of the *d* orbitals. For a given transition-metal atom, the valence $(n - 1)d$ orbitals are smaller than the corresponding valence *ns* and *np* orbitals. In quantum mechanical terms, the $(n - 1)d$ orbital wave functions drop off more rapidly as we move away from the nucleus than do the *ns* and *np* orbital

*Note that the radii defined in this way, often referred to as metallic radii, differ somewhat from the bonding atomic radii defined in Section 7.3.

⚠ GO FIGURE

In which transition-metal ion of this group are the 3*d* orbitals completely filled?

▲ **Figure 23.4 Aqueous solutions of transition metal ions.** Left to right: Co^{2+}, Ni^{2+}, Cu^{2+}, and Zn^{2+}. The counterion is nitrate in all cases.

wave functions. This characteristic feature of the *d* orbitals limits their interaction with orbitals on neighboring atoms, but not so much that they are insensitive to surrounding atoms. As a result, electrons in these orbitals behave sometimes like valence electrons and sometimes like core electrons. The details depend on location in the periodic table and the atom's environment.

When transition metals are oxidized, *they lose their outer s electrons before they lose electrons from the d subshell.* ∞ (Section 7.4) The electron configuration of Fe is $[Ar]3d^6 4s^2$, for example, whereas that of Fe^{2+} is $[Ar]3d^6$. Formation of Fe^{3+} requires loss of one 3*d* electron, giving $[Ar]3d^5$. Most transition-metal ions contain partially occupied *d* subshells, which are responsible in large part for three characteristics:

1. Transition metals often have more than one stable oxidation state.
2. Many transition-metal compounds are colored, as shown in ▲ **Figure 23.4**.
3. Transition metals and their compounds often exhibit magnetic properties.

⚠ GO FIGURE

For which of the ions shown in this figure are the 4*s* orbitals empty? For which ions are the 3*d* orbitals empty?

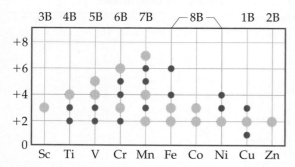

▲ **Figure 23.5 Nonzero oxidation states of the period 4 transition metals.**

◄ **Figure 23.5** shows the common nonzero oxidation states for the period 4 transition metals. The +2 oxidation state, which is common for most transition metals, is due to the loss of the two outer 4*s* electrons. This oxidation state is found for all these elements except Sc, where the 3+ ion with an [Ar] configuration is particularly stable.

Oxidation states above +2 are due to successive losses of 3*d* electrons. From Sc through Mn the maximum oxidation state increases from +3 to +7, equaling in each case the total number of 4*s* plus 3*d* electrons in the atom. Thus, manganese has a maximum oxidation state of $2 + 5 = +7$. As we move to the right beyond Mn in Figure 23.5, the maximum oxidation state decreases. This decrease is due in part to the attraction of *d* orbital electrons to the nucleus, which increases faster than the attraction of the *s* orbital electrons to the nucleus as we move left to right across the periodic table. In other words, in each period the *d* electrons become more corelike as the atomic number increases. By the time we get to zinc, it is not possible to remove electrons from the 3*d* orbitals through chemical oxidation.

In the transition metals of periods 5 and 6, the increased size of the $4d$ and $5d$ orbitals makes it possible to attain maximum oxidation states as high as $+8$, which is achieved in RuO_4 and OsO_4. In general, the maximum oxidation states are found only when the metals are combined with the most electronegative elements, especially O, F, and in some cases Cl.

Give It Some Thought

Why does Ti^{5+} not exist?

Magnetism

The spin an electron possesses gives the electron a *magnetic moment*, a property that causes the electron to behave like a tiny magnet. In a *diamagnetic* solid, defined as one in which all the electrons in the solid are paired, the spin-up and spin-down electrons cancel one another. ∞ (Section 9.8) Diamagnetic substances are generally described as being nonmagnetic, but when a diamagnetic substance is placed in a magnetic field, the motions of the electrons cause the substance to be very weakly repelled by the magnet. In other words, these supposedly nonmagnetic substances do show some very faint magnetic character in the presence of a magnetic field.

A substance in which the atoms or ions have one or more unpaired electrons is *paramagnetic*. ∞ (Section 9.8) In a paramagnetic solid, the electrons on one atom or ion do not influence the unpaired electrons on neighboring atoms or ions. As a result, the magnetic moments on the atoms or ions are randomly oriented and constantly changing direction, as shown in ▶ Figure 23.6(a). When a paramagnetic substance is placed in a magnetic field, however, the magnetic moments tend to align parallel to one another, producing a net attractive interaction with the magnet. Thus, unlike a diamagnetic substance, which is weakly repulsed by a magnetic field, a paramagnetic substance is attracted to a magnetic field.

When you think of a magnet, you probably picture a simple iron magnet. Iron exhibits **ferromagnetism**, a form of magnetism much stronger than paramagnetism. Ferromagnetism arises when the unpaired electrons of the atoms or ions in a solid are influenced by the orientations of the electrons in neighboring atoms or ions. The most stable (lowest-energy) arrangement is when the spins of electrons on adjacent atoms or ions are aligned in the same direction, as in Figure 23.6(b). When a ferromagnetic solid is placed in a magnetic field, the electrons tend to align strongly in a direction parallel to the magnetic field. The attraction to the magnetic field that results may be as much as one million times stronger than that for a paramagnetic substance.

When a ferromagnet is removed from an external magnetic field, the interactions between the electrons cause the ferromagnetic substance to maintain a magnetic moment. We then refer to it as a *permanent magnet* (▶ Figure 23.7).

The only ferromagnetic transition metals are Fe, Co, and Ni, but many alloys also exhibit ferromagnetism, which is in some cases stronger than the ferromagnetism of the pure metals. Particularly powerful ferromagnetism is found in compounds containing both transition metals and lanthanide metals. Two of the most important examples are $SmCo_5$ and $Nd_2Fe_{14}B$.

Two additional types of magnetism involving ordered arrangements of unpaired electrons are depicted in Figure 23.6. In materials that exhibit **antiferromagnetism** [Figure 23.6(c)], the unpaired electrons on a given atom or ion align so that their spins are oriented in the direction opposite the spin direction on neighboring atoms.

GO FIGURE

Describe how the representation shown for the paramagnetic material would change if the material were placed in a magnetic field.

Paramagnetic; spins random; spins do align if in a magnetic field

(a)

Ferromagnetic; spins aligned parallel to each other

(b)

Antiferromagnetic; spins align in opposite directions and cancel each other

(c)

Ferrimagnetic; unequal spins align in opposite directions but do not completely cancel each other

(d)

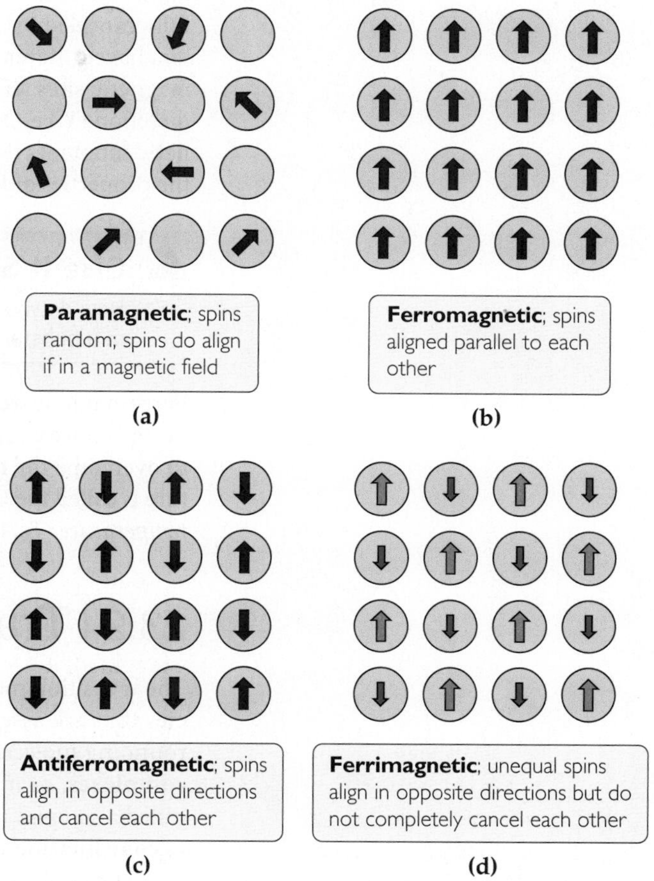

▲ Figure 23.6 **The relative orientation of electron spins in various types of magnetic substances.**

▲ Figure 23.7 **A permanent magnet.** Permanent magnets are made from ferromagnetic and ferrimagnetic materials.

This means that the spin-up and spin-down electrons cancel each other. Examples of antiferromagnetic substances are chromium metal, FeMn alloys, and such transition-metal oxides as Fe_2O_3, $LaFeO_3$, and MnO.

A substance that exhibits **ferrimagnetism** [Figure 23.6(d)] has both ferromagnetic and antiferromagnetic characteristics. Like an antiferromagnet, the unpaired electrons align so that the spins in adjacent atoms or ions point in opposite directions. However, unlike an antiferromagnet, the net magnetic moments of the spin-up electrons are not fully canceled by the spin-down electrons. This can happen because the magnetic centers have different numbers of unpaired electrons ($NiMnO_3$), because the number of magnetic sites aligned in one direction is larger than the number aligned in the other direction ($Y_3Fe_5O_{12}$), or because both these conditions apply (Fe_3O_4). Because the magnetic moments do not cancel, the properties of ferrimagnetic materials are similar to the properties of ferromagnetic materials.

 Give It Some Thought

How do you think spin–spin interactions of unpaired electrons on adjacent atoms in a substance are affected by the interatomic distance?

Ferromagnets, ferrimagnets, and antiferromagnets all become paramagnetic when heated above a critical temperature. This happens when the thermal energy is sufficient to overcome the forces determining the spin directions of the electrons. This temperature is called the *Curie temperature*, T_C, for ferromagnets and ferrimagnets and the *Néel temperature*, T_N, for antiferromagnets.

23.2 | Transition-Metal Complexes

The transition metals occur in many interesting and important molecular forms. Species that are assemblies of a central transition-metal ion bonded to a group of surrounding molecules or ions, such as $[Ag(NH_3)_2]^+$ and $[Fe(H_2O)_6]^{3+}$, are called **metal complexes**, or merely *complexes*.* If the complex carries a net charge, it is generally called a *complex ion*. ∞ (Section 17.5) Compounds that contain complexes are known as **coordination compounds**.

The molecules or ions that bond to the metal ion in a complex are known as **ligands** (from the Latin word *ligare*, "to bind"). There are two NH_3 ligands bonded to Ag^+ in the complex ion $[Ag(NH_3)_2]^+$, for instance, and six H_2O ligands bonded to Fe^{3+} in $[Fe(H_2O)_6]^{3+}$. Each ligand functions as a Lewis base and so donates a pair of electrons to form the ligand–metal bond. ∞ (Section 16.11) Thus, every ligand has at least one unshared pair of valence electrons. Four of the most frequently encountered ligands,

illustrate that most ligands are either polar molecules or anions. In forming a complex, the ligands are said to *coordinate* to the metal.

 Give It Some Thought

Is the interaction between an ammonia ligand and a metal cation a Lewis acid–base interaction? If so, which species acts as the Lewis acid?

---

*Most of the coordination compounds we examine in this chapter contain transition-metal ions, although ions of other metals can also form complexes.

The Development of Coordination Chemistry: Werner's Theory

Because compounds of the transition metals are beautifully colored, the chemistry of these elements fascinated chemists even before the periodic table was introduced. During the late 1700s through the 1800s, the many coordination compounds that were isolated and studied had properties that were puzzling in light of the bonding theories prevailing at the time. ▼ **Table 23.3**, for example, lists a series of $CoCl_3-NH_3$ compounds that have strikingly different colors. Note that the third and fourth species have different colors even though the originally assigned formula was the same for both, $CoCl_3 \cdot 4\,NH_3$.

The modern formulations of the compounds in Table 23.3 are based on various lines of experimental evidence. For example, all four compounds are strong electrolytes ∞ (Section 4.1) but yield different numbers of ions when dissolved in water. Dissolving $CoCl_3 \cdot 6\,NH_3$ in water yields four ions per formula unit ($[Co(NH_3)_6]^{3+}$ plus three Cl^- ions), whereas $CoCl_3 \cdot 5\,NH_3$ yields only three ions per formula unit ($[Co(NH_3)_5Cl]^{2+}$ and two Cl^- ions). Furthermore, the reaction of the compounds with excess aqueous silver nitrate leads to the precipitation of different amounts of $AgCl(s)$. When $CoCl_3 \cdot 6\,NH_3$ is treated with excess $AgNO_3(aq)$, 3 mol of $AgCl(s)$ precipitate per mole of complex, which means all three Cl^- ions in the complex can react to form $AgCl(s)$. By contrast, when $CoCl_3 \cdot 5\,NH_3$ is treated with excess $AgNO_3(aq)$, only 2 mol of $AgCl(s)$ precipitate per mole of complex, telling us that one of the Cl^- ions in the complex does not react. These results are summarized in Table 23.3.

In 1893 the Swiss chemist Alfred Werner (1866–1919) proposed a theory that successfully explained the observations in Table 23.3. In a theory that became the basis for understanding coordination chemistry, Werner proposed that any metal ion exhibits both a primary valence and a secondary valence. The *primary valence* is the oxidation state of the metal, which is +3 for the complexes in Table 23.3. ∞ (Section 4.4) The *secondary valence* is the number of atoms bonded to the metal ion, which is also called the **coordination number**. For these cobalt complexes, Werner deduced a coordination number of 6 with the ligands in an octahedral arrangement around the Co^{3+} ion.

Werner's theory provided a beautiful explanation for the results in Table 23.3. The NH_3 molecules are ligands bonded to the Co^{3+} ion (through the nitrogen atom as we will see later); if there are fewer than six NH_3 molecules, the remaining ligands are Cl^- ions. The central metal and the ligands bound to it constitute the **coordination sphere** of the complex.

In writing the chemical formula for a coordination compound, Werner suggested using square brackets to signify the makeup of the coordination sphere in any given compound. He therefore proposed that $CoCl_3 \cdot 6\,NH_3$ and $CoCl_3 \cdot 5\,NH_3$ are better written as $[Co(NH_3)_6]Cl_3$ and $[Co(NH_3)_5Cl]Cl_2$, respectively. He further proposed that the chloride ions that are part of the coordination sphere are bound so tightly that they do not dissociate when the complex is dissolved in water. Thus, dissolving $[Co(NH_3)_5Cl]Cl_2$ in water produces a $[Co(NH_3)_5Cl]^{2+}$ ion and two Cl^- ions.

Werner's ideas also explained why there are two forms of $CoCl_3 \cdot 4\,NH_3$. Using Werner's postulates, we write the formula as $[Co(NH_3)_4Cl_2]Cl$. As shown in **Figure 23.8**, there are two ways to arrange the ligands in the $[Co(NH_3)_4Cl_2]^+$ complex, called the *cis* and *trans* forms. In the cis form, the two chloride ligands occupy adjacent vertices of the octahedral arrangement. In *trans*-$[Co(NH_3)_4Cl_2]^+$ the two chlorides are opposite

Table 23.3 Properties of Some Ammonia Complexes of Cobalt(III)

Original Formulation	Color	Ions per Formula Unit	"Free" Cl^- Ions per Formula Unit	Modern Formulation
$CoCl_3 \cdot 6\,NH_3$	Orange	4	3	$[Co(NH_3)_6]Cl_3$
$CoCl_3 \cdot 5\,NH_3$	Purple	3	2	$[Co(NH_3)_5Cl]Cl_2$
$CoCl_3 \cdot 4\,NH_3$	Green	2	1	*trans*-$[Co(NH_3)_4Cl_2]Cl$
$CoCl_3 \cdot 4\,NH_3$	Violet	2	1	*cis*-$[Co(NH_3)_4Cl_2]Cl$

GO FIGURE

Is there another way to arrange the chloride ions in the $[Co(NH_3)_4Cl_2]^+$ ion besides the two shown in this figure?

Two Cl on same side of metal ion

cis isomer

Two Cl on opposite sides of metal ion

trans isomer

▲ **Figure 23.8 Isomers of [Co(NH₃)₄Cl₂]⁺.** The cis isomer is violet, and the trans isomer is green.

each other. It is this difference in positions of the Cl ligands that leads to two compounds, one violet and one green.

The insight Werner provided into the bonding in coordination compounds is even more remarkable when we realize that his theory predated Lewis's ideas of covalent bonding by more than 20 years! Because of his tremendous contributions to coordination chemistry, Werner was awarded the 1913 Nobel Prize in Chemistry.

SAMPLE EXERCISE 23.1 | Identifying the Coordination Sphere of a Complex

Palladium(II) tends to form complexes with coordination number 4. A compound has the composition $PdCl_2 \cdot 3\,NH_3$. **(a)** Write the formula for this compound that best shows the coordination structure. **(b)** When an aqueous solution of the compound is treated with excess $AgNO_3(aq)$, how many moles of $AgCl(s)$ are formed per mole of $PdCl_2 \cdot 3\,NH_3$?

SOLUTION

Analyze We are given the coordination number of Pd(II) and a chemical formula indicating that the complex contains NH_3 and Cl^-. We are asked to determine **(a)** which ligands are attached to Pd(II) in the compound and **(b)** how the compound behaves toward $AgNO_3$ in aqueous solution.

Plan (a) Because of their charge, the Cl^- ions can be either in the coordination sphere, where they are bonded directly to the metal, or outside the coordination sphere, where they are bonded ionically to the complex. The electrically neutral NH_3 ligands must be in the coordination sphere, if we assume four ligands bonded to the Pd(II) ion. **(b)** Any chlorides in the coordination sphere do not precipitate as AgCl.

Solve

(a) By analogy to the ammonia complexes of cobalt(III) shown in Figure 23.7, we predict that the three NH_3 are ligands attached to the Pd(II) ion. The fourth ligand around Pd(II) is one chloride ion. The second chloride ion is not a ligand; it serves only as a *counterion* (a noncoordinating ion that balances charge) in the compound. We conclude that the formula showing the structure best is $[Pd(NH_3)_3Cl]Cl$.

(b) Because only the non-ligand Cl^- can react, we expect to produce 1 mol of $AgCl(s)$ per mole of complex. The balanced equation is

$$[Pd(NH_3)_3Cl]Cl(aq) + AgNO_3(aq) \longrightarrow [Pd(NH_3)_3Cl]NO_3(aq) + AgCl(s)$$

This is a metathesis reaction ∞ (Section 4.2) in which one of the cations is the $[Pd(NH_3)_3Cl]^+$ complex ion.

> **Practice Exercise 1**
>
> When the compound $RhCl_3 \cdot 4 NH_3$ is dissolved in water and treated with excess $AgNO_3(aq)$ one mole of $AgCl(s)$ is formed for every mole of $RhCl_3 \cdot 4 NH_3$. What is the correct way to write the formula of this compound? **(a)** $[Rh(NH_3)_4Cl_3]$, **(b)** $[Rh(NH_3)_4Cl_2]Cl$, **(c)** $[Rh(NH_3)_4Cl]Cl_2$, **(d)** $[Rh(NH_3)_4]Cl_3$, **(e)** $[RhCl_3](NH_3)_4$.
>
> **Practice Exercise 2**
>
> Predict the number of ions produced per formula unit when the compound $CoCl_2 \cdot 6 H_2O$ dissolves in water to form an aqueous solution.

The Metal–Ligand Bond

The bond between a ligand and a metal ion is a Lewis acid–base interaction. ∞ (Section 16.11) Because the ligands have available pairs of electrons, they can function as Lewis bases (electron-pair donors). Metal ions (particularly transition-metal ions) have empty valence orbitals, so they can act as Lewis acids (electron-pair acceptors). We can picture the bond between the metal ion and ligand as the result of their sharing a pair of electrons initially on the ligand:

$$Ag^+(aq) + 2 \underset{\underset{H}{|}}{\overset{\overset{H}{|}}{:N}}{-}H(aq) \longrightarrow \left[H{-}\underset{\underset{H}{|}}{\overset{\overset{H}{|}}{N}}{:}Ag{:}\underset{\underset{H}{|}}{\overset{\overset{H}{|}}{N}}{-}H \right]^+ (aq) \qquad [23.1]$$

The formation of metal–ligand bonds can markedly alter the properties we observe for the metal ion. A metal complex is a distinct chemical species that has physical and chemical properties different from those of the metal ion and ligands from which it is formed. As one example, ▼ **Figure 23.9** shows the color change that occurs when aqueous solutions of NCS^- (colorless) and Fe^{3+} (yellow) are mixed, forming $[Fe(H_2O)_5NCS]^{2+}$.

▲ **GO FIGURE**

Write a balanced chemical equation for the reaction depicted in this figure.

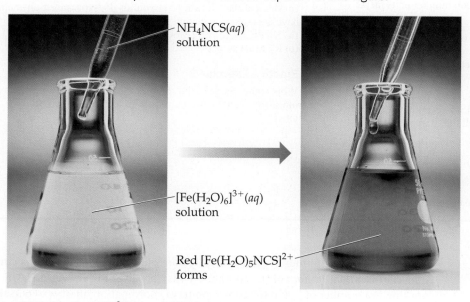

NH$_4$NCS(aq) solution

[Fe(H$_2$O)$_6$]$^{3+}$(aq) solution

Red [Fe(H$_2$O)$_5$NCS]$^{2+}$ forms

▲ **Figure 23.9 Reaction of $Fe^{3+}(aq)$ and $NCS^-(aq)$.**

Complex formation can also significantly change other properties of metal ions, such as their ease of oxidation or reduction. Silver ion, for example, is readily reduced in water,

$$Ag^+(aq) + e^- \longrightarrow Ag(s) \qquad E° = +0.799 \text{ V} \qquad [23.2]$$

but the $[Ag(CN)_2]^-$ ion is not so easily reduced because complexation by CN^- ions stabilizes silver in the +1 oxidation state:

$$[Ag(CN)_2]^-(aq) + e^- \longrightarrow Ag(s) + 2\,CN^-(aq) \quad E° = -0.31 \text{ V} \quad [23.3]$$

Hydrated metal ions are complexes in which the ligand is water. Thus, $Fe^{3+}(aq)$ consists largely of $[Fe(H_2O)_6]^{3+}$. ∞ (Section 16.11) It is important to realize that ligands can undergo reaction. For example, we saw in Figure 16.16 that a water molecule in $[Fe(H_2O)_6]^{3+}(aq)$ can be deprotonated to yield $[Fe(H_2O)_5OH]^{2+}(aq)$ and $H^+(aq)$. The iron ion retains its oxidation state; the coordinated hydroxide ligand, with a 1− charge, reduces the complex charge to 2+. Ligands can also be displaced from the coordination sphere by other ligands, if the incoming ligands bind more strongly to the metal ion than the original ones. For example, ligands such as NH_3, NCS^-, and CN^- can replace H_2O in the coordination sphere of metal ions.

Charges, Coordination Numbers, and Geometries

The charge of a complex is the sum of the charges on the metal and on the ligands. In $[Cu(NH_3)_4]SO_4$ we can deduce the charge on the complex ion because we know that the charge of the sulfate ion is 2−. Because the compound is electrically neutral, the complex ion must have a 2+ charge, $[Cu(NH_3)_4]^{2+}$. We can then use the charge of the complex ion to deduce the oxidation number of copper. Because the NH_3 ligands are uncharged molecules, the oxidation number of copper must be +2:

$$+2 + 4(0) = +2$$
$$[Cu(NH_3)_4]^{2+}$$

SAMPLE EXERCISE 23.2 | Determining the Oxidation Number of a Metal in a Complex

What is the oxidation number of the metal in $[Rh(NH_3)_5Cl](NO_3)_2$?

SOLUTION

Analyze We are given the chemical formula of a coordination compound and asked to determine the oxidation number of its metal atom.

Plan To determine the oxidation number of Rh, we need to figure out what charges are contributed by the other groups. The overall charge is zero, so the oxidation number of the metal must balance the charge due to the rest of the compound.

Solve The NO_3 group is the nitrate anion, which has a 1− charge. The NH_3 ligands carry zero charge, and the Cl is a coordinated chloride ion, which has a 1− charge. The sum of all the charges must be zero:

$$x + 5(0) + (-1) + 2(-1) = 0$$
$$[Rh(NH_3)_5Cl](NO_3)_2$$

The oxidation number of rhodium, x, must therefore be +3.

Practice Exercise 1

In which of the following compounds does the transition-metal have the highest oxidation number? **(a)** $[Co(NH_3)_4Cl_2]$, **(b)** $K_2[PtCl_6]$, **(c)** $Rb_3[MoO_3F_3]$, **(d)** $Na[Ag(CN)_2]$, **(e)** $K_4[Mn(CN)_6]$.

Practice Exercise 2

What is the charge of the complex formed by a platinum(II) metal ion surrounded by two ammonia molecules and two bromide ions?

Recall that the number of atoms directly bonded to the metal atom in a complex is the *coordination number* of the complex. Thus, the silver ion in $[Ag(NH_3)_2]^+$ has a coordination number of 2, and the cobalt ion has a coordination number of 6 in all four complexes in Table 23.3.

Some metal ions have only one observed coordination number. The coordination number of chromium(III) and cobalt(III) is invariably 6, for example, and that of platinum(II) is always 4. For most metals, however, the coordination number is different for different ligands. In these complexes, the most common coordination numbers are 4 and 6.

The coordination number of a metal ion is often influenced by the relative sizes of the metal ion and the ligands. As the ligand gets larger, fewer of them can coordinate to the metal ion. Thus, iron(III) is able to coordinate to six fluorides in $[FeF_6]^{3-}$ but to only four chlorides in $[FeCl_4]^-$. Ligands that transfer substantial negative charge to the metal also produce reduced coordination numbers. For example, six ammonia molecules can coordinate to nickel(II), forming $[Ni(NH_3)_6]^{2+}$, but only four cyanide ions can coordinate to this ion, forming $[Ni(CN)_4]^{2-}$.

The most common coordination geometries for coordination complexes are shown in ▶ **Figure 23.10**. Complexes in which the coordination number is 4 have two geometries—tetrahedral and square planar. The tetrahedral geometry is the more common of the two and is especially prevalent among nontransition metals. The square planar geometry is characteristic of transition-metal ions with eight d electrons in the valence shell, such as platinum(II) and gold(III). Complexes with a coordination number of 6 almost always have an octahedral geometry. Even though the octahedron can be drawn as a square with one ligand above and another below the plane, all six vertices are equivalent.

23.3 | Common Ligands in Coordination Chemistry

The ligand atom that binds to the central metal ion in a coordination complex is called the **donor atom** of the ligand. Ligands having only one donor atom are called **monodentate ligands** (from the Latin, meaning "one-toothed"). These ligands are able to occupy only one site in a coordination sphere. Ligands having two donor atoms are **bidentate ligands** ("two-toothed"), and those having three or more donor atoms are **polydentate ligands** ("many-toothed"). In both bidentate and polydentate species, the multiple donor atoms can simultaneously bond to the metal ion, thereby occupying two or more sites in a coordination sphere. Table 23.4 gives examples of all three types of ligands.

Because they appear to grasp the metal between two or more donor atoms, bidentate and polydentate ligands are also known as **chelating agents** (pronounced "KEE-lay-ting"; from the Greek *chele*, "claw").

One common chelating agent is the bidentate ligand *ethylenediamine*, denoted en:

$$H_2\ddot{N} \overset{\displaystyle CH_2-CH_2}{\diagup\qquad\diagdown} \ddot{N}H_2$$

in which each donor nitrogen atom has one nonbonding electron pair. These donor atoms are sufficiently far apart to allow both of them to bond to the metal ion in adjacent positions. The $[Co(en)_3]^{3+}$ complex ion, which contains three ethylenediamine ligands in the octahedral coordination sphere of cobalt(III), is shown in **Figure 23.11**. Notice that in the image on the right the en is written in a shorthand notation as two nitrogen atoms connected by an arc.

The ethylenediaminetetraacetate ion, $[EDTA]^{4-}$, is an important polydentate ligand that has six donor atoms. It can wrap around a metal ion using all six donor atoms, as shown in Figure 23.12, although it sometimes binds to a metal using only five of its donor atoms.

GO FIGURE

In the drawings on the right-hand side, what does the solid wedge connecting atoms represent? What does the dashed wedge connecting atoms represent?

A tetrahedron has four triangular faces and four equivalent vertices.

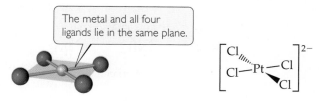

Tetrahedral geometry

The metal and all four ligands lie in the same plane.

Square planar geometry

An octahedron has eight triangular faces and six equivalent vertices.

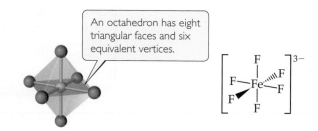

Octahedral geometry

▲ Figure 23.10 **Common geometries of coordination complexes.** In complexes having coordination number 4, the geometry is typically either tetrahedral or square planar. In complexes having coordination number 6, the geometry is nearly always octahedral.

 Give It Some Thought

Both H_2O and ethylenediamine (en) have two nonbonding pairs of electrons. Why cannot water act as a bidentate ligand?

Table 23.4 Some Common Ligands

Ligand Type	Examples

Monodentate

$H_2\ddot{O}:$ Water $:\ddot{\overset{..}{F}}:^-$ Fluoride ion $[:C\equiv N:]^-$ Cyanide ion $[:\ddot{\overset{..}{O}}-H]^-$ Hydroxide ion

$:NH_3$ Ammonia $:\ddot{\overset{..}{Cl}}:^-$ Chloride ion $[:\ddot{S}=C=\ddot{N}:]^-$ Thiocyanate ion $[:\ddot{\overset{..}{O}}-N=\ddot{O}:]^-$ Nitrite ion
                                  └─or─┘   └or┘

Bidentate

$\begin{array}{c} H_2C-CH_2 \\ H_2\ddot{N} \qquad \ddot{N}H_2 \end{array}$
Ethylenediamine (en)

Bipyridine (bipy or bpy)

Ortho-phenanthroline (*o*-phen)

$\left[\begin{array}{c} :\ddot{O}: \quad\quad :\ddot{O}: \\ \backslash\;\;\;\;\;/ \\ C - C \\ /\;\;\;\;\;\backslash \\ :\ddot{O}: \quad\quad :\ddot{O}: \end{array}\right]^{2-}$ Oxalate ion

$\left[\begin{array}{c} :O: \\ \| \\ C \\ /\;\;\;\backslash \\ :\ddot{O}: \quad :\ddot{O}: \end{array}\right]^{2-}$ Carbonate ion

Polydentate

$\begin{array}{c} H_2C-CH_2 \quad CH_2-CH_2 \\ H_2\ddot{N} \qquad \ddot{N}H \qquad\quad \ddot{N}H_2 \end{array}$
Diethylenetriamine

$\left[\begin{array}{c} :O: \qquad :O: \qquad :O: \\ \| \qquad\quad \| \qquad\quad \| \\ :\ddot{O}-P-\ddot{O}-P-\ddot{O}-P-\ddot{O}: \\ | \qquad\quad | \qquad\quad | \\ :\ddot{O}: \qquad :\ddot{O}: \qquad :\ddot{O}: \end{array}\right]^{5-}$
Triphosphate ion

$\left[\begin{array}{c} :O: \qquad\qquad\qquad\qquad\qquad\qquad :O: \\ \| \qquad\qquad\qquad\qquad\qquad\qquad \| \\ :\ddot{O}-C-CH_2 \qquad\qquad\quad CH_2-C-\ddot{O}: \\ \qquad\quad \ddot{N}-CH_2-CH_2-\ddot{N} \\ :\ddot{O}-C-CH_2 \qquad\qquad\quad CH_2-C-\ddot{O}: \\ \| \qquad\qquad\qquad\qquad\qquad\qquad \| \\ :O: \qquad\qquad\qquad\qquad\qquad\qquad :O: \end{array}\right]^{4-}$
Ethylenediaminetetraacetate ion (EDTA^{4-})

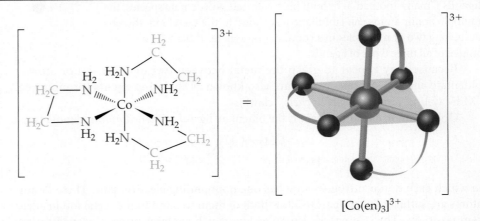

▲ **Figure 23.11 The [Co(en)$_3$]$^{3+}$ ion.** The ligand is ethylenediamine.

In general, the complexes formed by chelating ligands (that is, bidentate and polydentate ligands) are more stable than the complexes formed by related monodentate ligands. The equilibrium formation constants for $[Ni(NH_3)_6]^{2+}$ and $[Ni(en)_3]^{2+}$ illustrate this observation:

$$[Ni(H_2O)_6]^{2+}(aq) + 6\,NH_3(aq) \rightleftharpoons [Ni(NH_3)_6]^{2+}(aq) + 6\,H_2O(l)$$
$$K_f = 1.2 \times 10^9 \qquad [23.4]$$

$$[Ni(H_2O)_6]^{2+}(aq) + 3\,en(aq) \rightleftharpoons [Ni(en)_3]^{2+}(aq) + 6\,H_2O(l)$$
$$K_f = 6.8 \times 10^{17} \qquad [23.5]$$

▲ **Figure 23.12 The complex ion [Co(EDTA)]$^-$.** The ligand is the polydentate ethylenediaminetetraacetate ion, whose full representation is given in Table 23.2. This representation shows how the two N and four O donor atoms coordinate to cobalt.

Although the donor atom is nitrogen in both instances, $[Ni(en)_3]^{2+}$ has a formation constant that is more than 10^8 times larger than that of $[Ni(NH_3)_6]^{2+}$. This trend of generally larger formation constants for bidentate and polydentate ligands, known as the **chelate effect**, is examined in the "A Closer Look" essay on page 1010.

Chelating agents are often used to prevent one or more of the customary reactions of a metal ion without removing the ion from solution. For example, a metal ion that interferes with a chemical analysis can often be complexed and its interference thereby removed. In a sense, the chelating agent hides the metal ion. For this reason, scientists sometimes refer to these ligands as *sequestering agents*.

Phosphate ligands, such as sodium tripolyphosphate, $Na_5[OPO_2OPO_2OPO_3]$, are used to sequester Ca^{2+} and Mg^{2+} ions in hard water so that these ions cannot interfere with the action of soap or detergents.

Chelating agents are used in many prepared foods, such as salad dressings and frozen desserts, to complex trace metal ions that catalyze decomposition reactions. Chelating agents are used in medicine to remove toxic heavy metal ions that have been ingested, such as Hg^{2+}, Pb^{2+}, and Cd^{2+}. One method of treating lead poisoning, for example, is to administer $Na_2Ca(EDTA)$. The EDTA chelates the lead, allowing it to be removed from the body via urine.

 Give It Some Thought

> Cobalt(III) has a coordination number of 6 in all its complexes. Is the carbonate ion a monodentate or bidentate ligand in the $[Co(NH_3)_4(CO_3)]^+$ ion?

Metals and Chelates in Living Systems

Ten of the 29 elements known to be necessary for human life are transition metals. ∞ (Section 2.7, "Elements Required by Living Organisms") These ten elements—V, Cr, Mn, Fe, Co, Ni, Cu, Zn, Mo, and Cd—form complexes with a variety of groups present in biological systems.

Although our bodies require only small quantities of metals, deficiencies can lead to serious illness. A deficiency of manganese, for example, can lead to convulsive disorders. Some epilepsy patients have been helped by the addition of manganese to their diets.

Among the most important chelating agents in nature are those derived from the *porphine* molecule (▶ Figure 23.13). This molecule can coordinate to a metal via its four nitrogen donor atoms. Once porphine bonds to a metal ion, the two H atoms on the nitrogens are displaced to form complexes called **porphyrins**. Two important porphyrins are *hemes*, in which the metal ion is Fe(II), and *chlorophylls*, with a Mg(II) central ion.

Figure 23.14 shows a schematic structure of myoglobin, a protein that contains one heme group. Myoglobin is a *globular protein*, one that folds into a compact, roughly spherical shape. Myoglobin is found in the cells of skeletal muscle, particularly in seals, whales, and porpoises. It stores oxygen in cells, one molecule of O_2 per myoglobin, until it is needed for metabolic activities. Hemoglobin, the protein that transports oxygen in human blood, is made up of four heme-containing subunits, each of which is very similar to myoglobin. One hemoglobin can bind up to four O_2 molecules.

In both myoglobin and hemoglobin, the iron is coordinated to the four nitrogen atoms of a porphyrin and to a nitrogen atom from the protein chain (**Figure 23.15**). In hemoglobin, the sixth position around the iron is occupied either by O_2 (in oxyhemoglobin, the bright red form) or by water (in deoxyhemoglobin, the purplish red form). (The oxy form is the one shown in Figure 23.15.)

 GO FIGURE

What is the coordination number of the metal ion in heme b? In chlorophyll a?

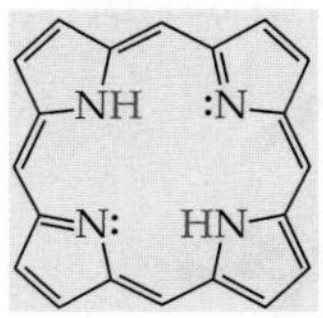

Porphine

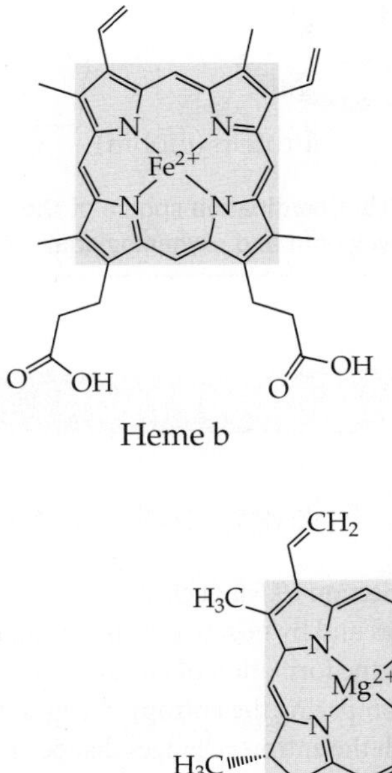

Heme b

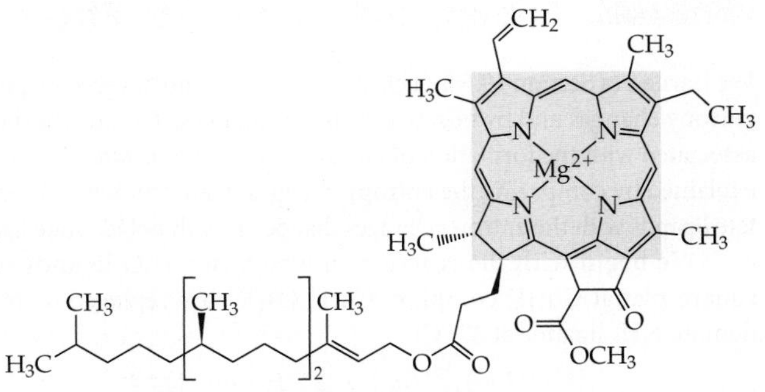

Chlorophyll a

▲ Figure 23.13 **Porphine and two porphyrins, heme b and chlorophyll a.** Fe(II) and Mg(II) ions replace the two blue H atoms in porphine and bond with all four nitrogens in heme b and chlorophyll a, respectively.

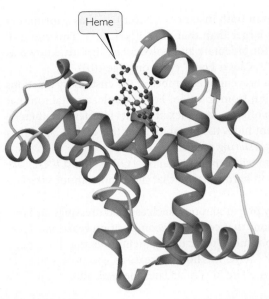

▲ **Figure 23.14 Myoglobin.** This ribbon diagram does not show most of the atoms.

▲**GO FIGURE**

What is the coordination number of iron in the heme unit shown here? What is the identity of the donor atoms?

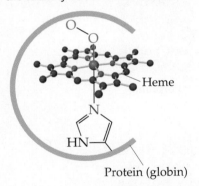

▲ **Figure 23.15 Coordination sphere of the hemes in oxymyoglobin and oxyhemoglobin.**

Carbon monoxide is poisonous because the equilibrium binding constant of human hemoglobin for CO is about 210 times greater than that for O_2. As a result, a relatively small quantity of CO can inactivate a substantial fraction of the hemoglobin in the blood by displacing the O_2 molecule from the heme-containing subunit. For example, a person breathing air that contains only 0.1% CO takes in enough CO after a few hours to convert up to 60% of the hemoglobin (Hb) into COHb, thereby reducing the blood's normal oxygen-carrying capacity by 60%.

Under normal conditions, a nonsmoker breathing unpolluted air has about 0.3 to 0.5% COHb in her or his blood. This amount arises mainly from the production of small quantities of CO in the course of normal body chemistry and from the small amount of CO present in clean air. Exposure to higher concentrations of CO causes the COHb level to increase, which in turn leaves fewer Hb sites to which O_2 can bind. If the level of COHb becomes too high, oxygen transport is effectively shut down and death occurs. Because CO is colorless and odorless, CO poisoning occurs with very little warning. Improperly ventilated combustion devices, such as kerosene lanterns and stoves, thus pose a potential health hazard.

A Closer Look

Entropy and the Chelate Effect

We learned in Section 19.5 that chemical processes are favored by positive entropy changes and by negative enthalpy changes. The special stability associated with the formation of chelates, called the *chelate effect*, can be explained by comparing the entropy changes that occur with monodentate ligands with the entropy changes that occur with polydentate ligands.

We begin with the reaction in which two H_2O ligands of the square-planar Cu(II) complex $[Cu(H_2O)_4]^{2+}$ are replaced by monodentate NH_3 ligands at 27 °C:

$$[Cu(H_2O)_4]^{2+}(aq) + 2\,NH_3(aq) \rightleftharpoons$$
$$[Cu(H_2O)_2(NH_3)_2]^{2+}(aq) + 2\,H_2O(l)$$
$$\Delta H° = -46\text{ kJ}; \quad \Delta S° = -8.4\text{ J/K}; \quad \Delta G° = -43\text{ kJ}$$

The thermodynamic data tell us about the relative abilities of H_2O and NH_3 to serve as ligands in this reaction. In general, NH_3 binds more tightly to metal ions than does H_2O, so this substitution reaction is exothermic ($\Delta H < 0$). The stronger bonding of the NH_3

ligands also causes the $[Cu(H_2O)_2(NH_3)_2]^{2+}$ ion to be more rigid, which is probably the reason $\Delta S°$ is slightly negative.

We can use Equation 19.20, $\Delta G° = -RT \ln K$, to calculate the equilibrium constant of the reaction at 27 °C. The result, $K = 3.1 \times 10^7$, tells us that the equilibrium lies far to the right, favoring replacement of H_2O by NH_3. For this equilibrium, therefore, the enthalpy change, $\Delta H° = -46$ kJ, is large enough and negative enough to overcome the entropy change, $\Delta S° = -8.4$ J/K.

Now let's use a single bidentate ethylenediamine (en) ligand in our substitution reaction:

$$[Cu(H_2O)_4]^{2+}(aq) + en(aq) \rightleftharpoons$$
$$[Cu(H_2O)_2(en)]^{2+}(aq) + 2\,H_2O(l)$$
$$\Delta H° = -54\text{ kJ}; \quad \Delta S° = +23\text{ J/K}; \quad \Delta G° = -61\text{ kJ}$$

The en ligand binds slightly more strongly to the Cu^{2+} ion than two NH_3 ligands, so the enthalpy change here (-54 kJ) is slightly more negative than for $[Cu(H_2O)_2(NH_3)_2]^{2+}$ (-46 kJ). There is a big difference in the entropy change, however: $\Delta S°$ is -8.4 J/K for the NH_3 reaction but $+23$ J/K for the en reaction. We can explain the positive $\Delta S°$ value using concepts discussed in Section 19.3. Because a single en ligand occupies two coordination sites, two molecules of H_2O are released when one en ligand bonds. Thus, there are three product molecules in the reaction but only two reactant molecules. The greater number of product molecules leads to the positive entropy change for the equilibrium.

The slightly more negative value of $\Delta H°$ for the en reaction (-54 kJ versus -46 kJ) coupled with the positive entropy change leads to a much more negative value of $\Delta G°$ (-61 kJ for en, -43 kJ for NH_3) and thus a larger equilibrium constant: $K = 4.2 \times 10^{10}$.

We can combine our two equations using Hess's law ∞ (Section 5.6) to calculate the enthalpy, entropy, and free-energy changes that occur for en to replace ammonia as ligands on Cu(II):

$$[Cu(H_2O)_2(NH_3)_2]^{2+}(aq) + en(aq) \rightleftharpoons$$
$$[Cu(H_2O)_2(en)]^{2+}(aq) + 2\,NH_3(aq)$$

$$\Delta H° = (-54\text{ kJ}) - (-46\text{ kJ}) = -8\text{ kJ}$$
$$\Delta S° = (+23\text{ J/K}) - (-8.4\text{ J/K}) = +31\text{ J/K}$$
$$\Delta G° = (-61\text{ kJ}) - (-43\text{ kJ}) = -18\text{ kJ}$$

Notice that at 27 °C, the entropic contribution ($-T\Delta S°$) to the free-energy change, $\Delta G° = \Delta H° - T\Delta S°$ (Equation 19.12), is negative and greater in magnitude than the enthalpic contribution ($\Delta H°$). The equilibrium constant for the NH_3–en reaction, 1.4×10^3, shows that the replacement of NH_3 by en is thermodynamically favorable.

The chelate effect is important in biochemistry and molecular biology. The additional thermodynamic stabilization provided by entropy effects helps stabilize biological metal–chelate complexes, such as porphyrins, and can allow changes in the oxidation state of the metal ion while retaining the structural integrity of the complex.

Related Exercises: 23.31, 23.32, 23.96, 23.98

The **chlorophylls**, which are porphyrins that contain Mg(II) (Figure 23.13), are the key components in the conversion of solar energy into forms that can be used by living organisms. This process, called **photosynthesis**, occurs in the leaves of green plants:

$$6\,CO_2(g) + 6\,H_2O(l) \longrightarrow C_6H_{12}O_6(aq) + 6\,O_2(g) \qquad [23.6]$$

The formation of 1 mol of glucose, $C_6H_{12}O_6$, requires the absorption of 48 mol of photons from sunlight or other sources of light. Chlorophyll-containing pigments in the leaves of plants absorb the photons. Figure 23.13 shows that the chlorophyll molecule has a series of alternating, or *conjugated*, double bonds in the ring surrounding the metal ion. This system of conjugated double bonds makes it possible for chlorophyll to absorb light strongly in the visible region of the spectrum. As ▶ Figure 23.16 shows, chlorophyll is green because it absorbs red light (maximum absorption at 655 nm) and blue light (maximum absorption at 430 nm) and transmits green light.

Photosynthesis is nature's solar energy–conversion machine, and thus all living systems on Earth depend on photosynthesis for continued existence.

 Give It Some Thought

What property of the porphine ligand makes it possible for chlorophyll to play a role in plant photosynthesis?

GO FIGURE

Which peak in this curve corresponds to the lowest-energy transition by an electron in a chlorophyll molecule?

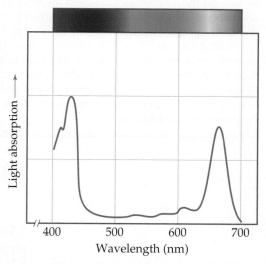

▲ Figure 23.16 **The absorption of sunlight by chlorophyll.**

Chemistry and Life

The Battle for Iron In Living Systems

Because living systems have difficulty assimilating enough iron to satisfy their nutritional needs, iron-deficiency anemia is a common problem in humans. Chlorosis, an iron deficiency in plants that makes leaves turn yellow, is also commonplace.

Living systems have difficulty assimilating iron because most iron compounds found in nature are not very soluble in water. Microorganisms have adapted to this problem by secreting an iron-binding compound, called a *siderophore*, that forms an extremely stable water-soluble complex with iron(III). One such complex is *ferrichrome* (**Figure 23.17**). The iron-binding strength of a siderophore is so great that it can extract iron from iron oxides.

When ferrichrome enters a living cell, the iron it carries is removed through an enzyme-catalyzed reaction that reduces the strongly bonding iron(III) to iron(II), which is only weakly complexed by the siderophore (**Figure 23.18**). Microorganisms thus acquire iron by excreting a siderophore into their immediate environment and then taking the resulting iron complex into the cell.

In humans, iron is assimilated from food in the intestine. A protein called *transferrin* binds iron and transports it across the intestinal wall to distribute it to other tissues in the body. The normal adult body contains about 4 g of iron. At any one time, about 3 g of this iron is in the blood, mostly in the form of hemoglobin. Most of the remainder is carried by transferrin.

A bacterium that infects the blood requires a source of iron if it is to grow and reproduce. The bacterium excretes a siderophore into the

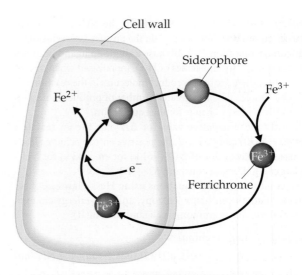

▲ **Figure 23.17 Ferrichrome.**

blood to compete with transferrin for iron. The equilibrium constants for forming the iron complex are about the same for transferrin and siderophores. The more iron available to the bacterium, the more rapidly it can reproduce and thus the more harm it can do.

Several years ago, New Zealand clinics regularly gave iron supplements to infants soon after birth. However, the incidence of certain bacterial infections was eight times higher in treated than in untreated infants. Presumably, the presence of more iron in the blood than absolutely necessary makes it easier for bacteria to obtain the iron needed for growth and reproduction.

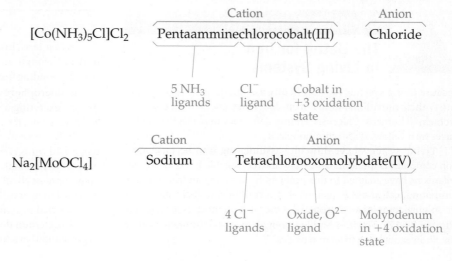

▲ **Figure 23.18 The iron-transport system of a bacterial cell.**

In the United States, it is common medical practice to supplement infant formula with iron sometime during the first year of life. However, iron supplements are not necessary for infants who breast-feed because breast milk contains two specialized proteins, lactoferrin and transferrin, which provide sufficient iron while denying its availability to bacteria. Even for infants fed with infant formulas, supplementing with iron during the first several months of life may be ill-advised.

For bacteria to continue to multiply in the blood, they must synthesize new supplies of siderophores. Synthesis of siderophores in bacteria slows, however, as the temperature is increased above the normal body temperature of 37 °C and stops completely at 40 °C. This suggests that fever in the presence of an invading microbe is a mechanism used by the body to deprive bacteria of iron.

Related Exercise: 23.74

23.4 | Nomenclature and Isomerism in Coordination Chemistry

When complexes were first discovered, they were named after the chemist who originally prepared them. A few of these names persist, as, for example, with the dark red substance $NH_4[Cr(NH_3)_2(NCS)_4]$, which is still known as Reinecke's salt. Once the structures of complexes were more fully understood, it became possible to name them in a more systematic manner. Let's use two substances to illustrate how coordination compounds are named:

	Cation	Anion
$[Co(NH_3)_5Cl]Cl_2$	Pentaamminechlorocobalt(III)	Chloride

| | 5 NH_3 ligands | Cl^- ligand | Cobalt in +3 oxidation state |

	Cation	Anion
$Na_2[MoOCl_4]$	Sodium	Tetrachlorooxomolybdate(IV)

| | 4 Cl^- ligands | Oxide, O^{2-} ligand | Molybdenum in +4 oxidation state |

1. **In naming complexes that are salts, the name of the cation is given before the name of the anion.** Thus, in $[Co(NH_3)_5Cl]Cl_2$ we name the $[Co(NH_3)_5Cl]^{2+}$ cation and then the Cl^-.

2. **In naming complex ions or molecules, the ligands are named before the metal. Ligands are listed in alphabetical order, regardless of their charges. Prefixes that give the number of ligands are not considered part of the ligand name in determining alphabetical order.** Thus, the $[Co(NH_3)_5Cl]^{2+}$ ion is pentaamminechlorocobalt(III). (Be careful to note, however, that the metal is written first in the chemical formula.)

3. **The names of anionic ligands end in the letter *o*, but electrically neutral ligands ordinarily bear the name of the molecules** (▲ Table 23.5). Special names are used for H_2O (aqua), NH_3 (ammine), and CO (carbonyl). For example, $[Fe(CN)_2(NH_3)_2(H_2O)_2]^+$ is the diamminediaquadicyanoiron(III) ion.

4. **Greek prefixes (*di-, tri-, tetra-, penta-, hexa-*) are used to indicate the number of each kind of ligand when more than one is present. If the ligand contains a Greek prefix (for example, *ethylenediamine*) or is polydentate, the alternate prefixes *bis-, tris-, tetrakis-, pentakis-, and hexakis-* are used and the ligand name is placed in parentheses.** For example, the name for $[Co(en)_3]Br_3$ is tris(ethylenediamine)-cobalt(III) bromide.

5. **If the complex is an anion, its name ends in -*ate*.** The compound $K_4[Fe(CN)_6]$ is potassium hexacyanoferrate(II), for example, and the ion $[CoCl_4]^{2-}$ is tetrachlorocobaltate(II) ion.

6. **The oxidation number of the metal is given in parentheses in Roman numerals following the name of the metal.**

Three examples for applying these rules are

$[Ni(NH_3)_6]Br_2$	Hexaamminenickel(II) bromide
$[Co(en)_2(H_2O)(CN)]Cl_2$	Aquacyanobis(ethylenediamine)cobalt(III) chloride
$Na_2[MoOCl_4]$	Sodium tetrachlorooxomolybdate(IV)

Table 23.5 Some Common Ligands and Their Names

Ligand	Name in Complexes	Ligand	Name in Complexes
Azide, N_3^-	Azido	Oxalate, $C_2O_4^{2-}$	Oxalato
Bromide, Br^-	Bromo	Oxide, O^{2-}	Oxo
Chloride, Cl^-	Chloro	Ammonia, NH_3	Ammine
Cyanide, CN^-	Cyano	Carbon monoxide, CO	Carbonyl
Fluoride, F^-	Fluoro	Ethylenediamine, en	Ethylenediamine
Hydroxide, OH^-	Hydroxo	Pyridine, C_5H_5N	Pyridine
Carbonate, CO_3^{2-}	Carbonato	Water, H_2O	Aqua

SAMPLE EXERCISE 23.3 Naming Coordination Compounds

Name the compounds **(a)** $[Cr(H_2O)_4Cl_2]Cl$, **(b)** $K_4[Ni(CN)_4]$.

SOLUTION

Analyze We are given the chemical formulas for two coordination compounds and assigned the task of naming them.

Plan To name the complexes, we need to determine the ligands in the complexes, the names of the ligands, and the oxidation state of the metal ion. We then put the information together following the rules listed in the text.

Solve

(a) The ligands are four water molecules—tetraaqua—and two chloride ions—dichloro. By assigning all the oxidation numbers we know for this molecule, we see that the oxidation number of Cr is +3:

$$+3 + 4(0) + 2(-1) + (-1) = 0$$

$$[Cr(H_2O)_4Cl_2]Cl$$

Thus, we have chromium(III). Finally, the anion is chloride. The name of the compound is tetraaquadichlorochromium(III) chloride.

(b) The complex has four cyanide ion ligands, CN^-, which means tetracyano, and the oxidation state of the nickel is zero:

$$4(+1) + 0 + 4(-1) = 0$$
$$K_4[Ni(CN)_4]$$

Because the complex is an anion, the metal is indicated as nickelate(0). Putting these parts together and naming the cation first, we have potassium tetracyanonickelate(0).

Practice Exercise 1

What is the name of the compound $[Rh(NH_3)_4Cl_2]Cl$? **(a)** Rhodium(III) tetraamminedichloro chloride, **(b)** Tetraammoniadichlororhodium(III) chloride, **(c)** Tetraamminedichlororhodium(III) chloride, **(d)** Tetraamminetrichlororhodium(III), **(e)** Tetraamminedichlororhodium(II) chloride.

Practice Exercise 2

Name the compounds **(a)** $[Mo(NH_3)_3Br_3]NO_3$, **(b)** $(NH_4)_2[CuBr_4]$. **(c)** Write the formula for sodium diaquabis(oxalato)ruthenate(III).

Isomerism

When two or more compounds have the same composition but a different arrangement of atoms, we call them **isomers**. ∞ (Section 2.9) Here we consider two main kinds of isomers in coordination compounds: **structural isomers** (which have different bonds) and **stereoisomers** (which have the same bonds but different ways in which the ligands occupy the space around the metal center). Each of these classes also has subclasses, as shown in ▼ Figure 23.19.

Structural Isomerism

Many types of structural isomerism are known in coordination chemistry, including the two named in Figure 23.19: linkage isomerism and coordination-sphere isomerism. **Linkage isomerism** is a relatively rare but interesting type that arises when a particular ligand is capable of coordinating to a metal in two ways. The nitrite ion, NO_2^-, for example, can coordinate to a metal ion through either its nitrogen or one of its oxygens

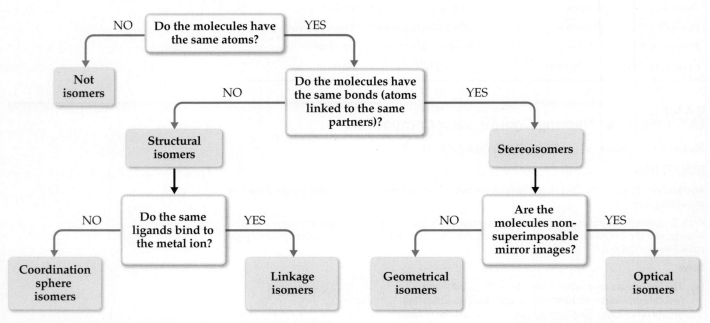

▲ Figure 23.19 **Forms of isomerism in coordination compounds.**

(▶ Figure 23.20). When it coordinates through the nitrogen atom, the NO_2^- ligand is called *nitro*; when it coordinates through the oxygen atom, it is called *nitrito* and is generally written ONO^-. The isomers shown in Figure 23.20 have different properties. The nitro isomer is yellow, for example, whereas the nitrito isomer is red.

Another ligand capable of coordinating through either of two donor atoms is thiocyanate, SCN^-, whose potential donor atoms are N and S.

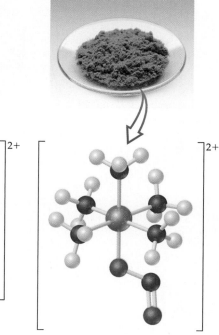

▲ **GO FIGURE**
What is the chemical formula and name for each of the complex ions in this figure?

Give It Some Thought

Can the ammonia ligand engage in linkage isomerism? Explain.

Coordination-sphere isomers are isomers that differ in which species in the complex are ligands and which are outside the coordination sphere in the solid. For example, three isomers have the formula $CrCl_3(H_2O)_6$. When the ligands are six H_2O and the chloride ions are in the crystal lattice (as counterions), we have the violet compound $[Cr(H_2O)_6]Cl_3$. When the ligands are five H_2O and one Cl^-, with the sixth H_2O and the two Cl^- out in the lattice, we have the green compound $[Cr(H_2O)_5Cl]Cl_2 \cdot H_2O$. The third isomer, $[Cr(H_2O)_4Cl_2]Cl \cdot 2\,H_2O$, is also a green compound. In the two green compounds, either one or two water molecules have been displaced from the coordination sphere by chloride ions. The displaced H_2O molecules occupy a site in the crystal lattice.

Nitro isomer
Bonding via ligand N atom

Nitrito isomer
Bonding via ligand O atom

▲ **Figure 23.20 Linkage isomerism.**

Stereoisomerism

Stereoisomers have the same chemical bonds but different spatial arrangements. In the square-planar complex $[Pt(NH_3)_2Cl_2]$, for example, the chloro ligands can be either adjacent to or opposite each other (▼ Figure 23.21). (We saw an earlier example of this type of isomerism in the cobalt complex of Figure 23.8, and we will return to that complex in a moment.) This form of stereoisomerism, in which the arrangement of the atoms is different but the same bonds are present, is called **geometric isomerism**. The isomer on the left in

▲ **GO FIGURE**
Which of these isomers has a nonzero dipole moment?

● = N   ● = Cl   ○ = H   ○ = Pt

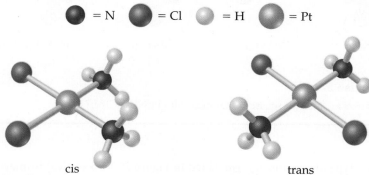

cis
Cl ligands adjacent to each other
NH_3 ligands adjacent to each other

trans
Cl ligands on opposite sides of central atom
NH_3 ligands on opposite sides of central atom

▲ **Figure 23.21 Geometric isomerism.**

Figure 23.21, with like ligands in adjacent positions, is the cis isomer, and the isomer on the right, with like ligands across from one another, is the trans isomer.

Geometric isomers generally have different physical properties and may also have markedly different chemical reactivities. For example, *cis*-[Pt(NH₃)₂Cl₂], also called *cisplatin*, is effective in the treatment of testicular, ovarian, and certain other cancers, whereas the trans isomer is ineffective. This is because cisplatin forms a chelate with two nitrogens of DNA, displacing the chloride ligands. The chloride ligands of the trans isomer are too far apart to form the N-Pt-N chelate with the nitrogen donors in DNA.

Geometric isomerism is also possible in octahedral complexes when two or more different ligands are present, as in the cis and trans tetraamminedichlorocobalt(III) ion in Figure 23.8. Because all the corners of a tetrahedron are adjacent to one another, cis–trans isomerism is not observed in tetrahedral complexes.

SAMPLE EXERCISE 23.4 Determining the Number of Geometric Isomers

The Lewis structure :C≡O: indicates that the CO molecule has two lone pairs of electrons. When CO binds to a transition-metal atom, it nearly always does so by using the C lone pair. How many geometric isomers are there for tetracarbonyldichloroiron(II)?

SOLUTION

Analyze We are given the name of a complex containing only monodentate ligands, and we must determine the number of isomers the complex can form.

Plan We can count the number of ligands to determine the coordination number of the Fe and then use the coordination number to predict the geometry. We can then either make a series of drawings with ligands in different positions to determine the number of isomers or deduce the number of isomers by analogy to cases we have discussed.

Solve The name indicates that the complex has four carbonyl (CO) ligands and two chloro (Cl⁻) ligands, so its formula is Fe(CO)₄Cl₂. The complex therefore has a coordination number of 6, and we can assume an octahedral geometry. Like [Co(NH₃)₄Cl₂]⁺ (Figure 23.8), it has four ligands of one type and two of another. Consequently, there are two isomers possible: one with the Cl⁻ ligands across the metal from each other, *trans*-[Fe(CO)₄Cl₂], and one with the Cl⁻ ligands adjacent to each other, *cis*-[Fe(CO)₄Cl₂].

Comment It is easy to overestimate the number of geometric isomers. Sometimes different orientations of a single isomer are incorrectly thought to be different isomers. If two structures can be rotated so that they are equivalent, they are not isomers of each other. The problem of identifying isomers is compounded by the difficulty we often have in visualizing three-dimensional molecules from their two-dimensional representations. It is sometimes easier to determine the number of isomers if we use three-dimensional models.

Practice Exercise 1

Which of the following molecules does not have a geometric isomer?

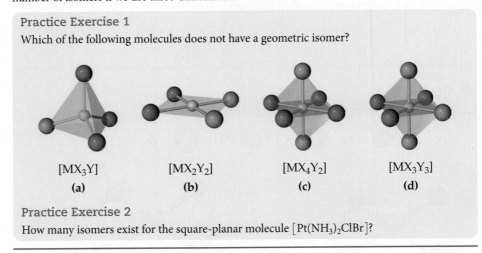

| [MX₃Y] | [MX₂Y₂] | [MX₄Y₂] | [MX₃Y₃] |
| (a) | (b) | (c) | (d) |

Practice Exercise 2

How many isomers exist for the square-planar molecule [Pt(NH₃)₂ClBr]?

The second type of stereoisomerism listed in Figure 23.19 is **optical isomerism**. Optical isomers, called **enantiomers**, are mirror images that cannot be superimposed on each other. They bear the same resemblance to each other that your left hand bears to your right hand. If you look at your left hand in a mirror, the image is identical to your right hand (▶ Figure 23.22). No matter how hard you try, however, you cannot

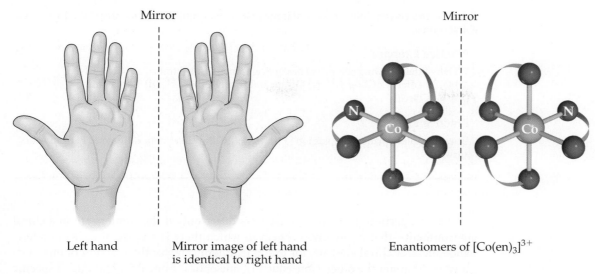

Enantiomers of $[Co(en)_3]^{3+}$

Left hand   Mirror image of left hand
                   is identical to right hand

▲ **Figure 23.22 Optical isomerism.**

superimpose your two hands on each other. An example of a complex that exhibits this type of isomerism is the $[Co(en)_3]^{3+}$ ion. Figure 23.22 shows the two enantiomers of this complex and their mirror-image relationship. Just as there is no way that we can twist or turn our right hand to make it look identical to our left, so also there is no way to rotate one of these enantiomers to make it identical to the other. Molecules or ions that are not superimposable on their mirror image are said to be **chiral** (pronounced KY-rul).

SAMPLE EXERCISE 23.5 Predicting Whether a Complex Has Optical Isomers

Does either *cis*-$[Co(en)_2Cl_2]^+$ or *trans*-$[Co(en)_2Cl_2]^-$ have optical isomers?

SOLUTION

Analyze We are given the chemical formula for two geometric isomers and asked to determine whether either one has optical isomers. Because en is a bidentate ligand, we know that both complexes are octahedral and both have coordination number 6.

Plan We need to sketch the structures of the cis and trans isomers and their mirror images. We can draw the en ligand as two N atoms connected by an arc. If the mirror image cannot be superimposed on the original structure, the complex and its mirror image are optical isomers.

Solve The trans isomer of $[Co(en)_2Cl_2]^+$ and its mirror image are:

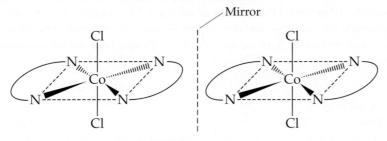

Notice that the mirror image of the isomer is identical to the original. Consequently *trans*-$[Co(en)_2Cl_2]^+$ does not exhibit optical isomerism.

The mirror image of the cis isomer cannot be superimposed on the original:

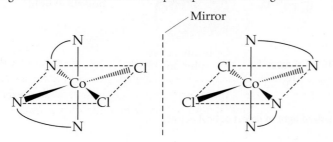

Thus, the two cis structures are optical isomers (enantiomers). We say that *cis*-$[Co(en)_2Cl_2]^+$ is a chiral complex.

> **Practice Exercise 1**
>
> Which of the following complexes has optical isomers? **(a)** Tetrahedral $[CdBr_2Cl_2]^{2-}$, **(b)** Octahedral $[CoCl_4(en)]^{2-}$, **(c)** Octahedral $[Co(NH_3)_4Cl_2]^{2+}$, **(d)** Tetrahedral $[Co(NH_3)BrClI]^-$.
>
> **Practice Exercise 2**
>
> Does the square-planar complex ion $[Pt(NH_3)(N_3)ClBr]^-$ have optical isomers? Explain your answer.

The properties of two optical isomers differ only if the isomers are in a chiral environment—that is, an environment in which there is a sense of right- and left-handedness. A chiral enzyme, for example, might catalyze the reaction of one optical isomer but not the other. Consequently, one optical isomer may produce a specific physiological effect in the body, with its mirror image producing either a different effect or none at all. Chiral reactions are also extremely important in the synthesis of pharmaceuticals and other industrially important chemicals.

Optical isomers are usually distinguished from each other by their interaction with plane-polarized light. If light is polarized—for example, by being passed through a sheet of polarizing film—the electric-field vector of the light is confined to a single plane (▼ Figure 23.23). If the polarized light is then passed through a solution containing one optical isomer, the plane of polarization is rotated either to the right or to the left. The isomer that rotates the plane of polarization to the right is **dextrorotatory**; it is the dextro, or *d*, isomer (Latin *dexter*, "right"). Its mirror image rotates the plane of polarization to the left; it is **levorotatory** and is the levo, or *l*, isomer (Latin *laevus*, "left"). The $[Co(en)_3]^{3+}$ isomer on the right in Figure 23.22 is found experimentally to be the *l* isomer of this ion. Its mirror image is the *d* isomer. Because of their effect on plane-polarized light, chiral molecules are said to be **optically active**.

 Give It Some Thought

What is the similarity and what is the difference between the *d* and *l* isomers of a compound?

When a substance with optical isomers is prepared in the laboratory, the chemical environment during the synthesis is not usually chiral. Consequently, equal amounts of the two isomers are obtained, and the mixture is said to be **racemic**. A racemic mixture does not rotate polarized light because the rotatory effects of the two isomers cancel each other.

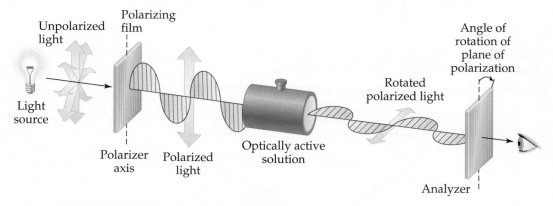

▲ Figure 23.23 **Using polarized light to detect optical activity.**

23.5 | Color and Magnetism in Coordination Chemistry

Studies of the colors and magnetic properties of transition-metal complexes have played an important role in the development of modern models for metal–ligand bonding. We discussed the various types of magnetic behavior of the transition metals in Section 23.1, and we discussed the interaction of radiant energy with matter in Section 6.3. Let's briefly examine the significance of these two properties for transition-metal complexes before we develop a model for metal–ligand bonding.

Color

In Figure 23.4, we saw the diverse range of colors seen in salts of transition-metal ions and their aqueous solutions. In general, the color of a complex depends on the identity of the metal ion, on its oxidation state, and on the ligands bound to it. ▼ **Figure 23.24**, for instance, shows how the pale blue color characteristic of $[Cu(H_2O)_4]^{2+}$ changes to deep blue-violet as NH_3 ligands replace the H_2O ligands to form $[Cu(NH_3)_4]^{2+}$.

For a substance to have color we can see, it must absorb some portion of the spectrum of visible light. ∞ (Section 6.1) Absorption happens, however, only if the energy needed to move an electron in the substance from its ground state to an excited state corresponds to the energy of some portion of the visible spectrum. ∞ (Section 6.3) Thus, the particular energies of radiation a substance absorbs dictate the color we see for the substance.

When an object absorbs some portion of the visible spectrum, the color we perceive is the sum of the unabsorbed portions, which are either reflected or transmitted by the object and strike our eyes. (Opaque objects *reflect* light, and transparent ones *transmit* it.) If an object absorbs all wavelengths of visible light, none reaches our eyes and the object appears black. If it absorbs no visible light, it is white if opaque or colorless if transparent. If it absorbs all but orange light, the orange light is what reaches our eye and therefore is the color we see.

▲ GO FIGURE

Is the equilibrium binding constant of ammonia for Cu(II) likely to be larger or smaller than that of water for Cu(II)?

$[Cu(H_2O)_4]^{2+}(aq)$ $NH_3(aq)$ $[Cu(NH_3)_4]^{2+}(aq)$

▲ Figure 23.24 **The color of a coordination complex changes when the ligand changes.**

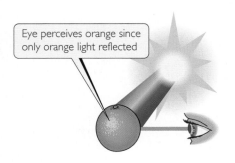

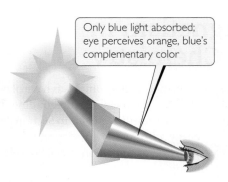

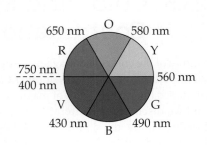

▲ **Figure 23.25 Two ways of perceiving the color orange.** An object appears orange either when it reflects orange light to the eye (left), or when it transmits to the eye all colors except blue, the complement of orange (middle). Complementary colors lie opposite to each other on an artist's color wheel (right).

An interesting phenomenon of vision is that we also perceive an orange color when an object absorbs only the blue portion of the visible spectrum and all the other colors strike our eyes. This is because orange and blue are **complementary colors**, which means that the removal of blue from white light makes the light look orange (and the removal of orange makes the light look blue).

Complementary colors can be determined with an artist's color wheel, which shows complementary colors on opposite sides (▲ Figure 23.25).

The amount of light absorbed by a sample as a function of wavelength is known as the sample's **absorption spectrum**. The visible absorption spectrum of a transparent sample can be determined using a spectrometer, as described in the "A Closer Look" box on page 582. The absorption spectrum of the ion $[Ti(H_2O)_6]^{3+}$ is shown in ▶ Figure 23.26. The absorption maximum is at 500 nm, but the graph shows that much of the yellow, green, and blue light is also absorbed. Because the sample absorbs all of these colors, what we see is the unabsorbed red and violet light, which we perceive as purple (the color purple is classified as a tertiary color located between red and violet on an artists color wheel).

SAMPLE EXERCISE 23.6 | Relating Color Absorbed to Color Observed

The complex ion *trans*-$[Co(NH_3)_4Cl_2]^+$ absorbs light primarily in the red region of the visible spectrum (the most intense absorption is at 680 nm). What is the color of the complex ion?

SOLUTION

Analyze We need to relate the color absorbed by a complex (red) to the color observed for the complex.

Plan For an object that absorbs only one color from the visible spectrum, the color we see is complementary to the color absorbed. We can use the color wheel of Figure 23.25 to determine the complementary color.

Solve From Figure 23.25, we see that green is complementary to red, so the complex appears green.

Comment As noted in Section 23.2, this green complex was one of those that helped Werner establish his theory of coordination (Table 23.3). The other geometric isomer of this complex, *cis*-$[Co(NH_3)_4Cl_2]^+$, absorbs yellow light and therefore appears violet.

Practice Exercise 1

A solution containing a certain transition-metal complex ion has the absorption spectrum shown here.

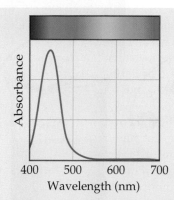

What color would you expect a solution containing this ion to be? **(a)** violet, **(b)** blue, **(c)** green, **(d)** orange, **(e)** red.

Practice Exercise 2

A certain transition-metal complex ion absorbs at 695 nm. Which color is this ion most likely to be—blue, yellow, green, or red?

Magnetism of Coordination Compounds

Many transition-metal complexes exhibit paramagnetism, as described in Sections 9.8 and 23.1. In such compounds the metal ions possess some number of unpaired electrons. It is possible to experimentally determine the number of unpaired electrons per metal ion from the measured degree of paramagnetism, and experiments reveal some interesting comparisons.

Compounds of the complex ion $[Co(CN)_6]^{3-}$ have no unpaired electrons, for example, but compounds of the $[CoF_6]^{3-}$ ion have four unpaired electrons per metal ion. Both complexes contain Co(III) with a $3d^6$ electron configuration. ⚬⚬ (Section 7.4) Clearly, there is a major difference in the ways in which the electrons are arranged in these two cases. Any successful bonding theory must explain this difference, and we present such a theory in the next section.

 Give It Some Thought

What is the electron configuration for **(a)** the Co atom and **(b)** the Co^{3+} ion? How many unpaired electrons does each possess? (See Section 7.4 to review electron configurations of ions.)

23.6 | Crystal-field Theory

Scientists have long recognized that many of the magnetic properties and colors of transition-metal complexes are related to the presence of *d* electrons in the metal cation. In this section, we consider a model for bonding in transition-metal complexes, **crystal-field theory**, that accounts for many of the observed properties of these substances.* Because the predictions of crystal-field theory are essentially the same as those obtained with more advanced molecular-orbital theories, crystal-field theory is an excellent place to start in considering the electronic structure of coordination compounds.

The attraction of a ligand to a metal ion is essentially a Lewis acid–base interaction in which the base—that is, the ligand—donates a pair of electrons to an empty orbital on the metal ion (▶ Figure 23.27). Much of the attractive interaction between the metal ion and the ligands is due, however, to the electrostatic forces between the positive charge on the metal ion and negative charges on the ligands. An ionic ligand, such as Cl^- or SCN^-, experiences the usual cation–anion attraction. When the ligand is a neutral molecule, as in the case of H_2O or NH_3, the negative ends of these polar molecules, which contain an unshared electron pair, are directed toward the metal ion. In this case, the attractive interaction is of the ion–dipole type. ⚬⚬ (Section 11.2) In either case, the ligands are attracted strongly toward the metal ion. Because of the metal–ligand electrostatic attraction, the energy of the complex is lower than the combined energy of the separated metal ion and ligands.

Although the metal ion is attracted to the ligand electrons, the metal ion's *d* electrons are repulsed by the ligands. Let's examine this effect more closely, specifically the case in which the ligands form an octahedral array around a metal ion that has coordination number 6.

In crystal-field theory, we consider the ligands to be negative points of charge that repel the negatively charged electrons in the *d* orbitals of the metal ion. The energy diagram in **Figure 23.28** shows how these ligand point charges affect the energies of the *d* orbitals. First we imagine the complex as having all the ligand point charges uniformly distributed on the surface of a sphere centered on the metal ion. The *average* energy of the metal ion's *d* orbitals is raised by the presence of this uniformly charged sphere. Hence, the energies of all five *d* orbitals are raised by the same amount.

 GO FIGURE

How would this absorbance spectrum change if you decreased the concentration of the $[Ti(H_2O)_6]^{3+}$ in solution?

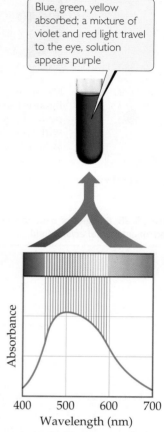

Blue, green, yellow absorbed; a mixture of violet and red light travel to the eye, solution appears purple

▲ **Figure 23.26 The color of $[Ti(H_2O)_6]^{3+}$** A solution containing the $[Ti(H_2O)_6]^{3+}$ ion appears purple because, as its visible absorption spectrum shows, the solution does not absorb light from the violet and red ends of the spectrum. That unabsorbed light is what reaches our eyes.

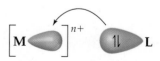

▲ **Figure 23.27 Metal–ligand bond formation.** The ligand acts as a Lewis base by donating its nonbonding electron pair to an empty orbital on the metal ion. The bond that results is strongly polar with some covalent character.

---

*The name *crystal field* arose because the theory was first developed to explain the properties of solid crystalline materials. The theory applies equally well to complexes in solution, however.

▲ **GO FIGURE**

Which *d* orbitals have lobes that point directly toward the ligands in an octahedral crystal field?

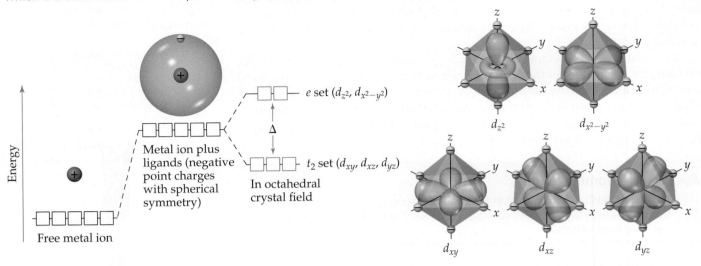

▲ **Figure 23.28 Energies of *d* orbitals in a free metal ion, a spherically symmetric crystal field, and an octahedral crystal field.**

This energy picture is only a first approximation, however, because the ligands are not distributed uniformly on a spherical surface and, therefore, do not approach the metal ion equally from every direction. Instead, we envision the six ligands approaching along *x*-, *y*-, and *z*-axes, as shown on the right in Figure 23.28. This arrangement of ligands is called an *octahedral crystal field*. Because the metal ion's *d* orbitals have different orientations and shapes, they do not all experience the same repulsion from the ligands and, therefore, do not all have the same energy under the influence of the octahedral crystal field. To see why, we must consider the shapes of the *d* orbitals and how their lobes are oriented relative to the ligands.

Figure 23.28 shows that the lobes of the d_{z^2} and $d_{x^2-y^2}$ orbitals are directed *along* the *x*-, *y*-, and *z*-axes and so point directly toward the ligand point charges. In the d_{xy}, d_{xz}, and d_{yz} orbitals, however, the lobes are directed *between* the axes and so do not point directly toward the charges. The result of this difference in orientation—$d_{x^2-y^2}$ and d_{z^2} lobes point directly toward the ligand charges; d_{xy}, d_{xz}, and d_{yz} lobes do not—is that the energy of the $d_{x^2-y^2}$ and d_{z^2} orbitals is higher than the energy of the d_{xy}, d_{xz}, and d_{yz} orbitals. This difference in energy is represented by the red boxes in the energy diagram of Figure 23.28.

It might seem like the energy of the $d_{x^2-y^2}$ orbital should be different from that of the d_{z^2} orbital because the $d_{x^2-y^2}$ has four lobes pointing at ligands and the d_{z^2} has only two lobes pointing at ligands. However, the d_{z^2} orbital does have electron density in the *xy* plane, represented by the ring encircling the point where the two lobes meet. More advanced calculations show that two orbitals do indeed have the same energy in the presence of the octahedral crystal field.

Because their lobes point directly at the negative ligand charges, electrons in the metal ion's d_{z^2} and $d_{x^2-y^2}$ orbitals experience stronger repulsions than those in the d_{xy}, d_{xz}, and d_{yz} orbitals. As a result, the energy splitting shown in Figure 23.28 occurs. The three lower-energy *d* orbitals are called the t_2 set of orbitals, and the two higher-energy ones are called the *e* set.* The energy gap Δ between the two sets is often called the *crystal-field splitting energy*.

---

*The labels t_2 for the d_{xy}, d_{xz}, and d_{yz} orbitals and *e* for the d_{z^2} and $d_{x^2y^2}$ orbitals come from the application of a branch of mathematics called *group theory* to crystal-field theory. Group theory can be used to analyze the effects of symmetry on molecular properties.

Crystal-field theory helps us account for the colors observed in transition-metal complexes. The energy gap Δ between the e and t_2 sets of d orbitals is of the same order of magnitude as the energy of a photon of visible light. It is therefore possible for a transition-metal complex to absorb visible light that excites an electron from a lower-energy (t_2) d orbital into a higher-energy (e) d orbital. In $[\text{Ti(H}_2\text{O)}_6]^{3+}$, for example, the Ti(III) ion has an $[\text{Ar}]3d^1$ electron configuration. (Recall from Section 7.4 that when determining the electron configurations of transition-metal ions, we remove the s electrons first.) Ti(III) is thus called a d^1 *ion*. In the ground state of $[\text{Ti(H}_2\text{O)}_6]^{3+}$, the single $3d$ electron resides in an orbital in the t_2 set (▶ **Figure 23.29**). Absorption of 495-nm light excites this electron up to an orbital in the e set, generating the absorption spectrum shown in Figure 23.26. Because this transition involves exciting an electron from one set of d orbitals to the other, we call it a ***d-d* transition**. As noted earlier, the absorption of visible radiation that produces this *d-d* transition causes the $[\text{Ti(H}_2\text{O)}_6]^{3+}$ ion to appear purple.

▲ Give It Some Thought

Why are compounds of Ti(IV) colorless?

The magnitude of the crystal-field splitting energy and, consequently, the color of a complex depend on both the metal and the ligands. For example, we saw in Figure 23.4 that the color of $[\text{M(H}_2\text{O)}_6]^{2+}$ complexes changes from reddish-pink when the metal ion is Co^{2+}, to green for Ni^{2+}, to pale blue for Cu^{2+}. If we change the ligands in the $[\text{Ni(H}_2\text{O)}_6]^{2+}$ ion the color also changes. $[\text{Ni(NH}_3)_6]^{2+}$ has a blue-violet color, while $[\text{Ni(en)}_3]^{2+}$ is purple (▼ **Figure 23.30**). In a ranking called the **spectrochemical series**, ligands are arranged in order of their abilities to increase splitting energy, as in this abbreviated list:

$$\longrightarrow \text{Increasing } \Delta \longrightarrow$$

$$\text{Cl}^- < \text{F}^- < \text{H}_2\text{O} < \text{NH}_3 < \text{en} < \text{NO}_2^-\text{(N-bonded)} < \text{CN}^-$$

▲ GO FIGURE

How would you calculate the energy gap between the t_2 and e orbitals from this diagram?

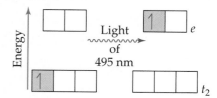

▲ Figure 23.29 **The *d-d* transition in $[\text{Ti(H}_2\text{O)}_6]^{3+}$ is produced by the absorption of 495-nm light.**

▲ GO FIGURE

If you were to use a ligand L that was a stronger field ligand than ethylenediamine, what color would you expect the $[\text{NiL}_6]^{2+}$ complex ion to have?

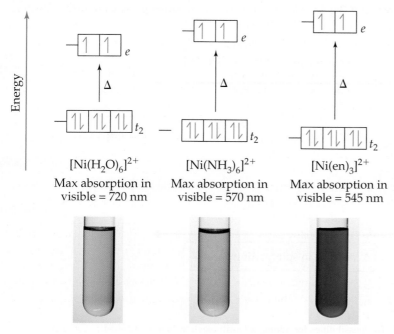

▲ Figure 23.30 **Effect of ligand on crystal-field splitting.** The greater the crystal-field strength of the ligand, the greater the energy gap Δ it causes between the t_2 and e sets of the metal ion's orbitals. This shifts the wavelength of the absorption maximum to shorter values.

The magnitude of Δ increases by roughly a factor of 2 from the far left to the far right of the spectrochemical series. Ligands at the low-Δ end of the spectrochemical series are termed *weak-field ligands*; those at the high-Δ end are termed *strong-field ligands*.

Let's take a closer look at the colors and crystal field splitting as we vary the ligands for the series of Ni^{2+} complexes discussed above. Because the Ni atom has an $[Ar]3d^84s^2$ electron configuration, Ni^{2+} has the configuration $[Ar]3d^8$ and therefore is a d^8 ion. The t_2 set of orbitals holds six electrons, two in each orbital, while the last two electrons go into the e set of orbitals. Consistent with Hund's rule, each e orbital holds one electron and both electrons have the same spin. ∞ (Section 6.8)

As the ligand changes from H_2O to NH_3 to ethylenediamine the spectrochemical series tells us that the crystal field, Δ, exerted by the six ligands should increase. When there is more than one electron in the d orbitals, interactions between the electrons make the absorption spectra more complicated than the spectrum shown for $[Ti(H_2O)_6]^{3+}$ in Figure 23.26, which complicates the task of relating changes in Δ with color. With d^8 ions like Ni^{2+} three peaks are observed in the absorption spectra. Fortunately, for Ni^{2+} complexes we can simplify the analysis because only one of these three peaks falls in the visible region of the spectrum.* Because the energy separation Δ is increasing, the wavelength of the absorption peak should shift to a shorter wavelength. ∞ (Section 6.3) In the case of $[Ni(H_2O)_6]^{2+}$ the absorption peak in the visible part of the spectrum reaches a maximum near 720 nm, in the red region of the spectrum. So the complex ion takes the complementary color—green. For $[Ni(NH_3)_6]^{2+}$ the absorption peak reaches its maximum at 570 nm near the boundary between orange and yellow. The resulting color of the complex ion is a mixture of the complementary colors—blue and violet. Finally for $[Ni(en)_3]^{2+}$ the peak shifts to an even shorter wavelength, 540 nm, which lies near the boundary between green and yellow. The resulting color purple is a mixture of the complimentary colors red and violet.

Electron Configurations in Octahedral Complexes

Crystal-field theory helps us understand the magnetic properties and some important chemical properties of transition-metal ions. From Hund's rule, we expect electrons to always occupy the lowest-energy vacant orbitals first and to occupy a set of degenerate (same-energy) orbitals one at a time with their spins parallel. ∞ (Section 6.8) Thus, if we have a d^1, d^2, or d^3 octahedral complex, the electrons go into the lower-energy t_2 orbitals, with their spins parallel. When a fourth electron must be added, we have the two choices shown in ▼ Figure 23.31: The electron can either go into an e orbital, where

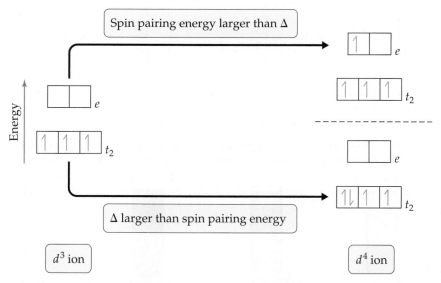

▲ **Figure 23.31 Two possibilities for adding a fourth electron to a d^3 octahedral complex.** Whether the fourth electron goes into a t_2 orbital or into an e orbital depends on the relative energies of the crystal-field splitting energy and the spin-pairing energy.

---

*The other two peaks fall in the infrared (IR) and ultraviolet (UV) regions of the spectrum. For $[Ni(H_2O)_6]^{2+}$ the IR peak is found at 1176 nm and the UV peak at 388 nm.

it will be the sole electron in the orbital, or become the second electron in a t_2 orbital. Because the energy difference between the t_2 and e sets is the splitting energy Δ, the energy cost of going into an e orbital rather than a t_2 orbital is also Δ. Thus, the goal of filling lowest-energy available orbitals first is met by putting the electron in a t_2 orbital.

There is a penalty for doing this, however, because the electron must now be paired with the electron already occupying the orbital. The difference between the energy required to pair an electron in an occupied orbital and the energy required to place that electron in an empty orbital is called the **spin-pairing energy**. The spin-pairing energy arises from the fact that the electrostatic repulsion between two electrons that share an orbital (and so must have opposite spins) is greater than the repulsion between two electrons that are in different orbitals and have parallel spins.

In coordination complexes, the nature of the ligands and the charge on the metal ion often play major roles in determining which of the two electron arrangements shown in Figure 23.31 is used. In $[CoF_6]^{3-}$ and $[Co(CN)_6]^{3-}$, both ligands have a $1-$ charge. The F^- ion, however, is on the low end of the spectrochemical series, so it is a weak-field ligand. The CN^- ion is on the high end and so is a strong-field ligand, which means it produces a larger energy gap Δ than the F^- ion. The splittings of the d-orbital energies in these two complexes are compared in ▶ Figure 23.32.

Cobalt(III) has an $[Ar]3d^6$ electron configuration, so both complexes in Figure 23.32 are d^6 complexes. Let's imagine that we add these six electrons one at a time to the d orbitals of the $[CoF_6]^{3-}$ ion. The first three go into the t_2 orbitals with their spins parallel. The fourth electron could pair up in one of the t_2 orbitals. The F^- ion is a weak-field ligand, however, and so the energy gap Δ between the t_2 set and the e set is small. In this case, the more stable arrangement is the fourth electron in one of the e orbitals. By the same energy argument, the fifth electron goes into the other e orbital. With all five d orbitals containing one electron, the sixth must pair up, and the energy needed to place the sixth electron in a t_2 orbital is less than that needed to place it in an e orbital. We end up with four t_2 electrons and two e electrons.

Figure 23.32 shows that the crystal-field splitting energy Δ is much larger in the $[Co(CN)_6]^{3-}$ complex. In this case, the spin-pairing energy is smaller than Δ, so the lowest-energy arrangement is the six electrons paired in the t_2 orbitals.

The $[CoF_6]^{3-}$ complex is a **high-spin complex**; that is, the electrons are arranged so that they remain unpaired as much as possible. The $[Co(CN)_6]^{3-}$ ion is a **low-spin complex**; that is, the electrons are arranged so that they remain paired as much as possible while still following Hund's rule. These two electronic arrangements can be readily distinguished by measuring the magnetic properties of the complex. Experiments show that $[CoF_6]^{3-}$ has four unpaired electrons and $[Co(CN)_6]^{3-}$ has none. The absorption spectrum also shows peaks corresponding to the different values of Δ in these two complexes.

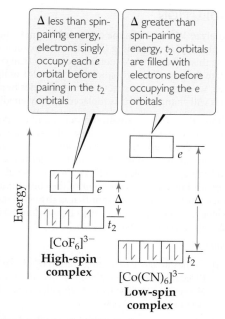

▲ **Figure 23.32 High-spin and low-spin complexes.** The high-spin $[CoF_6]^{3-}$ ion has a weak-field ligand and so a small Δ value. The low-spin $[Co(CN)_6]^{3-}$ ion has a strong-field ligand and so a large Δ value.

 Give It Some Thought

In octahedral complexes, for which d electron configurations is it possible to have high-spin and low-spin arrangements with different numbers of unpaired electrons?

In the transition metal ions of periods 5 and 6 (which have $4d$ and $5d$ valence electrons), the d orbitals are larger than in the period 4 ions (which have only $3d$ electrons). Thus, ions from periods 5 and 6 interact more strongly with ligands, resulting in a larger crystal-field splitting. Consequently, metal ions in periods 5 and 6 are invariably low spin in an octahedral crystal field.

SAMPLE EXERCISE 23.7 The Spectrochemical Series, Crystal Field Splitting, Color, and Magnetism

The compound hexaamminecobalt(III) chloride is diamagnetic and orange in color with a single absorption peak in its visible absorption spectrum. **(a)** What is the electron configuration of the cobalt(III) ion? **(b)** Is $[Co(NH_3)_6]^{3+}$ a high-spin complex or a low-spin complex? **(c)** Estimate the wavelength where you expect the absorption of light to reach a maximum? **(d)** What color and magnetic behavior would you predict for the complex ion $[Co(en)_3]^{3+}$?

SOLUTION

Analyze We are given the color and magnetic behavior of an octahedral complex containing Co with a +3 oxidation number. We need to use this information to determine its electron configuration, its spin state (low-spin or high-spin), and the color of light it absorbs. In part (d), we must use the spectrochemical series to predict how its properties will change if NH_3 is replaced by ethylenediamine (en).

Plan (a) From the oxidation number and the periodic table we can determine the number of valence electrons for Co(III) and from that we can determine the electron configuration. (b) The magnetic behavior can be used to determine whether this compound is a low-spin or high-spin complex. (c) Since there is a single peak in the visible absorption spectrum the color of the compound should be complementary to the color of light that is absorbed most strongly. (d) Ethylenediamine is a stronger field ligand than NH_3 so we expect a larger Δ for $[Co(en)_3]^{3+}$ than for $[Co(NH_3)_6]^{3+}$.

Solve

(a) Co has an electron configuration of $[Ar]4s^2 3d^7$ and Co^{3+} has three fewer electrons than Co. Because transition-metal ions always lose their valence shell s electrons, the electron configuration of Co^{3+} is $[Ar]3d^6$.

(b) There are six valence electrons in the d orbitals. The filling of the t_2 and e orbitals for both high-spin and low-spin complexes is shown below. Because the compound is diamagnetic we know all of the electrons must be paired up, which allows us to determine that $[Co(NH_3)_6]^{3+}$ is a low-spin complex.

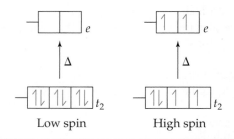

Low spin   High spin

(c) We are told that the compound is orange and has a single absorption peak in the visible region of the spectrum. The compound must therefore absorb the complementary color of orange, which is blue. The blue region of the spectrum ranges from approximately 430 nm to 490 nm. As an estimate we assume that the complex ion absorbs somewhere in the middle of the blue region, near 460 nm.

(d) Ethylenediammine is higher in the spectrochemical series than ammonia. Therefore, we expect a larger Δ for $[Co(en)_3]^{3+}$. Because Δ was already greater than the spin-pairing energy for $[Co(NH_3)_6]^{3+}$, we expect $[Co(en)_3]^{3+}$ to be a low-spin complex as well, with a d^6 configuration, so it will also be diamagnetic. The wavelength at which the complex absorbs light will shift to higher energy. If we assume a shift in the absorption maximum from blue to violet, the color of the complex will become yellow.

Practice Exercise 1

Which of the following octahedral complex ions will have the fewest number of unpaired electrons? (a) $[Cr(H_2O)_6]^{3+}$, (b) $[V(H_2O)_6]^{3+}$, (c) $[FeF_6]^{3-}$, (d) $[RhCl_6]^{3-}$, (e) $[Ni(NH_3)_6]^{2+}$.

Practice Exercise 2

Consider the colors of the ammonia complexes of Co^{3+} given in Table 23.3. Based on the change in color would you expect $[Co(NH_3)_5Cl]^{2+}$ to have a larger or smaller value of Δ than $[Co(NH_3)_6]^{3+}$? Is this prediction consistent with the spectrochemical series?

Tetrahedral and Square-Planar Complexes

Thus far we have considered crystal-field theory only for complexes having an octahedral geometry. When there are only four ligands in a complex, the geometry is generally tetrahedral, except for the special case of d^8 metal ions, which we will discuss in a moment.

The crystal-field splitting of d orbitals in tetrahedral complexes differs from that in octahedral complexes. Four equivalent ligands can interact with a central metal ion most effectively by approaching along the vertices of a tetrahedron. In this geometry the lobes of the two e orbitals point toward the edges of the tetrahedron, exactly in between the ligands (▶ Figure 23.33). This orientation keeps the $d_{x^2-y^2}$ and d_{z^2} as far from the ligand point charges as possible. Consequently, these two d orbitals experience less repulsion from the ligands and lie at lower energy than the other three d orbitals. The three t_2 orbitals do not point directly at the ligand point charges, but they do come closer to the ligands than the e set and as a result they experience more repulsion and are higher in energy. As we see in Figure 23.33, the splitting of d orbitals in a tetrahedral geometry is the opposite of what we find for the octahedral geometry, namely the e orbitals are now *below* the t_2 orbitals. The crystal-field splitting energy Δ is much smaller for tetrahedral complexes than it is for comparable octahedral complexes, in part because there are fewer ligand point charges in the tetrahedral geometry, and in part because neither set of orbitals have lobes that point directly at the ligand point charges. Calculations show that for the same metal ion and ligand set, Δ for the tetrahedral complex is only four-ninths as large as for the octahedral complex. For this reason,

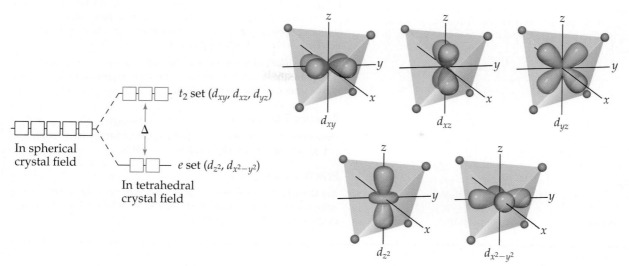

▲ **Figure 23.33 Energies of the *d* orbitals in a tetrahedral crystal field.** The splitting of the *e* and t_2 sets of orbitals is inverted with respect to the splitting associated with an octahedral crystal field. The crystal field splitting energy Δ is much smaller than it is in an octahedral crystal field.

all tetrahedral complexes are high spin; the crystal-field splitting energy is never large enough to overcome the spin-pairing energies.

In a square-planar complex, four ligands are arranged about the metal ion such that all five species are in the *xy* plane. The resulting energy levels of the *d* orbitals are illustrated in ▶ Figure 23.34. Note in particular that the d_{z^2} orbital is considerably lower in energy than the $d_{x^2-y^2}$ orbital. To understand why this is so, recall from Figure 23.28 that in an octahedral field the d_{z^2} orbital of the metal ion interacts with the ligands positioned above and below the *xy* plane. There are no ligands in these two positions in a square-planar complex, which means that the d_{z^2} orbital experiences less repulsion and so remains in a lower-energy, more stable state.

Square-planar complexes are characteristic of metal ions with a d^8 electron configuration. They are nearly always low spin, with the eight *d* electrons spin-paired to form a diamagnetic complex. This pairing leaves the $d_{x^2-y^2}$ orbital empty. Such an electronic arrangement is particularly common among the d^8 ions of periods 5 and 6, such as Pd^{2+}, Pt^{2+}, Ir^+, and Au^{3+}.

 Give It Some Thought

Why is the energy of the d_{xz} and d_{yz} orbitals in a square-planar complex lower than that of the d_{xy} orbital?

▲ **GO FIGURE**

For which *d* orbital(s) do the lobes point directly at the ligands in a square-planar crystal field?

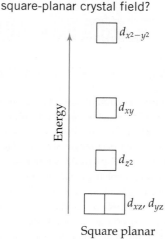

▲ **Figure 23.34 Energies of the *d* orbitals in a square-planar crystal field.**

Populating *d* Orbitals in Tetrahedral and Square-Planar Complexes

Nickel(II) complexes in which the metal coordination number is 4 can have either square-planar or tetrahedral geometry. $[NiCl_4]^{2-}$ is paramagnetic, and $[Ni(CN)_4]^{2-}$ is diamagnetic. One of these complexes is square planar, and the other is tetrahedral. Use the relevant crystal-field splitting diagrams in the text to determine which complex has which geometry.

SOLUTION

Analyze We are given two complexes containing Ni^{2+} and their magnetic properties. We are given two molecular geometry choices and asked to use crystal-field splitting diagrams from the text to determine which complex has which geometry.

Plan We need to determine the number of *d* electrons in Ni^{2+} and then use Figure 23.33 for the tetrahedral complex and Figure 23.34 for the square-planar complex.

Solve Nickel(II) has the electron configuration $[Ar]3d^8$. Tetrahedral complexes are always high spin, and square-planar complexes are almost always low spin. Therefore, the population of the d electrons in the two geometries is

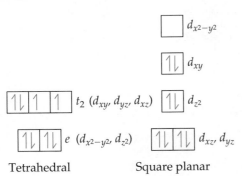

Tetrahedral   Square planar

The tetrahedral complex has two unpaired electrons, and the square-planar complex has none. We know from Section 23.1 that the tetrahedral complex must be paramagnetic and the square planar must be diamagnetic. Therefore, $[NiCl_4]^{2-}$ is tetrahedral, and $[Ni(CN)_4]^{2-}$ is square planar.

Comment Nickel(II) forms octahedral complexes more frequently than square-planar ones, whereas d^8 metals from periods 5 and 6 tend to favor square-planar coordination.

Practice Exercise 1

How many unpaired electrons do you predict for the tetrahedral $[MnCl_4]^{2-}$ ion? **(a)** 1, **(b)** 2, **(c)** 3, **(d)** 4, **(e)** 5.

Practice Exercise 2

Are there any diamagnetic tetrahedral complexes containing transition metal ions with partially filled d orbitals? If so what electron count(s) leads to diamagnetism?

Crystal-field theory can be used to explain many observations in addition to those we have discussed. The theory is based on electrostatic interactions between ions and atoms, which essentially means ionic bonds. Many lines of evidence show, however, that the bonding in complexes must have some covalent character. Therefore, molecular-orbital theory ∞ (Sections 9.7 and 9.8) can also be used to describe the bonding in complexes, although the application of molecular-orbital theory to coordination compounds is beyond the scope of our discussion. Crystal-field theory, although not entirely accurate in all details, provides an adequate and useful first description of the electronic structure of complexes.

A Closer Look

Charge-Transfer Color

In the laboratory portion of your course, you have probably seen many colorful transition-metal compounds. Many of these compounds are colored because of d-d transitions. Some colored complexes, however, including the violet permanganate ion, MnO_4^-, and the yellow chromate ion, CrO_4^{2-} ▼ **Figure 23.35**, derive their color from a different type of excitation involving the d orbitals.

The permanganate ion strongly absorbs visible light, with a maximum absorption at 565 nm. Because violet is complementary to yellow, this strong absorption in the yellow portion of the visible spectrum is responsible for the violet color of salts and solutions of the ion. What is happening during this absorption of light? The MnO_4^- ion is a complex

of Mn(VII). Because Mn(VII) has a d^0 electron configuration, the absorption cannot be due to a d-d transition because there are no d electrons to excite! That does not mean, however, that the d orbitals are not involved in the transition. The excitation in the MnO_4^- ion is due to a *charge-transfer transition*, in which an electron on one oxygen ligand is excited into a vacant d orbital on the Mn ion (▶ **Figure 23.36**). In essence, an electron is transferred from a ligand to the metal, so this transition is called a *ligand-to-metal charge-transfer (LMCT) transition*.

An LMCT transition is also responsible for the color of the CrO_4^{2-}, which is a d^0 Cr(VI) complex.

Also shown in Figure 23.35 is a salt of the perchlorate ion (ClO_4^-). Like MnO_4^-, ClO_4^- is tetrahedral and has its central atom in the +7 oxidation state. However, because the Cl atom does not have low-lying

KMnO₄

K₂CrO₄

KClO₄

▲ **Figure 23.35 The colors of compounds can arise from charge-transfer transitions.** $KMnO_4$ and K_2CrO_4 are colored due to ligand-to-metal charge-transfer transitions in their anions. The perchlorate anion in $KClO_4$ has no occupied d orbitals and its charge-transfer transition is at higher energy, corresponding to ultraviolet absorption; therefore it appears white.

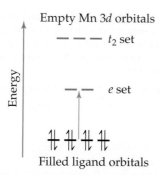

Empty Mn 3*d* orbitals

- - - - t_2 set

Energy

— — *e* set

⇡⇣ ⇡⇣ ⇡⇣ ⇡⇣
Filled ligand orbitals

▲ **Figure 23.36 Ligand-to-metal charge-transfer transition in MnO₄⁻.** As shown by the blue arrow, an electron is excited from a nonbonding pair on O into one of the empty *d* orbitals on Mn.

d orbitals, exciting a Cl electron requires a more energetic photon than does MnO₄⁻. The first absorption for ClO₄⁻ is in the ultraviolet portion of the spectrum, so all the visible light is transmitted and the salt appears white.

Other complexes exhibit charge-transfer excitations in which an electron from the metal atom is excited to an empty orbital on a ligand. Such an excitation is called a *metal-to-ligand charge-transfer (MLCT) transition*.

Charge-transfer transitions are generally more intense than *d-d* transitions. Many metal-containing pigments used for oil painting, such as cadmium yellow (CdS), chrome yellow (PbCrO₄), and red ochre (Fe₂O₃), have intense colors because of charge-transfer transitions.

Related Exercises: 23.82, 23.83

SAMPLE
INTEGRATIVE EXERCISE Putting Concepts Together

The oxalate ion has the Lewis structure shown in Table 23.4. **(a)** Show the geometry of the complex formed when this ion complexes with cobalt(II) to form $[Co(C_2O_4)(H_2O)_4]$. **(b)** Write the formula for the salt formed when three oxalate ions complex with Co(II), assuming that the charge-balancing cation is Na^+. **(c)** Sketch all the possible geometric isomers for the cobalt complex formed in part (b). Are any of these isomers chiral? Explain. **(d)** The equilibrium constant for the formation of the cobalt(II) complex produced by coordination of three oxalate anions, as in part (b), is 5.0×10^9, and the equilibrium constant for formation of the cobalt(II) complex with three molecules of *ortho*-phenanthroline (Table 23.4) is 9×10^{19}. From these results, what conclusions can you draw regarding the relative Lewis base properties of the two ligands toward cobalt(II)? **(e)** Using the approach described in Sample Exercise 17.14, calculate the concentration of free aqueous Co(II) ion in a solution initially containing 0.040 *M* oxalate (*aq*) and 0.0010 *M* Co^{2+}(*aq*).

SOLUTION

(a) The complex formed by coordination of one oxalate ion is octahedral:

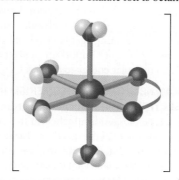

(b) Because the oxalate ion has a charge of 2−, the net charge of a complex with three oxalate anions and one Co^{2+} ion is 4−. Therefore, the coordination compound has the formula $Na_4[Co(C_2O_4)_3]$.

(c) There is only one geometric isomer. The complex is chiral, however, in the same way the $[Co(en)_3]^{3+}$ complex is chiral (Figure 23.22). The two mirror images are not superimposable, so there are two enantiomers:

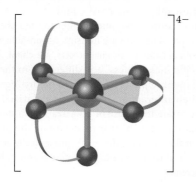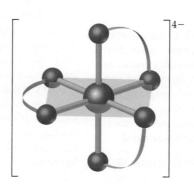

(d) The *ortho*-phenanthroline ligand is bidentate, like the oxalate ligand, so they both exhibit the chelate effect. Thus, we conclude that *ortho*-phenanthroline is a stronger Lewis base toward Co^{2+} than oxalate. This conclusion is consistent with what we learned about bases in Section 16.7, that nitrogen bases are generally stronger than oxygen bases. (Recall, for example, that NH_3 is a stronger base than H_2O.)

(e) The equilibrium we must consider involves 3 mol of oxalate ion (represented as Ox^{2-}).

$$Co^{2+}(aq) + 3\,Ox^{2-}(aq) \rightleftharpoons [Co(Ox)_3]^{4-}(aq)$$

The formation-constant expression is

$$K_f = \frac{[[Co(Ox)_3]^{4-}]}{[Co^{2+}][Ox^{2-}]^3}$$

Because K_f is so large, we can assume that essentially all the Co^{2+} is converted to the oxalato complex. Under that assumption, the final concentration of $[Co(Ox)_3]^{4-}$ is 0.0010 M and that of oxalate ion is $[Ox^{2-}] = (0.040) - 3(0.0010) = 0.037\ M$ (three Ox^{2-} ions react with each Co^{2+} ion). We then have

$$[Co^{2+}] = xM, [Ox^{2-}] \cong 0.037\ M, [[Co(Ox)_3]^{4-}] \cong 0.0010\ M$$

Inserting these values into the equilibrium-constant expression, we have

$$K_f = \frac{(0.0010)}{x(0.037)^3} = 5 \times 10^9$$

Solving for x, we obtain $4 \times 10^{-9}\ M$. From this, we see that the oxalate has complexed all but a tiny fraction of the Co^{2+} in solution.

Chapter Summary and Key Terms

THE TRANSITION METALS (SECTION 23.1) Metallic elements are obtained from **minerals**, which are solid inorganic compounds found in nature. **Metallurgy** is the science and technology of extracting metals from the earth and processing them for further use. Transition metals are characterized by incomplete filling of the *d* orbitals. The presence of *d* electrons in transition elements leads to multiple oxidation states. As we proceed through the transition metals in a given row of the periodic table, the attraction between the nucleus and the valence electrons increases more markedly for *d* electrons than for *s* electrons. As a result, the later transition elements in a period tend to have lower oxidation states.

The atomic and ionic radii of period 5 transition metals are larger than those of period 4 metals. The transition metals of periods 5 and 6 have comparable atomic and ionic radii and are also similar in other properties. This similarity is due to the **lanthanide contraction**.

The presence of unpaired electrons in valence orbitals leads to magnetic behavior in transition metals and their compounds. In **ferromagnetic**, **ferrimagnetic**, and **antiferromagnetic** substances, the unpaired electron spins on atoms in a solid are affected by spins on neighboring atoms. In a ferromagnetic substance the spins all point in the same direction. In an antiferromagnetic substance the spins point in opposite directions and cancel one another. In a ferrimagnetic substance the spins point in opposite directions but do not fully cancel. Ferromagnetic and ferrimagnetic substances are used to make permanent magnets.

TRANSITION-METAL COMPLEXES (SECTION 23.2) **Coordination compounds** are substances that contain **metal complexes**. Metal complexes contain metal ions bonded to several surrounding anions or molecules known as **ligands**. The metal ion and its ligands make up the **coordination sphere** of the complex. The number of atoms attached to the metal ion is the **coordination number** of the metal ion. The most common coordination numbers are 4 and 6; the most common coordination geometries are tetrahedral, square planar, and octahedral.

COMMON LIGANDS IN COORDINATION CHEMISTRY (SECTION 23.3) Ligands that occupy only one site in a coordination sphere are called **monodentate ligands**. The atom of the ligand that bonds to the metal ion is the **donor atom**. Ligands that have two donor atoms are **bidentate ligands**. **Polydentate ligands** have three or more donor atoms. Bidentate and polydentate ligands are also called **chelating agents**. In general, chelating agents form more stable complexes than do related monodentate ligands, an observation known as the **chelate effect**. Many biologically important molecules, such as the **porphyrins**, are complexes of chelating agents. A related group of plant pigments known as **chlorophylls** are important in **photosynthesis**, the process by which plants use solar energy to convert CO_2 and H_2O into carbohydrates.

NOMENCLATURE AND ISOMERISM IN COORDINATION CHEMISTRY (SECTION 23.4) In naming coordination compounds, the number and type of ligands attached to the metal ion are specified, as is the oxidation state of the metal ion. **Isomers** are compounds with the same composition but different arrangements of atoms and therefore different properties. **Structural isomers** differ in the bonding arrangements of the ligands. **Linkage isomerism** occurs when a ligand can coordinate to a metal ion through different donor atoms. **Coordination-sphere isomers** contain different ligands in the coordination sphere. **Stereoisomers** are isomers with the same chemical bonding arrangements but different spatial arrangements of ligands. The most common forms of stereoisomerism are **geometric isomerism** and **optical isomerism**. Geometric isomers differ from one another in the relative locations of donor atoms in the coordination sphere; the most common are cis–trans isomers. Geometric isomers differ from one another in their chemical and physical properties. Optical isomers are nonsuperimposable mirror images of each other. Optical isomers, or **enantiomers**, are **chiral**, meaning that they have a specific "handedness" and differ only in the presence of a chiral environment. Optical isomers can be distinguished from one another by their interactions with plane-polarized light; solutions of one isomer rotate the plane of polarization to the right (**dextrorotatory**), and solutions of its mirror image rotate the plane to the left (**levorotatory**). Chiral molecules, therefore, are **optically active**. A 50–50 mixture of two optical isomers does not rotate plane-polarized light and is said to be **racemic**.

COLOR AND MAGNETISM IN COORDINATION CHEMISTRY (SECTION 23.5) A substance has a particular color because it either reflects or transmits light of that color or absorbs light of the **complementary color**. The amount of light absorbed by a sample as a function of wavelength is known as its **absorption spectrum**. The light absorbed provides the energy to excite electrons to higher-energy states.

It is possible to determine the number of unpaired electrons in a complex from its degree of paramagnetism. Compounds with no unpaired electrons are diamagnetic.

CRYSTAL-FIELD THEORY (SECTION 23.6) Crystal-field theory successfully accounts for many properties of coordination compounds, including their color and magnetism. In crystal-field theory, the interaction between metal ion and ligand is viewed as electrostatic. Because some d orbitals point directly at the ligands whereas others point between them, the ligands split the energies of the metal d orbitals. For an octahedral complex, the d orbitals are split into a lower-energy set of three degenerate orbitals (the t_2 set) and a higher-energy set of two degenerate orbitals (the e set). Visible light can cause a

d-d **transition**, in which an electron is excited from a lower-energy d orbital to a higher-energy d orbital. The **spectrochemical series** lists ligands in order of their ability to increase the split in d-orbital energies in octahedral complexes.

Strong-field ligands create a splitting of d-orbital energies that is large enough to overcome the **spin-pairing energy**. The d electrons then preferentially pair up in the lower-energy orbitals, producing a **low-spin complex**. When the ligands exert a weak crystal field, the splitting of the d orbitals is small. The electrons then occupy the higher-energy d orbitals in preference to pairing up in the lower-energy set, producing a **high-spin complex**.

Crystal-field theory also applies to tetrahedral and square-planar complexes, which leads to different d-orbital splitting patterns. In a tetrahedral crystal field, the splitting of the d orbitals results in a higher-energy t_2 set and a lower-energy e set, the opposite of the octahedral case. The splitting in a tetrahedral crystal field is much smaller than that in an octahedral crystal field, so tetrahedral complexes are always high-spin complexes.

Learning Outcomes   After studying this chapter, you should be able to:

- Describe the periodic trends in radii and oxidation states of the transition-metal ions, including the origin and effect of the lanthanide contraction. (Section 23.1)

- Determine the oxidation number and number of d electrons for metal ions in complexes. (Section 23.2)

- Identify common ligands and distinguish between chelating and nonchelating ligands. (Section 23.3)

- Name coordination compounds given their formula and write their formula given their name. (Section 23.4)

- Recognize and draw the geometric isomers of a complex. (Section 23.4)

- Recognize and draw the optical isomers of a complex. (Section 23.4)

- Use crystal-field theory to explain the colors and to determine the number of unpaired electrons in a complex. (Sections 23.5 and 23.6)

Exercises

Visualizing Concepts

23.1 The three graphs below show the variation in radius, effective nuclear charge, and maximum oxidation state for the transition metals of period 4. In each part below identify which property is being plotted. [Section 23.1]

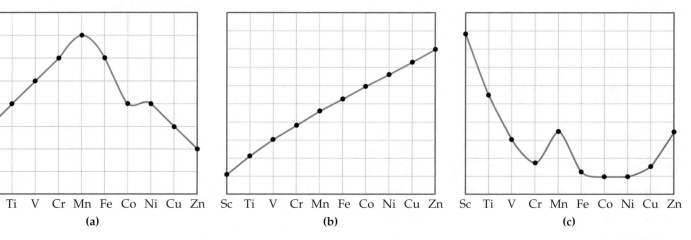

(a)   (b)   (c)

23.2 Draw the structure for Pt(en)Cl$_2$ and use it to answer the following questions: **(a)** What is the coordination number for platinum in this complex? **(b)** What is the coordination geometry? **(c)** What is the oxidation state of the platinum? **(d)** How many unpaired electrons are there? [Sections 23.2 and 23.6]

23.3 Draw the Lewis structure for the ligand shown here.
(a) Which atoms can serve as donor atoms? Classify this ligand
as monodentate, bidentate, or polydentate. **(b)** How many of
these ligands are needed to fill the coordination sphere in an
octahedral complex? [Section 23.2]

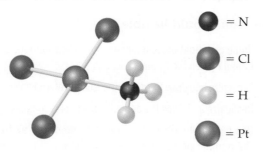

$$NH_2CH_2CH_2NHCH_2CO_2{}^-$$

23.4 The complex ion shown here has a $1-$ charge. Name the
complex ion. [Section 23.4]

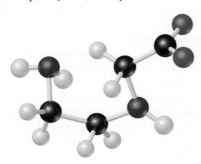

- = N
- = Cl
- = H
- = Pt

23.5 There are two geometric isomers of octahedral complexes of
the type MA_3X_3, where M is a metal and A and X are monodentate ligands. Of the complexes shown here, which are
identical to (1) and which are the geometric isomers of (1)?
[Section 23.4]

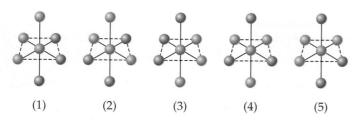

(1)   (2)   (3)   (4)   (5)

23.6 Which of the complexes shown here are chiral? [Section 23.4]

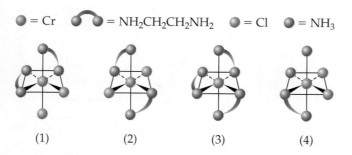

- = Cr
- = $NH_2CH_2CH_2NH_2$
- = Cl
- = NH_3

(1)   (2)   (3)   (4)

23.7 The solutions shown here each have an absorption spectrum
with a single absorption peak like that shown in Figure 23.26.
What color does each solution absorb most strongly?
[Section 23.5]

23.8 Which of these crystal-field splitting diagrams represents:
(a) a weak-field octahedral complex of Fe^{3+}, **(b)** a strong-field
octahedral complex of Fe^{3+}, **(c)** a tetrahedral complex of Fe^{3+},
(d) a tetrahedral complex of Ni^{2+}? (The diagrams do not indicate the relative magnitudes of Δ.) [Section 23.6]

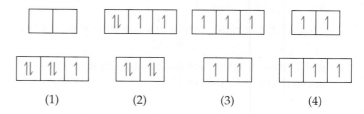

(1)   (2)   (3)   (4)

23.9 In the linear crystal field shown here, the negative charges are
on the z-axis. Using Figure 23.28 as a guide, predict which of
the following choices most accurately describes the splitting
of the d orbitals in a linear crystal field? [Section 23.6]

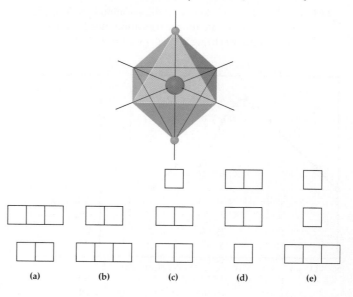

(a)   (b)   (c)   (d)   (e)

23.10 Two Fe(II) complexes are both low spin but have different ligands. A solution of one is green and a solution of the other is
red. Which solution is likely to contain the complex that has
the stronger-field ligand? [Section 23.6]

The Transition Metals (Section 23.1)

23.11 The lanthanide contraction explains which of the following periodic trends? **(a)** The atomic radii of the transition metals first decrease and then increase when moving horizontally across each period. **(b)** When forming ions the transition metals lose their valence *s* orbitals before their valence *d* orbitals. **(c)** The radii of the period 5 transition metals (Y–Cd) are very similar to the radii of the period 6 transition metals (Lu–Hg).

23.12 Which periodic trend is responsible for the observation that the maximum oxidation state of the transition-metal elements peaks near groups 7B and 8B? **(a)** The number of valence electrons reaches a maximum at group 8B. **(b)** The effective nuclear charge increases on moving left across each period. **(c)** The radii of the transition-metal elements reaches a minimum for group 8B and as the size of the atoms decreases it becomes easier to remove electrons.

23.13 For each of the following compounds determine the electron configuration of the transition metal ion. **(a)** TiO, **(b)** TiO_2, **(c)** NiO, **(d)** ZnO.

23.14 Among the period 4 transition metals (Sc–Zn), which elements do not form ions where there are partially filled $3d$ orbitals?

23.15 Write out the ground-state electron configurations of **(a)** Ti^{3+}, **(b)** Ru^{2+}, **(c)** Au^{3+}, **(d)** Mn^{4+}.

23.16 How many electrons are in the valence *d* orbitals in these transition-metal ions? **(a)** Co^{3+}, **(b)** Cu^+, **(c)** Cd^{2+}, **(d)** Os^{3+}.

23.17 Which type of substance is attracted by a magnetic field, a diamagnetic substance or a paramagnetic substance?

23.18 Which type of magnetic material cannot be used to make permanent magnets, a ferromagnetic substance, an antiferromagnetic substance, or a ferrimagnetic substance?

23.19 What kind of magnetism is exhibited by this diagram:

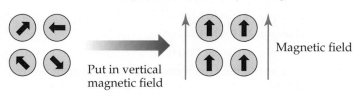

23.20 The most important oxides of iron are magnetite, Fe_3O_4, and hematite, Fe_2O_3. **(a)** What are the oxidation states of iron in these compounds? **(b)** One of these iron oxides is ferrimagnetic, and the other is antiferromagnetic. Which iron oxide is likely to show which type of magnetism? Explain.

Transition-Metal Complexes (Section 23.2)

23.21 **(a)** Using Werner's definition of valence, which property is the same as oxidation number, *primary valence* or *secondary valence*? **(b)** What term do we normally use for the other type of valence? **(c)** Why can the NH_3 molecule serve as a ligand but the BH_3 molecule cannot?

23.22 **(a)** What is the meaning of the term *coordination number* as it applies to metal complexes? **(b)** Give an example of a ligand that is neutral and one that is negatively charged. **(c)** Would you expect ligands that are positively charged to be common? Explain. **(d)** What type of chemical bonding is characteristic of coordination compounds? Illustrate with the compound

$Co(NH_3)_6Cl_3$. **(e)** What are the most common coordination numbers for metal complexes?

23.23 A complex is written as $NiBr_2 \cdot 6 NH_3$. **(a)** What is the oxidation state of the Ni atom in this complex? **(b)** What is the likely coordination number for the complex? **(c)** If the complex is treated with excess $AgNO_3(aq)$, how many moles of AgBr will precipitate per mole of complex?

23.24 Crystals of hydrated chromium(III) chloride are green, have an empirical formula of $CrCl_3 \cdot 6 H_2O$, and are highly soluble. **(a)** Write the complex ion that exists in this compound. **(b)** If the complex is treated with excess $AgNO_3(aq)$, how many moles of AgCl will precipitate per mole of $CrCl_3 \cdot 6 H_2O$ dissolved in solution? **(c)** Crystals of anhydrous chromium(III) chloride are violet and insoluble in aqueous solution. The coordination geometry of chromium in these crystals is octahedral as is almost always the case for Cr^{3+}. How can this be the case if the ratio of Cr to Cl is not 1:6?

23.25 Indicate the coordination number and the oxidation number of the metal for each of the following complexes:

(a) $Na_2[CdCl_4]$

(b) $K_2[MoOCl_4]$

(c) $[Co(NH_3)_4Cl_2]Cl$

(d) $[Ni(CN)_5]^{3-}$

(e) $K_3[V(C_2O_4)_3]$

(f) $[Zn(en)_2]Br_2$

23.26 Indicate the coordination number and the oxidation number of the metal for each of the following complexes:

(a) $K_3[Co(CN)_6]$

(b) $Na_2[CdBr_4]$

(c) $[Pt(en)_3](ClO_4)_4$

(d) $[Co(en)_2(C_2O_4)]^+$

(e) $NH_4[Cr(NH_3)_2(NCS)_4]$

(f) $[Cu(bipy)_2I]I$

Common Ligands in Coordination Chemistry (Section 23.3)

23.27 **(a)** What is the difference between a monodentate ligand and a bidentate ligand? **(b)** How many bidentate ligands are necessary to fill the coordination sphere of a six-coordinate complex? **(c)** You are told that a certain molecule can serve as a tridentate ligand. Based on this statement, what do you know about the molecule?

23.28 For each of the following polydentate ligands, determine (i) the maximum number of coordination sites that the ligand can occupy on a single metal ion and (ii) the number and type of donor atoms in the ligand: **(a)** ethylenediamine (en), **(b)** bipyridine (bipy), **(c)** the oxalate anion ($C_2O_4^{2-}$), **(d)** the 2− ion of the porphine molecule (Figure 23.13), **(e)** $[EDTA]^{4-}$.

23.29 Polydentate ligands can vary in the number of coordination positions they occupy. In each of the following, identify the polydentate ligand present and indicate the probable number of coordination positions it occupies:

(a) $[Co(NH_3)_4(o\text{-phen})]Cl_3$

(b) $[Cr(C_2O_4)(H_2O)_4]Br$

(c) $[Cr(EDTA)(H_2O)]^-$

(d) $[Zn(en)_2](ClO_4)_2$

23.30 Indicate the likely coordination number of the metal in each of the following complexes:

(a) $[Rh(bipy)_3](NO_3)_3$

(b) $Na_3[Co(C_2O_4)_2Cl_2]$

(c) $[Cr(o\text{-phen})_3](CH_3COO)_3$

(d) $Na_2[Co(EDTA)Br]$

23.31 **(a)** What is meant by the term *chelate effect*? **(b)** What thermodynamic factor is generally responsible for the chelate effect? **(c)** Why are polydentate ligands often called *sequestering agents*?

23.32 *Pyridine* (C_5H_5N), abbreviated py, is the molecule

(a) Why is pyridine referred to as a monodentate ligand? **(b)** For the equilibrium reaction

$$[Ru(py)_4(bipy)]^{2+} + 2\ py \rightleftharpoons [Ru(py)_6]^{2+} + bipy$$

what would you predict for the magnitude of the equilibrium constant? Explain your answer.

23.33 True or false? The following ligand can act as a bidentate ligand?

23.34 When silver nitrate is reacted with the molecular base *ortho*-phenanthroline, colorless crystals form that contain the transition-metal complex shown below. **(a)** What is the coordination geometry of silver in this complex? **(b)** Assuming no oxidation or reduction occurs during the reaction, what is charge of the complex shown here? **(c)** Do you expect any nitrate ions will be present in the crystal? **(d)** Write a formula for the compound that forms in this reaction. **(e)** Use the accepted nomenclature to write the name of this compound.

Nomenclature and Isomerism in Coordination Chemistry (Section 23.4)

23.35 Write the formula for each of the following compounds, being sure to use brackets to indicate the coordination sphere:

(a) hexaamminechromium(III) nitrate

(b) tetraamminecarbonatocobalt(III) sulfate

(c) dichlorobis(ethylenediamine)platinum(IV) bromide

(d) potassium diaquatetrabromovanadate(III)

(e) bis(ethylenediamine)zinc(II) tetraiodomercurate(II)

23.36 Write the formula for each of the following compounds, being sure to use brackets to indicate the coordination sphere:

(a) tetraaquadibromomanganese(III) perchlorate

(b) bis(bipyridyl)cadmium(II) chloride

(c) potassium tetrabromo(*ortho*-phenanthroline)-cobaltate (III)

(d) cesium diamminetetracyanochromate(III)

(e) tris(ethylenediamine)rhodium(III) tris(oxalato)-cobaltate(III)

23.37 Write the names of the following compounds, using the standard nomenclature rules for coordination complexes:

(a) $[Rh(NH_3)_4Cl_2]Cl$

(b) $K_2[TiCl_6]$

(c) $MoOCl_4$

(d) $[Pt(H_2O)_4(C_2O_4)]Br_2$

23.38 Write names for the following coordination compounds:

(a) $[Cd(en)Cl_2]$

(b) $K_4[Mn(CN)_6]$

(c) $[Cr(NH_3)_5(CO_3)]Cl$

(d) $[Ir(NH_3)_4(H_2O)_2](NO_3)_3$

23.39 Consider the following three complexes:

(Complex 1) $[Co(NH_3)_4Br_2]Cl$

(Complex 2) $[Pd(NH_3)_2(ONO)_2]$

(Complex 3) $[V(en)_2Cl_2]^+$,

Which of the three complexes can have **(a)** geometric isomers, **(b)** linkage isomers, **(c)** optical isomers, **(d)** coordination-sphere isomers?

23.40 Consider the following three complexes:

(Complex 1) $[Co(NH_3)_5SCN]^{2+}$

(Complex 2) $[Co(NH_3)_3Cl_3]^{2+}$

(Complex 3) $CoClBr \cdot 5NH_3$

Which of the three complexes can have **(a)** geometric isomers, **(b)** linkage isomers, **(c)** optical isomers, **(d)** coordination-sphere isomers?

23.41 A four-coordinate complex MA_2B_2 is prepared and found to have two different isomers. Is it possible to determine from this information whether the complex is square planar or tetrahedral? If so, which is it?

23.42 Consider an octahedral complex MA_3B_3. How many geometric isomers are expected for this compound? Will any of the isomers be optically active? If so, which ones?

23.43 Sketch all the possible stereoisomers of **(a)** tetrahedral $[Cd(H_2O)_2Cl_2]$, **(b)** square-planar $[IrCl_2(PH_3)_2]^-$, **(c)** octahedral $[Fe(o\text{-phen})_2Cl_2]^+$.

23.44 Sketch all the possible stereoisomers of **(a)** $[Rh(bipy)(o-phen)_2]^{3+}$, **(b)** $[Co(NH_3)_3(bipy)Br]^{2+}$, **(c)** square-planar $[Pd(en)(CN)_2]$.

Color and Magnetism in Coordination Chemistry; Crystal-Field Theory (Sections 23.5 and 23.6)

23.45 **(a)** If a complex absorbs light at 610 nm, what color would you expect the complex to be? **(b)** What is the energy in Joules of a photon with a wavelength of 610 nm? **(c)** What is the energy of this absorption in kJ/mol?

23.46 **(a)** A complex absorbs photons with an energy of 4.51×10^{-19} J. What is the wavelength of these photons? **(b)** If this is the only place in the visible spectrum where the complex absorbs light, what color would you expect the complex to be?

23.47 Identify each of the following coordination complexes as either diamagnetic or paramagnetic:

 (a) $[ZnCl_4]^{2-}$

 (b) $[Pd(NH_3)_2Cl_2]$

 (c) $[V(H_2O)_6]^{3+}$

 (d) $[Ni(en)_3]^{2+}$

23.48 Identify each of the following coordination complexes as either diamagnetic or paramagnetic:

 (a) $[Ag(NH_3)_2]^+$

 (b) square planar $[Cu(NH_3)_4]^{2+}$

 (c) $[Ru(bipy)_3]^{2+}$

 (d) $[CoCl_4]^{2-}$

23.49 In crystal-field theory, ligands are modeled as if they are point negative charges. What is the basis of this assumption, and how does it relate to the nature of metal–ligand bonds?

23.50 The lobes of which d orbitals point directly between the ligands in **(a)** octahedral geometry, **(b)** tetrahedral geometry?

23.51 **(a)** Sketch a diagram that shows the definition of the *crystal-field splitting energy* (Δ) for an octahedral crystal field. **(b)** What is the relationship between the magnitude of Δ and the energy of the *d-d* transition for a d^1 complex? **(c)** Calculate Δ in kJ/mol if a d^1 complex has an absorption maximum at 545 nm.

23.52 As shown in Figure 23.26, the *d-d* transition of $[Ti(H_2O)_6]^{3+}$ produces an absorption maximum at a wavelength of about 500 nm. **(a)** What is the magnitude of Δ for $[Ti(H_2O)_6]^{3+}$ in kJ/mol? **(b)** How would the magnitude of Δ change if the H_2O ligands in $[Ti(H_2O)_6]^{3+}$ were replaced with NH_3 ligands?

23.53 The colors in the copper-containing minerals malachite (green) and azurite (blue) come from a single *d-d* transition in each compound. **(a)** What is the electron configuration of the copper ion in these minerals? **(b)** Based on their colors in which compound would you predict the crystal field splitting Δ is larger?

23.54 The color and wavelength of the absorption maximum for $[Ni(H_2O)_6]^{2+}$, $[Ni(NH_3)_6]^{2+}$, and $[Ni(en)_3]^{2+}$ are given in Figure 23.30. The absorption maximum for the $[Ni(bipy)_3]^{2+}$ ion occurs at about 520 nm. **(a)** What color would you expect

for the $[Ni(bipy)_3]^{2+}$ ion? **(b)** Based on these data, where would you put bipy in the spectrochemical series?

23.55 Give the number of (valence) d electrons associated with the central metal ion in each of the following complexes: **(a)** $K_3[TiCl_6]$, **(b)** $Na_3[Co(NO_2)_6]$, **(c)** $[Ru(en)_3]Br_3$, **(d)** $[Mo(EDTA)]ClO_4$, **(e)** $K_3[ReCl_6]$.

23.56 Give the number of (valence) d electrons associated with the central metal ion in each of the following complexes: **(a)** $K_3[Fe(CN)_6]$, **(b)** $[Mn(H_2O)_6](NO_3)_2$, **(c)** $Na[Ag(CN)_2]$, **(d)** $[Cr(NH_3)_4Br_2]ClO_4$, **(e)** $[Sr(EDTA)]^{2-}$.

23.57 A classmate says, "A weak-field ligand usually means the complex is high spin." Is your classmate correct? Explain.

23.58 A classmate says, "A strong-field ligand means that the ligand binds strongly to the metal ion." Is your classmate correct? Explain.

23.59 For each of the following metals, write the electronic configuration of the atom and its 2+ ion: **(a)** Mn, **(b)** Ru, **(c)** Rh. Draw the crystal-field energy-level diagram for the d orbitals of an octahedral complex, and show the placement of the d electrons for each 2+ ion, assuming a strong-field complex. How many unpaired electrons are there in each case?

23.60 For each of the following metals, write the electronic configuration of the atom and its 3+ ion: **(a)** Fe, **(b)** Mo, **(c)** Co. Draw the crystal-field energy-level diagram for the d orbitals of an octahedral complex, and show the placement of the d electrons for each 3+ ion, assuming a weak-field complex. How many unpaired electrons are there in each case?

23.61 Draw the crystal-field energy-level diagrams and show the placement of d electrons for each of the following: **(a)** $[Cr(H_2O)_6]^{2+}$ (four unpaired electrons), **(b)** $[Mn(H_2O)_6]^{2+}$ (high spin), **(c)** $[Ru(NH_3)_5(H_2O)]^{2+}$ (low spin), **(d)** $[IrCl_6]^{2-}$ (low spin), **(e)** $[Cr(en)_3]^{3+}$, **(f)** $[NiF_6]^{4-}$.

23.62 Draw the crystal-field energy-level diagrams and show the placement of electrons for the following complexes: **(a)** $[VCl_6]^{3-}$, **(b)** $[FeF_6]^{3-}$ (a high-spin complex), **(c)** $[Ru(bipy)_3]^{3+}$ (a low-spin complex), **(d)** $[NiCl_4]^{2-}$ (tetrahedral), **(e)** $[PtBr_6]^{2-}$, **(f)** $[Ti(en)_3]^{2+}$.

23.63 The complex $[Mn(NH_3)_6]^{2+}$ contains five unpaired electrons. Sketch the energy-level diagram for the d orbitals, and indicate the placement of electrons for this complex ion. Is the ion a high-spin or a low-spin complex?

23.64 The ion $[Fe(CN)_6]^{3-}$ has one unpaired electron, whereas $[Fe(NCS)_6]^{3-}$ has five unpaired electrons. From these results, what can you conclude about whether each complex is high spin or low spin? What can you say about the placement of NCS^- in the spectrochemical series?

Additional Exercises

23.65 The *Curie temperature* is the temperature at which a ferromagnetic solid switches from ferromagnetic to paramagnetic, and for nickel, the Curie temperature is 354 °C. Knowing this, you tie a string to two paper clips made of nickel and hold the paper clips near a permanent magnet. The magnet attracts the paper clips, as shown in the photograph on the left. Now you heat one of the paper clips with a cigarette lighter, and the clip drops (right photograph). Explain what happened.

23.66 Explain why the transition metals in periods 5 and 6 have nearly identical radii in each group.

23.67 Based on the molar conductance values listed here for the series of platinum(IV) complexes, write the formula for each complex so as to show which ligands are in the coordination sphere of the metal. By way of example, the molar conductances of 0.050 M NaCl and BaCl$_2$ are 107 ohm^{-1} and 197 ohm^{-1}, respectively.

Complex	Molar Conductance (ohm^{-1})* of 0.050 M Solution
Pt(NH$_3$)$_6$Cl$_4$	523
Pt(NH$_3$)$_4$Cl$_4$	228
Pt(NH$_3$)$_3$Cl$_4$	97
Pt(NH$_3$)$_2$Cl$_4$	0
KPt(NH$_3$)Cl$_5$	108

*The ohm is a unit of resistance; conductance is the inverse of resistance.

23.68 **(a)** A compound with formula RuCl$_3 \cdot$ 5 H$_2$O is dissolved in water, forming a solution that is approximately the same color as the solid. Immediately after forming the solution, the addition of excess AgNO$_3$(aq) forms 2 mol of solid AgCl per mole of complex. Write the formula for the compound, showing which ligands are likely to be present in the coordination sphere. **(b)** After a solution of RuCl$_3 \cdot$ 5 H$_2$O has stood for about a year, addition of AgNO$_3$(aq) precipitates 3 mol of AgCl per mole of complex. What has happened in the ensuing time?

23.69 Sketch the structure of the complex in each of the following compounds and give the full compound name:
(a) cis-[Co(NH$_3$)$_4$(H$_2$O)$_2$](NO$_3$)$_2$
(b) Na$_2$[Ru(H$_2$O)Cl$_5$]
(c) trans-NH$_4$[Co(C$_2$O$_4$)$_2$(H$_2$O)$_2$]
(d) cis-[Ru(en)$_2$Cl$_2$]

23.70 **(a)** Which complex ions in Exercise 23.69 have a mirror plane? **(b)** Will any of the complexes have optical isomers?

23.71 The molecule dimethylphosphinoethane [(CH$_3$)$_2$PCH$_2$CH$_2$P(CH$_3$)$_2$, which is abbreviated dmpe] is used as a ligand for some complexes that serve as catalysts. A complex that contains this ligand is Mo(CO)$_4$(dmpe). **(a)** Draw the Lewis structure for dmpe, and compare it with ethylenediamine as a coordinating ligand. **(b)** What is the oxidation state of Mo in Na$_2$[Mo(CN)$_2$(CO)$_2$(dmpe)]? **(c)** Sketch the structure of the [Mo(CN)$_2$(CO)$_2$(dmpe)]$^{2-}$ ion, including all the possible isomers.

23.72 Although the cis configuration is known for [Pt(en)Cl$_2$], no trans form is known. **(a)** Explain why the trans compound is not possible. **(b)** Would NH$_2$CH$_2$CH$_2$CH$_2$CH$_2$NH$_2$ be more likely than en (NH$_2$CH$_2$CH$_2$NH$_2$) to form the trans compound? Explain.

23.73 The acetylacetone ion forms very stable complexes with many metallic ions. It acts as a bidentate ligand, coordinating to the metal at two adjacent positions. Suppose that one of the CH$_3$ groups of the ligand is replaced by a CF$_3$ group, as shown here:

Trifluoromethyl acetylacetonate (tfac)

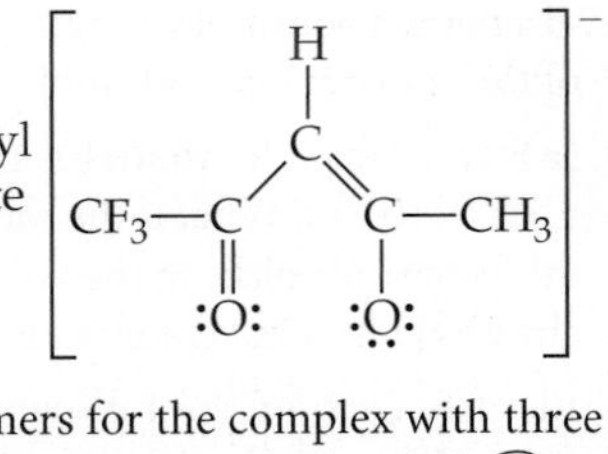

Sketch all possible isomers for the complex with three tfac ligands on cobalt(III). (You can use the symbol ◕ ◯ to represent the ligand.)

23.74 Which transition metal atom is present in each of the following biologically important molecules: **(a)** hemoglobin, **(b)** chlorophylls, **(c)** siderophores.

23.75 Carbon monoxide, CO, is an important ligand in coordination chemistry. When CO is reacted with nickel metal the product is [Ni(CO)$_4$], which is a toxic, pale yellow liquid. **(a)** What is the oxidation number for nickel in this compound? **(b)** Given that [Ni(CO)$_4$] is diamagnetic molecule with a tetrahedral geometry, what is the electron configuration of nickel in this compound? **(c)** Write the name for [Ni(CO)$_4$] using the nomenclature rules for coordination compounds.

23.76 Some metal complexes have a coordination number of 5. One such complex is Fe(CO)$_5$, which adopts a trigonal bipyramidal geometry (see Figure 9.8). **(a)** Write the name for Fe(CO)$_5$, using the nomenclature rules for coordination compounds. **(b)** What is the oxidation state of Fe in this compound? **(c)** Suppose one of the CO ligands is replaced with a CN$^-$ ligand, forming [Fe(CO)$_4$(CN)]$^-$. How many geometric isomers would you predict this complex could have?

23.77 Which of the following objects is chiral: **(a)** a left shoe, **(b)** a slice of bread, **(c)** a wood screw, **(d)** a molecular model of Zn(en)Cl$_2$, **(e)** a typical golf club?

23.78 The complexes [V(H$_2$O)$_6$]$^{3+}$ and [VF$_6$]$^{3-}$ are both known. **(a)** Draw the d-orbital energy-level diagram for V(III) octahedral complexes. **(b)** What gives rise to the colors of these complexes? **(c)** Which of the two complexes would you expect to absorb light of higher energy?

[23.79] One of the more famous species in coordination chemistry is the Creutz–Taube complex:

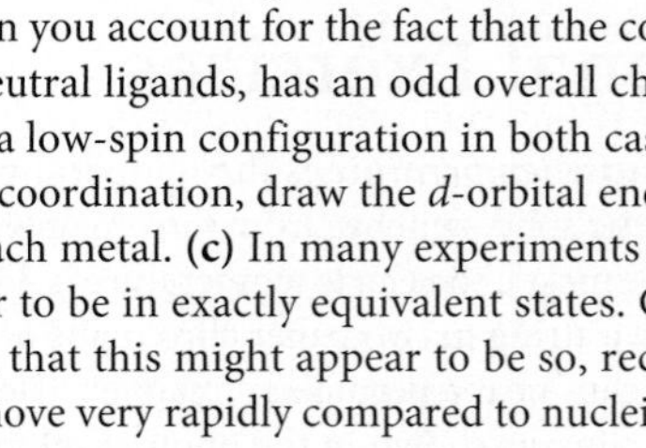

It is named for the two scientists who discovered it and initially studied its properties. The central ligand is pyrazine, a planar six-membered ring with nitrogens at opposite sides. **(a)** How can you account for the fact that the complex, which has only neutral ligands, has an odd overall charge? **(b)** The metal is in a low-spin configuration in both cases. Assuming octahedral coordination, draw the d-orbital energy-level diagram for each metal. **(c)** In many experiments the two metal ions appear to be in exactly equivalent states. Can you think of a reason that this might appear to be so, recognizing that electrons move very rapidly compared to nuclei?

23.80 Solutions of [Co(NH$_3$)$_6$]$^{2+}$, [Co(H$_2$O)$_6$]$^{2+}$ (both octahedral), and [CoCl$_4$]$^{2-}$ (tetrahedral) are colored. One is pink, one is blue, and one is yellow. Based on the spectrochemical series and remembering that the energy splitting in tetrahedral complexes is normally much less than that in octahedral ones, assign a color to each complex.

23.81 Oxyhemoglobin, with an O_2 bound to iron, is a low-spin Fe(II) complex; deoxyhemoglobin, without the O_2 molecule, is a high-spin complex. **(a)** Assuming that the coordination environment about the metal is octahedral, how many unpaired electrons are centered on the metal ion in each case? **(b)** What ligand is coordinated to the iron in place of O_2 in deoxyhemoglobin? **(c)** Explain in a general way why the two forms of hemoglobin have different colors (hemoglobin is red, whereas deoxyhemoglobin has a bluish cast). **(d)** A 15-minute exposure to air containing 400 ppm of CO causes about 10% of the hemoglobin in the blood to be converted into the carbon monoxide complex, called carboxyhemoglobin. What does this suggest about the relative equilibrium constants for binding of carbon monoxide and O_2 to hemoglobin? **(e)** CO is a strong-field ligand. What color might you expect carboxyhemoglobin to be?

[23.82] Consider the tetrahedral anions VO_4^{3-} (orthovanadate ion), CrO_4^{2-} (chromate ion), and MnO_4^- (permanganate ion). **(a)** These anions are *isoelectronic*. What does this statement mean? **(b)** Would you expect these anions to exhibit *d-d* transitions? Explain. **(c)** As mentioned in "A Closer Look" on charge-transfer color, the violet color of MnO_4^- is due to a *ligand-to-metal charge transfer* (LMCT) transition. What is meant by this term? **(d)** The LMCT transition in MnO_4^- occurs at a wavelength of 565 nm. The CrO_4^{2-} ion is yellow. Is the wavelength of the LMCT transition for chromate larger or smaller than that for MnO_4^-? Explain. **(e)** The VO_4^{3-} ion is colorless. Do you expect the light absorbed by the LMCT to fall in the UV or the IR region of the electromagnetic spectrum? Explain your reasoning.

23.83 **(a)** Given the colors observed for VO_4^{3-} (orthovanadate ion), CrO_4^{2-} (chromate ion), and MnO_4^- (permanganate ion) (see Exercise 23.82), what can you say about how the energy separation between the ligand orbitals and the empty *d* orbitals changes as a function of the oxidation state of the transition metal at the center of the tetrahedral anion?

[23.84] The red color of ruby is due to the presence of Cr(III) ions at octahedral sites in the close-packed oxide lattice of Al_2O_3. Draw the crystal-field splitting diagram for Cr(III) in this environment. Suppose that the ruby crystal is subjected to high pressure. What do you predict for the variation in the wavelength of absorption of the ruby as a function of pressure? Explain.

23.85 In 2001, chemists at SUNY-Stony Brook succeeded in synthesizing the complex *trans*-$[Fe(CN)_4(CO)_2]^{2-}$, which could be a model of complexes that may have played a role in the origin of life. **(a)** Sketch the structure of the complex. **(b)** The complex is isolated as a sodium salt. Write the complete name of this salt. **(c)** What is the oxidation state of Fe in this complex? How many *d* electrons are associated with the Fe in this complex? **(d)** Would you expect this complex to be high spin or low spin? Explain.

[23.86] When Alfred Werner was developing the field of coordination chemistry, it was argued by some that the optical activity he observed in the chiral complexes he had prepared was because of the presence of carbon atoms in the molecule. To disprove this argument, Werner synthesized a chiral complex of cobalt that had no carbon atoms in it, and he was able to resolve it into its enantiomers. Design a cobalt(III) complex that would be chiral if it could be synthesized and that contains no carbon atoms. (It may not be possible to synthesize the complex you design, but we will not worry about that for now.)

23.87 Generally speaking, for a given metal and ligand, the stability of a coordination compound is greater for the metal in the +3 rather than in the +2 oxidation state (for metals that form stable +3 ions in the first place). Suggest an explanation, keeping in mind the Lewis acid–base nature of the metal–ligand bond.

23.88 Many trace metal ions exist in the blood complexed with amino acids or small peptides. The anion of the amino acid glycine (gly),

$$H_2NCH_2\overset{\displaystyle O}{\overset{\|}{C}}{-}O^-$$

can act as a bidentate ligand, coordinating to the metal through nitrogen and oxygen atoms. How many isomers are possible for **(a)** $[Zn(gly)_2]$ (tetrahedral), **(b)** $[Pt(gly)_2]$ (square planar), **(c)** $[Co(gly)_3]$ (octahedral)? Sketch all possible isomers. Use the symbol to represent the ligand.

23.89 The coordination complex $[Cr(CO)_6]$ forms colorless, diamagnetic crystals that melt at 90 °C. **(a)** What is the oxidation number of chromium in this compound? **(b)** Given that $[Cr(CO)_6]$ is diamagnetic, what is the electron configuration of chromium in this compound? **(c)** Given that $[Cr(CO)_6]$ is colorless, would you expect CO to be a weak-field or strong-field ligand? **(d)** Write the name for $[Cr(CO)_6]$ using the nomenclature rules for coordination compounds.

Integrative Exercises

[23.90] Metallic elements are essential components of many important enzymes operating within our bodies. *Carbonic anhydrase*, which contains Zn^{2+} in its active site, is responsible for rapidly interconverting dissolved CO_2 and bicarbonate ion, HCO_3^-. The zinc in carbonic anhydrase is tetrahedrally coordinated by three neutral nitrogen-containing groups and a water molecule. The coordinated water molecule has a pK_a of 7.5, which is crucial for the enzyme's activity. **(a)** Draw the active site geometry for the Zn(II) center in carbonic anhydrase, just writing "N" for the three neutral nitrogen ligands from the protein. **(b)** Compare the pK_a of carbonic anhydrase's active site with that of pure water; which species is more acidic?

(c) When the coordinated water to the Zn(II) center in carbonic anhydrase is deprotonated, what ligands are bound to the Zn(II) center? Assume the three nitrogen ligands are unaffected. **(d)** The pK_a of $[Zn(H_2O)_6]^{2+}$ is 10. Suggest an explanation for the difference between this pK_a and that of carbonic anhydrase. **(e)** Would you expect carbonic anhydrase to have a deep color, like hemoglobin and other metal-ion containing proteins do? Explain.

23.91 Two different compounds have the formulation $CoBr(SO_4) \cdot 5\, NH_3$. Compound A is dark violet, and compound B is red-violet. When compound A is treated with $AgNO_3(aq)$, no reaction occurs, whereas compound B reacts with $AgNO_3(aq)$ to form a white precipitate. When compound A is treated with $BaCl_2(aq)$, a white precipitate is formed, whereas compound B has no reaction with $BaCl_2(aq)$. (a) Is Co in the same oxidation state in these complexes? (b) Explain the reactivity of compounds A and B with $AgNO_3(aq)$ and $BaCl_2(aq)$. (c) Are compounds A and B isomers of one another? If so, which category from Figure 23.19 best describes the isomerism observed for these complexes? (d) Would compounds A and B be expected to be strong electrolytes, weak electrolytes, or nonelectrolytes?

23.92 A manganese complex formed from a solution containing potassium bromide and oxalate ion is purified and analyzed. It contains 10.0% Mn, 28.6% potassium, 8.8% carbon, and 29.2% bromine by mass. The remainder of the compound is oxygen. An aqueous solution of the complex has about the same electrical conductivity as an equimolar solution of $K_4[Fe(CN)_6]$. Write the formula of the compound, using brackets to denote the manganese and its coordination sphere.

23.93 The $E°$ values for two low-spin iron complexes in acidic solution are as follows:

$$[Fe(o\text{-phen})_3]^{3+}(aq) + e^- \rightleftharpoons$$
$$Fe(o\text{-phen})_3]^{2+}(aq) \quad E° = 1.12 \text{ V}$$

$$[Fe(CN)_6]^{3-}(aq) + e^- \rightleftharpoons$$
$$Fe(CN)_6]^{4-}(aq) \quad E° = 0.36 \text{ V}$$

(a) Is it thermodynamically favorable to reduce both Fe(III) complexes to their Fe(II) analogs? Explain. (b) Which complex, $[Fe(o\text{-phen})_3]^{3+}$ or $[Fe(CN)_6]^{3-}$, is more difficult to reduce? (c) Suggest an explanation for your answer to (b).

23.94 A palladium complex formed from a solution containing bromide ion and pyridine, C_5H_5N (a good electron-pair donor), is found on elemental analysis to contain 37.6% bromine, 28.3% carbon, 6.60% nitrogen, and 2.37% hydrogen by mass. The compound is slightly soluble in several organic solvents; its solutions in water or alcohol do not conduct electricity. It is found experimentally to have a zero dipole moment. Write the chemical formula, and indicate its probable structure.

23.95 (a) In early studies it was observed that when the complex $[Co(NH_3)_4Br_2]Br$ was placed in water, the electrical conductivity of a 0.05 M solution changed from an initial value of 191 ohm^{-1} to a final value of 374 ohm^{-1} over a period of an hour or so. Suggest an explanation for the observed results. (See Exercise 23.67 for relevant comparison data.) (b) Write a balanced chemical equation to describe the reaction. (c) A 500-mL solution is made up by dissolving 3.87 g of the complex. As soon as the solution is formed, and before any change in conductivity has occurred, a 25.00-mL portion of the solution is titrated with 0.0100 M $AgNO_3$ solution. What volume of $AgNO_3$ solution do you expect to be required to precipitate the free $Br^-(aq)$? (d) Based on the response you gave to part (b), what volume of $AgNO_3$ solution would be required to titrate a fresh 25.00-mL sample of $[Co(NH_3)_4Br_2]Br$ after all conductivity changes have occurred?

23.96 The total concentration of Ca^{2+} and Mg^{2+} in a sample of hard water was determined by titrating a 0.100-L sample of the water with a solution of $EDTA^{4-}$. The $EDTA^{4-}$ chelates the two cations:

$$Mg^{2+} + [EDTA]^{4-} \longrightarrow [Mg(EDTA)]^{2-}$$
$$Ca^{2+} + [EDTA]^{4-} \longrightarrow [Ca(EDTA)]^{2-}$$

It requires 31.5 mL of 0.0104 M $[EDTA]^{4-}$ solution to reach the end point in the titration. A second 0.100-L sample was then treated with sulfate ion to precipitate Ca^{2+} as calcium sulfate. The Mg^{2+} was then titrated with 18.7 mL of 0.0104 M $[EDTA]^{4-}$. Calculate the concentrations of Mg^{2+} and Ca^{2+} in the hard water in mg/L.

23.97 Carbon monoxide is toxic because it binds more strongly to the iron in hemoglobin (Hb) than does O_2, as indicated by these approximate standard free-energy changes in blood:

$$Hb + O_2 \longrightarrow HbO_2 \quad \Delta G° = -70 \text{ kJ}$$
$$Hb + CO \longrightarrow HbCO \quad \Delta G° = -80 \text{ kJ}$$

Using these data, estimate the equilibrium constant at 298 K for the equilibrium

$$HbO_2 + CO \rightleftharpoons HbCO + O_2$$

[23.98] The molecule *methylamine* (CH_3NH_2) can act as a monodentate ligand. The following are equilibrium reactions and the thermochemical data at 298 K for reactions of methylamine and en with $Cd^{2+}(aq)$:

$$Cd^{2+}(aq) + 4 CH_3NH_2(aq) \rightleftharpoons [Cd(CH_3NH_2)_4]^{2+}(aq)$$
$$\Delta H° = -57.3 \text{ kJ}; \quad \Delta S° = -67.3 \text{ J/K}; \quad \Delta G° = -37.2 \text{ kJ}$$
$$Cd^{2+}(aq) + 2 en(aq) \rightleftharpoons [Cd(en)_2]^{2+}(aq)$$
$$\Delta H° = -56.5 \text{ kJ}; \quad \Delta S° = +14.1 \text{ J/K}; \quad \Delta G° = -60.7 \text{ kJ}$$

(a) Calculate $\Delta G°$ and the equilibrium constant K for the following *ligand exchange* reaction:

$$[Cd(CH_3NH_2)_4]^{2+}(aq) + 2 en(aq) \rightleftharpoons$$
$$[Cd(en)_2]^{2+}(aq) + 4 CH_3NH_2(aq)$$

Based on the value of K in part (a), what would you conclude about this reaction? What concept is demonstrated? (b) Determine the magnitudes of the enthalpic ($\Delta H°$) and the entropic ($-T\Delta S°$) contributions to $\Delta G°$ for the ligand exchange reaction. Explain the relative magnitudes. (c) Based on information in this exercise and in the "A Closer Look" box on the chelate effect, predict the sign of $\Delta H°$ for the following hypothetical reaction:

$$[Cd(CH_3NH_2)_4]^{2+}(aq) + 4 NH_3(aq) \rightleftharpoons$$
$$[Cd(NH_3)_4]^{2+}(aq) + 4 CH_3NH_2(aq)$$

23.99 The value of Δ for the $[CrF_6]^{3-}$ complex is 182 kJ/mol. Calculate the expected wavelength of the absorption corresponding to promotion of an electron from the lower-energy to the higher-energy d-orbital set in this complex. Should the complex absorb in the visible range?

[23.100] A Cu electrode is immersed in a solution that is 1.00 M in $[Cu(NH_3)_4]^{2+}$ and 1.00 M in NH_3. When the cathode is a standard hydrogen electrode, the emf of the cell is found to be +0.08 V. What is the formation constant for $[Cu(NH_3)_4]^{2+}$?

[23.101] The complex $[Ru(EDTA)(H_2O)]^-$ undergoes substitution reactions with several ligands, replacing the water molecule with the ligand. In all cases, the ruthenium stays in the +3 oxidation state and the ligands use a nitrogen donor atom to bind to the metal.

$$[Ru(EDTA)(H_2O)]^- + L \longrightarrow [Ru(EDTA)L]^- + H_2O$$

The rate constants for several ligands are as follows:

Ligand, L	k ($M^{-1}s^{-1}$)
Pyridine	6.3×10^3
SCN^-	2.7×10^2
CH_3CN	3.0×10

(a) One possible mechanism for this substitution reaction is that the water molecule dissociates from the Ru(III) in the rate-determining step, and then the ligand L binds to Ru(III) in a rapid second step. A second possible mechanism is that L approaches the complex, begins to form a new bond to the Ru(III), and displaces the water molecule, all in a single concerted step. Which of these two mechanisms is more consistent with the data? Explain. **(b)** What do the results suggest about the relative donor ability of the nitrogens of the three ligands toward Ru(III)? **(c)** Assuming that the complexes are all low spin, how many unpaired electrons are in each?

Design an Experiment

Following a procedure found in a scientific paper you go into the lab and attempt to prepare crystals of dichlorobis(ethylenediamine)cobalt(III) chloride. The paper states that this compound can be made by reacting $CoCl_2 \cdot 6 H_2O$, an excess of ethylenediamine, O_2 from the air (which acts as an oxidizing agent), water, and concentrated hydrochloric acid. At the end of the reaction you filter off the solution and are left with a green, crystalline product. **(a)** What experiment(s) could you perform to confirm that you have prepared $[CoCl_2(en)_2]Cl$ and not $[Co(en)_3]Cl_3$? **(b)** How could you verify that cobalt was present as Co^{3+} and determine the spin state of the cobalt complex in your product? **(c)** How many geometric isomers exist for $[CoCl_2(en)_2]Cl$? How could you determine if your product contains a single geometric isomer or a mixture of geometric isomers? **(d)** If the product does contain a single geometric isomer how would you determine which one was present? (*Hint: You may find the information in Table 23.3 helpful.*)

24

The Chemistry of Life: Organic and Biological Chemistry

We are all familiar with how chemical substances can influence our health and behavior. Aspirin, also known as acetylsalicylic acid, relieves aches and pains. Cocaine, whose full chemical name is methyl (1*R*,2*R*,3*S*,5*S*)-3-(benzoyloxy)-8-methyl-8-azabicyclo[3.2.1]octane-2-carboxylate, is a plant-derived substance that is used in clinical situations as an anesthetic, but also is used illegally to experience extreme euphoria (a "high").

Understanding how these molecules exert their effects, and developing new molecules that can target disease and pain, is an enormous part of the modern chemical enterprise. This chapter is about the molecules, composed mainly of carbon, hydrogen, oxygen, and nitrogen, that bridge chemistry and biology.

More than 16 million carbon-containing compounds are known. Chemists make thousands of new compounds every year, about 90% of which contain carbon. The study of compounds whose molecules contain carbon constitutes the branch of chemistry known as **organic chemistry**. This term arose from the eighteenth-century belief that organic compounds could be formed only by living (that is, organic) systems. This idea was disproved in 1828 by the German chemist Friedrich Wöhler when he synthesized urea (H_2NCONH_2), an organic substance found in the urine of mammals, by heating ammonium cyanate (NH_4OCN), an inorganic (nonliving) substance.

▶ **BRAIN ON COCAINE.** These positron-emitting tomography (PET) scans of the human brain show how rapidly glucose is metabolized in different parts of the brain. The top row of images show the brain of a normal person; the bottom two rows of images show the brain of a person who has taken cocaine, after 10 days and after 100 days (red = high glucose metabolism, yellow = medium, blue = low). Notice that glucose metabolism is suppressed in the brain of the person who has taken cocaine.

WHAT'S AHEAD

24.1 GENERAL CHARACTERISTICS OF ORGANIC MOLECULES We begin with a review of the structures and properties of organic compounds.

24.2 INTRODUCTION TO HYDROCARBONS We consider *hydrocarbons*, compounds containing only C and H, including the hydrocarbons called *alkanes*, which contain only single C—C bonds. We also look at *isomers*, compounds with identical compositions but different molecular structures.

24.3 ALKENES, ALKYNES, AND AROMATIC HYDROCARBONS We next explore hydrocarbons with one or more C=C bonds, called *alkenes*, and those with one or more C≡C bonds, called *alkynes*. *Aromatic* hydrocarbons have at least one planar ring with delocalized π electrons.

24.4 ORGANIC FUNCTIONAL GROUPS We recognize that a central organizing principle of organic chemistry is the *functional group*, a group of atoms at which most of the compound's chemical reactions occur.

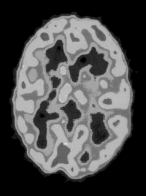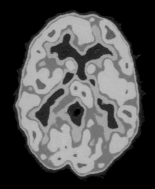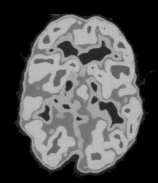

NORMAL

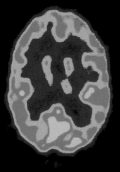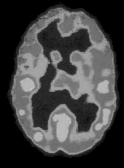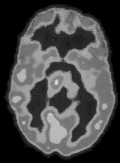

COCAINE ABUSER (10 DA)

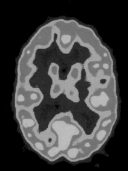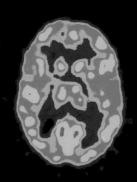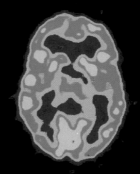

COCAINE ABUSER (100 DA)

24.5 CHIRALITY IN ORGANIC CHEMISTRY We learn that compounds with nonsuperimposable mirror images are *chiral* and that chirality plays important roles in organic and biological chemistry.

24.6 INTRODUCTION TO BIOCHEMISTRY We introduce the chemistry of living organisms, known as *biochemistry*, *biological chemistry*, or *chemical biology*. Important classes of compounds that occur in living systems are *proteins*, *carbohydrates*, *lipids*, and *nucleic acids*.

24.7 PROTEINS We learn that proteins are polymers of *amino acids* linked with *amide* (also called *peptide*) bonds. Proteins are used by organisms for structural support, as molecular transporters and as catalysts for biochemical reactions.

24.8 CARBOHYDRATES We observe that carbohydrates are sugars and polymers of sugars used primarily as fuel by organisms (glucose) or as structural support in plants (cellulose).

24.9 LIPIDS We find that lipids are a large class of molecules used primarily for energy storage in organisms.

24.10 NUCLEIC ACIDS We learn that nucleic acids are polymers of *nucleotides* that contain an organism's genetic information. *Deoxyribonucleic acid* (DNA) and *ribonucleic acid* (RNA) are polymers composed of nucleotides.

The study of the chemistry of living species is called *biological chemistry*, *chemical biology*, or **biochemistry**. In this chapter, we present some of the elementary aspects of both organic chemistry and biochemistry.

24.1 | General Characteristics of Organic Molecules

What is it about carbon that leads to the tremendous diversity in its compounds and allows it to play such crucial roles in biology and society? Let's consider some general features of organic molecules and, as we do, review principles we learned in earlier chapters.

The Structures of Organic Molecules

Because carbon has four valence electrons ($[He]2s^2 2p^2$), it forms four bonds in virtually all its compounds. When all four bonds are single bonds, the electron pairs are disposed in a tetrahedral arrangement. ∞ (Section 9.2) In the hybridization model, the carbon $2s$ and $2p$ orbitals are then sp^3 hybridized. ∞ (Section 9.5) When there is one double bond, the arrangement is trigonal planar (sp^2 hybridization). With a triple bond, it is linear (sp hybridization). Examples are shown in ▼ **Figure 24.1**.

Almost every organic molecule contains C—H bonds. Because the valence shell of H can hold only two electrons, hydrogen forms only one covalent bond. As a result, hydrogen atoms are always located on the *surface* of organic molecules whereas the C—C bonds form the *backbone*, or *skeleton*, of the molecule, as in the propane molecule:

$$
\begin{array}{ccccccc}
 & H & & H & & H & \\
 & | & & | & & | & \\
H & - & C & - & C & - & C & - & H \\
 & | & & | & & | & \\
 & H & & H & & H &
\end{array}
$$

 GO FIGURE

What is the geometry around the bottom carbon atom in acetonitrile?

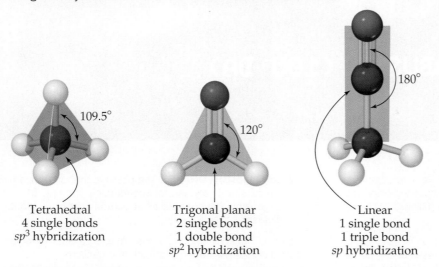

Tetrahedral
4 single bonds
sp^3 hybridization

Trigonal planar
2 single bonds
1 double bond
sp^2 hybridization

Linear
1 single bond
1 triple bond
sp hybridization

▲ Figure 24.1 **Carbon geometries.** The three common geometries around carbon are tetrahedral as in methane (CH_4), trigonal planar as in formaldehyde (CH_2O), and linear as in acetonitrile (CH_3CN). Notice that in all cases each carbon atom forms four bonds.

The Stabilities of Organic Substances

Carbon forms strong bonds with a variety of elements, especially H, O, N, and the halogens. ∞ (Section 8.8) Carbon also has an exceptional ability to bond to itself, forming a variety of molecules made up of chains or rings of carbon atoms. Most reactions with low or moderate activation energy ∞ (Section 14.5) begin when a region of high electron density on one molecule encounters a region of low electron density on another molecule. The regions of high electron density may be due to the presence of a multiple bond or to the more electronegative atom in a polar bond. Because of their strength (the C—C single bond enthalpy is 348 kJ/mol, the C—H bond enthalpy is 413 kJ/mol ∞ Table 8.4) and lack of polarity, both C—C single bonds and C—H bonds are relatively unreactive. To better understand the implications of these facts, consider ethanol:

$$
\begin{array}{ccc}
& \text{H} & \text{H} \\
& | & | \\
\text{H}-&\text{C}-\text{C}&-\text{O}-\text{H} \\
& | & | \\
& \text{H} & \text{H}
\end{array}
$$

The differences in the electronegativity values of C (2.5) and O (3.5) and of O and H (2.1) indicate that the C—O and O—H bonds are quite polar. Thus, many reactions of ethanol involve these bonds while the hydrocarbon portion of the molecule remains intact. A group of atoms such as the C—O—H group, which determines how an organic molecule reacts (in other words, how the molecule *functions*), is called a **functional group**. The functional group is the center of reactivity in an organic molecule.

Give It Some Thought

Which bond is most likely to be the location of a chemical reaction: C=N, C—C, or C—H?

Solubility and Acid–Base Properties of Organic Substances

In most organic substances, the most prevalent bonds are carbon–carbon and carbon–hydrogen, which are not polar. For this reason, the overall polarity of organic molecules is often low, which makes them generally soluble in nonpolar solvents and not very soluble in water. ∞ (Section 13.3) Organic molecules that are soluble in polar solvents are those that have polar groups on the molecule surface, such as glucose and ascorbic acid (▶ Figure 24.2). Organic molecules that have a long, nonpolar part bonded to a polar, ionic part, such as the stearate ion shown in Figure 24.2, function as *surfactants* and are used in soaps and detergents. ∞ (Section 13.6) The nonpolar part of the molecule extends into a nonpolar medium such as grease or oil, and the polar part extends into a polar medium such as water.

Many organic substances contain acidic or basic groups. The most important acidic organic substances are the carboxylic acids, which bear the functional group —COOH. ∞ (Sections 4.3 and 16.10) The most important basic organic substances are amines, which bear the —NH₂, —NHR, or —NR₂ groups, where R is an organic group made up of carbon and hydrogen atoms. ∞ (Section 16.7)

As you read this chapter, you will find many concept links (∞) to related materials in earlier chapters. *We strongly encourage you to follow these links and review the earlier material.* Doing so will enhance your understanding and appreciation of organic chemistry and biochemistry.

 GO FIGURE

How would replacing OH groups on ascorbic acid with CH₃ groups affect the substance's solubility in (**a**) polar solvents and (**b**) nonpolar solvents?

Glucose (C₆H₁₂O₆)

Ascorbic acid (HC₆H₇O₆)

Stearate (C₁₇H₃₅COO⁻)

▲ Figure 24.2 **Organic molecules that are soluble in polar solvents.**

24.2 | Introduction to Hydrocarbons

Because carbon compounds are so numerous, it is convenient to organize them into families that have structural similarities. The simplest class of organic compounds is the *hydrocarbons*, compounds composed of only carbon and hydrogen. The key structural feature of hydrocarbons (and of most other organic substances) is the presence of stable carbon–carbon bonds. Carbon is the only element capable of forming stable, extended chains of atoms bonded through single, double, or triple bonds.

Hydrocarbons can be divided into four types, depending on the kinds of carbon–carbon bonds in their molecules. ▼ Table 24.1 shows an example of each type.

Alkanes contain only single C—C bonds. **Alkenes**, also known as *olefins*, contain at least one C=C double bond, and **alkynes** contain at least one C≡C triple bond. In **aromatic hydrocarbons** the carbon atoms are connected in a planar ring structure, joined by both σ and delocalized π bonds between carbon atoms. ∞ (Section 8.6) Benzene (C_6H_6) is the best-known example of an aromatic hydrocarbon.

Each type of hydrocarbon exhibits different chemical behaviors, as we will see shortly. The physical properties of all four types, however, are similar in many ways. Because hydrocarbon molecules are relatively nonpolar, they are almost completely insoluble in water but dissolve readily in nonpolar solvents. Their melting points and boiling points are determined by dispersion forces. ∞ (Section 11.2) As a result, hydrocarbons of very low molecular weight, such as $C_2H_6(bp = -89\,°C)$, are gases at room temperature; those of moderate molecular weight, such as $C_6H_{14}(bp = 69\,°C)$, are liquids; and those of high molecular weight, such as $C_{22}H_{46}(mp = 44\,°C)$, are solids.

▶ Table 24.2 lists the ten simplest alkanes. Many of these substances are familiar because they are used so widely. Methane is a major component of natural gas. Propane is the major component of bottled gas used for home heating and cooking in areas where natural gas is not available. Butane is used in disposable lighters and in fuel canisters for gas camping stoves and lanterns. Alkanes with 5 to 12 carbon atoms per molecule are used to make gasoline. Notice that each succeeding compound in Table 24.2 has an additional CH_2 unit.

Table 24.1   The Four Hydrocarbon Types with Molecular Examples

Type			Example
Alkane	Ethane	CH_3CH_3	
Alkene	Ethylene	$CH_2{=}CH_2$	
Alkyne	Acetylene	$CH{\equiv}CH$	
Aromatic	Benzene	C_6H_6	

Table 24.2 First Ten Members of the Straight-Chain Alkane Series

Molecular Formula	Condensed Structural Formula	Name	Boiling Point (°C)
CH_4	CH_4	Methane	−161
C_2H_6	CH_3CH_3	Ethane	−89
C_3H_8	$CH_3CH_2CH_3$	Propane	−44
C_4H_{10}	$CH_3CH_2CH_2CH_3$	Butane	−0.5
C_5H_{12}	$CH_3CH_2CH_2CH_2CH_3$	Pentane	36
C_6H_{14}	$CH_3CH_2CH_2CH_2CH_2CH_3$	Hexane	68
C_7H_{16}	$CH_3CH_2CH_2CH_2CH_2CH_2CH_3$	Heptane	98
C_8H_{18}	$CH_3CH_2CH_2CH_2CH_2CH_2CH_2CH_3$	Octane	125
C_9H_{20}	$CH_3CH_2CH_2CH_2CH_2CH_2CH_2CH_2CH_3$	Nonane	151
$C_{10}H_{22}$	$CH_3CH_2CH_2CH_2CH_2CH_2CH_2CH_2CH_2CH_3$	Decane	174

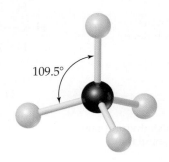

The formulas for the alkanes given in Table 24.2 are written in a notation called *condensed structural formulas*. This notation reveals the way in which atoms are bonded to one another but does not require drawing in all the bonds. For example, the structural formula and the condensed structural formulas for butane (C_4H_{10}) are

$$\begin{array}{c}\;\;H\;\;\;\;H\;\;\;\;H\;\;\;\;H\\ \;\;|\;\;\;\;\;|\;\;\;\;\;|\;\;\;\;\;|\\ H-C-C-C-C-H\\ \;\;|\;\;\;\;\;|\;\;\;\;\;|\;\;\;\;\;|\\ \;\;H\;\;\;\;H\;\;\;\;H\;\;\;\;H\end{array}$$

$$H_3C-CH_2-CH_2-CH_3$$

or

$$CH_3CH_2CH_2CH_3$$

 Give It Some Thought

How many C—H and C—C bonds are formed by the middle carbon atom of propane?

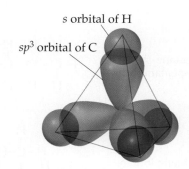

▲ **Figure 24.3 Bonds about carbon in methane.** This tetrahedral molecular geometry is found around all carbons in alkanes.

Structures of Alkanes

According to the VSEPR model, the molecular geometry about each carbon atom in an alkane is tetrahedral. ∞∞ (Section 9.2) The bonding may be described as involving sp^3-hybridized orbitals on the carbon, as pictured in ▶ Figure 24.3 for methane. ∞∞(Section 9.5)

Rotation about a carbon–carbon single bond is relatively easy and occurs rapidly at room temperature. To visualize such rotation, imagine grasping either methyl group of the propane molecule in ▶ Figure 24.4 and rotating the group relative to the rest of the molecule. Because motion of this sort occurs rapidly in alkanes, a long-chain alkane molecule is constantly undergoing motions that cause it to change its shape, something like a length of chain that is being shaken.

Structural Isomers

The alkanes in Table 24.2 are called *straight-chain* or *linear hydrocarbons* because all the carbon atoms are joined in a continuous chain. Alkanes consisting of four or more carbon atoms can also form *branched chains*, and when they do, they are called *branched-chain hydrocarbons*. (The branches in organic molecules are often called *side chains*.) **Table 24.3,** for example, shows all the straight-chain and branched-chain alkanes containing four and five carbon atoms.

Compounds that have the same molecular formula but different bonding arrangements (and hence different structures) are called **structural isomers**. Thus, C_4H_{10} has two structural isomers and C_5H_{12} has three. The structural isomers of a given alkane differ slightly from one another in physical properties, as the melting and boiling points in Table 24.3 indicate.

▲ **Figure 24.4 Rotation about a C—C bond occurs easily and rapidly in all alkanes.**

Table 24.3 **Isomers of C_4H_{10} and C_5H_{12}**

Systematic Name (Common Name)	Structural Formula	Condensed Structural Formula	Space-filling Model	Melting Point (°C)	Boiling Point (°C)
Butane (*n*-butane)	H–C–C–C–C–H (with H's)	$CH_3CH_2CH_2CH_3$		−138	−0.5
2-Methylpropane (isobutane)	CH_3–CH–CH_3 structure	CH_3—CH—CH_3 with CH_3		−159	−12
Pentane (*n*-pentane)	H–C–C–C–C–C–H (with H's)	$CH_3CH_2CH_2CH_2CH_3$		−130	+36
2-Methylbutane (isopentane)	branched structure	CH_3—CH—CH_2—CH_3 with CH_3		−160	+28
2,2-Dimethylpropane (neopentane)	branched structure	CH_3—C—CH_3 with two CH_3		−16	+9

The number of possible structural isomers increases rapidly with the number of carbon atoms in the alkane. There are 18 isomers with the molecular formula C_8H_{18}, for example, and 75 with the molecular formula $C_{10}H_{22}$.

 Give It Some Thought

What evidence can you cite to support the fact that although isomers have the same molecular formula they are in fact different compounds?

Nomenclature of Alkanes

In the first column of Table 24.3, the names in parentheses are called the *common names*. The common name of the isomer with no branches begins with the letter *n* (indicating the "normal" structure). When one CH_3 group branches off the major chain, the common name of the isomer begins with *iso-*, and when two CH_3 groups branch off, the common name begins with *neo-*. As the number of isomers grows, however, it becomes impossible to find a suitable prefix to denote each isomer by a common name. The need for a systematic means of naming organic compounds was recognized as early as 1892, when an organization called the International Union of Chemistry met in Geneva to formulate rules for naming organic substances. Since that time the task of updating the rules for naming compounds has fallen to the International Union of Pure and Applied Chemistry (IUPAC). Chemists everywhere, regardless of their nationality, subscribe to a common system for naming compounds.

The IUPAC names for the isomers of butane and pentane are the ones given first in Table 24.3. These systematic names, as well as those of other organic compounds, have three parts to them:

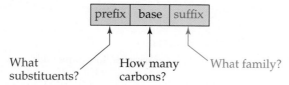

The following steps summarize the procedures used to name alkanes, which all have names ending with -*ane*. We use a similar approach to write the names of other organic compounds.

1. **Find the longest continuous chain of carbon atoms, and use the name of this chain (given in Table 24.2) as the base name.** Be careful in this step because the longest chain may not be written in a straight line, as in the following structure:

$$CH_3-\overset{2}{CH}-\overset{1}{CH_3}$$
$$\underset{3}{CH_2}-\underset{4}{CH_2}-\underset{5}{CH_2}-\underset{6}{CH_3}$$

2-Methyl*hexane*

Because the longest continuous chain contains six C atoms, this isomer is named as a substituted hexane. Groups attached to the main chain are called *substituents* because they are substituted in place of an H atom on the main chain. In this molecule, the CH_3 group not enclosed by the blue outline is the only substituent in the molecule.

2. **Number the carbon atoms in the longest chain, beginning with the end nearest a substituent.** In our example, we number the C atoms beginning at the upper right because that places the CH_3 substituent on C2 of the chain. (If we had numbered from the lower right, the CH_3 would be on C5.) The chain is numbered from the end that gives the lower number to the substituent position.

3. **Name each substituent.** A substituent formed by removing an H atom from an alkane is called an **alkyl group**. Alkyl groups are named by replacing the -*ane* ending of the alkane name with -*yl*. The methyl group (CH_3), for example, is derived from methane (CH_4) and the ethyl group (C_2H_5) is derived from ethane (C_2H_6). ▶ Table 24.4 lists six common alkyl groups.

4. **Begin the name with the number or numbers of the carbon or carbons to which each substituent is bonded.** For our compound, the name 2-methylhexane indicates the presence of a methyl group on C2 of a hexane (six-carbon) chain.

5. **When two or more substituents are present, list them in alphabetical order.** The presence of two or more of the same substituent is indicated by the prefixes *di*- (two), *tri*- (three), *tetra*- (four), *penta*- (five), and so forth. The prefixes are ignored in determining the alphabetical order of the substituents:

$$\overset{7}{CH_3}$$
$$CH_3-\overset{5}{CH}-\overset{6}{CH_2}$$
$$\overset{4}{CH}-\overset{3}{CH}-CH_2CH_3$$
$$CH_3|\overset{2}{CH}-CH_3$$
$$\overset{1}{CH_3}$$

3-Ethyl-2,4,5-trimethylheptane

Table 24.4 Condensed Structural Formulas and Common Names for Several Alkyl Groups

Group	Name
CH_3-	Methyl
CH_3CH_2-	Ethyl
$CH_3CH_2CH_2-$	Propyl
$CH_3CH_2CH_2CH_2-$	Butyl
CH_3 $\mid$ $HC-$ $\mid$ CH_3	Isopropyl
CH_3 $\mid$ CH_3-C- $\mid$ CH_3	*tert*-Butyl

SAMPLE EXERCISE 24.1 | Naming Alkanes

Give the systematic name for the following alkane:

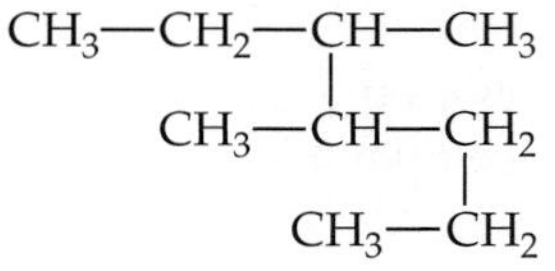

$$CH_3—CH_2—CH—CH_3$$
$$CH_3—CH—CH_2$$
$$CH_3—CH_2$$

SOLUTION

Analyze We are given the condensed structural formula of an alkane and asked to give its name.

Plan Because the hydrocarbon is an alkane, its name ends in *-ane*. The name of the parent hydrocarbon is based on the longest continuous chain of carbon atoms. Branches are alkyl groups, named after the number of C atoms in the branch and located by counting C atoms along the longest continuous chain.

Solve The longest continuous chain of C atoms extends from the upper left CH_3 group to the lower left CH_3 group and is seven C atoms long:

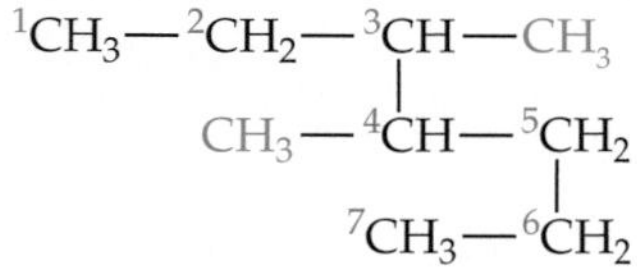

$$^1CH_3—^2CH_2—^3CH—CH_3$$
$$CH_3—^4CH—^5CH_2$$
$$^7CH_3—^6CH_2$$

The parent compound is thus heptane. There are two methyl groups branching off the main chain. Hence, this compound is a dimethyl-heptane. To specify the location of the two methyl groups, we must number the C atoms from the end that gives the lower two numbers to the carbons bearing side chains. This means that we should start numbering at the upper left carbon. There is a methyl group on C3 and one on C4. The compound is thus 3,4-dimethylheptane.

Practice Exercise 1

What is the proper name of this compound?

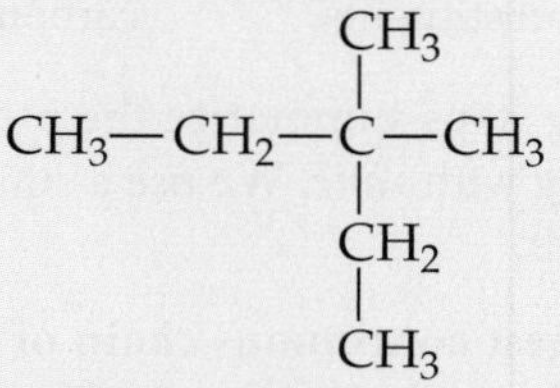

$$CH_3$$
$$CH_3—CH_2—C—CH_3$$
$$CH_2$$
$$CH_3$$

(a) 3-ethyl-3-methylbutane, (b) 2-ethyl-2-methylbutane, (c) 3,3-dimethylpentane, (d) isoheptane, (e) 1,2-dimethyl-neopentane.

Practice Exercise 2

Name the following alkane:

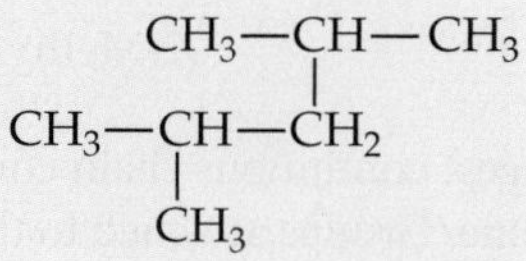

$$CH_3—CH—CH_3$$
$$CH_3—CH—CH_2$$
$$CH_3$$

SAMPLE EXERCISE 24.2 | Writing Condensed Structural Formulas

Write the condensed structural formula for 3-ethyl-2-methylpentane.

SOLUTION

Analyze We are given the systematic name for a hydrocarbon and asked to write its condensed structural formula.

Plan Because the name ends in *-ane*, the compound is an alkane, meaning that all the carbon–carbon bonds are single bonds. The parent hydrocarbon is pentane, indicating five C atoms (Table 24.2). There are two alkyl groups specified, an ethyl group (two carbon atoms, C_2H_5) and a methyl group (one carbon atom, CH_3). Counting from left to right along the five-carbon chain, the name tells us that the ethyl group is attached to C3 and the methyl group is attached to C2.

Solve We begin by writing five C atoms attached by single bonds. These represent the backbone of the parent pentane chain:

$$C—C—C—C—C$$

We next place a methyl group on the second C and an ethyl group on the third C of the chain. We then add hydrogens to all the other C atoms to make four bonds to each carbon:

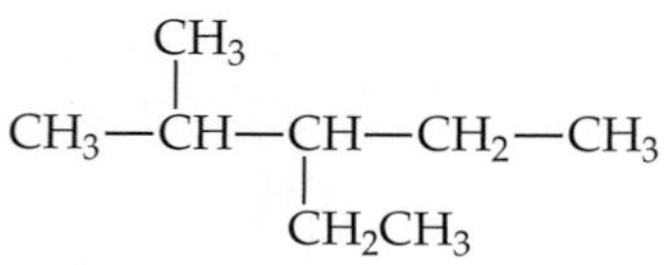

$$CH_3$$
$$CH_3—CH—CH—CH_2—CH_3$$
$$CH_2CH_3$$

The formula can be written more concisely as

$$CH_3CH(CH_3)CH(C_2H_5)CH_2CH_3$$

where the branching alkyl groups are indicated in parentheses.

Practice Exercise 1

How many hydrogen atoms are in 2,2-dimethylhexane?
(a) 6, (b) 8, (c) 16, (d) 18, (e) 20.

Practice Exercise 2

Write the condensed structural formula for 2,3-dimethylhexane.

Cycloalkanes

Alkanes that form rings, or cycles, are called **cycloalkanes**. As ▼ Figure 24.5 illustrates, cycloalkane structures are sometimes drawn as *line structures*, which are polygons in which each corner represents a CH_2 group. This method of representation is similar to that used for benzene rings. ∞ (Section 8.6) (Remember from our benzene discussion that in aromatic structures each vertex represents a CH group, not a CH_2 group.)

Carbon rings containing fewer than five carbon atoms are strained because the C—C—C bond angles must be less than the 109.5° tetrahedral angle. The amount of strain increases as the rings get smaller. In cyclopropane, which has the shape of an equilateral triangle, the angle is only 60°; this molecule is therefore much more reactive than propane, its straight-chain analog.

 Give It Some Thought

Are the C—C bonds cyclopropane weaker than those in cyclohexane?

Reactions of Alkanes

Because they contain only C—C and C—H bonds, most alkanes are relatively unreactive. At room temperature, for example, they do not react with acids, bases, or strong oxidizing agents. Their low chemical reactivity, as noted in Section 24.1, is due primarily to the strength and lack of polarity of C—C and C—H bonds.

Alkanes are not completely inert, however. One of their most commercially important reactions is *combustion* in air, which is the basis of their use as fuels. ∞ (Section 3.2) For example, the complete combustion of ethane proceeds according to this highly exothermic reaction:

$$2\ C_2H_6(g) + 7\ O_2(g) \longrightarrow 4\ CO_2(g) + 6\ H_2O(l) \qquad \Delta H° = -2855\ kJ$$

 GO FIGURE

The general formula for straight-chain alkanes is C_nH_{2n+2}. What is the general formula for cycloalkanes?

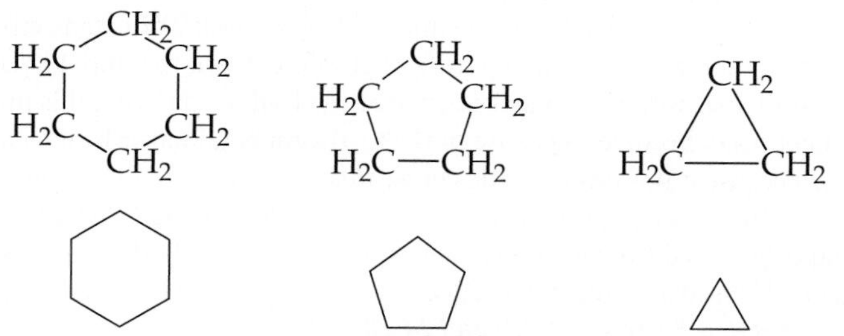

Cyclohexane	Cyclopentane	Cyclopropane
Each vertex represents one CH_2 group	Five vertices = five CH_2 groups	Three vertices = three CH_2 groups

▲ Figure 24.5 **Condensed structural formulas and line structures for three cycloalkanes.**

Chemistry Put to Work

Gasoline

Petroleum, or crude oil, is a mixture of hydrocarbons plus smaller quantities of other organic compounds containing nitrogen, oxygen, or sulfur. The tremendous demand for petroleum to meet the world's energy needs has led to the tapping of oil wells in such forbidding places as the North Sea and northern Alaska.

The usual first step in the *refining*, or processing, of petroleum is to separate it into fractions on the basis of boiling point (▼ Table 24.5). Because gasoline is the most commercially important of these fractions, various processes are used to maximize its yield.

Gasoline is a mixture of volatile alkanes and aromatic hydrocarbons. In a traditional automobile engine, a mixture of air and gasoline vapor is compressed by a piston and then ignited by a spark plug. The burning of the gasoline should create a strong, smooth expansion of gas, forcing the piston outward and imparting force along the driveshaft of the engine. If the gas burns too rapidly, the piston receives a single hard slam rather than a strong, smooth push. The result is a "knocking" or "pinging" sound and a reduction in the efficiency with which energy produced by the combustion is converted to work.

The *octane number* of a gasoline is a measure of its resistance to knocking. Gasolines with high octane numbers burn more smoothly and are thus more effective fuels (▶ Figure 24.6). Branched alkanes and aromatic hydrocarbons have higher octane numbers than straight-chain alkanes. The octane number of gasoline is obtained by comparing its knocking characteristics with those of isooctane (2,2,4-trimethylpentane) and heptane. Isooctane is assigned an octane number of 100, and heptane is assigned 0. Gasoline with the same

▲ Figure 24.6 **Octane rating.** The octane rating of gasoline measures its resistance to knocking when burned in an engine. The octane rating of the gasoline in the foreground is 89.

knocking characteristics as a mixture of 91% isooctane and 9% heptane, for instance, is rated as 91 octane.

The gasoline obtained by fractionating petroleum (called *straight-run* gasoline) contains mainly straight-chain hydrocarbons and has an octane number around 50. To increase its octane rating, it is subjected to a process called *reforming*, which converts the straight-chain alkanes into branched-chain ones.

Cracking is used to produce aromatic hydrocarbons and to convert some of the less-volatile fractions of petroleum into compounds suitable for use as automobile fuel. In cracking, the hydrocarbons are mixed with a catalyst and heated to 400–500 °C. The catalysts used are either clay minerals or synthetic Al_2O_3–SiO_2 mixtures. In addition to forming molecules more suitable for gasoline, cracking results in the formation of such low-molecular-weight hydrocarbons as ethylene and propene. These substances are used in a variety of reactions to form plastics and other chemicals.

Adding compounds called either *antiknock agents* or octane enhancers increases the octane rating of gasoline. Until the mid-1970s the principal antiknock agent was tetraethyl lead, $(C_2H_5)_4Pb$. It is no longer used, however, because of the environmental hazards of lead and because it poisons catalytic converters. ∞ (Section 14.7, "Catalytic Converters") Aromatic compounds such as toluene ($C_6H_5CH_3$) and oxygenated hydrocarbons such as ethanol (CH_3CH_2OH) are now generally used as antiknock agents.

Related Exercises: 24.19 and 24.20

Table 24.5 Hydrocarbon Fractions from Petroleum

Fraction	Size Range of Molecules	Boiling-Point Range (°C)	Uses
Gas	C_1 to C_5	−160 to 30	Gaseous fuel, production of H_2
Straight-run gasoline	C_5 to C_{12}	30 to 200	Motor fuel
Kerosene, fuel oil	C_{12} to C_{18}	180 to 400	Diesel fuel, furnace fuel, cracking
Lubricants	C_{16} and up	350 and up	Lubricants
Paraffins	C_{20} and up	Low-melting solids	Candles, matches
Asphalt	C_{36} and up	Gummy residues	Surfacing roads

24.3 | Alkenes, Alkynes, and Aromatic Hydrocarbons

Because alkanes have only single bonds, they contain the largest possible number of hydrogen atoms per carbon atom. As a result, they are called *saturated hydrocarbons*. Alkenes, alkynes, and aromatic hydrocarbons contain carbon–carbon multiple bonds (double,

GO FIGURE

How many isomers are there for propene, C_3H_6?

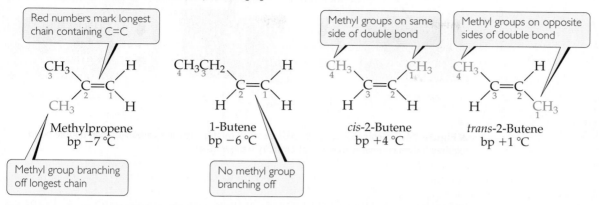

▲ **Figure 24.7 The alkene C_4H_8 has four structural isomers.**

triple, or delocalized π bonds). As a result, they contain less hydrogen than an alkane with the same number of carbon atoms. Collectively, they are called *unsaturated hydrocarbons*. On the whole, unsaturated molecules are more reactive than saturated ones.

Alkenes

Alkenes are unsaturated hydrocarbons that contain at least one $C=C$ bond. The simplest alkene is $CH_2=CH_2$, called ethene (IUPAC) or ethylene (common name), which plays important roles as a plant hormone in seed germination and fruit ripening. The next member of the series is $CH_3-CH=CH_2$, called propene or propylene. Alkenes with four or more carbon atoms have several isomers. For example, the alkene C_4H_8 has the four structural isomers shown in ▲ **Figure 24.7**. Notice both their structures and their names.

The names of alkenes are based on the longest continuous chain of carbon atoms that contains the double bond. The chain is named by changing the ending of the name of the corresponding alkane from *-ane* to *-ene*. The compound on the left in Figure 24.7, for example, has a double bond as part of a three-carbon chain; thus, the parent alkene is propene.

The location of the double bond along an alkene chain is indicated by a prefix number that designates the double-bond carbon atom that is nearest an end of the chain. The chain is always numbered from the end that brings us to the double bond sooner and hence gives the smallest-numbered prefix. In propene, the only possible location for the double bond is between the first and second carbons; thus, a prefix indicating its location is unnecessary. For butene (Figure 24.7), there are two possible positions for the double bond, either after the first carbon (1-butene) or after the second carbon (2-butene).

 Give It Some Thought

How many distinct locations are there for a double bond in a five-carbon linear chain?

If a substance contains two or more double bonds, the location of each is indicated by a numerical prefix, and the ending of the name is altered to identify the number of double bonds: diene (two), triene (three), and so forth. For example, $CH_2=CH-CH_2-CH=CH_2$ is 1,4-pentadiene.

The two isomers on the right in Figure 24.7 differ in the relative locations of their methyl groups. These two compounds are **geometric isomers**, compounds that have the same molecular formula and the same groups bonded to one another

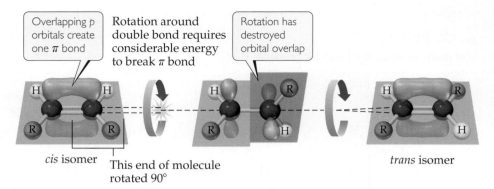

cis isomer — This end of molecule rotated 90° — *trans* isomer

▲ Figure 24.8 **Geometric isomers exist because rotation about a carbon–carbon double bond requires too much energy to occur at ordinary temperatures.**

but differ in the spatial arrangement of these groups. ∞ (Section 23.4) In the cis isomer the two methyl groups are on the same side of the double bond, whereas in the trans isomer they are on opposite sides. Geometric isomers possess distinct physical properties and can differ significantly from each other in their chemical behavior.

Geometric isomerism in alkenes arises because, unlike the C—C bond, the C=C bond resists twisting. Recall from Section 9.6 that the double bond between two carbon atoms consists of a σ and a π bond. ▲ Figure 24.8 shows a cis alkene. The carbon–carbon bond axis and the bonds to the hydrogen atoms and to the alkyl groups (designated R) are all in a plane, and the p orbitals that form the π bond are perpendicular to that plane. As Figure 24.8 shows, rotation around the carbon–carbon double bond requires the π bond to be broken, a process that requires considerable energy (about 250 kJ/mol). Because rotation does not occur easily around the carbon–carbon bond, the cis and trans isomers of an alkene cannot readily interconvert and, therefore, exist as distinct compounds.

SAMPLE
EXERCISE 24.3 Drawing Isomers

Draw all the structural and geometric isomers of pentene, C_5H_{10}, that have an unbranched hydrocarbon chain.

SOLUTION

Analyze We are asked to draw all the isomers (both structural and geometric) for an alkene with a five-carbon chain.

Plan Because the compound is named pentene and not pentadiene or pentatriene, we know that the five-carbon chain contains only one carbon–carbon double bond. Thus, we begin by placing the double bond in various locations along the chain, remembering that the chain can be numbered from either end. After finding the different unique locations for the double bond, we consider whether the molecule can have cis and trans isomers.

Solve There can be a double bond after either the first carbon (1-pentene) or second carbon (2-pentene). These are the only two possibilities because the chain can be numbered from either end. Thus, what we might erroneously call 3-pentene is actually 2-pentene, as seen by numbering the carbon chain from the other end:

$$\overset{1}{C}=\overset{2}{C}-\overset{3}{C}-\overset{4}{C}-\overset{5}{C}$$

$$\overset{1}{C}-\overset{2}{C}=\overset{3}{C}-\overset{4}{C}-\overset{5}{C}$$

$$\overset{1}{C}-\overset{2}{C}-\overset{3}{C}=\overset{4}{C}-\overset{5}{C} \quad \text{renumbered as} \quad \overset{5}{C}-\overset{4}{C}-\overset{3}{C}=\overset{2}{C}-\overset{1}{C}$$

$$\overset{1}{C}-\overset{2}{C}-\overset{3}{C}-\overset{4}{C}=\overset{5}{C} \quad \text{renumbered as} \quad \overset{5}{C}-\overset{4}{C}-\overset{3}{C}-\overset{2}{C}=\overset{1}{C}$$

Because the first C atom in 1-pentene is bonded to two H atoms, there are no cis–trans isomers. There are cis and trans isomers for 2-pentene, however. Thus, the three isomers for pentene are

$$CH_2{=}CH{-}CH_2{-}CH_2{-}CH_3$$

1-Pentene

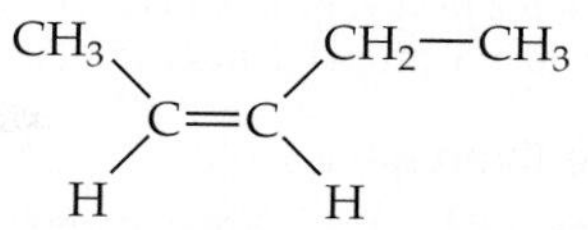

cis-2-Pentene

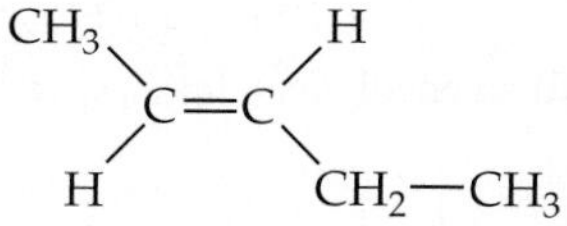

trans-2-Pentene

(You should convince yourself that *cis*-3-pentene is identical to *cis*-2-pentene and *trans*-3-pentene is identical to *trans*-2-pentene. However, *cis*-2-pentene and *trans*-2-pentene are the correct names because they have smaller numbered prefixes.)

Practice Exercise 1

Which compound does not exist?
(a) 1,2,3,4,5,6,7-octaheptaene,   **(b)** *cis*-2-butane,   **(c)** *trans*-3-hexene,
(d) 1-propene,   **(e)** *cis*-4-decene.

Practice Exercise 2

How many straight-chain isomers are there of hexene, C_6H_{12}?

Alkynes

Alkynes are unsaturated hydrocarbons containing one or more C≡C bonds. The simplest alkyne is acetylene (C_2H_2), a highly reactive molecule. When acetylene is burned in a stream of oxygen in an oxyacetylene torch, the flame reaches about 3200 K. Because alkynes in general are highly reactive, they are not as widely distributed in nature as alkenes; alkynes, however, are important intermediates in many industrial processes.

Alkynes are named by identifying the longest continuous chain containing the triple bond and modifying the ending of the name of the corresponding alkane from *-ane* to *-yne*, as shown in Sample Exercise 24.4.

SAMPLE EXERCISE 24.4 | **Naming Unsaturated Hydrocarbons**

Name the following compounds:

(a)

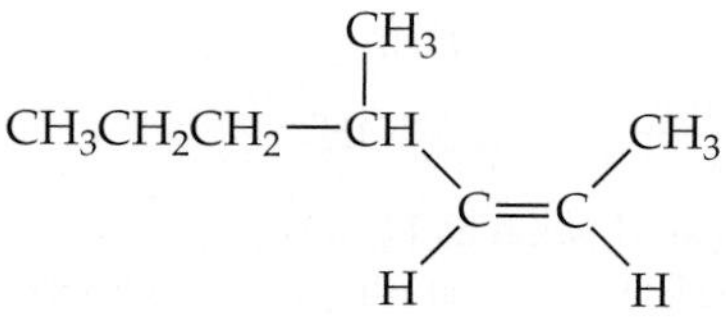

(b) 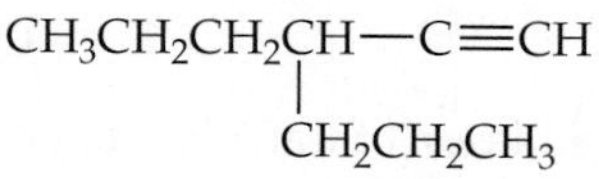 $CH_3CH_2CH_2CH{-}C{\equiv}CH$
              $CH_2CH_2CH_3$

SOLUTION

Analyze We are given the condensed structural formulas for an alkene and an alkyne and asked to name the compounds.

Plan In each case, the name is based on the number of carbon atoms in the longest continuous carbon chain that contains the multiple bond. In the alkene, care must be taken to indicate whether cis–trans isomerism is possible and, if so, which isomer is given.

Solve

(a) The longest continuous chain of carbons that contains the double bond is seven carbons long, so the parent hydrocarbon is heptene. Because the double bond begins at carbon 2 (numbering from the end closer to the double bond), we have 2-heptene. With a methyl group at carbon atom 4, we have 4-methyl-2-heptene. The geometrical configuration at the double bond is cis (that is, the alkyl groups are bonded to the double bond on the same side). Thus, the full name is 4-methyl-*cis*-2-heptene.

(b) The longest continuous chain containing the triple bond has six carbons, so this compound is a derivative of hexyne. The triple bond comes after the first carbon (numbering from the right), making it 1-hexyne. The branch from the hexyne chain contains three carbon atoms, making it a propyl group. Because this substituent is located on C3 of the hexyne chain, the molecule is 3-propyl-1-hexyne.

Practice Exercise 1

If a compound has two carbon–carbon triple bonds and one carbon–carbon double bond, what class of compound is it?
(a) an eneyne,   (b) a dieneyne,   (c) a trieneyne,   (d) an enediyne,   (e) an enetriyne.

Practice Exercise 2

Draw the condensed structural formula for 4-methyl-2-pentyne.

Addition Reactions of Alkenes and Alkynes

The presence of carbon–carbon double or triple bonds in hydrocarbons markedly increases their chemical reactivity. The most characteristic reactions of alkenes and alkynes are **addition reactions**, in which a reactant is added to the two atoms that form the multiple bond. A simple example is the addition of elemental bromine to ethylene to produce 1,2,-dibromoethane:

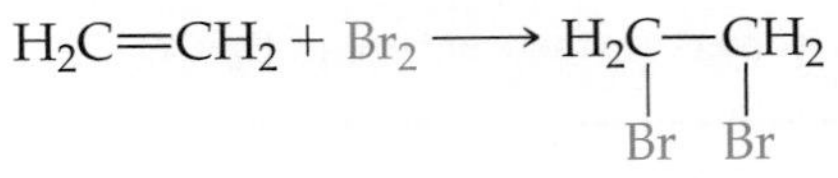

$$H_2C{=}CH_2 + Br_2 \longrightarrow \underset{\underset{Br \quad Br}{|\quad\;\;|}}{H_2C{-}CH_2} \qquad [24.1]$$

The π bond in ethylene is broken and the electrons that formed the bond are used to form two σ bonds to the two bromine atoms. The σ bond between the carbon atoms is retained.
    Addition of H_2 to an alkene converts it to an alkane:

$$CH_3CH{=}CHCH_3 + H_2 \xrightarrow{\text{Ni, 500 °C}} CH_3CH_2CH_2CH_3 \qquad [24.2]$$

The reaction between an alkene and H_2, referred to as *hydrogenation*, does not occur readily at ordinary temperatures and pressures. One reason for the lack of reactivity of H_2 toward alkenes is the stability of the H_2 bond. To promote the reaction, the reaction temperature must be raised (500 °C), and a catalyst (such as Ni) is used to assist in rupturing the H—H bond. We write such conditions over the reaction arrow to indicate they must be present in order for the reaction to occur. The most widely used catalysts are finely divided metals on which H_2 is adsorbed. ∞ (Section 14.7)
    Hydrogen halides and water can also add to the double bond of alkenes, as in these reactions of ethylene:

$$CH_2{=}CH_2 + HBr \longrightarrow CH_3CH_2Br \qquad [24.3]$$

$$CH_2{=}CH_2 + H_2O \xrightarrow{H_2SO_4} CH_3CH_2OH \qquad [24.4]$$

The addition of water is catalyzed by a strong acid, such as H_2SO_4.
    The addition reactions of alkynes resemble those of alkenes, as shown in these examples:

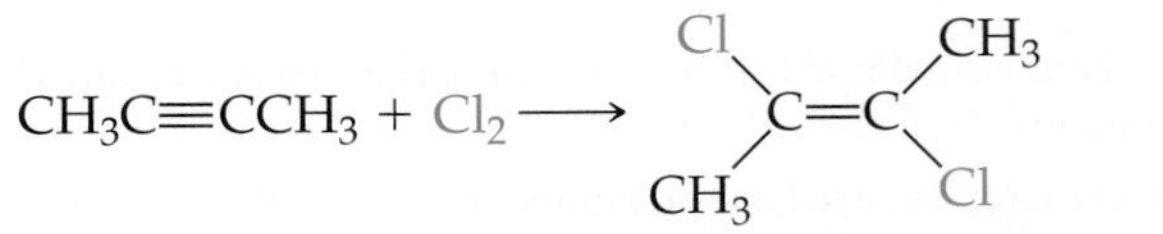

$$CH_3C{\equiv}CCH_3 + Cl_2 \longrightarrow \underset{\underset{CH_3}{}}{\overset{Cl}{}}C{=}C\underset{\underset{Cl}{}}{\overset{CH_3}{}} \qquad [24.5]$$

2-Butyne   *trans*-2,3-Dichloro-2-butene

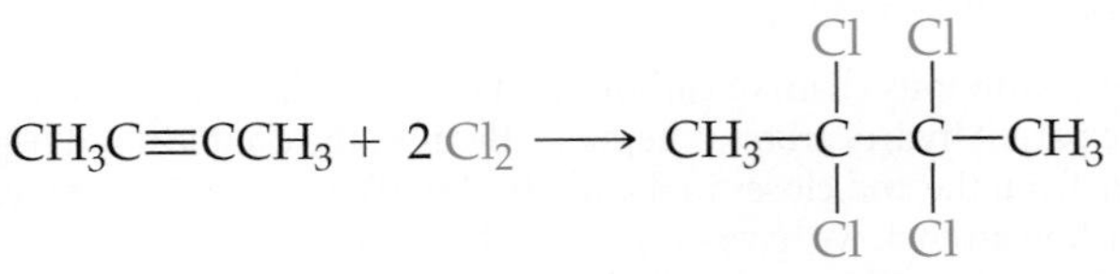

$$CH_3C{\equiv}CCH_3 + 2\,Cl_2 \longrightarrow CH_3{-}\underset{\underset{Cl}{|}}{\overset{\overset{Cl}{|}}{C}}{-}\underset{\underset{Cl}{|}}{\overset{\overset{Cl}{|}}{C}}{-}CH_3 \qquad [24.6]$$

2-Butyne   2,2,3,3-Tetrachlorobutane

SAMPLE EXERCISE 24.5 Predicting the Product of an Addition Reaction

Write the condensed structural formula for the product of the hydrogenation of 3-methyl-1-pentene.

SOLUTION

Analyze We are asked to predict the compound formed when a particular alkene undergoes hydrogenation (reaction with H_2) and to write the condensed structural formula of the product.

Plan To determine the condensed structural formula of the product, we must first write the condensed structural formula or Lewis structure of the reactant. In the hydrogenation of the alkene, H_2 adds to the double bond, producing an alkane.

Solve The name of the starting compound tells us that we have a chain of five C atoms with a double bond at one end (position 1) and a methyl group on C3:

$$CH_2{=}CH{-}\overset{\displaystyle CH_3}{\overset{|}{CH}}{-}CH_2{-}CH_3$$

Hydrogenation—the addition of two H atoms to the carbons of the double bond—leads to the following alkane:

$$CH_3{-}CH_2{-}\overset{\displaystyle CH_3}{\overset{|}{CH}}{-}CH_2{-}CH_3$$

Comment The longest chain in this alkane has five carbon atoms; the product is therefore 3-methylpentane.

Practice Exercise 1

What product is formed from the hydrogenation of 2-methylpropene? **(a)** propane, **(b)** butane, **(c)** 2-methylbutane, **(d)** 2-methylpropane, **(e)** 2-methylpropyne.

Practice Exercise 2

Addition of HCl to an alkene forms 2-chloropropane. What is the alkene?

A Closer Look

Mechanism of Addition Reactions

As the understanding of chemistry has grown, chemists have advanced from simply cataloging reactions that occur to explaining *how* they occur, by drawing the individual steps of a reaction based upon experimental and theoretical evidence. The sum of these steps is termed a *reaction mechanism*. ∞ (Section 14.6)

The addition reaction between HBr and an alkene, for instance, is thought to proceed in two steps. In the first step, which is rate determining ∞ (Section 14.6), HBr attacks the electron-rich double bond, transferring a proton to one of the double-bond carbons. In the reaction of 2-butene with HBr, for example, the first step is

$$CH_3CH{=}CHCH_3 + HBr \longrightarrow \left[\overset{\delta+}{CH_3CH}{\cdots}CHCH_3 \right]$$

with H and Br$^{\delta-}$ below.

$$\longrightarrow \overset{+}{CH_3CH}{-}CH_2CH_3 + Br^- \qquad [24.7]$$

The electron pair that formed the π bond is used to form the new C—H bond.

The second, faster step is addition of Br$^-$ to the positively charged carbon. The bromide ion donates a pair of electrons to the carbon, forming the C—Br bond:

$$\overset{+}{CH_3CH}{-}CH_2CH_3 + Br^- \longrightarrow \left[\overset{\delta+}{CH_3CH}{-}CH_2CH_3 \,\, \underset{Br^{\delta-}}{\vdots} \right]$$

$$\longrightarrow \underset{Br}{\overset{|}{CH_3CHCH_2CH_3}} \qquad [24.8]$$

Because the rate-determining step involves both the alkene and the acid, the rate law for the reaction is second order, first order in the alkene and first order in HBr:

$$\text{Rate} = -\frac{\Delta[CH_3CH{=}CHCH_3]}{\Delta t} = k[CH_3CH{=}CHCH_3][HBr]$$

$$[24.9]$$

The energy profile for the reaction is shown in **Figure 24.9**. The first energy maximum represents the transition state in the first step, and the second maximum represents the transition state in the second step. The energy minimum represents the energies of the intermediate species, $\overset{+}{CH_3CH}{-}CH_2CH_3$ and Br$^-$.

GO FIGURE

What features of an energy profile allow you to distinguish between an intermediate and a transition state?

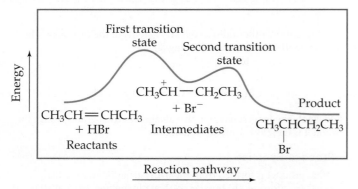

▲ Figure 24.9 **Energy profile for addition of HBr to 2-butene.** The two maxima tell you that this is a two-step mechanism.

To show electron movement in reactions like these, chemists often use curved arrows pointing in the direction of electron flow. For the addition of HBr to 2-butene, for example, the shifts in electron positions are shown as

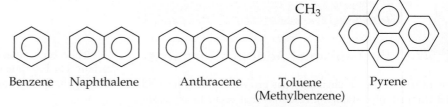

Aromatic Hydrocarbons

The simplest aromatic hydrocarbon, benzene (C_6H_6), is shown in ▼ Figure 24.10 along with some other aromatic hydrocarbons. Benzene is the most important aromatic hydrocarbon, and most of our discussion focuses on it.

Benzene   Naphthalene   Anthracene   Toluene   Pyrene
                                                    (Methylbenzene)

▲ Figure 24.10 **Line formulas and common names of several aromatic compounds.** The aromatic rings are represented by hexagons with a circle inscribed inside to denote delocalized π bonds. Each vertex represents a carbon atom. Each carbon is bound to three other atoms—either three carbons, or two carbons and a hydrogen—so that each carbon has the requisite four bonds.

Stabilization of π Electrons by Delocalization

If you draw one Lewis structure for benzene, you draw a ring that contains three CC double bonds and three single CC bonds. ∞ (Section 8.6) You might therefore expect benzene to resemble the alkenes and to be highly reactive. Benzene and the other aromatic hydrocarbons, however, are much more stable than alkenes because the π electrons are delocalized in the π orbitals. ∞ (Section 9.6)

We can estimate the stabilization of the π electrons in benzene by comparing the energy required to form cyclohexane by adding hydrogen to benzene, to cyclohexene (one double bond) and to 1,4-cyclohexadiene (two double bonds):

$+ \ 3 \ H_2 \longrightarrow$ $\Delta H° = -208 \ \text{kJ/mol}$

$+ \ H_2 \longrightarrow$ $\Delta H° = -120 \ \text{kJ/mol}$

$+ \ 2 \ H_2 \longrightarrow$ $\Delta H° = -232 \ \text{kJ/mol}$

From the second and third reactions, it appears that the energy required to hydrogenate each double bond is roughly 118 kJ/mol for each bond. Benzene contains the equivalent of three double bonds. We might expect, therefore, the energy required to hydrogenate benzene to be about 3 times −118, or −354 kJ/mol, if benzene behaved as though it were "cyclohexatriene," that is, if it behaved as though it had three isolated double bonds in a ring. Instead, the energy released is 146 kJ less than this, indicating that benzene is more stable than would be expected for three double bonds. The difference of 146 kJ/mol between the expected heat of hydrogenation (−354 kJ/mol) and the observed heat of hydrogenation, (−208 kJ/mol) is due to stabilization of the π electrons through delocalization in the π orbitals that extend around the ring. Chemists call this stabilization energy the *resonance energy*.

Substitution Reactions

Although aromatic hydrocarbons are unsaturated, *they do not readily undergo addition reactions*. The delocalized π bonding causes aromatic compounds to behave quite differently from alkenes and alkynes. Benzene, for example, does not add Cl_2 or Br_2 to its double bonds under ordinary conditions. In contrast, aromatic hydrocarbons undergo **substitution reactions** relatively easily. In a substitution reaction, one hydrogen atom of the molecule is removed and replaced (substituted) by another atom or group of atoms. When benzene is warmed in a mixture of nitric and sulfuric acids, for example, one of the benzene hydrogens is replaced by the nitro group, NO_2:

$$+ \text{HNO}_3 \xrightarrow{\text{H}_2\text{SO}_4} + \text{H}_2\text{O} \qquad [24.10]$$

Benzene   Nitrobenzene

More vigorous treatment results in substitution of a second nitro group into the molecule:

$$+ \text{HNO}_3 \xrightarrow{\text{H}_2\text{SO}_4} + \text{H}_2\text{O} \qquad [24.11]$$

There are three isomers of benzene that contain two nitro groups—the 1,2- or *ortho*- isomer, the 1,3- or *meta*-isomer, and the 1,4- or *para*-isomer of dinitrobenzene:

ortho-Dinitrobenzene	*meta*-Dinitrobenzene	*para*-Dinitrobenzene
1,2-Dinitrobenzene	1,3-Dinitrobenzene	1,4-Dinitrobenzene
mp 118 °C	mp 90 °C	mp 174 °C

In the reaction of Equation 24.11, the principal product is the *meta* isomer.

Bromination of benzene, carried out with FeBr$_3$ as a catalyst, is another substitution reaction:

Benzene   Bromobenzene   [24.12]

In a similar substitution reaction, called the *Friedel–Crafts reaction*, alkyl groups can be substituted onto an aromatic ring by reacting an alkyl halide with an aromatic compound in the presence of AlCl$_3$ as a catalyst:

$$\text{Benzene} + CH_3CH_2Cl \xrightarrow{AlCl_3} \text{Ethylbenzene} + HCl \qquad [24.13]$$

Benzene   Ethylbenzene

▲ **Give It Some Thought**

When the aromatic hydrocarbon naphthalene, shown in Figure 24.10, reacts with nitric and sulfuric acids, two compounds containing one nitro group are formed. Draw the structures of these two compounds.

24.4 | Organic Functional Groups

The C═C double bonds of alkenes and C≡C triple bonds of alkynes are just two of many functional groups in organic molecules. As noted earlier, these functional groups each undergo characteristic reactions, and the same is true of all other functional groups. Each kind of functional group often undergoes the same kinds of reactions in every molecule, regardless of the size and complexity of the molecule. Thus, the chemistry of an organic molecule is largely determined by the functional groups it contains.

▶ Table 24.6 lists the most common functional groups. Notice that, except for C═C and C≡C, they all contain either O, N, or a halogen atom, X.

We can think of organic molecules as being composed of functional groups bonded to one or more alkyl groups. The alkyl groups, which are made of C—C and C—H single bonds, are the less reactive portions of the molecules. In describing general features of organic compounds, chemists often use the designation R to represent any alkyl group: methyl, ethyl, propyl, and so on. Alkanes, for example, which contain no functional group, are represented as R—H. Alcohols, which contain the functional group —OH, are represented as R—OH. If two or more different alkyl groups are present in a molecule, we designate them R, R′, R″, and so forth.

Alcohols

Alcohols are compounds in which one or more hydrogens of a parent hydrocarbon have been replaced by the functional group —OH, called either the

Table 24.6 Common Functional Groups

Functional Group	Compound Type	Suffix or Prefix	Example Structural Formula	Ball-and-stick Model	Systematic Name (common name)
$C=C$ (alkene group)	Alkene	*-ene*	H–C=C–H (with H's)		Ethene (Ethylene)
$-C\equiv C-$	Alkyne	*-yne*	$H-C\equiv C-H$		Ethyne (Acetylene)
$-\overset{\|}{\underset{\|}{C}}-\ddot{O}-H$	Alcohol	*-ol*	$H-\overset{H}{\underset{H}{C}}-\ddot{O}-H$		Methanol (Methyl alcohol)
$-\overset{\|}{\underset{\|}{C}}-\ddot{O}-\overset{\|}{\underset{\|}{C}}-$	Ether	*ether*	$H-\overset{H}{\underset{H}{C}}-\ddot{O}-\overset{H}{\underset{H}{C}}-H$		Dimethyl ether
$-\overset{\|}{\underset{\|}{C}}-\ddot{X}:$ (X = halogen)	Alkyl halide or haloalkane	*-ide*	$H-\overset{H}{\underset{H}{C}}-\ddot{C}l:$		Chloromethane (Methyl chloride)
$-\overset{\|}{\underset{\|}{C}}-\ddot{N}-$	Amine	*-amine*	$H-\overset{H}{\underset{H}{C}}-\overset{H}{\underset{H}{C}}-\ddot{N}-H$		Ethylamine
$-\overset{:O:}{\overset{\|\|}{C}}-H$	Aldehyde	*-al*	$H-\overset{H}{\underset{H}{C}}-\overset{:O:}{\overset{\|\|}{C}}-H$		Ethanal (Acetaldehyde)
$-\overset{\|}{\underset{\|}{C}}-\overset{:O:}{\overset{\|\|}{C}}-\overset{\|}{\underset{\|}{C}}-$	Ketone	*-one*	$H-\overset{H}{\underset{H}{C}}-\overset{:O:}{\overset{\|\|}{C}}-\overset{H}{\underset{H}{C}}-H$		Propanone (Acetone)
$-\overset{:O:}{\overset{\|\|}{C}}-\ddot{O}-H$	Carboxylic acid	*-oic acid*	$H-\overset{H}{\underset{H}{C}}-\overset{:O:}{\overset{\|\|}{C}}-\ddot{O}-H$		Ethanoic acid (Acetic acid)
$-\overset{:O:}{\overset{\|\|}{C}}-\ddot{O}-\overset{\|}{\underset{\|}{C}}-$	Ester	*-oate*	$H-\overset{H}{\underset{H}{C}}-\overset{:O:}{\overset{\|\|}{C}}-\ddot{O}-\overset{H}{\underset{H}{C}}-H$		Methyl ethanoate (Methyl acetate)
$-\overset{:O:}{\overset{\|\|}{C}}-\ddot{N}-$	Amide	*-amide*	$H-\overset{H}{\underset{H}{C}}-\overset{:O:}{\overset{\|\|}{C}}-\ddot{N}-H$		Ethanamide (Acetamide)

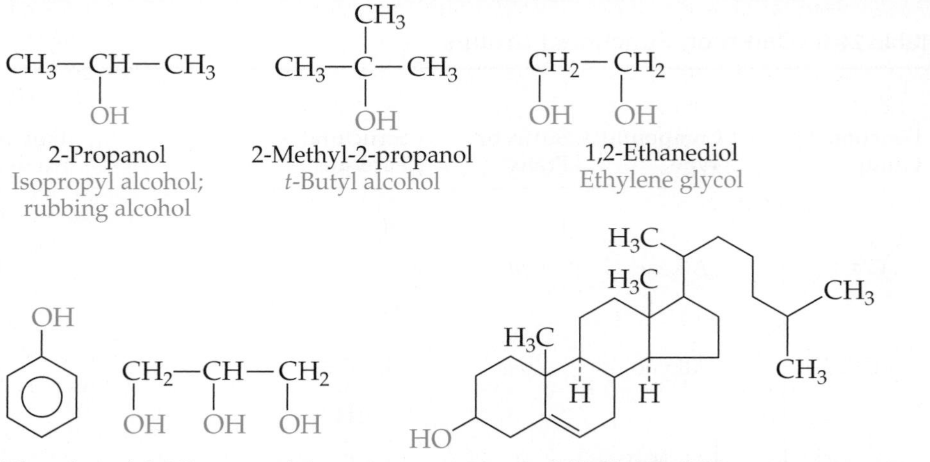

$$CH_3-\underset{\underset{OH}{|}}{CH}-CH_3 \qquad CH_3-\underset{\underset{OH}{|}}{\overset{\overset{CH_3}{|}}{C}}-CH_3 \qquad \underset{\underset{OH}{|}}{CH_2}-\underset{\underset{OH}{|}}{CH_2}$$

2-Propanol   2-Methyl-2-propanol   1,2-Ethanediol
Isopropyl alcohol;   *t*-Butyl alcohol   Ethylene glycol
rubbing alcohol

OH
          $$\underset{\underset{OH}{|}}{CH_2}-\underset{\underset{OH}{|}}{CH}-\underset{\underset{OH}{|}}{CH_2}$$

Phenol   1,2,3-Propanetriol   2,15-dimethyl-14-
          Glycerol; glycerin   (1,5-dimethylhexyl)tetracyclo[8.7.0.0^{2,7}.0^{11,15}]heptacos-
          7-en-5-ol
          Cholesterol

▲ **Figure 24.11 Condensed structural formulas of six important alcohols.** Common names are given in blue.

hydroxyl group or the *alcohol group*. Note in ▲ **Figure 24.11** that the name for an alcohol ends in *-ol*. The simple alcohols are named by changing the last letter in the name of the corresponding alkane to *-ol*—for example, ethan*e* becomes ethan*ol*. Where necessary, the location of the OH group is designated by a numeric prefix that indicates the number of the carbon atom bearing the OH group.

The O—H bond is polar, so alcohols are more soluble in polar solvents than are hydrocarbons. The —OH functional group can also participate in hydrogen bonding. As a result, the boiling points of alcohols are higher than those of their parent alkanes.

◄ **Figure 24.12** shows several commercial products that consist entirely or in large part of an alcohol.

The simplest alcohol, methanol (methyl alcohol), has many industrial uses and is produced on a large scale by heating carbon monoxide and hydrogen under pressure in the presence of a metal oxide catalyst:

$$CO(g) + 2\,H_2(g) \xrightarrow[400\,°C]{200-300\,atm} CH_3OH(g) \qquad [24.14]$$

Because methanol has a very high octane rating as an automobile fuel, it is used as a gasoline additive and as a fuel in its own right.

Ethanol (ethyl alcohol, C_2H_5OH) is a product of the fermentation of carbohydrates such as sugars and starches. In the absence of air, yeast cells convert these carbohydrates into ethanol and CO_2:

$$C_6H_{12}O_6(aq) \xrightarrow{yeast} 2\,C_2H_5OH(aq) + 2\,CO_2(g) \qquad [24.15]$$

In the process, the yeast cells derive energy necessary for growth. This reaction is carried out under carefully controlled conditions to produce beer, wine, and other beverages in which ethanol (called just "alcohol" in everyday language) is the active ingredient.

The simplest polyhydroxyl alcohol (an alcohol containing more than one OH group) is 1,2-ethanediol (ethylene glycol, $HOCH_2CH_2OH$), the major ingredient in automobile antifreeze. Another common polyhydroxyl alcohol is 1,2,3-propanetriol (glycerol, $HOCH_2CH(OH)CH_2OH$), a viscous liquid that dissolves readily in water and is used in cosmetics as a skin softener and in foods and candies to keep them moist.

Phenol is the simplest compound with an OH group attached to an aromatic ring. One of the most striking effects of the aromatic group is the greatly increased acidity of the OH group. Phenol is about 1 million times more acidic in water than a nonaromatic alcohol. Even so, it is not a very strong acid ($K_a = 1.3 \times 10^{-10}$). Phenol is used industrially to make plastics and dyes, and as a topical anesthetic in throat sprays.

▲ **Figure 24.12 Everyday alcohols.** Many of the products we use every day—from rubbing alcohol to hair spray and antifreeze—are composed either entirely or mainly of alcohols.

Cholesterol, shown in Figure 24.11, is a biochemically important alcohol. The OH group forms only a small component of this molecule, so cholesterol is only slightly soluble in water (2.6 g/L of H_2O). Cholesterol is a normal and essential component of our bodies; when present in excessive amounts, however, it may precipitate from solution. It precipitates in the gallbladder to form crystalline lumps called *gallstones*. It may also precipitate against the walls of veins and arteries and thus contribute to high blood pressure and other cardiovascular problems.

Ethers

Compounds in which two hydrocarbon groups are bonded to one oxygen are called **ethers**. Ethers can be formed from two molecules of alcohol by eliminating a molecule of water. The reaction is catalyzed by sulfuric acid, which takes up water to remove it from the system:

$$CH_3CH_2-OH + H-OCH_2CH_3 \xrightarrow{H_2SO_4} CH_3CH_2-O-CH_2CH_3 + H_2O$$

$$[24.16]$$

A reaction in which water is eliminated from two substances is called a *condensation reaction.* ∞ (Sections 12.8 and 22.8)

Both diethyl ether and the cyclic ether tetrahydrofuran, shown below, are common solvents for organic reactions. Diethyl ether was formerly used as an anesthetic (known simply as "ether" in that context), but it had significant side effects.

$$CH_3CH_2-O-CH_2CH_3$$

$$\begin{array}{cc} CH_2 & CH_2 \\ | & | \\ CH_2 & CH_2 \\ & O \end{array}$$

Diethyl ether   Tetrahydrofuran (THF)

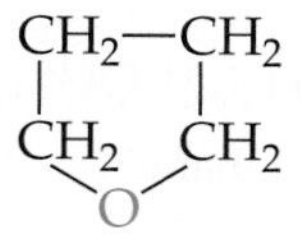

Aldehydes and Ketones

Several of the functional groups listed in Table 24.6 contain the **carbonyl group**, $C=O$. This group, together with the atoms attached to its carbon, defines several important functional groups that we consider in this section.

In **aldehydes**, the carbonyl group has at least one hydrogen atom attached:

$$H-\overset{\overset{\displaystyle O}{\|}}{C}-H \qquad CH_3-\overset{\overset{\displaystyle O}{\|}}{C}-H$$

Methanal   Ethanal
Formaldehyde   Acetaldehyde

In **ketones**, the carbonyl group occurs at the interior of a carbon chain and is therefore flanked by carbon atoms:

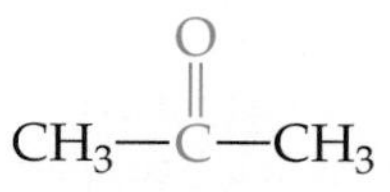

$$CH_3-\overset{\overset{\displaystyle O}{\|}}{C}-CH_3 \qquad CH_3-\overset{\overset{\displaystyle O}{\|}}{C}-CH_2CH_3$$

Propanone   2-Butanone   Testosterone
Acetone   Methyl ethyl ketone

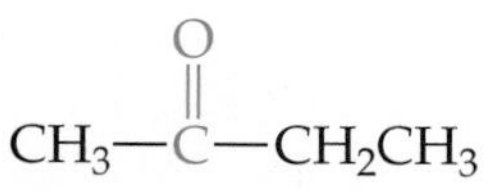

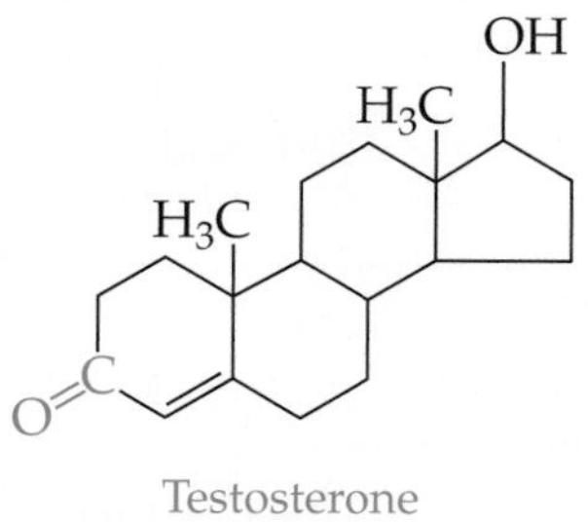

The systematic names of aldehydes contain -*al* and that ketone names contain -*one*. Notice that testosterone has both alcohol and ketone groups; the ketone functional group dominates the molecular properties. Therefore, testosterone is considered a ketone first and an alcohol second, and its name reflects its ketone properties.

Many compounds found in nature contain an aldehyde or ketone functional group. Vanilla and cinnamon flavorings are naturally occurring aldehydes. Two isomers of the ketone carvone impart the characteristic flavors of spearmint leaves and caraway seeds.

Ketones are less reactive than aldehydes and are used extensively as solvents. Acetone, the most widely used ketone, is completely miscible with water, yet it dissolves a wide range of organic substances.

Carboxylic Acids and Esters

Carboxylic acids contain the *carboxyl* functional group, often written COOH. ∞ (Section 16.10) These weak acids are widely distributed in nature and are common in citrus fruits. ∞ (Section 4.3) They are also important in the manufacture of polymers used to make fibers, films, and paints. ▼ Figure 24.13 shows the formulas of several carboxylic acids.

The common names of many carboxylic acids are based on their historical origins. Formic acid, for example, was first prepared by extraction from ants; its name is derived from the Latin word *formica*, "ant."

Carboxylic acids can be produced by oxidation of alcohols. Under appropriate conditions, the aldehyde may be isolated as the first product of oxidation, as in the sequence

$$\underset{\text{Ethanol}}{CH_3CH_2OH} + (O) \longrightarrow \underset{\text{Acetaldehyde}}{CH_3\overset{\displaystyle O}{\overset{\|}{C}H}} + H_2O \qquad [24.17]$$

$$\underset{\text{Acetaldehyde}}{CH_3\overset{\displaystyle O}{\overset{\|}{C}H}} + (O) \longrightarrow \underset{\text{Acetic acid}}{CH_3\overset{\displaystyle O}{\overset{\|}{C}OH}} \qquad [24.18]$$

where (O) represents any oxidant that can provide oxygen atoms. The air oxidation of ethanol to acetic acid is responsible for causing wines to turn sour, producing vinegar.

▲ **GO FIGURE**

Which of these substances have both a carboxylic acid functional group and an alcohol functional group?

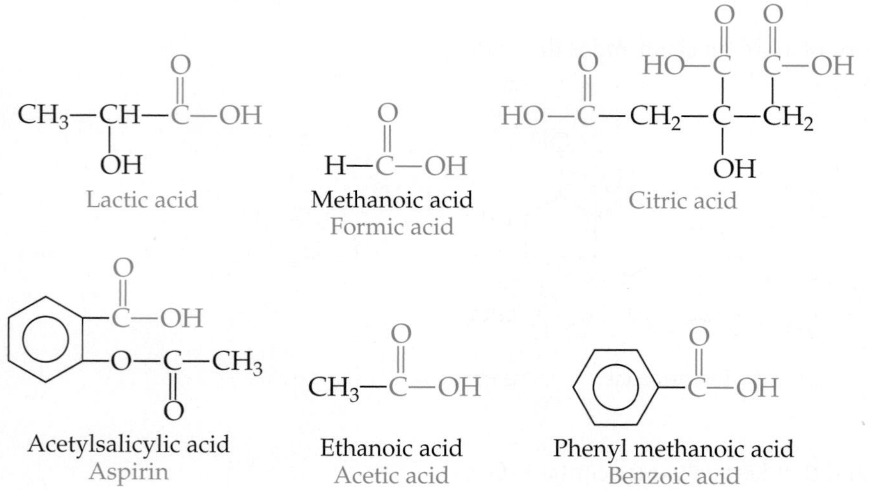

Lactic acid

Methanoic acid
Formic acid

Citric acid

Acetylsalicylic acid
Aspirin

Ethanoic acid
Acetic acid

Phenyl methanoic acid
Benzoic acid

▲ **Figure 24.13 Structural formulas of common carboxylic acids.** The monocarboxylic acids are generally referred to by their common names, given in blue type.

The oxidation processes in organic compounds are related to those oxidation reactions we studied in Chapter 20. Instead of counting electrons, the number of C—O bonds is usually considered to show the extent of oxidation of similar compounds. For example, methane can be oxidized to methanol, then formaldehye (methanal), then formic acid (methanoic acid):

$$
\underset{\text{Methane}}{\overset{\displaystyle H}{\underset{\displaystyle H}{H-C-H}}}
\quad
\underset{\text{Methanol}}{\overset{\displaystyle H}{\underset{\displaystyle H}{H-C-OH}}}
\quad
\underset{\text{Formaldehyde}}{\overset{\displaystyle O}{\overset{\|}{H-C-H}}}
\quad
\underset{\text{Formic acid}}{\overset{\displaystyle O}{\overset{\|}{H-C-O-H}}}
$$

From methanol to formic acid, the number of C—O bonds increases from 0 to 3 (double bonds are counted as two). If you were to calculate the oxidation state of carbon in these compounds, it would range from -4 in methane (if H's are counted as $+1$) to $+2$ in formic acid, which is consistent with carbon being oxidized. The ultimate oxidation product of any organic compound, then, is CO_2, which is indeed the product of combustion reactions of carbon-containing compounds (CO_2 has 4 C—O bonds, and C has the oxidation state $+4$).

Give It Some Thought

What chemical process is happening when formic acid is converted back to methane?

Aldehydes and ketones can be prepared by controlled oxidation of alcohols. Complete oxidation results in formation of CO_2 and H_2O, as in the burning of methanol:

$$CH_3OH(g) + \tfrac{3}{2}O_2(g) \longrightarrow CO_2(g) + 2\,H_2O(g)$$

Controlled partial oxidation to form other organic substances, such as aldehydes and ketones, is carried out by using various oxidizing agents, such as air, hydrogen peroxide (H_2O_2), ozone (O_3), and potassium dichromate ($K_2Cr_2O_7$).

Give It Some Thought

Write the condensed structural formula for the acid that would result from oxidation of the alcohol

$$
\begin{array}{c}
CH_2-CHOH \\
\diagup \qquad \diagdown \\
CH_2 \qquad CH_2 \\
\diagdown \quad \diagup \\
CH_2
\end{array}
$$

Acetic acid can also be produced by the reaction of methanol with carbon monoxide in the presence of a rhodium catalyst:

$$CH_3OH + CO \xrightarrow{\text{catalyst}} CH_3-\overset{\displaystyle O}{\overset{\|}{C}}-OH \qquad\qquad [24.19]$$

This reaction is not an oxidation; it involves, in effect, the insertion of a carbon monoxide molecule between the CH_3 and OH groups. A reaction of this kind is called *carbonylation*.

Carboxylic acids can undergo condensation reactions with alcohols to form esters:

$$\underset{\text{Acetic acid}}{CH_3-\overset{\displaystyle O}{\overset{\|}{C}}-OH} + \underset{\text{Ethanol}}{HO-CH_2CH_3} \longrightarrow \underset{\text{Ethyl acetate}}{CH_3-\overset{\displaystyle O}{\overset{\|}{C}}-O-CH_2CH_3} + H_2O$$

$$[24.20]$$

Esters are compounds in which the H atom of a carboxylic acid is replaced by a carbon-containing group:

$$-\overset{\overset{\displaystyle O}{\|}}{C}-O-\overset{|}{\underset{|}{C}}-$$

Give It Some Thought
What is the difference between an ether and an ester?

The name of any ester consists of the name of the group contributed by the alcohol followed by the name of the group contributed by the carboxylic acid, with the *-ic* replaced by *-ate*. For example, the ester formed from ethyl alcohol, CH_3CH_2OH, and butyric acid, $CH_3(CH_2)_2COOH$, is

$$CH_3CH_2CH_2\overset{\overset{\displaystyle O}{\|}}{C}-OCH_2CH_3$$

Ethyl butyrate

Notice that the chemical formula generally has the group originating from the acid written first, which is opposite of the way the ester is named. Another example is isoamyl acetate, the ester formed from acetic acid and isoamyl alcohol. Isoamyl acetate smells like bananas or pears.

$$(CH_3)_2CHCH_2CH_2-O-\overset{\overset{\displaystyle O}{\|}}{C}-CH_3$$

Isoamyl   Acetate

Many esters such as isoamyl acetate have pleasant odors and are largely responsible for the pleasing aromas of fruit.

An ester treated with an acid or a base in aqueous solution is *hydrolyzed*; that is, the molecule is split into an alcohol and a carboxylic acid or its anion:

$$CH_3CH_2-\overset{\overset{\displaystyle O}{\|}}{C}-O-CH_3 + OH^- \longrightarrow$$

Methyl propionate

$$CH_3CH_2-\overset{\overset{\displaystyle O}{\|}}{C}-O^- + CH_3OH$$

Propionate   Methanol   [24.21]

The **hydrolysis** of an ester in the presence of a base is called **saponification**, a term that comes from the Latin word for soap, *sapon*. Naturally occurring esters include fats and oils, and in making soap an animal fat or a vegetable oil is boiled with a strong base. The resultant soap consists of a mixture of salts of long-chain carboxylic acids (called fatty acids), which form during the saponification reaction.

SAMPLE EXERCISE 24.6 | Naming Esters and Predicting Hydrolysis Products

In a basic aqueous solution, esters react with hydroxide ion to form the salt of the carboxylic acid and the alcohol from which the ester is constituted. Name each of the following esters, and indicate the products of their reaction with aqueous base.

(a) 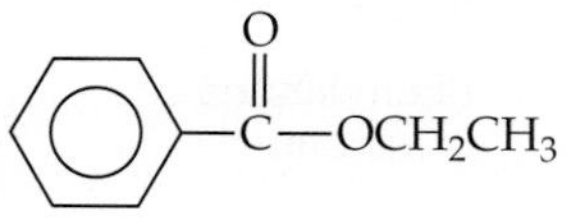   (b) 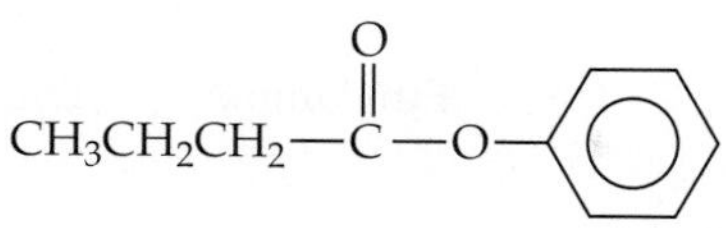 $CH_3CH_2CH_2 - \overset{\overset{\displaystyle O}{\|}}{C} - O -$

SOLUTION

Analyze We are given two esters and asked to name them and to predict the products formed when they undergo hydrolysis (split into an alcohol and carboxylate ion) in basic solution.

Plan Esters are formed by the condensation reaction between an alcohol and a carboxylic acid. To name an ester, we must analyze its structure and determine the identities of the alcohol and acid from which it is formed. We can identify the alcohol by adding an OH to the alkyl group attached to the O atom of the carboxyl (COO) group. We can identify the acid by adding an H to the O atom of the carboxyl group. We have learned that the first part of an ester name indicates the alcohol portion and the second indicates the acid portion. The name conforms to how the ester undergoes hydrolysis in base, reacting with base to form an alcohol and a carboxylate anion.

Solve

(a) This ester is derived from ethanol (CH_3CH_2OH) and benzoic acid (C_6H_5COOH). Its name is therefore ethyl benzoate. The net ionic equation for reaction of ethyl benzoate with hydroxide ion is

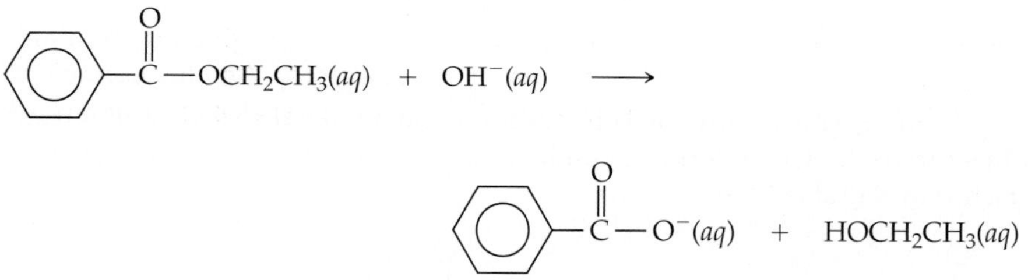

The products are benzoate ion and ethanol.

(b) This ester is derived from phenol (C_6H_5OH) and butanoic acid (commonly called butyric acid) $(CH_3CH_2CH_2COOH)$. The residue from the phenol is called the phenyl group. The ester is therefore named phenyl butyrate or phenyl butanoate. The net ionic equation for the reaction of phenyl butyrate with hydroxide ion is

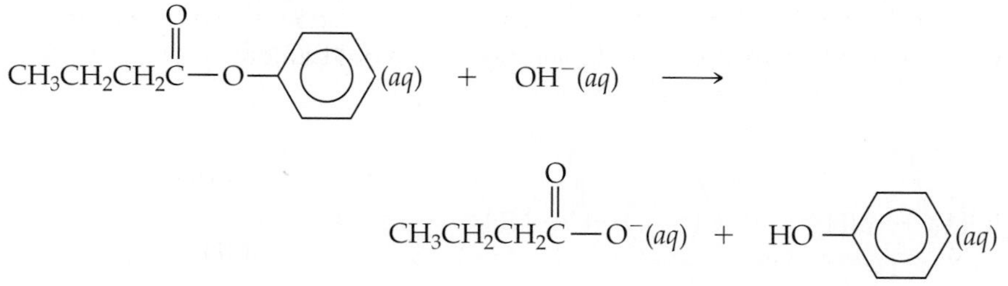

The products are butyrate ion and phenol.

Practice Exercise 1

For the generic ester RC(O)OR', which bond will hydrolyze under basic conditions?
(a) the R—C bond   (b) the C=O bond   (c) the C—O bond   (d) the O—R' bond
(e) more than one of the above

Practice Exercise 2

Write the condensed structural formula for the ester formed from propyl alcohol and propionic acid.

Amines and Amides

Amines are compounds in which one or more of the hydrogens of ammonia (NH_3) are replaced by an alkyl group:

$$CH_3CH_2NH_2 \qquad (CH_3)_3N \qquad \!-NH_2$$

Ethylamine   Trimethylamine   Phenylamine
Aniline

Amines are the most common organic bases. ∞ (Section 16.7) As we saw in the Chemistry Put to Work box in Section 16.8, many pharmaceutically active compounds are complex amines:

Cocaine   Morphine   Codeine

An amine with at least one H bonded to N can undergo a condensation reaction with a carboxylic acid to form an **amide**, which contains the carbonyl group ($C{=}O$) attached to N (Table 24.6):

$$CH_3\overset{\overset{\textstyle O}{\|}}{C}{-}OH \ + \ H{-}N(CH_3)_2 \longrightarrow CH_3\overset{\overset{\textstyle O}{\|}}{C}{-}N(CH_3)_2 \ + \ H_2O \qquad [24.22]$$

We may consider the amide functional group to be derived from a carboxylic acid with an NRR′, NH_2 or NHR′ group replacing the OH of the acid, as in these examples:

$$CH_3\overset{\overset{\textstyle O}{\|}}{C}{-}NH_2 \qquad \!-\overset{\overset{\textstyle O}{\|}}{C}{-}NH_2 \qquad CH_3\overset{\overset{\textstyle O}{\|}}{C}{-}NH$$

Ethanamide   Phenylmethanamide   N-(4-hydroxyphenyl)ethanamide
Acetamide   Benzamide   Acetaminophen

The amide linkage

$$R{-}\overset{\overset{\textstyle O}{\|}}{C}{-}\underset{\underset{\textstyle H}{|}}{N}{-}R'$$

where R and R′ are organic groups, is the key functional group in proteins, as we will see in Section 24.7.

24.5 | Chirality in Organic Chemistry

A molecule possessing a nonsuperimposable mirror image is termed **chiral** (Greek *cheir*, "hand"). ∞ (Section 23.4) *Compounds containing carbon atoms with four different attached groups are inherently chiral.* A carbon atom with four different attached groups is called a *chiral center*. For example, consider 2-bromopentane:

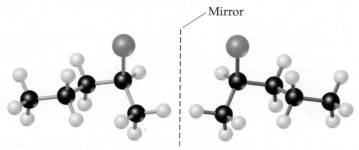

All four groups attached to C2 are different, making that carbon a chiral center. ▼ Figure 24.14 illustrates the nonsuperimposable mirror images of this molecule. Imagine moving the molecule shown to the left of the mirror over to the right of the mirror. If you then turn it in every possible way, you will conclude that it cannot be superimposed on the molecule shown to the right of the mirror. Nonsuperimposable mirror images are called either *optical isomers* or *enantiomers*. ∞ (Section 23.4) Organic chemists use the labels *R* and *S* to distinguish the two forms. We need not go into the rules for deciding on the labels.

▲ GO FIGURE

If you replace Br with CH_3, will the compound be chiral?

Mirror

▲ **Figure 24.14 The two enantiomeric forms of 2-bromopentane.** The mirror-image isomers are not superimposable on each other.

The two members of an enantiomer pair have identical physical properties and identical chemical properties when they react with nonchiral reagents. Only in a chiral environment do they behave differently from each other. One interesting property of chiral substances is that their solutions may rotate the plane of polarized light, as explained in Section 23.4.

Chirality is common in organic compounds. It is often not observed, however, because when a chiral substance is synthesized in a typical reaction, the two enantiomers are formed in precisely the same quantity. The resulting mixture is called a *racemic mixture*, and it does not rotate the plane of polarized light because the two forms rotate the light to equal extents in opposite directions. ∞ (Section 23.4)

Many drugs are chiral compounds. When a drug is administered as a racemic mixture, often only one enantiomer has beneficial results. The other is often inert, or nearly so, or may even have a harmful effect. For example, the drug (*R*)-albuterol (▶ Figure 24.15) is a bronchodilator used to relieve the symptoms of asthma. The enantiomer (*S*)-albuterol is not only ineffective as a bronchodilator but also actually counters the effects of (*R*)-albuterol. As another example, the nonsteroidal analgesic ibuprofen is a chiral molecule usually sold as the racemic mixture. However, a preparation consisting of just the more active enantiomer, (*S*)-ibuprofen (Figure 24.16), relieves pain and reduces inflammation more rapidly than the racemic mixture. For this reason, the chiral version of the drug may in time come to replace the racemic one.

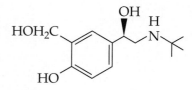

▲ Figure 24.15 (*R*)-Albuterol. This compound, which acts as a bronchodilator in patients with asthma, is one member of an enantiomer pair. The other member, (*S*)-albuterol, has the OH group pointing down, and does not have the same physiological effect.

Give It Some Thought

What are the requirements on the four groups attached to a carbon atom in order that it be a chiral center?

▲ **Figure 24.16 (S)-Ibuprofen.** For relieving pain and reducing inflammation, the ability of this enantiomer far outweighs that of the (R) isomer. In the (R) isomer, the positions of the H and CH_3 group on the far-right carbon are switched.

24.6 | Introduction to Biochemistry

The functional groups discussed in Section 24.4 generate a vast array of molecules with very specific chemical reactivities. Nowhere is this specificity more apparent than in *biochemistry*—the chemistry of living organisms.

Before we discuss specific biochemical molecules, we can make some general observations. Many biologically important molecules are quite large, because organisms build biomolecules from smaller, simpler substances readily available in the biosphere. The synthesis of large molecules requires energy because most of the reactions are endothermic. The ultimate source of this energy is the Sun. Animals have essentially no capacity for using solar energy directly, and so depend on plant photosynthesis to supply the bulk of their energy needs. ∞ (Section 23.3)

In addition to requiring large amounts of energy, living organisms are highly organized. In thermodynamic terms, this high degree of organization means that the entropy of living systems is much lower than that of the raw materials from which the systems formed. Thus, living systems must continuously work against the spontaneous tendency toward increased entropy. ∞ (Section 19.3)

In the "Chemistry and Life" essays that appear throughout this text, we have introduced some important biochemical applications of fundamental chemical ideas. The remainder of this chapter will serve as only a brief introduction to other aspects of biochemistry. Nevertheless, you will see some patterns emerging. Hydrogen bonding, ∞ (Section 11.2), for example, is critical to the function of many biochemical systems, and the geometry of molecules ∞ (Section 9.1) can govern their biological importance and activity. Many of the large molecules in living systems are polymers ∞ (Section 12.8) of much smaller molecules. These **biopolymers** can be classified into three broad categories: proteins, polysaccharides (carbohydrates), and nucleic acids. Lipids are another common class of molecules in living systems, but they are usually large molecules, not biopolymers.

24.7 | Proteins

Proteins are macromolecules present in all living cells. About 50% of your body's dry mass is protein. Some proteins are structural components in animal tissues; they are a key part of skin, nails, cartilage, and muscles. Other proteins catalyze reactions, transport oxygen, serve as hormones to regulate specific body processes, and perform other tasks. Whatever their function, all proteins are chemically similar, being composed of smaller molecules called *amino acids*.

Amino Acids

An **amino acid** is a molecule containing an amine group, $-NH_2$, and a carboxylic acid group, $-COOH$. The building blocks of all proteins are *α-amino acids*, where the α (alpha) indicates that the amino group is located on the carbon atom immediately adjacent to the carboxylic acid group. Thus, there is always one carbon atom between the amino group and the carboxylic acid group.

The general formula for an α-amino acid is represented by

One of about 20 different groups

$$H_2N-\overset{\displaystyle R}{\underset{\displaystyle H}{C}}-COOH \quad \text{or} \quad {}^{+}H_3N-\overset{\displaystyle R}{\underset{\displaystyle H}{C}}-COO^{-}$$

α carbon

The doubly ionized form, called a *zwitterion*, usually predominates at near-neutral pH values. This form is a result of the transfer of a proton from the carboxylic acid group to the amine group. ∞ (Section 16.10: "The Amphiprotic Behavior of Amino Acids")

Amino acids differ from one another in the nature of their R groups. Twenty-two amino acids have been identified in nature, and ▶ **Figure 24.17** shows the 20 of these 22 that are found in humans. Our bodies can synthesize 11 of these 20 amino acids in

GO FIGURE

Which group of amino acids has a net positive charge at pH 7?

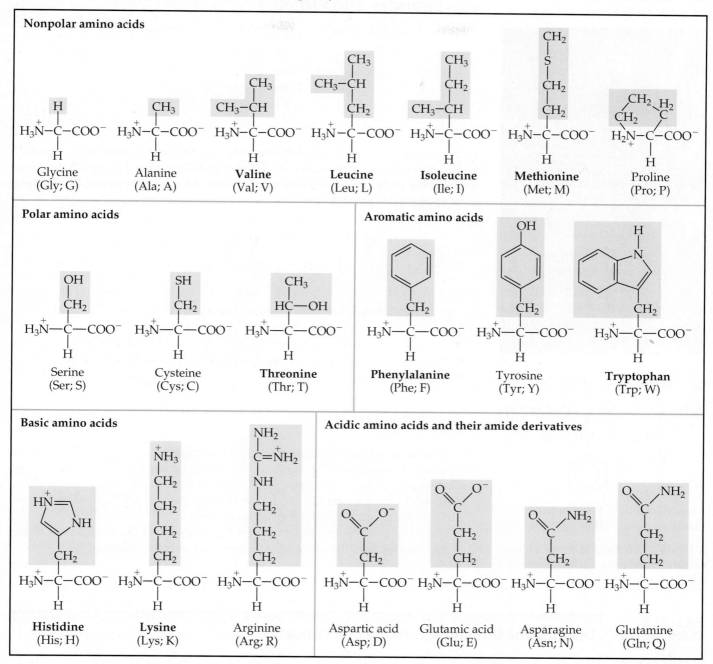

▲ **Figure 24.17 The 20 amino acids found in the human body.** The blue shading shows the different R groups for each amino acid. The acids are shown in the zwitterionic form in which they exist in water at near-neutral pH values. The amino acid names shown in bold are the nine essential ones.

sufficient amounts for our needs. The other 9 must be ingested and are called *essential amino acids* because they are necessary components of our diet.

The α-carbon atom of the amino acids, which is the carbon between the amino and carboxylate groups, has four different groups attached to it. The amino acids are thus chiral (except for glycine, which has two hydrogens attached to the central carbon). For historical reasons, the two enantiomeric forms of amino acids are often distinguished by the labels D (from the Latin *dexter*, "right") and L (from the Latin *laevus*, "left"). Nearly all the chiral amino acids found in living organisms have the

GO FIGURE

How many chiral carbons are there in one molecule of aspartame?

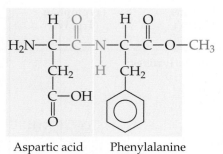

Aspartic acid   Phenylalanine
(Asp)   (Phe)

▲ **Figure 24.18 Sweet stuff.** The artificial sweetener aspartame is the methyl ester of a dipeptide.

L configuration at the chiral center. The principal exceptions are the proteins that make up the cell walls of bacteria, which can contain considerable quantities of the D isomers.

Polypeptides and Proteins

Amino acids are linked together into proteins by amide groups (Table 24.6):

$$R-\overset{\overset{\displaystyle O}{\|}}{C}-\underset{\underset{\displaystyle H}{|}}{N}-R$$

[24.23]

Each amide group is called a **peptide bond** when it is formed by amino acids. A peptide bond is formed by a condensation reaction between the carboxyl group of one amino acid and the amino group of another amino acid. Alanine and glycine, for example, form the dipeptide glycylalanine:

Glycine (Gly; G)   Alanine (Ala; A)

Glycylalanine (Gly–Ala; GA)

The amino acid that furnishes the carboxyl group for peptide-bond formation is named first, with a -*yl* ending; then the amino acid furnishing the amino group is named. Using the abbreviations shown in Figure 24.18, glycylalanine can be abbreviated as either Gly-Ala or GA. In this notation, it is understood that the unreacted amino group is on the left and the unreacted carboxyl group on the right.

The artificial sweetener *aspartame* (◄ Figure 24.18) is the methyl ester of the dipeptide formed from the amino acids aspartic acid and phenylalanine.

SAMPLE EXERCISE 24.7 | Drawing the Structural Formula of a Tripeptide

Draw the structural formula for alanylglycylserine.

SOLUTION

Analyze We are given the name of a substance with peptide bonds and asked to write its structural formula.

Plan The name of this substance suggests that three amino acids—alanine, glycine, and serine—have been linked together, forming a *tripeptide*. Note that the ending -*yl* has been added to each amino acid except for the last one, serine. By convention, the sequence of amino acids in peptides and proteins is written from the nitrogen end to the carbon end: The first-named amino acid (alanine, in this case) has a free amino group and the last-named one (serine) has a free carboxyl group.

Solve We first combine the carboxyl group of alanine with the amino group of glycine to form a peptide bond and then the carboxyl group of glycine with the amino group of serine to form another peptide bond:

Amino group ───────────→ Carboxyl group

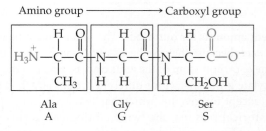

Ala   Gly   Ser
A   G   S

We can abbreviate this tripeptide as either Ala-Gly-Ser or AGS.

Practice Exercise 1

How many nitrogen atoms are in the tripeptide Arg-Asp-Gly?
(a) 3, (b) 4, (c) 5, (d) 6, (e) 7.

Practice Exercise 2

Name the dipeptide and give the two ways of writing its abbreviation.

$$\underset{\text{HOCH}_2}{\overset{\text{H}\quad\text{O}}{\text{H}_3\overset{+}{\text{N}}-\text{C}-\text{C}}}-\underset{\text{H}}{\text{N}}-\underset{\underset{\text{COOH}}{\text{CH}_2}}{\overset{\text{H}\quad\text{O}}{\text{C}-\text{C}}}-\text{O}^-$$

Polypeptides are formed when a large number of amino acids (>30) are linked together by peptide bonds. Proteins are linear (that is, unbranched) polypeptide molecules with molecular weights ranging from about 6000 to over 50 million amu. Because up to 22 different amino acids are linked together in proteins and because proteins consist of hundreds of amino acids, the number of possible arrangements of amino acids within proteins is virtually limitless.

Protein Structure

The sequence of amino acids along a protein chain is called its **primary structure** and gives the protein its unique identity. A change in even one amino acid can alter the biochemical characteristics of the protein. For example, sickle-cell anemia is a genetic disorder resulting from a single replacement in a protein chain in hemoglobin. The chain that is affected contains 146 amino acids. The substitution of an amino acid with a hydrocarbon side chain for one that has an acidic functional group in the side chain alters the solubility properties of the hemoglobin, and normal blood flow is impeded. (Section 13.6, "Sickle-Cell Anemia")

Proteins in living organisms are not simply long, flexible chains with random shapes. Rather, the chains self-assemble into structures based on the intermolecular forces we learned about in Chapter 11. This self-assembling leads to a protein's **secondary structure,** which refers to how segments of the protein chain are oriented in a regular pattern, as seen in **Figure 24.19.**

One of the most important and common secondary structure arrangements is the **α**(alpha)-**helix**. As the α-helix of Figure 24.19 shows, the helix is held in position by hydrogen bonds between amide H atoms and carbonyl O atoms. The pitch of the helix and its diameter must be such that (1) no bond angles are strained and (2) the N—H and C=O functional groups on adjacent turns are in proper position for hydrogen bonding. An arrangement of this kind is possible for some amino acids along the chain but not for others. Large protein molecules may contain segments of the chain that have the α-helical arrangement interspersed with sections in which the chain is in a random coil.

The other common secondary structure of proteins is the **β** (beta) **sheet.** Beta sheets are made of two or more strands of peptides that hydrogen-bond from an amide H in one strand to a carbonyl O in the other strand (Figure 24.19).

⚠ Give It Some Thought

If you heat a protein to break the intramolecular hydrogen bonds, will you maintain the α-helical or β-sheet structure?

Proteins are not active biologically unless they are in a particular shape in solution. The process by which the protein adopts its biologically active shape is called **folding**. The shape of a protein in its folded form—determined by all the bends, kinks, and sections of rodlike α-helical, β-sheet, or flexible coil components—is called the **tertiary structure**.

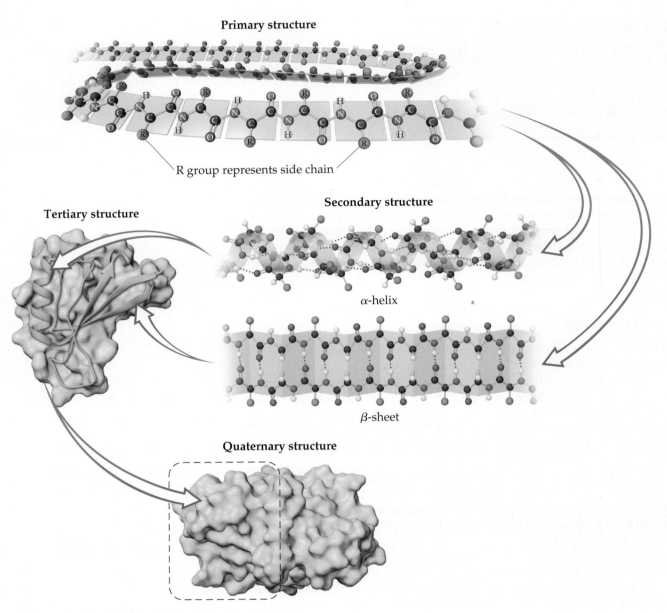

Primary structure

R group represents side chain

Secondary structure

α-helix

β-sheet

Tertiary structure

Quaternary structure

▲ Figure 24.19 **The four levels of structure of proteins.**

Globular proteins fold into a compact, roughly spherical shape. Globular proteins are generally soluble in water and are mobile within cells. They have nonstructural functions, such as combating the invasion of foreign objects, transporting and storing oxygen (hemoglobin and myoglobin), and acting as catalysts. The *fibrous proteins* form a second class of proteins. In these substances the long coils align more or less in parallel to form long, water-insoluble fibers. Fibrous proteins provide structural integrity and strength to many kinds of tissue and are the main components of muscle, tendons, and hair. The largest known proteins, in excess of 27,000 amino acids long, are muscle proteins.

The tertiary structure of a protein is maintained by many different interactions. Certain foldings of the protein chain lead to lower energy (more stable) arrangements than do other folding patterns. For example, a globular protein dissolved in aqueous solution folds in such a way that the nonpolar hydrocarbon portions are tucked within the molecule, away from the polar water molecules. Most of the more polar acidic and basic side chains, however, project into the solution, where they can interact with water molecules through ion–dipole, dipole–dipole, or hydrogen-bonding interactions.

Some proteins are assemblies of more than one polypeptide chain. Each chain has its own tertiary structure, and two or more of these tertiary subunits may aggregate into a larger functional macromolecule. The way the tertiary subunits are arranged is called the **quaternary structure** of the protein (Figure 24.19). For example, hemoglobin, the oxygen-carrying protein of red blood cells, consists of four tertiary subunits. Each subunit contains a component called a heme with an iron atom that binds oxygen as depicted in Figure 23.15. The quaternary structure is maintained by the same types of interactions that maintain the tertiary structure.

One of the most fascinating current hypotheses in biochemistry is that misfolded proteins can cause infectious disease. These infectious misfolded proteins are called *prions*. The best example of a prion is the one thought to be responsible for mad cow disease, which can be transmitted to humans.

24.8 | Carbohydrates

Carbohydrates are an important class of naturally occurring substances found in both plant and animal matter. The name **carbohydrate** ("hydrate of carbon") comes from the empirical formulas for most substances in this class, which can be written as $C_x(H_2O)_y$. For example, **glucose**, the most abundant carbohydrate, has the molecular formula $C_6H_{12}O_6$, or $C_6(H_2O)_6$. Carbohydrates are not really hydrates of carbon; rather, they are polyhydroxy aldehydes and ketones. Glucose, for example, is a six-carbon aldehyde sugar, whereas *fructose*, the sugar that occurs widely in fruit, is a six-carbon ketone sugar (▶ Figure 24.20).

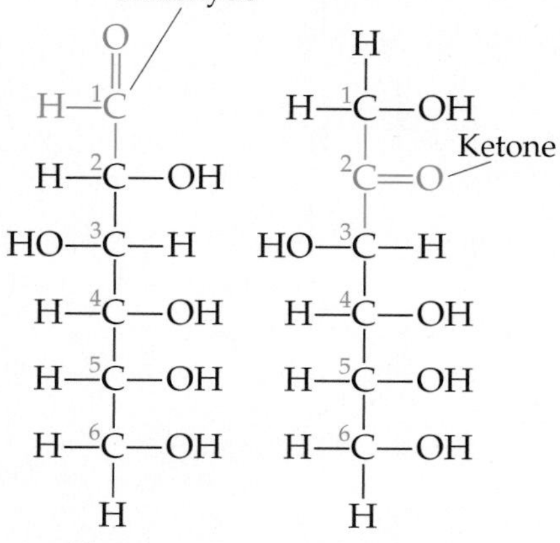

The glucose molecule, having both alcohol and aldehyde functional groups and a reasonably long and flexible backbone, can form a six-member-ring structure, as shown in ▼ Figure 24.21. In fact, in an aqueous solution only a small percentage of the glucose molecules are in the open-chain form. Although the ring is often drawn as if it were planar, the molecules are actually nonplanar because of the tetrahedral bond angles around the C and O atoms of the ring.

Figure 24.21 shows that the ring structure of glucose can have two relative orientations. In the α form the OH group on C1 and the CH_2OH group on C5 point in opposite directions, and in the β form they point in the same direction. Although the difference between the α and β forms might seem small, it has enormous biological consequences, including the vast difference in properties between starch and cellulose.

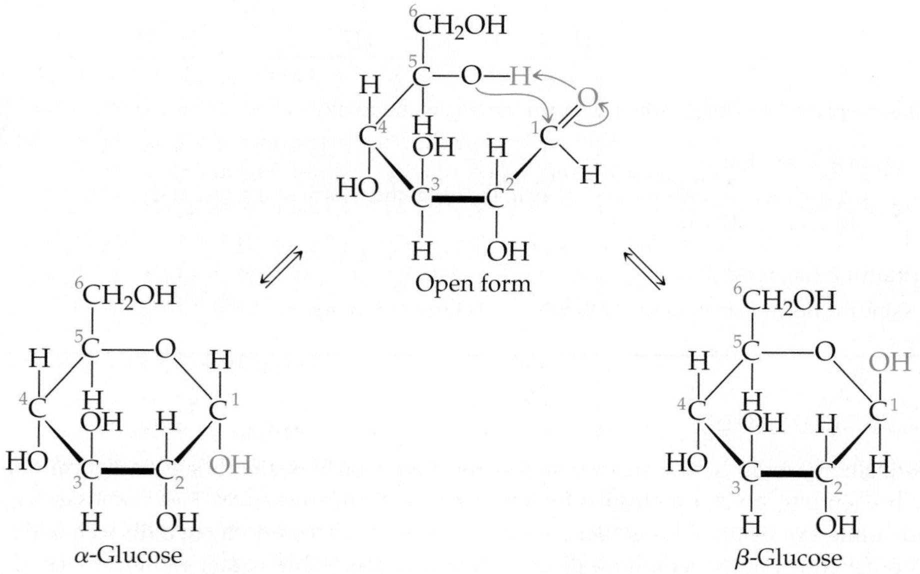

▲ Figure 24.20 Linear structure of the carbohydrates glucose and fructose.

▲ Figure 24.21 Cyclic glucose has an α form and a β form.

Fructose can cyclize to form either five- or six-member rings. The five-member ring forms when the C5 OH group reacts with the C2 carbonyl group:

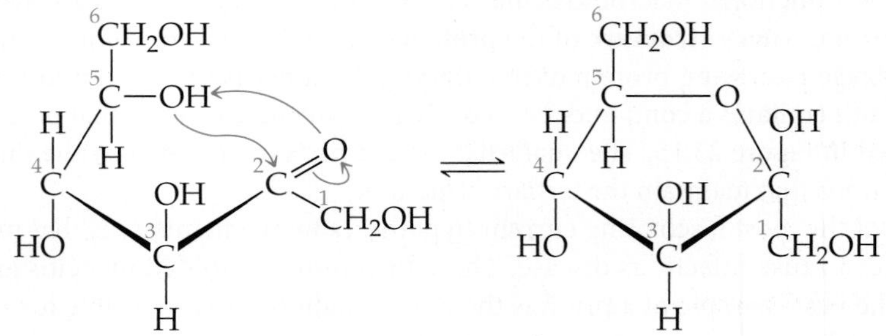

The six-member ring results from the reaction between the C6 OH group and the C2 carbonyl group.

SAMPLE
EXERCISE 24.8 **Identifying Functional Groups and Chiral Centers in Carbohydrates**

How many chiral carbon atoms are there in the open-chain form of glucose (Figure 24.20)?

SOLUTION

Analyze We are given the structure of glucose and asked to determine the number of chiral carbons in the molecule.

Plan A chiral carbon has four different groups attached (Section 24.5). We need to identify those carbon atoms in glucose.

Solve Carbons 2, 3, 4, and 5 each have four different groups attached to them:

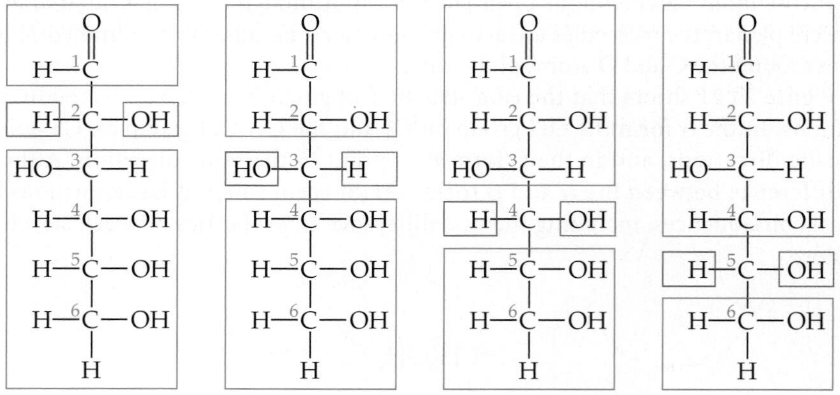

Thus, there are four chiral carbon atoms in the glucose molecule.

Practice Exercise 1
How many chiral carbon atoms are there in the open-chain form of fructose (Figure 24.20)?
(a) 0, (b) 1, (c) 2, (d) 3, (e) 4

Practice Exercise 2
Name the functional groups present in the beta form of glucose.

Disaccharides

Both glucose and fructose are examples of **monosaccharides**, simple sugars that cannot be broken into smaller molecules by hydrolysis with aqueous acids. Two monosaccharide units can be linked together by a condensation reaction to form a **disaccharide**. The structures of two common disaccharides, *sucrose* (table sugar) and *lactose* (milk sugar), are shown in ▶ Figure 24.22.

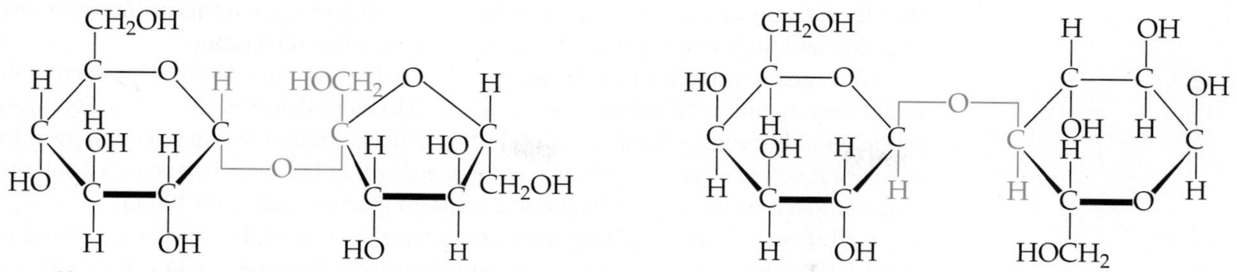

▲ Figure 24.22 **Two disaccharides.**

The word *sugar* makes us think of sweetness. All sugars are sweet, but they differ in the degree of sweetness we perceive when we taste them. Sucrose is about six times sweeter than lactose, slightly sweeter than glucose, but only about half as sweet as fructose. Disaccharides can be reacted with water (hydrolyzed) in the presence of an acid catalyst to form monosaccharides. When sucrose is hydrolyzed, the mixture of glucose and fructose that forms, called *invert sugar,** is sweeter to the taste than the original sucrose. The sweet syrup present in canned fruits and candies is largely invert sugar formed from hydrolysis of added sucrose.

Polysaccharides

Polysaccharides are made up of many monosaccharide units joined together. The most important polysaccharides are starch, glycogen, and cellulose, all three of which are formed from repeating glucose units.

Starch is not a pure substance. The term refers to a group of polysaccharides found in plants. Starches serve as a major method of food storage in plant seeds and tubers. Corn, potatoes, wheat, and rice all contain substantial amounts of starch. These plant products serve as major sources of needed food energy for humans. Enzymes in the digestive system catalyze the hydrolysis of starch to glucose.

Some starch molecules are unbranched chains, whereas others are branched. ▼ Figure 24.23(a) illustrates an unbranched starch structure. Notice, in particular, that

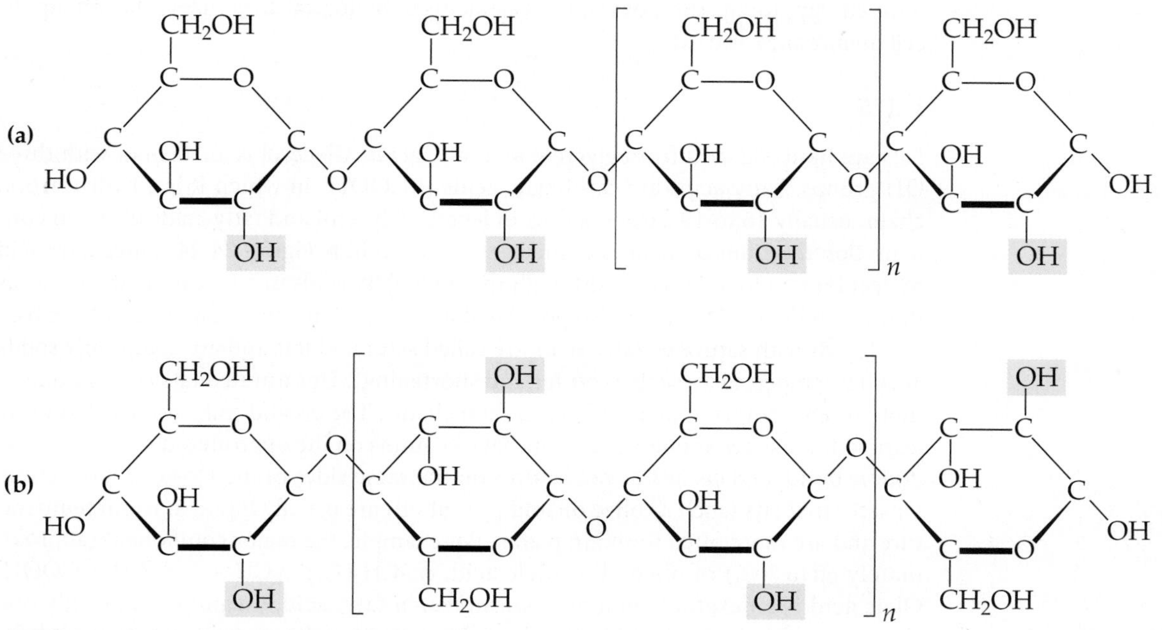

▲ Figure 24.23 **Structures of (a) starch and (b) cellulose.**

---

*The term *invert sugar* comes from the fact that rotation of the plane of polarized light by the glucose–fructose mixture is in the opposite direction, or inverted, from that of the sucrose solution.

the glucose units are in the α form with the bridging oxygen atoms pointing in one direction and the CH_2OH groups pointing in the opposite direction.

Glycogen is a starch-like substance synthesized in the animal body. Glycogen molecules vary in molecular weight from about 5000 to more than 5 million amu. Glycogen acts as a kind of energy bank in the body. It is concentrated in the muscles and liver. In muscles, it serves as an immediate source of energy; in the liver, it serves as a storage place for glucose and helps to maintain a constant glucose level in the blood.

Cellulose [Figure 24.23(b)] forms the major structural unit of plants. Wood is about 50% cellulose; cotton fibers are almost entirely cellulose. Cellulose consists of an unbranched chain of glucose units, with molecular weights averaging more than 500,000 amu. At first glance, this structure looks very similar to that of starch. In cellulose, however, the glucose units are in the β form with each bridging oxygen atom pointing in the same direction as the CH_2OH group in the ring to its left.

Because the individual glucose units have different relationships to one another in starch and cellulose, enzymes that readily hydrolyze starches do not hydrolyze cellulose. Thus, you might eat a pound of cellulose and receive no caloric value from it even though the heat of combustion per unit mass is essentially the same for both cellulose and starch. A pound of starch, in contrast, would represent a substantial caloric intake. The difference is that the starch is hydrolyzed to glucose, which is eventually oxidized with the release of energy. However, enzymes in the body do not readily hydrolyze cellulose, so it passes through the digestive system relatively unchanged. Many bacteria contain enzymes, called cellulases, that hydrolyze cellulose. These bacteria are present in the digestive systems of grazing animals, such as cattle, that use cellulose for food.

 Give It Some Thought

Which type of linkage, α or β, would you expect to join the sugar molecules of glycogen?

24.9 | Lipids

Lipids are a diverse class of nonpolar biological molecules used by organisms for long-term energy storage (fats, oils) and as elements of biological structures (phospholipids, cell membranes, waxes).

Fats

Fats are lipids derived from glyercol and fatty acids. Glycerol is an alcohol with three OH groups. Fatty acids are carboxylic acids (RCOOH) in which R is a hydrocarbon chain, usually 16 to 19 carbon atoms in length. Glycerol and fatty acids undergo condensation reactions to form ester linkages as shown in ▶ **Figure 24.24**. Three fatty acid molecules join to a glycerol. Although the three fatty acids in a fat can be the same, as they are in Figure 24.24, it is also possible that a fat contains three different fatty acids.

Lipids with saturated fatty acids are called saturated fats and are commonly solids at room temperature (such as butter and shortening). Unsaturated fats contain one or more double bonds in their carbon–carbon chains. The cis and trans nomenclature we learned for alkenes applies: Trans fats have H atoms on the opposite sides of the $C{=}C$ double bond, and cis fats have H atoms on the same sides of the $C{=}C$ double bond. Unsaturated fats (such as olive oil and peanut oil) are usually liquid at room temperature and are more often found in plants. For example, the major component (approximately 60 to 80%) of olive oil is oleic acid, *cis*-$CH_3(CH_2)_7CH{=}CH(CH_2)_7COOH$. Oleic acid is an example of a *monounsaturated* fatty acid, meaning it has only one carbon–carbon double bond in the chain. In contrast, *polyunsaturated* fatty acids have more than one carbon–carbon double bond in the chain.

For humans, trans fats are not nutritionally required, which is why some governments are moving to ban them in foods. How, then, do trans fats end up in our food? The process that converts unsaturated fats (such as oils) into saturated fats (such as

GO FIGURE

What structural features of a fat molecule cause it to be insoluble in water?

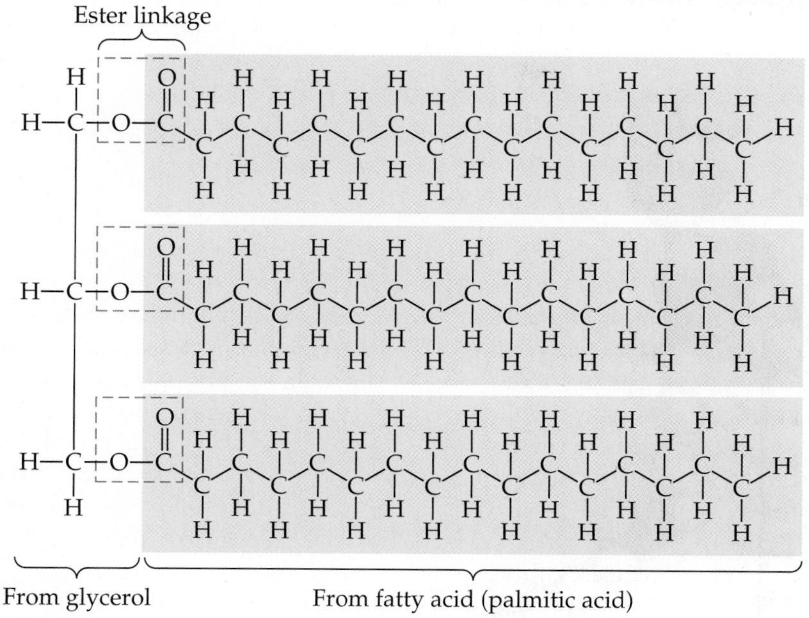

Ester linkage

From glycerol   From fatty acid (palmitic acid)

▲ Figure 24.24 Structure of a fat.

shortening) is hydrogenation. ∞ (Section 24.3) The by-products of this hydrogenation process include trans fats.

Some of the fatty acids essential for human health must be available in our diets because our metabolism cannot synthesize them. These essential fatty acids are ones that have the carbon–carbon double bonds either three carbons or six carbons away from the $-CH_3$ end of the chain. These are called omega-3 and omega-6 fatty acids, where *omega* refers to the last carbon in the chain (the carboxylic acid carbon is considered the first, or alpha, one).

Phospholipids

Phospholipids are similar in chemical structure to fats but have only two fatty acids attached to a glycerol. The third alcohol group of glycerol is joined to a phosphate group (Figure 24.25). The phosphate group can also be attached to a small charged or polar group, such as choline, as shown in the figure. The diversity in phospholipids is based on differences in their fatty acids and in the groups attached to the phosphate group.

In water, phospholipids cluster together with their charged polar heads facing the water and their nonpolar tails facing inward. The phospholipids thus form a bilayer that is a key component of cell membranes (Figure 24.26).

24.10 | Nucleic Acids

Nucleic acids are a class of biopolymers that are the chemical carriers of an organism's genetic information. **Deoxyribonucleic acids (DNAs)** are huge molecules whose molecular weights may range from 6 to 16 million amu. **Ribonucleic acids (RNAs)** are smaller molecules, with molecular weights in the range of 20,000 to 40,000 amu. Whereas DNA is found primarily in the nucleus of the cell, RNA is found mostly outside the nucleus in the *cytoplasm*, the nonnuclear material enclosed by the cell membrane. DNA stores the genetic information of the cell and specifies which proteins the cell can synthesize. RNA carries the information stored by DNA out of the cell nucleus into the cytoplasm, where the information is used in protein synthesis.

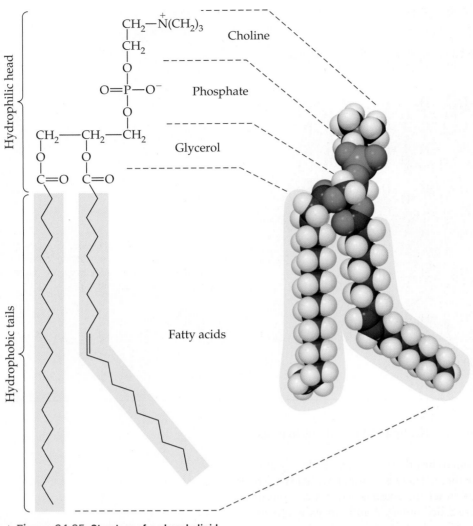

▲ Figure 24.25 **Structure of a phospholipid.**

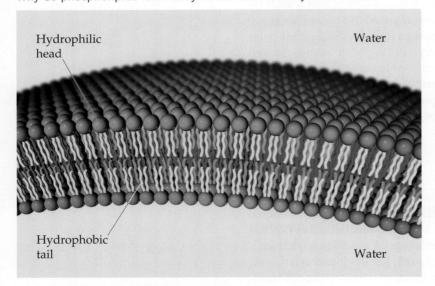

▲ **GO FIGURE**

Why do phospholipids form bilayers but not monolayers in water?

▲ Figure 24.26 **The cell membrane.** Living cells are encased in membranes typically made of phospholipid bilayers. The bilayer structure is stabilized by the favorable interactions of the hydrophobic tails of the phospholipids, which point away from both the water inside the cell and the water outside the cell, while the charged head groups face the two water environments.

The monomers of nucleic acids, called **nucleotides**, are formed from a five-carbon sugar, a nitrogen-containing organic base, and a phosphate group. An example is shown in ▶ Figure 24.27.

The five-carbon sugar in RNA is *ribose*, and that in DNA is *deoxyribose*:

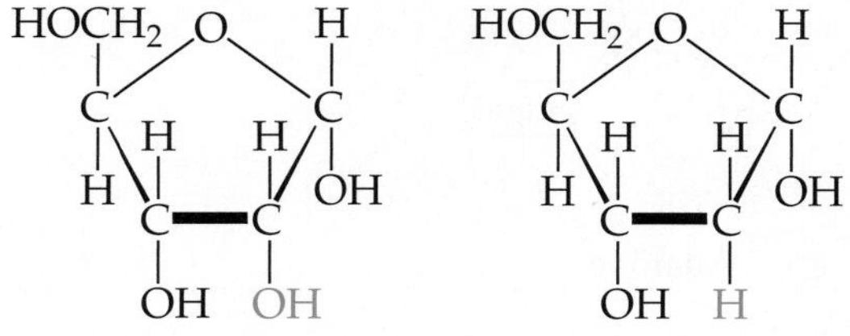

Ribose   Deoxyribose

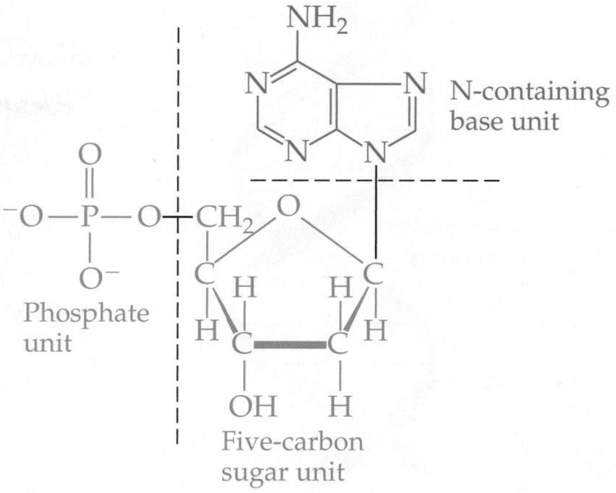

▲ **Figure 24.27 A nucleotide.** Structure of deoxyadenylic acid, the nucleotide formed from phosphoric acid, the sugar deoxyribose, and the organic base adenine.

Deoxyribose differs from ribose only in having one fewer oxygen atom at carbon 2.

There are five nitrogen-containing bases in nucleic acids:

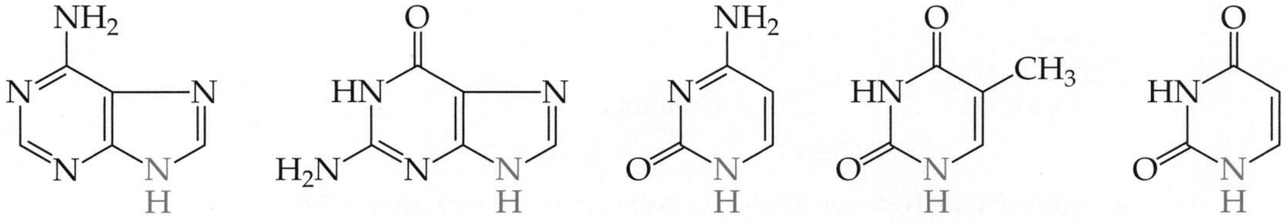

Adenine (A)   Guanine (G)   Cytosine (C)   Thymine (T)   Uracil (U)
DNA   DNA   DNA   DNA   RNA
RNA   RNA   RNA

The first three bases shown here are found in both DNA and RNA. Thymine occurs only in DNA, and uracil occurs only in RNA. In either nucleic acid, each base is attached to a five-carbon sugar through a bond to the nitrogen atom shown in color.

The nucleic acids RNA and DNA are *polynucleotides* formed by condensation reactions between a phosphoric acid OH group on one nucleotide and a sugar OH group on another nucleotide. Thus, the polynucleotide strand has a backbone consisting of alternating sugar and phosphate groups with the bases extending off the chain as side groups (▶ Figure 24.28).

The DNA strands wind together in a **double helix** (Figure 24.29). The two strands are held together by attractions between bases (represented by T, A, C, and G). These attractions involve dispersion forces, dipole–dipole forces, and hydrogen bonds. ∞ (Section 11.2) As shown in **Figure 24.30**, the structures of thymine and adenine make them perfect partners for hydrogen bonding. Likewise, cytosine and guanine form ideal hydrogen-bonding partners. We say that thymine and adenine are *complementary* to each other and cytosine and guanine are *complementary* to each other. In the double-helix structure, therefore, each thymine on one strand is opposite an adenine on the other strand, and each cytosine is opposite a guanine. The double-helix structure with complementary bases on the two strands is the key to understanding how DNA functions.

The two strands of DNA unwind during cell division, and new complementary strands are constructed on the unraveling strands (**Figure 24.31**). This process results in two identical double-helix DNA structures, each containing one strand from the original structure and one new strand. This replication allows genetic information to be transmitted when cells divide.

The structure of DNA is also the key to understanding protein synthesis, the means by which viruses infect cells, and many other problems of central importance to modern biology. These themes are beyond the scope of this book. If you take courses in the life sciences, however, you will learn a good deal about such matters.

▲ **GO FIGURE**

Is DNA positively charged, negatively charged, or neutral in aqueous solution at pH 7?

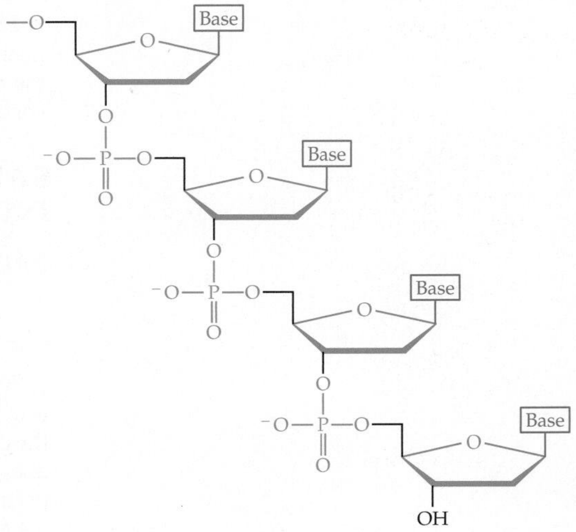

▲ **Figure 24.28 A polynucleotide.** Because the sugar in each nucleotide is deoxyribose, this polynucleotide is DNA.

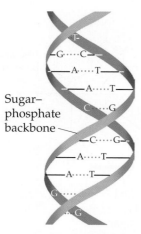

▲ Figure 24.29 **The DNA double helix,** showing the sugar–phosphate backbone as a pair of ribbons and dotted lines to indicate hydrogen bonding between the complementary bases.

▲ **GO FIGURE**

Which pair of complementary bases, AT or GC, would you expect to bind more strongly?

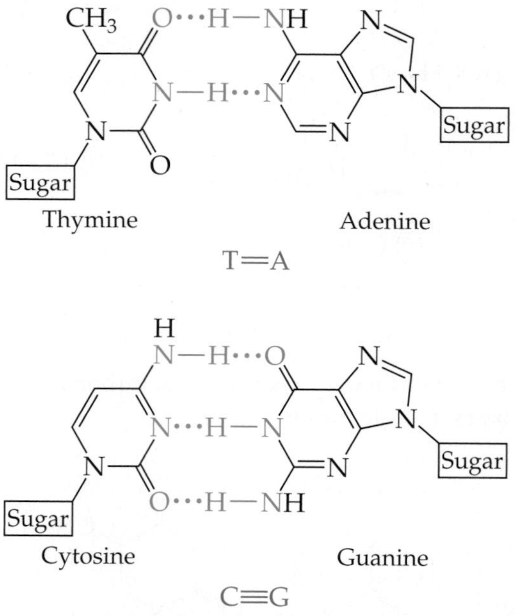

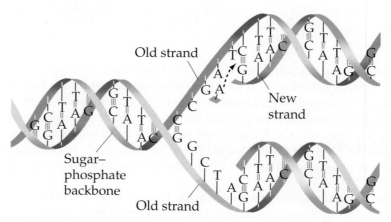

Thymine   Adenine

$$T=A$$

Cytosine   Guanine

$$C\equiv G$$

▲ Figure 24.30 **Hydrogen bonding between complementary bases in DNA.**

▲ Figure 24.31 **DNA replication.** The original DNA double helix partially unwinds, and new nucleotides line up on each strand in complementary fashion. Hydrogen bonds help align the new nucleotides with the original DNA chain. When the new nucleotides are joined by condensation reactions, two identical double-helix DNA molecules result.

SAMPLE INTEGRATIVE EXERCISE | Putting Concepts Together

Pyruvic acid,

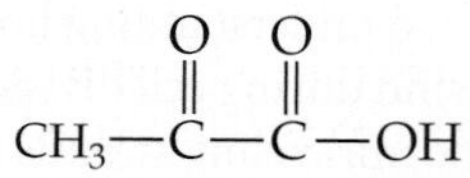

$$CH_3-\overset{\overset{\displaystyle O}{\|}}{C}-\overset{\overset{\displaystyle O}{\|}}{C}-OH$$

is formed in the body from carbohydrate metabolism. In muscles, it is reduced to lactic acid in the course of exertion. The acid-dissociation constant for pyruvic acid is 3.2×10^{-3}. **(a)** Why does pyruvic acid have a larger acid-dissociation constant than acetic acid? **(b)** Would you expect pyruvic acid to exist primarily as the neutral acid or as dissociated ions in muscle tissue, assuming a pH of 7.4 and an initial acid concentration of $2 \times 10^{-4}\ M$? **(c)** What would you predict for the solubility properties of pyruvic acid? Explain. **(d)** What is the hybridization of each carbon atom in pyruvic acid? **(e)** Assuming H atoms as the reducing agent, write a balanced chemical equation for the reduction of pyruvic acid to lactic acid (Figure 24.13). (Although H atoms do not exist as such in biochemical systems, biochemical reducing agents deliver hydrogen for such reductions.)

SOLUTION

(a) The acid-dissociation constant for pyruvic acid should be somewhat greater than that of acetic acid because the carbonyl function on the α-carbon atom of pyruvic acid exerts an electron-withdrawing effect on the carboxylic acid group. In the C—O—H bond system, the electrons are shifted from H, facilitating loss of the H as a proton. ∞ (Section 16.10)

(b) To determine the extent of ionization, we first set up the ionization equilibrium and equilibrium-constant expression. Using HPv as the symbol for the acid, we have

$$HPv \rightleftharpoons H^+ + Pv^-$$

$$K_a = \frac{[H^+][Pv^-]}{[HPv]} = 3.2 \times 10^{-3}$$

Let $[Pv^-] = x$. Then the concentration of undissociated acid is $2 \times 10^{-4} - x$. The concentration of $[H^+]$ is fixed at 4.0×10^{-8} (the antilog of the pH value). Substituting, we obtain

$$3.2 \times 10^{-3} = \frac{(4.0 \times 10^{-8})(x)}{(2 \times 10^{-4} - x)}$$

Solving for x, we obtain

$$x = [Pv^-] = 2 \times 10^{-4}\, M$$

This is the initial concentration of acid, which means that essentially all the acid has dissociated. We might have expected this result because the acid is quite dilute and the acid-dissociation constant is fairly high.

(c) Pyruvic acid should be quite soluble in water because it has polar functional groups and a small hydrocarbon component. We would predict it would be soluble in polar organic solvents, especially ones that contain oxygen. In fact, pyruvic acid dissolves in water, ethanol, and diethyl ether.

(d) The methyl group carbon has sp^3 hybridization. The carbon of the carbonyl group has sp^2 hybridization because of the double bond to oxygen. Similarly, the carboxylic acid carbon is sp^2 hybridized.

(e) The balanced chemical equation for this reaction is

$$\underset{\displaystyle CH_3\overset{\displaystyle \overset{O}{\|}}{C}COOH}{} + 2\,(H) \longrightarrow \underset{\displaystyle CH_3\overset{\displaystyle \overset{OH}{|}}{\underset{\displaystyle H}{C}}COOH}{}$$

Essentially, the ketonic functional group has been reduced to an alcohol.

Strategies in Chemistry

What Now?

If you are reading this box, you have made it to the end of our text. We congratulate you on the tenacity and dedication that you have exhibited to make it this far!

As an epilogue, we offer the ultimate study strategy in the form of a question: What do you plan to do with the knowledge of chemistry that you have gained thus far in your studies? Many of you will enroll in additional courses in chemistry as part of your required curriculum. For others, this will be the last formal course in chemistry that you will take. Regardless of the career path you plan to take—whether it is chemistry, one of the biomedical fields, engineering, the liberal arts, or another field—we hope that this text has increased your appreciation of the chemistry in the world around you. If you pay attention, you will be aware of encounters with chemistry on a daily basis, from food and pharmaceutical labels to gasoline pumps, sports equipment to news reports.

We have also tried to give you a sense that chemistry is a dynamic, continuously changing science. Research chemists synthesize new compounds, develop new reactions, uncover chemical properties that were previously unknown, find new applications for known compounds, and refine theories. The understanding of biological systems in terms of the underlying chemistry has become increasingly important as new levels of complexity are uncovered. Solving the global challenges of sustainable energy and clean water require the work of many chemists. We encourage you to participate in the fascinating world of chemical research by taking part in an undergraduate research program. Given all the answers that chemists seem to have, you may be surprised at the large number of questions that they still find to ask.

Finally, we hope you have enjoyed using this textbook. We certainly enjoyed putting so many of our thoughts about chemistry on paper. We truly believe it to be the central science, one that benefits all who learn about it and from it.

Chapter Summary and Key Terms

GENERAL CHARACTERISTICS OF ORGANIC COMPOUNDS (INTRO-DUCTION AND SECTION 24.1) This chapter introduces **organic chemistry**, which is the study of carbon compounds (typically compounds containing carbon–carbon bonds), and **biochemistry**, which is the study of the chemistry of living organisms. We have encountered many aspects of organic chemistry in earlier chapters. Carbon forms four bonds in its stable compounds. The C — C single bonds and the C — H bonds tend to have low reactivity. Those bonds that have a high electron density (such as multiple bonds or bonds with an atom of high electronegativity) tend to be the sites of reactivity in an organic compound. These sites of reactivity are called **functional groups**.

INTRODUCTION OF HYDROCARBONS (SECTION 24.2) The simplest types of organic compounds are hydrocarbons, those composed of only carbon and hydrogen. There are four major kinds of hydrocarbons: alkanes, alkenes, alkynes, and aromatic hydrocarbons. **Alkanes** are composed of only C — H and C — C single bonds. **Alkenes** contain one or more carbon–carbon double bonds. **Alkynes** contain one or more carbon–carbon triple bonds. **Aromatic hydrocarbons** contain cyclic arrangements of carbon atoms bonded through both σ and delocalized π bonds. Alkanes are saturated hydrocarbons; the others are unsaturated.

Alkanes may form straight-chain, branched-chain, and cyclic arrangements. Isomers are substances that possess the same molecular formula but differ in the arrangements of atoms. In **structural isomers**, the bonding arrangements of the atoms differ. Different isomers are given different systematic names. The naming of hydrocarbons is based on the longest continuous chain of carbon atoms in the structure. The locations of **alkyl groups**, which branch off the chain, are specified by numbering along the carbon chain.

Alkanes with ring structures are called **cycloalkanes**. Alkanes are relatively unreactive. They do, however, undergo combustion in air, and their chief use is as sources of heat energy produced by combustion.

ALKENES, ALKYNES, AND AROMATIC HYDROCARBONS (SECTION 24.3) The names of alkenes and alkynes are based on the longest continuous chain of carbon atoms that contains the multiple bond, and the location of the multiple bond is specified by a numerical prefix. Alkenes exhibit not only structural isomerism but geometric (*cis-trans*) isomerism as well. In **geometric isomers**, the bonds are the same, but the molecules have different geometries. Geometric isomerism is possible in alkenes because rotation about the C = C double bond is restricted.

Alkenes and alkynes readily undergo **addition reactions** to the carbon–carbon multiple bonds. Additions of acids, such as HBr, proceed via a rate-determining step in which a proton is transferred to one of the alkene or alkyne carbon atoms. Addition reactions are difficult to carry out with aromatic hydrocarbons, but **substitution reactions** are easily accomplished in the presence of catalysts.

ORGANIC FUNCTIONAL GROUPS (SECTION 24.4) The chemistry of organic compounds is dominated by the nature of their functional groups. The functional groups we have considered are

R — O — H   Alcohol

$$R-\overset{\overset{\textstyle O}{\|}}{C}-H$$
Aldehyde

$$\text{>C=C<}$$
Alkene

— C ≡ C —   Alkyne

$$R-\overset{\overset{\textstyle O}{\|}}{C}-N\overset{R'\,(\text{or H})}{\diagdown}$$
Amide

$$R-\overset{R'}{\underset{}{N}}-R''\,(\text{or H})$$
Amine

$$R-\overset{\overset{\textstyle O}{\|}}{C}-O-H$$
Carboxylic acid

$$R-\overset{\overset{\textstyle O}{\|}}{C}-O-R'$$
Ester

R — O — R'   Ether

$$R-\overset{\overset{\textstyle O}{\|}}{C}-R'$$
Ketone

R, R', and R" represent hydrocarbon groups—for example, methyl (CH_3) or phenyl (C_6H_5).

Alcohols are hydrocarbon derivatives containing one or more OH groups. **Ethers** are formed by a condensation reaction of two molecules of alcohol. Several functional groups contain the **carbonyl** (C = O) **group**, including **aldehydes, ketones, carboxylic acids, esters,** and **amides**. Aldehydes and ketones can be produced by the oxidation of certain alcohols. Further oxidation of the aldehydes produces carboxylic acids. Carboxylic acids can form esters by a condensation reaction with alcohols, or they can form amides by a condensation reaction with amines. Esters undergo **hydrolysis** (**saponification**) in the presence of strong bases.

CHIRALITY IN ORGANIC CHEMISTRY (SECTION 24.5) Molecules that possess nonsuperimposable mirror images are termed **chiral**. The two nonsuperimposable forms of a chiral molecule are called *enantiomers*. In carbon compounds, a chiral center is created when all four groups bonded to a central carbon atom are different, as in 2-bromobutane. Many of the molecules occurring in living systems, such as the amino acids, are chiral and exist in nature in only one enantiomeric form. Many drugs of importance in human medicine are chiral, and the enantiomers may produce very different biochemical effects. For this reason, synthesis of only the effective enantiomers of chiral drugs has become a high priority.

INTRODUCTION TO BIOCHEMISTRY; PROTEINS (SECTIONS 24.6 AND 24.7) Many of the molecules that are essential for life are large natural polymers that are constructed from smaller molecules called monomers. Three of these **biopolymers** are considered in this chapter: proteins, polysaccharides (carbohydrates), and nucleic acids.

Proteins are polymers of **amino acids**. They are the major structural materials in animal systems. All naturally occurring proteins are formed from 22 amino acids, although only 20 are common. The amino acids are linked by **peptide bonds**. A **polypeptide** is a polymer formed by linking many amino acids by peptide bonds.

Amino acids are chiral substances. Usually only one of the enantiomers is found to be biologically active. Protein structure is determined by the sequence of amino acids in the chain (its **primary structure**), the intramolecular interactions within the chain (its **secondary structure**), and the overall shape of the complete molecule (its **tertiary structure**). Two important secondary structures are the α-**helix** and the β-**sheet**. The process by which a protein assumes its biologically active tertiary structure is called **folding**. Sometimes several proteins aggregate together to form a **quaternary structure**.

CARBOHYDRATES AND LIPIDS (SECTIONS 24.8 AND 24.9) **Carbohydrates**, which are polyhydroxy aldehydes and ketones, are the major structural constituents of plants and are a source of energy in both plants and animals. **Glucose** is the most common **monosaccharide**, or simple sugar. Two monosaccharides can be linked together by means of a condensation reaction to form a **disaccharide**. **Polysaccharides** are complex carbohydrates made up of many monosaccharide units joined together. The three most important polysaccharides are **starch**, which is found in plants; **glycogen**, which is found in mammals; and **cellulose**, which is also found in plants.

Lipids are compounds derived from glycerol and fatty acids and include fats and **phospholipids**. Fatty acids can be saturated, unsaturated, cis, or trans depending on their chemical formulas and structures.

NUCLEIC ACIDS (SECTION 24.10) Nucleic acids are biopolymers that carry the genetic information necessary for cell reproduction; they also control cell development through control of protein synthesis. The building blocks of these biopolymers are **nucleotides**. There are two types of nucleic acids, **ribonucleic acids (RNA)** and **deoxyribonucleic acids (DNA)**. These substances consist of a polymeric backbone of alternating phosphate and ribose or deoxyribose sugar groups with organic bases attached to the sugar molecules. The DNA polymer is a double-stranded helix (**double helix**) held together by hydrogen bonding between matching organic bases situated across from one another on the two strands. The hydrogen bonding between specific base pairs is the key to gene replication and protein synthesis.

Learning Outcomes After studying this chapter, you should be able to:

- Distinguish among alkanes, alkenes, alkynes, and aromatic hydrocarbons. (Section 24.2)

- Draw hydrocarbon structures based on their names and name hydrocarbons based on their structures. (Sections 24.2 and 24.3)

- Predict the products of addition reactions and substitution reactions. (Section 24.3)

- Draw the structures of the functional groups: alkene, alkyne, alcohol, haloalkane, carbonyl, ether, aldehyde, ketone, carboxylic acid, ester, amine, and amide. (Section 24.4)

- Predict the oxidation products of organic compounds. (Section 24.4)

- Understand what makes a compound chiral and be able to recognize a chiral substance. (Section 24.5)

- Recognize the amino acids and understand how they form peptides and proteins via amide bond formation. (Section 24.7)

- Understand the differences among the primary, secondary, tertiary, and quaternary structures of proteins. (Section 24.7)

- Explain the difference between α-helix and β-sheet peptide and protein structures. (Section 24.7)

- Distinguish between starch and cellulose structures. (Section 24.8)

- Classify molecules as saccharides or lipids based on their structures. (Sections 24.8 and 24.9)

- Explain the difference between a saturated and unsaturated fat. (Section 24.9)

- Explain the structure of nucleic acids and the role played by complementary bases in DNA replication. (Section 24.10)

Exercises

Visualizing Concepts

24.1 All the structures shown here have the molecular formula C_8H_{18}. Which structures are the same molecule? (*Hint:* One way to answer this question is to determine the chemical name for each structure.) [Section 24.2]

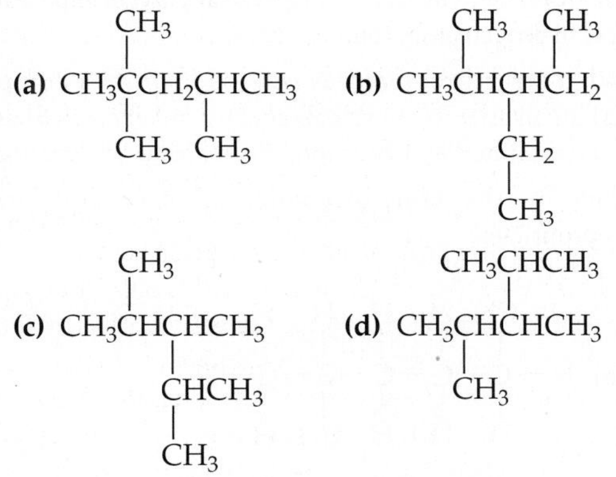

24.2 Which of these molecules is unsaturated? [Section 24.3]

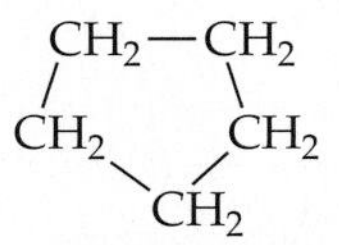

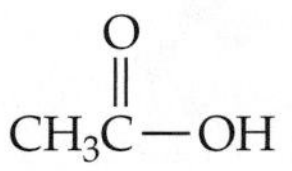

24.3 **(a)** Which of these molecules most readily undergoes an addition reaction? **(b)** Which of these molecules is aromatic? **(c)** Which of these molecules most readily undergoes a substitution reaction? [Section 24.3]

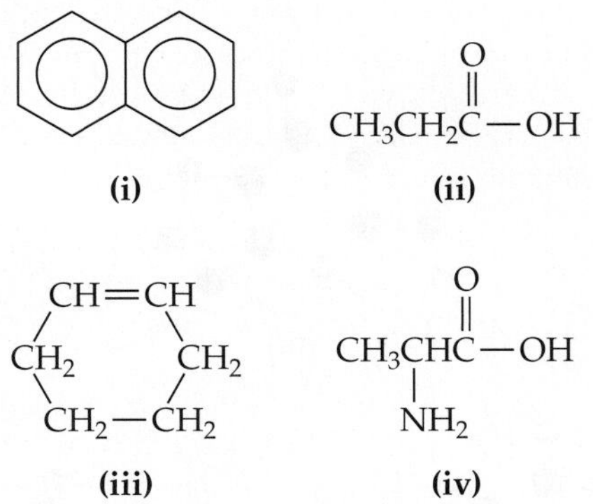

24.4 **(a)** Which of these compounds would you expect to have the highest boiling point? Which of the factors that determines boiling points described in Section 11.2 primarily accounts for this highest boiling point? **(b)** Which of these compounds is the most oxidized? (c) Which of these compounds, if any, is an ether? (d) Which of these compounds, if any, is an ester? (e) Which of these compounds, if any, is a ketone? [Section 24.4]

$$CH_3\overset{\displaystyle O}{\overset{\|}{C}}H \qquad CH_3CH_2OH \qquad CH_3C\equiv CH \qquad HCOCH_3$$

(i)   **(ii)**   **(iii)**   **(iv)**

24.5 Which of these compounds has an isomer? In each case where isomerism is possible, identify the type or types of isomerism. [Sections 24.2, 24.4]

$$\underset{\underset{NH_3^+}{|}}{CH_3CHCHC}\overset{\underset{O}{\|}}{\underset{CH_3}{|}}-O^-$$

(a)

(b) benzene ring with C—OH group at top and Cl at bottom left

(a)   Cl **(b)**

$$CH_3CH_2CH{=}CHCH_3 \qquad CH_3CH_2CH_3$$

(c)   **(d)**

24.6 From examination of the molecular models i–v, choose the substance that (a) can be hydrolyzed to form a solution containing glucose, (b) is capable of forming a zwitterion, (c) is one of the four bases present in DNA, (d) reacts with an acid to form an ester, (e) is a lipid. [Sections 24.6–24.10]

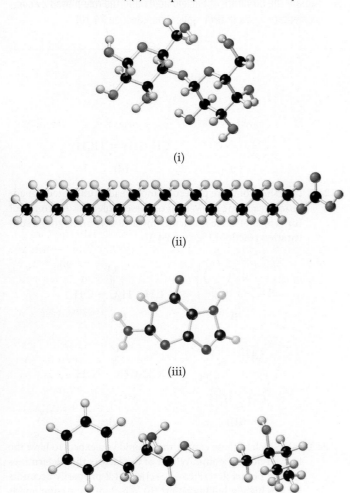

(i)

(ii)

(iii)

(iv)   **(v)**

Introduction to Organic Compounds; Hydrocarbons (Sections 24.1 and 24.2)

24.7 Indicate whether each statement is true or false. (a) Butane contains carbons that are sp^2 hybridized. (b) Cyclohexane is another name for benzene. (c) The isopropyl group contains three sp^3-hybridized carbons. (d) Olefin is another name for alkyne.

24.8 Indicate whether each statement is true or false. (a) Pentane has a higher molar mass than hexane. (b) The longer the linear alkyl chain for straight-chain hydrocarbons, the higher the boiling point. (c) The local geometry around the alkyne group is linear. (d) Propane has two structural isomers.

24.9 Predict the ideal values for the bond angles about each carbon atom in the following molecule. Indicate the hybridization of orbitals for each carbon.

$$CH_3CCCH_2COOH$$

24.10 Identify the carbon atom(s) in the structure shown that has (have) each of the following hybridizations: (a) sp^3, (b) sp, (c) sp^2.

$$N\equiv C-CH_2-CH_2-CH{=}CH-\underset{\underset{H}{\overset{\overset{C=O}{|}}{|}}}{CHOH}$$

24.11 Is ammonia an organic molecule? Explain.

24.12 Considering the comparative values of C—H, C—C, C—O, and C—Cl bond enthalpies (Table 8.4), predict whether compounds containing C—O and C—Cl bonds are more or less reactive than simple alkane hydrocarbons.

24.13 Indicate whether each statement is true or false. (a) Alkanes do not contain any carbon–carbon multiple bonds. (b) Cyclobutane contains a four-membered ring. (c) Alkenes contain carbon–carbon triple bonds. (d) Alkynes contain carbon–carbon double bonds. (e) Pentane is a saturated hydrocarbon but 1-pentene is an unsaturated hydrocarbon. (f) Cyclohexane is an aromatic hydrocarbon. (g) The methyl group contains one less hydrogen atom than methane.

24.14 What structural features help us identify a compound as (a) an alkane, (b) a cycloalkane, (c) an alkene, (d) an alkyne, (e) a saturated hydrocarbon, (f) an aromatic hydrocarbon?

24.15 Give the the name or condensed structural formula, as appropriate:

(a)
$$H-\underset{\underset{H}{|}}{\overset{\overset{CH_3}{|}}{C}}-\underset{\underset{H}{|}}{\overset{\overset{H}{|}}{C}}-\underset{\underset{H}{|}}{\overset{\overset{H}{|}}{C}}-\underset{\underset{CH_3}{|}}{\overset{\overset{H}{|}}{C}}-\underset{\underset{H}{|}}{\overset{\overset{H}{|}}{C}}-H$$

(b) $CH_3CH_2CH_2CH_2CH_2CH_2\underset{\underset{\underset{CH_3}{|}}{\underset{|}{CH_2}}}{\overset{\overset{CH_3}{|}}{C}}CH_2CHCH_3$

(c) 2-methylheptane
(d) 4-ethyl-2,3-dimethyloctane
(e) 1,2-dimethylcyclohexane

24.16 Give the name or condensed structural formula, as appropriate:

    (a) 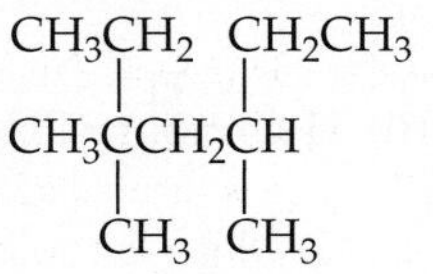
$$CH_3CH_2CH_2CH_2$$
$$CH_3CCH_2CH$$
$$CH_3 \quad CH_3$$

    (b) 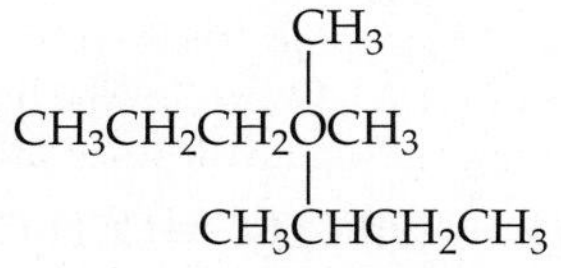
$$CH_3$$
$$CH_3CH_2CH_2OCH_3$$
$$CH_3CHCH_2CH_3$$

    (c) 2,5,6-trimethylnonane
    (d) 3-propyl-4,5-dimethylundecane
    (e) 1-ethyl-3-methylcyclohexane

24.17 Give the name or condensed structural formula, as appropriate:

    (a) 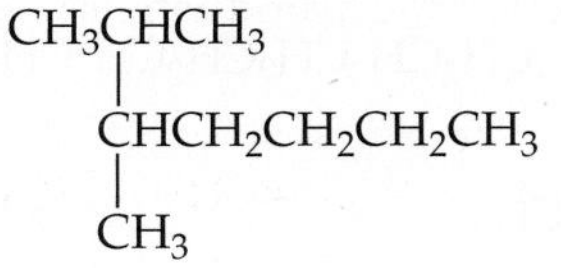
$$CH_3CHCH_3$$
$$CHCH_2CH_2CH_2CH_3$$
$$CH_3$$

    (b) 2,2-dimethylpentane
    (c) 4-ethyl-1,1-dimethylcyclohexane
    (d) $(CH_3)_2CHCH_2CH_2C(CH_3)_3$
    (e) $CH_3CH_2CH(C_2H_5)CH_2CH_2CH_2CH_3$

24.18 Give the name or condensed structural formula, as appropriate:

    (a) 3-phenylpentane
    (b) 2,3-dimethylhexane
    (c) 2-ethyl-2-methylheptane
    (d) $CH_3CH_2CH(CH_3)CH_2CH(CH_3)_2$

    (e) 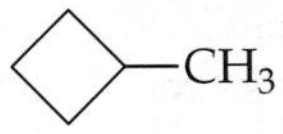 —CH_3

24.19 What is the octane number of a mixture that is 35% heptane and 65% isooctane?

24.20 Describe two ways in which the octane number of a gasoline consisting of alkanes can be increased.

Alkenes, Alkynes, and Aromatic Hydrocarbons (Section 24.3)

24.21 (a) Why are alkanes said to be saturated? (b) Is C_4H_6 a saturated hydrocarbon? Explain.

24.22 (a) Is the compound $CH_3CH=CH_2$ saturated or unsaturated? Explain. (b) What is wrong with the formula $CH_3CH_2CH=CH_3$?

24.23 Give the molecular formula of a hydrocarbon containing five carbon atoms that is (a) an alkane, (b) a cycloalkane, (c) an alkene, (d) an alkyne.

24.24 Give the molecular formula of a hydrocarbon containing six carbon atoms that is (a) a cyclic alkane, (b) a cyclic alkene, (c) a linear alkyne, (d) an aromatic hydrocarbon.

24.25 Enediynes are a class of compounds that include some antibiotic drugs. Draw the structure of an "enediyne" fragment that contains six carbons in a row.

24.26 Give the general formula for any cyclic alkene, that is, a cyclic hydrocarbon with one double bond.

24.27 Write the condensed structural formulas for two alkenes and one alkyne that all have the molecular formula C_6H_{10}.

24.28 Draw all the possible noncyclic structural isomers of C_5H_{10}. Name each compound.

24.29 Name or write the condensed structural formula for the following compounds:
    (a) *trans*-2-pentene
    (b) 2,5-dimethyl-4-octene
    (c)

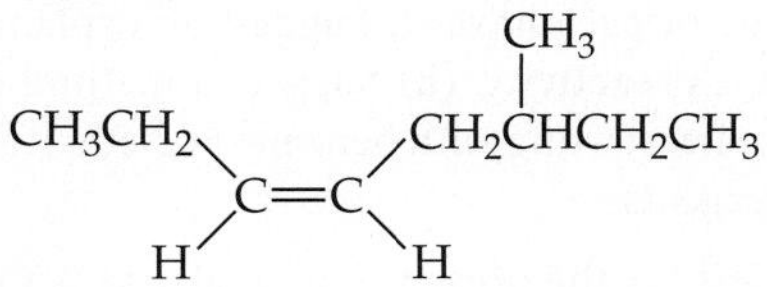

$$CH_3CH_2 \qquad\qquad CH_3$$
$$C=C \qquad CH_2CHCH_2CH_3$$
$$H \qquad\quad H$$

    (d)

    (e) 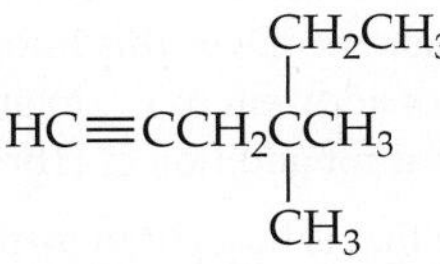
$$CH_2CH_3$$
$$HC\equiv CCH_2CCH_3$$
$$CH_3$$

24.30 Name or write the condensed structural formula for the following compounds:

    (a) 4-methyl-2-pentene
    (b) *cis*-2,5-dimethyl-3-hexene
    (c) *ortho*-dimethylbenzene
    (d) $HC\equiv CCH_2CH_3$
    (e) *trans*-$CH_3CH=CHCH_2CH_2CH_2CH_3$

24.31 Indicate whether each statement is true or false. (a) Two geometric isomers of pentane are *n*-pentane and neopentane. (b) Alkenes can have cis and trans isomers around the CC double bond. (c) Alkynes can have cis and trans isomers around the CC triple bond.

24.32 Draw all structural and geometric isomers of butene and name them.

24.33 Indicate whether each of the following molecules is capable of geometrical isomerism. For those that are, draw the structures: (a) 1,1-dichloro-1-butene, (b) 2,4-dichloro-2-butene, (c) 1,4-dichlorobenzene, (d) 4,4-dimethyl-2-pentyne.

24.34 Draw the three distinct geometric isomers of 2,4-hexadiene.

24.35 **(a)** True or false: Alkenes undergo addition reactions and aromatic hydrocarbons undergo substitution reactions. **(b)** Using condensed structural formulas, write the balanced equation for the reaction of 2-pentene with Br_2 and name the resulting compound. Is this an addition or a substitution reaction? **(c)** Write a balanced chemical equation for the reaction of Cl_2 with benzene to make *para*-dichlorobenzene in the presence of $FeCl_3$ as a catalyst. Is this an addition or a substitution reaction?

24.36 Using condensed structural formulas, write a balanced chemical equation for each of the following reactions: **(a)** hydrogenation of cyclohexene, **(b)** addition of H_2O to *trans*-2-pentene using H_2SO_4 as a catalyst (two products), **(c)** reaction of 2-chloropropane with benzene in the presence of $AlCl_3$.

24.37 **(a)** When cyclopropane is treated with HI, 1-iodopropane is formed. A similar type of reaction does not occur with cyclopentane or cyclohexane. Suggest an explanation for cyclopropane's reactivity. **(b)** Suggest a method of preparing ethylbenzene, starting with benzene and ethylene as the only organic reagents.

24.38 **(a)** One test for the presence of an alkene is to add a small amount of bromine, which is a red-brown liquid, and look for the disappearance of the red-brown color. This test does not work for detecting the presence of an aromatic hydrocarbon. Explain. **(b)** Write a series of reactions leading to *para*-bromoethylbenzene, beginning with benzene and using other reagents as needed. What isomeric side products might also be formed?

24.39 The rate law for addition of Br_2 to an alkene is first order in Br_2 and first order in the alkene. Does this information suggest that the mechanism of addition of Br_2 to an alkene proceeds in the same manner as for addition of HBr? Explain.

24.40 Describe the intermediate that is thought to form in the addition of a hydrogen halide to an alkene, using cyclohexene as the alkene in your description.

24.41 The molar heat of combustion of gaseous cyclopropane is -2089 kJ/mol; that for gaseous cyclopentane is -3317 kJ/mol. Calculate the heat of combustion per CH_2 group in the two cases, and account for the difference.

24.42 The heat of combustion of decahydronaphthalene ($C_{10}H_{18}$) is -6286 kJ/mol. The heat of combustion of naphthalene ($C_{10}H_8$) is -5157 kJ/mol. (In both cases $CO_2(g)$ and $H_2O(l)$ are the products.) Using these data and data in Appendix C, calculate the heat of hydrogenation and the resonance energy of naphthalene.

Functional Groups and Chirality (Sections 24.4 and 24.5)

24.43 **(a)** Which of the following compounds, if any, is an ether? **(b)** Which compound, if any, is an alcohol? **(c)** Which compound, if any, would produce a basic solution if dissolved in water? (Assume solubility is not a problem). **(d)** Which

compound, if any, is a ketone? **(e)** Which compound, if any, is an aldehyde?

(i) $H_3C{-}CH_2{-}OH$

(ii) (structure with N-H: $H_3C{-}\overset{H}{N}{-}CH_2CH{=}CH_2$)

(ii) 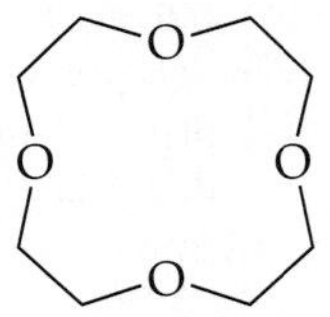 (cyclic ether structure with two O atoms)

(iv) 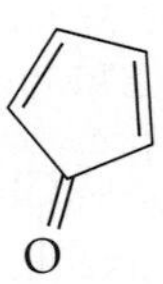 (cyclopentadienone ring with O)

(v) $CH_3CH_2CH_2CH_2CHO$

(vi) $CH_3C{\equiv}CCH_2COOH$

24.44 Identify the functional groups in each of the following compounds:

(a) 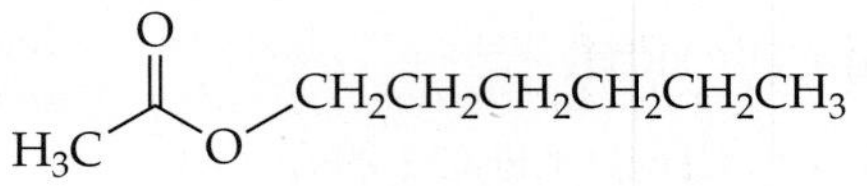
structure: H_3C ester O ${-}CH_2CH_2CH_2CH_2CH_2CH_3$

(b) 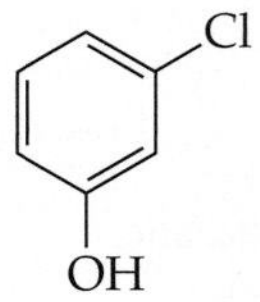
aromatic ring with Cl and OH substituents

(c) 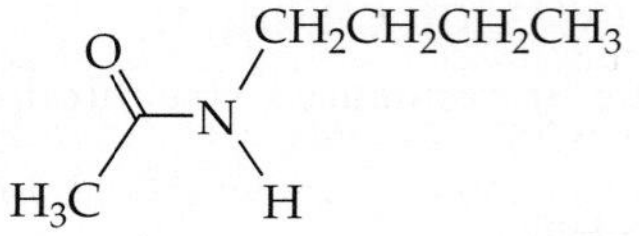
H_3C amide with N${-}CH_2CH_2CH_2CH_3$, N${-}H$

(d) 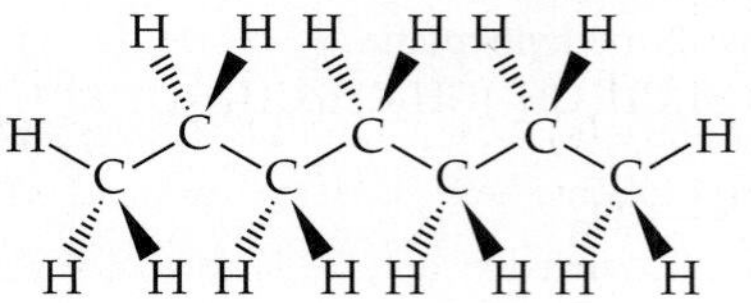
hydrocarbon chain structure

(e) 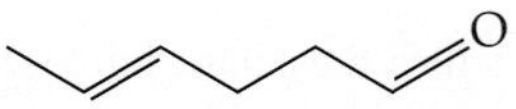
alkene-aldehyde structure

(f) 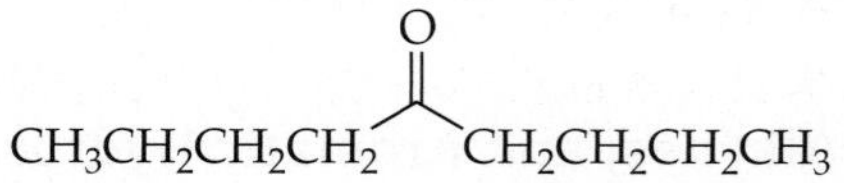
$CH_3CH_2CH_2CH_2$ ketone $CH_2CH_2CH_2CH_3$

24.45 Draw the molecular structure for **(a)** an aldehyde that is an isomer of acetone, **(b)** an ether that is an isomer of 1-propanol.

24.46 **(a)** Give the empirical formula and structural formula for a cyclic ether containing four carbon atoms in the ring. **(b)** Write the structural formula for a straight-chain compound that is a structural isomer of your answer to part (a).

24.47 The IUPAC name for a carboxylic acid is based on the name of the hydrocarbon with the same number of carbon atoms. The ending *-oic* is appended, as in ethanoic acid, which is the IUPAC name for acetic acid. Draw the structure of the following acids: **(a)** methanoic acid, **(b)** pentanoic acid, **(c)** 2-chloro-3-methyldecanoic acid.

24.48 Aldehydes and ketones can be named in a systematic way by counting the number of carbon atoms (including the carbonyl carbon) that they contain. The name of the aldehyde or ketone is based on the hydrocarbon with the same number of carbon atoms. The ending -*al* for aldehyde or -*one* for ketone is added as appropriate. Draw the structural formulas for the following aldehydes or ketones: **(a)** propanal, **(b)** 2-pentanone, **(c)** 3-methyl-2-butanone, **(d)** 2-methylbutanal.

24.49 Draw the condensed structure of the compounds formed by condensation reactions between **(a)** benzoic acid and ethanol, **(b)** ethanoic acid and methylamine, **(c)** acetic acid and phenol. Name the compound in each case.

24.50 Draw the condensed structures of the compounds formed from **(a)** butanoic acid and methanol, **(b)** benzoic acid and 2-propanol, **(c)** propanoic acid and dimethylamine. Name the compound in each case.

24.51 Write a balanced chemical equation using condensed structural formulas for the saponification (base hydrolysis) of **(a)** methyl propionate, **(b)** phenyl acetate.

24.52 Write a balanced chemical equation using condensed structural formulas for **(a)** the formation of butyl propionate from the appropriate acid and alcohol, **(b)** the saponification (base hydrolysis) of methyl benzoate.

24.53 Pure acetic acid is a viscous liquid, with high melting and boiling points (16.7 °C and 118 °C) compared to compounds of similar molecular weight. Suggest an explanation.

24.54 *Acetic anhydride* is formed from two acetic acid molecules, in a condensation reaction that involves the removal of a molecule of water. Write the chemical equation for this process, and show the structure of acetic anhydride.

24.55 Write the condensed structural formula for each of the following compounds: **(a)** 2-pentanol, **(b)** 1,2-propanediol, **(c)** ethyl acetate, **(d)** diphenyl ketone, **(e)** methyl ethyl ether.

24.56 Write the condensed structural formula for each of the following compounds: **(a)** 2-ethyl-1-hexanol, **(b)** methyl phenyl ketone, **(c)** *para*-bromobenzoic acid, **(d)** ethyl butyl ether, **(e)** *N, N*-dimethylbenzamide.

24.57 How many chiral carbons are in 2-bromo-2-chloro-3-methylpentane? **(a)** 0, **(b)** 1, **(c)** 2, **(d)** 3, **(e)** 4 or more.

24.58 Does 3-chloro-3-methylhexane have optical isomers? Why or why not?

Introduction to Biochemistry; Proteins (Sections 24.6 and 24.7)

24.59 **(a)** Draw the chemical structure of a generic amino acid, using R for the side chain. **(b)** When amino acids react to form proteins, do they do so via substitution, addition, or condensation reactions? **(c)** Draw the bond that links amino acids together in proteins. What is this called?

24.60 Indicate whether each statement is true or false. **(a)** Tryptophan is an aromatic amino acid. **(b)** Lysine is positively charged at pH 7. **(c)** Asparagine has two amide bonds.

(d) Isoleucine and leucine are enantiomers. **(e)** Valine is probably more water-soluble than arginine.

24.61 Draw the two possible dipeptides formed by condensation reactions between histidine and aspartic acid.

24.62 Write a chemical equation for the formation of methionyl glycine from the constituent amino acids.

24.63 **(a)** Draw the condensed structure of the tripeptide Gly-Gly-His. **(b)** How many different tripeptides can be made from the amino acids glycine and histidine? Give the abbreviations for each of these tripeptides, using the three-letter and one-letter codes for the amino acids.

24.64 **(a)** What amino acids would be obtained by hydrolysis of the following tripeptide?

$$\underset{\displaystyle (CH_3)_2CH}{H_2NCHC} \overset{\displaystyle O}{\underset{\displaystyle H_2COH}{NHCHC}} \overset{\displaystyle O}{\underset{\displaystyle H_2CCH_2COH}{NHCHCOH}}$$

(b) How many different tripeptides can be made from glycine, serine, and glutamic acid? Give the abbreviation for each of these tripeptides, using the three-letter codes and one-letter codes for the amino acids.

24.65 Indicate whether each statement is true or false. **(a)** The sequence of amino acids in a protein, from the amine end to the acid end, is called the primary structure of the protein. **(b)** Alpha helix and beta sheet structures are examples of quaternary protein structure. **(c)** It is impossible for more than one protein to bind to another and make a higher order structure.

24.66 Indicate whether each statement is true or false: **(a)** In the alpha helical structure of proteins, hydrogen bonding occurs between the side chains (R groups). **(b)** Dispersion forces, not hydrogen bonding, holds beta sheet structures together.

Carbohydrates and Lipids (Sections 24.8 and 24.9)

24.67 Indicate whether each statement is true or false: **(a)** Disaccharides are a type of carbohydrate. **(b)** Sucrose is a monosaccharide. **(c)** All carbohydrates have the formula $C_nH_{2m}O_m$.

24.68 **(a)** Are α-glucose and β-glucose enantiomers? **(b)** Show the condensation of two glucose molecules to form a disaccharide with an α linkage. **(c)** Repeat part (b) but with a β linkage.

24.69 **(a)** What is the empirical formula of cellulose? **(b)** What is the monomer that forms the basis of the cellulose polymer? **(c)** What bond connects the monomer units in cellulose: amide, acid, ether, ester, or alcohol?

24.70 **(a)** What is the empirical formula of starch? **(b)** What is the monomer that forms the basis of the starch polymer? **(c)** What bond connects the monomer units in starch: amide, acid, ether, ester, or alcohol?

24.71 The structural formula for the linear form of D-mannose is

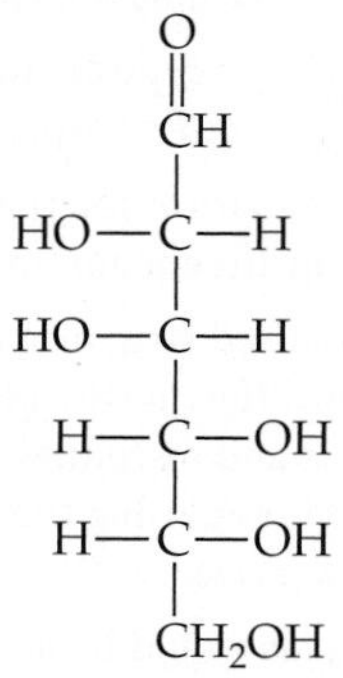

(a) Is this molecule a sugar? (b) How many chiral carbons are present in the molecule? (c) Draw the structure of the six-member-ring form of this molecule.

24.72 The structural formula for the linear form of galactose is

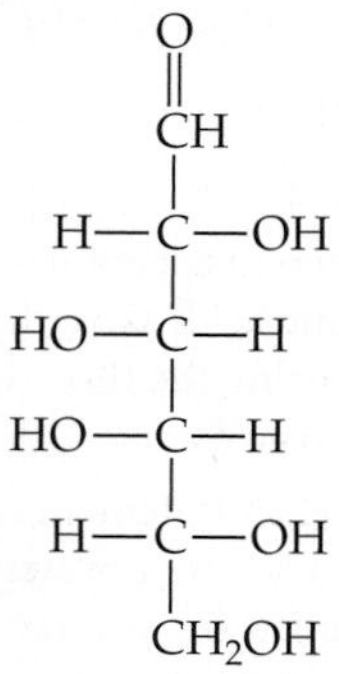

(a) Is this molecule a sugar? (b) How many chiral carbons are present in the molecule? (c) Draw the structure of the six-member-ring form of this molecule.

24.73 Indicate whether each statement is true or false: (a) Fat molecules contain amide bonds. (b) Phospholipids can be zwitterions. (c) Phospholipids form bilayers in water in order to have their long hydrophobic tails interact favorably with each other, leaving their polar heads to the aqueous environment.

24.74 Indicate whether each statement is true or false: (a) If you use data from Table 8.4 on bond enthalpies, you can show that the more C—H bonds a molecule has compared to C—O and O—H bonds, the more energy it can store. (b) Trans fats are saturated. (c) Fatty acids are long-chain carboxylic acids. (d) Monounsaturated fatty acids have one CC single bond in the chain, while the rest are double or triple bonds.

Nucleic Acids (Section 24.10)

24.75 Adenine and guanine are members of a class of molecules known as *purines*; they have two rings in their structure. Thymine and cytosine, on the other hand, are *pyrimidines*, and have only one ring in their structure. Predict which have larger dispersion forces in aqueous solution, the purines or the pyrimidines.

24.76 A nucleoside consists of an organic base of the kind shown in Section 24.10, bound to ribose or deoxyribose. Draw the structure for deoxyguanosine, formed from guanine and deoxyribose.

24.77 Just as the amino acids in a protein are listed in the order from the amine end to the carboxylic acid end (the primary structure or *protein sequence*), the bases in nucleic acids are listed in the order 5′ to 3′, where the numbers refer to the position of the carbons in the sugars (shown here for deoxyribose):

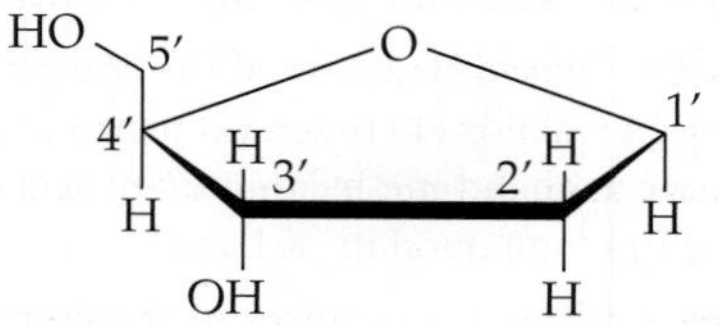

The base is attached to the sugar at the 1′ carbon. The 5′ end of a DNA sequence is a phosphate of an OH group, and the 3′ end of a DNA sequence is the OH group. What is the DNA sequence for the molecule shown here?

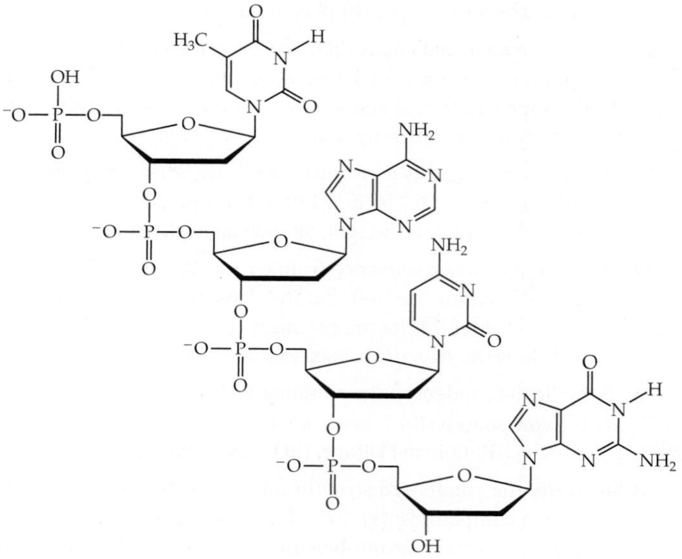

24.78 When samples of double-stranded DNA are analyzed, the quantity of adenine present equals that of thymine. Similarly, the quantity of guanine equals that of cytosine. Explain the significance of these observations.

24.79 Imagine a single DNA strand containing a section with the following base sequence: 5′-GCATTGGC-3′. What is the base sequence of the complementary strand? (The two strands of DNA will come together in an *antiparallel* fashion; that is, 5′-TAG-3′ will bind to 3′-ATC-5′.)

24.80 Which statement best explains the chemical differences between DNA and RNA? (a) DNA has two different sugars in its sugar–phosphate backbone, but RNA only has one. (b) Thymine is one of the DNA bases, whereas RNA's corresponding base is thymine minus a methyl group. (c) The RNA sugar–phosphate backbone contains fewer oxygen atoms than DNA's backbone. (d) DNA forms double helices but RNA cannot.

Additional Exercises

24.81 Draw the condensed structural formulas for two different molecules with the formula C_3H_4O.

24.82 How many structural isomers are there for a five-member straight carbon chain with one double bond? For a six-member straight carbon chain with two double bonds?

24.83 (a) Draw the condensed structural formulas for the cis and trans isomers of 2-pentene. (b) Can cyclopentene exhibit cis–trans isomerism? Explain. (c) Does 1-pentyne have enantiomers? Explain.

24.84 If a molecule is an "ene-one," what functional groups must it have?

24.85 Identify each of the functional groups in these molecules:

(a)

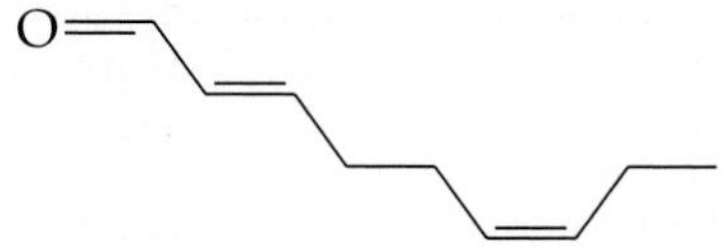

(Responsible for the odor of cucumbers)

(b)

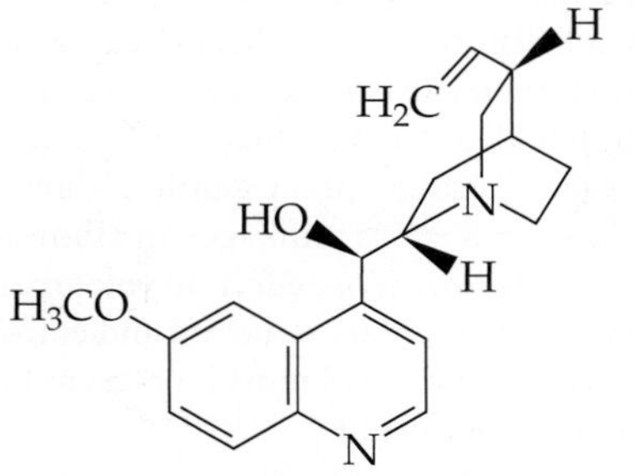

(Quinine — an antimalarial drug)

(c)

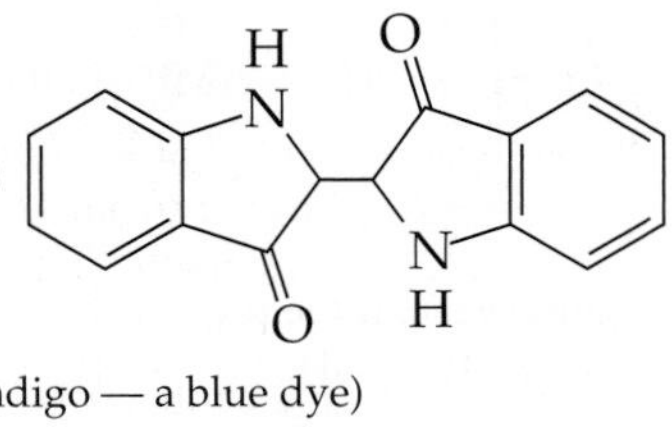

(Indigo — a blue dye)

(d)

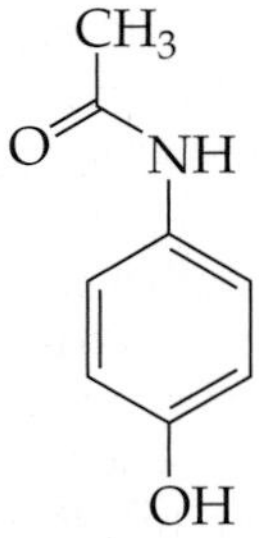

(Acetaminophen — aka Tylenol)

24.86 For the molecules shown in 24.85, (a) Which one(s) of them, if any, would produce a basic solution if dissolved in water? (b) Which one(s) of them, if any, would produce an acidic solution if dissolved in water? (c) Which of them is the most water-soluble?

24.87 Write a condensed structural formula for each of the following: (a) an acid with the formula $C_4H_8O_2$, (b) a cyclic ketone with the formula C_5H_8O, (c) a dihydroxy compound with the formula $C_3H_8O_2$, (d) a cyclic ester with the formula $C_5H_8O_2$.

24.88 Carboxylic acids generally have pK_a's of ~5, but alcohols have pK_a's of ~16. (a) Write the acid-dissociation chemical equation for the generic alcohol ROH in water. (b) Which compounds will produce more acidic solutions upon dissolution in water, acids or alcohols? (c) Suggest an explanation for the difference in pK_a's for acids compared to alcohols.

[24.89] Indole smells terrible in high concentrations but has a pleasant floral-like odor when highly diluted. Its structure is

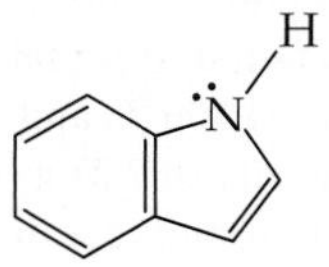

The molecule is planar, and the nitrogen is a very weak base, with $K_b = 2 \times 10^{-12}$. Explain how this information indicates that the indole molecule is aromatic.

24.90 Locate the chiral carbon atoms, if any, in each molecule:

(a) 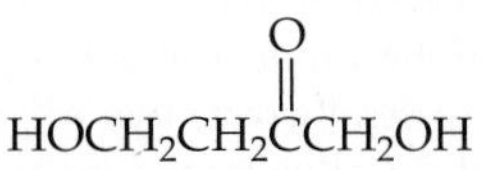 $HOCH_2CH_2\overset{\overset{\textstyle O}{\|}}{C}CH_2OH$

(b) $HOCH_2\overset{\overset{\textstyle OH}{|}}{C}H\underset{\underset{\textstyle O}{\|}}{C}CH_2OH$

(c) $HO\underset{}{\overset{\overset{\textstyle O}{\|}}{C}}\underset{\underset{\textstyle NH_2}{|}}{C}H\overset{\overset{\textstyle CH_3}{|}}{C}HC_2H_5$

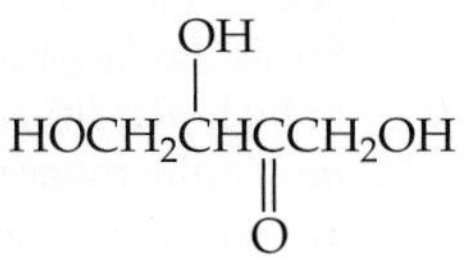

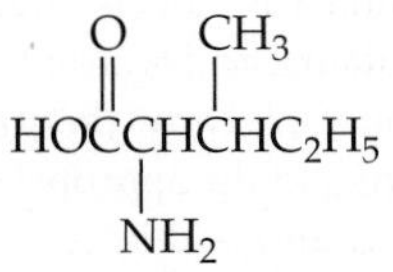

24.91 Which of the following peptides have a net positive charge at pH 7? (a) Gly-Ser-Lys, (b) Pro-Leu-Ile, (c) Phe-Tyr-Asp.

24.92 Glutathione is a tripeptide found in most living cells. Partial hydrolysis yields Cys-Gly and Glu-Cys. What structures are possible for glutathione?

24.93 Monosaccharides can be categorized in terms of the number of carbon atoms (pentoses have five carbons and hexoses have six carbons) and according to whether they contain an aldehyde (*aldo-* prefix, as in aldopentose) or ketone group (*keto-* prefix, as in ketopentose). Classify glucose and fructose in this way.

24.94 Can a DNA strand bind to a complementary RNA strand? Explain.

Integrative Exercises

24.95 Explain why the boiling point of ethanol (78 °C) is much higher than that of its isomer, dimethyl ether (−25 °C), and why the boiling point of CH_2F_2 (−52 °C) is far above that of CH_4 (−128 °C).

[24.96] An unknown organic compound is found on elemental analysis to contain 68.1% carbon, 13.7% hydrogen, and 18.2% oxygen by mass. It is slightly soluble in water. Upon careful oxidation it is converted into a compound that behaves chemically like a ketone and contains 69.7% carbon, 11.7% hydrogen, and 18.6% oxygen by mass. Indicate two or more reasonable structures for the unknown.

24.97 An organic compound is analyzed and found to contain 66.7% carbon, 11.2% hydrogen, and 22.1% oxygen by mass. The compound boils at 79.6 °C. At 100 °C and 0.970 atm, the vapor has a density of 2.28 g/L. The compound has a carbonyl group and cannot be oxidized to a carboxylic acid. Suggest a structure for the compound.

24.98 An unknown substance is found to contain only carbon and hydrogen. It is a liquid that boils at 49 °C at 1 atm pressure. Upon analysis it is found to contain 85.7% carbon and 14.3% hydrogen by mass. At 100 °C and 735 torr, the vapor of this unknown has a density of 2.21 g/L. When it is dissolved in hexane solution and bromine water is added, no reaction occurs. What is the identity of the unknown compound?

24.99 The standard free energy of formation of solid glycine is −369 kJ/mol, whereas that of solid glycylglycine is −488 kJ/mol. What is $\Delta G°$ for the condensation of glycine to form glycylglycine?

24.100 A typical amino acid with one amino group and one carboxylic acid group, such as serine, can exist in water in several ionic forms. **(a)** Suggest the forms of the amino acid at low pH and at high pH. **(b)** Amino acids generally have two pK_a values, one in the range of 2 to 3 and the other in the range of 9 to 10. Serine, for example, has pK_a values of 2.19 and 9.21. Using species such as acetic acid and ammonia as models, suggest the origin of the two pK_a values. **(c)** Glutamic acid is an amino acid that has three pK_a's: 2.10, 4.07, and 9.47. Draw the structure of glutamic acid, and assign each pK_a to the appropriate part of the molecule. **(d)** An unknown amino acid is titrated with strong base, producing the following titration curve. Which amino acids are likely candidates for the unknown?

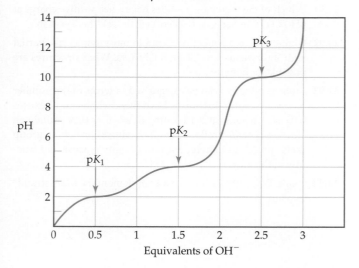

[24.101] The protein ribonuclease A in its native, or most stable, form is folded into a compact globular shape:

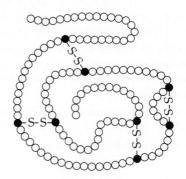

Native ribonuclease A

(a) Does the native form have a lower or higher free energy than the denatured form, in which the protein is an extended chain? **(b)** What is the sign of the system's entropy change in going from the denatured to the folded form? **(c)** In the native form, the molecule has four —S—S— bonds that bridge parts of the chain. What effect do you predict these four linkages to have on the free energy and entropy of the native form relative to the free energy and entropy of a hypothetical folded structure that does not have any —S—S— linkages? Explain. **(d)** A gentle reducing agent converts the four —S—S— linkages in ribonuclease A to eight —S—H bonds. What effect do you predict this conversion to have on the tertiary structure and entropy of the protein? **(e)** Which amino acid must be present for —SH bonds to exist in ribonuclease A?

24.102 The monoanion of adenosine monophosphate (AMP) is an intermediate in phosphate metabolism:

$$\text{A}-\text{O}-\overset{\overset{\textstyle O^-}{|}}{\underset{\underset{\textstyle O}{\|}}{\text{P}}}-\text{OH} = \text{AMP}-\text{OH}^-$$

where A = adenosine. If the pK_a for this anion is 7.21, what is the ratio of $[\text{AMP}-\text{OH}^-]$ to $[\text{AMP}-\text{O}^{2-}]$ in blood at pH 7.4?

Design an Experiment

Quaternary structures of proteins arise if two or more smaller polypeptides or proteins associate with each other to make a much larger protein structure. The association is due to the same hydrogen bonding, electrostatic, and dispersion forces we have seen before. Hemoglobin, the protein used to transport oxygen molecules in our blood, is an example of a protein that has quaternary structure. Hemoglobin is a tetramer; it is made of four smaller polypeptides, two "alphas" and two "betas." (These names do not imply anything about the number of alpha-helices or beta sheets in the individual polypeptides.) Design a set of experiments that would provide sound evidence that hemoglobin exists as a tetramer and not as one enormous polypeptide chain.

APPENDIX

Mathematical Operations

A.1 | Exponential Notation

The numbers used in chemistry are often either extremely large or extremely small. Such numbers are conveniently expressed in the form

$$N \times 10^n$$

where N is a number between 1 and 10, and n is the exponent. Some examples of this *exponential notation*, which is also called *scientific notation*, follow.

1,200,000 is 1.2×10^6 (read "one point two multi ten to the sixth power")

0.000604 is 6.04×10^{-4} (read "six point zero four times ten to the negative fourth power")

A positive exponent, as in the first example, tells us how many times a number must be multiplied by 10 to give the long form of the number:

$$1.2 \times 10^6 = 1.2 \times 10 \times 10 \times 10 \times 10 \times 10 \times 10 \quad \text{(six tens)}$$
$$= 1,200,000$$

It is also convenient to think of the *positive exponent* as the number of places the decimal point must be moved to the *left* to obtain a number greater than 1 and less than 10. For example, if we begin with 3450 and move the decimal point three places to the left, we end up with 3.45×10^3.

In a related fashion, a negative exponent tells us how many times we must divide a number by 10 to give the long form of the number.

$$6.04 \times 10^{-4} = \frac{6.04}{10 \times 10 \times 10 \times 10} = 0.000604$$

It is convenient to think of the *negative exponent* as the number of places the decimal point must be moved to the *right* to obtain a number greater than 1 but less than 10. For example, if we begin with 0.0048 and move the decimal point three places to the right, we end up with 4.8×10^{-3}.

In the system of exponential notation, with each shift of the decimal point one place to the right, the exponent *decreases* by 1:

$$4.8 \times 10^{-3} = 48 \times 10^{-4}$$

Similarly, with each shift of the decimal point one place to the left, the exponent *increases* by 1:

$$4.8 \times 10^{-3} = 0.48 \times 10^{-2}$$

Many scientific calculators have a key labeled EXP or EE, which is used to enter numbers in exponential notation. To enter the number 5.8×10^3 on such a calculator, the key sequence is

$$\boxed{5}\ \boxed{\cdot}\ \boxed{8}\ \boxed{\text{EXP}}\ (\text{or}\ \boxed{\text{EE}}\ )\ \boxed{3}$$

On some calculators the display will show 5.8, then a space, followed by 03, the exponent. On other calculators, a small 10 is shown with an exponent 3.

To enter a negative exponent, use the key labeled $+/-$. For example, to enter the number 8.6×10^{-5}, the key sequence is

$$\boxed{8}\;\boxed{\cdot}\;\boxed{6}\;\boxed{\text{EXP}}\;\boxed{+/-}\;\boxed{5}$$

When entering a number in exponential notation, do not key in the 10 if you use the EXP or EE button.

In working with exponents, it is important to recall that $10^0 = 1$. The following rules are useful for carrying exponents through calculations.

1. **Addition and Subtraction** In order to add or subtract numbers expressed in exponential notation, the powers of 10 must be the same.

$$(5.22 \times 10^4) + (3.21 \times 10^2) = (522 \times 10^2) + (3.21 \times 10^2)$$
$$= 525 \times 10^2 \quad \text{(3 significant figures)}$$
$$= 5.25 \times 10^4$$
$$(6.25 \times 10^{-2}) - (5.77 \times 10^{-3}) = (6.25 \times 10^{-2}) - (0.577 \times 10^{-2})$$
$$= 5.67 \times 10^{-2} \quad \text{(3 significant figures)}$$

When you use a calculator to add or subtract, you need not be concerned with having numbers with the same exponents because the calculator automatically takes care of this matter.

2. **Multiplication and Division** When numbers expressed in exponential notation are multiplied, the exponents are added; when numbers expressed in exponential notation are divided, the exponent of the denominator is subtracted from the exponent of the numerator.

$$(5.4 \times 10^2)(2.1 \times 10^3) = (5.4)(2.1) \times 10^{2+3}$$
$$= 11 \times 10^5$$
$$= 1.1 \times 10^6$$
$$(1.2 \times 10^5)(3.22 \times 10^{-3}) = (1.2)(3.22) \times 10^{5+(-3)} = 3.9 \times 10^2$$
$$\frac{3.2 \times 10^5}{6.5 \times 10^2} = \frac{3.2}{6.5} \times 10^{5-2} = 0.49 \times 10^3 = 4.9 \times 10^2$$
$$\frac{5.7 \times 10^7}{8.5 \times 10^{-2}} = \frac{5.7}{8.5} \times 10^{7-(-2)} = 0.67 \times 10^9 = 6.7 \times 10^8$$

3. **Powers and Roots** When numbers expressed in exponential notation are raised to a power, the exponents are multiplied by the power. When the roots of numbers expressed in exponential notation are taken, the exponents are divided by the root.

$$(1.2 \times 10^5)^3 = (1.2)^3 \times 10^{5\times3}$$
$$= 1.7 \times 10^{15}$$
$$\sqrt[3]{2.5 \times 10^6} = \sqrt[3]{2.5} \times 10^{6/3}$$
$$= 1.3 \times 10^2$$

Scientific calculators usually have keys labeled x^2 and $\sqrt{x}$ for squaring and taking the square root of a number, respectively. To take higher powers or roots, many calculators have y^x and $\sqrt[x]{y}$ (or INV y^x) keys. For example, to perform the operation $\sqrt[3]{7.5 \times 10^{-4}}$ on such a calculator, you would key in 7.5×10^{-4}, press the $\sqrt[x]{y}$ key (or the INV and then the y^x keys), enter the root, 3, and finally press $=$. The result is 9.1×10^{-2}.

SAMPLE EXERCISE 1 Using Exponential Notation

Perform each of the following operations, using your calculator where possible:

(a) Write the number 0.0054 in standard exponential notation.
(b) $(5.0 \times 10^{-2}) + (4.7 \times 10^{-3})$
(c) $(5.98 \times 10^{12})(2.77 \times 10^{-5})$
(d) $\sqrt[4]{1.75 \times 10^{-12}}$

SOLUTION

(a) Because we move the decimal point three places to the right to convert 0.0054 to 5.4, the exponent is -3:

$$5.4 \times 10^{-3}$$

Scientific calculators are generally able to convert numbers to exponential notation using one or two keystrokes; frequently "SCI" for "scientific notation" will convert a number into exponential notation. Consult your instruction manual to see how this operation is accomplished on your calculator.

(b) To add these numbers longhand, we must convert them to the same exponent.

$$(5.0 \times 10^{-2}) + (0.47 \times 10^{-2}) = (5.0 + 0.47) \times 10^{-2} = 5.5 \times 10^{-2}$$

(c) Performing this operation longhand, we have

$$(5.98 \times 2.77) \times 10^{12-5} = 16.6 \times 10^{7} = 1.66 \times 10^{8}$$

(d) To perform this operation on a calculator, we enter the number, press the $\sqrt[x]{y}$ key (or the INV and y^x keys), enter 4, and press the = key. The result is 1.15×10^{-3}.

Practice Exercise

Perform the following operations:
(a) Write 67,000 in exponential notation, showing two significant figures.
(b) $(3.378 \times 10^{-3}) - (4.97 \times 10^{-5})$
(c) $(1.84 \times 10^{15})(7.45 \times 10^{-2})$
(d) $(6.67 \times 10^{-8})^3$

A.2 | Logarithms

Common Logarithms

The common, or base-10, logarithm (abbreviated log) of any number is the power to which 10 must be raised to equal the number. For example, the common logarithm of 1000 (written log 1000) is 3 because raising 10 to the third power gives 1000.

$$10^3 = 1000, \text{ therefore, } \log 1000 = 3$$

Further examples are

$$\log 10^5 = 5$$
$$\log 1 = 0 \quad \text{Remember that } 10^0 = 1$$
$$\log 10^{-2} = -2$$

In these examples the common logarithm can be obtained by inspection. However, it is not possible to obtain the logarithm of a number such as 31.25 by inspection. The logarithm of 31.25 is the number x that satisfies the following relationship:

$$10^x = 31.25$$

Most electronic calculators have a key labeled LOG that can be used to obtain logarithms. For example, on many calculators we obtain the value of log 31.25 by entering 31.25 and pressing the LOG key. We obtain the following result:

$$\log 31.25 = 1.4949$$

Notice that 31.25 is greater than 10 (10^1) and less than 100 (10^2). The value for log 31.25 is accordingly between log 10 and log 100, that is, between 1 and 2.

Significant Figures and Common Logarithms

For the common logarithm of a measured quantity, the number of digits after the decimal point equals the number of significant figures in the original number. For example, if 23.5 is a measured quantity (three significant figures), then $\log 23.5 = 1.371$ (three significant figures after the decimal point).

Antilogarithms

The process of determining the number that corresponds to a certain logarithm is known as obtaining an *antilogarithm*. It is the reverse of taking a logarithm. For example, we saw previously that $\log 23.5 = 1.371$. This means that the antilogarithm of 1.371 equals 23.5.

$$\log 23.5 = 1.371$$
$$\text{antilog } 1.371 = 23.5$$

The process of taking the antilog of a number is the same as raising 10 to a power equal to that number.

$$\text{antilog } 1.371 = 10^{1.371} = 23.5$$

Many calculators have a key labeled 10^x that allows you to obtain antilogs directly. On others, it will be necessary to press a key labeled INV (for *inverse*), followed by the LOG key.

Natural Logarithms

Logarithms based on the number e are called natural, or base e, logarithms (abbreviated ln). The natural log of a number is the power to which e (which has the value 2.71828…) must be raised to equal the number. For example, the natural log of 10 equals 2.303.

$$e^{2.303} = 10, \text{therefore } \ln 10 = 2.303$$

Your calculator probably has a key labeled LN that allows you to obtain natural logarithms. For example, to obtain the natural log of 46.8, you enter 46.8 and press the LN key.

$$\ln 46.8 = 3.846$$

The natural antilog of a number is e raised to a power equal to that number. If your calculator can calculate natural logs, it will also be able to calculate natural antilogs. On some calculators there is a key labeled e^x that allows you to calculate natural antilogs directly; on others, it will be necessary to first press the INV key followed by the LN key. For example, the natural antilog of 1.679 is given by

$$\text{Natural antilog } 1.679 = e^{1.679} = 5.36$$

The relation between common and natural logarithms is as follows:

$$\ln a = 2.303 \log a$$

Notice that the factor relating the two, 2.303, is the natural log of 10, which we calculated earlier.

Mathematical Operations Using Logarithms

Because logarithms are exponents, mathematical operations involving logarithms follow the rules for the use of exponents. For example, the product of z^a and z^b (where z is any number) is given by

$$z^a \cdot z^b = z^{(a+b)}$$

Similarly, the logarithm (either common or natural) of a product equals the *sum* of the logs of the individual numbers.

$$\log ab = \log a + \log b \qquad \ln ab = \ln a + \ln b$$

For the log of a quotient,

$$\log(a/b) = \log a - \log b \qquad \ln(a/b) = \ln a - \ln b$$

APPENDIX A Mathematical Operations

Using the properties of exponents, we can also derive the rules for the logarithm of a number raised to a certain power.

$$\log a^n = n \log a \qquad \ln a^n = n \ln a$$

$$\log a^{1/n} = (1/n)\log a \quad \ln a^{1/n} = (1/n)\ln a$$

pH Problems

One of the most frequent uses for common logarithms in general chemistry is in working pH problems. The pH is defined as $-\log[H^+]$, where $[H^+]$ is the hydrogen ion concentration of a solution. ∞ (Section 16.4) The following sample exercise illustrates this application.

SAMPLE EXERCISE 2 | Using Logarithms

(a) What is the pH of a solution whose hydrogen ion concentration is 0.015 M?
(b) If the pH of a solution is 3.80, what is its hydrogen ion concentration?

SOLUTION

(1) We are given the value of $[H^+]$. We use the LOG key of our calculator to calculate the value of $\log[H^+]$. The pH is obtained by changing the sign of the value obtained. (Be sure to change the sign *after* taking the logarithm.)

$$[H^+] = 0.015$$

$$\log[H^+] = -1.82 \quad \text{(2 significant figures)}$$

$$pH = -(-1.82) = 1.82$$

(2) To obtain the hydrogen ion concentration when given the pH, we must take the antilog of $-$pH.

$$pH = -\log[H^+] = 3.80$$

$$\log[H^+] = -3.80$$

$$[H^+] = \text{antilog}(-3.80) = 10^{-3.80} = 1.6 \times 10^{-4}\ M$$

Practice Exercise

Perform the following operations: (a) $\log(2.5 \times 10^{-5})$, (b) $\ln 32.7$, (c) antilog -3.47, (d) $e^{-1.89}$.

A.3 | Quadratic Equations

An algebraic equation of the form $ax^2 + bx + c = 0$ is called a *quadratic equation*. The two solutions to such an equation are given by the quadratic formula:

$$x = \frac{-b \pm \sqrt{b^2 - 4ac}}{2a}$$

Many calculators today can calculate the solutions to a quadratic equation with one or two keystrokes. Most of the time, x corresponds to the concentration of a chemical species in solution. Only one of the solutions will be a positive number, and that is the one you should use; a "negative concentration" has no physical meaning.

SAMPLE EXERCISE 3 | Using the Quadratic Formula

Find the values of x that satisfy the equation $2x^2 + 4x = 1$.

SOLUTION

To solve the given equation for x, we must first put it in the form

$$ax^2 + bx + c = 0$$

and then use the quadratic formula. If

$$2x^2 + 4x = 1$$

then

$$2x^2 + 4x - 1 = 0$$

Using the quadratic formula, where $a = 2, b = 4$, and $c = -1$, we have

$$x = \frac{-4 \pm \sqrt{4^2 - 4(2)(-1)}}{2(2)}$$

$$= \frac{-4 \pm \sqrt{16 + 8}}{4} = \frac{-4 \pm \sqrt{24}}{4} = \frac{-4 \pm 4.899}{4}$$

The two solutions are

$$x = \frac{0.899}{4} = 0.225 \quad \text{and} \quad x = \frac{-8.899}{4} = -2.225$$

If this was a problem in which x represented a concentration, we would say $x = 0.225$ (in the appropriate units), since a negative number for concentration has no physical meaning.

A.4 | Graphs

Often the clearest way to represent the interrelationship between two variables is to graph them. Usually, the variable that is being experimentally varied, called the *independent variable*, is shown along the horizontal axis (*x*-axis). The variable that responds to the change in the independent variable, called the *dependent variable*, is then shown along the vertical axis (*y*-axis). For example, consider an experiment in which we vary the temperature of an enclosed gas and measure its pressure. The independent variable is temperature and the dependent variable is pressure. The data shown in ▶ Table A.1 can be obtained by means of this experiment. These data are shown graphically in ▶ Figure A.1. The relationship between temperature and pressure is linear. The equation for any straight-line graph has the form

$$y = mx + b$$

where *m* is the slope of the line and *b* is the intercept with the *y*-axis. In the case of Figure A.1, we could say that the relationship between temperature and pressure takes the form

$$P = mT + b$$

where *P* is pressure in atm and *T* is temperature in °C. As shown in Figure A.1, the slope is 4.10×10^{-4} atm/°C, and the intercept—the point where the line crosses the *y*-axis—is 0.112 atm. Therefore, the equation for the line is

$$P = \left(4.10 \times 10^{-4} \frac{\text{atm}}{\text{°C}}\right) T + 0.112 \text{ atm}$$

Table A.1 Interrelation between Pressure and Temperature

Temperature (°C)	Pressure (atm)
20.0	0.120
30.0	0.124
40.0	0.128
50.0	0.132

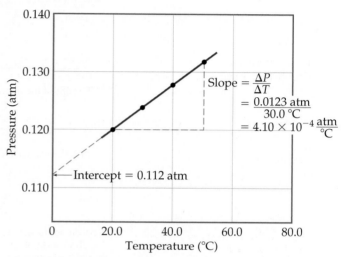

▲ **Figure A.1** A graph of pressure versus temperature yields a straight line for the data.

A.5 | Standard Deviation

The standard deviation from the mean, *s*, is a common method for describing precision in experimentally determined data. We define the standard deviation as

$$s = \sqrt{\frac{\sum_{i=1}^{N}(x_i - \bar{x})^2}{N - 1}}$$

where *N* is the number of measurements, $\bar{x}$ is the average (also called the mean) of the measurements, and x_i represent the individual measurements. Electronic calculators with built-in statistical functions can calculate *s* directly by inputting the individual measurements.

A smaller value of *s* indicates a higher precision, meaning that the data are more closely clustered around the average. The standard deviation has statistical significance. If a large number of measurements is made, 68% of the measured values is expected to be within one standard deviation of the average, assuming only random errors are associated with the measurements.

SAMPLE EXERCISE 4 Calculating an Average and Standard Deviation

The percent carbon in a sugar is measured four times: 42.01%, 42.28%, 41.79%, and 42.25%. Calculate **(a)** the average and **(b)** the standard deviation for these measurements.

SOLUTION

(a) The average is found by adding the quantities and dividing by the number of measurements:

$$\bar{x} = \frac{42.01 + 42.28 + 41.79 + 42.25}{4} = \frac{168.33}{4} = 42.08$$

(b) The standard deviation is found using the preceding equation:

$$s = \sqrt{\frac{\sum\limits_{i=1}^{N}(x_i - \bar{x})^2}{N - 1}}$$

Let's tabulate the data so the calculation of $\sum\limits_{i=1}^{N}(x_i - \bar{x})^2$ can be seen clearly.

Percent C	Difference between Measurement and Average, $(x_i - \bar{x})$	Square of Difference, $(x_i - x)^2$
42.01	$42.01 - 42.08 = -0.07$	$(-0.07)^2 = 0.005$
42.28	$42.28 - 42.08 = 0.20$	$(0.20)^2 = 0.040$
41.79	$41.79 - 42.08 = -0.29$	$(-0.29)^2 = 0.084$
42.25	$42.25 - 42.08 = 0.17$	$(0.17)^2 = 0.029$

The sum of the quantities in the last column is

$$\sum_{i=1}^{N}(x_i - \bar{x})^2 = 0.005 + 0.040 + 0.084 + 0.029 = 0.16$$

Thus, the standard deviation is

$$s = \sqrt{\frac{\sum\limits_{i=1}^{N}(x_i - \bar{x})^2}{N - 1}} = \sqrt{\frac{0.16}{4 - 1}} = \sqrt{\frac{0.16}{3}} = \sqrt{0.053} = 0.23$$

Based on these measurements, it would be appropriate to represent the measured percent carbon as 42.08 ± 0.23.

Properties of Water

Density:	0.99987 g/mL at 0 °C
	1.00000 g/mL at 4 °C
	0.99707 g/mL at 25 °C
	0.95838 g/mL at 100 °C
Heat (enthalpy) of fusion:	6.008 kJ/mol at 0 °C
Heat (enthalpy) of vaporization:	44.94 kJ/mol at 0 °C
	44.02 kJ/mol at 25 °C
	40.67 kJ/mol at 100 °C
Ion-product constant, K_w:	1.14×10^{-15} at 0 °C
	1.01×10^{-14} at 25 °C
	5.47×10^{-14} at 50 °C
Specific heat:	2.092 J/g-K = 2.092 J/g·°C for ice at −3 °C
	4.184 J/g-K = 4.184 J/g·°C for water at 25 °C
	1.841 J/g-K = 1.841 J/g·°C for steam at 100 °C

Vapor Pressure (torr) at Different Temperatures

T(°C)	P	T(°C)	P	T(°C)	P	T(°C)	P
0	4.58	21	18.65	35	42.2	92	567.0
5	6.54	22	19.83	40	55.3	94	610.9
10	9.21	23	21.07	45	71.9	96	657.6
12	10.52	24	22.38	50	92.5	98	707.3
14	11.99	25	23.76	55	118.0	100	760.0
16	13.63	26	25.21	60	149.4	102	815.9
17	14.53	27	26.74	65	187.5	104	875.1
18	15.48	28	28.35	70	233.7	106	937.9
19	16.48	29	30.04	80	355.1	108	1004.4
20	17.54	30	31.82	90	525.8	110	1074.6

Thermodynamic Quantities for Selected Substances at 298.15 K (25 °C)

Substance	ΔH_f° (kJ/mol)	ΔG_f° (kJ/mol)	S° (J/mol-K)	Substance	ΔH_f° (kJ/mol)	ΔG_f° (kJ/mol)	S° (J/mol-K)
Aluminum				$C_4H_{10}(g)$	−124.73	−15.71	310.0
$Al(s)$	0	0	28.32	$C_4H_{10}(l)$	−147.6	−15.0	231.0
$AlCl_3(s)$	−705.6	−630.0	109.3	$C_6H_6(g)$	82.9	129.7	269.2
$Al_2O_3(s)$	−1669.8	−1576.5	51.00	$C_6H_6(l)$	49.0	124.5	172.8
Barium				$CH_3OH(g)$	−201.2	−161.9	237.6
$Ba(s)$	0	0	63.2	$CH_3OH(l)$	−238.6	−166.23	126.8
$BaCO_3(s)$	−1216.3	−1137.6	112.1	$C_2H_5OH(g)$	−235.1	−168.5	282.7
$BaO(s)$	−553.5	−525.1	70.42	$C_2H_5OH(l)$	−277.7	−174.76	160.7
Beryllium				$C_6H_{12}O_6(s)$	−1273.02	−910.4	212.1
$Be(s)$	0	0	9.44	$CO(g)$	−110.5	−137.2	197.9
$BeO(s)$	−608.4	−579.1	13.77	$CO_2(g)$	−393.5	−394.4	213.6
$Be(OH)_2(s)$	−905.8	−817.9	50.21	$CH_3COOH(l)$	−487.0	−392.4	159.8
Bromine				**Cesium**			
$Br(g)$	111.8	82.38	174.9	$Cs(g)$	76.50	49.53	175.6
$Br^-(aq)$	−120.9	−102.8	80.71	$Cs(l)$	2.09	0.03	92.07
$Br_2(g)$	30.71	3.14	245.3	$Cs(s)$	0	0	85.15
$Br_2(l)$	0	0	152.3	$CsCl(s)$	−442.8	−414.4	101.2
$HBr(g)$	−36.23	−53.22	198.49	**Chlorine**			
Calcium				$Cl(g)$	121.7	105.7	165.2
$Ca(g)$	179.3	145.5	154.8	$Cl^-(aq)$	−167.2	−131.2	56.5
$Ca(s)$	0	0	41.4	$Cl_2(g)$	0	0	222.96
$CaCO_3(s, \text{calcite})$	−1207.1	−1128.76	92.88	$HCl(aq)$	−167.2	−131.2	56.5
$CaCl_2(s)$	−795.8	−748.1	104.6	$HCl(g)$	−92.30	−95.27	186.69
$CaF_2(s)$	−1219.6	−1167.3	68.87	**Chromium**			
$CaO(s)$	−635.5	−604.17	39.75	$Cr(g)$	397.5	352.6	174.2
$Ca(OH)_2(s)$	−986.2	−898.5	83.4	$Cr(s)$	0	0	23.6
$CaSO_4(s)$	−1434.0	−1321.8	106.7	$Cr_2O_3(s)$	−1139.7	−1058.1	81.2
Carbon				**Cobalt**			
$C(g)$	718.4	672.9	158.0	$Co(g)$	439	393	179
$C(s, \text{diamond})$	1.88	2.84	2.43	$Co(s)$	0	0	28.4
$C(s, \text{graphite})$	0	0	5.69	**Copper**			
$CCl_4(g)$	−106.7	−64.0	309.4	$Cu(g)$	338.4	298.6	166.3
$CCl_4(l)$	−139.3	−68.6	214.4	$Cu(s)$	0	0	33.30
$CF_4(g)$	−679.9	−635.1	262.3	$CuCl_2(s)$	−205.9	−161.7	108.1
$CH_4(g)$	−74.8	−50.8	186.3	$CuO(s)$	−156.1	−128.3	42.59
$C_2H_2(g)$	226.77	209.2	200.8	$Cu_2O(s)$	−170.7	−147.9	92.36
$C_2H_4(g)$	52.30	68.11	219.4	**Fluorine**			
$C_2H_6(g)$	−84.68	−32.89	229.5	$F(g)$	80.0	61.9	158.7
$C_3H_8(g)$	−103.85	−23.47	269.9	$F^-(aq)$	−332.6	−278.8	−13.8

Substance	ΔH_f° (kJ/mol)	ΔG_f° (kJ/mol)	S° (J/mol-K)	Substance	ΔH_f° (kJ/mol)	ΔG_f° (kJ/mol)	S° (J/mol-K)
$F_2(g)$	0	0	202.7	$MnO(s)$	−385.2	−362.9	59.7
$HF(g)$	−268.61	−270.70	173.51	$MnO_2(s)$	−519.6	−464.8	53.14
Hydrogen				$MnO_4^-(aq)$	−541.4	−447.2	191.2
$H(g)$	217.94	203.26	114.60	**Mercury**			
$H^+(aq)$	0	0	0	$Hg(g)$	60.83	31.76	174.89
$H^+(g)$	1536.2	1517.0	108.9	$Hg(l)$	0	0	77.40
$H_2(g)$	0	0	130.58	$HgCl_2(s)$	−230.1	−184.0	144.5
Iodine				$Hg_2Cl_2(s)$	−264.9	−210.5	192.5
$I(g)$	106.60	70.16	180.66	**Nickel**			
$I^-(g)$	−55.19	−51.57	111.3	$Ni(g)$	429.7	384.5	182.1
$I_2(g)$	62.25	19.37	260.57	$Ni(s)$	0	0	29.9
$I_2(s)$	0	0	116.73	$NiCl_2(s)$	−305.3	−259.0	97.65
$HI(g)$	25.94	1.30	206.3	$NiO(s)$	−239.7	−211.7	37.99
Iron				**Nitrogen**			
$Fe(g)$	415.5	369.8	180.5	$N(g)$	472.7	455.5	153.3
$Fe(s)$	0	0	27.15	$N_2(g)$	0	0	191.50
$Fe^{2+}(aq)$	−87.86	−84.93	113.4	$NH_3(aq)$	−80.29	−26.50	111.3
$Fe^{3+}(aq)$	−47.69	−10.54	293.3	$NH_3(g)$	−46.19	−16.66	192.5
$FeCl_2(s)$	−341.8	−302.3	117.9	$NH_4^+(aq)$	−132.5	−79.31	113.4
$FeCl_3(s)$	−400	−334	142.3	$N_2H_4(g)$	95.40	159.4	238.5
$FeO(s)$	−271.9	−255.2	60.75	$NH_4CN(s)$	0.4	—	—
$Fe_2O_3(s)$	−822.16	−740.98	89.96	$NH_4Cl(s)$	−314.4	−203.0	94.6
$Fe_3O_4(s)$	−1117.1	−1014.2	146.4	$NH_4NO_3(s)$	−365.6	−184.0	151
$FeS_2(s)$	−171.5	−160.1	52.92	$NO(g)$	90.37	86.71	210.62
Lead				$NO_2(g)$	33.84	51.84	240.45
$Pb(s)$	0	0	68.85	$N_2O(g)$	81.6	103.59	220.0
$PbBr_2(s)$	−277.4	−260.7	161	$N_2O_4(g)$	9.66	98.28	304.3
$PbCO_3(s)$	−699.1	−625.5	131.0	$NOCl(g)$	52.6	66.3	264
$Pb(NO_3)_2(aq)$	−421.3	−246.9	303.3	$HNO_3(aq)$	−206.6	−110.5	146
$Pb(NO_3)_2(s)$	−451.9	—	—	$HNO_3(g)$	−134.3	−73.94	266.4
$PbO(s)$	−217.3	−187.9	68.70	**Oxygen**			
Lithium				$O(g)$	247.5	230.1	161.0
$Li(g)$	159.3	126.6	138.8	$O_2(g)$	0	0	205.0
$Li(s)$	0	0	29.09	$O_3(g)$	142.3	163.4	237.6
$Li^+(aq)$	−278.5	−273.4	12.2	$OH^-(aq)$	−230.0	−157.3	−10.7
$Li^+(g)$	685.7	648.5	133.0	$H_2O(g)$	−241.82	−228.57	188.83
$LiCl(s)$	−408.3	−384.0	59.30	$H_2O(l)$	−285.83	−237.13	69.91
Magnesium				$H_2O_2(g)$	−136.10	−105.48	232.9
$Mg(g)$	147.1	112.5	148.6	$H_2O_2(l)$	−187.8	−120.4	109.6
$Mg(s)$	0	0	32.51	**Phosphorus**			
$MgCl_2(s)$	−641.6	−592.1	89.6	$P(g)$	316.4	280.0	163.2
$MgO(s)$	−601.8	−569.6	26.8	$P_2(g)$	144.3	103.7	218.1
$Mg(OH)_2(s)$	−924.7	−833.7	63.24	$P_4(g)$	58.9	24.4	280
Manganese				$P_4(s, red)$	−17.46	−12.03	22.85
$Mn(g)$	280.7	238.5	173.6	$P_4(s, white)$	0	0	41.08
$Mn(s)$	0	0	32.0	$PCl_3(g)$	−288.07	−269.6	311.7

Substance	ΔH_f° (kJ/mol)	ΔG_f° (kJ/mol)	S° (J/mol-K)	Substance	ΔH_f° (kJ/mol)	ΔG_f° (kJ/mol)	S° (J/mol-K)
$PCl_3(l)$	−319.6	−272.4	217	**Sodium**			
$PF_5(g)$	−1594.4	−1520.7	300.8	$Na(g)$	107.7	77.3	153.7
$PH_3(g)$	5.4	13.4	210.2	$Na(s)$	0	0	51.45
$P_4O_6(s)$	−1640.1	—	—	$Na^+(aq)$	−240.1	−261.9	59.0
$P_4O_{10}(s)$	−2940.1	−2675.2	228.9	$Na^+(g)$	609.3	574.3	148.0
$POCl_3(g)$	−542.2	−502.5	325	$NaBr(aq)$	−360.6	−364.7	141.00
$POCl_3(l)$	−597.0	−520.9	222	$NaBr(s)$	−361.4	−349.3	86.82
$H_3PO_4(aq)$	−1288.3	−1142.6	158.2	$Na_2CO_3(s)$	−1130.9	−1047.7	136.0
Potassium				$NaCl(aq)$	−407.1	−393.0	115.5
$K(g)$	89.99	61.17	160.2	$NaCl(g)$	−181.4	−201.3	229.8
$K(s)$	0	0	64.67	$NaCl(s)$	−410.9	−384.0	72.33
$K^+(aq)$	−252.4	−283.3	102.5	$NaHCO_3(s)$	−947.7	−851.8	102.1
$K^+(g)$	514.2	481.2	154.5	$NaNO_3(aq)$	−446.2	−372.4	207
$KCl(s)$	−435.9	−408.3	82.7	$NaNO_3(s)$	−467.9	−367.0	116.5
$KClO_3(s)$	−391.2	−289.9	143.0	$NaOH(aq)$	−469.6	−419.2	49.8
$KClO_3(aq)$	−349.5	−284.9	265.7	$NaOH(s)$	−425.6	−379.5	64.46
$K_2CO_3(s)$	−1150.18	−1064.58	155.44	$Na_2SO_4(s)$	−1387.1	−1270.2	149.6
$KNO_3(s)$	−492.70	−393.13	132.9	**Strontium**			
$K_2O(s)$	−363.2	−322.1	94.14	$SrO(s)$	−592.0	−561.9	54.9
$KO_2(s)$	−284.5	−240.6	122.5	$Sr(g)$	164.4	110.0	164.6
$K_2O_2(s)$	−495.8	−429.8	113.0	**Sulfur**			
$KOH(s)$	−424.7	−378.9	78.91	$S(s, \text{rhombic})$	0	0	31.88
$KOH(aq)$	−482.4	−440.5	91.6	$S_8(g)$	102.3	49.7	430.9
Rubidium				$SO_2(g)$	−296.9	−300.4	248.5
$Rb(g)$	85.8	55.8	170.0	$SO_3(g)$	−395.2	−370.4	256.2
$Rb(s)$	0	0	76.78	$SO_4^{2-}(aq)$	−909.3	−744.5	20.1
$RbCl(s)$	−430.5	−412.0	92	$SOCl_2(l)$	−245.6	—	—
$RbClO_3(s)$	−392.4	−292.0	152	$H_2S(g)$	−20.17	−33.01	205.6
Scandium				$H_2SO_4(aq)$	−909.3	−744.5	20.1
$Sc(g)$	377.8	336.1	174.7	$H_2SO_4(l)$	−814.0	−689.9	156.1
$Sc(s)$	0	0	34.6	**Titanium**			
Selenium				$Ti(g)$	468	422	180.3
$H_2Se(g)$	29.7	15.9	219.0	$Ti(s)$	0	0	30.76
Silicon				$TiCl_4(g)$	−763.2	−726.8	354.9
$Si(g)$	368.2	323.9	167.8	$TiCl_4(l)$	−804.2	−728.1	221.9
$Si(s)$	0	0	18.7	$TiO_2(s)$	−944.7	−889.4	50.29
$SiC(s)$	−73.22	−70.85	16.61	**Vanadium**			
$SiCl_4(l)$	−640.1	−572.8	239.3	$V(g)$	514.2	453.1	182.2
$SiO_2(s, \text{quartz})$	−910.9	−856.5	41.84	$V(s)$	0	0	28.9
Silver				**Zinc**			
$Ag(s)$	0	0	42.55	$Zn(g)$	130.7	95.2	160.9
$Ag^+(aq)$	105.90	77.11	73.93	$Zn(s)$	0	0	41.63
$AgCl(s)$	−127.0	−109.70	96.11	$ZnCl_2(s)$	−415.1	−369.4	111.5
$Ag_2O(s)$	−31.05	−11.20	121.3	$ZnO(s)$	−348.0	−318.2	43.9
$AgNO_3(s)$	−124.4	−33.41	140.9				

Aqueous Equilibrium Constants

Table D.1 Dissociation Constants for Acids at 25 °C

Name	Formula	K_{a1}	K_{a2}	K_{a3}
Acetic acid	CH_3COOH (or $HC_2H_3O_2$)	1.8×10^{-5}		
Arsenic acid	H_3AsO_4	5.6×10^{-3}	1.0×10^{-7}	3.0×10^{-12}
Arsenous acid	H_3AsO_3	5.1×10^{-10}		
Ascorbic acid	$H_2C_6H_6O_6$	8.0×10^{-5}	1.6×10^{-12}	
Benzoic acid	C_6H_5COOH (or $HC_7H_5O_2$)	6.3×10^{-5}		
Boric acid	H_3BO_3	5.8×10^{-10}		
Butanoic acid	C_3H_7COOH (or $HC_4H_7O_2$)	1.5×10^{-5}		
Carbonic acid	H_2CO_3	4.3×10^{-7}	5.6×10^{-11}	
Chloroacetic acid	$CH_2ClCOOH$ (or $HC_2H_2O_2Cl$)	1.4×10^{-3}		
Chlorous acid	$HClO_2$	1.1×10^{-2}		
Citric acid	$HOOCC(OH)(CH_2COOH)_2$ (or $H_3C_6H_5O_7$)	7.4×10^{-4}	1.7×10^{-5}	4.0×10^{-7}
Cyanic acid	$HCNO$	3.5×10^{-4}		
Formic acid	$HCOOH$ (or $HCHO_2$)	1.8×10^{-4}		
Hydroazoic acid	HN_3	1.9×10^{-5}		
Hydrocyanic acid	HCN	4.9×10^{-10}		
Hydrofluoric acid	HF	6.8×10^{-4}		
Hydrogen chromate ion	$HCrO_4^-$	3.0×10^{-7}		
Hydrogen peroxide	H_2O_2	2.4×10^{-12}		
Hydrogen selenate ion	$HSeO_4^-$	2.2×10^{-2}		
Hydrogen sulfide	H_2S	9.5×10^{-8}	1×10^{-19}	
Hypobromous acid	$HBrO$	2.5×10^{-9}		
Hypochlorous acid	$HClO$	3.0×10^{-8}		
Hypoiodous acid	HIO	2.3×10^{-11}		
Iodic acid	HIO_3	1.7×10^{-1}		
Lactic acid	$CH_3CH(OH)COOH$ (or $HC_3H_5O_3$)	1.4×10^{-4}		
Malonic acid	$CH_2(COOH)_2$ (or $H_2C_3H_2O_4$)	1.5×10^{-3}	2.0×10^{-6}	
Nitrous acid	HNO_2	4.5×10^{-4}		
Oxalic acid	$(COOH)_2$ (or $H_2C_2O_4$)	5.9×10^{-2}	6.4×10^{-5}	
Paraperiodic acid	H_5IO_6	2.8×10^{-2}	5.3×10^{-9}	
Phenol	C_6H_5OH (or HC_6H_5O)	1.3×10^{-10}		
Phosphoric acid	H_3PO_4	7.5×10^{-3}	6.2×10^{-8}	4.2×10^{-13}
Propionic acid	C_2H_5COOH (or $HC_3H_5O_2$)	1.3×10^{-5}		
Pyrophosphoric acid	$H_4P_2O_7$	3.0×10^{-2}	4.4×10^{-3}	2.1×10^{-7}
Selenous acid	H_2SeO_3	2.3×10^{-3}	5.3×10^{-9}	
Sulfuric acid	H_2SO_4	Strong acid	1.2×10^{-2}	
Sulfurous acid	H_2SO_3	1.7×10^{-2}	6.4×10^{-8}	
Tartaric acid	$HOOC(CHOH)_2COOH$ (or $H_2C_4H_4O_6$)	1.0×10^{-3}		

Table D.2 Dissociation Constants for Bases at 25 °C

Name	Formula	K_b
Ammonia	NH_3	1.8×10^{-5}
Aniline	$C_6H_5NH_2$	4.3×10^{-10}
Dimethylamine	$(CH_3)_2NH$	5.4×10^{-4}
Ethylamine	$C_2H_5NH_2$	6.4×10^{-4}
Hydrazine	H_2NNH_2	1.3×10^{-6}
Hydroxylamine	$HONH_2$	1.1×10^{-8}
Methylamine	CH_3NH_2	4.4×10^{-4}
Pyridine	C_5H_5N	1.7×10^{-9}
Trimethylamine	$(CH_3)_3N$	6.4×10^{-5}

Table D.3 Solubility-Product Constants for Compounds at 25 °C

Name	Formula	K_{sp}	Name	Formula	K_{sp}
Barium carbonate	$BaCO_3$	5.0×10^{-9}	Lead(II) fluoride	PbF_2	3.6×10^{-8}
Barium chromate	$BaCrO_4$	2.1×10^{-10}	Lead(II) sulfate	$PbSO_4$	6.3×10^{-7}
Barium fluoride	BaF_2	1.7×10^{-6}	Lead(II) sulfide*	PbS	3×10^{-28}
Barium oxalate	BaC_2O_4	1.6×10^{-6}	Magnesium hydroxide	$Mg(OH)_2$	1.8×10^{-11}
Barium sulfate	$BaSO_4$	1.1×10^{-10}	Magnesium carbonate	$MgCO_3$	3.5×10^{-8}
Cadmium carbonate	$CdCO_3$	1.8×10^{-14}	Magnesium oxalate	MgC_2O_4	8.6×10^{-5}
Cadmium hydroxide	$Cd(OH)_2$	2.5×10^{-14}	Manganese(II) carbonate	$MnCO_3$	5.0×10^{-10}
Cadmium sulfide*	CdS	8×10^{-28}	Manganese(II) hydroxide	$Mn(OH)_2$	1.6×10^{-13}
Calcium carbonate (calcite)	$CaCO_3$	4.5×10^{-9}	Manganese(II) sulfide*	MnS	2×10^{-53}
Calcium chromate	$CaCrO_4$	4.5×10^{-9}	Mercury(I) chloride	Hg_2Cl_2	1.2×10^{-18}
Calcium fluoride	CaF_2	3.9×10^{-11}	Mercury(I) iodide	Hg_2I_2	$1.1 \times 10^{-1.1}$
Calcium hydroxide	$Ca(OH)_2$	6.5×10^{-6}	Mercury(II) sulfide*	HgS	2×10^{-53}
Calcium phosphate	$Ca_3(PO_4)_2$	2.0×10^{-29}	Nickel(II) carbonate	$NiCO_3$	1.3×10^{-7}
Calcium sulfate	$CaSO_4$	2.4×10^{-5}	Nickel(II) hydroxide	$Ni(OH)_2$	6.0×10^{-16}
Chromium(III) hydroxide	$Cr(OH)_3$	6.7×10^{-31}	Nickel(II) sulfide*	NiS	3×10^{-20}
Cobalt(II) carbonate	$CoCO_3$	1.0×10^{-10}	Silver bromate	$AgBrO_3$	5.5×10^{-13}
Cobalt(II) hydroxide	$Co(OH)_2$	1.3×10^{-15}	Silver bromide	$AgBr$	5.0×10^{-13}
Cobalt(II) sulfide*	CoS	5×10^{-22}	Silver carbonate	Ag_2CO_3	8.1×10^{-12}
Copper(I) bromide	$CuBr$	5.3×10^{-9}	Silver chloride	$AgCl$	1.8×10^{-10}
Copper(II) carbonate	$CuCO_3$	2.3×10^{-10}	Silver chromate	Ag_2CrO_4	1.2×10^{-12}
Copper(II) hydroxide	$Cu(OH)_2$	4.8×10^{-20}	Silver iodide	AgI	8.3×10^{-17}
Copper(II) sulfide*	CuS	6×10^{-37}	Silver sulfate	Ag_2SO_4	1.5×10^{-5}
Iron(II) carbonate	$FeCO_3$	2.1×10^{-11}	Silver sulfide*	Ag_2S	6×10^{-51}
Iron(II) hydroxide	$Fe(OH)_2$	7.9×10^{-16}	Strontium carbonate	$SrCO_3$	9.3×10^{-10}
Lanthanum fluoride	LaF_3	2×10^{-19}	Tin(II) sulfide*	SnS	1×10^{-26}
Lanthanum iodate	$La(IO_3)_3$	7.4×10^{-14}	Zinc carbonate	$ZnCO_3$	1.0×10^{-10}
Lead(II) carbonate	$PbCO_3$	7.4×10^{-14}	Zinc hydroxide	$Zn(OH)_2$	3.0×10^{-16}
Lead(II) chloride	$PbCl_2$	1.7×10^{-5}	Zinc oxalate	ZnC_2O_4	2.7×10^{-8}
Lead(II) chromate	$PbCrO_4$	2.8×10^{-13}	Zinc sulfide*	ZnS	2×10^{-25}

*For a solubility equilibrium of the type $MS(s) + H_2O(l) \rightleftharpoons M^{2+}(aq) + HS^-(aq) + OH^-(aq)$

Standard Reduction Potentials at 25 °C

Half-Reaction	E° (V)	Half-Reaction	E° (V)
$Ag^+(aq) + e^- \longrightarrow Ag(s)$	+0.799	$2\,H_2O(l) + 2\,e^- \longrightarrow H_2(g) + 2\,OH^-(aq)$	−0.83
$AgBr(s) + e^- \longrightarrow Ag(s) + Br^-(aq)$	+0.095	$HO_2^-(aq) + H_2O(l) + 2\,e^- \longrightarrow 3\,OH^-(aq)$	+0.88
$AgCl(s) + e^- \longrightarrow Ag(s) + Cl^-(aq)$	+0.222	$H_2O_2(aq) + 2\,H^+(aq) + 2\,e^- \longrightarrow 2\,H_2O(l)$	+1.776
$Ag(CN)_2^-(aq) + e^- \longrightarrow Ag(s) + 2\,CN^-(aq)$	−0.31	$Hg_2^{2+}(aq) + 2\,e^- \longrightarrow 2\,Hg(l)$	+0.789
$Ag_2CrO_4(s) + 2\,e^- \longrightarrow 2\,Ag(s) + CrO_4^{2-}(aq)$	+0.446	$2\,Hg^{2+}(aq) + 2\,e^- \longrightarrow Hg_2^{2+}(aq)$	+0.920
$AgI(s) + e^- \longrightarrow Ag(s) + I^-(aq)$	−0.151	$Hg^{2+}(aq) + 2\,e^- \longrightarrow Hg(l)$	+0.854
$Ag(S_2O_3)_2^{3-}(aq) + e^- \longrightarrow Ag(s) + 2\,S_2O_3^{2-}(aq)$	+0.01	$I_2(s) + 2\,e^- \longrightarrow 2\,I^-(aq)$	+0.536
$Al^{3+}(aq) + 3\,e^- \longrightarrow Al(s)$	−1.66	$2\,IO_3^-(aq) + 12\,H^+(aq) + 10\,e^- \longrightarrow I_2(s) + 6\,H_2O(l)$	+1.195
$H_3AsO_4(aq) + 2\,H^+(aq) + 2\,e^- \longrightarrow H_3AsO_3(aq) + H_2O(l)$	+0.559	$K^+(aq) + e^- \longrightarrow K(s)$	−2.925
$Ba^{2+}(aq) + 2\,e^- \longrightarrow Ba(s)$	−2.90	$Li^+(aq) + e^- \longrightarrow Li(s)$	−3.05
$BiO^+(aq) + 2\,H^+(aq) + 3\,e^- \longrightarrow Bi(s) + H_2O(l)$	+0.32	$Mg^{2+}(aq) + 2\,e^- \longrightarrow Mg(s)$	−2.37
$Br_2(l) + 2\,e^- \longrightarrow 2\,Br^-(aq)$	+1.065	$Mn^{2+}(aq) + 2\,e^- \longrightarrow Mn(s)$	−1.18
$2\,BrO_3^-(aq) + 12\,H^+(aq) + 10\,e^- \longrightarrow Br_2(l) + 6\,H_2O(l)$	+1.52	$MnO_2(s) + 4\,H^+(aq) + 2\,e^- \longrightarrow Mn^{2+}(aq) + 2\,H_2O(l)$	+1.23
$2\,CO_2(g) + 2\,H^+(aq) + 2\,e^- \longrightarrow H_2C_2O_4(aq)$	−0.49	$MnO_4^-(aq) + 8\,H^+(aq) + 5\,e^- \longrightarrow Mn^{2+}(aq) + 4\,H_2O(l)$	+1.51
$Ca^{2+}(aq) + 2\,e^- \longrightarrow Ca(s)$	−2.87	$MnO_4^-(aq) + 2\,H_2O(l) + 3\,e^- \longrightarrow MnO_2(s) + 4\,OH^-(aq)$	+0.59
$Cd^{2+}(aq) + 2\,e^- \longrightarrow Cd(s)$	−0.403	$HNO_2(aq) + H^+(aq) + e^- \longrightarrow NO(g) + H_2O(l)$	+1.00
$Ce^{4+}(aq) + e^- \longrightarrow Ce^{3+}(aq)$	+1.61	$N_2(g) + 4\,H_2O(l) + 4\,e^- \longrightarrow 4\,OH^-(aq) + N_2H_4(aq)$	−1.16
$Cl_2(g) + 2\,e^- \longrightarrow 2\,Cl^-(aq)$	+1.359	$N_2(g) + 5\,H^+(aq) + 4\,e^- \longrightarrow N_2H_5^+(aq)$	−0.23
$2\,HClO(aq) + 2\,H^+(aq) + 2\,e^- \longrightarrow Cl_2(g) + 2\,H_2O(l)$	+1.63	$NO_3^-(aq) + 4\,H^+(aq) + 3\,e^- \longrightarrow NO(g) + 2\,H_2O(l)$	+0.96
$ClO^-(aq) + H_2O(l) + 2\,e^- \longrightarrow Cl^-(aq) + 2\,OH^-(aq)$	+0.89	$Na^+(aq) + e^- \longrightarrow Na(s)$	−2.71
$2\,ClO_3^-(aq) + 12\,H^+(aq) + 10\,e^- \longrightarrow Cl_2(g) + 6\,H_2O(l)$	+1.47	$Ni^{2+}(aq) + 2\,e^- \longrightarrow Ni(s)$	−0.28
$Co^{2+}(aq) + 2\,e^- \longrightarrow Co(s)$	−0.277	$O_2(g) + 4\,H^+(aq) + 4\,e^- \longrightarrow 2\,H_2O(l)$	+1.23
$Co^{3+}(aq) + e^- \longrightarrow Co^{2+}(aq)$	+1.842	$O_2(g) + 2\,H_2O(l) + 4\,e^- \longrightarrow 4\,OH^-(aq)$	+0.40
$Cr^{3+}(aq) + 3\,e^- \longrightarrow Cr(s)$	−0.74	$O_2(g) + 2\,H^+(aq) + 2\,e^- \longrightarrow H_2O_2(aq)$	+0.68
$Cr^{3+}(aq) + e^- \longrightarrow Cr^{2+}(aq)$	−0.41	$O_3(g) + 2\,H^+(aq) + 2\,e^- \longrightarrow O_2(g) + H_2O(l)$	+2.07
$Cr_2O_7^{2-}(aq) + 14\,H^+(aq) + 6\,e^- \longrightarrow 2\,Cr^{3+}(aq) + 7\,H_2O(l)$	+1.33	$Pb^{2+}(aq) + 2\,e^- \longrightarrow Pb(s)$	−0.126
$CrO_4^{2-}(aq) + 4\,H_2O(l) + 3\,e^- \longrightarrow$ $Cr(OH)_3(s) + 5\,OH^-(aq)$	−0.13	$PbO_2(s) + HSO_4^-(aq) + 3\,H^+(aq) + 2\,e^- \longrightarrow$ $PbSO_4(s) + 2\,H_2O(l)$	+1.685
$Cu^{2+}(aq) + 2\,e^- \longrightarrow Cu(s)$	+0.337	$PbSO_4(s) + H^+(aq) + 2\,e^- \longrightarrow Pb(s) + HSO_4^-(aq)$	−0.356
$Cu^{2+}(aq) + e^- \longrightarrow Cu^+(aq)$	+0.153	$PtCl_4^{2-}(aq) + 2\,e^- \longrightarrow Pt(s) + 4\,Cl^-(aq)$	+0.73
$Cu^+(aq) + e^- \longrightarrow Cu(s)$	+0.521	$S(s) + 2\,H^+(aq) + 2\,e^- \longrightarrow H_2S(g)$	+0.141
$CuI(s) + e^- \longrightarrow Cu(s) + I^-(aq)$	−0.185	$H_2SO_3(aq) + 4\,H^+(aq) + 4\,e^- \longrightarrow S(s) + 3\,H_2O(l)$	+0.45
$F_2(g) + 2\,e^- \longrightarrow 2\,F^-(aq)$	+2.87	$HSO_4^-(aq) + 3\,H^+(aq) + 2\,e^- \longrightarrow H_2SO_3(aq) + H_2O(l)$	+0.17
$Fe^{2+}(aq) + 2\,e^- \longrightarrow Fe(s)$	−0.440	$Sn^{2+}(aq) + 2\,e^- \longrightarrow Sn(s)$	−0.136
$Fe^{3+}(aq) + e^- \longrightarrow Fe^{2+}(aq)$	+0.771	$Sn^{4+}(aq) + 2\,e^- \longrightarrow Sn^{2+}(aq)$	+0.154
$Fe(CN)_6^{3-}(aq) + e^- \longrightarrow Fe(CN)_6^{4-}(aq)$	+0.36	$VO_2^+(aq) + 2\,H^+(aq) + e^- \longrightarrow VO^{2+}(aq) + H_2O(l)$	+1.00
$2\,H^+(aq) + 2\,e^- \longrightarrow H_2(g)$	0.000	$Zn^{2+}(aq) + 2\,e^- \longrightarrow Zn(s)$	−0.763

ANSWERS TO SELECTED EXERCISES

Chapter 1

1.1 (a) Pure element: i (b) mixture of elements: v, vi (c) pure compound: iv (d) mixture of an element and a compound: ii, iii **1.3** This kind of separation based on solubility differences is called *extraction*. The insoluble grounds are then separated from the coffee solution by *filtration*. **1.5** (a) The aluminum sphere is lightest, then nickel, then silver. (b) The platinum sphere is smallest, then gold, then lead. **1.7** (a) 7.5 cm; two significant figures (sig figs) (b) 72 mi/hr (inner scale, two significant figures) or 115 km/hr (outer scale, three significant figures) **1.9** Arrange the conversion factor so that the given unit cancels and the desired unit is in the correct position. **1.11** 464 jelly beans. The mass of an average bean has 2 decimal places and 3 sig figs. The number of beans then has 3 sig figs, by the rules for multiplication and division. **1.13** (a) Heterogeneous mixture (b) homogeneous mixture (heterogeneous if there are undissolved particles) (c) pure substance (d) pure substance. **1.15** (a) S (b) Au (c) K (d) Cl (e) Cu (f) uranium (g) nickel (h) sodium (i) aluminum (j) silicon **1.17** C is a compound; it contains both carbon and oxygen. A is a compound; it contains at least carbon and oxygen. B is not defined by the data given; it is probably also a compound because few elements exist as white solids. **1.19** Physical properties: silvery white; lustrous; melting point $= 649\,°C$, boiling point $= 1105\,°C$; density at $20\,°C = 1.738\,g/cm^3$; pounded into sheets; drawn into wires; good conductor. Chemical properties: burns in air; reacts with Cl_2. **1.21** (a) Chemical (b) physical (c) physical (d) chemical (e) chemical **1.23** (a) Add water to dissolve the sugar; filter this mixture, collecting the sand on the filter paper and the sugar water in the flask. Evaporate water from the flask to recover solid sugar. (b) Allow the mixture to settle so that there are two distinct layers. Carefully pour off most of the top oil layer. After the layers reform, use a dropper to remove any remaining oil. Vinegar is in the original vessel and oil is in a second container. **1.25** (a) 1×10^{-1} (b) 1×10^{-2} (c) 1×10^{-15} (d) 1×10^{-6} (e) 1×10^{6} (f) 1×10^{3} (g) 1×10^{-9} (h) 1×10^{-3} (i) 1×10^{-12} **1.27** (a) 22 °C (b) 422.1 °F (c) 506 K (d) 107 °F (e) 1644 K (f) −459.67 °F **1.29** (a) 1.62 g/mL. Tetrachloroethylene, 1.62 g/mL, is more dense than water, 1.00 g/mL; tetrachloroethylene will sink rather than float on water. (b) 11.7 g **1.31** (a) Calculated density $= 0.86\,g/mL$. The substance is probably toluene, density $= 0.866\,g/mL$. (b) 40.4 mL ethylene glycol (c) 1.11×10^{3} g nickel **1.33** 32 Pg **1.35** Exact: (b), (d), and (f) **1.37** (a) 3 (b) 2 (c) 5 (d) 3 (e) 5 (f) 1 **1.39** (a) 1.025×10^{2} (b) 6.570×10^{2} (c) 8.543×10^{-3} (d) 2.579×10^{-4} (e) -3.572×10^{-2} **1.41** (a) 17.00 (b) 812.0 (c) 8.23×10^{3} (d) 8.69×10^{-2} **1.43** 5 significant figures

1.45 (a) $\dfrac{1 \times 10^{-3}\ m}{1\ mm} \times \dfrac{1\ nm}{1 \times 10^{-9}\ m}$ (b) $\dfrac{1 \times 10^{-3}\ g}{1\ mg} \times \dfrac{1\ kg}{1000\ g}$

(c) $\dfrac{1000\ m}{1\ km} \times \dfrac{1\ cm}{1 \times 10^{-2}\ m} \times \dfrac{1\ in.}{2.54\ cm} \times \dfrac{1\ ft}{12\ in.}$ (d) $\dfrac{(2.54)^3\ cm^3}{1^3\ in.^3}$

1.47 (a) 54.7 km/hr (b) 1.3×10^{3} gal (c) 46.0 m (d) 0.984 in/hr **1.49** The volume of the box is $1.52 \times 10^{4}\ cm^3$. The uncertainty in the calculated volume is $\pm 0.4 \times 10^{4}\ cm^3$ cm. **1.51** (a) 4.32×10^{5} s (b) 88.5 m (c) \$0.499/L (d) 46.6 km/hr (e) 1.420 L/s (f) 707.9 cm³ **1.53** (a) 1.2×10^{2} L (b) 5×10^{2} mg (c) 19.9 mi/gal (2×10^{1} mi/gal for 1 significant figure) (d) 1.81 kg **1.55** 64 kg air **1.57** \$6 × 10⁴ **1.61** 8.47 g O; the *law of constant composition* **1.63** (a) Set I, 22.51; set II, 22.61. Based on the average, set I is more accurate. (b) The average deviation for both set I and set II is 0.02. The two sets display the same precision. **1.65** (a) Volume (b) area (c) volume (d) density (e) time (f) length (g) temperature **1.68** Substances (c), (d), (e), (g) and (h) are pure or nearly pure. **1.69** (a) 1.13×10^{5} quarters (b) 6.41×10^{5} g (c) $\$2.83 \times 10^{4}$ (d) 5.74×10^{8} stacks **1.73** The most dense liquid, Hg, will sink; the least dense, cyclohexane, will float; H_2O will be in the middle. **1.76** density of solid $= 1.63\,g/mL$ **1.79** (a) Density of peat $= 0.13\,g/cm^3$, density of soil $= 2.5\,g/cm^3$. It is not correct to say that peat is "lighter" than topsoil. Volumes must be specified in order to compare masses. (b) Buy 16 bags of peat (more than 15 are needed). (Results to 1 significant figure are not meaningful.) **1.83** The inner diameter of the tube is 1.71 cm. **1.85** The separation is successful if two distinct spots are seen on the paper. To quantify the characteristics of the separation, calculate a reference value for each spot: distance traveled by spot/distance traveled by solvent. If the values for the two spots are fairly different, the separation is successful. **1.88** (a) Volume $= 0.050$ mL (b) surface area $= 12.4\ m^2$ (c) 99.99% of the mercury was removed. (c) The spongy material weighs 17.7 mg after exposure to mercury.

Chapter 2

2.1 (a) The path of the charged particle bends because the particle is repelled by the negatively charged plate and attracted to the positively charged plate. (b) $(-)$ (c) increase (d) decrease **2.4** The particle is an ion. $^{32}_{16}S^{2-}$ **2.6** Formula: IF_5; name: iodine pentafluoride; the compound is molecular. **2.8** Only $Ca(NO_3)_2$, calcium nitrate, is consistent with the diagram. **2.10** (a) In the presence of an electric field, there is electrostatic attraction between the negatively charged oil drops and the positively charged plate as well as electrostatic repulsion between the negatively charged oil drops and the negatively charged plate. These electrostatic forces oppose the force of gravity and change the rate of fall of the drops. (b) Each individual drop has a different number of electrons associated with it. If the combined electrostatic forces are greater than the force of gravity, the drop moves up. **2.11** Postulate 4 of the atomic theory states that the relative number and kinds of atoms in a compound are constant, regardless of the source. Therefore, 1.0 g of pure water should always contain the same relative amounts of hydrogen and oxygen, no matter where or how the sample is obtained. **2.13** (a) 0.5711 g O/1 g N; 1.142 g O/1 g N; 2.284 g O/1 g N; 2.855 g O/1 g N (b) The numbers in part (a) obey the *law of multiple proportions*. Multiple proportions arise because atoms are the indivisible entities combining, as stated in Dalton's atomic theory. **2.15** (i) Electric and magnetic fields deflected the rays in the same way they would deflect negatively charged particles. (ii) A metal plate exposed to cathode rays acquired a negative charge. **2.17** (a) Most of the volume of an atom is empty space in which electrons move. Most alpha particles passed through this space. (b) The few alpha particles that hit the massive, positively charged gold nuclei were strongly repelled and deflected back in the direction they came from. (c) Because the Be nuclei have a smaller volume and a smaller positive charge than the Au nuclei, fewer alpha particles will be scattered and fewer will be strongly back scattered. **2.19** (a) 0.135 nm; 1.35×10^{2} or 135 pm (b) 3.70×10^{6} Au atoms (c) $1.03 \times 10^{-23}\ cm^3$ **2.21** (a) Proton, neutron, electron (b) proton $= 1+$, neutron $= 0$, electron $= 1-$ (c) The neutron is most massive. (The neutron and proton have very similar masses.) (d) The electron is least massive. **2.23** (a) Not isotopes (b) isotopes (c) isotopes **2.25** (a) Atomic number is the number of protons in the nucleus of an atom. Mass number is the total number of nuclear particles, protons plus neutrons, in an atom. (b) mass number **2.27** (a) ^{40}Ar: 18 p, 22 n, 18 e (b) ^{65}Zn: 30 p, 35 n, 30 e (c) ^{70}Ga: 31 p, 39 n, 31 e (d) ^{80}Br: 35 p, 45 n, 35 e (e) ^{184}W: 74 p, 110 n, 74 e (f) ^{243}Am: 95 p, 148 n, 95e

2.29

Symbol	^{79}Br	^{55}Mn	^{112}Cd	^{222}Rn	^{207}Pb
Protons	35	25	48	86	82
Neutrons	44	30	64	136	125
Electrons	35	25	48	86	82
Mass no.	79	55	112	222	207

2.31 (a) $^{196}_{78}$Pt (b) $^{84}_{36}$Kr (c) $^{75}_{33}$As (d) $^{24}_{12}$Mg **2.33** (a) $^{12}_{6}$C (b) Atomic weights are average atomic masses, the sum of the mass of each naturally occurring isotope of an element times its fractional abundance. Each B atom will have the mass of one of the naturally occurring isotopes, while the "atomic weight" is an average value. **2.35** 63.55 amu **2.37** (a) In Thomson's cathode ray tube, the charged particles are electrons. In a mass spectrometer, the charged particles are positively charged ions (cations). (b) The x-axis label is atomic weight, and the y-axis label is signal intensity. (c) The Cl^{2+} ion will be deflected more. **2.39** (a) average atomic mass $= 24.31$ amu

(b)

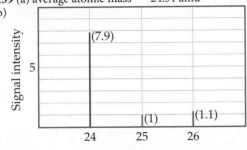

Atomic weight (amu)

2.41 (a) Cr, 24 (metal) (b) He, 2 (nonmetal) (c) P, 15 (nonmetal) (d) Zn, 30 (metal) (e) Mg, 12 (metal) (f) Br, 35 (nonmetal) (g) As, 33 (metalloid) **2.43** (a) K, alkali metals (metal) (b) I, halogens (nonmetal) (c) Mg, alkaline earth metals (metal) (d) Ar, noble gases (nonmetal) (e) S, chalcogens (nonmetal) **2.45** An empirical formula shows the simplest mole ratio of elements in a compound. A molecular formula shows the exact number and kinds of atoms in a molecule. A structural formula shows which atoms are attached to which. **2.47** From left to right: molecular, N_2H_4, empirical, NH_2; molecular, N_2H_2, empirical, NH, molecular and empirical, NH_3 **2.49** (a) $AlBr_3$ (b) C_4H_5 (c) C_2H_4O (d) P_2O_5 (e) C_3H_2Cl (f) BNH_2 **2.51** (a) 6 (b) 10 (c) 12

2.53

(a)

C_2H_6O, H—C—O—C—H (with H atoms)

(b)

C_2H_6O, H—C—C—O—H (with H atoms)

(c)

CH_4O, H—C—O—H (with H atoms)

(d) PF_3, F—P—F (with F below)

2.55

Symbol	^{59}Co^{3+}	^{80}Se^{2-}	^{192}Os^{2+}	^{200}Hg^{2+}
Protons	27	34	76	80
Neutrons	32	46	116	120
Electrons	24	36	74	78
Net Charge	3+	2−	2+	2+

2.57 (a) Mg^{2+} (b) Al^{3+} (c) K^+ (d) S^{2-} (e) F^- **2.59** (a) GaF_3, gallium(III) fluoride (b) LiH, lithium hydride (c) AlI_3, aluminum iodide (d) K_2S, potassium sulfide **2.61** (a) $CaBr_2$ (b) K_2CO_3 (c) $Al(CH_3COO)_3$ (d) $(NH_4)_2SO_4$ (e) $Mg_3(PO_4)_2$

2.63

Ion	K^+	NH_4^+	Mg^{2+}	Fe^{3+}
Cl^-	KCL	NH_4Cl	$MgCl_2$	$FeCl_3$
OH^-	KOH	NH_4OH	$Mg(OH)_2$	$Fe(OH)_3$
CO_3^{2-}	K_2CO_3	$(NH_4)_2CO_3$	$MgCO_3$	$Fe_2(CO_3)_3$
PO_4^{3-}	K_3PO_4	$(NH_4)_3PO_4$	$Mg_3(PO_4)_2$	$FePO_4$

2.65 Molecular: (a) B_2H_6 (b) CH_3OH (f) NOCl (g) NF_3. Ionic: (c) $LiNO_3$ (d) Sc_2O_3 (e) CsBr (f) Ag_2SO_4 **2.67** (a) ClO_2^- (b) Cl^- (c) ClO_3^- (d) ClO_4^- (e) ClO^- **2.69** (a) calcium, 2+; oxide, 2− (b) sodium, 1+; sulfate, 2− (c) potassium, 1+; perchlorate, 1− (d) iron, 2+, nitrate, 1− (e) chromium, 3+; hydroxide, 1− **2.71** (a) lithium oxide (b) iron(III) chloride (ferric chloride) (c) sodium hypochlorite (d) calcium sulfite (e) copper(II) hydroxide (cupric hydroxide) (f) iron(II) nitrate (ferrous nitrate) (g) calcium acetate (h) chromium(III) carbonate (chromic carbonate) (i) potassium chromate (j) ammonium sulfate **2.73** (a) $Al(OH)_3$ (b) K_2SO_4 (c) Cu_2O (d) $Zn(NO_3)_2$ (e) $HgBr_2$ (f) $Fe_2(CO_3)_3$ (g) NaBrO **2.75** (a) Bromic acid (b) hydrobromic acid (c) phosphoric acid (d) HClO (e) HIO_3 (f) H_2SO_3 **2.77** (a) Sulfur hexafluoride (b) iodine pentafluoride (c) xenon trioxide (d) N_2O_4 (e) HCN (f) P_4S_6 **2.79** (a) $ZnCO_3$, ZnO, CO_2 (b) HF, SiO_2, SiF_4, H_2O (c) SO_2, H_2O, H_2SO_3 (d) PH_3 (e) $HClO_4$, Cd, $Cd(ClO_4)_2$ (f) VBr_3 **2.81** (a) A hydrocarbon is a compound composed of the elements hydrogen and carbon only.

(b)

H—C—C—C—C—C—H (pentane structural formula with all H atoms)

molecular and empirical formulas, C_5H_{12}
2.83 (a) A functional group is a group of specific atoms that are constant from one molecule to the next. (b) —OH

(c)

H—C—C—C—C—C—OH (pentanol structural formula with all H atoms)

2.85 (a, b)

H—C—C—C—Cl   H—C—C—C—H (with Cl on middle carbon)

1-chloropropane   2-chloropropane

2.88 (a) 2 protons, 1 neutron, 2 electrons (b) tritium, ^{3}H, is more massive. (c) A precision of 1×10^{-27} g would be required to differentiate between ^{3}H$^+$ and ^{3}He$^+$. **2.90** Arrangement A, 4.1×10^{14} atoms/cm^2 (b) Arrangement B, 4.7×10^{14} atoms/cm^2 (c) The ratio of atoms going from arrangement B to arrangement A is 1.2 to 1. In three dimensions, arrangement B leads to a greater density for Rb metal. **2.94** (a) $^{16}_8$O, $^{17}_8$O, $^{18}_8$O (b) All isotopes are atoms of the same element, oxygen, with the same atomic number, 8 protons in the nucleus and 8 electrons. We expect their electron arrangements to be the same and their chemical properties to be very similar. Each has a different number of neutrons, a different mass number, and a different atomic mass. **2.96** (a) $^{69}_{31}$Ga, 31 protons, 38 neutrons; $^{71}_{31}$Ga, 31 protons, 40 neutrons (b) $^{69}_{31}$Ga, 60.3%, $^{71}_{31}$Ga, 39.7%. **2.99** (a) 5 significant figures (b) An electron is 0.05444% of the mass of an ^{1}H atom. **2.104** (a) nickel(II) oxide, 2+ (b) manganese(IV) oxide, 4+ (c) chromium(III) oxide, 3+ (d) molybdenium(VI) oxide, 6+ **2.107** (a) Perbromate ion (b) selenite ion (c) AsO_4^{3-} (d) $HTeO_4^-$ **2.110** (a) Potassium nitrate (b) sodium carbonate (c) calcium oxide (d) hydrochloric acid (e) magnesium sulfate (f) magnesium hydroxide

Chapter 3

3.1 Equation (a) best fits the diagram. **3.3** (a) NO_2 (b) No, because we have no way of knowing whether the empirical and molecular formulas are the same. NO_2 represents the simplest ratio of atoms in a molecule but not the only possible molecular formula. **3.5** (a) $C_2H_5NO_2$ (b) 75.0 g/mol (c) 225 g glycine . (d) Mass %N in glycine is 18.7%.
3.7

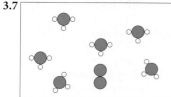

$N_2 + 3 H_2 \longrightarrow 2 NH_3$. Eight N atoms (4 N_2 molecules) require 24 H atoms (12 H_2 molecules) for complete reaction. Only 9 H_2 molecules are available, so H_2 is the limiting reactant. Nine H_2 molecules (18 H atoms) determine that 6 NH_3 molecules are produced. One N_2 molecule is in excess. **3.9** (a) Conservation of mass (b) Subscripts in chemical formulas should not be changed when balancing equations, because changing the subscript changes the identity of the compound (*law of constant composition*). (c) $H_2O(l)$, $H_2O(g)$, $NaCl(aq)$, $NaCl(s)$ **3.11** (a) $2 CO(g) + O_2(g) \longrightarrow 2 CO_2(g)$
(b) $N_2O_5(g) + H_2O(l) \longrightarrow 2 HNO_3(aq)$
(c) $CH_4(g) + 4 Cl_2(g) \longrightarrow CCl_4(l) + 4 HCl(g)$
(d) $Zn(OH)_2(s) + 2HNO_3(aq) \longrightarrow Zn(NO_3)_2(aq) + 2H_2O(l)$
3.13 (a) $Al_4C_3(s) + 12 H_2O(l) \longrightarrow 4Al(OH)_3(s) + 3CH_4(g)$
(b) $2 C_5H_{10}O_2(l) + 13 O_2(g) \longrightarrow 10 CO_2(g) + 10 H_2O(g)$
(c) $2 Fe(OH)_3(s) + 3 H_2SO_4(aq) \longrightarrow Fe_2(SO_4)_3(aq) + 6 H_2O(l)$
(d) $Mg_3N_2(s) + 4H_2SO_4(aq) \longrightarrow 3MgSO_4(aq) + (NH_4)_2SO_4(aq)$
3.15 (a) $CaC_2(s) + 2 H_2O(l) \longrightarrow Ca(OH)_2(aq) + C_2H_2(g)$
(b) $2 KClO_3(s) \overset{\Delta}{\longrightarrow} 2 KCl(s) + 3 O_2(g)$
(c) $Zn(s) + H_2SO_4(aq) \longrightarrow ZnSO_4(aq) + H_2(g)$
(d) $PCl_3(l) + 3 H_2O(l) \longrightarrow H_3PO_3(aq) + 3 HCl(aq)$
(e) $3 H_2S(g) + 2 Fe(OH)_3(s) \longrightarrow Fe_2S_3(s) + 6 H_2O(g)$
3.17 (a) NaBr (b) solid (c) 2 **3.19** (a) $Mg(s) + Cl_2(g) \longrightarrow MgCl_2(s)$
(b) $BaCO_3(s) \overset{\Delta}{\longrightarrow} BaO(s) + CO_2(g)$
(c) $C_8H_8(l) + 10 O_2(g) \longrightarrow 8 CO_2(g) + 4 H_2O(l)$
(d) $C_2H_6O(g) + 3 O_2(g) \longrightarrow 2 CO_2(g) + 3 H_2O(l)$
3.21 (a) $2 C_3H_6(g) + 9 O_2(g) \longrightarrow 6 CO_2(g) + 6 H_2O(g)$
    combustion
(b) $NH_4NO_3(s) \longrightarrow N_2O(g) + 2 H_2O(g)$ decomposition
(c) $C_5H_6O(l) + 6 O_2(g) \longrightarrow 5 CO_2(g) + 3 H_2O(g)$ combustion
(d) $N_2(g) + 3 H_2(g) \longrightarrow 2 NH_3(g)$ combination
(e) $K_2O(s) + H_2O(l) \longrightarrow 2 KOH(aq)$ combination
3.23 (a) 63.0 amu (b) 158.0 amu (c) 310.3 amu (d) 60.1 amu
(e) 235.7 amu (f) 392.3 amu (g) 137.5 amu **3.25** (a) 16.8% (b) 16.1%
(c) 21.1% (d) 28.8% (e) 27.2% (f) 26.5% **3.27** (a) 79.2% (b) 63.2%
(c) 64.6% **3.29** (a) 1×10^{-14} mol of people (b) atomic mass units, amu
(c) grams per mole, g/mol **3.31** 23 g Na contains 1 mol of atoms; 0.5 mol
H_2O contains 1.5 mol atoms; 6.0×10^{23} N_2 molecules contain 2 mol of
atoms. **3.33** 4.37×10^{25} kg (assuming 160 lb has 3 significant figures).
One mole of people weighs 7.31 times as much as Earth. **3.35** (a) 35.9 g
$C_{12}H_{22}O_{11}$ (b) 0.75766 mol $Zn(NO_3)_2$ (c) $6.0 \times 10^{17} CH_3CH_2OH$
molecules (d) 2.47×10^{23} N atoms **3.37** (a) 0.373 g $(NH_4)_3PO_4$
(b) 5.737×10^{-3} mol Cl^- (c) 0.248 g $C_8H_{10}N_4O_2$ (d) 387 g cholesterol/mol **3.39** (a) Molar mass = 162.3 g (b) 3.08×10^{-5} mol
allicin (c) 1.86×10^{19} allicin molecules (d) 3.71×10^{19} S atoms
3.41 (a) 2.500×10^{21} H atoms (b) 2.083×10^{20} $C_6H_{12}O_6$ molecules
(c) 3.460×10^{-4} mol $C_6H_{12}O_6$ (d) 0.06227 g $C_6H_{12}O_6$
3.43 3.2×10^{-8} mol C_2H_3Cl/L; 1.9×10^{16} molecules/L
3.45 (a) C_2H_6O (b) Fe_2O_3 (c) CH_2O **3.47** (a) $CSCl_2$ (b) C_3OF_6
(c) Na_3AlF_6 **3.49** 31 g/mol **3.51** (a) Empirical formula, CH; molecular formula, C_8H_8 (b) empirical formula, $C_4H_5N_2O$; molecular formula, $C_8H_{10}N_4O_2$ (c) empirical formula and molecular formula,

$NaC_5H_8O_4N$ **3.55** (a) C_7H_8 (b) The empirical and molecular formulas are $C_{10}H_{20}O$. **3.57** Empirical formula, C_4H_8O; molecular formula, $C_8H_{16}O_2$ **3.59** $x = 10$; $Na_2CO_3 \cdot 10 H_2O$
3.61 (a) 2.40 mol HF (b) 5.25 g NaF (c) 0.610 g Na_2SiO_3
3.63 (a) $Al(OH)_3(s) + 3 HCl(aq) \longrightarrow AlCl_3(aq) + 3 H_2O(l)$
(b) 0.701 g HCl (c) 0.855 g $AlCl_3$; 0.347 g H_2O
(d) Mass of reactants = 0.500 g + 0.701 g = 1.201 g;
mass of products = 0.855 g + 0.347 g = 1.202 g. Mass
is conserved, within the precision of the data.
3.65 (a) $Al_2S_3(s) + 6 H_2O(l) \longrightarrow 2 Al(OH)_3(s) + 3 H_2S(g)$
(b) 14.7 g $Al(OH)_3$ **3.67** (a) 2.25 mol N_2 (b) 15.5 g NaN_3 (c) 548 g NaN_3
3.69 (a) 5.50×10^{-3} mol Al (b) 1.47 g $AlBr_3$ **3.71** (a) The *limiting reactant* determines the maximum number of product moles resulting from a chemical reaction; any other reactant is an *excess reactant*.
(b) The limiting reactant regulates the amount of products because it is completely used up during the reaction; no more product can be made when one of the reactants is unavailable. (c) Combining ratios are molecule and mole ratios. Since different molecules have different masses, comparing initial masses of reactants will not provide a comparison of numbers of molecules or moles. **3.73** (a) 2255 bicycles (b) 50 frames left over, 305 wheels left over (c) the handle-bars **3.75** (a) $2 C_2H_5OH + 6 O_2 \longrightarrow 4 CO_2 + 6 H_2O$ (b) C_2H_5OH limits (c) If the reaction goes to completion, there will be four molecules of CO_2, six molecules of H_2O, zero molecules of C_2H_5OH and one molecule of O_2. **3.77** NaOH is the limiting reactant; 0.925 mol Na_2CO_3 can be produced; 0.075 mol CO_2 remains.
3.79 (a) $NaHCO_3$ is the limiting reactant. (b) 0.524 g CO_2 (c) 0.238 g citric acid remains **3.81** 0.00 g $AgNO_3$ (limiting reactant), 1.94 g Na_2CO_3, 4.06 g Ag_2CO_3, 2.50 g $NaNO_3$ **3.83** (a) The theoretical yield is 60.3 g C_6H_5Br. (b) 70.1% yield **3.85** 28 g S_8 actual yield
3.87 (a) $C_2H_4O_2(l) + 2 O_2(g) \longrightarrow 2 CO_2(g) + 2 H_2O(l)$
(b) $Ca(OH)_2(s) \longrightarrow CaO(s) + H_2O(g)$
(c) $Ni(s) + Cl_2(g) \longrightarrow NiCl_2(s)$ **3.91** (a) 8×10^{-20} g Si
(b) 2×10^3 Si atoms (with 2 significant figures, 1700 Si atoms)
(c) 1×10^3 Ge atoms (with 2 significant figures, 1500 Ge atoms)
3.95 $C_8H_8O_3$ **3.99** (a) 1.19×10^{-5} mol NaI
(b) 8.1×10^{-3} g NaI **3.103** 7.5 mol H_2 and 4.5 mol N_2 present
initially **3.106** 6.46×10^{24} O atoms **3.107** (a) 88 kg CO_2
(b) $4 \times 10^2 (400)$ kg CO_2 **3.110** (a) $S(s) + O_2(g) \longrightarrow SO_2(g)$;
$SO_2(g) + CaO(s) \longrightarrow CaSO_3(s)$ (b) 7.9×10^7 g CaO
(c) 1.7×10^8 g $CaSO_3$

Chapter 4

4.1 Diagram (c) represents Li_2SO_4 **4.5** $BaCl_2$ **4.7** (b) NO_3^- and
(c) NH_4^+ will always be spectator ions. **4.13** (a) False. Electrolyte solutions conduct electricity because *ions* are moving through the solution.
(b) True. Because ions are mobile in solution, the added presence of uncharged molecules does not inhibit conductivity. **4.15** Statement
(b) is most correct. **4.17** (a) $FeCl_2(aq) \longrightarrow Fe^{2+}(aq) + 2 Cl^-(aq)$
(b) $HNO_3(aq) \longrightarrow H^+(aq) + NO_3^-(aq)$
(c) $(NH_4)_2SO_4(aq) \longrightarrow 2 NH_4^+(aq) + SO_4^{2-}(aq)$
(d) $Ca(OH_4)_2(aq) \longrightarrow Ca^{2+}(aq) + 2 OH^-(aq)$
4.19 HCOOH molecules, H^+ ions, and $HCOO^-$ ions;
$HCOOH(aq) \rightleftharpoons H^+(aq) + HCOO^-(aq)$ **4.21** (a) Soluble
(b) insoluble (c) soluble (d) soluble (e) soluble **4.23** (a) $Na_2CO_3(aq) +$
$2 AgNO_3(aq) \longrightarrow Ag_2CO_3(s) + 2 NaNO_3(aq)$ (b) No precipitate
(c) $FeSO_4(aq) + Pb(NO_3)_2(aq) \longrightarrow PbSO_4(s) +$
$Fe(NO_3)_2(aq)$ **4.25** (a) K^+, SO_4^{2-} (b) Li^+, NO_3^- (c) NH_4^+, Cl^-
4.27 Only Pb^{2+} could be present. **4.29** (a) True (b) false (c) false
(d) true (e) false **4.31** The 0.20 M HI(aq) is most acidic.
4.33 (a) False. H_2SO_4 is a diprotic acid; it has two ionizable hydrogen atoms. (b) False. HCl is a strong acid. (c) False. CH_3OH is a molecular nonelectrolyte. **4.35** (a) Acid, mixture of ions and molecules (weak electrolyte) (b) none of the above, entirely molecules (nonelectrolyte)
(c) salt, entirely ions (strong electrolyte) (d) base, entirely ions
(strong electrolyte) **4.37** (a) H_2SO_3, weak electrolyte (b) C_2H_5OH,

nonelectrolyte (c) NH_3, weak electrolyte (d) $KClO_3$, strong electrolyte (e) $Cu(NO_3)_2$, strong electrolyte
4.39 (a) $2\ HBr(aq) + Ca(OH)_2(aq) \longrightarrow CaBr(aq) + 2\ H_2O(l)$; $H^+(aq) + OH^-(aq) \longrightarrow H_2O(l)$;
(b) $Cu(OH)_2(s) + 2\ HClO_4(aq) \longrightarrow Cu(ClO_4)_2(aq) + 2\ H_2O(l)$; $Cu(OH)_2(s) + 2\ H^+(aq) \longrightarrow 2\ H_2O(l) + Cu^{2+}(aq)$
(c) $Al(OH)_3(s) + 3\ HNO_3(aq) \longrightarrow Al(NO_3)_3(aq) + 3\ H_2O(l)$; $Al(OH)_3(s) + 3\ H^+(aq) \longrightarrow 3\ H_2O(l) + Al^{3+}(aq)$
4.41 (a) $CdS(s) + H_2SO_4(aq) \longrightarrow CdSO_4(aq) + H_2S(g)$; $CdS(s) + 2\ H^+(aq) \longrightarrow H_2S(g) + Cd^{2+}(aq)$
(b) $MgCO_3(s) + 2\ HClO_4(aq) \longrightarrow Mg(ClO_4)_2(aq) + H_2O(l) + CO_2(g)$; $MgCO_3(s) + 2\ H^+(aq) \longrightarrow H_2O(l) + CO_2(g) + Mg^{2+}(aq)$
4.43 (a)
$MgCO_3(s) + 2\ HCl(aq) \longrightarrow MgCl_2(aq) + H_2O(l) + CO_2(g)$;
$MgCO_3(s) + 2\ H^+(aq) \longrightarrow Mg^{2+}(aq) + H_2O(l) + CO_2(g)$;
$MgO(s) + 2\ HCl(aq) \longrightarrow MgCl_2(aq) + H_2O(l)$;
$MgO(s) + 2\ H^+(aq) \longrightarrow Mg^{2+}(aq) + H_2O(l)$;
$Mg(OH)_2(s) + 2\ H^+(aq) \longrightarrow Mg^{2+}(aq) + 2\ H_2O(l)$
(b) We can distinguish magnesium carbonate, $MgCO_3(s)$, because its reaction with acid produces $CO_2(g)$, which appears as bubbles. The other two compounds are indistinguishable, because the products of the two reactions are exactly the same. **4.45** (a) False (b) true **4.47** Metals in region A are most easily oxidized. Nonmetals in region D are least easily oxidized. **4.49** (a) $+4$ (b) $+4$ (c) $+7$ (d) $+1$ (e) $+3$ (f) -1
4.51 (a) $N_2 \longrightarrow 2\ NH_3$, N is reduced; $3\ H_2 \longrightarrow 2\ NH_3$, H is oxidized (b) $Fe^{2+} \longrightarrow Fe$, Fe is reduced; $Al \longrightarrow Al^{3+}$, Al is oxidized (c) $Cl_2 \longrightarrow 2\ Cl^-$, Cl is reduced; $2\ I^- \longrightarrow I_2$, I is oxidized (d) $S^{2-} \longrightarrow SO_4^{2-}$, S is oxidized; $H_2O_2 \longrightarrow H_2O$, O is reduced **4.53** (a) $Mn(s) + H_2SO_4(aq) \longrightarrow MnSO_4(aq) + H_2(g)$; $Mn(s) + 2\ H^+(aq) \longrightarrow Mn^{2+}(aq) + H_2(g)$
(b) $2\ Cr(s) + 6\ HBr(aq) \longrightarrow 2\ CrBr_3(aq) + 3\ H_2(g)$; $2\ Cr(s) + 6\ H^+(aq) \longrightarrow 2\ Cr^{3+}(aq) + 3\ H_2(g)$
(c) $Sn(s) + 2\ HCl(aq) \longrightarrow SnCl_2(aq) + H_2(g)$; $Sn(s) + 2\ H^+(aq) \longrightarrow Sn^{2+}(aq) + H_2(g)$
(d) $2\ Al(s) + 6\ HCOOH(aq) \longrightarrow 2\ Al(HCOO)_3(aq) + 3\ H_2(g)$; $2\ Al(s) + 6\ HCOOH(aq) \longrightarrow 2\ Al^{3+}(aq) + 6\ HCOO^-(aq) + 3\ H_2(g)$
4.55 (a) $Fe(s) + Cu(NO_3)_2(aq) \longrightarrow Fe(NO_3)_2(aq) + Cu(s)$
(b) NR (c) $Sn(s) + 2\ HBr(aq) \longrightarrow SnBr_2(aq) + H_2(g)$
(d) NR (e) $2\ Al(s) + 3\ CoSO_4(aq) \longrightarrow Al_2(SO_4)_3(aq) + 3\ Co(s)$
4.57 (a) i. $Zn(s) + Cd^{2+}(aq) \longrightarrow Cd(s) + Zn^{2+}(aq)$; ii. $Cd(s) + Ni^{2+}(aq) \longrightarrow Ni(s) + Cd^{2+}(aq)$ (b) The elements chromium, iron and cobalt more closely define the position of cadmium in the activity series. (c) Place an iron strip in $CdCl_2(aq)$. If $Cd(s)$ is deposited, Cd is less active than Fe; if there is no reaction, Cd is more active than Fe. Do the same test with Co if Cd is less active than Fe or with Cr if Cd is more active than Fe. **4.59** (a) Intensive; the ratio of amount of solute to total amount of solution is the same, regardless of how much solution is present. (b) The term 0.50 mol-HCl defines an amount (~ 18 g) of the pure substance HCl. The term 0.50 M HCl is a ratio; it indicates that there is 0.50 mol of HCl solute in 1.0 liter of solution. **4.61** (a) 1.17 M $ZnCl_2$ (b) 0.158 mol H^+ (c) 58.3 mL of 6.00 M NaOH **4.63** 16 g $Na^+(aq)$ **4.65** BAC of $0.08 = 0.02\ M$ CH_3CH_2OH (alcohol) **4.67** (a) 316 g ethanol (b) 401 mL ethanol **4.69** (a) 0.15 M K_2CrO_4 has the highest K^+ concentration. (b) 30.0 mL of 0.15 M K_2CrO_4 has more K^+ ions. **4.71** (a) 0.25 M Na^+, 0.25 M NO_3^- (b) $1.3 \times 10^{-2}\ M$ Mg^{2+}, $1.3 \times 10^{-2}\ M$ SO_4^{2-} (c) 0.0150 M $C_6H_{12}O_6$ (d) 0.111 M Na^+, 0.111 M Cl^-, 0.0292 M NH_4^+, 0.0146 M CO_3^{2-} **4.73** (a) 16.9 mL 14.8 M NH_3 (b) 0.296 M NH_3 **4.75** (a) Add 21.4 g $C_{12}H_{22}O_{11}$ to a 250-mL volumetric flask, dissolve in a small volume of water, and add water to the mark on the neck of the flask. Agitate thoroughly to ensure total mixing. (b) Thoroughly rinse, clean, and fill a 50-mL buret with the 1.50 M $C_{12}H_{22}O_{11}$. Dispense 23.3 mL of this solution into a 350-mL volumetric container, add water to the mark, and mix thoroughly. **4.77** 1.398 M CH_3COOH **4.79** (a) 20.0 mL of 0.15 M HCl (b) 0.224 g KCl (c) The KCl reagent is virtually free relative to the HCl solution. The KCl analysis is more cost-effective. **4.81** (a) 38.0 mL of 0.115 M $HClO_4$ (b) 769 mL of 0.128 M HCl

(c) 0.408 M $AgNO_3$ (d) 0.275 g KOH **4.83** 27 g $NaHCO_3$ **4.85** (a) Molar mass of metal hydroxide is 103 g/mol. (b) Rb^+ **4.87** (a) $NiSO_4(aq) + 2\ KOH(aq) \longrightarrow Ni(OH)_2(s) + K_2SO_4(aq)$ (b) $Ni(OH)_2$ (c) KOH is the limiting reactant. (d) 0.927 g $Ni(OH)_2$ (e) 0.0667 M $Ni^{2+}(aq)$, 0.0667 M $K^+(aq)$, 0.100 M $SO_4^{2-}(aq)$ **4.89** 91.39% $Mg(OH)_2$ **4.91** (a) $U(s) + 2ClF_3(g) \longrightarrow UF_6(g) + Cl_2(g)$ (b) This is not a metathesis reaction. (c) It is a redox reaction. **4.95** (a) $Al(OH)_3(s) + 3H^+(aq) \longrightarrow Al^{3+}(aq) + 3H_2O(l)$ (b) $Mg(OH)_2(s) + 2H^+(aq) \longrightarrow Mg^{2+}(aq) + 2H_2O(l)$ (c) $MgCO_3(s) + 2H^+(aq) \longrightarrow Mg^{2+}(aq) + H_2O(l) + CO_2(g)$ (d) $NaAl(CO_3)(OH)_2(s) + 4H^+(aq) \longrightarrow Na^+(aq) + Al^{3+}(aq) + 3H_2O(l) + CO_2(g)$ (e) $CaCO_3(s) + 2H^+(aq) \longrightarrow Ca^{2+}(aq) + H_2O(l) + CO_2(g)$ [In (c), (d) and (e), one could also write the equation for formation of bicarbonate, e.g., $MgCO_3(s) + H^+(aq) \longrightarrow Mg^{2+} + HCO_3^-(aq)$.] **4.99** 12.1 g $AgNO_3$ **4.103** (a) 2.055 M $Sr(OH)_2$ (b) $2\ HNO_3(aq) + Sr(OH)_2(aq) \longrightarrow Sr(NO_3)_2(aq) + 2\ H_2O(l)$ (c) 2.62 M HNO_3 **4.109** (a) $Mg(OH)_2(s) + 2\ HNO_3(aq) \longrightarrow Mg(NO_3)_2(aq) + 2\ H_2O(l)$ (b) HNO_3 is the limiting reactant. (c) 0.130 mol $Mg(OH)_2$, 0 mol HNO_3, and 0.00250 mol $Mg(NO_3)_2$ are present. **4.113** (a) $+5$ (b) silver arsenate (c) 5.22 % As

Chapter 5

5.1 (a) As the book falls, potential energy decreases and kinetic energy increases. (b) The initial potential energy of the book and its total kinetic energy at the instant before impact are both 71 J, assuming no transfer of energy as heat. (c) A heavier book falling from the same shelf has greater kinetic energy when it hits the floor. **5.4** (a) (iii) (b) none of them (c) all of them **5.7** (a) The sign of w is ($+$). (b) The sign of q is ($-$). (c) The sign of w is positive and the sign of q is negative, so we cannot absolutely determine the sign of ΔE. It is likely that the heat lost is much smaller than the work done on the system, so the sign of ΔE is probably positive. **5.10** (a) $N_2(g) + O_2(g) \longrightarrow 2NO(g)$. Since $\Delta V = 0$, $w = 0$. (b) $\Delta H = \Delta H_f = 90.37$ kJ. The definition of a formation reaction is one where elements combine to form one mole of a single product. The enthalpy change for such a reaction is the enthalpy of formation. **5.13** An object can possess energy by virtue of its motion or position. Kinetic energy depends on the mass of the object and its velocity. Potential energy depends on the position of the object relative to the body with which it interacts. **5.15** (a) 1.9×10^5 J (b) 4.6×10^4 cal (c) As the automobile brakes to a stop, its speed (and hence its kinetic energy) drops to zero. The kinetic energy of the automobile is primarily transferred to friction between brakes and wheels and somewhat to deformation of the tire and friction between the tire and road. **5.17** 1 Btu $= 1054$ J **5.19** (a) The *system* is the well-defined part of the universe whose energy changes are being studied. (b) A *closed system* can exchange heat but not mass with its surroundings. (c) Any part of the universe not part of the system is called the surroundings. **5.21** (a) Gravity; work is done because the force of gravity is opposed and the pencil is lifted. (b) Mechanical force; work is done because the force of the coiled spring is opposed as the spring is compressed over a distance. **5.23** (a) In any chemical or physical change, energy can be neither created nor destroyed; energy is conserved. (b) The *internal energy (E)* of a system is the sum of all the kinetic and potential energies of the system components. (c) Internal energy of a closed system increases when work is done on the system and when heat is transferred to the system. **5.25** (a) $\Delta E = -0.077$ kJ, endothermic (b) $\Delta E = -22.1$ kJ, exothermic **5.27** (a) Since no work is done by the system in case (2), the gas will absorb most of the energy as heat; the case (2) gas will have the higher temperature. (b) In case (2) $w = 0$ and $q = 100$ J. In case (1) energy will be used to do work on the surroundings ($-w$), but some will be absorbed as heat ($+q$). (c) ΔE is greater for case (2) because the entire 100 J increases the internal energy of the system rather than a part of the energy doing work on the surroundings. **5.29** (a) A *state function* is a property that depends only on the physical state (pressure, temperature, etc.) of the system, not on the route used to get to the current state. (b) Internal energy is a state function; heat is not

a state function. (c) Volume is a state function. The volume of a system depends only on conditions (pressure, temperature, amount of substance), not the route or method used to establish that volume. **5.31** -51 J **5.33** (a) ΔH is usually easier to measure than ΔE because at constant pressure, $\Delta H = q_p$. The heat flow associated with a process at constant pressure can easily be measured as a change in temperature, while measuring ΔE requires a means to measure both q and w. (b) H is a static quantity that depends only on the specific conditions of the system. q is an energy *change* that, in the general case, does depend on how the change occurs. We can equate change in enthalpy, ΔH, with heat, q_p, only for the specific conditions of constant pressure and exclusively P-V work. (c) The process is endothermic. **5.35** (a) We must know either the temperature, T, or the values of P and ΔV in order to calculate ΔE from ΔH. (b) ΔE is larger than ΔH. (c) Since the value of Δn is negative, the quantity $(-P\Delta V)$ is positive. We add a positive quantity to ΔH to calculate ΔE, so ΔE must be larger. **5.37** $\Delta E = 1.47$ kJ; $\Delta H = 0.824$ kJ
5.39 (a)
$$C_2H_5OH(l) + 3\,O_2(g) \longrightarrow 3\,H_2O + 2\,CO_2(g), \quad \Delta H = -1235 \text{ kJ}$$
(b) $C_2H_5OH(l) + 3\,O_2(g)$

$$\Delta H = -1235 \text{ kJ}$$

$$3\,H_2O(g) + 2\,CO_2(g)$$

5.41 (a) $\Delta H = -142.3$ kJ/mol $O_3(g)$ (b) $2\,O_3(g)$ has the higher enthalpy. **5.43** (a) Exothermic (b) -87.9 kJ heat transferred (c) 15.7 g MgO produced (d) 602 kJ heat absorbed **5.45** (a) -29.5 kJ (b) -4.11 kJ (c) 60.6 J **5.47** (a) $\Delta H = 726.5$ kJ (b) $\Delta H = -1453$ kJ (c) The exothermic forward reaction is more likely to be thermodynamically favored. (d) Vaporization is endothermic. If the product were $H_2O(g)$, the reaction would be more endothermic and would have a less negative ΔH. **5.49** (a) J/mol-°C or J/mol-K (b) J/g-°C or J/g-K (c) To calculate heat capacity from specific heat, the mass of the particular piece of copper pipe must be known. **5.51** (a) 4.184 J/g-K (b) 75.40 J/mol-°C (c) 774 J/°C (d) 904 kJ **5.53** (a) 2.66×10^3 J (b) It will require more heat to increase the temperature of one mole of octane, $C_8H_{18}(l)$, by a certain amount than to increase the temperature of one mole of water, $H_2O(l)$, by the same amount. **5.55** $\Delta H = -44.4$ kJ/mol NaOH **5.57** $\Delta H_{rxn} = -25.5$ kJ/g $C_6H_4O_2$ or -2.75×10^3 kJ/mol $C_6H_4O_2$ **5.59** (a) Heat capacity of the complete calorimeter $= 14.4$ kJ/°C (b) 7.56 °C **5.61** Hess's law is a consequence of the fact that enthalpy is a state function. Since ΔH is independent of path, we can describe a process by any series of steps that adds up to the overall process. ΔH for the process is the sum of ΔH values for the steps. **5.63** $\Delta H = -1300.0$ kJ **5.65** $\Delta H = -2.49 \times 10^3$ kJ **5.67** (a) *Standard conditions* for enthalpy changes are $P = 1$ atm and some common temperature, usually 298 K. (b) *Enthalpy of formation* is the enthalpy change that occurs when a compound is formed from its component elements. (c) *Standard enthalpy of formation*, ΔH_f°, is the enthalpy change that accompanies formation of one mole of a substance from elements in their standard states. **5.69** (a) $\frac{1}{2}N_2(g) + O_2(g) \longrightarrow NO_2(g)$, $\Delta H_f^\circ = 33.84$ kJ (b) $S(s) + \frac{3}{2}O_2(g) \longrightarrow SO_3(g)$, $\Delta H_f^\circ = -395.2$ kJ (c) $Na(s) + \frac{1}{2}Br_2(l) \longrightarrow NaBr(s)$, $\Delta H_f^\circ = -361.4$ kJ (d) $Pb(s) + N_2(g) + 3\,O_2(g) \longrightarrow Pb(NO_3)_2(s)$, $\Delta H_f^\circ = -451.9$ kJ **5.71** $\Delta H_{rxn}^\circ = -847.6$ kJ **5.73** (a) $\Delta H_{rxn}^\circ = -196.6$ kJ (b) $\Delta H_{rxn}^\circ = 37.1$ kJ (c) $\Delta H_{rxn}^\circ = -976.94$ kJ (d) $\Delta H_{rxn}^\circ = -68.3$ kJ **5.75** $\Delta H_f^\circ = -248$ kJ **5.77** (a) $C_8H_{18}(l) + \frac{25}{2}O_2(g) \longrightarrow$ $8\,CO_2(g) + 9\,H_2O(g)$ (b) $\Delta H_f^\circ = -259.5$ kJ **5.79** (a) $C_2H_5OH(l) + 3\,O_2(g) \longrightarrow 2\,CO_2(g) + 3\,H_2O(g)$ (b) $\Delta H_{rxn}^\circ = -1234.8$ kJ (c) 2.11×10^4 kJ/L heat produced (d) 0.071284 g CO_2/kJ heat emitted **5.81** (a) *Fuel value* is the amount of energy produced when 1 g of a substance (fuel) is combusted. (b) 5 g of fat (c) These products of metabolism are expelled as waste via the alimentary tract, $H_2O(l)$ primarily in urine and feces, and

$CO_2(g)$ as gas when breathing. **5.83** (a) 108 or 1×10^2 Cal/serving (b) Sodium does not contribute to the calorie content of the food because it is not metabolized by the body. **5.85** 59.7 Cal **5.87** (a) $\Delta H_{comb} = -1850$ kJ/mol C_3H_4, -1926 kJ/mol C_3H_6, -2044 kJ/mol C_3H_8 (b) $\Delta H_{comb} = -4.616 \times 10^4$ kJ/kg C_3H_4, -4.578×10^4 kJ/kg C_3H_6, -4.635×10^4 kJ/kg C_3H_8 (c) These three substances yield nearly identical quantities of heat per unit mass, but propane is marginally higher than the other two. **5.89** 1.0×10^{12} kg $C_6H_{12}O_6$/yr **5.91** (a) 469.4 m/s (b) 5.124×10^{-21} J (c) 3.086 kJ/mol **5.93** The spontaneous air bag reaction is probably exothermic, with $-\Delta H$ and thus $-q$. When the bag inflates, work is done by the system, so the sign of w is also negative. **5.97** $\Delta H = 38.95$ kJ; $\Delta E = 36.48$ kJ **5.99** 1.8×10^4 bricks **5.102** (a) $\Delta H_{rxn}^\circ = -353.0$ kJ (b) 1.2 g Mg needed **5.106** (a) $\Delta H^\circ = -631.3$ kJ (b) 3 mol of acetylene gas has greater enthalpy. (c) Fuel values are 50 kJ/g $C_2H_2(g)$, 42 kJ/g $C_6H_6(l)$. **5.109** If all work is used to increase the man's potential energy, the stair climbing uses 58 Cal and will not compensate for the extra order of 245 Cal fries. (More than 58 Cal will be required to climb the stairs because some energy is used to move limbs and some will be lost as heat.) **5.112** (a) 1.479×10^{-18} J/molecule (b) 1×10^{-15} J/photon. The X-ray has approximately 1000 times more energy than is produced by the combustion of 1 molecule of $CH_4(g)$. **5.114** (a) ΔH° for neutralization of the acids is HNO_3, -55.8 kJ; HCl, -56.1 kJ; NH_4^+, -4.1 kJ. (b) $H^+(aq) + OH^-(aq) \longrightarrow H_2O(l)$ is the net ionic equation for the first two reactions. $NH_4^+(aq) + OH^-(aq) \longrightarrow NH_3(aq) + H_2O(l)$ (c) The ΔH° values for the first two reactions are nearly identical, -55.8 kJ and -56.1 kJ. Since spectator ions do not change during a reaction and these two reactions have the same net ionic equation, it is not surprising that they have the same ΔH°. (d) Strong acids are more likely than weak acids to donate H^+. Neutralization of the two strong acids is energetically favorable, while the third reaction is barely so. NH_4^+ is likely a weak acid. **5.116** (a) $\Delta H^\circ = -65.7$ kJ (b) ΔH° for the complete molecular equation will be the same as ΔH° for the net ionic equation. Since the overall enthalpy change is the enthalpy of products minus the enthalpy of reactants, the contributions of spectator ions cancel. (c) ΔH_f° for $AgNO_3(aq)$ is -100.4 kJ/mol.

Chapter 6

6.2 (a) 0.1 m or 10 cm (b) No. Visible radiation has wavelengths much shorter than 0.1 m. (c) Energy and wavelength are inversely proportional. Photons of the longer 0.1-m radiation have less energy than visible photons. (d) Radiation with $\lambda = 0.1$ m is in the low-energy portion of the microwave region. The appliance is probably a microwave oven. **6.6** (a) Increase (b) Decrease (c) The light from the hydrogen discharge tube is a line spectrum, so not all visible wavelengths will be in our "hydrogen discharge rainbow." Starting on the inside, the rainbow will be violet, then blue and blue-green. After a gap, the final band will be red. **6.9** (a) 1 (b) p (c) (ii) **6.13** (a) Meters (b) 1/second (c) meters/second **6.15** (a) True (b) False. Ultraviolet light has shorter wavelengths than visible light. (c) False. X-rays travel at the same speed as microwaves. (d) False. Electromagnetic radiation and sound waves travel at different speeds. **6.17** Wavelength of X-rays < ultraviolet < green light < red light < infrared < radio waves **6.19** (a) 3.0×10^{13} s^{-1} (b) 5.45×10^{-7} m $= 545$ nm. (c) The radiation in (b) is visible; the radiation in (a) is not. (d) 1.50×10^4 m **6.21** 4.61×10^{14} s^{-1}; red. **6.23** (iii) **6.25** (a) 1.95×10^{-19} J (b) 4.81×10^{-19} J (c) 328 nm **6.27** (a) $\lambda = 3.3$ μm, $E = 6.0 \times 10^{-20}$ J; $\lambda = 0.154$ nm, $E = 1.29 \times 10^{-15}$ J (b) The 3.3-μm photon is in the infrared region and the 0.154-nm photon is in the X-ray region; the X-ray photon has the greater energy. **6.29** (a) 6.11×10^{-19} J/photon (b) 368 kJ/mol (c) 1.64×10^{15} photons (d) 368 kJ/mol **6.31** (a) The $\sim 1 \times 10^{-6}$ m radiation is in the infrared portion of the spectrum. (b) 8.1×10^{16} photons/s **6.33** (a) $E_{min} = 7.22 \times 10^{-19}$ J (b) $\lambda = 275$ nm (c) $E_{120} = 1.66 \times 10^{-18}$ J. The excess energy of the 120-nm photon is converted into the kinetic energy of the emitted electron. $E_k = 9.3 \times 10^{-19}$ J/electron. **6.35** When applied to atoms, the notion of quantized energies means that only certain values of ΔE are allowed. These are represented by the lines in the emission spectra of excited atoms. **6.37** (a) Emitted (b) absorbed (c) emitted **6.39** (a) $E_2 = -5.45 \times 10^{-19}$ J; $E_6 = -0.606 \times 10^{-19}$ J; $\Delta E = 4.84 \times 10^{-19}$ J; $\lambda = 410$ nm (b) visible, violet **6.41** (a) (ii) (b) $n_i = 3$, $n_f = 2$; $\lambda = 6.56 \times 10^{-7}$ m; this is the red line at 656 nm. $n_i = 4$, $n_f = 2$; $\lambda = 4.86 \times 10^{-7}$ m; this is the blue-green line at 486 nm. $n_i = 5$, $n_f = 2$; $\lambda = 4.34 \times 10^{-7}$ m; this is the blue-violet line at 434 nm. **6.43** (a) Ultraviolet region (b) $n_i = 7$, $n_f = 1$ **6.45** The order of increasing frequency of light absorbed is: $n = 4$ to $n = 9$; $n = 3$ to $n = 6$; $n = 2$ to $n = 3$; $n = 1$ to $n = 2$ **6.47** (a) $\lambda = 5.6 \times 10^{-37}$ m (c) $\lambda = 2.3 \times 10^{-13}$ m (d) $\lambda = 1.51 \times 10^{-11}$ m **6.49** 3.16×10^3 m/s **6.51** (a) $\Delta x \geq 4 \times 10^{-27}$ m (b) $\Delta x \geq 3 \times 10^{-10}$ m **6.53** (a) The uncertainty principle states that there is a limit to how precisely we can simultaneously know the position and momentum (a quantity related to energy) of an electron. The Bohr model states that electrons move about the nucleus in precisely circular orbits of known radius and energy. This violates the uncertainty principle. (b) De Broglie stated that electrons demonstrate the properties of both particles and waves and that each moving particle has a wave associated with it. A wave function is the mathematical description of the matter wave of an electron. (c) Although we cannot predict the exact location of an electron in an allowed energy state, we can determine the probability of finding an electron at a particular position. This statistical knowledge of electron location is the *probability density* and is a function of ψ^2, the square of the wave function ψ. **6.55** (a) $n = 4$, $l = 3, 2, 1, 0$ (b) $l = 2$, $m_l = -2, -1, 0, 1, 2$ (c) $m_l = 2$, $l \geq 2$ or $l = 2, 3$ or 4 **6.57** (a) $3p$: $n = 3$, $l = 1$ (b) $2s$: $n = 2$, $l = 0$ (c) $4f$: $n = 4$, $l = 3$ (d) $5d$: $n = 5$, $l = 2$ **6.59** (a) $2, 1, 0, -1, -2$ (b) $\frac{1}{2}, -\frac{1}{2}$ **6.61** (a) impossible, $1p$ (b) possible (c) possible (d) impossible, $2d$ **6.63**

(a) z (b) z (c) x

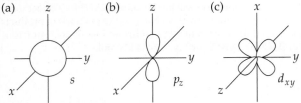

s p_z d_{xy}

6.65 (a) The hydrogen atom $1s$ and $2s$ orbitals have the same overall spherical shape, but the $2s$ orbital has a larger radial extension and one more node than the $1s$ orbital. (b) A single $2p$ orbital is directional in that its electron density is concentrated along one of the three Cartesian axes of the atom. The $d_{x^2-y^2}$ orbital has electron density along both the x- and y-axes, while the p_x orbital has density only along the x-axis. (c) The average distance of an electron from the nucleus in a $3s$ orbital is greater than for an electron in a $2s$ orbital. (d) $1s < 2p < 3d < 4f < 6s$ **6.67** (a) In the hydrogen atom, orbitals with the same principal quantum number, n, have the same energy. (b) In a many-electron atom, for a given n value, orbital energy increases with increasing l value: $s < p < d < f$. **6.69** (a) There are two main pieces of experimental evidence for electron "spin." The Stern-Gerlach experiment shows that atoms with a single unpaired electron interact differently with an inhomogeneous magnetic field. Examination of the fine details of emission line spectra of multi-electron atoms reveals that each line is really a close pair of lines. Both observations can be rationalized if electrons have the property of spin.

(b)   (c)

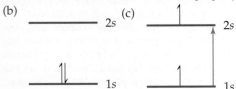

6.71 (a) 6 (b) 10 (c) 2 (d) 14 **6.73** (a) "Valence electrons" are those involved in chemical bonding. They are part or all of the outer-shell electrons listed after the core. (b) "Core electrons" are inner-shell

electrons that have the electron configuration of the nearest noble-gas element. (c) Each box represents an orbital. (d) Each half-arrow in an orbital diagram represents an electron. The direction of the half-arrow represents electron spin. **6.75** (a) Cs, $[\text{Xe}]6s^1$ (b) Ni, $[\text{Ar}]4s^23d^8$ (c) Se, $[\text{Ar}]4s^23d^{10}4p^4$ (d) Cd, $[\text{Kr}]5s^24d^{10}$ (e) U, $[\text{Rn}]5f^36d^17s^2$ (f) Pb, $[\text{Xe}]6s^24f^{14}5d^{10}6p^2$ **6.77** (a) Be, 0 unpaired electrons (b) O, 2 unpaired electrons (c) Cr, 6 unpaired electrons (d) Te, 2 unpaired electrons **6.79** (a) The fifth electron would fill the $2p$ subshell before the $3s$. (b) Either the core is [He], or the outer electron configuration should be $3s^23p^3$. (c) The $3p$ subshell would fill before the $3d$. **6.81** (a) $\lambda_A = 3.6 \times 10^{-8}$ m, $\lambda_B = 8.0 \times 10^{-8}$ m (b) $\nu_A = 8.4 \times 10^{15}$ s^{-1}, $\nu_B = 3.7 \times 10^{15}$ s^{-1} (c) A, ultraviolet; B, ultraviolet **6.84** (a) 8.93×10^3 mi/hr (b) 35.0 min **6.86** 1.6×10^{18} photons **6.91** (a) The Paschen series lies in the infrared. (b) $n_i = 4$, $\lambda = 1.87 \times 10^{-6}$ m; $n_i = 5$, $\lambda = 1.28 \times 10^{-6}$ m; $n_i = 6$, $\lambda = 1.09 \times 10^{-6}$ m **6.95** $\lambda = 10.6$ pm **6.99** (a) The nodal plane of the p_z orbital is the xy-plane. (b) The two nodal planes of the d_{xy} orbital are the ones where $x = 0$ and $y = 0$. These are the yz- and xz-planes. (c) The two nodal planes of the $d_{x^2-y^2}$ orbital are the ones that bisect the x- and y-axes and contain the z-axis. **6.102** (a) Br: $[\text{Ar}]4s^23d^{10}4p^5$, 1 unpaired electron (b) Ga: $[\text{Ar}]4s^23d^{10}4p^1$, 1 unpaired electron (c) Hf: $[\text{Xe}]6s^24f^{14}5^2$, 2 unpaired electrons (d) Sb: $[\text{Kr}]5s^24d^{10}5p^3$, 3 unpaired electrons (e) Bi: $[\text{Xe}]6s^24f^{14}5d^{10}6p^3$, 3 unpaired electrons (f) Sg: $[\text{Rn}]7s^25f^{14}6d^4$, 4 unpaired electrons **6.105** (a) 1.7×10^{28} photons (b) 34 s **6.109** (a) Bohr's theory was based on the Rutherford nuclear model of the atom: a dense positive charge at the center and a diffuse negative charge surrounding it. Bohr's theory then specified the nature of the diffuse negative charge. The prevailing theory before the nuclear model was Thomson's plum pudding model: discrete electrons scattered about a diffuse positive charge cloud. Bohr's theory could not have been based on the Thomson model of the atom. (b) De Broglie's hypothesis is that electrons exhibit both particle and wave properties. Thomson's conclusion that electrons have mass is a particle property, while the nature of cathode rays is a wave property. De Broglie's hypothesis actually rationalizes these two seemingly contradictory observations about the properties of electrons.

Chapter 7

7.2 The largest brown sphere is Br$^-$, the intermediate blue one is Br, and the smallest red one is F. **7.5** (a) The bonding atomic radius of A, r_A, is $d_1/2$; $r_x = d_2 - (d_1/2)$. (b) The length of the X—X bond is $2r_x$ or $2d_2 - d_1$. **7.8** (a) X + 2F$_2 \longrightarrow$ XF$_4$ (b) X in the diagram has about the same bonding radius as F, so it is likely to be a nonmetal. **7.9** (a) The results are 2, 8, 18, 32. (b) The atomic numbers of the noble gases are 2, 10, 18, 36, 54 and 86. The differences between sequential pairs of these atomic numbers is 8, 8, 18, 18 and 32. These differences correspond to the results in (a). They represent the filling of new subshells when moving across the next row of the periodic chart. (c) The Pauli exclusion principle is the source of the "2". **7.11** In general, elements are discovered according to their ease of isolation in elemental form. **7.13** (a) *Effective nuclear charge*, Z_{eff} is a representation of the average electrical field experienced by a single electron. It is the average environment created by the nucleus and the other electrons in the molecule, expressed as a net positive charge at the nucleus. (b) Going from left to right across a period, effective nuclear charge increases. **7.15** (a) For both Na and K, $Z_{\text{eff}} = 1$. (b) For both Na and K, $Z_{\text{eff}} = 2.2$. (c) Slater's rules give values closer to the detailed calculations: Na, 2.51; K, 3.49. (d) Both approximations give the same value of Z_{eff} for Na and K; neither accounts for the gradual increase in Z_{eff} moving down a group. (e) Following the trend from detailed calculations, we predict a Z_{eff} value of approximately 4.5. **7.17** The $n = 3$ electrons in Kr experience a greater effective nuclear charge and thus have a greater probability of being closer to the nucleus. **7.19** (a) Atomic radii are determined by measuring distances between atoms in various situations. (b) Bonding radii are calculated from the internuclear separation of two atoms joined by a covalent chemical bond. Nonbonding radii are calculated from the internuclear separation between two gaseous atoms that collide and move apart but do not bond. (c) For a given element, the nonbonding radius is always larger than the bonding radius. (d) If a free atom reacts to become part of a covalent molecule, its radius changes from nonbonding to bonding and the atom gets smaller. **7.21** (a) 1.37 Å (b) The distance between W atoms will decrease. **7.23** From the sum of the atomic radii, As—I = 2.58 Å. This is very close to the experimental value of 2.55 Å. **7.25** (a) Cs > K > Li (b) Pb > Sn > Si (c) N > O > F **7.27** (a) False (b) true (c) false **7.29** Ga^{3+}: none; Zr^{4+}: Kr; Mn^{7+}: Ar; I^-: Xe; Pb^{2+}: Hg **7.31** (a) Na^+ (b) F^-, $Z_{eff} = 7$; Na^+, $Z_{eff} = 9$ (c) S = 4.15; F^-, $Z_{eff} = 4.85$; Na^+, $Z_{eff} = 6.85$ (d) For isoelectronic ions, as nuclear charge (Z) increases, effective nuclear charge (Z_{eff}) increases and ionic radius decreases. **7.33** (a) Cl < S < K (b) $K^+ < Cl^- < S^{2-}$ (c) The neutral K atom has the largest radius because the n-value of its outer electron is larger than the n-value of valence electrons in S and Cl. The K^+ ion is smallest because, in an isoelectronic series, the ion with the largest Z has the smallest ionic radius. **7.35** (a) O^{2-} is larger than O because the increase in electron repulsions that accompanies addition of an electron causes the electron cloud to expand. (b) S^{2-} is larger than O^{2-} because for particles with like charges, size increases going down a family. (c) S^{2-} is larger than K^+ because the two ions are isoelectronic and K^+ has the larger Z and Z_{eff}. (d) K^+ is larger than Ca^{2+} because the two ions are isoelectronic and Ca^{2+} has the larger Z and Z_{eff}. **7.37** $Al(g) \longrightarrow Al^+(g) + 1e^-$; $Al^+(g) \longrightarrow Al^{2+}(g) + 1e^-$; $Al^{2+}(g) \longrightarrow Al^{3+}(g) + 1e^-$. The process for the first ionization energy requires the least amount of energy. **7.39** (a) The outer electron in Li has a smaller n-value and is closer to the nucleus. More energy is needed to overcome the greater attraction of the Li electron for the nucleus. (b) Sc: $[Ar] 4s^2 3d^1$; Ti: $[Ar]4s^2 3d^2$. The fourth ionization of titanium involves removing a $3d$ valence electron, while the fourth ionization of Sc requires removing a $3p$ electron from the [Ar] core. The core electron has a greater Z_{eff} and requires more energy to remove. (c) Li: $[He]2s^1$; Be: $[He]2s^2$. The second ionization of Be involves removing a $2s$ valence electron, while the second ionization of Li requires removing a $1s$ core electron. The core electron is closer to the nucleus and requires more energy to remove. **7.41** (a) The smaller the atom, the larger its first ionization energy. (b) Of the nonradioactive elements, He has the largest and Cs has the smallest first ionization energy. **7.43** (a) Cl (b) Ca (c) K (d) Ge (e) Sn **7.45** (a) Co^{2+},$[Ar]3d^7$ (b) Sn^{2+},$[Kr]5s^2 4d^{10}$ (c) Zr^{4+}, [Kr], noble gas configuration (d) Ag^+,$[Kr]4d^{10}$ (e) S^{2-}, $[Ne]3s^2 3p^6$, noble gas configuration **7.47** Ni^{2+},$[Ar]3d^8$;Pd^{2+},$[Kr]4d^8$;Pt^{2+},$[Xe]4f^{14}5d^8$ **7.49** (a) Positive, endothermic, values for ionization energy and electron affinity mean that energy is required to either remove or add electrons. Valence electrons in Ar experience the largest Z_{eff} of any element in the third row, resulting in a large, positive ionization energy. When an electron is added to Ar, the n = 3 electrons become core electrons that screen the extra electron so effectively that Ar^- has a higher energy than an Ar atom and a free electron. This results in a large positive electron affinity. (b) kJ/mol **7.51** Electron affinity of Br: $Br(g) + 1e^- \longrightarrow Br^-(g)$; $[Ar]4s^2 3d^{10}4p^5 \longrightarrow [Ar]4s^2 3d^{10}4p^6$; electron affinity of Kr: $Kr(g) + 1e^- \longrightarrow Kr^-(g)$; $[Ar]4s^2 3d^{10}4p^6 \longrightarrow [Ar]4s^2 3d^{10}4p^6 5s^1$. Br^- adopts the stable electron configuration of Kr; the added electron experiences essentially the same Z_{eff} and stabilization as the other valence electrons and electron affinity is negative. In Kr^- ion, the added electron occupies the higher energy $5s$ orbital. A $5s$ electron is farther from the nucleus, effectively shielded by the spherical Kr core and not stabilized by the nucleus; electron affinity is positive. **7.53** (a) Ionization energy (I_1) of Ne: $Ne(g) \longrightarrow Ne^+(g) + 1e^-$; $[He]2s^2 2p^6 \longrightarrow [He]2s^2 2p^5$; electron affinity ($E_1$) of F: $F(g) + 1e^- \longrightarrow F^-(g)$;$[He]2s^2 2p^5 \longrightarrow [He]2s^2 2p^6$. (b) I_1 of Ne is positive; E_1 of F is negative. (c) One process is apparently the reverse of the other, with one important difference. Ne has a greater Z and Z_{eff}, so we expect I_1 for Ne to be somewhat greater in magnitude and opposite in sign to E_1 for F. **7.55** (a) Decrease (b) Increase (c) The smaller the first ionization energy of an element, the greater the metallic character of that element. The trends in (a) and (b) are the opposite of the trends in ionization energy. **7.57** Agree. When forming ions, all metals form cations. The only nonmetallic

element that forms cations is the metalloid Sb, which is likely to have significant metallic character. **7.59** Ionic: SnO_2, Al_2O_3, Li_2O, Fe_2O_3; molecular: CO_2, H_2O. Ionic compounds are formed by combining a metal and a nonmetal; molecular compounds are formed by two or more nonmetals. **7.61** (a) An *acidic oxide* dissolved in water produces an acidic solution; a *basic oxide* dissolved in water produces a basic solution. (b) Oxides of nonmetals, such as SO_3, are acidic; oxides of metals, such as CaO, are basic. **7.63** (a) Dichlorine heptoxide (b) $2 Cl_2(g) + 7 O_2(g) \longrightarrow 2 Cl_2O_7(l)$ (c) A boiling point of 81 °C is not unexpected for a large molecule like Cl_2O_7. (d) Cl_2O_7 is an acidic oxide, so it will be more reactive to base, OH^-. (e) The oxidation state of Cl in Cl_2O_7 is +7; the corresponding electron configuration for Cl is $[He]2s^2 2p^6$ or [Ne]. **7.65** (a) $BaO(s)$ $H_2O(l) \longrightarrow Ba(OH)_2(aq)$ (b) $FeO(s) + 2 HClO_4(aq) \longrightarrow Fe(ClO_4)_2(aq) + H_2O(l)$ (c) $SO_3(g) + H_2O(l) \longrightarrow H_2SO_4(aq)$ (d) $CO_2(g) + 2 NaOH(aq) \longrightarrow Na_2CO_3(aq) + H_2O(l)$ **7.67** (a) Ca is more reactive because it has a lower ionization energy than Mg. (b) K is more reactive because it has a lower ionization energy than Ca. **7.69** (a) $2 K(s) + Cl_2(g) \longrightarrow 2 KCl(s)$ (b) $SrO(s) + H_2O(l) \longrightarrow Sr(OH)_2(aq)$ (c) $4 Li(s) + O_2(g) \longrightarrow 2 Li_2O(s)$ (d) $2 Na(s) + S(l) \longrightarrow Na_2S(s)$ **7.71** (a) Both classes of reaction are redox reactions where either hydrogen or the halogen gains electrons and is reduced. The product is an ionic solid, where either hydride ion, H^-, or a halide ion, X^-, is the anion. (b) $Ca(s) + F_2(g) \longrightarrow CaF_2(s)$; $Ca(s) + H_2(g) \longrightarrow CaH_2(s)$. Both products are ionic solids containing Ca^{2+} and the corresponding anion in a 1:2 ratio. **7.73** (a) Br, $[Ar]4s^2 4p^5$; Cl, $[Ne]3s^2 3p^5$ (b) Br and Cl are in the same group, and both adopt a 1− ionic charge. (c) The ionization energy of Br is smaller than that of Cl, because the $4p$ valence electrons in Br are farther from to the nucleus and less tightly held than the $3p$ electrons of Cl. (d) Both react slowly with water to form HX + HOX. (e) The electron affinity of Br is less negative than that of Cl, because the electron added to the $4p$ orbital in Br is farther from the nucleus and less tightly held than the electron added to the $3p$ orbital of Cl. (f) The atomic radius of Br is larger than that of Cl, because the $4p$ valence electrons in Br are farther from the nucleus and less tightly held than the $3p$ electrons of Cl. **7.75** (a) The term *inert* was dropped because it no longer described all the Group 8A elements. (b) In the 1960s, scientists discovered that Xe would react with substances having a strong tendency to remove electrons, such as F_2. Thus, Xe could not be categorized as an "inert" gas. (c) The group is now called the noble gases. **7.77** (a) $2 O_3(g) \longrightarrow 3 O_2(g)$ (b) $Xe(g) + F_2(g) \longrightarrow XeF_2(g)$; $Xe(g) + 2 F_2(g) \longrightarrow XeF_4(s)$; $Xe(g) + 3 F_2(g) \longrightarrow XeF_6(s)$ (c) $S(s) + H_2(g) \longrightarrow H_2S(g)$ (d) $2 F_2(g) + 2 H_2O(l) \longrightarrow 4 HF(aq) + O_2(g)$ **7.79** Up to Z = 82, there are three instances where atomic weights are reversed relative to atomic numbers: Ar and K; Co and Ni; Te and I. **7.81** (a) 5+ (b) 4.8+ (c) Shielding is greater for $3p$ electrons, owing to penetration by $3s$ electrons, so Z_{eff} for $3p$ electrons is less than that for $3s$ electrons. (d) The first electron lost is a $3p$ electron because it has a smaller Z_{eff} and experiences less attraction for the nucleus than a $3s$ electron does. **7.85** (a) The As—Cl distance is 2.24 Å. (b) The predicted As—Cl bond length is 2.21 Å **7.87** (a) Chalcogens, −2; halogens, −1. (b) The family with the larger value is: atomic radii, chalcogens; ionic radii of the most common oxidation state, chalcogens; first ionization energy, halogens; second ionization energy, halogens **7.91** C:$1s^2 2s^2 2p^2$. I_1 through I_4 represent loss of the $2p$ and $2s$ electrons in the outer shell of the atom. The values of $I_1 − I_4$ increase as expected. I_5 and I_6 represent loss of the $1s$ core electrons. These $1s$ electrons are much closer to the nucleus and experience the full nuclear charge, so the values of I_5 and I_6 are significantly greater than $I_1 − I_4$. **7.96** (a) Cl^-, K^+ (b) Mn^{2+},Fe^{3+} (c) Sn^{2+},Sb^{3+} **7.99** (a) For both H and the alkali metals, the added electron will complete an ns subshell, so shielding and repulsion effects will be similar. For the halogens, the electron is added to an np subshell, so the energy change is likely to be quite different. (b) True. The electron configuration of H is $1s^1$. The single $1s$ electron experiences no repulsion from other electrons and feels the full unshielded nuclear charge. The outer electrons of all other elements that form compounds are shielded by a spherical inner core of electrons and are less strongly attracted to the nucleus, resulting in larger bonding atomic radii. (c) Both H and the halogens have large ionization energies. The relatively large effective nuclear charge experienced by np electrons of the halogens is similar to the unshielded nuclear charge experienced by the H $1s$ electron. For the alkali metals, the ns electron being removed is effectively shielded by the core electrons, so ionization energies are low. (d) ionization energy of hydride, $H^-(g) \longrightarrow H(g) + 1e^-$ (e) electron affinity of hydrogen, $H(g) + 1 e^- \longrightarrow H^-(g)$. The value for the ionization energy of hydride is equal in magnitude but opposite in sign to the electron affinity of hydrogen. **7.102** The most likely product is (i). **7.107** Electron configuration, $[Rn]7s^2 5f^{14} 6d^{10} 7p^5$; first ionization energy, 805 kJ/mol; electron affinity, -235 kJ/mol; atomic size, 1.65 Å; common oxidation state, -1. **7.110** (a) Li, $[He]2s^1$; $Z_{eff} \approx 1+$ (b) $I_1 \approx 5.45 \times 10^{-19}$ J/atom ≈ 328 kJ/mol (c) The estimated value of 328 kJ/mol is less than the Table 7.4 value of 520 kJ/mol. Our estimate for Z_{eff} was a lower limit; the [He] core electrons do not perfectly shield the $2s$ electron from the nuclear charge. (d) Based on the experimental ionization energy, $Z_{eff} = 1.26$. This value is greater than the estimate from part (a) but agrees well with the "Slater" value of 1.3 and is consistent with the explanation in part (c). **7.113** (a) Mg_3N_2 (b) $Mg_3N_2(s) + 3 H_2O(l) \longrightarrow 3 MgO(s) + 2 NH_3(g)$; the driving force is the production of $NH_3(g)$. (c) 17% Mg_3N_2 (d) $3 Mg(s) + 2 NH_3(g) \longrightarrow Mg_3N_2(s) + 3 H_2(g)$. NH_3 is the limiting reactant and 0.46 g H_2 is formed. (e) $\Delta H^{\circ}_{rxn} = -368.70$ kJ

Chapter 8

8.1 (a) Group 4A or 14 (b) Group 2A or 2 (c) Group 5A or 15 **8.4** (a) Ru (b) $[Kr]5s^2 4d^6$. **8.7** (a) Four (b) In order of increasing bond length: $3 < 1 < 2$ (c) In order of increasing bond enthalpy: $2 < 1 < 3$ **8.9** (a) False (b) A nitrogen atom has 5 valence electrons. (c) The atom (Si) has 4 valence electrons. **8.11** (a) Si, $1s^2 2s^2 2p^6 3s^2 3p^2$ (b) four (c) The $3s$ and $3p$ electrons are valence electrons.

8.13 (a) ·Äl· (b) :B̈r· (c) :Är· (d) ·Sr

8.15 M̈g + :Ö: $\longrightarrow$ Mg^{2+} + $\left[:\ddot{O}:\right]^{2-}$

(b) two (c) Mg loses electrons. **8.17** (a) AlF_3 (b) K_2S (c) Y_2O_3 (d) Mg_3N_2 **8.19** (a) Sr^{2+}, $[Ar]4s^2 3d^{10} 4p^6 = [Kr]$, noble-gas configuration (b) Ti^{2+}, $[Ar]3d^2$ (c) Se^{2-}, $[Ar]4s^2 3d^{10} 4p^6 = [Kr]$, noble-gas configuration (d) Ni^{2+}, $[Ar]3d^8$ (e) Br^-, $[Ar]4s^2 3d^{10} 4p^6 = [Kr]$, noble-gas configuration (f) Mn^{3+}, $[Ar]3d^4$ **8.21** (a) Endothermic (b) $NaCl(s) \longrightarrow Na^+(g) + Cl^-(g)$ (c) Salts like CaO that have doubly charged ions will have larger lattice energies compared with salts that have singly charged ions **8.23** (a) The slope of this line is 3.32×10^3 kJ/cation charge (b) The slope of this line is 829 kJ/(cation charge)^2 (c) The data points fall much closer to the line on the graph of lattice energy versus the square of the cation charge. The trend of lattice energy versus the square of the cation charge is more linear. (d) The lattice energy of TiC should be 1.34×10^4 kJ **8.25** (a) K–F, 2.71 Å; Na–Cl, 2.83 Å; Na–Br, 2.98 Å; Li–Cl, 2.57 Å (b) $LiCl > KF > NaCl > NaBr$ (c) From Table 8.2: LiCl, 1030 kJ; KF, 808 kJ; NaCl, 788 kJ; NaBr, 732 kJ. The predictions from ionic radii are correct. **8.27** Statement (a) is the best explanation. **8.29** The lattice energy of $RbCl(s)$ is $+692$ kJ/mol. **8.31** (a) The bonding in (iii) and (iv) is likely to be covalent. (b) Covalent, because it is a gas at room temperature and below.

8.33

:C̈l· + :C̈l· + :C̈l· + :C̈l· + ·S̈i· $\longrightarrow$:C̈l—Si—C̈l:
with :Cl: above and :Cl: below Si

(a) 4 (b) 7 (c) 8 (d) 8 (e) 4 **8.35** (a) Ö=Ö (b) four bonding electrons (two bonding electron pairs) (c) An O=O double bond is shorter than an O—O single bond. The greater the number of shared electron

pairs between two atoms, the shorter the distance between the atoms. **8.37** Statement (b) is false. **8.39** (a) Mg (b) S (c) C (d) As **8.41** The bonds in (a), (c), and (d) are polar. The more electronegative element in each polar bond is (a) F (c) O (d) I. **8.43** (a) The calculated charge on H and Br is 0.12e. (b) Decrease **8.45** (a) SiF_4, molecular, silicon tetrachloride; LaF_3, ionic, lanthanum(III) fluoride (b) $FeCl_2$, ionic, iron(II) chloride; $ReCl_6$, molecular (metal in high oxidation state), rhenium hexachloride. (c) $PbCl_4$, molecular (by contrast to the distinctly ionic RbCl), lead tetrachloride; RbCl, ionic, rubidium chloride

8.47 (a)

H—Si—H with H above and H below (tetrahedral)

(b) :C≡O:

(c) :F̈—S̈—F̈:

(d) :Ö—S—Ö—H with :Ö: above and :Ö: below, and H below bottom O

(e) $\left[:\ddot{O}—\ddot{C}l—\ddot{O}:\right]^-$

(f) H—N̈—Ö—H with H below N

8.49 Statement (b) is most true. Keep in mind that when it is necessary to place more than an octet of electrons around an atom in order to minimize formal charge, there may not be a "best" Lewis structure. **8.51** Formal charges are shown on the Lewis structures; oxidation numbers are listed below each structure.

(a) Ö=C=S̈ (with -1 on S)
$\quad$ 0 $\quad$ 0 $\quad$ 0
O, -2; C, $+4$; S, -2

(b) :Ö: (with -1)
$\quad$ 0 :C̈l—S—C̈l: 0
$\qquad\qquad +1$
S, $+4$; Cl, -1; O, -2

(c) $\left[-1 :\ddot{O}—Br—\ddot{O}: -1 \right]^{1-}$ with :Ö:$^{-1}$ above Br ($+2$)
Br, $+5$; O, -2

(d) 0 H—Ö—C̈l—Ö: -1
$\qquad\quad 0 \quad +1$
Cl, $+3$; H, $+1$; O, -2

8.53 (a) $\left[\ddot{O}=N—\ddot{O}:\right]^- \longleftrightarrow \left[:\ddot{O}—N=\ddot{O}\right]^-$

(b) O_3 is isoelectronic with NO_2^-; both have 18 valence electrons. (c) Since each N—O bond has partial double-bond character, the N—O bond length in NO_2^- should be shorter than an N—O single bond but longer than an N=O double bond. **8.55** The more electron pairs shared by two atoms, the shorter the bond. Thus, the C—O bond lengths vary in the order $CO < CO_2 < CO_3^{2-}$. **8.57** (a) No. Two equally valid Lewis structures can be drawn for benzene.

The concept of resonance dictates that the true description of bonding is some hybrid or blend of these two Lewis structures. The most obvious blend of these two resonance structures is a molecule with six equivalent C—C bonds with equal lengths. (b) Yes. This model predicts a uniform C—C bond length that is shorter than a single bond but longer than a double bond. (b) No. The C—C bonds in benzene have some double bond character but they are not full double bonds. They are longer than isolated C=C double bonds. **8.59** (a) True (b) False (c) True (d) False **8.61** Assume that the dominant structure is the one that minimizes formal charge. Following this guideline, only ClO^- obeys the octet rule. ClO, ClO_2^-, ClO_3^-, and ClO_4^- do not obey the octet rule.

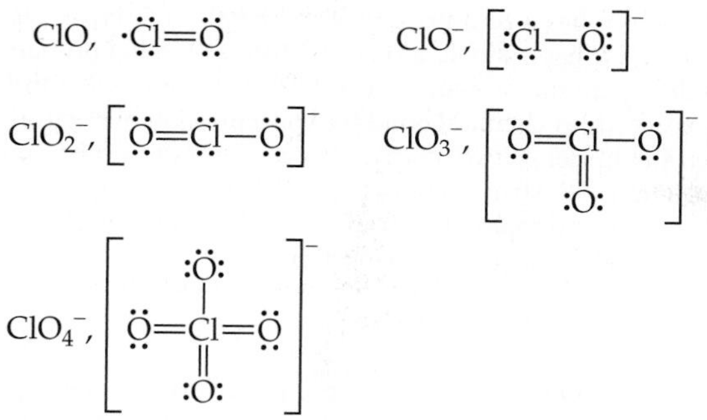

ClO, $\cdot\ddot{\text{C}}\text{l}\!=\!\ddot{\text{O}}$ ClO⁻, $\left[\,:\!\ddot{\text{C}}\text{l}\!-\!\ddot{\text{O}}:\,\right]^{-}$

ClO₂⁻, $\left[\,\ddot{\text{O}}\!=\!\ddot{\text{C}}\text{l}\!-\!\ddot{\text{O}}:\,\right]^{-}$ ClO₃⁻, $\left[\,\ddot{\text{O}}\!=\!\ddot{\text{C}}\text{l}\!-\!\ddot{\text{O}}\,\right]^{-}$ with double-bonded :O: below

ClO₄⁻, $\left[\,\ddot{\text{O}}\!=\!\text{Cl}\!=\!\ddot{\text{O}}\,\right]^{-}$ with :Ö: above and :Ö: below

8.63 (a) H—P̈—H with H below

(b) H—Al—H with H below

(b) Does not obey the octet rule. Central Al has only 6 electrons

(c) $\left[\,:\text{N}\!\equiv\!\text{N}\!-\!\ddot{\text{N}}:\,\right]^{-} \longleftrightarrow \left[\,:\ddot{\text{N}}\!-\!\text{N}\!\equiv\!\text{N}:\,\right]^{-} \longleftrightarrow$

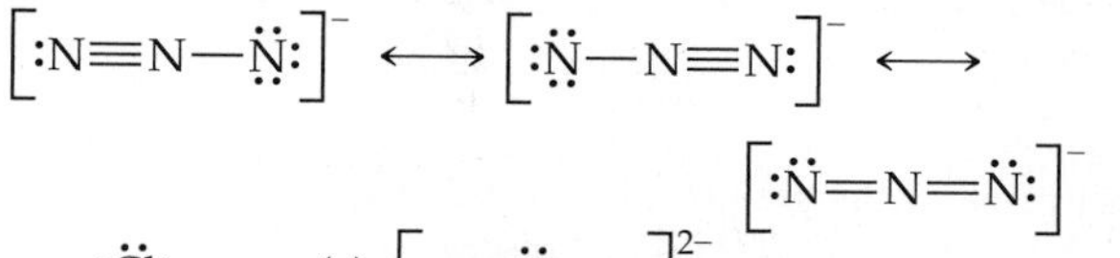

$\left[\,:\ddot{\text{N}}\!=\!\text{N}\!=\!\ddot{\text{N}}:\,\right]^{-}$

(d) :C̈l: :C̈l—C—H with H below

(e) $\left[\,\text{SnF}_6\,\right]^{2-}$ octahedral with six :F̈: around central Sn

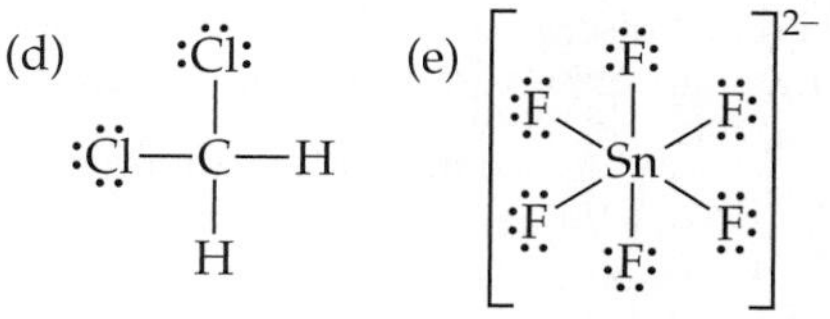

(e) Does not obey octet rule. Central Sn has 12 electrons.
8.65 (a) :C̈l—Be—C̈l: with formal charges 0 0 0

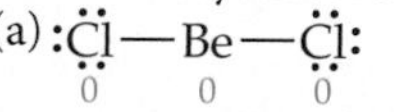

This structure violates the octet rule.

(b) C̈l=Be=C̈l ⟷ :C̈l—Be≡Cl ⟷ Cl≡Be—C̈l: with formal charges shown below (|1 |2 |1 ; 0 |2 |2 ; |2 |2 0)

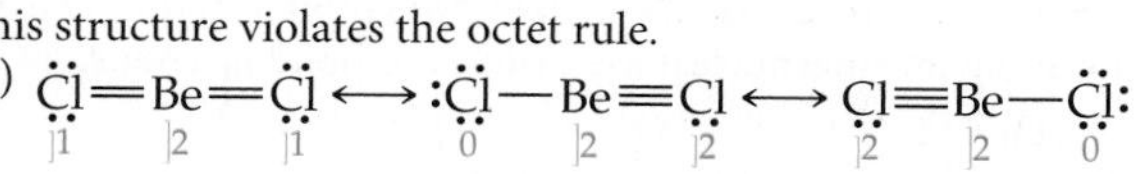

(c) Formal charges are minimized on the structure that violates the octet rule; this form is probably dominant. **8.67** Three resonance structures for HSO₃⁻ are shown here. Because the ion has a 1− charge, the sum of the formal charges of the atoms is −1.

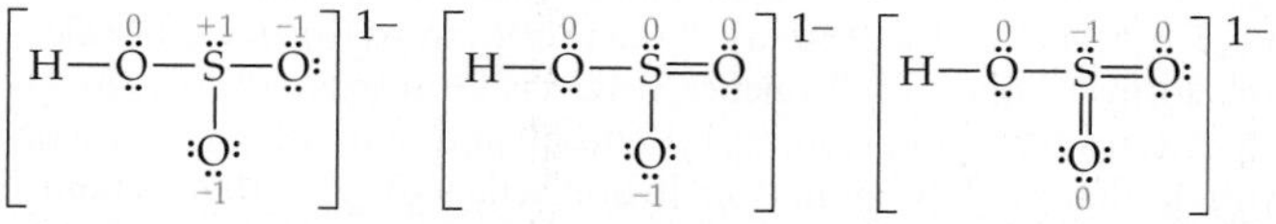

$\left[\,\text{H}\!-\!\ddot{\text{O}}\!-\!\overset{+1}{\text{S}}\!-\!\ddot{\text{O}}:\,\right]^{1-}$ with :O: below (formal charges 0 +1 −1, −1) $\left[\,\text{H}\!-\!\ddot{\text{O}}\!-\!\text{S}\!=\!\ddot{\text{O}}\,\right]^{1-}$ with :O: below (0 0 0, −1) $\left[\,\text{H}\!-\!\ddot{\text{O}}\!-\!\text{S}\!=\!\ddot{\text{O}}\,\right]^{1-}$ with double bond :O: below (0 −1 0, 0)

The structure with no double bonds obeys the octet rule for all atoms, but does not lead to minimized formal charges. The structures with one and two double bonds both minimize formal charge but do not obey the octet rule. Of these two, the structure with one double bond is preferred because the formal charge is localized on the more electronegative oxygen atom. **8.69** (a) $\Delta H = -304$ kJ (b) $\Delta H = -82$ kJ (c) $\Delta H = -467$ kJ **8.71** (a) $\Delta H = -321$ kJ (b) $\Delta H = -103$ kJ (c) $\Delta H = -203$ kJ **8.73** (a) ΔH calculated from bond enthalpies is −97 kJ; the reaction is exothermic. (b) The ΔH calculated from ΔH_f° values is −92.38 kJ. **8.75** The average Ti—Cl bond enthalpy is 430 kJ/mol. **8.77** (a) Seven elements in the periodic table have Lewis symbols with single dots. They are in Group 1A, the alkali metals. This includes hydrogen, whose placement is problematic, and francium, which is radioactive. **8.81** The charge on M is likely to be 3+. The range of lattice energies for ionic compounds with the general formula MX and a charge of 2+ on the metal is 3-4 × 10³ kJ/mol. The lattice energy of 6 × 10³ kJ/mol indicates that the charge on M must be greater than 2+. **8.85** (a) B—O. The most polar bond will be formed by the two elements with the greatest difference in electronegativity. (b) Te—I. These elements have the two largest covalent radii among this group. (c) TeI₂ The octet rule is satisfied for all three atoms. (d) P₂O₃. Each P atom needs to share 3 e⁻ and each O atom 2 e⁻ to achieve an octet. And B₂O₃. Although this is not a purely ionic compound, it can be understood in terms of gaining

and losing electrons to achieve a noble-gas configuration. If each B atom were to lose 3 e⁻ and each O atom were to gain 2 e⁻, charge balance and the octet rule would be satisfied. **8.90** (a) +1 (b) −1 (c) +1 (assuming the odd electron is on N) (d) 0 (e) +3 **8.95** (a) False. The B—A=B structure says nothing about the nonbonding electrons in the molecule. (b) True **8.98** (a) $\Delta H = 7.85$ kJ/g nitroglycerine (b) $4\,\text{C}_7\text{H}_5\text{N}_3\text{O}_6(s) \longrightarrow 6\,\text{N}_2(g) + 7\,\text{CO}_2(g) + 10\,\text{H}_2\text{O}(g) + 21\,\text{C}(s)$ **8.101** (a) Ti²⁺, [Ar]3d²; Ca, [Ar]4s². (b) Ca has no unpaired electrons and Ti²⁺ has two. (c) In order to be isoelectronic with Ca²⁺, Ti would have a 4+ charge. **8.107** (a) Azide ion is N₃⁻. (b) Resonance structures with formal charges are shown.

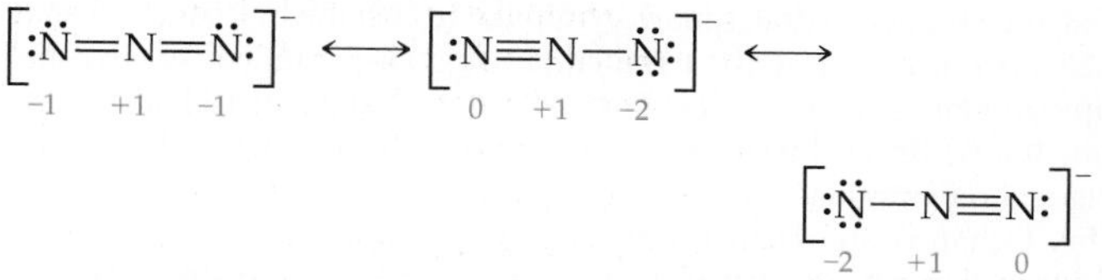

$\left[\,:\ddot{\text{N}}\!=\!\text{N}\!=\!\ddot{\text{N}}:\,\right]^{-}$ (−1 +1 −1) ⟷ $\left[\,:\text{N}\!\equiv\!\text{N}\!-\!\ddot{\text{N}}:\,\right]^{-}$ (0 +1 −2) ⟷

$\left[\,:\ddot{\text{N}}\!-\!\text{N}\!\equiv\!\text{N}:\,\right]^{-}$ (−2 +1 0)

(c) The structure with two double bonds minimizes formal charges and is probably the main contributor. (d) The N—N distances will be equal and have the approximate length of a N—N double bond, 1.24 Å. **8.112** (a) D(Br—Br)(l) = 223.6 kJ; D(Br—Br)(g) = 193 kJ (b)   D(C—Cl)(l) = 336.1 kJ;   D(C − Cl)(g) = 328 kJ (c) D(O—O)(l) = 192.7 kJ; D(O—O)(g) = 146 kJ (d) Breaking bonds in the liquid requires more energy than breaking bonds in the gas phase. Bond dissociation in the liquid phase can be thought of in two steps, vaporization of the liquid followed by bond dissociation in the gas phase. The greater bond dissociation enthalpy in the liquid phase is due to the contribution from the enthalpy of vaporization.

Chapter 9

9.1 Removing an atom from the equatorial plane of the trigonal bipyramid in Figure 9.3 creates a seesaw shape. **9.3** (a) Two electron-domain geometries, linear and trigonal bipyramidal (b) one electron-domain geometry, trigonal bipyramidal (c) one electron-domain geometry, octahedral (c) one electron-domain geometry, octahedral (d) one electron domain geometry, octahedral (e) one electron domain geometry, octahedral (f) one electron-domain geometry, trigonal bipyramidal (This triangular pyramid is an unusual molecular geometry not listed in Table 9.3. It could occur if the equatorial substituents on the trigonal bipyramid were extremely bulky, causing the nonbonding electron pair to occupy an axial position.) **9.5** (a) Zero energy corresponds to two separate, noninteracting Cl atoms. This infinite Cl—Cl distance is beyond the right extreme of the horizontal axis on the diagram. (b) According to the valence bond model, valence orbitals on the approaching atoms overlap, allowing two electrons to mutually occupy space between the two nuclei and be stabilized by two nuclei rather than one. (c) The Cl—Cl distance at the energy minimum on the plot is the Cl—Cl bond length. (d) At interatomic separations shorter than the bond distance, the two nuclei begin to repel each other, increasing the overall energy of the system. (e) The y-coordinate of the minimum point on the plot is a good estimate of the Cl—Cl bond energy or bond strength. **9.11** (a) i, Two s atomic orbitals; two p atomic orbitals overlapping end to end; two p atomic orbitals overlapping side to side (b) i, σ-type MO; ii, σ-type MO; iii, π-type MO (c) i, antibonding; ii, bonding; iii, antibonding (d) i, the nodal plane is between the atom centers, perpendicular to the interatomic axis and equidistant from each atom. ii, there are two nodal planes; both are perpendicular to the interatomic axis. One is left of the left atom and the second is right of the right atom. iii, there are two nodal planes; one is between the atom centers, perpendicular to the interatomic axis and equidistant from each atom. The second contains the interatomic axis and is perpendicular to the first. **9.13** (a) Yes. The stated shape defines the bond angle and the bond length tells the size. (b) No. Atom A could have 2, 3, or 4 nonbonding electron pairs. **9.15** A molecule with tetrahedral molecular geometry has an atom at each vertex of the tetrahedron. A trigonal-pyramidal molecule has one vertex of the tetrahedron occupied by a nonbonding electron pair rather than an atom. **9.17** (a) The number of electron domains in a molecule or ion is the number of bonds (double and triple bonds count as one domain) plus the number of nonbonding electron pairs. (b) A *bonding electron domain* is a region between two bonded atoms that contains one or more pairs of bonding electrons. A *nonbonding electron domain* is localized on a single atom and contains one pair of nonbonding electrons. **9.19** (a) No effect on molecular shape (b) 1 nonbonding pair on P influences molecular shape (c) no effect (d) no effect (e) 1 nonbonding pair on S influences molecular shape **9.21** (a) 2 (b) 1 (c) none (d) 3 **9.23** (a) Tetrahedral, tetrahedral (b) trigonal bipyramidal, T-shaped (c) octahedral, square pyramidal (d) octahedral, square planar **9.25** (a) Linear, linear (b) tetrahedral, trigonal pyramidal (c) trigonal bipyramidal, seesaw (d) octahedral, octahedral (e) tetrahedral, tetrahedral (f) linear, linear **9.27** (a) i, trigonal planar; ii, tetrahedral; iii, trigonal bipyramidal (b) i, 0; ii, 1; iii, 2 (c) N and P (d) Cl (or Br or I). This T-shaped molecular geometry arises from a trigonal-bipyramidal electron-domain geometry with 2 nonbonding domains. Assuming each F atom has 3 nonbonding domains and forms only single bonds with A, A must have 7 valence electrons and be in or below the third row of the periodic table to produce these electron-domain and molecular geometries. **9.29** (a) 1, less than $109.5°$; 2, less than $109.5°$ (b) 3, different than $109.5°$; 4, less than $109.5°$ (c) 5, $180°$ (d) 6, slightly more than $120°$; 7, less than $109.5°$; 8, different than $109.5°$ **9.31** Each species has four electron domains around the N atom, but the number of nonbonding domains decreases from two to zero, going from NH_2^- to NH_4^+. Since nonbonding domains exert greater repulsive forces on adjacent domains, the bond angles expand as the number of nonbonding domains decreases. **9.33** (a) Although both ions have 4 bonding electron domains, the 6 total domains around Br require octahedral domain geometry and square-planar molecular geometry, while the 4 total domains about B lead to tetrahedral domain and molecular geometry. (b) The angles will vary as $H_2O > H_2S > H_2Se$. The less electronegative the central atom, the larger the nonbonding electron domain, and the greater the effect of repulsive forces on adjacent bonding domains. The less electronegative the central atom, the greater the deviation from ideal tetrahedral angles. **9.35** A bond dipole is the asymmetric charge distribution between two bonded atoms with unequal electronegativities. A molecular dipole moment is the three-dimensional sum of all the bond dipoles in a molecule. **9.37** (a) Yes. The net dipole moment vector points along the $Cl—S—Cl$ angle bisector. (b) No, $BeCl_2$ does not have a dipole moment. **9.39** (a) For a molecule with polar bonds to be nonpolar, the polar bonds must be (symmetrically) arranged so that the bond dipoles cancel. Any nonbonding pairs about the central atom must be arranged so that they also cancel. In most cases, nonbonding electron domains will be absent from the central atom. (b) AB_2, linear electron domain and molecular geometry, trigonal bipyramidal electron domain geometry and linear molecular geometry; AB_3, trigonal planar electron domain and molecular geometry; AB_4, tetrahedral electron domain and molecular geometry, octahedral electron domain geometry and square planar molecular geometry **9.41** (a) IF (d) PCl_3 and (f) IF_5 are polar. **9.43** (a) Lewis structures

Molecular geometries

Polar   Nonpolar   Polar

(b) The middle isomer has a zero net dipole moment. (c) C_2H_3Cl has only one isomer, and it has a dipole moment. **9.45** (a) *Orbital overlap* occurs when valence atomic orbitals on two adjacent atoms share the same region of space. (b) A chemical bond is a concentration of electron density between the nuclei of two atoms. This concentration can take place because orbitals on the two atoms overlap. **9.47** (a) $H — Mg — H$, linear electron domain and molecular geometry (b) The linear electron-domain geometry in MgH_2 requires sp hybridization.
(c)

9.49 (a) B, $[He]2s^2 2p^1$. One $2s$ electron is "promoted" to an empty $2p$ orbital. The $2s$ and two $2p$ orbitals that each contain one electron are hybridized to form three equivalent hybrid orbitals in a trigonal-planar arrangement. (b) sp^2
(c)

(d) A single $2p$ orbital is unhybridized. It lies perpendicular to the trigonal plane of the sp^2 hybrid orbitals. **9.51** (a) sp^2 (b) sp^3 (c) sp (d) sp^3 **9.53** Left, no hybrid orbitals discussed in this chapter form angles of $90°$ with each other; p atomic orbitals are perpendicular to each other. center, $109.5°$, sp^3; right, $120°$, sp^2

9.55 (a)

(b)

(c) A σ bond is generally stronger than a π bond because there is more extensive orbital overlap. (d) No. Overlap of two s orbitals results in electron density along the internuclear axis, while a π bond has none.
9.57 (a)

(b) sp^3, sp^2, sp (c) nonplanar, planar, planar (d) 7σ, 0π; 5σ, 1π; 3σ, 2π **9.59** (a) 18 valence electrons (b) 16 valence electrons form σ bonds. (c) 2 valence electrons form π bonds. (d) No valence electrons are nonbonding. (e) The left and central C atoms are sp^2 hybridized; the right C atom is sp^3 hybridized. **9.61** (a) $\sim 109.5°$ about the leftmost C, sp^3; $\sim 120°$ about the right-hand C, sp^2 (b) The doubly bonded O can be viewed as sp^2, and the other as sp^3; the nitrogen is sp^3 with bond angles less than $109.5°$. (c) nine σ bonds, one π bond **9.63** (a) In a localized π bond, the electron density is concentrated between the two atoms forming the bond. In a delocalized π bond, the electron density is spread over all the atoms that contribute p orbitals to the network. (b) The existence of more than one resonance form is a good indication that a molecule will have delocalized π bonding. (c) delocalized
9.65 (a)

(b) sp^2 (c) Yes, there is one other resonance structure. (d) The C and two O atoms have p_π orbitals. (e) There are four electrons in the π system of the ion. **9.67** (a) Linear (b) The two central C atoms each have trigonal planar geometry with $\sim 120°$ bond angles about them. The C and O atoms lie in a plane with the H atoms free to rotate in and out of this plane. (c) The molecule is planar with $\sim 120°$ bond angles about the two N atoms. **9.69** (a) Hybrid orbitals are mixtures of atomic orbitals from a single atom and remain localized on that atom. Molecular orbitals are combinations of atomic orbitals from two or more atoms and are delocalized over at least two atoms. (b) Each MO can hold a maximum of two electrons. (c) Antibonding molecular orbitals can have electrons in them.

9.71 (a)

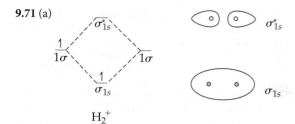

H_2^+

(b) There is one electron in H_2^+. (c) σ_{1s}^1 (d) BO $= \frac{1}{2}$ (e) Fall apart. If the single electron in H_2^+ is excited to the σ_{1s}^* orbital, its energy is higher than the energy of an H $1s$ atomic orbital and H_2^+ will decompose into a hydrogen atom and a hydrogen ion. (f) Statement (i) is correct.

9.73

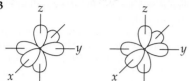

(a) 1 σ bond (b) 2 π bonds (c) 1 σ^* and 2 π^* **9.75 (a)** When comparing the same two bonded atoms, bond order and bond energy are directly related, while bond order and bond length are inversely related. When comparing different bonded nuclei, there are no simple relationships. (b) Be_2 is not expected to exist; it has a bond order of zero and is not energetically favored over isolated Be atoms. Be_2^+ has a bond order of 0.5 and is slightly lower in energy than isolated Be atoms. It will probably exist under special experimental conditions. **9.77 (a, b)** Substances with no unpaired electrons are weakly repelled by a magnetic field. This property is called *diamagnetism*. (c) O_2^{2-}, Be_2^{2+} **9.79 (a)** B_2^+, $\sigma_{2s}^2\sigma_{2s}^{*2}\pi_{2p}^1$, increase (b) Li_2^+, $\sigma_{1s}^2\sigma_{1s}^{*2}\sigma_{2s}^1$, increase (c) N_2^+, $\sigma_{2s}^2\sigma_{2s}^{*2}\pi_{2p}^4\sigma_{2p}^1$, increase (d) Ne_2^{2+}, $\sigma_{2s}^2\sigma_{2s}^{*2}\sigma_{2p}^2\pi_{2p}^4\pi_{2p}^{*4}$, decrease **9.81** CN, $\sigma_{2s}^2\sigma_{2s}^{*2}\sigma_{2p}^2\pi_{2p}^3$, bond order = 2.5; CN^+, $\sigma_{2s}^2\sigma_{2s}^{*2}\sigma_{2p}^2\pi_{2p}^2$, bond order = 2.0; CN^-, $\sigma_{2s}^2\sigma_{2s}^{*2}\sigma_{2p}^2\pi_{2p}^4$, bond order = 3.0. (a) CN^- (b) CN, CN^+ **9.83 (a)** $3s$, $3p_x$, $3p_y$, $3p_z$ (b) π_{3p} (c) 2 (d) If the MO diagram for P_2 is similar to that of N_2, P_2 will have no unpaired electrons and be diamagnetic. **9.87 (a)** PF_4^-, BrF_4^- and ClF_4^- (b) AlF_4^- (c) BrF_4^- (d) PF_4^- and ClF_4^+

9.90

Molecule	Electron-Domain Geometry	Hybridization of Central Atom	Dipole Moment? Yes or No
CO_2	Linear	sp	No
NH_3	Tetrahedral	sp^3	Yes
CH_4	Tetrahedral	sp^3	No
BH_3	Trigonal planar	sp^2	No
SF_4	Trigonal bipyramidal	Not applicable	Yes
SF_6	Octahedral	Not applicable	No
H_2CO	Trigonal planar	sp^2	Yes
PF_5	Trigonal bipyramidal	Not applicable	No
XeF_2	Trigonal bipyramidal	Not applicable	No

9.91 (a) 2 σ bonds, 2 π bonds (b) 3 σ bonds, 4 π bonds (c) 3 σ bonds, 1 π bond (d) 4 σ bonds, 1 π bond

9.98

(a) The molecule is nonplanar. (b) Allene has no dipole moment. (c) The bonding in allene would not be described as delocalized. The π electron clouds of the two adjacent C=C are mutually perpendicular,

so there is no overlap and no delocalization of π electrons. **9.101 (a)** All O atoms have sp^2 hybridization. (b) The two σ bonds are formed by overlap of sp^2 hybrid orbitals, the π bond is formed by overlap of atomic p orbitals, one nonbonded pair is in a p atomic orbital and the other five nonbonded pairs are in sp^2 hybrid orbitals. (c) unhybridized p atomic orbitals (d) four, two from the π bond and two from the nonbonded pair in the p atomic orbital **9.104** The silicon analogs would have the same hybridization as the C compounds. Silicon, which is in the row below C, has a larger bonding atomic radius and atomic orbitals than C. The close approach of Si atoms required to form strong, stable π bonds in Si_2H_4 and Si_2H_2 is not possible and these Si analogs do not readily form. **9.108** $\sigma_{2s}^2\sigma_{2s}^{*2}\pi_{2p}^4\sigma_{2p}^1\pi_{2p}^{*1}$ (a) Paramagnetic (b) The bond order of N_2 in the ground state is 3; in the first excited state it has a bond order of 2. Owing to the reduction in bond order, N_2 in the first excited state has a weaker N—N bond. **9.114 (a)** $2\ SF_4(g) + O_2(g) \longrightarrow 2\ OSF_4(g)$
(b)

:O:
‖
:F—S—F:
:F: :F:

(c) $\Delta H = -551$ kJ, exothermic (d) The electron-domain geometry is trigonal bipyramidal. The O atom can be either equatorial or axial. (e) Because F is more electronegative than O, S–F bonding domains are smaller than the S=O domain. The structure that minimizes S=O repulsions has O in the equatorial position and is more likely. **9.118** ΔH from bond enthalpies is 5364 kJ; ΔH from ΔH_f° values is 5535 kJ. The difference in the two results, 171 kJ/mol C_6H_6, is due to the resonance stabilization in benzene. Because the π electrons are delocalized, the molecule has a lower overall energy than that predicted using bond enthalpies for localized C—C and C=C bonds. Thus, the amount of energy actually required to decompose 1 mole of $C_6H_6(g)$, represented by the Hess' law calculation, is greater than the sum of the localized bond enthalpies.

9.122 (a)

H—C—N=C=Ö: ⟷ H—C—N≡C—Ö:

(The structure on the right does not minimize formal charges and will make a minor contribution to the true structure.)
(b)

180°
109° N=C=Ö
H   120°
109° C
H   109°
109° H

Both resonance structures predict the same bond angles.
(c) The two extreme Lewis structures predict different bond lengths. These bond length estimates assume that the structure minimizing formal charge makes a larger contribution to the true structure. C—O, 1.28 Å; C=N, 1.33 Å; C—N, 1.43 Å; C—H, 1.07 Å (d) The molecule will have a dipole moment. The C=N and C=O bond dipoles are opposite each other, but they are not equal. And, there are nonbonding electron pairs that are not directly opposite each other and will not cancel.

Chapter 10

10.1 It would be much easier to drink from a straw on Mars. When a straw is placed in a glass of liquid, the atmosphere exerts equal pressure inside and outside the straw. When we drink through a straw, we withdraw air, thereby reducing the pressure on the liquid inside. If only 0.007 atm is exerted on the liquid in the glass, a very small reduction in pressure inside the straw will cause the liquid to rise. **10.3** At the same temperature, volume and the lower pressure, the container would have half as many particles as at the higher pressure. **10.5** For a fixed amount of ideal gas at constant volume, if the pressure is doubled, the temperature also doubles. **10.7** (a) $P_{red} < P_{yellow} < P_{blue}$ (b) $P_{red} = 0.28$ atm; $P_{yellow} = 0.42$ atm; $P_{blue} = 0.70$ atm **10.09** (a) Curve B is helium. (b) Curve B corresponds to the higher temperature. (c) The root mean square speed is highest. **10.11** The $NH_4Cl(s)$ ring will form at location A. **10.13** (a) A gas is much less dense than a liquid. (b) A gas is much more compressible than a liquid. (c) All mixtures of gases are homogenous. Similar liquid molecules form homogeneous mixtures, while very dissimilar molecules form heterogeneous mixtures. (d) Both gases and liquids conform to the shape of their container. A gas also adopts the volume of its container, while a liquid maintains its own volume. **10.15** (a) 1.8×10^3 kPa (b) 18 atm (c) 2.6×10^2 lb/in.^{2} **10.17** (a) 10.3 m (b) 2.1 atm **10.19** (a) 0.349 atm (b) 265 mm Hg (c) 3.53×10^4 Pa (d) 0.353 bar (e) 5.13 psi **10.21** (a) $P = 773.4$ torr (b) $P = 1.018$ atm (c) The pressure in Chicago is greater than standard atmospheric pressure, and so it makes sense to classify this weather system as a "high-pressure system." **10.23** (i) 0.31 atm (ii) 1.88 atm (iii) 0.136 atm **10.25** The action in (c) would double the pressure **10.27** (a) Boyle's Law, $PV =$ constant or $P_1V_1 = P_2V_2$, at constant V, $P_1/P_2 = 1$; Charles' Law, $V/T =$ constant or $V_1/T_1 = V_2/T_2$, at constant V, $T_1/T_2 = 1$; then $P_1/T_1 = P_2/T_2$ or $P/T =$ constant. Amonton's law is that pressure and Kelvin temperature are directly proportional at constant volume. (b) 34.7 psi **10.29** (a) STP stands for standard temperature, 0 °C (or 273 K), and standard pressure, 1 atm. (b) 22.4 L (c) 24.5 L (d) 0.08315 L-bar/mol-K **10.31** Flask A contains the gas with a molar mass of 30 g/mol and flask B contains the gas with a molar mass of 60 g/mol.

10.33

P	V	n	T
200 atm	1.00 L	0.500 mol	48.7 K
0.300 atm	0.250 L	3.05×10^{-3} mol	27 °C
650 torr	11.2 L	0.333 mol	350 K
10.3 atm	585 mL	0.250 mol	295 K

10.35 8.2×10^2 kg He **10.37** (a) 5.15×10^{22} molecules (b) 6.5 kg air **10.39** (a) 91 atm (b) 2.3×10^2 L **10.41** $P = 4.9$ atm **10.43** (a) 29.8 g Cl_2 (b) 9.42 L (c) 501 K (d) 2.28 atm **10.45** (a) $n = 2 \times 10^{-4}$ mol O_2 (b) The roach needs 8×10^{-3} mol O_2 in 48 h, approximately 100% of the O_2 in the jar. **10.47** For gas samples at the same conditions, molar mass determines density. Of the three gases listed, (c) Cl_2 has the largest molar mass. **10.49** (c) Because the helium atoms are of lower mass than the average air molecule, the helium gas is less dense than air. The balloon thus weighs less than the air displaced by its volume. **10.51** (a) $d = 1.77$ g/L (b) molar mass $= 80.1$ g/mol **10.53** molar mass $= 89.4$ g/mol **10.55** 4.1×10^{-9} g Mg **10.57** (a) 21.4 L CO_2 (b) 40.7 L O_2 **10.59** 0.402 g Zn **10.61** (a) When the stopcock is opened, the volume occupied by $N_2(g)$ increases from 2.0 L to 5.0 L. $P_{N_2} = 0.40$ atm (b) When the gases mix, the volume of $O_2(g)$ increases from 3.0 L to 5.0 L. $P_{O_2} = 1.2$ atm (c) $P_t = 1.6$ atm **10.63** (a) $P_{He} = 1.87$ atm, $P_{Ne} = 0.807$ atm, $P_{Ar} = 0.269$ atm, (b) $P_t = 2.95$ atm **10.65** $X_{CO_2} = 0.00039$ **10.67** $P_{CO_2} = 0.305$ atm, $P_t = 1.232$ atm **10.69** $P_{CO_2} = 0.9$ atm **10.71** 2.5 mole %O_2 **10.73** $P_t = 2.47$ atm **10.75** (a) Decrease (b) increase (c) decrease **10.77** The two common assumptions are: (i) the volume of gas molecules is negligible relative to the container volume; and (ii) attractive and repulsive forces among molecules are negligible. **10.79** The root-mean-square speed of WF_6 is approximately 8 times slower than that of He. **10.81** (a) Average kinetic energy of the molecules increases. (b) Root mean square speed of the molecules increases. (c) Strength of an average impact with the container walls increases. (d) Total collisions of molecules with walls per second increases. **10.83** (a) In order of increasing speed and decreasing molar mass: HBr $<$ NF_3 $<$ SO_2 $<$ CO $<$ Ne (b) $u_{NF_3} = 324$ m/s (c) The most probable speed of an ozone molecule in the stratosphere

is 306 m/s. **10.85** Effusion is the escape of gas molecules through a tiny hole. Diffusion is the distribution of a gas throughout space or throughout another substance. **10.87** The order of increasing rate of effusion is $^2H^{37}Cl < {}^1H^{37}Cl < {}^2H^{35}Cl < {}^1H^{35}Cl$. **10.89** As_4S_6 **10.91** (a) Non-ideal-gas behavior is observed at very high pressures and low temperatures. (b) The real volumes of gas molecules and attractive intermolecular forces between molecules cause gases to behave nonideally. **10.93** Ar ($a = 1.34, b = 0.0322$) will behave more like an ideal gas than CO_2 ($a = 3.59, b = 0.427$) at high pressures. **10.95** (a) $P = 4.89$ atm (b) $P = 4.69$ atm (c) Qualitatively, molecular attractions are more important as the amount of free space decreases and the number of molecular collisions increases. Molecular volume is a larger part of the total volume as the container volume decreases. **10.97** From the value of b for Xe, the nonbonding radius is 2.72 Å. From Figure 7.7, the bonding atomic radius of Xe is 1.40 Å. We expect the bonding radius of an atom to be smaller than its nonbonding radius, but our claculated value is nearly twice as large. **10.99** $V = 3.1$ mm^3 **10.101** $P = 0.43$ mm Hg **10.103** (a) 13.4 mol $C_3H_8(g)$ (b) 1.47×10^3 mol $C_3H_8(l)$ (c) The ratio of moles liquid to moles gas is 110. Many more molecules and moles of liquid fit in a container of fixed volume because there is much less space between molecules in the liquid phase. **10.105** (a) Molar mass of the unknown gas is 100.4 g/mol (b) We assume that the gases behave ideally, and that P, V and T are constant. **10.107** (a) 0.00378 mol O_2 (b) 0.0345 g C_8H_8 **10.109** 42.2 g O_2 **10.111** (a) Molar mass of the unknown gas is 50.46 g/mol (b) The ratio d/P varies with pressure because of the finite volumes of gas molecules and attractive intermolecular forces. **10.113** $T_2 = 687$ °C **10.115** (a) More significant (b) less significant **10.117** (a) At STP, argon atoms occupy 0.0359% of the total volume. (b) At 200 atm pressure and 0 °C, argon atoms occupy 7.19% of the total volume. **10.119** (a) The molecular formula of cyclopropane is C_3H_6. (b) Although the molar masses of Ar and C_3H_6 are similar, we expect intermolecular attractions to be more significant for the more complex C_3H_6 molecules, and that C_3H_6 will deviate more from ideal behavior at the conditions listed. If the pressure is high enough for the volume correction in the van der Waals equation to dominate behavior, the larger C_3H_6 molecules definitely deviate more than Ar atoms from ideal behavior. (c) Cyclopropane would effuse through a pinhole slower than methane, because it has the greater molar mass. **10.121** (a) 44.58% C, 6.596% H, 16.44% Cl, 32.38% N (b) $C_8H_{14}N_5Cl$ (c) Molar mass of the compound is required in order to determine molecular formula when the empirical formula is known. **10.123** (a) $NH_3(g)$ remains after reaction. (b) $P = 0.957$ atm (c) 7.33 g NH_4Cl **10.125** (a) $:\ddot{O}\!-\!\ddot{Cl}\!-\!\ddot{O}:$ (b) ClO_2 is very reactive because it is an odd-electron molecule. Adding an electron both pairs the odd electron and completes the octet of Cl. ClO_2 has a strong tendency to gain an electron and be reduced. (c) $\left[:\ddot{O}\!-\!\ddot{Cl}\!-\!\ddot{O}:\right]^-$ (d) The bond angle is approximately 170°. (e) 11.2 g ClO_2 **10.127** (a) $P_{IF_3} = 0.515$ atm (b) $X_{IF_5} = 0.544$ (c)

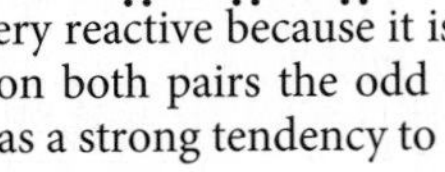

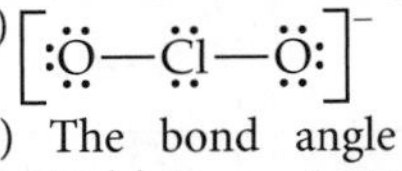

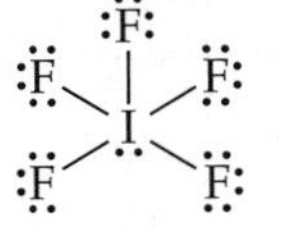

(d) Total mass in the flask is 20.00 g; mass is conserved.

Chapter 11

11.1 (a) The diagram best describes a liquid. (b) In the diagram, particles are close together, mostly touching, but there is no regular arrangement or order. This rules out a gaseous sample, where the particles are far apart, and a crystalline solid, which has a regular repeating structure in all three directions. **11.4** (a) In its final state, methane is a gas at 185 °C. **11.6** The stronger the intermolecular forces, the higher the boiling point of a liquid. Propanol, $CH_3CH_2CH_2OH$, has hydrogen bonding and the higher boiling point. **11.9** (a) Solid < liquid < gas (b) gas < liquid < solid (c) Matter in the gaseous state is most easily compressed because particles are far apart and there is much empty space. **11.11** (a) It increases. Kinetic energy is the energy of motion. As melting occurs, the motion of atoms relative to each other increases. (b) It increases somewhat. The density of liquid lead is less than the density of solid lead. The smaller density means a greater sample volume and average distance between atoms in three dimensions. **11.13** (a) The molar volumes of Cl_2 and NH_3 are nearly the same because they are both gases. (b) On cooling to 160 K, both compounds condense from the gas phase to the solid-state, so we expect a significant decrease in the molar volume. (c) The molar volumes are 0.0351 L/mol Cl_2 and 0.0203 L/mol NH_3 (d) Solid-state molar volumes are not as similar as those in the gaseous state, because most of the empty space is gone and molecular characteristics determine properties. $Cl_2(s)$ is heavier, has a longer bond distance and weaker intermolecular forces, so it has a significantly larger molar volume than $NH_3(s)$. (e) There is little empty space between molecules in the liquid state, so we expect their molar volumes to be closer to those in the solid state than those in the gaseous state. **11.15** (a) London dispersion forces (b) dipole–dipole forces (c) hydrogen bonding **11.17** (a) SO_2, dipole–dipole and London dispersion forces (b) CH_3COOH, London dispersion, dipole–dipole, and hydrogen bonding (c) H_2S, dipole–dipole and London dispersion forces (but not hydrogen bonding) **11.19** (a) *Polarizability* is the ease with which the charge distribution in a molecule can be distorted to produce a transient dipole. (b) Sb is most polarizable because its valence electrons are farthest from the nucleus and least tightly held. (c) in order of increasing polarizability: CH_4 < SiH_4 < $SiCl_4$ < $GeCl_4$ < $GeBr_4$ (d) The magnitudes of London dispersion forces and thus the boiling points of molecules increase as polarizability increases. The order of increasing boiling points is the order of increasing polarizability given in (c). **11.21** (a) H_2S (b) CO_2 (c) GeH_4 **11.23** Both rodlike butane molecules and spherical 2-methylpropane molecules experience dispersion forces. The larger contact surface between butane molecules facilitates stronger forces and produces a higher boiling point. **11.25** (a) A molecule must contain H atoms, bound to either N, O, or F atoms, in order to participate in hydrogen bonding with like molecules. (b) CH_3NH_2 and CH_3OH **11.27** (a) Replacing a hydroxyl hydrogen with a CH_3 group eliminates hydrogen bonding in that part of the molecule. This reduces the strength of intermolecular forces and leads to a lower boiling point. (b) $CH_3OCH_2CH_2OCH_3$ is a larger, more polarizable molecule with stronger London dispersion forces and thus a higher boiling point.

11.29

Physical Property	H_2O	H_2S
Normal boiling point, °C	100.00	−60.7
Normal melting point, C	0.00	−85.5

(a) Based on its much higher normal melting point and boiling point, H_2O has much stronger intermolecular forces. (b) H_2O has hydrogen bonding, while H_2S has dipole–dipole forces. Both molecules have London dispersion forces. **11.31** SO_4^{2-} has a greater negative charge than BF_4^-, so ion–ion electrostatic attractions are greater in sulfate salts and they are less likely to form liquids. **11.33** (a) As temperature increases, surface tension decreases; they are inversely related. (b) As temperature increases, viscosity decreases; they are inversely related. (c) The same attractive forces that cause surface molecules to be difficult to separate (high surface tension) cause molecules elsewhere in the sample to resist movement relative to each other (high viscosity). **11.35** (a) $CHBr_3$ has a higher molar mass, is more polarizable, and has stronger dispersion forces, so the surface tension is greater. (b) As temperature increases, the viscosity of the oil decreases because the average kinetic energy of the molecules increases. (c) Adhesive forces between polar water and nonpolar car wax are weak, so the large surface tension

of water draws the liquid into the shape with the smallest surface area, a sphere. (d) Adhesive forces between nonpolar oil and nonpolar car wax are similar to cohesive forces in oil, so the oil drops spread out on the waxed car hood. **11.37** (a) The three molecules have similar structures and experience the same types of intermolecular forces. As molar mass increases, the strength of dispersion forces increases and the boiling points, surface tension, and viscosities all increase. (b) Ethylene glycol has an — OH group at both ends of the molecule. This greatly increases the possibilities for hydrogen bonding; the overall intermolecular attractive forces are greater and the viscosity of ethylene glycol is much greater. (c) Water has the highest surface tension but lowest viscosity because it is the smallest molecule in the series. There is no hydrocarbon chain to inhibit their strong attraction to molecules in the interior of the drop, resulting in high surface tension. The absence of an alkyl chain also means the molecules can move around each other easily, resulting in the low viscosity. **11.39** (a) Melting, endothermic (b) evaporation, endothermic (c) deposition, exothermic (d) condensation, exothermic **11.41** Melting does not require separation of molecules, so the energy requirement is smaller than for vaporization, where molecules must be separated. **11.43** 2.3×10^3 g H_2O **11.45** (a) 39.3 kJ (b) 60 kJ **11.47** (a) The critical pressure is the pressure required to cause liquefaction at the critical temperature. (b) As the force of attraction between molecules increases, the critical temperature of the compound increases. (c) All the gases in Table 11.6 can be liquefied at the temperature of liquid nitrogen, given sufficient pressure. **11.49** Properties (c) intermolecular attractive forces, (d) temperature and (e) density of the liquid affect vapor pressure of a liquid. **11.51** (a) CBr_4 < $CHBr_3$ < CH_2Br_2 < CH_2Cl_2 < CH_3Cl < CH_4. (b) CH_4 < CH_3Cl < CH_2Cl_2 < CH_2Br_2 < $CHBr_3$ < CBr_4 (c) By analogy to attractive forces in HCl, the trend will be dominated by dispersion forces, even though four of the molecules are polar. The order of increasing boiling point is the order of increasing molar mass and increasing strength of dispersion forces. **11.53** (a) The temperature of the water in the two pans is the same. (b) Vapor pressure does not depend on either volume or surface area of the liquid. At the same temperature, the vapor pressures of water in the two containers are the same. **11.55** (a) Approximately 48 °C (b) approximately 340 torr (c) approximately 17 °C (d) approximately 1000 torr **11.57** (a) The critical point is the temperature and pressure beyond which the gas and liquid phases are indistinguishable. (b) The line that separates the gas and liquid phases ends at the critical point because conditions beyond the critical temperature and pressure, there is no distinction between gas and liquid. In experimental terms a gas cannot be liquefied at temperatures higher than the critical temperature, regardless of pressure. **11.59** (a) $H_2O(g)$ will condense to $H_2O(s)$ at approximately 4 torr; at a higher pressure, perhaps 5 atm or so, $H_2O(s)$ will melt to form $H_2O(l)$. (b) At 100 °C and 0.50 atm, water is in the vapor phase. As it cools, water vapor condenses to the liquid at approximately 82 °C, the temperature where the vapor pressure of liquid water is 0.50 atm. Further cooling results in freezing at approximately 0 °C. The freezing point of water increases with decreasing pressure, so at 0.50 atm the freezing temperature is very slightly above 0 °C. **11.61** (a) 24 K (b) Neon sublimes at pressures less than the triple point pressure, 0.43 atm. (c) No **11.63** (a) Methane on the surface of Titan is likely to exist in both solid and liquid forms. (b) As pressure decreases upon moving away from the surface of Titan, $CH_4(l)$ (at −178 °C) will vaporize to $CH_4(g)$, and $CH_4(s)$ (at temperatures below −180 °C) will sublime to $CH_4(g)$. **11.65** In a nematic liquid crystalline phase, molecules are aligned along their long axes, but the molecular ends are not aligned. Molecules are free to translate in all dimensions, but they cannot tumble or rotate out of the molecular plane, or the order of the nematic phase is lost and the sample becomes an ordinary liquid. In an ordinary liquid, molecules are randomly oriented and free to move in any direction. **11.67** The presence of polar groups or nonbonded electron pairs leads to relatively strong dipole–dipole interactions between molecules. These are a significant part of the orienting forces necessary for liquid crystal formation. **11.69** Because order is maintained in at least one dimension, the molecules in a liquid-crystalline phase are not totally free to change orientation. This makes the liquid-crystalline phase more resistant to flow, more viscous, than the isotropic liquid. **11.71** Melting provides kinetic energy sufficient to disrupt molecular alignment in one dimension in the solid, producing a smectic phase with ordering in two dimensions. Additional heating of the smectic phase provides kinetic energy sufficient to disrupt alignment in another dimension, producing a nematic phase with one-dimensional order. **11.73** (a) Decrease (b) increase (c) increase (d) increase (e) increase (f) increase (g) increase **11.76** (a) The cis isomer has stronger dipole-dipole forces; the trans isomer is nonpolar. (b) The cis isomer boils at 60.3 °C and the trans isomer boils at 47.5 °C. **11.78** When a halogen is substituted for H in benzene, molar mass, polarizability and strength of dispersion forces increase; the order of increasing molar mass is the order of increasing boiling points for the first three compounds. C_6H_5OH experiences hydrogen bonding, the strongest force between neutral molecules, so it has the highest boiling point. **11.81** A plot of number of carbon atoms versus boiling point indicates that the boiling point of C_8H_{18} is approximately 130 °C. The more carbon atoms in the hydrocarbon, the longer the chain, the more polarizable the electron cloud, the higher the boiling point. **11.83** (a) Evaporation is an endothermic process. The heat required to vaporize sweat is absorbed from your body, helping to keep it cool. (b) The vacuum pump reduces the pressure of the atmosphere above the water until atmospheric pressure equals the vapor pressure of water and the water boils. Boiling is an endothermic process, and the temperature drops if the system is not able to absorb heat from the surroundings fast enough. As the temperature of the water decreases, the water freezes. **11.88** At low Antarctic temperatures, molecules in the liquid crystalline phase have less kinetic energy due to temperature, and the applied voltage may not be sufficient to overcome orienting forces among the ends of molecules. If some or all of the molecules do not rotate when the voltage is applied, the display will not function properly.

11.92

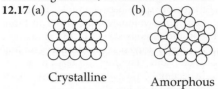

(i) **M** = 44   (ii) **M** = 72   (iii) **M** = 123

(iv) **M** = 58   (v) **M** = 123   (vi) **M** = 60

(a) Molar mass: Compounds (i) and (ii) have similar rodlike structures. The longer chain in (ii) leads to greater molar mass, stronger London dispersion forces, and higher heat of vaporization. (b) Molecular shape: Compounds (iii) and (v) have the same chemical formula and molar mass but different molecular shapes. The more rodlike shape of (v) leads to more contact between molecules, stronger dispersion forces, and higher heat of vaporization. (c) Molecular polarity: Compound (iv) has a smaller molar mass than (ii) but a larger heat of vaporization, which must be due to the presence of dipole–dipole forces. (d) Hydrogen bonding interactions: Molecules (v) and (vi) have similar structures. Even though (v) has larger molar mass and dispersion forces, hydrogen bonding causes (vi) to have the higher heat of vaporization. **11.96** $P(\text{benzene vapor}) = 98.7$ torr

Chapter 12

12.1 The orange compound is more likely to be a semiconductor and the white one an insulator. The orange compound absorbs light in the visible spectrum (orange is reflected, so blue is absorbed), while the white compound does not. This indicates that the orange compound has a lower energy electron transition than the white one. Semiconductors have lower energy electron transitions than insulators. **12.3** (a) The structure is hexagonal close-packed. (b) The coordination number, CN, is twelve. (c) CN(1) = 9, CN(2) = 6. **12.5** Fragment (b) is more likely to give rise to electrical conductivity. Arrangement (b) has a delocalized π system, in which electrons are free to move. Mobile electrons are required for electrical conductivity. **12.7** We expect linear polymer (a), with ordered regions, to be more crystalline and to have a higher melting point than branched polymer (b). **12.9** Statement (b) **12.11** (a) Hydrogen bonding, dipole-dipole forces, London dispersion forces (b) covalent chemical bonds (c) ionic bonds (d) metallic bonds **12.13** (a) Ionic (b) metallic (c) covalent-network (It could also be characterized as ionic with some covalent character to the bonds.) (d) molecular (e) molecular (f) molecular **12.15** Because of its relatively high melting point and properties as a conducting solution, the solid must be ionic.

12.17 (a)   (b)

Crystalline   Amorphous

12.19

Two-dimensional structure	(i)	(ii)
(a) unit cell	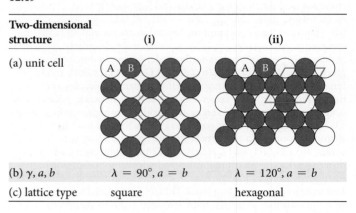	
(b) γ, a, b	$\lambda = 90°, a = b$	$\lambda = 120°, a = b$
(c) lattice type	square	hexagonal

12.21 (a) Tetragonal **12.23** (e) Triclinic and rhombohedral **12.25** (b) 2 **12.27** (a) Primitive hexagonal unit cell (b) NiAs **12.29** Potassium. A body-centered cubic structure has more empty space than a face-centered cubic one. The more empty space, the less dense the solid. We expect the element with the lowest density, potassium, to adopt the body-centered cubic structure. **12.29** (a) Structure types A and C have equally dense packing and are more densely packed than structure type B. (b) Structure type B is least densely packed. **12.33** (a) The radius of an Ir atom is 1.355 Å. (b) The density of Ir is 22.67 g/cm³ **12.35** (a) 4 Al atoms per unit cell (b) coordination number = 12 (c) $a = 4.04$ Å or 4.04×10^{-8} cm (d) density = 2.71 g/cm³ **12.35** An *alloy* contains atoms of more than one element and has the properties of a metal. In a *solution alloy* the components are randomly dispersed. In a *heterogeneous alloy* the components are not evenly dispersed and can be distinguished at a macroscopic level. In an *intermetallic compound* the components have interacted to form a compound substance, as in Cu_3As **12.39** Statement (b) is false. **12.41** (a) True (b) false (c) false **12.43** (a) Nickel or palladium, substitutional alloy (b) copper, substitutional alloy (c) silver, substitutional alloy **12.45** (a) True (b) false (c) false (d) false

12.47

(a) Six AOs require six MOs (b) zero nodes in the lowest energy orbital (c) five nodes in highest energy orbital (d) two nodes in the HOMO (e) three nodes in the LUMO (f) The HOMO-LUMO energy gap for the six-atom diagram is smaller than the one for the four-atom diagram. In general, the more atoms in the chain, the smaller the HOMO-LUMO energy gap. **12.49** (a) Ag is more ductile. Mo has stronger metallic bonding, a stiffer lattice, and is less susceptible to distortion. (b) Zn is more ductile. Si is a covalent-network solid with a stiffer lattice than metallic Zn. **12.51** Moving from Y to Mo, the number of valence electrons, occupancy of the bonding band, and strength of metallic bonding increase. Stronger metallic bonding requires more energy to break bonds and mobilize atoms, resulting in higher melting points. **12.53** (a) $SrTiO_3$ (b) Six (c) Each Sr atom is coordinated to a total of twelve O atoms in the eight unit cells that contain the Sr atom. **12.55** (a) $a = 4.70$ Å (b) 2.69 g/cm^3 **12.57** (a) 7.711 g/cm^3 (b) We expect Se^{2-} to have a larger ionic radius than S^{2-}, so HgSe will occupy a larger volume and the unit cell edge will be longer. (c) The density of HgSe is 8.241 g/cm^3. The greater mass of Se accounts for the greater density of HgSe. **12.59** (a) Cs^+ and I^- have the most similar radii and will adopt the CsCl-type structure. The radii of Na^+ and I^- are somewhat different; NaI will adopt the NaCl-type structure. The radii of Cu^+ and I^- are very different; CuI has the ZnS-type structure. (b) CsI, 8; NaI, 6; CuI, 4 **12.61** (a) 6 (b) 3 (c) 6 **12.63** (a) False (b) true **12.65** (a) Ionic solids are much more likely to dissolve in water. (b) Covalent-network solids can become electrical conductors via chemical substitution. **12.67** (a) CdS (b) GaN (c) GaAs **12.69** Ge or Si (Ge is closer to Ga in bonding atomic radius.) **12.71** (a) A 1.1 eV photon corresponds to a wavelength of 1.1×10^{-6} m (b) According to the figure, Si can absorb all wavelengths in the visible portion of the solar spectrum. (c) Si absorbs wavelengths less than 1,100 nm. This corresponds to approximately 80–90% of the total area under the curve. **12.73** The emitted light has a wavelength of 560 nm; its color is yellow-green. **12.75** The band gap is approximately 1.85 eV, which corresponds to a wavelength of 672 nm. **12.77** (a) A monomer is a small molecule with low molecular mass that can be joined together to form a polymer. They are the repeating units of a polymer. (b) ethene (also known as ethylene) **12.79** Reasonable values for a polymer's molecular weight are 10,000 amu, 100,000 amu, and 1,000,000 amu.

12.81

$$CH_3-\overset{\overset{O}{\|}}{C}-\boxed{O-H} \ + \ \boxed{H}-O-CH_2-CH_3 \longrightarrow$$

Acetic acid   Ethanol

$$CH_3-\overset{\overset{O}{\|}}{C}-O-CH_2CH_3 \ + \ H_2O$$

Ethyl acetate

If a dicarboxylic acid and a dialcohol are combined, there is the potential for propagation of the polymer chain at both ends of both monomers.

12.83

(a) [structure: 1-chloroethene / vinyl chloride]

(b) [structures: a diamine $H_2N-(CH_2)_6-NH_2$ and a diacid $HO-\overset{O}{C}-(CH_2)_4-C-OH$]

(c) [structures: ethylene glycol $HO-CH_2-CH_2-OH$ and terephthalic acid]

12.85

[structures: isophthalic acid HOOC—C6H4—COOH and benzene-1,3-diamine H_2N—C6H4—NH_2] and

12.87 (a) Flexibility of the molecular chains causes flexibility of the bulk polymer. Flexibility is enhanced by molecular features that inhibit order, such as branching, and diminished by features that encourage order, such as cross-linking or delocalized π electron density. (b) Less flexible **12.89** Low degree of crystallinity **12.91** If a solid has nanoscale dimensions of 1–10 nm, there may not be enough atoms contributing atomic orbitals to produce continuous energy bands of molecular orbitals. **12.93** (a) False. As particle size decreases, the band gap increases. (b) False. As particle size decreases, wavelength decreases. **12.95** 2.47×10^5 Au atoms **12.97** Statement (b) is correct. **12.103** 3 Ni atoms, 1 Al atom; 6 Nb atoms, 2 Sn atoms; 1 Sm atom, 5 Co atoms. In each case, the atom ratio matches the empirical formula given in the exercise. **12.105** (a) White tin is a metal, while gray tin is a semiconductor. Delocalized electrons in the close-packed structure of white tin lead to metallic properties, while gray tin has the covalent-network structure characteristic of other Group IV semiconductors. (b) Metallic white tin has the longer bond distance because the valence electrons are shared with twelve nearest neighbors rather than being localized in four bonds as in gray tin. **12.107** (a) Zinc sulfide, ZnS (b) Covalent (c) In the solid, each Si is bound to four C atoms in a tetrahedral arrangement, and each C is bound to four Si atoms in a tetrahedral arrangement, producing an extended three-dimensional network. SiC is high-melting because melting requires breaking covalent Si—C bonds, which takes a huge amount of thermal energy. It is hard because the three-dimensional lattice resists any deformation that would weaken the Si—C bonding network. **12.113** The distance between planes in crystalline silicon is 3.13 Å. **12.119** (a) 109° (b) 120° (c) atomic p orbitals **12.123** (a) 2.50×10^{22} Si atoms (To 1 sig fig, the result is 2×10^{22} Si atoms.) (b) 1.29×10^{-3} mg P ($1.29 \ \mu$g P) **12.125** 99 Si atoms

Chapter 13

13.1 (a) < (b) < (c) **13.3** (a) No (b) The ionic solid with the smaller lattice energy will be more soluble in water. **13.7** Vitamin B_6 is more water soluble because of its small size and capacity for extensive hydrogen-bonding interactions. Vitamin E is more fat soluble. The long, rodlike hydrocarbon chain will lead to strong dispersion forces among vitamin E and mostly nonpolar fats. **13.9** (a) Yes, the *molarity* changes with a change in temperature. Molarity is defined as moles solute per unit volume of solution. A change of temperature changes solution volume and molarity. (b) No, *molality* does not change with change in temperature. Molality is defined as moles solute per kilogram of solvent. Temperature affects neither mass nor moles.

13.11 The volume inside the balloon will be 0.5 L, assuming perfect osmosis across the semipermeable membrane. **13.13** (a) False (b) false (c) true **13.15** (a) Dispersion (b) hydrogen bonding (c) ion–dipole (d) dipole–dipole **13.17** Very soluble. (b) ΔH_{mix} will be the largest negative number. In order for ΔH_{soln} to be negative, the magnitude of ΔH_{mix} must be greater than the magnitude of $(\Delta H_{solute} + \Delta H_{solvent})$. **13.19** (a) ΔH_{solute} (b) ΔH_{mix} **13.21** (a) ΔH_{soln} is nearly zero. Since the solute and solvent experience very similar London dispersion forces, the energy required to separate them individually and the energy released when they are mixed are approximately equal. $\Delta H_{solute} + \Delta H_{solvent} \approx -\Delta H_{mix}$. (b) The entropy of the system increases when heptane and hexane form a solution. From part (a), the enthalpy of mixing is nearly zero, so the increase in entropy is the driving force for mixing in all proportions. **13.23** (a) Supersaturated (b) Add a seed crystal. A seed crystal provides a nucleus of prealigned molecules, so that ordering of the dissolved particles (crystallization) is more facile. **13.25** (a) Unsaturated (b) saturated (c) saturated (d) unsaturated **13.27** (a) We expect the liquids water and glycerol to be miscible in all proportions. The —OH groups of glycerol facilitate strong hydrogen bonding similar to that in water; like dissolves like. (b) Hydrogen bonding, dipole-dipole forces, London dispersion forces **13.29** Toluene, $C_6H_5CH_3$, is the best solvent for nonpolar solutes. Without polar groups or nonbonding electron pairs, it forms only dispersion interactions with itself and other molecules. **13.31** (a) Dispersion interactions among nonpolar $CH_3(CH_2)_{16}$— chains dominate the properties of stearic acid, causing it to be more soluble in nonpolar CCl_4 (b) Dioxane can act as a hydrogen bond acceptor, so it will be more soluble than cyclohexane in water. **13.33** (a) CCl_4 is more soluble because dispersion forces among nonpolar CCl_4 molecules are similar to dispersion forces in hexane. (b) C_6H_6 is a nonpolar hydrocarbon and will be more soluble in the similarly nonpolar hexane. (c) The long, rodlike hydrocarbon chain of octanoic acid forms strong dispersion interactions and causes it to be more soluble in hexane. **13.35** (a) A sealed container is required to maintain a partial pressure of $CO_2(g)$ greater than 1 atm above the beverage. (b) Since the solubility of gases increases with decreasing temperature, more $CO_2(g)$ will remain dissolved in the beverage if it is kept cool. **13.37** $S_{He} = 5.6 \times 10^{-4}\,M$, $S_{N_2} = 9.0 \times 10^{-4}\,M$ **13.39** (a) 2.15% Na_2SO_4 by mass (b) 3.15 ppm Ag **13.41** (a) $X_{CH_3OH} = 0.0427$ (b) 7.35% CH_3OH by mass (c) 2.48 m CH_3OH **13.43** (a) $1.46 \times 10^{-2}\,M$ $Mg(NO_3)_2$ (b) 1.12 M $LiClO_4 \cdot 3H_2O$ (c) 0.350 M HNO_3 **13.45** (a) 4.70 m C_6H_6 (b) 0.235 m NaCl **13.47** (a) 43.01% H_2SO_4 by mass (b) $X_{H_2SO_4} = 0.122$ (c) 7.69 m H_2SO_4 (d) 5.827 M H_2SO_4 **13.49** (a) $X_{CH_3OH} = 0.227$ (b) 7.16 m CH_3OH (c) 4.58 M CH_3OH **13.51** (a) 0.150 mol $SrBr_2$ (b) 1.56×10^{-2} mol KCl (c) 4.44×10^{-2} mol $C_6H_{12}O_6$ **13.53** (a) Weigh out 1.3 g KBr, dissolve in water, dilute with stirring to 0.75 L. (b) Weigh out 2.62 g KBr, dissolve it in 122.38 g H_2O to make exactly 125 g of 0.180 m solution. (c) Dissolve 244 g KBr in water, dilute with stirring to 1.85 L. (d) Weigh 10.1 g KBr, dissolve it in a small amount of water, and dilute to 0.568 L. **13.55** 71% HNO_3 by mass **13.57** (a) 3.82 m Zn (b) 26.8 M Zn **13.59** (a) 0.046 atm (b) $1.8 \times 10^{-3}\,M$ CO_2 **13.61** F (a) False (b) true (c) true (d) false **13.63** The vapor pressure of both solutions is 17.5 torr. Because these two solutions are so dilute, they have essentially the same vapor pressure. Generally, the less concentrated solution, the one with fewer moles of solute per kilogram of solvent, will have the higher vapor pressure. **13.65** (a) $P_{H_2O} = 186.4$ torr (b) 78.9 g $C_3H_8O_2$ **13.67** (a) $X_{Eth} = 0.2812$ (b) $P_{soln} = 238$ torr (c) X_{Eth} in vapor $= 0.472$ **13.69** (a) Because NaCl is a strong electrolyte, one mole of NaCl produces twice as many dissolved particles as one mole of the molecular solute $C_6H_{12}O_6$. Boiling-point elevation is directly related to total moles of dissolved particles, The 0.10 m NaCl has more dissolved particles so its boiling point is higher than the 0.10 m $C_6H_{12}O_6$. (b) In solutions of strong electrolytes like NaCl, ion pairing reduces the effective number of particles in solution, decreasing the change in boiling point. The actual boiling point is then lower than the calculated boiling point for a 0.10 m solution. **13.71** 0.050 m LiBr $<$ 0.120 m glucose $<$ 0.050 m $Zn(NO_3)_2$ **13.73** (a) $T_f = -115.0\,°C$, $T_b = 78.7\,°C$ (b) $T_f = -67.3\,°C$, $T_b = 64.2\,°C$ (c) $T_f = -0.4\,°C$, $T_b = 100.1\,°C$

(d) $T_f = -0.6\,°C$, $T_b = 100.2\,°C$ **13.75** 167 g $C_2H_6O_2$ **13.77** $\Pi = 0.0168$ atm $= 12.7$ torr **13.79** The approximate molar mass of adrenaline is 1.8×10^2 g. **13.81** Molar mass of lysozyme $=1.39 \times 10^4$ g **13.83** (a) $i = 2.8$ (b) The more concentrated the solution, the greater the ion pairing and the smaller the measured value of i. **13.85** (a) No. In the gaseous state, particles are far apart and intermolecular attractive forces are small. When two gases combine, all terms in Equation 13.1 are essentially zero and the mixture is always homogeneous. (b) To determine whether Faraday's dispersion is a true solution or a colloid, shine a beam of light on it. If light is scattered, the dispersion is a colloid. **13.87** Choice (d), $CH_3(CH_2)_{11}$ COONa, is the best emulsifying agent. The long hydrocarbon chain will interact with the hydrophobic component, while the ionic end will interact with the hydrophilic component, as well as stabilize the colloid. **13.89** (a) No. The hydrophobic or hydrophilic nature of the protein will determine which electrolyte at which concentration will be the most effective precipitating agent. (b) Stronger. If a protein has been "salted out", protein-protein interactions are stronger than protein-solvent interactions and solid protein forms. (c) The first hypothesis seems plausible, since ion-dipole interactions among electrolytes and water molecules are stronger than dipole-dipole and hydrogen bonding interactions between water and protein molecules. But, we also know that ions adsorb on the surface of a hydrophobic colloid; the second hypothesis also seems plausible. If we could measure the charge and adsorbed water content of protein molecules as a function of salt concentration, then we could distinguish between these two hypotheses. **13.91** The periphery of the BHT molecule is mostly hydrocarbon-like groups, such as —CH_3. The one —OH group is rather buried inside and probably does little to enhance solubility in water. Thus, BHT is more likely to be soluble in the nonpolar hydrocarbon hexane, C_6H_{14}, than in polar water. **13.94** (a) $k_{Rn} = 7.27 \times 10^{-3}$mol/L-atm (b) $P_{Rn} = 1.1 \times 10^{-4}$ atm; $S_{Rn} = 8.1 \times 10^{-7}\,M$ **13.98** (a) 2.69 m LiBr (b) $X_{LiBr} = 0.0994$ (c) 81.1% LiBr by mass **13.100** $X_{H_2O} = 0.808$; 0.0273 mol ions; 0.0136 mol NaCl; 0.798 g NaCl **13.103** (a) $-0.6\,°C$ (b) $-0.4\,°C$ **13.106** (a), CF_4, $1.7 \times 10^{-4}\,m$; $CClF_3$, $9 \times 10^{-4}\,m$; CCl_2F_2, $2.3 \times 10^{-2}\,m$; $CHClF_2$, $3.5 \times 10^{-2}\,m$ (b) Molality and molarity are numerically similar when kilograms solvent and liters solution are nearly equal. This is true when solutions are dilute and when the density of the solvent is nearly 1g/mL, as in this exercise. (c) Water is a polar solvent; the solubility of solutes increases as their polarity increases. Nonpolar CF_4 has the lowest solubility and the most polar fluorocarbon, $CHClF_2$, has the greatest solubility in H_2O. (d) The Henry's law constant for $CHClF_2$ is 3.5×10^{-2} mol/L-atm. This value is greater than the Henry's law constant for $N_2(g)$ because $N_2(g)$ is nonpolar and of lower molecular mass than $CHClF_2$. **13.109**

(a) cation (g) + anion (g) + solvent

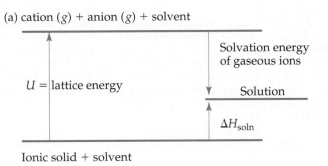

Ionic solid + solvent

(b) Lattice energy (U) is inversely related to the distance between ions, so salts with large cations like $(CH_3)_4N^+$ have smaller lattice energies than salts with simple cations like Na^+. Also the —CH_3 groups in the large cation are capable of dispersion interactions with nonpolar groups of the solvent molecules, resulting in a more negative solvation energy of the gaseous ions. Overall, for salts with larger cations, lattice energy is smaller (less positive), the solvation energy of the gaseous ions is more negative, and ΔH_{soln} is less endothermic. These salts are more soluble in polar nonaqueous solvents. **13.113** (a) $P_t = 330$ torr (b) Weak hydrogen bonds between the two molecules prevent molecules from escaping to the vapor phase and result in a lower than ideal vapor pressure for the solution. There is no hydrogen bonding in the pure liquids. (c) Exothermic. According to Coulomb's law, electrostatic attractive forces lead to an overall lowering of the energy of the system. $\Delta H_{soln} < 0$. (d) No. Chloromethane has only one chlorine atom withdrawing electron density from the central carbon atom, the three C—H bonds are less polarized, and interactions with acetone are weaker. There is probably some vapor pressure lowering, but much less than for chloroform and acetone.

Chapter 14

14.1 The rate of the combustion reaction in the cylinder depends on the surface area of the droplets in the spray. The smaller the droplets, the greater the surface area exposed to oxygen, the faster the combustion reaction. In the case of a clogged injector, larger droplets lead to slower combustion. Uneven combustion in the various cylinders can cause the engine to run roughly and decrease fuel economy. **14.3** Equation (iv) (b) rate $= -\Delta[B]/\Delta t = \frac{1}{2}\Delta[A]/\Delta t$ **14.9** (1) Total potential energy of the reactants (2) E_a, activation energy of the reaction (3) ΔE, net energy change for the reaction (4) total potential energy of the products **14.12** (a) $NO_2 + F_2 \longrightarrow NO_2F + F; NO_2 + F \longrightarrow NO_2F$ (b) $2NO_2 + F_2 \longrightarrow 2NO_2F$ (c) F (atomic fluorine) is the intermediate (d) rate $= k[NO_2][F_2]$ **14.15** (a) Net reaction: $AB + AC \longrightarrow BA_2 + C$ (b) A is the intermediate. (c) A_2 is the catalyst. **14.17** (a) *Reaction rate* is the change in the amount of products or reactants in a given amount of time. (b) Rates depend on concentration of reactants, surface area of reactants, temperature, and presence of catalyst. (c) No. The stoichiometry of the reaction (mole ratios of reactants and products) must be known to relate rate of disappearance of reactants to rate of appearance of products.

14.19

Time (min)	Mol A	(a) Mol B	[A] (mol/L)	Δ[A] (mol/L)	(b) Rate (M/s)
0	0.065	0.000	0.65		
10	0.051	0.014	0.51	−0.14	2.3×10^{-4}
20	0.042	0.023	0.42	−0.09	1.5×10^{-4}
30	0.036	0.029	0.36	−0.06	1.0×10^{-4}
40	0.031	0.034	0.31	−0.05	0.8×10^{-4}

(c) $\Delta[B]_{avg}/\Delta t = 1.3 \times 10^{-4} M/s$

14.21 (a)

Time (s)	Time Interval (s)	Concentration (M)	ΔM	Rate (M/s)
0		0.0165		
2,000	2,000	0.0110	−0.0055	28×10^{-7}
5,000	3,000	0.00591	−0.0051	17×10^{-7}
8,000	3,000	0.00314	−0.00277	9.3×10^{-7}
12,000	4,000	0.00137	−0.00177	4.43×10^{-7}
15,000	3,000	0.00074	−0.00063	2.1×10^{-7}

(b) The average rate of reaction is $1.05 \times 10^{-6} M/s$ (c) The average rate between $t = 2,000$ and $t = 12,000$ s ($9.63 \times 10^{-7} M/s$) is greater than the average rate between $t = 8,000$ and $t = 15,000$ s ($3.43 \times 10^{-7} M/s$). (d) From the slopes of the tangents to the graph, the rates are $12 \times 10^{-7} M/s$ at 5000 s, $5.8 \times 10^{-7} M/s$ at 8000 s. **14.23** (a) $-\Delta[H_2O_2]/\Delta t = \Delta[H_2]/\Delta t = \Delta[O_2]/\Delta t$ (b) $-\frac{1}{2}\Delta[N_2O]/\Delta t = \frac{1}{2}\Delta[N_2]/\Delta t = \Delta[O_2]/\Delta t$

(c) $-\Delta[N_2]/\Delta t = -1/3\Delta[H_2]/\Delta t = -1/2\Delta[NH_3]/\Delta t$ (d) $-\Delta[C_2H_5NH_2]/\Delta t = \Delta[C_2H_4]/\Delta t = \Delta[NH_3]/\Delta t$ **14.25** (a) $-\Delta[O_2]/\Delta t = 0.24$ mol/s; $\Delta[H_2O]/\Delta t = 0.48$ mol/s (b) P_{total} decreases by 28 torr/min. **14.27** (a) If [A] doubles, there is no change in the rate or the rate constant. (b) The reaction is zero order in A, second order in B, and second order overall. (c) units of $k = M^{-1}s^{-1}$ **14.29** (a) Rate $= k[N_2O_5]$ (b) Rate $= 1.16 \times 10^{-4} M/s$ (c) When the concentration of N_2O_5 doubles, the rate doubles. (d) When the concentration of N_2O_5 is halved, the rate doubles. **14.31** (a, b) $k = 1.7 \times 10^2 M^{-1}s^{-1}$ (c) If $[OH^-]$ is tripled, the rate triples. (d) If $[OH^-]$ and $[CH_3Br]$ both triple, the rate increases by a factor of 9. **14.33** (a) Rate $= k[OCl^-][I^-]$ (b) $k = 60 M^{-1}s^{-1}$. (c) Rate $= 6.0 \times 10^{-5} M/s$ **14.35** (a) Rate $= k[BF_3][NH_3]$ (b) The reaction is second order overall. (c) $k_{avg} = 3.41 M^{-1}s^{-1}$ (d) $0.170 M/s$ **14.37** (a) Rate $= k[NO]^2[Br_2]$ (b) $k_{avg} = 1.2 \times 10^4 M^{-2}s^{-1}$ (c) $\frac{1}{2}\Delta[NOBr]/\Delta t = -\Delta[Br_2]/\Delta t$ (d) $-\Delta[Br_2]/\Delta t = 8.4 M/s$ **14.39** (a) $[A]_0$ is the molar concentration of reactant A at time zero. $[A]_t$ is the molar concentration of reactant A at time t. $t_{1/2}$ is the time required to reduce $[A]_0$ by a factor of 2. k is the rate constant for a particular reaction. (b) A graph of ln[A] versus time yields a straight line for a first-order reaction. (c) On a graph of ln[A] versus time, the rate constant is the (–slope) of the straight line. **14.41** (a) $k = 3.0 \times 10^{-6}s^{-1}$ (b) $t_{1/2} = 3.2 \times 10^4$ s **14.43** (a) $P = 30$ torr (b) $t = 51$ s **14.45** Plot $(\ln P_{SO_2Cl_2})$ versus time, $k = -$slope $= 2.19 \times 10^{-5}s^{-1}$ **14.47** (a) The plot of $1/[A]$ versus time is linear, so the reaction is second order in [A]. (b) $k = 0.040 M^{-1}$ min^{-1} (c) $t_{1/2} = 38$ min **14.49** (a) The plot of $1/[NO_2]$ versus time is linear, so the reaction is second order in NO_2. (b) $k = $ slope $= 10 M^{-1}s^{-1}$ (c) rate at $0.200 M = 0.400 M/s$, rate at $0.100 M = 0.100 M/s$; rate at $0.050 M = 0.025 M/s$ **14.51** (a) The energy of the collision and the orientation of the molecules when they collide determine whether a reaction will occur. (b) At a higher temperature, there are more total collisions and each collision is more energetic. (c) The rate constant usually increases with an increase in reaction temperature. **14.53** $f = 4.94 \times 10^{-2}$. At 400 K, approximately 1 out of 20 molecules has this kinetic energy. **14.55** (a)

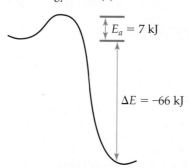

(b) $E_a(\text{reverse}) = 73$ kJ **14.57** (a) False (b) false (c) true **14.59** Reaction (b) is fastest and reaction (c) is slowest. **14.61** (a) $k = 1.1$ s^{-1} (b) $k = 13$ s^{-1} (c) The method in parts (a) and (b) assumes that the collision model and thus the Arrhenious equation describe the kinetics of the reactions. That is, activation energy is constant over the temperature range under consideration. **14.63** A plot of ln k versus $1/T$ has a slope of -5.71×10^3; $E_a = -R(\text{slope}) = 47.5$ kJ/mol. **14.65** (a) An *elementary reaction* is a process that occurs as a single event; the order is given by the coefficients in the balanced equation for the reaction. (b) A *unimolecular* elementary reaction involves only one reactant molecule; a *bimolecular* elementary reaction involves two reactant molecules. (c) A *reaction mechanism* is a series of elementary reactions that describes how an overall reaction occurs and explains the experimentally determined rate law. (d) A *rate-determining step* is the slow step in a reaction mechanism. It limits the overall reaction rate. **14.67** (a) Unimolecular, rate $= k[Cl_2]$ (b) bimolecular, rate $= k[OCl^-][H_2O]$ (c) bimolecular, rate $= k[NO][Cl_2]$ **14.69** (a) Two intermediates, B and C. (b) three transition states (c) C $\longrightarrow$ D is fastest. (d) ΔE is positive. **14.71** (a) $H_2(g) + 2ICl(g) \longrightarrow I_2(g) + 2HCl(g)$ (b) HI is the intermediate. (c) If the first step is slow,

the observed rate law is rate $= k[H_2][ICl]$. **14.73** (a) The two-step mechanism is consistent with the data, assuming that the second step is rate determining. (b) No. The linear plot guarantees that the overall rate law will include $[NO]^2$. Since the data were obtained at constant $[Cl_2]$, we have no information about reaction order with respect to $[Cl_2]$. **14.75** (a) A catalyst is a substance that changes (usually increases) the speed of a chemical reaction without undergoing a permanent chemical change itself. (b) A homogeneous catalyst is in the same phase as the reactants, while a hetereogeneous catalyst is in a different phase. (c) A catalyst has no effect on the overall enthalpy change for a reaction, but it does affect activation energy. It can also affect the frequency factor.
14.77

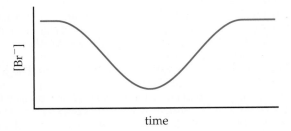

14.79 (a) Multiply the coefficients in the first reaction by 2 and sum. (b) $NO_2(g)$ is a catalyst. (c) This is a homogeneous catalysis. **14.81** (a) Use of chemically stable supports makes it possible to obtain very large surface areas per unit mass of the precious metal catalyst because the metal can be deposited in a very thin, even monomolecular, layer on the surface of the support. (b) The greater the surface area of the catalyst, the more reaction sites, the greater the rate of the catalyzed reaction. **14.83** To put two D atoms on a single carbon, it is necessary that one of the already existing C—H bonds in ethylene be broken while the molecule is adsorbed, so that the H atom moves off as an adsorbed atom and is replaced by a D atom. This requires a larger activation energy than simply adsorbing C_2H_4 and adding one D atom to each carbon. **14.85** Carbonic anhydrase lowers the activation energy of the reaction by 42 kJ. **14.87** (a) The catalyzed reaction is approximately 10,000,000 times faster at 25 °C (b) The catalyzed reaction is 180,000 times faster at 125 °C. **14.91** (a) Rate $= 4.7 \times 10^{-5} M/s$ (b, c) $k = 0.84\ M^{-2}s^{-1}$ (d) If the [NO] is increased by a factor of 1.8, the rate would increase by a factor of 3.2. **14.95** (a) The reaction is second order in NO_2. (b) If $[NO_2]_0 = 0.100\ M$ and $[NO_2]_t = 0.025\ M$, use the integrated form of the second-order rate equation to solve for t. $t = 48\ s$ **14.99** (a) The half-life of ^{241}Am is 4.3×10^2 yr, that of ^{125}I is 63 days (b) ^{125}I decays at a much faster rate. (c) 0.13 mg of each isotope remains after 3 half-lives. (d) The amount of ^{241}Am remaining after 4 days is 1.00 mg. The amount of ^{125}I remaining after 4 days is 0.96 mg. **14.103** (a) The plot of $1/[C_5H_6]$ versus time is linear and the reaction is second order. (b) $k = 0.167\ M^{-1}\ s^{-1}$ **14.107** (a) When the two elementary reactions are added, $N_2O_2(g)$ appears on both sides and cancels, resulting in the overall reaction. $2NO(g) + H_2(g) \longrightarrow N_2O(g) + H_2O(g)$ (b) First reaction, $-[NO]/\Delta t = k[NO]^2$, second reaction, $-[H_2]/\Delta t = k[H_2][N_2O_2]$ (c) N_2O_2 is the intermediate. (d) Because $[H_2]$ appears in the rate law, the second step must be slow relative to the first. **14.110** (a) $Cl_2(g) + CHCl_3(g) \longrightarrow HCl(g) + CCl_4(g)$ (b) $Cl(g)$, $CCl_3(g)$ (c) reaction 1, unimolecular; reaction 2, bimolecular; reaction 3, bimolecular (d) Reaction 2 is rate determining. (e) Rate $= k[CHCl_3][Cl_2]^{1/2}$. **14.115** The enzyme must lower the activation energy by 22 kJ in order to make it useful.
14.120 (a) $k = 8 \times 10^7\ M^{-1}s^{-1}$

(b) $:N≡\ddot{O}:$

$:\ddot{O}=\ddot{N}-\ddot{F}: \longleftrightarrow (:\ddot{O}-\ddot{N}=\ddot{F})$

(c) NOF is bent with a bond angle of approximately 120°

(d)

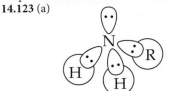

(e) The electron-deficient NO molecule is attracted to electron-rich F_2 so the driving force for formation of the transition state is greater than simple random collisions.
14.123 (a)

(b) A reactant that is attracted to the lone pair of electrons on nitrogen will produce a tetrahedral intermediate. This can be a moiety with a full, partial or even transient positive charge.

Chapter 15

15.1 $k_f > k_r$ (b) The equilibrium constant is greater than one. **15.3** K is greater than one. **15.5** $A(g) + B(g) \rightleftharpoons AB(g)$ has the larger equilibrium constant. **15.6** From the smallest to the largest equilibrium constant, (c) < (b) < (a). **15.7** Statement (b) is true. **15.9** Two B atoms should be added to the diagram. **15.11** K_c decreases as T increases, so the reaction is exothermic. **15.13** (a) $K_p = K_c = 8.1 \times 10^{-3}$. (b) At equilibrium, the partial pressure of A is greater than the partial pressure of B. **15.15** (a) $K_c = [N_2O][NO_2]/[NO]^3$; homogeneous (b) $K_c = [CS_2][H_2]^4/[CH_4][H_2S]^2$; homogeneous (c) $K_c = [CO]^4/[Ni(CO)_4]$; heterogeneous (d) $K_c = [H^+][F^-]/[HF]$; homogeneous (e) $K_c = [Ag^+]^2/[Zn^{2+}]$; heterogeneous (f) $K_c = [H^+][OH^-]$; homogeneous (g) $K_c = [H^+]^2[OH^-]^2$; homogeneous **15.17** (a) Mostly reactants (b) mostly products **15.19** (a) True (b) false (c) false **15.21** $K_p = 1.0 \times 10^{-3}$ **15.23** (a) The equilibrium favors NO and Br_2 at this temperature. (b) $K_c = 77$ (c) $K_c = 8.8$ **15.25** (a) $K_p = 0.541$ (b) $K_p = 3.42$ (c) $K_c = 281$ **15.27** $K_c = 0.14$ **15.29** (a) $K_p = P_{O_2}$ (b) $K_c = [Hg(solv)]^4[O_2(solv)]$ **15.31** $K_c = 10.5$ **15.33** (a) $K_p = 51$ (b) $K_c = 2.1 \times 10^3$ **15.35** (a) $[H_2] = 0.012\ M$, $[N_2] = 0.019\ M$, $[H_2O] = 0.138\ M$ (b) $K_c = 653.7 = 7 \times 10^2$ **15.37** (a) $P_{CO_2} = 4.10$ atm, $P_{H_2} = 2.05$ atm, $P_{H_2O} = 3.28$ atm, (b) $P_{CO_2} = 3.87$ atm, $P_{H_2} = 1.82$ atm, $P_{CO} = 0.23$ atm (c) $K_p = 0.11$ (d) $K_c = 0.11$ **15.39** $K_c = 2.0 \times 10^4$ **15.41** (a) To the right (b) The concentrations used to calculate Q must be equilibrium concentrations. **15.45** (a) $Q = 1.1 \times 10^{-8}$, the reaction will proceed to the left. (b) $Q = 5.5 \times 10^{-12}$, the reaction will proceed to the right. (c) $Q = 2.19 \times 10^{-10}$, the mixture is at equilibrium. **15.45** $P_{Cl_2} = 5.0$ atm **15.47** $[Br_2] = 0.00767\ M$, $[Br] = 0.00282\ M$, 0.0451 g $Br(g)$ **15.49** $[I] = 2.10 \times 10^{-5} M$, $[I_2] = 1.43 \times 10^{-5} M$, 0.0362 g I_2 **15.51** $[NO] = 0.002\ M$, $[N_2] = [O_2] = 0.087\ M$ **15.53** The equilibrium pressure of $Br_2(g)$ is 0.416 atm. **15.55** (a) $[Ca^{2+}] = [SO_4^{2-}] = 4.9 \times 10^{-3}\ M$ (b) A bit more than 1.0 g $CaSO_4$ is needed in order to have some undissolved $CaSO_4$ (s) in equilibrium with 1.4 L of saturated solution. **15.57** $[IBr] = 0.223\ M$, $[I_2] = [Br_2] = 0.0133\ M$ **15.59** (a) $P_{CH_3I} = P_{HI} = 0.422$ torr, $P_{CH_4} = 104.7$ torr, $P_{I_2} =$ torr **15.61** (a) Shift equilibrium to the right (b) decrease the value of K (c) shift equilibrium to the left (d) no effect (e) no effect (f) shift equilibrium to the right **15.63** (a) No effect (b) no effect (c) no effect (d) increase equilibrium constant (e) no effect **15.65** (a) $\Delta H° = -155.7$ kJ (b) The reaction is exothermic, so the equilibrium constant will decrease with increasing temperature. (c) Δn does not equal zero, so a change in volume at constant temperature will affect the fraction of products in the equilibrium mixture. **15.67** An increase in pressure favors formation of ozone. **15.71** $K_p = 24.7$; $K_p = 24.7$; $K_c = 3.67 \times 10^{-3}$ **15.74** (a) $P_{Br_2} = 1.61$ atm, $P_{NO} = 0.628$ atm, $P_{NOBr} = 0.179$ atm; $K_c = 0.0643$ (b) $P_t = 0.968$ atm (c) 10.49 g NOBr **15.77** At equilibrium, $P_{IBr} = 0.21$ atm, $P_{I_2} = P_{Br_2} = 1.9 \times 10^{-3}$ atm **15.80** $K_p = 4.33$, $K_c = 0.0480$ **15.83** $[CO_2] = [H_2] = 0.264\ M$, $[CO] = [H_2O] = 0.236\ M$ **15.87** (a) 26% of the CCl_4 is converted to C and Cl_2. (b) $P_{CCl_4} =$

1.47 atm, $P_{Cl_2} = 1.06$ atm **15.91** $Q = 8 \times 10^{-6}$. $Q > K_p$, so the system is not at equilibrium; it will shift left to attain equilibrium. A catalyst that speeds up the reaction and thereby promotes the attainment of equilibrium would decrease the CO concentration in the exhaust. **15.93** At equilibrium, $[H_4IO_6^-] = 0.0015\ M$ **15.97** At 850 °C, $K_p = 14.1$; at 950 °C, $K_p = 73.8$; at 1050 °C, $K_p = 2.7 \times 10^2$; at 1200 °C, $K_p = 1.7 \times 10^3$. Because K increases with increasing temperature, the reaction is endothermic.

Chapter 16

16.1 (a) HCl, the H^+ donor, is the Brønsted–Lowry acid. NH_3, the H^+ acceptor, is the Brønsted–Lowry base. (b) HCl, the electron pair acceptor, is the Lewis acid. NH_3, the electron pair donor, is the Lewis base. **16.3** (a) True. (b) False. Methyl orange turns yellow at a pH slightly greater than 4, so solution B could be at any pH greater than 4. (c) True. **16.5** (a) HY is a strong acid. There are no neutral HY molecules in solution, only H^+ cations and Y^- anions. (b) HX has the smallest K_a value. It has most neutral acid molecules and fewest ions. (c) HX has fewest H^+ and highest pH. **16.7** (a) Curve C (b) For a weak acid, the percent ionization is inversely related to acid concentration; curve C depicts this relationship. It is the only curve that shows a decrease in percent ionization as acid concentration increases. **16.9** (a) Basic (b) Phenylephrine hydrochloride is a salt, while phenylephrine is a neutral molecule. The cation in the salt has an additional H atom bound to the N of phenylephrine. The anion is chloride. **16.11** (a) Molecule (b) is more acidic because its conjugate base is resonance-stabilized and the ionization equilibrium favors the more stable products. (b) Increasing the electronegativity of X increases the strength of both acids. As X becomes more electronegative and attracts more electron density, the O—H bond becomes weaker, more polar, and more likely to be ionized. An electronegative X group also stabilizes the anionic conjugate bases by delocalizing the negative charge. The equilibria favor products and the values of K_a increase. **16.13** (a) The Arrhenius definition of an acid is confined to aqueous solution; the Brønsted–Lowry definition applies to any physical state. (b) HCl is the Brønsted–Lowry acid; NH_3 is the Brønsted–Lowry base. **16.15 (a)** (i) IO_3^- (ii) NH_3 **(b)** (i) OH^- (ii) H_3PO_4
16.17

Acid	+	Base	⇌	Conjugate Acid	+	Conjugate Base
(a) $NH_4^+(aq)$		$CN^-(aq)$		$HCN(aq)$		$NH_3(aq)$
(b) $H_2O(l)$		$(CH_3)_3N(aq)$		$(CH_3)_3NH^+(aq)$		$OH^-(aq)$
(c) $HCOOH(aq)$		$PO_4^{3-}(aq)$		$HPO_4^{2-}(aq)$		$HCOO^-(aq)$

16.19 (a) Acid: $HC_2O_4^-(aq) + H_2O(l) \rightleftharpoons C_2O_4^{2-}(aq) + H_3O^+(aq)$; Base: $HC_2O_4^-(aq) + H_2O(l) \rightleftharpoons H_2C_2O_4(aq) + OH^-(aq)$. (b) $H_2C_2O_4$ is the conjugate acid of $HC_2O_4^-$. $C_2O_4^{2-}$ is the conjugate base of $HC_2O_4^-$ **16.21** (a) CH_3COO^-, weak base; CH_3COOH, weak acid (b) HCO_3^-, weak base; H_2CO_3 weak acid (c) O_2^-, strong base; OH^-, strong base (d) Cl^-, negligible base; HCl, strong acid (e) NH_3, weak base; NH_4^+, weak acid **16.23** (a) HBr (b) F^- **16.25** (a) $OH^-(aq) + OH^-(aq)$, the equilibrium lies to the right. (b) $H_2S(aq) + CH_3COO^-(aq)$, the equilibrium lies to the right. (c) $HNO_3(aq) + OH^-(aq)$, the equilibrium lies to the left. **16.27** Statement (ii) is correct. **16.29** (a) $[H^+] = 2.2 \times 10^{-11}\ M$, basic (b) $[H^+] = 1.1 \times 10^{-6}\ M$, acidic (c) $[H^+] = 1.0 \times 10^{-8}\ M$, basic **16.31** $[H^+] = [OH^-] = 3.5 \times 10^{-8}\ M$ **16.33** (a) $[H^+]$ changes by a factor of 100. (b) $[H^+]$ changes by a factor of 3.2 **16.35**

$[H^+]$	$[OH^-]$	pH	pOH	Acidic or Basic
$7.5 \times 10^{-3}\ M$	$1.3 \times 10^{-12}\ M$	2.12	11.88	acidic
$2.8 \times 10^{-5}\ M$	$3.6 \times 10^{-10}\ M$	4.56	9.44	acidic
$5.6 \times 10^{-9}\ M$	$1.8 \times 10^{-6}\ M$	8.25	5.75	basic
$5.0 \times 10^{-9}\ M$	$2.0 \times 10^{-6}\ M$	8.30	5.70	basic

16.37 $[H^+] = 4.0 \times 10^{-8}\ M$, $[OH^-] = 6.0 \times 10^{-7}\ M$, pOH = 6.22 **16.39** (a) Acidic (b) The range of possible integer pH values for the solution is 4-6. (c) Methyl violet, thymol blue, methyl orange and methyl red would help determine the pH of the solution more precisely. **16.41** (a) True (b) true (c) false **16.43** (a) $[H^+] = 8.5 \times 10^{-3}\ M$, pH = 2.07 (b) $[H^+] = 0.0419\ M$, pH = 1.377 (c) $[H^+] = 0.0250\ M$, pH = 1.602 (d) $[H^+] = 0.167\ M$, pH = 0.778 **16.45** (a) $[OH^-] = 3.0 \times 10^{-3}\ M$, pH = 11.48 (b) $[OH^-] = 0.3758\ M$, pH = 13.5750 (c) $[OH^-] = 8.75 \times 10^{-5}\ M$, pH = 9.942 (d) $[OH^-] = 0.17\ M$, pH = 13.23 **16.47** $3.2 \times 10^{-3}\ M$ NaOH **16.49** (a) $HBrO_2(aq) \rightleftharpoons H^+(aq) + BrO_2^-(aq)$, $K_a = [H^+][BrO_2^-]/[HBrO_2]$; $HBrO_2(aq) + H_2O(l) \rightleftharpoons H_3O^+(aq) + BrO_2^-(aq)$ $K_a = [H_3O^+][BrO_2^-]/[HBrO_2]$ (b) $C_2H_5COOH(aq) \rightleftharpoons H^+(aq) + C_2H_5COO^-(aq)$ $K_a = [H^+][C_2H_5COO^-]/[C_2H_5COOH]$; $C_2H_5COOH(aq) + H_2O(l) \rightleftharpoons H_3O^+(aq) + C_2H_5COO^-(aq)$ $K_a = [H_3O^+][C_2H_5COO^-]/[C_2H_5COOH]$ **16.51** $K_a = 1.4 \times 10^{-4}$ **16.53** $[H^+] = [ClCH_2COO^-] = 0.0110\ M$, $[ClCH_2COOH] = 0.089\ M$, $K_a = 1.4 \times 10^{-3}$ **16.55** $0.089\ M$ CH_3COOH **16.57** $[H^+] = [C_6H_5COO^-] = 1.8 \times 10^{-3}\ M$, $[C_6H_5COOH] = 0.048\ M$ **16.59** (a) $[H^+] = 1.1 \times 10^{-3}\ M$, pH = 2.95 (b) $[H^+] = 1.7 \times 10^{-4}\ M$, pH = 3.76 (c) $[OH^-] = 1.4 \times 10^{-5}\ M$, pH = 9.15 **16.61** $[H^+] = 2.0 \times 10^{-2}\ M$, pH = 1.71 **16.63** (a) $[H^+] = 2.8 \times 10^{-3}\ M$, 0.69% ionization (b) $[H^+] = 1.4 \times 10^{-3}\ M$, 1.4% ionization (c) $[H^+] = 8.7 \times 10^{-4}\ M$, 2.2% ionization **16.65** (a) $[H^+] = 5.1 \times 10^{-3}\ M$, pH = 2.30. The approximation that the first ionization is less than 5% of the total acid concentration is not valid; the quadratic equation must be solved. The $[H^+]$ produced from the second and third ionizations is small with respect to that present from the first step; the second and third ionizations can be neglected when calculating the $[H^+]$ and pH. (c) $[C_6H_5O_7^{3-}]$ is much less than $[H^+]$. **16.67** (a) $HONH_3^+$ (b) When hydroxylamine acts as a base, the nitrogen atom accepts a proton. (c) In hydroxylamine, O and N are the atoms with nonbonding electron pairs; in the neutral molecule both have zero formal charges. Nitrogen is less electronegative than oxygen and more likely to share a lone pair of electrons with an incoming (and electron-deficient) H^+. The resulting cation with the +1 formal charge on N is more stable than the one with the +1 formal charge on O. **16.69** (a) $(CH_3)_2 NH(aq) + H_2O(l) \rightleftharpoons (CH_3)_2 NH_2^+(aq) + OH^-(aq)$; $K_b = [(CH_3)_2NH_2^+][OH^-]/[(CH_3)_2NH]$ (b) $CO_3^{2-}(aq) + H_2O(l) \rightleftharpoons HCO_3^-(aq) + OH^-(aq)$; $K_b = [HCO_3^-][OH^-]/[(CO_3^{2-})]$ (c) $HCOO^-(aq) + H_2O(l) \rightleftharpoons HCOOH(aq) + OH^-(aq)$; $K_b = [HCOOH][OH^-]/[HCOO^-]$ **16.71** From the quadratic formula, $[OH^-] = 6.6 \times 10^{-3}\ M$, pH = 11.82. **16.73** (a) $[C_{10}H_{15}ON] = 0.033\ M$, $[C_{10}H_{15}ONH^+] = [OH^-] = 2.1 \times 10^{-3}\ M$ (b) $K_b = 1.4 \times 10^{-4}$ **16.75** (a) For a conjugate acid/conjugate base pair such as $C_6H_5OH/C_6H_5O^-$, K_b for the conjugate base can always be calculated from K_a for the conjugate acid, so a separate list of K_b values is not necessary. (b) $K_b = 7.7 \times 10^{-5}$ (c) Phenolate is a stronger base than NH_3. **16.77** (a) Acetic acid is stronger. (b) Hypochlorite ion is the stronger base. (c) For CH_3COO^-, $K_b = 5.6 \times 10^{-10}$; for ClO^-, $K_b = 3.3 \times 10^{-7}$. **16.79** (a) $[OH^-] = 6.3 \times 10^{-4}\ M$, pH = 10.80 (b) $[OH^-] = 9.2 \times 10^{-5}\ M$, pH = 9.96 (c) $[OH^-] = 3.3 \times 10^{-6}\ M$, pH = 8.52 **16.81** $4.5\ M$ $NaCH_3COO$ **16.83** (a) Acidic (b) acidic (c) basic (d) neutral (e) acidic **16.85** K_b for the anion of the unknown salt is 1.4×10^{-11}; K_a for the conjugate acid is 7.1×10^{-4}. The conjugate acid is HF and the salt is NaF. **16.87** (a) HNO_3 is a stronger acid because it has one more nonprotonated oxygen atom and thus a higher oxidation number on N. (b) For binary hydrides, acid strength increases going down a family, so H_2S is a stronger acid than H_2O. (c) H_2SO_4 is a stronger acid because H^+

is much more tightly held by the anion HSO_4^-. (d) For oxyacids, the greater the electronegativity of the central atom, the stronger the acid, so H_2SO_4 is the stronger acid. (e) CCl_3COOH is stronger because the electronegative Cl atoms withdraw electron density from other parts of the molecule, which weakens the O—H bond and stabilizes the anionic conjugate base. Both effects favor increased ionization and acid strength. **16.89** (a) BrO^- (b) BrO^- (c) HPO_4^{2-} **16.91** (a) True (b) False. In a series of acids that have the same central atom, acid strength increases with the number of nonprotonated oxygen atoms bonded to the central atom. (c) False. H_2Te is a stronger acid than H_2S because the H—Te bond is longer, weaker, and more easily ionized than the H—S bond. **16.93** Yes. Any substance that fits the narrow Arrhenius definition will fit the broader Brønsted and Lewis definitions of a base. **16.95** (a) Acid, $Fe(ClO_4)_3$ or Fe^{3+}, base, H_2O (b) Acid, H_2O; base, CN^- (c) Acid, BF_3; base, $(CH_3)_3N$ (d) Acid, HIO; base, NH_2^- **16.97** (a) Cu^{2+}, higher cation charge (b) Fe^{3+}, higher cation charge (c) Al^{3+}, smaller cation radius, same charge **16.101** $K = 3.3 \times 10^7$ **16.106** pH = 7.01 (not 5.40, from the typical calculation, which does not make sense.) Usually we assume that $[H^+]$ and $[OH^-]$ from the autoionization of water do not contribute to the overall $[H^+]$ and $[OH^-]$. However, for acid or base solute concentrations less than $1 \times 10^{-6} M$, the autoionization of water produces significant $[H^+]$ and $[OH^-]$ and we must consider it when calculating pH. **16.109** (a) $pK_b = 9.16$ (b) pH = 3.07 (c) pH = 8.77 **16.113** Nicotine, protonated; caffeine, neutral base; strychnine, protonated; quinine, protonated **16.116** $6.0 \times 10^{13} H^+$ ions **16.119** (a) To the precision of the reported data, the pH of rainwater 40 years ago was 5.4, no different from the pH today. With extra significant figures, $[H^+] = 3.61 \times 10^{-6} M$, pH = 5.443 (b) A 20.0-L bucket of today's rainwater contains 0.02 L (with extra significant figures, 0.0200 L) of dissolved CO_2. **16.123** Rx 1, $\Delta H = 104$ kJ; Rx 2, $\Delta H = -32$ kJ. Reaction 2 is exothermic while reaction 1 is endothermic. For binary acids with heavy atoms (X) in the same family, the longer and weaker the H—X bond, the stronger the acid (and the more exothermic the ionization reaction). **16.126** (a) $K(i) = 5.6 \times 10^3, K(ii) = 10$ (b) Both (i) and (ii) have $K > 1$ so both could be written with a single arrow.

Chapter 17

17.1 The middle box has the highest pH. For equal amounts of acid HX, the greater the amount of conjugate base X^-, the smaller the amount of H^+ and the higher the pH. **17.7** (a) The red curve corresponds to the more concentrated acid solution. (b) On the titration curve of a weak acid, pH = pK_a at the volume halfway to the equivalence point. Reading the pK_a values from the two curves, the red curve has the smaller pK_a and the larger K_a. **17.09** (a) The right-most diagram represents the solubility of $BaCO_3$ as HNO_3 is added. (b) The left-most diagram represents the solubility of $BaCO_3$ as Na_2CO_3 is added. (c) The center diagram represents the solubility of $BaCO_3$ as $NaNO_3$ is added. **17.13** Statement (a) is most correct. **17.15** (a) $[H^+] = 1.8 \times 10^{-5} M$, pH = 4.73 (b) $[OH^-] = 4.8 \times 10^{-5} M$, pH = 9.68 (c) $[H^+] = 1.4 \times 10^{-5} M$, pH = 4.87 **17.17** (a) 4.5% ionization (b) 0.018% ionization **17.19** Only solution (a) is a buffer. **17.21** (a) pH = 3.82 (b) pH = 3.96 **17.23** (a) pH = 5.26
(b) $Na^+(aq) + CH_3COO^-(aq) + H^+(aq) + Cl^-(aq) \longrightarrow$
$CH_3COOH(aq) + Na^+(aq) + Cl^-(aq)$
(c) $CH_3COOH(aq) + Na^+(aq) + OH^-(aq) \longrightarrow$
$CH_3COO^-(aq) + H_2O(l) + Na^+(aq)$
17.25 (a) pH = 1.58 (b) 36 g NaF **17.27** (a) pH = 4.86 (b) pH = 5.0 (c) pH = 4.71 **17.29** (a) $[HCO_3^-]/[H_2CO_3] = 11$ (b) $[HCO_3^-]/[H_2CO_3] = 5.4$ **17.31** 360 mL of 0.10 M HCOONa, 640 mL of 0.10 M HCOOH **17.33** (a) Curve B (b) pH at the approximate equivalence point of curve A = 8.0, pH at the approximate equivalence point of curve B = 7.0 (c) For equal volumes of A and B, the concentration of acid B is greater, since it requires a larger volume of base to reach the equivalence point. (d) The pK_a value of the weak acid is approximately 4.5 **17.35** (a) False (b) true (c) true **17.37** (a) Above pH 7 (b) below pH 7 (c) at pH 7 **17.39** The second color change of thymol blue is in the correct pH range to show the equivalence point

of the titration of a weak acid with a strong base. **17.41** (a) 42.4 mL NaOH soln (b) 35.0 mL NaOH soln (c) 29.8 mL NaOH soln **17.43** (a) pH = 1.54 (b) pH = 3.30 (c) pH = 7.00 (d) pH = 10.69 (e) pH = 12.74 **17.45** (a) pH = 2.78 (b) pH = 4.74 (c) pH = 6.58 (d) pH = 8.81 (e) pH = 11.03 (f) pH = 12.42 **17.47** (a) pH = 7.00 (b) $[HONH_3^+] = 0.100 M$, pH = 3.52 (c) $[C_6H_5NH_3^+] = 0.100 M$, pH = 2.82 **17.49** (a) True (b) false (c) false (d) true **17.51** (a) The concentration of undissolved solid does not appear in the solubility product expression because it is constant. (b) $K_{sp} = [Ag^+][I^-]$; $K_{sp} = [Sr^{2+}][SO_4^{2-}]$; $K_{sp} = [Fe^{2+}][OH^-]^2$; $K_{sp} = [Hg_2^{2+}][Br^-]^2$ **17.53** (a) $K_{sp} = 7.63 \times 10^{-9}$ (b) $K_{sp} = 2.7 \times 10^{-9}$ (c) 5.3×10^{-4} mol $Ba(IO_3)_2/L$ **17.55** $K_{sp} = 2.3 \times 10^{-9}$ **17.57** (a) 7.1×10^{-7} mol $AgBr/L$ (b) 1.7×10^{-11} mol $AgBr/L$ (c) 5.0×10^{-12} mol $AgBr/L$ **17.59** (a) The amount of $CaF_2(s)$ on the bottom of the beaker increases. (b) The $[Ca^{2+}]$ in solution increases. (c) The $[F^-]$ in solution decreases. **17.61** (a) 1.4×10^3 g $Mn(OH)_2/L$ (b) 0.014 g/L (c) 3.6×10^{-7} g/L **17.63** More soluble in acid: (a) $ZnCO_3$ (b) ZnS (d) AgCN (e) $Ba_3(PO_4)_2$ **17.65** $[Ni^{2+}] = 1 \times 10^{-8} M$ **17.67** (a) 9.1×10^{-9} mol AgI per L pure water (b) $K = K_{sp} \times K_f = 8 \times 10^4$ (c) 0.0500 mol AgI per L 0.100 M NaCN **17.69** (a) $Q < K_{sp}$; no $Ca(OH)_2$ precipitates (b) $Q < K_{sp}$; no Ag_2SO_4 precipitates **17.71** pH = 11.5 **17.73** AgI will precipitate first, at $[I^-] = 4.2 \times 10^{-13} M$. **17.75** AgCl will precipitate first. **17.77** The first two experiments eliminate group 1 and 2 ions (Figure 17.23). The absence of insoluble phosphate precipitates in the filtrate from the third experiment rules out group 4 ions. The ions that might be in the sample are those from group 3, Al^{3+}, Fe^{3+}, Cr^{3+}, Zn^{2+}, Ni^{2+}, Mn^{2+}, or Co^{2+}, and from group 5, NH_4^+, Na^+, or K^+. **17.79** (a) Make the solution acidic with 0.2 M HCl; saturate with H_2S. CdS will precipitate; ZnS will not. (b) Add excess base; $Fe(OH_3)(s)$ precipitates, but Cr^{3+} forms the soluble complex $Cr(OH)_4^-$. (c) Add $(NH_4)_2HPO_4$; Mg^{2+} precipitates as $MgNH_4PO_4$; K^+ remains soluble. (d) Add 6 M HCl; precipitate Ag^+ as $AgCl(s)$; Mn^{2+} remains soluble. **17.81** (a) Base is required to increase $[PO_4^{3-}]$ so that the solubility product of the metal phosphates of interest is exceeded and the phosphate salts precipitate. (b) K_{sp} for the cations in group 3 is much larger; to exceed K_{sp}, a higher $[S^{2-}]$ is required. (c) They should all redissolve in strongly acidic solution. **17.83** pOH = pK_b + log{$[BH^-]/[B]$} **17.89** (a) 6.5 mL glacial acetic acid, 5.25 g CH_3COONa **17.91** (a) The molecular weight of the acid is 94.6 g/mol. (b) $K_a = 1.4 \times 10^{-3}$ **17.97** (a) A person breathing normally exhales $CO_2(g)$. Rapid breathing causes excess $CO_2(g)$ to be removed from the blood. This causes both equilibria in Equation 17.10 to shift right, reducing $[H^+]$ in the blood and raising blood pH. (b) When a person breaths into a paper bag, the gas in the bag contains more CO_2 than ambient air. When the person inhales gas from the bag, a greater-than-normal partial pressure of $CO_2(g)$ in the lungs shifts the equilibria left, increasing $[H^+]$ and lowering blood pH. **17.102** The solubility of $Mg(OH)_2$ in 0.50 M NH_4Cl is 0.11 mol/L **17.108** $[OH^-] \geq 1.0 \times 10^{-2} M$ or pH $\geq$ 12.02 **17.111** (a) The molecular weight of the acid is 94.5 g/mol. (b) $K_a = 1.3 \times 10^{-5}$ (c) From Appendix D, butanoic acid is the closest match for K_a and molar mass, but the agreement is not exact. **17.113** (a) At the conditions given for the stomach, 99.7% of the aspirin is present as neutral molecules. **17.117** (a) $[Ag^+]$ from AgCl is 1.4×10^3 ppb or 1.4 ppm. (b) $[Ag^+]$ from AgBr is 76 ppb. (c) $[Ag^+]$ from AgI is 0.98 ppb. AgBr would maintain $[Ag^+]$ in the correct range.

Chapter 18

18.1 (a) A volume greater than 22.4 L (b) No. The relative volumes of one mole of an ideal gas at 50 km and 85 km depend on the temperature and pressure at the two altitudes. From Figure 18.1, the gas will occupy a much larger volume at 85 km than at 50 km. (c) We expect gases to behave most ideally in the thermosphere, around the stratopause and in the troposphere at low altitude. **18.3** (a) A = troposphere, 0–10 km; B = stratosphere, 12–50 km; C = mesosphere, 50–85 km (b) Ozone is a pollutant in the troposphere and filters UV radiation in the stratosphere. (c) The troposphere (d) Only region C in the diagram is involved in an aurora borealis, assuming a narrow "boundary" between the stratosphere and mesosphere at 50 km. (e) The concentration of water vapor is greatest near Earth's surface in region A and decreases with altitude. Water's single bonds are susceptible to photodissociation in regions B and C, so its concentration is likely to be very low in these regions. The relative concentration of CO_2, with strong double bonds, increases in regions B and C, because it is less susceptible to photodissociation. **18.5** The Sun **18.7** $CO_2(g)$ dissolves in seawater to form $H_2CO_3(aq)$. The basic pH of the ocean encourages ionization of $H_2CO_3(aq)$ to form $HCO_3^-(aq)$ and $CO_3^{2-}(aq)$. Under the correct conditions, carbon is removed from the ocean as $CaCO_3(s)$ (sea shells, coral, chalk cliffs). As carbon is removed, more $CO_2(g)$ dissolves to maintain the balance of complex and interacting acid-base and precipitation equilibria. **18.9** Above ground, evaporation of petroleum gases at the well head and waste ponds, as well as evaporation of volatile organic compounds from the ponds are potential sources of contamination. Below ground, petroleum gases and fracking liquid can migrate into groundwater, both deep and shallow aquifers. **18.11** (a) Its temperature profile (b) troposphere, 0 to 12 km; stratosphere, 12 to 50 km; mesosphere, 50 to 85 km; thermosphere, 85 to 110 km **18.13** (a) The partial pressure of O_3 is 3.0×10^{-7} atm (2.2×10^{-4} torr). (b) 7.3×10^{15} O_3 molecules/1.0 L air **18.15** 8.6×10^{16} CO molecules/1.0 L air **18.17** (a) 570 nm (b) visible electromagnetic radiation **18.19** (a) *Photodissociation* is cleavage of a bond such that two neutral species are produced. *Photoionization* is absorption of a photon with sufficient energy to eject an electron, producing an ion and the ejected electron. (b) Photoionization of O_2 requires 1205 kJ/mol. Photodissociation requires only 495 kJ/mol. At lower elevations, high-energy short-wavelength solar radiation has already been absorbed. Below 90 km, the increased concentration of O_2 and the availability of longer-wavelength radiation cause the photodissociation process to dominate. **18.21** Ozone depletion reactions, which involve only O_3, O_2, or O (oxidation state = 0), do not involve a change in oxidation state for oxygen atoms. Reactions involving ClO and one of the oxygen species with a zero oxidation state do involve a change in the oxidation state of oxygen atoms. **18.23** (a) A chlorofluorocarbon is a compound that contains chlorine, fluorine, and carbon, while a hydrofluorocarbon is a compound that contains hydrogen, fluorine, and carbon. An HFC contains hydrogen in place of the chlorine present in a CFC. (b) HFCs are potentially less harmful than CFCs because their photodissociation does not produce Cl atoms, which catalyze the destruction of ozone. **18.25** (a) The C—F bond requires more energy for dissociation than the C—Cl bond and is not readily cleaved by the available wavelengths of UV light. (b) Chlorine is present as chlorine atoms and chlorine oxide molecules, Cl and ClO, respectively. **18.27** (a) Methane, CH_4 arises from decomposition of organic matter by certain microorganisms; it also escapes from underground gas deposits. (b) SO_2 is released in volcanic gases and also is produced by bacterial action on decomposing vegetable and animal matter. (c) Nitric oxide, NO, results from oxidation of decomposing organic matter and is formed in lightning flashes.

18.29 (a)

$H_2SO_4(aq) + CaCO_3(s) \longrightarrow CaSO_4(s) + H_2O(l) + CO_2(g)$

(b) The $CaSO_4(s)$ would be much less reactive with acidic solution, since it would require a strongly acidic solution to shift the relevant equilibrium to the right: $CaSO_4(s) + 2H^+(aq) \rightleftharpoons Ca^{2+}(aq) + 2HSO_4^-(aq)$. $CaSO_4$ would protect $CaCO_3$ from attack by acid rain, but it would not provide the structural strength of limestone. **18.31** (a) Ultraviolet (b) 357 kJ/mol (c) The average C—H bond energy from Table 8.4 is 413 kJ/mol. The C—H bond energy in CH_2O, 357 kJ/mol, is less than the "average" C—H bond energy.

(d)

$$\overset{\displaystyle :\ddot{O}:}{\underset{\displaystyle H-C-H}{\parallel}} + h\nu \longrightarrow \overset{\displaystyle :\ddot{O}:}{\underset{\displaystyle H-C\cdot}{\parallel}} + H\cdot$$

18.33 Incoming and outgoing energies are in different regions of the electromagnetic spectrum. CO_2 is transparent to incoming visible radiation but absorbs outgoing infrared radiation. **18.35** $0.099\ M$ Na^+

18.37 (a) 3.22×10^3 gH_2O (b) The final temperature is 43.4 °C. **18.39** 4.361×10^5 g CaO **18.41** (a) *Groundwater* is freshwater (less than 500 ppm total salt content) that is under the soil; it composes 20% of the world's freshwater. (b) An *aquifer* is a layer of porous rock that holds groundwater. **18.43** The minimum pressure required to initiate reverse osmosis is greater than 5.1 atm. **18.45** (a) $CO_2(g)$, HCO_3^-, $H_2O(l)$, SO_4^{2-}, NO_3^-, HPO_4^{2-}, $H_2PO_4^-$ (b) $CH_4(g)$, $H_2S(g)$, $NH_3(g)$, $PH_3(g)$ **18.47** 25.1 g O_2 **18.49** $Mg^{2+}(aq) + Ca(OH)_2(s) \longrightarrow Mg(OH)_2(s) + Ca^{2+}(aq)$ **18.51** 0.42 mol $Ca(OH)_2$, 0.18 mol Na_2CO_3 **18.53** $4\,FeSO_4(aq) + O_2(aq) + 2\,H_2O(l) \longrightarrow 4\,Fe^{3+}(aq) + 4\,OH^-(aq) + 4\,SO_4^{2-}(aq);$ $Fe^{3+}(aq) + 3HCO_3^-(aq) \longrightarrow Fe(OH)_3(s) + 3\,CO_2(g)$ **18.55** (a) *Trihalomethanes* are the by-products of water chlorination; they contain one central carbon atom bound to one hydrogen and three halogen atoms.

(b)

$$\overset{\displaystyle H}{\underset{\displaystyle Cl}{Cl-\overset{|}{\underset{|}{C}}-Cl}} \qquad \overset{\displaystyle H}{\underset{\displaystyle Cl}{Cl-\overset{|}{\underset{|}{C}}-Br}}$$

18.57 The fewer steps in a process, the less waste is generated. Processes with fewer steps require less energy at the site of the process and for subsequent cleanup or disposal of waste. **18.59** (a) H_2O (b) It is better to prevent waste than to treat it. Atom economy. Less hazardous chemical synthesis and inherently safer for accident prevention. Catalysis and design for energy efficiency. Raw materials should be renewable. **18.61** (a) Water as a solvent, by criteria 5, 7, and 12. (b) Reaction temperature of 500 K, by criteria 6, 12, and 1. (c) Sodium chloride as a by-product, according to criteria 1, 3, and 12. **18.66** Multiply Equation 18.7 by a factor of 2; then add it to Equation 18.9. 2 Cl(g) and 2 ClO(g) cancel from each side of the resulting equation to produce Equation 18.10. **18.69** (a) A CFC has C—Cl bonds and C—F bonds. In an HFC, the C—Cl bonds are replaced by C—H bonds. (b) Stratospheric lifetime is significant because, the longer a halogen-containing molecule exists in the stratosphere, the greater the likelihood that it will encounter light with energy sufficient to dissociate a carbon-halogen bond. Free halogen atoms are the bad actors in ozone destruction. (c) HFCs have replaced CFCs because it is infrequent that light with energy sufficient to dissociate a C—F bond will reach an HFC molecule. F atoms are much less likely than Cl atoms to be produced by photodissociation in the stratosphere. (d) The main disadvantage of HFCs as replacements for CFCs is that they are potent greenhouse gases. **18.71** (a) $CH_4(g) + 2\,O_2(g) \longrightarrow CO_2(g) + 2\,H_2O(g)$ (b) $2\,CH_4(g) + 3\,O_2(g) \longrightarrow 2\,CO_2(g) + 4\,H_2O(g)$ (c) 9.5 L of dry air **18.75** 7.1×10^8 m^2 **18.77** (a) CO_3^{2-} is a relatively strong Brønsted–Lowry base and produces OH^- in aqueous solution. If $[OH^-(aq)]$ is sufficient for the reaction quotient to exceed K_{sp} for $Mg(OH)_2$, the solid will precipitate. (b) At these ion concentrations, $Q > K_{sp}$ and $Mg(OH)_2$ will precipitate. **18.81** (a) 2.5×10^7 ton CO_2, 4.2×10^5 ton SO_2 (b) 4.3×10^5 ton $CaSO_3$

18.85 (a) $H-\ddot{O}-H \longrightarrow H\cdot + \cdot\ddot{O}-H$

(b) 258 nm (c) The overall reaction is $O_3(g) + O(g) \longrightarrow 2O_2(g)$. $OH(g)$ is the catalyst in the overall reaction because it is consumed and then reproduced. **18.87** The enthalpy change for the first step is -141 kJ, for the second step, -249 kJ, for the overall reaction, -390 kJ. **18.90** (a) Rate $= k[O_3][H]$ (b) $k_{avg} = 1.13 \times 10^{44}\ M^{-1}\ s^{-1}$ **18.94** (a) Process (i) is greener because it involves neither the toxic reactant phosgene nor the by-product HCl. (b) Reaction (i): C in CO_2 is linear with *sp* hybridization; C in $R-N=C=O$ is linear with *sp* hybridization; C in the urethane monomer is trigonal planar with sp^2 hybridization. Reaction (ii): C in $COCl_2$ is trigonal planar with sp^2 hybridization; C in $R-N=C=O$ is linear with *sp* hybridization; C in the urethane monomer is trigonal planar with sp^2 hybridization. (c) The greenest way to promote formation of the isocyanate is to remove by-product, either water or HCl, from the reaction mixture.

Chapter 19

19.1 (a)

(b) $\Delta H = 0$ for mixing ideal gases. ΔS is positive because the disorder of the system increases. (c) The process is spontaneous and therefore irreversible. (d) Since $\Delta H = 0$, the process does not affect the entropy of the surroundings. **19.4** Both ΔH and ΔS are positive. The net change in the chemical reaction is the breaking of five blue-blue bonds. Enthalpies for bond breaking are always positive. There are twice as many molecules of gas in the products, so ΔS is positive for this reaction. **19.7** (a) At 300 K, $\Delta G = 0$, and the system is at equilibrium. (b) The reaction is spontaneous at temperatures above 300 K. **19.10** (a) True (b) false (c) false (d) false (e) true **19.11** Spontaneous: a, b, c, d; nonspontaneous: e **19.13** (a) Yes. If ΔS is large and positive, ΔG can be negative even if ΔH is positive. (b) Yes. Phase changes are examples of this. **19.15** (a) Endothermic (b) above 100 °C (c) below 100 °C (d) at 100 °C **19.17** (a) False (b) true (c) false (d) false **19.19** (a) No. Temperature is a state function, so a change in temperature does not depend on pathway. (b) No. An isothermal process occurs at constant temperature. (c) No. ΔE is a state function. **19.21** (a) An ice cube can melt reversibly at the conditions of temperature and pressure where the solid and liquid are in equilibrium. (b) We know that melting is a process that increases the energy of the system even though there is no change in temperature. ΔE is not zero for the process. **19.23** (a) True (b) true (c) false **19.25** (a) Entropy increases. (b) 89.2 J/K **19.27** (a) False (b) true (c) false **19.29** (a) Positive ΔS (b) $\Delta S = 1.02$ J/K (c) Temperature need not be specified to calculate ΔS, as long as the expansion is isothermal. **19.31** (a) Yes, the expansion is spontaneous. (b) As the ideal gas expands into the vacuum, there is nothing for it to "push back," so no work is done. Mathematically, $w = -P_{ext}\Delta V$. Since the gas expands into a vacuum, $P_{ext} = 0$ and $w = 0$. (c) The "driving force" for the expansion of the gas is the increase in entropy. **19.33** (a) An increase in temperature produces more available microstates for a system. (b) A decrease in volume produces fewer available microstates for a system. (c) Going from liquid to gas, the number of available microstates increases. **19.35** (a) ΔS is positive. (b) S of the system clearly increases in 19.11 (b) and (e); it clearly decreases in 19.11 (c). The entropy change is difficult to judge in 19.11 (a) and definition of the system in (d) is problematic. **19.37** S increases in (a) and (c); S decreases in (b). **19.39** (a) False (b) true (c) false (d) true **19.41** (a) $Ar(g)$ (b) $He(g)$ at 1.5 atm (c) 1 mol of $Ne(g)$ in 15.0 L (d) $CO_2(g)$ **19.43** (a) $\Delta S < 0$ (b) $\Delta S > 0$ (c) $\Delta S < 0$ (d) $\Delta S \approx 0$

19.45 (a)

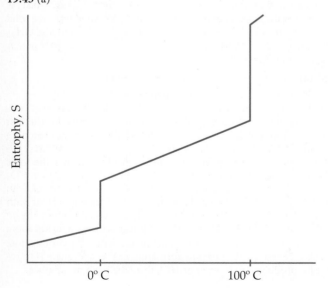

(b) Boiling water, at 100 °C, has a much larger entropy change than melting ice at 0 °C. **19.47** (a) $C_2H_6(g)$ (b) $CO_2(g)$ **19.49** (a) $Sc(s)$, 34.6 J/mol-K; $Sc(g)$, 174.7 J/mol-K. In general, the gas phase of a substance has a larger $S°$ than the solid phase because of the greater volume and motional freedom of the molecules. (b) $NH_3(g)$, 192.5 J/mol-K; $NH_3(aq)$, 111.3 J/mol-K. Molecules in the gas phase have more motional freedom than molecules in solution. (c) 1 mol of $P_4(g)$, 280 J/K; 2 mol of $P_2(g)$, 2(218.1) = 436.2 J/K. More particles have a greater motional energy (more available microstates). (d) C (diamond), 2.43 J/mol-K; C (graphite), 5.69 J/mol-K. The internal entropy in graphite is greater because there is translational freedom among planar sheets of C atoms, while there is very little freedom within the covalent-network diamond lattice. **19.51** For elements with similar structures, the heavier the atoms, the lower the vibrational frequencies at a given temperature. This means that more vibrations can be accessed at a particular temperature, resulting in greater absolute entropy for the heavier elements. **19.53** (a) $\Delta S° = -120.5$ J/K. $\Delta S°$ is negative because there are fewer moles of gas in the products. (b) $\Delta S° = +176.6$ J/K. $\Delta S°$ is positive because there are more moles of gas in the products. (c) $\Delta S° = +152.39$ J/K. $\Delta S°$ is positive because the product contains more total particles and more moles of gas. (d) $\Delta S° = +92.3$ J/K. $\Delta S°$ is positive because there are more moles of gas in the products. **19.55** (a) Yes. $\Delta G = \Delta H - T\Delta S$ (b) No. If ΔG is positive, the process is nonspontaneous. (c) No. There is no relationship between ΔG and rate of reaction. **19.57** (a) Exothermic (b) $\Delta S°$ is negative; the reaction leads to a decrease in disorder. (c) $\Delta G° = -9.9$ kJ (d) If all reactants and products are present in their standard states, the reaction is spontaneous in the forward direction at this temperature. **19.59** (a) $\Delta H° = -537.22$ kJ, $\Delta S° = 13.7$ J/K, $\Delta G° = -541.40$ kJ, $\Delta G° = \Delta H° - T\Delta S° = -541.31$ kJ (b) $\Delta H° = -106.7$ kJ, $\Delta S° = -142.2$ kJ, $\Delta G° = -64.0$ kJ, $\Delta G° = \Delta H° - T\Delta S° = -64.3$ kJ (c) $\Delta H° = -508.3$ kJ, $\Delta S° = -178$ kJ, $\Delta G° = -465.8$ kJ, $\Delta G° = \Delta H° - T\Delta S° = -455.1$ kJ. The discrepancy in $\Delta G°$ values is due to experimental uncertainties in the tabulated thermodynamic data. (d) $\Delta H° = -165.9$ kJ, $\Delta S° = 1.4$ kJ, $\Delta G° = -166.2$ kJ, $\Delta G° = \Delta H° - T\Delta S° = -166.3$ kJ **19.61** (a) $\Delta G° = -140.0$ kJ, spontaneous (b) $\Delta G° = +104.70$ kJ, nonspontaneous (c) $\Delta G° = +146$ kJ, nonspontaneous (d) $\Delta G° = -156.7$ kJ, spontaneous **19.63** (a) $2\,C_8H_{18}(l) + 25\,O_2(g) \longrightarrow 16\,CO_2(g) + 18\,H_2O(l)$ (b) Because $\Delta S°$ is positive, $\Delta G°$ is more negative than $\Delta H°$. **19.65** (a) The forward reaction is spontaneous at low temperatures but becomes nonspontaneous at higher temperatures. (b) The reaction is nonspontaneous in the forward direction at all temperatures. (c) The forward reaction is nonspontaneous at low temperatures but becomes spontaneous at higher temperatures. **19.67** $\Delta S > 60.8$ J/K **19.69** (a) $T = 330$ K (b) nonspontaneous **19.71** (a) $\Delta H° = 155.7$ kJ, $\Delta S° = 171.4$ kJ. Since $\Delta S°$ is positive, $\Delta G°$ becomes more negative with increasing temperature. (b) $\Delta G° = 19$ kJ. The reaction is not spontaneous under standard conditions at 800 K (c) $\Delta G° = -15.7$ kJ. The reaction is spontaneous under standard conditions at 1000 K. **19.73** (a) $T_b = 79$ °C (b) From the *Handbook of Chemistry and Physics*, 74th Edition, $T_b = 80.1$ °C. The values are remarkably close; the small difference is due to deviation from ideal behavior by $C_6H_6(g)$ and experimental uncertainty in the boiling point measurement and the thermodynamic data. **19.75** (a) $C_2H_2(g) + \frac{5}{2}O_2(g) \longrightarrow 2\,CO_2(g) + H_2O(l)$ (b) -1299.5 kJ of heat produced/mol C_2H_2 burned (c) $w_{max} = -1235.1$ kJ/mol C_2H_2 **19.77** (a) ΔG decreases; it becomes more negative. (b) ΔG increases; it becomes more positive. (c) ΔG increases; it becomes more positive. **19.79** (a) $\Delta G° = -5.40$ kJ (b) $\Delta G = 0.30$ kJ **19.81** (a) $\Delta G° = -16.77$ kJ, $K = 870$ (b) $\Delta G° = 8.0$ kJ, $K = 0.039$ (c) $\Delta G° = -497.9$ kJ, $K = 2 \times 10^{87}$ **19.83** $\Delta H° = 269.3$ kJ, $\Delta S° = 0.1719$ kJ/K (a) $P_{CO_2} = 6.0 \times 10^{-39}$ atm (b) $P_{CO_2} = 1.6 \times 10^{-4}$ atm **19.85** (a) $HNO_2(aq) \rightleftharpoons H^+(aq) + NO_2^-(aq)$ (b) $\Delta G° = 19.1$ kJ (c) $\Delta G = 0$ at equilibrium (d) $\Delta G = -2.7$ kJ **19.87** (a) The thermodynamic quantities T, E, and S are state functions. (b) The quantities q and w depend on the path taken. (c) There is only one *reversible* path between states. (d) $\Delta E = q_{rev} + w_{max}$, $\Delta S = q_{rev}/T$. **19.91** (a) 16 arrangements (b) 1 arrangement (c) The gas will spontaneously adopt the state with the most possible arrangements for the molecules, the state with maximum disorder. **19.96** (a) For all three compounds listed, there are fewer moles of gaseous products than reactants in the formation reaction, so we expect ΔS_f° to be negative. If $\Delta G_f^\circ = \Delta H_f^\circ - T\Delta S_f^\circ$ and ΔS_f° is negative, $-T\Delta S_f^\circ$ is positive and ΔG_f° is more positive than ΔH_f°. (b) In this reaction, there are more moles of gas in products, ΔS_f° is positive, $-T\Delta S_f^\circ$ is negative and ΔG_f° is more negative than ΔH_f°. **19.101** (a) For the oxidation of glucose in the body, $\Delta G^\circ = -2878.8$ kJ, $K = 5 \times 10^{504}$; for the anaerobic decomposition of glucose, $\Delta G^\circ = -228.0$ kJ, $K = 9 \times 10^{39}$ (b) A greater maximum amount of work at standard conditions can be obtained from the oxidation of glucose in the body, because ΔG° is much more negative. **19.104** (a) $\Delta G = 8.77$ kJ (b) $w_{min} = 8.77$ kJ. In practice, a larger than minimum amount of work is required. **19.107** (a) Acetone, $\Delta S_{vap}^\circ = 88.4$ J/mol-K; dimethyl ether, $\Delta S_{vap}^\circ = 86.6$ J/mol-K; ethanol, $\Delta S_{vap}^\circ = 110$ J/mol-K; octane, $\Delta S_{vap}^\circ = 86.3$ J/mol-K; pyridine, $\Delta S_{vap}^\circ = 90.4$ J/mol-K. Ethanol does not obey Trouton's rule. (b) Hydrogen bonding (in ethanol and other liquids) leads to more ordering in the liquid state and a greater than usual increase in entropy upon vaporization. Liquids that experience hydrogen bonding are probably exceptions to Trouton's rule. (c) Owing to strong hydrogen bonding interactions, water probably does not obey Trouton's rule. $\Delta S_{vap}^\circ = 109.0$ J/mol-K. (d) ΔH_{vap} for $C_6H_5Cl \approx 36$ kJ/mol **19.114** (a) For any given total pressure, the condition of equal moles of the two gases can be achieved at some temperature. For individual gas pressures of 1 atm and a total pressure of 2 atm, the mixture is at equilibrium at 328.5 K or 55.5 °C. (b) 333.0 K or 60 °C (c) 374.2 K or 101.2 °C (d) The reaction is endothermic, so an increase in the value of K as calculated in parts (a)–(c) should be accompanied by an increase in T.

Chapter 20

20.1 In this analogy, the electron is analogous to the proton (H^+). Redox reactions can be viewed as electron-transfer reactions, just as acid-base reactions can be viewed as proton-transfer reactions. Oxidizing agents are themselves reduced; they gain electrons. A strong oxidizing agent would be analogous to a strong base. **20.3** (a) The process represents oxidation (b) The electrode is the anode. (c) When a neutral atom loses a valence electron, the radius of the resulting cation is smaller than the radius of the neutral atom. **20.7** (a) The sign of ΔG° is positive. (b) The equilibrium constant is less than one. (c) No. An electrochemical cell based on this reaction cannot accomplish work on its surroundings. **20.10** (a) Zinc is the anode. (b) The energy density of the silver oxide battery is most similar to the nickel-cadmium battery. The molar masses of the electrode materials and the cell potentials for these two batteries are most similar. **20.13** (a) *Oxidation* is the loss of electrons. (b) Electrons appear on the products' side (right side). (c) The *oxidant* is the reactant that is reduced. (d) An *oxidizing agent* is the substance that promotes oxidation; it is the oxidant. **20.15** (a) True (b) false (c) true **20.17** (a) I, +5 to 0; C, +2 to +4 (b) Hg, +2 to 0; N, -2 to 0 (c) N, +5 to +2; S, -2 to 0 **20.19** (a) No oxidation-reduction (b) Iodine is oxidized; chlorine is reduced. (c) Sulfur is oxidized; nitrogen is reduced. **20.21** (a) $TiCl_4(g) + 2 Mg(l) \longrightarrow Ti(s) + 2 MgCl_2(l)$ (b) Mg(l) is oxidized; $TiCl_4(g)$ is reduced. (c) Mg(l) is the reductant; $TiCl_4(g)$ is the oxidant. **20.23** (a) $Sn^{2+}(aq) \longrightarrow Sn^{4+}(aq) + 2 e^-$, oxidation (b) $TiO_2(s) + 4 H^+(aq) + 2 e^- \longrightarrow Ti^{2+}(aq) + 2 H_2O(l)$, reduction (c) $ClO_3^-(aq) + 6 H^+(aq) + 6 e^- \longrightarrow Cl_2(aq) + 3 H_2O(l)$, reduction (d) $N_2(g) + 8 H^+(aq) + 6 e^- \longrightarrow 2 NH_4^+(aq)$, reduction (e) $4 OH^-(aq) \longrightarrow O_2(g) + 2 H_2O(l) + 4 e^-$, oxidation (f) $SO_3^{2-}(aq) + 2 OH^-(aq) \longrightarrow SO_4^{2-}(aq) + H_2O(l) + 2 e^-$, oxidation (g) $N_2(g) + 6 H_2O(l) + 6 e^- \longrightarrow 2 NH_3(g) + 6 OH^-(aq)$, reduction **20.25** (a) $Cr_2O_7^{2-}(aq) + I^-(aq) + 8 H^+(aq) \longrightarrow 2 Cr^{3+}(aq) + IO_3^-(aq) + 4 H_2O(l)$; oxidizing agent, $Cr_2O_7^{2-}$; reducing agent, I^- (b) $4 MnO_4^-(aq) + 5 CH_3OH(aq) + 12 H^+(aq) \longrightarrow 4 Mn^{2+}(aq) + 5 HCO_2H(aq) + 12 H_2O(aq)$; oxidizing agent,

MnO_4^-; reducing agent, CH_3OH (c) $I_2(s) + 5 OCl^-(aq) + H_2O(l) \longrightarrow 2 IO_3^-(aq) + 5 Cl^-(aq) + 2 H^+(aq)$; oxidizing agent, OCl^-; reducing agent, I_2 (d) $As_2O_3(s) + 2 NO_3^-(aq) + 2 H_2O(l) + 2 H^+(aq) \longrightarrow 2 H_3AsO_4(aq) + N_2O_3(aq)$; oxidizing agent, NO_3^-; reducing agent, As_2O_3 (e) $2 MnO_4^-(aq) + Br^-(aq) + H_2O(l) \longrightarrow 2 MnO_2(s) + BrO_3^-(aq) + 2 OH^-(aq)$, oxidizing agent, MnO_4^-; reducing agent, Br^- (f) $Pb(OH)_4^{2-}(aq) + ClO^-(aq) \longrightarrow PbO_2(s) + Cl^-(aq) + 2 OH^-(aq) + H_2O(l)$; oxidizing agent, $ClO^-(aq)$; reducing agent, $Pb(OH)_4^{2-}$ **20.27** (a) The reaction $Cu^{2+}(aq) + Zn(s) \longrightarrow Cu(s) + Zn^{2+}(aq)$ is occurring in both figures. In Figure 20.3 the reactants are in contact, while in Figure 20.4 the oxidation half-reaction and reduction half-reaction are occurring in separate compartments. In Figure 20.3 the flow of electrons cannot be isolated or utilized; in Figure 20.4 electrical current is isolated and flows through the voltmeter. (b) Na^+ cations are drawn into the cathode compartment to maintain charge balance as Cu^{2+} ions are removed. **20.29** (a) Fe(s) is oxidized, $Ag^+(aq)$ is reduced. (b) $Ag^+(aq) + e^- \longrightarrow Ag(s)$; $Fe(s) \longrightarrow Fe^{2+}(aq) + 2 e^-$. (c) Fe(s) is the anode, Ag(s) is the cathode. (d) Fe(s) is negative; Ag(s) is positive. (e) Electrons flow from the Fe electrode ($-$) toward the Ag electrode ($+$). (f) Cations migrate toward the Ag(s) cathode; anions migrate toward the Fe(s) anode. **20.31** *Electromotive force*, emf, is the potential energy difference between an electron at the anode and an electron at the cathode of a voltaic cell. (b) One *volt* is the potential energy difference required to impart 1 J of energy to a charge of 1 coulomb. (c) *Cell potential*, E_{cell}, is the emf of an electrochemical cell. **20.33** (a) $2 H^+(aq) + 2 e^- \longrightarrow H_2(g)$ (b) $H_2(g) \longrightarrow 2 H^+(aq) + 2 e^-$ (c) A *standard* hydrogen electrode, SHE, has components that are at standard conditions, 1 M $H^+(aq)$ and $H_2(g)$ at 1 atm. (c) The platinum foil in a SHE serves as an inert electron carrier and a solid reaction surface. **20.35** (a) $Cr^{2+}(aq) \longrightarrow Cr^{3+}(aq) + e^-$; $Tl^{3+}(aq) + 2 e^- \longrightarrow Tl^+(aq)$ (b) $E_{red}^\circ = 0.78$ V (c)

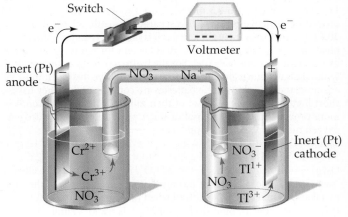

Switch · Voltmeter · Inert (Pt) anode · Inert (Pt) cathode · NO_3^- · Na^+ · Cr^{2+} · Cr^{3+} · NO_3^- · NO_3^- · Tl^{1+} · NO_3^- · Tl^{3+}

Movement of cations

Movement of anions

20.37 (a) $E^\circ = 0.823$ V (b) $E^\circ = 1.89$ V (c) $E^\circ = 1.211$ V. (d) $E^\circ = 0.62$ V **20.39** (a) $3 Ag^+(aq) + Cr(s) \longrightarrow 3 Ag(s) + Cr^{3+}(aq)$, $E^\circ = 1.54$ V (b) Two of the combinations have essentially equal E° values: $2 Ag^+(aq) + Cu(s) \longrightarrow 2 Ag(s) + Cu^{2+}(aq)$, $E^\circ = 0.462$ V; $3 Ni^{2+}(aq) + 2 Cr(s) \longrightarrow 3 Ni(s) + 2 Cr^{3+}(aq)$, $E^\circ = 0.46$ V **20.41** (a) Anode, Sn(s); cathode, Cu(s). (b) The copper electrode gains mass as Cu is plated out, and the tin electrode loses mass as Sn is oxidized. (c) $Cu^{2+}(aq) + Sn(s) \longrightarrow Cu(s) + Sn^{2+}(aq)$. (d) $E^\circ = 0.473$ V. **20.43** (a) Mg(s) (b) Ca(s) (c) $H_2(g)$ (d) $BrO_3^-(aq)$ **20.45** (a) $Cl_2(aq)$, strong oxidant (b) $MnO_4^-(aq)$, acidic, strong oxidant (c) Ba(s) strong reductant (d) Zn(s), reductant

20.47 (a) $Cu^{2+}(aq) < O_2(g) < Cr_2O_7^{2-}(aq) < Cl_2(g) < H_2O_2(aq)$ (b) $H_2O_2(aq) < I_2(aq) < Sn^{2+}(aq) < Zn(s) < Al(s)$ **20.49** Al and $H_2C_2O_4$ **20.51** (a) $2\,Fe^{2+}(aq) + S_2O_6^{2-}(aq) + 4\,H^+(aq) \longrightarrow 2\,Fe^{3+}(aq) + 2\,H_2SO_3(aq); 2\,Fe^{2+}(aq) + N_2O(aq) + 2\,H^+(aq) \longrightarrow 2\,Fe^{3+}(aq) + N_2(g) + H_2O(l); Fe^{2+}(aq) + VO_2^+(aq) + 2\,H^+(aq) \longrightarrow Fe^{3+}(aq) + VO^{2+}(aq) + H_2O(l)$ (b) $E° = -0.17\,V, \Delta G° = 33\,kJ; E° = -2.54\,V, \Delta G° = 4.90 \times 10^2\,kJ; E° = 0.23\,V, \Delta G° = -22\,kJ$ (c) $K = 1.8 \times 10^{-6} = 10^{-6}; K = 1.2 \times 10^{-86} = 10^{-86}; K = 7.8 \times 10^3 = 8 \times 10^3$ **20.53** $\Delta G° = 21.8\,kJ, E°_{cell} = -0.113\,V$ **20.55** (a) $E° = 0.16\,V, K = 2.54 \times 10^5 = 3 \times 10^5$ (b) $E° = 0.277\,V, K = 2.3 \times 10^9$ (c) $E° = 0.45\,V, K = 1.5 \times 10^{75} = 10^{75}$ **20.57** (a) $K = 9.8 \times 10^2$ (b) $K = 9.5 \times 10^5$ (c) $K = 9.3 \times 10^8$ **20.59** $w_{max} = -130\,kJ/mol\,Sn$ **20.61** (a) In the Nernst equation, $Q = 1$ if all reactants and products are at standard conditions. (b) Yes. The Nernst equation is applicable to cell emf at nonstandard conditions, so it must be applicable at temperatures other than 298 K. Values of $E°$ at temperatures other than 298 K are required. And, there is a variable for T in the second term. If the short-hand form of the equation, Equation 20.18, is used, a coefficient other than 0.0592 is required. **20.63** (a) E decreases (b) E decreases (c) E decreases (d) no effect **20.65** (a) $E° = 0.48\,V$ (b) $E = 0.53\,V$ (c) $E = 0.46\,V$ **20.67** (a) $E° = 0.46\,V$ (b) $E = 0.37\,V$ **20.69** (a) The compartment with $[Zn^{2+}] = 1.00 \times 10^{-2}\,M$ is the anode. (b) $E° = 0$ (c) $E = 0.0668\,V$ (d) In the anode compartment $[Zn^{2+}]$ increases; in the cathode compartment $[Zn^{2+}]$ decreases **20.71** $E° = 0.763\,V$, pH $= 1.6$ **20.73** (a) $464\,g\,PbO_2$ (b) 3.74×10^5 C of charge transferred **20.75** (a) The anode (b) $E° = 0.50\,V$ (c) The emf of the battery, 3.5 V, is exactly the standard cell potential calculated in part (b). (d) At ambient conditions, $E \approx E°$, so log $Q \approx 1$. Assuming that the value of $E°$ is relatively constant with temperature, the value of the second term in the Nernst equation is approximately zero at 37 °C, and $E \approx 3.5\,V$. **20.77** (a) The cell emf will have a smaller value. (b) NiMH batteries use an alloy such as $ZrNi_2$ as the anode material. This eliminates the use and disposal problems associated with Cd, a toxic heavy metal. **20.79** (a) When the battery is fully charged, the mole ratio of the elements in the cathode is: 0.5 mol Li^+ to 0.5 mol Co^{3+} to 0.5 mol Co^{4+} to 2 mol O^{2-}. (b) 4.9×10^3 C of electricity delivered on full discharge of a fully charged battery **20.81** The main advantage of a fuel cell is that fuel is continuously supplied, so that it can produce electrical current for a time limited only by the amount of available fuel. For the hydrogen-oxygen fuel cell, this is also a disadvantage because volatile and explosive hydrogen must be acquired and stored. Alkaline batteries are convenient, but they have a short lifetime, and the disposal of their zinc and manganese solids is more problematic than disposal of water produced by the hydrogen-oxygen fuel cell. **20.83** (a) anode: $Fe(s) \longrightarrow Fe^{2+}(aq) + 2\,e^-$; cathode: $O_2(g) + 4\,H^+(aq) + 4\,e^- \longrightarrow 2\,H_2O(l)$ (b) $2\,Fe^{2+}(aq) + 3\,H_2O(l) + 3\,H_2O(l) \longrightarrow Fe_2O_3 \cdot 3\,H_2O(s) + 6\,H^+(aq) + 2\,e^-$; $O_2(g) + 4\,H^+(aq) + 4\,e^- \longrightarrow 2\,H_2O(l)$ **20.85** (a) Mg is called a "sacrificial anode" because it has a more negative $E°_{red}$ than the pipe metal and is preferentially oxidized when the two are coupled. It is sacrificed to preserve the pipe. (b) $E°_{red}$ for Mg^{2+} is $-2.37\,V$, more negative than most metals present in pipes, including Fe and Zn. **20.87** Under acidic conditions, air (O_2) oxidation of $Zn(s)$, 1.99 V; $Fe(s)$, 1.67 V; and $Cu(s)$, 0.893 V are all spontaneous. When the three metals are in contact, Zn will act as a sacrificial anode for both Fe and Cu, but after the Zn is depleted, Fe will be oxidized (corroded). **20.89** (a) *Electrolysis* is an electrochemical process driven by an outside energy source. (b) By definition, electrolysis reactions are nonspontaneous. (c) $2\,Cl^-(l) \longrightarrow Cl_2(g) + 2\,e^-$ (d) When an aqueous solution of NaCl undergoes electrolysis, sodium metal is not formed because H_2O is preferentially reduced to form $H_2(g)$. **20.91** (a) 236 g $Cr(s)$ (b) 2.51 A **20.93** (a) 4.0×10^5 g Li (b) The minimum voltage required to drive the electrolysis is $+4.41\,V$. **20.95** Gold is less active than copper and thus more difficult to oxidize. When crude copper is refined by electrolysis, Cu is oxidized from the crude anode, but any metallic gold present in the crude copper is not oxidized, so it accumulates near the anode, available for collection. **20.97** (a) $2\,Ni^+(aq) \longrightarrow Ni(s) + Ni^{2+}(aq)$

(b) $3\,MnO_4^{2-}(aq) + 4\,H^+(aq) \longrightarrow 2\,MnO_4^-(aq) + MnO_2(s) + 2\,H_2O(l)$ (c) $3\,H_2SO_3(aq) \longrightarrow S(s) + 2\,HSO_4^-(aq) + 2\,H^+(aq) + H_2O(l)$ (d) $Cl_2(aq) + 2\,OH^-(aq) \longrightarrow Cl^-(aq) + ClO^-(aq) + H_2O(l)$ **20.99** (a) $E° = 0.627\,V$, spontaneous (b) $E° = -0.82\,V$, nonspontaneous (c) $E° = 0.93\,V$, spontaneous (d) $E° = 0.183\,V$, spontaneous **20.103** $K = 1.6 \times 10^6$ **20.106** The ship's hull should be made negative. The ship, as a negatively charged "electrode," becomes the site of reduction, rather than oxidation, in an electrolytic process. **20.109** 3×10^4 kWh required **20.114** (a) $E° = 0.028\,V$ (b) cathode: $Ag^+(aq) + e^- \longrightarrow Ag(s)$; anode: $Fe^{2+}(aq) \longrightarrow Fe^{3+}(aq) + e^-$ (c) $\Delta S° = 148.5\,J$. Since $\Delta S°$ is positive, $\Delta G°$ will become more negative and $E°$ will become more positive as temperature is increased. **20.117** K_{sp} for AgSCN is 1.0×10^{-12}.

Chapter 21

21.1 (a) ^{24}Ne; outside; reduce neutron-to-proton ratio via β decay (b) ^{32}Cl; outside; increase neutron-to-proton ratio via positron emission or orbital electron capture (c) ^{108}Sn; outside; increase neutron-to-proton ratio via positron emission or orbital electron capture (d) ^{216}Po; outside; nuclei with $Z \geq 84$ usually decay via α emission. **21.6** (a) 7 min (b) 0.1 min^{-1} (c) 30% (3/10) of the sample remains after 12 min. (d) $^{88}_{41}Nb$ **21.7** (a) $^{10}_5B$, $^{11}_5B$; $^{12}_6C$, $^{13}_6C$; $^{14}_7N$, $^{15}_7N$; $^{16}_8O$, $^{17}_8O$, $^{18}_8O$; $^{19}_9F$ (b) $^{14}_6C$ (c) $^{11}_6C$, $^{13}_7N$, $^{15}_8O$, $^{18}_9F$ (d) $^{11}_6C$ **21.9** (a) 24 protons, 32 neutrons (b) 81 protons, 112 neutrons (c) 18 protons, 20 neutrons **21.11** (a) ^1_0n (b) ^4_2He or α (c) $^0_0\gamma$ or γ **21.13** (a) $^{90}_{37}Rb \longrightarrow {}^{90}_{38}Sr + {}^0_{-1}e$ (b) $^{72}_{34}Se + {}^0_{-1}e$ (orbital electron) $\longrightarrow {}^{72}_{33}As$ (c) $^{76}_{36}Kr \longrightarrow {}^{76}_{35}Br + {}^0_1e$ (d) $^{226}_{88}Ra \longrightarrow {}^{222}_{86}Rn + {}^4_2He$ **21.15** (a) $^{211}_{82}Pb \longrightarrow {}^{211}_{83}Bi + {}^0_{-1}\beta$ (b) $^{50}_{25}Mn \longrightarrow {}^{50}_{24}Cr + {}^0_1e$ (c) $^{179}_{74}W + {}^0_{-1}e \longrightarrow {}^{179}_{73}Ta$ (d) $^{230}_{90}Th \longrightarrow {}^{226}_{88}Ra + {}^4_2He$ **21.17** 7 alpha emissions, 4 beta emissions **21.19** (a) Positron emission (for low atomic numbers, positron emission is more common than electron capture) (b) beta emission (c) beta emission (d) beta emission **21.21** (a) $^{40}_{19}K$, radioactive, odd proton, odd neutron; $^{39}_{19}K$, stable, 20 neutrons is a magic number (b) $^{208}_{83}Bi$, radioactive, odd proton, odd neutron; $^{209}_{83}Bi$, stable, 126 neutrons is a magic number (c) $^{65}_{28}Ni$, radioactive, high neutron-to-proton ratio; $^{58}_{28}Ni$ stable, even proton, even neutron **21.23** (a) ^4_2He (b) $^{40}_{20}Ca$ (c) $^{126}_{82}Pb$ **21.25** The alpha particle, ^4_2He, has a magic number of both protons and neutrons, while the proton is an odd proton, even neutron particle. Alpha is a very stable emitted particle, which makes alpha emission a favorable process. The proton is not a stable emitted particle, and its formation does not encourage proton emission as a process. **21.27** Protons and alpha particles are positively charged and must be moving very fast to overcome electrostatic forces that would repel them from the target nucleus. Neutrons are electrically neutral and not repelled by the nucleus. **21.29** (a) $^{252}_{98}Cf + {}^{10}_5B \longrightarrow 3\,{}^1_0n + {}^{259}_{103}Lr$ (b) $^2_1H + {}^3_2He \longrightarrow {}^4_2He + {}^1_1H$ (c) $^1_1H + {}^{11}_5B \longrightarrow 3\,{}^4_2He$ (d) $^{122}_{53}I \longrightarrow {}^{122}_{54}Xe + {}^0_{-1}e$ (e) $^{59}_{26}Fe \longrightarrow {}^0_{-1}e + {}^{59}_{27}Co$ **21.31** (a) $^{238}_{92}U + {}^4_2He \longrightarrow {}^{241}_{94}Pu + {}^1_0n$ (b) $^{14}_7N + {}^4_2He \longrightarrow {}^{17}_8O + {}^1_1H$ (c) $^{56}_{26}Fe + {}^4_2He \longrightarrow {}^{60}_{29}Cu + {}^0_{-1}e$

21.33 (a) True. The decay rate constant and half-life are inversely related. (b) False. If X is not radioactive, its half-life is essentially infinity. (c) True. Changes in the amount of A would be substantial and measurable over the 40-year time frame, while changes in the amount of X would be very small and difficult to detect. **21.35** When the watch is 50 years old, only 6% of the tritium remains. The dial will be dimmed by 94%. **21.37** The source must be replaced after 2.18 yr or 26.2 months; this corresponds to August 2015. **21.39** (a) 1.1×10^{11} alpha particles emitted in 5.0 min (b) 9.9 mCi

21.41 $k = 1.21 \times 10^{-4}\,yr^{-1}; t = 4.3 \times 10^3\,yr$

21.43 $k = 5.46 \times 10^{-10}\,yr^{-1}; t = 3.0 \times 10^9\,yr$ **21.45** The energy released when one mole of Fe_2O_3 reacts is $8.515 \times 10^5\,J$. The energy released when one mole of ^4_2He is formed from protons and neutrons is $2.73 \times 10^{12}\,J$. This is 3×10^6 or 3 million times as much energy as the thermite reaction.

21.47 $\Delta m = 0.2414960$ amu, $\Delta E = 3.604129 \times 10^{-11}$ J/^{27}Al nucleus required, 8.044234×10^{13} J/100 g ^{27}Al   **21.49** (a) Nuclear mass: ^{2}H, 2.013553 amu; ^{4}He, 4.001505 amu; ^{6}Li, 6.0134771 amu (b) nuclear binding energy: ^{2}H, 3.564×10^{-13} J; ^{4}He, 4.5336×10^{-12} J; ^{6}Li, 5.12602×10^{-12} J. (c) binding energy/nucleon: ^{2}H, 1.782×10^{-13} J/nucleon; ^{4}He, 1.1334×10^{-12} J/nucleon; ^{6}Li, 8.54337×10^{-13} J/nucleon. This trend in binding energy/nucleon agrees with the curve in Figure 21.12. The anomalously high calculated value for ^{4}He is also apparent on the figure. **21.51** (a) 1.71×10^{5} kg/d (b) 2.1×10^{8} g ^{235}U **21.53** (a) ^{59}Co; it has the largest binding energy per nucleon, and binding energy gives rise to mass defect. **21.55** (a) NaI is a good source of iodine because iodine is a large percentage of its mass; it is completely dissociated into ions in aqueous solution, and iodine in the form of I$^{-}(aq)$ is mobile and immediately available for biouptake. (b) A Geiger counter placed near the thyroid immediately after ingestion will register background, then gradually increase in signal until the concentration of iodine in the thyroid reaches a maximum. Over time, iodine-131 decays, and the signal decreases. (c) The radioactive iodine will decay to 0.01% of the original amount in approximately 82 days. **21.57** (a) Characteristics (ii) and (iv) are required for a fuel in a nuclear power plant. (b) ^{235}U **21.59** The *control rods* in a nuclear reactor regulate the flux of neutrons to keep the reaction chain self-sustaining and also to prevent the reactor core from overheating. They are composed of materials such as boron or cadmium that absorb neutrons. **21.61** (a) $^{2}_{1}$H + $^{2}_{1}$H $\longrightarrow$ $^{3}_{2}$He + $^{1}_{0}$n
(b) $^{239}_{92}$U + $^{1}_{0}$n $\longrightarrow$ $^{133}_{51}$Sb + $^{98}_{41}$Nb + 9 $^{1}_{0}$n
21.63 (a) $\Delta m = 0.006627$ g/mol; $\Delta E = 5.956 \times 10^{11}$ J = 5.956×10^{8} kJ/mol $^{1}_{1}$H (b) The extremely high temperature is required to overcome electrostatic charge repulsions between the nuclei so that they can come together to react. **21.65** (a) Boiling water reactor (b) fast breeder reactor (c) gas-cooled reactor **21.67** Hydrogen abstraction: RCOOH + OH $\longrightarrow$ RCOO + H$_2$O; deprotonation: RCOOH + OH^{-} $\longrightarrow$ RCOO^{-} + H$_2$O. Hydroxyl radical is more toxic to living systems because it produces other radicals when it reacts with molecules in the organism. Hydroxide ion, OH^{-}, on the other hand, will be readily neutralized in the buffered cell environment. The acid–base reactions of OH^{-} are usually much less disruptive to the organism than the chain of redox reactions initiated by OH radical.
21.69 (a) 5.3×10^{8} dis/s, 5.3×10^{8} Bq
(b) 6.1×10^{2} mrad, 6.1×10^{-3} Gy
(c) 5.8×10^{3} mrem, 5.8×10^{-2} Sv
21.72 $^{210}_{82}$Pb **21.74** (a) $^{36}_{17}$Cl $\longrightarrow$ $^{36}_{18}$Ar + $^{0}_{-1}$e (b) ^{35}Cl and ^{37}Cl both have an odd number of protons but an even number of neutrons. ^{36}Cl has an odd number of protons and neutrons, so it is less stable than the other two isotopes. **21.76** (a) $^{6}_{3}$Li + $^{56}_{28}$Ni $\longrightarrow$ $^{62}_{31}$Ga
(b) $^{40}_{20}$Ca + $^{248}_{96}$Cm $\longrightarrow$ $^{147}_{62}$Sm + $^{141}_{54}$Xe (c) $^{88}_{38}$Sr + $^{84}_{36}$Kr $\longrightarrow$
$^{116}_{46}$Pd + $^{56}_{28}$Ni (d) $^{40}_{20}$Ca + $^{238}_{92}$U $\longrightarrow$ $^{70}_{30}$Zn + 4 $^{1}_{0}$n + 2 $^{102}_{41}$Nb
21.80 $k = 9.2 \times 10^{-13}$ s^{-1}; $N = 9.7 \times 10^{20}$ Pu atoms; rate = 8.9×10^{8} counts/s **21.84** ^{7}Be, 8.612×10^{-13} J/nucleon; ^{9}Be, 1.035×10^{-12} J/nucleon; ^{10}Be: 1.042×10^{-12} J/nucleon. The binding energies/ nucleon for ^{9}Be and ^{10}Be are very similar; that for ^{10}Be is slightly higher. **21.90** 1.4×10^{4} kg C$_8$H$_{18}$

Chapter 22

22.1 (a) C$_2$H$_4$, the structure on the left, is the stable compound. Carbon can form strong multiple bonds to satisfy the octet rule, while silicon cannot. (b) The geometry about the central atoms in C$_2$H$_4$ is trigonal planar. **22.3** Molecules (b) and (d) will have the seesaw structure shown in the figure. **22.6** The graph shows the trend in (c) density for the group 6A elements. Going down the family, atomic mass increases faster than atomic volume (radius), and density increases. **22.9** The compound on the left, with the strained three-membered ring, will be the most generally reactive. The larger the deviation from ideal bond angles, the more strain in the molecule and the more generally reactive it is. **22.11** Metals: (b) Sr, (c) Mn, (e) Na; nonmetals: (a) P, (d) Se, (f) Kr; metalloids: none. **22.13** (a) O (b) Br (c) Ba (d) O (e) Co (f) Br **22.15** (a) N is too small a central atom to fit five fluorine atoms, and it does not have available d orbitals, which can help accommodate more

than eight electrons. (b) Si does not readily form π bonds, which are necessary to satisfy the octet rule for both atoms in the molecule. (c) As has a lower electronegativity than N; that is, it more readily gives up electrons to an acceptor and is more easily oxidized.
22.17 (a) NaOCH$_3$(s) + H$_2$O(l) $\longrightarrow$ NaOH(aq) + CH$_3$OH(aq)
(b) CuO(s) + 2 HNO$_3$(aq) $\longrightarrow$ Cu(NO$_3$)$_2$(aq) + H$_2$O(l)
(c) WO$_3$(s) + 3 H$_2$(g) $\longrightarrow$ W(s) + 3 H$_2$O(g)
(d) 4 NH$_2$OH(l) + O$_2$(g) $\longrightarrow$ 6 H$_2$O(l) + 2 N$_2$(g)
(e) Al$_4$C$_3$(s) + 12 H$_2$O(l) $\longrightarrow$ 4 Al(OH)$_3$(s) + 3 CH$_4$(g)
22.19 (a) $^{1}_{1}$H, protium; $^{2}_{1}$H, deuterium; $^{3}_{1}$H, tritium (b) in order of decreasing natural abundance: protium > deuterium > tritium (c) Tritium is radioactive. (d) $^{3}_{1}$H $\longrightarrow$ $^{3}_{2}$He + $^{0}_{-1}$e **22.21** Like other elements in group 1A, hydrogen has only one valence electron and its most common oxidation number is +1.
22.23 (a) Mg(s) + 2 H^{+}(aq) $\longrightarrow$ Mg^{2+}(aq) + H$_2$(g)
(b) C(s) + H$_2$O(g) $\xrightarrow{1100\,°C}$ CO(g) + 3 H$_2$(g)
(c) CH$_4$(g) + H$_2$O(g) $\xrightarrow{1100\,°C}$ CO(g) + 3 H$_2$(g)
22.25 (a) NaH(s) + H$_2$O(l) $\longrightarrow$ NaOH(aq) + H$_2$(g)
(b) Fe(s) + H$_2$SO$_4$(aq) $\longrightarrow$ Fe^{2+}(aq) + H$_2$(g) + SO$_4^{2-}$(aq)
(c) H$_2$(g) + Br$_2$(g) $\longrightarrow$ 2 HBr(g)
(d) 2 Na(l) + H$_2$(g) $\longrightarrow$ 2 NaH(s)
(e) PbO(s) + H$_2$(g) $\xrightarrow{\Delta}$ Pb(s) + H$_2$O(g) **22.27** (a) Ionic (b) molecular (c) metallic **22.29** Vehicle fuels produce energy via combustion reactions. The combustion of hydrogen is very exothermic and its only product, H$_2$O, is a non-pollutant. **22.31** Xenon has a lower ionization energy than argon; because the valence electrons are not as strongly attracted to the nucleus, they are more readily promoted to a state in which the atom can form bonds with fluorine. Also, Xe is larger and can more easily accommodate an expanded octet of electrons. **22.33** (a) Ca(OBr)$_2$, Br, +1 (b) HBrO$_3$, Br, +5 (c) XeO$_3$, Xe, + 6 (d) ClO$_4^{-}$, Cl, +7 (e) HIO$_2$, I, +3 (f) IF$_5$; I, +5; F, -1 **22.35** (a) iron(III) chlorate, Cl, +5; (b) chlorous acid, Cl, +3 (c) xenon hexafluoride, F, -1 (d) bromine pentafluoride; Br, +5; F, -1 (e) xenon oxide tetrafluoride, F, -1 (f) iodic acid, I, +5; **22.37** (a) van der Waals intermolecular attractive forces increase with increasing number of electrons in the atoms. (b) F$_2$ reacts with water: F$_2$(g) + H$_2$O(l) $\longrightarrow$ 2 HF(g) + O$_2$(g). That is, fluorine is too strong an oxidizing agent to exist in water. (c) HF has extensive hydrogen bonding. (d) Oxidizing power is related to electronegativity. Electronegativity and oxidizing power decrease in the order given.
22.39 (a) 2 HgO(s) $\xrightarrow{\Delta}$ 2 Hg(l) + O$_2$(g)
(b) 2 Cu(NO$_3$)$_2$(s) $\xrightarrow{\Delta}$ 2 CuO(s) + 4 NO$_2$(g) + O$_2$(g)
(c) PbS(s) + 4 O$_3$(g) $\longrightarrow$ PbSO$_4$(s) + 4 O$_2$(g)
(d) 2 ZnS(s) + 3 O$_2$(g) $\longrightarrow$ 2 ZnO(s) + 2 SO$_2$(g)
(e) 2 K$_2$O$_2$(s) + 2 CO$_2$(g) $\longrightarrow$ 2 K$_2$CO$_3$(s) + O$_2$(g)
(f) 3 O$_2$(g) $\xrightarrow{h\nu}$ 2 O$_3$(g) **22.41** (a) acidic (b) acidic (c) amphoteric (d) basic **22.43** (a) H$_2$SeO$_3$, Se, +4 (b) KHSO$_3$, S, +4 (c) H$_2$Te, Te, -2 (d) CS$_2$, S, -2 (e) CaSO$_4$, S, +6 (f) CdS, S, -2 (g) ZnTe, Te, -2
22.45 (a)
2 Fe^{3+}(aq) + H$_2$S(aq) $\longrightarrow$ 2 Fe^{2+}(aq) + S(s) + 2 H^{+}(aq)
(b) Br$_2$(l) + H$_2$S(aq) $\longrightarrow$ 2 Br^{-}(aq) + S(s) + 2 H^{+}(aq)
(c) 2 MnO$_4^{-}$(aq) + 6 H^{+}(aq) + 5 H$_2$S(aq) $\longrightarrow$ 2 Mn^{2+}(aq) + 5 S(s) + 8 H$_2$O(l) (d) 2 NO$_3^{-}$(aq) + H$_2$S(aq) + 2 H^{+}(aq) $\longrightarrow$ 2 NO$_2$(aq) + S(s) + 2 H$_2$O(l)
22.47

(a)

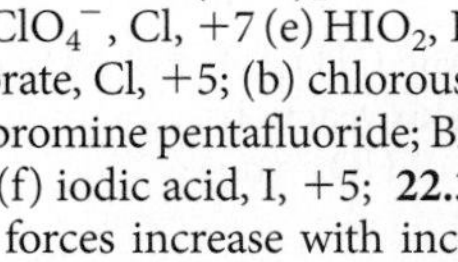

Trigonal pyramidal

(b)

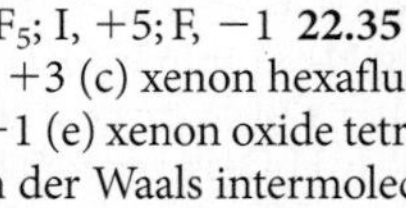

Bent (free rotation around S–S bond)

(c)

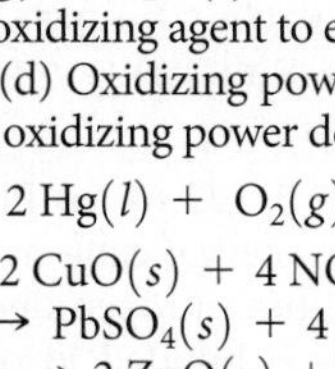

Tetrahedral
(around S)

22.49 (a) $SO_2(s) + H_2O(l) \rightleftharpoons H_2SO_3(aq) \rightleftharpoons H^+(aq) + HSO_3^-(aq)$
(b) $ZnS(s) + 2 HCl(aq) \longrightarrow ZnCl_2(aq) + H_2S(g)$
(c) $8 SO_3^{2-}(aq) + S_8(s) \longrightarrow 8 S_2O_3^{2-}(aq)$
(d) $SO_3(aq) + H_2SO_4(l) \longrightarrow H_2S_2O_7(l)$ **22.51** (a) $NaNO_2, +3$
(b) $NH_3, -3$ (c) $N_2O, +1$ (d) $NaCN, -3$ (e) $HNO_3, +5$ (f) $NO_2, +4$
(g) $N_2, 0$ (h) $BN, -3$

22.53 (a) $:\ddot{O}=\ddot{N}-\ddot{O}-H \longleftrightarrow :\ddot{O}-\ddot{N}=\ddot{O}-H$
The molecule is bent around the central oxygen and nitrogen atoms; the four atoms need not be coplanar. The right-most form does not minimize formal charges and is less important in the actual bonding model. The oxidation state of N is +3.

(b)
$$\left[:\ddot{N}=N=\ddot{N}:\right]^- \longleftrightarrow \left[:N\equiv N-\ddot{\ddot{N}}:\right]^- \longleftrightarrow \left[:\ddot{\ddot{N}}-N\equiv N:\right]^-$$

The molecule is linear. The oxidation state of N is $-1/3$.

(c)
$$\left[\begin{array}{c} H \quad H \\ | \quad | \\ H-N-N: \\ | \quad | \\ H \quad H \end{array}\right]^+$$

The geometry is tetrahedral around the left nitrogen, trigonal pyramidal around the right. The oxidation state of N is -2.

(d)
$$\left[\begin{array}{c} :\ddot{O}: \\ | \\ :\ddot{O}-N=\ddot{O} \end{array}\right]^-$$

The ion is trigonal planar; it has three equivalent resonance forms. The oxidation state of N is +5.

22.55 (a) $Mg_3N_2(s) + 6 H_2O(l) \longrightarrow 3 Mg(OH)_2(s) + 2 NH_3(aq)$
(b) $2 NO(g) + O_2(g) \longrightarrow 2 NO_2(g)$, redox reaction
(c) $N_2O_5(g) + H_2O(l) \longrightarrow 2 H^+(aq) + 2 NO_3^-(aq)$
(d) $NH_3(aq) + H^+(aq) \longrightarrow NH_4^+(aq)$
(e) $N_2H_4(l) + O_2(g) \longrightarrow N_2(g) + 2 H_2O(g)$, redox reaction

22.57 (a) $HNO_2(aq) + H_2O(l) \longrightarrow NO_3^-(aq) + 2 e^-$
(b) $N_2(g) + H_2O(l) \longrightarrow N_2O(aq) + 2 H^+(aq) + 2 e^-$

22.59 (a) $H_3PO_3, +3$ (b) $H_4P_2O_7, +5$ (c) $SbCl_3, +3$ (d) $Mg_3As_2, +5$
(e) $P_2O_5, +5$ (f) $Na_3PO_4, +5$ **22.61** (a) Phosphorus is a larger atom than nitrogen, and P has energetically available $3d$ orbitals, which participate in the bonding, but nitrogen does not. (b) Only one of the three hydrogens in H_3PO_2 is bonded to oxygen. The other two are bonded directly to phosphorus and are not easily ionized. (c) PH_3 is a weaker base than H_2O so any attempt to add H^+ to PH_3 in the presence of H_2O causes protonation of H_2O. (d) The P_4 molecules in white phosphorus have more severely strained bond angles than the chains in red phosphorus, causing white phosphorus to be more reactive.
22.63 (a) $2 Ca_3PO_4(s) + 6 SiO_2(s) + 10 C(s) \longrightarrow$
$P_4(g) + 6 CaSiO_3(l) + 10 CO(g)$
(b) $PBr_3(l) + 3 H_2O(l) \longrightarrow H_3PO_3(aq) + 3 HBr(aq)$
(c) $4 PBr_3(g) + 6 H_2(g) \longrightarrow P_4(g) + 12 HBr(g)$
22.65 (a) HCN (b) $Ni(CO)_4$ (c) $Ba(HCO_3)_2$ (d) CaC_2
(e) K_2CO_3 **22.67** (a) $ZnCO_3(s) \xrightarrow{\Delta} ZnO(s) + CO_2(g)$
(b) $BaC_2(s) + 2 H_2O(l) \longrightarrow Ba^{2+}(aq) + 2 OH^-(aq) + C_2H_2(g)$
(c) $2 C_2H_2(g) + 5 O_2(g) \longrightarrow 4 CO_2(g) + 2 H_2O(g)$
(d) $CS_2(g) + 3 O_2(g) \longrightarrow CO_2(g) + 2 SO_2(g)$
(e) $Ca(CN)_2(s) + 2 HBr(aq) \longrightarrow CaBr_2(aq) + 2 HCN(aq)$
22.69

(a) $2 CH_4(g) + 2 NH_3(g) + 3 O_2(g) \xrightarrow{800°C} 2 HCN(g) + 6 H_2O(g)$
(b) $NaHCO_3(s) + H^+(aq) \longrightarrow CO_2(g) + H_2O(l) + Na^+(aq)$
(c) $2 BaCO_3(s) + O_2(g) + 2 SO_2(g) \longrightarrow 2 BaSO_4(s) + 2 CO_2(g)$
22.71 (a) $H_3BO_3, +3$ (b) $SiBr_4, +4$ (c) $PbCl_2, +2$
(d) $Na_2B_4O_7 \cdot 10 H_2O, +3$ (e) $B_2O_3, +3$ (f) $GeO_2, +4$
22.73 (a) Tin (b) carbon, silicon, and germanium (c) silicon
22.75 (a) Tetrahedral (b) Metasilicic acid will probably adopt the single-strand silicate chain structure shown in Figure 22.33 (b). The Si

to O ratio is correct and there are two terminal O atoms per Si that can accommodate the two H atoms associated with each Si atom of the acid. **22.77** (a) Two Ca^{2+} (b) Two OH^- **22.79** (a) Diborane has bridging H atoms linking the two B atoms. The structure of ethane has the C atoms bound directly, with no bridging atoms. (b) B_2H_6 is an electron-deficient molecule. The 6 valence electron pairs are all involved in B—H sigma bonding, so the only way to satisfy the octet rule at B is to have the bridging H atoms shown in Figure 22.35. (c) The term *hydridic* indicates that the H atoms in B_2H_6 have more than the usual amount of electron density for a covalently bound H atom.
22.81 (a) False (b) true (c) false (d) true (e) true (f) true (g) false
22.84 (a) SO_3 (b) Cl_2O_5 (c) N_2O_3 (d) CO_2 (e) P_2O_5 **22.88** (a) PO_4^{3-}, +5; NO_3^-, +5, (b) The Lewis structure for NO_4^{3-} would be:

$$\left[\begin{array}{c} :\ddot{O}: \\ | \\ :\ddot{O}-N-\ddot{O}: \\ | \\ :\ddot{O}: \end{array}\right]^{3-}$$

The formal charge on N is +1 and on each O atom is -1. The four electronegative oxygen atoms withdraw electron density, leaving the nitrogen deficient. Since N can form a maximum of four bonds, it cannot form a π bond with one or more of the O atoms to regain electron density, as the P atom in PO_4^{3-} does. Also, the short N—O distance would lead to a tight tetrahedron of O atoms subject to steric repulsion. **22.92** (a) $1.94 \times 10^3 \text{ g } H_2$ (b) $2.16 \times 10^4 \text{ L } H_2$ (c) $2.76 \times 10^5 \text{ kJ}$
22.94 (a) $-285.83 \text{ kJ/mol } H_2$; $-890.4 \text{ kJ/mol } CH_4$
(b) $-141.79 \text{ kJ/g } H_2$; $-55.50 \text{ kJ/g } CH_4$ (c) $1.276 \times 10^4 \text{ kJ/m}^3 \text{ } H_2$; $3.975 \times 10^4 \text{ kJ/m}^3 \text{ } CH_4$ **22.96** (a) $[HClO] = 0.036M$ (b) pH = 1.4
22.99 (a) $SO_2(g) + 2 H_2S(aq) \longrightarrow 3 S(s) + 2 H_2O(l)$ or
$8 SO_2(g) + 16 H_2S(aq) \longrightarrow 3 S_8(s) + 16 H_2O(l)$
(b) $4.0 \times 10^3 \text{ mol} = 9.7 \times 10^4 \text{ L } H_2S$ (c) $1.9 \times 10^5 \text{ g S produced}$
22.101 The average bond enthalpies are H—O, 463 kJ; H—S, 367 kJ; H—Se, 316 kJ; H—Te, 266 kJ. The H—X bond enthalpy decreases steadily in the series. The origin of this effect is probably the increasing size of the orbital from X with which the hydrogen $1s$ orbital must overlap. **22.105** Dimethylhydrazine produces 0.0369 mol gas per gram of reactants, while methylhydrazine produces 0.0388 mol gas per gram of reactants. Methylhydrazine has marginally greater thrust.
2.107 (a)
$3 B_2H_6(g) + 6 NH_3(g) \longrightarrow 2 (BH)_3(NH)_3(l) + 12 H_2(g)$;
$3 LiBH_4(s) + 3 NH_4Cl(s) \longrightarrow 2 (BH)_3(NH)_3(l) + 9 H_2(g) + 3 LiCl(s)$
(b)

$$\begin{array}{c} H \\ | \\ H{-}B{\diagup}^{N}{\diagdown}B{-}H \\ \end{array} \longleftrightarrow \longleftrightarrow$$

(c) $2.40 \text{ g } (BH)_3(NH)_3$

Chapter 23

23.2

$$\begin{array}{c} N \diagdown \quad \diagup Cl \\ \quad Pt \\ N \diagup \quad \diagdown Cl \end{array}$$

(a) Coordination number is 4 (b) coordination geometry is square planar (c) oxidation state is +2. **23.6** Molecules (1), (3), and (4) are chiral because their mirror images are not superimposible on the original molecules. **23.8** (a) Diagram (4) (b) diagram (1) (c) diagram (3) (d) diagram (2) **23.11** The lanthanide contraction explains trend (c). The lanthanide contraction is the name given to the decrease in atomic size due to the build-up in effective nuclear charge as we move through the lanthanides and beyond them. This effect offsets the expected increase in atomic size going from period 5 to period 6 transition elements. **23.13** (a) Ti^{2+}, $[Ar]3d^2$ (b) Ti^{4+}, $[Ar]$ (c) Ni^{2+}, $[Ar]3d^8$ (d) Zn^{2+}, $[Ar]3d^{10}$ **23.15** (a) Ti^{3+}, $[Ar]3d^1$ (b) Ru^{2+}, $[Kr]4d^6$ (c) Au^{3+}, $[Xe]4f^{14}5d^8$ (d) Mn^{4+}, $[Ar]3d^3$ **23.17** (a) The unpaired electrons in a paramagnetic material cause it to be weakly attracted into a magnetic field. **23.19** The diagram shows a material with misaligned spins that become aligned in the direction of an applied magnetic field. This is a paramagnetic material. **23.21** (a) Primary valence is roughly the same as oxidation number. Oxidation number is a broader term than ionic charge, but Werner's complexes contain metal ions where cation charge and oxidation number are equal. (b) Coordination number is the modern term for secondary valence. (c) NH_3 can serve as a ligand because it has an unshared electron pair, while BH_3 does not. Ligands are the Lewis base in metal–ligand interactions. As such, they must possess at least one unshared electron pair. **23.23** (a) +2 (b) 6 (c) 2 mol $AgBr(s)$ will precipitate per mole of complex. **23.25** (a) Coordination number = 4, oxidation number = +2 (b) 5, +4 (c) 6, +3 (d) 5, +2 (e) 6, +3 (f) 4, +2 **23.27** (a) A monodentate ligand binds to a metal via one atom, a bidentate ligand binds through two atoms. (b) Three bidentate ligands fill the coordination sphere of a six-coordinate complex. (c) A tridentate ligand has at least three atoms with unshared electron pairs in the correct orientation to simultaneously bind one or more metal ions. **23.29** (a) *Ortho*-phenanthroline, *o*-phen, is bidentate (b) oxalate, $C_2O_4^{2-}$, is bidentate (c) ethylenediaminetetraacetate, EDTA, is pentadentate (d) ethylenediamine, en, is bidentate. **23.31** (a) The term chelate effect refers to the special stability associated with formation of a metal complex containing a polydentate (chelate) ligand relative to a complex containing only monodentate ligands. (b) The increase in entropy, $+\Delta S$ associated with the substitution of a chelating ligand for two or more monodentate ligands generally gives rise to the *chelate effect*. Chemical reactions with $+\Delta S$ tend to be spontaneous, have negative ΔG and large values of K. (c) Polydentate ligands are used as sequestering agents to bind metal ions and prevent them from undergoing unwanted chemical reactions without removing them from solution. **23.33** False. The ligand is not typically a bidentate ligand. The entire molecule is planar and the benzene rings on either side of the two N atoms inhibit their approach in the correct orientation for chelation. **23.35** (a) $[Cr(NH_3)_6](NO_3)_3$ (b) $[Co(NH_3)_4CO_3]_2SO_4$ (c) $[Pt(en)_2Cl_2]Br_2$ (d) $K[V(H_2O)_2Br_4]$ (e) $[Zn(en)_2][HgI_4]$ **23.37** (a) tetraamminedichlororhodium(III) chloride (b) potassium hexachlorotitanate(IV) (c) tetrachlorooxomolybdenum(VI) (d) tetraaqua(oxalato)platinum (IV) bromide **23.39** (a) Complexes 1, 2 and 3 can have geometric isomers; they each have cis-trans isomers. (b) Complex 2 can have linkage isomers, owing to the presence of the nitrite ligand. (c) The cis geometric isomer of complex 3 can have optical isomers (d) Complex 1 can have coordination sphere isomers. It is the only complex with a counter ion that can also be a ligand. **23.41** Yes. No structural or stereoisomers are possible for a tetrahedral complex of the form MA_2B_2. The complex must be square planar with cis and trans geometric isomers. **23.43** (a) One isomer (b) trans and cis isomers with 180° and 90° $Cl-Ir-Cl$ angles, respectively (c) trans and cis isomers with 180° and 90° $Cl-Fe-Cl$ angles, respectively. The cis isomer is optically active. **23.45** (a) Blue (b) $E = 3.26 \times 10^{-19}$ J/photon (c) $E = 196$ kJ/mol **23.47** (a) Diamagnetic. Zn^{2+}, $[Ar]3d^{10}$. There are no unpaired electrons. (b) Diamagnetic. Pd^{2+}, $[Kr]4d^8$. Square planar complexes with 8 d-electrons are usually diamagnetic, especially with a heavy metal center like Pd. (c) Paramagnetic. V^{3+}, $[Ar]3d^2$. The two d-electrons would be unpaired in any of the d-orbital energy level diagrams. **23.49** Most of the attraction between a metal ion and a ligand is electrostatic. Whether the interaction is ion–ion or ion–dipole, the ligand is strongly attracted to the metal center and can be modeled as a point negative charge.

23.51 (a) ____ ____ $d_{x^2y^2}, d_{z^2}$

Δ (with up arrow)

____ ____ ____ d_{xy}, d_{xz}, d_{yz}

(b) The magnitude of Δ and the energy of the *d-d* transition for a d^1 complex are equal. (c) $\Delta = 220$ kJ/mol **23.53** (a) Cu^{2+}, $[Ar]3d^9$. Since the two minerals are colored, the copper ions in them must be in the +2 oxidation state. (b) Azurite will probably have the larger Δ. It absorbs orange visible light, which has shorter wavelengths than the red light absorbed by malachite. **23.55** (a) Ti^{3+}, d^1, (b) Co^{3+}, d^6 (c) Ru^{3+}, d^5 (d) Mo^{5+}, d^1 (e) Re^{3+}, d^4 **23.57** Yes. A weak-field ligand leads to a small Δ value and a small d-orbital splitting energy. If the splitting energy of a complex is smaller than the energy required to pair electrons in an orbital, the complex is high spin. **23.59** (a) Mn, $[Ar]4s^23d^5$; Mn^{2+}, $[Ar]3d^5$; 1 unpaired electron (b) Ru, $[Kr]5s^14d^7$; Ru^{2+}, $[Kr]4d^6$; 0 unpaired electrons (c) Rh, $[Kr]5s^14d^8$; Rh^{2+}, $[Kr]4d^7$; 1 unpaired electron **23.61** All complexes in this exercise are six-coordinate octahedral.

(a) d^4, high spin

(b) d^5, high spin

(c) d^6, low spin

(d) d^5, low spin

(e) d^3

(f) d^8

23.63 high spin

23.67 $[Pt(NH_3)_6]Cl_4$; $[Pt(NH_3)_4Cl_2]Cl_2$; $[Pt(NH_3)_3Cl_3]Cl$; $[Pt(NH_3)_2]Cl_4$; $K[Pt(NH_3)Cl_5]$

23.71

```
    H        H  H        H
    |        |  |        |
H—C—P—C—C—P—C—H
    |        |  |        |
    H        H  H        H
    H—C—H  H—C—H
        |        |
        H        H
```

Both dmpe and en are bidentate ligands, binding through P and N, respectively. Because phosphorus is less electronegative than N, dmpe is a stronger electron pair donor and Lewis base than en. Dmpe creates a stronger ligand field and is higher on the spectrochemical series. Structurally, dmpe occupies a larger volume than en. M–P bonds are longer than M—N bonds and the two —CH_3 groups on each P atom in dmpe create more steric hindrance than the H atoms on N in en. (b) The oxidation state of Mo is zero. (c) The symbol $\widehat{P\,P}$ represents the bidentate dmpe ligand.

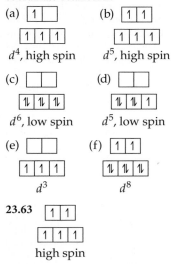

optical isomers

23.74 (a) Iron (b) magnesium (c) iron **23.76** (a) Pentacarbonyliron(0) (b) The oxidation state of iron must be zero. (c) Two. One isomer has CN in an axial position and the other has it in an equatorial position.
23.78 (a)

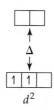

d^2

(b) These complexes are colored because the crystal-field splitting energy, Δ, is in the visible portion of the electromagnetic spectrum. Visible light with $\lambda = hc/\Delta$ is absorbed by the complex, promoting one of the d-electrons into a higher-energy d-orbital. The remaining wavelengths are reflected or transmitted; the combination of these wavelengths is the color we see. (c) $[V(H_2O)_6]^{3+}$ will absorb light with higher energy because it has a larger Δ than $[VF_6]^{3-}$. H_2O is in the middle of the spectrochemical series and causes a larger Δ than F^-, a weak-field ligand. **23.80** $[Co(NH_3)_6]^{3+}$, yellow; $[Co(H_2O)_6]^{2+}$, pink; $[CoCl_4]^{2-}$, blue

23.85 (a)

$$\left[\begin{array}{c} O \\ \| \\ C \\ NC \diagdown \diagup CN \\ Fe \\ NC \diagup \diagdown CN \\ C \\ \| \\ O \end{array} \right]^{2-}$$

(b) sodium dicarbonyltetracyanoferrate(II) (c) $+2$, 6 d-electrons (d) We expect the complex to be low spin. Cyanide (and carbonyl) are high on the spectrochemical series, which means the complex will have a large Δ splitting, characteristic of low-spin complexes. **23.91** (a) Yes, the oxidation state of Co is $+3$ in both complexes. (b) Compound A has SO_4^{2-} outside the coordination sphere and coordinated Br^-, so it forms a precipitate with $BaCl_2(aq)$ but not $AgNO_3(aq)$. Compound B has Br^- outside the coordination sphere and coordinated SO_4^{2-}, so it forms a precipitate with $AgNO_3(aq)$ but not $BaCl_2(aq)$. (c) Compounds A and B are coordination sphere isomers. (d) Both compounds are strong electrolytes. **23.94** The chemical formula is $[Pd(NC_5H_5)_2Br_2]$. This is an electrically neutral square-planar complex of Pd(II), a nonelectrolyte whose solutions do not conduct electricity. Because the dipole moment is zero, it must be the trans isomer. **23.96** 47.3 mg Mg^{2+}/L, 53.4 mg Ca^{2+}/L **23.99** $\Delta E = 3.02 \times 10^{-19}$ J/photon, $\lambda = 657$ nm. The complex will absorb in the visible around 660 nm and appear blue-green.

Chapter 24

24.1 Molecules (c) and (d) are the same molecule. **24.7** (a) False (b) false (c) true (d) false **24.9** Numbering from the right on the condensed structural formula, C1 has trigonal-planar electron-domain geometry, 120° bond angles, and sp^2 hybridization; C2 and C5 have tetrahedral electron-domain geometry, 109° bond angles, and sp^3 hybridization; C3 and C4 have linear electron-domain geometry, 180° bond angles, and sp hybridization. **24.11** NH_3 and CO are not typical organic molecules. NH_3 contains no carbon atoms. Carbon monoxide contains a C atom that does not form four bonds. **24.13** (a) True (b) true (c) false (d) false (e) true (f) false (g) true **24.15** (a) 2-methylhexane (b) 4-ethyl-2,4-dimethyldecane (c) $CH_3CH_2CH_2CH_2CH_2CH(CH_3)_2$ (d) $CH_3CH_2CH_2CH_2CH(CH_2CH_3)CH(CH_3)CH(CH_3)_2$
(e)

H H
H—C—C—CH₃
H—C—C—H
H—C—C—CH₃
H—C—C—H
H H

or

CH₂—CH₃
H₂C—CH
H₂C—CH
CH₂—CH₃

24.17 (a) 2,3-dimethylheptane (b) $CH_3CH_2CH_2C(CH_3)_3$

(c)
H₃C   CH₃
  \ /
   C
  / \
H₂C   CH₂
 | |
H₂C   CH₂
  \   /
   CH
    |
   CH₂CH₃

(d) 2,2,5-trimethylhexane (e) methylcyclobutane **24.19** 65
24.21 (a) Alkanes are said to be saturated because they cannot undergo addition reactions, such as those characteristic of carbon–carbon double bonds. (b) No. The compound C_4H_6 does not contain the maximum possible number of hydrogen atoms and is unsaturated.
24.23 (a) $CH_3CH_2CH_2CH_2CH_3$, C_5H_{12}
(b)
CH₂
H₂C   CH₂ C_5H_{10}
H₂C—CH₂
,

(c) $CH_2{=}CHCH_2 CH_2 CH_3$, $C_5 H_{10}$
(d) $HC{\equiv}CCH_2CH_2CH_3$, C_5H_8 **24.25** One possible structure is $CH{\equiv}C{-}CH{=}CH{-}C{\equiv}CH$ **24.27** There are at least 46 structural isomers with the formula C_6H_{10}. A few of them are

$CH_3CH_2CH_2CH_2C{\equiv}CH$ $CH_3CH_2CH_2C{\equiv}CCH_3$

H H
| |
$CH_3C{=}C{-}CH_2CH{=}CH_2$

H
|
$CH_3C{=}C{-}CH_2CH{=}CH_2$
|
H

HC=CH   CH=C—CH₃   CH=CH—CH₃
  \ CH₂   /   \   /   \
H₂C | H₂C   CH₂   H₂C   C—H
  \ CH₂   \   /   \   /
   CH₂   CH₂   CH₂

24.29 (a)
H   CH₃
 \   /
  C=C
 /   \
CH₃CH₂ H

(b)
       CH₃   CH₃
        | |
$CH_3CH_2CH_2{-}C{=}CH{-}CH_2{-}CH{-}CH_3$

(c) *cis*-6-methyl-3-octene (d) *para*-dibromobenzene (e) 4,4-dimethyl-1-hexyne **24.31** (a) True (b) true (c) false **24.33** (a) No

(b)
H   Cl   ClH₂C   Cl
 \   /   \   /
  C=C   C=C
 /   \   /   \
ClH₂C CH₃   H   CH₃

(c) no (d) no **24.35** (a) True

(b) $CH_3CH_2CH{=}CH{-}CH_3 + Br_2 \longrightarrow$
2-pentene

$CH_3CH_2CH(Br)CH(Br)CH_3$
2, 3-dibromopentane

This is an addition reaction.

(c)

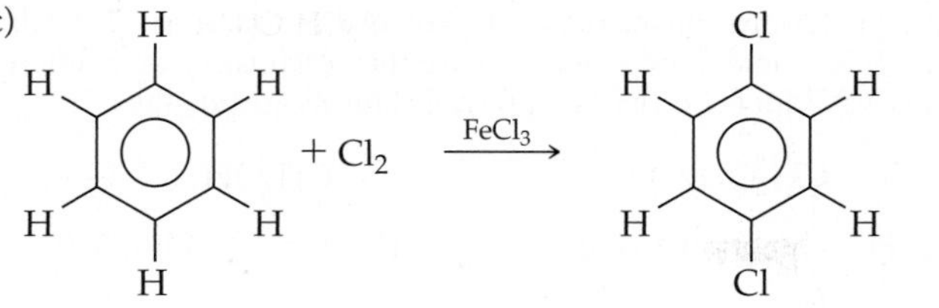

$$C_6H_6 + Cl_2 \xrightarrow{FeCl_3} C_6H_4Cl_2$$

This is an substitution reaction.

24.37 The $60°\,C—C—C$ angles in the cyclopropane ring cause strain that provides a driving force for reactions that result in ring opening. There is no comparable strain in the five-membered or six-membered rings. (b) $C_2H_4(g) + HBr(g) \longrightarrow CH_3CH_2Br(l)$; $C_6H_6(l) + CH_3CH_2Br(l) \xrightarrow{AlCl_3} C_6H_5CH_2CH_3(l) + HBr(g)$ **24.39** Yes, this information suggests (but does not prove) that the reactions proceed in the same manner. That the two rate laws are first order in both reactants and second order overall indicates that the activated complex in the rate-determining step in each mechanism is bimolecular and contains one molecule of each reactant. This is usually an indication that the mechanisms are the same, but it does not rule out the possibility of different fast steps or a different order of elementary steps. **24.41** $\Delta H_{comb}/\text{mol } CH_2$ for cyclopropane = 696.3 kJ, for cyclopentane = 663.4 kJ. $\Delta H_{comb}/CH_2$ group for cyclopropane is greater because C_3H_6 contains a strained ring. When combustion occurs, the strain is relieved and the stored energy is released. **24.43** (a) (iii) (b) (i) (c) (ii) (d) (iv) (e) (v) **24.45** (a) Propionaldehyde (or propanal):

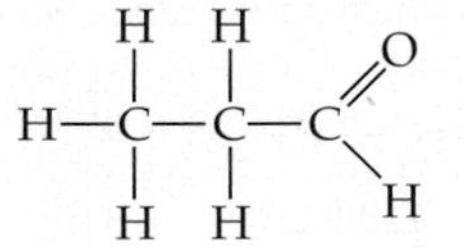

(b) ethylmethyl ether:

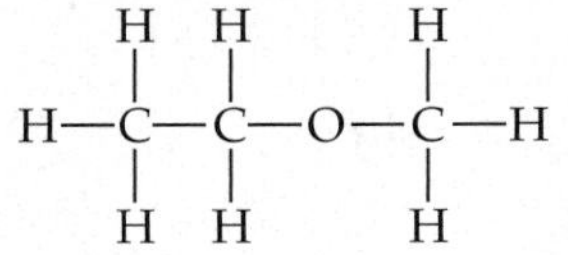

24.47 (a)

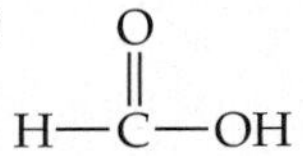

(b)

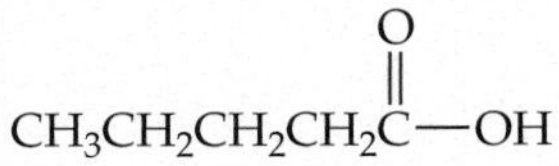

$$CH_3CH_2CH_2CH_2\overset{\displaystyle O}{\overset{\displaystyle \|}{C}}—OH$$

or

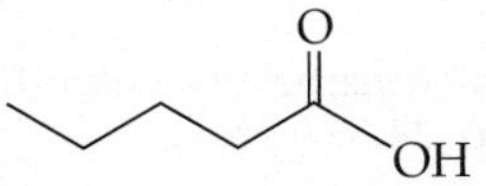

(c)

$$CH_3CH_2CH_2CH_2CH_2CH_2CH_2\overset{\displaystyle CH_3}{\underset{\displaystyle |}{CH}}—\overset{\displaystyle Cl}{\underset{\displaystyle |}{C}}—\overset{\displaystyle O}{\overset{\displaystyle \|}{C}}—OH$$

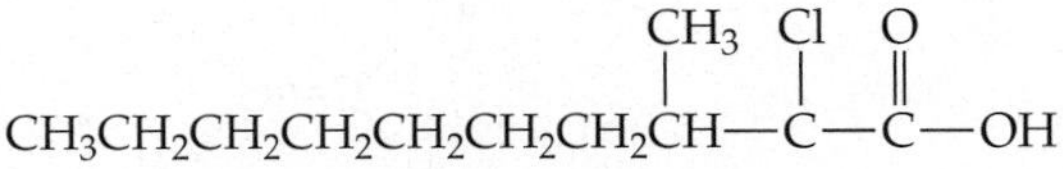

or

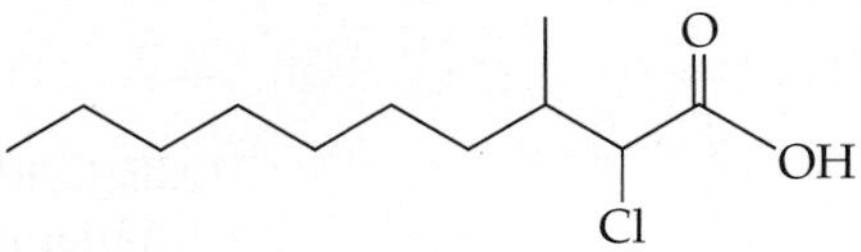

24.49

(a)

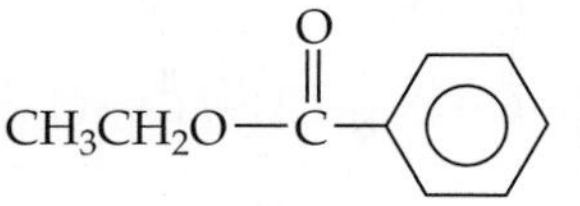

Ethylbenzoate

(b) $CH_3\overset{\displaystyle H}{\underset{\displaystyle |}{N}}—\overset{\displaystyle O}{\overset{\displaystyle \|}{C}}CH_3$

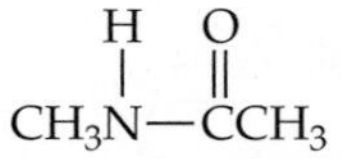

N-methylethanamide or N-methylacetamide

(c)

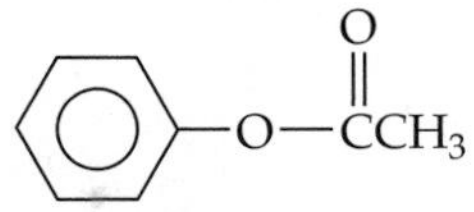

Phenylacetate

24.51

(a) $CH_3CH_2\overset{\displaystyle O}{\overset{\displaystyle \|}{C}}—O—CH_3 + NaOH \longrightarrow \left[CH_3CH_2C\overset{\displaystyle O}{\underset{\displaystyle O}{<}} \right]^-$

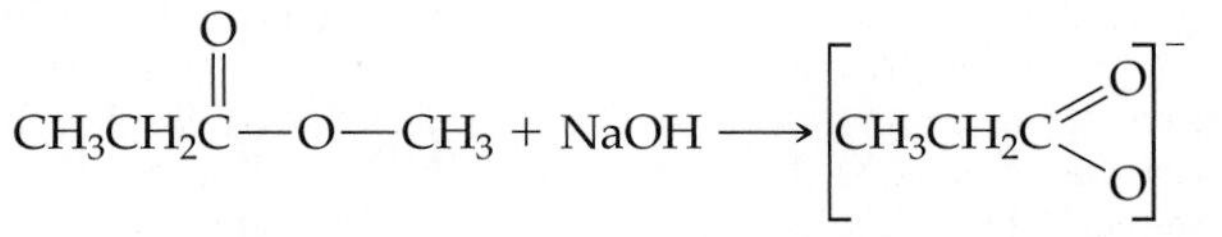

$+ Na^+ + CH_3OH$

(b) $CH_3\overset{\displaystyle O}{\overset{\displaystyle \|}{C}}—O—$$+ NaOH \longrightarrow \left[CH_3C\overset{\displaystyle O}{\underset{\displaystyle O}{<}} \right]^- + Na^+$

$+$

24.53 High melting and boiling points are indicators of strong intermolecular forces in the bulk substance. The presence of both $—OH$ and $—C=O$ groups in pure acetic acid leads us to conclude that it will be a strongly hydrogen-bonded substance. That the melting and boiling points of pure acetic acid are both higher than those of water, a substance we know to be strongly hydrogen-bonded, supports this conclusion. **24.55** (a) $CH_3CH_2CH_2CH(OH)CH_3$ (b) $CH_3CH(OH)CH_2OH$

(c)

$$CH_3\overset{\displaystyle O}{\overset{\displaystyle \|}{C}}OCH_2CH_3$$

(d)

(e) $CH_3OCH_2CH_3$ **24.57** (c) There are two chiral carbon atoms in the molecule.

24.59 (a)

$$H_2N—\overset{\displaystyle R}{\underset{\displaystyle H}{C}}—COOH$$

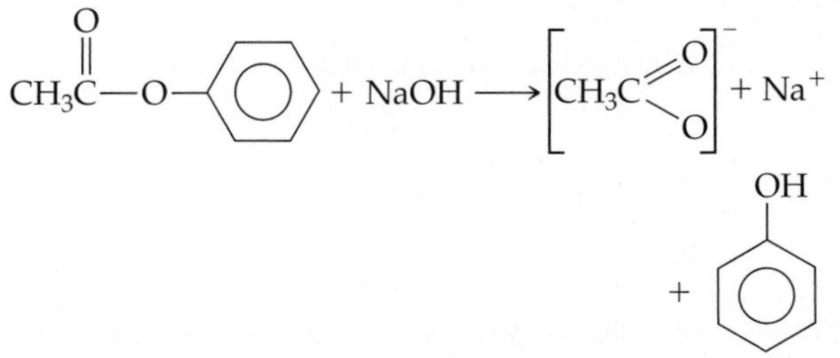

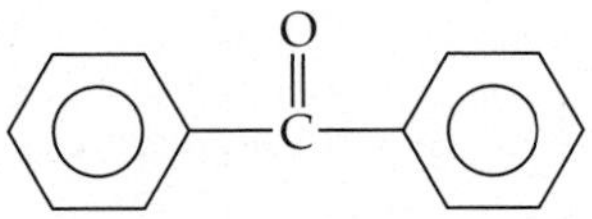

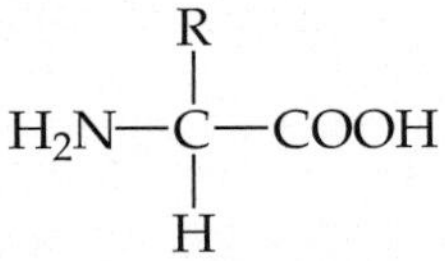

(b) In protein formation, amino acids undergo a condensation reaction between the amino group of one molecule and the carboxylic acid group of another to form the amide linkage. (c) The bond that links amino acids in proteins is called the peptide bond. It is shown in bold in the figure below.

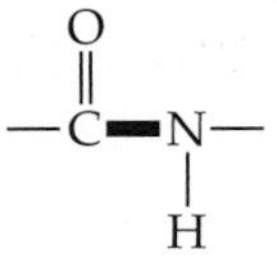

24.61

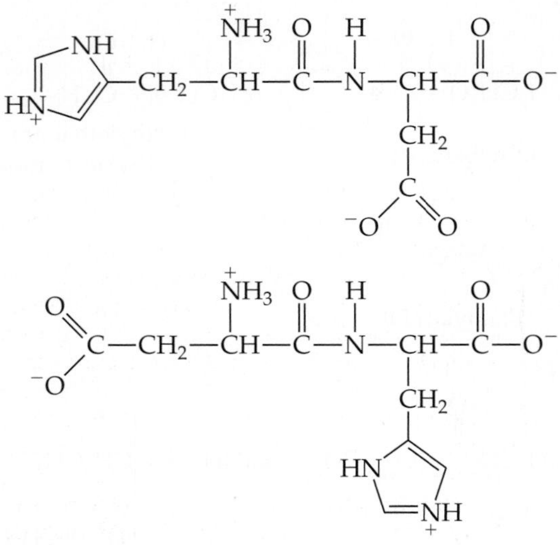

24.63 (a)

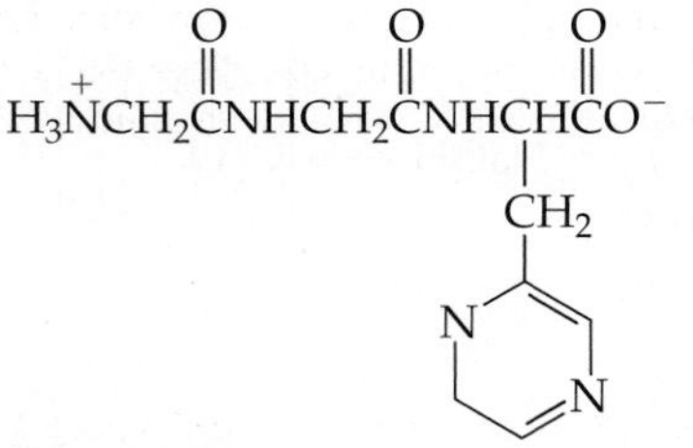

$$\underset{+}{H_3NCH_2CNHCH_2CNHCHCO^-}$$

(with three C=O groups and a CH₂ substituent connected to a pyrazine ring)

(b) Three tripeptides are possible: Gly-Gly-His, GGH; Gly-His-Gly, GHG; His-Gly-Gly, HGG **24.65** (a) True (b) false (c) false **24.67** (a) True (b) false (c) true   **24.69** (a) The empirical formula of cellulose is $C_6H_{10}O_5$. (b) The six-membered ring form of glucose is the monomer unit that is the basis of the polymer cellulose. (c) Ether linkages connect the glucose monomer units in cellulose. **24.71** (a) Yes. (b) Four.

In the linear form of mannose, the aldehydic carbon is C1. Carbon atoms 2, 3, 4, and 5 are chiral because they each carry four different groups. (c) Both the α (left) and β (right) forms are possible.

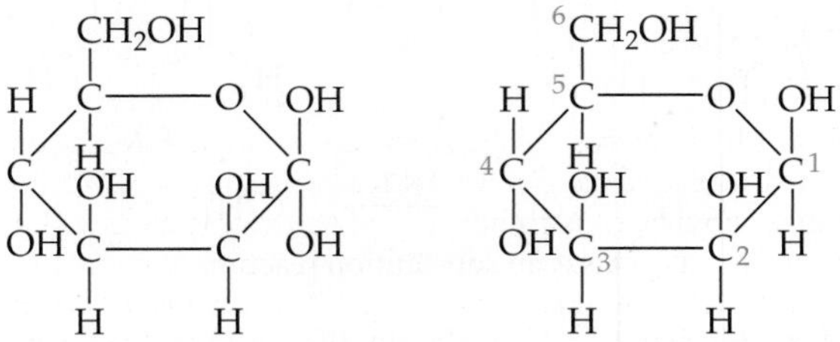

24.73 (a) False (b) true (c) true **24.75** *Purines*, with the larger electron cloud and molar mass, will have larger dispersion forces than *pyrimidines* in aqueous solution. **24.77** 5′−TACG−3′ **24.79** The complimentary strand for 5′−GCATTGGC−3′ is 3′−CGTAACCG−5′.

24.81

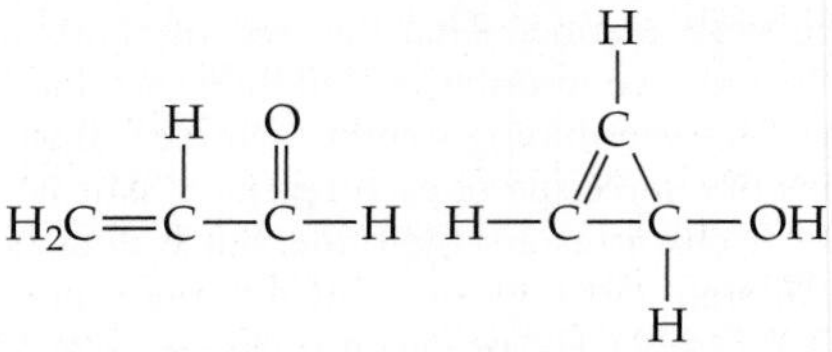

24.90 (a) None (b) The carbon bearing the secondary —OH has four different groups attached, and is thus chiral. (c) The carbon bearing the —NH₂ group and the carbon bearing the CH₃ group are both chiral. **24.95** In both cases, stronger intermolecular forces lead to the higher boiling point. Ethanol contains O—H bonds, which form strong intermolecular hydrogen bonds, while dimethyl ether experiences only weak dipole–dipole and dispersion forces. The heavier and polar CH_2F_2 experiences dipole–dipole and stronger dispersion forces, while CH_4 experiences only weaker dispersion forces. **24.99** $\Delta G° = 13$ kJ

ANSWERS TO GIVE IT SOME THOUGHT

Chapter 1

page 5 (a) 100 (b) atoms **page 10** Water is composed of two types of atoms: hydrogen and oxygen. Hydrogen is composed only of hydrogen atoms, and oxygen is composed only of oxygen atoms. Therefore, hydrogen and oxygen are elements and water is a compound. **page 12** Density is an intensive property. Because it is measured per unit of volume, it is independent of how much of the material is present. **page 13** (a) Chemical change: Carbon dioxide and water are different compounds than sugar. (b) Physical change: Water in the gas phase becomes water in the solid phase (frost). (c) Physical change: Gold in the solid state becomes liquid and then resolidifies. **page 16** The candela, cd. **page 19** 10^3 **page 19** $2.5 \times 10^2 \ m^3$ is, because it has units of length to the third power. **page 22** (b) Mass of a penny **page 23** 154.9 ± 0.1 lbs. The average deviation is about 0.1 lb **page 25** the 18 m measurement needs to be made to greater accuracy, so that is also has three significant figures **page 28** Use all digits given in the conversion factor. Conversion factors may be exact and then have "infinite" significant digits (for example, 2.54 cm = 1 inch exactly). Usually, your answer will have its number of significant digits limited by those of the quantities given in the problem.

Chapter 2

page 43 (a) The law of multiple proportions. (b) The second compound must contain two oxygen atoms for each carbon atom (that is, twice as many carbon atoms as the first compound). **page 44** This observation shows that the cathode rays are comprised of the same entity present in all the elements used as cathode, not something characteristic of the individual elements. **page 47** Most α particles pass through the foil without being deflected because most of the volume of the atoms that comprise the foil is empty space. **page 48** (a) The atom has 15 electrons because atoms have equal numbers of electrons and protons. (b) The protons reside in the nucleus of the atom. **page 51** Any single atom of chromium must be one of the isotopes of that element. The isotope mentioned has a mass of 52.94 amu and is probably ^{53}Cr. The atomic weight differs from the mass of any particular atom because it is the *average* atomic mass of the naturally occurring isotopes of the element. **page 55** (a) Cl, (b) third period and group 7A, (c) 17, (d) nonmetal **page 56** B_2H_6 and $C_4H_2O_2$ can be only molecular formulas; their empirical formulas would be BH_3 and C_2HO. SO_2 and CH could be empirical formulas or molecular formulas. No formula can be *only* an empirical formula; there could always be a molecule of that composition **page 57** (a) C_2H_6, (b) CH_3, (c) Probably the ball-and-stick model because the angles between the sticks indicate the angles between the atoms **page 61** No. The formula does not contain any information regarding the nature of the bonds between the constituent atoms. **page 64** (a) Chromium exhibits differing charges in its compounds, so we need to identify what charge it has in this case. In contrast, calcium is always 2+ in its compounds (b) It tells us that the ammonium ion is a cation; that is, positively charged. **page 65** The endings convey the number of oxygen atoms bound to the nonmetallic element **page 65** Borate by extension of Nitrate $\longrightarrow$ Carbonate $\longrightarrow$ should be BO_3^{3-}. Silicate should be SiO_4^{4-}. **page 67** (a) $CaHCO_3$, (b) $KHSO_4$, LiH_2PO_4 **page 68** Iodic acid, by analogy to the relationship between the chlorate ion and chloric acid **page 69** No, it contains three different elements. **page 70**

Chapter 3

page 82 Each $Mg(OH)_2$ has 1 Mg, 2 O, and 2 H; thus, 3 $Mg(OH)_2$ represents 3 Mg, 6 O, and 6 H. **page 88** The product is an ionic compound involving Na^+ and S^{2-}, and its chemical formula is therefore Na_2S. **page 94** (a) A mole of glucose. By inspecting their chemical formulas we find that glucose has more atoms of H and O than water and in addition it also has C atoms. Thus, a molecule of glucose has a greater mass than a molecule of water. (b) They both contain the same number of molecules because a mole of each substance contains 6.02×10^{23} molecules. **page 99** No, chemical analysis cannot distinguish compounds that have different molecular formulas but the same empirical formula. **page 102** There are experimental uncertainties in the measurements. **page 103** 3.14 mol because $2 \ mol \ H_2 \cong 1 \ mol \ O_2$ based on the coefficients in the balanced equation **page 104** Two mol of H_2 are consumed for every mol of O_2 that reacts, so 3.14 mol H_2 would be consumed. **page 105** To obey the law of conservation of mass the mass of the products consumed must equal the mass of reactants formed, therefore the mass of H_2O produced is $1.00 \ g + 3.59 \ g - 3.03 \ g = 1.56 \ g \ H_2O$.

Chapter 4

page 126 (a) $K^+(aq)$ and $CN^-(aq)$ (b) $Na^+(aq)$ and $ClO_4^-(aq)$ **page 127** NaOH because it is the only solute that is a strong electrolyte **page 131** $Na^+(aq)$ and $NO_3^-(aq)$ **page 133** Three. Each COOH group will partially ionize in water to form $H^+(aq)$. **page 134** Only soluble metal hydroxides are classified as strong bases and $Al(OH)_3$ is insoluble. **page 138** $SO_2(g)$ **page 141** (a) -3, (b) $+5$ **page 144** (a) Yes, nickel is below zinc in the activity series so $Ni^{2+}(aq)$ will oxidize $Zn(s)$ to form $Ni(s)$ and $Zn^{2+}(aq)$. (b) No reaction will occur because the $Zn^{2+}(aq)$ ions cannot be further oxidized. **page 146** The second solution is more concentrated, 2.50 M, than the first solution, which has a concentration of 1.00 M. **page 150** The concentration is halved to 0.25 M.

Chapter 5

page 168 (a) No. The potential energy is lower at the bottom of the hill. (b) Once the bike comes to a stop, its kinetic energy is zero, just as it was at the top of the hill. **page 169** Open system. Humans exchange matter and energy with their surroundings. **page 173** Endothermic **page 175** The balance (current state) does not depend on the ways the money may have been transferred into the account or on the particular expenditures made in withdrawing money from the account. It depends only on the net total of all the transactions. **page 176** No. If ΔV is zero, then the expression $w = -P\Delta V$ is also zero. **page 177** ΔH is positive; the fact that the flask (part of the surroundings) gets cold means that the system is absorbing heat, meaning that q_P is positive. (See Figure 5.8.) Because the process occurs at constant pressure, $q_p = \Delta H$. **page 179** No. Because only half as much matter is involved, the value of ΔH would be $\frac{1}{2}(-483.6 \ kJ) = -241.8 \ kJ$. **page 182** Hg($l$). Rearranging

Equation 5.22 gives $\Delta T = \dfrac{q}{C_s \times m}$. When q and m are constant

for a series of substances, then $\Delta T = \dfrac{constant}{C_s}$. Therefore, the

element with the smallest C_s in Table 5.2 has the largest ΔT, Hg(l). **page 187** (a) The sign of ΔH changes. (b) The magnitude of ΔH doubles. **page 191** No. Because $O_3(g)$ is not the most stable form of oxygen at 25 °C, 1 atm $[O_2(g)$ is], ΔH_f° for $O_3(g)$ is not necessarily zero. In Appendix C we see that it is 142.3 kJ/mol. **page 196** Fats, because they have the largest fuel value of the three

Chapter 6

page 214 No. Both visible light and X-rays are forms of electromagnetic radiation. They therefore both travel at the speed of light, *c*. Their differing ability to penetrate skin is due to their different energies, which we will discuss in the next section.   **page 217** The notes on a piano go in "jumps"; for example, one can't play a note between B and C on a piano. In this analogy, a violin is continuous—in principle, one can play any note (such as a note halfway between B and C).   **page 218** No. The energy of the emitted electron will equal the energy of the photon minus the work function.   **page 219** Wave-like behavior.   **page 220** The energies in the Bohr model have only certain specific allowed values, much like the positions on the steps in Figure 6.6.   **page 221** A larger orbit means the electron is farther from the nucleus and therefore experiences less attraction. Thus, the energy is higher.   **page 222** ΔE is negative for an emission. The negative sign yields a positive number that corresponds to the energy of the emitted photon.   **page 222** This will be an absorption so $1/\lambda = \Delta E / hc$.   **page 224** Yes, all moving objects produce matter waves, but the wavelengths associated with macroscopic objects, such as the baseball, are too small to allow for any way of observing them.   **page 226** The small size and mass of subatomic particles. The term $h/4\pi$ in the uncertainty principle is a very small number that becomes important only when considering extremely small objects, such as electrons.   **page 227** In the first statement, we know exactly where the electron is. In the second statement, we are saying that we know the probability of the electron being at a point, but we don't know exactly where it is. The second statement is consistent with the Uncertainty Principle.   **page 228** Bohr proposed that the electron in the hydrogen atom moves in a well-defined circular path around the nucleus (an orbit). In the quantum-mechanical model, the motion of the electron is not well-defined and we must use a probabilistic description. An orbital is a wave function related to the probability of finding the electron at any point in space.   **page 230** The energy of an electron in the hydrogen atom is proportional to $-1/n^2$, as seen in Equation 6.5. The difference between $-1/(2)^2$ and $-1/(1)^2$ is much greater than the difference between $-1/(3)^2$ and $-1/(2)^2$.   **page 235** No. Based on what we have learned, all we know is that both the 4s and 3d orbitals are higher in energy than the 3s orbital. In most atoms, the 4s is actually lower in energy than the 3d, as shown in Figure 6.25.   **page 240** The 6s orbital, which starts to hold electrons at element 55, Cs.   **page 245** We can't conclude anything! Each of the three elements has a different valence electron configuration for its $(n-1)d$ and ns subshells: For Ni, $3d^8 4s^2$; for Pd, $4d^{10}$; and for Pt, $5d^9 6s^1$.

Chapter 7

page 259 Co and Ni and Te and I are other pairs of elements whose atomic weights are out of order compared to their atomic numbers.   **page 262** The 2p electron in a Ne atom would experience a larger Z_{eff} than the 3s electron in Na, due to the greater shielding of the 3s electron of the Na atom by all the 2s and 2p electrons.   **page 264** These trends work against each other: Z_{eff} increasing would imply that the valence electrons are pulled tighter in to make the atom smaller, while orbital size "increasing" would imply that atomic size would also increase. The orbital size effect is larger: As you go down a column in the periodic table, atomic size generally increases.   **page 268** It is harder to remove another electron from Na^+, so the process in Equation 7.3 would require more energy and, hence, shorter-wavelength light (see Sections 6.1 and 6.2).   **page 269** I_2 for a carbon atom corresponds to ionizing an electron from C^+, which has the same number of electrons as a neutral B atom. Z_{eff} will be greater for C^+ than for B, so I_2 for a carbon atom will be greater than I_1 for a boron atom.   **page 271** The same.   **page 273** The magnitudes are the same, but they have opposite signs.   **page 274** No. The oxidation state of As will be positive when combined with Cl, and negative when combined with Mg.   **page 277** Because the melting point is so low, we would expect a molecular rather than ionic compound. Thus, A is more likely to be P than Sc because PCl_3 is a compound of two nonmetals and is

therefore more likely to be molecular.   **page 280** Its low first ionization energy.   **page 282** In the acidic environment of the stomach, metal carbonates can react to give carbonic acid, which decomposes to water and carbon dioxide gas. Thus, calcium carbonate is much more soluble in acidic solution than it is in neutral water.   **page 284** The longest wavelength of visible light is about 750 nm (Section 6.1). We can assume that this corresponds to the lowest energy of light (since $E = hc/\lambda$) needed to break bonds in hydrogen peroxide. If we plug in 750 nm for λ, we can calculate the minimum energy to break one O—O bond in one molecule of hydrogen peroxide, in joules. If we multiple by Avogadro's number, we can calculate how many joules it would take to break a mole of O—O bonds in hydrogen peroxide (which is the number one normally finds).   **page 285** The halogens all have ground-state electron configurations that are $ns^2 np^5$; sharing an electron with only one other atom generally makes a stable compound.   **page 286** Based on the trends in the table, we might expect the radius to be about 1.5 Å, and the first ionization energy to be about 900 kJ/mol. In fact, its bonding radius is indeed 1.5 Å, and the experimental ionization energy is 920 kJ/mol.

Chapter 8

page 300 No. Cl has seven valence electrons. The first and second Lewis symbols are both correct—they both show seven valence electrons, and it doesn't matter which of the four sides has the single electron. The third symbol shows only five electrons and is incorrect.   **page 302** CaF_2 is an ionic compound consisting of Ca^{2+} and F^- ions. When Ca and F_2 react to form CaF_2, each Ca atom loses two electrons to form a Ca^{2+} ion and each fluorine atom in F_2 takes up an electron, forming two F^- ions. Thus, we can say that each Ca atom transfers one electron to each of two fluorine atoms.   **page 303** No. This reaction corresponds to the lattice energy for KCl, which is a large positive number. Therefore, the reaction will cost energy, not release energy.   **page 306** Rhodium, Rh   **page 307** Weaker. In both H_2 and H_2^+ the two H atoms are principally held together by the electrostatic attractions between the nuclei and the electron(s) concentrated between them. H_2^+ has only one electron between the nuclei whereas H_2 has two and this results in the H — H bond in H_2 being stronger.   **page 309** Triple bond. CO_2 has two C—O double bonds. Because the C—O bond in carbon monoxide is shorter, it is likely to be a triple bond.   **page 309** Electron affinity measures the energy released when an isolated atom gains an electron to form a 1− ion and has units of energy. Electronegativity has no units, and is the ability of the atom in a molecule to attract electrons to itself within that molecule.   **page 311** Polar covalent. The difference in electronegativity between S and O is 3.5 − 2.5 = 1.0. Based on the examples of F_2, HF, and LiF, the difference in electronegativity is great enough to introduce some polarity to the bond but not sufficient to cause a complete electron transfer from one atom to the other.   **page 312** IF. Because the difference in electronegativity between I and F is greater than that between Cl and F, the magnitude of Q should be greater for IF. In addition, because I has a larger atomic radius than Cl, the bond length in IF is longer than that in ClF. Thus, both Q and r are larger for IF and, therefore, $\mu = Qr$ will be larger for IF.   **page 313** Smaller dipole moment for C—H The magnitude of Q should be similar for C—H and H—I bonds because the difference in electronegativity for each bond is 0.4. The C—H bond length is 1.1 Å and the H—I bond length is 1.6 Å. Therefore $\mu = Qr$ will be greater for H—I because it has a longer bond (larger r).   **page 315** OsO_4. The data suggest that the yellow substance is a molecular species with its low melting and boiling points. Os in OsO_4 has an oxidation number of +8 and Cr in Cr_2O_3 has an oxidation number of +3. In Section 8.4, we learn that a compound with a metal in a high oxidation state should show a high degree of covalence and OsO_4 fits this situation.   **page 318** There is probably a better choice of Lewis structure than the one chosen. Because the formal charges must add up to 0 and the formal charge on the F atom is +1, there must be another atom that has a formal charge of −1. Because F is the most electronegative element, we don't expect it to carry a positive formal charge.   **page 320** Yes. There are two resonance structures for ozone that each contribute equally to the overall description of the molecule. Each O—O bond is therefore an average of a single bond and a double bond, which is a "one-and-a-half" bond. ***page 321*** As "one-and-a-third" bonds. There are three resonance structures, and each of the three N—O bonds is single in two of those structures and double in the third. Each bond in the actual ion is an average of these: $(1 + 1 + 2)/3 = 1\frac{1}{3}$. ***page 322*** No, it will not have multiple resonance structures. We can't "move" the double bonds, as we did in benzene, because the positions of the hydrogen atoms dictate specific positions for the double bonds. We can't write any other reasonable Lewis structures for the molecule. ***page 323*** The formal charge of each atom is shown here:

$$\overset{..}{\text{N}}=\overset{..}{\text{O}} \qquad \overset{..}{\text{N}}=\overset{..}{\text{O}}$$
F.C. 0   0   −1 +1

The first structure shows each atom with a zero formal charge and therefore it is the dominant Lewis structure. The second one shows a positive formal charge for an oxygen atom, which is a highly electronegative atom, and this is not a favorable situation. ***page 326*** The atomization of ethane produces $2\,\text{C}(g) + 6\,\text{H}(g)$. In this process, six C—H bonds and one C—C bond are broken. We can use $6D(\text{C—H})$ to estimate the amount of enthalpy needed to break the six C—H bonds. The difference between that number and the enthalpy of atomization is an estimate of the bond enthalpy of the C—C bond, $D(\text{C—C})$. ***page 327*** H_2O_2. From Table 8.4, the bond enthalpy of the O—O single bond in H_2O_2 (146 kJ/mol) is much lower than that of the O=O bond in O_2 (495 kJ/mol). The weaker bond in H_2O_2 is expected to make it more reactive than O_2.

Chapter 9

page 346 Removal of two opposite atoms from an octahedral arrangement would lead to square-planar shape. ***page 347*** No, the molecule does not satisfy the octet rule because there are ten electrons around the atom A. Each of the atoms B does satisfy the octet rule. There are four electron domains around A: Two single bonds, one double bond, and one nonbonding electron domain around A. ***page 349*** Each counts as a single electron domain around the central atom. ***page 352*** Yes. Because there are three equivalent dominant resonance structures, each of which puts the double bond between the N and a different O, the average structure has the same bond order for all three N—O bonds. Thus, each electron domain is the same and the angles are predicted to be 120°. ***page 352*** In a square planar arrangement of electron domains, each domain is 90° from two other domains (and 180° from the third domain). In a tetrahedral arrangement, each domain is 109.5° from the other three domains. Domains always try to minimize the number of 90° interactions, so the tetrahedral arrangement is favored. ***page 357*** No. Although the bond dipoles are in opposite directions, they will not have the same magnitude because the C—S bond is not as polar as the C—O bond. Thus, the sum of the two vectors will not equal zero and the molecule is polar. ***page 361*** They are both oriented perpendicular to the F—Be—F axis. ***page 362*** None. All of the 2p orbitals are used in constructing the sp^3 hybrid orbitals. ***page 367*** There are three electron domains about each N atom, so we expect sp^2 hybridization at each of the N atoms. The H—N—N angles should therefore be roughly 120°, and the molecule is not expected to be linear. In order for the π bond to form, all four atoms would have to be in the same plane. ***page 373*** sp hybridization. ***page 376*** The excited molecule would fall apart. It would have an electron configuration $\sigma_{1s}^1\sigma_{1s}^{*1}$ and therefore a bond order of 0. ***page 377*** Yes, it will have a bond order of $\frac{1}{2}$. ***page 383*** No. If the σ_{2p} MO were lower in energy than the π_{2p} MOs we would expect the last two electrons to go into the π_{2p} MOs with the same spin, which would cause C_2 to be paramagnetic.

Chapter 10

page 401 No. The heaviest gas in Table 10.1 is SO_2, whose molar mass, 64 g/mol, is less than half that of Xe, whose molar mass is

131 g/mol. ***page 402*** 1.4×10^3 lbs ***page 402*** (a) 745 mm Hg, (b) 0.980 atm, (c) 99.3 kPa, (d) 0.993 bar ***page 405*** It would be halved. ***page 406*** No because the absolute temperature is not halved, it only decreases from 373 K to 323 K. ***page 409*** 28.2 cm ***page 413*** Less dense because water has a smaller molar mass, 18 g/mol, than N_2, 28 g/mol. ***page 416*** The partial pressure of N_2 is not affected by the introduction of another gas, but the total pressure will increase. ***page 420*** Slowest HCl < O_2 < H_2 fastest. ***page 422*** $u_{rms}/u_{mp} = \sqrt{3/2}$. This ratio will not change as the temperature changes and it will be the same for all gases. ***page 426*** (a) Decrease, (b) have no effect. ***page 427*** (b) 100 K and 5 atm ***page 428*** The negative deviation is due to attractive intermolecular forces.

Chapter 11

page 446 $H_2O(g)$; during boiling, energy is provided to overcome intermolecular forces between H_2O molecules allowing the vapor to form. ***page 448*** CH_4 < CCl_4 < CBr_4. Because all three molecules are nonpolar, the strength of dispersion forces determines the relative boiling points. Polarizability increases in order of increasing molecular size and molecular weight, CH_4 < CCl_4 < CBr_4; hence, the dispersion forces and boiling points increase in the same order. ***page 452*** Mainly hydrogen bonds, which hold the individual H_2O molecules together in the liquid. ***page 452*** $Ca(NO_3)_2$ in water, because calcium nitrate is a strong electrolyte that forms ions and water is a polar molecule with a dipole moment. Ion–dipole forces cannot be present in a CH_3OH/H_2O mixture because CH_3OH does not form ions. ***page 456*** (a) Both viscosity and surface tension decrease with increasing temperature because of the increased molecular motion. (b) Both properties increase as the strength of intermolecular forces increases. ***page 459*** Melting (or fusion), endothermic ***page 461*** The intermolecular attractive forces in H_2O are much stronger than those in H_2S because H_2O can form hydrogen bonds, The stronger intermolecular forces results in a higher critical temperature and pressure. ***page 463*** CCl_4. Both compounds are nonpolar; therefore, only dispersion forces exist between the molecules. Because dispersion forces are stronger for the larger, heavier CBr_4, it has a lower vapor pressure than CCl_4. The substance with the larger vapor pressure at a given temperature is more volatile.

Chapter 12

page 484 Tetragonal. There are two three-dimensional lattices that have a square base with a third vector perpendicular to the base, tetragonal and cubic, but in a cubic lattice the a, b, and c lattice vectors are all of the same length. ***page 487*** Ionic solids are composed of ions. Oppositely-charged ions slipping past each other could create electrostatic repulsions, so ionic solids are brittle. ***page 491*** The packing efficiency decreases as the number of nearest neighbors decreases. The structures with the highest packing efficiency, hexagonal and cubic close packing, both have atoms with a coordination number of 12. Body-centered cubic packing, where the coordination number is 8, has a lower packing efficiency, and primitive cubic packing, where the coordination number is 6, has a lower packing efficiency still. ***page 492*** Interstitial, because boron is a small nonmetal atom that can fit in the voids between the larger palladium atoms ***page 498*** Gold, Au should have more electrons in antibonding orbitals. Tungsten, W, lies near the middle of the transition metal series where the bands arising from the d orbitals and the s orbital are approximately half-filled. This electron count should fill the bonding orbitals and leave the antibonding orbitals mostly empty. Because both elements have similar numbers of electrons in the bonding orbitals but tungsten has fewer electrons in antibonding orbitals, it will have a higher melting point. ***page 499*** No. In a crystal the lattice points must be identical. Therefore, if an atom lies on top of a lattice point, then the same type of atom must lie on all lattice points. In an ionic compound there are at least two different types of atoms, and only one can lie on the lattice points. ***page 501*** Four. The empirical formula of potassium oxide is K_2O. Rearranging Equation 12.1 we can determine the potassium coordination number to be anion coordination number × (number of anions per formula unit/number of cations per formula unit) = 8(1/2) = 4.   ***page 511*** A condensation polymer. The presence of both —COOH and —NH$_2$ groups allow molecules to react with one another forming C—N bonds and eliminating H$_2$O.   ***page 512*** As the vinyl acetate content increases more side chain branching occurs which inhibits the formation of crystalline regions thereby lowering the melting point. ***page 515*** No. The emitted photons have energies that are similar in energy to the band gap of the semiconductor. If the size of the crystals is reduced into the nanometer range, the band gap will increase. However, because 340-nm light falls in the UV region of the electromagnetic spectrum, increasing the energy of the band gap will only shift the light deeper into the UV.

Chapter 13

page 532 Entropy increases as the ink molecules disperse into water.   ***page 533*** The lattice energy of NaCl(*s*) must be overcome to separate Na$^+$ and Cl$^-$ ions and disperse them into a solvent. C$_6$H$_{14}$ is nonpolar. Interactions between ions and nonpolar molecules tend to be very weak. Thus, the energy required to separate the ions in NaCl is not recovered in the form of ion–C$_6$H$_{14}$ interactions. ***page 535*** (a) Separating solvent molecules from each other requires energy and is therefore endothermic. (b) Forming the solute–solvent interactions is exothermic.   ***page 537*** The added solute provides a template for the solid to begin to crystallize from solution, and a precipitate will form.   ***page 540*** The solubility in water would be considerably lower because there would no longer be hydrogen bonding with water, which promotes solubility.   ***page 543*** Dissolved gases become less soluble as temperature increases, and they come out of solution, forming bubbles below the boiling point of water.   ***page 545*** 230 ppm (1 ppm is 1 part in 10^6); 2.30 × 10^5 ppb (1 ppb is 1 part in 10^9).   ***page 546*** For dilute aqueous solutions the molality will be nearly equal to the molarity. Molality is the number of moles of solute per kilogram of solvent, whereas molarity is the number moles of solute per liter of solution. Because the solution is dilute, the mass of solvent is essentially equal to the mass of the solution. Furthermore, a dilute aqueous solution will have a density of 1.0 kg/L. Thus, the number of liters of solution and the number of kilograms of solvent will be essentially equal.   ***page 549*** The lowering of the vapor pressure depends on the total solute concentration (Equation 13.11). One mole of NaCl (a strong electrolyte) provides 2 mol of particles (1 mol of Na$^+$ and 1 mol Cl), whereas one mole of (a nonelectrolyte) provides only 1 mol particles.   ***page 552*** Not necessarily; if the solute is a strong or weak electrolyte, it could have a lower molality and still cause an increase of 0.51 °C. The total molality of all the particles in the solution is 1 m.   ***page 555*** The 0.20-m solution is hypotonic with respect to the 0.5-m solution. (A hypotonic solution will have a lower concentration and hence a lower osmotic pressure.)   ***page 557*** They would have the same osmotic pressure because they have the same concentration of particles. (Both are strong electrolytes that are 0.20 *M* in total ions.)   ***page 560*** No, hydrophobic groups would face outward to make contact with hydrophobic lipids.   ***page 561*** The smaller droplets carry negative charges because of the embedded stearate ions and thus repel one another.

Chapter 14

page 577 Increasing the partial pressure increases the number of collisions between molecules. For any reaction that depends on collisions (which is nearly all of them), we would expect the rate to increase with increasing partial pressure.   ***page 580*** You can see visually that the slope of the line connecting 0 s and 600 s is smaller than the slope at 0 s and larger than the slope at 600 s. The order from fastest to slowest is therefore (ii) > (i) > (iii).   ***page 583*** Reaction rate is the change in the concentration of a reactant (disappearance) or

product (appearance) over a time interval. The rate law governs how the reaction rate depends on the concentrations of the reactants. The rate constant is part of the rate law.   ***page 583*** Generally no. Rate always has units of *M/s*. The units of the rate constant depend on the specific rate law, as we shall see throughout this chapter.   ***page 584*** (a) Second order in [NO] and first order in [H$_2$]; (b) The rate increases more when we double [NO]. Doubling [NO] will increase the rate fourfold whereas doubling [H$_2$] will double the rate.   ***page 585*** No. The two reactions could have different rate laws. Only if they have the same rate law and the same value of *k* will they have the same rate.   ***page 592*** After 3 half-lives, the concentration will be 1/8 of its original value, so 1.25 g of the substance remains.   ***page 593*** As seen in Equation 14.15, the half-life of a first-order reaction is independent of initial concentration. By contrast, the half-life of a second-order reaction depends on initial concentration (Equation 14.17). ***page 596*** According to the figure, the energy barrier is lower in the forward direction than in the reverse. Thus, more molecules will have energy sufficient to cross the barrier in the forward direction. The forward rate will be greater   ***page 596*** No. If B can be isolated, it can't correspond to the top of the energy barrier. There would be transition states for each of the individual reactions above ***page 600*** Bimolecular.   ***page 603*** To determine the rate, we need to know the elementary reactions that add up to give the balanced equation   ***page 605*** The likelihood of three molecules colliding at exactly the same time is vanishingly small.   ***page 609*** A heterogeneous catalyst is in a different phase than the reactants and is therefore fairly easy to remove from the mixture. The removal of a homogeneous catalyst can be much more difficult as it exists in the same phase as the reactants.   ***page 611*** Yes. Like other catalysts, enzymes speed up reactions by lowering the activation energy, which is the energy of the transition state.

Chapter 15

page 632 (a) The rates of the forward and reverse reactions. (b) Greater than 1   ***page 632*** When the concentrations of reactants and products are no longer changing   ***page 635*** It does not depend on starting concentrations.   ***page 635*** Units of moles/L are used to calculate K_c; units of partial pressure are used to calculate K_p.   ***page 636*** Yes, $K_c = K_p$ when the number of moles of gaseous products and the number of moles of gaseous reactants are equal.   ***page 637*** 0.00140 ***page 639*** $K_c = 91$ ***page 639*** It is cubed.   ***page 642*** $K_p = P_{H_2O}$ ***page 644*** $K_c = [NH_4^+][OH^-]/[NH_3]$ ***page 652*** (a) to the right, (b) to the left   ***page 654*** (bottom) It will shift to the left, the side with a larger number of moles of gas.   ***page 655*** As the temperature increases, a larger fraction of molecules in the liquid phase have enough energy to overcome their inter molecular attractions and go into the vapor; the evaporation process is endothermic. Raising the temperature of an endothermic reaction shifts the equilibrium to the right, increasing the amount of gas present.   ***page 658*** No, the presence of a catalyst can accelerate the reaction but does not alter the value of K, which is what limits the amount of product produced.

Chapter 16

page 672 The H$^+$ ion for acids and the OH$^-$ ion for bases. ***page 674*** CH$_3$NH$_2$ is the base because it accepts a H$^+$ from H$_2$S as the reaction moves from the left-hand to the right-hand side of the equation.   ***page 677*** As the conjugate base of a strong acid, we would classify ClO$_4^-$ as having negligible basicity.   ***page 681*** pH is defined as −log[H$^+$]. This quantity will become negative if the H$^+$ concentration exceeds 1 *M*, which is possible. Such a solution would be highly acidic.   ***page 683*** pH = 14.00 − 3.00 = 11.00. This solution is basic because pH > 7.0.   ***page 685*** Both NaOH and Ba(OH)$_2$ are soluble hydroxides. Therefore, the hydroxide concentrations will be 0.001 *M* for NaOH and 0.002 *M* for Ba(OH)$_2$. Because the Ba(OH)$_2$ solution has a higher [OH$^-$], it is more basic and

has a higher pH.   **page 686** Because CH_3^- is the conjugate base of a substance that has negligible acidity, CH_3^- must be a strong base. Bases stronger than OH^- abstract H^+ from water molecules: $CH_3^- + H_2O \longrightarrow CH_4 + OH^-$.   **page 687** Oxygen.
page 691 $pH < 7$. We must consider the $[H^+]$ that is due to the autoionization of water. The additional $[H^+]$ from the very dilute acid solution will make the solution acidic.
page 694 $HPO_4^{2-}(aq) \rightleftharpoons H^+(aq) + PO_4^{3-}(aq)$
page 700 4.   **page 702** Nitrate is the conjugate base of nitric acid, HNO_3. The conjugate base of a strong acid does not act as a base, so NO_3^- ions will not affect the pH. Carbonate is the conjugate base of hydrogen carbonate, HCO_3^-, which is a weak acid. The conjugate base of a weak acid acts as a weak base, so CO_3^{2-} ions will increase the pH.   **page 708** $HBrO_3$. For an oxyacid, acidity increases as the electronegativity of the central ion increases, which would make $HBrO_2$ more acidic than HIO_2. Acidity also increases as the number of oxygens bound to the central atom increases, which would make $HBrO_3$ more acidic than $HBrO_2$. Combining these two relationships we can order these acids in terms of increasing acid-dissociation constant, $HIO_2 < HBrO_2 < HBrO_3$.   **page 709** The carboxyl group, —COOH.   **page 710** It must have an empty orbital that can interact with the lone pair on a Lewis base.

Chapter 17

page 729 (top) The Cl^- ion is the only spectator ion. The pH is determined by the equilibrium

$$NH_3(aq) + H_2O(l) \rightleftharpoons OH^-(aq) + NH_4^+(aq).$$

page 729 (bottom) HNO_3 and NO_3^-. To form a buffer we need comparable concentrations of a weak acid and its conjugate base. HNO_3 and NO_3^- will not form a buffer because HNO_3 is a strong acid and the NO_3^- ion is merely a spectator ion.   **page 731** (a) The OH^- of NaOH (a strong base) reacts with the acid member of the buffer (CH_3COOH), abstracting a proton. Thus, $[CH_3COOH]$ decreases and $[CH_3COO^-]$ increases. (b) The H^+ of HCl (a strong acid) reacts with the base member of the buffer $[CH_3COO^-]$. Thus, $[CH_3COO^-]$ decreases and $[CH_3COOH]$ increases.   **page 735** HClO would be more suitable for a pH = 7.0 buffer solution. To make a buffer we would also need a salt containing ClO^-, such as NaClO.   **page 740** The pH = 7. The neutralization of a strong base with a strong acid gives a salt solution at the equivalence point. The salt contains ions that do not change the pH of water.   **page 744** The conjugate base of the weak acid is the majority species in solution at the equivalence point, and this conjugate base reacts with water in a K_b reaction to produce OH^-. Therefore, the pH at the equivalence point for a weak acid/strong base titration is greater than 7.00.   **page 746** The nearly vertical portion of the titration curve at the equivalence point is smaller for a weak acid–strong base titration; as a result fewer indicators undergo their color change within this narrow range.   **page 746** The following titration curve shows the titration of 25 mL of Na_2CO_3 with HCl, both with 0.1 M concentrations. The overall reaction between the two is

$$Na_2CO_3(aq) + HCl(aq) \longrightarrow 2\,NaCl(aq) + CO_2(g) + H_2O(l)$$

The initial pH (sodium carbonate in water only) is near 11 because CO_3^{2-} is a weak base in water. The graph shows two equivalence points, **A** and **B**. The first point, **A**, is reached at a pH of about 9:

$$Na_2CO_3(aq) + HCl(aq) \longrightarrow NaCl(aq) + NaHCO_3(aq)$$

HCO_3^- is weakly basic in water and is a weaker base than the carbonate ion. The second point, **B**, is reached at a pH of about 4:

$$NaHCO_3(aq) + HCl(aq) \longrightarrow NaCl(aq) + CO_2(g) + H_2O(l)$$

H_2CO_3, a weak acid, forms and decomposes to carbon dioxide and water.

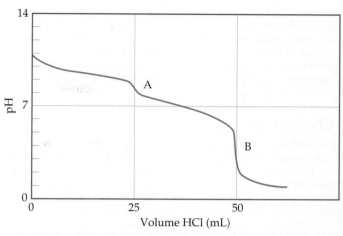

page 749 AgCl. Because all three compounds produce the same number of ions, their relative stabilities correspond directly to the K_{sp} values, with the compound with the largest K_{sp} value being the most soluble.   **page 759** Amphoteric substances are insoluble in water but dissolve in the presence of sufficient acid or base. Amphiprotic substances can both donate and accept protons.   **page 763** The solution must contain one or more of the cations in group 1 of the qualitative analysis scheme, Ag^+, Pb^{2+} or Hg_2^{2+}.

Chapter 18

page 777 We would expect that helium is relatively more abundant at the higher elevation, because Earth's gravitational field would exert a greater downward force on the heavier argon atoms. The effect, however, would be small. (new)   **page 780** In photoionization, a molecule loses an electron upon illumination with radiant energy. In photodissociation, illumination causes rupture of a chemical bond.   **page 781** Because those molecules do not absorb light at those wavelengths.   **page 783** Yes; Cl is neither a product nor a reactant in the overall reaction, and its presence does speed the reaction up.   **page 786** SO_2 in the atmosphere reacts with oxygen to form SO_3. SO_3 in the atmosphere reacts with water in the atmosphere to form H_2SO_4, sulfuric acid. The sulfuric acid dissolves in water droplets that fall to Earth, causing "acid rain" that has a pH of 4 or so.   **page 787** NO_2 photodissociates to NO and O; the O atoms react with O_2 in the atmosphere to form ozone, which is a key ingredient in photochemical smog.
page 790 Higher humidity means there is more water in the air. Water absorbs infrared light, which we feel as heat. After sundown, the ground that has been warmed earlier in the day reradiates heat out. In locations with higher humidity, this energy is absorbed somewhat by the water and in turn is reradiated to some extent back to the Earth, resulting in warmer temperatures compared to a low-humidity location.   **page 791** You can see from the phase diagram that to pass from the solid to the gaseous state we need to be below water's critical point. Therefore, to sublime water we need to be below 0.006 atm. A wide range of temperatures will work for sublimation at this pressure—the most environmentally relevant ones are $-50\,°C$ to $100\,°C$.   **page 795** The pollutants are capable of being oxidized (either directly by reaction with dissolved oxygen or indirectly by the action of organisms such as bacteria).   **page 799** With a catalyst, the reaction is always faster, therefore costing less energy to run. In addition, with a catalyst the reaction may occur readily at a lower temperature, also costing less energy.   **page 800** Fossil fuel combustion puts a great deal more CO_2 in the atmosphere right now than any supercritical use of CO_2. Compared to other (halogenated organic) solvents, supercritical CO_2 is far less toxic to life. In addition, the CO_2 used already exists, so we are not making more of it, simply putting some of what exists to an environmentally friendly use. Using CO_2 as a solvent or a reactant in industrial processes is a reasonable choice for environmental

sustainability. *page 801* Improve efficiency of synthesis of the starting reagent. Improve efficiency of isolating and recycling the acetone and phenol. Improve the activity and cost of production of the catalyst. Work toward using room temperature and room pressure; use water as a solvent if possible; use O_2 as the oxidizing agent instead of hydrogen peroxide if possible. *page 802* sp before reaction; sp^2 after reaction.

Chapter 19

page 815 No, nonspontaneous processes can occur so long as they receive some continuous outside assistance. Examples of nonspontaneous processes with which we may be familiar include the building of a brick wall and the electrolysis of water to form hydrogen gas and oxygen gas. *page 817* No. Just because the system is restored to its original condition doesn't mean that the surroundings have likewise been restored to their original condition, so it is not necessarily reversible. *page 819* ΔS depends not merely on q but on q_{rev}. Although there are many possible paths that could take a system from its initial to final state, there is always only one reversible isothermal path between two states. Thus, ΔS has only one particular value regardless of the path taken between states. *page 821* Because rusting is a spontaneous process, ΔS_{univ} must be positive. Therefore, the entropy of the surroundings must increase, and that increase must be larger than the entropy decrease of the system. *page 823* $S = 0$, based on Equation 19.5 and the fact that $\ln 1 = 0$. *page 824* No. Argon atoms are not attached to other atoms, so they cannot undergo vibrational motion. *page 827* It must be a perfect crystal at 0 K (third law of thermodynamics), which means it has only a single accessible microstate. *page 831* ΔS_{surr} always increases. For simplicity, assume that the process is isothermal. The change in entropy of the surroundings in an isothermal process is $\Delta S_{surr} = \dfrac{-q_{sys}}{T}$. Because the reaction is exothermic, $-q_{sys}$ is a positive number. Thus, ΔS_{surr} is a positive number and the entropy of the surroundings increases. *page 832* (a) In any spontaneous process the entropy of the universe increases. (b) In any spontaneous process operating at constant temperature, the free energy of the system decreases. *page 834* It indicates that the process to which the thermodynamic quantity refers has taken place under standard conditions, as summarized in Table 19.2. *page 837* Above the boiling point, vaporization is spontaneous, and $\Delta G < 0$. Therefore, $\Delta H - T\Delta S < 0$, and $\Delta H < T\Delta S$. Therefore $T\Delta S$ is greater in magnitude. *page 841* No. K is the ratio of product concentration to reactant concentration at equilibrium; K could be very small but is not zero.

Chapter 20

page 859 Oxygen is first assigned an oxidation number of -2. Nitrogen must then have a $+3$ oxidation number for the sum of oxidation numbers to equal -1, the charge of the ion. *page 862* No. Electrons should appear in the two half-reactions but cancel when the half-reactions are added properly. *page 869* Yes. A redox reaction with a positive standard cell potential is spontaneous under standard conditions. *page 871* 1 atm pressure of $Cl_2(g)$ and 1 M concentration of $Cl^-(aq)$ *page 873* Ni *page 877* Sn^{2+} *page 893* Al, Zn. Both are easier to oxidize than Fe.

Chapter 21

page 911 The mass number decreases by 4. *page 913* Only the neutron, as it is the only neutral particle listed. *page 918* From Figure 21.4 we can see that each of these four elements has only one stable isotope, and from their atomic numbers we see that they each have an odd number of protons. Given the rarity of stable isotopes with odd numbers of neutrons and protons, we expect that each isotope will possess an even number of neutrons. From their atomic weights we see that this is the case: F (10 neutrons), Na (12 neutrons), Al (14 neutrons), and P (16 neutrons). *page 920* No. Electric and magnetic fields are only effective at accelerating charged particles and a neutron is not charged. *page 923* It doubles as well. The number of disintegrations per second is proportional to the number of atoms or the radioactive isotope. *page 924* No; changing the mass would not change the half-life as shown in Equation 21.20. *page 926* Alpha radiation, which has the smallest relative penetrating power (Table 21.1). *page 927* Any process that depends on the mass of the molecule, such as the rate of gaseous effusion (Section 10.8) *page 931* (top) The values in Table 21.7 only reflect the mass of the nucleus, while the atomic mass is the sum of the mass of the nucleus and the electrons. So the atomic mass of iron-56 is 26 $\times$ m_e larger than the nuclear mass. *page 931* (bottom) No. Stable nuclei having mass numbers around 100 are the most stable nuclei. They could not form a still more stable nucleus with an accompanying release of energy. *page 941* The absorbed dose is equal to 0.10 J $\times$ $(1$ rad$/1 \times 10^{-2}$ J$) = 10$ rads. The effective dose is calculated by multiplying the absorbed dose by the relative biological effectiveness (RBE) factor, which is 10 for alpha radiation. Thus, the effective dosage is 100 rems.

Chapter 22

page 955 No. There is a triple bond in N_2. P does form triple bonds, as it would have to in order to form P_2. *page 957* H^-, hydride. *page 958* $+1$ for everything except H_2, for which the oxidation state of H is 0. *page 964* 0 for Cl_2; -1 for Cl^-; $+1$ for ClO^- *page 966* They should both be strong, since the central halogen is in the $+5$ oxidation state for both of them. We need to look up the redox potentials to see which ion, BrO_3^- or ClO_3^-, has the larger reduction potential. The ion with the larger reduction potential is the stronger oxidizing agent. BrO_3^- is the stronger oxidizing agent on this basis ($+1.52$ V standard reduction potential in acid compared to $+1.47$ V for ClO_3^-). *page 968* HIO_3 *page 972* $SO_3(g) + H_2O(l) \longrightarrow H_2SO_4(l)$ *page 976* (a) $+5$ (b) $+3$ *page 978* In PF_3, phosphorus in the $+3$ oxidation state. The reaction with water produces the oxyacid with phosphorus in that oxidation state, H_3PO_3. Thus, the reaction is analogous to Equation 22.53, which also shows a trihalide reacting with water. *page 982* $CO_2(g)$ *page 987* Silicon is the element, Si. Silica is SiO_2. Silicones are polymers that have an O—Si—O backbone and hydrocarbon groups on the Si. *page 988* $+3$

Chapter 23

page 999 Sc has the largest radius. *page 1001* Because titanium only has 4 valence electrons you would have to remove a core electron to create a Ti^{5+} ion. *page 1002* The larger the distance, the weaker the spin–spin interactions. *page 1002* Yes, it is a Lewis acid–base interaction; the metal ion is the Lewis acid (electron pair acceptor). *page 1007* The nonbonding electron pairs in H_2O are both located on the same atom, which makes it impossible for both pairs to be donated to the same metal atom. To act as a chelating agent the nonbonding electron pairs need to be on different atoms that are not connected to each other. *page 1009* Bidentate. *page 1011* The porphine ligand contains a conjugated system of pi bonds (alternating single and double bonds), which allows it to absorb visible light. *page 1015* No, ammonia cannot engage in linkage isomerism—the only atom that can coordinate to a metal is the nitrogen. *page 1018* Both isomers have the same chemical formulas and the same donor atoms on the ligands bonding to the metal ion. The difference is that the d isomer has a right-handed "twist" and the l isomer has a "left-handed" twist. *page 1021* (a) Co is $[Ar] 4s^2 3d^7$. (b) Co^{3+} is $[Ar] 3d^6$. The Co atom has 3 unpaired electrons and the Co^{3+} ion has 4 unpaired electrons, assuming all five d orbitals have the same energy. *page 1023* Because Ti(IV) ions have an $[Ar] 3d^0$ electron configuration there are no d-d transitions that can absorb photons of visible light. *page 1025* d^4, d^5, d^6, d^7. *page 1027* Because the ligands are located in the xy plane, electrons in a d_{xy} orbital, which has its lobes in the xy plane, feel more repulsion than electrons in the d_{xz} and d_{yz} orbitals.

Chapter 24

page 1043 C═N, because it is a polar double bond. C—H and C—C bonds are relatively unreactive. (The CN double bond does not have to fully break to be reactive.)   ***page 1045*** Two C—H bonds and two C—C bonds   ***page 1046*** The isomers have different properties, as seen in Table 24.3 (e.g., different melting points and different boiling points).   ***page 1049*** Yes, since cyclopropane is more strained.   ***page 1051*** Only two of the four possible C═C bond sites are distinctly different in the linear chain of five carbon atoms with one double bond.   ***page 1058***

page 1063 Reduction, which is the reverse of oxidation.
page 1063 $CH_3CH_2CH_2CH_2COOH$ ***page 1064*** The generic formula of an ether is ROR′, while an ester is RCOOR′. If you carbonylate or oxidize an ether, you can get an ester.   ***page 1067*** All four groups must be different from one another.   ***page 1071*** No. Breaking the hydrogen bonds between N—H and O═C groups in a protein by heating causes the α-helix structure to unwind and the β-sheet structure to separate.   ***page 1076*** The alpha form of the C—O—C linkage. Glycogen serves as a source of energy in the body, which means that the body's enzymes must be able to hydrolyze it to sugars. The enzymes work only on polysaccharides having the α linkage.

ANSWERS TO GO FIGURE

Chapter 1

Figure 1.1 Aspirin. It contains 9 carbon atoms. **Figure 1.4** Vapor (gas) **Figure 1.5** Molecules of a compound are composed of more than one type of atom, and molecules of an element are composed of only one type of atom. **Figure 1.6** Earth is rich in silicon and aluminum; the human body is rich in carbon and hydrogen **Figure 1.7** The relative volumes are in direct proportion t to the number of molecules **Figure 1.14** The separations are due to physical processes of adsorption of the materials onto the column **Figure 1.17** True **Figure 1.18** 1000 **Figure 1.22** The darts would be scattered widely (poor precision) but their average position would be at the center (good accuracy).

Chapter 2

Figure 2.3 We know the rays travel from the cathode because the rays are deflected by the magnetic field as though coming from the left. **Figure 2.4** The electron beam would be deflected downward because of repulsion by the upper negative plate and attraction toward the positive plate. **Figure 2.5** No, the electrons are of negligible mass compared with an oil drop. **Figure 2.7** The beta rays, whose path is diverted away from the negative plate and toward the positive plate, consist of electrons. Because the electrons are much less massive than the alpha particles, their motion is affected more strongly by the electric field. **Figure 2.9** The beam consists of alpha particles, which carry a +2 charge. **Figure 2.10** 10^{-2} pm **Figure 2.13** Based on the periodic trend, we expect that elements that precede a nonreactive gas, as F does, will also be reactive nonmetals. The elements fitting this pattern are H and Cl. **Figure 2.15** The metals are in the form of solid blocks, or relatively large pieces, as opposed to powders. They have a metallic sheen that is missing from the nonmetals. The nonmetals more readily form powders, and are more varied in color and consistency than metals. **Figure 2.17** The ball-and-stick model. **Figure 2.18** The elements are in the following groups: Ag^+ is 1B, Zn^{2+} is 2B, and Sc^{3+} is 3B. Sc^{3+} has the same number of electrons as Ar (element 18). **Figure 2.21** It is a difference in a physical property, color. **Figure 2.22** Removing one O atom from the perbromate ion gives the bromate ion, BrO_3^-.

Chapter 3

Figure 3.4 There are two CH_4 and four O_2 molecules on the reactant side, which contain 2 C atoms, 8 H atoms and 8 O atoms. The number of each type of atom remains the same on the product side as it must. **Figure 3.8** The flame gives off heat and therefore the reaction must release heat. **Figure 3.9** As shown, 18.0 g H_2O = 1 mol H_2O = 6.02 × 10^{23} molecules H_2O. Thus, 9.00 g H_2O = 0.500 mol H_2O = 3.01 × 10^{23} molecules H_2O. **Figure 3.12** (a) The molar mass of CH_4, 16.0 g CH_4/1 mol CH_4. (b) Avogadro's number, 1 mol CH_4/6.02 × 10^{23} formula units CH_4, where a formula unit in this case is a molecule. **Figure 3.17** If the amount of H_2 is doubled then O_2 becomes the limiting reactant. In that case (7 mol O_2) × (2 mol H_2O/1 mol O_2) = 14 mol H_2O would be produced.

Chapter 4

Figure 4.3 NaCl(*aq*) **Figure 4.4** K^+ and NO_3^- **Figure 4.9** Two moles of hydrochloric acid are needed to react with each mole of $Mg(OH)_2$. **Figure 4.12** Two. Each O atom becomes an O^{2-} ion. **Figure 4.13** One, based on the reaction stoichiometry. **Figure 4.14** Because the Cu(II) ion produces a blue color in aqueous solution. **Figure 4.18** The volume needed to reach the end point if $Ba(OH)_2(aq)$ were used would be one-half the volume needed for titration with NaOH(*aq*), because there are two hydroxide ions for every barium ion.

Chapter 5

Figure 5.1 In the act of throwing, the pitcher transfers energy to the ball, which then becomes kinetic energy of the ball. For a given amount of energy E transferred to the ball, Equation 5.1 tells us that the speed of the ball is $v = \sqrt{2E/m}$ where m is the mass of the ball. Because a baseball has less mass than a bowling bowl, it will have a higher speed for a given amount of energy transferred. **Figure 5.2** When she starts going uphill, kinetic energy is converted to potential energy and her speed decreases. **Figure 5.3** The electrostatic potential energy of two oppositely charged particles is negative (Equation 5.2). As the particles become closer, the electrostatic potential energy becomes even more negative—that is, it decreases. **Figure 5.4** Yes, the system is still closed—matter can't escape the system to the surroundings unless the piston is pulled completely out of the cylinder. **Figure 5.5** If $E_{final} = E_{initial}$, then $\Delta E = 0$. **Figure 5.6**

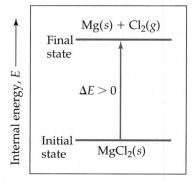

Figure 5.7 $\Delta E = 50 \text{ J} + (-85 \text{ J}) = -35 \text{ J}$ **Figure 5.10** The battery is doing work on the surroundings, so $w < 0$. **Figure 5.11** We need to know whether Zn(*s*) or HCl(*aq*) is the limiting reagent of the reaction. If it is Zn(*s*), then the addition of more Zn will lead to the generation of more $H_2(g)$ and more work will be done. **Figure 5.16** Endothermic—heat is being added to the system to raise the temperature of the water. **Figure 5.17** Two cups provide more thermal insulation so less heat will escape the system. **Figure 5.18** The stirrer ensures that all of the water in the bomb is at the same temperature. **Figure 5.20** The condensation of 2 $H_2O(g)$ to 2 $H_2O(l)$ **Figure 5.21** Yes, ΔH_3 would remain the same as it is the enthalpy change for the process $CO(g) + \frac{1}{2} O_2(g) \longrightarrow CO_2(g)$. **Figure 5.23** Grams of fat

Chapter 6

Figure 6.3 Wavelength = 1.0 m, frequency = 3.0 × 10^8 cycles/s. **Figure 6.4** Longer by 3 to 5 orders of magnitude (depending on what part of the microwave spectrum is considered). **Figure 6.5** The hottest area is the white or yellowish white area in the center. **Figure 6.7** The energy comes from the light shining on the surface. **Figure 6.12** The $n = 2$ to $n = 1$ transition involves a larger energy change than the $n = 3$ to $n = 2$ transition. (Compare the space differences between the states in the figure.) If the $n = 3$ to $n = 2$ transition produces visible light, the $n = 2$ to $n = 1$ transition must produce radiation of greater energy. The infrared radiation has lower frequency and, hence, lower energy than visible light, whereas ultraviolet radiation has higher frequency and greater energy. Thus, the $n = 2$ to $n = 1$ transition is more likely to produce ultraviolet radiation. **Figure 6.13** The $n = 4$ to $n = 3$ transition involves a smaller energy difference and will therefore emit light of longer wavelength. **Figure 6.17** The region of highest electron density is where the density of dots is highest, which is near the nucleus.

Figure 6.18 The fourth shell ($n = 4$) would contain four subshells, labeled 4s, 4p, 4d, and 4f.   **Figure 6.19** There would be four maxima and three nodes.   **Figure 6.23** (a) The intensity of the color indicates that the probability of finding the electron is greater at the interior of the lobes than on the edges. (b) $2p_x$.   **Figure 6.24** The d_{z^2} orbital has two large lobes that look like those of a p orbital, but also has a "doughnut" around the middle.   **Figure 6.25** The 4d and 4f subshells are not shown.   **Figure 6.26** In this pictorial view of spin, there are only two directions in which the electron can spin   **Figure 6.31** Osmium

Chapter 7

Figure 7.1 These three metals do not readily react with other elements, especially oxygen, so they are often found in nature in the elemental form as metals (such as gold nuggets)   **Figure 7.4** Yes, because of the peak near the nucleus in the 2s curve there is a probability, albeit small, that a 2s electron will be closer to the nucleus than a 1s electron.   **Figure 7.7** Bottom and left   **Figure 7.8** They get larger, just like the atoms do.   **Figure 7.10** 900 kJ/mol   **Figure 7.11** There is more electron–electron repulsion in the case of oxygen because two electrons have to occupy the same orbital.   **Figure 7.12** The added electron for the Group 4A elements leads to a half-filled np^3 configuration. For the Group 5A elements, the added electron leads to an np^4 configuration, so the electron must be added to an orbital that already has one electron in it.   **Figure 7.13** They are opposite: As ionization energy increases, metallic character decreases, and vice versa.   **Figure 7.15** Anions are above and to the right of the line; cations are below and to the left of the line.   **Figure 7.16** No. The Na^+ and NO_3^- ions will simply be spectator ions. The H^+ ions of an acid are needed in order to dissolve NiO.   **Figure 7.17** No. As seen in the photo, sulfur crumbles as it is hit with a hammer, typical of a solid nonmetal.   **Figure 7.21** Because Rb is below K in the periodic table, we expect Rb to be more reactive with water than K.   **Figure 7.23** Lilac (see Figure 7.22).   **Figure 7.25** The bubbles are due to $H_2(g)$. This could be confirmed by carefully testing the bubbles with a flame—there should be popping as the hydrogen gas ignites.   **Figure 7.26** Water does not decompose on sitting the way that hydrogen peroxide does.   **Figure 7.27** A regular octagon.   **Figure 7.28** I_2 is a solid whereas Cl_2 is a gas. Molecules are more closely packed together in a solid than they are in a gas, as will be discussed in detail in Chapter 11.

Chapter 8

Figure 8.1 We would draw the chemical structure of sugar molecules (which have no charges and have covalent bonding between atoms in each molecule) and indicate weak intermolecular forces (especially hydrogen bonding) between sugar molecules.   **Figure 8.2** Yes, the same sort of reaction should occur between any of the alkali metals and any of the elemental halogens.   **Figure 8.3** Cations have a smaller radius than their neutral atoms and anions have a larger radius. Because Na and Cl are in the same row of the periodic table, we would expect Na^+ to have a smaller radius than Cl^-, so we would infer that the larger green spheres represent the chloride ions and the smaller purple spheres represent the sodium ions.   **Figure 8.4** The distance between ions in KF should be larger than that in NaF and smaller than that in KCl. We would thus expect the lattice energy of KF to be between 701 and 910 kJ/mol.   **Figure 8.6** The repulsions between the nuclei would decrease, the attractions between the nuclei and the electrons would decrease, and the repulsions between the electrons would be unaffected.   **Figure 8.7** The electronegativity decreases with increasing atomic number.   **Figure 8.9** μ will decrease   **Figure 8.10** The bonds are not polar enough to cause enough excess electron density on the halogen atom to lead to a red shading.   **Figure 8.12** The lengths of the bonds of the outer O atoms to the inner O atom are the same.   **Figure 8.13** Yes. The electron densities on the left and right parts of the molecule are the same, indicating that resonance has made the two O — O bonds equivalent to one another.   **Figure 8.14** The dashed bonds represent the "half bonds" that result when the two resonance structures are averaged.   **Figure 8.15** Exothermic   **Figure 8.17** It should be halfway between the values for single and double bonds, which we can estimate to be about 280 kJ/mol from the graph.

Chapter 9

Figure 9.1 The radii of the atoms involved (see Section 7.3).   **Figure 9.3** Octahedral.   **Figure 9.4** No. We will get the same bent-shaped geometry regardless of which two atoms we remove.   **Figure 9.7** The electron pair in the bonding domain is attracted to two nuclei, whereas the nonbonding pair is attracted to only one nucleus.   **Figure 9.8** 90°.   **Figure 9.9** The C — O — H bond involving the right O because of the greater repulsions due to the nonbonding electron domains. The angle should be less than the ideal value of 109.5°.   **Figure 9.10** Zero. Because they are equal in magnitude but opposite in sign, the vectors cancel upon addition.   **Figure 9.13** The Cl 3p orbital extends farther in space than does the H 1s orbital. Thus, the Cl 3p orbital can achieve an effective overlap with a H 1s orbital at a longer distance.   **Figure 9.14** At very short internuclear distances the repulsion between the nuclei causes the potential energy to rise rapidly.   **Figure 9.16** The hybrid orbitals have to overlap with the 2p orbitals of the F atoms. The large lobes of the hybrid orbitals will lead to a much more effective overlap.   **Figure 9.17** 120°.   **Figure 9.18** The p_z orbital.   **Figure 9.19** No. All four hybrids are equivalent and the angles between them are all the same, so we can use any of the two to hold the nonbonding pairs.   **Figure 9.20** Because the P 3p orbitals are larger than the N 2p orbitals, we would expect somewhat larger lobes on the hybrid orbitals in the right-most drawing. Other than that, the molecules are entirely analogous.   **Figure 9.23** They have to lie in the same plane in order to allow the overlap of the π orbitals to be effective in forming the π bond.   **Figure 9.24** Acetylene should have the higher carbon-carbon bond energy because it has a triple bond as compared to the double bond in ethylene.   **Figure 9.26** It has six C — C σ and six C — H σ bonds.   **Figure 9.32** Zero. A node, by definition, is the place where the value of the wave function is zero.   **Figure 9.33** The energy of σ_{1s} would rise (but would still be below the energy of the H 1s atomic orbitals).   **Figure 9.34** The two electrons in the σ_{1s} MO.   **Figure 9.35** σ_{1s}^* and σ_{2s}^*.   **Figure 9.36** The end-on overlap in the σ_{2p} MO is greater than the sideways overlap in the π_{2p}.   **Figure 9.42** The σ_{2p} and π_{2p} MOs have switched order.   **Figure 9.43** F_2 contains four more electrons than N_2. These electrons go into the antibonding π_{2p}^* orbitals, thus lowering the bond order.   **Figure 9.45** N_2 is diamagnetic so it would not be attracted to the magnetic field. The liquid nitrogen would simply pour through the poles of the magnet without "sticking."   **Figure 9.46** 11. All the electrons in the $n = 2$ level are valence-shell electrons.

Chapter 10

Figure 10.2 It will increase.   **Figure 10.4** Decreases   **Figure 10.5** 1520 torr or 2 atm   **Figure 10.6** Linear   **Figure 10.9** one   **Figure 10.10** Chlorine, Cl_2.   **Figure 10.13** About one-sixth   **Figure 10.14** O_2 has the largest molar mass, 32 g/mol, and H_2 has the smallest, 2.0 g/mol.   **Figure 10.16** n, moles of gas   **Figure 10.20** True   **Figure 10.22** It would increase.

Chapter 11

Figure 11.2 The density in a liquid is much closer to a solid than it is to a gas.   **Figure 11.3** The distance within the molecule (the covalent bond distance) represented by the solid black line is smaller than the intermolecular distance represented by the red dotted line.   **Figure 11.5** The halogens are diatomic molecules and have much greater size and mass, and therefore greater polarizability, than the noble gases, which are monatomic.   **Figure 11.8** They stay roughly the same because the molecules have roughly the same molecular weights. Thus, the change in boiling point moving left to right is due mainly to the increasing dipole-dipole attractions.

Figure 11.9 Both compounds are nonpolar and incapable of forming hydrogen bonds. Therefore, the boiling point is determined by the dispersion forces, which are stronger for the larger, heavier SnH_4. **Figure 11.10** The non-hydrogen atom must possess a nonbonding electron pair.   **Figure 11.11** There are four electron pairs surrounding oxygen in a water molecule. Two of the electron pairs are used to make covalent bonds to hydrogen within the H_2O molecule, while the other two are available to make hydrogen bonds to neighboring molecules. Because the electron-pair geometry is tetrahedral (four electron domains around the central atom), the $H — O \cdots H$ bond angle is approximately 109°.   **Figure 11.13** The O atom is the negative end of the polar H_2O molecule; the negative end of the dipole is attracted to the positive ion.   **Figure 11.14** There are no ions present, but when we ask, "Are polar molecules present?" we make a distinction between the two molecules because SiH_4 is nonpolar and SiH_2Br_2 is polar.   **Figure 11.19** Wax is a hydrocarbon that cannot form hydrogen bonds. Therefore, coating the inside of tube with wax will dramatically decrease the adhesive forces between water and the tube and change the shape of the water meniscus to an inverted U-shape. Neither wax nor glass can form metallic bonds with mercury so the shape of the mercury meniscus will be qualitatively the same, an inverted U-shape.   **Figure 11.20** Because energy is a state function, the energy to convert a gas to a solid is the same regardless of whether the process occurs in one or two steps. Thus, the energy of deposition equals the energy of condensation plus the energy of freezing.   **Figure 11.21** Because we are dealing with a state function, the energy of going straight from a solid to a gas must be the same as going from a solid to a gas through an intermediate liquid state. Therefore, the heat of sublimation must be equal to the sum of the heat of fusion and the heat of vaporization: $\Delta H_{sub} = \Delta H_{fus} + \Delta H_{vap}$.   **Figure 11.22** The temperature of the liquid water is increasing.   **Figure 11.24** Increases, because the molecules have more kinetic energy as the temperature increases and can escape more easily   **Figure 11.25** All liquids including ethylene glycol reach their normal boiling point when their vapor pressure is equal to atmospheric pressure, 760 torr.   **Figure 11.27** It must be lower than the temperature at the triple point.

Chapter 12

Figure 12.5 There is not a centered square lattice, because if you tile squares and put lattice points on the corners and the center of each square it would be possible to draw a smaller square (rotated by 45°) that only has lattice points on the corners. Hence a "centered square lattice" would be indistinguishable from a primitive square lattice with a smaller unit cell.   **Figure 12.12** Face-centered cubic, assuming similar size spheres and cell edge lengths, since there are more atoms per volume for this unit cell compared to the other two.   **Figure 12.13** A hexagonal lattice   **Figure 12.15** The solvent is the majority component and the solute the minority component. Therefore, there will be more solvent atoms than solute atoms.   **Figure 12.17** The samarium atoms sit on the corners of the unit cell so there is only $8 \times (1/8) = 1$ Sm atom per unit cell. Eight of the nine cobalt atoms sit on faces of the unit cell, and the other sits in the middle of the unit cell so there are $8 \times (1/2) + 1 = 5$ Co atoms per unit cell.   **Figure 12.19** P_4, S_8, and Cl_2 are all molecules, because they have strong chemical bonds between atoms and have well-defined numbers of atoms per molecule.   **Figure 12.21** In the fourth period, vanadium and chromium have very similar melting points. Molybdenum and tungsten have the highest melting points in the fifth and sixth periods, respectively. All of these elements are located near the middle of the period where the bonding orbitals are mostly filled and the antibonding orbitals mostly empty.   **Figure 12.22** The molecular orbitals become more closely spaced in energy.   **Figure 12.23** Potassium has only one valence electron per atom $(4s^1)$. Therefore we expect the $4s$ band to be approximately half full. If we fill the $4s$ band halfway a small amount of electron density might leak over and start to fill the $3d$ orbitals as well. The $4p$ orbitals should be empty.   **Figure 12.24** Ionic substances cleave because the nearest neighbor interactions switch from attractive to repulsive if the atoms slide so that ions of like

charge (cation–cation and anion–anion) touch each other. Metals don't cleave because the atoms are attracted to all other atoms in the crystal through metallic bonding.   **Figure 12.25** No, ions of like charge do not touch in an ionic compound because they are repelled from one another. In an ionic compound the cations touch the anions.   **Figure 12.27** In NaF there are four Na^+ ions $(12 \times 1/4)$ and four F^- ions $(8 \times 1/8 + 6 \times 1/2)$ per unit cell. In MgF_2 there are two Mg^{2+} ions $(8 \times 1/8 + 1)$ and four F^- ions $(4 \times 1/2 + 2)$ per unit cell. In ScF_3 there is one Sc^{3+} ion $(8 \times 1/8)$ and three F^- ions $(12 \times 1/4)$ per unit cell.   **Figure 12.28** The intermolecular forces are stronger in toluene, as shown by its higher boiling point. The molecules pack more efficiently in benzene, which explains its higher melting point, even though the intermolecular forces are weaker.   **Figure 12.30** The band gap for an insulator would be larger than the one for a semiconductor.   **Figure 12.31** If you doubled the amount of doping in panel (b), the amount of blue shading the conduction band would also double.   **Figure 12.44** Decrease. As the quantum dots get smaller, the band gap increases and the emitted light shifts to shorter wavelength.   **Figure 12.47** Each carbon atom in C_{60} is bonded to three neighboring carbon atoms through covalent bonds. Thus, the bonding is more like graphite, where carbon atoms also bond to three neighbors, than diamond, where carbon atoms bond to four neighbors.

Chapter 13

Figure 13.1 Gas molecules move in constant random motion.   **Figure 13.2** Opposite charges attract. The electron-rich O atom of the H_2O molecule, which is the negative end of the dipole, is attracted to the positive Na^+ ion.   **Figure 13.3** The negative end of the water dipole (the O) is attracted to the positive Na^+ ion, whereas the positive end of the dipole (the H) is attracted to the negative Cl^- ion.   **Figure 13.4** For exothermic solution processes the magnitude of ΔH_{mix} will be larger than the magnitude of $\Delta H_{solute} + \Delta H_{solvent}$ **Figure 13.6** 237.6 g/mol, which includes the waters of hydration.   **Figure 13.7** If the solution wasn't supersaturated, solute would not crystallize from it.   **Figure 13.12** If the partial pressure of a gas over a solution is doubled, the concentration of gas in the solution would double.   **Figure 13.13** The slopes increase as the molecular weight increases. The larger the molecular weight, the greater the polarizability of the gas molecules, leading to greater intermolecular attractive forces between gas molecules and water molecules.   **Figure 13.15** Looking at where the solubility curves for KCl and NaCl intersect the 80 °C line, we see that the solubility of KCl is about 51 g/100 g H_2O, whereas NaCl has a solubility of about 39 g/100 g H_2O. Thus, KCl is more soluble than NaCl at this temperature.   **Figure 13.16** N_2 has the same molecular weight as CO but is nonpolar, so we can predict that its curve will be just below that of CO.   **Figure 13.22** The water will move through the semipermeable membrane toward the more concentrated solution. Thus, the liquid level in the left arm will increase.   **Figure 13.23** Water will move toward the more concentrated solute solution, which is inside the red blood cells, causing them to undergo hemolysis.   **Figure 13.26** The negatively charged groups both have the composition $—CO_2^-$.   **Figure 13.28** Recall the rule that likes dissolve likes. The oil drop is composed of nonpolar molecules, which interact with the nonpolar part of the stearate ion via dispersion forces.

Chapter 14

Figure 14.2 No. The surface area of a steel nail is much smaller than that for the same mass of steel wool, so the reaction with O_2 would not be as vigorous. Depending on how hot it is, it might not burn at all.   **Figure 14.3** Our first guess might be half way between the values at 20 s and 40 s, namely 0.41 mol A. However, we also see that the change in the number of moles of A between 0 s and 20 s is greater than that between 20 s and 40 s—in other words, the rate of conversion gets smaller as the amount of A decreases. So we would guess that the change from 20 s to 30 s is greater than the change from 30 s to 40 s, and we would estimate that the number of moles of A is between 0.41 and 0.30 mol.   **Figure 14.4** The instantaneous rate decreases as the reaction proceeds. *Figure 14.8* The reaction is first order. *Figure 14.10* At the beginning of the reaction when both plots are linear or nearly so. *Figure 14.13* The reaction that causes the light occurs more slowly at lower temperatures than at higher ones. *Figure 14.14* No, it will not turn down. The rate constant increases monotonically with increasing temperature because the kinetic energy of the colliding molecules continues to increase. *Figure 14.16* He would not need to hit the ball as hard; that is, less kinetic energy would be required if the barrier were lower. *Figure 14.17* As shown, the magnitude of energy needed to overcome the energy barrier is greater than the magnitude of the energy change in the reaction. *Figure 14.18* It would be more spread out, the maximum of the curve would be lower and to the right of the maximum of the red curve, and a greater fraction of molecules would have kinetic energy greater than E_a than for the red curve. *Figure 14.20* As shown, it is easier to convert to products because the activation energy is lower in that direction. *Figure 14.21* It can't be determined. If the amount of backup at the two toll plazas is roughly the same in the two scenarios, the time to get from point 1 to point 3 will be about the same. *Figure 14.22* The color is characteristic of molecular bromine, Br_2, which is present in appreciable quantities only in the middle photo. *Figure 14.23* The intermediate is at the valley in the middle of the blue curve. The transition states are the peaks, one for the red curve and two for the blue curve. *Figure 14.26* Grinding increases the surface area, exposing more of the catalase to react with the hydrogen peroxide. *Figure 14.27* Substrates must be held more tightly so that they can undergo the desired reaction. Products are released from the active site.

Chapter 15

Figure 15.1 The same because once the system reaches equilibrium the concentrations of NO_2 and N_2O_4 stop changing. *Figure 15.2* Greater than 1 *Figure 15.6* The boxes would be approximately the same size. *Figure 15.7* It would decrease. To reestablish equilibrium the concentration of $CO_2(g)$ would need to return to its previous value. The only way to do that would be for more $CaCO_3$ to decompose to produce enough $CO_2(g)$ to replace what was lost. *Figure 15.9* High pressure and low temperature, 500 atm and 400 °C in this figure. *Figure 15.10* Nitrogen must react with some of the added hydrogen to create ammonia and restore equilibrium. *Figure 15.14* (a) the energy difference between the initial state and the transition state. *Figure 15.15* About 5×10^{-4}

Chapter 16

Figure 16.2 Hydrogen bonds. *Figure 16.4* $O^{2-}(aq) + H_2O(l) \longrightarrow 2\,OH^-(aq)$. *Figure 16.5* Basic. The mixture of the two solutions will still have $[H^+] < [OH^-]$. *Figure 16.6* Lemon juice. It has a pH of about 2 whereas black coffee has a pH of about 5. The lower the pH, the more acidic the solution. *Figure 16.8* Phenolphthalein changes from colorless, for pH values less than 8, to pink for pH values greater than 10. A pink color indicates pH > 10. *Figure 16.9* Bromothymol blue would be most suitable because it changes pH over a range that brackets pH = 7. Methyl red is not sensitive to pH changes when pH > 6, while phenolphthalein is not sensitive to pH changes when pH < 8, so neither changes color at pH = 7. *Figure 16.12* Yes. The equilibrium of interest is $H_3CCOOH \rightleftharpoons H^+ + H_3CCOO^-$. If the percent dissociation remained constant as the acid concentration increased, the concentration of all three species would increase at the same rate. However, because there are two products and only one reactant, the product of the concentrations of products would increase faster than the concentration of the reactant. Because the equilibrium constant is constant, the percent dissociation decreases as the acid concentration increases. *Figure 16.13* The acidic hydrogens belong to carboxylate (—COOH) groups, whereas the fourth proton bound to oxygen is part of a hydroxyl (—OH) group. In organic acids, like citric acid, the acidic protons are almost always part of a carboxylate group.

Figure 16.14 The nitrogen atom in hydroxylamine accepts a proton to form NH_3OH^+. As a general rule, nonbonding electron pairs on nitrogen atoms are more basic than nonbonding electron pairs on oxygen atoms. *Figure 16.16* The range of pH values is so large that we can't show the effects using a single indicator (see Figure 16.8). *Figure 16.18* Yes. HI is a strong acid, which is consistent with the trends shown in the figure. *Figure 16.19* The HOY molecule on the left because it is a weak acid. Most of the HOY molecules remain undissociated.

Chapter 17

Figure 17.6 The pH will increase upon addition of the base. *Figure 17.7* 25.00 mL. The number of moles of added base needed to reach the equivalence point remains the same. Therefore, by doubling the concentration of added base the volume needed to reach the equivalence point is halved. *Figure 17.9* The volume of base needed to reach the equivalence point would not change because this quantity does not depend on the strength of the acid. However, the pH at the equivalence point, which is greater than 7 for a weak acid–strong base titration, would decrease to 7 because hydrochloric acid is a strong acid. *Figure 17.11* The pH at the equivalence point increases (becomes more basic) as the acid becomes weaker. The volume of added base needed to reach the equivalence point remains unchanged. *Figure 17.12* Yes. Any indicator that changes color between pH = 3 and pH = 11 could be used for a strong acid–strong base titration. Methyl red changes color between pH values of approximately 4 and 6. *Figure 17.22* ZnS and CuS would both precipitate on addition of H_2S, preventing separation of the two ions. *Figure 17.23* Yes. CuS would precipitate in step 2 on addition of H_2S to an acidic solution, while the Zn^{2+} ions remained in solution.

Chapter 18

Figure 18.1 About 85 km, at the mesopause. *Figure 18.3* The atmosphere absorbs a significant fraction of solar radiation. *Figure 18.4* The peak value is about 5×10^{12} molecules per cm^3. If we use Avogadro's number to convert molecules to moles, and the conversion factor of $1000\ cm^3 = 1000\ mL = 1\ L$, we find that the concentration of ozone at the peak is 8×10^{-9} mole/L. *Figure 18.7* The major sources of sulfur dioxide emission are located in the eastern half of the United States, and prevailing winds carry emissions in an eastward direction. *Figure 18.9* $CaSO_3(s)$ *Figure 18.11* $(168\ W/m^2)/(342\ W/m^2) = 0.49$; that is, about 50% of the total. (Keep in mind that the IR radiation absorbed by Earth's surface is not directly solar radiation.) *Figure 18.13* The increasing slope corresponds to an increasing *rate* of addition of CO_2 to the atmosphere, probably as a result of ever-increasing burning of fossil fuels worldwide. *Figure 18.15* Evaporation from sea water, evaporation from freshwater; evaporation and transpiration from land. *Figure 18.16* The variable that most affects the density of water in this case is the change in temperature. As the water grows colder, its density increases. *Figure 18.19* Water is the chemical species that is crossing the membrane, not the ions. The water flows oppositely to the normal flow in osmosis because of the application of high pressure. *Figure 18.20* To reduce concentrations of dissolved iron and manganese, remove H_2S and NH_3, and reduce bacterial levels.

Chapter 19

Figure 19.1 Yes, the potential energy of the eggs decreases as they fall. *Figure 19.2* Because the final volume would be less than twice the volume of Flask A, the final pressure would be greater than 0.5 atm. *Figure 19.3* The freezing of liquid water to ice is exothermic. *Figure 19.4* To be truly reversible, the temperature change δT must be infinitesimally small. *Figure 19.8* There are two other independent rotational motions of the H_2O molecule:

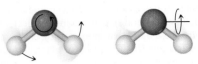

Figure 19.9 Ice, because it is the phase in which the molecules are held most rigidly **Figure 19.11** The decrease in the number of molecules due to the formation of new bonds. **Figure 19.12** During a phase change, the temperature remains constant but the entropy change can be large as molecules increase their degrees of freedom and motion. **Figure 19.13** Based on the three molecules shown, the addition of each C increases $S°$ by 40–45 J/mol-K. Based on this observation, we would predict that $S°(C_4H_{10})$ would 310–315 J/mol-K. Appendix C confirms that this is a good prediction: $S°(C_4H_{10}) = 310.0$ J/mol-K. **Figure 19.14** Spontaneous **Figure 19.15** If we plot progress of the reaction versus free energy, equilibrium is at a minimum point in free energy, as shown in the figure. In that sense, the reaction runs "downhill" until it reaches that minimum point.

Chapter 20

Figure 20.1 (a) The bubbling is caused by the hydrogen gas formed in the reaction. (b) The reaction is exothermic, and the heat causes the formation of steam. **Figure 20.2** The permanganate, MnO_4^-, is reduced, as the half-reactions in the text show. The oxalate ion, $C_2O_4^{2-}$, acts as the reducing agent. **Figure 20.3** The blue color is due to $Cu^{2+}(aq)$. As this ion is reduced, forming $Cu(s)$, its concentration decreases and the blue color fades. **Figure 20.4** The Zn is oxidized and, therefore, serves as the anode of the cell. **Figure 20.5** The electrical balance is maintained in two ways: Anions migrate into the half-cell, and cations migrate out. **Figure 20.9** As the cell operates, H^+ is reduced to H_2 in the cathode half-cell. As H^+ is depleted, the positive Na^+ ions are drawn into the half-cell to maintain electrical balance in the solution. **Figure 20.11** Yes. **Figure 20.13** The variable n is the number of moles of electrons transferred in the process. **Figure 20.14** The $Ni^{2+}(aq)$ ions and the cations in the salt bridge migrate toward the cathode. The $NO_3^-(aq)$ ions and the anions in the salt bridge migrate toward the anode. **Figure 20.19** The cathode consists of $PbO_2(s)$. Because each oxygen has an oxidation state of -2, lead must have an oxidation state of $+4$ in this compound. **Figure 20.20** Zn **Figure 20.21** Co^{3+}. The oxidation number increases as the battery charges **Figure 20.24** $O_2(g) + 4 H^+ + 4 e^- \longrightarrow 2 H_2O(g)$ **Figure 20.25** The oxidizing agent of $O_2(g)$ from the air **Figure 20.29** 0 V

Chapter 21

Figure 21.2 From Figure 21.2 we see that the belt of stability for a nucleus containing 70 protons lies at approximately 102 neutrons. **Figure 21.4** Only three of the elements with an even number of protons have fewer than three isotopes: He, Be, and C. Note that these three elements are the lightest elements that have an even atomic number. Because they are so light, any change in the number of neutrons will change the neutron/proton ratio significantly. This helps to explain why they do not have more stable isotopes. None of the elements in Figure 21.4 that have an odd number of protons have more than two stable isotopes. **Figure 21.6** 6.25 g. After one half-life, the amount of the radioactive material will have dropped to 25.0 g. After two half-lives, it will have dropped to 12.5 g. After three half-lives, it will have dropped to 6.25 g. **Figure 21.7** Plants convert $^{14}CO_2$ to ^{14}C-containing sugars via photosynthesis. When mammals eat the plants, they metabolize the sugars, thereby incorporating ^{14}C in their bodies. **Figure 21.8** Gamma rays. Both X rays and gamma rays consist of high-energy electromagnetic radiation, whereas alpha and beta rays are streams of particles. **Figure 21.9** Ionization energy. Detection depends on the ability of the radiation to cause ionization of the gas atoms. **Figure 21.13** The mass numbers are equal on both sides. Remember that this does not mean that mass is conserved—mass is lost

during the reaction, which appears as energy released. **Figure 21.14** 16. Each of the eight neutrons would split another uranium-235 nucleus, releasing two more neutrons. **Figure 21.15** Critical without being supercritical so that the release of energy is controlled. **Figure 21.19** Because large quantities of water are needed to condense the secondary coolant once it passes through the turbine. **Figure 21.23** Alpha rays are less dangerous when outside the body because they cannot penetrate the skin. However, once inside the body they can do great harm to any cells that are nearby.

Chapter 22

Figure 22.5 Beaker on the right is warmer. **Figure 22.6** HF is the most stable, SbH_3 the least stable. **Figure 22.8** More soluble in CCl_4 — the colors are deeper. **Figure 22.9** CF_2 **Figure 22.10** Redox reactions: The halides are being oxidized. **Figure 22.13** No **Figure 22.15** Based on this structure—yes, it would have a dipole moment. In fact, if you look it up, hydrogen peroxide's dipole moment is larger than water's! **Figure 22.20** They have been converted into water. **Figure 22.21** Formally they could both be $+2$. If we consider that the central sulfur is like SO_4^{2-}, however, then the central sulfur would be $+6$, like SO_4^{2-}, and then the terminal sulfur would be -2. **Figure 22.22** Nitrite **Figure 22.23** Longer. (There is a triple bond in N_2.) **Figure 22.25** The NO double bond **Figure 22.27** In both compounds the electron domains around the P atoms are tetrahedral. In P_4O_{10} all the electron domains around the P atoms are bonding. In P_4O_6 one of the electron domains about each P atom is nonbonding. **Figure 22.32** The minimum temperature should be the melting point of silicon; the temperature of the heating coil should not be so high that the silicon rod starts to melt outside the zone of the heating coil.

Chapter 23

Figure 23.3 No. The radius decreases first and then flattens out before increasing on moving past group 8B, while the effective nuclear charge increases steadily on moving left to right across the transition metal series. **Figure 23.4** Zn^{2+}. **Figure 23.5** The 4s orbitals are always empty in transition metal ions, so all of the ions shown in this table have empty 4s orbitals. The 3d orbitals are only empty for those ions that have lost all of their valence electrons: Sc^{3+}, Ti^{4+}, V^{5+}, Cr^{6+} and Mn^{7+}. **Figure 23.6** The electron spins would tend to align with the direction of the magnetic field. **Figure 23.8** No. If you start with a chloride ion on one vertex of the octahedron and then generate structures by placing the second chloride ion on any of the other five vertices you will get one complex that is the *trans* isomer and four complexes that are equivalent to the *cis* isomer shown in this figure. **Figure 23.9** $[Fe(H_2O)_6]^{3+}(aq) + SCN^-(aq) \longrightarrow [Fe(H_2O)_5(SCN)]^{2+}(aq) + H_2O(l)$ **Figure 23.10** The solid wedge represents a bond coming out of the plane of the page. The dashed wedge represents a bond that is going into the plane of the page. **Figure 23.13** Four for both (assuming no other ligands coordinate to the metal). **Figure 23.15** The coordination number is 6. Five nitrogen atoms (four from the heme and one from the protein) and one oxygen atom (from the O_2 molecule) coordinate iron. **Figure 23.16** The peak at 660 nm. **Figure 23.20** $[Fe(NH_3)_5(NO_2)]^{2+}$ and pentaamminenitroiron(III) ion for the complex on the left. $[Fe(NH_3)_5(ONO)]^{2+}$ and pentaamminenitritoiron(III) ion for the complex on the right. **Figure 23.21** The *cis* isomer. **Figure 23.24** Larger, since ammonia can displace water. **Figure 23.26** The peak would stay in the same position in terms of wavelength, but its absorbance would decrease. **Figure 23.28** $d_{x^2-y^2}$ and d_{z^2}. **Figure 23.29** Convert the wavelength of light, 495 nm, into energy in joules using the relationship $E = hc/\lambda$. **Figure 23.30** The absorption peak would shift to shorter wavelengths absorbing green light and the color of the complex ion would become red. An even larger shift would move the absorption peak into the blue region of the spectrum and the color would become orange. **Figure 23.34** Only the $d_{x^2-y^2}$ orbital points directly at the ligands.

Chapter 24

Figure 24.1 Tetrahedral   *Figure 24.2* The OH group is polar whereas the CH_3 group is nonpolar. Hence, adding CH_3 will (a) reduce the substance's solubility in polar solvents and (b) increase its solubility in nonpolar solvents.   *Figure 24.5* C_nH_{2n}, because there are no CH_3 groups, each carbon has two hydrogens.   *Figure 24.7* Just one   *Figure 24.9* Intermediates are minima and transition states are maxima on energy profiles.   *Figure 24.13* Both lactic acid and citric acid   *Figure 24.14* No, because there are not four different groups around any carbon   *Figure 24.17* Those labeled "basic amino acids," which have basic side groups that are protonated at pH 7   *Figure 24.18* Two.   *Figure 24.24* The long hydrocarbon chains, which are nonpolar   *Figure 24.26* The polar parts of the phospholipids seek to interact with water whereas the nonpolar parts seek to interact with other nonpolar substances and to avoid water.   *Figure 24.28* Negatively charged, because of the charge on the phosphate groups.   *Figure 24.30* GC because each base has three hydrogen bonding sites, whereas there are only two in AT.

ANSWERS TO SELECTED PRACTICE EXERCISES

Chapter 1

Sample Exercise 1.1
Practice Exercise 2: It is a compound because it has constant composition and can be separated into several elements.

Sample Exercise 1.2
Practice Exercise 2: (a) 10^{12} pm, (b) 6.0 km, (c) 4.22×10^{-3} g, (d) 0.00422 g.

Sample Exercise 1.3
Practice Exercise 2: (a) 261.7 K, (b) 11.3 °F

Sample Exercise 1.4
Practice Exercise 2: (a) 8.96 g/cm³, (b) 19.0 mL, (c) 340 g.

Sample Exercise 1.5
Practice Exercise 2: Five, as in the measurement 24.995 g, the uncertainty being in the third decimal place.

Sample Exercise 1.6
Practice Exercise 2: No. The number of feet in a mile is a defined quantity and is therefore exact, but the distance represented by one foot is not exact, although it is known to high accuracy.

Sample Exercise 1.7
Practice Exercise 2: (a) four, (b) two, (c) three.

Sample Exercise 1.8
Practice Exercise 2: 9.52 m/s (three significant figures).

Sample Exercise 1.9
Practice Exercise 2: No. Even though the mass of the gas would then be known to four significant figures, the volume of the container would still be known to only three.

Sample Exercise 1.10
Practice Exercise 2: 804.7 km

Sample Exercise 1.11
Practice Exercise 2: 12 km/L.

Sample Exercise 1.12
Practice Exercise 2: 1.2×10^4 ft.

Sample Exercise 1.13
Practice Exercise 2: 832 g

Chapter 2

Sample Exercise 2.1
Practice Exercise 2: (a) 154 pm, (b) 1.3×10^6 C atoms

Sample Exercise 2.2
Practice Exercise 2: (a) 56 protons, 56 electrons, and 82 neutrons, (b) 15 protons, 15 electrons, and 16 neutrons.

Sample Exercise 2.3
Practice Exercise 2: $^{208}_{82}\text{Pb}$

Sample Exercise 2.4
Practice Exercise 2: 28.09 amu

Sample Exercise 2.5
Practice Exercise 2: Na, atomic number 11, is a metal; Br, atomic number 35, is a nonmetal.

Sample Exercise 2.6
Practice Exercise 2: B_5H_7

Sample Exercise 2.7
Practice Exercise 2: 34 protons, 45 neutrons, and 36 electrons

Sample Exercise 2.8
Practice Exercise 2: (a) 3+, (b) 1−

Sample Exercise 2.9
Practice Exercise 2: (a) Rb is from group 1, and readily loses one electron to attain the electron configuration of the nearest noble gas element, Kr. (b) Nitrogen and the halogens are all nonmetallic elements, which form molecular compounds with one another. (c) Krypton, Kr, is a noble gas element and is chemically inactive except under special conditions. (d) Na and K are both from group 1 and adjacent to one another in the periodic table. They would be expected to behave very similarly. (e) Calcium is an active metal and readily loses two electrons to attain the noble gas configuration of Ar.

Sample Exercise 2.10
Practice Exercise 2: (a) Na_3PO_4, (b) $ZnSO_4$, (c) $Fe_2(CO_3)_3$

Sample Exercise 2.11
Practice Exercise 2: BrO^- and BrO_2^-

Sample Exercise 2.12
Practice Exercise 2: (a) ammonium bromide, (b) chromium(III) oxide, (c) cobalt(II) nitrate

Sample Exercise 2.13
Practice Exercise 2: (a) HBr, (b) H_2CO_3

Sample Exercise 2.14
Practice Exercise 2: (a) $SiBr_4$, (b) S_2Cl_2, (c) P_2O_6.

Sample Exercise 2.15
Practice Exercise 2: No, they are not isomers because they have different molecular formulas. Butane is C_4H_{10}, whereas cyclobutane is C_4H_8.

Chapter 3

Sample Exercise 3.1
Practice Exercise 2: (a) $C_2H_4 + 3\,O_2 \longrightarrow 2\,CO_2 + 2\,H_2O$. (b) Nine O_2 molecules

Sample Exercise 3.2
Practice Exercise 2: (a) 4, 3, 2; (b) 2, 6, 2, 3; (c) 1, 2, 1, 1, 1

Sample Exercise 3.3
Practice Exercise 2: (a) $HgS(s) \longrightarrow Hg(l) + S(s)$, (b) $4\,Al(s) + 3\,O_2(g) \longrightarrow 2\,Al_2O_3(s)$

Sample Exercise 3.4
Practice Exercise 2: $C_2H_5OH(l) + 3\,O_2(g) \longrightarrow 2\,CO_2(g) + 3\,H_2O(g)$

Sample Exercise 3.5
Practice Exercise 2: (a) 78.0 amu, (b) 32.0 amu, (c) 211.0 amu

Sample Exercise 3.6
Practice Exercise 2: 16.1%

Sample Exercise 3.7
Practice Exercise 2: 1 mol H_2O (6×10^{23} O atoms) $< 3 \times 10^{23}$ molecules O_3 (9×10^{23} O atoms) $<$ 1 mol CO_2 (12×10^{23} O atoms)

Sample Exercise 3.8
Practice Exercise 2: (a) 9.0×10^{23}, (b) 2.71×10^{24}

Sample Exercise 3.9
Practice Exercise 2: 164.1 g/mol

Sample Exercise 3.10
Practice Exercise 2: 55.5 mol H_2O.

Sample Exercise 3.11
Practice Exercise 2: **(a)** 6.0 g, **(b)** 8.29 g.

Sample Exercise 3.12
Practice Exercise 2: **(a)** 4.01×10^{22} molecules HNO_3,
(b) 1.20×10^{23} atoms O

Sample Exercise 3.13
Practice Exercise 2: C_4H_4O

Sample Exercise 3.14
Practice Exercise 2: **(a)** CH_3O, **(b)** $C_2H_6O_2$

Sample Exercise 3.15
Practice Exercise 2: **(a)** C_3H_6O, **(b)** $C_6H_{12}O_2$

Sample Exercise 3.16
Practice Exercise 2: 1.77 g

Sample Exercise 3.17
Practice Exercise 2: 26.5 g

Sample Exercise 3.18
Practice Exercise 2: **(a)** Al, **(b)** 1.50 mol, **(c)** 0.75 mol Cl_2

Sample Exercise 3.19
Practice Exercise 2: **(a)** $AgNO_3$, **(b)** 1.59 g, **(c)** 1.39 g, **(d)** 1.52 g Zn

Sample Exercise 3.20
Practice Exercise 2: **(a)** 105 g Fe, **(b)** 83.7%

Chapter 4

Sample Exercise 4.1
Practice Exercise 2: **(a)** 6, **(b)** 12, **(c)** 2, **(d)** 9

Sample Exercise 4.2
Practice Exercise 2: **(a)** insoluble, **(b)** soluble, **(c)** soluble

Sample Exercise 4.3
Practice Exercise 2: **(a)** $Fe(OH)_3$, **(b)** $Fe_2(SO_4)_3(aq) + 6\,LiOH(aq)$ $\longrightarrow 2\,Fe(OH)_3(s) + 3\,Li_2SO_4(aq)$

Sample Exercise 4.4
Practice Exercise 2: $3\,Ag^+(aq) + PO_4^{3-}(aq) \longrightarrow Ag_3PO_4(s)$

Sample Exercise 4.5
Practice Exercise 2: The diagram would show 10 Na^+ ions, 2 OH^- ions, 8 Y^- ions, and 8 H_2O molecules.

Sample Exercise 4.6
Practice Exercise 2: $C_6H_6O_6$ (nonelectrolyte) < CH_3COOH (weak electrolyte, existing mainly in the form of molecules with few ions) < $NaCH_3COO$ (strong electrolyte that provides two ions, Na^+ and CH_3COO^-) < $Ca(NO_3)_2$ (strong electrolyte that provides three ions, Ca^{2+} and $2\,NO_3^-$)

Sample Exercise 4.7
Practice Exercise 2:
(a) $H_3PO_3(aq) + 3\,KOH(aq) \longrightarrow 3\,H_2O(l) + K_3PO_3(aq)$,
(b) $H_3PO_3(aq) + 3\,OH^-(aq) \longrightarrow 3\,H_2O(l) + PO_3^{3-}(aq)$.
(H_3PO_3 is a weak acid and therefore a weak electrolyte, whereas KOH, a strong base, and K_3PO_3, an ionic compound, are strong electrolytes.)

Sample Exercise 4.8
Practice Exercise 2: **(a)** +5, **(b)** −1, **(c)** +6, **(d)** +4, **(e)** −1

Sample Exercise 4.9
Practice Exercise 2:
(a) $Mg(s) + CoSO_4(aq) \longrightarrow MgSO_4(aq) + Co(s)$;
$Mg(s) + Co^{2+}(aq) \longrightarrow Mg^{2+}(aq) + Co(s)$,
(b) Mg is oxidized and Co^{2+} is reduced.

Sample Exercise 4.10
Practice Exercise 2: Zn and Fe

Sample Exercise 4.11
Practice Exercise 2: 0.278 M

Sample Exercise 4.12
Practice Exercise 2: 0.030 M

Sample Exercise 4.13
Practice Exercise 2: **(a)** 1.1 g, **(b)** 76 mL

Sample Exercise 4.14
Practice Exercise 2: **(a)** 0.0200 L = 20.0 mL, **(b)** 5.0 mL, **(c)** 0.40 M

Sample Exercise 4.15
Practice Exercise 2: **(a)** 0.240 g, **(b)** 0.400 L

Sample Exercise 4.16
Practice Exercise 2: 0.210 M

Sample Exercise 4.17
Practice Exercise 2: **(a)** 1.057×10^{-3} mol MnO_4^-,
(b) 5.286×10^{-3} mol Fe^{2+}, **(c)** 0.2952 g, **(d)** 33.21%

Chapter 5

Sample Exercise 5.1
Practice Exercise 2: **(a)** 1.4×10^{-20} J, **(b)** 8.4×10^3 J

Sample Exercise 5.2
Practice Exercise 2: +55 J

Sample Exercise 5.3
Practice Exercise 2: 0.69 L-atm = 70 J

Sample Exercise 5.4
Practice Exercise 2: In order to solidify, the gold must cool to below its melting temperature. It cools by transferring heat to its surroundings. The air around the sample would feel hot because heat is transferred to it from the molten gold, meaning the process is exothermic. (You may notice that solidification of a liquid is the reverse of the melting we analyzed in the exercise. As we will see, reversing the direction of a process changes the sign of the heat transferred.)

Sample Exercise 5.5
Practice Exercise 2: −14.4 kJ

Sample Exercise 5.6
Practice Exercise 2: **(a)** 4.9×10^5 J,
(b) 11 K decrease = 11 °C decrease

Sample Exercise 5.7
Practice Exercise 2: −68,000 J/mol = −68 kJ/mol

Sample Exercise 5.8
Practice Exercise 2: **(a)** −15.2 kJ/g, **(b)** −1370 kJ/mol

Sample Exercise 5.9
Practice Exercise 2: +1.9 kJ

Sample Exercise 5.10
Practice Exercise 2: −304.1 kJ

Sample Exercise 5.11
Practice Exercise 2: $C(graphite) + 2\,Cl_2(g) \longrightarrow CCl_4(l)$;
$\Delta H_f^\circ = -106.7$ kJ/mol.

Sample Exercise 5.12
Practice Exercise 2: −1367 kJ

Sample Exercise 5.13
Practice Exercise 2: −156.1 kJ/mol

Sample Exercise 5.14
Practice Exercise 2: **(a)** 15 kJ/g, **(b)** 100 min

Chapter 6

Sample Exercise 6.1
Practice Exercise 2: The expanded visible-light portion of Figure 6.4 tells you that red light has a longer wavelength than blue light. The lower wave has the longer wavelength (lower frequency) and would be the red light.

Sample Exercise 6.2
Practice Exercise 2: **(a)** $1.43 \times 10^{14}\,s^{-1}$, **(b)** 2.899 m

Sample Exercise 6.3
Practice Exercise 2: **(a)** $3.11 \times 10^{-19}\,J$, **(b)** 0.16 J, **(c)** 4.2×10^{16} photons

Sample Exercise 6.4
Practice Exercise 2: **(a)** $\Delta E < 0$, photon emitted, **(b)** $\Delta E > 0$, photon absorbed

Sample Exercise 6.5
Practice Exercise 2: $7.86 \times 10^2\,m/s$

Sample Exercise 6.6
Practice Exercise 2: **(a)** $5p$; **(b)** 3; **(c)** $1, 0, -1$

Sample Exercise 6.7
Practice Exercise 2: **(a)** $1s^2 2s^2 2p^6 3s^2 3p^2$, **(b)** two

Sample Exercise 6.8
Practice Exercise 2: group 4A

Sample Exercise 6.9
Practice Exercise 2: **(a)** $[Ar]4s^2 3d^7$ or $[Ar]3d^7 4s^2$, **(b)** $[Kr]5s^2 4d^{10}5p^1$ or $[Kr]4d^{10}5s^2 5p^1$

Chapter 7

Sample Exercise 7.1
Practice Exercise 2: P—Br

Sample Exercise 7.2
Practice Exercise 2: C < Be < Ca < K

Sample Exercise 7.3
Practice Exercise 2: S^{2-}

Sample Exercise 7.4
Practice Exercise 2: Cs^+

Sample Exercise 7.5
Practice Exercise 2: Ca

Sample Exercise 7.6
Practice Exercise 2: Al lowest, C highest

Sample Exercise 7.7
Practice Exercise 2: **(a)** >

Sample Exercise 7.8
Practice Exercise 2:
$CuO(s) + H_2SO_4(aq) \longrightarrow CuSO_4(aq) + H_2O(l)$

Sample Exercise 7.9
Practice Exercise 2: $P_4O_6(s) + 6\,H_2O(l) \longrightarrow 4\,H_3PO_3(aq)$

Sample Exercise 7.10
Practice Exercise 2: $2K(s) + S(s) \longrightarrow K_2S(s)$

Chapter 8

Sample Exercise 8.1
Practice Exercise 2: ZrO_2

Sample Exercise 8.2
Practice Exercise 2: Mg^{2+} and N^{3-}

Sample Exercise 8.3
Practice Exercise 2: They both show 8 valence electrons; methane has 4 bonding pairs and neon has 4 nonbonding pairs

Sample Exercise 8.4
Practice Exercise 2: Se—Cl

Sample Exercise 8.5
Practice Exercise 2: **(a)** F, **(b)** 0.11−

Sample Exercise 8.6
Practice Exercise 2: **(a)** 20, **(b)**

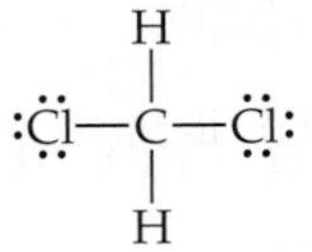

Sample Exercise 8.7
Practice Exercise 2: **(a)** $[:N\equiv O:]^+$, **(b)**

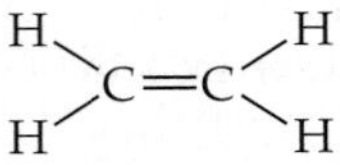

Sample Exercise 8.8
Practice Exercise 2: **(a)** $[:\ddot{O}—\ddot{Cl}—\ddot{O}:]^-$ **(b)**

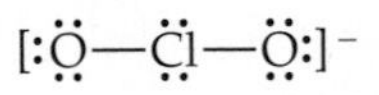

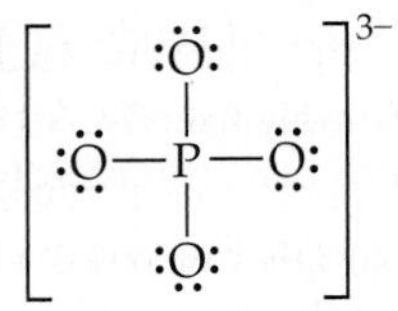

Sample Exercise 8.9
Practice Exercise 2:

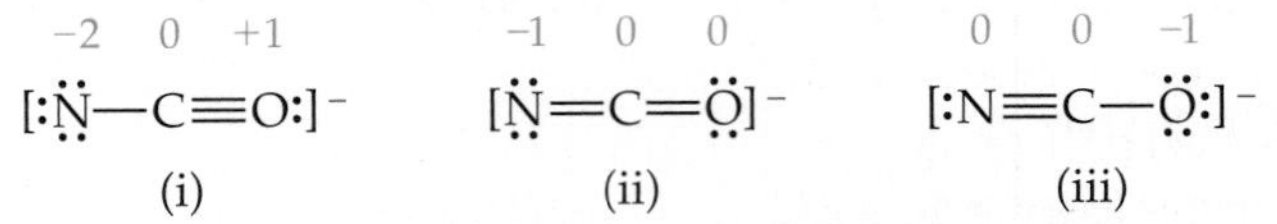

(b) Structure (iii), which places a negative charge on oxygen, the most electronegative element in the ion, is the dominant Lewis structure.

Sample Exercise 8.10

Practice Exercise 2:

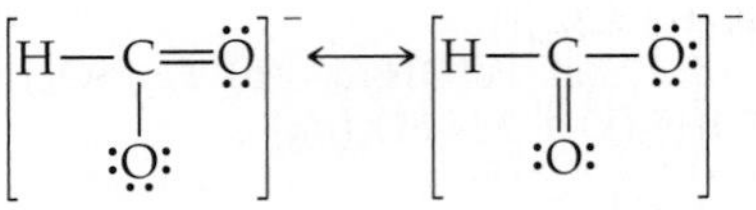

Sample Exercise 8.11
Practice Exercise 2:
(a) C, **(b)** $:\ddot{F}—\ddot{X}e—\ddot{F}:$

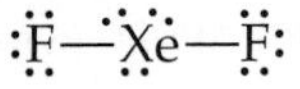

Sample Exercise 8.12
Practice Exercise 2: −86 kJ

Chapter 9

Sample Exercise 9.1
Practice Exercise 2: **(a)** tetrahedral, bent; **(b)** trigonal planar, trigonal planar

Sample Exercise 9.2
Practice Exercise 2: **(a)** trigonal bipyramidal, T-shaped; **(b)** trigonal bipyramidal, trigonal bipyramidal.

Sample Exercise 9.3
Practice Exercise 2: 109.5°, 180°

Sample Exercise 9.4
Practice Exercise 2: **(a)** polar because polar bonds are arranged in a seesaw geometry, **(b)** nonpolar because polar bonds are arranged in a tetrahedral geometry

Sample Exercise 9.5
Practice Exercise 2: tetrahedral, sp^3

Sample Exercise 9.6
Practice Exercise 2: (a) approximately 109° around the left C and 180° around the right C; (b) sp^3, sp; (c) five σ bonds and two π bonds

Sample Exercise 9.7
Practice Exercise 2: SO_2 and SO_3, as indicated by the presence of two or more resonance structures involving π bonding for each of these molecules.

Sample Exercise 9.8
Practice Exercise 2: $\sigma_{1s}^2 \sigma_{1s}^{*1}$; $\frac{1}{2}$

Sample Exercise 9.9
Practice Exercise 2: (a) diamagnetic, 1; (b) diamagnetic, 3

Chapter 10

Sample Exercise 10.1
Practice Exercise 2: (D) 1.6 M

Sample Exercise 10.2
Practice Exercise 2: 807.3 torr

Sample Exercise 10.3
Practice Exercise 2: 5.30×10^3 L

Sample Exercise 10.4
Practice Exercise 2: 2.0 atm

Sample Exercise 10.5
Practice Exercise 2: 3.83×10^3 m³

Sample Exercise 10.6
Practice Exercise 2: 27 °C

Sample Exercise 10.7
Practice Exercise 2: 5.9 g/L

Sample Exercise 10.8
Practice Exercise 2: 29.0 g/mol

Sample Exercise 10.9
Practice Exercise 2: 14.8 L

Sample Exercise 10.10
Practice Exercise 2: 2.86 atm

Sample Exercise 10.11
Practice Exercise 2: 1.0×10^3 torr N_2, 1.5×10^2 torr Ar, and 73 torr CH_4

Sample Exercise 10.12
Practice Exercise 2: (a) increases, (b) no effect, (c) no effect

Sample Exercise 10.13
Practice Exercise 2: 1.36×10^3 m/s

Sample Exercise 10.14
Practice Exercise 2: $r_{N_2}/r_{O_2} = 1.07$

Sample Exercise 10.15
Practice Exercise 2: (a) 7.472 atm, (b) 7.181 atm

Chapter 11

Sample Exercise 11.1
Practice Exercise 2: chloramine, $NHCl$

Sample Exercise 11.2
Practice Exercise 2: (a) CH_3CH_3 has only dispersion forces, whereas the other two substances have both dispersion forces and hydrogen bonds, (b) CH_3CH_2OH

Sample Exercise 11.3
Practice Exercise 2: -20.9 kJ $- 33.4$ kJ $- 6.09$ kJ $= -60.4$ kJ

Sample Exercise 11.4
Practice Exercise 2: about 340 torr (0.45 atm)

Sample Exercise 11.5
Practice Exercise 2: (a) -162 °C; (b) It sublimes whenever the pressure is less than 0.1 atm; (c) The highest temperature at which a liquid can exist is defined by the critical temperature. So we do not expect to find liquid methane when the temperature is higher than -80 °C.

Sample Exercise 11.6
Practice Exercise 2: Because rotation can occur about carbon–carbon single bonds, molecules whose backbone consists predominantly of C–C single bonds are too flexible; the molecules tend to coil in random ways and, thus, are not rodlike.

Chapter 12

Sample Exercise 12.1
Practice Exercise 2: 0.68 or 68%

Sample Exercise 12.2
Practice Exercise 2: $a = 4.02$ Å and density $= 4.31$ g/cm³

Sample Exercise 12.3
Practice Exercise 2: smaller.

Sample Exercise 12.4
Practice Exercise 2: A group 5A element could be used to replace Se.

Chapter 13

Sample Exercise 13.1
Practice Exercise 2: $C_5H_{12} < C_5H_{11}Cl < C_5H_{11}OH < C_5H_{10}(OH)_2$ (in order of increasing polarity and hydrogen-bonding ability)

Sample Exercise 13.2
Practice Exercise 2: 1.0×10^{-5} M

Sample Exercise 13.3
Practice Exercise 2: 90.5 g of NaOCl

Sample Exercise 13.4
Practice Exercise 2: 0.670 m

Sample Exercise 13.5
Practice Exercise 2: (a) 9.00×10^{-3}, (b) 0.505 m

Sample Exercise 13.6
Practice Exercise 2: (a) 10.9 m, (b) $X_{C_3H_8O_3} = 0.163$, (c) 5.97 M

Sample Exercise 13.7
Practice Exercise 2: 0.290

Sample Exercise 13.8
Practice Exercise 2: -65.6 °C

Sample Exercise 13.9
Practice Exercise 2: 0.048 atm

Sample Exercise 13.10
Practice Exercise 2: 110 g/mol

Sample Exercise 13.11
Practice Exercise 2: 4.20×10^4 g/mol

Chapter 14

Sample Exercise 14.1
Practice Exercise 2: 1.8×10^{-2} M/s

Sample Exercise 14.2
Practice Exercise 2: 1.1×10^{-4} M/s

Sample Exercise 14.3
Practice Exercise 2: (a) 8.4×10^{-7} M/s, (b) 2.1×10^{-7} M/s

Sample Exercise 14.4
Practice Exercise 2: $2 = 3 < 1$

Sample Exercise 14.5
Practice Exercise 2: **(a)** 1, **(b)** $M^{-1}s^{-1}$.

Sample Exercise 14.6
Practice Exercise 2: **(a)** rate $= k[NO]^2[H_2]$, **(b)** $k = 1.2\,M^{-2}\,s^{-1}$
(c) rate $= 4.5 \times 10^{-4}\,M/s$

Sample Exercise 14.7
Practice Exercise 2: 51 torr

Sample Exercise 14.8
Practice Exercise 2: $[NO_2] = 1.00 \times 10^{-3}\,M$

Sample Exercise 14.9
Practice Exercise 2: **(a)** 0.478 yr $= 1.51 \times 10^7$ s, **(b)** it takes two
half-lives, $2(0.478\ yr) = 0.956$ yr

Sample Exercise 14.10
Practice Exercise 2: $2 < 1 < 3$ because, if you approach the barrier
from the right, the E_a values are 40 kJ/mol for reverse reaction 2,
25 kJ/molfor reverse reaction 1, and 15 kJ/mol for reverse reaction 3.

Sample Exercise 14.11
Practice Exercise 2: $2 \times 10^{13}\,s^{-1}$

Sample Exercise 14.12
Practice Exercise 2: **(a)** Yes, the two equations add to yield the equation for the reaction. **(b)** The first elementary reaction is unimolecular, and the second one is bimolecular. **(c)** $Mo(CO)_5$.

Sample Exercise 14.13
Practice Exercise 2: **(a)** Rate $= k[NO]^2[Br_2]$, **(b)** No, because termolecular reactions are very rare.

Sample Exercise 14.14
Practice Exercise 2: Because the rate law conforms to the molecularity
of the first step, the first step must be the rate-determining step. The
second step must be much faster than the first one.

Sample Exercise 14.15

Practice Exercise 2: $[Br] = \left(\dfrac{k_1}{k_{-1}}[Br_2]\right)^{1/2}$

Chapter 15
Sample Exercise 15.1

Practice Exercise 2: **(a)** $K_c = \dfrac{[HI]^2}{[H_2][I_2]}$, **(b)** $K_c = \dfrac{[CdBr_4^{2-}]}{[Cd^{2+}][Br^-]^4}$

Sample Exercise 15.2
Practice Exercise 2: 0.335

Sample Exercise 15.3
Practice Exercise 2: At the lower temperature because K_p is larger at
the lower temperature

Sample Exercise 15.4

Practice Exercise 2: $\dfrac{(54.0)^3}{1.04 \times 10^{-4}} = 1.51 \times 10^9$

Sample Exercise 15.5
Practice Exercise 2:

(a) $K_c = \dfrac{[Cr^{3+}]}{[Ag^+]^3}$, **(b)** $K_p = \dfrac{(P_{H_2})^4}{(P_{H_2O})^4}$

Sample Exercise 15.6
Practice Exercise 2: $H_2(g)$

Sample Exercise 15.7
Practice Exercise 2: 1.79×10^{-5}

Sample Exercise 15.8
Practice Exercise 2: 33

Sample Exercise 15.9
Practice Exercise 2: $Q_p = 16$; $Q_p > K_p$, and so the reaction will proceed from right to left, forming more SO_3.

Sample Exercise 15.10
Practice Exercise 2: 1.22 atm

Sample Exercise 15.11
Practice Exercise 2: $P_{PCl_5} = 0.967$ atm, $P_{PCl_3} = P_{Cl_2} = 0.693$ atm

Sample Exercise 15.12
Practice Exercise 2: **(a)** right, **(b)** left, **(c)** right, **(d)** left

Sample Exercise 15.13
Practice Exercise 2: $\Delta H° = 508.3$ kJ; the equilibrium constant will
increase with increasing temperature.

Chapter 16
Sample Exercise 16.1
Practice Exercise 2: H_2SO_3, HF, HPO_4^{2-}, HCO^+

Sample Exercise 16.2
Practice Exercise 2: $O^{2-}(aq) + H_2O(l) \longrightarrow OH^-(aq) + OH^-(aq)$.
The OH^- is both the conjugate acid of O^{2-} and the conjugate base
of H_2O.

Sample Exercise 16.3
Practice Exercise 2: **(a)** left, **(b)** right

Sample Exercise 16.4
Practice Exercise 2: **(a)** basic, **(b)** neutral, **(c)** acidic

Sample Exercise 16.5
Practice Exercise 2: **(a)** $5 \times 10^{-9}\,M$, **(b)** $1.0 \times 10^{-7}\,M$,
(c) $7.1 \times 10^{-9}\,M$

Sample Exercise 16.6
Practice Exercise 2: **(a)** 3.42, **(b)** $[H^+] = 5.3 \times 10^{-9}\,M$, so pH $= 8.28$

Sample Exercise 16.7
Practice Exercise 2: $[H^+] = 6.6 \times 10^{-10}$

Sample Exercise 16.8
Practice Exercise 2: 0.0046 M

Sample Exercise 16.9
Practice Exercise 2: **(a)** $7.8 \times 10^{-3}\,M$, **(b)** $2.4 \times 10^{-3}\,M$

Sample Exercise 16.10
Practice Exercise 2: 1.5×10^{-5}

Sample Exercise 16.11
Practice Exercise 2: 2.7%

Sample Exercise 16.12
Practice Exercise 2: 3.41

Sample Exercise 16.13
Practice Exercise 2: **(a)** 3.9%, **(b)** 12%

Sample Exercise 16.14
Practice Exercise 2: **(a)** pH $= 1.80$, **(b)** $[C_2O_4^{2-}] = 6.4 \times 10^{-5}\,M$

Sample Exercise 16.15
Practice Exercise 2: Methylamine (because it has the larger K_b value of
the two amine bases in the list)

Sample Exercise 16.16
Practice Exercise 2: 0.12 M

Sample Exercise 16.17
Practice Exercise 2: **(a)** $PO_4^{3-}(K_b = 2.4 \times 10^{-2})$,
(b) $K_b = 7.9 \times 10^{-10}$

Sample Exercise 16.18
Practice Exercise 2: **(a)** $Fe(NO_3)_3$, **(b)** KBr, **(c)** CH_3NH_3Cl,
(d) NH_4NO_3

Sample Exercise 16.19
Practice Exercise 2: acidic

Sample Exercise 16.20
Practice Exercise 2: (a) HBr, (b) H_2S, (c) HNO_3, (d) H_2SO_3

Chapter 17

Sample Exercise 17.1
Practice Exercise 2: 3.42

Sample Exercise 17.2
Practice Exercise 2: $[HCOO^-] = 9.0 \times 10^{-5}$, pH $= 1.00$

Sample Exercise 17.3
Practice Exercise 2: 4.42

Sample Exercise 17.4
Practice Exercise 2: 4.62 (using quadratic).

Sample Exercise 17.5
Practice Exercise 2: 0.13 M

Sample Exercise 17.6
Practice Exercise 2: (a) 4.68, (b) 1.70

Sample Exercise 17.7
Practice Exercise 2: (a) 10.30, (b) 3.70

Sample Exercise 17.8
Practice Exercise 2: (a) 4.20, (b) 9.26

Sample Exercise 17.9
Practice Exercise 2: (a) 8.21, (b) 5.28

Sample Exercise 17.10
Practice Exercise 2: (a) $K_{sp} = [Ba^{2+}][CO_3^{2-}] = 5.0 \times 10^{-9}$,
(b) $K_{sp} = [Ag^+]^2[SO_4^{2-}] = 1.5 \times 10^{-5}$

Sample Exercise 17.11
Practice Exercise 2: 1.6×10^{-12}

Sample Exercise 17.12
Practice Exercise 2: 9×10^{-6} mol/L

Sample Exercise 17.13
Practice Exercise 2: 4.0×10^{-10} M

Sample Exercise 17.14
Practice Exercise 2: (a) $CuS(s) + H^+(aq) \rightleftharpoons Cu^{2+}(aq) + HS^-(aq)$,
(b) $Cu(N_3)_2(s) + 2H^+(aq) \rightleftharpoons Cu^{2+}(aq) + 2HN_3(aq)$

Sample Exercise 17.15
Practice Exercise 2: 1×10^{-16} M

Sample Exercise 17.16
Practice Exercise 2: Yes, CaF_2 precipitates because Q $= 4.6 \times 10^{-7}$ is larger than $K_{sp} = 3.9 \times 10^{-11}$

Sample Exercise 17.17
Practice Exercise 2: $Cu(OH)_2$ precipitates first, beginning when $[OH^-] > 9.8 \times 10^{-10}$ M. $Mg(OH)_2$ begins to precipitate when $[OH^-] > 1.9 \times 10^{-5}$ M.

Chapter 18

Sample Exercise 18.1
Practice Exercise 2: 3.0×10^{-3} torr

Sample Exercise 18.2
Practice Exercise 2: 127 nm

Chapter 19

Sample Exercise 19.1
Practice Exercise 2: No, the reverse process is spontaneous at this temperature.

Sample Exercise 19.2
Practice Exercise 2: -163 J/K

Sample Exercise 19.3
Practice Exercise 2: no

Sample Exercise 19.4
Practice Exercise 2: (a) 1 mol of $SO_2(g)$ at STP, (b) 2 mol of $NO_2(g)$ at STP

Sample Exercise 19.5
Practice Exercise 2: 180.39 J/K

Sample Exercise 19.6
Practice Exercise 2: $\Delta G° = -14.7$ kJ; the reaction is spontaneous.

Sample Exercise 19.7
Practice Exercise 2: -800.7 kJ

Sample Exercise 19.8
Practice Exercise 2: More negative

Sample Exercise 19.9
Practice Exercise 2: (a) $\Delta H° = -196.6$ kJ, $\Delta S° = -189.6$ J/K;
(b) $\Delta G° = -120.8$ kJ

Sample Exercise 19.10
Practice Exercise 2: 330 K

Sample Exercise 19.11
Practice Exercise 2: -26.0 kJ/mol

Sample Exercise 19.12
Practice Exercise 2: $\Delta G° = -106.4$ kJ/mol, $K = 4 \times 10^{18}$

Chapter 20

Sample Exercise 20.1
Practice Exercise 2: $Al(s)$ is the reducing agent; $MnO_4^-(aq)$ is the oxidizing agent.

Sample Exercise 20.2
Practice Exercise 2: $Cu(s) + 4 H^+(aq) + 2 NO_3^-(aq) \longrightarrow Cu^{2+}(aq) + 2 NO_2(g) + 2 H_2O(l)$

Sample Exercise 20.3
Practice Exercise 2: $2 Cr(OH)_3(s) + 6 ClO^-(aq) \longrightarrow 2 CrO_4^{2-}(aq) + 3 Cl_2(g) + 2 OH^-(aq) + 2 H_2O(l)$

Sample Exercise 20.4
Practice Exercise 2: (a) The first reaction occurs at the anode and the second reaction at the cathode. (b) Zinc is oxidized to form Zn^{2+} as the reaction proceeds and therefore the zinc electrode loses mass. (c) The platinum electrode is not involved in the reaction and none of the products of the reaction at the cathode are solids, so the mass of platinum electrode does not change. (d) The platinum cathode is positive.

Sample Exercise 20.5
Practice Exercise 2: -0.40 V

Sample Exercise 20.6
Practice Exercise 2: 2.20 V

Sample Exercise 20.7
Practice Exercise 2:
(a) $Co \longrightarrow Co^{2+} + 2 e^-$; (b) $+0.499$ V

Sample Exercise 20.8
Practice Exercise 2: $Al(s) > Fe(s) > I^-(aq)$

Sample Exercise 20.9
Practice Exercise 2: Reactions (b) and (c) are spontaneous.

Sample Exercise 20.10
Practice Exercise 2: $+77$ kJ/mol

Sample Exercise 20.11
Practice Exercise 2: It would increase.

Sample Exercise 20.12
Practice Exercise 2: pH $= 4.23$ (using data from Appendix E to obtain $E°$ to three significant figures)

Sample Exercise 20.13
Practice Exercise 2: **(a)** The anode is the half-cell in which $[Zn^{2+}] = 3.75 \times 10^{-4}\,M$.   **(b)** The emf is 0.105 V.

Sample Exercise 20.14
Practice Exercise 2: **(a)** 30.2 g of Mg,   **(b)** $3.97 \times 10^3\,s$

Chapter 21

Sample Exercise 21.1
Practice Exercise 2: $^{212}_{84}Po$

Sample Exercise 21.2
Practice Exercise 2: $^{15}_{8}O \longrightarrow {}^{15}_{7}N + {}^{0}_{+1}e$

Sample Exercise 21.3
Practice Exercise 2:
(a) α emission,   **(b)** β^- emission

Sample Exercise 21.4
Practice Exercise 2: $^{16}_{8}O(p, \alpha){}^{13}_{7}N$

Sample Exercise 21.5
Practice Exercise 2: 3.12%

Sample Exercise 21.6
Practice Exercise 2: 2200 yr

Sample Exercise 21.7
Practice Exercise 2: 15.1%

Sample Exercise 21.8
Practice Exercise 2: $-3.19 \times 10^{-3}\,g$

Chapter 22

Sample Exercise 22.1
Practice Exercise 2: **(a)** Cs,   **(b)** Cl,   **(c)** C,   **(d)** Sb

Sample Exercise 22.2
Practice Exercise 2: $NaH(s) + H_2O(l) \longrightarrow NaOH(aq) + H_2(g)$

Sample Exercise 22.3
Practice Exercise 2: trigonal bipyramidal, linear

Sample Exercise 22.4
Practice Exercise 2: $2\,Br^-(aq) + Cl_2(aq) \longrightarrow Br_2(aq) + 2\,Cl^-(aq)$

Sample Exercise 22.5
Practice Exercise 2: $NaI(s) + H_3PO_4(l) \longrightarrow NaH_2PO_4(s) + HI(g)$

Sample Exercise 22.6
Practice Exercise 2: $5\,N_2O_4(l) + 4\,N_2H_3CH_3(l) \longrightarrow 9\,N_2(g) + 4\,CO_2(g) + 12\,H_2O(g)$

Sample Exercise 22.7
Practice Exercise 2: $6-$

Chapter 23

Sample Exercise 23.1
Practice Exercise 2: three: $[Co(H_2O)_6]^{2+}$ and two Cl^-

Sample Exercise 23.2
Practice Exercise 2: zero

Sample Exercise 23.3
Practice Exercise 2: **(a)** triamminetribromomolybdenum(IV) nitrate, **(b)** ammonium tetrabromocuprate(II),   **(c)** $Na[Ru(H_2O)_2(C_2O_4)_2]$

Sample Exercise 23.4
Practice Exercise 2: two

Sample Exercise 23.5
Practice Exercise 2: No, because the complex is flat. This complex ion does, however, have geometric isomers (for example, the Cl and Br ligands could be cis or trans).

Sample Exercise 23.6
Practice Exercise 2: green

Sample Exercise 23.7
Practice Exercise 2: $[Co(NH_3)_5Cl]^{2+}$ is purple which means it must absorb light near the boundary between the yellow and green regions of the spectrum. $[Co(NH_3)_6]^{3+}$ is orange which means it absorbs blue light. Because yellow-green have a lower energy (longer wavelength) than blue photons Δ is smaller for $[Co(NH_3)_5Cl]^{2+}$. This prediction is consistent with the spectrochemical series because Cl^- ligands are at the low end of the spectrochemical series. Thus replacing NH_3 with Cl^- should make Δ smaller.

Sample Exercise 23.8
Practice Exercise 2: No it is not possible to have a diamagnetic complex with partially filled d orbitals if the complex is high spin, and all tetrahedral complexes are high spin.

Chapter 24

Sample Exercise 24.1
Practice Exercise 2: 2,4-dimethylpentane

Sample Exercise 24.2
Practice Exercise 2:

$$CH_3CH\overset{\displaystyle CH_3}{|}\!-\!CHCH_2CH_2CH_3 \quad \text{or} \quad CH_3CH(CH_3)CH(CH_3)CH_2CH_2CH_3$$

Sample Exercise 24.3
Practice Exercise 2: five (1-hexene, *cis*-2-hexene, *trans*-2-hexene, *cis*-3-hexene, *trans*-3-hexene)

Sample Exercise 24.4
Practice Exercise 2:

$$CH_3\!-\!C\!\equiv\!C\!-\!\underset{\underset{\displaystyle CH_3}{|}}{CH}\!-\!CH_3$$

Sample Exercise 24.5
Practice Exercise 2: propene

Sample Exercise 24.6
Practice Exercise 2:

$$CH_3CH_2\overset{\displaystyle O}{\overset{\|}{C}}\!-\!O\!-\!CH_2CH_2CH_3$$

Sample Exercise 24.7
Practice Exercise 2: serylaspartic acid; Ser-Asp, SD

Sample Exercise 24.8
Practice Exercise 2: alcohol, ether

GLOSSARY

absolute zero The lowest attainable temperature; 0 K on the Kelvin scale and $-273.15\ ^\circ\text{C}$ on the Celsius scale. (Section 1.4)

absorption spectrum A pattern of variation in the amount of light absorbed by a sample as a function of wavelength. (Section 23.5)

accuracy A measure of how closely individual measurements agree with the correct value. (Section 1.5)

acid A substance that is able to donate a H^+ ion (a proton) and, hence, increases the concentration of $\text{H}^+(aq)$ when it dissolves in water. (Section 4.3)

acid-dissociation constant (K_a) An equilibrium constant that expresses the extent to which an acid transfers a proton to solvent water. (Section 16.6)

acidic anhydride (acidic oxide) An oxide that forms an acid when added to water; soluble nonmetal oxides are acidic anhydrides. (Section 22.5)

acidic oxide (acidic anhydride) An oxide that either reacts with a base to form a salt or with water to form an acid. (Section 22.5)

acid rain Rainwater that has become excessively acidic because of absorption of pollutant oxides, notably SO_3, produced by human activities. (Section 18.2)

actinide element Element in which the $5f$ orbitals are only partially occupied. (Section 6.8)

activated complex (transition state) The particular arrangement of atoms found at the top of the potential-energy barrier as a reaction proceeds from reactants to products. (Section 14.5)

activation energy (E_a) The minimum energy needed for reaction; the height of the energy barrier to formation of products. (Section 14.5)

active site Specific site on a heterogeneous catalyst or an enzyme where catalysis occurs. (Section 14.7)

activity The decay rate of a radioactive material, generally expressed as the number of disintegrations per unit time. (Section 21.4)

activity series A list of metals in order of decreasing ease of oxidation. (Section 4.4)

addition polymerization Polymerization that occurs through coupling of monomers with one another, with no other products formed in the reaction. (Section 12.8)

addition reaction A reaction in which a reagent adds to the two carbon atoms of a carbon–carbon multiple bond. (Section 24.3)

adsorption The binding of molecules to a surface. (Section 14.7)

alcohol An organic compound obtained by substituting a hydroxyl group ($-\text{OH}$) for a hydrogen on a hydrocarbon. (Sections 2.9 and 24.4)

aldehyde An organic compound that contains a carbonyl group ($\text{C}=\text{O}$) to which at least one hydrogen atom is attached. (Section 24.4)

alkali metals Members of group 1A in the periodic table. (Section 7.7)

alkaline earth metals Members of group 2A in the periodic table. (Section 7.7)

alkanes Compounds of carbon and hydrogen containing only carbon–carbon single bonds. (Sections 2.9 and 24.2)

alkenes Hydrocarbons containing one or more carbon–carbon double bonds. (Section 24.2)

alkyl group A group that is formed by removing a hydrogen atom from an alkane. (Section 25.3)

alkynes Hydrocarbons containing one or more carbon–carbon triple bonds. (Section 24.2)

alloy A substance that has the characteristic properties of a metal and contains more than one element. Often there is one principal metallic component, with other elements present in smaller amounts. Alloys may be homogeneous or heterogeneous. (Section 12.3)

alpha decay A type of radioactive decay in which an atomic nucleus emits an alpha particle and thereby transforms (or "decays") into an atom with a mass number 4 less and atomic number 2 less. (Section 21.1)

alpha (α) helix A protein structure in which the protein is coiled in the form of a helix with hydrogen bonds between $\text{C}=\text{O}$ and $\text{N}-\text{H}$ groups on adjacent turns. (Section 24.7)

alpha particles Particles that are identical to helium-4 nuclei, consisting of two protons and two neutrons, symbol ^4_2He or $^4_2\alpha$. (Section 21.1)

amide An organic compound that has an NR_2 group attached to a carbonyl. (Section 24.4)

amine A compound that has the general formula R_3N, where R may be H or a hydrocarbon group. (Section 16.7)

amino acid A carboxylic acid that contains an amino ($-\text{NH}_2$) group attached to the carbon atom adjacent to the carboxylic acid ($-\text{COOH}$) functional group. (Section 24.7)

amorphous solid A solid whose molecular arrangement lacks the regularly repeating long-range pattern of a crystal. (Section 12.2)

amphiprotic Refers to the capacity of a substance to either add or lose a proton (H^+). (Section 16.2)

amphoteric oxides and hydroxides Oxides and hydroxides that are only slightly soluble in water but that dissolve in either acidic or basic solutions. (Section 17.5)

angstrom A common non-SI unit of length, denoted Å, that is used to measure atomic dimensions: $1\text{Å} = 10^{-10}$ m. (Section 2.3)

anion A negatively charged ion. (Section 2.7)

anode An electrode at which oxidation occurs. (Section 20.3)

antibonding molecular orbital A molecular orbital in which electron density is concentrated outside the region between the two nuclei of bonded atoms. Such orbitals, designated as σ^* or π^*, are less stable (of higher energy) than bonding molecular orbitals. (Section 9.7)

antiferromagnetism A form of magnetism in which unpaired electron spins on adjacent sites point in opposite directions and cancel each other's effects. (Section 23.1)

aqueous solution A solution in which water is the solvent. (Chapter 4: Introduction)

aromatic hydrocarbons Hydrocarbon compounds that contain a planar, cyclic arrangement of carbon atoms linked by both σ and delocalized π bonds. (Section 24.2)

Arrhenius equation An equation that relates the rate constant for a reaction to the frequency factor, A, the activation energy, E_a, and the temperature, T: $k = Ae^{-E_a/RT}$. In its logarithmic form it is written $\ln k = -E_a/RT + \ln A$. (Section 14.5)

atmosphere (atm) A unit of pressure equal to 760 torr; $1\ \text{atm} = 101.325$ kPa. (Section 10.2)

atom The smallest representative particle of an element. (Sections 1.1 and 2.1)

atomic mass unit (amu) A unit based on the value of exactly 12 amu for the mass of the isotope of carbon that has six protons and six neutrons in the nucleus. (Sections 2.3 and 3.3)

atomic number The number of protons in the nucleus of an atom of an element. (Section 2.3)

atomic radius An estimate of the size of an atom. See bonding atomic radius. (Section 7.3)

atomic weight The average mass of the atoms of an element in atomic mass units (amu); it is numerically equal to the mass in grams of one mole of the element. (Section 2.4)

autoionization The process whereby water spontaneously forms low concentrations of $\text{H}^+(aq)$ and $\text{OH}^-(aq)$ ions by proton transfer from one water molecule to another. (Section 16.3)

Avogadro's hypothesis A statement that equal volumes of gases at the same temperature and pressure contain equal numbers of molecules. (Section 10.3)

Avogadro's law A statement that the volume of a gas maintained at constant temperature and pressure is directly proportional to the number of moles of the gas. (Section 10.3)

Avogadro's number (N_A) The number of ^{12}C atoms in exactly 12 g of ^{12}C; it equals $6.022 \times 10^{23}\ \text{mol}^{-1}$. (Section 3.4)

band An array of closely spaced molecular orbitals occupying a discrete range of energy. (Section 12.4)

band gap The energy gap between a fully occupied band called a valence band and an empty band called the conduction band. (Section 12.7)

band structure The electronic structure of a solid, defining the allowed ranges of energy for electrons in a solid. (Section 12.7)

bar A unit of pressure equal to 10^5 Pa. (Section 10.2)

base A substance that is an H^+ acceptor; a base produces an excess of $OH^-(aq)$ ions when it dissolves in water. (Section 4.3)

base-dissociation constant (K_b) An equilibrium constant that expresses the extent to which a base reacts with solvent water, accepting a proton and forming $OH^-(aq)$. (Section 16.7)

basic anhydride (basic oxide) An oxide that forms a base when added to water; soluble metal oxides are basic anhydrides. (Section 22.5)

basic oxide (basic anhydride) An oxide that either reacts with water to form a base or reacts with an acid to form a salt and water. (Section 22.5)

battery A self-contained electrochemical power source that contains one or more voltaic cells. (Section 20.7)

becquerel The SI unit of radioactivity. It corresponds to one nuclear disintegration per second. (Section 21.4)

Beer's law The light absorbed by a substance (A) equals the product of its extinction coefficient (ε), the path length through which the light passes (b), and the molar concentration of the substance (c): $A = \varepsilon bc$. (Section 14.2)

beta emission A nuclear decay process where a beta particle is emitted from the nucleus; also called beta decay. (Section 21.1)

beta particles Energetic electrons emitted from the nucleus, symbol $_{-1}^{0}e$ or β^-. (Section 21.1)

beta sheet A structural form of protein in which two strands of amino acids are hydrogen-bonded together in a zipperlike configuration. (Section 24.7)

bidentate ligand A ligand in which two linked coordinating atoms are bound to a metal. (Section 23.3)

bimolecular reaction An elementary reaction that involves two molecules. (Section 14.6)

biochemistry The study of the chemistry of living systems. (Chapter 24: Introduction)

biodegradable Organic material that bacteria are able to oxidize. (Section 18.4)

body-centered lattice A crystal lattice in which the lattice points are located at the center and corners of each unit cell. (Section 12.2)

bomb calorimeter A device for measuring the heat evolved in the combustion of a substance under constant-volume conditions. (Section 5.5)

bond angles The angles made by the lines joining the nuclei of the atoms in a molecule. (Section 9.1)

bond dipole The dipole moment that is due to unequal electron sharing between two atoms in a covalent bond. (Section 9.3)

bond enthalpy The enthalpy change, ΔH, required to break a particular bond when the substance is in the gas phase. (Section 8.8)

bonding atomic radius The radius of an atom as defined by the distances separating it from other atoms to which it is chemically bonded. (Section 7.3)

bonding molecular orbital A molecular orbital in which the electron density is concentrated in the internuclear region. The energy of a bonding molecular orbital is lower than the energy of the separate atomic orbitals from which it forms. (Section 9.7)

bonding pair In a Lewis structure a pair of electrons that is shared by two atoms. (Section 9.2)

bond length The distance between the centers of two bonded atoms. (Section 8.3)

bond order The number of bonding electron pairs shared between two atoms, minus the number of antibonding electron pairs: bond order = (number of bonding electrons − number of antibonding electrons)/2. (Section 9.7)

bond polarity A measure of the degree to which the electrons are shared unequally between two atoms in a chemical bond. (Section 8.4)

boranes Covalent hydrides of boron. (Section 22.11)

Born–Haber cycle A thermodynamic cycle based on Hess's law that relates the lattice energy of an ionic substance to its enthalpy of formation and to other measurable quantities. (Section 8.2)

Boyle's law A law stating that at constant temperature, the product of the volume and pressure of a given amount of gas is a constant. (Section 10.3)

Brønsted–Lowry acid A substance (molecule or ion) that acts as a proton donor. (Section 16.2)

Brønsted–Lowry base A substance (molecule or ion) that acts as a proton acceptor. (Section 16.2)

buffer capacity The amount of acid or base a buffer can neutralize before the pH begins to change appreciably. (Section 17.2)

buffered solution (buffer) A solution that undergoes a limited change in pH upon addition of a small amount of acid or base. (Section 17.2)

calcination The heating of an ore to bring about its decomposition and the elimination of a volatile product. For example, a carbonate ore might be calcined to drive off CO_2. (Section 23.2)

calorie A unit of energy; it is the amount of energy needed to raise the temperature of 1 g of water by 1 °C from 14.5 °C to 15.5 °C. A related unit is the joule: 1 cal = 4.184 J. (Section 5.1)

calorimeter An apparatus that measures the heat released or absorbed in a chemical or physical process. (Section 5.5)

calorimetry The experimental measurement of heat produced in chemical and physical processes. (Section 5.5)

capillary action The process by which a liquid rises in a tube because of a combination of adhesion to the walls of the tube and cohesion between liquid particles. (Section 11.3)

carbide A binary compound of carbon with a metal or metalloid. (Section 22.9)

carbohydrates A class of substances formed from polyhydroxy aldehydes or ketones. (Section 24.8)

carbon black A microcrystalline form of carbon. (Section 22.9)

carbonyl group The C=O double bond, a characteristic feature of several organic functional groups, such as ketones and aldehydes. (Section 24.4)

carboxylic acid A compound that contains the —COOH functional group. (Sections 16.10 and 24.4)

catalyst A substance that changes the speed of a chemical reaction without itself undergoing a permanent chemical change in the process. (Section 14.7)

cathode An electrode at which reduction occurs. (Section 20.3)

cathode rays Streams of electrons that are produced when a high voltage is applied to electrodes in an evacuated tube. (Section 2.2)

cathodic protection A means of protecting a metal against corrosion by making it the cathode in a voltaic cell. This can be achieved by attaching a more easily oxidized metal, which serves as an anode, to the metal to be protected. (Section 20.8)

cation A positively charged ion. (Section 2.7)

cell potential The potential difference between the cathode and anode in an electrochemical cell; it is measured in volts: 1 V = 1 J/C. Also called electromotive force. (Section 20.4)

cellulose A polysaccharide of glucose; it is the major structural element in plant matter. (Section 24.8)

Celsius scale A temperature scale on which water freezes at 0° and boils at 100° at sea level. (Section 1.4)

chain reaction A series of reactions in which one reaction initiates the next. (Section 21.7)

changes of state Transformations of matter from one state to a different one, for example, from a gas to a liquid. (Section 1.3)

charcoal A form of carbon produced when wood is heated strongly in a deficiency of air. (Section 22.9)

Charles's law A law stating that at constant pressure, the volume of a given quantity of gas is proportional to absolute temperature. (Section 10.3)

chelate effect The generally larger formation constants for polydentate ligands as compared with the corresponding *monodentate* ligands. (Section 23.3)

chelating agent A polydentate ligand that is capable of occupying two or more sites in the coordination sphere. (Section 23.3)

chemical bond A strong attractive force that exists between atoms in a molecule. (Section 8.1)

chemical changes Processes in which one or more substances are converted into other substances; also called chemical reactions. (Section 1.3)

chemical equation A representation of a chemical reaction using the chemical formulas of the reactants and products; a balanced chemical equation contains equal numbers of atoms of each element on both sides of the equation. (Section 3.1)

chemical equilibrium A state of dynamic balance in which the rate of formation of the products of a reaction from the reactants equals the rate of formation of the reactants from the products; at equilibrium the concentrations of the reactants and products remain constant. (Section 4.1; Chapter 15: Introduction)

chemical formula A notation that uses chemical symbols with numerical subscripts to convey the relative proportions of atoms of the different elements in a substance. (Section 2.6)

chemical kinetics The area of chemistry concerned with the speeds, or rates, at which chemical reactions occur. (Chapter 14: Introduction)

chemical nomenclature The rules used in naming substances. (Section 2.8)

chemical properties Properties that describe a substance's composition and its reactivity; how the substance reacts or changes into other substances. (Section 1.3)

chemical reactions Processes in which one or more substances are converted into other substances; also called chemical changes. (Section 1.3)

chemistry The scientific discipline that studies the composition, properties, and transformations of matter. (Chapter 1: Introduction)

chiral A term describing a molecule or an ion that cannot be superimposed on its mirror image. (Sections 23.4 and 24.5)

chlorofluorocarbons Compounds composed entirely of chlorine, fluorine, and carbon. (Section 18.3)

chlorophyll A plant pigment that plays a major role in conversion of solar energy to chemical energy in photosynthesis. (Section 23.3)

cholesteric liquid crystalline phase A liquid crystal formed from flat, disc-shaped molecules that align through a stacking of the molecular discs. (Section 11.7)

coal A naturally occurring solid containing hydrocarbons of high molecular weight, as well as compounds containing sulfur, oxygen, and nitrogen. (Section 5.8)

colligative property A property of a solvent (vapor-pressure lowering, freezing-point lowering, boiling-point elevation, osmotic pressure) that depends on the total concentration of solute particles present. (Section 13.5)

collision model A model of reaction rates based on the idea that molecules must collide to react; it explains the factors influencing reaction rates in terms of the frequency of collisions, the number of collisions with energies exceeding the activation energy, and the probability that the collisions occur with suitable orientations. (Section 14.5)

colloids (colloidal dispersions) Mixtures containing particles larger than normal solutes but small enough to remain suspended in the dispersing medium. (Section 13.6)

combination reaction A chemical reaction in which two or more substances combine to form a single product. (Section 3.2)

combustion reaction A chemical reaction that proceeds with evolution of heat and usually also a flame; most combustion involves reaction with oxygen, as in the burning of a match. (Section 3.2)

common-ion effect A shift of an equilibrium induced by an ion common to the equilibrium. For example, added Na_2SO_4 decreases the solubility of the slightly soluble salt $BaSO_4$, or added NaF decreases the percent ionization of HF. (Section 17.1)

complementary colors Colors that, when mixed in proper proportions, appear white or colorless. (Section 23.5)

complete ionic equation A chemical equation in which dissolved strong electrolytes (such as dissolved ionic compounds) are written as separate ions. (Section 4.2)

complex ion (complex) An assembly of a metal ion and the Lewis bases (ligands) bonded to it. (Section 17.5)

compound A substance composed of two or more elements united chemically in definite proportions. (Section 1.2)

compound semiconductor A semiconducting material formed from two or more elements. (Section 12.7)

concentration The quantity of solute present in a given quantity of solvent or solution. (Section 4.5)

concentration cell A voltaic cell containing the same electrolyte and the same electrode materials in both the anode and cathode compartments. The emf of the cell is derived from a difference in the concentrations of the same electrolyte solutions in the compartments. (Section 20.6)

condensation polymerization Polymerization in which molecules are joined together through condensation reactions. (Section 12.8)

condensation reaction A chemical reaction in which a small molecule (such as a molecule of water) is split out from between two reacting molecules. (Sections 12.6 and 22.8)

conduction band A band of molecular orbitals lying higher in energy than the occupied valence band and distinctly separated from it. (Section 12.7)

conjugate acid A substance formed by addition of a proton to a Brønsted–Lowry base. (Section 16.2)

conjugate acid–base pair An acid and a base, such as H_2O and OH^-, that differ only in the presence or absence of a proton. (Section 16.2)

conjugate base A substance formed by the loss of a proton from a Brønsted–Lowry acid. (Section 16.2)

continuous spectrum A spectrum that contains radiation distributed over all wavelengths. (Section 6.3)

conversion factor A ratio relating the same quantity in two systems of units that is used to convert the units of measurement. (Section 1.6)

coordination compound A compound containing a metal ion bonded to a group of surrounding molecules or ions that act as ligands. (Section 23.2)

coordination number The number of adjacent atoms to which an atom is directly bonded. In a complex the coordination number of the metal ion is the number of donor atoms to which it is bonded. (Sections 12.37 and 24.2)

coordination sphere The metal ion and its surrounding ligands. (Section 23.2)

coordination-sphere isomers Structural isomers of coordination compounds in which the ligands within the coordination sphere differ. (Section 23.4)

copolymer A complex polymer resulting from the polymerization of two or more chemically different monomers. (Section 12.8)

core electrons The electrons that are not in the outermost shell of an atom. (Section 6.8)

corrosion The process by which a metal is oxidized by substances in its environment. (Section 20.8)

covalent bond A bond formed between two or more atoms by a sharing of electrons. (Section 8.1)

covalent-network solids Solids in which the units that make up the three-dimensional network are joined by covalent bonds. (Section 12.1)

critical mass The amount of fissionable material necessary to maintain a nuclear chain reaction. (Section 21.7)

critical pressure The pressure at which a gas at its critical temperature is converted to a liquid state. (Section 11.4)

critical temperature The highest temperature at which it is possible to convert the gaseous form of a substance to a liquid. The critical temperature increases with an increase in the magnitude of intermolecular forces. (Section 11.4)

crystal-field theory A theory that accounts for the colors and the magnetic and other properties of transition-metal complexes in terms of the splitting of the energies of metal ion d orbitals by the electrostatic interaction with the ligands. (Section 23.6)

crystal lattice An imaginary network of points on which the repeating motif of a solid may be imagined to be laid down so that the structure of the crystal is obtained. The motif may be a single atom or a group of atoms. Each lattice point represents an identical environment in the crystal. (Section 12.2)

crystalline solid (crystal) A solid whose internal arrangement of atoms, molecules, or ions possesses a regularly repeating pattern in any direction through the solid. (Section 12.2)

crystallization The process in which molecules, ions, or atoms come together to form a crystalline solid. (Section 13.2)

cubic close packing A crystal structure where the atoms are packed together as close as possible, and the close-packed layers of atoms adopt a three-layer repeating pattern that leads to a face-centered cubic unit cell. (Section 12.3)

curie A measure of radioactivity: 1 curie $= 3.7 \times 10^{10}$ nuclear disintegrations per second. (Section 21.4)

cycloalkanes Saturated hydrocarbons of general formula C_nH_{2n} in which the carbon atoms form a closed ring. (Section 24.2)

Dalton's law of partial pressures A law stating that the total pressure of a mixture of gases is the sum of the pressures that each gas would exert if it were present alone. (Section 10.6)

d–d transition The transition of an electron in a transition-metal compound from a lower-energy d orbital to a higher-energy d orbital. (Section 23.6)

decomposition reaction A chemical reaction in which a single compound reacts to give two or more products. (Section 3.2)

degenerate A situation in which two or more orbitals have the same energy. (Section 6.7)

delocalized electrons Electrons that are spread over a number of atoms in a molecule or a crystal rather than localized on a single atom or a pair of atoms. (Section 9.6)

density The ratio of an object's mass to its volume. (Section 1.4)

deoxyribonucleic acid (DNA) A polynucleotide in which the sugar component is deoxyribose. (Section 24.10)

desalination The removal of salts from seawater, brine, or brackish water to make it fit for human consumption. (Section 18.4)

deuterium The isotope of hydrogen whose nucleus contains a proton and a neutron: $_1^2H$. (Section 22.2)

dextrorotatory, or merely dextro or d A term used to label a chiral molecule that rotates the plane of polarization of plane-polarized light to the right (clockwise). (Section 23.4)

diamagnetism A type of magnetism that causes a substance with no unpaired electrons to be weakly repelled from a magnetic field. (Section 9.8)

diatomic molecule A molecule composed of only two atoms. (Section 2.6)

diffusion The spreading of one substance through a space occupied by one or more other substances. (Section 10.8)

dilution The process of preparing a less concentrated solution from a more concentrated one by adding solvent. (Section 4.5)

dimensional analysis A method of problem solving in which units are carried through all calculations. Dimensional analysis ensures that the final answer of a calculation has the desired units. (Section 1.6)

dipole A molecule with one end having a partial negative charge and the other end having a partial positive charge; a polar molecule. (Section 8.4)

dipole–dipole force A force that becomes significant when polar molecules come in close contact with one another. The force is attractive when the positive end of one polar molecule approaches the negative end of another. (Section 11.2)

dipole moment A measure of the separation and magnitude of the positive and negative charges in polar molecules. (Section 8.4)

dispersion forces Intermolecular forces resulting from attractions between induced dipoles. Also called London dispersion forces. (Section 11.2)

displacement reaction A reaction in which an element reacts with a compound, displacing an element from it. (Section 4.4)

donor atom The atom of a ligand that bonds to the metal. (Section 23.2)

doping Incorporation of a hetero atom into a solid to change its electrical properties. For example, incorporation of P into Si. (Section 12.7)

double bond A covalent bond involving two electron pairs. (Section 8.3)

double helix The structure for DNA that involves the winding of two DNA polynucleotide chains together in a helical arrangement. The two strands of the double helix are complementary in that the organic bases on the two strands are paired for optimal hydrogen bond interaction. (Section 24.10)

dynamic equilibrium A state of balance in which opposing processes occur at the same rate. (Section 11.5)

effective nuclear charge The net positive charge experienced by an electron in a many-electron atom; this charge is not the full nuclear charge because there is some shielding of the nucleus by the other electrons in the atom. (Section 7.2)

effusion The escape of a gas through an orifice or hole. (Section 10.8)

elastomer A material that can undergo a substantial change in shape via stretching, bending, or compression and return to its original shape upon release of the distorting force. (Section 12.6)

electrochemistry The branch of chemistry that deals with the relationships between electricity and chemical reactions. (Chapter 20: Introduction)

electrolysis reaction A reaction in which a nonspontaneous redox reaction is brought about by the passage of current under a sufficient external electrical potential. The devices in which electrolysis reactions occur are called electrolytic cells. (Section 20.9)

electrolyte A solute that produces ions in solution; an electrolytic solution conducts an electric current. (Section 4.1)

electrolytic cell A device in which a nonspontaneous oxidation–reduction reaction is caused to occur by passage of current under a sufficient external electrical potential. (Section 20.9)

electromagnetic radiation (radiant energy) A form of energy that has wave characteristics and that propagates through a vacuum at the characteristic speed of 3.00×10^8 m/s. (Section 6.1)

electrometallurgy The use of electrolysis to reduce or refine metals. (Section 20.9)

electromotive force (emf) A measure of the driving force, or *electrical pressure*, for the completion of an electrochemical reaction. Electromotive force is measured in volts: $1\ V = 1\ J/C$. Also called the cell potential. (Section 20.4)

electron A negatively charged subatomic particle found outside the atomic nucleus; it is a part of all atoms. An electron has a mass $1/1836$ times that of a proton. (Section 2.3)

electron affinity The energy change that occurs when an electron is added to a gaseous atom or ion. (Section 7.5)

electron capture A mode of radioactive decay in which an inner-shell orbital electron is captured by the nucleus. (Section 21.1)

electron configuration The arrangement of electrons in the orbitals of an atom or molecule. (Section 6.8)

electron density The probability of finding an electron at any particular point in an atom; this probability is equal to ψ^2, the square of the wave function. Also called the probability density. (Section 6.5)

electron domain In the VSEPR model, a region about a central atom in which an electron pair is concentrated. (Section 9.2)

electron-domain geometry The three-dimensional arrangement of the electron domains around an atom according to the VSEPR model. (Section 9.2)

electronegativity A measure of the ability of an atom that is bonded to another atom to attract electrons to itself. (Section 8.4)

electronic charge The negative charge carried by an electron; it has a magnitude of 1.602×10^{-19} C. (Section 2.3)

electronic structure The arrangement of electrons in an atom or molecule. (Chapter 6: Introduction)

electron-sea model A model for the behavior of electrons in metals. (Section 12.4)

electron shell A collection of orbitals that have the same value of n. For example, the orbitals with $n = 3$ (the $3s$, $3p$, and $3d$ orbitals) comprise the third shell. (Section 6.5)

electron spin A property of the electron that makes it behave as though it were a tiny magnet. The electron behaves as if it were spinning on its axis; electron spin is quantized. (Section 6.7)

element A substance consisting of atoms of the same atomic number. Historically defined as a substance that cannot be separated into simpler substances by chemical means. (Sections 1.1 and 1.2)

elemental semiconductor A semiconducting material composed of just one element. (Section 12.7)

elementary reaction A process in a chemical reaction that occurs in a single step. An overall chemical reaction consists of one or more elementary reactions or steps. (Section 14.6)

empirical formula A chemical formula that shows the kinds of atoms and their relative numbers in a substance in the smallest possible whole-number ratios. (Section 2.6)

enantiomers Two mirror-image molecules of a chiral substance. The enantiomers are nonsuperimposable. (Section 23.4)

endothermic process A process in which a system absorbs heat from its surroundings. (Section 5.2)

energy The capacity to do work or to transfer heat. (Section 5.1)

energy-level diagram A diagram that shows the energies of molecular orbitals relative to the atomic orbitals from which they are derived. Also called a **molecular-orbital diagram**. (Section 9.7)

enthalpy A quantity defined by the relationship $H = E + PV$; the enthalpy change, ΔH, for a reaction that occurs at constant pressure is the heat evolved or absorbed in the reaction: $\Delta H = q_p$. (Section 5.3)

enthalpy of formation The enthalpy change that accompanies the formation of a substance from the most stable forms of its component elements. (Section 5.7)

enthalpy of reaction The enthalpy change associated with a chemical reaction. (Section 5.4)

entropy A thermodynamic function associated with the number of different equivalent energy states or spatial arrangements in which a system may be found. It is a thermodynamic state function, which means that once we specify the conditions for a system—that is, the temperature, pressure, and so on—the entropy is defined. (Section 19.2)

enzyme A protein molecule that acts to catalyze specific biochemical reactions. (Section 14.7)

equilibrium constant The numerical value of the equilibrium-constant expression for a system at equilibrium. The equilibrium constant is most usually denoted by K_p for gas-phase systems or K_c for solution-phase systems. (Section 15.2)

equilibrium-constant expression The expression that describes the relationship among the concentrations (or partial pressures) of the substances present in a system at equilibrium. The numerator is obtained by multiplying the concentrations of the substances on the product side of the equation, each raised to a power equal to its coefficient in the chemical equation. The denominator similarly contains the concentrations of the substances on the reactant side of the equation. (Section 15.2)

equivalence point The point in a titration at which the added solute reacts completely with the solute present in the solution. (Section 4.6)

ester An organic compound that has an OR group attached to a carbonyl; it is the product of a reaction between a carboxylic acid and an alcohol. (Section 24.4)

ether A compound in which two hydrocarbon groups are bonded to one oxygen. (Section 24.4)

exchange (metathesis) reaction A reaction between compounds that when written as a molecular equation appears to involve the exchange of ions between the two reactants. (Section 4.2)

excited state A higher energy state than the ground state. (Section 6.3)

exothermic process A process in which a system releases heat to its surroundings. (Section 5.2)

extensive property A property that depends on the amount of material considered; for example, mass or volume. (Section 1.3)

face-centered lattice A crystal lattice in which the lattice points are located at the faces and corners of each unit cell. (Section 12.2)

Faraday constant (*F*) The magnitude of charge of one mole of electrons: 96,500 C/mol. (Section 20.5)

***f*-block metals** Lanthanide and actinide elements in which the 4*f* or 5*f* orbitals are partially occupied. (Section 6.9)

ferrimagnetism A form of magnetism in which unpaired electron spins on different-type ions point in opposite directions but do not fully cancel out. (Section 23.1)

ferromagnetism A form of magnetism in which unpaired electron spins align parallel to one another. (Section 23.1)

first law of thermodynamics A statement that energy is conserved in any process. One way to express the law is that the change in internal energy, ΔE, of a system in any process is equal to the heat, q, added to the system, plus the work, w, done on the system by its surroundings: $\Delta E = q + w$. (Section 5.2)

first-order reaction A reaction in which the reaction rate is proportional to the concentration of a single reactant, raised to the first power. (Section 14.4)

fission The splitting of a large nucleus into two smaller ones. (Section 21.6)

folding The process by which a protein adopts its biologically active shape. (Section 24.7)

force A push or a pull. (Section 5.1)

formal charge The number of valence electrons in an isolated atom minus the number of electrons assigned to the atom in the Lewis structure. (Section 8.5)

formation constant For a metal ion complex, the equilibrium constant for formation of the complex from the metal ion and base species present in solution. It is a measure of the tendency of the complex to form. (Section 17.5)

formula weight The mass of the collection of atoms represented by a chemical formula. For example, the formula weight of NO_2 (46.0 amu) is the sum of the masses of one nitrogen atom and two oxygen atoms. (Section 3.3)

fossil fuels Coal, oil, and natural gas, which are presently our major sources of energy. (Section 5.8)

fracking The practice in which water laden with sand and other materials is pumped at high pressure into rock formations to release natural gas and other petroleum materials. (Section 18.4)

free energy (Gibbs free energy, *G*) A thermodynamic state function that gives a criterion for spontaneous change in terms of enthalpy and entropy: $G = H - TS$. (Section 19.5)

free radical A substance with one or more unpaired electrons. (Section 21.9)

frequency The number of times per second that one complete wavelength passes a given point. (Section 6.1)

frequency factor (*A*) A term in the Arrhenius equation that is related to the frequency of collision and the probability that the collisions are favorably oriented for reaction. (Section 14.5)

fuel cell A voltaic cell that utilizes the oxidation of a conventional fuel, such as H_2 or CH_4, in the cell reaction. (Section 20.7)

fuel value The energy released when 1 g of a substance is combusted. (Section 5.8)

functional group An atom or group of atoms that imparts characteristic chemical properties to an organic compound. (Section 24.1)

fusion The joining of two light nuclei to form a more massive one. (Section 21.6)

galvanic cell See **voltaic cell**. (Section 20.3)

gamma radiation Energetic electromagnetic radiation emanating from the nucleus of a radioactive atom. (Section 21.1)

gas Matter that has no fixed volume or shape; it conforms to the volume and shape of its container. (Section 1.2)

gas constant (*R*) The constant of proportionality in the ideal-gas equation. (Section 10.4)

geometric isomerism A form of isomerism in which compounds with the same type and number of atoms and the same chemical bonds have different spatial arrangements of these atoms and bonds. (Sections 23.4 and 24.4)

Gibbs free energy A thermodynamic state function that combines enthalpy and entropy, in the form $G = H - TS$. For a change occurring at constant temperature and pressure, the change in free energy is $\Delta G = \Delta H - T\Delta S$. (Section 19.5)

glass An amorphous solid formed by fusion of SiO_2, CaO, and Na_2O. Other oxides may also be used to form glasses with differing characteristics. (Section 22.10)

glucose A polyhydroxy aldehyde whose formula is $CH_2OH(CHOH)_4CHO$; it is the most important of the monosaccharides. (Section 24.8)

glycogen The general name given to a group of polysaccharides of glucose that are synthesized in mammals and used to store energy from carbohydrates. (Section 24.7)

Graham's law A law stating that the rate of effusion of a gas is inversely proportional to the square root of its molecular weight. (Section 10.8)

gray (Gy) The SI unit for radiation dose corresponding to the absorption of 1 J of energy per kilogram of biological material; 1 Gy = 100 rads. (Section 21.9)

green chemistry Chemistry that promotes the design and application of chemical products and processes that are compatible with human health and that preserve the environment. (Section 18.5)

greenhouse gases Gases in an atmosphere that absorb and emit infrared radiation (radiant heat), "trapping" heat in the atmosphere. (Section 18.2)

ground state The lowest-energy, or most stable, state. (Section 6.3)

group Elements that are in the same column of the periodic table; elements within the same group or family exhibit similarities in their chemical behavior. (Section 2.5)

Haber process The catalyst system and conditions of temperature and pressure developed by Fritz Haber and coworkers for the formation of NH_3 from H_2 and N_2. (Section 15.2)

half-life The time required for the concentration of a reactant substance to decrease to half its initial value; the time required for half of a sample of a particular radioisotope to decay. (Sections 14.4 and 21.4)

half-reaction An equation for either an oxidation or a reduction that explicitly shows the electrons involved, for example, $Zn^{2+}(aq) + 2\,e^- \longrightarrow Zn(s)$. (Section 20.2)

halogens Members of group 7A in the periodic table. (Section 7.8)

hard water Water that contains appreciable concentrations of Ca^{2+} and Mg^{2+}; these ions react with soaps to form an insoluble material. (Section 18.4)

heat The flow of energy from a body at higher temperature to one at lower temperature when they are placed in thermal contact. (Section 5.1)

heat capacity The quantity of heat required to raise the temperature of a sample of matter by 1 °C (or 1 K). (Section 5.5)

heat of fusion The enthalpy change, ΔH, for melting a solid. (Section 11.4)

heat of sublimation The enthalpy change, ΔH, for vaporization of a solid. (Section 11.4)

heat of vaporization The enthalpy change, ΔH, for vaporization of a liquid. (Section 11.4)

Henderson–Hasselbalch equation The relationship among the pH, pK_a, and the concentrations of acid and conjugate base in an aqueous solution: $pH = pK_a + \log \dfrac{[\text{base}]}{[\text{acid}]}$. (Section 17.2)

Henry's law A law stating that the concentration of a gas in a solution, Sg, is proportional to the pressure of gas over the solution: $S_g = kP_g$. (Section 13.3)

Hess's law The heat evolved in a given process can be expressed as the sum of the heats of several processes that, when added, yield the process of interest. (Section 5.6)

heterogeneous alloy An alloy in which the components are not distributed uniformly; instead, two or more distinct phases with characteristic compositions are present. (Section 12.3)

heterogeneous catalyst A catalyst that is in a different phase from that of the reactant substances. (Section 14.7)

heterogeneous equilibrium The equilibrium established between substances in two or more different phases, for example, between a gas and a solid or between a solid and a liquid. (Section 15.4)

hexagonal close packing A crystal structure where the atoms are packed together as closely as possible. The close-packed layers adopt a two-layer repeating pattern, which leads to a primitive hexagonal unit cell. (Section 12.3)

high-spin complex A complex whose electrons populate the d orbitals to give the maximum number of unpaired electrons. (Section 23.6)

hole A vacancy in the valence band of a semiconductor, created by doping. (Section 12.7)

homogeneous catalyst A catalyst that is in the same phase as the reactant substances. (Section 14.7)

homogeneous equilibrium The equilibrium established between reactant and product substances that are all in the same phase. (Section 15.4)

Hund's rule A rule stating that electrons occupy degenerate orbitals in such a way as to maximize the number of electrons with the same spin. In other words, each orbital has one electron placed in it before pairing of electrons in orbitals occurs. (Section 6.8)

hybridization The mixing of different types of atomic orbitals to produce a set of equivalent hybrid orbitals. (Section 9.5)

hybrid orbital An orbital that results from the mixing of different kinds of atomic orbitals on the same atom. For example, an sp^3 hybrid results from the mixing, or hybridizing, of one s orbital and three p orbitals. (Section 9.5)

hydration Solvation when the solvent is water. (Section 13.1)

hydride ion An ion formed by the addition of an electron to a hydrogen atom: H^-. (Section 7.7)

hydrocarbons Compounds composed of only carbon and hydrogen. (Section 2.9)

hydrogen bonding Bonding that results from intermolecular attractions between molecules containing hydrogen bonded to an electronegative element. The most important examples involve OH, NH, and HF. (Section 11.2)

hydrolysis A reaction with water. When a cation or anion reacts with water, it changes the pH. (Sections 16.9 and 24.4)

hydronium ion (H_3O^+) The predominant form of the proton in aqueous solution. (Section 16.2)

hydrophilic Water attracting. The term is often used to describe a type of colloid. (Section 13.6)

hydrophobic Water repelling. The term is often used to describe a type of colloid. (Section 13.6)

hypothesis A tentative explanation of a series of observations or of a natural law. (Section 1.3)

ideal gas A hypothetical gas whose pressure, volume, and temperature behavior is completely described by the ideal-gas equation. (Section 10.4)

ideal-gas equation An equation of state for gases that embodies Boyle's law, Charles's law, and Avogadro's hypothesis in the form $PV = nRT$. (Section 10.4)

ideal solution A solution that obeys Raoult's law. (Section 13.5)

immiscible liquids Liquids that do not dissolve in one another to a significant extent. (Section 13.3)

indicator A substance added to a solution that changes color when the added solute has reacted with all the solute present in solution. The most common type of indicator is an acid–base indicator whose color changes as a function of pH. (Section 4.6)

instantaneous rate The reaction rate at a particular time as opposed to the average rate over an interval of time. (Section 14.2)

insulators Materials that do not conduct electricity. (Section 12.7)

intensive property A property that is independent of the amount of material considered, for example, density. (Section 1.3)

interhalogens Compounds formed between two different halogen elements. Examples include IBr and BrF_3. (Section 22.4)

intermediate A substance formed in one elementary step of a multistep mechanism and consumed in another; it is neither a reactant nor an ultimate product of the overall reaction. (Section 14.6)

intermetallic compound A homogeneous alloy with definite properties and a fixed composition. Intermetallic compounds are stoichiometric compounds that form between metallic elements. (Section 12.3)

intermolecular forces The short-range attractive forces operating between the particles that make up the units of a liquid or solid substance. These same forces also cause gases to liquefy or solidify at low temperatures and high pressures. (Chapter 11: Introduction)

internal energy The total energy possessed by a system. When a system undergoes a change, the change in internal energy, ΔE, is defined as the heat, q, added to the system, plus the work, w, done on the system by its surroundings: $\Delta E = q + w$. (Section 5.2)

interstitial alloy An alloy in which smaller atoms fit into spaces between larger atoms. The larger atoms are metallic elements and the smaller atoms are typically nonmetallic elements. (Section 12.3)

ion Electrically charged atom or group of atoms (polyatomic ion); ions can be positively or negatively charged, depending on whether electrons are lost (positive) or gained (negative) by the atoms. (Section 2.7)

ion–dipole force The force that exists between an ion and a neutral polar molecule that possesses a permanent dipole moment. (Section 11.2)

ionic bond A bond between oppositely charged ions. The ions are formed from atoms by transfer of one or more electrons. (Section 8.1)

ionic compound A compound composed of cations and anions. (Section 2.7)

ionic hydrides Compounds formed when hydrogen reacts with alkali metals and also the heavier alkaline earths (Ca, Sr, and Ba); these compounds contain the hydride ion, H^-. (Section 22.2)

ionic solids Solids that are composed of ions. (Section 12.1)

ionization energy The energy required to remove an electron from a gaseous atom when the atom is in its ground state. (Section 7.4)

ionizing radiation Radiation that has sufficient energy to remove an electron from a molecule, thereby ionizing it. (Section 21.9)

ion-product constant For water, K_w is the product of the aquated hydrogen ion and hydroxide ion concentrations: $[H^+][OH^-] = K_w = 1.0 \times 10^{-14}$ at 25 °C. (Section 16.3)

irreversible process A process that cannot be reversed to restore both the system and its surroundings to their original states. Any spontaneous process is irreversible. (Section 19.1)

isoelectronic series A series of atoms, ions, or molecules having the same number of electrons. (Section 7.3)

isomers Compounds whose molecules have the same overall composition but different structures. (Sections 2.9 and 23.4)

isothermal process One that occurs at constant temperature. (Section 19.1)

isotopes Atoms of the same element containing different numbers of neutrons and therefore having different masses. (Section 2.3)

joule (J) The SI unit of energy, 1 kg-m^2/s^2. A related unit is the calorie: 4.184 J = 1 cal. (Section 5.1)

Kelvin scale The absolute temperature scale; the SI unit for temperature is the kelvin. Zero on the Kelvin scale corresponds to −273.15 °C. (Section 1.4)

ketone A compound in which the carbonyl group ($C=O$) occurs at the interior of a carbon chain and is therefore flanked by carbon atoms. (Section 24.4)

kinetic energy The energy that an object possesses by virtue of its motion. (Section 5.1)

kinetic-molecular theory A set of assumptions about the nature of gases. These assumptions, when translated into mathematical form, yield the ideal-gas equation. (Section 10.7)

lanthanide contraction The gradual decrease in atomic and ionic radii with increasing atomic number among the lanthanide elements, atomic numbers 57 through 70. The decrease arises because of a gradual increase in effective nuclear charge through the lanthanide series. (Section 23.1)

lanthanide (rare earth) element Element in which the 4f subshell is only partially occupied. (Sections 6.8 and 6.9)

lattice energy The energy required to separate completely the ions in an ionic solid. (Section 8.2)

lattice points Points in a crystal all of which have identical environments. (Section 12.2)

lattice vectors The vectors a, b, and c that define a crystal lattice. The position of any lattice point in a crystal can be represented by summing integer multiples of the lattice vectors. (Section 12.2)

law of constant composition A law that states that the elemental composition of a pure compound is always the same, regardless of its source; also called the **law of definite proportions**. (Section 1.2)

law of definite proportions A law that states that the elemental composition of a pure substance is always the same, regardless of its source; also called the **law of constant composition**. (Section 1.2)

law of mass action The rules by which the equilibrium constant is expressed in terms of the concentrations of reactants and products, in accordance with the balanced chemical equation for the reaction. (Section 15.2)

Le Châtelier's principle A principle stating that when we disturb a system at chemical equilibrium, the relative concentrations of reactants and products shift so as to partially undo the effects of the disturbance. (Section 15.7)

levorotatory, or merely levo or l A term used to label a chiral molecule that rotates the plane of polarization of plane-polarized light to the left (counterclockwise). (Section 24.4)

Lewis base An electron-pair donor. (Section 16.11)

Lewis structure A representation of covalent bonding in a molecule that is drawn using Lewis symbols. Shared electron pairs are shown as lines, and unshared electron pairs are shown as pairs of dots. Only the valence-shell electrons are shown. (Section 8.3)

Lewis symbol (electron-dot symbol) The chemical symbol for an element, with a dot for each valence electron. (Section 8.1)

ligand An ion or molecule that coordinates to a metal atom or to a metal ion to form a complex. (Section 23.2)

limiting reactant (limiting reagent) The reactant present in the smallest stoichiometric quantity in a mixture of reactants; the amount of product that can form is limited by the complete consumption of the limiting reactant. (Section 3.7)

line spectrum A spectrum that contains radiation at only certain specific wavelengths. (Section 6.3)

linkage isomers Structural isomers of coordination compounds in which a ligand differs in its mode of attachment to a metal ion. (Section 23.4)

lipid A nonpolar molecule derived from glycerol and fatty acids that is used by organisms for long-term energy storage. (Section 24.9)

liquid Matter that has a distinct volume but no specific shape. (Section 1.2)

liquid crystal A substance that exhibits one or more partially ordered liquid phases above the melting point of the solid form. By contrast, in nonliquid crystalline substances the liquid phase that forms upon melting is completely unordered. (Section 11.7)

lock-and-key model A model of enzyme action in which the substrate molecule is pictured as fitting rather specifically into the active site on the enzyme. It is assumed that in being bound to the active site, the substrate is somehow activated for reaction. (Section 14.7)

low-spin complex A metal complex in which the electrons are paired in lower-energy orbitals. (Section 23.6)

magic numbers Numbers of protons and neutrons that result in very stable nuclei. (Section 21.2)

main-group elements Elements in the s and p blocks of the periodic table. (Section 6.9)

mass A measure of the amount of material in an object. It measures the resistance of an object to being moved. In SI units, mass is measured in kilograms. (Section 1.4)

mass defect The difference between the mass of a nucleus and the total masses of the individual nucleons that it contains. (Section 21.6)

mass number The sum of the number of protons and neutrons in the nucleus of a particular atom. (Section 2.3)

mass percentage The number of grams of solute in each 100 g of solution. (Section 13.4)

mass spectrometer An instrument used to measure the precise masses and relative amounts of atomic and molecular ions. (Section 2.4)

matter Anything that occupies space and has mass; the physical material of the universe. (Section 1.1)

matter waves The term used to describe the wave characteristics of a moving particle. (Section 6.4)

mean free path The average distance traveled by a gas molecule between collisions. (Section 10.8)

metal complex An assembly of a metal ion and the Lewis bases bonded to it. (Section 23.2)

metallic bond Bonding, usually in solid metals, in which the bonding electrons are relatively free to move throughout the three-dimensional structure. (Section 8.1)

metallic character The extent to which an element exhibits the physical and chemical properties characteristic of metals, for example, luster, malleability, ductility, and good thermal and electrical conductivity. (Section 7.6)

metallic elements (metals) Elements that are usually solids at room temperature, exhibit high electrical and heat conductivity, and appear lustrous. Most of the elements in the periodic table are metals. (Sections 2.5 and 12.1)

metallic hydrides Compounds formed when hydrogen reacts with transition metals; these compounds contain the hydride ion, H^-. (Section 22.2)

metallic solids Solids that are composed of metal atoms. (Section 12.1)

metalloids Elements that lie along the diagonal line separating the metals from the nonmetals in the periodic table; the properties of metalloids are intermediate between those of metals and nonmetals. (Section 2.5)

metallurgy The science of extracting metals from their natural sources by a combination of chemical and physical processes. It is also concerned with the properties and structures of metals and alloys. (Section 23.1)

metathesis (exchange) reaction A reaction in which two substances react through an exchange of their component ions: $AX + BY \longrightarrow AY + BX$. Precipitation and acid–base neutralization reactions are examples of metathesis reactions. (Section 4.2)

metric system A system of measurement used in science and in most countries. The meter and the gram are examples of metric units. (Section 1.4)

microstate The state of a system at a particular instant; one of many possible energetically equivalent ways to arrange the components of a system to achieve a particular state. (Section 19.3)

mineral A solid, inorganic substance occurring in nature, such as calcium carbonate, which occurs as calcite. (Section 23.1)

miscible liquids Liquids that mix in all proportions. (Section 13.3)

mixture A combination of two or more substances in which each substance retains its own chemical identity. (Section 1.2)

molal boiling-point-elevation constant (K_b) A constant characteristic of a particular solvent that gives the increase in boiling point as a function of solution molality: $\Delta T_b = K_b m$. (Section 13.5)

molal freezing-point-depression constant (K_f) A constant characteristic of a particular solvent that gives the decrease in freezing point as a function of solution molality: $\Delta T_f = -K_f m$. (Section 13.5)

molality The concentration of a solution expressed as moles of solute per kilogram of solvent; abbreviated m. (Section 13.4)

molar heat capacity The heat required to raise the temperature of one mole of a substance by $1\ ^\circ C$. (Section 5.5)

molarity The concentration of a solution expressed as moles of solute per liter of solution; abbreviated M. (Section 4.5)

molar mass The mass of one mole of a substance in grams; it is numerically equal to the formula weight in atomic mass units. (Section 3.4)

mole A collection of Avogadro's number (6.022×10^{23}) of objects; for example, a mole of H_2O is 6.022×10^{23} H_2O molecules. (Section 3.4)

molecular compound A compound that consists of molecules. (Section 2.6)

molecular equation A chemical equation in which the formula for each substance is written without regard for whether it is an electrolyte or a nonelectrolyte. (Section 4.2)

molecular formula A chemical formula that indicates the actual number of atoms of each element in one molecule of a substance. (Section 2.6)

molecular geometry The arrangement in space of the atoms of a molecule. (Section 9.2)

molecular hydrides Compounds formed when hydrogen reacts with nonmetals and metalloids. (Section 22.2)

molecularity The number of molecules that participate as reactants in an elementary reaction. (Section 14.6)

molecular orbital (MO) An allowed state for an electron in a molecule. According to molecular-orbital theory, a molecular orbital is entirely analogous to an atomic orbital, which is an allowed state for an electron in an atom. Most bonding molecular orbitals can be classified as σ or π, depending on the disposition of electron density with respect to the internuclear axis. (Section 9.7)

molecular-orbital diagram A diagram that shows the energies of molecular orbitals relative to the atomic orbitals from which they are derived; also called an **energy-level diagram**. (Section 9.7)

molecular-orbital theory A theory that accounts for the allowed states for electrons in molecules. (Section 9.7)

molecular solids Solids that are composed of molecules. (Sections 12.1 and 12.6)

molecular weight The mass of the collection of atoms represented by the chemical formula for a molecule. (Section 3.3)

molecule A chemical combination of two or more atoms. (Sections 1.1 and 2.6)

mole fraction The ratio of the number of moles of one component of a mixture to the total moles of all components; abbreviated X, with a subscript to identify the component. (Section 10.6)

momentum The product of the mass, m, and velocity, v, of an object. (Section 6.4)

monodentate ligand A ligand that binds to the metal ion via a single donor atom. It occupies one position in the coordination sphere. (Section 23.3)

monomers Molecules with low molecular weights, which can be joined together (polymerized) to form a polymer. (Section 12.8)

monosaccharide A simple sugar, most commonly containing six carbon atoms. The joining together of monosaccharide units by condensation reactions results in formation of polysaccharides. (Section 24.8)

nanomaterial A solid whose dimensions range from 1 to 100 nm and whose properties differ from those of a bulk material with the same composition. (Section 12.1)

natural gas A naturally occurring mixture of gaseous hydrocarbon compounds composed of hydrogen and carbon. (Section 5.8)

nematic liquid crystalline phase A liquid crystal in which the molecules are aligned in the same general direction, along their long axes, but in which the ends of the molecules are not aligned. (Section 11.7)

Nernst equation An equation that relates the cell emf, E, to the standard emf, E°, and the reaction quotient, Q: $E = E^\circ - (RT/nF) \ln Q$. (Section 20.6)

net ionic equation A chemical equation for a solution reaction in which soluble strong electrolytes are written as ions and spectator ions are omitted. (Section 4.2)

neutralization reaction A reaction in which an acid and a base react in stoichiometrically equivalent amounts; the neutralization reaction between an acid and a metal hydroxide produces water and a salt. (Section 4.3)

neutron An electrically neutral particle found in the nucleus of an atom; it has approximately the same mass as a proton. (Section 2.3)

noble gases Members of group 8A in the periodic table. (Section 7.8)

node Points in an atom at which the electron density is zero. For example, the node in a 2s orbital is a spherical surface. (Section 6.6)

nonbonding pair In a Lewis structure a pair of electrons assigned completely to one atom; also called a lone pair. (Section 9.2)

nonelectrolyte A substance that does not ionize in water and consequently gives a nonconducting solution. (Section 4.1)

nonionizing radiation Radiation that does not have sufficient energy to remove an electron from a molecule. (Section 21.9)

nonmetallic elements (nonmetals) Elements in the upper right corner of the periodic table; nonmetals differ from metals in their physical and chemical properties. (Section 2.5)

nonpolar covalent bond A covalent bond in which the electrons are shared equally. (Section 8.4)

normal boiling point The boiling point at 1 atm pressure. (Section 11.5)

normal melting point The melting point at 1 atm pressure. (Section 11.6)

nuclear binding energy The energy required to decompose an atomic nucleus into its component protons and neutrons. (Section 21.6)

nuclear disintegration series A series of nuclear reactions that begins with an unstable nucleus and terminates with a stable one; also called a radioactive series. (Section 21.2)

nuclear model Model of the atom with a nucleus containing protons and neutrons and with electrons in the space outside the nucleus. (Section 2.2)

nuclear transmutation A conversion of one kind of nucleus to another. (Section 21.3)

nucleic acids Polymers of high molecular weight that carry genetic information and control protein synthesis. (Section 24.10)

nucleon A particle found in the nucleus of an atom. (Section 21.1)

nucleotide Compounds formed from a molecule of phosphoric acid, a sugar molecule, and an organic nitrogen base. Nucleotides form linear polymers called DNA and RNA, which are involved in protein synthesis and cell reproduction. (Section 24.10)

nucleus The very small, very dense, positively charged portion of an atom; it is composed of protons and neutrons. (Section 2.2)

octet rule A rule stating that bonded atoms tend to possess or share a total of eight valence-shell electrons. (Section 8.1)

optical isomerism A form of isomerism in which the two forms of a compound (stereoisomers) are nonsuperimposable mirror images. (Section 23.4)

optically active Possessing the ability to rotate the plane of polarized light. (Section 23.4)

orbital An allowed energy state of an electron in the quantum mechanical model of the atom; the term *orbital* is also used to describe the spatial distribution of the electron. An orbital is defined by the values of three quantum numbers: n, l, and m_l (Section 6.5)

organic chemistry The study of carbon-containing compounds, typically containing carbon–carbon bonds. (Section 2.9; Chapter 24: Introduction)

osmosis The net movement of solvent through a semipermeable membrane toward the solution with greater solute concentration. (Section 13.5)

osmotic pressure The pressure that must be applied to a solution to stop osmosis from pure solvent into the solution. (Section 13.5)

Ostwald process An industrial process used to make nitric acid from ammonia. The NH_3 is catalytically oxidized by O_2 to form NO; NO in air is oxidized to NO_2; HNO_3 is formed in a disproportionation reaction when NO_2 dissolves in water. (Section 22.7)

overall reaction order The sum of the reaction orders of all the reactants appearing in the rate expression when the rate can be expressed as rate $= k[A]^a[B]^b\ldots$. (Section 14.3)

overlap The extent to which atomic orbitals on different atoms share the same region of space. When the overlap between two orbitals is large, a strong bond may be formed. (Section 9.4)

oxidation A process in which a substance loses one or more electrons. (Section 4.4)

oxidation number (oxidation state) A positive or negative whole number assigned to an element in a molecule or ion on the basis of a set of formal rules; to some degree it reflects the positive or negative character of that atom. (Section 4.4)

oxidation–reduction (redox) reaction A chemical reaction in which the oxidation states of certain atoms change. (Section 4.4; Chapter 20: Introduction)

oxidizing agent, or oxidant The substance that is reduced and thereby causes the oxidation of some other substance in an oxidation–reduction reaction. (Section 20.1)

oxyacid A compound in which one or more OH groups, and possibly additional oxygen atoms, are bonded to a central atom. (Section 16.10)

oxyanion A polyatomic anion that contains one or more oxygen atoms. (Section 2.8)

ozone The name given to O_3, an allotrope of oxygen. (Section 7.8)

paramagnetism A property that a substance possesses if it contains one or more unpaired electrons. A paramagnetic substance is drawn into a magnetic field. (Section 9.8)

partial pressure The pressure exerted by a particular gas in a mixture. (Section 10.6)

particle accelerator A device that uses strong magnetic and electrostatic fields to accelerate charged particles. (Section 21.3)

parts per billion (ppb) The concentration of a solution in grams of solute per 10^9 (billion) grams of solution; equals micrograms of solute per liter of solution for aqueous solutions. (Section 13.4)

parts per million (ppm) The concentration of a solution in grams of solute per 10^6 (million) grams of solution; equals milligrams of solute per liter of solution for aqueous solutions. (Section 13.4)

pascal (Pa) The SI unit of pressure: $1 \text{ Pa} = 1 \text{ N/m}^2$. (Section 10.2)

Pauli exclusion principle A rule stating that no two electrons in an atom may have the same four quantum numbers (n, l, m_l, and m_s). As a reflection of this principle, there can be no more than two electrons in any one atomic orbital. (Section 6.7)

peptide bond A bond formed between two amino acids. (Section 24.7)

percent ionization The percent of a substance that undergoes ionization on dissolution in water. The term applies to solutions of weak acids and bases. (Section 16.6)

percent yield The ratio of the actual (experimental) yield of a product to its theoretical (calculated) yield, multiplied by 100. (Section 3.7)

period The row of elements that lie in a horizontal row in the periodic table. (Section 2.5)

periodic table The arrangement of elements in order of increasing atomic number, with elements having similar properties placed in vertical columns. (Section 2.5)

petroleum A naturally occurring combustible liquid composed of hundreds of hydrocarbons and other organic compounds. (Section 5.8)

pH The negative log in base 10 of the aquated hydrogen ion concentration: $pH = -\log[H^+]$. (Section 16.4)

pH titration curve A graph of pH as a function of added titrant. (Section 17.3)

phase change The conversion of a substance from one state of matter to another. The phase changes we consider are melting and freezing (solid $\rightleftharpoons$ liquid), sublimation and deposition, and vaporization and condensation (liquid $\rightleftharpoons$ gas). (Section 11.4)

phase diagram A graphic representation of the equilibria among the solid, liquid, and gaseous phases of a substance as a function of temperature and pressure. (Section 11.6)

phospholipid A form of lipid molecule that contains charged phosphate groups. (Section 24.9)

photochemical smog A complex mixture of undesirable substances produced by the action of sunlight on an urban atmosphere polluted with automobile emissions. The major starting ingredients are nitrogen oxides and organic substances, notably olefins and aldehydes. (Section 18.2)

photodissociation The breaking of a molecule into two or more neutral fragments as a result of absorption of light. (Section 18.2)

photoelectric effect The emission of electrons from a metal surface induced by light. (Section 6.2)

photoionization The removal of an electron from an atom or molecule by absorption of light. (Section 18.2)

photon The smallest increment (a quantum) of radiant energy; a photon of light with frequency ν has an energy equal to $h\nu$. (Section 6.2)

photosynthesis The process that occurs in plant leaves by which light energy is used to convert carbon dioxide and water to carbohydrates and oxygen. (Section 23.3)

physical changes Changes (such as a phase change) that occur with no change in chemical composition. (Section 1.3)

physical properties Properties that can be measured without changing the composition of a substance, for example, color and freezing point. (Section 1.3)

pi (π) bond A covalent bond in which electron density is concentrated above and below the internuclear axis. (Section 9.6)

pi (π) molecular orbital A molecular orbital that concentrates the electron density on opposite sides of an imaginary line that passes through the nuclei. (Section 9.8)

Planck constant (h) The constant that relates the energy and frequency of a photon, $E = h\nu$. Its value is 6.626×10^{-34} J-s. (Section 6.2)

plastic A material that can be formed into particular shapes by application of heat and pressure. (Section 12.8)

polar covalent bond A covalent bond in which the electrons are not shared equally. (Section 8.4)

polarizability The ease with which the electron cloud of an atom or a molecule is distorted by an outside influence, thereby inducing a dipole moment. (Section 11.2)

polar molecule A molecule that possesses a nonzero dipole moment. (Section 8.4)

polyatomic ion An electrically charged group of two or more atoms. (Section 2.7)

polydentate ligand A ligand in which two or more donor atoms can coordinate to the same metal ion. (Section 23.3)

polymer A large molecule of high molecular mass, formed by the joining together, or polymerization, of a large number of molecules of low molecular mass. The individual molecules forming the polymer are called monomers. (Sections 12.1 and 12.8)

polypeptide A polymer of amino acids that has a molecular weight of less than 10,000. (Section 24.7)

polyprotic acid A substance capable of dissociating more than one proton in water; H_2SO_4 is an example. (Section 16.6)

polysaccharide A substance made up of many monosaccharide units joined together. (Section 24.8)

porphyrin A complex derived from the porphine molecule. (Section 23.3)

positron A particle with the same mass as an electron but with a positive charge, $_{+1}^{0}e$, or β^{+}. (Section 21.1)

positron emission A nuclear decay process where a positron, a particle with the same mass as an electron but with a positive charge, symbol $_{+1}^{0}e$, or β^{+} is emitted from the nucleus. (Section 21.1)

potential energy The energy that an object possesses as a result of its composition or its position with respect to another object. (Section 5.1)

precipitate An insoluble substance that forms in, and separates from, a solution. (Section 4.2)

precipitation reaction A reaction that occurs between substances in solution in which one of the products is insoluble. (Section 4.2)

precision The closeness of agreement among several measurements of the same quantity; the reproducibility of a measurement. (Section 1.5)

pressure A measure of the force exerted on a unit area. In chemistry, pressure is often expressed in units of atmospheres (atm) or torr: 760 torr = 1 atm; in SI units pressure is expressed in pascals (Pa). (Section 10.2)

pressure–volume (PV) work Work performed by expansion of a gas against a resisting pressure. (Section 5.3)

primary cell A voltaic cell that cannot be recharged. (Section 20.7)

primary structure The sequence of amino acids along a protein chain. (Section 24.7)

primitive lattice A crystal lattice in which the lattice points are located only at the corners of each unit cell. (Section 12.2)

probability density (ψ^2) A value that represents the probability that an electron will be found at a given point in space. Also called **electron density**. (Section 6.5)

product A substance produced in a chemical reaction; it appears to the right of the arrow in a chemical equation. (Section 3.1)

property A characteristic that gives a sample of matter its unique identity. (Section 1.1)

protein A biopolymer formed from amino acids. (Section 24.7)

protium The most common isotope of hydrogen. (Section 22.2)

proton A positively charged subatomic particle found in the nucleus of an atom. (Section 2.3)

pure substance Matter that has a fixed composition and distinct properties. (Section 1.2)

pyrometallurgy A process in which heat converts a mineral in an ore from one chemical form to another and eventually to the free metal. (Section 23.2)

qualitative analysis The determination of the presence or absence of a particular substance in a mixture. (Section 17.7)

quantitative analysis The determination of the amount of a given substance that is present in a sample. (Section 17.7)

quantum The smallest increment of radiant energy that may be absorbed or emitted; the magnitude of radiant energy is $h\nu$. (Section 6.2)

quaternary structure The structure of a protein resulting from the clustering of several individual protein chains into a final specific shape. (Section 24.7)

racemic mixture A mixture of equal amounts of the dextrorotatory and levorotatory forms of a chiral molecule. A racemic mixture will not rotate the plane of polarized light. (Section 23.4)

rad A measure of the energy absorbed from radiation by tissue or other biological material; 1 rad = transfer of 1×10^{-2} J of energy per kilogram of material. (Section 21.9)

radial probability function The probability that the electron will be found at a certain distance from the nucleus. (Section 6.6)

radioactive Possessing **radioactivity**, the spontaneous disintegration of an unstable atomic nucleus with accompanying emission of radiation. (Section 2.2; Chapter 21: Introduction)

radioactive decay chain A series of nuclear reactions that begins with an unstable nucleus and terminates with a stable one. Also called **nuclear disintegration series**. (Section 21.2)

radioisotope An isotope that is radioactive; that is, it is undergoing nuclear changes with emission of radiation. (Section 21.1)

radionuclide A radioactive nuclide. (Section 21.1)

radiotracer A radioisotope that can be used to trace the path of an element in a chemical system. (Section 21.5)

Raoult's law A law stating that the partial pressure of a solvent over a solution, $P_{solution}$, is given by the vapor pressure of the pure solvent, $P^{\circ}_{solvent}$, times the mole fraction of a solvent in the solution, $X_{solvent}$: $P_{solution} = X_{solvent}P^{\circ}_{solvent}$. (Section 13.5)

rare earth element See **lanthanide element**. (Sections 6.8 and 6.9)

rate constant A constant of proportionality between the reaction rate and the concentrations of reactants that appear in the rate law. (Section 14.3)

rate-determining step The slowest elementary step in a reaction mechanism. (Section 14.6)

rate law An equation that relates the reaction rate to the concentrations of reactants (and sometimes of products also). (Section 14.3)

reactant A starting substance in a chemical reaction; it appears to the left of the arrow in a chemical equation. (Section 3.1)

reaction mechanism A detailed picture, or model, of how the reaction occurs; that is, the order in which bonds are broken and formed and the changes in relative positions of the atoms as the reaction proceeds. (Section 14.6)

reaction order The power to which the concentration of a reactant is raised in a rate law. (Section 14.3)

reaction quotient (Q) The value that is obtained when concentrations of reactants and products are inserted into the equilibrium expression. If the concentrations are equilibrium concentrations, $Q = K$; otherwise, $Q \neq K$. (Section 15.6)

reaction rate A measure of the decrease in concentration of a reactant or the increase in concentration of a product with time. (Section 14.2)

redox (oxidation–reduction) reaction A reaction in which certain atoms undergo changes in oxidation states. The substance increasing in oxidation state is oxidized; the substance decreasing in oxidation state is reduced. (Section 4.4; Chapter 20: Introduction)

reducing agent, or reductant The substance that is oxidized and thereby causes the reduction of some other substance in an oxidation–reduction reaction. (Section 20.1)

reduction A process in which a substance gains one or more electrons. (Section 4.4)

rem A measure of the biological damage caused by radiation; rems = rads × RBE. (Section 21.9)

renewable energy sources Energy such as solar energy, wind energy, and hydroelectric energy derived from essentially inexhaustible sources. (Section 5.8)

representative (main-group) element An element from within the s and p blocks of the periodic table (Figure 6.29). (Section 6.9)

resonance structures (resonance forms) Individual Lewis structures in cases where two or more Lewis structures are equally good descriptions of a single molecule. The resonance structures in such an instance are "averaged" to give a more accurate description of the real molecule. (Section 8.6)

reverse osmosis The process by which water molecules move under high pressure through a semipermeable membrane from the more concentrated to the less concentrated solution. (Section 18.4)

reversible process A process that can go back and forth between states along exactly the same path; a system at equilibrium is reversible if equilibrium can be shifted by an infinitesimal modification of a variable such as temperature. (Section 19.1)

ribonucleic acid (RNA) A polynucleotide in which ribose is the sugar component. (Section 24.10)

root-mean-square (rms) speed (μ) The square root of the average of the squared speeds of the gas molecules in a gas sample. (Section 10.7)

rotational motion Movement of a molecule as though it is spinning like a top. (Section 19.3)

salinity A measure of the salt content of seawater, brine, or brackish water. It is equal to the mass in grams of dissolved salts present in 1 kg of seawater. (Section 18.3)

salt An ionic compound formed by replacing one or more hydrogens of an acid by other cations. (Section 4.3)

saponification Hydrolysis of an ester in the presence of a base. (Section 24.4)

saturated solution A solution in which undissolved solute and dissolved solute are in equilibrium. (Section 13.2)

scientific law A concise verbal statement or a mathematical equation that summarizes a wide range of observations and experiences. (Section 1.3)

scientific method The general process of advancing scientific knowledge by making experimental observations and by formulating hypotheses, theories, and laws. (Section 1.3)

secondary cell A voltaic cell that can be recharged. (Section 20.7)

secondary structure The manner in which a protein is coiled or stretched. (Section 24.7)

second law of thermodynamics A statement of our experience that there is a direction to the way events occur in nature. When a process occurs spontaneously in one direction, it is nonspontaneous in the reverse direction. It is possible to state the second law in many different forms, but they all relate back to the same idea about spontaneity. One of the most common statements found in chemical contexts is that in any spontaneous process the entropy of the universe increases. (Section 19.2)

second-order reaction A reaction in which the overall reaction order (the sum of the concentration-term exponents) in the rate law is 2. (Section 14.4)

semiconductor A material that has electrical conductivity between that of a metal and that of an insulator. (Section 12.7)

sigma (σ) bond A covalent bond in which electron density is concentrated along the internuclear axis. (Section 9.6)

sigma (σ) molecular orbital A molecular orbital that centers the electron density about an imaginary line passing through two nuclei. (Section 9.7)

significant figures The digits that indicate the precision with which a measurement is made; all digits of a measured quantity are significant, including the last digit, which is uncertain. (Section 1.5)

silica Common name for silicon dioxide. (Section 22.4)

silicates Compounds containing silicon and oxygen, structurally based on SiO_4 tetrahedra. (Section 22.10)

single bond A covalent bond involving one electron pair. (Section 8.3)

SI units The preferred metric units for use in science. (Section 1.4)

smectic liquid crystalline phase A liquid crystal in which the molecules are aligned along their long axes and arranged in sheets, with the ends of the molecules aligned. There are several different kinds of smectic phases. (Section 12.8)

solid Matter that has both a definite shape and a definite volume. (Section 1.2)

solubility The amount of a substance that dissolves in a given quantity of solvent at a given temperature to form a saturated solution. (Sections 4.2 and 13.2)

solubility-product constant (solubility product) (K_{sp}) An equilibrium constant related to the equilibrium between a solid salt and its ions in solution. It provides a quantitative measure of the solubility of a slightly soluble salt. (Section 17.4)

solute A substance dissolved in a solvent to form a solution; it is normally the component of a solution present in the smaller amount. (Section 4.1)

solution A mixture of substances that has a uniform composition; a homogeneous mixture. (Section 1.2)

solution alloy A homogeneous alloy, where two or more elements are distributed randomly and uniformly throughout the solid. (Section 12.3)

solvation The clustering of solvent molecules around a solute particle. (Section 13.1)

solvent The dissolving medium of a solution; it is normally the component of a solution present in the greater amount. (Section 4.1)

specific heat (C_s) The heat capacity of 1 g of a substance; the heat required to raise the temperature of 1 g of a substance by 1 °C. (Section 5.5)

spectator ions Ions that go through a reaction unchanged and that appear on both sides of the complete ionic equation. (Section 4.2)

spectrochemical series A list of ligands arranged in order of their abilities to split the d-orbital energies (using the terminology of the crystal-field model). (Section 23.6)

spectrum The distribution among various wavelengths of the radiant energy emitted or absorbed by an object. (Section 6.3)

spin magnetic quantum number (m_s) A quantum number associated with the electron spin; it may have values of $+\frac{1}{2}$ or $-\frac{1}{2}$. (Section 6.7)

spin-pairing energy The energy required to pair an electron with another electron occupying an orbital. (Section 23.6)

spontaneous process A process that is capable of proceeding in a given direction, as written or described, without needing to be driven by an outside source of energy. A process may be spontaneous even though it is very slow. (Section 19.1)

standard atmospheric pressure Defined as 760 torr or, in SI units, 101.325 kPa. (Section 10.2)

standard emf, also called the standard cell potential ($E°$) The emf of a cell when all reagents are at standard conditions. (Section 20.4)

standard enthalpy change ($\Delta H°$) The change in enthalpy in a process when all reactants and products are in their stable forms at 1 atm pressure and a specified temperature, commonly 25 °C. (Section 5.7)

standard enthalpy of formation ($\Delta H_f°$) The change in enthalpy that accompanies the formation of one mole of a substance from its elements, with all substances in their standard states. (Section 5.7)

standard free energy of formation ($\Delta G_f°$) The change in free energy associated with the formation of a substance from its elements under standard conditions. (Section 19.5)

standard hydrogen electrode (SHE) An electrode based on the half-reaction $2\,H^+(1\,M) + 2\,e^- \longrightarrow H_2(1\,atm)$. The standard electrode potential of the standard hydrogen electrode is defined as 0 V. (Section 20.4)

standard molar entropy ($S°$) The entropy value for a mole of a substance in its standard state. (Section 19.4)

standard reduction potential ($E°_{red}$) The potential of a reduction half-reaction under standard conditions, measured relative to the standard hydrogen electrode. A standard reduction potential is also called a standard electrode potential. (Section 20.4)

standard solution A solution of known concentration. (Section 4.6)

standard temperature and pressure (STP) Defined as 0 °C and 1 atm pressure; frequently used as reference conditions for a gas. (Section 10.4)

starch The general name given to a group of polysaccharides that acts as energy-storage substances in plants. (Section 24.8)

state function A property of a system that is determined by its state or condition and not by how it got to that state; its value is fixed when temperature, pressure, composition, and physical form are specified; P, V, T, E, and H are state functions. (Section 5.2)

states of matter The three forms that matter can assume: solid, liquid, and gas. (Section 1.2)

stereoisomers Compounds possessing the same formula and bonding arrangement but differing in the spatial arrangements of the atoms. (Section 23.4)

stoichiometry The relationships among the quantities of reactants and products involved in chemical reactions. (Chapter 3: Introduction)

stratosphere The region of the atmosphere directly above the troposphere. (Section 18.1)

strong acid An acid that ionizes completely in water. (Section 4.3)

strong base A base that ionizes completely in water. (Section 4.3)

strong electrolyte A substance (strong acids, strong bases, and most salts) that is completely ionized in solution. (Section 4.1)

structural formula A formula that shows not only the number and kinds of atoms in the molecule but also the arrangement (connections) of the atoms. (Section 2.6)

structural isomers Compounds possessing the same formula but differing in the bonding arrangements of the atoms. (Sections 23.4 and 24.2)

subatomic particles Particles such as protons, neutrons, and electrons that are smaller than an atom. (Section 2.2)

subshell One or more orbitals with the same set of quantum numbers n and l. For example, we speak of the $2p$ subshell ($n = 2, l = 1$), which is composed of three orbitals ($2p_x, 2p_y$, and $2p_z$). (Section 6.5)

substitutional alloy A homogeneous (solution) alloy in which atoms of different elements randomly occupy sites in the lattice. (Section 23.6)

substitution reactions Reactions in which one atom (or group of atoms) replaces another atom (or group) within a molecule; substitution reactions are typical for alkanes and aromatic hydrocarbons. (Section 24.3)

substrate A substance that undergoes a reaction at the active site in an enzyme. (Section 14.7)

supercritical mass An amount of fissionable material larger than the critical mass. (Section 21.7)

supersaturated solution A solution containing more solute than an equivalent saturated solution. (Section 13.2)

surface tension The intermolecular, cohesive attraction that causes a liquid to minimize its surface area. (Section 11.3)

surroundings In thermodynamics, everything that lies outside the system that we study. (Section 5.1)

system In thermodynamics, the portion of the universe that we single out for study. We must be careful to state exactly what the system contains and what transfers of energy it may have with its surroundings. (Section 5.1)

termolecular reaction An elementary reaction that involves three molecules. Termolecular reactions are rare. (Section 14.6)

tertiary structure The overall shape of a large protein, specifically, the manner in which sections of the protein fold back upon themselves or intertwine. (Section 24.7)

theoretical yield The quantity of product that is calculated to form when all of the limiting reagent reacts. (Section 3.7)

theory A tested model or explanation that satisfactorily accounts for a certain set of phenomena. (Section 1.3)

thermochemistry The relationship between chemical reactions and energy changes. (Chapter 5: Introduction)

thermodynamics The study of energy and its transformation. (Chapter 5: Introduction)

thermonuclear reaction Another name for fusion reactions; reactions in which two light nuclei are joined to form a more massive one. (Section 21.8)

thermoplastic A polymeric material that can be readily reshaped by application of heat and pressure. (Section 12.8)

thermosetting plastic A plastic that, once formed in a particular mold, is not readily reshaped by application of heat and pressure. (Section 12.8)

third law of thermodynamics A law stating that the entropy of a pure, crystalline solid at absolute zero temperature is zero: $S(0 \text{ K}) = 0$. (Section 19.3)

titration The process of reacting a solution of unknown concentration with one of known concentration (a standard solution). (Section 4.6)

torr A unit of pressure (1 torr = 1 mm Hg). (Section 10.2)

transition elements (transition metals) Elements in which the d orbitals are partially occupied. (Section 6.8)

transition state (activated complex) The particular arrangement of reactant and product molecules at the point of maximum energy in the rate-determining step of a reaction. (Section 14.5)

translational motion Movement in which an entire molecule moves in a definite direction. (Section 19.3)

transuranium elements Elements that follow uranium in the periodic table. (Section 21.3)

triple bond A covalent bond involving three electron pairs. (Section 8.3)

triple point The temperature at which solid, liquid, and gas phases coexist in equilibrium. (Section 11.6)

tritium The isotope of hydrogen whose nucleus contains a proton and two neutrons. (Section 22.2)

troposphere The region of Earth's atmosphere extending from the surface to about 12 km altitude. (Section 18.1)

Tyndall effect The scattering of a beam of visible light by the particles in a colloidal dispersion. (Section 13.6)

uncertainty principle A principle stating there is an inherent uncertainty in the precision with which we can simultaneously specify the position and momentum of a particle. This uncertainty is significant only for particles of extremely small mass, such as electrons. (Section 6.4)

unimolecular reaction An elementary reaction that involves a single molecule. (Section 14.6)

unit cell The smallest portion of a crystal that reproduces the structure of the entire crystal when repeated in different directions in space. It is the repeating unit or building block of the crystal lattice. (Section 12.2)

unsaturated solution A solution containing less solute than a saturated solution. (Section 13.2)

valence band A band of closely spaced molecular orbitals that is essentially fully occupied by electrons. (Section 12.7)

valence-bond theory A model of chemical bonding in which an electron-pair bond is formed between two atoms by the overlap of orbitals on the two atoms. (Section 9.4)

valence electrons The outermost electrons of an atom; those that occupy orbitals not occupied in the nearest noble-gas element of lower atomic number. The valence electrons are the ones the atom uses in bonding. (Section 6.8)

valence orbitals Orbitals that contain the outer-shell electrons of an atom. (Chapter 7: Introduction)

valence-shell electron-pair repulsion (VSEPR) model A model that accounts for the geometric arrangements of shared and unshared electron pairs around a central atom in terms of the repulsions between electron pairs. (Section 9.2)

van der Waals equation An equation of state for nonideal gases that is based on adding corrections to the ideal-gas equation. The correction terms account for intermolecular forces of attraction and for the volumes occupied by the gas molecules themselves. (Section 10.9)

vapor Gaseous state of any substance that normally exists as a liquid or solid. (Section 10.1)

vapor pressure The pressure exerted by a vapor in equilibrium with its liquid or solid phase. (Section 11.5)

vibrational motion Movement of the atoms within a molecule in which they move periodically toward and away from one another. (Section 19.3)

viscosity A measure of the resistance of fluids to flow. (Section 11.3)

volatile Tending to evaporate readily. (Section 11.5)

voltaic (galvanic) cell A device in which a spontaneous oxidation–reduction reaction occurs with the passage of electrons through an external circuit. (Section 20.3)

vulcanization The process of cross-linking polymer chains in rubber. (Section 12.6)

watt A unit of power; 1 W = 1 J/s. (Section 20.5)

wave function A mathematical description of an allowed energy state (an orbital) for an electron in the quantum mechanical model of the atom; it is usually symbolized by the Greek letter ψ. (Section 6.5)

wavelength The distance between identical points on successive waves. (Section 6.1)

weak acid An acid that only partly ionizes in water. (Section 4.3)

weak base A base that only partly ionizes in water. (Section 4.3)

weak electrolyte A substance that only partly ionizes in solution. (Section 4.1)

work The movement of an object against some force. (Section 5.1)

PHOTO AND ART CREDITS

INDEX

Note: Page references followed by *n* indicate contents available in endnotes. Page references followed by *f* indicate contents available in figures.